Handbook of
Thin Film Materials

Handbook of Thin Film Materials

Volume 2
Characterization and Spectroscopy of Thin Films

Edited by

Hari Singh Nalwa, M.Sc., Ph.D.
Stanford Scientific Corporation
Los Angeles, California, USA

Formerly at
Hitachi Research Laboratory
Hitachi Ltd., Ibaraki, Japan

ACADEMIC PRESS

A Division of Harcourt, Inc.

San Diego   San Francisco   New York   Boston   London   Sydney   Tokyo

ACADEMIC PRESS
A division of Harcourt, Inc.
525 B Street, Suite 1900, San Diego, CA 92101-4495, USA
http://www.academicpress.com

Academic Press
Harcourt Place, 32 Jamestown Road, London, NW1 7BY, UK
http://www.academicpress.com

Library of Congress Catalog Card Number: 00-2001090614
International Standard Book Number, Set: 0-12-512908-4
International Standard Book Number, Volume 2: 0-12-512910-6

Printed in the United States of America
01 02 03 04 05 06 07 MB 9 8 7 6 5 4 3 2 1

1002666974

To my children
Surya, Ravina, and Eric

Preface

Thin film materials are the key elements of continued technological advances made in the fields of electronic, photonic, and magnetic devices. The processing of materials into thin-films allows easy integration into various types of devices. The thin film materials discussed in this handbook include semiconductors, superconductors, ferroelectrics, nanostructured materials, magnetic materials, etc. Thin film materials have already been used in semiconductor devices, wireless communication, telecommunications, integrated circuits, solar cells, light-emitting diodes, liquid crystal displays, magneto-optic memories, audio and video systems, compact discs, electro-optic coatings, memories, multilayer capacitors, flat-panel displays, smart windows, computer chips, magneto-optic disks, lithography, microelectromechanical systems (MEMS) and multifunctional protective coatings, as well as other emerging cutting edge technologies. The vast variety of thin film materials, their deposition, processing and fabrication techniques, spectroscopic characterization, optical characterization probes, physical properties, and structure-property relationships compiled in this handbook are the key features of such devices and basis of thin film technology.

Many of these thin film applications have been covered in the five volumes of the *Handbook of Thin Film Devices* edited by M. H. Francombe (Academic Press, 2000). The *Handbook of Thin Film Materials* is complementary to that handbook on devices. The publication of these two handbooks, selectively focused on thin film materials and devices, covers almost every conceivable topic on thin films in the fields of science and engineering.

This is the first handbook ever published on thin film materials. The 5-volume set summarizes the advances in thin film materials made over past decades. This handbook is a unique source of the in-depth knowledge of deposition, processing, spectroscopy, physical properties, and structure–property relationship of thin film materials. This handbook contains 65 state-of-the-art review chapters written by more than 125 world-leading experts from 22 countries. The most renowned scientists write over 16,000 bibliographic citations and thousands of figures, tables, photographs, chemical structures, and equations. It has been divided into 5 parts based on thematic topics:

Volume 1: Deposition and Processing of Thin Films
Volume 2: Characterization and Spectroscopy of Thin Films
Volume 3: Ferroelectric and Dielectric Thin Films
Volume 4: Semiconductor and Superconductor Thin Films
Volume 5: Nanomaterials and Magnetic Thin Films

Volume 1 has 14 chapters on different aspects of thin film deposition and processing techniques. Thin films and coatings are deposited with chemical vapor deposition (CVD), physical vapor deposition (PVD), plasma and ion beam techniques for developing materials for electronics, optics, microelectronic packaging, surface science, catalytic, and biomedical technological applications. The various chapters include: methods of deposition of hydrogenated amorphous silicon for device applications, atomic layer deposition, laser applications in transparent conducting oxide thin film processing, cold plasma processing in surface science and technology, electrochemical formation of thin films of binary III–V compounds, nucleation, growth and crystallization of thin films, ion implant doping and isolation of GaN and related materials, plasma etching of GaN and related materials, residual stresses in physically vapor deposited thin films, Langmuir–Blodgett films of biological molecules, structure formation during electrocrystallization of metal films, epitaxial thin films of intermetallic compounds, pulsed laser deposition of thin films: expectations and reality and b″-alumina single-crystal films. This volume is a good reference source of information for those individuals who are interested in the thin film deposition and processing techniques.

Volume 2 has 15 chapters focused on the spectroscopic characterization of thin films. The characterization of thin films using spectroscopic, optical, mechanical, X-ray, and electron microscopy techniques. The various topics in this volume include: classification of cluster morphologies, the band structure and orientations of molecular adsorbates on surfaces by angle-resolved electron spectroscopies, electronic states in GaAs-AlAs short-period superlattices: energy levels and symmetry, ion beam characterization in superlattices, *in situ* real time spectroscopic ellipsometry studies: carbon-based materials and metallic TiN_x thin films growth, *in situ* Faraday-modulated fast-nulling single-wavelength ellipsometry of the growth of semiconductor, dielectric and metal thin films, photocurrent spectroscopy of thin passive films, low frequency noise spectroscopy for characterization of polycrystalline semiconducting thin films and polysilicon thin film transistors, electron energy loss spectroscopy for surface study, theory of low-energy electron diffraction and photoelectron spectroscopy from ultra-thin films, *in situ* synchrotron structural studies of the growth of oxides and metals, operator formalism in polarization nonlinear optics and spectroscopy of polarization inhomogeneous media, secondary ion mass spectrometry (SIMS) and its application to thin films characterization, and a solid state approach to Langmuir monolayers, their phases, phase transitions and design.

Volume 3 focuses on dielectric and ferroelectric thin films which have applications in microelectronics packaging, ferroelectric random access memories (FeRAMs), microelectromechanical systems (MEMS), metal–ferroelectric–semiconductor field-effect transistors (MFSFETs), broad band wireless communication, etc. For example, the ferroelectric materials such as barium strontium titanate discussed in this handbook have applications in a number of tunable circuits. On the other hand, high-permittivity thin film materials are used in capacitors and for integration with MEMS devices. Volume 5 of the *Handbook of Thin Film Devices* summarizes applications of ferroelectrics thin films in industrial devices. The 12 chapters on ferroelectrics thin films in this volume are complimentary to Volume 5 as they are the key components of such ferroelectrics devices. The various topics include electrical properties of high dielectric constant and ferroelectrics thin films for very large scale integration (VLSI) integrated circuits, high permittivity (Ba, Sr)TiO$_3$ thin films, ultrathin gate dielectric films for Si-based microelectronic devices, piezoelectric thin films: processing and properties, fabrication and characterization of ferroelectric oxide thin films, ferroelectric thin films of modified lead titanate, point defects in thin insulating films of lithium fluoride for optical microsystems, polarization switching of ferroelecric crystals, high temperature superconductor and ferroelectrics thin films for microwave applications, twinning in ferroelectrics thin films: theory and structural analysis, and ferroelectrics polymers Langmuir–Blodgett films.

Volume 4 has 13 chapters dealing with semiconductor and superconductor thin film materials. Volumes 1, 2, and 3 of the *Handbook of Thin Film Devices* summarize applications of semiconductor and superconductors thin films in various types of electronic, photonic and electro-optics devices such as infrared detectors, quantum well infrared photodetectors (QWIPs), semiconductor lasers, quantum cascade lasers, light emitting diodes, liquid crystal and plasma displays, solar cells, field effect transistors, integrated circuits, microwave devices, SQUID magnetometers, etc. The semiconductor and superconductor thin film materials discussed in this volume are the key components of such above mentioned devices fabricated by many industries around the world. Therefore this volume is in coordination to Volumes 1, 2, and 3 of the *Handbook of Thin Film Devices*. The various topics in this volume include; electrochemical passivation of Si and SiGe surfaces, optical properties of highly excited (Al, In)GaN epilayers and heterostructures, electical conduction properties of thin films of cadmium compounds, carbon containing heteroepitaxial silicon and silicon/germanium thin films on Si(001), germanium thin films on silicon for detection of near-infrared light, physical properties of amorphous gallium arsenide, amorphous carbon thin films, high-T_c superconducting thin films, electronic and optical properties of strained semiconductor films of group V and III-V materials, growth, structure and properties of plasma-deposited amorphous hydrogenated carbon–nitrogen films, conductive metal oxide thin films, and optical properties of dielectric and semiconductor thin films.

Volume 5 has 12 chapters on different aspects of nanostructured materials and magnetic thin films. Volume 5 of the *Handbook of Thin Film Devices* summarizes device applications of magnetic thin films in permanent magnets, magneto-optical recording, microwave, magnetic MEMS, etc. Volume 5 of this handbook on magnetic thin film materials is complimentary to Volume 5 as they are the key components of above-mentioned magnetic devices. The various topics covered in this volume are; nanoimprinting techniques, the energy gap of clusters, nanoparticles and quantum dots, spin waves in thin films, multi-layers and superlattices, quantum well interference in double quantum wells, electro-optical and transport properties of quasi-two-dimensional nanostrutured materials, magnetism of nanoscale composite films, thin magnetic films, magnetotransport effects in semiconductors, thin films for high density magnetic recording, nuclear resonance in magnetic thin films, and multilayers, and magnetic characterization of superconducting thin films.

I hope these volumes will be very useful for the libraries in universities and industrial institutions, governments and independent institutes, upper-level undergraduate and graduate students, individual research groups and scientists working in the field of thin films technology, materials science, solid-state physics, electrical and electronics engineering, spectroscopy, superconductivity, optical engineering, device engineering nanotechnology, and information technology, everyone who is involved in science and engineering of thin film materials.

I appreciate splendid cooperation of many distinguished experts who devoted their valuable time and effort to write excellent state-of-the-art review chapters for this handbook. Finally, I have great appreciation to my wife Dr. Beena Singh Nalwa for her wonderful cooperation and patience in enduring this work, great support of my parents Sri Kadam Singh and Srimati Sukh Devi and love of my children, Surya, Ravina and Eric in this exciting project.

Hari Singh Nalwa
Los Angeles, CA, USA

Contents

Chapter 1. CLASSIFICATION OF CLUSTER MORPHOLOGIES

Nan Li, Martin Zinke-Allmang

Chapter 2. BAND STRUCTURE AND ORIENTATION OF MOLECULAR ADSORBATES ON SURFACES BY ANGLE-RESOLVED ELECTRON SPECTROSCOPIES

P. A. Dowben, Bo Xu, Jaewu Choi, Eizi Morikawa

Chapter 3. SUPERHARD COATINGS IN C−B−N SYSTEMS: GROWTH AND CHARACTERIZATION

Arun K. Sikder, Ashok Kumar

Chapter 4. ATR SPECTROSCOPY OF THIN FILMS

Urs Peter Fringeli, Dieter Baurecht, Monira Siam, Gerald Reiter, Michael Schwarzott, Thomas Bürgi, Peter Brüesch

Chapter 5.   ION-BEAM CHARACTERIZATION IN SUPERLATTICES

Z. Zhang, J. R. Liu, Wei-Kan Chu

Chapter 6.   IN SITU AND REAL-TIME SPECTROSCOPIC ELLIPSOMETRY STUDIES: CARBON BASED AND METALLIC TiN$_x$ THIN FILMS GROWTH

S. Logothetidis

Chapter 7. *IN SITU* FARADAY-MODULATED FAST-NULLING SINGLE-WAVELENGTH ELLIPSOMETRY OF THE GROWTH OF SEMICONDUCTOR, DIELECTRIC, AND METAL THIN FILMS

J. D. Leslie, H. X. Tran, S. Buchanan, J. J. Dubowski, S. R. Das, L. LeBrun

Chapter 8. PHOTOCURRENT SPECTROSCOPY OF THIN PASSIVE FILMS

F. Di Quarto, S. Piazza, M. Santamaria, C. Sunseri

Chapter 9. ELECTRON ENERGY LOSS SPECTROSCOPY FOR SURFACE STUDY

Takashi Fujikawa

Chapter 10. THEORY OF LOW-ENERGY ELECTRON DIFFRACTION AND PHOTOELECTRON SPECTROSCOPY FROM ULTRA-THIN FILMS

Jürgen Henk

Chapter 11.   *IN SITU* SYNCHROTRON STRUCTURAL STUDIES OF THE GROWTH OF OXIDES AND METALS

A. Barbier, C. Mocuta, G. Renaud

Chapter 12.   OPERATOR FORMALISM IN POLARIZATION-NONLINEAR OPTICS AND SPECTROSCOPY OF POLARIZATION-INHOMOGENEOUS MEDIA

I. I. Gancheryonok, A. V. Lavrinenko

Chapter 13. SECONDARY ION MASS SPECTROMETRY AND ITS APPLICATION TO THIN FILM CHARACTERIZATION

Elias Chatzitheodoridis, George Kiriakidis, Ian Lyon

Chapter 14. A SOLID-STATE APPROACH TO LANGMUIR MONOLAYERS, THEIR PHASES, PHASE TRANSITIONS, AND DESIGN

Craig J. Eckhardt, Tadeusz Luty

Chapter 15.   SOLID STATE NMR OF BIOMOLECULES

Akira Naito, Miya Kamihira

About the Editor

Dr. Hari Singh Nalwa is the Managing Director of the Stanford Scientific Corporation in Los Angeles, California. Previously, he was Head of Department and R&D Manager at the Ciba Specialty Chemicals Corporation in Los Angeles (1999–2000) and a staff scientist at the Hitachi Research Laboratory, Hitachi Ltd., Japan (1990–1999). He has authored over 150 scientific articles in journals and books. He has 18 patents, either issued or applied for, on electronic and photonic materials and devices based on them.

He has published 43 books including *Ferroelectric Polymers* (Marcel Dekker, 1995), *Nonlinear Optics of Organic Molecules and Polymers* (CRC Press, 1997), *Organic Electroluminescent Materials and Devices* (Gordon & Breach, 1997), *Handbook of Organic Conductive Molecules and Polymers*, Vols. 1–4 (John Wiley & Sons, 1997), *Handbook of Low and High Dielectric Constant Materials and Their Applications*, Vols. 1–2 (Academic Press, 1999), *Handbook of Nanostructured Materials and Nanotechnology*, Vols. 1–5 (Academic Press, 2000), *Handbook of Advanced Electronic and Photonic Materials and Devices*, Vols. 1–10 (Academic Press, 2001), *Advanced Functional Molecules and Polymers*, Vols. 1–4 (Gordon & Breach, 2001), *Photodetectors and Fiber Optics* (Academic Press, 2001), *Silicon-Based Materials and Devices*, Vols. 1–2 (Academic Press, 2001), *Supramolecular Photosensitive and Electroactive Materials* (Academic Press, 2001), *Nanostructured Materials and Nanotechnology*–Condensed Edition (Academic Press, 2001), and *Handbook of Thin Film Materials*, Vols. 1–5 (Academic Press, 2002). The *Handbook of Nanostructured Materials and Nanotechnology* edited by him received the 1999 Award of Excellence in Engineering Handbooks from the Association of American Publishers.

Dr. Nalwa is the founder and Editor-in-Chief of the *Journal of Nanoscience and Nanotechnology* (2001–). He also was the founder and Editor-in-Chief of the *Journal of Porphyrins and Phthalocyanines* published by John Wiley & Sons (1997–2000) and serves or has served on the editorial boards of *Journal of Macromolecular Science-Physics* (1994–), *Applied Organometallic Chemistry* (1993–1999), *International Journal of Photoenergy* (1998–) and *Photonics Science News* (1995–). He has been a referee for many international journals including *Journal of American Chemical Society, Journal of Physical Chemistry, Applied Physics Letters, Journal of Applied Physics, Chemistry of Materials, Journal of Materials Science, Coordination Chemistry Reviews, Applied Organometallic Chemistry, Journal of Porphyrins and Phthalocyanines, Journal of Macromolecular Science-Physics, Applied Physics, Materials Research Bulletin*, and *Optical Communications*.

Dr. Nalwa helped organize the First International Symposium on the Crystal Growth of Organic Materials (Tokyo, 1989) and the Second International Symposium on Phthalocyanines (Edinburgh, 1998) under the auspices of the Royal Society of Chemistry. He also proposed a conference on porphyrins and phthalocyanies to the scientific community that, in part, was intended to promote public awareness of the *Journal of Porphyrins and Phthalocyanines*, which he founded in 1996. As a member of the organizing committee, he helped effectuate the First International Conference on Porphyrins and Phthalocyanines, which was held in Dijon, France in 2000. Currently he is on the organizing committee of the BioMEMS and Smart Nanostructures, (December 17–19, 2001, Adelaide, Australia) and the World Congress on Biomimetics and Artificial Muscles (December 9–11, 2002, Albuquerque, USA).

Dr. Nalwa has been cited in the *Dictionary of International Biography, Who's Who in Science and Engineering, Who's Who in America,* and *Who's Who in the World.* He is a member of the American Chemical Society (ACS), the American Physical Society (APS), the Materials Research Society (MRS), the Electrochemical Society and the American Association for the Advancement of Science (AAAS). He has been awarded a number of prestigious fellowships including a National Merit Scholarship, an Indian Space Research Organization (ISRO) Fellowship, a Council of Scientific and Industrial Research (CSIR) Senior fellowship, a NEC fellowship, and Japanese Government Science & Technology Agency (STA) Fellowship. He was an Honorary Visiting Professor at the Indian Institute of Technology in New Delhi.

Dr. Nalwa received a B.Sc. degree in biosciences from Meerut University in 1974, a M.Sc. degree in organic chemistry from University of Roorkee in 1977, and a Ph.D. degree in polymer science from Indian Institute of Technology in New Delhi in 1983. His thesis research focused on the electrical properties of macromolecules. Since then, his research activities and professional career have been devoted to studies of electronic and photonic organic and polymeric materials. His endeavors include molecular design, chemical synthesis, spectroscopic characterization, structure-property relationships, and evaluation of novel high performance materials for electronic and photonic applications. He was a guest scientist at Hahn-Meitner Institute in Berlin, Germany (1983) and research associate at University of Southern California in Los Angeles (1984–1987) and State University of New York at Buffalo (1987–1988). In 1988 he moved to the Tokyo University of Agriculture and Technology, Japan as a lecturer (1988–1990), where he taught and conducted research on electronic and photonic materials. His research activities include studies of ferroelectric polymers, nonlinear optical materials for integrated optics, low and high dielectric constant materials for microelectronics packaging, electrically conducting polymers, electroluminescent materials, nanocrystalline and nanostructured materials, photocuring polymers, polymer electrets, organic semiconductors, Langmuir-Blodgett films, high temperature-resistant polymer composites, water-soluble polymers, rapid modeling, and stereolithography.

List of Contributors

Numbers in parenthesis indicate the pages on which the author's contribution begins.

A. BARBIER (527)
CEA/Grenoble, Département de Recherche Fondamentale sur la Matière Condensée
SP2M/IRS, 38054 Grenoble Cedex 9, France

DIETER BAURECHT (191)
Institute of Physical Chemistry, University of Vienna, Althanstrasse 14/UZA II,
A-1090 Vienna, Austria

PETER BRÜESCH (191)
ABB Management Ltd., Corporate Research (CRB), CH-5405 Baden-Dattwil, Switzerland and
Departement de Physique, Ecole Polytechnique, Federale de Lausanne (EPFL), PH-Ecublens,
CH-1015 Lausanne, Switzerland

S. BUCHANAN (331)
Waterloo Digital Electronics Division of WDE Inc., 279 Weber St. N., Waterloo, Ontario,
N2J 3H8, Canada

THOMAS BÜRGI (191)
Laboratory of Technical Chemistry, Swiss Federal Institute of Technology, ETH Zentrum,
CH-8092 Zürich, Switzerland

ELIAS CHATZITHEODORIDIS (637)
IESL/FORTH, Crete, Greece

JAEWU CHOI (61)
Center for Advanced Microstructures and Devices, Louisiana State University,
Baton Rouge, Louisiana, USA

WEI-KAN CHU (231)
Department of Physics, Texas Center for Superconductivity, University of Houston,
Houston, Texas, USA

S. R. DAS (331)
Institute for Microstructural Sciences, National Research Council, Ottawa,
Ontario, K1A 0R6, Canada

P. A. DOWBEN (61)
Department of Physics and Astronomy and the Center for Materials Research and Analysis,
Behlen Laboratory of Physics, University of Nebraska, Lincoln, Nebraska, USA

F. DI QUARTO (373)
Dipartimento di Ingegneria Chimica dei Processi e dei Materiali, Università di Palermo,
Viale delle Scienze, 90128 Palermo, Italy

J. J. DUBOWSKI (331)
Institute for Microstructural Sciences, National Research Council, Ottawa, Ontario,
K1A 0R6, Canada

CRAIG J. ECKHARDT (685)
Department of Chemistry, Center for Materials Research and Analysis,
University of Nebraska–Lincoln, Lincoln, Nebraska, USA

URS PETER FRINGELI (191)
Institute of Physical Chemistry, University of Vienna, Althanstrasse 14/UZA II,
A-1090 Vienna, Austria

TAKASHI FUJIKAWA (415)
Graduate School for Science, Chiba University, Chiba 263-8522, Japan

I. I. GANCHERYONOK (597)
Department of Physics, Belarusian State University, F. Skariny av. 4, Minsk 2220080, Belarus

JÜRGEN HENK (479)
Max-Planck-Institut für Mikrostrukturphysik, Halle/Saale, Germany

MIYA KAMIHIRA (735)
Department of Life Science, Himeji Institute of Technology, 3-2-1 Kouto, Kamigori,
Hyogo 678-1297, Japan

GEORGE KIRIAKIDIS (637)
IESL/FORTH, Crete, Greece

ASHOK KUMAR (115)
Center for Microelectronics Research, College of Engineering, University of South Florida,
Tampa, Florida, USA

A. V. LAVRINENKO (597)
Department of Physics, Belarusian State University, F. Skariny av. 4, Minsk 2220080, Belarus

L. LEBRUN (331)
Institute for Microstructural Sciences, National Research Council, Ottawa, Ontario,
K1A 0R6, Canada

J. D. LESLIE (331)
Department of Physics, University of Waterloo, Waterloo, Ontario, N2L 3G1, Canada

NAN LI (1)
Department of Physics and Astronomy, The University of Western Ontario, London, Ontario,
N6A 3K7, Canada

J. R. LIU (231)
Department of Physics, Texas Center for Superconductivity, University of Houston,
Houston, Texas, USA

S. LOGOTHETIDIS (277)
Aristotle University of Thessaloniki, Physics Department, Solid State Physics Section,
GR-54006, Thessaloniki, Greece

TADEUSZ LUTY (685)
Institute of Physical and Theoretical Chemistry, Technical University of Wrocław,
Wrocław, Poland

IAN LYON (637)
Manchester University, England

C. MOCUTA (527)
CEA/Grenoble, Département de Recherche Fondamentale sur la Matière Condensée
SP2M/IRS, 38054 Grenoble Cedex 9, France

EIZI MORIKAWA (61)
Center for Advanced Microstructures and Devices, Louisiana State University,
Baton Rouge, Louisiana, USA

AKIRA NAITO (735)
Department of Life Science, Himeji Institute of Technology, 3-2-1 Kouto, Kamigori,
Hyogo 678-1297, Japan

S. PIAZZA (373)
Dipartimento di Ingegneria Chimica dei Processi e dei Materiali, Università di Palermo,
Viale delle Scienze, 90128 Palermo, Italy

G. RENAUD (527)
CEA/Grenoble, Département de Recherche Fondamentale sur la Matière Condensée
SP2M/IRS, 38054 Grenoble Cedex 9, France

GERALD REITER (191)
Institute of Physical Chemistry, University of Vienna, Althanstrasse 14/UZA II,
A-1090 Vienna, Austria

M. SANTAMARIA (373)
Dipartimento di Ingegneria Chimica dei Processi e dei Materiali, Università di Palermo,
Viale delle Scienze, 90128 Palermo, Italy

MICHAEL SCHWARZOTT (191)
Institute of Physical Chemistry, University of Vienna, Althanstrasse 14/UZA II,
A-1090 Vienna, Austria

MONIRA SIAM (191)
Institute of Physical Chemistry, University of Vienna, Althanstrasse 14/UZA II,
A-1090 Vienna, Austria

ARUN K. SIKDER (115)
Center for Microelectronics Research, College of Engineering, University of South Florida,
Tampa, Florida, USA

C. SUNSERI (373)
Dipartimento di Ingegneria Chimica dei Processi e dei Materiali, Università di Palermo,
Viale delle Scienze, 90128 Palermo, Italy

H. X. TRAN (331)
Department of Physics, University of Waterloo, Waterloo, Ontario, N2L 3G1, Canada

BO XU (61)
Department of Physics and Astronomy and the Center for Materials Research and Analysis,
Behlen Laboratory of Physics, University of Nebraska, Lincoln, Nebraska, USA

Z. ZHANG (231)
Department of Physics, Texas Center for Superconductivity, University of Houston,
Houston, Texas, USA

MARTIN ZINKE-ALLMANG (1)
Department of Physics and Astronomy, The University of Western Ontario, London, Ontario,
N6A 3K7, Canada

Handbook of Thin Film Materials

Edited by H.S. Nalwa

Volume 3. FERROELECTRIC AND DIELECTRIC THIN FILMS

Volume 4. SEMICONDUCTOR AND SUPERCONDUCTING THIN FILMS

Volume 5. NANOMATERIALS AND MAGNETIC THIN FILMS

Chapter 1

CLASSIFICATION OF CLUSTER MORPHOLOGIES

Nan Li, Martin Zinke-Allmang

Department of Physics and Astronomy, The University of Western Ontario, London, Ontario, N6A 3K7, Canada

Contents

1. STRUCTURES DEVELOPING DURING FILM GROWTH

1.1. Early Growth Morphology

The formation of clusters and islands on surfaces is a widespread phenomenon since these structures are often thermodynamically favored over uniform films. Thus, it is important to identify and to characterize any nonhomogeneity in the growth of heterostructures, whether obtained intentionally or unintentionally. In this chapter, we provide a wide range of clustering structures which allow the reader to either identify the processes which led to an observed morphology or to plan the growth such that a particular cluster morphology is obtained. Throughout the chapter, the term *cluster* is used for three-dimensional structures and the term *island* is used for two-dimensional structures.

1.1.1. Thin Film Growth Modes

Clustering is a phase separation process governed by nonequilibrium thermodynamics concepts. On surfaces, it can be initiated by any type of nonequilibrium situation during an experiment. Typical processes causing clustering to occur are il-

lustrated in Figure 1 schematically: A thin film system reaches a metastable or instable region of its phase diagram either through temperature variation or through variation of the concentration of surface components. The figure shows an idealized miscibility gap for a two component system. The two components in a thin film system can be two chemical species of a partially miscible alloy film or, more typically, a single component partially wetting the surface. The coexistence curve (solid line) separates stable and nonstable structures and the spinodal line (dashed line) separates metastable, outer regions and instable, inner regions of the miscibility gap. The shaded area near the coexistence curve indicates the regime where the classical nucleation theory applies [1, 2]. Pathways into the coexistence regime are indicated by arrows *a* and *b* based on temperature changes and arrows *c* and *d* based on concentration changes.

Temperature quenches transfer a film system fast into the two-phase region under mass conservation conditions. Through the choice of the initial concentration, both instable and metastable regions can be reached. The mass conservation condition implies that the system evolution develops toward an Ostwald ripening mechanism [3], with an initial nucleation stage for off-critical quenches close to the coexistence curve (as shown with arrow *a*) and an initial spinodal decomposition

Handbook of Thin Film Materials, edited by H.S. Nalwa
Volume 2: Characterization and Spectroscopy of Thin Films

ISBN 0-12-512910-6/$35.00

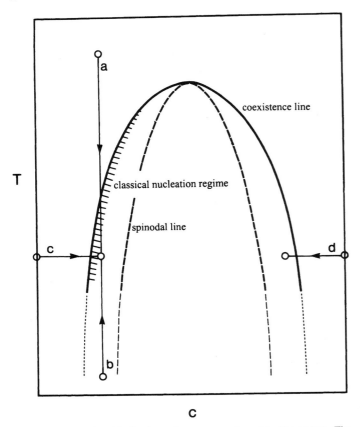

c

Fig. 1. Sketch of an idealized coexistence curve for a thin film system. The system can be brought into the two-phase coexistence regime by a temperature quench (arrow a), by a fast sample heating from a frozen state (arrow b), through a thin film deposition process (arrow c), and through a desorption process (arrow d). Reprinted with permission from [20], © 1999, Elsevier Science.

for critical quenches toward regimes closer to the center of the miscibility gap. In experimental thin film applications, this approach is suitable for systems with full miscibility of the phases at high temperatures below the desorption limit.

Most films with thicknesses in excess of a single monolayer or systems with a propensity toward three-dimensional clustering cannot be studied this way. Desorption typically establishes an upper temperature limit at which a single, miscible phase is not yet reached [4]. For these systems, pathway b in Figure 1 represents an alternative. A deposition of the film at temperatures, at which any morphological changes are kinetically frozen, is followed by a sudden increase in temperature. Subsequent structural development again occurs under mass conservation. Systems evolving under mass conservation along pathways a and b are discussed in Section 2.

During thin film growth, the pathways of interest in Figure 1 are those at constant temperature but with changing concentration on the surface. Arrows c and d define nonmass conserved systems which lead asymptotically to coalescence and, particularly for two-dimensional films, percolation. Since deposition (i) occurs at a slower timescale than temperature quenches and (ii) the miscibility gap is entered through the classical nucleation regime, usually nucleation is observed to initiate the phase

separation. Spinodal decomposition should only be observed [2] if the phase separation is sufficiently slow that the deposition process can cross over the spinodal line before phase separation starts. Despite this fact, a few experimental studies suggest that spinodal may be observable for surface alloy systems [5, 6].

Systems entering the two-phase coexistence regime along a decreasing concentration (arrow d in Fig. 1) have been studied as well. While this process does not describe a typical approach toward a thin film, there are systems where film mass reducing processes are either easier to accomplish experimentally or are technologically interesting. The latter includes sputtering processes where the evolution of clustered structures depends on the ion beam configuration (relevant, e.g., to secondary ion mass spectroscopy (SIMS) and transmission electron microscopy (TEM) sample preparation) [7]. Desorption plays an important role in compound semiconductors, many of which are thermally instable at epitaxial film growth temperatures. Studies of the thermal decomposition of such wafers in ultrahigh vacuum establish the evolution of a film structure under non-mass conserved conditions as one component is continuously lost to the vacuum with the other component steadily increasing on the surface. Processes leading to coalescence dominated growth behavior are discussed in Section 1.3.

The conditions under which a clustered morphology is thermodynamically favored for a given thin film growth system are illustrated in Figure 2. In the figure, the thermodynamic state of a clustered morphology is compared with a uniform layer. The ordinate of the plot is the excess chemical potential for the clustered film and the abscissa is the number of atoms in a deposit. The particular curve is given for a film layer growing commensurately on a substrate with a nonzero misfit below the roughening transition [8–11]. The misfit introduces a strain energy in the film which increases linearly with film thickness. We distinguish two cases of the film atom-substrate interaction strength relative to the interaction strength between film atoms, W, with $W = E_s/E_f$. E_s is the bond energy between film and substrate atoms and where E_f is the bond energy between two film atoms. Thus, the substrate interaction dominates for $W > 1$ and the interaction between film atoms is stronger for $W < 1$.

The energy axis is a relative axis between the two states, clustered and uniform film. It is given by $(E(n) - E(0) - \mu_c n)$ with μ_c being the chemical potential of a bulk piece of the material forming the film, and n is the number of monolayers in the film. Thus, when the slope of a curve in the plot is positive, $\mu_{film} < \mu_c$ and it is energetically favorable to form bulk of the film material, i.e., clusters.

We now assume a growth experiment where a random deposition of film atoms leads to a completion of one full layer after another (which is a simplification as second layer formation may start before the previous layer is complete). This growth of a single-phase film would lead for the case $W > 1$ to the oscillations shown in Figure 2, where unpaired bonds, edge- and corner-arrangements cause the increase in energy between complete layers. Whenever the system can alternatively lower its energy through a phase separation, the real growth follows the alternate route. In the current model, this is the case be-

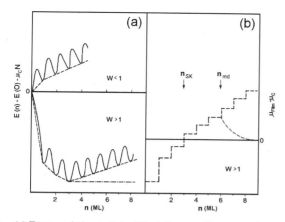

Fig. 2. (a) Energy relative to that of the bulk crystalline state, plotted versus the film thickness for a film-substrate interaction strength larger ($W > 1$) and smaller ($W < 1$) than the film–film interaction. The dashed line represents the minimum energy configuration for a single-phase system, the solid line includes two-phase regimes between full layers. (b) Chemical potentials relative to the bulk chemical potential for the same system. n_{SK} labels the Stranski–Krastanov layer thickness and n_{md} labels the critical thickness for misfit dislocations.

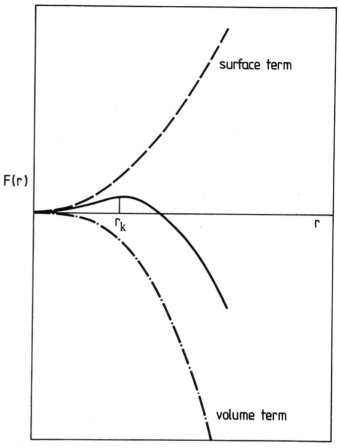

Fig. 3. Sketch illustrating the Gibbs free energy differential for a cluster on a surface. Reprinted with permission from [181], © 1992, Elsevier Science.

tween the completion of full layers where a two-phase regime can be traversed (represented by the common-tangent of two subsequent energy minima) by growing the new layer in form of large two-dimensional islands, avoiding unpaired bonds, edge- and corner-arrangements (dashed line to $n = 3$ ML).

Three-dimensional clustering can be established in Figure 2 based on the same argument. It occurs when the chemical potential μ_{film} exceeds μ_C, i.e., the slope of the common-tangent curve becomes positive (dashed line beyond $n = 3$ ML). Instead of following the dashed line toward higher energy, the system follows the dashed–dotted line, defining a wetting transition [12]. The resulting film morphology, consisting of a few uniform layers and clustering of subsequently deposited material is referred to as a Stranski–Krastanov (SK) growth condition [13].

Based on the thermodynamic equilibrium arguments applied to Figure 2, two growth models are distinguished: if the structure grows with the formation of three-dimensional clusters on the bare substrate, the system is called a Volmer–Weber system [14] and if the growth results initially in a uniform layer the system is either a Frank–van der Merwe or a Stranski–Krastanov system. The latter is distinguished by the occurrence of clusters beyond a given coverage above one monolayer, as illustrated in Figure 2. This coverage is called the SK-layer thickness.

During the growth of a film, however, clustering usually does not occur immediately after exceeding the SK-layer thickness due to a kinetic barrier toward clustering. The reason for this observation is illustrated in Figure 3. The figure shows the Gibbs free energy of a three-dimensional cluster as a function of its radius (assuming for simplicity a spherical cluster shape). At small radii, the energy gain due to the formation of bulk material, which is essentially proportional to the cube of the radius times the difference between the chemical potential of bulk film material and the chemical potential of free adatoms on the sub-

strate, $(\mu_C - \mu_{ad}) \cdot r_C^3$, is exceeded by the energy loss due to the surface of the cluster, which is essentially proportional to the square of the cluster radius, γr_C^2, with γ the surface tension of the cluster material. With increasing radius, the volume term begins to dominate, establishing a critical radius in Figure 3 with a maximum of the Gibbs free energy $F_C(r)$.

Applying the condition $dF_C(r)/dr = 0$ allows us to calculate the critical cluster size for which the increasing surface energy and the decreasing volume free energy are balanced. This critical radius can be interpreted in two ways, (i) in experiments with variable deposition rates and variable free adatom concentrations, the critical radius identifies the size of a cluster which needs to form before it can become stable with the addition of a single adatom. Such clusters may require cluster sizes of the order of several tens of nanometers to be energetically favored. (ii) For a cluster with a given radius r a vapor pressure can be calculated such that this cluster grows and decomposes with the same probability. This leads to the Gibbs–Thomson equation defining the equilibrium concentration of a cluster of radius r [15–19],

$$c(r) = c_\infty \exp\left(\frac{2\gamma v_C}{rkT}\right) \cong c_\infty\left(1 + \frac{2\gamma v_C}{rkT}\right) \qquad (1)$$

where c_∞ is the equilibrium concentration for an infinitely large cluster, γ is the surface tension, v_C is the atomic volume of the cluster material, and k is the Boltzmann constant. The second equation in Eq. (1) applies when $r \geq 5$ nm. For large clusters, the exponent in Eq. (1) vanishes and $c(r)$ approaches c_∞. In this form, the Gibbs–Thomson equation is applied to cluster growth under zero deposition rate conditions in part 2 of this chapter.

Thus, uniform layer growth often continues beyond the SK-layer thickness, forming coherently strained but metastable layers. The decay of such metastable structures occurs along different kinetic pathways, separately discussed earlier: The traditional model allows a metastable uniform film to grow to a thickness where local regions start to exceed a critical thickness for misfit dislocations. In these regions, homogeneous strain relaxes. Gettering mobile adatoms at these sites results in the nucleation of clusters and a decay of the metastability by approaching the bulk chemical potential. Nucleation mechanisms and the resulting cluster distributions are discussed in Section 1.1.2. Another mechanism possible for approaching the bulk chemical potential is through spinodal decomposition [20]. This mechanism is expected for states deep in the miscibility gap shown in Figure 1. The formation of well-defined clusters by nucleation is replaced in these cases by long-wavelength instabilities in the supersaturation leading to a network of denser and more dilute areas. This perturbation then increases, forming sharp concentration steps. Spinodal decomposition is seldomly seen in thin film growth situations as indicated already when discussing Figure 1. We still discuss it briefly in Section 1.1.3 as a few examples of the resulting morphologies can be given.

A third kinetic pathway was identified. Before the supersaturation reaches the onset point of nucleation of relaxed clusters, coherently strained clusters can form in some systems [21–23]. These clusters still represent an energetically more favorable state than a uniform, strained film as the corrugation of the surface allows for some strain relief. However, these structures are only metastable as a fully relaxed cluster is the energetically most favorable state for all systems [3, 20]. Since the coherent clusters have received a lot of interest lately, a more detailed discussion is provided in a separate section (Section 1.2).

When studying clustered structures, it should also be noted that at all nonzero temperatures a free concentration of adatoms must be present between the clusters. This is already implicitly indicated in the Gibbs–Thomson equation (Eq. (1)) where each cluster on a surface requires a specific adatom concentration to be in thermodynamic equilibrium. However, in any experimental system, there is a range of cluster sizes on the surface, and most of them are not in thermodynamic equilibrium. Thus, this concentration cannot be derived from Eq. (1) alone, although this equation indicates which two contributions to the free concentration have to be included: (i) the temperature-dependent solubility term c_∞ and (ii) the additional contribution associated with the exponential factor for a representative cluster in the cluster size distribution. The second contribution depends obviously on the actual cluster size distribution and thus on the growth processes leading to the cluster morphology of interest. In general, however, this contribution is a small correction (in the range below 1%) of the contribution due to c_∞. This term is discussed in the present context as it establishes a free adatom concentration in the range of up to a few tenths of a monolayer for all systems discussed in this chapter.

For a two-phase system with both phases in thermodynamic equilibrium, in our case the dense cluster phase and the dilute adatom phase on a surface, thermodynamics requires that the chemical potential between both phases is balanced [24], $\mu_C(T) = \mu_{ad}(T)$, where the index ad refers the free adatom phase and C refers to the dense cluster phase. If we neglect for simplicity the dependence on the number of atoms in a cluster, we derive from the balance condition of the chemical potential a Clausius–Clapeyron type of equation [25],

$$\ln\big(c_\infty(T)\big) \propto -\frac{E_f}{kT} \qquad (2)$$

where E_f is the energy of formation of a cluster (i.e., the energy difference for a film atom between being a free adatom and an integral part of a larger cluster). The temperature dependence of c_∞ has been verified [25] in a direct measurement of the adatom concentration as a function of temperature using scanning Auger microscopy in which the surface concentration between clusters is directly determined.

1.1.2. Nucleation

The time evolution of cluster formation is divided into three stages [20]: nucleation, early stage growth, and late stage growth, as sketched in Figure 4. Random nucleation and spinodal decomposition (see Section 1.1.3) are the dominant processes in the first stage. While nucleation is continuing, the first nuclei start to grow, capturing atoms from the supersaturated adatom phase. When the supersaturation is mostly reduced the nucleation process ceases. Clusters will continue to grow but other processes dominate, such as ripening, i.e., growth of larger clusters at the expense of smaller clusters which dissolve (Lifshitz–Slyozov–Wagner (LSW) model [26, 27], discussed in Section 2.2), and coalescence, i.e., clusters growing into each other [28, 29] (see Section 1.3).

The early stage of growth is a transition stage between nucleation and late stage growth regime. In the early stage, the development of the cluster morphology depends primarily on the deposition rate. If deposition ceases, cluster nucleation ends and the clusters grow individually from the surrounding supersaturation which far exceeds the equilibrium concentration c_∞ while growth becomes a global phenomenon in the late stage with the entire cluster distribution interacting. If deposition continues during the early stages of cluster growth, the system asymptotically approaches a coalescence state. During the formation of precoalescence morphologies, i.e., structures where the merging of clusters does not yet dominate the morphology, coherent clustering may lead to novel structures which are discussed in Section 1.2.

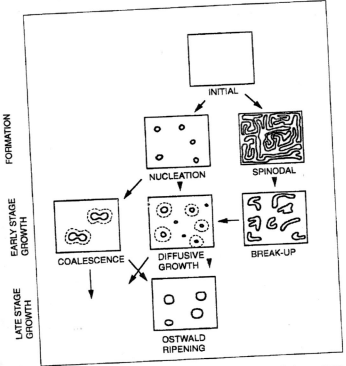

Fig. 4. Sketch of the different stages of phase formation and phase separation. Indicated are random nucleation and spinodal decomposition in the cluster formation regime, diffusive growth, coalescence, and breakup of spinodal networks in the early stage and coalescence and Ostwald ripening in the late stage. Reprinted with permission from [181], © 1992, Elsevier Science.

1.1.2.1. Fundamental Concepts of Nucleation

Essentially, two approaches for nucleation and early stage growth have to be distinguished, (i) a model based on analytical formulations requiring usually restrictive assumptions such as a strict separation of the different stages of growth and (ii) a model based on kinetic rate equations with fewer assumptions, allowing the inclusion of growth modifications such as variable deposition rates or competing early stage coalescence [30, 31]. The kinetic rate equation approach has been chosen most frequently in the literature since it allows studing a wider range of cases. Thus, the concepts used in analytical models are only summarized briefly and the kinetic rate equations model is discussed in more detail.

In this section, we assume that a system, consisting of a homogenous adlayer on a substrate surface, is brought into a two-phase coexistence regime by a sudden change in supersaturation, e.g., by a temperature quench or during a deposition process. We also assume that an energy barrier exists toward formation of instabilities, i.e., a small fluctuation in the surface concentration does usually decay, i.e., the adlayer can exist as a metastable configuration.

In the thermodynamic limit, the energetic situation for a randomly formed small aggregation of atoms is described by the total Gibbs free energy. For a specific critical radius in Figure 3, the positive contribution due to surface increase and the negative bulk contribution is balanced. After a cluster of critical radius has formed, the addition of one further atom stabilizes this cluster, i.e., then the cluster does not decay anymore. The main quantity besides the critical radius to be determined is the nucleation rate J which is the number of stable clusters formed per unit time and unit area on the substrate. The nucleation rate of stable clusters is given by the areal density of critical clusters, and the rate at which these clusters gain an additional atom. The nucleation rate is then deduced from the principle of detailed balance with the key equation given in the form [32],

$$J = N(r_C)N(1)\delta_{in}(r_C) \qquad (3)$$

where $N(1)$ is the monomer adatom concentration, $N(r_C)$ is the areal density of clusters of critical radius, and δ_{in} is a collision factor describing the likelihood of a monomer to enter a cluster. The three factors in Eq. (3) are obtained as follows:

(1) $N(1)$ is calculated from the deposition rate and certain assumptions about the time constant for adatoms to remain in the free concentration as opposed to joining a cluster at low temperatures or possibly desorbing at higher temperatures.

(2) The number density of critical nuclei $N(r_C)$ is estimated by analyzing Figure 3 quantitatively. It is assumed that the densities of all subcritical clusters reach a steady state.

(3) The collision factor contains two major contributions: direct impingement from the vapor phase by the deposition process and surface diffusion of monomers (neglecting mobility of larger clusters). The diffusion mechanism dominates in most experimental cases [33, 34]. The collision factor then depends exponentially on the activation energy for diffusion.

For condensation in most three-dimensional systems, e.g., rain clouds, the critical cluster contains more than 100 atoms and an analysis based on the continuum thermodynamic model is justified [35]. On surfaces however, critical clusters often contain only a few atoms [36] or even a single atom, as discussed later in greater detail. Atomic level quantities, such as the binding energies of atoms at different sites on the surface or at the cluster, have to be used replacing macroscopic thermodynamic quantities [37]. The applicability of this approach suffers, however, from the same restrictions mentioned above. A qualitative advantage of the atomic approach is given for nucleation of crystalline nuclei, when Eq. (3) is modified to reflect the equilibrium shape of the cluster. In the atomic approach, a sequence of nucleation steps toward this shape can be described.

Still, both approaches cannot deal with major factors in experimental studies, e.g., a spatial gradient of the free adatom concentration $N(1)$ including diffusion effects on the surface. To rectify these shortcomings, an independent model was developed by Zinsmeister [31] and Venables [30], called the kinetic rate equations approach. Frankl and Venables also gave the first comprehensive discussion of the applicability of these equations in the analysis of nucleation experiments [38].

In the rate equation model for each cluster size (represented by j atoms in the cluster), the change of number of the clus-

ters is written as a function of all processes contributing to the change of this number. To keep the model transparent, surface mobility is restricted to single atoms in this section. Further, coalescence phenomena are neglected and are discussed later. The kinetic rate equations are then given by

$$
\frac{dN(1)}{dt} = R - \frac{N(1)}{\tau_a} - 2N(1)^2 \delta_{in}(1) + N(2)\delta_{out}(2)
$$
$$
+ \sum_{j=2}^{\infty} [\delta_{out}(j) - N(1)\delta_{in}(j)]N(j)
$$
$$
\frac{dN(2)}{dt} = N(1)[\delta_{in}(1)N(1) - \delta_{in}(2)N(2)]
$$
$$
+ [\delta_{out}(3)N(3) - \delta_{out}(2)N(2)] \qquad (4)
$$
$$
\cdots
$$
$$
\frac{dN(j)}{dt} = N(1)[\delta_{in}(j-1)N(j-1) - \delta_{in}(j)N(j)]
$$
$$
+ [\delta_{out}(j+1)N(j+1) - \delta_{out}(j)N(j)]
$$
$$
\cdots
$$

where δ_{in} is the collision factor for monomer capture and δ_{out} is the collision factor for the release of a monomer; both factors are cluster size-dependent proportionality factors. The first equation describes the change in the number of single atoms, which is equivalent to changes of the free adatom concentration. It contains five terms: the first term is the deposition rate R, the second term represents losses due to reevaporation (where τ_a is a time constant for evaporation), and the third and fourth terms describe loss and gain due to charge in the number of dimers. The first term in the bracket of the sum corresponds to capture of single atoms by clusters and the second term corresponds to release of single atoms from a cluster. Each of these rates is given by the product of the number of relevant species and the collision factor which depends on the size of the cluster. The second and all following equations give the number change of clusters with j atoms, losses (negative terms) due to capture or release of single atoms, and gains (positive terms) due to growth of the next smaller cluster or decomposition of the next larger cluster.

Nucleation for the case of a constant deposition rate allowed Venables [39] to simplify Eq. (4) by separating three groups of formulas: a formula for single adatoms, a formula combining all stable clusters, and a formula for the nonstable clusters ($j \leq i^*$, where i^* represents the number of atoms in a cluster of critical radius). Assuming a steady state for these sizes, the simplified set of kinetic rate equations reads:

$$
\frac{dN(1)}{dt} = R - \frac{N(1)}{\tau_a} - \frac{d(N_x j_x)}{dt}
$$
$$
\frac{dN(j)}{dt} = 0 \quad \text{for } 1 < j < i^* \qquad (5)
$$
$$
\frac{dN_x}{dt} = N(1)\delta_{in}(i^*)N(i^*) - 2N_x \frac{d\Phi_C}{dt}
$$

with $N_x j_x$ the total number of atoms in stable clusters and Φ_C the fraction of the substrate surface area covered by clusters. The last term in Eq. (5) allows for random coalescence events, which are often found to terminate the nucleation stage [40]. Note that the first term of the last formula is equal to the nucleation rate as given in Eq. (3).

A typical nucleation process at sufficiently high temperatures proceeds in three stages based on Eq. (5), where initially the deposition rate R dominates on the right-hand side of the formula. This transient regime is characterized by $N(1) = Rt$ and $N_x = \Phi_C = 0$. The next regime is characterized by steady-state conditions for the concentration of single adatoms. In most experimental systems, it is very quickly reached [41, 42], and most nuclei form in this regime. Finally, coalescence starts, since the clusters are growing. At the same time, the nucleation rate decreases sharply.

1.1.2.2. Applications of the Fundamental Nucleation Concepts

We first summarize a few examples where experimental data were analyzed based on Eq. (5). In a detailed analysis for several metal overlayers on metals and silicon, the maximum number of stable clusters reached in the nucleation regime, the surface diffusion and desorption activation energies and the energy of formation of a critical cluster were determined [43–45]. For the quantitative analysis, a two-dimensional nearest neighbor pair bond model is used with the pair binding energy and the activation energy of surface diffusion as free parameters. The model requires the determination of all critical nuclei sizes possible based on the criterion of a minimum nucleation density. These numbers of atoms in critical clusters are defined when an additional atom forms three new bonds to the island periphery. The difference in how the two activation energies depend on the critical cluster size as a function of temperature allows a separate calculation of the activation energies in a single Arrhenius-type measurement [44]. Figure 5 gives an example for the system Ag/Si(100). Shown is the areal density of stable clusters as a function of $1/T$ for two deposition rates. The numbers attached along the fitted curves indicate the number of atoms in a critical cluster at each temperature interval as found in the data analysis.

Note that there are some specific assumptions required which limit the generality of this technique [44]. For example, for Ag overlayer structures in Figure 5 the following assumptions were made: (i) all critical clusters are two-dimensional islands of known shape with a hexagonal bond structure, (ii) the energetic difference of Ag in islands and Ag monomers is described by a nearest neighbor pair bond model neglecting altering contributions from the substrate as expected for covalent binding adlayers or three-dimensional critical clusters, and (iii) defect induced, heterogenous nucleation, and island mobility are excluded.

Table I presents activation energies for clustering obtained by this data analysis technique, for Ag on Si(111) and Si(100), Ag on Mo(100), and Ag on W(110). The table also contains data obtained from cluster growth studies as described in the

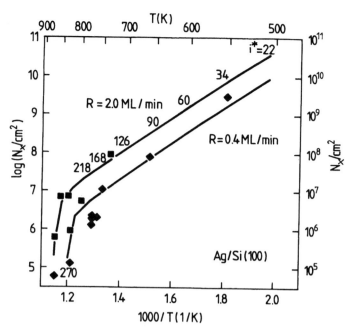

Fig. 5. Nucleation densities of stable clusters as a function of temperature of Ag on Si(001) for two different deposition rates R. The additional numbers indicated give the corresponding critical cluster size, i^*, in the respective temperature interval based on a two-dimensional hexagonal model of the islands as discussed in the text. Technique: SEM. Reprinted with permission from [43], © 1984, Elsevier Science.

Table I. Activation Energies for Clustering on Surface, E_c, from Nucleation and Cluster Growth Experiments

System	E_c (eV)	Method
Sn/Si(111) [313]	0.32 ± 0.04	RBS/SEM
Sn/Si(001) [313]	1.0 ± 0.2	RBS/SEM
Ge/Si(001) [284]	1.0 ± 0.1	RBS/TEM
Ga/Si(111) [313]	0.49 ± 0.05	RBS/REM
Ga/Si(001) [313]	0.8 ± 0.07	RBS/REM
Ga/Si(001) [313] 4° miscut	0.8 ± 0.07	RBS/REM
Ga/As/Si(111) [314]	1.23 ± 0.05	RBS/REM
GaAs/Si(001) [315]	0.7 ± 0.4	TEM/SEM
GaAs/Si(001) [315]	1.0 ± 0.1	TEM
Ag/Si(001) [43, 44]	1.0	SEM
Ag/Si(111) [43]	0.65	SEM
Ag/Si(111) [46]	0.6	b-SEI
Ga/GaAs(001) [316]	1.15 ± 0.2	RBS/REM
Ga/GaAs(001) [50]	1.3 ± 0.1	RHEED
Sn/GaAs(001) [317]	1.8 ± 0.3	TEM
O/W(110) [3]	0.16 ± 0.04	RHEED
Ag/W(110) [45]	0.65 ± 0.03	SEM
Ag/Mo(001) [318]	0.9	SEM

(Reprinted from Surface Science Reports, 16, M. Zinke-Allmang, L. C. Feldman, and M. H. Grabow, p. 377, © 1992, with permission from North-Holland Elsevier Science.)

second part of this chapter. The measurements on both Si(111) and Si(100) surfaces allow for a direct comparison of different surface geometries. The resulting clusters for Si(100) are smaller indicating a different affinity to coarsening in the nucleation regime. The equilibrium shapes of clusters on both surfaces also differ with irregular shaped flat clusters of Ag on Si(111) (aspect ratio 0.01–0.04) and a more uniform shape on Si(100) with an aspect ratio of 0.3–0.6.

The same type of studies have been extended to nucleation after microarea deposition. Biased secondary electron imaging (b-SEI) analysis of a 5-ML Ag patch diameter on Si(111) [46, 47] is shown in Figure 6 as a function of temperature at a given deposition rate. The patch size measurement replaces the evaluation of the maximum number of stable clusters, $N_x(T)$. The high temperature regime is described by a steady-state diffusion process limited by the time τ_a, i.e., the average time for a deposited atom to desorb. Temperature properties of patch size growth in this regime are therefore dominated by the desorption and diffusion energies, $E_a - E_d$. The low temperature regime is dominated by nucleation and subsequent capture of adatoms by the islands where the activation energy can be written as $E_d - 3E_b$, again using a two-dimensional hexagonal model for the critical cluster shapes. Note that these type of measurements can be directly compared to surface diffusion measurements based on scanning Auger microscopy (SAM) techniques for postdeposit annealing [48, 49].

Nucleation of two-dimensional islands on surfaces during growth is of relevance in homoepitaxial growth on stepped semiconductor surfaces. In this case, nucleation on terraces is in competition with step flow growth, i.e., that all atoms are accumulated at steps causing these steps to sweep over the surface during growth. The transition between two-dimensional nucleation and step growth is believed to be the origin of the destruction of reflection high-energy electron diffraction (RHEED) oscillations and therefore is used to estimate surface self-diffusion coefficients [50]. These phenomena are investigated for low-temperature deposition on vicinal Si(100) miscut by 2° toward the ⟨110⟩ direction, which leads to stable double-height steps [51]. In a scanning tunneling microscopy (STM) study it was shown (i) that the critical cluster at room temperature is a monomer with the smallest stable cluster a dimer, (ii) that N_x is a function of the deposition rate R and decreases with temperature at a given value of R, (iii) that anisotropic island shapes result, their size increasing with temperature, (iv) that nucleation is not dominated by a defect decoration mechanism, and (v) that island coarsening is not observed below 520 K.

Examples of the anisotropic island shapes are shown in Figure 7 for annealing temperatures between room temperature and 350°C. Each STM scan represents an area of 36 × 36 nm. Shape anisotropy usually indicate anisotropic surface diffusion of adatoms, e.g., due to anisotropic surface reconstruction, or

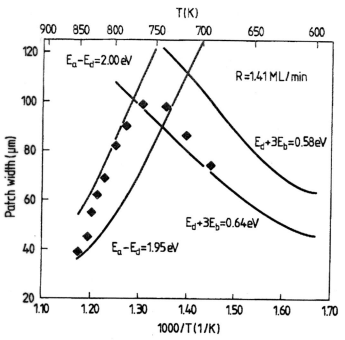

Fig. 6. Patch width as a function of substrate temperature during deposition of 5-ML Ag on Si(111) at $R = 1.41$ ML/min. Model calculations for the indicated parameter values are superimposed. Technique: b-SEI. Reprinted with permission from [46], © 1990, Kluwer Academic/Plenum Publishers.

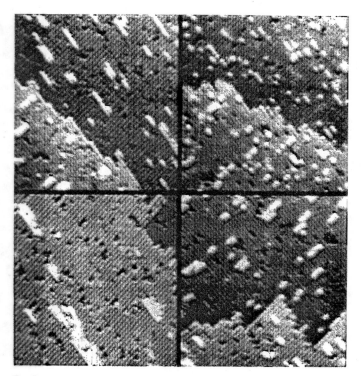

Fig. 7. Micrographs of Si coarsening on Si(100). Top left: deposition at room temperature; top right: 4 min 245°C; bottom left: additional 2 min 310°C; bottom right: additional 2 min 350°C. Terraces are of single atomic height. Island shapes are highly anisotropic. Scale 360 × 360 Å each panel. Technique: STM. Reprinted with permission from [312], © 1989, Elsevier Science.

anisotropic attachment probabilities for monomers approaching the island from specific directions.

1.1.2.3. Developments in Nucleation Theory

Research has focused on the validity of the basic assumptions of the nucleation theory. Theoretical and simulation studies have been undertaken to study deviations arising from alternative assumptions in simplifying the rate equations. In this section, modifications to the kinetic rate equations model are summarized. These modifications focus on (i) systems where monomers represent the critical island (irreversible nucleation), (ii) systems where a single adatom already is stable (surface alloying), (iii) heteroepitaxial systems where defects and steps lead to irreversible nucleation on zero- and one-dimensional subspaces of the surface, (iv) systems where even stable islands are subject to detachment of monomers and the critical island size, which is a central concept in the classical theory, becomes a less useful, and (v) systems with irreversible nucleation of mobile clusters.

(i) *Irreversible Nucleation with Monomers as Critical Islands* ($i^* = 1$). The case $i^* = 1$, i.e., where a monomer already is a critical island is labeled "irreversible nucleation" [52–57] as it does not involve a quasi-steady state between subcritical islands ($j \leq i^*$). Instead, the rate of change of the number of stable islands is dominated by random walk diffusion of monomers when stable islands are assumed immobile. When limiting the discussion to low island densities on the surface (modeled as point islands), i.e., the regime preceding the termination of nucleation due to coalescence events, a set of rate equations is formulated which are quite similar to Eq. (5). However, the new model does not require a formula for intermediate island sizes (second rate equation) as a dimer is already stable. This leads conceptually to a new approach since the change in number of stable islands does not further result from the rate at which monomers are added to critical sized islands but results from the rate of monomers combining with atoms of the deposition flux and due to diffusive encounter of monomers. Bartelt and Evans [52, 54–57] and Bartelt, Tringides, and Evans [53] used a scaling analysis and Monte Carlo simulations to obtain the properties of the system evolving two dimensionally on a perfect crystalline substrate. This simulation approach provides access to the island size distributions which cannot be obtained from the rate equations approach.

The scaling analysis shows that the number of stable islands increases initially as $N_x \propto t^3$ and continues to increase slower at later times with $N_x \propto t^{1/3}$. The number of stable islands further decreases for diffusion rates increasingly dominating over the deposition rate. A scaling is observed with the asymptotic form $N_x \propto (D/R)^{-1/3}$ where R is the deposition rate and D is the diffusion coefficient. The average size of an island $\langle j \rangle$ is related to N_x and depends on D/R in the simulations as $\langle j \rangle \propto (D/R)^{1/3}$. Due to the dependence on diffusion, variations of the results for isotropic and anisotropic diffusion are expected [55, 58–60].

The time dependence of the number of monomers in the irreversible nucleation case is in agreement with low-temperature results for the rate equations in the form of Eq. (5) below the onset of coalescence. At early times, the monomer density increases as $n \propto t$ and decreases later as $n \propto t^{-1/3}$. The island size distribution scales in the range $10^4 \leq D/R \leq 10^8$ as [54],

$$N(j) \propto \frac{\theta}{\langle j \rangle^2} \mathcal{F}\left(\frac{j}{\langle j \rangle}\right) \qquad (6)$$

where $\langle j \rangle$ is the number of atoms in an average sized island. The same scaling form had been found for late stage coalescence growth [61, 62]. The size distributions show a peak at finite size with a steep scale toward larger sizes and a decrease toward smaller sizes extrapolating to a finite number of islands of size zero. The size distributions are broader and have a less distinct peak height than predicted by the rate equations [63] due to neglecting island–island interactions in the rate equation model.

An interesting finding for the irreversible nucleation case is that spatial distributions of island–island neighbor distances are nonrandom [56]. This is due to the competition for monomers which leads to depletions in the number of close island pairs for small island–island distances. Island size distributions should also depend on anisotropy effects in diffusion and sticking as well as on uniaxial strain. It has been suggested that in a strained submonolayer system the total strain can be reduced through formation of small islands and that therefore a uniform distribution of islands forms which do not grow due to increasing detachment from larger islands [65]. This is discussed later when reversible nucleation is considered, i.e., when even stable islands are allowed to release monomers.

Unstrained systems with significant anisotropy were studied by Nosho et al. [66]. In a lattice gas Monte Carlo simulation, (i) the activation energies for monomer migration on the substrate, (ii) the edge diffusion activation energies around an island (the primary cause for island shape anisotropy), and (iii) the interaction energy with neighboring monomers or islands were included with variable values in perpendicular directions. Figure 8 shows rescaled size distributions for a strain-free system with diffusion activation energies varying by 20%, interaction energies varying by 50%, and the attachment probability varying by a factor of 10 for two perpendicular directions. Each of the three panels shows the data collapse for surface coverages between 10 and 25%. The top panel is the scaled size distribution and does not deviate significantly from size distributions of isotropic systems. The second and third panels show the island length in the two perpendicular directions with the inset indicating the absolute lengths for each coverage.

In the simulations, similar island size distributions were obtained for point islands [53] and for square-shaped islands [54, 57]. The advantage to proceed with square-shaped islands is that their finite size allows extending the simulations to larger areal coverages. The model is still restrictive as the islands are assumed to reach the square shape instantaneously during growth. As expected, the coalescence regime can be reached.

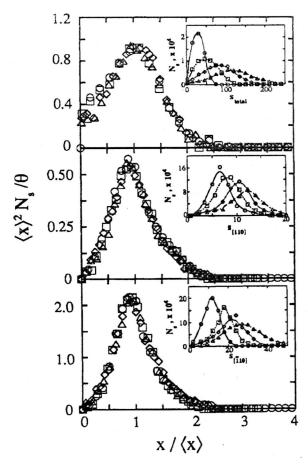

Fig. 8. Island size distributions for irreversible nucleation with anisotropic surface diffusion, anisotropic monomer interaction, and significant difference in monomer attachment rates to islands. Top panel shows scaled size distributions for coverages between 10 and 25%, the two bottom panels show the length distributions in two perpendicular directions. The insets give the simulation data before scaling. Reprinted with permission from [66], © 1996, Elsevier Science.

The size distribution changes significantly to a bimodal distribution with one peak at size zero in the range between 30 and 50% coverage (Fig. 9). The 50% coverage curve is similar to size distributions predicted for late stage coalescence by Family and Meakin as discussed in more detail in Section 1.3 [61].

In a kinetic Monte Carlo simulation [67], the same process was studied allowing for the reshaping of the growing islands. The method is based on a Monte Carlo simulation with pre-set energy barriers for all relevant fundamental processes (such as surface diffusion and edge diffusion) with these barriers used as parameters. The island shape is an important parameter since the assumption of a compact island is often not sufficient when shape instabilities occur [68]. When edge diffusion is hindered, like on metal (111) faces, fractal shapes occur. However, even when the edge diffusion is overcome, growth instability prevails [69, 70]. Usually, the fractal instability is favored at high deposition rates [90]. The simulation by Bales and Chrzan [67] confirms the results of Bartelt et al. for the time dependence of the monomer concentration, the total number of stable islands

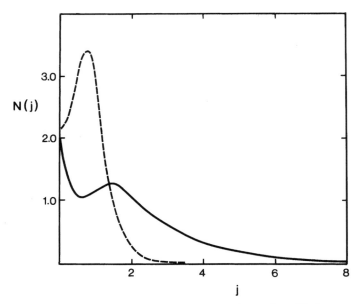

Fig. 9. Island size distributions for a simulation of irreversible, diffusion mediated two-dimensional nucleation with $D/F = 10^6$ at 30% (dashed line) and 50% (solid line) coverage. Size j and number of islands at size j, $N(j)$, are given in arbitrary units. Reprinted with permission from [54], © 1993, Elsevier Science.

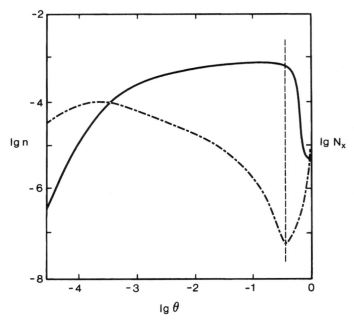

Fig. 10. Logarithmic monomer density (dash-dotted line) and density of stable islands (solid line) for irreversible nucleation ($i^* = 1$) as a function of logarithmic coverage. The data represent a simulation for $D/F = 10^8$, i.e., slow deposition with a constant rate where the coverage dependence is equivalent to the time evolution. The dashed vertical line indicates the onset of coalescence. Reprinted with permission from [75], © 1996, Elsevier Science.

on the surface, and the island size distribution at coverages of up to 15% (i.e., below the coalescence limit). A comparison with the rate equation model leads to good agreement for all averaged properties of the systems while the island size distributions are broadened.

Amar and Family [72] studied irreversible and classical nucleation for spatially extended islands with the critical island size varying between $i^* = 1$ and $i^* = 3$. Island size distributions were obtained from kinetic Monte Carlo simulations. For $i^* > 1$, detachment of monomers was allowed from island sites with less than a certain number of atoms. For $i^* = 1$, the number of neighbors necessary to prevent detachment is one. In addition, an enhanced edge diffusion of atoms with just one neighboring atom was included to allow for reshaping of islands. The resulting island size distributions become narrower for larger critical island sizes and approach zero toward small island sizes when $i^* > 1$. The size distribution for the irreversible nucleation case differs from the results in the preceding studies in that the number of islands with small sizes is significantly reduced. Two subsequent studies, by Mulheran and Blackman [73] and Bartelt and Evans [63] confirmed these results independently based on a Voronoi cell construction.

In a separate kinetic Monte Carlo study [74], the transition temperatures for the increase of the size of the critical island were determined. Deviations were observed between the simulations and the rate equation predictions for metal (111) faces with a triangular lattice and a transition from $i^* = 1$ to $i^* = 2$ as well as for metal (100) faces with square lattice and a transition from $i^* = 1$ to $i^* = 3$. The rate equations tend to underestimate the transition temperature due to neglecting spatial island correlation effects.

The kinetic Monte Carlo simulations were extended to areal coverages up to 100%. Figure 10 shows the evolution of the monomer density (dash-dotted line) and the density of stable islands (solid line) for $D/R = 10^8$ [75, 76]. After a short initial stage (note the logarithmic scale for the coverage θ) where the monomer density dominates, the monomer and the stable island densities develop in good agreement with the point island model. As the coverage increases, the direct aggregation of deposited monomers to islands becomes more likely. Above coverages of 30% coalescence and percolation start leading to a sudden decrease of the number of stable islands. The concurrent recovery of the concentration of monomers is not found in the classical studies. This increase of the monomer concentration is linked to second-layer growth. The sudden onset of the coalescence and percolation regime allows defining a threshold θ_c as indicated by the dashed vertical line in Figure 10.

(ii) *Irreversible Nucleation for $i^* = 0$.* Significantly different predictions for the island size distribution result for $i^* = 0$. This case has been treated [72, 77] as a systematic extension of the simulations for $i^* = 1$ to $i^* = 3$. $i^* = 0$ requires spontaneous freezing of single monomers. The frozen monomer is a stable island of size one and grows into a larger island if mobile monomers get frozen with increased probability when they encounter the stable monomer island. Processes described by this model are metal alloying and surfactant mediated growth of semiconductors. In both studies, simulations were done for $D/R \geq 10^9$ and for coverages in the range of $4\% \leq \theta \leq 30\%$. Scaled size distributions (ordinate given in the

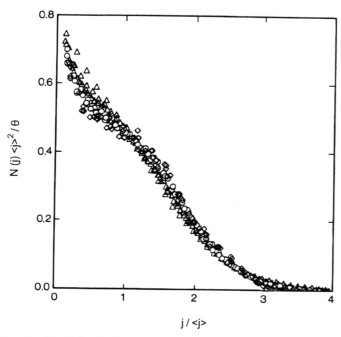

Fig. 11. Island size distribution for $i^* = 0$ for $4 \leq \theta \leq 30\%$ and $D/F \geq 10^9$ from Monte Carlo simulations. (Data taken from Amar and Family [75]. Reprinted with permission from [75], © 1996, Elsevier Science.)

form $N(j)\langle j\rangle^2/\theta)$ are shown in Figure 11. The distributions are distinguished because no peak occurs at finite cluster size.

An interesting intermediate case between $i^* = 1$ and $i^* = 0$ emerges for vicinal surfaces in homoepitaxial systems. When steps are present, an additional sink for monomers is created through irreversible attachment of mobile monomers. Bales [78] studied the case of irreversible nucleation ($i^* = 1$) with both the kinetic Monte Carlo simulations and the rate equation models with a new parameter representing the competition between monomers attaching to the steps and monomers attaching to existing islands or forming new islands. As a function of this new parameter, a crossover from scaling is observed toward a new scaling form in the limit of strong step influence (i.e., monomer depletion due to attachment to steps). Resulting size distributions are quite similar to the case discussed earlier for $i^* = 0$. This may arise from the similarity of the effects of monomer exchange with substrate atoms [77] and the attachment of a monomer to a step. In both cases, the monomer is immobilized and leaves behind a stable structure object to further monomer attachment.

(iii) *Irreversible Nucleation in Reduced Dimensional Subspaces* ($i^* = 1$). The influence of imperfections of the substrate surface, such as steps or random point defects, can vary the nucleation dynamics significantly. Nucleation in heteroepitaxy is affected differently as monomers do not disappear completely when incorporated in a step.

Mulheran and Blackman [79, 80] studied the case of irreversible nucleation on a stepped surface. Monomers cannot detach from steps after diffusive attachment. Subsequent diffusion along the step is possible and leads to nucleation when

a second monomer is encountered or leads to aggregation if a larger island is met. This reduces the irreversible nucleation model to one dimension. Monte Carlo simulations and rate equation calculations have been performed for point islands and circular shaped, extended islands. Their island size distributions show that the use of a finite island shape in the calculations has a noticeable effect. Spatial correlations are demonstrated in the form of nonrandom nearest neighbor distance distributions. Nucleation is not expected to be spatially random if cluster formation is diffusion controlled since the probability to nucleate a new island between existing islands is proportional to the square of the monomer concentration at each position which in turn is determined by the diffusion profile.

Heteroepitaxial nucleation at random point defects has been investigated by Venables [64] and by Heim et al. [81]. The point defects are introduced as traps with an areal density and a trapping energy. A local equilibrium in the form of a Langmuir adsorption isotherm is established on the surface between the density of monomers in traps and monomers on the perfect surface because adsorbed monomers block further monomers from adsorbing at the same site. For large values of the trapping energy, all defects are decorated with a monomer and nucleation is strongly favored at the defect site. A consequence of this model is that the nucleation density is temperature independent over a wide range of substrate temperatures. Depending on the trapping energy relative to competing activation energies, such as the pair bond energy and the surface diffusion activation energy, this temperature interval may coincide with typical substrate temperatures in nucleation experiments.

(iv) *Reversible Nucleation with Monomer Detachment from Stable Islands.* Ratsch et al. [82] studied the opposite case where two-dimensional island formation is reversible. This scenario is particularly interesting as it contradicts the assumption of a well-defined value for i^* as all islands can decompose. The reversibility of islanding is introduced in the Monte Carlo simulation by steadily decreasing a single pair bond energy barrier E_{pair} for atom detachment from islands. Varying the pair bond energy by a factor of 3 leads immediately to a deviation from the expected behavior from the kinetic rate equations (Eq. (5)) for N_x as the exponent of the term D/R varies continuously between 1/3 and 3/5, the discrete values expected for $i^* = 1$ and $i^* = 3$. This behavior has been interpreted as a crossover between two different stable island sizes [72, 83] but has been used to question the concept of a discrete i^* value by Ratsch et al. [84]. Figure 12 illustrates the resulting size distributions. The data set represents six different coverages between 7.5 and 25%. Thus, reversible nucleation leads to sharper size distributions with a significant reduction of the number of islands of sizes near zero.

A slightly more complicated system was studied by Nosho et al. [66] including strain in their model of a system with significant anisotropy. The strain is incorporated through a modification of the monomer attachment probability in one orthogonal surface direction. The strain coefficient, 0, is given such that no islands wider than 20 monomer units in this di-

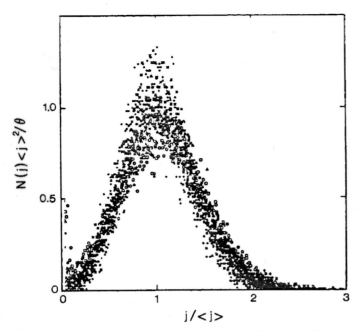

Fig. 12. Island size distributions collapsed for six areal coverages (between 7.5 and 25%) for reversible nucleation. Reversibility is included by varying the pair binding energy in the Monte Carlo simulations. Reprinted with permission from [82], © 1994, American Physical Society.

rection are allowed. This additional strain leads to stronger anisotropy of the island shape. The additional strain effect modifies the size distribution more significantly than the energy anisotropies alone when compared with the isotropic system. Further, scaling is not achieved in the direction of the uniaxial stress where no broadening of the length distribution occurs with increasing coverage.

(v) *Irreversible Nucleation with Mobile Islands.* In many systems, island mobility, particularly dimer mobility at low temperatures where $i^* = 1$, is expected due to low barriers for twisting motion on (100) metal surfaces where one monomer moves around the other in the dimer configuration [85]. Other mechanisms of cluster motion include the motion of dislocations across the island which leads to a significant increase in island mobility at certain sizes [86]. Modifications to the properties of irreversible nucleation have been studied in several simulations [85, 87–89] based on the initial rate equation study by Villain et al. Villain et al. [58] limited the mobility to dimers and derived a modified scaling form for the stable island density where $N_x \propto (D_1/R)^{-1/3}$ is replaced by $N_x \propto (D_1 D_2/R^2)^{-2/5}$, with D_1 the diffusion rate of monomers and D_2 the diffusion rate of dimers. However, for this type of scaling to occur, the rates for three fundamental processes have to be compared: (i) if the dimer immobilization through incorporation of mobile monomers dominates, the dimer mobility does not contribute significantly to the process and scaling of immobile dimers of the irreversible nucleation model is reproduced. (ii) If the dissociation rate of dimers dominates, then the classical nucleation scenario as described by the kinetic rate equations (5) for $i^* > 1$ applies [39, 90, 91]. (iii) The novel

scaling is only expected when both of these rates are smaller than the rate at which dimers aggregate with larger islands as a result of their diffusive motion [89].

For irreversible nucleation with significant dimer mobility, several Monte Carlo simulations have been performed to obtain island size distributions. For hopping rates between 1 and 100% of the monomer hopping rate, size distributions become narrower with a more distinct peak near the average size. In addition, in some studies [88] a separate peak toward small island sizes is reported. Furman and Biham [87] extended the studies to systems with mobile trimers finding an even narrower size distribution.

1.1.2.4. Novel Experimental Aspects in Nucleation

Experimental systems are grouped by substrate material, including (i) element semiconductors, (ii) compound semiconductors, (iii) metals, and (iv) insulators.

(i) *Semiconductor Surfaces: Silicon and Germanium.* Only a few studies connect experimental nucleation data to kinetic Monte Carlo simulations or the rate equation model on semiconductor surfaces. The complexity of these systems leads to a large number of parameters which limit the usefulness of data simulation by numerical methods [64]. In addition to the basic activation energies of adsorption–desorption, diffusion, and island formation which are required in the rate equation model, several new processes play a significant role, including crossing of step edges, diffusion along island edges, reconstructions of the substrate and adlayer islands. Only some of these processes can be studied with STM [92, 93].

The homoepitaxial system Si/Si(100) is of particular interest due to technological relevance. Mo et al. [59, 60] analyzed the number density of Si islands in STM to obtain surface diffusion data for this system. In the temperature range 350 K ≤ T ≤ 500 K, island coalescence did not interfere with the data and dimers are stable, but larger islands showed a significant anisotropy of the two-dimensional island shape (long needle-like structures of mostly dimer-length width). The experimental data were analyzed using a Monte Carlo simulation with random walk diffusion and random deposition with an activation energy of diffusion of $E_d = 0.67 \pm 0.08$ eV and a preexponential factor of $10^{-4} \leq D_0 \leq 10^{-2}$ cm²/s. To model the data properly, a combination of anisotropy of diffusion and attachment to stable islands has to be taken into account. Anisotropic bonding (sticking limited to the ends of islands) is found to affect the model less than anisotropic diffusion (jump rate along substrate dimers 1000 times higher than across the rows, i.e., quasi-one-dimensional diffusion). This anisotropy also alters the relation between island density and diffusion rate [58].

Two other complications arise in the homoepitaxy of silicon. The first problem is associated with steps. Monomers can attach to steps like to large islands. This leads to denuded zones near step edges [60] with widths depending on several factors including Ehrlich–Schwoebel energy barriers [94]. A second complication is the lack of observations of small islands which are slightly larger than dimers. This draws in question whether a dimer is indeed the smallest stable island and establishes the kinetic pathway to larger islands. For a SiGe alloy layer on Si(100) chains of adatom pairs have been observed which are terminated at both ends by buckled dimers [95]. The adatom pairs are not dimers as the distance between the two atoms in the pair is too large. These structures are metastable and convert into stable dimer chains in a transition with a positive activation energy. This process is not affected by isolated dimers concurrently observed on the surface.

Direct experimental evidence exists for critical two-dimensional island sizes of the order of 650 dimers for Si(100) homoepitaxy at 925 K (typical epitaxy growth temperature). In a low-energy electron microscopy experiment (LEEM). Theis and Tromp [96] studied the step-terrace exchange kinetics of a surface growing at $R = 0.0017$ ML/s to $R = 0.033$ ML/s with chemical vapor deposition of silane and disilane. They found nucleation at this growth temperature to occur much closer to thermal equilibrium than previously assumed, with the deposition rate varying the intrinsic monomer concentration on the surface by less than 2%. The data can be extrapolated to lower temperatures at the same deposition rate and can predict a minimal stable island size of a dimer for temperatures below 775 K for $R = 0.017$ ML/s.

Heteronucleation on silicon is often complicated due to chemical interaction with the highly reactive, clean silicon surface, e.g., in form of silicide formation for many metals. Temperature thresholds for chemical processes vary and may interfere with the nucleation experiment. Cobalt, as an example, clusters as a metal on Si(100) at temperatures below 650 K [97] but forms silicide at higher temperatures. At 595 K, reactive epitaxy of cobalt on Si(111) leads to monolayer thick islands of triangular shape which nucleate at the faulted side of the Si 7 × 7 reconstruction and grow in length in multiples of the 7 × 7 cell size [98].

Two studies on Ge(111) surfaces [99, 100] extend previous studies of Ag on Si(111) [101]. The experiments are based on the widening of a silver deposit through a 20 × 100-μm mask, studied by biased secondary electron imaging (b-SEI) and scanning Auger microscopy (SAM). Nucleation starts after a uniform, reconstructed $\sqrt{3}$ structure of 1-ML thickness is formed. The nucleation density as a function of temperature for Ag/Ge(111) in the temperature range from 525 to 725 K is shown in Figure 13 (full symbols) in comparison with a measurement for Ag/Si(111) (open symbols [102]). Both data sets were obtained for similar deposition rates close to $R = 1.3$ ML/min. Note that the analysis of the Ag/Si(111) data led to large critical nucleus sizes, with up to several hundred atoms.

To extract the microscopic activation energies involved in the nucleation process, the broadening at the edges of the deposited patch have been observed. In this regime, an instability of the $\sqrt{3}$ structure for Ag/Ge(111) is observed with the formation of a new (4 × 4) reconstruction with a thickness between 0.25 and 0.375 ML [100]. The patch width for a 6.7-ML deposit varies with temperature as shown in Figure 14 (full circles for

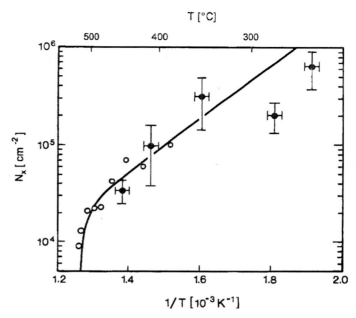

Fig. 13.   Nucleation density for Ag/Ge(111) as a function of inverse deposition temperature (full circles), compared with data for Ag/Si(111) (open circles). The simulation (solid line) is based on a pair bond energy of 0.05 eV, on activation energy for diffusion of 0.4 eV, and on activation energy for adsorption of 2.55 eV. Reprinted with permission from [99], © 1996, Elsevier Science.

$\sqrt{3}$ structure and open circles for 4 × 4 structure, compared with data for a 5-ML deposition of Ag/Si(111), open squares). These data can be modeled with the rate equation model. In addition using a two-dimensional diffusion equation, reasonable values for the activation energy of surface diffusion, the pair binding energies, and the activation energies for adsorption are obtained. The activation energy for adsorption is higher for Ag/Ge(111) which results in a higher monomer concentration.

(ii) *Semiconductor Surfaces: Compound Semiconductors.* Systematic nucleation studies are rare on III–V semiconductor surfaces since most films are grown by chemical vapor deposition (CVD) which is incompatible with most *in situ* analytical tools. Often, studies focus on practical aspects, for example, the control of the island density with ion presputtering to achieve smoother growth surfaces. Wang et al. [103] showed that significantly increased nucleation densities are obtained for Ge on *in situ* cleaved GaAs(110) when the substrate is exposed to a 2-keV Xe beam prior to Ge deposition at 695 K.

The studies selected in this section focus on specific properties of nucleation on GaAs surfaces. Nucleation, coalescence, and layer-by-layer growth for the system Ge on cleaved GaAs(110) was studied with STM by Yang, Luo, and Weaver [104, 105]. The first article focuses on nucleation at 695 K with a deposition rate of roughly $R = 1$ ML/min and a total deposition of 0.2 and 0.4 ML. For both final coverages, the nearest neighbor distance distributions are compared with random simulations in Figure 15. Although the random simulation does not take areal exclusions into account to prevent overlapping of islands, the deviation from the simulation is sufficiently

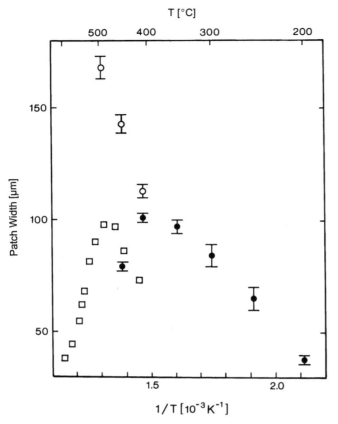

Fig. 14. Width of a deposited patch of Ag on Ge(111) (open and full circles) and Ag on Si(111) (open squares). The Ge(111) data split in two branches with the open circles linked to the 4 × 4 reconstruction and the solid circles linked to the $\sqrt{3}$ surface structure. The Ag coverage for Ge(111) is 6.7 ML and the coverage for Si(111) is 5 ML. Reprinted with permission from [99, 102], © 1992, Elsevier Science.

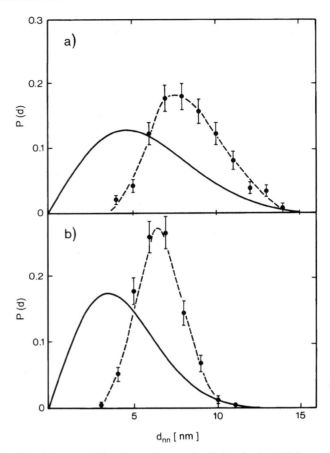

Fig. 15. Nearest neighbor cluster distance distribution for (a) 0.2-ML coverage and (b) 0.4-ML coverage of Ge on cleaved GaAs(001) at 695 K. The dashed line connects the data points and shows a significant depletion of shorter cluster distances in comparison to a random simulation (solid line). The random simulation did not include an areal exclusion provision to avoid cluster overlap. Reprinted with permission from [104], © 1992, American Physical Society.

large to indicate that nucleation does not occur random, rather with depletion zones around existing clusters. This is due to the fact that an adsorbed monomer attaches easier to an existing nearby cluster instead of nucleating a new cluster with other free monomers. These data do not allow an unambiguous measurement of the depletion zone width, however, as parameters such as the mobility of small clusters are not known. The depletion zone around clusters is easier to be seen for clusters nucleating at substrate steps. Preferred nucleation at steps leads to increased cluster densities at the steps with corresponding depletion zones of the order of 10-nm width on the adjacent terraces. Nucleation at steps is only kinetically favored since cluster ripening during postnucleation annealing at 825 K leads to a reduction of the cluster density at steps [104].

Homoepitaxial nucleation has been studied for metalorganic vapor-phase epitaxy (MOVPE) for GaAs by Kasu and Kobayashi [106]. The islands have anisotropic shapes (factor of 2 longer in $\langle 110 \rangle$ direction) due to a difference in the attachment probability for mobile monomers of a factor of 3 for the two perpendicular directions. In addition, a variation of the width of denuded zones near step edges was observed, indicating a significant role of steps with different attachment probabilities

(favoring the descending steps for monomer incorporation by a factor of 10–300).

Bressler-Hill et al. [107] studied island scaling for InAs growth on an As rich GaAs(001)-(2 × 4) surface at 725 K. Total coverages ranged from $\theta = 0.15$ ML to $\theta = 0.35$ ML. Data analysis was done on areas with low step density and parallel to the $\langle 110 \rangle$ direction as steps along $\langle 110 \rangle$ (B-type steps) influence the island shape while steps perpendicular to $\langle 110 \rangle$ (A-type steps) have no influence. The island size distribution, obtained from STM images, is shown in Figure 16. The size distribution clearly disagrees with the simulation in Figure 8 for an unstrained system, but agreement is also not reached for strained system simulations.

If the same system is grown with chemical vapor deposition (CVD) techniques [108], stable island densities are much smaller than those predicted by the standard nucleation models [91]. Stoyanov [109] predicted such reduced nucleation densities with a modified model based on a two-species nucleation. This model predicts a dependence on the indium flux and the partial arsenic pressure which is observed in the experiments for the InAs island density on GaAs(001).

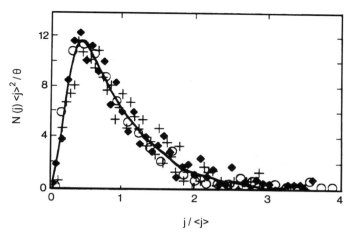

Fig. 16. Island size distributions for nucleation of InAs on GaAs(001) at 725 K studied by STM. The data represent varying coverages (0.15 ML $\leq \theta \leq$ 0.35 ML). The solid lines are smooth fits to the data. Reprinted with permission from [107], © 1995, American Physical Society.

(iii) *Metal Substrates.* Cluster formation has been studied on a wide range of metal surfaces. In most cases, good agreement with the kinetic Monte Carlo simulations and the solutions of the rate equation model has been found.

A data set frequently compared to theoretical models is Fe on Fe(001) whiskers [110]. Fe nucleation occurs irreversibly at room temperature with $i^* = 1$. Scaling of the island size distribution is demonstrated in Figure 17. The island size distribution is shown for the temperature range of 293 to 480 K with $10^6 \leq D/F \leq 10^9$ in Figure 17a. In Figure 17b, the island size distribution is shown for two higher temperatures (at 575 K with $D/F = 8 \times 10^9$ and at 630 K with $D/F = 1.7 \times 10^{10}$). The inset of Figure 17b shows an Arrhenius plot for the stable cluster density, indicating a transition of i^* with $i^* = 1$ for $T \leq 520$ K and $i^* = 3$ above.

These data should be compared with simulations for irreversible nucleation at the lower temperatures ($T \leq 480$ K) and the classical rate equation model for $i^* > 1$ at the higher temperatures, assuming $i^* = 2$ at 520 K and $i^* = 3$ at $T \geq 575$ K [72, 110], as well as with simulations for reversible nucleation [84]. Due to the scatter of the size distribution, a definite choice of the appropriate model is however not possible.

Bott et al. [111] studied the homoepitaxial system Pt/Pt(111) with STM. A comparison with kinetic Monte Carlo simulations is undertaken to provide an independent test of the rate equation model. Figure 18 presents this comparison for the saturation island density versus inverse substrate temperature. The experimental data were obtained with a deposition rate of $R = 6.6 \pm 0.7 \times 10^{-4}$ ML/s and a temperature quench to 20 K immediately following deposition. The kinetic Monte Carlo simulations require just two free parameters, the diffusion coefficient and the diffusion prefactor. The best agreement with the data is obtained for $E_d = 0.26$ eV and $v_0 = 5 \times 10^{12}$ Hz. A further Monte Carlo simulation result from Bales and Chrzan [67] is included in the figure as a dash-dotted line. The two numerical solutions for the rate equations model (dashed lines)

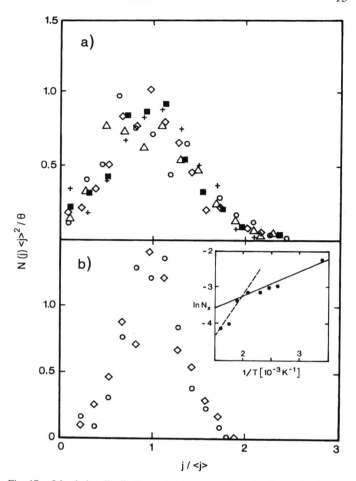

Fig. 17. Island size distribution and island separation distribution for Fe on Fe(001) whiskers. (a) Low-temperature size distributions for $T \leq 480$ K and (b) high-temperature data for $T \geq 575$ K. The inset of (b) shows the stable cluster density as a function of substrate temperature. The solid and the dashed lines indicate $i^* = 1$ at low temperatures and $i^* = 3$ at high temperatures. Reprinted with permission from [52, 53, 56], © 1999, Elsevier Science.

are based on $i^* = 1$ with immobile stable islands and negligible evaporation. These simulations are limited by the accuracy of assumptions for the collision factors δ_{in} [76, 112–115]. The experimental data are also somewhat uncertain as STM tip effects on the mean displacement of surface atoms were observed [111].

Brune et al. [114] investigated Ag on Pt(111) at low temperatures (50–120 K) with STM. The nucleation density was measured as a function of coverage in the range $\theta \leq 14\%$. The variation of coverage leads to an order of magnitude difference in nucleation density. Figure 19 shows the nucleation density as a function of coverage at 75 K. The dashed line indicates the low coverage limit for a model with $i^* = 1$. The solid line is the best fit using the classical rate equation model with the collision terms δ_{in} obtained from the gradient of the monomer concentration at the island surface [39]. The good agreement between rate equation model and experimental data indicates the applicability of the classical rate equation model in this case. The monomer diffusion activation energy for Ag on Pt(111) is de-

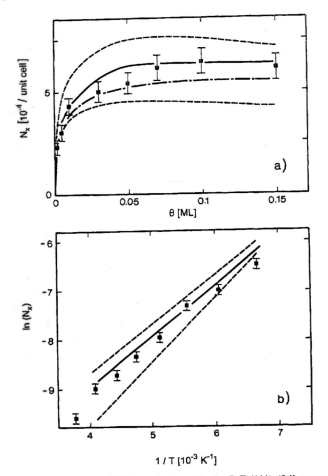

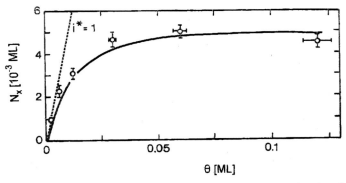

Fig. 19. Ag/Pt(111) nucleation STM images on island density as a function of coverage at 75 K with lines for the low coverage limit for $i^* = 1$ (dashed line) and full data simulation (solid line) based on the classical rate equation model. Reprinted with permission from [114], © 1994, American Physical Society.

Fig. 18. Comparison of STM nucleation data for Pt/Pt(111) (full squares) with kinetic Monte Carlo simulation (solid line) and numerical solutions of rate equation model with $i^* = 1$ (dashed lines) using $E_d = 0.26$ eV and $\nu_0 = 5 \times 10^{12}$ Hz. The rate equation model curves are based on two different assumptions for the capture cross-sections σ. The plots show the saturation island density at 0.1-ML coverage as a function of substrate temperature. Reprinted with permission from [111], © 1996, American Physical Society.

rived as $E_d = 0.16 \pm 0.01$ eV. A molecular dynamic simulation study for the same system was unfortunately limited to substrate temperatures above 400 K [116]. In that regime, the system behaves differently as the deposited atoms are always embedded in a metallic environment as suggested by photoemission measurements [117].

The Ag on the Pt(111) system was further studied by Brune et al. [118] in comparison with Ag on Ag(111) and a single monolayer of Ag on Pt(111), focusing on the influence of strain on the nucleation process. An Ag monolayer on Pt(111) is under 4.2% compressive strain leading to a significant number of dislocations in the film. Dislocations and strain have a significant effect on the nucleation properties of Ag adatoms, e.g., a lowering of the surface diffusion activation barrier (40% compared to Ag/Ag(111)) as derived from the saturation island density as a function of temperature from STM data. In general, reduction of diffusion barriers are expected for compressive strain while tensile strain increases the diffusion barrier. In a

subsequent study, Röder et al. [119] varied the temperature between 130 and 300 K. The temperature increase leads to an increased layer thickness for the initial two-dimensional layer. This behavior is also attributed to the compressive strain of the Ag film which reduces the energy barrier toward interlayer mass transport (Ehrlich–Schwoebel barrier).

The dependence of the average island spacing on the deposition rate has been studied for Cu/Cu(100) by high-resolution low-energy electron diffraction (HRLEED) [120]. The average island spacing, $\langle d \rangle$, is related to the island density, N_x, in the form $\langle d \rangle \propto \sqrt{N_x}$ for two-dimensional nuclei. Thus, the classical rate equation approach predicts $\langle d \rangle \propto i^*/2(i^* + 2)$. The experimental data for island separation versus inverse deposition flux, $1/R$, are shown in Figure 20 for two substrate temperatures and two coverages. In the double logarithmic representation, an $i^* = $ constant behavior is reflected in linear curve segments. For low deposition rates, the low temperature data agree with a slope of 0.165 ± 0.015 indicating irreversible nucleation with $i^* = 1$, while the high temperature data are fitted with a slope of 0.29 ± 0.01 leading to $i^* = 3$ [121, 122].

Another mechanism is proposed for the system Fe/Cu(100) [123–125] studied the room temperature by STM. A monotonously decreasing cluster size distribution with cluster size was found. This suggests a model with $i^* = 0$. The mechanism apparently involves the exchange of an Fe monomer with a Cu substrate atom leading to an embedded Fe inclusion. This mechanism had been proposed initially by Kellogg and Feibelman [126] for Pt on Pt(100). Subsurface growth of nuclei is also observed for Cu on Pb(111) [127].

That monomer exchange with the substrate does not automatically cause $i^* = 0$ as shown in an HRLEED study for Fe on Au(001) [128]. Submonolayer Fe deposition at 315 K leads to a direct exchange with the substrate, displaying a 1×1 LEED pattern for coverages above 0.2 ML [129]. The island size distributions for coverages in the range 0.15 ML $\leq \theta \leq 0.6$ ML were compared with simulations by Amar, Family, and Lam [76] leading to the conclusion that $i^* = 1$ best describes the data. Coalescence terminates nucleation only above 0.6 ML, which has been established for this system through a steep in-

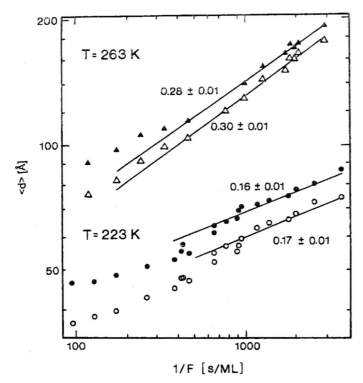

Fig. 20. Average island spacing as a function of inverse deposition rate for nucleation on Cu(100) at two temperatures ($T = 263$ K: open and solid triangles, and $T = 223$ K: open and solid circles) and two coverages ($\theta = 0.3$-ML open symbols and $\theta \approx 0.7$-ML solid symbols). The straight line segments are simulations using Venables' rate equation model for $i^* = 1$ at the lower temperature and $i^* = 3$ at the higher temperature. Deviations from the classical behavior occur at high deposition rates. Reprinted with permission from [121], © 1992, American Physical Society.

crease in the average island size and a corresponding decrease in the density of stable clusters.

A metal system with strong anisotropy effects is Au on Au(100) due to a surface reconstruction with a dense quasi-hexagonal top layer on the square bulk lattice [130]. The density of rectangular islands is found to vary with the deposition flux in the form predicted by the classical rate equation model, i.e., $N_x \propto (D/R)^{-\chi}$ but with the exponent given in the form $\chi = 0.37 \pm 0.03$. Assuming isotropic diffusion, only $i^* = 1$ with $\chi = 1/3$ would be in agreement with the data. The anisotropy of the morphology would still have to be explained with anisotropic edge diffusion and monomer attachment rates. However, Monte Carlo simulations with these parameters predict absolute values for the stable nuclei density of a factor of almost 10 higher than the experimental data [54, 57]. Assuming strongly anisotropic diffusion, $i^* = 1$ leads to $\chi = 1/4$ which does not agree with the experimental data either. Thus, $i^* \geq 2$ is required and $i^* = 3$ leads to the best agreement with the stable island density in the experiment.

(iv) *Insulator Surfaces.* On insulators, defect trapping becomes a major issue in nucleation studies. Heim et al. [81] investigated nucleation of Fe and Co on $CaF_2(111)$ films grown on Si(111) in an ultrahigh vacuum scanning transmis-

sion electron microscope (UHVSTEM). The nucleation density is independent of the deposition temperature in the range from 295 to 575 K and decreases with coverages larger than 20% due to coalescence. The density of nuclei is much higher than for metal systems. These observations have been simulated with the rate equation model modified to include point defects on the surface [64]. The simulations suggest $i^* = 1$ for the studied temperature regime with i^* increasing as incomplete condensation occurs at the highest temperatures. The trap density is roughly 1% of fluor-binding sites on the surface and must interact with the monomers chemically due to the large trapping energy (e.g., fluor vacancies or oxygen–hydroxyl groups). Deposition of Ag on the same surface under the same conditions led to lower nucleation densities in agreement with the standard rate equation nucleation model, i.e., not requiring trapping sites on the surface or allowing only for sites with detrapping energies below 0.1 eV. Point defects in combination with mobility of small islands also affect the spatial island distribution in the Au/NaCl(100) system [131].

1.1.3. Spinodal Decomposition

The nucleation models discussed before apply only to a shallow regime inside the coexistence curve, i.e., the dashed area in Figure 1. At temperatures close to the critical point, the correlation length of density fluctuations starts to diverge and the classical picture of a cluster as a homogenous and sharply confined atom aggregation becomes meaningless. However, no universal model exists for the entire range.

Cluster formation through spinodal decomposition in a mean-field approach is usually illustrated using the Cahn–Hilliard theory [132, 133]. A qualitative discussion of the changes, which are expected when a system crosses over from the nucleation to the spinodal decomposition regime, is given based on Figure 21 [36]. The first panel shows the development of a system toward equilibrium in the metastable limit and the last panel shows the same process in the unstable limit. The second and third panels study the situation adjacent to the spinodal line where the Cahn–Hilliard model predicts diverging critical cluster sizes. In the figure, the local density profiles and the morphological representations are compared. For a classical cluster (Fig. 21a) with the concentration of the system c just inside the coexistence line of Figure 1, the density drops from the value in the well-defined cluster, c_C to the free adatom concentration c_∞ at a sharp edge over a length defined as a coexistence length λ_{coex}. Far in the spinodal regime (Fig. 21d), density fluctuations in the length regime of the order of λ_{coex} occur. Close to the spinodal, the correlation length for the critical cluster (Fig. 21b) and the wavelength of the critical fluctuation diverge.

The reshaping of the cluster surface toward the equilibrium shape may dominate the processes in the spinodal regime [134]. For a two-dimensional unstable fluid of Lennard–Jones atoms, Amar, Sullivan, and Mountain [135] observed spinodal decay of the supersaturation by isothermal molecular dynamics simulations. Characteristic of the end of the spinodal regime is the time dependence of the size of the largest cluster crossing a

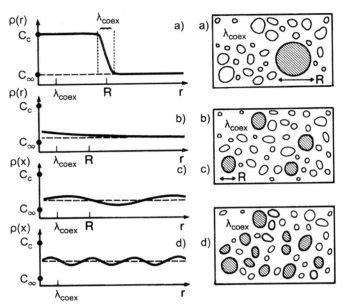

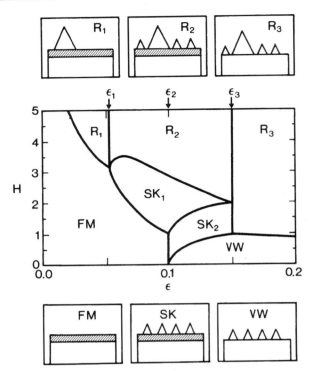

Fig. 21. Schematic of the density profile of thermal fluctuations against which the system is unstable according to the Cahn–Hilliard theory. (a) At densities close to the coexistence curve the unstable fluctuation is a spherical critical cluster. (b) At densities close to the spinodal curve the critical cluster becomes very diffuse with the density in the interior only slightly higher than outside, but the radius diverges. (c) At densities slightly higher than the spinodal density the unstable fluctuation is weak. (d) Far in the unstable region all fluctuations are unstable and grow. Snapshots of the corresponding density fluctuations are shown on the right side. Reprinted with permission from [36], © 1976, Taylor & Francis Ltd.

Fig. 22. Equilibrium phase diagram in a function of the coverage h and the misfit ε. The small panels on the top and the bottom illustrate the morphology of the surface in the six growth modes. The small empty clusters indicate the presence of stable s, while the large ones refer to ripened clusters. The phases are separated by the following phase boundary lines: $H_{c1}(\varepsilon)$: FM-R_1, FM–SK$_1$; $H_{c2}(\varepsilon)$: SK$_1$-R_2; $H_{c3}(\varepsilon)$: SK$_2$-SK$_1$; $H_{c4}(\varepsilon)$: VW–SK$_2$, VW–R_3. The parameters used to obtain the phase diagram are $a = 1$, $C = 40E_0$, $\Phi_{AA} = E_0$, $\Phi_{AB} = 1.27E_0$, $g = 0.7$, $p = 4.9$, $b = 10$, and $\gamma = 0.3$. Reprinted with permission from [143], © 1997, American Physical Society.

maximum. This is due to the wave type of fluctuations forming highly interconnected and thus large clusters. Breaking up of this network leads to smaller clusters which are closer to the equilibrium shape. The authors report an $r \propto t^{1/5}$ dependence of the average cluster size in the spinodal regime for two-dimensional clusters after eliminating short-range order effects.

1.2. Coherent Clustering Morphologies

Coherent clustering in heteroepitaxial systems was first observed at the beginning of the 1990s in the growth of Ge on Si(001) [20, 22], and was later confirmed for other systems (InGaAs [23, 136] and InAs [137] on GaAs(001), and GeSi on Si(001) [138]). Although it affects only a few systems up to now, the coherent clustering attracts a great deal of attention because of the new physical concepts raised, including materials self organization. Also important are the potential applications for growth of nanoscaled structures, e.g., quantum dots (QDs) for electronic and optoelectronic devices [139, 140]. With self-organized clustering, we expect to take advantage of fabricating damage-free QD structures *in situ* without the help of expensive lithographic processes, possibly allowing a higher sample throughput [141]. A significant number of theoretical studies have been published discussing the coherent clustering mechanism based on energetic and kinetic models. In experimental investigations, details of the growth have been explored, e.g.,

the controlling of the size, the shape, and the spatial distributions of QDs.

For heteroepitaxial systems, there are three equilibrium growth modes as discussed in Section 1.1: (i) the Frank–van der Merwe (FM) growth mode (two-dimensional layer by layer growth), (ii) the Volmer–Weber (VW) mode (formation of three-dimensional clusters on the bare substrate), and (iii) the intermediate Stranski–Krastanow (SK) mode (two-dimensional layer growth followed by three-dimensional cluster formation) [13, 14, 142]. A more detailed phase diagram has been developed in a study [143] using an equilibrium theory, distinguishing the main growth modes as a function of the layer thickness and the lattice constant misfit between film and substrate (see Fig. 22). The key features of this phase diagram are: (i) at a lattice misfit ϵ, which is smaller than a critical value ϵ_2 and the equivalent film thickness H (in monolayers) which is smaller than a critical value $H_{c1}(\epsilon)$, the heteroepitaxial system will undergo FM or pseudomorphic growth, with the latter defining a wetting layer for which the lattice mismatch between the epitaxial layer and the substrate is completely accommodated by elastic strain; (ii) for $\epsilon_1 < \epsilon < \epsilon_3$ and H smaller than another critical value $H_{c2}(\epsilon)$, the system grows in the SK mode growth; (iii) for a system with ϵ larger than a certain value ϵ_2, the system grows in the VW mode; (iv) with H larger than certain critical values $H_c(\epsilon)$, cluster ripening occurs in either cases with or without the existence of the wetting layer. (All the critical values shown here are determined through a minimization of the system energy.)

Based on this equilibrium picture, coherent clustering is found to be closely related to the SK growth mode. Three-dimensional clusters appear in both cases on the supersaturated wetting layer as a means of reducing the strain built in the wetting layer as the epilayer thickness reaches a critical value. The critical value is associated with the extension to which the lattice mismatch between the deposited material and the substrate can be accommodated fully by elastic strain (pseudomorphic epilayer). Subsequent deposition leads to a supersaturated epilayer. Growth often continues layer-by-layer, however, as the increasing strain energy in the film disfavors the uniform epilayer energetically, but kinetic barriers prevent the immediate formation of relaxed clusters.

The propensity of the metastable film toward a stabilized structure causes the system to relax to a state of lower total free energy along any possible kinetic path; i.e., any strain relief mechanism for the accumulated strain energy has to be considered. The conventional mechanism is now called the classical SK model, as distinguished by Seifert et al. [144]. When the strain energy is high enough to overcome the energy barrier for the introduction of misfit dislocations, nucleation and growth of three-dimensional relaxed clusters occurs across the surface. Coherent clustering is introduced as an alternative pathway for the strain relief, and it will be observed in systems where the kinetic threshold for the process occurs at a lower film thickness than the threshold for the formation of dislocations. The mechanism is based on a roughening of the film under strain with a subsequent diffusive aggregation of monomers at elevated points representing a lower total strain energy. This transition from two-dimensional layer to three-dimensional coherent clustering is defined [145] as a morphological phase transition, and the origin of the total strain relief in this process was identified by Eaglesham and Cerullo [21] as illustrated in Figure 23. Figure 23a and b schematically compare two possible structures of a commensurate cluster on a lattice mismatched substrate. The sketches indicate two sources of strain relief; first, partial lateral relaxation within the cluster, and second, elastic local adjustment of the substrate lattice parameter. The latter process may accommodate an equivalent amount of strain energy as the cluster itself [146]. Figure 23c is a cross-sectional TEM micrograph of a large coherent Ge cluster on Si(100) grown at 775 K with a deposition rate of 6×10^{-3} nm/s. As shown, the substrate is a bend below the defect-free cluster.

1.2.1. Equilibrium Theories

Ratsch and Zangwill [147] developed an equilibrium theory based on minimizing the total energy of an adlayer system. They used a rigid crystalline substrate and they describe the film such that both elastic and plastic strain can be accommodated (Frenkel–Kontorova model). This model is particularly suited

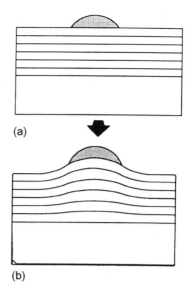

(a)

(b)

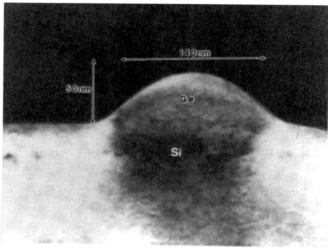

(c)

Fig. 23. (a) and (b) Sketch of a commensurate cluster on a crystalline, lattice mismatched surface. The structure in (a) allows for a partial lattice relaxation within the cluster and therefore represents a system with a lower energy than a strained film of uniform thickness. An additional, local lattice adjustment of the substrate below the cluster can lower the energy of the system further (b). Both effects are observed in cross-sectional TEM for Ge on Si(100) grown at 775 K with a deposition rate of 0.006 nm/s (c). The Ge island is 50-nm high and 140 nm in diameter, representing a coherent island near the threshold for misfit generation. Note the strong strain contrast around the island. Reprinted with permission from [21], © 1990, American Physical Society.

to compare uniform films with coherent and relaxed clusters. For diamond-like materials, the analysis results in a phase diagram allowing for coherent clusters. The stability window for coherent clusters depends on lattice mismatch, surface energy, and total number of particles in the film; in Figure 24 the phase diagram is shown as a function of misfit and total number of atoms in the film with a fixed dimer binding energy of 1.7 eV. A minimum of 2% lattice mismatch is required for the coherent cluster phase (CC), and this phase narrows significantly for misfits larger than 5% due to dominance of the formation of

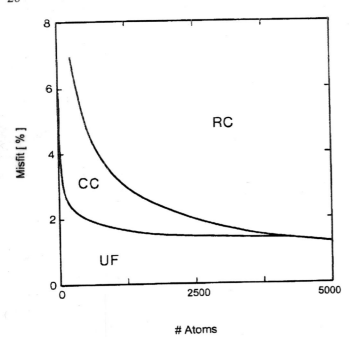

Atoms

Fig. 24. Phase diagram for thin crystalline films grown on a diamond lattice with a dimer binding energy of 1.7 eV. As a function of misfit and a total number of atoms in the film (coverage), three distinct phases have been identified, a coherent cluster phase (CC), a uniform film phase (UF), and a relaxed cluster phase (RC) [192], with only the latter two considered in classical Stranski–Krastanov growth of lattice mismatched systems. Reprinted with permission from [20], © 1999, Elsevier Science.

relaxed clusters (RC). For a misfit of 3%, the coherent cluster phase disappears for binding energies larger than 2.0 eV where the uniform film phase (UF) is stable. There is also an upper limit of the number of atoms in the system for a stable coherent cluster phase.

The picture developed by Ratsch and Zangwill is more detailed since the actual surface energy of relaxed clusters is taken into account, which favors the stability of uniform films or coherent clusters for smaller lattice mismatch or small particle numbers in the film. The authors note however that their model required several simplifications such as neglecting strain nonuniformity in the substrate surface. Thus, only a qualitative comparison with experimental data should be possible.

Shchukin et al. [148, 149] treat the energetics of coherent clustering based on the evaluation of the change of the total energy of a heterophase system, with a uniformly strained film of thickness H_a on a (001) substrate, due to formation of a single cluster. This energy change is written in the form,

$$\Delta E_{isl} = \Delta E_{fac} + E_{ed} + \Delta E_{el} \qquad (7)$$

where ΔE_{fac} is the surface energy of a tilted facet, E_{ed} is the short-range energy of edges, and ΔE_{el} is the change of the strain energy due to an elastic relaxation. With the facet energy included of the minimum of ΔE_{isl} corresponds to an cluster bounded by low index facets, e.g., a pyramid or a prism, essentially defining the shape of the coherent clusters. From the equilibrium elasticity theory [150], the dependence of ΔE_{el} on

the size of an cluster, L, is determined

$$\Delta E_{el} = -f_1(\phi_0)\lambda\epsilon^2 L^3 - f_2(\phi_0)\epsilon\tau L^2$$
$$- f_3(\phi_0)\tau^2/\lambda L \ln\left(\frac{L}{2\pi a}\right) \qquad (8)$$

where f_1, f_2, and f_3 are coefficients which depend on the facet tilt angle ϕ, λ is the elastic modulus, $\phi_0 = \tan^{-1}(k/l)$ with k and l Miller indices for a $(k0l)$ facet. The formula shows that ΔE_{el} also depends on the strain field factors due to the lattice mismatch, $\epsilon = \Delta a/a$, and the discontinuity of the intrinsic surface stress tensor, τ_{ij} at the edges. Equation (8) represents a relation between the elastic energy and the cluster size. A square-based pyramid, or "quantum dot," is energetically more favorable than an elongated prism, or "quantum wire," because it provides for a larger elastic relaxation.

Substituting Eq. (8) in Eq. (7) and using an additional term for the elastic interaction between clusters (anticipating high cluster densities) the cluster size-dependent energy density (per unit surface area) is found

$$E(L) = E_0[-2\chi^{-2}|\ln(e^{1/2}\chi) + 2\alpha e^{-1/2}\chi^{-1} + 4\beta e^{-3/4}\chi^{-3/2}] \qquad (9)$$

where E_0 is the characteristic energy per unit area, $\chi = L/L_0$ is the dimensionless length, and L_0 is the characteristic length. α and β are parameters containing the surface energy and the cluster interaction energy, respectively. The first term in Eq. (9) represents the contribution from the edges to the elastic relaxation energy, the second and third terms represent the contribution from the surface and the cluster interactions. Equation (9) gives a representative expression of the energy density for a coherent clustering system that has widely been cited by later studies. A phase diagram results from this formula based on the energy minimum for variables α and β. A key result of this model is that a minimum of the total energy exists at a finite cluster size L when the elastic relaxation energy at cluster edges exceeds the surface energy. In this case, clusters resume their optimum size and do not have a propensity to grow.

Another equilibrium theoretical model with two-dimensional platelets as a precursor for three-dimensional coherent SK clustering was presented by Priester and Lannoo [151, 152]. This model also provides an explanation for the initial nucleation and narrow size distribution of the coherent clusters. For distant and noninteracting islands, the authors applied the mass action law to obtain a thermal equilibrium distribution in relation to the reduced energy of the system (the energy per atom in the island). This allows islands to grow or to decompose. An absolute minimum of the reduced energy should exist in the equilibrium distribution for physically interesting cases of thermal equilibrium. Based on continuous elasticity theory, the total system energy was studied with a valence force field approach to include the elastic properties of the system. Separately, the surface energy is included. The study allows one to estimate the reduced energy for three-dimensional clusters versus the cluster sizes, in the form,

$$\epsilon(n) = \epsilon_{ps} - A_r + A_s n^{-1/3} \qquad (10)$$

where ϵ_{ps} is the energy per atom of the corresponding pseudomorphic layer, A_r and A_s are two positive constants representing the relaxation and surface energies respectively, and n is the number of atoms in the cluster. By testing this expression with experimental data for heteroepitaxial growth on GaAs, the authors found that the reduced energy does not vary with the surface coverage for clusters larger than 1000 atoms. This led them to conclude that direct formation of three-dimensional clusters cannot explain the existence of a critical coverage and the narrow distribution in the three-dimensional cluster sizes. For a two-dimensional system with noninteracting square platelets, they further determined the reduced energy from classical elasticity theory as a function of the platelet size in the form,

$$\epsilon(n) = \epsilon_{ps} - B_r n^{-1/2} \ln(n^{1/2}) + B_s n^{-1/2} \qquad (11)$$

where B_r and B_s are positive constants representing the relaxation and surface energies, respectively. Based on this expression, a Gaussian size distribution is obtained for the two-dimensional platelets in cases where the equilibrium can be reached, and the resulting distribution curve, together with the two-dimensional and three-dimensional clustering energy, are shown in Figure 25 as a function of clusters size. At coverage smaller than a critical value and cluster sizes smaller than 20,000 atoms, the two-dimensional platelets have a lower energy thus grow as stable prenuclei, and this remains the case as long as the two-dimensional platelets do not interact; when the coverage reaches a critical value, three-dimensional clusters have lower energy and become more stable than the two-dimensional platelets. Thus, the authors expect that the islands undergo a spontaneous transition from two-dimensional platelets to coherent three-dimensional clusters. They conclude further that the corresponding size distribution of the three-dimensional clusters must be narrow as well, like that of the two-dimensional platelets.

Several groups used Monte Carlo techniques to identify the kinetic pathways for coherent SK mode clustering. Such studies were initiated by Orr et al. [153] who first discussed how to incorporate elasticity more realistically in Monte Carlo simulations. In addressing the question of the kinetic pathway, Ratsch et al. [154] simulated thin film growth with Monte Carlo methods as a random deposition on a square lattice. The effects of lattice misfit in the heteroepitaxial system was included through strain relaxation at the edges of clusters and through a reduction of the detachment barrier with an increasing number of lateral neighbors. Neither vacancies nor overhangs were allowed in the direction perpendicular to the substrate surface, thus excluding misfit dislocations and limiting the model to a comparison of uniform film versus coherent clustering. Monomers had decreasing mobility with an increasing number of neighbor atoms to allow for surface diffusion.

Results of the model simulations showed that up to a coverage of $\theta = 0.25$ ML, two-dimensional islanding is observed with the clusters growing in size, covering an increasing fraction of the surface. At higher coverage, three-dimensional clusters form. In this stage, the areal coverage increases only

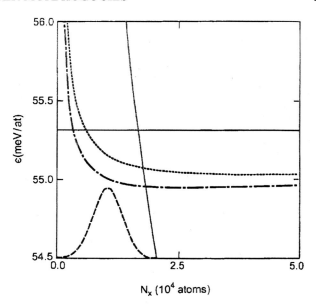

Fig. 25. Representation of the relative variations of three-dimensional (full line) and two-dimensional (dotted and dashed lines) reduced energy curves versus the number of atoms in the island, referred to the corresponding ideally flat reduced energy (horizontal full line). The two-dimensional dashed line corresponds to a coverage lower than the critical coverage, and the two-dimensional dotted line to a coverage greater than the critical coverage. The three-dimensional curve does not depend on the coverage. The chain-dotted curve gives the size distribution for noninteracting two-dimensional platelets at 500°C. Reprinted with permission from [51], © 1995, American Physical Society.

slightly and the additional material deposited (up to 0.5 ML) is used to build the clusters up. This growth in the third dimension continues and clusters of about four layers of thickness are obtained at 1-ML coverage. During the growth of the three-dimensional clusters, some two-dimensional islands dissolve.

Seifert et al. [144, 155] use a picture closer to classical nucleation theory for the transition from the two-dimensional supersaturated phase to three-dimensional clusters. Stress plays an important role in their model as the first nuclei, formed through film thickness fluctuations, lead to a lateral strain profile as illustrated in Figure 26. Within the cluster, the local strain energy density is lowered due to the partial relaxation associated with the finite lateral size of the cluster. In turn, a maximum strain exists at the edge of the cluster which may accelerate the decomposition of the supersaturated layer in the vicinity of the cluster. These strain field predictions have been quantified in a combined analytical and computational two-dimensional continuum model of partially spherical, coherent clusters by Johnson and Freund [156]. The enhanced strain at the cluster edge leads to forced growth of the nuclei with high initial growth rates. The resulting clusters have kinetically controlled shapes, i.e., are bound by low index planes which grow slowest. Like in classical nucleation studies, the temperature does influence the diffusion process and the growth rate of facets, which alter the nucleation density.

The transition to three-dimensional clusters should occur on a timescale much shorter than the deposition process and thus

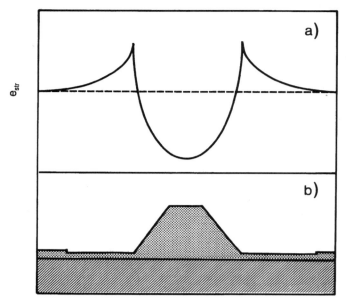

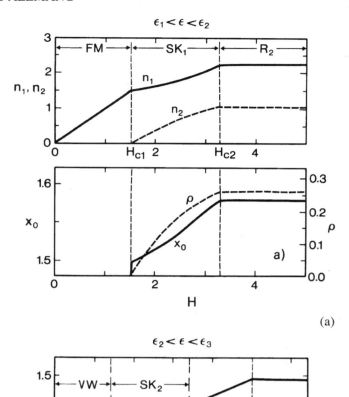

Fig. 26. Sketch of a coherent cluster nucleated within a two-dimensional, supersaturated wetting layer. The top graph shows the local strain energy density profile across the coherent cluster, with strain relaxation inside the cluster and a strain increase at the cluster edge. Reprinted with permission from [144], © 1996, Elsevier Science.

does not directly depend on the deposition rate. However, the deposition rate has been shown to play a major role in experimental studies. This may be explained by reaching a higher supersaturation with a higher deposition rate which in turn may shorten the time period between onset and completion of the nucleation process [155].

In the model given by Shchukin et al. [148], as discussed above, the existence of the wetting layer was neglected in the total energy, thus (i) the actual growth mode, (ii) the cluster sizes and the densities, and (iii) the wetting layer thickness as a function of the deposited material cannot be predicted. These quantities can be measured experimentally, however, with great accuracy. With this argument, Daruka and Barabasi [143] present an equilibrium model for strained heteroepitaxial growth to address these problems. The model incorporates the growth of the wetting layer, dislocation-free cluster formation, and ripening, which results in the phase diagram of strained heteroepitaxial systems as shown in Figure 22. For such a system where H monolayers have been deposited, of which n_1 monolayers are incorporated in the wetting layer and the rest of the material ($n_2 = H - n_1$) is distributed in the clusters, this model expresses the energy density of the system as

$$u(H, n_1, n_2, \epsilon) = E_{\mathrm{ml}}(n_1) + n_2 E_{\mathrm{isl}} + (H - n_1 - n_2)E_{\mathrm{rip}} \quad (12)$$

where $E_{\mathrm{ml}}(n_1)$ is the energy contribution of the strained wetting layer (given as an integral over the binding and the elastic energy densities), E_{isl} is the corresponding contribution of the clusters (in the form of Eq. (7) as given by Shchukin et al. [148]), and E_{rip} is the ripening energy which is obtained from the limit of infinitely large clusters. Minimization of Eq. (12)

Fig. 27. (a) Wetting film thickness (n_1), cluster coverage (n_2) (top), cluster size (χ_0), and cluster density (ρ) (bottom) as a function of H for $\varepsilon = 0.08$. At H_{c1}, there is a transition from the FM to the SK_1 phase, the cluster size jumping discontinuously. In the R_2 phase, present for $H > H_{c2}$, ripening takes place; (b) same as (a) but for $\varepsilon = 0.12$. At H_{c4}, there is a transition from the VW to the SK_2 phase followed by the SK_1 phase at H_{c3}. Finally, at H_{c2} the system reaches the R_2 phase. Reprinted with permission from [143], © 1997, American Physical Society.

with respect to n_1, n_2, and χ (the normalized cluster size as defined in Eq. (9)) determines the equilibrium properties of the system. From this minimization, different growth modes (or phases) are found as shown in Figure 22. In addition the phase diagram, the model also predicts some criteria for the formation of coherent clusters and ripening as shown in Figure 27.

For strain induced clustering, a critical strain ϵ_1 exists such that for any value $\epsilon > \epsilon_1$ stable clusters are possible. On the other hand, for any ϵ value there exists an upper limit to the number of layers in the film, H_{c2}, beyond which ripening occurs. Thus, the stability of the coherent clusters is sensitively dependent on the coverage. To grow stable clusters, thin deposition layers must be used. For small coverage and small values of ϵ, the formation of the pseudomorphic wetting layer should become a general feature of strained layer formation. The obtained phase diagram shows that the critical thickness of the wetting layer for initial nucleation of three-dimensional coherent clusters decreases with increasing lattice misfit ϵ, in agreement with experimental measurements [143, 157]. Uniform clustering occurs after the critical thickness has been reached, and in the vicinity of the critical thickness, H_{c1}, the cluster density increases linearly with $(H - H_{c1})$. The equilibrium model also shows that unlike the cluster density, the equilibrium cluster size does not increase continuously near H_{c1}, but jumps discontinuously from zero, again in agreement with the results of experimental observations [144, 158].

From the above studies, we emphasize that (i) equilibrium theories are successful in describing the energetics of the heteroepitaxial systems with a two-dimensional to three-dimensional transition; (ii) the classical continuity theory is a good approximation for strained epilayer growth in the initial SK stage and for the coherent cluster growth; (iii) the origin of coherent clustering is attributed to the mismatch strain induced by the lattice misfit between the film and the substrate material, causing the strain energy in the initial two-dimensional wetting layer to overcome the barrier for formation of three-dimensional coherent clusters which is observed as a two-dimensional to three-dimensional SK mode transition. Two-dimensional platelets are possibly a precursor stage; (iv) the surface facets of the coherent clusters, their edges, and the elastic strain within the cluster establish additional energies which determine the shapes and the sizes of the three-dimensional coherent cluster.

1.2.2. Kinetics Considerations

The foregoing discussion indicates that coherent SK mode clustering in heteroepitaxial growth is based on the assumption of a thermodynamic equilibrium: the free energy of a "macroscopic" three-dimensional cluster competing with that of an epitaxial film. On the other hand, kinetic processes such as thermal fluctuation and diffusion, the existence of possible intermediate phases, and mass transfer in the two-dimensional to three-dimensional transition, must also contribute to the formation of coherent cluster morphologies. Various studies have been conducted [147, 159–163] to address the kinetics issues of the coherent clustering morphologies in heteroepitaxial growth.

In thin film growth, surface diffusion is critically important since it allows the deposited material to be redistributed that may significantly affect the final morphology of the film. For the initial stage of the coherent clustering, such surface diffusion effect may accelerate or delay the nucleation process by allowing sufficient or insufficient material to reach particular positions on the surface. After the initial nucleation, the competition between formation of new nuclei and growth of an existing nucleus may also be altered. Thus, diffusion can affect the critical thickness in coherent clustering and alter the cluster size and density distributions. Snyder, Mansfield, and Orr [159] presented an early work on the kinetics of the coherent clustering morphologies with an experimental investigation and model study for the $In_xGa_{1-x}As/GaAs(001)$ system. In the study, they demonstrated a temperature dependence of the critical thickness for the coherent cluster formation. STM and RHEED measurements showed that at 520°C, the critical thickness for the two-dimensional to three-dimensional transition increases with decreasing lattice misfit. At a fixed lattice misfit, the critical thickness increases with decreasing temperature, and the transition from the two-dimensional wetting layer to the three-dimensional clusters becomes more gradual as shown in Figure 28. Thus, a suppression of the two-dimensional to three-dimensional transition occurs at lower temperatures. In addition, they observed that planar films grown at moderate and low temperatures are unstable against cluster formation at subsequent high-temperature annealing, in agreement with the theoretical prediction that the strain-relieved three-dimensional cluster morphology is the lower energy configuration. A model incorporating the surface diffusion was proposed in which the kinetic diffusion barriers delay or hinder the morphological transformation to coherent clusters at lower temperature. This model uses the surface diffusion length $\sqrt{D\tau}$ as a characteristic length on the two-dimensional surface $L = \sqrt{D\tau}$, where τ is taken as the time to deposit a monolayer. The volume of the deposited material, $V = Ld$, either in the form of a uniform film with thickness d or in a cluster, raises the system energy due to the strain and the surface energies, which is expressed as [159],

$$\frac{E}{V} = \kappa \epsilon^2 f(x) + \gamma_{FV} \sqrt{\frac{x}{V}} \qquad (13)$$

where κ is the bulk modulus, ϵ is the misfit, γ_{FV} is the surface tension, x is the aspect ratio of the cluster height to width (with $x = 0$ for a uniform film), and $f(x)$ is an empirical function incorporating the strain relaxation. In this equation, it is assumed that elastic strain is built in both the substrate and the clusters. Minimizing this system energy with respect to the cluster aspect ratio and comparing it to the energy of the uniform film, a critical thickness criterion is obtained ($d_C \cong \gamma_{FV}^2/\kappa^2\epsilon^4 L$), at which the aspect ratio jumps from zero to a finite value, thus predicting the formation of coherent clusters. From this study, it is clear that the critical thickness is inversely dependent of the diffusion length, and consequently, dependent of the growth temperature in the same fashion. Further, with decreasing misfit or increasing surface tension, the onset for coherent clustering increases, which is consistent with the energetic models.

In addition to the surface diffusion, mass transfer in the two-dimensional to three-dimensional transition of the coherent clustering has been reported by Ramachandran et al. [160]. Considering the experimental results for both Ge/Si [138, 161] and InGaAs/GaAs [136, 137] systems, a partial

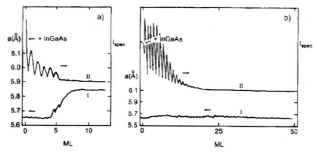

Fig. 28. RHEED data characterizing (a) the high-temperature ($T \cong 520°C$) and (b) the low-temperature ($T \cong 320°C$) growth of $In_{0.5}Ga_{0.5}As$ (3.5% lattice mismatch) on GaAs(001). Both the intensity oscillations of the specular beam (top curve) and the corresponding surface lattice constant (bottom curve) are plotted as a function of epilayer thickness. Reprinted with permission from [159], © 1992, American Physical Society.

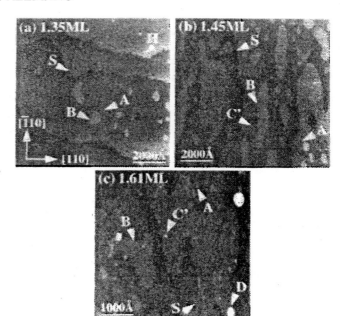

Fig. 29. *In situ* UHVSTM images for InAs depositions on GaAs(001) of (a) 1.35 ML, (b) 1.45 ML, and (c) 1.61 ML. The observed structural features are labeled as follows: A-small two-dimensional islands (1 ML ~ 0.3-nm high, lateral size ≤20 nm), B-large two-dimensional islands (lateral size ≥50 nm), C-small quasi-three-dimensional clusters (0.6–1.2-nm high), D-three-dimensional clusters (~2–4-nm high), S-steps of 1-ML high, and H-holes of 1-ML deep. Note the completely two-dimensional morphology in (a), the appearance of small quasi-three-dimensional clusters in (b), and the presence of all features A–D in (c). Reprinted with permission from [160], © 1997, American Institute of Physics.

breakup of the two-dimensional layer in the formation of three-dimensional clusters is expected. The authors conducted a systematic study of the two-dimensional to three-dimensional morphology change for InAs on GaAs(100) focusing on the material transfer in the process. Using *in situ* STM to obtain the micrographs in Figure 29, they observed four main structures (small (≤20 nm) and large (≥50 nm) two-dimensional islands, small quasi-three-dimensional and three-dimensional clusters) in the morphological evolution from a purely two-dimensional surface to the final three-dimensional structures as the InAs coverage varies (in the range of 1.35–2.18 ML). By plotting the density (Figure 30a) and total volume (Figure 30b) of each type of the structures versus the surface coverage of InAs, they found that the two-dimensional to three-dimensional transition occurs over a range of coexisting two-dimensional islands, small quasi-three-dimensional and three-dimensional clusters. The quasi-three-dimensional clusters only exist as a mediate structure in the transition regime but play a significant role in the material redistribution. They also found that after the formation of the first three-dimensional clusters and below a critical coverage, the total volume of each of the three structures (two-dimensional islands, small quasi-three-dimensional, and three-dimensional clusters) increase with increasing coverage. This indicates that three-dimensional structures initially form as a result of the local strain-driven kinetics but not as a straight shape transition from two-dimensional platelets to three-dimensional clusters, contrary to the equilibrium model prediction given by Priester and Lannoo [151, 152]. Beyond the critical coverage, the total volume of two-dimensional islands decreases accompanying with the increase of the volume of the three-dimensional clusters, indicating a pure mass transfer from two-dimensional islands to three-dimensional clusters. The analysis further shows that with a subsequent increase of the coverage, the small quasi-three-dimensional clusters gradually disappear with their mass redistributed into the three-dimensional clusters, indicating a ripening-like growth of the three-dimensional clusters.

Another kinetic process identified in coherent clustering is related to the so-called cooperative nucleation as studied by Jes-

son et al. [162]. In an experimental investigation of the strain induced two-dimensional to three-dimensional transition in the $Si_xGe_{1-x}/Si(100)$ system, they recorded with atomic force microscopy (AFM) the formation of a ripple morphology through a gentle postdeposition annealing step at about 590°C of a 5-nm thick alloy layer which was grown at a lower temperature. The annealed surface region shows a well-defined and continuous ripple geometry shown in Figure 31a. At a surface region annealed at 570°C, isolated ripple domains, consisting of alternating clusters and pits, were observed with intermittent planar regions of the strained layer as shown in Figure 31b. The latter morphology indicates a direct transformation from a strained planar film to a ripple morphology, while the continuous ripple in Figure 31a is considered to be the coalesced stage of a cooperative nucleation process involving clusters and pits. As the elastic interaction between clusters and pits is negative (they attract each other), the nucleation of clusters and pits in a cooperative manner may reduce the activation barrier for this type of nucleation. With a simple model, the authors estimated the energy change, ΔG_N, which is associated with the nucleation of N pits and clusters for various simultaneous and sequential configurations. The resulting nucleation energy barrier is shown in Figure 32 with the data normalized with the energy barrier ΔG_1 for nucleation of an cluster with $N = 1$. The figure indicates that for simultaneous nucleation of a domain consisting of N pits and clusters the activation barrier increases linearly with N.

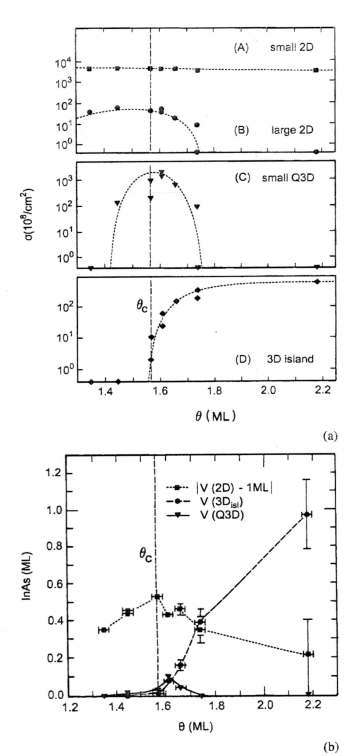

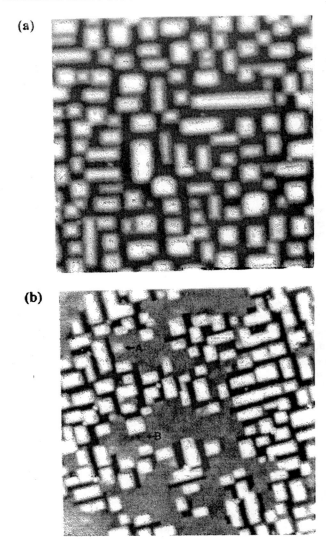

Fig. 31. AFM image of a 5-nm thick $Si_{0.5}Ge_{0.5}$ alloy layer grown on Si(001) and annealed for 5 min at (a) 590°C where a regular surface ripple pattern is formed at the center of the wafer, and (b) 570°C where the cooperative nucleation of surface ripple domains is captured. Clusters (A) and pits (B) nucleate adjacent to each other as indicated, with substantial planar regions of the film remaining in (b). Note that in both images, the cluster and the pits are bounded by (501) facets. Reprinted with permission from [162], © 1996, American Physical Society.

Fig. 30. (a) Density of various InAs structural features on GaAs(001) as a function of InAs delivery, θ. The open symbols indicate a density of $<4 \times 10^7/cm^2$. (b) Total amount of InAs contained in the two-dimensional features, small quasi-three-dimensional, and three-dimensional clusters, denoted as V(two-dimensional)-1 ML, V(quasi-three-dimensional), and V(three-dimensional), respectively. Note the decrease in V(two-dimensional) beyond $\theta = 1.57$ ML, establishing mass transfer to three-dimensional features. Reprinted with permission from [160], © 1997, American Institute of Physics.

Thus the nucleation rate is proportional to $\exp(-\Delta G_N^*/kT)$, i.e., a considerably reduced rate in comparison to nucleation of an cluster of $N = 1$. On the other hand, the activation barrier for subsequent nucleation of a pit adjacent to an existing stable cluster decreases and is approximately constant if the cluster nucleates adjacent to domains of several pits and clusters. Therefore, the authors believe that after the first nucleation event involving an individual pit or cluster, the formation of the ripple morphology proceeds on the surface with a reduced activation barrier via the sequential nucleation of clusters and pits, as observed experimentally. The sequential nucleation is favored over simultaneous nucleation since in the latter case the system gains the elastic interaction energy and surface en-

ergy at the same time for both the pits and clusters while in the former case the system gains the elastic interaction energy only at the expense of the surface energy for either the cluster or the pit, which lowers the barriers.

A similar cooperative nucleation kinetics has been observed by Goldfarb et al. [163] in gas source MBE growth of Ge on Si(100) using GeH$_4$. In the initial stage of the growth, two-dimensional ($M \times N$) patches formed and gradually coalesced, forming a two-dimensional wetting layer with hut-shaped pits on the surface. In subsequent growth at 620 K, they observed new clusters to nucleate at the positions of the hut pits as shown by the STM results in Figure 33. To explain the cluster formation mechanism, the authors incorporated $\langle 100 \rangle$ steps at the hut pit edges as the low barrier kinetic pathway for initial nucleation of the three-dimensional clusters [22], confirming the cooperative nucleation model as presented by Jesson et al. [162]. These results demonstrate that kinetics plays a significant role in the formation of coherent clustering morphologies.

1.2.3. Size and Spatial Distribution Control: Self-Organized Process

The surprisingly uniform size distributions found in coherent SK mode clustering [20, 22, 23, 136–138, 164] are the main reasons why this growth process has been termed "self-organized" and attracts a great deal of interest. With the intrinsic properties of the Stranski–Krastanov growth, ordered spatial distributions of the coherent clustering are also expected in theoretical and experimental investigations. These studies try to identify (i) what is the origin of the uniformity of the cluster size and the spatial distribution in coherent SK mode clustering and (ii) how can energetic and kinetic processes affect the size and the spatial distributions.

In the energy-based approach to coherent SK clustering as given by Shchukin et al. [148], a finite cluster size is deduced as a result of the existence of the total energy minimum when the elastic relaxation energy at cluster edges exceeds the surface energy. Thus, this model predicts a uniform optimum size for the coherent clusters and a stable array of coherent clusters (which do not undergo ripening) at the optimum size. A comparison of this model with experimental observations on a quasi-periodic square lattice of InAs coherent clusters on GaAs(001) shows good agreement. In the equilibrium thermodynamic theory given by Priester and Lannoo [151, 152], in which a two-dimensional platelet precursor allowed very narrow cluster size distributions are believed to be the result of a spontaneous transition from the two-dimensional platelets to three-dimensional clusters. The size distribution of the three-dimensional clusters then reflects that of the platelets, which has a narrow Gaussian distribution.

Seifert et al. [144] addressed, for the transition from a two-dimensional supersaturated phase to a three-dimensional clustering, the relative stability and uniformity in size and shape of the SK clusters based on classical nucleation theory. The authors described, as shown in Figure 26, that as soon as the first clusters are formed, a lateral strain profile is generated

across the cluster, which promotes the reorganization of material on the surface. The clusters form potential minimums on the surface that attract nearby available material to grow on the clusters, until a quasi-equilibrium with the material in the wetting layer is reached. Near the edges of the clusters the increased strain field produces a potential maximum that act as a barrier for diffusion of surface adatoms toward the clusters. With such a barrier, smaller clusters are favored to grow faster than larger ones, allowing the system to reach a uniform cluster size distribution.

This simple picture was confirmed in an investigation by Chen and Washburn [165] based on experimental results for the Ge/Si system. Stress field calculations are used within the framework of linear elastic theory to interpret the results. From an analytic solution of the stress field in a two-dimensional "mound" on a strained semi-infinite substrate [166], the boundary of the mound is defined in Figure 34. Comparing the experimentally observed structure of the Ge cluster from Figure 34b to this model, the authors concluded that the corresponding planar strain obtained from the elastic theory can be used as an approximation to the strain for a three-dimensional cluster. The calculated strain ϵ_{xx} is shown as a contour plot of the absolute strain value as included in Figure 34c. In Figure 34d, the calculated surface tangential strain ϵ_s along the surface of the system is shown. The contour plot clearly shows that the strain ϵ_{xx} caused by lattice mismatch in the cluster is partially relaxed at the cost of inducing an extra strain in the substrate and increasing the strain in the wetting layer near the cluster edge. The tangential strain along the system surface of the system, ϵ_s, is very similar to the local strain energy density around the coherent cluster as suggested by Seifert et al. [144].

With the analytic solution of the elastic theory, Chen and Washburn [165] further deduced the energy barrier in a first-order approximation for subsequently deposited adatoms to diffuse and then to attach to a previously nucleated cluster of radius R, predicting a cluster growth rate shown in Figure 35. Since the height of the energy barrier for adatoms to diffuse to an existing cluster is proportional to R^{2n}, it prevents further growth of larger clusters but favors smaller clusters to increase their size, resulting in a uniform cluster size distribution. In addition, since the energy barrier is also proportional to ϵ_0^2, smaller coherent clusters should be observed for a larger lattice mismatch. For subsequently deposited adatoms to diffuse and to attach to previously formed clusters, an increase of the growth temperature is needed to allow the adatoms to overcome the energy barrier at the cluster edges. These predictions are in agreement with the experimental results [22, 159, 167]. With continuous deposition and growth, further increase of the cluster sizes lead to a monotonous increase of the strain concentration at the cluster edges. When this strain concentration exceeds a critical value and allows the system free energy to overcome the misfit dislocation barrier, the deposited adatoms may occupy a crystallographically "wrong" but less strained position then misfit dislocations are formed around the cluster edge. This process may further relax the strain and may reduce

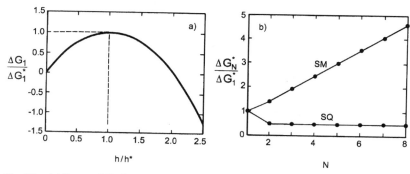

Fig. 32. (a) Energy ΔG_1 for the nucleation of the two-dimensional island ($N = 1$) as a function of height h. To attain a stable geometry, the islands must first overcome the energy barrier ΔG_1^*, which occurs at height h^*. (b) Energy barrier ΔG_N^* associated with the simultaneous (SM) or the sequential (SQ) nucleation of N clusters and pits. Reprinted with permission from [162], © 1996, American Physical Society.

the energy barrier at the cluster edges, leading to resumption of rapid growth of the clusters.

Based on these considerations, the uniformity of the cluster size and spatial distributions in the coherent clustering can be understood in the following way. The lattice misfit induces strain which builds up in the initial two-dimensional epilayer and later in the three-dimensional coherent clusters in a heteroepitaxial system. This strain not only raises an additional energy for the formation of the dislocation free clusters, but also introduces an additional structural factor into the free energy of the system which promotes the ordering in the size and spatial distributions of the coherent clustering morphologies. In a microscopic description, the lattice misfit induced strain produces a potential barrier at the edges of initially formed coherent clusters. Through this barrier, the growth of the clusters is predominately influenced by the system structure thus leading to specific final size distributions.

1.2.4. Shape Transitions: The Role of Additional Issues

Previous actions we have shown that the structural transition of a planar thin film to a three-dimensional cluster morphology is a critical issue to heteroepitaxial growth. In these structural transitions, coherent clustering provides for a pathway for relaxation of structural mismatch, leading to a uniform distribution of cluster sizes. In the described models, the geometry of the critical nucleus and the final clusters plays an important role since all energy terms depend on the geometric configuration. It was already noticed in the first observations of coherent clusters [20, 22] that they can be classified based on their shapes and the corresponding uniform size distribution. Mo et al. [22] reported the small metastable clusters with well-defined shapes of pyramidal huts and distinguished {501} facets on their surface, besides the larger macroscopic clusters, in their study of the SK mode growth of Ge on Si(100) surface as illustrated in Figure 36. This observation has been confirmed by several later experimental studies [168] of the Ge/Si system. Well-defined cluster shapes have also been observed in heteroepitaxial growth of the InGaAs/GaAs systems. Shape related effects have been incorporated into theoretical

studies by Tersoff and Tromp [169] and Tersoff and LeGoues [170] and Jesson et al. [171]. Tersoff et al. treated the growing cluster as a flat top pyramid representing facets and calculated the particular surface and the strain energies corresponding to this cluster shape. Although they did not reach a conclusive result in terms of a particular shape configuration of the coherent clusters, their minimized energy expression involving the geometric relation to stable coherent cluster configurations proved important in studies on shape transitions in the coherent clustering [172]. Jesson et al. studied the strain relaxation mechanism in heteroepitaxial growths in terms of the evolution of surface cusps with a sinusoidal shape. They concluded that strain causes the sinusoidal cusp to grow higher eventually leading to strain relaxation by nucleation of half-loop dislocations at the cusp edges.

In a further study, Chen et al. [173] presented a more detailed model for critical nucleus shapes which predicts the energetic and kinetic pathway of the two-dimensional to three-dimensional transition in coherent clustering. Based on their experimental results, they showed that the initially observed three-dimensional $Si_{0.5}Ge_{0.5}$ clusters do not involve the {501} facets that commonly appear in the Si–Ge heteroepitaxial systems. That is possibly related to the role of atomic steps in the nucleation kinetics. To clarify this point, the authors used surface step interactions in their kinetic model and demonstrated that the resulting kinetic pathway is consistent with the energetics of interacting steps and depends on the relative importance of facets and step energies. With ex situ AFM, an inclination angle θ of the cluster faces was measured as shown in Figure 37. Many smaller but stable clusters have a pyramidal shape with θ significantly below the {501} facet angle of 11°, that indicates a two-dimensional to three-dimensional transition prior to the formation of {501} facets. The cluster faces with small inclination angles are composed of stepped terraces as shown in Figure 38. The energetics of the interacting steps rather than the surface energy of a particular stable facet is believed to determine the kinetic pathway of cluster nucleation.

From these data, the free energy change for the formation of the three-dimensional cluster from a two-dimensional epilayer is calculated. Using ΔG_r as the cluster strain relaxation energy,

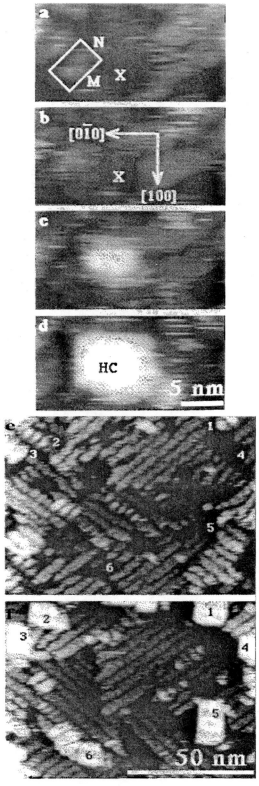

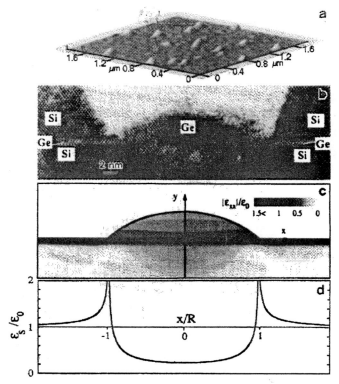

Fig. 34. (a) An AFM image of Ge clusters grown on a Si substrate: (b) A cross-section TEM image showing a Ge cluster on a 6-ML-thick Ge wetting layer on an Si substrate: (c) The boundary of Ling's "mound" and a contour diagram showing the calculated strain ε_{xx} in the system: (d) The variation of surface strain ε_s along the system surface. Reprinted with permission from [165], © 1996, American Physical Society.

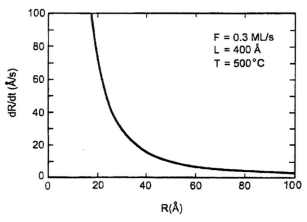

Fig. 35. Cluster growth rate versus cluster radius for a Ge cluster grown on an Si substrate. Reprinted with permission from [165], © 1996, American Physical Society.

Fig. 33. Hut cluster nucleation on void (marked "X") $\langle 100 \rangle$-type edges. (a) The initial void, (b) formation of the $\langle 100 \rangle$-edges, (c) nucleation, and (d) nucleus growing into a hut cluster (HC). Such hut nucleation is observed on typical surface of a Ge wetting layer prior (e) and after (f) hut cluster nucleation at 620 K. Note the correspondence between the 1–6 cluster locations and the 1–6 nucleation sites at surface irregularities in (e). Reprinted with permission from [163], © 1997, American Physical Society.

ΔG_s as the surface energy, the total Gibbs free energy is written as $\Delta G = \Delta G_s - \Delta G_r$. Further using the simple structural model of Figure 38, the surface free energy of a pyramidal cluster for [100] type monoatomic steps can be quantified. Taking the strain energy term from Tersoff and Tromp [169] and Tersoff and LeGoues [170] a contour map of the Gibbs free energy is drawn and shown in Figure 37. The data show a saddle point

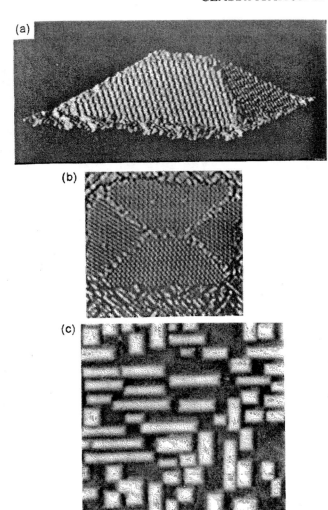

Fig. 36. STM images of Ge hut clusters on Si(001). (a) Perspective plot of a single Ge "hut" cluster. (b) Curvature-mode gray-scale plot. The crystal structure on all four facets as well as the dimer rows in the two-dimensional Ge layer around the cluster are visible. The two-dimensional layer dimer rows are 45° to the axis of the cluster. (c) Hut clusters have rectangular or square bases, in two orthogonal orientations, corresponding to ⟨100⟩ directions in the substrate. Reprinted with permission from [22], © 1990, American Physical Society.

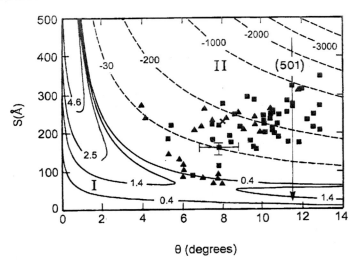

Fig. 37. A distribution of experimentally observed cluster geometries characterized by half-base size s and inclination angle θ. The energy contours $\Delta G(s, \theta)$ in units of electron volts are fitted to the smallest stable clusters, yielding $s_c = 60$ Å and $\theta_c \approx 7°$. Positive and negative energies are represented by solid and dashed lines, respectively. Reprinted with permission from [173], © 1997, American Physical Society.

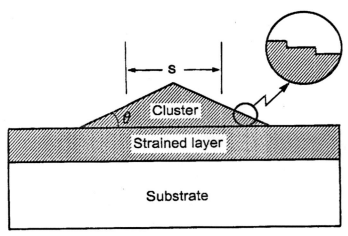

Fig. 38. Cross section of a square-based pyramidal cluster geometry. The three-dimensional cluster is considered as a vertical stack of monolayer islands. Reprinted with permission from [173], © 1997, American Physical Society.

corresponding to a minimum energy barrier. The saddle point represents the most likely pathway for cluster nucleation and defines the critical nucleus shape. The smallest observed clusters in the system appear to be consistent with this view. Over the saddle point, rapid cluster growth with the scatter in clusters shape and size is observed which is considered to be due to steep gradients in the energy surface and to variations in the local kinetics.

In the subsequent growth of the coherent clusters, further shape transitions can occur as studied by Kamins et al. [174] in CVD growth of Ge clusters on Si(100) at substrate temperatures of 550 and 600°C and with Ge coverages up to 20 equivalent ML. A bimodal distribution of the coherent cluster sizes and distinct cluster shapes were observed. Figure 39 shows two entirely different size–shape classes, one composed of clusters of small volume in pyramid shape with {501} facets, and the other composed of clusters of larger volume in dome shape with multiple facets. Both shapes coexist in the bimodal distribution as illustrated in the histogram of Figure 40. The major dome facets are [175] (113) and (102) planes, which form angles of 25.2 and 26.6°, respectively, with the (001) substrate plane. It is observed that both size–shape classes are stable over some range of surface coverages, even for annealing at 550°C for up to 80 min.

In addition to these results, a third structure has been identified and called "superdomes" as shown in Figure 41. A phase diagram for the three different cluster shapes is given in Figure 42 [176]. These results indicate a rather complex interplay of structures that replaces the concept of an idealized SK mode growth and affects the understanding of self-organized coherent

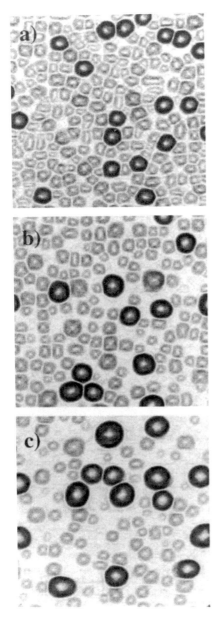

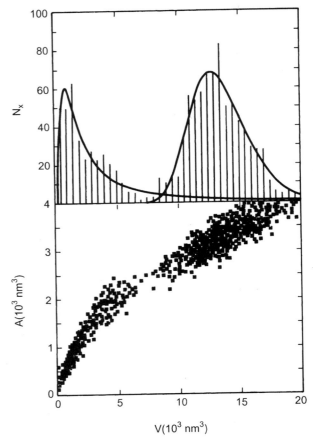

Fig. 40. Histogram of over 1000 nanocrystal volumes determined from topographs (lateral resolution of 2.4 nm) for a surface covered with 10 equivalent ML Ge (top) and the corresponding scatter plot of the surface area of the nanocrystals as a function of their volume (bottom). The scatter plot reveals the two families of nanocrystal shapes. Reprinted with permission from [168], © 1998, American Association for the Advancement of Science.

Fig. 39. AFM images of eight equivalent ML Ge films after annealing times of (a) 0 s, (b) 30 s, and (c) 4800 s. The scan direction is [100]. The gray scale in the images was determined by calculating the projection of the local surface normal onto the growth plane. Darker regions correspond to steeper facets. Reprinted with permission from [175], © 1998, American Physical Society.

growth processes and control of the final cluster sizes. Ribeiro et al. [168] presented an explanation for the bimodal distribution and the shape transition of Ge clusters on Si(001) from pyramids to domes based on the energetic model given by Shchukin et al. [148]. Assuming the surface diffusion is fast compared with the incident flux of Ge, the ensemble of coherent clusters can be considered to be close to equilibrium. Then, the cluster size distribution has a Boltzmann form [176],

$$P(n) = \frac{n_0}{n} P_0(n_0) \exp\left(\frac{(\Delta E(n)/n - \Delta E_0(n_0)/n_0)}{k_\mathrm{B} T} \right) \quad (14)$$

where $P(n)$ is the probability density of finding a cluster with n atoms, $\Delta E(n_0)/n_0$ is the minimum energy per atom of the cluster with n_0 atoms which can be obtained from Eq. (9), and T is the substrate temperature. With this expression, least-square fits to the experimental data are possible for the volume distributions of pyramids and domes shown in Figure 40, and the volume dependence of the pyramid and dome free energies can be obtained. With these results, a qualitative representation of these cluster energies is drawn and illustrated in Figure 43, in which two separate energy minimums are found corresponding to stable pyramids and stable domes. From the observations, it is concluded that the largest pyramid is 20% larger than the smallest dome. Further, the existence of an activation barrier is suggested which is associated with the energy for removing a sufficient number of atoms from a pyramid to form an intermediate structure which can grow into a dome. This allows adding a reaction coordinate to the energy plot in Figure 43. The shape transition from pyramid to dome is explained with an activation energy barrier and an additional assumption that two-dimensional Ge islands exist on top of the wetting layer where they act as a reservoir.

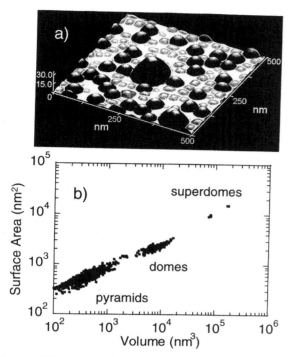

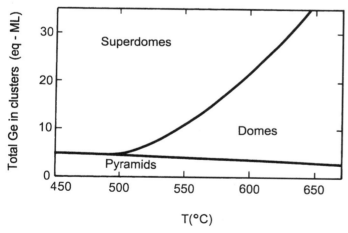

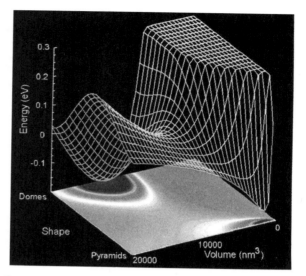

Fig. 41. (a) AFM topograph of a 13 equivalent ML film of Ge deposited onto Si(001) and annealed for 30 min at 550°C. The gray scale is keyed to the local facet angle with respect to the substrate plane, with darker shades corresponding to steeper angles. Pyramids, domes, and superdomes are readily recognized by both their size and their shading. (b) Scatter plot showing the exposed surface area of the clusters versus their volumes on a 1-μm area for the same sample as in (a). Each cluster shape forms a family of points on the graph, allowing the members of each family to be identified for analysis. Reprinted with permission from [176], © 1998, American Chemical Society.

Fig. 42. Shape diagram for Ge nanocrystals on Si(001) determined from experimental data on annealed samples at 8 and 13 ML equivalent coverage at 550°C and 6 and 11 ML equivalent coverage at 600°C. The horizontal axis represents the temperature at which the equilibrium of the cluster distribution is achieved. The boundaries represent the conditions for which two different shapes contain equal amounts of Ge. Any of the three shapes can be found in any of the regions at equilibrium, but the relative amounts of Ge in each shape can vary by many orders of magnitude, making it extremely unlikely to find certain shapes in some regions of the map. Reprinted with permission from [176], © 1998, American Chemical Society.

Fig. 43. A model free-energy surface for Ge nanocrystals on Si(001), plotted with reference to a pseudomorphic two-dimensional island on top of the wetting layer. The shape axis is a reaction coordinate that includes an activation energy barrier to account for the rearrangement of atoms to change the size of the base of the pyramid to that of an equal-volume dome. The saddle point represents the transition state of the shape change, occurring at a volume that is between those of the largest pyramid and the smallest dome. Reprinted with permission from [168], © 1998, American Association for the Advancement of Science.

An alternative model to the coherent cluster shape transition was proposed by Ross, Tersoff, and Tromp [172] based on their observations of the coarsening process which leads to the bimodal distribution of the coherent cluster sizes. With increasing volume, a shape transition from small size pyramids to larger size domes is accompanied by an abrupt change in chemical potential, which does not require a minimum energy configuration: the dependence of the chemical potential on the shape is sufficient to explain the bifurcation and the narrowing of the size distribution. They obtained the temporal evolution of the cluster histogram from real time TEM of Ge clusters as shown in Figure 44. The data show that a coarsening process occurred after nucleation. This coarsening process results in the development of a bimodal size distribution as seen before [168–170], where the larger clusters have a rather narrow distribution. The evolution of individual clusters is shown in Figure 45: within a time period of ~80 s, many clusters in the observed area vanished, while some others developed into two bands of clusters centered at two different diameters. The gray scale histogram also shows that the larger clusters increased their sizes at the expense of small clusters in both their number and size. This indicates a coexistence of the cluster shapes. The observed peak positions of the bimodal size distribution move continuously with time, consistent with the model description given by Ribeiro et al. [168]. Thus, the chemical potential of the clusters is then estimated using an approximated energy of a shallow pyramid [169] which is given as

$$\mu = \frac{\partial E}{\partial V} = \frac{2}{3} \alpha^{4/3} V^{-1/3} - \alpha \tag{15}$$

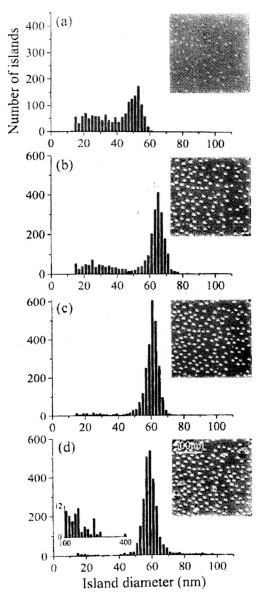

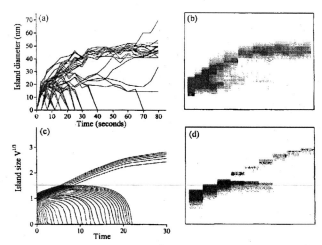

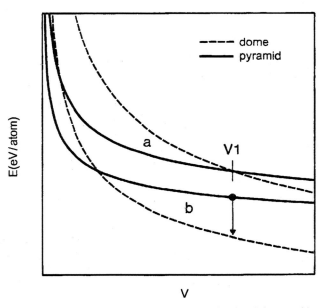

Fig. 44. Images (inset) and histograms obtained during Ge deposition from 2×10^{-7} torr digermane at a substrate temperature of 640°C. Images were obtained (a) 21 s, (b) 51 s, (c) 98 s, and (d) 180 s after "nucleation" (i.e., the time at which distinct strain contrast is first seen). The low contrast from the smallest clusters is responsible for the cutoff at about 15-nm diameter. In (d), the inset histogram shows the very large clusters making up the tail of the distribution. Reprinted with permission from [172], © 1998, American Physical Society.

Fig. 45. (a) The evolution of every cluster within an area of 0.25 μm^2. The cluster showing rapid growth near the end of the sequence has probably formed a dislocation. (b) The same data plotted as a gray scale histogram to show the two branches of cluster size. (c) Simulation showing the fate of clusters with different initial sizes. (d) Histogram of the size distribution in the simulation as a function of time. The model does not include dislocated clusters and therefore at large times it predicts a higher growth rate in strained clusters than is observed experimentally. Reprinted with permission from [172], © 1998, American Physical Society.

Fig. 46. Energy per atom (a) and chemical potential μ (b) of clusters with two types of facets with values of α in the ratio 1 : 2. The latter is calculated from Eq. (15). The shape transition occurs at volume $V1$, where the energy curves cross, at which point there is a discontinuity in μ. Reprinted with permission from [172], © 1998, American Physical Society.

where α is the facet angle and V is the cluster volume. From Eq. (15) it is seen that for clusters with different shapes (different values of α) the volume dependence of the chemical potential differs. Figure 46 illustrates a qualitative sketch of these volume dependences for the pyramids and the domes with α assumed to have the ratio 1 : 2. With increasing volume, the cluster energy crosses over from one volume dependent curve to another at the cross-point V_1 as the cluster assumes the shape with the lowest energy. At this point, the chemical potential undergoes a discontinuous transition.

An additional experimental and model study on the same issue for the SiGe cluster shape transition was given by Floro et al. [177]. The authors measured the surface morphological evolution from pure pyramids to pure dome configuration with sequential scanning electron microscopy (SEM) images. When increasing the equivalent Ge layer thickness, they observed a transition from pure (501) faceted huts to domes at about 13 nm. By measuring the film stress [178] and mean spatial distribution of the clusters [179], they found a correlation between cluster impingement and the shape transformation. With these results, they derived an expression for the areal energy density of the coherent clustering morphology as a function of cluster volume, in which the elastic interaction between clusters is incorporated. The authors concluded that elastic interactions between clusters can significantly reduce the equilibrium transition volume and may also modify the activation barrier for the transition.

1.3. Coalescence and Percolation

A system continues to evolve toward a coalescence morphology when the experimental conditions past the nucleation regime favor merging of clusters. This requires either the mobile clusters or a nonmass conserved system, e.g., with continuous deposition or desorption of monomers. Beysens, Knobler, and Schaffar [180] have characterized three different processes in the coalescence regime, (i) nucleation and individual cluster growth without reduction in the cluster density (transient pre-coalescence regime with detailed discussion in Sections 1.1 and 1.2), (ii) cluster merging, leading to sudden reductions in the substrate area covered by clusters, and (iii) secondary nucleation and cluster growth on freshly exposed fractions of the surface.

As the areal coverage increases in mass nonconserved systems, clusters merge frequently and the cluster size distribution changes significantly. In addition to the merging of clusters due to continuous deposition of material (called static coalescence), an alternative mechanism has to be considered, coalescence due to collision of mobile clusters, which may also occur in mass-conserved systems (called dynamic coalescence) [181]. When clusters merge in nonmass-conserved systems, the fraction of substrate surface covered by clusters is suddenly reduced, allowing secondary nucleation and growth of new clusters to take place. Beysens, Knobler, and Schaffar [180] state that the morphology in these areas evolves in the same way as in the early transient stage.

1.3.1. Fundamental Concepts in Coalescence Growth

In the late stage of coalescence growth, the dynamics of the global cluster distribution and the timescale on which clusters reach their equilibrium shape becomes relevant. Two extremes are usually considered, (i) clusters attaching to each other without any reshaping and (ii) clusters recovering the equilibrium fast in comparison to cluster growth. These processes are discussed in detail for three-dimensional systems, see, e.g., [182, 183].

Theoretical coalescence models usually focus on the case with shape preservation as it is of greater fundamental interest in the context of self-similarity [184]. In experiments, both limiting cases are observed. When clusters do not reshape, it is important to distinguish results from different experimental techniques: microscopic methods, e.g., analyze the areal cluster density with merged clusters as a single cluster [8] while diffraction techniques may treat a merged cluster as several clusters. This has been discussed for Pb on Cu(001) by Li, Vidali, and Biham [185]. In this system, commensurate Pb islands have a high degeneracy and diffract incoherently in most cases.

1.3.1.1. Dynamic Coalescence

We discuss coalescence of mobile clusters first. Cluster mobility has to be defined carefully during late stage coalescence growth since clusters reshape after merging, leading to some degree of motion of the cluster. For immobile clusters, the material motion is limited to reestablishing the equilibrium shape. This does not lead to a change of the center of mass of the cluster, unless additional constraints such as nonplanar substrate structures interfere with the process. Even in the latter case, the cluster should come to rest immediately.

Cluster mobility favors coalescence growth even for mass-conserved systems. Theoretical studies by Binder and Stauffer [186] and Siggia [187] predicted that cluster mobility due to Brownian motion leads to a significant role of coalescence in the clustering process for volume fractions as low as 10% [188]. The microscopic mechanisms leading to Brownian cluster motion are discussed by Khare, Bartelt, and Einstein [189] and Khare and Einstein [190]. The authors proposed three processes which may limit the growth rate of islands, (i) edge diffusion, where mass transport is confined to the periphery of the island, (ii) terrace diffusion, where monomers are rapidly evaporated from the island and their motion is limited by diffusion, and (iii) evaporation–condensation limited island diffusion, where the attachment and detachment barriers for monomers at the island periphery govern the island motion. In all three cases, the island diffusion coefficient depends on the island radius in a power law form.

Cluster mobility can also result when in an otherwise immobile cluster dislocation lines sweep across the cluster. This has been studied by Hamilton [86] showing that two-dimensional islands have enhanced mobility for certain sizes. A third, very system specific mechanism of cluster mobility is discussed by Święch, Bauer, and Mundschau [191] for Au on Si(111). A very high cluster mobility above 600 K is the result of the formation of a liquid eutectic alloy at the cluster-substrate interface.

Four early cluster size distribution predictions for this case are shown in Figure 47 in double-logarithmic representation [37, 192–194]. All four distributions agree on a unimodal distribution with the cluster number density for small cluster sizes usually vanishing. The assumptions for the four curves vary slightly, with Venables et al. (Fig. 47a) assuming a constant deposition rate; a zero deposition rate used in the study of Ruckenstein et al. (Fig. 47b), using three-dimensional clusters on a surface with an effective diffusion rate of a cluster independent of cluster size (dashed line) or varying as $1/r^2$ with r being the radius of the cluster (solid line); Gruber et al. (Fig. 47c) studying a three-dimensional system with an effective cluster diffusion rate proportional to $1/r^2$.

DYNAMIC COALESCENCE

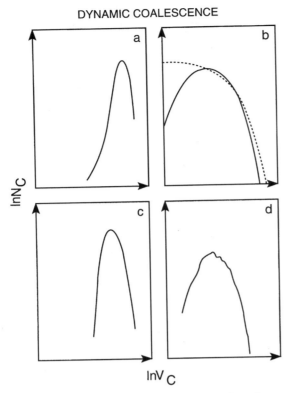

Fig. 47. Theoretical cluster size distribution for dynamic coalescence, i.e., growth of mobile clusters into each other upon impact. Predictions taken from (a) Venables, Spiller, and Hanbücken [37] with $R = $ const, $d_S = 2$, and a nonzero cluster diffusion; (b) Ruckenstein and Pulvermacher [192] with $R = 0$, $d_S = 2$, $d_C = 3$, the diffusion of clusters $\propto 1/r^2$ (solid line) and $\propto$ const (dashed line); (c) Gruber [193] with $R = 0$, $d_S = d_C = 3$, the diffusion of clusters $\propto 1/r^4$; and (d) Meakin [194] with $R = 0$, $d_S = 3$, $d_C = 1$, the diffusion of clusters inversely proportional to the cluster size. Reprinted with permission from [181], © 1992, Elsevier Science.

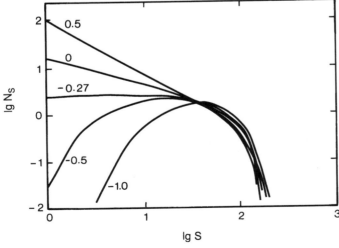

Fig. 48. Cluster size distribution for island growth by dynamic coalescence with cluster diffusivity proportional to cluster size S as $D(S) = D_0 S^{\gamma}$. The five distributions correspond to the range of exponent γ with $-1.0 \le \gamma \le 0.5$ as indicated. Reprinted with permission from [195], © 1985, American Physical Society.

Meakin, Vicsek, and Family [195] studied dynamic coalescence for two-dimensional islands on surfaces with variable cluster diffusivity in a Monte Carlo simulation. Figures 47d and 48 show five cluster size distributions in double-logarithmic representation. The distribution depends on the mobility of larger clusters, which is varied by choosing the coefficient γ in the diffusivity relation $D(S) = D_o S^{\gamma}$ where S is the size of the cluster, i.e., small values of γ correspond to significant mobility of larger clusters. A qualitative change from a unimodal, bell-shaped distribution for highly mobile clusters to a monotonous decreasing size distribution for low diffusivity of large clusters is observed [195, 196]. The same system has been studied by Sholl and Skodje [197].

Steyer et al. [198] report simulations allowing for monomer and cluster mobility, where a random motion was allowed per time step inversely proportional to the radius of the cluster. The simulation leads to a bimodal size distribution, which is similar to the distributions for immobile clusters discussed as follows. In a second simulation, the authors extended the simulation to include competing cluster growth through the attachment of monomers from the vapor phase. The resulting size distribution varied little from the previous case. Steyer et al. find further that the areal coverage in simulations with mobile clusters remains lower than in the simulations for immobile clusters and that the radius of an average, larger cluster grows with a power law with exponent of 0.48. The authors suggest that this exponent approaches $1/3$, the value for early transient growth as clusters become large enough to neglect cluster mobility.

Even though mobility of small clusters has been observed in several studies, systematic experimental studies of dynamic coalescence are rare. A major problem is the dependence of the cluster mobility on the cluster size; as this function is usually not known, a comparison with theoretical models is limited. An example is Rh clustering on various substrates (mica, aluminum oxide, and NaCl) where the mobility of clusters appears to contribute to the observed cluster size distributions which maintained however a weak bimodality [199].

A more detailed STM study for two-dimensional islands in the size range from 100 to 700 atoms is reported for Ag on Ag(100) [200, 201]. The island diffusion coefficients at room temperature are shown in Figure 49 as a function of island size. The diffusivity varies insignificantly with size. The mechanism of island motion seems to be evaporation and condensation of monomers [200], in agreement with a Monte Carlo simulation by Sholl and Skodje [202]. In this postdeposition system, cluster growth can occur either via Ostwald ripening or through dynamic coalescence. For each disappearing small island, the mass transfer to neighboring islands is followed. Ripening processes, as discussed in detail in the second part of the current chapter, lead to a gradual growth of several neighboring clusters while coalescence leads to the sudden growth of a single neighboring cluster. At medium coverages (0.1 to 0.3 ML) dynamic coalescence dominates while both processes compete below 0.1-ML coverage. This result is mainly due to the significant mobility of Ag islands at 300 K [200].

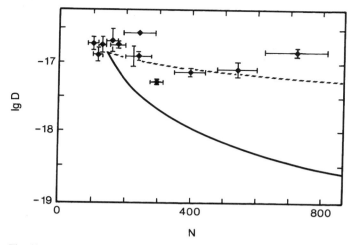

Fig. 49. Diffusion coefficient of Ag islands on Ag(100) at room temperature as a function of the number of atoms N in the island. The dashed line is drawn for $D(N) = N^{-0.5}$, the solid line is drawn for $D(N) = N^{-1.75}$ with both lines arbitrarily passing a common value at $N \approx 150$. Reprinted with permission from [200], © 1994, American Physical Society.

Another study, where mobility of small clusters is suspected, is reported by Søndergård et al. [203] for Sn on 5-nm SiO$_x$ films on amorphous carbon. Sn forms a Volmer–Weber structure on this substrate at 375 K which has been observed with *ex situ* TEM for nominal Sn layer thicknesses ranging from 0.75 to 50 nm. Clusters grown at a constant deposition rate have a bimodal size distribution above 5-nm film thickness. The late stage growth rate agrees well with the static coalescence model of Beysens, Knobler, and Schaffar [180] with an exponent of 0.92 observed for large clusters. However, the exponent at low coverages is significantly higher than the predicted value of 1/3 for static coalescence. The authors [203] suggest that mobility of small clusters, as introduced by Meakin [204] may lead to the deviation.

1.3.1.2. Static Coalescence

Figure 50 shows four cluster size distributions for static coalescence, again showing a double-logarithmic representation of the cluster number density as a function of the cluster volume [28, 37, 40, 205]. Venables et al. predict a bimodal distribution based on the kinetic equations discussed in Section 1.1 in Figure 50a. Cluster size distributions are also obtained in computer simulations by Family and Meakin for a model of random addition of small droplets onto a surface, with coalescence under mass and shape preservation of the clusters (Fig. 50d). The asymptotic distribution shows a power-law decay for small sizes superimposed on a monodispersed, bell-shaped distribution peaked at the mean cluster size. The two other distributions by Vincent (Fig. 50b) and Jayanth (Fig. 50c) differ qualitatively in that they predict unimodal distributions. Note that Jayanth's study treats a three-dimensional case and Vincent's distribution was obtained with a variable deposition rate.

Viovy, Beysens, and Knobler predict a power-law behavior for an average sized cluster (or island) [206]. For three-

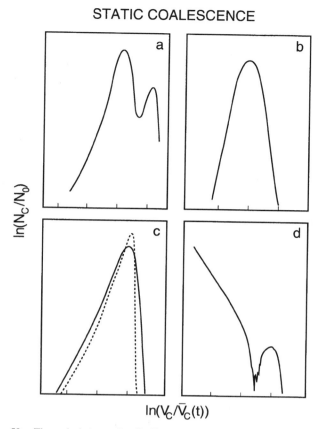

Fig. 50. Theoretical cluster size distribution for static coalescence, i.e., growth of immobile clusters into each other when their perimeter lines touch. Predictions taken from (a) Venables, Spiller, and Hanbücken [37] with $R = $ const, $d_S = 2$; (b) Vincent [40] with $R \neq$ const, $d_S = 2$, $d_C = 3$; (c) Jayanth and Nash [205] with $R = 0$, $d_S = d_C = 3$; and (d) Family and Meakin [61] with $R = $ const, $d_S = 2$, $d_C = 3$. The dashed line in plot (c) corresponds to the Ostwald ripening distribution. Reprinted with permission from [181], © 1992, Elsevier Science.

dimensional clusters on a surface a linear time dependence of the average cluster radius follows. The analysis is based on scaling law arguments using the assumption of scale invariance of the growth process. This result is confirmed by Family and Meakin in computer simulations [28]. The scaling exponent does not vary whether direct impingement or surface diffusion of monomers to the cluster is dominant. In the scaling analysis and the computer simulations, the deposition rate is considered to be constant and the total mass accumulated on the surface can be used to define the timescale. In an earlier study, a monotonously decreasing deposition rate is shown to result in different exponents. Rogers, Elder, and Desai [207] and Viovy, Beysens, and Knobler [208] showed that the power-law exponent for transient, diffusive mass transport limited growth (x), and the overall power-law exponent for growth including cluster merging (y) are related for three-dimensional clusters in the form $y = 3x$. All power-law predictions should apply as long as the total surface coverage remains less than 55% (the jamming limit where percolation growth becomes dominant) [62, 180, 209].

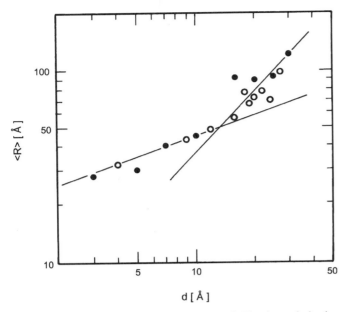

Fig. 51. Cluster radius growth rate for Ag on a-C. The timescale is given as the average film thickness, which is an equivalent measure due to a constant deposition rate in the experiment. The two symbols indicate different stabilizing films deposited after growth and prior to *ex situ* microscopic analysis (open circles for an SiO cap layer and solid circles for an a-C cap layer). Reprinted with permission from [180], © 1990, American Physical Society.

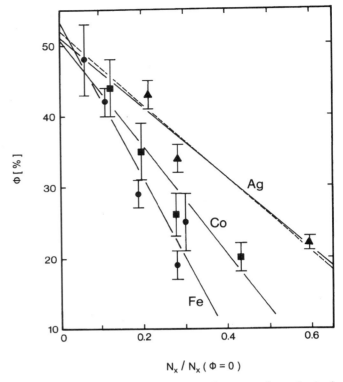

Fig. 52. Plot of experimental areal coverage data versus cluster density for Ag, Co, and Fe on a 10-nm CaF2 film on Si(111). Growth temperatures varied between 295 and 575 K. The intercept at zero cluster density represents the jamming limit. The cluster density is given relative to the extrapolated cluster density at zero areal coverage. The dashed line corresponds to a coalescence-based model by Vincent [40]. Only the Ag data match the model quantitatively. Reprinted with permission from [81], © 1996, American Institute of Physics.

Family and Meakin studied the cluster size distributions for coalescence growth [61, 62, 210]. In Monte Carlo simulations, objects of size one were randomly added to a surface and merged if they touched a previously formed cluster. This model does not allow monomer diffusion [198]. The cluster size distribution in Figure 50d emphasizes a power-law behavior of the decrease of cluster density with size for smaller sizes and a bimodal peak which develops slowly for larger cluster sizes as a constriction between the power-law decrease and the peak becomes more distinct. The first feature of the size distribution to approach the asymptotic self-similar distribution is the power-law decay at small cluster sizes, which is proportional to R^{-5}.

The power-law prediction for the growth of the radius of an average cluster has been tested experimentally [180] using electron microscopy data from Ag on amorphous carbon (a-C) [211]. In the experiments, the Ag deposition rate was kept constant and the deposition time was varied to obtain different coverages. Thus, the average film thickness d represents a timescale. A protective a-C or SiO layer has been deposited *in situ* to stabilize the film. Figure 51 shows the growth rate of an average cluster, $\langle R \rangle$. Until the film thickness reaches 1.5 nm, transient cluster growth dominates with a power-law exponent of 0.38. For later times, growth accelerates due to merging of clusters leading to a power-law exponent of 1.08. Note however the limited range of data points for the late stage as the jamming limit is reached at a film thickness of about 2.5 nm [180]. The power-law exponents are in good agreement with the theoretical values of 1/3 and 1, respectively, [209].

For areal coverages exceeding 20% for Ag, Co, and Fe on CaF$_2$(111), reductions in the areal density of clusters were ob-

served which are consistent with the onset of coalescence [81]. A quantitative comparison is made using a theoretical concept introduced by Vincent [212] linking the cluster density and the areal coverage. Vincent showed that the surface fraction covered by clusters depends on the cluster density and the jamming limit where only one cluster remains. Figure 52 shows the experimental data of Heim et al. [81] which follow the linear dependence predicted by Vincent. However, only the Ag data match Vincent's model (dashed line) quantitatively, with the Co and the Fe data deviating because clusters do not resume their equilibrium shape.

Cluster size distributions during late stage coalescence growth have been obtained for Ga on GaAs(001) at 950 K and for In on InP(001) at 775 K. For both systems, coalescence growth of the metal component was observed since the growth temperature is above the decomposition temperature of the semiconductor, leading to a rapid loss of the group V species [213]. The experimental data for Ga/GaAs(001) are shown in Figure 53 as vertical bars. Note in particular the good agreement with the model prediction of the power-law decrease with size for small clusters [61]. The experimental exponent of 4.8 ± 0.3 matches the theoretical value of 5 well. The bimodal character of the experimental size distribution is less distinct than the theoretical prediction. This indicates that the experimental data are

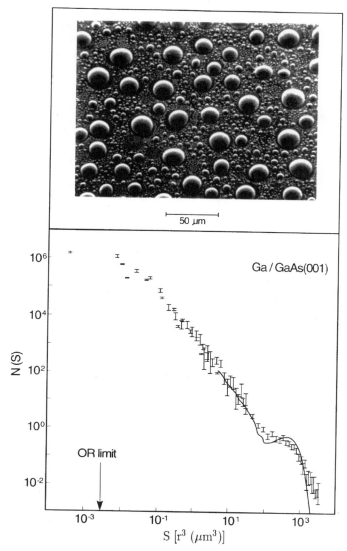

Fig. 53. SEM micrograph and cluster size distribution for Ga on GaAs(001) annealed for 5 min at 935 K. The size distribution is shown double–logarithmic with the size corresponding to the third power of the radius. The solid line is a Monte Carlo simulation of diffusion-less coalescence growth [61]. The experimental distribution is given as error bars. The decrease at smaller cluster sizes follows a power-law $N \propto \gamma^{-\alpha}$ with $a = 4.8 \pm 0.3$ (theoretical value $\alpha = 5$ [61]). Reprinted with permission from [213], © 1992, American Physical Society.

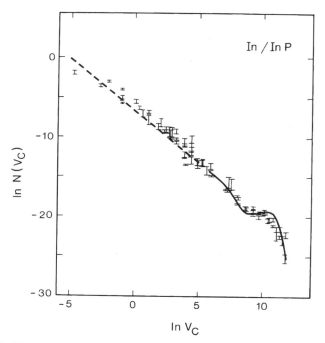

Fig. 54. Experimental cluster size distribution for In on InP(001) annealed at 775 K for 10 min. Data are shown double logarithmic as a function of cluster volume [214]. The solid line is a Monte Carlo simulation for diffusionless coalescence [61] which is extended to smaller cluster sizes based on the power law relation discussed in the theoretical model (dashed line). Reprinted with permission from [61], © 1994, American Physical Society.

taken early in late stage coalescence regime and that additional effects such as surface diffusion interfere with the growth [180, 198].

The In/InP(001) cluster size distribution after annealing of InP at 775 K for 10 min is shown in Figure 54 as a function of the cluster volume V_C [214]. The data again agree well with the simulations for static coalescence (solid line) [61]. In the plot, the simulation curve is extrapolated to smaller cluster sizes (dashed line). The bimodal character is again less distinct.

Yang, Luo, and Weaver studied the coalescence of Ag [215] and Ge [105] on cleaved GaAs(110) surfaces. In the Ge study, deposition of 1–10 ML was followed at 695 K with STM.

Lateral overlayer growth led to static coalescence with an insignificant change in cluster height. The lack of growth in height is the result of the total energy of a cluster which has a flat minimum at six layers. Coalescing of clusters already start at deposition of 0.6 to 0.8 ML, indicated by a reduction of the cluster density. Clusters do not reshape after they merge, rather they form elongated networks. Thus, a self-similar description of the coalescence regime for these samples is not expected. In the study of Ag/GaAs(110), a hydrogen terminated surface was used at 300 K. Ag crystallite coalescence was observed for coverages starting at about 0.5 ML with crystallites bound by {111} facets. In the range between 0.5 and 1 ML, a reduction of the cluster density was observed but the cluster shape was preserved. The necessary Ag transport within the merging cluster is favored by a weak bond energy to the hydrogen terminated substrate and a large cohesive energy of Ag. Above 1-ML coverage, cluster shapes become distorted as clusters interact through competition for monomers.

During coalescence growth clusters do not order spatially. Figure 55 shows the spatial distribution for Ga clusters on GaAs(001) grown at 935 K for 5 min [216]. For the evaluation of nearest neighbor cluster distances in a broad distribution, only clusters above a certain size limit can be included, indicated by the shaded area in the cluster size distribution of the sample (see the inset of Fig. 55). The cutoff includes all clusters in the peak region of the bimodal distribution. Variation of the cutoff size does not alter the observed distribution. The solid

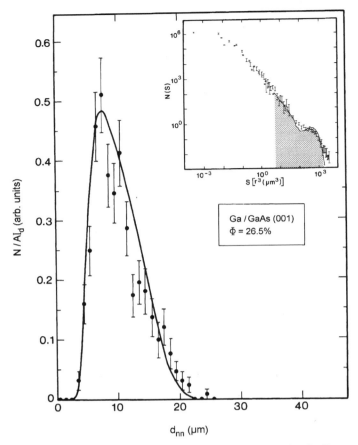

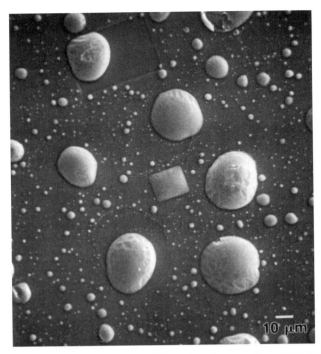

Fig. 55. Normalized nearest neighbor cluster distance distribution for Ga coalescence on GaAs(001) at 935 K. The graph is based on 594 clusters which represent the larger clusters in the cluster size distribution (shaded area in the inset). The cutoff size corresponds to a radius of 16% of the maximum cluster radius. The solid line is a random simulation for the same cluster density with areal exclusion based on the experimental cluster size distribution. The areal coverage of Ga clusters is 26. 5% for the clusters included in the main plot. Reprinted with permission from [216], © 1997, Elsevier Science.

Fig. 56. SEM micrograph of InP(001) surface after annealing at 775 K for 10 min in ultrahigh vacuum. Note dominant fraction of spherical shaped In clusters with roughly 10% clusters associated with rectangular cluster shapes (e.g., rectangular cluster at the center and rectangular depressions near the top left and the top right). Reprinted with permission from [214], © 1994, American Physical Society.

line is a Monte Carlo simulation for a random placement of clusters with the same areal density using the experimental size distribution for an areal exclusion provision. The close agreement with the experimental data indicates that the coalescing clusters are randomly positioned on the surface.

1.3.2. Concurrent Processes in Coalescence Growth and Their Role for Self-Similarity

Statistical self-similarity is a fundamental phenomena which dominates the late stages of phase separation. It has analytically first been established for ripening processes by Lifshitz and Slyozov [217]. Mullins [184] generalized the concept by applying it to other phase separation processes which cannot be described analytically. An example is static coalescence growth, where Family and Meakin's Monte Carlo simulations [61] demonstrate self-similarity.

As self-similarity is applied to a wider range of growth phenomena, it is necessary to define the conditions under which it

holds in detail. Mullins [184] listed the preservation of the cluster shape in cases where sufficient atomic diffusion in and on clusters can occur. A system excluded through that condition is Pt on Al_2O_3 in oxygen atmospheres [218] where the change in cluster shape is associated with a change in the dominant growth process.

A more precise interpretation of this condition is needed if cyclic violations of the cluster shape preservation are considered. First observations of shape oscillations were reported by Tersoff, Denier van der Gon, and Tromp [219] who showed with low-energy electron microscopy (LEEM) that small, facetted three-dimensional Ag clusters on Si(111) at 500 K changed repeatedly between rounded and sharp corners. This effect results from a nucleation-inhibited growth mechanism of the facets and is likely limited to clusters in an early stage of growth, i.e., before self-similar behavior is expected, thus not affecting Mullins' condition.

A second study focused on the late stage coalescence in the system In on InP(001) annealed at 775 K for 10 min. A micrograph of morphological details of this sample is shown in Figure 56. Indium clusters are liquid during growth. The micrograph highlights the following features: (i) about 90% of the clusters are partially spherical with diameters of up to 30 μm; (ii) a fraction of about 10% of the clusters are associated with one or more rectangular features. These features are either clusters with a rectangular contact area to the substrate, e.g., in the center of the micrograph, or are rectangular depressions with a partially spherical cluster located at one of the corners. The clusters follow a shape cycle where in a first step subsurface depressions below clusters form. The formation of rectangular shapes from partial spheres is due to preferential etching of high index crystallographic planes, leading to a depression confined by (100), (111)A, and (111)B surfaces [220]. The cluster in the depression follows the development of a rectangular based shape until further elongation of the rectangular-base becomes energetically unfavorable. At this point, the cluster partially detaches from the depression and retracts to an equilibrium spherical shape. In combination with the self-similar evolution of the cluster size distribution in Figure 54, the In/InP(100) study shows that a self-similar cluster evolution may be maintained even with cyclic cluster shape variations.

1.3.3. Jamming Limit and Percolation Growth

As coverage increases, a coalescing system reaches the jamming limit which is the onset of percolation. The jamming limit is defined as the maximum areal coverage for placing separate clusters on a surface [221]. When jamming occurs during a phase separation process, several physical parameters abruptly change such as the conductivity of a metallic film. In Monte Carlo simulations, the jamming limit [222] is reached for an areal coverage of 54.7% independent of the area of the objects added to the system. However, random sequential adsorption leads to fractal clusters which are only expected in low-temperature experiments where no reshaping via edge diffusion or adsorption–desorption of monomers from the island occurs [223].

Percolation is clearly distinguished from coalescence since the cluster density decreases rapidly with a cluster-network spanning across the surface. Simulations of the formation of a percolating structure based on the rate equations model [39] is complicated as cluster shapes cannot be incorporated but are important for determining the jamming limit. Therefore, most simulations are based on Monte Carlo simulations [223].

Percolation has been studied for increasing coverage in the submonolayer regime for dentritic island growth where only free monomers are mobile [75, 76]. In the simulations, the threshold for percolation, θ_j, depends on the ratio of the diffusion and the deposition rate (D/R) with a minimum value of 56% at $D/R \approx 10^4$ (Fig. 57). At lower values of D/R, first layer percolation coverage remains just above 55%, but significant second layer growth leads to higher total coverages (thresholds up to 70% for $D/R = 10$). Such second layer growth prior to completion of the first coalescing layer has been reported for the first two monolayer depositions of Ge on Si(111) by Voigtländer, Bonzel, and Ibach [224]. If D/R is larger than 10^4, the onset of coalescence is delayed as clusters grow larger with a smaller number density, leading to an increasing percolation threshold (up to 65% coverage at $D/R \approx 10^9$).

Blacher et al. [225] have presented experimental data showing the progressive transition of a coalescence morphology (for

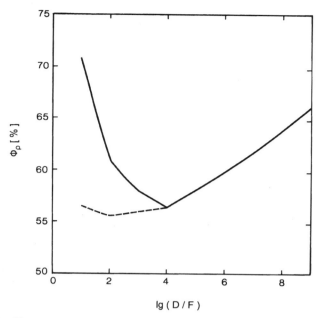

Fig. 57. Jamming limit for submonolayer epitaxial film growth as a function of ratio of diffusion and deposition rate, D/F. The minimum near $D/F \approx 10^4$ results from a significant second-layer monomer concentration at lower monomer mobility (with the dashed line indicating just the first-layer coverage at the jamming limit) and an increase of the coverage for higher mobility due to a delayed onset of coalescence. Reprinted with permission from [75], © 1996, Elsevier Science.

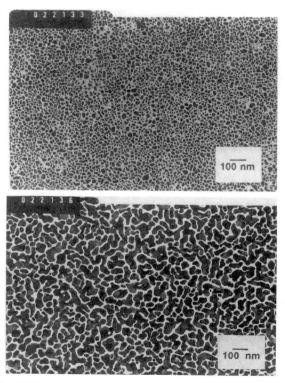

Fig. 58. TEM micrographs of Au coalescence and percolation morphologies on amorphous, polished borosilicate glass. (a) Au deposition at room temperature with final areal coverage of 45% resulting in separate clusters of irregular shape, (b) with final areal coverage of 70% showing highly interconnected Au clusters near the jamming limit. Reprinted with permission from [225], © 1999, American Institute of Physics.

a system without cluster reshaping) toward a percolating structure. Figure 58 shows Au granular films on an amorphous,
polished borosilicate glass deposited at room temperature with
a deposition rate of less than 0.1 nm/s for medium (45%,
Fig. 58a) and high (70%, Fig. 58b) areal coverage. The earlier
morphology shows isolated clusters while the later structure is
highly interconnected.

2. STRUCTURES DEVELOPING AFTER COMPLETED DEPOSITION

Cluster growth in a mass-conserved system, i.e., when no further material is added to the system, evolves asymptotically
toward an Ostwald ripening morphology. However, the asymptotic behavior of the evolution of the system dominates the
morphology only after a transitional period where local effects
determine the cluster structures. The literature distinguishes
therefore an early stage and a late stage, where the early stage
growth corresponds to the transient period.

In the present section, we first provide the underlying theory
for Ostwald ripening, based on the Lifshitz–Slyozov–Wagner
model (LSW theory), modified for surface systems. After this
theoretical introduction, early stage growth is discussed in Section 2.2 and late stage phenomena are presented in Section 2.3.

2.1. Lifshitz–Slyozov–Wagner Model for Clustering on Surfaces

2.1.1. Basic Equations

The term *Ostwald ripening* defines materials system ripening
processes, e.g., the ageing of precipitates in a solution which
Ostwald was the first to interpret in the proper fashion. The
growth of precipitates (coarsening) in a solution is driven by
the Gibbs–Thomson effect which states that a cluster of larger
radius is more stable than a smaller cluster, i.e., that the larger
cluster requires a smaller free concentration in the surrounding
solute to be thermodynamically balanced (monomer loss and
addition rates are equal). The first complete analytical treatment
of this growth model was developed by Lifshitz and Slyozov
and extended by Wagner (referred to as the LSW theory). The
LSW model [217, 226] is generally accepted to describe Ostwald ripening in a uniform three-dimensional solution under
mass conservation for the limit of negligible volume fraction of
the minority phase.

The analytical description of the radial and the time dependence of the cluster size distribution, $f(r, t)$, giving the number
of clusters in the radius interval $(r, r + dr)$ per unit surface area
is based on three main concepts, (i) the Gibbs–Thomson effect,
(ii) a mass transport mechanism (e.g., surface diffusion), and
(iii) the mass conservation.

The driving force for cluster growth is the Gibbs–Thomson
effect (Eq. (1)), describing the solute concentration of the precipitate material as a function of the radius of the precipitate
cluster. Since in most cases the inequality $2\gamma v_C / rkT \leq 0.3$
applies, the second equation of Eq. (1) in Section 1.1 is used

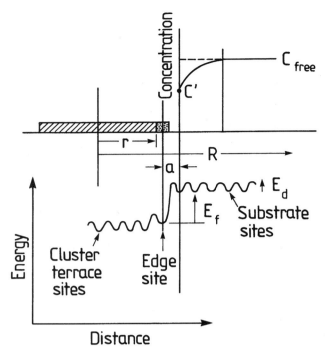

Fig. 59. Superposition of concentration-distance and energy-distance diagrams for two-dimensional clusters on a surface, showing the different energy
levels involved in cluster growth.

to develop the LSW theory. The radii which satisfy this condition vary for different systems, e.g., for metals melting below
1000 K the cutoff radius is estimated as $r \geq 9$ nm. Note that water and organic liquids have much lower surface tensions than
metals and therefore smaller cutoff radii.

In the first step, the Gibbs–Thomson effect is combined with
the formula for the mass transport between clusters. The microscopic processes defining the mass transport mechanisms are
sketched in Figure 59 [227]. When an adatom leaves or enters
a cluster, an energy barrier has to be passed which is given by
E_d, the activation energy for surface diffusion, and E_f, which
is the energy of formation of a cluster, i.e., the energy difference
for an adatom to be in the cluster or outside of the cluster. This
energy determines the ratio of the concentration of adatoms at
cluster edge sites to the theoretical equilibrium concentration of
adatoms given by the Gibbs–Thomson effect. Two mass transport mechanisms, (i) surface diffusion and (ii) passing of the
surface barrier, are operative at the same time with an equal
flux of adatoms crossing both barriers to get a steady-state condition for the mass transport. For all limiting conditions, the
surface diffusion mechanism eventually dominates if the radii
of the clusters grow. Wynblatt and Gjostein [227] quantified
this statement correlating the diffusion coefficient and the kinetic rate constant for the processes in Figure 59.

In the surface diffusion limit, the difference of the actual
concentration at the cluster surface and the average free adatom
concentration per unit area between the clusters is the driving
force for the mass transport. In the three-dimensional case, the
growth rate of a cluster is obtained by coupling the Gibbs–

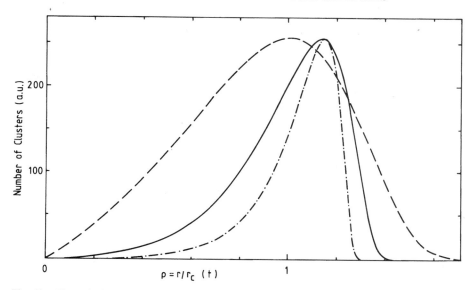

Fig. 60. Theoretical cluster size distributions for ripening processes with different power laws for the growth of the critical cluster with time: $t^{1/2}$ (dashed line), $t^{1/3}$ (solid line), and $t^{1/4}$ (dashed–dotted line). Reprinted with permission from [181], © 1992, Elsevier Science.

Thomson effect and the solution of Fick's law of diffusion for a time-dependent concentration field, $\partial c(r, t)/\partial t = D\nabla^2 c(r, t)$. The diffusion equation has a finite steady-state solution in the form $\nabla^2 c(r) = 0$, leading to $c(r) = A + B/r$ for a single spherical source or sink. With boundary conditions taken (i) at the cluster surface from the Gibbs–Thomson effect, i.e., $c(r) = c(R)$ at $r = R$, and (ii) far from the cluster as a mean-field value, i.e., $c(r) = c_{\text{free}}$ at $r = \infty$, the molar transfer to or from the cluster is equal to the change in the size of the cluster:

$$4\pi RD\big(c_{\text{free}} - c(R)\big) = \frac{1}{v_M}\frac{d}{dt}\left(\frac{4\pi}{3}R^3\right) \qquad (16)$$

In the last step, this formula is combined with the equation of continuity which represents the conservation of material in the system. The model as set up earlier leads to two predictions, (i) a power law for the critical cluster in the form $R_c(t) \propto t^{1/3}$ and (ii) a narrow cluster size distribution shown as the solid line in Figure 60 [217]. The number density of clusters is plotted in the figure versus the radius of the cluster, where the radius axis is renormalized with the critical radius to eliminate the time dependence. The figure shows that most clusters are near the critical size with a steep cutoff toward larger, growing clusters and a tail in the distribution toward small cluster sizes.

The two predictions of the LSW theory are exact for the limit of a zero volume fraction of the minority phase, which corresponds to negligible cluster–cluster interactions. For finite volume fractions, a number of studies predict the same power law but broader and more symmetric cluster size distributions [228–231]. The results of these studies suggest that the diffusive exchange between neighboring clusters depends on their relative positions, an aspect not included in the LSW theory, which is a mean-field model.

The case of island growth on a surfaces (two-dimensional–two-dimensional) or cluster growth on surfaces (three-dimen-sional–two-dimensional) has to be addressed carefully as the phase separation of islands or clusters with diffusion in two dimensions cannot be treated as a simple two-dimensional analogy to the LSW theory (three-dimensional/three-dimensional case). Diffusion has to be dealt with differently since there is no steady-state solution in two dimensions with a finite far-field concentration level as the solution in cylindrical coordinates has the form $c(r) = A + B\ln(r)$ which diverges for $r \to \infty$.

The problem of diverging diffusion fields has been approached in two different ways. Rogers and Desai [232] implemented the proper nonsteady-state solutions into the dynamics. Major simplifications are required in these calculations, including a mean-field assumption ($\lim \Phi \to 0$ where Φ is the volume fraction of the minority phase). In a second approach, the steady-state solution of the diffusion equation is recovered by starting with a finite areal coverage of islands to cut off the diverging concentration field. Unfortunately, the most frequently cited approach by Chakraverty [233] does not treat the cutoff satisfactorily. A more consistent treatment was introduced by Marqusee [234] and Marder [235], starting with the time-dependent form of Fick's law. Marqusee has modified Fick's law to include additional source and sink terms to represent neighboring islands,

$$\frac{\partial c(r, t)}{\partial t} = D\nabla^2 c(r, t) - \frac{D}{\xi^2}c(r, t) + S \qquad (17)$$

with the second term on the right-hand side representing a sink term in the concentration field, containing the screening length ξ. The screening length allows representing the effect of the large number of differently sized islands on the surface with a single quantity as sketched in Figure 61. The screening length represents the distance at which the diffusive interaction with neighboring islands is strong enough so that the concentration gradient around the central island is obscured.

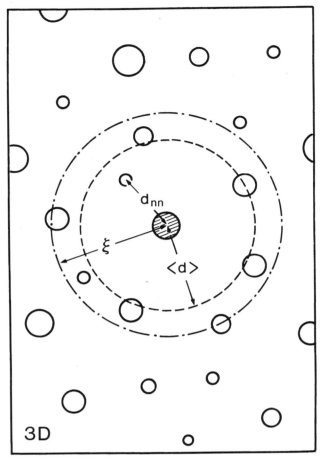

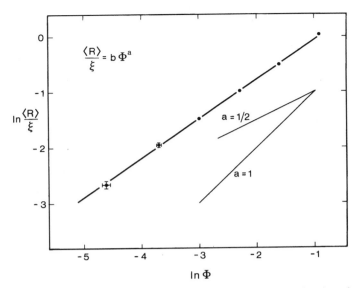

Fig. 62. The screening length as derived self-consistently as a function of areal fraction Φ and the average radius $\langle R \rangle$ by Marqusee [234]. The data are replotted as $\langle R \rangle / \xi$ versus Φ in double-logarithmic representation [242] and are fitted with $\langle R \rangle / \xi = b\Phi^a$ where $a = 0.73 \pm 0.03 \approx 3/4$ and $b = 1.95 \pm 0.15 \approx 2$. As the figure shows, the data exclude $a = 1$ or $a = 1/2$. Reprinted with permission from [20], © 1999, Elsevier Science.

Fig. 61. Sketch of the screening length concept to replace individual cluster–cluster interactions. The screening length ξ corresponds to the distance at which the contribution from neighboring clusters to the monomer concentration is large enough that the influence of the concentration gradient around the central cluster is screened. The screening length is of the order of the next nearest neighbor cluster distance, d_{nn}, or collectively, of the average nearest cluster distance, $\langle d \rangle$. Reprinted with permission from [20], © 1999, Elsevier Science.

tion theory for the local concentration fields to solve the many body problem without an *a priori* assumption of a screening length, (iii) Yao et al. [238–240], and (iv) Akaiwa and Meiron [241].

The screening length is derived numerically as a function of the areal fraction of clusters covering the surface, Φ, and the average cluster radius $\langle R \rangle$. The data are plotted as $\langle R \rangle / \xi$ versus Φ in double-logarithmic representation in Figure 62 [242]. From the figure follows $\langle R \rangle / \xi = b\Phi^a$, with $a = 0.73 \pm 0.03 \approx 3/4$ and $b = 1.95 \pm 0.15 \approx 2$. The data clearly exclude $a = 1$ or $a = 1/2$. Thus, the following relation is obtained for two-dimensional–two-dimensional systems,

$$\xi \approx \frac{\langle R \rangle}{2\Phi^{3/4}} \approx \frac{\langle d \rangle}{2\pi^{3/4}} \sqrt{\frac{\langle d \rangle}{\langle R \rangle}} \qquad (18)$$

where $\langle d \rangle$ is the average distance between islands as sketched in Figure 61. The time dependence of ξ follows $\xi \propto t^{1/3}$ since $\langle R \rangle \propto t^{1/3}$ [234] and the average cluster distance is connected to $\langle R \rangle$ through the mass conservation with $\langle R \rangle^2 N_x = \text{const}$ and $\langle d \rangle^2 \propto 1/N_x$, i.e., $\langle R \rangle \propto \langle d \rangle$, where N_x is the areal density of clusters on the surface.

However, the quantitative form of the screening length remains an open issue. Marqusee's approach, introducing the screening length in the equation of diffusion as a sink term averaged over all clusters, is not derived from first principles and thus the degree of approximation cannot be predicted. In addition, the concept is inherently a finite areal fraction model, which cannot be extended to the case $\Phi = 0$. This justifies a simpler approach by Ardell [243] with the *a priori* assumption that $\xi = 1/2\langle d \rangle$. This model still contains the essential features of the screening concept. With this approach, the same power

The last term in Marqusee's diffusion equation represents a source term which also depends on the island distribution in space and size. The latter two terms are initially unknown functions since they are arbitrarily introduced to recover a steady-state diffusion field. The source and the sink terms are connected with each other, however, as they have to be balanced far from the central island. This leads to a modified form of the steady-state diffusion equation, $(\nabla^2 - \xi^{-2})(c(r) - c_{\text{free}}) = 0$, where the divergence is eliminated due to the screening length term. The diffusion field $c(r)$ and the screening length ξ can be found in a self-consistent manner [234] and can lead to the same power law of island growth as in the three-dimensional–three-dimensional case, $R_c \propto t^{1/3}$. The 1/3 power law for the two-dimensional–two-dimensional case has further been confirmed in several analytical and numerical studies, including work by (i) Zheng and Gunton [236] who included island–island correlations of higher order (beyond the screening length concept), (ii) Hayakawa and Family [237] who used perturba-

law was predicted, while a simpler analytical expression for the cluster size distribution followed. Interestingly, Ardell's size distribution function agrees closely with the result by Rogers and Desai [232] solving the nonstationary problem for $\Phi = 0$. In Rogers and Desai's study, however, the cluster growth law does not follow power-law behavior but is given in the form $R_c \propto (t/\ln t)^{1/3}$.

A comprehensive and satisfactory description of the ripening model for three-dimensional clusters on surfaces does not yet exist. In early studies of diffusion limited growth, preceding the reevaluation of the two-dimensional–two-dimensional case by Marqusee [234], the divergence in the steady-state solution of the two-dimensional diffusion equation had been circumvented with unphysical assumptions [233, 244, 245]. Nevertheless, these studies suggested that a new power law should result in the form $R_c \propto t^{1/4}$. Viñals and Mullins [246] revisited this problem, linking scaling exponents and self-similarity arguments. They suggested that the 1/4 power law may only approximately apply for short periods and scaling should not be applicable due to logarithmic correction terms. The same result was obtained through an extension of Marqusee's calculations and Ardell's simpler model to the three-dimensional–two-dimensional case [247]. Essentially, no power law exists since the equation of continuity and the mass conservation condition cannot be satisfied simultaneously if a power-law scaling is assumed. Further, it is not possible to determine the proper cluster size distribution function without an appropriate scaling form for the critical cluster radius. Power-law behavior can still be obtained under certain approximations, e.g., for small areal fractions. Logarithmic correction terms have also been demonstrated for the simpler case of the growth of a single, three-dimensional cluster embedded in a surface supersaturation [248]. Using a quasi-static approximation, the radius grows proportional to $[t/\ln(t)]^{1/3}$.

Zheng and Bigot [249] reported a Monte Carlo study based on mass exchange between two isolated hemispherical clusters. The simulations led to a power-law exponent of 0.233 ± 0.003, close to the suggested value of 1/4. Atwater and Yang [250] numerically studied the three-dimensional–two-dimensional system for the case where monomer attachment and detachment limits the mass transport. They found an asymptotic growth exponent of 1/3 for a wide range of initial cluster size distributions. This power exponent corresponds to an exponent of 1/4 for the diffusion limited case based on self-similarity arguments [181].

The cluster size distribution predicted for the three-dimensional–two-dimensional case is shown as the dash-dotted line in Figure 60 [251]. Note that the distribution is even narrower than the distribution predicted for the LSW-theory case. These size distribution functions are universal; i.e., they do not depend on the initial distribution or on other initial values if the appropriate scaled parameters are used. This allows for a comparison of various experimental data (e.g., cluster size distributions evaluated by electron microscopy or cluster height measurements by ion scattering techniques) directly with the theoretical prediction. As a result, the normalized cluster size distribution $f(r, t)$

is statistically called self-similar [252]. For the morphological configurations observed during a film growth experiment, self-similarity means that sequential size distributions cannot be distinguished after rescaled with the average cluster size. Mullins and Vinals [253] showed for statistically self-similar cluster size distributions that the power-law exponents of the growth of an average sized cluster can be predicted based on a few general assumptions. This is true independent of the initial size distribution.

2.2. Early Stage Phase Separation Morphologies

The early stage of initial cluster growth begins with the formation of the first stable nuclei. Then, the driving force already is the Gibbs–Thomson effect which thermodynamically favors larger clusters. Early during the deposition, the supersaturation and the nucleation rate decrease. The growth of nuclei in this stage is dominated by local effects, e.g., (a) formation of a concentration profile around the cluster and (b) coalescence when two clusters grow into each other due to their size increase or due to impact if clusters are mobile.

2.2.1. Theoretical Concepts Linking the Early Stages to Nucleation and Late Stage Growth

A first model to link the nucleation regime directly to the late stage growth regime is discussed by Langer et al. They extend the equation of continuity as used in the description of the late stage by a term which allows the continuation of nucleation into the early stage. This approach has two advantages over the nucleation theory:

(1) Cluster growth is immediately allowed for any emerging nucleus. This includes completion times of the nucleation process in the theoretical model which is sometimes necessary to understand experimental results.
(2) The variable nucleation rate in the model addresses more complex quench experiments with a changing supersaturation due to incorporation of adatoms in the cluster phase by nucleation events. Such an extension reaches beyond systems with a constant supersaturation defined by a nonzero deposition rate.

Under simplifying assumptions, two regimes are distinguished: for initial supersaturations close to the coexistence curve with large activation energies for the formation of clusters, the nucleation rate is small and the completion occurs when a small number of nuclei have grown to rather large sizes. In addition, an incubation time is needed for the first nuclei to appear after the quench is expected. For deeper quenches to higher supersaturations with small activation energies for the formation of clusters, rapid nucleation is predicted where almost no cluster growth of the stable clusters occurs concurrently. Thus, near critical clusters are maintained to the end of the nucleation phase.

Theoretically, a more satisfactory approach is a direct link of the rate equations of Section 1.1 and the Lifshitz–Slyozov–Wagner theory developed in Section 2.1. Many studies in the transition regime toward late stage ripening apply either the rate equation concept or the late stage concepts, assuming that the basic features of nucleation or ripening can be extrapolated. This approach is validated later [254, 255].

The starting points are the classic rate equations (Eq. (4)) with two specific conditions, (i) monomers are the only mobile species and (ii) the cluster volume fraction is sufficiently low to exclude cluster coalescence. Then, the number of clusters with j atoms, $N(j)$, changes only through monomer attachment–detachment with the rate equation for ($j \geq 2$) given in the form,

$$\frac{dN(j)}{dt} = n\big[\sigma_{\text{in},j-1}N(j-1) - \sigma_{\text{in},j}N(j)\big]$$
$$+ \big[\sigma_{\text{out},j+1}N(j+1) - \sigma_{\text{out},j}N(j)\big] \quad (19)$$

where σ_{in} is the capture cross section and σ_{out} is the release cross section for monomers. In the first step, the discrete differences are rewritten as continuous differential terms, since clusters contain very large numbers of atoms in the late stage ($j \gg 1$):

$$\sigma_{j-1}N(j-1) - \sigma_j N(j) \equiv -\frac{\partial}{\partial j}\big(\sigma_j N(j)\big) \quad (20)$$

The partial differentials indicate that the variables also depend on other parameters, e.g., the number density of clusters on time. With this relation, the rate equation reads

$$\frac{\partial N(j)}{\partial t} = -n\frac{\partial}{\partial j}\big(\sigma_{\text{in},j}N(j)\big) + \frac{\partial}{\partial j}\big(\sigma_{\text{out},j}N(j)\big) \quad (21)$$

The number density of monomers, n, depends explicitly on time and the critical cluster size, but not on any specific value of j. Thus, n can be integrated into the differential term:

$$\frac{\partial N(j)}{\partial t} = \frac{\partial}{\partial j}\big[N(j)(-n\sigma_{\text{in},j} + \sigma_{\text{out},j})\big] \quad (22)$$

Explicit functional forms for the monomer concentration and the cross-section terms exist for systems in late stage ripening. n is connected to the free monomer concentration far from clusters in the form $n = c_{\text{free}}(t) = c_{GT}(j_c(t))$. j_c denotes the number of atoms in a cluster in equilibrium with the free monomer concentration. The index GT (for Gibbs–Thomson) indicates that cluster size and equilibrium concentration of clusters are linked. Quantitatively, the Gibbs–Thomson effect is given by $c_{GT}(R) = c_\infty[1 + (2(v_M/RkT)]$ where c_∞ is the equilibrium solubility at infinite cluster radius, γ is the surface tension, and v_M is the atomic volume.

The Gibbs–Thomson effect is also used to relate the two cross-section terms, σ_{in} and σ_{out},

$$\sigma_{\text{out},j} = c_{GT}(j)\sigma_{\text{in},j} \quad (23)$$

representing a microscopic balance for each cluster size j. Substituting the formulas for n and σ in the rate equation leads to

$$\frac{\partial N(j)}{\partial t} = \frac{\partial}{\partial j}\big[N(j)\sigma_{\text{in},j}\big\{-c_{GT}\big(j_c(t)\big) + c_{GT}(j)\big\}\big] \quad (24)$$

At this point, an explicit form for $\sigma_{\text{in},j}$ has to be introduced [255]. Assuming that monomer attachment to larger clusters limits the mass transport (Wagner's model [226]), the cross-section term is written as $\sigma_{\text{in},j} = \sigma_0 S(j)$. σ_0 is a constant for the net-attachment–detachment rate per surface site on the cluster and $S(j)$ is the total number of sites on the cluster surface. The independent variable is changed from the number of atoms in a cluster, j, to the radius of the cluster, R, using $j = (4\pi/3)R^3/v_M$ and $(\partial/\partial j) = (v_M/4\pi R^2)(\partial/\partial R)$. The distribution function transforms to the radial size distribution $f(R,t)$ with $N(j)dj = N(j)(4\pi/v_M)R^2 dR = f(R,t)dR$. Using the Gibbs–Thomson relation this leads to

$$\frac{\partial f(R,t)}{\partial t} = \frac{\partial}{\partial R}\left[v_M\sigma_0 f(R,t)c_\infty\frac{2\gamma v_M}{kT}\left(\frac{1}{R} - \frac{1}{R_c}\right)\right]$$
$$= \beta\frac{\partial}{\partial R}\left[f(R,t)\left(\frac{1}{R} - \frac{1}{R_c}\right)\right] \quad (25)$$

with

$$\beta = \frac{2\gamma v_M^2 c_\infty \sigma_0}{kT} \quad (26)$$

This equation is equivalent to the respective equations in the LSW model [181, 226].

2.2.2. Unique Features of the Early Stages Leading to Ripening Patterns

(i) *Nucleation in the Postdeposition Stage*. Most nucleation studies are based on finite values for the deposition rate R. In these cases, we saw that nucleation terminates abruptly in the transition to late stage coalescence growth. However, a system can enter a two-phase coexistence regime in Figure 1 while $R = 0$ along arrows a and b. A sudden temperature quench is experimentally feasible, however often limited to systems of low interest in thin film growth, such as surface gas contaminations. Following path b is complicated since uncontrolled nucleation may occur in experiments and an ill-defined initial nonequilibrium state has to be considered in theoretical models.

Li, Rojo, and Sander [256] applied kinetic Monte Carlo simulation techniques to submonolayer epitaxial systems of point-size islands without a monomer flux arriving at the surface. The conserved system ($\theta = \text{const}$) is established with a predetermined number of randomly distributed monomers on the surface. An off-critical temperature quench along arrow a in Figure 1 initiates the phase separation. The mass conservation defines the length scale on the surface by $\theta^{-1/2}$. At time $t = 0$, the monomers are released and perform a temperature-dependent random walk. In the simulations, critical island sizes were varied between $i^* = 1$ and $i^* = 3$ for initial coverages between 2 and 20%. The obtained island size distributions are scaled and shown smoothed in Figure 63 for $i^* = 1$ and $i^* = 2$. The distributions are significantly different from those obtained in Section 1.1, particularly the distribution for $i^* = 1$ does not have a peak at a finite cluster size. The data were further analyzed for spatial correlations of the islands. For all three critical

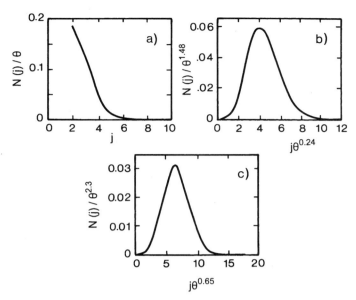

Fig. 63. Scaled island size distributions from kinetic Monte Carlo simulations of nucleation with $F = 0$ for (a) $i^* = 1$, (b) $i^* = 2$, and (c) $i^* = 3$. Initial coverages vary between 2 and 20%. Reprinted with permission from [256], © 1997, American Physical Society.

island sizes, secondary nucleation near an existing island is suppressed, most effectively for the largest critical islands. No peak in the island–island correlation function was observed near the average island distance.

Tomellini [257] tried to develop a model for a system following arrow b in Figure 1 to interpret experimental data for the system Cu/Ni(100) by Müller et al. [258]. A priori assumptions have to be made to extend the rate equation approach beyond the nucleation regime for submonolayer coverages, including a specific growth rate for stable islands in the early stages following nucleation. Using the growth power law for late stage processes allows only a qualitative approach since the applicability of the Gibbs–Thomson formula for small islands is questionable [259] and experiments indicate that the late stage power law does not hold [260]. In addition, other effects may complicate the description particularly in this regime, e.g., a limited dissipation of latent heat of cluster formation can lead to overheated clusters which delays growth [261].

For Cu on Ni(100), Müller et al. [258] observed that nucleation and island growth continue in the postdeposition regime if the experiment was done with low diffusion rates, i.e., temperatures below $T = 160$ K, or at high deposition rates, $R > 2 \times 10^{-3}$ ML/s at 145 K with $D/R < 10^4$. In these cases, the evolution of the morphology is dominated by the monomer concentration at the time of termination of the deposition process and the island density is independent of the substrate temperature, resembling distributions for $i^* = 0$ in Section 1.1.

(ii) *Direct Ripening.* The reduction of the initial supersaturation is the main effect during early stage growth. Direct diffusional growth slows down and the area around the clusters defined by the diffusion length Λ start to overlap. As the to-

tal overlap approaches the entire surface, the late stage growth regime starts as described in the next section. For a short intermediate time interval, a growth process becomes possible which is called "direct ripening" [262]. This process is characterized by a local interaction between two neighboring clusters of slightly different size. The smaller cluster starts to decompose to maintain the concentration gradient toward the larger cluster due to the Gibbs–Thomson effect. The appearance of such processes is critically dependent on the initial conditions, especially the fraction of spatially close cluster pairs. The process is part of the early stage growth since it is a local effect.

Hirth [263] used this picture to describe nonrandom spatial distributions of clusters on the substrate surface, i.e., that very close clusters are less likely to survive into the late stage. Physically, this shows up in SEM or TEM pictures as a denuded zone around larger clusters. These zones may be due to a diffusive dissolution of small clusters in the neighborhood of a larger cluster (direct ripening). Alternatively, a reduced probability of nucleation due to the decreased supersaturation around large clusters may lead to the denuded zones (direct growth from the supersaturated phase).

(iii) *Coalescence.* Although coalescence, i.e., the growth of clusters into each other, is normally considered part of the late stage growth, it may also be important in the early stages. This process depends strongly on the cluster density on the surface and becomes dominant when the fraction of the surface covered by the cluster increases. This is the case during the nucleation regime, when the number of nuclei increases or when continuous deposition of monomers steadily increases the average size of existing clusters.

A semiquantitative argument on the relative contribution of single atoms (ripening) or stable clusters (coalescence) in the early growth regime is given by Kern [264] by comparing the monomer concentration on the surface and the number of stable clusters when nucleation reaches the maximum density. The latter value is usually about $10^9–10^{11}$ cm^{-2}. In the particular study, the monomer concentration is estimated from the difference of the heat of adsorption and evaporation, e.g., for Au on KCl(100) or C(0001) (graphite) as 10^3 monomers/cm^2 at 1100 K and for Au on MgO(100) or Si(111) as 10^{11} monomers/cm^2 at 1100 K. Thus, a lower monomer concentration favors coalescence growth. For Ge on Si(100) at typical heteroepitaxial growth temperatures (550°C), the monomer concentration is estimated by molecular dynamic simulations to be of the order of 0.1 ML [265] which is confirmed by an experimental scanning Auger microscopic study [266].

2.3. Late Stage Ripening Morphologies

2.3.1. Experimental Tests of the Basic Ripening Theory on Surfaces

2.3.1.1. The Two-Dimensional–Two-Dimensional Case

A detailed experimental study for two-dimensional–two-dimensional ripening has been presented by Krichevsky and Stavans

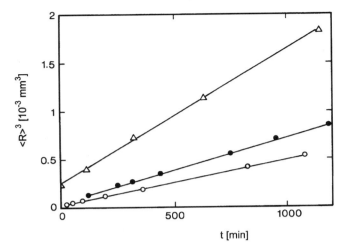

Fig. 64. Dependence of the third power of the radius of succinonitrile solid droplet growth in its melt near 320 K in a quasi-two-dimensional arrangement between two glass slides. Data represent different volume fractions, $\Phi = 19\%$ (open circles), $\Phi = 40\%$ (solid circles), and $\Phi = 54\%$ (triangles). Reprinted with permission from [268], © 1995, American Physical Society.

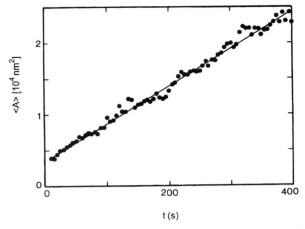

Fig. 65. Time dependence of the average island area during ripening of 0.1-ML Si on Si(001). The plot shows the growth of an average sized island on a terrace at 945 K. Reprinted with permission from [270], © 1996, American Physical Society.

[267, 268]. A thin layer of succinonitrile with solid droplets surrounded by their melt was prepared between glass slides 25 μm apart. Since the droplet sizes exceeded 25 μm, the system essentially behaves two dimensional. The solid fraction can be controlled with an impurity in a coexistence region near 320 K. Volume fractions of the solid phase were varied between $\Phi = 13\%$ and $\Phi = 54\%$. For all concentrations, the growth power-law $R \propto t^{\alpha}$ was determined with exponent $\alpha = 0.335 \pm 0.005$ (Fig. 64).

The predicted power law was also studied for an amphiphiles mixture forming a monolayer at an air–water interface (Langmuir film) [269]. For $\Phi = 25\%$, the components were miscible at 292 K but become immiscible through a rapid mechanical expansion (10% areal increase). The analysis of the average domain area as a function of time leads to an exponent of $\alpha = 0.28 \pm 0.01$ slightly below the expected value at $\alpha = 1/3$.

Step edge attachment–detachment limited ripening has been observed for 0.1-ML Si on Si(001) at 945 and at 1135 K [270, 271]. By imaging the evolution of individual islands with low energy electron microscopy (LEEM [272]) the time dependence of the average island area (Fig. 65) was followed. The solid line illustrates a linear dependence, corresponding to $\alpha = 0.5$, in agreement with the exponent expected [181].

Another case of attachment–detachment controlled island growth has been observed by Cavalleri et al. [273, 274]. During the formation of a uniform thiol layer (H_2S derivates with an organic (CH_3-$[CH_2]_x$-) ligand) on an Au(111) surface, erosion of the gold surface led to thiol covered, nanometer sized depressions of monolayer depth. The annealing process of these holes at about 635 K was followed by STM. The holes coarsen in an Ostwald ripening mechanism with α varying between $\alpha = 0.48$ and $\alpha = 0.52$ (error margin $\leq \pm 0.1$) for three chain lengths with $x = 5$, 9, and 17. The interface barrier dominated process is linked to breaking and forming interchain bonds between the

thiol molecules. This interpretation is supported by a faster annealing for molecules with shorter organic ligands.

Observation of deviating power laws requires special considerations. For example, Ernst, Fabre, and Lapujoulade [275] observed low-temperature islanding in the system Cu on Cu(100) with a 0.5-ML coverage using helium atom beam scattering. Deposition at 100 K was followed by ripening in the temperature range of 212–266 K for annealing times of 5–165 min. Due to the high coverage, strongly interlinked structures are expected on the surface. The time evolution of the characteristic length scale at four temperatures in this interval showed slopes that varied between 0.17 and 0.25. These values imply an asymptotic exponent of 1/4 which can only be explained with diffusion limited ripening along the edges of the interlocked morphology as predicted by Mullins based on scaling arguments [184].

2.3.1.2. Three-Dimensional–Two-Dimensional Case

The 1/4 power law for a three-dimensional–two-dimensional system was first experimentally observed for the system Sn on Si(111) and Si(100) [260, 276]. The difficulty to establish a power law in such experiments is illustrated by comparing linearizations with a third- and a fourth-power assumption for the cluster height versus time. Figure 66 shows such a comparison for 22.3- and 3.8-ML Sn on Si(100) at 795 K [277]. The fourth-power law agrees with the data better if a transient regime for cluster sizes up to radii of 100 nm is included. This transient regime coincides with estimates of the transition from early stage growth to late stage growth for this system [260]. Figure 67 shows a similar measurement for 2.9-ML Sn on Si(111) at 525 K. The same conclusions apply, supported in this figure with a double-logarithmic plot in the inset of the figure. The inset illustrates that the fourth-power law is indeed the best fit to the data after an initial period representing early stage growth [277]. The activation energies for clustering as obtained from such measurements have been included in Table I.

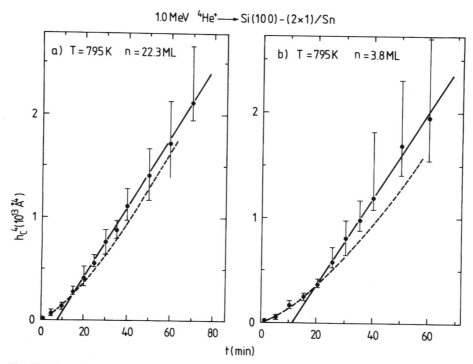

Fig. 66.   Power-law dependence of the height of clusters as a function of time at 795 K for two different coverages on Si(001) 2×1, (a) 22.3-ML and (b) 3.8-ML equivalent coverage of Sn. The solid line corresponds to the fourth-power dependence of the cluster height and the dashed line to the cube of the cluster height. Technique: ion scattering (RBS). Reprinted with permission from [260], © 1991, Elsevier Science.

Rogers et al. [278] measured the growth exponent for a liquid–liquid (succinonitrile–water) two-phase system at 315 K. The experiment was carried out in a quartz test cell with clustering of partial spherical caps on the walls of the cell. Clustering was observed for times up to 4 months. Figure 68 shows in double-logarithmic representation the average cluster radius as a function of time. The late stage is described by a power law with coefficient $\alpha = 0.247$ (solid line in Fig. 68) in good agreement with the theoretical value of $1/4$.

2.3.2. Finite Areal Fraction and Volume Fraction Effects

Although the power-law exponent of cluster growth is not affected by the volume or the areal fraction of clusters, two other main properties depend on the volume fraction as has been shown for three-dimensional–three-dimensional systems: (i) the cluster size distribution function broadens and becomes more symmetric with increasing volume fraction, and (ii) the absolute growth rate of clusters increases by a factor of 2 to 3 between $\Phi = 0\%$ and $\Phi = 50\%$ [181, 228–231].

2.3.2.1. Two-Dimensional–Two-Dimensional Systems: Analogy to Three-Dimensional–Three-Dimensional Systems

Areal fraction effects have been predicted for two-dimensional–two-dimensional systems by Marqusee [234]. Both aspects have been investigated rigorously during the last few years due

to the related problems with the diffusion process in two dimensions.

Yao et al. [238–240] compared two-dimensional–two-dimensional and three-dimensional–three-dimensional systems in a series of mean-field analytical and numerical studies of many droplet systems with screening effects. The results are summarized in Figure 69 for the absolute cluster growth rate and in Figure 70 for the size distributions. Figure 69 compares absolute rates of cluster growth, $K(\Phi)$, defined by $R_c = [K(\Phi)t]^{1/3}$, which is the suitable power law for both three-dimensional–three-dimensional and two-dimensional–two-dimensional systems. Note that the data are shown as absolute values [234, 236, 238, 239, 243]. The data agree on an increase of cluster growth rates with increasing volume fraction. Marder [235] attributed this to the increased direct ripening [279, 280] between neighboring clusters which accelerates the growth of the larger cluster in a binary pair. Figure 70 illustrates the changes of the normalized size distribution for various areal fractions in the range from $\Phi = 0\%$ to $\Phi = 8.5\%$ for two-dimensional–two-dimensional systems. For comparable volume fractions Φ the changes of the cluster size distributions are comparable with three-dimensional–three-dimensional predictions, with the shown two-dimensional–two-dimensional distribution broadening slightly stronger.

Cluster size distributions are quantitatively compared using the different moments of the distributions. Most frequently, the standard deviation and the skewness (a measure for the

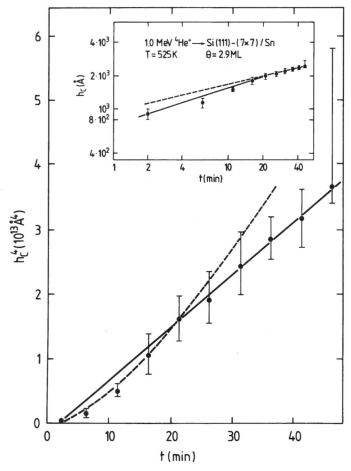

Fig. 67. RBS measurement of the cluster height as a function of time for 2.9-ML equivalent coverage of Sn on Si(111) deposited at room temperature and held at 525 K. The solid line corresponds to the fourth-power dependence of the cluster height and the dashed line corresponds to the cube of the cluster height. The inset shows the same data in double-logarithmic form. Technique: RBS. Reprinted with permission from [260], © 1991, Elsevier Science.

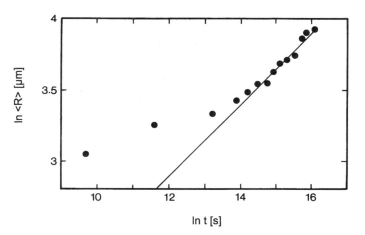

Fig. 68. Double-logarithmic representation of the clustering in a binary liquid mixture of water and succinonitrile (clusters are succinonitrile rich phase) on a quartz surface at 315 K. The linear regression to the late stage yields a power-law coefficient of $\alpha = 0.247$ in good agreement with the theoretical value of $\alpha = 1/4$. Reprinted with permission from [278], © 1994, Minerals, Metals, and Materials Society.

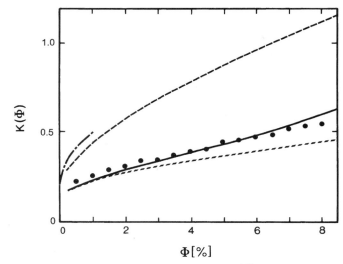

Fig. 69. Plots of ripening rates K, in $R_c = (Kt)^{1/3}$, versus areal coverage fraction in a two-dimensional–two-dimensional system. The data are taken from Ardell [243] (long dashed line), Marqusee [234] (short dashed line), Zheng and Gunton [236] (dash-dotted line), and Yao et al. [238–240] (solid line and solid circles). Reprinted with permission from [20], © 1999, Elsevier Science.

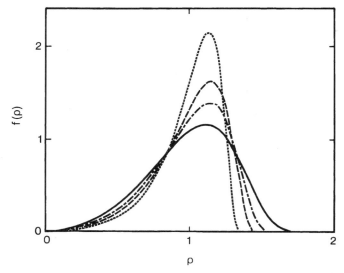

Fig. 70. Scaled normalized size distributions for two-dimensional–two-dimensional system with variable areal–volume fractions. The short dashed line corresponds to $\Phi = 0\%$, the long dashed line corresponds to $\Phi = 1\%$, the dash-dotted line corresponds to $\Phi = 4\%$ and the solid line corresponds to $\Phi = 8.5\%$. Reprinted with permission from [239], © 1993, American Physical Society.

asymmetry of the distribution) are reported while the kurtosis (a measure of the symmetric deviation from a Gaussian shape) and higher moments carry too large error margins in experiments. Moments for two-dimensional–two-dimensional and three-dimensional–three-dimensional systems are shown in Figure 71 (standard deviation) and Figure 72 (skewness). The relative increase of the standard deviation is shown twice as a function of Φ, with a linear Φ axis in Figure 71a and as a logarithmic Φ axis in Figure 71b to highlight variations at small

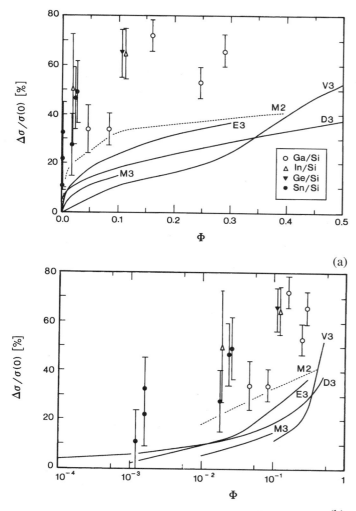

(a)

(b)

Fig. 71. Plots of the relative increase of the standard deviation of the cluster size distribution (in percent) as a function of the volume–areal fraction Φ with (a) the Φ axis linear and (b) the Φ axis logarithmic to emphasize low coverages. The experimental data are for three-dimensional–two-dimensional ripening in Ga/Si (open circles), In/Si (open triangles), Ge/Si (solid inverse triangles), and Sn/Si (solid circles) [283]. The solid lines are theoretical predictions for three-dimensional–three-dimensional systems, taken from Enomoto, Tokuyama, and Kawasaki (E3 [229]), Davies, Nash, and Stevens (D3 [231]), Marqusee and Ross (M3 [228]) and Voorhees and Glicksman (V3 [230]). The dashed line (M2) is a prediction for two-dimensional–two-dimensional systems from Marqusee. Reprinted with permission from [20], © 1999, Elsevier Science.

values. The two-dimensional–two-dimensional data (dashed line) are taken from Marqusee and the three-dimensional–three-dimensional data (solid lines) from Enomoto, Tokuyama, and Kawasaki [229], Voorhees and Glicksman [230], Marqusee and Ross [228], and Davies, Nash, and Stevens [231].

Three experimental groups reported cluster size distributions for the two-dimensional–two-dimensional case. Krichevsky and Stavans [267, 268] analyzed ripening of solid succinonitrile islands in the solute at different areal coverages. The size distributions for 13% areal coverage and for 40% areal coverage show increasing broadening with areal fraction, but still

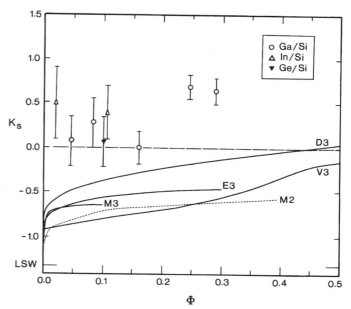

Fig. 72. Skewness of cluster size distribution for ripening systems as a function of areal–volume fraction Φ. Notations of experimental and theoretical data are the same as in Figure 71. Reprinted with permission from [283], © 1996, Elsevier Science.

negative skewness. The authors attribute the broadening to correlated growth rates of neigboring clusters (direct ripening [279, 280]).

Seul, Morgan, and Sire [269] and Morgan and Seul [281] studied a binary mixture of two amphiphiles at a water–air interface. When brought into an immiscible state, coarsening was observed. The size distribution was studied for areal coverages between 11 and 15% at different times during the phase separation. Although data scatter strongly, a broadening and a significant reduction of the skewness in comparison with zero areal fraction models is observed.

Theis, Bartelt, and Tromp [271] and Bartelt, Theis, and Tromp [270] studied silicon homoepitaxy on Si(001) with LEEM. The power law for island growth suggests an attachment–detachment dominated mass transport in this case. This conclusion is confirmed by the island size distribution as shown in Figure 73 at two different times during the coarsening evolution. The width of the experimental distributions match with the predicted width for the detachment limited case (solid line, taken from a two-dimensional grain growth model [282]). The skewness of the theoretical curve, however, is more negative than the experimental value.

2.3.2.2. Three-Dimensional–Two-Dimensional Systems: Qualitative Differences

Cluster–cluster interactions should act between three-dimensional clusters on a surface (three-dimensional–two-dimensional system) in the same fashion as discussed in the previous section for two-dimensional–two-dimensional and three-dimensional–three-dimensional systems. Thus, the same devia-

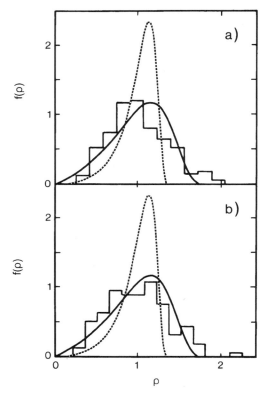

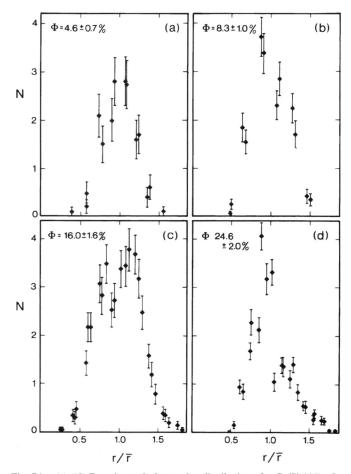

Fig. 73. Normalized island size distributions for Si on Si(100) at 945 K for two different times, (a) 50 s and (b) 100 s. The experimental data are shown as histograms and are compared with two theoretical distributions, mean-field predictions for ripening with diffusion limited mass transport (dashed line) and with attachment–detachment limited mass transport (solid line). Reprinted with permission from [270], © 1996, American Physical Society.

Fig. 74. (a)–(d) Experimental cluster size distributions for Ga/Si(111) after annealing at 810 K for 30 min for initial coverages of (a) 3.4 ML, (b) 7.2 ML, (c) 17.5 ML, and (d) 49 ML. The areal cluster coverage of the final morphology is indicated in each panel. Reprinted with permission from [283], © 1996, Elsevier Science.

tions should occur from mean-field predictions as a function of an areal fraction. This has been tested in several studies.

Two computer simulation studies generally confirmed the trends discussed in the previous section. Zheng and Bigot [249] used solutions for pairwise intercluster diffusion in a Monte Carlo study at low areal coverages which resulted in broader size distributions (based on a comparison to the cluster size distribution given by Chakraverty [233] which is however questionable due to an inappropriate formulation of the screening length) and maintained a negative skewness. Atwater and Yang [250] used computer simulations to study the case where the mass transport between clusters is limited by monomer attachment–detachment. Size distributions were obtained for areal coverages between 0 and 50% and show an increasing broadening and a negative skewness at all coverages.

Cluster size distributions have been analyzed in a sequence of experiments on clean semiconductor surfaces [283]. Figure 74 shows a set of cluster size distributions at various areal coverages for Ga ripening at 810 K for 30 min on Si(111). The size distributions and areal coverages were obtained with SEM and range from $\Phi = 5$–25%. Figure 75 shows a cluster size distribution obtained in the same way for 21.8-ML In on Si(111) at 650 K with annealing for 35 min, leading to an areal coverage of $11.3 \pm 0.5\%$. To emphasize the smaller density of large

clusters, the data set in Figure 75 is shown with a logarithmic number density of clusters. Note that all distributions either are nearly symmetric or have a positive skewness.

The corresponding moments of the cluster size distributions are shown in Figures 71 and 72. Note that the first three metals are liquid at the ripening temperatures but Ge ripens as solid clusters [284]. The three-dimensional–two-dimensional experimental data show significantly larger broadening than the three-dimensional–three-dimensional and two-dimensional–two-dimensional simulations. However, this conclusion depends on the choice of the reference size distribution at $\Phi = 0$ for the three-dimensional–two-dimensional case and may change if the used distribution [233] has to be corrected. The skewness data, on the other hand, show qualitative differences with the simulations with positive values already at areal coverages below 10%.

Two interpretations have been proposed for these deviations. Initially, coalescence or memory effects of coalescence were discussed [283]. However, clusters are rather distant from each other in the late stage and are even located at

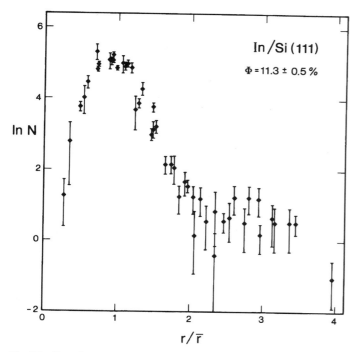

Fig. 75. Experimental cluster size distributions for 21.8-ML In/Si(111) after annealing at 650 K for 35 min. The areal cluster coverage of the final morphology is indicated in the panel. Reprinted with permission from [283], © 1996, Elsevier Science.

larger than random distances due to partial spatial ordering. In three-dimensional–three-dimensional simulations Tokuyama and Enomoto [285] showed that coalescence occurs infrequently for volume fraction between 10 and 20% and is not observed below 10%. Akaiwa and Meiron [286] illustrated further for two-dimensional–two-dimensional systems in numerical simulations that even at 50% areal coverage ripening without significant coalescence is possible if the shape restriction on the islands is lifted.

A second explanation has been suggested, studying a new mechanism based on the unique dimensionality of the three-dimensional–two-dimensional system. In this model, the cluster shape relaxes and the monomer exchange with the free concentration occurs only at the contact line on the substrate. If the shape relaxation occurs on a timescale slower or comparable to ripening growth, the apparent radius of the cluster at the substrate contact line is different from the radius determined from volume and Young–Dupré contact angle. A growing cluster appears to be too large as monomers are not removed fast enough from the periphery area and a decomposing cluster appears to be too small as detached atoms are not replaced fast enough from upper areas of the cluster. This causes changes to the mass-conservation condition and the Gibbs–Thomson concentration in the formalism reported by Marqusee [234]. The changes are sufficient to explain a broadening and a shift of the skewness toward positive values for the cluster size distribution, although absolute magnitudes for changes have not been calculated yet [247].

2.3.3. Screening Effects

The introduction of the screening length not only is a mathematical method to simplify the ripening problem with two-dimensional mass transport (or the ripening problem in three dimensions at finite volume fraction) but has real physical implications, e.g., for systems with concentration gradients.

(i) *Theoretical Concepts.* We demonstrate the relevance of the screening length for the early development of open systems (i.e., before significant mass transport occurred across the interface) based on the treatment of three-dimensional systems by Beenakker and Ross [287]. The model system consists of two infinite half-spaces (cells) in diffusive contact, where one cell contains a phase separating system in late stage Ostwald ripening. The other cell contains a fixed monomer concentration which acts as a thermodynamic reservoir. The description of the ripening process in the respective cells differs in two ways from the LSW model: (i) the far-field value for the concentration field near the interface is changed to a constant value c_o representing the fixed concentration of the reservoir in the adjacent cell, and (ii) relative positions of clusters have to be introduced. Beenakker and Ross follow the LSW calculations but modify the equation of continuity in the form,

$$\frac{1}{v_M} \frac{d}{dt} \left(\frac{4}{3} \pi R_i^3 \right) = 4\pi D \sum_{j=1}^{\infty} R_j \left(\delta_{ij} - (1 - \delta_{ij}) g(R_i, r_i, r_j) \right)$$
$$\times \left(c_o - c(R_j) \right) \tag{27}$$

where the right-hand side represents the source strength for the ith cluster. The source strength depends on the difference of the concentration at the cluster surface (given by the Gibbs–Thomson effect) and the far-field concentration c_o, multiplied with a transport matrix. This matrix has two terms, the first delta function represents the contribution of the central cluster to the growth rate and the second term describes the interaction with all other clusters where g is a complex function of the radius of the ith cluster, R_i, and the positions r_i and r_j of the ith and the jth cluster, respectively.

To solve this problem, again requires the introduction of a screening length ξ in the following form for three-dimensional systems,

$$\xi = \left(4\pi \langle R \rangle \frac{N}{V} \right)^{-1/2} \approx \langle d \rangle \sqrt{\frac{\langle d \rangle}{4\pi \langle R \rangle}} \approx \frac{\langle R \rangle}{\sqrt{3\Phi}} \tag{28}$$

The screening length is first given as a function of the average radius of the clusters, $\langle R \rangle$, and the average distance between clusters, $\langle d \rangle$, i.e., $\xi = \xi(\langle R \rangle, \langle d \rangle)$. The cluster density N/V is inversely proportional to the volume per cluster using $V/N \cong \langle d \rangle^3$, which is a good approximation for spatially ordered clusters. Alternatively, the screening length is given as a function of average cluster radius and volume fraction of the minority phase, Φ, i.e., $\xi = \xi(\langle R \rangle, \Phi)$, with the volume fraction determined as $\Phi = (4/3)\pi \langle R \rangle^3 (N/V)$. The latter is the form explicitly provided by Beenakker and Ross [287].

The result for the screening length can be compared to the two-dimensional result discussed in the previous section. Marquee's self-consistent derivation of the screening length [234] led to the relation $\langle R \rangle / \xi = 2\Phi^{3/4}$ (Fig. 62). The same dependence of the screening length follows in both cases, however, when the screening length is written as a function of average radius and average nearest neighbor distance, $\xi = \xi(\langle R \rangle, \langle d \rangle)$ [242].

The mass transfer through the open interface leads to a zone of width $L(t)$ at the interface where the evolution of the coarsening cell depends on the reservoir parameters. The interface width is of the order of the screening length, $L(t) \approx \xi$, and thus depends on time as $L(t) \approx t^{1/3}$. This result has been confirmed in a numerical simulation where the material flux across the interface in a three-dimensional system was found to be proportional to $t^{-2/3}$ [288].

(ii) *Experimental Tests of the Screening Concept for Open Systems.* An open surface system can be established by shadowing part of the substrate with a mask during deposition [64, 293]. The system Ag/Ge(111) was studied with this approach [97] to measure Ag surface diffusion. As the deposited patch expands, it forms uniform Stranski–Krastanov layers. On these layers, three-dimensional clusters form. The experiments have been simulated with a combination of the rate equation models and the diffusion equation. Due to the large number of variables a full data fit is not feasible. Complicated monomer profiles and cluster patterns occur near the initial patch edge. An increase in the monomer concentration near the interface is in agreement with the simulations by Lacasta, Sancho, and Yeung [288]. The width of the boundary zone scales roughly as $N_x^{-1/2}$, in qualitative agreement with the model by Beenakker and Ross [287].

Sn on Si(111) was studied with partial masking during deposition [289–291] in the range of 5–50 ML and temperatures between 745 and 775 K. Ostwald ripening had previously been established for postdeposition annealing in this temperature range [260]. Concentration profiles across the interface were determined by scanning Auger microscopy and cluster size and densities were determined by electron microscopy with Figure 76 showing a sample deposited with a partially masked surface followed by postdeposition annealing for 120 min at 675 K. A comparison with estimated values for the diffusion lengths for the used growth conditions showed that the interface broadens significantly less than the diffusion length predicts, but that the broadening agrees well with estimates based on the screening length [290, 291].

An open, three-dimensional system is often established in thin film growth or ion beam surface modification if diffusion perpendicular to the growth front is possible. An intensively investigated system is the immiscible system silicon–cobalt where phase separation dynamics is expected due to the existence of a thermodynamically favored chemical compound, cobalt disilicide. It is commonly observed that the initial cobalt depth profile collapses during annealing into a buried $CoSi_2$ layer (mesotaxy) which bears a potential for applications due to its metallic properties [292, 293].

Fig. 76. Sn/Si(100) micrograph of a sample deposited with a partial mask, where the white spots are the Sn clusters formed during postdeposition annealing for 120 min at 675 K. Clusters have an average radius of 1.68 ± 0.05 μm, which is a size well within the late stage ripening regime [181]. The difference in cluster density across a roughly horizontal line corresponds to a initial deposition of 47.7-ML Sn in the upper part and 4.9-ML Sn in the lower part. Reprinted with permission from [289], © 1994, National Research Council of Canada.

A data set focusing on the role of the screening effect during $CoSi_2$ mesotaxy is shown in Figure 77 [291, 294, 295]. Four ion scattering depth profiles near the surface of low-dose implantation of 400-keV Co are shown, with each spectrum displaying an as-implanted profile and a profile obtained after annealing at 1275 K for 10 min. Varied is the dose of the implanted cobalt, ranging from 5% of the critical dose down to 1% [296, 297]. The profiles shown in Figure 77 therefore consist of spatially separated clusters. For doses above 3%, a secondary peak occurs near the surface which resists the collapse of the cobalt profile. This peak can be explained by the screening effect which becomes more dominant as the cluster density increases. When late stage clustering is reached, the zone near the surface with the initially lower cobalt concentration has smaller clusters (see the discussion of volume fraction effects on clustering rates in the next section), i.e., this zone is unstable against dissolution in favor of the larger clusters in the peak zone. If the density of clusters near the surface is sufficiently high, however, the screening length becomes significantly shorter than the width of the zone and cluster dissolution is slowed down.

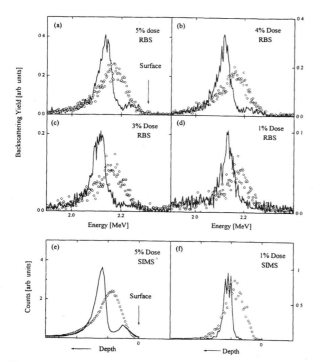

Fig. 77. Cobalt depth profiles (ion scattering data in panels (a)–(d)) for different implantation doses before (open circles) and after rapid thermal annealing at 1275 K for 10 min (solid lines). The dose is given in percent of the critical dose for formation of a coherent buried film. Note that the secondary peak near the surface is not present for doses less than 3% of the critical dose. Reprinted with permission from [295], © 1998, National Research Council of Canada.

2.3.4. Cluster Size and Spatial Ordering

(i) *Simulation Models.* In this section, we focus on spatial ordering during clustering, i.e., the formation of a nonrandom distribution of the distances and the relative orientation of clusters on the surface. While issues regarding the size distribution have been discussed intensively in the literature, only a few studies address spatial ordering. The importance of spatial distributions for a complete description has first been noted for three-dimensional–three-dimensional systems [298–300]. Akaiwa and Voorhees [301] made systematic numerical simulations and found that the multiparticle diffusion problem required a multipole expansion beyond the monopole term for volume fractions larger than 10%. At higher volume fractions, the higher order terms lead to an effective mobility of clusters with a significant effect on the spatial distributions. These effects include large depletion zones between small clusters but not between large clusters and a reduced probability to find small clusters near larger clusters.

The first group to study spatial ordering in two-dimensional–two-dimensional systems were Sagui and Desai [302–306]. They considered a late stage ripening system where a long-range repulsive interaction competes with the short-range attractive force due to surface tension. Two different approaches have been studied, based on a Langevin description [302, 303, 305, 307] and on a Ginzburg–Landau free-energy description [304, 306]. Both methods lead to comparable results. If the

long-range repulsive force is a dominant contribution, two noncoarsening structures emerge. If islands are sufficiently mobile, a highly ordered hexagonal structure results, or a frozen disordered structure in which islands have negligible mobility. If the repulsive interaction is weaker, a coarsening regime is reached where islands continue to grow. A power law behavior was not reached in Sagui and Desai's studies [304]. When the islands order hexagonally, a monodisperse size distribution is found. When mobility is reduced and the islands do not grow, their size distribution does not fully collapse as the correlation between growth and position is hindered. This frozen state leads to a bell-shaped size distribution which lies between the fully ordered and the ripening distribution. As the repulsive interaction between clusters becomes weaker, the size distribution resembles the classical LSW distribution more closely.

(ii) *Experimental Observations of Ordering.* Two-dimensional–two-dimensional systems were first studied for nonrandom ordering. Data by Krichevsky and Stavans [267, 268] and by Seul, Morgan, and Sire [269] and Morgan and Seul [281] have been used in several theoretical studies for comparison. The data suggest that a ring of neighboring islands increasingly orders as the areal fraction is increased and indicate that a strong anticorrelation occurs where islands with large areas tend to border preferentially to islands with small areas.

Ordering has also been observed for three-dimensional–two-dimensional systems. Anton, Harsdorff, and Möller [131] studied Au on cleaved NaCl surfaces, but the study focused on a very early regime where late stage conditions are not yet established. Ordering during late stage Ostwald ripening was studied for metal overlayers on Si(111) [216, 308, 309]. Since clustering during late stage ripening depends strongly on the areal fraction of clusters, data were obtained for low areal coverages ($1\% < \Phi < 3\%$) for the system Sn/Si(111) and for high areal coverages ($\Phi = 10$–11%) for the system In/Si(111). The cluster size distributions for both systems indicate strong cluster–cluster interactions. Figure 78a shows a SEM micrograph for Sn clustering on Si(111) with equivalent coverage of 47.7 ML, annealed at 675 K for 120 min with final areal coverage of $\Phi = 2.4\%$ [216]. For qualitative comparison, the micrograph is supplemented in Figure 78b by a computer simulated random distribution for the same areal density of clusters. In comparison, the cluster distances appear more uniform in the experimental data. To quantify this observation, Figure 79 shows the corresponding nearest neighbor distance distribution for the micrograph in Figure 78a. The experimental distribution is compared with two simulations, the same number of point clusters (negligible radius) and the same number of clusters of uniform radius $R = R_{max}$ with R_{max} the maximum cluster radius of the experimental distribution. Figure 79 shows that spatial distributions deviate significantly from random for ripening dominated growth.

In Figure 80, nearest neighbor distance distributions have been compiled for a larger number of experimental data with a wide range of morphological conditions, including a variation of an order of magnitude in average cluster–cluster distances

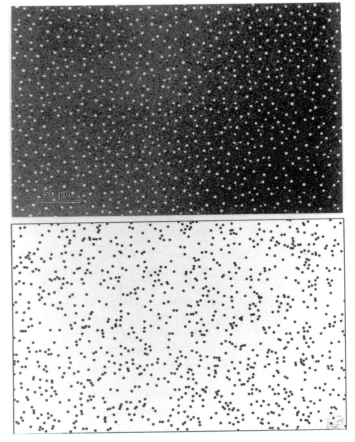

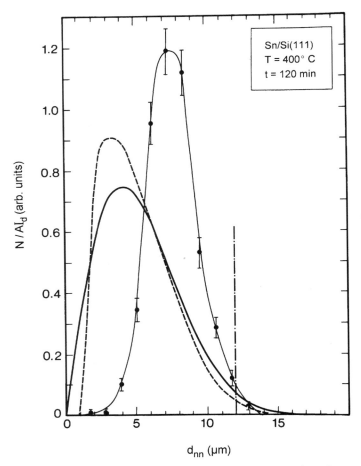

Fig. 78.   (a) SEM micrograph of Sn/Si(111) clustering of 47.7 ML at 675 K for 120 min. The areal coverage is $\Phi = 2.4\%$. (b) Computer simulation of random cluster distribution with same areal cluster density. In the simulation, overlapping of clusters is excluded. Reprinted with permission from [216], © 1997, Elsevier Science.

Fig. 79.   Nearest neighbor distance distribution for the same sample as shown in Figure 78. The experimental data are given as full symbols, connected with a thin solid line to guide the eye. Three theoretical curves are included for comparison: (i) a random simulation for point clusters (thick solid line), (ii) a random simulation for clusters with the same radius as the largest clusters in the micrograph of Figure 78a (dashed line), and (iii) a fully ordered, hexagonal arrangement with maximum cluster distances (dash-dotted vertical line). Reprinted with permission from [216], © 1997, Elsevier Science.

and more than 1 magnitude in areal coverage [308]. The collapse of the data onto a single curve suggests that late stage ripening morphologies develop with a self-similar spatial distribution [184]. As self-similarity holds in this system for both, the cluster size distribution [283, 290] and the cluster spatial distribution, the ripening mechanism must also be responsible for the ordering processes.

2.4. Transition between Ripening and Coalescence

The evolution of a mass-conserved systems has been discussed in terms of the Ostwald ripening concept. Coalescence has been shown to develop quite distinct morphologies for mass-nonconserved systems. In this section, we focus on the transition from one asymptotic growth evolution to the other by gradually increasing the deposition rate R.

The entire range of processes between ripening and coalescence dominated growth, including the transition between both regimes where the deposition rate is smaller than diffusive mass transfer between clusters, has been studied by Carlow, Barel, and Zinke-Allmang [309] and Carlow [310]. The study is based

on Sn and In/Si(111) for ripening and on Sn/Si(111) for the intermediate regime, established by a finite Sn deposition rate during substrate annealing. The Sn deposition rate was sufficiently low that a perturbation was given to first-order ripening conditions with the Sn flux; i.e., coalescence represents the asymptotic growth regime but the transient time toward coalescence is long compared to the time to complete the experiment.

Intermediate morphologies were obtained at Sn deposition rates of 0.1 nm/min with a total deposition for 120 min at 635 K. A novel feature is the propensity toward pairwise grouping of clusters. The cluster size distribution is narrower than that for samples grown under mass conservation at the same areal coverage. The nearest neighbor distributions are significantly broadened compared to those obtained with ripening conditions and are close to a random distribution. Thus the grouping of clusters compensates for the small nearest neighbor distances missing in the partially ordered late stage systems. The small deposition flux reduces the cluster–cluster interaction, reducing

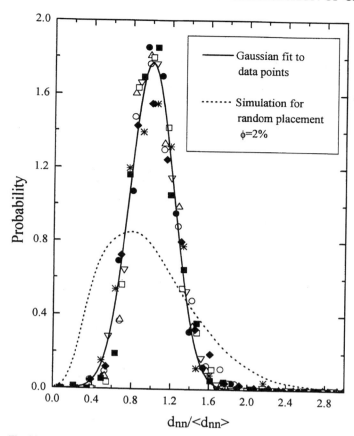

Fig. 80.   Rescaled nearest neighbor distance distributions for eight Sn/Si(111) samples using varying preparation parameters. The final areal coverages range from 0.12 to 2.4%. The solid line is a common fit to the experimental data. The dashed line is a computer simulated random distribution for a cluster areal coverage of 2%. Reprinted with permission from [308], © 1997, American Physical Society.

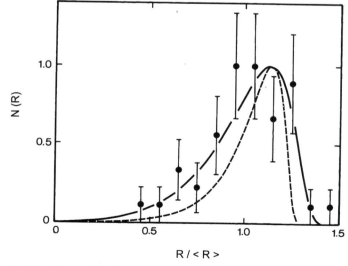

Fig. 81.   Cluster size distribution of 3.4 ± 0.1-ML Ge on Si(111) grown at 725 K with a deposition rate of 0.12 nm/min [311]. The solid line is the three-dimensional–three-dimensional LSW distribution and the dashed distribution is the three-dimensional–two-dimensional distribution predicted by Chakraverty. Reprinted with permission from [233], © 1996, Elsevier Science.

the rate at which clusters decompose at near proximity (local ripening), effectively recovering the LSW mean-field conditions except for a minor violation of mass conservation.

The narrowing of cluster size distributions during growth with a small deposition flux has also been reported by Deelman, Thundat, and Schowalter [311] for Ge growth on Si(111) and Si(100). The deposition rates were in the range of 0.12 nm/min to 0.3 nm/min and substrate temperatures were varied between 675 and 975 K. Figure 81 shows an experimental distribution in comparison with the LSW distribution and the distribution proposed by Chakraverty [233]. Note that both distributions are not suitable for this system since the LSW distribution was derived for three-dimensional–three-dimensional systems and Chakraverty's distribution was derived with an improper screening length. However, the close agreement with the experimental data indicate that the grown distribution is significantly narrower than the comparable ripening distributions for the same areal fraction [311]. This result is in agreement with earlier observations for the same system [181].

REFERENCES

1. K. Binder, in "Alloy Phase Stability" (G. M. Stocks and A. Gonis, Eds.), p. 233. Kluwer, Norwell, MA, 1989.
2. K. Binder, in "Materials Science and Technology: Phase Transformations in Materials" (R. W. Cahn, P. Haasen, and E. J. Kramer, Eds.), Vol. 5, p. 405. VCH, Weinheim, Germany, 1991.
3. M. C. Tringides, P. K. Wu, and M. G. Lagally, *Phys. Rev. Lett.* 59, 315 (1987).
4. M. Zinke-Allmang and L. C. Feldman, *Surf. Sci.* 191, L749 (1987).
5. G. R. Carlow, T. D. Lowes, M. Grunwell, and M. Zinke-Allmang, *Mater. Res. Soc. Symp. Proc.* 382, 419 (1995).
6. K. Yaldram and K. Binder, *Acta Metall. Mater.* 39, 707 (1991).
7. G. R. Calow, *Scaning Microsc.* 11, 947 (1997).
8. M. Bienfait, J. L. Seguin, J. Suzanne, E. Lerner, J. Krim, and J. G. Dash, *Phys. Rev. B* 29, 983 (1984).
9. M. H. Grabow and G. H. Gilmer, in "Semiconductor-based Heterostructures: Interfacial Structure and Stability" (M. L. Green, J. E. E. Baglin, G. Y. Chin, H. W. Deckman, W. Mayo, and D. Narasinham, Eds.), p. 3. Metallurgical Soc., Warrendale, PA, 1986.
10. D. A. Huse, *Phys. Rev. B* 29, 6985 (1984).
11. R. Pandit, M. Schick, and M. Wortis, *Phys. Rev. B* 26, 5112 (1982).
12. K. Binder, *Phys. Rev. B* 37, 1745 (1988).
13. I. N. Stranski and L. Krastanov, *Sitz. Akad. Wissenschaft., Wien* 146, 797 (1938).
14. M. Volmer and A. Weber, *Z. Phys. Chem. (Leipzig)* 119, 277 (1926).
15. W. Ostwald, *Z. Phys. Chem.* 34, 495 (1900).
16. M. L. Dundon and E. Mack, Jr., *J. Am. Chem. Soc.* 45, 2479 (1923).
17. W. D. Kingery, H. K. Bowen, and D. R. Uhlmann, "Introduction to Ceramics," Chaps. 5 and 9. Wiley, New York, 1976.
18. H. Freundlich, "Fortschritte der Kolloidchemie," Steinkopff, Dresden, 1926; translated in "New Conceptions in Colloidal Chemistry," Methuen, London, 1926.
19. G. W. Greenwood, *Acta Metall.* 4, 243 (1956).
20. M. Zinke-Allmang, *Thin Solid Films* 346, 1 (1999).
21. D. J. Eaglesham and M. Cerullo, *Phys. Rev. Lett.* 64, 1943 (1990).
22. Y.-W. Mo, D. E. Savage, B. S. Swartzentruber, and M. G. Lagally, *Phys. Rev. Lett.* 65, 1020 (1990).
23. S. Guha, A. Madhukar, and K. C. Rajkumar, *Appl. Phys. Lett.* 57, 2110 (1990).
24. C. H. P. Lupis, "Chemical Thermodynamics of Materials," Chaps. 1–3 and 13. Elsevier Science, New York, 1983.

25. M. Zinke-Allmang, L. C. Feldman, and M. H. Grabow, *Surf. Sci.* 200, L427 (1988).

26. I. M. Lifshitz and V. V. Slyozov, *J. Phys. Chem. Solids* 19, 35 (1961); *Sov. Phys. JETP* 35, 331 (1959).

27. C. Wagner, *Z. Elektrochem.* 65, 581 (1961).

28. F. Family and P. Meakin, *Phys. Rev. Lett.* 61, 428 (1988).

29. M. von Smoluchowski, *Z. Phys. Chem. (Leipzig)* 92, 129 (1917); *Phys. Z.* 17, 585 (1916).

30. J. A. Venables, *Philos. Mag.* 27, 697 (1973).

31. G. Zinsmeister, *Vacuum* 16, 529 (1966); translation from: R. Niedermayer and J. Mayer, Eds., "Basic Problems in Thin Film Physics," p. 33. Vandenhoeck and Ruprecht, Goettingen, 1966.

32. J. Frenkel, *Z. Phys.* 26, 117 (1924); "Kinetic Theory of Liquids," p. 42. Dover, New York, 1955.

33. R. A. Sigsbee, *J. Appl. Phys.* 42, 3904 (1971).

34. L. Yang, C. E. Birchenall, G. M. Pound, and M. T. Simnad, *Acta Metall.* 2, 462 (1954).

35. J. P. Hirth and G. M. Pound, "Condensation and Evaporation—Nucleation and Growth Kinetics," Pergamon, New York, 1963.

36. K. Binder and D. Stauffer, *Adv. Phys.* 25, 343 (1976).

37. J. A. Venables, G. D. T. Spiller, and M. Hanbücken, *Rep. Prog. Phys.* 47, 399 (1984).

38. D. R. Frankl and J. A. Venables, *Adv. Phys.* 19, 409 (1970).

39. J. A. Venables, *Philos. Mag.* 27, 697 (1973).

40. R. Vincent, *Proc. R. Soc. (London) A* 321, 53 (1971).

41. G. Zinsmeister, *Thin Solid Films* 2, 497 (1968).

42. J. L. Katz, in "Interfacial Aspects of Phase Transitions" (B. Mutaftschiev, Ed.), p. 261. Reidel, Dordrecht, 1982.

43. M. Hanbücken, M. Futamoto, and J. A. Venables, *Surf. Sci.* 147, 433 (1984).

44. J. A. Venables, *J. Vac. Sci. Technol. B* 4, 870 (1986).

45. G. W. Jones, J. M. Marcano, J. K. Norskov, and J. A. Venables, *Phys. Rev. Lett.* 65, 3317 (1990).

46. J. A. Venables, T. Doust, J. S. Drucker, and M. Krishnamurthy, in "Kinetics of Ordering and Growth at Surfaces" (M. G. Lagally, Ed.), p. 437. Plenum, New York, 1990.

47. T. Doust, F. L. Metcalfe, and J. A. Venables, *Ultramicroscopy* 31, 116 (1989).

48. H.-J. Gossmann and G. J. Fisanick, *Scanning Microsc.* 4, 543 (1990).

49. M. Henzler, *J. Phys. C* 16, 1543 (1983).

50. J. H. Neave, P. J. Dobson, B. A. Joyce, and J. Zhang, *Appl. Phys. Lett.* 47, 101 (1985).

51. Y.-W. Mo, R. Kariotis, B. S. Swartzentruber, M. B. Webb, and M. G. Lagally, *J. Vac. Sci. Technol. A* 8, 201 (1990).

52. M. C. Bartelt and J. W. Evans, *Phys. Rev. B* 46, 12,675 (1992).

53. M. C. Bartelt, M. C. Tringides, and J. W. Evans, *Phys. Rev. B* 47, 13,891 (1993).

54. M. C. Bartelt and J. W. Evans, *Surf. Sci.* 298, 421 (1993).

55. M. C. Bartelt and J. W. Evans, *Europhys. Lett.* 21, 99 (1993).

56. J. W. Evans and M. C. Bartelt, *Surf. Sci.* 284, L437 (1993).

57. J. W. Evans and M. C. Bartelt, *J. Vac. Sci. Technol. A* 12, 1800 (1994).

58. J. Villain, A. Pimpinelli, L. Tang, and D. Wolf, *J. Phys. I* 2, 2107 (1992).

59. Y.-W. Mo, J. Kleiner, M. B. Webb, and M. G. Lagally, *Phys. Rev. Lett.* 66, 1998 (1991).

60. Y.-W. Mo, J. Kleiner, M. B. Webb, and M. G. Lagally, *Surf. Sci.* 268, 275 (1992).

61. F. Family and P. Meakin, *Phys. Rev. Lett.* 61, 428 (1988).

62. F. Family and P. Meakin, *Phys. Rev. A* 40, 3836 (1989).

63. M. C. Bartelt and J. W. Evans, *Phys. Rev. B* 54, R17,359 (1996).

64. J. A. Venables, *Physica A* 239, 35 (1997).

65. C. Ratsch, A. Zangwill, and P. Šmilauer, *Surf. Sci.* 314, L937 (1994).

66. B. Z. Nosho, V. Bressler-Hill, S. Varma, and W. H. Weinberg, *Surf. Sci.* 364, 164 (1996).

67. G. S. Bales and D. C. Chrzan, *Phys. Rev. B* 50, 6057 (1994).

68. W. W. Mullins and R. F. Sekerka, *J. Appl. Phys.* 34, 323 (1963).

69. A. Pimpinelli, J. Villain, and D. E. Wolf, *J. Phys. I* 3, 447 (1993).

70. T. Michely, M. Hohage, and G. Comsa, in "Surface Diffusion: Atomistic and Collective Processes." NATO ASI Series B: Physics (M. C. Tringides, Ed.), Vol. 360, p. 125. Plenum, New York, 1997.

71. Z. Zhang, X. Chen, and M. G. Lagally, *Phys. Rev. Lett.* 73, 1829 (1994).

72. J. G. Amar and F. Family, *Phys. Rev. Lett.* 74, 2066 (1995).

73. P. A. Mulheran and J. A. Blackman, *Phys. Rev. B* 53, 10,261 (1996).

74. J. G. Amar and F. Family, *Surf. Sci.* 382, 170 (1997).

75. J. G. Amar and F. Family, *Thin Solid Films* 272, 208 (1996).

76. J. G. Amar, F. Family, and P.-M. Lam, *Phys. Rev. B* 50, 8781 (1994).

77. A. Zangwill and E. Kaxiras, *Surf. Sci.* 326, L483 (1995).

78. G. S. Bales, *Surf. Sci.* 356, L439 (1996).

79. J. A. Blackman and P. A. Mulheran, *Phys. Rev. B* 54, 11681 (1996).

80. P. A. Mulheran and J. A. Blackman, *Surf. Sci.* 376, 403 (1997).

81. K. R. Heim, S. T. Coyle, G. G. Hembree, J. A. Venables, and M. R. Scheinfein, *J. Appl. Phys.* 80, 1161 (1996).

82. C. Ratsch, A. Zangwill, P. Šmilauer, and D. D. Vvedensky, *Phys. Rev. Lett.* 72, 3194 (1994).

83. M. C. Bartelt, L. S. Perkins, and J. W. Evans, *Surf. Sci.* 344, L1193 (1995).

84. C. Ratsch, P. Šmilauer, A. Zangwill, and D. D. Vvedensky, *Surf. Sci.* 329, L599 (1995).

85. S. Liu, L. Bönig, and H. Metiu, *Phys. Rev. B* 52, 2907 (1995).

86. J. C. Hamilton, *Phys. Rev. Lett.* 77, 885 (1996).

87. I. Furman and O. Biham, *Phys. Rev. B* 55, 7917 (1997).

88. L. Kuipers and R. E. Palmer, *Phys. Rev. B* 53, R7646 (1996).

89. M. C. Bartelt, S. Günther, E. Kopatzki, R. J. Behm, and J. W. Evans, *Phys. Rev. B* 53, 4099 (1996).

90. J. A. Venables, *Surf. Sci.* 299–300, 798 (1993).

91. J. A. Venables, G. D. T. Spiller, and M. Hanbücken, *Rep. Prog. Phys.* 47, 399 (1984).

92. M. G. Lagally, *Jpn. J. Appl. Phys.* 32, 1493 (1993).

93. U. Köhler, *Microsc. Microanal. Microstruct.* 5, 247 (1994).

94. Y.-W. Mo and M. G. Lagally, *Surf. Sci.* 248, 313 (1991).

95. M. G. Lagally, private communication, presented at the CpiP'98 Conference in Barbados.

96. W. Theis and R. M. Tromp, *Phys. Rev. Lett.* 76, 2770 (1996).

97. T. D. Lowes and M. Zinke-Allmang, *Scanning Microsc.* 11, 947 (1997).

98. P. A. Bennett, S. A. Parikh, and D. G. Cahill, *J. Vac. Sci. Technol. A* 11, 1680 (1993).

99. F. L. Metcalfe and J. A. Venables, *Surf. Sci.* 369, 99 (1996).

100. J. A. Venables, F. L. Metcalfe, and A. Sugawara, *Surf. Sci.* 371, 420 (1997).

101. J. A. Venables, T. Doust, J. S. Drucker, and M. Krishnamurthy, in "Kinetics of Ordering and Growth at Surfaces." NATO ASI Series, (M. G. Lagally, Ed.), p. 437. Plenum, New York, 1990.

102. G. Raynerd, T. N. Doust, and J. A. Venables, *Surf. Sci.* 261, 251 (1992).

103. X.-S. Wang, J. Brake, R. J. Pechman, and J. H. Weaver, *Appl. Phys. Lett.* 68, 1660 (1996).

104. Y.-N. Yang, Y. S. Luo, and J. H. Weaver, *Phys. Rev. B* 46, 15,387 (1992).

105. Y.-N. Yang, Y. S. Luo, and J. H. Weaver, *Phys. Rev. B* 46, 15,395 (1992).

106. M. Kasu and N. Kobayashi, *J. Cryst. Growth* 170, 246 (1997).

107. V. Bressler-Hill, S. Varma, A. Lorke, B. Z. Nosho, P. M. Petroff, and W. H. Weinberg, *Phys. Rev. Lett.* 74, 3209 (1995).

108. R. E. Welser and L. J. Guido, *Appl. Phys. Lett.* 68, 912 (1996).

109. S. Stoyanov, *Appl. Phys. A* 50, 349 (1990).

110. J. A. Stroscio and D. T. Pierce, *Phys. Rev. B* 49, 8522 (1994).

111. M. Bott, M. Hohage, M. Morgenstern, T. Michely, and G. Comsa, *Phys. Rev. Lett.* 76, 1304 (1996).

112. M. J. Stowell, *Philos. Mag.* 26, 349 (1972).

113. G. Zinsmeister, *Thin Solid Films* 7, 51 (1971).

114. H. Brune, H. Röder, C. Boragno, and K. Kern, *Phys. Rev. Lett.* 73, 1955 (1994).

115. C. Ratsch, A. Zangwill, P. Šmilauer, and D. D. Vvedensky, *Phys. Rev. Lett.* 72, 3194 (1994).

116. P. Blandin, C. Massobrio, and P. Ballone, *Phys. Rev. B* 49, 16,637 (1994).

117. B. Schmiedeskamp, B. Kessler, B. Vogt, and U. Heinzmann, *Surf. Sci.* 223, 465 (1989).

118. H. Brune, K. Bromann, H. Röder, K. Kern, J. Jacobsen, P. Stoltze, K. Jacobsen, and J. Nørskov, *Phys. Rev. B* 52, R14,380 (1995).
119. H. Röder, K. Bromann, H. Brune, and K. Kern, *Surf. Sci.* 376, 13 (1997).
120. J.-K. Zuo, J. F. Wendelken, H. Dürr, and C.-L. Liu, *Phys. Rev. Lett.* 72, 3064 (1994).
121. H.-J. Ernst, F. Fabre, and J. Lapujoulade, *Phys. Rev. B.* 46, 1929 (1992).
122. S. V. Ghaisas and S. Das Sarma, *Phys. Rev. B* 46, 7308 (1992).
123. D. D. Chambliss, R. J. Wilson, and S. Chiang, *J. Vac. Sci. Technol. B* 10, 1993 (1992).
124. K. E. Johnson, D. D. Chambliss, R. J. Wilson, and S. Chiang, *J. Vac. Sci. Technol. A* 11, 1654 (1993).
125. D. D. Chambliss and K. E. Johnson, *Phys. Rev. B* 50, 5012 (1994).
126. G. L. Kellogg and P. J. Feibelman, *Phys. Rev. Lett.* 64, 3143 (1990).
127. C. Nagl, E. Platzgummer, M. Schmid, P. Varga, S. Speller, and W. Heiland, *Surf. Sci.* 352–354, 540 (1996).
128. Q. Jiang and G.-C. Wang, *Surf. Sci.* 324, 357 (1995).
129. Y.-L. He and G.-C. Wang, *Phys. Rev. Lett.* 71, 3834 (1993).
130. S. Günther, E. Kopatzki, M. C. Bartelt, J. W. Evans, and R. J. Behm, *Phys. Rev. Lett.* 73, 553 (1994).
131. R. Anton, M. Harsdorff, and A. Möller, *Phys. Status Solidi A* 146, 269 (1994).
132. J. W. Cahn and J. E. Hilliard, *J. Chem. Phys.* 28, 258 (1958); 31, 688 (1959).
133. K. Kawasaki, *J. Stat. Phys.* 12, 365 (1975).
134. T. M. Rogers and R. C. Desai, *Phys. Rev. B* 39, 11,956 (1989).
135. F. F. Abraham, S. W. Koch, and R. C. Desai, *Phys. Rev. Lett.* 49, 923 (1982).
136. C. W. Snyder, B. G. Orr, D. Kessler, and L. M. Sander, *Phys. Rev. Lett.* 66, 3032 (1991).
137. J. M. Moison, F. Houzay, F. Barthe, L. Leprince, E. Andre, and O. Vatel, *Appl. Phys. Lett.* 64, 196 (1994); D. Leonard, K. Pond, and P. M. Petroff, *Phys. Rev. B* 50, 11,687 (1994).
138. F. Iwaraki, M. Tomitori, and O. Nishikawa, *Surf. Sci. Lett.* 253, L411 (1991); J. Knall and K. B. Pethica, *Surf. Sci.* 265, 156 (1992).
139. D. Leonard, M. Krishnamurthy, C. M. Reaves, S. P. Denbaars, and P. M. Petroff, *Appl. Phys. Lett.* 63, 3203 (1993); J. M. Moison, F. Houzay, F. Barthe, and L. Leprince, *Appl. Phys. Lett.* 64, 196 (1994); A. Madhukar, Q. Xie, P. Chen, and A. Konkar, *Appl. Phys. Lett.* 64, 2727 (1994).
140. N. Kirstaedter, N. N. Ledentsov, M. Grundmann, D. Bimberg, V. M. Ustinov, S. S. Ruvimov, M. V. Maximov, P. S. Kop'ev, Zh. I. Alferov, U. Richter, P. Werner, U. Goesele, and J. Heydenriech, *Electron. Lett.* 30, 1416 (1994); Zh. I. Alferov, *Phys. Scr.* 68, 32 (1996); N. N. Ledentsov, V. M. Ustinov, V. A. Shchukin, P. S. Kop'ev, Zh. I. Alferov, and D. Bimberg, *Fiz. Tekh. Poluprovoddn.* 32, 385 (1998) [*Semiconductors* 32, (1998)].
141. See, e.g., M. Krishnamurthy, J. S. Drucker, and J. A. Venables, *J. Appl. Phys.* 69, 6461 (1991); D. Leonard, M. Krishnamurthy, S. Fafard, J. L. Merz, and P. M. Petroff, *J. Vac. Sci. Technol. B* 12, 1063 (1994).
142. F. C. Frank and J. H. Van der Merwe, *Proc. R. Soc. London, Ser. A* 198, 205 (1949).
143. I. Daruka and A.-L. Barabasi, *Phys. Rev. Lett.* 79, 3708 (1997).
144. W. Seifert, N. Carlsson, M. Miller, M.-E. Pistol, L. Samuelson, and L. R. Wallenberg, *Prog. Cryst. Growth Charact.* 33, 423 (1996).
145. C. Weisbuch and B. Vinter, "Quantum Semiconductor Structures: Fundamentals and Applications," Academic Press, New York, 1991.
146. D. J. Eaglesham and R. Hull, *Mater. Sci. Eng. B* 30, 197 (1995).
147. C. Ratsch and A. Zangwill, *Surf. Sci.* 293, 123 (1993).
148. V. A. Shchukin, N. N. Ledentsov, P. S. Kop'ev, and D. Bimberg, *Phys. Rev. Lett.* 75, 2968 (1995).
149. V. A. Shchukin, A. I. Borovkov, N. N. Ledentsov, and D. Bimberg, *Phys. Rev. B* 51, 10,104 (1995); V. A. Shchukin, A. I. Borovkov, N. N. Ledentsov, and P. S. Kop'ev, *Phys. Rev. B* 51, 17,767 (1995).
150. V. A. Shchukin and D. Bimberg, *Appl. Phys. A* 67, 687 (1998).
151. C. Priester and M. Lannoo, *Phys. Rev. Lett.* 75, 93 (1995).
152. C. Priester and M. Lannoo, *Appl. Surf. Sci.* 104–105, 495 (1996).
153. B. G. Orr, D. Kessler, C. W. Snyder, and L. Sander, *Europhys. Lett.* 19, 33 (1992).
154. C. Ratsch, P. Šmilauer, D. D. Vvedensky, and A. Zangwill, *J. Phys. I* 6, 575 (1996).
155. W. Seifert, N. Carlsson, J. Johansson, M.-E. Pistol, and L. Samuelson, *J. Cryst. Growth* 170, 39 (1997).
156. H. T. Johnson and L. B. Freund, *J. Appl. Phys.* 81, 6081 (1997).
157. G. Abstreiter, P. Schittenhelm, C. Engel, E. Silveira, A. Zrenner, D. Meertens, and W. Jager, *Semicond. Sci. Technol.* 11, 1521 (1996).
158. P. M. Petroff and G. Medeiros-Ribeiro, *MRS Bull.* 21, 50 (1996); T. I. Kamins, E. C. Carr, R. S. Williams, and S. R. Rosner, *J. Appl. Phys.* 81, 211 (1997).
159. C. W. Snyder, J. F. Mansfield, and B. G. Orr, *Phys. Rev. B* 46, 9551 (1992).
160. T. R. Ramachandran, R. Heitz, P. Chen, and A. Madhukar, *Appl. Phys. Lett.* 70, 640 (1997).
161. H. Sunamura, N. Shiraki, and S. Fukatsu, *Appl. Phys. Lett.* 66, 3024 (1995); P. Schittenhelm, M. Gail, J. Brunner, J. F. Nutzel, and G. Abstreiter, *Appl. Phys. Lett.* 67, 1292 (1995).
162. D. E. Jesson, K. M. Chen, S. J. Pennycook, T. Thundat, and R. J. Warmack, *Phys. Rev. Lett.* 77, 1330 (1996).
163. I. Goldfarb, P. T. Hayden, J. H. G. Owen, and G. A. D. Briggs, *Phys. Rev. Lett.* 78, 3959 (1997).
164. D. E. Jesson, K. M. Chen, S. J. Pennycook, T. Thundat, and R. J. Warmack, *Science* 268, 1161 (1995).
165. Y. Chen and J. Washburn, *Phys. Rev. Lett.* 77, 4046 (1996).
166. C. Ling, *J. Math. Phys.* 26, 284 (1948).
167. B. Voigtlander and A. Zinner, *Appl. Phys. Lett.* 63, 3055 (1993); F. K. LeGoues, M. C. Reuter, J. Tersoff, M. Hammer, and R. M. Tromp, *Phys. Rev. Lett.* 73, 300 (1994); S. Guha, A. Madhukar, and K. C. Rajkumar, *Appl. Phys. Lett.* 57, 2111 (1990); M. Sopanen, H. Lipsanen, and J. Ahopelto, *Appl. Phys. Lett.* 67, 3768 (1995); Y. Chen, X. W. Lin, Z. L. Weber, J. Washburn, E. R. Weber, A. Sasaki, A. Wakahara, and T. Hasegawa, *Phys. Rev. B* 52, 16,581 (1995).
168. G.-M. Ribeiro, A. M. Bratkovski, T. I. Kamins, D. A. A. Ohlberg, and R. S. Williams, *Science* 279, 353 (1998).
169. J. Tersoff and R. M. Tromp, *Phys. Rev. Lett.* 70, 2782 (1993).
170. J. Tersoff and F. K. LeGoues, *Phys. Rev. Lett.* 72, 3570 (1994).
171. D. E. Jesson, S. J. Pennycook, J.-M. Baribeau, and D. C. Houghton, *Phys. Rev. Lett.* 71, 1744 (1993).
172. F. M. Ross, J. Tersoff, and R. M. Tromp, *Phys. Rev. Lett.* 80, 984 (1998).
173. K. M. Chen, D. E. Jesson, S. J. Pennycook, T. Thundat, and R. J. Warmack, *Phys. Rev. B* 56, R1700 (1997).
174. T. I. Kamins, E. C. Carr, R. S. Willaims, and S. J. Rosner, *J. Appl. Phys.* 81, 211 (1997).
175. G.-M. Ribeiro, T. I. Kamins, D. A. A. Ohlberg, and R. S. Williams, *Phys. Rev. B* 58, 3533 (1998).
176. R. S. Williams, G.-M. Ribeiro, T. I. Kamins, and D. A. A. Ohlberg, *J. Phys. Chem. B* 102, 9605 (1998).
177. J. A. Floro, G. A. Lucadamo, E. Chason, L. B. Freund, M. Sinclair, R. D. Twesten, and R. Q. Hwang, *Phys. Rev. Lett.* 80, 4717 (1998).
178. J. A. Floro, E. Chason, R. D. Twestern, R. Q. Hwang, and L. B. Freund, *Phys. Rev. Lett.* 79, 3946 (1997); J. A. Floro, E. Chason, S. R. Leee, R. D. Twestern, R. Q. Twang, and L. B. Freund, *J. Electon. Mater.* 26, 983 (1997).
179. E. Chason, M. Sinclair, J. A. Floro, and G. A. Lucadamo, to be published.
180. D. A. Beysens, C. M. Knobler, and H. Schaffar, *Phys. Rev. B* 41, 9814 (1990).
181. M. Zinke-Allmang, L. C. Feldman, and M. H. Grabow, *Surf. Sci. Rept.* 16, 377 (1992).
182. D. A. Beysens, P. Guenoun, P. Sibille, and A. Kumar, *Phys. Rev. E* 50, 1299 (1994).
183. D. A. Beysens, *Physica A* 239, 329 (1997).
184. W. W. Mullins, *J. Appl. Phys.* 59, 1341 (1986).
185. W. Li, G. Vidali, and O. Biham, *Phys. Rev. B* 48, 8336 (1993).
186. K. Binder and D. Stauffer, *Phys. Rev. Lett.* 33, 1006 (1974).
187. E. D. Siggia, *Phys. Rev. A* 20, 595 (1979).
188. D. A. Beysens and Y. Jayalakshmi, *Physica A* 213, 71 (1995).
189. S. V. Khare, N. C. Bartelt, and T. L. Einstein, *Phys. Rev. Lett.* 75, 2148 (1995).

190. S. V. Khare and T. L. Einstein, *Phys. Rev. B* 54, 11,752 (1996).
191. W. Święch, E. Bauer, and M. Mundschau, *Surf. Sci.* 253, 283 (1991).
192. E. Ruckenstein and B. Pulvermacher, *J. Catal.* 29, 224 (1973).
193. E. E. Gruber, *J. Appl. Phys.* 38, 243 (1967).
194. P. Meakin, in "Phase Transitions and Critical Phenomena" (C. Domb and J. L. Lebowitz, Eds.), p. 335. Academic Press, London, 1988.
195. P. Meakin, T. Vicsek, and F. Family, *Phys. Rev. B* 31, 564 (1985).
196. F. Family, in "On Clusters and Clustering, from Atoms to Fractals" (P. J. Reynolds, Ed.), p. 323. Elsevier, Amsterdam, 1993.
197. D. S. Sholl and R. T. Skodje, *Physica A* 231, 631 (1996).
198. A. Steyer, P. Guenoun, D. Beysens, and C. M. Knobler, *Phys. Rev. A* 44, 8271 (1991).
199. K. Mašek, V. Matolín, and M. Gillet, *Thin Solid Films* 260, 252 (1995).
200. J.-M. Wen, S.-L. Chang, J. W. Burnett, J. W. Evans, and P. A. Thiel, *Phys. Rev. Lett.* 73, 2591 (1994).
201. J.-M. Wen, J. W. Evans, M. C. Bartelt, J. W. Burnett, and P. A. Thiel, *Phys. Rev. Lett.* 76, 652 (1996).
202. D. S. Sholl and R. T. Skodje, *Phys. Rev. Lett.* 75, 3158 (1995).
203. E. Søndergård, R. Kofman, P. Cheyssac, and A. Stella, *Surf. Sci.* 364, 467 (1996).
204. P. Meakin, *Physica A* 165, 1 (1990).
205. C. S. Jayanth and P. Nash, *J. Mater. Sci.* 24, 3041 (1989).
206. J. L. Viovy, D. Beysens, and C. M. Knobler, *Phys. Rev. A* 37, 4965 (1988).
207. T. M. Rogers, K. R. Elder, and R. C. Desai, *Phys. Rev. A* 38, 5303 (1988).
208. J. L. Viovy, D. Beysens, and C. M. Knobler, *Phys. Rev. A* 37, 4965 (1988).
209. P. Meakin, *Phys. Scr. T* 44, 31 (1992).
210. P. Meakin and F. Family, *J. Phys. A* 22, L225 (1989).
211. G. Desrousseaux, A. Carlan, B. Robrieux, H. Schaffar, J. Trompette, J. P. Dussaulcy, and R. Faure, *J. Phys. Chem. Solids* 46, 929 (1985).
212. R. Vincent, *Proc. R. Soc. London Ser. A* 321, 53 (1971).
213. M. Zinke-Allmang, L. C. Feldman, and W. van Saarloos, *Phys. Rev. Lett.* 68, 2358 (1992).
214. T. D. Lowes and M. Zinke-Allmang, *Phys. Rev. B* 49, 16,678 (1994).
215. Y.-N. Yang, Y. S. Luo, and J. H. Weaver, *Phys. Rev. B* 45, 3606 (1992).
216. R. Barel, G. R. Carlow, M. Zinke-Allmang, Y. Wu, and T. Lookman, *Physica A* 239, 53 (1997).
217. I. M. Lifshitz and V. V. Slyozov, *J. Phys. Chem. Solids* 19, 35 (1961); *Sov. Phys. JETP* 35, 331 (1959).
218. J. T. Wetzel, L. D. Roth, and J. K. Tien, *Acta Metall.* 32, 1573 (1984).
219. J. Tersoff, A. W. Denier van der Gon, and R. M. Tromp, *Phys. Rev. Lett.* 70, 1143 (1993).
220. W. Y. Lum and A. R. Clawson, *J. Appl. Phys.* 50, 5296 (1979).
221. E. L. Hinrichsen, J. Feder, and T. Jossang, *J. Stat. Phys.* 44, 793 (1986).
222. B. S. Shklovskii and A. L. Efros, "Electronic Properties of Doped Semiconductors," Springer-Verlag, Berlin, 1994.
223. P. Bruschi, P. Cagnoni, and A. Nannini, *Phys. Rev. B* 55, 7955 (1997).
224. B. Voigtländer, H. P. Bonzel, and H. Ibach, *Z. Phys. Chem.* 198, 189 (1997).
225. S. Blacher, F. Brouers, P. Gadenne, and J. Lafait, *J. Appl. Phys.* 74, 207 (1993).
226. C. Wagner, *Z. Elektrochem.* 65, 581 (1961).
227. P. Wynblatt and N. A. Gjostein, in "Progress in Solid State Chemistry" (J. O. McCaldin and G. Somorjai, Eds.), p. 21. Pergamon, Oxford, U.K., 1975.
228. J. A. Marqusee and J. Ross, *J. Chem. Phys.* 80, 536 (1984).
229. Y. Enomoto, M. Tokuyama, and K. Kawasaki, *Acta Metall.* 34, 2119 (1986).
230. P. W. Voorhees and M. E. Glicksman, *Acta Metall.* 32, 2001, 2013 (1984).
231. C. K. L. Davies, P. Nash, and R. N. Stevens, *Acta Metall.* 28, 179 (1980).
232. T. M. Rogers and R. C. Desai, *Phys. Rev. B* 39, 11,956 (1989).
233. B. K. Chakraverty, *J. Phys. Chem. Solids* 28, 2401 (1967).
234. J. A. Marqusee, *J. Phys. Chem.* 81, 976 (1984).
235. M. Marder, *Phys. Rev. A* 36, 858 (1987).
236. Q. Zheng and J. D. Gunton, *Phys. Rev. A* 39, 4848 (1989).
237. H. Hayakawa and F. Family, *Physica A* 163, 491 (1990).
238. J. H. Yao, K. R. Elder, H. Guo, and M. Grant, *Phys. Rev. B* 45, 8173 (1992).
239. J. H. Yao, K. R. Elder, H. Guo, and M. Grant, *Phys. Rev. B* 47, 14,110 (1993).
240. J. H. Yao, K. R. Elder, H. Guo, and M. Grant, *Physica A* 204, 770 (1994).
241. N. Akaiwa and D. I. Meiron, *Phys. Rev. E* 51, 5408 (1995).
242. M. Zinke-Allmang, G. R. Carlow, and S. Yu. Krylov, *Physica A* 261, 115 (1998).
243. A. J. Ardell, *Phys. Rev. B* 41, 2554 (1990).
244. P. Wynblatt and N. A. Gjostein, *Acta Metall.* 24, 1165 (1976); P. Wynblatt, *Acta Metall.* 24, 1175 (1976).
245. C. V. Thompson, *Acta Metall.* 36, 2929 (1988).
246. J. Viñals and W. W. Mullins, *J. Appl. Phys.* 83, 621 (1998).
247. K. Shorlin, S. Yu. Krylov, and M. Zinke-Allmang, *Physica A* 261, 248 (1998).
248. P. L. Krapivsky, *Phys. Rev. E* 47, 1199 (1993).
249. X. Zheng and B. Bigot, *J. Phys. II* 4, 743 (1994).
250. H. A. Atwater and C. M. Yang, *J. Appl. Phys.* 67, 6202 (1990).
251. R. D. Vengrenovich, *Acta Metall.* 30, 1079 (1982).
252. W. W. Mullins, *J. Appl. Phys.* 59, 1341 (1986).
253. W. W. Mullins and J. Vinals, *Acta Metall.* 37, 991 (1989).
254. K. Binder, *Phys. Rev. B* 15, 4425 (1977).
255. H. Xia and M. Zinke-Allmang, *Physica A* 261, 176 (1998).
256. J. Li, A. G. Rojo, and L. M. Sander, *Phys. Rev. Lett.* 78, 1747 (1997).
257. M. Tomellini, *Appl. Surf. Sci.* 99, 67 (1996).
258. B. Müller, L. Nedelmann, B. Fischer, H. Brune, and K. Kern, *Phys. Rev. B* 54, 17,858 (1996).
259. B. Krishnamachari, J. McLean, B. Cooper, and J. Sethna, *Phys. Rev. B* 54, 8899 (1996).
260. M. Zinke-Allmang and L. C. Feldman, *Appl. Surf. Sci.* 52, 357 (1991).
261. S. A. Kukushkin and A. V. Osipov, *J. Phys. Chem. Solids* 56, 211 (1995).
262. E. Ruckenstein and D. B. Dadyburjor, *Thin Solid Films* 55, 89 (1978).
263. J. P. Hirth, *J. Cryst. Growth* 17, 63 (1972).
264. R. Kern, in "Interfacial Aspects of Phase Transitions" (B. Mutaftschiev, Ed.), p. 203. Reidel, Dordrecht, 1982.
265. M. Zinke-Allmang, L. C. Feldman, and M. H. Grabow, *Surf. Sci.* 200, L427 (1988).
266. H.-J. Gossmann and G. J. Fisanick, *Scanning Microsc.* 4, 543 (1990).
267. O. Krichevsky and J. Stavans, *Phys. Rev. Lett.* 70, 1473 (1993).
268. O. Krichevsky and J. Stavans, *Phys. Rev. E* 52, 1818 (1995).
269. M. Seul, N. Y. Morgan, and C. Sire, *Phys. Rev. Lett.* 73, 2284 (1994).
270. N. C. Bartelt, W. Theis, and R. M. Tromp, *Phys. Rev. B* 54, 11,741 (1996).
271. W. Theis, N. C. Bartelt, and R. M. Tromp, *Phys. Rev. Lett.* 75, 3328 (1995).
272. E. Bauer, *Appl. Surf. Sci.* 92, 20 (1996).
273. O. Cavalleri, A. Hirstein, and K. Kern, *Surf. Sci.* 340, L960 (1995).
274. O. Cavalleri, A. Hirstein, J.-P. Bucher, and K. Kern, *Thin Solid Films* 284–285, 392 (1996).
275. H.-J. Ernst, F. Fabre, and J. Lapujoulade, *Phys. Rev. Lett.* 69, 458 (1992).
276. M. Zinke-Allmang, *Nucl. Instrum. Methods B* 64, 113 (1992).
277. M. Zinke-Allmang and L. C. Feldman, *Appl. Surf. Sci.* 52, 357 (1992).
278. J. R. Rogers, J. P. Downey, W. K. Witherow, B. R. Facemire, D. O. Frazier, V. E. Fradkov, S. S. Mani, and M. E. Glicksman, *J. Electron. Mater.* 23, 999 (1994).
279. J. P. Hirth, *J. Cryst. Growth* 17, 63 (1972).
280. E. Ruckenstein and D. B. Dadyburjor, *Thin Solid Films* 55, 89 (1978).
281. N. Y. Morgan and M. Seul, *J. Phys. Chem.* 99, 2088 (1995).
282. M. Hillert, *Acta Metall.* 13, 227 (1965).
283. R. Barel, Y. Mai, G. R. Carlow, and M. Zinke-Allmang, *Appl. Surf. Sci.* 104–105, 669 (1996).
284. M. Zinke-Allmang, L. C. Feldman, S. Nakahara, and B. A. Davidson, *Phys. Rev. B* 39, 7848 (1989).
285. M. Tokuyama and Y. Enomoto, *Int. J. Thermophys.* 15, 1145 (1994).
286. N. Akaiwa and D. I. Meiron, *Phys. Rev. E* 54, R13 (1996).
287. C. W. J. Beenakker and J. Ross, *J. Chem. Phys.* 83, 4710 (1985).
288. A. M. Lacasta, J. M. Sancho, and C. Yeung, *Europhys. Lett.* 27, 291 (1994).
289. G. R. Carlow and M. Zinke-Allmang, *Can. J. Phys.* 72, 812 (1994).
290. G. R. Carlow and M. Zinke-Allmang, *Surf. Sci.* 328, 311 (1995).

291. G. R. Carlow, D. D. Perovi, and M. Zinke-Allmang, *Appl. Surf. Sci.* 130–132, 704 (1998).

292. A. E. White, K. T. Short, R. C. Dynes, J. P. Garno, and J. M. Gibson, *Appl. Phys. Lett.* 50, 95 (1987).

293. A. E. White, K. T. Short, Y.-F. Hsieh, and R. Hull, *Mater. Sci. Eng. B* 12, 107 (1992).

294. M. Zinke-Allmang, in "Surface Diffusion: Atomistic and Collective Processes," NATO ASI Series B: Physics (M. C. Tringides, Ed.), Vol. 360, p. 389. Plenum, New York, 1997.

295. G. R. Carlow and M. Zinke-Allmang, *Can. J. Chem.* 76, 1737 (1998).

296. Y.-F. Hsieh, R. Hull, A. E. White, and K. T. Short, *J. Appl. Phys.* 70, 7354 (1991).

297. Y.-F. Hsieh, R. Hull, A. E. White, and K. T. Short, *Appl. Phys. Lett.* 58, 122 (1991).

298. M. Tokuyama and K. Kawasaki, *Physica A* 123, 386 (1984).

299. M. Marder, *Phys. Rev. Lett.* 55, 2953 (1985).

300. C. W. J. Beenakker, *Phys. Rev. A* 33, 4482 (1986).

301. N. Akaiwa and P. W. Voorhees, *Phys. Rev. E* 49, 3860 (1994).

302. C. Sagui and R. C. Desai, *Phys. Rev. Lett.* 71, 3995 (1993).

303. C. Sagui and R. C. Desai, *Phys. Rev. E* 49, 2225 (1994).

304. C. Sagui and R. C. Desai, *Phys. Rev. E* 52, 2822 (1995).

305. C. Sagui and R. C. Desai, *Phys. Rev. E* 52, 2807 (1995).

306. C. Sagui and R. C. Desai, *Phys. Rev. Lett.* 74, 1119 (1995).

307. C. Roland and R. C. Desai, *Phys. Rev. B* 42, 6658 (1990).

308. G. R. Carlow and M. Zinke-Allmang, *Phys. Rev. Lett.* 78, 4601 (1997).

309. G. R. Carlow, R. J. Barel, and M. Zinke-Allmang, *Phys. Rev. B* 56, 12,519 (1997).

310. G. R. Carlow, *Physica A* 239, 65 (1997).

311. P. W. Deelman, T. Thundat, and L. J. Schowalter, *Appl. Surf. Sci.* 104–105, 510 (1996).

312. M. G. Lagally, R. Kariotis, B. S. Swartzentruber, and Y.-W. Mo, *Ultramicroscopy* 31, 87 (1989).

313. M. Minke-Allmang, L. C. Feldman, and S. Nakahara, *Appl. Phys. Lett.* 51, 975 (1987).

314. M. Minke-Allmang, L. C. Feldman, and S. Nakahara, *Appl. Phys. Lett.* 52, 144 (1988).

315. D. K. Biegelsen, F. A. Ponce, A. J. Smith, and J. C. Tranmontana, *J. Appl. Phys.* 61, 1856 (1987).

316. M. Minke-Allmang, L. C. Feldman, and S. Nakahara, in "Kinetics of Ordering and Growth at Surfaces" (M. G. Lagally, Ed.), p. 455. Plenum, New York, 1990.

317. J. J. Harris, B. A. Joyce, J. P. Gowers, and J. H. Neave, *Appl. Phys. A* 28, 63 (1982).

318. K. Hartig, A. P. Janssen, and J. A. Venables, *Surf. Sci.* 74, 69 (1978).

Chapter 2

BAND STRUCTURE AND ORIENTATION OF MOLECULAR ADSORBATES ON SURFACES BY ANGLE-RESOLVED ELECTRON SPECTROSCOPIES

P. A. Dowben, Bo Xu

Department of Physics and Astronomy and the Center for Materials Research and Analysis, Behlen Laboratory of Physics, University of Nebraska, Lincoln, Nebraska, USA

Jaewu Choi, Eizi Morikawa

Center for Advanced Microstructures and Devices, Louisiana State University, Baton Rouge, Louisiana, USA

Contents

Handbook of Thin Film Materials, edited by H.S. Nalwa
Volume 2: Characterization and Spectroscopy of Thin Films
Copyright © 2002 by Academic Press
All rights of reproduction in any form reserved.

ISBN 0-12-512910-6/$35.00

1. INTRODUCTION

Band structure is the energy dependence of electron states as a function of wave vector. This is the result of the combination (in the most simple *ansatz*: a linear combination) of atomic orbitals in a periodic lattice. Similarly, molecular orbitals can also combine to form a band structure [1–4]. Though molecular orbitals are highly localized in molecular overlayers and molecular crystals, there is sufficient interaction between adjacent molecules to result in dispersion, as in many examples of crystalline molecular overlayers and thin films. This is the subject of this review.

As such, this subject is complementary to reviews on the application of electron spectroscopies, principally photoemission, for determining molecular orientation from dipole and symmetry selection rules [1–4]. We are extending a recent review of molecular orientation studies using photoemission [4] to include experimental molecular electronic band structure determinations.

The band structure of the occupied (valence band) and unoccupied (conduction band) molecular orbitals can be measured using angle-resolved photoemission spectroscopy (ARPES or ARUPS) and angle-resolved inverse photoemission spectroscopy (ARIPES), respectively. It is important that the molecular overlayer or crystal be ordered if electronic band structure is to be clearly and accurately determined. Such ordering includes well-defined molecular orientation(s).

The dipole and symmetry selection rules in photoemission and the principles for mapping the experimental band structure have long been applied to the study of small molecule overlayers (CO and N_2). There has been a trend to extend the use of photoemission to study not only the extramolecular band structure but also the intramolecular band structure which becomes important with increasingly larger molecular systems [5]. This is in addition to the characterization of the binding sites and orientation of increasingly larger molecules [4]. As we shall show, the rules learned with the smaller molecules seem to be readily applicable to larger molecular systems.

2. CRYSTALLINITY AND THE RECIPROCAL LATTICE

Band structure is experimentally determined in terms of the reciprocal space lattice — the reciprocal lattice of the diffraction pattern. Due to the translational periodicity parallel to the surface, elastically photoemitted or backscattered electrons can appear only in directions defined by the surface parallel momentum ($k_\parallel^n = \mathbf{g}_{hk} + k_\parallel^o$) where $\mathbf{g}_{hk}$ is a reciprocal lattice vector parallel to the surface as defined by the symmetry of the surface lattice. So the diffraction condition is the two-dimensional Laue condition [6–9]

$$\Delta\mathbf{q} = k_\parallel^n - k_\parallel^o = \mathbf{g}_{hk} \qquad (1)$$

where $\mathbf{g}_{hk} = h\mathbf{a}_1^* + k\mathbf{a}_2^*$. The diffraction patterns are representative of the reciprocal lattice which can be related to the real lattice using the orthogonal properties

$$\mathbf{a}_1^* \bullet \mathbf{a}_1 = \mathbf{a}_2^* \bullet \mathbf{a}_2 = 2\pi \qquad (2)$$

and

$$\mathbf{a}_1^* \bullet \mathbf{a}_2 = \mathbf{a}_2^* \bullet \mathbf{a}_1 = 0 \qquad (3)$$

The reciprocal surface lattice structure, determining the Brillouin zone, can be described in terms of the underlayer reciprocal lattice structure or bulk structure or some other reference lattice (with reciprocal lattice vectors $\mathbf{b}_1^*$ and $\mathbf{b}_2^*$ denoted as $\mathbf{b}^*$ using matrix notation) by a transformation matrix $\mathfrak{R}$ as

$$\mathbf{a}^* = \mathfrak{R}\mathbf{b}^* \qquad (4)$$

The real space translation vectors can then be related to the real space reference lattice by $\mathbf{a} = \widetilde{\mathfrak{R}}^{-1}\mathbf{b}$, where $\widetilde{\mathfrak{R}}^{-1}$ is the transpose inverse of $\mathfrak{R}$.

The absence of a low energy electron diffraction (LEED) pattern does not necessarily mean that a wave vector dependent band structure cannot be observed. Even in the absence of a LEED pattern, the local order is sometimes sufficient for the observation of a wave vector dependent electronic structure. Further, if there is textured growth, resulting in a cylindrically averaged crystal structure, this type of disorder, even if playing a dominant role in the overlayer thin film, does not preclude band structure. In the latter case, the band structure is also a cylindrical average of the Brillouin zone and dispersion can still be a substantial factor. Nonetheless, a key element of band structure, the Brillouin zone, is defined by the reciprocal lattice.

3. EXPERIMENTAL MEASUREMENTS OF ELECTRONIC BAND STRUCTURE

3.1. Photoemission

Typically, the electronic structure of the valence band (the occupied bands) is studied using angle-resolved ultraviolet photoemission spectroscopy (ARPES or ARUPS). This has long been a basic experimental method for studying the occupied electronic structure of atoms and molecules in both the gas phase and solids [1–3, 10–17]. The photoemission process can be expressed in terms of the absorption of the incident photon (with

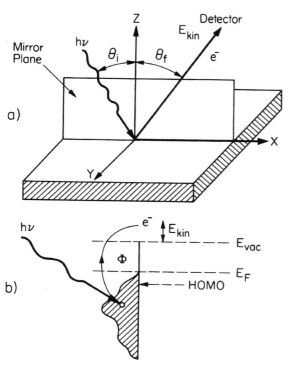

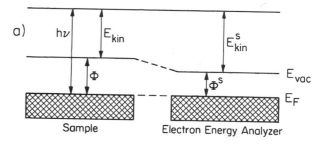

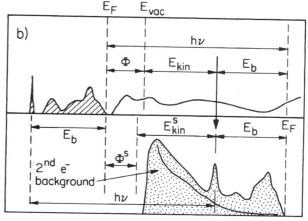

Fig. 1. The schematic diagram of ultraviolet photoemission spectroscopy. (a) The photoelectron with kinetic energy E_{kin}, and emission angle θ_f is shown with respect to the surface normal and the incidence photon with energy $h\nu$ and angle θ_i. (b) Schematic diagram of the emission process of the photoelectron with respect to energy.

Fig. 2. (a) The relationship of energy reference levels between the photoelectron emitter and the photoelectron detector. (b) The relationship of the electronic structure with the measured energy distribution curve using an electron energy analyzer with work function Φ_s.

energy $h\nu$) by matter and then emission of an electron with kinetic energy E_{kin}. For solids, this is schematically shown in Figure 1a. From the principle of energy conservation, the relationship between the energy of an electron in the initial state E_i and in its final state E_f is given by

$$E_f - E_i = h\nu \qquad (5a)$$

While the penetration depth of the photon depends on the wavelength and the material, it is usually much larger than 100 Å and is certainly on a much longer length scale than the mean free path of the electron. Additionally, in order for it to be detected, the electron excited by the incident photon must escape from the solid and therefore must have sufficient energy to overcome the work function barrier Φ as shown in Figure 1b. Using the Fermi level as an energy reference, the kinetic energy of the emitted electron, E_{kin}, can be written as

$$E_{kin} = h\nu - E_b - \Phi \qquad (6a)$$

where Φ is the work function of the solid sample and E_b is the binding energy as shown in Figure 2.

The electrons emitted from the solid can be analyzed by an electron energy analyzer and counted as a function of the photoelectron kinetic energy, although typically the data are plotted as a function of binding energy with respect to the Fermi level or chemical potential. As shown in Figure 2a, the measured kinetic energy of the emitted electron, E_{kin}^s, using an electron energy analyzer, is slightly different from the kinetic energy of

the directly emitted electron from the sample. This is due to the difference of the work function between the sample and the spectrometer. When the sample is in electrical contact with the electron energy analyzer, to align the Fermi levels, the measured kinetic energy of the emitted electron is

$$E_{kin}^s = h\nu - E_b - \Phi_s \qquad (6b)$$

where Φ_s is the work function of the electron energy analyzer. As a result, the binding energy of the emitted electron is defined as

$$E_b = h\nu - E_{kin}^s - \Phi_s \qquad (6c)$$

Energy distribution curves of the occupied states are then obtained by counting the detected photoelectrons as a function of their kinetic energy, with a fixed incident photon energy.

3.2. Inverse Photoemission

Inverse photoemission spectroscopy [18–27] has been traditionally employed to study the unoccupied electronic structure of solids. In the inverse photoemission process, an incident electron loses a discrete amount of energy and falls from a continuum state to an unoccupied bound or continuum state by means of the emission of a photon, as schematically shown in Figure 3. The inverse photoemission process is an electron spectroscopy which can be undertaken in one of two different modes. Either the photon energy detected is held fixed and the

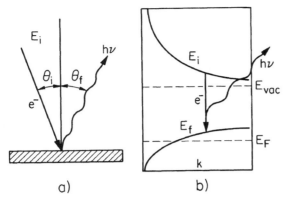

Fig. 3. Inverse photoemission process. An electron with kinetic energy E_i is incident on the sample with the incidence angle θ_i and a photon is emitted from the surface with energy $h\nu$ and emission angle θ_f, as shown in geometrical space (a) and in energy space (b).

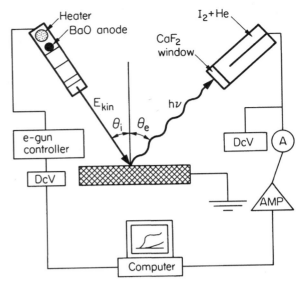

Fig. 4. Schematic diagram of the inverse photoemission spectroscopy apparatus with a modified Zipf electron gun and an iodine–helium gas filled Geiger–Müller tube with CaF_2 window.

electron energy is swept (isochromatic mode) or the electron energy is held fixed and a monochromator is used to determine the energies for the inverse photoemission transitions from the variation in light intensity as a function of emitted photon energy.

The latter technique is more powerful, but the count rates are low and the approach more difficult, so more commonly, the photon energy detector is designed to accept a narrow range of photon energies and the incident electron energy is changed. A Geiger–Müller tube is typically used for photon detection within a narrow photon energy range. The Geiger–Müller tube consists of a central bar (positive) and a concentric outer cylinder (negative). The photon energy band-pass filter is formed by the window transmission which falls as a function of increasing $h\nu$, and the ionization cross-section of the active gas which rises as a function of increasing $h\nu$ [27]. The most typical gas/window combination is I_2/CaF_2 or I_2/SrF_2 because of the high counting rate and reliable performance, although the typical helium–I_2 mixture used is corrosive. The performance of the iodine based Geiger–Müller tube is relatively difficult to assess because the photoyield depends on the vapor pressure of I_2, which is strongly dependent on temperature. The detector band-pass characteristics are somewhat irregular but nonetheless possess a sharp function centered on $h\nu = 9.7$ eV, with about a 400 meV full width at half maximum.

If an electron with well-defined energy and momentum is directed onto the sample (as in Figure 4), the electron can undergo radiative transitions from its initial state with energy E_i to a lower lying unoccupied state with a final state energy E_f. The photons emitted in this process are analyzed in the Geiger–Müller detector. The energy of the final state is given by energy conservation as the difference between the known incident electron energy $E_i = E_{kin}$ and the measured photon energy $h\nu$,

$$E_i - E_f - h\nu = 0 \tag{5b}$$

4. SYMMETRY AND SELECTION RULES

The photoelectron intensity is proportional to the differential cross-section, by absorption of one photon with radiation field $\mathbf{A}$, between two eigenstates denoted by Ψ_i and Ψ_f and described by the Golden Rule [1–5, 14–16, 28–32],

$$\left(\frac{d\sigma}{d\Omega}\right)_{PES} \propto \left|\langle\Psi_f|\mathbf{A}\bullet\mathbf{p}+\mathbf{p}\bullet\mathbf{A}|\Psi_i\rangle\right|^2\delta(E_f-E_i-h\nu) \tag{7a}$$

where two photon processes or second order effects are neglected and $\mathbf{p}$ is the momentum operator. Mathematically, we choose a gauge to make the scalar potential zero, and thus simplify Eq. (7a) and the other variations to Eq. (7) below. The perturbation to the Hamiltonian, $H' = \frac{1}{2}(\mathbf{A} \cdot \mathbf{p} + \mathbf{p} \cdot \mathbf{A})$, is usually further simplified (at least for smaller molecules) by assuming that $\mathbf{A} \cdot \mathbf{p} + \mathbf{p} \cdot \mathbf{A} \propto 2\mathbf{A} \cdot \mathbf{p}$ or that the spatial variation of the light field is small on the atomic or molecular scale. This simplification also requires that we neglect surface effects, a somewhat suspect assumption because the dielectric response changes dramatically across the surface barrier [32]. The radiation field $\mathbf{A}$ can be written as a plane wave in free space as either a classical vector potential Eq. (8a) or a quantum radiation field Eq. (8b) [32, 33], as follows:

$$\mathbf{A}(x, t) = \mathbf{A}_o e^{(-i\omega t+i\mathbf{q}\bullet\mathbf{r})} \tag{8a}$$

or

$$\mathbf{A}(\mathbf{x}, t) = V_p^{-1/2} \sum_{\mathbf{q}} \sum_{\alpha} \frac{\hbar c}{\nu} \left(a_{\mathbf{q},\alpha}(t)\hat{\varepsilon}^\alpha e^{i\mathbf{q}\bullet\mathbf{x}} + a_{\mathbf{q},\alpha}^+(t)\hat{\varepsilon}^\alpha e^{-i\mathbf{q}\bullet\mathbf{x}}\right) \tag{8b}$$

In ultraviolet photoemission spectroscopy, the wavelength of the incident photon (100 to 1000 Å) is much greater than the linear dimension of the atom so we can use the dipole approximation. The resultant differential cross-section can be described in three representations, momentum, and force forms, of the matrix elements of the dipole operator in a semiclassical treatment as

$$\left(\frac{d\sigma}{d\Omega}\right)_{PES} \propto \left|\langle\psi_f|\mathbf{A}\bullet\mathbf{p}|\psi_i\rangle\right|^2\delta(E_f - E_i - h\nu) \quad (7b)$$

$$\left(\frac{d\sigma}{d\Omega}\right)_{PES} \propto \left|\langle\psi_f|\nabla V\bullet\mathbf{A}|\psi_i\rangle\right|^2\delta(E_f - E_i - h\nu) \quad (7c)$$

and

$$\left(\frac{d\sigma}{d\Omega}\right)_{PES} \propto \left|\mathbf{A}\bullet\langle\psi_f|\nabla V_{bulk}|\psi_i\rangle\right.$$
$$+ \mathbf{A}\bullet\langle\psi_f|\nabla V_{surface}|\psi_i\rangle\Big|^2$$
$$\times \delta(E_f - E_i - h\nu) \quad (7d)$$

or quantum mechanically by

$$\left(\frac{d\sigma}{d\Omega}\right)_{PES} = \frac{e^2}{4\pi^2}\frac{\mathbf{k}}{m}\frac{1}{\hbar^2 c\nu}\left|\langle\psi_f, n_{\mathbf{q},\alpha} - 1|\hat{\varepsilon}\bullet\mathbf{p}|\psi_i, n_{\mathbf{q},\alpha}\rangle\right|^2$$
$$(7e)$$

where $\mathbf{k}$ is the wave vector of the electron. The cross-section depends on the initial state, the final state (whose parity is always even), the light polarization, and the dipole moment field of the sample. Based on this dependence, the symmetry of the initial state wave function can be obtained from experiment.

Since the physical processes of inverse photoemission are so strongly related to those of photoemission, we can use the same Hamiltonian and quantum radiation field. For photoemission, the radiation field $\mathbf{A}$ can be treated classically, and the analysis of the data will largely be the same as if a proper quantum treatment has been undertaken [33] because the electronic system is, from the outset, affected by the perturbing electromagnetic field. However, for inverse photoemission, we have an excited electron system but no electromagnetic field at the outset. Hence the matrix elements of the dipole operator in the semiclassical expression are zero. So with inverse photoemission, it is essential to quantize the electromagnetic field in order to allow for the spontaneous emission of light. That is why it is essential to replace the classical vector potential $\mathbf{A}$ by the quantum radiation field operator $\mathbf{A}(\mathbf{x}, t)$ of Eq. (7d) [33]. For inverse photoemission, the initial state consists of an electron in a continuum state with no photons, $|\psi_i, n_{h\nu} = 0\rangle$, and the final state consists of an electron in a bound state and a photon with wave vector $\mathbf{q}$, $|\psi_f, n_{h\nu}(\mathbf{q}) = 1\rangle$:

$$\left(\frac{d\sigma}{d\Omega}\right)_{IPES} = \frac{\nu e^2}{m\hbar c^3}\frac{1}{\hbar|k|}\left|\langle\psi_f, n_{\mathbf{q},\alpha} = 1|\hat{\varepsilon}\bullet\mathbf{p}|\psi_i, n_{\mathbf{q},\alpha} = 0\rangle\right|^2$$
$$(7f)$$

The symmetries of the states are labeled according to the group representation (Tables I–IV for several different point group symmetries). Any linear combination of states, which may include several different rectangular representations within the same group representation, can contribute to one state of that symmetry. This is also complicated by the fact that each lattice site also has a contribution; hence the wave vector is also important.

Table I. Character Table and Group Representation for States at the $\overline{\Gamma}$ and $\overline{M}$, C_{4v}, High Symmetry Points of a C_{4v} Surface

C_{4v}	e	$2C_4$	C_2	$2\sigma_v$	$2\sigma_d$	$\overline{\Gamma}$ top, $\overline{M}$ top	Center	$\overline{M}$ center	$a_1 \leftrightarrow b_2$ $a_2 \leftrightarrow b_2$
a_1 (Δ_1)	1	1	1	1	1	s, p_z	$d_{3z^2-r^2}$		d_{xy}
a_2 (Δ_2)	1	1	1	−1	−1				$d_{x^2-y^2}$
b_1 (Δ_3)	1	−1	1	1	−1		$d_{x^2-y^2}$		
b_2 (Δ_4)	1	−1	1	−1	1		d_{xy}	s, p_z	$d_{3z^2-r^2}$
e (Δ_5)	2	0	−2	0	0	p_x, p_y	$d_{xz,yz}$	p_x, p_y	$d_{xz,yz}$

Table II. Character Table and Group Representation for States at the $\overline{X}C_{2v}$ High Symmetry Point of a C_{4v} Surface

C_{2v}	e	C_2	$\sigma_v(xy)$	$\sigma_v'(yz)$	Top		Center	$a_1 \leftrightarrow b_2$ $a_2 \leftrightarrow b_2$
a_1	1	1	1	1	s, p_z	$d_{3z^2-r^2}$, $d_{x^2-y^2}$	p_x	d_{xz}
a_2	1	1	−1	−1		d_{xy}	p_y	d_{yz}
b_1	1	−1	1	−1	p_x	d_{xz}	s, p_z	$d_{3z^2-r^2}$, $d_{x^2-y^2}$
b_2	1	−1	−1	1	p_y	d_{yz}		d_{xy}

Table III. Character Table and Group Representation for States along $\overline{\Sigma}C_{1h}$ High Symmetry Direction of a C_{4v} Surface

C_{1h}	e	σ_h	Top		Center
a	1	1	s, p_z, p_{x+y}	$d_{3z^2-r^2}, d_{xy}$	$d_{(x+y)z}$
a'	1	−1	p_{x-y}	$d_{x^2-y^2}$	$d_{(x-y)z}$

Table IV. Character Table and Group Representation for States along $\overline{\Delta}C_{1h}$ High Symmetry Direction of a C_{4v} Surface

C_{1h}	e	σ_h	Top		Center	
a	1	1	s, p_z, p_x	$d_{3z^2-r^2}, d_{xz}$	s, p_z, p_x	$d_{3z^2-r^2}, d_{xz}$
a'	1	−1	p_y	d_{xy}, d_{yz}	p_y	d_{xy}, d_{yz}

To use the symmetry selection rules effectively, we need to know the symmetry (group representation) of each molecular orbital, the orientation of the light vector potential and the emission (incidence) angle of the detected photoelectrons (incident electrons). The better the extent of light polarization, and the better the angular resolution of the photoelectron energy analyzer, or incident electron source, the better the determination of the molecular orientation. Schematically, the geometry considerations for angle-resolved photoemission of surfaces are summarized in Figure 1. When using the highly plane polarized light available from a synchrotron, p-polarized light typically requires the incident polar angle θ_i to be at a large incidence angle to place the light vector potential **A** along the surface normal, while s-polarized light typically uses a geometry where the incident polar angle θ_i is at an angle close to zero to place the light vector potential **A** parallel with the surface. In the notation of Figure 1, odd or "forbidden" geometry for an upright diatomic molecule would be with the azimuthal angle ϕ at 90° (as opposed to setting ϕ equal to zero or "even" geometry), θ_i at an angle close to zero (s-polarized light), and θ_f at a nonzero angle.

Applying the basic concepts of photoemission and inverse photoemission selection rules, dictated by Eqs. (7) and (8), is relatively straightforward. Using a diatomic molecule as an example, bonded upright along the surface normal, if the final photoemitted electron is detected along the molecular axis, the final state wave function is of σ (or typically the a_1, fully symmetric, group representation element) symmetry. Molecular orbitals of π (or typically group representations b_1 and b_2 or e_1 and e_2) symmetry have no amplitude along the molecular axis. This means that if the polarization of the incident light is arranged so as to have the vector potential **A** along the molecular axis, **A** · **p** has σ symmetry (i.e., the fully symmetric group representation element, typically a_1). For photoemission to be allowed with this polarization and detection geometry, the initial state must also be of σ (or typically the a_1, fully symmetric, group representation) symmetry to couple to the allowed final state. With the vector potential perpendicular to the molecular axis, **A** · **p** has $\pi_{x,y}$ (or typically group representations b_1 and b_2 or e_1 and e_2) symmetry, and only $\pi_{x,y}$ (or group representations b_1 and b_2 or e_1 and e_2) symmetry initial states are allowed with the electron energy analyzer detecting only those photoelectrons with wave functions symmetric to all symmetry operations of the molecule. There is another photoemission geometry worth bearing in mind. With the photoelectron detector in a plane perpendicular to the plane defined by the molecular axis and the incident light vector potential, only odd symmetry states can be detected, noting also that odd π_y symmetry states can exist but not σ odd symmetry states.

Not all points within the Brillouin zone will share the same symmetry. Thus it must be understood that the conditions that apply to the symmetry and selection rules, dictated by Eqs. (7) and (8), change depending on the position within the Brillouin zone, as well as on the local binding site. For the fourfold symmetric surface at different symmetry points in the Brillouin zone, this complexity imposed on the group representations is summarized in Tables I–IV. For example, with a fourfold symmetric surface, with Brillouin zone shown in Figure 5, the a_1 (or Δ_1 including atomic orbitals with the rectangular representations of s, p_z, $d_{z^2-r^2}$) and e (or Δ_5 including atomic orbitals with the rectangular representations of p_x, p_y, d_{xz}, d_{yz}) molecular or atomic orbitals can be observed depending on the light polarization, but orbitals with a_2, b_1, and b_2 symmetry cannot.

As has been indicated, highly polarized light can be very useful in determining both the molecular orientation [1–5, 17] and the symmetry [1–5, 14–18, 28, 19, 30–32] of a given state.

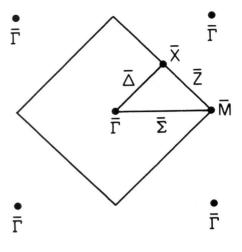

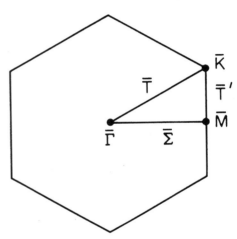

Fig. 5. The high symmetry points and lines of the surface Brillouin zone of a C_{4v} or fourfold symmetric surface (top) and a C_{6v} or sixfold symmetric surface (bottom).

This typically requires utilization of a synchrotron source for angle-resolved photoemission and perhaps use of a polarizer in combination with the photon detector for inverse photoemission. The changes in light polarization mean that different orbitals are accessed in photoemission and inverse photoemission as summarized in Table V, with the additional considerations of the angular momentum restrictions noted below. For example, it should be noted that in the a_1 representation the orbitals $d_{x^2-y^2}$ (in C_{2v}) and $d_{x^2+y^2}$ (in C_{4v} and C_{6v}), are suppressed in p-polarized light (**A** perpendicular to the surface). This is because of the selection rule that $\langle \pm m_1(\text{final state})| \pm m_1(\text{perturbing Hamiltonian})| \pm m_1(\text{initial state})\rangle = 0$. Since $m_1 = 2$ for $d_{x^2-y^2}$ (in C_{2v}) and $d_{x^2+y^2}$ (in C_{4v} and C_{6v}) and $m_1(\text{perturbing Hamiltonian}) = 0$ with p-polarized light, finding a final state (or an initial state) with which to access $d_{x^2-y^2}$ or $d_{x^2+y^2}$ with p-polarized light is difficult.

Off-normal emission angles (or incidence angles) of the photoelectron (incident electron) in photoemission (inverse pho-

Table V. Relationship between the Orientation of the Vector Potential of the Incident (or Emitted) Light and the States Observed in Photoemission (or Inverse Photoemission) for the Normal Emission Photoelectron (or Incidence Electron) Collection Geometry

Component of the vector potential **A**	C_{2v}	C_{4v}	C_{6v}
A perpendicular to the surface	s	s	s
	p_z	p_z	p_z
	p_x	$d_{3z^2-r^2}$	$d_{3z^2-r^2}$
	$d_{3z^2-r^2}$	(A_1)	
			(A_1)
	(A_1)		
A parallel with the surface	p_x	p_x	p_x
	p_y	p_y	p_y
	d_{xz}	d_{xz}	d_{xz}
	d_{yz}	d_{yz}	d_{yz}
	(B_1, B_2)	(E)	
			(E_1)

Note: The symmetry points are given for center of the surface Brillouin zone of a twofold symmetric surface, a fourfold symmetric surface such a (100) face of a bcc or fcc crystal, and a sixfold symmetric surface such as a hcp (0001) or fcc (111) face.

toemission) sample different parts of the Brillouin zone. As a result, the changing wave vector affects both the point group symmetry and the position in the Brillouin zone. Thus band structure can play a role in determining the binding energy (because of dispersion) as well as affect the photoemission (inverse photoemission) matrix elements. These complications of band structure aside, as we shall see in this review, molecular orientation and symmetry play a dominant role in the application of selection rules.

As we shall see, there are a number of similarities between the bonding of large molecules to metal surfaces and molecular CO adsorption. A surprising number of adsorbate molecules tend to bond with the molecular axis along the surface normal (although not all), making it possible to compare, by analogy at the very least, almost all molecular adsorbates with CO and N_2. The rules and principles just described, developed one to two decades ago for studying diatomic molecules like CO and N_2, are the same rules we apply to much larger molecules today. In part, this review is an effort to emphasize that valence band photoemission, inverse photoemission symmetry, and selection rules are applicable to a wide variety of molecular adsorbates as well as polymeric thin films.

In resonant photoemission, or indeed in any resonant excited electron process, selection rules are also obeyed [34–42]. In both resonant photoemission and near edge X-ray absorption fine structure (NEXAFS or XANES) spectroscopies a core exciton is formed—a core electron is excited to an unoccupied state near the Fermi energy, forming a transient excited state. The final state is identical to the direct photoemission process and thus leads to a resonant "enhancement" of the valence band

photoemission features. This resonant process can be probed not only by photoemission but also by fluorescent decay, Auger de-excitation, and Auger coincidence spectroscopies as well as by monitoring the sample current.

To some extent, the symmetry of the unoccupied states (those involved in the transient excited states) can be probed [34–39, 41]. This is a consequence of the requirement that the term

$$M(\varphi_f) \sim \left| \langle \varphi_i | H' | \varphi_p \rangle \langle \varphi_p \varphi_q | V(r) | \varphi_q \varphi_f \rangle \right| \times \delta(E_f - E_i - h\nu) \quad (9)$$

must be nonzero, where φ_p, φ_q are intermediate valence excited states. The second term in Eq. (9) (the Auger process) involves two electrons instead of one [as in Eqs. (7)] and usually has little influence on the symmetry selection in most experiments [34] because the core exciton lifetime is usually long compared to the excitation lifetime.

Since the dipole selection rules are most easily applied to the excitation to determine the intermediate (core exciton) excited state as well as the final state, the typical one electron transition requires that $\Delta l = \pm 1$, $\Delta j = \pm 1, 0$, and $\Delta s = 0$. Typically, angular momentum increases so that in most cases $\Delta l = 1$; in other words an excitation of a core p state to an unoccupied d state is favored over an excitation to an s state [43].

For excitations to an unoccupied orbital from a core s orbital, for example, excitations from a carbon $1s$, nitrogen $1s$, or oxygen $1s$ core (as is typical in molecular adsorbate NEXAFS studies [44]), the dipole selection rules are quite restrictive. With p-polarized light, only the $a_1(s) \rightarrow a_1(p_z)$ excitations are possible in C_{2v}, C_{4v}, and C_{6v}. With s-polarized light, only the $a_1(s) \rightarrow e(p_{x,y})$ excitations in C_{4v}, the $a_1(s) \rightarrow e_1(p_{x,y})$ excitations in C_{6v}, and the $a_1(s) \rightarrow b_{1,2}(p_{x,y})$ excitations in C_{2v} are allowed.

As might be expected, the selection rules from a core p shell to the unoccupied molecular orbitals are somewhat more complex. In C_{6v}, with p-polarized light, both $a_1(p_z) \rightarrow a_1(s, d_{z^2})$ and $e_1(p_{x,y}) \rightarrow e_1(d_{xz}, d_{yz})$ excitations are possible. With s-polarized light, $a_1(p_z) \rightarrow e_1(d_{xz}, d_{yz})$, $e_1(p_{x,y}) \rightarrow a_1(s, d_{z^2}, d_{x^2+y^2})$ and $e_2(d_{x^2-y^2}, d_{xy}) \rightarrow a_2$ excitations are possible. In C_{4v}, with p-polarized light, both $a_1(p_z) \rightarrow a_1(s, d_{z^2})$ and $e(p_{x,y}) \rightarrow e(d_{xz}, d_{yz})$ excitations are possible. With s-polarized light, $a_1(p_z) \rightarrow e_1(d_{xz}, d_{yz})$, $e(p_{x,y}) \rightarrow a_1(s, d_{z^2})$ and $e(p_{x,y}) \rightarrow b_1(d_{x^2-y^2})$, $b_2(d_{xy})$ excitations are possible. Detailed analysis of the transition matrix in the atomic limit [41, 45] suggests that pure p-polarized light excites p electrons into unoccupied d_{z^2} states more than into d_{xz}, d_{yz} states and the reverse occurs with pure s-polarized light in most point group symmetries.

5. BAND DISPERSION

Band dispersion originates from the wave vector dependence of the overall wave function containing contributions from each lattice site. Band dispersion is the variation of orbital binding energy with wave vector. The bandwidth is the energy difference between the energy minimum and the energy maximum for a dispersing band across the Brillouin zone. This variation in orbital binding energy with wave vector is a consequence of the periodic potential [1–3, 5, 16, 17, 30, 32, 33, 46, 47]. In addition, the wave vector plays an important role in determining the appropriate irreducible representation and changes in the wave vector can change the point group symmetry, as we have already noted.

In one dimension, the periodic, symmetry corrected, wave function is the linear combination of wave functions on each lattice site n as

$$\Psi_k = \sum_n e^{ikna}\xi_n \tag{10}$$

At $\mathbf{k} = 0$, the total wave function is $\Psi_{k=0} = \sum_n \xi_n$. This is generally what a chemist would refer to as a bonding configuration. At the Brillouin zone boundary ($\mathbf{k} = \pi/a$) the linear combination is $\Psi_{k=\pi/a} = \sum_n(-1)^n\xi_n$ (an antibonding configuration). This wave vector dependence does not just affect the overall wave function but clearly has obvious consequences for the binding energy [1, 2, 17, 46, 47]. Similar constructions are possible in two and three dimensions, even with very complicated wave function combinations originating from each lattice site.

Momentum conservation can be used to determine the energy band dispersion relation with respect to wave vector. Since the momentum of the incident (or emitted) photon is negligible in the ultraviolet photoemission (inverse photoemission) range and the momentum of the emitted photoelectron (or incident electron) is relatively large, we may neglect the momentum of the incident (or emitted) light. The wave vector component parallel to the surface ($k_\parallel$) is conserved across the vacuum solid interface and can be related to the kinetic energy of the emitted (or incident) electron as

$$E_{\text{kin}} = \frac{\hbar^2}{2m}\big(k_\parallel^2 + k_\perp^2\big) \tag{11}$$

where $k_\perp$ in this expression is the value after emission, which is different from $k_\perp$ in the crystal. The parallel momentum can be derived as follows from the kinetic energy and the emission angle:

$$k_\parallel = \sqrt{\frac{2m}{\hbar^2}E_{\text{kin}}}\sin(\theta) = 0.51198\sqrt{E_{\text{kin}}/\text{eV}}\sin(\theta)\ \text{Å}^{-1} \tag{12}$$

The perpendicular component of the crystal wave vector ($k_\perp$) is not conserved across the solid vacuum interface, because of the truncation of the crystal at the surface. This perpendicular component of the wave vector in the crystal along the surface normal (which is not pertinent to overlayers that conserve two dimensionality of state) is

$$k_\perp = \left(\frac{2m}{\hbar^2}\big\{E_{\text{kin}}(\cos(\theta))^2 + U_o\big\}\right)^{1/2} \tag{13}$$

where θ is the emission angle of the photoelectron or incident angle of the electron in inverse photoemission and U_o is the inner potential of the solid. The dispersion of $k_\perp$ is important for long chain molecules oriented along the surface normal, however, and is, of course, important for films thick enough to exhibit a three-dimensional band structure.

While molecules are generally considered as existing in real space with the well-defined positions of balls (atoms) and sticks (bonds), band structure is clearer in momentum space and is best defined in terms of the reciprocal lattice. It is, after all, the reciprocal lattice that defines the Brillouin zone. The electron wave vector places the position of the photoemission measurement or inverse photoemission event in the Brillouin zone. For a molecular overlayer or a thin film that is only a molecular monolayer or a few molecular layers thick, two-dimensionality of state is conserved. Under these conditions, the two-dimensional Brillouin zone is sufficient and Eq. (7) can be written in terms of the electron wave vector parallel with the film as

$$\left(\frac{d\sigma}{d\Omega}\right)_{\text{PES}} \propto \big|\langle\Psi_f|\mathbf{A}\bullet\mathbf{p} + \mathbf{p}\bullet\mathbf{A}|\Psi_i\rangle\big|^2$$
$$\times \delta(E_f - E_i - h\nu)\delta\big(k_\parallel^f - \mathbf{g}_{hk} - k_\parallel^i\big) \tag{14}$$

since momentum parallel to the surface is conserved in angle-resolved electron spectroscopies [48].

The Brillouin zone center is the point in reciprocal space with the highest symmetry—so symmetry and selection rules are at their most restrictive. With increasing wave vector, even along a high symmetry direction, the symmetry is reduced, though at the Brillouin zone edge, the symmetry can increase again. Orbitals hybridize and the number of group representation elements is reduced away from the high symmetry points in the Brillouin zone. More molecular orbitals can be observed in both photoemission and inverse photoemission as the symmetry is reduced, although light polarization continues to play a role since the symmetry selection rules are not completely absent along a high symmetry direction. At the high symmetry points on the Brillouin zone edge, $\overline{M}$ or $\overline{X}$ (for the fourfold symmetric surface, whose Brillouin zone is schematically represented in Figure 5), as with the Brillouin zone center, the group symmetry increases and the group representation that applies also changes. The number of group representation elements and corresponding molecular orbital symmetries increases, and the number of orbitals that can be observed in photoemission or inverse photoemission decreases.

The influence of the wave vector on the overall configuration of the wave function for a fourfold symmetric lattice is schematically described in Figure 6 (using the labeling of the Brillouin zone in Figure 5). This is done for both Δ_1 or a_1 (rectangular representations of s, p_z, $d_{z^2-r^2}$) and e or Δ_5 (rectangular representations of p_x, p_y, d_{xz}, d_{yz}) bands. Of course, in practice, band dispersion is a smoothly varying function of the wave vector. Note that even or odd symmetry with respect to the mirror plane, established by the orientation of the vector potential and the plane of emission (or incidence), now has an influence on the part of the wave function that is sampled. One can preferentially sample b_1 or b_2 orbitals as one changes the wave vector

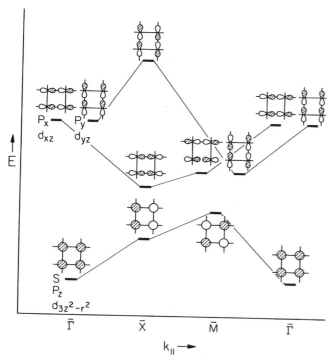

Fig. 6. The schematic band structure of a fourfold symmetric lattice containing p_x, p_y, and s or p_z atomic orbitals. Adapted from [1, 2, 32, 47].

along the mirror plane or perpendicular to the mirror plane. The band dispersion and collection geometry now both matter (as indicated above).

6. CARBON MONOXIDE MONOLAYERS

The diatomic molecules, in particular CO, have been more extensively studied by angle-resolved photoemission than have any other molecular adsorbates [1, 3, 4]. Molecular CO adsorption has become the paradigm for molecular adsorption and adsorbed molecular CO was the first application of the use of angle-resolved photoemission and light polarization dependent near edge X-ray adsorption fine structure (NEXAFS or XANES) to determine molecular orientation. Gadzuk [49, 50] and others [51–54] calculated the relative angular dependence of the photoemission intensities for "model surface molecules" in the mid 1970s. The early angle-resolved photoemission studies (and some later studies as well) used the emission angle dependence of the mostly unpolarized helium discharge lamp [55–60]. The calculation of Davenport [53] provided the foundation for looking at the angle-resolved partial cross-sections in photoemission by changing the incident light polarization specifically for oriented CO and experiments soon followed [54, 61–77] as summarized in Table VI.

Determining the orientation through the use of dipole and symmetry selection rules has sometimes been more successful than have more general diffraction techniques such as LEED. An early LEED study of CO on Ni(100) determined

an orientation of 34° off the surface normal [78]. From angle-resolved photoemission, a clear and defensible case was made that the CO bonded with the molecular axis along the surface normal and was not tilted [69]. In LEED intensity versus voltage analysis, one must account for CO domain size and for surface regions absent of CO. This can make the analysis of the LEED data inadvertently coverage dependent. No similar problems exist with using photoemission to determine molecular orientation. Based on high resolution electron energy loss spectroscopy (HREELS) measurements, probing the vibrational modes, Hoffman and de Paola argued for CO tilted or oriented parallel to the surface with low coverages of CO coadsorbed with potassium on Ru(100) [79, 80]. Again, angle-resolved photoemission provides a far more convincing case that the CO bonds with the molecular axis along the surface normal [63–65, 70].

Recent work provides compelling angle-resolved photoemission evidence for CO bonded to Fe(110) tilted with respect to the surface normal [71] although not all angle-resolved studies concur with this conclusion. Jensen and Rhodin argued for CO bonded to the surface with the molecular axis along the surface normal [73, 74]. CO on Fe(111) also seems to adopt a tilted configuration [81].

CO on Fe(100) is also seen to be tilted from NEXAFS [82] and photoelectron diffraction [83] and is suggested by shifts of the photoemission binding energies [84]. Two different bonding configurations for CO on Cr(110) have been postulated. One "lying down" [85, 86], and the other "upright" [86].

There is a less convincing case for CO tilted on the surface of Cu(100) [87] and this postulate does not agree with all measurements [62, 66]. Generally CO bonds upright, with the molecular axis parallel with the surface normal. At least this seems to be the case on Ni(100) [60, 67–69, 88], Ni(111) [55, 59], Cu(100) [62, 66], Ru(100) [58, 63–65, 70, 75, 76], Co(0001) [61], Ir(111) [72], NiAl(110) [77], NiAl(111) [77], Mo(110) [89], and Pt(111) [55, 56]. Only on iron and chromium surfaces and a few other select surfaces [e.g., Pt(110)] are there strong deviations from this behavior.

The binding energies summarized in Table VI are complicated by the fact that CO molecular binding energies are dependent upon both coverage and emission angle. The emission angle dependence is due to the formation of band structure from the hybridization of the molecular orbitals of CO on Ag(111) [90], Co(0001) [61], Cu(111) [91], Fe(110) [71, 73, 74], Ir(111) [72, 92], Ni(100) [60], Ni(110) [93, 94], Ni(111) [59], Os(0001) [94], Pd(100) [95], Pd(111) [96, 97], Pt(111) [98], Pt(110) [98], and Ru(0001) [58, 63, 64]. This hybridization of CO molecular orbitals is shown schematically, in real space, for a hexagonal lattice of CO in Figure 7 and is usually perturbed by the metal substrate as indicated in Figure 8.

The band dispersion resulting from the hybridization of CO molecular orbitals can be quite significant and the bandwidth, not unexpectedly, is larger the smaller the lattice constant [61, 63, 64, 72, 74, 94, 95, 99]. With the smaller CO–CO lattice constants some bandwidths are of the order of an eV or

DOWBEN ET AL.

Table VI.   Summary of Photoemission Studies of Adsorbed CO Molecular Orbitals and Bonding Configurations

Substrate	Orientation	1π Binding energy	5σ Binding energy	4σ Binding energy	Remarks
Ni(100)	upright [67]	7.8 [67]	8.3 [67]	10.8 [67]	lamp source [60], polarized light [67–69], existence of π orbital confirmed using odd geometry
	upright [68]	7.9 [68]	8.2 [68]	11.2 [68]	
	upright [69]	7.5 [69]	8.3 [69]	10.8 [69]	
	upright [60]	8 [60]	8 [60]	11 [60]	
Ni(110)	tilted [93]	in even geom. 7.0 [93] in 'y' odd geom. 7.7 [93]	in even geom. 7.7 [93] in odd geom. 8.5 [93]	in even geom. 10.7 [93] in odd geom. 11 [93]	polarized light [93], existence of π orbital confirmed using odd geometry
Ni(111)	upright [59]	7.1 [59]	8.7 [59]	11.7 [59]	lamp source [55, 59]
	upright [55]	8.1 [55]	8.1 [55]	10.8 [55]	
Cr(110)	flat [86]	7.6 [86]	7.6 [86]	11.8 [86]	polarized light but angle integrated [86]
Cu(100)	upright [62]	7.5 [62]	7.5 [62]	11.7 [62]	lamp source [87], polarized light [62, 66], existence of π orbital confirmed using odd geometry
	upright [66]	8.5 [66]	8.9 [66]	11.8 [66]	
	tilted [87]	8.5 [87]	6.7 [87]	12.7 [87]	
Ru(100)	upright [58]	7.7 [58]	7.7 [58]	10.5 [58]	polarized lamp source [58, 75, 76], polarized light [63–65, 70], existence of π orbital confirmed using odd geometry
	upright [75]	8 [75]	8 [75]	11 [75]	
	upright [65]	7.5 [65]	7.6 [65]	10.8 [65]	
	upright [76]	7.3 [76]	8.5 [76]	11.5 [76]	
	upright [64]	7.6 [64]	8.4 [64]	11.05 [64]	
	upright [63]	8 [63]	8 [63]	10.6 [63]	
	upright [70]	7.55 [70]	8.4 [70]	11.05 [70]	
Fe(110)	canted [71]	7.0 [71]	7.5 [71]	10.5 [71]	polarized light [71, 74]
	upright [74]	6.8 [74]	8.3 [74]	11.0 [74]	
Fe(111)	canted [81]	6.8 [81]	8.2 [81]	11.0 [81]	polarized light [81]
Ir(111)	upright [72]	8.8 [72]	9.1 [72]	11.7 [72]	polarized light [72]
Co(0001)	upright [61]	6.6 [61]	8.2 [61]	10.8 [61]	polarized light [61], existence of π orbital confirmed using odd geometry
Pt(111)	upright [55]	9.2 [55]	9.2 [55]	11.7 [55]	lamp source [55, 56]
	upright [56]	8.3 [56]	9.2 [56]	11.7 [56]	
Pt(110)	canted [56, 60]	8.2 [56]	9.25 [56]	11.7 [56]	lamp source [56, 60]
NiAl(110)	upright [77]	9.1 [77]	9.1 [77]	11.6 [77]	polarized light [77]
NiAl(111)	upright [77]	8.4 [77]	8.4 [77]	11.5 [77]	polarized light [77]
Ru(0001) + K	upright [75]	6.3 [75]	8 [75]	11.5 [75]	polarized lamp source [75, 76], polarized light [70, 65], existence of π orbital confirmed using odd geometry
	upright [65]	6.5 [65]	8.4 [65]	11.5 [65]	
	upright [76]	6.3 [76]	7.9 [76]	11.5 [76]	
	upright [70]	6.5 [70]	8.4 [70]	11.45 [70]	
Cu(100) + K	upright [65]	8.5 [65]	8.5 [65]	11.7 [65]	polarized light [65, 66] existence of π orbital confirmed using odd geometry
	upright [66]	7.9 [66]	9.0 [66]	11.6 [66]	

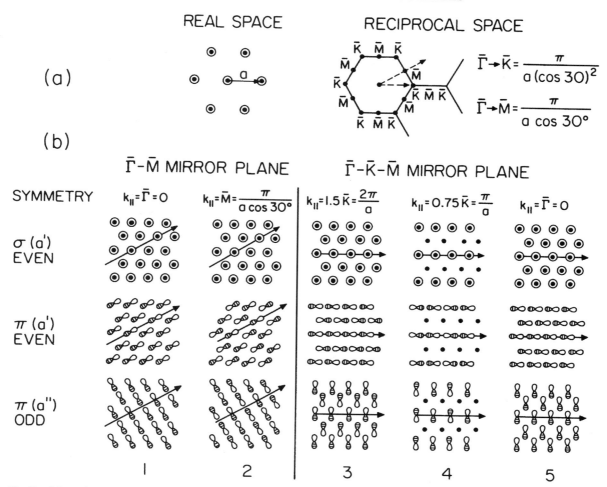

Fig. 7.   Schematic representation of the wave functions (in real space) for a hexagonal CO overlayer. Depicted are the real and reciprocal lattices of the hexagonal structure (a) and wave functions for σ and π states at various points in the Brillouin zone. The arrows indicate the direction of the wave vector $k_\parallel$. Reprinted with permission from [61], copyright 1983, American Institute of Physics.

so. For example, in CO adsorbed on Co(0001), with a CO–CO spacing of 3.29 Å, the bandwidth is 0.8 eV for the 5σ band and 1.1 eV for the 1π orbitals, as shown in Figure 9 [61]. Similarly, CO overlayers form a close packed structure on Ru(0001) with a CO–CO spacing of about 3.5 Å and observed bandwidths are about 0.9 eV [64] to 0.7 eV [63] for the 5σ band, 0.5 eV [63] for the 1π band, and 0.5 eV [64] to 0.4 eV [63] for the 4σ orbitals, but the less dense CO overlayer lattice on Ru(0001) has a smaller bandwidth. This general observation [61, 63, 64, 74, 94, 99] that the small CO–CO lattice spacings have the largest bandwidth is somewhat summarized in Figure 10, adapted from [99].

In the case of CO adsorbed on Cu(111), there is an anomalously small band dispersion of 0.15 eV for the 4σ orbital compared with a value of about 0.36 eV expected from theory. A shift of intensity to the satellite lines [91, 99–101] that depends upon the wave vector $k_\parallel$ is postulated to be responsible [91]. The "correct" band dispersion can be recovered using a sum rule that accounts for the shift in intensity between the screened and unscreened final states of a molecular orbital and their respective band shifts as a function of $k_\parallel$ [91].

In spite of all of these complications, there is general consensus that on many surfaces, CO bonds with the axis along the surface normal with the carbon atom closest to the surface, for most surfaces [Fe(110), Cr(110), Pt(110), Ni(110), and Fe(100) may be among the few exceptions]. While much of this work may seem old and "dusty" from the perspective of today, nonetheless, the work with CO adsorption did lay down a foundation for the use of angle-resolved photoemission. Angle-resolved photoemission studies of CO adsorption continue to be carried out and the coadsorption of molecular CO with other molecules as overlayers and within multilayer thin films is also attracting interest [2, 102, 103].

7. MOLECULAR NITROGEN

Molecular nitrogen adsorption is very similar to that of CO. Although the molecule is symmetric in the gas phase (unlike CO), the N_2 molecule is isoelectronic to that of CO (14 electrons/molecule). We can compare the $1\pi_u$, $3\sigma_g$, and $2\sigma_u$ molecular orbitals of N_2 to the 1π, 5σ, and 4σ, respectively, of CO, as

ONE ELECTRON SCHEME OF AN ADSORBATE

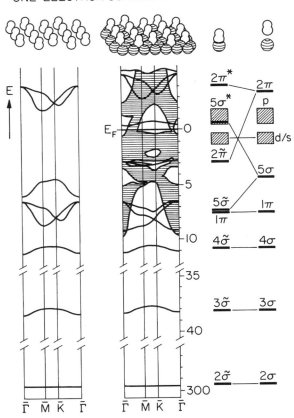

Fig. 8. The schematic changes in the CO band structure upon adsorption on a transition metal surface. The one electron level scheme is shown at right. The band structure of a CO monolayer is shown at the extreme left, and the two-dimensional band structure on a metal is shown in the middle. The band structure of the metal substrate, projected on the surface, is schematically shown as the hatched area. Adapted from [99].

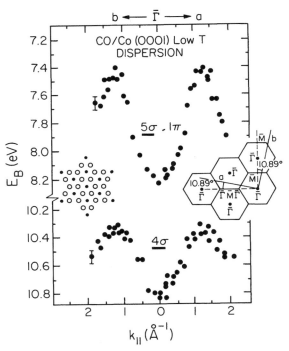

Fig. 9. The dispersion of the bands form from the 4σ, 5σ, and 1π molecular orbitals of the $(2\sqrt{3} \times 2\sqrt{3})R30°$ CO overlayer on Co(0001) along the a and b directions as indicated in the inset schematic of the Brillouin zone. The LEED pattern is also shown as an inset. Reprinted with permission from [61], copyright 1983, American Institute of Physics.

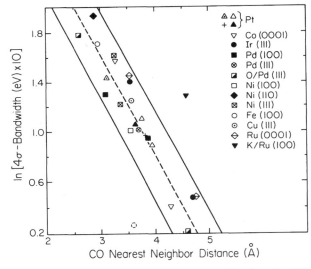

Fig. 10. The 4σ molecular orbital band dispersion plotted against CO–CO spacing for CO on various surfaces. Adapted from [99].

indicated in Figure 11. With bonding orientations that place the molecular axis along the surface, the surface breaks the symmetry of the nitrogen molecule so that nitrogen even more resembles CO. This was noted from a very, very early stage of molecular nitrogen adsorption so that, for convenience, the CO molecular orbital assignment is used for N_2. When N_2 adsorbs with the molecular axis along the surface normal or nearly so, the photoemission spectra of N_2, now with the symmetry of the molecule broken, are thus very similar to those of CO [104–107]. In fact, in XPS, the N $1s$ core level spectra show the nitrogen atoms in adsorbed N_2 to be inequivalent [108–115]. The light incidence angle dependence of adsorbed nitrogen, as seen in Figure 12, is representative of both CO and N_2 bonded to a surface with the molecular axis parallel to the surface normal.

As with CO, N_2 bonds with the molecular axis along the surface normal on many surfaces including Ni(100) [116], Ni(110) [117], Ru(100) [64], and W(110) [118]. These angle-resolved photoemission results, indicating an upright bonding configuration, are consistent with other measurements such as infrared adsorption in the case of Ni(110) [119] and

Ni(111) [120], HREELS in the case of Ru(100) [64, 121, 122], Ru(10$\bar{1}$0) [123], and Ni(110) [117], and NEXAFS and extended X-ray adsorption fine structure (EXAFS) spectroscopies in the case of Ni(100) [115, 124]. In the case of nickel surfaces, the upright vertical bonding of molecular nitrogen is expected from theoretical cluster calculations [125] as well.

The canted bonding configurations for N_2 adsorption (as with CO) occur on iron surfaces, with indications of a canted

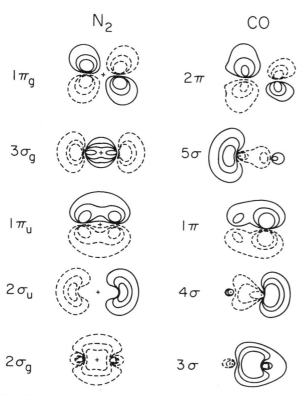

Fig. 11. The wave functions for N_2 and CO. The carbon end of the CO is at the left. The binding energies with respect to the vacuum level are given in eV. Adapted from [4, 117].

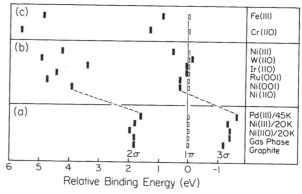

Fig. 13. The relative photoemission binding energies of the occupied molecular orbitals of N_2 with repect the N_2 1π molecular orbital for (a) weak chemisorption systems (often misnamed as physisorbed), (b) moderately chemisorbed molecular nitrogen systems, and (c) strongly chemisorbed molecular nitrogen systems. Adapted from [127, 128].

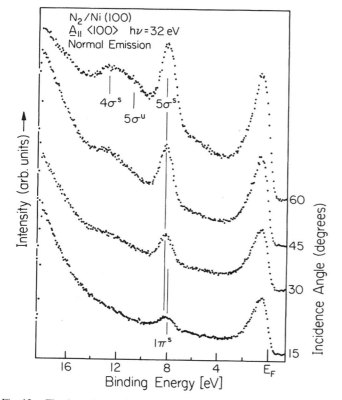

Fig. 12. The dependence of the photoemission spectra on the light incidence angle for a $c(2 \times 2)$ N_2 overlayer on Ni(100). The photoelectrons were collected normal to the surface. Adapted from [116].

bonding configuration on Fe(111) in angle-resolved photoemission [81] and electron energy loss spectroscopy [126]. This π-bonded canted or in-plane bonding configuration may also occur for the strongly chemisorbed molecular nitrogen bonding to Cr(110) [127, 128] and Cr(111) [129] as well as for the very weakly chemisorbed nitrogen bonding to graphite [130, 131]. The weakly chemisorbed molecular nitrogen on Pd(111) is seen to be more randomly oriented [117].

The difference between the strong molecular nitrogen chemisorption systems and the weak molecular nitrogen chemisorption systems is that the greater the chemisorption bond [127, 128], the greater the molecular orbital $2\sigma_u$ and $3\sigma_g$ binding energies relative to the $1\pi_u$ molecular orbital, as schematically indicated in Figure 13. This increase in the $3\sigma_g$ binding energies relative to the 1π molecular orbital (or the 5σ binding energies relative to the 1π molecular orbital in the case of molecular CO) is consistent [132] with the Blyholder model [133] of electron donation from the $3\sigma_g$ orbital of N_2 (5σ of CO, NO) to the metal substrate and back donation from the metal to the $1\pi_g$ molecular orbital of N_2 (2π molecular orbital of CO, NO). Such a bonding configuration would tend to favor linear (upright) bonding orientations for diatomic molecules (CO, N_2, NO, CN) so affected. This would be particularly true if the surface of the transition metal is dominated by a d_{z^2} state, so that diatomic σ orbital ($3\sigma_g$ orbital of N_2, 5σ of CO, NO) and metal bonding orbitals share the same group representation. *Ab initio* calculations for Cr–CO [134] and Cr–N_2 [132, 135] model systems indicate that the charge transfer to the molecular adsorbate is greater when the molecule is tilted (or nearly flat in the case of N_2), leading to a greater chemisorption bond with the metal substrate but a weakening of the intramolecular bond.

As with CO, the valence band (occupied molecular orbital) binding energies, outlined in Figure 13, are complicated by the fact that there is both coverage dependence and emission angle dependence to the N_2 molecular orbital binding energies [90, 116]. Band dispersion, created from hybridization of the nitro-

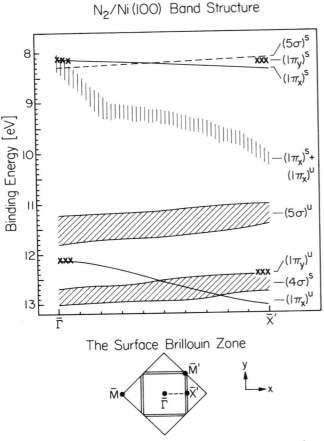

Fig. 14. The band dispersion of the N_2 molecular orbitals from $\overline{\Gamma}$ to $\overline{X}'$ of the reduced surface Brillouin zone. The character of the bands is indicated at the right margin. The features with considerable uncertainty in binding energies are indicated by the broad bands. The "weighted" band derived from the k-dependent intensities of the screened $1\pi_x$ and unscreened $1\pi_x$ is indicated. The surface Brillouin zone is also indicated. The double line indicates the surface Brillouin zone formed by the $c(2 \times 2)$ N_2 overlayer on Ni(100). All dispersion data are taken along the symmetry direction indicated by the dashed line. Adapted from [116].

gen molecular orbitals, has been observed for Ni(100) [116] and graphite [90]. Like the CO case, the satellite photoemission features in N_2 adsorption complicate the band structure [107, 116, 117], and sum rules [116], like those applied to CO, can indicate a greater dispersion of the nitrogen molecular orbital bands than might be apparent from a more "single-particle" interpretation of the wave vector dependent photoemission spectra. Thus a shift of spectral weight from the screened photoemission final state to the unscreened final state can increase the bandwidth from a few hundred meV to several eV [116]. Such an example is shown for molecular nitrogen on Ni(100) in Figure 14.

8. NITROSYL BONDING

The study of NO has been often undertaken as a comparison with CO and N_2. NO and NO_x adsorption has also attracted attention within the surface science community because of the industrial significance of the reduction of NO_x to N_2 and O_2 as well as of the oxidation of ammonia to NO. With the three 1π, 5σ, and 4σ prominent molecular orbitals in angle-resolved photoemission, molecular NO adsorption is similar to the isoelectronic pair CO and N_2. However, NO has one more electron than CO and N_2 and is a less stable adsorbate. The additional electron occupies the antibonding 2π molecular orbital in the isolated molecule and this acts to reduce any tendency for "back donation" of electrons from the metal substrate and to enhance the tendency to fragment upon adsorption. This additional electron also tends to favor molecular dimerization in molecular crystal/condensates [136–138] and at surfaces [139–150].

Still there are very strong similarities among the three molecular adsorbates. There are far fewer valence band angle-resolved photoemission studies of NO compared to the other diatomic low-Z molecules and the assignment of bonding orientation is complicated by the fact that the bonding configuration (upright or tilted) is often coverage dependent to a greater degree than occurs for CO and N_2. There is the additional complexity that, as noted, NO dimers, $(NO)_2$, often form and this can be a ready route to N_2O formation [151]. As a further complexity, using nitrosyl vibrational frequencies to assign bonding orientations appears to be often misleading [151].

The complexity is apparent in the controversy surrounding NO bonding on Ni(111). The valence band angle-resolved photoemission study of NO on Ni(111) (using linearly polarized synchrotron radiation [152]) found that the molecular axis is along the surface normal at 120 K, like CO [55, 59, 60, 67–69, 88] and N_2 [115–117, 119, 120, 124] on a variety of nickel surfaces. The similarity in the photoemission selection rules is apparent in the comparison of the light incidence angle dependence for NO on Ni(111) (Figure 15) and N_2 on Ni(100) (Figure 11). Not in complete agreement with this ARUPS result, a HREELS study [153] and a later reflection absorption infrared spectroscopy (RAIRS) study [154] suggested the existence of two states: a bent NO in the twofold site and an upright state in a twofold bridge site. In agreement with the angle-resolved photoemission, but in contradiction with the studies of the adsorbate vibrational modes, electron stimulated desorption ion angular distribution (ESDIAD) [155], surface extended X-ray absorption (SEXAFS) [156], NEXAFS [157], and photoelectron diffraction [158] found only an upright (and no bent or tilted) NO. This is consistent with cluster calculations [159, 160] for NO on Ni(111).

This problem in assigning the vibrational modes, observed with NO on Ni(111), also occurs with NO on Pt(111), where again two sets of vibrational modes were observed in RAIRS [161] and HREELS [162, 163]. While the X-ray photoelectron spectroscopy core level binding energies are consistent with two binding sites for Pt(111) [164], this is not the case for Ni(111) [165]. Both LEED [166] and density functional theory calculations [167] support a single upright molecular NO species on Pt(111). Potential problems with assigning the binding site and geometry from vibrational spectra are also suggested [168] for Pd(111) and Rh(111).

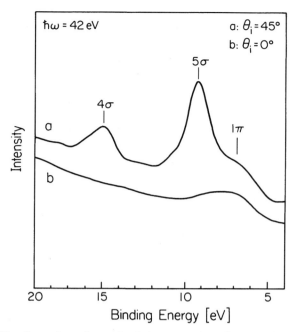

Fig. 15. Comparison of normal emission spectra for 45° (a) and 0° (b) light incidence angles, containing $p+s$-polarized light and s-polarized light, respectively, for NO on Ni(111). Reprinted with permission from [152], copyright 1989, Elsevier Science.

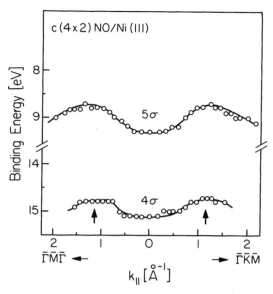

Fig. 16. The two-dimensional band dispersion for the 4σ and 5σ levels of the $c(4 \times 2)$ NO overlayer on Ni(111) along two different symmetry directions. The arrows indicate the average radius of the adsorbate surface Brillouin zone (1.12 Å^{-1}). Reprinted with permission from [152], copyright 1989, Elsevier Science.

In an early NEXAFS study [124], photoemission selection rules were used to compare the orientation of NO on Ni(100) with that of CO and N_2. All three adsorbates were found to bond to the surface with the molecular axis along the surface normal. This result is consistent with cluster calculations [160]. NO adsorption on Ni(100), formed from the dissociation of NO_2, appears to be tilted away from the surface normal in HREELS [169] as well as bonding in the upright configuration, but there is the complexity of coadsorbed oxygen atoms from the NO_2 dissociation. A later NEXAFS [169] is generally consistent with the earlier NEXAFS study [124]. Valence band angle-resolved photoemission measurements demonstrate that, like CO and N_2, NO is also bonded on Ru(0001) [170] with a molecular axis along the surface normal by HREELS [171] and LEED [172].

By way of contrast, angle-resolved photoemission studies of NO on Pd(111) [173] suggested a canted configuration with the molecular axis 20° to 25° off the surface normal. This behavior is also comparable to the other light diatomic molecular adsorbates such as N_2, which bonds with multiple orientations on Pd(111) and not with the molecular axis along the surface normal [117].

While formation of NO dimers can confuse the assignment of binding site and orientation of NO, the reactivity of NO also presents problems on some surfaces such as on Ag(111) where NO is particularly reactive. Angle-resolved photoemission found that NO is bonded in an upright configuration at 150 K on Ag(111) [174]. In contrast, a study of the vibrational modes by HREELS suggested that adsorption of NO leads to both dissociative and molecular adsorption on Ag(111) [175]

and so to molecular NO in both bent and upright configurations. On the basis of evidence supporting the presence of N_2O, So et al. [175] suggested that the route to N_2O occurred through dissociatively adsorbed NO combining with molecular NO. Theoretical calculations suggested [176] that some of the vibrational features could be assigned to NO bound to the surface with the oxygen toward the surface—although there is no compelling evidence to support this molecular bonding configuration. More recent RAIRS vibrational mode studies [148] provided evidence of multilayer $(NO)_2$ formation on Ag(111). The monolayer, left after desorption of the additional multilayers, also consisted of $(NO)_2$ dimers on Ag(111), with the N—N bond parallel with the surface [148]. N_2O formed from the dimers pairs adsorbed on Ag(111) at 70 to 90 K [148]. A similar sequence of adsorbed species, leading to the formation of N_2O, appears to occur on Cu(111) in a synchrotron RAIRS study [146]. In this latter case, the N_2O bonded to the copper surface through the oxygen and the O—N—N molecular axis was postulated to be parallel with the surface.

The dispersion of NO overlayers is quite similar to that for CO and N_2, as seen in Figure 16 for NO on Ni(111) [152]. The dispersion of the 4σ and 5σ molecular orbitals is consistent with the upright configuration of NO on Ni(111) and the bonding character of these orbitals.

9. MOLECULAR OXYGEN

With one more electron than NO and two more electrons than the isoelectronic pair of CO and N_2, it seems likely, based upon the Blyholder model [133], that electron donation from molecular oxygen to a metal substrate must occur from the $1\pi_g$ molecular orbital of O_2 with back donation from the metal to the higher lying unoccupied O_2 σ orbital. Such a bonding configuration with the substrate would favor an O—O bonding orientation with the molecular axis parallel with the surface. This does seem to be generally the case in most O_2 molecular adsorption studies.

In an angle-resolved photoemission study of molecular oxygen, strong light polarization of the molecular orbitals for two different molecular adsorption states on Ag(110) was observed [177]. As seen in Figure 17, the molecular adsorption state (often called the physisorbed state and sometimes a precursor state) formed with adsorption below 40 K exhibits a strong suppression of the $1\pi_g$ and $1\pi_u$ states in s-polarized light—a clear indication of the molecular axis being oriented parallel with the surface. In Figure 18, the situation is similar for the irreversibly chemisorbed O_2 (formed by annealing molecular oxygen to 40 K or by undertaking molecular adsorption at a higher temperature, e.g., 80 K), but there is little suppression of the $1\pi_u$. Prince et al. have explained their data [177] by noting that if the C_{2v} symmetry of the surface is preserved (or even if a lower symmetry applies), following molecular adsorption, then for both molecular adsorption states the molecular axis must be parallel with the surface. If the molecular axis is along the surface normal, then the $1\pi_g$ O_2 molecular orbital is of $b_1 + b_2$ symmetry and is observable in s-polarized light, the $1\pi_u$ O_2 molecular orbital is also of $b_1 + b_2$ symmetry and is observable in s-polarized light, the $3\sigma_g$ O_2 molecular orbital is of a_1 symmetry and is observable in p-polarized light, and the $2\sigma_u$ O_2 molecular orbital is also of a_1 symmetry and is observable in p-polarized light, and this is inconsistent with the experimental data. If the molecular axis is parallel with the surface then the $1\pi_g$ O_2 molecular orbital is of $a_2 + b_1$ symmetry (or $a + b$ in C_2) and is observable in s-polarized light with C_{2v} symmetry and is observable in both s- and p-polarized light with a C_2 symmetry. The $1\pi_u$ is of $a_1 + b_2$ symmetry (or $a + b$ in C_2) and is observable in both s- and p-polarized light. The $3\sigma_g$ is of a_1 symmetry (or a in C_2) and is observable in p-polarized light. The $2\sigma_u$ is b_1 symmetry (or b in C_2) and is observable in s-polarized light. Both molecular adsorption states on Ag(110) are attributed to a C_{2v} configuration with the molecular axis parallel with the surface [177]. This assignment is consistent with the HREELS data [177, 178] and the NEXAFS data for O_2 on Ag(110) [179] which place the molecular axis along the surface $\langle 1\bar{1}0\rangle$ azimuth.

On Pt(111), the O_2 molecular axis also lies largely parallel with the surface as shown by NEXAFS [179, 180], HREELS [181–184], and scanning tunneling microscopy (STM) [185, 186]. Although the STM image of molecular O_2 shows a molecule that is slightly canted out of the plane of the surface [185, 186], theory [187] indicates that one of the molecularly adsorbed states of O_2 is in the plane of the surface while the other molecular adsorption state is slightly canted (consistent with the STM).

Similarly, molecular O_2 adsorbs in two bonding configurations with the molecular axis largely parallel with the surface on Cu(111) [188], consistent with theory [189, 252]. On a

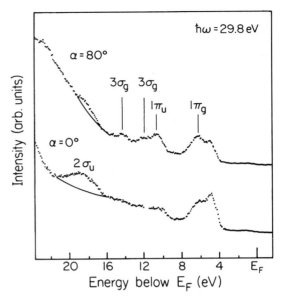

Fig. 17. Light incidence angle dependence of NO on Ag(110) at 25 K at 80° (p-polarized) and 0° (s-polarized) light incidence angles and normal emission. Reprinted with permission from [177], copyright 1986, Elsevier Science.

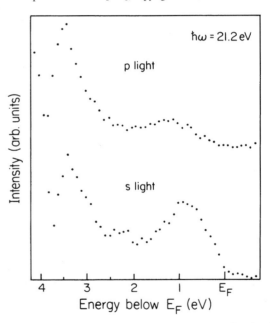

Fig. 18. Polarized (helium I) light photoemission spectra of molecular oxygen of chemisorbed molecular oxygen on Ag(110) at a light incidence angle 45° along the $\langle 001\rangle$ azimuth and normal emission. Reprinted with permission from [177], copyright 1986, Elsevier Science.

stepped Pt(133), molecular O_2 adopts several different orientations along the surface but again, the molecular axis is largely in plane and parallel with the surface as determined from NEXAFS [190].

Distinct from the Pt, Ag, and Cu surfaces, from X-ray diffraction [191], the low coverage phase of molecular O_2 on graphite appears to have the molecular axis parallel with the surface, consistent with photoemission [192], but in the higher coverage phase molecular O_2 on graphite appears to adopt a configuration with the molecular axis along the surface normal. On many other surfaces, oxygen adsorption is initially dissociative followed by molecular adsorption. This is much like the adsorption of the molecular halogens.

10. DI-HALOGEN ADSORPTION

Unlike the low Z diatomic species, molecular halogen adsorption occurs on metal surfaces only following the formation of a dissociatively chemisorbed halogen overlayer, even at substrate temperatures well below room temperature [193]. This adsorption process is schematically illustrated in Figure 19. These are large Z diatomic molecular species and the valence band spectra, as seen in Figure 20, are more complex than the low Z diatomic species—both the $1\pi_g$ and $1\pi_u$ molecular orbitals exhibit spin–orbit splitting.

In spite of the complexities, the molecular orientation of Br_2 [194] and I_2 [195] were determined using polarization dependent angle-resolved photoemission in much the same way as was undertaken with CO, N_2, and NO. One example is molecular iodine adsorption on a chemisorbed iodine overlayer on

Fe(110), as illustrated by the series of spectra as a function of light incidence angle in Figure 20. In fact, as seen in Figure 20, the iodine $1\pi_{g3/2}$ (1.7 eV), $1\pi_{g1/2}$ (2.2 eV), $1\pi_{u3/2}$ (3.2 eV), and $1\pi_{u1/2}$ (4.2 eV) orbitals are suppressed as compared to the $2\sigma_u$ (5.4 eV) in p-polarized light (large incidence angles). Similar behavior is observed for Br_2 molecularly adsorbed on a chemisorbed bromine layer in Figure 21 [194], where two different bonding states, both with the molecular axis normal to the surface, are observed. This confirms the preferred upright molecular bonding orientation of the molecular halogens [193–195]. We believe that the different polarization dependence that can be at lower coverages, in angle-resolved photoemission, may be indicative of a bonding configuration of the molecular axis parallel with the surface for I_2 though not Br_2. Although this is a somewhat modified interpretation of the published data [195], there is evidence of a molecular axis parallel with the surface for very low coverages of molecular iodine on Fe(110), as may be the case for low coverages of molecular I_2 on FeI_2.

For the most part, Br_2 and I_2 bond with the molecular axis along the surface normal and in this respect they very closely resemble CO, N_2, and NO. Molecular I_2 adsorption [195] further resembles O_2 adsorption in that at low molecular adsorption coverages, the molecule bonds with the molecular axis either canted well away for the surface normal or parallel with the surface. In the case of molecular Br_2 adsorption on Ni(100),

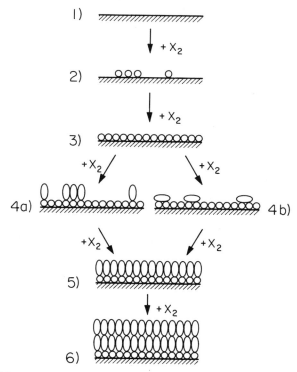

Fig. 19. A schematic representation of halogen adsorption on low index faces of iron, nickel, and iron iodide. The initial adsorption adsorption on the clean surfaces of iron and nickel (but not iron iodide) is dissociative (1–3) and subsequent adsorption is molecular (4–6). For I_2 on iron iodide [and possibly Fe(110)] the initial adsorption appears to have the molecular axis parallel with the surface, while for Br_2 the adsorption appears to be generally with the molecular axis along the surface normal. At higher coverages, the molecular axis is along the surface normal, though in the case for multilayer adsorption of Br_2, the evidence could be argued to favor a reorientation with a molecular axis either in both upright and canted configurations. Reprinted with permission from [193], copyright 1987, CRC Press.

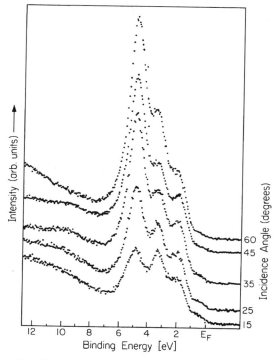

Fig. 20. Light incidence angle dependence of linearly polarized light for molecular I_2 adsorbed on Fe(110) at 110 K. The photon energy is 24 eV, the vector potential A component is parallel along the $\langle 1\bar{1}0 \rangle$ direction, and the photoelectrons are collected normal to the surface. Increasing light incidence angle with respect to the surface normal contains increasingly p-polarized light. Reprinted with permission from [195], copyright 1985, Elsevier Science.

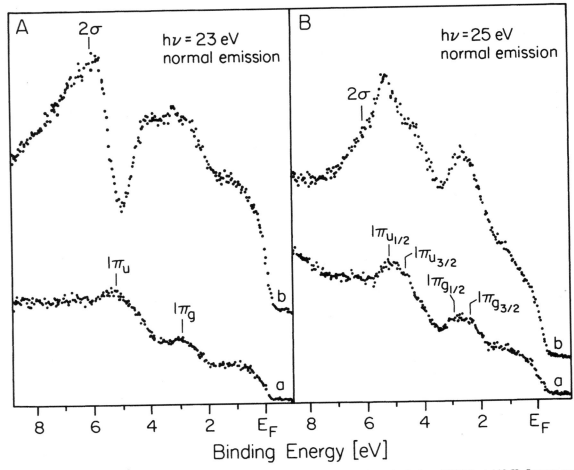

Fig. 21. Light incidence angle dependent photoemission spectra of molecular Br$_2$ adsorbed on Ni(100) at 110 K. Spectra are shown for normal emission and light incidence angles of (a) 15° with respect to the surface normal (s-polarized light) and (b) 45° with respect to the surface normal (s + p-polarzied light), for the initial metastable molecular state (A) and the chemisorbed molecular state (B). Reprinted with permission from [194], copyright 1985, Elsevier Science.

a metastable state is observed that converts to a more stable molecularly chemisorbed state [194], much like molecular O$_2$ adsorption on many surfaces.

11. AMMONIA

The valence band of adsorbed ammonia is dominated by the $2a_1$, $1e$, and $3a_1$ molecular orbitals. The $3a_1$ orbital is representative of the nitrogen lone pair electrons, which are donated, to a significant extent, to the surface in most ammonia molecular adsorption systems. The $2a_1$ orbital has most of the oscillator strength from the nitrogen $2s$ while the $1e$ ammonia molecular orbital is dominated by the N—H bonding electrons.

For ammonia on Ni(111) [196], angle-resolved photoemission showed that the $3a_1$ molecular orbital intensity increases with greater p-polarization (the increase of the vector potential **A** along the surface normal). This indicates that the molecular axis is along the surface normal and the likely bonding scenario is with the nitrogen toward the surface. Angle-resolved photoemission exhibited azimuthal variations in the cross-section

of the $1e$ molecular orbital [197], again consistent with an orientation with nitrogen bonding toward the surface and hydrogens away. Since the $1e$ molecular orbital is dominated by the N—H bonding electrons this suggests that there are specific N—H bond orientations with respect the substrate surface crystal directions [197]. This ammonia bonding orientation is consistent with ESDIAD studies of NH$_3$ and NH$_3$ coadsorption on Ni(111) [198–200] and with both the RAIRS [201] and HREELS [202] derived vibrational modes, but the ESDIAD studies for NH$_3$ on Ni(111) do not show the strong azimuthal dependence, in the absence of a coadsorbate, exhibited in angle-resolved photoemission for both NH$_3$ on Ni(111) [197] and Ir(111) [203].

Also using the ammonia $1e$ molecular orbital, dominated by the N—H bonding electrons, Purtell et al. [203] were also able to show threefold symmetry to the azimuthal intensity in angle-resolved photoemission, as seen in Figure 22. As in the case of NH$_3$ on Ni(111), this strongly indicates a preferential bonding orientation of ammonia on Ir(111) with the molecular axis along the surface normal.

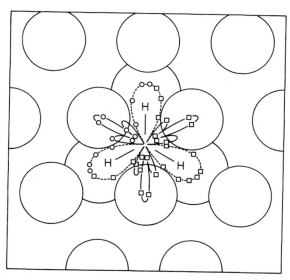

Fig. 22. The azimuthal plot of the $1e$ molecular orbital of NH_3 photoemission signal on Ir(111). The angle of incidence is for $s + p$-polarized light and the emission angle is 45°, with the unmodulated component of the signal subtracted. The photon energy is 25 eV. The difference signal of the NH_3 $1e$ orbital is superimposed upon the Ir(111) surface.

Using the light polarization dependence in angle-resolved photoemission, a similar bonding orientation was assigned to molecular ammonia adsorption on Ni(110) [204] where again the ammonia bonds to the surface through the nitrogen lone pair. This is also consistent with ESDIAD [198, 199] and HREELS [202, 205]. With photoelectron diffraction, ammonia is seen to occupy the top site on the Cu(110) rows, again bonding the nitrogen to the copper through the lone pair [206].

Both ESDIAD [207] and HREELS vibrational studies [208, 209] confirm the adsorption of ammonia bonding through the nitrogen lone pair on Ru(100), consistent with Ni(111), Ir(111), Ni(110), and Cu(110), just discussed, as well as with the HREELS vibrational mode studies on Pt(111) [210], Fe(110) [211, 212], Ag(110) [213], and Ag(311) [214].

Even for compound surfaces the ammonia adsorption appears to occur through the nitrogen lone pair. A crude photoelectron diffraction found evidence of ammonia clustering, providing indirect evidence for the ammonia bonding through the nitrogen with the hydrogen away from the surface on Ni(100) with a chemisorbed nitrogen adlayer [215]. Similarly, HREELS is consistent with the nitrogen bonding toward the nickel surface through the lone pair [216]. While the photoemission results for ammonia adsorption on ZnO and CuCl surfaces are not compelling, the self consistent field (SCF-Xα) calculations also suggest ammonia bonding to the metal again through the nitrogen lone pair [217].

12. WATER

The adsorption of water and the formation of ice has been extensively studied on many different surfaces, both metals as well as oxides and insulators. Water typically adsorbs on metal surfaces with the oxygen close to the surface and the hydrogens farther away with the C_{3v} molecular axis along the surface normal in a similar configuration to ammonia [218–220]. With over 100 publications on water and hydroxy adsorption each year, we cannot do justice to water and hydroxy adsorption within the scope of this review and therefore it is one of the few adsorbates to which we purposefully give short shrift.

13. NO$_2$, SO$_2$, CO$_2$

The triatomic species NO_2, SO_2, and CO_2 all have attracted interest because of their role in atmospheric chemistry [221]. CO_2 is rather more inert than NO_2 and SO_2. All of these triatomic adsorbates adopt several different bonding configurations often complicated with coadsorbed atomic oxygen so that there are really few generalizations possible.

Sulfur dioxide, in particular, has become of increasingly investigated because of its key role in air pollution and in airborne particulates [222]. Generally SO_2 dissociates on metal surfaces forming either SO_3 plus atomic sulfur or SO plus atomic oxygen. On silver, palladium, and platinum (at low temperatures) SO_2 adsorption is molecular. While HREELS suggested that the SO_2 on Ag(110) is tilted [223], NEXAFS has the SO_2 molecular axis on Ag(110) placed along the surface normal [224, 225], with the sulfur bonded via a site atop the silver, consistent with SCF-molecular orbital calculations [226] and HREELS vibrational mode studies [223]. On Pd(100), SO_2 adsorption is also molecular below 240 K and the vibrational HREELS spectra [227, 228] are consistent with the SO_2 plane parallel with the surface normal, but the O—O axis is not parallel with the surface and so the molecule is effectively tilted in the plane (of the molecule-surface normal). On Pt(111) a similar bonding configuration is believed to occur for molecular SO_2 adsorption, but perhaps with more tilt [229]. These two configurations for SO_2 on silver as compared with SO_2 on palladium and platinum are schematically shown in Figure 23.

Analysis of the surface extended X-ray adsorption fine structure (surface-EXAFS) [230] indicates that there are several SO_2 adsorption geometries on Cu(100) for the SO_2 plus coadsorbed atomic oxygen, while, according to surface-EXAFS (SEXAFS), the SO_2 species lies flat [222, 230]. Generally, the molecules are bonded through the S and generally with one oxygen. NEXAFS places the C_{3v} axis of SO_3 (formed from adsorbed SO_2) along the surface normal, consistent with STM [231], while EXAFS places the oxygen toward the surface and the sulfur away [231]. NEXAFS studies of SO_3 and SO_2 on Cu(111) [232] also place the oxygen toward the surface and the sulfur away, with the C_{3v} and C_{2v} molecular axis along the surface normal. This molecular bonding orientation for SO_3 is the mirror image of the bonding exhibited by ammonia and water, but it is quite similar to the bonding configuration of the similar formate (HCOO) molecule on Cu(100) [233, 234].

On the more reactive nickel surfaces, SO_2 adsorption is largely molecular, with a small amount of dissociation at the lower coverages On Ni(110) angle-resolved photoemission

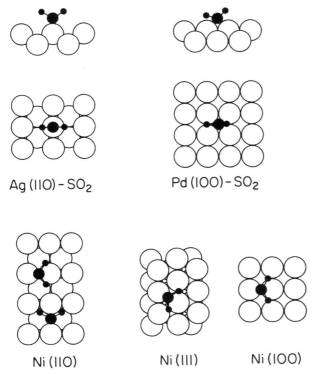

Fig. 23. The schematic SO_2 bonding geometry on Ag(110), Pd(100), and the various faces of nickel. Reprinted with permission from [222], copyright 1987, American Institute of Physics.

placed the O—S—O plane perpendicular to the surface and along the $\langle 100 \rangle$ azimuth [235]. Unfortunately this appears to disagree with other measurements, including those also based upon dipole selection rules. NEXAFS [236, 237] indicates that the molecular plane is in the plane of the surface with little azimuthal dependence and this is consistent with SEXAFS [236, 237]. The SO_2 bonding orientation on Ni(111) [238, 239] and Ni(100) [239] is qualitatively similar, and SO_2 is placed with the molecular plane parallel with the surface.

CO_2 adsorption is largely linear on Pd(111) in the plane of the surface, as determined by angle-resolved photoemission [240] and high resolution electron energy loss [241], with similar HREELS results obtained for CO_2 on Pd(100) [242].

On Ni(110), the molecular axis of CO_2 is mostly in the plane of the surface, at low temperatures, as determined by angle-resolved photoemission [243, 244], although some tilt is expected. A bent anionic CO_2 species is observed at higher temperatures on Ni(110) [243]. These results are supported by HREELS [243, 245] and NEXAFS [246]. The CO_2 structure, at higher temperature, does resemble a formate species with the oxygen bonding toward the surface [233], as discussed later.

As on Ni(110), angle-resolved photoemission identifies a linear CO_2 species on Fe(111) is oriented with the molecular axis parallel with the surface [247], consistent with the HREELS for CO_2 on Fe(111) and Fe(110) [248]. A minority fraction is a bent anionic CO_2 species that does not desorb as

easily as the linear species and is stable to higher temperatures [247, 249].

NO_2 adopts bonding configurations between that of CO_2 and SO_2. The NO_2 bonds with the nitrogen down on Pt(111) [250, 251] in bent configuration, like SO_2 on Pd(100), as determined by HREELS [250, 251]. On Ni(100) [169], the case for such a bonding configuration is even more convincing with confirmation from angle-resolved photoemission, HREELS, and NEXAFS. On Au(111), the molecular NO_2 bonds with the oxygen toward the surface [253]. Like NO, NO_2 dimerizes to form N_2O_4, not surprising given that NO_2 and N_2O_4 coexist in equilibrium [254, 255], making the weak chemisorption of NO_2 a somewhat complex picture.

The complexities of SO_2, NO_2, and CO_2 are not limited to a wide variety of bonding configurations and the reader is directed to recent review for further discussion. The surface chemistries of both SO_2 [222] and CO_2 [249] adsorption have been recently reviewed.

14. FORMATE

As noted previously, CO_2 adsorption on some surfaces can resemble the adsorption of formate (HCO_2), although the O—C—O axis is typically bent considerably in formate. On Cu(110), angle-resolved photoemission [256, 257], NEXAFS [258, 259], and photoelectron diffraction [260] are consistent with the more detailed NEXAFS studies [257] that place the O—C—O axis parallel (not necessarily linear) with the surface and parallel with the $\langle 1\bar{1}0 \rangle$ azimuthal direction with the molecular plane perpendicular to the surface (hydrogen up). This is consistent with density functional theory, which suggests that the bridge site is favored but the O—C—O bond is bent [261]. This bonding appears to be similar on Cu(100) [260]. NEXAFS studies of the adsorbed formate orientation also suggest some preferential azimuthal orientation on Cu(100) [233, 234] consistent with theory [261, 262] and photoelectron diffraction [263].

On Ag(110), the NEXAFS data indicate that the molecular plane is, however, tilted [264], but HREELS data [265, 266] for formate on Ni(110) suggest the formate molecular plane is along the surface normal, with the oxygen bonding toward the surface, much like formate on Cu(110).

15. METHANOL (METHOXY), METHANETHIOL (THIOLATE), AND RELATED SPECIES

On many metal surfaces, the adsorption of methanol results in the formation of a species (methoxy) with an oxygen–metal substrate bond. The C—O molecular axis is largely (but not always) along the surface normal with the oxygen toward the surface and the hydrogen away.

Vibrational spectra of the methoxy species on Cu(100) [267, 268] suggest a C—O molecular axis parallel to the surface normal. This result is in agreement with the NEXAFS results for the methoxy species on Cu(100) [180, 269]. RAIRS spectra for methoxy on Cu(111) (consistent with later photoelectron diffraction [270, 271] studies and model calculations [272]) were interpreted to indicate that again methoxy was upright [273] and the authors reinterpreted earlier RAIRS vibration data [274] for methoxy on Cu(100) as consistent with an upright configuration. Definitive evidence for an upright methoxy species on Cu(100) is also provided by angle-resolved photoemission [269]. Only an early NEXAFS/SEXAFS study suggested a picture in which methoxy on Cu(100) is tilted away from the surface normal [233]. HREELS also suggests that ethanol (ethanoxide) is also upright on Cu(100) [267], although nonlinear.

While methoxy on Cu(111) and Cu(100) adopts an upright configuration, on Cu(110), methoxy is tilted away from the surface normal. Vibrational RAIRS [275] and HREELS [276, 277] spectra of methoxy on Cu(110) find modes indicative of a canted configuration. For most other surfaces—Mo(110) [278], Mo(100) [279], Rh(111) [280], Ni(100) [281], and Fe(100) [282]—the vibrational spectra are consistent with the molecular axis being upright along the surface normal.

Angle-resolved photoemission studies of methoxy on Ni(111) [283] showed a strong increase in the $2e$ (5.4 eV) molecular orbitals with increasingly p-polarized light and off normal emission, while the methoxy $1e$ molecular orbital (9.7 eV), which overlaps in binding energy with the $5a_1$ orbital, exhibited no such photoemission geometry dependence. This confirms that, like CO molecular adsorbates on nickel surfaces, the methoxy fragment bonds with the molecular axis along the surface normal. This is confirmed by photoelectron diffraction [284, 285] and NEXAFS [286]. Model calculations [287] of the vibrational spectra [288] also suggest an upright configuration for methoxy on Ni(111), although the original HREELS results did suggest a slightly tilted configuration [288]. Similar model calculations also suggest an upright methoxy configuration on Ni(100) [289].

Methanethiol (methyl mercaptan) is a sulfur-containing species that resembles methanol. On Cu(111), Cu(410), Cu(110), Cu(100), and Ni(100), both CH_3SH and $(CH_3S)_2$ react to form the common surface species, CH_3S (methyl mercaptide [290–297]), although later studies suggest multiple species resulting from methanethiol adsorption on Cu(111) [298, 299]. Angle-resolved photoemission shows a largely substrate band, induced with adsorption, becoming symmetry forbidden at normal emission [292]. This is taken to indicated that the CH_3S methanethiol species is upright and bonding occurs in the threefold hollow site, consistent with the previous angle-resolved photoemission study [293], which placed the C—S bond along the surface normal. This is also largely consistent with later normal incidence standing wave experiments which have the intact thiol (CH_3SH) adsorbing on Cu(111) with the C—S axis tilted away from the surface normal and coadsorbing with the thiolate (CH_3S) which has the C—S axis placed along the surface normal [298]. SEXAFS studies of the thiolate on Cu(111) [300] also place the sulfur as bonding to the copper with the molecular axis along the surface normal. NEXAFS [301] has the C—S molecular axis tilted, although, in view of the recent standing

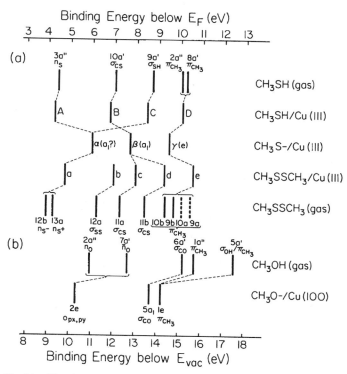

Fig. 24. The observed energy levels for gaseous, condensed, and chemisorbed methane thiolate, together with $(CH_3S)_2$ on Cu(111) compared to the methanol (gas phase) and methoxy on Cu(100) binding energies. The energy levels are referenced to the Fermi level or the vacuum level as appropriate. Reprinted with permission from [290], copyright 1987, Elsevier Science.

wave X-ray data, this result might refer to the intact thiol rather than to the thiolate.

Angle-resolved photoemission studies [293] have been undertaken of the thiolate (CH_3S) on Cu(110) and Ni(100). The results for Cu(110) are consistent with NEXAFS [301]. The angle-resolved photoemission [293] and photoelectron diffraction [302] also are consistent with the C—S axis of the thiolate being normal to the surface.

The orientation of the molecular axis of thiolate (CH_3S) on Cu(100) appears tilted in HREELS [303] and NEXAFS [301] experiments. Analysis of the photoelectron diffraction [304] has the thiolate C—S bond tilted away from the surface normal with the methyl group tilted toward the bridge site on Ni(111). This tilted species on Cu(100) and Ni(111) differs from the upright configuration observed for thiolate on other copper and nickel surfaces, just mentioned. The thiolate also appears, in both NEXAFS and HREELS, as a tilted species on Pt(111) [305]. In spite of these differences in different substrates, the thiolate adsorption is very similar to the methoxy adsorption, as indicated in comparison of the methoxy and thiolate photoemission results on copper in Figure 24.

The close nitrogen analog, acetonitrile (CH_3CN), appears to be different in bonding configuration to thiolate, methoxy, and ethylidene, while methyl isocyanide (CH_3NC) on Pt(111) adopts a terminal bonding configuration with the molecular N≡C bond upright or slightly tilted [306, 307]. With acetonitrile on Pt(111) [306, 307], HREELS vibrational studies suggest that the imine C=N lies close to parallel with the surface. Vibrational HREELS studies of acetonitrile on Cu(111) [308] and Ag(110) [309] suggest the adsorbed species are little different from the gas phase species and the NEXAFS places the linear molecule molecular axis parallel with the Ag(110) surface as well [310].

The ethylidene (CH₃C) adsorbs in a fashion similar to the thiolate (the sulfur analog) and methoxy (oxygen analog) with the molecular axis generally along the surface and adopting C_{3v} symmetry where possible. The ethylidene adsorption should, in principle, be compared with the adsorption of thiolate and methoxy surface species, but, it should be noted, ethylidene is often formed with adsorption of acetylene and ethylene at room temperature, which we discuss in the next section. The ethylidene species was "first" discovered in a HREELS study of ethylene adsorption on Pt(111) at room temperature [311] and the placement of the molecular axis, determined by LEED [312, 313], was consistent with later vibrational mode studies [314–316]. The adsorbed ethylidene species has also been investigated by angle-resolved photoemission on Pt(111) [317] and Pd(111) [318] and on Fe(100) [319] by HREELS and in all cases, again, the molecular axis is placed along the surface normal.

16. ETHYLENE AND ACETYLENE

The interaction of ethylene and acetylene with the surface has long been expected to be more substantial than ethane [320–322] and therefore more likely to exhibit ordering and band structure. Several different bonding configurations for molecular ethylene and acetylene have been postulated (as indicated schematically for acetylene on Ni(110) in Figure 25 [323]) generally with the C—C bond parallel with the surface.

Angle-resolved photoemission studies, in agreement with NEXAFS [323], demonstrate that acetylene on Ni(110) bonds

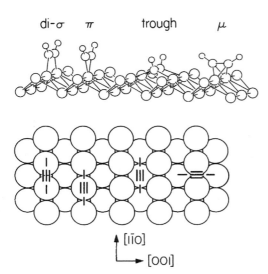

Fig. 25. The schematic view of the di-σ, π, trough, and bridge μ adsorption sites for acetylene adsorbed on Ni(110). Reprinted with permission from [323], copyright 1995, American Institute of Physics.

with the molecular axis parallel to the surface with the C—C bond along the $\langle 1\bar{1}0 \rangle$ azimuth [323]. Symmetry and selection rules dictate the same bonding configuration occurs for ethylene on Ni(110) [324–327], again consistent with NEXAFS [324] and vibrational mode studies from HREELS studies [328]. The problem is that such a molecular orientation for both acetylene on Ni(110) and ethylene on Ni(110) is consistent with three possible binding sites [323–327]. This is a good illustration of the limitations of angle-resolved photoemission—the symmetry of these different sites is essentially the same—so a clear distinction between the sites is not really possible from valence band angle-resolved photoemission alone.

Similar behavior is seen for acetylene and ethylene on other nickel surfaces. On Ni(111), vibration mode HREELS experiments [329] place acetylene and ethylene [329, 330] with the molecular axis parallel with the surface and, in combination with LEED [329], suggest strong azimuthal orientation of the adsorbed molecules. This is consistent with density functional calculations [331] and photoelectron diffraction [332]. For Ni(100), the symmetry of the molecule appeared to be slightly broken in an early HREELS vibrational study [333], suggesting that the ethylene is slightly tilted with respect to the surface.

On Pt(111) as well, the molecular axis of both acetylene and ethylene, as determined by angle-resolved photoemission, is also placed [317] parallel with the surface. This is consistent with HREELS and RAIRS vibrational loss studies [311, 315, 334–338]. Results that have been obtained from vibrational studies of ethylene on Pd(111) [339, 340] again suggest a molecular axis largely parallel with the surface. From the observed vibrational modes, it appears that the nature of the ethylene bonding (π-bonding, as indicated in Figure 25) on Pd(111) [339, 340], Pd(100) [341], Pd(110) [342–344], and Cu(100) [345] differs from most other surfaces ("di-σ bonding," as indicated in Figure 25) including Fe(111) [346], Fe(110) [347], Fe(100) [319], Ni(110) [328], Ni(111) [329, 330], Ni(100) [333, 348, 349], Ru(100) [350], and Pt(100) [351]. For Pt(111) both di-σ bonding [311, 315] and π bonding [337, 338] have been postulated. From HREELS, it appears that C_2H_4 bonds as both a π bonding species and as a di-σ bonding species Pt₃Cu(111) [352]. The strong preferential bonding orientation for ethylene on Pd(110) along the $\langle 1\bar{1}0 \rangle$ azimuth, with the C—C axis parallel with the surface, is evident in both NEXAFS [344] and STM [353].

On Si(100), NEXAFS [354] also places acetylene and ethylene with the molecular axis parallel with the surface and parallel with the 2 × 1 dimer rows. C_2H_4 is only weakly bonded to the As-terminated GaAs(100) surface, but again the C—C bond is parallel with the surface [355]. The differences between ethylene adsorption on Si(100) and GaAs(100) are consistent with theoretical predictions [356].

Surprisingly, the linear molecule acetylene appears by photoelectron diffraction [332] to be in a distorted configuration, with the C—C axis parallel with the surface on Ni(111). While this is understandable due to the strength of the substrate interaction with the adsorbed acetylene, this appears, in NEX-

AFS experiments, to be also the case on Cu(111) [357, 358] and Pd(111) [358, 359] where the interaction is expected to be much weaker. Thus, while the C—C axis is parallel with the surface in photoelectron diffraction [357] and NEXAFS [359], the acetylene appears to be distorted, in spite of the weak substrate adsorbate interaction [358, 360]. Vibrational studies, however, have the acetylene tilted on Pd(110) [361] and Pd(111) [343] although, as noted, this is not consistent with NEXAFS [344].

The band structure can to some extent help determine the bonding site. In fact, the extensive dispersion exhibited by both acetylene and ethylene overlayers is impressive—with the experimentally mapped bands extending into multiple Brillouin zones. The possible extended and reduced Brillouin zone schemes for the $c(2 \times 4)$ structure formed by ethylene on Ni(110) [324] are shown in Figure 26. In fact, few of these structures fit the experimental ethylene on Ni(110) band structure seen in Figure 27 [324], but the fit with the expected band structure of the "geometry II" ethylene structure of Figure 26 is marked. The extensive band dispersion of the molecular orbitals of ethylene on Ni(110) is shared by acetylene on Ni(110) [323]. Again the fit to theory is impressive, as seen in Figure 28. The

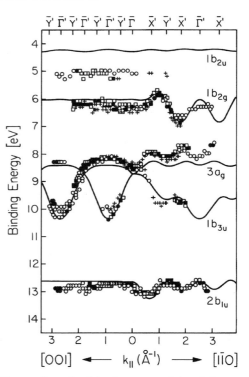

Fig. 27. The two-dimensional band structure of the ethylene overlayer on Ni(110) determined from angle-resolved photoemission. Shown is the calculated band structure for geometry II (see Figure 26) shown as a solid line. Reprinted with permission from [324], copyright 1992, American Physical Society.

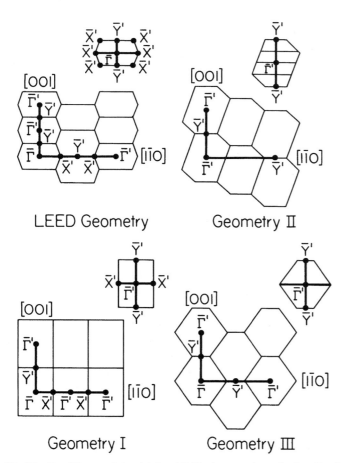

Fig. 26. Possible extended and reduced Brillouin zone scheme for the $c(2 \times 4)$ ethylene structure on Ni(110). The experimental k-paths (wave vector dependence) along the $\langle 100 \rangle$ and $\langle 1\bar{1}0 \rangle$ azimuths of the Ni(110) substrate are indicated. Reprinted with permission from [324], copyright 1992, American Physical Society.

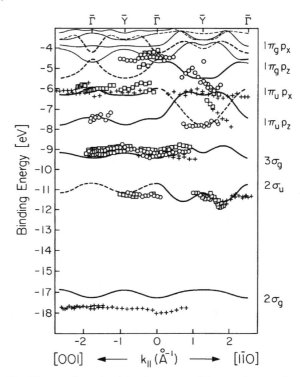

Fig. 28. The two-dimensional band structure of the acetylene overlayer on Ni(110) determined from angle-resolved photoemission. Shown is the calculated band structure for the trough geometry (see Figure 25) shown as a solid line. Reprinted with permission from [323], copyright 1995, American Institute of Physics.

distinction between a π bonding species and a di-σ bonding species, however, on the basis of photoemission alone, seems to be rife with difficulties, judging from recent NiC$_2$H$_4$ cluster calculations [362].

17. CYANOGEN AND CN

While quite different, other linear molecules with a similar bonding configuration to acetylene and ethylene are the cyano (CN) containing molecules, although the data available on the amines, imines, and nitriles are far less than those for the alkane, alkene, and alkyne counterparts. While CN$^-$ is isoelectronic to CO, the bonding configuration for CN is quite different from what is generally observed for CO.

The adsorption of cyanogen, C$_2$N$_2$, on Pd(110) has been studied by angle-resolved photoemission [363, 364] and found to be π bonded with the molecular axis parallel with the surface and with the molecule oriented along the $\langle 100 \rangle$ azimuth direction. This bonding configuration is consistent with the "flat" adsorption geometry for cyanogen on Pd(111) and Pd(100) deduced from HREELS measurements [365–368] and NEXAFS [368]. There are complications, nonetheless, as there is persistent evidence of a high coverage "tilted" molecular orientation phase [366].

On Ni(110), the C$_2$N$_2$ molecule was first thought to appear tilted in angle-resolved photoemission [369, 370], although a possible preferential orientation along the $\langle 100 \rangle$ direction was acknowledged [369]. This tilt may be a result of the fact that while cyanogen adopts a $c(2 \times 2)$ configuration on both Pd(110) and Ni(110), the lattice spacing is a little smaller on Ni(110) (about 10%). In later angle-resolved photoemission experiments, the data were thought sufficient to find that the C$_2$N$_2$ exhibited good azimuthal orientation along the $\langle 1\bar{1}0 \rangle$ direction with the molecular axis parallel with the surface [371]. This was believed to be consistent with theory, although a low barrier to other configurations was acknowledged [371]. Other theory placed the molecular axis parallel to the surface and with preferential orientation along $\langle 100 \rangle$ direction (orthogonal to the $\langle 1\bar{1}0 \rangle$ trough) [372]. Recent HREELS studies found that the angle-resolved photoemission results could all be reconciled, since the multilayer, when annealed to 120 K, was shown to be a combination of CN oriented along the $\langle 1\bar{1}0 \rangle$ direction and a tilted C$_2$N$_2$ that exhibited good azimuthal orientation along $\langle 100 \rangle$ direction (orthogonal to the $\langle 1\bar{1}0 \rangle$ trough) [373].

Annealing cyanogen results in the formation of a CN layer [294, 363, 373]. Angle-resolved photoemission studies of CN on Pd(110) [374] find that the molecular axis is parallel with the surface and along the $\langle 1\bar{1}0 \rangle$ direction. The degeneracy of the π-orbital is lifted in the angle-resolved photoemission of Pd(111) [375] and Ni(110) [369], as is also the case for CN on Pd(110). While not as conclusive, these results are consistent with HREELS data for Pd(100) [365–367], Pd(111) [367, 368], Cu(111) [376, 377], and Ni(110) [373], as well as the more definitive NEXAFS for CN on Pd(111) [368]. The "flat" (parallel with the surface) or slightly tilted away from parallel

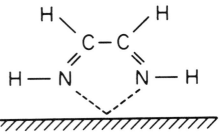

Fig. 29. A likely bonding configuration for the hydrogenated cyanogen species (CNH$_2$)$_2$. Adapted from [294].

with the surface bonding configurations are not entirely consistent with existing calculations, however [378–380].

The other simple cyanogen species, HCN, has been seen in angle-resolved photoemission on Cu(110) [381] and Pd(111) [382]. The angle-resolved photoemission seems consistent with high resolution electron energy loss [368] and NEXAFS [368] results for HCN on Pd(111), although HREELS has the hydrogen somewhat tilted with respect the C—N axis that is parallel with the surface. The diaminothylene (CNH$_2$)$_2$ has been observed in FT-RAIRS and a bonding configuration like those in Figure 29 is likely (a bridge bonding configuration) on Pd(111) [383], as suggested earlier [294]. Some of the larger cyanogen species (acetonitrile and methyl isocyanide) have previously been discussed in the context of methanol and methane thiole.

18. BENZENE, PYRIDINE, AND SMALL AROMATICS

Like ethylene and acetylene, benzene adsorbs molecularly on Ni(110) and bonds "flat," i.e., with the molecular C—C bonds parallel with the surface, as observed by NEXAFS [384], STM [385], photoelectron diffraction [386], and angle-resolved photoemission [2, 384, 387, 388]. A similar molecular orientation [2, 4, 294, 387, 389–392] has been found for benzene on Pd(100) [393, 394], Pd(111) [395–398], Pt(111) [399–401], Ag(111) [402, 403], Rh(111) [392, 395, 396, 404–411], Ir(111) [412], Al(111) [413], Ni(111) [325, 387, 414–417], Ni(100) [418], Cu(110) [385, 418], Cu(111) [419], Ru(100) [170, 420, 421], and Os(100) [422–425].

While replete with molecular orbitals (some of which are schematically shown in Figure 30), the adsorption of benzene in the "flat" configuration [2, 384, 387] leads to a reduction in symmetry from C_{6v} to C_{2v} on Ni(110) in angle-resolved photoemission. There appears to be preferential orientation along the $\langle 001 \rangle$ direction [384, 387]. A specific binding site for benzene on Ni(110) [2, 325, 384, 387] and Cu(110) [426] cannot be uniquely determined from photoemission alone. The bonding configuration in photoemission is consistent with the scanning tunneling microscopy images of benzene on Ni(110) and Cu(110) which place the benzene in a hollow site parallel with the surface and bridging the close packed rows [385]. Photoelectron diffraction [386] confirms the STM binding of benzene on Ni(110) over a hollow site as well, again, as placing

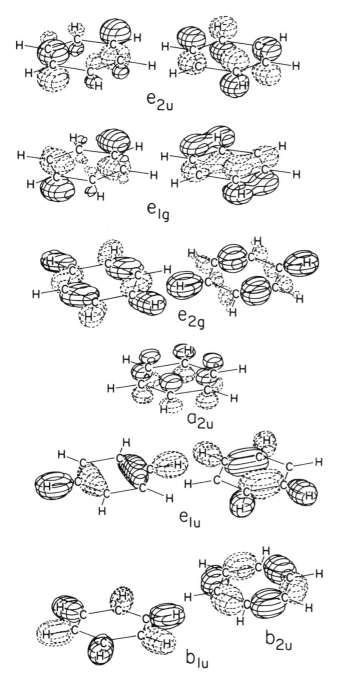

Fig. 30. A schematic representation of the benzene molecular orbitals. Figure adapted from [387] and W. L. Jorgensen and L. Salem, "The Organic Chemist's Book of Molecular Orbitals," p. 257. Academic Press, San Diego, 1973.

distortion of the benzene planar ring structure, as suggested by theory [428].

The degeneracy of the benzene $2e_{1u}$ orbitals (schematically shown in Figure 30) is lifted in photoemission, resulting in the formation of two bands, the b_1 and b_2 bands (at 7.5 and 7.9 eV binding energy) on Ni(110) [384]. The same is true of the $1e_{1g}$ orbitals (b_1 and b_2 bands at 4.3 and 4.6 eV) and the $2e_{2g}$ orbitals (a_1 and a_2 bands at 5.5 and 6.1 eV) [384] also shown in Figure 30. The reduction of the symmetry to C_{3v} for benzene on Ni(111) [414] may result in two $2e_{2g}$ orbitals (at 5.9 and 6.5 eV binding energy). For benzene on Ni(111), the preferential orientation apparent in angle-resolved photoemission [387, 414] has been confirmed by photoelectron diffraction [416]. Low energy electron diffraction structural studies [417] suggest that benzene on Ni(111) is also placed with the molecular plane parallel with the surface, although there is a slight molecular distortion. (This distortion is somewhat similar to that postulated for benzene on Ni(110) [386].) There is buckling of the Ni(111) substrate induced by the chemisorption of benzene [417]. When the Ni(111) surface is covered by only a monolayer of copper, the angle-resolved photoemission [429] suggests that the adsorbed benzene loses azimuthal orientation and becomes much more weakly bound, like benzene on Cu(111) [419]. Similarly, benzene on Al(111) is also weakly bound, with the molecular plane seen to be parallel with the surface in angle-resolved photoemission and HREELS [413]. The C_{6v} molecular symmetry of benzene is preserved on Al(111) [413] and, in spite of some preferential orientation for adsorbed benzene on Rh(111), the C_{6v} molecular symmetry also appears to be preserved in angle-resolved photoemission [392]. Angle-resolved photoemission suggests a strong preferential orientation for benzene on Rh(111) [392, 406] and Pt(111) [399], but it is clear that almost undetectable amounts of coadsorbed CO can induce a preferential orientation for benzene on the very flat surfaces like Rh(111) [392].

However, there is a fairly convincing case to be made [430] that the Jahn–Teller effect in the photoemission process itself leads to a "double peak structure" for the $2e_{2g}$ from adsorbed benzene. This lowering of symmetry in the photoemission process is also observed in NEXAFS in the core excitation to the $1e_{2u}$ unoccupied molecular orbitals (shown in Figure 30) for benzene on Pt(110) [389], on Mo(110) [431], and in condensed molecular films [432].

For benzene on Pd(110) [364, 391], the selection rules applied to the angle-resolved photoemission measurements suggest that, instead of lying "flat," the adsorption geometry is slightly tilted. A tilt geometry of 10° to 20° was proposed [391], reducing the overall symmetry to C_S. This result is fairly compelling in spite of early HREELS data that suggest a "flat" orientation [433, 434]. Similarly, benzene on Pt(110) (reconstructed to the 1×2 structure) is also seen to be tilted, with the overall symmetry of C_S, by angle resolved photoemission [389]. With NEXAFS [389], a tilt angle of about 30° in the $\langle 001 \rangle$ direction is deduced, placing the benzene almost planar with the (111) facets on this platinum surface, as shown in Figure 31. This tilted orientation is a consequence of the

the molecular plane parallel with the surface. This compelling further refinement of the benzene adsorption site is not, however, in agreement with the expected theoretical binding site of the benzene bonding atop a nickel atom [427]. In spite of the compelling STM [385] and NEXAFS [418] results, benzene on Cu(110) appeared to be tilted out of the plane of the surface in HREELS and angle-resolved photoemission [425]. These conflicting results for benzene on Cu(110) may be reconciled by

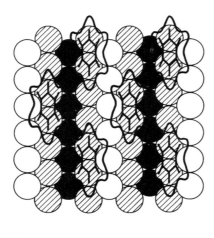

Fig. 31. The proposed bonding configuration for benzene in the 4×2 structure on Pt(110) 1×2. The adsorption sites are chosen somewhat arbitrarily, but the tilt is indicated. Reprinted with permission from [389], copyright 1998, Elsevier Science.

surface structure and is not like the tilting of phenylacetylene on Pt(111) (the phenyl ring is tilted by 34° to 37° out of the plane of the surface as observed in NEXAFS) caused by the end group [435].

The reduction of symmetry in the photoemission final state can make unequivocal assignment of the bonding configuration difficult: without angle-resolved measurements, assigning benzene adsorption on Si(111) to a Diels–Alder like addition [436] on the basis of the photoemission bands alone is uncertain. Benzene chemisorbed on Si(100)-2 × 1 appears to lose much of its aromatic character in infrared adsorption and NEXAFS [437]. The majority of the benzene, in the simple overlayer, chemisorbs to the surface with a Diels–Alder like configuration, resembling that of 1,4-cyclohexadiene tilted with respect to the surface by about 30° [437].

The closely related molecule pyridine does not lie flat as readily as benzene. Angle-resolved photoemission measurements show quite clearly that the molecule bonds through the N atom on Ir(111) [412]. The molecule stands up (i.e., upright), occupying a relatively high point group symmetry in this bonding configuration, with the molecular axis along the surface normal [412]. Angle resolved photoemission results for pyridine on Pd(111) suggest a tilted configuration (neither "flat" nor "upright" in bonding configuration) with the molecule interacting with the surface through both the nitrogen lone pair electrons and the π electrons [397, 438]. While angle-resolved photoemission band studies have been undertaken for pyridine on Cu(111) [439], the assignment of the bands has been questioned [438, 440].

Pyridine on Pd(110) appears from angle-resolved photoemission to bond in a "flat" configuration [441]. This is quite different from many of the results for the pyridine bonding configuration on Cu(110), where the pyridine is again observed, by photoelectron diffraction [442], to bond atop the copper atoms and to have its molecular plane tilted with respect to the surface. Angle-resolved photoemission from Pd(110) has also provided unusual results for benzene, as just noted. Surprisingly, while benzene generally bonds with the molecular plane parallel with most surfaces, benzene has also been observed to bond on Pd(110) in a tilted configuration [364, 391].

The slightly larger benzotriazole ($C_6H_5N_3$) is found, by NEXAFS, to adsorb (at submonolayer coverages) with the molecular plane nearly perpendicular to the Cu(100) surface [443], a more "upright" configuration than even pyridine. The preferential orientation of the molecule in the multilayer is much closer to a configuration with the molecular plane parallel with the surface [443]. The related benzimidazole ($C_7H_6N_2$) and 1-methyl benzotriazole ($C_7H_7N_3$) do not appear to have a strong preferential orientation on Cu(100) [443]. Dimethyl pyridine, as well as 1,3 dimethylbenzene, adsorb on Pd(111) with the molecular plane parallel with the surface [397]. The absence of a strong preferential orientation for benzimidazole and 1-methyl benzotriazole and the "flat" bonding configuration of dimethyl pyridine may be due to steric hindering of the bonding with the surface through a protruding N lone pair that may be characteristic of both pyridine and benzotriazole bonding. The unsaturated bonds of phenyl (C_6H_5) and benzyne (C_6H_4) [398, 422] are also probably responsible for the slightly tilted bonding configuration of these species [425].

The ligand cyclopentadienyl (C_5H_5) has also been studied by angle-resolved photoemission on Rh(111) [444]. Much like benzene on Rh(111) [392], the molecular plane is parallel with the surface and the C_{5v} symmetry of the molecule dominating the photoemission, although there is some reduction symmetry toward C_s [444]. Consistent with this result, cyclopentadienyl rings have been imaged by scanning tunneling microscopy on Ag(100) and exhibit high mobility across the Ag(100) surface (like benzene) providing every indication that the molecular plane is parallel with the surface [445]. If ever investigated in future studies, we would expect the *nido*-cage carborane ligands (similar to the *closo*-carboranes discussed in the next section), because of their similar electronic structure to cyclopentadienyl, to bond in a similar fashion to the surface.

The intermolecular interaction of adsorbed benzene is strongly evident in the dispersion of the $2a_{1g}$ (σ_{CH}) band for Ni(111) [415], Ni(110) [384, 387, 388], Rh(111) [392], and Os(0001) [423] as seen in Figure 32 for benzene on Ni(110). As with the simpler CO overlayers, the benzene molecular orbital bandwidth (the extent of dispersion) can be related to the nearest neighbor lattice constant, as seen in Figure 33. The exception seems to be Pd(110) [364], where the lattice spacing might lead one to conclude that the bandwidth should be about 1/2 eV but in fact is about 0.23 eV, perhaps because the benzene is adsorbed in a tilted configuration.

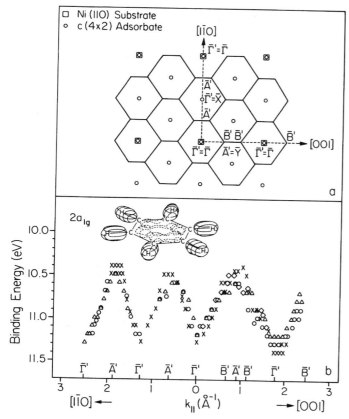

Fig. 32. The Brillouin zone formed for the benzene $c(4 \times 2)$ overlayer structure on Ni(110) is shown at the top (a). The experimental two-dimensional band structure for the benzene $2a_{1g}$ molecular orbital (b) along the high symmetry directions as indicated at the top. Reprinted with permission from [384], copyright 1991, Elsevier Science.

19. CARBORANES

Two different main group cage molecules have been studied by angle-resolved photoemission, in an effort to determine a preferential bonding orientation [446, 447]. Both carboranes $C_2B_4H_6$ [446] and $C_2B_{10}H_{12}$ [447] exhibit preferential bonding orientations that not only resemble each other, but also resemble the smaller molecular adsorbates like CO as well as larger molecules like methoxy and benzotriazole.

Valence band angle resolved photoemission provides very clear evidence that the initial adsorption of the small borane molecule, *nido*-2,3-diethyl-2,3-dicarbahexaborane, $(C_2H_5)_2$ $C_2B_4H_6$, on Si(111) occurs with partial dissociation of the cluster molecule [446, 448]. The dissociation results in the loss of the ethyl groups [446, 448]. Schematically, the parent molecule and fragment are shown along with their respective molecular orbitals in Figure 34. The carborane cage fragment, $C_2B_4H_6$, adsorbs molecularly with the molecular axis and the basal, C_2B_3, plane of the cage parallel with the surface normal [446].

As with many molecular adsorbates, bonding of the carboranes to the metal surface perturbs the molecular orbitals, and some to a greater extent than others. From a comparison of the observed binding energies of the molecular orbitals of the adsorbed species with simple model calculations [446, 448], it is possible to determine that the molecule bonds to the surface through the carbon(s), C2 and C3 in Figure 34. This is clearly similar to the "upright" bonding configuration of CO on many metal surfaces in that the molecular axis is along the surface normal and the dipole is pointed toward the surface. Subsequent adsorption of *nido*-2,3-diethyl-2,3-dicarbahexaborane on the oriented dicarbahexaborane is molecular but with no preferred bonding orientation that could be ascertained from angle-resolved photoemission.

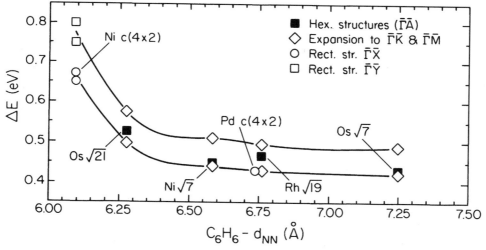

Fig. 33. Comparison of the $2a_{1g}$ band dispersions (bandwidths) as a function of benzene nearest neighbor distances, as described in the text. The nearest neighbor distance for benzene on Pd(110) is indicated, but the dispersion (0.23 eV) does not fit on the scale of the figure. Adapted from [392].

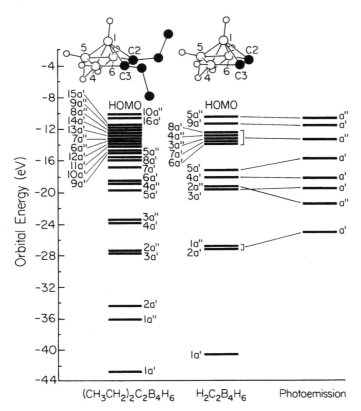

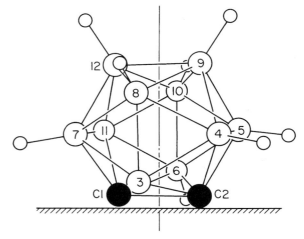

Fig. 35. The schematic representation of the bonding and orientation of orthocarborane on Cu(100) at low coverages [447, 451].

Fig. 34. The schematic representation of *nido*-2,3-(C₂H₅)₂-2,3-C₂B₄H₆ and *nido*-2,3-C₂B₄H₈ carboranes with the energy levels calculated using modified neglect of differential overlap technique (taken from [446]). The photoemission results are shifted by 7.3 eV to account for the work function and screening effects and are included for comparison. The open circles are boron, the small open circles are hydrogen, and the filled circles are carbon. For clarity, not all hydrogens are shown. Adapted from [446, 448].

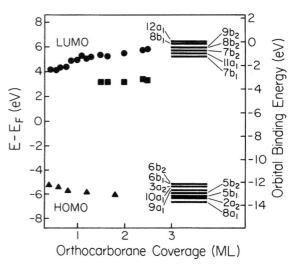

Fig. 36. The experimental binding energies of HOMO (▲) and LUMO (●) and an exopolyhedra state (■), reference to the substrate Fermi level, for orthocarborane adsorbed on Cu(100) as a function of coverage. Comparison with theoretical orbital energies is indicated at right for the free molecule [Adam P. Hitchcock, A. T. Wen, Sunwoo Lee, J. A. Glass, J. T. Spencer, and P. A. Dowben, *J. Phys. Chem.* 97, 8171 (1993); A. P. Hitchcock, S. G. Urquhart, A. T. Wen, A. L. D. Kilcoyne, T. Tyliszczak, E. Rühl, N. Kosugi, J. D. Bozek, J. T. Spencer, D. N. McIlroy, and P. A. Dowben, *J. Phys. Chem. B* 101, 3483 (1997); S. Lee, D. Li, S. M. Cendrowski-Guillaume, P. A. Dowben, F. Keith Perkins, S. P. Frigo, and R. A. Rosenberg, *J. Vac. Sci. Technol. A* 10, 2299 (1992)]. Adapted from [451].

The larger carborane cage molecule *closo*-1,2-dicarbadodecaborane (C₂B₁₀H₁₂), commonly referred to as orthocarborane, bonds to Cu(100) with a molecular bonding orientation [447] quite similar to that of dicarbahexaborane on Si(111). This bonding orientation also has the molecular axis along the surface normal. From the shifts in the molecular orbitals from the gas phase, we can infer that the molecule bonds with the carbon(s) toward the metal substrate, as shown schematically in Figure 35. The partial dehydrogenation of this molecular adsorbate may occur as indicated in Figure 35, but this cannot be determined from photoemission or STM. (The molecule appears to adsorb largely "intact" on Si(111) in scanning tunneling microscopy [449].) For both of the carborane cage molecules presented here as examples, photoemission cannot provide too much insight into the possible dehydrogenation of the molecules upon adsorption, and HREELS measurements are clearly indicated to complement the angle-resolved photoemission studies. If there is partial dehydrogenation, then this species bonds in a fashion similar to ortho-xylene with dehydrogenation of the methyl groups (on the benzene ring) [450], benzotriazole [443], or methoxy.

As with *nido*-2,3-diethyl-2,3-dicarbahexaborane, (C₂H₅)₂-C₂B₄H₆, adsorption on Si(111), the *closo*-1,2-dicarbadodeca-

borane (C₂B₁₀H₁₂), or orthocarborane loses preferential bonding orientation with multilayer adsorption. There is also very little perturbation of both the occupied and unoccupied molecular orbitals from theory with the condensation of several molecular layers, as indicated in Figure 36 for orthocarborane.

The lateral interactions between the adsorbed molecules in multilayer adsorption are believed to be weak but not completely absent. While very slight, the orthocarborane molecular orbitals do exhibit some wave vector dependence of the molecular orbitals for both one and two molecularly adsorbed layers,

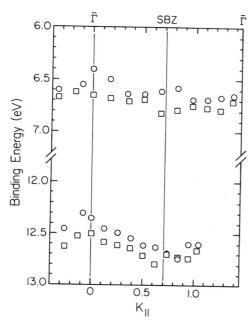

Fig. 37. The dispersion of the orthocarborane, on Cu(100), induced 6.5 and 12.5 eV molecular orbital bands for one monolayer (○) and two monolayers (□). Reprinted with permission from [451], copyright 1995, American Institute of Physics.

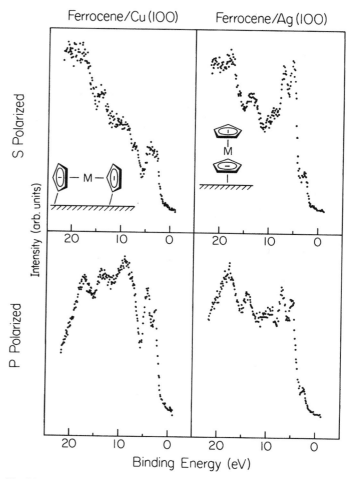

Fig. 38. The incident light polarization dependence of molecular ferrocene adsorbed on Cu(100) and Ag(100). The photoelectrons were collected normal to the surface. s-Polarized light is with an incidence angle of 34°, while p-polarized light is with an incidence angle of 70°. Adapted from [457].

in spite of the absence of long range order in the molecular adsorbate layer. This "band" dispersion is indicative of lateral interactions [451]. In this respect, this cage molecule is similar to many of the molecular adsorbates discussed, although the dispersion is quite small ($\approx$200 meV), as seen in Figure 37.

In some sense, the carborane cage molecule closo-1,2-dicarbadodecaborane ($C_2B_{10}H_{12}$) could be thought of as a small version of the fullerenes, the buckyball C_{60}, and related species discussed at the end of this review.

20. METALLOCENES

Metallocene (i.e., MCp_2, where Cp is C_5H_5 in the η^5 configuration with the metal center, and M is a transition metal) adsorption on surfaces has been an area of increasing interest, in part because of the utility of metallocenes for the selective area chemical vapor deposition of metals [452–456]. Of primary interest are the mechanisms for dissociation and fragmentation, but bonding orientation has also been explored.

The substrate has a surprisingly strong influence on the bonding orientation of metallocenes. The adsorption of molecular ferrocene (Fe(C_5H_5)$_2$) on Ag(100) [457], Cu(100) [457], and Mo(112) [458] has been recently studied by angle-resolved photoemission. As seen in Figure 38, there is a polarization dependence of the molecular orbitals and this polarization dependence differs from Cu(100) to Ag(100). There is a strong enhancement of the $4e_{2g}$, $6e_{1u}$, and $4e_{1g}$ orbitals in p-polarized light for ferrocene on Cu(100). This does not occur on Ag(100). This is interpreted as demonstrating that the ferrocene molecular axis is parallel to the surface for molecular ferrocene adsorption on Cu(100), while the molecular axis is normal to the surface for ferrocene adsorption on Ag(100).

These postulated differences in bonding orientation have been confirmed by independent measurements. Scanning tunneling microscopy images, as shown in Figure 39, show that ferrocene does indeed adsorb with a bonding orientation that places the molecular axis parallel to the Cu(100) surface [457]. HREELS measurements show strong dipolar a_{2u} vibrational modes in specular scattering but not in off-specular scattering for ferrocene molecularly adsorbed on Ag(100) [457, 459, 460]. In particular, the two dipole active modes $\nu_{as}(M(Cp)_2)$ and π(CH), at 60.4 and 93.2 meV, respectively, are observed with great intensity only in the dipole scattering geometry (specular scattering) as seen in Figure 40. This observation is consistent with the bonding orientation in which the ferrocene molecule adsorbs with its molecular axis normal to the Ag(100) surface [457, 459, 460], as indicated in the inset to Figure 38.

While angle-resolved photoemission places the ferrocene Cp–Fe–Cp molecular axis parallel with the surface on Cu(100) [457] and Mo(112) [458], with a very strong azimuthal orien-

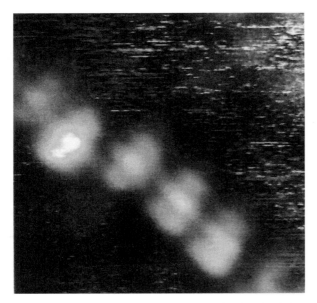

Fig. 39. A constant-current STM image (4 nm by 4 nm) of ferrocene molecules adsorbed on Cu(100) following adsorption at room temperature. The relative orientation of the molecule on the Cu(100) surface is along the ⟨110⟩ direction. Adapted from [457].

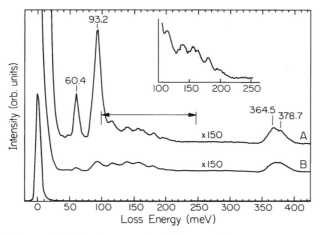

Fig. 40. HREELS for ferrocene adsorbed on Ag(100) at 110 K. The spectra were taken following 2 langmuirs exposure. Both the specular (A) and off-specular (B, $\Delta\theta = 9°$) spectra are shown with the same scale but magnified by 150 as compared to the specular elastic peak. The inset shows the features in the range from 100 to 250 meV. The intensities of the two dipole active modes $\nu_{as}(M(Cp)_2)$ and $\pi(CH)$, at 60.4 and 93.2 meV, respectively, are greatly suppressed in the off-specular geometry while the peaks at 364.5 and 378.7 meV are not. Adapted from [457, 459].

tation for ferrocene on Mo(112) [458], the general bonding orientation for metallocenes is with the molecular axis along the surface normal. This "upright" orientation has been observed for ferrocene on graphite [461] by HREELS, as well as for ferrocene on Ag(100) [457, 459, 460], though the preferential "upright" orientation is lost in very thick ferrocene films [457, 460].

Similarly, nickelocene is also seen, by HREELS, to adopt the "upright" orientation with the molecular axis along the surface normal on Ag(100) [462–464]. Angle-resolved photoemission

results were initially taken to suggest that the nickelocene was tilted away from the surface normal [465], like cobaltocene on Cu(111) [466] and nickelocene on Si(111) [467]. More detailed studies suggest that there are two adsorption states [468] for nickelocene on Ag(100)—a low coverage state consistent with HREELS and a high coverage state consistent with the angle-resolved photoemission. The initial adsorption of nickelocene on Ni(100) [463], NiO(100)/Ni(100) [463], and initial cobaltocene adsorption on Cu(111) [466] appears to be dissociative, thus complicating assignment of the molecular bonding configuration of metallocenes on these surfaces.

21. PHTHALOCYANINES AND PORPHYRINS

A quantitative approach to determine molecule orientation from angle-resolved photoemission has been examined, first, for simple molecules adsorbed on crystal surfaces, as indicated by the earlier sections of this review and early reviews [1, 4, 294]. As far as large organic molecules are concerned, the first effort was undertaken by Koch and co-workers [469] in 1983 in determining the molecular orientation of lead-phthalocyanine (Pb-$C_{32}H_{16}N_8$). The angular distribution of the photoelectrons from the uppermost π band was theoretically analyzed. Although Richardson [470] later reanalyzed the data of Koch and co-workers, both results nonetheless indicated a bonding orientation with flat configuration on the crystal surface.

In the model calculation by Koch and co-workers [469], the initial and final states of the photoemission process were approximated by a single p_z atomic orbital (point emitter model) and a free-electron plane-wave, respectively. Although the challenge was the quantitative analysis of angle-resolved photoemission intensities for a large molecule, the analysis of Permien et al. (Koch and co-workers) [469] was criticized due to the use of a plane-wave for the final state. Richardson [470] later reanalyzed their data by employing a spherical–harmonic expansion for the final state and found better agreement with the experimental results. This theoretical method has been applied to analyze the molecular orientation of absorbed benzene on a Pd(100) surface [394] and pyridine on a Cu(110) surface [439]. In those theoretical approaches, the molecule was treated as a single–point emitter and interference between coherent photoelectron waves emitted from the atoms constituting the molecule was completely neglected. These calculations are too simplified to apply into large and complex organic molecules. As a result, the intensity of the molecular orbitals derived from ARUPS has continually been discussed qualitatively in terms of symmetry analysis because of the difficulties in calculating accurate quantitative descriptions of the photoelectron intensity.

The next theoretical development in emission-dependent photoemission intensities from large molecules was realized in the molecular orientation study of thin films of bis(1,2,5-thiadiazolo)-p-quinobis(1,3-dithiole) on graphite performed by Hasegawa and co-workers [471] in 1993. Briefly, they applied the independent-atomic-center (IAC) approximation, formulated by Grobman [472], in conjunction with molecular orbital calculations to calculate angular dependent photoemission intensities. The photoelectron intensity $I_n(\mathbf{R})$ from the nth molecular orbital at the detector position $\mathbf{R}$ is expressed as

$$I_n(\mathbf{R}) \propto \left| A_{\text{tot}}^n(\mathbf{R}) \right|^2 \quad (15)$$

where $A_{\text{tot}}^n(\mathbf{R})$ is the photoelectron wave function at $\mathbf{R}$ represented by

$$
\begin{aligned}
\left| A_{\text{tot}}^n(\mathbf{R}) \right| = {} & \sum_a \sum_{X_a} D_a C_{X_a}^n e^{-i\mathbf{k}_n \mathbf{R}_a} \sum_L Y_L^*(\hat{\mathbf{R}}) M_{LX_a} \\
& + \sum_a \sum_{b \neq a} \sum_{X_a} D_b C_{X_a}^n e^{-i\mathbf{k}_n \mathbf{R}_b} \sum_L \sum_{L'} Y_{L'}^*(\hat{\mathbf{R}}) \\
& \times t_b^{l'}(k_n) G_{L'L}(\mathbf{R}_b - \mathbf{R}_a) M_{LX_a} \quad (16)
\end{aligned}
$$

where D_a is the phenomenological damping factor for the photoelectron wave along $\mathbf{R}$ from atom a to the surface due to inelastic processes, $C_{X_a}^n$ is the nth molecular orbital coefficient of the Slater-type atomic orbital X_a, $K_n (= \mathbf{k}_n\hat{\mathbf{R}})$ is the wave vector of a photoelectron, $\mathbf{R}_a$ is the position of atom a, $\hat{\mathbf{R}}$ is the unit vector along $\mathbf{R}$, and M_{LX_a} represents the matrix element including the phase shift and the radial integral. The terms $t_b^{l'}(k_n)$ and $G_{L'L}(\mathbf{R}_b - \mathbf{R}_a)$ are the single-scattering vertex and the free electron propagator, respectively. The first term of Eq. (16) is defined as the independent-atomic-center/molecular orbital (IAC/MO) approximation, where the initial state is expressed by using molecular orbital calculation, the photoelectron wave function is approximated by a coherent sum of the waves emitted from atomic orbitals that build up the molecular orbital (IAC approximation), and the self-scattering due to the residual hole upon the photoemission is taken into account for the final state. Although the independent-atomic-center/molecular orbital calculations only explain the photoelectron angular distribution reasonably well for large organic molecules [473, 474], an appropriate choice of experimental conditions, such as light polarization and analyzer position, is essential to minimize contributions due to the single/multiple scattering of photoelectrons by surrounding atoms which are completely ignored in the IAC/MO calculation. When the kinetic energy of photoelectrons is relatively low, as in the case of ultraviolet photoemission, the scattering due to surrounding atoms is substantial and cannot be ignored. The second term of Eq. (16) describes the scattering in which the photoelectron waves scattered singly by the atoms surrounding the independent-atomic-center atoms are included in the calculation for the final states. Eq. (16) corresponds to the single-scattering approximation combined with the molecular orbital calculation (SS/MO).

Ueno et al. [475] studied the orientation of the metal-free phthalocyanine (H_2-$C_{32}H_{16}N_8$) on the cleaved MoS$_2$ surface by analyzing the photoelectron angular distribution quantitatively with the IAC approximation [471]. They found that the observed photoelectron take-off-angle dependence could be explained well by an angular distribution calculated for the "flat"

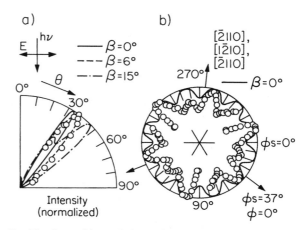

Fig. 41. The observed (open circles) and calculated photoelectron angular distributions for the HOMO band of Cu-phthalocyanine. (a) The tilt angle (β) dependencies calculated by the SS/MO approximation are shown for $\beta = 0°$ (—), $\beta = 6°$ (- - - -), and $\beta = 15°$ (—·—). (b) Comparison between the calculated azimuthal angle (ϕ) dependence for $\beta = 0°$ and the observed one. Reprinted with permission from [476], copyright 1999, American Institute of Physics.

orientation, i.e., with the molecular plane parallel with the surface. Later, a more detailed investigation was performed by employing LEED in addition to the ARUPS technique in order to determine a full structure of the copper–phthalocyanine (Cu–C$_{32}$H$_{16}$N$_8$) and the H$_2$-phthalocyanine monolayer films [476]. The electron take-off angle as well as the azimuthal angle dependencies of the top π band intensity (as shown in Figure 41), within the context of the single-scattering approximation/molecular orbital calculation (SS/MO) [471], confirmed again that the phthalocyanine molecule lay flat on the MoS$_2$ surface. Moreover, the azimuthal orientation of the molecules (angle between molecular axis and surface crystal axis of MoS$_2$) was found to be about $-7°$, $-37°$, and $-67°$ with respect to the three equivalent surface crystal axes of the MoS$_2$, as indicated in Figure 42. From the LEED measurements, it was determined that the molecules form a square lattice with the lattice constant of 13.7 Å [476], again as indicated in Figure 42. These results for copper–phthalocyanine are consistent with scanning tunneling microscopy studies which have imaged copper–phthalocyanine on Cu(100) and Si(111) [477], in a packing arrangement much like that shown in Figure 42a. Both copper–phthalocyanine and cobalt–phthalocyanine have been imaged on Au(111) as well [478].

The orientation of molecules in ultrathin films (0.5 to 5 monolayers) of chloroaluminum phthalocyanine (ClAlPc) on MoS$_2$ has been studied by ARUPS and LEED together with Penning ionization electron spectroscopy or metastable quenching spectroscopy (MQS) [479]. In Penning ionization electron spectroscopy, or MQS, the kinetic energy of electrons (like photoelectrons) ejected from the target molecule as a result of collisions with helium metastable atoms is analyzed. Since the metastable atoms do not penetrate into inner molecular layers, only the outermost surface layer or electron orbital can be distinctively probed. For the ClAlPc molecule, the Cl atom bonded to the center Al atom of the phthalocyanine ring pro-

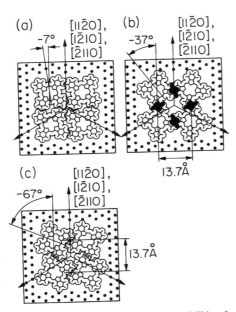

Fig. 42. The experimentally determined three possibilities for the surface structures of Cu-phthalocyanine and H$_2$-phthalocyanine monolayers on MoS$_2$. The molecular structures are shown with molecular van der Waals radii. The arrows indicate the surface crystal axes of MoS$_2$ and the black points indicate surface sulfur atoms. Azimuthal orientations of the molecule are $-7°$, $-37°$, and $-67°$ for (a), (b), and (c), respectively. The overlap of van der Waals radius between neighboring molecules is indicated by shaded area. The structure (a) gives the smallest van der Waals overlap. Adapted from [476].

trudes from the phthalocyanine ring. A change in the molecular orientation due to thermal annealing was observed. At 1 monolayer equivalence of the ClAlPc deposited at room temperature, islands are formed on the substrate surface with the Cl atoms in an "up and down" mixed configuration. Upon annealing the overlayer film up to 100°C, a uniform monolayer of the ClAlPc is formed where the molecules are oriented flat to the substrate with the majority of the Cl atoms protruding outside the monolayer film surface. Moreover, the structure of the film depends upon the film thickness [479]. At 5 monolayers equivalent, the AlClPc molecules are tilted with an inclination angle of about 10° with respect to the substrate. These molecular orientation changes in the ClAlPc thin films were further characterized by low-energy electron transmission spectroscopy [480]. The ClAlPc deposited on HOPG (highly oriented pyrolytic graphite) surfaces was studied by high-resolution electron energy loss spectroscopy [481]. The ClAlPc molecules in the as-grown monolayer were found to lie flat on the HOPG surface.

The shape of titanyl phthalocyanine (OTiPc) is similar to that of the ClAlPc, but instead an oxygen atom projects from the phthalocyanine ring. The dependence of film thickness and temperature on the molecular orientation in OTiPc ultrathin films (0.2–3 monolayer equivalents) on graphite was studied by Penning ionization electron spectroscopy/MQS and ultraviolet photoemission [482]. In the submonolayers, the majority of the molecules are oriented flat on the substrate surface with the oxygen atoms protruding outside the film. When coverage approaches one monolayer, the number of molecules having the oxygen atom directed toward the substrate increases. By annealing the films, a uniformly oriented monolayer is created in which the OTiPc molecules are oriented with the oxygen atoms directing upward (away from the surface).

Different molecular orientations in a monolayer film on two substrate surfaces (graphite and MoS$_2$) were also realized in the investigation performed by Penning ionization electron spectroscopy/MQS [483]. At room temperature, ClAlPc molecules aggregate and form islands on the MoS$_2$ surface, whereas the molecules spread out over the surface, wetting and almost covering the graphite surface.

ARUPS measurements were used to investigate the molecular orientation of Langmuir–Blodgett (LB) grown films (eight layers) of copper-tetrakis(n-butoxylcarbonyl) phthalocyanine [484]. The photoelectron emission-angle measurements suggested that the phthalocyanine molecules are oriented with the phthalocyanine rings perpendicular to the substrate surface— much different from the results for most phthalocyanines. This conclusion is consistent with the results obtained by polarized UV-visible absorption measurements on the LB grown films [485, 486].

The phthalocyanine molecules generally lie flat, with the molecular plane parallel with the surface, on a substrate surface until monolayer coverage is established. A center metal atom is of little influence in determining bonding orientation; the geometrical shape of the molecule largely influences the molecular orientation.

In this regard, porphyrins may differ slightly from the phthalocyanines. Evaporated films of zinc 5,10,15,20-tetraphenylporphyrin (ZnTPP) and 5,10,15,20-tetraphenylporphyrin (H$_2$-TPP) on Ag substrates were investigated by polarized NEXAFS/XANES spectroscopy [487, 488]. The analysis of the polarization dependence of NEXAFS/XANES peak intensities (the transition from the N 1s core level to the π^* LUMO) revealed that ZnTPP molecules in the film deposited on Ag substrates at 367 K have a high degree of orientation with the central macrocylic plane inclined by 28° $\pm$ 10°. On the other hand, ZnTPP films evaporated on room-temperature substrates and H$_2$-TPP films evaporated on Ag substrates at both room temperature and 367 K showed little polarization dependence which suggests that the molecules in these films are oriented rather more randomly. On the other hand, the porphyrin based Cu-tetra-3,5-di-tertiary-butyl-phenyl porphyrin seems to lie with the molecular plane parallel with the Cu(100) surface and the Au(110) surface, with a strong azimuthal orientation along the $\langle 010 \rangle$ direction of the Cu(100) surface in scanning tunneling microscopy images [489].

22. LARGE AROMATIC HYDROCARBONS AND ORGANIC SPECIES

In many respects the large aromatic compounds behave much like their pyridene and benzene building blocks. It is common for the planar molecules to lie "flat" with the molecular plane parallel with the surface. The long chain organics are frequently dominated by the end groups. The large organics like 17,19-hexatriacondieyne [490] as well as smaller species like phenylactyelene [435] and cyclopentadiene [491] are building blocks for larger polymers and in some cases will polymerize on a surface [490, 491]. Such molecules are of interest not only for their own sake but also because of the interest in polymers, as discussed later.

The bonding orientations of naphthalene and some other aromatic hydrocarbons on Ag(111) and/or Cu(100) substrates were investigated by NEXAFS using synchrotron radiation [492]. By measuring the polarization dependence of the $C_{1S} \rightarrow \pi^*$ transition, naphthalene molecules were found to be oriented with an almost flat configuration (a tilt angle of $0° \pm 30°$) with respect to the Ag(111) substrate surface.

Naphthalene-dicarboxylicacid-anhydride (NDCA), which is essentially half a perylene-tetracarboxylicacid-dianhydride (PTCDA) molecule (discussed later), was investigated on clean and oxygen $p(2 \times 2)$ precovered Ni(111) surfaces by NEXAFS [493, 494]. On clean Ni(111), NDCA lies flat on the surface and bonds via the naphthalene π system, whereas on the O-precovered substrate an upright orientation with bonding via the anhydride group is observed. The orientation and chemical bonding of big organic adsorbates depend strongly on the properties of the surface.

A preferential orientation of anthracene on a metal crystal surface was investigated by both NEXAFS [492] and ARUPS [495]. By measuring the polarization dependence of the $C_{1S} \rightarrow \pi^*$ transition in the NEXAFS spectra, flat molecular configurations of anthracene on Ag(111) (a tilt angle of $10 \pm 10°$) and on Cu(100) were concluded. With one more ring, naphthacene (tetracene) exhibits a slightly altered bonding configuration. NEXAFS [492], with a dipole symmetry analysis, concluded that naphthacene molecules orient with a slightly canted ($27 \pm 10°$) configuration on the Ag(111) surface, but with an upright configuration ($87 \pm 10°$) on Cu(100). The upright orientation of naphthalene on Cu(100) was also determined by angle-resolved photoemission [496].

Theoretical simulations of the valence band spectrum have been undertaken for oriented thin films of naphthacene [474, 497] and have been compared with the observed data on cleaved HOPG, as seen in Figure 43. The quantitative analysis of the photoelectron take-off angle dependence of the full valence band spectra were simulated completely with not only the experimental peak binding energies but also photoemission intensities in agreement with theory. It was determined that the molecular orientation in the thin films (thickness about 24 Å) is quite similar to that in the single crystal, when the a–b plane of the naphthacene single crystal is placed parallel to the substrate, hence corresponding to an upright molecular configuration for the overlayer.

Almost flat bonding configurations ($19 \pm 10°$ and $20 \pm 10°$) in perylene thin films were suggested by NEXAFS for Ag(111) and Cu(100) surfaces, respectively [492]. Most recently, the insulating α-perylene single crystal was studied using NEXAFS and ARUPS techniques [498]. Some compensation techniques

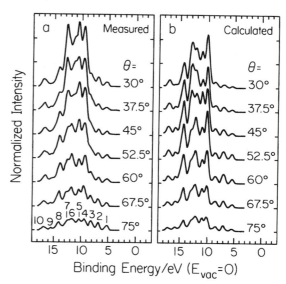

Fig. 43. The comparison between the measured and calculated photoelectron take-off-angle dependencies of the valence band spectra for the naphthacene thin films. (a) The measured spectra after background subtraction. (b) The calculated spectra by the SS/MO approximation. The mean free path of the photoelectron 5.0 Å was used in the calculation. Adapted from [497].

were used to overcome the influence of sample charging. The spectra, so obtained for α-perylene single crystals, display several sharp peaks whose intensities strongly depend on the polarization angle of the incident synchrotron radiation. From the angular dependence, an average angle of inclination of the molecular planes relative to the (001) cleavage plane of $85 \pm 5°$ was derived.

The PTCDA molecule is "built" around perylene but is different from the perylene molecule in that it contains reactive groups. Since PTCDA was found to form excellent ordered multilayers with high stability [499, 500], it has gained increasing interest as a promising material for organic devices. Ordered adsorbate layers of perylene and PTCDA on various single crystal surfaces, such as Ag(111), Ni(111), and Si(111), were studied using NEXAFS [501, 502]. The angular dependence of the NEXAFS data reveals that the interaction between the substrate and the adsorbed molecule again plays an important role in determining the molecule orientation. For weak chemisorption or perhaps physisorption, such as PTCDA/Ag(111) and perylene/Si(111), a coplanar adsorption geometry at the interface is achieved through bonding via the molecular π-system. This weak interaction of perylene and PTCDA on Si(111) and the planar configuration have been confirmed by angle-resolved photoemission [503]. The planar geometry of PTCDA on Ag(111) is consistent with the STM images and LEED of monolayers on Ag(111) and Ag(110) which also confirm a strong azimuthal orientation dependence [504]. If the bonding at the interface is too strong and involves the reactive group(s), a tilted or bent adsorption geometry may occur, such as for PTCDA on Ni(111) and Si(111). Angle-resolved photoemission was carried out for determining molecular orientation of PTCDA thin films on MoS$_2$ [505]. The take-off angle dependence of the photoelectron intensity from the highest π band was analyzed with the theoretical SS/MO calculation method (discussed in the section on phthalocyanines and porphyrins), and a flat configuration was concluded for the molecular orientation. Furthermore, new bandgap states due to reaction at the organic/metal interface were observed. The reaction between the molecule and metal atoms was considered to occur at the C=O groups in the molecule [506, 507]. On the Si(111) 7×7 surface, both NEXAF and angle-resolved photoemission suggest that PTCDA has the molecular plane tilted slightly (perhaps some 20–30°) away from the surface [502].

The molecular orientation of naphthalene-tetracarboxylicacid-dianhydride (NTCDA) was also investigated by NEXAFS [508–510]. Some details of the interaction between NTCDA and the substrate can be derived from NEXAFS data. Whereas for Ag substrates the π^*-resonances attributed to the naphthalene cores are mainly affected by the substrate interaction, the anhydride groups are more involved in the molecular bonding to Cu substrates and even more for Ni substrates. NTCDA monolayers bond strongly to Ag(111) and Cu(100) surfaces via the conjugated π-system of the naphthalene "molecular core," which leads to a orientation of the molecular plane parallel with respect to the substrate. For larger coverages, however, the interaction between molecule and substrate is negligible, leading to possible misorientation at the higher coverages. The molecular orientation can be altered by varying the substrate temperature during overlayer film preparation: the molecule is observed to be parallel to the substrate at low substrate temperature and perpendicular at higher temperature [509]. The bonding orientation of coronene on Ag(111) is also quite similar to that of perylene—a flat configuration with a tilt of $16 \pm 10°$ [492].

In addition, the core-hole effects in NEXAFS for many of these large aromatic hydrocarbons have been discussed in terms of *ab initio* molecular calculations [511, 512]. NEXAFS spectra of chrysene, perylene, and coronene were measured and the results were analyzed in detail using *ab initio* MO calculations. The observed spectra showed a large deviation from the calculated density of unoccupied states, indicating the presence of a large core-hole effect. The observed spectra were simulated well by theoretical calculations that took this effect into account by the improved "virtual orbital" method.

ARUPS, infrared reflection absorption (IR-RAS), and X-ray diffraction were applied to p-sexiphenyl to investigate its valence band structure and molecular bonding orientation [513, 514]. It was found by ARUPS, using dipole selection rules to the light incidence angle as indicated in Figure 44, that the molecules in the films evaporated on a heated Ag substrate are highly oriented with the molecular axis/benzene plane parallel to the surface normal.

The technique of measuring band structure, as a result of intramolecular periodicity, can be undertaken easily by changing the photon energy. This approach requires long molecules with many repeating units and with the molecular axis oriented along the surface normal [5]. Photoemission from the oriented films

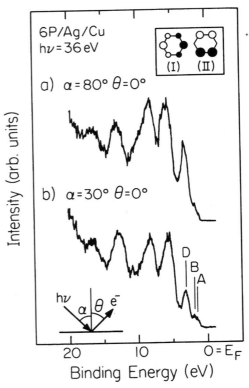

Fig. 44. Photon incidence angle (α) dependence of normal-emission ARUPS spectra of vacuum-evaporated p-sexiphenyl film on a heated Ag/Cu substrate at 423 K for $h\nu = 36$ eV. Each spectrum is normalized by the intensity of secondary electrons. In the insert, molecular orbital patterns of the two HOMOs of benzene are shown [514].

of p-sexiphenyl appear to exhibit wave number ($\mathbf{k}$) conservation (intramolecular energy band dispersion) [514], implying that the wave number $k_\perp$ along the molecular axis serves as a reasonably good quantum number even in a system of only six repeating units, as indicated in Figure 45. Similar studies of intramolecule energy band mapping were performed for the π-band of a urethane-derived polydiacetylene [515]. Ultrathin films of oriented p-sexiphenyl on a GaAs(001) wafer were also studied by angle-resolved photoemission with the intent of obtaining the intramolecular band structure [516], using photons in the energy range between 20 and 60 eV. The thickness of the deposited films played a very important role in the orientation of p-sexiphenyl molecules. For layers having a thickness of about 35 Å, the molecules were oriented perpendicular to the substrate surface, but for films of about 300 Å thickness the molecules lost their "upright" orientation. From the experimental photoelectron spectra, the full one-dimensional valence band structure of oriented p-sexiphenyl layers was determined. While valence states between 2 and 5 eV show only weak dispersion, the bands between 5 and 11 eV show a clear momentum dispersion, which indicates the high degree of order in the investigated sexiphenyl layers.

α-Sexithienyl (6T), also a six-repeating-unit molecule similar to p-sexiphenyl, is a model molecule (an oligomer) of polythiophene, which has potential device applications, for exam-

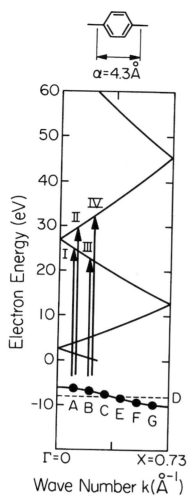

Fig. 45. The energy-band dispersion relation $E = E(k)$ in the Brillouin zone along the molecular axis (Γ-X) for p-sexiphenyl and poly(p-phenylene) with a D_{2h} symmetry (unit-cell constant $a = 4.3$ Å). The solid line and filled circles A–G below the vacuum level ($E = 0$) show the occupied π orbitals delocalized over the molecule of poly(p-phenylene) and p-sexiphenyl, respectively, derived from one of the HOMOs of benzene. The broken line (D) shows localized levels derived from the other HOMOs of benzene. Above the vacuum level, free-electron-like final-state bands with a free-electron-like energy dispersion are shown with an inner potential of 5.5 eV [514].

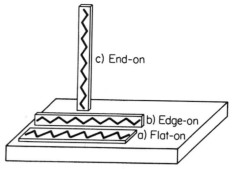

Fig. 46. Schematic view of the three typical orientation models for n-alkane: (a) flat-on orientation, (b) edge-on orientation, and (c) end-on orientation. The zigzag indicates the alkyl-chain axes. Adapted from [526].

a high degree of orientation for thin layers (<3 ML). However, for thicker layers ($\geqslant$3 ML), this orientation order gradually decreases. This is consistent with the Fourier transform IR-RAS results on Ag(111) [520].

A series of end-capped oligothiophene (ECnT) monolayers with different chain lengths n ($n = 3$–6) were vapor deposited onto the Ag(111) surface and found to form highly ordered superstructures with flat-lying molecules [521, 522] by using LEED and STM. However, the absence of angle-resolved photoemission and NEXAFS work on such systems leaves many issues open for future research.

Molecular orientation changes upon a thin film solid–liquid transition were investigated by polarized NEXAFS in pentacontane (n-CH$_3$(CH$_2$)$_{48}$CH$_3$) film evaporated on Cu substrate [523, 524]. The low vapor pressure of the long-chain n-alkane, in the liquid phase, permits such measurements to be performed under UHV conditions. At room temperature (as deposited), the pentacontane molecules appear to be placed in an almost flat configuration with respect to the substrate surface. With increasing temperature, the molecules gradually reorient upright, and, at an elevated temperature just below the bulk melting point of pentacontane, the molecular axis is oriented almost perpendicular to the substrate surface. At the bulk melting point, this ordered structure is still preserved. However, at temperatures well above the bulk melting point (in the liquid phase), the molecular orientation becomes quite random. Upon cooling, the film surface is first crystallized and then bulk crystallization followed. In the (re)crystallized film, the molecules remain oriented in the upright configuration and this is maintained down to room temperature.

The long-chain alkane tetratetracontane (n-C$_{44}$H$_{90}$) has been studied by LEED combined with ARUPS on Cu(100) [525–527] and Au(111) [527]. The observed LEED patterns indicated that the molecules lie with its zigzag chain axis parallel to the $\langle 110 \rangle$ direction for both substrates. Two additional possible orientations (schematically indicated in Figure 46) having the $-$C$-$C$-$C$-$ plane of the molecule parallel (flat-on) or perpendicular (edge-on) to the substrate surface were examined. These flat-on and edge-on orientations have been realized in n-alkanes investigated by Firment et al. [528] on Pt(111), in n-C$_{44}$H$_{90}$ on Cu(111) by Dudde et al. [529], and on Au, Ag, and

ple, in field-effect transistors [517]. The molecular orientation of evaporated α-sexithienyl (6T) films on Au, Ag, and Cu were studied by polarized NEXAFS and IR-RAS [518]. It was found that the α-sexithienyl (6T) molecules on Ag and Cu substrates are highly oriented, with their molecular axis inclined by about 70° with respect to the substrate surface. By way of contrast, the α-sexithienyl (6T) molecules on Au exhibited only a little polarization dependence, indicating that the molecules are oriented nearly randomly.

The orientation of quaterthiophene (4T) molecules, deposited on the Ag(111) surface, was investigated for different film thicknesses [519]. NEXAFS data taken at the S $2p$ absorption edge for a 4T film show the molecular planes must preferentially be oriented parallel to the substrate surface with

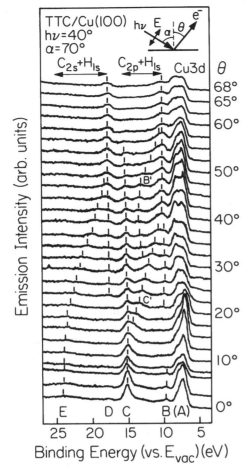

Fig. 47.   Take-off-angle θ dependence of the ARUPS spectra for an ultrathin film of tetratetrapentacontane on Cu (100). Adapted from [526].

Pb substrates by Seki et al. [530]. By comparing the measured take-off-angle dependence of the photoemission spectra (Figure 47) with theoretical calculations (Figure 48), it was concluded that the tetratetrapentacontane molecules are oriented in the flat-on configuration with respect to the substrate surface.

Intramolecular energy band dispersion was also observed in tetratetracontane, as observed by angle-resolved photoemission, from the emission angle dependence of the binding energies in angle-resolved photoemission, as indicated in Figure 47. The dispersion is substantial in these long alkyl chain compounds [531–533], as indicated in Figure 49. The observation of such dispersion from the emission dependence is also consistent with orientation of the molecules in the flat-on configuration.

The linear hexatriaconate and Cd-arachidate, like the long chain 1-octanethiol, are oriented normal to the surface of Ag(111) [532]. Intramolecular band structure down the chain of hexatriaconate on Ag(111) (as shown in Figure 50) as well as that for mercaptan-22 (22-alkane thiol) on Ag(111) and Au(111) have been (again) obtained from the dispersion with changing photon energy ($k_\perp$) at normal emission [532], which

is made possible by the upright configuration of the adsorbed molecules. The shift in the critical points of the band structure suggests that the alkane thiol is tilted about 15° off the surface normal [532]. The long chain 1-octanethiol, on the other hand, is seen by standing wave X-ray scattering and NEXAFS to be oriented along the Cu(111) [534] normal like the smaller thiolates.

Bis(1, 2, 5-thiadiazole)-p-quinbis(1,3-dithiole)($C_4 H_4 S_6 N_4$) (BTQBT) is another planar organic molecule, as schematically shown in Figure 51. This planar molecule is a novel single-component organic semiconductor [535], and thus its electronic structure of such molecules is of great importance (particularly since a transistor based on a thiophene oligomer has been fabricated [536]). The resistivity of a single crystal is remarkably low ($1.2 \times 10^3 \ \Omega$ cm) and also shows a Hall effect, which is rarely observable in organic semiconductors [537]. Photoemission studies [538] of the electronic structure of BTQBT indicated that there is only a small energy difference of 0.3 eV between the Fermi level and the top of the valence band, and this is considered to be part of the origin of the high conductivity in BTQBT. The strong intermolecular interaction of BTQBT has been investigated by ARUPS [539], by probing the photon energy dependence, as seen in Figure 52. Intermolecular energy band dispersion (as opposed to the intramolecular band dispersion just discussed) in a single-component organic molecular crystal was observed for the highest occupied molecular orbital (HOMO) and the next highest occupied molecular orbital (NHOMO) π bands [539], as indicated in Figure 53. The total HOMO bandwidth of 0.4 eV was obtained for the dispersion along the perpendicular direction from one plane of molecules to the next (as indicated in Figure 51). In the context of all the previous discussion, this is extraordinarily large for organic solids and explains the large hole mobility of 4 cm^2/V s observed in this material. This extramolecular band structure should be compared with the extensive band structure observed for the one-dimensional organic conductor tetrathiafuvalene-tetracyanoquinodimethane discussed in the next section.

The photoemission angular dependence of BTQBT thin films was measured by Hasegawa et al. [471] to investigate the molecular orientation on a HOPG substrate surface. The photoemission angular distribution was analyzed [471] with the IAC/MO approximation (discussed in the previous section on phthalocyanines and porphyrins). From this comparison of experiment with theory, it is concluded that the BTQBT molecules in the thin film (∼30 Å thickness) lie nearly flat (inclination angle ⩽10°) with respect to the substrate surface (as indicated schematically in Figure 51). In addition, ultrathin films of BTQBT deposited on a MoS$_2$ surface were studied by ARUPS [540]. The observed angular distributions were better explained by SS/MO, considering intramolecularly scattered waves than by calculations using the IAC/MO approximation. Further, with the help of the LEED measurements, the full structure of the thin film, the two-dimensional lattice, and the full molecular orientations at the lattice points were proposed. These results [540] suggest strong azimuthal orientation of the

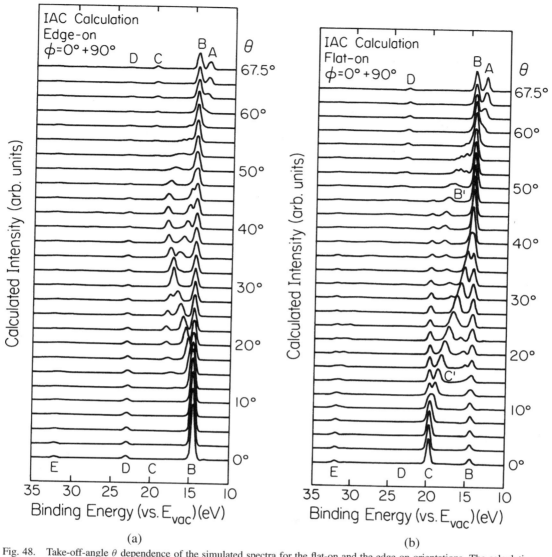

Fig. 48. Take-off-angle θ dependence of the simulated spectra for the flat-on and the edge-on orientations. The calculations were performed for a dotriacontane molecule with the IAC/MO approximation. Adapted from [526].

molecules, with the molecular plane, again, parallel with the surface.

The smaller diphenyl-carbonate is seen, by NEXAFS [541], to be oriented with the phenyl rings parallel with the Ag(111) surface and the carbonate group tilted by some 20° out of the plane. Stronger interactions are seen with an Mg overlayer on Ag(111) [541]. The even smaller 2,5-dimethyl-N,N'-dicyanoquinonediime (DMe-DCNQI) also adsorbed on Ag(111) with the molecular plane parallel with the surface for up to several monolayers [542], consistent with the CN and benzene building blocks and the LEED and STM derived structures for DMe-DCNQI on Ag(110) [543]. The related charge transfer salts Cu(DMe-DCNQI)$_2$ and Na(DMe-DCNQI)$_2$ result in the (DMe-DCNQI) molecules tilted considerably out of the plane of the surface (though a less likely scenario is a random orientation) [542].

23. ORGANIC POLYMERS AND "ONE" DIMENSIONAL CONDUCTORS

To date, polymer surfaces have been investigated mainly by XPS [5, 544] and only more recently by valence band photoemission [5, 545, 546], which provides some information on the chemical groups present at the film surfaces. For molecular orientation, however, second-harmonic generation and other optical methods [547], along with neutron scattering and X-ray diffraction, seem to be the more dominant experimental approaches. This is mainly due to the fact that molecules are considered to be randomly oriented at polymer surfaces. The polymer thin film samples made by traditional solvent casting and spin-coating methods are polymorphous, containing multiple phases and incompletely oriented crystallites (if crystalline at all), limiting the detail and accuracy of scientific studies, not to mention making surface chemistry studies of the

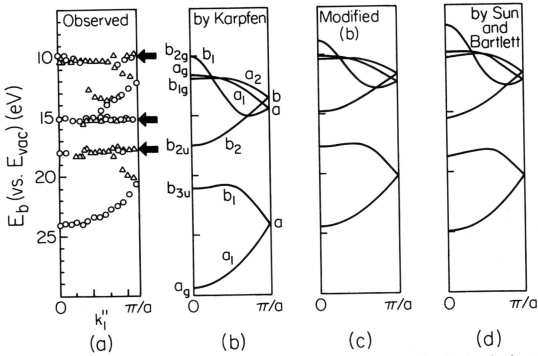

Fig. 49. (a) Experimentally obtained one-dimensional intramolecular energy-band structure plotted in the reduced zone scheme. (b) Calculated band structure for all-trans polyethylene from A. Karpfen, *J. Chem. Phys.* 75, 238 (1981). (c) Modified scale of (a). (d) Calculated band structure for all-trans polyethylene from J. Sun and R. J. Bartlett, *Phys. Rev. Lett.* 77, 3669 (1996). Adapted from [526].

polymer quite difficult to interpret [544]. In words of David Briggs [544]: "there are very few definitive studies of the polymer surface structure–property relationships" and he has noted that even recent surface science and surface chemistry textbooks make no mention of polymers. Regarding the electronic structure of some polymers, three excellent review overviews are available [545–547]. In addition, NEXAFS spectroscopy applications to polymers have been outlined [546, 548]. The subsections below cover only a limited number of polymer systems.

The conventional wisdom has been that in most polymer systems the surface differs from the bulk because there is a different distribution of end groups or copolymer at the surface compared with the bulk [544]. Furthermore, as polymers are not typically chemically pure, additives, prepolymers, and crosslinking agents segregate to the surface [544, 549–555]. New techniques for depositing polymers, such as LB film growth techniques and film transfer using tribological (rubbing) techniques, can produce highly ordered polymer films. With single component systems, band structure measurements are now possible, both in angle-resolved photoemission and angle-resolved inverse photoemission.

23.1. Polystyrene

Random molecular orientation in a polymer was first realized by ARUPS in measuring the take-off-angle dependence of pho-

toemission intensity from polystyrene thin films [556]. The observed angular dependence was successfully explained by the asymmetry parameter (β) under the assumption that the pendant phenyl groups are randomly oriented throughout the thin film. An important message from this work is that an inhomogeneous photoelectron angular distribution is expected to be observed even from a "perfectly" randomly oriented system.

More recently, the polystrene thin film surface was investigated again [557] with a quantitative analysis of angular distribution by the SS/MO method (discussed in the section on phthalocyanines and porphyrins), as indicated in Figure 54. The photoelectron take-off-angle dependence was found to be clearly different between the 100 and 270 Å thickness films. The analysis indicated that the orientation of the phenyl pendant groups at the surface of 100 Å thick film is random, while the phenyl groups at the surface of the 270 Å thick film are tilted at rather larger angles. This implies that the molecular planes of the pendant groups become more perpendicularly oriented with increasing film thickness.

In addition, an enhancement of the whole valence band emission at the core excitation region was studied by XANES for polystyrene [558].

23.2. Poly(2-vinylnaphthalene)

The orientation of pendant naphthalene groups in poly(2-vinylnaphthalen) has been studied by ARUPS and NEXAFS

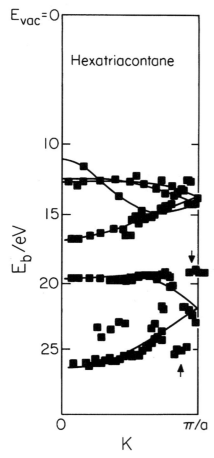

Fig. 50. The experimental band structure for hexatriacontane together with the modified theoretical calculations for polyethylene from A. Karpfen, *J. Chem. Phys.* 75, 238 (1981). Adapted from [532].

tation [561, 562]. The rubbing process is commonly used in the manufacture of well-oriented polymer films. By analyzing the incident angle dependence of NEXAFS at carbon and fluorine K-edges, it was concluded that such films were made of highly oriented chains. In the evaporated PTFE film, the fluorocarbon chains are almost parallel to the surface and are uniaxially realigned along the rubbing direction by the mechanical rubbing process, consistent with atomic force microscopy experiments [563] and other NEXAFS studies [564]. In evaporated PFT films, the fluorocarbon chains are oriented normal to the surface and are hardly changed in their alignment direction by the mechanical rubbing process. This is probably due to the fact that PTFE has a generally lower crystallinity and much longer chains than PFT, and hence the realignment of the chains along the surface and parallel with the "rubbing" direction can induced by the rubbing process and remains stable.

Theoretical calculations [565] of the NEXAFS spectra confirm that orientational ordering can be determined conclusively, and the calculated spectra are in agreement with the experimental NEXAFS studies of highly oriented films of PTFE [564].

The model compound for PTFE, perfluorotetracosane (n-CF$_3$(CF$_2$)$_{22}$CF$_3$), can be oriented (as just indicated) with the chain backbone oriented along the surface normal, as indicated by angle-resolved photoemission [566]. In such a configuration, the photon energy dependence in angle-resolved photoemission can be used to recover the intramolecular band dispersion, similar to those studies undertaken for a number of long-chain organic species discussed in the previous section. The experimental band dispersion [566] is qualitatively similar to the band structure calculated for PTFE in the planar zigzag configuration by Seki et al. [567]. Nonetheless, the dispersion of the highest occupied molecular orbital is more than several eV and larger than expected.

23.4. Polyvinylidene Fluoride–Trifluoroethylene Copolymers

The Teflon™-like polymer polyvinylidene fluoride (PVDF) (shown in Figure 55) and its copolymers with trifluoroethylene (TrFE) and tetrafluoroethylene are a rich system for the study of ferroelectricity [568–570] and are discussed extensively in another chapter in this series. Through a variation of the Langmuir–Blodgett technique, copolymer polyvinylidene fluoride with trifluoroethylene (PVDF-TrFE) films can be grown crystalline with the chains parallel with the substrate surface [570–572]. There is no evidence that the surface of (PVDF-TrFE70:30) differs from the bulk on a molecular level [571]. The surface, nonetheless, is fundamentally different from the bulk [572]. Surface structures and a surface structure phase transition have been identified that are distinct from the known bulk ferroelectric–paraelectric phase transition of crystalline copolymer films of vinylidene fluoride (70%) with trifluoroethylene (30%) [572, 573]. The temperature-dependent changes in the surface structure are accompanied by the physical rotation of the polar group (CH$_2$–CF$_2$) [572]. The surface

[559, 560]. Although NEXAFS is a powerful tool for investigating molecular orientation, the technique is not as straightforward when molecules are oriented either randomly or at the magic angle (as just indicated). ARUPS, with sophisticated theoretical analysis, has an advantage over NEXAFS in determining molecular orientations in such a disordered systems. For the naphthalene pendant groups at the polymer surface, the photoemission angular distribution calculated for a three-dimensional isotropic random orientation model was found in good agreement with the angle-resolved photoemission measurements [559], suggesting that a majority of the naphthalene pendant groups are tilted at rather large angles with respect to the polymer surface.

23.3. Poly(tetrafluoroethylene)

Poly(tetrafluoroethylene) (CF$_2$)$_n$ (or PTFE), known by the trade name Teflon™, is one of the polymers widely used in a variety of industrial applications. Polarized NEXAFS studies were performed on (PTFE) oligomers and its model compound perfluorotetracosane (PFT) (n-CF$_3$(CF$_2$)$_{22}$CF$_3$) to investigate the effect of mechanical rubbing on the molecular orien-

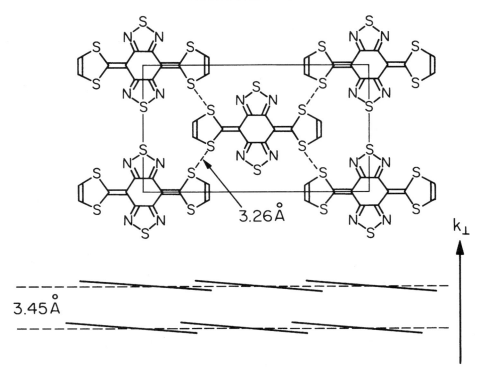

Fig. 51. Bis(1,2,5-thiadiazolo)-p-quinobis(1,3-dithiole) molecules in single crystal. The network sheet is formed by short S–S contacts of 3.26 Å. The spacing between network sheets is 3.45 Å, and each molecule is inclined by ~5° in the sheet. The energy-band dispersion was observed along $k_\perp$ [539].

phase transition near 20°C is particularly interesting as it involves the same conformational changes [572, 573] as the bulk transition near 80°C, but is different in several important aspects. The surface transition is accompanied by a large shift of the Fermi level (that makes the high-temperature phase nearly metallic [573, 574]), a work function, and dipole reorientation [572–575] and by a dynamic Jahn–Teller distortion [576]. The surface transition is first order [575] just like the bulk transition but, unlike the bulk transition [571], is strongly first order. In addition, the surface has band structure and preserves surface crystallographic symmetries that are inconsistent with a proper ferroelectric transition [575]. Therefore, despite the similarities with the well-known bulk ferroelectric transition, the precise nature of the surface transition is at present unknown.

From the band structure of the surface [572, 573], LEED, and scanning tunneling microscopy (STM) [572, 573], it is also very clear that the surface is reconstructed [572], with a packing arrangement different from the bulk, as indicated in Figure 56.

Unlike poly(tetrafluoroethylene), these crystalline copolymer films of vinylidene fluoride (70%) with trifluoroethylene (30%) have little band dispersion in the occupied part of the molecular orbitals. On the other hand, the unoccupied molecular orbitals exhibit extensive and considerable intramolecular band dispersion, as seen in Figure 57 [572, 573]. The unoccupied band structure is seen to change significantly as the temperature increases to above the surface (but not the bulk) ferroelectric transition [572, 573] as seen in Figure 57.

23.5. Polyimides

Rubbed polyimide films are widely used for liquid crystal displays, while polyimides are a common component of circuit boards and are generally used in the "packaging" of electronics. Unidirectionally rubbed (usually performed mechanically with a cloth) polyimide induces uniaxial alignment of the molecules, and the optical axis (molecular long axis) often makes a relatively small pretilt angle (0°–15°) from the polymer-coated substrate. It is widely believed that the mechanically rubbed polymer is subjected to local structural changes, such as realignment of the polymer chain. Complete understanding, however, is still lacking. NEXAFS was applied to the study of two different types of polyimides, biphenyl-3,3′,4,4′-tetracarboxylic dianhydride and pyromellitic dianhydride [577, 578]. It was concluded that the surface structure of the polyimides strongly affects the liquid crystal pretilt angle.

23.6. Polythiophenes

NEXAFS studies of poly-3-alkylthiophenes [579, 580] indicated that the polymeric chains, namely the hetero-aromatic rings, lie flat on platinum substrate surfaces, unlike the generally more or less upright structures formed in "self-assembled" films of thiols and thiophenes [581–583], as well as the smaller adsorbed thiols and thiolates previously discussed. A strong interaction between the antibonding π orbitals of the polymer and the electron states of the metal was observed.

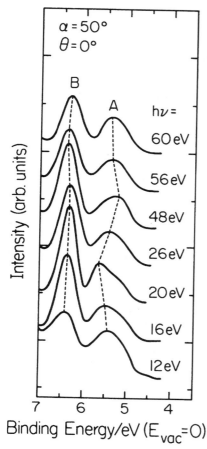

Fig. 52. Photon energy dependence (energy-band dispersion) of photoelectron spectra at normal emission for bis(1,2,5-thiadiazolo)-p-quinobis(1,3-dithiole) thin films. The HOMO and NHOMO bands are labeled A and B [539].

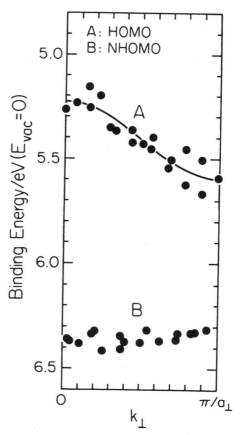

Fig. 53. Intermolecular energy-band dispersion for bis(1,2,5-thiadiazolo)-p-quinobis(1,3-dithiole) thin films in the reduced zone scheme, where $a_\perp$ is 3.4 Å [539].

23.7. Organometallic Polymers

The molecular orientation of rodlike organometallic polymers $[-Me(Pn-Bu_3)_2-C{\equiv}C-C_6H_4-C_6H_4-C{\equiv}C-]_n$, Me=Pt, Pd, in thick films (~1 μm) has also been determined by NEXAFS spectroscopy [584]. Quantitative analysis suggested that the organic moiety of the polymer ($-C{\equiv}C-C_6H_4-C_6H_4-C{\equiv}C-$) is oriented at about 40° with respect to the surface normal, while the two benzene rings are nearly coplanar, and the plane containing the metal and the phosphines (Pn–Bu$_3$) is probably aligned with the plane of the benzene rings. Finding a well-ordered molecular orientation in such a thick polymer film is surprising. It was found, nonetheless, that the polymers containing two different transition metals, Pt and Pd, show the same molecular structure and orientation.

23.8. Tetrathiafuvalene-Tetracyanoquinodimethane

Tetracyanoquinodimethane (TCNQ) molecules form dimers with other functionals with about one electron per dimer [585, 586]. Some combinations of other organic groups with TCNQ are one-dimensional conductors which may undergo a Peierls metal to nonmetal transition with decreasing temperature.

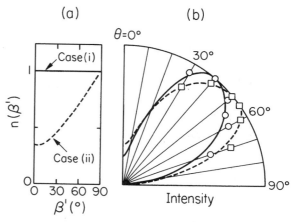

Fig. 54. Calculated take-off angular (θ) dependence of the HOMO band intensity for multiply oriented phenyl groups for polystyrene compared with experiment [557]. (a) Number-weight distributions of the molecules of tilt angle β' [$n(\beta')$] used to obtain θ dependencies shown in (b). Case (i) corresponds to a randomly oriented system of pendant phenyl groups, and case (ii) corresponds to an orientation system of pendant phenyl groups where $n(\beta')$ increases with β'. (b) Calculated θ dependencies for the two systems with different multiple molecular orientations, cases (i) and (ii). The observed results (open circles: 100-Å-thick film, open squares: 270-Å-thick film) are also shown for comparison. The calculations were performed by assuming azimuthal disorder of the phenyl groups.

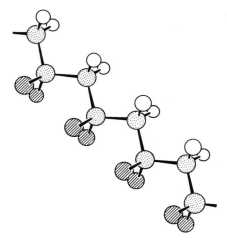

Fig. 55. Structure of PVDF in the all-trans *TTTT* conformation of the ferro-electric β phase; carbons are gray, fluorines are striped, and hydrogen is white.

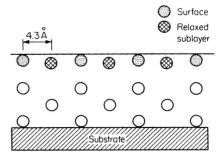

Fig. 56. A schematic side view of the reconstruction of the crystalline PVDF-TrFE copolymer Langmuir–Blodgett film surface. Reprinted with permission from [572], copyright 2000, American Physical Society.

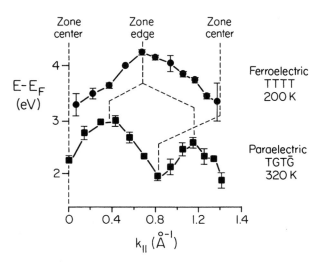

Fig. 57. The band dispersion of the LUMO states in the ferroelectric phase (●) at 200 K and in the "surface" high temperature phase (■) at 320 K along the chain symmetry direction of the Brillouin zone. The wave vector of the Brillouin zone edge is indicated as are the Brillouin zone center(s). The dashed lines indicate the changing positions of the center of the Brillouin zone edge and the Brillouin zone centers across the surface ferroelectric transition. Adapted from [572, 573].

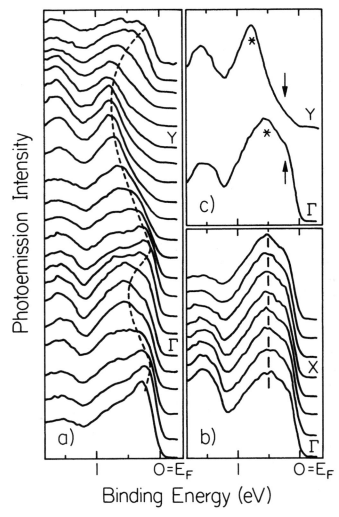

Fig. 58. The angle-resolved photoemission spectra at 150 K of TTF-TCNQ along (a) and perpendicular to the chain direction (b). Γ to Y are the Brillouin zone center and Brillouin zone boundary, respectively. In panel (c) are selected spectra, with the background subtracted, showing the main dispersive peak (marked by *) and an emission band not accounted for by theory marked with an arrow. Reprinted with permission from [592], copyright 1998, American Physical Society.

The Peierls transition in tetrathiafuvalene-tetracyanoquinodimethane (TTF-TCNQ) is anticipated to occur at 54 K [585, 587–589]. TTF-TCNQ contains segregated stacks of the planar TTF and TCNQ molecules and it is the charge transfer from TTF to TCNQ that leads to the partially occupied bands and the metallic character [590].

Band structure mappings of TTF-TCNQ, using angle-resolved photoemission, have been undertaken and band dispersion along the chains was observed [590–592], as seen in the Γ to Y direction in Figure 58. Little or no dispersion was observed in the direction perpendicular to the chains (the Γ to X direction in Figure 58). These band mapping results provide some indication of a change in the density of states near the Fermi level, at the anticipated Fermi wave vector k_F (Figure 58), as the temperature is altered from above to below the temperature of the anticipated Peierls transition [590–594]. In some ways, these temperature dependent changes to the band structure, along the one-dimensional chains, are qualitatively similar to those of the polyvinylidene fluoride–trifluoroethylene copolymer crystals, just discussed, which exhibit not only a change in the density of states but also a massive change in band structure across a surface ferroelectric transition, associated with a change in metallicity.

24. BAND STRUCTURE OF BUCKYBALL FILMS ON METALS AND SEMICONDUCTORS

While interest in the fullerene (carbon clusters) has now been extended to carbon nanotubes, in this section we will focus on fullerenes. The several possible hybridizations of the carbon $2s$ and $2p$ orbitals lead to several different carbon structures dominated by different dimensionalities, such as diamond [a three-dimensional (3D) carbon network], graphite (a largely 2D network), graphene tubules (a 1D carbon network) and fullerenes (0D). In the "classic" fullerene, C_{60}, each carbon atom has two single bonds along adjacent sides of a pentagon and one double bond between two adjoining hexagons as shown in Figure 59 (in the figure: red bonds around each pentagonal face are single bonds; yellow bonds are double bonds). This is the buckminster fullerene (buckyball) with icosahedral structural symmetry (I_h). Buckminster fullerene C_{60} was first synthesized using pulsed-laser evaporation by Smalley and co-workers in 1985 [595], but the mass production, using arc-discharge methods, has now made large quantities of purified C_{60} and other fullerenes available (there are a number of stable even-numbered clusters of carbon atoms in the range C_{40}–C_{100} [596]). The C_{60} icosahedral fullerene has 120 distinct symmetry operations, as indicated in the character table in Table VII.

In a sense, the C_{60} fullerene can be thought as electronically related to the smaller icosahedral *closo*-carborane cage molecule *closo*-1,2-dicarbadodecaborane ($C_2B_{10}H_{12}$), commonly referred to as orthocarborane, discussed in a previous section. As with the *closo*-carborane [597–599], there

have been many studies on the alkali doped C_{60} and related fullerenes and the associated thin films [600–604]. Consistent with the tenor of this review thus far, we only consider the electronic band structure and orientation of the undoped adsorbed fullerenes.

Also like orthocarborane, the electronic structure of the isolated fullerenes has been investigated theoretically using *ab initio* methods [605]. There are similarities (both form *n*-type semiconductorlike molecular films) but unlike the *closo*-carborane, C_{60} more resembles graphite and has a much smaller experimental highest occupied molecular orbital to lowest unoccupied molecular orbital (HOMO–LUMO) gap. Graphite is a semimetal, due to the weak coupling between two graphite layers, with a zero energy gap as the interlayer interaction goes to zero. Fullerene is a semiconductor with a ≈ 2.3 eV band gap [606]. The theoretical HOMO h_u–LUMO t_{1u} gap of the fullerene is about 1.5 eV [607] while the experimental HOMO–LUMO gap, for the condensed film of C_{60}, is about 3.5 eV [606], as shown in Figure 60.

The electronic structure of the solid C_{60} films has been studied on a number of different semiconductor and metals sub-

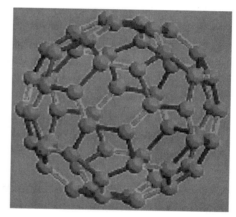

Fig. 59. Geometrical structure of C_{60}, the buckminster fullerene (red bonds around each pentagonal face are single bonds; yellow bonds are double bonds).

Table VII. Character Table for the Icosahedral Group (I_h) Representing the C_{60} Symmetry

I_h	E	$12C_5$	$12C_5^2$	$20C_3$	$15C_2$	i	$12S_{10}$	$12S_{10}^3$	$20S_6$	$15\sigma_v$	Top	Center
A_g	1	1	1	1	1	1	1	1	1	1		$x^2+y^2+z^2$
T_{1g}	3	β	α	0	-1	3	β	α	0	-1	(R_x, R_y, R_z)	
T_{2g}	3	α	β	0	-1	3	α	β	0	-1		
G_g	4	-1	-1	1	0	4	-1	-1	-1	0		
H_g	5	0	0	-1	1	5	0	0	1	1		xy, yz, zx
A_u	1	1	1	1	1	-1	-1	-1	-1	-1		
T_{1u}	3	β	α	0	-1	-3	β	$-\beta$	0	1		
T_{2u}	3	α	β	0	-1	-3	α	$-\alpha$	0	1		
G_u	4	-1	-1	1	0	-4	1	1	-1	0	(T_x, T_y, T_z)	(x, y, z)
H_u	5	0	0	-1	1	-5	0	0	1	-1		

$\alpha = -2\cos 72°$, $\beta = -\cos 144°$.

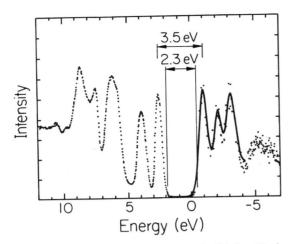

Fig. 60. Valence and conduction band spectra of solid C_{60}. The band gap is 2.3 eV while the HOMO–LUMO gap is 3.5 eV. Reprinted with permission from [606], copyright 1992, American Physical Society.

strates, such as Si (100) [608], Si(111)-(7 × 7) [609–612], GaAs [613, 614]. Cu(111) [615, 616], Au(110) [617–621], Au(111) [607, 617, 622], Ag(100) [623, 624], Ag(110) [620, 621], Ag(111) [625], and polycrystalline noble metals [626, 627] using photoemission, inverse photoemission, and scanning tunneling microscopy. The details of the interactions between the substrate and the adsorbed C_{60} have also been probed using photoemission and inverse photoemission spectroscopy [606, 628–630].

The charge transfer from substrate to C_{60} has been observed using photoemission spectroscopy and has been estimated to be about 0.8 electrons per C_{60} molecule on Au(111), and 1.6 electrons on Cu(111) [622]. The charge transfer from polycrystalline noble metal substrates is in a similar range: 1.8 electrons per C_{60} molecule on a Cu substrate, 1.7 on Ag, and 1.0 on Au, with an uncertainty of about 0.2 electrons [627]. This is a substantial charge transfer from substrate to the C_{60} molecules, considering the inert character of C_{60}. It is the charge transfer, however, that is believed to induce an increase of the density of states near Fermi level [627], partly filling the LUMO band for very thin C_{60} films, as shown in Figure 61. The density of states near Fermi level on the occupied side disappears with increasing coverage, as the LUMO band cannot easily be filled in the top layers of the multilayer film, as shown in Figure 62 [623]. Charge transfer to C_{60} is directly related to the metallicity changes of the C_{60} film and to its superconducting behavior (under some conditions) [631–641]. Potassium-doped C_{60} shows superconductivity at 18 K [631]. The charge transferred from the potassium fills the unoccupied t_{1g} and t_{1u} states but leaves the t_{1u} molecular band mostly unoccupied [637].

There are few significant differences in electronic structure between the gas phase and the condensed phase at high coverage (at least above one monolayer). This has been attributed to the strong molecular character of C_{60} and to localized electronic states with short mean free paths. Solid C_{60} crystal is an insulator, with (as we have mentioned) a ~2.3 eV electronic band gap [606]. Core level electronic structure studies

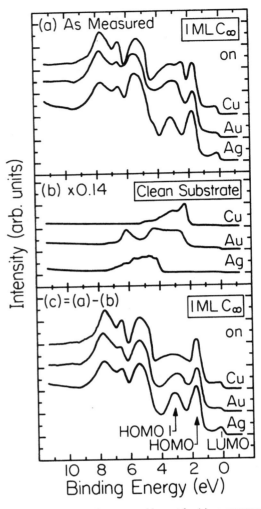

Fig. 61. Undoped C_{60} monolayer on noble metals: (a) as-measured photoemission spectra of the overlayer on polycrystalline Cu, Au, and Ag; (b) photoemission spectra of clean polycrystalline Cu, Au, and Ag multiplied by 0.14; (c) net photoemission spectra of the overlayer on polycrystalline Cu, Au, and Ag, taken as the difference between (a) and (b). Reprinted with permission from [627], copyright 1998, American Physical Society.

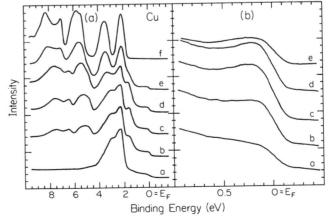

Fig. 62. The He I UPS spectra of varying thickness of C_{60} on Cu. Curve (a), bulk Cu; curve (b), 0.25 ML C_{60}; curve (c), 0.5 ML; curve (d), 1 ML; curve (e), 2 ML; curve (f), bulk C_{60}. (a) shows full UPS spectra; (b) shows near Fermi-edge region. Reprinted with permission from [626], copyright 1992, American Physical Society.

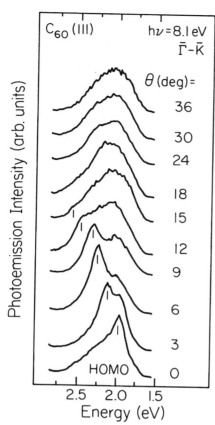

Fig. 63. The electronic band dispersion of the HOMO bands from crystalline $C_{60}(111)$ as a function of the emission angle using 8.1 eV photon energy. Reprinted with permission from [644], copyright 1994, American Physical Society.

25. CONCLUSION

Many of the very large molecular adsorbates are seen to exhibit the orientational order and band structure behavior(s) of the smaller "simple" molecules. Certainly, the application of symmetry selection rules is quite similar. The greater abundance of molecular orbitals with the larger molecular adsorbates requires greater energy resolution in order for the various molecular orbitals to be distinctly identified but there are indeed limits to the amount of information gained from increasing resolution—there will be finite widths to photoemission from molecular adsorbates due to lifetime broadening, Franck–Condon factors, and the Jahn–Teller effects in the photoelectron excitation process. In fact the more serious limitation is phonon broadening, a prosaic problem that is addressed, in part, by undertaking photoemission from samples at very low temperatures.

As the molecular size increases, ordered molecular overlayers have an increasingly smaller Brillouin zone. To probe the lateral interactions between adsorbate molecules through the dispersion of the molecular orbitals with changing electron wave vector **k**, one must either increase the angular resolution or reduce the photon energy. Reducing the photon energy does help but tends to restrict studies to molecular orbitals with smaller and smaller binding energies. It is also clear that, for increasingly large molecules, there is a general trend toward smaller extramolecular dispersions (though there are exceptions), which, if they are to be accurately mapped out, require (again) better energy resolution. The molecular orbitals of adsorbates do combine to form bands, much as the atomic orbitals form bands in crystalline solids; as the molecule size increases, intramolecular band dispersion increases. Both extramolecular and intramolecular band structures have been observed in sufficiently ordered molecular thin films. Band structure has been observed in both angle-resolved photoemission and angle-resolved inverse photoemission by changing $k_{\parallel}$ (incidence or emission angle for molecules oriented along the surface) or changing $k_{\perp}$ (for molecules with orientations along the surface normal).

In spite of the complexities of working with molecular systems, often the band structure of molecular systems is easier to understand with simple models than is the band structure of more conventional solid state systems. The observations of band structure in molecular adlayers and thin films provide new opportunities for the study of band structure in solid state physics. Increasingly, conventional topics of solid state physics now include molecular systems.

New directions for study include the coadsorption of different adsorbates, and this can introduce new molecular orientations and molecular crystal packing arrangements. While much of the early work with molecular adsorption focused upon adsorption on metal substrates, adsorption on semiconductor surfaces is increasingly important and indeed new vistas for research are being introduced with the study of molecular adsorption on molecular thin film surfaces.

on gas-phase and solid C_{60} [642] indicate that the solid state effects (the effect of condensation and formation of a film) are small. Electron correlation effects are mostly determined by intramolecular interactions instead of intermolecular interactions. The localized electronic character has been well documented in angle-resolved photoemission spectroscopy studies [643–645]. Nonetheless, a very small band dispersion of HOMO, ~400 to ~600 meV, has been observed [643, 644], as shown in Figure 63. Very recently, the band structure has been also been measured in oriented carbon nanotubes, using angle-resolved photoemission spectroscopy [646].

The effect of molecular orientation on electronic structure and charge distribution has been studied theoretically [647]. It has been difficult, however, to identify the preferential molecular orientation because, in addition to the complexities of the HOMO and LUMO states and in addition to the electronic convolution with the states of the STM tip, the molecules tend to "spin" in the surface layer above the room temperature. Recently, the molecular orientation of C_{60} on Si(111) has been studied successfully by using a combination of low temperature scanning tunneling microscopy and local density approximation [648]. The molecular orientation depends on the site and STM tip bias [648].

Acknowledgments

The authors thank Neil Boag, Friedrich Hensel, Kazuhiko Seki, and Michael Grunze for their help and encouragement in preparing this manuscript and to Brian Robertson, Hans-Peter Steinrück and Tony Caruso for their help in proofing the manuscript (though any errors that remain are ours). This work was supported by the National Science Foundation, the Office of Naval Research, AFOSR Contract F30602-98-C-0105 to the W. M. Keck Center for Molecular Electronics, Subcontract 3531141, the Nebraska Research Initiative, and the donors of the Petroleum Research Fund administered by the ACS.

REFERENCES

1. E. W. Plummer and W. Eberhardt, *Adv. Chem. Phys.* 49, 533 (1982).
2. H.-P. Steinrück, *Vacuum* 45, 715 (1994).
3. N. V. Richardson and A. M. Bradshaw, "Electron Spectroscopy: Theory, Techniques and Applications" (C. R. Brundle and A. D. Baker, Eds.), Vol. 4, p. 153. Academic Press, San Diego, 1984.
4. P. A. Dowben, *Z. Phys. Chem.* 202, 227 (1997).
5. W. R. Salaneck, S. Stafström, and J.-L. Brédas, "Conjugated Polymer Surfaces and Interfaces: Electronic and Chemical Structure of Interfaces for Polymer Light Emitting Devices." Cambridge Univ. Press, Cambridge, UK, 1996.
6. J. B. Pendry, "Low Energy Electron Diffraction." Academic Press, London/New York, 1974.
7. L. J. Clarke, "Surface Crystallography — An Introduction to Low Energy Electron Diffraction." Wiley, New York, 1985.
8. K. Heinz, *Rep. Prog. Phys.* 58, 637 (1995).
9. P. J. Estrup and E. G. McRae, *Surf. Sci.* 25, 1 (1971).
10. R. E. Ballard, "Photoelectron Spectroscopy and Molecular Orbital Theory." Hilger, Bristol, 1978.
11. M. Cardona and L. Ley, "Photoemission in Solids I," Topics in Applied Physics, Vol. 26. Springer-Verlag, Berlin/Heidelberg, 1978.
12. L. Ley and M. Cardona, "Photoemission in Solids II," Topics in Applied Physics, Vol. 27. Springer-Verlag, Berlin/Heidelberg, 1979.
13. D. W. Turner, C. Baker, A. D. Baker, and C. R. Brundle, "Molecular Photoelectron Spectroscopy," Chap. 2. Wiley, New York, 1970.
14. J. Braun, *Rep. Prog. Phys.* 59, 1267 (1996).
15. J. Hermanson, *Solid State Commun.* 22, 9 (1977).
16. W. Eberhardt, "Applications of Synchrotron Radiation," Springer Series in Surface Sciences, Vol. 35. Springer-Verlag, Berlin/Heidelberg, 1995.
17. S. D. Kevan (Ed.), "Angle-Resolved Photoemission." Elsevier, Amsterdam, 1992.
18. P. M. G. Allen et al., *Solid State Commun.* 55, 701 (1985).
19. D. Funnemann et al., *J. Phys. E* 19, 554 (1986).
20. P. O. Gartland and B. J. Slagsvold, *Phys. Rev. B* 12, 4047 (1975).
21. N. V. Smith, *Rep. Prog. Phys.* 51, 1227 (1988).
22. F. J. Himpsel, *Surf. Sci. Rep.* 12, 1 (1990).
23. N. Memmel, *Prog. Surf. Sci.* 42, 75 (1993).
24. B. Reihl and R. R. Schlittler, *Phys. Rev. Lett.* 52, 1826 (1984).
25. P. M. Echenique and M. E. Uranga, *Surf. Sci.* 247, 125 (1991).
26. J. C. Fuggle and J. E. Inglesfield, "Unoccupied Electronic States" Topics in Applied Physics, Vol. 69, Springer-Verlag, Berlin/New York, 1992.
27. V. Dose, *Appl. Phys.* 14, 117 (1977).
28. N. V. Richardson, *Surf. Sci.* 126, 337 (1983).
29. M. Scheffler, K. Kambe, and F. Forstmann, *Solid State Commun.* 25, 93 (1978).
30. P. A. Dowben, M. Onellion, and Y. J. Kime, *Scanning Microscopy* 2, 177 (1988).
31. W. Eberhardt and F. J. Himpsel, *Phys. Rev. B* 21, 5572 (1980).
32. J. E. Inglesfield and E. W. Plummer, in "Angle-Resolved Photoemission" (S. D. Kevan, Ed.). Elsevier, Amsterdam, 1992.
33. J. J. Sakurai, "Advanced Quantum Mechanics." Addition–Wesley, Reading, MA, 1980.
34. G. Wendin, "Breakdown of the One Electron Picture in Photoelectron Spectra." Structure and Bonding, Vol. 45. Springer-Verlag, New York, 1981.
35. P. A. Dowben, D. Li, J. Zhang, and M. Onellion, *J. Vac. Sci. Technol. A* 13, 1549 (1995).
36. P. A. Dowben, D. LaGraffe, and M. Onellion, *J. Phys. Cond. Mat.* 1, 6571 (1989).
37. G. J. Lapeyre and J. Anderson, *Phys. Rev. Lett.* 35, 117 (1975).
38. C. T. Chen, L. H. Tjeng, J. Kwo, H. L. Kao, P. Rudolf, F. Sette, and R. M. Fleming, *Phys. Rev. Lett.* 68, 2543 (1992).
39. H. Ueba, *Surf. Sci.* 242, 266 (1991).
40. E. E. Koch, "Handbook on Synchrotron Radiation." North-Holland, New York, 1983.
41. A. Bianconi, S. Della Longa, C. Li, M. Pompa, A. Congiu-Castellano, D. Urdon, A. M. Flank, and P. Lagarde, *Phys. Rev. B* 44, 10126 (1991).
42. E. Tamura, J. van Ek, M. Fröba, and J. Wong, *Phys. Rev. Lett.* 74, 4899 (1995).
43. G. Schutz, M. Knulle, R. Wienke, W. Wilhelm, W. Wagner, P. Kienle, and R. Frahm, *Z. Phys. B* 73, 67 (1988).
44. M. N. Piancastelli, *J. Electron Spectrosc. Rel. Phenom.* 100, 167 (1999).
45. P. G. Burke, N. Chandra, and F. A. Gianturo, *J. Phys. B* 5, 2212 (1972).
46. R. Hoffman, *J. Phys. Cond. Mat.* 5, A1 (1993).
47. R. Hoffman, "Solids and Surfaces: A Chemist's View of Bonding in Extended Structures." VCH, Weinheim/New York, 1988.
48. W. E. Spicer, *Phys. Rev.* 112, 114 (1958).
49. J. W. Gadzuk, *Phys. Rev. B* 10, 5030 (1974).
50. J. W. Gadzuk, *Surf. Sci.* 53, 132 (1975).
51. A. Liebsch, *Phys. Rev. B* 13, 544 (1976).
52. S. Y. Tong and A. R. Lubensky, *Phys. Rev. Lett.* 39, 498 (1977).
53. J. W. Davenport, *Phys. Rev. Lett.* 36, 945 (1976).
54. T. Gustafsson, *Surf. Sci.* 94, 593 (1980).
55. G. Apai, P. S. Wehner, R. S. Williams, J. Stöhr, and D. A. Shirley, *Phys. Rev. Lett.* 37, 1497 (1976).
56. P. Hofmann, S. R. Bare, N. V. Richardson, and D. A. King, *Solid State Commun.* 42, 645 (1982).
57. P. Hofmann, S. R. Bare, and D. A. King, *Surf. Sci.* 117, 245 (1982).
58. J. C. Fuggle, M. Steinkilberg, and D. Menzel, *Chem. Phys.* 11, 307 (1975).
59. P. M. Williams, P. Butcher, J. Wood, and K. Jacobi, *Phys. Rev. B* 14, 3215 (1976).
60. K. Horn, A. M. Bradshaw, and K. Jacobi, *Surf. Sci.* 72, 719 (1978).
61. F. Greuter, D. Heskett, E. W. Plummer, and H. J. Freund, *Phys. Rev. B* 27, 7117 (1983).
62. C. C. Allyn, T. Gustafsson, and E. W. Plummer, *Solid State Commun.* 24, 531 (1977).
63. P. Hofmann, J. Gossler, A. Zartner, M. Glanz, and D. Menzel, *Surf. Sci.* 161, 303 (1985).
64. D. Heskett, E. W. Plummer, R. A. de Paola, W. Eberhardt, F. M. Hoffmann, and H. R. Moser, *Surf. Sci.* 164, 490 (1985).
65. W. Eberhardt, F. M. Hoffmann, R. dePaola, D. Heskett, I. Strathy, E. W. Plummer, and H. R. Moser, *Phys. Rev. Lett.* 54, 1856 (1985).
66. D. Heskett, I. Strathy, E. W. Plummer, and R. A. de Paola, *Phys. Rev. B* 32, 6222 (1985).
67. C. L. Allyn, T. Gustafsson, and E. W. Plummer, *Chem. Phys. Lett.* 47, 127 (1977).
68. R. J. Smith, J. Anderson, and G. J. Lapeyre, *Phys. Rev. Lett.* 37, 1081 (1976).
69. C. L. Allyn, T. Gustafsson, and E. W. Plummer, *Solid State Commun.* 28, 85 (1978).
70. D. Heskett, E. W. Plummer, R. A. de Paola, and W. Eberhardt, *Phys. Rev. B* 33, 5171 (1986).
71. T. Maruyama, Y. Sakisaka, H. Kato, Y. Aiura, and H. Yanashima, *Surf. Sci.* 304, 281 (1994).

72. C. W. Seabury, E. S. Jensen, and T. N. Rhodin, *Solid State Commun.* 37, 383 (1981).

73. E. Jensen and T. N. Rhodin, *J. Vac. Sci. Technol.* 18, 470 (1981).

74. E. Jensen and T. N. Rhodin, *Phys. Rev. B* 27, 3338 (1983).

75. J. J. Weimer, E. Umbach, and D. Menzel, *Surf. Sci.* 159, 83 (1985).

76. J. J. Weimer and E. Umbach, *Phys. Rev. B* 30, 4863 (1984).

77. J. M. Mundenar, R. H. Gaylord, S. C. Lui, E. W. Plummer, D. M. Zehner, W. K. Ford, and L. G. Sneddon, *Mat. Res. Soc. Symp. Proc.* 83, 59 (1987); R. H. Gaylord, Ph.D. Thesis, University of Pennsylvania, 1987, and Ref. [64].

78. S. Andersson and J. B. Pendry, *Surf. Sci.* 71, 75 (1978).

79. F. M. Hoffman and R. A. de Paola, *Phys. Rev. Lett.* 52, 1697 (1984).

80. R. A. de Paola, J. Hrbek, and F. M. Hoffman, *J. Chem. Phys.* 82, 2484 (1985).

81. H. J. Freund, R. Bartos, R. P. Messmer, M. Grunze, H. Kuhlenbeck, and M. Neumann, *Surf. Sci.* 185, 187 (1987).

82. D. W. Moon, S. Cameron, F. Zaera, W. Eberhardt, R. Carr, S. L. Bernasek, J. L. Gland, and D. J. Dwyer, *Surf. Sci.* 180, L123 (1987).

83. R. S. Saki, G. S. Herman, M. Yamada, J. Osterwald, and C. S. Fadley, *Phys. Rev. Lett.* 63, 283 (1989).

84. C. Benndorf, B. Nieberand, and B. Krüger, *Surf. Sci.* 177, L907 (1986).

85. N. D. Shinn and T. E. Madey, *Phys. Rev. Lett.* 53, 2481 (1984).

86. N. D. Shinn, *J. Vac. Sci. Technol. A* 4, 1351 (1986).

87. J. E. Demuth and D. E. Eastman, *Solid State Commun.* 18, 1497 (1976).

88. J. Stöhr, K. Baberschke, R. Jaeger, T. Treicher, and S. Brennan, *Phys. Rev. Lett.* 47, 381 (1981).

89. D. A. Outka and J. Stöhr, *J. Chem. Phys.* 88, 3539 (1988).

90. D. Schmeisser, F. Grueter, W. Plummer, and H.-J. Freund, *Phys. Rev. Lett.* 54, 2095 (1985).

91. H.-J. Freund, W. Eberhardt, D. Heskett, and E. W. Plummer, *Phys. Rev. Lett.* 50, 768 (1983).

92. C. W. Seabury, T. N. Rhodin, M. M. Traum, R. Benbow, and Z. Hurych, *Surf. Sci.* 97, 363 (1980).

93. H. Kuhlenbeck, M. Neumann, and H.-J. Freund, *Surf. Sci.* 173, 194 (1986).

94. H. Kuhlenbeck, H. B. Saalfeld, U. Buskotte, M. Neumann, H.-J. Freund, and E. W. Plummer, *Phys. Rev. B* 39, 3475 (1989).

95. K. Horn, A. M. Bradshaw, K. Hermann, and I. P. Batra, *Solid State Commun.* 31, 257 (1979).

96. R. Miranda, K. Wandelt, D. Rieger, and R. D. Schell, *Surf. Sci.* 139, 430 (1984).

97. G. Odörfer, E. W. Plummer, H.-J. Freund, H. Kuhlenbeck, and M. Neumann, *Surf. Sci.* 198, 331 (1988).

98. D. Rieger, R. D. Schell, and W. Steinmann, *Surf. Sci.* 143, 157 (1984).

99. H.-J. Freund and M. Neumann, *Appl. Phys. A* 47, 3 (1988).

100. R. P. Messmer, S. H. Lamson, and D. R. Salahub, *Phys. Rev. B* 25, 3576 (1982).

101. C. Mariani, H-U. Middelmann, M. Iwan, and K. Horn, *Chem. Phys. Lett.* 93, 308 (1982).

102. Zhi Xu, L. Hanley, and J. T. Yates, Jr., *J. Chem. Phys.* 96, 1621 (1992).

103. Zhi Xu, J. T. Yates, Jr., L. C. Wang, and H. J. Kreutzer, *J. Chem. Phys.* 96, 1628 (1992).

104. P. S. Bagus, C. R. Brundle, K. Hermann, and D. Menzel, *J. Electron Spectrosc. Rel. Phenom.* 20, 253 (1980).

105. P. S. Bagus, K. Hermann, and M. Stell, *J. Vac. Sci. Technol.* 18, 435 (1981).

106. K. Hermann, P. S. Bagus, C. R. Brundle, and D. Menzel, *Phys. Rev. B* 24, 7025 (1981).

107. C. R. Brundle, P. S. Bagus, D. Menzel, and K. Hermann, *Phys. Rev. B* 24, 7041 (1981).

108. J. C. Fuggle and D. Menzel, *Vakuum-Tech.* 27, 130 (1978).

109. J. C. Fuggle and D. Menzel, *Surf. Sci.* 79, 1 (1979).

110. J. C. Fuggle, E. Umbach, D. Menzel, K. Wandelt, and C. R. Brundle, *Solid State Commun.* 27, 65 (1978).

111. E. Umbach, *Surf. Sci.* 117, 482 (1982).

112. D. Saddei, H. J. Freund, and G. Hohlneicher, *Surf. Sci.* 102, 359 (1981).

113. M. Golze, M. Grunze, R. K. Driscoll, and W. Hirsch, *Appl. Surf. Sci.* 6, 464 (1980).

114. N. Mårtensson and A. Nilsson, *J. Electron Spectrosc. Rel. Phenom.* 52, 1 (1990).

115. E. J. Moler, S. A. Kellar, W. R. A. Huff, Z. Hussain, Y. Zheng, E. A. Henderson, Y. F. Chen, and D. A. Shirley, *Chem. Phys. Lett.* 264, 502 (1997).

116. P. A. Dowben, Y. Sakisaka, and T. N. Rhodin, *Surf. Sci.* 147, 89 (1984).

117. K. Horn, J. DiNardo, W. Eberhardt, H.-J. Freund, and E. W. Plummer, *Surf. Sci.* 118, 465 (1982).

118. E. Umbach, A. Schichl, and D. Menzel, *Solid State Commun.* 36, 93 (1980).

119. M. Grunze, R. K. Driscoll, G. N. Burland, J. C. L. Gornish, and J. P. Prichard, *Surf. Sci.* 89, 381 (1979).

120. J. Yoshinobu, R. Zenobi, J. Xu, Zhi Xu, and J. T. Jates, Jr., *J. Chem. Phys.* 95, 9393 (1991).

121. A. B. Anton, N. R. Avery, B. H. Toby, and W. H. Weinberg, *J. Electron Spectrosc. Rel. Phenom.* 29, 181 (1983).

122. H. Shi and K. Jacobi, *Surf. Sci.* 278, 281 (1992).

123. M. Gruyters and K. Jacobi, *Surf. Sci.* 336, 314 (1995).

124. J. Stöhr and R. Jaeger, *Phys. Rev. B* 26, 4111 (1982).

125. Y. Wu and P.-L. Cao, *Surf. Sci. Lett.* 179, L26 (1987).

126. M. Grunze, M. Golze, W. Hirschwald, H.-J. Fruend, H. Pulm, U. Seip, N. C. Tsai, G. Ertl, and J. Küppers, *Phys. Rev. Lett.* 53, 850 (1984).

127. N. D. Shinn, *Phys. Rev. B* 41, 9771 (1990).

128. N. D. Shinn and K.-L. Tsang, *J. Vac. Sci. Technol. A* 8, 2449 (1990).

129. Y. Fukuda and M. Nagoshi, *Surf. Sci.* 203, L651 (1988).

130. R. D. Diehl, M. F. Toney, and S. C. Fain, *Phys. Rev. Lett.* 48, 177 (1982).

131. R. D. Diehl and S. C. Fain, *Surf. Sci.* 125, 116 (1983).

132. C. A. Taft, T. C. Guimarães, A. C. Pavão, and W. A. Lester, Jr., *Int. Rev. Phys. Chem.* 18, 163 (1999).

133. G. Blyholder, *J. Phys. Chem.* 68, 2772 (1964).

134. A. C. Pavão, B. L. Hammond, M. M. Soto, W. A. Lester, Jr., and C. A. Taft, *Surf. Sci.* 323, 340 (1995).

135. T. C. Guimarães, A. C. Pavão, C. A. Taft, and W. A. Lester, Jr., *Phys. Rev. B* 60, 11789–11794 (1999).

136. B. P. Tonner, C. M. Kao, E. W. Plumber, T. C. Caves, R. P. Messmer, and W. R. Salaneck, *Phys. Rev. Lett.* 51, 1378 (1983).

137. C. Y. Ng, P. W. Tiedemann, B. H. Mahan, and Y. T. Lee, *J. Chem. Phys.* 66, 3985 (1977).

138. S. G. Kukolich, *J. Mol. Spectrosc.* 98, 80 (1983).

139. R. J. Behm and C. R. Brundle, *J. Vac. Sci. Technol. A* 2, 1040 (1984).

140. C. J. Nelin, P. S. Bagus, R. J. Behm, and C. R. Brundle, *Chem. Phys. Lett.* 105, 58 (1984).

141. K. T. Queeney and C. M. Friend, *J. Chem. Phys.* 107, 6432 (1997).

142. K. T. Queeney and C. M. Friend, *Surf. Sci.* 414, L957 (1998).

143. K. T. Queeney and C. M. Friend, *J. Phys. Chem. B* 102, 9251 (1998).

144. K. T. Queeney, S. Pang, and C. M. Friend, *J. Chem. Phys.* 109, 8058 (1998).

145. M. Pérez-Jigato, D. A. King, and A. Yoshimori, *Chem. Phys. Lett.* 300, 639 (1999).

146. P. Dumas, M. Suhren, Y. J. Chabal, C. J. Hirschmugel, and G. P. Williams, *Surf. Sci.* 371, 200 (1997).

147. W. A. Brown, R. K. Sharma, D. A. King, and S. Haq, *J. Phys. Chem. B* 100, 12559 (1996).

148. W. A. Brown, P. Gardner, and D. A. King, *J. Phys. Chem. B* 99, 7065 (1995).

149. W. A. Brown, P. Gardner, M. Pérez-Jigato, and D. A. King, *J. Chem. Phys.* 102, 7277 (1995).

150. A. Ludviksson, C. Huang, H. J. Jänsch, and R. M. Martin, *Surf. Sci.* 284, 328 (1993).

151. W. A. Brown and D. A. King, *J. Phys. Chem. B* 104, 2578 (2000).

152. H.-P. Steinrück, C. Schneider, P. A. Heimann, T. Pache, E. Umbach, and D. Menzel, *Surf. Sci.* 208, 136 (1989).

153. S. Lehwald, J. T. Yates, and H. Ibach, "Proceedings of the ECOSS-3" (D. A. Degras and M. Costa, Eds.), 1980, p. 221.

154. W. Erley, *Surf. Sci.* 205, L771 (1988).

155. F. P. Netzer and T. E. Madey, *Surf. Sci.* 110, 251 (1981).

156. S. Aminopirooz, A. Schmaltz, L. Becker, and J. Haase, *Phys. Rev. B* 45, 6337 (1992).

157. F. Bozso, J. Arias, C. P. Hanraham, J. T. Yates, Jr., R. M. Martin, and H. Metiu, *Surf. Sci.* 141, 591 (1984).

158. M. C. Asensio, D. P. Woodruff, A. W. Robinson, K.-M. Schindler, P. Gardner, D. Richen, A. M. Bradshaw, J. Consea, and A. R. González-Elipe, *Chem. Phys. Lett.* 192, 259 (1992).

159. K. M. Neyman and N. Rösch, *Surf. Sci.* 307–309, 1193 (1994).

160. A. Schichl and N. Rösch, *Surf. Sci.* 137, 261 (1984).

161. B. E. Hayden, *Surf. Sci.* 131, 419 (1983).

162. J. L. Gland and B. Sexton, *Surf. Sci.* 94, 355 (1980).

163. H. Ibach and S. Lehwald, *Surf. Sci.* 76, 1 (1978).

164. M. Kiskinova, G. Pirug, and H. P. Bonzel, *Surf. Sci.* 136, 285 (1984).

165. M. J. Breitschafter, E. Umbach, and D. Menzel, *Surf. Sci.* 109, 495 (1981).

166. N. Materer, A. Barbieri, D. Gardin, U. Starke, J. D. Batteas, M. A. van Hove, and G. A. Somorjai, *Surf. Sci.* 303, 319 (1994).

167. Q. Ge and D. A. King, *Chem. Phys. Lett.* 285, 15 (1998).

168. D. Loffreda, D. Simion, and P. Sautet, *Chem. Phys. Lett.* 291, 15 (1998).

169. H. Geisler, G. Odörfer, G. Illing, R. Jaeger, H.-J. Freund, G. Watson, E. W. Plummer, M. Neuber, and M. Neumann, *Surf. Sci.* 234, 237 (1990).

170. P. A. Heimann, P. Jacob, T. Pache, H.-P. Steinrück, and D. Menzel, *Surf. Sci.* 210, 282 (1989).

171. H. Conrad, R. Scala, W. Stenzel, and R. Unwin, *Surf. Sci.* 145, 1 (1984); G. E. Thomas and W. H. Weinberg, *Phys. Rev. Lett.* 41, 1181 (1978).

172. M. Stichler and D. Menzel, *Surf. Sci.* 391, 47 (1997).

173. E. Miyazaki, I. Kojima, M. Orita, K. Sawa, N. Sanada, K. Edamoto, T. Miyahara, and H. Kato, *J. Electron Spectrosc. Rel. Phenom.* 43, 139 (1987).

174. K. Edamoto, S. Maechama, E. Miyazaki, T. Migahana, and H. Kato, *Surf. Sci.* 204, L739 (1988).

175. S. K. So, R. Franchy, and W. Ho, *J. Chem. Phys.* 91, 5701 (1989).

176. P. S. Bagus and F. Illias, *Chem. Phys. Lett.* 224, 576 (1994).

177. K. C. Prince, G. Paolucci, and A. M. Bradshaw, *Surf. Sci.* 175, 101 (1986).

178. C. Backx, C. P. M. de Groot, and P. Biloen, *Surf. Sci.* 104, 300 (1981).

179. D. A. Outka, J. Stöhr, W. Jark, P. Stevens, J. Solomon, and R. J. Madix, *Phys. Rev. B* 35, 4119 (1987).

180. J. Stöhr, J. L. Gland, W. Eberhardt, D. Outka, R. J. Madix, F. Sette, R. J. Koestner, and U. Döbler, *Phys. Rev. Lett.* 51, 2414 (1983).

181. P. D. Nolan, B. R. Lutz, P. L. Tanaka, J. E. Davis, and C. B. Mullins, *J. Chem. Phys.* 111, 3696 (1999).

182. N. R. Avery, *Chem. Phys. Lett.* 96, 371 (1983).

183. H. Steininger, S. Lehwald, and H. Ibach, *Surf. Sci.* 123, 1 (1982).

184. J. L. Gland, B. A. Sexton, and G. B. Fischer, *Surf. Sci.* 95, 587 (1980).

185. B. C. Stipe, M. A. Rezaei, W. Ho, S. Gao, M. Persson, and B. I. Lundqvist, *Phys. Rev. Lett.* 78, 4410 (1997).

186. B. C. Stipe, M. A. Rezaei, and W. Ho, *Science* 279, 1907 (1998).

187. A. Eichler and J. Hafner, *Phys. Rev. Lett.* 79, 4481 (1997).

188. T. Sueyoshi, T. Sasaki, and Y. Iwasawa, *Surf. Sci.* 365, 310 (1996).

189. G. Pacchioni and P. S. Bagus, in "Elementary Reaction Steps in Heterogeneous Catalysis" (R. Joyner and R. van Santen, Eds.), NATO ASI Ser. C. Kluwer Academic, Dordrecht/Norwell, MA, USA, 1993.

190. M. Sano, Y. Seimiya, Y. Ohno, T. Matsushima, S. Tanaka, and M. Kamada, *Surf. Sci.* 421, 386 (1999).

191. P. A. Heiney, P. W. Stephens, S. G. J. Mochrie, J. Akimitsu, and R. J. Birgeneau, *Surf. Sci.* 125, 539 (1983).

192. W. Eberhardt and E. W. Plummer, *Phys. Rev. Lett.* 47, 1476 (1981).

193. P. A. Dowben, *CRC Critical Rev. Solid State Mater. Sci.* 13, 191 (1987).

194. P. A. Dowben, D. Mueller, T. N. Rhodin, and Y. Sakisaka, *Surf. Sci.* 155, 567 (1985).

195. D. Mueller, T. N. Rhodin, and P. A. Dowben, *Surf. Sci.* 164, 271 (1985).

196. C. W. Seabury, T. N. Rhodin, R. J. Purtell, and R. P. Merrill, *Surf. Sci.* 93, 117 (1980).

197. M. W. Kang, C. H. Li, S. Y. Tong, C. W. Seabury, T. N. Rhodin, R. J. Purtell, and R. P. Merrill, *Phys. Rev. Lett.* 47, 931 (1981).

198. A.-M. Lanzillotto, M. P. Dresser, M. D. Alvey, and J. T. Yates, *Surf. Sci.* 191, 15 (1987).

199. M. J. Dresser, A.-M. Lanzilloto, M. D. Alvey, and J. T. Yates, Jr., *Surf. Sci.* 191, 1 (1987).

200. T. E. Madey, J. E. Houston, C. W. Seabury, and T. N. Rhodin, *J. Vac. Sci. Technol.* 18, 476 (1981).

201. Zhi Xu, L. Hanley, and J. T. Yates, Jr., *J. Chem. Phys.* 96, 1621 (1992).

202. G. B. Fisher and G. E. Mitchell, *J. Electron Spectrosc. Rel. Phenom.* 29, 253 (1983).

203. R. J. Purtell, R. P. Merrill, C. W. Seabury, and T. N. Rhodin, *Phys. Rev. Lett.* 44, 1279 (1980).

204. K. Jacobi, E. S. Jensen, T. N. Rhodin, and R. P. Merrill, *Surf. Sci.* 108, 397 (1981).

205. I. C. Bassignana, K. Wagemann, J. Küppers, and G. Ertl, *Surf. Sci.* 175, 22 (1987).

206. C. J. Hirschmugl, K.-M. Schindler, O. Schaff, V. Fernandez, A. Theobald, Ph. Hofmann, A. M. Bradshaw, R. Davis, N. A. Booth, D. P. Woodruff, and V. Fritzsche, *Surf. Sci.* 352–354, 232 (1996).

207. C. Benndorf and T. E. Madey, *Surf. Sci.* 135, 164 (1983).

208. J. E. Parmeter, Y. Wang, C. B. Mullins, and W. H. Weinberg, *J. Chem. Phys.* 88, 5225 (1988).

209. Y. Zhou, Z.-M. Liu, and J. M. White, *J. Phys. Chem.* 97, 4182 (1993).

210. B. A. Sexton and G. E. Mitchell, *Surf. Sci.* 99, 523 (1980).

211. W. Erley and H. Ibach, *Surf. Sci.* 119, L357 (1982).

212. W. Erley and H. Ibach, *J. Electron Spectrosc. Rel. Phenom.* 29, 263 (1982).

213. J. L. Gland, B. A. Sexton, and G. E. Mitchell, *Surf. Sci.* 115, 623 (1982).

214. S. T. Ceyer and J. T. Yates, Jr., *Surf. Sci.* 155, 584 (1985).

215. M. Grunze, P. A. Dowben, and C. R. Brundle, *Surf. Sci.* 128, 311 (1983).

216. M.-C. Wiu, C. M. Truong, and D. W. Goodman, *J. Phys. Chem.* 97, 4182 (1993).

217. J. Lin, P. M. Jones, M. D. Lowery, R. R. Gay, S. L. Cohen, and E. I. Solomon, *Inorg. Chem.* 31, 686 (1992).

218. P. A. Thiel and T. E. Madey, *Surf. Sci. Rep.* 7, 211 (1987).

219. C. Nöbl, C. Benndorf, and T. E. Madey, *Surf. Sci.* 157, 29 (1985).

220. C. Benndorf and T. E. Madey, *Surf. Sci.* 194, 63 (1988).

221. J. G. Calvert, F. Su, J. W. Bottonheim, and O. P. Strausz, *Atmos. Environ.* 12, 197 (1987).

222. J. Haase, *J. Phys. Cond. Mat.* 9, 3647 (1997).

223. J. L. Solomon, R. J. Madix, W. Wurth, and J. Stöhr, *J. Phys. Chem.* 95, 3687 (1991).

224. D. A. Outka, R. J. Madix, G. B. Fisher, and C. DiMaggio, *Langmuir* 2, 406 (1986).

225. A. Gutiérrez-Sosa, J. F. Walsh, C. A. Muryun, G. Finetti, G. Thornton, A. W. Robinson, S. D'Addato, and S. P. Frigo, *Surf. Sci.* 364, L519 (1996).

226. J. A. Rodriguez, *Surf. Sci.* 226, 101 (1990).

227. M. L. Burke and R. J. Madix, *Surf. Sci.* 194, 223 (1988).

228. M. L. Burke and R. J. Madix, *J. Vac. Sci. Technol. A* 6, 789 (1988).

229. Y.-M. Sun, D. Sloan, D. J. Alberas, M. Kovar, Z.-J. Sun, and J. M. White, *Surf. Sci.* 319, 34 (1994).

230. N. Pangher, L. Wilde, M. Polcik, and J. Haase, *Surf. Sci.* 372, 211 (1997).

231. T. Nakahashi, S. Terada, T. Yokoyama, H. Hamamatsu, Y. Kitajima, M. Sakano, F. Matsui, and T. Ohta, *Surf. Sci.* 373, 1 (1997).

232. G. J. Jackson, S. M. Driver, D. P. Woodruff, N. Abrams, R. G. Jones, M. T. Butterfield, M. D. Crapper, B. C. C. Cowie, and V. Formoso, *Surf. Sci.* 459, 231 (2000).

233. D. A. Outka, R. J. Madix, and J. Stöhr, *Surf. Sci.* 164, 235 (1985).

234. J. Stöhr, D. A. Outka, R. J. Madix, and U. Döbler, *Phys. Rev. Lett.* 54, 1256 (1985).

235. P. Zebisch, M. Weinelt, and H.-P. Steinrück, *Surf. Sci.* 295, 295 (1993).

236. S. Terada, A. Imanishi, T. Yokoyama, S. Takenaka, Y. Kitajima, and T. Ohta, *Surf. Sci.* 336, 55 (1995).

237. L. Wilde, M. Polcik, J. Haase, B. Brena, D. Cocco, G. Comelli, and G. Paolucci, *Surf. Sci.* 405, 215 (1998).

238. G. J. Jackson, J. Lüdecke, S. M. Driver, D. P. Woodruff, R. G. Jones, A. Chan, and B. C. C. Cowie, *Surf. Sci.* 389, 223 (1997).

239. T. Yokoyama, S. Terada, S. Yagi, A. Imanishi, S. Takenaka, Y. Kitajima, and T. Ohta, *Surf. Sci.* 324, 25 (1995).

240. J. Wambach, G. Odörfer, H. J. Fruend, H. Kuhlenbeck, and M. Neumann, *Surf. Sci.* 209, 159 (1989).

241. S. Wohlrab, D. Ehrlich, J. Wambach, H. Kulenbeck, and H.-J. Freund, *Surf. Sci.* 220, 243 (1989).

242. A. Berkó and F. Solymosi, *Surf. Sci.* 171, L498 (1986).

243. B. Bartos, H.-J. Freund, H. Kuhelbeck, M. Neumann, H. Linder, and K. Müller, *Surf. Sci.* 179, 59 (1987).

244. B. Bartos, H.-J. Fruend, H. Kuhlenbeck, and M. Neumann, in "Kinetics of Interface Reactions" (M. Grunze and H.-J. Kreuzer, Eds.), Springer Series in Surface Science, Vol. 8, p. 164. Springer-Verlag, Berlin/New York, 1987.

245. H. Linder, D. Rupprecht, L. Hammer, and K. Muller, *J. Electron Spectrosc. Rel. Phenom.* 44, 59 (1987).

246. G. Illig, D. Heskett, E. W. Plumber, H.-J. Freund, J. Sommers, Th. Linder, A. M. Bradshaw, M. Buskotte, M. Neumann, U. Starke, K. Heinz, P. L. deAndres, D. Saldin, and J. B. Pendry, *Surf. Sci.* 220, 1 (1989).

247. H.-J. Freund, B. Bartos, H. Behner, G. Wedler, H. Kuhlenbeck, and M. Neumann, *Surf. Sci.* 180, 550 (1987).

248. H. Behner, W. Spiess, G. Wedler, D. Borgmann, and H.-J. Fruend, *Surf. Sci.* 184, 335 (1987).

249. H.-J. Fruend and M. W. Roberts, *Surf. Sci. Rep.* 25, 225 (1996).

250. M. E. Bartram, R. G. Windham, and B. E. Koel, *Surf. Sci.* 184, 57 (1987).

251. M. E. Bartram, R. G. Windam, and B. E. Koel, *Langmuir* 4, 240 (1988).

252. R. A. Van Santen and M. Neurock, *Catal. Rev. Sci. Eng.* 37, 557 (1995).

253. M. E. Bartram and B. E. Koel, *Surf. Sci.* 213, 137 (1989).

254. A. J. Vosper, *J. Chem. Soc. A* 625 (1970).

255. M. Herman, J. C. van Craen, J. van der Auwera, and G. W. Hills, *Chem. Phys. Lett.* 115, 445 (1985).

256. P. Hofmann and D. Menzel, *Surf. Sci.* 191, 353 (1987).

257. Th. Linder, J. Somers, A. M. Bradshaw, and G. P. Williams, *Surf. Sci.* 185, 75 (1987).

258. A. Pushman, J. Haase, M. D. Crapper, C. E. Riley, and D. P. Woodruff, *Phys. Rev. Lett.* 54, 2250 (1985).

259. M. D. Crapper, C. E. Riley, D. P. Woodruff, A. Puschmann, and J. Haase, *Surf. Sci.* 171, 1 (1986).

260. D. P. Woodruff, C. F. McConville, A. L. D. Kilcoyne, Th. Linder, J. Somers, M. Surman, G. Paolucci, and A. M. Bradshaw, *Surf. Sci.* 201, 228 (1989).

261. J. R. B. Gomer and J. A. N. F. Gomes, *Surf. Sci.* 432, 279 (1999).

262. M. Casarin, G. Granozzi, M. Sambi, E. Tondello, and A. Vittadini, *Surf. Sci.* 309, 95 (1994).

263. L. S. Caputi, G. Chiarello, M. G. Lancellotti, G. A. Rizzi, M. Sambi, and G. Granozzi, *Surf. Sci.* 291, L756 (1993).

264. P. A. Steven, R. J. Madix, and J. Stöhr, *Surf. Sci.* 230, 1 (1990).

265. T. S. Jones, M. R. Ashton, and N. V. Richardson, *J. Chem. Phys.* 90, 7564 (1989).

266. T. S. Jones and N. V. Richardson, *Surf. Sci.* 211/212, 377 (1989).

267. B. A. Sexton, *Surf. Sci.* 88, 299 (1979).

268. B. A. Sexton, *Appl. Phys. A* 26, 1 (1981).

269. Th. Linder, J. Somers, A. M. Bradshaw, A. L. D. Kilcoyne, and D. P. Woodruff, *Surf. Sci.* 203, 333 (1988).

270. K. M. Schindler, P. Hofmann, V. Fritzsche, S. Bao, S. Kulkarni, A. M. Bradshaw, and D. P. Woodruff, *Phys. Rev. Lett.* 71, 2054 (1993).

271. P. Hofmann, K. M. Schindler, S. Bao, V. Fritzsche, D. E. Ricken, A. M. Bradshaw, and D. P. Woodruff, *Surf. Sci.* 304, 74 (1994).

272. K. Hermann and C. Meyer, *Surf. Sci.* 277, 377 (1992).

273. M. A. Chesters and E. M. Cash, *Spectrochim. Acta A* 43, 1625 (1987).

274. R. Ryberg, *Phys. Rev. B* 31, 2545 (1985).

275. A. Peremans, F. Maseri, J. Darville, and J.-M. Giles, *J. Vac. Sci. Technol. A* 8, 3224 (1990).

276. B. A. Sexton, A. E. Hughes, and N. R. Avery, *Appl. Surf. Sci.* 22/23, 404 (1985).

277. B. A. Sexton, A. E. Hughes, and N. R. Avery, *Surf. Sci.* 155, 366 (1985).

278. P. Uvdal, M. K. Weldon, and C. M. Friend, *Phys. Rev. B* 50, 12258 (1984).

279. S. L. Miles, L. Bernasek, and J. L. Gland, *J. Phys. Chem.* 87, 1626 (1983).

280. C. Houtman and M. A. Barteau, *Langmuir* 6, 1558 (1990).

281. H. Yang, J. L. Whitten, J. S. Huberty, and R. J. Madix, *Surf. Sci.* 375, 268 (1997).

282. M. R. Albert, J.-P. Lu, S. L. Bernasek, and D. J. Dwyer, *Surf. Sci.* 221, 197 (1989).

283. J. L. Erskine and A. M. Bradshaw, *Chem. Phys. Lett.* 72, 260 (1980).

284. O. Schaff, G. Hess, V. Fritzsche, V. Fernandez, K. M. Schindler, A. Theobald, P. Hofmann, A. M. Bradshaw, R. Davis, and D. P. Woodruff, *Surf. Sci.* 333, 201 (1995).

285. O. Schaff, G. Hess, V. Fernandez, K. M. Schindler, A. Theobald, P. Hofmann, A. M. Bradshaw, V. Fritzsche, R. Davis, and D. P. Woodruff, *J. Electron Spectrosc. Rel. Phenom.* 75, 117 (1995).

286. K. Amemiya, Y. Kitajima, Y. Yonamoto, S. Terada, H. Tsukabayashi, T. Yokoyama, and T. Ohta, *Phys. Rev. B* 59, 2307 (1999).

287. H. Yang, J. L. Whitten, and C. M. Friend, *Surf. Sci.* 313, 295 (1994).

288. J. E. Demuth and H. Ibach, *Chem. Phys. Lett.* 60, 395 (1979).

289. H. Yang, J. L. Whitten, J. S. Huberty, and R. J. Madix, *Surf. Sci.* 375, 268 (1997).

290. S. Bao, C. F. McConville, and D. P. Woodruff, *Surf. Sci.* 187,133 (1987).

291. H. J. Kuhr, W. Ranke, and J. Finster, *Surf. Sci.* 178, 171 (1986).

292. G. L. Nyberg, T. Gengenbach, and J. Liesegang, *Physica Scripta* 41, 517 (1990).

293. S. Bao, C. F. McConville, and D. P. Woodruff, *Surf. Sci.* 187, 133 (1987).

294. F. P. Netzer and M. G. Ramsey, *Critical Rev. Solid State. Mat. Sci.* 17, 397 (1992).

295. B. A. Sexton and R. J. Madix, *Surf. Sci.* 105, 177 (1981).

296. M. S. Kariapper, C. Fisher, D. P. Woodruff, B. C. C. Cowie, and R. G. Jones, *J. Phys. Cond. Mat.* 12, 2153 (2000).

297. P. Hofmann, S. R. Bare, N. V. Richardson, and D. A. King, *Surf. Sci.* 133, L459 (1983).

298. G. L. Jackson, D. P. Woodruff, R. G. Jones, N. K. Singh, A. S. Y. Chan, B. C. C. Cowie, and V. Formoso, *Phys. Rev. Lett.* 84, 119 (2000).

299. M. S. Kariapper, G. F. Grom, G. L. Jackson, C. F. McConville, and D. P. Woodruff, *J. Phys. Cond. Mat.* 10, 8661 (1998).

300. N. P. Prince, D. L. Semour, D. P. Woodruff, R. G. Jones, and W. Walter, *Surf. Sci.* 215, 566 (1989).

301. D. L. Seymour, S. Bao, C. F. McConville, M. D. Crapper, D. P. Woodruff, and R. G. Jones, *Surf. Sci.* 189/190, 529 (1987).

302. D. R. Mullins, T. Tang, X. Chen, V. Shneerson, D. K. Saldin, and W. T. Tysoe, *Surf. Sci.* 372, 193 (1997).

303. B. A. Sexton and G. L. Nyberg, *Surf. Sci.* 165, 251 (1986).

304. D. R. Mullins, D. R. Huntley, T. Tang, D. K. Saladin, and W. T. Tysoe, *Surf. Sci.* 380, 468 (1997).

305. R. J. Koestner, J. Stöhr, J. L. Gland, E. B. Kollin, and F. Sette, *Chem. Phys. Lett.* 120, 285 (1985).

306. B. A. Sexton and N. R. Avery, *Surf. Sci.* 129, 21 (1983).

307. N. R. Avery and T. W. Matheson, *Surf. Sci.* 143, 110 (1984).

308. W. Erley, *J. Electron Spectrosc. Rel. Phenom.* 44, 65 (1987).

309. A. J. Capote, A. V. Hanza, N. D. S. Canning, and R. J. Madix, *Surf. Sci.* 175, 445 (1986).

310. P. A. Stevens, R. J. Madix, and J. Stöhr, *J. Chem. Phys.* 91, 4338 (1989).

311. H. Ibach and S. Lehwald, *J. Vac. Sci. Technol.* 15, 407 (1978).

312. L. L. Kesmodel, L. H. Dubois, and G. A. Somorjai, *Chem. Phys. Lett.* 56, 267 (1978).

313. L. L. Kesmodel, L. H. Dubois, and G. A. Somorjai, *J. Chem. Phys.* 70, 2180 (1979).

314. P. Skinner, M. W. Howard, I. A. Oxton, S. F. A. Kettle, D. B. Powell, and N. Sheppard, *J. Chem. Soc. Faraday Trans. II* 77, 1203 (1981).

315. H. Steininger, H. Ibach, and S. Lehwald, *Surf. Sci.* 117, 685 (1982).

316. M. A. Chesters and E. M. McCash, *Surf. Sci.* 187, L639 (1987).

317. M. R. Albert, L. G. Sneddon, W. Eberhardt, F. Grueter, T. Gustafsson, and E. W. Plumber, *Surf. Sci.* 120, 19 (1982).

318. D. R. Lloyd and F. P. Netzer, *Surf. Sci.* 129, L249 (1983).

319. W.-H. Hung and S. L. Bernasek, *Surf. Sci.* 339, 272 (1995).

320. J. E. Demuth and D. E. Eastman, *Phys. Rev. Lett.* 32, 1123 (1974).

321. D. E. Eastman and J. E. Demuth, *Jpn. Appl. Phys. Suppl.* 2, 827 (1974).

322. J. E. Demuth and D. E. Eastman, *Phys. Rev. B* 13, 1523 (1976).

323. M. Weinelt, W. Huber, P. Zebisch, H.-P. Steinrück, P. Ulbricht, U. Birkenheuer, J. C. Boettger, and N. Rösch, *J. Chem. Phys.* 102, 9709 (1995).

324. M. Weinelt, W. Huber, P. Zebisch, H.-P. Steinrück, B. Reichert, U. Birkenheuer, and N. Rösch, *Phys. Rev. B* 46, 1675 (1992).

325. H.-P. Steinrück, *Appl. Phys. A* 59, 517 (1994).

326. M. Weinelt, W. Huber, P. Zebisch, H.-P. Steinrück, M. Pabst, and N. Rösch, *Surf. Sci.* 271, 539 (1992).

327. M. Weinelt, P. Zebisch, W. Huber, M. Pabst, U. Birkenheuer, B. Reichert, N. Rösch, and H.-P. Steinrück, "Proceedings Symposium on Surface Science" (M. Alnot, J. J. Ehrhardt, C. Launois, B. Mutaftschiev, and M. R. Tempère, Eds.), La Plagne, Savoie, France, p. 276.

328. C. E. Anson, B. J. Bandy, M. A. Chesters, B. Keiller, I. A. Oxton, and N. Sheppard, *J. Electron Spectrosc. Rel. Phenom.* 29, 315 (1983).

329. L. Hammer, T. Hertlein, and K. Müller, *Surf. Sci.* 178, 693 (1986).

330. S. Lehwald and H. Ibach, *Surf. Sci.* 89, 425 (1979).

331. A. Fahmi and T. A. van Santen, *Surf. Sci.* 371, 53 (1997).

332. S. Bao, P. Hofmann, K. M. Schindler, V. Fritsche, A. M. Bradshaw, D. P. Woodruff, C. Casado, and M. C. Asensio, *Surf. Sci.* 307–309, 722 (1994).

333. S. Lehwald, H. Ibach, and H. Steininger, *Surf. Sci.* 117, 342 (1982).

334. H. Ibach, H. Hopster, and B. Sexton, *Appl. Phys.* 14, 21 (1977).

335. H. Ibach and H. Hopster, *Appl. Surf. Sci.* 1, 1 (1977).

336. A. M. Baro and H. Ibach, *J. Chem. Phys.* 74, 4194 (1981).

337. J. Kubota, S. Ichihara, J. Kondo, K. Domen, and C. Hirose, *Langmuir* 12, 1926 (1996).

338. J. Kubota, S. Ichichara, J. N. Kondo, K. Domen, and C. Hirose, *Surf. Sci.* 357–358, 634 (1996).

339. J. A. Gates and L. L. Kesmodel, *Surf. Sci.* 120, L461 (1982).

340. J. A. Gates and L. L. Kesmodel, *Surf. Sci.* 124, 68 (1982).

341. B. Tardy and J. C. Bertolini, *J. Chem. Phys.* 82, 407 (1985).

342. M. A. Chesters, G. S. McDougall, M. E. Pemble, and N. Sheppard, *Appl. Surf. Sci.* 22/23, 369 (1985).

343. J. Yoshinobu, T. Sekitani, M. Onchi, and M. Nishijima, *J. Electron Spectrosc. Rel. Phenom.* 54/55, 697 (1990).

344. H. Okuyama, S. Ichihara, H. Ogasawara, H. Kato, T. Komeda, M. Kawai, and J. Yoshinobu, *J. Chem. Phys.* 112, 5948 (2000).

345. C. Nyberg, C. G. Tengstal, S. Andersson, and M. W. Holmes, *Chem. Phys. Lett.* 87, 87 (1982).

346. U. Seip, M.-C. Tsai, J. Küppers, and G. Ertl, *Surf. Sci.* 147, 65 (1984).

347. W. Erley, A. M. Baro, and H. Ibach, *Surf. Sci.* 120, 273 (1982).

348. F. Zaera and R. B. Hall, *Surf. Sci.* 180, 1 (1987).

349. F. Zaera and R. B. Hall, *J. Phys. Chem.* 91, 4318 (1987).

350. M. A. Barteau, J. Q. Broughton, and D. Menzel, *Appl. Surf. Sci.* 19, 92 (1984).

351. G. H. Hatzikos and R. I. Masel, *Surf. Sci.* 185, 479 (1987).

352. C. Becker, T. Pelster, M. Tanemura, J. Breitbach, and K. Wandelt, *Surf. Sci.* 433–435, 822 (1999).

353. S. Ichihara, J. Yoshinobu, H. Ogasawara, M. Nantoh, M. Kawai, and K. Domen, *J. Electron Spectrosc. Rel. Phenom.* 88–91, 1003 (1998).

354. F. Matsui, H. W. Yeom, A. Imanishi, K. Isawa, I. Matsuda, and T. Ohta, *Surf. Sci.* 401, L413 (1998).

355. Y. Chen, J. Schmidt, L. Siller, J. C. Barnard, and R. E. Palmer, *Phys. Rev. B* 58, 1177 (1998).

356. C. M. Goringe, L. J. Clarke, H. M. Lee, M. C. Payne, I. Stich, J. A. White, M. J. Gillan, and A. P. Sutton, *J. Phys. Chem. B* 101, 1498 (1997).

357. S. Bao, K.-M Schindler, Ph. Hofmann, V. Fritzsche, A. M. Bradshaw, and D. P. Woodruff, *Surf. Sci.* 291, 295 (1993).

358. A. Clotet and G. Pacchioni, *Surf. Sci.* 346, 91 (1996).

359. H. Hofmann, F. Zaera, R. M. Ormerod, R. M. Lambert, J. M. Yao, D. K. Saldin, L. P. Wang, D. W. Bennett, and W. T. Tysoe, *Surf. Sci.* 268, 1 (1992).

360. K. Hermann and M. Witko, *Surf. Sci.* 337, 205–214 (1995).

361. J. Yoshinobu, T. Sekitani, M. Onchi, and M. Nishijima, *J. Phys. Chem.* 94, 4269 (1990).

362. M. Ohno and W. von Niessen, *Phys. Rev. B* 55, 5466 (1997).

363. M. G. Ramsey, G. Rosina, F. P. Netzer, H. B. Saalfeld, and D. R. Lloyd, *Surf. Sci.* 217, 140 (1989).

364. F. P. Netzer, *Vacuum* 41, 49 (1990).

365. M. E. Kordesch, W. Stenzel, and H. Conrad, *J. Electron Spectrosc. Rel. Phenom.* 39, 89 (1986).

366. K. Besenthal, G. Chiarello, M. E. Kordesch, and H. Conrad, *Surf. Sci.* 178, 667 (1986).

367. M. E. Kordesch, W. Stenzel, and H. Conrad, *Surf. Sci.* 186, 601 (1987).

368. M. E. Kordesch, Th. Linder, J. Somers, H. Conrad, A. M. Bradshaw, and G. P. Williams, *Spectrochim. Acta* 43A, 11567 (1987).

369. M. G. Ramsey, D. Steinmüller, F. P. Netzer, N. Rösch, L. Ackermann, and G. Rosina, *Surf. Sci.* 260, 163–174 (1992).

370. N. Rösch, Th. Fox, F. P. Netzer, M. G. Ramsey, and D. Steinmüller, *J. Chem. Phys.* 94, 3276 (1991).

371. M. G. Ramsey, D. Steinmüller, F. P. Netzer, S. Kostlmeier, J. Lauber, and N. Rösch, *Surf. Sci.* 309, 82 (1994).

372. L. Ackermann and N. Rösch, *Chem. Phys.* 168, 259 (1992).

373. R. I. R. Blyth, I. Kardinal, F. P. Netzer, D. Chrysostomou, and D. R. Lloyd, *Surf. Sci.* 415, 227 (1998).

374. M. G. Ramsey, G. Rosina, F. P. Netzer, H. B. Saalfeld, and D. R. Lloyd, *Surf. Sci.* 218, 317 (1989).

375. J. Somers, M. E. Kordesch, R. Hemmen, Th. Linder, H. Conrad, and A. M. Bradshaw, *Surf. Sci.* 198, 400 (1988).

376. M. E. Kordesch, W. Stenzel, H. Conrad, and M. J. Weaver, *J. Am. Chem. Soc.* 109, 1878 (1987).

377. M. E. Kordesch, W. Feng, W. Stenzel, M. J. Weaver, and H. Conrad, *J. Electron Spectrosc. Rel. Phenom.* 44, 149 (1987).

378. K. Hermann, W. Müller, and P. S. Bagus, *J. Electron Spectrosc. Rel. Phenom.* 39, 107 (1986).

379. M. R. Philpott, P. S. Bagus, C. J. Nelin, and H. Seki, *J. Electron Spectrosc. Rel. Phenom.* 45, 169 (1987).

380. X.-Y. Zhou, D.-H. Shi, and P.-L. Cao, *Surf. Sci.* 223, 393 (1989).

381. M. Connolly, T. McCabe, D. R. Lloyd, and E. Taylor, in "Structure and Reactivity of Surfaces" (C. Morterra, A. Zecchina, and G. Costa, Eds.), p. 30. Elsevier, New York, 1989.

382. J. Somers, M. E. Kordesch, R. Hemmen, Th. Linder, H. Conrad, and A. M. Bradshaw, *Surf. Sci.* 198, 400 (1988).

383. P. Mills, D. Jentz, and M. Trenary, *J. Am. Chem. Soc.* 119, 9002 (1997).

384. W. Huber, M. Weinelt, P. Zebisch, and H.-P. Steinrück, *Surf. Sci.* 253, 72 (1991).

385. M. Doering, H.-P. Rust, B. G. Riner, and A. M. Bradshaw, *Surf. Sci.* 410, L736 (1998).

386. J. H. Kang, R. L. Toomes, J. Robinson, D. P. Woodruff, O. Schaff, R. Terborg, R. Lindsay, P. Baemgartel, and A. M. Bradshaw, *Surf. Sci.* 448, 23 (2000).

387. H.-P. Steinrück, *J. Phys. Cond. Mat.* 8, 6465 (1996).

388. M. G. Ramsey, D. Steinmüller, F. P. Netzer, T. Schedel, A. Santianello, and D. R. Lloyd, *Surf. Sci.* 251/252, 979 (1991).

389. P. Zebisch, M. Stichler, P. Trischberger, M. Weinelt, and H. P. Steinrück, *Surf. Sci.* 396, 61 (1998).

390. A. M. Bradshaw, *Current Opinion Solid State Mater. Sci.* 2, 530 (1997).

391. F. P. Netzer, G. Rangelov, G. Rosina, H. B. Saalfeld, M. Neumann, and D. R. Lloyd, *Phys. Rev. B* 37, 10399 (1988).

392. M. Neuber, F. Schneider, C. Zubrägel, and M. Neumann, *J. Phys. Chem.* 99, 9160 (1995).

393. P. Hoffman, K. Horn, and A. M. Bradshaw, *Surf. Sci.* 105, L260 (1981).

394. G. L. Nyberg and N. V. Richardson, *Surf. Sci.* 85, 335 (1979).

395. H. Othani, M. A. van Hove, and G. A. Somorjai, *J. Phys. Chem.* 92, 3974 (1988).

396. A. Barbieri, M. A. van Hove, and G. A. Somorjai, *Surf. Sci.* 306, 261 (1994).

397. F. P. Netzer and J. U. Mack, *J. Chem. Phys.* 79, 1017 (1983).

398. G. D. Waddill and L. L. Kesmodel, *Phys. Rev. B* 31, 4940 (1985).

399. J. Somers, M. E. Bridge, D. R. Lloyd, and T. McCabe, *Surf. Sci.* 181, L167 (1987).

400. D. F. Ogletree, M. A. van Hove, and G. A. Somorjai, *Surf. Sci.* 183, 1 (1987).

401. A. Wander, G. Held, R. Q. Hwang, G. S. Blackman, M. L. Xu, P. de Andres, M. A. van Hove, and G. A. Somorjai, *Surf. Sci.* 249, 21 (1991).

402. R. Dudde, K.-H. Frank, and E. E. Koch, *Surf. Sci.* 225, 267 (1990).

403. P. Yannoulis, R. Dudde, K. H. Frank, and E. E. Koch, *Surf. Sci.* 217, 103 (1989).

404. M. Neumann, J. U. Mack, E. Bertel, and F. P. Netzer, *Surf. Sci.* 155, 629 (1985).

405. E. Bertel, G. Rosina, and F. P. Netzer, *Surf. Sci.* 172, L515 (1986).

406. F. P. Netzer, G. Rosina, E. Bertel, and H. Saalfeld, *Surf. Sci.* 184, L397 (1987).

407. G. Rosina, G. Rangelov, E. Bertel, H. Saalfeld, and F. P. Netzer, *Chem. Phys. Lett.* 140, 200 (1987).

408. C. T. Kao, C. M. Mate, G. S. Blackman, B. E. Bent, M. A. van Hove, and G. A. Somorjai, *J. Vac. Sci. Technol. A* 6, 786 (1988).

409. S. Chiang, R. J. Wilson, C. M. Mate, and H. Othani, *J. Microsc.* 152, 567 (1988).

410. R. F. Lin, G. S. Blackman, M. A. van Hove, and G. A. Somorjai, *Acta Crystallogr. B* 43, 368 (1987).

411. M. A. van Hove, R. F. Lin, and G. A. Somorjai, *J. Am. Chem. Soc.* 108, 2532 (1986).

412. J. U. Mack, E. Bertel, and F. P. Netzer, *Surf. Sci.* 159, 265 (1985).

413. R. Duschek, F. Mittendorfer, R. I. R. Blyth, F. P. Netzer, J. Hafner, and M. G. Ramsey, *Chem. Phys. Lett.* 318, 43 (2000).

414. W. Huber, H.-P. Steinrück, T. Pache, and D. Menzel, *Surf. Sci.* 217, 103 (1991).

415. W. Huber, P. Zebisch, T. Bornemann, and H.-P. Steinrück, *Surf. Sci.* 258, 16 (1991).

416. O. Schaff, V. Fernandez, P. Hofmann, K.-M. Schindler, A. Theobald, V. Fritsche, A. M. Bradshaw, R. Davids, and D. P. Woodruff, *Surf. Sci.* 348, 89 (1996).

417. G. Held, M. P. Bessent, S. Titmuss, and D. A. King, *J. Chem. Phys.* 105, 11305 (1996).

418. M. Weinelt, N. Wassdahl, T. Wiell, O. Karis, J. Hasselström, P. Bennich, A. Nilsson, J. Stöhr, and M. Samant, *Phys. Rev. B* 58, 7351 (1998).

419. C. H. Patterson and R. M. Lambert, *J. Phys. Chem.* 92, 1266 (1988).

420. P. Jacob and D. Menzel, *Surf. Sci.* 201, 503 (1988).

421. C. Stellwag, G. Held, and D. Menzel, *Surf. Sci.* 325, L379 (1995).

422. H. H. Graen, M. Neuber, M. Neumann, G. Illig, H.-J. Freund, and F. P. Netzer, *Surf. Sci.* 223, 33 (1989).

423. H. H. Graen, M. Neuber, M. Neaumann, G. Odörfer, and H. J. Freund, *Europhys. Lett.* 12, 173 (1990).

424. F. P. Netzer, H. H. Graen, H. Kuhlenbeck, and M. Neumann, *Chem. Phys. Lett.* 133, 49 (1987).

425. H. H. Graen, M. Neumann, J. Wambach, and H.-J. Freund, *Chem. Phys. Lett.* 165, 137 (1990).

426. J. R. Lomas, C. R. Baddeley, M. S. Tikhov, and R. M. Lambert, *Chem. Phys. Lett.* 263, 591 (1996).

427. F. A. Grimm and D. R. Huntley, *J. Phys. Chem.* 97, 3800 (1993).

428. L. G. M. Pettersson, H. Agren, Y. Luo, and L. Triguero, *Surf. Sci.* 408, 1 (1998).

429. H. Koschel, H. Held, P. Trischberger, W. Widdra, and H. P. Steinrück, *Surf. Sci.* 437, 125 (1999).

430. J. Eiding, W. Domcke, W. Huber, and H.-P. Steinrück, *Chem. Phys. Lett.* 180, 133 (1991).

431. A. C. Liu and C. M. Friend, *J. Chem. Phys.* 89, 4396 (1988).

432. Y. Ma, F. Sette, G. Meigs, S. Modesti, and C. T. Chen, *Phys. Rev. Lett.* 63, 2044 (1989).

433. G. L. Nyberg, S. R. Bare, P. Hofmann, D. A. King, and M. Surman, *Appl. Surf. Sci.* 22/23, 392 (1985).

434. M. Surman, S. R. Bare, P. Hofmann, and D. A. King, *Surf. Sci.* 179, 243 (1987).

435. G. Polzonetti, V. Carravetta, M. V. Russo, G. Contini, P. Parent, and C. Laffon, *J. Electron Spectrosc. Rel. Phenom.* 98–99, 175 (1999).

436. M. Carbone, M. N. Piancastelli, R. Zanoni, G. Comtet, G. Dujardin, and L. Hellner, *Surf. Sci.* 407, 275 (1998).

437. M. J. Kone, A. V. Teplyakov, J. G. Lyubovitsky, and S. F. Bent, *Surf. Sci.* 411, 286 (1998).

438. F. P. Netzer and J.-U. Mack, *Chem. Phys. Lett.* 95, 492 (1983).

439. B. J. Bandy, D. R. Lloyd, and N. V. Richardson, *Surf. Sci.* 89, 344 (1979).

440. G. L. Nyberg, *Surf. Sci.* 95, L273 (1980).

441. F. P. Netzer, G. Rangelov, G. Rosina, and H. B. Saalfeld, *J. Chem. Phys.* 89, 3331 (1988).

442. T. Giessel, O. Schaff, R. Lindsay, P. Baumgartel, M. Polcik, A. M. Bradshaw, A. Koebbel, T. McCabe, M. Bridge, D. R. Lloyd, and D. P. Woodruff, *J. Chem. Phys.* 110, 9666 (1999).

443. J. F. Walsh, H. S. Dhariwal, A. Gutierrez-Sosa, P. Finetti, C. A. Muryn, N. B. Brookes, R. J. Oldman, and G. Thornton, *Surf. Sci.* 415, 423 (1998).

444. F. P. Netzer, G. Rosina, E. Bertel, and H. B. Saalfeld, *J. Electron Spectrosc. Rel. Phenom.* 46, 373 (1988).

445. W. W. Pai, Z. Y. Zhang, J. D. Zhang, and J. F. Wendelken, *Surf. Sci.* 393, L106 (1997).

446. S. Lee, D. Li, P. A. Dowben, F. K. Perkins, M. Onellion, and J. T. Spencer, *J. Am. Chem. Soc.* 113, 8444 (1991).

447. H. Zeng, D. Byun, J. Zhang, G. Vidali, M. Onellion, and P. A. Dowben, *Surf. Sci.* 313, 239 (1994).

448. F. K. Perkins, M. Onellion, S. Lee, D. Li, J. Mazurowski, and P. A. Dowben, *Appl. Phys. A* 54, 442 (1992).

449. J. M. Carpinelli, E. W. Plummer, D. Byun, and P. A. Dowben, *J. Vac. Sci. Technol. B* 13, 1203 (1995).

450. D. E. Wilk, C. D. Stanners, Y. R. Shen, and G. A. Somorjai, *Surf. Sci.* 280, 298 (1993).

451. J. Zhang, D. N. McIlroy, P. A. Dowben, H. Zeng, G. Vidali, D. Heskett, and M. Onellion, *J. Phys. Cond. Mat.* 7, 7185–7194 (1995).

452. N. M. Boag and P. A. Dowben, "Metallized Plastics 4: Fundamental and Applied Aspects" (K. L. Mittal, Ed.). Plenum, New York, 1997.

453. J. T. Spencer, "Progress in Inorganic Chemistry" (K. D. Karlin, Ed.), Vol. 41, p. 145, 1994.

454. P. A. Dowben, J. T. Spencer, and G. T. Stauf, *Mater. Sci. Eng. B* 2, 297 (1989).

455. A. A. Zinn, L. Brandt, H. D. Kaesz, and R. F. Hicks, in "The Chemistry of Metal CVD" (T. Kodas and M. Hampden-Smith, Eds.), p. 329. VCH, New York, 1994.

456. D. Welipitiya, C. Waldfried, C. N. Borca, P. A. Dowben, N. M. Boag, H. Jiang, I. Gobulukoglu, and B. W. Robertson, *Surf. Sci.* 418, 466 (1998).

457. C. Waldfried, D. Welipitiya, C. W. Hutchings, H. S. V. de Silva, G. A. Gallup, P. A. Dowben, W. W. Pai, J. Zhang, J. F. Wendelken, and N. M. Boag, *J. Phys. Chem. B* 101, 9782 (1997).

458. P. A. Dowben, C. Waldfried, T. Komesu, D. Welipitiya, T. McAvoy, and E. Vescovo, *Chem. Phys. Lett.* 283, 44 (1998).

459. D. Welipitiya, P. A. Dowben, J. Zhang, W. W. Pai, and J. F. Wendelken, *Surf. Sci.* 367, 20 (1996).

460. C. M. Woodbridge, D. L. Pugmire, R. C. Johnson, N. M. Boag, and M. A. Langell, *J. Phys. Chem. B* 104, 3085 (2000).

461. P. J. Durston and R. E. Palmer, *Surf. Sci.* 400, 277 (1998).

462. D. L. Pugmire, C. M. Woodbridge, and M. A. Langell, *Surf. Sci.* 411, L844 (1998).

463. D. L. Pugmire, C. M. Woodbridge, S. Root, and M. A. Langell, *J. Vac. Sci. Technol. A* 17, 1581 (1999).

464. D. L. Pugmire, C. M. Woodbridge, N. M. Boag, and M. A. Langell, *Surf. Sci.* 472, 155 (2001).

465. D. Welipitiya, C. N. Borca, C. Waldfried, C. Hutchings, L. Sage, C. M. Woodbridge, and P. A. Dowben, *Surf. Sci.* 393, 34 (1997).

466. J. Choi and P. A. Dowben, in preparation.

467. D. L. Pugmire, Ph.D. Thesis, University of Nebraska, 2000.

468. C. N. Borca, D. Welipitiya, P. A. Dowben, and N. M. Boag, *J. Phys. Chem. B* 104, 1047 (2000).

469. T. Permien, R. Engelhardt, C. A. Feldman, and E. E. Koch, *Chem. Phys. Lett.* 98, 527 (1983).

470. N. V. Richardson, *Chem. Phys. Lett.* 102, 390 (1983).

471. S. Hasegawa, S. Tanaka, Y. Yamashita, H. Inokuchi, H. Fujimoto, K. Kamiya, K. Seki, and N. Ueno, *Phys. Rev. B* 48, 2596 (1993).

472. W. G. Grobman, *Phys. Rev. B* 17, 4573 (1978).

473. K. Kamiya, M. Momose, A. Kitamura, Y. Harada, N. Ueno, T. Miyazaki, S. Hasegawa, H. Inokuchi, S. Narioka, H. Ishii, and K. Seki, *Mol. Cryst. Liq. Cryst.* 267, 211 (1995).

474. N. Ueno, *J. Electron Spectrosc. Rel. Phenom.* 78, 345 (1996).

475. N. Ueno, K. Suzuki, S. Hasegawa, K. Kamiya, K. Seki, and H. Inokuchi, *J. Chem. Phys.* 99, 7169 (1993).

476. K. K. Okudaira, S. Hasegawa, H. Ishii, K. Seki, Y. Harada, and N. Ueno, *J. Appl. Phys.* 85, 6453 (1999).

477. P. H. Lippel, R. J. Wilson, M. D. Miller, Ch. Wöll, and S. Chiang, *Phys. Rev. Lett.* 62, 171 (1989).

478. X. Lu, W. Hipps, X. D. Wang, and U. Mazur, *J. Am. Chem. Soc.* 118, 7198 (1996).

479. M. Aoki, S. Masuda, Y. Einaga, K. Kamiya, A. Kitamura, M. Momose, N. Ueno, Y. Harada, T. Miyazaki, S. Hasegawa, H. Inokuchi, and K. Seki, *J. Electron Spectrosc. Rel. Phenom.* 76, 259 (1996).

480. N. Ueno, Y. Azuma, T. Yokota, M. Aoki, K. K. Okudaira, and Y. Harada, *Jpn. J. Appl. Phys.* 36, 5731 (1997).

481. Y. Azuma, T. Yokota, S. Kera, M. Aoki, K. K. Okudaira, Y. Harada, and N. Ueno, *Thin Solid Films* 327–329, 303 (1998).

482. S. Kera, A. Abduaini, M. Aoki, K. K. Okudaira, N. Ueno, Y. Harada, Y. Shirota, and T. Tsuzuki, *Thin Solid Films* 327–329, 278 (1998).

483. S. Kera, A. Abduaini, M. Aoki, K. K. Okudaira, N. Ueno, Y. Harada, Y. Shirota, and T. Tsuzuki, *J. Electron Spectrosc. Rel. Phenom.* 88–91, 885 (1998).

484. N. Ueno, K. Kamiya, K. Ogawa, H. Yonehara, M. Takahashi, H. Nakahara, K. Seki, K. Sugita, K. Fukuda, and H. Inokuchi, *Thin Solid Films* 210/211, 678 (1992).

485. K. Ogawa, S. Kinoshita, H. Yonehara, H. Nakahara, and K. Fukuda, *J. Chem. Soc. Chem. Comm.* 478 (1989).

486. K. Ogawa, H. Yonehara, T. Shoji, S. Kinoshita, and E. Maekawa, *Thin Solid Films* 178, 439 (1989).

487. S. Narioka, M. Sei, H. Ishii, S. Hasegawa, Y. Ouchi, T. Ohta, and K. Seki, *Trans. Mater. Res. Soc. Jpn.* 15A, 631 (1994).

488. S. Narioka, H. Ishii, Y. Ouchi, T. Yokoyama, T. Ohta, and K. Seki, *J. Phys. Chem.* 99, 1332 (1995).

489. J. K. Gimzewski, T. A. Jung, M. T. Cuberes, and R. R. Schlittler, *Surf. Sci.* 386, 101 (1997).

490. H. Ozaki, T. Magara, and Y. Mazaki, *J. Electron Spectrosc. Rel. Phenom.* 88–91, 867 (1998).

491. F. P. Falko, *Chem. Phys. Lett.* 146, 566 (1988).

492. P. Yannoulis, R. Dudde, K. H. Frank and E. E. Koch, *Surf. Sci.* 189/190, 519 (1987).

493. J. Taborski, V. Wüstenhagen, P. Väterlein, and E. Umbach, *Chem. Phys. Lett.* 239, 380 (1995).

494. E. Umbach, C. Seidel, J. Taborski, R. Li, and A. Soukopp, *Phys. Status Solidi B* 192, 389 (1995).

495. P. Yannoulis, K. H. Frank, and E. E. Koch, *Surf. Sci.* 241, 325 (1991).

496. P. Yannoulis, E. E. Koch, and M. Lähdeniemi, *Surf. Sci.* 192, 299 (1987).

497. S. Hasegawa, H. Inokuchi, K. Seki, and N. Ueno, *J. Electron Spectrosc. Rel. Phenom.* 78, 391 (1996).

498. U. Zimmermann, G. Schnitzler, V. Wustenhagen, N. Karl, R. Dudde, E. E. Koch, and E. Umbach, *Mol. Cryst. Liquid Cryst.* 339, 231 (2000).

499. S. R. Forrest, P. E. Burrows, E. I. Haskal, and F. F. So, *Phys. Rev. B* 49, 11309 (1994).

500. A. Schmidt, T. J. Schuerlein, G. E. Collins, and N. R. Armstrong, *J. Phys. Chem.* 99, 11770 (1995).

501. J. Taborski, P. Väterlein, H. Dietz, U. Zimmermann, and E. Umbach, *J. Electron Spectrosc. Rel. Phenom.* 75, 129 (1997).

502. E. Umbach, *Progr. Surf. Sci.* 35, 113 (1991).

503. M. Jung, U. Baston, G. Schnitzler, M. Kaiser, J. Papst, T. Porwol, H. J. Freund, and E.Umbach, *J. Mol. Struct.* 293, 239 (1993).

504. K. Glöckler, C. Seidel, A. Soukopp, M. Sokolowski, E. Umbach, M. Böhringer, R. Berndt, and W.-D. Schneider, *Surf. Sci.* 405, 1 (1998).

505. Y. Azuma, T. Hasebe, T. Miyamae, K. K. Okudaira, Y. Harada, K. Seki, E. Morikawa, V. Saile, and N. Ueno, *J. Synchrotron Rad.* 5, 1044 (1998).

506. Y. Hirose, W. Chen, E. I. Haskal, S. R. Forrest, and A. Kahn, *Appl. Phys. Lett.* 64, 3482 (1994).

507. Y. Hirose, A. Kahn, V. Aristov, P. Soukiassian, V. Bulovic, and S. R. Forrest, *Phys. Rev. B* 54, 13748 (1996).

508. D. Gador, C. Buchberger, and R. Fink, *Europhys. Lett.* 41, 231 (1998).

509. D. Gador, C. Buchberger, R. Fink, and E. Umbach, *J. Electron Spectrosc. Rel. Phenom.* 96, 11 (1998).

510. D. Gador, Y. Zou, C. Buchberger, M. Bertram, R. Fink, and E. Umbach, *J. Electron Spectrosc. Rel. Phenom.* 101–103, 523 (1999).

511. H. Oji, R. Mitsumoto, E. Ito, H. Ishii, Y. Ouchi, K. Seki, and N. Kosugi, *J. Electron Spectrosc. Rel. Phenom.* 78, 379 (1996).

512. H. Oji, R. Mitsumoto, E. Ito, H. Ishii, Y. Ouchi, K. Seki, T. Yokoyama, T. Ohta, and N. Kosugi, *J. Chem. Phys.* 109, 10410 (1998).

513. H. Ishii, S. Narioka, K. Edamatsu, K. Kamiya, S. Hasegawa, T. Ohta, N. Ueno, and K. Seki, *J. Electron Spectrosc. Rel. Phenom.* 76, 553 (1995).

514. S. Narioka, H. Ishii, K. Edamatsu, K. Kamiya, S. Hasegawa, T. Ohta, N. Ueno, and K. Seki, *Phys. Rev. B* 52, 2362 (1995).

515. W. R. Salaneck, M. Fahlman, C. Lapersonne-Meyer, J.-L. Fave, M. Schott, M. Loegdlund, and J. L. Bredas, *Synth. Met.* 67, 309 (1994).

516. H. Schurmann, N. Koch, A. Vollmer, S. Schrader, and M. Neumann, *Synth. Met.* 111, 591 (2000).

517. A. Tsumura, H. Koezuka, and T. Ando, *Appl. Phys. Lett.* 49, 1210 (1986).

518. T. Okajima, S. Narioka, S. Tanimura, K. Hamano, T. Kurata, Y. Uehara, T. Araki, H. Ishii, Y. Ouchi, K. Seki, T. Ogama, and H. Koezuka, *J. Electron Spectrosc. Rel. Phenom.* 78, 383 (1996).

519. A. Soukopp, C. Seidel, R. Li, M. Bässler, M. Sokolowski, and E. Umbach, *Thin Solid Films* 284–285, 343 (1996).

520. R. Li, P. Bäuerle, and E. Umbach, *Surf. Sci.* 331–333, 100 (1995).

521. A. Soukopp, K. Glockler, P. Kraft, S. Schmitt, M. Sokolowski, E. Umbach, E. Mena-Osteritz, P. Bauerle, and E. Hadicke, *Phys. Rev. B* 58, 13882 (1998).

522. A. Soukopp, K. Glöckler, P. Bäuerle, M. Sokolowski, and E. Umbach, *Adv. Mater.* 8, 902 (1996).

523. Y. Yamamoto, R. Mitsumoto, T. Araki, Y. Ouchi, H. Ishii, K. Seki, N. Ueno, and Y. Takanishi, *J. Phys. IV France* 7 (C2), 709 (1997).

524. Y. Yamamoto, R. Mitsumoto, E. Ito, T. Araki, Y. Ouchi, K. Seki, and Y. Takanishi, *J. Electron Spectrosc. Rel. Phenom.* 78, 367 (1996).

525. D. Yoshimura, H. Ishii, Y. Ouchi, E. Ito, T. Miyamae, S. Hasegawa, N. Ueno, and K. Seki, *J. Electron Spectrosc. Rel. Phenom.* 88–91, 875 (1998).

526. D. Yoshimura, H. Ishii, Y. Ouchi, E. Ito, T. Miyamae, S. Hasegawa, K. K. Okudaira, N. Ueno, and K. Seki, *Phys. Rev. B* 60, 9046 (1999).

527. H. Ishii, E. Morikawa, S. J. Tang, D. Yoshimura, E. Ito, K. Okudaira, T. Miyamae, S. Hasgawa, P. T. Sprunger, N. Ueno, K. Seki, and V. Saile, *J. Electron Spectrosc. Rel. Phenom.* 101–103, 559 (1999).

528. L. E. Firment and G. A. Somorjai, *J. Chem. Phys.* 66, 2901 (1977).

529. R. Dudde and B. Reihl, *Chem. Phys. Lett.* 196, 91 (1992).

530. K. Seki, E. Ito, and H. Ishii, *Synth. Met.*, in press.

531. K. Seki, N. Ueno, O. Uif, R. Karlson, R. Engelhardt, and E. E. Koch, *Chem. Phys.* 105, 247 (1986).

532. Ch. Zubägel, F. Schneider, M. Neumann, G. Hähner, Ch. Wöll, and M. Gunze, *Chem. Phys. Lett.* 219, 127 (1994).

533. N. Ueno, K. Seki, N. Sato, H. Fujimoto, T. Kuramouchi, K. Sugita, and H. Inokuchi, *Phys. Rev. B* 41, 1176 (1990).

534. H. Rieley, G. K. Kendall, A. Chan, R. G. Jones, J. Lüdecke, D. P. Woodruff, and B. C. C. Cowie, *Surf. Sci.* 392, 143 (1997).

535. Y. Yamashita, S. Tanaka, K. Imaeda, and H. Inokuchi, *Chem. Phys. Lett.* 7, 1213 (1991).

536. G. Horowitz, X. Peng, D. Fichou, and F. Ganier, *J. Appl. Phys.* 67, 528 (1990).

537. K. Imaeda, Y. Yamashita, Y. Li, T. Mori, H. Inokuchi, and M. Sano, *J. Mater. Chem.* 2, 115 (1992).

538. H. Fujimoto, K. Kamiya, S. Tanaka, T. Mori, Y. Yamashita, H. Inokuchi, and K. Seki, *Chem. Phys.* 165, 135 (1992).

539. S. Hasegawa, T. Mori, K. Imaeda, S. Tanaka, Y. Yamashita, H. Inokuchi, H. Fujimoto, K. Seki, and N. Ueno, *J. Chem. Phys.* 100, 6969 (1994).

540. N. Ueno, A. Kitamura, K. K. Okudaira, T. Miyamae, Y. Harada, S. Hasegawa, H. Ishii, H. Inokuchi, T. Fujikawa, T. Miyazaki, and K. Seki, *J. Chem. Phys.* 107, 2079 (1997).

541. D. Gador, C. Buchberger, A. Soukopp, M. Sokolowski, R. Fink, and E. Umbach, *J. Electron Spectrosc. Rel. Phenom.* 101–103, 529 (1999).

542. M. Bässler, R. Fink, C. Heske, J. Müller, P. Väterlein, J. U. von Schütz, and E. Umbach, *Thin Solid Films* 284–285, 234 (1996).

543. C. Seidel, H. Kopf, and H. Fuchs, *Phys. Rev. B* 60, 14341 (1999).

544. D. Briggs, "Surface Analysis of Polymers by XPS and Static SIMS." Cambridge Solid State Science Series (D. R. Clarke, S. Suresh, and I. M. Ward, Eds.). Cambridge Univ. Press, Cambridge, UK, 1998.

545. W. R. Salaneck, R. H. Friend, and J. L. Brédas, *Phys. Rep.* 319, 231 (1999).

546. K. Seki, H. Ishii, and Y. Ouchi, in "Chemical Applications of Synchrotron Radiation" (T. K. Sham, Ed.), Advanced Series of Physical Chemistry. World Scientific, Singapore.

547. K. Seki, in "Optical Techniques to Characterize Polymer Systems" (L. L. and M. Cardona, Eds.), pp. 261–298. Springer-Verlag, Berlin, 1979.

548. D. A. Outka and J. Stöhr, *Chem. Phys. Solid Surf.* 7, 201 (1988).

549. M. Geoghegan, F. Boue, A. Menelle, F. Abel, T. Russ, H. Ermer, R. Brenn, and D. G. Bucknall, *J. Phys. Cond. Mat.* 12, 5129 (2000).

550. B. N. Dev, A. K. Das, S. Dev, D. W. Schubert, M. Stamm, and G. Materlik, *Phys. Rev. B* 61, 8462 (2000).

551. S. D. Kim, E. M. Boczar, A. Klein, and L. H. Sperling, *Langmuir* 16, 1279 (2000).

552. A. Budkowski, *Adv. Polymer Sci.* 148, 1 (1999).

553. A. Budkowski, J. Rysz, F. Scheffold, and J. Klein, *Europhys. Lett.* 43, 404 (1998).

554. R. A. L. Jones, E. J. Kramer, M. H. Rafailovich, J. Sokolov, and S. A. Schwarz, *Phys. Rev. Lett.* 62, 280 (1989).

555. I. Schmidt and K. Binder, *J. Phys.* 46, 1631 (1985).

556. N. Ueno, W. Gaedeke, E. E. Koch, R. Engelhardt, and R. Dudde, *J. Electron Spectrosc. Rel. Phenom.* 36, 143 (1985).

557. N. Ueno, Y. Azuma, M. Tsutsui, K. Okudaira, and Y. Harada, *Jpn. J. Appl. Phys.* 37, 4979 (1998).

558. J. Kikuma and B. P. Tonner, *J. Electron Spectrosc. Rel. Phenom.* 82, 41 (1996).

559. E. Morikawa, V. Saile, K. K. Okudaira, Y. Azuma, K. Meguro, Y. Harada, K. Seki, S. Hasegawa, and N. Ueno, *J. Appl. Phys.* 112, 10476 (2000).

560. K. K. Okudaira, E. Morikawa, D. A. Hite, S. Hasegawa, H. Ishii, M. Imamura, H. Shimada, Y. Azuma, K. Meguro, Y. Harada, V. Saile, K. Seki, and N. Ueno, *J. Electron Spectrosc. Rel. Phenom.* 101–103, 389 (1999).

561. K. Nagayama, R. Mitsumoto, T. Araki, Y. Ouchi, and K. Seki, *Physica B* 208–209, 419 (1995).

562. K. Nagayama, M. Mei, R. Mitsumoto, E. Ito, T. Araki, H. Ishii, Y. Ouchi, K. Seki, and K. Kondo, *J. Electron Spectrosc. Rel. Phenom.* 78, 375 (1996).

563. G. Beamson, D. T. Clark, D. E. Deegan, N. W. Hayes, D. S. L. Law, J. R. Rasmusson, and W. R. Salaneck, *Surf. Interface Anal.* 24, 204 (1996).

564. G. Ziegler, T. Schedelniedrig, G. Beamson, D. T. Clark, W. R. Salaneck, H. Sotobayashi, and A. M. Bradshaw, *Langmuir* 10, 4399 (1994).

565. H. Ågren, V. Carravetta, O. Vahtras, and L. G. M. Pettersson, *Phys. Rev. B* 51, 17848 (1995).

566. T. Miyamae, S. Hasegawa, D. Yoshimura, H. Ishii, N. Ueno, and K. Seki, *J. Chem. Phys.* 112, 3333 (2000).

567. K. Seki, H. Tanaka, T. Ohta, Y. Aoki, A. Imamura, H. Fujimoto, H. Yamamoto, and H. Inokuchi, *Phys. Scripta* 41, 167 (1990).

568. A. J. Lovinger, *Macromol.* 16, 1529 (1983).

569. T. Furukawa, *Phase Transitions* 18, 143 (1989).

570. L. M. Blinov, V. M. Fridkin, S. P. Palto, A. V. Bune, P. A. Dowben, and S. Ducharme, *Uspekhi Fizicheskikh Nauk* 170, 247 (2000); *Phys. Uspekhi* 43, 243 (2000).

571. A. V. Bune, V. M. Fridkin, S. Ducharme, L. M. Blinov, S. P. Palto, A. Sorokin, S. G. Yudin, and A. Zlatkin, *Nature* 391, 874 (1998).

572. J. Choi, C. N. Borca, P. A. Dowben, A. Bune, M. Poulsen, S. Pebley, S. Adenwalla, S. Ducharme, L. Robertson, V. M. Fridkin, S. P. Palto, N. Petukhova, and S. G. Yudin, *Phys. Rev. B* 61, 5760 (2000).

573. J. Choi, P. A. Dowben, S. Ducharme, V. M. Fridkin, S. P. Palto, N. Petukhova, and S. G. Yudin, *Phys. Lett. A* 249, 505 (1998).

574. J. Choi, P. A. Dowben, S. Pebley, A. Bune, S. Ducharme, V. M. Fridkin, S. P. Palto, and N. Petukhova, *Phys. Rev. Lett.* 80, 1328 (1998).

575. J. Choi, S.-J. Tang, P. T. Sprunger, P. A. Dowben, V. M. Fridkin, A. V. Sorokin, S. P. Palto, N. Petukhova, and S. G. Yudin, *J. Phys. Cond. Mat.* 12, 4735 (2000).

576. J. Choi, P. A. Dowben, C. N. Borca, S. Adenwalla, A. V. Bune, S. Ducharme, V. M. Fridkin, S. P. Palto, and N. Petukhova, *Phys. Rev. B* 59, 1819 (1999).

577. Y. Ouchi, I. Mori, M. Sei, E. Ito, T. Araki, H. Ishii, K. Seki, and K. Kondo, *Physica B* 208–209, 407 (1995).

578. I. Mori, T. Araki, H. Ishii, Y. Ouchi, K. Seki, and K. Kondo, *J. Electron Spectrosc. Rel. Phenom.* 78, 371 (1996).

579. G. Tourillon, D. Guay, A. Fontaine, R. Garrett, and G. P. Williams, *Faraday Discuss. Chem. Soc.* 89 (1990).

580. A. P. Hitchcock, G. Tourillon, R. Garrett, G. P. Williams, C. Mahatsekake, and C. Andrien, *J. Phys. Chem.* 96, 5987 (1992).

581. G. Hahner, M. Kinzler, C. Thummler, Ch. Wöll, and M. Grunze, *J. Vac. Sci. Technol. A* 10, 2758 (1992).

582. M. Zharnikov, S. Frey, H. Rong, Y. J. Yang, K. Heister, M. Buck, and M. Grunze, *Phys. Chem. Chem. Phys.* 2, 3359 (2000).

583. A. J. Pertsin and M. Grunze, *Langmuir* 10, 3668 (1994).

584. M. V. Russo, G. Infante, G. Polzonetti, G. Contini, G. Tourillon, Ph. Parent, and C. Laffon, *J. Electron Spectrosc. Rel. Phenom.* 85, 53 (1997).

585. D. Jérome and H. J. Schultz, *Adv. Phys.* 31, 299 (1982).

586. M. J. Rice, V. M. Yarstev, and J. S. Jacobsen, *Phys. Rev. B* 21, 3437 (1980).

587. J. P. Pouget, S. K. Khanna, F. Denoyer, R. Comés, A. F. Garito, and A. J. Heeger, *Phys. Rev. Lett.* 37, 437 (1976).

588. F. Denoyer, R. Comés, A. F. Garito, and A. J. Heeger, *Phys. Rev. Lett.* 35, 445 (1975).

589. S. Kagoshima, H. Azai, K. Kajima, and R. Ishigoro, *J. Phys. Soc. Jpn.* 39, 1143 (1975).

590. M. Grioni and J. Voit, *J. Phys. IV* 10, 91 (2000).

591. F. Zwick, M. Grioni, M. Onellion, and G. Margaritondo, *Helv. Phys. Acta* 71, S13 (1998).

592. F. Zwick, D. Jérome, G. Margaritondo, M. Onellion, J. Voit, and M. Grioni, *Phys. Rev. Lett.* 81, 2974 (1998).

593. F. Zwick, M. Grioni, M. Onellion, L. K. Montgomery, and G. Margaritondo, *Physica B* 265, 160 (1999).

594. F. Zwick, H. Berger, M. Grioni, G. Margaritondo, and M. Onellion, *J. Phys. IV* 9, Pr10 (2000).

595. H. W. Kroto, J. R. Heath, S. C. O'Brien, R. F. Curl, and R. E. Smalley, *Nature (London)* 318, 162 (1985).

596. H. W. Kroto, A. W. Allaf, and S. P. Balm, *Chem. Rev.* 91, 1213 (1991).

597. D. N. McIlroy, J. Zhang, P. A. Dowben, P. Xu, and D. Heskett, *Surf. Sci.* 328, 47 (1995).

598. D. N. McIlroy, J. Zhang, P. A. Dowben, and D. Heskett, *Mater. Sci. Eng. A* 217/218, 64 (1996).

599. D. N. McIlroy, C. Waldfried, T. McAvoy, J. Choi, P. A. Dowben, and D. Heskett, *Chem. Phys. Lett.* 264, 168 (1997).

600. M. Schluter, M. Lannoo, M. Needels, G. A. Baraff, and D. Tomanek, *Phys. Rev. Lett.* 68, 526 (1992).

601. S. Hino, K. Matsumoto, S. Hasegawa, K. Kamiya, H. Inokuchi, T. Morikawa, T. Takahashi, K. Seki, K. Kikuchi, S. Suzuki, I. Ikemoto, and Y. Achiba, *Chem. Phys. Lett.* 190, 169 (1992).

602. P. J. Bennig, D. M. Poirier, T. R. Ohno, Y. Chen, M. B. Jost, F. Stepniak, G. H. Kroll, J. H. Weaver, J. Fure, and R. E. Smalley, *Phys. Rev. B* 45, 6899 (1992).

603. S. Hino, K. Matsumoto, S. Hasegawa, H. Inokuchi, T. Morikawa, T. Takahashi, K. Seki, K. Kikuchi, S. Suzuki, I. Ikemoto, and Y. Achiba, *Chem. Phys. Lett.* 197, 38 (1992).

604. S. Hino, K. Matsumoto, S. Hasegawa, K. Iwasaki, K. Yakushi, T. Morikawa, T. Takahashi, K. Seki, K. Kikuchi, I. Ikemoto, and Y. Achiba, *Phys. Rev. B* 48, 8418 (1993).

605. G. E. Scuseria, *Science* 271, 942 (1996).

606. R. W. Lof, M. A. van Veenendaal, B. Koopmans, H. T. Jonkman, and G. A. Sawatzky, *Phys. Rev. Lett.* 68, 3924 (1992).

607. Y. Zhang, X. Gao, and M. J. Weaver, *J. Phys. Chem.* 96, 510 (1992).

608. D. Chen and D. Sarid, *Surf. Sci.* 329, 206 (1995).

609. D. Chen and D. Sarid, *Phys. Rev. B* 49, 7612 (1994).

610. D. Chen, J. Chen, and D. Sarid, *Phys. Rev. B* 50, 10905 (1994).

611. K. Sakamoto, D. Kondo, Y. Ushimi, A. Kimura, A. Kakizaki, and S. Suto, *Surf. Sci.* 438, 248 (1999).

612. K. Sakamoto, D. Kondo, Y. Ushimi, M. Harada, A. Kimura, A. Kakizaki, and S. Suto, *J. Electron Spectrosc. Rel. Phenom.* 101–103, 413 (1999).

613. T. R. Ohno, Y. Chen, S. E. Harvey, G. H. Froll, J. H. Weaver, R. E. Haufler, and R. E. Smalley, *Phys. Rev. B* 44, 13747 (1991).

614. J. H. Weaver, *Acc. Chem. Res.* 25, 143 (1992).

615. K. Motai, T. Hashizume, H. Shinohara, Y. Saito, H. W. Pickering, Y. Nishina, and T. Sakuri, *Jpn. J. Appl. Phys.* 32, L450 (1993).

616. K.-D. Tsuei, J. Y. Yuh, C.-T. Tzeng, R.-Y. Chu, S.-C. Chung, and K.-L. Tsang, *Phys. Rev. B* 56, 15412 (1997).

617. E. I. Altman and R. J. Colton, *Surf. Sci.* 279, 49 (1992).

618. M. R. C. Hunt, P. Rudolf, and S. Modesti, *Phys. Rev. B* 55, 7882 (1997).

619. A. J. Maxwell, P. A. Bruhwiler, A. Nilsson, and N. Mårtensson, *Phys. Rev. B* 49, 10717 (1994).

620. D. Purdie, H. Bernhoff, and B. Reihl, *Surf. Sci.* 364, 279 (1996).

621. J. Kovac, G. Scarel, O. Sakho, and M. Sancrotti, *J. Electron Spectrosc. Rel. Phenom.* 72, 71 (1995).

622. C.-T. Tzeng, W.-S. Lo, J.-Y. Yuh, R.-Y. Chu, and K.-D. Tsuei, *Phys. Rev. B* 61, 2263 (2000).

623. A. Goldoni and G. Paolucci, *Surf. Sci.* 437, 353 (1999).

624. C. Cepek, L. Giovanelli, M. Sancrotti, G. Costantini, C. Boragno, and U. Valbusa, *Surf. Sci.* 454–456, 766 (2000).

625. G. K. Wertheim and D. N. E. Buchanan, *Phys. Rev. B* 50, 11070 (1994).

626. S. J. Chase, W. S. Bacsa, M. G. Mitch, L. J. Pilione, and J. S. Lannin, *Phys. Rev. B* 46, 7873 (1992).

627. B. W. Hoogenboom, R. Hesper, L. H. Tjeng, and G. A. Sawatzky, *Phys. Rev. B* 57, 11939 (1998).

628. J. L. Martins, N. Troullier, and J. H. Weaver, *Chem. Phys. Lett.* 180, 457 (1991).

629. M. B. Jost, P. J. Benning, D. M. Poirier, J. H. Weaver, L. P. F. Chibante, and R. E. Smalley, *Chem. Phys. Lett.* 184, 423 (1991).

630. J. H. Weaver, *J. Phys. Chem. Solids* 53, 1433 (1992).

631. A. F. Hebard, M. J. Rosseinsky, R. C. Haddon, D. W. Murphy, S. H. Glarum, T. T. M. Palstra, A. P. Ramirez, and A. R. Kortan, *Nature (London)* 350, 600 (1991).

632. P. J. Benning, J. L. Martins, J. H. Weaver, L. P. F. Chibante, and R. E. Smalley, *Science* 252, 1417 (1991).

633. G. K. Wertheim, J. E. Rowe, D. N. E. Buchanan, B. E. Chaban, A. F. Hebard, A. R. Kortan, A. V. Makhija, and R. C. Haddon, *Science* 252, 1419 (1991).

634. J. L. Martins and N. Troullier, *Phys. Rev. B* 46, 1766 (1992).

635. O. A. Bruhwiler, A. J. Maxwell, A. Nilsson, N. Martensson, and O. Gunnarsson, *Phys. Rev. B* 48, 18296 (1993).

636. T. Takhashi, S. Suzuki, T. Morikawa, H. K. Yoshida, S. Hasegawa, H. Inkuchi, K. Seki, K. Kikuchi, S. Suzuki, K. Ikemoto, and Y. Achiba, *Phys. Rev. Lett.* 68, 1232 (1992).

637. L. Q. Jiang and B. E. Koel, *Phys. Rev. Lett.* 72, 140 (1994).

638. J. P. Lu, *Phys. Rev. B* 49, 5687 (1994).

639. S. Satpathy, V. P. Antropov, O. K. Anderson, O. Jepsen, O. Gunnarsson, and A. I. Liechtenstein, *Phys. Rev. B* 46, 1773 (1992).

640. P. J. Benning, F. Stepniak, and J. H. Weaver, *Phys. Rev. B* 48, 9086 (1993).

641. W. Andreoni, P. Giannozzi, and M. Parrinello, *Phys. Rev. Lett.* 72, 848 (1994).

642. S. Jrummacher, M. Biermann, M. Neeb, A. Liebsch, and W. Eberhardt, *Phys. Rev. B* 48, 8424 (1993).

643. G. Gensterblum, J.-J. Pireaux, P. A. Thiry, R. Caudano, T. Bushapls, R. L. Johnson, G. Lelay, V. Aristov, R. Gunter, A. Taleb-Ibrahimi, G. Indlekofer, and Y. Petroff, *Phys. Rev. B* 48, 14756 (1993).

644. P. J. Benning, C. G. Olson, D. W. Lynch, and J. H. Weaver, *Phys. Rev. B* 50, 11239 (1994).

645. P. He, S. Bao, C. Yu, and Y. Xu, *Surf. Sci.* 328, 287 (1995).

646. J. Choi, Y. C. Choi, S. M. Lee, Y. H. Lee, and J. C. Jiang, submitted for publication.

647. B.-L. Gu, Y. Maruyama, J.-Z. Yu, K. Ohno, and Y. Kawazoe, *Phys. Rev. B* 49, 16202 (1994).

648. J. G. Hou, J. Yang, H. Wang, Q. Li, C. Zeng, H. Lin, W. Bing, D. M. Chen, and Q. Zhu, *Phys. Rev. Lett.* 83, 3001 (1999).

Chapter 3

SUPERHARD COATINGS IN C−B−N SYSTEMS: GROWTH AND CHARACTERIZATION

Arun K. Sikder, Ashok Kumar

Center for Microelectronics Research, College of Engineering, University of South Florida, Tampa, Florida, USA

Contents

1. INTRODUCTION

Hard and wear-resistant coatings are an important segment in the US and world economy; for instance, machine tools are a $15 billion per year industry worldwide. Hard and superhard, wear-resistant, and corrosion-resistant coatings are frequently used in industry to improve product quality and life. They allow products to survive in adverse environments where they are subject to high wear, abrasion or corrosion [1, 2]. Coatings are also vital to advancements in magnetic recording technology where worldwide annual revenues are close to $100 billion (including media and equipment) [3]. The United States accounts for about 40% of these revenues, the largest portion coming from magnetic hard-disk systems. Many applications of these hard coatings require well-controlled properties. For mechanical applications this necessitates control of hardness, surface roughness, and adhesion to substrates. The choice of what hard coating to use for a particular application depends on a number

Handbook of Thin Film Materials, edited by H.S. Nalwa
Volume 2: Characterization and Spectroscopy of Thin Films

ISBN 0-12-512910-6/$35.00

of factors including in part: (i) relative coating/substrate thermal expansion, (ii) chemical stability of coating, (iii) coating hardness, and (iv) required deposition temperature.

Diamond and cubic boron nitride (cBN), together with some ternary compounds from the C—B—N triangle [4], are generally classified as intrinsic "superhard" materials [5]. The materials whose hardness and other mechanical properties are determined by their microstructure are called extrinsic. To understand what that means, realize that an uncoated carbide has Vickers hardness number 1800. A titanium nitride is in the vicinity of 2100 to 2200, whereas cBN goes to 5500 and there is a big leap to diamond, which is at 9500. Essentially, diamond and cBN are the two hardest materials known for cutting applications. If diamond is the hardest known material, then why bother to search for new hard materials in the industry and what is the reason for simulating the new materials by researchers? Reason behind this is the limitations of each material in way they behave for specific purposes. Figure 1 shows the theoretical prediction of ultrahard materials and most of them are in practical reality.

The hardest known material, diamond, is made out of the same thing as graphite, a very soft material, which is carbon. The difference between the two is in the way that carbon atoms are arranged. Diamond consists of tightly interlinked carbon atoms with covalent bonding. Graphite, by contrast, has good linkages within the horizontal layers but poor binding between vertical layers and hence it is soft and lubricating. cBN is similar in linkage arrangement to diamond. The boron atom is close to carbon on the periodic table. However, there is nitrogen involved in cBN, which makes the material somewhat trickier to work in terms of developing the process of taking a vapor of boron and nitride atoms, bombarding it with energetic ions, and ending up with a useful thin film of cBN. Now carbon and iron have an affinity. Carbon and iron, under the right conditions, form steel. So ferrous machining applications—and of all machining that takes place, some 70% is of ferrous materials—are

not particularly good for diamond tools due to fast chemical wear of the tools. This reaction does not occur between cBN and ferrous materials. This is why enormous of interest has been shown on cBN, even though it is not as hard as diamond. For machining of hypereutectic aluminum alloys (with >12.2% Si), used in automobile industries, diamond coated tools are preferred. Therefore, selection of materials overcoat depends on the materials to be machined and also the work environment.

Another hard material in the ternary system B—C—N is being investigated extensively, which is carbon nitride. Sung and Sung predicted in 1984 that C_3N_4 should have hardness comparable to that of diamond. A considerable amount of interest has been generated in producing crystalline carbon nitride since Liu and Cohen [6, 7] derived a formula which confirmed the prediction of Sung and Sung. Study of C_3N_4 was motivated from a empirical model [8] for the bulk modulus of tetrahedral solids which indicates that short bond lengths and low ionicity are favorable for achieving large bulk moduli. Since the C—N bond satisfies these conditions, tetrahedral C—N solids were suggested by as candidates for new low compressibility solids. In spite of many research efforts around the world, reports of successful synthesis of β-C_3N_4 have been few and usually unsubstantiated.

In this chapter we will discuss the basic properties, growth, and characterization of intrinsic superhard materials: diamond, cBN, and carbon nitride films. We also discuss the basic principles of growth techniques and characterization tools in order to understand more the specific purposes. A large number of articles came out regarding the different issues of these novel materials, but to our knowledge there are no reports which give a detailed study of all three materials in the C—B—N system.

2. DIAMOND

2.1. Introduction

Diamond is the hardest material known and has the highest thermal conductivity among all known materials. Combined with these important properties, diamond has very low thermal expansion and high electrical resistance. Because of its hardness, diamond is far more effective and efficient than other competing materials used for abrasive, cutting, shaping, or finishing tools. Its very high thermal conductivity makes it ideal for spreading and conducting heat out of compact, high power, high-speed electronic packages.

Though the initial work in the synthesis of diamond at low temperature and pressure was carried out in the 1950s [9–11], the activities in the area of chemical vapor deposited (CVD) diamond picked up really in the early 1980s [12]. This was due to the advances made in the CVD techniques. A variety of CVD techniques such as microwave plasma CVD (MPCVD) [13, 14], hot filament CVD (HFCVD) [15, 16], oxyacetylene torch CVD [17, 18], dc plasmas [19], arc discharges [20, 21], plasma jet CVD [22], etc. have been used for the synthesis of diamond films. It is also noted that MPCVD and HFCVD, which have remained the leading techniques, have been used widely by a

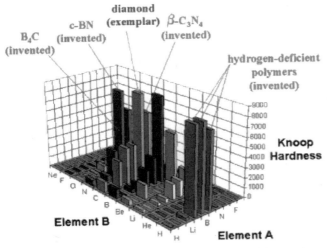

Fig. 1.   Theoretical prediction of ultrahard materials; most of them are in practical reality (from www.imagination-engines.com/newpcai.htm).

Table I.   CVD Diamond vs Competing Materials

Properties of diamond		Comparison with competitor	Possible applications
Vicker's hardness (kg mm^{-2})	12,000–15,000	Hardest known material	Drill bits, polishing materials, cutting tools, sintered or brazed diamond components
Coefficient of friction	0.1 (in air)	More than Teflon	Wear resistance coatings on lenses, bearings, tools or hard disks, sliding parts
Young's modulus (N m^{-2})	1.2×10^{12}	Twice the value of alumina; highest mechancal strength	Stiff membranes for lithography masks, radiation windows, micromechanical applications
Sound propagation velocity (km s^{-1})	18.2	1.6 times the value of alumina at room temperature	
Chemical inertness	Inert	Resistant to all chemicals	Coatings for reactor vessels, sample containers for analytical instruments
Range of high transmittance (μm)	0.22–2.5 and >6	In IR orders of magitude lower than other materials	UV-VIS-IR windows and coatings, spectrometery sample containers, microwave windows, optical interference filters, optical waveguide
Refractive index	2.41	1.6× value of silica	
Bandgap (eV)	5.45	1.1 & 1.43 for Si &GaAs	Passive and active electronics, high power, high frequency semiconducting devices, wide range thermistors, hot transistors, Shottkey diodes, short wavelength lasers, γ-ray detectors
Electron/hole mobility (cm A s^{-1})	1900/1600	1500/1600 for Si 8500/400 for GaAs	
Dielectric constant	5.5	11 & 12.5 for Si &GaAs	
Thermal conductivity (W cm^{-1} K^{-1})	20	At room temperature 4× the value of copper	Highly efficient, insulating heat sinks for high power ULSI electronics
Luminescence (μm)	0.44 0.52 (B-doped)	Blue luminescence Scarce β-SiC	Blue of green LEDs, detectors

number of workers both in laboratories and in industries. These films are expected to be used in a variety of applications from cutting tools to wear-resistant parts, and from electronic to optical applications. One advantage of CVD diamond technology over high-pressure technology is low cost and its ability to coat on any shape.

Diamond formed by sp^3 hybridized carbon atoms is a unique structure in nature. Its unique properties make it suitable for a variety of commercial applications. Not only is it used prominently as a gemstone, but it is also a very useful industrial material. Table I compares the properties of CVD diamond with the other competing materials. The diamond coated cutting tools, abrasive wheels, polycrystalline diamond inserts, and diamond heat sinks are a few products used routinely in industry. The properties of CVD diamond are almost at par with its natural counterpart. There are, however, several technical problems yet to be solved. For instance, in tribological applications adhesion of diamond to nondiamond materials is still a serious problem [23–25]. For optical and electronic applications, defects and impurity-free diamond coatings are desired.

2.1.1. Chemistry and Physics of Diamond

Diamond is an allotrope of carbon with a tetrahedrally bonded structure. The C atoms in diamond structure form strong covalent bonds resulting from the hybridization of the outer shell s and p electrons. Other solids crystallizing in diamond structure

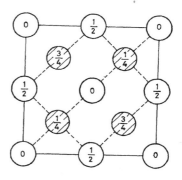

Fig. 2.   Projected cubic cell of the diamond structure showing the atomic arrangements. The open circles are part of a fcc lattice and the shaded circles in the same lattice are part of another lattice displaced by one-fourth of the body diagonal of the fcc lattice. The fractions inside the circle are the heights of the atoms above the base in units of a cube edge. It can be considered as a fcc lattice with a basis of two atoms at 0 0 0; ¼, ¼, ¼.

are Ge, Si, gray tin, etc. The diamond crystal structure can be defined as two interpenetrating face-centered cubic (fcc) Bravais lattices, displaced along the body diagonal by one-fourth of its length [26]. It can also be described as a primitive basis of atoms at 000; ¼, ¼, ¼ associated with each lattice point [27] of a fcc lattice as shown in Figure 2. The unit cell of diamond is shown in Figure 3 in which the shaded circles represent one of the two fcc lattices. The lattice constant of diamond is 3.567 Å. Comparison of structural peoperties of diamond and graphite is shown in Table II.

Table II.   Structural Properties of Cubic and Hexagonal Diamond and Graphite

	Cubic	Hexagonal	Graphite
Point group of atoms	Td	Td	
Space group	Oh^7; $Fd3$	D^4_{-6h}; $P6_{3/mmc}$	D^4_{6H}
Unit cell constants	Cubic	$a = 0.252$ nm	
	0.357	$c = 0.412$ nm	
C–C nearest neighbor distance	0.154 nm	0.154 nm	
Bonding	sp^3	sp^3	sp^2
Atoms located at	(001)(½ ½0)(0½ ½)	(000) (003/8)	
	(½0½)(¼¼¼)(¾¾¼)	(1/32/32/3)	
	(¼¾¾)(¾¼¾)		
General crystal	Zinc blend type	2H-wurtzite	
Density	3.52×10^3 kg m^{-3}	3.52×10^3 kg m^{-3}	2.30×10^3 kg m^{-3}

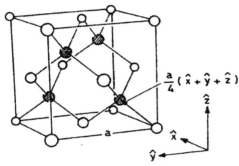

Fig. 3.   Unit cell of diamond showing the tetrahedral bond arrangements of sp^3 carbon.

Table III. d-Spacing (in Å) of Diffracting Rings in Polycrystalline Diamond Thin Films

No. of measured d-spacing	Cubic diamond	3C- hexagonal diamond	2H- hexagonal ring diamond	6H
1	2.06	2.06 (111)	2.06	2.06
2	1.89	1.77 (200)	1.92	1.93
3	1.52	—	1.50	—
4	1.30	—	—	1.37
5	1.27	1.26 (220)	1.26	1.26
6	1.20	—	1.17	1.16
7	1.04	—	1.07	1.07
8	0.97	—	1.05	—

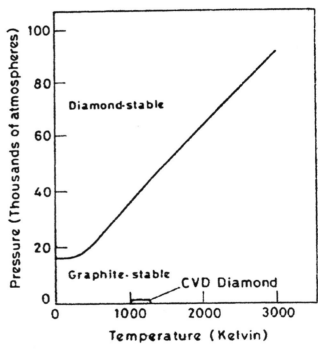

Fig. 4.   Carbon phase diagram showing the stable conditions for diamond and graphite. The narrow range of the conditions of CVD diamond is also shown in the diagram.

Diamonds may also exist in a series of polytypes [28] (Table III). The two extreme cases of the series are cubic diamond and the 2H hexagonal form of diamond, which is also known as Loansdaleite. The other crystalline forms of carbon are graphite in sp^2 hybridization, carbyne in sp^1 hybridization, and fullerenes in a combination of sp^2 and sp^3 hybridization [29]. The phase diagram of carbon (Fig. 4) shows that diamond is a stable phase, thermodynamically, only at thousands of atmospheric pressure and high temperatures. The phase diagram can, therefore, broadly be split in two regions: the high pressure region where diamond is a stable phase, and the low pressure region where graphite is a stable phase of carbon. The techniques based on the later region attracted considerable attention over the last few years. In this technique diamond can be nucleated on the nondiamond substrates and an oriented or polycrystalline thin film can be deposited by dissociating the precursor gases at low pressure [30]. These polycrystalline films show character-

Table IV.   Comparison of the Properties of the CVD, the Natural, and the PCD Polycrystall Diamond

Property	CVD diamond	Single crystal diamond	PCD
Density ($\times 10^3$ kg m^3)	3.52	3.52	4.12
Hardness (GPa)	85–100	50–100	50
Fracture toughness (MPa m exp 0.5)	5.5	3.4	8.81
Young's modulus (GPa)	1000–1100	1000–1100	776
Poisson's ratio	0.07	0.07	0.07
Tensile strength (MPa)	400–800 (growth–nucleation)	1050–3000 (orientation dependent)	1260
Transverse rupture strength (GPa)	1.3	2.9	1.2
Compressive strength (GPa)	16.0	9.0	7.60
Thermal conductivity at 20°C (W m K)	500–2200	600–2200	560
Thermal conductivity at 200°C (W m K)	500–1100	600–1100	200
Thermal diffusivity (cm^2 s)	2.8–11.6	5.5–11.6	2.7
Thermal expansion coef.			
100°C–250°C (ppm K)	1.21	1.21	4.2
500°C (ppm K)	3.84	3.84	—
100°C (ppm K)	4.45	4.45	6.3

istics similar to those of natural or synthetic diamond and have high potential for various applications (Table IV).

2.1.2. CVD Diamond

Generally a carbon precursor gas (CH_4, C_2H_2, etc.) with large concentration of hydrogen (H_2) is used for the deposition of diamond films [31]. Perfection in diamond growth is achieved only on some particular substrates. In the CVD process the ratio of CH_4/H_2, substrate temperature (T_s), and deposition pressure (P_d) are the three most important parameters which control the film quality. Other parameters vary depending on the system and technique used. Production of atomic hydrogen in the gas phase and atomic H concentration on the substrate surface are very important for obtaining good quality diamond films. High atomic H concentration on the growing surface helps to etch graphite preferentially [32] and stabilizes the sp^3 phase of carbon as diamond [33].

2.2. Deposition of Diamond Films

2.2.1. Chemical Nature of the Substrate

Growth of diamond from the vapor phase is highly dependent on the chemical nature of the substrates [34, 35] and the pretreatment of the substrate surface, which will be discussed in the next section. Normally the carbide-forming nature of the

Table V.   Classification of Substrate Materials

Carbon substrate interactions	Substrate materials
Little or no solubility or reaction	Diamond, graphite, carbon, Cu, Ag, Au, Sn, Pb, etc.
C-diffusion only, C dissolves in MeC mixed crystals	Pt, Pd, Rh, etc.
Carbide formation	
Metallic	Ti, Zr, Hf, V, Nb, Ta, Cr, Mo, W, Fe, Co, Ni (metastable)
Covalent	B, Si, etc.
Ionic	Al, Y, rare earth metals, etc.

Reproduced with permission from [38], copyright 1993, Elsevier Science.

substrate is seen to be suitable for diamond growth [36]. It has also been used in noncarbide forming substrates to grow diamond for several applications with the help of either surface pretreatment or suitable buffer layers [37]. Table V shows the classification of the substrate materials in terms of carbon–substrate interactions [38]. A large number of substrates are being used for vapor phase diamond grown, such as Mo, W, WC, Ta, Cr, Co, Pt, Au, Al, Cu, Ni, Fe, stainless steel (SS), NiAl, $FeSi_2$, Ti, Ti−2Al−1.5Mn, Ti−6Al−4V, TiN, TiC, Si, SiC, Si_3N_4, silica, SiAlON, MgO, Al_2O_3, cBN, Y−ZrO_2, etc. [35]. Among all these substrates Si is most commonly used substrate in CVD diamond synthesis due to its nature of carbide formation and similar crystal structure with diamond.

2.2.2. Substrate Pretreatment

Substrate pretreatment is another important criteria to control the nucleation, morphology, orientation, and surface roughness. Surface roughening or seeding with diamond powder [39] or nondiamond powders [40] as well as graphite flakes, application of buffer layers, and substrate biasing are commonly used techniques to increase the nucleation density and decrease the incubation time. Ion implantation [41], pulsed laser modification of substrate surfaces [42], carburization of substrate surfaces [43–45], and chemical etching of substrate surfaces [46] are few other treatments used to increase the nucleation density and adhesion of the films to the substrate.

The most commonly used method of substrate pretreatment to increase the nucleation is scratching by diamond powder or paste. Nucleation density and surface damage of the substrate strongly depend on the size of the diamond particles used for the pretreatment method. In this process of substrate pretreatment small fragments of diamond particles are embedded in the substrate materials and directly act as the nucleation centers for diamond growth. It is also shown that a rough surface is more suitable for nucleation than the mirror polished substrates. This is shown by number of researchers by using abrasive particles (SiC, BN, Al$_2$O$_3$, MoB, etc.) other than diamond [47–49]. Possibly the mechanism for enhancement of nucleation of diamond on substrate surfaces scratched by other abrasives is due to the formation of microscopic sharp edges. These sites are suitable for diamond growth due to the following reasons [50, 51]: (i) minimization of interfacial energy by the formation of diamond nuclei on sharp convex surfaces, (ii) breaking of a certain number of surface bonds and presence of more dangling bonds at sharp edges favoring the chemisorption of nucleating species, (iii) strain field effects, and (iv) most rapid carbon saturation at sharp edges [52, 53]. The third possible mechanism is that scratching produces nonvolatile graphitic particles through local pyrolysis of adsorbed hydrocarbons. These graphitic clusters would subsequently be hydrogenated in the atomic hydrogen environment under the typical CVD conditions to form the precursor molecules. Finally, the removal of surface oxides is also suggested to be a possible operating mechanism of nucleation enhancement by scratching [49, 54, 55].

Regarding the effect of various abrasives (Table VI) on diamond nucleation, it is found that polishing with diamond has the most pronounced effect, and the effect decreases in the order diamond> cBN> SiC [56]. Diamond nucleation densities after scratching pretreatments range typically from 10^5 to 10^{10} cm^{-2} as compared to <10^5 cm^{-2} on untreated surfaces. Scratching of the substrate causes serious surface damage and hence the application of the coatings for optical and electronic materials needs alternate nucleation techniques. For this purpose substrate was seeded with submicron diamond powders by different means and by this method one can grow epitaxial or oriented diamond on a preferred substrate surface [57–59].

2.2.3. Deposition Parameters

Depending on the growth parameters discussed above, incorporation of impurities in the films varies over a wide range of concentrations. Major impurities in CVD diamond are nondiamond carbon, hydrogen (H), nitrogen (N), oxygen (O), and other metallic impurities intrinsic to the deposition techniques. As the deposition (P_d) varies in the chamber nondiamond carbon, H and paramagnetic defects (dangling bonds) vary substantially in the deposits depending on the growth techniques.

P_d and the substrate temperature (T_s) are the two most important parameters in diamond deposition by the CVD method. A good quality diamond film having the least amount of nondiamond impurities can be grown in a narrow range of T_s (700–1000°C) and P_d (5–100 Torr). The required T_s in the CVD apparatus can be achieved either by using the same input power which activates the gas phase or by using a separate heater for the substrate. Both arrangements have certain advantages and disadvantages. The design of the former system in which the substrate is self-heated is easier compared to the latter. On the other hand if the external heater is inside the chamber, it would be an additional source of contamination of the films. The disadvantage of the self-heated CVD system is that the deposition parameters, viz., input power, T_s, and P_d, become interdependent.

2.2.4. Deposition Method

A large number of deposition techniques have been employed for successful deposition of diamond from vapor phase. Deposition techniques are mainly divided into two major categories, namely, CVD and physical vapor deposition (PVD). Though few attempts have been made using PVD techniques, CVD proved its dominance for diamond deposition. In CVD, the source of carbon is an energetically excited or activated gas phase. This gas phase contains a mixture of molecules, atoms, radicals, ions, and electrons which oscillate randomly in three dimensions and will condense as diamond only upon surfaces possessing suitable characteristics. PVD techniques typically involve the evaporation or sublimation of carbon from a solid source into the gas phase. Evaporated carbon is chemically and energetically altered so that it condenses as diamond on the surface of a substrate placed in its trajectory.

A large variety of carbon-containing gas species have been employed to synthesize diamond using CVD techniques. These include methane, aliphatic and aromatic hydrocarbons, alcohols, ketones, amines, ethers, and carbon monoxides. In addition to carbon containing carriers, the gas phase usually must contain nondiamond carbon etchants and diamond phase stabilizing agents, such as hydrogen, oxygen, or chlorine and fluorine atoms. For example, the most commonly used methane/hydrogen system, more than 97–99 vol% of hydrogen

Table VI. Size and Physical Properties of Diamond and Various Ceramic Compounds Used for Scratching Pretreatment

Abrasive material	Particle size (μm)	Density (kg m^{-3})	Hardness (kg mm^{-2})	Crystal structure
Diamond	0.25–40 (ultrasonic)	3515	5700–10,400	Cubic/hexagonal
	0.25–15			
Oxides				
Al$_2$O$_3$	0.3–1	3970	2000	Hexagonal
ZrO$_2$	0.1–0.3	5560	1019	Cubic
SiO$_2$		2320	790	Hexagonal
Borides				
TiB$_2$	2–10	4530	3370	Hexagonal
CrB	5–20	6110	1250	Orthorhombic
ZrB$_2$	5–15	6090	2252	Hexagonal
NbB$_2$	0.5–1	7000	2600	Hexagonal
MoB	1–5	8670	2350	Tetragonal
LaB$_6$	10–40	4720	2770	Cubic
TaB$_2$	0.5–1	12,620	2500	Hexagonal
WB	1–5	15,730	3700	Tetragonal
Carbides				
B$_4$C	10–20	2510	2750	Rhombohedral
SiC	5–20	3220	2550	Hexagonal
TiC	10–30	4920	3170	Cubic
V$_8$C$_7$	20–40	5480	2480	Cubic
Cr$_3$C$_2$	2–10	6740	1800	Orthorhombic
ZrC	2–10	6660	2950	Cubic
NbC	10–30	7820	2170	Cubic
Mo$_2$C	0.5–3	9180	1499	Hexagonal
TaC	10–20	14,400	1720	Cubic
WC	20–30	15,770	1716	Hexagonal
Nitrides				
BN	8–12	3480	4530	Cubic
AlN	1–10	3260	1200	Hexagonal
Si$_3$N$_4$	0.2–1	3180	2100	Hexagonal
TiN	3–15	5440	2050	Cubic
VN	1–5	6100	1310	Cubic
Cr$_2$N	0.5–3	6510	1571	Hexagonal
ZrN	3–15	7350	1670	Cubic
NbN	5–30	8310	1461	Cubic
TaN	1–5	14,360	2416	Hexagonal
Silicides				
Ti Si$_2$	10–30	4040	892	Orthorhombic
Cr Si$_2$	10–50	4980	1131	Hexagonal
Zr Si$_2$	10–30	4860	1063	Orthorhombic
Nb Si$_2$	10–40	5660	1050	Hexagonal
Mo Si$_2$	10–40	6240	1200	Tetragonal
Ta Si$_2$	10–60	9100	1407	Hexagonal
WSi$_2$	5–15	9800	1074	Tetragonal

is necessary for pure growth of diamond phase. Among all the CVD techniques MPCVD and HFCVD techniques are the most commonly used due to their simplicity and easy commercialization. Growth rate and comparative process technical data are summarized in Table VII.

2.2.4.1. Plasma Assisted CVD

Microwave Plasma CVD. From its invention in the early 1980s [62], the technique has found a lot of success because of its simplicity, flexibility, and the early commercial avail-

Table VII.   Characteristics of Different Diamond CVD Depositions and Technical Data

Method	Rate (μm h^{-1})	Area (cm^2)	Quality (Raman)[a]	Substrate material	Advantage	Disadvantage
Hot filament	0.4–40	100–400	+++	Si, Mo, silica, Al$_2$O$_3$, etc.	Simple, large area	Contamination, stability
DC discharge (low P)	<0.1	70	+	Si, Mo, silica, Al$_2$O$_3$, etc.	Simple, area	Quality, rate
DC discharge (medium P)	20–250	<2	+++	Si, Mo, Al$_2$O$_3$, etc.	Rate, quality	Area
DC plasma jet	10–930	2–100	+++	Mo, Si, W, Ta, Cu, Ni, Ti, stainless steel	Highest Rate, quality	Stability, Homogeneity,
RF (low P)	<0.1	1–10	-/+	Si, Mo, Al$_2$O$_3$, BN, Ni	Scale up	Quality, rate, contamination,
RF (thermal, 1 atm)	30–500	3–78	+++	Mo	Rate	Stability, Homogeneity,
Microwave (0.9–2.45 GHz)	1 (low P) 30 (high P)	40	+++	Si, Mo, silica, WC, etc.	Quality, area, stabilty	Rate
Microwave (ECR 2.45 GHz)	0.1	<40	-/+	Si	Area, low P	Quality, rate, coat, contamination
Flame Combustion	30–200	1–100	+++	Si, Mo, Al$_2$O$_3$, TiN	Simple rate	Stability, uniformity

[a] = poor quality, +++ = excellent quality.

Reproduced with permission from [61], Noyes Publishing.

ability of reactors from several companies. Figure 5 shows the schematic diagram of a MPCVD reactor. In this technique process gases are introduced into a reactor chamber, which contains the substrate to be coated. Microwave power is then coupled into the chamber through a dielectric window or tube in order to create a discharge. The chamber is an integral part of an electromagnetic cavity in which the microwave electric field profiles are such that the discharge location can be reproducibly controlled. Typically, the substrate to be coated with diamond is immersed into the plasma within this cavity. Plasma is formed by the electron impact on the feed gas resulting in atomic hydrogen and carbon containing species. Plasma energy (measured in terms of electron temperature) typically depends on the chamber pressure which is controlled by controlling the pumping speed of the roughing pump.

A movable substrate holder assembly is normally used in the systems to facilitate an independent variation of substrate temperature (T_s). This enables us to overcome the problem of interdependence of various parameters. For example, various T_s can be achieved at the same microwave power in MPCVD apparatus by keeping the substrate at various positions in the plasma column. Substrates inside the chamber rest on a graphite holder and T_s is measured by either an optical pyrometer or Chromel–Alumel thermocouple. A bypass line is made normally across the valve in the fore-line where a needle valve is used to control the pressure inside the chamber during deposition.

Initially, a base vacuum has to be created in the chamber and hydrogen flushing is used to remove the residual contamination

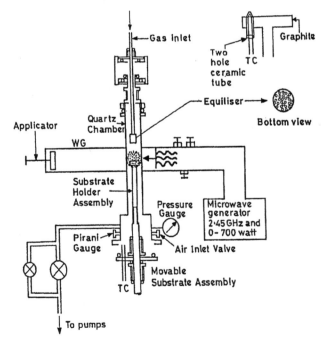

Fig. 5.   Schematic diagram of MPCVD growth chamber. Reproduced with permission from [84].

in the gas phase. Once hydrogen flushing is over a fixed hydrogen flow is set (typically 100–500 sccm) in the chamber through a mass flow controller. After a few minutes a microwave generator of the frequency 2.45 GHz and power 0–60 kW is used for generating plasma. The microwave energy is supplied to the gas load inside the chamber via the waveguide. One end of the waveguide is connected to the microwave generator while the other is terminated by a plunger which is used to shift the plasma ball and also to get the maximum impedance match with the help of the three stub tuners. Hydrogen plasma is created and the plasma treatment of the substrate is carried out until the required T_s is achieved. Separate substrate heating and cooling arrangements are common to the deposition system. The deposition starts with the flow of methane or other carbon containing gas. T_s is controlled by controlling the input power or the heating or cooling arrangement of the substrate holder assembly. After the deposition, the methane flow is terminated first and the H_2 is continued for 2–5 minutes. Subsequently the microwave power is switched off and the substrate is cooled down to 150°C in H_2. Again the chamber is evacuated after closing the flow of H_2 and the substrate is removed when T_s comes down to room temperature. This is a typical operation of the plasma CVD process. Several researchers have carried out plasma diagnostics measurements in order to understand the gas phase chemistry and growth process and substrate plasma interactions [63].

Electron cyclotron resonance (ECR) with a MPCVD chamber is a variation in the plasma enhanced CVD [64]. When a magnetic field is coupled to the microwaves, plasmas can be altered in intensity and position. However, if the pressure in the chamber is held below 10 mTorr, then the mean free path of electrons in the plasma will be long enough to oscillate in electron cyclotron resonance if the appropriate magnetic field is applied. At 2.45 GHz, the ECR is impossible if the pressure exceeds ~10 mTorr; a process at such high pressures should be called magnetically enhanced MPCVD. Diamond films can be grown by both magnetically enhanced MPCVD as well as genuine ECR conditions [65, 66].

Direct Current Plasma Assisted CVD (DC PACVD). In DC plasma CVD, DC voltage (~1 kV) is applied to the substrate or an auxiliary electrode. Intermediate electrodes may be used to guide the plasma and alter plasma position and properties. A typical schematic diagram of the DC PACVD system is shown in Figure 6. In this process electrical energy is converted to thermal and kinetic energy of a flowing gas mixture by an electric arc discharge. Normally, substrates are mounted on the anode of the deposition system as putting the substrates on the cathode results in deposition of graphite or nondiamond phase of carbon rather than diamond [67, 68]. In this technique larger substrate plates may be used, with coated areas nearly 70 cm^2 [61]. The diamond deposition area is limited by the electrodes and the DC power supply. DC plasma CVD has the ability to synthesize high quality diamond with a commercially viable production rate. Kurihara et al. [69] have developed a DC plasma jet process entitled the DIA-JET, which is comprised of a gas injection nozzle composed of a rod cathode concentric with a tube anode. The gas mixture is CH$_4$ and H$_2$ with a carrier gas, Ar or He, passing between the cathode and anode and sprayed out form an orifice in the anode. The DIA-JET process could produce freestanding diamond plates

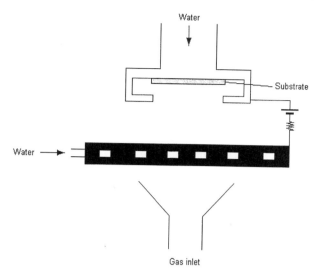

Fig. 6.   Schematic diagram of DC plasma assisted CVD method.

$10 \times 10 \times 2$ mm^3. DIA-JET is also known as DC thermal plasma CVD. A few groups also coupled this DC plasma method along with the HFCVD process and found an increased deposition rate [70, 71]. The highest known deposition rate, 930 μm h^{-1}, has been achieved using the DC plasma jet CVD by Ohtake et al. [72] and similar results are repeatedly produced by Matsumoto et al. [73]. Norton in United States grows routinely > 4 inch diameter wafers using DC plasma jet deposition.

Radio Frequency Plasma Assisted CVD (RF PACVD).   A RF plasma is glow discharge plasma, generated in a gas at relatively low pressure (~1 Torr) by high-frequency electric field at a frequency of 13.56 MHz. A typical RF PECVD system developed by Matsumoto [74] is shown in Figure 7. An advantage of RF plasmas is that they can be easily generated over much larger areas than microwave plasmas. Since ion bombardment from the plasma can create diamondlike carbon (DLC) films, RF PACVD is more likely to produce DLC coatings rather than diamond [75]. Low deposition rates (<0.1 μm h^{-1}) are also a drawback of this method. At atmospheric pressure, a thermal plasma can be created by RF induction as demonstrated by Matsumoto in 1987 [76]. Very high growth rates, 300–500 μm h^{-1}, have been achieved by Matsumoto and also by Koshino et al. [67]. Large are as with high deposition rates are reported by several other groups [77–83].

2.2.4.2. Hot Filament CVD

The schematic representation of the HFCVD apparatus is shown in Figure 8. The CVD reactor consists of a cylindrical stainless steel chamber. W, Ta, Mo, or Re, heated to about 1800–2400°C, is used as a heating element which serves as the source of energy to dissociate carbon containing gas and hydrogen. In this process, 2–10% hydrogen is dissociated into atomic H and methane undergoes pyrolysis reactions leading to the formation of radicals such as CH$_3$ and CH$_2$, and stable species

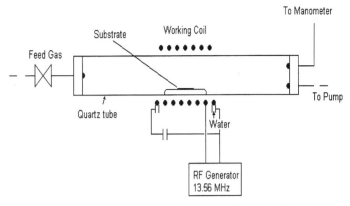

Fig. 7. Schematic diagram of a RF PACVD system. Reproduced with permission from [74].

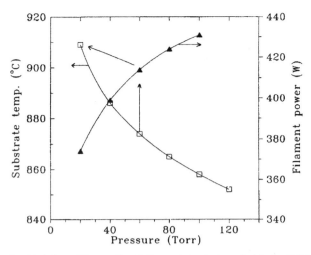

Fig. 9. Variation of the quality of diamond sheets deposited in the HFCVD chamber with pressure and T_s. A sheet deposited at 900°C at 40 Torr (marked with an arrow) measures one on the quality scale. The qualities of the rest are marked relative to this. Reproduced with permission from [84].

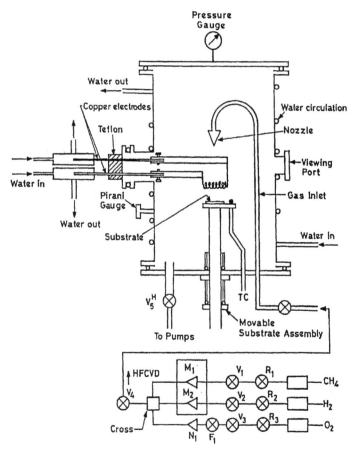

Fig. 8. Schematic diagram of the HFCVD apparatus. Reproduced with permission from [84].

such as C_2H_2, C_2H_4, and C_2H_6. Diamond is deposited on a substrate mounted at a distance of typically 0.4 to 2 cm from the heated filament and kept at temperature 600 to 900°C either by the radiation from the filament or by a separate substrate heating cooling arrangement. T_s and filament temperature are measured using an optical pyrometer and a thermocouple.

To find the exact substrate surface temperature it is advisable to measure the T_s in an independent experiment under the same

deposition conditions with the help of a tiny Chromel–Alumel thermocouple and a two color optical pyrometer. Deposition rates typically vary from 0.2 to 15 μm h^{-1} and can be increased by using oxygen containing gas in the gas phase [67].

The basic parameters to consider in HFCVD techniques are filament materials, filament carburization, substrate filament distance, chamber pressure, and substrate temperature. Carburization is done in a higher methane concentration in the gas phase till the filament current gets saturated in a particular filament power. Fixing the substrate filament distance one can optimize the process at different temperatures and pressures. It was seen that [84] keeping the T_s the same one can grow diamond with varied quality. Figure 9 shows the effect of variation of pressure and T_s on the quality of the films. The maximum value on the quality scale has been chosen to be one, which corresponds to the samples in which there is no evidence of nondiamond impurities according to Raman spectroscopy. For instance, the sheets deposited at 40 Torr and 900°C measure up to one on the quality scale. However, it may be prudent to add that the samples measuring one on the quality scale may not be completely free of sp^2 bonded carbon. The quality of the rest of the samples can be marked with respect to this normalized value. It is also to be noted that while increasing the pressure in the chamber with varying pumping speed T_s goes down due to higher heat losses at constant filament power. Also the recombination of atomic hydrogen increases substantially with pressure in the gas phase, leading to a reduced recombination effect at the substrate surface which further reduces T_s. To minimize this interdependence, the filament power is varied along with the pressure to keep T_s constant. Figure 10 shows the variation of T_s while changing pressure at constant filament power. In this figure the Y-axis on the right is taken as the power axis. It is evident that at 20 Torr, T_s is 910°C with the filament power at 374 W. However, at 60 Torr, T_s has come down to 873°C

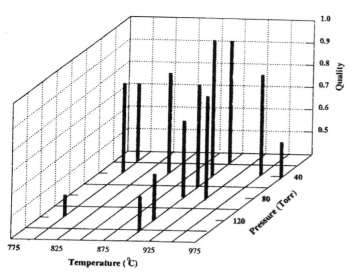

Fig. 10. Variation of T_s while changing deposition pressure. To keep the T_s at 910°C for all P_d, the filament poser is varied according to the curve plotted between the filament power and pressure. Pressure axis is the same for both the curves. Reproduced with permission from [84].

at the same power. It is then essential to increase the power to 413 W to get the T_s back to 910°C.

2.2.4.3. Flame CVD

Combustion provides gas phase activation through the highly exothermic chemical reaction between acetylene and oxygen. This process can produce diamond with high quality at atmospheric pressure. Hirose and co-workers at the Nippon Institute of Technology first demonstrated the growth of diamond by combustion [87]. The results were then confirmed by Hanssen et al. [88]. A schematic diagram of a simple combustion synthesis apparatus is illustrated in Figure 11a. In Hirose's experiments and those that followed [89–93], an oxyacetylene torch was typically used in a 1 : 1 mixture of C_2H_2 and O_2. In a flame CVD process, a small premixed flame issues from a nozzle, surrounded by a diffusion flame where excess fuel and CO continue to be oxidized. While hydrogen, oxygen, and acetylene are burned, diamond forms on the deposition substrate positioned in the reducing part of the flame at a substrate temperature of about 800–1000°C and a gas temperature of about 2000°C in the immediate vicinity of the substrate surface. As substrate is kept in direct contact with the flame it is very important to cool it actively. Normally substrates mounted on a metal holder which is screwed into a water cooled copper base (Fig. 11a). Three regions can be observed visually in the flame: the inner cone, the acetylene feather, and the outer flame. In Figure 11b, and c, the components of the flame and corresponding growth regions are shown schematically [86]. The focus of most combustion reactor designs is the optimization of the flame nozzle for production of a uniform flame nozzle and the growth of uniform film. Extensive studies on the mechanisms of diamond nucleation and growth, modification of the flame

method for large area uniform deposition, and improvement of quality of the films have extensively studied by many groups [88–90, 94–96]. Frenklach et al. [97] placed a flame in a microwave plasma to demonstrate the homogeneous nucleation of diamond particles. They also synthesized hexagonal polytypes of diamond, including a novel 6H polytype. A 15R polytype of diamond is also reported by Kapil et al. [98]. The advantages of the flame CVD are the simple technique, low cost, high growth rate, and good quality. Major drawbacks of this technique are poor uniformity and low deposition area. Tzeng et al. [99] have developed a method for large area coatings using multiple burners in combination with a rotating substrate holder.

2.2.4.4. Other Methods for Diamond Deposition

In spite of the variety of techniques described in earlier sections, development of new deposition methods continues. CVD techniques provide the capability to coat various shapes and produce many forms of diamond, such as fine powders, protective coatings, epitaxial and whisker growth, and coating of particles. Although there are many ways to produce diamond films and many applications, much needs to be accomplished before the full potential of diamond CVD is realized. This section provides a brief overview of other methods, or combinations of the above discussed methods, developed for growing diamond.

Spitsyn and Deryagin were the first to report pyrolysis of halogens, such as CBr_4 and CI_4, in a process for depositing diamond on diamond at temperatures from 800°C to 1000°C and pressures of 3×10^{-6} mm of mercury [100]. Several researchers have used halogens in conventional deposition techniques along with hydrocarbon and H_2 gas mixture [101–120]. Addition of halogen in the gas phase increases the deposition rate by abstracting hydrogen from the growing diamond surface and providing more open sites for further diamond growth.

A novel technique based on atomic layer epitaxy (ALE) [121] was developed by Hukka and co-workers [122, 123]. To achieve layer by layer growth across an entire substrate, a growth process has been chosen that is self- limiting after one atomic layer of growth. Growth can continue as long as the original surface is still exposed; once that surface is covered with one atomic layer of new material, the process stops and a new cycle must be used to begin the process again. Diamond is well suited for the ALE process because the surface structure is only stable if it is predominantly terminated with noncarbon atoms [124, 125]. Thus, cycling between two deposition chemistries, each of which results in a growth surface terminated by a different element, could be engineered to proceed by ALE. Hukka et al. [126] has used a $CHCl_3$ and H_2 mixture into a stream of atomic fluorine at pressures of 10^{-3} to 10^{-2} Torr. Beginning with a hydrogen terminating surface, a halocarbon gas is introduced to the growth chamber. A dangling bond is terminated with a halocarbon. The halocarbon then reconstructs such that the diamond surface grows by a carbon atom that is terminated by a halogen. When this process has gone to completion, the

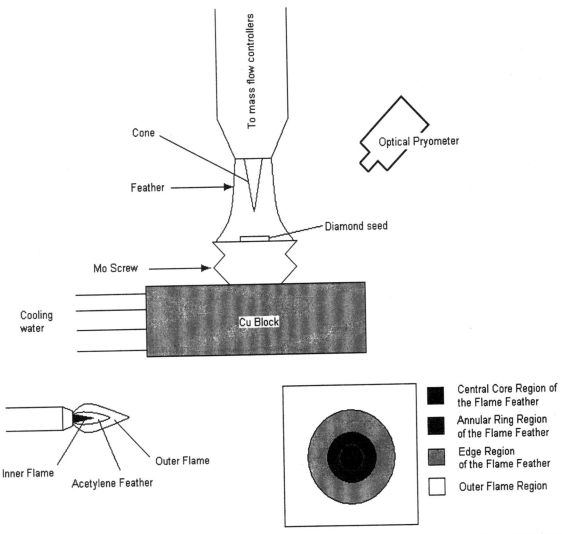

Fig. 11. Schematic diagram of (a) combustion activated CVD system, (b) the flame structure of oxyacetylene torch, (c) side and top view of growth regions. Reproduced with permission from [85, 86], copyright 1993, Kluwer Academic Publishers.

substrate surface is terminated by halogens instead of hydrogens. Hydrogen is then introduced into the system to react with the halogens to form gaseous hybrids and terminate the gangling bonds with hydrogen. When this part of the cycle has gone to completion, the substrate surface is hydrogen terminated and the process can be repeated with the introduction of the halocarbon. This process has been used for homoepitaxy [126] and for growth on Cu [127].

Laser assisted growth of diamond is used by many researchers. Laser-assisted techniques use the laser to couple photon energy into the reactants and the substrate. Fedoseev et al. [128, 129] reported using a CO_2 laser to pyrolyze small particles of graphite of carbon black to form diamond powder. Molian and Waschek reported a CO_2 laser CVD process in which a gas mixture of 2% CH_4 in balanced H_2 was pyrolyzed to deposit diamond on GaAs or Si substrate [130]. Katahama et al. reported diamond deposition on Si substrate using 193 nm irradiation from an ArF laser to activate a 0.5–2% C_2H_2 in bal-

ance H_2 [131, 132]. But upon reinvestigation of this founding it was confirmed that the deposit was a graphite layer [133].

QQC, Inc. has developed a novel diamond deposition process which does not require a vacuum, hydrogen, substrate heating, or surface pretreatment [134] and thereby offers great potential for coating aluminum, plastics, and other low-melting materials. This process uses a combination of lasers, ranging from the UV to the IR. Plasma formed by these lasers can create a wide variety of chemically reactive species as well as very hot electrons, which produce a unique crystal growth environment. The process utilizes three types of lasers: two UV excimer lasers, an IR CO_2 laser, and an IR Nd:YAG laser. Each laser emits a beam which is directed through an opening of a nozzle toward the substrate surface. CO_2 and N_2 are used as carbon source and shielding gas, respectively.

Pulsed laser deposition has been used for diamond deposition from graphite, or carbon target. But technologies dependent on ion ablation of graphite of other PVD for depositing diamond produce basically nondiamond carbon along with a small fraction of diamond [135–137]. Few reports show the formation of diamond using nondiamond targets [138–143]. PVD methods thus are not viable for producing diamond for commercial applications.

2.2.4.5. Epitaxy and Oriented Diamond Deposition

Homoepitaxial diamond growth has been demonstrated by many researchers [144, 145]. But heteroepitaxial growth of diamond is very important for utilizing its unique properties in different applications. The primary difficulty inherent in this issue is the small number of materials with suitable crystal structures and lattice constants (Ni, Cu, Fe, cBN, BeO, SiC, W, Mo, etc.). Several authors [57, 146–149] have studied the growth of epitaxial diamond on nondiamond substrates. But epitaxial film growth is less promising than the oriented diamond films grown by bias enhancement or other techniques.

A polycrystalline diamond surface on nondiamond substrate is typically rough because it consists of {100} and {111} facets. The smooth surface that many electronic applications require may be obtained by postdeposition polishing; such techniques add considerable expense to the final product and contribute stress, defects, and contamination to the film surface. Also conventional surface pretreatments to grow diamond damage the substrate surface and hence hinder a lot of applications, such as diamond electronics and optical applications. So it would be desirable to produce nucleation sites without damaging the underlying substrate. One method for increasing nucleation without damaging the substrate material is bias enhanced nucleation (BEN) [150, 151]. The mechanism responsible for texture development in polycrystalline films in general can be described by the principle of evolutionary selection described by van der Drift [152]. The two assumptions upon which evolutionary selection is based, (i) a crystallite morphology that is independent of crystallite orientation and (ii) an absence of secondary nucleation, are met in the growth of diamond particles and films [153]. Evolutionary selection occurs when some fraction of the grains in a growing polycrystalline film is oriented such that the crystal direction of most rapid growth is normal to the substrate surface. This favorable orientation allows these grains to dominate the film morphology as the film grows. Textured films have been demonstrated in diamond for <100>, [153–155], <111> [155], and <110> [156–159] growth directions. Wild et al. [153] have given the growth parameter depending on the relative growth of {100} and {111} faces: $\alpha = V_{100}/V_{111} \cdot \sqrt{3}$, where V_{hkl} refers to the growth rate on the {hkl} plane. The resulting crystal morphologies corresponding to $1 \leq \alpha \leq 3$ are shown in Figure 12. The direction of the fastest growth for each crystal shape in Figure 12 is indicated by the arrow [153]. This idea can be explained along with the diamond heteroepitaxial growth sequence on β-SiC (001) and is shown schematically in Figure 13 [153]. In order to obtain [001]-oriented diamond films after BEN (Fig. 13a), the direction of fastest growth must be [001]; that is, the α-parameter must be close to or above 3. The alignment of the direction of

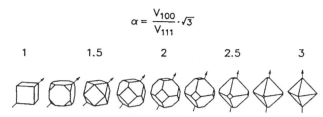

$$\alpha = \frac{V_{100}}{V_{111}} \cdot \sqrt{3}$$

Fig. 12. Idiomorphic crystal shapes for different values of the growth parameter α. The arrows indicate the direction of fastest growth. Reproduced with permission from [153].

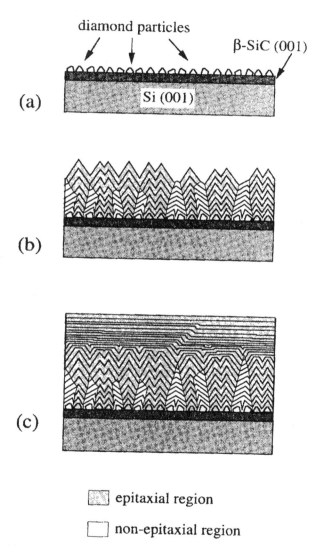

Fig. 13. A schematic of the growth of heteroepitaxial diamond on β-SiC(001). (a) BEN. (b) [001] fast growth with $\alpha \approx 3.0$. (c) Smooth growth of the (001) surface with $\alpha < 3.0$. Reproduced with permission from [160], copyright 1993, publisher Elsevier Science. Reproduced with permission from [161], copyright 1995, Springer-Verlag.

fastest growth leads to an alignment of [001] directions perpendicular to the substrate surface (Fig. 13b). At this stage, the surface has a rough structure, with pyramide like shapes. Subsequently, the α-parameter is decreased to below 3 and as a result, [001] facets appear on the tips of these pyramids and grow larger rapidly. By this technique a smooth surface can be realized (Fig. 13c).

The BEN technique, pioneered by the team of Yugo et al in 1990 [162], is conceptually straightforward—an electrical bias is applied to an untreated electrically conducting substrate, under similar conditions to CVD diamond growth, for a short period (a few minutes to several hours) prior to actual deposition. This treatment was found to produce a vast increase in nucleation density from 10^3–10^5 cm^{-2} (pristine Si) [163, 164] to $>10^{10}$ cm^{-2} [164, 165] (BEN). Typically, for comparison, manual abrasion can achieve a nucleation density of $\sim 10^7$ cm^{-2} [165]. The bias applied to the substrate is usually negative DC of ~ 200 V magnitude, with respect to an anode (which in many cases is the grounded CVD chamber walls). The advantage of the process is that it allows the user to nucleate diamond without mechanically damaging the substrate, as is the case for manual abrasion, for example. In principle, conventional methods described in earlier sections can be used to grow textured or oriented diamond films. Among them the MPCVD technique [166–168] is generally used because of the presence of ions and the ability to control the bias between the substrate and the plasma at relatively high pressure. The HFCVD technique is also used by several researchers for bias enhanced nucleation and growth of oriented diamond films on different substrates [169–173].

BEN is generally carried out under similar conditions to diamond CVD, but prior to the deposition. Typically, an enriched carbon feed is used, often ~ 2–5% CH$_4$/H$_2$ [164, 165, 174, 175], but sometimes as high as 15% [176, 177]. Other studies found it beneficial to instead use a prebias carburization step, where a short period (~ 1 hour) of CVD diamond growth was carried out on the otherwise untreated wafer under comparatively CH$_4$ rich conditions (2–5%), before application of the bias voltage [178, 179].

The diamond crystallites nucleated during BEN are sometimes oriented with the texture of the SiC layer, and hence the Si wafer underneath [178–180]. This causes diamond crystals formed during BEN to align with the crystallographic planes of the Si substrate, allowing growth of oriented diamond particles and films. Oriented, polycrystalline diamond films have one particular crystal facet parallel to the substrate surface and hence can be made very smooth. This would be beneficial for many applications of CVD diamond technology as it allows growth of smooth diamond surfaces (optics applications), or heteroepitaxial single crystal growth (microelectronics applications).

2.2.4.6. Nanocrystalline Diamond Deposition

In general surface scratching or substrate biasing are applied for nucleation of diamond on nondiamond substrates. The first method produces normally randomly oriented diamond films with high surface roughness. The second method though could produce oriented smooth diamond films, but in both cases the films are comprised of micron-sized crystallites and sizes increase with the increase of the thickness of the coatings. A thick diamond films therefore has a low density of grain boundary and a rough surface. This is because of the columnar nature of growth by the CVD method. If it is possible to interrupt the grain growth process by a continuous secondary nucleation process, diamond films with decreased grain size should be formed [181]. It is well known that the grain size of a film strongly affects its properties; this can be attributed to the grain boundary density [182]. Nanocrystalline diamond films with high grain boundary density have attracted enormous interest due their to fascinating mechanical, electrical, and optical properties [183]. Due to the negative electron affinity diamond is an optimum candidate for field electron emission (FEE), which has potential applications in areas such as flat panel displays and microelectronic devices [184]. Reduction in grain size may increase the conducting pathways; it is possible to improve diamond FEE by depositing size controlled diamond films. Several articles relating growth and characterization of nanodiamond have been published [185], but control growth for specific applications needs more careful research on this important material. Gruen et al. reported the nucleation and growth of nanocrystalline diamond film on scratched Si (100) from H$_2$/Ar plasma using fullerenes as carbon precursors [186, 187]. With the addition of N$_2$ in the precursor gases a nanocrystalline phase of diamond can be produced [188]. Schaller et al. described the surface properties of nanodiamond films deposited by electrophoresis on flat Si (001) using nanometersized diamond particles. Gu and Jiang [181] developed a new method to synthesize nanocrystalline diamond films. In their MPCVD reactor a negative bias is applied for initial nucleation and films are grown with bombardment of H$^+$ ion of different energies. Scanning electron microscopy (SEM) and Raman spectroscopy reveal the signature of nanophase diamond in their growth technique. Field emission properties of nanocrystalline diamond have been studied by many authors who observed a threshold as low as <3 V/μm [184, 189–191]. Using fullerenes, which are spherical molecules of pure carbon containing 60 carbon atoms, Argonne National Laboratory developed a nobel method to grow nanocrystalline diamond [192]. Fullerene powder is vaporized and introduced into an argon plasma, causing the spheres to fragment into two-atom carbon molecules (dimers). They also showed that dimers can be formed by introducing CH$_4$ into an argon plasma without adding any H$_2$. The plasma forms a ball over a wafer of silicon or other substrate materials on which a fine grained diamond powder formed. As the carbon dimers settle out of the plasma, onto the substrate, they arrange themselves into films of small diamond crystals about 3–5 nm in diameter.

Ultrananocrystalline diamonds produced by Argonne National Laboratory researchers and others have superior mechanical, tribological, and thermal properties suitable for the rapidly expanding field of microelectromechanical systems (MEMS) technology. MEMS is a manufacturing strategy that integrates miniature mechanical devices and semiconductor microcircuitry on a silicon chip. The result is a chip that not only thinks but also senses and acts. In the new market for MEMS, many potential applications for these microscopic devices are not, in fact, practical because the properties of the material currently used, silicon, are not suitable. Table VIII shows the markets and possible products for MEMS [192]. The exceptional physical properties of diamond (hardness, wear resistance, low coefficient of friction comparable with that of Teflon, and thermal and chemical stability) would expand the range of applications for microdevices.

2.2.5. Nucleation Mechanism

It is very important to understand the fundamental phenomena related to the nucleation and growth of this technologically important material. In order to grow single crystal diamond which is necessary for diamond electronics and protective coatings on soft and low melting point materials it is necessary to understand the initial growth mechanism of diamond. Several groups have studied this important issues of diamond growth [193, 194]. Though emphasis of most studies on nucleation and growth of diamond has been placed on the heterogeneous formation of diamond particles and the crystallization and deposition of diamond films on substrate surfaces, there is evidence of showing that diamond can nucleate homogeneously in the gas phase also [195, 196].

Diamond can grow homoepitaxially on a diamond surface without nucleation problems. Also diamond easily nucleates on cBN due to the identical crystal structure. In the conventional growth process in CVD of polycrystalline diamond films there are several states: (a) incubation period, (b) 3-D nucleation of individual crystallites, (c) coalescence of individual crystallites, and (d) growth of continuous film. Normally diamond does not nucleate directly on a nondiamond substrate surface. Always an intermediate layer forms on the substrate surface before diamond starts nucleation.

The nature of the intermediate layer normally depends on the chemical nature of the substrates. An a-C layer normally

grows on the substrates like Cu which does not react in the typical CVD atmosphere [197]. Diamond can nucleate without pretreatment on foreign substrates which form their carbides (Si, Mo, Ta, W). It is established that diamond nucleates on Si through the β-SiC buffer layer [198]. Due to the catalytic effect and high diffusivity of carbon in Fe, Co, and Ni diamond nucleate through a intermediate layer of graphite. This also happens with the Pt and sometimes with Si and Cu substrates.

2.3. Characterization of Diamond Films

The growth of CVD diamond films has been optimized so much that the properties of the films are reaching very close to those of their natural counterpart. However, the defects and impurities which get incorporated in the films due to the nature of the low pressure deposition process make it difficult to attain extreme perfection in the films. Often various properties of the films are controlled by these defects and impurities. To exploit the potential of CVD diamond fully, it is necessary to investigate the nature of the defects and impurities in the films and understand their origin. In particular, nondiamond carbon along with hydrogen and related impurities has very serious influence on the characteristics of the films [199]. In addition to hydrogenous impurities various other defects and centers, related to boron, oxygen, nitrogen, silicon, etc., have also been identified [200]. There have been several reports [201–216] on the identification of the various defects which are unique to CVD diamond. Similarly the grain boundaries, vacancies and other related structural defects also play an important role in controlling the electrical and optical characteristics of the films [199, 217, 218]. Figure 14 presents a schematic representation of the various kinds of possible defect centers in diamond.

Table VIII. Markets and Possible Products for MEMS

Markets	Possible products
Aerospace systems	Miniaturized power for aerospace and communications application
Information processing	Vibration sensors
Communication systems	Acceleration sensors
Auto and transportation	Gyroscopes
Consumer products	Precision positioners
Medical and biological technologies	Flow-control values
Biochemical analysis	Pressure transducers
Manufacturing automation	Microwave signal generators
Environmental control	Precision instruments for laser-optic surgery

Reproduced with permission from [192] http://www.techtransfer.anl.gov/techour/diamoandmems.html.

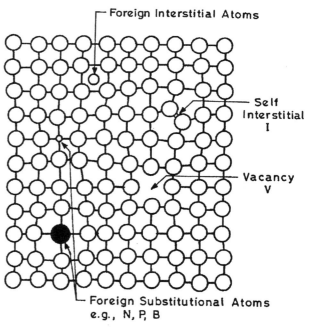

Fig. 14. Schematic representation of various types of possible defect centers in diamond. Reproduced with permission from [280], copyright 1995, Kluwer Academic Publishers.

Table IX.   Techniques for Characterization of CVD Diamond Films

Technique	Basic physical process		Type of information
Vibrational spectroscopy	Raman spectroscopy	Inelastic light scattering with lattice vibration which changes the polarizability	Carbon bonding (sp^2/sp^3); structural perfection stress state, crystalline (domain) size
	Infrared spectroscopy	Inelastic light scattering with lattice vibration which changes the existing dipole movement	Chemical bonding (C—C, C—H, C—Si), adsorption coefficient, refractive index, impurity
Electron spectroscopy	X-Ray photoelectron spectroscopy (XPS)	Excitation and emission of electrons by incident X-rays	Surface species and chemical bonding (C, C—O, Si—C; surface electronic state
	Auger electron spectroscopy	Creation of inner hole by incident electrons beams which is subsequently filled by an outer electron accompanied by the release of Auger electron	Carbon bonding (sp^2/sp^3) surface electronic structure
Electron microscopy	Transmission electron microscopy (TEM)	High energy (50–200 keV) electron beam passing through a thin sample and forming an image, the contrast of which is determined by the crystallographic orientation or the atomic mass	Internal structural defects (twins, stacking faults, dislocations, secondary phases interfaces, etc.)
	SEM	Scanning and focusing an electron beam (5–30 keV) over a surface and sensing the secondary electrons emitted to form an image	Surface morphology
Diffraction methods	X-ray diffraction (XRD)	X-ray plastic scattering with atoms (Bragg diffraction)	Phase (structure) identification, lattice constant measurement, crystal orientation or texture, residual stress
	Electron selected area diffraction	High energy (50–200 keV) electron elastic scattering with atoms (Bragg diffraction)	Phase (structure) identification, lattice constant measurement, crystal orientation or texture, esp. when the sample is thin or both the size and amount are small
	Low energy electron diffraction	Low energy (10–200 keV) electrons backscattered on the surface (Bragg diffraction surface specific)	Surface structure reconstruction
	Reflection high energy electron diffraction	High energy (30–100 keV) electron elastic scattering with atoms (Bragg diffraction, glazing incidence, and emergence angles, surface specific)	Surface structure
Others	Secondary ion mass spectroscopy	Scanning and focusing an electron beam (5–30 keV) over a surface and sensing the secondary electrons emitted to form an image	Surface morphology
	Luminescence spectroscopy	Excitation of electrons by photon or electron radiation resulting in emission of photons	Electronic state, point defects
	Rutherford scattering spectroscopy	Collision of He^+ ions (1–3 MeV) with "target" atoms leading to energy transfer	Quantitative atomic composition (including H), depth profile
	Spectroscopic ellipsometry	Ellipsometry angles of psuedomorphic dielectric functions are modeled by linear regression analysis and multiplayer optical techniques to deduce photon energy independent structural parameters.	Microstructure (volume fraction of sp^2/sp^3 bonded carbon), layer thickness, densities, refractive index

Reproduced with permission from [219].

A number of characterization techniques have been employed to monitor the impurities and defects in diamond films. Table IX provides the brief description of characterization techniques along with the information they can produce.

2.3.1. Scanning Electron Microscopy

SEM is one of the most widely used techniques to characterize thin film surface morphology. Cross-sectional SEM has also been employed for interface studies and measurement of thickness of the films. It operates by scanning a focused electron beam over a surface and sensing the secondary electrons emitted from the surface. The electron beam can be focused on a very small spot (5–20 nm) and the beam size is the detrimental factor of the microscopy resolution. SEM is an easier and simpler technique than TEM where sample preparations are often very difficult. In addition, the energy dispersive X-ray (EDX) mode of operation allows one to perform elemental identification and distribution in the film. It is often necessary to coat a diamond film surface with a thin (~10 nm) layer of gold or carbon for avoiding charging effects. Working at low electron voltage is always useful for avoiding the charging effect while working with insulating films. It has been recognized that diamond films produced on nondiamond substrates are comprised of different surface morphologies depending on the deposition conditions and substrate chemical nature. Typical surface morphologies of the sheets grown at different deposition pressures (P_d) in a HFCVD reactor with 0.8% CH_4 in balanced hydrogen are shown in Figure 15 [84]. Substrate temperature was kept fixed at 890°C for the entire deposition. It can be seen

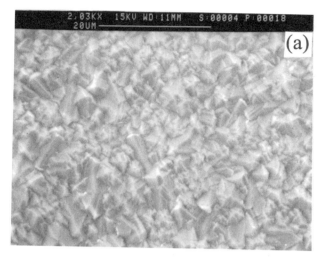

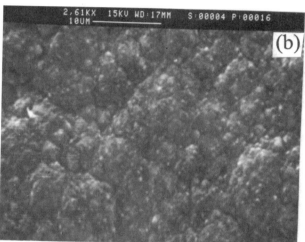

Fig. 15. SEM micrographs of diamond sheets grown in the HFCVD chamber at different pressures: (a) 20 Torr, (b) 120 Torr. For all the CH_4 deposition, was 0.8%, in balanced H_2 with $T_s = 890°C$. Reproduced with permission from [84].

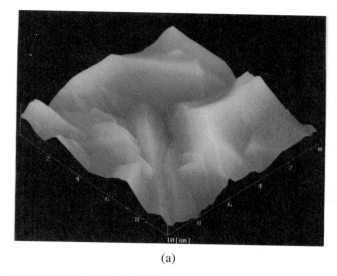

(a)

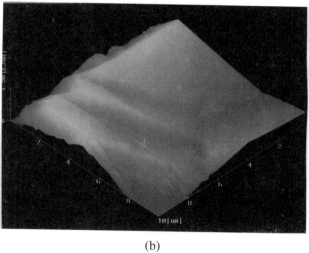

(b)

Fig. 16. Typical AFM image scanned in an area of $10 \times 10 \mu m^2$ of the growing surface of diamond sheets grown in HFCVD chamber at (a) 20 Torr, (b) 40 Torr, (c) 60 Torr, and (d) 80 Torr. Reproduced with permission from [84].

from the Figure 15a that sheets grown at lower P_d are comprised of sharp crystallites. Crystals are randomly oriented but dominated by {111} and {220} textured grains. It is observed from Figure 15b that a further increase in P_d makes the crystallite spherulitical and particle sizes become submicron. A higher methane concentration affects the film morphology similar to the films grown at higher P_d. Several researchers have studied the variation of morphology of the films grown at different deposition conditions with the help of SEM [220–223].

2.3.2. Atomic Force Microscopy (AFM)

AFM is primarily based upon surface imaging by monitoring variation in attractive/repulsive atomic forces such as van der Waal forces, Born repulsion, and electrostatic and magnetic forces, between adjacent surfaces. It basically consists of a cantilever-type sensor that responds to a force (range of $\sim 10^{-8}$–10^{-13} N) and a detector that measures that sensor's re-

sponse, i.e., position/movement of tip mounted on cantilever. The output of sensor thus provides a force map and hence the image of sample surface under consideration. AFM has been extensively used to investigate the surface morphology, roughness, and initial stage of nucleation [224–226]. The typical surface morphology of diamond films grown in a HFCVD chamber is shown in Figure 16 [84].

2.3.3. Transmission Electron Microscopy

TEM is a powerful structural characterization technique. It employs an accelerated electron beam associated with a much shorter wavelength, which enables the operator to obtain fundamentally better resolutions and substantially higher useful magnifications. The electron beams can be deflected and focused by electrostatic or electromagnetic lenses in complete analogy to the focusing of light waves by glass lenses. TEM

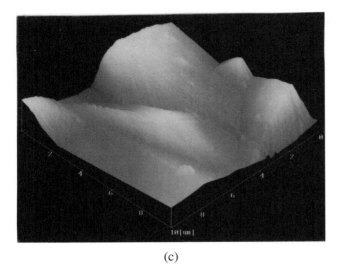

(c)

(d)

Fig. 16. (Continued).

has very high magnifications (in the range of 8000–300,000×) associated with a high resolution (0.2–0.3 nm using 100 kV acceleration voltage), allowing very detailed study of localized regions of the sample. Selected area diffraction allows detailed correlation between the image and the electron diffraction pattern. Specifically, TEM is capable of detecting and analyzing various internal structural defects, such as lattice imperfections of staking faults, twins and dislocations, very small precipitates of second phase inclusions, grain boundaries, and interfaces. The basic requirement for a TEM is the thinning the sample down to 50–300 nm, depending upon their atomic weights, in order to make them electron transparent. A diamond sample can be thinned by passing a stream of dry O_2 at atmospheric pressure over their surface at 750°C. CVD diamond samples are usually thinned by mechanical polishing and ion milling. Extensive TEM investigation of structural defects in CVD diamond is reported in the literature [227–233]. Some typical TEM micrographs, showing different defect structures, are shown in Figure 17 [229].

2.3.4. X-Ray Diffraction

XRD is a very important experimental technique that has long been used to address all issues related to the crystal structure of bulk solids and thin films, including lattice constants, identification of unknown materials, orientation of single crystals, and preferred orientation of polycrystals, defects, stresses, and particle sizes. XRD methods have advantages because they are nondestructive and do not require elaborate sample preparation (like TEM) or film removal from the substrate. Principles of XRD are based on Bragg's diffraction law [234]: $2d \sin \theta = n\lambda$ where d is the interplaner spacing, θ is the diffraction angle, n is an integer and, λ is the incident wavelength. An X-ray diffractometer usually consists of an X-ray source, a goniometer, a counter tube, and counting electronics. Results are obtained in the form of a chart of peak intensity vs 2θ. For CVD diamond films, the XRD method has been employed for lattice parameter measurement, texture determination, and other crystalline phases of carbon. To obtain a pattern characteristic of polycrystalline diamond requires that the films are comprised of crystallites of the order of ~10 nm or larger. Due to the high symmetry of the diamond, cubic crystal system, structure factor gives very few allowable diffracting planes ({111}, {220}, {311}, {400}, and {331}). Typical XRD patterns of {111} and {100} surface dominated films are shown in Figure 18 [238]. Several researchers have employed the XRD method to study surface morphology, lattice spacings, and texture analysis [229, 230, 235–237].

2.3.4.1. Line Broadening

Figure 19 shows a high resolution diffraction pattern recorded in the region of the (111) peak using narrower slits and an angular step of 0.02° [84]. The diffractometer resolution function was also obtained for these settings using the silicon X-ray diffraction internal d-spacing standard (NIST 640 B) powder. Note that the diffraction peak for the 100 Torr sample is broader than that of the 20 Torr sample. These samples were grown in a HFCVD chamber [84]. The true full width at half maxima (FWHM) of the (111) peak was obtained from the measured FWHM after correcting for the intrinsic resolution of the diffractometer assuming a Gaussian profile. The true FWHM thus obtained was used for estimating the particle size in various samples using the Scherrer formula,

$$B = 0.9\lambda/D \cos \theta,$$

where B = broadening of the diffraction line measured at half its maximum intensity (radians) and D = diameter of the particle. The widths of the diffraction peak for the good crystalline samples are found to be very close to the diffraction limited resolution. Hence the particle sizes for these samples are expected to be in the micrometer range and hence the Scherrer formula is not applicable. The particle size of the film grown at 100 Torr is estimated to be in the range of ~30 nm [239].

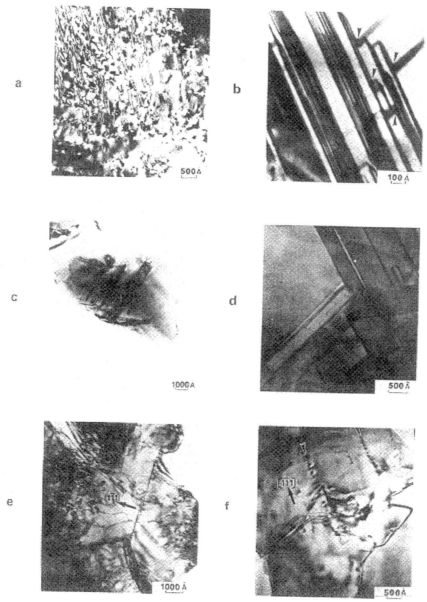

Fig. 17. Bright field TEM images of typical defects in CVD diamond: (a) and (b) {111} stacking faults, (c) and (d) {111} twins, and (e) and (f) dislocations. Arrows in image (b) indicate interactions between stacking faults, and arrows in image (f) indicate aligned end-on dislocations. Reproduced with permission from [229], Noyes Publishing.

2.3.5. Raman Spectroscopy

Raman spectroscopy has been the most widely used tool for the examination of the types of bonding present in carbon films. It is a complementary technique to IR spectroscopy. The Raman signals correspond to inelastically scattered light resulting from the radiative emission of dipoles induced by the electric field of the incident light and coupled with atomic vibrations. The Raman efficiency depends on the polarizability and electron–phonon interaction of the solid. Using Raman spectroscopy, diamond, which has a phonon density of states very different from other carbon phases, can be detected without any ambiguity. The Raman signal is very sensitive to short-range disorder and subsequently, it can reveal different forms of amorphous carbon i.e., a-C or a-C:H, and graphite. The position of various Raman bands, their FWHM values, observed in CVD diamond films, and their corresponding assignments are given in Table X.

Diamond belongs to the face centered cubic lattice with space group $Fd3m$. There are two carbon atoms in the primitive cell and it has a single triply degenerate first order phonon with symmetry T_{2g}. From the selection rules for the factor group O_h, this mode is expected to be Raman active only; the diamond structure has no first order infrared absorption. Raman

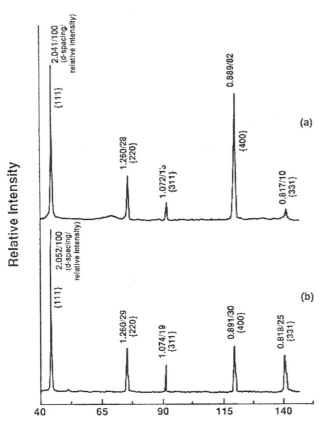

Fig. 18.   Typical XRD patterns for diamond films with (a) {100} facets; and (b) {111} facets on the surface. Reproduced with permission from [238], Noyes Publishing.

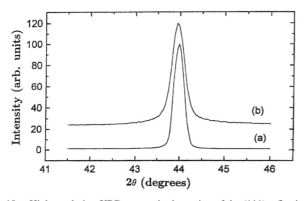

Fig. 19.   High resolution XRD pattern in the region of the (111) reflection of diamond phase. Note that the width of the diffraction peak for the 100 Torr sample (b) is more than that of the 20 Torr sample (a) grown in the HFCVD chamber. Reproduced with permission from [84].

Table X.   Observed Raman Bands in CVD Diamond Films

Position (cm^{-1})	FWHM (cm^{-1})	Assignments	Modes
1130–1160	30–60	a-diamond	
1180–1250	70–130	Diamond polytypes	
1330–1340	4–15	Cubic diamond	T_{2g}
1340–1380	120–200	Graphitic band D	E_g
1430–1540	70–130	Mixed phase	
1550–1600	130–200	Graphite band G	

The presence of disorder or small crystallite size gives rise to a Raman peak at 1355 cm^{-1}, known as the graphitic D-band. In CVD diamond the position of the graphitic G-band appears at slightly lower wave numbers than the actual position (1581 cm^{-1}) observed for highly oriented pyrolytic graphite. In highly disordered graphitic carbons, which may contain tetrahedrally as well as trigonally bound C (sp^3 and sp^2 hybridization, respectively), the width, position, and relative position of the D- and G-bands can vary significantly. It has been established that the particle size of microcrystalline graphite can be calculated from the intensity ratio of the D-band and G-band [241].

In many diamond films, one observes not only a sharp diamond peak at 1332 cm^{-1} but also a broad band centered at $\sim$ 1500 cm^{-1}, commonly known as the band due to nondiamond carbon. The broad band is a superposition of the G-band, D-band, and mixed band (M-band). The position and intensity of this band depends on the deposition conditions and the wavelength of the exciting photon [247]. Though origin of the M-band is not clearly understood, the most commonly accepted assignment is that it is due to the highly disordered C phase (diamond like carbon), consisting of sp^3 and sp^2 hybridized carbon. Nanocrystalline (crystallite sizes of 1–100 nm) diamond samples show additional features including a broad peak centered at 1133 cm^{-1}. The peak arises from the effects of small size or disorder in the tetrahedral carbon network similar to the explanation suggested for the microcrystalline graphitic peak at 1355 cm^{-1}. Raman active bands for various diamond polytypes have also been predicted theoretically and their presence has been confirmed in CVD diamond [242].

Laser Raman spectroscopy is also useful in determining the presence of sp^3 bonded clusters in the presence of an amorphous carbon background. This is particularly important for so-called amorphic diamond films. These are hard carbon films which contain a high proportion of sp^3-bonded carbon and may be said to exhibit short-range order but no long-range order. Compared to diamondlike films, which can contain up to 30% hydrogen, amorphic diamond films contain very little, or no, hydrogen. Raman spectra of laser ablated amorphic diamond films generally show a broad band centered at $\sim$1500 cm^{-1} with a large tail towards lower wave numbers. This should be contrasted with the Raman spectrum from DLC, which typically displays a band peaking at around 1555 cm^{-1}.

spectrum of diamond is distinct from that of graphite, which appears at 1581 cm^{-1}, known as the graphitic G-band (Raman active mode is E_{2g}). The value of the Raman scattering coefficient for diamond (6×10^{-7} cm^{-1} sr^{-1}) is $\sim$50 times less than that for graphite (307×10^{-7} cm^{-1} sr^{-1}) [240]. Thus even a very small concentration of the sp^2 phase can be easily detected using Raman spectroscopy.

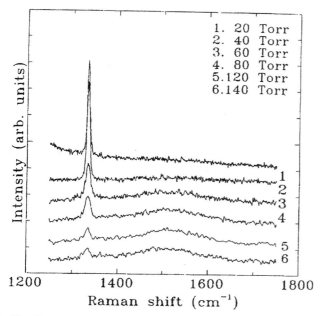

Fig. 20. Typical Raman spectra recorded on the sheets grown at different pressures with 0.8% CH_4 in balance H_2 and $T_s = 890°C$. All the spectra show the characteristic Raman line at 1332 cm^{-1} corresponding to crystalline diamond. Spectra also show a broad band around 1500 cm^{-1} assigned to the nondiamond phase of carbon. Reproduced with permission from [84].

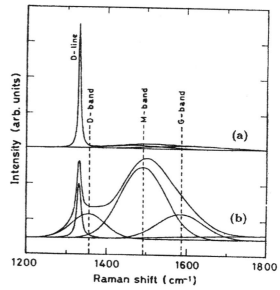

Fig. 21. Typical deconvoluted Raman spectra; (a) sheet grown at 20 Torr and (b) sheet grown at 120 Torr. Reproduced with permission from [84].

Broadening of the 1332 cm^{-1} diamond Raman line has been reported by many authors [243–245], which was earlier assigned to strains. Ager et al. [246] have carried out detailed Raman investigations of CVD diamond films and found a symmetric broadening and blue shift of the 1332 cm^{-1} line which could not be accounted for using the Gaussian confinement model [246]. Hence scattering from defects and compressive strains were invoked to explain the observed behavior. Although there have been some studies on nanocrystalline diamond, lineshape analysis of the Raman spectrum of nanocrystalline diamond coexisting with a-C has not been carried out.

Laser Raman spectra are normally recorded in the range 1000–1700 cm^{-1} with a step size of 2 cm^{-1}. To resolve the fine structure of the Raman diamond line, in the range 1280–1380 cm^{-1}, a step size of 0.5 cm^{-1} can be used. An Ar$^+$ laser ($\lambda = 488$ nm) with 50 mW power was used for recording the spectra shown in Figure 20 [84]. A sharp Raman line at 1332.5 cm^{-1} is observed in all the sheets grown at different P_d in HFCVD chamber [84]. This implies that all the sheets contain crystalline diamond [241, 245]. A broad band corresponding to the nondiamond impurities [246] also appears at around 1500 cm^{-1} in the sheets. The nondiamond carbon components in the sheets increase systematically with the increase in P_d in the HFCVD chamber. Fitting of the spectra can be done with three Gaussian peaks and one Lorentzian peak in order to identify the different phases of carbon that exist in the sheets.

Two typical fitted Raman spectra are shown in Figure 21 and they consist of D (crystalline diamond), graphitic D-band, M (mixed phase of sp^3 and sp^2 hybridized carbon), and graphitic G-band phases [84]. An increase of nondiamond carbon in the

sheets with P_d can be explained with the rate of etching and growth of sp^2 and sp^3 bonded carbon in the CVD environment. Atomic H is known to desorp H from the growing surface and stabilize a sp^3 precursor for further growth. Therefore, a continuous and sufficient supply of impingement flux density of atomic hydrogen (IFDH) on the growing surface is required in the CVD process for depositing diamond. Insufficient IFDH will leave a few C—H bonds intact and hence hydrogen may get incorporated in the diamond lattice. Inside the diamond lattice, the termination of the sp^3 carbon bond with H may give rise to the sp^2 bonding in the surrounding environment. This implies that the bonded H in the diamond films, which is a result of insufficient IFDH, will give rise to more hydrogenated carbon impurities. In the HFCVD process the dissociation of H_2 into H atoms takes place in the vicinity of the hot filament. The recombination of H atoms occurs during their movement toward the substrate. The recombination rate will increase as the mean free path of H atoms decreases with increase in P_d. This will result in lower IFDH at higher P_d and higher nondiamond carbon as well as higher H concentrations in the sheets. This hypothesis, although a simplified picture of a complex situation, can be used to explain the well-observed correlation between the H content and the nondiamond carbon impurities reported by several groups [199, 248, 249].

Wagner et al. [247] used resonant Raman scattering to study the incorporation of sp^2 bonded carbon in polycrystalline diamond films. They found that the broad band at 1500 cm^{-1} corresponding to the a-C phase is shifted to a higher wave number side with the increase of incident photon energy. This shift was interpreted in terms of scattering from π bonded carbon clusters which is resonantly enhanced for photon energies approaching the π^* resonance of sp^2 bonded carbon. At the same time no shift was observed in the Raman diamond line. Robertson and O'Reilly [250] have shown that the most stable configuration of sp^2 sites is in the form of clusters of fused six-fold rings. It is likely that in the HFCVD chamber at high P_d deposited films graphitic inclusions are present in large concentrations as proposed by Robertson and O'Reilly. Knight and White [241] also pointed out that the graphitic carbon in the diamond films has the smallest particle size and is the most disordered form of carbon in diamond films.

Figure 22 shows that the linewidth of the diamond Raman line is high in the sheets grown at higher P_d in HFCVD chamber. Further it can be seen from Figure 22 that the diamond line is symmetric for low pressure grown samples, while for high pressure samples it has marked asymmetry on the low frequency side. As expected the FWHM values of diamond lines of low pressure deposited sheets are higher than those of natural diamond. This may be related to the scattering of phonons at defects, impurities, and grain boundaries in CVD diamond.

It is well established that Raman lines associated with an optical phonon in a small particle can exhibit asymmetric broadening [251, 252] similar to that found in the high P_d deposited sheets. The asymmetry arises because of the contribution of phonons away from the Brillouin-zone center to the Raman scattering process. This happens because of the relaxation of the $q = 0$ selection rule in the case of a small particle. Richter [251] proposed a Gaussian confinement model for the calculation of confined-phonon lineshape and evaluated the effect of particle size on the linewidth of nanocrystalline Si. A Gaussian confinement model alone could not define the asymmetry of the Raman line observed by Arora et al. [239]. An alternative discrete phonon confinement model by Arora et al. has also been proposed to explain the observed asymmetry of the diamond Raman line [239].

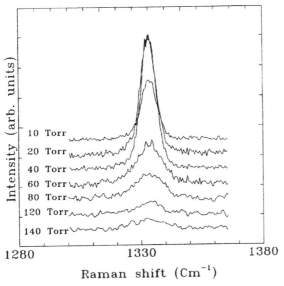

Fig. 22. Raman spectra in the region of 1300–1365 cm^{-1}, recorded with a step size of 0.5 cm^{-1} of the diamond sheets grown at different P_d in the HFCVD chamber. Reproduced with permission from [84].

2.3.6. Fourier Transform Infrared Spectroscopy (FTIR)

IR spectroscopy is based on inelastic molecular light-scattering arising from the interaction of photons with lattice vibrations or phonons. The selection rule for IR active vibration modes is that the vibration produces a finite change in the existing dipole moment, as opposed to the change of polarizability for Raman active vibration modes. An IR spectrum can be measured either by dispersive or interferometric methods [253, 254]. The dispersive method involves dispersing radiation from an incoherent source via a prism or grating, which requires monochromator entrance and exit slits. FTIR samples all the frequencies of radiation from the source using an interferometric method resulting in improved signal-to-noise ratios.

Measurements of the concentration of H atoms in diamond films have been carried out by various groups [255–257] and elastic recoil detection analysis (ERDA), nuclear reaction analysis (NRA), nuclear magnetic resonance (NMR), and infrared spectroscopy (IR) techniques have been used for this purpose. However, the concentrations obtained by various techniques do not show good agreement. Hydrogen concentration as measured using NMR is generally much lower than that obtained in IR techniques [257]. ERDA and NRA measurements of H concentration also show similar results, i.e., higher concentration than from NMR studies. Recent ERDA experiments [256] have indicated that the H content near the substrate/film, interface may be significantly higher than the front surface of the films. ERDA is a nondestructive analytical nuclear technique widely used for the identification and depth profiling of light elements in thin films [258].

Highly transparent diamond films with smooth surfaces are extremely desirable for optical applications due to their excellent optical transmission from the far IR to UV region of the electromagnetic spectrum. This property makes it an ideal material for IR windows and optical coatings. Polycrystalline diamond, however, contains grain boundaries and defects which are not present in single crystal stones. The defects usually occur due to the impurities present which can cause absorption in the regions of interest. The impurities include hydrogen, oxygen, nitrogen, etc. FTIR is an extremely powerful analytical technique for both qualitative and quantitative analysis. It is a nondestructive, simple, and quicker technique for analyzing hydrogen in CVD diamond. FTIR can also be used to identify specific hydrocarbon groups responsible for absorption. CVD diamond, for IR window applications, requires low absorption in the range 2–12 μm. Increased absorption in this region is associated with a similar increase in absorption in the CH-stretch region which indicates an increase in hydrogen concentration in the films. Even a small concentration of H ($<1\%$) in diamond sheets deteriorates the properties of the sheets greatly. H in polycrystalline diamond films influences the electrical [259], thermal [260], optical [261], field emission [262], etc. severely. Therefore, determination of the H content (bonded and unbonded) and studying its influence on the properties of the sheets are important tasks.

Atomic hydrogen (H) in the growth environment is a key factor in most viable diamond deposition techniques [263]. Many roles have been proposed for H atoms in CVD diamond film growth [264, 265]. For example, gas phase H atoms produce condensable carbon radicals by reaction with hydrocarbons [266]. H atoms incident on the surface abstract hydrogen to produce vacant sites [267] and surface radicals [265, 268, 269], refill vacant sites by absorption [265, 268–270], and etch the graphite surface [271]. Adsorbed hydrogen stabilizes the diamond surface structure [264, 272]. Due these reasons more than 99% of H_2 is fed in the gas mixture for any typical CVD route of diamond growth, and so chance of hydrogen getting incorporated as an impurity in the diamond deposit is highly likely. There are three components of hydrogenous impurities: (i) bonded to carbon atoms in the diamond lattice, (ii) unbonded hydrogen probably located at grain boundaries and interstitial sites, and (iii) molecular hydrogen (H_2) which is the stable configuration in the absence of defect sites to which the hydrogen can bond. Bonded hydrogen with host carbon atoms gives rise to the IR absorption mostly in the CH stretch region. Figure 23 shows the description and symmetries of the C—H IR vibration modes of CH_n groups.

Intrinsic two-phonon absorption in natural as well as in polycrystalline diamond films occurs between 1333 and 2666 cm^{-1}. Diamond with defects shows additional absorption in the symmetry forbidden one-phonon region below 1333 cm^{-1} and in the three-phonon region between 2750 and 3150 cm^{-1}, known as the CH stretch region. The one-phonon region (due to nitrogen and structural defects) is normally weak in CVD diamond and the three-phonon region is masked by the high intensity CH stretch region in polycrystalline diamond films. Due to the growth of diamond films in a H atmosphere, the main absorption naturally takes place in the CH stretching region. There are a number of absorptions, many of which can be related to oxygen- and nitrogen-containing groups, including O—CH_3 and N—CH_3. Figure 24 shows the CH stretching region after deconvolution of a typical diamond film grown in a HVCVD chamber. Table XI shows the various IR absorptions observed in CVD diamond.

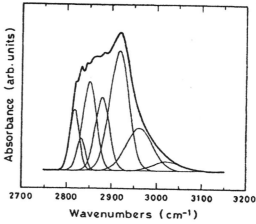

Fig. 24. Deconvolution of the CH stretching region of a typical diamond films grown in the HFCVD chamber showing different mode of vibrations. Reproduced with permission from [84].

Table XI. CH Stretching Frequencies and Corresponding Modes of Vibration Observed in CVD Diamond

Wave number (cm^{-1})	Mode of vibration
1030	O—CH_3, deformation
1130–1150	N—CH_3, O—CH_3, rock
1220	substitutional nitrogen
1250	C—N, stretch
	C—N
1332	nitrogen–vacancy pair
1350	disallowed one-phonon mode
1375	
1450	substitutional N
1500	N—CH_3, scissor
1610	sp^3, CH_3 deformation
1740	O—CH_3, N—CH_3, deformation
2820	sp^3, CH_2 scissor
2832	C=C, aromatic stretch
2850	C=C, stretch, isolated
2880	C—O, stretch
2920	N—CH_3
2960	O—CH_3
3025	sym. sp^3 CH_2
	sym. sp^3 CH_3
	sym. sp^3 CH_3
	sp^2 CH

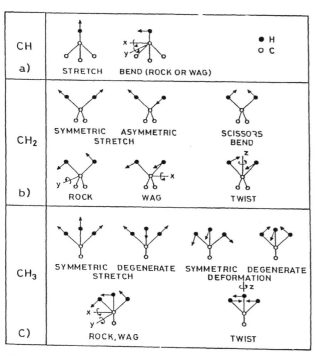

Fig. 23. Description and symmetries of the C—H IR vibration modes of CH_n groups.

As H is one of the key impurities in CVD diamond to control several properties it is worth trying to calculate the amount of total hydrogen and to estimate the bonded and unbonded contributions. The integrated absorbance of each band can be used to estimate the hydrogen concentration in that particular mode. The number of hydrogen atoms contributing to the stretching absorption band is proportional to the integrated area under the particular stretching mode. Different groups have chosen different values for the proportionality constant (A_n). In all the studies, constant values of A_n corresponding to different hybridization states of carbon, which was suggested by Dischler et al. [273], have been used. Brodsky et al. investigated silicon–hydrogen bonds in amorphous hydrogeneted silicon (a-Si:H) and showed that the absorption strengths of all possible Si—H vibrations are the same [274]. At the same time they also added a caveat that this is not true for C—H vibrations. This discrepancy is solely due to the fact that the oscillator strength corresponding to a particular mode of vibration is not independent of the structure and environment of that particular groups. Therefore, it is essential to know the exact number of bonded H contributing to a particular mode in order to calculate the exact value of A_n and to compare the concentration of bonded H measured using a standard sample (like a polymethyl methacrylate sample).

2.3.7. Photoluminescence (PL) and Cathodoluminescence (CL) Spectroscopy

For characterization of the optical center in diamond and other materials luminescence spectroscopy is a very useful and accurate technique. PL involves an excitation of incident photon, absorption, and photon emission. CL is the electron excitation of electrons in the sample to higher energy levels and subsequent radiative decay of electrons from the excited high energy states to some lower energy states, resulting in the emission of photons. The frequency of these emitted photons is determined by the particular electronic excitation process, and a characteristic decay time is associated with the excitation and emission. PL is often excited by an intense laser beam which can be the same as that used to stimulate Raman scattering as described earlier. CL is excited by a high energy electron beam (typically 50 keV) which can often be equipped with an electron microscope.

Both CL and PL have been used extensively to characterize natural as well as CVD diamond [275–277]. Table XII gives the

Table XII. Major Peaks Corresponding to CL and PL in Diamond

Position	Name	Description
>5.5 eV (<225 nm)	Band edge	Indirect gap, observable only in relatively defect-free diamond
5.260 eV (236 nm)	Band-B nitrogen	Four n atoms on substitutional sites symmetrically surrounding a vacancy
4.640 eV (267 nm)		Oxygen related
4.582 eV (270 nm)	5 RL, system	Interstitial carbon atoms
3.750 eV (331 nm)		Oxygen related
3.188 eV (389 nm)		Interstitial carbon atom and a nitrogen atom
3.150 eV (394 nm)	ND1	Negatively charged vacancies
Broad peak between 3.1 and 2.5 eV (400–496 nm)	Band A nitrogen	A nearest neighbor pair of substituional nitrogen atoms
2.985 eV (415 nm)	N_3 system	Three nitrogen atoms of (111) plane bonded to common carbon atom, or three nitrogen atoms surrounding a vacancy
2.560 eV (484 nm)		Defects involving Ni impurities
2.526 eV (491 nm)		Defects involving slip planes
2.499 eV (496 nm)	H_4 system	Vacancies trapped at nitrogen B-centers
2.463 eV (503 nm)	H_3 system	Vacancies trapped at nitrogen A-centers
2.305 eV (538 nm)		Associated with H_3 system, observed only in Cl
2.156 eV (575 nm)		A vacancy and a nitrogen atom often found in the flame deposited diamond
1.945 eV (637 nm)	NV system	A vacancy trapped in subsituitonal nitrogen atoms, often found in flame deposit diamond
1.681 eV (738nm)		Defects involving Si impurities and associated vacancies, observed mostly in CVD diamond
1.673 eV (741 nm)	GRI system	Neutral vacancies
1.400 eV (886 nm)		Defects involving Ni impurities observed predominantly from (111) surface
1.267 eV (979 nm)	H_2	Negatively charged H_3 center

major peaks corresponding to CL and PL observed in diamond [278, 279]. Luminescence spectra normally contain different features and they are characteristic to different optical centers present in the sample. More than 100 optical centers have been observed in diamond. Most are seen in all diamonds while some are seen only in CVD diamond films.

2.3.8. Electron Paramagnetic Resonance Spectroscopy

Electron paramagnetic resonance (EPR) is a powerful technique to study the paramagnetic point efects, their interactions with the lattice, and their mutual interactions. When an external magnetic field is applied to the system, the magnetic moment of the electron is either aligned with or opposite to the magnetic field. If a second alternating magnetic field is applied using a high frequency microwave-resonant cavity, the microwaves may induce transitions between the two states of the unpaired electron. When the microwave energy ($h\nu$) coincides with the energy level separation between the two electron states ($E_{1/2} - E_{-1/2}$), resonance absorption takes place. The resulting spectrum of the absorbed microwave energy as a function of the magnetic field is analyzed to determine the mechanisms associated with the interaction of the unpaired electron with the external magnetic field and its environment.

In diamond synthesis under low pressure conditions graphite is thermodynamically more stable than diamond. It is also argueable that growth is a nonequilibrium process and therefore equilibrium considerations are not relevant. Thus, a number of groups [281–283] discussed diamond growth entirely as a kinetic process. It was observed [284] that a qualitative correlation between diamond with a certain concentration of defects is thermodynamically stable over graphite. Besides any role the defects may play in the stabilization of diamond, their study is important because defects affect the mechanical, thermal, optical, and electronic properties of the material. In many cases, defects in the films are undesirable; however, certain mechanical properties, such as toughness, could improve with certain types of defects. Therefore, the ability to control the defects during film growth and to characterize their nature after film growth is an important area of study.

Paramagnetic defect centers related to nitrogen and nickel have been studied extensively in natural and high pressure grown synthetic diamond [285–287]. Electron spin resonance studies have also been carried out on neutron and electron irradiated diamond [288]. Several authors have also reported EPR signals in diamond films grown by the CVD method [289–292]. CVD diamond films show a characteristic symmetric spectrum with g-values in the range 2.0027(2) with a bulk concentration of spins ranging between 10^{16} and 10^{20} cm^{-3} [293]. It is due to three types of defects: sp^3, and sp^2 carbon defects and a dangling bond H-center, where a dangling bond is hyperfine-coupled to an adjacent proton. This latter contribution reveals the presence of H in the films. A satellite structure often seen in CVD films is attributed to the carbon dangling bond H-center [218, 294]. This center has not been detected in natural or synthetic diamond.

2.3.9. Positron Annihilation Spectroscopy

Positrons emitted from a radioactive source such as ^{22}NaCl are injected into the solid, wherein they thermalize and annihilate with one of the electrons in the medium resulting in the emission of two photons of energy 0.511 MeV. The annihilation rate (inverse of lifetime) is determined by the electron density at the site of the positron. Positron lifetime measurements are carried out by measuring the time delay between the 1.28 MeV gamma ray which signals the birth of the positron and the 0.511 MeV annihilation photons using ultrafast timing methods derived from the momentum of the electron–positron pair.

Depending on the material and the defects in it, the positron annihilates with different lifetimes. Some of the positrons can annihilate in the bulk of the diamond sample. This will correspond to a lifetime of nearly 105 ps observed in natural as well as in CVD diamond. Some of the positrons can be trapped by vacancy-type defects. These positrons will annihilate slower than those residing in the bulk due to the lower electron density. Positrons trapped by monovacancies and divacancies have a lifetime of 155 and 185 ps, respectively [295]. The amount of positrons trapped is proportional to the concentration of the defects. Positrons that annihilate in vacancy clusters show a life of the order of nearly 300 ps. When enough free space is available for the positrons, a bound state between a positron and an electron can form. This bound state is called positronium (Ps), and when formed, exhibits a lifetime typically in excess of 1000 ps. The Ps can be promoted by vacancy clusters or by the empty spaces between crystallites in polycrystalline materials.

For CVD diamond, incorporation of vacancies can be significant due to the nonequilibrium nature of the growth process and the presence of many other defects such as dislocations, grain boundaries, and planar defects [296]. The nature of the vacancies in the diamond lattice is complicated and not well understood. Most of the vacancies are coupled with other defects such as impurities and grain boundaries. Positron Annihilation Spectroscopy has been used for studying defects in the diamond lattice [295, 297, 298].

Figure 25 shows the raw data of the lifetime spectra for the two diamond films grown at 20 and 100 Torr, respectively, in the HFCVD chamber [84]. The spectrum recorded on HF81 (100 Torr) is much more extended compared to the spectrum recorded on HF80 (20 Torr). This indicates that sheet HF81 is more defective and contains a larger concentration of vacancy clusters. Table XIII gives the positron lifetime and relative intensity data obtained for the two sheets [84]. The table also includes the positron lifetime data on various types of natural diamond. It may be noted that all natural and synthetic diamond data are characterized by two different lifetime components, τ_1 and τ_2, respectively. Intensity I_1 corresponding to τ_1 is substantially larger than intensity of τ_2 components. In contrast a set of three different lifetimes τ_1, τ_2 τ_3 are calculated from the PA spectra recorded on the samples grown in HFCVD system. Moreover the intensities I_1 and I_2 are nearly equal in CVD diamond sheets (~47%) and are significantly higher than intensity I_3 corresponding to lifetime τ_3.

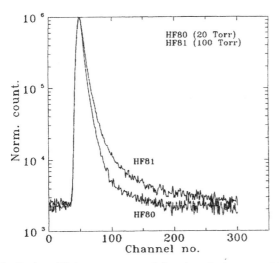

Fig. 25. Positron lifetime spectra recorded on two sheets grown at 20 and 100 Torr, respectively. 0.8% CH_4 in balance H_2 and $T_s = 890°C$ were used for both the depositions in HFCVD chamber. Reproduced with permission from [84].

Table XIII. Positron Lifetime Results Obtained by Sikder et al. [84] on Diamond Sheets Grown at Different Deposition Pressure and Results for Three Other Natural and Synthetic Diamond Samples

Sample	τ_1 (ps)	τ_2 (ps)	τ_3 (ps)	I_1 (%)	I_2 (%)	I_3 (%)
Natural type Ia	112	400		88	4	
Synthetic type Ib	105	430		97	3	
Natural type IIa	100	380		91.5	8.5	
HF80	122	271	1750	47	50	3
HF81	178	350	1170	48	43	9

Slight deviation in τ_1 may be related to the modification of τ_1 due to the contributions from monovacancy and positronium formation in CVD diamond. The lifetimes related to monovacancy and positronium formation are 150 and 125 ps, respectively. τ_1 in HF81 is 178 ps and is much higher than that in HF80. This value is very close to the value of τ_1 observed in graphite. In defective films normally graphitic inclusion is more common in CVD diamond. τ_2 for the two types of sheets are 271 and 350 ps, respectively. This is attributed to positron annihilation in vacancy clusters and grain boundaries. Smaller τ_2 for the HF80 sample suggests a smaller size of vacancy clusters in a less defective sample. Long-lived components of lifetime τ_3 correspond to the annihilation of the positron in positronium formation. The positronium formation is clearly preferred in the HF81 sample.

2.4. Diamond Deposition on Cutting Tool Materials

Diamond is the hardest known material with the highest thermal conductivity. Along with these, the properties suitable for hard coatings for tribological applications, such as low friction coefficient, chemical inertness, and good abrasive resistance, make

it the most attractive material for the cutting tool industry. Cutting tools need two important properties: wear resistance and toughness. Ceramic materials with high wear resistance generally have low toughness. The material with the highest known hardness, single crystal diamond, can not fully be used due to its easy cleavage. The two basic ways of utilizing the extreme properties of diamond in hard wear-resistant tools are: (a) a composite polycrystalline diamond (PCD) with metal binder (mostly Co is used) and (b) a thin (typically 5–15 μm thick) polycrystalline diamond film grown on a suitable base materials. However, because of higher costs of PCD, its use as a cutting tool is limited to only certain applications where higher precision and better surface finish is required. With the advent of CVD diamond technology it is now possible to coat the surface of cutting tools with diamond thin film and enhance the tool life manyfold. In this section we discuss the basic problems in low pressure diamond deposition on common tool materials and the steps being adopted to overcome them. Adhesion of CVD diamond to the substrate material is the main hindrance in exploiting its unique properties as a hard wear-resistant coating. The most widely used cutting tool materials are stainless steel (SS), high speed steel (HSS) and tungsten carbide–cobalt (WC–Co). Both the steel and WC–Co contain elements like iron (Fe) and cobalt (Co), which are very good solvents for carbon (Table XIV). Due to high solubility, diamond nucleates on these materials through a nondiamond layer, which is one of the main reasons for poor adhesion. The catalytic effect of Fe, Co, and the high solubility of carbon into them results in a highly unsuitable base for CVD diamond growth [299–305]. Surface modifications or the introduction of suitable buffer layers compatible with both substrates and CVD diamond are the only solutions for growing diamond on them.

It has been mentioned that growth of CVD diamond is perfected on only a few carbide forming substrates. Severe adhesion problems come into the picture when diamond is to be deposited on substrates which do not form carbide in the typical CVD environment. Also the chemical nature of the substrates has a significant effect on the diamond deposition. The thermal expansion coefficient of diamond (0.8×10^{-6} K^{-1} at 293 K) differs largely from those of various materials and hence is another major problem for better adhesion to the substrates. Low temperature diamond growth will result in better adhesion than conventional high temperature (800–900°C) growth. Needless to say, though large number of reports have been published on low temperature diamond synthesis, the technique has not yet been perfected.

2.4.1. Diamond Coatings on WC–Co Substrate

For machining nonferrous metals such Al, Cu, and bronze, most ceramic tools are chemically incompatible. However, diamond tools work very well in machining these materials as well as graphite, wood products, glass fiber-reinforced plastics, and light metal composites. Due to the increasing demand of low silicon Al alloy in automotive industries, the machining of them with higher precision is essential. Diamond coated tool inserts

Table XIV. Relevant Physical and Chemical Properties of Diamond and Group VIII Transition Metals

Materials	Electron configuration	Crystal structure	Lattice constant (20°C, °A)	Melting point (°C)	Coefficient of thermal expansion (20°C, 10^{-6} K^{-1})	Hydrogen solubility (1 atm, ppm)	Me–C eutectic MP. (°C, wt% C)
Diamond	[He]2p^2s^2	fcc	3.56	3830	0.8	—	—
Ni	[Ar]3$d^6 4s^2$	fcc	1453	1453	13.3	200 (420°C) 500 (900°C)	1326 (2.2 %)
Co	[Ar]3$d^7 4s^2$	hep (α-Co) <417°C; fcc (β-CO) >417°C	$a = 2.51$ $c = 4.07$ 3.55	1495	12.5	65 (700°C) 134 (900°C)	1320 (2.6%)
Fe	[Ar] 3$d^6 4s^2$	bcc (a-Fe) <912°C; fcc (γ)-Co >417°C	2.87	1536	12.6	120 (800°C) 250 (1000°C)	1147 (4.3%)
Cr	[Ar]3$d^5 4s^1$	bcc	2.88	1875	6.2	13 (400°C) 44 (800°C)	1534 (3.6%)

are the best choice for machining this high performance material. WC–Co is the most suitable material due to its hardness, ductility, and stiffness for depositing diamond films to make wear-resistant cutting tools. Diamond coated tools can be produced either brazing free-standing diamond sheets onto carrier substrates (Fig. 26), or *in situ* CVD diamond coatings (typically <50 μm) on the base materials (Fig. 27) [306]. In the first method a thick diamond layer is deposited on well-suited dummy substrates (Si, Mo, etc). After laser cutting to the desired shape free-standing diamond sheets are produced to braze with the actual tool materials (mainly WC–Co).

Diamond coatings deposited by CVD on WC–Co generally suffer from poor adhesion due to mismatch of thermal expansion coefficients leading to large thermal stresses at the interface [307, 308], and more importantly the interaction between the carbon and thin layer of cobalt on the surface of the tool [309]. Binder metal Co is not only between the WC grains but also thinly covers the grains at the surface. The carbon dissolution into Co during typical CVD conditions first extends the incubation period for diamond nucleation [310, 311] and then enhances the accumulation of graphite at the diamond–carbide interface [312, 313]. The graphitic interface layer weakens the bond between diamond and carbide and degrades the quality of diamond deposit [314]. So it is essential

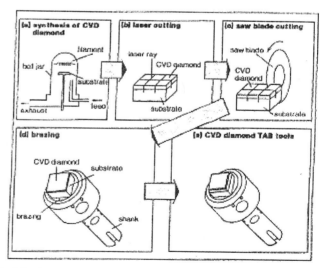

Fig. 26. Schematic diagram of the process steps of making brazed cutting tools. Reproduced with permission from [306], Noyes Publishing.

to get a reasonable bonding with tools to diamond coatings either by removing the surface cobalt [314–317] or by using a suitable Co diffusion barrier [317–320] which should provide

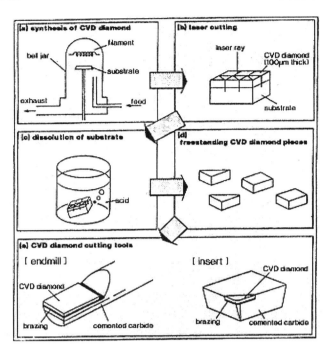

Fig. 27. Schematic diagram of the process steps of an *in situ* CVD diamond coated TAB tool. Reproduced with permission from [306], Noyes Publishing.

a good surface to grow diamond. Replacing Co with Cu also shows the improvement of adhesion of diamond films on carbide substrate [321]. Forming stable compound as interfacial layers has also been tried to improve the adhesion [322–324]. A number of pretreatment procedures aim at a depletion of the Co at the hard metal substrate surface by selective chemical etching with acids such as HNO_3, HCl, or CH_3COOH [299–303]. Another etching pretreatment consists of two steps. First, WC is removed by Murakami solution ($K_3[Fe(CN)_6]$ in KOH and H_2O), leaving an excess of Co on the hard metal surface. Subsequently, the Co is treated by a H_2SO_4/H_2O_2 solution. It is found that combination of both etching steps can give improved adhesion results [302, 303]. Using other binders in WC tool materials can be used to overcome the adhesion problems [325]. In another approach WC-Co substrate is heat treated in a protective atmosphere as a pretreatment of the substrate [326]. This produces WC grain at the surface, increases the roughness without forming substrate porosity, which is the basic problem in Co chemical etching, and virtually eliminates surface Co by evaporation. Different methods have been employed to achieve diamond coatings on tool materials in the laboratory as well as in the industries. Among them HFCVD, MPCVD, flame CVD, and arc discharge plasma CVD are used extensively due to their versatile nature.

2.4.2. Diamond Deposition on Steel Substrate

The deposition of diamond films on steel is possible now because of the recent advances made in the CVD techniques. High quality diamond films, if deposited successfully on steel, may revolutionize the usefulness of steel in modern industry. For in-

stance, with this technology one can think of coating steel ball bearings, sheets, and cutting tools with diamond and many other similar applications.

It was already mentioned that the catalytic effect of Fe and high solubility of carbon into stainless steel (SS) makes it most unsuitable base for CVD diamond growth [305]. Therefore, the direct deposition of CVD diamond on steel substrates has so far resulted in poor quality films. Surface modification by ion implantation also could not inhibit carbon diffusion into the bulk and hence does not provide a suitable solution for growth of diamond on steel substrates [327]. A number of attempts [328–331] have been made to develop good quality films using suitable buffer layers prior to diamond deposition. The quality of diamond films produced, however, remains poor. Sikder and Misra [332] have developed a buffer layer based on a composite Ni–diamond coating which is found suitable for deposition of high quality diamond films with good adhesion on SS. Here a brief description of the buffer layer and its influence on the properties of CVD diamond are described.

Ni has only a 4% lattice mismatch and can be an ideal candidate for the deposition of epitaxial diamond films. However, the strong catalytic effect of Ni on hydrocarbon decomposition [333] and graphite formation [334, 335] at low pressure inhibit the diamond growth. In spite of this, several groups have made attempts to grow diamond epitaxially on single crystal Ni [328–331]. In the following section we discuss the growth of diamond on SS substrates by the HFCVD technique using different Ni based buffer layers [84].

2.4.2.1. Substrate Preparation

Mirror polished SS substrates (grade 304) of 12 mm × 12 mm × 1 mm size were used as substrates. Substrates were cleaned and pretreated throughly as described below to improve the adhesion of the buffer layer.

- ultrasonically cleaned by acetone and tetrachloride ethylene,
- dipped in hot (60–70°C) NaOH (30% v/v) solution for 10 min,
- rinsed in deionized (DI) water,
- anodic treatment in dilute (30% v/v) H_2SO_4 with a SS cathode,
- rinsed in DI water.

2.4.2.2. Buffer Layer

A strike Ni layer of thickness 1–2 μm was deposited on surface cleaned substrates prior to any buffer layer deposition for better adhesion. Three different buffer layers have been experimented based on Ni and Ni–diamond composite type coatings in three successive steps:

1. Ni–diamond composite coatings,
2. Ni/Ni–diamond composite coatings, and
3. modified Ni–diamond composite coatings.

2.4.2.3. Composite Electroplating

The concept of composite material is to combine certain assets of the components or members of the composite systems and suppress the shortcomings of the individual members in order to give the newly synthesized composite material improved and useful properties. Three basic types of composite materials can be distinguished, viz., dispersion-strengthened, particle-reinforced, and fiber-reinforced or whisker-reinforced materials. In this study, diamond particles have been used in the Ni matrix to make particle reinforced-composite coatings. In the rest of the chapter this will be referred to as Ni–diamond composite coating. One of the main objectives of this study was to develop a suitable buffer layer which not only prevents the interdiffusion of Fe and C atoms but also becomes conducive to diamond growth in a CVD environment. Further, it was anticipated that Ni acts as a diffusion barrier and diamond particles will strengthen the layer as well as act as nucleation sites for CVD diamond growth.

Figure 28 shows a schematic diagram of the composite plating bath [84]. A standard Watt's type of electrolyte solution is used for Ni plating. A porous container containing diamond slurry (particle size of diamond was 8–20 μm) was used as a cathode chamber. The porosity of the container was such so as to avoid the mixing of diamond particles in the whole electrolyte solution. The suspension of the diamond particles was achieved by means of a magnetic stirrer. Commercial grade Ni plate is used as a anode. A typical SEM micrograph of Ni–diamond composite coated SS substrate is shown in Figure 29a [84]. Area coverage of the diamond is particles estimated from the micrograph as 65–70% and can be controlled depending on the applications.

Diamond particles embedded in the Ni matrix are irregular in shape. No overplating can be seen. This is highly desirable because the diamond grains covered with Ni can no longer act as nucleation centers for CVD diamond growth. Also the performance of electroplated tools is reduced by overplating.

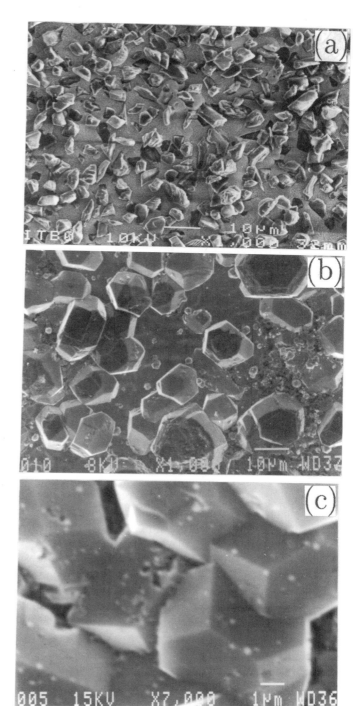

Fig. 29. SEM micrographs (a) showing the diamond particles are embedded in the Ni matrix, (b) discontinuous CVD diamond deposited on Ni–diamond composite coated substrate, and (c) continuous film on Ni–diamond composite coated SS substrate. Reproduced with permission from [84].

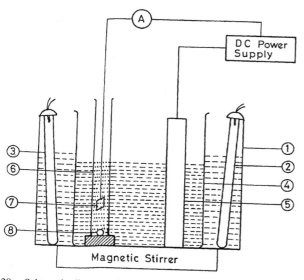

Fig. 28. Schematic diagram of composite electrodeposition setup. Numbers are described as: (1) hot water bath, (2,3) glass immersion heater, (4) deposition bath, (5) Ni anode, (6) cathode chamber, (7) sample, and (8) magnetic needle. Reproduced with permission from [84].

Figure 29b shows the typical surface morphology of the diamond grown on the composite coated substrate. The initially irregular-shaped particles are converted to regular cuboctahedron shapes. An increase in growth time and area coverage of diamond particles results in continuous diamond films. In Figure 29c the micrograph of a typical continuous diamond film grown on Ni–diamond composite coated substrate is shown.

XRD patterns of the composite coated substrate before and after CVD are shown in Figure 30a [84]. As can be seen from the figure, before CVD both Ni and diamond peaks are present. After CVD, diamond (111), (220), and (311) peaks cannot be distinguished due to the overlap with the Ni peaks as the latter reflections are shifted toward the lower 2θ position. This is well supported by the shift of the Ni (200) peak where no diamond peak appears. The downshifting of the Ni peaks gives an indication for Ni lattice expansion. This might be caused by the diffusion of the carbon from the gas phase resulting in the formation of Ni−C solid solution. However, as Ni matrix contains diamond seeds the carbon atoms from the seeds might also dissolve in Ni. To investigate this further, bare Ni coated substrates were subjected to CVD with 0.3% CH$_4$ in balance H$_2$ for different times. We found that all the Ni peaks in XRD pattern were shifted after CVD to lower 2θ values.

Figure 30b shows the typical XRD pattern of the bare Ni coated substrates before and after CVD with 0.3% CH$_4$ for 10 h. This indicates an expansion in the Ni unit cell due to the carbon diffusion in the CVD environment forming a Ni−C solid solution. Similar expansions in the Co lattice have been observed by Kubelka et al. [307]. The expanded Ni unit cell matches closely with the diamond and may be responsible for the epitaxial growth of diamond on Ni observed by others [335–339]. Observed saturation in lattice expansion may be related to the solubility limit of the carbon in Ni. Based on the expanded volume of the Ni lattice, it is estimated that the maximum carbon concentration [340] in the Ni−C solid solution may be ~4 wt% [341]

Carbon concentration in the Ni lattice was calculated form the following basic density equation of the solid solution [340] and is shown in Table XV [84]. Alternately, hydrogen can also lead to the lattice expansion in a similar way as carbon. However, under the present CVD conditions the solubility of H$_2$ is negligible [335]. To confirm the incorporation of carbon in the Ni lattice, XPS measurements were carried out on the bare Ni coated SS substrates after CVD [84]. Basics of XPS measurement are discussed in the carbon nitride section.

Figure 31 shows the deconvoluted XP spectra of the samples treated with CVD for different times [84]. Deconvolution of the C(1s) spectra using a standard fitting program shows three distinct transitions at 284, 284.7, and 286 eV along with a less intense broad peak at higher binding energy (Table XVI). The latter can be avoided with tail corrections and taking the proper Gaussian and Lorentzian ratio [342]. The three main peaks can respectively be assigned to carbidic phase (due to Ni−C solid solution), pure carbon (diamond, nondiamond, graphite), and hydrogenated/hydrogen terminated carbon. In the case of Ni(2p$_{3/2}$) spectra, the transition can be deconvoluted into three

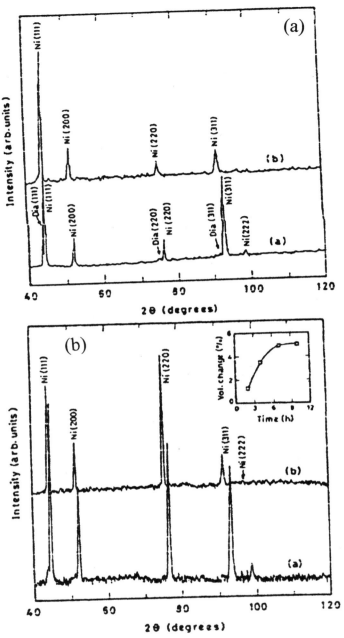

Fig. 30. XRD patterns show the shift of Ni peaks after CVD (a) on a Ni–diamond composite coated substrate and (b) on a bare Ni coated (without diamond seeding) substrate. Inset shows the percentage volume change of Ni unit cell with growth time. Reproduced with permission from [84].

peaks. The peaks at 853 and 858.8 eV, respectively are due to the main and satellite peaks of metallic Ni [342]. The peak at 854 eV can be attributed to Ni−C solid solution which is in accordance with the C(1s) peak assignment. In Figure 32b, the ratio of the area of the carbon to Ni in the Ni−C region $(A_{\text{C:Ni}})/(A_{\text{Ni:C}})$ is plotted. It is evident from the figure that the ratio which is proportional to the carbon concentration in the Ni−C solid solution increases and approaches saturation with the deposition time. This observation is well supported by the lattice expansion studies (cf. Fig. 32a) [341].

Table XV.   Unit Cell Expansion of Ni after CVD (0.3% CH$_4$) on Ni Coated SS [84]

Time (h)	Before CVD		After CVD		Volume change (%)	Carbon conc. (wt%)
	a (Å)	V_u (Å^3)	a (Å)	V_u (Å^3)		
2	3.517 (2)	43.52 (7)	3.532 (1)	44.06(3)	1.24	1.2
4	3.522 (1)	43.69(3)	3.563(1)	45.22(5)	3.5	3.7
7	3.518(1)	43.55(4)	3.575 (4)	45.70(2)	4.95	4.7
10	3.516 (1)	43.46(5)	3.574(2)	45.67(8)	5.09	4.6

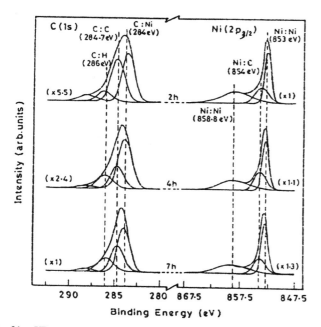

Fig. 31.   XP spectra of the bare Ni-coated SS substrates subjected to CVD for 2, 4, 7 h while keeping other conditions same. Treatment was done at 900°C, 320 W (microwave power), and 30 Torr pressure in 0.3% CH$_4$ in balance H$_2$. Reproduced with permission from [84]

Table XVI.   XPS Data of the Bare Ni Coated SS Substrates after CVD [84]

Time (h)	Peak positions (eV)		$A_{C:Ni}/A_{Ni:C}$
	C(1s)	Ni(2$p_{3/2}$)	
2	283.7, 284.8, 286.2	852.9, 853.9, 858.4	0.13
4	283.9, 284.8, 286.1	853, 853.9, 857.6	0.49
7	284, 284.7, 286	853, 854, 858.8	0.85

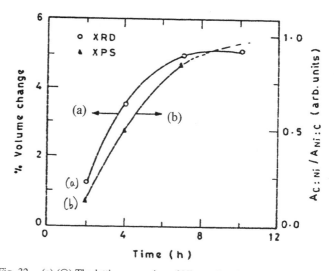

Fig. 32.   (a) (○) The lattice expansion of Ni as a function of deposition time and (b) (▲) ratios of the areas under the Ni−C curves in C (1s) and Ni (2$p_{3/2}$) regions as a function of deposition time. Reproduced with permission from [84].

The lack of diamond nucleation on the substrates (Fig. 33a) exposed to 0.3% CH$_4$ in balance H$_2$ for 10 h may also be due to the dissolution of the carbon from the gas phase resulting in an insufficient concentration of diamond precursors. It naturally follows that if the substrate is exposed to higher CH$_4$ percentage the carbon dissolution may approach saturation relatively quickly and sufficient precursors will be present for diamond nucleation. This conjecture has been verified with the observation that diamond starts nucleating rather rapidly at higher methane concentrations (Fig. 33b) [341].

A sharp peak at 1334 cm^{-1} (small upshift due to instrument) corresponds to good quality crystalline diamond (Fig. 34a). A broad band at 1500 cm^{-1} indicates the presence of nondiamond carbon in the film. The intensity of the broad band increases with even a small increase of methane concentration in the gas phase. There is no distinct graphitic peak at 1580 cm^{-1} in any of the films grown on either Ni–diamond composite coated or bare Ni coated substrates. This may be suppressed due to the postannealing treatment of the substrates

in the hydrogen atmosphere [343]. Electron Dispersive X-ray results show that a small amount of Fe is still present in the diamond layers coated on the Ni–diamond composite coatings. The diamond particles which are trapped at the early stage of the composite electrodeposition will be very close to the SS substrate. At the time of CVD, Fe may segregate to the surface through the diamond–Ni interfaces. This may be the reason for the observed Fe on the surface which may be one of the main causes for the nondiamond formation [84]. A thick Ni (∼12 μm) layer deposited prior to composite coatings may stop Fe diffusion significantly and may help in improving the quality of the films.

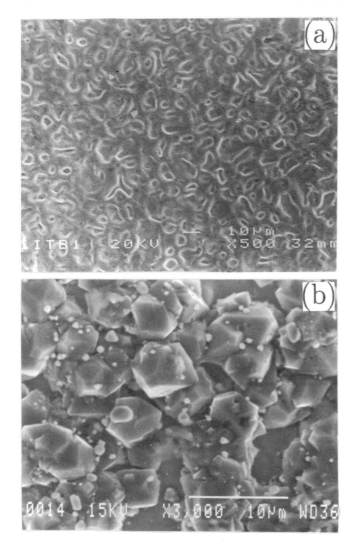

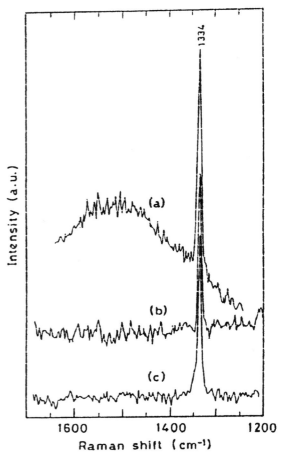

Fig. 34.   Comparison of Raman spectra recorded on CVD diamond deposited on buffer layered SS substrate: (a) Ni–diamond composite coated substrate, (b) Ni/Ni–diamond composite coated substrate, and (c) modified Ni–diamond composite coated substrate. As the modified layer is suppressing the adverse effects of Ni, a very high quality of diamond is growing on these surfaces. Reproduced with permission from [84].

Fig. 33.   SEM micrographs showing the surface morphology of Ni coated SS substrates after (a) 10 h CVD treatment with 0.3% CH$_4$ and (b) 2 h CVD treatment with 1.6% CH$_4$ in balanced H$_2$. No diamond nucleation is seen in Figure 32a. Reproduced with permission from [84].

2.4.2.4. Ni/Ni–Diamond Composite Layer

Keeping this in mind, a set of samples was prepared with Ni and Ni–diamond bilayers [84]. The idea of giving an intermediate Ni layer prior to the Ni–diamond composite layer is that the Ni layer will prevent Fe diffusion completely and diamond particles on the top will help for nucleation as discussed in the earlier section. This bilayer-type buffer layer helps to improve the quality of the CVD diamond significantly. It was found that crystallinity of the diamond grains has improved considerably with the intermediate Ni coating, over the films grown without the Ni intermediate layer [84]. Raman spectroscopy shows a sharp peak at 1334 cm^{-1} without any nondiamond components (Fig. 34b). EDX analysis shows almost non existence of Fe on the surface of the substrates after CVD. Though crystalline quality was improved considerably with the Ni/Ni–diamond buffer layer, adhesion to the substrate was poor.

2.4.2.5. Modified Ni–Diamond Composite Layer

In the Ni–diamond composite coatings part of the Ni matrix was exposed to CVD atmosphere and had an adverse influence on the adhesion of the diamond crystallites with the metal matrix after CVD. Therefore, it is essential to hide the Ni matrix by modifying it with some means or covering the exposed Ni with some other matrix. In this work we modified the Ni layer with a suitable coating on Ni–diamond composite coating [344]. This layer not only suppressed the Fe diffusion in the surface but also created an excellent atmosphere to grow state-of-the-art diamond films on buffer layered SS substrates. Hiding the Ni layer from the CVD atmosphere not only helps to grow clean sharp faceted diamond but also the adhesion of the crystallites improves greatly. Figure 35 shows the typical surface morphology of the diamond grown on modified Ni–diamond composite coated substrate. It can be seen that crystallite facets are much more sharper and cleaner. A sharp Raman peak without humps is an indication of very good crystalline diamond (Fig. 34c).

Fig. 35.   SEM of diamond films grown on modified buffer layered SS substrate. Reproduced with permission from [84].

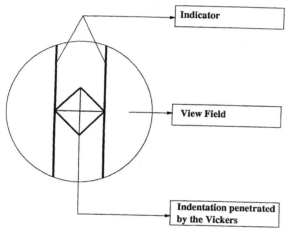

Fig. 36.   Schematic digram of Vicker's indentation measurement.

Simple scratch testing on the samples shows that the diamond grown on the modified Ni–diamond composite coated substrates is adhering much more strongly than any other buffer layers. Microhardness testing was done in a Shimadzu microhardness tester (HMV-2000). Vickers hardness was determined based on the test load used when penetrating a Vickers indenter on the test piece surface and the indentation area calculated from the indented diagonal length. Figure 36 shows the schematic diagram of the indentation measurement procedure. Measuring the diagonal length, Vickers hardness can be calculated from the expression HV $= 1.854F/d_i$, where HV $=$ Vickers hardness, $F =$ test load (kgf) and $d_i =$ mean of the indentation diagonal length (mm). Microhardness testing shows no indentation under 200 gm of load on the modified layer after CVD, whereas under this condition indentation marks were clearly visible on other composite coated samples prior to and after CVD.

2.5. Conclusions

Extensive research has been done on the growth of diamond films and an in-depth understanding of the vapor growth at low temperature and pressure has been reached. Still a few issues have not been perfected yet. These are

- growth at low temperature which will help to grow diamond on low melting substrates,
- coating with large area conformity,
- nanocrystalline diamond coatings for smoother films,
- oriented diamond films for smooth surface,
- heteroepitaxial diamond growth,
- advancement toward single crystal diamond growth.

Characterization of diamond films is now very much perfected in order to know the surface morphology, roughness, crystalline quality, defects, and physical properties. More techniques have to be employed to investigate the growth mechanism in order to resolve some of the growth problems described earlier for using the potential of this novel material fully.

Though diamond coated tools are on the market, adhesion problems on some extensively used materials (e.g., iron based materials) still persist. For wear-resistant hard coating on the tool materials, a major problem comes in the way the adhesion of the coatings with the substrates. Substrate pretreatment and intermediate buffer layers are techniques to overcome the adhesion problem. The method described here to grow good quality diamond on iron base materials may prove to be a way to avoid the adverse effect of iron on the deposition of diamond on them. This coating can also be employed to improve the electroplated tools and other tool materials where direct deposition of CVD diamond with sufficient adhesion is nearly impossible.

3. CUBIC BORON NITRIDE

3.1. Introduction

Boron nitride (specially in cubic form, cBN) thin films are of great interest because of their excellent physical, chemical, thermal, electrical, optical, and mechanical properties. cBN is the second hardest known material [345, 346] and has the largest band gap among IV and III–V materials [346]. It can be doped to both p- and n-type semiconductors [347, 348], whereas n-type doping has been difficult for diamond. In fact it has been shown that a cBN p–n junction grown at high pressure and high temperature can operate at 530°C [349]. It has very high temperature oxidation resistance and the ability to machine ferrous-based materials (superior than diamond). Because of its high bandgap [350] ($Eg \approx 6$ eV) and good thermal conductivity, cBN also has the potential for the same high temperature and high power electronic applications similar to diamond films. Significant ion irradiation during BN film growth is currently necessary to form the sp^3 bonded cubic phase over other phases (mainly the hexagonal phase) [351, 352]. In the 1990s this was one of the most studied materials and it is still being investigated extensively. Thus available literature is extensive, covering many different aspects of cBN and other phases of boron nitride, such as microstructure [353–355, 364], stress and strain fields [365–367], effect of ion bombardment [368, 369], substrate effects [370–372, 372], surface morphology [373, 374], growth mechanism [365, 368, 369, 375–385], mechanical properties [361, 386], and others [387–389]. Several review papers have been published that discuss the issues of growth, phase stability, characterization techniques, and several other issues of cBN [390–395].

3.2. Phases of Boron Nitride

It was indicated from the pressure–temperature phase diagram that hexagonal BN (hBN) is thermodynamically stable at ambient temperatures and pressures, whereas cBN is only stable at high pressures [396, 397]. It was also indicated the cBN can be stable in ambient conditions [398–400]. In contrast to the phase transition of graphite to diamond there are significant kinetic barriers hindering the direct transition from sp^2 to sp^3 bonding under ambient conditions. Similarities in stacking sequence under pressure cause well-ordered hBN to transform to wuttzite BN, whereas rhombohedral BN transforms to cBN [401]. Table XVII shows the structural data for different phases of BN and Figure 37 shows the structures of different phases of BN [402].

3.2.1. Hexagonal Boron Nitride (hBN)

Boron nitride is a compound isoelectronic with carbon and, like carbon, can possess sp^2 and sp^3 bonded phases. The most common phase of boron nitride is the hexagonal phase (hBN). This phase is structurally similar to graphite, and like graphite, it is comprised of sp^2 bonds forming planer hexagonal networks that are stacked along the c-axis in an AA′A . . . configuration. The in plane and c-axis lattice constants of hBN are $a = 2.50$ Å and $c = 6.66$ Å, respectively [403] which are very similar to the values of graphite ($a = 2.41$ Å and $c = 6.70$ Å) [404]. The c-axis lattice constants for hBN and graphite are equal to twice the interplaner separation due to the stacking. While graphite is a semimetal, hBN is an insulator with a direct bandgap energy of 5.2 eV. Because of the weak interplanar bonding in both graphite and hBN, the planes can slide easily against each other which makes the materials excellent lubricants.

3.2.2. Rhombohedral Boron Nitride (rBN)

Another phase of BN with sp^2 bonding has a rhombohedral (or trigonal) primitive unit cell (rBN). This phase is physically similar to the carbon polytype called rombohedral graphite. In this structure, BN atoms are also arranged in a planar hexagonal network, but the hexagons are stacked along the c-axis in an A B C A . . . configuration. Although it is usually referred to as a hexagonal structure, the primitive unit cell is a rhombus. The lattice constants are $a = 2.498$ Å and $c = 9.962$ Å [405] and are comparable to that of rhombohedral graphite, $a = 2.456$ Å and $c = 10.044$ Å [406]. Physical properties are little known of rBN since it does not exist in crystals of adequate size for many experiments and since rBN seldom exists as a pure phase without the presence of hBN. rBN powders are usually produced by mixing turbostratic hBN with KCN and heating the mixture to over 1000°C [405].

3.2.3. Turbostratic Boron Nitride (tBN)

As with graphitic carbon, sp^2 bonded BN often is found in a disordered, turbostatic form (tBN). This is the form of sp^2 bonded material most commonly observed in boron nitride thin films. For a turbostratic structure, the two-dimensional in-plane order of the hexagonal basal planes is largely retained, but these planes are stacked in a random sequence and with random rotation about the c-axis. tBN produces a broad and diffuse diffraction pattern that is distinct from that for h- and rBN [407].

3.2.4. Cubic Boron Nitride (cBN)

The second most common phase of BN has the cubic zinc-blend structure (cubic BN or cBN). Like diamond, cBN is formed in equilibrium at high temperatures and pressures from hexagonal material and is comprised of sp^3 bonds. While in diamond the bonding among the atoms is entirely covalent, the bonding in cBN is partially ionic (~60%) [408]. cBN has a crystal structure similar to diamond. The lattice constant for cBN ($a = 3.62$ Å) is very similar to that of diamond ($a = 3.56$ Å).

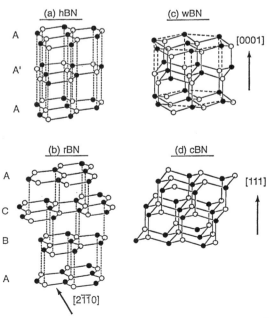

Fig. 37. Structure of the sp^3 bonded phases cBN and wBN and the sp^2 bonded phases hBN and rBN. Reproduced with permission from [402], copyright 1997, Elsevier Science.

Table XVII. Structural Data for the Boron Nitride Phases

Phase	a (Å)	c (Å)	Space group	Atom position
hBN[20]	2.5043	6.6562	$P6_3/mmc$ (194)	B: (0,0,0), (2/3,1/3,1/2) N: (2/3, 1/3,0), (0,0,1/2)
rBN[23]	2.5042	9.99	$R3m$ (160)	B: (0,0,0), (2/3, 1/3, 1/3), (1/2,2/3,2/3) N: (2/3,1/3,0), (1/3,2/3,1/3), (0,0,2/3)
cBN[1,26]	3.6153		$F43m$ (216)	B: (0,0,0), (1/2,1/2,0), (0,1/2,1/2), (1/2,0,1/2) N:(1/4, 1/4, 1/4), (3/4,3/4,1/4), (1/4,3/4,3/4), (3/4,1/4,3/4)
wBN[26,27]	2.5505	4.210	$P6_3mc$ (186)	B: (0,0,0), (1/3,2/3,1/2) N: (0,0,3/8), (1/3,2/3,7/8)

Reproduced with permission from [402], copyright 1997, Elsevier Science.

3.2.5. Wurtzite Boron Nitride (wBN)

Both the stacking sequences for hBN and cBN can vary and produce sp^2 and sp^3 bonded polytypes that are metastable. Another phase of BN has the wurtzite structure (wBN) and is similar to lonsdaleite or hexagonal diamond. wBN consists of a two-layer (A A′ A...) stacking of (0002) planes. The lattice constants for wBN are $a = 2.52$ Å and $c = 4.228$ Å [409] which are comparable to that of lonsdaleite, $a = 2.52$ Å and $c = 4.12$ Å [410]. Not much is known regarding the physical properties of wBN since single phases materials of high purity have not yet been produced. Figure 37 shows crystal structures of BN in different forms.

3.3. Cubic Boron Nitride Synthesis

It is difficult to grow cBN films in pure phase without ion bombardment. Thus the deposition processes differ primarily in the sources of B and N and how the ions are generated and transported. Figure 38 shows a simplified schematic representation of the two basic processes: ion-assisted deposition and plasma-assisted deposition [402].

3.3.1. Electron Beam Evaporation/Ion Plating

Ion-assisted electron beam evaporation, sometimes referred to as ion plating, is a very effective technique used in early attempts and it has the ability to grow a very pure phase of cBN [411]. In most cases ions from sources, like Kaufman-type sources, bombard the growing surface of film with little or no bias applied to the substrate [412–419]. Some processes involve striking a plasma discharge and extracting and accelerating ions to the biased substrate surface [411, 418, 420–422]. Several studies used processes of a hybrid nature [423].

3.3.2. Ion Assisted Pulsed Laser Deposition

In this process cBN films are synthesized by combining a deposition flux generated by ablating a BN target with a pulsed laser source with an ion flux from a Kaufman source [424–427]. This process has several advantages for the growth of cBN

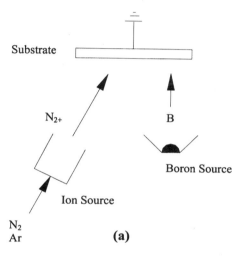

(a)

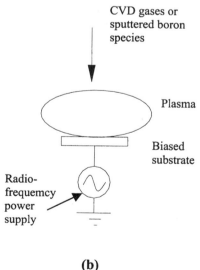

(b)

Fig. 38. Schematic summarizing microstructural schematic illustrations of two basic techniques for cBN deposition. In ion-assisted deposition (a), an ion flux from an ion source is combined with a deposition flux from a source such as a boron evaporator. In plasma-assisted deposition (b), the substrate is typically RF biased and ions from a plasma discharge are accelerated toward the substrate. Reproduced with permission from [402], copyright 1997, Elsevier Science.

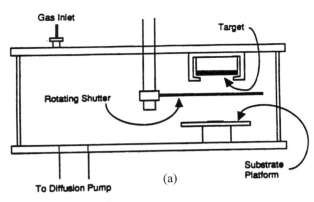

(a)

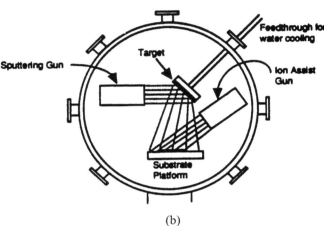

(b)

Fig. 39.   (a) Schematic diagram of a sputtering system and (b) schematic diagram of a dual ion-beam sputtering system.

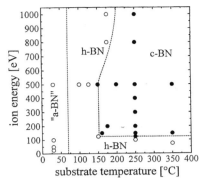

Fig. 40.   Formation of cBN grown by MSIBD technique as function of ion energy and temperature. Reproduced with permission from [447], copyright 1997, American Physical Society.

3.3.4. Plasma Assisted CVD

Plasma assisted CVD has also been used extensively to grow cBN films using different gas combinations, like B_2H_6 in N_2 [438–441] or NH_3 [442], BH_3–NH_3 in H_2 [443], $NaBH_4$ in NH_3 [444], $HBN(CH_3)_3$ in N_2 [445], $B_3H_3N_3(CH_3)_3$ in N_2 [446], etc. The reason for using a more toxic gas like B_2H_6 in most of the cases rather than the less toxic and less explosive $B_3H_3N_3(CH_3)_3$ is that less substrate temperature is necessary to promote cBN formation ($\leq 400°C$) [438, 439, 441, 446]. Disadvantages of CVD process are the necessary high substrate temperatures (this generates high stress in the films during the cooling stage after deposition), the use of more contamination sources due to more variables, and lower deposition rate. Advantages of the CVD process are that it is easily scalable for industrial use and has better conformity.

3.3.5. Other Techniques

Mass separated ion beam deposition (MSIBD) is another useful technique which gives high cBN content in the deposited films [447–450]. In this process, singly ionized B^+ and N^+ provide the ion irradiation and the deposition flux without other additional ions and neutral atoms. The technique has the advantage of allowing for precise, independent control of the ion-bombardment parameters of all species without any unwanted residual inert gas in the films. Sharp threshold values of 125 eV ion energy and 150°C substrate temperature, were found in almost phase pure cBN formation (Fig. 40) [447]. Because of reduced film sputtering, the cBN formation window extended from about 125 to 1000 eV, which is wider than the typical windows of other energetic techniques.

In another method a low density supersonic plasma jet fed by BCl_3 and N_2 was impinged on a substrate biased between -60 and -90 eV in order to grow cBN [451]. Cathodic arc evaporation which produced a large fraction of evaporated ionized atoms is also being used to produce cBN with a high deposition rate. As this technique needed a conducting target, heated boron sources (make it conducting) are being used to grow nearly phase pure cBN [452, 453]. Deposition of cBN is

though it has limited potential for industrial scale-up and it incorporates particle into the film during deposition. Advantages include separate ion and deposition sources and the ability to use multicomponent target [402].

3.3.3. Sputtering

DC or rf sputtering has been used extensively in the study of cBN films [428–431]. The success largely results from techniques that enhance ion bombardment during film growth. Figure 39 shows a schematic of a sputtering system. Substrate biasing, external magnets, or an extra ion source are normally used along with the conventional sputtering methods [432–435]. RF magnetron sputtering is used when insulating targets such as hBN are used. Several people used B_4C targets (with N_2 added to the ambient) to deposit cBN films [436–438]. Boron is another potential target material as it becomes sufficiently conductive above 800°C [429]. Sputtering is a well-established technique in industries and so it is easily scalable. However, a disadvantage of sputtering for research studies is the inherent coupling of the ion flux and deposition flux, which makes it more difficult to independently vary, control, and measure these important parameters [402].

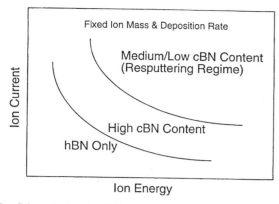

Fig. 41. Schematic showing cBN content as a function of ion current energy at constant ion mass and deposition rate [411]. The transition to low/medium cBN content with increased ion current and energy is now known to result from increased sputtering, which gives a thinner film and ultimately a no-growth condition. Reproduced with permission from [402], copyright 1997, Elsevier Science.

Table XVIII. Summary of cBN Film Growth on Non-Si Substrate

Substrate	Charac. technique	≈CBN content
a-C/Si	FTIR	Very high
Al/Si	XRD	Low
Al	FTIR	Low/med
Al	FTIR	Low
Ag	FTIR	Low
Au	FTIR	Medium
B/Si	FTIR	Very high
Cu	FTIR	None
Cu	FTIR	Medium
Diamond	FTIR	High
Diamond	FTIR	Very high
Mo	FTIR	High
Mo	FTIR	High
Ni/Si	XRD	High
Ni	FTIR	Medium
Ni	FTIR	High
Ni	XRD	Very high
Nb	FTIR	High
Pt/Al$_2$O$_3$	FTIR	Medium
β-SiC/Si	FTIR	Very high
Steel	FTIR	High
Ta	FTIR	High
TiN/WC-Co	XRD	High
TiN/WC-Co	FTIR	High
TiN/MgO	FTIR	High/ Very high
WC	FTIR	Very high
WC-Co	FTIR	High

Key: Very high: >85% cBN, high: ~60–85% cBN, medium: ~35–60% cBN, low: ~5–35% cBN, none: <5% cBN.

Reproduced with permission from [402], copyright 1997, Elsevier Science.

being reported by many other techniques, e.g., hot filament assisted evaporation of boron [454], metal–organic CVD without the use of a plasma [455], and laser ablation with an energetic ion source [456–459].

3.3.6. Key Parameters for Controlling cBN Phase

Major parameters which play key roles in the formation of cBN phase dominantly over other phases are (i) ion bombardment, (ii) the chemical nature of substrates, and (iii) substrate temperature. Most of the deposition techniques used highly energetic ion bombardment in order to stabilize the cBN phase. It is shown by Inagawa et al. [411] that cBN film formation only occurred in a specific range of ion current and substrate bias (Fig. 41). They observed that the process-parameter boundaries, as well as the maximum cBN percentage attained, were influenced by the ratio of Ar$^+$ to N$_2$ ions. Several other studies also showed the narrow window of ion energies for the deposition of cBN phase [420, 460–462]. Normally ion mass m, ion energy E, ion flux J (ions cm^{-2} s^{-1}) and deposition flux a (atoms cm^{-2} s^{-1}) are the key parameters to control the ion bombardment on the deposition surface. Extensive work has been done on the effect of ion bombardment during the growth of cBN by different deposition techniques [414, 415, 429, 434, 460], [462–469].

Normally in the PVD process the nature of substrates does not play as significant a role as it does in the CVD process. Mostly, the deposition of cBN has been performed in the energetic PVD process with Si used as substrate. Several studies show that the highest cBN-content films form on hard, covalent substrates, as indicated by Mirkarimi et al. [402]. Table XVIII shows the results of cBN film formation on nonsilicon substrates [402]. It was found that a lower content of cBN forms on soft metal substrate (Al, Ag) than on the hard metal substrates (Nb, Ta, Ni). Also it is difficult to obtain cBN films on insulating substrate, but using a rf substrate bias would be the way to overcome this obstacle [469]. Surface roughness may also play significant role for the formation of the initial layer of cBN films [421].

In contrast to the narrow window of ion energies, cBN films can be grown in a wide range of substrate temperatures [425]. Temperature dependent growth of cBN has been studied by many researchers [414, 425, 469, 470]. Though room temperature deposition of cBN films has also been reported [471, 472], the actual temperature must be much higher than room temperature due to energetic particle bombardment, as suggested by Mirkarimi et al. [402]. It was found that there is a threshold temperature below which only tBN films form instead of cBN and that threshold temperature depends on the ion-to-atom ratio [425, 462, 469, 470]. This threshold temperature might be important for only the initial layer formation for cBN growth [473]. Also the microstructure of cBN films varies with the substrate temperature [425, 468, 474]. A higher substrate temperature can help to relieve the stress in the films generated during the growth [434, 436].

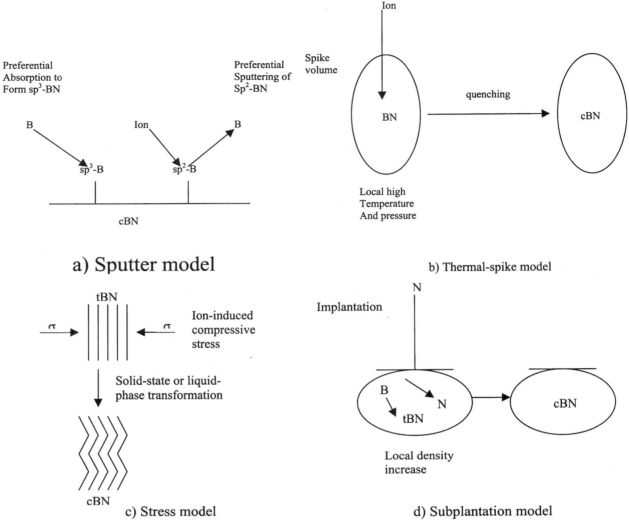

Fig. 42. Simplified schematic representations of (a) the sputter model, (b) the thermal-spike model, (c) the stress model, and (d) the subplantation model of cBN formation. Reproduced with permission from [402], copyright 1997, Elsevier Science.

3.4. Growth Mechanism

Mikarimi [402] has discussed several growth mechanisms which are briefly discussed in the following with schematics of each model shown in Figure 42 [402].

3.4.1. Sputter Model

Here growth of cBN occurs by preferential sputtering of cBN relative to graphitic BN, as proposed by Reinke and co-workers [464, 475, 476] (Fig. 42a). Reinke et al. [476] show that over 60% of the incident boron atoms are sputtered. The validity of this model has been investigated by many researchers [391, 463, 467, 468, 477, 478]. Though this model describes the process with relative etching of hBN and cBN, it does not consider the phase stabilization of cBN at the same condition which happens in the growth of diamond films at the typical CVD atmosphere.

3.4.2. Thermal Spike Model

Although near the end of its trajectory in a solid, an ion will not have sufficient energy to displace atoms but ion energy will be dissipated into phonons in what has been termed a thermal spike [402] (Fig. 42b). These thermal spikes can result in very high temperatures (several thousand degrees Celsius) and pressures (upto 10 GPa) experienced locally over a very small period of time (order of picoseconds). If the energy within the spike region is very rapidly quenched, kinetics will favor formation of the metastable phase, cBN [479]. Ronning et al. recently proposed a cylindrical thermal spike model to describe the formation of diamondlike carbon phases and extended that for the explanation of the formation of cBN films by ion assisted beam deposition [395].

3.4.3. Stress Model

Due to the ion bombardment for the promotion of the cBN phase over the hBN phase a significant amount of compressive stress is generated in the grown films [480]. According to McKenzie et al. [481] cBN forms because the ion-induced stress places the BN material in the cBN stable region of the thermodynamic phase diagram. There a boundary of the film's stress (4–5 GPa) was found, over which the cBN phase stabilizes over hBN. It was observed that the stress quickly rises to a very large value during the initial growth of cBN and then falls off with deposition time [419]. Substrate must be strong enough to withhold the high stress and this might be the reason hard substrates are better than those soft ones for the growth of cBN films [482].

3.4.4. Subplantation Model

Lifshitz et al. [483] proposed this model to describe the formation of sp^3 hybridized carbon by ion irradiation. If ions of sufficient energy penetrate below the surface of sp^2 hybridized carbon, ions will displace more sp^2 carbon than sp^3 carbon, resulting in net increase of sp^3 carbon (Fig. 42d). Several researchers commented on the validity of this model for the growth of cBN formation and proposed a modified subplantation model also [425, 477, 484–486].

3.4.5. Nanoarc Model

Recently Collazo-Davila et al. [487] showed the atomic scale description of the ion-bombardment-induced nucleation of cBN. In this study they started with hBN (99.5% purity, Advanced Ceramics Corp.) and a prepared sample on a 1000-mesh gold grid for TEM analysis. Samples were then examined with HRTEM with and without exposure to 50–100 A cm^{-2} (300 keV) for 10 min [487]. These irradiation conditions produced 5–10 dpa which is a range that is comparable to the estimated 5 dpa occurring for ion-induced cBN formation. The structures that are seen after intense electron irradiation were referred to as "nanoarches" (Fig. 43a). They postulated that for nucleation of cBN, the bond strain in the nanoarches lowers the barrier to sp^3 hybridization. Further deposition of boron or nitrogen atoms on an arch will release some of the strain energy and finally extend the sp^3 bonding network which will help to form a small cBN crystal. Figure 43 shows the steps in this nucleation process [487]. A schematic representation of the growth step is shown in Figure 44 [487]. First, the ion bombardment creates the textured tBN layer (Fig. 44a) through compressive stress mechanism. The compressive stress is important for compressing the sp^2 bonded sheets by ~5% to obtain a good lattice match to cBN. In the second step the model shows the formation of nanoarches which is the key element for the nucleation. Finally, the nanoarches grow, coalesce, and form the naoncrystalline cBN film (Fig. 44c). The nucleation model was consistent with their experimental reports on the formation of cBN films [487].

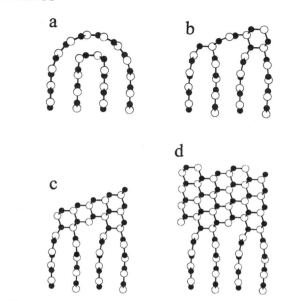

Fig. 43. Conversion of a nanoarch to a cBN nucleus. Reproduced with permission from [487], copyright 1999, Elsevier Science.

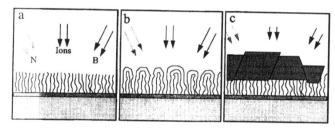

Fig. 44. Depiction of cBN groth process based on nanoarch nucleation mechanism: (a) formation of textured t-BN layer, (b) formation of nanoarches, and (c) cBN nucleation and growth. Reproduced with permission from [487], copyright 1999, Elsevier Science.

3.5. Characterization Techniques

Characterization of BN films is nontrivial, and conclusive phase indentification requires applications of several complementary techniques [488]. Characterization is made more complicated due to their small grain size and also often cBN forms along with the mixture of other phases. A number of experimental techniques have been used to characterize the bonding and microstructure of the BN films. Out of those FTIR, TEM, and electron energy loss spectroscopy (EELS) are found to be very effective. In this section we discuss some of the characterization tools used to analyze the boron nitride films and give a brief description of the basic principle of the EELS technique.

3.5.1. FTIR Spectroscopy and Raman Spectroscopy

IR and Raman spectroscopies are very sensitive to the bonding of BN films. FTIR is one of the most used characterizations of BN films. Vibrational properties of single crystal cBN and highly crystalline hBN are well known while those of rBN and wBN are not well investigated (Table XIX) due to unavailability of large single crystals of them. As vibrational frequencies are

Table XIX. Optical Vibrational Modes of cBN, wBn, hBN, and rBN

Crystal structure	Activity	Mode symmetry		Measured frequencies (cm⁻¹)	
				Infrared (IR)	Raman (R)
CBN	R, IR	F_2	TO	1065 ~ 1050	1056
			LO	1340 ~ 1310	1304
WBN	R,IR	A_1	TO		950
			LO	1090	1015
	R,IR	A_1	TO	1120	1050
			LO	1230	1295
	R	E_2			
	R	E_2			
	silent	B_1			
	silent	B_2			
HBN	IR	A_{2u}	TO	783	
			LO	828	
	silent	B_{1g}			
	silent	B_{1g}			
	IR	E_{1u}	TO	1367	
			LO	1610	
	R	E_{2g}			52
	R	E_{2g}			1366–1370
RBN	IR	A_{2u}	TO	790	
			LO		
	IR	A_{2u}	TO		1367
			LO		

Reproduced with permission from [402], copyright 1997, Elsevier Science.

bond sensitive it is straightforward to distinguish sp^2 and sp^3 hybridized BN, but difficulties arise in separating the phases with the same hybridization. For cBN, the single, triply degenerate phonon is both IR and Raman active and is split into TO and LO components at ~1060 and 1310 cm⁻¹, respectively. A typical FTIR spectrum taken on BN film with high cBN fraction and a Raman spectrum taken on sintered bulk cBN samples are shown in Figure 45a and b, respectively [402]. For hBN, two IR-active phonons have TO frequencies at about ~780 and ~1370 cm⁻¹, respectively, and a typical IR spectrum of it is shown in Figure 45c. The 1370 cm⁻¹ mode is a stretching of the B–N bond within the basal plane and the 780 cm⁻¹ mode is a bending of the B–N–B bond between the basal planes. rBN has two vibration modes, both of which are IR and Raman active and they are closely related to the IR-active modes of hBN. As in most of the cases in which Si is being used as a sub-

strate for cBN formation, care must be taken to interpret the IR peak of cBN films. This is due to the appearance of intense IR absorption associated with SiO_x between 1100 and 1050 cm⁻¹, depending on the oxygen content [489] (Fig. 45d). The common practice for phase quantification is that, the volume fraction of cBN is equal to the ratio of I_{1060} and total of I_{1060} and I_{1370} [460, 482, 490]. I_{1060} and I_{1370} are the normalized reflected or transmitted intensities of IR absorptions at 1060 and 1370 cm⁻¹, respectively. Shifting of IR peaks due to the high compressive stress and poor crystallinity has been discussed by many researchers [413, 420, 425, 491, 492].

3.5.2. X-Ray and Electron Diffraction

Small crystallites, crystallographic disorder, overlap of reflections from multiple phases, and low scattering intensities make diffraction techniques very complicated. The expected diffraction peaks and d-spacings for the cBN are shown in Table XX [402]. The strongest peak is the {111} reflection with $d = 2.09$ Å. Several reflections are very weak and some are rarely seen. This is the reason conventional θ–2θ geometries have been relatively unsuccessful in characterizing cBN thin films. Grazing angle XRD is proved to have better counting statistics, while transmission electron diffraction is the best. Normally the presence of secondary graphitic phases significantly complicated the analysis of cBN films. Crystalline hBN is rarely seen in BN films; instead the turbostratic modification is dominant.

3.5.3. TEM

TEM is the most useful tool for microanalysis of boron nitride films. Both dark field TEM and HRTEM have been used extensively for phase identification, microstructural investigation, and interface studies [364, 381, 385, 493]. Dark field TEM methods are useful for determining the phase distribution, grain size and morphology in cBN films. As the ordered regions of both the sp^2 and sp^3 hybridized phases in BN films are nanocrystalline dimensions, HRTEM has been a very useful tools for studying cBN films. For cBN structure, the most commonly identified fringes correspond to the {111} planes ($d = 2.088$ Å) and for graphitic boron nitride material, the basal plane fringes ($d \approx 3.3$ Å) are easily resolved. The sensitivity of HRTEM is very high and hence care must be taken in crystal orientation, beam tilt, objective lens focus, and aberrations [424, 494]. From different characterization techniques, and especially from TEM microanalysis, it is found that cBN forms on the graphitic boron nitride layer which forms on an amorphous layer attached to the substrate [412]. A typical HRTEM picture which shows the distinct layered structure of cBN formation is shown in Figure 46 [412]. It has been seen that BN films consists of a 20 Å thick amorphous layer, a 20–50 Å thick oriented graphitic BN layer, and the top polycrystalline cBN layer. It has been also seen that the seed layer consists of tBN instead of an amorphous layer [425, 495, 496]. It is not clear whether the growth of an amorphous layer at the beginning of the cBN film

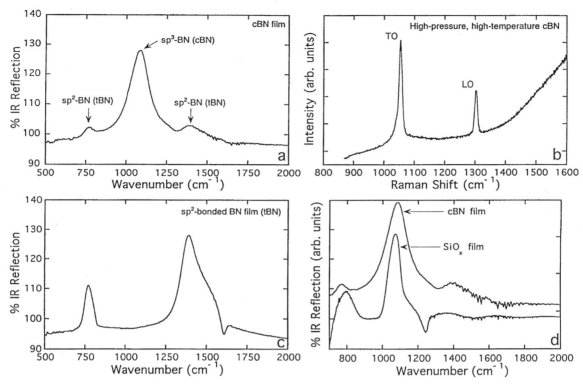

Fig. 45.   Vibrational spectroscopy of BN phases. (a) FTIR spectrum for an 100 nm thick film with high cBN fraction taken in reflection and ratioed to the spectrum of an uncoated silicon substrate. (b) Raman spectrum of a commercial, sintered compact of cBN grains synthesized at high temperature and pressure. Obtained using 514.5 nm laser. (c) FTIR spectrum for an sp^2 bonded BN film (tBN) taken in reflection and ratioed to the spectrum of an uncoated silicon substrate. (d) Comparison of FTIR spectra from a mainly cBN film and a SiO_x film, the latter being the oxide seal coat of the back side of a commercial Si (100) wafer. Both spectra taken in reflection and are ratioed to the spectrum of a clean silicon substrate. Reproduced with permission from [402], copyright 1997, Elsevier Science.

growth is necessary for the nucleation of tBN or the crystalline hBN layer prior to cBN layer [448, 497–499]. Orientation of this graphitic BN layer strongly affects the orientation of the cBN layer [481, 500]. It is also found using different surface sensitive techniques that the top layer of the cBN films consists of a layer of sp^2 bonded BN. Mirkarimi summarized the microstructure of the cBN films on a substrate schematically in Figure 47 [402]. Grain size of the cBN films are found to be very small (few nanometers to 100 nm), which is one of the major problem in growth of cBN films. Grains also contain high twin and stacking fault densities [501, 502].

3.5.4. Electron Energy Loss Spectroscopy

EELS is a well-used microanalytical tool for analysis of electronic, chemical, and crystallographical structures in materials. This technique is useful because of its ability to detect light elements such as carbon and its ability to differentiate the types of bonding in a polymorphic material. The incident electrons which interact via electrostatic forces with the valence electrons in the conduction bands produce plasmon excitation. The collected electrons contribute to the low-loss peaks (5–50 eV region). Interaction of the incident electrons with the inner core electrons may be inelastic, and the energy loss is much greater and will produce much higher energy loss edges (core loss).

Table XX.   Diffraction Line Expected for cBN

Hlk	d (Å)	I/I_o (400 keV electrons)	I/I_o (JCPDS 25-1033)
111	2.09	100.0	100
200	1.81	0.8	2
220	1.28	49	6
311	1.1	26	3
222	1.04	0.5	1
400	0.9	5	1
331	1.83	8	3
420	0.81	0.4	
422	0.74	9	
333.115	0.7	5	

Reproduced with permission from [402], copyright 1997, Elsevier Science.

The mean free path for such interactions is ~ 1 μm. So a sample which is ~ 100 nm thick will experience few inner shell excitations. EELS offers a higher spatial resolution than many other techniques such as Auger electron spectroscopy (AES) and X-ray emission spectroscopy. The energy at which the edge occurs is approximately the binding energy of the correspond-

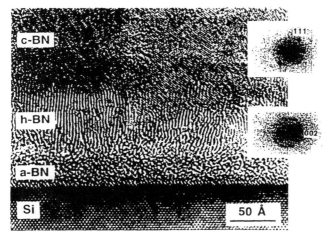

Fig. 46.   HRTEM image of a cBN film in cross-section showing the different layers in the growth of cBN film. Reproduced with permission from [497], copyright 1994, publisher ??.

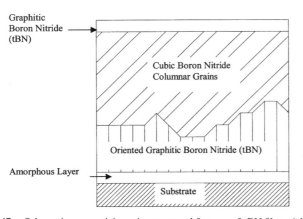

Fig. 47.   Schematic summarizing microstructural features of cBN films. A layered structure is typically found with a layer of oriented tBN forming before the cBN crystallites form. An amorphous layer is commonly observed between the substrate and the tBN layer. The cBN layer is capped by an sp^2 bonded BN layer. Reproduced with permission from [402], copyright 1997, Elsevier Science.

ing atomic shell, which is dependent on the atomic number of the element. Going from the conducting phase graphite to the insulating diamond phase introduces an energy-band gap, raising the first available empty state by several eV and thus increasing the ionization threshold energy. The core-loss spectra also contain near-edge energy-loss fine structures due to an electron being excited from an inner shell to unoccupied bond states around its nucleus before being excited to the vacuum energy level and leaving the atom ionized. The strength of the fine structure depends on the transition probability to each available level and the density of unoccupied states of that level. Consequently, the position and shape depend on the energy band structure, i.e., the type of bonding of the elements. These fine structures also contain information about the atomic environment such as the number of neighboring atoms, bonding angles, and local electronic states, and the fine structure helps to identify which allotrope of carbon is present. In sp^2 carbon, a small

$1s$-π^* peak can be observed at the rise of the ionization edge. This arises from transitions of carbon K-shell electrons to the π^* bound states below the ionization threshold. Fine structures in the 6–8 eV region (known as the π-π^* peak) due to the π-electron transitions also give an indication to the presence of double bonds, i.e., sp^2 carbon.

EELS normally attached with TEM provides useful information regarding the bonding character of the boron nitride polymorphs [503]. Peaks located at about 188, 283, and 402 eV correspond to k edges of boron, carbon, and nitrogen, respectively [504]. In the low loss regime, the energy of the plasmon peak exhibits a significant shift between hBn ($E_p = 21.5$ eV) and cBN ($E_p = 30.1$ eV), reflecting the large difference in density between the two phases. In the high-energy-loss regime, analyzing the near-edge fine structure of boron and nitrogen K ionization edges, sp^2 and sp^3 bonded material can be distinguished and can also quantified the fraction of sp^2 bonded materials on a local scale.

Several other characterization techniques, such as AES, XPS, Rutherford backscattering, and near-edge X-ray absorption fine-structure spectroscopy, have been used to analyze the grown boron nitride films. Since surfaces of BN films are normally different than the bulk, and the electron or ion beam often modifies the phase of the BN films, characterization of BN films with these techniques are used very cautiously.

3.6. Mechanical Properties

Before going to the discussion of mechanical properties of boron nitride films a discussion of the basic theory of the nanoindentation technique to measure hardness and modulus for thin films is appropriate. Indentation behavior varies depending on type of materials. For example, the indentation of most metals causes an almost entirely plastic deformation, while for many ceramics the deformation is more elastic. In metals displacement at maximum load is close to the displacement after the load is removed. Due to an elastic recovery, indenter displacement after the load is removed is considerably smaller for the ceramic. Due to this, a large error is involved in microhardness testing. In nanoindentation testing, though elastic recovery can be accounted for, it makes the analysis more complicated.

3.6.1. Nanoindentaion

To evaluate the mechanical properties, nanoindentation over a small scale has been used extensively in recent years [505]. This is a depth sensing indentation at low loads and is a well-established technique for the investigation of localized mechanical behavior of materials. The displacement and load resolution can be as low as 0.02 nm and 50 nN, respectively. A typical load vs. displacement curve showing contact depth (h_c), and maximum depth (h_t) after unloading is shown in Figure 48. Hardness and Young's modulus of elasticity are derived from the experimental indentation data by an analytical method

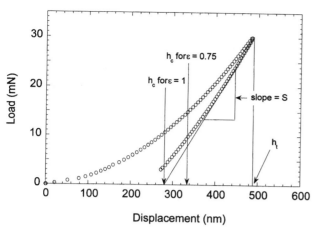

Fig. 48. Typical curve showing the loading and unloading as a function of indenter penetration depth.

Table XXI. Mechanical Properties of Bulk cBN

Hardness H	45–55 GPA	Vickers—polycrystalline
"	$\approx$55 GPa	Berkovitch/nanoindentation polycrystalline
"	40–60 GPa	Knoop–polycrystalline
"	30–93 GPa	Knoop–single crystal (depending upon orientation)
Young's modulus E	800–900 GPa	Various methods polycrystalline
Bulk modulus B	370–385 GPa	Pressure dependence of lattice parameter
"	400 GPa	Brillouin scattering single crystal
C_{11}	820 GPa	"
C_{12}	190 GPa	"
C_{44}	480 GPa	"
Fracture toughness K	2.8 MPa m$^{1/2}$	Single crystal
"	3.5–5.0 MPa m$^{1/2}$	Polycrystalline

using a number of simplifications [506]. Contact depth h_c can be calculated by:

$$h_c = h_t - \frac{\varepsilon F}{S} \qquad (1)$$

where h_t is maximum depth of penetration including elastic deformation of the surface under load, F is the maximum force, and $\varepsilon = 0.75$ is a geometrical constant associated with the shape of the Berkovitch indenter [506]. Once h_c is determined, the projected area A of actual contact can be calculated from the cross-sectional shape of the indenter along its length. S is the stiffness which can be derived experimentally from

$$S = \frac{dF}{dh} = \frac{2}{\sqrt{\pi}} E_r \sqrt{A} \qquad (2)$$

where E_r is the reduced modulus. Hardness is then calculated from the simple relation

$$H = F/A \qquad (3)$$

The reduced modulus E_r is normally defined as

$$\frac{1}{E_r} = \frac{1 - \nu^2}{E} + \frac{1 - \nu_i}{E_i} \qquad (4)$$

where E and ν are Young's modulus and Poisson's ratio for the sample and E_i and ν_i are the same for the indenter, respectively. During nanoindentation on thin films, the lesser the indentation depth (10–20% of the film thickness), the lesser the substrate will affect the measurement. Again indenter size effects can be avoided using a larger indentation depth [507].

Due to the poor crystallinity [508, 509] and large compressive film stress [510, 511] generated mainly due to high ion irradiation, hard wear-resistant coating applications are presently more viable than electronic applications. It is then very important to investigate the mechanical properties carefully. Mechanical properties of bulk cBN are given in Table XXI [402]. Hardness values of 40–60 GPa, measured using Vickers or Knoop microhardness testing, have been reported

[432, 472, 512–516]. In most of the studies high hardness values were obtained for films on soft substrates in which the indent depth was typically a large fraction (50% or more) of the film thickness. As Mirkarimi et al. [517] mentioned, high indentation depths were used in those experiments because of the constraint of low thickness of the films and minimum load used in indentation techniques. Several groups [518, 519] have used a nanoindentation technique which allows for shallower indent depths and hence constraints in minimum load problem can be overcome. There is a need for a sufficiently thick film in order to have a reasonable indent depth, which is needed to initiate deformation beyond the elastic regime, as well as to minimize the indentation size effect.

Mirkarimi et al. [517] performed nanoindentation on cBN films of various thicknesses and indentation depth was taken to be 100 nm. A typical loading/unloading curve on a 700 nm cBN films is shown in Figure 49a and Figure 49b shows the same on bulk cBN [517]. The results obtained on the films along with some bulk values are shown in Table XXII. The lower modulus and elastic recovery values were attributed to substrate effects [517].

3.7. Some Issues on the Formation of cBN Films

3.7.1. Phase Purity

It has been observed that before nucleation of cBN there is always amorphous or tBN layer formed at the interface. Kester et al. [412] suggested that once cBN nucleation takes place, it grows as a single phase. It is also been suggested that graphitic BN may exist at the boundaries of individual crystallites [520]. In order to utilize the potential of this novel material it is

Table XXII.   Mechanical Properties of cBN Films and Selected References

Sample	Sputter target used	Film thickness (nm)	Indent depth (nm)	Mean H (Gpa)	Mean E (Gpa)	% Recovery[a]
CBN Film	B_4C	700	100	68 ± 11	530 ± 60	80%
CBN Film	B_4C	700	135	57 ± 13	420 ± 50	80%
CBN Film	B_4C	550-	100	69 ± 6	360 ± 20	80%
CBN Film	B_4C	550	135	59 ± 5	305 ± 15	80%
CBN Film	B_4C	400	100	55 ± 8	320 ± 20	80%
CBN Film	B_4C	450	100	58 ± 14	330 ± 25	80%
CBN Film	BN	450-	100	55 ± 29	290 ± 55	80%
TBN Film[b]	B_4C	700+	100	2 ± 0.5	50 ± 10	—
Bulk cBN[c]	N/A	Bulk	100	53 ± 14	850 ± 150	65%

[a] % recovery = (depth at maximum load–depth after load removed)/depth at maximum load.

[b] Deposited under the same conditions as the cBN films except at a lower ion flux.

[c] Bulk, polycrystalline, translucent cBN, manufactured by Sumitomo, Inc.

Reproduced with permission from [517], copyright 1997, American Institute of Physics.

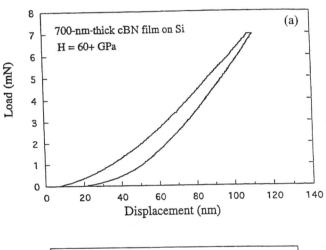

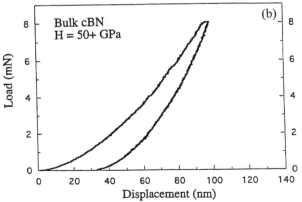

Fig. 49.   Nanoindentation loading/unloading curve for (a) 700 nm thick cBN film and (b) bulk cBN. Reproduced with permission from [517], copyright 1997, American Institute of Physics.

essential to grow a phase pure cBN layer both homo- and heteroepitaxially.

3.7.2. Adhesion

As ion-assisted deposition of cBN films generates a high amount of intrinsic compressive stress the adhesion of the films on the substrate is limited by the thickness of the films. It was observed that films >3000 Å thick delaminate after some period of time in air, while films >5000 Å thick delaminate during deposition [516]. Delamination of the films from substrates occurs more severely in the presence of a high-humidity environment. It has been suggested that water may react with BN at the interfacial soft BN layer and change the volume while making the end product $(B_2O_3)_2(OH)$ at the interface [499]. This plays a major role in the poor adhesion along with the large compressive stress. This problem of delamination is also normally substrate independent [470, 520], except diamond substrate [470].

3.7.3. Buffer Layer

The adhesion problems come due to the interfacial soft layer and the high compressive stress in the films. Using an intermediate layer (buffer layer) in between film and substrate has been proven to enhance the adhesion manyfold. Using a boron layer and a graded BN_x layer prior to the cBN layer can improve the film's adhesion with the substrate [441]. Inagawa et al. [516] investigated a number of graded layers consisting of $B_xN_xZ_{1-x-y}$ (where Z = C, Si, Ge, Al, Fe, Ti, and Cr) toward the improvement of adhesion. They could deposit cBN films of 1.5 μm using $B_xN_xSi_{1-x-y}$ interlayers. Several other buffer layers have been used to improve the adhesion and quality of the films [421, 521–523].

3.7.4. Ion Energy in Growth

In order to stabilize the cubic phase over other competitive phases, the importance of ion bombardment of the substrate surface during the deposition has been proven. Though it helps to stabilize the cBN phase, but it also incorporates a large amount of compressive stress, which is the limiting factor for growing a thick cBN layer. In CVD processes using additional gas with the conventional precursors it may be possible to stabilize the cBN phase at lower ion energies. The formation of a cBN phase in the CVD method is via a chemical route [524]. The role of hydrogen, cholorine, and florine in the gas phase is not well studied. Usually substrate biasing guides the species (ions) on the substrate surface and this effectively acts as an ion bombardment. Biasing is found to be related to the H/F ratio in the gas phase. Zhang and Matsumoto [525] found that a critical bias voltage is needed for the formation of cBN in their DC jet plasma CVD method using a gas mixture of $Ar-N_2-BF_3-H_2$. They also found that fluorine acts as an effective etchant, which can preferentially etch hBN phase, and hydrogen addition is necessary for the formation of solid BN from the gas phase. A critical bias voltage has also been observed for cBN formation and its value decreased as the H/F ratio decreased. Adverse effects of ion bombardment may be avoided by using chemical methods to grow thick cBN films.

3.8. Conclusions

Cubic boron nitride can now be deposited by number of physical or chemical vapor deposition techniques. Success in growing cBN films in most of the techniques is dependent on the ion bombardment on the growing surface. Films grown by different techniques contain more than 90% cBN. However, growing cBN at larger areas, thicknesses, and larger grain sizes has not yet been perfected. Deposition of cBN at low pressure and temperature is very much process dependent and has a narrow window in the variation of process parameters. High ion energy, which is being proved to be necessary for the formation of cBN films, is the cause of high intrinsic stress in the films. This again restricts the deposition of thicker films. The following issues need to be discussed in order to improve the growth of cBN films with better perfection:

- growth of cBN with low energy ion bombardment,
- relation with substrate temperature and ion bombardment,
- conversion of graphitic BN into cBN, which might open the window of post-treatments effects,
- addition of buffer layers suitable to both substrate and cBN,
- CVD of cBN with the use of catalytic gas in the conventional precursor gases.

Lot of different growth mechanisms are being discussed but none of them could describe the deposition of cBN completely. Among them the cylindrical thermal spike model and the nanoarc model may help to explain the growth mechanism completely. Large numbers of characterization tools have been used to investigate the cBN films and the most important of them are the FTIR and TEM analysis. One has to take enough care to avoid the artifacts which may be produced during the measurements. Though few companies are producing the cBN coatings on the tool materials, there are several issues that need

to be solved in order to utilize this novel superhard material in thin film form.

4. CARBON NITRIDE THIN FILMS

4.1. Introduction

A considerable amount of interest has been generated on producing crystalline carbon nitride since Liu and Cohen [6, 7] predicted a carbon nitride, β-C_3N_4, with a structure similar to β-Si_3N_4. Study of C_3N_4 was motivated by an empirical model [8] for the bulk modulus of tetrahedral solids which indicates that short bond lengths and low ionicity are favorable for achieving large bulk moduli. Since the $C-N$ bond satisfies these conditions, tetrahedral $C-N$ solids were suggested by as candidates for new low compressibility solids [7]. According to their calculations it was predicted that zinc-blende $C-N$ compounds would be unstable but β-C_3N_4 can have bulk moduli comparable to that of diamond, which has the largest known bulk modulus [7]. Due to a shorter bond, β-C_3N_4 may possess extremely high hardness comparable to or greater than that of diamond. These interesting properties attract researchers to synthesize this hypothetical material by means of a variety of techniques. Carbon nitrides with crystalline as well as amorphous microstructures are also of great engineering interest [526, 527] as a potential material for microelectronic devices and for optical, magnetic, and tribological applications [526, 528–533].

The computer hard-disk industry is currently working on achieving an areal density of 100 Gbits/in^2 [534]. To achieve such a density the magnetic spacing (spacing between the magnetic read/write head and and the magnetic recording layer on the disk) has to be less than 10 nm. This only leaves ~2 nm for the protective overcoat. Diamondlike carbon or CN_x coatings (5–10 nm thick) may be used to improve the tribological performance with improved corrosion resistance (Fig. 50) [535].

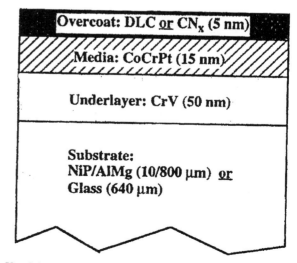

Fig. 50. Schematic of the different hard disk components and their layer thickness. Reproduced with permission from [535], copyright 1999, Institute of Physics.

Table XXIII.   Space Group, Lattice Parameters (a and c), Mass Densities ρ and Bulk Moduli B of the Theoretically Predicted α- and β-C_3N_4 Phases

Space group								
β-C_3N_4								
Space group	$P6_3/m$				$P3$		$P6_3/m$	
a (Å)	6.44	6.41	6.37	6.47	6.35	6.40	6.42	6.39
c (Å)	2.47	2.40	2.40	2.45	2.46	2.40		2.40
ρ (g cm^{-3})					3.56			
B (GPa)	427	437	421	450	250	451	557	
α-C_3N_4								
Space group					$P3_1c$			
a (Å)					6.35	6.47	6.46	6.45
c (Å)					4.46	4.71		4.70
ρ (g cm^{-3})					3.78			3.61
B (GPa)					189	425	567	

The hardest material known at present is diamond, with a hardness of about 100 GPa, while the second hardest material, cBN, has a hardness of only 50 GPa. Liu and Cohen calculated [536, 537] the properties of the hypothetical compound β-C_3N_4 (space group $P3$), which has a structure similar to tetrahedral β-Si_3N_4 (space group $P6_3/m$). Because the sp^2-sp^3 hybridized bonds between the tetrahedrally bonded C—N are shorter than the bond length of C—C in diamond, it is predicted that the bulk modulus will be larger than diamond. The three-dimensional tetrahedrally bonded structure would thus have a high isotropic hardness. They indicated that the material would have mechanical and thermal properties similar to diamond. On top of that, the strength, hardness, and high corrosion and wear resistance of Si_3N_4 (which the β-C_3N_4 structure has been based on) suggest that the hypothetical C_3N_4 compound could have improved structural properties.

Further calculations have also predicted that C_3N_4 can exist in two other types of structures, a zinc-blend-like cubic structure (space group $P = 43m$) and a graphitelike phase with rhombohedral stacking (space group $R3m$) [538]. Both the zinc-blend-like and β-C_3N_4 phases have similar structures, with each C atom having four N neighbors and each N atom bonded to three C atoms. The angles between the C—N—C bonds, however, are different for both the structures. The bond angle for the zinc-blend structure is close to 109.47° and for β-C_3N_4, 120°. The local coordination for this structure is similar to that in the hexagonal phase, except that the N atoms form sp^3 bonds rather than sp^2 bonds. The graphitelike phase consists of holey graphitelike sheets with rhombohedral stacking order (ABCABC...). Each of the C atoms has three N atoms as neighbors (threefold coordinated), as does one of the four N atoms in each unit cell. The other three N atoms are only twofold coordinated, having only two C neighbors with one C atom missing. The interlayer bonding for the sheet structures is expected then to be weak. Tables XXIII and XXIV summarize the structural parameters for different phases of carbon nitride [539].

From the enthalapy–pressure (H–P) data it was found that the Willemite and alpha C_3N_4 phases are the only stable phases. The alpha phase is stable from low pressures up to 81 GPa and the Willemite phase was found to be stable between 81 GPa and the highest simulated pressure ($\sim$300 GPa). The transition between the lower pressure alpha phase and the Willemite phase occurs as the pressure–volume term in the enthalpy becomes increasingly important at higher pressures. Since the Willemite structure has a lower volume configuration, at higher pressures the pressure–volume term will be less significant compared to that of the alpha, and above 81 GPa, the enthalpy of Willemite is smaller. The beta and defect zinc-blende phases were found not to be the stable phases throughout the entire pressure range. At higher pressures the Willemite structure is preferred over the defect zinc-blende structure because of the higher energy configuration of the lone pair electrons occuring in defect zinc blende. Since hardness depends on shear modulus, it would need the shear for the alpha phase to make further conclusions about its hardness. The Willemite structure has a bulk modulus higher than the one for diamond but its shear modulus is much lower than that of diamond and therefore its hardness is expected to be much lower.

4.2. Phases of Carbon Nitride

There is considerable discussion as to which of the four is the most stable, as well as to the precise values of their bulk moduli. The most recent values are 448, 496, 425, and 437 GPa for the defect zinc blende, cubic, α, and β forms of C_3N_4, respectively [540]. These values are similar to, or greater than, that for diamond at 443 GPa. Similarly, the velocity of sound in β-C_3N_4 has been estimated to be high, $\sim$$10^6$ cm s^{-1}, meaning that the material should have a high thermal conductivity. The bandgaps of all of the high density C_3N_4 materials are expected

Table XXIV. Structural Parameters (Space Group, Formula Units/Cell Z, Lattice Constants a, b, c), Mass Densities, and Bulk Moduli of the Theoretically Predicted Cubic, Pseudocubic, and Graphitic C_3N_4 Phases

Face centered cubic C_3N_4						
Space group	$I\bar{4}3d$			$I\bar{4}3d$		
Z	4			4		
a (Å)				4.67		
ρ (g cm^{-3})	5.40			3.89		
B (GPa)	496					
Pseudocubic C_3N_4						
Space group	$P\bar{4}3m$	$P\bar{4}2m$		$P\bar{4}3m$		
a (Å)	3.43	3.42	3.41	3.42		
ρ (g cm^{-3})				3.82		
B (GPa)	425	448	556			
c-C_3N_4						
Space group						
a (Å)					6.87	
ρ (g cm^{-3})					3.77	
B (GPa)					396	
Graphitic C_3N_4[a]						
Space group	$P\bar{3}m1$	$P\bar{6}m2$		$P\bar{6}m2$ & $R3m$	$P2mm$	
Lattice		Rhombohedral				Orthorhombic
Z	3		3	2 & 1[b]		2 & 1[c]
a (Å)	4.11	2	4.09	4.37 & 4.11		410
b (Å)		4.74			4.70	
c (Å)				6.69 & 4.11		3.2 & 6.4
α (°)	70.5	6.72		90 & 70.38		
ρ (g cm^{-3})				2.35 & 2.56		

[a] Stacking orders: AA, AB, or ABC.

[b] Depends on the space group.

[c] AA and AB stacking.

to be in the range 3–4 eV. The cohesive energy calculated for all of the C_3N_4 compounds indicates that they should be, at least, metastable. Furthermore, it is clear that the low-bulk-modulus graphitelike structures are the most stable, and that the other high density forms have similar but lower stabilities.

4.2.1. Alpha Structure

According to the data collected, the alpha structure is the most stable of the structures investigated. In theory, this is supported by examining the lone pair interactions present in the structure. It is a trend that less of these interactions serve to lower the energy of the structure, thus increasing the stability. The decrease in these interactions is due to fewer nonplanar NC_3 groups present in the alpha structure, hence the lower energy. When examining the hardness of the alpha phase, the modulii approach, but are not greater than, that of diamond. Figure 51 shows the crystal structure of different phases of carbon nitride.

4.2.2. Beta Structure

Calculations for the beta structure compare very well with experimental data. β-Si_3N_4 is softer than β-C_3N_4 because the N—N bond is longer and therefore there is not so much resistance to compression due to lone pairs (N—N bond changes 18% as opossed to 11% change for C_3N_4). The coordination tetrahedra are highly deformed at high pressures.

4.2.3. Willemite Structure

C_3N_4 in the Willemite structure has a very high bulk modulus (474 GPa). The structure does not move around too much under hydrostatic compression. This can be seen from the graphs that are included. The bond lengths linearly decrease, and the C—C and N—N atomic distances do as well. The C—C atomic distance, however, decreases faster than the N—N atomic distance. This is probably due to lone pair repulsion. This conclusion can be reached with the aid of the bond angle plot for hydrostatic compression. The C—N—C angles stay the same, but the N—C—N angles decrease with increasing pressure. The two

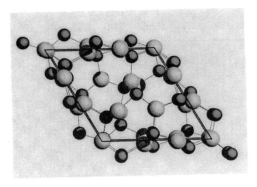

(a)Alpha C3N4 (black is nitrogen)

(b) Beta C3N4

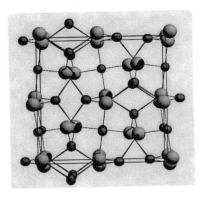

(c)Cubic-C3N4

Fig. 51.   Crystal structure of different phases of carbon nitride: (a) α-C$_3$N$_4$, (b) β-C$_3$N$_4$, (c) cubic-C$_3$N$_4$, (d) pseudocubic C$_3$N$_4$, and (e) graphitic C$_3$N$_4$.

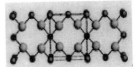

(d) Pseudocubic-C3N4

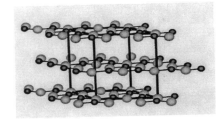

(e) Graphitic C3N4

Fig. 51.   (Continued).

nitrogens are trying to get away from each other due to lone pair repulsion. Keep in mind that this effect is slight and only noticeable at very high pressures. Si$_3$N$_4$ in the Willemite structure has a much lower bulk modulus (266 GPa). The C—C atomic distance decreases under hydrostatic pressure but the N—N distance decreases only slightly and then actually increases, allowing the nitrogen atoms to slide over one another. This has the effect of decreasing the C—N—C bond angle while the N—C—N bond angle stays constant. Under hydrostatic compression of Si$_3$N$_4$, the C—N—C bond angle change does not allow as much energy. The shear modulus of C$_3$N$_4$ in the Willemite structure is relatively low (266 GPa). The N—C—N bond angle changes quite a lot compared to the C—N—C bond angle. This suggests that the lone pairs act to help the nitrogen atoms glide over one another, lowering resistance to shear.

4.2.4. Defect Zinc-Blende Structure

The C$_3$N$_4$ zinc-blende structure was found to have a bulk modulus of 425.8 GPa and other calculations have determined a value of around 448 GPa. With this variation, the true value is somewhere in the range of diamond (443 GPa). The high bulk modulus of the defect zinc-blende structure is primarily due to strong interactions of the lone pair electrons which are all directed toward one another and strongly resist compression. The shear modulus for the defect zinc-blende structure was significantly lower (159.6 GPa) compared to that of diamond (534.7 GPa). The low shear resistance difference is again due to the presence of the lone pair electrons which move away from each other under a shearing motion, thus reducing their interaction. The deformation of the C$_3$N$_4$ defect zinc-blende structure is depicted by the bond length and bond angles vs strain/pressure graphs in the various deformation regimes. Since the shear modulus has been found to more accurately represent hardness properties, the defect zinc blende does not appear to have a hardness on the same level as diamond. Furthermore, from the $H-P$ curves, the C3N4 defect zinc blende was shown not to be stable at any pressure; this could prevent the material from being a realistic consideration, unless a processing technique is developed which is capable of capturing the structure in a metastable state.

4.3. Carbon Nitride Synthesis

Another sp^3 structure that is of current interest is C$_3$N$_4$. The main attraction to this elusive material is its hardness which was predicted to be comparable to that of diamond. Many methods have been used to synthesize this material and it is found that carbon nitride is metastable and can exist in various structures. Therefore, carbon in sp^3 and sp^2 form can be found simultaneously in the films. The common observation is that the deposited CN_x films are spatially inhomogeneous. Evidence of such structures can only be found on average at a micron scale. Predicted to have mechanical properties similar to diamond, carbon nitride compound has captured the attention of many researchers since 1989. So far, however, there is a little evidence of successful growth of crystalline single phase C_3N_4. Here we discuss some of the deposition techniques used by many researchers.

4.3.1. High Pressure Pyrolysis and Explosive Shock

Attempts have been made [541] to synthesize CN_x by a combustion of various organic precursors, such as mixtures of tetrazole and its sodium salt, or pyrolysis of polymeric precursors. Similar to diamond synthesis, the resultant materials have been subjected to explosive shock compression to synthesize the required structure. Also a-CN, produced by PECVD, has also been treated with a similar method and triple bonded carbon clusters joined by nitrogen bridging atoms were found. The other starting materials gave a-C, well-ordered diamond crystallites, or, at best, carbon structures with bridging nitrogen bonds [542]. This indicates that a considerable amount of nitrogen loss occurred during the compression stage. In the work of Wixom [543], using shock wave compression technology with nitrogen containing precursor and typical conditions of diamond and boron nitride synthesis, material was produced with low overall nitrogen content which contains some well-ordered diamond. In another work [544], synthesis of a new crystalline phase was reported from a mixture of carbon and nitrogen heated to 2000–2500 K at 30 ± 5 GPa. The phase obtained in this work has an X-ray diffraction pattern compatible with cubic symmetry but does not match with theoretically expected patterns for CN_x phase [545]. In their work Dymont et al. [546] prepared bulk crystalline CN_x phase from a precursor containing carbon, nitrogen, and hydrogen. Precursor was synthesized from a electrochemical process. Syntheses of the CN_x phase in block powder form from this precursor were performed at pressures up to 7 GPa and the temperature was varied between 300 and 600 K. X-ray diffraction spectra showed that the precursor is successively changed from completely amorphous carbon–nitride to crystalline carbon–nitride.

4.3.2. Carbon Source with Nitrogen Bombardment

Basically this involved the implantation of nitrogen (incident energies of 3–60 eV) into graphite at temperature of <400°C, with or without annealing up to 1000°C [547]. High nitrogen content may be found in as-grown films with high energy, but nitrogen content was drastically reduced upon annealing, indicating that little stable compound formation had occurred. There is an energy barrier found for the formation of sp^3 phase of carbon [548]. No crystalline CN formation has been observed. Another process involves chemical/physical sputtering,

or arc evaporation of graphite and the transport of C and CN species to a substrate. Veprek et al. [549] found amorphous deposits with C_3N_4 composition at medium pressure, high power, and high temperature; 0.2 Torr, 3 kW, and 800°C. In a DC arc jet, using high temperature, power, and pressure (60 Torr), a small crystallite embedded in an amorphous matrix was found and was confirmed by XRD and Raman spectroscopy [550].

Yen and Chou [551] reported the use of a high pressure DC arc jet to make CN deposits. In this system, the arc was struck between a thoriated tungsten cone-shaped cathode and the inside wall of a graphite anode, used as a carbon source. The system was operated at high power, ~400 W, and pressure, 60 Torr, thus ensuring that the ion energy was very low. The substrate temperature depended on the substrate–arc distance and typically varied from 600 to 800°C. The N/C ratio depended on the temperature used, and at the highest temperature it was ~0.58. Diffraction measurements showed the existence of small crystallites, within a carbon-rich matrix, whose lattice spacings and intensities closely agreed with those of β-C_3N_4. The c/a value at 0.39 was larger than the theoretical value and some of the observed d-spacings could not identified. However, Raman analysis gave five small peaks whose positions were in good agreement with simulated spectra generated from the spectra of β-Si_3N_4 multiplied by the scaling factor proposed by Yen and Chen [551]. This work presents some of the most convincing evidence for the formation of crystalline C_3N_4.

In order to produce low energy carbon atoms in a nitrogen or ammonia atmosphere, or with nitrogen ion bombardment at a biasing of 150–600 V, the cathode arc technique with and without filtering is one of the methods used [552, 553]. Pressures been used in all the studies are less than 50 mTorr while substrate temperatures were varied from room temperature to 1000°C. Films were amorphous and decomposed with nitrogenoss at higher annealing temperature (>800°C). Properties of the films are not very sensitive with the nitrogen arrival or the bias. Evaporation of carbon with nitrogen ion bombardment is another technique to grow CN_x [554]. This is a high vacuum technique while substrate temperatures are kept at room temperature. The N/C impingement rate on the substrate surface appeared to be very sensitive for film properties but ion energy has less effect. However, high energy ion bombardment deteriorates the film's mechanical and optical properties. Freeman sources have also been used to generate alternating C and N beams where both ion energies (from 5 to 100 eV) and the relative arrival rates were controlled [548]. Mostly crystallite graphite is observed. The sp^3 C content was measured as a function of the incident ion energy, and two maxima at ~30 and <1 eV were seen. Chemical sputtering of the carbon and CN by nitrogen was found to be of great importance. Significantly, bombardment by nitrogen ions with energies greater than ~30 eV was seen to restrain the amount of this element in the deposit. Specifically, nitrogen bombardment was seen to remove deposited material through the formation of CN, with an etch rate of ~0.5 carbon atoms per incident N_2^+ ion. This process was estimated to limit the maximum nitrogen content to N/C ~ 1.86; however, experimental evidence indicates that the maximum may be closer to 0.67 [555]. Todorov et al. suggested that an additional mechanism involving the promotion of the formation of either molecular nitrogen or the low-boiling-point compound C_2N_2 ($-26°C$) may also take place at high nitrogen contents [556].

Kohzaki et al. used electron-beam evaporation of carbon but at a fixed low nitrogen-ion energy of 200 eV and substrate temperatures of <100, 300, and 500°C [557]. Here the N/C ratio decreased toward saturation, at a value of ~0.6, as the relative nitrogen arrival rate increased with little dependence on the substrate temperature, although the FTIR spectra showed that the 2200 cm^{-1} peak (CN) decreased in intensity, at an approximately constant N/C value, as the substrate temperature increased. However, the XPS measurements indicated almost no change in the bonding configuration. Columnar structure at the highest substrate temperature was also reported but films were not crystalline [558].

In one study polyethylene was evaporated at 300°C through a nozzle in such a way that the accompanying adiabatic, supersonic expansion caused condensation into clusters [559]. These were then ionized by electron bombardment and accelerated toward the silicon substrate held at 300°C. Nitrogen pressures from 3×10^{-5} to 3×10^{-4} Torr were used. No composition data were provided but FTIR absorption bands in the 1000–1600 cm^{-1} range were seen to change as the nitrogen pressure increased. XPS and Raman spectroscopy studies showed the indication of formation of crystalline carbon nitride.

Another work by Veprek involved chemical sputtering of graphite by an intense, high frequency nitrogen plasma to generate large amounts of CN radicals in an excess of excited nitrogen molecules and atoms [560]. Increasing the distance between the graphite source and the substrate could decrease the relative concentration of CN to nitrogen. It is being claimed that the material might contain rhombohedral C_3N_4.

A novel method used by Yen involved various combinations of explosive evaporation of carbon, surface condensation of nitrogen, and mixing of these layers by either intense ion or electron bombardment [561]. Different approaches were used as follows: (i) Nitrogen was condensed on a graphite source electrode; then both carbon and nitrogen were explosively evaporated by electron impact. (ii) Nitrogen was condensed on the substrate at ~80 K; then carbon was explosively evaporated on top with the carbon-ion energy being controlled through biasing. (iii) Carbon was evaporated onto the substrate, nitrogen was condensed on top, and then the two were mixed by using an intense pulse of energy-controlled electrons. (iv) In a reversal of (iii), nitrogen was condensed, carbon evaporated on top, and the two were mixed by using a pulse of electrons. The characteristics of the deposits were independent of the type of substrate (silicon, titanium, copper, nickel, and graphite). The nitrogen content could be varied up to a maximum of 59%, $N/C = 1.44$, and 1 μm films could easily be formed using multiple experiments. The XRD data of the $N/C = 1.33$ material showed that this was crystalline with a few strong peaks whose d-values agreed with those for β-C_3N_4. For all of the four regimes the threshold energy density to cause the amorphous-to-crystalline

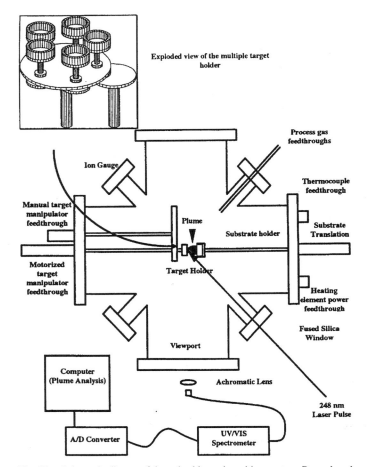

Fig. 52. Schematic digram of the pulsed laser deposition system. Reproduced with permission from [578c], copyright 1998, Elsevier Science.

transition was reported to be 10^4 J cm^{-3} for electrons and 5×10^4 J cm^{-3} for ions (this was expressed as an energy density because of the different penetration depths for electrons and ions) [562]. The hardness of the films depended on the processing type used and was greatest, 20–30 GPa, for regimes III and IV. Similarly, under these conditions, the friction coefficient was quite low at 0.1. One of the important aspects of this study was that chemical sputtering was avoided by using thermal condensation of the carbon and nitrogen.

4.3.3. Laser Processing

Laser processing is mainly ablation of a pure graphite target in a nitrogen or ammonia atmosphere, with or without ion bombardment of the substrate, or a secondary discharge [563, 564, 564–567]. Figure 52 shows the schematic diagram of a laser ablation system. Excimer lasers are normally operated in the UV or near UV; however, some studies have used Nd:YAG lasers operating at 532 nm. Operating pressures in the chamber are varied from 4 to ~100 mTorr and temperatures are generally <500°C.

The laser-plasma plume does generate large quantities of N_2^+ but little atomic nitrogen. The nitrogen molecules absorbed on the graphite are decomposed but the principal ablated species are C_x ($x = 1$–4) and CN. The relative quantities of these depend on the ambient gas pressure, which nitrogen-containing gas is used, the laser fluence, and its wavelength. In nitrogen, large amounts of CN are observed but in ammonia, C_2 dominates and there is little CN production [568, 569]. Therefore, the species incident on the substrate consist, principally, of ions and radicals of C_x, CN and N_2, or NH. As described in other sections of this chapter, it has been seen that N_2 species are inefficient at introducing nitrogen in the deposit and molecular nitrogen ions can remove nitrogen by chemical sputtering.

Several studies using a pulsed laser deposition technique reported the formation of crystalline phase of carbon nitride by varying the basic deposition conditions, additional arrangements, or using different laser sources [570–577]. Most deposits to date have been amorphous forms of carbon nitride with $N/C < 1$. There have been some reports of the formation of small crystals using laser ablation, with diffraction data containing some of the peaks calculated for β-C_3N_4 [578, 577]. However, it is frequently found that some peaks are absent, others cannot be indexed, and the relative intensities often do not agree with the calculated data.

Niu et al. described Nd:YAG ablation of graphite but with an intense low energy (1 eV) atomic-nitrogen beam incident on the substrate [578a]. The N/C ratio was directly proportional to the atomic-nitrogen flux, maximum value 0.82, and was independent of the substrate temperature up to 600°C. The deposit was thermally stable up to 800°C and small crystals were observed using electron diffraction, with six d-values in close agreement with those predicted for β-C_3N_4. This work is similar to that by Xu et al., who used a novel atomic-nitrogen source with the laser ablation [578b].

4.3.4. Reactive Sputtering

In this method usually a graphite target is being sputtered with a DC or RF power source using N_2 or NH_3 gas mixtures [578d, 578e]. A schematic of a Penning-type sputtering method is shown in Figure 53 which has been used by Cameron et al. to grow CN_x films [579]. Here two opposite cathodes have a magnetic field perpendicular to the cathode surfaces. Secondary electrons emitted during sputtering are constrained to reside within the volume between the cathodes for an extended period. This makes possible the creation of an intense discharge which is sustainable down to low pressures. In sputtering substrate temperature was varied from ambient to 600°C and a little or no deposition was found at higher temperature. Normally amorphous films were deposited and they decomposed at higher temperature annealing. Addition of small amount of hydrogen in the gas mixture drastically reduced the deposition rate [580]. Kin Man Yu et al. [581] reported the formation of β-C_3N_4 crystals along with graphite inclusions in their RF sputtering experiment using pure N_2 with a substrate temperature of ~600°C. The XRD pattern of their films could not conclude the formation of crystalline phase of C_3N_4.

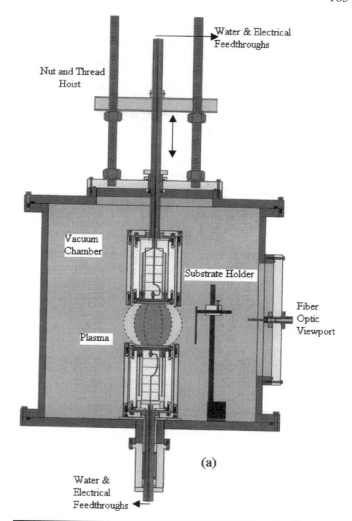

Fig. 53. (a) Schematic of the Penning-type sputtering chamber; (b) a photograph of the plasma within the vacuum chamber. Reproduced with permission from [579], copyright 2000, Elsevier Science.

4.3.4.1. Sputtering with Ion Bombardment

Experiments were done for the formation of CN using substrate biasing and dual beam ion sputtering and substrate bombardment [582–584]. It was observed that if only N_2 is present then chemical sputtering releases CN while physical sputtering gives C. Therefore low energy bombardment promotes chemical etching of C=N, but if hydrogen is present then volatile CNH is created for substrate biasing >30 V [583]. Increasing biasing results in a decrease in deposition rate. It is also seen that film formation occurs if the ion to atom impingement ratio is less than some critical value which depends on the ion energy and probably on the N/C arrival ratio, as well as the substrate temperature and the binding energy of CN to the substrate material.

4.3.5. Plasma Enhanced Chemical Vapor Deposition (PECVD)

The plasma enhanced method of growing CN has been studied extensively. Usually RF or DC plasma decomposition of CH_4 or other carbon containing precursors mixed with N_2 or NH_3 has used a gas pressure of 0.4 to 10 Torr [585, 586]. A deposit grown in this method loses its nitrogen content while annealed at >800°C and no deposition was observed at >600°C. In general the deposition rate, refractive index, stress, resistivity, and hardness decrease as the nitrogen concentration increases above ~2%. The nitrogen concentration of the films depends on the nitrogen concentration in the gas mixture and NH_3 is more effective than N_2 at introducing nitrogen in the deposit. Nitrogen containing precursors is proved to help to increase the nitrogen incorporation in the films but no significant improvement was observed in the structure of the deposit [587]. Crystalline deposit of CNSi was found using microwave decomposition of a $CH_4/NH_3/H_2$ mixture at high pressure (80 Torr), high power (3.5 kW), and high substrate temperature (1000–1200°C) [597]. Though crystallites containing ~5% Si were found, the hexagonal structure was still significantly different from both α- and β-C_3N_4. The interesting conclusion is that if no silicon was present, no deposit was found, but conversely if silicon was present then a deposit could be formed even on substrates other than silicon.

4.3.6. Filament Assisted CVD

This is a modified version of the method used to grow diamond films. In this case either CH_4 mixed with N_2 or $NH_3 + H_2$ is used as a precursor. Here either CH_4 concentration was very low, ~1% or a high H_2 dilution, 98%, was used, with the NH_3 to CH_4 ratio being either 1 or 5 [588, 589]. Substrate temperature was typically 800–900°C and chamber pressure can be varied from 1 to 15 Torr. Crystalline deposits have frequently been obtained with no amorphous phase. Occasionally crystalline structures are found which are mixture of different phases of CN_x. It is not clear about the role of hydrogen in the gas phase. It might be playing the same role as in diamond growth, and plentiful amounts of atomic hydrogen and nitrogen are required to create C_3N_4 [590]. This work indicates that there are at least two regimes, one with high hydrogen dilution and the other with high nitrogen dilution, which can be used to grow crystalline C_3N_4.

4.3.7. Other Methods

Sun et al. used the electrochemical method to synthesize amorphous CN_x films [591]. Normally an electrochemical deposition method is used to deposit metallic films. Recently this method was used to grow diamondlike carbon films [592]. In this method they used a silicon substrate of about $8 \times 6 \times 0.3$ mm mounted on the positive graphite electrode after cleaning the native oxide layer. Analytically pure acetonitrile was used as the electrolyte and the negatively biased graphite electrode was kept 2 to 3 mm away from the positively biased Si substrate. Current density was in the range 5–60 mA cm^{-2} while applied voltage varied in the range 200 to 1000 V. Substrate temperature was kept at 25°C by water cooling the deposition system. Deposited films were yellow-brown color and inert to both strong acids and alkalis and did not dissolve in organic solvents. Films mainly consisted of C and N with a sp^2 carbon, C=N, and C≡N bonding configuration. The N to C concentration ratio is estimated to be about 0.1 in their films and they showed distinct photoluminescence property.

Fu et al. [593] used a solvothermal synthetic route to prepare polycrystalline carbon nitride. In this study polycrystalline β-C_3N_4 was produced by the solvothermal reaction of 1,3,5-trichlorotriazine ($C_3N_3Cl_3$) and lithium nitride (Li_3N), using benzene as the solvent. The process is expressed as $C_3N_3Cl_3 + Li_3N \rightarrow C_3N_4 + 3LiCl$. Produced powder was characterized using XRD, FTIR, XPS, and TEM. Results indicate the formation of polycrystalline β-C_3N_4 in the powder. This might be a useful route to synthesize metastable bulk C_3N_4.

It was found that nitrogen ion implantation into Si can form a silicon nitride layer of composition similar to Si_3N_4 [594]. A number of studies show the formation of only sp^2 bonded C−N network with nitrogen implantation in carbon materials [595, 596]. Some studies also show that Si incorporation plays a very crucial role in promoting the crystal growth of the carbon nitride [597]. Takahiro et al. [598] studied the effect of silicon preimplantation on glassy carbon plates on the formation of carbon nitride by nitrogen ion implantation. It was found that the carbon nitride layer with a Si concentration of 5–8 at.% is amorphous and has sp^2 bonded carbon structure with a small amount of C≡N bonds. XPS measurement suggests the existence of local C=N−Si(C_nN_{3-n}) arrangements in the nitride layer, which might help to increase the possible bonding sites for N atoms and result in the higher saturation level of the implanted atoms. A carbon nitride compound with defect zinc-blende structure was synthesized by chemical precursor route [599]. Samples were prepared by heating N,N-diethyl-1,4-phenylene-diammonium sulphate, $C_{10}H_{18}N_2O_4S$, in the presence of N_2 atmosphere with a SeO_2 catalyst. High resolution electron microscopy, electron nanodiffraction, and electron energy loss measurements identified the synthesized material as cubic zinc-blende with C_3N_4 composition.

In order to grow good crystalline carbon nitride several researchers have used hybrid techniques, which are the combination of different techniques and principles [600–610].

4.4. Characterization of Carbon Nitride Films

As the complexity of CN_x films is very high and it is very difficult to grow in pure single phase form, characterization of the films is very important in order to identify the phase, structure, and composition correctly. A large number of techniques have been used for characterization of CN_x films. In the diamond section we discussed most of the characterization techniques. Here we will discuss the basics of some techniques that may be used to characterize CN_x films.

4.4.1. Raman Spectroscopy and FTIR

The basic principles of Raman spectroscopy and FTIR were already described in the diamond section. Here we discuss these two characterization tools which have been used to characterize the bonding hybridization, nature and quantitative information of C, N and other elements in the CN_x complex. Detection of bonding hybridizations (sp, sp^2 and sp^3) by Raman and IR spectroscopy depends on the selection rules applied to the molecular vibrations. The total zone-center optic modes can be given as the following irreducible representation for graphite [611]:

$$\Gamma = A_{2u} + 2B_{2g} + E_{1u} + 2E_{2g}$$

The A_{2u} and E_{1u} modes are IR active and are observed at 867 and 1588 cm^{-1}, respectively [612]. The $2E_{2g}$ modes are Raman active and are observed at 42 and 1581 cm^{-1} [613]. E-symmetry modes exhibit in-plane atomic displacements while A and B symmetry modes have out-of-plane displacements. For amorphous carbon the E_{2g} mode is the symmetric Raman active G-band (usually 1584 cm^{-1}) and E_{1u} is weakly IR active (usually 1588 cm^{-1}). A Raman active D-band (usually 1360 cm^{-1}) appears due to bond angle disorder in the sp^2 graphitelike microdomains. Most of the studies show the Raman spectra which consist of D- and G-bands along with some other broad peaks which might be related to the amorphous carbon.

FTIR has been used extensively to study carbon nitride but there is considerable discussion concerning the analysis of the spectra. Absorption bands in the infrared can give a great deal of qualitative and sometimes quantitative information. The incorrect use of group frequency tables can be misleading when trying to establish the sp^2 content, such as assigning the C=N stretching mode to what is, in fact, the bending mode of aliphatic CH_2 or CH_3 groups. The C=N frequency is reduced upon conjugation but is seldom found below 1500 cm^{-1} in the IR and is relatively weak [614], unless coupled with another vibration. Furthermore, when the bond is part of a heterocycle, the vibration is difficult to distinguish from the C=C vibrations [615].

At high nitrogen contents a peak at ~2200 cm^{-1} is often seen and is accepted to be related to CN, but even here the situation is not simple. All of the following components have absorption bands in this region: −N=C=N−, −C=N=N−, C=C=N−, C−N$^+$C−, C−CN and N−CN. It should be noted that, except for the nitriles, all these groups are linear, rigid, and continuous (i.e., nonterminal), and therefore impose conditions

on the structure of the material which may preclude the formation of high density C_3N_4 materials [562]. Most researchers have assumed that the number of sp bonds is low on the basis of infrared spectra alone. This is not a good assumption, especially for films with high nitrogen content, since the absorption intensity of carbonitriles is highly dependent on their chemical environment, a fact often ignored in the CN literature [562].

Various broad absorption bands are found in the range 1250–1600 cm^{-1}. Those at ~1570 and 1370 cm^{-1} are normally considered to be related to the G and D peaks seen in Raman, where the incorporation of nitrogen in the carbon matrix makes the bands IR-active [616]. Different CN and CC single- and double-bond absorption bands are also found in this region. Furthermore, bands due to the different C_3N_4 structures are predicted in this range [617]. Therefore interpretation and association of the IR spectra is not straightforward. The exact position of the IR peaks also depends on the residual stress, and since it is expected that high density forms of carbon nitride should contain considerable stress, analogous to the cases of diamondlike carbon and cubic boron nitride, such stress-related effects further complicate the precise analysis by FTIR.

4.4.2. X-Ray Diffraction (XRD)

XRD is one of the most important techniques used to investigate the crystalline phase of carbon nitride. Several studies have claimed to synthesize the crystalline form of carbon nitrides, and XRD patterns were used for their identification. Crystal structure images were taken with TEM in some cases. However, their identification in most of the cases has been ambiguous. This is partly because the crystals were always deposited in small amounts and, in many cases, composed of small crystallites embedded in amorphous phases. It is therefore important to study the XRD pattern very carefully in order to identify crystalline phases correctly. Matsumoto et al. [618] simulated XRD patterns for the five proposed structures of C_3N_4 with RIETAN [619] using the lattice constants and structure parameters reported by Teter and Hemley [620], and they are shown in Figure 54a–e.

Table XXV summarizes their peak positions along with other parameters. Matsumoto et al. reviewed extensively and found that the quality of the XRD patterns provided by many groups varies widely. Many of the studies that claimed to produce crystalline CN are questionable [621–629]. Out of many reports only a few have shown the XRD results which are relatively similar to the simulated data [630–639].

4.4.3. Auger Electron Spectroscopy

Rutherford back scattering (RBS), AES, and photon induced X-ray emission can be used to identify the composition of the films and also the C : N ratio. Because the films are thin (<300 nm), AES and XPS spectroscopy studies are applicable as they are good surface analysis techniques due to their high surface sensitivity. The spatial resolution of AES is of the order of 2 μm. In this technique a low electron beam (≤10 keV)

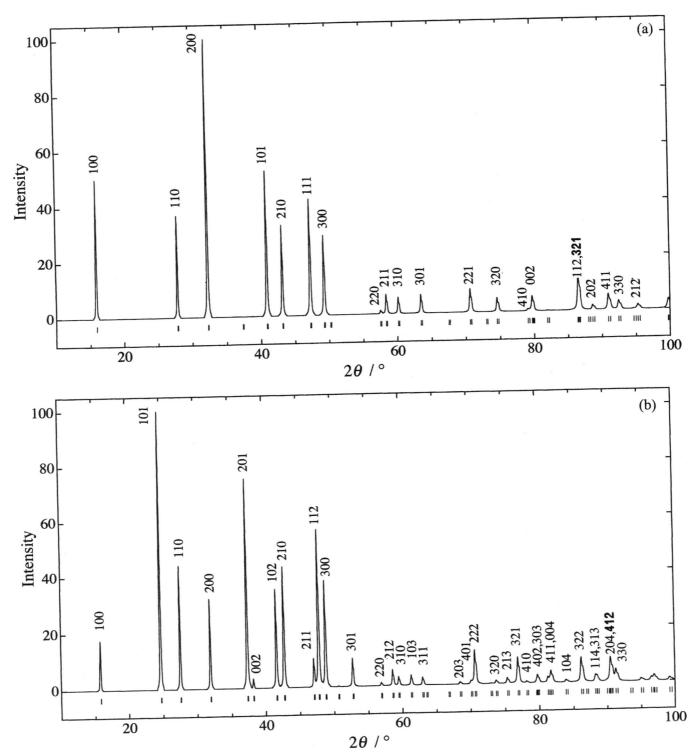

Fig. 54.   Simulated X-ray diffraction patterns (Cu Kα$_1$ + Cu Kα$_2$) for the five proposed C$_3$N$_4$ structures: (a) β-C3N4, (b) α-C$_3$N$_4$, (c) cubic C$_3$N$_4$, (d) pseudocubic C$_3$N$_4$, and (e) graphitic C$_3$N$_4$. Reproduced with permission from [618], copyright 1999, Elsevier Science.

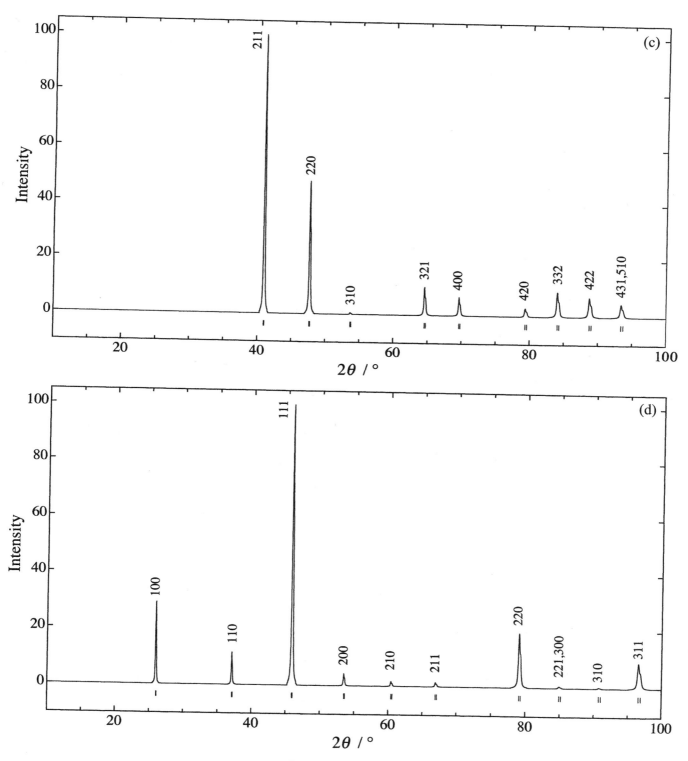

Fig. 54.   (Continued.)

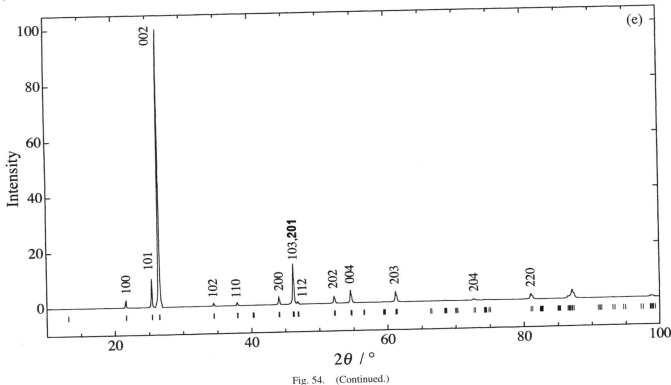

Fig. 54.   (Continued.)

is bombarded on the sample. It can detect any element that is above He in the periodic table, thus making it suitable in the study of C—N—Si—O systems. AES is also capable of depth profiling, which is ideal for the composition study of the CN_x films as a function of depth. Due to the nature of the film/substrate that is being studied, the Auger transitions that will occur will be based on the K-shell ionization ($KL_{2,3}L_{2,3}$ transitions). The kinetic energy of the escaped electron is given by

$$E_{KL1L2,3} = E_K - E_{L1} - E^*_{L2,3}$$

where E_i are the binding energies of the ith atomic energy levels. $E^*_{L2,3}$ is the binding energy of the $L_{2,3}$ level in the presence of a hole in the L_1 level. As sets of atomic binding energies are not the same for any two elements, the analysis of the Auger energies leads to elemental identification. Similar to EELS, the different phases of an element due to the chemical shift in the binding energy of the ionized core level will produce varying lineshapes. The fine structures or lineshapes of the peaks will provide information about the valence band density of states (hence, the electronic structure) and the chemical environment of the sample. The diamond spectrum contains more fine structures on the lower energy side of the main Auger peak than either graphite or amorphous carbon. The more damage there is to the lattice structure, the more smeared out the fine structures will be. Insulators, however, tend to cause some problems in proper calibration of the spectra. Because the emitted electrons constituting the secondary electron emission have low energies, they are easily attracted back to the sample even if the sample

has been charged minimally positive due to the loss of electrons. This would cause the elemental peaks to shift down in energy. The situation is reversed if the sample is charged negatively.

4.4.4. X-Ray Photoelectron Spectroscopy

Though XPS may not have as high a spatial resolution as AES, it is less destructive than AES. XPS is also less affected by the nature of the sample (insulating or conducting) since the electron current densities are much lower. The incident X-rays give rise to photoelectron emission according to the equation

$$E_k = h\nu - E_B - \Phi$$

where E_k is the measured electron kinetic energy, $h\nu$ is the energy of the excitation radiation, E_B is the binding energy of the electron in the sample, and Φ is the work function. This kinetic energy is unlike the kinetic energy of an Auger electron, whose value is independent of the excitation radiation. Again, similar to EELS and AES, XPS is able to identify the atoms of the same element under different environments by the lineshape. However, many of the binding energy shifts are small (~0.2 eV) and the linewidths of the peaks are 1–2 eV which makes the shift difficult to detect. A disadvantage in using XPS and AES is that reference materials are needed for proper calibration of the data. On top of that, XPS has a further disadvantage. The binding energies of some common elements have been reported in the literature to vary, even when referenced to the gold $Au\,4f_{7/2}$ line of 84 eV. The graphite C $1s$ line has been reported to occur at energies from 284.5 to as much as 285.2 eV. This makes

Table XXV.   Diffraction Angles (for Cu Kα_1 O, Lattice Spacings, and Relative Intensities Calculated for the Five Proposed Structures of C_3N_4

β-C_3N_4

$2\theta(°)$	$d(Å)$	I/I_1	hkl
15.97	5.544	50.87	100
27.85	3.201	36.36	110
32.27	2.772	100.00	200
40.88	2.206	62.34	101
43.14	2.095	40.84	210
47.25	1.922	56.40	111
49.27	1.848	40.01	300
50.19	1.816	0.01	201
57.54	1.600	2.01	220
58.37	1.580	11.34	211
60.13	1.538	9.79	310
63.44	1.465	11.90	301
67.53	1.386	0.00	400
70.65	1.332	17.13	221
72.98	1.295	0.07	311
74.55	1.272	10.91	320
79.09	1.210	1.53	410
79.70	1.202	12.39	002
79.81	1.201	0.62	401
81.95	1.175	0.41	102
86.39	1.125	1.21	112
86.50	1.124	35.25	321
88.01	1.109	0.18	500
88.61	1.103	5.08	202
90.92	1.081	16.50	411
92.43	1.067	9.61	330
94.65	1.048	0.65	420
95.25	1.043	5.53	212
99.71	1.008	6.39	302
99.82	1.007	7.15	501
101.35	0.996	0.82	510

α-C_3N_4

$2\theta(°)$	$d(Å)$	I/I_1	hkl
15.81	5.600	18.95	100
24.68	3.604	100.00	101
27.57	3.233	45.53	110
31.94	2.800	33.37	200
37.33	2.407	87.93	201
38.19	2.355	4.06	002
41.57	2.171	44.45	102
42.68	2.117	55.69	210
47.03	1.931	14.45	211
47.74	1.904	79.77	112
48.74	1.867	54.57	300
50.61	1.802	0.40	202
52.70	1.735	15.63	301
56.91	1.617	1.96	220
58.59	1.574	9.82	212
59.46	1.553	5.47	310
61.27	1.512	6.51	103
62.96	1.475	5.31	311
63.55	1.463	0.05	302
66.76	1.400	0.02	400
68.46	1.369	1.81	203
70.06	1.342	2.22	401
70.61	1.333	25.96	222
72.90	1.297	0.43	312
73.68	1.285	2.80	320
75.31	1.261	4.77	213
76.85	1.239	22.19	321
78.15	1.222	1.63	410
79.60	1.203	1.26	402
79.75	1.201	6.59	303
81.26	1.183	5.05	411
81.72	1.177	10.69	004
83.91	1.152	2.43	104
86.15	1.128	24.91	322
86.90	1.120	0.21	500
88.25	1.106	6.97	114
88.48	1.104	2.77	313
89.97	1.090	0.71	501
90.42	1.085	2.90	204
90.49	1.085	23.79	412
91.24	1.078	12.53	330

Cubic C_3N_4

$2\theta(°)$	$d(Å)$	I/I_1	hkl
40.92	2.203	100.00	211
47.62	1.908	53.91	220
53.66	1.707	0.83	310
64.55	1.442	15.57	321
69.62	1.349	11.29	400
79.32	1.207	6.03	420
84.04	1.151	19.33	332
88.72	1.102	16.68	422
93.39	1.059	12.36	431, 510

Pseudocubic C_3N_4

$2\theta(°)$	$d(Å)$	I/I_1	hkl
26.01	3.423	21.48	100
37.11	2.421	9.63	110
45.88	1.976	100.00	111
53.49	1.712	4.94	200
60.42	1.531	2.15	210
66.90	1.398	2.20	211
79.06	1.210	35.02	220
84.92	1.141	1.34	221, 300
90.73	1.083	0.87	310
96.54	1.032	22.35	311

Graphitic C_3N_4

$2\theta(°)$	$d(Å)$	I/I_1	hkl
21.62	4.107	2.64	100
25.40	3.504	10.17	101
26.50	3.360	100.00	002
34.46	2.601	1.18	102
37.92	2.371	1.29	110
44.07	2.053	3.78	200
46.12	1.967	1.22	103
46.19	1.964	18.94	201
46.86	1.937	1.14	112
52.16	1.752	3.87	202
54.58	1.680	7.24	004
59.39	1.555	0.15	104
59.51	1.552	0.13	210
61.18	1.514	6.10	203
61.24	1.512	0.64	211
66.27	1.409	0.17	212
68.38	1.371	0.27	114
68.49	1.369	0.18	300
72.65	1.300	1.04	204
74.17	1.277	0.20	105
74.28	1.276	0.33	213
74.83	1.268	0.27	302
81.05	1.186	4.84	220
85.01	1.140	0.08	214
85.11	1.139	0.04	310
86.46	1.125	1.78	205
86.62	1.123	0.21	311
86.90	1.120	1.38	006
87.10	1.118	7.93	222
90.93	1.081	0.04	106
91.14	1.079	0.06	312
93.08	1.061	0.16	304
97.23	1.027	0.23	400
98.59	1.016	0.17	215
98.70	1.015	0.17	313
98.75	1.015	1.34	401
99.03	1.013	0.11	116

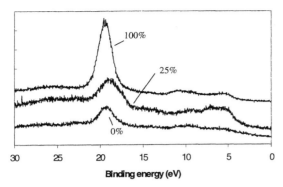

Fig. 55. Valence band XPS spectra of films grown at 0, 25, and 100% nitrogen partial pressure (N_{pp}). Reproduced with permission from [640], copyright 1999, Elsevier Science.

an exact identification of the type of bonding present difficult. Most often carbon (or graphite) formed as a contaminant on the surface is used as reference despite the associated uncertainties. The binding energies of the peaks must also be regarded with caution due to the fact that the energies are also dependent on the thickness of the layer, the method used for referencing, and the accuracy of the energy calibration of the spectrometer on which it is determined. Furthermore, comparison between results from different laboratories may not be meaningful unless the spectrometers are properly calibrated.

In many cases, the XPS spectrum consists of a number of overlapping chemically shifted peaks, of different shapes and intensities. One method of analyzing the spectrum is to fit it with functions. The most common functions are Gaussian and Lorentzian. The basic shape of an XPS peak is Lorentzian but may be modified by factors like instrumental and phonon broadening to give a Gaussian contribution. The final peak shape is usually asymmetrical due to various loss processes. In this current study, the XPS lineshapes are represented by mixed Gaussian–Lorentzian functions in the form of pseudo-Voigt functions.

A typical valence band (VB) XPS spectra taken on carbon nitride films, grown with different nitrogen concentrations, is shown in Figure 55 [640]. A peak located in the region 17.5–22.5 eV is related to $s-p$ hybridized states. This gives an indication that films are polymerlike in nature. For the film grown at 100% N_{pp}, the $s-p$ hybridized peak dominates the VB spectra due to the fact that the nature of nitrogen incorporation changes when $N_{pp} > 25\%$. As nitrogen is increasing in the films, a band located at ~4–7 eV appears, due to C $2p$ and N $2p$ electrons associated to π bonds in aromatic rings and probably N lone-pair electrons. A band also emerges at 8–12 eV, which is normally associated with C—N and C—C σ-states. No structure can be seen in the interval 0–4 eV, which is normally associated with C—C π bonds due to $2p$ electrons.

4.4.5. Rutherford Backscattering Spectroscopy

RBS is commonly used to identify elements in thin films or surface layers, to measure the concentrations of the constituents,

and also to estimate the density distribution in depth. This technique involves a sample or target being bombarded with very energetic (in the MeV range) particles and detecting the scattered particles. Due to the high energy of the incident particles, the scattering would involve the nuclei of both the incident ion and the target atom. The energy of the scattered ions or particles carries the compositional, scattering depth, and structural information of the target material, thus identifying the elements present. Some of the high energy incident ions are capable of traveling a certain depth into the target material undergoing inelastic collisions before being scattered. These ions would carry the depth information required for a depth profile determination of any given constituent of the sample. The quantitation of the RBS spectra requires a standard for energy calibration. Using the fitting program RUMP, the calibrated parameters of the standard sample can then be used to quantify the unknown CN_x films. Although RBS is capable of 1 μm in lateral resolution, it has poorer depth resolution compared with AES.

4.5. Mechanical and Tribological Properties

Although pure crystalline phases of carbon nitride are hardly found so far, mechanical and tribological behaviors of CN_x films were studied by many researchers using nanoindentation and friction and wear testing [535, 641–651]. Amorphous CN_x films show attractive mechanical properties, and they are characterized by high hardness (up to 65 GPa) [652], high eleastic recovery (up to 85%) [652, 653], good adhesion to substrates [654], and low friction coefficient and wear [655]. High hardness, good adhesion, and low friction and wear make this material very important in industrial applications [656]. The magnetic recording industry may switch from hard amorphous carbon films to hard amorphous CN_x films simply by substituting argon for nitrogen sputtering gas.

Wei et al. [649] deposited β-C_3N_4 films on Si and WC cermet substrates by unbalanced magnetron sputtering using a graphite target in an Ar/N_2 plasma. Structural analysis was performed using FTIR, XPS, and TEM analytical techniques. The dry friction and wear characteristics at room temperature were determined using a ball-on-disk tribometer. The friction coefficient decreases with higher N/C ratio films. For the films with a N/C ratio of 0.4, the coefficient of friction increased from an initially low value to a higher stable value of about 0.075 (after 1000 rev, i.e., 62.8 m). The friction coefficient of CN_x films can be decreased during dry sliding conditions [642]. This is due to the fact that sliding-induced heat may be accumulated on local contact areas and cause a gradual destabilization of C—N bonds in the sp^3 tetrahedral structure. Now removal of N atoms can trigger the transformation of sp^3 structure into a graphite-like sp^2 structure and hence the decrease in friction coefficient. The wear rate of the films decreases with the increase of N/C ratio due to the lower friction coefficient.

The nanoindentation technique is also being used to determine the hardness and modulus of the films [657–659]. It is a depth sensing measurement of hardness and modulus of the films. Typical loading/unloading curves obtained for different

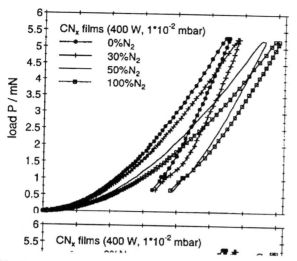

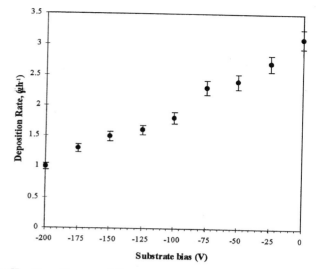

Fig. 56. Typical load-displacement curves observed in CN_x films. Comparison shows that the hardness is reduced with increasing N_2 partial pressure applied during films deposition. Reproduced with permission from [657], copyright 1999, American Institute of Physics.

Fig. 57. Deposition rate of films as a function of substrate bias. Reproduced with permission from [660], copyright 1999, Elsevier Science.

Table XXVI. Hardness and Modulus Values of CN_x Films Grown at Different Nitrogen Concentrations

N_2/Ar ratio (%)	at.% N_2	ρ (g/cm^3)	H (GPa)	E (Gpa)
0	0	1.4	16.0	175
30	26	1.2	12.5	130
50	26	1.1	10.0	110
100	33	1.0	9.5	90

Reproduced with permission from [657], copyright 1999, Materials Research Society.

CN_x films grown with different N_2/Ar concentrations in the gas phase in a magnetron sputtering chamber are shown in Figure 56 [657]. The penetration depth of less than 200 nm out of a film thickness of 800 nm is a good approximation of avoiding the substrate effects. Hardness and modulus values obtained from the analysis of loading/unloading curves are shown in Table XXVI along with other data [657]. Hardness and modulus values are very low in these films due to the low density of the films.

4.6. Some Interesting Results on Carbon Nitride

Extensive studies on growth and characterization of carbon nitride films were accomplished by Cameron et al. [640, 660–663]. In this section we review their work a little elaborately. Films were grown using DC magnetron sputtering using an opposite target Penning-type geometry [657] (discussed in the sputtering section). Graphite of 99.95% purity was used as a target while an Ar/N_2 mixture of 99.999% purity with variable mixture was used as sputtering gas. Pressure during deposition was 1×10^{-3} mbar. Partial pressure was varied from 0 to 100%, keeping the total pressure constant. Films were grown mostly on polished Si substrates. Adhesion of films on tool steel sub-

strates was also studied [643]. It was seen that substrate bias and nitrogen partial pressure during the deposition are the two most parameters affecting the film's structural and physical properties [660, 661].

The proportion of sp^3 bonding in the deposited CN_x films, which is important in the growth of crystalline CN films, can be influenced by controlling the energy of the ions which bombard the substrate during the growth. The mechanisms involved in this ion induced growth process are described by the densification model of Robertson [664] and the induced stress models of Davis [665] and McKenzie et al. [666]. Again sp^3 fraction is normally found maximum in a small window of ion energies (typically 50–250 eV) [660]. Outside this range the sp^3 fraction is reduced due to either the ion energy being too low to cause significant change to the structure or being too high such that the desired bond type is disrupted. Chowdhury et al. investigated the deposition rate, film composition, and structural properties by varying the ion energies by controlling the substrate bias during deposition in sputtering [660]. The most obvious effect of the substrate bias changes was a monotonic decrease in deposition rate as the bias became more negative (Fig. 57). Growth rate declines by approximately a factor of three between 0 and -200 V bias. Four different mechanism can cause of this decline: (i) preferential sputtering and a change in the film stoichiometry, (ii) a densification of the film, (iii) reduced target sputtering and (iv) chemical sputtering of the films. RBS measurements on the films do not show significant change in the nitrogen content in the films (Fig. 58). Therefore the first option may not be correct. Although they did not measure the density of the films, it is not not credible that a variation of the scale necessary to account for the thickness changes could occur without a major effect on the other properties, in particular the hardness (Fig. 59). Hence the second option is also not valid. Substrate bias does not change significantly the target potential and hence the sputtering rate would not have changed, which can change the deposition rate. The

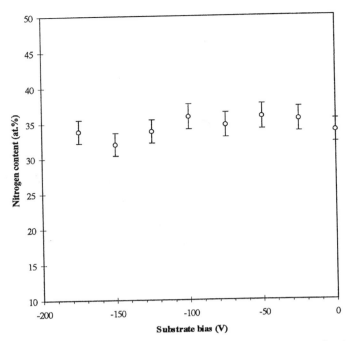

Fig. 58. Nitrogen content of films as a function of substrate bias. Reproduced with permission from [660], copyright 1999, Elsevier Science.

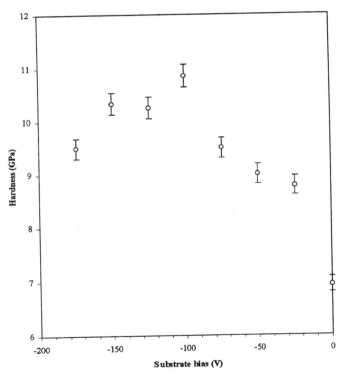

Fig. 59. Hardness of films as a function of substrate bias. Reproduced with permission from [660], copyright 1999, Elsevier Science.

most likely explanation is that the increased nitrogen ion energy at the substrate stimulated a desorption of carbon and nitrogen in the ratio in which they were present in the film [667]. Valence band XPS results on these films at different substrate

biases indicate that sp^3 bonding is increasing in the films during biasing in the range −75 to 150 V [660]. This is might be the reason for the higher hardness values obtained in this range (Figure 59). It might be necessary to study further this biasing range with changing substrate temperatures which they did not change during their experiments (∼325°C). It was found from RBS measurements that while the nitrogen content of the sputtering gas is varied from 0 to 100%, the atomic concentration of nitrogen in the films does not vary much once the N_2 partial is above 25% (Fig. 60).

Structural properties due to the nitrogen partial pressure are being investigated using XPS, ultraviolet photoelectron spectroscopy (UPS), EELS, and EPR studies. Deconvoluted XPS and UPS spectra of the CN_x films are shown in Figures 61 and 62. In XPS data, in comparison to the structures of diamond, the large peaks at ∼19 eV binding energy are due to mixed s–p electronic states and the peak at ∼14 eV to p states. The lower energy peaks are usually due to the σ and π bonding electrons [668, 669]. Both UPS and XPS showed similar trends as the N_2 partial pressure increased from 0 to 100%. At 0% the UPS spectrum shows a main peak at 9.7 eV (σ bonding electron) and three other states due to π electron states. The larger peak, called π_1, at ∼8.0 eV is due to the C=C π bond in aromatic rings. Others are due to sp bonded carbon atoms ($\pi_2 = 5.5$ eV) and lone pair electrons from some residual nitrogen in the films ($\pi_3 = 2.6$ eV). XPS also shows similar results in the films with 0% N_2. Interesting results were observed in their films with higher N_2 partial pressure. In both XPS and UPS results on their 25% N_2 spectra show larger π peaks than σ peaks. This is certainly because of nitrogen induced increase of sp^2 bonding at the expense of the sp^3 bonding [670]. Clearly resolved π_3 peaks, in both UPS and XPS are assigned to lone pair electrons on N atoms which are bonded to C atoms in a nondoping configuration such as C≡N [670, 671]. A major change in the bonding occurs in the films with 100% N_2 in the gas phase, although there has been little change in the total N concentration in the films. Both XPS and UPS spectra indicate the reduction of sp^2 bonding and hence an increase in the sp^3 fraction. This is an important indication of the possible formation of crystalline β-C_3N_4 [672].

EELS (Fig. 63) of their same samples confirms the similarity between the 0 and 100% films and the difference of the 25% films which show both a broader π-π^* and a much larger π-σ peak which would indicate a higher density of π states.

Electron paramagnetic resonance (also known as EPR) results on a range of CN_x films (with different N_2 partial pressures) are shown in Figure 64. Spin concentration is initially decreased up to 50% for N_2 films and is then increased again. The g-factor also starts decreasing at the 50% level. This result is consistent with the idea [662] of the N bonding to C in a C≡N nondoping configuration which has the effect of reducing the unpaired electrons because of the termination of the carbon dangling bonds with nitrogen atoms. An increase in the spin density with higher N_2 partial pressure is a indication of changing bonding structure.

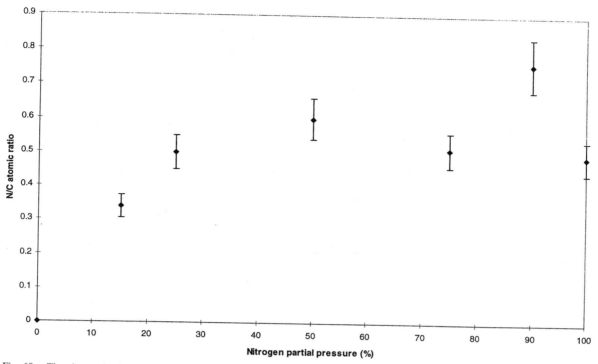

Fig. 60. The nitrogen/carbon ratio of carbon nitride films as a function of the nitrogen partial pressure during deposition. Reproduced with permission from [662], copyright 2000, Elsevier Science.

The activation energy calculated for the temperature dependent resistivity measurements on the films grown at different N_2 pressures showed that it is increasing with nitrogen content (Fig. 65) [661]. The temperature-dependent resistivity of the highly conducting films (0 at.% of N, $\rho \approx 1 \times 10^{-2}$ Ω cm) exhibits only a small variation resistivity in the temperature range. The relatively high conductivity and low gap of the carbon nitride films suggest an electronic structure having a large number of n-type band tail states with a structure which consists mostly of sp^2 hybridized aromatic rings. The increased activation energy and resistivity for nitrogen containing films suggest a reduction of density of states in the mobility gap and perhaps removal of defect states near the conduction band edge [661].

In another study [662] structural studies were performed on the CN_x films containing different nitrogen at.% percent calculated from the RBS measurement. Figure 66 presents the Raman (a) and IR (b) spectra of CN_x films containing different N concentrations. Spectra show several features due to vibrations of bonds between C and N or the same element. The out-of-plane vibrations of C—C at 700 cm^{-1} are becoming more intense as the nitrogen incorporation in the films increases. This is a indication of nitrogen induced stabilization of C—C bonding [17, 125]. The Raman active D- and G-bands (region 1350–1650 cm^{-1}) are becoming very prominent with increasing nitrogen in the films. Kaufman et al. [673] concluded that nitrogen substitution is responsible for the symmetry breaking of the E_{2g} mode and intensity of D- and G-bands corresponding to sp^2 hybridization. The band at ~2200 cm^{-1} is due to the stretching vibration of C≡N bonding, while

at higher wave numbers (~2900 cm^{-1}) C—H stretching band is observed. The presence of this stretching band is due to residual water vapor in the chamber during deposition [663]. The E_{2g} symmetry mode become IR active due to nitrogen incorporation. Unlike Raman spectra D- and G-band overlaps in IR spectra, the C≡N stretching band at ~2200 cm^{-1} in IR spectra is much stronger than Raman. The N_2 molecule is not IR or Raman active as it has neither net dipole moment nor a change in polarizability during the vibration. But it can be Raman/IR active while it replaces carbon atoms from the six-fold carbon ring. The formation of a N—N stretching band is unlikely here, as there is no evidence of absorption at 1150–1030 cm^{-1}. As nitrogen incorporation increases in CN compounds, the E_{2g} symmetry mode becomes more and more IR and Raman active. Increase in intensity of the 1350–1650 cm^{-1} band region is caused by the Raman active in-plane or out-of-plane N=N stretching vibration due to excess nitrogen incorporation. This N=N band overlaps with the C=N stretching band. It was also shown that over the range 25–44 at.% N there is no systematic variation of the IR absorption coefficient. This is an indication of bonding of N with N instead of C. A close looking of the Raman spectra in the region 1100–1800 cm^{-1} indicate a new feature in between D- and G-bands. This new peak is being assigned by N band (1400–1500 cm^{-1}). Weich et al. [674] also derived the possibility of the formation of nitrogen–nitrogen bonding within the CN compound. This was also supported by the fact that the growth continues to increase for films with higher nitrogen content.

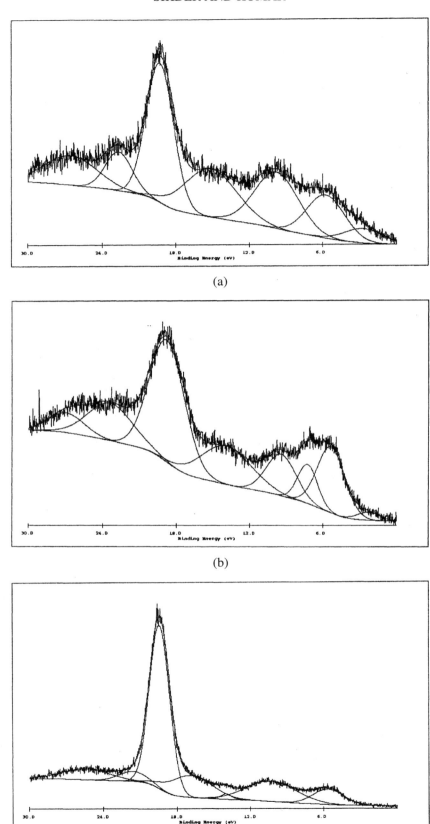

(a)

(b)

(c)

Fig. 61.   Valence band XPS spectra resolved into Gaussian peaks: (a) 0%, (b) 25%, and (c) 100%
N_2 partial pressure. Reproduced with permission from [662], copyright 2000, Elsevier Science.

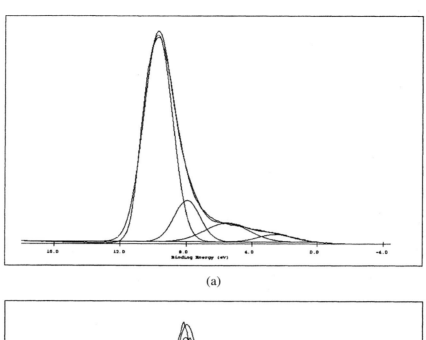

(a)

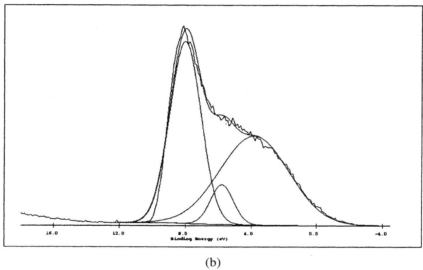

(b)

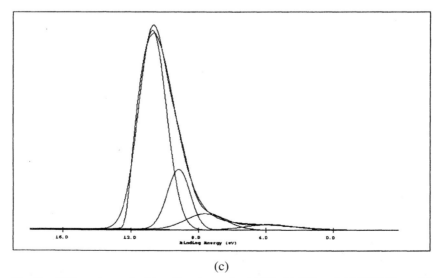

(c)

Fig. 62. UPS spectra resolved into Gaussian peaks: (a) 0%, (b) 25%, and (c) 100% N_2 partial pressure. Reproduced with permission from [662], copyright 2000, Elsevier Science.

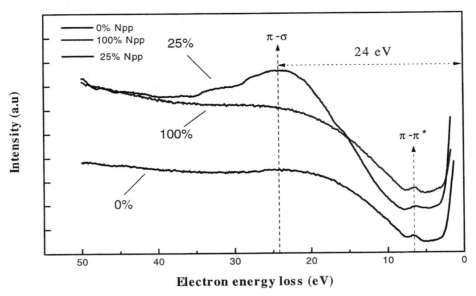

Fig. 63.   EELS data on films grown with 0%, 25%, and 100% N_2 partial pressure. Reproduced with permission from [662], copyright 2000, Elsevier Science.

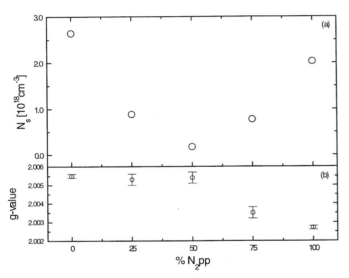

Fig. 64.   ESR measurements on films with varying N_2 partial pressure. Reproduced with permission from [662], copyright 2000, Elsevier Science.

Annealing behavior of the sample containing 33 at.% nitrogen is being investigated by Raman and IR and valence band XPS spectroscopes. Annealing is important in order to study the phase stability with respect to temperature. Raman spectra of as-grown and annealed samples (Fig. 67a) show the increase of D-band intensity. This is due to the fact that as annealing progresses the microdomains grow in size or number, making a large contribution to the D-band. Due to annealing, outdiffusion of nitrogen and disruption of bonds occur. This disruption does not affect the N=N bond as much it affects C−C and C≡N bonds. This indicates that sp^2 bonding in the CN compound is the most stable phase. These results were backed by the valence band XPS results, where the sp hybridized feature remains un-

affected after annealing but the $2s$ state is severely affected. IR absorbance of the same samples (Fig. 67b) shows broadly the same results except that the sp^2 band becomes more IR active. This is due to the fact that annealing favors the symmetry breaking mechanism. After annealing, the D- and G-bands are shifted due to the change in sp^2 domain size. The N=N band also shifts because the annealing process allows the bonded nitrogen atoms to be repositioned within the ring, giving rise to a wide range of shifting of the N=N stretching band.

Thin CN_x films were deposited on crystalline silicon substrates by e-beam evaporation of graphite and simultaneous nitrogen ion bombardment. The films were deposited at an e-beam power of 0.5–1.5 kW and N^+ ion energies of 300–400 eV while substrate temperature was varied from 800 to 1150°C. Films were characterized by Raman and XPS spectroscopy. In the Raman spectrum D- and G-bands were seen in all the films. Two additional peaks at 1230 and 1466 cm^{-1} were also seen in the films grown at lower temperatures and low ion energies. These peaks are sensitive to nitrogenization changes and are identified as being due to nanocrystalline diamond species dispersed in the amorphous carbon layer [675, 676]. Decrease of this phase with higher nitrogen energy and deposition temperature is due to the formation of CN_x phase by substituting carbon atoms with nitrogen atoms [677]. XPS analysis of the deposited films shows during the growth process that the ratio of sp^2 to sp^3 bonds changes and the number of C−N bonds increases with increasing deposition energy and substrate temperature.

Ni substrate has been studied extensively for deposition of diamond due to its close matching of lattice constants with diamond and chemical affinity of Ni and carbon. Single crystal Ni substrates have also been studied to grow carbon nitride on it. Yan et al. [678] reported the synthesis of crystalline α- and β-phase of C_3N_4 on Ni (100) substrate using a HFCVD method with nitrogen and methane. In another studies crystalline car-

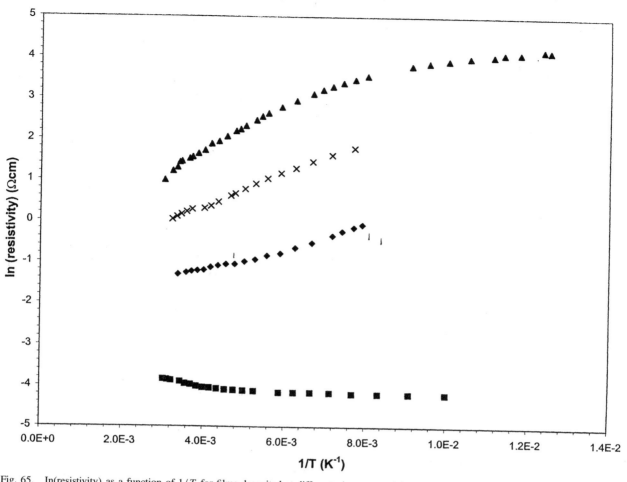

Fig. 65. ln(resistivity) as a function of $1/T$ for films deposited at different nitrogen partial pressures: (■) 0%, (◆) 25%, (×) 50%, and (▲) 100%. Reproduced with permission from [661], copyright 1999, Elsevier Science.

bon nitride films were grown on polycrystalline Ni substrate using a RF bias assisted HFCVD technique [679]. Large cystallites (∼10 μm) are seen in the SEM picture and XRD reveals the formation of a crystalline phase of carbon nitride, but no Raman signal is found.

Much literature has come out recent days regarding the formation of crystalline or amorphous carbon nitride. Here we mention the brief description of the deposition technique and obtained results. CN thin films were synthesized on Si substrate using electron beam evaporation of carbon and simultaneous nitrogen ion bombardment [680]. Clusters of C_xN_y phase with $x > y$ are formed along with the sp^3 phase consisting β-C_3N_4, as indicated by XPS, Raman, and FTIR analysis. Formation of $\beta-C_3N_4$ has been reported on the Si, Ta, Mo, and Pt substrates using the MPCVD method with a methane and N_2 gas mixture [681]. Extensive XRD results are shown in support of their results. Graphitic-C_3N_4 films were grown on Si and a highly pyrolytic grahite substrate by a MPCVD system using a methane and N_2 gas mixture [682]. Films were characterized by scanning tunneling microscopy and suggested the formation of graphitic-C_3N_4. Hexagonal beta carbon nanocrystals were observed in the CN_x films deposited on Si substrates by the RF

PECVD technique using ammonia and ethylene gas mixture followed by rapid thermal annealing to 1000°C [683]. Amorphous carbon nitride nanotips were synthesized on silicon to be used as a electron emitters by the ECR-CVD method in which a negative bias was applied to the graphite substrate holder and a mixture of C_2H_2, N_2, and Ar was used as precursor [684]. An onset emission field as low as 1.5 V μm and a current density of up to 0.1 mA cm² at 2.6 V μm were observed. Crystallization behavior of amorphous carbon nitride films has been studied by Xiao et al. [685]. CN_x films were first prepared by DC reactive magnetron sputtering and then films were heat treated under protective nitrogen. Results showed that heat treatment above 1100°C could induce the transition from an amorphous to a crystalline state of carbon nitride films. Carbon nitride thin films and nanopowders were produced by CO_2 laser pyrolysis of sensitized $NH_3-N_2O-C_2H_2$ reactant gas mixtures [686]. Data obtained by XRD, XPS, and TEM analysis on power suggested the presence of CN bonded phases and are taken as an indication of the formation of α- and β-C_3N_4.

Several other studies also reported the formation of either crystalline or amorphous carbon nitride films on different substrates [687–689, 700].

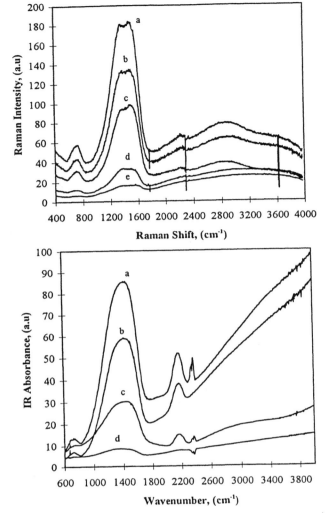

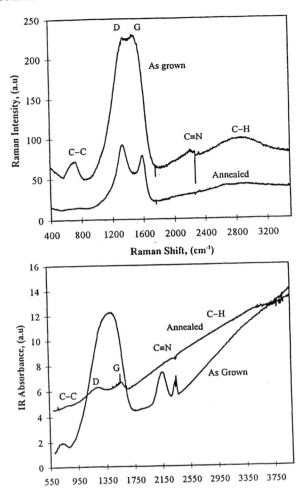

Fig. 66. (a) Raman spectra as a function of relative nitrogen concentration. Data were translated but not rescaled (a = 43.3 at.%, b = 37.6 at.%, c = 33.7 at.%, d = 25.2 at.%, e = 0 at.% N). (b) IR spectra as a function of relative nitrogen concentration. Data were translated but not rescaled (a = 43.3 at.%, b = 37.6 at.%, c = 33.7 at.%, d = 25.2 at.% nitrogen). Reproduced with permission from [663], copyright 1999, Elsevier Science.

4.7. Conclusions

It is seen form the literature that very little definite evidence for the existence of crystalline carbon nitride is being shown. XRD and TEM analyses suggest the probability of the formation of disordered polytypic diamond in the presence of nitrogen, which might be interesting in connection to diamond synthesis [618]. In some cases crystalline phases are being identified in their films but that is in a small area or in a mixture of an amorphous carbon–nitrogen mixture. The following important comments are made by Muhl and Méndez [562] in their recent review on the formation and characterization of carbon nitride films:

- Substrate temperatures $\geq$ 800°C may be used for the process with gaseous precursors. Such temperatures should

Fig. 67. (a) Raman spectra, (b) IR spectra of the as-grown sample containing 33.2 at.% N before and after annealing at 600°C (23.2 at.% N). Reproduced with permission from [663], copyright 1999, Elsevier Science.

help inhibit the formation of polymeric and inorganic CN_x compounds.
- Atomic nitrogen and CN are probably preferable as precursors rather than molecular nitrogen and hydrocarbon ions and radicals. The use of a CN precursor is likely to help prevent the formation of stable carbon-ring structures.
- Ion energies below 10 eV are necessary at low pressure synthesis to avoid chemical sputtering and the concurrent reduction in nitrogen content.
- For gas phase processes, the use of medium to high pressures helps to ensure that the kinetic energy of the incident neutrals is kept below 10 eV. Use of high to ultrahigh predeposition vacuums is recommendable to minimize the residual contamination.
- Substrate effects are complex but important; reasonable adhesion between the carbon nitride and the substrate is needed but without compound formation as in the case of silicon. There are indications that there are advantages in using nickel, titanium, or Si_3N_4 coated substrates. Different

buffer layers may be used in order to promote crystalline formation of carbon nitride.

- Various characterization techniques should be employed in parallel in order to identify phases correctly.
- Measured nonoptimum N/C ratios do not mean necessarily that crystalline C_3N_4 does not exist within the deposit.
- Low sp^1 bonding cannot be assumed only by the presence of small peaks in the 2000 to 2200 cm^{-1} region of FTIR spectra.
- Effects of the chemical environment cannot be ignored in XPS or other techniques that are sensitive to the energies of valence electrons.

Further efforts are imperative for synthesizing materials that can unambiguously be identified as crystalline carbon nitride.

Acknowledgments

Work leading to this chapter was partially funded by a grant under the NSF Career Award 9983535 and NSF DMI grant 0096255. The authors thank the Center for Microelectronics Research, College of Engineering at the University of South Florida, and its Director, Dr. M. Anthony, for the support and the internal release time, which made this chapter possible. The help received from Jean-Paul Deeb, F. Giglio, Ismail Irfan, P. Zantye, and T. Vestagaarden in the preparation of this manuscript is also greatly acknowledged. One of the authors (A.K.S.) would like to thank his doctoral supervisor (Professor D. S. Misra, IIT Bombay, India) for his kind help and affection during his thesis work; a part of the thesis work has been presented in this chapter.

REFERENCES

1. "Hard Coatings Based on Borides, Carbides and Nitrides: Synthesis, Characterization and Applications" (A. Kumar, Y.-W. Chung, and R. W. Chia, Eds.). TMS, 1998.
2. "Surface Engineering: Science and Technology I" (A. Kumar, Y.-W. Chung, J. J. Moore, and J. E. Smugeresky, Eds.). TMS, 1999.
3. "Magnetic Recording Technology" (C. D. Mee and E. D. Daniel, Eds.). McGraw–Hill, New York, 1996.
4. "Phase diagram of Ternary Boron Nitride and Silicon Nitride Systems" (P. Rogl and J. C. Schuster, Eds.). ASM International, Metals Park, OH, 1992.
5. S. Veprek, J. Vac. Sci. Technol. A17, 2401 (1999).
6. A. Y. Liy and M. L. Cohen, Science 245, 841 (1989).
7. A. Y. Liy, M. L. Cohen, Phys. Rev, 41(15), 10727 (1990).
8. M. L. Cohen, Phys. Rev 32(12), 7988 (1985).
9. F. P. Bundy, H. T. Hall, H. M. Strong, and R. H. Wentorf, Nature 176, 51 (1955).
10. P. W. Bridgman, Sci. Am. 193, 42 (1955).
11. W. G. Eversole, U.S. Patent 3,630,677, 1961.
12. S. Matsumoto, Y. S. M. Kamo, and N. Setaka, Jpn. J. Appl. Phys. 21, L183 (1982).
13. M. Kamo, Y. Sato, S. Matsumoto, and N. Setaka, J. Crystal Growth 62, 642 (1983).
14. A. Badzian, T. Badzian, R. Roy, R. Meisser, and K. E. Spear, Mater. Res. Bull. 23, 531 (1988).
15. S. Matsumoto, Y. Sato, M. Tsutsumi, and N. Setaka, J. Mater. Sci. 17, 3106 (1982).
16. K. Okano, N. Naruki, Y. Akiba, T. Kurosu, M. Ida, and Y. Hirose, Jpn. J. Appl. Phys. 27, L153 (1988).
17. R. S. YaJamanchi and K. S. Harshavardhan, J. Appl. Phys. 68, 5941 (1990).
18. Y. Tzeng, R. Phillips, C. Cutshaw, and Srivinyunon, App. Phys. Lett. 58, 2645 (1991).
19. K. Suzuki, A. Sawabe, H. Yasuda, and T. Inuzuka, Appl. Phys. Lett. 50, 728 (1987).
20. F. Akatsuka, Y. Hirose, and K. Komaki, Jpn. J. Appl. Phys. 27, L1600 (1988).
21. F. Zhang, Y. Zhang, Y. Yang, G. Chen, and X. Jiang, Appl. Phys. Lett. 57, 1467 (1990).
22. K. Kurihara, K. Sasaki, M. Kararada, and N. Koshimo, Appl. Phys. Lett. 52, 437 (1988).
23. P. XiLing and G. ZhaoPing, Thin Solid Films 239, 47 (1994).
24. J. Karner, M. Pedrazzini, I. Reineck, M. E. Sjostrand, and E. Bergmann, Mater. Sci. Eng. A 209, 405 (1996).
25. T. Leyendecker, O. Lemmer, A. Jurgens, S. Esser, and J. Ebberink, Surf. Coatings Technol. 48, 261 (1991).
26. N. W. Ashcroft and N. D. Mermin, "Solid State Physics." Saunders, Philadelphia, 1976.
27. C. Kittel, "Introduction to Solid State Physics." Wiley Eastern Ltd., New Delhi, 1986.
28. K. E. Spear, A. W. Phelps, and W. B. White, J. Mater. Res. 5, 2277 (1990).
29. J. C. Angus, Thin Solid Films 216, 126 (1992).
30. R. E. Clausing, L. L. Horton, J. C. Angus, and P. Koidl, "Diamond and Diamond-like Films and Coatings." Plenum, New York, 1991.
31. W. A. Yarbrough and R. Messier, Science 247, 688 (1990).
32. J. C. Angus, H. A. Will, and W. S. Stanko, J. Appl. Phys. 39, 2915 (1968).
33. T. R. Anthony, Vacuum 41, 1356 (1990).
34. B. V. Spitzyn, L. L. Bouilov, and B. V. Derjaguin, J. Crystal Growth 52, 219 (1981); B. V. Spitzyn and B. V. Derjaguin, In "Problems of Physics and Technology of Wide-Gap Semiconductors," pp. 22–34. Akad. Nauk SSSR, Leningrad, 1979.
35. S. Matsumoto, Y. Sato, M. Kamo, and N. Setaka, Jpn. J. Appl. Phys. 21, L183 (1982).
36. S. J. Harris, J. Appl. Phys. 56, 2298 (1990).
37. W. Banholzer, Surf. Coat. Technol. 53, 1 (1992).
38. R. Haubner and B. Lux, Diamond Relat. Mater. 2(9), 1277 (1993).
39. P. K. Bachmann, W. Drawe, D. Knight, R. Weimer, and R. Messier, "Spring MRS Meeting Extended Abstracts, Symposium D" (M. W. Geis, G. H. Johnson, and A. R. Badzian, Eds.), pp. 99–102. MRS, Pittsburgh, PA, 1988.
40. H. Liu and D. S. Dandy, "Diamond Chemical Vapor Deposition, Nucleation and Early Growth Stages," p. 40. Noyes, Park Ridge, NJ.
41. T. P. Ong, F. Xiong, R. P. H. Chang, and C. W. White, J. Mater. 7, 2429 (1992).
42. J. Narayan, V. P. Godbole, O. Matcra, and R. K. Singh, J. Appl. Phys. 71, 966 (1992).
43. S. D. Wolter, B. R. Stoner, J. T. Glass, P. J. Ellis, D. S. Buhaenko, C. E. Jenkins, and P. Southworth, Appl. Phys. Lett. 62, l215 (1993). e
44. D. Michau, B.Tanguy, G. Demazeau, M. Couzi, and R. Cavagnat, Diamond Relat. Mater. 2, 19 (1993).
45. K. Kobayashi, T. Nakano, N. Mutsukura, and Y. Machi, Vacuum, 44, 1 (1993).
46. U. Yoshikawa, Y. Kaneko, C. F. Yang, H. Tokura, and M. Kamo, J. Jpn. Soc. Precision Eng. 54, 1703 (1988); M, Yoshikawa, Y. Kaneko, C. F. Yang, and H. Tokura, J. Jpn. Soc. Precision Eng. 55, 155 (1989).
47. G. F. Zhang, X. Zheng, and Z. T. Liu, J. Cryst, Growth 133, 117 (1993).
48. R. A. Bauer, N. U Sbrockey, and W. E. Brower, Jr., J. Mater. Res. 8, 2858 (1993).
49. K. Kobayashi, N. Mutsukura, and Y. Machi, Mater. Manufacturing Processes 7, 395 (1992).
50. P. A. Dennig, R Shiomi, D. A. Stevenson, and N. M. Johnson, Thin Solid films 212, 63 (1992).

51. P. A. Dennig, and D., A. Stevenson, *Appl. Phys. Lett.* 59, 1562 (1991).

52. B. Lux, and R. Haubner, "Diamond and Diamond-like Filmds and Coatings." *In* (R. E. Clausing, L. L. Horton, J. C. Angus, and P. Koidl, Eds.), Plenum Press, New York, 1991, pp. 579–609.

53. J. C. Angus, Y. Wang, and M. Sunkara, *Ann. Rev. Mater. Sci.* 21, 221 (1991).

54. H. Liu and D. S. Dandy, "Diamond Chemical Vapor Deposition, Nucleation and Early Growth Stages," p. 97. Noyes, Park Ridge, NJ.

55. P. Ascarelli, and S. Fontana, *Appl. Surf. Sci.* 64, 307 (1993).

56. W. A. Yarbrough, *Appl. Diamond Relat. Mater.* 25 (1991).

57. P. C. Yang, W. Zhu, and J. T. Glass, *J. Mater. Res.* 8, 1773 (1993).

58. M. W. Geis, A. Argoitia, J. C. Angus, G. H. M. Ma, J. T. Glass, J. Butler, C. J. Robinson, and R. Pryor, *Appl. Phys. Lett.* 58, 2485 (1991).

59. M. W. Geis, *Appl. Phys. Lett.* 55, 550 (1989).

60. H. Liu and D. S. Dandy, "Diamond Chemical Vapor Deposition, Nucleation and Early Growth Stages," p. 100. Noyes, Park Ridge, NJ.

61. H. Liu and D. S. Dandy, "Diamond Chemical Vapor Deposition, Nucleation and Early Growth Stages," p. 15. Noyes, Park Ridge, NJ.

62. M. Kamo, Y. Sato, S. Matsumoto, and N. Setaka, *J. Cryst, Growth* 62, 642 (1983).

63. T. Sharda, D. S. Misra, D. K. Avasthi, and G. K. Mehta, *Solid State Commun.* 98, 879 (1996).

64. R. F. Davis, "Diamond Film And Coatings Development, Properties and Applications," p. 162. Noyes Publishing, Park Ridge, NJ.

65. H. Kawarada, K. S. Mar, and A. Hiraki, *Jap. J. Appl. Phys.* 26, L1 (1987).

66. J. Wei, H. Kawarada, J. Suzuki, K. Yanagihara, K. Numata, and A. Hiraki, *Electrochem. Soc. Proc.* 393, 89 (1989).

67. P. K. Bachmann, and R. Messier, *C&EN* 67, 24 (1989).

68. J. C. Angus, and C. C. Hayman, *Science* 241, 4868, 913 (1988).

69. K. Kudhara, K. Sasaid, M. Kawarada, and N. Koshino, *Appl. Phys. Lett.* 52, 437 (1988); K. Kudhara, K. Sasaid, M. Kawarada, A. Teshima, and K. Koshino, *SPIE Proc.* 1146, 28 (1990).

70. N. Fujimori, A. lkegaya, T. Imal, K. Fukushima, and N. Ota, *Electrochem. Soc. Proc.* 465, 89 (1989).

71. Y. H. Lee, P. D. Richard, K. J. Bachmann, J. T. Glass, *Appl. Phys. Lett.* 56, 620 (1990).

72. N. Ohtake and M. Yoshikawa, *J. Electrochem. Soc.* 137, 717 (1990); N. Ohtake and M. Yoshikawa, *Thin Solid Films* 212, 112 (1992).

73. S. Matsumoto, Y. Manabe, and Y. Hibino, *J. Mater. Sci.* 27, 5905 (1992).

74. S. Matsumoto, *J. Mater. Sci. Lett.* 4, 600 (1985).

75. H. O. Pierson, "Graphite." Noyes, Park Ridge, NJ, 1993.

76. S. Matsumoto, M. Hino, and T. Kobayashi, *Appl. Phys. Lett.* 51, 737 (1987).

77. S. Matsumoto and K. Okada, *Oyo Buturi* 61, 726 (1992).

78. X. Chen and J. Narayan, *J. Appl. Phys.* 74, 4168 (1993).

79. Z. P. Lu, J. Heberlein, and E. Pfender, *Plasma Proc.* 12, 55 (1992).

80. A. Hirata, and M. Yoshikawa, *Diamond Relat. Mater.* 2, 1402 (1993).

81. K. Eguchi, S. Yata, and T. Yoshida, *Appl. Phy. Lett.* 64, 58(1994).

82. Watanabe, T. Matsushita, and K. Sasahara, *Jpn. J. Appl. Phys.* 31, 1428 (1992).

83. M. Kohzaki, K. Uchida, K. Higuchi, and S. Noda, *Jpn. J. Appl. Phys.* 32, L438 (1993).

84. A. K. Sikder, "Electrical and Structural Characterization of Diamond Films and Growth of Diamond Films on Stainless Steel for Tool Applications." Doctoral Thesis, IIT Bombay, India, 1999.

85. C. M. Marks, H. R. Burris, J. Grun and K. A. Snail, *J. Appl. Phys.* 73, 755 (1993).

86. X. H. Wang, W. Zhu, J. von Windheirn, and J. T. Glass. *Cryst. Growth* 129, 45 (1993).

87. I. Y. Hirose, and N. Kondo, "Japan Applied Physics Spring Meeting," 1988.

88. L. M. Hanssen, W. A. Carrington, J. E. Butler, and K. A. Snail, *Mater. Lett.* 7, 289 (1988).

89. P. G. Kosky, and D. S. McAtee, *Mater. Lett.* 8, 369 (1989).

90. M. A. Cappelli, and P. H. Paul, *J. Appl. Phys.* 67, 2596 (1990).

91. K. V. Ravi, and C. A. Koch, *Appl. Phys. Lett.* 57, 348 (1990).

92. Y. Tseng, C. Cutshaw, R. Phillips, and T. Srivinyunon, *Appl. Phys. Lett.* 56, 134 (1990).

93. D. B. Oakes, and J. E. Butler, *J. Appl. Phys.* 69, 2602 (1991).

94. Y. Matsui, A. Yuuki, M. Sahara, and Y. Hirose, *Jpn. J. Appl. Phys.* 28, 1718 (1989).

95. Y. Matsui, H. Yabe, and Y. Hirose, *Jpn. J. Appl. Phys.* 29, 1552 (1990).

96. Y. Matsui, H. Yabe, T. Sigimoto, and Y. Hirose, *Diamond Relat. Mater.* 1, 19 (1991).

97. R. F. Davis, "Diamond Film And Coatings Development, Properties and Applications," p. 182. Noyes, Park Ridge, NL.

98. R. Kapil, B. R. Mehta, V. D. Vankar, *Thin Solid Films* 312, 106 (1998).

99. P. K. Bachman, W. Drawe, D. Knight, R. Weimer and R. Meisser, *In* "Diamond and Diamond-like Materials Synthesis" (M.W. Geis, G. H. Johnson and A. R. Badzian, Eds.), pp. 99-102. MRS, Pittsburgh, 1988.

100. B. V. Spitsyn and B. V. Deryagin, USSR Patent 399134, 1980.

101. R. A. Rudder, G. C. Hudson, J. B. Posthill, R. E. Thomas, and R. J. Markunas, *Appl. Phys. Lett.* 59, 791 (1991).

102. B. J. Bai, C. J. Chu, D. E. Patterson, R. H. Hauge, and J. L. Margrave, *J. Mater. Res.* 8, 233 (1993).

103. N. J. Komplin, B. J. Bai, C. J. Chu, J. L. Margrave, and R. H. Hauge, *In* "Third International Symposium on Diamond Materials" (J. P. Dismukes and K. V. Ravi, Eds.), p. 385. The Electrochemical Society Processing Series, 1993.

104. J.-J. Wu and F. C.-N. Hong, *J. Appl. Phys.* 81, 3647 (1997)

105. F. C-N. Hong, G. T. Lang, D. Chang, and S. C. Yu, *In* "Applications of Diamond Films and Related Materials" (Y. Tzeng, M.Yoshikawa, M. Murakawa, and A. Feldman, Eds.), pp. 577–82. Elsevier, Amsterdam, 1991.

106. F. C. Hong, G. T. Lang, J. J. Wu, D. Chang, and J. C. Hsieh, *Appl. Phys. Lett.* 63, 3149 (1993).

107. C. A. Rego, R. S. Tsang, P. W. May, M. N. R. Ashfold, and K. N. Rosser, *J. Appl. Phys.* 79, 7264 (1996).

108. R. S. Tsang, C. A. Rego, P. W. May, J. Thumin, M. N. R. Ashfold, K. N. Rosser, C. M. Younes, and M. J. Holt, *Diamond Rel. Mater.* 5, 359 (1996).

109. E. J. Corat, V. J. Trava-Airoldi, N. F. Leite, A. F. V. Pena, and V. Baranauskas, *Mater. Res. Soc. Symp. Proc.* 349, 439 (1994).

110. E. J. Corat, R. C. Mendes de Barros, V. J. Trava-Airoldi, N. G. Ferreira, N. F. Leite, and K. lha, *Diamond and Relat. Mater.* 6, 1172 (1997).

111. V. J. Trava-Airoldi, B. N. Nobrega, E. J. Corat, E. Delbosco, N. F. Leite, and V. Baranauskas, *Vacuum* 46, 5 (1995).

112. I. Schmidt, F. Hentschel, and C. Benndorf, *Diamond Rel. Mater.* 5, 1318 (1996).

113. M. Kadono, T. Inoue, A. Midyadera, and S. Yamazaki, *Appl. Phys. Lett.* 61, 772 (1992).

114. H. Maeda, M. hie, T. Hino, K. Kusakabe, and S. Morooka, *Diamond Rel. Mater.* 3, 1072 (1994).

115. R. Ramesharm, R. F. Askew, and B. H. Loo, *In* "Third International Symposium on Diamond Materials" (J. P. Dismukes and K. V. Ravi, Eds.), p. 394. The Electrochemical Society Processing Series, 1993.

116. K. J. Grannen and R. P. H. Chang, *J. Mater. Res.* 9, 2154 (1994).

117. K. J. Grannen, D. V. Tsu, R. J. Meflunas, and R. P. H. Chang, *Appl. Phys. Lett.* 59, 745 (1991).

118. C. A. Fox, M. C. McMaster, W. L. Hsu, M. A. I Kelly, and S. B. Hagstrorn, *Appl. Phys. Lett.* 67, 2379 (1995).

119. T. Kotaki, Y. Amada, K. Harada, and O. Matsumoto, *Diamond Rel. Mater.* 342 (1993).

120. T. Kotaki, N. Horii, H. Isono, and O. Matsumoto, *J. Electrochern. Soc.* 143, 2003 (1996).

121. T. Suntola, *Mater. Sci. Rep.* 4, 261 (1989).

122. T. I. Hukka, R. E. Rawles and M. P. D'Evelyn, *Thin Solid Films* 225, 212 (1993)

123. R., Gat, T. I. Hukka, R. E. Rawles, and M. P. D'Evelyn, *Ann. Tech. Conf. Soc. Vac. Coat.* 353, (1993).

124. B. B. Pate, et al., *Physica B* 117/118, 783 (1983).

125. B. B. Pate, *Surf. Sci.* 165, 83 (1986).

126. T. 1. Hukka, R. E. Rawles and M. P. D'Evelyn, *Thin Solid Films* 225, 212 (1996).

127. R. Gat, T. I. Hukka and M. P. D'Evelyn, *In* "Diamond Materials" (H. P. Dismukes, K. V. Ravi, K. E. Spear, B. Lux, and N. Setaka, Eds.). The Electrochemical Society, 1993.

128. D. V. Fedoseev, I. G. Varshavskaya, A. V. lavent'ev, and B. V. Deryagin, *Powder Technol.* 44, 125 (1985).

129. D. V. Fedoseev, and B. V. Deryagin, *Mater. Res. Soc. Symp. Proc.* 439 (1989).

130. P. A. Molian and A. Waschek, *J. Mater. Sci.* 28, 1733 (1993).

131. K. Kitahama, K. Hirata, H. Nakamatsu, S. Kawai, N. Fujimori, T. Imai, H. Yoshino, and A. Doi, *Appl. Phys. Lett.* 49, 634 (1986).

132. K. Kitaharna, K. Hirata, H, Nakamatsu, S. Kawai, N. Fujimori, and T. Imai, *Mater. Res. Soc. Symp. Proc.* 75, (1987).

133. K. Kitahama, *Appl. Phys. Lett.* 53, 1812 (1988).

134. P. Mistry, M. C. Turchan, S. Liu, G. O. Granse, T. Baurmann, and M. G. Shara, *Innov. Mater. Res.* 1, 193 (1996).

135. D. L. Pappas, K. L. Saenger, J. Bruley, Y. V. Kra,kow, J. J. Cuomo, T. Gu, and R. W. Collins, *J. Appl. Phys.* 71, 5675 (1992).

136. M. B. Guseva, V. G. Babaev, V. V. Khvostov, Z. K. Valioullova, A. Yu. Bregadze, A. N. Obraztsov, and A. E. Alexenko, *Diamond Rel. Mater.* 3, 328 (1994).

137. J. Seth, R. Padiyath, D. H. Rasmussen, and S. V. Babu, *Appl. Phys. Lett.* 63, 473 (1993).

138. M. C. Polo, J. Cifre, G. Sanchez, R. Aguiar, M. Varela, and J. Esteve, *Appl. Phys. Lett.* 67, 485 (1995)

139. M. C. Polo, 1. Cifre, G. Sanchez, R. Aguiar, M. Yarela, and J. Esteve, *Diamond Rel. Mater.* 4, 780 (1995)

140. P. A. Molian, B. Janvrin, and A. M. Molian, *J. Mater. Sci.* 30, 4751 (1995).

141. M. E. Kozlov, P. Fons, H.-A. Durand, K. Nozaki, M. Tokumoto, K. Yase, and N. Minami, *J. Appl. Phys.* 80, 1182 (1996).

142. J. Wagner, C. Wild, and P. Koidl, *Appl. Phys. Lett.* 59, 779 (1991).

143. S. Ltsch and H. Hiraoka, *J. Min. Met. Mater. Soc.* 46, 64 (1994).

144. B. V. Spitzyn, L. L. Bouilov, and B. V. Derjaguin, *J. Crystal Growth* 52, 219 (1981); B. V. Spitzyn and B. V. Derjaguin, *Akad. Nauk SSSR, Leningrad* 22 (1979).

145. R. J. Graham, J. B. Posthill, R. A. Rudder, and R. I. Markunas, *Appl. Phys. Lett.* 59, 2463 (1991).

146. W. A. Yarbrough, *J. Vacuum Sci. Technol. A* 9, 1145 (1991).

147. D. N. Belton, and S. J. Schmieg, *Thin Solid Films*, 212, 68 (1992).

148. A.-R. Badzian and T. Badzian, *In* "Proc. MRS Fall Meeting" (T. M. Bessmann, B. M. Gallois, and J. Warren, Eds.), Vol. 250B, 339–349. Boston, MA, 1991.

149. Y. Sato, I. Yashima, H. Fujita, T. Ando, and M. Kamo, *In* "New Diamond Science and Technology" (R. Messier, J. T. Glass, I. E. Butler, and R. Roy, Eds.), pp. 371–376. MRS, Pittsburgh, PA, 1991.

150. S. Yugo, T. Kanai, T. Kimura, and T. Muto, *Appl. Phys. Lett.*, 58, 1036 (1991).

151. B. R. Stoner, G.-H. M. Ma, D. S. Wolter, and J. T. Glass, *Phys. Rev. B* 45, 11067 (1992).

152. A. van der Drift, *Phillips Res. Rep.* 22, 267 (1967).

153. C. Wild, P. Koidl, W. MOller-Sebert, H. Walcher, R. Kohl. N. Herres, R. Locher, R. Samlenski and R. Brenn, *Diamond Relat. Mater.* 2, 158 (1993).

154. R. E. Clausing, L. Heatherly, E. D. Specht and K. L. More, *In* "New Diamond Science and Technology, R. Messier" (J. T. Glass, J. E. Butler and R. Roy, Eds.), p. 575. Materials Research Society, 1991.

155. H. J. Fiffler, M. Rosler, M. Hartweg, R. Zachai, X. Jiang, and C. P. Klages, (1993).

156. J. E. Butler and R. L. Woodin, *Philos. Trans. Roy Soc. London Ser. A* 342, 209 (1993).

157. K. Kobashi, K. Nishimura, K. Miyata, K. Kumagai and A. Nakaue, *J. Mater. Res.* 5, 2469 (1990).

158. F. G. Celii, J. White D. and A. J. Purdes, *Thin Solid Films* 212, 140 (1992).

159. L. Chow, A. Homer, and H. Sakoui, *J. Mater. Res.* 7, 1606 (1992).

160. C. Wild, P. Koidl, W. Muller-Sebert, H. Walcher, R. Kohl, N. Herres, R. Locher, R. Samlenski, and R. Brenn, *Diamond Rel. Mater.* 2, 158 (1993).

161. T. Suesada, N. Nakamura, H. Nagasawa, and H. Kawarada, *Jpn. J. Appl. Phys.* 34, 4898 (1995).

162. E. Burton, *Diamond London* (1978).

163. J. C. Angus, *Thin Solid Films* 216, 126 (1992).

164. S. P. McGinnis, M. A. Kelly, and S. B. Hagstrom, *Appl. Phys. Lett.*, 66 3117 (1995).

165. J. E. Field, "The Properties of Natural and Synthetic Diamond." Academic Press, London, 1992.

166. X. Jiang, C.-P. Klages, R. Zachai, M. Hartweg, and H.-J. Fosser, *Appl. Phys. Lett.* 62, 26 (1993).

167. H. Yagi, K. Hosina, A. Hatta, T. Ito, T. Sasaki, and A. Hiraki, *J. Jpn. Appl. Phys.* 36, L507 (1997).

168. X. Jiang, K. Shiffmann, C.-P. Klages, D. Wittorf, C. L. Jia, and K. Urban, *J. Appl. Phys.* 83, 5 (1998).

169. G. Popovici, C. H. Chao, M. A. Prelas, E. J. Charlson and J. M. Meese, *J. Mater. Res.* 10, 8 (1995).

170. O. Chena and Z. Lin, *J. Appl. Phys.* 80, (1996).

171. W. Zhu, F. R. Sivazlian, B. R. Stoner and J. T. Glass *J. Mater. Res.* 10, 2 (1995).

172. F. Stubhan, M. Ferguson, and H.-J. Fusser, R. J. Behm, *Appl. Phys. Lett.* 66, 15 (1995).

173. X. Li, Y. Hayashi, and S. Nishino, *Jpn. J. Appl. Phys.* 36, 5197 (1997).

174. W. Kulisch, L. Ackermann, and B. Sobisch, *Phys. Stat. Sol. (a)* 154, 155 (1996).

175. H. P. Lorenz, W. Schilling, *Diamond Relat. Mater.* 6, 6 (1997).

176. J. Robertson, J. Gerber, S. Sattel, M. Weiler, K. Jung, and H. Ehrhardt, *Appl. Phys. Lett.* 66, 3287 (1995).

177. S. Sattel, J. Gerber, and H. Ehrhardt, *Phys. Stat. Sol. (a)* 154, 141 (1996).

178. S. D. Wolter, B. R. Stoner, I T. Glass, P. J. Ellis, D. S. Buhhaenko, C. E. Jenkins, and P. Southworth, *Appl. Phys. Lett.* 62, 1215 (1993).

179. B. X. Wang, W. Zhu, J, Ahn, and H. S. Tan, *J. Cryst. Growth* 151, 319 (1995).

180. P. John, D. K. Milne, P. G. Roberts, M G. Jubber, M. Liehr, and J. I. B. Wilson, *J. Mater. Res.* 9, 3083 (1994).

181. C. Z. Gu and X. Jiang, *J. Appl. Phys.* 88, 4 (2000).

182. S. Chattopadhyay, P. Ayyub, V. R. Palkar, and M. Multani, *Phys. Rev. B* 52, 177 (1995).

183. D. A Gruen, *MRS Bull.* 9, 32 (1998).

184. Kehui Wu, E. G. Wang, and Z. X. Cao, *J. Appl. Phys.* 88, 5 (2000).

185. B. Hong, J. Lee, R. W. Collins, Y. Kuang, W. Drawl, R. Messier, T. T. Tsong, and Y. E. Strausser, *Diamond Relat. Mater.* 6, 55 (1997).

186. D. M. Gruen, S. Liu, A. R. Krauss, J. Luo, and X. Pan, *Appl. Phys. Lett.* 64, 1502 (1994).

187. D. Zhou, D. M. Gruen, L. C. Qin, T. G. McCauley, and A. R. Krauss, *J. Appl. Phys.* 84, 1981 (1998).

188. S. A. Catledge and Y. K. Vohra, *J. Appl. Phys.* 86, 698 (1999).

189. O. Groning, M. Kuttel, P. Groning, and L. Schlapbach, *J. Vac. Sci. Technol. B* 17, 5 (1999).

190. W. Zhu, G. P. Kochanski, and S. Jin, *Science* 282, 1471 (1998).

191. Zhou, A. R. Krauss, L. C. Qin, T. G. McCauley, D. M. Gruen, T. D. Corrigan, R. P. H. Chang, and H. Gnaser, *J. Appl. Phys.* 82, 4546 (1997).

192. http://www.techtransfer.anl.gov/techtour/diamondmems.html.

193. T. Sharda, D. S. Misra, D. K. Avasthi, and G. K. Metha, *Solid State Commun.* 98, 879 (1996).

194. Y. N. K. Tamaki, Y. Watanabe, and S. Hirayama, *J. Mater. Res.* 10, 431 (1995).

195. W. Howard, D. Huang, J. Yuan, M. Frenklach, K. E. Spear, R. Koba, and A. W. Phelps, *J. Appl. Phys.* 68, 1247 (1990).

196. M. Frenklach, W. Howard, D. Huang, J. Yuan, K. E. Spear, and R. Koba, *Appl. Phys. Lett.* 59, 546 (1991).

197. J. Singh, *J. Mater. Sci.* 29, 2761 (1993).

198. A. R. Badzian, and T. Badzian, *Surf. Coat. Technol.* 36, 283 (1991).

199. K. M. McNamara, B. E. Williams, K. K. Gleason, and B. E. Scruggs, *J. Appl. Phys.* 76, 2466 (1994).

200. T. Sharda, A. K. Sikder, D. S. Misra, A. T. Collins, S. Bhargava, H. D. Bist, P. Veluchamy, H. Minoura, D. Kabiraj, D. K. Avasthi, and P. Selvam, *Diamond Relat. Mater.* 7, 250 (1998).

201. P. K. Bachmann, A. T. Collins, and M. Seal, "Diamond 1992." Elsevier Sequoia, S.A., 1993.

202. J. Ruan, W. J. Choyke, and W. D. Paxtlow, *J. Appl. Phys.* 69, 6632 (1991).

203. J. Ruan, K. Kobashi, and W. J. Choyke, *Appl. Phys. Lett.* 60, 1884 (1992).

204. L. H. Robins, L. P. Cook, E. N. Farabaugh, and A. Feldman, *Phys. Rev. B* 39, 13367 (1989).

205. A. T. Collins, M. Kamo, and Y. Sato, *J. Phys. D* 22, 1402 (1989).

206. A. T. Collins, M. Kamo, and Y. Sato, *J. Mater. Res.* 5, 2507 (1990).

207. R. J. Graham, T. D. Moustakas, and M. M. Disko, *J. Appl. Phys.* 69, 3212 (1991).

208. A. T. Collins, *Diamond Relat. Mater.* 1, 457 (1992).

209. A. T. Collins, L. Allers, C. J. H. Wort, and G. A. Scarsbrook, *Diamond Relat. Mater.* 3, 932 (1994).

210. Y. L. Khong, A. T. Collins, and L. Allers, *Diamond Relat. Mater.* 3, 1023 (1994).

211. J. Ruan, W. J. Choyke, and K. Kobashi, *Appl. Phys. Lett.* 62, 1379 (1993).

212. Y. H. Singh, D. H. Rich, and F. S. Pool, *J. Appl. Phys.* 71, 6036 (1992).

213. X. Yang, A. V. Barnes, M. M. Albert, R. G. Albridge, J. T. McKinley, N. H. Tolk, and J. L. Davidson, *J. Appl. Phys.* 77, 1758 (1995).

214. Z. Zhang, K. Liao, and W. Wang, *Thin Solid Films* 269, 108 (1995).

215. G. Z. Cao, J. J. Schermer, W. J. P. van Enckevort, W. A. L. M. Elst, and L. J. Giling, *J. Appl. Phys.* 79, 1357 (1996).

216. T. Feng and B. D. Schwartz, *J. Appl. Phys.*, 73 1415 (1993).

217. X. Jiang and C. P. Klages, *Phys. Stat. Sol.* 154, 175 (1996).

218. H. Jia, J. Shinar, D. P. Lang, and M. Pruski, *Phys. Rev. B* 48, 17595 (1993).

219. R. F. Davis, "Diamond Film And Coatings Development, Properties and Applications," p. 246. Noyes, Park Ridge, NJ.

220. B. V. Spitsyn, L. L. Bouilov, and B. V. Deryagin, *J. Cryst. Growth* 52, 219 (1981).

221. K. Kobashi, K. Nishimura, Y. Kawate, and T. Horiuchi, *Phys. Rev. B* 38, 4067 (1988).

222. R. Haubner, and B. Lux, *Int. J. Refract. Hard Metals* 6, 210 (1987).

223. W. Zhu, A. R. Badzian, and R. Messier, *Proc. SPIE* 1325, 187 (1990).

224. G. Sanchez, J. Servat, P. Gorostiza, M. C. Polo, W. L. Wang, and J. Esteve, *Diamond Relat. Mater.* 5, 592 (1996).

225. X. Jiang, K Schiffmann, A. Westphal, and C.-P. Klages, *Diamond Relat. Mater.* 4, 155 (1995).

226. E. Johansson and J.-O. Carlsson, *Diamond Relat. Mater.* 4, 155 (1995).

227. B. E. Williams and J. T. Glass, *J. Mater. Res.* 4, 373 (1989).

228. H. Kawarada, K. S. Mar, J. Suzuki, Y. Ito, H. Mori, H. Fujita, and A. Hirald, *Jpn. J. AppL. Phys.* 26, 1903 (1987).

229. W. Zhu, A. R. Badzian, and R. Messier, *J. Mater. Res.* 4, 659 (1989).

230. R. E. Clausing, L. Heatherly, and K. L. More, *Surf. Coat. Technol.* 39/40, 199 (1989).

231. J. L. Kaae, P. K. Gantzel, J. Chin, and W. P. West, *J. Mater. Res.* 5, 1480 (1990).

232. A. V. Hetherington, C. J. H. Wort, and P. Southworth, *J. Mater. Res.* 5, 1591 (1990).

233. J. Narayan, *J. Mater. Res.* 5, 2414 (1990).

234. W. L. Bragg, *In* "The Crystalline State" (W. H. Bragg and W. L. Bragg, Eds.), Vol. I, pp. 12–21. Bell, London, 1933.

235. C. Wild, N. Herres, and P. Koidl, *J. Appl. Phys.* 68, 973 (1990).

236. E. D. Specht, R. E. Clausing, and L. Heatherly, *J. Mater. Res.* 5, 2351 (1990).

237. R. E. Clausing, L. Heatherly, E. D. Specht, and K. L. More, *In* "New Diamond Science and Technology" (R. Messier, J. T. Glass, J. E. Butler and R. Roy, Eds.), pp. 575–580. Materials Research Society, Pittsburgh, PA, 1991.

238. R. F. Davis, "Diamond Film And Coatings Development, Properties and Applications," p. 300. Noyes, Park Ridge, NJ.

239. A. K. Arora, T. R. Ravindran and G. L. N. Reddy, A. K. Sikder and D. S. Misra, *Diamond Relat. Mater.* 10, 1477 (2001).

240. N. Wada and S. A. Solin, *Physica* 105B, 353 (1981).

241. D. S. Knight and W. B. White, *J. Mater. Res.* 4, 385 (1989).

242. K. E. Spear, A. W. Phelps, and W. B. White, *J. Mater. Res.* 5, 2277 (1990).

243. L. Abello, G. Lucazeau, B. Andre, and T. Priem, *Diamond Relat. Mater.* 1, 512 (1992).

244. P. V. Huong, *Mater. Sci. Eng. B* 11, 235 (1992).

245. M. Yoshikawa, G. Katagiri, H. Ishida, A. Ishitani, M. Ono, and K. Matsumura, *Appl. Phys. Lett.* 55, 2608 (1989).

246. J. W. Ager III, D. K. Veirs, and G. M. Rosenblatt, *Phys. Rev. B* 43, 6491 (1991).

247. J. Wagner, M. Ramsteiner, C. Wild, and P. Koidl, *Phys. Rev. B* 40, 1817 (1989).

248. H. Windischmann, G. F. Epps, Y. Cong, and R. W. Collins, *J. Appl. Phys.* 69, 2231 (1991).

249. W. Wanlu, G. Kejun, G. Jinying, and L. Aimin, *Thin Solid Films* 215, 174 (1992).

250. J. Robertson, *Adv. Phys.* 35, 317 (1986).

251. H. Ricter, Z. P. Wang, and L. Ley, *Solid State Commun.* 39, 625 (1981).

252. I. H. Campbell and P. Fauchet, *Solid State Commun.* 58, 739 (1986).

253. N. B. Colthup, L. H. Daly, and S. E. Wiberley, "Introduction to Infrared & Raman Spectroscopy." Academic Press, New York, 1975.

254. P. R. Griffiths, "Chemical Infrared Fourier Transform Spectroscopy." Wiley, New York, 1975.

255. T. Sharda, D. S. Misra, and D. K. Avasthi, *Vacuum* 47, 1259 (1996).

256. G. Dollinger, A. Bergmaier, C. M. Frey, M. Roesler, and H. Verhoeven, *Diamond Relat. Mater.* 4, 591 (1995).

257. K. M. McNamara, K. K. Gleason, and C. J. Robinson, *J. Vac. Sci. Technol. A* 10, 3143 (1992).

258. D. K. Avasthi, *Vacuum* 48, 1011 (1997).

259. M. I. Landstrass and K. V. Ravi, *Appl. Phys. Lett.* 55, 1391 (1989).

260. K. Baba, Y. Aikawa, and N. Shohata, *J. Appl. Phys.* 69, 7313 (1991).

261. K. M. McNamara, D. H. Levy, K. K. Gleason, and C. J. Robinson, *Appl. Phys. Lett.* 60, 580 (1992).

262. G. T. Mearini, I. L. Krainsky, J. J. A. Dayton, Y. Wang, C. A. Zorman, J. C. Angus, and R. W. Hoffman, *Appl. Phys. Lett.* 65, 2702 (1994).

263. F. G. Celli and J. E. Butler, *Appl. Phys. Lett.* 54, 1031 (1989).

264. J. E. Butler and R. L. Woodwin, *Trans. Roy Soc. London A Sect.* 342, 209 (1993).

265. T. R. Anthony, *Vacuum* 41, 1356 (1990).

266. F. G. Celli and J. E. Butler, *Annu. Rev. Chem.* 42, 643 (1991).

267. K. E. Spear, *J. Am. Ceram. Soc.* 72, 171 (1989).

268. M. Page and D. W. Brenner, *J. Am. Chem. Soc.* 96, 3270 (1991).

269. D. Huang and M. Frenklavh, *Phys. Chem.* 96, 1868 (1992).

270. B. B. Pate, M. H. Hecht, C. Binns, I. Lindau, and W. E. Spicer, *J. Vac. Sci. Technol.* 21, 364 (1982).

271. J. C. Angus, H. A. Will, and W. S. Stanko, *J. Appl. Phys.* 39, 2915 (1968).

272. W. A. Yarbrough and R. Messier, *Science* 247, 688 (1990).

273. B. Dischler, A. Bubenzer, and P. Koidl, *Solid State Commun.* 48, 105 (1983).

274. A H. Brodsky, M. Cardona, and J. J. Cuomo, *Phys. Rev. B* 162, 3556 (1977).

275. D. M. Bibby and P. A. Thrower, Dekker, New York, 1982.

276. J. Ruan, W. J. Choyke, and W. D. Partlow, *Appl. Phys. Lett.* 58, 295 (1991).

277. Y. Mori, N. Eimori, H. Kozuka, Y. Yokota, J. Moon, J. S. Ma, T. Ito, and A. Hiraki, *Appl. Phys. Lett.* 60, 47 (1992).

278. G. Davies, Dekker, New York, 1977.

279. C. D. Clark, A. T. Collins, and G. S. Woods, "The Properties of Natural and Synthetic Diamond," Academic Press, London, 1992.

280. L. S. Pan and D. R. Kania, "Diamond Electronic Properties and Applications," p. 201. Kluwer Academic, Dordrecht/Norwell, MA, 1995.

281. A. R. Badzian and R. DeVries, *Mater. Res. Bull.* 23, 385 (1988).

282. M. Tsuda, M. Nakajima, and S. Oikawa, *J. Am. Chem. Soc.* 108, 5780 (1986).

283. M. Frenklach and K. Spear, *J. Mater. Res.* 3, 133 (1988).

284. Y. B. Yam and T. D. Moustakas, *Nature* 342, 786 (1989).

285. M. Fanciulli, S. Jin, and T. D. Moustakas, *Physica B* 229, 27 (1996).

286. M. H. Nazare, P. W. Mason, G. D. Watkins, and H. Kanda, *Phys. Rev. B* 51, 16741 (1995).

287. E. Rohrer, C. F. O. Graeff, R. Janssen, C. E. Nebel, and M. Stutzmann, *Phys. Rev. B* 54, 7874 (1996).

288. E. A. Faulkner and J. N. Lomer, *Philos. Mag. B* 7, 1995 (1962).

289. M. Fanciulli and T. D. Moustakas, *Phys. Rev. B* 48, 14982 (1993).

290. Y. von Kaenel, J. Stiegler, E. Blank, O. Chauvet, C. Hellwig, and K. Plamann, *Phys. Stat. Sol.* 154, 219 (1996).

291. Y. Show, T. Izumi, M. Deguchi, M. Kitabatake, T. Hirao, Y. Morid, A. Hatta, T. Ito, and A. Hiraki, *Nucl. Instrum. Methods B* 127-128, 217 (1997).

292. S. L. Holder, L. G. Rowan, and J. J. Krebs, *Appl. Phys. Lett.* 64, 1091 (1994).

293. D. J. Keeble and B. Ramakrishnan, *Appl. Phys. Lett.* 69, 3836 (1996).

294. D. F. Talbot-Ponsonby, M. E. Newton, J. M. Baker, G. A. Scarsbrook, R. S. Sussmann, and C. J. H. Wort, *J. Phys. Condensed Mater.* 8, 837 (1996).

295. S. Dannefaer, T. Bretagnon, and D. Kerr, *Diamond Relat. Mater.* 2, 1479 (1993).

296. S. Dannefaer, W. Zhu, T. Bretagnon, and D. Kerr, *Phys. Rev. B* 53, 1979 (1996).

297. R. N. West, *Adv. Phys.* 22, 263 (1973).

298. N. Amrane, B. Soudini, N. Amrane, and H. Aourag, *Mater. Sci. Eng. B* 40, 119 (1996).

299. T.H. Huang, C.T. Kuo, C.S. Chang, C.T. Kao, and H.Y. Wen, *Diamond Relat. Mater.* 1, 594 (1992).

300. K. Shibuki, M. Yagi, K. Saijo, and S. Takatsu, *Surf. Coat. Technol.* 36, 295 (1988).

301. A. K. Mehlmann, S. F. Dimfeld, and Y. Avigal, *Diamond Rel. Mater.* 1, 600 (1992).

302. M.G. Peters and R.H. Cummings, European Patent 0519 587 Al, 1992.

303. X.X Pan, Doctoral Thesis, Vienna University of Technology, 1989.

304. X. Chen and J. Narayan, *J. Appl. Phys.* 74, 4168 (1993).

305. A. K. Sikder, T. Sharda, D. S. Misra, D. Chandrasekaram, and P. Selvam, *Diamond Relat. Mater.* 7, 1010 (1998).

306. T. Yashiki, T. Nakamura, N. Fujimori, and T. Nakai, Int. Conf. On Metallurgical Coatings and Thin Films (ICMCTF), San Diego, California April 22-26.

307. S. Kubelka, R. Haubner, B. Lux, R. Steiner, G. Stingeder, and M. Grasserbauer, *Diamond Relat. Mater.* 3, 1360 (1994).

308. C. T. Kuo, T. Y. Yen, and T. H. Huang, *J. Mater. Res.* 5, 2515 (1990).

309. B. Lux and R. Haubner, *Int. J. Refractory, Met. Hard Mater.* 8, 158 (1989).

310. M. Murakawa, S. Takeuchi, H. Miyazawa, and Y. Hirose, *Surf. Coat. Technol.* 36, 303 (1988).

311. S. Soderberg, *Vacuum* 41, 1317 (1990).

312. R. C. McCune, R. E. Chase, and W. R. Drawl, *Surf. Coat. Technol.* 39/40, 223 (1999).

313. S. Soderberg, A. Gerendas, and M. Sjostrand, *Vacuum* 41, 2515 (1990).

314. B. S. Park, Y. J. Baik, K. R. Lee, K. Y. Eun, and D. H. Kim, *Diamond Relat. Mater.* 2, 910 (1993).

315. H. Suzuki, H. Matsubara, and N. Horie. *J. Jpn. Soc. Powder Metallurgy* 33, 262 (1986).

316. Y. Saito. K. Sato, S. Matsuda, and H. Koinuma. *J. Water. Sci. B* 2937 (1991).

317. M. I. Nesladek, J. Spinnewyn, C. Asinari, R. Lebout, and R. Lorent, *Diamond Relat. Mater.* 3, 98 (1993).

318. K. Saijo, M. Yagi, K. Shibiki, and S. Takatsu, *Surf. Coat. Technol.* 43/44, 30 (1990).

319. T. Isozaki, Y. Saito, A. Masuda, K. Fukumoto, M. Chosa, T. Ito, E. J. Oles, A. Inspektor, and C. E. Bauer, *Diamond Relat. Mater.* 2, 1156 (1993).

320. S. Kubelka, R. Haubner, B. Lux, R. Steiner, G. Stingeder, and M. Grasserbauer, *Diamond Relat. Mater.* 3, 1360 (1994).

321. M. Nesladek, K. Vandierendonck, C. Quaeyhaegens, M. Kerkhofs, and L. M. Stals, *Thin Solid Films* 270, 184 (1995).

322. S. Kubelka, R. Haubner, B. Lux, R. Steiner, G. Stingeder, and M. Grasserbauer, *Diamond Rel. Mater.* 3, 1360 (1994).

323. W. G. Moffatt, New York.

324. J. Kamer, E. Bergmann, M. Pedrazzini, I. Reineck, and M. Sjostrand, Int. Patent. WO 95/12009, 1995.

325. Volker Weihnacht, W. D. Fan, K. Jagannadham, and J. Narayan, *J. Mater. Res.* 11, 9 (1996).

326. E. J. Oles, A. Inspektor, and C. E. Bauer, "The New Diamond-Coated Carbide Cutting Tools." Corp. Technol. Center. PA.

327. C. Tsai, J. Nelson, W. W. Gerberich, J. Heberlein, and E. Pfender, *J. Mater. Res.* 7, 1967 (1992).

328. A. Fayer and O. G. A. Hoffman, *Appl. Phys. Lett.* 67, 2299 (1995).

329. J. Narayan, V. P. Godbole, G. Matera, and R. K. Singh, *J. Appl. Phys.* 71, 966 (1992).

330. E. Heidarpour and Y. Namba, *J. Mater. Res.* 8, 2840 (1993).

331. C. Tsai, J. C. Nelson, W. W. Gerberich, D. Z. Liu, J. Heberiein, and E. Pfender, *Thin Solid Films* 237, 181 (1994).

332. A. K. Sikder, D. S. Misra, D. Singhbal, S. Chakravorty, *Surface Coatings Technol.* 114, 230 (1999).

333. H. Yang and J. L. Whitten, *J. Chem. Phys.* 91, 126 (1989).

334. E. Johansson, P. Skytt, J. O. Carlsson, N. Wassdahl, and J. Nordgren, *J. Appl. Phys.* 79, 7248 (1996).

335. M. Yudasaka, K. Tasaka, R. Kikuchi, Y. Ohki, and S. Yoshimura, *J. Appl. Phys.* 81, 7623 (1997).

336. W. Zhu, P. C. Yang, J. T. Glass, and F. Arezzo, *J. Mater. Res.* 10, 1455 (1995).

337. D. N. Belton and S. J. Schmieg, *J. Appl. Phys.* 66, 4223 (1989).

338. P. C. Yang, W. Zhu, and J. T. Glass, *Mater. Res.* 9, 1063 (1994).

339. W. Zhu, P. C. Yang, and J. T. Glass, *Appl. Phys. Lett.* 63, 1640 (1993).

340. Y. Sato, H. Fujita, T. Ando, T. Tanaka, and M. Kamo, *Philos. Trans. Roy Soc. London A* 342, 225 (1993).

341. B. D. Cullity, "Elements of X-Ray Diffraction." Addison–Wesley, Reading, MA, 1956.

342. A. K. Sikder, T. Sharda, D. S. Misra, D. Chandrasekaram, P. Veluchamy, H. Minoura, and P. Selvam, *J. Mater. Res.* 14, 1148 (1999).

343. D. N. Belton and S. J. Schmieg, *Thin Solid Films* 212, 68 (1992).

344. P. C. Yang, W. Zhu, and J. T. Glass, *J. Mater. Res.* 8, 1773 (1993)

345. H. Holleck, *J. Vac. Sci. Technol. A* 4, 2661 (1986).

346. L. Vel, G. Demazeau, and J. Etourneau, *Mater. Sci. Eng. B* 10, 149 (1991).

347. R. H. Wentorf Jr., *J. Chem. Phys.* 36, 1990 (1962).

348. O. Mishima, In *"Synthesis and Properties of Boron Nitride,* (J. J. Pouch and S. A. Alterovitz, Eds.), Vol. 54/55, pp. 313-328. Trans Tech. Publications LTD, Brookfield, 1990.

349. O. Mishima, J. Tanaka, and O. Funkunaga, *Science* 238, 181 (1987).

350. L. Vel, G. Demazeau, and J. Etourneau, *Mater. Sci. Eng. B* 10, 149 (1991).

351. L. B. Hackenberger, L. J. Pilione, and R. Messier, In "Science and Technology of Thin Films" (F. Matacotta and G. Ottaviani, Eds.). World Scientific, Singapore, 1996.

352. P. Widmayer, P. Ziemann, S. Ulrich and H. Ehrhardt, *Diamond Relat. Mater.* 6, 621 (1997).

353. D. R. McKenzie, W. G.Sainty, and D. Green, *Mater. Sci. Forum* 54/55, 193 (1990).

354. D. R. McKenzie, D. J. H. Cockayne, D. A. Muller, M. Murakawa, S. Miyake, and P. Fallon, *J. Appl. Phys.* 70, 3007 (1991).

355. D. J. Kester, K. S. Ailey, R. F. Davis, and K. L. More, *J. Mater. Res.* 8, 1213 (1993).

356. D. L. Medlin, T. A. Friedman, P. B. Mirkarimi, M. J. Mills, and K. F. McCarty, *Phys. Rev. B* 50, 7884 (1994).

357. D. L. Medlin, T. A. Friedman, P. B. Mirkarimi, P. Rez, M. J. Mills, and K. F. McCarty, *J. Appl. Phys.* 76, 295 (1994).

358. D. J. Kester, K. S. Ailey, R. F. Davis, and D. J. Lichtenwalner, *J. Vac. Sci. Technol.* 12, 3074 (1994).

359. S. Watanabe, S. Mayake, W. Zhou, Y. Ikuhara, T. Suzuki, and M. Murakawa, *Appl. Phys. Lett.* 66 2490 (1995).

360. W. L. Zhou, Y. Ikuhara, M. Murakawa, S. Watanabe, and T. Suzuki, *Appl. Phys. Lett.* 66, 2490 (1995).

361. P. B. Mirkarimi, D. L. Medlin, K. F. McCarty, D. C. Dibble, W. M. Clift, J. A. Knapp, and J. C. Barbour, *J. Appl. Phys.* 82, 1617 (1997).

362. Y. Yamada-Takamura, O. Tsuda, H. Ichinose, and T. Yoskida, *Phys. Rev. B* 59, 352 (1999).

363. D. Zhang, D. M. Davalle, W. L. O'Brien, and D. N. McIlroy, *Surface Sci.* 461, 16 (2000).

364. W. L. Wang, K. J. Liao, S. X. Wang, and Y. W. Sun, *Thin Solid Films* 368, 283 (2000).

365. D. R. McKenzie, W. D. McFall, W. G. Sainty, C. A. Davis, and R. E. Collins, *Diamond Relat. Mater.* 2, 970 (1993).

366. G. F. Cardinale, D. L. Medlin, P. B. Mirkarimi, K. F. McCarty, and D. G. Howitt, *J. Vac. Sci. Technol.* 15, 196 (1997).

367. M. Zeitler, S. Seinz, and B. Rauschenbach, *J. Vac. Sci. Technol.* 17, 597 (1999).

368. D. J. Kester and R. Messier, *J. Appl. Phys.* 72, 504 (1992).

369. P. B. Mirkarimi, D. L. Medlin, K. F. McCarty, T. A. Friedman, G. F. Cardinale, D. G. Howitt, E. J. Klaus, and W. G. Wolfer, *J. Mater. Res.* 9, 2925 (1994).

370. P. B. Mirkarimi, D. L. Medlin, K. F. McCarty, and J. C. Barbour, *Appl. Phys. Lett.* 66, 2813 (1995).

371. P. B. Mirkarimi, D. L. Medlin, K. F. McCarty, G. F. Cardinale, D. K. Ottensen, and H. A. Johnsen, *J. Vac. Sci. Technol.* 14, 251 (1996).

372. Z. F. Zhou, I. Bello, V. Kremnican, M. K. Fung, K. H. Lai, K. Y. Li, C. S. Lee, and S. T. Lee. *Thin Solid Films* 368, 292 (2000).

373. T. Ichiki, S. Amagi, and T. Yoshida, *J. Appl. Phys.* 79, 4381 (1996).

374. T. Takami, I. Kusunoki, K. Suzuki, K.P. Loh, I. Sakaguchi, M. Nishitani-Gamo, T. Taniguchi, and T. Ando, *Appl. Phys. Lett.* 73, 2733 (1998).

375. R. A. Roy, P. Catania, K. L. Saenger, J. J. Cuomo, and R. L. Lossy, *J. Vac. Sci. Technol. B* 11, 1921 (1993).

376. Y. Lfshitz, S. R. Kasi, and J. W. Rabalais, *Phys. Rev. B* 41, 10468 (1990).

377. W. Dworkschak, K. Jung, and H. Ehrhardt, *Thin Solid Films* 254, 65 (1995).

378. J. Robertson, J. Gerber, S. Sattel, M. Weiler, K. Jung, and H. Ehrhardt, *Appl. Phys. Lett.* 66, 3287 (1995).

379. S. Uhlmann, T. Frauenheim, and U. Stephan, *Phys. Rev. B* 51, 4541 (1995).

380. S. Reinke, M. Kuhr, W. Kulisch, and R. Kassing, *Diamond Relat. Mater.* 4, 272 (1995).

381. L. Jiang, A. G. Fitzgerald, M. J. Rose, A. Lousa, and S. Gimeno, *Appl. Surface Sci.* 167, 89 (2000).

382. C. Hu, S. Kotake, Y. Suzuki, and M. Senoo, *Vacuum* 59, 748 (2000).

383. K. P.Loh, I. Sakaguchi, M. Nishitani-Gamo, T. Taniguchi, and T. Ando, *Phys. Rev. B* 57, 7266 (1998).

384. J. Deng, B. Wang, L. Tan, H. Yan, and G. Chen, *Thin Solid Films* 368, 312 (2000).

385. X. Zhang, H. Yan, B. Wang, G. Chen, and S. P. Wong, *Mater. Lett.* 43, 148 (2000).

386. J. Vilcarromero, M. N. P. Carreno, and I. Pereyra, *Thin Solid Films* 373, 273 (2000).

387. K. P. Loh, I. Sakaguchi, M. Nishitani-Gamo, S. Tagawa, T. Sugino, and T. Ando, *Appl. Phys. Lett.* 74, 28 (1999).

388. K. P. Loh, I. Sakaguchi, M. Nishitani-Gamo, T. Taniguchi, and T. Ando, *Appl. Phys. Lett.* 72, 3023(1998).

389. X. W. Zhang, Y. J. Zou, H. Yan, B. Wang, G. H. Chen, and S. P. Wong, *Mater. Lett.* 45, 111 (2000).

390. S. P. S. Arya, and A. D'Amico, *Thin Solid Films* 157, 267 (1988).

391. L. B. Hackenberger, L. J. Pilione, and R. Messier, *In* "Science and Technology of Thin Films" (F. Matacotta and G. Ottaviani, Eds.). World Scientific, Singapore, 1996.

392. W. Kulisch and S. Reinke, *Diamond Films Technol.* 7, 105 (1997).

393. F. Richter, *In* "Diamond Materials IV Electrochemical Society Proc." (K. V. Ravi, K. E. Spear, J. L. Davidson, R. H. Hadge, and J. P. Dismukes, Eds.), Vol. 94/95. The Electrochemical Society, Pennington, NJ, 1995.

394. T. Yoshida, *Diamond Relat. Mater.* 5, 501 (1996).

395. C. Ronning, H. Feldermann, and H. Hofsass, *Diamond Relat. Mater.* 9, 1767 (2000)

396. F. P. Bundy and R. H. Wentorf, *J. Chem. Phys.* 38, 1144 (1963).

397. F. R. Corrigan and F. P. Bundy, *J. Chem. Phys.* 63, 3812 (1975).

398. V. L. Solozhenko, *Thermochimica Acta* 218, 221 (1993).

399. S. Nakano, O. Funkunaga, *Diamond Relat. Mater.* 2, 1409 (1993).

400. S. Bohr, R. Haubner, and B. Lux, *Diamond Relat. Mater.* 4, 714 (1995).

401. T. Sato, T. Ishii, and N. Setaka, *Commun. Am. Ceram. Soc.* 65, C162 (1982).

402. P. B. Mirkarimi et al., *Mater. Sci. Eng.* R21, 47 (1997).

403. JCPDS 34-421.

404. JCPDS 41-1487.

405. T. Sato, *Proc. Japan Acad. Ser. B* 61, 459 (1985).

406. JCPDS 26-1079.

407. J. Thomas Jr., N. E. Weston, and T. E. O'Connor, *J. Amer. Chem. Soc.* 84, 4619 (1963).

408. J. A. Van Vechten, *Phys. Rev.* 187, 1007 (1969).

409. JCPDS 26-773.

410. JCPDS 19-268.

411. K. Inagawa, K. Watanabe, H. Ohsone, K. Saitoh, and A. Itoh, *J. Vac. Sci. Technol. A* 5, 2696 (1987).

412. D. J. Kester, K. S. Ailey, R. F. Davis, and K. L. More, *J. Mater. Res.* 8, 1213 (1993).

413. L. B. Hackenberger, L. J. Pilione, R. Messier, and G. P. Lamaze, *J. Vac. Sci. Technol. A* 12, 1569 (1994).

414. D. J. Kester and R. Messier, *J. Appl. Phys.* 72, 504 (1992).

415. D. Bouchier, G. Sene, M. A. Djouadi, and P. Moller, *Nucl. Instum. Methods B* 89, 369 (1994).

416. M. Sueda, T. Kobayashi, H. Tsukamoto, T. Rokkaku, S. Morimoto, Y. Fukaya, N. Yamashita, and T. Wada, *Thin Solid Films* 228, 97 (1993).

417. T. Wada and N. Yamashita, *J. Vac. Sci. Technol. A* 10, 515 (1992).

418. T. Ikeda, T. Satou, and H. Satoh, *Surf. Coat. Technol.* 50, 33 (1991).

419. S. Watanabe, S. Miyake, and M. Murakawa, *Surf. Coat. Technol.* 49, 406 (1991).

420. D. R. McKenzie, W. D. McFall, W. G. Sainty, C. A. Davis, and R. E. Collins, *Diamond Relat. Mater.* 2, 970 (1993).

421. T. Ikeda, Y. Kawate, and Y. Hirai, *J. Vac. Sci. Technol. A* 8, 3168 (1990).

422. M. Murakawa and S. Watanabe, *Surf. Coat. Technol.* 43/44, 128 (1990).

423. M. Lu, A. Bousetta, R. Sukach, A. Bensaoula, K. Walters, K. Eipers-Smith, and A. Schultz, *Appl. Phys. Lett.* 64, 1514 (1994).

424. S. Mineta, M. Kolrata, N. Yasunaga, and Y. Kikuta, *Thin Solid Films* 189, 125 (1990).

425. T. A. Friedmann, P. B. Mirkarimi, D. L. Medlin, K. F. McCarty, E. J. Klaus, D. Boehme, H. A. Johnsen, M. J. Mills, and D. K. Ottesen, *J. Appl. Phys.* 76, 3088 (1994).

426. S. Turan and K. M. Knowles, *Phys. Status Solidi* 150, 227 (1995).

427. A.K. Ballal, L. Salamanca-Riba, G. L. Doll, C. A. Taylor II, and R. Clarke, *J. Mater. Res.* 7, 1618 (1992).

428. C. R. Aita, *In* "Synthesis and Properties of Boron Nitride" (J. J. Pouch, S. A. Alterovitz, Eds.), Materials Science Forum, Vol. 54/55. Trans Tech, Brookfield, 1990.

429. R. Ganzetti and W. Gissler, *Mater. Manuf. Processes* 9, 507 (1994).

430. J. Hahn, M. Friedrich, R. Pintaske, M. Schaller, N. Kahl, D. R. T. Zahn, and F. Richter, *Diamond Relat. Mater.* 5, 1103 (1996).

431. M. Friedrich, J. Hahn, S. Laufer, F. Richter, H. J. Hinneberg, and D. R. T. Zahn, *In* "Diamond Materials IV Electrochemical Society Proc." (K. V. Ravi, K. E. Spear, J. L. Davidson, R. H. Hauge, and J. P. Dismukes, Eds.), Vol. 94/95. The Electrochemical Society, Pennington, NJ, 1995.

432. V. Y. Kulikovsky, L. R. Shaginyan, V. M. Vereschaka, and N. G. Hatyenko, *Diamond Relat. Mater.* 4, 113 (1995).

433. M. Mieno and T. Yoshida, *Surf. Coat. Technol.* 52, 87 (1992).

434. S. Kidner, C. A. Taylor II, and R. Clarke, *Appl. Phys. Lett.* 64, 1859 (1994).

435. L. F. Allard, A. K. Datye, T. A. Nolan, S. L. Mahan, and R. T. Paine, *Ultramicroscopy* 37, 153 (1991).

436. S. Nishiyama, N. Kuratani, A. Ebe, and K. Ogata, *Nucl. Instrum. Meth. B* 80/81, 1485 (1993).

437. T. Ichiki, T. Momose, and T. Yoshida, *J. Appl. Phys.* 75, 1330 (1994).

438. M. Okamoto, Y. Utsumi, and Y. Osaka, *Plasma Sources Sci. Technol.* 2, 1 (1993).

439. W. Dworschak, K. Jung, and H. Ehrhardt, *Thin Solid Films* 245, 65 (1995).

440. A. Chayahara, H. Yokoyama, T. Imura, and Y. Osaka, *Appl. Surf. Sci.* 33/34, 561 (1988).

441. M. Okamoto, H. Yokoyama, and Y. Osaka, *Jpn. J. Appl. Phys.* 29, 930 (1990).

442. K. J. Lao and W. L. Wang, *Phys. Stat. Sol.* 147, K9 (1995).

443. H. Saitoh, T. Hirose, H. Matsui, Y. Hirotsu, and Y. Ichinose, *Surf. Coat. Technol.* 39/40, 265 (1989).

444. H. Saitoh and W. A. Yarbrough, *Appl. Phys. Lett.* 58, 2230 (1991).

445. S. Reinke, M. Kuhr and W. Kulisch, *Surf. Coat. Technol.* 74/75, 806 (1995).

446. A. Weber, U. Bringmann, R. Nikulski, and C. P. Klages, *Diamond. Relat. Mater.* 2, 201 (1993).

447. H. Hofsass, H. Feldermann, M. Sebastian, and C. Ronning, *Phys. Rev. B* 55, 13230 (1997).

448. H. Hofsass, C. Ronning, U. Griesmeier, M. Gross, S. Reinke, and M. Kuhr, *Appl. Phys. Lett.* 67, 46 (1995).

449. T. Ichiki, S. Amagi, and T. Yoshida, *J. Appl. Phys.* 79, 4381(1996).

450. M. F. Plass, W. Fukarek, A. Kolitsch, M. Mader, and W. Moller, *Phys. Stat. Sol. A* 155, K1 (1996).

451. D. H. Berns and M. A. Capelli, *Appl. Phys. Lett.* 68, 2711 (1996).

452. G. Krannich, F. Richter, J. Hahn, R. Pintaske, V. B. Filippov, and Y. Paderno, *Diamond Relat. Mater.* 6, 1005 (1997).

453. G. Krannich, F. Richter, J. Hahn, R. Pintaske, M. Friedrich, S. Schmidbauer, and D. R. T. Zahn, *Surf. Coat. Technol.* 90, 178 (1996).

454. S. Manorama, G. N. Chaudhari, and V. J. Rao, *J. Phys. D* 26, 1793 (1993).

455. A. R. Phani, S. Roy, and V. J. Rao, *Thin Solid Films* 258, 21 (1995).

456. P. B. Mirkarimi, D. L. Medlin, T. F. Friedmann, M. J. Mills, and K. F. McCarty, *Phys. Rev. B* 50, 7884 (1994).

457. P. B. Mirkarimi, D. L. Medlin, T. F. Friedmann, M. J. Mills, and K. F. McCarty, *Phys. Rev. B* 50, 8907 (1994).

458. C. A. Davis, K. M. Knowles, and G. A. J. Araratunga, *Surf. Coat. Technol.* 76, 316 (1995).

459. D. A. Muller, Y. Tzou, R. Rah, and J. Silcox, *Nature* 366, 725 (1993).

460. H. Yokoyama, M. Okamoto, and Y. Osaka, *Jpn. J. Appl. Phys.* 30, 344 (1991).

461. T. Ikeda, *Appl. Phys. Lett.* 61, 786 (1992).

462. N. Tanabe, T. Hayashi, and M. Iwaki, *Diamond Relat. Mater.* 1, 151 (1992).

463. P. B. Mirkarimi, D. L. Medlin, T. F. Friedmann, W. G. Wolfer, E. J. Klaus, G. F. Cardinale, D. G. Howitt, and K. F. McCarty, *J. Mater. Res.* 9, 2925 (1994).

464. S. Reinke, M. Kuhr, and W. Kulisch, *Surf. Coat. Technol.* 74/75, 723 (1995).

465. G. Rene, D. Bouchier, S. Ilias, M. A. Djouadi, J. Pascallon, V. Stambouli, P. Moller, and G. Hug, *Diamond Relat. Mater.* 5, 530 (1996).

466. D. Litvinov, C. A. Taylor II, and R. Clarke, *Diamond Relat. Mater.* 7, 360 (1998).

467. D. R. McKenzie, W. D. McFall, H. Smith, B. Higgins, R. W. Boswell, A. Durandet, B. W. James, and L. S. Falconer, *Nucl. Instrum. Meth. B* 106, 90 (1995).

468. J. Hahn, F. Richter, R. Pintaske, M. Roder, E. Schneider, and T. Welzel, *Surf. Coat. Technol.* 92, 129 (1996).

469. P. B. Mirkarimi, D. L. Medlin, J. C. Barbour, andK. F. McCarty, *Appl. Phys. Lett.* 66, 2813 (1995).

470. D. J. Kester, K. S. Ailey, D. J. Lichtenwaler, and R. F. Davis, *J. Vac. Sci. Technol. A* 12, 3074 (1994).

471. P. X. Yan, S. Z. Yang, B. Li, and X. S. Chen, *J. Cryst. Growth* 148, 232 (1995).

472. W. L. Lin, Z. Xia, Y. L. Liu, and Y .C. Fen, *Mater. Sci Eng. B* 7, 107 (1990).

473. J. J. Cuomo, J. P. Doyle, J. Bruley, and J. C. Liu, *Appl. Phys. Lett.* 58, 466 (1991).

474. K. F. McCarty, P. B. Mirkarimi, D. L. Medlin, and T. F. Friedmann, *Diamond Relat. Mater* 5, 1519 (1996).

475. S. Reinke, M. Kuhr, and W. Kulisch, *Diamond Relat. Mater.* 3, 341 (1994).

476. S. Reinke, M. Kuhr, W. Kulisch, and R. Kassing, *Diamond Relat. Mater.* 4, 272 (1995).

477. J. Robertson, *Diamond Relat. Mater.* 5, 519 (1996).

478. J. Malherbe, *Crit. Rev. Sol. State* 19, 55 (1994).

479. C. Weissmantel, K. Bewilogua, D. Dietrich, H. J. Hinneberg, S. Klose, W. Nowick, and G. Reisse, *Thin Solid Films* 72, 19 (1980).

480. S. Reinke, M. Kuhr, and W. Kulisch, *Diamond Relat. Mater.* 5, 508 (1996).

481. D. R. McKenzie, D. J. H. Cockayne, D. A. Muller, M. Murakawa, S. Miyake, S. Wantanabe, and P. Fallon, *J. Appl. Phys.* 70, 3007 (1991).

482. P. B. Mirkarimi, K. F. McCarty, G. F. Cardinale, D. L. Medlin, D. K. Ottensen, and H. A. Johnsen, *J. Vac. Sci. Technol. A* 14, 251 (1996).

483. Y. Lifshitz, S. R. Kasi, and J. W. Rabelais, *Phys. Rev. Lett.* 62, 1290 (1987).

484. J. Robertson, *Diamond Relat. Mater.* 3, 984 (1993).

485. C. A. Davis, *Thin Solid Films* 226, 30 (1993).

486. M. Nastasi and J. W. Mayer, *Mater. Sci. Rep.* 6, 1 (1991).

487. C. Collazo-Davila, E. Bengu, L. D. Marks, and M. Kirk, *Diamond Relat. Mater.* 8, 1091 (1999).

488. D. R. McKenzie, W. G. Sainty, and D. Green, *In* "Synthesis and Properties of Boron Nitride" (J. J. Pouch and S. A. Alterovitz, Eds.), Materials Science Forum, Vol. 54/55. Trans Tech Publications, Brookfield, 1990.

489. D. M. Back, *In* "Physics of Thin Films," Vol. 15, p. 287. Academic Press, Boston, 1991.

490. S. Jager, K. Bewilogua, and C. P. Klages, *Thin Solid Films* 245, 50 (1994).

491. S. Fahy, *Phys. Rev B* 51, 12873 (1995).

492. S. Fahy, C. A. Taylor II, and R. Clarke, *Phys. Rev B* (1997).

493. M. Terauchi, M. Tanaka, K. Suzuki, A. Ogino, and K. Kimura, *Chem. Phys. Lett.* 324, 359 (2000).

494. D. J. Smith, W. O. Saxton, M. A. O'Keefe, G. J. Wood, and W. M. Stobbs, *Ultramicroscopy* 11, 263 (1983).

495. P. B. Mirkarimi, D. L. Medlin, P. Rez, T. F. Friedmann, M. J. Mills, and K. F. McCarty, *J. Appl. Phys.* 76, 295 (1994).

496. M. P. Johansson, L. Hultman, S. Daaud, K. Bewilogua, H. Luthje, A. Schutze, S. Kouptsidas, and G. S. A. M. Theunissen, *Thin Solid Films* 287, 193 (1996).

497. D. J. Kester, K. S. Ailey, and R. F. Davis, *Diamond Relat. Mater.* 3, 332 (1994).

498. H. Hofsass, C. Ronning, U. Griesmeier, M. Gross, S. Reinke, M. Kuhr, J. Zweck, and R. Fischer, *Nucl. Instrum. Meth. B* 106, 153 (1990).

499. D. G. Howitt, P. B. Mirkarimi, D. L. Medlin, K. F. McCarty, E. J. Klaus, G. F. Cardinale, and W. M. Clift, *Thin Solid Films* 253, 130 (1994).

500. A. K. Ballal, L. Salamanca-Riba, C. A. Taylor II, and G. L. Doll, *Thin Solid Films* 224, 46 (1993).

501. D. G. Rickerby, P. N. Gibson, W. Gissler, and J. Haupt, *Thin Solid Films* 209, 155 (1992).

502. W. L. Zhou, Y. Ikuhara, and T. Suzuki, *Appl. Phys. Lett.* 67, 3551 (1995).

503. H. K. Schmid, *Microsc. Microanal. Microstruct.* 6, 99 (1995).

504. Y. Bando, K. Kurashima, and S. Nakano, *J. European Ceramic Soc.* 16, 379 (1996).

505. M. R. McGurk and T. F. Page, *J. Mater. Res.* 14, 2283 (1999).

506. W. C. Oliver and G. M. Pharr, *J. Mater. Res.* 7, 1564 (1992).

507. P. M. Sargent, *In* "Microindentation Techniques in Science and Engineering" (P. Blau and B. Lawn, Eds.), p. 160. ASTM, Ann Arbor, 1995.

508. D. J. Kester, K. S. Ailey, R. F. Davis, and K. L. More, *J. Mater. Res.* 8, 1213 (1993).

509. D. L. Medlin, T. A. Friedmann, P. B. Mirkarimi, P. Rez, K. F. McCarty, and M. J. Mills, *J. Appl. Phys.* 76, 295 (1994).

510. D. R. McKenzie, W. D. McFall, W. G. Sainty, C. A. Davis, and R. E. Collins, *Diamond Relat. Mater.* 2, 970 (1993).

511. G. F.Cardinale, D. G. Howitt, K. F. McCarty, D. L. Medlin, P. B. Mirkarimi, and N. R. Moody, *Diamond Relat. Mater.* 5, 1295 (1996)

512. Y. Andoh, S. Nishiyama, H. Kirimura, T. Mikami, K.Ogata, and F. Fujimoto, *Nucl. Instrum. Methods B* 59/60, 276 (1991).

513. S. Nishiyama, N. Kuratani, A. Ebe, and K. Ogata, *Nucl. Instrum. Methods B* 80/81, 1485 (1993).

514. W. L. Lin, Z. Xia, Y. L. Liu and Y. C. Fen, *Mater. Sci Eng. B* 7, 107 (1990).

515. T. Wada and N. Yamashita, *J. Vac. Sci. Technol. A* 10, 515 (1992).

516. K. Inagawa, K. Watanabe, K. Saitoh, Y. Yuchi, and A. Itoh, *Surf. Coat. Technol.* 39/40, 253 (1989).

517. P. B. Mirkarimi, D. L. Medlin, K. F. McCarty, D. C. Dibble, and W. M. Clift, *J. Appl. Phys.* 82, 1617 (1997).

518. H. Luthje, K. Bewilogua, S. Daaud, M. Johansson, and L. Hultman, *Thin Solid Films* 257, 40 (1995).

519. M. P. Johansson, H. Sjostrom, and L. Hultman, *Vacuum* 53, 441 (1999).

520. M. Murakawa, S. Watanabe, W. L. Zhou, Y. Ikuhara, and T. Suzuki, *Appl. Phys. Lett.* 66, 2490 (1995).

521. M. Murakawa and S. Watanabe, *Surf. Coat. Technol.* 43/44, 145 (1990).

522. H. Kohzuki, Y. Okuno, and M. Motoyama, *Diamond Films Technol.* 5, 95 (1995).

523. P. B. Mirkarimi, unpublished work, (1996).

524. I. Konyashin, J. Bill, and F. Aldinger, *Chem. Vapor Deposition* 3, 239 (1997).

525. W. J. Zhang and S. Matsumoto, *Chem. Phys. Lett.* 330, 243 (2000).

526. T. W. Scharf, R. D Ott, D. Yang, and J. A. Baranard, *J. Appl. Phys.* 85, 3142 (1999).

527. E. C. Cutiongco, D. Li, Y. C. Chung, and C. S. Bhatia, *J. Tribol.* 118, 543 (1996).

528. G. Jungnickel, P. K. Stich, and T. Frauenheim, *Phys. Rev. B* 57, R661 (1998).

529. K. Ogata, J. F. D. Chubaci, and F. Fujimoto, *J. Appl. Phys.* 76, 3791 (1994).

530. J. Robertson and C. A. Davis, *Diamond Relat.* 4, 441 (1995).

531. Z. J. Zang, J. Hung, S. Fan, and C. M. Lieber, *Mater. Sci. Eng. A* 209 5 (1996).

532. Z. J. Zang, J. Hung, S. Fan, and C. M. Lieber, *Appl. Phys. Lett.* 68, 2639 (1996).

533. V. Hajek, K. Rusnaka, J. Vlacek, L. Martinu, and H. M. Hawthorne, *Wear* 213, 80 (1997).

534. E. Murdock, R. Simmons, and R. Davidson, *IEEE Trans. Magn.* 28, 3078 (1992).

535. T. W. Scharf, R. D. Ott, D. Yang, and J. A. Baranard, *J. Appl. Phys.* 85, 3142 (1999).

536. A. M. Liu and M. L. Cohen, *Science* 245, 841 (1989).

537. A. M. Liu and M. L. Cohen, *Phys. Rev. B* 41, 10727 (1990).

538. A. M. Liu and R. M. Wentzcovitch, *Phys. Rev. B* 50, 10362 (1990).

539. T. Malkow, *Mater. Sci. Eng. A* 292, 112 (2000).

540. D. M. Teter, *MRS Bull.* 23, 22 (1998).

541. M. R. Wixom, *J. Am. Ceram.* 73, 1973 (1990).

542. M. B. Guseva et al., *Diamond Relat. Mater.* 6, 640 (1997).

543. M. R. Wixom, *J. Am. Ceram. Soc* 73 , 1973 (1990).

544. J. H. Nguyen and R. Jeanolz, *Mater. Sci. Eng. A* 209, 23 (1996).

545. D. M. Teter, and R. J. Hemley, *Science* 271, 53 (1996).

546. V. P. Dymont, E. M. Nekrashevich, and I. M. Starchenko, *Solid State Commun.* 111, 443 (1999).

547. A. Hoffman et al. *Surf. Coat. Technol.* 68/69, 616 (1995).

548. K. J. Boyd et al. *J. Vas. Sci. Technol. A* 13, 2110 (1995).

549. S. Veprek, J, Weidmann and F. Glatz, *J. Vas. Sci. Technol. A* 13, 2914 (1995).

550. Tyan-Ywan Yen and Chang-Pin Chou, *Appl. Phys. Lett.* 67, 2801 (1995).

551. T.-Y. Yen and C.-P. Chou, *Appl. Phys. Lett.* 67, 2801 (1995).

552. C. Spaeth, M. Kuhn, U. Kreissig, and F. Richter, *Diamond Relat. Mater.* 6, 626 (1997).

553. D. G. McCulloch and A. R. Merchant, *Thin Solid Films* 290-291, 99 (1996).

554. K. Ogata, J. F. Diniz Chubaci and F. Fujimoto, *J. Appl. Phys.* 76, 3791 (1994).

555. P. Hammer, W. Gissler, *Diamond Relat. Mater.* 5, 1152 (1996).

556. S. S. Todorov et al., *J. Vac. Sci. Technol. A* 12, 3192 (1994).

557. M. Kohzaki et al., *Jpn. J. Appl. Phys.* 362313 (1997).

558. M. Kohzaki et al., *Thin Solid Films* 308/309, 239 (1997).

559. H. W. Lu, X. R. Zou, J. Q. Xie, and J. Y. Feng, *J. Phys. D* 31, 363 (1998).

560. S. Veprek, J. Weidmann, and F. Glatz, *J. Vac. Sci. Technol. A* 13, 2914 (1995).

561. S. A. Korenev et al., *Thin Solid Films* 308/309, 233 (1997).

562. S. Muhl and J. M. Mendez, *Diamond Relat. Mater.* 8, 1809 (1999)

563. J. Bulir et al., *Thin Solid Films* 292, 318 (1997).

564. R. Gonzalez et al., *Appl. Surf. Sci.* 109/110, 380 (1997)

565. X.-A. Zhao et al., *Appl. Phys. Lett.* 66, 2652 (1995)

566. C. W. Ong et al., *Thin Solid Films* 280, 1 (1996).

567. Z. J. Zhang, S. Fan, J. Huang, and C. M. Lieber, *Appl. Phys. Lett.* 68, 3582 (1995)

568. E. Aldea et al., *Jpn. J. Appl. Phys.* 36, 4686 (1997).

569. S. Acquaviva et al., *Appl. Surf. Sci.* 109/110, 408 (1997).

570. Z. M. Ren et al., *Phys. Rev. B* 51, 5274 (1995).

571. D. J. Johnson, Y. Chen, Y. He, and R. H. Prince, *Diamond Relat Mater.* 6, 1799 (1997).

572. A. A. Voevodin and M. S. Donley, *Surf. Coat. Technol.* 82, 199 (1996).

573. P. Merel et al., *Appl. Phys. Lett.* 71, 3814 (1997).

574. J. Hu, P. Yang, and C. M. Lieber, *Phys. Rev. B* 57, 3185 (1998).

575. A. K. Sharma et al., *Appl. Phys. Lett.* 69, 3489 (1996).

576. S. A. Uglov, V. E. Shub, A. A. Beloglazov, and V. I. Konov, *Appl. Surf. Sci.* 92, 656 (1996).

577. Y. Chen, L. Guo, F. Chen, and E. G. Wang, *J. Phys. Condens. Matter* 8, L685 (1996).

578. (a) C. Niu, Y. Z. Lu, and C. M. Lieber, *Science* 261, 334 (1993).
(b) N. Xu et al., *J. Phys. D* 30, 1370 (1997).
(c) A. Kumar, U. Ekanayake, and J. S. Kapat, *Surf. Coat. Technol.* 102, 113 (1998).
(d) L. Wan and R. F. Egerton, *Thin Solid Films* 279, 34 (1996).
(e) H. Sjostrom et al., *J. Mater. Res.* 11, 981 (1996).

579. R. Kurt, R. Sanjines, A. Karimi, and F. Levy, *Diamond Relat. Mater.* 9, 566 (2000).

580. B. C. Holloway et al., *Thin Solid Films* 290-291, 94 (1996).

581. Kin Man Yu et al., *Phys. Rev. B* 49, 5034 (1994).

582. H. Sjostrom et al., *J. Mater. Res.* 11, 981 (1996)

583. P. Hammer and W. Gissler, *Diamond Relat. Mat er.* 5, 1152 (1996)

584. N. Axen et al., *Surf and Coat. Tech.* 81, 262 (1996)

585. P. Wood, T. Wydeven and O. Tsuji, *Thin Solid Films* 258, 151 (1995)

586. F. Lfreire Jr, and D. F. Franceschini, *Thin Solid Films* 293, 236 (1997)

587. M. M. Lacerda et al., *Diamond Relat. Mater.* 6, 631 (1997)

588. Y. Zhang, Z. Zhou, and H. Li, *Appl. Phys. Lett.* 68, 634 (1996)

589. H. K. Woo et al., *Diamond Relat. Mater.* 6, 635 (1997)

590. P. H. Fang, *Appl. Phys. Lett.* 69, 136 (1996).

591. J. Sun, Y. Zhang, X. He, W. Liu, C. S. Lee, and S. T. Lee, *Mater. Lett.* 38, 98 (1999).

592. V. P. Novikov and V. P. Dymont, *Appl. Phys. Lett.* 70, 200 (1997).

593. Qiang Fu, Chuan-Bao Cao, and He-Sun Zhu, *Chem. Phys. Lett.* 314, 223 (1999).

594. K. H. Park, B. C. Kim, and H. Kang, *J. Chem. Phys.* 97, 2742 (1992).

595. A. R. Merchant, D. G. McCulloch, D. R. McKenzie, Y. Yin, L. Hall, and E. G. Gerstner, *J. Appl. Phys.* 79, 6914 (1996).

596. F. Link, H. Baumann, A. Markwitz, E. F. Krimmel, and K. Bethge, *Nucl. Instrum. Methods B* 113, 235 (1996).

597. D. M. Bhusari, C. K. Chen, K. H. Chen, T. J. Chuang, L. C. Chen, and M. C. Lin, *J. Mater. Res.* 12, 322 (1997).

598. K. Takahiro, H. Habazaki, S. Nagata, M. Kishimoto, S. Yamaguchi, F. Nishiyama, and S. Nimori, *Nucl. Instrum. Methods B* 152, 301 (1999).

599. J. Martin-Gi, M. Sarikaya, M. Jose-Yacaman, and A. Rubio, *J. Appl. Phys.* 81, 2555 (1997).

600. X. R. Zou, H. W. Lu, J. Q. Xie, and J. Y. Feng, *Thin Solid Films* 345, 208 (1999).

601. D. Marton, K. J. Boyd, A. H. Al-Bayati, S. S. Todorov, and J. W. Rabalais, *Phys. Rev. Lett.* 73, 118 (1994).

602. I. Bertoti, A. Mohai, A. Toth, and B. Zelei, *Nucl. Instrum. Methods B* 148, 645 (1999).

603. J. Wang, S. F. Durrant, and M. A. B. Moraes, *J. Non-Crytst. Solids* 262, 216 (2000).

604. M. Balaceanu, E. Grigore, F. Truica-Marasescu, D. Pantelica, F. Negoita, G. Pavelescu, and F. Ionescu, *Nucl. Instrum. Methods B* 161–163, 1002 (2000).

605. M. Bai, K. Kato, N. Umehara, Y. Miyake, J. Xu, and H. Tokisue, *This Solid Films* 376, 170 (2000).

606. N. Tajima, H. Saze, H. Sugimura, and O. Taki, *Vacuum* 59 567 (2000).

607. K. Suenaga, M. Yudasaka, C. Colliex, S. Iijima, *Chem. Phys. Lett.* 316, 365 (2000).

608. F.-R. Weber and H. Oechsner, *Thin Solid Films* 355–356, 73 (1999).

609. C. Popov, M. F. Plass, R. Kassing, W. Kulisch, *Thin Solid Films* 355–356, 406 (1999).

610. D. Y. Lee, Y. H. Kim, I. K. Kim, and H. K. Buik, *Thin Solid Films* 355–356, 239 (1999).

611. C. Niu, Y. z. Lu and C. M. Lierber, *Science* 261, 334 (1993).

612. K. M. Yu, M. L. Cohen, E. E. Haller, W. L. Hansen, A. Y. Liy and I. C. Wu, *Phys. Rev. B* 49, 5034 (1994).

613. M. Y. Chen, X. Lin, V. P. Dravid, Y. W. Chung, M. S. Wong, and W. D. Sproul, *Surf. Coat. Tech.* 54/55, 360 (1992).

614. C. Sandorfy, *In* "The Chemistry of Functional Groups" (S. Patai, Ed.). The Chemistry of the Carbon-Nitrogen Double Bonds, Interscience Publishers, London, 1970.

615. L. J. Bellamy, "The Infrared Spectra of Complex Molecules." Chapman and Hall, London, 1975.

616. J . H. Kauffman, S. Metin, and D. D. Saperstein, *Phys. Rev. B* 39, 3053 (1989).

617. J. Widany et al., *Diamond Relat. Mater.* 5, 1031 (1996).

618. S. Matsumoto, E.-Qxie, and F. Izumi, *Diamond Relat. Mater.* 8, 1175 (1999).

619. F. Izumi, Rietveld analysis programs RIETAN and PREMOS and special applications, *In* "The Rietveld Method" (R. Y. Young, Ed.), Chapter 13. Oxford University Press, Oxford, 1995.

620. D. M. Teter, and R. J. Hemley, *Science* 271, 53 (1996).

621. Z. M. Ren, Y. C. Du, Z. F. Ying, F. M. Li, J. Lin, Y. Z. Ren, and X. F. Zong, *Nucl. Instrum. Methods Pkys. Res. B* 117, 249 (1996).

622. L. A. Bursil, J. L. Peng, V. N. Gurarie, A. N. Orlov, and S. Prawer, *J. Mater. Res.* 10, 2277 (1995).

623. D. Li, XW. Lin, S. C. Cheng, V. P. Dravid, Y. W. Chung, M. S. Wong, and W. D. Sproul, *Appl. Phys. Lett.* 68, 1211 (1996).

624. A. Badzian and T. Badzian, *Diamond Relat. Mater.* 5, 1051 (1996).

625. J. H. Nguyen and R. Jeanloz, *Mater. Sci. Eng. A* 209, 23 (1996).

626. T. Werninghaus, D. R. T. Zahn, E. G. Wang, and Y. Chen, *Diamond Relat. Mater.* 7, 52 (1998).

627. M. Kawaguchi, Y. Tokimatsu, K. Nozaki, Y. Kaburagi, and Y. Hishiyama, *Chem. Lett.* 1997, 1003 (1997).

628. Y. S. Jin, Y. Matsuda, and H. Fujiyama, *Jpn J. Appl. Phys.* 37, 4544 (1998).

629. M. L. De Giorgi, G. Leggieri, A. Luches, M. Martino, A. Perrone, A. Zocco, G. Barucca, G. Majni, E. Gyorgy, I. N. Mihailescu, and M. Popescu, *Appl. Surf. Sci.* 27, 481 (1998).

630. C. Niu, Y. Z. Lu, and C. M. Lieber, *Science* 261, 334 (1993).

631. K. M. Yu, M. L. Cohen, E. E. Haller, W. L. Hansen, A. L. Hansen, A. Y. Liu, and I. C. Wu, *Phys. Rev. B* 49, 5034 (1994).

632. J. Narayan, J. Reddy, N. Biunno, S. M. Kanetkar, P. Tiwari, and N. Parikh, *Mater. Sci. Eng. B* 26, 49 (1994).

633. J. Szmidt, A. Werbowy, K. Zdunek, A. Sokolowska, J. Konwerska-Hrabowska, and S. Mitura, *Diamond Relat. Mater.* 5, 564 (1996).

634. X. W. Su, H. W. Song, F. Z. Cui, and W. Z. Li, *J. Phys. Condens. Matter* 7, L517 (1995).

635. T. Y. Yen and C. P. Chou, *Appl. Phys. Lett.* 67, 2801 (1995).

636. T. Y. Yen and C. P. Chou, *Solid State Commun.* 95, 281 (1995).

637. S. Muhl, A. Gaona-Couto, J. M. Me'ndeza, S. Rodil, G. Gonzalez, A. Merkulov, and R. Asomoza, *Thin Solid Films* 308-309, 228 (1997).

638. Y. A. Li, S. Xu, H. S. Li, and W. Y. Luo, *J. Mater. Sci. Lett.* 17, 31 (1998).

639. S. Xu, H. S. Li, Y. A. Li, S. Lee, and C. H. A. Huan, *Chem. Phys. Lett.* 287, 731 (1998).

640. M. A. Monclus, D. C. Cameron, A. K. M. S. Chowdhury, R. Barkley, and M. Collins, *Thin Solid Films* 355–356, 79 (1999).

641. V. Hajek, K. Rusank, J. Vlcek, L. Martinu, and H. M. Hawthorne, *Wear* 213, 80 (1997).

642. Y. Fu, N. L. Loh, J. Wei, B. Yan, and P. Hing, *Wear* 237, 12 (2000).

643. A. K. M. S. Chowdhury, D. C. Cameron, and M. S. J. Hashmi, *Surf. Coat. Technol.* 118–119, 46 (1999).

644. H. Q. Lou, N. Axen, R. E. Somekh, and I. M. Hutchings, *Diamond Relat. Mater.* 5, 1202 (1996).

645. Dong F. Wang, Koji Kato, *J. Tribo.* 33, 115 (2000).

646. K. Kato, H. Koide, and N. Umehara, *Wear* 238, 40 (2000).

647. T. Hayashi, A. Matsumuro, M. Muramatsu, M. Kohzaki, and K. Yamaguchi, *Thin Solid Films* 376, 152 (2000).

648. W. Precht, M. Pancielejko, and A. Czyzniewski, *Vacuum* 53, 109 (1999).

649. J. Wei, P. Hing, and Z. Q. Mo, *Wear* 225–229, 1141 (1999).

650. M. Y. Chen, X. Lin, V. P. Dravid, Y. W. Chung, M. C. Wong, and W. D. Sproul, *Tribol. Trans.* 36, 491 (1993).

651. V. Haek, K. Rusnak, J. Vlcek, L. Martinu, and H. M. Hawthorne, *Wear* 213, 80 (1997).

652. H. Sjbström, S. Stafströni, M. Bohman, and J.-E. Sundgren, *Phys. Rev. Lett.* 75, 1336 (1995).

653. H. Sjbström, L. Hultman, J.-E. Sundgren, S. V. Hainsworth, T. F. Page, and G. S. A. M. Theunissen, *J. Vac. Sci. Technol. A* 14, 56 (1996).

654. V. Hajek, K. Rusnak, J. Vlcek. L. Martinu, and H. M. Hawthorne, *Wear* 213, 80 (1997).

655. A. Khurshudov, K. Kato, and S. Daisuke, *J. Vac. Sci. Technol. A* 14, 2935 (1996).

656. W. D. Sproul from BIRL Industrial Research Laboratory, Northwestern University, Evanston, Illinois, private communication.

657. M. J. Murphy, J. Monaghan, M. Tyrrel, R. Walsh, D. C. Cameron, A. K. M. S. Chowdhury, M. Monclus, and S. J. Hashmi, *J. Vac. Sci. Technol. A* 17, 62 (1999).

658. W.-C. Chan, M.-K. Fung, K.-H. Lai, I. Bello, S.-T. Lee, and C.-S. Lee, *J. Non. Crys. Solids* 254, 180 (1999).

659. M. Gioti, S. Logothetidis, C. Charitidis, and H. Lefakis, *Vacuum* 53, 53 (1999).

660. A. K. M. S. Chowdhury, D. C. Cameron, and M. A. Moclus, *Thin Solid Films* 355–356, 85 (1999).

661. A. K. M. S. Chowdhury, D. C. Cameron, and M. A. Moclus, *Thin Solid Films* 341,94 (1999).

662. A. K. M. S. Chowdhury, D. C. Cameron, M. A. Moclus, and R. Barklie, *Surf. Coat. Technol.* 131, 488 (2000).

663. A. K. M. S. Chowdhury, D. C. Cameron, and M. S. J. Hashmi, *Thin Solid Films* 332, 62 (1999).

664. J. Robertson, *Diamond Relat. Mater.* 2, 984 (1993).

665. C. A. Davis, *Thin Solid Films* 226, 30 (1993).

666. D. R. McKenzie, D. Muller, and B. A. Pailthorpe, *Phys. Rev. Lett.* 67, 773 (1991).

667. H. Sjbström, M. Johansson, T. F. Page, L. R. Wallenberg, S. V. Hainsworth, L. Hultman, J. E. Sundgren, and L. Ivanov, *Thin Solid Films* 246, 103 (1994).

668. V. G. Aleshin, and Yu. N. Kucherenko, *J. Electron Spectrosc.* 8, 411 (1976).

669. V. V. Nemoshkalenko, V. G. Aleshin, and Yu. N. Kucherenko, *Sol. State. Commun.* 20, 1155 (1976).

670. S. R. P. Silva, J. Robertson, G. A. J. Amaratunga et al., *J. Appl. Phys.* 81, 2626 (1997).

671. M. C. dos Santos, and F. Alvarez, *Phys. Rev. B* 58, 13918 (1998).

672. A. K. M. S. Chowdhury, D. C. Cameron, M. S. J. Hashmi, and J. M. Gregg, *J. Mater. Res.* 14, 2359 (1999).

673. J. H. Kaufman, S. Metin, and D. D. Saperstein, *Phys. Rev. B* 39, 13053 (1989).

674. F. Weich, J. Widany, and Th. Frauenheim, *Phys. Rev. Lett.* 78, 3326 (1997).

675. M. Zarrabian, N. Fourches-Coulon, G. Turban, C. Marhic, and M. Lancin, *Appl. Phys. Lett.* 70, 2535 (1997).

676. M. Zarrabian, N. Fourches-Coulon, G. Turban, M. Lancin, C. Marhic, and M. Lancin, *Diamond. Relat. Mater.* 6, 542 (1997).

677. S. Battacharyya, C. Cardinaud, and G. Turban, *J. Appl. Phys.* 83. 4491 (1998).

678. Y. Chen, L. Guo, and G. Wang, *Philos. Maga. Lett.* 75, 155 (1997).

679. Y. Zhang, Z. Zhou, and H. Li, *Appl. Phys. Lett.* 68, 634 (1996).

680. P. Petrov, D. Dimitrov, G. Beshkov, V. Krastev, S. Nemska, and Ch. Georgiev, *Vacuum* 52, 501(1999).

681. Y. S. Gu, Y. P. Zhang, Z. J. Duan, X. R. Chang, Z. Z. Tian, D. X. Shi, L. P. Ma, X. F. Zhang, and L. Yuan, *Mat. Sci. Eng. A* 271, 206 (1999).

682. L. P. Ma, Y. S. Gu, Z. J. Duan, L. Yuan, and S. J. Pang, *Thin Solid Films* 349, 10 (1999).

683. S. F. Lim, A. T. S. Wee, J. Lin, D. H. C. Chua, and C. H. A. Huan, *Chem. Phys. Lett.* 306, 53 (1999).

684. X. W. Liu, C. H. Lin, L. T. Chao, and H. C. Shih, *Mater. Lett.* 44, 304 (2000).

685. X.-C. Xiao, W.-H. jiang, L.-X. Song, J.-F. Tian, and X.-F. Hu, *Chem. Phys. Lett.* 310, 240 (1999).

686. R. Alexandrescu, R. Cireasa, C. S. Cojocaru, A. Crunteanu, I. Morjan, F. Vasiliu, and A. Kumar, *Surf. Eng.* 15, 230 (1999).

687. Z. Zhang, H. Guo, G. Zhong, F. Yu, Q. Xiong, and X. Fan, *Thin Solid Films* 346, 96 (1999).

688. D. X. shi, X. F. Zhang, L. Yuan, Y. S. Gu, Y. P. Zhang. Z. J. Duan, X. R. Chang, Z. Z. Tian, and N. X. Chen, *Appl. Surf. Sci.* 148, 50 (1999).

689. S. Acquaviva, E. D. Anna, M. L. De Giorgi, M. Fernandez, G. Leggieri, A. Luches, A. Zocco, and G. Majni, *Appl. Surf. Sci.* 154–155, 369 (2000).

690. P. Gonzalez, R. Soto, B. Leon, M. Perez-Amor, and T. Szorenyi, *Appl. Surf. Sci.* 154–155, 454 (2000).

691. I. Alexandrou, I. Zergioti, G. A. J. Amaratunga, M. J. F. Healy, C. J. Kiely, P. Hatto, M. Velegrakis, and C. Fotakis, *Mater. Lett.* 39, 97 (1999).

692. X. C. Xiao, Y. W. Li, W. H. Jiang, L. X. Song, and X. F. Hu, *J. Chem Phys. Solids* 61, 915 (2000).

693. S. Acquaviva, A. Perrone, A. Zocco, A. Klini, and C. Fotakis, *Thin Solid Films* 373, 266 (2000).

694. Y. P. Zhang, Y. S. Gu, X. R. Chang, Z. Z. Tian, D. X. Shi, and X. F. Zhang, *Mater. Sci. Eng. B* 78, 11 (2000).

695. L. Y. Feng, R. Z. Min, N. H. Qiao, H. Zi. Feng, D. S. H. Chan, L. T. Seng, C. S. Yin, K. Gamini, C. Geng, and L. Kun, *Appl. Surf. Sci.* 138–139, 494 (2000).

696. K. Yamamoto, Y. Koga, S. Fujiwara, F. Kokai, J. I. Kleiman, and K. K. Kim, *Thin Solid Films* 339, 38 (1999)..

697. S. L. Sung, C. H. Tseng, F. K. Chiang, X. J. Guo, X. W. Liu. and H. C. Shih, *Thin Solid Films* 340, 169 (1999).

698. Y. K. Yap, Y. Mori, S. Kida, T. Aoyama, and T. Sasaki, *J. Cryst. Growth* 198/199, 1028 (1999).

699. Y. Kusano, C. Christou, Z. H. Barber, J. E. Evetts. and I. M. Hutchings, *Thin Solid Films* 355–356, 117 (1999).

700. M. E. Ramsey, E. Poindexter, J. S. Pelt, J. Marin, and S. M. Durbin, *Thin Solid Films* 360, 82 (2000).

Chapter 4

ATR SPECTROSCOPY OF THIN FILMS

Urs Peter Fringeli, Dieter Baurecht, Monira Siam, Gerald Reiter,
Michael Schwarzott

*Institute of Physical Chemistry, University of Vienna, Althanstrasse 14/UZA II,
A-1090 Vienna, Austria*

Thomas Bürgi

*Laboratory of Technical Chemistry, Swiss Federal Institute of Technology, ETH Zentrum,
CH-8092 Zürich, Switzerland*

Peter Brüesch

*ABB Management Ltd., Corporate Research (CRB), CH-5405 Baden-Dattwil, Switzerland
and Departement de Physique, Ecole Polytechnique, Federale de Lausanne (EPFL),
PH-Ecublens, CH-1015 Lausanne, Switzerland*

Contents

1. INTRODUCTION

Since the invention of internal reflection as a spectroscopic tool by Harrick [1] and Fahrenfort [2], many applications have been reported in the meantime. The first general review was given by Harrick in his famous book [3] which was followed by a supplement [4]. Infrared surface techniques are also discussed in Ref. [5]. Further books should be mentioned among the literature. Application of internal reflection spectroscopy in a variety of fields has been reviewed in the book by Mirabella [6], while the book by Urban [7] treats internal reflection spectroscopy applied to polymers. Finally, attention should be drawn to Ref. [8] and for readers interested in biological systems attention should be drawn to the book by Gremlich and Yan [9]. In this chapter,

Handbook of Thin Film Materials, edited by H.S. Nalwa
Volume 2: Characterization and Spectroscopy of Thin Films

ISBN 0-12-512910-6/$35.00

we aim to give a comprehensive introduction to basic theory of attenuated total reflection (ATR) spectroscopy paralleled by experimental examples to enable the reader to choose an adequate approximation for quantitative analysis of the ATR spectra. It is of general interest to use Harrick's *thin film approximation* and the concept of *effective thickness* as long as possible, since working with exact mathematical expressions turns out to be a lavish expenditure. However, it is even so important to be able to check for the limits of approximate analytical methods.

The introduction into basic ATR theory is followed by a discussion of orientation measurements which also results in the means of the calculation of surface concentrations of oriented samples. Two techniques for enhanced background compensation are presented, too. The single-beam-sample-reference (SBSR) technique renders a Fourier transform infrared (FTIR) single-beam spectrometer into a pseudodouble-beam instrument, enabling the measurement of sample and reference data with only little time delay. The second technique deals with modulated excitation (ME) spectroscopy. Each sample that enables periodic external stimulation by a variation of a parameter such as temperature, pressure, concentration, electric field, light flux, etc. can be investigated by ME spectroscopy. Periodic stimulation will result in a periodic response in the infrared from only that part of the system that has been affected by ME. Phase sensitive detection enables a narrow band detection of modulated responses which leads to a high selectivity and a low noise level, i.e., a high signal-to-noise ratio which is 1 or 2 magnitudes better than achieved by conventional difference spectroscopy. Moreover, if the frequency of ME is adapted to the kinetic response of the stimulated system, phase shifts with respect to the stimulation and the amplitude damping give valuable information with respect to relaxation times and reaction schemes.

Heterogeneous catalysis on distinct metal surfaces is shown to be accessible by ATR spectroscopy, too, despite relatively high reflection losses due to high absorption indices of metals. The latter fact, however, requires the application of the exact ATR theory for quantitative analysis.

2. FUNDAMENTAL THEORY

In this section a review and a discussion of the theory for attenuated total reflection (ATR) is given. The starting points are Maxwell's equations and the materials equations for isotropic absorbing media. Electromagnetic waves, refractive index, dielectric constant, and the intensity of plane waves are introduced. A short discussion of the Kramers–Kronig relation is also given. We then consider in detail reflection and refraction in isotropic transparent and absorbing media and we derive the appropriate Fresnel equations and reflectivities for *s*- and *p*-polarized light. Finally, we discuss ATR for two bulk media as well as for stratified media. The formalism is illustrated by considering the system Ge/H$_2$O and the fictitious single-layer system Ge/HCl/H$_2$O. Approximate relations for relatively weakly absorbing systems are worked out which express the

normalized reflectivities in terms of effective thicknesses and absorption coefficients. For general references, the reader is referred to the literature [3, 10–20].

2.1. Plane Electromagnetic Wave in an Absorbing Medium

2.1.1. Maxwell's Equations and Material Relations

Maxwell's equations in an isotropic medium are

$$\text{curl}\,\mathbf{E} = -\frac{\partial \mathbf{B}}{\partial t} \tag{1}$$

$$\text{curl}\,\mathbf{H} = \frac{\partial \mathbf{D}}{\partial t} + \mathbf{j} \tag{2}$$

$$\text{div}\,\mathbf{D} = \rho \tag{3}$$

$$\text{div}\,\mathbf{B} = 0 \tag{4}$$

$\mathbf{E}$ and $\mathbf{D}$ are the electric field and the electric displacement, while $\mathbf{H}$ and $\mathbf{B}$ are the magnetic field and the magnetic induction, respectively. ρ is the charge density and $\mathbf{j}$ is the current density. Equation (1) is Faraday's law of induction. Equation (2) expresses the dependence of the magnetic field on the displacement current density $\partial \mathbf{D}/\partial t$, or rate-of-change of the electric field, and on the conduction current density $\mathbf{j}$, or rate of motion of charge. Equation (3) is the equivalent of Coulomb's law and Eq. (4) states that there are no sources of magnetic fields except currents [13]. The materials equations are relations linking the Maxwell fields $\mathbf{E}$ and $\mathbf{H}$ with the induced dielectric polarization $\mathbf{P}$, magnetization $\mathbf{M}$, and current density $\mathbf{j}$, thereby also defining $\mathbf{D}$ and $\mathbf{B}$. For the time being, we disregard nonlinear optical effects observed in strong electric and magnetic fields. For sufficiently small field strengths, the material equations are linear and in isotropic media these are

$$\mathbf{D} = \varepsilon\mathbf{E} = \varepsilon_0\varepsilon_r\mathbf{E} = \varepsilon_0\mathbf{E} + \mathbf{P} \tag{5}$$

$$\mathbf{B} = \mu\mathbf{H} = \mu_n\mu_r\mathbf{H} = \mu_n(\mathbf{H} + \mathbf{M}) \tag{6}$$

$$\mathbf{j} = \sigma\mathbf{E} \tag{7}$$

The parameters describing the material properties are the real part of the dielectric constant ε_r (relative permittivity), the magnetic susceptibility μ_r (relative permeability), and the electric conductivity σ. In anisotropic media, Eqs. (5)–(7) become more complicated with ε_r, μ_r and σ being tensors rather than scalar quantities. The international system of units (SI) is used throughout this Chapter. Table I shows the definitions of the quantities in the equations together with the appropriate SI units. To the preceding equations, we can add the relations,

$$\varepsilon = \varepsilon_0\varepsilon_r \tag{8}$$

$$\mu = \mu_0\mu_r \tag{9}$$

$$\varepsilon_0 = \frac{1}{\mu_0 c^2} \tag{10}$$

where ε_0 and μ_0 are the permittivity and the permeability of free space, respectively, and c is the velocity of light in vacuum (Table II).

Table I.   Definitions and SI Units of Important Quantities

Symbol	Physical Quantity	SI Unit	Symbol for SI Unit
E	Electric field	Volts per meter	V m^{-1}
D	Electric displacement	Coulombs per square meter	C m^{-2}
H	Magnetic field	Amperes per meter	A m^{-1}
B	Magnetic induction	Tesla	T
j	Electric current density	Amperes per square meter	A m^{-2}
ρ	Electric charge density	Coulomb per cubic meter	C m^{-3}
σ	Electric conductivity	Siemens per meter	S m^{-1}
ε	Permittivity	Farads per meter	F m^{-1}
μ	Permeability	Henries per meter	H m^{-1}
P	Polarization	Coulombs per square meter	C m^{-2}
M	Magnetization	Amperes per meter	A m^{-1}

Table II.   Fundamental Constants c, ε_0, and μ_0

Symbol	Physical Quantity	Value
c	Speed of light in vacuum	$2.997925 \cdot 10^8$ m s^{-1}
ε_0	Permittivity of vacuum	$8.8541853 \cdot 10^{-12}$ F m^{-1}
μ_0	Permeability of vacuum	$4\pi \cdot 10^{-7}$ H m^{-1}

In the following, we assume that there is no charge density in the medium, i.e., $\rho = 0$, and from Eq. (3) it follows,

$$\text{div } \mathbf{D} = 0 \tag{11}$$

In addition, for isotropic media, there is no spatial variation in ε_r. Thus,

$$\text{div } \mathbf{E} = 0 \tag{12}$$

From Eqs. (1) and (2), it is possible to eliminate **E** or **H**. Forming the curl of Eq. (1), differentiating Eq. (2) with respect to t, using the vector identity,

$$\text{curl curl } \mathbf{E} = \text{grad div } \mathbf{E} - \nabla^2 \mathbf{E} \tag{13}$$

as well as Eqs. (5) and (6) yields

$$\nabla^2 \mathbf{E} = \varepsilon\mu \frac{\partial^2 \mathbf{E}}{\partial t^2} + \mu\sigma \frac{\partial \mathbf{E}}{\partial t} \tag{14}$$

In a similar way, one obtains

$$\nabla^2 \mathbf{H} = \varepsilon\mu \frac{\partial^2 \mathbf{H}}{\partial t^2} + \mu\sigma \frac{\partial \mathbf{H}}{\partial t} \tag{15}$$

2.1.2. *Plane Waves, Complex Refractive Index, and Dielectric Constant*

We look for a solution of Eq. (14) in the form of a plane-polarized, plane harmonic wave, and we choose the complex form of this wave, the physical meaning being associated with the real part of the expression. For a wave propagating along the x axis with velocity v, we write

$$\mathbf{E} = \mathbf{E}_0 e^{i\omega(x/v-t)} \tag{16}$$

Substituting Eq. (16) into Eq. (14) gives

$$\omega^2/v^2 = \omega^2\varepsilon\mu + i\omega\mu\sigma \tag{17}$$

This equation shows that in a conducting medium ($\sigma \neq 0$), the velocity v is a complex quantity. In a vacuum, we have $\sigma = 0$, $\varepsilon = \varepsilon_0$, $\mu = \mu_0$, and $v = c$, resulting in

$$c = (\mu_0\varepsilon_0)^{-1/2} \tag{18}$$

which is identical to Eq. (10).

Multiplying Eq. (17) with $c^2 = 1/\varepsilon_0\mu_0$ and dividing through by ω^2, one obtains

$$\frac{c^2}{v^2} = \varepsilon_r\mu_r + i\frac{\mu_r\sigma}{\omega\varepsilon_0} \tag{19}$$

where c/v is clearly a dimensionless parameter of the medium, which we denote by $\hat{n}$,

$$\hat{n}^2 = \varepsilon_r\mu_r + i\frac{\mu_r\sigma}{\omega\varepsilon_0} \tag{20}$$

This implies that $\hat{n}$ is of the form,

$$\hat{n} = \frac{c}{v} = n + ik \tag{21}$$

There are two possible values of $\hat{n}$ from Eq. (20), but for physical reasons we choose that which gives a positive value of n. $\hat{n}$ is known as the complex refractive index, n as the real part of the refractive index (or often simply as the refractive index, because $\hat{n}$ is real in an ideal dielectric material) and k is known as the extinction coefficient. In the literature, the so-called attenuation index κ is often used which is related to k by $k = n\kappa$ or $\hat{n} = n(1 + i\kappa)$. In the latter equation, the minus sign is often used which corresponds to the choice of $i\omega(t - x/v)$ in the exponent of Eq. (16) for the electric field.

At this point, we introduce the complex dielectric constant $\hat{\varepsilon}$, defined by

$$\hat{\varepsilon} = \varepsilon' + i\varepsilon'' = \frac{\hat{n}^2}{\mu_r} \qquad (22)$$

From Eqs. (20) and (21), it follows

$$\varepsilon' = \varepsilon_r = \frac{n^2 - k^2}{\mu_r} \qquad (23)$$

$$\varepsilon'' = \frac{\sigma}{\omega\varepsilon_0} = \frac{2nk}{\mu_r} \qquad (24)$$

For nonmagnetic materials $\mu_r = 1$ and solving Eqs. (23) and (24) for n and k gives

$$n = \frac{1}{\sqrt{2}} \left(|\varepsilon| + \varepsilon' \right)^{1/2} \qquad (25)$$

$$k = \frac{1}{\sqrt{2}} \left(|\varepsilon| + \varepsilon' \right)^{1/2} \qquad (26)$$

where

$$|\varepsilon| = \left(\varepsilon'^2 + \varepsilon''^2 \right)^{1/2} \qquad (27)$$

is the modulus of ε.

Using Eq. (21), the electric field given by Eq. (16) can now be written in the form,

$$\mathbf{E} = \mathbf{E}_0 \, e^{-(2\pi k/\lambda)} \, e^{i[(2\pi n/\lambda)x - \omega t]} \qquad (28)$$

where we have introduced the wavelength λ in free space, $\lambda = 2\pi c/\omega$. An analogous expression holds for $\mathbf{H}$. From Eq. (28), the significance of k emerges as being a measure of the absorption in the medium: the distance $\lambda/2\pi k$ is that in which the amplitude of the wave falls to $1/e$ of its initial value. Equation (28) represents a plane-polarized wave propagating along the x axis. For a similar wave propargating in a direction given by the direction cosines (α, β, γ), the expression becomes

$$\mathbf{E} = \mathbf{E}_0 \, e^{i[(2\pi/\lambda)\hat{n}(\alpha x + \beta y + \gamma z) - \omega t]} \qquad (29)$$

where $\alpha = \cos(\mathbf{s}, \mathbf{e}_x)$, $\beta = \cos(\mathbf{s}, \mathbf{e}_y)$, $\gamma = \cos(\mathbf{s}, \mathbf{e}_z)$ and

$$\mathbf{s} = \alpha\mathbf{e}_x + \beta\mathbf{e}_y + \gamma\mathbf{e}_z = (\alpha, \beta, \gamma) \qquad (30)$$

is a unit vector in the direction of propagation, and $\mathbf{e}_x$, $\mathbf{e}_y$, $\mathbf{e}_z$ are unit vectors along the x, y, and z axes, respectively. Defining the vector $\mathbf{r} = (x, y, z)$ and the complex wave vector,

$$\mathbf{q} = \frac{2\pi}{\lambda}\hat{n}\mathbf{s} = \mathbf{q}' + i\mathbf{q}'' \qquad (31)$$

where

$$\mathbf{q}' = |\mathbf{q}'| = \frac{2\pi}{\lambda}n = \frac{\omega}{c}n, \qquad \mathbf{q}'' = |\mathbf{q}''| = \frac{2\pi}{\lambda}k = \frac{\omega}{c}k \quad (32)$$

and λ is the vacuum wavelength, the electric field can be written in the form,

$$\mathbf{E} = \mathbf{E}_0 e^{i[\mathbf{q}\mathbf{r} - \omega t]} \qquad (33)$$

From Eqs. (2), (5), (7), (10), and (28), together with Eq. (20) it follows,

$$\text{curl } \mathbf{H} = (\sigma - i\omega\varepsilon_0\varepsilon_r)\mathbf{E} = -i\omega\frac{\hat{n}^2}{c^2\mu_0\mu_r}\mathbf{E} \qquad (34)$$

On the other hand, using a plane-wave representation for $\mathbf{H}$ analogous to Eq. (28) one finds

$$\text{curl } \mathbf{H} = \begin{vmatrix} \mathbf{e}_x & \mathbf{e}_y & \mathbf{e}_z \\ \partial/\partial x & \partial/\partial y & \partial/\partial z \\ H_x & H_y & H_z \end{vmatrix} = i\omega\frac{\hat{n}}{c}(\mathbf{s} \times \mathbf{H}) \qquad (35)$$

Comparison of Eqs. (34) and (35) gives

$$\mathbf{s} \times \mathbf{H} = -\frac{\hat{n}}{c\mu_0\mu_r}\mathbf{E} \qquad (36)$$

and from Eq. (36) it follows,

$$\frac{\hat{n}}{c\mu_0\mu_r}(\mathbf{s} \times \mathbf{E}) = \mathbf{H} \qquad (37)$$

For this type of wave, therefore, $\mathbf{E}, \mathbf{H}$, and $\mathbf{s}$ are mutually perpendicular and form a right-handed set. The quantity $\hat{n}/(c\mu_0\mu_r)$ has the dimension of an admittance and is known as the characteristic optical admittance of the medium, written Y:

$$Y = \frac{\hat{n}}{c\mu_0\mu_r} \qquad (38)$$

For a vacuum, we have $\hat{n} = n = 1$, $\mu_r = 1$ and from Eq. (10) it follows that its admittance is given by

$$Y_V = (\varepsilon_0/\mu_0)^{1/2} = 2.6544 \cdot 10^{-3} \text{ S} \qquad (39)$$

At optical frequencies $\mu_r = 1$, and we can write

$$Y = \hat{n} \cdot Y_V \qquad (40)$$

and

$$\mathbf{H} = Y(\mathbf{s} \times \mathbf{E}) = \hat{n}Y_V(\mathbf{s} \times \mathbf{E}) = \hat{n}Y_V\frac{\mathbf{q} \times \mathbf{E}}{q} \qquad (41)$$

2.1.3. Intensity of Electromagnetic Waves

Electromagnetic waves transport energy and it is this energy which is observed. The instantaneous rate of flow of energy across the unit area is given by the Pointing vector,

$$\mathbf{P} = \mathbf{E} \times \mathbf{H} \qquad (42)$$

Since $\mathbf{E}$ is perpendicular to $\mathbf{H}$, the magnitude of $\mathbf{P}$ is $P = E \cdot H$, where E and H are the complex and time-dependent field strengths. The intensity of the electromagnetic wave is defined as the time average of the real part of E and H: $I = \langle \text{Re}(E) \cdot \text{Re}(H) \rangle_t$. It can be shown [12] that this average can be written in the form $(1/2) \text{Re}(E \cdot tH^*)$, where H^* is the complex conjugate of H. Hence, the intensity of the electromagnetic wave is given by

$$I = \tfrac{1}{2} \text{Re}(E \cdot H^*) \qquad (43)$$

where E and H are complex scalar quantities.

It is important to note that the electric and magnetic vectors in Eq. (42) should be the total resultant field due to all waves which are involved. From Eq. (41), it follows that $H = \hat{n} Y_V E$ and we obtain $I = (1/2) \operatorname{Re}(E\hat{n}^* Y_V E^*) = (1/2) \operatorname{Re}(Y^*)EE^*$. Using Eqs. (21) and (40), one obtains $I = (1/2)nY_V EE^*$. The product EE^* is obtained from Eq. (29) and the final expression for the intensity is given by

$$I = \tfrac{1}{2} n Y_V E_0^2 e^{-(4\pi/\lambda)k(\alpha x + \beta y + \gamma z)} \tag{44}$$

The unit of I is W m^{-2}.

The expression $(\alpha x + \beta y + \gamma z)$ is simply the distance traveled along the direction of propagation, and thus the intensity drops to $1/e$ of its initial value in a distance given by $\lambda/4\pi k$. The inverse of the distance is defined as the absorption coefficient α, that is

$$\alpha = 4\pi k/\lambda = 4\pi \tilde{\nu} k \tag{45}$$

where $\tilde{\nu}$ is the frequency in wave numbers (in units of cm^{-1}).

2.1.4. Kramers–Kronig Relation

The real and imaginary parts of the complex dielectric constant $\hat{\varepsilon} = \varepsilon' + i\varepsilon''$ or the complex refractive index $\hat{n} = n + ik$ are not quite independent of one another. They are linked by dispersion relations, also called the Kramers–Kronig relations [15]. Equation (5) shows that the dielectric displacement D is linearly related to the electric field E: $D = \varepsilon E$. When a linear relation of this kind is considered as a function of frequency, that is, as a function of time, it must satisfy the requirements of causality: there must be no displacement until after the application of the field E. It is well known that this condition requires that the real and imaginary parts of $\varepsilon(\omega)$ or $\hat{n}(\omega)$ should satisfy the Kramers–Kronig relation. For $\varepsilon'(\omega)$ and $\varepsilon''(\omega)$, these relations are

$$\varepsilon'(\omega) = 1 + \frac{2}{\pi} P \int_0^\infty \frac{\omega'\varepsilon''(\omega')}{\omega'^2 - \omega^2} \, d\omega' \tag{46}$$

$$\varepsilon''(\omega) = -\frac{2\omega}{\pi} P \int_0^\infty \frac{\varepsilon(\omega') - 1}{\omega'^2 - \omega^2} \, d\omega' \tag{47}$$

where P stands for the Cauchy principle value [15].

We are interested in the Kramers–Kronig relations modified for the analysis of vibrational transitions in solids and liquids. The resonance absorptions for light by vibrational transitions are then generally well separated in frequency from the optical processes associated with electronic transitions which take place at much higher frequencies (in the visible or UV part of the spectrum). The integration in Eq. (46) can therefore be divided in two parts, namely,

$$\varepsilon'(\omega) - 1 = \frac{2}{\pi} P \int_0^{\omega_c} \frac{\omega'\varepsilon''(\omega')}{\omega'^2 - \omega^2} \, d\omega'$$
$$+ \frac{2}{\pi} P \int_{\omega_c}^\infty \frac{\omega'\varepsilon''(\omega')}{\omega'^2 - \omega^2} \, d\omega' \tag{48}$$

Since we are evaluating $\varepsilon'(\omega)$ for $\omega \ll \omega_c$ where ω_c is well above the vibrational frequencies, but well below the elec-

tronic frequencies, we can neglect ω in the second integral. The second term is therefore constant and can approximately be replaced by $\varepsilon_\infty - 1$ where ε_∞ is the "high frequency" optical dielectric constant associated with electronic transitions. The Kramers–Kronig relation modified for vibrational transitions is therefore,

$$\varepsilon'(\omega) - \varepsilon_\infty = \frac{2}{\pi} P \int_0^{\omega_c} \frac{\omega'\varepsilon''(\omega')}{\omega'^2 - \omega^2} \, d\omega' \tag{49}$$

The inverse relation corresponding to Eq. (49) is

$$\varepsilon''(\omega) = -\frac{2\omega}{\pi} P \int_0^{\omega_c} \frac{\varepsilon'(\omega) - \varepsilon_\infty}{\omega'^2 - \omega^2} \, d\omega' \tag{50}$$

The static dielectric constant ε_{st} can be obtained from Eq. (49) by setting $\omega = 0$ resulting in

$$\int_0^{\omega_c} \frac{\varepsilon''(\omega)}{\omega} \, d\omega = \frac{\pi}{2}(\varepsilon_{st} - \varepsilon_\infty) \tag{51}$$

Relation (51) constitutes a sum rule which provides further information and checks for experimental data.

2.2. Reflection and Refraction

2.2.1. Isotropic, Nonabsorbing Media

In this section, we consider the reflection and the refraction of electromagnetic waves at a plane interface between dielectrics, characterized by the refractive indices n_1 and n_2 [11]. The coordinate system and symbols, appropriate to the problem are shown in Figure 1. The media above and below the plane $z = 0$ have permeabilities and refractive indices μ_{r1}, n_1 and μ_{r2}, n_2, respectively.

From Eqs. (23) and (32), we have $n_1 = (c/\omega)q_i = \sqrt{\mu_{r1}\varepsilon_{r1}}$ and $n_2 = (c/\omega)q_t = \sqrt{\mu_{r2}\varepsilon_{r2}}$, where $q_i = q'_i$ and $q_t = q_t$ for nonabsorbing media. A plane wave with wave vector $\mathbf{q}_i$ and frequency ω is incident from medium μ_{r1}, n_1 at an angle of incidence θ_i. The refracted and reflected waves have wave vectors $\mathbf{q}_t$ and $\mathbf{q}_r$, respectively, and $\mathbf{e}_z$ is a unit vector directed

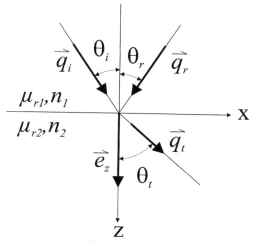

Fig. 1. The incident wave $\mathbf{q}_i$ strikes the plane interface between different media, giving rise to a reflected wave $\mathbf{q}_r$ and a transmitted (refracted) wave $\mathbf{q}_t$.

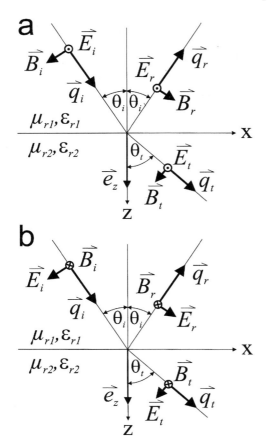

Fig. 2. (a) and (b) Reflection and refraction for s-polarized (a) and p-polarized (b) light. All the electric fields in (a) are shown directed toward the viewer. All the magnetic fields in (b) are shown directed away from the viewer.

from medium μ_{r1}, n_1 to μ_{r2}, n_2. θ_t is the angle of the refracted (transmitted) wave and θ_r that of the reflected wave.

Using Eqs. (6), (18), (33), and (37), the three waves shown in Figure 2 are as follows

incident:

$$\mathbf{E}_i = \mathbf{E}_{0i} e^{i(\mathbf{q}_i \mathbf{r} - \omega t)}$$

$$\mathbf{B}_i = \sqrt{\mu_1 \varepsilon_1} \frac{\mathbf{q}_i \times \mathbf{E}_i}{q_i} \qquad (52)$$

refracted:

$$\mathbf{E}_t = \mathbf{E}_{0t} e^{i(\mathbf{q}_t \mathbf{r} - \omega t)}$$

$$\mathbf{B}_t = \sqrt{\mu_2 \varepsilon_2} \frac{\mathbf{q}_t \times \mathbf{E}_t}{q_t} \qquad (53)$$

reflected:

$$\mathbf{E}_r = \mathbf{E}_{0r} e^{i(\mathbf{q}_r \mathbf{r} - \omega t)}$$

$$\mathbf{B}_r = \sqrt{\mu_1 \varepsilon_1} \frac{\mathbf{q}_r \times \mathbf{E}_r}{q_r} \qquad (54)$$

where $\mu_1 = \mu_0 \mu_{r1}$, $\varepsilon_1 = \varepsilon_0 \varepsilon_{r1}$ and similarly for μ_2 and ε_2.

The wave vectors have the magnitudes,

$$|\mathbf{q}_i| = |\mathbf{q}_r| = q_i = \frac{\omega}{c}\sqrt{\mu_{r1}\varepsilon_{r1}} = \frac{\omega}{c}n_1$$

$$|\mathbf{q}_t| = q_t = \frac{\omega}{c}\sqrt{\mu_{r2}\varepsilon_{r2}} = \frac{\omega}{c}n_2 \qquad (55)$$

The boundary conditions at $z = 0$ must be satisfied at all points on the plane at all times, implying that the spatial (and time) variation of all fields must be the same at $z = 0$. Consequently, we must have the phase factors all equal at $z = 0$,

$$(\mathbf{q}_i \mathbf{r})_{z=0} = (\mathbf{q}_r \mathbf{r})_{z=0} = (\mathbf{q}_t \mathbf{r})_{z=0} \qquad (56)$$

independent of the nature of the boundary conditions. Equation (56) contains the kinematic aspects of reflection and refraction, in particular it follows that all three wave vectors must lie in a plane.

From Eq. (31) and Figure 1, we have

$$(\mathbf{q}_i \mathbf{r})_{z=0} = \frac{\omega}{c}n_1\big[(\sin\theta_i, 0, \cos\theta_i)(x, y, z)\big]_{z=0}$$

$$= \frac{\omega}{c}n_1\big[x\sin\theta_i + z\cos\theta_i\big]_{z=0} = \frac{\omega}{c}n_1 x\sin\theta_i \quad (57)$$

and similar expressions for $(\mathbf{q}_r \mathbf{r})_{z=0}$ and $(\mathbf{q}_t \mathbf{r})_{z=0}$.

We therefore find

$$\theta_r = \theta_i \qquad (58)$$

and Snell's law,

$$n_1 \sin\theta_i = n_2 \sin\theta_t \qquad (59)$$

The dynamic properties are contained in the boundary conditions, that is, normal components of $\mathbf{D}$ and $\mathbf{B}$ are continuous; tangential components of $\mathbf{E}$ and $\mathbf{H}$ are continuous. For a proof of this condition, the reader is referred to [11]. For the normal components of $\mathbf{D}$ and $\mathbf{B}$ the scalar products $\mathbf{D}\mathbf{e}_z$ and $\mathbf{B}\mathbf{e}_z$ of the three waves must be compared, while for the tangential components of $\mathbf{E}$ and $\mathbf{H}$ the vector products $\mathbf{E} \times \mathbf{e}_z$ and $\mathbf{H} \times \mathbf{e}_z$ are involved. Using Eqs. (52)–(55) and $\mathbf{D} = \varepsilon\mathbf{E}$, $\mathbf{H} = \mathbf{B}/\mu$, the boundary conditions at $z = 0$ are [11],

$$\varepsilon_{r1}(\mathbf{E}_{0i} + \mathbf{E}_{0r})\mathbf{e}_z = \varepsilon_{r2}\mathbf{E}_{0t}\mathbf{e}_z$$

$$(\mathbf{q}_i \times \mathbf{E}_{0i} + \mathbf{q}_r \times \mathbf{E}_{0r})\mathbf{e}_z = (\mathbf{q}_t \times \mathbf{E}_{0t})\mathbf{e}_z$$

$$(\mathbf{E}_{0i} + \mathbf{E}_{0r}) \times \mathbf{e}_z = \mathbf{E}_{0t} \times \mathbf{e}_z \qquad (60)$$

$$\frac{1}{\mu_{r1}}(\mathbf{q}_i \times \mathbf{E}_{0i} + \mathbf{q}_r \times \mathbf{E}_{0r}) \times \mathbf{e}_z = \frac{1}{\mu_{r2}}(\mathbf{q}_t \times \mathbf{E}_{0t}) \times \mathbf{e}_z$$

In applying these boundary conditions, it is convenient to consider two separate cases, one in which the incident plane wave is linearly polarized with its polarization vector perpendicular to the plane of incidence (plane defined by $\mathbf{q}$ and $\mathbf{e}_z$), and the other in which the polarization vector is parallel to the plane of incidence.

2.2.1.1. E Perpendicular to the Plane of Incidence

We first consider the electric field perpendicular to the plane of incidence (s polarization, from German senkrecht), as shown in

Figure 2a. All the electric fields are shown directed toward the viewer. The orientations of the **B** vectors are chosen to give a positive flow of energy in the direction of the wave vectors (see Eq. (42)).

Since for s-polarized light (Fig. 2a) the electric fields are all parallel to the surface, the first condition in Eq. (60) yields nothing. Evaluation of the third and fourth conditions in Eq. (60) gives

$$E_{0i} + E_{0r} = E_{0t}$$

$$\sqrt{\frac{\varepsilon_{r1}}{\mu_{r1}}}(E_{0i} - E_{0r})\cos\theta_i = \sqrt{\frac{\varepsilon_{r2}}{\mu_{r2}}}E_{0t}\cos\theta_t \tag{61}$$

while the second, using Snell's law (Eq. (59)), duplicates the third. The relative amplitudes of the refracted (transmitted) and reflected waves can be found from Eqs. (61),

$$\tau_s = \frac{E_{0t}}{E_{0i}} = \frac{2\cos\theta_i}{\cos\theta_i + \mu_{12}\sqrt{n_{21}^2 - \sin^2\theta_i}}$$

$$\rho_s = \frac{E_{0r}}{E_{0i}} = \frac{\cos\theta_i - \mu_{12}\sqrt{n_{21}^2 - \sin^2\theta_i}}{\cos\theta_i + \mu_{12}\sqrt{n_{21}^2 - \sin^2\theta_i}} \tag{62}$$

In calculating τ_s and ρ_s, we have used Snell's law (59), $\cos\theta_t = (1 - \sin^2\theta_t)^{1/2}$, and we have introduced the ratios of the permeabilities and refractive indices, namely,

$$\mu_{12} = \frac{\mu_1}{\mu_2}, \qquad n_{21} = \frac{n_2}{n_1} \tag{63}$$

2.2.1.2. E Parallel to the Plane of Incidence

If the electric field is parallel to the plane of incidence, as shown in Figure 2b, the boundary conditions involved are normal **D**, tangential **E**, and tangential **H**, the first, third, and fourth condition in Eq. (60). The tangential **E** and **H** continues demand that

$$(E_{0i} - E_{0r})\cos\theta_i = E_{0t}\cos\theta_t$$

$$\sqrt{\frac{\varepsilon_{r1}}{\mu_{r1}}}(E_{0i} + E_{0r}) = \sqrt{\frac{\varepsilon_{r2}}{\mu_{r2}}}E_{0t} \tag{64}$$

The normal **D** continuous condition, plus Snell's law merely duplicate the second of the previous equations. The relative amplitudes of refracted (transmitted) and reflected fields are therefore,

$$\tau_p = \frac{2n_{21}\cos\theta_i}{\mu_{12}n_{21}^2\cos\theta_i + \sqrt{n_{21}^2 - \sin^2\theta_i}}$$

$$\rho_p = \frac{\mu_{12}n_{21}^2\cos\theta_i - \sqrt{n_{21}^2 - \sin^2\theta_i}}{\mu_{12}n_{21}^2\cos\theta_i + \sqrt{n_{21}^2 - \sin^2\theta_i}} \tag{65}$$

For normal incidence ($\theta_i = 0$) and $\mu_1 = \mu_2$, it becomes unnecessary to distinguish between s and p polarization and one obtains

$$\tau(\theta_i = 0) = \frac{2n_1}{n_2 + n_1}$$

$$\rho(\theta_i = 0) = \frac{n_2 - n_1}{n_2 + n_1} \tag{66}$$

2.2.2. Isotropic Absorbing Media

A derivation of the Fresnel relation from Maxwell's equations for absorbing media described by complex refractive indices $\hat{n}_1 = n_1 + ik_1$ and $\hat{n}_2 = n_2 + ik_2$ gives the result that Eqs. (62) and (65) are still valid, provided the real refractive indices n_1 and n_2 are replaced by $\hat{n}_1$ and $\hat{n}_2$, respectively.

In the following, we consider the important case where medium 1 is transparent with $\hat{n}_1 = n_1$ and medium 2 is absorbing with $\hat{n}_2 = n_2 + ik_2$. For this case, the reflection coefficients r_s and r_p are complex and are given by

$$r_s = \frac{n_1\cos\theta_i - [(n_2 + ik_2)^2 - n_1^2\sin^2\theta_i]^{1/2}}{n_1\cos\theta_i + [(n_2 + ik_2)^2 - n_1^2\sin^2\theta_i]^{1/2}} \tag{67}$$

$$r_p = \frac{(n_2 + ik_2)^2\cos\theta_i - n_1[(n_2 + ik_2)^2 - n_1^2\sin^2\theta_i]^{1/2}}{(n_2 + ik_2)^2\cos\theta_i + n_1[(n_2 + ik_2)^2 - n_1^2\sin^2\theta_i]^{1/2}} \tag{68}$$

and we have assumed that $\mu_1 = \mu_2$ which is usually a good approximation at optical frequencies (in nonmagnetic materials $\mu_1 = \mu_2 = 1$). The corresponding reflectivities are given by

$$R_s = |r_s|^2 = r_s r_s^* \tag{69}$$

$$R_p = |r_p|^2 = r_p r_p^* \tag{70}$$

2.3. Total Reflection and Attenuated Total Reflection

2.3.1. Two Bulk Media

2.3.1.1. Isotropic Transparent Media

We consider the phenomenon of total internal reflection. The word internal implies that the incident and reflected waves propagate in a medium of larger refractive index than the refracted wave, i.e., $n_1 > n_2$. According to Snell's law, $\sin\theta_t = n_{12}\sin\theta_i$ ($n_{12} = n_1/n_2$) and if $n_1 > n_2$ then $\theta_t > \theta_i$. The critical incident angle θ_{ic} is defined by the condition $\theta_t = \pi/2$ or by

$$\sin\theta_{ic} = n_{21} = n_2/n_1 \tag{71}$$

For waves incident at angles $\theta_i = \theta_{ic}$, the refracted wave is propagated parallel to the surface ($\theta_t = \pi/2$). There can be no energy flow across the surface. Hence, at that angle of incidence there must be total reflection.

What happens if $\theta_i > \theta_{ic}$? In this case, according to Snell's law $\sin\theta_t > 1$, hence $\sin\theta_i > n_{21}$ or $n_{21}^2 - \sin^2\theta_i < 0$ and the square root occurring in the Fresnel coefficients (62) and (65) becomes purely imaginary, i.e., $(n_{21}^2 - \sin^2\theta_i)^{1/2} = i(\sin^2\theta_i - n_{21}^2)^{1/2}$. We then obtain

$$\rho_s = \frac{\cos\theta_i - i(\sin^2\theta_i - n_{21}^2)^{1/2}}{\cos\theta_i + i(\sin^2\theta_i - n_{21}^2)^{1/2}} \tag{72}$$

and

$$\rho_p = \frac{n_{21}^2 \cos\theta_i - i(\sin^2\theta_i - n_{21}^2)^{1/2}}{n_{21}^2 \cos\theta_i + i(\sin^2\theta_i - n_{21}^2)^{1/2}} \qquad (73)$$

From these equations, it follows that $|\rho_s| = |\rho_p| = 1$, indicating that the reflection is total when n_{21} is real.

For angles larger than θ_{ic}, it follows that

$$\cos\theta_t = \sqrt{1 - \sin^2\theta_t} = in_{12}\sqrt{\sin^2\theta_i - n_{21}^2} \qquad (74)$$

that is, $\cos\theta_t$ is purely imaginary while $\sin\theta_t = n_{12}\sin\theta_i$ is real. θ_t is therefore a complex angle with a real sine and a purely imaginary cosine.

The meaning of these complex quantities becomes clear when we consider the wave $\mathbf{E}_t$ as defined by Eq. (53). We have $\mathbf{q}_t\mathbf{r} = q_t(\sin\theta_t, 0, \cos\theta_t)(x, y, z) = q_t(x\sin\theta_t + z\cos\theta_t)$. Using Eq. (74) and $\sin\theta_t = n_{12}\sin\theta_i$, the propagation factor for the refracted wave becomes

$$e^{i\mathbf{q}_t\mathbf{r}} = e^{-q_t n_{12}\sqrt{\sin^2\theta_i - n_{21}^2}\cdot z} \cdot e^{iq_t n_{12}\sin\theta_i \cdot x} \qquad (75)$$

This shows explicitly that for $\theta_i > \theta_{ic}$, the refracted wave is propagated only parallel to the surface (second factor in Eq. (75)), but is attenuated exponentially beyond the interface (first factor in Eq. (75)). The attenuation occurs within a fraction of the wavelength λ, except at $\theta_i = \theta_{ic}$. This becomes clear if in Eq. (75) we replace q_t by its value given by Eq. (55), namely, $q_t = (\omega/c)n_2 = (2\pi/\lambda)n_2$. The exponentially decaying factor in Eq. (75) can then be written as $\exp(-z/d_p)$, where

$$d_p = \frac{\lambda}{2\pi n_1\sqrt{\sin^2\theta_i - n_{21}^2}} \qquad (76)$$

is the depth of penetration of the evanescent wave, defined as the distance required for the electric field amplitude to fall to e^{-1} of its value at the surface. Note that in Eq. (76), λ is the wavelength in vacuum and $\lambda_1 = \lambda/n_1$ is the wavelength in the denser medium.

Even though fields exist on the other side of the surface, in the optically rarer medium, there is no energy flow through the surface. Hence, total internal reflection occurs for $\theta_i > \theta_{ic}$. This can be verified by calculating the time-averaged normal component, $\langle e_z \mathbf{P}\rangle_t$ of the complex Pointing vector $\mathbf{P}$. Referring to Figure 2 we have for the transmitted wave,

$$\langle e_z\mathbf{P}_t\rangle_t = \frac{1}{2}\,\mathrm{Re}\big[\mathbf{n}\cdot(\mathbf{E}_t \times \mathbf{H}_t^*)\big] \qquad (77)$$

Now, $\mathbf{E}_t \times \mathbf{H}_t^* = E_t H_t^* \mathbf{s}_t$, where $\mathbf{s}_t$ is a unit vector in the direction of the transmitted wave. According to Eq. (41), $\mathbf{H}_t = Y_t(\mathbf{s}_t \times \mathbf{E}_t) = Y_t E_t \mathbf{e}$ where $\mathbf{e}$ is a unit vector in the direction of $\mathbf{H}_t$ and $Y_t = n_2 Y_V$ is real for a transparent rarer medium. Therefore, $H_t^* = Y_t E_t^*$ and using $\mathbf{e}_z\mathbf{s}_t = \cos\theta_t$ we obtain

$$\langle e_z\mathbf{P}_t\rangle_t = \frac{1}{2}n_2 Y_V\,\mathrm{Re}\big[\cos\theta_t|E_{t0}|^2\big] \qquad (78)$$

However, since $|E_{t0}|^2$ is real and since according to Eq. (74) $\cos\theta_t$ is purely imaginary for $\theta_i > \theta_{ic}$ it follows that $\langle e_z\mathbf{P}_t\rangle_t = 0$. This is clearly not true if the rarer medium is absorbing, since in this case, both, $Y_t = \hat{n}_2 Y_V$ and $\cos\theta_t$ are complex quantities

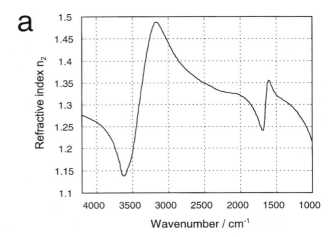

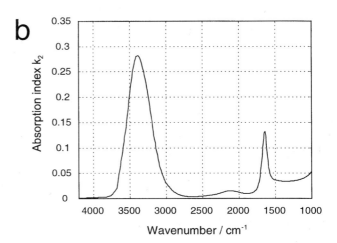

Fig. 3. (a) and (b) Refractive index n_2 (a) and extinction coefficient k_2 (b) of liquid water at 300 K, as a function of wave number $\hat{\nu}$.

and $\mathrm{Re}[\cos\theta_t Y_t^*|E_{t0}|^2] \neq 0$. Part of the incident wave is then absorbed by the rarer medium and we are dealing with attenuated total reflection (ATR).

2.3.1.2. Absorbing Rarer Medium

If the rarer medium is absorbing, the reflectivities can still be calculated from Eqs. (72) and (73) by replacing n_2 with the complex refraction index $\hat{n}_2 = n_2 + ik_2$. The resulting expressions are identical with those given by Eqs. (67) and (68) which are valid both for $\theta_i < \theta_{ic}$ and for $\theta_i > \theta_{ic}$. It turns out that for θ_i well above θ_{ic}, measured and calculated reflectivities resemble transmission measurements, i.e., they exhibit a behavior described by $k_2(\omega)$. On the other hand, if $\theta_i < \theta_{ic}$, the spectra resemble the mirror image of the dispersion of $n_2(\omega)$ (see [3, pp. 69–75 and 243–250]).

As an example, Figure 4 contains the calculated ATR spectra of water for a single total reflection using a Ge-ATR element with refractive index $n_1 = 4.0$ and an angle of incidence $\theta_i = 45°$; the calculation is based on the relations $R_s = |\rho_s|^2$

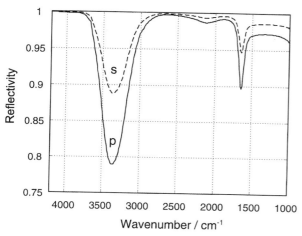

Fig. 4. Calculated ATR spectra of liquid water at 300 K in contact with a Ge-ATR element ($n_1 = 4.0$) for s-polarized (dashed line) and p-polarized (solid line) light (single total reflection at an angle of incident of $\theta_i = 45°$).

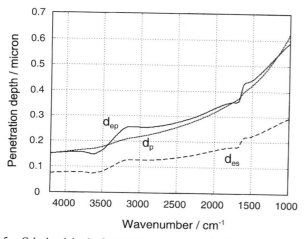

Fig. 5. Calculated depth of penetration d_p (short dashed line), effective thicknesses d_{es} (long dashed line), and d_{ep} (solid line) for s- and p-polarizations, respectively, for the Ge/H$_2$O system described in Figure 4. The weak structure is due to the dispersion of n_2 as shown in Figure 3a.

and $R_p = |\rho_p|^2$ where ρ_s and ρ_p are given by Eqs. (72), (73). The experimentally determined input spectra of water, $n_2(\tilde{v})$ and $k_2(\tilde{v})$, as a function of $\tilde{v}$ ($\tilde{v} = 1/\lambda = v/c = \omega/2\pi c$) are shown in Figure 3a and 3b, respectively. The absorption peaks of $k_2(\tilde{v})$ near $\tilde{v} = 3300$ and $1650\,\text{cm}^{-1}$ are due to the symmetric and asymmetric stretching and H$-$O$-$H bending modes, respectively. The depth of penetration d_p is shown in Figure 5; it is calculated using Eq. (76) with $n_1 = 4.0$ and $n_2 = n_2(\tilde{v})$; the weak structures are due to the dispersion of n_2. An identical curve is obtained if absorption is included in the calculation of $d_p(\tilde{v})$; this is due to the fact that $(k_2/n_1)^2 \gg 1$. It should also be noted that the depth of penetration is independent of the state of polarization of the infrared light.

While the expressions $R_s = |\rho_s|^2$ and $R_p = |\rho_p|^2$ with ρ_s and ρ_p defined by Eqs. (72), (73) give an exact expression for the reflectivities as illustrated in Figure 4, it is desirable to establish approximate expressions for R_s and R_p which give

some physical insight into the coupling of the evanescent wave with the rarer absorbing medium. The calculation is based again on Eqs. (72), (73) (with n_{21} replaced by $\hat{n}_2/n_1 = (n_2 + ik_2)/n_1$) and retaining only terms linear in k_2. We are therefore dealing with the "low absorption" approximation. Terms such as $(k_2/n_1)^2$ are therefore neglected which is usually a good approximation for Ge and Si in contact with a weak or moderately weak absorber (see later). A straightforward but lengthy calculation gives the following result for the approximate reflectivities R_{as} and R_{ap}:

$$R_{as} = 1 - d_{es} \cdot \alpha \tag{79}$$
$$R_{ap} = 1 - d_{ep} \cdot \alpha \tag{80}$$

where $\alpha = (2\omega/c)k_2$ is the absorption coefficient of the rarer medium (see Eq. (45)), while d_{es} and d_{ep} are the effective thicknesses for isotropic media for perpendicular and parallel polarization, respectively,

$$d_{es} = \frac{(\lambda/n_1)n_{21}\cos\theta_i}{\pi(1 - n_{21}^2)(\sin^2\theta_i - n_{21}^2)^{1/2}} \tag{81}$$

$$d_{ep} = \frac{(\lambda/n_1)n_{21}\cos\theta_i(2\sin^2\theta_i - n_{21}^2)}{\pi(1 - n_{21}^2)[(1 + n_{21}^2)\sin^2\theta_i - n_{21}^2](\sin^2\theta_i - n_{21}^2)^{1/2}} \tag{82}$$

A detailed discussion of these effective thicknesses is given in the excellent book by Harrick [3] (note that there is an error in sign in Harrick's formula (27) at p. 43, which has been corrected in our Eq. (82)). The relations (81) and (82) are valid for a single reflection and for sufficiently low absorptions, i.e., for $\alpha d < 0.1$ in transmission experiments of thin films of thickness d with transmissions $T = I/I_0 = \exp(-\alpha d) \approx 1 - \alpha d$. The effective thicknesses d_{es} and d_{ep} represent a measure for the strength of interaction of the evanescent field with the rarer medium. It should be noted that according to Eqs. (81), (82) $d_{es} = 1/2d_{ep}$ for $\theta_i = 45°$. Figure 5 shows d_{es} and d_{ep} for the system Ge/H$_2$O for $\theta_i = 45°$, $n_1 = 4$, and $n_2(\tilde{v})$ according to Figure 3a, along with the depth of penetration d_p.

We have performed calculations of the approximate reflectivities R_{as} and R_{ap} defined by Eqs. (79), (80) and compared them with the exact reflectivities $R_s = |\rho_s|^2$ and $R_p = |\rho_p|^2$ shown in Figure 4 which have been obtained from the generalized Fresnel coefficients given by Eqs. (72), (73) (in which n_2 is replaced by $\hat{n}_2 = n_2 + ik_2$). Calculations have been done for the systems Ge/H$_2$O, Si/H$_2$O, and ZnSe/H$_2$O. For the relatively weak H$-$O$-$H bending absorption near $1640\,\text{cm}^{-1}$, the relative errors of the peak heights are 3, 4.7, and 16% for s polarization, and 6, 9, and 30% for p polarization. On the other hand, for the strong O$-$H absorption near $3300\,\text{cm}^{-1}$ the corresponding errors are 9, 12, and 47% for s polarization, and 14, 23, and 87% for p polarization. This illustrates the fact that for absolute reflectivities as defined by Eqs. (79), (80) the "weak absorption approximation" is good for Ge/H$_2$O, acceptable for Si/H$_2$O, but completely breaks down for ZnSe/H$_2$O. However, as discussed later, the weak absorption approximation is much better for suitably normalized reflectivities.

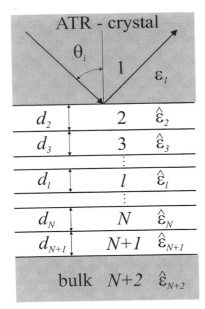

Fig. 6. N layers: ($l = 2, 3, \ldots, N + 1$) are sandwiched between the ATR-crystal (1) and the outer bulk-phase ($N + 2$). The ATR crystal is considered to be transparent in the interesting range of wave numbers, all other refractive indices are complex. The thickness of the lth layer is d_l.

2.3.2. Stratified Media

2.3.2.1. General Formalism

In the following, we consider N layers sandwiched between two bulk phases 1 and $N + 2$; phase 1 is transparent with $\varepsilon_1 = n_1^2$ and phase $N + 2$ is absorbing with complex refraction index $\hat{n}_{N+2} = n_{N+2} + ik_{N+2}$ and dielectric constant $\hat{\varepsilon}_{N+2} = \hat{n}_{N+2}^2$. The N layers are numbered by $l = 2, 3, \ldots, N + 1$ and the thickness of layer l is d_l and its complex dielectric constant is $\hat{\varepsilon}_l = \hat{n}_l^2 = (n_l + ik_l)^2$. The situation is shown in Figure 6.

In many cases, the thicknesses d_l of the thin films are much less than the wavelength λ. It is then possible to derive simple, linear-approximation relations for the optical changes of reflectivity. These relations are valid for weakly and moderate weakly thin absorbing layers (i.e., for $\alpha d < 0.1$) but not for thin metallic layers (see what follows). In the following, we assume $d_l \ll \lambda$ for all N layers, $l = 2, 3, \ldots, N + 1$, and we quote the final expressions for the normalized reflectivities R_{ns} and R_{np} for s- and p-polarized light without proof. The interested reader is referred to the literature for more detailed information [17]. If u denotes the polarization s or p, then the normalized reflectivity R_{nu} is defined as

$$R_{nu} = R_u(1, 2, \ldots, N + 2)/R_u(1, N + 2) \quad (83)$$

where $R_u(1, 2, \ldots, N + 2)$ is the reflectivity of the whole system shown in Figure 6, and $R_u(1, N + 2)$ is the reference reflectivity that is measured in the absence of the N layers, i.e., if the ATR-crystal 1 is in direct contact with the bulk-phase $N + 2$. In practice, it is usually not possible to measure the absolute reflectance $R_u(1, 2, \ldots, N + 2)$ of a sample inclosed in an experimental cell owing to unknown energy losses caused

by window reflections, ambient absorption, diffuse scattering, etc. However, since these loss factors tend to be constant, the reflectance ratio, R_{nu}, can be measured accurately. By definition, R_{nu} may be smaller or larger than 1, depending on whether $R_u(1, 2, \ldots, N + 2)$ is smaller or larger than $R_u(1, N + 2)$. If all layers $2, 3, \ldots, N + 1$ are identical and equal to the bulk-phase $N + 2$, then $R_u(1, 2, \ldots, N + 2) = R_u(1, N + 2)$ and $R_{nu} = 1$. The general expressions for $d_l \ll \lambda$ are [17]

$$R_{ns} = 1 - (8\pi/\lambda)n_1 \cos\theta_i \sum_{l=2}^{N+1} d_l \, \mathrm{Im}\left[\frac{\hat{\varepsilon}_l - \hat{\varepsilon}_{N+2}}{\varepsilon_1 - \hat{\varepsilon}_{N+2}}\right] \quad (84)$$

$$R_{np} = 1 - (8\pi/\lambda)n_1 \cos\theta_i \sum_{l=2}^{N+1} d_l$$
$$\times \mathrm{Im}\left[\frac{\hat{\varepsilon}_l - \hat{\varepsilon}_{N+2}}{\varepsilon_1 - \hat{\varepsilon}_{N+2}} \frac{1 - \varepsilon_1(\hat{\varepsilon}_l^{-1} + \hat{\varepsilon}_{N+2}^{-1})\sin^2\theta_i}{1 - \varepsilon_1(\varepsilon_1^{-1} + \hat{\varepsilon}_{N+2}^{-1})\sin^2\theta_i}\right] \quad (85)$$

where λ is the wavelength in vacuum and Im stands for the imaginary part of the expression in parentheses.

It should be noted that Eqs. (84) and (85) are valid only if the overall thickness $d = \sum d_l$ of the layers remains very much less than the vacuum wavelength λ. In addition, the layers must be weak or moderately weak absorbers such that $\alpha d < 0.1$, where α is the mean absorption coefficient of the N layers. Metallic layers having very large values of the absorption coefficient are excluded, even for thicknesses as small as 50–100 Å. Furthermore, it should be emphasized that the sum occurring in Eqs. (84) and (85) implies that R_{ns} and R_{np} are independent on the ordering of the surface films, i.e., the linear-approximation relations neglect the phase changes of the light beam during its traversal through the layers and accounts only for the attenuation of its electric field amplitude due to absorption and for the phase changes on reflection. As a result, it is usually not possible to distinguish the optical effects due to the individual layers; only the integral absorption loss can be detected in surface reflection spectroscopy experiments.

2.3.2.2. The Single-Layer System: General

Figure 7 shows the system considered: The layer with thickness $d = d_2$ and refractive index $\hat{n}_2 = n_2 + ik_2$ is sandwiched between the transparent bulk-phase 1 (ATR crystal) with refractive index n_1 and the bulk-phase 3 with refractive index $\hat{n}_3 = n_3 + ik_3$. It is assumed that $d \ll \lambda$. For the one-layer system ($N = 1$), the normalized reflectivities defined by Eqs. (84) and (85) are given by [17]

$$R_{ns} = 1 - 8\pi n_1 \frac{d}{\lambda} \cos\theta_i \cdot \mathrm{Im}\left[\frac{\hat{\varepsilon}_2 - \hat{\varepsilon}_3}{\varepsilon_1 - \hat{\varepsilon}_3}\right] \quad (86)$$

$$R_{np} = 1 - 8\pi n_1 \frac{d}{\lambda} \cos\theta_i$$
$$\times \mathrm{Im}\left[\frac{\hat{\varepsilon}_2 - \hat{\varepsilon}_3}{\varepsilon_1 - \hat{\varepsilon}_3} \frac{1 - \varepsilon_1(\hat{\varepsilon}_2^{-1} + \hat{\varepsilon}_3^{-1})\sin^2\theta_i}{1 - \varepsilon_1(\varepsilon_1^{-1} + \hat{\varepsilon}_3^{-1})\sin^2\theta_i}\right] \quad (87)$$

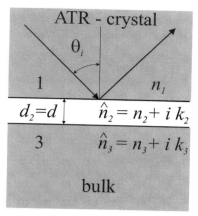

Fig. 7. The single-layer system consists of only one layer between the ATR-crystal (1) and the outer bulk-phase (3). The refractive indices of the layer and the outer bulk medium are complex.

2.3.2.3. The Single-Layer System: Approximation for Weak Absorption

Although Eqs. (86), (87) can be worked out easily by computer calculations using complex quantities, they are physically not appealing, especially for p-polarized light. It is therefore again desirable to work out simplified relations for the case of weak absorption ($\alpha d < 0.1$) which express the normalized reflectivities in terms of absorption coefficients and effective thicknesses. We have done such a calculation by retaining only linear terms in k_2 and k_3 of Eqs. (86) and (87). In addition, we have performed the calculation for a large n_1 ($n_1 = 4$ for Ge) and for comparable values of n_2 and n_3, i.e., $n_2 = n_3 + \Delta n$, where Δn is relatively small for all frequencies. Using these approximations, we find the following relations for the normalized reflectivities R_{nu} ($u = s$- or p-polarized light),

$$R_{nu} = 1 - d_{eu2} \cdot \alpha_2 + d_{eu3} \cdot \alpha_3 \qquad (88)$$

where $\alpha_2 = 4\pi k_2 \tilde{\nu}$ and $\alpha_3 = 4\pi k_3 \tilde{\nu}$ are the absorption coefficients of the layer 2 and the bulk-phase 3, respectively, and where d_{eu2} and d_{eu3} are effective thickness of layer 2 and phase 3, respectively, in the state of polarization u. We find

$$d_{eu2} = \frac{4 n_{21} d \cos\theta_i}{1 - n_{31}^2} G_u \qquad (89)$$

$$d_{eu3} = \frac{4 n_{31} d \cos\theta_i}{1 - n_{31}^2} G_u \qquad (90)$$

where

$$G_s = 1 \qquad (91)$$

$$G_p = \frac{1 - (n_{12}^2 + n_{13}^2)\sin^2\theta_i}{\cos^2\theta_i - n_{13}^2\sin^2\theta_i} \qquad (92)$$

Note that both d_{eu2} and d_{eu3} are proportional to the geometrical film thickness d. If the condition $n_2 \approx n_3$ is not satisfied, the general form of Eq. (88) is still valid, but the expressions for d_{ep2} and d_{ep3} are modified (see the Appendix).

For $n_1 = 4.0$ (Ge), $n_2 = 1.5$, and $n_3 = 1.33$ we have $G_p = 2.01$ at $\theta_i = 30°$, 1.76 at $\theta_i = 45°$, 1.70 at $\theta_i = 60°$, and 1.68 at $\theta_i = 80°$. For $n_1 = 4.0$ and $n_2 = n_3 = 1.33$ we have $G_p = 2.33$ at $\theta_i = 30°$, 2.00 at $\theta_i = 45°$, 1.92 at $\theta_i = 60°$, and 1.89 at $\theta_i = 80°$; in this case $G_p = 2.00$ at $45°$, independent of the values of n_1 and n_2.

To illustrate the single-layer model, we consider the infrared optical response of a Helmholtz–Gouy–Chapman layer (HGC layer) formed at the surface of a polarized Ge-ATR crystal, immersed in pure water. Due to self-dissociation, hydronium ions, H_3O^+, and hydroxide ions, OH^-, with number densities of $6.0 \cdot 10^{13}$ cm^{-3} are present in the bulk water at 300 K. If the Ge-ATR crystal is cathodically polarized to about -1 V, the HGC layer consists of hydrated hydronium ions, and our computer simulations show that the maximum number density of hydronium ions at the outer Helmholtz plane near 3 Å is of the order of 10^{21}–10^{22} cm^{-3} and decreases rapidly to less than 10^{20} cm^{-3} at 20 Å. Since infrared spectroscopic data of the HGC layer are not yet available, we consider an approximate system in which the HGC layer is replaced by a fictitious 20-Å thick 10-N HCl layer which may be thought to be adsorbed at the surface of the Ge crystal. The number density of hydronium ions of this layer is $6 \cdot 10^{21}$ cm^{-3} and its pH value is about -1, values which are comparable to the corresponding values of the HGC layer. On the other hand, the HGC layer is positively charged (implying div $\mathbf{D} = \rho \neq 0$, i.e., Eq. (11) is not satisfied), while the HCl layer is electrically neutral and div $\mathbf{D} = 0$. Thus, we consider the following system: phase 1 is a Ge-ATR crystal with $n_1 = 4.0$ and $\theta_i = 45°$, the bulk-phase 3 is pure water with $\hat{n}_3 = n_3 + i k_3$, where the frequency dependence of n_3 and k_3 in the infrared region is shown in Figure 3a and b, respectively. The layer 2 is a fictitious 20-Å thick 10-N HCl layer with $\hat{n}_2 = n_2 + i k_2$, where k_2 has been evaluated from experimental transmission data [21]. Although n_2 is expected to be slightly different from n_3, we have adopted the approximation $n_2 = n_3$ for all wave numbers. Since k_2 is only slightly larger than k_3, the optical contrast of the layer with respect to the bulk background phase will be very small, leading to reflectivity changes in the per thousand range (see Fig. 8a); we are confident, however, that such small signals can be observed by using field-modulation ATR spectroscopy [16, 22]. Such small reflectivity changes are also expected for the HGC layer.

For $n_2 = n_3 = n$, the relations (89)–(92) simplify to

$$d_{es2} = d_{es3} = d_{es} = \frac{4 d n_1 n \cos\theta_i}{n_1^2 - n^2} \qquad (93)$$

$$d_{ep2} = d_{ep3} = d_{ep} = d_{es} G_p \qquad (94)$$

$$G_p = \frac{n^2 - 2 n_1^2 \sin^2\theta_i}{n^2\cos^2\theta_i - n_1^2\sin^2\theta_i} \qquad (95)$$

and the normalized reflectivities R_{nu} ($u = s, p$) are given by

$$R_{nu} = 1 - d_{eu}(\alpha_2 - \alpha_3) \qquad (96)$$

The latter equation is particularly simple and shows that the decisive quantity which controls the reflectivity changes is the difference of the absorption coefficients α_2 and α_3 of the thin

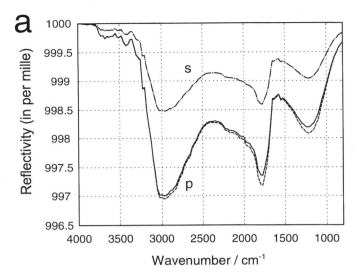

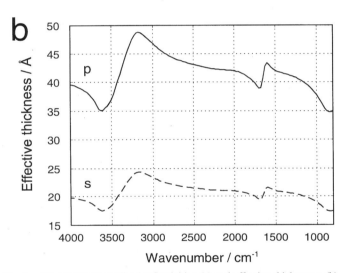

Fig. 8. (a) and (b) Normalized reflectivities (a) and effective thicknesses (b) for the single-layer system as described in the text. The 20-Å thick layer consists of a fictitious 10-N HCl solution which is sandwiched between the Ge-ATR crystal and pure water. (a) Calculated normalized reflectivity of p-polarized light using exact calculation (solid line) and approximation based on Eq. (88) (dashed line). In the case of s-polarized light, the exact and approximate calculations coincide (dashed–dotted line). The absorption near 1200 cm^{-1} is assigned to the inversion mode of the H_3O^+ ion. (b) Effective thicknesses d_{es} and d_{ep} for s- (dashed line) and p-polarized (solid line) light, respectively. For the present situation with $n_2 = n_3 = n$ and $\theta_i = 45°$, $d_{ep} = 2d_{es}$. The structure in d_{es} and d_{ep} is due to the dispersion of n.

film and the backing bulk phase. The simple relation (96) deserves some notes: remember that by definition (Eq. (83)), the normalized reflectivity R_{nu} is the ratio of two reflectivities and as such can be smaller or larger than 1. If $\alpha_2 > \alpha_3$, then $R_{nu} < 1$ and there are some absorption minima in R_{nu} as a function of $\tilde{\nu}$. On the other hand, if $\alpha_2 < \alpha_3$ then $R_{nu} > 1$ and absorption maxima occur in R_{nu} as a function of $\tilde{\nu}$. If $\alpha_2 = \alpha_3$, then $R_u(1, 2, 3) = R_u(1, 3)$ and $R_{nu} = R_u(1, 2, 3)/R_u(1, 3) = 1$ (see Eq. (83)). Furthermore, $R_{nu} = 1$ if $\alpha_2 \neq \alpha_3$ and the geometrical film thickness d and hence d_{eu} tends to zero. For the

particular case of a transparent bulk-phase 3 ($\alpha_3 = 0$), $R_{nu} = 1 - d_{eu}\alpha_2$ with absorption minima, while for a transparent layer 2 ($\alpha_2 = 0$), $R_{nu} = 1 + d_{eu}\alpha_3$ with absorption maxima.

Figure 8a shows R_{ns} and R_{np} for the system Ge/HCl/H_2O, both calculated on the basis of the original expressions (86) and (87) as well as using the approximate relations (93)–(96). The approximate relations give excellent results. Figure 8b illustrates the effective thicknesses d_{es} and d_{ep}; again, d_{ep} is twice as large as d_{es} for $\theta_i = 45°$, reflecting the fact that the coupling of the p-polarized light is stronger than that of the s-polarized light.

Referring to Figure 8a it is remarkable that the "weak absorption" approximation yields normalized reflectivities in close agreement with those obtained from the exact relations, even in the case of aqueous systems which have strong O—H stretching absorption bands as shown in Figure 3b. Computer simulations also show that for the present case study, the weak absorption approximation for the normalized reflectivities is a very good approximation for layer thicknesses d up to at least 100 nm for which $d/\lambda = 0.0166$ at $\lambda = 6$ μm ($\tilde{\nu} = 1666$ cm^{-1}). In addition, excellent agreement is obtained not only for the system Ge/HCl/H_2O, but also for Si/HCl/H_2O and ZnSe/HCl/H_2O.

The weak absorption approximation appears to be interesting for three reasons. First, as illustrated before, its range of validity extends into the region of moderately strong or even strong absorption with acceptable errors. Second, the approximate relations (88)–(92) are much simpler than the original relations (86) and (87) and can easily be generalized to the N-layer system. Third, the description is physically appealing because it is based on the concepts of effective thicknesses and absorption coefficients. Writing the reflectivities in terms of products $d_{el}\alpha_l - d_{e,N+2}\alpha_{N+2}$ for layer l of the N-layer system (see Fig. 6) allows the separation of the geometrical and dispersive factors (contained in the effective thicknesses) from the absorption properties (continued in the αs). The effective thicknesses are only weakly frequency dependent (Fig. 8b) and in a rough approximation they can be treated as constants. On the other hand, the absorption coefficients depend strongly on frequency and they give rise to the peaks in the normalized reflectivities.

We finally quote an approximate expression for the absolute reflectivity $R_u(1, 2, 3) = R(1, 3) \cdot R_{nu}$ of the single-layer system in the polarization state $u = (s, p)$. R_{nu} is given by Eqs. (88)–(92), while $R(1, 3)$ is given by Eqs. (79)–(82) where in Eqs. (79), (80) α is replaced by α_3 and in Eqs. (81), (82) n_{21} must be replaced by n_{31} and we write $R(1, 3) = 1 - \tilde{d}_{eu3}\alpha_3$. Hence, $\tilde{d}_{eu3}$ is defined by Eqs. (81), (82) with n_{21} replaced by n_{31}. Note that $\tilde{d}_{eu3}$ is independent on the geometrical thickness d of the layer but rather depends on the wavelength λ, while according to Eqs. (89), (90) both, d_{eu2} and d_{eu3} depend on d. Retaining only linear terms in α_2 and α_3, one obtains

$$R_u(1, 2, 3) \approx 1 - d_{eu2}\alpha_2 - \tilde{d}_{eu3}\left(1 - \frac{d_{eu3}}{\tilde{d}_{eu3}}\right)\alpha_3 \qquad (97)$$

The ratio $r_u = d_{eu3}/\tilde{d}_{eu3}$ is proportional to d/λ and for $d/\lambda \ll 1$, r_u is small compared to 1. As an example, we put $n_1 = 4.0$, $n_2 = n_3 = 1.33$, $\theta_i = 45°$, $d = 100$ Å, and $\lambda = 6$ μm and we find $r_s = r_p = 0.052$. We then obtain

$$R_u(1, 2, 3) \approx 1 - d_{eu2}\alpha_2 - \tilde{d}_{eu3}\alpha_3 \qquad (98)$$

Equation (98) is identical with Harrick's Eq. (4) in his book [3] at p. 282; it is valid only if $d/\lambda \ll 1 (r_u \ll 1)$ and for low absorptions, i.e., $\alpha d_e \leq 0.1$. The more general Eq. (97) reflects the fact that the effective thickness $\Delta_{eu3} = \tilde{d}_{eu3} - d_{eu3}$ of phase 3 depends not only on the properties of phase 3 but also on the thickness of layer 2: the larger d, the smaller is Δ_{eu3} and hence the absorption contribution from phase 3. Equation (97) is valid for $r_u < 1$. Furthermore, as for the relations (79) and (80), computer calculations show that the approximations (97) and (98) for the absolute reflectivities are good for Ge/HCl/H_2O, acceptable for Si/HCl/H_2O, but unacceptable for ZnSe/HCl/H_2O.

2.3.2.4. Approximation for Weak Absorption: N Layers

Referring to Figure 6 we find the following approximate relations for the normalized reflectivities R_{nu} in the polarization state $u(= s, p)$,

$$R_{nu} \approx 1 - \sum_{l=2}^{N+1} \left[d_{eu}(l, l)\alpha_l - d_{eu}(l, N+2)\alpha_{N+2} \right] \qquad (99)$$

where α_l is the absorption coefficient of layer l, α_{N+2} is the absorption coefficient of the bulk-phase $N + 2$, and where the effective thicknesses are given by

$$\begin{aligned} d_{eu}(l, l) &= f G_{ul} d_l n_l \\ d_{eu}(l, N+2) &= f G_{ul} d_l n_{N+2} \end{aligned} \qquad (100)$$

In theses relations, d_l is the geometrical thickness of layer l and

$$\begin{aligned} f &= \frac{4n_1 \cos\theta_i}{n_1^2 - n_{N+2}^2} \\ G_{sl} &= 1 \\ G_{pl} &= \frac{1 - (n_{1l}^2 + n_{1,N+2}^2)\sin^2\theta_i}{\cos^2\theta_i - n_{1,N+2}^2\sin^2\theta_i} \end{aligned} \qquad (101)$$

It is easily verified that for $N = 1$ the relations are identical to Eqs. (88)–(92). Note that according to Eqs. (97), (98) the normalized reflectivities depend on the differences contained in the square brackets; they are a measure of the optical contrast between layer l and the substrate $N + 2$. The relations (99)–(101) are obtained if in Eqs. (84), (85) only linear terms in the extinction coefficients k are retained. Furthermore, these are valid only if $(n_l^2 - n_{N+2}^2)/(n_1^2 - n_{N+2}^2) \ll G_{pl}$, a condition which is satisfied for many materials studied with Ge-ATR crystals ($n_1 = 4.0$), but usually not for metallic films. If the foregoing condition is not satisfied, the general form of Eq. (99) is still valid, but the expressions for $d_{ep}(l, l)$ and $d_{ep}(l, N+2)$ are modified (see the Appendix).

2.3.2.5. Systems Containing Metallic Layers

As an example, we consider the system illustrated in Figure 7 in which layer 2 is a metallic layer with thickness d and complex dielectric constant $\hat{\varepsilon}_2$.

We are interested in the normalized ATR reflectivity,

$$R_{nu}(1, 2, 3) = \frac{R_u(1, 2, 3)}{R_u(1, 2, \text{air})} \qquad (102)$$

where $u = (s, p)$ is the polarization state, $R_u(1, 2, 3)$ is the absolute reflectivity of the system considered, and $R_u(1, 2, \text{air})$ is the reference reflectivity which is measured by replacing the bulk-phase 3 by air (or by vacuum). As mentioned in Sections 2.3.2.1 and 2.3.2.3, it is not possible to evaluate the appropriate Fresnel coefficients on the basis of the linear-approximation relations (86), (87) if metallic layers are present, even if $d/\lambda \ll 1$. This is due to the large absorption coefficient k_2 (or large values of $\alpha_2 d$). This gives rise to appreciable phase changes and absorbing loss by multiply-reflected light in the metallic layer. We must therefore resort to the exact Fresnel coefficients $r_u(1, 2, 3)$ and $r_u(1, 2, \text{air})$ which are given by [17],

$$r_u(1, 2, 3) = \frac{r_u(1, 2) + r_u(2, 3)e^{2i\beta}}{1 + r_u(1, 2)r_u(2, 3)e^{2i\beta}} \qquad (103)$$

$$r_u(1, 2, \text{air}) = \frac{r_u(1, 2) + r_u(2, \text{air})e^{2i\beta}}{1 + r_u(1, 2)r_u(2, \text{air})e^{2i\beta}} \qquad (104)$$

where

$$r_s(j, k) = \frac{\zeta_j - \zeta_k}{\zeta_j + \zeta_k} \qquad r_p(j, k) = \frac{\eta_j - \eta_k}{\eta_j + \eta_k} \qquad (105)$$

are the Fresnel reflection coefficients of the interface of two contiguous phases j and k and

$$\zeta_j = \left(\hat{\varepsilon}_j - \varepsilon_1 \sin^2\theta_i\right)^{1/2} \qquad \eta_j = \zeta_j/\hat{\varepsilon}_j \qquad (106)$$

and

$$\beta = 2\pi \zeta_2 d/\lambda \qquad (107)$$

The absolute reflectivities are now given by

$$\begin{aligned} R_u(1, 2, 3) &= \left| r_u(1, 2, 3) \right|^2 \\ R_u(1, 2, \text{air}) &= \left| r_u(1, 2, \text{air}) \right|^2 \end{aligned} \qquad (108)$$

from which the desired normalized reflectivity $R_{nu}(1, 2, 3)$ can be calculated using Eq. (102).

In the following, we consider the system Ge/Pt/H_2O, i.e., a Ge-ATR crystal (phase 1) with $\varepsilon_1 = n_1^2$, $n_1 = 4.01$ with an evaporated Pt layer of thickness d and dielectric constant $\hat{\varepsilon}_2$ (phase 2), in contact with bulk water with dielectric constant $\hat{\varepsilon}_3$ (phase 3). The angle of incidence is θ_i and we consider the bending mode of liquid water with n_3 and k_3 as illustrated in Figure 3. In the limited wave number range considered, 1400 cm$^{-1} \leq \tilde{\nu} \leq 1900$ cm^{-1}, $\hat{\varepsilon}_2 = (n_2 + ik_2)^2$ is taken to be independent of $\tilde{\nu}$ with $n_2 = 5.71$ and $k_2 = 23.35$ [23].

Before presenting the results of the calculations, we show that for metallic films the linear-approximation relations (86), (87) break down. This can be seen by substituting $\hat{\varepsilon}_2 = \varepsilon_{21} +$

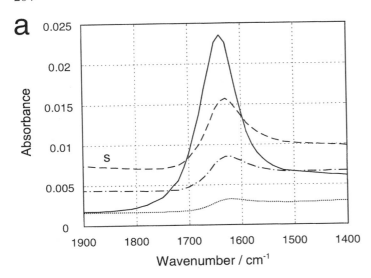

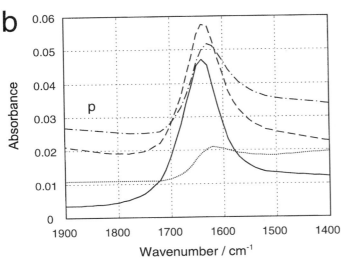

Fig. 9. (a) and (b) Calculated absorbance $A_u(\tilde{\nu}) = -\log_{10}[R_u(1,2,3)/R_u(1,2,\text{air})]$ for s-polarized (a) and p-polarized (b) light for the system Ge/Pt/H_2O for different thicknesses of Pt. Parameters: $n_1 = 4.01$ (Ge), $\hat{n}_2 = 5.71 + 23.35i$ (Pt at 1670 cm^{-1}), $\theta_i = 45°$; the semi-infinite third phase of water was characterized by the wavelength-dependent complex refractive index n_3 depicted in Figure 3 (water bending mode) for the calculation of $R_u(1,2,3)$ and by $n_3 = 1.0$ (vacuum or air) for the reference $R_u(1,2,\text{air})$. The thickness of the Pt layer is $d_{Pt} = 0$ (solid line), $d_{Pt} = 5$ nm (dashed line), $d_{Pt} = 10$ nm (dashed–dotted line), and $d_{Pt} = 20$ nm (dotted line). As the thickness of the Pt layer increases, the field in phase 3 an hence the absorbance of the water bending mode decrease.

$i\varepsilon_{22}$ in ζ_2 given by Eq. (106); observing that $|\varepsilon_{21}| = |n_2^2 - k_2^2| \gg \varepsilon_1 \sin^2\theta_i$ for metals one finds $2i\beta \approx -4\pi k_2 d/\lambda + 4\pi n_2 d/\lambda$ for the exponent in Eqs. (103), (104). Due to the large value of k_2, a linear development of the form $\exp(2i\beta) \approx 1 + 2i\beta$ is not accurate, even for very small values of d/λ (e.g., for $d = 10$ nm, $\lambda = 6$ μm). Figure 9 shows the absorbance $A_s(\tilde{\nu})$ and $A_p(\tilde{\nu})$, of the bending mode of water,

$$A_u(\tilde{\nu}) = -\log_{10}[R_{nu}(1,2,3)] \qquad (109)$$

for different thicknesses of the Pt layer ($d = 0, 5, 10, 20$ nm)

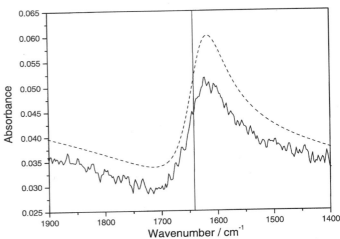

Fig. 10. Calculated absorbance (dashed line) for unpolarized light (composed of 60% p- and 40% s-polarized light) for water over a ZnSe IRE ($n_1 = 2.43$) coated with 10-nm Pt. The wave number-dependent complex refraction index of bulk Pt was used for the calculation [23]. Angle of incidence $\theta_i = 45°$. The solid line is the measured spectrum for the same conditions. The vertical line indicates the band center for bulk water. For the optical constants of water a damped harmonic oscillator representation was used. [60]. Reproduced by permission of The Royal Society of Chemistry on behalf of the PCCP Owner Societies.

and for $\theta_i = 45°$. McIntyre [17] has also given relations for the normalized reflectivity changes of the two-layer system,

ATR (1) / metal film (2) / adsorbate layer (3) / absorbing ambient (4)

i.e., expressions for $R_{nu}(1,2,3,4) = R_u(1,2,3,4)/R_u(1,4)$. Such relations, although complicated, are of great importance for many applications, e.g., for electrochemical systems involving a Pt film (2) evaporated on the ATR (1), an adsorbate film (3), and water as ambient (4).

As a last example, we consider the experimental absorbance of the bending mode with the calculated absorbance as shown in Figure 10. A 10-nm thick Pt film with refractive index $\hat{n}_2 = n_2 + ik_2$ has been evaporated on a ZnSe-ATR crystal with $n_1 = 2.43$. Water with refractive index $\hat{n}_3 = n_3 + ik_3$ is in contact with the Pt film; the system is shown in Figure 7. The experiment has performed with unpolarized light (single reflection) and the angle of incidence was 45°. In contrast to the calculations shown in Figure 9 in which $\hat{n}_2$ was taken to be independent of $\tilde{\nu}$, the calculated absorbance A_{nat} is based on a frequency-dependent complex refractive index $\hat{n}_2$ obtained from a curve fitting to the data given in [23]. For $\hat{n}_3(\tilde{\nu})$ of the bending mode of water, a damped harmonic oscillator representation has been used. The absorbance for unpolarized light ("natural" light in the spectroscopic experiment is approximately composed of 40% s-polarized and 60% p-polarized light in the interesting range of wave numbers) is defined by $A_{nat} = -\log_{10}(R_{nat}/R_{0,nat})$, where $R_{nat} = 0.4R_s + 0.6R_p$ and $R_{0,nat} = 0.4R_{0,s} + 0.6R_{0,p}$; here, R_s and R_p are the absolute reflectivities of the system ZnSe/Pt/H_2O, while $R_{0,s}$ and $R_{0,p}$ are the absolute reflectivities of the system ZnSe/Pt–air.

It is gratifying that the main characteristics of the experimental curve are very well reproduced by the model calculations.

3. ORIENTATION MEASUREMENTS

3.1. Uniaxial Alignment

In this section, tools are provided for the analysis of spectra measured with polarized light. We use the approach described by Zbinden [24]. Considering thin films, one often encounters the situation of partial molecular orientation along the axis perpendicular to the support, which means according to Figure 11 alignment along the z axis. Explicit solutions are derived for the special case that the molecules exhibit free rotation around the molecular axis as well as an isotropic arrangement around the z axis. For more general cases and alternative approaches, the reader is referred to the book by Michl and Thulstrup [25].

3.1.1. Coordinate Systems and Vector Transformations

The general situation is described by Figure 11. The unit vector $\mathbf{m}_{i,k}$ denotes the direction of the kth transition dipole moment associated with the ith functional group of a molecule which is aligned along the t axis. The local orientation axis of the corresponding group is denoted by a_i. There are three relevant coordinate systems to be considered: First, the group-fixed system $\{x_{ai}, y_{ai}, z_{ai}\}$ where the z axis coincides with the ith group axis a_i. This coordinate system enables the description of the directions of group transition dipole moments relative to the molecular structure of the respective functional group by the angles $\phi_{ik,a}$ and $\theta_{ik,a}$. Information on these angles are available for many so-called group vibrations. In the case of more complex vibrational modes, the application of normal coordinates analysis is required [26]. Second, a reorientation of the ith functional group, e.g., due to a conformational change of the molecule will be described in the molecule-fixed coordinate system $\{x_t, y_t, z_t\}$. Generally, the origin of this system is set into the center of gravity of the undistorted molecule, and the coordinate axes are coinciding with the principal axes. The relating parameters between the coordinate systems are the Eulerian angles $\phi_{ai}, \theta_{ai}, \psi_{ai}$, and ϕ_t, θ_t, ψ_t, respectively. The latter establish the connection between the molecule-fixed coordinate system $\{x_t, y_t, z_t\}$ and the laboratory coordinate system $\{x, y, z\}$ in which experimental data are obtained ("observer's system"). Unfortunately, there exist different definitions of the Eulerian angles in literature. In this context, we have used the definitions given in Ref. [27]. All indicated rotations have to be performed counterclockwise. To generate the molecule-fixed system from the laboratory coordinate system starting from coinciding axes, one has first to rotate around z resulting in an angle ϕ_t between the x and x_t axis. In the next step, rotation about x_t by the angle θ_t, is performed followed by a final rotation about z_t by the angle ψ_t. An analogous procedure relates the molecule- and group-fixed coordinate systems t and a_i. Experimental data are related to the laboratory system (observer's

system) l, $\{x, y, z\}$. The components of the unit vector $\mathbf{m}_{i,k}$ which are experimentally accessible are then given by

$$m_{ik,x} = \sin\theta_{ik}\cos\phi_{ik}$$
$$m_{ik,y} = \sin\theta_{ik}\sin\phi_{ik} \qquad (110)$$
$$m_{ik,z} = \cos\theta_{ik}$$

The corresponding components in the t and a systems are obtained by orthogonal transformations [27]. The aim is now to express the components given by Eq. (110) as a function of molecular structure and molecular orientation. For that purpose, we start in the a_i coordinate system where $\mathbf{m}_{i,k}$ has the components $m_{ik,a,x}$, $m_{ik,a,y}$, $m_{ik,a,z}$. For the sake of simplicity, the index i, denoting the ith functional group, is omitted from now on. Thus, transformation of $\mathbf{m}_k$ from the a into the t coordinate system is performed by the matrix A according to

$$A = \begin{pmatrix} a_{11} & a_{12} & a_{13} \\ a_{21} & a_{22} & a_{23} \\ a_{31} & a_{32} & a_{33} \end{pmatrix}$$

$$= \begin{pmatrix} \cos\psi_a\cos\phi_a - \cos\theta_a\sin\phi_a\sin\psi_a \\ \cos\psi_a\sin\phi_a + \cos\theta_a\cos\phi_a\sin\psi_a \\ \sin\psi_a\sin\theta_a \end{pmatrix.$$

$$\begin{matrix} -\sin\psi_a\cos\phi_a - \cos\theta_a\sin\phi_a\cos\psi_a \\ -\sin\psi_a\sin\phi_a + \cos\theta_a\cos\phi_a\cos\psi_a \\ \cos\psi_a\sin\theta_a \end{matrix}$$

$$\left.\begin{matrix} \sin\theta_a\sin\phi_a \\ -\sin\theta_a\cos\phi_a \\ \cos\theta_a \end{matrix}\right) \qquad (111)$$

Furthermore, denoting the matrix converting the coordinates of $\mathbf{m}_k$ from the t system to the laboratory system (l) by T one obtains

$$T = \begin{pmatrix} t_{11} & t_{12} & t_{13} \\ t_{21} & t_{22} & t_{23} \\ t_{31} & t_{32} & t_{33} \end{pmatrix}$$

$$= \begin{pmatrix} \cos\psi_t\cos\phi_t - \cos\theta_t\sin\phi_t\sin\psi_t \\ \cos\psi_t\sin\phi_t + \cos\theta_t\cos\phi_t\sin\psi_t \\ \sin\psi_t\sin\theta_t \end{pmatrix.$$

$$\begin{matrix} -\sin\psi_t\cos\phi_t - \cos\theta_t\sin\phi_t\cos\psi_t \\ -\sin\psi_t\sin\phi_t + \cos\theta_t\cos\phi_t\cos\psi_t \\ \cos\psi_t\sin\theta_t \end{matrix}$$

$$\left.\begin{matrix} \sin\theta_t\sin\phi_t \\ -\sin\theta_t\cos\phi_t \\ \cos\theta_t \end{matrix}\right) \qquad (112)$$

It should be noted once more that the matrices (111) and (112) hold for each functional group i with orientation axis a_i. For the sake of simplicity, the index i had been omitted.

The components of $\mathbf{m}_k$ in the laboratory system may now be expressed by the following transformation,

$$\mathbf{m}_k = T \cdot A \cdot \mathbf{m}_{k,a} \qquad (113)$$

Equation (113) holds for a distinct orientation of $\mathbf{m}_k$ in space. The expressions for the components $m_{k,x}, m_{k,y}, m_{k,z}$ are rather

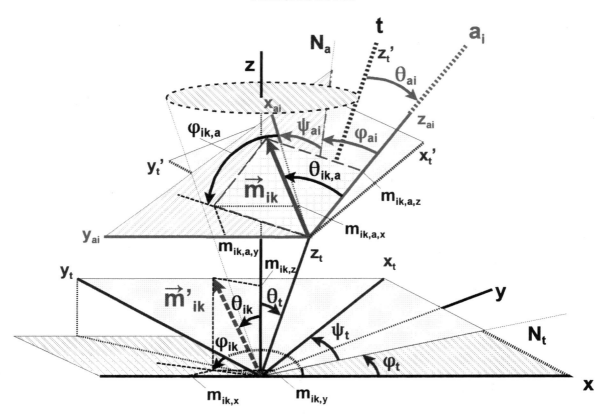

Fig. 11.   Uniaxial orientation of molecules about the z axis of the laboratory coordinate system $\{x, y, z\}$. The direction of the transition moment of the kth molecular vibration located in the ith functional group of the molecule is defined by the angles ϕ_k and θ_k in the laboratory coordinate system and by the angles $\phi_{ik,a}$ and $\theta_{ik,a}$ in the group-fixed coordinate system $\{x_{ai}, y_{ai}, z_{ai}\}$. The orientation axis of the ith group is denoted by a_i. The a_i system is related to the molecule-fixed coordinate system $\{x_t, y_t, z_t\}$ by the Eulerian angles ϕ_{ai}, θ_{ai}, and ψ_{ai} (the definition of Eulerian angles follows Ref. [27]). Both the molecular axis t and the group axis a_i coincide with the respective z axes, z_t and z_{ai}. N_t and N_a are the nodal lines. For the sake of clarity, the $\{x_t, y_t, z_t\}$ system and the transition moment direction unit vector $\mathbf{m}_{i,k}$ are redrawn after shifts along the z axis and denoted by $\mathbf{m}_{i,k}$. Finally, the $\{x_t, y_t, z_t\}$ system is related to the laboratory-fixed coordinate system (observer's system) $\{x, y, z\}$ by the Eulerian angles ϕ_t, θ_t, and ψ_t.

complicated. The simplest expression result for $m_{k,z}$ is given in Eq. (114) as an example to visualize the superposition of conformational and orientational effects. Then,

$$
\begin{aligned}
\mathbf{m}_{k,z} = & \big[\sin\psi_t \sin\theta_t(\cos\psi_a \cos\phi_a - \cos\theta_a \sin\phi_a \sin\psi_a) \\
& + \cos\psi_t \sin\theta_t(\cos\psi_a \sin\phi_a + \cos\theta_a \cos\phi_a \sin\psi_a) \\
& + \cos\theta_t \sin\psi_a \sin\theta_a\big]\mathbf{m}_{k,a,x} \\
& + \big[\sin\psi_t \sin\theta_t(-\sin\psi_a \cos\phi_a - \cos\theta_a \sin\phi_a \cos\psi_a) \\
& + \cos\psi_t \sin\theta_t(-\sin\psi_a \sin\phi_a + \cos\theta_a \cos\phi_a \cos\psi_a) \\
& + \cos\theta_t \cos\psi_a \sin\theta_a\big]\mathbf{m}_{k,a,y} \\
& + (\sin\psi_t \sin\theta_t \sin\theta_a \sin\phi_a - \cos\psi_t \sin\theta_t \sin\theta_a \cos\phi_a \\
& + \cos\theta_t \cos\theta_a)\mathbf{m}_{k,a,z}
\end{aligned}
\tag{114}
$$

Terms with subscript a, as shown by Figure 11, depend on the molecular conformation, whereas terms with subscript t depend on molecular orientation. The situation as described by Eq. (114) holds for a single molecule or for an oriented crystalline ultrastructure (OCU) where all molecules have the same orientation and conformation. In a real system, however, molecules are not oriented exactly alike, but exhibit angular flexibilities. As a consequence, optical measurements give information about the ensemble mean of system parameters. To take account of this fact, averaging over all possible angular regions must be performed by the use of angular probability density functions.

3.1.2. Probability Density

As already mentioned, we distinguish two types of flexibility, a conformational one described in the a system $\{x_{ai}, y_{ai}, z_{ai}\}$ and an orientational one described in the t system $\{x_t, y_t, z_t\}$. The latter deals with the molecular orientation with respect to the laboratory system $\{x, y, z\}$ and is described by the *probability density* or the *orientation distribution function* $f_t(\Omega_t)$. Denoting $f_t(\Omega_t) \cdot d\Omega_t$ the probability that a given molecule is within the limits Ω_t and $\Omega_t + d\Omega_t$ where Ω_t stands for the Eulerian angles ϕ_t, θ_t, ψ_t. Correspondingly, we denote the *conformational distribution function* $f_{ai}(\Omega_{ai})$. $f_{ai}(\Omega_{ai}) \cdot d\Omega_{ai}$ is the probability that the ith functional group is oriented within Ω_{ai} and $\Omega_{ai} + d\Omega_{ai}$. Here, Ω_{ai} stands for the Eulerian angles $\phi_{ai}, \theta_{ai}, \psi_{ai}$. The probability density is normalized; i.e., the integral over all possible values of Ω must be equal to 1. Then,

$$\int_\Omega f(\Omega)\,d\Omega = 1 \qquad (115)$$

For an isotropic sample, $f(\Omega)$ is equal to 1. The average value of an arbitrary angular-dependent quantity $p(\Omega)$ is then given by

$$\langle p \rangle = \int_\Omega p(\Omega) f(\Omega)\,d\Omega \qquad (116)$$

If two or more probability density functions are associated with a given system as in our case (see Fig. 11), the overall probability density is given by the product of the involved density functions. Moreover, if n equal functional groups are considered at different positions in a molecule then the average of $p(\Omega)$ is determined by

$$\langle p \rangle = \sum_{i=1}^{n}\langle p_i \rangle = \int_{\Omega_t}\sum_{i=1}^{n}\int_{\Omega_{ai}} p(\Omega_{ai}, \Omega_t)\cdot f_{ai}(\Omega_{ai})$$
$$\times f_t(\Omega_t)\,d\Omega_{ai}\,d\Omega_t \quad (117)$$

The magnitude of $d\Omega$ is given by

$$d\Omega_{ai} = \frac{1}{8\pi^2}\sin\theta_{ai}\,d\phi_{ai}\,d\theta_{ai}\,d\psi_{ai}$$

and $\qquad\qquad\qquad\qquad\qquad\qquad (118)$

$$d\Omega_t = \frac{1}{8\pi^2}\sin\theta_t\,d\phi_t\,d\theta_t\,d\psi_t$$

leading to the normalization conditions,

$$\int_{\phi_{ai}=0}^{2\pi}\int_{\theta_{ai}=0}^{\pi}\int_{\psi_{ai}=0}^{2\pi} f_{ai}(\phi_{ai},\theta_{ai},\psi_{ai})\frac{1}{8\pi^2}$$
$$\times \sin\theta_{ai}\,d\phi_{ai}\,d\theta_{ai}\,d\psi_{ai} = 1$$

and $\qquad\qquad\qquad\qquad\qquad\qquad (119)$

$$\int_{\phi_t=0}^{2\pi}\int_{\theta_t=0}^{\pi}\int_{\psi_t=0}^{2\pi} f_t(\phi_t,\theta_t,\psi_t)\frac{1}{8\pi^2}\sin\theta_t\,d\phi_t\,d\theta_t\,d\psi_t = 1$$

The general normalization condition for the coupled a and t systems as presented in Figure 11 is

$$\int_{\phi_t=0}^{2\pi}\int_{\theta_t=0}^{\pi}\int_{\psi_t=0}^{2\pi}\int_{\phi_{ai}=0}^{2\pi}\int_{\theta_{ai}=0}^{\pi}\int_{\psi_{ai}=0}^{2\pi} f_{ai}(\phi_{ai},\theta_{ai},\psi_{ai})$$
$$\times f_t(\phi_t,\theta_t,\psi_t)\frac{1}{64\pi^4}\sin\theta_{ai}\sin\theta_t$$
$$\times d\phi_{ai}\,d\theta_{ai}\,d\psi_{ai}\,d\phi_t\,d\theta_t\,d\psi_t = 1 \qquad (120)$$

Uniaxial orientation means isotropy around the z axis which reduces Eq. (120) to

$$\int_{\theta_t=0}^{\pi}\int_{\psi_t=0}^{2\pi}\int_{\phi_{ai}=0}^{2\pi}\int_{\theta_{ai}=0}^{\pi}\int_{\psi_{ai}=0}^{2\pi} f_{ai}(\phi_{ai},\theta_{ai},\psi_{ai})f_t(0,\theta_t,\psi_t)$$
$$\times\frac{1}{32\pi^3}\sin\theta_{ai}\sin\theta_t\,d\phi_{ai}\,d\theta_{ai}\,d\psi_{ai}\,d\theta_t = 1 \qquad (121)$$

Furthermore, allowing free rotation around the molecular axis (t axis) results in the normalization condition,

$$\int_{\theta_t=0}^{\pi}\int_{\phi_{ai}=0}^{2\pi}\int_{\theta_{ai}=0}^{\pi}\int_{\psi_{ai}=0}^{2\pi} f_{ai}(\phi_{ai},\theta_{ai},\psi_{ai})f_t(0,\theta_t,0)$$
$$\times\frac{1}{16\pi^2}\sin\theta_{ai}\sin\theta_t\,d\phi_{ai}\,d\theta_{ai}\,d\psi_{ai}\,d\theta_t = 1 \qquad (122)$$

Finally, consider the special case of a stiff molecule exhibiting only orientational flexibility. In this case, the angles ϕ_{ai}, θ_{ai}, ψ_{ai} assume distinct values ϕ_{ai0}, θ_{ai0}, ψ_{ai0} and the corresponding conformational distribution function becomes

$$f_{ai}(\phi_{ai},\theta_{ai},\psi_{ai}) = 2\pi\delta(\phi_{ai}-\phi_{ai0})\cdot\frac{2}{\sin\theta_{ai0}}\delta(\theta_{ai}-\theta_{ai0})$$
$$\times 2\pi\delta(\psi_{ai}-\psi_{ai0}) \qquad (123)$$

where δ denotes the Dirac delta function. Introducing Eq. (123) into (Eq. 122) results in the corresponding normalization condition,

$$\int_{\theta_t=0}^{\pi} f_t(0,\theta_t,0)\frac{1}{2}\sin\theta_t\,d\theta_t = 1 \qquad (124)$$

3.1.3. Light Absorption and Dichroic Ratio

The probability of electric dipole light absorption is proportional to the square of the scalar product between the electric field and the transition dipole moment according to

$$\Delta I_{ik} \propto (\mathbf{E}\cdot\mathbf{m}_{ik})^2$$
$$= |\mathbf{E}|^2\cdot|\mathbf{m}|_{ik}\cdot^2\cdot\cos^2(\mathbf{E},\mathbf{m}_{ik}) \qquad (125)$$
$$= (E_x m_{ik,x} + E_y m_{ik,y} + E_z m_{ik,z})^2$$

Obviously, the intensity of light absorption depends on the mutual orientation of the electric field $\mathbf{E}$ and the unit vector in the direction of the transition dipole moment $\mathbf{m}_{i,k}$. Equation (125) is the basis for orientation measurements. $m_{ik,x}$, $m_{ik,y}$, and $m_{ik,z}$ denote the components of the unit vector in the direction of the transition dipole moment in the laboratory coordinate system, see Figure 11. In the case of ATR spectroscopy, $\mathbf{E}$ denotes the electric field of the evanescent wave. According to Section 2, parallel polarized incident light (p) results in the x and z components of the electric field of the evanescent wave, while perpendicular polarized incident light (s) produces the y component. It is usual to evaluate orientation measurements in terms of dimensionless relative intensities to get rid of physical and molecular constants, such as the magnitude of the transition moment. Introducing the so-called dichroic ratio R, which is the absorbance ratio of the kth vibrational mode obtained from spectra measured with parallel and perpendicular polarized incident light.

$$R_{ik} = \frac{A_{ik,p}}{A_{ik,s}} = \frac{d_{ei,p}}{d_{ei,s}}$$
$$= \frac{E_x^2 m_{ik,x}^2 + E_z^2 m_{ik,z}^2 + 2E_x E_z m_{ik,x} m_{ik,z}}{E_y^2 m_{ik,y}^2} \qquad (126)$$

Introducing Eq. (110) into Eq. (126) results in the dichroic ratio as depending on the angles ϕ_{ik} and Θ_{ik}. In case of a rigid system, a so-called oriented crystalline ultrastructure (OCU) [28], these angles may be determined from p- and s-polarized spectra, provided the ith functional group exhibits at least two vibrational modes (k and $k+1$) with known mutual orientation of the transition moments, e.g., the symmetric and asymmetric vibrations of a methylene group. Most practical systems, however, are not of OCU type, i.e., measured absorbances reflect the average over many different position of the molecules of the system.

3.1.4. Ultrastructure and Order Parameter

Deviation from OCU will now be described by the use of the conformation and orientation distribution function $f_a(\Omega_a)$ and $f_t(\Omega_t)$, respectively (see Section 3.1.2 and Fig. 11). The angle $\theta_{ik,a}$ is determined by the structure and the vibrational mode of the ith functional group. In a methylene group, $\theta_{ik,a} = 90°$ for both, the symmetric ν_s (CH$_2$) and asymmetric ν_{as} (CH$_2$) stretching vibration. However, $\phi_{ik,a}$ behaves different for these two modes. Denoting ν_s (CH$_2$) by the index $k = 1$ and ν_{as} (CH$_2$) by $k = 2$, it follows from spectroscopic reasons that $\phi_{i2,a} = \phi_{i1,a} + 90°$. Variation of $\phi_{ik,a}$ means rotation of the ith group around the a_i axis. This may be partial or total, the latter leading to isotropy about this axis, the former being described by a probability density function which contains terms of the potential energy resulting from torsion about chemical bonds (e.g., gauche transconversion) and other contributions of intra- and intermolecular interaction. Ultrastructures as introduced in Ref. [28] can now be defined as follows: Referring to Figure 11, *oriented crystalline ultrastructure* (OCU) means that all angles are distinct, *microcrystalline ultrastructure* (MCU) means isotropic ϕ_t (isotropic arrangement of molecules around the z axis), all other angles are distinct. Finally, *liquid crystalline ultrastructure* (LCU) is defined as isotropic Ψ_t (free rotation about the molecular axis (t axis)), isotropic ϕ_t (isotropic arrangement of molecules around the z axis), and fluctuating θ_t about a mean value $\bar{\theta}_t$. The mean dichroic ratio of the kth overlapping vibration resulting from n equal functional groups which are at different positions in a molecule can thus be expressed in the following form,

$$\langle R_k \rangle = \frac{\langle A_{k,p} \rangle}{\langle A_{k,s} \rangle} = \frac{\langle d_{e,k,p} \rangle}{\langle d_{e,k,s} \rangle}$$
$$= \left(E_x^2 \sum_{i=1}^{n} \langle m_{ik,x}^2 \rangle + E_z^2 \sum_{i=1}^{n} \langle m_{ik,z}^2 \rangle \right. \qquad (127)$$
$$\left. + 2E_x E_z \sum_{i=1}^{n} \langle m_{ik,x} m_{ik,z} \rangle \right) \left(E_y^2 \sum_{i=1}^{n} \langle m_{ik,y}^2 \rangle \right)^{-1}$$

$\langle m_{ik,x}^2 \rangle$, $\langle m_{ik,y}^2 \rangle$, and $\langle m_{ik,z}^2 \rangle$ denote the averaged squares of the components of the unit vector of the kth molecular vibration resulting from the ith functional group expressed in the laboratory fixed coordinate system $\{x, y, z\}$. They may be calculated

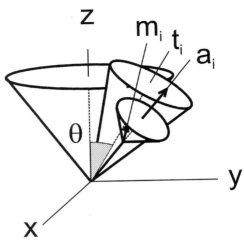

Fig. 12. Superposition of three uniaxial distributions. m_i is the instantaneous orientation of the unit vector of the transition moment. a_i is the instantaneous orientation of the segmental molecular axis. t_i is the tilted average orientation of the molecular axis. Angles: θ_t is the tilt angle between the z axis and t_i, θ_{ai} is the angles the between t_i and a_i, θ is the angle between a_i and the transition moment m_i.

by Eq. (113) and averaged by Eq. (117) using the corresponding distribution functions. In general, such operations are only feasible by numerical means.

3.1.5. Order Parameter Associated with a Liquid Crystalline Ultrastructure

An explicit analytical solution is now discussed for the special case that the three angles ϕ_t, ψ_t and $\phi_{ik,a}$ are isotropic, which means free rotation about the a_i, t_i, and z axes. In this case, the distribution of the transition moments of a given vibration forms the surface of a cone about the respective axis as depicted in Figure 12. A detailed discussion of this model is given in Ref. [29]. Insertion of Eq. (110) into (127) results in

$$\langle R_k \rangle = \frac{\langle A_{k,p} \rangle}{\langle A_{k,s} \rangle} = \frac{\langle d_{e,k,p} \rangle}{\langle d_{e,k,s} \rangle}$$
$$= \left(E_x^2 \sum_{i=1}^{n} \langle \sin^2\theta_{ik}\cos^2\phi_{ik} \rangle + E_z^2 \sum_{i=1}^{n} \langle \cos^2\theta_{ik} \rangle \right.$$
$$\left. + 2E_x E_z \sum_{i=1}^{n} \langle \sin\theta_{ik} \cos\phi_{ik} \cos\theta_{ik} \rangle \right)$$
$$\times \left(E_y^2 \sum_{i=1}^{n} \langle \sin^2\theta_{ik}\sin^2\phi_{ik} \rangle \right)^{-1} \qquad (128)$$

Assuming isotropy about the z axis leads to a probability density according to

$$f(\phi_{ik}, \theta_{ik}) = f(0, \theta_{ik}) = f(0) \cdot f(\theta_{ik}) \quad \text{with } f(0) = 1 \quad (129)$$

Thus, one obtains from Eq. (128) the following expression for the dichroic ratio [29],

$$\langle R_k \rangle = \frac{\langle A_{k,p} \rangle}{\langle A_{k,s} \rangle} = \frac{\langle d_{e,k,p} \rangle}{\langle d_{e,k,s} \rangle} = \frac{E_x^2}{E_y^2} + 2\frac{E_z^2}{E_y^2}\frac{\sum_{i=1}^n \langle \cos^2\theta_{ik} \rangle}{n - \sum_{i=1}^n \langle \cos^2\theta_{ik} \rangle}$$
$$= \frac{E_x^2}{E_y^2} + 2\frac{E_z^2}{E_y^2}\frac{\overline{\langle \cos^2\theta_k \rangle}}{1 - \overline{\langle \cos^2\theta_k \rangle}} \tag{130}$$

In practice, Eq. (130) has turned out to be a valuable tool for orientation measurements. Experimentally, the mean dichroic ratio $\langle R_k \rangle$ is directly accessible via the measurement of the mean absorbance of the kth vibrational band $\langle A_{k,p} \rangle$ and $\langle A_{k,s} \rangle$ with parallel (p) and perpendicular(s) polarized light, respectively. Obviously, the mean square of the cosine of the angle between the direction of the transition moment and the z axis, $\langle \cos^2\theta_{ik} \rangle$ plays a significant role. Three special cases are now considered, namely, perfect alignment of $\mathbf{m}_{ik}$ along the z axis, second, perfect alignment of $\mathbf{m}_{ik}$ parallel to the x, y plane, and third, isotropic distribution of $\mathbf{m}_{ik}$ in space. According to Eq. (124), $\langle \cos^2\theta_{ik} \rangle$ has to be calculated as

$$\langle \cos^2\theta_{ik} \rangle = \int_0^\pi f(\theta_{ik})\cos^2\theta_{ik}\frac{1}{2}\sin\theta_{ik}\,d\theta_{ik} \tag{131}$$

In the first case of a perfect alignment along an axis forming a fixed angle $\theta_{ik,0}$ with the z axis, one gets the corresponding distribution function from Eq. (123). Insertion into Eq. (131) leads to

$$\langle \cos^2\theta_{ik} \rangle = \int_0^\pi \frac{2}{\sin\theta_{ik,0}}\delta(\theta_{ik} - \theta_{ik,0})\cos^2\theta_{ik}\frac{1}{2}\sin\theta_{ik}\,d\theta_{ik}$$
$$= \cos^2\theta_{ik,0} \tag{132}$$

For the special orientations $\theta_{ik,0} = 0$, i.e., perfect alignment along the z axis, and $\theta_{ik,0} = 90°$, i.e., perfect alignement parallel to the xy-plane, it follows

$$\langle \cos^2\theta_{ik,0} \rangle_z = 1 \quad \text{and} \quad \langle \cos^2\theta_{ik,0} \rangle_{xy} = 0 \tag{133}$$

The probability density for the isotropic case is $f(\theta_{ik}) = 1$. Thus, resulting from Eq. (131),

$$\langle \cos^2\theta_{ik} \rangle_{iso} = \int_0^\pi \cos^2\theta_{ik}\frac{1}{2}\sin\theta_{ik}\,d\theta_{ik} = \frac{1}{3} \tag{134}$$

Now, introducing the segmental order parameter S_{seg} in the form,

$$S_{seg}(i,k) = \frac{3}{2}\langle \cos^2\theta_{ik} \rangle - \frac{1}{2} \tag{135}$$

one obtains from Eqs. (133)–(135) for the special cases discussed earlier,

$$S_{seg,z}(i,k) = \frac{3}{2}\langle \cos^2\theta_{ik} \rangle_z - \frac{1}{2} = 1$$
$$S_{seg,xy}(i,k) = \frac{3}{2}\langle \cos^2\theta_{ik} \rangle_{xy} - \frac{1}{2} = -\frac{1}{2} \tag{136}$$
$$S_{seg,iso}(i,k) = \frac{3}{2}\langle \cos^2\theta_{ik} \rangle_{iso} - \frac{1}{2} = 0$$

As expressed, e.g., by Eq. (130) a superposition of n absorption bands of equal functional groups i at different positions in the

molecule results in arithmetic averaging over n order parameters according to

$$\overline{S}_{seg}(k) = \frac{1}{n}\sum_{i=1}^n S_{seg}(i,k)$$
$$= \frac{3}{2n}\sum_{i=1}^n \langle \cos^2\theta_{ik} \rangle - \frac{1}{2} = \frac{3}{2}\overline{\langle \cos^2\theta_k \rangle} - \frac{1}{2} \tag{137}$$

$\overline{S}_{seg}(k)$ is a measure for the averaged fluctuation of the transition moments $\mathbf{m}_{ik}$ as resulting from the kth molecular vibration. This quantity is experimentally accessible, via dichroic analysis. Solving Eq. (130) for $\overline{\langle \cos^2\theta_k \rangle}$ results in

$$\overline{\langle \cos^2\theta_k \rangle} = \frac{E_x^2 - \langle R_k \rangle \cdot E_y^2}{E_x^2 - \langle R_k \rangle \cdot E_y^2 - 2E_z^2} \tag{138}$$

Considering now the coupled motion composed of structural and orientational fluctuations in the simple framework of a liquid crystalline ultrastructure, it becomes obvious from Figure 12 that the treatment of the a, t, and l systems is analogous, the latter having just been discussed in the context with the segmental order parameter. Analogous to Eq. (135) one can define order parameters associated to the angles θ_t and $\theta_{ik,a}$ for the orientational part,

$$S_{mol} = \frac{3}{2}\langle \cos^2\theta_t \rangle - \frac{1}{2} \tag{139}$$

and for the conformational part,

$$S_{conf}(i) = \frac{3}{2}\langle \cos^2\theta_{ai} \rangle - \frac{1}{2} \tag{140}$$

The same procedure may also be applied to the angle $\theta_{ik,a}$, the deviation of the transition moment from the a_i axis, although this angle remains unaltered in most cases, i.e., averaging is not necessary. Then,

$$S_{vibr}(ik) = \frac{3}{2}\langle \cos^2\theta_{ik,a} \rangle - \frac{1}{2} \tag{141}$$

$S_{seg}(i,k)$ contains the averaged information on orientational and conformational flexibility. In the case of the LCU model which is depicted in Figure 12, it can be expressed as a product of the order parameters defined by Eqs. (139)–(141) resulting in

$$S_{seg}(i,k) = S_{mol}(k) \cdot S_{conf}(i) \cdot S_{vibr}(ik)$$
$$= \left(\frac{3}{2}\langle \cos^2\theta_t \rangle - \frac{1}{2}\right) \cdot \left(\frac{3}{2}\langle \cos^2\theta_{ai} \rangle - \frac{1}{2}\right)$$
$$\times \left(\frac{3}{2}\langle \cos^2\theta_{ik,a} \rangle - \frac{1}{2}\right) \tag{142}$$

To get the relevant parameter for molecular fluctuation, one has to solve Eq. (142) for $S_{mol}(k)$. Furthermore, averaging over the n equal functional groups which are at different positions in the same molecule is required since the experimental data result in general average information due to band overlapping. Selective measurements would require selective isotopic modification of a molecule. Thus, it follows for the general case,

$$S_{mol}(k) = \frac{\overline{S}_{seg}(k)}{\overline{S}_{conf}\overline{S}_{vibr}(k)} \tag{143}$$

The index k stands for the kth vibrational mode of the molecule. The fluctuation of the molecular axis t (see Fig. 11) should be independent of k, thus the results obtained by means of different absorption bands in the same spectrum should vary only within the range of the experimental error, provided Figure 12 is a good approximation of the real system. As already mentioned, $\overline{S}_{\mathrm{seg}}(k)$ is directly accessible by experiment. However, $\overline{S}_{\mathrm{conf}}$ and $\overline{S}_{\mathrm{vibr}}(k)$ depend on molecular structure and vibration, respectively. To evaluate them, at least empirical knowledge about group vibrations is required. This approach works well for a limited number of functional groups [28]. A more general approach, however, is possible by means of molecular modeling and normal mode analysis. By these means, it should be possible to calculate the conformational and vibrational distribution functions. On the other hand, data obtained via polarized ATR spectra could also be used in the reverse direction, i.e., to fit parameters and to set constraints in theoretical calculations. An example for a structural analysis of lipid mono- and bilayers using the approach just described is given in Ref. [29]. For a comprehensive discussion of the concept of order parameters and orientation factors, the reader is referred to Ref. [25].

3.1.6. Calculation of Surface Concentrations of an Oriented Layer

Calculation of the surface concentration which we denote by Γ is based on the Lambert–Beer's law. Adaptation for the ATR technique requires the introduction of a hypothetical thickness d_e, the so-called "effective thickness," introduced for the first time by Harrick [3]. Considering a real ATR experiment with a sample of thickness d, then d_e denotes a theoretical thickness of this sample which would result in the same absorbance in a corresponding transmission experiment. As discussed in Section 2, the effective thickness is different for parallel (p, //) and perpendicular (s, ⊥) polarized incident light, even for an isotropic sample (Eqs. (81), (82) and (A.2), (A.4), (A.7)). In a first view, this is a quite astonishing situation which, however, was shown to result from the fact that s-polarized light consists only of the electric field component in the y direction, while the p-polarized light consists of field components in the x and z direction. The expressions for d_e as derived in Section 2 and in the Appendix hold for isotropic samples. Therefore, they are denoted by $d_{e,u,\mathrm{iso}}$ from now on, where u stands for s (⊥) and p (//). Considering axial symmetry around the z axis, it follows from Eqs. (127)–(130),

$$
\begin{aligned}
\langle R_k \rangle &= \frac{\langle A_{k,p} \rangle}{\langle A_{k,s} \rangle} = \frac{\langle d_{e,k,p} \rangle}{\langle d_{e,k,s} \rangle} = \frac{\langle d_{e,k,x} \rangle + \langle d_{e,k,z} \rangle}{\langle d_{e,k,y} \rangle} \\
&= \frac{E_x^2 \sum_{i=1}^{n} \langle m_{ik,x}^2 \rangle + E_z^2 \sum_{i=1}^{n} \langle m_{ik,z}^2 \rangle}{E_y^2 \sum_{i=1}^{n} \langle m_{ik,y}^2 \rangle} \\
&= \frac{E_x^2 \overline{\langle m_{k,x}^2 \rangle} + E_z^2 \overline{\langle m_{k,z}^2 \rangle}}{E_y^2 \overline{\langle m_{k,y}^2 \rangle}}
\end{aligned}
\tag{144}
$$

In the isotropic case, the probability to absorb light by interaction of the transition dipole moment with the electric fields in the x, y, and z directions is the same and equals 1/3 (see also Eq. (134)). Thus, Eq. (144) results in

$$
\begin{aligned}
\langle R_k \rangle_{\mathrm{iso}} &= \frac{\langle A_{k,p} \rangle_{\mathrm{iso}}}{\langle A_{k,s} \rangle_{\mathrm{iso}}} = \frac{\langle d_{e,p} \rangle_{\mathrm{iso}}}{\langle d_{e,s} \rangle_{\mathrm{iso}}} \\
&= \frac{\langle d_{e,x} \rangle_{\mathrm{iso}} + \langle d_{e,z} \rangle_{\mathrm{iso}}}{\langle d_{e,y} \rangle_{\mathrm{iso}}} = \frac{E_x^2 + E_z^2}{E_y^2}
\end{aligned}
\tag{145}
$$

Note that the dichroic ratio obtained from an isotropic sample differs from 1 in the ATR measurement, but equals 1 in a corresponding transmission experiment. It follows that the axial mean effective thicknesses $\langle d_{e,k,x} \rangle$, $\langle d_{e,k,y} \rangle$, and $\langle d_{e,k,z} \rangle$ of a partially oriented sample are given by

$$
\begin{aligned}
\langle d_{e,k,x} \rangle &= 3\overline{\langle m_{k,x}^2 \rangle} \langle d_{e,x} \rangle_{\mathrm{iso}} \\
\langle d_{e,k,y} \rangle &= 3\overline{\langle m_{k,y}^2 \rangle} \langle d_{e,y} \rangle_{\mathrm{iso}} \\
\langle d_{e,k,z} \rangle &= 3\overline{\langle m_{k,z}^2 \rangle} \langle d_{e,z} \rangle_{\mathrm{iso}}
\end{aligned}
\tag{146}
$$

The factor 3 has to be introduced because in the isotropic case $\overline{\langle m_{k,x}^2 \rangle}_{\mathrm{iso}} = \overline{\langle m_{k,y}^2 \rangle}_{\mathrm{iso}} = \overline{\langle m_{k,z}^2 \rangle}_{\mathrm{iso}} = 1/3$. The aim is now to replace the mean squares of transition moment unit vectors by quantities which are experimentally directly accessible. It follows from Eqs. (130) and (144),

$$
\overline{\langle m_{k,x}^2 \rangle} = \overline{\langle m_{k,y}^2 \rangle} = \frac{1}{2}\left(1 - \overline{\langle \cos^2 \theta_k \rangle}\right) \quad \text{and} \quad \overline{\langle m_{k,z}^2 \rangle} = \overline{\langle \cos^2 \theta_k \rangle}
\tag{147}
$$

Now, introducing Eqs. (138) and (147) into Eq. (146) results in

$$
\begin{aligned}
\langle d_{e,k,x} \rangle &= \frac{3}{2}\left(1 - \frac{E_x^2 - \langle R_k \rangle \cdot E_y^2}{E_x^2 - \langle R_k \rangle \cdot E_y^2 - 2E_z^2}\right) \langle d_{e,x} \rangle_{\mathrm{iso}} \\
\langle d_{e,k,y} \rangle &= \frac{3}{2}\left(1 - \frac{E_x^2 - \langle R_k \rangle \cdot E_y^2}{E_x^2 - \langle R_k \rangle \cdot E_y^2 - 2E_z^2}\right) \langle d_{e,y} \rangle_{\mathrm{iso}} \\
\langle d_{e,k,z} \rangle &= 3\frac{E_x^2 - \langle R_k \rangle \cdot E_y^2}{E_x^2 - \langle \cdot R_k \rangle E_y^2 - 2E_z^2} \langle d_{e,z} \rangle_{\mathrm{iso}}
\end{aligned}
\tag{148}
$$

with

$$
\begin{aligned}
\langle d_{e,k,p} \rangle &= \langle d_{e,k,x} \rangle + \langle d_{e,k,z} \rangle \\
&= 3\Bigg[\frac{1}{2}\langle d_{e,x} \rangle_{\mathrm{iso}} + \frac{E_x^2 - \langle R_k \rangle \cdot E_y^2}{E_x^2 - \langle R_k \rangle \cdot E_y^2 - 2E_z^2} \\
&\quad \times \left(\langle d_{e,z} \rangle_{\mathrm{iso}} - \frac{1}{2}\langle d_{e,x} \rangle_{\mathrm{iso}}\right)\Bigg]
\end{aligned}
\tag{149}
$$

and

$$
\langle d_{e,k,s} \rangle = \langle d_{e,k,y} \rangle = \frac{3}{2}\left(1 - \frac{E_x^2 - \langle R_k \rangle \cdot E_y^2}{E_x^2 - \langle R_k \rangle \cdot E_y^2 - 2E_z^2}\right) \langle d_{e,y} \rangle_{\mathrm{iso}}
$$

For the isotropic case, $(E_x^2 - \langle R_k \rangle_{\mathrm{iso}} \cdot E_y^2)/(E_x^2 - \langle R_k \rangle_{\mathrm{iso}} E_y^2 - 2E_z^2) = 1/3$, thus fulfilling the condition $\langle d_{e,k,p} \rangle_{\mathrm{iso}} = \langle d_{e,k,x} \rangle_{\mathrm{iso}} + \langle d_{e,k,z} \rangle_{\mathrm{iso}}$ and $\langle d_{e,k,s} \rangle_{\mathrm{iso}} = \langle d_{e,k,y} \rangle_{\mathrm{iso}}$. The final step leading from effective thicknesses to concentrations and surface concentrations is easily done by applying Lambert–Beer's law, resulting in the volume concentration of a given species as determined by the kth molecular vibration. So,

$$c = \frac{\int_{\text{band }k} \langle A_{k,u} \rangle \, d\tilde{v}}{nN \langle d_{e,k,u} \rangle \int_{\text{band }k} \varepsilon_k(\tilde{v}) \, d\tilde{v}} \quad (150)$$

$\int_{\text{band }k} \langle A_{k,u} \rangle \, d\tilde{v}$ is the integrated absorbance where u stands again for $(p, //)$- and $(s, \perp)$-polarized incident light. n is the number of equal functional groups per molecule and N denotes the number of active internal reflections. $\langle d_{e,k,u} \rangle$ is the effective thickness relevant for the kth vibration measured with $(p, //)$- and $(s, \perp)$-polarized incident light. Note that in our notation $\langle d_{e,k,u} \rangle$ depends on the orientation of the transition moment of the kth vibration. Only $\langle d_{e,u} \rangle_{\text{iso}}$ is independent of the vibrational mode considered. Finally, $\int_{\text{band }k} \varepsilon_k(\tilde{v}) d\tilde{v}$ denotes the integrated molar absorption coefficient of the kth vibration. It should be noted that Eq. (150) also holds for wave number-specific spectroscopic data, such as peak absorbance and peak molar absorption coefficient according to

$$c = \frac{\langle A_{k,u} \rangle}{nN \langle d_{e,k,u} \rangle \varepsilon_k(\tilde{v})} \quad (151)$$

In the case of thin layers ($d < d_p$), it is often adequate to quantify in terms of surface concentration Γ instead of volume concentration c. The relation is given by

$$\Gamma = cd \quad (152)$$

d denotes the geometrical thickness of the sample. Thus, Γ indicates the number of particles in the unit volume projected to the unit area.

4. SPECIAL EXPERIMENTAL TECHNIQUES

In the following two sections, we describe two experimental techniques which have turned out to considerably enhance the quality of background compensation and in the case of modulated excitation (ME) spectroscopy to enable time-resolved measurements. In the latter case, the measurement of phase lags of the system response with respect to the external stimulation, as well as the frequency depends of the amplitude, is the principal means for kinetic analysis and for the evaluation of the underlying reaction scheme.

4.1. Single-Beam-Sample-Reference Technique

Most FTIR spectrometers are working in the single-beam (SB) mode. As a consequence, a single channel reference spectrum has to be stored for later conversion of single channel sample spectra into transmittance and absorbance spectra. This technique favors inaccuracy due to drifts resulting from the instrument or from the sample as well as disturbance by atmospheric absorptions. To eliminate these unwanted effects to a great extent, a new type of ATR attachment has been constructed, converting a single beam instrument into a pseudodouble beam instrument. The principal features of this attachment are depicted in Figure 13. As usual, a convergent IR beam enters the

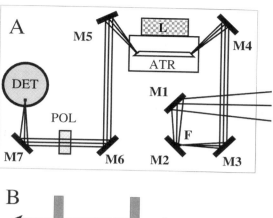

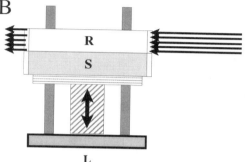

Fig. 13. Single-beam-sample-reference (SBSR) ATR attachment. (A) The focus in the sample compartment is displaced to the position F by the planar mirrors M1 and M2. The off-axis parabolic mirror M3 produces a parallel beam with a diameter of one centimeter, i.e., half of the height of the IRE. The cylindrical mirror M4 focuses the light to the entrance face of the IRE. M5 which has the same shape as M4 reconverts to parallel light passing via the planar mirror M6 through the polarizer POL and being focused to the detector DET by the off-axis parabolic mirror M7. (B) Alternative change from sample to reference and vice versa is performed by computer-controlled lifting and lowering of the ATR cell body. Reproduced from [41] by permission of the American Institute of Physics, © 1998.

sample compartment with a focal point in the middle. This focal point is now displaced by the planar mirrors M1 and M2 to the new position F, whereas the off-axis parabolic mirror M3 performs a conversion of the divergent beam into a parallel beam with fourfold reduced cross section. This beam is focused to the entrance face of a trapezoidal internal reflection elemen (IRE) by a cylindrical mirror M4. Therefore, the ray propagation in the IRE is still parallel to the direction of light propagation (x axis), enabling a subdivision of the large IRE surfaces (x, y plane) in perpendicular direction (y axis) to the light propagation. One half of the IRE is then used for the sample (S) and the other one for the reference (R). Both, S and R, were encapsulated by flow-through cuvettes, independently accessible by liquid or gaseous flow through. This principle is referred to as the *single-beam-sample-reference (SBSR)* technique. A computer-controlled lift moves the cell platform alternatively up and down aligning the sample and reference cuvettes with the IR beam, respectively. Thus, SBSR absorbance spectra are calculated from sample and reference single channel spectra which have been measured with very short mutual time delay. Figure 14 shows the results of a series of hydrogen deuterium (HD) exchange measurements performed in the

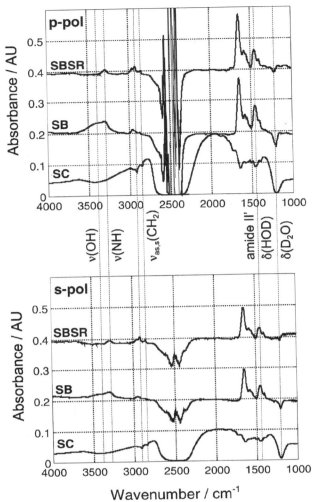

Fig. 14. Comparison of single-beam-sample-reference (SBSR) pseudodouble-beam technique with conventional single-beam (SB) technique. Very low energy in the 2500 and 1200 cm^{-1} region resulted from stretching $\nu(D_2O)$ and bending $\delta(D_2O)$ absorptions of liquid D_2O, respectively, as shown by a single channel (SC) spectrum. The supported bilayers in the sample and reference cuvettes consisted of a dipalmitoylphosphatidic acid (DPPA) LB monolayer and a cardiolipin (CL) adsorbed monolayer. Both membranes exhibited the same age, since the LB layer covered the whole width of the IRE, and the CL adsorption from vesicles occurred synchronously by two independent equal circuits. In a second step, creatine kinase (CK) was adsorbed from a circulating solution in the sample channel. Therefore, the absorbance spectra shown in this figure reflect adsorbed CK as well as any other differences between the S and R channels. Obviously, there are more detectable differences in the SB mode than in the SBSR mode, because SBSR reflects the actual difference between S and R, while SB shows the difference between the actual sample spectrum and a stored (older) reference spectrum. In this case, the partly different results obtained in the SBSR and SB mode result predominately from a slow uptake of H_2O vapor by the circulating D_2O solutions. This leads to overlapping of $\delta(HDO)$ and amide II' of CK, as well as an overcompensation of D_2O absorption bands ($\sim$2500 and 1200 cm^{-1}). It should be noted that the slight overcompensation of $\delta(D_2O)$ at 1200 cm^{-1} is significant, since it reflects the reduced water content in the sample (13) cuvette due to displacement by the CK layer. Ge IRE, angle of incidence, $\theta = 51°$, number of active internal reflections, $N = 36.7$. The refractive indices were: $n_1 = 4.0$, $n_2 = 1.45$, and $n_3 = 1.30$. Reproduced from [41] by permission of the American Institute of Physics, © 1998.

SBSR mode with the enzyme creatine kinase (CK). The enzyme was adsorbed from H_2O buffer to a dipalmitoyl phosphatidic acid/cardiolipin (DPPA/CL) supported bilayer. The conventional SB spectrum reflects the whole history of the sample, whereas the SBSR spectrum reflects the sample state when compared with a reference of the same age. Therefore, the SB spectrum contains the partially deuterated water (HDO) produced by slight H_2O contamination during the experiment in addition to the spectrum of CK. The former obscures the shape of $\nu(NH)$ and amide II' bands, which is an obvious disadvantage of the SB mode. For HD exchange, a D_2O buffer solution was circulated through the sample and reference cuvette of the ATR cell for 3 days. As a consequence, slight contamination of D_2O by atmospheric H_2O could not be avoided in the course of this long-time experiment. The resulting HDO gave rise to absorption bands near 3400 and 1450 cm^{-1} interfering with the NH stretching ($\nu(NH)$) of nonexchanged amide protons, and with amide II' of the deuterated amide groups of the protein, respectively. Since sample and reference contamination by hydronium ions is approximately the same due to equal treatment of the circulating D_2O buffer solutions, the HDO absorption bands ($\nu(OH)$ and $\delta(HDO)$) are compensated to a major extent, as demonstrated by Figure 14, trace SBSR. The SBSR trace represents predominantly membrane-bound CK in a partially deuterated state. The sample consisted of a DPPA/CL/CK assembly, and the reference of a DPPA/CL supported bilayer. Since a sequence of SBSR spectra consists of two independent sequences of single channel spectra, the collected data may by analyzed in the SB mode as well. Doing this by using the single channel spectrum of the sample channel measured in the SBSR mode before CK adsorption (DPPA/CL in a D_2O environment in S and R) as a reference and a corresponding single channel spectrum after about 12 h of CK exposure to the D_2O buffer as a single channel sample spectrum. The resulting SB absorbance spectrum is also presented in Figure 14, trace SB. Thus, SBSR and SB spectra shown had exactly the same experimental conditions. To make the best use of SBSR data, it is recommended to analyze the data by both modes, SBSR and SB since an unwanted synchronous breakdown of sample and reference assembly, e.g., by hydrolysis of a polymer matrix existing in the S and R channels, or by equal loss of lipid molecules from a supported bilayer, would be obscured in the SBSR mode, but unambiguously detected in the SB mode.

4.2. Modulated Excitation Spectroscopy

Modulation spectroscopy or modulated excitation (ME) spectroscopy can always be applied if a system admits a periodic alteration of its state by the variation of an external parameter, such as temperature (T), pressure (p), concentration (c), electric field (E), electric potential (ψ), radiant power (Φ), or mechanical force (F). A schematic is shown in Figure 15. The response of the system to an ME will also be periodic, exhibiting the same frequency as the stimulation. In the case of a non linear system, the response to a sinusoidal stimulation will also contain multiples of the fundamental frequency. After an initial

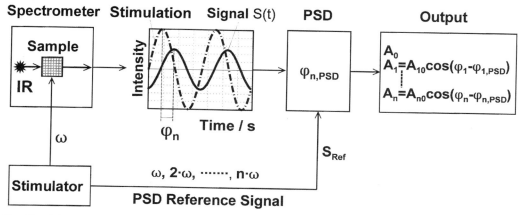

Fig. 15. Schematic for modulated excitation (ME) spectroscopy. A periodic excitation is exerted on the sample with frequency ω. The sample response $S(t)$, as sensed by IR radiation, contains the frequency ω and higher harmonics at wavelengths that are significant for those parts of the sample that have been affected by the stimulation. Selective detection of the periodic sample responses is performed by phase sensitive detection (PSD), resulting in the dc output A_n of fundamentals ω ($n = 1$) and their harmonics $n\omega$ ($n = 2, 3, \ldots$), as well as the phase shifts ϕ_n between the nth harmonic and the stimulation. This phase shift is indicative of the kinetics of the stimulated process and of the underlying chemical reaction scheme. Since the PSD output A_n ($n = 1, 2, \ldots, n$; frequency $n \cdot \omega$) is proportional to $\cos(\phi_n - \phi_{n,\mathrm{PSD}})$, absorption bands featuring the same phase shift ϕ_n are considered to be correlated, i.e., to be representative of a population consisting of distinct molecules or molecular parts. $\phi_{n,\mathrm{PSD}}$ is the operator controlled PSD phase setting. Because of the cosine dependence, different populations will have their absorbance maxima at different $\phi_{n,\mathrm{PSD}}$ settings, thus enabling selective detection. Moreover, since in the case that $0.1 < \omega \cdot \tau_i < 10$ (τ_i denotes the ith relaxation time of the system), ϕ_n becomes ω dependent, $\phi_n = \phi_n(\omega)$. The spectral information can then be spread in the $\phi_{n,\mathrm{PSD}}\omega$ plane resulting in a significantly enhancement of resolution with respect to standard difference spectroscopy. Reproduced from [22] by permission of Marcel Dekker, © 2000.

period of stimulation, the system will reach the stationary state, which is characterized by periodic alterations around a constant mean which corresponds to the equilibrium state of the system at the mean value of the parameter. In the case of incomplete reversibility, e.g., the existence of an irreversible exit in the reaction scheme, the signal amplitudes of the initial components and of the intermediate species will decline, as the system is approaching its final state.

Phase sensitive detection (PSD) is used for the evaluation of amplitudes and phase lags of the periodic system response. In a simple view, PSD applied to data from a spectroscopic ME experiment results in a special kind of difference spectra between excited and nonexcited states. Let us consider a system, which is stimulated by a sinusoidally oscillating external parameter. During one half-wave, there is excitation followed by relaxation in the other. In the stationary state, this alteration between excitation and relaxation may be repeated as many times as necessary to obtain a good signal-to-noise (S/N) ratio of the modulation spectra. Moreover, it should be noted that PSD is a narrow band detection technique, i.e., noise contributes only from a frequency range which is close to the stimulation frequency ω. Since the periodic system response is evaluated automatically within each period of stimulation, instabilities of the spectrometer, the environment, and the sample are much better compensated than with conventional techniques, where a reference spectrum has to be measured separately. As a consequence, ME technique generally leads to high quality background compensation with a low noise level resulting in

enhanced sensitivity by at least 1 order of magnitude. So far, ME spectroscopy appears as a special type of difference spectroscopy. This is true if the frequency of stimulation is slow compared to the kinetics of the response of the stimulated system. However, if one or more relaxation times of the externally excited process fulfil the condition $0.1 < \omega \cdot \tau_i < 10$, where ω denotes the angular frequency of stimulation and τ_i is the ith relaxation time of the system, significant phase lags ϕ_i between stimulation and sample responses will occur. As will be derived in [22], phase lag and relaxation time are related by $\phi_i = \mathrm{atan}(-\omega \cdot \tau_i)$. This phenomenon is paralleled by the damping of the response amplitudes A_i. Both are significant for the underlying reaction scheme and the associated rate constants of the stimulated process [22, 30]. In this case, selectivity of ME spectroscopy, e.g., with respect to single components in heavily overlapping absorption bands, is significantly higher than that achievable by normal difference spectroscopy. The reason is the typical dependence of phase lags ϕ_i and amplitudes A_i on the modulation frequency ω. If a set of absorption bands of a modulation spectrum exhibits the same phase lag ϕ_i, it is considered a correlated population. Such a population consists, e.g., of molecules or parts of them that are involved in the same reaction step. The assignment of a group of absorption bands in a modulation spectrum to a population is considered to be validated if upon changing the stimulation frequency ω all these bands exhibit further on the same dependence with respect to phase lag $\phi_i(\omega)$ and amplitude $A_i(\omega)$. Moreover, the dependence of phase lag and amplitude on ω may be calculated based on a given reaction scheme. Analytical expressions for a simple reversible reaction $A_1 = A_2$ are given for demonstration of the principles. Obviously, ME spectroscopy enables a very rigorous test of the significance of a reaction scheme, since consistence of experimental data with theory derived from a given reaction scheme must hold on over the whole frequency range of stimulation.

Theoretically, as shown in Ref. [22], modulation spectroscopy and relaxation spectroscopy [31, 32] have the same information content regarding kinetic parameters. It should be noted, that due to the parameter ω modulation spectroscopy enables the collection of an arbitrary number of independent experiments, which is of special importance in the critical frequency range $0.1 < \omega \cdot \tau_i < 10$. No doubt that this approach is significantly more time consuming than a relaxation experiment which collects the whole information with "one shot." For that purpose a broad electronic bandwidth is required, rendering a considerably higher noise level than in a modulation experiment. Thus, the latter enables a more rigorous validation of experimental data since the system response at any frequency must be consistent with the reaction scheme underlying the kinetic analysis. The price to pay for this advantage is, besides a longer experimental phase, a more complicated theoretical approach for the evaluation of kinetic parameters. In this section, we report on temperature modulated excitation (T-ME) of poly-L-lysine. For more details and examples on concentration modulated excitation (c-ME), electric field modulated excitation (E-ME), and UV-VIS modulated excitation (Φ-ME), the reader is referred to Ref. [22] whereas basic information on data acquisition and treatment, especially with FTIR instruments, is available in Ref. [30]. Finally, as an example, the modulation spectroscopic approach to kinetic data of the simple reversible reaction $A_1 \rightleftarrows A_2$ is discussed.

4.2.1. Temperature Modulated Excitation

Using the general symbol η to denote the external parameter used for ME, one obtains in case of a harmonic stimulation,

$$\eta(t) = \eta_0 + \Delta\eta_0 + \Delta\eta_1 \cos(\omega t + \theta) \quad (153)$$

$\Delta\eta_0$ denotes the offset of the stationary state with respect to the initial state, i.e., $\eta_0 + \Delta\eta_0$ is the average value of the parameter in the stationary state. $\Delta\eta_1$ is the modulation amplitude of the corresponding parameter. θ denotes the phase of the stimulation, a parameter which is under experimental control by the operator. The shape for harmonic stimulation with $\theta = \pi$ is shown in Figure 16. For an analytical description of the influence of the external parameter T on rate constants, one may use the approximation by Arrhenius,

$$k = Ae^{-E_a/RT} \quad (154)$$

By introducing Eq. (153) into Eq. (154) one obtains

$$k(t) = Ae^{-E_a/(R(T_0 + \Delta T_0 + \Delta T_1 \cos(\omega t + \theta)))} \quad (155)$$

For small perturbations, i.e., $\Delta T_0, \Delta T_1 << T_0$, Eq. (155) may be approximated by the linear part of a Taylor series expansion

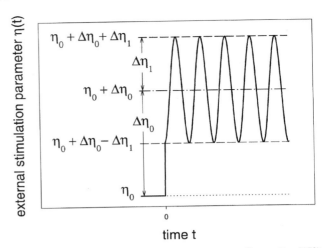

Fig. 16.   External harmonic modulated excitation according to Eq. (153). η denotes any external parameter. η_0 is the initial value at the beginning of the experiment. $\Delta\eta_0$ is the offset from the initial state to the stationary state, which corresponds to the equilibrium state at the parameter setting $\eta_0 + \Delta\eta_0$. $\Delta\eta_1$ denotes the modulation amplitude. θ is the phase angle, determining the onset of the stimulation. With $\theta = \pi$, stimulation starts at the minimum value of the parameter. Reproduced from [22] by permission of Marcel Dekker, © 2000.

Table III.   Signs of the Trigonometric Functions

Quadrant	Sin	Cos	Tan
I	+	+	+
II	+	−	−
III	−	−	+
IV	−	+	−

at T_0,

$$k(T(t)) = k(T_0) + k(T_0)\frac{T_A}{T_0^2}(\Delta T_0 + \Delta T_1 \cos(\omega t + \theta)) \quad (156)$$

$T_A = E_a/R$ will be referred to as the Arrhenius temperature. Under these conditions, Eq. (156) may be used to get the relevant rate equations. An example is given in Section 4.2.3. It should be noted that if the system undergoes a phase transition or is involved in a cooperative process this approach will still hold, however, more complicated reaction schemes have to be used. Cooperative phenomena have been described, e.g., by Hill [33] and Monod [34].

4.2.2. Other Types of Modulated Excitation

As mentioned earlier, ME may be performed by any external parameter which can influence the system by periodic variation. Analytical expressions describing the influence of pressure, electric field, concentration, and light flux on the rate constant are summarized in Ref. [22].

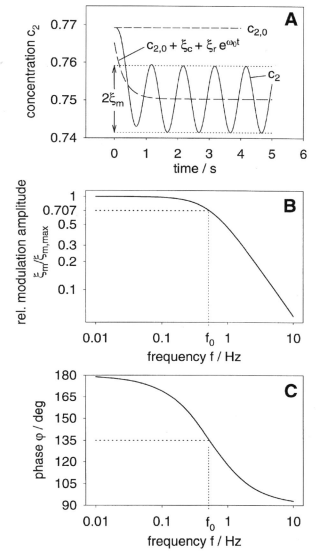

Fig. 17. Kinetics of a temperature modulated model system. The concentration c_2 of the product in reaction $A_1 \rightleftarrows A_2$ is calculated as a function of time using Eqs. (160) and (162) (A). The mean value of the concentration decreases exponentially caused by the temperature jump ΔT_0. After a few periods, the stationary state is reached, exhibiting the changes in the concentration due to the temperature modulation amplitude ΔT_1. With increasing frequency, the modulation amplitude ξ_m decreases assuming a slope of -6 db per octave at elevated frequency according to Eq. (162) (B), and the phase lag ϕ (C) decreases according to Eq. (163). At the frequency f_0, the amplitude ξ_m has decreased by a factor of $1/\sqrt{2}$ (-3 db). The phase lag at the 3-db point results in $\phi = 3\pi/4$ (C). The parameters used for calculation were: $E_{a,+} = 35,000$ J mol^{-1}, $E_{a,-} = 55,000$ J mol^{-1}, $k_+ = 2$ s^{-1}, $k_- = 0.6$ s^{-1}, $T_0 = 298$ K, $\Delta T_0 = 5$ K, $\Delta T_1 = 5$ K, $f = 1$ Hz and $\theta = -\pi$. Reproduced from [22] by permission of Marcel Dekker, © 2000.

4.2.3. Kinetic Analysis of a Reversible First-order Reaction

As an example of a kinetic analysis using modulation technique, we will demonstrate the procedures for the simplest case of a reversible first-order reaction of the type,

$$A_1 \underset{k_-}{\overset{k_+}{\rightleftarrows}} A_2 \qquad (157)$$

which may be typical for a simple conformational change, e.g., a cis-trans isomerization. The rate equations $\dot{c}_i(t)$ ($i = 1, 2$) are given by

$$\dot{c}_1(t) = -k_+ c_1(t) + k_- c_2(t)$$
$$\dot{c}_2(t) = k_+ c_1(t) - k_- c_2(t) \qquad (158)$$

where $c_i(t)$ denotes time-dependent concentrations of the species A_i and and k_+ and k_- are the rate constants of forward and backward reaction, respectively. Introducing the extent of reaction $\xi(t)$ as relevant a parameter which is related to concentrations according to

$$c_i(t) = c_{i,0} + \nu_i \xi(t) \qquad (159)$$

c_i and $c_{i,0}$ denote the actual and initial concentrations of the ith species, respectively, and ν_i is the stoichiometric number which is, according to convention, negative for reactants and positive for products [35]. Thus, it follows for the concentrations of A_1 and A_2,

$$c_1(t) = c_{1,0} - \xi(t) \qquad c_2(t) = c_{2,0} + \xi(t) \qquad (160)$$

The rate equation in terms of the extent of reaction follows from combining Eqs. (158) and (160),

$$\dot{\xi}(t) = -(k_+ + k_-)\xi(t) + k_+ c_{1,0} - k_- c_{2,0} \qquad (161)$$

Up to now, there was no external stimulation considered. To describe a temperature modulated excitation with small amplitudes, we introduce the temperature-dependent rate constants as given by Eq. (156) into Eq. (161). The following steps are depicted in detail in Ref. [22]. For the sake of shortness, we go directly to the solution for the modulated part of the extent of the reaction in the stationary state $\xi_{m,\text{stat}}(t)$, i.e., the situation as encountered in a typical experiment. Then,

$$\xi_{m,\text{stat}}(t) = \frac{|c_{1,0}\Delta k_+ - c_{2,0}\Delta k_-| \cdot \tau_m}{|\sqrt{1 + (\omega\tau_m)^2}|} \cos(\omega t + \theta + \phi) \quad (162)$$

$\tau_m = (\bar{k}_+ + \bar{k}_-)^{-1}$ is the relaxation time of the system. It should be noted, that the right quadrant for the phase angle ϕ can be determined only by considering the signs of two trigonometric functions indicated in Eq. (163), as depicted by Table III. So,

$$\cos\phi = \frac{\text{sign}(c_{1,0}\Delta k_+ - c_{2,0}\Delta k_-) \cdot (\bar{k}_+ + \bar{k}_-)}{|\sqrt{(\bar{k}_+ + \bar{k}_-)^2 + \omega^2}|}$$

$$= \frac{\text{sign}(c_{1,0}\Delta k_+ - c_{2,0}\Delta k_-)}{|\sqrt{1 + (\omega\tau_m)^2}|}$$

$$\sin\phi = -\frac{\text{sign}(c_{1,0}\Delta k_+ - c_{2,0}\Delta k_-) \cdot \omega}{|\sqrt{(\bar{k}_+ + \bar{k}_-)^2 + \omega^2}|} \qquad (163)$$

$$= -\frac{\text{sign}(c_{1,0}\Delta k_+ - c_{2,0}\Delta k_-) \cdot \omega\tau_m}{|\sqrt{1 + (\omega\tau_m)^2}|}$$

$$\tan\phi = -\frac{\omega}{\bar{k}_+ + \bar{k}_-} = -\omega\tau_m$$

As an illustration, the time-dependent concentration $c_2(t)$ of species A_2 (see Eq. (157)) was calculated and was shown in Figure 17A. The frequency dependence of the modulation amplitude $|\xi_m(\omega, \theta)|$ and the phase lag $\phi(\omega)$ are shown in Figure 17B and C. Obviously, kinetic rate constants can be determined from both the frequency dependence of the amplitude $|\xi_m(\omega, \theta)|$ and the phase lag ϕ, respectively. For details, the reader is referred to Refs. [22, 36].

5. APPLICATIONS

5.1. Model Biomembranes

5.1.1. Preparation of Immobilized Membrane Assemblies

Model bilayers for *in situ* ATR studies have been prepared according to the Langmuir–Blodgett (LB)–vesicle method [29]. The principal steps of this procedure are depicted in Figure 18. In practice, after the transfer of the first LB layer, the monolayer-coated internal reflection element (IRE) is mounted in an ATR flow through cuvette, as depicted by Figure 19. After spectroscopic examination of quantity and quality (ordering) of the LB monolayer, a vesicle solution of any phospholipid is circulated through the cuvette. This setup enables direct monitoring of the state of adsorption as shown by Figure 18. After about 30 min, the lipid bilayer is completed. A special washing procedure turned out to be necessary to detach loosely bound vesicle fragments. For details, the reader is referred to Refs. [16, 29, 37].

Typical spectra of a DPPA monolayer transferred by the Langmuir–Blodgett (LB) technique from the air–water interface to the germanium crystal as well as of a palmitoyl oleoyl phosphatidylcholine (POPC) monolayer attached to the DPPA layer are shown in Figure 20. Such bilayers can now be used for *in situ* membrane interaction studies with drugs [40, 41], pesticides [42], endotoxins [43], and proteins [37, 44]. An example of the formation of a planar lipid-protein assembly is presented later and is described in more detail in Ref. [37]. Mitochondrial creatine kinase (CK) is a highly ordered octamer. It has nearly a cubic shape with an edge of 93 Å [45] and features an accumulation of positive charges at two opposite sides of the cube. Probably, one of these sides binds to the membrane surface, predominantly by electrostatic interaction. The bilayer consisted of an inner DPPA monolayer and an outer cardiolipin (CL) monolayer. CL was used to achieve a bilayer membrane with a negative surface charge. Now, pumping CK through the ATR flow-through cuvette (Fig. 19) leads to a spontaneous adsorption of CK as schematically depicted in Figure 18 at the left hand. The observed surface coverage of about 60% and the kinetics of CK adsorption were in accordance with earlier results obtained by plasmon resonance [46]. A typical spectrum of membrane-bound CK is shown in Figure 14. A modified procedure had to be applied for membrane anchoring of alkaline phosphatase (AP), see Figure 18 right-hand side. In contrast to CK, AP had to be solubilized by a detergent (β-octyl glucoside

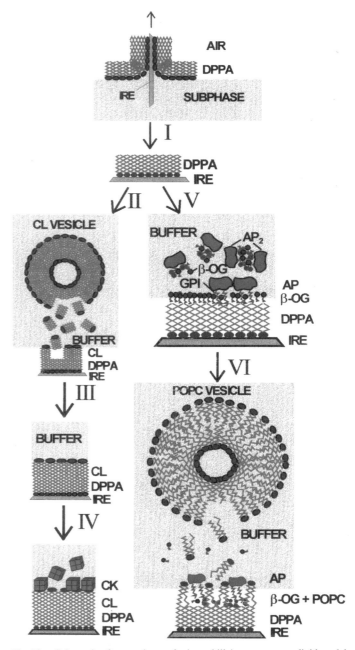

Fig. 18. Schematic of two pathways for immobilizing enzymes on lipid model membranes attached to an IRE plate. Path 1: Immobilization of mitochondrial creatine kinese (Mi-CK). (I) Transfer of the inner IRE-attached DPPA monolayer from the air–water interface of a film balance to an internal reflection element (IRE) by the Langmuir–Blodgett (LB) technique; (II) spontaneous adsorption of cardiolipin (CL) lipids from vesicles energetically driven by the reduction of the unfavorable high energy of the hydrophobic surface of the DPPA monolayer in contact with the aqueous environment; (III) completed asymmetric CL/DPPA-bilayer; (IV) adsorption of Mi-CK to the bilayer by electrostatic interactions. Path 2: Immobilization of alkaline phosphatase (AP). (I) and (II) as described before; (V) spontaneous adsorption of AP (solubilized by β-octyl glucoside (β-OG)) to the DPPA monolayer via its GPI (glycosyl-phosphatidylinositol) anchor; (VI) reconstitution of a bilayer-like system by passing POPC vesicles over the AP DPPA assembly. Reproduced from [37] by permission of the American Institute of Physics, © 1998.

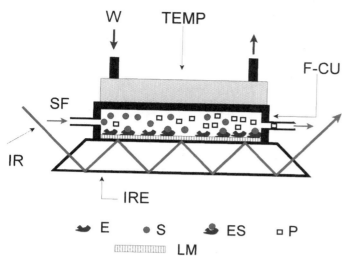

Fig. 19. Flow-through cuvette (F-CU) for *in situ* FTIR ATR spectroscopy. The IRE is coated by a supported bilayer (LM) to whom the enzyme (*E*) is immobilized. The substrate (*S*) flowing through the cuvette (SF) is enzymatically converted into the product (*P*). Reproduced from [41] by permission of the American Institute of Physics, © 1998.

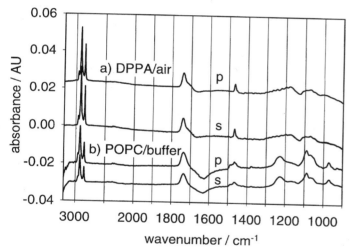

Fig. 20. Polarized FTIR ATR absorbance spectra of a DPPA monolayer attached by the polar head to a germanium internal reflection plate, and of a POPC monolayer exposing the polar head group to the aqueous environment (see Fig. 18). (a) Parallel (*p*) and vertical (*s*) polarized spectra of DPPA against air, transferred at 30 mN m^{-1} from an aqueous subphase (0.1 mmol L^{-1} CaCl$_2$) to a trapezoidal Ge plate; $\Gamma = (3.9 \pm 0.1) \times 10^{-10}$ mol cm^{-2}, corresponding to an area of 43 Å^2 per molecule, which is close to the value of 43.8 Å^2 as reported in [38]. Dichroic ratio $R = 0.93$, $S_{mol} = 0.99$ (reference: clean Ge plate; active internal reflections $N_{act} = 22.5$, integration between: 2867–2832 cm^{-1}). (b) Parallel (*p*) and vertical (*s*) polarized absorbance spectra of a POPC monolayer adsorbed from a vesicle solution to a DPPA monolayer; $\Gamma = (2.3 \pm 0.1) \times 10^{-10}$ mol cm^{-2}; mean molecular area $A_m = 74$ Å^2, in good agreement with 68 Å^2, as reported in [39, 74]. Dichroic ratio $R = 1.33$, $S_{mol} = 0.43$ (reference: DPPA monolayer in buffer; active internal reflections $N_{act} = 18.4$; integration between: 2869.9-2832 cm^{-1}); surface concentrations for (a) and (b) were calculated with the thin film approximation using the CH$_2$ stretching vibration ν_s(CH$_2$) at 2850 cm^{-1}. Integrated molar absorption coefficient $\int \varepsilon(\tilde{\nu}) \, d\tilde{\nu} = 5.22 \times 10^3$ m mol^{-1}, angle of incidence $\theta = 45°$. Refractive indices at 2850 cm^{-1}, $n_1 = 4$, $n_2 = 1.43$, $n_3 = 1$ (DPPA–air), and 1.41 (POPC–buffer), respectively.

(β-OG)). Pumping the AP solution through the flow cell containing the DPPA coated Ge IRE has led also to spontaneous adsorption, however, β-OG molecules took the place of lipid molecules in the outer membrane. In the presence of an excess of palmitoyl oleoyl phosphatidylcholine (POPC) vesicles, however, the detergent could be exchanged by the phospholipid [37].

5.1.2. Biomedical and Environmental Interaction Studies

5.1.2.1. Endotoxins

Lipopolysaccharides (LPSs, endotoxins) are complex lipid-linked carbohydrates which are found in the outer membranes of gram-negative bacteria. These negatively charged molecules are usually composed of a polymeric carbohydrate (O antigen), a short oligosaccharide (*R* core), and a fatty-acylated region (lipid A). Through its action on macrophages, LPS can trigger responses that are protective or injurious to the host [47]. To get more insight into the mechanisms of blood purification [48] and toxic action, FTIR ATR spectroscopy was used to investigate *in situ* the interactions of LPS from *Pseudomonas aeruginosa* (serotype 10) with different kinds of surfaces: a hydrophilic germanium (Ge) plate, a hydrophobic monolayer consisting of dipalmitoyl phosphatidic acid (DPPA), a positively charged cross-linked polymer consisting of γ-aminopropyltriethoxysilane (ATS) and, finally, a positively charged bilayer consisting of DPPA as an inner and a (1:1 molar) mixture of palmitoyl oleoyl phosphatidylcholine (POPC) and hexadecylpyridinium (HDPyr) as an outer leaflet. Since LPS exhibits a negative surface charge, one could expect that a positively charged surface would adsorb endotoxins [43]. As an example of these investigations, the interaction of LPS with an immobilized bilayer consisting of an inner DPPA layer and an outer positively charged layer of a 1:1 molar mixture of POPC and HDPyr, the latter being responsible for the positive surface charge, will be mentioned. LPS indeed has bound to the membrane, however, the attractive force between LPS and HDPyr was stronger than the retaining force of the membrane. During the interaction of about 30-min duration, practically 100% of the positively charged lipid was extracted from the membrane by the endotoxin. This process could be quantified in terms of surface concentration Γ determined according to Eq. (152) and mean orientation of the bisectrice of the HCH bond angle of the methylene groups of hydrocarbon chains according to Eq. (138), see also legend to Figure 21.

5.1.2.2. Chlorophenols

Weak organic acids are of increasing environmental concern since many pesticides and many pharmaceuticals contain acidic groups [49]. Among them, phenolic compounds substituted by chloride at various positions have been studied with respect to their toxic effects on the energy metabolism. Many chlorophenols are known as uncouplers of the chemiosmotic phosphorylation; i.e., they destroy the electrochemical proton

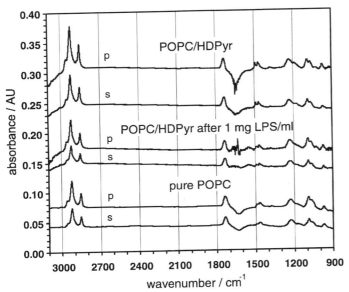

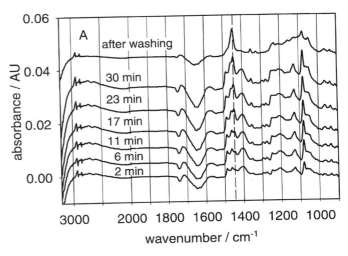

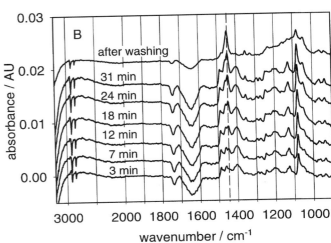

Fig. 21. Parallel (p) and vertical (s) polarized absorbance spectra of lipid monolayers with the polar headgroups facing the aqueous phase. Top: DPPA/(POPC:HDPyr; 1:1 molar) bilayer; 20-mM phosphate buffer pH 7.0, 100-mM NaCl; $T = 25$ °C; reference, DPPA monolayer in phosphate buffer. Dichroic ratio $R = 1.29$ (at 2853 cm^{-1}) resulting in according to Eq. (138) a mean value for the angle between the bisectrice of the methylene group and the z axis of $\overline{\theta} = 65.3°$ (note: a hydrocarbon chain in all-trans conformation, being aligned parallel to the z axis would result in $\overline{\theta} = 90°$, see also Fig. 11). Further general conditions were: angle of light incidence $\theta_i = 45°$; number of active internal reflections $N = 33$. Middle: DPPA/(POPC:HDPyr; 1:1 molar) bilayer after 30-min contact with 1-mg LPS ml^{-1}. Dichroic ratio $R = 1.59$, $\overline{\theta} = 59.1°$; the surface concentration as determined by Eq. (162) is found to be $\Gamma = 1.90 \times 10^{-10}$ mol cm^{-2}, corresponding to a surface coverage of about 80% by POPC. Note, practically all positively charged lipids (HD-Pyr) have been extracted by the endotoxin. This is manifested by the decrease of CH$_2$ stretching bands and to a significant enhancement of group fluctuation. Furthermore, the most typical HDPyr band at near 1500 cm^{-1} (aromatic C−C stretching) has vanished. Bottom: DPPA/POPC bilayer for comparison. DPPA monolayer as reference. Dichroic ratio $R = 1.35$, $\overline{\langle\cos^2\theta\rangle} = 0.1940$, $\overline{\theta} = 63.9°$. The spectra in the middle and at the bottom are quite similar, featuring pure POPC. LPS obviously removed only the positively charged HDPyr from the initial membrane.

Fig. 22. (A) and (B) Time-resolved IR ATR absorbance spectra of the adsorption of 2,4,5-trichlorophenol (TCP) to a DPPA/POPC bilayer. A TCP solution ($c_{TCP} = 2.0$ mmol L^{-1}) in 25 mmol L^{-1} potassium phosphate buffer pH 6.0 (c_{total} (K$^+$) 100 mmol L^{-1}) was pumped into a flow-through cell (Fig. 19) with a flow rate of 0.5 mL min^{-1} at 25 °C. For 0.5-h parallel (p) polarized and vertical (s) polarized ATR, spectra were measured in turn every 64 s. After about 1 h, the TCP solution was exchanged by a buffer and the adsorbate was washed for 15 min. Parallel (p) and vertical (s) polarized SBSR-ATR spectra were recorded. (A) parallel (p) polarized absorbance spectra of the washed adsorbate (top) and of the adsorption process arranged by increasing time from bottom to top. (B) vertical (s) polarized spectra of the washed adsorbate (top) and of the adsorption process arranged by increasing time from bottom to top. A dashed line marks a peak emerging with delay at 1446 cm^{-1}. Reference: DPPA/POPC bilayer against 25 mmol L^{-1} potassium phosphate buffer pH 6.0 (c_{total} (K$^+$) 100 mmol L^{-1}); Measurement conditions: Ge trapezoidal IRE, angle of incidence $\theta = 45°$, number of active internal reflections $N = 13.5$.

gradient of energy transducing membranes [50]. As an example of possible contributions of *in situ* FTIR ATR spectroscopy to such problems, we have studied the interaction of 2,4,5-trichlorophenol (TCP) with a DPPA/POPC supported bilayer membrane.

Three highly significant features observed in Figure 22A (parallel (p) polarized spectra) and Figure 22B (vertical (s) polarized spectra) should be mentioned. Looking first to the CH$_2$ stretching region near 2900 cm^{-1}, it is obvious that in the course of the adsorption process p-polarized spectra show increasingly sigmoidal shapes, while s-polarized spectra reflect increasingly negative bands also with slight tendency to a sigmoidal shape. Second, the C=O stretching mode of the ester groups of phospholipids near 1730 cm^{-1} shows negative bands in both polarizations during adsorption of TCP. In a first view, one would suggest lipid loss upon interaction with TCP. This, however, is not the case, since washing of the membrane after

about 1 h of TCP adsorption restores the major part of the lipid structure. A buffer flow through removes a considerable part of adsorbed TCP, but leaving a prominent band near 1450 cm^{-1} and a smaller, well-resolved band at 1352 cm^{-1}. This band was used to quantify the remaining tightly membrane-bound species. As noticed by visual inspection at 1450 cm^{-1} (Fig. 22) and analytically (Fig. 23), this species appears delayed with respect to the species quantified by means of the 1080 cm^{-1}

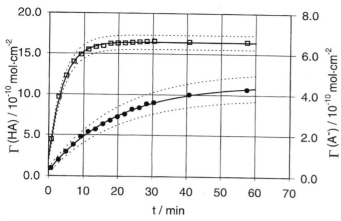

Fig. 23. Calculated surface concentrations Γ for phenol (HA) and phenoxide (A^-) vs time. Squares denote values for Γ(HA) and refer to the left-hand axis. Whereas, filled circles indicate values for $\Gamma(A^-)$, referring to the right-hand axis. The surface concentrations Γ were determined with the thin film approximation, using peak heights A_{max} at 1080 cm^{-1} for Γ(HA) and a molar absorption coefficient of $\varepsilon = 18.5 \pm 0.8$ m^2 mol^{-1}, as well as peak heights A_{max} at 1352 cm^{-1} for $\Gamma(A^-)$ with $\varepsilon = 15 \pm 2$ m^2 mol^{-1}. Data of peak heights were fitted with $f(t) = d + g(1 - \exp(-kt))$, resulting in functions with the following parameters ($g/k/d/Rsqr/s$): parallel polarization at 1352 cm^{-1}: 1.64×10^{-3} AU/0.048 min^{-1}/7.53 $\times 10^{-5}$ AU/0.98/6.458 $\times 10^{-5}$; vertical polarization at (1352 cm^{-1}): 8.41×10^{-4} AU/0.059 min^{-1}/8.99 $\times 10^{-5}$ AU/0.947/5.709 $\times 10^{-5}$; parallel polarization at (1080 cm^{-1}): 6.19×10^{-3} AU/0.249 min^{-1}/9.37 $\times 10^{-4}$ AU/0.983/2.495 $\times 10^{-4}$; vertical polarization at (1080 cm^{-1}): 3.45×10^{-3} AU/0.263 min^{-1}/4.82 $\times 10^{-4}$ AU/0.915/2.49 $\times 10^{-4}$. The surface concentration after 57 min resulted in $\Gamma_{HA} = 1.65 \times 10^{-9}$ mol cm^{-2} and $\Gamma_{A^-} = 0.43 \times 10^{-9}$ mol cm^{-2}. Dotted lines represent the standard deviation of $\Gamma(t)$, derived from standard deviations of the molar absorption coefficient and the error of the fit–data relation. A DPPA/POPC bilayer in contact with 25 mmol L^{-1} potassium phosphate buffer pH 6.0 (c_{total} (K$^+$) 100 mmol L^{-1}) was used as reference. Measurement conditions: Ge trapezoidal IRE, angle of incidence $\theta = 45°$, number of active internal reflections $N_{act} = 13.5$; refractive indices: $n_1 = 4$ (Ge), $n_2 = 1.50$ (membrane), $n_3 = 1.31$ (H$_2$O, 1352 cm^{-1}), and $n_3 = 1.26$ (H$_2$O, 1080 cm^{-1}).

band. The latter can be assigned to the neutral chlorophenol (AH) while the bands at 1450 and 1352 cm^{-1} are tentatively assigned to the phenolate ion [42]. As depicted by Figure 23, the uncharged species AH adsorbs about five times faster than the charged species A^-. Several observations favor the interpretation that in a first step AH adsorbs to the membrane and reorients in the adsorbed state forming in part a so-called heterodimer consisting of AHA$^-$ which binds more tightly to the membrane than the primary adsorbates. This finding is consistent with models proposed in the literature [50, 51].

Finally, it should be noted again, that the disturbances of the lipids appearing during TCP adsorption vanish for the major part upon replacing the TCP solution in the flow-through cell by buffer solution. Thus, lipid loss cannot serve for understanding this spectroscopic effect. A more reasonable explanation would be the postulation of some amounts of chlorophenolates in the loosely bound adsorbate which could result in conformational changes in the lipid membrane due to immobilized surface charges [42]. A similar effect was already observed upon controlled production of negative surface charges by de-

protonation of the carboxylic acid group of an arachidic acid bilayer attached to a Ge IRE by the LB technique [22, 52].

5.2. Heterogeneous Catalysis by Metals

The relevant chemical step in heterogeneous catalysis by metals takes place at the metal-gas or metal-liquid interface, depending on whether the reaction runs in a gas or liquid environment [53]. Fundamental research in these areas largely relies on methods able to probe these interfaces under reaction conditions. For the investigation of gas–solid interfaces, vibrational spectroscopy contributed much to the current understanding of processes at such interfaces due to the detailed information of the adsorbate layer that can be obtained. Despite their importance, catalytic processes at metal-liquid interfaces are much less studied and understood. One reason for this is the experimental difficulty to probe such interfaces *in situ*. As concerns vibrational spectroscopy, the challenge is discriminating small signals from the interface from large solution signals. ATR spectroscopy offers the possibility to probe metal-liquid interfaces with high sensitivity and is therefore an ideal tool for research in heterogeneous catalysis.

5.2.1. Preparation of Thin Metal Films

There are several possibilities to immobilize a solid catalyst onto an internal reflection element (IRE). Catalyst powders can be immobilized by simply dropping a suspension of the catalyst onto the IRE followed by drying or by dipping the IRE in a suspension of the catalyst (dip coating). On the other hand, model catalysts can be obtained by coating the IRE with a thin film of the active metal. Evaporation and sputtering are the most convenient ways to prepare such films. In sputtering, ions (usually Ar$^+$) are accelerated onto a sample of the desired metal. Small metal particles are thus detached from the sample and deposited onto the nearby IRE. Evaporation, on the other hand, is achieved by heating the metal either resistively or through an electron beam. Film thickness can be measured using an oscillating quartz microbalance. To obtain homogeneous films over the IRE, it is important that the distance between the source and the IRE is large.

5.2.2. Optical Properties of Thin Metal Films

Metals are very strong infrared absorbers with k values typically between 10 and 60 in the mid-IR. Nevertheless, thin metal films are transparent enough for investigations by ATR. The reflectivity of an internal reflection element coated with a metal film as a function of film thickness is shown in Figure 24 for the example of Pt on Ge. First, the reflectivity strongly drops with increasing film thickness and reaches a minimum. For the example shown in Figure 24, this minimum is observed slightly below 10 nm. When working in this region, the use of multiple reflections is not indicated since too much intensity is lost per reflection, resulting in an increased noise level. For even thicker films, the reflectivity rises again: The metal acts

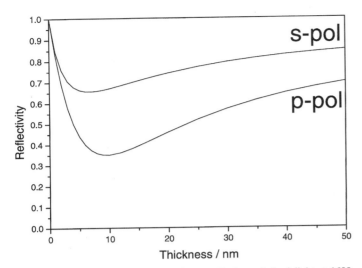

Fig. 24. Reflectivity for parallel and perpendicular polarized light at 1600 cm^{-1} for the system Ge/Pt–vacuum as a function of the thickness of the Pt film. For the calculations, the optical constants of bulk Pt were used. $n_{Ge} = 4.01$, $\hat{n}_{Pt} = 5.71 + 23.35i$. The angle of incidence is 45°. Calculations were performed according to the matrix formalism described by Hansen [18].

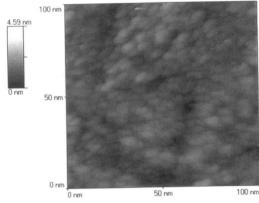

Fig. 25. STM image (100 × 100 nm) of a 1-nm Pt film electron beam evaporated on Ge. Evaporation conditions: Pressure 1.5 × 10^{-5} mbar, room temperature, 0.5 Å s^{-1} evaporation speed. From [69] by permission of the American Chemical Society, © 2001.

as a mirror. However, this situation is not desirable for ATR since the evanescent electromagnetic wave does not probe the outer surface of the metal film. Suitable metal films are typically 1- to 30-nm thick. Such metal films are very often not homogeneous but have island structure. The structure of the films (shape of the islands, diameter, density) depends on the metal, the substrate (internal reflection element), and various parameters during film preparation. For evaporated metal films, important parameters are temperature, presence of gases, and evaporation speed. As an example, Figure 25 shows an STM image of a 1-nm Pt film evaporated on Ge. The film consists of densely packed Pt islands of about 6-nm diameter.

Due to this island structure, the effective (space averaged) optical constants of thin metal films differ considerably from the bulk values of the corresponding metal (see, for example,

[54–56]). The effective optical constants of a metal film depend not only on the optical constants of the metal, but on several other parameters. According to the Maxwell-Garnett [57], the Bruggeman [58], and other effective medium theories [59], the optical constants of a composite film depend on the packing density of the particles, the optical constant of the surrounding, and the polarizability of the mutually interacting metal island particles. The polarizability itself is a complicated function of size and shape of the metal particles. For thin silver films at 590 nm, for example, it was found that the refractive index n strongly increased (by about 1 order of magnitude) whereas the absorption index k decreased when going from 10- to 1-nm thick films [54]. For thicker films, the optical constants approach bulk values. For silver films, this is achieved at around 10 nm [54]. As can be seen from Eqs. (20)–(27), the optical constants are closely related to electric conductivity. The conductivity of very thin films is different from bulk conductivity. For island films, the conductivity and hence the optical constants drastically change as the individual islands merge (percolation). For thicker metal films (10 nm and more), the bulk optical constants are a good first approximation for the film optical constants. This is demonstrated in Figure 10, where good agreement between experiment and simulation was achieved for the system ZnSe/10 nm Pt/ H$_2$O by using the bulk optical constants of Pt.

The optical constants of the thin metal film can have a pronounced influence on the line shape of an absorption band associated with an absorber above the metal film (i.e., bulk medium or adsorbate) [60]. Simulations show that strongly distorted lines are expected for strong absorbers even far away from the critical angle. As an illustration, Figure 26 shows the line shape calculated for a thin adsorbate film on a 20-nm metal film as a function of the optical constants of the metal film. The optical constants for the adsorbate film were calculated from a damped harmonic oscillator model,

$$\varepsilon(\tilde{\nu}) = \hat{n}_3^2(\tilde{\nu}) = n_e^2(\tilde{\nu}) + \frac{B}{\tilde{\nu}_0^2 - \tilde{\nu}^2 - i\gamma\tilde{\nu}} \qquad (164)$$

where $\tilde{\nu}$ [cm^{-1}] is the wave number, $\tilde{\nu}_0$ [cm^{-1}] is the wave number of the absorption band center, γ [cm^{-1}] is the damping constant, which gives the bandwidth, B [cm^{-2}] determines the intensity of the band, and n_e is the refractive index far away from resonance.

The properties of the adsorbate film are those of CO on Pt. For the middle spectrum, the complex refractive index ($\hat{n}_2 = 5 + 20i$) is close to the one of Pt at 2000 cm^{-1}. n and k were varied between 2.5 and 10 and between 10 and 40, respectively, in view of the fact that n and k vary drastically over the midinfrared region and from metal to metal [61].

Both n and k of the metal film have an influence on the line shape of the adsorbate. With the middle spectrum as the reference, Figure 26 shows that increasing n leads to a more distorted line shape, whereas increasing k leads to a less distorted line shape. Increasing k (more absorbing metal) also leads to a decreased absorbance. Without metallic film, the

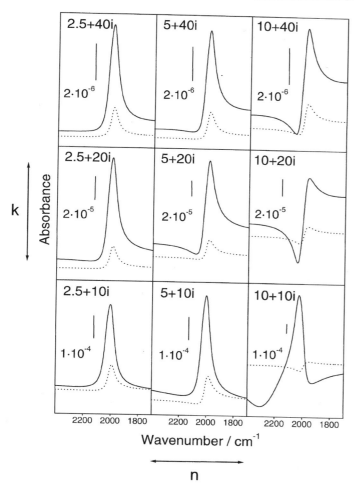

Fig. 26. Calculated ATR spectra (angle of incidence 45°) for a monolayer adsorbate on a 20-nm thick metal film in contact with a solvent as a function of the complex refractive index $\hat{n}_2$ of the metal film. Solid line: parallel polarized light, dotted line: perpendicular polarized light. The respective complex refractive index $\hat{n}_2$ is given at the top of each spectrum. The vertical bars indicate the scale for the absorbance, which differs for each spectrum. Parameters: $n_1 = 4.01$ (Ge), $n_4 = 1.4$ (organic solvent) $d_3 = 3$ Å, $n_e = 1.6$, $B = 280,000$ cm^{-2}, $\tilde{v}_0 = 2000$ cm^{-1}, $g = 60$ cm^{-1}. The parameters correspond to adsorbed CO. The calculations were performed using the formalism proposed by Hansen [18] and the results are given in absorbance $A = -\log_{10}(R/R_0)$, where R is the reflectivity of the system Ge/Pt–adsorbate–solvent and R_0 is the reflectivity of the system Ge/Pt–solvent. From [60] by permission of the Royal Society of Chemistry on behalf of the PCCP Owner Societies, © 2001.

pends on the structure of the metal film, especially the shape and the size of the islands and their spacing. Enhanced infrared absorption has been found for various metals (Au, Ag, Cu, Pb, Pt, Pd, Ni) but up to now strong enhancement was mainly reported for Au and Ag. For the latter, enhancement factors (i.e., the signal of a vibration in the presence of the metal film divided by the signal in the absence of the metal film) of 200 are not unusual [65]. For other metals, typical enhancement factors are below 10. The enhancement is restricted to near the metal film. Enhancement factors strongly decreased with an increasing number of deposited stearic acid Langmuir–Blodgett layers, demonstrating that only molecules near the metal give rise to enhanced absorption. However, the asymmetric CH$_3$ stretching also showed strong enhancement. Thus, the enhancement mechanism has a long-range nature of at least one monolayer [66]. Analyzing spectra of molecules for which the local adsorption geometry is known (e.g., p-nitrobenzoic acid), it was found that only modes giving rise to a dynamic dipole moment perpendicular to the local surface (totally symmetric modes) were enhanced. Therefore, for the surface enhancement the same selection rules were found as for infrared reflection absorption spectroscopy (IRRAS) measurements of adsorbates on metals. This indicates that the electric field near the metal particles of the film is locally perpendicular to the surface [65].

The large infrared enhancement together with the local nature of the enhancement is a big advantage for *in situ* investigation of processes at metal-liquid interfaces. The enhanced signal-to-noise ratio allows following relatively fast processes [67] and the short-range nature of the enhancement helps the discrimination of the signals from the bulk solution. Due to these advantages, surface enhanced infrared absorption (SEIRA) in the ATR configuration has been applied more and more for the investigations of electrochemical interfaces [67]. However, it should be restated that large enhancement factors have been found only for few metals.

5.2.3. Adsorption Studies on Pt

Carbon monoxide CO is the most widely used probe molecule for the characterization of catalytic metal surfaces. CO adsorption is therefore ideal to probe evaporated metal films. Since Pt is a reactive metal, samples of freshly deposited films are immediately covered by a resistant contamination layer, when exposed to air. Attempts to adsorb CO from the gas phase onto such films were not successful. One way to clean the Pt is by flowing a solvent, such as CH$_2$Cl$_2$, saturated with hydrogen over the sample. Doing this in an ATR flow-through cell the cleaning effect of the hydrogen can be monitored *in situ*. After this cleaning step, CO adsorption from the solvent can be monitored as shown in Figure 27. The observed band at around 2050 cm^{-1} is associated with the C—O stretching vibration of linearly bound CO (CO coordinated to one Pt atom). The series of spectra reveals that the frequency of the bands is a function of coverage; i.e., the frequency increases with increasing the coverage of CO on Pt. This phenomenon is well studied on single

absorption band is symmetric (not shown). Some thin metal island films have another property relevant to IR spectroscopy. Several research groups reported that infrared absorption was considerably enhanced for molecules in contact with the metal (surface enhanced infrared absorption, SEIRA). Since the first report [62], many investigations aimed at elucidating the origin of the enhancement, yet the phenomenon is not completely understood. Many findings support the view that the electric field around a metal particle within the metal film is enhanced due to a plasma resonance [63, 64]. Even though the enhancement mechanism is not completely understood, the experimental findings can be summarized as follows: The enhancement de-

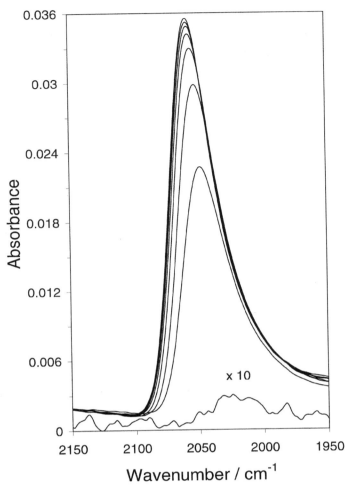

Fig. 27. ATR-IR spectra showing CO adsorption on a 1-nm Pt thin film evaporated on Ge after cleaning with hydrogen. CO was admitted to the sample by flowing CH$_2$Cl$_2$ saturated with a mixture of 0.5% CO in Ar over the sample. Spectra were recorded between 1 and 120 min after admitting CO. Ten active internal reflections were used. From [69] by permission of the American Chemical Society, © 2001.

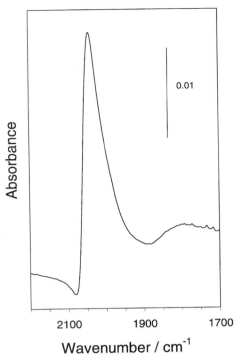

Fig. 28. ATR-IR spectra showing CO adsorption on a 1-nm Pt thin film evaporated on Ge. CO was admitted to the sample by flowing CH$_2$Cl$_2$ saturated with a mixture of 0.5% CO in Ar over the sample. Ten active internal reflections were used.

crystals and supported Pt particles [68]. It arises due to vibrational coupling between neighboring molecules when domains of CO are formed. The CO adsorption experiments therefore show that after hydrogen treatment at room temperature large "clean" domains exist on the Pt, even in the presence of a commercial solvent.

As mentioned earlier, simulations show (Fig. 26) that for a strong absorber above a metal film the line shape of an absorption band can be distorted. This is already demonstrated in Figure 10 for the system ZnSe/10 nm Pt/H$_2$O. Figure 28 shows that a distorted line shape is also found for CO on Pt (Ge/1 nm Pt/CO/CH$_2$Cl$_2$). The spectrum shows the C—O stretching vibration of linearly (bound to one Pt atom) and bridged (bound to two Pt atoms) CO at around 2050 and 1800 cm^{-1}, respectively. We have found that the line shape depended on the condition of the Pt film [69]. For example, distorted line shapes were found for Pt films on Ge, whereas hardly undistorted line shapes were

found for Pt films, for which an Al$_2$O$_3$ film was deposited between Ge and Pt.

Another common molecule to probe metal surfaces is pyridine. The vibrational spectrum of pyridine is well studied. With the use of the metal surface selection rule, detailed information about the orientation of pyridine on metal surfaces has been obtained. Pyridine adsorption onto Pt(111) under ultrahigh vacuum (UHV) conditions depends on coverage and temperature [70]. Between 140 and 250 K, pyridine adsorbs initially flat on the Pt(111) surface. With increasing coverage, the molecule shows some tilting with respect to the surface normal. At around room temperature, an α-pyridyl species is formed (dissociation of the C—H bond in α position to the N), which is bound in an upright position. The spectrum of the intact molecule and the α-pyridyl species are clearly different. Figure 29 shows an ATR spectrum, which was recorded after flowing a solution of 0.001 mol l^{-1} pyridine in hydrogen saturated CH$_2$Cl$_2$ over a 1-nm Pt film at room temperature, as the reference served the sample after it was treated ("cleaned") with hydrogen. Based on the mentioned UHV work, the sharp positive peaks can be assigned to pyridine, which is adsorbed intact onto the Pt in a tilted way. Under UHV conditions, the α-pyridyl species predominate at this temperature. This difference in behavior is assigned to the presence of solvent and hydrogen on the Pt. This example illustrates that the adsorption behavior in solution and in the presence of hydrogen can be quite different from the one observed in vacuum and highlights the importance of the *in situ* approach for the study of catalytic

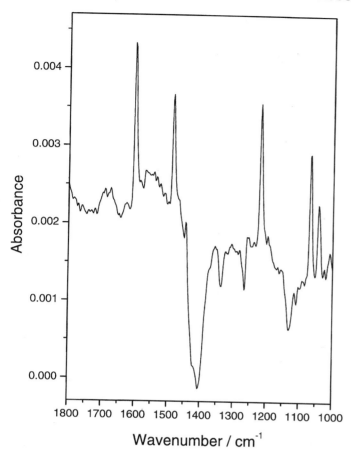

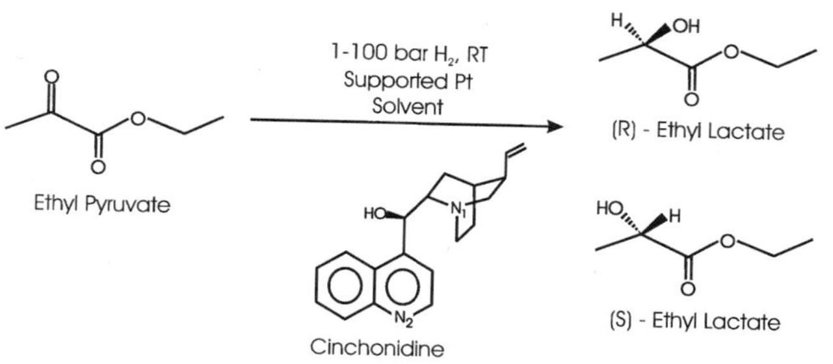

Fig. 30. Reaction scheme of the heterogeneous enantioselective hydrogenation of ethyl pyruvate to ethyl lactate over cinchonidine modified Pt. In the absence of cinchonidine, the reaction yields a racemic mixture of the two enantiomers of ethyl lactate. In the presence of cinchonidine, (R)-ethyl lactate is yielded in high enantiomeric excess (>95%).

tion of pyridine, at least a part of these decomposition products are displaced, leading to the negative bands.

5.2.4. Enantioselective Hydrogenation by Chirally Modified Pt

Asymmetric heterogeneous catalysis in which a catalyst is modified with one enantiomer of a chiral compound is a field of rapidly growing interest. The interest in such reactions arises from the insight that very often enantiomers have quite different properties in a chiral environment such as a metabolism. The wrong enantiomer has to be considered as ballast at best and, at worst, as a toxin, which can drastically overweight the useful effect of the right enantiomer. One of the few reactions which has been studied in some detail in the past is the enantioselective hydrogenation of activated carbonyl compounds such as ethyl pyruvate (EP) over cinchonidine (CD) modified Pt (Fig. 30) [72, 73]. A better understanding of the mechanism of enatiodifferentiation in this reaction is a prerequisite for rational catalyst and modifier improvements.

However, to date information on a molecular level is barely available, rendering the proposed mechanistic models highly speculative. Especially, the modification of the Pt surface by the chiral modifier is not well investigated mainly due to the experimental difficulties in studying such processes. An recent NEXAFS (near edge X-ray absorption fine structure) study on the adsorption of 10,11-dihydrocinchonidine on Pt(111) under UHV conditions showed that the quinoline moiety of the modifier is nearly parallel to the surface at 298 K, whereas at 323 K a tilting angle of 60° has been found [74].

Figure 31 shows a series of ATR spectra of cinchonidine (CD) adsorbed on a Pt/Al$_2$O$_3$ model catalyst recorded under different conditions. The spectra were measured by flowing solutions of CD in hydrogen saturated CH$_2$Cl$_2$ over the sample. The sample after hydrogen treatment ("cleaning") served as the reference. Spectra (a), (b), and (c) were recorded while flowing 0, 10^{-6} and 10^{-3} mol l^{-1} solutions of CD over the sample. Spectrum (d) was recorded in the presence of a neat solvent, but after a 10^{-3} mol l^{-1} solution was admitted to the sample. For comparison, spectrum (e) shows a scaled transmission

Fig. 29. ATR-IR spectrum of pyridine adsorbed on a Pt/Al$_2$O$_3$ model catalyst, prepared by evaporating 100-nm Al$_2$O$_3$ on a Ge internal reflection element, followed by evaporating 1-nm Pt. The spectrum was recorded *in situ* while flowing a solution of 10^{-3} mol L^{-1} pyridine in hydrogen saturated CH$_2$Cl$_2$ over the sample. The reference spectrum was recorded while flowing a neat solvent saturated with hydrogen over the sample. Ten active internal reflections were used.

processes at metal-liquid interfaces. There are also strong negative bands arising upon adsorption of pyridine. These bands are consistently observed also upon adsorption of other molecules from CH$_2$Cl$_2$. The band at 1267 cm^{-1} is associated with the strongest CH$_2$Cl$_2$ band and arises from incomplete compensation. The other bands at around 1400 cm^{-1} and at 1338 and 1131 cm^{-1} are explained by the chemistry of CH$_2$Cl$_2$ on Pt. CH$_2$Cl$_2$ decomposes on Pt, yielding CH$_X$ fragments, which, in the presence of hydrogen, are hydrogenated to methane [71]. The broadband at 1400 cm^{-1} is associated with the C—H bending modes of CH$_X$ fragments. Furthermore, these CH$_X$ fragments can recombine to yield C$_2$ hydrocarbons and ultimately C$_2$H$_6$ (ethane) [71]. The two bands at 1131 and 1338 cm^{-1} match well with the most intensive bands of C—CH$_3$ (ethylidyne), the most stable C$_2$ hydrocarbon on Pt under hydrogenation conditions. The spectrum in Figure 29 therefore shows that Pt is covered by decomposition products in the presence of CH$_2$Cl$_2$ solvent and hydrogen. Upon adsorp-

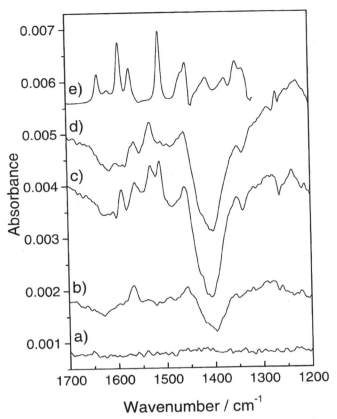

Fig. 31. (a)–(d) ATR-IR spectra of cinchonidine (CD) adsorbed on a Pt/Al$_2$O$_3$ model catalyst. The ATR spectra were recorded upon flowing a solution of CD in hydrogen saturated CH$_2$Cl$_2$ solvent over the sample. (a) neat solvent, (b) 10^{-6} mol L^{-1} CD, (c) 10^{-3} mol L^{-1} CD. Spectrum (d) was recorded by first flowing a 10^{-3} mol L^{-1} CD solution over the sample, followed by a neat solvent. Spectrum (e) is a scaled transmission spectrum of a CD solution for comparison. The reference spectrum for the ATR was recorded before admitting the CD solution to the sample. Ten active internal reflections were used.

Relative intensities of vibrational bands of adsorbed molecules on metal surfaces strongly depend on the local orientation of the molecules with respect to the metal surface. Therefore, the different adsorption modes observed exhibit different orientation on the Pt surface. Some of the CD molecules are irreversibly adsorbed on the Pt, as can be seen from Figure 31d. These species exhibit a strong band at 1530 cm^{-1}, which is missing in the solution spectrum. This indicates a strong interaction between the guinotine ring and the Pt surfaces. As for the example of pyridine adsorption, negative bands are observed when CD is adsorbed due to the displacement of solvent decomposition products from the Pt-CH$_2$Cl$_2$ interface. CD has to compete with these decomposition products for adsorption sites on Pt. This example shows that ATR is able to yield detailed information about the modification of metal surfaces under reactive conditions [76].

5.3. Temperature Modulated Excitation of a Hydrated Poly-L-Lysine Film

A poly-(L)-lysine (PLL) film cast on an ATR plate and hydrated with D$_2$O (80% relative humidity, 28 °C) was exposed to a periodic temperature variation of $\Delta T/2 = \pm 2$ °C at the mean value of $T = 28$ °C. The results obtained after phase sensitive detection (PSD) are shown in Figure 32. Part (A) shows the stationary spectrum and part (B) shows phase-resolved spectra of the system response with the fundamental frequency ω. The numbers indicated on the spectra denote the phase difference between the modulated excitation and the phase setting at the phase sensitive detector (PSD). The ME spectra shown in Figure 32B may be expressed by Eq. (165) (see also Refs. [22, 77]). Then,

$$\Delta A(\tilde{\nu}, \phi_{PSD}) = \kappa' \sum_{i=1}^{N} \Delta A_{0i}(\tilde{\nu}) \cos(\phi_i - \phi_{PSD}) \quad (165)$$

$\Delta A_{0i}(\tilde{\nu})$ is the ith component spectrum in which each band has the same phase angle ϕ_i. Consequently, this set of bands may be considered to be correlated, i.e., to belong to a population of molecules or functional groups featuring the same kinetic response to the external stimulation. In such a population, all absorbance bands exhibit the same periodic dependence on the PSD phase setting ϕ_{PSD}. The amplitudes become maximum for $(\phi_i - \phi_{PSD}) = 0°$, minimum (negative) for $(\phi_i - \phi_{PSD}) = 180°$, and zero for $(\phi_i - \phi_{PSD}) = 90°$ or $270°$. Obviously, ϕ_{PSD} can be used to sense the phase angle ϕ_i of a population of absorption bands, because ϕ_{PSD} is a parameter under experimental control. The most accurate determination of ϕ_i is obtained by performing a line shape analysis of the phase-resolved spectra shown in Figure 32B, followed by fitting each component according to Eq. (165), as performed Ref. [77]. The first impression on comparing Figure 32A with Figure 32B is that modulation spectra are significantly better resolved. The spectral resolution was 4 cm^{-1} for both stationary and modulation spectra. However, overlap is drastically reduced in the latter, because they contain only absorption bands from species that

spectrum of CD in solution. In the transmission spectrum, the bands above 1500 cm^{-1} are associated with ring vibrations of the quinoline moiety of CD, with the exception of the band at 1636 cm^{-1}, which is associated with the stretching vibration of the C=C double bond. The bands at 1450 cm^{-1} and below are associated mainly with the quinuclidine part of the molecule. The spectrum of the quinoline ring vibrations of adsorbed CD is clearly different from the one measured in solution. Not only the intensity of the bands is different, but some bands shift in frequency. This shows that the CD is anchored through the quinoline part of the molecule. The stretching vibration of the C=C bond cannot be observed in any of the spectra, indicating that this part of the molecule is hydrogenated, in accord with catalytic experiments, where 10,11-dihydrocinchonidine can be detected in the solution [75]. In addition, the spectrum of adsorbed CD strongly changes with solution concentration. Furthermore, time resolved experiments reveal changing relative signal intensities during the first stages of adsorption. All this indicates that the adsorption mode of CD is coverage dependent and that CD can be adsorbed on Pt in different ways.

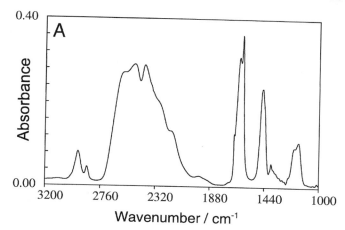

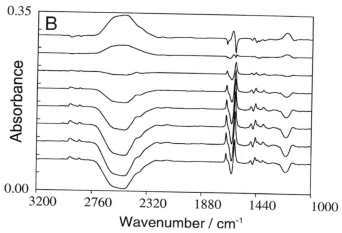

Fig. 32. (A) and (B) Parallel polarized temperature-modulated excitation (T-ME) FTIR spectra of a poly-(L)-lysine (PLL) deuterobromide film hydrated with D$_2$O (80% relative humidity, 28 °C). The film was deposited on a CdTe ATR plate. A rectangular temperature stimulation was applied with a period of 14.7 min ($\omega = 0.427$ min^{-1}) at $\bar{T} = 28$ °C ± 2 °C. Angle of incidence: $\theta = 45°$, mean number of internal reflections: $N = 9$–10. (A) Stationary part of the T-ME-IR spectrum of PLL. (B) Set of phase-resolved T-ME-IR spectra after phase sensitive detection (PSD) at phase settings $\phi_{PSD} = 0$–157.5° (phase resolution 22.5°) with respect to the T stimulation. $\phi_{PSD} = 0°$ means in-phase with temperature switching from 26 to 30 °C. Heat transfer from the thermostats to the sample resulted in an additional phase lag of $\phi_T = 25°$. Reproduced from [77] by permission of the American Chemical Society, © 1996.

have been affected by the external stimulation. Furthermore, Figure 32B shows that not only the intensity but also the shape of phase-resolved spectra is changing with ϕ_{PSD} setting. This is an unambiguous indication of the existence of populations of conformational states featuring different phase angles ϕ_i. Extraction of these populations according to Eq. (165) enabled the assignment of transient species in the amide I$'$ and II$'$ regions. Attention should be drawn to a correlation between CH$_2$ stretching and the secondary structure of PLL. The weak absorption bands at 2865 and 2935 cm^{-1} result from symmetric and antisymmetric stretching of the CH$_2$ groups of the lysine side chains. They are displaced by approximately 3 cm^{-1} toward lower wave numbers with respect to the corresponding bands in the stationary state (Fig. 32A). This finding is indica-

tive for a conformational change of a hydrocarbon chain from gauche defects into trans conformations [78]. These bands are correlated with the formation of antiparallel β-pleated sheet structure (amide I$'$ bands at 1614 and 1685 cm^{-1}). We conclude therefore, that the conversion of PLL from an α helix to a β sheet is paralleled by a conformational change of the side chain from a bent to an extended structure.

5.4. Aqueous Solutions

So far, most experimental data shown in this chapter have dealt with immobilized systems. However, FTIR ATR spectroscopy can also be used to get high quality spectra from compounds which are dissolved in a strongly absorbing solvent, such as water. The sensitivity achievable with advanced compensation techniques like the SBSR technique (see Section 4.1) enables differentiation between bulk water and hydration water, i.e., water bound to a molecule or an ion. Spectra of hydration shells are very specific, they even enable distinguishing non-IR-active ions such as sodium and potassium due to different absorption bands of bound water. In this section, we want to show spectra of hydrochloric acid (HCl) and sodium hydroxide (NaOH) at different concentrations. Measurement of such a series had two aims, first, test of the feasibility of the compensation of absorption bands resulting from hydration water, and second, search and validation of the adequate quantitative analytical approach as derived in Section 2.

5.4.1. Influence of Hydration on Water Compensation

The compensation of water is a common problem in biological FTIR spectroscopy, since lipids as well as proteins and nucleic acids have charged and hydrophilic groups which may be strongly hydrated. Taking, e.g., the amide I vibration of polypeptides or proteins, a vibration which is often used for secondary structure determination. This band is generally centered around 1650 cm^{-1} and overlaps therefore directly with the H$_2$O bending band near 1640 cm^{-1}. Access to the pure amide I band shape is a prerequisite for a reliable vibrational analysis. Subtracting the bulk water spectrum must fail since the spectrum of water in the hydration shell is characteristic of the hydrated group or ion [16]. The complexity of OH stretching and bending modes in H$_2$O solutions is impressively demonstrated by the spectra of HCl and NaOH, Figures 33 and 34.

The FTIR ATR spectrum of pure water was used as a reference spectrum throughout this series. Therefore, one should expect a certain overcompensation at wave numbers of bulk water absorption. Negative bands appear indeed near 3400 and 1640 cm^{-1}, i.e., in the H$_2$O stretching and bending regions. However, new more intense bands lead to predominantly positive absorbances. In case of HCl (Fig. 33), typical bands of the hydronium ion appear in the region of 2600–2700 cm^{-1} featuring symmetric and asymmetric H$_3$O$^+$ stretching. The asymmetric bending of H$_3$O$^+$ is expected near 1665 cm^{-1} and the symmetric bending near 1125 cm^{-1}.

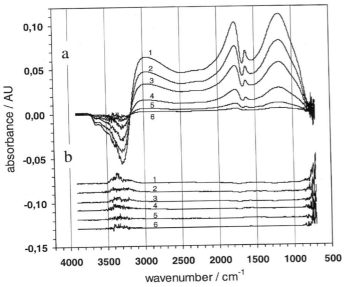

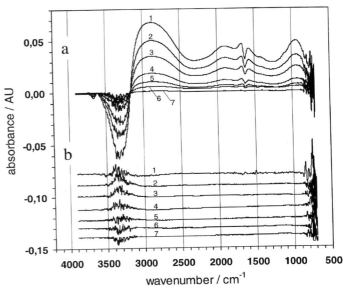

Fig. 33. Stationary difference spectra of aqueous hydrochloric acid (HCl) solutions. (a) Parallel polarized HCL spectra measured at concentrations of 768 mM, pH 0.18 (1); 576 mM, pH 0.30 (2); 384 mM, pH 0.45 (3); 192 mM, pH 0.71 (4); 77 mM, pH 1.09 (5); 31 mM, pH 1.49 (6). (b) Isotropy test by means of dichroic difference spectra according to $A^* = A_p - R_{iso} \cdot A_s$ (with A_p: parallel polarized ATR absorbance spectra; A_s: vertical polarized ATR absorbance spectra and R_{iso}: Dichroic ratio of an isotropic bulk sample. $R_{iso} = 2, 0$ at $\theta = 45°$. The numbers 1–6 coincide with (a). For an isotropic sample A^* should equal the baseline as long as the weak absorber model holds. Deviations occur in regions of strong absorbance such as the stretching region of water near 3400 cm^{-1} and in regions where oriented adsorbates contribute, which however, is not the case here. $T = 22\ °C$. Ge IRE with angle of incidence $\theta = 45°$. Mean number of active internal reflections $N = 13, 5$. Reference: pure water.

Fig. 34. Stationary difference spectra of aqueous sodium hydroxide (NaOH) solutions. (a) Parallel polarized NaOH spectra measured at concentrations of 676 mM, pH 13.05' (1); 487 mM, pH 13.04 (2); 338 mM, pH 13.01 (3); 175 mM, pH 12.86 (4); 96 mM, pH 12.70 (5); 47 mM, pH 12.50 (6); 13 mM, pH 11.98 (7). (b) Isotropy test by means of dichroic difference spectra analogues to Figure 1(b). The numbers 1–7 coincide with (a). $T = 22\ °C$. Ge IRE with angle of incidence $\theta = 45°$. Mean number of active internal reflections $N = 13, 5$. Reference: pure water. Note that measured pH values near and above pH 13 are not reliable due to the alkaline error of the pH glass electrode.

The spectrum of sodium hydroxide (Fig. 34) differs significantly from that of hydrochloric acid. However, one may conclude from Figures 33 and 34 that in standard application perturbations by hydronium and hydroxyl ions may be neglected in the pH range between 3 and 11. This finding is consistent with the experience that the detectivity limit of FTIR ATR is in the 1-mM range.

5.4.2. Adequate Approach for Quantitative Analysis of the ATR Spectra

A comprehensive overview of ATR theory is given in Section 2 of this chapter. It spans from exact quantitative calculations to approximate expressions. Our work confirms conclusions based on experimental data drawn in earlier work (see the Appendix of Ref. [16]) pointing out that the concept of *effective thickness* [3] is applicable for decadic absorption coefficients $\alpha = \varepsilon \cdot c \leq 2000$ cm^{-1}, provided germanium is taken as IRE at an angle of incidence $\theta \geq 45°$ and wavelength-dependent refractive indices are used. Most applications of ATR spectroscopy to organic compounds and aqueous solutions are within this limit. The latter is demonstrated by traces *b* of Figures 33 and 34 which represent the so-called dichroic difference spectra D^* [79] calculated from corresponding *p*- and *s*-polarized spectra

according to

$$D^*(\tilde{\nu}) = A_p(\tilde{\nu}) - A_s(\tilde{\nu}) R_{iso} \qquad (166)$$

$D^*(\tilde{\nu})$ is zero, i.e., the baseline, for an isotropic substance, as can be concluded from Eq. (145) (for explicit expressions of relative electric field components see, e.g., Ref. [16]). In the framework of the *weak absorber approximation* (see Section 2 and Appendix), it follows that the dichroic ratio of an isotropic medium $R_{iso} = 2.0$ for $\theta = 45°$ angle of incidence, irrespective of refractive indices. Spectra of HCl and NaOH fulfil this condition excellently, thus proving the applicability of this approximation. A further experimental proof of the validity of the *weak absorber approximation* in this case is the test whether experimental ATR data fulfil the Lambert–Beer relation.

As demonstrated by Figure 35, absorption bands of HCl, NaCl, and NaOH result in a linear relation between absorbance or integrated absorbance and concentration, thus fulfilling the Lambert–Beer condition.

6. CONCLUSIONS

In this chapter, a comprehensive overview has been given with respect to the applicability of infrared ATR spectroscopy to thin films as well as to bulk solutions. It has turned out that this technique is optimum for *in situ* studies of chemical and biochemical processes at interfaces and in solution. Several techniques have been mentioned for the modification of the

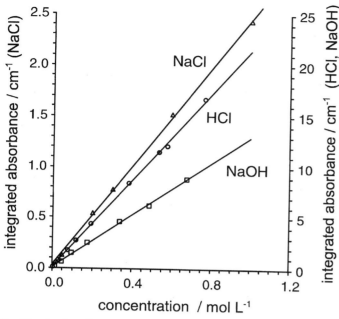

Fig. 35. Lambert–Beer plot of integrated absorbances of the hydration shell of NaCl, HCl, and NaOH. The graphs express a linear dependence between the integrated absorbances of the solvate and the electrolyte concentration of NaCl (triangles), HCl (circles), and NaOH (squares), respectively. NaCl exhibits a clearly resolved band at 1639 cm^{-1} (not shown), which could be used for integration (integration limits: 1708; 1553 cm^{-1}). For HCl, an overall integration between 1757 and 1167 cm^{-1} was performed. NaOH spectra were evaluated in a similar region from 1672 to 950 cm^{-1}. Areas were multiplied by -1. Results of HCl measurements in 0.1-M NaCl (filled circles) demonstrate the independence of the acid effect, i.e., electrostatic interaction of NaCl with HCl is not detectable with FTIR hydration spectroscopy.

surface internal reflection elements (IRE), such as the immobilization of model biomembrane assemblies as well as coating of the IRE by metal films. Fundamental theory derived from Maxwell's equations paralleled by experiments gives evidence that in most applications the *concept of effective thickness* in combination with the *weak absorber approximation*, both introduced originally by Harrick, can be used for quantitative analysis without introducing significant errors. This enables the application of the Lambert–Beers law in the same way as in transmission spectroscopy. However, unambiguous evidence is also given that the validity of the *weak absorber approximation* breaks down, as soon as the IRE is coated by a thin metal film. In this case, a more advanced theory as discussed in Sections 2 and 5.2 has to be applied for quantitative analysis. Finally, it has been demonstrated that high quality spectra can be obtained under conditions of very strong background interferences, provided adequate experimental tools, such as SBSR- and ME-techniques are applied [80].

APPENDIX: WEAK-ABSORPTION APPROXIMATION

In Sections 2.3.2.3 and 2.3.2.4, we derived simple expressions for the normalized reflectivities R_{nu} $(u = s, p)$ in terms of effective thicknesses and absorption coefficients. As mentioned

in these sections, the relations (89)–(92), (100), and (101) are, however, only valid if the quantity $(n_l^2 - n_{N+2}^2)/(n_1^2 - n_{N+2}^2) \ll 1$ for each layer l. This condition is well satisfied for systems such as Ge/HCl/H$_2$O for which excellent results are obtained (see Fig. 8). For cases where this condition is not satisfied it is, however, possible to derive more general relations for R_{nu}. The general structure of R_{nu} remains the same, but the expressions for the effective thicknesses for p-polarized light are different. For the single-layer system ($N = 1$), i.e., ATR crystal–layer 2–bulk 3, we find

$$R_{nu} = 1 - d_{eu2}\alpha_2 + d_{eu3}\alpha_3 \tag{A1}$$

where

$$d_{es2} = \frac{4n_{21}d\cos\theta_i}{1 - n_{31}^2}, \qquad d_{es3} = \frac{4n_{31}d\cos\theta_i}{1 - n_{31}^2}\gamma \tag{A2}$$

with

$$\gamma = \frac{1 - n_{21}^2}{1 - n_{31}^2} \tag{A3}$$

and

$$d_{ep2} = d_{es2}G_{p2}, \qquad d_{ep3} = d_{es3}G_{p3} \tag{A4}$$

$$G_{p2} = \frac{C_2}{C_1}, \qquad G_{p3} = \left(\frac{C_3}{C_1} + \frac{C_4}{C_1^2}\right)\frac{1}{\gamma} \tag{A5}$$

and

$$
\begin{aligned}
C_1 &= (1 + n_{31}^2)\sin^2\theta_i - n_{31}^2 \\
C_2 &= (1 + n_{32}^4)\sin^2\theta_i - n_{31}^2 \\
C_3 &= \left[(1 + n_{32}^2)\sin^2\theta_i - n_{31}^2\right]\gamma \\
C_4 &= (1 - n_{23}^2)(n_{32}^2 - n_{31}^2)\sin^4\theta_i
\end{aligned}
\tag{A6}
$$

The effective thicknesses d_{es2} and d_{ep2} agree with the corresponding quantities given by Harrick [3, p. 51, Eqs. (31) and (32)] and by Fringeli [16, p. 280, Eq. (13)]. Unfortunately, a print error just occurred in this equation. The correct notation of Eq. (13) in [16] is

$$
\begin{aligned}
d_{eu2} &= \frac{n_{21}d_p}{2\cos\theta}\cdot\left(1 - \exp\left(-\frac{2d}{d_p}\right)\right)E_{02,u}^{r}{}^2 \\
&\approx \frac{n_{21}d}{\cos\theta}E_{02,u}^{r}{}^2
\end{aligned}
\tag{A7}
$$

where $E_{02,u}^{r}{}^2$ denotes the square of the relative electric field component in medium 2 as obtained from Eqs. (8)–(10) on p. 279 in Ref. [16]. Finally, it should be noted, that using Eqs. (84) and (85), the relations (A1)–(A6) can easily be generalized for N-layer systems.

REFERENCES

1. N. J. Harrick, *Phys. Rev. Lett.* **4**, 224 (1960).
2. J. Fahrenfort, *Spectrochim. Acta* **17**, 698 (1961).
3. N. J. Harrick, "Internal Reflection Spectroscopy," Interscience, New York, 1967.

4. F. M. Mirabella, Jr. and N. J. Harrick, "Internal Reflection Spectroscopy: Review and Supplement," Harrick Sci. Corp., Ossining, NY, 1985.

5. J. R. Durig, (Ed.), "Chemical, Biological and Industrial Applications of Infrared Spectroscopy," Wiley, Chichester/New York, 1985.

6. F. M. Mirabella, Jr., "Internal Reflection Spectroscopy, Theory and Applications," Dekker, New York/Basel, 1993.

7. M. W. Urban, "Attenuated Total Reflectance Spectroscopy of Polymers," Am. Chem. Soc., Washington, DC, 1996.

8. J. C. Lindon, G. E. Tranter, and J. L. Holmes, "Encyclopedia of Spectroscopy and Spectrometry," Academic Press, London/San Diego, 2000.

9. H.-U. Gremlich and B. Yan, "Infrared and Raman Spectroscopy of Biological Materials," Dekker, New York/Basel, 2001.

10. M. Born and E. Wolf, "Principles of Optics," 3rd ed. Pergamon, Oxford, U.K., 1979.

11. J. D. Jackson, "Classical Electrodynamics," 2nd ed. Wiley, New York, 1975.

12. J. A. Shahon, Electromagnetic Theory, McGraw-Hill, New York/London, 1994.

13. H. A. Macleod, "Thin-film Optical Filters," 2nd ed. Hilger, Bristol, U.K., 1986.

14. J. R. Wait, "Electromagnetic Waves in Stratified Media," Pergamon, Oxford, U.K., 1970.

15. P. Brüesch, "Phonons: Theory and Experiments II." Springer Series in Solid State Sciences, Vol. 65. Springer-Verlag, Berlin/Heidelberg, 1986.

16. U. P. Fringeli, in "*In Situ* Infrared Attenuated Total Reflection Membrane Spectroscopy." Internal Reflection Spectroscopy, Theory and Applications (F. M. Mirabella, Jr., Ed.), pp. 255–324. Dekker, New York, 1992.

17. J. D. E. McIntyre, in "Optical Properties of Solids, New Developments" (B. O. Seraphin, Ed.), pp. 555–630. North-Holland, Amsterdam, 1976.

18. W. N. Hansen, *J. Opt. Soc. Am.* 58, 3, 380 (1968).

19. W. N. Hansen, in "Advances in Electrochemistry and Electrochemical Engineering." Optical Techniques in Electrochemistry (R. H. Muller, Ed.). Vol. 9, pp. 1–60. Wiley, New York, 1973.

20. J. D. E. McIntyre, in "Advances in Electrochemistry and Electrochemical Engineering." Optical Techniques in Electrochemistry (R. H. Muller, Ed.), Vol. 9, pp. 61–166. Wiley, New York, 1973.

21. P. A. Giguère and S. Turrel, *Can. J. Chem.* 54, 3477 (1976).

22. U. P. Fringeli, D. Baurecht, and H. H. Günthard, in "Infrared and Raman Spectroscopy of Biological Materials" (H. U. Gremlich and B. Yan, Eds.), pp. 143–192. Dekker, New York/Basel, 2000.

23. B. W. Johnson, B. Pettinger, and K. Doblhofer, *Ber. Bunsenges. Phys. Chem.* 97, 412 (1993).

24. R. Zbinden, "Infrared Spectroscopy of High Polymers," pp. 166–233. Academic Press, New York/London, 1964.

25. J. Michl and E. W. Thulstrup, "Spectroscopy with Polarized Light," pp. 171–221. VCH, New York, 1986.

26. E. B. Wilson, J. C. Decius, and P. C. Cross, "Molecular Vibrations," McGraw-Hill, New York, 1955.

27. H. Goldstein, "Classical Mechanics," Addison-Wesley, Reading, MA, 1959.

28. U. P. Fringeli and Hs. H. Günthard, in "Infrared Membrane Spectroscopy." Molecular Biology, Biochemistry and Biophysics (E. Grell, Ed.), Vol. 31, pp. 270–332. Springer-Verlag, Heidelberg, 1981.

29. P. Wenzl, M. Fringeli, J. Goette, and U. P. Fringeli, *Langmuir* 10, 4253 (1994).

30. D. Baurecht and U. P. Fringeli, in press.

31. M. Eigen and L. De Maeyer, in: "Techniques of Organic Chemistry" (S. L. Friess, E. S. Lewis, and A. Weissberger, Eds.), Vol. 8, Part 2, pp. 895–1054. Wiley-Interscience, New York, 1963.

32. H. Strehlow and W. Knoche, "Fundamentals of Chemical Relaxation," VCH, Weinheim/New York, 1977.

33. T. L. Hill, "Thermodynamics for Chemists and Biologists," Addison-Wesley, Reading, MA, 1968.

34. J. Monod, J. Wyman, and P. Changeux, *J. Mol. Biol.* 12, 88 (1965).

35. I. Mills, T. Cvitas, K. Homann, N. Kallay, and K. Kuchitsu, "Quantities, Units and Symbols in Physical Chemistry," p. 38. Blackwell Sci., Oxford, U.K., 1988.

36. Hs. H. Günthard, *Ber. Bunsenges. Phys. Chem.* 78, 1110 (1974).

37. M. Siam, G. Reiter, D. Baurecht, and U. P. Fringeli, in "Interaction of Two Different Types of Membrane Proteines with Model Membranes Investigated by FTIR ATR Spectroscopy," Fourier Transform Spectroscopy: 11th International Conference, AIP Conference Proceedings 430, (J. A. deHaseth, Ed.), pp. 336–339. Am. Inst. of Phys., Woodbury, New York, 1998.

38. R. A. Demel, C. C. Yin, B. Z. Lin, and H. Hauser, *Chem. Phys. Lipids* 60, 209 (1992).

39. R. W. Evans, M. A. Willams, and J. Tinoco, *Biochem. J.* 245, 455 (1987).

40. M. Schöpflin, U. P. Fringeli, and X. Perlia, *J. Am. Chem. Soc.* 109, 2375 (1987).

41. U. P. Fringeli, J. Goette, G. Reiter, M. Siam, and D. Baurecht, in "Structural Investigations of Oriented Membrane Assemblies by FTIR-ATR Spectroscopy," Fourier Transform Spectroscopy: 11th International Conference, AIP Conference Proceedings 430, (J. A. deHaseth, Ed.), pp. 729–747. Am. Inst. of Phys., Woodbury, New York, 1998.

42. M. Siam, G. Reiter, R. W. Hunziker, B. I. Escher, and U. P. Fringeli, in preparation.

43. G. Reiter, M. Siam, W. Gollneritsch, D. Falkenhagen, and U. P. Fringeli, submitted for publication.

44. L. K. Tamm and S. A. Tatulian, *Q. Rev. Biophys.* 30, 365 (1997).

45. K. Fritz-Wolf, Th. Schnyder, Th. Wallimann, and W. Kapsch, *Nature* 381, 341 (1996).

46. O. Stachowiak, M. Dolder, and Th. Wallimann, *Biochemistry* 35, 15,522 (1996).

47. M. J. Sweet and D. A. Hume, *J. Leukocyte Biol.* 60, 8 (1996).

48. C. Weber, H. K. Stummvoll, S. Passon, and D. Falkenhagen, *Int. J. Artif. Organs* 21, 335 (1998).

49. B. Halling-Sørensen, B. Nielsen, S. N. Lansky, F. Ingerslev, H. C. H. Lützhøft, and S. E. Jørgensen, *Chemosphere* 36, 357 (1998).

50. B. I. Escher, M. Snozzi, and R. P. Schwarzenbach, *Environ. Sci. Technol.* 30, 3071 (1996).

51. B. I. Escher, R. Hunziker, and R. P. Schwarzenbach, *Environ. Sci. Technol.* 33, 560 (1999).

52. D. Baurecht and U. P. Fringeli, in preparation.

53. G. A. Somorjai, "Introduction to Surface Chemistry and Catalysis," Wiley, New York, 1994.

54. K. Ishiguro, *J. Phys. Soc. Jpn.* 6, 71 (1951).

55. E. David, *Z. Phys.* 114, 389 (1939).

56. T. Wakamatsu, K. Kato, and F. Kaneko, *J. Mod. Opt.* 43, 2217 (1996).

57. J. C. Maxwell-Garnett, *Philos. Trans. R. Soc. A* 203, 385 (1904).

58. D. A. Bruggeman, *Ann. Phys. (Leipzig)* 24, 636 (1935).

59. G. A. Niklasson and C. G. Granqvist, *J. Appl. Phys.* 55, 3382 (1984).

60. T. Bürgi, *Phys. Chem. Chem. Phys.* 3, 2127 (2001).

61. E. D. Palik, "Handbook of Optical Constants of Solids." Academic Press, New York, 1985.

62. A. Hartstein, J. R. Kirtley, and J. C. Tsang, *Phys. Rev. Lett.* 45, 201 (1980).

63. M. Osawa and M. Ikeda, *J. Phys. Chem.* 95, 9914 (1991).

64. A. Wokaun, *Mol. Phys.* 56, 1 (1895).

65. M. Osawa, K.-I. Ataka, K. Yoshii, and Y. Nishikawa, *Appl. Spectrosc.* 47, 1497 (1993).

66. T. Kamata, A. Kato, J. Umemura, and T. Takenaka, *Langmuir* 3, 1150 (1987).

67. M. Osawa, *Bull. Chem. Soc. Jpn.* 70, 2861 (1997).

68. R. P. Eischens and W. A. Pliskin, *Adv. Catal.* 9, 1 (1958).

69. D. Ferri, T. Bürgi, and A. Baiker, *J. Phys. Chem. B* 105, 3187 (2001).

70. S. Haq and D. A. King, *J. Phys. Chem.* 100, 16,957 (1996).

71. E. Toukoniity, P. Mäki-Arvela, A. N. Villela, A. K. Neyestanaki, T. Salmi, R. Leino, R. Sjöholm, E. Laine, J. Väyrynen, T. Ollonqvist, and P. J. Kooyman, *Catal. Today.* 60, 175 (2000).

72. A. Baiker, *J. Mol. Catal. A: Chem.* 115, 473 (1997).

73. H. U. Blaser and M. Müller, *Stud. Surf. Sci. Catal.* 59, 73 (1991).

74. T. Evans, A. P. Woodhead, A. Gutierrez-Soza, G. Thornton, T. J. Hall, A. A. Davis, N. A. Young, P. B. Wells, R. J. Oldman, O. Plashkevych, O. Vahtras, H. Agren, and V. Carravetta, *Surf. Sci.* 436, L691 (1999).

75. J. T. Wehrli, A. Baiker, D. M. Monti, and H. U. Blaser, *J. Mol. Catal.* 49, 195 (1989).

76. D. Ferri, T. Bürgi, and A. Baiker, in press.

77. M. Müller, R. Buchet, and U. P. Fringeli, *J. Phys. Chem.* 100, 10,810 (1996).

78. H. H. Mantsch and R. N. McElhaney, *Chem. Phys. Lipids* 57, 213 (1991).

79. U. P. Fringeli, P. Leutert, H. Thurnhofer, M. Fringeli, and M. M. Burger, *Proc. Natl. Acad. Sci. U.S.A.* 83, 1315 (1986).

80. Equipment for single-beam-sample-reference (SBSR) and modulated excitation (ME) spectroscopy is available from OPTISPEC, Rigishasse 5, CH-8173 Neerash, Switzerland, (OPTISPEC@bluewin.ch).

Chapter 5

ION-BEAM CHARACTERIZATION IN SUPERLATTICES

Z. Zhang, J. R. Liu, Wei-Kan Chu

Department of Physics, Texas Center for Superconductivity, University of Houston, Houston, Texas, USA

Contents

1. INTRODUCTION

This chapter is devoted to the application of the ion-beam analysis technique to the characterization of the superlattice. Ion-beam analysis technique is one of the most important analysis techniques for materials characterization, especially for thin film (including superlattice) characterization. It normally includes four techniques: (1) Rutherford backscattering spec-

trometry (RBS), (2) ion-beam channeling effect, (3) particle induced X-ray emission (PIXE), and (4) nuclear reaction analysis (NRA). As for the characterization in superlattices, literature search indicates that among the four techniques Rutherford backscattering spectrometry and ion-beam channeling effect have been most extensively used. Therefore, this chapter concentrates on the application of RBS and ion-beam channeling effect.

Handbook of Thin Film Materials, edited by H.S. Nalwa
Volume 2: Characterization and Spectroscopy of Thin Films

ISBN 0-12-512910-6/$35.00

Rutherford backscattering spectrometry can provide very useful information about the surface region of the materials, such as (1) the composition of a multielemental sample, (2) the thickness of a thin film, (3) the identity of an unknown element in the materials, (4) the depth profile of elements in a sample with depth resolution as high as a few nanometers, (5) the stopping power of ions (usually light ions) in materials, etc. When combined with the ion-beam channeling effect, RBS can be used to characterize the crystalline materials, providing useful information such as (1) the crystalline quality of crystals or epitaxial films, (2) defects or radiation damage and their depth distribution, (3) the lattice location of impurity atoms, (4) the surface structure of crystals, (5) the kinetics of the solid phase epitaxial regrowth, (6) the distortion of the lattice, and (7) the strains in superlattices or heterostructures (e.g., quantum well), etc.

This chapter first discusses the basic principles, the basic concepts, and the basic calculations of the Rutherford backscattering spectrometry (Section 2). The applications of RBS to the characterization of thin films is described in detail, including the determination of the thickness and the composition of the thin films, the identification of unknown elements and the mass resolution, the depth profiling and the depth resolution, etc. In Section 3 (ion-beam channeling technique), first the theoretical description of the channeling effect, i.e., continuum model, established by Lindhard is introduced. The basic concepts of the channeling effect, such as channeling minimum yield ($\chi_{\min}$), critical angle ($\psi_{1/2}$), surface peak, transverse energy (E_{tr}), transverse continuum potentials ($U_{\mathrm{tr}}(r)$), and dechanneling, etc. are discussed in detail. Application of the channeling effect to the analysis of defects, such as dislocations, are also discussed. Sections 4 and 5 concentrate on the applications of the ion-beam technique to superlattices. Section 4 presents five examples, including the studies on the interdiffusion (thermal stability), the structural change of the superlattice, the determination of the thickness and the composition of the superlattice, and the studies on the ion irradiation effects in superlattices. Section 5 gives detailed discussions on the characterization of strains in the superlattice. Three major methods for the determination of strains are described, i.e., axial dechanneling analysis channeling angular scan analysis, and planar dechanneling analysis.

The advantages of the RBS and the channeling analysis are fast, simple, and nondestructive. Normally, RBS measurement can be done within 5–15 min which is much faster than transmission electron microscopy (TEM) and secondary ion mass spectrometry (SIMS). Ion-beam channeling analysis may take a longer time because alignment of the crystal takes extra time. However, since ion-beam channeling analysis is based on the direct ion-solid collision, it provides a direct real space probe to the crystalline materials. This advantage makes the ion-beam channeling technique a unique and very powerful tool in the characterization of superlattices, especially in terms of the structural distortion and the strain measurements.

2. RUTHERFORD BACKSCATTERING SPECTROMETRY (RBS)

2.1. Introduction

Rutherford backscattering spectrometry (RBS) [1] is the method that characterizes the surface region of materials by using the large-angle elastic Coulomb scattering between the probing ions and the target atoms. This method is called Rutherford backscattering spectrometry because Rutherford and his students Geiger and Marsden performed the first backscattering experiment in the early twentieth century, in an effort to verify the atomic model. In modern conventional RBS analysis, a collimated beam of energetic ions (normally He$^+$ ions) generated by an accelerator (Fig. 1a and b) impinges on the sample. The ions that are backscattered from the sample are registered in a particle detector (normally an Si detector) and the signals (electrical pulses) generated in the detector by the backscattered ions are amplified, shaped, and finally analyzed by a multichannel analyzer (MCA) (Fig. 2). As mentioned in Section 1, RBS is capable of providing versatile information about the surface region of the materials. Therefore, it becomes one of the most frequently used techniques for characterizing the surface regions of various solids. The strengths of the RBS are that it is fast, simple, easy, and basically nondestructive. The weaknesses of the RBS are not being sensitive to light elements, having poor mass resolution for heavy elements, and not being able to provide chemical information.

Several related ion-beam techniques have been developed to overcome the weaknesses of the conventional RBS. (1) Non-Rutherford elastic scattering. In this technique, the energy of the probing ions (normally ^{4}He$^+$ or ^{1}H$^+$ ions) is raised so that in addition to the Coulomb force, the nuclear force is also involved in the scattering. This often results in enhancement of the scattering cross section (the probability of the scattering) in some energy regions for light elements, such as C, N, and O. Thus, the detection sensitivities to those light elements are increased. (2) High-energy heavy-ion RBS. In this technique, high-energy heavy ions (much heavier than He ions) are used as the probing ions. This technique improves the mass resolution for the heavy elements as well as the depth resolution [2]. (3) RBS combined with time-of-flight (TOF) technique. This method improves both the depth resolution and the mass resolution. Depth resolution as good as 2 nm and mass resolution less than 1 amu have been demonstrated [3–5]. (4) By using an electrostatic analyzer or a 90° sector magnetic spectrometer as a detector instead of an Si detector, the depth resolution (δt) can be further improved to the monolayer level as reported, for instance, by Hunter et al. [6], Smeenk et al. [7], Carstanjen [8], and Kimura et al. ($\delta t \sim 0.3$–0.5 nm) [9, 10].

The application of these ion-beam techniques in characterizing superlattices are discussed later in this chapter. To obtain more details on the Rutherford backscattering spectrometry, the readers are referred to the reference book by Chu, Mayer, and Nicolet [1] and the handbooks edited by Mayer and Rimini [11] and by Tesmer et al. [12].

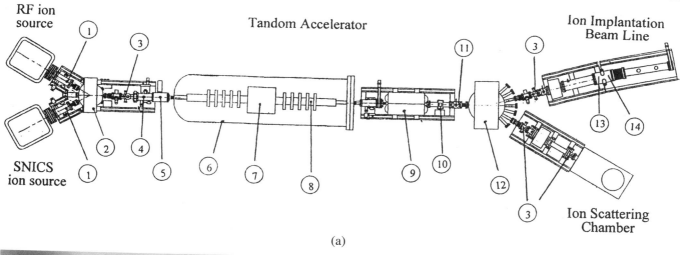

(a)

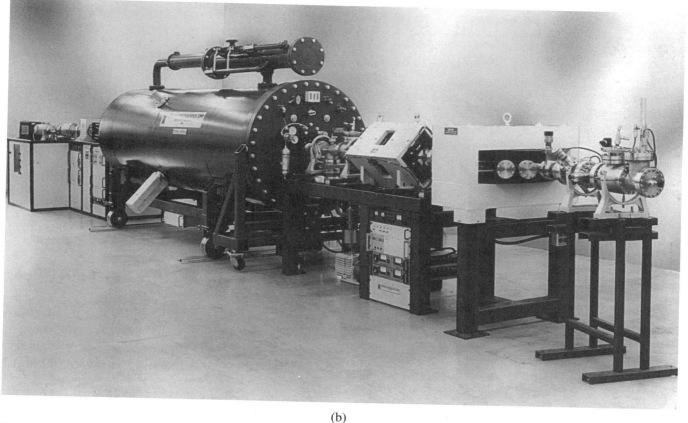

(b)

Fig. 1. (a) Schematics of a Tandom accelerator: (1) Y-steerer, (2) injector magnet, (3) slits, (4) X- and Y-steerer, (5) Enzel lens, (6) high pressure tank, (7) high voltage terminal, (8) acceleration tube, (9) quadrupole lens, (10) magnetic deflector, (11) Faraday cup, (12) switching magnet, (13) X- and Y-scanner, (14) X-deflector. (Adapted from drawing by National Electrostatics Corp., USA.) (b) 5SDH-2 1.7-MV Tandom accelerator with dual ion source injectors (protons, helium, and heavy-ion beams) manufactured by National Electrostatics Corp., Wisconsin, USA (photo from National Electrostatic Corp.).

2.2. Instrumentation

To perform conventional RBS measurement, three basic pieces of equipment are needed: (1) an ion accelerator, (2) a scattering chamber, (3) and a particle detection system including a multi-channel analyzer (MCA) (Figs. 1 and 2).

The ion accelerator generates well-collimated monoenergetic ion-beams. Modern small accelerators dedicated to ion-beam analysis can generate protons, $^4He^+$ ions, and other heavy ions with energy ranging from a few hundred kiloelectron-volts (keV) to a few tens of megaelectron-volts (MeV). In conventional RBS, usually 1 to 2-MeV $^4He^+$ ions are used because in this energy range the scatterings are mostly pure Coulomb scattering and the stopping power data are very well known. $^4He^+$ ions are preferred to Li, B, or C ions because $^4He^+$ ions

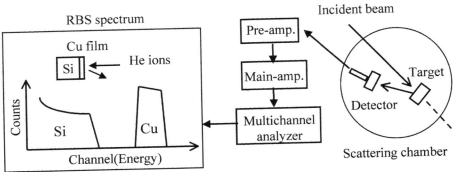

Fig. 2.   Schematics of scattering chamber, electronics, and RBS spectrum of Cu film on an Si substrate.

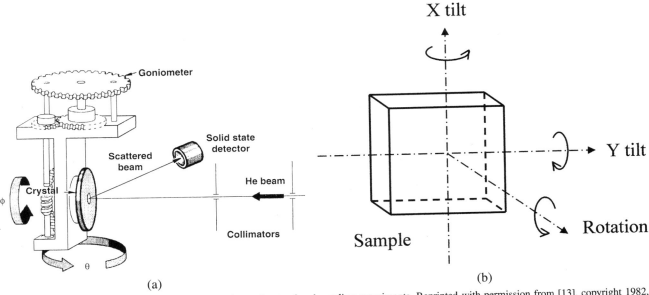

Fig. 3.   (a) Schematic of a goniometer and an experimental setup for channeling experiments. Reprinted with permission from [13], copyright 1982, Academic Press. (b) Schematically showing the three movements of the goniometer: X, Y tilt, and rotation.

generate less radiation damage in an Si detector, which is the dominant particle detector used in conventional RBS analysis.

The ion-beam generated by the accelerator is then introduced through beam lines into the scattering chamber and impinges on the sample. The sample is usually mounted on a goniometer, which is a mechanical device that can tilt, rotate, and translate the sample (Fig. 3 [13]). In the scattering chamber, a particle detector is set at a certain angle (typically 160–170°) to the incident direction of the beam. In most cases, the Si detector is preferred to other types of detectors because it is easy to operate, is compact, inexpensive, has a lifetime of many years, and has a reasonably good energy resolution (13–15 keV). The backscattered ions that enter the Si detector create electron-hole pairs. These electron-hole pairs are collected by the electrode and form electrical pulses. One ion creates one electrical pulse in the detector. The amplitude of the pulse is proportional to the energy of the ion. The pulses from the detector are further amplified and shaped by the preamplifier and the main amplifier. Then, they are sent to the multichannel analyzer (MCA)(Fig. 2). MCA is an electronic system that converts the pulses to digital

signals, sorts, and displays the signals on the screen in the form of counts/channel vs channel, which is called the RBS energy spectrum. Figure 4 is a schematic of a typical RBS spectrum of a thin elemental film grown on a light substrate. The horizontal ordinate is the channel number, which is proportional to the energy of the ions and can be converted into an energy scale or a depth scale. The element of the thin film is assumed to be heavier than the substrate. Therefore, the energy of the ions backscattered from the film is higher than that scattered from the substrate. The RBS signals of the thin film thus appear in the high-energy portion of the spectrum and the RBS signals of the substrate appear in the low-energy portion. In the following sections, we discuss how to extract useful information from an RBS spectrum.

2.3. Basic Concepts and Basic Calculations in RBS Analysis

An RBS spectrum often appears complicated. However, only four correlated observables are important. They are (1) surface energy, E_s; (2) energy width of the thin film spectrum,

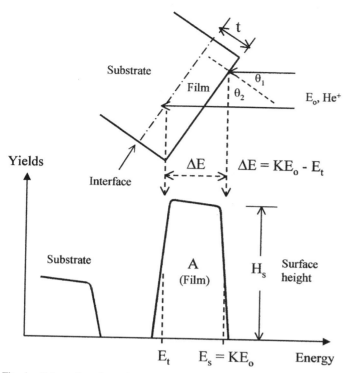

Fig. 4. Schematics of a typical RBS spectrum of a thin film deposited on a lighter substrate. The four basic elements that consist of an RBS spectrum are (1) surface energy E_s, (2) surface height H_s, (3) energy width, ΔE, and (4) area (total counts), A.

ΔE; (3) surface spectrum height, H_s; (4) total counts (area), A (Fig. 4). To relate the RBS spectrum to the properties of the sample, three basic parameters are needed. They are (1) kinematic factor, K, (2) scattering cross-section, $\sigma(E, \theta)$, (3) stopping cross-section, $\varepsilon(E)$. These four observables of an RBS spectrum and the three basic parameters form the foundation of the RBS analysis.

2.3.1. Kinematic Factor K and Surface Energy E_s

Assuming that before scattering, the probing ion has energy E_0. After scattering, its energy is reduced to E_1. The change of the energy in one single scattering event is characterized by the kinematic factor K,

$$K = E_1/E_0 \tag{1}$$

The value of the K factor depends on the mass of the probing ions, M_1, the mass of the target atoms, M_2, and the scattering angle, θ. It can be calculated from

$$K = \left[\frac{(M_2^2 - M_1^2 \sin^2 \theta)^{1/2} + M_1 \cos \theta}{M_2 + M_1} \right]^2$$
$$= \left\{ \frac{[1 - (M_1/M_2)^2 \sin^2 \theta]^{1/2} + (M_1/M_2) \cos \theta}{1 + (M_1/M_2)} \right\}^2 \tag{2}$$

Several reference books and handbooks offer tabulated K values for various ion-target combinations and various scattering angles [1, 11, 12].

In Figure 4, the initial energy of the incident $^4He^+$ ions is E_0. The angle between the incident direction and the sample normal is θ_1 and the angle between the detector and the sample normal is θ_2, so the scattering angle is $\theta = \pi - (\theta_1 + \theta_2)$. The energy of the ions that scattered at the surface of the sample is called the surface energy of the RBS spectrum, denoted as E_s, as shown in Figure 4. E_s can be calculated from the initial incident energy E_0 and the kinematic factor K,

$$E_s = K E_0 \tag{3}$$

2.3.2. Identifying Unknown Elements in the Sample

The K factor provides a tool for identifying unknown elements. The K factor can be determined experimentally from the surface energy of the RBS spectrum using Eq. (3). On the other hand, the K factor can be calculated from Eq. (2). By comparing the experimentally determined K value with the calculated one, it is possible to determine the mass of the target atom, M_2, i.e., the identity of the unknown target atom.

2.3.3. Mass Resolution, δM_2

The accuracy of the identification of unknown elements depends on the mass resolution of the RBS analysis system. When the mass of element A is close to the mass of element B ($M_A \sim M_B$), the difference between their surface energies, ($E_{sA} - E_{sB}$), will be small since they have almost the same K factor (Eqs. (2) and (3)). RBS may have difficulties in distinguishing elements A and B because the RBS may not be able to resolve the surface energy E_{sA} from E_{sB}. In this case, we say that the mass resolution is poor. The mass resolution can be obtained by differentiating Eq. (3),

$$\Delta E_s = E_0(dK/dM_2)\, dM_2 \tag{4}$$

where ΔE_s is the difference in the surface energies between two elements and the dM_2 is the mass difference. The minimum energy difference that can be resolved is limited by the energy resolution of the RBS analysis system, δE. Mass resolution δM_2 is thus defined as

$$\delta M_2 = \delta E / \left[E_0/(dK/dM_2) \right] \tag{5}$$

The derivative of the K factor, dK/dM_2, depends on the mass of the target atoms (see Ref. [1], pp. 186–193). Light elements have larger dK/dM_2, thus smaller δM_2, i.e., RBS has higher mass resolution for identifying light elements. While, for the heavy elements, the dK/dM_2 is small, thus RBS has poor mass resolution for identifying heavy elements. It is often difficult or even impossible to identify heavy unknown elements by conventional RBS measurement. The mass resolution, δM_2, also depends on the energy resolution, δE, of the RBS analysis system (Eq. (5)). δE mainly consists of three components. A detailed discussion on δE is given later in the discussion of the depth resolution.

There are several ways to improve the mass resolution for heavy elements: (1) Increase the initial incident energy, E_0. (2) Use a larger scattering angle, θ, because dK/dM_2 is larger for larger θ. (3) Use detectors with better energy resolution, such as an electrostatic analyzer. (4) Use heavier ions as probing particles because the dK/dM_2 of heavier probing ions is larger.

2.3.4. Energy Width of the Thin Film, ΔE

A thin elemental film deposited on a substrate has a surface and an interface (Fig. 4). The energy of the ions scattered at the surface is E_s (surface energy) and E_t is the energy of the ions scattered at the interface that is recorded in the detector. The difference between E_s and E_t is called the energy width of the film, denoted as ΔE,

$$\Delta E = E_s - E_t = K E_0 - E_t \tag{6}$$

Simple calculation [1] gives following relation between thickness t and energy width ΔE

$$\Delta E = [S] t \tag{7}$$

$$[S] = \left[\frac{K}{\cos\theta_1} \left(\frac{dE}{dz} \right)_{\text{in, av}} + \frac{1}{\cos\theta_2} \left(\frac{dE}{dx} \right)_{\text{out, av}} \right] t \tag{8}$$

where K is the kinematic factor, $(dE/dx)_{\text{in, av}}$ is the average value of the stopping power along the inward path of the incident ion, $(dE/dx)_{\text{out, av}}$ is the average value of the stopping power along the outward path of the scattered ion, and t is the thickness of the film. $[S]$ is usually called the energy loss factor or the S factor.

2.3.5. Stopping Power, dE/dx

As seen in Eq. (8), the ability of the RBS technique to measure the thickness or the depth stems from the energy loss of the probing particles in the samples. The stopping power, dE/dx, is the energy loss per unit distance traveled by the ions. The accuracy of the stopping power data is crucial for the thickness measurement or the depth profiling. To a good approximation, stopping power dE/dx, can be divided into two components: electronic stopping, $(dE/dx)_e$ and nuclear stopping, $(dE/dx)_n$ [14, 15]. The electronic stopping, $(dE/dx)_e$, is the energy lost in the collision with the electrons of the target atom and the nuclear stopping, $(dE/dx)_n$, is the energy lost to the target atom as a whole. In the energy range used in conventional RBS analysis using $^4\text{He}^+$ ions as probing particles (few keV to few MeV), where the velocity of the $^4\text{He}^+$ ions (V) is much larger than the average orbital velocity of the electron (v_o), the electron stopping, $(dE/dx)_e$, is much larger than the nuclear stopping, $(dE/dx)_n$, and the Bethe–Bloch formula applies for estimating the value of the electronic stopping [16, 17]. Bethe–Bloch formula has the general form,

$$\left(\frac{dE}{dx} \right)_e = \frac{4\pi Z_1^2 e^4 Z_2 N}{m v^2} \ln\left(\frac{2m v^2}{I} \right) \tag{9}$$

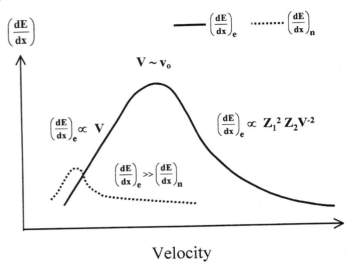

Fig. 5. Schematic of the general features of stopping power. $(dE/dx)_e$ is the electronic stopping power, $(dE/dx)_n$ is the nuclear stopping power, V is the velocity of the projectile, v_o is the average orbital velocity of the electrons, Z_1 and Z_2 are the atomic number of the projectile and the target atom, respectively.

where Z_1 and Z_2 are the atomic number of the probing ions and the target atoms, respectively, e is the electron charge, m is the electron mass, v is the velocity of the probing ions, N is the atomic density of the target, and I is the mean excitation potential. Bloch had shown that the mean excitation potential, I, is proportional to Z_2, $I = C Z_2$, where $C \sim 10$ eV [17]. The Bethe–Bloch formula (Eq. (9)) shows that the $(dE/dx)_e$ increases with atomic number Z_1 and Z_2, but decreases with the increase of the velocity of the probing ions. More sophisticated theoretical treatments on stopping power have been developed by Lindhard and Scharff [18], Lindhard, Scharff, and Schiott [19], Firsov [20], and others. It was experimentally observed that the electron stopping, $(dE/dx)_e$, oscillates with atomic number, Z_1 and Z_2. This so-called Z_1 and Z_2 oscillations, which were accounted for by using the Hartree–Fock–Slater atomic model [21, 22].

With the development of the RBS analysis, many nonconventional RBS techniques were established. One of them is the low-energy (1 to 30-keV) backscattering technique. Within such a low energy region, the nuclear stopping, $(dE/dx)_n$, is not negligible as compared with the electron stopping, $(dE/dx)_e$. Many theoretical descriptions have been developed for nuclear stopping, $(dE/dx)_n$ (e.g., Lindhard, Nielsen, and Scharff [23], Winterbon, Sigmund, and Sanders [24], Ziegler, Biersack, and Littmark [25, 26], and Biersack and Haggmark [27]). The general feature of the stopping power vs the velocity of the ions is presented in Figure 5.

2.3.6. Stopping Cross-Section, $\varepsilon(E)$

Stopping power, dE/dx, involves the density of the sample (Eq. (9)). Since the density of a thin film sample often is not precisely known, the accurate values of dE/dx in thin films are difficult to be determined. Therefore, it is necessary to use the stopping cross-section, $\varepsilon(E)$, to replace dE/dx. The stopping cross-section, $\varepsilon(E)$, is the energy loss per atom and does not involve the density of the materials. So,

$$\varepsilon = \frac{1}{N}\left(\frac{dE}{dx}\right) \qquad (10)$$

where N is the volume density of the target atoms. The values of the stopping cross-section, $\varepsilon(E)$, in various materials can be found in several references. For RBS applications, the current wildly accepted sources for tabulated values of the $\varepsilon(E)$ in various elemental materials are (1) computer code TRIM [25–27] and (2) Handbook by Mayer and Rimini [11] and by Tesmer et al. [12]. For more details about the energy loss of charged particles through matters, readers are referred to Chapt. 2 in Ref. [12] and the references therein.

2.3.7. Determine the Thickness of the Thin Film Using Energy Width, ΔE

By introducing the stopping cross-section, $\varepsilon(E)$, energy width ΔE (Eqs. (6) and (7)) now becomes

$$\Delta E = KE_0 - E_t = [S]t = [\varepsilon(E)]Nt \qquad (11)$$

where $[\varepsilon(E)]$ is usually called the stopping cross-section factor or, the ε factor. Then,

$$[\varepsilon(E)] = [S]/N = \left[\frac{K}{\cos\theta_1}\varepsilon(E)_{\text{in, av}} + \frac{1}{\cos\theta_2}\varepsilon(E)_{\text{out, av}}\right] \qquad (12)$$

where $\varepsilon(E)_{\text{in, av}}$ and $\varepsilon(E)_{\text{out, av}}$ are the average stopping cross section along the inward path and the outward path, respectively. There are several approximations to estimate their values [1]. The most popular and also the simplest approximation is the surface approximation. In the surface approximation, $\varepsilon(E)_{\text{in, av}}$ is evaluated at initial incident energy E_0 and $\varepsilon(E)_{\text{out, av}}$ is evaluated at surface energy E_s ($E_s = KE_0$, Eq. (3)). Then,

$$\varepsilon(E)_{\text{in, av}} \approx \varepsilon(E_0)_{\text{in, av}} \qquad (13)$$

$$\varepsilon(E)_{\text{out, av}} \approx \varepsilon(KE_0)_{\text{out, av}} \qquad (14)$$

From Eq. (11), it can be seen that RBS measurement only provides the information about the area density, (Nt). Only when the density N is known, then can the physical thickness t be determined.

In addition to determining the thickness of thin films, Eqs. (11) and (12) can be used for depth profiling. If we consider the thickness t in Eq. (11) as the depth in bulk materials, Eq. (11) actually links the energy of the backscattered ions to the depth, thus providing the depth scale to convert the energies (channel number) in the RBS spectrum to depth, t.

2.3.8. Depth Resolution, δt

For depth profiling, the capability of resolving two features that are close in distance is important. The minimum distance resolvable, δt, is called the depth resolution. δt is related to the

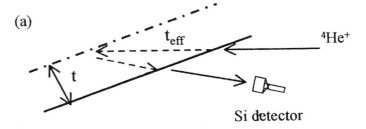

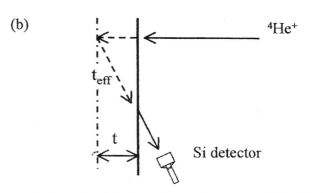

Fig. 6. Schematics of glancing-angle scattering geometry. (a) Double glancing geometry: both the incident beam and the detector are at glancing angles to the surface of the sample. (b) Single glancing geometry: the detector is at a glancing angle to the surface, but the incident beam impinges around the normal of the surface. This geometry is often used in channeling experiments.

minimum resolvable energy difference (energy resolution), δE, between the scattered particles, through

$$\delta t = \delta E/[S(E)] = \delta E/([\varepsilon(E)]N) \qquad (15)$$

This equation can be obtained by differentiating both sides of Eqs. (7) or (11). δE is the energy resolution of the RBS detection system. It consists of three major components: (1) The energy spread of the incident ion beam, δE_{beam}. Modern accelerators can normally provide an ion beam with relative energy spread, $\delta E_{\text{beam}}/E_0$, less than 0.1%; (2) the energy resolution of the detectors, δE_{det}, which depends on the type of the detectors used. For Si detectors, the δE_{det} ranges from 11 to 20 keV. For an electrostatic analyzer, δE_{det} could be as low as 0.5 keV; (3) energy loss straggling, Ω. As seen later, the energy loss straggling, Ω, increases with the depth. Therefore, the depth resolution, δt, becomes poorer and poorer with the depth.

Several techniques have been developed to improve the depth resolution. One handy technique is to tilt the sample so that the incident beam impinges the sample at a glancing angle (Fig. 6a) or to set the detector at a glancing angle to the surface of the sample (Fig. 6b). In both cases, the total path length that a probing particle must traverse is greatly increased; this increase, in turn, results in a great increase in the effective depth resolution. This technique has been extensively used in characterizing the superlattice [28]. Another method to improve the depth resolution is to use an electrostatic analyzer instead of an Si detector. For 2-MeV ^{4}He$^+$ ions, the depth resolution for an Si detector set at conventional scattering geometry ($\theta = 160$–$170°$) is 20–30 nm. When the Si detector is set at glancing geometry ($\theta < 100°$) (Fig. 6), it is possible to improve the depth resolution to 5–8 nm (see the study carried out by Barradas et al. on the Si/SiGe superlattice [28]). However, using an electrostatic analyzer as a particle detector, a depth resolution as small as monolayer thickness has been achieved (see the study carried out by Hunter et al. [6]).

As mentioned before, the depth resolution, δt, depends closely on energy loss straggling, Ω. In addition, the mass resolution also depends on the energy loss straggling, Ω. Therefore, it is necessary to discuss the energy loss straggling in more detail.

2.3.9. Energy Loss Straggling, Ω

Particles traveling in the matter lose energy through the collisions with electrons and atoms. This is a statistical process. Particles passing through the same distance in the matter do not necessarily loss the same amount of energy. Thus, the energy of the incident particles spreads with the depth. In the first order of approximation, such an energy spread can be approximately described by a Gaussian distribution. The standard deviation of this distribution is called energy loss straggling, normally denoted as Ω. Energy loss straggling, Ω, includes (1) electronic energy loss straggling, Ω_e, which is related to the energy loss in the collisions with the electrons, and (2) nuclear energy loss straggling, Ω_n, which is related to the energy loss in the collisions with atoms. For protons and ^{4}He$^+$ ions in the energy range used in the conventional RBS analysis, the nuclear energy loss straggling, Ω_n, is negligible compared with the electronic energy loss straggling, Ω_e, i.e., $\Omega = \Omega_e + \Omega_n \sim \Omega_e$. Bohr derived an expression for Ω_e at depth t [29] for high-energy light particles,

$$\Omega_B^2 \text{ (keV}^2) = 0.26 Z_1^2 Z_2 N t (10^{18} \text{ at. cm}^{-2}) \quad (16)$$

where Z_1 and Z_2 are the atomic numbers of the probing ions and the target atoms, respectively, N is the atomic density of the sample, and t is the depth. Bohr's expression shows that the energy loss straggling increases with the depth but is independent of the incident energy for high-energy light ions. Lindhard and Scharff [30] extended Bohr's expression (Eq. (16)) to a lower energy region. So,

$$\frac{\Omega^2}{\Omega_B^2} = 0.5 L(x) \qquad E > 75 Z_2 \text{ (keV amu}^{-1})$$

$$\Omega^2/\Omega_B^2 = \frac{\Omega^2}{\Omega_B^2} = 1 \qquad E \leq 75 Z_2 \text{ (keV amu}^{-1}) \quad (17)$$

$$L(x) = 1.36 x^{1/2} - 0.016 x^{3/2}$$

$$x = E \text{ (keV amu}^{-1})/(25 Z_2)$$

2.3.10. Bragg's Rule

For a compound sample, for instance $A_m B_n$, the stopping cross-section, $\varepsilon^{AB}(E)$, may be obtained by a linear combination of the stopping powers of the individual elements [31],

$$\varepsilon^{AB}(E) = m \varepsilon^A(E) + n \varepsilon^B(E) \quad (18)$$

where, m and n are the atomic fraction of the elements A and B in the compound, respectively. $\varepsilon^A(E)$ and $\varepsilon^B(E)$ are the stopping cross sections for elements A and B, respectively. Equation (18) is usually called Bragg's rule [31].

Bragg's rule assumes that the interaction between the ions and the target atoms is independent of the physical and the chemical state of the medium. However, chemical bonding and physical state, to a certain extent, do affect the energy loss. Deviation from Bragg's rule has been observed. Especially around the maximum of the stopping power, where the velocity of the ions is comparable to the average orbital velocity of the electrons (Fig. 5), deviations as large as 10–20% have been reported (Ref. [12] and the references therein). The effects of physical state and chemical bonding on the stopping power in compounds have been reviewed by Thwaites [32, 33] and Bauer [34]. The deviation from Bragg's rule for the heavy ions has been studied by Herault et al. [35]. Zielger and Manoyan have developed a CAB-model (Cores and Bondes model) to allow for the effect of the chemical bonding [36].

2.3.11. Surface Spectrum Height, H_s and Scattering Cross-Section, $\sigma(E, \theta)$

To determine the stoichiometric ratio of a compound, it is necessary to know the surface spectrum height H_s and the scattering cross-section, $\sigma(E, \theta)$.

The surface spectrum height, H_s, refers to the counts of the front edge of the spectrum (Fig. 4). The surface height, H_s, is proportional to the $\sigma(E, \theta)$ (scattering cross section), ω (the solid angle of the detector subtended to the beam spot on the sample surface), and Q (dose, total number of the incident ions),

$$H_s = \sigma(E_0, \theta) \omega Q \eta / ([\varepsilon(E_0)] \cos \theta_1) \quad (19)$$

where E_0 is the initial incident energy and η is the energy per channel, i.e., the energy width corresponding to one channel of the spectrum. In Eq. (19), the stopping cross-section factor $[\varepsilon(E)]$ is evaluated at E_0 (surface approximation).

When the scattering between the probing ions and the target atoms is pure Coulomb scattering, the scattering cross-section, $\sigma(E, \theta)$, is given by the Rutherford scattering cross section,

$$\sigma(E, \theta) = \left(\frac{Z_1 Z_2 e^2}{4E} \right)^2 \frac{4}{\sin^4 \theta}$$

$$\times \frac{\{[1 - ((M_1/M_2) \sin \theta)^2]^{1/2} + \cos \theta\}^2}{[1 - ((M_1/M_2) \sin \theta)^2]^{1/2}} \quad (20)$$

where Z_1 and Z_2 are the atomic numbers for the probing ions and the target atoms, respectively, e is the electronic charge, M_1 and M_2 are the masses for the probing ions and the target atoms, respectively, and θ is the scattering angle. Equation (20) is given in centimeter-gram-second (cgs) units. In RBS analysis, normally the unit for energy is electron voltage (eV), therefore, it is convenient to express e^2 in units of $e^2 = 1.44 \times 10^{-13}$ MeV cm $= 1.44$ eV nm.

From Eq. (20), it is obvious that RBS analysis is more sensitive to the heavier elements than to the light elements since $\sigma(E, \theta)$ is proportional to Z_2^2. The detection sensitivity can be improved by reducing the incident energy, E, however, at the cost of reducing mass resolution (δM_2 increases, Eq. (5)).

2.3.12. Total Counts (Area), A

In an RBS spectrum, the integrated counts of a peak are called area, denoted as A. The area of a peak, A_i, of the ith element in a multielement sample is proportional to the dose of the incident probing ions Q, the solid angle of the detector ω, the scattering cross-section $\sigma_i(E, \theta)$, and the area density of the ith element ($N_i t$), where N_i is the density of the ith element and t is the physical thickness of the compound. We have

$$A_i = \sigma_i(E, \theta)\omega Q(N_i t)/\cos\theta_1 \qquad (21)$$

2.3.13. Determine the Thickness (Nt) by the Peak Area, A

When the materials to be characterized are extremely thin, limited by the depth resolution of the detection system, the corresponding RBS spectrum is just a small peak. An example is shown in Figure 7, which shows that with the decrease of the

thickness of the film, the energy width of the RBS spectrum of the film decreases. As the thickness of the film reduces to less than the depth resolution, the spectrum changes from a rectangular shape to a small peak. In such a case, the method that uses energy width ΔE to determine the thickness (Nt) of a thin film (described in Section 2.3.7 and Fig. 7) cannot be applied because it is difficult to define energy width ΔE of a small peak. In this case, the peak area, A_i, can be used to determine the thickness ($N_i t$). From Eq. (21),

$$(N_i t) = A_i \cos\theta_1 / [\sigma_i(E, \theta)\omega Q] \qquad (22)$$

Usually, accurate values of Q and ω are difficult to measure. To avoid difficulty in measuring the Q and ω, a standard sample with known thickness ($N_{st} t$) is often used to calibrate the value of the ($Q\omega$). Measuring the RBS spectrum on the standard sample under the same experimental conditions as that for the sample under investigation, obtains

$$(N_{st} t) = A_{st} \cos\theta_1 / [\sigma_{st}(E, \theta)\omega Q] \qquad (23)$$

where A_{st} and $\sigma_{st}(E, \theta)$ are the peak area and the scattering cross section of the standard sample, respectively, ($N_{st} t$) is the thickness of the standard sample. Comparing Eq. (22) with Eq. (23), the $\cos\theta_1$, Q, and ω are canceled and give the thickness of the thin film, ($N_i t$),

$$(N_i t) = (N_{st} t) A_i \sigma_{st}(E, \theta) / [A_{st}\sigma_i(E, \theta)] \qquad (24)$$

2.3.14. Determine the Stoichiometric Ratio of a Compound

There are mainly two methods to determine the stoichiometric ratio of a compound; one uses the spectrum surface height, H_s, and the other uses the peak area (integrated counts of a peak), A.

2.3.14.1. Surface Height Method

Surface height of the RBS spectrum, H_s, can be used to determine the stoichiometric ratio of a compound sample. Take a uniform compound consisting of two elements, $A_m B_n$, as an example. Figure 8 is the schematic RBS spectrum for this compound. From Eq. (19), the surface heights for elements A and B, under surface approximation, are

$$H_{s,A} = \sigma_A(E_0, \theta)\omega Q m\eta / ([\varepsilon_A^{AB}(E_0)]) \qquad (25)$$

$$H_{s,B} = \sigma_B(E_0, \theta)\omega Q n\eta / ([\varepsilon_B^{AB}(E_0)]) \qquad (26)$$

where E_0 is the initial incident energy. The stoichiometric ratio m/n is given by comparing Eqs. (25) and (26),

$$m/n = \left(\frac{H_{s,A}}{H_{s,B}}\right)\left(\frac{\sigma_B(E_0, \theta)}{\sigma_A(E_0, \theta)}\right)\left(\frac{[\varepsilon_A^{AB}(E_0)]}{[\varepsilon_B^{AB}(E_0)]}\right) \qquad (27)$$

In Eq. (27), the calculation of the stopping cross-section ratio ($[\varepsilon_A^{AB}(E_0)]/[\varepsilon_B^{AB}(E_0)]$) involves the knowledge of ratio m/n. This problem can be solved by an iterative process. In most cases, ($[\varepsilon_A^{AB}(E_0)]/[\varepsilon_B^{AB}(E_0)]$) is close to unity. In the first-order approximation, take ($[\varepsilon_A^{AB}(E_0)]/[\varepsilon_B^{AB}(E_0)]$) as unity and calculate the stoichiometric ratio m/n from Eq. (27). Then, use this first-order ratio m/n to calculate

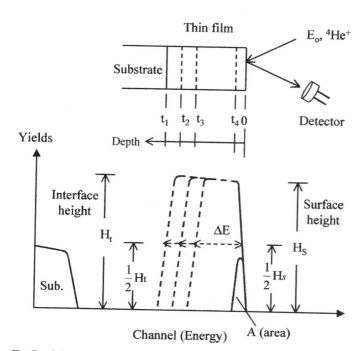

Fig. 7. Schematics of RBS spectra for thin films with various thickness. H_s is the surface spectrum height and H_t is the interface spectrum height. The energy width ΔE reduces with a decrease of the thickness. When the thickness of a film is too small, such as t_4, the proper value of the energy width ΔE cannot be defined from the spectrum.

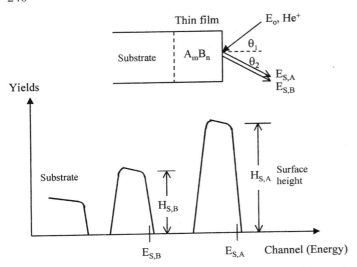

Element	$H_{1,0}^{AB...D}$ (counts)	σ_R^1 (1 MeV, 170°) ($\times 10^{-24}$ cm^2/sr)	Atomic percent
O	365 ± 30	0.2965	63.8 ± 1.6
Na	125 ± 15	0.5993	10.8 ± 1.0
Si	410 ± 15	0.9905	21.5 ± 1.3
K	70 ± 7	1.861	2.0 ± 0.2
Zn	175 ± 6	4.701	1.9 ± 0.1
Cd	9 ± 1	12.09	0.039 ± 0.004

Source: J. R. Tesmer et al., reprinted with permission from Chap. 4 in [12], copyright 1995, Materials Research Society.

Fig. 8.   Schematic RBS spectrum for a compound film, $A_m B_n$. Element A is assumed heavier than B, therefore RBS signals of element A appear in the higher energy portion and RBS signals of B is in the lower energy portion. The surface height and the surface energy for element A is $H_{S,A}$ and $E_{S,A}$, respectively. For element B, it is $H_{S,B}$ and $E_{S,B}$.

height H_s is only meaningful for the near-surface region. For a nonuniform compound, computer simulation should be used to obtain the depth profile. When the compound sample is a thin film, the surface height method may not apply if the film is too thin to define the surface height (Fig. 7), then the peak area method should be used.

2.3.14.2. Peak Area Method

The stoichiometric ratio, m/n, of a compound thin film, $A_m B_n$, can be calculated from the ratio of the peak areas. From Eq. (21), obtain

$$\frac{m}{n} = \frac{(N_A t)}{(N_B t)} = \frac{(A_A \sigma_B(E, \theta))}{(A_B \sigma_A(E, \theta))} \qquad (28)$$

where, A_A and A_B are the peak areas of elements A and B, respectively. It should be noted that the stoichiometric ratio given by Eq. (28) is the average value of the stoichiometric ratio in the film. For a thick film [e.g., Fig. 8], although the surface height method is applicable, the peak area method may offer more accurate results because the peak area can be read more accurately from the RBS spectrum than the surface height and the stopping power is not involved in the calculation (Eq. 28). An example of the peak area method is shown in Figure 10, which is the 3-MeV ^{4}He$^+$ ions RBS spectrum of a Y-Ba-Cu-O thin film grown on an Si substrate. The atomic fractions of Y, Ba, and Cu were calculated using the peak area method.

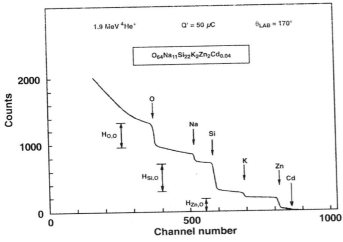

Fig. 9.   The 1.9-MeV He ion RBS spectrum of a multielement ceramic sample. The stoichiometry shown was determined from the surface heights of the elements. Reprinted with permission from [12], copyright 1995, Materials Research Society.

the $([\varepsilon_A^{AB}(E_0)]/[\varepsilon_B^{AB}(E_0)])$. After a few steps of iteration, a stable ratio m/n can be reached. An example of RBS measurement for a stoichiometric ratio in a ceramic glass sample ($O_{64}Na_{11}Si_{22}K_2Zn_2Cd_{0.04}$) using the surface height method is shown in Figure 9 [12]. The corresponding analysis results obtained from the RBS spectrum are listed in Table I. The large uncertainties in the atomic fractions are due to both the large errors in reading the surface heights from the spectrum and the large statistical errors of the surface heights. Later, it is shown that by using the peak area method, experimental errors can be reduced significantly.

If the elements in the compound are not uniformly distributed, the stoichiometric ratio m/n obtained by using surface

2.4. Non-Rutherford Elastic Scattering

Depending on the type and the energy of the probing ions and the type of the target atoms, the elastic scattering may not always follow the Rutherford scattering cross section. At a lower incident energy, due to partial screening of the nuclear charges by electron shells, the actual scattering cross sections may be less than the Rutherford scattering cross section. While at a higher energy, due to the involvement of short-range nuclear forces, the actual scattering cross sections may exhibit resonant structures. Figure 11 shows the non-Rutherford scattering cross sections of ^{4}He$^+$ ions on oxygen atoms in the energy range from 2.0 to 4.0 MeV [37]. It can be seen that when the energy of the $^4He^+$ ions exceeds 2.4 MeV, the scattering cross section departs from the Rutherford scattering cross section. The sharp resonance peak around 3.05 MeV is about 15 times larger than the Rutherford cross section. This sharp resonance is often used by researchers for depth profiling of oxygen. Non-Rutherford scattering cross sections for light elements are often larger than the Rutherford scattering cross section and are frequently used for improving the detection sensitivities to light elements. Useful non-Rutherford cross-section data for light elements analysis can be found in the handbook edited by Tesmer et al. [12].

3. ION-BEAM CHANNELING TECHNIQUE

3.1. Ion-Beam Channeling Effect in Single Crystals

Many samples studied in material research have a crystallographic structure, for instance, single crystals, epitaxial thin films, or man-made superlattices. As mentioned in the previous sections, the RBS combined with the channeling effect can provide abundant useful information on some important properties of crystalline samples.

When the incident direction of a probing particle is aligned with or at a small angle to a major crystallographic axis or plane of the crystal, due to the collective Coulomb repulsion from the target atoms, the incident particle will have lower probability to closely approach the target atoms. This is because before the incident particles could get close to any of the target atoms, for instance atom A in Figure 12a, it had

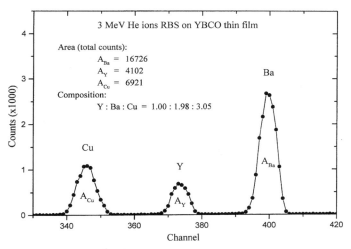

Fig. 10. The 3-MeV $^4He^+$ ions RBS spectrum of a YBCO thin film deposited on Si. (The RBS signals from Si were not shown.) The composition of the film is determined from the peak areas (total counts) and is Y:Ba:Cu = 1.00:1.98:3.05.

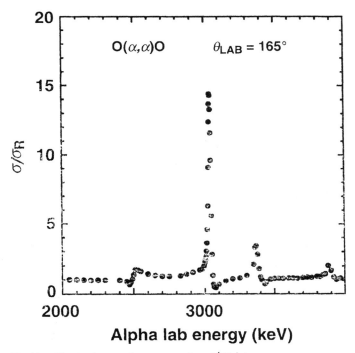

Fig. 11. The elastic scattering cross-section of $^4He^+$ ions on oxygen atoms at a scattering angle of 165°. The scattering of the $^4He^+$ ions follows the Rutherford scattering until about 2.4 MeV. Above 2.4 MeV, the scattering becomes non-Rutherford. A sharp resonance appears around 3.05 MeV, with a scattering cross-section about 15 times of the Rutherford scattering cross-section. Reprinted with permission from [12], copyright 1995, Materials Research Society.

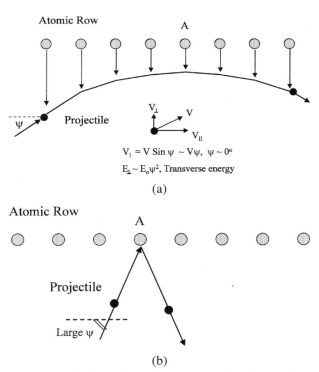

Fig. 12. (a) When the incident angle, ψ, is small, the incident particle only experiences a series of small-angle scattering. The incident particle could not get close to any target atoms because before it could get close to one atom (e.g., atom A), it had already been deflected away by the atoms proceeding this atom. (Adapted from Andersen's Lecture Notes on Channeling, 1980.) (b) When the incident angle, ψ, is large, the incident particle has large transverse energy, E_{tr}. It is able to get very close to the target atoms and have large-angle scattering with the target atom. (Adapted from Andersen's Lecture Notes on Channeling, 1980.)

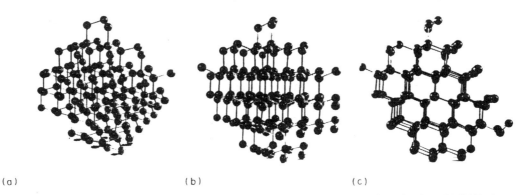

Fig. 13. Models of the lattice atoms with diamond structure, viewed along (a) random direction, (b) {110} planar direction, and (c) ⟨110⟩ axial direction. Reprinted with permission from [1], copyright 1978, Academic Press.

already been deflected away by the atoms proceeding the atom A in the atomic row through a series of correlated small-angle scatterings (Fig. 12a). The large-angle scattering events (i.e., scattering events with small impact parameters) such as backscattering events are suppressed because they require having a close encounter with the target atoms. From the energy point of view, when the incident direction of the particles is almost parallel to the axis or the plane, the kinetic energy of the particles in the transverse direction will be small (Fig. 12a). Therefore, the collective repulsive potential of the target atoms can easily keep the incident particles from closely approaching the target atoms. The incident particles that impinge into the crystals along or at a small angle to the axis or the plane and that only experience small-angle scattering (scattering with large impact parameter) are called channeled particles. In the literature, normally, the axial or planar directions are called channeling directions. For channeled particles, not only the backscattering events are suppressed but also the nuclear reaction events and the inner shell ionization events are suppressed because both events require having a close encounter (small impact parameter collision) with the target nuclei. On the other hand, when the probing particles impinge into crystals at a large incident angle to the axis or the plane, the kinetic energy of the particles in the transverse direction is much larger. In this case, the incident particles can easily overcome the collective repulsive potential in the close vicinity of the target atoms and can get very close to the target atoms (Fig. 12b). Thus, the probability of having close encounter events (backscattering, nuclear reaction, inner shell electron ionization etc.) with the target atoms are high. Such incident directions are called random direction and the particles impinging the crystals along the random direction are called random particles. The definitions of the channeling direction and the random direction can be illustrated in Figure 13 [1], which presents a model of lattice atoms. Figure 13a shows the random distribution of the atoms as viewed along the random direction. Figure 13b and c shows the regular arrangement of the atoms as viewed along the planar and axial channeling directions, respectively. When the incident direction switches from the random direction to the channeling direction, the probabilities of backscattering, nuclear reaction, and inner shell ionization will reduce dramatically. Normally, it reduces

20–40 times. Because the channeled particles are prohibited from getting close to the target atoms, they are always moving in the regions where the electron density is lower. Therefore, the electronic energy loss of channeled particles is less than that of random particles. In other words, channeled particles penetrate deeper than random particles. All the phenomena discussed so far, i.e., the suppression of the close encounter events and the reduction of the energy loss, are called the ion-beam channeling effect.

Ion-beam channeling effect was first observed around the beginning of the 1960s independently by Davies and Sims [38] in ion range measurement in single crystal Al and by Robinson and Oen [39] in computer simulation of the penetration of Cu ions through a Cu single crystal. Shortly after, Lindhard established the first theoretical model–continuum model for the channeling effect [40]. Then, a few years later, Barrett accomplished detailed studies on the Monte Carlo simulation of the channeling effect [41]. In most of the reference books [13, 42, 43] and handbooks [11, 12] dealing with the channeling effect, the theoretical calculations presented to readers are mainly based on Lindhard's continuum model and Barrett's Monte Carlo simulation. Since the 1970s, more studies on the channeling effect using high-energy electrons and high-energy (up to gigaelectron volts (GeV)) ions have been carried out. However, these studies are out of the scope of this chapter. In this chapter, we concentrate on the channeling effect of the $^4\mathrm{He}^+$ ions in crystals with energy ranging from few keV to few MeV.

3.2. Basic Concepts in Ion-Beam Channeling Effect

In this section, several important concepts in channeling effect are discussed. They are (1) transverse continuum potential, $U_{\mathrm{tr}}(r)$, (2) transverse energy, E_{tr}, and spatial distributions of the channeled particles, (3) critical angle, $\psi_{1/2}$, (4) channeling minimum yield, $\chi_{\min}$, (5) surface peak, and (6) dechanneling.

3.2.1. Transverse Continuum Potential, $U_{\mathrm{tr}}(r)$

As already mentioned, a channeled particle will only experience a series of correlated large impact parameter collisions (small-angle scatterings) with the target atoms in the atomic rows (or planes). In a perfect single crystal, an uncorrelated single large-angle scattering event with a single target atom will never occur to a channeled particle because whenever it moves close to one target atom, it must also come close to the proceeding atoms in the row as well as the atoms behind (Fig. 12a). When a channeled particle passes by an atomic row, instead of colliding with a single target atom, it always collides with many target atoms in the atomic row. Lindhard's continuum model [40] showed that for a channeled particle, such " particle-many atoms" collision can be approximated by " particle-atomic row" collision, which can be described by the interaction between the channeled particle and a transverse potential, $U_{tr}(r)$. This transverse potential can be obtained by averaging the potentials of individual target atoms in an atomic row,

$$U_{tr}(r) = \left(\frac{1}{d}\right) \int_{-\infty}^{+\infty} V\left(\sqrt{z^2 + r^2}\right) dz \qquad (29)$$

where r is the perpendicular distance of the channeled particle to the atomic row, z is the distance traveled along the atomic row, d is the spacing between the atoms in the atomic row, $V[(r^2 + z^2)^{1/2}] = V(R)$ is the interatomic potential between the incident particle and the target atom. In the channeling effect study, since normally only large impact parameter collision are involved, the Thomas–Fermi type of screening potentials are used as interatomic potential, $V(R)$,

$$V(R) = \frac{Z_1 Z_2 e^2}{R} \Phi\left(\frac{R}{a}\right) \qquad (30)$$

where R is the separation distance, Z_1 and Z_2 are the atomic number of the incident particle and the target atom, respectively, e is the electronic charge, $\Phi(R/a)$ is the Thomas–Fermi type of screening function, and a is the Thomas–Fermi screening radius. Following the usage in [1] and [11], the screening radius a is given by [44],

$$a = 0.8853a_o\left(Z_1^{1/2} + Z_2^{1/2}\right)^{-2/3} \qquad (31)$$

$$a \approx 0.47(Z_2)^{-1/3} \ (\text{\AA}) \qquad Z_1 \ll Z_2 \qquad (32)$$

where a_o is the Bohr radius, $a_o = 0.5292$ Å. For the Thomas–Fermi screening function, $\Phi(R/a)$, an analytical approximation given by Moliéré [45] is often used for calculating transverse continuum potentials,

$$\Phi\left(\frac{R}{a}\right) = \sum_{3}^{i=1} \alpha_i \exp\left(\beta_i \frac{R}{a}\right) \qquad (33)$$

where a is the screening radius (Eqs. (31) and (32)) and α_i and β_i are given by

$$\{\alpha_i\} = \{0.35, 0.55, 0.10\} \qquad (34)$$

$$\{\beta_i\} = \{0.3, 1.20, 6\} \qquad (35)$$

Ziegler, Biersack, and Littmark (ZBL) extended an earlier study made by Wilson, Haggmark, and Biersack [46] and developed a new interatomic potential [25–27, 47]. They calculated the interatomic potentials of hundreds of randomly selected ion-atom

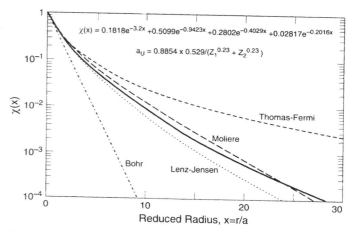

Fig. 14. The ZBL universal screening function, $\chi(x)$ (solid line) (Eqs. (36)–(39)), with its argument, x, being defined as $x = r/a_u$, where a_u is the universal screening length given in Eq. (37). Reprinted with permission from [25], copyright 1985, Pergamon.

pairs using the Hartree–Fock–Slater atomic model (for estimation of the electron density). From the calculated hundreds of potentials of individual ion-atom pairs, an optimal fitting curve, which is called the universal potential, was obtained by least-square fitting and by adjusting the screening radius. The proposed universal potential consists of four exponential terms,

$$\Phi\left(\frac{R}{a}\right) = \sum_{3}^{i=1} c_i \exp\left(b_i \frac{R}{a}\right) \qquad (36)$$

where a is the ZBL screening radius [47],

$$a = 0.8853a_o\left(Z_1^{0.23} + Z_2^{0.23}\right)^{-1} \qquad (37)$$

and c_i and b_i are given by

$$\{c_i\} = \{0.1818, 0.5099, 0.2802, 0.02817\} \qquad (38)$$

$$\{b_i\} = \{3.2, 0.9423, 0.4028, 0.2016\} \qquad (39)$$

Figure 14 shows the universal potential (Eqs. (36)–(39)) proposed by Ziegler, Biersack, and Littmark [25]. For comparison, the Bohr potential, the Lenz–Jensen potential, the Moliéré potential, and the Thomas–Fermi potential are also shown in Figure 14. The Lenz–Jensen potential and the Moliéré potential are close to the universal potential. For more information about the interatomic potentials, readers are referred to the monographs by Terrens [48] and by Nastasi, Mayer, and Hirvonen [49].

The transverse potential given by (Eq. (29)) is usually called the axial continuum potential, which is an averaged potential produced by the atoms in the atomic row. Equation (29) is a static continuum potential because it neglected the thermal vibration of the atoms. A more realistic potential is the thermally averaged continuum potential, which takes into account the displacement of the atoms due to thermal vibration. Figure 15 is the axial continuum potentials calculated by Andersen and Feldman [50] using the Moliéré potential (Eq. (33)) as the interatomic potential. It shows the axial thermal average potentials

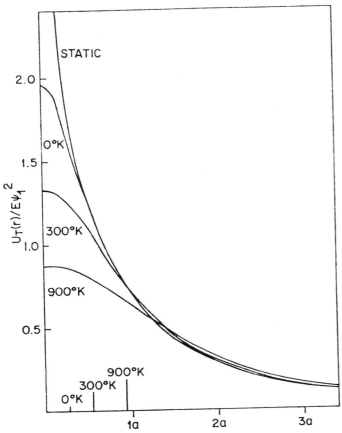

Fig. 15. Thermally averaged axial continuum potentials, $U_T(r)$, for the $\langle 110 \rangle$ atomic row in W. r is the distance from the atomic row. As comparison, a static axial continuum potential is also shown. Reprinted with permission from [50], copyright 1970, American Physical Society.

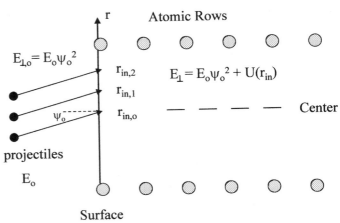

Fig. 16. The transverse energy, E_{tr}, of the channeled particle in the crystal is defined as the sum of the initial transverse kinetic energy, $E_{tr,0}$ ($= E_0\Psi^2$) and the initial transverse potential energy, $U(r_{in})$, which is acquired at the entrance of the sample. (Adapted from Andersen's Lecture Notes on Channeling, 1980.)

continuum model defines that the initial transverse energy, E_{tr}, of a channeled particle at the entrance point is the sum of the initial transverse kinetic energy, $E\psi_0^2$, and the initial transverse potential energy, $U_{tr}(r_{in})$,

$$E_{tr} = E\psi_0^2 + U_{tr}(r_{in}) \tag{40}$$

The transverse energy E_{tr} has two interesting features: (1) the value of the transverse energy, E_{tr}, depends on the entrance point, r_{in}. Even the incident particles have the same incident energy E_0 and same incident angle ψ_0, they may have different transverse energy if they cross the surface at different entrance points and thus acquire different transverse potential energy, $U_{tr}(r_{in})$. The particles with the entrance points near the atomic row have higher transverse energy than the particles with the entrance points near the center of the channel; (2) Lindhard [40] showed that, to a good approximation, the transverse energy, E_{tr}, is conserved (if energy losses are neglected).

3.2.3. Transverse Spatial Distributions of the Channeled Particles across the Channel

The open space between the atomic rows (or atomic planes) is usually called the channel (Fig. (13)). Within the transverse plane of a channel, the areas that are accessible to the channeled particles are determined by their transverse energy (E_{tr}). Channeled particles can only access the areas where their transverse energy is equal to or larger than the potential $U_{tr}(r)$ ($E_{tr} \geq U_{tr}(r)$). The central region of an axial channel has the lowest potential, therefore it is accessible to all the channeled particles. While for the region near the atomic rows the potential is high, only a few particles with higher transverse energy are able to enter. Therefore, a uniform parallel beam of particles incident along an axial channeling direction ($\psi = 0$) will be transformed into a flux of channeled particles that are highly concentrated near the center of the channel. This is the so-called "flux peaking." The flux peaking was first observed by Andersen et al. in the lattice location of Yb in Si [54]. As an ex-

and the axial static potential for protons in the W $\langle 100 \rangle$ channel. When r goes to 0, the static potential goes to an unrealistic value (infinity), while the thermal average potential gives a realistic finite value. For large r (far away from the atomic row), there is no significant difference between the static potential and the thermally averaged potential.

Similarly, planar transverse continuum potentials can be calculated [40]. The planar continuum potential is weaker than the axial continuum potential.

Based on the continuum potential, the trajectories of the channeled particles (see, for example, [51]), the channeling angular scan (channeling dip), and many other channeling parameters can be calculated [52, 53].

3.2.2. Transverse Energy of a Channeled Particle, E_{tr}

Transverse energy is one of the important concepts in Lindhard's continuum model [40]. Assuming that there is a beam of particles with initial energy E_0 impinging at a small-angle ψ_0 to an axial channel (Fig. 16), the initial transverse kinetic energy of the incident particles before entering the crystal, is $E_{tr} = E_0 \sin^2 \psi_0 \sim E\psi_0^2$. At the surface of the sample, the incident particles acquire transverse potential energy $U_{tr}(r_{in})$. The r_{in} is the distance from the entrance point to the atomic row. The

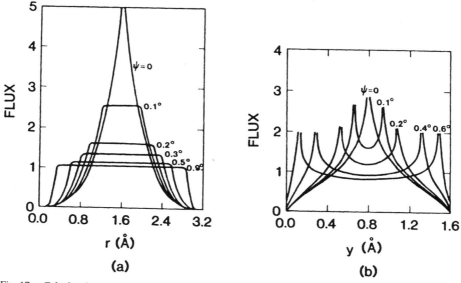

Fig. 17. Calculated transverse spatial distributions of channeled particles in a tungsten crystal as a function of the incident angle ψ for 1-MeV ^{4}He$^+$ ions in (a) $\langle 100 \rangle$ axial channel and (b) (100) planar channel. Reprinted with permission from [13], copyright 1982, Academic Press.

ample, Figure 17a shows the distribution of 1-MeV He ions in the $\langle 100 \rangle$ channel of a tungsten crystal [13]. In axial channeling, the flux peaking effect reduces as the incident angle increases. In the case of planar channeling, the transverse spatial distribution of the channeled particles differs from that in the axial case: when the incident angle is small, the channeled particles tend to gather near the central region, while when the incident angle is large they tend to gather near the atomic planes (Fig. 17b) [13].

It should be pointed out that the calculation of transverse spatial distribution shown in Figure 17 is based on the assumption that the statistical equilibrium has been reached. This not only means that the incident particles are uniformly distributed over their accessible area in the transverse plane, but also means that at any point along the channel, the transverse space distribution of the channeled particles is the same. It is generally accepted that the incident particles need to travel a few hundred nanometers before the statistical equilibrium is reached [40, 55, 56]. It is important to note that before reaching statistical equilibrium, the transverse spatial distribution varies with the depth along the channel. The trajectories of the incident particles oscillate with the depth in the channels like waves. In the case of planar channeling, the trajectories oscillate between two neighboring atomic planes (a schematic of such waves is shown in Fig. 18). Since such oscillations are close to regular harmonic oscillation, they can be approximately characterized by a wavelength, λ [13]. Typically $\lambda = 10$–40 nm depending on the incident energy and the type of the sample.

3.2.4. Channeling Critical Angle, $\Psi_{1/2}$

To illustrate the concept of critical angle, let us perform a virtual angular scan. Assume that there is a 2-MeV He ion beam impinging around the $\langle 100 \rangle$ axis of an Si crystal. The Si crystal

Trajectory of the channeled particles

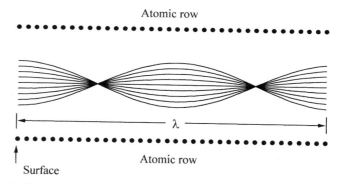

Fig. 18. Schematically shows the oscillation of the trajectories of the channeled particles along a planar channel before the statistical equilibrium is reached. The trajectories are approximated by sine waves.

is mounted on a goniometer (Fig. 3) so that the incident angle of the He ions to the Si crystal can be varied by rocking the Si sample. Assume that the He ion beam scans through a series of incident angles in steps of $0.2°$, starting from the random direction at one side of the $\langle 100 \rangle$ axis, decreasing step by step to the $\langle 100 \rangle$ axial direction and then increasing to the random direction at the other side of the $\langle 100 \rangle$ axis. At each angle, the RBS spectrum is taken. Figure 19a is the schematic of the recorded spectra (for a clearer view, not all the spectra are displayed). Near the surface region, an energy window is opened and the total counts within this window are taken from each spectrum. Plotting these total counts against the incident angle gives angular distribution of the counts (Fig. 19b). In Figure 19b the total counts have been normalized by the counts taken from the random spectrum. Such angular distribution is called a channeling angular scan or a channeling dip. The half value of the FWHM

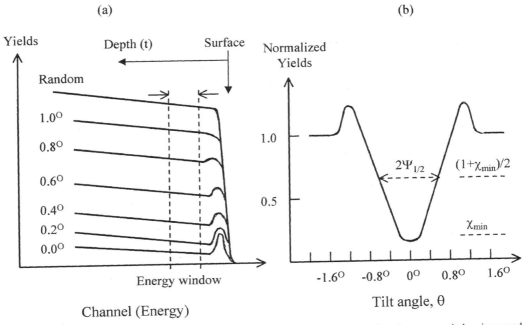

Fig. 19.   (a) Schematic of the ^{4}He$^+$ ions RBS spectra taken at various incident angles. An energy window is opened near the surface of the sample. The total counts within this window are taken from each spectrum and are plotted against the incident angle, as shown in (b). (b) Schematic of a channeling angular scan (channeling dip). The RBS counts have been normalized by the random spectrum. The channeling critical angle, $\Psi_{1/2}$, is defined as half of the full width at half maximum (FWHM).

(full width at half maximum) of the channeling dip is called the critical angle, normally denoted as $\Psi_{1/2}$. For axial channeling, Lindhard's continuum model [40] shows that the critical angle $\Psi_{1/2}$ is related to the atomic number Z_1 (probing particle) and Z_2 (target atom), energy of the probing particle E and the lattice spacing d through

$$\Psi_{1/2} = C\Psi_1 \qquad (41)$$

where Ψ_1 is called the characteristic angle,

$$\Psi_1 = \sqrt{\frac{2Z_1 Z_2 e^2}{Ed}} \qquad (42)$$

The constant C depends on the thermal vibrational amplitude of the lattice. Barrett's Monte Carlo computer simulation [41] gives

$$C = 0.85 F_{RS}(\xi) \qquad (43)$$

where $F_{RS}(\xi)$ is a function of thermal vibration of the lattice atoms and the value of $F_{RS}(\xi)$ is given in Figure 20 [41]. ξ is given by

$$\xi = 1.2 u_1/a \qquad (44)$$

where a is the Thomas–Fermi screening radius (Eqs. (31) and (32)) and u_1 is the one-dimensional root mean square (rms) thermal vibration amplitude, which can be calculated from Debye theory of thermal vibration,

$$u_1 = 1.21 \left\{ \frac{[(\phi(x)/x)] + 1/4}{M_2 \Theta_D} \right\}^{1/2} \quad \text{(nm)} \qquad (45)$$

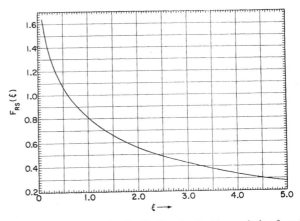

Fig. 20.   Graph of the function Frs(ξ). Reprinted with permission from [41], copyright 1971, American Physical Society.

where M_2 is the atomic weight of the target atom, Θ_D is the Debye temperature (in degrees Kelvin), $x = \Theta_D/T$, T is the temperature of the sample (in degrees Kelvin), and $\phi(x)$ is the Debye function. Figure 21 presents the tabulated values (taken from Ref. [57]) and the graph (taken from Gemmel [58]) of the Debye function. The values of u_1 for C, Al, Si, Ge, and W are listed in Table II [1, 11] along with the measured and calculated critical angles, $\Psi_{1/2}$, for axial channeling.

Critical angle $\Psi_{1/2}$ is very sensitive to the lattice distortions (static or dynamic). By measuring the critical angle of an axis, the atomic displacement perpendicular to this axis, as small as 0.001 nm, can be detected.

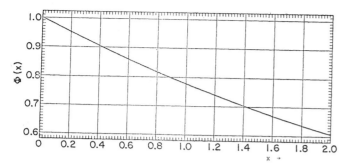

Debye Function

x	$\frac{1}{x}\int_0^e \frac{t\,dt}{e^4-1}$	x	$\frac{1}{x}\int_0^e \frac{t\,dt}{e^4-1}$
0.0	1.000000	2.6	0.528375
0.1	0.975278	2.8	0.502682
0.2	0.951111	3.0	0.480435
0.3	0.927498	3.2	0.459555
0.4	0.904437	3.4	0.439962
0.5	0.881927	3.6	0.421580
0.6	0.859964	3.8	0.404332
0.7	0.838545	4.0	0.388148
0.8	0.817665	4.2	0.372958
0.9	0.979820	4.4	0.358696
1.0	0.777505	4.6	0.345301
1.1	0.758213	4.8	0.332713
1.2	0.739438	5.0	0.320876
1.3	0.721173	5.5	0.294240
1.4	0.703412	6.0	0.271260
1.6	0.669366	6.5	0.251331
1.8	0.637235	7.0	0.233948
2.0	0.606947	7.5	0.218698
2.2	0.578427	8.0	0.205239
2.4	0.551596	8.5	0.193294
		9.0	0.182633
		9.5	0.173068
		10.0	0.164443

Fig. 21. Graph and table of the Debye function. Reprinted with permission from [11], copyright 1977, Academic Press. Originally, graph from D. S. Gemmel. Reprinted with permission from [58], copyright 1974, American Physical Society, and table from "Handbook of Mathematical Functions" (M. Abramowitz and I. A. Segun, Eds.), p. 998. Dover, New York, 1965. Reprinted with permission from [57], copyright 1965, Dover Publishing.

3.2.5. Channeling Minimum Yield, χ_{min}

Let us go back to Figure 19a. It shows a series of RBS spectra corresponding to different incident angles around an axial direction. Among them, the spectrum taken along the axial direction ($\theta = 0$) is called the aligned spectrum while the spectrum taken along the random direction is called the random spectrum. In the near surface region, the ratio of the counts of the aligned spectrum to that of the random spectrum is called the minimum yield, often denoted as χ_{min},

$$\chi_{min} = \frac{Y_s\ (\text{aligned})}{Y_s\ (\text{random})} \qquad (46)$$

When such a ratio is taken at depth t, it is denoted as $\chi(t)$. The channeling minimum yield, χ_{min}, increases with temperature and decreases with the incident energy [40, 41]. χ_{min} also increases with the thickness of the surface imperfect layer (damage caused by polishing, contamination, or oxides, etc.). Both χ_{min} and $\chi(t)$ are important parameters characterizing the crystalline quality of single crystals or epitaxial films. The lower the χ_{min} and $\chi(t)$ are, the better the crystalline quality is. For most good crystals or epifilms, the χ_{min} is between 2–4%.

3.2.6. Surface Peak in Channeling Aligned Spectrum

Let us look at the aligned spectrum ($\theta = 0°$) in Figure 19a. In this spectrum, a little "hill" can be seen at the surface. This little hill is called the surface peak. For most crystals, the surface layer (usually 2 to 5-nm thick) is full of imperfections, including the damage caused by mechanical polishing, the oxides, and the contamination, etc. Because the atoms in imperfect layers are randomly distributed, the incident probing particles have high scattering yields when passing through such layers. After passing through the imperfect surface layer, the probing particles enter a region where the sample has good crystalline structure, the scattering yield thus drops due to the channeling effect. It should be noted that even though there is no imperfect surface layer lying on the top of the crystal, the channeling aligned spectrum will still have a surface peak because the incident particles will have larger angle scattering with the outermost atoms of the atomic rows.

3.2.7. Dechanneling

The RBS counts in an aligned spectrum ($\theta = 0°$) increase with the depth t (see Fig. 19a). One of the reasons for such increase is dechanneling. A channeled particle moving through the crystal experiences various types of scattering: (1) multiple scattering with thermally vibrating target atoms, (2) multiple scattering with electrons, and (3) scattering with defects such as displaced atoms and dislocations. Due to these scatterings, for some of the channeled particles, the incident angles to the atomic row (or atomic plane) gradually increase with the depth. Finally, the incident angles become larger than the critical angle $\Psi_{1/2}$ and these channeled particles are dechanneled (becoming random particles). This process is called dechanneling. Figure 22 schematically shows the dechanneling of the channeled incident particles. A dechanneled particle behaves like a random particle. It has a high probability of having close encounter events with any target atoms.

Dechanneling is very important in characterizing the strained superlattice (SSL). In a strained superlattice, along the inclined axial (or planar) direction, due to the strain the atomic rows (or atomic planes) have a kink (angle shift) at the interface between the layers of the superlattice or between the superlattice and the substrate (Fig. 23). When a channeled particle impinging along such inclined axial direction passes through the interface, its incident angle to the atomic row will have a sudden change, $\delta\Psi$. By adjusting the energy of the particle, the change of the incident angle, $\delta\Psi$, can be larger than the critical angle, $\Psi_{1/2}$,

Table II. Data of d (Lattice Constant), Θ_D (Debye Temperature), u_1 (One-dimensional Thermal Vibration Amplitude), $\Psi_{1/2}$ (Critical Angle) for Single Crystals of C (Diamond), Al, Si, Ge, and W

Name Structure	Z_2	M_2	N (at. cm^{-2} $\times 10^{22}$)	a (Å) (= $0.47Z_2^{-1/3}$)	u_1 (Å) Θ_D (K)	(293 K)	d_0 (Å)	$\Psi_{1/2}$ (Å)[a] Calculated	Measured
C face-centered cubic (fcc) (dia.)	6	12.1	1.13	0.258	2000	0.04	3.567	0.75	0.75
Al fcc	13	26.98	6.02	0.199	390	0.105	4.050	0.45[b]	0.4[b]
Si fcc (dia.)	14	28.09	4.99	0.194	543	0.075	5.431	0.73	0.75
Ge fcc (dia.)	32	72.59	4.42	0.148	290	0.085	5.657	0.93	0.95
W body-centered cubic	74	183.85	6.32	0.112	310	0.050	3.165	0.83[c]	0.85[c]

Adapted from Chu, Mayer, and Nicolet [1], copyright 1978, Academic Press.

Source: Taken from D. S. Gemmel [58] and J. W. Mayer and E. Rimini [11].

[a] Values are for 1.0-MeV He along $\langle 110 \rangle$ unless otherwise specified.

[b] For 1.4-MeV He along $\langle 110 \rangle$.

[c] For 3.0-MeV H along $\langle 111 \rangle$.

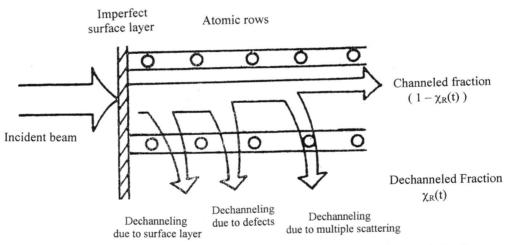

Fig. 22. Illustration of the factors that result in dechanneling. Channeled particles may be dechanneled by (1) scattering in the imperfect surface layer, (2) scattering with defects, and (3) scattering with electrons and thermally vibrating target atoms.

so that the particle that is channeled in one layer of the superlattice immediately becomes dechanneled in the next layer. Thus, a sudden and tremendous increase in the scattering yield will be observed in the channeling spectrum. Combined with computer simulation, information about strains in the superlattice can be extracted from such a sudden increase of scattering yield. In most cases, the planar channeling effect is not as useful as the axial channeling effect because planar channeling is weaker than the axial channeling. The critical angles of the planar channeling are two to three times smaller than that of axial channeling. However, in the analysis of the strains of the superlattices, planar channeling is as important as axial channeling.

3.3. Experimental Procedure for Aligning the Axis of the Crystal to the Incident Beam

To perform channeling analysis, the first thing that needs to be done is to align the axis or planes of the crystal to the incident probing ion beam. The alignment is accomplished with the aid of the goniometer. A goniometer is a mechanical device that can tilt or rotate the sample (see Fig. 3). The axis of the lattice is the intersection of planes. Therefore, to find an axis, the first step is to locate the planes. The procedure usually used to locate the planes is as follows: (1) set the Y tilt angle at $+5°$ (assuming that the normal of the sample is close to $0°$), then scan X from -5 to $+5°$ in steps of $0.2°$. At each step, take the

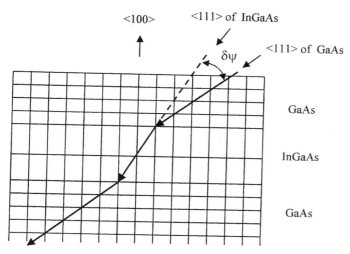

Fig. 23. Schematic of the ⟨100⟩ oriented GaAs/InGaAs/GaAs superlattice (only one period is shown). The InGaAs lattice is elongated to match the lattice constant of GaAs lattice, thus an angular shift, $\delta\psi$, in the inclined ⟨111⟩ direction occurs when going from the GaAs layer into the InGaAs layer.

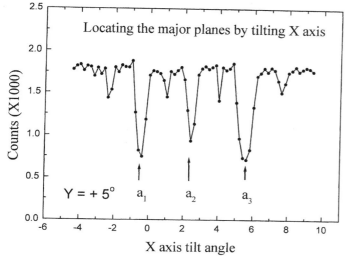

Fig. 24. Plot of the RBS counts (within the preset energy window) against the X tilt angles at $Y = 5°$. Whenever the incident beam hit the planes, the RBS counts reduce. Three major planes appear at X tilt angles, a_1, a_2, and a_3.

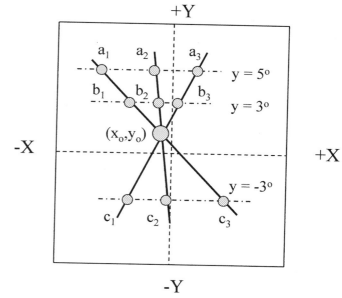

Fig. 25. Schematic of the procedure for locating the axis. The incident beam across three major planes at X tilt angles, a_1, a_2, a_3, b_1, b_2, b_3, and c_1, c_2, c_3. The intersection of the three planes give the coordinators of the axis, (x_o, y_o).

tilt $= x_o$, and the incident beam is now aligned with the axis of the sample.

3.4. Analysis of Defects in Single Crystals

Single crystals often contain defects. The defects seen frequently in crystals are displaced atoms, point defects (interstitials and vacancies), or point defect clusters, impurity atoms, strains, dislocations, amorphous islands, mosaic spread, stacking faults, twins and voids, etc. The RBS-channeling technique is a powerful tool to characterize the defects in single crystals. Various information about the defects can be obtained: (1) the depth profiles of the defects [1, 59]; (2) the identity of the defects (e.g., dislocations [60–63]); (3) the regrowth of the amorphized layers [64–66]; (4) radiation damages; (5) strains in superlattices [67–70] and epitaxial heterostructures [71–74] and so on. This section discusses the application of the RBS-channeling technique to the characterization of defects in single crystals. First, some basic concepts are introduced. Then, as an example of application, the characterization of dislocation is discussed.

3.4.1. Dechanneling Fraction in a Defect-Free Crystal, $\chi_v(t)$

Imagine that a well-collimated probing beam impinges along a major axis onto a perfect single crystal containing no defects. After entering the crystal, not all the incident particles become channeled particles. Some of the incident particles will be deflected into random direction becoming dechanneled particles by (1) the scattering with the atoms in the imperfect surface layer (oxides, contamination, or amorphous phase etc.), (2) the multiple scattering with the thermally vibrating host atoms,

RBS counts within a pre-fixed energy window (normally such a window is set near the surface region but excluding the surface peak). During the scan, when the probing beam hits a plane, the RBS counts will drop. Record the X tilt angles, a_i, at which the minimum counts appear (Fig. 24). Scanning X from -5 to $+5°$, the beam may hit two to three planes (Fig. 24); (2) repeat step (1) at $Y = +3$ and $-3°$; (3) plot the recorded points a_i, b_i, and c_i where the beam hits the planes (minimum counts occur) on a graphic paper (Fig. 25). Connect the points with three straight lines in an appropriate manner. These lines represent the planes. The intersection of the lines, (x_o, y_o), is the coordinate of the axis (Fig. 25). Normally, fine scans around the x_o and y_o over a small-angle range are necessary to obtain the precise location of the axis. Finally, set the goniometer at Y tilt $= y_o$ and the X

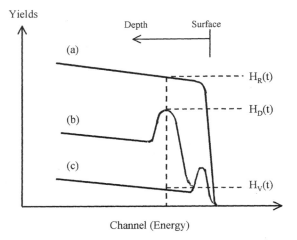

Fig. 26.   Schematics of (a) the random spectrum, (b) the aligned spectrum of a partially damaged Si crystal, and (c) the aligned spectrum of a virgin Si crystal.

and (3) the multiple scattering with the electrons (Fig. 22). The dechanneled particles are random particles. If the normalized yield of an aligned spectrum (incident angle $\theta = 0°$) is $\chi_v(t) = H_V(t)/H_R(t)$ (Fig. 26), then at depth t the percentage of the dechanneled particles (dechanneling fraction) should be $\chi_v(t)$ because in a defect-free crystal all the backscattering counts are from the dechanneled particles. The percentage of the remaining channeled particle should be $(1 - \chi_v(t))$. Normally in defect-free crystals, in the near-surface region, the $\chi_v(t \sim 0)$ (i.e., χ_{min}) is only 2–4%. The $\chi_v(t)$ increases with depth as more and more channeled particles are being dechanneled.

3.4.2. Dechanneling Fraction in a Defected Crystal, $\chi_R(t)$

If the crystal contains defects, the scatterings with the defects will also cause the dechanneling of the channeled particles. Assuming that at depth t the probability for a channeled particle to be dechanneled by the defects is $P(t)$, then the contribution from defects to the dechanneling will be $(1 - \chi_v(t))P(t)$. Assuming that the different types of scatterings that cause dechanneling have no correlation with each other, then the total dechanneling fraction in a defected crystal, $\chi_R(t)$, should be given by

$$\chi_R(t) = \chi_V(t) + (1 - \chi_V(t))P(t) \qquad (47) \text{ [41, 75–77]}$$

In a defected crystal, the percentage of the channeled particles is $(1 - \chi_R(t))$.

3.4.3. Direct Scattering and Indirect Scattering

When the incident angle to the axis $\psi = 0$, channeled particles are concentrated near the central region of the channel (see Fig. 17). Therefore, channeled particles have very low probability to have **direct** backscattering with the host atoms that are sitting on the regular lattice sites. However, the channeled particles have high probability to have **direct** backscattering with

displaced host atoms (defects). As for dechanneled particles, since they are random particles, there is always a high probability of having backscattering with any host atoms, no matter that they are regularly sited atoms or displaced atoms. The scattering between dechanneled particles and host atoms is called **indirect** scattering, because it only occurs after a channeled particle is dechanneled.

Based on the preceding discussion, the backscattering yield of an aligned spectrum (incident angle $\psi = 0°$) taken from a defected crystal at depth t, $H_D(t)$, should be the sum of **indirect scattering** (the scatterings between the dechanneled particles and all host atoms) and **direct scattering** (the scatterings between the channeled particles and the displaced host atoms (defects)) (Fig. 26),

$$H_D(t) = \chi_R(t)H_R(t) + (1 - \chi_R(t))H_R(t)\left(f\frac{n_D}{N}\right)$$
$$(48) \text{ [13, 75–77]}$$

where $H_R(t)$ is the RBS yield of the random spectrum at depth t, $n_D(t)$ is the concentration of the defects, N is the density of the host atoms, and f is the defect scattering factor. The defect scattering factor, f, is introduced because the amount of the direct scattering depends on the type of the defects [75]. For instance, for randomly distributed interstitials, all the interstitials contribute to the direct scattering, i.e., $f = 1$. However, for dislocations, the direct scattering is nearly zero as compared with dechanneling, i.e., $f \sim 0$ [41, 76]. Equation (47) can be normalized by the random spectrum (divided by $H_R(t)$),

$$\chi_D(t) = \chi_R(t) + (1 - \chi_R(t))\left(f\frac{n_D}{N}\right) \qquad (49) \text{ [13, 75–77]}$$

where $\chi_D(t) = H_D(t)/H_R(t)$. Equations (47) and (49) are the basic formulas in the RBS-channeling analysis of the defects. $\chi_D(t)$ can be measured experimentally, but $\chi_R(t)$ cannot be directly measured because it depends on the dechanneling probability $P(t)$. The calculation of $P(t)$ depends on the theoretical models and, is discussed later for several special cases. Once having $P(t)$, the depth profile of the defects, $n_D(t)$, can be calculated from the measured channeling aligned spectrum using an iterative procedure (see, for instance, [1] for more details).

3.4.4. Dechanneling Probability, $P(t)$

The dechanneling probability, $P(t)$, depends on the type of the defects. In the following, the $P(t)$ for three types of defects that are commonly seen in the ion implanted semiconductor crystals will be discussed: (1) low-density randomly displaced atoms, (2) high-density randomly displaced atoms, and (3) dislocations.

3.4.4.1. Low-Density Randomly Displaced Atoms

When the density of the displaced atoms is low, normally the single-scattering model [75] is used for calculating the dechanneling probability, $P(t)$. In the single-scattering model, the dechanneling of the channeled particles is assumed to be due to the single scattering with the displaced atoms. The $P(t)$ is given by

$$p(t) = 1 - \exp\left[-\int_0^t \sigma_D n_D(x)\, dx\right] \quad (50)$$

where σ_D is the dechanneling cross section, which is the integration of the Rutherford cross section (θ integrated from $\psi_{1/2}$ to π and ϕ integrated from 0 to π),

$$\sigma_D = \iint \sigma_R\, d\Omega = \frac{\pi d^2}{4}\frac{\Psi_1^4}{\Psi_{1/2}^2} \quad (51)$$

Substitute $P(t)$ (Eq. (50)) and σ_D (Eq. (51)) into Eqs. (47) and (49) and after rearrangement, they give

$$-\ln\left[\frac{1-\chi_D(t)}{1-\chi_V(t)}\right] = \int \sigma_D(E) n_D(x)\, dx \quad (52)$$

Usually, $\{-\ln[(1-\chi_D(t))/(1-\chi_v(t))]\}$ is called the dechanneling parameter. Using surface approximation, $E \sim E_0$, Eq. (52) can be simplified as

$$-\ln\left[\frac{1-\chi_D(t)}{1-\chi_V(t)}\right] = \sigma_D(E_0) \int n_D(x)\, dx \quad (53)$$

Therefore, when the displaced atoms are located near the surface and the concentration is low, the dechanneling parameter is proportional to the reciprocal of the incident energy E_0 (Eq. (51)),

$$-\ln\left[\frac{1-\chi_D(t)}{1-\chi_V(t)}\right] \propto \frac{1}{E_0} \quad (54)$$

This relation is often used to identify the randomly displaced atoms [63].

3.4.4.2. High-Density Randomly Displaced Atoms

When the concentration of the displaced atoms is high, the small-angle multiple scattering plays a major role in the dechanneling. Here, we follow the calculation given by Meyer [78], Sigmund and Winterbon [79], Luguijo and Mayer [80], and Thompson and Galvin [81]. The calculated probability, $P(t)$, for a channeled particle scattered at depth t by an angle larger than the channeling critical angle, $\psi_{1/2}$, is given by

$$P(t) = P(\theta_C, m_D) \quad (55)$$

where θ_C is the reduced critical angle and m_D is the reduced thickness defined by

$$\theta_C = aE\Psi_{1/2}/(2Z_1 Z_2 e^2) \quad (56)$$

$$m_D = \pi a^2 \int n_D(x)\, dx \quad (57)$$

where a is the Thomas–Fermi screening radius, E is the energy of the particle, e is the electronic charge, Z_1 and Z_2 are the atomic numbers for the incident particle and the host atom, respectively. Figure 27 presents the calculated dechanneling probability, $P(\theta_C, m_D)$ (Lugujjo and Mayer [80]) and reduced thickness, m_D (Thompson and Galvin [81]).

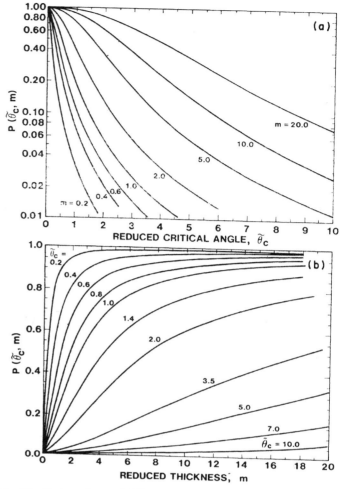

Fig. 27. Dechanneling probability (using the Meyer treatment) vs reduced angle for various reduced thickness. Adapted from [13], copyright 1982, Academic Press. Originally (a) from E. Lugujjo and J. W. Mayer. Reprinted with permission from [80], copyright 1973, American Physical Society. (b) from calculation by M. D. Thompson and G. J. Galvin [81]. Reprinted with permission from the author.

Usually, when the reduced thickness $m_D < 0.2$, the single-scattering model is preferred while when $m_D > 0.2$, the multiple-scattering model gives better results.

3.4.5. Analysis of the Dislocations

3.4.5.1. Dechanneling by Dislocations

Dislocations are common defects in crystals. Generally, there are three types of dislocation: edge dislocation, screw dislocation, and dislocation loops. Take edge dislocation as an example. Figure 28 is the schematics of an edge dislocation. The major structural character of an edge dislocation is that at the center of the dislocation half of the atomic plane is missing. The edge of the remaining half of the atomic plane is called the dislocation line. The atomic planes that surround the dislocation line are bent. In a bent channel, either axial channel or planar channel, the probability for a channeled particle to be deflected

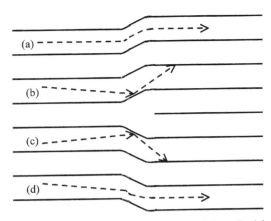

Fig. 28. Illustration of dechanneling by dislocation defects. Particles (a) and (d) are not dechanneled because they are far away from the core of the dislocation where the channels are bent less. Particles (b) and (c) are dechanneled because they are close to the core of the dislocation where the channels are bent more.

into a random direction (dechanneling) is much higher than the probability of having direct backscattering with the host atoms. Assuming that the density of the dislocation at depth t is $l(t)$ (projected length of the dislocation line per unit depth per unit area), from Eq. (49), obtains

$$\chi_D(t) = \chi_R(t) + \left(1 - \chi_R(t)\right)\left[f\frac{l(t)}{N}\right] \qquad (58)$$

For dislocations, the defect scattering factor, $f \sim 0$ because the direct backscattering is negligible as compared with the dechanneling. Thus,

$$\chi_D(t) \approx \chi_R(t) \qquad (59)$$

Moving in the distorted region around the dislocation line, the channeled particles are dechanneled by a single collision with the wall of the channels. This is similar to the single-scattering dechanneling mechanism. Thus, from Eqs. (47) and (50),

$$\chi_D(t) \approx \chi_R(t) = \chi_V(t) + \left(1 - \chi_V(t)\right)$$
$$\times \left\{1 - \exp\left[-\int \lambda(E)l(z)\,dz\right]\right\} \qquad (60)$$

where $\lambda(E)$ is the dechanneling cross section for dislocation and $l(z)$ is the density of the dislocation at depth z. $\lambda(E)$ normally is called the dechanneling diameter. In ion implanted crystals, the dislocations are the common secondary defects after annealing. Also, dislocations are often generated as a mean for strained superlattice to relax. The dislocations in ion implanted samples are often distributed within a narrow region. Since the energy loss is negligible in a narrow region, the $\lambda(E)$ is approximately a constant and can be taken out of the integration in Eq. (60). Thus, from Eq. (60),

$$\lambda(E)\int l(z)\,dz = \lambda(E)L(t) = -\ln\left[\frac{(1 - \chi_D(t))}{(1 - \chi_V(t))}\right] \qquad (61)$$

where $L(t)$ is the total length of the projected dislocation lines within the unit area integrated from the surface of the sample to

depth t. The right side of Eq. (61) is the dechanneling parameter. Differentiating both sides of Eq. (61) gives the depth profile of the dislocation, $l(t)$,

$$l(t) = \left(\frac{-1}{\lambda}\right)\frac{d}{dt}\left\{\ln\left[\frac{(1 - \chi_D(t))}{(1 - \chi_V(t))}\right]\right\} \qquad (62)$$

The dechanneling diameter, $\lambda(E)$, can be calculated using theoretical models. Thus, from Eq. (62), the depth profile of the dislocation can be obtained from the RBS-channeling spectrum.

3.4.5.2. Calculation of the Dechanneling Diameter, $\lambda(E)$—Quèrè Model [60, 61]

When a channeled particle moves from a straight channel into a bent channel around the dislocation line (Fig. 28), its transverse energy, E_{tr}, increases. Whether this channeled particle will be dechanneled (deflected by the wall into a random direction) or not depends on the balance between the centrifugal force of the particle and the centripetal force provided by the transverse potential of the channel ($U_{\mathrm{tr}}(r)$). If the radius of the curvature of the bent channel is too small so that the centrifugal force is larger than the centripetal force, the channeled particle that enters this bent channel will be deflected into a random direction. Based on the balance between centrifugal force and centripetal force combined with the dislocation theory, Quèrè has successfully developed a model for the calculation of the dechanneling diameter of the dislocation, $\lambda(E)$[60, 61]. In Quèrè's model, the $\lambda(E)$ for the axial case and the planar case are as follows.

For the axial case,

$$\lambda_{\mathrm{ax}}(E) = \left[\frac{\alpha b d E}{2Z_1 Z_2 e^2}\right]^{1/2} \qquad (63)$$

where $\alpha = 12.5$, for straight screw dislocation, $\alpha = 4.5$ for straight edge dislocation, $\alpha = 7.2$, for screw and edge mixed dislocation, E is the energy of the incident particle, b is the Burger vector, d is the lattice spacing of the atomic row, Z_1 and Z_2 are the atomic number of the incident particle and host atom, respectively, and e is the electronic charge.

For the planar case,

$$\lambda_P(E) = \left[\frac{Eb}{\beta 2Z_1 Z_2 e^2 N_P}\right]^{1/2} \qquad (64)$$

where $\beta = 8.6$, for straight screw dislocation, $\beta = 3.2$, for straight edge dislocation, $\beta = 5$, for screw and edge mixed dislocation, N_P is the atomic planar density of the host atoms.

The Quèrè model predicts that the dechanneling diameter, $\lambda(E)$, is proportional to the square root of the incident energy, $E^{1/2}$. This has been confirmed by the experiment carried out by Picraux et al. [62, 82, 83]. The sample studied by Picraux et al. [82] was an Si single crystal implanted with phosphorous ions and annealed at 1050°C. The TEM microphoto shows that a network of misfit edge dislocations was formed around the tail of the P concentration profile (Fig. 29a [82]). The 2.5-MeV $^4\mathrm{He}^+$ ions RBS spectrum along the (110) plan and the ⟨111⟩ axis are shown in Figure 29b [82]. The sudden increase of the dechanneling yields around a depth of 450 nm in both the (110)

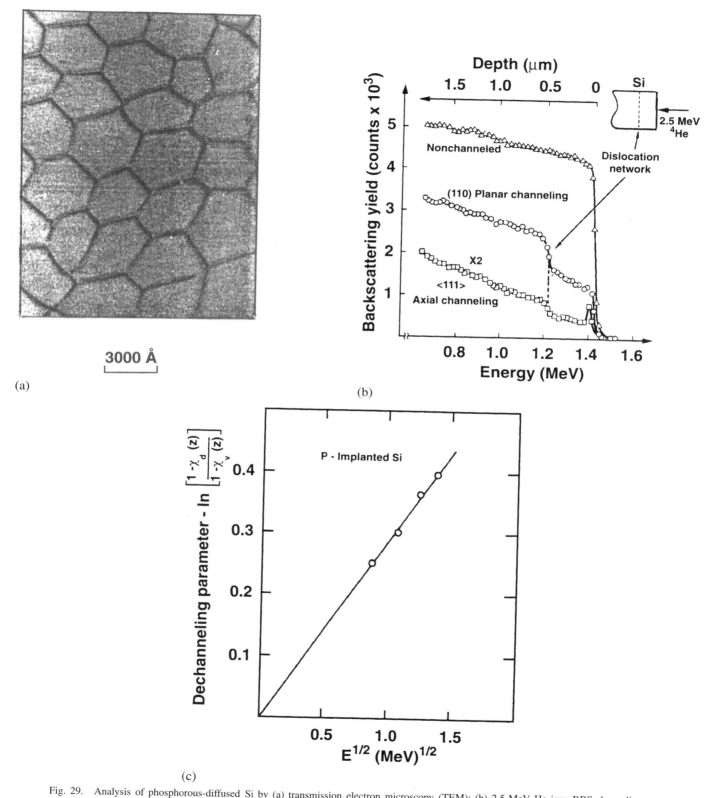

Fig. 29. Analysis of phosphorous-diffused Si by (a) transmission electron microscopy (TEM); (b) 2.5-MeV He ions RBS-channeling aligned spectra along the {110} plane and the ⟨111⟩ axis; and (c) energy dependence of the planar dechanneling. A misfit dislocation network is observed at a depth of 450 nm. (From S. I. Picraux et al. [82].)

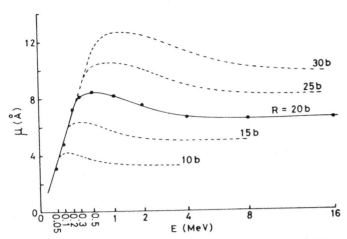

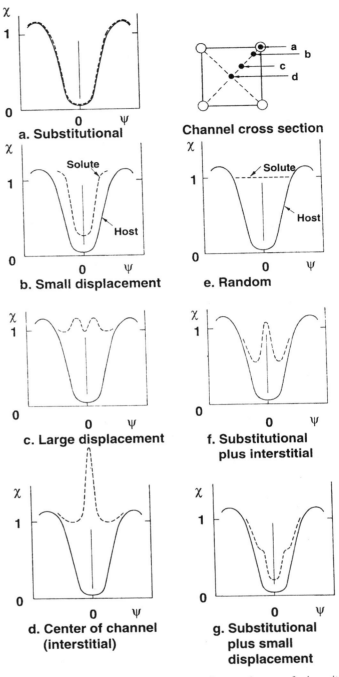

Fig. 30. Calculated energy dependence of the dechanneling width μ of dislocation loops by the Kudo model. The solid line was obtained from the numerical calculation for a dislocation loop with radius $R = 20b$ (b is the Burger vector of the dislocation loops). The broken lines were obtained from the solid line by using a scaling function defined by Kudo. Reprinted with permission from [85], copyright 1978, American Physical Society.

and $\langle 111 \rangle$ directions was due to the dechanneling by dislocations. Obviously the planar channeling is more sensitive to the dislocation than the axial channeling. The energy dependence of the dechanneling parameter, $\{-\ln[(1 - \chi_D(z))/(1 - \chi_V(z))]\}$, for the (110) planar direction was measured and is displayed in Figure 29c [82]. It shows that the dechanneling caused by dislocations is proportional to the square root of the energy as predicted by the Quèrè model [60, 61].

3.4.5.3. Calculation of the Dechanneling Diameter, $\lambda(E)$—Kudo Model [84–86]

The dechanneling by dislocations has also been studied by Kudo [84–86] using an approach which is different from the Quèrè model [60, 61]. Kudo applied the continuum model to the distorted crystal and established the differential equation for the trajectories of the channeled particles in distorted channels. The dechanneling probability was obtained by solving this differential equation. Figure 30 shows the energy dependence of the dechanneling diameter, $\lambda(E)$, for the dislocation loops predicted by the Kudo model [85]. For low-energy He ions, the $\lambda(E)$ of the dislocation loops is linear with the square root of the energy. For high-energy ^{4}He$^+$ ions, the $\lambda(E)$ becomes constant. Kudo's prediction on the dechanneling of dislocation loops was confirmed by the experiment carried out by Picraux and Follstaedt [83].

3.5. Examples of the Applications of RBS-Channeling Technique in Materials Analysis

3.5.1. Lattice Location of Impurity Atoms

In the following, two examples will be presented: (a) Impurity Sn atoms occupy the substitutional site in SiC (6H) and (b) Er atoms occupy the tetrahedral interstitial site in GaAs.

Fig. 31. Characteristic profiles of the channeling angular scans for impurity atoms residing at various lattice sites in the single crystals. Reprinted with permission from [90], copyright 1983, Academic Press.

3.5.1.1. Impurity Sn Atoms Occupy the Substitutional Site in SiC (6H)

For impurity atoms occupying the substitutional site, the shape and the critical angle, $\Psi_{1/2}$, of their channeling dip will be identical to that of the host atoms [87] (see Fig. 31). One example is the lattice location of the hot-implanted Sn atoms in single crystal SiC (6H) studied by Petersen et al. [88]. The

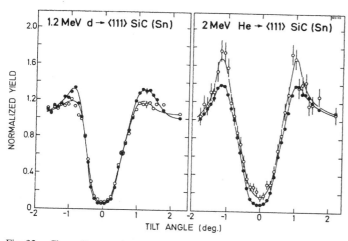

Fig. 32. Channeling angular scans along the ⟨111⟩ direction of Sn implanted (at ~550°C) 6H SiC and postannealed at 1120°C in vacuum. (a) The 1.2-MeV d+ ions scattered from Si atoms (solid circles) and proton yield from the ^{12}C (d, p) ^{13}C reaction (open circles). (b) 2-MeV ^{4}He+ ions scattered from Si atoms (solid circles) and Sn atoms (open circles). Reprinted with permission from [88], copyright 1981, Kluwer.

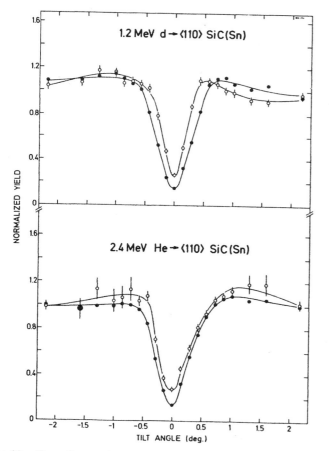

Fig. 33. Channeling angular scans along the ⟨110⟩ direction of Sn implanted (at ~550°C) 6H SiC and postannealed at 1120°C in vacuum. (a) The 1.2-MeV d+ ions scattered from Si atoms (solid circles) and proton yield from the ^{12}C (d, p) ^{13}C reaction (open circles). (b) The 2.4-MeV ^{4}He+ ions scattered from Si atoms (solid circles) and Sn atoms (open circles). Reprinted with permission from [88], copyright 1981, Kluwer.

Table III. The Measured $\Psi_{1/2}$ for Si and C of SiC (6H) Single Crystal, and Impurity Sn along ⟨110⟩ (Separated Atomic Rows) and ⟨111⟩ (Mixed Atomic Rows) Axial Directions

	⟨111⟩		⟨110⟩	
	1.2-MeV d+	2-MeV He+	1.2-MeV d+	2.4-MeV He+
Si	0.48(2)°	0.51(2)°	0.30(2)°	0.30(2)°
C	0.48(2)°		0.21(2)°	
Sn		0.50(3)°		0.28(2)°

Source: Reprinted with permission from J. W. Peterson et al. [88], copyright 1981, Kluwer.

2-MeV ^{4}He+ ions RBS were used for angular scans on Si and Sn atoms and 1.2-MeV deuterium induced ^{12}C (d, p) ^{13}C nuclear reaction was used for an angular scan on carbon atoms. Channeling angular scans around the ⟨110⟩ and ⟨111⟩ axis are shown in Figures 32 and 33. The measured $\Psi_{1/2}$ are listed in Table III. Along the ⟨111⟩ direction, the Si and C atoms are sitting on the same atomic row, so the critical angles of the channeling dip for Si and for C are expected to be the same. This was confirmed by the measurement, which shows that along the ⟨111⟩ direction $\Psi_{1/2}^{Si} = \Psi_{1/2}^{C} = 0.48°$ [Fig. 32]. However, along the ⟨110⟩ direction, the Si and the C are sitting on separate atomic rows, therefore the critical angle for Si and for C should be different. This was also confirmed by measurement, which shows that $\Psi_{1/2}^{Si} = 0.30°$ and $\Psi_{1/2}^{C} = 0.21°$, differ by 30% [Fig. 33]. Along the ⟨111⟩ direction, within the experimental error, the critical angle for impurity Sn, $\Psi_{1/2}^{Sn}$ (0.50°) is the same as that for Si rows, $\Psi_{1/2}^{Si}$ (0.51°), showing that the impurity Sn atoms are sitting along the ⟨111⟩ direction. However, along the ⟨110⟩ direction, $\Psi_{1/2}^{Sn}$ (0.28°) is, within the experimental error, in agreement with $\Psi_{1/2}^{Si}$ (0.30°), but is different from $\Psi_{1/2}^{C}$ (0.21°) by 25%. This was interpreted by Petersen et al. as that the most Sn atoms were sitting on the Si sites. The substitutional fraction of the Sn atoms that were sitting on the Si sites was estimated as 85% [88] using the expression proposed by Merz et al. [87],

$$f = \frac{(1 - \chi_{min}^{imp.})}{(1 - \chi_{min}^{host})} \qquad (65)$$

where f is the substitutional fraction of the impurity atoms, $\chi_{min}^{imp.}$ is the channeling minimum yield of the impurity atoms, and χ_{min}^{host} is the channeling minimum yield of the host atoms.

3.5.1.2. Er Atoms Occupy the Tetrahedral Interstitial Sites in GaAs

In the following, the study on the impurity atoms occupying the interstitial site, carried out by Alves et al. [89], is introduced. The sample studied was Er doped (2×10^{20} at. cm^{-3}) GaAs epitaxial film grown by MBE on a (100) semi-insulating GaAs substrate. The χ_{min} near the surface was 3.8% in the

⟨110⟩ direction and 4.5% in the ⟨111⟩ direction, showing that the film had good crystalline quality. The channeling angular scans around the ⟨110⟩ and ⟨111⟩ axial and the (111) planar directions are shown in Figure 34. The results of the computer simulations are also shown in Figure 34. Channeling angular scan around the ⟨110⟩ axis was performed in the (110) plane. The measured channeling angular scan (Fig. 34a) for GaAs is a dip but for Er atoms it is a peak around the center of the channel, indicating that the Er atoms are sitting around the center of the ⟨110⟩ channel. In the ⟨111⟩ direction, both Er and GaAs angular scans show a strong dip and the Er angular scan matches with the GaAs angular scan (Fig. 34b), indicating that the Er atoms are sitting along the ⟨111⟩ atomic rows. In the (111) planar direction, the angular scan for Er atoms shows a broad peak (Fig. 34c). Computer simulation carried out by Alves et al. showed that the best fit to the three measured scans for Er atoms

can be achieved when about 95% of Er atoms assume the tetrahedral interstitial site.

The profile of the channeling angular scan (channeling dip) varies with the lattice sites. Figure 31 [90] schematically shows the characteristic profiles of the channeling angular scans for impurity atoms residing at various sites in the single crystal.

3.5.2. Study the Ion-Beam Induced Epitaxial Crystallization

Ion-beam induced epitaxial crystallization (IBIEC) has been studied in many materials, such as Si [91, 92], GaAs [92–94], InP [94], Ge-Si alloys [95–97], NiS_2 [98], SiC [99], and Al [100]. In all the IBIEC researches, RBS and channeling techniques have been extensively used to study the kinetics of the regrowth processes. As an example, the study on the activation energy of crystallization of Si induced by light ions irradiation carried out by Kinomura et al. [92] is presented here. The sample used was (100)-oriented P-type Si crystals, which have about 0.2-μm preamorphized layers at the surface. To study the IBIEC of the preamorphized surface layers, the samples were irradiated by B^+, C^+, and O^+ ions with energy up to 3 MeV. As a comparison, the IBIEC induced by heavy ions of Au^+ and Ge^+ were also studied. During the irradiation, the samples were heated to various temperatures ranging from 150 to 400°C. Figure 35 [92] displays the epitaxial regrowth of the preamorphized layers induced by the irradiation of 0.7-MeV B^+ ions. It shows that the regrowth proceeds in a layer-by-layer fashion and the thickness of the regrown layer increases with the temperature. From Figure 35, the regrowth rate (thickness of the regrown layer/dose) can be determined. The activation energy of the regrowth process can be obtained from the Arrhenius plot, as shown in Figure 36 [92]. Data for IBIEC induced by the irradiation of 3-MeV Au^+ and Ge^+ are also shown for

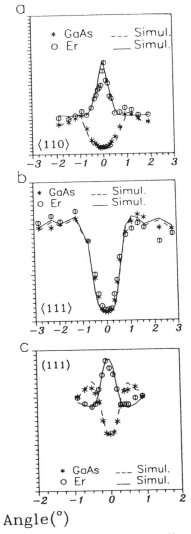

Fig. 34. Channeling angular scans for GaAs: E_r (2×10^{20} at. cm^{-3}) (a) across the ⟨110⟩ axis along the (110) plane, (b) across the ⟨111⟩ axis, and (c) across the (111) plane. The solid and dashed lines are the results of Monte Carlo simulation. Reprinted with permission from [89], copyright 1993, Elsevier Science.

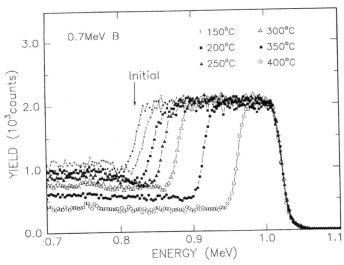

Fig. 35. The 1.5-MeV $^4He^+$ ions RBS-channeling spectra of Si single crystals annealed by IBIEC process using 0.7-MeV B^+ irradiation to a dose of 2×10^{17} cm^{-2}. The spectrum plotted with small dots is from the control sample without annealing and B ions irradiation (initial amorphous thickness). Reprinted with permission from [92], copyright 1999, Elsevier Science.

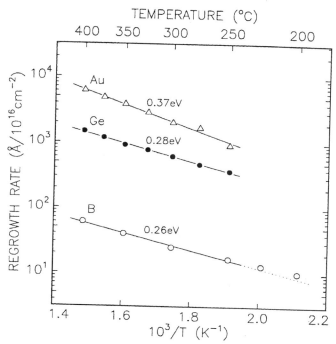

Fig. 36. Arrhenius plot of regrowth rates vs annealing temperatures for 0.7-MeV B ions irradiation (1×10^{13} cm^{-2} s^{-1}) compared with regrowth rates for 3-MeV Au ions and Ge ions irradiation (2×10^{12} cm^{-2} s^{-1}). Reprinted with permission from [92], copyright 1999, Elsevier Science.

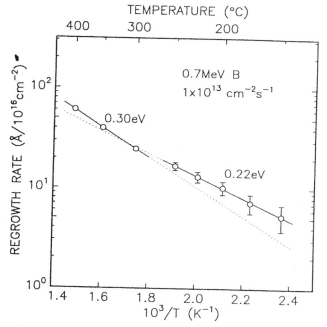

Fig. 37. Arrhenius plot of regrowth rates vs annealing temperatures for 0.7-MeV B ions irradiation (1×10^{13} cm^{-2} s^{-1}). Two activation energies for regrowth were observed in the temperature range of 150–400°C. Reprinted with permission from [92], copyright 1999, Elsevier Science.

comparison. Figure 36 shows that the activation energy of the IBIEC is mass dependent. Although Figure 36 only shows a single activation energy for 0.7-MeV B$^+$ ions, the Arrhenius plot (Fig. 37) obtained in a wider temperature range suggests

that the 0.7 B$^+$ induced regrowth has two activation energies, indicating several types of defect may be operative depending on irradiation conditions [92].

3.5.3. Characterization of Strain Reduction in Si/Si$_{1-x}$Ge$_x$ Heterostructure

Strained Si$_{1-x}$Ge$_x$ layers epitaxially grown on Si are promising materials for optical and electronic devices [101]. The band structure of the Si$_{1-x}$Ge$_x$ layers is related to the strains in the layers [102]. Akane et al. [103] have shown that by adding carbon atoms into Si$_{1-x}$Ge$_x$, the strains in the Si$_{1-x}$Ge$_x$ can be reduced in a controllable manner. The Si$_{1-x}$Ge$_x$C$_y$ alloy layers studied by Akane et al. were grown on a (100)-oriented Si substrate by the gas-source molecular beam epitaxy process (GSMBE). The amount of the carbon in the alloys was controlled by varying the partial pressure of MMGe gas (monomethylgermane: CH$_3$GeCH$_3$). For partial pressures PMMGe = 2.4×10^{-6}, 4×10^{-6}, and 7×10^{-6} Torr, the corresponding carbon composition y in ternary Si$_{1-x-y}$Ge$_x$Cy layers are 0.004, 0.01, and 0.014, respectively. Due to the strain in the Si$_{1-x}$Ge$_x$ layer, which arises from the larger atomic size of Ge atoms as compared to that of Si atoms, the angular offset between the inclined ⟨110⟩ axis of the strained layer and the inclined ⟨110⟩ axis of the substrate is expected. By adding carbon atoms into the Si$_{1-x}$Ge$_x$ layers, the strain will be reduced because carbon atom is much smaller than germanium atom. The reduction of the strain is reflected in the reduction of the angular offset, as shown in Figure 38 [103]. Figure 38 displays the channeling angular scans around the inclined ⟨110⟩ axis of the Si$_{1-x-y}$Ge$_x$Cy/Si (100) heterostructure for the samples containing various carbon contents. It shows that the angular offset, thus the strain, in the heteroepitaxial layer decreases with the increase of the carbon content, indicating that adding carbon atoms can control the strain in the Si$_{1-x}$Ge$_x$ layer.

4. APPLICATION OF ION-BEAM TECHNIQUES TO SUPERLATTICE (I)

RBS and channeling techniques have been extensively used in the characterization of the superlattices. The information about the superlattice that can be provided by RBS and the channeling analysis includes (1) interdiffusion and thermal stability [104–107], (2) structural analysis [108–112], (3) thickness and composition [2, 107, 111, 113–117], (4) quality of the epitaxy [113, 114, 118–120], (5) radiation damage and epitaxial regrowth [121–123], (6) interface [124–126], (7) stopping power [109, 110], and (8) strains (see references in Section 5). In the following, some interesting examples are introduced.

4.1. Interdiffusion in Superlattice Studied by RBS-Channeling

The thermal stability of the superlattice is a critical issue for device application because the high temperature processing steps

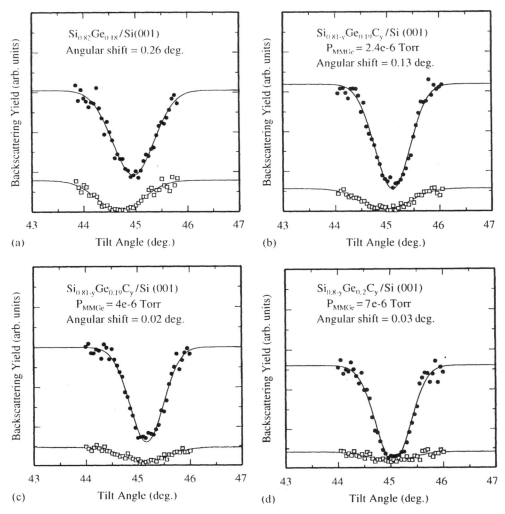

Fig. 38.   Channeling dips for the inclined ⟨110⟩ axis of the $Si_{1-x-y}Ge_xC_y$/Si(001) heterostructures. The filled circles and the open circles correspond to channeling dips of Si in the substrate and correspond to channeling dips of Ge in the stained epi-$Si_{1-x-y}Ge_xC_y$ layer, respectively. The angular shifts of the strained layer reduced with the increase of the carbon concentration. Reprinted with permission from [103], copyright 1999, Elsevier Science.

are inevitably involved in the device fabrication processes. The high temperature processing steps will result in three by-effects: (1) interdiffusion, (2) relaxation of the strains, and (3) formation of the secondary defects, such as dislocations. These by-effects will negatively affect the properties of the superlattice and need to be characterized appropriately. RBS and channeling analysis have been successfully used to characterize these by-effects. One excellent example is the study carried out by Hollander et al. on the Si/$Si_{1-x}Ge_x$ superlattice [104–106].

The samples studied by Hollander et al. [104–106] are five-period Si/$Si_{1-x}Ge_x$ strained superlattice with a period of 20 nm and Ge concentration between $x = 0.2$ and $x = 0.7$. Both asymmetrically strained and symmetrically strained superlatives [104] were studied. Rapid thermal annealing (RTA) was carried out with the temperatures between 900 and 1125°C. RBS and channeling analysis were performed using 1.4-MeV He ions with the Si detector set at 170° to the beam incident direction. When studying the Ge interdiffusion,

a glancing geometry was used to improve the depth resolution.

Before RTA and after RTA, RBS measurements were performed. For annealing temperature of 975°C the corresponding random RBS spectra are shown in Figure 39 [106]. The Ge signals appear in the high-energy portion (0.8–1.2 MeV) and the Si signals corresponding to the Si in the superlattice appear in the low-energy portion (0.6–0.8 MeV). Both signals have five peaks confirming that the superlattice has five periods. The measured RBS random spectra carry three types of information: (1) stoichiometry, (2) thickness, and (3) interdiffusion. The three methods for calculating the stoichiometry and the thickness described in Section 2, i.e., the methods based on total area A, based on spectrum height H, and based on energy width ΔE, are not suitable for the spectra shown in Figure 39 because the RBS signals (peaks) overlap with each other, making the determination of A, H, or ΔE impossible. Therefore, computer simulation was used to obtain the thickness of the period and

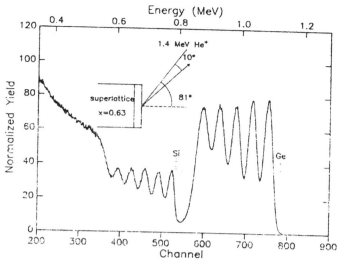

Fig. 39. The 1.4-MeV $^4He^+$ ions gracing incidence random RBS spectrum taken from a Si/Si$_{0.37}$Ge$_{0.63}$ superlattice, showing that the superlattice consisted of five periods. Reprinted with permission from [104], copyright 1992, American Physical Society.

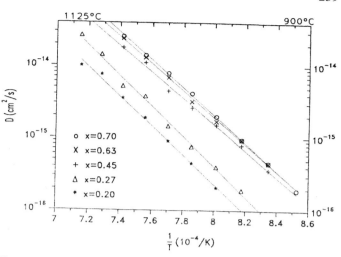

Fig. 40. Arrhenius plot of the diffusion coefficients of a group of symmetrically strained Si/Si$_{1-x}$Ge$_x$ superlattices with Ge concentration varies from $x = 0.2$ to $x = 0.7$. Diffusion of Ge in the Si/Si$_{1-x}$Ge$_x$ superlattice significantly increases with Ge concentration. Reprinted with permission from [104], copyright 1992, American Physical Society.

the stoichiometry of the Si$_{1-x}$Ge$_x$ alloy. Based on the changes of the Ge concentration after annealing, Hollander et al. calculated the interdiffusion coefficients of the Ge vs temperatures using a Fourier algorithm [127]. In a periodic structure with period length H, the diffusion equation,

$$\frac{\partial c}{\partial t} = D\nabla^2 c \tag{66}$$

is solved by a Fourier series,

$$c = \frac{1}{2} + \sum_m \beta_m e^{-(2\pi m/H)^2 Dt} \sin\frac{2\pi m}{H}z \tag{67}$$

with

$$\beta_m = 2/(\pi m) \qquad m = 1, 3, 5, \ldots$$
$$= 0 \quad \text{otherwise} \tag{68}$$

The annealing time t is known, the period length H and the Ge concentration can be determined from the RBS spectra, therefore, the interdiffusion coefficients can be calculated from Eq. (67) and Eq. (68). Figure 40 [104] shows the Arrhenius plot of the diffusion coefficients obtained from symmetrically strained superlattices with various Ge concentrations. It shows that the diffusion of the Ge in the Si/Si$_{1-x}$Ge$_x$ strained superlattice increases with the Ge concentration. Similar behavior was also observed in an asymmetrically strained Si/Si$_{1-x}$Ge$_x$ strained superlattice. Combining the RBS results with other considerations, Hollander et al. suggested a diffusion mechanism by interstitials.

The information about the strain relaxation and the defects in the Si/Si$_{1-x}$Ge$_x$ strained superlattice were obtained by channeling measurement. Figure 41 [104] shows the random and channeling aligned spectra along the $\langle 100 \rangle$ axis of the asymmetrically strained Si/Si$_{1-x}$Ge$_x$ superlattices with Ge

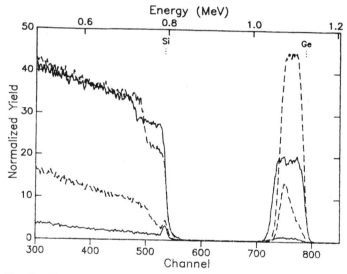

Fig. 41. The 1.4-MeV He ions random and aligned spectra (along the $\langle 100 \rangle$ axis) of the asymmetrically strained Si/Si$_{0.73}$Ge$_{0.27}$ superlattice (----) and Si/Si$_{0.37}$Ge$_{0.63}$ superlattice (- - - -). Significant dechanneling observed in the Si/Si$_{0.37}$Ge$_{0.63}$ superlattice, indicating the formation of mismatch dislocations. For the Si/Si$_{0.73}$Ge$_{0.27}$ superlattice, the low χ_{min} value shows a high crystalline perfection. Reprinted with permission from [104], copyright 1992, American Physical Society.

concentration $x = 0.27$ and $x = 0.63$ (before annealing). For the sample with higher Ge concentration, $x = 0.63$, the channeling aligned spectrum (dashed line) along the $\langle 100 \rangle$ direction shows a high χ_{min} value and a sudden increase of dechanneling throughout the superlattice, indicating the presence of high-density dislocations within the superlattice. This is because at Ge concentration $x = 0.63$ the critical thickness has been exceeded, thus mismatch dislocations were induced to relieve the strain. The channeling angular scan along the (100)

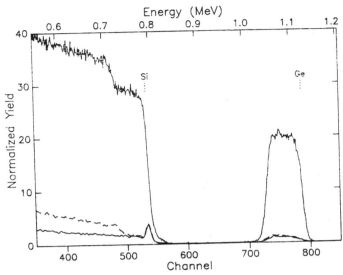

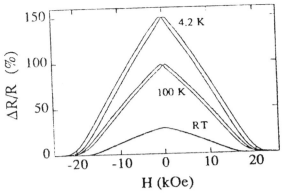

Fig. 43.   Magnetoresistance loops at three different temperatures for a (100)-oriented 50-period [Fe(1.4 nm)/Cr(0.8 nm)] superlattice prepared by magnetron sputtering. Reprinted with permission from [108], copyright 1997, Elsevier Science.

Fig. 42.   The 1.4-MeV He ions random and aligned spectra (along the ⟨100⟩ axis) of an asymmetrically strained Si/Si₀.₇₃Ge₀.₂₇ superlattice before (- - - -) and after RTA (- - - -). The increase of the channeling yield after RTA indicates the formation of crystal defects at the substrate-superlattice interface. Reprinted with permission from [104], copyright 1992, American Physical Society.

plane through an inclined ⟨110⟩ axis was performed to measure the angular shift with respect to the substrate. It showed a large discrepancy between the measured angular shift and the theoretical value, confirming that the strain had been relaxed in this sample. However, for the sample with lower Ge concentration, $x = 0.27$, the ⟨100⟩ aligned spectrum (solid line) shows a low χ_{min} value (4%) and no significant increase of dechanneling within the superlattice, indicating a low dislocation density. The channeling angular scan around the inclined ⟨110⟩ direction of this sample also shows that the measured angular shift of the inclined ⟨110⟩ axis, $-0.75 \pm 0.05°$, is in good agreement with the theoretical value, $-0.85°$, as estimated using a biaxial strain model [128], suggesting a nearly coherent growth. However, after 1075° RTA for 25 s, thermally induced dislocation formation as well as the strain relaxation occurred. This is revealed by the sudden increase of the dechanneling at the interface between the superlattice and the substrate in the ⟨100⟩ channeling aligned spectra shown in Figure 42 [104]. From this example, it can be seen that just using the RBS and channeling technique, abundant information such as interdiffusion coefficients, the activation energies, strain relaxation, defects formation, and the influences of the strains and dislocation on the interdiffusion, can be obtained.

4.2. Study the Structural Change in Fe/Cr GMR Superlattice

Ruders et al. studied the structural changes in the Fe/Cr superlattice using RBS-channeling technique [108]. The Fe/Cr superlattice is a so-called GMR (giant-magnetoresistance) material [129]. The magnetoresistance loops measured at 4.2, 100 K, and room temperature (Fig. 43 [130]) shows a strong

temperature dependence. The GMR occurs in the Fe/Cr superlattice as a result of antiferromagnetic interlayer coupling. However, the nature of the coupling of the Fe layers strongly depends on the thickness of the Cr layer [131]. With the increasing thickness of the Cr layer, the coupling of the Fe layers oscillates between ferromagnetic and antiferromagnetic, exhibiting a period of 1.8 nm. The oscillatory coupling is attributed to the formation of a static spin density wave in the Cr layer. The aim of the study of Ruders et al. is to search for any abnormal structural change that might be responsible for the strong temperature dependence seen in Figure 43. Such structural changes could be either a temperature-dependent lattice strain, or a subtle periodic lattice distortion accompanying the spin density wave that communicates the coupling information.

Any structural change in a crystal will manifest itself in the channeling minimum yield, χ_{min}, as well as in the critical angle, $\Psi_{1/2}$, of the channeling angular scan. The critical angle, $\Psi_{1/2}$, is more sensitive to the structural change than the minimum yield, χ_{min}. As demonstrated in this study and in other studies [132, 133], structural changes as small as few picometers (m^{-12}) can be detected by ion channeling measurement. The χ_{min} of the axial channeling is described by the Lindhard continuum model [40] and by Barrett's computer simulation [41] in

$$\chi_{min} = 18.8 N \, du_1^2 \left(1 + \varsigma^{-2}\right)^{1/2} \quad (69)$$

where N is the number of the crystal atoms per unit volume (in cubic angstroms), d is the atomic spacing along the atomic rows, u_1 is the one-dimensional root mean square (rms) of the thermal vibration amplitude (in angstroms) and

$$\varsigma = 126 u_1 / (\Psi_{1/2} d) \quad (70)$$

where $\Psi_{1/2}$ is the channeling critical angle and is given in degrees by Eq. (42). The one-dimensional root mean square (rms) of the thermal vibration amplitude, u_1, can be obtained from Figure 20 and Eq. (45). It should be pointed out that from Eqs. (69), (41) and (42), a smooth and nearly linear dependence on the temperature for the χ_{min} and the $\Psi_{1/2}$ are expected when temperature varies from 100 to 300 K. Deviation from

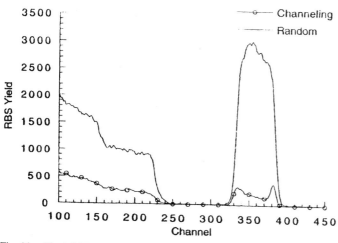

Fig. 44. The 1.5-MeV He ions RBS random and the $\langle 100 \rangle$-channeling spectra from a 14-period $\langle 100 \rangle$-oriented [Fe(1.4 nm)/Cr(6.2 nm)] superlattice grown on an MgO single crystal substrate. The χ_{min} is less than 5%. Reprinted with permission from [108], copyright 1997, Elsevier Science.

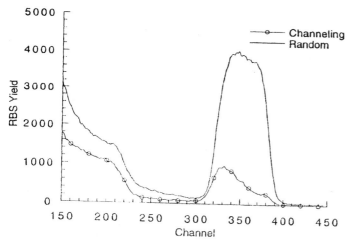

Fig. 45. The 1.5-MeV ^{4}He$^+$ ions RBS random and the $\langle 111 \rangle$-channeling spectra from the same superlattice as in Figure 44. The χ_{min} is higher than that in the $\langle 100 \rangle$ direction but is less than 10%. Reprinted with permission from [108], copyright 1997, Elsevier Science.

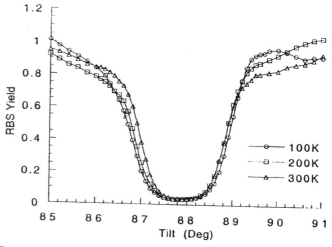

Fig. 46. Channeling angular scans taken across the $\langle 100 \rangle$ axis of a 14-period $\langle 100 \rangle$-oriented [Fe(1.4 nm)/Cr(6.2 nm)] superlattice at temperatures of 150, 200, and 300 K. The scans were plotted using the total counts within a window from channels 345 to 375 (see Fig. 44). Reprinted with permission from [108], copyright 1997, Elsevier Science.

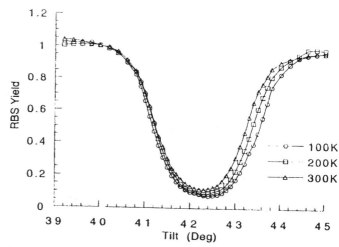

Fig. 47. Channeling angular scans taken across the inclined $\langle 111 \rangle$ axis of the 14-period $\langle 100 \rangle$-oriented [Fe(1.4 nm)/Cr(6.2 nm)] superlattice at temperatures of 150, 200, and 300 K. The scans were plotted using the total counts within a window from channels 348 to 374 (see Fig. 45). Reprinted with permission from [108], copyright 1997, Elsevier Science.

this smooth and nearly linear dependence could be the indication of an abnormal structural change.

The sample studied by Ruders et al. was a (100)-oriented [Fe(1.4 nm)/Cr(6.2 nm)] superlattice (14 periods) grown on a single crystal MgO with a 10-nm Cr buffer layer. The sample was cooled by a closed-cycle refrigeration system. A 1.5-MeV well-collimated (angular divergence <0.05°) ^{4}He$^+$ ion beam was used. The backscattered ^{4}He$^+$ ions were recorded in an Si detector placed at 138° to the incident beam. The channeling aligned spectrum along the $\langle 100 \rangle$ and $\langle 111 \rangle$ axes, together with the random spectrum, are presented in Figures 44 and 45, respectively. The $\langle 100 \rangle$ channeling minimum yield χ_{min} is less than 5%, indicating high crystalline quality of the superlattice. In the inclined $\langle 111 \rangle$ direction, the χ_{min} is less than 10%. To search for abnormal structural change, Ruders et al. performed

channeling angular scans around the $\langle 100 \rangle$ and $\langle 111 \rangle$ axes at several temperatures ranging from 100 to 300 K. Three scans along the $\langle 100 \rangle$ axis (plotted using the total RBS counts between channels 345 and 375) are displayed in Figure 46. Three scans along the $\langle 111 \rangle$ axis (plotted using the total RBS counts between channels 328 and 368) are displayed in Figure 47. The critical angles, $\Psi_{1/2}$, and the minimum yields, χ_{min}, obtained from all of the scans are plotted in Figures 48 and 49, respectively. It can be seen that the critical angle, $\Psi_{1/2}$, and the χ_{min} changes smoothly and linearly with the temperature, which is expected from a normal Debye behavior of the lattice thermal vibration. Hence, no abnormal structural changes were observed in the Fe(1.4 nm)/Cr(6.2 nm) superlattice when

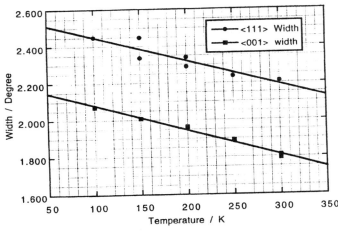

Fig. 48. FWHM of the ⟨100⟩ and ⟨111⟩ channeling angular scans (Figs. 46 and 47) from the 14-period ⟨100⟩-oriented [Fe(1.4 nm)/Cr(6.2 nm)] superlattice as a function of temperature. Reprinted with permission from [108], copyright 1997, Elsevier Science.

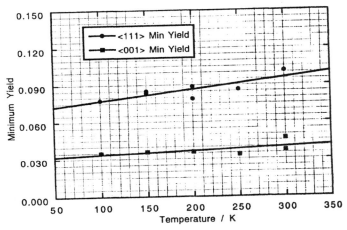

Fig. 49. Minimum yields, χ_{min}, for the ⟨100⟩ and ⟨111⟩ scans (Figs. 46 and 47) from the 14-period ⟨100⟩-oriented [Fe(1.4 nm)/Cr(6.2 nm)] superlattice as a function of temperature. Reprinted with permission from [108], copyright 1997, Elsevier Science.

temperature varied from 100 to 300 K. Although no abnormal structural changes that can be linked to the lattice strain or the distortion were observed, the ion channeling study carried out by Ruders et al. provides an excellent example about the characterization of the structural change in the superlattice using the RBS-channeling technique.

4.3. Characterization of Superlattice Containing Sb δ-Layers by Medium Energy RBS

The semiconductor n-i-p-i superlattice with thin doped δ-layers is a promising material for novel device applications due to its quantum size effects and the two-dimensional carrier gas [134, 135]. Generally speaking, integral properties of such a superlattice depend on the following parameters: (1) the thickness of the δ-layers, (2) the dopant activation (dopant lattice location), and (3) the dopant redistribution during the growth process and

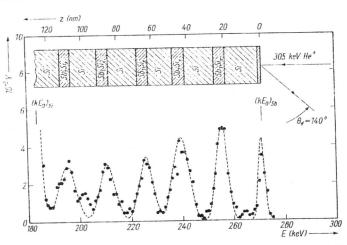

Fig. 50. The 305-keV ^{4}He$^+$ ions RBS random spectrum of a five-period Si superlattice with an Sb δ-layer (Y is the RBS counts and Z is the depth). The dashed line is the computer simulated spectrum. $(KE_o)_{Sb}$ is the surface position of the Sb element and $(KE_o)_{Si}$ is the surface position of the Si element. Reprinted with permission from [107], copyright 1990, John Wiley & Sons.

the following thermal treatments [107]. Therefore, characterization of those parameters is crucial for growing and processing such types of the superlattices. Obviously, for characterization of the superlattices containing impurity doped δ-layers, a technique with high depth resolution is required. Secondary ion mass spectrometry (SIMS) is a good tool for depth profiling that requires high depth resolution. However, it is a destructive method and is not capable of providing the information about the crystalline quality of the superlattices and the lattice locations of the dopant. The RBS technique combined with the ion channeling effect not only can provide high depth resolution comparable with that of SIMS but also can provide the information about the crystalline quality of the superlattices and the lattice locations of the dopant. In addition, RBS and channeling techniques are nondestructive methods. Lenkeit et al. have successfully applied RBS and channeling techniques to the characterization of the Si superlattices with Sb doped δ-layers [107]. To achieve high depth resolution, Lenkeit et al. used medium-energy ^{4}He$^+$ ions (305 keV) as the probing particles and a cylindrical electrostatic analyzer as the detector. Depth resolution as high as 0.8 nm has been achieved in the near-surface region.

The samples investigated are five-period Si superlattices with Sb doped δ-layers [Fig. 50], grown either on (100) Si or on (111) Si substrates. The superlattices were grown by MBE at low temperature (200°C), and then annealed at 600–800°C for amorphous layers to recrystallize into crystals. The concentration of the Sb in the δ-layers ranges from 0.4×10^{14} cm^{-2} to 1.4×10^{14} cm^{-2}. To achieve high depth resolution in the RBS measurement, ^{4}He$^+$ ions with energy as low as 305 keV (to obtain higher energy loss) were used as probing particles. To further improve the depth resolution, a cylindrical electrostatic analyzer was used to record the backscattered ^{4}He$^+$ ions, which has an energy resolution of $\delta E/E_0 = 7.6 \times 10^{-3}$, 20 times smaller than that of the Si detector ($\delta E/E_0 =$

12 keV/305 keV $= 3.9 \times 10^{-2}$). That allows a very high depth resolution of 0.8 nm in the near-surface region of Si. In the RBS-channeling measurement for checking the crystalline quality, 1.7-MeV $^4He^+$ ions were used as the probing particles.

Figure 50 shows a typical experimental (solid circles) and simulated (dashed line) random RBS spectra of 305-keV $^4He^+$ ions on a five-period Sb δ-doped superlattice [107]. Six Sb peaks are clearly resolved. The first Sb peak located near channel 272 is from the Sb that segregated to the surface. The other five Sb peaks correspond to the five Sb doped δ-layers. To determine the thickness of the δ-layers and the concentration of the Sb from the RBS spectra, computer simulation has to be performed. Lenkeit et al. developed a computer program that is capable of simulating RBS spectra for $^1H^+$ or $^4He^+$ ions in the energy range from 100 keV to 2 MeV and for sam-

ples consisting of 30 layers and containing a maximum of 10 elements. The following important information were obtained by RBS-channeling measurement: (1) the thickness of the δ-layer, the concentration of Sb, and the depth of the δ-layers, which were determined by the comparison between the measured RBS spectrum and the simulated one (see Fig. 50); (2) the channeling analysis done with 1.7-MeV $^4He^+$ ions showed that the superlattices had χ_{min} as low as 3.2% near-surface region, indicating that Sb δ-doped Si superlattices have very high crystalline quality (see Figure 51 [107]); (3) the superlattices investigated were thermally stable as no broadening of the Sb profiles in the δ-layers were observed after 700°C (60 min) annealing (Fig. 52 [107]); (4) although Sb diffusion in the superlattice did not take place during the annealing process from 600 to 800°C, segregation of Sb towards the surface did occur as revealed in Figure 53 [107], indicating enhanced diffusion near the surface.

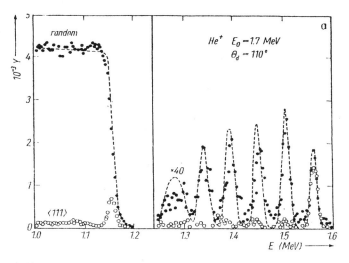

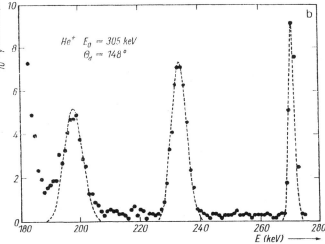

Fig. 51. RBS-channeling analysis on the Si superlattice with an Sb δ-layer: (a) 1.7-MeV $^4He^+$ ions RBS random (filled circles) and the ⟨111⟩ channeling (open circles) spectra measured with an Si surface barrier detector; (b) 305-keV $^4He^+$ ions RBS random spectrum measured with an electrostatic analyzer. Dashed lines are the computer-simulated spectra. Reprinted with permission from [107], copyright 1990, John Wiley & Sons.

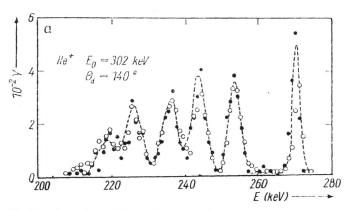

Fig. 52. Comparison of the random spectra for superlattice number 6 measured before annealing (filled circles) and after annealing (700°C, 60 min) (open circles). No measurable difference was observed. The dashed line is the computer-simulated spectrum. Reprinted with permission from [107], copyright 1990, John Wiley & Sons.

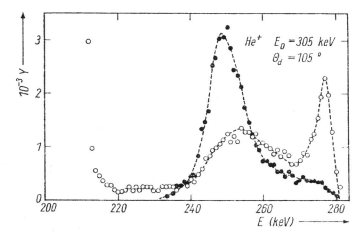

Fig. 53. The change of the random spectra of Sb buried under an amorphous Si top layer ($d = 21$ nm). Filled circle: measured before annealing. Open circles: measured after annealing (700°C, 60 min). Part of the Sb in the first δ-layer segregated to the surface. Reprinted with permission from [107], copyright 1990, John Wiley & Sons.

In summary, Lenkeit et al. demonstrated that by using medium-energy $^4He^+$ ions, such as 305-keV $^4He^+$, and a detector with high-energy resolution, such as an electrostatic analyzer, a depth resolution as high as 0.8 nm could be achieved and many important parameters of the superlattices with doped δ-layers can be determined. A few years later, Kimura et al. performed a similar analysis on an Sb δ doped Si single-layer structure using 300-keV $^4He^+$ ions and a 90° sector magnetic spectrometer with energy resolution of $\delta E/E \sim 0.1\%$ [9]. Successive atomic layers were resolved, indicating that monolayer resolution was achieved.

4.4. Characterization of Shallow Superlattices by TOF-MEIS Technique

This section discusses the characterization of shallow superlattices, i.e., superlattices that are located in the region less than 50 nm below the surface. Several techniques are frequently used for analysis of shallow structures, such as X-ray diffraction (XRD), Auger spectroscopy (AES), X-ray photoelectron spectroscopy (XPS), and secondary ion mass spectrometry (SIMS). XRD does not have the high depth resolution that can resolve the layered structure of the superlattices. AES and XPS sample only the top 1–2 nm of the materials, therefore, to obtain the structural and compositional information of the shallow superlattices layer stripping is required. Secondary ion mass spectrometry (SIMS) has the high depth resolution that can satisfy the needs of depth profiling in superlattices, but before the sample can be studied a significant layer of the material has to be consumed until the sputtering process reaches the equilibrium. On the other hand, Time-of-flight technique (TOF) combined with ion scattering technique offers high depth resolution and high mass resolution that are suitable for the characterization of the shallow superlattices. A depth resolution as high as 2–3 nm and mass resolution less than 1 amu has been achieved by TOF systems (combined with medium- or high-energy ion scattering) [113, 114, 136–138]. Sugden et al. [113] and McConville et al. [114] have demonstrated that time-of-flight combined with medium-energy ion scattering spectroscopy (TOF-MEIS) is capable of resolving the structure of a subsurface superlattice located at a depth as shallow as 15 nm.

The samples investigated by Sugden et al. and McConville et al. were MBE or CVD grown $Si/Si_{0.78}Ge_{0.22}$ three-period superlattices on (100) Si substrates, with each layer being only 2.5-nm thick. The TOF-MEIS system is shown schematically in Figure 54 [113, 114]. The $^4He^+$ ions were generated by a Duoplasmatron ion source with energy in the range of 5–30 keV. An electrostatic switching device chops the $^4He^+$ ions into pulses with a duration of 10–50 ns. After passing a start detector, the pulsed $^4He^+$ ions are directed onto the sample. The backscattered $^4He^+$ ions are recorded in a microchannel plate detector (MCP), which also serves as a stop detector.

Figure 55 is the time-of-flight spectra for the three-period (2.5:2.5 nm) $Si/Si_{0.78}Ge_{0.22}$ superlattice taken with 21-keV $^4He^+$ ions [113, 114]. Both channeling aligned and random

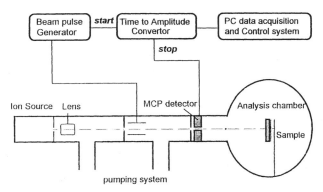

Fig. 54. Schematic of the time-of-flight medium-energy ion scattering (TOF-MEIS) spectrometer and data collection system. The ion source generates a 21-keV $^4He^+$ ion beam, which is directed to the sample through a small hole in the center of the microchannel plate (MCP) detector. The backscattered $^4He^+$ ions and neutrals are recorded in a time-flight mode by the MCP detector. Reprinted with permission from [113], copyright 1995, American Institute of Physics. Reprinted with permission from [114], copyright 1996, Elsevier Science.

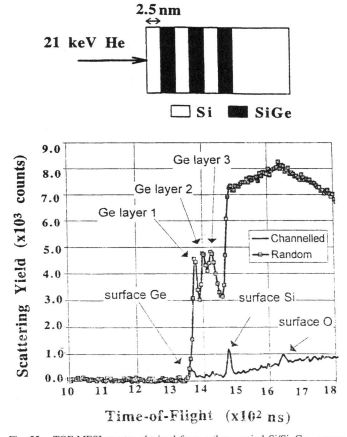

Fig. 55. TOF-MESI spectra obtained from a three-period Si/Si_xGe_x grown by MBE with each layer being only 2.5-nm thick. Three Ge peaks are clearly resolved. The χ_{min} of the aligned spectrum near the surface is about 3–4%. The small oxygen peak is from the native silicon oxide on the surface. Reprinted with permission from [113], copyright 1995, American Institute of Physics. Reprinted with permission from [114], copyright 1996, Elsevier Science.

spectrum are shown. The horizontal scale is the time of the flight, which can be converted into depth. Figure 55 provides the following useful information: (1) The crystalline quality of the superlattice. Near the surface, the χ_{min} value is as low as 3–4%. The large reduction in the channeling yields compared with random yield shows that the $Si/Si_{0.78}Ge_{0.22}$ superlattice has a high crystalline quality; (2) depth resolution. Three Ge peaks, seen at flight times, 1375, 1400, and 1425 ns, respectively, are clearly resolved, indicating the depth resolution to be about 3 nm; (3) the thickness of the period. The separation of the Ge peaks is about 25 ns, which equates to an energy difference of 1400 eV. Using the stopping power in silicon, Sugden et al. and McConville et al. estimated that the thickness of the period is about 6 nm, which is close to the anticipated Ge layer separation of 5 nm; (4) the segregation of the Ge to the surface. On the channeling aligned spectrum, there are three small peaks. The first peak at 1360 ns is the signals from the Ge atoms residing at the surface, indicating the possibility that some of the Ge atoms had segregated to the surface through the Si capping layer.

In summary, the study carried out by Sugden et al. and McConville et al. [113, 114] has shown that the TOF-MEIS technique is capable of providing the structural and compositional information in a shallow region not accessible to other conventional surface analysis techniques. The depth resolution (~3 nm) achieved by the TOF-MEIS system designed by Sugden et al. and McConville et al. is not as good as the depth resolution achieved by the medium-energy ion scattering system (MEIS) used by Lenkeit et al. [107], which used a bulky electrostatic analyzer. However, the TOF-MEIS system used by Sugden et al. and McConville et al. is compact and small, thus is more suitable for commercial applications. It can be easily attached to a commercial MBE or CVD growth chamber to provide *in situ* real-time diagnostic analysis for the growth processes.

4.5. Characterization of Ion Effects in Superlattices

Ion effects, such as ion implantation induced compositional and structural disorder, ion induced strain relaxation, and ion-mixing induced disordering, provide unique ways to tailor the electronic properties of the superlattices. To develop those unique techniques, it is crucial to have accurate characterization on the compositional and structural disordering, strain relaxation, radiation damage, and damage recovery in the superlattices. RBS and channeling techniques are among the techniques that are capable of providing such services [139–142].

Picraux et al. [139] have studied the effect of ion implantation on the structural and compositional properties of the III–V compound superlattices using RBS and channeling techniques. One of the compounds studied was the $In_{0.2}Ga_{0.8}As/GaAs$ strained-layer superlattice (SLS) with a layer thickness of 10–25 nm (less than the critical thickness). Ion implantation of 75-keV Be^+, 75-keV N^+, 150-keV Si^+, and 50 to 250-keV Zn^+ was carried out at room temperature. The structural and compositional disorders induced by the ion implantation were

characterized using the 2 to 3-MeV $^4He^+$ ion RBS and channeling technique.

4.5.1. Low Dose, Low-Energy Zn Ion Implantation and Disorder Annealing

The RBS random and channeling spectra in Figure 56a show the effect of the 50-keV Zn^+ ions shallow implantation and 30-min post-annealing on the $In_{0.2}Ga_{0.8}As/GaAs$ SLS. The main information provided by the RBS-channeling spectra are (1) 50-keV Zn^+ ion implantation with a low dose of 2.5×10^{14} cm^{-2} only partially damaged the superlattice since the RBS yields in the $\langle 100 \rangle$ aligned spectrum are much less than that in the random spectrum (Fig. 56a), (2) the $\langle 100 \rangle$ aligned spectrum taken after 600°C post-annealing is almost the same as that of the virgin sample (Fig. 56a), indicating that the 600°C annealing could effectively remove the implantation induced damage, (3) along the inclined $\langle 110 \rangle$ direction, the strong dechanneling observed in the channeling spectra is due to the strain in the superlattice (Fig. 56b). Furthermore, there is no significant difference in the $\langle 110 \rangle$ dechanneling before and after Zn^+ ion implantation plus 600°C annealing (Fig. 56b), showing that there is no detectable loss of strain in the structure, (4) the RBS signals of the In atoms have not been changed after Zn^+ ion implantation to a low dose of 2.5×10^{14} cm^{-2} or after 600°C post-annealing (Fig. 56c), suggesting that no compositional disordering was induced.

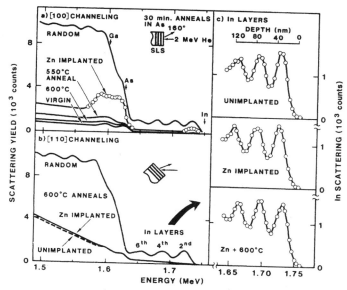

Fig. 56. The 2-MeV $^4He^+$ ion RBS and channeling aligned spectra for the $In_{0.2}Ga_{0.8}As/GaAs$ superlattice (21-nm layers) after 50 keV, 2×10^{14} cm^{-2}, Zn^+ ion implantation plus 550°C 30 min^{-1} or 600°C 30 min^{-1} annealing. (a) The $\langle 100 \rangle$ channeling aligned spectra and random spectrum. (b) The inclined $\langle 110 \rangle$ channeling aligned spectra and random spectrum. (c) Indium RBS signals along a random direction for unimplanted, Zn^+ ion implanted, and Zn^+ ion implanted plus 600°C annealing. Reprinted with permission from [139], copyright 1985, Elsevier Science.

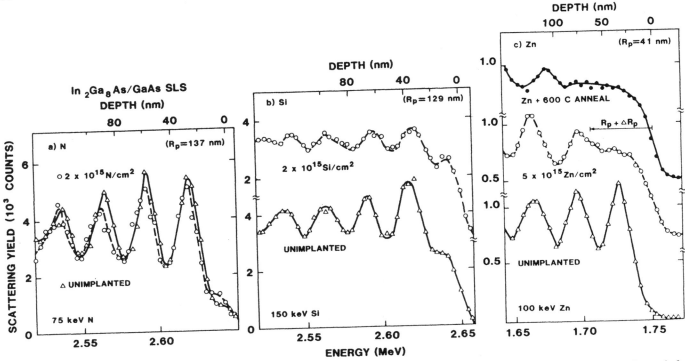

Fig. 57. The 3-MeV ^{4}He$^+$ ion RBS random spectra showing the indium RBS signals of the first four layers of In$_{0.2}$Ga$_{0.8}$As/GaAs SLS. (a) Before and after 75-keV N$^+$ ion implantation to 2×10^{15} cm^{-2} for 15-nm layers. (b) Before and after 150-keV Si$^+$ ion implantation to 2×10^{15} cm^{-2} for 15-nm layers. (c) Before and after 100-keV Zn$^+$ ion implantation to 5×10^{15} cm^{-2} for 21-nm layers and after 600°C 60 min^{-1} annealing. $R_p = 41$ nm is the projected range of the 100-keV Zn$^+$ ions in the sample. ΔR_p is the range straggling. Reprinted with permission from [139], copyright 1985, Elsevier Science.

4.5.2. High Dose Ion Implantation

One could expect that ion-beam mixing would occur when the superlattices were irradiated by ions with a larger mass and higher doses. Figure 57 [139] shows the In signals in the RBS spectra taken under different irradiation conditions. (1) Under 75-keV N$^+$ ion irradiation to a dose of 2×10^{15} cm^{-2}, no significant difference in the In signal oscillations was observed before and after irradiation (Fig. 57a), indicating that the intermixing has not been induced. (2) Under 75-keV Si$^+$ ion irradiation to a dose of 2×10^{15} cm^{-2}, some intermixing clearly occurred since the In signal peaks became smaller and wider after Si$^+$ irradiation (Fig. 57b). However, after being annealed at 600°C, no significant difference in the RBS spectra was observed between the as-implanted and 600°C annealed sample (spectra not shown), showing that no alloying had occurred. (3) Under 100-keV Zn$^+$ ion irradiation to a dose of 5×10^{15} cm^{-2}, the oscillation of the indium signals for the second and fourth layers almost disappeared (Fig. 57c), indicating that significant intermixing had occurred. Upon annealing to 600°C for 60 min, the oscillation for the second and fourth layers completely disappeared, showing that alloying was induced by the 100-keV Zn$^+$ ion irradiation at a dose level of 5×10^{15} cm^{-2}.

The example presented above shows that abundant information about the effect of ion irradiation on superlattice, such as radiation damage and annealing, strain relaxation, composition change, structural change and interdiffusion etc., can be obtained by RBS and channeling techniques.

5. APPLICATION OF ION-BEAM TECHNIQUES TO SUPERLATTICE (II)

5.1. Kink Angle $\triangle \Psi$ and Strain in Superlattice

Superlattice structures consist of alternating single crystalline layers of different materials ordered in a near perfect crystal arrangement. These layers differ in composition but have the same type of crystalline structure. The resulting modulated structure is not only of fundamental interest in the investigation of quantum mechanical phenomena but also has important application in electronic and optical devices. The superlattices by room temperature epitaxy exhibit a high degree of crystal perfection and a minimum of interdiffusion and chemical reactions. The periodic structures with sharp interfaces provide a square potential well for electrons, which exhibit a unique electronic and optical property [143]. When the lattice mismatch is not very small, dislocations or other defects may occur at the interfaces. When the mismatch of the lattice constant of these layers is small (in the order of 0.5–5%) and if the layer thickness is less than a certain critical thickness, the misfit of the lattice constant can be accommodated by the elastic strain. Under epitaxial growth, the in-plane lattice constant remains the same, giving rise to alternating compressive and tensile stresses in the layers. Due to the Poisson effect, these stresses cause tetragonal distortions along the growth direction with the suppression and the expansion of the lattice constant perpendicular to the plane, $a_{\perp 1}$ and $a_{\perp 2}$ (Fig. 58). Thus, unique man-made multilayer composite structures can be formatted as strained-

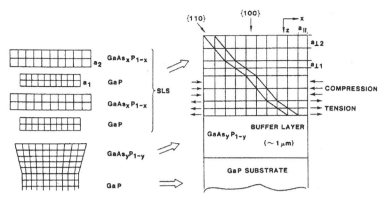

Fig. 58. Schematic of a strained-layer superlattice (SLS) sample and the lattice distortion due to mismatch (a); the distorted channel along the $\langle 110 \rangle$ direction. Reprinted with permission from [68], copyright 1983, Elsevier Science.

layer superlattices (SLS). The electronic and optical properties of the SLS are closely related to the thickness and the composition of the layer, the sharpness of the interface, and the magnitude of the elastic strain in the layers. One example of an SLS is a $GaAs_x P_{1-x}$/GaP strained-layer superlattice grown on substrate GaP through a buffer layer of $GaAs_y P_{1-y}$/GaP is shown in Figure 58 [68]. The epitaxial growth direction was along the $\langle 100 \rangle$ axis with the same in-plane lattice constant $a_\parallel$ in both $GaAs_x P_{1-x}$ and GaP layers in the $\{100\}$ plane. In the direction of the growth (perpendicular to the $\{100\}$ plane), due to the Poisson effect, the lattice constant in GaP layers $a_\parallel$ contracts to $a_{\perp 1}$, whereas the lattice constant in $GaAs_x P_{1-x}$ layers a_2 expands to $a_{\perp 2}$. This contract and expansion causes a tetragonal distortion occurring along the $\langle 100 \rangle$ direction. The distortion gives an abrupt kink angle of $\Delta \Psi$ at the interfaces between $GaAs_x P_{1-x}$ and GaP layers along the $\langle 110 \rangle$ direction. The magnitude of the kink angle $\Delta \Psi$ can be calculated by

$$\Delta \Psi = \tan^{-1}(a_{\perp 1}/a_\parallel) - \tan^{-1}(a_{\perp 2}/a_\parallel) \sim (a_{\perp 2} - a_{\perp 1})/2a_\parallel \tag{71}$$

Typical value of $\Delta \Psi$ is 0.1–1.0° for the lattice mismatch of 0.5–5%. This tetragonal distortion and the periodical kink along the $\langle 110 \rangle$ direction will affect the behavior of channeled ions in the SLS structure. The kink angle $\Delta \Psi$ itself is a direct measure of the relative strain between the strained layers. In studies of the strained-layered superlattice (SLSs) RBS-channeling is an excellent tool for measuring the strain. Several methods for measuring the strain in SLS structures using RBS-channeling have been studied and developed by Saris et al. [67], Chu et al. [68, 69, 144–149], Picraux et al.[70, 150–154], Ellison et al. [154], Barret [156, 157] and others [158, 159]. Three major methods are discussed in this section, they are:

(1) axial dechanneling analysis;
(2) angular scan analysis;
(3) planar catastrophic dechanneling analysis.

5.2. Axial Dechanneling Analysis of Strain

Figure 58a [68] shows the layer structure of an $GaAs_x P_{1-x}$/GaP strained superlattice grown in the $\langle 100 \rangle$ direction. When an

ion-beam for RBS/Channeling is aligned along the $\langle 100 \rangle$ direction, where the atomic rows are straight, the ion-beam can pass through the interfaces between the InGaAs and GaAs with little dechanneling (Fig. 58b). On the other hand, along the inclined direction such as the $\langle 110 \rangle$ direction, the atomic rows are not straight, but have a small kink of $\Delta \Psi$ at each interface. When the analyzing ion-beam impinges along such inclined direction, at each interface, the transverse energy of the ions experience a sudden change due to the presence of the kinks at interfaces. Some of the ions may receive a significant increase in their transverse energy so that their minimum approaches to the atomic row exceed the critical impact parameter. These ions thus have large angle scattering with the atoms in the atomic rows and become dechanneled (Fig. 58c). This enhanced dechanneling is a good measure of the kink angle $\Delta \Psi$. Figure 59 shows RBS-channeling aligned spectra along the $\langle 100 \rangle$ and $\langle 110 \rangle$ directions of an InAs/GaSb strained-layer superlattice [145]. The random spectrum is also shown above the aligned spectrum. The oscillation in the random spectra comes from the different elements of the layers of 20 periods InAs/GaSb with 41 nm/41 nm for each period (Fig. 59b). Dashed smooth curves are the corresponding spectra for single crystal GaSb. The RBS-channeling spectrum along the $\langle 100 \rangle$ growth direction shows normal low dechanneling with a little oscillation due to the different elements in InAs/GaSb. The overall dechanneling is comparable to that of a single crystal GaSb. Along the inclined $\langle 110 \rangle$ direction, the aligned RBS-channeling spectrum shows very high dechanneling compared to the aligned spectrum of the single crystal GaSb due to the existence of kinks (strain) at the interface in the InAs/GaSb superlattice (Fig. 59c). The enhanced dechanneling along the $\langle 110 \rangle$ direction relative to the normal dechanneling along the $\langle 100 \rangle$ direction is a clear indication of the strain. This enhanced dechanneling is a complicated function of many parameters, such as the thickness of the surface layer, composition of the layers, the energy of the incident ions, and the geometry of the experiment. The advantage of the axial dechanneling technique is that it provides the depth-resolved measure of the strain. But quantitative analysis of the strain requires computer simulation [155–159].

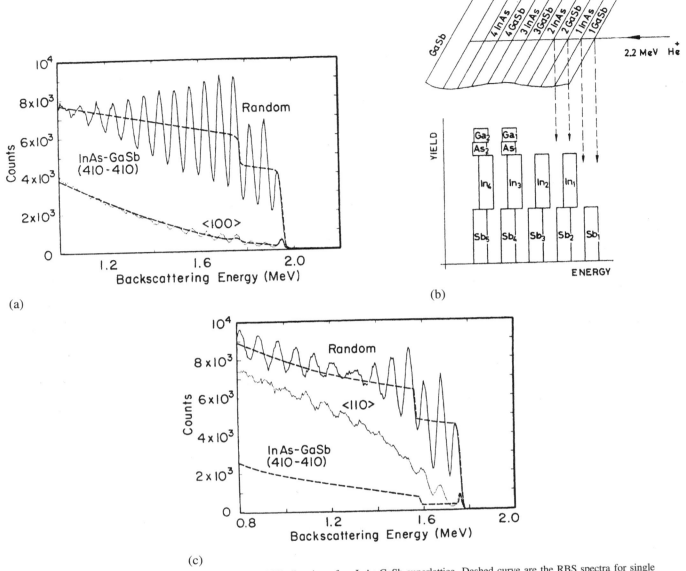

Fig. 59. (a) RBS-channeling spectra along the ⟨100⟩ direction of an InAs-GaSb superlattice. Dashed curve are the RBS spectra for single crystal GaSb. (b) Schematic of the origin of the oscillatory yield of an RBS random spectrum in (a). (c) RBS-channeling spectra along the ⟨110⟩ direction. Abnormally high dechanneling is observed. Reprinted with permission from [145], copyright 1982, American Physical Society.

5.3. Strain Measurement by Channeling Angular Scan Analysis

RBS-channeling spectra are depth sensitive or, for the superlattice, layer sensitive. The counts within energy windows for different layers will give information of the specific layers. For strain analysis, one has to perform a channeling angular scan around the inclined direction. A number of spectra are taken around the inclined ⟨110⟩ direction. The angular scan of the top layer will locate the centroid of the dip for the top layer and will give the accurate determination of the direction of the top layer. The approach of an absolute measurement of the strain is to determine precisely the direction of the inclined crystal axis in the top layer relative to a reference direction. The reference direction can be determined from the average direction of

the angular scans (averaged over several layers deep in the SLS structure).

As an example, Figure 60a shows the angular scan of an SLS sample at four different depths for four different layers [69, 160]. The sample consists of (30 nm/30 nm) GaAs/AlSb periodic structure with a total of 10 periods grown on a GaSb substrate along the [100] direction (top part of Fig. 60b [69, 160]). The four angular scans were obtained from 52 RBS spectra taken at 52 different angles across the inclined ⟨110⟩ direction. The angular position of the centroid of the dips at various depths shown in Figure 60b. The centroids were obtained from the dip minimum of the angular scan shown in Figure 60a. A clear oscillation of the angular shift can be seen from the angular shift-depth dependence with oscillation damp-

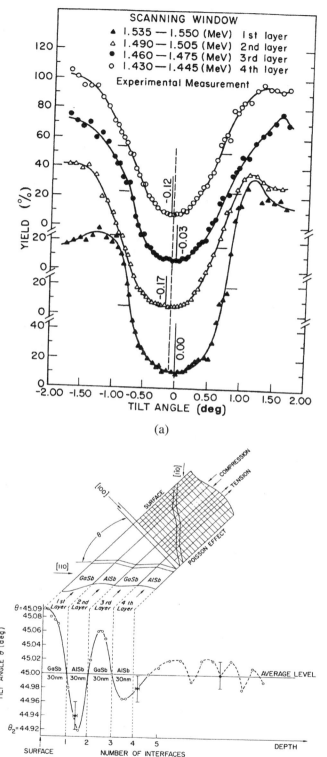

Fig. 60. (a) Angular scan by setting an energy window from the first to the fourth layer from 52 spectra run at different angles on a GaSb/AlSb superlattice across the inclined {110} direction. (b) The top portion shows the sample schematics of the GaSb/AlSb superlattice. The lower portion shows the angular position of the dips in (a). Reprinted with permission from C. K. Pan et al. [149], copyright 1983, American Physical Society. Reprinted with permission from W. K. Chu et al. [69], copyright 1983, American Physical Society.

ing with depth (low part of Fig. 60b [69, 160]). The top part in Figure 60b shows the structure of the SLS. The observed difference between the best channeling direction for the first two layers, $0.17 \pm 0.03°$, gives a lower limit for the strain. This measurement also can be compared with the Monte Carlo simulation and the real strain value can be obtained from the best agreement between the experiment and the simulation. However, care needs to be taken with the beam steering effect by the surface layer. The steering effect could significantly distort the shape of the angular scan of the strained layer, making the interpretation of the angular scan difficult.

Planar channeling angular scan also can be used to measure the strains. Due to the tetragonal distortion along the growth direction, for an SLS grown along the $\langle 100 \rangle$ direction, the atomic planes are not straight when looking along the inclined planar direction such as the {110} planar direction. There is a small-angle change from layer to layer at each interface. Hence, similar to the axial case, an angular scan across the inclined planar direction also can provide information about the strain. Davies et al. gives the details of the strain measurement by planar angular scan in Ref. [73, 74]. The influence of the steering effect is more serious in planar channeling than axial channeling. Detailed discussion on steering effect is given at the end of this section.

This angular scan technique can measure the strain from a large value down to about $0.03°$ or $\sim 0.06\%$ misfit.

5.4. Strain Measurement by Planar Dechanneling Analysis

The third and most sensitive method of analysis of the strain in SLS is to use the resonance match between the superlattice period and the planar channeling trajectory wavelength of the probing ions [146–148, 151, 153, 155, 156].

When an ion beam is channeled along a planar channel direction, it will be steered back and forth between the planes giving rise to a focusing effect. The focusing effect of the planar channel on the probing ions was first observed through the oscillation character of the planar aligned energy spectra by Abel et al. [159]. Figure 61 shows the planar RBS-channeling spectrum with an oscillation for 1.2-MeV He ions entering GaP {110} planes [69, 160]. The interpretation of the measured oscillation is very simple based on a harmonic planar potential. The wavelength of the ion-beam can be expressed as [160],

$$\lambda = 2\pi (2E/\alpha)^{1/2} \tag{72}$$

where E is the energy of the projectile and α is the force constant of the harmonic potential $U(x) = 1/2\alpha x^2$ with $x = 0$ at the center between two neighboring atomic planes. The steering potential in a real crystal is anharmonic, so the real focus will be blurred and only an effective wavelength or a mean wavelength can be defined. In practice, the effective wavelength can be measured from the energy shifts between the oscillating peaks or the valleys, and then convert this energy shift in to depth by the energy loss of the ions in the materials. For example, in Figure 61, the effective wavelength of a 1.2-MeV He ion in [110] planes of a GaP crystal is defined as $\lambda/2 = 48$ nm.

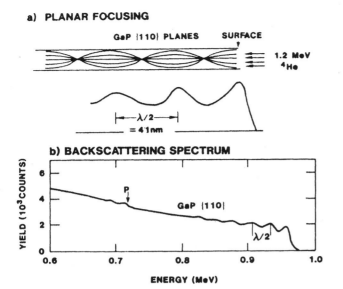

a) PLANAR FOCUSING

b) BACKSCATTERING SPECTRUM

Fig. 61. Planar channeling of 1.2-MeV He ions from {110} planes of a GaP single crystal. (a) Illustration of focusing and defocusing is used to explain the dips and the peaks and the definition of the effective wavelength on a RBS measurement given in (b). From W. K. Chu [158].

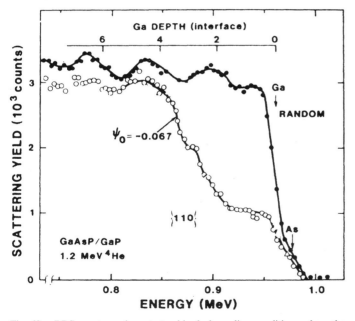

Fig. 62. RBS spectra under catastrophic dechanneling conditions where the half-wavelength of the 1.2-MeV He ions (48 nm) matches the resonant condition. From W. K. Chu [158].

For a parallel beam, all trajectories are matched in phase upon entering the crystal. When the half-wavelength of the channeled ions, $\lambda/2$, is matched to the path length of the ions along an inclined plane for a pair of superlattice layers, a resonance phenomenon called catastrophic dechanneling occurs.

Figure 62 shows an example of catastrophic dechanneling in a 1.2-MeV RBS-channeling spectrum for $GaAs_xP_{1-x}/GaP$ (x close to 0.12) SLS near the {110} plane for random direction and aligned direction [158]. The incident angle relative to the

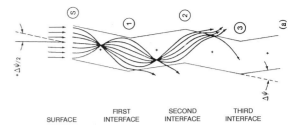

Fig. 63. Computer-simulated trajectories of 1.2-MeV channeled He ions in planar channels of an SLS structure with incident angle $+\Delta\psi/2$ respect to the orientation of the surface layer, where $\Delta\psi$ is the kink angle. Adapted from S. I. Picraux et al. [153].

surface plane is $\psi_0 = -\Delta\psi/2 = -0.067°$. The oscillation in the random spectrum is due to the absence of As in the second, fourth, and succeeding even layers and the energy difference between adjacent oscillation peaks gives the measured thickness of 58 nm per a pair of $GaAs_xP_{1-x}$ and GaP layers. The path length, L, that the particles traverse in one layer along the {110} direction is 29 nm $\div \sin 45° = 41$ nm. The incident energy of the He ions is chosen to be 1.2 MeV to give an effective wavelength of about 82 nm which matches the resonant condition, $\lambda/2 = L$. The sharp increase in dechanneling after the second interface and the third interface is observed (Fig. 62). Here, $\Delta\psi$ is the alternating tilt angle (kink angle) between the planes at each interface.

To understand the catastrophic dechanneling, we refer to a simple modified harmonic model [155] in which the planar continuum potential is approximated by a simple harmonic potential, $U(x) = kx^2$. This simplified harmonic potential was assumed to calculate the ion trajectories. The match of the parameters, $\lambda/2 = L$, is the necessary condition for catastrophic dechanneling. The calculated trajectories are shown in Figure 63 [153]. Due to the steering of the planar potential, the trajectories of the particles are shaped with alternative focal points and waist planes. When a simple harmonic potential is used in calculation, the trajectories have sharp focal points. With more realistic potentials such as Moliéré continuum potential, a blurred focal point will be observed, but the general picture is the same. In the first layer, the focal point is near the center of the channel and little dechanneling occurs. At the first interface, all the particles experience an angle shift, $\Delta\Psi$, relative to the second layer. At the second interface, the particles undergo another angle shift. This angle shift moves the focal point of the channeled particles onto the channel wall in the third layer, causing an abrupt dechanneling. By the time the rest of the channeled particles reach the fourth layer, they are completely dechanneled.

Catastrophic dechanneling is easily understood by the phase plane description shown in Figure 64. The horizontal axis x/x_c represents the position in the planar channel and the vertical axis ψ/ψ_c represents the angle of the probing ions. In the modified-harmonic model, all ions have the same wavelength and move on circles in the phase plane. When the beam moves in the planar channel, the line representing the beam rotates on the phase plane within a circle. When the line on the phase plane moves outside the circle, the beam in the planar channel hit the channel wall and is dechanneled. The position and the angle of the projectiles at the surface can be represented by the horizontal line marked as 0 on the phase plane diagram. A uniform beam of ions enters the surface at an angle ψ_0. Upon reaching the first interface, the line has rotated $\theta = \pi$ to the line 1. When the parallel ions move across the first interface, they experience an instantaneous change of the angle $-\Delta\psi$ due to the SLS lattice distortion. On the phase diagram, this is equivalent to the dashed line. Subsequent shifts in the horizontal axis by $-\Delta\psi$ of alternating sign after each π rotation results in the ions moving to large radii. When the line is outside the circle, the ions hit the channel wall and get dechanneled.

The catastrophic dechanneling is very attractive because the sharp dechanneling occurred at well-defined depth. Picraux et al. showed that by evaluating the incident angle

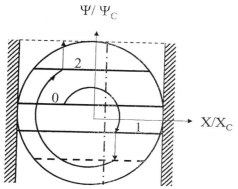

Fig. 64. The phase plane diagram of the planar channeled particles.

(ψ_0)–dechanneling depth (T_{ch}) dependence of the catastrophic dechanneling accurate value of the strain can be obtained [151]. Picraux et al. has studied the ψ_0–T_{ch} dependence in the inclined {110} planes of GaAs$_{1-x}$P$_x$/GaP (48.2/48.2 nm) SLS for 1.2 MeV He$^+$ ions. The results are shown in Figure 65 [151]. The incident angle, ψ_0, is measured relative to the first layer. The dechanneling depth in Figure 65 is defined as the depth where the 85% increase in dechanneling is occurred. The theoretical calculations are also presented in Figure 65 using both modified harmonic potential (dashed line) and static Moliéré potential (solid line). The comparison between the experiment and calculation shows that the best agreement is achieved when the kink angle $\Delta\psi = 0.153°$ and the critical impact parameter r_c is chosen as 1.25 times of the Thomas–Fermi screening radius a_T, here $a_T = 0.01375$ nm. Figure 65 shows that the calculated angular dependence using modified harmonic potential is close to that using detailed Moliéré calculation. However, the calculation using modified harmonic potential is simpler and faster. Further more, the averaged slope (dotted line) of the modified harmonic results is simply $1/\Delta\psi$ [152]. The ψ_0–T_{ch} dependence is a step-like structure. As pointed out by Picraux et al. [151], the position of the steps depends very sensitively on the kink angle $\Delta\psi$. A 0.01° decrease $\Delta\psi$ can cause a 0.06° shift of the step position, indicating that accuracy better than 0.01° in $\Delta\psi$ determination is possible. The channeling critical angle of the planar channeling is smaller than that of axial channeling. Therefore, for larger kink angle, it may be necessary to use axial channeling for the determination of the strains in SLS. But for smaller kink angles, planar catastrophic dechanneling is the most accurate method among the three methods discussed so far for measuring the strains in superlattice.

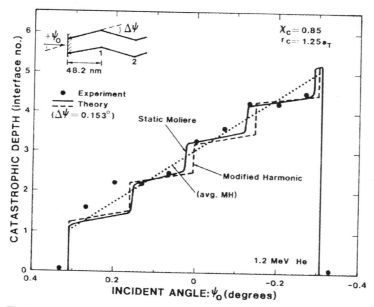

Fig. 65. Experimental angular dependence of planar catastrophic dechanneling and its depth vs incident angle dependence (solid circles) along with theoretical calculations based on Moliéré potential (solid line), modified harmonic potential (dashed line), and averaged modified-harmonic value (dotted line) given for $\Delta\psi = 0.153°$. Reprinted with permission from [151], copyright 1986, Elsevier Science.

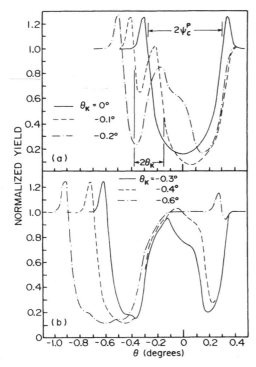

Fig. 66. The variation of double-dip feature with kink angle θ_k at fixed surface layer thickness. $\theta = 0°$ corresponds to the best channeling direction in the surface layer. Computer simulation of a 2 MeV He$^+$ {110} planar angular scan in the strained layer. Reprinted with permission from [74], copyright 1989, American Institute of Physics.

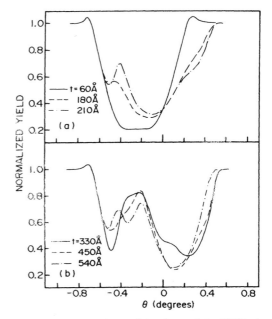

Fig. 67. Showing the dependence of the shape of the {110} planar angular scans of the strained layer on the thickness of the surface layer. $\theta = 0°$ corresponds to the best channeling direction in the surface layer. Computer simulation of a 1 MeV He$^+$ {110} planar angular scan in the strained layer, with θ_k fixed at $-0.26°$. Reprinted with permission from [74], copyright 1989, American Institute of Physics.

5.5. Steering Effect

This section provides detailed discussion on the steering effect. In the early studies on the ion beam channeling in superlattices, it has been noticed that the shape of the channeling angular scan of the underlying strained layer would be influenced by the history of the channeled particles in the surface cap layer [69, 149, 151]. Further detailed studies on this problem were carried out by several groups [73, 74, 160–162]. These studies showed that when the channeled particles in the surface layer are steered into the channeling direction of the underlying strained layer, the channeling behavior of the particles in the surface layer would interfere with their channeling behavior in the underlying strained layer. The shape of the angular scan of the underlying strained layer will be significantly distorted. When certain conditions are met, such as the surface layer is thin or the kink angle θ_k is comparable with the channeling critical angle Ψ_c, the distortion of the angular scan could be so large that the determination of the kink angle become complicated and ambiguous. As examples, Figure 66 shows the dependence of the distortion of off-normal {110} planar angular scan on the kink angle θ_k and Figure 67 shows the dependence of the distortion on the thickness of the surface layer [73]. The discussion presented in the following is mainly based on the studies carried out by Davies and Stevens and their co-workers [73, 74]. The samples used in their studies are InGaAs strained layers grown

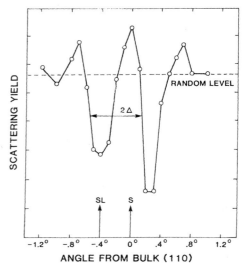

Fig. 68. Off-normal {110} planar angular scan of the indium scattering yields (InGaAs strained layer) in sample #192: surface layer (GaAs)–36 nm; strained layer (In$_{0.1}$Ga$_{0.9}$As)–22 nm. Incident beam: 2 MeV He$^+$; kink angle $\theta_k = 0.38°$; $\Psi_c = 0.22°$. Double-dip feature occurred because $2\Psi_c > \theta_k > \Psi_c$. The angular difference between the two left-hand sides gives reasonable estimate of $2\theta_k$ (shown as 2Δ in the figure). Reprinted with permission from [73], copyright 1989, American Vacuum Society.

on ⟨100⟩ GaAs substrate and capped by a thin GaAs epi-layers. The thickness of the surface capping layers (S-layers) and underlying strained layers (SL-layers) are ranging from 25 nm to 36 nm and from 22 nm to 25 nm respectively.

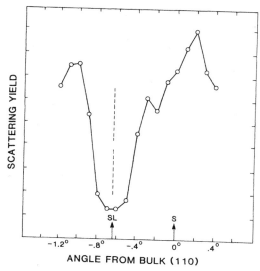

Fig. 69. Off-normal {110} planar angular scan of the indium scattering yields (InGaAs strained layer) in sample #108: surface layer (GaAs)–25 nm; strained layer (In$_{0.17}$Ga$_{0.83}$As)–25 nm. Incident beam: 2 MeV He$^+$; kink angle θ_k = 0.65°; Ψ_c = 0.22°. Normal single-dip angular scan observed. The mid-point of the scan agrees with the orientation of the strained layer. Reprinted with permission from [73], copyright 1989, American Vacuum Society.

5.5.1. Planar Channeling

The influence of the steering effect in planar channeling depends on the thickness of the surface layer t (or path length L_p), wavelength λ, (roughly is $\lambda = 2d/\Psi_p^s$), value of the kink angle θ_k, and the critical angle Ψ_p^s. Follow the reference [74] here the critical angle Ψ_p^s is defined as the half width at the random level. A superscript "S" is used to distinguish it from the normal definition of planar channeling critical angle Ψ_p. To simplify the problem, Davies et al. [73] considered an ideal case in which the beam energy had been tuned such that the path length L_p in the surface layer is exactly $\lambda/2$. This means that the channeled ions will undergo one reflection off the atomic rows in passing through the surface layer. Under this ideal condition, only two cases need to be considered: (1) the kink angle θ_k is comparable to the critical angle Ψ_p^s, and (2) the kink angle θ_k is much less than critical angle Ψ_p^s.

Case (1), θ_k is comparable to Ψ_p^s (either $\theta_k > \Psi_p^s$ or $\theta_k < \Psi_p^s$). The resulting channeling angular scan of the strained layer about the off-normal axis is a double-dip structure (Fig. 68 [74]). Davies et al. has established the factor that the angular difference between the two left-hand sides of the dips should give a reasonable estimate for $2\theta_k$ (Fig. 68 [74]). This has been confirmed by several other groups [71, 161, 163]. It should be noted that for InGaAs layer grown on GaAs substrate, the InGaAs layers are under tension so the kink angle θ_k always has negative values. However, when the strained layer is under compression, the kink angle θ_k always has a positive value. In this case, the kink angle would be obtained from the two right-hand sides of the dips.

Case (2), $\theta_k > 2\Psi_p^s$. In this case, the He$^+$ ions that are channeled in underlying strained layer must be in random direction

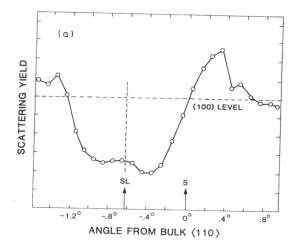

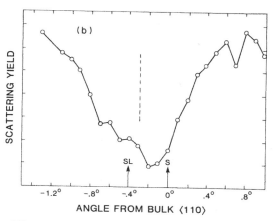

Fig. 70. Off-normal axial ⟨110⟩ angular scan of the indium scattering yields (InGaAs strained layer) using 2 MeV He$^+$ ions. (a) Sample #108: surface layer (GaAs)–25 nm; strained layer (In$_{0.17}$Ga$_{0.83}$As)–25 nm. (b) Sample #192: surface layer (GaAs)–36 nm; strained layer (In$_{0.1}$Ga$_{0.9}$As)–22 nm. Reprinted with permission from [73], copyright 1989, American Vacuum Society.

when they are traversing the surface layer and vice versa. Therefore, the scattering of the He$^+$ ions in the surface layer will not interfere with their channeling behavior in the strained layer. Thus the mid-point of the angular scan of the strained layer gives the correct value of the kink angle (Fig. 69).

5.5.2. Axial Channeling

The steering effect is less serious in axial channeling as compared with that in planar channeling. Axial scan about the off-normal axis of the InGaAs strained layer (SL-layer) from two significantly different samples are shown in Fig. 70 [73]. In each case, the correct ⟨110⟩ orientations of the SL layer and the surface layer (S) are indicated by arrows.

For sample #108, the path length L_p in the surface layer (35 nm) is smaller than the characteristic interaction distance d/Ψ_c (40 nm for 2 MeV He$^+$ in GaAs) between the channeled ions and the ⟨110⟩ atomic rows, the mid point (dashed line) of the measured axial scan gives the correct SL orientation

(Fig. 70a), indicating that most of the He$^+$ ions do not suffer significant steering in traversing the surface layer. In other words, most channeled ions do not have reflection in traversing the surface layer.

But for sample #192, the path length L_p in the surface layer (54 nm) is much larger than d/Ψ_c. A major fraction of the channeled ions undergo at least one reflection off the atomic rows in traversing the surface layer and thus have been shifted from their incident direction before entering the strained layer (SL). The resulting angular scan is now much more asymmetric than that in sample #108 (Fig. 70a) and the mid-point of the scan deviates significantly from the correct orientation of the strained layer (Fig. 70b).

From these two examples, Davies et al. [73] concluded that (1) the steering on the channeled ions in traversing the surface layer can cause significant distortion of the shape of axial angular scan of the underlying strained layer if the thickness of the surface layer is larger than the d/Ψ_c. In this case, the accurate value of the kink angle can not be obtained by the axial channeling angular scan; (2) to reduce the influence of the steering effect on the kink angle measurement, it is necessary to tune the energy of the ions in such way that the characteristic interaction distance d/Ψ_c is never less than the thickness of the surface layer or to make the thickness of the surface layer less than d/Ψ_c.

REFERENCES

1. W. K. Chu, J. W. Mayer, and M.-A. Nicolet, "Rutherford Backscattering Spectrometry," Academic Press, Orlando, FL, 1978.
2. M. Dobeli, A. Lombao, D. Vetterli, and M. Suter, *Nucl. Instrum. Methods B* 89, 174 (1994).
3. J. P. Thomas, M. Fallavier, and A. Ziani, *Nucl. Instrum. Methods B* 15, 443 (1986).
4. T. M. Stanescu, J. D. Meyer, H. Baumann, and K. Bethge, *Nucl. Instrum. Methods B* 50, 167 (1990).
5. R. A. Weller, K. McDonald, D. Pedersen, and J. A. Keenan, *Nucl. Instrum. Methods B* 118, 556 (1996).
6. D. Hunter, O. Meyer, J. Reiner, and G. Linker, *Nucl. Instrum. Methods B* 118, 578 (1996).
7. R. G. Smeenk, R. M. Tromp, H. H. Kersten, A. J. H. Boerboom, and F. W. Saris, *Nucl. Instrum. Methods* 195, 581 (1982).
8. H. D. Carstanjen, *Nucl. Instrum. Methods B* 136–138, 1183 (1998).
9. K. Kimura, K. Ohshima, K. Nakajima, Y. Fujii, M. Mannami, and H.-J. Gossmann, *Nucl. Instrum. Methods B* 99, 472 (1995).
10. K. Kimura, K. Nakajima, and H. Imura, *Nucl. Instrum. Methods B* 140, 397 (1998).
11. J. W. Mayer and E. Rimini, "Ion Beam Handbook for Material Analysis," Academic Press, New York, 1977.
12. J. R. Tesmer, M. Nastasi, J. C. Barbour, C. J. Maggiore, and J. W. Mayer, "Handbook of Modern Ion Beam Materials Analysis," Materials Research Soc., Pittsburgh, 1995.
13. L. C. Feldman, J. W. Mayer, and S. T. Picraux, "Materials Analysis by Ion Channeling," Academic Press, New York, 1982.
14. N. Bohr, *Philos. Mag.* 25, 10 (1913).
15. N. Bohr, *Philos. Mag.* 30, 581 (1915).
16. H. A. Bethe, *Z. Phys.* 76, 293 (1932); H. A. Bethe, *Ann. Phys.* 5, 325 (1930); H. A. Bethe and W. Heitler, *Proc. R. Soc. London A* 146, 83 (1934).
17. F. Bloch, *Ann. Phys.* 16, 287 (1933); *Z. Phys.* 81, 363 (1933).
18. J. Lindhard and M. Scharff, *Mat. Fys. Medd. K. Dan. Vidensk. Selsk.* 27, (1953).
19. J. Lindhard, M. Scharff, and H. E. Schiott, *Mat. Fys. Medd. K. Dan. Vidensk. Selsk.* 34, (1963).
20. O. B. Firsov, *Zh. Eksp. Teor. Fiz.* 36, 1517 (1959); *Zh. Eksp. Teor. Fiz.* 32, 1464 (1957); *Zh. Eksp. Teor. Fiz.* 33, 696 (1957); *Zh. Eksp. Teor. Fiz.* 34, 447 (1958).
21. W. K. Chu and D. Powers, *Phys. Rev. A* 13, 2057 (1976).
22. W. K. Chu and D. Powers, *Phys. Lett.* 40A, 23 (1972); W. K. Chu and D. Powers, *Phys. Lett.* 38A, 267 (1972).
23. J. Lindhard, V. Nielsen, and M. Scharff, *Mat. Fys. Medd. K. Dan. Vidensk. Selsk.* 36, (1968).
24. K. B. Winterbon, P. Sigmund, and J. B. Sanders, *Mat. Fys. Medd. K. Dan. Vidensk. Selsk.* 37, (1970); K. B. Winterbon, *Radiat. Eff.* 13, 215 (1972).
25. J. F. Ziegler, J. P. Biersack, and U. Littmark, "The Stopping and Range of Ions in Solids," Pergamon, New York, 1985.
26. J. F. Ziegler, "Helium: Stopping Powers and Ranges in All Elemental Matters," Pergamon, New York, 1978.
27. J. P. Biersack and L. G. Haggmark, *Nucl. Instrum. Methods* 174, 257 (1980).
28. N. P. Barradas, C. Jeynes, O. A. Mironov, P. J. Phillips, and E. H. C. Parker, *Nucl. Instrum. Methods B* 139, 239 (1998).
29. N. Bohr, *Mat. Fys. Medd. K. Dan. Vidensk. Selsk.* 18, (1948).
30. J. Lindhard and M. Scharff, *Mat. Fys. Medd. K. Dan. Vidensk. Selsk.* 27, (1953).
31. W. H. Bragg and R. Kleeman, *Philos. Mag.* 10, S318 (1905); 10, 5318 (1905).
32. D. J. Thwaites, *Nucl. Instrum. Methods B* 12, 84 (1985).
33. D. J. Thwaites, *Nucl. Instrum. Methods B* 27, 293 (1985).
34. P. Bauer, *Nucl. Instrum. Methods B* 45, 673 (1985).
35. J. Herault, R. Bimbot, H. Ganvin, B. Kubico, R. Anne, G. Bastin, and F. Hubert, *Nucl. Instrum. Methods B* 61, 156 (1985).
36. J. F. Ziegler and J. M. Manoyan, *Nucl. Instrum. Methods B* 35, 215 (1988).
37. Feng et al. private communications to R. P. Cox, J. A. Leavit, and L. C. McIntyre, Jr., Appendix 7 in [12].
38. J. A. Davies and G. A. Sims, *Can. J. Chem.* 39, 601 (1961); J. A. Davies, J. Friensen, and J. D. McIntyre, *Can. J. Chem.* 38, 1526 (1960).
39. M. T. Robinson and O. S. Oen, *Appl. Phys. Lett.* 2, 30 (1963); M. T. Robinson and O. S. Oen, *Phys. Rev.* 132, 2385 (1963).
40. J. Lindhard, *Mat. Fys. Medd. K. Dan. Vidensk. Selsk.* 34, (1965).
41. J. H. Barrett, *Phys. Rev. B* 3, 1527 (1971).
42. J. U. Andersen, "Lecture Notes on Channeling," Institute of Physics, Aarhus University, Denmark, 1980.
43. D. V. Morgan, Ed., "Channeling: Theory, Observations and Applications," Wiley-Interscience, New York, 1973.
44. O. B. Firsov, *Sov. Phys. JETP* 6, 534 (1958).
45. G. Moliéré, *Z. Naturforschung A* 2, 133 (1947).
46. W. D. Wilson, L. G. Haggmark, and J. P. Biersack, *Phys. Rev.* 15, 2458 (1977).
47. J. P. Biersack and J. F. Zirgler, *Nucl. Instrum. Methods* 194, 93 (1982).
48. I. M. Terrens, "Interatomic Potentials," Academic Press, San Diego, 1972.
49. M. Nastasi, J. Mayer, and J. Hirvonen, "Ion-Solid Interactions: Fundamentals and Applications," Cambridge Univ. Press, Cambridge, U.K., 1996.
50. J. U. Andersen and L. C. Feldman, *Phys. Rev. B* 1, 2063 (1970).
51. J. U. Andersen, N. G. Chechenin, Yu. A. Timoshnikov, and Z. H. Zhang, *Radiat. Eff.* 83, 91 (1984).
52. J. U. Andersen, *Mat. Fys. Medd. K. Dan. Vidensk. Selsk.* 36, (1967).
53. S. T. Picraux, W. L. Brown, and W. M. Gibson, *Phys. Rev. B* 6, 1382 (1972).
54. J. U. Andersen, O. Andreasen, J. A. Davies, and E. Uggerhøj, *Radiat. Eff.* 7, 25 (1971).
55. B. A. Davidson, L. C. Feldman, J. Bevk, and J. P. Mannaerts, *Appl. Phys. Lett.* 50, 135 (1986).
56. R. W. Fearick, *Nucl. Instrum. Methods B* 164, 88 (2000).

57. M. Abramowitz and I. A. Segun, Eds., in "Handbook of Mathematics Functions," p. 998. Dover, New York, 1965.

58. D. S. Gemmel, *Rev. Mod. Phys.* 46, 129 (1974).

59. J. E. Westmoreland, J. W. Mayer, F. H. Eisen, and B. Welch, *Radiat. Eff.* 6, 161 (1970).

60. Y. Quèrè, *Phys. Status Solidi A* 30, 713 (1968);

61. Y. Quèrè, *Ann. Phys.* 5, 105 (1968).

62. S. T. Picraux, E. Remini, G. Foti, and S. U. Camisano, *Phys. Rev. B* 18, 2078 (1978).

63. K. Yasuda, M. Nastasi, K. E. Sickafus, C. J. Maggiore, and N. Yu, *Nucl. Instrum. Methods B* 136–138, 499 (1998).

64. Z. Atzmon, M. Eizenberg, Y. Shacham-Diamand, and J. W. Mayer, *J. Appl. Phys.* 75, 3936 (1994).

65. Z. Zhang, I. Rusakova, and W. K. Chu, *Nucl. Instrum. Methods B* 136–138, 404 (1998).

66. N. Yu, T. W. Simpson, P. C. McIntyre, M. Nastasi, and I. U. Mitchell, *Appl. Phys. Lett.* 67, 924 (1995).

67. F. W. Saris, W. K. Chu, C. A. Chang, R. Ludeke, and L. Esaki, *Appl. Phys. Lett.* 37, 931 (1980).

68. W. K. Chu, J. A. Ellison, S. T. Picraux, R. M. Biefeld, and G. C. Osboun, *Nucl. Instrum. Methods* 218, 81 (1983).

69. W. K. Chu, C. K. Pan, and C.-A. Chang, *Phys. Rev. B* 28, 4033 (1983).

70. S. T. Picraux, L. R. Dawson, G. C. Osboun, R. M. Biefeld, and W. K. Chu, *Appl. Phys. Lett.* 43, 1020 (1983).

71. D. W. Hetherington, P. F. Hinrichsen, R. A. Masut, D. Morris, and A. P. Roth, *Can. J. Phys.* 69, 378 (1991).

72. G. W. Fleming, R. Brenn, E. C. Larkins, S. Burkner, G. Bender, M. Baeumler, and J. D. Ralston, *J. Cryst. Growth* 143, 29 (1994).

73. J. A. Davies, B. J. Robinson, J. L. E. Stevens, D. A. Thompson, and J. Zhao, *Vacuum* 39, 73 (1989).

74. J. L. E. Stevens, B. J. Robinson, J. A. Davies, D. A. Thompson, and T. E. Jackman, *J. Appl. Phys.* 65, 1510 (1989).

75. E. Bøgh, *Can. J. Phys.* 46, 653 (1968).

76. K. L. Merkel, P. P. Pronko, D. S. Gemmell, R. C. Mikkelson, and J. R. Wrobel, *Phys. Rev. B* 8, 1002 (1973).

77. F. H. Eison, in "Channeling: Theory, Observation and Applications" (D. V. Morhan, Ed.), Chap. 14, p. 415. Wiley-Interscience, New York, 1973.

78. L. Meyer, *Phys. Status Solidi B* 44, 253 (1971).

79. P. Sigmund and W. K. Winterbon, *Nucl. Instrum. Methods* 119, 541 (1974); *Nucl. Instrum. Methods* 121, 491 (1975).

80. E. Lugujjo and J. W. Mayer, *Phys. Rev. B* 7, 1782 (1973).

81. M. D. Thompson and G. J. Galvin, private communication to Feldman et al., Chap. 6.

82. S. T. Picraux, D. M. Follstaedt, P. Baeri, S. U. Camisano, G. Foti, and E. Remini, *Radiat. Eff.* 49, 75 (1980).

83. S. T. Picraux and D. M. Follstaedt (1982), in "Material Analysis by Ion Channeling" by L. C. Feldman, J. W. Mayer, and S. T. Picraux, Academic Press, New York, 1982, p. 103.

84. H. Kudo, *J. Phys. Soc. Jpn.* 40, 1645 (1945).

85. H. Kudo, *Phys. Rev. B* 18, 5995(1978).

86. H. Kudo, *Nucl. Instrum. Methods* 170, 129 (1970).

87. L. Z. Merz, L. C. Feldman, D. W. Mingay, and W. M. Augustyniak, in "Ion Implantation in Semiconductors" (L. Ruge and J. Graul, Eds.), p. 182. Springer-Verlag, Berlin, 1971.

88. J. W. Petersen, J. U. Andersen, S. Damgaard, F. G. Lu, I. Stensgaard, J. T. Tang, G. Weyer, and Z. H. Zhang, *Hyperfine Interact.* 10, 989 (1981).

89. E. Alves, M. F. da Silva, K. R. Evans, C. R. Jones, A. A. Melo, and J. C. Soares, *Nucl. Instrum. Methods B*, 80–81, 180 (1993).

90. L. M. Howe, M. L. Swanson, and J. A. Davies, "Methods of Experimental Physics," Vol. 21, p. 275. Academic Press, New York.

91. R. G. Elliman, J. S. Williams, W. L. Brown, A. Leiberich, D. M. Maher, and R. V. Knoell, *Nucl. Instrum. Methods B* 19–20, 423 (1987).

92. A. Kinomura, A. Chayahara, N. Tsubouchi, Y. Horino, and J. S. Williams, *Nucl. Instrum. Methods B* 148, 370 (1999).

93. S. T. Johnson, J. S. Williams, E. Nygreen, and R. G. Elliman, *J. Appl. Phys.* 64, 6567 (1988).

94. M. C. Ridgway, S. T. Johnson, and R. G. Elliman, *Nucl. Instrum. Methods B* 59–60, 353 (1991).

95. N. Kobayashi, *Thin Solid Films* 270, 307 (1995).

96. R. G. Elliman, M. C. Ridgway, J. S. Williams, and J. C. Bean, *Appl. Phys. Lett.* 55, 843 (1989).

97. A. J. Yu, J. W. Mayer, D. J. Eaglesham, and J. M. Poate, *Appl. Phys. Lett.* 54, 2342 (1989).

98. M. C. Ridgway, R. G. Elliman, and J. S. Williams, *Appl. Phys. Lett.* 56, 2117 (1990).

99. N. Kobayashi, D. H. Zhu, M. Hasegawa, H. Katsumata, Y. Tanaka, N. Hayashi, Y. Makita, H. Shibata, and S. Uekusa, *Nucl. Instrum. Methods B* 127–128, 350 (1997).

100. N. Yu and M. Nastasi, *Nucl. Instrum. Methods B* 106, 579 (1995).

101. S. C. Jain, "Germanium, Silicon Strained Layers and Heterostructures," pp. 69–93. Academic Press, New York, 1994.

102. C. C. Van d Walle and R. M. Martin, *Phys. Rev. B* 8, 5621 (1986).

103. T. Akane, M. Sano, H. Okumura, Y. Tubo, T. Ishikawa, and S. Matsumoto, *J. Cryst. Growth* 203, 80 (1999).

104. B. Hollander, R. Butz, and S. Mantl, *Phys. Rev. B* 46, 6975 (1992).

105. B. Hollander, S. Mantl, B. Stritzker, and R. Butz, *Superlattices Microstruct.* 9, 415 (1991).

106. B. Hollander, S. Mantl, B. Stritzker, and R. Butz, *Nucl. Instrum. Methods B* 59–60, 994 (1991).

107. K. Lenkeit, A. Pirrwitz, A. I. Nikiforov, B. Z. Kanter, S. I. Stenin, and V. P. Popov, *Phys. Status Solidi A* 121, 523 (1990).

108. F. Ruders, L. E. Rehn, P. M. Baldo, E. E. Fullterton, and S. D. Bader, *Nucl. Instrum. Methods B* 121, 30 (1997).

109. K. Lenkeit, R. Flagmeyer, and R. Grotzschel, *Nucl. Instrum. Methods B* 66, 453 (1992).

110. K. Lenkeit and A. Pirwitz, *Nucl. Instrum. Methods B* 68, 253 (1992).

111. K. Lenkeit, *Superlattices Microstruct.* 9, 203 (1991).

112. G. W. Flemig, R. Brenn, E. C. Larkins, S. Bürkner, G. Bender, M. Baeumler, and J. D. Ralston, *J. Cryst. Growth* 143, 29 (1994).

113. S. Sugden, C. J. Sofield, T. C. Q. Noakes, R. A. A. Kubiak, and C. F. McConville, *Appl. Phys. Lett.* 65, 2849 (1995).

114. C. F. McConville, T. C. Q. Noakes, S. Sugden, P. K. Hucknell, and C. J. Sofield, *Nucl. Instrum. Methods B*, 118, 573 (1996).

115. D. Love, D. Endisch, T. W. Simpson, T. D. Lowes, I. V. Mitchel, and J.-M. Baribeau, *Mater. Res. Soc. Symp. Proc.* 379, 461 (1995).

116. T. Laursen, D. Chandrasekhar, D. J. Smith, J. W. Mayer, E. T. Croke, and A. T. Hunter, *Thin Solid Films* 308–309, 358 (1997).

117. O. A. Mironov, P. J. Phillips, E. H. C. Parker, M. G. Dowsett, N. P. Barradas, C. Jeynes, V. Mironov, V. P. Gnezdilov, V. Ushakov, and V. V. Eremenko, *Thin Solid Films* 306, 307 (1997).

118. C. Lin, P. L. F Hemment, C. W. M. Chen, J. Li, W. Zhu, R. Ni, G. Zhou, and S. C. Zou, *Phys. Status Solidi A* 132, 419 (1992).

119. G. Q. Zhao, Y. H. Ren, Z. Y. Zhou, X. J. Zhang, G. L. Zhou, and C. Sheng, *Solid State Commun.* 67, 661 (1988).

120. J. G. Beery, B. K. Laurich, C. J. Maggiore, D. L. Smith, K. Elcess, C. G. Fonstad, and C. Mailhiot, *Appl. Phys. Lett.* 54, 233 (1989).

121. A. Pathak, S. V. S. Nageshwara Rao, and A. M. Siddiqui, *Nucl. Instrum. Methods B* 161–163, 487 (2000).

122. T. Xu, P. Zhu, J. Zhou, D. Li, B. Gong, Y. Wan, and S. Mu, *Nucl. Instrum. Methods B* 90 392 (1994).

123. E. A. Dobisz, M. Fatemi, H. B. Dietrich, A. W. McCormick, and J. P. Harbison, *Appl. Phys. Lett.* 59, 1338 (1991).

124. R. Flagmeyer, *Superlattices Microstruct.* 9, 181 (1991).

125. L. V. Melo, I. Trindade, M. From, P. P. Freitas, N. Teixeira, M. F. da Silva, and J. C. Soares, *J. Appl. Phys.* 70, 7370 (1991).

126. T. Taguchi, Y. Kawakami, and Y. Yamada, *Physica B* 191, 23 (1993).

127. K. Roll and W. Reill, *Thin Solid Films* 89, 221 (1982).

128. M. Murakami, *CRC Crit. Rev. Solid State Mater. Sci.* 11, 317 (1983).

129. M. N. Baibich, J. M. Broto, A. Fert, F. Nguyen Van Dau, F. Petroff, P. Etienne, B. Greuzet, A. Friederich, and J. Chazelas, *Phys. Rev. Lett.* 61, 2471 (1988).

130. E. E. Fullerton, M. J. Conover, J. E. Mattson, C. H. Sowers, and S. D. Bader, *Phys. Rev. B* 48, 15755 (1993).

131. S. S. P. Parkin, N. More, and K. P. Roche, *Phys. Rev. Lett.* **64**, 2304 (1990).

132. R. P. Sharma, T. Venkatesan, Z. H. Zhang, J. R. Liu, R. Chu, and W. K. Chu, *Phys. Rev. Lett. B* **77**, 4624 (1996).

133. W. K. Chu, "Laser and Electron Beam Processing of Electronic Materials" (C. L. Anderson, G. C. Heller, and G. A. Rozgonyi, Eds.), p. 361. Electrochemical Soc., Pennington, NJ, 1980.

134. G. H. Dohler, *IEEE J. Quantum Electron.* **22**, 1682 (1986).

135. L. H. Yang, R. F. Gallup, and C. Y. Fong, *Phys. Rev. B* **39**, 3795 (1989).

136. J. P. Thomas, M. Fallavier, and A. Ziani, *Nucl. Instrum. Methods B* **15**, 443 (1986).

137. T. M. Stanescu, J. D. Meyer, H. Baumann, and K. Bethge, *Nucl. Instrum. Methods B* **50**, 167 (1990).

138. R. A. Weller, K. McDonald, D. Pedersen, and J. A. Keenan, *Nucl. Instrum. Methods B* **118**, 556 (1996).

139. S. T. Picraux, G. W. Arnold, D. R. Myers, L. R. Dawson, R. M. Biefeld, I. J. Fritz, and T. E. Zipperian, *Nucl. Instrum. Methods B* **7–8**, 453 (1985).

140. J. Li, Q. Z. Hong, G. Vizkelethy, J. W. Mayer, C. Cozzolino, W. Xia, B. Zhu, S. N. Hsu, S. S. Lau, B. Hollander, R. Butz, and S. Mantl, *Nucl. Instrum. Methods B* **59–60**, 989 (1991).

141. D. R. Myers, S. T. Picraux, B. L. Doyle, G. W. Arnold, L. R. Dawson, and R. M. Biefeld, *J. Appl. Phys.* **60**, 3631 (1986).

142. D. R. Myers, G. W. Arnold, T. E. Zipperian, L. R. Dawson, R. M. Biefeld, I. J. Fritz, and C. E. Barnes, *J. Appl. Phys.* **60**, 1131 (1986).

143. L. Esaki, and R. Tsu, *IBM J. Res. Develop.* **14**, 61(1970).

144. W. K. Chu, *Phys. Rev. A* **13**, 2057 (1976).

145. W. K. Chu, F. W. Saris, C.-A. Chang, R. Ludeke, and L. Esaki, *Phys. Rev. B* **26**, 1999 (1982).

146. W. K. Chu, J. A. Ellison, S. T. Picraux, R. M. Biefeld, and G. C. Osbourn, *Phys. Rev. Lett.* **52**, 125 (1984).

147. W. K. Chu, W. R. Allen, J. A. Ellison, and S.T. Picraux, *Nucl. Instrum. Methods B* **13**, 39 (1986).

148. W. K. Chu, W. R. Allen, S. T. Picraux, and J. A. Ellison, *Phys. Rev. B* **42**, 5932 (1990).

149. C. K. Pan, D. C. Zheng, T. G. Finstad, W. K. Chu, V. S. Speriosu, and M.-A. Nicolet, *Phys. Rev. B* **31**, 1270 (1983).

150. S. T. Picraux, W. K. Chu, W. R. Allen, and J. A. Ellison, *Nucl. Instrum. Methods B* **15**, 57 (1986).

151. S. T. Picraux, W. K. Chu, W. R. Allen, and J. A. Ellison, *Nucl. Instrum. Methods B* **15**, 306 (1986).

152. S. T. Picraux, W. R. Allen, R. M. Biefeld, J. A. Ellison, and W. K. Chu, *Phys. Rev. Lett.* **54**, 2355 (1985).

153. S. T. Picraux, R. M. Biefeld, W. R. Allen, W. K. Chu, and J. A. Ellison, *Phys. Rev. B* **38**, 11086 (1988).

154. J. A. Ellison, S. T. Picraux, W. R. Allen, and W. K. Chu, *Phys. Rev. B* **37**, 7295 (1988).

155. J. H. Barret, *Appl. Phys. Lett.* **49**, 482 (1982).

156. J. H. Barret, *Phys. Rev. B* **28**, 2328 (1983).

157. D. Y. Han, W. R. Allen, W. K. Chu, J. H. Barret, and S. T. Picraux, *Phys. Rev. B* **35**, 4159 (1987).

158. W. K. Chu, in "Nuclear Physics Applications on Materials Science" (E. Recknagel and J. C. Soares, Eds.), pp. 117–132. Kluwer Academic, Dordrecht/Norwell, MA, 1988.

159. F. Abel, G. Amsel, M. Bruneaux, and C. Cohen, *Phys. Lett.* **42A**, 165 (1972).

160. S. Hashimoto, Y.-Q. Peng, W. M. Gibson, L. J. Schowalter, and B. D. Hunt, *Nucl. Instrum. Methods B* **13**, 45 (1986).

161. C. Wu, S. Yin, J. Zhang, G. Xiao, J. Liu, and P. Zhu, *J. Appl. Phys.* **68**, 2100 (1990).

162. K. Lenkeit and A. Pirrwitz, *Nucl. Instrum. Methods B* **67**, 217 (1992).

163. A. Kozancecki, J. Kaczanowski, B.J. Sealy, and W.P. Gillin, *Nucl. Instrum. Methods B* **118**, 640 (1996).

Chapter 6

IN SITU AND REAL-TIME SPECTROSCOPIC ELLIPSOMETRY STUDIES: CARBON BASED AND METALLIC TiN$_x$ THIN FILMS GROWTH

S. Logothetidis

Aristotle University of Thessaloniki, Physics Department, Solid State Physics Section, GR-54006, Thessaloniki, Greece

Contents

1. INTRODUCTION

The increasing demands of performance specifications and the related sophistication of manufacturing processes have provided a strong incentive to develop highly sensitive diagnostics that can monitor—in situ and, more importantly, in real time—thin film properties and processing. However, for control process applications, any in situ and real-time probe must satisfy severe requirements, such as to be noninvasive, nonperturbing, contactless, and fast enough. Diffraction techniques like reflec-

Handbook of Thin Film Materials, edited by H.S. Nalwa
Volume 2: Characterization and Spectroscopy of Thin Films

ISBN 0-12-512910-6/$35.00

tion high-energy electron diffraction have been used widely to monitor growth of materials under ultrahigh-vacuum conditions by molecular beam epitaxy. Traditional surface probes based on either electrons or ions, however, cannot be extended to the high pressures and reactive environments associated with the chemical vapor deposition and the physical vapor deposition techniques, which are extensively used in thin film and material processing.

Optical techniques based on specular reflection can satisfy the previously mentioned requirements. Among them, ellipsometry measures the change in polarization of a polarized beam of light after nonnormal reflection from a material to be studied. The polarization state of the polarized light can be characterized by two parameters (e.g., the relative amplitude and the relative phase) of the electric field. This provides ellipsometry an advantage over other optical techniques, since from it one can directly calculate two quantities, e.g., the complex dielectric function $[\tilde{\varepsilon}(\omega) = \varepsilon_1 + i\varepsilon_2]$ of a bulk absorbing material at a given photon energy ω (which is related to the electronic, vibrational, structural, and morphological characteristics of the material) or the film thickness d and the real part of dielectric function ε_1 ($= n^2$, the refractive index) in transparent films. In contrast, reflectance, for example, provides only one parameter: the ratio of the reflected to the incident intensity of the electric field. As a consequence, spectroscopic ellipsometry (SE) has extensively been used for thin film growth monitoring [1–3]. SE can provide a lot of information about thin films, such as the optical gap, electronic structure, physical structure (crystalline and amorphous phases), composition, and stoichiometry, as well as their density and thickness. Moreover, SE analysis can be extended to multilayer systems, can be used in situ to identify the various stages of the film growth (nucleation, coalescence, and surface roughness evolution [2, 3]), and is very sensitive to the surface and bulk properties of the materials [4–6]. Since over all types of materials the absorption coefficient varies from 1 to 2×10^6 cm^{-1}, the penetration depth of light from the infrared (IR) to the deep ultraviolet (UV) range varies from $\sim 10^8$ to 0.5×10^2 Å, a fact that enables SE to be used as a probe for screening not only bulk, but also surface material properties. In metals with a penetration depth in the IR-to-UV range below about 200 Å, SE can be used as a probe to screen the bulk and surface properties with high sensitivity. In semiconductors, SE can probe different depths of the material from the near infrared (NIR) to the visible (Vis) energy region. In particular, in the UV range the penetration depth is reduced to the order of 100 Å or below, and then ellipsometry can provide their surface electronic properties and characterize their surfaces and interfaces.

Whatever the configuration or approach used, SE has generally been applied in the last fifteen years in the near-UV–Vis range. However, the optical response in the UV–Vis range is sensitive neither to the vibrational properties of materials nor to the electronic properties of the wide-bandgap semiconductors (e.g. GaN [4, 5], SiC [6]) and the insulating materials (e.g. SiN$_x$ [7], SiO$_2$). The former can be a strong limitation when dealing with complex compounds, such as thin films of hydrogenated

amorphous carbon (a-C:H) [2, 8] and carbon-related materials such as carbon nitride (CN$_x$) [9] and boron nitride (BN). A detailed understanding, for example, of hydrogen and nitrogen incorporation from the gas phase into the film and the dependence of the vibrational properties upon the local environment and the CN$_x$ and BN film bonding configuration and composition requires the extension of the spectroscopic measurements into the IR, and this has been done the last few years [2, 8–15]. Finally, the availability of the synchrotron radiation sources has given SE the ability to overcome the limitation of conventional light sources to energies below 6.5 eV, and (together with the development of the new polarized devices) to be extended in the deep UV energy region, where the strong absorption of wide-bandgap and insulating materials takes place [4–7, 16–18].

Moreover, as a consequence of recent advances in optical instrumentation, fast real-time SE monitoring (in the UV–Vis range) is becoming compatible with most of the kinetics involved in thin film processing techniques [2, 3]. Thus, this fast and sensitive optical technique makes possible feedback control of the thickness and the refractive index of transparent films [19–21] or the thickness and the composition in absorbing films [22, 23].

Various applications of SE are presented here, with particular emphasis on in situ and real-time studies of thin film processing. More precisely, this review is organized as follows. The fundamentals of ellipsometry in bulk and thin films and the application of SE to the most-studied material, silicon (which is also used widely as a substrate for thin films) are briefly described in Section 2. Section 3 provides an overview of the effective-medium theory and microscopic surface roughness in an attempt to understand and quantify the information about the properties and growth stages of thin films during and after preparation. Among the various ellipsometric techniques available, rotating analyzer ellipsometry (RAE) at low frequency (≈ 20 Hz) and phase-modulated spectroscopic ellipsometry (PMSE) at high frequency (50 kHz) are the most important. These techniques, together with the extension of ellipsometry to IR and deep UV energy are presented in Section 4. Then selected examples of applications of both techniques to thin film processing are described in two sections. In particular, Section 5 is devoted first to a detailed in situ spectroscopic study of the optical and electronic properties of a-C:H films in the energy region 1.5 to 10 eV, combining Vis–UV and synchrotron radiation ellipsometry. After that are described real-time ellipsometry and spectroscopic ellipsometry studies in the UV–Vis energy range on hydrogen-free amorphous carbon (a-C) films during deposition by magnetron sputtering techniques. The results obtained by these studies are discussed and reviewed as a case study in optimizing the deposition parameters and in producing films with desired properties for specific applications. Moreover, a very fast PMSE technique at certain wavelengths is introduced, which has been applied to investigate the initial stages of growth of carbon-based (a-C and CN$_x$) materials and their differences in growth when they are produced by various magnetron sputtering techniques. Finally, recent applications of IR ellipsometry are reviewed, where the high sensitivity of this technique to film properties such as vibrational modes, bonding configuration, and composition of carbon nitride films is underlined.

Section 6 describes the in situ and real-time monitoring of metallic TiN thin films by PMSE in the UV–Vis range, from which information about the electronic structure, stoichiometry, oxidation, and structural and morphological characteristics of the films are obtained. The procedure to mesure the TiN_x film's stoichiometry and its thickness during deposition is described analytically; it has been applied in real-time monitoring of TiN_x film deposition by a very fast deposition technique—unbalanced magnetron sputtering—in an attempt to test the acquisition speed and software analysis limits of the current SE technology. It is shown that real-time ellipsometry, with fast recording capability, is a promising technique to provide feedback control of the growth of absorbing thin films.

2. THEORETICAL BACKGROUND OF ELLIPSOMETRY FOR STUDY OF THE PROPERTIES OF MATERIALS

2.1. Ellipsometry in Bulk Materials and Macroscopic Dielectric Function

The interaction between a material and a polarized light beam, incident at an angle θ and reflected from the material's surface, modifies the polarization of the beam. Ellipsometry is conducted in order to obtain information about a system that modifies the state of polarization of the specular reflection. This

is achieved by measuring the initial (incident) and final (reflected) states of polarization of the beam.

An ellipsometric arrangement is shown in Figure 1. A well-collimated monochromatic beam from a suitable light source is passed through a polarizer to produce light of known controlled linear polarization. This is reflected by the surface of the sample under investigation, and is passed through the second polarizer (the analyzer) of the ellipsometer, to be analyzed with respect to the new polarization state established by the reflection, and finally the intensity of the linearly polarized light is detected and transformed to raw data through the electronics and computing devices. This last part of the ellipsometric arrangement determines the accuracy and the speed of acquisition of the information obtained about the investigated sample.

In order to understand how ellipsometry works, the differences between it and other optical techniques, and the differences among various ellipsometric techniques, we are going to outline some results of the electromagnetic theory. Consider first the propagation of an electromagnetic plane wave through a nonmagnetic medium, which can be described by the electric field vector $\vec{E}$; in the simplest case this is a plane wave given by the following expression:

$$\vec{E} = \vec{E}_i e^{i(kz - \omega t)} \tag{1}$$

For oblique incidence, plane waves are typically referred to a local coordinate system (x, y, z), where z is the direction of propagation of light with wave number k, and x and y define the plane where the transverse electromagnetic wave oscillates. That is, the latter are the directions (see Fig. 1) parallel (p) and perpendicular (s) to the plane of incidence, respectively (these

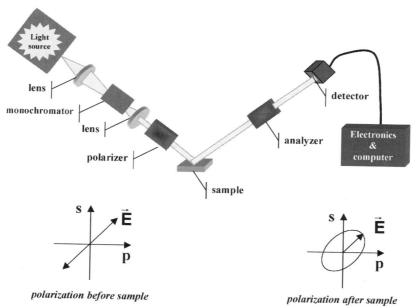

polarization before sample

polarization after sample

Fig. 1. Schematic diagram of an ellipsometer with the incident monochromatic beam linearly polarized. Here p and s denote polarization parallel and perpendicular to the plane of incidence (defined by the normal to the sample and the incident beam), respectively. The detector, electronics, and computing devices collect the information about the sample and transform it to raw data.

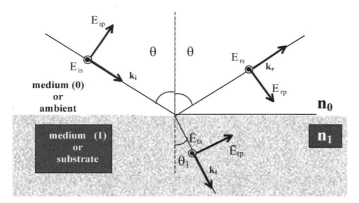

Fig. 2. Oblique reflection and transmission of a plane electromagnetic wave at the sharp interface between two media 0 and 1 with refractive indexes n_0 and $\tilde{n}_1$, respectively. The electric field components E_p and E_s, parallel (p) and perpendicular (s) to the plane of incidence, and the wave vector for the incident (i), reflected (r), and transmitted (t) waves are shown. θ and θ_1 are the angles of incidence and refraction.

two directions are the two optical eigenaxes of the material under study). The complex electric field amplitudes E_p and E_s represent the projections of the plane wave $\vec{E}$ along x and y. Therefore, the quantity $\vec{E}_i$ in Eq. (1) does not only carry information about the amplitude of the plane wave $\vec{E}$ when it is propagated in vacuum, but also carries information about its polarization as well. Namely,

$$\vec{E}_i = E_{ix}\hat{x} + E_{iy}\hat{y} \qquad (2)$$

In addition, the amplitude k of the wave vector during the propagation of the wave in matter is in general a complex number given by the dispersion expression $k = \tilde{n}\omega/c$. Here $\tilde{n}(\omega)$ is the *refractive index*, which is in general a complex quantity, and it is related to *dispersion* and *absorption* of the radiation by the medium:

$$\tilde{n}(\omega) = n + i\kappa \qquad (3)$$

The complex dielectric function $\tilde{\varepsilon}(\omega)$ ($= \varepsilon_1 + i\varepsilon_2$) is the quantity directly related to the material properties, and is connected to the refractive index through the following equation:

$$\tilde{\varepsilon}(\omega) = \varepsilon_1 + i\varepsilon_2 \equiv \tilde{n}^2(\omega) = (n + i\kappa)^2$$
$$\Rightarrow \varepsilon_1 = n^2 - \kappa^2, \quad \varepsilon_2 = 2n\kappa \qquad (4)$$

Equation (1) for an electromagnetic wave transmitted in a dispersive and absorbing material can be rewritten as follows:

$$\vec{E}_i = \left(E_{ix}\hat{x} + E_{iy}\hat{y}\right)e^{i\omega nz/c}e^{-\omega\kappa z/c}e^{-i\omega t} \qquad (5)$$

with the absorption coefficient $\alpha = 2\omega\kappa/c$ ($= 4\pi\kappa/\lambda$) representing the depth of the wave in the material, and λ being the wavelength in vacuum. When the electromagnetic wave is reflected by the smooth surface of the material (see Fig. 2), the polarization of the outgoing wave can be represented as

$$\vec{E}_r = \left(\tilde{r}_p E_{0x}\hat{x} + \tilde{r}_s E_{0y}\hat{y}\right) \qquad (6)$$

In this approximation, the interaction of the electromagnetic wave with the material is described by the two complex Fres-

nel reflection coefficients $\tilde{r}_p$ and $\tilde{r}_s$. These reflection coefficients describe the influence of the material on the p and s electric field components referred to the plane of incidence, and characterize the interface between two media—e.g., the ambient (medium 0), and the material studied (medium 1)—and are given by the expressions

$$\tilde{r}_p = \frac{\tilde{E}_{r,p}}{\tilde{E}_{i,p}} = \frac{|\tilde{E}_{r,p}| \cdot e^{i\theta_{r,p}}}{|\tilde{E}_{i,p}| \cdot e^{i\theta_{i,p}}} = \left|\frac{\tilde{E}_{r,p}}{\tilde{E}_{i,p}}\right| \cdot e^{i(\theta_{r,p}-\theta_{i,p})}$$
$$= |\tilde{r}_p|e^{i\delta_p} \qquad (7)$$

$$\tilde{r}_s = \frac{\tilde{E}_{r,s}}{\tilde{E}_{i,s}} = \frac{|\tilde{E}_{r,s}| \cdot e^{i\theta_{r,s}}}{|\tilde{E}_{i,s}| \cdot e^{i\theta_{i,s}}} = \left|\frac{\tilde{E}_{r,s}}{\tilde{E}_{i,s}}\right| \cdot e^{i(\theta_{r,s}-\theta_{i,s})}$$
$$= |\tilde{r}_s|e^{i\delta_s} \qquad (8)$$

The Fresnel reflection coefficients on the interface between two media i and j, e.g., medium 0 and medium 1 in Fig. 2, with refractive indices $\tilde{n}_i$ and $\tilde{n}_j$, respectively, are given by the following expressions:

$$\tilde{r}_{ij,p} = \frac{\tilde{n}_j \cos\theta_i - \tilde{n}_i \cos\theta_j}{\tilde{n}_j \cos\theta_i + \tilde{n}_i \cos\theta_j} \qquad (9)$$

$$\tilde{r}_{ij,s} = \frac{\tilde{n}_i \cos\theta_i - \tilde{n}_j \cos\theta_j}{\tilde{n}_i \cos\theta_i + \tilde{n}_j \cos\theta_j} \qquad (10)$$

where the incident (θ_i) and refracted (θ_j) angles are related through *Snell's law* $\tilde{n}_i \sin\theta_i = \tilde{n}_j \sin\theta_j$.

When the light beam does not penetrate medium 1, due either to its high absorption coefficient or to its infinite thickness, as shown in Fig. 2, we are dealing with a two-phase (ambient–substrate) system, or a bulk material surrounding by medium 0. In this case the ratio of the p to the s Fresnel reflection coefficients called the *complex reflection ratio*, is the quantity measured directly by ellipsometry, and it is given by the expression

$$\tilde{\rho} = \frac{\tilde{r}_p}{\tilde{r}_s} = \left|\frac{\tilde{r}_p}{\tilde{r}_s}\right|e^{i(\delta_p-\delta_s)} = \tan\Psi e^{i\Delta} \qquad (11)$$

that characterizes any bulk material. In this expression Ψ and Δ are the ellipsometric angles, and for a bulk material take values $0° < \Psi < 45°$ and $0° < \Delta < 180°$. From an ellipsometric measurement the complex reflection ratio $\tilde{\rho}$ is estimated through the calculation of the amplitude ratio $\tan\Psi$ and the phase difference Δ. From these two quantities one can extract all the other optical constants of the material. For example, the complex dielectric function of a bulk material with smooth surfaces is directly calculated by the following expression [24]:

$$\tilde{\varepsilon}(\omega) = \varepsilon_1 + i\varepsilon_2 = \tilde{\eta}^2(\omega)$$
$$= \tilde{\varepsilon}_0 \sin^2\theta \left\{1 + \left[\frac{1-\tilde{\rho}(\omega)}{1+\tilde{\rho}(\omega)}\right]^2 \tan^2\theta\right\} \qquad (12)$$

where θ is the angle of incidence of the beam, and $\tilde{\varepsilon}_0$ the dielectric constant of the ambient medium (for the case of air $\tilde{\varepsilon}_0 = \tilde{n}_0 = 1$).

As an example, Fig. 3 shows the real (ε_1) and the imaginary (ε_2) parts of the dielectric function $\tilde{\varepsilon}(\omega)$ of the crystalline

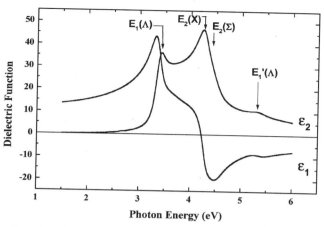

Fig. 3. The real (ε_1) and imaginary (ε_2) parts of the dielectric function of c-Si, measured by ellipsometry, versus the photon energy. The arrows denote the energies where the interband transitions of crystalline Si take place.

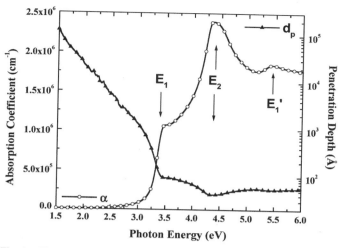

Fig. 4. The absorption coefficient α (open circles) and the penetration depth d_p (solid triangles) of light in c-Si versus photon energy, in the energy region 1.5–6.0 eV. Arrows denote the energies of the interband transitions where strong electronic absorption takes place.

Si (c-Si), one of the most well-known and studied semiconducting materials, in the energy region 1.5–6 eV, measured by spectroscopic ellipsometry. The dielectric function of c-Si shows considerable structure in the form of peaks and shoulders. These structures arise from electronic transitions from the filled valence bands to the empty conduction band. Such structures are due to the electronic (interband) transitions between valence and conduction bands in crystalline material (with long-range order); they are smoothed out by disorder and disappear in amorphous materials. The latter exhibit only short-range order.

What we can learn from the $\tilde{\varepsilon}(\omega)$ spectrum of c-Si is that the material starts to absorb strongly above 3 eV, where the direct interband electronic transitions take place and contribute to the dielectric function. These interband transitions, which characterize the crystalline material, take place in c-Si mainly in the near UV range. In particular, in c-Si the electronic transitions occur in the energy regions around 3.37, 4.27, 4.47, and 5.3 eV and correspond to the so-called critical points in the Brillouin zone: $E_1(\Lambda)$, $E_2(X)$, $E_2(\Sigma)$, and $E_1'(\Lambda)$, respectively [25]. These interband electronic transitions are referred to in the classical description as resonances in the wave absorption by matter; they are replaced in amorphous material (due to the lack of symmetry and long-range order) with an average electronic absorption that is characterized by a strong absorption at an energy that is characteristic of each material (see for example the description of amorphous silicon in what follows).

In addition, the real part $\varepsilon_1(\omega)$ and the imaginary part $\varepsilon_2(\omega)$ of the dielectric function are strongly related through the well-known Kramers–Kronig relation (based on the principle of the causality [26]):

$$\varepsilon_1(\omega) = 1 + \frac{2}{\pi} P \int_0^\infty \frac{\omega' \varepsilon_2(\omega')}{\omega'^2 - \omega^2} d\omega' \qquad (13a)$$

$$\varepsilon_2(\omega) = -\frac{2\omega}{\pi} P \int_0^\infty \frac{\varepsilon_1(\omega') - 1}{\omega'^2 - \omega^2} d\omega' \qquad (13b)$$

where P means the principal value of the integral at the material's electronic resonance ($\omega' = \omega$) and

$$\varepsilon_1(\omega = 0) = 1 + \frac{2}{\pi} P \int_0^\infty \frac{\varepsilon_2(\omega')}{\omega'} d\omega' \qquad (13c)$$

is the static dielectric function (the ratio of the material's dielectric constant to that of vacuum, $\varepsilon_0 = 1$), which describes all losses over the whole electromagnetic spectrum in the material due to electronic absorption. For example, the value measured for c-Si is $\varepsilon_1(\omega = 0) = 11.3$, and the value calculated with Eq. (13c) (assuming $\omega' \approx 4$ eV, $\varepsilon_2(\omega) \approx 40$, and $\Delta\omega' \approx 1.6$ eV) is $\varepsilon_1(\omega = 0) = 11.5$. That is, in c-Si this quantity is proportional to the electronic absorption in the near UV range, and no other strong electronic absorption is expected to contribute above this energy range.

If the photon energy is below the fundamental bandgap in a semiconducting material, the sample absorption coefficient is actually zero or very small. When the photon energy is increased from below to above the bandgap, typically the absorption coefficient α increases rapidly to values as large as or larger than 10^4 cm^{-1}. As a result the sample becomes opaque to photon energies above the bandgap unless its thickness is very small. Therefore, it is necessary to thin down the sample to about $d_p = \alpha^{-1}$ in order to detect much transmitted light. The quantity d_p is known as the optical penetration depth, and for the case of c-Si both the absorption coefficient and d_p in the energy region 1.5–6 eV are shown in Fig. 4. For semiconductors α is typically of the order 10^4–10^6 cm^{-1} above the absorption edge. That is, light with photon energy close to the first absorption edge will probe only a thin layer, about 1 µm or less thick, at the top of the material. As a result, the measured dielectric function will be very sensitive to the presence of surface contaminants such as oxides or even ambient pollutants at higher photon energies. Then the measured dielectric function is that of a *composite*, consisting of a surface layer and the bulk mate-

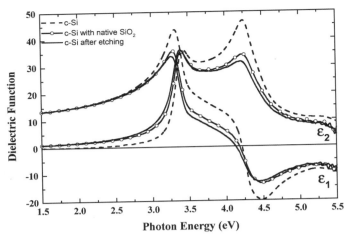

Fig. 5. The real (ε_1) and imaginary (ε_2) parts of the dielectric function of: the c-Si (dashed lines) from an atomically clean surface, a c-Si wafer (open circles) covered by an native silicon oxide overlayer of 20 Å, and the same Si wafer (solid curves) after dry etching to remove the oxide layer by ion bombardment.

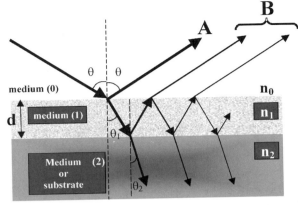

Fig. 6. The propagation of a light beam at oblique incidence in an optical system consisting of three media with reflective indexes n_0, $\tilde{n}_1$, and $\tilde{n}_2$ and parallel interfaces. Medium 1 is a film with thickness d (of the order or less of the penetration depth of the light beam) on a substrate (medium 2), θ is the angle of incidence, and θ_1 and θ_2 are the angles of refraction in the film and substrate, respectively. The intensity of the electric field of beam A is given by $\tilde{r}_{01i}$, and that of beam B by the second term of Eq. (14a).

rial, and is referred to as the *pseudo-dielectric function* $\langle \varepsilon(\omega) \rangle$. As an example of the sensitivity of the dielectric function to material surface quality, we compare in Figure 5 the dielectric function of an abrupt and atomically clean surface of c-Si with that of an oxidized c-Si (e.g. a crystalline Si wafer) surface, $\approx$20-Å native SiO_2 surface, and one obtained after bombardment of the Si wafer with low-energy ions (etching) to remove the native oxide and other contaminants. Note that ion bombardment of the Si wafer also induces an amorphization of the top c-Si to a depth of a few monolayers. The largest differences in the dielectric function due to these surface effects (native oxide and amorphization of c-Si) occur around 4.2-eV photon energy. This is because in this energy region the absorption coefficient of c-Si (see e.g. Fig. 4) is very high, $\approx$2.2 × 10^6 cm^{-1}, with a corresponding penetration depth of about 50 Å.

2.2. Ellipsometry in Thin Film Systems

In bulk materials—medium 1 (or the substrate) in Fig. 2—the transmitted beam is totally absorbed and is not reflected by the back side of the sample. This means that the penetration depth of light is smaller than the substrate thickness. When the penetration depth of light is larger than the medium thickness, which corresponds to the case of a thin film grown on a bulk substrate (see e.g. Fig. 6), then the light is back-reflected at the film–substrate interface, is transmitted again through the film (medium 1), and finally goes out to medium 0. In Figure 6, in addition to the first reflected beam A (which corresponds to $\tilde{r}_{01p}$ and $\tilde{r}_{01s}$, the reflected intensities of the electric fields of the p and s components), there are also shown the reflected beams B, which are originated by the multiple reflections of the light at the interfaces of three media. The situation that is described by this figure is referred as three-phase (ambient–film–substrate) system. The contribution of the secondary multiple reflections B in the optical response of the complex three-phase system is

determined by the total Fresnel reflection coefficients $\tilde{R}_p$ and $\tilde{R}_s$, which are given by the following expressions [24]:

$$\tilde{R}_p = \tilde{r}_{01p} + \frac{(1 - \tilde{r}_{01p})\tilde{r}_{12}e^{i2\beta}}{1 + \tilde{r}_{01p}\tilde{r}_{12p}e^{i2\beta}} \quad (14a)$$

or

$$\tilde{R}_p = \frac{\tilde{r}_{01p} + \tilde{r}_{12p}e^{i2\beta}}{1 + \tilde{r}_{01p}\tilde{r}_{12p}e^{i2\beta}} \quad (14b)$$

and

$$\tilde{R}_s = \frac{\tilde{r}_{01s} + \tilde{r}_{12s}e^{i2\beta}}{1 + \tilde{r}_{01s}\tilde{r}_{12s}e^{i2\beta}} \quad (15)$$

where r_{01i} and r_{12i} are the *Fresnel* reflection coefficients for the interfaces between media 0 and 1 and between media 1 and 2, respectively, and β is a phase angle defined by

$$\beta = 2\pi \left(\frac{d}{\lambda} \right) \sqrt{\tilde{n}_1^2 - n_0^2 \sin^2 \theta} \quad (16)$$

The complex reflection ratio of the total Fresnel reflection coefficients in analogy to Eq. (11) is

$$\tilde{\rho} = \frac{\tilde{R}_p}{\tilde{R}_s} \quad (17)$$

Equation (12) is also applied for the calculation of the dielectric function of the three-phase (ambient–film–substrate) system, referred to as the *pseudo-dielectric function* $\langle \varepsilon(\omega) \rangle$, which carries information on the substrate and the film dielectric functions and the film thickness d, too. See for example the discussion and Figures 4 and 5 of the previous section. The effect of the multiple reflections on the measured $\langle \tilde{\varepsilon}(\omega) \rangle$ spectrum is more pronounced at the energies where the film is transparent (Im $\tilde{\varepsilon}_1 = 0$) and when the thickness of the film is $d \geq \lambda/4n_1$.

Therefore, interference fringes with high modulation amplitude versus energy or wavelength will appear.

By substituting R_p and R_s from (14) and (15), Eq. (17) is recast in the following quadratic form in terms of the exponential variable $X = e^{i2\beta}$:

$$\alpha_1(\tilde{n}_1, \rho)X^2 + \alpha_2(\tilde{n}_1, \rho)X + \alpha_3(\tilde{n}_1, \rho) = 0 \qquad (18)$$

When the film in the three-phase (ambient–film–substrate) system is transparent ($\tilde{n}_1 = n_1$), its thickness d is given by

$$d = i\left[\frac{\lambda}{4\pi(n_1^2 - \sin^2\theta)^{1/2}}\right] \log X \qquad (19)$$

The appropriate root X of Eq. (18) is a periodic function of the thickness d and the angle of incidence θ, and leads to a positive real film thickness d. For example, as the film thickness d is increased (e.g., during the growth of a transparent thin film) starting from the point $X = 1$ (when $d = 0$), X passes through 1 every time the film thickness sweeps through a thickness period

$$D = \frac{\lambda}{[2(n_1^2 - \sin^2\theta)^{1/2}]} \qquad (20)$$

If the film thickness is such that $d \ll 4\pi/\lambda$, it can be shown from Eqs. (12) and (13)–(17), with a first-order approximation, that

$$\langle\tilde{\varepsilon}\rangle = \tilde{\varepsilon}_2 + \frac{4i\pi d}{\lambda}(\tilde{\varepsilon}_2 - \sin^2\theta)\frac{\tilde{\varepsilon}_2(\tilde{\varepsilon}_2 - \tilde{\varepsilon}_1)(\tilde{\varepsilon}_1 - 1)}{\tilde{\varepsilon}_1(\tilde{\varepsilon}_2 - 1)} \qquad (21)$$

where $\langle\tilde{\varepsilon}\rangle$ is the pseudo-dielectric constant defined as the simple transformation of the ellipsometric ratio $\tilde{\rho}$ according to Eq. (12) when a very thin layer is deposited on top of a bulk material with dielectric function $\tilde{\varepsilon}_2$. The difference between $\langle\tilde{\varepsilon}\rangle$ and the dielectric function of the clean substrate, $\tilde{\varepsilon}_2$, is proportional to the ratio d/λ.

An excellent example of the meaning of this expression it is shown in Fig. 5, where the dielectric function $\tilde{\varepsilon}_2$ of atomically clean c-Si is compared with those of a silicon wafer covered with a 20-Å native oxide layer and of c-Si with an amorphous layer $\approx$5 Å thick on top of it. Notice in Fig. 5 that in the energy region 1.5–2.5 eV, where c-Si ($\tilde{\varepsilon}_2 \approx 14 + i0$) and SiO$_2$ ($\tilde{\varepsilon}_1 \approx 4 + i0$) are transparent, the calculated pseudo-dielectric function, $\langle\tilde{\varepsilon}\rangle \approx 14 + i1.25$, exhibits a large apparent absorptive (imaginary) part, whereas its real part has been changed only slightly.

3. MACROSCOPIC DIELECTRIC FUNCTION, EFFECTIVE-MEDIUM THEORY, AND MICROSCOPIC SURFACE ROUGHNESS

The dielectric response of a microscopically heterogeneous but macroscopically homogeneous bulk material that consists of a random mixture of separate phases depends on the relative compositions (volume fractions) and the shape distributions of the separate regions [27]. Shapes enter because of screening effects (the accumulation of charge at the boundaries between phases).

Also, screening effects are most effective for disk-shaped inclusions oriented perpendicular to the electric field, and least effective for needle-shaped inclusions oriented parallel to the field. The description of the dielectric response of microscopically heterogeneous materials, including both compositional and shape aspects, is the objective of *effective-medium theory* (EMT).

The key to understanding the optical properties of heterogeneous materials is to recognize that the dielectric function $\tilde{\varepsilon}$ for any material should be calculated via a two-step process. An externally applied electric field causes displacements $\Delta\vec{r}_i$ of discrete charges q_i or charge densities $\rho(\vec{r})$. These screening charges give rise to corrections to the field on a microscopic scale. The first step is to solve this local-field problem self-consistently for the microscopic field $\vec{e}(\vec{r})$ and polarization $\vec{p}(\vec{r}) = \rho(\vec{r})\Delta\vec{r}$ at every point in space. In an isotropic material and for a set of discrete changes in a volume Ω when the wavelength of light, λ, is far greater that the microstructural dimensions, what can be actually observed are the macroscopic averages $\vec{E} = \langle\vec{e}(\vec{r})\rangle$ and $\vec{P} = \langle\vec{p}(\vec{r})\rangle = (4\pi/\Omega)\sum_i q_i \Delta\vec{r}_i$. Then the measured macroscopic dielectric function $\tilde{\varepsilon}(\omega)$ is given by $\vec{D} = \varepsilon\vec{E} = \vec{E} + 4\pi\vec{P}$, with $\vec{D} = \langle\vec{dr}\rangle = \langle\vec{e}(\vec{r}) + 4\pi\vec{p}(\vec{r})\rangle$ the macroscopic displacement field. That is, the displacements clearly depend on the bonding properties and the local fields, while the averaging process is represented by the sum and volume normalization. The normalization suggests that optical measurements can be a contactless means of determining the sample density. Thus, $\tilde{\varepsilon}$ is a macroscopic average quantity, and whether a dielectric description makes sense depends on whether the averaging process itself makes sense.

The derivation of an EMT expression begins with a model microstructure. As an example we first suppose a microstructure where all boundaries are parallel to the applied field. Then $\vec{e}(\vec{r})$ $(= \vec{E})$ is everywhere constant, and the spatial average over $\vec{d}(\vec{r})$ reduces simply to a spatial average over $\tilde{\varepsilon}(\vec{r})$. For a two-phase composite material this leads to an EMT expression for the dielectric function given by

$$\langle\tilde{\varepsilon}\rangle = f_i\tilde{\varepsilon}_i + f_j\tilde{\varepsilon}_j, \qquad f_i + f_j = 1 \qquad (22)$$

which is a linear average and corresponds to no screening, where $\tilde{\varepsilon}_i$ and $\tilde{\varepsilon}_j$ are the dielectric functions of the constituents, and f_i and f_j are the corresponding volume fractions. As a second example we suppose that all boundaries are perpendicular to $\vec{E}$; then $\vec{d}(\vec{r})$ $(= \vec{D})$ is everywhere constant, and the spatial average over $\vec{e}(\vec{r})$ becomes

$$\langle\tilde{\varepsilon}\rangle^{-1} = f_i/\tilde{\varepsilon}_i + f_j/\tilde{\varepsilon}_j, \qquad f_i + f_j = 1 \qquad (23)$$

which corresponds to maximum screening. According to this discussion, all values of the dielectric function $\langle\tilde{\varepsilon}\rangle$ of a composite material for any composition and internal shape distribution must fall within these limits.

The description of the macroscopic dielectric response of microscopically heterogeneous materials such as bulk films, including both compositional and shape aspects, is (as already mentioned) the objective of an EMT. A composite film has an effective dielectric function depending on the shape and size of the constituent fractions, as well as on the orientation of the individual grains. Therefore, EMTs allow the macroscopic dielectric response $\tilde{\varepsilon}$ of a microscopically heterogeneous material to be deduced from the dielectric functions $\tilde{\varepsilon}_i$ of its constituents and the wavelength-independent parameters f_i. Thus, EMTs appear as a basic tool in material characterization by optical means that can be used if the separate regions are small compared to the wavelength of the light but large enough so that the individual component dielectric functions are not distorted by size effects. All EMTs, as already discussed, should obey the quasi-static approximation and can be represented by [28]

$$\frac{\tilde{\varepsilon} - \tilde{\varepsilon}_h}{\tilde{\varepsilon} + \kappa \tilde{\varepsilon}_h} = \sum_i f_i \frac{\tilde{\varepsilon}_i - \tilde{\varepsilon}_h}{\tilde{\varepsilon}_i + \kappa \tilde{\varepsilon}_h}, \qquad \sum_i f_i = 1 \qquad (24)$$

where κ represents the screening effect (for example, for spheres $\kappa = 2$) and ε_h is the host dielectric function. In the case of Maxwell–Garnett (MG) theory [29], the host material is the main component. However, an EMT that is more suitable and generally applicable to thin films is the Bruggeman effective-medium theory (BEMT) [30]. The BEMT is self-consistent with $\varepsilon = \varepsilon_h$, and in the small-sphere limit ($\kappa = 2$) of the Lorentz–Mie theory [31], Eq. (24) gives

$$\sum_{i=1} \frac{\tilde{\varepsilon}_i - \tilde{\varepsilon}}{\tilde{\varepsilon}_i + 2\tilde{\varepsilon}} f_i = 0, \qquad \sum_i f_i = 0 \qquad (25)$$

There has not yet appeared a theory analogous to the Lorenz–Mie theory for spheres, for the case of nonspherical particles; instead an expression analogous to Eq. (25) is used [31]. The parameter κ in Eq. (22) is replaced by the screening parameter k_j. This quantity represents the accumulation of charge at the boundaries between separate phases and is related to the Lorenz depolarization factor q_j, for the direction along the electric field, by the expression

$$k_i = (1 - q_j)/q_j \qquad (26)$$

In the Wiener limits (see [27]), for any composition and microstructure the depolarization factor varies within the limits $0 < q_j < 1$. However, if the composite film is macroscopically isotropic in two or three dimensions, Bergman–Milton limits [32, 33] restrict the depolarization factor to $q = 1/2$ and $q = 1/3$, respectively. The fact that the parameter q is allowed to vary gives the EMT flexibility for better representation. This factor incorporates the effect of the shape of the individual regions that constitute the microstructure.

An important and very general manifestation of inhomogeneity in films occurs as microscopic roughness on sample surfaces. The presence of surface roughness with correlation on the order of the wavelength of light will change the specular reflection properties of surfaces. Two physical processes are responsible for this effect [34]: (a) part of the incident light is scattered away from the specular direction, and (b) part of it may be converted into surface plasmon excitations. However, if the scale of surface irregularities is much less than the wavelength λ, then the electromagnetic wave neither will be scatter nor can form plasmons, and the surface region must presented

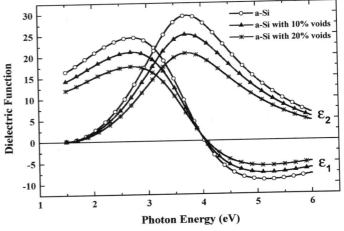

Fig. 7. The real (ε_1) and imaginary (ε_2) parts of the dielectric function of a-Si (open circles), and the effect of bulk density decrease (voids) on the a-Si dielectric function for 10% (solid triangles) and 20% (asterisks) void fraction.

as an effective medium consisting of part substrate and part ambient material. According to Eq. (12), the sample surface will exhibit a low dielectric function and change its absolute $\varepsilon(\omega)$ value in a similar way to that of an oxide overlayer (see for example Fig. 5).

As an example, according to the BEMT, in a composite material consisting of amorphous silicon (a-Si) and voids with volume fractions $1 - f_v$ and f_v, respectively, the measured dielectric function of the material, $\langle \varepsilon \rangle$ is given by

$$\frac{\varepsilon_\alpha - \langle \varepsilon \rangle}{\varepsilon_\alpha + 2\langle \varepsilon \rangle}(1 - f_v) + \frac{1 - \langle \varepsilon \rangle}{1 + 2\langle \varepsilon \rangle} f_v = 0 \qquad (27)$$

where ε_α ($= \varepsilon_{1\alpha} + i\varepsilon_{2\alpha}$) is the dielectric function of a-Si. In a first approximation, assuming $f_v \ll 1$, Eq. (27) becomes

$$\langle \varepsilon \rangle = \varepsilon_\alpha \left(1 - \tfrac{3}{2} f_v\right) \qquad (28)$$

which shows that voids in the material reduce the dielectric function $\langle \varepsilon \rangle$, both real and imaginary parts, everywhere uniformly. This behavior is illustrated in Figure 7 for the dielectric function of a-Si in the energy region 1.5–6.0 eV. This approximation works fairly well for the changes observed in the imaginary part of the dielectric function of an amorphous material (e.g., a-Si:H [35], a-C:H [36]), and can be used for the estimation of the hydrogen incorporated in its matrix during preparation or to study the modifications that take place during thermal postannealing. For example, the reduction in void fraction (bulk density decrease) shown in Fig. 7 for a-Si can be estimated from the changes of the maximum $\varepsilon_{2\alpha}$ value, $\langle \Delta \varepsilon_2 \rangle = (\langle \varepsilon_2 \rangle - \varepsilon_{2\alpha})$, with the formula derived from Eq. (28): $\delta f_v = -\tfrac{2}{3} \langle \Delta \varepsilon_2 \rangle / \varepsilon_{2\alpha}$. Notice that in Figure 7 the dielectric function (both the real part $\varepsilon_{1\alpha}$ and the imaginary part $\varepsilon_{2\alpha}$) of a-Si decreases uniformly and that the energy where the maximum $\varepsilon_{2\alpha}$ occurs does not shift.

In a similar way a material with microscopic surface roughness will exhibit a lower dielectric function or a density deficit, and its top surface region can be treated with an EMT consisting of the bulk film material and a void fraction. However, the

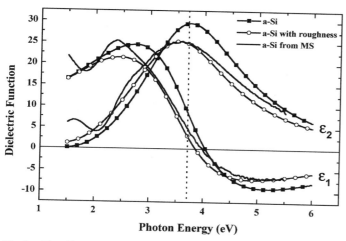

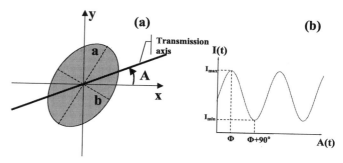

Fig. 9. The basic principle of RAE: (a) The polarization state of the reflected beam, in general, is elliptical and is defined by the long and short semiaxes and the azimuthal angle A. (b) The rotation of the transmission axis of an analyzer having an azimuth A causes sinusoidal time modulation of intensity $I(t)$.

Fig. 8. The effect of the microscopic surface roughness (open circles) on the dielectric function $\varepsilon(\omega)$ of a perfect a-Si (solid squares) material and on the dielectric function of a hydrogen-free a-Si film (solid line) with thickness about 5000 Å prepared by magnetron sputtering (MS).

effect of this density deficit on microscopic surface roughness (see Fig. 8) and on the measured dielectric function $\varepsilon(\omega)$ will be different than that of the uniform decrease of the bulk density shown in Fig. 7. For example, the microscopic surface roughness in a-Si (solid squares) induces in the measured dielectric function (open circles) a red shift as shown in Figure 8. That is, the maximum ε_2-value in a-Si appears at about 3.8 eV, and is shifted to about 3.5 eV when its surface exhibits microscopic roughness. In Figure 8 is also presented the pseudo-dielectric function of an amorphous Si film, about 5000 Å thick, free of hydrogen, prepared by magnetron sputtering, where due to the low adatom mobility of the deposited material an rms surface roughness of ≈ 10 Å is formed. Again the maximum ε_2-value in the dielectric function of this film appears at ≈ 3.5 eV, and its absolute value corresponds to a bulk a-Si material with $\approx 10\%$ voids. The multiple reflections in the dielectric function of the a-Si film below ≈ 3 eV is due to the low absorption coefficient (below 10^5 cm^{-1}) in this energy region.

4. ELLIPSOMETRIC TECHNIQUES AND EXTENSION OF ELLIPSOMETRY FROM THE IR TO THE DEEP UV RANGE

Two different experimental techniques, based on the polarization modulation of the light beam, are commonly used to perform SE measurements. Rotating element (e.g. analyzer) ellipsometers (RAEs) offer the advantages of conceptual simplicity and wavelength insensitivity [37] and can be combined with detector arrays (see for example Fig. 1) to provide real-time spectroscopic capability [3]. Their main difficulty is related to the low-frequency modulation provided by the mechanical rotation (20–100 Hz). This can be a strong limitation for in situ and real-time measurements dealing with weak signals, because of the presence of low-frequency experimental noise (e.g., due to rotary pumps); it can be combated with another

low-frequency modulation technique, such as Fourier transform spectrometry in the IR, as discussed later. On the other hand, phase-modulated spectroscopic ellipsometers (PMSEs) use photoelastic devices to perform the polarization modulation [2, 38, 39]. The main advantage of this technique is the use of a modulation device that is at least two orders of magnitude faster (30–50 kHz) than the mechanical rotation of a polarizer. Thus the PMSE technique can be easily adapted to real-time applications, where time resolution as short as 1 ms can be achieved. Both the RAE and PMSE techniques can be combined with detector arrays to provide real-time spectroscopic capabilities [2, 3, 20].

4.1. Rotating Analyzer Spectroscopic Ellipsometry

In *rotating analyzer ellipsometry* [37], the modulation of the beam reflected from the sample's surface is achieved by mechanical rotation of the analyzer. Although RAE measurements are of high precision, they are characterized by a low speed in data collection and acquisition, due to the relatively low frequency of the analyzer rotation (≈ 50 Hz).

The basic principal of RAE is illustrated in Figure 9. In the general case the light reflected from the sample has an elliptical polarization, which is defined by the long and short axes of the ellipse and the azimuth Φ, as measured in an orthogonal x–y Cartesian system. The beam passes through the analyzer, which is rotating with a constant speed. The intensity of the light beam is detected with a photomultiplier or CCD detector. If we take into account that the analyzer passes only the component that is parallel to the analyzer's transmission axis, the intensity has a modulation equal to twice the frequency of the rotation. This is because in a full revolution of the analyzer it takes on its maximum ($I_{\max}$ for $A = \Phi$) and its minimum ($I_{\min}$ for $A = \Phi \pm 90°$) twice. Thus, the sinusoidal time modulation of the light intensity $I(t)$ is given by the following expression:

$$I(t) = I_0(1 + a\cos 2A + b\sin 2A) \tag{29}$$

where I_0 is the mean value of the intensity, and a, b the normalized Fourier coefficients of the amplitude and phase of the ac component of the sinusoidal signal presented in Figure 9. The time dependence of the analyzer's azimuth angle is given by

$$A = A(t) = 2\pi f_0 t + \theta_0 \qquad (30)$$

where f_0 is the rotation frequency of the transmission axis, and θ_0 is a constant phase factor. The Fourier coefficients of Eq. (29) are related to the optical properties of the material under investigation. The complex reflection ratio $\tilde{\rho}$ then is calculated through the equation

$$\tilde{\rho} = \tan(P - P_s) \frac{\frac{1}{\tan(\Phi - A_s)} - ie}{1 + \frac{ie}{\tan(\Phi - A_s)}} = \tan\Psi\, e^{i\Delta} \qquad (31)$$

where P is the azimuth of the polarizer, and Φ and e are the azimuth and the ratio of long to short semiaxes of the ellipse, respectively. The constants P_s and A_s are the deviations from the zero position of the two polarizing components of the ellipsometric setup—the polarizer and the analyzer, respectively—from the plane of incidence, and can be determined through a calibration procedure [37].

4.2. Phase-Modulated Spectroscopic Ellipsometry

The phase modulation ellipsometric setup [2, 39, 40] consists of the light source, the polarizer, the modulator, the sample, the analyzer, and the detection system, as shown in Fig. 10. The modulation polarization is provided by a photoelastic material, consisting in the UV–Vis range of a fused silica block sandwiched between piezoelectric quartz crystals oscillating at the frequency ω ($\approx$50 kHz), which generates a periodic phase shift $\delta(t)$ between two orthogonal components of the transmitted electric field. In a first-order approximation, $\delta(t) = \alpha \sin \omega t$, where α is the modulation amplitude, which is proportional to V_m/λ, V_m being the excitation voltage applied to the photoelastic device.

As in RAE, the sinusoidal time modulation of the light intensity takes the general form [2, 38–40]

$$I(t) = I\big[I_0 + I_s \sin\delta(t) + I_c \cos\delta(t)\big] \qquad (32)$$

where I_0, I_s, and I_c are trigonometric functions of the ellipsometric angles Ψ and Δ, and for an ideal photoelastic modulator

$$\sin\delta(t) = 2J_1(\alpha)\sin\omega t + 2J_3(\alpha)\sin 3\omega t + \cdots \quad (33a)$$
$$\cos\delta(t) = 2J_0(\alpha) + 2J_2(\alpha)\cos 2\omega t + \cdots \quad (33b)$$

with $J_n(\alpha)$ being the nth Bessel function of α.

In configuration measurements, the polarizer and modulator are oriented in such a way that $P - M = \pm 45°$. It can be shown that suitable choices of the azimuthial angles A and M allow simple determination of the ellipsometric angles (Ψ, Δ) from I_0, I_s, and I_c [2, 38–40]. The most useful configurations for the azimuthial setting $P - M = 45°$ are

$$M = 0°, \qquad A = +45°$$
$$I_0 = 1$$
$$I_s = \sin 2\Psi \sin\Delta \qquad\qquad (34)$$
$$I_c = \sin 2\Psi \cos\Delta$$

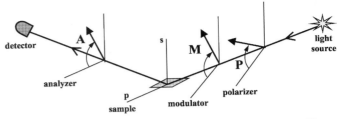

Fig. 10.   Arrangement for a typical phase-modulated ellipsometer. The azimuthal angles P, M, and A of the three optical elements—the polarizer, the modulator (a photoelastic device), and the analyzer—are referred to the plane of incidence.

and

$$M = +45°, \qquad A = +45°$$
$$I_0 = 1$$
$$I_s = \sin 2\Psi \sin\Delta \qquad\qquad (35)$$
$$I_c = \cos\Delta$$

Then combining Eqs. (13), (16), and (17) leads to a simple linear relation between the harmonics S_0, S_ω and $S_{2\omega}$ determined from numerical signal processing and the useful quantities I_0, I_s, and I_c:

$$\begin{pmatrix} S_0 \\ S_\omega \\ S_{2\omega} \end{pmatrix} = I \begin{pmatrix} 1 & 0 & J_0(\alpha) \\ 0 & 2T_1 J_1(\alpha) & 0 \\ 0 & 0 & 2T_2 J_2(\alpha) \end{pmatrix} \begin{pmatrix} I_0 \\ I_s \\ I_c \end{pmatrix} \qquad (36)$$

where T_1 and T_2 are the attenuation coefficients of the detection system at the frequencies ω and 2ω, respectively.

A typical PMSE optical setup, in the Vis–UV range, is presented in Figure 11. Optical fibers are introduced in both arms of the ellipsometer, in order to increase its compactness. Since the light beam goes through a polarizer before the modulation and through an analyzer before being detected, PMSE is insensitive to any polarization effect due to the optical fibers. The light source is a high-pressure Xe arc lamp, and a mechanical shutter is used in order to evaluate the continuous background. The polarizer and the photoelastic modulator are located in the excitation head. The light beam spot size on the sample is usually reduced to the order 1–3 mm^2. A second optical holder contains the analyzer. The polarizer, the modulator, and the analyzer are mounted on 1-arcmin precision rotators. The excitation head and the optical holder with the analyzer can be easily mounted and demounted from such an ellipsometric setup for use, for example, in a deposition system for in situ real-time ellipsometric measurements.

The reflected light beam can be analyzed by a grating monochromator providing full spectrum capability. In the simplest setup a photomultiplier is used as detector in the wavelength range 225 to 830 nm. In addition, several detectors (for example, 16 photomultipliers, as will be discussed in Sections 5 and 6) can be adapted to a spectrograph through optical fibers, providing real-time spectroscopic capability [22]. Each channel then can be measured at 1-kHz frequency, without dead time between two consecutive acquisitions. Using digital parallel signal processing, a spectrum consisting of at least 16 photon

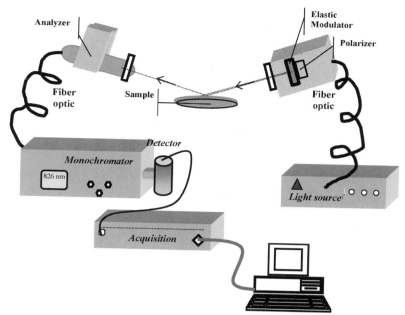

Fig. 11. An optical setup of a phase-modulated spectroscopic ellipsometer (PMSE) in the Vis–UV energy region. The light source (Xe arc lamp), the polarizer, and the elastic modulator constitute the excitation head, whereas the analyzer, monochromator, and detector constitute the detection head. The acquisition board and the computer are necessary to collect and transform the information into raw data for further analysis.

energies between 1.5 and 4.3 eV can be recorded in less than 100 ms [22, 23].

4.3. Ellipsometry in the IR and Deep UV Energy Regions

The difficulty in using spectroscopic ellipsometry to study the vibrational properties of materials in the infrared energy region is mainly because of the lack of intensity in this energy range, and the lack of high-quality optical components [2, 10–12, 41, 42]. These limitations, however, have successfully been overcome during the last ten years, using IR phase-modulated ellipsometers [2, 15] based on two different approaches. In the first one, a high-intensity IR source was implemented in conventional PMSE based on dispersive spectroscopy [43, 44]. In the second one, which will be described below, PMSE is combined with Fourier transform (FT) spectroscopy (FTPME) [15] and exhibits good sensitivity. Similarly, FTIR ellipsometers based on the RAE technique have been reported [10–12, 41, 42]. However, they are not well adapted to real-time applications, since the ellipsometric spectra are obtained from a combination of several interferograms recorded at different analyzer orientations. This difficulty is mainly due to the low modulation frequency of the rotated polarizer (20–100 Hz), which overlaps the interferogram frequencies, themselves well below 1 kHz.

The basic principle of the FTPME technique [2, 15], which is a double modulation technique, is based on conventional FT spectrometers that record the detected signal as a function of the optical path difference x between the two interfering beams.

The detected intensity in this configuration can take the form

$$I(x, t) = \int_0^\infty I(\sigma, t) \frac{1 + \cos 2\pi\sigma x}{2} \, d\sigma \qquad (37)$$

where σ is the wave number ($\sigma = 1/\lambda$) and $I(\sigma, t)$ is given by an expression similar to Eq. (32):

$$I(\sigma, t) = S_0(\sigma) + S_\omega(\sigma)\sin\omega t + S_{2\omega}(\sigma)\cos 2\omega t + \cdots \quad (38)$$

The Fourier analysis of the signal modulated at the frequency ω for each retardation x enables the determination of the interferograms $I_i(x)$ corresponding to the dc (S_0), fundamental (S_ω), and second harmonic ($S_{2\omega}$). As a result Eq. (37) yields

$$I_i(x) = \int_0^\infty S_i(\sigma) \frac{1 + \cos 2\pi\sigma x}{2} \, d\sigma, \qquad i = 0, \omega, 2\omega \quad (39)$$

Using numerical methods, the interferograms are obtained from the recording of N points at $x_k \, (= k\Delta x)$; then the arrays $S_i(\sigma_n)$, with $\sigma_n = n/(N\Delta_x)$, are obtained from the inversion of Eq. (39):

$$S_i(\sigma_n) = \frac{1}{\sqrt{N}} \sum_{k=-N/2+1}^{N/2} I_i(k\,\Delta_x) e^{i2\pi nk/N} \qquad (40)$$

where I_i is the interferogram after subtraction of the mean value and apodization by a triangle function. In FTPME the data recording system is modified to make I_i symmetric and the quantities S_ω and $S_{2\omega}$ real trigonometric functions with signs

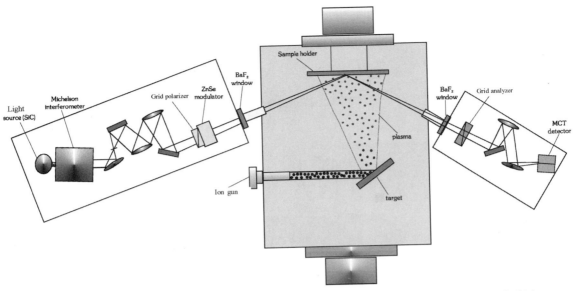

Fig. 12. An optical setup for a Fourier transform infrared phase-modulated ellipsometer (FTPME) adapted to an ultrahigh-vacuum (UHV) deposition system for in situ and/or real-time ellipsometry measurements during deposition of thin films. The SiC light source, the Michelson interferometer, the grid polarizer, and the ZnSe photoelastic modulator constitute the excitation head, which is mounted on the UHV chamber with a BaF_2 window. The detection head consists of a grid analyzer and the InSb and HgCdTe detectors. An ion gun and the target material for the preparation of thin films are also shown.

depending on σ. From $I_0(\sigma)$, $I_\omega(\sigma)$, and $I_{2\omega}(\sigma)$ the ratios

$$R_\omega = \frac{I_\omega(\sigma)}{I_0(\sigma)} = \frac{S_\omega(\sigma)}{S_0(\sigma)} \quad \text{and} \quad R_{2\omega} = \frac{I_{2\omega}(\sigma)}{I_{0h}(\sigma)} = \frac{S_{2\omega}(\sigma)}{S_0(\sigma)} \tag{41}$$

can be obtained, which are functions of the ellipsometric angles Ψ and Δ.

The optical setup for FTPME, which is schematically shown in Figure 12, is also based on the conventional polarizer–modulator–sample–analyzer ellipsometer configuration. The incident beam is provided by a conventional IR spectrometer, which contains a SiC light source and a Michelson interferometer. The parallel IR light beam coming out of the spectrometer is focused on an ∼1-cm² area on the sample surface by means of mirrors. Before reaching the sample, the IR beam passes through the IR grid polarizer and the ZnSe photoelastic modulator at 37 kHz. After reflection from the sample, the beam goes through the analyzer (IR grid polarizer) and is focused on the sensitive area of a photovoltaic detector: InSb (above 1850 cm⁻¹) or HgCdTe (MCT). The spectral range of the ellipsometer is limited by the detector specifications. At present it ranges from 900 to 4000 cm⁻¹. The resolution of the spectra can be varied from 1 to 128 cm⁻¹. An ellipsometric spectrum can be recorded in less than 2 s, which corresponds to a single scan of the interferometer. By increasing the integration time to a few minutes, the precision of Ψ and Δ can be improved by more than one order of magnitude. Thus a resolution of the order of 0.01° or better can be achieved for both angles Ψ and Δ [9].

Figure 13 shows a representative example of FTPME data collected in the range 1100–1900 cm⁻¹ with a resolution of 8 cm⁻¹ after the deposition of a BN film about 1600 Å thick

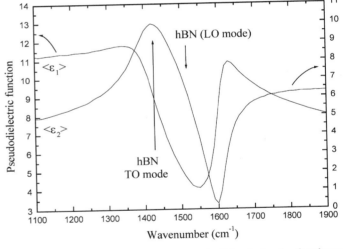

Fig. 13. Real and imaginary parts of the pseudo-dielectric function for a boron nitride (BN) film (≈1600 Å) deposited by rf magnetron sputtering on a c-Si substrate. The arrows denote the positions of the TO and LO vibration modes characteristic of the hexagonal BN (hBN).

prepared by rf magnetron sputtering onto a c-Si substrate. The real ($\langle\varepsilon_1\rangle$) and imaginary ($\langle\varepsilon_2\rangle$) parts of the pseudo-dielectric function show strong IR absorption bands in the range 1400–1600 cm⁻¹, which are characteristic of the TO and LO vibrational modes of the hexagonal BN materials. In Section 5 is discussed in detail the ability of FTPME to investigate the bonding structure and the vibrational properties of CN_x films.

In the case of wide-bandgap semiconductors and insulating materials in which the strong absorption due to interband

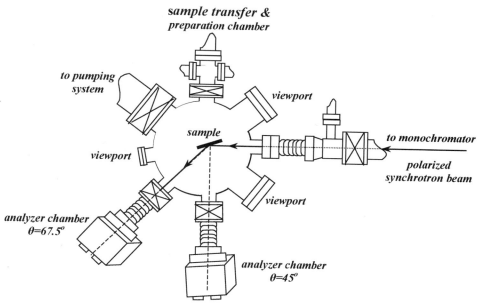

Fig. 14. Schematic diagram of the VUV ellipsometric setup. On the main chamber (base pressure 10^{-10} mbar), which is mounted directly in the polarized synchrotron beam coming out of the monochromator, are mounted the analyzer chamber at either of two angles (67.5° and 45°) and the preparation chamber.

electronic transitions occur at high photon energies, conventional light sources, such as the commonly used Xe lamps, are insufficient for an optical study, because do not emit at energies above 6.5 eV. The best light source for spectroscopic ellipsometry in the vacuum ultraviolet (VUV) region is without doubt monochromatized synchrotron radiation (SR) because of its tunability, high intensity, and natural collimation. It is also highly polarized, although the degree of linear polarization is not as high as that can be achieved in the visible region with conventional polarizers.

SR extends over a broad continuum from IR to X-ray regions, so a monochromator has to be used to filter out a narrow bandwidth. On this principle was designed a VUV ellipsometer [4–7, 16] at the BESSY I Synchrotron Radiation Laboratory in Berlin. The light from the storage ring was monochromatized by a 2-m Seya Namioka monochromator, which nominally covered the energy region 4.0–40 eV using two gratings. Although the light from the monochromator is highly linearly polarized in the orbital plane, it may still contain a component of unpolarized light at the level of a few percent. The proportion of these components could vary as the monochromator was scanned, and this should be taken into account (especially above 10 eV) when evaluating the data, for example by means of full Mueller matrix notation to describe the electromagnetic beam [24].

The SR ellipsometer, which is a RAE, consists mainly of two vacuum chambers: (i) the main UHV chamber (base pressure 10^{-10} mbar) around the sample manipulator, which is mounted directly in the synchrotron beam, and (ii) the analyzer chamber, which is adapted to an angle of incidence of 67.5° or 45°, depending on the type of material and its optical response in this energy region. In order to tilt the plane of the beam polarization with respect to the plane of incidence (which is usually ac-

complished by turning the polarizer), the whole ellipsometer is rotated about the axis of the incoming synchrotron light beam. As shown in Figure 14, the sample under study can be mounted on the precision manipulator at the center of the main UHV chamber. Samples can be prepared in the main UHV chamber, or are introduced via a rapid load–lock and exchanged with a transfer system. The analyzer chambers contain the rotating analyzers and the detectors. The angular orientation of the analyzers is measured with an angle encoder mounted directly on the rotating drum. The photodiode detectors are mounted on a linear-motion feedthrough perpendicular to the axis of the analyzers.

A representative example of the application of VUV ellipsometry in the study of wide-bandgap semiconductors is presented in Figure 15, where are plotted the dielectric functions of cubic and hexagonal gallium nitride (GaN) films [4, 5]. The arrows designate the locations of structures in the experimental spectra. At energies below 3 eV, GaN films are transparent, and thus the spectra are dominated by interference fringes originating from multiple reflection of the incident light beam at the film–substrate interface. From this figure we can see that strong electronic absorption in GaN materials due to the interband transitions occurs at high energies, above 6 eV, and we can appreciate the usefulness of VUV ellipsometry in materials study. In Section 5 we are going to discuss in more detail the investigation, by SR ellipsometry, of amorphous carbon and diamond [18] materials.

What we have achieved by the extension of spectroscopic ellipsometry to the deep UV energy region is the ability to study in detail the electronic structure of materials of high scientific and technological interest, such as the wide-bandgap semiconductors [4–6, 18], the d-bands of III–V semiconductors [17],

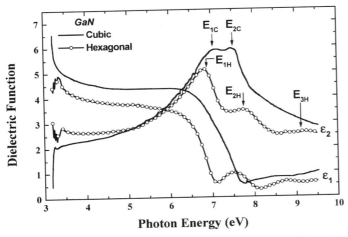

Fig. 15. The real (ε_1) and imaginary (ε_2) parts of the dielectric function of cubic (solid lines) and hexagonal (symbols) GaN films in the energy region 3–9.5 eV. The arrows indicate the locations of interband transitions, as assigned to the observed structures (see [4]).

and the optical properties and the electronic structures of insulating films [7]—as well as surface phenomena, since in this energy region the penetration depth of light in most materials is very small and ellipsometry can be used as a surface technique to probe their surface and bonding properties.

5. ELLIPSOMETRIC STUDIES OF CARBON-BASED THIN FILMS

Carbon is unique among the elements of group IV in that it exhibits both semimetallic (graphite) and insulating (diamond) characteristics. Carbon atoms can form structures with twofold (sp^1 hybridization), threefold (sp^2 hybridization) and fourfold (sp^3 hybridization) coordination. This property gives rise to an enormous variety of solid carbon phases, both crystalline and amorphous [45]. The desired properties of diamond have attracted scientific and technological interest in developing carbon films that exhibit properties close to those of diamond [46].

In general, carbon films exhibit a combination of properties similar to those of bulk crystalline diamond that makes them ideal for use as an overcoat in thin-film structures [47]. Their hardness, wear and corrosion resistance, and low friction coefficient enable their application in magnetic recording media [45, 48, 49], whereas their chemical inertness and infrared transparency favor their further use in magnetooptic [50] and optoelectronic [51] devices. The wish to grow carbon films with properties close to that of diamond has led to the development of various deposition techniques in the last decade [47, 52–64]. The unique capability of carbon to form two kinds of bonding configurations, namely, the planar sp^2 and the tetrahedral sp^3, that can be intermixed in various relative amounts in the same specimen, results in materials possessing diverse properties [65]. Therefore, the deposited films are termed either *diamond-like* or *graphite-like* by virtue of certain criteria, such

as the prevailing bonding configuration, the estimated film hardness, and the value of the optical gap.

In various growth techniques the source materials for the deposition of carbon atoms are hydrocarbon gases, and in these cases the incorporation of hydrogen in the film, often in significant fractions, is inherent. Such films are usually denoted a-C:H. The role of hydrogen in carbon films is not yet well understood, but it is believed to stabilize the tetrahedral bonding in the specimens, which is agreed to be the origin of the diamond-like properties. Recent works suggest that hydrogen saturates carbon π-bonds and thus converts sp^2 sites to sp^3 ones [66]. In any case, hydrogen might merely reinforce the diamond-like character, since nowadays it is clear that hydrogen-free amorphous carbon films can also possess diamond-like properties [67, 68], a fact that will be discussed in detail in this section. Hydrogen is also associated with the existence of defects, which are thought to localize the existing π-electrons in small sp^2 clusters and thus make the material insulating [69]. Thus it is of interest to investigate the effect of hydrogen on the properties of a-C:H films.

Several difficulties are involved in studying the dielectric function of carbon-based materials. First, it requires light sources in the ultraviolet region, where these materials exhibit their strong absorption. This calls for vacuum ultraviolet spectroscopy, e.g., based on synchrotron radiation sources. Conventional lamps present problems in the energy region above 6.5 eV. Mainly because of this, we do not yet know the exact dielectric function and interband transitions of diamond and of crystalline graphite [70–72] above 10 eV. Second, no one has measured the dielectric function of amorphous, fully sp^3 or sp^2 carbon-bonded materials over the whole energy region. Finally, the response of a-C and CN_x materials prepared as thin films is affected by that of the substrate, especially at low photon energies, where the penetration depth of light is large enough. In general the films are composite materials consisting of differently coordinated carbon sites [62, 72–74].

In this section is described the optical and electronic characterization of representative a-C:H films with spectroscopic ellipsometry in the extended energy region from NIR to VUV (1.5 to 10 eV), the latter using synchrotron radiation spectroscopic ellipsometry (SRSE). We aim to demonstrate first how the investigation of the optical properties of relatively thick (above 3000 Å) a-C:H films in the Vis to VUV energy regions can reveal their diamond-like or graphite-like character, and to describe the optical properties or the dielectric response in this energy region. In this regard we also study, by in situ SE, the effect of thermal annealing, which is known to cause hydrogen effusion, on the optical properties of the films and the subsequent gradual graphitization of the materials [36]. For this purpose we have performed SE measurements in conjunction with in situ annealing experiments on a-C:H films prepared by various physical vapor deposition techniques and studied the obtained pseudo-dielectric function $\langle \varepsilon(\omega) \rangle$. The information obtained by SE through data analysis of the pseudo-dielectric function $\langle \varepsilon(\omega) \rangle$ of a-C:H films enable us to determine the bulk dielectric function $\varepsilon(\omega)$ of sp^3- and sp^2-

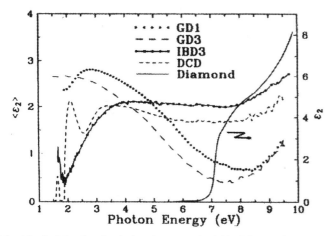

Fig. 16. Left-hand scale: the imaginary part $\langle\varepsilon_2(\omega)\rangle$ of the pseudo-dielectric function of various a-C:H films prepared with different deposition techniques in the photon energy region 1.5–10 eV. Right-hand scale: the corresponding $\varepsilon_2(\omega)$ data for bulk crystalline diamond (solid line). Notice the multiple reflections in films IBD3 and DCD in the visible energy region and the low $\varepsilon_2(\omega)$ values around 7–8 eV in the GD1 and GD3 films. (After [36].)

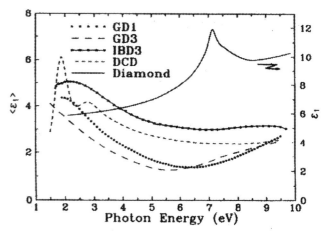

Fig. 17. As in Figure 16, but for the real part $\langle\varepsilon_1(\omega)\rangle$ of the pseudo-dielectric function. Notice the low $\varepsilon_1(\omega)$ values in GD1 and GD3 film in the energy region 5–7 eV. (After [36].)

bonded carbon materials, which can be used in studying the measured $\langle\varepsilon(\omega)\rangle$ of a-C thin films free of hydrogen during deposition by applying the BEMT. As a second example we present the ability to investigate and characterize the growth, using the magnetron sputtering technique, of a-C thin films by in situ spectroscopic ellipsometry and real-time ellipsometry. When needed, we also introduce results obtained from complementary techniques to confirm those obtained by ellipsometry, such as the film thickness and density of the deposited a-C films as measured by X-ray reflectometry (XRR). In addition, we demonstrate real-time measurements by means of multiwavelength (spectroscopic) ellipsometry to monitor the progress of high-deposition-rate sputtering in industrial scale applications as a first step towards control and characterization during film and component production. Finally, in this section results of the newly developed FTPME spectroscopic ellipsometry are presented in a study of the bonding configuration, the vibrational properties, and the electronic structure of CN_x films [9] prepared with the magnetron sputtering technique.

5.1. The Dielectric Function of a-C:H Films in the Energy Region 1.5 to 10 eV

The various deposition techniques developed nowadays prepare a-C:H films that are reported to be mixtures of sp^2- and sp^3-bonded carbon atoms and hence are to some extent intermediate between graphite and diamond. Since graphite is a semimetal and diamond an insulator, the lineshape of the dielectric function of a-C:H films depends decisively on the graphite- or diamond-like nature of the films. A simple look at the pseudo-dielectric $\langle\varepsilon(\omega)\rangle$ spectrum of a film reveals if it is graphite- or diamond-like, as can be seen, for example, in Figures 16 and 17. In Figure 16 we present the imaginary parts $\langle\varepsilon_2(\omega)\rangle$ of the pseudo-dielectric function spectra of some a-C:H films,

and in Figure 17 the corresponding real parts $\langle\varepsilon_1(\omega)\rangle$. These films were prepared by different physical vapor deposition techniques such as glow discharge (GD) [36], ion beam deposition (IBD) [36, 75], and dc magnetron sputtering (DCD) [36]. They have thicknesses above 6000 Å, which means that the measured dielectric function above about 4 eV is a bulk property. Solid lines in these figures designate the corresponding $\varepsilon(\omega)$ data for bulk crystalline diamond [18, 76] and correspond to the right-hand axis scale. As can be seen in Figure 16, the $\langle\varepsilon_2(\omega)\rangle$ spectra of the a-C:H films exhibit profound differences, which can be attributed to the existence or not of an optical gap and a minimum at ≈ 7.5 eV.

In detail, in the lower-energy part of the spectrum, some films exhibit attenuating interference fringes, which are due to the multiple reflection of the incident light beam at the film–substrate interface and are followed by a more or less rapid rise of the imaginary part $\varepsilon_2(\omega)$ due to the beginning of interband electronic transitions from the valence to the conduction band (i.e., across the optical gap). On the other hand, other films do not exhibit interference fringes in the energy region of the measurements, and they have an imaginary part $\varepsilon_2(\omega)$ of high absolute value at low energies, which implies the absence of an optical gap or at most the existence of a gap at very low energies, below 1 eV. The former films resemble the optical response of diamond (solid line in Figs. 16 and 17) and therefore can be called diamond-like, whereas the latter resemble the opposite extreme and can be called graphite-like.

The dip of $\varepsilon_1(\omega)$ that occurs between 5 and 7 eV [in the fundamental gap region of films with a thickness of the order of 6000 Å, the measured $\langle\varepsilon(\omega)\rangle$ and $\varepsilon(\omega)$ have the same value] is another feature characteristic of the film. This is because the energy at which ε_1 exhibits a minimum and ε_2 exhibits low absolute values corresponds to a characteristic excitation of the carbon-based material, called a *plasmon*. This is seen in comparing, for example, the spectra of films GD1 and GD3 with those of IBD3 and DCD. This dip does not always exist in amorphous carbon films and may be shallow when it does (see, e.g., the spectra of films IBD3 and DCD in Figs. 16 and 17, respectively). It can be seen from Figures 16 and 17 that films that do not exhibit an optical gap around or above 1.5 eV show a profound dip in their $\langle \varepsilon_1(\omega) \rangle$ spectra at around 6 eV. This dip is accompanied by a similarly pronounced minimum in the $\langle \varepsilon_2(\omega) \rangle$ spectra that is situated at slightly higher energies around 7.5 eV, as follows from the Kramers–Kronig relation (13) between $\varepsilon_1(\omega)$ and $\varepsilon_2(\omega)$. The existence and sharpness of these two features are distinctive for the existence of diamond character in an a-C:H material and can be used as a further criterion for the characterization of such materials, being complementary to the optical gap criterion that is described in the previous paragraph. Later in this section, we will see that the absence of an optical gap around 1–2 eV and the minimum in $\varepsilon_2(\omega)$ at ≈ 7.5 eV in this type of materials are due to the onset and the saturation of the $\pi-\pi^*$ electronic transitions of graphite, respectively.

5.2. Graphitization of a-C:H Films during Annealing

The structural change of amorphous carbon from diamond-like into graphite-like with thermal annealing results in radical changes in its optical behavior. In the following the information obtained by ex situ and in situ ellipsometry on the dielectric function will be discussed, and evidence presented on the changes of the a-C:H material during annealing.

Annealing of a-C:H films at high temperatures modifies the dielectric function of films having diamond-like character. In order to study these modifications we performed [36] persistent annealing experiments at higher temperature. In detail, the GD2 and DCD a-C:H films, with thickness ≈ 1 and 0.5 µm, respectively, were gradually heated to 675 °C in vacuum conditions at a temperature rate of about 6 °C/min; the annealing process was repeatedly interrupted in order to measure the pseudo-dielectric function spectroscopically at intermediate temperatures. Including the time dedicated to these measurements, it took around 4 h to perform the experiment and reach the final temperature, 675 °C. There the films remained for 30 min before being finally measured. After that the films were gradually cooled back to room temperature, and ellipsometric measurements of $\langle \varepsilon(\omega) \rangle$ at intermediate temperatures were also taken.

In situ measurements of the dielectric function were performed at fixed energy 4.5 eV. This energy value was selected in this type of experiment for two reasons. First, because of the thickness of the films, the measured $\varepsilon(\omega)$ is a bulk property and is not affected by the silicon substrate. Second, around this energy the dielectric function $\varepsilon_2(\omega)$ of diamond-like a-C:H films exhibits a broad plateau (see Fig. 16) and thus a shift of the dielectric function solely due to the temperature dependence of the valence and conduction bands should not affect the values of $\varepsilon_2(\omega)$. In other words, reversible modifications are not expected to have a significant effect on $\varepsilon_2(\omega)$, whereas nonreversible changes, such as the enhancement of the graphite-like character that occurs at temperatures above 500 °C, are anticipated to alter its line shape and absolute values.

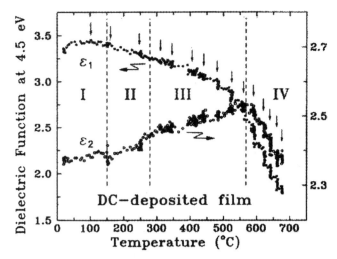

Fig. 18. The real (ε_1) and imaginary (ε_2) parts of the dielectric function at fixed photon energy 4.5 eV of the dc-deposited (DCD) a-C:H film against temperature during thermal annealing up to 675 °C. The arrows indicate the temperatures at which the annealing was interrupted for spectroscopic measurements of the dielectric function. (After [36].)

Typical results are shown in Figure 18 for the DCD film, where the values of ε_1 and ε_2 at 4.5 eV are plotted against temperature and the arrows designate the temperatures at which heating was interrupted to measure the dielectric function spectroscopically. The time to obtain such a spectrum was approximately 6 min. The deviations observed in ε_1 and ε_2 during that time are attributed to the structural modifications, hydrogen effusion, and rearrangements in the material that meanwhile occur, and to uncertainty in the actual temperature attained (± 5 °C). Concerning the dependence of ε_1 and ε_2 on temperature that is shown in Figure 18, one can distinguish four regimes. In regime I ($T < 150$ °C) the dielectric function remains almost constant. In regime II (150 °C $< T <$ 280 °C), ε_1 decreases gradually with increasing temperature [indicating mostly the temperature dependence and to a lesser extent the occurrence of structural modifications in the film, such as reduction of density deficiencies (voids) and local imperfections], whereas of ε_2 remains almost constant.

The thermal stability of a-C:H films is determined by the loss of hydrogen during thermal annealing and the subsequent bond changes. The simplest reaction that occurs is the elimination of hydrogen from adjacent sites and the formation of olefinic groups [66]. Hydrogen is reported to effuse over a broad temperature range above 300 °C in the case of soft [66, 77] as well as hard [78] a-C:H films, and thermal evolution studies show that the onset of effusion depends strongly on the actual growth parameters, but in most cases takes place between 300 and 400 °C [66, 77]. In our case, the rate of fall (rise) of ε_1 (ε_2) increases in regime III of Figure 18 (280 °C $< T <$ 550 °C), which therefore must be associated with the beginning of hydrogen effusion from the film. The drastic changes in ε_1 and ε_2 at temperatures above 550 °C (regime IV, Fig. 18) show that above this temperature the major modifications turn the material into a graphite-like substance dominated by hydrogen

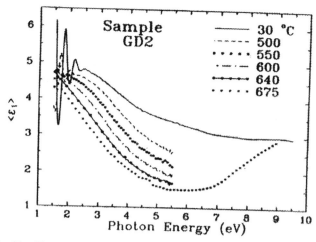

Fig. 19. The real part $\langle\langle\varepsilon_1\rangle\rangle$ of the pseudo-dielectric function of the glow-discharge-deposited (GD2) a-C:H film in the photon energy region 1.5–10 eV at selected temperatures during thermal annealing at 675 °C. (After [36].)

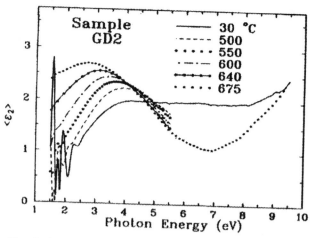

Fig. 20. As in Figure 19, but for the imaginary part $\langle\langle\varepsilon_2\rangle\rangle$ of the pseudo-dielectric function. (After [36].)

effusion. The absolute value of ε_1 shrinks by 50% to a value of 1.75 at 6 75 °C, and this implies (a) the formation of a pronounced minimum in $\varepsilon_1(\omega)$, which is centered at the energy of the π-electron plasmon resonance, and (b) a considerable change in the electronic structure of the as-grown a-C:H materials, as will be discussed in the next paragraph. A hydrogen effusion maximum at temperatures around and above 600 °C is also reported in some works [66, 77, 79], and this may be related to the breaking of strong C–H bonds, in contrast to the effusion of hydrogen at 300–400 °C, which may originate from weak C–H bonds located at polymeric or void surface sites in a manner similar to what is found for a-Si:H [35, 66, 80]. In any case, the evolution of H_2 molecules rather than H atoms is found to be favored, which indicates that the network of all a-C:H films prepared to date is open [66] and thus associated with a large void component and low dielectric function [compare, for example, the $\varepsilon_1(\omega)$ values at ≈ 2 eV in diamond and a-C:H films in Fig. 17]. Hence the full transformation of the material must originate from the strong C–H bonds.

The real and imaginary parts of the pseudo-dielectric function of the GD2 film during the thermal annealing are presented at selected temperatures in Figures 19 and 20. The measurements were taken during the heating of the sample, as described in the previous paragraph. From Figures 19 and 20, one can see the drastic change in the pseudo-dielectric function line shape and absolute values during the annealing, especially at temperatures around and above 600 °C. The interference fringes that are dominant in $\langle\varepsilon(\omega)\rangle$ below 2.5 eV at low temperatures are gradually red-shifted and attenuated in strength, which implies the lowering of the fundamental gap and the subsequent transformation of the semiconducting behavior of the film into a graphite-like one. The resulting dielectric function spectrum (dotted lines in Figs. 19 and 20) does not resemble the initial one (solid lines in Figs. 19 and 20), and this must be associated with the effusion of hydrogen and the prevalance of sp^2 character and the $\pi-\pi^*$ electronic transitions characterizing graphite,

as we are going to discuss. During cooling back to room temperature, the shape and absolute values of the dielectric function remain almost unaffected.

5.3. Dielectric Function of Amorphous Carbon Materials with Various Bonding

It is known that the $\pi \rightarrow \pi^*$ electronic transitions of crystalline graphite [36, 72] occur around 4.5 eV, and a strong absorption due to the $\sigma \rightarrow \sigma^*$ electronic transitions occurs in the range 14–20 eV. In diamond the indirect bandgap is at ≈ 5.3 eV, the first direct gap is at ≈ 7.2 eV, and the strong electronic absorption due to the $\sigma \rightarrow \sigma *$ transitions occurs at ≈ 12 eV [18]. Thus, the experimental dielectric spectra of carbon films are influenced in the Vis–UV energy region mainly by the optical transitions of sp^2 sites. Based on what we know from the crystalline-silicon–amorphous-silicon counterpart [81, 82], the strong electronic absorption in films that are predominantly sp^3-bonded is expected to shift to lower energies (well below the 12 eV of diamond), accompanied by peak broadening and a drop in magnitude, due to the loss of the long-range order of the sp^3-bonded carbon. On the other hand, in composite carbon films, contain both sp^3 and sp^2 carbon bonds, the strong absorption is expected to vary within the energy region 9–15 eV [83]. Thus, for a reliable description of the dielectric response of a-C films, all the contributions of the sp^2 sites ($\pi \rightarrow \pi^*$ and $\sigma \rightarrow \sigma^*$ transitions) and the sp^3 sites (low- and high-energy absorption of $\sigma \rightarrow \sigma^*$ transitions) should be taken into account [84].

Based on what is presented in the previous paragraphs, we further analyzed the measured pseudo-dielectric functions of the GD2 a-C:H film, as grown and annealed at 675 °C, by means of dielectric function models describing amorphous materials (for more details see for example [83, 84]), in order to get their bulk dielectric function in the energy region 1.5–10 eV. This is because in the as-grown GD2 film the sp^3 carbon bonded character prevails, whereas after annealing at 675 °C the sp^2

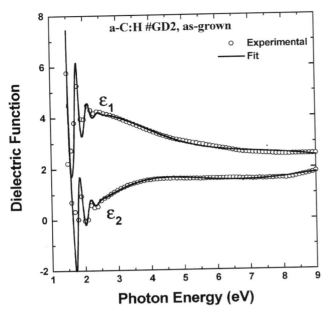

Fig. 21. The experimental $\langle\varepsilon(\omega)\rangle$ of the as-grown GD2 a-C:H film (open circles) and the corresponding fitted one (solid lines) using two Tauc–Lorentz oscillators.

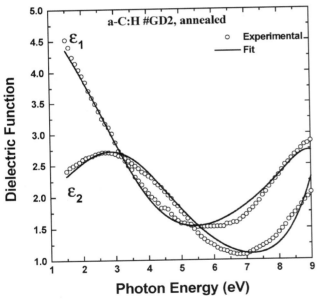

Fig. 22. The experimental $\langle\varepsilon(\omega)\rangle$ of the annealed GD2 a-C:H film (open circles) and the corresponding fitted one (solid lines) using two Tauc–Lorentz oscillators.

$\varepsilon_2(\omega)$ obtained from the Lorentz oscillator model. Thus, the TL model provides the ability to determine the fundamental optical gap ω_g of the electronic transitions, in addition to the energy ω_0, the broadening C, and the strength A of each Lorentz oscillator [74]. The energy ω_0 of this model corresponds more or less to the well-known Penn gap where the strong absorption in a material takes place. The imaginary part $\varepsilon_2(\omega)$ of the dielectric function $\varepsilon(\omega)$ in the TL oscillator model is described by the following expressions [85]:

$$\varepsilon_2(\omega) = \frac{A\omega_0 C(\omega - \omega_g)^2}{(\omega^2 - \omega_0^2)^2 + C^2\omega^2} \cdot \frac{1}{\omega}, \qquad \omega > \omega_g \quad (42)$$

$$\varepsilon_2(\omega) = 0, \qquad \omega \le \omega_g \quad (43)$$

and the real part $\varepsilon_1(\omega)$ is obtained by the Kramers–Kronig integration

$$\varepsilon_1(\omega) = \varepsilon_0 + \frac{2}{\pi} P \int_{\omega_g}^{\infty} \frac{\omega' \varepsilon_2(\omega')}{\omega'^2 - \omega^2} \, d\omega' \quad (44)$$

with ε_0 being equal to one, larger when account is taken of the existence of electronic transitions beyond the measured energy region of 10 eV.

In Figures 21 and 22 are also plotted, with solid lines, the respective calculated dielectric functions from the described modeling procedure using two Tanc–Lorentz oscillators. By using the best-fit parameters (A, ω_g, ω_0, C, and ε_0) of these two Tanc–Lorentz oscillators, the bulk dielectric functions of the as-grown and annealed a-C:H film were calculated. The calculated dielectric functions $\varepsilon(\omega)$ in the energy region up to 6 eV are referred hereafter to mainly sp^2-bonded carbon (as grow) and mainly sp^3-bonded carbon (annealed), and their real and imaginary parts are presented in Figure 23a and b, respectively. In the same figure are also plotted the real and imaginary parts of the calculated $\varepsilon(\omega)$ for a hydrogen-free a-C film, referred as a-D ("amorphous diamond"), taken form [57]. This film was prepared using pulsed ArF-excimer-laser ablation of graphite at room temperature, and according to [57] exhibits fully (100%) "amorphous diamond" character.

The dielectric functions $\varepsilon(\omega)$ presented in Figure 23a and b can be used as reference dielectric functions in a fitting procedure to analyze the measured $\langle\varepsilon(\omega)\rangle$ of an amorphous carbon film by means of EMT, assuming that the film is a composite material consisting of sp^2- and sp^3-bonded carbon.

5.4. Real-Time Ellipsometry to Optimize the Growth of Sputtered a-C Films

Amorphous hydrogen-free carbon (a-C) films with controllable sp^3 content have found an increasing number of applications in improving nanoscale components such as those used in magnetic hard disk and wear-resistant coatings, and they are promising for semiconductor devices and optical film applications [45, 46]. Although a vast effort has been devoted in the last few years to controlling the mechanisms of development and the properties of a-C films, many questions yet need to be answered concerning their mechanical properties and thermal

prevails, and the corresponding dielectric functions can be used as references for various studies. Figures 21 and 22 show the $\langle\varepsilon(\omega)\rangle$ of the as-grown and annealed GD2 film, respectively. In order to evaluate the spectral dependence of the $\varepsilon(\omega)$ of the mentioned films, we fitted the experimental $\langle\varepsilon(\omega)\rangle$ by using the modified Tauc–Lorentz (TL) model [83–85]. In the TL model the imaginary part $\varepsilon_2(\omega)$ of the dielectric function is determined by multiplying the Tauc joint density of states [86] by

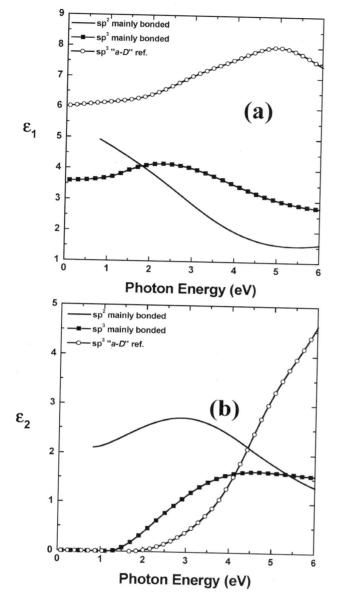

Fig. 23.   (a) The real and (b) the imaginary part of the reference dielectric functions of the mainly sp²-bonded (solid lines) and mainly sp³-bonded (solid squares) carbon materials and the fully amorphous diamond (a-D) from [57] in the energy region up to 6 eV.

may modify the subplantation process and the relaxation mechanism that explains the growth of ta-C films when energetic C^+ species are used. The ion bombardment induces local rearrangements of the film near surface atoms, resulting in very dense or even new, metastable structures rich in sp³ bonds. High density (ρ) and compressive stress are the main characteristics of highly sp³-bonded films, and it has been proposed that the densification [87] or the stress [88] controls the sp³ C–C bonding. The existence has also been reported of a relatively narrow ion energy window in which the formation of the sp³ bonds is favored [87, 88].

Among the other deposition techniques, sputtering is distinguished by its ability not only to control the deposition parameters and to tailor the films' properties, but also to be easily applied on an industrial scale [56, 73, 83, 89, 90]. Nevertheless, the development of a-C films by sputtering has been a challenge. Indeed, due to the statistical character of the ion energies and fluxes in sputtering, it was generally believed at the beginning of the last decade that the development of a-C films rich in sp³ bonds could not be achieved. Recently, however, sp³-rich a-C films have been sputtered by applying an external negative bias voltage to the substrate during deposition [91]. Within the context of this work will be presented the application of in situ spectroscopic ellipsometry and real-time ellipsometry monitoring in order to investigate the growth of sputtered a-C films, and to study the effect of bias voltage (during Ar^+ ion bombardment) as the main deposition parameter for tailoring the film properties in these processes.

The experimental setup is presented in Figure 24, which shows the deposition system and the ellipsometer used to measure Ψ and Δ either at fixed photon energy or over a photon energy range from 1.5 to 5.5 eV. The instrument consists of a Xe light source and a fiber optic transferring the light from a polarizer assembly to an elastic modulator. The light passes through an UHV strain-free optical window and is reflected at 70.4° angle of incidence from the from the sample (substrate or thin film); then it is linearly polarized by the analyzer and, after passing through the grating monochromator, is detected, and the information is transferred to the PC through an acquisition board.

Amorphous and polycrystalline films inevitably exhibit nucleation and growth phenomena and roughness on the film surface, in general much greater than for a monolayer. For a semitransparent film that ultimately grows to opacity, one typically compares the trajectory (or trajectories) in the (Ψ, Δ) plane (the plot of single-photon-energy ellipsometric angles) as a function of time during film deposition, with the corresponding trajectory assuming perfect uniform growth. Under these assumptions, discrepancies in the experimental data from the optical model can be considered as being due to the deviations from the perfect layer-by-layer process. Then either more sophisticated optical models of film growth can be constructed (for example, a nucleation process consisting of cluster formation, coalescence, and surface roughness evolution), or additional experimental results obtained from SE or real-time SE can be used to understand the discrepancies from perfect

stability and their potential for specific applications. a-C films with highly sp³-bonded carbon (ta-C) have been deposited with energetic carbon ions (C^+) produced by several techniques [47, 52–58]. Most of these techniques provide well-defined deposition conditions (i.e., the deposited species and their energies are known and controllable); however, they are difficult to use in industrial applications. Recently, it has been reported that dense and highly sp³-bonded a-C films may also be deposited by MS, including MS with intense ion plating (MS/IP) [59, 60], unbalanced MS [61], and negatively substrate-biased MS [62–64].

The statistical character of the sputtering technique and the use of heavy Ar^+ ions bombarding the film during deposition

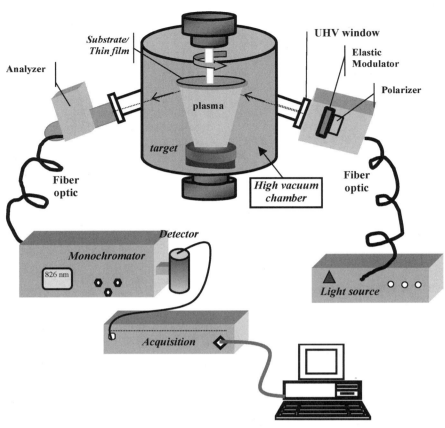

Fig. 24.   A schematic diagram of the deposition system for the magnetron sputtering technique with
the help of a phase-modulated ellipsometer, working in the energy region 1.5–5.5 eV (see Fig. 11).
Details on the magnetron sputtering technique and the plasma are also shown in Figure 25.

growth. The resulting simulations can then be compared with the experimental (Ψ, Δ) data in order to understand the discrepancies from perfect growth. Once these discrepancies are understood, the deduced features can be studied as a function of deposition conditions to obtain insights into the film growth process. In general, because of the complexities associated with the growth process and the film properties at the initial stages of growth, ellipsometry at one photon energy is not sufficient to establish unique growth models. Recent advances in ellipsometric instrumentation, however, have been stimulated in the last few years by the ability to solve these more complex problems, and have culminated in the development of real-time spectroscopic ellipsometry as a more powerful in situ probe of thin film growth, as we will discuss in the next subsections.

Therefore, the real-time monitoring of the growth of transparent films can be accomplished by measuring the ellipsometric angles Ψ and Δ [62, 92]. According to this procedure, when the deposited film is transparent, the experimental Ψ–Δ trajectory will display a closed cycle with a thickness period D depending on the refractive index n of the grown material and the wavelength λ of the light used. The thickness period is given by the expression [see Eq. (16)] $D = (m\lambda/2)(n^2 - \sin^2\theta)^{-1/2}$, where m is an integer [62].

Real-time ellipsometry at fixed energies was first used to monitor the deposition of a-C films and to define the optimum deposition conditions, e.g., where a-C films rich in sp^3 sites are grown. A schematic representation of the rf magnetron sputtering technique and the main deposition parameters affecting the a-C films' growth are shown in Figure 25. Since highly sp^3-bonded a-C films exhibit optical transparency in the visible energy region, the procedure applied was to monitor the growth of transparent films by measuring in real time the ellipsometric angles Ψ, Δ at fixed energies in this region. That is, when the deposited a-C film is transparent, the real-time Ψ–Δ trajectory will display a closed cycle with a thickness period depending on the refractive index of the grown material and the photon energy of the light used [62].

By using this procedure and tuning the deposition parameters such as the Ar partial pressure P_{Ar} and the target voltage V_T (Figs. 26 and 27), as well as the target-to-substrate distance δ and an external negative bias voltage V_b applied to the substrate, we succeeded in depositing a-C films with high sp^3 content. In particular, the real-time Ψ–Δ trajectories presented in Figs. 26 and 27 correspond to the tuning of P_{Ar} and V_T values, respectively. The results in these figures clearly reveal that the optimum conditions are obtained when the Ar partial pressure $P_{Ar} \approx 1.9 \times 10^{-2}$ mbar, the target voltage $V_T \approx 160$ V,

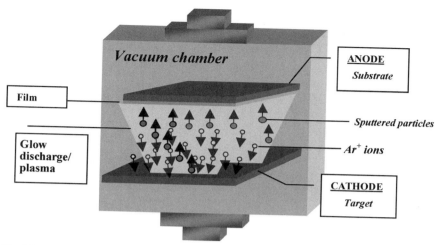

Fig. 25. A schematic representation of a magnetron sputtering deposition configuration. For amorphous carbon films an rf magnetron cathode was used with controllable rf voltage V_T, inert working gas (Ar) partial pressure $P_{Ar} \sim 10^{-2}$ mbar, and distance between the graphite target and the substrate.

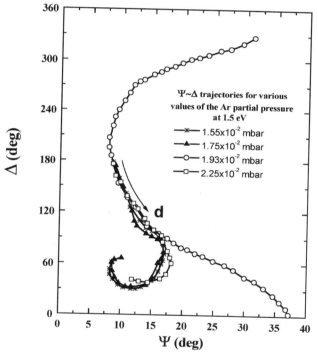

Fig. 26. Real-time examination of a-C films deposited on c-Si at various P_{Ar} at fixed target voltage applied on the target ($V_T = 160$ V), through the monitoring of the ellipsometric angles Ψ and Δ at the photon energy 1.5 eV. The evolution of film thickness is shown by the arrow.

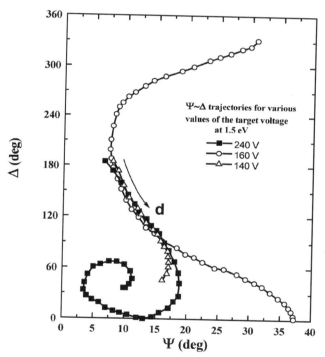

Fig. 27. As in Figure 26, but for various target voltages V_T applied to the graphite target at fixed Ar partial pressure ($P_{Ar} = 1.9 \times 10^{-2}$ mbar).

and the substrate-to-target distance $\delta \approx 65$ mm, leading to the development of films almost transparent in the visible energy region. In particular, a-C films developed with these deposition conditions exhibit an sp^3 content as high as 30%. An increase of sp^3 content to 55% can be decomplished by applying an external negative bias voltage V_b to the substrate during deposition. This bias controls mainly the energy, and to a lesser extent the flux of Ar^+ ions bombarding the growing film surface.

In Figure 28 are illustrated the experimental $\Psi-\Delta$ trajectories obtained during the deposition of two representative a-C films deposited by rf MS. This is an example showing how we can get the optimum deposition conditions to promote the formation of sp^3 carbon bonding and the transparency of the a-C films. The a-C film #a was deposited in sequential thin layers, and the (Ψ, Δ) values were recorded at 1.5 eV (solid line) and 2.3 eV (dotted line). The $\Psi-\Delta$ trajectory obtained at 1.5 eV during film growth displays a closed cycle evolving from the

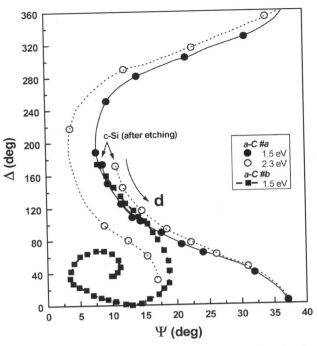

Fig. 28. Real-time kinetic measurements of the ellipsometric angles Ψ and Δ at fixed photon energies during the deposition of two a-C films on c-Si substrates. Film #a is transparent in the NIR–Vis energy region, revealing its sp^3-carbon-bonded character, while film #b is opaque, having sp^3 character.

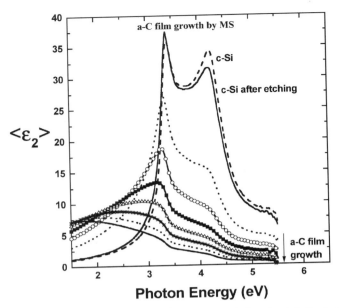

Fig. 29. The imaginary parts of the pseudo-dielectric function $\langle \varepsilon(\omega) \rangle$ of an a-C film deposited in successive layers on c-Si substrate to a final thickness ≈ 300 Å, for a total deposition time $t = 18$ min. Also plotted is the $\varepsilon_2(\omega)$ of c-Si before (dashed lines) and after (solid lines) ion etching (to remove the native SiO_2 overlayer).

(Ψ, Δ) setting of the c-Si substrate, where the film thickness is zero, and provides an indication that the deposited film is transparent. The corresponding Ψ–Δ trajectory obtained at 2.3 eV does not pass from the (Ψ, Δ) setting of the bare Si substrate (after etching) and indicates that the deposited film in this energy is semitransparent. That is, film #a is transparent in the NIR–Vis energy region and thus reveals its sp^3-carbon-bonded character. In Figure 28 are also presented the (Ψ, Δ) values that were recorded at 1.5 eV (dashed line) for an a-C film #b prepared under different deposition conditions than the optimum ones. In contrast to film #a, they exhibit high absorption at 1.5 eV and the sp^2 character of a-C films.

5.5. In Situ Spectroscopic Ellipsometry to Study the Optical Properties and Bonding of a-C Films

The growth of the MS a-C films was accomplished first in successively deposited very thin layers. Upon completion of each layer, the $\langle \varepsilon(\omega) \rangle$ spectrum was measured in situ in the energy range 1.5–5.5 eV. An example of the measured $\langle \varepsilon(\omega) \rangle$ of the a-C-film–Si-substrate system, after deposition of several thin layers, is presented in Figure 29. In particular, in this figure are shown the imaginary parts of the dielectric function of the c-Si substrate, that of c-Si after ion etching (to remove the SiO_2 native oxide induces due to ion bombardment in the formation of an a-Si overlayer), and those of the very thin a-C film on the Si substrate at successive deposition times, until a total time of 18 min.

The question now is how to analyze and interpret the pseudo-dielectric $\langle \varepsilon(\omega) \rangle$ spectra shown in Figure 29 in order to get information about the a-C film's composition, its thickness, and the growth mechanisms at various stages of the deposition. This is commonly done with a least-squares linear regression analysis (fitting procedure). The quantity used to describe the agreement between the experimental data and the modeling process is the unbiased estimator x^2 of the measured pseudo-dielectric function $\langle \varepsilon(\omega) \rangle^m$ or the pseudoreflection ratio $\langle \rho(\omega) \rangle^m$, defined through the expression

$$
\begin{aligned}
x^2 &\left(d, f_{sp^3}, f_{sp^2}, \dots\right) \\
&= \frac{1}{N - p - 1} \sum_i^N \left[\left(\langle \varepsilon(\omega) \rangle_{\mathrm{R}i}^m - \langle \varepsilon(\omega, d, f_{sp^3}, f_{sp^2}, \dots) \rangle|_{\mathrm{R}i}^c \right)^2 \right. \\
&\left. + \left(\langle \varepsilon(\omega) \rangle_{\mathrm{Im}\, i}^m - \langle \varepsilon(\omega, d, f_{sp^3}, f_{sp^2}, \dots) \rangle|_{\mathrm{Im}\, i}^c \right)^2 \right] \quad (45)
\end{aligned}
$$

In this expression, $\langle \varepsilon(\omega, d, f_{sp^3}, f_{sp^2}, \dots) \rangle^c$ represents the calculated dielectric function of the a-C-film–Si-substrate system, N the number of experimental points or spectral points, the subscript R (Im) the real (imaginary) part of $\langle \varepsilon(\omega) \rangle$, and p the number of the energy-independent (unknown) parameters describing the whole system. The 95% confidence limits on the deduced unknown parameter values together with the unbiased estimator are used to determine the applicability of different model processes. That is, the fitting procedure depends on all geometrical characteristics of the film–substrate system (see Fig. 6) and the relations (14)–(17) that are described in Section 2. Moreover, what is needed in order to extract the information about the film is a model to describe the dielectric

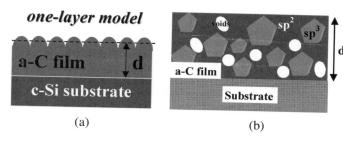

Fig. 30. (a) A schematic representation of the one-layer model with an effective thickness d that represents the islands at the initial stages of growth or the surface roughness in later stages, all consisting of the same carbon phases. (b) The microstructure of the composite a-C film with an average thickness d, consisting of sp^2 and sp^3 carbon phases and voids, grown on a bulk substrate according to the one-layer model.

function of the bulk film, and that is going to be discussed in the following.

In modeling the amorphous thin films in the nucleation stage (initial stages) of growth, when the films consist of isolated clusters on the substrate that increase in size with thickness or time, we can use an EMT to calculate the dielectric function $\tilde{\varepsilon}_i$ ($= \varepsilon_{1i} + i\varepsilon_{2i}$) of the bulk film. In the simplest case, when the deposited material includes only one carbon phase, it is assumed in the model that the film consists of a volume fraction f_i of material, which simulates the nuclei, with dielectric function the same as that of the bulk material; and a volume fraction f_v ($= 1 - f_i$) of free space, which simulates the voids between the clusters. In the case that the amorphous carbon film includes more than one carbon phase, we can construct the model in a similar way. The only difference is that the analysis is more complicated and attention must be paid to the reference dielectric functions of carbon phases. In either case, this procedure is referred as the *one-layer model* (see Fig. 30a). It provides an average over the thickness of the film and simulates the inhomogeneity that is developed in the nucleation and coalescence processes. At a thickness depending on the size of the nuclei (say 10 Å) the nuclei make contact (coalescence), and a a two-layer model can better describe the experimental data. In the two-layer model the top layer (of thickness d_s) simulates the nuclei that may have been formed in the early stages of growth, as well as the surface roughness on the film in the later stages. The underlying layer simulates a bulk like layer (of thickness d_b), and its dielectric function is assumed to be thickness-independent. However, the two-layer model for a-C films introduces a large number of unknown parameters, since several carbon phases and the thicknesses d_s and d_b have to be taken into account, thus reducing the accuracy and reliability of the fitting procedure. Thus, in the following we are going to analyze the $\langle \varepsilon(\omega) \rangle$ of a-C-film–Si-substrate system based on the one-layer model.

According to the above, the a-C film composition was obtained through the analysis of the measured $\langle \varepsilon(\omega) \rangle$ by means of the one-layer model and the BEMT. In this analysis (see for example Figs. 6 and 30b) we assume that the film, with an effective thickness d (see Fig. 30a), is a composite mate-

rial (e.g., consisting of sp^3 and sp^2 components and voids) in combination with the three-phase (air–a-C-film–c-Si-substrate) model [27, 62].

According to the BEMT the bulk dielectric function $\varepsilon(\omega)$ of a composite a-C material is given by Eq. (23):

$$\sum_i f_i \frac{\varepsilon_i + \varepsilon}{\varepsilon_i + 2\varepsilon} = 0, \qquad i = sp^2, \ sp^3, \ voids$$

where f_i is the volume fraction of each component. Since in the BEMT analysis we need reference dielectric functions of sp^3 and sp^2 carbon phases, we used for the sp^3 component that of an amorphous, fully tetrahedrally bonded material [57], and for the sp^2 component that of an amorphous graphite-like material [36] (see Fig. 23a and b). We have estimated that the latter reference contains a fraction 10–15% sp^3 sites. The results from BEMT analysis are expected to underestimate the sp^3 fraction by a corresponding amount.

By this fitting procedure one can obtain the volume fractions of the sp^2 and sp^3 components, as well as the film thickness. In Figures 31 and 32 are shown the experimental (open circles) and the fitted (solid lines) $\langle \varepsilon(\omega) \rangle$ of a-C films of thicknesses 50 and 300 Å, respectively, for comparison with those presented in Fig. 29 and obtained during the a-C film deposition. We estimate that the error in calculating the film thickness with the BEMT analysis is of the order of 2–3%. The film thickness of the a-C films calculated by SE was confirmed by XRR and transmission electron microscopy in cross sectional geometry (XTEM). In Figure 33, we present for comparison the results for the thickness of the a-C films shown in Figure 29 obtained by SE data analysis, plotted against those obtained by the XRR technique. Details on the XRR technique and how one can get the thickness, density, and interfacial properties of a thin-film–substrate system will be presented in Section 6 [93]. The agreement on the a-C film thicknesses between the two techniques is excellent, showing that the one-layer model and the reference dielectric functions used in the above SE analysis are quite reasonable.

The evolution of the sp^2 and sp^3 bond fractions of two representative a-C films prepared by MS with two different bias voltages $V_b = -20$ and $+10$ V is shown in Figure 34. It is evident that the negative V_b (ion bombardment during deposition) enhances the sp^3 formation with respect to the positive V_b (floating bias, without ion bombardment). More specifically, in the a-C film prepared with $V_b = -20$ V, with the increase of film thickness the sp^2 and sp^3 fractions beyond the initial stages of growth (≈ 80 Å) are constant at about 60% and 40%, respectively, indicating the uniform deposition of the film. However, in the a-C film deposited with $V_b = +10$ V, the sp^2 (sp^3) fraction increases (decreases) with increasing film thickness. The reason is the existence of an intense positive Ar$^+$ ion bombardment of the growing film surface when a negative bias is applied onto the substrate. The energy E_i of the bombarding ions is controlled by the applied negative bias V_b on the substrate through the relation $E_i = E_p + e|V_b|$, where E_p is the discharge energy and in our case is ≈ 30 eV [91]. When the energy E_i is above ≈ 30 eV, the ion bombardment induces local

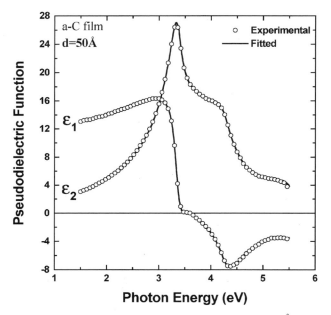

Fig. 31. The measured (open circles) $\langle \varepsilon(\omega) \rangle$ of an a-C film ≈ 50 Å thick deposited on c-Si, and the curves fitted by using the BEMT (solid lines).

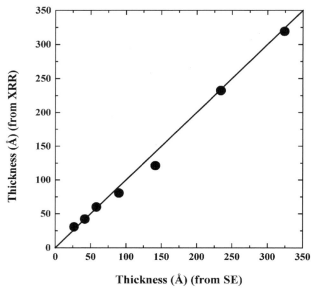

Fig. 33. Comparison of the thickness of a-C films measured by spectroscopic ellipsometry (SE) and by X-ray reflectometry (XRR). The difference between the thicknesses measured by the two techniques is, in general, below 5%.

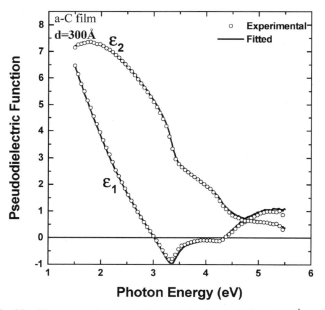

Fig. 32. The measured (open circles) $\langle \varepsilon(\omega) \rangle$ of an a-C film ≈ 300 Å thick deposited on c-Si, and the curves fitted by using the BEMT (solid lines).

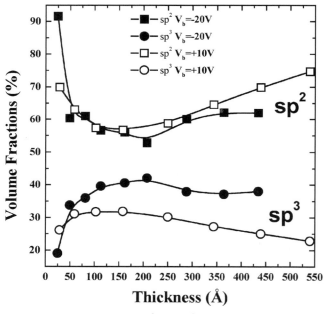

Fig. 34. The evolution of the sp^2 and sp^3 bond fractions of two representative a-C films prepared by MS when applying onto the substrate with $V_b = -20, +10$ V.

rearrangement of the film near surface atoms, resulting in very dense (i.e., rich in sp^3-bonded carbon) or even new metastable carbon structures [73, 74, 83].

Figure 35 shows the thicknesses deduced from the above analysis for three a-C films deposited at different bias voltages V_b, versus the deposition time. For $V_b = +10$ and -40 V there is an almost linear dependence of the film thickness with deposition rates ≈ 18 and 14.5 Å/min, respectively. The film deposited at high negative $V_b = -160$ V exhibits an almost quadratic dependence with a deposition rate ≈ 15 Å/min. That

is, when the external negative bias (or the Ar^+ ion energy E_i) increases above a critical value, the development of the sp^3 sites and the deposition rate are reduced, possibly because of the preferential etching of the sp^2 sites by the Ar^+ ion bombardment. In addition, above this critical energy (or V_b) carbon structures other than the sp^2 and sp^3 may be developed. This point will be addressed and discussed below. Since the a-C films are composite materials and they were produced with the coexistence of at least two different carbon bonded sites and some

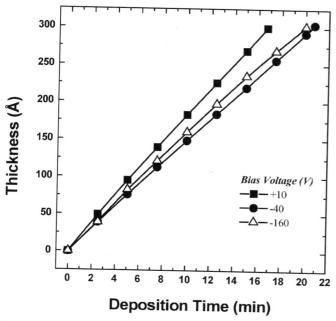

Fig. 35. Dependence of a-C film thickness on deposition time at various bias voltages V_b applied to the substrate during deposition.

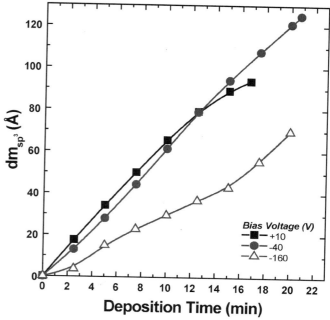

Fig. 36. sp^3 mass thickness during the a-C film growth, plotted as a function of deposition time.

amount of voids (free space), an analysis combining the results on composition (Fig. 34) with those on thickness (Fig. 35) during the film growth is expected to provide more information on the film growth.

Thus, for a better understanding of the deposition mechanisms of a-C films with respect to the applied V_b, we calculated the mass thickness of sp^3 (dm_{sp^3}) and sp^2 (dm_{sp^2}) components, according to the following expressions:

$$dm_{sp^3} = f_{sp^3}d$$
$$dm_{sp^2} = f_{sp^2}d \qquad (46)$$

where f_{sp^i} is the volume fraction of the i, component (see Fig. 34), and d the calculated film thickness (see Fig. 35). In Figures 36 and 37 the quantities dm_{sp^3} and dm_{sp^2}, respectively, are plotted versus the deposition time. Starting from the film deposited with $V_b = +10$ V (film type I), we can see an almost linear dependence for dm_{sp^2} and a square root one for dm_{sp^3}. The latter exhibits a tendency to saturation above ≈ 14-min deposition time, suggesting that the sp^3 formation is no longer favored. In the film deposited with $V_b = -40$ V (film type II), the sp^2 and sp^3 phases coexist throughout the film growth, more or less at a constant percentage, and as a result dm_{sp^3} and dm_{sp^2} exhibit almost linear growth. Finally, the dm_{sp^3} and dm_{sp^2} of the film deposited with $V_b = -160$ V (film type III) show a complicated growth behavior. In particular, the sp^2 carbon sites exhibit a tendency for saturation above ≈ 18 min, whereas at the same time the growth rate of sp^3 sites increases. In order to get quantitative results on the above observations we fitted the data presented in Figures 36 and 37 with the following expression:

$$dm_{sp^i} = \alpha t^n \qquad (47)$$

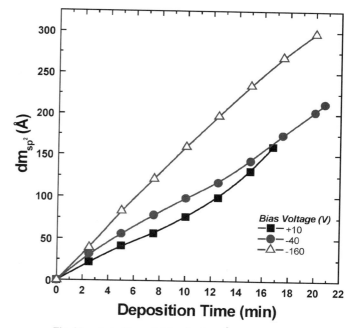

Fig. 37. As in Figure 36, but for the sp^2 mass thickness.

In Table I are presented the results of the fitting and some remarks on the growth of very thin a-C films. In addition, we have to mention here some results obtained from X-ray reflectometry and atomic force microscopy studies [94]. According to them, the a-C films deposited with negative biases are atomically smooth, and those deposited with positive biases exhibit a rms surface roughness of the order of 5 Å for films ≈ 500 Å thick. This means that the a-C films developed with negative biases do not have to be analyzed with a two-layer model.

Table I. The Parameters α and n Obtained by Fitting the Combined Results on Composition and Thickness of a-C Films Shown in Figures 36 and 37, Respectively, with the Expression $dm_{sp^i} = \alpha t^n$

V_b (V)	dm_{sp^3} α (Å/min)	n	dm_{sp^2} α (Å/min)	n	Remarks
+10	8.3	0.83	3.2	1.4	Saturation in the formation of the sp^3 phase, which is not formed above $\approx$200–300 Å.
−40	6	1.01	10	1	Both sp^2 and sp^3 phases coexist during growth, and a-C film is homogeneously grown.
−160	1.9	1.2	20.2	0.9	Saturation in the formation of the sp^2 phase above $\approx$400 Å.

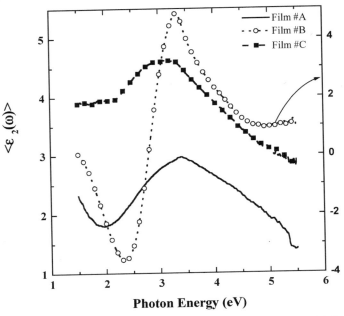

Fig. 39. The imaginary part of the measured pseudo-dielectric function $\langle \varepsilon(\omega) \rangle$ versus photon energy for three a-C films #A, #B, and #C grown with the same thickness $\approx$110 nm.

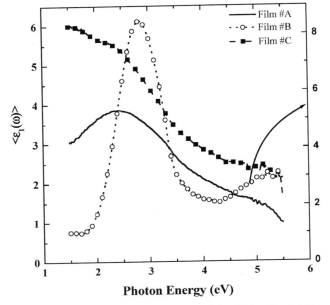

Fig. 38. The real part of the measured pseudo-dielectric function $\langle \varepsilon(\omega) \rangle$ versus photon energy for three a-C films #A, #B, and #C grown with low, medium, and high Ar$^+$ ion energies, respectively, and the same thickness $\approx$110 nm.

Furthermore, we have to examine the carbon phases and the geometrical model assumed in the above discussion, especially in the case of films deposited with high negative biases. In particular, the reference $\varepsilon(\omega)$ for sp^3 and sp^2 phases when used alone in BEMT analysis can fail to fully describe the dielectric response of these films, because of the formation of (i) a new form of carbon with different dielectric response than the sp^2- and sp^3-carbon-bonded materials [74, 83], as we will discuss below, and (ii) a SiC polycrystalline phase at the initial stages of growth ($\sim$1 nm) [95–97].

Having discussed the in situ SE monitoring of the very thin a-C films during deposition and the procedure to get quantitative results about the films' growth, we proceed to the description of optical properties of thicker a-C films. Figures 38 and 39 display the real ($\langle \varepsilon_1 \rangle$) and the imaginary part ($\langle \varepsilon_2 \rangle$) of the pseudo-dielectric function of the three representative types

of a-C films. In particular, these films were grown with low (#A), medium (#B), and high (#C) Ar$^+$ ion energies, to the same thickness of about 110 nm, and with the other deposition parameters in the rf MS process identical [74]. As we can see in these figures, the films exhibit profound differences that can be ascribed to their optical transparency or opacity, which are directly related to the large or small numbers of sp^3 sites in the a-C films, respectively. In detail, the optical transparency of the film induces attenuating interference fringes, which are due to the multiple reflections of the incident light beam at the film–substrate interface. This effect is pronounced in film #B, whereas in films #A and #C the effect of back reflection is reduced. That is, films #A and #C are more absorbing than film #B. Also, film #C exhibits higher absolute $\langle \varepsilon_2 \rangle$ values in the whole energy region and is the more absorbing one. Thus, if we take into account their different optical responses, this can be used as a criterion to classify the a-C films into three types. In type I belong the films grown with positive or zero bias voltage V_b applied to the substrate during deposition. In types II and III belong the films grown with $-100 \leq V_b < 0$ V and $-200 \leq V_b < -100$ V, or Ar$^+$ ion energies $130 > E_i \geq 30$ eV and $230 \geq E_i > 130$ eV, respectively.

In order to extract the bulk dielectric function $\varepsilon(\omega)$ of the a-C films from the measured $\langle \varepsilon(\omega) \rangle$ we used an inverse fitting procedure [24, 74]. In this mathematical inversion fitting procedure all the quantities that are involved in the three-phase (air–a-C-film–Si-substrate) system are known except the bulk dielectric function $[\varepsilon(\omega) = \varepsilon_1 + i\varepsilon_2]$ of the a-C film, which is deduced with a Newton method. More specifically, the pseudo-dielectric function $\langle \varepsilon(\omega) \rangle$ of the a-C-film–Si-substrate system and the dielectric function $\varepsilon(\omega)$ of the Si substrate (after ion etching) are

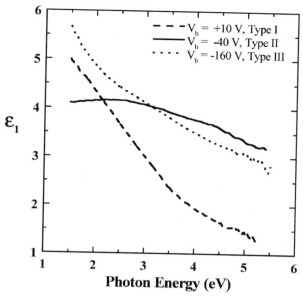

Fig. 40. The real part $\varepsilon_1(\omega)$ for three representative a-C films of type I, II, and III deposited by magnetron sputtering at different bias voltages V_b applied to the substrate during growth.

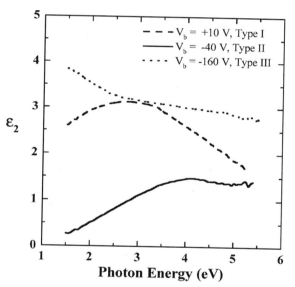

Fig. 41. The imaginary part $\varepsilon_2(\omega)$ of three representative a-C films of type I, II, and III deposited by magnetron sputtering at different bias voltages V_b.

measured (known) quantities, and the film thickness d is known (for example, calculated by SE, XRR, and XTEM) with high accuracy too. Then Eq. (18) is solved at each photon energy for the two unknown quantities ε_1 and ε_2 of the film.

In Figures 40 and 41, the real and imaginary parts, respectively, of the bulk dielectric functions obtained through the inverse fitting procedure for the films of types I and II are presented. The bulk $\varepsilon(\omega)$, for a-C films of type III, is measured directly from an a-C film ≈ 180 nm thick, where no contribution of the Si substrate occurs, since these films are very absorbing in the energy region 1.5–5.5 eV. The results of this analysis are

in agreement with those found with the Lorentz oscillator description [74]. Films of type I exhibit high absolute ε_2-values in the range 1.5–4.0 eV, an optical gap below 0.5 eV, and a fast reduction in $\varepsilon_2(\omega)$ above 4.5 eV, the characteristic behavior of graphite as discussed in Sections 5.1–5.3. The low absolute values in $\varepsilon_2(\omega)$ over the whole energy range 1.5–5.5 eV in films of type II, and the appearance of an optical band gap ≈ 1.5 eV, are in agreement with the fact that these films contain larger percentages of sp^3 sites. On the other hand, $\varepsilon(\omega)$ in films of type III exhibits high absorption over the whole spectral range and an interband transition at ≈ 1.4 eV [74, 83]. The optical response of these films is rather unexpected if we assume that the C–C bonding configurations in the material are only the well-known sp^2 and sp^3 ones. Additional studies on the films by XRR and stress measurements showed that these films are denser with lower stress and sp^3 content than the ones of type II [73]. Nevertheless, a complete analysis of the amorphous carbon films, with regard to their the electronic and optical properties versus their different bonding configurations in the energy region from IR to deep UV, will be presented in a forthcoming publication [84].

5.6. Multiwavelength Real-Time Ellipsometry to Study the Kinetics of the Growth of Carbon-Based Films Deposited with Sputtering Techniques

Real-time monitoring is essential for the production of modern engineering materials. Particularly, nondestructive optical access to a surface in vacuum is important for real-time monitoring, control, and characterization during film development, providing reduction in production time and increase of production yield. If the data collected in real time can be also interpreted in real time, then it becomes possible to control materials' characteristics through a closed-loop adjustment of process variables, such as an external bias voltage applied to the substrate during deposition, which controls the bonding and composition of the carbon-based materials. From a technological point of view, this means that one can assess process reproducibility and quickly attain the desired process variables. Even if the real-time interpretation is not possible and only one characteristic of the thin film is obtained, the information deduced in a postprocess analysis of the real-time measurements can be applied to a better understanding of the process.

In situ spectroscopic ellipsometry (SE) is a surface-sensitive optical technique used to monitor deposition rates and composition of films during deposition and can be used in a complementary manner with other postdeposition techniques such as Auger electron spectroscopy (AES) and X-ray photoelectron spectroscopy (XPS) [98]. In this subsection we will present a real-time SE monitoring process by multiwavelength ellipsometry (MWE) of a-C film deposition by closed field unbalanced magnetron sputtering (CFUBMS) [22, 23], which is a technique achieving high deposition rates and is applied on an industrial scale. The high deposition rate of a-C by CFUBMS is used to test the acquisition speed and software analysis limits of the currently available SE MWE technology.

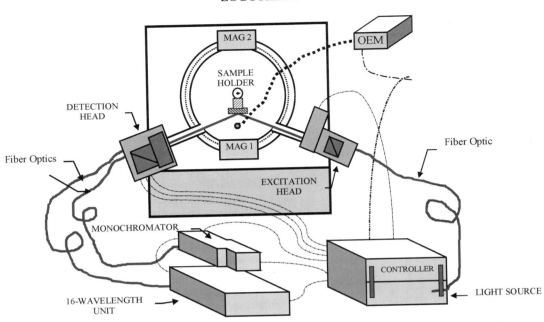

Fig. 42.   A schematic representation of the high-vacuum closed field unbalanced magnetron sputtering (CFUBMS) system, the magnetrons (MAG 1, MAG 2) and the sample holder, and the attached multiwavelength ellipsometer, with the excitation and detection heads, the light sources, the 16-wavelength unit, and the monochromator. The monochromator allows the whole ellipsometric unit to be used as a spectroscopic ellipsometer in the energy region 1.5–5.5 eV as well. A fiber optic in front MAG 1 is used for optical emission spectroscopy (OEM).

The deposition experiments were performed in a high-vacuum system with base pressure better than 10^{-5} Torr by using the CFUBMS technique. The deposition system and the ellipsometers used for these experiments as well as their integration are shown in Figure 42. The high-vacuum deposition system consists of two or four high-field-strength magnetrons with graphite targets, mounted on the vertical walls of the cylindrical chamber. Plasma etching of c-Si and pre-sputtering were performed prior to carbon and carbon nitride thin film deposition. The partial pressures of working gas (Ar) and reactive gas (N_2) were 3 and 1.6 mTorr, respectively. The real-time measurements were performed with a 16-wavelength phase-modulated ellipsometer, which was attached to the CFUBMS vacuum chamber.

The ellipsometer provides also the possibility of SE measurements in the energy region 1.5–5.5 eV. With the 16-wavelength ellipsometer are obtained simultaneously the ellipsometric angles Ψ and Δ for the 16 different wavelengths of the film–substrate system in continuous, distinct, timed steps. The 16 wavelengths are distributed in the Vis–UV energy region from 1.52 to 4.19 eV. The dielectric function $\varepsilon(\omega)$ of the system is calculated directly from Ψ and Δ through the complex reflection ratio $\tilde{\rho}$ ($= \tan \Psi \, e^{i\Delta}$). In the case of thin films the measured quantity is the pseudo-dielectric function $\langle \varepsilon(\omega) \rangle$ that includes the effect of the substrate. The speed of the real-time measurements depended on the integration time (IT) used for the simultaneous acquisition of the 16 wavelengths. The smaller the IT value, the closer the monitoring of the phenomena. In order to monitor the CFUBMS processes, $10 \leq \text{IT} \leq 125$ ms must be used, and subsequently the sampling time (ST) for ev-

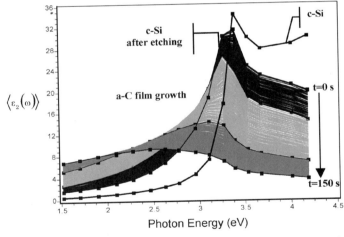

Fig. 43.   Real-time 16-wavelength spectra of $\langle \varepsilon_2(\omega) \rangle$ during a-C film growth on a c-Si substrate for 150-s total deposition time. The MWE spectra of the substrate before and after etching are also plotted.

ery step measurement must be at least slightly longer than 10 and 125 ms, respectively. The IT = 10 ms is set by the accuracy and the reproducibility of the MWE measurements. The results presented in this work were obtained with IT = 62.5 ms, providing a total time of 1 s for each 16-wavelength spectrum. In Figure 42 is shown the use of the 16-wavelength ellipsometer with the high-vacuum CFUBMS deposition chamber, which is used for real-time MWE and in situ SE measurements. The ellipsometers have the same excitation and detection head, and the angle of incidence is 70.05° [22, 23].

The real-time monitoring of thin film optical properties is achieved with the spectral response of dielectric function at 16 wavelengths in the energy region 1.52–4.19 eV. Figure 43 presents the evolution of the imaginary part $\langle\varepsilon_2(\omega)\rangle$ versus photon energy during the deposition of an a-C film grown by unbalanced magnetron sputtering for 150-s total deposition time. Figure 44 presents $\langle\varepsilon_2(\omega)\rangle$ measured by the same procedure during deposition of CN_x films grown with the same technique. The multiwavelength spectra were collected with IT = 0.25 s and ST = 0.50 s for the first 30 s of deposition. After the first 30 s, in order to reduce the number of spectra, the IT and ST were set at 0.25 and 1 s, respectively.

In continuing the discussion on the modeling of the initial stages of growth in Section 5.5, we should mention here that at these stages of a-C and CN_x film growth on a Si

substrate, phenomena such as interdiffusion, chemical reactions, and SiC formation [95, 96] also take place, depending on the deposition conditions. On the other hand, the volume and the rate of data collection by real-time MWE are very high and yield more or less enough information about these phenomena. The problem is how to handle the SE data and how to model them. For example, the simplified one-layer model we used in the data analysis in Section 5.5 may not be able to provide all the details, but at least provides reliable information on whether or not inhomogeneous film growth takes place. A more sophisticated model of analysis is required to describe quantitatively the a-C-film–Si and the CN_x–Si substrate systems. However, a more sophisticated model increases the free parameters, a fact that affects its accuracy and reliability. That is beyond of the scope of this work; instead (i) describe the phenomena at the a-C–Si interface qualitatively, leaving details about the a-C–Si interface phenomena under ion bombardment during deposition and layer-by-layer growth to [95–99], and (ii) discuss and compare the a-C film growth mechanisms in different magnetron sputtering techniques and the differences between a-C and CN_x films prepared by CFUBMS, in an attempt to optimize the techniques and to tailor the films' properties for specific applications.

Thus, for the analysis of measured $\langle\varepsilon(\omega)\rangle$ spectra we applied the BEMT in combination with the three-phase (air–composite-film–c-Si substrate) model [27, 62], assuming that the composite film, of thickness d, consists of sp^2 and sp^3 phases (see for example the discussion in Section 5.3) and voids. Two representative examples of this analysis are shown in Figure 45, where are plotted two experimental MWE $\langle\varepsilon(\omega)\rangle$ spectra, among those collected during deposition, at times $t = 15$ and 150 s of Figure 43, together with the corresponding fitted spectra (solid lines). By the described analysis of the measured $\langle\varepsilon(\omega)\rangle$ we calculated the thickness, deposition rate, and composition

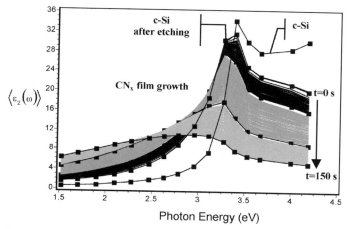

Fig. 44. Real-time 16-wavelength spectra of $\langle\varepsilon_2(\omega)\rangle$ during CN_x film growth on a c-Si substrate for 150-s total deposition time. The MWE spectra of the substrate before and after etching are also plotted.

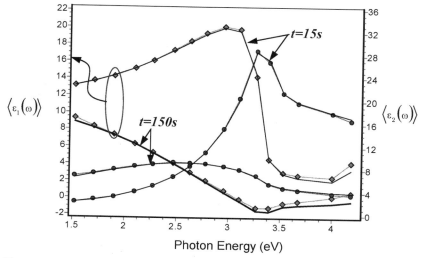

Fig. 45. The measured real [$\langle\varepsilon_1(\omega)\rangle$, closed diamonds] and imaginary [$\langle\varepsilon_2(\omega)\rangle$, closed circles] parts of the pseudo-dielectric function during growth of an a-C film at deposition times $t = 15$ and 150 s, and the fitted results (solid lines) based on the analysis with one-layer model in combination with BEMT.

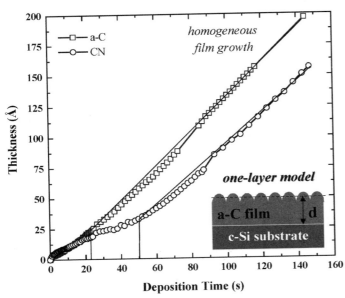

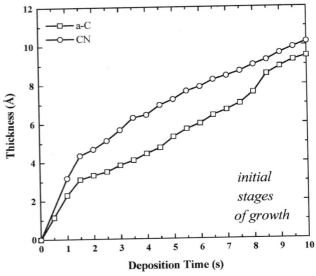

Fig. 46. Evolution of thickness of a-C and CN$_x$ films, grown by closed field unbalanced magnetron sputtering (CFUBMS) under identical conditions, obtained from real-time MWE data analysis, based on the one-layer model (see inset) and assuming that the grown films, with an average thickness d, consist of sp^2 and sp^3 carbon phases and voids.

Fig. 47. Evolution of thickness of a-C and CN$_x$ films, as in Figure 46, but at the very initial stages of growth.

(not presented here) of representative (i) a-C and CN$_x$ films deposited by CFUBMS and (ii) a-C films deposited by planar dc MS with dc bias voltage pulses applied to the substrate (for more details see [23]).

The a-C and CN$_x$ films were deposited by CFUBMS using identical deposition conditions [23]. The only difference between them is the introduction of nitrogen gas (partial pressure $P_{N_2} = 3 \times 10^{-3}$ Torr) and the equivalent reduction of the argon partial pressure in the chamber when depositing the CN$_x$ film.

The results for the best-fit evolution of the thickness d of the a-C and CN$_x$ films are given in Figures 46 and 47. In Figure 47, the scale is magnified to show the details at the very initial stages of growth. The results of Figure 47 provide quantitative signatures of the nucleation processes of a-C and CN$_x$ materials. In a-C the nuclei increase in size to ≈ 10 Å in the first 7 s and then make contact until 20 s, and for $t > 20$ s the bulk layer thickness increases linearly with time. In CN$_x$ the nuclei initially increase more abruptly in size, to ≈ 25 Å in the first 35 s. The coalescence stage lasts until 60 s, and for $t > 60$ s we obtain the linear time dependence of film thickness that signifies bulk homogeneous CN$_x$ layer deposition.

The differences deduced in the deposition rates between the two CFUBMS materials deposited under identical conditions are due to the presence of nitrogen in the growth of CN$_x$ films. In the stage of homogeneous film growth presence of the nitrogen leads to reduction of the deposition rate, mainly due to the preferential sputtering under the ion bombardment at the growing film surface. At the initial stages of growth the main differences are due to the chemical reactions of the nitrogen species with the Si substrate and their mobility at the growing film surface, which dissociate the sputtered carbon clusters and

affect the bonding structure and bonding configuration. More details about the formation and bonding structure and configurations of CN$_x$ films will be presented in Section 5.7.

A floating bias and a pulsed negative dc bias voltage ($V_b = -50$ V) were applied to the Si substrate for the deposition of a-C films by dc MS. The results for the best-fit evolution of thickness are presented in Figure 48a, whereas in Figure 48b they are magnified to show the details at the initial stages of growth. The nucleation and coalescence stages are not completely different in the two a-C films [23] as in the case of rf-magnetron-sputtered films presented in Section 5.5, because the applied negative dc bias here is low. Even so, when a negative dc bias $V_b = -50$ V is applied, the nucleus size is smaller and the duration of nucleus contact is shorter than with floating bias. This is more or less expected, due to the contribution of Ar$^+$ ion bombardment to the growth process when the negative bias is applied. Consequently, the stage of homogeneous film growth, identified by the linear time dependence of thickness (Fig. 48a), is characterized by a lower deposition rate.

In analyzing the MWE data [99] the process of the film evolution can be better described by the ideas of the two-layer model shown in Figure 49. Accordingly, we performed a linear regression of the thickness (according to one-layer model) calculated by MWE versus time deposition rate during homogeneous film growth, that is, while there is a clear steady state and a linear dependence of thickness on time (e.g., for the films of Fig. 48a for time beyond 250 s). The linear regression provides a deposition rate of a homogeneous layer (Fig. 49) that has the same properties during growth; only its thickness varies. On the other hand, a thin nucleation layer (or surface layer), where the C adatoms are sticking on the Si or a-C surface, is formed on top of the homogeneous layer. The thickness of this layer is determined by subtracting from the curve of total thickness versus time the contribution of the homogeneous layer. This analysis discriminates between nucleation and the

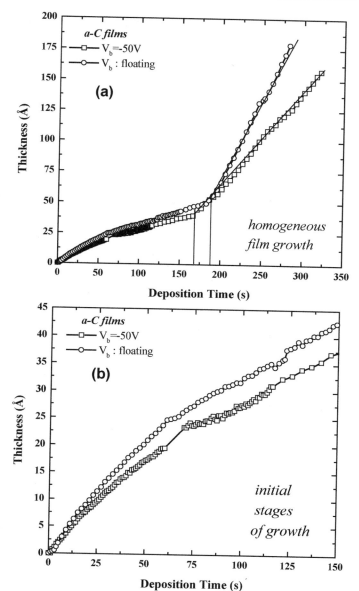

Fig. 48. Evolution of thickness of a-C films, grown by dc magnetron sputtering at different bias voltages V_b applied to the substrate, obtained from multiwavelength ellipsometry data analysis (a) at different stages of growth, and (b) at the initial stages of growth.

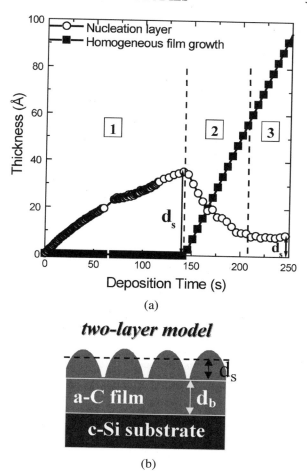

Fig. 49. (a) Evolution of thicknesses (d_s and d_b) of an a-C film deposited on a c-Si substrate with dc magnetron sputtering and pulsed dc biasing of the substrate, obtained by fitting the two-layer model. The three stages of growth—the nucleation (1), the coalescence (2), and homogeneous film growth (3)—are shown. (b) A schematic representation of the two-layer model with the nucleation layer (surface layer) d_s and the homogeneous bulk layer d_b, both consisting of the same carbon phases.

homogeneous layer, as shown in Figure 49a. At the very initial stages of growth (region 1), there is only the development of the nucleation layer and corresponding growth of a-C nuclei before they make contact with each other and cover the whole Si substrate. As the size of the carbon islands increases, the thickness d_s of this nucleation layer gradually increases to a maximum $d_s \approx 36$ Å at about 150 s. At the second stage of film growth (Fig. 49a, region 2), the Si substrate has been completely covered by the a-C film, there is no interaction of the deposited C species with the Si, and the homogeneous layer starts to form. At this stage the thickness d_s of the nucleation layer decreases to a final value ≈ 10 Å as the deposited C species fill the free

space between the a-C islands. In the steady state (region 3), the development of the homogeneous layer dominates the film growth. The open space between islands has been filled, and a very thin nucleation (surface roughness) layer exists due to the formation of smaller islands by the deposited species, which act as nucleation sites during deposition. These islands are formed continuously, contributing to the development of the homogeneous layer. By this type of analysis, based on the one-layer model and using the ideas of the two-layer model [99], we can find the differences (in the nucleation, coalescence, and homogeneous growth stages) between the a-C films grown by dc MS and those grown with pulsed dc bias, and between a-C and CN_x films deposited by CFUBMS.

5.7. Study of Carbon Nitride Films with FTIR Spectroscopic Ellipsometry

Continuous technological progress creates an increasing demand on the properties and functionality of materials. Improvements of the mechanical, optical, thermal, chemical, and electrical features of thin films, especially hard coatings, are highly desirable in modern industry. Carbon nitride (CN_x) thin films are among the good candidates for several technological applications. The theoretical prediction of the β-C_3N_4 phase [100, 101] and its interesting properties initiated vast efforts by several groups for its synthesis. Carbon nitride is now considered as an outstanding material, characterized by high hardness, low friction coefficient, chemical inertness, and variable optical bandgap. Its use is spreading to a wide variety of applications such as wear-resistant coatings, hard coatings, protective optical coatings, protective coatings on magnetic disk drives, cutting tools, and hard barriers against corrosion, and as a novel semiconductor material. This wide range of valuable application has led to an intensive investigation of CN_x films using several techniques, such as magnetron sputtering, ion-beam deposition, laser ablation, and plasma-enhanced chemical vapor deposition CVD [102]. However, there has been no clear evidence of the formation of crystalline stoichiometric β-C_3N_4 except for the existence of small crystallites of this phase, embedded in an amorphous carbon matrix [103–105]. The difficulty in its production arises from the large amount of N (57 at. %) that must be incorporated in the films. Nevertheless, nitrogen-deficient (≈ 20–40 at. %) carbon nitride films are found to exhibit interesting properties, comparable to other carbon-based materials. These properties have been attributed to the bending and cross-linking of the graphite planes through sp^3-coordinated bonds, leading to the formation of a fullerene-like microstructure through the formation of three-dimensional molecular structures [106–109].

Fourier transform IR spectroscopic ellipsometry (FTPME) can be used for the study of carbon nitride materials, since the analysis of the characteristic bands in the infrared can give a great deal of both qualitative and quantitative information about their bonding structure. This can significantly contribute to the understanding of the mechanisms that take place during the formation of the different bonding structures between carbon and nitrogen atoms. The excellent diamond-like properties that these materials exhibit are due to the sp^3 bonding fraction, which varies considerably with the synthesis parameters. Furthermore, during the growth of CN_x thin films by sputtering, there exists an intense positive ion bombardment on the growing film surface [110]. The energy E_i of the bombarding ions is controlled by the negative bias voltage V_b applied to the substrate through the relation $E_i = E_p + e|V_b|$, where E_p is the discharge energy. The use of FTPME can help to establish a definite correlation between the growth conditions (V_b) and the films' bonding structure, which straightforwardly reflects their optical, mechanical, and electrical properties as well as their thermal stability [9].

The role of nitrogen atoms in determining the degree of diamond-like character is complicated, because N has the ability to form several bonding configurations with C [9]. These include the sp^3-hybridized (tetrahedral, C—N), sp^2-hybridized (trigonal, C=N), and chain-terminating sp^1-hybridized (linear, —C≡N and —N≡C) bonds. The vibration frequency of each

bonding structure depends on the atom species and the bonding configuration, but also on its neighboring environment. Although the identification of bonding structures of different hybridization is possible, that is not the case for different bonding structures with the same hybridization state, which contribute to the FTPME spectra in overlapping vibration bands that are difficult to distinguish. Particularly, the existence of sp^3-hybridized C—N bonds in the films gives rise to an characteristic band in the wave-number region 1212–1270 cm^{-1}, while the existence of sp^2-hybridized C=N bonds is evidenced by a vibration mode appeared at ≈ 1530 cm^{-1} [9, 111–113]. The sp^2 C=C bonds are normally IR-inactive, since the C=C bond is a nonpolar bond. However, the contribution of this bond to the FTPME spectra is the result of N incorporation in the graphitic rings, which tends to destabilize the rings' planar geometry [114], rendering the sp^2 C=C bonds IR-active [115]. Furthermore, the existence of sp^2 C=N bonds in linear chains is evidenced by a characteristic band appearing in the wave-number region between 1650 and 1680 cm^{-1} [112]. The sp^1-hybridized bonding structures between carbon and nitrogen atoms are characterized by low IR responses. The vibration frequencies of the terminating C≡N stretching modes associated with both nitrile (—C≡N) and isonitrile (—N≡C) structures greatly depend on the type of component bonded to these structures. The contributions of the above structures to the FTPME spectra corresponds to ≈ 2250 and ≈ 2150 cm^{-1}, respectively [9, 111–113].

In the context of this discussion, the optical response of CN_x thin films deposited by reactive rf magnetron sputtering (see for example Figs. 24 and 25) on c-Si (100) substrates was studied [9, 116]. The films were deposited using a graphite target in a deposition chamber with a base pressure better than 1×10^{-7} mbar at room temperature (RT), by applying different V_b's on the substrate, film #A ($V_b = -20$ V), and film #B ($V_b = -250$ V). The substrates were located 65 mm above the target and coated using a sputtering power of 100 W. The partial pressure of the sputtering gas was 4×10^{-3} mbar. Both films have thicknesses ≈ 4500 Å. The FTPME measurements were performed in the wave-number range 900–3500 cm^{-1} at a resolution of 8 cm^{-1}, using the FTIR phase-modulated-ellipsometer, which has been described in Section 4.3. Figure 50 shows the imaginary part $\langle \varepsilon_2(\omega) \rangle$ of the pseudo-dielectric function for the two CN_x films measured, together with the characteristic bands corresponding to the various carbon–nitrogen bonding structures.

From Figure 50 it can be seen that the contribution of the various carbon–nitrogen bonding structures is different in the measured pseudo-dielectric functions of the two films. For example, the contribution of the sp^2-hybridized C=C bonds and sp^2-hybridized C=N bonds contained in linear chains is greater in film #A than in film #B. These differences in the IR response of the CN_x films are a result of the nitrogen distribution in the films and among the possible bonding configurations with carbon atoms. The mechanisms governing N incorporation in the a-C network are affected by the energy E_i, controlled by the applied V_b, of the positive ions (N$^+$, N^{2+}, C$_n^+$, (CN)$_n^+$, etc.) that bombard the growing film surface during the films' growth

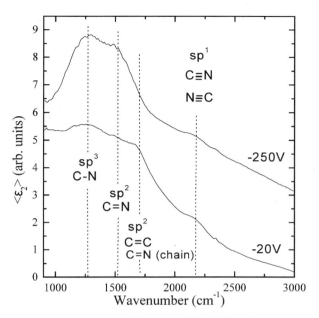

Fig. 50. Imaginary parts $\langle \varepsilon_2 \rangle$ of the pseudo-dielectric function for two carbon nitride (CN_x) films deposited by rf magnetron sputtering onto c-Si substrates by applying different bias voltages (film #A: $V_b = -20$ V; film #B: $V_b = -250$ V). The various sp^3, sp^2, and sp^1 hybridized carbon–nitrogen and C=C bonds are shown. The vertical dotted lines denote the positions of the characteristic band frequencies of the various carbon–nitrogen bonding structures.

[110]. The high-energy ion bombardment during deposition enhances the chemical reactions between the different species as well as their mobility at the growing film surface. As a result, the high-energy gas ions dissociate the sputtered carbon clusters, affecting the nitrogen distribution among the possible bonding configurations with carbon atoms, and thus the films' bonding structure.

The optical response of the CN_x films measured in the infrared spectral region is a direct result of the distribution of nitrogen atoms among the mentioned bonding configurations with carbon atoms. The bonding vibrations that are excited by the electric field of the IR beam are modeled, to a first approximation, using a damped harmonic oscillator (Lorentz model). The contribution of each oscillator to the measured pseudo-dielectric function is evidenced by a maximum in the imaginary part $\langle \varepsilon_2 \rangle$ together by an inflection in the real part $\langle \varepsilon_1 \rangle$. The study of the imaginary part is preferable, since the contribution of the vibrational modes is more pronounced in this representation. The effect of the several vibration modes on the complex dielectric function is described by the expression

$$\tilde{\varepsilon}(\omega) = \varepsilon_\infty + \sum_i \frac{f_i \omega_{0i}^2}{\omega_{0i}^2 - \omega^2 + i\Gamma_i \omega} \quad (48)$$

where ω is the energy of the light and ω_{0i} is the absorption energy of the ith vibration mode. The constants f_i and Γ_i are the oscillator strength and the damping (broadening) of the specific vibration mode, respectively. Considering a specific bond between certain atoms where the effective charge e^*, the reduced mass m, and the absorption energy ω_0 have fixed values, the os-

cillator strength is analogous to the concentration N_i (number of oscillators per unit volume) of the specific bond:

$$f_i \omega_{0i}^2 = \frac{4\pi N_i e^{*2}}{m} \quad (49)$$

The quantity ε_∞, described by the relation

$$\varepsilon_\infty = 1 + \frac{\omega_p^2}{\omega_0^2} \quad (50)$$

is the static dielectric constant and represents the contribution to the dielectric function of the electronic transition that occurs at an energy ω_0 in the NIR–Vis–UV energy region. The plasma energy ω_p is given by

$$\omega_p^2 = \frac{4\pi N_e e^2}{m^*} \quad (51)$$

where e and m^* are the electron charge and effective mass, respectively. N_e is the density of electrons [117, 118], which is associated with the existence of voids and structural imperfections and is a measure of the density of the film.

Figure 51 shows the real ($\langle \varepsilon_1 \rangle$) and imaginary ($\langle \varepsilon_2 \rangle$) parts of the pseudo-dielectric function for two CN_x films #A and #B and the best-fit simulations resulting from the fitting analysis of the experimental data, represented by solid lines [9]. For the fitting analysis we assumed a layer of thickness d on top of a Si substrate (bulk), while the optical response of the layer was modeled by three oscillators.

It is evident from the above discussion that the overlapping of the characteristic bands corresponding to the different carbon–nitrogen bonding structures has to be overcome in order to enhance the contribution of each bonding structure to the spectra. This can be performed by calculating the first derivative $d\langle \varepsilon(\omega) \rangle/d\omega$ of the pseudo-dielectric function $\langle \varepsilon(\omega) \rangle$, which allows the enhancement of the contribution of each bonding structure around the absorption bands [9]. The expressions for the real and imaginary parts of the first derivative of $\varepsilon(\omega)$ can be calculated through Eq. (48) and are given by

$$\frac{d\langle \varepsilon_1(\omega) \rangle}{d\omega} = -\sum_i \frac{2f_i \omega_{0i}^2 \omega \cdot [\Gamma_i^2 \omega_{0i}^2 - (\omega_{0i}^2 - \omega^2)^2]}{[(\omega_{0i}^2 - \omega^2)^2 + \Gamma_i^2 \omega^2]^2} \quad (52)$$

$$\frac{d\langle \varepsilon_2(\omega) \rangle}{d\omega} = -\sum_i \frac{f_i \Gamma_i \omega_{0i}^2 [(\omega_{0i}^2 - \omega^2)(\omega_{0i}^2 + 3\omega^2) - \Gamma_i^2 \omega^2]}{[(\omega_{0i}^2 - \omega^2)^2 + \Gamma_i^2 \omega^2]^2} \quad (53)$$

Thus, in order to obtain quantitative information about the vibrational as well as the electronic properties from the measured IR spectra, two types of analysis can be performed [9]. In the first analysis the experimental $\langle \varepsilon(\omega) \rangle$ data were analyzed based on Eq. (48) and assuming a CN_x layer of thickness d on top of a bulk substrate (c-Si) (see for example the results in Fig. 51), providing information about the film's bonding structure and its electronic properties as well as its thickness. In the second analysis, by calculating the first derivative and analyzing the experimental data by using Eqs. (52) and (53), assuming a bulk CN_x material around the region of the absorption bands,

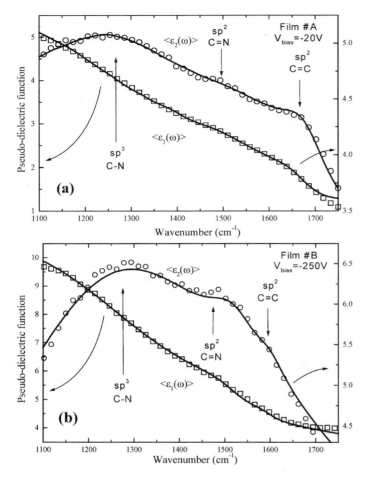

Fig. 51. Real ($\langle \varepsilon_1 \rangle$) and imaginary ($\langle \varepsilon_2 \rangle$) parts of the pseudo-dielectric function for two carbon nitride (CN_x) films deposited by rf magnetron sputtering onto c-Si substrates by applying different bias voltages (film #A: $V_b = -20$ V; film #B; $V_b = -250$ V). The solid lines represent the best-fit simulations resulting from the fitting analysis of the experimental data based on the theoretical model of $\varepsilon(\omega)$ described by Eq. (48) and assuming a CN_x film of thickness d on top of a c-Si substrate.

Table II. Results of Fitting the First Derivative of the Pseudo-dielectric Function $\langle \varepsilon(\omega) \rangle$ for Films #A and #B

Film	V_b (V)	ε_∞	Bond	F	ω_0 (cm^{-1})	γ (cm^{-1})
#A	-20	4.2	C$-$N	1.66	1338	750
			C$=$N	0.03	1524	177
			C$=$C	0.28	1693	322
			$-$C$\equiv$N	0.003	2233	104
			$-$N$\equiv$C	0.027	2161	193
#B	-250	6.8	C$-$N	2.6	1330	709
			C$=$N	0.123	1540	233
			C$=$C	0.11	1645	274
			$-$C$\equiv$N	0.008	2210	153
			$-$N$\equiv$C	0.021	2145	201

one can obtain information about the vibrational properties of the films. Combining the results from the above types of analyses with those of the dielectric function in the NIR–Vis–UV spectral region a detailed study of the bonding mechanisms, the electronic properties, and the microstructure of the films is completed.

In Figure 52 the first derivative of the real parts $\langle \varepsilon_1 \rangle$ of the pseudo-dielectric functions for films #A and #B are shown along with the simulated spectra. Chemical bonds, for example, for the sp^3 C$-$N, sp^2 C$=$N, and sp^2 C$=$N bonds can be identified by extrema in $d\langle \varepsilon_1(\omega) \rangle / d\omega$ much better than in the case of the direct $\langle \varepsilon(\omega) \rangle$ spectra as in Figures 50 and 51.

The observed differences in $d\langle \varepsilon \rangle / d\omega$ spectra between the CN_x films #A and #B prove that the bonding structure of the films is significantly affected by the energy of the ion bombardment during deposition. The characteristic band around 1300 cm^{-1} is attributed to the C$-$N bonds, while the one around 1530 cm^{-1} is assigned to the stretching vibration of

the sp^2-hybridized C$=$N bonds contained in aromatic rings [9, 111–113]. Since the stretching vibration frequencies of the sp^2-hybridized C$=$N bonds contained in linear chains and of the C$=$C bonds are reported in the region 1650–1680 cm^{-1} and at 1680 cm^{-1}, respectively [112], their contributions overlap the characteristic band at around 1680 cm^{-1}, making its separation difficult. The C$=$C vibration mode is normally IR-forbidden, since the C$=$C bond is a nonpolar bond. However, the contribution of this bond to the FTPME spectra is the result of nitrogen atom incorporation into the amorphous carbon network, where the excess of electrons coming from the nitrogen atoms induces charge variations in the C$=$C neighborhood and renders these bonds IR-active [115]. Also, the appearance of a small amount of the chain-terminating nitrile ($-$C$\equiv$N) and isonitrile ($-$N$\equiv$C) groups is obvious in all spectra as a broad peak centered at around 2200 cm^{-1}. The results of the fitting analysis on the experimental data concerning films #A and #B are summarized in Table II.

It is clear, based on the strength of the oscillators f_i, that in film #B, which is characterized by high-energy ion bombardment during deposition, the formation of sp^3 C$-$N as well as sp^2 C$=$N bonds is favored, even though the IR response of the latter is variable and very dependent on the local symmetry of the graphite rings. The reduction of the sp^2 C$=$C bonds in film #B could be the result of the existence of graphitic rings with more than one N atom because otherwise the distribution of N to a large number of rings would induce a large-scale distortion in the graphite planar symmetry. Furthermore, as far as sp^1-hybridized $-$C$\equiv$N bonds are concerned, their formation is gradually suppressed in film #B [9].

The variation of the N concentration in the films, as measured by X-ray photoelectron spectroscopy (XPS), cannot explain the observed differences in the bonding structure between films #A and #B [9, 119]. Therefore, these differences must be the result of the N bonding and distribution in the films as affected by the applied V_b, which controls the energy of the ion bombardment during deposition. Particularly, low-energy

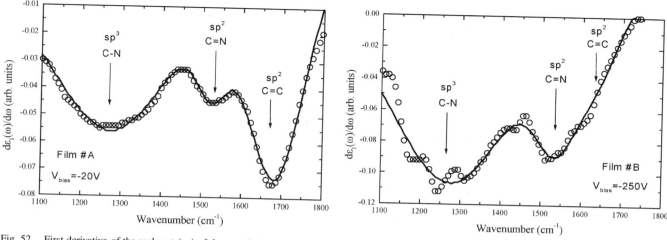

Fig. 52. First derivative of the real part $\langle \varepsilon_1 \rangle$ of the pseudo-dielectric function for two CN_x films deposited by rf magnetron sputtering onto c-Si substrates by applying different bias voltages (film #A: $V_b = -20$ V; film #B; $V_b = -250$ V) in the IR energy region 1100–1800 cm^{-1}, where the C—N, C=N, and C=C bonds are active. The solid lines represent the best-fit simulations resulting from the fitting analysis of the experimental data.

ion bombardment during deposition (film #A) promotes the homogeneous N distribution in a large number of graphitic rings and results in a large-scale distortion of the graphite plane symmetry. These distortions explain the enhanced contribution of the sp^2-hybridized C=C bond-stretching vibration in film #A's FTPME spectra.

In contrast, high-energy ion bombardment during deposition (film #B) promotes nonhomogeneous N distribution in local regions in which the nitrogen concentration is higher than that measured in the film. In these regions of increased N concentration, where a large number of graphitic rings contain more than one N atom, intense local distortions are induced, since the lengths of the C—N (1.47 Å) and C=N (1.22 Å) bonds are shorter than the lengths of C—C (1.54 Å) and C=C (1.33 Å) bonds. Relaxation of these distortions arises through the formation of pentagonal rings, which lead to local reduction of the distance between graphite planes and cross-linking with sp^3-coordinated bonds [106–109]. Thus, the formation of fullerene-like or C_3N_4 structures is expected. These three-dimensional bonding structures are capable of a high degree of bond angle deformation, resulting in highly elastic behavior, significantly improving the films' mechanical properties [119].

Additional support for the above discussion arises from the study of the static dielectric constants ε_s and ε_∞ as determined by analysis of the pseudo-dielectric function $\langle \varepsilon(\omega) \rangle$. Here ε_s describes the losses in the material in the whole electromagnetic region [9]. Figure 53 shows the evolution of ε_s and ε_∞ for a series of CN_x films of the same thickness (≈ 4500 Å) that were deposited by applying V_b in the range between +16 and −250 V [9]. From Figure 53 a clear trend is observed. In films grown with high negative V_b (high energy of the bombarding ions during deposition) the contribution of the IR-active bonds is enhanced. This is followed by an increase of ε_∞.

According to Eq. (50), the increase of ε_∞ suggests that either the kth electronic transition energy ω_{0k} decreases, or the plasma energy ω_{pk}, which is proportional to the density, increases. This

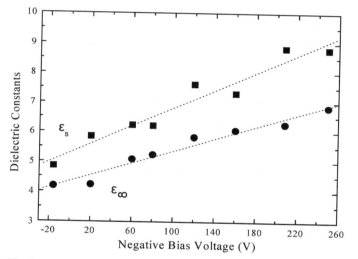

Fig. 53. Dependence of the static dielectric constants ε_s and ε_∞ of IR-active bond and electronic contributions as a function of applied V_b for a series of CN_x films of the same thickness (≈ 4500 Å), calculated through analysis of the pseudo-dielectric function $\langle \varepsilon(\omega) \rangle$ spectra. The dotted lines are guides for the eye.

is also demonstrated by considering the Kramers–Kronig integral

$$\varepsilon_1 (\omega = 0) = \varepsilon_\infty = 1 + \frac{1}{2\pi} \int_0^\infty \frac{\varepsilon_2(\omega') \, d\omega'}{\omega'} \quad (54)$$

where the static dielectric constant ε_∞ is inversely proportional to ω and proportional to $\varepsilon_2(\omega)$ (the density of states of electronic transitions), which is in turn proportional to the film's density.

The correlation between the variation of ε_∞ and the bonding structure of the films can be examined by studying their electronic properties by SE measurements in the NIR–Vis–UV spectral region (1.5 to 5.5 eV) [9]. The analysis of the pseudo-dielectric function $\langle \varepsilon(\omega) \rangle$ of the films in this energy region provides further insight through the study of the $\pi-\pi^*$ (attributed to sp^2 bonds) and $\sigma-\sigma^*$ (attributed to both sp^2 and sp^3 bonds) electronic transitions [9]. In particular, this analysis reveals that the energy of both $\pi-\pi^*$ and $\sigma-\sigma^*$ electronic transitions decreases in films grown with high-energy ion bombardment during deposition. This decrease is followed by an increase of their strength [higher $\varepsilon_2(\omega)$ values] as well as by an additional oscillator, at ≈ 1.5 eV, needed for describing the measured pseudo-dielectric function. This structure has been correlated, in the amorphous carbon films, with the formation of a dense carbon phase [74, 120], however, in the case of CN$_x$ films it is not clear yet whether it is correlated with carbon–carbon and/on carbon–nitrogen bonds too.

Finally, the ability of FTPME to perform a precise determination of thin film bonding structure has just been illustrated. FTPME is well adapted for the study of carbon-based materials, and by combining it with SE in the NIR–Vis–UV spectral region it can provide important information about the vibrational and electronic properties [9] of the materials under study. This capability constitutes a crucial advantage for studies in which information about the type as well as the amount of chemical bonds is needed. Furthermore, this technique can also probe the local structural order of thin films. Thus, taking into consideration its high sensitivity and capability for real-time measurements, it is anticipated that FTPME will be used extensively in the future for process monitoring as well as for probing the growth mechanisms of a wide variety of thin film materials.

6. OPTICAL CHARACTERIZATION AND REAL-TIME MONITORING OF THE GROWTH OF METALLIC TiN$_x$ FILMS

6.1. The Dielectric Function of TiN$_x$ Films

One of the most interesting and important categories of engineering materials are the metal nitrides, used mainly as hard, protective coatings. Especially, titanium nitride (TiN$_x$) thin films ([N]/[Ti] close to 1) with the NaCl crystalline structure are known to exhibit a unique combination of high hardness with excellent wear and corrosion resistance [121] and are widely used as coatings for cutting tools and wear parts [122]. Moreover, TiN$_x$ has lately gained much interest in different areas of Si device technology as a diffusion barrier in metallization schemes, in rectifying and ohmic contacts [123], and in Schottky barrier contacts [124]. TiN$_x$ exhibits a typical metallic optical response due to the intersection of the Ti 3d valence electron band by the Fermi energy level [125]. In addition to that response, which is due to intraband absorption by the Ti 3d electrons, interband absorption takes place in the visible energy region [126]. This behavior is strikingly similar to that of the noble metals, where the electronic interband transition is responsible for the characteristic color. In partiuclar, in stoichiometric TiN an interband transition occurs in the same energy region as in gold, and it therefore exhibits a gold like color. Figure 54a shows the dielectric function of a pure, stoichiometric polycrystalline TiN film of thickness about 1000 Å, in the

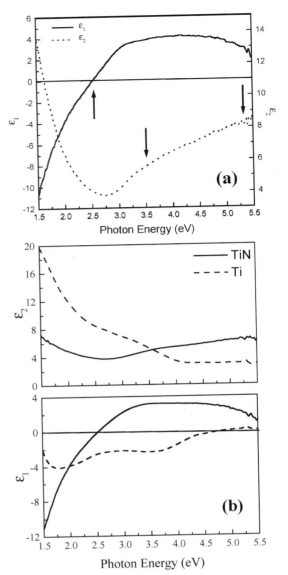

Fig. 54. The dielectric function (real part ε_1 and imaginary part ε_2) of: (a) A pure, stoichiometric polycrystalline TiN film of about 1000 Å. The characteristic features in the dielectric function of TiN are the screened plasma energy in $\varepsilon_1(\omega)$ and the energies of the interband absorption in $\varepsilon_2(\omega)$, all denoted by arrows. (b) TiN and pure Ti films, in the energy region 1.5–5.5 eV, measured by spectroscopic ellipsometry in vacuum after deposition. The high energy of the interband transitions of TiN distinguishes clearly the intraband (up to ≈ 2.5 eV) and interband (above ≈ 2.5 eV) absorption regimes. In pure Ti the intraband regime is not clear, due to the shift of the interband transitions to lower photon energy.

energy region 1.5 to 5.5 eV. Figure 54b shows for comparison the dielectric functions $\varepsilon(\omega)$ ($= \varepsilon_1 + i\varepsilon_2$) of bulk TiN and pure polycrystalline Ti films, both deposited by dc MS, in the same energy region.

In situ SE and MWE have been used to study and characterize TiN$_x$ films, and a detailed monitoring procedure by MWE has been developed in order to determine the film thickness and stoichiometry as well as other microstructural details during deposition [22, 126].

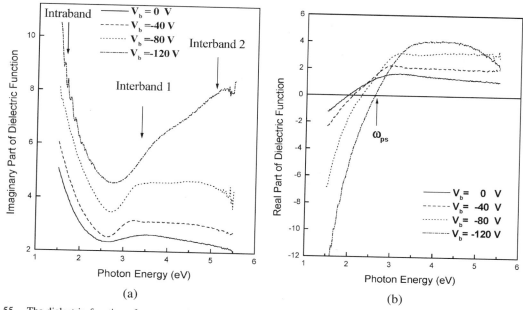

Fig. 55. The dielectric function of representative TiN$_x$ films deposited with different V_b: (a) imaginary ε_2, and (b) real part ε_1. Arrows in (a) denote the intraband contribution and the position of interband transitions, whereas in (b) the arrow denotes the position of ω_{ps}. The deposition conditions strongly influence both the intraband and interband absorption due to stoichiometry and microstructure variations. (After [129].)

Two sets of TiN$_x$ thin films were deposited on (001) Si wafers by dc reactive magnetron sputtering (conventional and unbalanced) from a Ti target (99.999% purity) up to a thickness of 1000 Å. The experimental deposition chambers were already described and are shown in Figures 24 and 42 (see the discussion of a-C in Section 5). The main difference between the sputter deposition of TiN and the deposition of a-C is in the reactive process, viz., for TiN the sputtering is done by using a mixture of working inert gas (Ar) and reactive gas (N) to promote the Ti—N chemical reaction. The reactive sputtering process is very sensitive and requires careful selection of process parameters as well as extreme stability of gas flows and pressures. It was found that there is a very narrow window of process parameters to produce stoichiometric TiN by reactive sputtering [127–129]. Here we briefly discuss the most important conditions and parameters affecting the properties and quality of the deposited films. The Si substrates were cleaned by standard surface cleaning procedures before entering the deposition chamber and then by a dry ion etching process, with very low-energy Ar$^+$ to remove the native SiO$_2$ and avoid substrate damage. In the conventional MS a gas mixture of Ar and N$_2$ was used at a flow rate of 15 sccm and 2.2 sccm, respectively, producing a total pressure of 8.3×10^{-3} mbar. The discharge power on the Ti target was 450 W, and the substrate-to-target distance 65 mm. The final structure, stoichiometry, and properties of TiN$_x$ thin films mainly depend on the N$_2$ flow and the bias voltage V_b applied to the substrate [127–129]. As an example, we report here on the latter dependence. In order to develop films with different stoichiometries, deposition was carried out at various values of bias voltage V_b applied to the substrate be-

tween 0 and −220 V and substrate temperatures from RT to 650 °C with fixed N$_2$ flow.

In situ SE spectra of the dielectric function were obtained with a phase-modulated spectroscopic ellipsometer mounted on the deposition system (see Fig. 24) and operated in the energy region from 1.5 to 5.5 eV with a step of 20 meV. SE measures directly the complex dielectric function $\varepsilon(\omega)$ of the materials and enables the investigation of their electronic structure [e.g., $\varepsilon_2(\omega)$ is related directly to the joint density of states for interband absorption] and their structural and morphological characteristics. Figure 55 shows the real and the imaginary parts of $\varepsilon(\omega)$ obtained from a number of fcc TiN$_x$ films deposited at four different bias voltages. It is clearly seen that $\varepsilon(\omega)$ is strongly dependent on the applied bias voltage. Of special interest, however, is the plasma energy ω_{pu} of the TiN$_x$, which depends strongly on the stoichiometry of the material [127–129]. FTPME measurements in the energy region 900–3500 cm^{-1} were performed using a Jobin–Yvon FTIR phase-modulated ellipsometer. The incident beam was initiated by an IR spectrometer including a glow bar source and a Michelson interferometer (see Section 4.3). The spectral resolution of the FTPME measurements was 16 cm^{-1}, and the average total number of measurements per spectrum was 600.

6.2. Optical Response of TiN$_x$ Films and Correlation with Stoichiometry

TiN, like all real metals, exhibits a dielectric function that can be analyzed by the Drude and Lorentz models. The Drude model, which was proposed for ideal metals, describes phenomenologically very well the intraband transitions that cor-

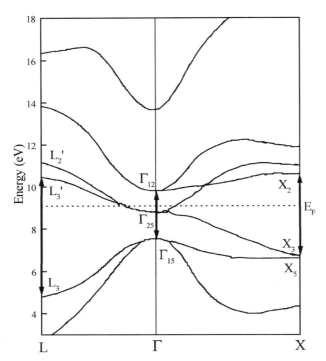

Fig. 56.   Band energy diagram for the bulk stoichiometric TiN (based on the band structure calculations of [130]). We have assigned the interband transitions (denoted by arrows) to the $\Gamma_{15} \rightarrow \Gamma_{12}$, $X_5 \rightarrow X_2$, and $L_3 \rightarrow L'_3$ points of the Brillouin zone, which correspond to ≈ 2.8, 3.9, and 6.5 eV calculated by the SE data analysis based on Lorentz oscillators.

respond to the optical excitation of an electron from below the Fermi energy to another state above the Fermi energy but within the same band. There is no threshold energy for such transitions; however, they can occur only in metals. In the case of TiN only the Ti 3d valence electrons contribute to the intraband absorption, since all the N and the other Ti electrons are bound below the Fermi energy level. Insulators do not have partially filled bands that would allow excitation of an electron from a filled state below the Fermi energy to an empty state within the same band. That is, of course, what makes an insulator nonconducting. On the other hand, the Lorentz model can describe classically the interband transitions, which are optical excitations of electrons to another band. The direct interband transitions involve only the excitation of an electron by a photon. Since the momentum of a photon is very small compared with the range of values of crystal momentum in the Brillouin zone, conservation of total crystal momentum of the electron plus photon means that the wave vector $\vec{k}$ for the electron is essentially unchanged in the reduced zone scheme.

Direct interband transitions have a threshold energy. It is the energy required for the transition from the Fermi level at $\vec{k}_0$ to the same state $\vec{k}_0$ in the next higher band. This threshold energy is analogous to that for the excitation of an electron across the bandgap in an insulator, i.e., even in metals, like TiN, interband transitions occur at certain photon energies. In particular, in crystalline TiN, based on the band structure in [130], we have assigned the interband transitions (see Fig. 56) to $\Gamma_{15} \rightarrow \Gamma_{12}$,

$X_5 \rightarrow X_2$, and $L_3 \rightarrow L'_3$ of the Brillouin zone, which correspond to strong interband absorption at ≈ 2.8, 3.9, and 6.5 eV [126]. There is also a weak transition at ≈ 2.2 eV assigned to $\Gamma'_{25} \rightarrow \Gamma_{12}$, which is hardly seen in the current SE data, due the Drude term and the $\Gamma_{15} \rightarrow \Gamma_{12}$ transition. This transition can only be discriminated from the Drude and the other interband contributions in the second-derivative spectra of the dielectric function and for TiN$_x$ films with large crystalline grains and low defect density [126]. In any case, this first interband transition should be responsible for the characteristic color of the TiN$_x$ films.

The optical properties depend strongly on the stoichiometry and the structural characteristics of the TiN$_x$ films as in all real metals. Assuming such an ideal metal (only free electrons contribute to its properties and the energy loss mechanisms due to scattering are very low), obtain

$$\tilde{\varepsilon}(\omega) = 1 - \frac{\omega_{pu}^2}{\omega^2 + i\omega\Gamma_D} \approx 1 - \frac{\omega_{pu}^2}{\omega^2 + \Gamma_D^2} \quad (55a)$$

so that, when $\Gamma_D^2 \ll \omega_{pu}^2 \approx \omega^2$,

$$\tilde{\varepsilon}\,(\omega \approx \omega_{pu}) = \varepsilon_1\,(\omega \approx \omega_{pu}) = 0 \quad (55b)$$

The energy ω_{pu} is of special interest and is called unscreened plasma energy of the material.

Since Γ_D ($\ll \omega_{pu}$) is very low in an ideal metal, the imaginary part ε_2 of the dielectric function around this energy ($\omega \approx \omega_{pu}$) approaches zero as

$$\varepsilon_2\,(\omega \approx \omega_{pu}) = \omega_{pu}^2 \Gamma_D/\omega^3 \approx \Gamma_D/\omega_{pu} \quad (56)$$

However, the existence of interband transitions (bonded electrons) at energies lower than ω_{pu} shifts the energy position where $\varepsilon_1 = 0$, and thus ω_{pu}, to lower energy. This latter energy is called the screened plasma energy ω_{ps}. The discrimination between ω_{pu} and ω_{ps} in TiN$_x$ can be achieved by using a model to describe the optical response, e.g. the SE data, including a Drude (free electron) and two Lorentz oscillator (bonded electron) contributions. In addition, the results of the SE analysis can be combined with the results obtained from X-ray photoelectron spectroscopy (XPS) and Auger electron spectroscopy (AES) [129] as well as from resistivity measurements [131] in order to study the TiN$_x$ stoichiometry and metallic character, respectively, in terms of the ω_{pu} and ω_{ps}. All these will be described in detail in the following subsections.

The unscreened plasma energy ω_{pu} depends on the concentration of the free electrons in the material and is defined through the relation [132]

$$\omega_{pu} = (4\pi Ne^2/m^*)^{1/2} \quad (57)$$

where N is the free electron (carrier) density, e is the electron charge, and m^* is the electron effective mass. Since ω_{pu} is directly correlated with the free electron density, it can be used to determine the metallic character and the stoichiometry of the TiN$_x$. Indeed, an explicit relation has been found between ω_{pu} and electrical resistivity, as we will discuss below. On the

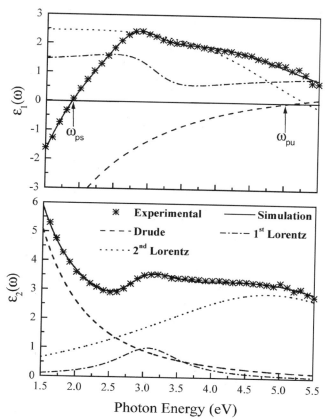

Fig. 57. The real (ε_1) and the imaginary (ε_2) part of the dielectric function of a 1000-Å-thick TiN$_x$ film deposited with $V_b = -40$ V and $T_d = $ RT. The experimental dielectric function (solid lines with asterisks) has been fitted with three terms: the Drude term (dotted lines) describing the optical response of free electrons, and two Lorentz oscillators located at 3.3 eV (dashed–dotted lines) and 5.2 eV (dashed lines) and corresponding to the TiN$_x$ interband transitions. Arrows on the graph of ε_1 denote the screened (ω_{ps}) and unscreened (ω_{pu}) plasma energies. The characteristics of the Drude term (ω_{pu} and Γ_D) and the energy position of the Lorentz oscillators are the fingerprints of the stoichiometry and microstructure differences of the TiN$_x$ films.

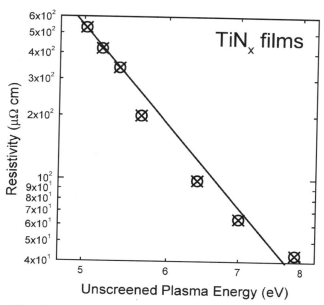

Fig. 58. Electrical resistivity versus ω_{pu}. The two quantities are correlated through an inverse square law similar to the one described by the Drude theory of a free electron gas. The inverse square law is clear evidence that all TiN$_x$ films are in general good metals following the free electron model and only quantitative differences exist regarding their metallic behavior.

other hand, ω_{ps} depends on both the free and the bound electrons in the material and is related with a more complicated manner with the TiN$_x$ film metallic character and stoichiometry [129]. However, this quantity is directly obtained from SE measurements without the need for analysis. For example, Figure 57 shows the $\varepsilon_1(\omega)$ of a TiN$_x$ film ≈ 1000 Å thick, deposited with $V_b = -40$ V at RT in the energy region 1.5–5.5 eV; here ω_{ps} (denoted by an arrow) is a directly measured quantity and is about 1.8 eV. In Figure 57 are also shown strong interband transitions in TiN$_x$ (with $x \approx 1.1$) that occur at about 3.0 and 4.5 eV. These energies differ from the 2.8-, 3.8-, and 6.5-eV energy differences in the $\Gamma_{15} \rightarrow \Gamma_{12}$, $X_5 \rightarrow X_2$, and $L_3 \rightarrow L_3'$ directions in the Brillouin zone, where the interband transitions take place in pure stoichiometric TiN films. These differences may be due to the difference in stoichiometry of this film which exhibits a cell size $a = 0.424$ (instead of 0.430 nm of the stoichiometric TiN), and to a different density of defects (vacancies at Ti lattice sites) [126, 133].

In order to calculate ω_{pu}, we used a model that describes the dielectric response including both the Drude and the interband absorption (Lorentz oscillator) contributions [132]:

$$\tilde{\varepsilon}(\omega) = \varepsilon_\infty - \frac{\omega_{pu}^2}{\omega^2 + i\Gamma_D\omega} + \sum_{j=1}^{2} \frac{f_j\omega_{0j}^2}{\omega_{0j}^2 - \omega^2 - i\gamma_j\omega} \quad (58)$$

Here ε_∞ is a background constant, larger than unity (varying from 2.1 to 2.3, depending on the deposition conditions) due to the contribution of the higher-energy interband transitions not taken into account. The Drude term is characterized by the unscreened plasma energy ω_{pu}, as discussed previously, and the damping Γ_D. In turn, Γ_D is due to scattering of electrons and, according to the free-electron theory, is the reciprocal of the relaxation time τ_D. The quantities Γ_D and τ_D are closely related to the electrical resistivity ρ of the metal and are influenced by the existence of grain boundaries, defects, phonons, and electron–electron interactions not taken into account in the free electron model. Each of the Lorentz oscillators is located at an energy position ω_{0j}, with strength f_j and damping (broadening) factor γ_j. The energy position, strength, and broadening of each individual oscillator depend on the deposition conditions, since the latter affect the stoichiometry and the microstructural characteristics of the deposited films.

Figure 57 describes the above procedure for a representative TiN$_x$ film. The fitting of both real and imaginary parts of the measured $\varepsilon(\omega)$ spectra with Eq. (58) is very precise. The contribution of the free electrons to the total dielectric function is described by the Drude term (dotted curve), from which we calculate $\omega_{pu} = 5$ eV (denoted with an arrow in Fig. 57) and $\Gamma_D = 1.3$ eV. In addition to the Drude contribution, we see in Figure 57 the two strong interband transitions in the Vis–UV spectral region, which are described by the corresponding Lorentz oscillators (dashed curves). The first oscillator, which is located at $\omega_{01} = 3.1$ eV, is well defined (low values of the broadening γ_i) with less strength f_i than the second one, in all cases. Although the broad second oscillator at $\omega_{02} = 5$ eV is acting as a smooth background in the $\varepsilon(\omega)$ in the Vis–UV region, it has a considerable effect on the ω_{ps} because of its high strength. In general, the two TiN_x Lorentz oscillators, in all examined films, are located in the ranges 3.1–3.9 and 5.0–6.5 eV, respectively. The experimental value of ω_{ps} (= 2 eV), denoted by an arrow in Figure 57, is far away from $\omega_{pu} = 5$ eV because it is strongly affected by the existence of interband transitions (Lorentz contributions) and thus in a way describes the response of all electrons (bound and free). Although ω_{ps} is affected by several factors—free electron density, interband absorption, grain size, etc.—exhibits an almost linear dependence on the film stoichiometry, as was found by combined SE and XPS–AES analysis; this will be discussed subsequently. On the other hand, changes of ω_{pu} may be attributed, according to Eq. (57), either to changes in free electron density or to changes of the electron effective mass [134]. In any case, the increase of ω_{pu} is strongly correlated with the enhancement of TiN_x metallic character expressed in terms of electrical resistivity. Indeed, experiment has confirmed the explicit relation between ω_{pu}, calculated by the above analysis, and the ex situ measured resistivity [134]. The resistivity follows an inverse square law, $\rho \approx 20.500/\omega_{pu}^2$, in ω_{pu}. The inverse square law is illustrated on a semilog scale in Figure 58 and is clear evidence that all TiN_x films, though of different stoichiometry and microstructure, are in general good metals following the free electron model, and only quantitative differences exist regarding their metallic behavior.

This inverse square law is an intrinsic property of fcc TiN_x films: it was observed for films deposited with various T_d and V_b, and thus describes the combined effect of these two deposition parameters on the electrical resistivity. In order to explain the inverse square law we should consider all the possible effects contributing to the resistivity decrease in terms of the free electron model, according to which ω_{pu} is closely related to electrical resistivity ρ. The relaxation time is associated with the film resistivity ρ through the relation (in cgs units) [135]

$$\tau_D = m^*/\rho N e^2 = 4\pi/\rho \omega_{pu}^2 \qquad (59)$$

From the fitting of the experimental points (Fig. 58) with a relation of the form $\rho = \alpha/\omega_{pu}^2$ we can roughly estimate the mean relaxation time for all TiN_x films deposited under different conditions and exhibiting different resistivity values. According to this analysis $\tau_D = 0.1 \times 10^{-14}$ s, which is a typical value for heavy transition metals with high resistivity (above 40 μΩ cm).

The above analysis is useful in the case of TiN_x films for applications in microelectronics, since it can provide a realistic estimation of the TiN_x resistivity during deposition, through its dependence on ω_{pu}. TiN_x films have been found to make very good rectifying or ohmic contacts on Si, depending on the deposition conditions. High T_d and V_b result in ohmic contacts with very low resistivity, while low T_d and V_b result in rectifying contacts (Schottky type) with barrier height $V = 0.5$ V and ideality factor close to 1 [124].

6.3. Study of the Stoichiometry of TiN_x Films by Spectroscopic Ellipsometry

Auger electron spectroscopy (AES) and X-ray photoelectron spectroscopy (XPS) have been used in the past for the characterization of this type of films [121, 136, 137]. However, the above and other analytical techniques used for compositional analysis suffer from the inherent difficulties that (1) they usually require the exposure of the TiN_x films to air, leading to absorption and reaction with oxygen, and (2) even worse, they are destructive. Optical nondestructive access to a surface in vacuum is important for real-time information, control, and characterization during film development. Techniques such as SE and MWE are already used to monitor deposition rates and composition of films during deposition [22, 23] and can be used in a complementary manner with other postdeposition techniques. Here we present an approach [127, 129] to study the stoichiometry and characterize TiN_x thin films by using either in situ or ex situ SE in combination with XPS and AES.

The electronic and optical properties of TiN_x films, as already discussed, depend on their actual nitrogen content and microstructure, and they are strongly affected by the deposition conditions (e.g. the bias voltage applied to the substrate during deposition). The real and imaginary parts of the dielectric functions of representative TiN_x films deposited with different V_b have been shown in Figure 55a and b. As we have already mentioned, of special interest is the adjustable screened plasma energy ω_{ps} of TiN_x, which depends strongly on the stoichiometry of the material [129]. The value of the plasma energy ω_{ps} based on the XPS results has been used as a criterion to determine in situ the stoichiometry of TiN_x films during deposition. That value was found to be about 2.6 eV when $x = 1$, and less than 2.6 eV for $x > 1$.

Figure 59 presents the values of the screened plasma energy ω_{ps} for TiN_x thin films that were deposited at various values of V_b. The ω_{ps}-values were extracted from the real part of the measured dielectric function spectra, as shown in Figs. 55b and 57, obtained from the TiN_x films. The results show that films prepared with V_b between −100 and −140 V are stoichiometric (region II in Fig. 59). Films deposited at $|V_b|$ values lower than 100 V (region I) deviate from stoichiometry and more than likely are overstoichiometric, as they exhibit a bronze color. The screened plasma energy exhibits strong, almost linear dependence on $|V_b|$ in the region from 0 to 100 V, with a slope $d\omega_{ps}/dV_b = 5$ meV/V. In the same region the stoichiometry changes from 1.1 to 1 (see Fig. 60), indicating that the ω_{ps} is a very sensitive probe of the TiN_x film stoichiometry. Finally, the slight decrease in ω_{ps} observed beyond −140 V (region III) may be attributed to the expected substantial increase in Ti sputtering in that range (−100 to −240 V) under the higher energy of the sputtering plasma ions [138].

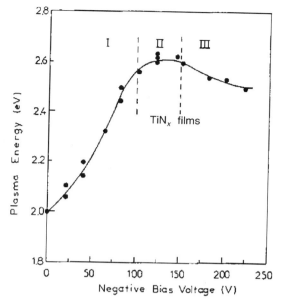

Fig. 59. The variation of the screened plasma energy ω_{ps} with the bias voltage V_b on the substrate during deposition. At 100 V a saturation value is reached, indicating that by applying higher bias the material does not change significantly with regard to stoichiometry and microstructure. (After [129].)

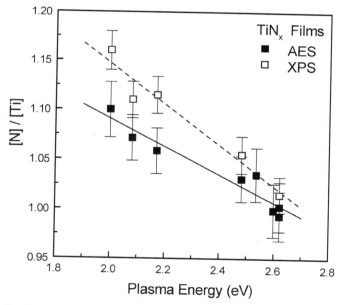

Fig. 60. The [N]/[Ti] ratios of TiN$_x$ films obtained by XPS (open squares) and AES (solid squares) versus the screened plasma energy ω_{ps}. The direct correlation of ω_{ps} with the film stoichiometry [N]/[Ti] enables the use of this feature of TiN$_x$ films for real-time ellipsometry monitoring of TiN$_x$ stoichiometry during deposition. (After [129]).

High-resolution XPS spectra from films deposited at low $|V_b|$, viz. $V_b = -25$ V, showed broad Ti $2p_{3/2}$ peaks with peak heights at binding energies $E_b = 455.4$ eV and about 458 eV. The former value corresponds to overstoichiometric TiN$_x$ (for example, for $V_b = -25$ V it was found that $x = 1.1$) and the latter is close to TiO$_2$ (the maximum position reported for TiO$_2$ is 458.5 eV) [139]. Spectra from films deposited at higher

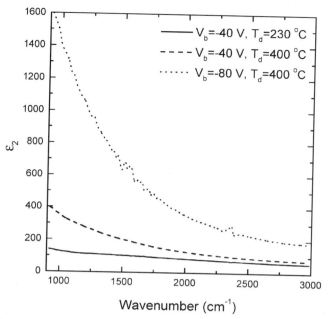

Fig. 61. The IR absorption of representative TiN$_x$ films. With increasing bias and substrate temperature the intraband absorption (measured in the IR region) becomes stronger, indicating the increase of conduction electron density and mobility. The conduction electron density and mobility increase as the TiN$_x$ films approach the stoichiometric TiN ($x = 1$), which can be consider similar to ideal metals.

$|V_b|$ values showed only sharper Ti $2p_{3/2}$ peaks situated at $E_b = 455.2$ eV. The XPS results suggest that oxidation (more than likely with formation of TiO$_2$) takes place in films developed at low $|V_b|$, whereas no oxidation occurs in films deposited at higher $|V_b|$, as we will discuss in Section 6.4.

Figure 60 shows the dependence of [N]/[Ti] ratios, estimated by XPS and AES, on the screened plasma energy ω_{ps}. The data presented in Figure 60 exhibit an almost linear relationship between [N]/[Ti] and ω_{ps}. The differences in [N]/[Ti] between the two techniques, especially in TiN$_x$ films developed with V_b below 60 V, are mainly due to the inherent difficulties of AES rather than to the calibration procedure used in analyzing the XPS measurements [139].

6.4. Electronic and Microstructural Features of TiN$_x$ Films

Having made a detailed discussion of the correlation of ω_{ps} and ω_{pu} with the intrinsic properties of the TiN$_x$ films, let us now consider the actual effect of the substrate temperature (T_d) and $|V_b|$ on the optical response of the TiN$_x$ films. FTPME was used to study the intraband absorption within the conduction band in the IR region. Figure 61 shows representative FTPME $\varepsilon_2(\omega)$ spectra of TiN$_x$ films in the range 950–3000 cm^{-1}. The IR intraband absorption gradually increases with both V_b and T_d. The maximum absorption, or the most intense metallic optical response, was observed in films deposited at high $|V_b| > 100$ V and $T_d = 400$ °C (dotted curve). Several causes, such as TiN$_x$ stoichiometry, vacancies at Ti lattice sites, grain size, and void

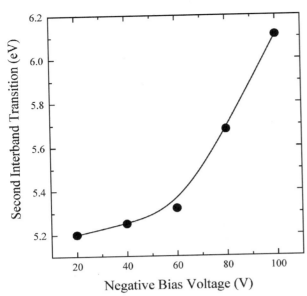

Fig. 62. The energy position of the second interband transition of TiN$_x$ films versus V_b. The energy position of this absorption is an indication either of stoichiometry variation changing the structure of the Brillouin zone, or of the changes of the lattice parameter with the bias voltage.

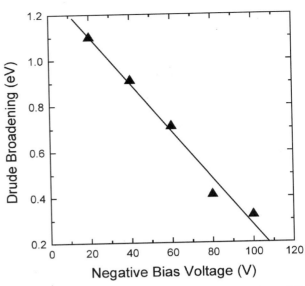

Fig. 63. The variation of Γ_D versus V_b. The linear behavior is not followed for $|V_b| > 100$ V; from that point a saturation value of ≈ 0.3 eV is reached, indicating that in this bias regime there is no further change in film stoichiometry and microstructure, in accordance with the behavior of the plasma energy.

content, are contributing to this behavior. In general, increase of negative bias V_b and of the substrate temperature T_d affects the adatom mobility, promotes the formation of larger grains, and leads to the elimination of voids and the growth of stoichiometric ($x = 1$) TiN$_x$. In particular, the film stoichiometry and crystal structure may affect the TiN$_x$ band structure, as has already been mentioned and will be discussed in detail in the following.

The effect of the deposition conditions on the band structure can be mainly described by the variations of the interband absorption characteristics (in a classical way by the Lorentz oscillators), viz. the energy position ω_{0j}, strength f_j, and damping (broadening) factor γ_j. All the Drude and Lorentz parameters are strongly affected by the electronic and microstructural features of TiN$_x$ films and therefore may be used to probe and study the TiN$_x$ microstructure with respect to the deposition conditions and film stoichiometry. In all cases, the first Lorentz oscillator is well defined (low broadening values γ_i), with less strength f_i than the second one, and is located between 3.0 and 3.8 eV; its location is considerably affected by the deposition conditions such as V_b and T_d. On the other hand, the second one is very sensitive to structural and compositional changes caused by changing the deposition conditions. Figure 62 shows the variation of the energy position of the second strong interband absorption of TiN$_x$ versus V_b. For high $|V_b|$, approaching 100 V, the grown TiN$_x$ approaches the stoichiometric TiN [129], suggesting that the second interband absorption of TiN is located at ≈ 6.5 eV. The lower energy of interband absorption that was detected for films deposited with low $|V_b|$ is an indication of the band-structure modifications of overstoichiometric ($x > 1$) TiN$_x$ films with respect to TiN. The modification of the band structure is attributed to changes of the crystal cell size of TiN$_x$

and to the Ti vacancy density, as we have already discussed, and therefore is directly related to the film microstructure and stoichiometry.

Intraband absorption is also affected by the film microstructure (grain size and void content) and composition. In addition to the unscreened plasma energy ω_{pu}, the intraband absorption is also described by the Drude broadening Γ_D, which is caused by the scattering of free electrons and is the reciprocal of the relaxation time τ_D [140, 141], according to the free-electron theory. Γ_D and τ_D are closely related to the electrical resistivity of the metal, as we have already discussed, and are influenced by the existence of grain boundaries, defects, phonons, and electron–electron interactions. Elevated deposition temperatures promote the adatom mobility and result in elimination of microvoids and dissociation of the weakly bonded excess nitrogen atoms. The negative V_b applied to the substrate induces Ar$^+$ ion bombardment of the film surface during deposition. This ion bombardment results in compositional (sputtering of the nitrogen excess) [129] and structural (larger grain size) modifications in the films. The energy transferred to the film from the bombarding ions is considerably higher than the thermal energy for deposition at $T_d < 650$ °C. Thus, $|V_b|$ has similar but stronger effect than T_d on the optical properties, structure, and composition of TiN$_x$ films. Figure 63 shows the variation of Γ_D versus V_b. The decrease in Γ_D is attributed to the higher free electron mobility, due mainly to the stoichiometry changes from overstoichiometric to stoichiometric TiN$_x$, and to a lesser extent to the increase of the TiN$_x$ grain size. For $|V_b| > 100$ V a saturation value of ≈ 0.3 eV is reached, as shown in Figure 63, and remains almost constant up to 200 V, similar to what we have found for the dependence of stoichiometry on V_b.

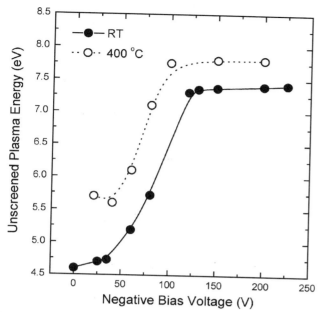

Fig. 64. The effect of V_b on ω_{pu}. There is an increase of ω_{pu} with $|V_b|$ until a saturation value is reached. Deposition at high T_d shifts the ω_{pu} values to higher energy.

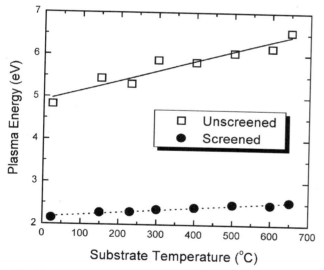

Fig. 65. The effect of deposition temperature on ω_{pu} and ω_{ps}. It is seen that ω_{pu} is more affected by T_d.

In Figure 64 the variation of ω_{pu} with $|V_b|$ is presented for substrate temperature T_d at RT and 400 °C. We see that ω_{pu} increases with $|V_b|$ from 4.5 eV to a saturation value of ≈ 7.5 eV for $|V_b| > 100$ V. Deposition of TiN$_x$ films at $T_d = 400$ °C results in a similar curve, but shifted to higher ω_{pu} values (up to 7.8 eV). The differences between the two curves (Fig. 64) are larger for low $|V_b|$. This happens because for higher $|V_b|$ the bombarding Ar$^+$ ion energy is much higher than the thermal energy and thus $|V_b|$ plays the dominant role in the deposition process. When the TiN$_x$ films are deposited at higher T_d, then relatively small changes with V_b are observed (see Fig. 65) in

both ω_{pu} and ω_{ps}. In addition, both quantities exhibit a linear dependence on T_d on keeping the other deposition conditions constant. However, the two lines in Figure 65 are not parallel, and ω_{pu} is strongly affected by T_d. By raising T_d from RT to 650 °C a 15% and a 35% increase in ω_{ps} and ω_{pu}, respectively, are observed.

The parameters of the Drude term, ω_{pu} and Γ_D, may be used to calculate the mean free path of free electrons in TiN$_x$. The mean free path of electrons is governed by the type of scattering sites in TiN, so it can be limited by the grain boundaries of crystallites [140–142] or by defects (mainly vacancies in Ti lattice sites) [126]. Combined SE and XRD studies have shown that for thin, stoichiometric films the grain boundaries play the dominant role in determining the mean free path, whereas in substoichiometric films the Ti vacancies also play a significant role. The mean free path of electrons in stoichiometric TiN corresponds to the mean crystalline grain size and thus is directly related to the film's microstructure [126]. In the case of substoichiometric TiN$_x$ ($x > 1$) the calculated mean free path of electrons can be smaller than the grain size, due to the existence of Ti vacancies in the crystallites that act as scattering centers. Nevertheless, the mean free path of electrons is expected to be close to the grain size and thus can be used as an estimate of grain size [126].

The mean free path L_g of the free electrons in TiN$_x$ can be calculated from the relaxation time. It be expressed [140, 141], in first approximation (assuming that the relaxation time associated with the bulk crystalline TiN is very small) [126], in terms of the relaxation time and the mean electron velocity as follows:

$$L_g = \tau_D v_F \tag{60}$$

where v_F is the velocity of electrons at the Fermi surface (ground state energy of electrons at $T = 0$ K). The velocity of electrons due to thermal energy is neglected, since it is two orders of magnitude smaller than the velocity at the Fermi surface.

A detailed calculation of τ_D for each particular case may be performed using the damping factor Γ_D through the relation (the units are shown in brackets):

$$\tau_D \, [\text{s}] = \frac{1}{\Gamma_D \, [1/\text{s}]} = \frac{0.66 \times 10^{-15}}{\Gamma_D \, [\text{eV}]} \tag{61}$$

The velocity at Fermi surface is calculated according to the free electron model through the following relations [131] in cgs units:

$$v_F = \frac{\hbar}{m^*}\left(3\pi^2 N\right)^{1/3} = \left(\frac{0.75\hbar^3\pi\omega_{pu}^2}{(m^*e)^2}\right)^{1/3} \tag{62}$$

From Eqs. (60)–(62) it is evident that the mean free path L_g may be calculated in terms of the quantities Γ_D and ω_{pu} obtained by SE data analysis. Figure 66 shows the calculated mean free path versus V_b and T_d.

The void content in the bulk TiN$_x$ films was estimated by BEMT, assuming that each of the deposited TiN$_x$ films consisted of stoichiometric TiN and voids. According to BEMT the dielectric function $\varepsilon(\omega)$ of a layer consisting of n different

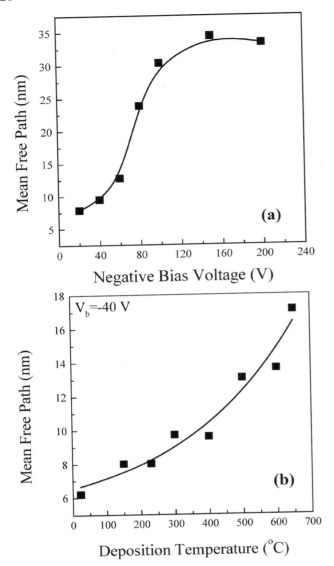

Fig. 66. The effect of (a) the bias voltage V_b applied to TiN$_x$ films deposited at $T_d = 400$ °C, and (b) deposition temperature T_d of TiN$_x$ films deposited with $V_b = -40$ V, on the mean free path L_g of the free electrons. There is an increase of L_g with $|V_b|$ until a saturation value is reached. Deposition at high T_d results in large L_g.

phases can be described by

$$\sum_{i=1}^{n} \frac{\varepsilon_i - \varepsilon}{\varepsilon_i + 2\varepsilon} f_i = 0 \qquad (63)$$

where ε_i and f_i are the dielecric function and volume fraction of constituent i (TiN or voids). This analysis suffers from the fact that TiN$_x$ films show differences in their second interband absorption when the films are deposited at various V_b. This is due to the fact that we use the same TiN reference dielectric function, which corresponds to stoichiometric TiN film. Therefore, we expect the accuracy in determining the void content to decrease when examining films deposited at low $|V_b|$. In order to estimate the accuracy of BEMT analysis, we have also calcu-

lated density and void content by XRR. We briefly describe here how one can calculate the density of the TiN$_x$ films from XRR measurements; more information can be found in [143, 144].

Information on the density, thickness, and surface roughness is deduced from XRR through the dependence of the reflection coefficient on the angle of incidence. XRR is based on the same principles as SE regarding the optical response of materials. The dielectric function of all materials in the X-ray region has the form [see Eqs. (55a) and (58) [126]

$$\varepsilon_1(\omega) = 1 - \frac{\omega_{pu}'^2}{\omega^2} \quad \text{and} \quad \varepsilon_2 \approx \frac{\omega_{pu}'^2 \Gamma_D'}{\omega^3} \approx 0$$

$$\text{for} \quad \omega \gg \omega_{pu} \quad \text{and} \quad \omega_{pu}'^2 \gg \Gamma_D'^2 \qquad (64)$$

where ω_{pu}' and Γ_D' are the effective plasma energy and broadening and take into account all free and bound electrons. Therefore, the refractive index n for X-rays, which is associated with $\varepsilon_1(\omega)$ through the relation

$$n(\omega) = \varepsilon_1^{1/2}(\omega) \approx 1 - \frac{\omega_{pu}'^2}{2\omega^2} \qquad (65)$$

will be less than 1 (not taking into account the negligible absorption term associated with ε_2). Total reflection of the X-ray beam occurs at the film surface. Its refractive index is smaller than that of the ambient air, which is closer to 1. The critical angle of total reflection, which is between 0.2 and 0.6° for most materials, can be expressed in terms of the refractive index n of the material:

$$\cos \theta_c = 1 - \frac{\theta_c^2}{2} = n \qquad (66)$$

From Eqs. (57) and (64)–(66) the mass density ρ_m of the material is correlated with the critical angle for total reflection:

$$\theta_c^2 = 2N_0 \left(\frac{e^2}{2\pi mc^2} \right) \left(\frac{Z\rho_m}{A} \right) \lambda^2 \qquad (67)$$

where N_0 is Avogadro's number, A the atomic mass, λ the X-ray wavelength, and Z the number of free electrons per atom.

The measured intensity in a specular reflectivity scan that is obtained from a thin film at incidence angles larger than the critical angle θ_c exhibits interference fringes that originate from the multiple reflection of X-rays at the film–substrate interface. These fringes, beyond the region for total reflection, are related to the thickness of the film and are very sensitive to the surface and interface roughness, which causes an overall reduction in the reflected intensity by scattering into nonspecular directions [145, 146]. The distribution of roughness is usually assumed to be a Gaussian one. The rms roughness, in each interface, is incorporated as a damping effect on the specular reflectivity R_0: $R_0 \exp(-q_z^2 \sigma^2)$, with q_z the perpendicular component of the momentum transfer vector. R_0 is calculated through the Fresnel diffraction theory [147] or equivalently the Ewald–von Laue dynamical diffraction theory.

The XRR measurements were analyzed using a Monte Carlo algorithm assuming a single-layer model of stoichiometric TiN on an atomically smooth c-Si(001) substrate. The maximum measured density was 5.7 g/cm^3 for a film deposited at 400 °C

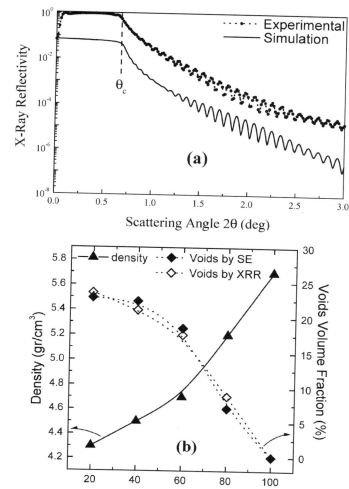

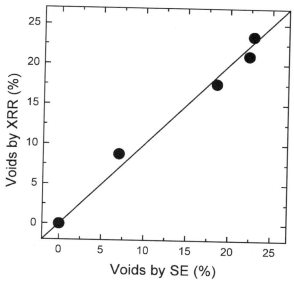

Fig. 68.   The validation of void content calculated by SE and XRR measurements. The two techniques are in very good agreement, confirming that by the SE (BEMT) analysis we can extract accurate quantitative results for the void fraction, whether the voids are microvoids at grain boundaries or Ti vacancies.

Fig. 67.   (a) XRR curve and the results of the Monte Carlo fitting for a stoichiometric TiN film 890 Å thick. The film density is calculated from the critical angle for total external reflection (θ_c), indicated by the dashed vertical line, while the film thickness is calculated from the interference fringes using Bragg's law. The film density was 5.7 g/cm³, which is 5% higher than the density of the bulk, unstrained, stoichiometric TiN and is attributed to the stress-reduced lattice parameter of the film. (b) The effect of V_b on the density and void concentration of TiN$_x$ films. The intense ion bombardment induced by V_b during deposition results in the elimination of voids and sputtering of the excess N atoms, which are responsible for the formation of vacancies in Ti lattice sites. Both mechanisms influence the density of TiN$_x$ films.

Figure 67b shows the evolution of density and void volume fraction in TiN$_x$ films with V_b. The density increase is attributed to the elimination of voids and vacancies and to changes in stoichiometry (or cell size) with V_b, as is shown by the agreement between the independent results of XRR and SE. A similar variation of the density and void volume was also found for T_d. However, the effect of V_b is much stronger (variation 25%) than that T_d (variation < 10% for deposition up to 400 °C). TiN$_x$ films exhibit columnar growth, with column size and intercolumnar spacing depending on V_b and T_s [148].

Figure 68 illustrates the comparison of the results of the two independent techniques concerning the void content in the TiN$_x$ film. Surprisingly, the two techniques are in good agreement, with a discrepancy below 2%.

At this point we should discuss the nature of the calculated void fraction. XRR directly measures the total electron density, and therefore it does not discriminate the microscopic voids (such as grain boundary areas or intercolumnar spacing) from the atomic-scale voids such as vacancies (holes of probed electrons). The huge amount of voids (30%) in the substoichiometric TiN$_x$ films measured by XRR cannot be exclusively attributed to microvoids; we need to consider the Ti vacancies as well. Transmission electron microscopy observations have shown some microvoids across the intercolumnar space at the initial stages of growth, but much less than the 30% measured by XRR, supporting the above arguments [148]. The existence of vacancies in the Ti sublattice, as suggested by XRR, and the small size of the grains are crucial structural properties, and they affect the penetration of ambient atmospheric oxygen and the oxidation mechanism [149] of such films, as we will discuss below. The interesting result here is that SE analysis by BEMT is in excellent agreement with the XRR, although it is expected

and $V_b = -100$ V. This value is 5% higher than the density of the bulk, unstrained, stoichiometric TiN [143]. Figure 67a shows the XRR experimental measurements (dots) and the results of the Monte Carlo fitting for a stoichiometric TiN film 890 Å thick. The void volume fraction was calculated using this maximum density as normalization constant ρ_{norm} through the relation

$$\text{voids \%} = 100\% \times \frac{\rho_{norm} - \rho_m^{exp}}{\rho_{norm}} \qquad (68)$$

where ρ_m^{exp} is the measured density of TiN$_x$ films.

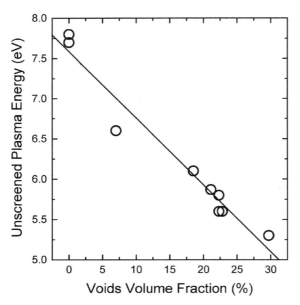

Fig. 69.   The variation of unscreened plasma energy ω_{pu} versus the void content (as defined in the text) in TiN$_x$ films.

that SE cannot probe such localized defects and the BEMT assumes constituent materials with dimensions comparable with the light wavelength [27, 32].

After calculating and validating the void content in the film, we investigated its relation to ω_{pu}. Figure 69 illustrates the direct correlation between void content and ω_{pu}. This linear behavior suggests that by changing the deposition conditions and the film stoichiometry and structure, a considerable, nonlinear decrease of electron effective mass m^* is required for Eq. (57) to be fulfilled.

6.5. Multiavelength Real-Time Ellipsometry of TiN$_x$ Films Prepared by Unbalanced Magnetron Sputtering

Real-time monitoring is essential for the production of modern engineering materials. Particularly, optical nondestructive access to a surface in vacuum is important for real-time monitoring, control, and characterization during film development, providing reduction in production process time and increase of production yield. In this subsection we will present a real-time SE monitoring process by MWE of TiN deposition with closed field unbalanced magnetron sputtering (CFUBMS), which is a technique achieving high deposition rate and is applied in industrial-scale applications. The high deposition rate of TiN by CFUBMS is used as a second example to test the acquisition speed and software analysis limits of the currently available SE–MWE technology.

The deposition experiments were performed in a vacuum system with base pressure better than 10^{-5} Torr by using the CFUBMS technique (see Fig. 42). The system consisted of two high-field-strength magnetrons with Ti targets, mounted on the vertical walls of a cylindrical chamber. Plasma etching of c-Si and presputtering were performed prior to Ti and TiN$_x$ thin

film deposition. The partial pressures of working gas (Ar) and reactive gas (N$_2$) were 3 and 1.6 mTorr, respectively. A fiber optic between magnetron 1 and the substrate was used for optical emission spectroscopy (OEM). The real-time measurements were performed with a 16-wavelength phase-modulated ellipsometer, which was attached to the vacuum chamber of CFUBMS. The ellipsometer provides also the possibility of spectroscopic ellipsometry (SE) measurements in the energy region 1.5–5.5 eV. With the 16-wavelength ellipsometer are obtained simultaneously the ellipsometric angles Ψ and Δ for the 16 different wavelengths of the system film–substrate system, in timed steps. The 16 wavelengths are distributed in the Vis–UV energy region 1.54–4.32 eV. The speed of the real-time measurements depended on the integration time (IT) used for the acquisition of each wavelength (for details about the 16-wavelength ellipsometer see Section 5.6). The results presented here were obtained with IT = 62.5 ms, providing a total time of 1 s for each 16-wavelength spectrum.

The monitoring of the growth of thin TiN$_x$ films was performed with the help of process parameters of dc magnetron sputtering, such as the negative dc bias voltage V_b applied to the substrate during plasma etching of the c-Si substrate, with various values from −200 to −500 V. Our studies are focused on: (i) the top layer amorphization of c-Si, which occurred during dry wet etching of silicon substrate [22], (ii) the constitutive ratio x, [N]/[Ti] during reactive unbalanced magnetron sputtering and thus the stoichiometry of the growing TiN$_x$ film, and (iii) the calculation of the TiN$_x$ film thickness during deposition. The TiN$_x$ film stoichiometry, as already discussed, is correlated with the plasma frequency ω_{ps}, the energy where $\varepsilon_1(\omega_{ps}) = 0$ [131]. OEM diagnostics control the adsorbed number of N$^+$ on Ti atoms by determining the optical intensity of the reactive plasma as a percentage of the initial optical plasma intensity of Ti$^+$ and Ar$^+$ ions. Thus, the constitutive ratio x of TiN$_x$ was controlled by the optical intensity value of reactive plasma, and its sensitivity was monitored with MWE.

The real-time monitoring of thin films' optical properties is accomplished with multiple spectra of the pseudo-dielectric function obtained at the 16 fixed energies within the spectral region 1.54–4.32 eV. Before the TiN$_x$ film deposition, plasma etching with high Ar$^+$ ion energy to clean the c-Si substrate from Si oxide native oxide was performed. During this procedure an amorphous Si (a-Si) layer is also formed. The amorphization depth and the etching rate of silicon substrate as a function of Ar$^+$ ion energy, which depends on the voltage V_b applied to the substrate, were studied [22]. Figure 70a shows the $\langle \varepsilon_1(\omega) \rangle$ spectra obtained during the c-Si substrate bombardment with Ar$^+$ ions. The measured pseudo-dielectric function $\langle \varepsilon(\omega) \rangle$ of the c-Si amorphization was analyzed with the Bruggeman effective-medium theory (BEMT) in combination with the three-phase (air–amorphous-layer–c-S substrate) model, assuming that the top of Si substrate layer of thickness d consists of either a-Si and voids, or a-Si, voids, and Ti [22].

Before the TiN deposition by CFUBMS, a Ti interlayer $\approx$1000 Å thick was deposited on top of amorphized silicon in order to enhance the adhesion of TiN to silicon, and the pseudo-

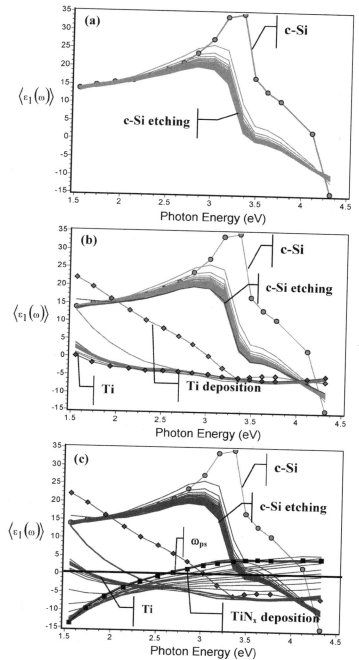

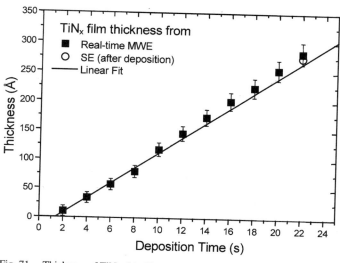

Fig. 71. Thickness of TiN$_x$ thin film (solid squares) measured during deposition with the CFUBMS technique by analyzing the MWE data. The open circles present the film thickness calculated from the in situ SE spectrum obtained after the film deposition, and the solid line is the result of a linear fit to MWE results (the deposition rate is 12.9 Å/s).

Fig. 70. Real-time monitoring of 16-wavelength $\langle \varepsilon_1 \rangle$ spectra obtained continuously during: (a) plasma etching of c-Si substrate with Ar$^+$ ion bombardment with a high voltage $V_b = -500$ V applied to the substrate; (b) Ti interlayer deposition up to $\approx$1000 Å (solid diamonds); and (c) TiN$_x$ film deposition up to $\approx$1000 Å (solid squares), where the position of unscreened plasma energy ω_{ps} for a stoichiometric TiN film is denoted.

dielectric function $\langle \varepsilon_1(\omega) \rangle$ during its deposition is shown in Figure 70b. Finally, Figure 70c presents the evolution of TiN$_x$ growth on top of the Ti interlayer. The stoichiometry of the growing TiN$_x$ film was inferred from the screened plasma frequency ω_{ps}. At the initial growth steps of TiN$_x$, the real part $\langle \varepsilon_1(\omega) \rangle$ of the pseudo-dielectric function is influenced only by

the Ti interlayer. The strong interband absorption occurring in TiN$_x$ at $\approx$3 eV, which causes a positive contribution to $\varepsilon_1(\omega)$, shifts the point at which $\varepsilon_1(\omega_{ps}) = 0$ to lower energies [131]. By observing the real-time MWE spectra, it seems that the evolution of TiN$_x$ growth leads to the formation of a stoichiometric thin film, and the final plasma frequency of the deposited TiN$_x$ is $\omega_{ps} = 2.6$ eV, as denoted in Fig. 70c. MWE with CFUBMS can be used to monitor optical properties of a TiN$_x$ thin film even from its initial growth steps.

An example that illustrates the sensitivity of the MWE technique during real-time measurements of thin film bulk properties is the following. The penetration depth of light limits the accuracy of MWE and SE in measuring the thickness of TiN$_x$ films when it is more than $\approx$600 Å (see Fig. 70c). Above this thickness, MWE and SE measure the bulk properties of the TiN$_x$. In TiN$_x$ thin film deposition (Fig. 70c), the N$_2$ content in the reactive plasma was controlled with OEM by setting the reactive plasma's optical emission intensity at 54% of the initial emission intensity in the Ar$^+$ ion plasma. This OEM diagnostic controlled the reactive plasma's optical emission intensity for stoichiometric adsorption of N$^+$ on Ti$^+$ ions during the TiN$_x$ growth. However, it was ascertained with real-time MWE monitoring that ω_{ps} was slightly varying during deposition of TiN$_x$ films thicker than 1000 Å. This is attributed to the higher sensitivity of MWE to any structural or optical modification of film during its deposition, a subject that will be discussed in detail elsewhere.

An example of real-time monitoring analysis of TiN$_x$ growth evolution is presented in Fig. 71, from which by analyzing the measured $\langle \varepsilon(\omega) \rangle$ spectra, the thickness and the deposition rate of a TiN$_x$ thin film can be estimated during deposition [22]. The film was prepared under identical conditions to the one shown in Figure 70. Growth of TiN$_x$ with CFUBMS results in high deposition rates, and thus fast acquisition times are required from MWE. For every measurement step of 2 s, an IT of ≈ 62.5 ms (a total time of ≈ 1 s is required to collect the 16-wavelength spectrum) was used for monitoring and calculating the film thickness. By analyzing the sequential multiwavelength spectra with the three-phase (air–TiN-film–Ti-interlayer) model, a deposition rate for TiN_x film of 12.9 Å/s is obtained, whereas the total thickness of TiN_x film after a deposition period of 11 steps is found to be 283 ± 8 Å. It should be noted here that the expected accuracy in determining the TiN_x film thickness (due to IT = 62.5 ms) at each step of measurement (every 2 s) is ≈ 6.5 Å. In order to compare the accuracy of real-time MWE in thin film thickness calculation, in situ SE measurements in the energy region 1.5–5.5 eV and ex situ XRR measurements were also conducted. By analyzing the in situ SE spectrum obtained in the same deposition time period, the thickness of TiN_x was found to be 275 ± 3 Å. Analysis of the XRR measurements obtained from the same specimen provides a thickness of 280 ± 4 Å and an rms surface roughness of ≈ 5 Å. A comparison between these results confirms the ability of MWE to determine accurately the thickness and the deposition rate of TiN_x produced by CFUBMS, a fast deposition technique.

6.6. Oxidation Study of TiN_x Films

In order to monitor the oxidation of bulk TiN_x films (films thicker than 1000 Å), real-time ellipsometric measurements were taken during exposure of the films to Ar, N_2, an Ar–O_2 mixture, and air at fixed energy [149]. The partial pressures (flow rates) of the gases were 15×10^{-3} mbar (30 sccm) for Ar, and 2.6×10^{-3} mbar (5 sccm) for N_2. In the Ar–O_2 mixture, the partial pressure (flow rate) was 14.9×10^{-3} mbar (30 sccm) for Ar, and 1.2×10^{-3} mbar (3 sccm) for O_2. All exposures to different atmospheres were performed at RT. Figure 72 presents successive $\varepsilon(\omega)$ spectra from a film deposited at $V_b = -25$ V, obtained after exposure to atmospheric air at RT for various periods of time. It is evident from Figure 72 that there is a shift in the value of $\varepsilon_1(\omega)$ to lower energies. The as-deposited TiN_x film exhibits a plasma energy $\omega_{ps} = 2.05$ eV that gradually decreases with exposure to air, reaching a value of $\omega_{ps} = 1.85$ eV after 215 h [149].

The value of the unscreened plasma energy is proportional to the square root of the density of free carriers, which are mainly due to the Ti content. As mentioned earlier, previous studies have shown that films deposited at low $|V_b|$ values are overstoichiometric with a weak bond between Ti and N atoms [128]. Thus, the reduction in plasma energy can be attributed to the formation of an oxide. It is well known that the reaction of titanium nitride with oxygen is thermodynamically favorable ($\Delta G^0 = -139$ kcal/mol), forming rutile (TiO_2) [150]. Thus, a stronger bond is expected between Ti and O atoms than between Ti and N, causing a reduction in plasma energy.

In an effort to clarify the process of oxidation of TiN_x films when exposed to air and to explain the $\varepsilon(\omega)$ behavior (Fig. 72), it was assumed that oxidation proceeds with the formation of a thin TiO_2 overlayer on the surface of the polycrystalline TiN_x.

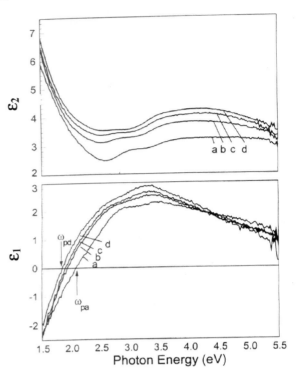

Fig. 72. The in situ real (ε_1) and imaginary (ε_2) parts of the dielectric function of a TiN_x film deposited at $V_b = -25$ V and exposed to air at RT. The $\varepsilon(\omega)$ spectra were obtained after exposure to air for (a) 0 h, (b) 0.1 h, (c) 7 h, (d) 215 h. The arrows denote the position of ω_{ps} in the as-deposited film (ω_{pd}) and after exposure to air for 215 h (ω_{pa}). The oxidation process and the formation of insulating TiO_2 in the films (even in small fraction) have considerable effect on both the interband and the intraband absorption of TiN_x films. (After [149].)

The calculated results obtained by this simple two-phase (TiO_2-overlayer–TiN_x-film) model for $\varepsilon_1(\omega)$, assuming various values of the thickness of the overlayer, are presented in Figure 73, together with the real part of the dielectric function of the as-deposited TiN_x films. It is clear that this possibility does not resemble the behavior of the experimental $\varepsilon_1(\omega)$ spectra versus time (Fig. 72). The plasma energy ω_{ps} (denoted by an arrow in Fig. 73) of the calculated $\varepsilon_1(\omega)$ spectra is unchanged, and large changes are observed in the spectra only above 3.5 eV, where the TiO_2 starts to absorb. Therefore, the SE analysis clearly suggests that surface oxidation cannot proceed for more than a few angstroms and most of the oxidation occurs in the internal structure of the film, possibly around the intercolumnar spacing of the fibrous, columnar structure of TiN_x.

In view of the above results, and assuming that during exposure to an oxygen atmosphere the film can be described as a composite material consisting of variable volume fractions of TiN_x and TiO_2, the measured $\varepsilon(\omega)$ can be simulated by using the Bruggeman effective-medium approximation. The volume fraction of TiO_2 in the film as a function of the exposure time obtained from this analysis is shown in Figure 74. The time dependence of the TiO_2 fraction exhibits exponential behavior, supporting that the oxidation is faster in the first few minutes after exposure to air.

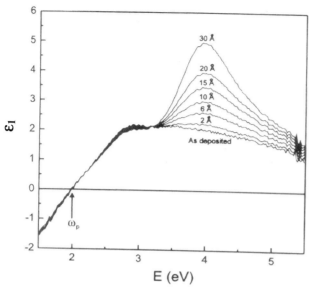

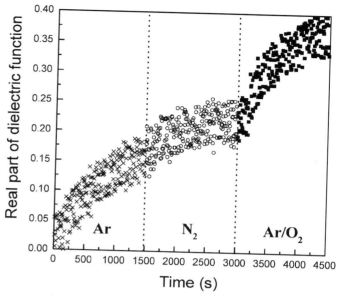

Fig. 73. The effect of a hypothetical TiO_2 overlayer with various values on thickness on the $\varepsilon_1(\omega)$ dielectric function of a TiN_x film (deposited with $V_b = -25$ V), calculated assuming the two-phase (TiO_2-overlayer–TiN_x-film) model. The arrow denotes the position of ω_{ps} of the as-deposited TiN_x film. The strong variations at ≈ 4 eV are attributed to absorption of TiO_2. The absence of such variation in the experimental spectra supports the formation of TiO_2 in the bulk of the film (around the intercolumnar spacing in the fibrous, columnar structure of TiN_x) and not the surface. (After [149]).

Fig. 75. Real-time ellipsometric measurements during exposure to various gases of a TiN_x film deposited at $V_b = -20$ V, leading to fast oxidation of TiN_x films in air. The real part (ε_r) measurements were obtained at 2-eV photon energy, which corresponds to the plasma energy of the as-deposited film.

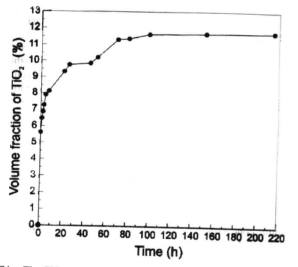

Fig. 74. The TiO_2 volume fraction, in a TiN_x film deposited with $V_b = -25$ V, versus the exposure time to air, calculated from the analysis of the $\varepsilon(\omega)$ spectra by using the Bruggeman effective-medium approximation. The saturation is reached at the time stage when the intercolumnar spacing is filled by TiO_2, contributing to the densification of the film. (After [149]).

Real-time ellipsometric measurements were performed at fixed photon energy every 100 ms. For these measurements, the energy that corresponds to the plasma energy ω_{ps} of the as-deposited TiN_x film was selected. In this energy region, large changes in the measured dielectric function are expected. Furthermore, in order to investigate the TiN_x oxidation under very low-oxygen atmospheres and to detect possible oxidation of the

film during deposition (for example, due to residual oxygen in the working and reacting gases), similar experiments were conducted in Ar, N_2, and O_2 gases. The $\varepsilon(\omega)$ spectrum was obtained immediately after deposition, and the plasma energy was determined. For example, for the TiN_x film deposited at $V_b = -20$ V, a plasma energy $\omega_{ps} \approx 2$ eV was determined. This photon energy was used for subsequent real-time ellipsometric measurements during exposure of the TiN_x film to different atmospheres. Figure 75 shows the results from these experiments. An increase in the real part ε_r of the dielectric function of the film can be observed, supporting the conclusion that ω_{ps} decreases and that the TiN_x film was oxidized even by the small residual oxygen in the deposition gases. Obviously the decrease in ω_{ps} during exposure to the Ar–O_2 mixture is more significant, due to the increased presence of oxygen.

Analysis of $\varepsilon(\omega)$ spectra obtained after exposure of TiN_x to the various process gases mentioned and to air, using BEMT, indicates an increase in the TiO_2 content. It is also found that the decrease of ω_{ps} versus time is associated with the increase of TiO_2 content. In Figure 76 ω_{ps} is plotted versus the volume fraction of TiN_x, for films deposited at various bias voltages and exposed for up to 220 h to atmospheric air. The open circles in the figure denote the SE spectra that were obtained during exposure to air. Large changes in ω_{ps} can be seen in films deposited at low V_b (−20 V), whereas in films deposited at $V_b = -120$ V, ω_{ps} remains almost constant. These graphs provide an overall view of how oxidation affects the bulk dielectric properties of the TiN_x films deposited under various bias voltages. The formation of titanium oxide when oxygen reacts with TiN is expected to cause a gradual distortion of the TiN_x lattice and a final change in its volume.

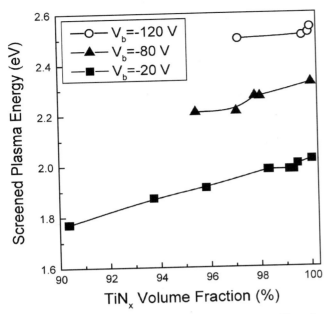

Fig. 76. The plasma energy of TiN$_x$ films deposited at various bias voltages versus the TiN$_x$ volume fraction after exposure to air for several hours open circles for $V_b = -120$ V, solid triangles for $V_b = -80$ V, and solid squares for $V_b = -20$ V. The volume fraction of each film was calculated through the measured dielectric function before and after exposure to air, using BEMT. (After [149].)

The results obtained in the present study support the following: (i) even though oxygen is absorbed by all TiN$_x$ films, oxidation occurs only in those deposited at low $|V_b|$ (overstoichiometric films with loosely bound nitrogen); (ii) the oxidation proceeds into the bulk structure of the TiN$_x$ films; and (iii) different processes of oxygen diffusion into the film at RT seem to take place during the oxidation of TiN$_x$. The initial fast diffusion of oxygen in the TiN$_x$ films is a result of the columnar structure of the films. The wide vertical columnar boundaries act as fast channels for oxygen to enter the film and saturate the intercolumnar spacing. Thus, it is plausible that oxygen diffusion and reaction to form an oxide may be responsible for the observed gradual increase in the level of the compressive stresses in the film [149]. AES studies have shown the presence of weakly bonded nitrogen in TiN$_x$ films deposited at low $|V_b|$ [149]. Since a higher density of defects and weaker Ti–N bonds are expected at the grain boundaries, oxygen can easily react at these sites to form titanium oxide. Oxidation of TiN$_x$ deposited at low $|V_b|$ can then take place even at very low partial pressure of oxygen, due to the reaction of oxygen with the weakly bonded Ti–N at grain boundaries.

7. SUMMARY AND CONCLUSIONS

Over the last ten years ellipsometry has been developed as a powerful probe to screen the preparation and properties of thin films. In ellipsometry, one measures the amplitude and phase of the complex reflection ratio $\tilde{\rho}$ that characterizes the change in polarization state incurred when polarized light is reflected from the surface of a material. This gives ellipsometry an advantage over other optical techniques, in that it can directly calculate two quantities, e.g., the complex dielectric function $\tilde{\varepsilon}(\omega) = \varepsilon_1 + i\varepsilon_2$ of a bulk absorbing material at a given photon energy ω, or the film thickness d and ε_1 ($= n^2$) in transparent films. Spectroscopic ellipsometry (SE) involves measuring $\tilde{\rho}$ (or $\tilde{\varepsilon}$) versus photon energy and has been recently extended from the IR to the deep UV energy region, with the adaptation of the Fourier transform technology and synchrotron radiation sources, respectively. Real-time ellipsometry measures $\tilde{\rho}$ continuously versus time during film growth, at one fixed photon energy or, recently, at many photon energies (multiwavelength ellipsometry, MWE), with a resolution in thickness at the monolayer level.

SE has been employed to determine the optical properties of bulk materials, static surfaces, and thin films, as well as the structure of complex, multilayered samples, by determining the dielectric function of the sample under investigation. From the dielectric function measured versus photon energy for a bulk material, its electronic, vibrational, structural, and morphological properties can be obtained. In addition, SE can characterize multilayered structures, with individual layers exhibiting complex microstructures, by applying an effective-medium theory and using the thicknesses and component volume fractions as free parameters. Thus, SE can provide a lot of information about thin films, such as electronic structure, vibrational modes, microstructure (crystalline and amorphous phases), bonding configuration, composition, and stoichiometry, as well as their density and thickness, and can be used in situ to identify the various stages of film growth (nucleation, coalescence, homogeneous, surface roughness evolution). Most of these uses are reviewed in this chapter.

On the other hand, the increasing demands on performance and the related sophistication of the deposition processes have provided a strong incentive to develop highly sensitive diagnostics that can monitor—not only in situ but, more importantly, in real time—thin film properties and processing. However, for control process applications, any in situ and real-time probe must satisfy severe requirements, such as to be noninvasive, nonperturbing, contactless, and (especially) fast enough. The recently developed real-time phase-modulated MWE has been advanced to monitor very fast thin film deposition processes. The key to the development of real-time MWE is the replacement of the monochromator and photomultiplier detector in the conventional instruments with a grating and several photomultipliers, each for one wavelength. In addition, instead of the earlier serial scanning instruments, the MWE is provided with parallel detection, allowing the simultaneous collection of raw data that are spread over the energy region 1.52–4.2 eV. Moreover, as a consequence of recent advances in optical instrumentation, real-time MWE (in the UV–Vis range) is becoming compatible with most of the kinetics involved in thin film processing. Thus, this fast and sensitive optical technique can make possible the feedback control of the thickness and the refractive index of transparent films or the thickness and the composition of absorbing films. The MWE has been applied here in studying the growth of the technologically important carbon based (a-C and CN_x) and TiN_x thin films prepared by various magnetron sputtering deposition techniques. We summarize in the following the most applications presented and reviewed in this work.

First, the optical and electronic characterization of representative amorphous hydrogenated carbon (a-C:H) films was accomplished with SE in the extended energy region from the near IR to the deep UV (1.5 to 10 eV), the latter using synchrotron radiation spectroscopic ellipsometry. We demonstrated how the dielectric function of relatively thick a-C:H films in this energy region can reveal their diamond- or graphite-like character. In this connection also were presented in situ SE studies of the effect of thermal annealing, which is known to cause hydrogen effusion, on the optical properties of the films and the subsequent gradual graphitization of the materials. For this purpose SE measurements were performed in conjunction with in situ annealing experiments on a-C:H films prepared by various techniques, and the obtained pseudo-dielectric function $\langle\varepsilon(\omega)\rangle$ was studied. The results and information obtained by SE through the data analysis of the pseudo-dielectric function of a-C:H films enabled us to measure the bulk dielectric function $\varepsilon(\omega)$ of sp^3- and sp^2-carbon-bonded materials, which can be used in studying the measured $\langle\varepsilon(\omega)\rangle$ of a-C thin films free of hydrogen during deposition by applying the Bruggeman effective-medium theory (BEMT).

Second, we demonstrated the ability to investigate and characterize the growth, using magnetron sputtering (MS), of a-C thin films free of hydrogen by in situ SE and real-time ellipsometry. When needed, we also introduced results obtained from complementary techniques to confirm those obtained by ellipsometry, such as film thickness and density of the deposited a-C films by X-ray reflectometry (XRR). In addition, we demonstrated real-time measurements by means of multiwavelength ellipsometry to monitor the process of high-deposition-rate sputtering in industrial-scale applications, as a first step towards control and characterization during film and component development. The different nucleation stages and the type of growth were studied by SE and MWE in terms of the time dependence of thickness and deposition rate. In all cases there is an initial nucleation stage, with low deposition rate and interaction of the deposited material with the Si substrate, before the homogeneous growth of the bulk film. The thickness of this nucleation layer depends on the type of growth of a-C films. SE and MWE have shown that a-C films deposited by rf MS on floating substrates and dc MS follow an island-type growth, while films deposited by rf MS and CFUBMS on biased substrates (with Ar^+ ion bombardment during deposition) follow layer-by-layer growth with almost atomically smooth surfaces. This difference is attributed to the variation of the ion flux to the film surface during deposition, with rf MS and CFUBMS on biased substrates being the techniques achieving the higher ion flux. Furthermore, the high deposition rate of a-C films by CFUBMS was used to test the acquisition speed and software analysis limits of the currently available SE–MWE technology.

Third, the ability of Fourier transform ellipsometry (FTPME) to perform precise determination of thin film bonding structure has been illustrated. FTPME is well adapted for the study of carbon nitride materials, providing important information about their vibrational properties. This constitutes a crucial advantage for studies in which information about the type as well as the amount of chemical bonds is needed. Furthermore, this technique can also probe the local structural order of thin films. Thus, taking into consideration its high sensitivity and capability for real-time measurements, it is anticipated that FTPME will be used extensively in the future for process monitoring as well as for probing the growth mechanisms of a wide variety of thin film materials. A detailed analysis of the bonding structure of sputtered CN_x films, as affected by the ion bombardment during deposition, has been performed by means of the FTRME technique. The results were discussed through the films' electronic behavior obtained by FTPME and SE measurements in the NIR–Vis–UV range. The conditions for the formation of hard and elastic three-dimensional carbon–nitrogen bonding structures were clearly correlated with the high-energy ion bombardment. These results can contribute to a further understanding of the bonding mechanisms and electronic modifications during N incorporation in the amorphous carbon matrix and can provide insight towards the production of CN_x films with desired properties, such as fullerene-like films and crystalline C_3N_4 structures.

Finally, SE and MWE were used to study the optical properties and to monitor the growth and oxidation of TiN_x films. Their optical properties depend strongly on their composition and structural characteristics. SE was proved a valuable technique that can estimate the stoichiometry and electrical resistivity of TiN_x films. Determination of the film composition was made through the screened plasma energy ω_{ps}, which is obtained directly from the measured $\varepsilon_1(\omega)$ spectrum, while the electrical resistivity was correlated with the screened (ω_{ps}) and unscreened (ω_{pu}) plasma energies. ω_{pu} was calculated by an optical model including the free and bonded electron contributions, and it is correlated with the electrical resistivity through an exponential decay law. Both ω_{pu} and ω_{ps} were found to be almost linearly related to the with TiN_x film stoichiometry and the bias voltage V_b applied to the substrate during deposition. Similarly, by raising the substrate temperature to 400 °C, 15% and 35% increases in ω_{ps} and ω_{pu}, respectively, were observed. However, $|V_b|$ has an even stronger effect on optical properties of TiN_x; for example, ω_{pu} increases with $|V_b|$ from 4.5 eV at $V_b = -20$ V to a saturation value of ≈ 7.5 eV for $|V_b| > 100$ V.

The detailed analysis of SE based on the free electron model has proved a great asset in determining the TiN_x films' stoichiometry, density, and grain size with respect to the main deposition conditions, which influence the growth mechanism of the films. The films mass density was calculated in situ by SE data analysis, and the results are confirmed by XRR. Moreover, the mean free path of electrons was calculated by SE and compared with the grain size as measured by X-ray diffraction and transmission electron microscopy, from which it was found that the main electron scattering sites are the grain boundaries and the Ti vacancies for stoichiometric and substoichiometric ($x \approx 1.1$) films, respectively. In addition, the interband electronic absorption in TiN_x films was studied in terms of the Lorentz oscillators. Two strong oscillators were identified and studied, and they are located at energies positions ≈ 3.2–3.9 and 5.2–6.2 eV, respectively. A third, weak electronic transition also exists at ≈ 2.7 eV. All the electronic transitions were assigned according to the band structure diagram of the bulk stoichiometric TiN. However, the energy position, strength, and broadening of the Lorentz oscillators, which simulate the interband transitions, were varied with respect to deposition parameters and were associated with TiN_x crystal cell size, stoichiometry, and grain orientation. The results of this analysis have shown that in situ SE may provide valuable information for the microstructure of metallic TiN_x films, and they were used to develop a real-time monitoring procedure of the TiN_x film growth by closed field unbalanced magnetron sputtering (CFUBMS), a technique with high deposition rate and industrial applications. Real-time monitoring is essential for the production of modern engineering materials, providing reduction in production process times and increase of production yield. The high deposition rate of TiN by CFUBMS was used to test the acquisition speed and software analysis limits of the currently available SE–MWE technology. MWE was proved capable of successfully monitoring not only the TiN_x growth, but the preceding Si etching and the Ti interlayer deposition as well. The MWE results were confirmed by conventional SE analysis and XRR.

The oxidation of TiN_x films grown by dc magnetron sputtering was also studied in films deposited under various bias voltages at RT. TiN_x films deposited with $V_b = -120$ V, which corresponds to stoichiometric TiN, are dense and stable when exposed to atmospheric air. TiN_x films deposited with $|V_b|$ below 80 V exhibit an excess of nitrogen and the presence of weakly bonded nitrogen, and are oxidized very fast at RT. Simulation of the dielectric function suggests that surface oxidation of TiN_x films cannot proceed further than a few angstroms. Analysis of the dielectric function spectra obtained during exposure of TiN_x films to air and various gases with an BEMT provides the TiO_2 volume fraction of the oxidized material. The weakly bonded nitrogen in TiN_x films and their columnar structure can explain the deep diffusion of oxygen through the grain boundaries.

Acknowledgments

The author would like to thank his collaborators M. Gioti, P. Patsalas, and A. Laskarakis for their contributions to this research. Part of this research was financially supported by the EU BRPR-CT96-0265 project and the General Secretariat of Research and Technology of Greece, under the EPET-333, YPER-97YP3-211, and PENED-99ED645 projects.

REFERENCES

1. A. C. Boccara, C. Pickering, and J. Rivory, Eds., "Spectroscopic Ellipsometry." Elsevier, Amsterdam, 1993.
2. B. Drevillon, in "Science and Technology of Thin Films" (F. C. Matacotta and C. Ottaviani, Eds.), p. 189. World Scientific Publishers, 1995.
3. R. W. Collins, I. An, V. Nguyen, Y. Li, and Y. Lu, in "Optical Characterization of Real Surfaces and Films" (M. H. Francombe and J. L. Vossen, Eds.), Vol. 19, p. 49. Academic Press, San Diego, 1994.
4. S. Logothetidis, J. Petalas, M. Cardona, and T. D. Moustakas, *Phys. Rev. B* 50, 18017 (1994).
5. S. Logothetidis, J. Petalas, M. Cardona, and T. D. Moustakas, *Mater. Sci. Engin. B* 29, 65 (1995).
6. S. Logothetidis and J. Petalas, *J. Appl. Phys.* 80, 1768 (1996).
7. S. Logothetidis, J. Petalas, A. Markwitz, and R. L. Johnson, *J. Appl. Phys.* 73, 8514 (1993).
8. T. Heitz, B. Drévillon, C. Gobet, and J. E. Bourée, *Phys. Rev. B* 58, 13957 (1998).
9. A. Laskarakis, S. Logothetidis, and M. Gioti, *Phys. Rev. B*, 15 August (2001).
10. A. Roeseler, "Infrared Spectroscopic Ellipsometry." Akademie Verlag, Berlin, 1990.
11. F. Ferrieu, *Rev. Sci. Instrum.* 60, 3212 (1989).
12. J. Bremer, O. Hunderi, K. Fanping, T. Skauli, and E. Wold, *Appl. Opt.* 31, 471 (1988).
13. R. T. Graf, F. Eng, J. L. Koening, and H. Ishida, *Appl. Spectrosc.* 40, 498 (1986).
14. N. Blayo, B. Drevillon, and R. Ossikovski, *SPIE Symp. Proc.* 1681, 116 (1992).
15. A. Canillas, E. Pascual, and B. Drevillon, *Rev. Sci. Instrum.* 64, 2153 (1993).
16. J. Barth, R. L. Johnson, S. Logothetidis, M. Cardona, D. Fuchs, and A. M. Bradshaw, "Soft X-Ray Optics and Technology." International Conference Proceedings, SPIE, Berlin, 1986.
17. J. Barth, R. L. Johnson, and M. Cardona, in "Handbook of Optical Constants of Solids II" (E. D. Palik, Ed.), Chapter 10, p. 213. Academic Press, 1991.
18. S. Logothetidis, J. Petalas, H. M. Polatoglou, and D. Fuchs, *Phys. Rev. B* 46, 4483 (1992).
19. D. E. Aspnes, W. E. Quinn, and S. Gregory, *Appl. Phys. Lett.* 57, 2707 (1990).
20. W. M. Duncan and S. A. Henck, *Appl. Surf. Sci.* 63, 9 (1993).
21. I. F. Wu, J. B. Dottelis, and M. Dagenais, *J. Vac. Sci. Technol. A* 11, 2398 (1993).
22. V. G. Kechagias, M. Gioti, S. Logothetidis, E. Benferhat, and D. Teer, *Thin Solid Films* 364, 213 (2000).
23. S. Logothetidis, M. Gioti, and P. Patsalas, *Diamond Relat. Mater.* 19, 117 (2001).
24. R. M. A. Azzam and N. M. Bashara, in "Ellipsometry and Polarized Light." North Holland, Amsterdam, 1977.
25. P. Lautenschlager, M. Garriga, L. Viña, and M. Cardona, *Phys. Rev. B* 36, 4821 (1987).
26. L. V. Keldysh, D. A. Kirzhnits, and A. A. Maradudin, Eds., "The Dielectric Function of Condensed Systems." North-Holland, 1989.
27. D. E. Aspnes, *Thin Solid Films* 89, 249 (1982), and references therein.
28. O. Hunderi, *Surf. Sci.* 96, 1 (1980), and references therein.
29. A. M. Brodsky and M. I. Urbakh, *Prog. Surf. Sci.* 15, 121 (1984).
30. D. A. G. Bruggeman, *Ann. Phys. (Leipzig)* 24, 636 (1935).
31. G. A. Niklasson and C. G. Granqvist, *J. Appl. Phys.* 55, 3382 (1984), and references therein.
32. D. J. Bergman, *Phys. Rev. Lett.* 44, 1285 (1980).
33. G. W. Milton, *Appl. Phys. Lett.* 37, 300 (1980).
34. S. Logothetidis, *J. Appl. Phys.* 65, 2416 (1989).
35. S. Logothetidis, G. Kiriakidis, and E. C. Paloura, *J. Appl. Phys.* 70, 2791 (1991).
36. S. Logothetidis, J. Petalas, and S. Ves, *J. Appl. Phys.* 79, 1040 (1996).
37. D. E. Aspnes and A. A. Studna, *Appl. Opt.* 14, 220 (1975).

38. S. N. Jasperson and S. E. Schnatterly, *Rev. Sci. Instrum.* 40, 761 (1969).

39. B. Drevillon, J. Perrin, R. Marbot, A. Violet, and J. L. Dalby, *Rev. Sci. Instrum.* 53, 969 (1982).

40. O. Acher, E. Bigan, and B. Drevillon, *Rev. Sci. Instrum.* 60, 65 (1989).

41. M. J. Dignam, B. Rao, and R. W. Stobie, *Surf. Sci.* 46, 308 (1976).

42. B. E. Hayden, W. Wyrobisch, W. Oppermann, S. Hachicha, P. Hofmàn, and A. M. Bradshaw, *Surf. Sci.* 109, 207 (1981).

43. N. Blayo and B. Drevillon, *J. Non-cryst. Solids* 137&138, 771 (1991).

44. R. Ossikovski, H. Shirai, and B. Drevillon, *Appl. Phys. Lett.* 64, 1815 (1994).

45. B. Bhushan, *Diamond Relat. Mater.* 8, 1985 (1999).

46. B. Bhushan, A. J. Kellock, N. H. Cho, and W. Ager III, *J. Mater. Res.* 7, 404 (1992).

47. D. R. Mckenzie, D. A. Muller, and B. A. Paithorpe, *Phys. Rev. Lett.* 67, 773 (1991).

48. H. Tsai and D. B. Bogy, *J. Vac. Sci. Technol. A* 5, 3287 (1987).

49. B. K. Gupta and B. Bhushan, *Wear* 190, 110 (1995).

50. J. Wagner and P. Lautenschlager, *J. Appl. Phys.* 59, 2044 (1986).

51. A. A. Smolin, V. G. Ralchenko, S. M. Pimerov, T. V. Kononenko, and E. N. Loubnin, *Appl. Phys. Lett.* 62, 3449 (1993).

52. P. J. Fallon, V. S. Veerasamy, C. A. Davis, J. Robertson, G. A. J. Amaratunga, W. I. Milne, and J. Koskinen, *Phys. Rev. B* 48, 4777 (1993).

53. S. R. P. Silva, S. Xu, B. X. Tay, H. S. Tan, and W. I. Milne, *Appl. Phys. Lett.* 69, 491 (1996).

54. Y. Lifshitz, S. R. Kasi, J. W. Rabalais, and W. Eckstein, *Phys. Rev. B* 41, 10468 (1990).

55. Y. Lifshitz, G. Lempert, and E. Grossman, *Phys. Rev. Lett.* 72, 2753 (1997).

56. J. J. Cuomo, D. L. Pappas, J. Bruley, J. P. Doyle, and K. L. Saenger, *J. Appl. Phys.* 70, 1706 (1991).

57. F. Xiong, Y. Y. Wang, and R. P. H. Chang, *Phys. Rev. B* 48, 8016 (1993).

58. B. André, F. Rossi, A. van Veen, P. E. Mijnarends, H. Schut, and M. P. Delplancke, *Thin Solid Films* 241, 171 (1994).

59. J. Schwan, S. Ulrich, H. Roth, H. Ehrhardt, S. R. P. Silva, J. Robertson, R. Samienski, and R. Brenn, *J. Appl. Phys.* 79, 1416 (1996).

60. J. Schwan, S. Ulrich, T. Theel, H. Roth, H. Ehrhardt, P. Becker, and S. R. P. Silva, *J. Appl. Phys.* 82, 6024 (1997).

61. E. Mounier, and Y. Pauleau, *Diamond Relat. Mater.* 6, 1182 (1997).

62. S. Logothetidis, *Appl. Phys. Lett.* 69, 158 (1996).

63. S. Logothetidis and M. Gioti, *Mater. Sci. Engin. B* 46, 119 (1997).

64. M. Gioti and S. Logothetidis, *Diamond Relat. Mater.* 7, 444 (1998).

65. G. Curro, F. Neri, G. Mondio, G. Compagnini, and G. Foti, *Phys. Rev. B* 49, 8411 (1994).

66. J. Robertson, *Prog. Solid State Chem.* 21, 199 (1992).

67. N. Savvides, *J. Appl. Phys.* 59, 4133 (1986).

68. V. Paillard, P. Melinon, V. Dupuis, A. Perez, J. P. Perez, G. Guiraud, J. Fornazero, and G. Panczer, *Phys. Rev. B* 49, 11433 (1994).

69. M. A. Tamor and C. H. Wu, *J. Appl. Phys.* 67, 1007 (1990).

70. R. Ahuja, S. Auluck, J. M. Wills, M. Alouani, B. Johansson, and O. Eriksson, *Phys. Rev. B* 55, 4999 (1997).

71. A. K. Solanki, A. Kashyap, T. Nautiyal, S. Auluck, and M. A. Khan, *Solid State Commun.* 100, 645 (1996).

72. A. B. Djurisic and E. H. Li, *J. Appl. Phys.* 85, 7404 (1999).

73. S. Logothetidis, M. Gioti, P. Patsalas, and C. Charitidis, *Carbon* 37, 765 (1999).

74. M. Gioti and S. Logothetidis, *Diamond Relat. Mater.* 8, 446 (1999).

75. V. Liebler, H. Baumann, and K. Bethge, *Diamond Relat. Mater.* 2, 584 (1993).

76. D. F. Edwards and H. R. Phillip, in "Handbook of Optical Constants of Solids" (E. D. Palik, Ed.), p. 665. Academic Press, Orlando, FL, 1985.

77. X. Jiang, W. Beyer, and K. Reichelt, *J. Appl. Phys.* 68, 1378 (1990).

78. B. Dischler, in "Amorphous Hydrogenated Carbon Films" (P. Koidl and P. Oelhafen, Eds.), *Proc. Euro. Mater. Res. Soc.* 17, 189, Les Editions de Physique, Paris, 1987.

79. C. Wild and P. Koidl, *Appl. Phys. Lett.* 51, 1506 (1987).

80. W. Beyer and H. Wagner, *J. Appl. Phys.* 53, 8745 (1982).

81. P. Lautenschlager, M. Garriga, and M. Cardona, *Phys. Rev. B* 36, 4821 (1987).

82. S. Logothetidis and G. Kiriakidis, *J. Appl. Phys.* 64, 2389 (1988).

83. M. Gioti, D. Papadimitriou, and S. Logothetidis, *Diamond Relat. Mater.* 9, 741 (2000).

84. S. Logothetidis and M. Gioti, to be published.

85. G. E. Jellison and F. A. Modine, *Appl. Phys. Lett.* 69, 371 (1996).

86. J. Tauc, R. Grigorovici, and A. Vaucu, *Phys. Stat. Sol.* 15, 627 (1966).

87. J. Robertson, *Diamond Relat. Mater.* 3, 361 (1994).

88. C. A. Davis, *Thin Solid Films* 226, 30 (1993).

89. S. Logothetidis and C. Charitidis, *Thin Solid Films* 353, 208 (1999).

90. S. Logothetidis, M. Gioti, C. Charitidis, P. Patsalas, J. Arvanitidis, and J. Stoemenos, *Appl. Surf. Sci.* 138–139, 244 (1999).

91. S. Logothetidis, *Int. J. Mod. Phys. B* 14, 113 (2000).

92. M. Kildemo and B. Drévillon, *Appl. Phys. Lett.* 67, 918 (1999).

93. S. Logothetidis and G. Stergioudis, *Appl. Phys. Lett.* 71, 2463 (1997).

94. S. Logothetidis, G. Stergioudis, and N. Vouroutzis, *Surf. Coat. Technol.* 100–101, 486 (1998).

95. P. C. Kelires, M. Gioti, and S. Logothetidis, *Phys. Rev. B* 59, 5074 (1999).

96. S. Logothetidis, P. Patsalas, M. Gioti, A. Galdikas, and L. Pranevicius, *Thin Solid Films* 376, 56 (2000).

97. P. Patsalas and S. Logothetidis, *J. Appl. Phys.* 88, 6346 (2000).

98. P. Patsalas, M. Handrea, S. Logothetidis, M. Gioti, S. Kennou, and W. Kautek, *Diamond Relat. Mater.* 10, 960 (2001).

99. S. Logothetidis, P. Patsalas, and M. Gioti, to be published (2001).

100. S. A. Y. Liu and M. L. Cohen, *Science* 245, 841 (1989).

101. A. Y. Liu and M. L. Cohen, *Phys. Rev. B* 41, 10727 (1990).

102. S. Muhl and J. M. Mendez, *Diamond Relat. Mater.* 8, 1809 (1999).

103. K. M. Yu, M. L. Cohen, E. E. Haller, W. L. Hansen, and A. Y. Liu, *Phys. Rev. B* 49, 2034 (1994).

104. C. Niu, Y. Z. Lu, and C. M. Lieber, *Science* 261, 334 (1993).

105. D. Marton, K. J. Boyd, A. H. Al-Bayati, S. S. Todorov, and J. W. Rabalais, *Phys. Rev. Lett.* 73, 118 (1994).

106. N. Hellgren, M. P. Johansson, E. Broitman, L. Hultman, and J.-E. Sundgren, *Phys. Rev. B* 59, 5162 (1999).

107. H. Sjöström, S. Stafstrom, M. Boman, and J.-E. Sundgren, *Phys. Rev. Lett.* 75, 1336 (1995).

108. H. Sjöström, L. Hultman, J.-E. Sundgren, S. V. Hainsworth, T. F. Page, and G. S. A. M. Theunissen, *J. Vac. Sci. Technol. A* 14, 56 (1996).

109. W. T. Zheng, E. Broitman, N. Hellgren, K. Z. Xing, I. Ivanov, H. Sjöström, L. Hultman, and J.-E. Sundgren, *Thin Solid Films* 308, 223 (1997).

110. R. Kaltofen, T. Sebald, J. Schulte, and G. Weise, *Thin Solid Films* 347, 31 (1999).

111. M. R. Wixom, *J. Am. Ceram. Soc.* 73, 1973 (1990).

112. D. Lin-Vien, N. B. Colthup, W. G. Fateley, and J. G. Grasselli, "The Handbook of Infrared and Raman Characteristic Frequencies of Organic Molecules." Academic Press, Boston, 1991.

113. Y. Liu, C. Jiaa, and H. Do, *Surf. Coat. Technol.* 115, 95 (1999).

114. M. C. dos Santos and F. Alvarez, *Phys. Rev. B* 58, 13918 (1998).

115. J. H. Kaufman, S. Metin, and D. D. Saperstein, *Phys. Rev. B* 39, 13053 (1989).

116. M. Gioti, S. Logothetidis, C. Charitidis, and H. Lefakis, *Vacuum* 53, 53 (1999).

117. J. Fink, Th. Müller-Heinzerling, J. Pflüger, B. Sheerer, B. Dischler, P. Koidl, A. Bubenzer, and R. E. Sach, *Phys. Rev. B* 30, 4713 (1984).

118. P. Kovarik, E. B. D. Bourdon, and R. H. Prince, *Phys. Rev. B* 48, 12123 (1993).

119. A. Laskarakis, S. Logothetidis, C. Charitidis, M. Gioti, Y. Panayiotatos, M. Handrea, and W. Kautek, *Diamond Relat. Mater.* 10, 1179 (2001).

120. S. Logothetidis, H. Lefakis, and M. Gioti, *Carbon* 36, 757 (1998).

121. E. I. Meletis, *J. Mater. Eng.* 11, 159 (1989).

122. T. Arai, H. Fujita, and M. Watanabe, *Thin Solid Films* 154, 387 (1987).

123. M. Wittmer, *J. Vac. Sci. Technol. A* 3, 1797 (1985).

124. C. A. Dimitriadis, S. Logothetidis, and I. Alexandrou, *Appl. Phys. Lett.* 66, 582 (1995).

125. L. Soriano, M. Abbate, H. Pen, P. Prieto, and J. M. Sanz, *Solid State Commun.* 102, 291 (1997).

126. P. Patsalas and S. Logothetidis, *J. Appl. Phys.*, submitted.

127. S. Logothetidis, I. Alexandrou, and A. Papadopoulos, *J. Appl. Phys.* 77, 1043 (1995).

128. A. A. Adjaottor, E. I. Meletis, S. Logothetidis, I. Alexandrou, and S. Kokkou, *Surf. Coat. Technol.* 76–77, 142 (1995).

129. S. Logothetidis, E. I. Meletis, and G. Kourouklis, *J. Mater. Res.* 14, 436 (1999).

130. K.-H. Hellwege and J. L. Olsen, in "Numerical Data and Functional Relationships in Science and Technology" (Landolt-Boernstein, Eds.), Group III, Vol. 13a. Springer-Verlag, Berlin, 1981.

131. P. Patsalas, S. Logothetidis, and C. A. Dimitriadis, *Mater. Res. Soc. Symp. Proc.* 569, 113 (1999).

132. F. Wooten, "Optical Properties of Solids." Academic Press, New York, 1972.

133. J. Hojo, O. Iwamoto, Y. Maruyama, and A. Kato, *J. Less-Common Met.* 53, 265 (1977).

134. P. Patsalas, C. Charitidis, S. Logothetidis, C. A. Dimitriadis, and O. Valassiades, *J. Appl. Phys.* 86, 5296 (1999).

135. N. W. Ashcroft, N. D. Mermin, "Solid State Physics." Saunders College Publishing, Orlando, FL, 1976.

136. W. D. Chen, H. Bender, W. Vandervorst, and H. E. Maes, *Surf. Interf. Anal.* 12, 151 (1988).

137. M. J. Vasile, A. B. Emerson, and F. A. Baiocchi, *J. Vac. Sci. Technol. A* 8, 99 (1990).

138. I. Petrov, L. Hultman, J.-E. Sudgren, and J. E. Greene, *J. Vac. Sci. Technol. A* 10, 265 (1992).

139. I. N. Michailescu, N. Chitica, L. C. Nistor, M. Popescu, V. S. Theodorescu, I. Ursu, A. Andrei, A. Barborica, A. Luches, M. Luisa De Giorgi, A. Perrone, B. Dubreil, and J. Hermann, *J. Appl. Phys.* 74(9), 5781 (1993).

140. U. Kreibig, *J. Phys. F* 4, 999 (1974).

141. H. V. Nguyen, I. An, and R. W. Collins, in "Optical Characterization of Real Surfaces and Films" (M. H. Francombe and J. L. Vossen, Eds.), Vol. 19, p. 127. Academic Press, San Diego, 1994.

142. S. Logothetidis, H. M. Polatoglou, and S. Ves, *Solid State Commun.* 68, 1075 (1988).

143. P. Patsalas, C. Charitidis, and S. Logothetidis, *Surf. Coat. Technol.* 125, 335 (2000).

144. P. Patsalas, C. Charitidis, and S. Logothetidis, *Appl. Surf. Sci.* 154–155, 256 (2000).

145. D. E. Savage, J. Kleiner, N. Schimke, Y. H. Phang, T. Jankowsky, J. Jacobs, R. Darious, and M. G. Lagally, *J. Appl. Phys.* 69, 1411 (1991).

146. J. B. Kotright, *J. Appl. Phys.* 70, 4286 (1991).

147. M. Born and E. Wolf, "Principles of Optics." Pergamon, New York, 1983.

148. B. Pecz, N. Frangis, S. Logothetidis, I. Alexandrou, P. B. Barna, and J. Stoemenos, *Thin Solid Films* 268, 57 (1995).

149. S. Logothetidis, E. I. Meletis, G. Stergioudis, and A. A. Adjaottor, *Thin Solid Films* 338, 304 (1999).

150. J. Desmaison, P. Lefort, and M. Billy, *Oxid. Met.* 13, 203 (1979).

Chapter 7

IN SITU FARADAY-MODULATED FAST-NULLING SINGLE-WAVELENGTH ELLIPSOMETRY OF THE GROWTH OF SEMICONDUCTOR, DIELECTRIC, AND METAL THIN FILMS

J. D. Leslie, H. X. Tran

Department of Physics, University of Waterloo, Waterloo, Ontario,
N2L 3G1, Canada

S. Buchanan

Waterloo Digital Electronics Division of WDE Inc., 279 Weber St. N., Waterloo, Ontario,
N2J 3H8, Canada

J. J. Dubowski, S. R. Das, L. LeBrun

Institute for Microstructural Sciences, National Research Council, Ottawa, Ontario,
K1A 0R6, Canada

Contents

Handbook of Thin Film Materials, edited by H.S. Nalwa
Volume 2: Characterization and Spectroscopy of Thin Films
Copyright © 2002 by Academic Press
All rights of reproduction in any form reserved.

ISBN 0-12-512910-6/$35.00

1. INTRODUCTION TO ELLIPSOMETRY

1.1. Ellipsometry and Fresnel Coefficients

Ellipsometry is an optical technique which can be used to characterize the thickness (Th_L) and the complex index of refraction n_L of one or more layers on a substrate of complex index of refraction n_s by studying the change in the state of polarization of light reflected off the layers on the substrate. The change in the state of polarization is obtained by using classical electromagnetism to calculate the Fresnel coefficients for the surface of that sample, which of course depends on the index of refraction and thickness of each layer on the substrate and the index of refraction of the substrate itself. Basically, the Fresnel coefficients describe the change of the reflected and transmitted light relative to the incident beam. Since Fresnel coefficients have been derived in a number of electromagnetic [1, 2] and thin film optics [3] textbooks as well as on the worldwide web [4], we will not repeat the derivations totally here. Instead, we just define the geometries and quantities involved, we sketch the derivations, and we give the relevant results.

The incident, reflected, and transmitted light at a single interface between two semi-infinite media is depicted in Figure 1. The two media are characterized by their refractive indices n_0 and n_1. These refractive indices can be complex numbers. If n_0 represents an ambient medium, which usually in a film growth case is a vacuum, then it has a value of 1. n_1 may represent the refractive index of an optically thick substrate, which normally is a complex number for a semiconductor or metal. The angles of incidence and refraction are denoted by ϕ_0 and ϕ_1, respectively. The superscript i, r, or t on E denotes the incident, reflected, or transmitted electric vector, respectively. Each electric vector can be decomposed into two components which are indicated by subscripts p and s. The subscript p represents the component of the electric vector which is parallel to the plane of incidence, while s represents the component of the electric vector which is perpendicular to the plane of incidence. (s comes

from the German senkrecht which means perpendicular.) The incident, reflected, and transmitted waves are assumed to be plane waves with the form $\mathbf{E} = \mathbf{E}_0 e^{i(\omega t - \mathbf{k} \cdot \mathbf{r})}$. The phase factors associated with the incident, reflected, and transmitted waves for this case are given by

$$\exp\left[i\left(\omega t - \frac{2\pi n_0 x \sin \phi_0}{\lambda} - \frac{2\pi n_0 z \cos \phi_0}{\lambda}\right)\right]$$

$$\exp\left[i\left(\omega t - \frac{2\pi n_0 x \sin \phi_0}{\lambda} + \frac{2\pi n_0 z \cos \phi_0}{\lambda}\right)\right] \quad (1)$$

$$\exp\left[i\left(\omega t - \frac{2\pi n_1 x \sin \phi_1}{\lambda} - \frac{2\pi n_1 z \cos \phi_1}{\lambda}\right)\right]$$

respectively, and λ is the wavelength in vacuum. The most interesting part in the argument of the phase factor is the z component, since the wave is assumed to propagate in the z direction, and thus the other parts of the argument will be left out throughout the discussion for convenience. At the boundary, where $z = 0$ is assumed, the components of the electric and magnetic vectors in the x and y directions for the first medium are given by

$$E_{0x} = \left(E_p^i - E_p^r\right) \cos \phi_0$$

$$E_{0y} = E_s^i + E_s^r$$

$$H_{0X} = \left(-E_s^i + E_s^r\right) n_0 \cos \phi_0 \quad (2)$$

$$H_{0Y} = \left(E_p^i + E_p^r\right) n_0$$

There is only a transmitted wave in the second medium, and the components of the electric and magnetic vectors are given by

$$E_{1x} = E_p^t \cos \phi_1$$

$$E_{1y} = E_s^t$$

$$H_{1x} = -n_1 E_p^t \cos \phi_1 \quad (3)$$

$$H_{1y} = n_1 E_p^t$$

By applying the standard boundary conditions for electromagnetic waves to the interface between two semi-infinite media depicted in Figure 1, the amplitudes of the reflected and transmitted electric vectors can be written in terms of the incident electric vectors and hence the Fresnel coefficients [1–4] can be derived and are given as

$$r_p = \frac{E_p^r}{E_p^i} = \frac{n_0 \cos \phi_1 - n_1 \cos \phi_0}{n_0 \cos \phi_1 + n_1 \cos \phi_0} \quad (4)$$

$$r_s = \frac{E_s^r}{E_s^i} = \frac{n_0 \cos \phi_0 - n_1 \cos \phi_1}{n_0 \cos \phi_0 + n_1 \cos \phi_1} \quad (5)$$

$$t_p = \frac{E_p^t}{E_p^i} = \frac{2 n_0 \cos \phi_0}{n_1 \cos \phi_0 + n_0 \cos \phi_1} \quad (6)$$

$$t_s = \frac{E_s^t}{E_s^i} = \frac{2 n_0 \cos \phi_0}{n_0 \cos \phi_0 + n_1 \cos \phi_1} \quad (7)$$

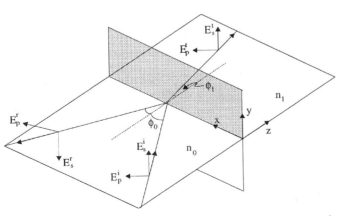

Fig. 1. Illustration of incident, reflected, and transmitted electric vectors for the p and s states of polarization at a single interface between two semi-infinite media characterized by n_0 and n_1.

Two assumptions have been made in the derivation of the Fresnel coefficients. The first assumption is that the electromagnetic waves are assumed to be plane waves. The second assumption is that the media are isotropic such that the refractive indices are the same in all directions. It should be remembered that the refractive indices in the Fresnel coefficients (4)–(7) are complex numbers, with the form $n - ik$, for the case of a semiconductor or a metal.

1.2. Reflection of Light by a Single Film

The reflections and the transmissions of light by a single film are shown in Figure 2. Figure 2a illustrates the method of summation while Figure 2b illustrates the method of resultant waves in the derivation of the reflection coefficients. The final expression for the reflection coefficients will be valid for either p or s polarization, and thus the subscripts p and s will be omitted during the derivation. Another assumption made here is that the film is nonabsorbing such that the refractive index is real. Even with this assumption, there is no loss in generality in the final expression of the reflection coefficients. For the case of an absorbing film, the same expression for the reflection coefficients for a nonabsorbing film can be used except that the complex number, $n - ik$, for the refractive index is used for this case instead of a real number, n, for the case of a nonabsorbing film.

1.3. Method of Summation

The multiple reflections and transmissions by interfaces of a single film are shown in Figure 2. The total reflected ampli-

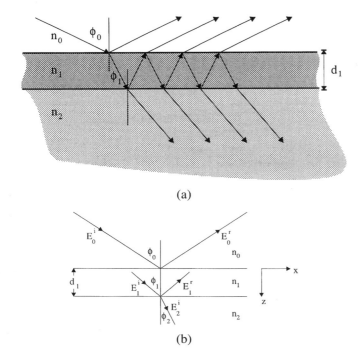

(a)

(b)

Fig. 2.   (a) The multiple reflections and transmissions of light in a single film are shown (method of summation) and (b) the incoming and outgoing electric wave vectors are shown (method of resultant waves).

tudes are obtained by adding up the contributions to the reflected amplitude due to each of the multiple rays within the film while taking into account the phase differences associated with the various rays. **Appendix A** goes through this method of summation in detail and shows that the total reflected amplitudes for s and p polarization of light can be written

$$R_s = \frac{r_{1s} + r_{2s}e^{-2i\delta_1}}{1 + r_{1s}r_{2s}e^{-i2\delta_1}} \qquad (8)$$

$$R_p = \frac{r_{1p} + r_{2p}e^{-2i\delta_1}}{1 + r_{1p}r_{2p}e^{-i2\delta_1}} \qquad (9)$$

where r_1 (r_2) is the Fresnel reflection coefficient at the n_0–n_1 (n_1–n_2) interface, respectively, and $e^{-i2\delta_1}$ is the phase factor associated with one complete traversal of the film and the phase difference,

$$2\delta_1 = \frac{4\pi}{\lambda}n_1 d_1 \cos\phi_1 \qquad (10)$$

1.4. Method of Resultant Waves

The method of summation becomes very awkward when one gets to a more complicated situation than a single layer on a substrate, whereas the method of resultant waves proves to be much more versatile. This method uses the vector sums of the reflected and transmitted waves and applies the appropriate boundary conditions at the interfaces to derive the reflection coefficients. **Appendix B** shows how the method of resultant waves is applied first to a single layer and then to a multilayer structure. Each layer in a multilayer structure can be represented by a characteristic matrix which links components of the electric and magnetic vectors in one layer to the preceding layer. We consider the case of m layers each of thickness d_m sandwiched in between two semi-infinite media n_0 and n_{m+1}. The theory is developed in terms of the tangential components of the electric and magnetic fields at each interface, so that $\overline{E}_j$, $\overline{H}_j$ are these components at the interface between the jth and $j + 1$th layers. The characteristic matrix of the jth layer is denoted by $M_j(d_j)$. **Appendix B** shows that the tangential components $\overline{E}_0$, $\overline{H}_0$ in the semi-infinite medium n_0 at the interface with the first layer may be expressed in terms of the tangential components $\overline{E}_{m+1}$, $\overline{H}_{m+1}$ in the semi-infinite medium n_{m+1} at the interface with the mth layer, as

$$\begin{bmatrix} \overline{E}_0 \\ \overline{H}_0 \end{bmatrix} = M \begin{bmatrix} \overline{E}_{m+1} \\ \overline{H}_{m+1} \end{bmatrix} \qquad (11)$$

where $M = M_1(d_1) * M_2(d_2) \cdots M_j(d_j) \cdots M_m(d_m)$. Of course, it is necessary to differentiate the state of polarization and this is accomplished by adding the superscript p or s to the matrix variable M and the subscript p or s to the other appropriate variables.

Appendix B shows that Eq. (11) can be rewritten in terms of the reflection coefficients, and thus one can derive that the reflection coefficient, r_p, for TM (Transverse Magnetic) waves (p polarization) can be written as

$$r_p = \frac{c_p - 1}{c_p + 1} \tag{12}$$

where c_p is an expression involving the components of M^p. Similarly, one can derive that the reflection coefficient, r_s, for TE (Transverse Electric) waves (s polarization) can be written as

$$r_s = \frac{c_s - 1}{c_s + 1} \tag{13}$$

where c_s is an expression involving the components of M^s.

The key fact is that the reflection coefficients can be calculated from Eqs. (12) and (13) by a computer program once the optical characteristics of the system being studied are input. Once the reflection coefficients are known, the ellipsometric parameters can be determined. The relation between the reflection coefficients and the ellipsometric parameters are discussed in the next section.

1.5. Principles of Ellipsometry and Null Ellipsometry

In general, a reflection of a light beam at the interface between two media will cause the state of polarization of the light beam to change. The change in the state of polarization can be described by the fundamental equation of ellipsometry, defined as

$$\rho = \frac{r_p}{r_s} = \tan \psi e^{i\Delta} \tag{14}$$

where $\Delta = \Delta_1 - \Delta_2$ and $\tan \psi = |r_p|/|r_s|$. Δ_1 is defined as the phase difference between the perpendicular component and the parallel component of the incoming wave while Δ_2 is defined as the phase difference between the perpendicular component and the parallel component of the outgoing wave. Equation (14) is the basis of defining ellipsometric data as the pair of values ψ and Δ, and the use of this data to obtain the refractive indices and layer thicknesses of the sample being studied.

The most accurate way of measuring ψ and Δ employs null ellipsometry. Usually, when linearly polarized light is incident on the surface of a sample, the sample substrate plus any layers on the substrate will cause the reflected light to become elliptically polarized as shown in Figure 3a. In null ellipsometry, the earlier process is reversed. Instead of having linearly polarized incident light, elliptically polarized light it used. The ellipticity of the incident light is selected so that it will cancel the ellipticity introduced by the surface. As a result, the light reflected off the sample is linearly polarized as shown in Figure 3b. The orientation of the resulting linearly polarized light can be determined by passing it through a polarizer (henceforth called the analyzer) and rotating the analyzer until a detector indicates no light is being received. In this condition, the detector indicates a null, hence the name "null ellipsometry." The orientation of the linearly polarized light is perpendicular to the direction of the polarization axis of the analyzer.

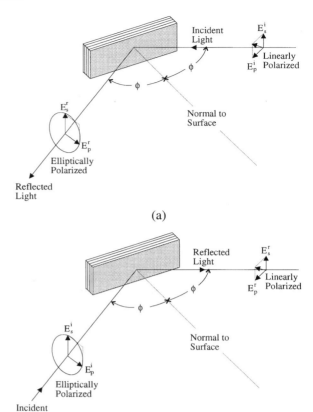

Fig. 3. (a) Illustrates that in general linearly polarized light incident on a sample will produce elliptically polarized reflected light, while (b) illustrates that if we select the correct elliptically polarized light incident on a surface, the ellipticity caused by the reflection can be canceled and we obtain linearly polarized light.

1.6. Operation of the *EXACTA 2000* Faraday-Modulated Single-Wavelength Ellipsometer

The *EXACTA 2000*, whose schematic diagram is shown in Figure 4, normally operates in the following way: A linearly polarized laser (L) and the first quarterwave plate (Q_1) produce circularly polarized light. The polarizer prism (PP) and the second quarterwave plate (Q_2) produce elliptically polarized light, whose Δ value is set by the polarizer angle P, to produce a linearly polarized reflected beam. The analyzer prism (AP), when rotated to the appropriate angle A, produces a null at the photodetector (D). The action of the Faraday rods and drive coils are described later. (**Appendix C** shows that the relation between the ellipsometric parameters Δ and ψ and the null ellipsometer variables P and A is just $\Delta = 2P + 90°$ and $\psi = A$. Here, P and A are angles of rotation of the axis of the polarizing direction of the polarizer and the analyzer from the plane of incidence, respectively, and the expressions assume that the fast axis of the second quarterwave plate is at $-45°$, and we are in the first zone.)

The polarizer (PP) and analyzer (AP) prisms are each mounted on precision rotary positioners which can be posi-

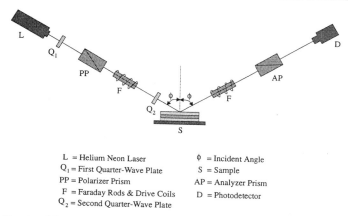

L = Helium Neon Laser
Q₁ = First Quarter-Wave Plate
PP = Polarizer Prism
F = Faraday Rods & Drive Coils
Q₂ = Second Quarter-Wave Plate

φ = Incident Angle
S = Sample
AP = Analyzer Prism
D = Photodetector

Fig. 4. Schematic of the **EXACTA 2000** Faraday-modulated self-nulling single-wavelength ellipsometer.

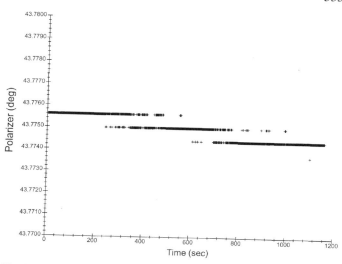

Fig. 5. Plot of the **EXACTA 2000** ellipsometer polarizer null readings versus time from an air-exposed single crystal silicon surface where each data point is the average value per second of the 10 readings per second, and data were taken for a total of 1159 s.

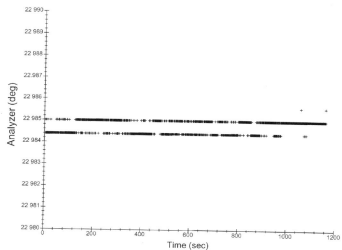

Fig. 6. Plot of the **EXACTA 2000** ellipsometer analyzer null readings versus time from an air-exposed single crystal silicon surface where each data point is the average value per second of the 10 readings per second, and data were taken for a total of 1159 s.

tioned with a resolution of 0.000625° under the control of stepping motors with their controllers. A Faraday rod in a Faraday drive coil (F) is in series optically with each of the polarizer and analyzer prisms. The Faraday rods and drive coils are used to slightly rotate the polarization of the light passing through the polarizer and the analyzer using quadrature modulation, and the detector signal is analyzed by a two-phase lock-in amplifier to provide feedback to the stepping motors rotating the polarizer and analyzer prisms to keep them at null. With this combination of Faraday modulation and mechanical rotation of the polarizer and analyzer prisms, we can obtain up to 10 polarizer and analyzer null readings per second over a very large range of polarizer and analyzer angles. With this approach, we can measure the polarizer (P) and analyzer (A) angles at null with a precision of better than $\pm 0.01°$ at 10 readings per second and better than $\pm 0.003°$, if we average the 10 readings over 1 s. Figures 5 and 6 show some drift measurements of P, A for a single crystal silicon wafer that shows that with good surfaces one can get even less scatter than is implied by the previous precision specifications. Figure 5 shows that the P value is drifting slightly over the 1159 s of the measurement, but that when averaged over 1 s the P value can be measured to better than $\pm 0.001°$. Figure 6 shows that the A value when averaged over 1 s can be measured to better than $\pm 0.001°$.

One can also operate the ellipsometer in an all-Faraday drive mode in which one can improve the precision of the polarizer and analyzer null angle determinations to better than $\pm 0.001°$[1] at one reading per second when P and A vary within a range of $\pm 0.25°$. The self-nulling mode via the motor control modules is switched off. Now, the polarization rotation function that was provided by the motor control modules, stepping motor drivers, rotary stages, and polarizing prisms is replaced by time integrating the lock-in amplifier signals and feeding them to the summing points in the Faraday rod coil driver amplifiers. Now, the voltage at the output of the integrator or the dc current in the Faraday rod coils represents the amount that the null angles

have rotated from the initial null positions. The conversion factor between angular and electrical variables, i.e., degrees per volt or degrees per ampere, is determined by measuring the P, A reading of a known surface, then backing P and A off by a set amount such as 0.25° from the values measured and determining the voltage at the output of the integrator or the dc current in the Faraday rod coils to bring the detector signal back to null.

1.7. What a Single-Wavelength Nulling Ellipsometer Can Measure

A single measurement of a single-wavelength nulling ellipsometer can yield only two pieces of information, i.e., the values of P and A that yield a null at the detector, which henceforth we will call just the P, A values. We will see that this has

---

[1]With a good surface such as a single crystal silicon substrate, the precision can be $\pm 0.0003°$.

a very significant effect on what ellipsometry can tell about a system consisting of a substrate with a sequence of layers on it. It will also be important to differentiate between *ex situ* and *in situ* measurements. *Ex situ* measurements are where the sample exists in completed form before the ellipsometric measurements are made. *In situ* measurements are where the ellipsometric measurements are being made while the sample is being fabricated, so that the ellipsometer can follow the time evolution of the fabrication of the sample. Of course, for a single-wavelength ellipsometer the index of refraction for the substrate or any layer is the value at the wavelength of the measurement, which in the present case is always 633 nm.

For the case of an **optically thick substrate**, a single ellipsometric measurement is sufficient to determine the index of refaction of the substrate. (By an **optically thick substrate**, we mean that no light reflected from the back surface of the substrate reaches the spot on the front surface where the light is being reflected. This could be due to the substrate being sufficiently thick that the light being reflected from the back surface is shifted laterally sufficiently to avoid being directed into the detector, or the back surface being sufficiently rough that no specular reflection occurs, or the layer is so absorbing that no reflected energy reaches the front surface.) For this case, the substrate is either characterized by n if it is nonabsorbing or by $n - ik$ if it is absorbing. The two values (P and A) obtained from a single ellipsometric measurement are sufficient to determine the unknown values (n or $n - ik$) for this case and thus we have sufficient information from a single ellipsometric measurement to characterize an optically thick substrate.

For the case of a single layer on top of an optically thick substrate (assuming the index of refraction of the substrate is known), a single ellipsometric measurement is or is not sufficient to characterize the layer depending on whether the layer is nonabsorbing or absorbing, respectively. A nonabsorbing layer is characterized by the index of refraction n and layer thickness d. Since the number of unknowns (2) for this case is the same as the number of data values (two—one for P and one for A), a single ellipsometric measurement can determine the value of n and d, assuming that the index of refraction of the substrate is known. If the layer is absorbing, then it will be characterized by three unknowns, $n - ik$ and d, then a single ellipsometric measurement does not have enough information to characterize this layer. Since this is an *ex situ* measurement, the only way to characterize an absorbing layer with a single-wavelength ellipsometer is to measure the P and A values at another angle of incidence or on another sample with a different thickness (assuming that the layer has the same $n - ik$ and the same substrate, then one has enough information to uniquely fit the data to a uniform layer model).

However, the best way for a single-wavelength ellipsometer to determine the index of refraction and thickness of such an absorbing layer would be to monitor the growth *in situ* of the absorbing layer on the optically thick substrate. By monitoring the growth, a number of different values of P and A can be measured as the thickness of the layer increases to its final value. The plot of A versus P traces out a trajectory which is characterized by the index of refraction and thickness of the layer. This trajectory is referred to throughout this chapter as a P–A trajectory. The curvature of the P–A trajectory depends on the value of the index of refraction and the displacement along the trajectory depends on the layer thickness. By doing the best fit of this P–A trajectory based on the fundamental ellipsometric equation defined in Eq. (14), the index of refraction and the thickness of the layer can be determined.

We have shown earlier that P and A are linearly related to the ellipsometric parameters Δ and ψ. We plot P and A in the P–A trajectory, rather than convert over to Δ and ψ for the following reasons. First, since we measure P and A directly, the plot of P and A show the actual experimental error that is being encountered in the measurement of P and A. If we convert P over into Δ and A into ψ, we make the error in Δ appear bigger than the error in ψ because the factor of 2 in the relation converting P into Δ. Second, since the operation of the **EXACTA 2000** ellipsometer involves measuring values of P and A, it is easier to think in terms of P and A rather than always to be converting P and A values into Δ and ψ values.

1.8. Advantages–Disadvantages of Using a Spectral Ellipsometer for *In Situ* Measurements

Spectral ellipsometers are often used in *ex situ* measurements, because one can use the ellipsometric information, Δ and ψ, obtained at different wavelengths to characterize the sample in terms of layers on a substrate. Such characterization is always somewhat model dependent, in that the user has to define the number of layers that are going to be used in the fit, and the user has to assume dispersion relations for the different layers, i.e., how the index of refraction of each layer is assumed to vary with wavelength. However, obviously, spectral ellipsometers provide more information for fitting than is available from varying the angle of incidence on a single-wavelength ellipsometer.

Spectral ellipsometers have less utility in doing *in situ* measurements. First, spectral ellipsometers tend to have a severe tradeoff on accuracy versus speed: the shorter the time spent in acquiring the ellipsometric parameters at each wavelength, the lower the accuracy in each measurement. In *ex situ* measurements, one can take additional time to obtain more accurate measurements. However, with an *in situ* measurement, the film growth is occurring at a rate determined by what is optimum for the deposition of the film, and cannot be slowed down to make it easier to get more accurate measurements. The only possible advantage that a spectral ellipsometer might have would be to operate it at only one wavelength that has been selected to give the maximum difference in index of refraction between adjacent layers to get a larger structure in the P–A trajectory. However, usually the technique used to measure the ellipsometric parameters in a spectral ellipsometer is sufficiently inaccurate that a Faraday-modulated fast-nulling ellipsometer can outperform it, even if the difference in index of refraction between adjacent layers is small.

1.9. Practical Considerations for *In Situ* Ellipsometry

In doing *in situ* ellipsometry, one normally has the sample inside some sample growth chamber, e.g., a magnetron sputtering (MS) system, a molecular beam epitaxy (MBE) system, a metal organic chemical vapor deposition (MOCVD) system, or even an electrolytic cell if the film growth involves anodic oxidation or electrodeposition. In all cases, the laser beam from the polarizer unit has to enter the growth chamber via an entrance window, strike the sample at the angle of incidence, reflect off the sample at the same angle as the angle of incidence and in the plane of incidence, exit the growth chamber via an exit window and enter the analyzer unit and strike the detector. The **EXACTA 2000** uses on-axis optics, so the alignment procedure is very simple, i.e., one uses a pinhole at the entrance to the analyzer and a pinhole at the detector to line up the laser beam when the polarizer unit is facing the analyzer unit. (Essentially, one is lining up the optical axis of the polarizer unit with the optical axis of the analyzer unit. The laser beam exiting the polarizer unit defines the optical axis of the polarizer unit. The pinholes at the entrance to the analyzer and at the detector define the optical axis of the analyzer unit.) In the operation of the ellipsometer, one needs a clear view of the sample from both the entrance window and the exit window, because unfortunately we cannot use any mirrors to redirect the beam, because such a mirror would change the state of polarization of the beam and of course this change would be different for different states of polarization of the beam hitting the mirror. So, we have some practical considerations to address regarding how we mount the polarizer and analyzer units and how we handle the entrance and exit windows.

1.9.1. External Beam Mount

If the sample can be moved and rotated in a **measurable** way from outside the growth chamber, then the process of aligning the sample relative to the ellipsometer is relatively straightforward. One adjusts the orientation of the sample until the incident beam is reflected directly back along the incident beam. The orientation of the sample then defines the direction of the incident beam. Then, one changes the orientation of the sample until the light beam incident on the sample is reflected through the exit window and into the entrance pinhole on the analyzer unit and strikes the center of the detector. Then, from the change in orientation of the sample, one can determine what the angle of incidence is.

However, usually the sample cannot be moved and rotated in a **measurable** way from outside the growth chamber. In such cases, we have found an external beam mount to be the easiest way to align the ellipsometer on a sample. Figure 7 shows a schematic of an external beam mount and shows that it consists of an external beam on which are two parallel shafts, that can be rotated through measurable amounts by the use of protractors on each shaft. The polarizer unit is mounted transverse to one shaft and the analyzer unit is mounted transverse to the other shaft. By rotating the polarizer and analyzer units until

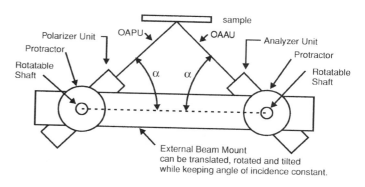

OAPU = Optical Axis of Polarizer Unit
OAAU = Optical Axis of Analyzer Unit

$\alpha = 90° - \phi$ ϕ = angle of incidence

sample
Polarizer Unit   OAPU   OAAU   Analyzer Unit
Protractor   Protractor
Rotatable Shaft α α Rotatable Shaft

External Beam Mount
can be translated, rotated and tilted
while keeping angle of incidence constant.

Rotatable Shafts are perpendicular to External Beam Mount.
Protractors are attached to Rotatable Shafts
to measure angle of rotation.
Polarizer and Analyzer units are rigidly attached to shafts
so they rotate in the plane perpendicular to the shafts.

Fig. 7. Schematic of an external beam mount.

they face each other by using the pinhole at the front of the analyzer unit and a pinhole in the detector location, one can put the ellipsometer into the straight-through aligned position. (Essentially, we are aligning the optical axis of the polarizer unit with the optical axis of the analyzer unit.) If one cannot do this, then the polarizer and analyzer units are not perpendicular to the shafts and at the same distance from the beam, or the two shafts are not parallel to each other. After aligning for the first two conditions, one can check for parallel shafts by rotating the polarizer unit and the analyzer unit to the same angle with respect to the external beam mount and demonstrating that one can find a position of a mirror so that the laser beam from the polarizer unit is reflected through the pinhole in the analyzer unit into the pinhole at the detector. Once the shafts have been aligned perpendicular to the beam and the polarizer unit and analyzer unit are perpendicular to the shafts, then the angle of incidence can be set by rotating each unit through $90° - \phi$ from the straight-through aligned position. At this point, the incident laser beam and the reflected laser beam are in the same plane of incidence with an angle of incidence of ϕ. Figure 7 has been drawn assuming $\phi = 45°$. Now, by moving and tilting the external beam mount one can ensure that the incident laser beam will hit the sample at the desired location and the reflected laser beam will pass out through the exit window and through the pinhole in the analyzer unit and will strike the center of the detector, while knowing that the external beam is maintaining the angle of incidence at ϕ, and keeping the incident laser beam and the reflected laser beam in the plane of incidence.

1.9.2. Effect of Windows

If the incident laser beam and the reflected laser beam pass through the windows normally, i.e., an angle of incidence of zero, then Eqs. (6) and (7) show that the Fresnel coefficients for both the *p* and *s* orientations are the same, and the windows have no effect on the polarization of the light passing through. Even if there are layers of deposits on the windows, as long as the angle of incidence is zero, the layers on the windows will have no effect on the polarization of the incident and reflected laser beams. If one is using an external beam mount, it is much easier to adjust the orientation of the windows so that the beams go through the windows at an angle of incidence of zero.

With the windows properly aligned so that the angle of incidence on the window is zero, the polarization states do not change for the incident and reflected laser beams that pass through the windows. However, the intensity of the incident and reflected beams will be lowered. This can affect the accuracy of ellipsometers that use the variation of intensity with angle of rotation (e.g., a rotating polarizer or a rotating analyzer ellipsometer), since as the deposited layers get thicker, the variation of intensity with rotation gets smaller. This is one of the advantages of a nulling ellipsometer such as the **EXACTA 2000**. In a nulling ellipsometer, as long as one can still detect a null, a drop in intensity has no affect on the accuracy. We have made measurements on growth chambers where the deposits on the windows were so thick that the intensity was cut to 1% of its initial value without any effect on the accuracy of the measurements.

Birefringence in the entrance and exit windows is another matter for concern, because such birefringence shifts and rotates the P–A trajectory from its ideal behavior and makes it harder to fit. There are essentially two approaches to this problem. First, one can employ windows that have low birefringence, such as the Bomco window. Second, one can use a software-based correction that removes the birefringence from the P, A data as it is being collected. The **EXACTA 2000** has software to make such a correction in real time. However, to do this one needs to know the amount of the birefringence of each window and the orientation of the fast axis in each window. One of the problems that we and others have encountered is that if one measures the birefringence of a window and the orientation of the fast axis, the process of bolting the window back onto the growth chamber changes the birefringence due to stress on the window. However, we finally discovered a simple solution to this problem, i.e., bolt the window to be measured onto a bellows-equipped port-aligner before measuring the birefringence in the ellipsometer. Then, do not tighten or loosen or otherwise adjust the bolts holding the window to the port-aligner. Then the port-aligner, with the window already attached and measured, can be bolted to the growth chamber without affecting the magnitude of the birefringence of the window. In addition, if the direction of the fast axis has been marked on the window, one can easily determine (or even set) the orientation of the fast axis with respect to the plane of incidence.

The birefringence of a window acts like a compensator, i.e., there is a fast axis and a slow axis, and the refractive index of the window is not isotropic. If the birefringence is small then it will only translate the P–A trajectory. However, if the birefringence is large, it not only causes the whole P–A trajectory to shift but will also cause it to tilt, which will make the P–A trajectory more difficult to fit. A Bomco window is a special type of window with a much smaller birefringence, such that the trans-

lation of the P–A trajectory is small, and the tilt is small enough not to be readily observable in the P–A trajectory. It should be remembered that birefringence of the windows causes an error in the ellipsometric data but if the effect of the birefringence is small, then it will not affect the accuracy of fitting the ellipsometric data to determine the thickness of layers on samples inside the chamber.

In theory [5], the error caused by birefringence of the windows can be determined. The ellipsometric data can be corrected by taking this error into account in the ellipsometric equation. In calculating the error, the window is treated as a compensator which is characterized by a retardation, t, and an angle of the fast axis, W, relative to the plane of incidence. If the retardation and angle of the fast axis relative to the plane of incidence for the entrance window are t_p and W_p, and for the exit window are t_a and W_a, then the errors in P and A can be predicted by the followings [5],

$$\Delta P = -\frac{1}{2}t_p \cos 2W_p - \frac{1}{2}\cos 2W_a$$
$$+ \frac{1}{2}t_a \sin 2W_a \cot 2A \tag{15}$$

$$\Delta A = -\frac{1}{2}t_p \sin 2W_p \sin\left(2P + \frac{\pi}{2}\right)\sin 2A \tag{16}$$

Hence, to calculate the errors, the retardations and the angles of the fast axis relative to the plane of incidence of the windows have to be determined.

The two birefringence parameters, the retardation and the angle of the fast axis relative to the plane of incidence, can be determined by using the ellipsometer in a straight-through mode. The positions of the plane of incidence of the windows when they are mounted on the growth chamber should be marked on the windows before they are taken off the growth chamber. Each window is placed between the polarizer and the analyzer unit of the ellipsometer such that the laser beam is incident at the center of the window. When the window is rotated at a constant rate, the plot of P versus time or A versus time will trace out a sinusoidal plot. The amplitude of this plot is the retardation and the position of the window where this maximum amplitude occurs should be marked on the window, because this is the position of the fast axis of the window. The angle of the fast axis relative to the plane of incidence is the angle between the mark of the plane of incidence and the mark of the fast axis on the window.

1.10. Simulations of P–A Trajectories

Let us examine the P–A trajectories that we can expect if we were doing *in situ* monitoring of the growth of a dielectric, semiconductor, or metal layer on a semiconductor substrate. The P–A trajectories that we will show are based on the simulation of the changes in ellipsometric parameters that would be taking place as the layer is being deposited. These simulations are obtained in a computer program as follows: The layer is divided into N sublayers of equal thickness, and the matrix M from Eq. (11) associated with each sublayer is generated for each polarization. By multiplying the matrices M associated

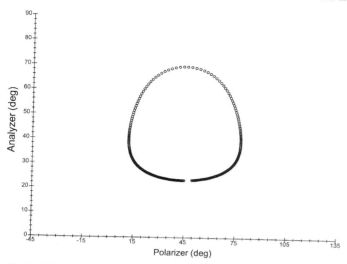

Fig. 8. Shows the simulated P–A trajectory of the deposition of 260 nm of a dielectric with an index of refraction at 633 nm of 1.46 on a single crystal silicon substrate for an angle of incidence of 60°.

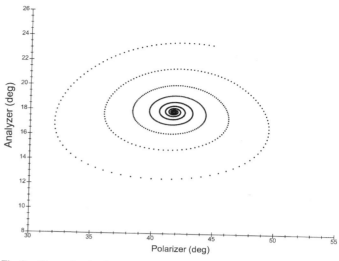

Fig. 9. Shows the simulated P–A trajectory of the deposition of 800 nm of the semiconductor CdTe with an index of refraction at 633 nm of $3.030 - i0.230$ on a single crystal silicon substrate at an angle of incidence of 60°.

with each sublayer together sublayer by sublayer as in Eq. (11) one can step through the layer finding the values of the resultant matrix M up to that sublayer in the layer. Then, by using Eqs. (12) and (13) one can calculate the value of r_s and r_p at that point in the deposition of the layer. By using the values of r_s and r_p in Eq. (14), one can then find the values of Δ and ψ at that point in the deposition of the layer. Then, using Eq. (C9) the value of P is found from Δ and using Eq. (C11) the value of A is found from ψ. Finally, one has the P–A trajectory for the layer that shows the values of P, A as the sequence of N sublayers is deposited to form the complete layer. Obviously, the same procedure can be done for a second, third, or more layer, and one can also handle situations where the properties of a layer varies from bottom to top. These simulations will show the general aspects of the P–A trajectories in each case. However, complications that might occur in real systems are not considered here, but are treated in Section 2 when we consider measurements on real systems.

Figure 8 shows the simulated P–A trajectory of the deposition of 260 nm of a dielectric on a single crystal silicon substrate for an angle of incidence of 60°. The index of refraction at 633 nm for single crystal silicon is $3.870 - i0.020$. If one uses this index to calculate the P, A values that we would measure for a single crystal silicon surface, then for an angle of incidence of 60°, we would obtain $P = 44.853°$ and $A = 23.314°$. (Note that these P, A values are dependent on what is chosen for the angle of incidence, and we would get significantly different P, A values if we had chosen the angle of incidence to be 45°.) In Figure 8, we have computed the P, A values that result as 260 increments of 1 nm of a dielectric of index of refraction $n = 1.46$ is added to the surface of the single crystal silicon surface. The P–A trajectory starts from $P = 44.853°$, $A = 23.314°$ (representing the bare silicon surface) and moves clockwise until it ends at $P = 49.617°$, $A = 23.431°$ when a layer of dielectric of index of refraction $n = 1.46$ and thickness 260 nm exists on the surface.

In Figure 8, note first that the spacing between dots varies as one goes along the trajectory, e.g., it is biggest on the top portion of the trajectory. This just means that the ellipsometer sensitivity to a 1-nm increment in the thickness of the dielectric varies as one goes along the trajectory as the thickness grows from zero to 260 nm of the dielectric. From Figure 8, it would appear that the P–A trajectory is going to form a closed loop, in that the end value of A is 23.431° and the starting value of A is 23.314°. If we were to continue the simulation for a thickness greater than 260 nm of dielectric, one would see that the P–A trajectory for a dielectric is indeed a closed loop and repeats itself over and over as the thickness of the dielectric is increased. This is why it is so hard in an *ex situ* measurement of a very thick layer to determine the thickness, because one does not know when a P, A value is measured, how many cycles of the P–A trajectory have taken place. However, if one is measuring the P–A trajectory *in situ* there is no confusion, because one sees how many times the P–A trajectory has cycled around the closed loop.

Figure 9 shows the simulated P–A trajectory of the deposition of 800 nm of the semiconductor CdTe with an index of refraction at 633 nm of $3.030 - i0.230$ on a single crystal silicon substrate at an angle of incidence of 60°. Once again the starting point of the P–A trajectory is $P = 44.853°$ and $A = 23.314°$, corresponding to a bare single crystal silicon substrate with an index of refraction of $3.870 - i0.020$. In Figure 9, we have computed the P, A values that result as 800 increments of 1 nm of CdTe with an index of refraction $= 3.0330 - i0.230$ is added to the surface of the single crystal silicon surface. The P–A trajectory starts from $P = 44.853°$, $A = 23.314°$ (representing the bare silicon surface) and moves counterclockwise until it ends at $P = 41.466°$, $A = 17.882°$ when an 800-nm layer of CdTe exists on the surface.

Figure 9 shows that the P–A trajectory is spirally inward heading for some "end-point." This behavior is typical of any absorbing material with a small absorption coefficient. If we kept the simulation going for CdTe thicknesses greater than 800 nm, we would gradually spiral in on the end point with $P = 42.711°$, $A = 17.901°$. If we consider this end point to correspond to a "substrate," and if we ask what $n - ik$ does the substrate have, we would find $n - ik = 3.030 - i0.230$, i.e., the index of refraction of CdTe. So this spiral into an end point is easy to interpret physically. If we cover a single crystal silicon substrate with a layer of an absorbing semiconductor such as CdTe, when the layer of CdTe gets thick enough we no longer see the underlying single crystal silicon surface, and the layer of CdTe just behaves like an infinitely thick CdTe layer, i.e., like a CdTe substrate.

In Figure 9, the points correspond to increments of 1 nm of CdTe. We see that the spacing of the points varies along the trajectory. There is a general tendency for the spacing to get closer and closer together as we get closer and closer to the end point, but along each cycle of the spiral we get some variation in the thickness where the spacing is bigger for part of the cycle. We later see that this variation **along a cycle** is due to a variation in the distance in P–A space of the current P, A point from the P, A value of the end point. The change in P, A values with an increment in thickness of the layer for a given cycle is greater (less) depending on whether this distance in P–A space is smaller (larger).

Figure 10 shows the simulated P–A trajectory of the deposition of a sequence of metal layers on a single crystal silicon substrate, i.e., the deposition of 70 nm of platinum (Pt) with an index of refraction at 633 nm of $2.000 - i5.000$ on top of the silicon substrate, then 70 nm of nickel (Ni) with an index of refraction at 633 nm of $1.970 - i3.720$ on top of the Pt layer,

and finally 70 nm of titanium (Ti) with an index of refraction at 633 nm of $3.230 - i3.620$ on top of the Ni layer. The simulation assumes an angle of incidence of 60° and a thickness increment of 1 nm for each layer. Once again, the starting point of the P–A trajectory is $P = 44.853°$ and $A = 23.314°$, corresponding to a bare single crystal silicon substrate with an index of refraction of $3.870 - i0.020$.

Figure 10 shows that the P, A values initially change very rapidly with every increment of 1 nm of Pt on the silicon substrate but then decrease rapidly as we approach an end point. Although there is some curvature to the P–A trajectory, there is no spiral as in the case of CdTe in Figure 9, because the absorption coefficient is so large. The P–A trajectory associated with the 70 nm of Pt on the silicon substrate starts rotating slightly clockwise and then rotates slightly counterclockwise as it reaches the end point. The P, A values associated with 70 nm of Pt on the silicon substrate are $P = 30.457°$, $A = 39.647°$. The reason that with 70 nm of Pt on the silicon substrate we appear to be reaching an end point is that a 70-nm Pt film is almost opaque, so the ellipsometer cannot see the underlying silicon substrate. If we ask for the P, A associated with the Pt $n - ik = 2.000 - i5.000$, we find $P = 30.450°$, $A = 39.632°$. Note how close these P, A values are to the P, A values after 70 nm of Pt on Si.

The next part of the P–A trajectory in Figure 10 is associated with the deposition of 70 nm of Ni on top of the Pt layer. The points show increments of 1 nm of Ni on top of the Pt layer. Once again, the P–A values initially change rapidly with increments of 1 nm of Ni but then the changes become smaller and smaller as another end point is approached. Once again, the P–A trajectory does not really spiral but instead curves slightly counterclockwise initially and then curves more sharply as the end point is reached. The P, A values associated with 70 nm of Ni on Pt are $P = 27.162°$, $A = 36.753°$. Once again, the reason that with 70 nm of Ni on the Pt layer we appear to be reaching an end point is that a 70-nm Ni film is almost opaque, so the ellipsometer cannot see the underlying Pt layer. If we ask for the P, A associated with Ni $n - ik = 1.970 - i3.720$, we find $P = 27.186°$, $A = 36.760°$. Note how close these P, A values are to the P, A values after 70 nm of Ni on Pt.

The next part of the P–A trajectory in Figure 10 is associated with the deposition of 70 nm of Ti on top of the Ni layer. The points show increments of 1 nm of Ti on top of the Ni layer. Once again, the P–A values initially change rapidly with increments of 1 nm of Ti but then the changes become smaller and smaller as another end point is approached. Once again, the P–A trajectory does not really spiral but instead curves slightly counterclockwise initially and then curves more sharply as the end point is reached. The P, A values associated with 70 nm of Ti on Ni are $P = 31.373°$, $A = 34.210°$. Once again, the reason that with 70 nm of Ti on the Ni layer we appear to be reaching an end point is that a 70-nm Ti film is almost opaque, so the ellipsometer cannot see the underlying Ni layer. If we ask for the P, A associated with Ti $n - ik = 3.230 - i3.620$, we find $P = 31.353°$, $A = 34.190°$. Note how close these P, A values are to the P, A values after 70 nm of Ti on Ni.

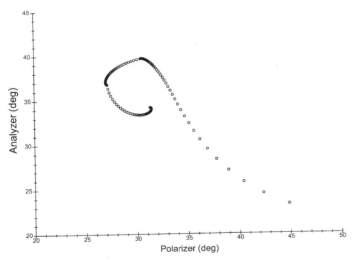

Fig. 10.   Shows the simulated P–A trajectory of the deposition of a sequence of metal layers on a single crystal silicon substrate, i.e., the deposition of 70 nm of platinum (Pt) with an index of refraction at 633 nm of $2.000 - i5.000$ on top of the silicon substrate, then 70 nm of nickel (Ni) with an index of refraction at 633 nm of $1.970 - i3.720$ on top of the Pt layer, and finally 70 nm of titanium (Ti) with an index of refraction at 633 nm of $3.230 - i3.620$ on top of the Ni layer. The simulation assumes an angle of incidence of 60° and a thickness increment of 1 nm for each layer.

The first thing that should be stressed about Figure 10 is that with this *in situ* monitoring we have followed the deposition of the Pt, Ni, and Ti layers until each in turn has become opaque. If one were given the completed sample as an *ex situ* sample, at most one would be able to determine the thickness of the top Ti layer. So, with this *in situ* monitoring approach we would be able to follow the deposition of many more layers of Pt, Ni, and Ti. The other point that should be emphasized is that if we had deposited the layers in a different sequence, we would have obtained a significantly different *P–A* trajectory, although the end points for each layer would still be the same and would be associated with the *P*, *A* values for an opaque layer of that material.

2. EXPERIMENTAL TECHNIQUES

2.1. Introduction

To show the use of *in situ* Faraday-modulated fast-nulling single-wavelength ellipsometry in monitoring the growth of semiconductor and metal films, we are going to present some results from the Ph.D. thesis of Tran [6]. Two different growth systems were used, i.e., a pulsed laser evaporation and epitaxy [7, 8] (PLEE) system, and a magnetron sputtering [9, 10] (MS) system. The work on ellipsometry with the PLEE system was a collaboration involving the University of Waterloo,

Waterloo Digital Electronics Division of WDE Inc., and Dr. J. J. Dubowski of the Institute for Microstructural Sciences, National Research Council of Canada. The PLEE system was used to grow II–VI compound ($Cd_{1-x}Mn_xTe$-CdTe) quantum well and superlattice structures and hence it was essential that this system be able to grow epitaxial layers. In fact, epitaxial $Cd_{1-x}Mn_xTe$-CdTe multilayer structures have been previously and successfully grown by this system [11]. The work on ellipsometry with the MS system was a collaboration involving the University of Waterloo, Waterloo Digital Electronics Division of WDE Inc., and Dr. S. R. Das and L. LeBrun of the Institute for Microstructural Sciences, National Research Council of Canada. The MS system was used to grow the metallic multilayer structures involving layers of platinum (Pt), nickel (Ni), and titanium (Ti) on single crystal silicon which were monitored by ellipsometry. The Waterloo Digital Electronics Division of WDE Inc. provided the two **EXACTA 2000** ellipsometers, one on each growth chamber, that were used for this study. Steve Buchanan of WDE Inc. provided invaluable support in connection with the alignment, the maintenance, and the use of these two ellipsometers.

2.2. The Pulsed Laser Evaporation and Epitaxy System

A schematic of the PLEE system equipped with an **EXACTA 2000** ellipsometer is shown in Figure 11. The details of the

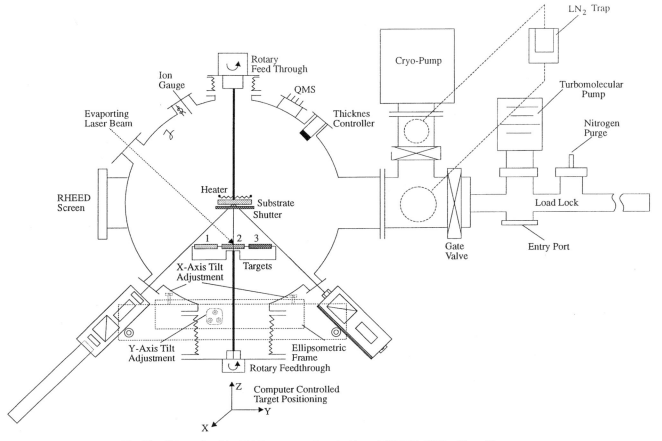

Fig. 11.   Schematic of the PLEE system equipped with an **EXACTA 2000** nulling ellipsometer.

evaporation system and the technique are described elsewhere [6, 8]. Briefly, the system consists of target holders and a sample holder inside a cryopumped vacuum chamber with a load lock. The base pressure of the chamber after bakeout was 1.5×10^{-8} Torr. XeCl and Nd:YAG lasers were used to ablate the high purity $Cd_{1-x}Mn_xTe$ and CdTe polycrystalline targets, respectively. The XeCl laser operating at 0.308 μm was triggered at a rate of 20 Hz and the Nd:YAG laser, operating at 1.06 μm, was triggered at a rate of 500 Hz. The two lasers were computer controlled to sequentially vaporize the high purity polycrystalline targets to produce $CdTe/Cd_{1-x}Mn_xTe$ quantum well and superlattice structures. The substrates used in these experiments were (111) GaAs and (100) CdZnTe. The substrates were mounted on an Mo block that could be rotated during deposition to minimize the lateral temperature gradient and film thickness nonuniformity. However, in this research all the substrates were stationary during the growth so that the growth could be monitored by the ellipsometer. The sample stage was not perfectly perpendicular to the axis of rotation and the rotation of the sample stage would have caused a variation in the incident angle of the light of the ellipsometer. Hence, the reflected light would not be aligned with the analyzer unit and the center of the photodetector and this would cause errors in the ellipsometric measurements. (However, in measurements on other systems we have demonstrated that one can make ellipsometric measurements on a rotating substrate as long as the surface of the sample is sufficiently close to perpendicular to the axis of rotation by locating the ellipsometer laser spot directly on the location of the axis of rotation on the sample.)

At the bottom of Figure 11 is shown the external beam mount that was used to align the ellipsometer on this PLEE system. The beam mount was custom made such that both the polarizer unit and the analyzer unit are aligned with the chamber windows which are 90° apart. Hence, the angle of incidence for the ellipsometric measurements had to be set at 45°. The external beam mount was bolted onto the PLEE system, and the alignment was achieved as follows. The beam mount allows both the polarizer unit and the analyzer unit to be rotated about the x axis such that the angle of incidence can be set at 45° in this case. It allows the whole of the ellipsometer (polarizer and analyzer units) to be tilted about the y axis with a three point suspension which is labeled as Y-axis tilt. It also allows the ellipsometer to be tilted about the x axis by adjusting the screws which are labeled as X-axis tilt. The whole external beam mount could be shifted in the z and y directions to position the ellipsometer beam on a particular point on the sample. This procedure of moving and tilting the external beam mount eliminated the need for an adjustable sample mount, which was not present in the PLEE system. Otherwise, to be able to align the reflected light so it left a particular point on the sample, passed through the pinhole on the front of the analyzer unit and struck the center of the photodetector, one would have to be able to tilt the sample about the x and y axes, translate the sample in the x–y plane and move the sample up and down along the z axis. Most of the ellipsometric data were taken using Bomco windows and no birefringence corrections were applied.

2.3. The Magnetron Sputtering System

A schematic of the MS system equipped with an **EXACTA 2000** ellipsometer is shown in Figure 12. It consists of a stainless steel growth chamber equipped with a load lock (not shown in the figure) for translating the substrates. A cryogenic pump is used to produce an ultrahigh vacuum in the growth chamber with a base pressure of 10^{-9} Torr, and a turbomolecular pump is used to keep the pressure as low as 10^{-7} Torr (both of these pumps are not shown in the diagram). The ultrahigh vacuum is not affected when the substrate is transferred from the load lock to the growth chamber. The substrate can be rotated and translated through the rotary feedthrough. The targets are mounted on stainless steel cylinders which can be slid into the entry ports and tightly secured. The MS system had to be opened to the atmosphere to take out the cylinders for target replacement.

In the experiments with the MS system, the substrates were single crystal (100) Si and the targets used for sputtering were platinum (Pt), nickel (Ni), and titanium (Ti). The targets were 1 in in diameter. The radio frequency (rf) power used for sputtering was 100 W for Pt and Ni, and 200 W for Ti. During sputtering, 0.01 Torr was maintained by controlling the flow rate of high purity argon using electronic mass flow controllers. The rf power was fed to the water-cooled cathode via a tuning network. The Si substrates were cleaned by the RCA method and were given a final dip into an HF solution before they were clamped on the substrate holder which was then loaded into the chamber. The Si substrates were kept under nitrogen atmosphere at all times after HF cleaning until loading into the load lock.

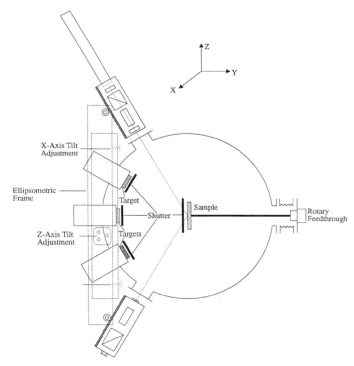

Fig. 12.   Schematic of the MS system equipped with an **EXACTA 2000** nulling ellipsometer.

Similar to the PLEE system, an external beam was used to mount the ellipsometer onto the MS system. The external beam and the ellipsometer are shown in Figure 12. Again, the external beam had similar functions to the one for the PLEE system such as setting the angle of incidence and tilting about the x and y axes. However, due to the configuration of the MS system, the ellipsometer is oriented vertically, and the windows of the chamber used for the ellipsometer are 120° apart and hence the angle of incidence of the ellipsometer was set at 60°. The whole external beam mount could be shifted in the z and x directions to position the ellipsometer beam on a particular point on the sample. For this system, only normal glass window view ports were used and no corrections were made for birefringence.

2.4. Fitting Procedure for Ellipsometric P–A Trajectories

In general, there are two steps involved in fitting an *in situ* ellipsometric P–A trajectory. The first step is to define an **optical model** for a physical structure of thin films of a sample corresponding to an *in situ* ellipsometric P–A trajectory and the second step is to fit this ellipsometric data with the defined model. For example, an ellipsometric P–A trajectory was obtained from monitoring the growth of a CdTe layer on a GaAs substrate. In fitting this experimental P–A trajectory, the obvious optical model would consist of a complex index of refraction of a GaAs substrate and a CdTe layer that was uniform with thickness, and a rough estimated thickness of the CdTe layer. Next, these parameters, the values of complex indices of refraction of GaAs and CdTe determined from other ellipsometric measurements or reported in the literature and the estimated thickness were input into the fitting program to calculate the P–A trajectory and to see how close it was to the experimental one. By varying these fitting parameters, eventually the best fit to the experimental P–A trajectory could be achieved and hence, the thickness of the layers in the sample could be determined. For convenience, this second step; i.e., the actual fitting process, involved in the fitting is referred to as a **fitting routine**. This **fitting routine** is the basis of all the fitting of the P–A trajectories presented in this chapter.

Actually, defining an optical model for a P–A trajectory is an essential step in the fitting and is not a trivial matter. The accuracy of the thickness and the index of refraction determined from fitting a P–A trajectory depends on how well an optical model can describe the physical structure of a sample. For example, one might assume that the sample consists of a film-free substrate, when in fact the sample is composed of a thin film of one material on top of a substrate of another. Assuming that the ellipsometer is operating properly, the measured P and A values will be correct, but the complex index of refraction resulting from the fit assuming no layer will be incorrect. In fact, for an optical model to accurately describe the P–A trajectory, one needs to know not only about the structure of the samples but also about all the physical effects occurring during the growth that might change the complex index of refraction and hence the P–A trajectory. Some of the known physical

effects that can cause changes in the complex index of refraction are: (1) strain induced in a layer, (2) a sudden relaxation in a layer, (3) diffusion, e.g., Mn from a CdMnTe layer into a CdTe layer, (4) surface roughness, (5) birefringence in the windows, (6) island formation in the initial stage of film growth, and (7) intermixing at the interface between the deposited layer and the underlying layer. The difficulty is to know which physical effects have occurred and how they can cause changes in the P–A trajectory.

Experimentally, the fitting procedure is carried out in the reverse order; fitting of a P–A trajectory is done first and then the fitting parameters are justified by defining an optical model corresponding to the known physical effects that might occur during the growth. In this way, one would know how the complex index of refraction varied along the trajectory. By identifying the possible physical effect(s) that might cause similar variations in the complex index of refraction, the optical model for the P–A trajectory can be established. However, it is not always easy for one to know which physical effects have occurred and how they can cause changes in the P–A trajectory, especially the first time that such anomalies in the P–A trajectory are observed.

Since we do not always know for sure the physical effects that have occurred during a growth, a certain fitting procedure has to be followed to obtain reliable results, especially for the thickness of a layer. Of course, if no physical effect occurred during the growth that resulted in a change in the complex index of refraction, then the trajectory can be fitted with a single index of refraction using the earlier fitting routine. Even if a P–A trajectory is characterized by a complex index of refraction that is varying slowly with thickness, it still can be fitted using the previous fitting routine, but modeling the overall layer as a number of sublayers each with its own index of refraction and thickness. To obtain a reliable thickness for the overall layer, the number of indices of refraction has to be kept to a minimum; i.e., one has to fit as large a portion of the P–A trajectory as possible for each value of complex index of refraction, and the variation of the complex index of refraction from its bulk or accepted value also has to be kept small. For a P–A trajectory characterized by a complex index of refraction that varies quickly with thickness, the trajectory can still be fitted with a number of layers, each with its own complex index of refraction, but each index of refraction can fit only a small portion of the P–A trajectory. The complication of fitting a small portion of a trajectory is that sometimes it does not have enough curvature to have a meaningful fit, i.e., it can be fitted with a number of possible indices of refraction. A previous article [12] has given detailed examples of how well the CdMnTe layers can be fitted in a CdTe-CdMnTe quantum well structure; later in this chapter in Figure 22 we give further proof of the validity of the fitting procedure. A final aid to fitting is that the meaningful complex index of refraction is the one that would give a thickness close to the value estimated from the growth rate for that portion of the trajectory and which is not much different from the bulk value of that material.

The preceding fitting procedure works well if the physical effects occurred at the surface of a layer while the layer is monitored by the ellipsometer. However, if the physical effect occurs in the layer then the complex index of refraction obtained from the fitting procedure described earlier will not have a direct physical connection to the properties of the portion of the layer being deposited at that instant. Nevertheless, the thickness determined for the overall layer can still be quite reliable. The present fitting program assumes that all the mechanisms that might cause changes in the complex index of refraction occur at the surface, and thus, it will not be able to simulate a mechanism that occurs within a layer. For example, assuming that a layer is not perfectly lattice matched to a substrate, when it first deposited on a substrate, it will be strained. This effect will be seen by the ellipsometer, and since it occurs at the surface it can be modeled by the fitting program. When the layer reaches its critical thickness, the next layer deposited on the surface will cause the layer beneath, i.e., the strained layer, to relieve its strain. The ellipsometer will be able to detect the change in the complex index of refraction due to strain relief in the layer and at the surface due to the deposited layer, but the fitting program can only simulate this part of the P–A trajectory as if the deposited layer at the surface has a continuous change in the complex index of refraction. Thus, the complex indices of refraction obtained from fitting this portion of the trajectory are not directly related to the physical effect causing the change. Nevertheless, the thickness determined from the fit is usually not seriously in error because the complex indices of refraction used in the fit were selected to give a thickness similar to the value expected by the growth rate.

To obtain accurate thicknesses of layers, other than having an accurate optical model, the calculated P–A trajectory has to be fitted to the experimental P–A trajectory as closely as possible. In general, this condition is difficult to achieve when the trajectory of a sample consists of more than one layer deposited on a substrate. For example, the calculated P–A trajectory that fits best to the experimental P–A trajectory of the first layer might not end where the experimental P–A trajectory of the second layer begins. Hence, the start of the calculated P–A trajectory of the second layer is not the same as the experimental one and that would introduce an error in the fitting of the second layer. This kind of error can be minimized, if not actually eliminated, if a **pseudosubstrate** approximation is used. Basically, a pseudosubstrate approximation allows one to represent any point on the trajectory where one wants to start the fitting as a substrate. Then, the subsequent trajectory of the layer can be fitted as if it is the first layer deposited on the substrate. This type of pseudosubstrate approximation has been used earlier by Aspnes [13] and Gilmore and Aspnes [14], who referred to it either as a "common pseudosubstrate approximation," a "virtual substrate," or a "virtual interface." Our main contribution to the development of the pseudosubstrate approximation has been to examine closely under what conditions this approximation is valid.

The pseudosubstrate approximation provides faster and more flexible fits. If one is interested only in the thickness of the last layer in an N layer structure, without using a pseudosubstrate approximation one would have to fit the P–A trajectory for all $N - 1$ layers starting from the bare substrate, before fitting the P–A trajectory of the Nth layer. With the pseudosubstrate approximation, one can start with the P, A value at the start of the deposition of the Nth layer, treat it as a pseudosubstrate and only fit the P–A trajectory of the last layer. Similarly, if one were depositing an additional layer on a substrate that already had N layers deposited on it, if one can use the pseudosubstrate approximation, one does not need to know the indices of refraction and thicknesses of the N layers already deposited to fit the P–A trajectory caused by the additional layer.

2.5. Validity of the Use of a Pseudosubstrate Approximation

In general, a pseudosubstrate approximation can only be applied to certain cases. In this section, the cases where a pseudosubstrate can and cannot be applied are discussed. We cannot provide an actual proof that a pseudosubstrate approximation, in general, can or cannot be applied to the cases that will be mentioned. However, we can justify our belief by demonstrating that the P–A trajectories, for a number of specific cases, simulated using the P, A values of the substrate and of the pseudosubstrate are equivalent. All of the structures considered in these simulations consist of a substrate and two layers on top of it. First, the P–A trajectory for the whole structure will be simulated. Next, the trajectory of the layer of interest, i.e., the second layer, in the structure will be simulated using the pseudosubstrate, which can be determined from the P, A values where the trajectory of the first layer ended or the second layer began, as if this second layer is being deposited on a substrate. If the P–A trajectory of the second layer simulated starting from the P, A values of the pseudosubstrate is the same as that starting from the P, A values of the substrate, then the pseudosubstrate approximation is valid to apply to this P–A trajectory. From now on, when the pseudosubstrate approximation is said to be valid to apply to a P–A trajectory, it means that the effect of the P–A trajectory of all the layers and the substrate that lie beneath the top layer (or the layer of interest) can be represented by a pseudosubstrate.

In general, the validity of using a pseudosubstrate depends on the types of materials of the underlying layers and the overlying layer. The underlying layers and the substrate can be any combination of different types of materials such as dielectric, semiconductor, and metal. It should be noted that the underlying layer next to the top layer (or the layer of interest) is assumed to have a thickness which is not thick enough to absorb all the light transmitted through it or else the pseudosubstrate in this case becomes in fact a real substrate. This implies that a pseudosubstrate can be used in any case as long as the underlying layer next to the layer of interest is thick enough to absorb all the light. This can be true if the underlying layer is either a semiconductor or a metal, but not in the case of a dielectric. A number of different combinations of types of underlying layer and types of overlying layer on types of substrates have been used to test for the validity of the pseudosubstrate approximation [6]. Here, we present just two examples of these simulations. The point on the $P–A$ trajectory denoted by S indicates the P, A values of the substrate and PS indicates the P, A values of the pseudosubstrate for the second layer. The simulated $P–A$ trajectory of the whole structure starting from the substrate is represented by "+" and the $P–A$ trajectory of the second layer using the pseudosubstrate is represented by the gray solid line.

Figure 13 shows the case of 100 nm of the **dielectric** SiO$_2$ on top of 100 nm of the **semiconducto**r CdTe on top of the **semiconductor** Si substrate. The simulated $P–A$ trajectory of

the top SiO$_2$ layer starting from the pseudosubstrate PS (solid gray line) is equivalent to the simulated $P–A$ trajectory of the top SiO$_2$ layer on the CdTe layer on top of a Si substrate (+). Therefore, the pseudosubstrate approximation **is valid** for a **dielectric** top layer (SiO$_2$) on an underlying **semiconductor** layer (CdTe) on top of a **semiconductor** substrate (Si).

Figure 14 shows the case of 100 nm of the **metal** Ti on top of 50 nm of the **dielectric** glass on top of the **metal** Ni substrate. The simulated $P–A$ trajectory of the top Ti layer starting from the pseudosubstrate PS (solid gray line) is not equivalent to the simulated $P–A$ trajectory of the top Ti layer on the glass layer on top of a Ni substrate (+). Therefore, the pseudosubstrate approximation is **not valid** for a **metal** layer (Ti) on a **dielectric** layer (glass) on top of a **metal** substrate (Ni).

From studying all possible combinations of types of overlying layer, underlying layer, and substrate, the following rules for when the pseudosubstrate can and cannot be applied have been obtained:

(1) As long as the underlying layers and the substrate are made of the same material, which is not a dielectric, a pseudosubstrate approach can be used no matter what type of material the overlying layer or the layer of interest is.

(2) A pseudosubstrate cannot be applied when the underlying layers and the substrate are dielectrics. The next three situations assume that the underlying layers and the substrate are not made of the same type of material.

(3) A pseudosubstrate cannot be applied to the case where the underlying layer next to the top layer (or layer of interest) is made of a dielectric.

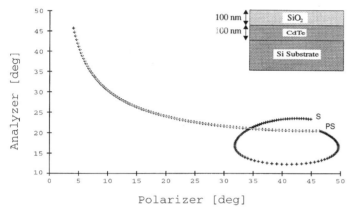

Fig. 13.   Showing the equivalence of the simulation of a $P–A$ trajectory of SiO$_2$ on CdTe layer on top of an Si substrate from the substrate S and from the pseudosubstrate PS. Simulation from $S+++$, and simulation from PS is a gray solid line.

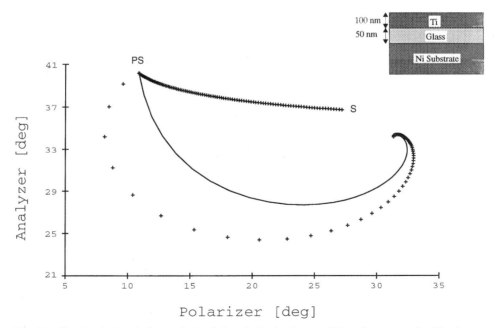

Fig. 14.   Showing the inequivalence of a simulation of a $P–A$ trajectory of Ti on glass on top of an Ni substrate from the substrate S and from the pseudosubstrate PS. Simulation from $S+++$, and simulation from PS is a gray solid line.

(4) If the overlying layer is a metal, then the underlying layers can be represented as a pseudosubstrate no matter what types of materials the underlying layers and the substrate are, as long as the underlying layer next to it is not a dielectric.

(5) If the overlying layer is a semiconductor or a dielectric, then a pseudosubstrate can or cannot be used to represent the underlying layers depending on how thick the underlying layer next to the overlying layer is and if the underlying layer next to the overlying layer material is not a dielectric.

Even though the discussion of the pseudosubstrate approximation is based on structures that consist of only two layers and a substrate, the discussion can be applied to structures that have many layers. **When one deals with cases where the underlying layers and the substrate are made of different types of materials, and the overlying layer or layer of interest is a semiconductor or a dielectric, one has to be careful in using a pseudosubstrate.** One has to make sure that it is valid to replace the underlying layers and the substrate beneath the layer of interest with a pseudosubstrate. Fortunately, all the layers and the substrates in the structures that are grown by the PLEE system are made of the same type of material, i.e., a semiconductor, hence a pseudosubstrate can be applied at any point on the trajectory of a structure. For samples produced by the MS system, two different types of materials were used, the substrate is the semiconductor Si while the deposited layers are metal. Any point on the trajectory of these samples is valid to be the point of a pseudosubstrate. One of the reasons is that the overlying metallic layer is independent of the type of material of the underlying layers as long as the underlying layer next to it is not a dielectric. The other reason is that a thick metallic layer is grown on the Si substrate before growing the subsequent metallic layers. Normally, the first layer grown on the Si substrate is so thick that it can be considered as a substrate. Thus, the whole structure can be considered as if it is made of the same type of material.

3. EXPERIMENTAL RESULTS

3.1. Experimental Results for PLEE System

The PLEE system was used to grow single epilayers of CdTe or CdMnTe, and CdMnTe-CdTe quantum well and superlattice structures. Two types of substrates, GaAs and CdZnTe, were used for growing these structures. The samples, which consisted of single layers of CdTe, CdMnTe, and CdMnTe-CdTe quantum well and superlattice structures, are identified with the first letters of the names of the samples as CT, CMT, and CCM, respectively. One of the reasons that we wanted to grow the epitaxial layers in the CCM samples is that we wanted to check that the ellipsometer was really measuring the true thicknesses of these structures by comparing the results with measurements made on the same samples by transmission electron microscopy (TEM).

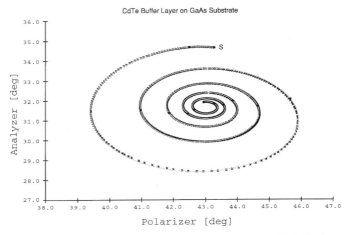

Fig. 15. *In situ* measurement of the $P–A$ trajectory of a CdTe buffer layer grown on a GaAs substrate, sample CCM306. CdTe $\times \times \times$.

It is difficult to grow epilayers and it is even more difficult to grow good quality epitaxial layers in the case of a large lattice mismatch between two different materials such as GaAs and CdTe. The difference in lattice constants between these two materials is approximately 14%. Normally, with this large of a lattice mismatch, a thick buffer layer has to be grown on the substrate such that it can mask the dislocations, caused by the misfit in the lattice constants at the interface between the GaAs and the CdTe, before growing a quantum well or superlattice structure. However, if the growth condition is not right, the buffer layer will not be able to mask the dislocations; i.e., the dislocations will extend up through the buffer layer into the structure layers and will degrade the quantum well or superlattice structure, no matter how thick the buffer layer is. For this reason, a number of thick single layers of CdMnTe were grown on GaAs substrates at different temperatures to determine the optimum growth temperature. CdMnTe was tried as a buffer layer instead of CdTe, since the lattice mismatch, although it is still large, between CdMnTe and GaAs is slightly smaller than between CdTe and GaAs. However, we did not succeed in growing any good quality epitaxial layers with GaAs substrates for making TEM measurements to obtain the thickness of the layers to compare with the thickness of the layers determined from fitting the ellipsometric results.

Figure 15 shows a typical experimental $P–A$ trajectory of a CdTe buffer layer on a GaAs substrate as part of the fabrication of sample CCM306. Before the deposition, the $P–A$ trajectory always started at a particular point denoted by S. Theoretically, this point represents the optical characteristics of the GaAs substrate and the complex index of refraction of this substrate can be determined once this point is measured. Experimentally, this point may not truly represent the optically thick GaAs substrate for two reasons. When a substrate is exposed to the atmosphere after degreasing, it always has a thin oxide on its surface. Ellipsometric measurement of the substrate under this condition would yield P, A values, which are not the true values of the bare substrate, but rather those associated with a thin oxide on

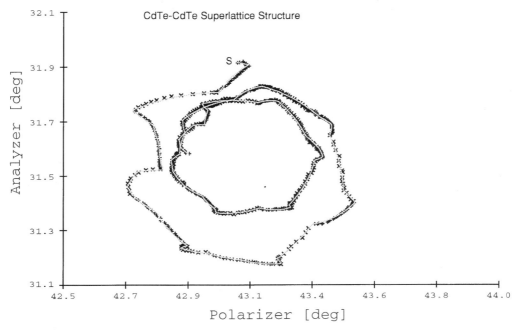

Fig. 16. *In situ* measurement of the *P–A* trajectory of a CdTe-CdMnTe quantum well structure grown on the CdTe buffer layer of Figure 15, sample CCM306, CdTe + + +, and CdMnTe **x x x**. Preliminary fit shown as a gray solid line.

the bare substrate. It is possible to eliminate the oxide on the substrate once it is loaded into the growth chamber by annealing it at high temperature. However, a side effect of removing the oxide this way is that the surface would be roughened and the resulting surface roughness would affect the true *P*, *A* values of the bare substrate.

For a semiconductor substrate, a thin (less than 20 Å) oxide on the substrate tends to mostly decrease the value of *P*, but produces little or no decrease in the value of *A*. If one interprets these *P* and *A* values as being due to a bare substrate, this shift in *P* due to the thin oxide produces a complex index of refraction of this "bare substrate" which will have a larger value of *k* than when the substrate is free of oxide. As long as the oxide on the substrate is not thick, which is normally about 50 Å or less for a substrate exposed to atmosphere, it will cause the *P*, *A* values to shift horizontally to lower *P*.

When CdTe was deposited on the GaAs substrate, the *P–A* trajectory shown in Figure 15 was traced out. Note that the *P–A* trajectory is reasonably large, extending approximately 7° in *P* and 6° in *A*; this is due to the large difference between the indices of refraction of GaAs ($3.920 - i0.210$) and CdTe ($3.030 - i0.230$), since the size of the *P–A* trajectory depends on the magnitude of this difference in indices. The shape of the *P–A* trajectory is determined by the complex index of refraction of the layer being deposited, in this case CdTe, while the distance along the trajectory is determined by the thickness of the layer being deposited. As we discussed in the case of the CdTe on GaAs simulation of Figure 9, since the CdTe is optically absorbing, the *P–A* trajectory will spiral in toward an end point, whose *P*, *A* values would be determined by the complex index of refraction of CdTe.

While the *P–A* trajectory of Figure 15 appears reasonably smooth, close inspection shows that there are deviations from the behavior that would be expected if the buffer layer had a single value of the complex index of refraction. When CdTe (or CdMnTe) is first deposited on a GaAs substrate, the *P–A* trajectory started out rather flat before it spiraled inward. This initial region is believed to be associated with the initial strain in the layer when it is first deposited on the substrate, which causes the complex index of refraction of the CdTe in this part of the layer to be different from the bulk value and hence the initial trajectory is rather flat and different from the rest of the trajectory. The slight tilt in the *P–A* trajectory of Figure 15 is due to a small amount of birefringence in the windows of the growth chamber.

Figure 16 shows the *P–A* trajectory of the CdTe-CdMnTe superlattice structure of sample CCM306 that was grown after the deposit of the buffer layer whose *P–A* trajectory was shown in Figure 15. The CdTe layers are shown with +++ and the CdMnTe layers are shown with × × ×. Note the small size of the *P–A* trajectory which extends approximately 0.8° in *P* and 0.7° in *A*. This is because the pseudosubstrate, represented by the end of the *P–A* trajectory associated with the buffer layer, has an effective index of refraction which is not much different than the indices of refraction of CdTe and CdMnTe, therefore the size of the *P–A* trajectory associated with the CdTe-CdMnTe quantum well structure is small. Experimentally, the *P–A* trajectory of a CdTe or CdMnTe layer is not smooth, sometimes its curvature even changes quickly within the layer and there is even a sharp break within the *P–A* trajectory of the same layer of CdTe or CdMnTe. Hence, the *P–A* trajectory of a CdTe or CdMnTe layer in a quantum well or superlattice

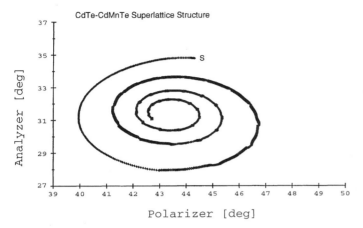

Fig. 17. *In situ* measurement of the *P–A* trajectory of a CdTe-CdMnTe superlattice structure grown on a thin CdTe buffer layer, sample CCM302, CdTe **x x x**, and CdMnTe + + +.

structure cannot be fitted with a single complex index of refraction. The nonuniformity of the complex index of refraction in an individual CdTe or CdMnTe layer is believed to be associated with the initial strain in the layer, the relief of the strain in the layer after the layer reaches its critical thickness, and the diffusion [15] of Mn from a CdMnTe layer into a CdTe layer, or a combination of these effects. Because we did not think that the quality of sample CCM306 was that good, we only did a preliminatry fit of the *P–A* trajectory of Figure 16.

Figure 17 shows the *P–A* trajectory associated with the deposition of the CdTe buffer layer on GaAs and the deposition of a CdTe-CdMnTe superlattice structure on the buffer layer involved in the fabrication of sample CCM302. Note that in Figure 17, the overall size of the *P–A* trajectory is quite large, i.e., 7° in *P* and 7° in *A*. The reason for this is that in sample CCM302 the buffer layer was made relatively thin (1000 Å), consequently the pseudosubstrate, represented by the end of the *P–A* trajectory associated with the buffer layer, has an effective index of refraction that is still substantially different from the indices of refraction of CdTe and CdMnTe, hence the size of the subsequent *P–A* trajectory is large. The complex index of refraction of CdMnTe $(3.040 - i0.180)$ is not much different from that of CdTe $(3.030 - i0.230)$, because the composition of Mn in CdMnTe is only 20%. Although the *P–A* trajectory associated with the CdTe-CdMnTe superlattice structure on the buffer layer has some small bumps on it, overall the *P–A* trajectory spirals in toward the end point as if there is no difference in the index of refraction between the CdTe and CdMnTe layers. The *P–A* trajectory associated with the CdTe-CdMnTe superlattice structure can be fitted quite well (neglecting the bumps) with the average of the indices of CdTe and CdMnTe. Since we know when we switch from depositing CdTe to depositing CdMnTe (and vice versa), we could use data such as this to determine the thicknesses of the various CdTe and CdMnTe layers, just by knowing how much of the *P–A* trajectory is associated with each layer. However, quantum wells and superlattice structures are not going to always occur on thin buffer layers and we felt

that we should concentrate on the more general case where the effective index of refraction at the pseudosubstrate at the end of the buffer layer is closer to that of CdTe and CdMnTe.

The initial anomalous trajectory of a CdTe layer on a GaAs substrate is not so apparent, as has been shown in Figure 15, and can only be seen when it is closely inspected on a blown up scale. However, one sees in Figure 18, when one is dealing with a substrate of CdZnTe, which is better lattice matched to CdTe or CdMnTe, that the anomalous initial trajectory due to interfacial effects is much more apparent. Note that in Figure 18 in the fabrication of sample CCM401 the overall size of the *P–A* trajectory of the CdTe buffer layer is quite small, i.e., 0.6° in *P* and 0.15° in *A*. This is due to the fact that the complex index of refraction of CdZnTe $(3.030 - i0.242)$ is much closer to that of CdTe $(3.030 - i0.230)$ and CdMnTe $(3.040 - i0.180)$. Since the size of the *P–A* trajectory depends on the small difference in these indices of refraction, it means that the *P–A* trajectory is very sensitive to small changes in these indices.

We believe that the portion of the trajectory between points *A* and *C* in Figure 18 is associated with the strain in the CdTe layer. When the deposited CdTe layer reaches its critical thickness, the layer starts to relieve its strain, and, as the CdTe continues to be deposited, the layer relaxes to an unstrained layer and the *P–A* trajectory reverts to a normal spiral toward its end point. The curvature of the initial anomalous *P–A* trajectory is changing too drastically and cannot be fitted with a single complex refractive index. Theoretically, the normal part of the *P–A* trajectory should be fitted with a single complex refractive index. However, in some samples such as CCM401 we believe that due to the effect of surface roughness, and the slow relaxation in the layer, the normal part of the *P–A* trajectory does not spiral symmetrically inward toward its end point. Sometimes, it spirals asymmetrically such that it intersects another part of the trajectory as observed in the *P–A* trajectory of Figure 18. Hence, it cannot be fitted with a single complex refractive index but rather with a number of complex refractive indices such that they decrease toward the bulk value of the deposited material, CdTe in this case. However, we shortly present the case of sample CCM410, which has the *P–A* trajectory of its buffer layer such that the normal part can be fitted with a single complex index of refraction.

Another point that should be noticed about Figure 18 is that the trajectory actually started at point *S*, and the trajectory moved initially approximately horizontally to the right to the point labeled *A* before it turned around to trace out the initial anomalous trajectory. This initial horizontal part of the trajectory has been observed in the trajectories of a number of samples. Normally, after thermally etching the oxide off the substrate, one leaves the sample in the sample holder with the shutter over it so that the lasers can be tested to see if they are working by vaporizing the targets. During this time, the surface might absorb some of the residue from the surroundings and hence shift the *P*, *A* value of the substrate to a new point, to the left as expected when an oxide is deposited on the substrate. It should be remembered that the base pressure is $\sim 5 \times 10^{-8}$ Torr, and so it is not an ultrahigh vacuum (UHV) and so surface con-

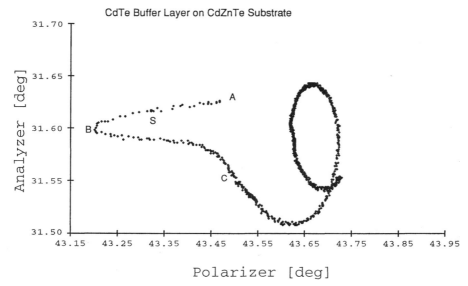

Fig. 18. *In situ* measurement of the *P–A* trajectory of a CdTe buffer layer grown on a CdZnTe substrate, sample CCM401.

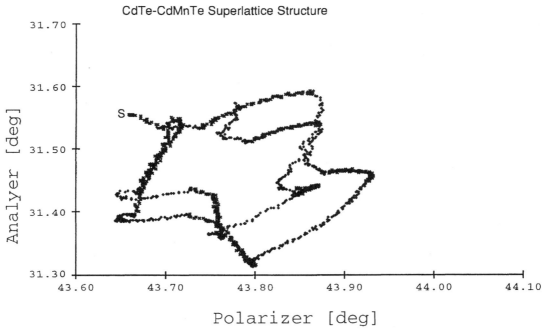

Fig. 19. *In situ* measurement of the *P–A* trajectory of a CdTe-CdMnTe superlattice structure grown on the CdTe buffer layer of Figure 18, sample CCM401, CdTe **x x x**, and CdMnTe **+ + +**.

tamination is possible. Probably, the contamination is due to physiadsorption, which means that the contamination is only weakly bonded to the surface. When the deposition starts, the vaporized CdTe or CdMnTe can sputter off the contamination that is weakly bonded to the surface, and the *P–A* trajectory shifts back to the right before it traces out a typical trajectory observed for other samples. This process is possible because the vaporized CdTe or CdMnTe molecules are highly energetic [16] and the contamination is weakly bonded.

Figure 19 shows the *P–A* trajectory for sample CCM401 of the CdTe-CdMnTe superlattice structure grown on the CdTe buffer layer grown on the CdZnTe substrate whose *P–A* trajectory has been shown in Figure 18. For a lattice-matched system of materials as in this case, a thin buffer layer is sufficient to produce a good quality superlattice structure, and the kinks between the parts of the *P–A* trajectory due to CdTe and CdMnTe are sharply defined. However, the fitting of the *P–A* trajectory of a superlattice is not easy. First, the size of the *P–A* trajec-

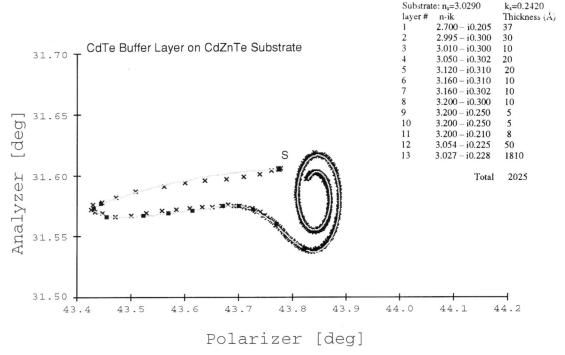

Substrate: n_s=3.0290		k_s=0.2420
layer #	n-ik	Thickness (Å)
1	2.700 − i0.205	37
2	2.995 − i0.300	30
3	3.010 − i0.300	10
4	3.050 − i0.302	20
5	3.120 − i0.310	20
6	3.160 − i0.310	10
7	3.160 − i0.302	10
8	3.200 − i0.300	10
9	3.200 − i0.250	5
10	3.200 − i0.250	5
11	3.200 − i0.210	8
12	3.054 − i0.225	50
13	3.027 − i0.228	1810
	Total	2025

Fig. 20. *In situ* measurement of the *P–A* trajectory of a CdTe buffer layer as it was deposited on a CdZnTe substrate, sample CCM410. Experimental data: CdTe **x x x**, and the theoretical fit is the gray solid line.

tory is small (0.3° in *P* and 0.3° in *A*) so the data have to be acquired with as little scatter as possible. In general, the trajectory of the CdTe or the CdMnTe behaves anomalously, when it first deposits on CdMnTe or CdTe, respectively. For example, when the CdTe is first deposited on the CdMnTe, the trajectory changes its direction to trace out half a loop producing a kink in its trajectory. After some thickness, the trajectory behaves normally and moves with a curvature that tends to spiral inward toward its end point. We will see that in general, the *P–A* trajectories in these CdTe and CdMnTe layers in the superlattice structure have to be fitted with two sublayers with slightly different indices of refraction.

Figure 20 shows the *P–A* trajectory of the CdTe buffer layer for sample CCM410. This is a typical trajectory of CdTe deposited on a CdZnTe substrate. The initial deposit of CdTe onto the substrate causes an anomalous *P–A* trajectory, until a sufficient thickness of CdTe has been deposited, whereupon the *P–A* trajectory reverts to a normal spiral toward the end point that would correspond to such a thick layer of uniform CdTe that the laser beam would no longer be able to see the underlying CdZnTe substrate. The thickness of this anomalous trajectory from fitting is 215 Å. With this close lattice match between CdZnTe (a_0 = 6.4658 Å [17]) and CdTe (a_0 = 6.4658 Å [18]), it is reasonable to associate the initial anomalous trajectory to strain in the initially deposited CdTe layer. After the initial strain in the layer, it behaves normally and the trajectory spirals into its end point. Because the birefringence in the windows is low (indicated by a small zone difference), the effect of surface roughness is small, and the layer could be fully relaxed,

the trajectory spirals into its end point symmetrically. Hence, this part of the trajectory can be fitted with a single complex index of refraction for a thickness of 1810 Å. This is the best trajectory of a buffer layer that we obtained in all the samples we fabricated. The fitted complex index of refraction for CdTe from this trajectory has a value close to that reported in the literature [19].

The *P–A* trajectories of the CdTe buffer layers of all the samples with CdZnTe substrates have the same shape, although the size of the initial anomalous trajectory is different from one sample to the next. This might be associated with the degree of strain in the initial layer. There are differences in the ways the trajectories revert from the initial anomalous trajectory to normal spirals toward their end points. Most of these trajectories did not spiral symmetrically inward, instead they shifted to the right as they spiral inward such as parts of their trajectories intersect with other parts. One possible explanation for this behavior is related to the degree of relaxation of the layer. It has been suggested in the literature [20] that the relaxation of the residual strain in the layer can be slow and thus could occur over a large thickness of the deposited CdTe layer. The decrease of the fitted complex indices of refraction of CdTe toward the value associated with bulk CdTe as the trajectory starts to spiral normally near the end point supports this explanation.

The *P–A* trajectory of the CdTe-CdMnTe superlattice structure grown on the buffer layer of sample CCM410 is shown in Figure 21. Note the small size of the *P–A* trajectory, i.e., 0.25° in *P* and 0.25° in *A*. There are kinks in the *P–A* trajectories within the same layer of CdTe and CdMnTe, and it is

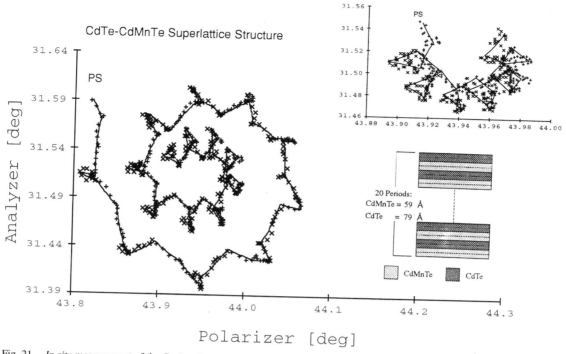

Fig. 21. *In situ* measurement of the *P–A* trajectory of a CdTe-CdMnTe superlattice structure grown on the CdTe buffer layer on a CdZnTe substrate of Figure 20, sample CCM410. Experimental data: CdTe × × ×, and CdMnTe + + +, and the theoretical fit is the black solid line.

Table I.   Summary of Fitting Parameters for CdTe-CdMnTe Superlattice of Sample CCM410

	CdMnTe			CdTe		
	Range of n	Range of k	Range of thickness (Å)	Range of n	Range of k	Range of thickness (Å)
First sublayer	$3.030 \rightarrow 3.050$	$0.180 \rightarrow 0.200$	$16 \rightarrow 34$	$2.990 \rightarrow 3.010$	$0.210 \rightarrow 0.233$	$26 \rightarrow 56$
Second sublayer	$3.030 \rightarrow 3.020$	$0.180 \rightarrow 0.200$	$28 \rightarrow 46$	$3.015 \rightarrow 3.040$	$0.200 \rightarrow 0.230$	$26 \rightarrow 54$
Layer			$52 \rightarrow 68$			$72 \rightarrow 90$
		Average thickness = 59 Å			Average thickness = 79 Å	

not possible to fit any layer of CdTe or CdMnTe with a single $n - ik$. Some parts of the trajectories of the CdTe layers, even though they do not have kinks within them, have curvatures that change so rapidly that they form small loops. However, it has been possible to fit over 16 periods of the superlattice with a model in which each layer consists of two sublayers, and the details of this fit have been published previously [12]. Table I gives a summary of this fit. The *P–A* trajectories of three other CdTe-CdMnTe superlattices grown on CdTe buffer layers on CdZnTe have been fitted successfully with the same two sublayer model [6].

The purpose of Table I is to give the reader an idea of the values used in the fit. In fact, the fitting parameters are more consistent than indicated by the range listed in the table. If one were to scan through a table of fitting parameters, one would see that the fitting parameters of a layer are consistent with other fitting parameters of the adjacent layers. The range of fitting can be misleading because the range could be large if there are one or two odd layers where the fitting parameters are quite different from the rest. Another point that should be noticed is that the range of thicknesses of the sublayers does not add to give the range of thickness of the layer. For example, from the range of thicknesses of the sublayers of CdMnTe for sample CCM410 in Table I, one might expect the range of thickness of the layer to be from $44(16 + 28)$ Å to $80(34 + 46)$ Å, but the range of thickness of the layer shown in the table is from 52 to 68 Å. The range of thicknesses of the sublayers can add up to the range of thickness of the layer only if the layer with the smallest thickness corresponds to the sum of the smallest thicknesses of both first and second sublayers, and if the layer with the largest thickness corresponds to the sum of the largest thicknesses of both first and second sublayers. In general, this is not the case; the smallest (largest) thickness of a layer might consist of the first and second sublayers with thicknesses that do not correspond to the smallest (largest) thicknesses of the ranges. The thickness accuracy of a layer in the CdTe-CdMnTe superlattice structure is $\pm 10\%$. However, for the buffer layer, the error introduced from interfacial effects on the P–A trajectory is small compared to the thickness of the buffer layer and thus it can be determined with an accuracy of $\pm 5\%$.

Another point that should be noted from the trajectory of the superlattice structure shown in Figure 21 is that the overall P–A trajectory of the superlattice structure spirals inward toward the center of the overall trajectory. If one determines the complex index of this center point, one will find that it has a value close to the average value of the complex indices of refraction of CdMnTe and CdTe. For example, the center point of the overall trajectory is approximately $43.95°$ in P and $31.51°$ in A, which corresponds to a complex index of refraction of $3.015 - i0.205$. This value is close to the average value, $3.018 - i0.204$, of the complex indices of refraction of CdTe ($3.026 - i0.223$) and CdMnTe ($3.009 - i0.185$). The complex indices of refraction of CdTe and CdMnTe before are determined from averaging the values of complex indices for the second sublayers of CdTe and CdMnTe from the table of fitting parameters that was used to generate Table I. The physical implication of the trajectory of a superlattice spiraling toward the end point, characterized by the average of complex indices of refraction of CdTe and CdMnTe, is that if the number of periods of the superlattice is large then the light will see the superlattice as a single layer with an average complex index of the two materials that make up the superlattice structure.

The trajectory inset in Figure 21 is the trajectory of the subsequent periods of CdTe-CdMnTe layers on top of the superlattice structure shown in the main plot of Figure 21. One can see that the size of this structure is small and the scattering of P, A values is comparable to the trajectory such that it is difficult to see the trajectory clearly. In fact, the last few periods of layers of this structure were not fitted because the trajectory between CdTe and CdMnTe layers could not be clearly differentiated. There were a number of sets of ellipsometric data of superlattice structures for which the last few periods of layers were not fitted because of the same reason. At the time these data were taken, we did not have the all-Faraday mode of operation of the ellipsometer, which would have reduced the scatter on these small structures by a factor of at least 3 and probably have made them fittable.

Since all our results on the CdTe-CdMnTe superlattices are based on our method of fitting the P–A trajectories with two sublayers, we think we should present evidence of how well this fit can be made and how clearly the evidence points toward two distinct sublayers. Figure 22a–d shows an enlarged view of the P–A trajectory associated with the 13th layer of CdMnTe in the CdTe-CdMnTe superlattice of sample CCM501 grown on a CdTe buffer layer on a CdZnTe substrate. Figure 22a shows the sensitivity of the fitting of the first sublayer of this CdMnTe layer to variations of 0.01 in n, while k is held constant at its best value. The three complex refractive indices used in the fit-

ting are $3.060 - i0.300$, $3.050 - i0.300$, and $3.070 - i0.300$. Even with this small change in n, the three fittings can be clearly differentiated such that one can say with confidence that $3.060 - i0.300$ gives the best fit. Figure 22b illustrates the sensitivity of the fitting of the first sublayer of this CdMnTe layer to variations of 0.005 in k while n is held constant at its best value. The three complex refractive indices used in the fitting are $3.060 - i0.300$, $3.060 - i0.295$, and $3.06 - i0.305$. Again, even with this small change in k, the three fittings can be clearly differentiated such that one can say with confidence that $3.060 - i0.300$ gives the best fit. However, the P–A trajectory for the rest of the 13th layer of CdMnTe takes such a sharp turn, that it is readily apparent that there is a second sublayer in this layer that cannot be fitted with the same index $3.060 - i0.300$. Figure 22c shows the sensitivity of the fitting of the second sublayer of this CdMnTe layer to variations of 0.01 in n with k held constant at its best value. The $n - ik$ for the three fittings are $3.000 - i0.284$, $2.990 - i0.284$, and $3.010 - i0.284$, and $3.000 - i0.284$ is clearly seen to give the best fit to the experimental data. However, Figure 22d shows that this second layer has less sensitivity to variations in the value of k, since fittings with $n - ik$ of $3.000 - i0.284$, $3.000 - i0.279$, and $3.000 - i0.289$ are not as clearly differentiated as the earlier three cases. When this happens, one can try using larger variations in k, to compensate for the lack of sensitivity, to try to get predicted P–A trajectories that differ equally from the data on either side of the experimental points so that one can choose the average of these values of k. If this does not work, then one has to use the value of k that has been obtained from an adjacent similar sublayer of the CdMnTe layer where there was enough k sensitivity to determine a value. Using these approaches, we were able to determine that the value of k should be 0.284 for this second sublayer. The same approaches have to be applied if a lack of sensitivity in determining n is encountered on some other part of the P–A trajectory.

One of the reasons for fabricating the CdTe-CdMnTe superlattices was so that we could check the thickness determined ellipsometrically with the thickness determined by transmission electron microscopy (TEM) for the same sample. Because of the high sample quality demands of TEM, layer thickness was determined only for samples CCM401, CCM501, and CCM502. The thickness of the buffer layer for sample CCM501 could not be determined due to a defect in the TEM specimen for that sample. The images of the layers in the CdTe-CdMnTe superlattice structures were obtained, with the CdMnTe layers appearing bright and the CdTe layers appearing dark. The thicknesses of the layers were measured off the image negatives using an Abbé comparator. The accuracy of the thickness measurement is $\pm 10\%$.

Table II shows a comparison of the layer thicknesses determined for samples CCM410, CCM501, and CCM502 by ellipsometry and TEM. The conclusion is that both techniques agree in their determination of the layer thicknesses of the CdTe and CdMnTe layers in these CdTe-CdMnTe superlattices within the experimental error of $\pm 10\%$ for each technique.

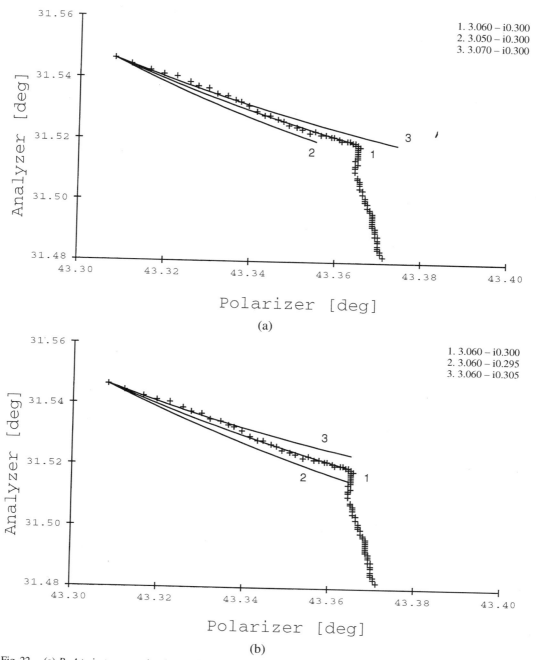

1. $3.060 - i0.300$
2. $3.050 - i0.300$
3. $3.070 - i0.300$

(a)

1. $3.060 - i0.300$
2. $3.060 - i0.295$
3. $3.060 - i0.305$

(b)

Fig. 22. (a) P–A trajectory associated with the 13th CdMnTe layer of the CdTe-CdMnTe superlattice structure of sample CCM501 grown on a CdTe buffer layer on a CdZnTe substrate. The first sublayer is fitted with variations of 0.01 in n (k is held constant at its best value). Experimental data: $+++$, and theoretical fits are the solid lines. (b) P–A trajectory associated with the 13th CdMnTe layer of the CdTe-CdMnTe superlattice structure of sample CCM501 grown on a CdTe buffer layer on a CdZnTe substrate. The first sublayer is fitted with variations of 0.005 in k (n is held constant at its best value). Experimental data: $+++$, and theoretical fits are the solid lines. (c) P–A trajectory associated with the 13th CdMnTe layer of the CdTe-CdMnTe superlattice structure of sample CCM501 grown on a CdTe buffer layer on a CdZnTe substrate. The second sublayer is fitted with variations of 0.01 in n (k is held constant at its best value). Experimental data: $+++$, and theoretical fits are the solid lines. (d) P–A trajectory associated with the 13th CdMnTe layer of the CdTe-CdMnTe superlattice structure of sample CCM501 grown on a CdTe buffer layer on a CdZnTe substrate. The second sublayer is fitted with variations of 0.005 in k (n is held constant at its best value). Experimental data: $+++$, and theoretical fits are the solid lines.

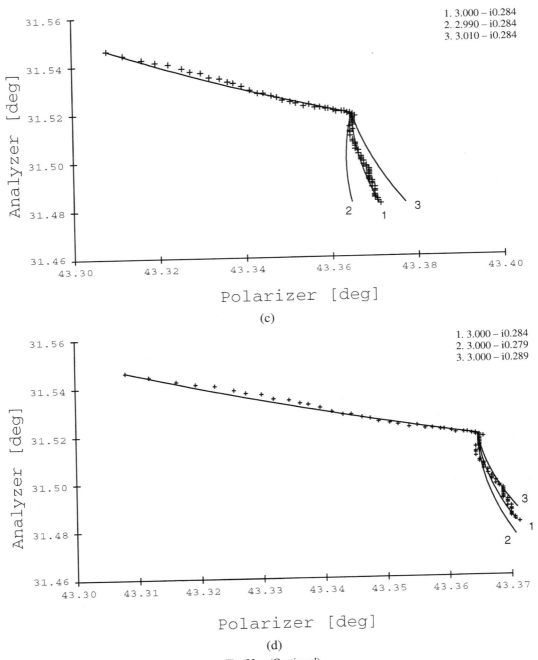

Fig. 22. (Continued).

Table II. Layer Thickness Determined by Null Ellipsometry and TEM

	Null ellipsometry				Transmission electron microscopy (TEM)			
Sample	Buffer layer (Å)	CdTe layer (Å)	CdMnTe layer (Å)	Period thickness (Å)	Buffer layer (Å)	CdTe layer (Å)	CdMnTe layer (Å)	Period thickness (Å)
CCM410	2025	79	59	138	2098	82	50	132
CCM501	2373	102	73	175	N/A	101	77	178
CCM502	1286	104	71	175	1315	101	69	170

3.2. Experimental Results for MS System

The MS system was used for depositing layers of Pt, Ni, and Ti on (100) single crystal silicon either as individual layers or as sequences of layers to produce metallic multilayer structures. The layer growths were at room temperature with a base pressure of approximately 10^{-9} Torr. Due to these growth conditions, all the grown metallic layers were expected to be polycrystalline. Samples EL2, EL3, and EL4 were metallic multilayer structures involving Pt, Ni, and Ti in various sequences. Samples X131A, X131B, and X131C involved depositing repeated periods of the same thicknesses of two metal layers, i.e., Ni-Ti, Pt-Ti, and Pt-Ni, respectively. Samples X131D and X132A through X132E involved individual layers of Pt, Ni, and Ti deposited on silicon for various thicknesses of the layer.

In situ ellipsometric measurements made during the deposition of samples, consisting of either single or multiple metallic layers produced by the MS system, yield *P–A* trajectories that are significantly different from those of semiconductor single layers or superlattice structures produced by the PLEE system. First, the size of the *P–A* trajectory involving the deposition of a metallic layer can be much larger than that associated with the deposition of semiconductor layers. Second, the trajectory does not spiral around its end point, but moves much more directly toward its end point, when the thickness that has been deposited is sufficient for the layer to become totally opaque. This is due to the fact that metallic layers are much more absorbing than semiconductor layers, or equivalently that the value of *k* for a metal is much larger than the value of *k* for a semiconductor. A typical *P–A* trajectory for a thick metallic layer is categorized into three regions: the initial anomalous region, the normal region, and the final anomalous region. The initial anomalous region is the result of the interface between the overlying layer and the underlying layer or the substrate. When a layer is sufficiently thick that it is not affected by the interface, it would be characterized by a complex index of refraction that corresponds to a bulk sample of the material. Due to the nature of the growth, which is predominantly a polycrystalline growth, the surface of the metallic layer becomes rough as the layer gets thicker. The effect of surface roughness has a major impact on the trajectory and is the cause of the final anomalous region.

The initial anomalous trajectory occurs at the interface between the material being deposited and the underlying layer or the substrate. When the anomaly is small, the initial anomalous trajectory can be fitted with a single layer with one value of $n - ik$, whereas when the anomaly is large, one needs to fit the initial anomalous trajectory with a number of sublayers each with a different value of $n - ik$. When the initial anomalous trajectory can be fitted with a single layer, the fitted refractive index has a value close to that of the bulk index for the deposited material. When the anomaly is large, the initial anomalous trajectory has to be fitted with a number of sublayers, whose refractive indices start at an $n - ik$ close to that of the underlayer and approach the bulk index of the material being deposited in the final sublayer. With this approach to the fitting, the thickness for that part of the layer associated with the initial anomalous trajectory is found to be consistent with that expected from the growth rate.

After the initial anomalous trajectory, the *P–A* trajectory behaves normally; i.e., the trajectory in this region can be fitted with the single value of $n - ik$ close to that of the bulk material being deposited. The process of fitting the trajectory in this region is trivial, since one only needs to vary the value of *n* and *k* until the best fit is obtained. In this normal region of the trajectory, an accurate thickness can be obtained and hence a good value of the growth rate can be determined. It is very useful to know the growth rate at which the layer is being deposited, because from it one can approximately determine the thickness in the initial anomalous region and in the region where surface roughness is important. This knowledge can help to narrow down the range of $n - ik$ which may be used for fitting these regions.

After the middle region, where the trajectory behaves normally, the trajectory gradually diverges from a normal trajectory or sometimes even produces a sharp break in the trajectory and moves in a different direction. A number of researchers in the literature have attributed this phenomenon to the effects of surface roughness. In general, surface roughness is believed to change (decrease or increase) the value of $n - ik$ from that associated with the bulk material. The surface becomes significantly rough only when the layer becomes very thick, and at the stage where the trajectory also becomes less sensitive to a change in thickness of the layer because the trajectory is close to its end point. In this final region where the effect of surface roughness is important, the *P–A* trajectory depends on both changes in thickness and refractive index, and thus it is difficult to fit the trajectory in this region to obtain a thickness with as small an uncertainty as in the normal *P–A* trajectory region where the index of refraction is constant. We have explored various approaches for finding the thickness of the metal film in this difficult final region.

One method to obtain the thickness of the final portion of the layer where surface roughness is important is by utilizing the growth rate. The growth rate can be determined from fitting the trajectory where it behaves normally as has been mentioned previously. Since the **EXACTA 2000** monitors the deposition in real time, the time that has elapsed between when the *P–A* trajectory first starts to deviate from normal behavior until the end of the trajectory for that layer can be determined easily. Hence, the thickness of that portion of the layer where surface roughness is important to the *P–A* trajectory can be determined from this ellipsometrically determined growth rate, on the assumption that the growth rate in this region is the same as that in the normal region of the trajectory.

Another method to determine the thickness in this region is to fit the *P–A* trajectory. We used this approach to derive our thickness for this portion of a layer and we used the growth rate thickness determination just to confirm that the fitting yields an appropriate thickness. The fitting approach that we used is described by the following iterative procedure.

1. We change the value of *k* from that of the refractive index of the normal region by a small amount and then we vary the value of *n* from that of the refractive index

of the normal region such that it would give a good fit to the first segment of this final anomalous *P–A* trajectory that corresponds to a sublayer with a small thickness having this $n - ik$. In fitting this region of the *P–A* trajectory where surface roughness is important, we believe that the change in n or k must be consistent, i.e., it must either be always decreased, increased, or kept constant. In other words, the values of n or k cannot be decreased in one part and then increased in another part of the trajectory. In general, the values of n or k should be decreased due to surface roughnes, but for the case of Pt the values of n and k both increased or k stayed constant. It is reasonable to assume that n or k changes by only a small amount for a small increase in thickness (50 or 100 Å) because it is believed that the roughness of the surface increases slowly with thickness.

2. We change the k value slightly more from its bulk value for the next segment of the *P–A* trajectory and then we repeat the process in 1 to obtain the best fit for that segment corresponding to the next sublayer with a small thickness having this $n - ik$. This process is repeated sublayer by sublayer to fit the rest of the trajectory.

The foregoing procedures were used to fit all the *P–A* trajectories shown in this section.

Figure 23a shows the complete *P–A* trajectory for sample EL2. Starting with a clean Si(100) substrate, an 810-Å layer of

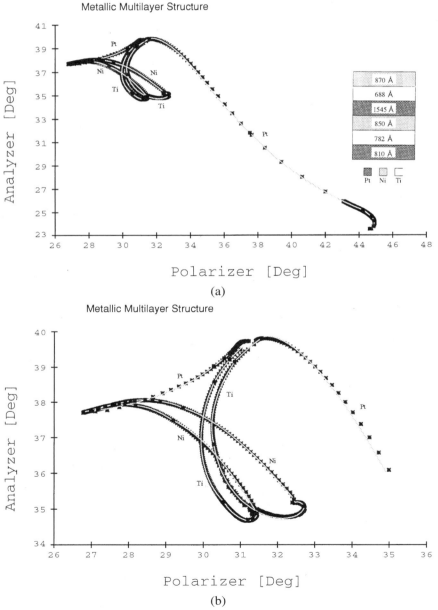

Fig. 23.   (a) *P–A* trajectory of all six layers in sample EL2. (b) *P–A* trajectory of the last five layers in sample EL2.

Table III.   The Fitting Parameters for Sample EL2, Figure 23a and b

Mat	L	$n - ik$	Th (Å)	Mat	L	$n - ik$	Th (Å)	Mat	L	$n - ik$	Th (Å)
$n_s = 3.913, k_s = 0.0518$				Ni	8	$1.85 - i3.95$	50	Pt	16	$2.143 - i5.235$	100
Pt	1	$4.60 - i0.80$	7		9	$1.83 - i3.90$	50		17	$2.150 - i5.245$	50
	2	$4.60 - i2.00$	5		10	$1.81 - i3.85$	50				**1545**
	3	$4.20 - i3.00$	3		11	$1.80 - i3.80$	50				
	4	$3.50 - i3.30$	5		12	$1.79 - i3.78$	50	$n_{ps} = 2.160, k_{ps} = 5.250$			
	5	$2.60 - i5.00$	60				**850**	Ti	1	$3.60 - i3.20$	12
	6	$2.23 - i5.38$	730						2	$2.80 - i4.16$	30
			810	$n_{ps} = 1.772, k_{ps} = 3.861$					3	$3.30 - i3.92$	150
				Pt	1	$2.00 - i5.10$	145		4	$3.25 - i3.84$	116
$n_{ps} = 2.232, k_{ps} = 5.239$					2	$2.05 - i5.10$	50		5	$3.185 - i3.840$	50
Ti	1	$3.60 - i3.20$	12		3	$2.09 - i5.10$	50		6	$3.180 - i3.838$	180
	2	$3.00 - i4.16$	120		4	$2.10 - i5.10$	50		7	$3.170 - i3.828$	150
	3	$3.54 - i4.03$	400		5	$2.11 - i5.10$	100				**688**
	4	$3.48 - i4.01$	250		6	$2.12 - i5.10$	100				
			782		7	$2.12 - i5.11$	100	$n_{ps} = 3.165, k_{ps} = 3.860$			
					8	$2.12 - i5.12$	100	Ni	1	$2.05 - i406$	190
$n_{ps} = 3.460, k_{ps} = 4.050$					9	$2.11 - i5.13$	100		2	$1.90 - i4.01$	160
Ni	1	$2.05 - i4.15$	200		10	$2.11 - i5.15$	100		3	$1.87 - i4.00$	100
	2	$1.99 - i4.14$	100		11	$2.11 - i5.17$	100		4	$1.835 - i3.950$	100
	3	$1.95 - i4.13$	50		12	$2.11 - i5.18$	100		5	$1.805 - i3.900$	100
	4	$1.93 - i4.09$	50		13	$2.12 - i5.20$	100		6	$1.795 - i3.850$	70
	5	$1.91 - i4.07$	100		14	$2.125 - i5.210$	100		7	$1.77 - i3.80$	80
	6	$1.88 - i4.05$	100		15	$2.132 - i5.235$	100		8	$1.745 - i3.750$	70
	7	$1.87 - i4.00$	50								**870**

Pt was deposited, followed by a 782-Å layer of Ti, followed by an 850-Å layer of Ni, followed by a 1545-Å layer of Pt, followed by a 688-Å layer of Ti, and followed by an 870-Å layer of Ni. Figure 23b shows an enlarged view of the P–A trajectory of the last five layers in sample EL2.

Table III shows the fitting parameters for sample EL2. Each fitting table is formatted in the following way: The n_s and k_s are the real and imaginary parts of the index of refraction of the Si substrate before the deposition. n_{ps} and k_{ps} are the real and imaginary parts of the index of refraction of a pseudosubstrate, if this was used in the fitting. The layer material is given as Ni, Pt, and Ti. Then, the sublayer number (L), and the complex index of refraction ($n - ik$) and the thickness (Th) of that sublayer is given in rows. The bold number after the last sublayer is the total thickness of the layer based on the sum of the thicknesses of the sublayers for that layer. The experimental P–A trajectories for Pt, Ti, and Ni are plotted in symbols *, ×, and +, respectively. The fit is the gray solid line. The points on a trajectory at which the fitting parameters were changed is indicated by a larger square dot. The schematic of the physical structure is also inset in the plot of the P–A trajectory to help one to visualize the physical meaning of the P–A trajectory.

Figure 24a shows the P–A trajectory for the first four layers of sample EL3. Starting with a clean Si(100) substrate denoted by S, an 880-Å layer of Ti was deposited, followed by a 978-Å layer of Ni, followed by a 1390-Å layer of Pt, followed by a 810-Å layer of Ti. The trajectory of this last Ti layer instead of

spiraling into the end point as the layer got thicker, suddenly produced a kink and moved in a different direction (down and to the left in the plot). Figure 24b shows the P–A trajectory for the next five layers of sample EL3. Starting at the end of the Ti layer deposited in Figure 24a denoted by PS, a 958-Å layer of Pt was deposited, followed by a 198-Å layer of Ti, followed by a 957-Å layer of Pt, followed by a 260-Å layer of Ti, followed by a 899-Å layer of Pt. The P–A trajectory of the Pt layer initially deposited on the Ti traced out a half flat loop and only when the Pt layer was thick enough did it start to behave normally. However, when the Pt layer became too thick, instead of spiraling into its end point, the trajectory changed direction and moved horizontally to the right. When Ti is deposited on Pt, the initial anomalous trajectory is not as apparent as when Pt is deposited on Ti. Table IV shows the fitting parameters for the layers shown in Figure 24a and b, respectively.

Figure 25(a) shows the P–A trajectory for the first three layers of sample EL4. Starting with a clean Si(100) substrate denoted by S, a 1450-Å layer of Ni was deposited, followed by a 920-Å layer of Pt, followed by a 946-Å layer of Ti. Again, toward the end of the trajectory of this Ti layer, the trajectory produced a kink and moved off in a different direction, to the left of the plot. Figure 25b shows the P–A trajectory for the next nine layers of sample EL4. Starting at the end of the Ti layer deposited in Figure 25a denoted by PS, a 91-Å layer of Ni was deposited, followed by an 80-Å layer of Pt, followed by a 116-Å layer of Ti, followed by an 84-Å layer of Ni, followed by a 78-Å layer of Pt, followed by a 116-Å layer of Ti,

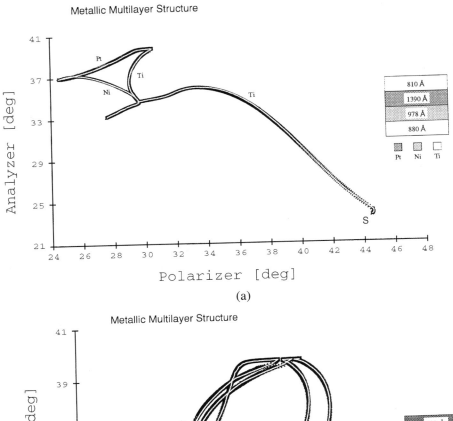

Fig. 24. (a) *P–A* trajectory of the first four layers in sample EL3. (b) *P–A* trajectory of the last five layers in sample EL3.

followed by a 94-Å layer of Ni, followed by a 76-Å layer of Pt, and then finally followed by a 595-Å layer of Ti. There was no significant initial anomaly in the *P–A* trajectory when Pt was deposited on Ni. We are not saying that the initial anomaly in the *P–A* trajectory when Pt is deposited on Ni is exactly zero, but if it does exist, it is too small in these experiments to be measurable. However, when Ni was deposited on Ti, the initial anomaly was large, with the *P–A* trajectory tracing out a small loop before it behaved normally. When Ti was deposited on Pt, the initial anomaly was again not apparent. As before, when the last Ti layer of this sample was thick, it produced a kink and moved in a different direction. Table V shows the fitting parameters for the layers shown in Figure 25a and b, respectively.

The *P–A* trajectories or tables of fitting data for samples X131A through X131D and X132A through X132E, which are in the Ph.D. thesis of Tran [6] will not be presented here, since they just support the same conclusions that can be made on the basis of samples EL2, EL3, and EL4. However, in the few cases where they give additional results, we will present them, and we will present the results of certain subsidiary experiments on some of these samples.

In the rest of this section, we first present general observations about the experimental results and the analysis. Next, we discuss models of the effective complex index of refraction to handle surface roughness, island formation on film growth, and intermixing of two different layers. Then, we discuss the result-

Table IV.   Fitting Parameters for sample EL3, Figure 24a and b

Mat	L	$n - ik$	Th (Å)	Mat	L	$n - ik$	Th (Å)	Mat	L	$n - ik$	Th (Å)
$n_s = 3.910, k_s = 0.0708$				Ni	1	$2.20 - i4.00$	8	Pt	10	$1.865 - i4.900$	100
Ti	1	$4.60 - i0.80$	2		2	$2.00 - i3.70$	20		11	$1.90 - i5.00$	100
	2	$4.50 - i1.50$	2		3	$1.80 - i3.55$	30		12	$1.93 - i5.00$	100
	3	$4.20 - i2.00$	3		4	$1.60 - i3.41$	400		13	$1.96 - i5.05$	100
	4	$3.40 - i3.80$	16		5	$1.583 - i3.38$	70		14	$1.99 - i5.08$	70
	5	$3.60 - i4.05$	280		6	$1.58 - i3.36$	100				**1390**
	6	$3.45 - i4.00$	50		7	$1.57 - i3.35$	100	$n_{ps} = 1.995, k_{ps} = 5.020$			
	7	$3.40 - i3.95$	50		8	$1.563 - i3.350$	100	Ti	1	$2.70 - i3.70$	20
	8	$3.35 - i3.90$	50		9	$1.553 - i3.345$	150		2	$3.00 - i3.74$	170
	9	$3.30 - i3.85$	50				**978**		3	$2.95 - i3.67$	50
	10	$3.23 - i3.80$	50	$n_{ps}=1.558, k_{ps}=3.359$					4	$2.90 - i3.58$	40
	11	$3.15 - i3.75$	50	Pt	1	$1.63 - i4.50$	120		5	$2.80 - i3.55$	70
	12	$3.06 - i3.70$	50		2	$1.72 - i4.50$	100		6	$2.77 - i3.40$	50
	13	$3.00 - i3.65$	50		3	$1.76 - i4.50$	100		7	$2.70 - i3.35$	90
	14	$2.90 - i3.60$	50		4	$1.77 - i4.60$	100		8	$2.66 - i3.30$	100
	15	$2.85 - i3.55$	50		5	$1.77 - i4.70$	100		9	$2.62 - i3.00$	90
	16	$2.80 - i3.50$	50		6	$1.79 - i4.72$	100		10	$2.56 - i3.00$	50
	17	$2.73 - i3.45$	50		7	$1.80 - i4.75$	100		11	$2.52 - i3.00$	80
			880		8	$1.815 - i4.800$	100				**810**
$n_{ps}=2.756, k_{ps}=3.582$					9	$1.84 - i4.85$	100				
				$n_{ps} = 1.923, k_{ps} = 4.963$				Ti	1	$2.98 - i4.00$	6
$n_{ps} = 2.470, k_{ps} = 3.085$				Ti	1	$3.00 - i4.00$	5		2	$3.00 - i3.30$	20
Pt	1	$3.50 - i4.00$	5		2	$3.00 - i3.10$	15		3	$2.95 - i3.71$	50
	2	$2.40 - i4.50$	4		3	$3.04 - i3.68$	178		4	$2.99 - i3.74$	184
	3	$2.00 - i4.50$	9				**198**				**260**
	4	$1.64 - i4.60$	100	$n_{ps} = 2.680, k_{ps} = 3.570$				$n_{ps} = 2.800, k_{ps} = 3.590$			
	5	$1.70 - i4.65$	50	Pt	1	$3.60 - i3.90$	13	Pt	1		15
	6	$1.73 - i4.70$	60		2	$3.20 - i4.70$	4		2		4
	7	$1.83 - i4.70$	80		3	$2.30 - i4.80$	10		3		250
	8	$1.82 - i4.75$	100		4	$1.935 - i4.940$	250		4		290
	9	$1.82 - i4.79$	100		5	$1.935 - i4.960$	150	$n_{ps} = 1.982, k_{ps} = 5.035$			
	10	$1.83 - i4.79$	100		6	$1.940 - i4.977$	150		5		200
	11	$1.845 - i4.870$	100		7	$1.955 - i5.010$	180		6		150
	12	$1.88 - i4.93$	100		8	$1.97 - i5.03$	200				**899**
	13	$1.91 - i4.97$	100				**957**				
	14	$1.93 - i5.00$	50	$n_{ps} = 1.970, k_{ps} = 5.030$							
			958								

ing fitting parameters and thicknesses determined. Finally, we discuss the physical implications of the $P–A$ trajectory analysis.

3.2.1. General Comments

1. When Ni, Pt, or Ti is deposited first on the Si(100) single crystal substrates, there is always an initial anomalous region of the $P–A$ trajectory, in that the index of refraction used to fit this initial sublayer of the metal being deposited is significantly different from that of the bulk index of refraction for that metal. The thickness of this initial sublayer depends on the metal:

it appears to be largest in the case of Pt, smaller in the case of Ni, and smallest in the case of Ti.

2. Anomalous $P–A$ trajectories are also sometimes observed at the interface between two metal layers. The occurrence of such an anomalous $P–A$ trajectory appears to depend strongly not only on the two metals involved, but also on the order in which they are deposited. For example, as shown in Figure 24b for sample EL3, when Pt is deposited on Ti the associated $P–A$ trajectory is initially anomalous for a certain thickness of deposited Pt before the subsequent deposited part of the Pt layer begins to behave

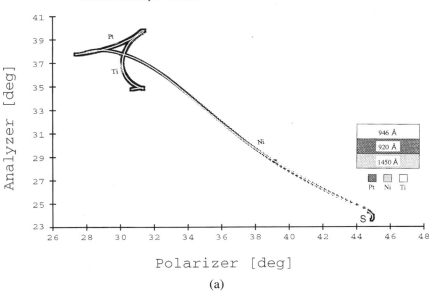

(a)

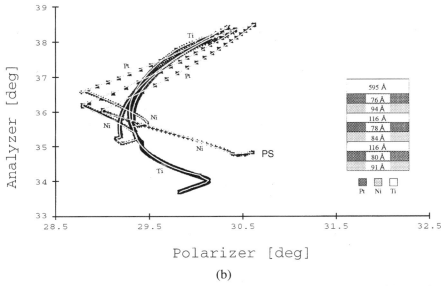

(b)

Fig. 25.   (a) *P–A* trajectory of the first three layers in sample EL4. (b) *P–A* trajectory of the last nine layers in sample EL4.

normally. On the other hand, as shown also in Figure 24b for sample EL3, an anomalous *P–A* trajectory is not apparent when Ti is deposited on Pt.

3. Theoretically, as the thickness of a metal overlayer deposited on an underlayer is increased to the point that all the light incident on the overlayer is absorbed in this layer, then the *P–A* trajectory should spiral into a final point, P_f, A_f, which is characteristic of the bulk index of refraction of that metal. Experimentally, when the metal overlayer is made very thick on an underlayer, the *P–A* trajectory instead of spiraling into this final point, sometimes diverged from its normal

trajectory. For some metals, the trajectory might even change direction or might produce a sharp break in the trajectory. A good illustration of a sharp break in the *P–A* trajectory for the same layer is shown in Figure 24a for the final layer of Ti on Pt in that figure.

4. The range of values on $n - ik$ needed to fit the *P–A* trajectory associated with a particular metal layer is dependent on the underlying layer or substrate. For example, the $n - ik$ needed to fit the first portion of the *P–A* trajectory for Ti when it is initially deposited on Si has a value close to that of Si. On the other hand, the $n - ik$ needed to fit the *P–A* trajectory for Ti when it is

Table V. Fitting Parameters for sample EL4, Figure 25a and b

Mat	L	n − ik	Th (Å)	Mat	L	n − ik	Th (Å)	Mat	L	n − ik	Th (Å)
$n_s = 3.9180, k_s = 0.0163$				Ni	13	1.825 − i3.910	160	Ti	1	3.10 − i3.80	30
Ni	1	4.50 − i1.10	3				1450		2	3.35 − i3.95	226
	2	4.40 − i1.70	3	$n_{ps} = 1.828, k_{ps} = 3.926$				$n_{ps} = 3.175, k_{ps} = 3.800$			
	3	3.70 − i2.50	4	Pt	1	2.06 − i5.15	120		3	3.172 − i3.790	100
	4	2.73 − i3.20	60		2	2.11 − i5.22	100		4	3.155 − i3.830	100
	5	2.15 − i4.22	300		3	2.13 − i5.25	100		5	3.135 − i3.800	100
	6	2.00 − i4.15	120		4	2.155 − i5.270	100		6	3.11 − i3.77	70
	7	1.94 − i4.05	100		5	2.165 − i5.280	100		7	3.07 − i3.72	50
	8	1.92 − i4.00	100		6	2.175 − i5.300	100		8	3.05 − i3.71	50
	9	1.90 − i4.00	100		7	2.190 − i5.320	100		9	3.025 − i3.710	70
	10	1.88 − i4.00	250		8	2.200 − i5.327	200		10	3.01 − i3.71	70
	11	1.865 − i3.950	100				920		11	3.00 − i3.70	80
	12	1.825 − i3.930	150	$n_{ps} = 2.218, k_{ps} = 5.287$							946
$n_{ps} = 2.975, k_{ps} = 3.735$				Ni	4	1.95 − i3.75	54	$n_{ps} = 2.300, k_{ps} = 3.930$			
Ni	1	2.70 − i3.00	9				84	Pt	1	1.92 − i5.20	76
	2	2.00 − i3.60	82	$n_{ps} = 2.360, k_{ps} = 3.858$				$n_{ps} = 2.300, k_{ps} = 4.660$			
			91	Pt	1	1.93 − i5.20	78	Ti	1	2.78 − i3.72	30
$n_{ps} = 2.430, k_{ps} = 3.850$				$n_{ps} = 2.330, k_{ps} = 4.670$					2	3.05 − i3.65	100
Pt	1	1.89 − i5.30	80	Ti	1	3.10 − i3.30	10		3	3.05 − i3.59	150
$n_{ps} = 2.330, k_{ps} = 4.670$					2	2.99 − i3.66	106		4	3.000 − i3.495	60
Ti	1	3.10 − i3.30	10				116		5	2.99 − i3.40	25
	2	3.14 − i3.50	106	$n_{ps} = 2.550, k_{ps} = 3.809$					6	2.94 − i3.40	60
			116	Ni	1	3.50 − i3.55	4		7	2.925 − i3.350	70
$n_{ps} = 2.577, k_{ps} = 3.680$					2	2.55 − i4.20	18		8	2.905 − i3.350	100
Ni	1	3.50 − i3.65	5		3	2.20 − i3.95	8				595
	2	2.55 − i4.10	13		4	2.00 − i3.83	64				
	3	2.20 − i3.90	12				94				

deposited on Pt has a value close to that reported in the literature for Ti. This suggests that Ti behaves normally when deposited on Pt and abnormally when deposited on Si.

5. The values of $n − ik$ needed to fit the $P–A$ trajectory of a deposited metal layer might either increase or decrease as the layer gets thicker. For example, in Figure 23b for sample EL2, when Ni was deposited on Ti the values of $n − ik$ needed to fit the $P–A$ trajectory of the Ni layer decrease as the thickness increases. On the other hand, in Figure 23b for sample EL2, when Pt was deposited on Ni the values of $n − ik$ needed to fit the $P–A$ trajectory of the Pt layer increase as the Pt layer became thicker. In some cases, only n changes while k stays constant.

3.2.2. Models of the Effective Complex Index of Refraction

A number of models used to calculate an effective refractive index have been proposed to handle surface roughness, island formation on film growth, and intermixing of two different layers. These models are based on simple effective medium theories [21, 22] that represent heterogeneous mixtures, which have the form,

$$\frac{N_e^2 - N_h^2}{N_e^2 + 2N_h^2} = f_1 \frac{N_1^2 - N_h^2}{N_1^2 + 2N_h^2} + f_2 \frac{N_2^2 - N_h^2}{N_2^2 + 2N_h^2} + \cdots \quad (17)$$

where N_e, N_h, N_1, N_2 are the complex indices for the effective medium, the host medium, and the inclusions of types 1, 2, etc. The f_1, f_2, etc. represent the volume fraction of inclusions 1, 2, etc. It should be noted that the "host" medium has not been defined. In fact, the primary difference in different models is the choice of the host medium.

The Lorentz–Lorenz model is good for calculating the effective complex index of refraction which involves roughness due to island formation film growth. In this model, $N_h = 1$ and Eq. (17) becomes

$$\frac{N_e^2 - 1}{N_e^2 + 2} = f_1 \frac{N_1^2 - 1}{N_1^2 + 2} + f_2 \frac{N_2^2 - 1}{N_2^2 + 2} + \cdots \quad (18)$$

However, the approximations involved in this model are only reasonable when the volume fraction is less than 20%.

A second model, i.e., the Maxwell–Garnett model, assumes that the inclusions are in a host background (other than vacuum). In the case of a single inclusion in a single host, then Eq. (17) becomes

$$\frac{N_e^2 - N_h^2}{N_e^2 + 2N_h^2} = f_1 \frac{N_1^2 - N_h^2}{N_1^2 + 2N_h^2} \qquad (19)$$

Again, this model is not a bad approximation if the inclusions make up a small fraction of the total volume. However, this model, as has been shown by Aspnes, Theeten, and Hottier [23], is not a good model to apply to the case of a rough surface where there is about as much inclusion as there is host.

The case of a rough surface is best described by the model suggested by Bruggeman [24], which is referred to as an effective medium approximation (EMA) model. In this model, the host is assumed to have an effective medium, i.e., $N_h = N_e$. With this approximation, Eq. (17) becomes

$$0 = f_1 \frac{N_1^2 - N_h^2}{N_1^2 + 2N_h^2} + f_2 \frac{N_2^2 - N_h^2}{N_2^2 + 2N_h^2} \qquad (20)$$

This is the effective medium approximation (EMA) equation for two materials. It can easily be extended to more materials in the obvious way by adding more terms.

In the references to a rough surface before, one has to be clear that what is meant is a surface which is microscopically rough. There are two types of surface roughness, macroscopic roughness and microscopic roughness. Macroscopic roughness occurs when the length scale of the irregularities on the surface is greater than the wavelength of the light used. The light that reflects from any sections of the surface of the sample that are parallel to the average plane of the surface will be reflected at the prescribed angle of incidence into the detector and light that reflects from sections of the surface of the sample that are not parallel to the average plane will be directed elsewhere and hence will not be detected. Overall, macroscopic roughness will just decrease the intensity of the reflected beam but will not affect the ellipsometric measurement of P and A values, and hence the complex refractive index will not be affected as long as there is enough light reflected into the detector to obtain a reasonable signal.

Microscopic roughness occurs when the length scale of the irregularities on the surface is smaller than the wavelength of the light such that the light interacts with the surface as a whole, rather than interacting with different facets of the surface individually. It is easy to tell experimentally when the surface is becoming microscopically rough, because the detector signal becomes increasingly sensitive to vibration and the reflected "spot" becomes diffuse over a larger and larger solid angle. In theory, the effect of microscopic roughness on the far-field radiation pattern can be approximated [26] by one or more layers of an "effective medium," which is embedded between a perfect substrate and a perfect ambient. The parameters for this model are the same as for any single film, i.e., the thickness and the complex index of refraction of the "effective film." The calcu-

lation of the different models of the effective complex index of refraction have been discussed earlier.

The effects of island formation on film growth and the intermixing of two different materials on the complex refractive index can be described by the models discussed previously. It has been mentioned that the effect of island formation in film growth is best described by the Lorentz–Lorenz model. The effect of intermixing, depending on the fraction of mixing, can be described by any of the three models before. If the fraction of intermixing is small, i.e., the inclusions occupy only a small volume of the total volume, then this effect is best described either by the Lorentz–Lorenz model or by the Maxwell–Garnet model. On the other hand, when the fraction of intermixing is large, i.e., there is as much inclusion as host, then it is best described by the effective medium approximation.

3.2.3. Discussion of the Fitting Parameters

The complex refractive indices for Si [19], Ti [26], Ni [19], and Pt [19] at a wavelength of 632.8 nm reported in the literature are $3.882 - i0.019$, $3.23 - i3.62$, $1.97 - i3.72$, and $2.390 - i4.236$, respectively. The average complex index of refraction of Si determined in this research is $3.911 - i0.024$. Comparing this index with the one reported in the literature, one observes that there is a small difference (0.029 for n and 0.005 for k) between the two indices. This small discrepancy might be due to the difference in surface treatment of the Si substrate, the effect of a thin oxide on the Si substrate due to it being exposed to atmosphere when it was being transferred to the load lock, and/or the small effect of birefringence in the windows.

In the discussion of the fitting parameters for Ti, Ni, and Pt we only consider the complex indices of refraction used to fit the normal parts of the P–A trajectories because they are the best to represent the index of a bulk sample of a material. In addition, the indices for the initial anomalous regions are only used for a small fraction of the total thickness of a layer. We believe that the surface roughness in an underlayer will affect the indices used to fit the normal parts of the trajectories in higher order layers. For example, the fifth layer of sample EL3 is Pt and the complex index of refraction used to fit the normal part of the trajectory of this layer is $1.64 - i4.60$, which is different from the bulk value of the index for Pt determined from this research.

The range of n and k used to fit the normal parts of the trajectories of the Ti layers was 3.00 to 3.60 and 3.73 to 4.05, respectively. For Ni, the range of the fitted n and k was 1.90 to 2.20 and 3.60 to 4.07, respectively. Finally for Pt, the range of fitted n and k was 2.00 to 2.50 and 5.00 to 5.38, respectively. Comparing the range of the fitted n and k of Ti and Ni to the values reported in the literature, we see that the values reported in the literature are within the range of the fitted n and k. From the tables of fitting parameters, one notices that most of the fitted n values for Ti are close to the reported values, but most of the fitted k values are around 4.00, which is higher than the reported value of 3.62. The index of refraction of Ti of $3.23 - i3.62$ from the literature [26] agrees favorably with the values obtained by

Smith and Mansfeld [27] of $3.16 - i3.54$ in ultrahigh vacuum and of $3.26 - i3.62$ in electrolyte using a single-wavelength ellipsometric technique. We believe that birefringence in the windows of the MS system is the most likely cause of the discrepancy in the k value for Ti. Similarly, for Ni, one finds that the fitted n values are close to the reported values while the fitted k values are slightly higher than the value reported of 3.74. The index of refraction given in the reference was determined from a Kramers–Kronig analysis of the absorptivity measured by a calorimetric technique on a single crystal of Ni at 4 K. It should be noted that the sample had annealed for 72 h at 1300°C, but the anneal produced no change in the absorptivity spectrum. One possible explanation for the discrepancy in the index of refraction of Ni is due to birefringence in the windows used in our measurements. The other possibility is that the index of refraction given in the reference was determined from an Ni substrate that had a thin oxide on the surface of the substrate which would affect the index of refraction that they measured. Birefringence in the windows is believed to be the major cause of the discrepancies in the indices of refraction of Ti and Ni between the reported values and the values determined in this research. However, we made these measurements before we had developed our technique for measuring and correcting for the birefringence in the windows, so we do not have any information on how large the birefringence was.

Finally, if we compare the range of fitted n and k for Pt with the value reported in the literature, we see that the fitted n value is close to the reported value, but the fitted k value is much higher than that reported. We believe that the effect of birefringence alone cannot be used to explain this large discrepancy. The reported $n - ik$ of Pt was determined from analyzing reflectance data by using the Kramers–Kronig technique. The reflectance data was obtained from Pt films evaporated by an electron gun at a high deposition rate of 100 Å/s. Thus, in addition to the effect of birefringence, it is possible that the difference in the growth techniques and growth rates of the Pt layer produced different surface conditions and hence caused a difference in the fitted $n - ik$ from that reported in the literature.

One other check on our fitting involves the relation between the expected growth rate of these metal films and the growth rate determined from our ellipsometric fits. The expected growth rates for Ti, Ni, and Pt on the MS system were 25, 55, and 95 Å/min, respectively. The average growth rates of Ti, Ni, and Pt, determined from the ellipsometric fits [6] were 25, 55, and 94 Å/min, respectively, which agrees with the expected growth rates and supports the validity of our fitting procedure.

3.2.4. Discussion of the Physical Implications of the P–A Trajectory Analysis

We have shown how the $P–A$ trajectories for these metal layers can be analyzed in terms of an optical model that distinguishes between an initial anomalous part of the $P–A$ trajectory, a normal part of the $P–A$ trajectory, and a final anomalous part of the $P–A$ trajectory. We have described how we have analyzed

the $P–A$ trajectory in terms of this optical model to derive the fitting parameters involving a number of sublayers, each with its own index of refraction, $n - ik$, and thickness, Th. What we do in this section is discuss the physical mechanisms that can explain the origins of the various regions of the $P–A$ trajectory. In some cases, there are alternative physical explanations for a particular region, that are consistent with the derived fitting parameters of the optical model. While we may feel that one explanation is more likely than another, we cannot always definitively choose which is the correct physical explanation.

As we mentioned in Section 3.2.1, an initial anomalous part of the $P–A$ trajectory is always observed when P, Ni, or Ti is initially being deposited on an Si substrate. This initial anomalous trajectory has to be fitted with a number of sublayers, each with its own value of $n - ik$ and thickness, Th, to obtain a thickness for this portion of the layer that is within the range expected from the growth rate. The fitted $n - ik$ for the first sublayer has a value slightly larger than that of Si and the fitted $n - ik$ values increase through the sublayers associated with this region until the final fitted $n - ik$ value is close to that of the refractive index of a bulk layer of the material being deposited. One physical mechanism that can be used to explain this phenomenon is intermixing between the material being deposited and the Si substrate. Initially, there is only a small amount of the metal being deposited onto the Si and obviously this intermixing layer would behave almost like Si. As more metal is deposited, this intermixing layer starts to be dominated by the deposited metal and thus the $n - ik$ is expected to approach the value of the bulk index of the deposited metal. This graded interface may be a combination of ion implantation initially of the metal into the Si, followed by diffusion of Si into the metal as a thicker layer of metal is deposited. In our optical measurements, we would see this as the growth of a series of sublayers with $n - ik$ varying between that of Si and that of the bulk metal. For a thick Si substrate, the ellipsometric measurements cannot distinguish between a graded layer being formed below the initial surface of the Si or above the initial surface of the Si, because it cannot detect whether the bulk Si wafer is being thinned.

The second possible physical mechanism which can be used to explain the initial anomalous $P–A$ trajectory is that the initial layer is discontinuous. We believe that this physical mechanism is most likely the origin of the initial anomaly in the $P–A$ trajectory when a metal layer is initially deposited on an Si substrate. This physical mechanism was suggested in papers by Yamamoto and Namioka [28], and Yamamoto and Arai [29]. They believe that the metal initially deposited on the substrate forms islands. As more metal is deposited, the metal remains in the form of separate islands that just keep getting larger until a certain average thickness is reached at which these islands start to coalesce to form a continuous film. They confirmed their belief by depositing a number of gold layers on different Si substrates with layer thickness ranging from very thin, i.e., when (in our terminology) the $P–A$ trajectory behaves anomalously, to thick enough such that the $P–A$ trajectory behaves normally. Then, TEM was used to examine the gold layers on the Si substrates.

They found that for a very thin layer, the islands are small and farther apart and these islands start to coalesce to form a continuous film when the gold layer becomes thicker. This physical mechanism also will give the fitted $n - ik$ of the first sublayer having a value close to that of Si, since the islands are initially far apart and cover only a small portion of the Si surface. The fitted $n - ik$ would also approach the value of the bulk index of refraction of the metal as the islands coalesce to form a continuous film.

This second physical mechanism can also explain the $n - ik$ of the sublayers used to fit the initial anomalous $P-A$ trajectory better than that of an intermixing mechanism for the following reason. For an intermixing mechanism, we expect the fitted values of n and k to decrease or increase from the value of Si to that of the deposited layer, depending on whether the deposited layer has n and k values that are less or greater than that of Si, respectively. This is what we observed in the fitted k values but not in the fitted n values for the initial anomalous trajectory. For example, when a Pt layer is initially deposited on Si, we expected the fitted n value to decrease from that of Si to that of Pt, but instead we observed the value of n to first increase to a value greater than that of Si before it quickly decreases to the value of Pt. This behavior of the fitted n value agrees with the predictions of the Lorentz–Lorenz model which is used to calculate the effective complex index of refraction due to island formation growth. It should be remembered that this model is correct only up to a volume fraction of 20%. This model indeed predicts [6] an increase in the n and k value greater than that Si for a deposited metal layer up to 20% of the volume fraction. For the volume fraction of a deposited layer over 20%, the EMA model is used to calculate the effective n and k. This model indeed predicts [6] a decrease in n from Si to the value of Pt, while k always increases from the value of Si to the value of Pt.

From the fitted $n - ik$ in Tables III–V it is apparent that the fitted $n - ik$ values for the same layer vary as the deposited layer gets thicker. This variation in $n - ik$ for the same layer is best explained by the surface roughness mechanism. Surface roughness can be modeled as a layer which consists of a large fraction of volume which is void of constituents (or normally it is referred to as void) and a small fraction of the volume occupied by the deposited and underlying metal, and so it has its own effective $n - ik$, which is best calculated by the EMA (effective medium approximation) model. According to this model, the value for the effective k should be decreasing as the surface becomes rougher but the effective n value is decreasing or increasing depending on the degree of roughness and the type of material.

In fact, the change in n and k caused by the surface roughness can explain why there can be a sharp break in the $P-A$ trajectory, with the $P-A$ trajectory suddenly moving in an opposite direction as in the case for Pt in Figure 24b. When a metal layer is thick, the $P-A$ trajectory is near the end point for that metal and so the $P-A$ trajectory is insensitive to an additional deposit of material with the same index of refraction. However, when the $P-A$ trajectory is at or close to the end point of the underlying metal, an additional layer that has a slightly different index of refraction, $n - ik$, from that of the underlying metal can cause the $P-A$ trajectory to deviate sharply in a new direction just as if we were depositing a new layer of different material. We have observed this extreme sensitivity of the $P-A$ trajectory near the end point of a layer as discussed in Section 3.1 for the $P-A$ trajectories associated with buffer layers and quantum wells. In this case, the added layer with a different index of refraction from the underlying layer can be considered to be due to surface roughness as predicted by the EMA model.

We proposed that the divergence of a $P-A$ trajectory or the final anomalous region of a trajectory is caused by surface roughness. The fitted $n - ik$ of the sublayers for this part of the $P-A$ trajectory is supported by the EMA model. Next, we show the physical evidence that the final part of the $P-A$ trajectory is indeed the result of surface roughness. In this investigation, we have produced samples consisting of single layers of Ti, Nt, and Pt on single crystal (100) Si. There were two samples produced for each material. The thicknesses of these two samples were monitored by the **EXACTA 2000** ellipsometer such that the thickness for one sample was large enough to produce a kink in the $P-A$ trajectory or a divergence of the $P-A$ trajectory from its end point, and the thickness for the other sample was small enough that a kink was not observed in the $P-A$ trajectory. Since there were three different metals, six samples were produced in this investigation. The surfaces of these samples were then scanned by an atomic force microscope (AFM). The mean roughness of these samples is shown in Table VI. From the results in Table VI, we can make the following observations. The surface did become rougher after the divergence of the $P-A$ trajectory from its end point (i.e., a large layer thickness) than before the kink was produced in the $P-A$ trajectory (i.e., a small layer thickness). Second, the rate of increase of surface roughness with thickness is highest for Ti, is intermediate for Ni, and is lowest for Pt. This corresponds to the final anomalous $P-A$ trajectory that is largest for Ti, is intermediate for Ni, and is smallest for Pt. From the fitting parameters of the $P-A$ trajectory for Ti layers on Si substrates of samples X131D and X132A, the fitted $n - ik = 3.58 - i3.90$ for the normal part of the trajectory is decreased to $n - ik = 2.45 - i2.90$ for the last sublayer used to fit the final part of the $P-A$ trajectory. In the process of fitting the anomalous $P-A$ trajectory for sample X131D, as one goes from $n - ik = 3.58 - i3.90$ at the start to $n - ik = 2.45 - i2.90$ at the end, one of the

Table VI.   Surface Roughness Measured by AFM

Samples	Layer thickness (Å)	Mean roughness (Ra) in Å
X131D (Ti layer)	1233	48.3
X132A (Ti layer)	313	9.45
X132B (Pt layer)	375	2.46
X132C (Pt layer)	1430	9.91
X132D (Ni layer)	1441	13.64
X132E (Ni layer)	471	3.64

sublayers was fitted with the value of $n - ik = 3.22 - i3.60$ which is very close to the value of $n - ik = 3.23 - i3.62$ given in the literature for Ti. This suggests that it is possible that the complex index of refraction for Ti reported in the literature might have been measured on a rough surface of Ti. Similarly, from the fitted parameters [6] of the P–A trajectory for Pt layers on Si substrates of samples X132B and X132C, the fitted $n - ik = 2.50 - i5.32$ for the normal part of the P–A trajectory decreases to $n - ik = 2.412 - i5.230$ when the surface becomes rough. Similarly, from the fitted parameters [6] of the P–A trajectory for Ni layers on Si substrates of samples X132D and X132E, the fitted $n - ik = 2.15 - i4.04$ for the normal part of the trajectory is decreasing to $n - ik = 1.885 - i3.720$ for the last sublayer used to fit the final anomalous part of the P–A trajectory. One of the sublayers in the final anomalous part of the P–A trajectory for sample X132D is fitted with $n - ik = 1.965 - i3.83$ which is close to the value of $n - ik = 1.97 - i3.72$ given in the literature for Ni. Again, this suggests that the discrepancy between the $n - ik$ value used to fit the normal part of the P–A trajectory for Ni and the value of $n - ik$ given in the literature for Ni might be because the value given in the literature was measured on a rough surface of Ni. The change in the fitted $n - ik$ from normal to the final anomalous part of the P–A trajectory is largest for Ti, is intermediate for Ni, and is smallest for Pt. Thus, the change of the fitted $n - ik$ from the normal to the final anomalous part of the trajectory is consistent with the surface roughness measured by the AFM for the three materials Ti, Ni, and Pt. This is very strong evidence that the final anomalous part of the P–A trajectory is the result of surface roughness.

Finally, we discuss the physical implications of our optical model results at the interface between an overlayer of one metal and an underlayer of another metal. When the anomaly is small in the initial anomalous P–A trajectory, such as is observed for Ti on Pt and Pt on Ni, the fitted $n - ik$ of the overlayer has a value close to that of the bulk index of refraction of the deposited material. This can be a good indication that there is a sharp interface between these two materials. For example, the first sublayer of the second Pt layer on the Ni layer of sample EL2 has an $n - ik$ close to that of the bulk index of refraction of Pt. It can be considered that for this layer there is no initial anomalous region: the first sublayer represents the normal region of the P–A trajectory; the rest of the sublayers are for fitting the final anomalous P–A trajectory. However, for an initial deposition of Pt on Ti and Ni on Ti, the initial anomaly in the P–A trajectory is large as shown in Figure 24b and Figure 25b. The fitted $n - ik$ for the sublayers in the optical model that characterize these initial anomalous P–A trajectories start with values that are close to that of Ti and approach that of the other metal at the end of the initial anomalous P–A trajectory. One can attempt to explain these initial anomalous trajectories involving an overlayer on an underlayer by either the intermixing mechanism or the island growth mechanism that were discussed for the case of metal deposited on Si. However, there are problems with both these mechanisms, since it is not immediately apparent how they can explain the significant difference in the

optical results with the different order of deposition of an interface. For example, if we attribute the initial anomalous trajectory at an interface between two metals to intermixing, we get intermixing for Pt deposited on Ti and Ni deposited on Ti, but no intermixing for Ti deposited on Pt and Pt deposited on Ni. If the intermixing is due to ion implantation and diffusion, it is not apparent why materials will only diffuse in one direction and will not diffuse in the other direction. Similarly, if we attribute the initial anomalous P–A trajectory to island formation, then we get island formation for Pt deposited on Ti and Ni deposited on Ti, but no island formation for Ti deposited on Pt and Pt deposited on Ni. Since if we attribute the initial anomalous trajectory on Si to island formation, then all three metals Pt, Ti, and Ni are observed to form islands when deposited on Si. It is not apparent why island formation is inhibited when Ti is deposited on Pt or Pt is deposited on Ni. Obviously, this is a topic in this study of metallic multilayer structures that requires more experimental and theoretical work.

4. CONCLUSIONS

We have shown that *in situ* single-wavelength ellipsometry with a fast-nulling Faraday-modulated ellipsometer like the **EX-ACTA 2000** has the capabilities to give much information about the deposition of semiconductor, metal, and dielectric films.

1. From fitting the P–A trajectory of CdTe-CdMnTe superlattice structures, we have found that the determined thickness of the layers is in agreement with the thickness obtained by TEM, within the experimental accuracy of $\pm 10\%$ for both techniques. We have also shown that the ellipsometrically determined deposition rate agrees with the expected deposition rate for magnetron sputtered Ti, Ni, and Pt films. Both of these results show that *in situ* ellipsometry and our approach to fitting the P–A trajectory gives a thickness for such semiconductor and metal layers in agreement with other standard measurement techniques.
2. We have shown that the P–A trajectories associated with CdTe-CdMnTe superlattice structures can only be fitted with a model whereby each CdTe or CdMnTe layer consists of two layers with slightly different indices of refraction.
3. We have shown that in the P–A trajectory associated with the deposition of a CdTe or CdMnTe buffer layer on a GaAs or CdZnTe substrate there is always an initial anomalous trajectory associated with initial strain in the layer. We have shown that in the case of a GaAs substrate, because the difference in the complex index of refraction between GaAs and CdTe or CdMnTe is so large, the initial anomalous trajectory amounts to only a slight initial flattening on an otherwise very large spiral P–A trajectory, and this initial anomalous part of the trajectory can be fitted

with a single $n - ik$. On the other hand, in the case of a CdZnTe substrate, because the difference in the complex index of refraction between CdZnTe and CdTe or CdMnTe is so small, the resulting $P-A$ trajectory is small, and the initial anomalous trajectory is quite noticeable and has to be fitted with a number of sublayers with different indices of refraction.

4. We have shown that we can measure the thickness of metallic multilayer structures, consisting of sequential layers of Pt, Ni, and Ti deposited on top of each other on a single crystal silicon substrate. We have shown that in general the $P-A$ trajectory associated with any layer can be divided up into three regions: the initial anomalous region, the normal region, and the final anomalous region.

5. The initial anomalous region of the $P-A$ trajectory of the first metallic layer on the single crystal Si substrate can be explained by either intermixing or island formation growth. However, the Lorentz–Lorenz model for the effective index of refraction in such a situation and previous work on Au layers by other researchers suggests that the initial anomalous region of the $P-A$ trajectory in our case is due to island formation growth.

6. The cause of the initial anomalous region of the $P-A$ trajectory at the interface between two metal layers has not yet been established. The $P-A$ trajectories show that some of these interfaces are sharp (i.e., no initial anomalous region) and some of the interfaces are diffuse (i.e., an initial anomalous region). We cannot yet explain why one obtains a sharp interface when Ti is deposited on Pt, Pt is deposited on Ni, or Ni is deposited on Pt, whereas one gets a diffuse interface when Pt or Ni is deposited on Ti.

7. The normal region of the $P-A$ trajectory for these metallic layers is associated with the deposition of a uniform layer with a single index of refraction. This normal region of a trajectory can be fitted very well to give an excellent value for the thickness of this region and hence an excellent value for the layer growth rate.

8. The final anomalous part of the trajectory is associated with surface roughness. We have shown that the effect of surface roughness on the complex index of refraction is described well by the EMA model. By direct measurement with an atomic force microscope, we have shown that the rate of increase of surface roughness with thickness is highest for Ti, is intermediate for Ni, and is lowest for Pt. The change of the fitted $n - ik$ from the normal to the final anomalous part of the trajectory is consistent with the surface roughness measured by the AFM for the three materials Ti, Ni, and Pt. This is very strong evidence that the final anomalous part of the $P-A$ trajectory is the result of surface roughness.

9. We have developed rules for the applicability of the pseudosubstrate approximation. With the help of this

approximation, the fit of a $P-A$ trajectory can be done faster, more efficiently, and accurately.

10. Finally, although we have not shown any experimental results for the case of dielectrics, only a simulation, *in situ* ellipsometry might have its greatest application in measuring the thickness of dielectric layers in precision optical filters. Such dielectric layers should have none of the complications that we have encountered with CdTe-CdMnTe superlattices or Pt/Ti/Ni multilayer structures, so *in situ* ellipsometry should be able to measure the thickness of such layers with great precision.

APPENDIX A: METHOD OF SUMMATION

The multiple reflections and transmissions by interfaces of a single thin film are shown in Figure 2. The Fresnel coefficients for propagation of light from n_0 to n_1 are denoted by r_1 and t_1 and from n_1 to n_2 by r_2 and t_2. The Fresnel coefficients for propagation of light from n_1 to n_0 are denoted by r_1' and t_1' and from n_2 to n_1 are denoted by r_2' and t_2'. When the incident beam strikes at the first interface, the beam is divided into reflected and transmitted parts with amplitudes r_1 and t_1, respectively. The transmitted part of the beam is then incident on the second interface and again is split into reflected and transmitted parts with amplitudes $t_1 r_2$ and $t_1 t_1'$, respectively. Such a division occurs each time the beam strikes the interfaces and the reflected and transmitted parts are further divided into other reflected and transmitted parts with smaller amplitudes. The process keeps repeating as shown in Figure 2a. The amplitudes of the successive beams reflected into the medium n_0 are thus given by r_1, $t_1 t_1' r_2$, $t_1 t_1' r_2^2 r_1'$, $t_1 t_1' r_2^3 r_1'^2$, $t_1 t_1' r_2^4 r_1'^3$, ... The amplitudes of the successive beams transmitted into the medium n_1 will not be presented here since only reflected coefficients are essential in null ellipsometry. Before summing up all the reflected amplitudes, the phase change of the beam due to its traversal in the film must be determined.

The path difference of the reflected beams between the two interfaces is shown in Figure A1. The refractive indices of the

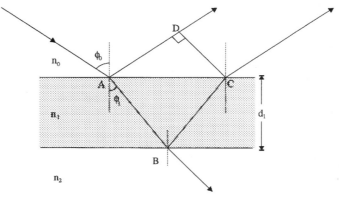

Fig. A1.   Illustration of optical path difference of the reflected beams emerging from a film in between two semi-infinite media n_0 and n_1.

media are n_0, n_1, and n_2 and the thickness of the film is denoted by d_1. The reflected beams from the interfaces shown in the figure imply that the optical path difference (**PD**) for these beams between the two interfaces should be **PD** $= n_1(AB + BC) - n_0 AD$. In addition, from Figure A1 the following relationships can be deduced:

$$AB = BC = d_1/\cos\phi_1$$

$$AC = 2AB\sin\phi_1 \quad\text{and}\quad AD = AC\sin\phi_0 \quad\text{(A1)}$$

By substituting the appropriate expressions given in (A1), the path difference can be written as

$$\mathbf{PD} = \frac{2d_1 n_1}{\cos\phi_1} - \frac{2d_1 n_0 \sin\phi_0 \sin\phi_1}{\cos\phi_1} \quad\text{(A2)}$$

Using Snell's law, $n_0 \sin\phi_0 = n_1 \sin\phi_1$, (A2) can be written as

$$\begin{aligned}
\mathbf{PD} &= \frac{2d_1 n_1}{\cos\phi_1} - \frac{2d_1 n_1 \sin^2\phi_1}{\cos\phi_1} \\
&= 2d_1 n_1 \left(\frac{1 - \sin^2\phi_1}{\cos\phi_1}\right) \\
&= 2d_1 n_1 \cos\phi_1 \quad\text{(A3)}
\end{aligned}$$

The phase difference, $2\delta_1$, of the beams due to its total traversal in the film is defined as $(2\pi/\lambda)\mathbf{PD}$ and thus is given by the following expression,

$$2\delta_1 = \frac{4\pi}{\lambda} n_1 d_1 \cos\phi_1 \quad\text{(A4)}$$

Using either Eq. (4) or (5), we have $r_1 = -r_1'$. The total reflected amplitude including the phase difference is then given by

$$\begin{aligned}
R &= r_1 + t_1 t_1' r_2 e^{-2i\delta_1} - t_1 t_1' r_1 r_2^2 e^{-4i\delta_1} + t_1 t_1' r_1^2 r_2^3 e^{-6i\delta_1} - \cdots \\
&= r_1 + t_1 t_1' r_2 e^{-2i\delta_1} \{1 - (r_1 r_2 e^{-2i\delta_1}) + (r_1 r_2 e^{-2i\delta_1})^2 \\
&\quad - (r_1 r_2 e^{-2i\delta_1})^3 + \cdots\} \quad\text{(A5)}
\end{aligned}$$

The second factor of the second term of Eq. (A5) is equivalent to the geometric series,

$$\frac{1}{1+x} = 1 - x + x^2 - x^3 + x^4 - \cdots \quad\text{(A6)}$$

Thus, the total reflected amplitude given by (A5) can be simplified and written as

$$R = r_1 + \frac{(1 - r_1^2) r_2 e^{-2i\delta_1}}{1 + r_1 r_2 e^{-2i\delta_1}} \quad\text{(A7)}$$

Conservation of energy implies that $t_1 t_1' = 1 - r_1$. This relation can be shown by using Eqs. (4) and (6) or Eqs. (5) and (7). Equation (A7) can be further simplified to

$$R = \frac{r_1 + r_2 e^{-2i\delta_1}}{1 + r_1 r_2 e^{-2i\delta_1}} \quad\text{(A8)}$$

The total reflected amplitudes for the s and p polarizations of light can thus be written

$$R_s = \frac{r_{1s} + r_{2s} e^{-2i\delta_1}}{1 + r_{1s} r_{2s} e^{-2i\delta_1}} \quad\text{(A9)}$$

$$R_p = \frac{r_{1p} + r_{2p} e^{-2i\delta_1}}{1 + r_{1p} r_{2p} e^{-2i\delta_1}} \quad\text{(A10)}$$

Equations (A9) and (A10) have been written in the main body of the chapter as Eqs. (8) and (9), respectively.

APPENDIX B: METHOD OF RESULTANT WAVES

B1. Single Layer on a Substrate

This method uses the vector sums of the reflected and transmitted waves and applies the appropriate boundary conditions at the interfaces to derive the reflection coefficients. A wave incident on a film of index n_1 between media of indices n_0 and n_2 is shown in Figure 2b. For convenience, all waves traveling toward medium n_2 are defined as incoming waves and are denoted with a superscript i while all waves traveling toward medium n_0 are called outgoing waves and are denoted with a superscript r. In medium n_0, the incident electric vector is defined as E_0^i and that of the reflected electric vector is E_0^r. The subscript on E is to indicate the type of medium. The electric vectors for the different media are clearly illustrated in Figure 2b. Inside the film, the sum of all the incoming and outgoing waves (in the previous method of summation) can be represented by the electric vectors E_1^i and E_1^r, respectively. In the third medium, there are only incoming waves and they sum to an electric vector E_2^i. In writing down the components of the electric and magnetic vectors, it is necessary to distinguish the plane of polarization of the waves and this is accomplished by adding the suffix p or s to the subscript.

Similar to the case of a single interface, the phase factors associated with the waves in the three media can be written as

$$\exp\left[i\left(\omega t - \frac{2\pi n_0 x \sin\phi_0}{\lambda} \pm \frac{2\pi n_0 z \cos\phi_0}{\lambda}\right)\right]$$
for medium n_0

$$\exp\left[i\left(\omega t - \frac{2\pi n_1 x \sin\phi_1}{\lambda} \pm \frac{2\pi n_1 z \cos\phi_1}{\lambda}\right)\right]$$
for medium n_1 $\quad$ (B1)

$$\exp\left[i\left(\omega t - \frac{2\pi n_2 x \sin\phi_2}{\lambda} \pm \frac{2\pi n_2 z \cos\phi_2}{\lambda}\right)\right]$$
for medium n_2

The upper sign indicates the wave is traveling toward n_0 while the lower sign indicates the wave is traveling toward n_2. Again, only z components of the arguments of the phase factors will appear in the discussion and $(2\pi n_m \cos\phi_m)/\lambda$ will be denoted by α_m. Based on the orientation defined in Figure 2b, the electric and magnetic vectors in the x and y directions for the three media can be written as

$$E_{0x} = \left(E_{0p}^i e^{-i\alpha_0 z} - E_{0p}^r e^{i\alpha_0 z}\right)\cos\phi_0$$

$$E_{0y} = E_{0s}^i e^{-i\alpha_0 z} + E_{0s}^r e^{i\alpha_0 z}$$

$$H_{0x} = \left(-E_{0s}^i e^{-i\alpha_0 z} + E_{0s}^r e^{i\alpha_0 z}\right)n_0\cos\phi_0 \tag{B2}$$

$$H_{0y} = \left(E_{0p}^i e^{-i\alpha_0 z} + E_{0p}^r e^{i\alpha_0 z}\right)n_0$$

$$E_{1x} = \left(E_{1p}^i e^{-i\alpha_1 z} - E_{1p}^r e^{i\alpha_1 z}\right)\cos\phi_1$$

$$E_{1y} = E_{1s}^i e^{-i\alpha_1 z} + E_{1s}^r e^{i\alpha_1 z}$$

$$H_{1x} = \left(-E_{1s}^i e^{-i\alpha_1 z} + E_{1s}^r e^{i\alpha_1 z}\right)n_1\cos\phi_1 \tag{B3}$$

$$H_{1y} = \left(E_{1p}^i e^{-i\alpha_1 z} + E_{1p}^r e^{i\alpha_1 z}\right)n_1$$

$$E_{2x} = E_{2p}^i e^{-i\alpha_2 z}\cos\phi_2$$

$$E_{2y} = E_{2s}^i e^{-i\alpha_2 z}$$

$$H_{2x} = -E_{2s}^i e^{-i\alpha_2 z}n_2\cos\phi_2 \tag{B4}$$

$$H_{2y} = E_{2p}^i e^{-i\alpha_2 z}n_2$$

At the interfaces $z = 0$ and $z = d_1$, the tangential components of the electric and magnetic vectors are continuous and by equating these tangential components the following relationships are obtained. Then,

$$\left(E_{0p}^i - E_{0p}^r\right)\cos\phi_0 = \left(E_{1p}^i - E_{1p}^r\right)\cos\phi_1$$

$$\left(E_{0p}^i + E_{0p}^r\right)n_0 = \left(E_{1p}^i + E_{1p}^r\right)n_1$$

$$z = 0 \tag{B5}$$

$$E_{0s}^i + E_{0s}^r = E_{1s}^i + E_{1s}^r$$

$$\left(-E_{0s}^i + E_{0s}^r\right)n_0\cos\phi_0 = \left(-E_{1s}^i + E_{1s}^r\right)n_1\cos\phi_1$$

$$z = 0 \tag{B6}$$

$$\left(E_{1p}^i e^{-i\alpha_1 d_1} - E_{1p}^r e^{i\alpha_1 d_1}\right)\cos\phi_1 = E_{2p}^i e^{-i\alpha_2 d_1}\cos\phi_2$$

$$\left(E_{1p}^i e^{-i\alpha_1 d_1} + E_{1p}^r e^{i\alpha_1 d_1}\right)n_1 = E_{2p}^i e^{-i\alpha_2 d_1}n_2$$

$$z = d_1 \tag{B7}$$

$$E_{1s}^i e^{-i\alpha_1 d_1} + E_{1s}^r e^{i\alpha_1 d_1} = E_{2s}^i e^{-i\alpha_2 d_1}$$

$$\left(-E_{1s}^i e^{-i\alpha_1 d_1} + E_{1s}^r e^{-i\alpha_1 d_1}\right)n_1\cos\phi_1 = E_{2s}^i e^{-i\alpha_2 d_1}n_2\cos\phi_2$$

$$z = d_1 \tag{B8}$$

It should be noted that the tangential components of the magnetic vectors are continuous across the interface only when the current densities of the media are zero. With some additional effort, we could use Eqs. (B5)–(B8) to derive equations equivalent to Eqs. (A9) and (A10). However, we want to move directly on to the main use of the method of resultant waves, i.e., dealing with multilayer structures. It can be seen from Eqs. (B5)–(B8) that the application of the boundary conditions to the interface couples the electric vectors in one medium to the electric vectors in the preceding medium. We will see that this fact will simplify the solution of a system consisting of multiple layers in such a way that the reflection coefficients can be determined efficiently, especially with the help of a computer program.

B2. Reflection Coefficients of Multilayer Structures

The method of resultant waves will be shown to give a simple and elegant solution to the problem of multilayer structures. The stack of multiple layers and their labels are shown in Figure B1. There are m layers in between the two semi-infinite media n_0 and n_{m+1}. The interface between the $(j-1)$th and jth layers is denoted by z_j. The thickness of a layer is in the z direction and in general the thickness of the jth layer is defined as $d_j = z_j - z_{j-1}$. The tangential components of the electric and magnetic vectors for the jth layer can be generalized from Eqs. (B3). Similar to the case of a single layer, there are two cases to be discussed, the case of the p state of polarization (TM waves) and the s state of polarization (TE waves).

For the case of TM waves, the electric and magnetic vectors can be written as

$$E_{jx} = \left(E_{jp}^i e^{-i\alpha_j z} - E_{jp}^r e^{i\alpha_j z}\right)\cos\phi_j$$

$$H_{jy} = \left(-E_{jp}^i e^{-i\alpha_j z} + E_{jp}^r e^{i\alpha_j z}\right)n_j \tag{B9}$$

where $\alpha_j = (2\pi n_j\cos\phi_j)/\lambda$. For simplicity, z_{j-1} is assumed to be zero and hence z_j is equal to d_j. By applying the boundary conditions for electric and magnetic vectors to the interfaces z_{j-1} and z_j, similar relations to Eqs. (B5) and (B7) can be ob-

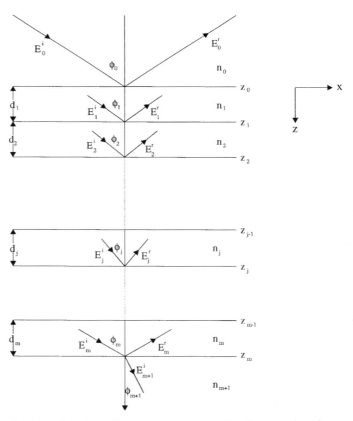

Fig. B1. Illustration of incoming and outgoing electric vectors in m layers between two semi-infinite media n_0 and n_{m+1}.

tained. These relations are

$$\left(E_{j-1p}^i - E_{j-1p}^r\right)\cos\phi_{j-1} = \left(E_{jp}^i - E_{jp}^r\right)\cos\phi_j$$
$$\left(E_{j-1p}^i + E_{j-1p}^r\right)n_{j-1} = \left(E_{jp}^i + E_{jp}^r\right)n_j$$
$$z_{j-1} = 0 \tag{B10}$$
$$\left(E_{jp}^i e^{-i\alpha_j d_j} - E_{jp}^r e^{i\alpha_j d_j}\right)\cos\phi_j = E_{j+1p}^i e^{-i\alpha_{j+1}d_1}\cos\phi_{j+1}$$
$$\left(E_{jp}^i e^{-i\alpha_j d_j} + E_{jp}^r e^{i\alpha_j d_j}\right)n_j = E_{j+1p}^i e^{-i\alpha_{j+1}d_1}n_{j+1}$$
$$z_j = d_j \tag{B11}$$

It should be noted that the left-hand side of Eq. (B10) are the tangential components of E and H at $z = z_{j-1}$ (i.e., 0) and the right-hand side of Eq. (B11) are the tangential components of E and H at $z = z_j$ (i.e., d_j). For convenience, we denote the tangential components of E and H at $z = z_{j-1}$ as $\overline{E}_{j-1}$ and $\overline{H}_{j-1}$ and at $z = z_j$ as $\overline{E}_j$ and $\overline{H}_j$. Using Eqs. (B10) and (B11), E_{jp}^i and E_{jp}^r can be eliminated. By using the notation $\overline{E}$ and $\overline{H}$ for the tangential components of E and H, respectively, at $z = z_{j-1}$ and $z = z_j$, and by writing $e^{\pm i\alpha_j d_j} = \cos\alpha_j d_j \pm i\sin\alpha_j d_j$, the following expressions can be derived

$$\overline{E}_{j-1} = (\cos\alpha_j d_j)\overline{E}_j + (iq_j\sin\alpha_j d_j)\overline{H}_j$$
$$\overline{H}_{j-1} = \left(i\frac{\sin\alpha_j d_j}{q_j}\right)\overline{E}_j + (\cos\alpha_j d_j)\overline{H}_j \tag{B12}$$

where $q_j = \cos\phi_j/n_j$.

For the case of TE waves, the tangential electric and magnetic vectors are

$$E_{jy} = \left(E_{js}^i e^{-i\alpha_j z} - E_{js}^r e^{i\alpha_j z}\right)$$
$$H_{jx} = \left(-E_{js}^i e^{-i\alpha_j z} + E_{js}^r e^{i\alpha_j z}\right)n_j\cos\phi_j \tag{B13}$$

Similar to the case of TM waves, by applying the boundary conditions to the interfaces the following expressions are obtained

$$E_{j-1s}^i + E_{j-1s}^r = E_{js}^i + E_{js}^r$$
$$\left(-E_{j-1s}^i + E_{j-1s}^r\right)n_{j-1}\cos\phi_{j-1} = \left(-E_{js}^i + E_{js}^r\right)n_j\cos\phi_j$$
$$z_{j-1} = 0 \tag{B14}$$
$$E_{js}^i e^{-i\alpha_j d_j} + E_{js}^r e^{i\alpha_j d_j} = E_{j+1s}^i e^{-i\alpha_{j+1}d_1}$$
$$\left(-E_{js}^i e^{-i\alpha_j d_j} + E_{js}^r e^{i\alpha_j d_j}\right)n_j\cos\phi_j$$
$$= \left(-E_{j+1s}^i e^{-i\alpha_{j+1}d_1} + E_{j+1s}^r e^{i\alpha_{j+1}d_1}\right)n_{j+1}\cos\phi_{j+1}$$
$$z_j = d_j \tag{B15}$$

Following the same notation and procedure as for the TM case, the following expressions can be obtained

$$\overline{E}_{j-1} = (\cos\alpha_j d_j)\overline{E}_j + \left(-\frac{i}{p_j}\sin\alpha_j d_j\right)\overline{H}_j$$
$$\overline{H}_{j-1} = (-ip_j\sin\alpha_j d_j)\overline{E}_j + (\cos\alpha_j d_j)\overline{H}_j \tag{B16}$$

where $p_j = n_j\cos\phi_j$.

Equations (B12) and (B16) show that $\overline{E}_{j-1}$, $\overline{H}_{j-1}$ are linearly dependent on $\overline{E}_j$, $\overline{H}_j$ and thus these equations can be

written in matrix form as

$$\begin{bmatrix} \overline{E}_{j-1} \\ \overline{H}_{j-1} \end{bmatrix} = \begin{bmatrix} \cos\alpha_j d_j & iq_j\sin\alpha_j d_j \\ i\dfrac{\sin\alpha_j d_j}{q_j} & \cos\alpha_j d_j \end{bmatrix} \begin{bmatrix} \overline{E}_j \\ \overline{H}_j \end{bmatrix}$$

(TM or p case) $\qquad$ (B17)

and

$$\begin{bmatrix} \overline{E}_{j-1} \\ \overline{H}_{j-1} \end{bmatrix} = \begin{bmatrix} \cos\alpha_j d_j & -i\dfrac{\sin\alpha_j d_j}{p_j} \\ -ip_j\sin\alpha_j d_j & \cos\alpha_j d_j \end{bmatrix} \begin{bmatrix} \overline{E}_j \\ \overline{H}_j \end{bmatrix}$$

(TE or s case) $\qquad$ (B18)

Equations (B17) and (B18) indicate that each layer in a multilayer structure can be represented by a characteristic matrix which links components of the electric and magnetic vectors in one layer to the preceding layer. Hence, the components $\overline{E}_0$, $\overline{H}_0$ may be expressed in terms of $\overline{E}_j$, $\overline{H}_j$ and the reflection coefficients can be determined. If the characteristic matrix in Eqs. (B17) and (B18) of the jth layer is denoted by $M_j(d_j)$, then for the case of m layers sandwiched in between two semi-infinite media n_0 and n_{m+1} as shown in Figure B1, the components $\overline{E}_0$, $\overline{H}_0$ may thus be expressed in terms of $\overline{E}_{m+1}$, $\overline{H}_{m+1}$ as

$$\begin{bmatrix} \overline{E}_0 \\ \overline{H}_0 \end{bmatrix} = M \begin{bmatrix} \overline{E}_{m+1} \\ \overline{H}_{m+1} \end{bmatrix} \tag{B19}$$

where $M = M_1(d_1) * M_2(d_2)\cdots M_j(d_j)\cdots M_m(d_m)$. It should be noted that there are only incoming waves in the semi-infinite media and thus the second terms on the right-hand side of Eqs. (B10), (B11), (B14), and (B15) vanish. Following the notation described earlier, $\overline{E}_0$ and $\overline{H}_0$ are the tangential components of E_0 and H_0 at $z = 0$ and $\overline{E}_{m+1}$ and $\overline{H}_{m+1}$ are the tangential components of E_{m+1} and H_{m+1} at $z = d_m$. Now, it is also necessary to differentiate the state of polarization and this is accomplished by adding the superscript p or s to the matrix variable M and the subscript p or s to the other appropriate variables. Hence, Eq. (B19) can be rewritten as

$$\begin{bmatrix} (E_{0p}^i - E_{0p}^r)\cos\phi_0 \\ (E_{0p}^i + E_{0p}^r)n_0 \end{bmatrix} = M^p \begin{bmatrix} E_{m+1p}^i e^{-i\alpha_{m+1}d_m}\cos\phi_{m+1} \\ E_{m+1p}^i e^{-i\alpha_{m+1}d_m}n_{m+1} \end{bmatrix}$$

(TM or p case) $\qquad$ (B20)

$$\begin{bmatrix} (E_{0s}^i + E_{0s}^r) \\ (-E_{0s}^i - E_{0s}^r)n_0\cos\phi_0 \end{bmatrix}$$
$$= M^s \begin{bmatrix} E_{m+1s}^i e^{-i\alpha_{m+1}d_m} \\ E_{m+1s}^i e^{-i\alpha_{m+1}d_m}n_{m+1}\cos\phi_{m+1} \end{bmatrix}$$

(TE or s case) $\qquad$ (B21)

One should remember that $r_p = E_{0p}^r/E_{0p}^i$, $t_p = (E_{m+1p}^i \times e^{-i\alpha_{m+1}d_m})/E_{0p}^i$, $r_s = E_{0s}^r/E_{0s}^i$, $t_s = (E_{m+1s}^i e^{-i\alpha_{m+1}d_m})/E_{0p}^i$. By appropriate substitutions of r_p and t_p, and r_s and t_s into Eqs. (B20) and (B21), these matrix equations can be written in compact form as shown

$$\begin{bmatrix} (1 - r_p) \cos \phi_0 \\ (1 + r_p)n_0 \end{bmatrix} = M^p \begin{bmatrix} t_p \cos \phi_{m+1} \\ t_p n_{m+1} \end{bmatrix}$$

(TM or p case) (B22)

$$\begin{bmatrix} (1 + r_s) \\ (1 - r_s)n_0 \cos \phi_0 \end{bmatrix} = M^s \begin{bmatrix} t_s \\ t_s n_{m+1} \cos \phi_{m+1} \end{bmatrix}$$

(TE or s case) (B23)

where M^p and M^s are the effective characteristic matrices for the case of TM waves (p polarization) and TE waves (s polarization), respectively. The matrix M^p and M^s can be written as

$$M^p = \begin{bmatrix} m_{11}^p & m_{12}^p \\ m_{21}^p & m_{22}^p \end{bmatrix} \quad \text{and} \quad M^s = \begin{bmatrix} m_{11}^s & m_{12}^s \\ m_{21}^s & m_{22}^s \end{bmatrix}$$ (B24)

By substituting M^p of Eq. (B24) into Eq. (B22) and by performing the matrix multiplication, Eq. (B22) can be written as two linear equations,

$$(1 - r_p) \cos \phi_0 = (m_{11}^p \cos \phi_{m+1} + m_{12}^p n_{m+1})t_p$$ (B25)

$$(1 + r_p)n_0 = (m_{21}^p \cos \phi_{m+1} + m_{22}^p n_{m+1})t_p$$ (B26)

Similarly, Eq. (B23) can be written as

$$(1 + r_s) = (m_{11}^s + m_{12}^s n_{m+1} \cos \phi_{m+1})t_s$$ (B27)

$$(1 - r_s)n_0 \cos \phi_0 = (m_{21}^s + m_{22}^s n_{m+1} \cos \phi_{m+1})t_s$$ (B28)

Next, the transmission coefficients will be eliminated and the expression for the reflection coefficients can be determined. For the case of TM waves (p polarization), Eqs. (B25) and (B26) are divided to eliminate the transmission coefficient and so obtain

$$\frac{(1 - r_p) \cos \phi_0}{(1 + r_p)n_0} = \frac{m_{11}^p \cos \phi_{m+1} + m_{12}^p n_{m+1}}{m_{21}^p \cos \phi_{m+1} + m_{22}^p n_{m+1}}$$ (B29)

Remembering that $q_0 = \cos \phi_0 / n_0$ and $q_{m+1} = \cos \phi_{m+1}/n_{m+1}$, Eq. (B29) can be rewritten as

$$\frac{(1 + r_p)}{(1 - r_p)} = q_0 \frac{q_{m+1}m_{21}^p + m_{22}^p}{q_{m+1}m_{11}^p + m_{12}^p}$$ (B30)

If the right-hand of Eq. (B30) is called c_p, then the reflection coefficient for TM waves (p polarization) can be written in terms of c_p as shown

$$r_p = \frac{c_p - 1}{c_p + 1}$$ (B31)

Similarly, the case for TE waves (s polarization) can be written as

$$\frac{(1 + r_s)}{(1 - r_s)p_0} = \frac{m_{11}^s + m_{12}^s p_{m+1}}{m_{21}^s + m_{22}^s p_{m+1}}$$ (B32)

where $p_0 = n_0 \cos \phi_0$ and $p_{m+1} = n_{m+1} \cos \phi_{m+1}$. If the right side of Eq. (B32) is called c_s/p_0, then the reflection coefficient

for TE waves (s polarization) in terms of c_s is given by

$$r_s = \frac{c_s - 1}{c_s + 1}$$ (B33)

With the forms given by Eqs. (B31) and (B33), the reflection coefficients can be calculated efficiently by a computer program once the optical characteristics of the system being studied are input. Once the reflection coefficients are known, the ellipsometric parameters can be easily derived as shown in Appendix C.

APPENDIX C: RELATION BETWEEN ELLIPSOMETRIC PARAMETERS Δ AND ψ AND THE NULL ELLIPSOMETER VARIABLES P AND A

The relation between the ellipsometer parameters, Δ and ψ, and the null ellipsometer variables, P and A, can be derived as follows. First, linearly polarized light from the laser is passed through the first quarterwave plate oriented so as to produce circularly polarized light. The circularly polarized light is then incident on a polarizer with its transmission axis oriented at angle P with respect to the plane of incidence, so as to produce linearly polarized light oriented at angle P. This linearly polarized light is shown in Figure C1a. It is next incident on the second quarterwave plate with its fast axis oriented at $-45°$ to the plane of incidence and can be decomposed into components with electric vectors parallel to the quarterwave plate axes. The wave plate introduces a 90° phase difference between these components. Hence, the emergent light of the second quarterwave plate is elliptically polarized with semimajor axis a and semiminor axis b. As the angle P is varied, a and b change, but the orientation of the ellipse remains fixed at 45°. The orientation of the elliptically polarized light is depicted as in Figure C1b. E_s^i and E_p^i, the s and p components of the light incident on the surface, are always equal in magnitude. Next, E_s^i and E_p^i are written in terms of the semimajor axis a and semiminor axis b. From Figure C1b, the following expression for E_s^i can be deduced

$$E_s^i = \frac{1}{\sqrt{2}}(a - ib)$$ (C1)

Similarly, the electric vector in the plane of incidence, E_p^i, can be written as

$$E_p^i = \frac{1}{\sqrt{2}}(a + ib)$$ (C2)

By dividing Eq. (C1) by Eq. (C2), the phase difference between E_s^i and E_p^i, which is defined in the previous text as Δ_1, can be determined and is given by

$$\frac{|E_s^i|}{|E_p^i|}e^{i\Delta_1} = \frac{a - ib}{a + ib} = \frac{b + ia}{-b + ia}$$ (C3)

Equation (C3) can be rewritten as

$$\frac{|E_s^i|}{|E_p^i|}e^{i\Delta_1} = \frac{\sqrt{b^2 + a^2}e^{i \tan^{-1}(a/b)}}{\sqrt{b^2 + a^2}e^{-i \tan^{-1}(a/b)}}$$ (C4)

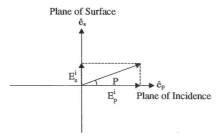

Plane of Surface
$\hat{e}_s$

E_s^i P
E_p^i Plane of Incidence

Output from the Polarizer Prism

(a)

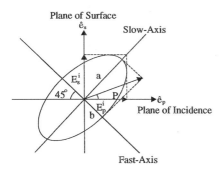

Plane of Surface
$\hat{e}_s$
                        Slow-Axis

E_s^i a
45°   P
         $\hat{e}_p$
E_p^i Plane of Incidence
b

Fast-Axis

Output from Quarter-wave Plate Q_2

(b)

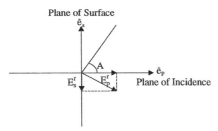

Plane of Surface
$\hat{e}_s$

         A
              $\hat{e}_p$
E_s^r E_p^r Plane of Incidence

Nulling output from Analyzer

(c)

Fig. C1. (a) The output of the light beam from the polarizer is shown. (b) The output of the light beam from the second quarterwave plate is shown. (c) The linearly polarized reflected light is shown.

and hence

$$\Delta_1 = 2\tan^{-1}\left(\frac{a}{b}\right) \tag{C5}$$

If the polarizer angle P is selected such that the ellipticity of the elliptically polarized incident light cancels out the ellipticity induced by the surface, then the reflected light is linearly polarized. The angle of polarization of this reflected light is determined by setting the analyzer to an angle A to produce a null. Hence, the electric vector of the reflected light must be at 90° to the analyzer's transmission axis. The orientations of the electric vector of the reflected light and analyzer's transmission axis are shown in Figure C1c. E_s^r and E_p^r are the s and p components, respectively, of the electric vector of the reflected light. After

the reflection, the components of the reflected light in the plane of incidence and the surface are in phase; i.e., the reflected light is linearly polarized and hence $\Delta_2 = 0$. One can then conclude that

$$\Delta = \Delta_1 - \Delta_2 = 2\tan^{-1}\left(\frac{a}{b}\right) \tag{C6}$$

or

$$\tan\left(\frac{\Delta}{2}\right) = \frac{a}{b} \tag{C7}$$

From Figure C1b, the ratio a to b in terms of P can be deduced and is given by

$$\frac{a}{b} = \frac{\sqrt{a^2 + b^2}\sin(\pi/4 + P)}{\sqrt{a^2 + b^2}\cos(\pi/4 + P)} = \tan\left(\frac{\pi}{4} + P\right) \tag{C8}$$

Comparing Eqs. (C7) and (C8) one can conclude that

$$\Delta = \frac{\pi}{2} + 2P \tag{C9}$$

It is clear from Figure C1c that $\tan A = E_p^r / E_s^r$. Since there is no phase difference between E_s^r and E_p^r and the magnitude of E_s^i and E_p^i are equal, we can write

$$\tan A = \frac{|E_p^r / E_p^i|}{|E_s^r / E_s^i|} = \frac{|r_p|}{|r_s|} \tag{C10}$$

The fundamental equation of ellipsometry shown in Eq. (14) defines that $\tan\psi = |r_p|/|r_s|$ and hence

$$\psi = A \tag{C11}$$

Thus, the ellipsometric parameters, Δ and ψ, are related to the measured angles at a null, P and A, in the simple linear way of Eqs. (C9) and (C11). Since $\Delta = \pi/2 + 2P$ and $\psi = A$, one can consider null ellipsometry as a direct method of measuring the ellipsometric parameters.

From Figure C1, we can see that there are several combinations of P and A that can give a null. For example, one of the possible polarizer angles that can give a reflected linearly polarized light is 90° relative to the P value shown in Figure C1. However, if we restrict the ranges of P and A, the various combinations can be reduced to two zones, normally referred to as zones 1 and 3. For zone 1, we have chosen the range of P between −45 and 135° and A between 0 and 90°. The ellipsometric parameters Δ and ψ related to P and A are given by Eqs. (C9) and (C11). For the zone 3, the ranges of P between −45 and 135° and A between −90 and 0° are chosen, and the Δ and ψ related to P and A can be shown to be

$$\Delta = 2P - \frac{\pi}{2} \quad \text{and} \quad \psi = -A \tag{C12}$$

Similarly, zones 2 and 4 can be derived when the fast axis of the second quarterwave plates of the nulling ellipsometer is oriented at +45° relative to the plane of incidence. In the measurements that are the basis for this chapter, all the ellipsometric data were collected in zone 1, and only occasionally did we switch into zone 3 to check how big the zone differences were for the same sample.

REFERENCES

1. M. Born and E. Wolf, "Principles of Optics," 5th ed., Pergamon, New York, 1975.
2. P. Lorrain, D. P. Corson, and F. Lorain, "Electromagnetic Fields and Waves," 3rd ed., Freeman, New York, 1988.
3. O. S. Heavens, "Optical Properties of Thin Solid Films," Academic Press, New York, 1955.
4. D. Desmet at www.tusc.net/~ddesmet.
5. R. M. A. Azzam and N. M. Bashara, "Ellipsometry and Polarized Light," pp. 206–207, North-Holland, Amsterdam, 1977.
6. H. X. Tran, *In Situ* Single-Wavelength Fast-Nulling Ellipsometric Measurements on CdTeCd$_{1-x}$Mn$_x$Te Quantum Well Structures Grown by Pulsed Laser Evaporation and Epitaxy and Multi-Layer Structures Consisting of Sequential Layers of Platinum, Titanium, and Nickel by Magnetron Sputtering, Ph.D. Thesis, Department of Physics, University of Waterloo, Waterloo, Ontario, Canada, 1996 (unpublished).
7. J. J. Dubowski and D. F. Williams, *Thin Solid Films* 117, 289 (1984).
8. J. J. Dubowski, D. F. Williams, J. M. Wrobel, P. B. Sewell, J. LeGeyt, C. Halpin, and D. Todd, *Can. J. Phys.* 67, 343 (1989).
9. S. R. Das, K. Rajan, P. van der Meer, and J. G. Cook, *Can. J. Phys.* 65, 864 (1987).
10. S. R. Das and J. G. Cook, *Thin Solid Films* 163, 409 (1988).
11. J. J. Dubowski, J. R. Thompson, S. J. Rolfe, and J. P. McCaffrey, *Superlattices Microstruct.* 9, 327 (1990).
12. H. X. Tran, J. D. Leslie, S. Buchanan, and J. J. Dubowski, *SPIE* 2403, 116 (1995).
13. D. E. Aspnes, *J. Opt. Soc. Am.* 10, 974 (1993).
14. W. Gilmore III and D. E. Aspnes, *Appl. Phys. Lett.* 66, 1617 (1995).
15. J. M. Wrobel, J. J. Dubowski, and P. Becla, *SPIE* 2403, 251 (1995).
16. J. J. Dubowski, *CHEMTRONICS* 3, 66 (1988).
17. J. J. Dubowski, A. P. Roth, Z. R. Wasilewski, and S. J. Rolfe, *Appl. Phys. Lett.* 59, 1591 (1991).
18. W. Gilmore III and D. E. Aspnes, *Appl. Phys. Lett.* 66, 1617 (1995).
19. E. D. Palik, "Handbook of Optical Constants of Solids," Academic Press, New York, 1985.
20. T. P. Pearsall, "Strained-Layer Superlattices: Physics," Chap. 1. Academic Press, New York, 1990.
21. C. G. Grandqvist and O. Hunderi, *Phys. Rev. B* 16, 3513 (1977).
22. R. Landauer, "Proceedings of the First Conference on Electrical Transport and Optical Properties of Inhomogeneous Media," American Institute of Physics Conference Proceedings (J. C. Garland and D. B. Tanner, Eds.), Vol. 40, Am. Inst. of Phys., New York, 1978.
23. D. E. Aspnes, J. B. Theeten, and F. Hottier, *Phys. Rev. B* 20, 3292 (1979).
24. D. A. G. Bruggeman, *Ann. Phys. (Leipzig)* 24, 636 (1935).
25. D. V. Sivukhin, *Zh. Eksp. Teor. Fiz.* 21, 367 (1951).
26. J. L. Ord, D. J. Desmet, and D. J. Beckstead, *J. Electrochem. Soc.* 136, 2178 (1989).
27. J. Smith and F. Mansfeld, *J. Electrochem. Soc.* 119, 663 (1972).
28. M. Yamamoto and T. Namioka, *App. Opt.* 31, 1612 (1992).
29. M. Yamamoto and A. Arai, *Thin Solid Films* 223, 268 (1993).

Chapter 8

PHOTOCURRENT SPECTROSCOPY OF THIN PASSIVE FILMS

F. Di Quarto, S. Piazza, M. Santamaria, C. Sunseri

*Dipartimento di Ingegneria Chimica dei Processi e dei Materiali, Università di Palermo,
Viale delle Scienze, 90128 Palermo, Italy*

Contents

1. INTRODUCTION

The formation of a protective film on a metallic surface is a key step in establishing passivity, with strong reduction of the corrosion rate of the underlying metal [1]. In many cases of practical importance passivity of metals is attained in the presence of very thin (a few nanometers thick) oxide films, whose physicochemical properties play an important role in determining the nature of possible chemical reactions occurring at the interface between the passive metal and the environment. For these reasons the physicochemical characterization of passive films and corrosion layers is a preliminary task for a deeper understanding of the corrosion behavior of metals and alloys. In many cases the identification of the passive films on metal and alloys requires the use of different in situ and ex situ techniques in order to get information on chemical composition, morphology, and the crystalline or disordered nature of the passivating layers.

Although useful information on the film composition can be gathered from ex situ techniques (Auger, ESCA, XPS, SIMS), they suffer some drawbacks, especially when investigating very thin films, owing to the risk of changing the structure and composition of the passive film on going from the potentiostatic control in solution to the vacuum. For this reason, a large agreement exists on the advantages of using in situ techniques in corrosion and passivity studies.

Besides more traditional (and mainly optical) techniques (differential and potential-modulated reflectance, ellipsometry, interferometry, Fourier transform infrared spectroscopy, Raman and Mössbauer spectroscopy), new in situ techniques have been introduced in recent years, like EXAFS, XANES, STM, and AFM, which are capable of providing useful information on the structure, composition and morphology of passive films. The use of these analytical techniques has improved our understanding of the structure and composition of passive films grown on metal and alloys. Information on the use of such techniques for

Handbook of Thin Film Materials, edited by H.S. Nalwa
Volume 2: Characterization and Spectroscopy of Thin Films

ISBN 0-12-512910-6/$35.00

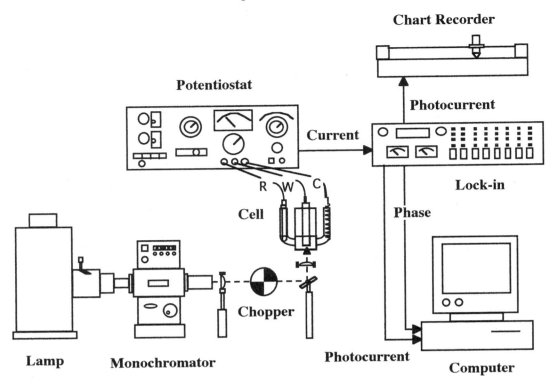

Fig. 1.   Experimental setup employed for the Photocurrent Spectroscopy experiments.

characterizing passive films on electrode surfaces can be obtained from the specialized literature reported in [2–13].

However, all these techniques suffer some limitations when applied to real corroding metal surfaces, owing to experimental difficulties or intrinsic limitations, so that a large demand exists for new in situ techniques that can be routinely used by electrochemists for getting useful information on the physicochemical properties of corrosion layers and passive films. The final goal is to better understand the role of solid state properties and chemical composition of passive films in determining the overall electrochemical behavior of passive electrodes.

Among the other optical methods, photocurrent spectroscopy (PCS) has attracted much attention in recent decades [14, 15]. PCS is a nondestructive technique based on the analysis of the electrochemical response (photocurrent) of the passive electrode–electrolyte interface under irradiation with photons of suitable energy and intensity. The choice of a potentiostatic control is preferred, in view of the preminent role of the electrode potential in the establishment of electrochemical equilibria involving different metal oxidation states and reactivity of the passive films [16].

The aims of this review are:

(a) to provide a theoretical background on the photoelectrochemistry of metal and semiconductor electrodes on which PCS relies, by focusing particularly on new features that are typical of the photoelectrochemical behavior of thin passive films and usually absent in the behavior of bulk crystalline semiconductors;

(b) to highlight the advantages of PCS in getting in situ information on the structure of the metal–passive-film–electrolyte systems as well as to show a more recent quantitative use of this technique in characterizing the composition of passive films.

Figure 1 displays the experimental setup usually employed in photoelectrochemical experiments, where the use of monochromatic light over a quite extended range of wavelengths (200–800 nm) is accessible by using an UV–Vis monochromator–lamp system. This optical setup is easily expandible to the near IR region. In order to improve the sensitivity of the signal detection, a lock-in amplifier coupled to a mechanical chopper allows one to scrutinize very thin films (1–2 nm thick) under illumination of low intensity, so minimizing the risk of modification of the passive films under scrutiny.

If we consider that in PCS the photocurrent response of the passive-film–electrolyte interface is directly related to the number of absorbed photons, it appears evident that such a technique is not demanding of surface finishing and thus allows one to monitor long-term corrosion processes, where large changes of surface reflectivity are expected owing to the formation of rough surfaces covered by corrosion products.

The main limitations to the use of PCS can be traced to the following aspects:

(1) the technique is able to scrutinize only passive films having semiconducting or insulating properties;

(2) surface layers having an optical bandgap larger than 5.5 eV or smaller than 1 eV require a special setup or

Table I.   Reported Values for the Optical Bandgap of Oxides [76, 108, 159]

Oxide	E_g (eV)	Oxide	E_g (eV)
MgO	8.70	Ta_2O_5	3.95
Al_2O_3	6.30	MnO	3.60
Ga_2O_3	4.80	Cr_2O_3	3.50
SnO	4.22	NiO	3.50
SnO_2	3.50	Nb_2O_5	3.35
Sb_2O_3	3.00	TiO_2	3.05–3.20
In_2O_3	2.80	ZnO	3.20
Bi_2O_3	2.80	WO_3	2.75
PbO orth.	2.75	CoO	2.60
PbO tetr.	1.90	CdO	2.50
Tl_2O_3	1.60	FeO	2.40
Y_2O_3	5.80	Fe_2O_3	1.90
MoO_3	2.90	HgO	1.90
V_2O_5	2.30	Cu_2O	1.86
HfO_2	5.20	CuO	1.40
ZrO_2	4.60	PdO	1.00

(1) energetics at the metal–passive-film–electrolyte interfaces (flat-band potential; location of conduction and valence band edges);
(2) electronic structure and (indirectly, through the optical bandgap values) chemical composition of passive films in situ and under controlled potential in long-lasting experiments;
(3) mechanisms of generation (geminate recombination effects) and transport of photocarriers (trap-limited mobility) in localized electronic states of amorphous materials at constant thickness;
(4) kinetics of growth of photoconducting films under illumination, and interference effects in absorbing materials at constant growth rate.

The first two aspects are of paramount importance in corrosion studies if we consider that both the ion transfer reactions (ITRs) and electron transfer reactions (ETRs) are controlled by the electronic properties of the passive film and the energetics at the metal–film and film–electrolyte interfaces [17–21]. We mention that both ITRs and ETRs are involved in determining the kinetics of growth and the breakdown processes of passive films. The onset of breakdown [22–24] determines the film thickness that can be reached during the anodization process, and thus establishes the limits of application of passive films as dielectric materials in electrolytic capacitors.

As for the last two aspects, they provide important information on the solid state properties of passive films as well as on the interaction between ionic and electronic transport processes during the film formation. This gives more insight into the nature of mobile species and defects formed during the growth of passivating layer, so allowing a deeper understanding of the microscopic precesses operating during the formation of the passive layers.

Finally, we mention that the presence of inhomogeneities in the structure of passive films can be evidenced by photoelectrochemical imaging with lateral resolution on the order of 5 μm, as reported by different authors [25–27]; this could help to understand the role of inclusions in the local corrosion of metal and alloys.

Most of the experimental data presented in the following pertain to the photoelectrochemical behavior of passive films having an amorphous or microcrystalline structure, and this requires an extension of interpretative models usually employed for crystalline materials. A short theoretical introduction to the photoelectrochemistry of crystalline semiconductors will be provided for readers not acquainted with the subject, in order to evidence the differences in photoelectrochemical behavior between passive films and bulk crystalline semiconductors. For a more extensive and detailed introduction to the principles of photoelectrochemistry of semiconductors, the reader can take advantage of classical books and workshop discussions published on the subject [28–35].

Theoretical interpretations of the experimental results will be presented on the basis of simple models developed initially for interpreting the photoelectrochemical behavior of passive

are experimentally not accessible in aqueous solutions;
(3) information on the structure and chemical composition of the layer is not directly accessible and usually requires a complementary investigation by other techniques.

The first two limitations are more apparent than real if we take into account the data of Table I, showing the optical bandgap values experimentally measured or inferred from data reported in the recent literature for amorphous or crystalline passive films grown on a large number of sp- and d-metal oxides. From such data it appears that, with the exception of very a few oxides of low electronegativity, the most common base metal oxides have bandgap values largely lying within the optical limits experimentally accessible by PCS. Moreover, with the exception of noble metals (Ir, Ru, etc.), which are covered by oxides only at high electrode potentials, most metals are thermodynamically unstable on immersion in aqueous solution, so that they are covered spontaneously by photoactive films having insulating or semiconducting properties.

The third limitation is the principal one: it is within the aim of this work to show that, although structural and compositional information is not directly accessible by PCS, such a technique can provide this type of information if a realistic interpretative model of the photoelectrochemical behavior of passive films is considered, by taking into account both the more complex electronic structure of amorphous materials and the complementary information accessible by other techniques in situ or ex situ.

Despite these limitations, we mention that, besides the obvious advantages related to the use of a relatively simple equipment, PCS can provide information on:

films on valve metals (Al, Ta, Zr, Nb, Ti, W). In order to show the ability of PCS to scrutinize also complex systems, we will discuss some relevant results pertaining to passive films grown on metals of technological interest (Fe, Cr, Ni). Finally, a more quantitative use of PCS for the compositional characterization of passive films on pure metals and alloys will be presented.

2. METAL–ELECTROLYTE AND SEMICONDUCTOR–ELECTROLYTE INTERFACES

2.1. The Structure of M/El and SC/El Interfaces at Equilibrium

The structure of the metal–electrolyte (M/El) and semiconductor–electrolyte (SC/El) interfaces has been the object of a long series of studies [36–44]. These concluded that the main differences between the structures of the two electrode–electrolyte interfaces can be traced to the different electronic properties of the electrodes [28–35]. A brief summary of the subject will be given, in order to introduce some concepts and to derive some equations for use in explaining the photoelectrochemical behavior of such interfaces.

The potential drop across a M/El interface, $\Delta\Phi_{m,el}$, is given by (see Fig. 2):

$$\Delta\Phi_{m,el} = \{\phi_m - \phi_{el}(-x_H)\} + \{\phi_{el}(-x_H) - \phi_{el}(-\infty)\} \quad (2.1.1)$$

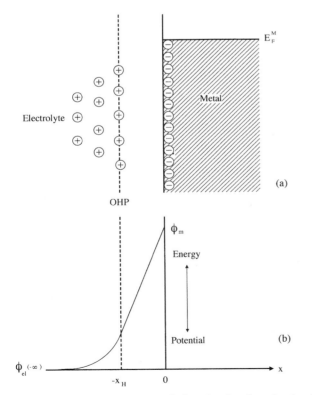

Fig. 2. (a) Schematic diagram of the metal–electrolyte interface, showing the charge excess at the phase boundaries; (b) potential profile across the interface. OHP is the outer Helmholtz plane. E_F^M represents the Fermi level of the metal.

where ϕ_m and ϕ_{el} represent the Galvani potential in the metal and the electrolytic phase, respectively, whilst the arguments within the parentheses are distances from the metal surface, assumed as the origin of the x-axis. A positive x-direction has been assumed from the surface towards the interior of the metal, so that $-x_H$ is the coordinate of the outer Helmholtz plane (OHP). In the absence of ions specifically adsorbed, this last represents the distance of closest approach of solvated ions in solution to the metal surface. The infinite distance into the solution means simply the bulk of the solution, where the Gouy diffuse layer vanishes. In previous work the Galvani potential in the bulk of the solution is usually assumed as the reference potential level, but as a physical quantity it is not accessible experimentally. This difficulty is a general one, and it is avoided in electrochemistry by using a reference electrode, with respect to which we can measure the Galvani potential drop at any single interface.

This implies that the measured electrode potential, U_e, is really the sum of three Galvani potential drops at the metal–solution, solution–reference-electrode, and reference-electrode–contact-metal interfaces [45]:

$$U_e = \Delta\Phi_{m,el} + \Delta\Phi_{el,ref} + \Delta\Phi_{ref,m'} \quad (2.1.2)$$

where the contact metal m' has the same physicochemical properties of the metal m.

If it is assumed that the Galvani potential drop at a reference electrode is constant, then any change in the electrode potential, and thus any modification in the charge distribution, can be attributed to the M/El interface under study. Moreover, at electrochemical equilibrium the potential drop at the M/El interface is entirely localized on the solution side, where the excess of ionic charge in the compact double layer (Helmholtz layer) and in the diffuse double layer (Gouy layer) counterbalances the excess of charge (opposite in sign) at the metal surface. In the absence of a net circulating current across the interface, an electrical equivalent circuit consisting of two capacitors in series accounts for the electrical behavior of such an interface [36–42]. This simple model of an interface has been used frequently in studies on the double layer structure at the M/El interface. More refined models of the metal–solution interface have been proposed, and they can be found in [43, 44].

The SC/El interface does not present significant differences from the M/El one on the solution side, but it differs drastically on the electrode side, as sketched in Figure 3. In this case a new term in the Galvani potential drop appears, so that we can write

$$\Delta\Phi_{sc,el} = \{\phi_{sc}(\infty) - \phi_{sc}(0)\} + \{\phi_{sc}(0) - \phi_{el}(-x_H)\} + \{\phi_{el}(-x_H) - \phi_{el}(-\infty)\} \quad (2.1.3)$$

where the first term on the right, $\Delta\Phi_{sc} = \phi_{sc}(\infty) - \phi_{sc}(0)$, represents the potential drop within the SC electrode. In contrast with the M/El interface, owing to the much lower concentration of mobile electrical carriers (electrons and holes for n-type and for p-type semiconductors, respectively), a space-charge region now appears in the interior of the SC phase (see Fig. 4).

In equilibrium conditions the potential drop inside the semiconductor can be calculated by solving the Poisson equation

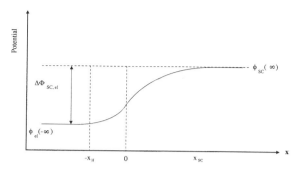

Fig. 3. Potential profile across the semiconductor–electrolyte interface.

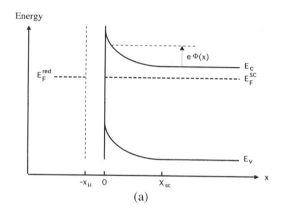

(a)

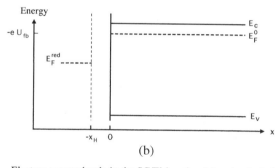

(b)

Fig. 4. Electron energy levels in the SC/El junction (a) under depletion conditions and (b) at the flat-band potential. $\Phi(x)$ represents the potential drop at a point x of the SC space-charge region. E_F^{SC} and E_F^{red} are the Fermi levels of the SC and the redox couple in solution, respectively. U_{fb} is the flat-band potential.

The hypothesis of absence of surface states at the SC/El interface has been made in Figures 3 and 4, in order to simplify the analysis. In this case and for not highly doped semiconductors, the total potential drop in the SC space charge accounts for the variation of electrode potential on going from the flat-band potential conditions ($\Delta\Phi_{sc} = 0$; see Fig. 4b) to depletion conditions ($\Delta\Phi_{sc} > 0$ for n-type SCs; see Fig. 4a) when a space-charge region, having width X_{sc}, is formed inside the semiconductor. In the case of p-type material a space-charge layer develops inside the SC for $\Delta\Phi_{sc} < 0$.

2.2. Determination of the Space-Charge Width in Crystalline SCs and the Mott–Schottky Equation

In order to calculate the dependence of X_{sc} on $\Delta\Phi_{sc}$, we need to solve the Poisson equation, which relates the Galvani potential drop at a point x of the space-charge region, $\Phi(x) = \phi_{sc}(x) - \phi_{sc}(\infty)$, to the charge density inside the semiconductor, $\rho(x)$:

$$\frac{d^2\Phi}{dx^2} = -\frac{\rho(x)}{\varepsilon\varepsilon_0} \qquad (2.2.1)$$

where ρ is the total density of charge (mobile and fixed) inside the space-charge region, ε is the dielectric constant of the SC, and ε_0 is the vacuum permittivity. The usual boundary conditions of zero electric field and zero Galvani potential in the bulk of the semiconductor [$d\Phi/dx(\infty) = 0$ and $\Phi_{sc}(\infty) = 0$] can be assumed. This choice implies that $\Phi_{sc}(x) = \phi_{sc}(x) - \phi_{sc}(\infty)$ has opposite sign to the electrochemical scale of potential.

The width of the space-charge region, X_{sc}, as well as the potential distribution inside the semiconductor under reverse polarization of the SC/El junction can be easily found by using the Schottky barrier model of the junction and by solving the Poisson equation (2.2.1) under the depletion approximation and the hypothesis of a homogeneously doped semiconductor with fully ionized donors N_d (for n-type SCs) or acceptors N_a (for p-type SCs) [28–30, 46]. The depletion approximation implies that the net charge density varies from the zero value of the bulk to the value $+eN_d$ (or $-eN_a$ for a p-type SC) at the depletion edge.

As a consequence of such a charge distribution, a linear variation of the electric field is obtained according to Gauss's law, which relates the electric field intensity at the surface of the SC, F_s, to the charge density:

$$F_s = \frac{eN_d X_{sc}}{\varepsilon\varepsilon_0} \qquad (2.2.2)$$

For an electric field varying linearly inside the space-charge region, the potential drop inside the SC can be calculated as

$$\Delta\Phi_{sc} = \frac{F_s X_{sc}}{2} = \frac{\varepsilon\varepsilon_0 F_s^2}{2eN_d} \qquad (2.2.3)$$

from which a dependence of F_s on $(\Delta\Phi_{sc})^{1/2}$ is derived. By relaxing the depletion approximation and taking into account the more gradual drop of the electron density at the

(see Section 2.2) under the same conditions used for an ideal metal–semiconductor (M/SC) Schottky barrier and by taking into account that the potential drop within the semiconductor is only a part of the total potential difference measured with respect to a reference electrode. Moreover, by taking into account that in the presence of a sufficiently concentrated (>0.1 M) electrolytic solution the potential drop in the Gouy layer is negligible, the equivalent electrical circuit of the interface in electrochemical equilibrium can be represented again by two capacitors in series: C_{sc} for the SC space-charge layer, and C_H, for the Helmholtz layer [28–30]. The value of C_{sc} changes with the width of the space-charge region within the semiconductor, X_{sc}, and it is a function of the total potential drop $\Delta\Phi_{sc}$ within the SC.

depletion edge, we get for F_s the expression [46]

$$F_s = \left(\frac{2eN_d}{\varepsilon\varepsilon_0}\right)^{1/2}\left(\Delta\Phi_{sc} - \frac{kT}{e}\right)^{1/2} \qquad (2.2.4)$$

showing that the depletion approximation differs from the exact one by the thermal voltage contribution kT/e.

The dependence of the width of the space-charge region on the potential drop $\Delta\Phi_{sc}$ can be obtained by using Eqs. (2.2.2) and (2.2.4) [46]:

$$X_{sc} = \left(\frac{2\varepsilon\varepsilon_0}{eN_d}\right)^{1/2}\left(\Delta\Phi_{sc} - \frac{kT}{e}\right)^{1/2} \qquad (2.2.5)$$

This expression can be used for deriving also the dependence of the space-charge capacitance on $\Delta\Phi_{sc}$ [29, 33]:

$$C_{sc} = \frac{\varepsilon\varepsilon_0}{X_{sc}} = \left(\frac{\varepsilon\varepsilon_0 eN_d}{2}\right)^{1/2}\left(\Delta\Phi_{sc} - \frac{kT}{e}\right)^{-1/2} \qquad (2.2.6)$$

This is the well-known Mott–Schottky (MS) equation, which can be employed to derive the flat-band potential U_{fb} of the SC/El junction. For this aim we need to relate the Galvani potential drop within the semiconductor to the measured electrode potential U_e.

By defining the flat-band potential of the SC/El junction as the electrode potential at which $\Delta\Phi_{sc} = 0$ (Fig. 4b), in the absence of surface states we can write at any other electrode potential

$$\Delta\Phi_{sc} = U_e - U_{fb} \qquad (2.2.7)$$

where U_e and U_{fb} are measured with respect to the same reference electrode. By substituting Eq. (2.2.7) in Eq. (2.2.6) we get the final form of the MS equation, usually employed for obtaining the flat-band potential and the energetics of the n-SC/El junction [29, 33]:

$$\left(\frac{1}{C_{sc}}\right)^2 = \frac{2}{\varepsilon\varepsilon_0 eN_d}\left(U_e - U_{fb} - \frac{kT}{e}\right) \qquad (2.2.8)$$

An analogous equation holds for a p-type material, with a minus sign in front of right-hand side and N_a instead of N_d. From the electrical equivalent circuit of the interface (see Section 2.1), it is obvious that the following relation holds in a concentrated electrolyte:

$$\frac{1}{C_{sc}} = \frac{1}{C_m} - \frac{1}{C_H} \qquad (2.2.9)$$

where C_m is the capacitance of the interface measured experimentally, and a constant value of $C_H = 30\ \mu F/cm^2$ is frequently used for aqueous solutions.

The determination of the flat-band potential is the first step in the location of the energy levels at the SC/El interface. Once U_{fb} is known, it is possible to locate the Fermi level of the SC with respect to the electrochemical scale by means of the relationship:

$$E_F^0(el) = -eU_{fb} \qquad (2.2.10)$$

It is also possible to relate the Fermi level of the SC to the *physical scale* (zero electron potential energy in the vacuum)

using the following equation:

$$E_F^0(vac) = -eU_{fb}(ref) + eU_{ref}(vac) \qquad (2.2.11)$$

where the first term on the right is the flat-band potential, measured with respect to a standard reference electrode, and the second one is the potential of the reference electrode with respect to the vacuum scale [33, 45]. In the case of a standard hydrogen reference electrode (SHE), the most commonly accepted value of $eU_{SHE}(vac)$ is -4.50 eV. Slightly different values have been reported by different authors, so that an uncertainty on the order of 0.2 eV remains when the energy levels are referred to the vacuum scale [35].

The location of the remaining energy levels of the junction is easily obtained by means of the usual relationships for n- and p-type semiconductors:

$$E_c = E_F^0 + kT\ln\left(\frac{N_c}{N_d}\right) \qquad \text{for n-type SC} \quad (2.2.12a)$$

$$E_v = E_F^0 - kT\ln\left(\frac{N_v}{N_a}\right) \qquad \text{for p-type SC} \quad (2.2.12b)$$

$$E_g = E_c - E_v \qquad (2.2.13)$$

where N_c and N_v are the effective densities of states at the bottom of the SC conduction band and at the top of the SC valence band, N_d and N_a are the donor and acceptor concentrations in the SC, E_c and E_v are the conduction and valence band edges, respectively, and E_g is the bandgap of the SC.

The rather simple MS method for locating the energy levels at the SC/El interface is very popular among electrochemists. However, we have to mention that, although the validity of Eq. (2.2.8) has been tested rigorously for several SC/El interfaces [47–56], since the seminal work of Dewald on single-crystal ZnO electrodes [47, 48], in many cases a misuse has been made of this equation, neglecting the limits under which it was derived. This is particularly true in the case of passive films, which are usually amorphous or strongly disordered and display a frequency dependence of the measured capacitance not in agreement with the hypotheses underlying Eq. (2.2.8). A more detailed discussion of the inadequacy of the MS analysis for interpreting the capacitance data of the amorphous SC/El interface can be found in Section 3. Here we summarize the main assumptions used for deriving Eq. (2.2.8) [54]:

(1) a crystalline SC electrode homogeneously doped in the depletion regime;
(2) a fully ionized single donor (or acceptor, for p-type SC) level;
(3) absence of deep-lying donor (acceptor) levels in the forbidden gap of the SC;
(4) negligible contribution of surface states and minority carriers to the measured capacitance;
(5) absence of electrochemical faradaic processes at the SC/El interface.

In the case of passive films it is rather improbable that the first four points are satisfied simultaneously. In our opinion, any model of the space-charge capacitance of passive films that assumes the existence of discrete donor (acceptor) levels in the forbidden gap without taking into account the dependence of the response of localized gap states to the frequency of the modulating ac signal, fails to provide the correct analysis of experimental data.

2.3. Photocurrent-vs-Potential Curves for Crystalline SC/El Junctions: The Gärtner–Butler Model

The first attempt to model the photocurrent behavior of a SC/El junction under illumination can be ascribed to M. A. Butler [57], who adapted a previous model developed by Gärtner for the M/SC Schottky barrier [58] to the case of a SC/El interface. A close similarity between the Schottky barrier junction and the semiconductor–liquid interface was assumed in the model. The quantitative fitting of the experimental photocurrent-vs-potential curves for an n-type single-crystal WO_3–electrolyte junction under monochromatic irradiation, performed through the Butler–Gärtner model, attracted the attention of many electrochemists, who recognized in the Butler–Gärtner mathematical treatment an alternative way to get the flat-band potential of the junction, circumventing some of the difficulties related to the MS analysis. In this subsection we will derive the theoretical expressions for the photocurrent in a crystalline SC/El junction as a function of the electrode potential in the frame of the Gärtner–Butler model. The results will also be used for getting the relationship between bandgap of photoelectrodes and the quantum efficiency of the junction at constant potential. The limits of the Gärtner–Butler model will be discussed briefly, and the new features of photoelectrochemical behavior of the passive-film–electrolyte junction will be discussed in Section 3.2.

In Figure 5 we report a schematic picture for a crystalline n-type SC/El interface under illumination. In the figure, Φ_0 (in $cm^{-2} s^{-1}$) is the photon flux entering the semiconductor (corrected for the reflection losses at the SC/El interface), which is absorbed following the Lambert–Beer law. According to this, the number of electron–hole pairs generated per second and unit volume at any distance from the SC surface, $g(x)$, is given by

$$g(x) = \Phi_0 \alpha e^{-\alpha x} \qquad (2.3.1)$$

where α, the light absorption coefficient of the semiconductor, is a function of the irradiating wavelength. It is assumed that each absorbed photon (having energy, $h\nu$, higher than the optical bandgap of the semiconductor photoelectrode, E_g) originates a free electron–hole pair.

In the Gärtner model the total photocurrent collected in the external circuit is calculated as the sum of two terms: a migration term I_{drift}, which takes into account the contribution of the minority carriers generated in the space-charge region, and a diffusion term I_{diff}, arising from the minority carriers entering the edge of the space-charge region from the field-free region ($x > X_{sc}$). No reflections of light at the rear interface is assumed, so that all the entering light is absorbed within the SC.

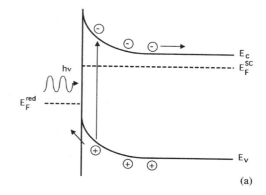

(a)

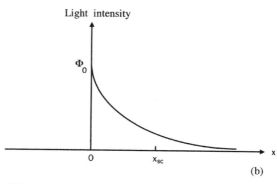

(b)

Fig. 5. Schematic representation of a crystalline n-type SC/El interface under illumination, showing (a) the electron–hole pair generation and (b) the change of light intensity due to absorption within the semiconductor.

Moreover, in the model it is assumed that minority carriers generated in the space-charge region of the SC do not recombine at all, owing to the presence of an electric field which separates efficiently the photogenerated carriers. The same assumption is made for the minority carriers arriving at the depletion edge from the bulk region of the SC. In order to calculate I_{diff}, Gärtner solved the transport equation, which for a n-type SC is

$$\frac{dp(x)}{dt} = D_p \frac{d^2 p(x)}{dx^2} - \frac{p - p_0}{t} + g(x) \qquad (2.3.2)$$

with the boundary conditions $p = p_0$ for $x \to \infty$ and $p = 0$ for $x = X_{sc}$. Here p is the hole concentration under illumination, p_0 the equilibrium concentration of holes in the bulk of the (not illuminated) SC, and D_p the diffusion coefficient for holes. The zero value of p at the boundary of the depletion region follows from the previous assumption that all the holes generated into the space-charge region are swept away without recombining. According to Gärtner, for the total photocurrent we can write

$$I_{ph} = I_{drift} + I_{diff}$$

$$= e \int_0^{X_{sc}} g(x)\, dx + e D_p \left(\frac{dp(x)}{dx} \right)_{x=X_{sc}} \qquad (2.3.3)$$

where e is the absolute value of the electronic charge. By solving Eq. (2.3.2) in the steady-state approximation to obtain the distribution of holes in the field-free region, and by substituting Eq. (2.3.1) for $g(x)$ in the integral of the drift term, we get finally the Gärtner equation for n-type semiconductor [58]:

$$I_{ph} = e\Phi_0 \frac{1 - \exp(-\alpha X_{sc})}{1 + \alpha L_p} + e p_0 \frac{D_p}{L_p} \qquad (2.3.4)$$

where L_p is the hole diffusion length. The same equation holds for p-type SCs, with D_n and L_n (the electron diffusion coefficient and diffusion length, respectively) instead of D_p and L_p, and n_0 (the electron equilibrium concentration) instead of p_0.

Equation (2.3.4) was further simplified by Butler for the case of wide-bandgap SCs (e.g., n-type WO_3), where the concentration of minority carriers in the bulk, p_0, is very small. In this case, by using Eq. (2.2.5) for X_{sc}, he derived the Gärtner–Butler equation for the photocharacteristics of a crystalline SC/El junction [57]:

$$I_{ph} = e\Phi_0 \left[1 - \frac{\exp\left(-\alpha X_{sc}^0 \sqrt{\Delta\Phi_{sc} - kT/e}\right)}{1 + \alpha L_p} \right] \qquad (2.3.5)$$

In this equation, X_{sc}^0 represents the space-charge width at the SC electrode at 1 V of band bending, and $\Delta\Phi_{sc} = U_e - U_{fb}$. It is easy to show [57] that if $\alpha X_{sc} \ll 1$ and $\alpha L_p \ll 1$, the photocurrent crossing the n-type SC/El interface can be written as

$$I_{ph} = e\Phi_0 \alpha X_{sc}^0 \left(U_e - U_{fb} - \frac{kT}{e} \right)^{1/2} \qquad (2.3.6)$$

Equation (2.3.6) predicts a quadratic dependence of the photocurrent on the electrode potential, which can be used for getting the flat-band potential. In fact, neglecting the termal voltage, a plot of I_{ph}^2 vs U_e will be a straight line that intercepts the voltage axis at the flat-band potential U_{fb}.

The following relationship holds between the absorption coefficient α and bandgap E_g of a material in the vicinity of the optical absorption threshold [33]:

$$\alpha = A \frac{(h\nu - E_g)^{n/2}}{h\nu} \qquad (2.3.7)$$

with n depending on the kind of the optical transition between occupied electronic states and vacant states of the semiconductor. From this and the Butler equation (2.3.6) it is possible to derive the following expression for the quantum yield:

$$\frac{I_{ph}}{e\Phi_0} = \eta h\nu = A(h\nu - E_g)^{n/2} \left(L_i + X_{sc}^0 \sqrt{|U_e - U_{fb}|} \right) \quad (2.3.8)$$

Here η represents the collection efficiency, E_g the optical threshold for the onset of photocurrent of the electrode, and $L_i = L_p$ or L_n for n-type or p-type SC, respectively. From Eq. (2.3.8) it is possible to get the optical bandgap of the material from the wavelength dependence of the photocurrent (usually referred as the photocurrent spectrum of the junction in the photoelectrochemical literature), by plotting $(\eta h\nu)^{2/n}$ vs $h\nu$ at constant electrode potential ($U_e - U_{fb} = $ const). We will

come back on this point when discussing the electronic structure of passive films.

Another interesting application of the Gärtner–Butler model for obtaining the diffusion length of the minority carriers was suggested later by Morrison [59]. In fact, for large-bandgap materials, inserting in Eq. (2.3.5) the relationship (2.2.6) between the space-charge width and the SC capacitance, and assuming $\alpha X_{sc} \ll 1$, at constant wavelength one has the following relationship:

$$I_{ph} = \text{const}\left(L_i + \frac{\varepsilon\varepsilon_0}{C_{sc}} \right) \qquad (2.3.9)$$

where ε is the relative dielectric constant of the SC, ε_0 the vacuum permittivity, and L_i the minority carrier diffusion length. From this relationship a straight line is expected for the plot of I_{ph} vs C_{sc}^{-1}, and the parameter L_i is given by

$$L_i = -\left. \frac{\varepsilon\varepsilon_0}{C_{sc}} \right|_{I_{ph}=0} \qquad (2.3.10)$$

We have to mention that, apart the initial assumption by Gärtner of an ideal Schottky barrier, several hypotheses underlie the use of Eqs. (2.3.4) to (2.3.10) for interpreting photoelectrochemical data. One of the most striking is the assumption of the absence of any kinetic control from the solution side, which implies the absence of any limitation in the supply of reagents from the solution as well as the absence of any recombination either in the space-charge region or at the surface of the semiconductor. The possible existence of kinetic control at the SC surface or within the space-charge region has been considered by other authors [60–65], who have shown clearly that:

(a) In the presence of strong surface recombination effects, the onset of photocurrent may occur at much higher band bending than that predicted by the Gärtner–Butler equation [60].

(b) A square root dependence of the photocurrent on the electrode potential is still compatible with the existence of some mechanism of recombination (first-order kinetics) within the space-charge region [63], so that the determination of the flat-band potential from a plot of the square of the photocurrent vs the electrode potential must be taken with some caution.

This last aspect is evidenced in Figure 6, where the presence of a kinetic control due to a recombination process within the SC space-charge region at lower electrode potentials leads to two straight lines with different slopes in the I_{ph}^2-vs-U_e plots [66]. As reported in the figure, a rather large potential range exists where the extrapolated quadratic dependence of the photocurrent on the electrode potential gives values more positive than the true flat-band potential of the junction [$U_{fb} = -0.05$ V (MSE) in 0.1 N H_3PO_4]. According to the previous considerations, a more reliable derivation of the flat-band potential of the n-WO_3–electrolyte junction was performed by extrapolating the I_{ph}^2-vs-U_e plots from the highest-potential region. Such a finding is in agreement with the theoretical treatment of the illuminated SC/El junction made by Albery

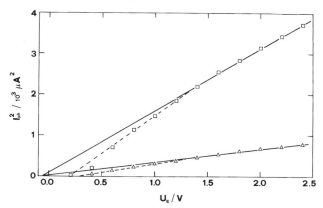

Fig. 6. Plots of I_{ph}^2 vs U_e recorded for a WO$_3$ film grown at 8 mA/cm^2 in 0.1 N H$_3$PO$_4$ electrolyte up to 100 V and crystallized for 3 h at 350°C under argon atmosphere. Electrode surface: 0.053 cm^2. Irradiating wavelengths: □, $\lambda = 300$ nm; △, $\lambda = 380$ nm. [From "Solar Energy Materials," Vol. 11, pp. 419–433, 1985; reprinted with permission from Elsevier Science.]

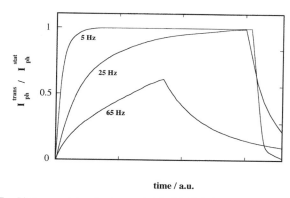

Fig. 7. Photocurrent transients recorded in 0.5 M H$_2$SO$_4$ solution during the chopping period of the lock-in at different frequencies for a TiO$_2$ film 250 nm thick. Electrode potential: 3.5 V (SCE).

et al. [63], who first stressed that a proportionality of I_{ph} to the square root of the potential, even in a rather large region of electrode potential, cannot be interpreted as support for the validity of Butler equation and of the assumptions made in deriving it. We will come back to these aspects when discussing an analogous behavior sometimes observed with photoconducting anodic films.

A model that takes into account surface recombination of the photogenerated minority carriers has been proposed by Wilson [60]. With respect to Gärtner's treatment, both the overall recombination rate S_r and the rate of minority carrier reaction at the SC/El interface, S_t, are considered. The two rates are dependent in different ways on the potential, which controls their competition. In agreement with experimental results on crystalline TiO$_2$ (rutile) electrodes, the Wilson model predicts S-shaped photocharacteristics in which I_{ph} saturates at high band bendings [60]. The effect of surface recombination can be approximated by multiplying the right-hand side of Eqs. (2.3.4) to (2.3.6) by a factor $S_t/(S_t + S_r)$ [66].

From our point of view the importance of the Butler model is to enable the use of photoelectrochemical data for determining the optical bandgap of photoelectrodes. In view of the limited usefulness of the Gärtner–Butler model for getting the flat-band potential of the SC/El junction (owing to the numerous assumptions just discussed), we believe that the most important result of the model is the use of Eq. (2.3.8) for getting information on the optical bandgap in photoelectrodes in a rather simple and direct way. This finding has opened an electrochemical route to the physicochemical characterization of photoactive layers on metal and alloys in contact with electrolytic solutions by means of photocurrent spectroscopy.

Before concluding this subsection we recall that all the preceding equations pertain to the steady-state value of the dc photocurrent. In this case the photocurrent circulating in the cell at a band bending $\Delta\Phi_{sc}$ is defined as

$$I_{ph}(\Delta\Phi_{sc}) = I_l(\Delta\Phi_{sc}) - I_d(\Delta\Phi_{sc}) \qquad (2.3.11)$$

where I_l and I_d are the currents under illumination and in the dark, respectively, measured in the external circuit at the same electrode potential, U_e.

When the photocurrent is measured by means of the lock-in technique (see Fig. 1), we are dealing with a periodically chopped photon flux, i.e., an illumination of the junction that is interrupted several times each second. In this case the lock-in measures a signal whose intensity depends on the ratio between the chopping angular frequency ω_c and the time constant of the electrical equivalent circuit of the junction, τ, including also the electrolyte resistance ($\tau = R_t C_t$, with R_t and C_t representing the total resistance and capacitance of the junction). Only for low enough chopping frequencies is the measured signal proportional to the steady-state value of the dc photocurrent, given by the previous relations, provided that the condition $\omega_c\tau \ll 1$ is obeyed. The effect of the chopping frequency on the recorded photocurrent signal is shown in Figure 7 for an anodic TiO$_2$ film. More details on the lock-in technique can be found in [67].

The transient behavior of the photocurrent as a function of time and frequency has been investigated by several researchers [68–73], through new techniques like intensity-modulated photocurrent spectroscopy (IMPS) and the frequency domain analysis of photoprocesses by transient transformation, which can be employed in the study of recombination processes and charge transfer kinetics at the SC/El junction. About the intensity- and frequency-modulated photocurrent spectroscopy techniques, mainly employed in the study of the charge transfer kinetics at the SC/El junction, the interested reader can find more details in [72].

3. THE PASSIVE-FILM–ELECTROLYTE INTERFACE

In order to understand the photoelectrochemical behavior of passive films, we must consider that their electrical behavior may span the whole range from insulating to metallic. However, with the exception of a very few noble metals, like Re and Ru, overwhelming evidence exists that anodic and/or cathodic photocurrents can be detected for almost any oxidized metal [14, 15, 74–76]. This follows from the presence on the metal surface of oxide (or hydroxide) layers having insulating or semiconducting properties. Indeed the widely known phenomenon of light-stimulated current at electrochemical interfaces dates back to the discovery of the Becquerel effect [77] at the very initial stages of the electrochemical science. However, on going from thick crystalline SC electrodes to very thin insulating or semiconducting corrosion films on metals, new experimental features are observed, which require the extension of old interpretative models and the introduction of new theoretical concepts in order to account for novel results not observed previously in bulk crystalline materials.

As a result of the research effort on passive films in recent decades, it is now generally accepted that the extreme variability of the physicochemical properties thickness (from 1–2 nm to micrometers), crystal structure (from crystalline to amorphous), morphology (barrier-type or porous films), and chemical composition (nonstoichiometry and hydration degree) are the most important parameters in determining the overall electrochemical behavior of passive films on metals [17, 21–23, 32, 78].

In the next two subsections we will focus on the influence of the disordered structure on the electronic and optical properties of passive films, and in the following one (Section 3.3) we will discuss the influence of both thickness and solid state properties of the films on their photoelectrochemical behavior.

3.1. Electronic Properties of Disordered Passive Films

The earlier studies of the electronic properties of passive films [79–81] were performed by means of differential capacitance and photoelectrochemical measurements, which in the case of crystalline material allow one to get quite easily the main information on the solid state properties (bandgap, E_g, and concentration of donors or acceptors for n- or p-type semiconducting films) and on the energetics (flat-band potential U_{fb}) of the passive-film–electrolyte junction. Unfortunately, the application of interpretative models valid for crystalline materials (Mott–Schottky theory for the space-charge capacitance and the Gärtner–Butler model for the photocurrent curves) to experimental data pertaining to amorphous or strongly disordered passive films gives misleading results [82, 83]. Two prominent examples are the strong frequency dependence of the capacitance and the change of shape of the photocurrent-vs-potential curves (photocharacteristics) with the wavelength of the incident light, often observed for the passive-film–electrolyte junction [84, 85]. Neither finding is explained in the frame of the theory for crystalline materials, outlined in the previous section. An example of the first effect is reported in Figure 8, showing the MS lines measured experimentally at different frequencies of the ac signal (10 Hz–3 kHz) for a WO$_3$ film in sulfuric acid solution. The sharp change both in the slope and in the intercept of the straight lines is a consequence of the flattening of the curves of capacitance vs electrode potential with increasing frequency (see below). At even higher frequencies the capacitance curve tends to become flat, so that the intercept of the MS plots shifts toward unrealistic, very negative potentials.

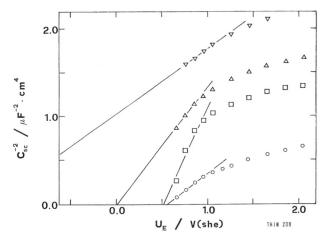

Fig. 8. Mott–Schottky plots at different frequencies for an amorphous film grown anodically on W in 0.5 M H$_2$SO$_4$ electrolyte up to 54 nm and recorded in the same solution. $\bigcirc$, $f = 10$ Hz; $\square$, $f = 100$ Hz; $\triangle$, $f = 1$ kHz; $\triangledown$, $f = 3$ kHz. [*AIChE J.* 38(2), 219–226 (1992); reprinted with permission from the American Institute of Chemical Engineering.]

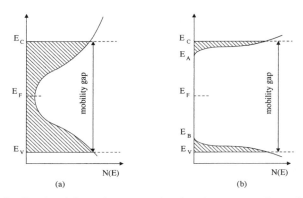

Fig. 9. Density of electronic states as a function of energy according to different amorphous semiconductor models: (a) Cohen–Fritzsche–Ovshinsky model; (b) Mott–Davis model. Hatched areas represent the localized gap states; E_c and E_v are the conduction and valence band mobility edges, respectively. [*Ber. Bunsenges. Phys. Chem.* 90, 549–555 (1986); reprinted with permission from VCH Verlagsgesellschaft mbH.]

For these reasons, an extension of previous theory to the case of disordered materials is necessary. Before discussing in detail the photocurrent behavior of the passive-film–electrolyte junction, it is necessary to highlight the main differences between the electronic properties of amorphous and crystalline materials [86–93].

In Figure 9 we have sketched two band structure models usually employed for insulating or semiconducting amorphous materials. We stress that usually amorphous materials retain the same short-range order as their crystalline counterparts and that the main differences arise out from the absence of the long-range order, typical of crystalline phases. It is now generally accepted by solid state physicists that the band structure model also retains its validity in the absence of the long-range lattice periodicity. This means that the long-range disorder perturbs but does not annihilate the band structure: its main effect is the presence of a finite density of states (DOS) within the so-called *mobility gap*, $E_c - E_v$, of the amorphous semiconductor (a-SC) or insulator. In Figure 9 two different distributions of localized electronic states within the mobility gap are sketched: the DOS distribution of Figure 9a, initially proposed by Cohen, Fritzsche, and Ovshinsky (CFO model) [88], considers the presence of defect states within the semiconductor that produce a continuous distribution of electronic states within the mobility gap (as in a-Si:H), whilst the DOS distribution of Figure 9b, due to Mott and Davis [90], can be attributed to an ideal amorphous material in which only long-range lattice disorder is taken into account. Other models have been suggested for explaining the behavior of different classes of amorphous materials, but they involve only minor modifications to those of Figure 9, when the existence of specific defects in the investigated material is considered. As evidenced in the Figure 9, the general features of the DOS of crystalline materials are preserved also for disordered phases, but some differences from the previous case are now evident. These can be attributed out to the following facts:

(a) The existence of a finite DOS within the mobility gap defined by the two sharp mobility edges, E_c and E_v, in the conduction band (CB) and in the valence band (VB), respectively.

(b) The free-electron-like DOS, $N(E) \propto E^{1/2}$, valid for energy levels close the CB and VB edges, is no longer valid below E_c or above E_v. In these energy regions the presence of a tail of states, decreasing exponentially [91] or linearly [90], has been suggested by different authors for explaining the optical properties of different amorphous materials.

(c) Different mechanisms of charge carrier transport are invoked in extended (above E_c or below E_v) or localized (within the mobility gap) electronic states. A free-carrier-like mechanism of transport is involved in the first case, whilst transport by hopping (thermally activated) is assumed in localized states.

These differences between the distributions of electronic states in crystalline and disordered materials have noticeable influence on both the impedance and the photoelectrochemical behavior of the a-SC/El junction. In order to highlight such differences in the impedance behavior of the junction, we summarize in the following the most important results obtained by us with three typical semiconducting amorphous passive films: a-WO$_3$, a-Nb$_2$O$_5$, and a-TiO$_2$ [83, 94–96]. For these systems the existence of deep electronic states in the mobility gap of the material influences both the shape of the space-charge region and the frequency response of the barrier to the modulating ac signal [97–103].

As for the first aspect, the potential distribution inside the a-SC can be obtained by solving the one-dimensional Poisson equation (2.2.1) with the usual boundary conditions. For mathematical simplicity, a constant DOS, $N(E) = N$, is assumed within the mobility gap of the a-SC, as shown in Figure 10. According to various authors [97–103], we get

$$\Phi(x) = \Delta\Phi_{sc} \exp\left(-\frac{x}{x_0}\right) \qquad (3.1.1)$$

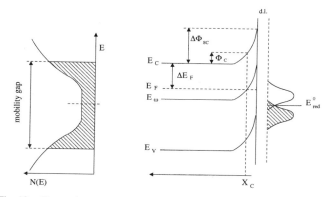

Fig. 10. Energetics at an n-type a-SC/El junction under reverse bias. A constant density of bandgap states is assumed (left side). E_{red}^0 is the redox level of the solution; d.l. is the Helmholtz double layer; $\Delta\Phi_{sc}$ is the total bandgap bending. For the other symbols see text. [*Ber. Bunsenges. Phys. Chem.* 90, 549–555 (1986); reprinted with permission from VCH Verlagsgesellschaft mbH.]

where $x_0 = (\varepsilon\varepsilon_0/e^2 N)$ is the screening length of the a-SC and N is the DOS in cm^{-3} eV^{-1}. Such an exponential dependence is obeyed as long as the Fermi energy extends through the barrier, i.e., the electron quasi-Fermi level is flat throughout the a-SC. At high band bending a parabolic distribution will appear in the deep depletion region of the a-SC, located near the interface region, followed by the exponential depletion region close to the bulk of the SC [85, 97–104].

A detailed analysis of the impedance behavior of the a-SC/El junction has been presented in [83, 85, 94–96], by taking into account the theoretical studies carried out by different authors on the amorphous silicon solid state Schottky barrier [97–103]. These last have shown that, in contrast with the case of the crystalline SC, the filled electronic states in the gap do not follow the ac signal instantaneously, but need a finite response time that depends on their energy position and can be much longer than the period of the imposed ac signal. In fact the relaxation time τ for the capture of emission of electrons from electronic states below E_F is assumed to follow the relationship

$$\tau = \tau_0 \exp\left(\frac{E_c - E}{kT}\right) \qquad (3.1.2)$$

where, at constant temperature, τ_0 is a constant characteristic of each material. According to Eq. (3.1.2), on decreasing the energy of the localized state in the gap, τ increases sharply, so that deep states (for which $\omega\tau \gg 1$) do not respond to the ac signal.

By assuming a full response for states satisfying the condition $\omega\tau \ll 1$ and a null response for states having $\omega\tau \gg 1$, a sharp cutoff energy level, E_ω separating states responding from those not responding to the signal, can be defined from the condition $\omega\tau = 1$. The location of the cutoff level is easily found by imposing $\omega\tau = 1$ for $E = E_\omega$ in Eq. (3.1.2), which gives

$$E_c - E_\omega = -kT \ln(\omega\tau_0) \qquad (3.1.3)$$

This condition occurs at some position within the barrier ($x = X_c$) that is a function of band bending and ac frequency (see

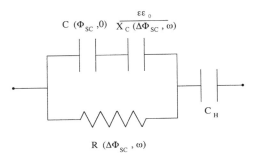

Fig. 11.   Equivalent electrical circuit employed for interpreting the impedance data on the a-SC–electrolyte junction. C_H is the Helmholtz capacitance. For the other symbols see text.

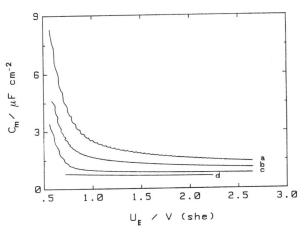

Fig. 12.   Measured parallel capacitance C_m vs electrode potential at different frequencies in 0.5 M H_2SO_4 solution for an a-WO$_3$ film grown in the same solution up to 54 nm. (a) $f = 3$ Hz; (b) $f = 10$ Hz; (c) $f = 100$ Hz; (d) $f = 3$ kHz. [AIChE J. 38(2), 219–226 (1992); reprinted with permission from the American Institute of Chemical Engineering.]

Fig. 10). This means that, at constant band bending, on changing the ac frequency, the levels that can follow the signal change too. On the other hand, at constant ac frequency, on changing the band bending, the deep levels of the depletion region lying below E_ω will not change in occupancy.

The intersection of the cutoff level with the Fermi level allows us to locate the corresponding point within the barrier X_c that separates two regions of the a-SC (see Fig. 10): the first (for $x > X_c$) where all electronic states fully respond to the ac signal, and the second (for $x < X_c$) where no states respond at all. The band bending at X_c is given by

$$\Phi_c = \Phi(X_c) = -kT \ln(\omega\tau_0) - \Delta E_F \qquad (3.1.4)$$

where $\Delta E_F = (E_c - E_F)_{bulk}$.

It was shown in [83, 94–96] that by assuming the model of an a-SC Schottky barrier to be valid also for the a-SC/El junction, the equivalent electrical circuit of the junction reported in Figure 11 can be obtained for concentrated electrolytic solutions. Once again, the main differences with respect to the equivalent electrical circuit of the junction for crystalline materials occur on the electrode side, which is now represented by a frequency-dependent resistance, $R(\Delta\Phi_{sc}, \omega)$, in parallel with two capacitors in series, the first of which, $C(\Phi_c, 0)$, is quite insensitive to both frequency and bias, whilst the second, $C(X_c) = \varepsilon\varepsilon_0/X_c(\Delta\Phi_{sc}, \omega)$, is frequency- as well as bias-dependent.

In the simplified case of a constant DOS within the mobility gap of the a-SC [$N(E) = N$; see Fig. 10], the analytical solutions for the admittance components of the junction can be derived for $\Delta\Phi_{sc} > \Phi_c$ and at not too high band bending ($\Delta\Phi_{sc} \leq E_g/2$ [83, 85, 94, 104]). The total parallel capacitance is given by

$$C(\Delta\Phi_{sc}, \omega) = \frac{(\varepsilon\varepsilon_0 e^2 N)^{1/2}}{1 + \ln(\Delta\Phi_{sc}/\Phi_c)} \qquad (3.1.5)$$

whilst the parallel conductance of the junction is given by

$$G(\Delta\Phi_{sc}, \omega) = \frac{\omega\pi}{2} \frac{kT}{e\Phi_c} \frac{(\varepsilon\varepsilon_0 e^2 N)^{1/2}}{[1 + \ln(\Delta\Phi_{sc}/\Phi_c)]^2} \qquad (3.1.6)$$

It has been shown that $G(\Delta\Phi_{sc}, \omega)$ has a spectroscopic character with respect to the distribution of electronic states, whilst variations in DOS cause only minor changes in the

$C(\Delta\Phi_{sc}, \omega)$-vs-potential plots [94–96]. With respect to the MS analysis, which allows one to get both the doping density of the SC and the flat-band potential of the junction from the same plot [see Eq. (2.2.8)], in the case of a-SC a longer fitting procedure is required in order to get the flat-band potential and to locate the mobility band edges. As reported in [94–96], the fitting must be carried out on both components of the admittance of the junction, after the correction of the measured quantities for the equivalent circuit shown in Figure 11, by imposing the condition that at all the frequencies employed for the ac signal both the $1/C_{sc}$-vs-$\Delta\Phi_{sc}$ and the $1/G_{sc}$-vs-$\Delta\Phi_{sc}$ plots give the same U_{fb}-value, within an assigned uncertainty (in our case, 0.05 V). Moreover, an additional constraint arises from Eq. (3.1.4), which predicts for Φ_c a variation of 59 mV over a decade of frequency at room temperature.

Figures 12–14 display the experimental capacitance and conductance curves recorded for a-WO$_3$ and a-Nb$_2$O$_5$ anodic films at various frequencies, whilst a fitting of the admittance components according to Eqs. (3.1.5) and (3.1.6) is shown in Figure 15 for an a-TiO$_2$ film; the results of the fitting procedure are reported in Table II. Despite the simplifying assumption of constant DOS, the very good agreement with the experimental data supports the theoretical analysis outlined.

The fact that the theory of a-SC Schottky junctions predicts a parabolic potential distribution inside the a-SC in the high-depletion regime [94–104] explains why different authors have observed experimentally linear MS plots for many disordered passive film–electrolyte junctions. However, the extrapolation of such MS plots to the potential axis gives flat-band potential values that are completely wrong, as can be easily inferred from the study of the a-TiO$_2$/El junction reported in [85, 96, 104].

More importantly, the theory of a-SC Schottky barrier allows us:

(a)   to explain, without any further assumption, the experimentally well-known frequency dependence of the capacitance values reported for amorphous semiconducting passive films;

(b) to locate correctly the CB mobility edge, E_c, through Eq. (3.1.4) [and not by means of the Eq. (2.2.8), which is valid for crystalline SCs but misleading when used for a-SCs];

(c) to calculate the DOS distribution around the Fermi level by means of the relationship [94, 96, 102]

$$N[E_F - e\Phi_c(\omega)] = \text{const}\frac{G_p}{\omega} \qquad (3.1.7)$$

where G_p is the parallel conductance and the constant is independent of the ac frequency.

Further support for the model for the impedance of the a-SC/El junction, outlined above, derives from new features

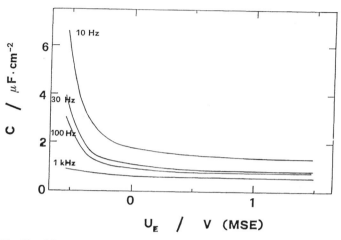

Fig. 13.   Measured parallel capacitance vs electrode potential at different frequencies in 0.5 M H_2SO_4 solution for an a-Nb_2O_5 film grown in the same solution up to 84 nm. [*Electrochim. Acta* 35(1), 99–107 (1990); reprinted with permission from Elsevier Science.]

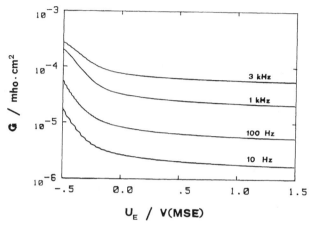

Fig. 14.   Parallel conductance vs electrode potential at different frequencies for the same film as in Figure 13. [*Electrochim. Acta* 35(1), 99–107 (1990); reprinted with permission from Elsevier Science.]

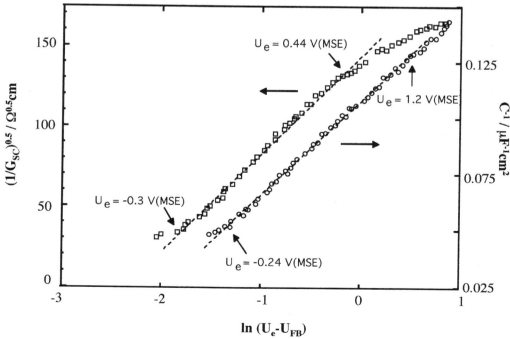

Fig. 15.   Fitting, according to Eqs. (3.1.5) and (3.1.6), of the impedance data recorded at $f = 3$ Hz for an a-TiO_2 film grown at 2 V/s in 0.5 M H_2SO_4 solution up to 20 V. Flat-band potential: -0.47 V (MSE). [*Ber. Bunsenges. Phys. Chem.* 101(6), 932–942 (1997); reprinted with permission from VCH Verlagsgesellschaft mbH.]

Table II.   Physical Parameters Derived for Three Anodic Oxide Films[a]

Oxide	Thickness (nm)	U_{fb} (SHE) (V)	Φ_c ($f = 3$ Hz) (mV)	ΔE_F (eV)	$N(E_F)$ (cm^{-3} eV^{-1})
a-WO$_3$	54	0.51 ± 0.04	174	0.44	2.0×10^{19}
a-Nb$_2$O$_5$	84	0.06 ± 0.04	186	0.43	1.5×10^{19}
a-TiO$_2$	83	0.10 ± 0.05	162	0.45	2.3×10^{20}

[a]From the fitting, according to Eqs. (3.1.5) and (3.1.6), of the impedance data in 0.5 M H$_2$SO$_4$. Raw data were corrected for the electrical equivalent circuit of Figure 11; a value of 30 μF/cm^2 has been used for C_H. [*AIChE J.* 38(2), 219–226 (1992); reprinted with permission from the American Institute of Chemical Engineering.]

observed in the photoelectrochemical behavior of amorphous passive films, which are strictly related to the electronic structure of the films.

3.2. Optical Absorption and Photoelectrochemical Response of the Metal–Passive-Film–Electrolyte Junction

Like the electronic properties, the optical properties of materials, are considerably affected by their amorphous nature. The main differences in the photocurrent response of disordered thin films with respect to the case of bulk crystalline semiconductors arise from the following facts:

(a) The optical bandgap of an amorphous material may coincide or not with that of the crystalline counterpart, depending on the presence of different types of defects that can modify the DOS distribution.

(b) In contrast with crystalline materials, the generation of free carriers by the absorption of photons having energy equal to or higher than the optical bandgap of the film may depend on the electric field, owing to the presence of an initial (geminate) recombination.

(c) The presence of reflecting metal–film and film–electrolyte interfaces makes possible the occurence of multiple reflections, even for photons having energy higher than the optical absorption threshold; this fact causes interference effects in the curves of photocurrent vs film thickness.

(d) The small thickness of the film makes possible optical excitation at the inner metal–film interface, which can inject photocarriers from the underlying metal into the VB or CB of the passive film (internal photoemission), or directly into the electrolyte (external photoemission) in the case of very thin films (1–2 nm thick).

In the following sub-subsections we will analyze the foregoing points in detail by comparing the theoretical expectations with the experimental results pertaining to different metal–passive film–electrolyte junctions.

3.2.1. *Optical Gap in Amorphous SCs*

As to the first point, we recall the relationship between the optical absorption coefficient and the opticalband gap E_g of a material, valid near the optical absorption threshold [33, 105]:

$$\alpha h\nu = A(h\nu - E_g)^{n/2} \qquad (2.3.7)$$

where, for crystalline materials, n can assume different values depending on the nature of the optical transitions between occupied electronic states of the VB and vacant states of the CB. The optical transitions at energies near the bandgap of a crystalline material may be direct or indirect. In the first case no intervention of other particles is required, apart the incident photon and the electron of the VB; in the second case the optical transition is assisted by lattice vibrations.

Assuming a parabolic DOS distribution [$N(E) \propto E^{1/2}$] near the band edges, in the case of direct transitions n assumes values equal to 1 or 3, depending on whether the optical transitions are allowed or forbidden in the quantum-mechanical sense [33]. In the case of indirect optical transitions the value of n in Eq. (2.3.7) is equal to 4 [33].

In the case of amorphous materials, owing to the relaxation of the k-conservation selection rule, "no intervention of phonons is invoked to conserve momentum and all energy required is provided by the incident photons" [105]. By assuming again a parabolic DOS distribution in the vicinity of the mobility edges of both the conduction and the valence band (above E_c and below E_v, with reference to Fig. 9), it has been shown [91] that the following relationship holds:

$$\alpha h\nu = \text{const}(h\nu - E_g^m)^2 \qquad (3.2.1)$$

where $E_g^m = E_c - E_v$ is the mobility gap of the a-SC. The exponent 2 is reminiscent of the indirect optical transitions in crystalline material, but now photons interact with the solid as a whole; this type of transition in amorphous materials is termed *nondirect*.

Because some tailing of states is theoretically foreseen for a-SC by every proposed model of DOS, E_g^m represents an extrapolated rather then a real zero in the density of states.

In the presence of a DOS distribution varying linearly with energy in the ranges $E_c - E_A$ and $E_B - E_v$ of Figure 9b, it is still possible to get a similar expression for the absorption coefficient [90]:

$$\alpha h\nu = \text{const}(h\nu - E_g^{opt})^2 \qquad (3.2.2)$$

where E_g^{opt} now represents the difference of energy $E_A - E_v$ or $E_c - E_B$ in Figure 9b, whichever is the smaller, and the constant assumes values close to 10^5 eV^{-1} cm^{-1}. The range of energy over which Eq. (3.2.2) should be valid is on the order of 0.3 eV or less [105].

In order to distinguish between these two different models of optical transitions, both giving the same dependence of the absorption coefficient on the photon energy, we will refer to the first one as the Tauc approximation for the calculation of the SC mobility gap and to the second one as the Mott–Davis approximation for the SC optical gap. When the plots of $(\alpha h\nu)^{0.5}$ vs $h\nu$

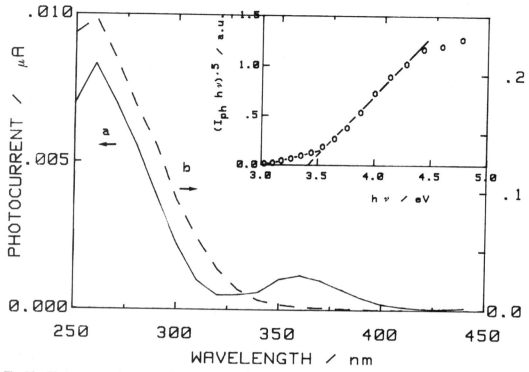

Fig. 16. Photocurrent action spectra for films grown anodically on Nb in 0.5 M H_2SO_4 electrolyte up to (a) 0 V and (b) +0.5 V (MSE). The spectra were recorded polarizing the electrode at the final formation voltage. Inset: determination of the optical bandgap from spectrum b. [*J. Electroanal. Chem.* 293, 69–84 (1990); reprinted with permission from Elsevier Science.]

display a linear region larger than 0.3 eV, it seems more correct to interpret the data on the basis of the Tauc model of optical transitions. In Figure 16 we report an interesting case of coexistence of both types of transitions for thin anodic films grown anodically on niobium [106], where the presence of a mobility gap on the order of 3.5 eV is observed in the high photon energy range, extending around 1 eV, followed by an optical gap (in the Mott sense) of around 3.05 eV over an energy interval of about 0.3 eV.

In the case of anodic films on valve metals, the so-called Urbach tail for the absorption coefficient is frequently observed for photon energies lower than the mobility gap; it follows the law

$$\alpha = \alpha_0 \exp\left(-\gamma \frac{E_0 - h\nu}{kT}\right) \qquad (3.2.3)$$

with γ and α_0 constant. Such a relationship, which has been found to hold for crystalline materials too, has been rationalized in the case of amorphous SCs by assuming an exponential distribution of localized states in the band edge tails [107]. In this situation the value of E_0 obtained from the $\ln \alpha$-vs-$h\nu$ (Urbach) plot should coincide with the mobility gap determined according to Eq. (3.2.1). A typical example is reported in Figure 17 for an a-MoO_3 film [108]: the value of E_g^m, equal to about 3.10 eV, is in good agreement with the value of E_0 ($\cong 3.15$ eV) derived form the Urbach plot. Other explanations have been suggested

by different authors for such behavior in the case of crystalline materials [105].

In interpreting the information provided by the optical gap values, it is important to recall the statement in Mott and Davis's book [105]: "A general rule appears to be that, if the local atomic order is not appreciably altered in the amorphous phase, the gaps in the two states (amorphous and crystalline) are not appreciably different." This rule works better for materials whose band structure is mainly determined by the nearest-neighbor overlap integral (tight-binding materials). In Table III we report the optical gap values determined by photocurrent spectroscopy for different passive films: for the amorphous anodic oxides on valve metals they are systematically larger than those of their crystalline counterparts. The difference $E_g^{amor} - E_g^{cryst}$ is in the range of 0.1–0.35 eV, in agreement with the extension of the localized state regions near the band edges due to the lattice disorder. Moreover, by taking into account that a value of E_0 nearly coincident with the mobility gap has been frequently derived from the Urbach plot, it seems quite reasonable to suggest for this a class of amorphous materials a band model similar to that shown in Figure 9b, with an exponentially decreasing DOS in the mobility gap of the films at energies lying below E_c and above E_v.

A mobility gap of the passive film lower than the bandgap of the crystalline counterpart must be interpreted as an indication

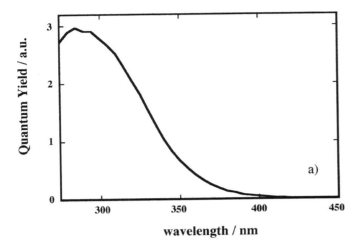

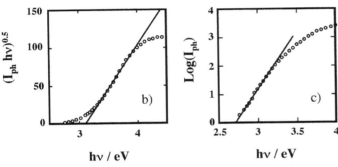

Fig. 17. (a) Photocurrent action spectrum recorded at +2 V (SCE) for a film grown up to 40 V on sputtered Mo in 0.1 M sodium acetate in a 2% v/v aqueous acetic acid solution. Lower part: determination of the optical bandgap, (b) assuming indirect transitions, and (c) from the Urbach tail. [*J. Electrochem. Soc.* 147, 1366–1375 (2000); reprinted with permission from The Electrochemical Society, Inc.]

Table III. Measured Optical Gap E_g^m for Passive Films on Pure Metals Compared with the Bandgap E_g of the Crystalline Counterpart

Phase	E_g^m (eV)	E_g (eV)	ΔE_{am}^a (eV)
ZrO_2	4.70–4.80	4.50	0.20–0.30
Ta_2O_5	3.95–4.05	3.85	0.10–0.20
Nb_2O_5	3.30–3.40	3.15	0.15–0.25
TiO_2	3.20–3.35	3.05^b–3.20^c	0.15–0.20
WO_3	2.95–3.05	2.75	0.20–0.30
MoO_3	2.95–3.10	2.90	0.05–0.20
Cr_2O_3	3.30–3.55	3.30	0.0–0.25
NiO	3.43	3.45–3.55	0
Cu_2O	1.86	1.86	0
Fe_2O_3	1.95	1.90	0.05

^{a}Difference between E_g^m and E_g (see text).

^{b}Rutile phase.

^{c}Anatase phase.

that there are differences in short-range order between the two phases. A different short-range order can lead to the formation of a defective structure, with a high density of localized states within the mobility gap as well as changes in the density of the passive film, which is known to affect also the value of the optical gap in amorphous materials [86, 87, 90, 91].

The experimental findings on passive anodic films and corrosion layers suggest that large differences in the bandgap values between the amorphous and crystalline counterparts must be due to a different chemical environment around the metallic cation or to the presence of large amount of defects within the passive films, which produce electronic states within the mobility gap. We will come back to these aspects in Section 4, where the use of the PCS technique in a more quantitative way will be discussed.

In order to derive the optical gap of passive films from the photocurrent spectra (see Figs. 16, 17), we need to relate the optical absorption coefficient α to the photocurrent measured experimentally. This goal will be accomplished in Section 3.3.

3.2.2. Geminate Recombination in Amorphous Films

A major aspect to take into account in the formulation of the transport equations of the photocarriers is related to the possible presence of geminate recombination effects in the generation of mobile photocarriers. This phenomenon occurrs generally in any material where the photogenerated carriers display very low mobility. In the case of amorphous materials, localized states are present below the CB and above the VB mobility edges as a consequence of lattice disorder. The mobility of carriers in these states is much lower than that in the extended states, so that the existence of initial recombination effects in amorphous materials is quite probable. In fact, during the thermalization time the electron–hole pairs do not cover a distance long enough to prevent recombination due to their mutual coulombic attraction. Owing to this insufficient separation, a certain fraction of the photogenerated carriers recombine before the transport process can separate them permanently. As a consequence, the efficiency of free photocarrier generation must be taken into account when dealing with amorphous materials, and it acts to lower the quantum yield in comparison with crystalline materials.

The clearest evidence for the presence of geminate recombination effects in passive films comes from the photocurrent-vs-potential curves (photocharacteristics) at constant film thickness recorded under suprabandgap illumination with light of different wavelengths. In the absence of initial recombination effects, no influence of the photon energy on the shape of the I_{ph}-vs-U_e plots is expected according to the simple Gärtner–Butler model valid for crystalline SCs. Moreover, the model predicts a linear dependence of I_{ph}^2 on the electrode potential [see Eq. (2.3.6)] as long as the surface and space-charge recombination rates are negligible, which has been confirmed experimentally at high electrode potentials. Even more complicated models for the photocurrent in crystalline SC/El junctions

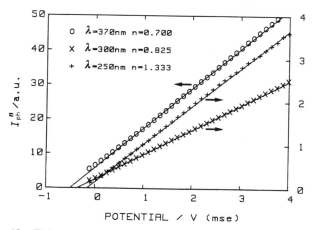

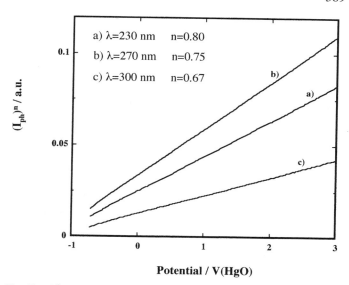

Fig. 18. Fitting, according to different power laws, of experimental curves of photocurrent vs potential for an anodic film grown on Ti at 2 V/s in 0.5 M H_2SO_4 electrolyte, recorded at different wavelengths. [*Electrochim. Acta* 38(1), 29–35 (1993); reprinted with permission from Elsevier Science.]

Fig. 19. I_{ph}^n vs electrode potential at different wavelengths for a Ta_2O_5 film grown at 8 mA/cm^{-2} in 0.1 M NaOH solution up to 10 V (MSE).

do not predict any change of the shape of the photocharacteristics with the wavelength. In fact, however, the energy of incident photons affects the shape of the photocharacteristics, recorded experimentally for different passive-film–electrolyte junctions, in such a way that the curves are fitted as power laws, I_{ph}^n vs U_e with n increasing for decreasing wavelength. Such behavior, depicted in Figure 18 for a thin anodic oxide on titanium, has been observed in amorphous semiconducting films on valve metals (a-WO$_3$, a-Nb$_2$O$_5$, a-TiO$_2$) [84, 85, 109–111].

In the case of insulating crystalline materials, one expects the photocurrent to depend linearity on the electrode potential, and the shape of the photocharacteristics to be independent of the photon energy, in the absence of trapping phenomena which could modify the electric field distribution [112]. In presence of trapping, phenomena sublinear behavior of the photocurrent has been predicted [113]. On the other hand, if trap-limited mobility of the photocarriers is taken into account, supralinear photocurrent behavior is expected only at very high electric fields—once again, independent of the photon energy [114]. The different behavior of amorphous materials is evidenced in Figure 19, displaying the photocharacteristics at various wavelengths for an insulating Ta$_2$O$_5$ anodic film [115]: a clear change of the supralinear shape with the wavelength is reflected in the different exponent n of the fitted power laws, all giving the same zero-photocurrent potential (about -1.3 V with respect to HgO).

A model for the initial recombination effects in amorphous materials, based on the treatment originally developed by Onsager for electrolytic solutions [116], has been proposed by Pai and Enck [117]. The model takes into account the three-dimensional aspect of the generation of photocarriers and yields a mathematical expression for the efficiency of free carrier generation, η_g, as a function of the electric field F and of the thermalization distance r_0. The latter quantity is the distance traveled by a photogenerated electron–hole pair during the thermalization time. According to these authors the efficiency of

generation is given by the following expression:

$$\eta_g(r_0, F) = \frac{kT}{eFr_0} \exp(-A) \exp\left(-\frac{eFr_0}{kT}\right)$$
$$\times \sum_{m=0}^{\infty} \frac{A^m}{m!} \sum_{n=0}^{\infty} \sum_{l=m+n+1}^{\infty} \left[\left(\frac{eFr_0}{kT}\right)^l \times \frac{1}{l!}\right]$$

$$(3.2.4)$$

where k is the Boltzmann constant, T the absolute temperature, e the electronic charge (in absolute value), and $A = e^2/4\pi\varepsilon\varepsilon_0 r_0 kT$. In Eq. (3.2.4) the only parameter the varies with the photon energy is the thermalization distance r_0, which is a function of the *excess energy* $h\nu - E_g$, representing the difference between the energy of incident photons and the optical gap of the material. This excess energy is dissipated by collisions with the constituents of the material during the thermalization time. It has been suggested that in Eq. (3.2.4) an average dielectric constant between the static and the high-frequency values should be used. Anyway, the value of ε affects the absolute value of η_g, but it does not change its functional dependences. This fact is evident in Figures 20 and 21, where the theoretical curves of η_g as a function of the electric field and the thermalization distance are plotted for two different values of the dielectric constant. From the figures it is clear that both r_0 and ε act in the same direction, because an increase in either tends to lower the coulombic attraction between the photogenerated electron–hole pair (Poole–Frenkel effect).

The mathematical structure of the Pai–Enck expression for η_g [Eq. (3.2.4)] is in agreement with a series of experimental results on passive films, showing the influence of electric field and photon energy on the shape of the photocurrent potential curves [84, 85, 109–111, 115]. According to these findings, we will use Eq. (3.2.4) for η_g, both for insulating (constant electric field) and for semiconducting amorphous films. In this last case

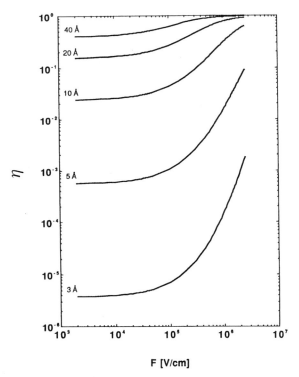

F [V/cm]

Fig. 20. Theoretical dependence of the efficiency of free carriers generation on the electric field, calculated from Eq. (3.2.4) for a material having $\varepsilon = 15$. Different curves correspond to different values of the thermalization length. [*Electrochim. Acta* 38(1), 29–35 (1993); reprinted with permission from Elsevier Science.]

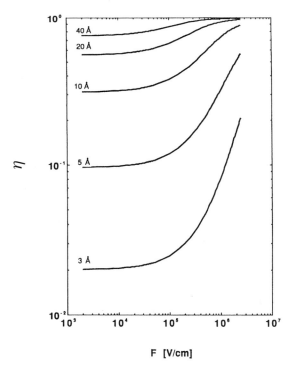

F [V/cm]

Fig. 21. Theoretical dependence of the efficiency of free carriers generation on the electric field, calculated from Eq. (3.2.4) for a material having $\varepsilon = 48$. Different curves correspond to different values of the thermalization length. [*Electrochim. Acta* 38(1), 29–35 (1993); reprinted with permission from Elsevier Science.]

an average value of the electric field within the space-charge region will be used in Eq. (3.2.4) for calculating the generation efficiency to be used in the expressions for the photocurrent vs electrode potential (see Section 3.3).

3.2.3. Interference Effects during the Growth of Passive Films

In our opinion, one of the most interesting findings about the photoelectrochemical behavior of thin films in comparison with bulk materials is the appearance of interference effects in the photocurrent measured as a function of the film thickness. The phenomenon of multiple reflections of unabsorbed light in thin films is very well known, and it has been used in the past for obtaining information on the change of thickness of passive films during anodization [118–120]. In this case the light reflected from the metal–film–electrolyte function is monitored as a function of the film thickness. Less frequently has been reported the presence of interference effects on the measured photocurrent when photoconducting films are illuminated with photons having energy higher than their optical bandgap.

Figure 22 shows the interference effects on the photocurrent measured during the anodic growth of an oxide film on tantalum metal under illumination with light having energy higher than the mobility gap of a-Ta_2O_5 (about 4.05 eV). Maxima and minima in the measured photocurrent as a function of the formation voltage (proportional to the film thickness) are very evident as long as the film thickness is much smaller than the light absorption length α^{-1}, so that the phenomenon is well observable for incident photons having energy slightly higher than the bandgap of the film. With increasing photon energy the interference effects disappear at progressively lower thicknesses, as expected for strongly absorbed light. Analogous findings have been observed with semiconducting oxide films both during the growth (variable thickness and nearly constant anodizing electric field) and with electrodes anodized up to different thicknesses and investigated at lower electrode potentials [121, 122]. In the latter case, the photocurrent values recorded at constant wavelength and electrode potential on passive films grown on niobium up to different formation potentials have been plotted against the film thickness: the nonmonotonic behavior of these curves was analogous to previous findings on anodic TiO_2 films [123].

The theoretical fitting of this phenomenon will be performed in the next section by taking into account multiple interference effects caused by the reflections of the light at the solution–oxide and oxide–metal interfaces (see Fig. 23), once the analytical expressions are known for the photocurrent in insulating or semiconducting amorphous films. It is worth mentioning here that the film thicknesses at which maxima and minima occur in the photocurrent intensity are essentially governed by the anodization ratio (expressed in angstroms per volt), which is the reciprocal of the anodizing electric field. In turn, the fitting of the curves reported in Figure 22 allows one to derive the kinetic parameters for the film growth. The possibility of obtaining the growth parameters of the films by simply recording the photocurrent during their growth opens a new

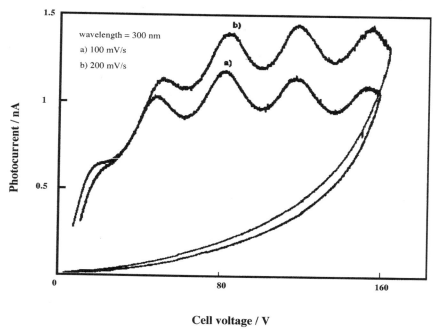

Fig. 22. Photocurrent intensity vs formation voltage for passive films growing tensiodynamically on Ta at different growth rates in 0.5 M H_2SO_4 solution, under monochromatic irradiation with $\lambda = 300$ nm.

Multiple internal reflections

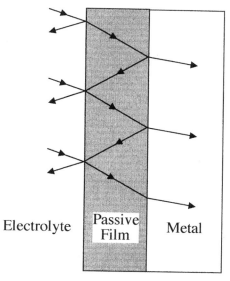

Fig. 23. Optical model for multiple internal reflections at the metal–passive-film–electrolyte junction. The light flux is partially transmitted and partially reflected at both the interfaces.

prospect for investigating the influence of the absorbed light on the kinetics of growth of passive films as well as on their solid state properties during growth [120, 121, 124].

Another interesting finding, reported in Figure 24, is the presence of interference effects in the recorded photocurrent during the anodization of electropolished aluminum electrodes

in organic-anion-containing solutions under illumination with photons having energy much lower than the bandgap of Al_2O_3 ($\cong$6.3 eV). The intriguing aspect of such experiments is the occurrence of the interference effects only in the presence of selected organic electrolytes, which favor the incorporation into the anodic oxide of species originating a band of defects within the forbidden gap of the a-Al_2O_3 film. Support for this interpretation comes from both the anodic photocurrent spectra and experiments performed successively in different electrolytic solutions containing or not containing the organic anions [125].

3.2.4. Photoemission Phenomena at the Metal–Passive-Film Interface

In this sub-subsection we discuss the role of the inner metal–film interface in the generation processes under illumination for thin passive films. In this case, regardless of the wavelength of the incident light, a large fraction of photons impinging the oxide–solution interface arrive at the metal–film interface, where the electrons in the metal surface can be excited to higher energy levels, leaving vacant states below the Fermi level of the metal. The fate of the excited states in the metal depends on the occurrence of different physical deactivation processes at that interface. Besides thermal deactivation by scattering of excited electrons by lattice vibrations, photoemission of excited photocarriers of the metal can be observed.

In the case of very thin passive films ($d_{ox} \leq 2$ nm), external photoemission processes become possible, by tunneling of the electrons or holes excited at the metal surface through the film. Although a hole photoemission process was suggested years ago in the case of a gold electrode covered with a very

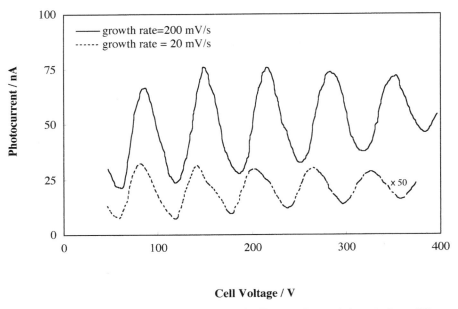

Fig. 24.   Photocurrent intensity vs formation voltage for films growing tensiodynamically at different rates on Al in 0.1 M ammonium tartrate electrolyte, under monochromatic irradiation at λ = 280 nm.

thin oxide [126], the photoemission of electrons directly from the metal to the ground state of liquid water has been observed more frequently through very thin oxide films covering metals [127–129]. When such an external photoemission process occurs in the absence of diffuse double layer effects or adsorbed large molecules, it has been shown that the emission photocurrent from the metal to the acceptor species in solution depends on both the photon energy and the electrode potential according to the so-called 5/2 power law [130]:

$$I_{ph} \propto (h\nu - h\nu_0 - eU_e)^{5/2} \qquad (3.2.5)$$

where U_e is the electrode potential measured with respect to a reference electrode and $h\nu_0$ is the photoelectric threshold at $U_e = 0$. From Eq. (3.2.5) it follows that the photoelectric threshold changes with the reference electrode. By extrapolating to $I_{ph} = 0$ the plot of $I_{ph}^{0.4}$ as a function of the electrode potential, under irradiation with light at a fixed wavelength, we can derive the photoelectric threshold $h\nu_0 = h\nu - eU^*(\text{ref})$, where U*(ref) is the intercept potential measured with respect to the reference electrode employed. From this value it is possible to derive the energy level of a hydrated electron in the electrolyte (the so-called "conduction band" of liquid water, for aqueous solutions) with respect to the vacuum level, using the relationship

$$E^c(H_2O) = h\nu_0 + eU_{ref}(\text{vac}) \qquad (3.2.6)$$

Equation (3.2.6) is analogous to Eq. (2.2.11) for locating the Fermi level of a semiconductor on the physical scale of energy. Again, $U_{ref}(\text{vac})$ is the potential of the reference electrode with respect to the vacuum scale. Moreover, from Eq. (3.2.5) it follows that a plot of $h\nu_0$ values obtained at different wavelengths vs $h\nu$ must have unit slope.

On the other hand, at constant electrode potential U_e the photoemission threshold can be obtained from the energy spectrum of the photoemission current [127, 129] by plotting $I_{ph}^{0.4}$ vs $h\nu$ (see Figs. 25, 26). In this case, the threshold energy value should shift by 1 eV/V on changing the electrode potential (see the inset of Fig. 26), according to Eq. (3.2.5). Once again, $E^c(H_2O)$ can be obtained by means of Eq. (3.2.6). Following this procedure, the location of the "conduction band" of liquid water, at -1.0 ± 0.1 eV with respect to the vacuum level, has been evaluated by us based on experimental data pertaining to Al, Zr, and Ni [127–129], in agreement with the analogous determination reported in [126].

The preceding relationships help to discriminate between a cathodic photocurrent due to external photoemission and those originating from band-to-band excitation in p-type semiconducting thin films, like those formed on Ni, Cr, and stainless steels (see Section 4). In fact, in this last case an optical bandgap value independent of (or very slightly dependent) on the electrode potential is expected.

For thicker films ($d_{ox} \geq 5$ nm), where the external photemission processes are forbidden owing to a very low probability of tunneling through the film, the possibility of internal photoemission due to the injection of photoexcited electrons (or holes) from the metal into the CB (or VB) of the passive film must be considered. In such a situation the internal photocurrent emission varies with the photon energy according to the so-called Fowler law [130, 131]:

$$I_{ph} = \text{const}(h\nu - E_{th})^2 \qquad (3.2.7)$$

where E_{th} is the internal photoemission threshold energy, which can be obtained from a plot of $I_{ph}^{0.5}$ vs the photon energy. This threshold is a measure of the distance in energy between the Fermi level of the metal and the edge of the film CB (electron photoemission) or VB (hole photoemission). The occurrence of electron or hole internal photemission in the case of insulating films is established by the direction of the electric field, and in turn by the electrode potential value with respect to the inversion photocurrent potential. In the absence of trapping effects, the inversion photocurrent potential can be used to determine the flat-band potential of insulating passive films.

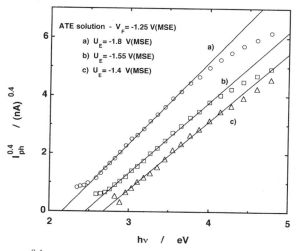

Fig. 25. $I_{ph}^{0,4}$ vs photon energy at different electrode potentials for a film grown on Al in 0.1 M ammonium tartrate electrolyte up to -1.25 V (MSE): (a) $U_e = -1.8$ V; (b) $U_e = -1.55$ V; (c) $U_e = -1.4$ V (MSE). [*J. Electrochem. Soc.* 140(11), 3146–3152 (1993); reprinted with permission from The Electrochemical Society, Inc.]

Obviously, Eqs. (3.2.5) and (3.2.7) can be referred also to the photoemission yield ($I_{ph}/e\Phi_0$).

In the case of insulating anodic films on valve metals, internal electron photemission processes are usually observed under cathodic polarization and under illumination with photons having energy lower than the optical bandgap of the film [127, 129]. In Figure 27 we report the determination of the internal photoemission threshold for an anodic oxide film grown on electropolished aluminum metal. It is interesting to note that, for anodic oxide films grown on Al irradiated with photons having energy lower than the bandgap of Al_2O_3 ($E_g \cong 6.3$ eV), the cathodic photoemission process is always observed, regardless of the anodizing conditions and the metal surface treatment, whilst an anodic photocurrent spectrum is recorded only for films anodized in solutions containing organic species [125] or when a hydrated layer is formed on the surface of the films [132]. Moreover, we mention that in the intermediate range of film thickness, between the very thin (≤ 2 nm) and thicker (≥ 5 nm) anodic films grown on aluminum, the presence of a variable cathodic photoemission threshold, due to the quantum size effects, has been hypothesized [127].

In Table IV we report the cathodic internal photoemission thresholds for a series of insulating oxide films grown on different valve metals; these allow one to locate the energy level of the conduction band of the oxide films with respect to the Fermi level of the underlying metal, once the work function of the metal is known. In the case of semiconducting films no internal photoemission is expected, owing to the absence of any electric field at the metal–film interface.

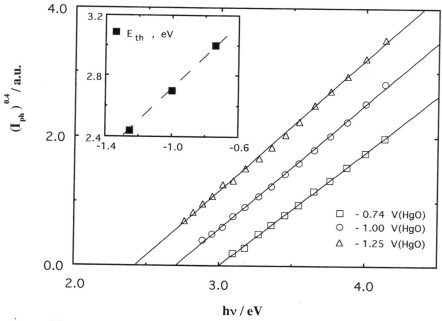

Fig. 26. $I_{ph}^{0,4}$ vs photon energy derived from the long-wavelength region of the photocurrent spectra recorded at different cathodic polarizations for a passive layer grown on a Zr electrode treated mechanically and immersed in 0.1 M NaOH solution at the corrosion potential. Inset: energy threshold for external photoemission as a function of the electrode potential. [*Electrochim. Acta* 41(16), 2511–2522 (1996); reprinted with permission from Elsevier Science.]

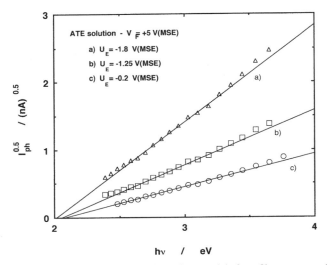

Fig. 27. Fowler plots at different electrode potentials for a film grown on Al in 0.1 M ammonium tartrate electrolyte up to +5 V (MSE). (a) $U_e = -1.8$ V; (b) $U_e = -1.25$ V; (c) $U_e = -0.2$ V (MSE). [*J. Electrochem. Soc.* 140(11), 3146–3152 (1993); reprinted with permission from The Electrochemical Society, Inc.]

Table IV.   Threshold Energy Values Derived from the Fowler Plots for Internal Photoemission at the Metal–Passive-Film Interface [14, 15, 108, 115, 127, 129, 146, 180]

Interface	$E_{th}{}^a$ (eV)
Y/Y$_2$O$_3$	2.10
Bi/Bi$_2$O$_3$	1.40
Pb/PbO	1.00
Zr/ZrO$_2$	1.80
Al/Al$_2$O$_3$	2.00
Ta/Ta$_2$O$_5$	1.50
Hf/HfO$_2$	2.15
Mo–Ta 79 at. %/anodic oxide	1.75
Al–Ta 34 at. %/anodic oxide	1.53

^{a}Uncertainty ±0.1 eV.

3.3. Photocurrent–Potential Curves in Passive-Film–Electrolyte Junctions

The differences in the electronic properties between crystalline and amorphous materials affect greatly the physical models for describing the photocurrent behavior of the passive-film–electrolyte junction. As discussed in Section 3.2, both the generation processes and the transport properties of the photogenerated carriers are strongly influenced by the band structure of the films. In the case of disordered semiconducting films, simplifying assumptions can be made on the contributions to the collected photocurrent arising from the different regions of the material. Moreover, the electric field distribution within the space-charge region also is changed with respect to that of crystalline SCs (Section 3.1); this fact influences the quantum yield of the passive-film–electrolyte junction. For in-

sulating films different electric field distributions within the film are predicted, depending on the occurrence of trapping phenomena of the photogenerated carriers. These facts must be considered in writing the equations for the collected photocurrent.

To clarify the differences between the photoelectrochemical behavior of passive films and that of bulk crystalline materials, we will discuss separately the expressions for the photocurrent for semiconducting and insulating films. The efficiency of photocarrier generation (Section 3.2.2) will be taken into account, in order to fit the experimental photocharacteristics recorded at constant film thickness under irradiation with varying energy of the incident photons. Finally, the influence of the film thickness on the photocurrent behavior of passive-film–electrolyte junctions will be taken into account, in order to explain quantitatively the interference effects on the photocurrent described in Section 3.2.

3.3.1. A Photocurrent Expression for the a-SC/El Junction

Due to the low mobility of carriers in amorphous materials, it is reasonable to assume that a negligible contribution to the measured photocurrent arises from the field-free region of the semiconductor. In this case it is quite easy to derive an expression for the migration term in the space-charge region of the a-SC, in a quite similar way to that followed by Gärtner.

As in the Butler model, we will assume the absence of kinetic control in the solution and a negligible recombination rate at surface of the semiconductor. The limits of validity of that assumption have been discussed in the literature for the case of crystalline SC/El junctions [61], and they will not be repeated here. Initially we will assume also an 100% efficiency of free carrier generation ($\eta_g = 1$) under illumination with light having energy higher than the SC mobility gap; then we will relax that assumption in discussing the behavior of particular systems.

In steady-state conditions the recombination of photogenerated carriers in the space-charge region can be taken into account by assuming that the probability for any carrier photogenerated at a position x to leave the space-charge region is given by [133]

$$P(x, \overline{F}) = \exp\left(-\frac{x}{L_d}\right) \quad (3.3.1)$$

where $\overline{F}$ is the mean electric field in the space-charge region of the a-SC, and L_d is the drift length of the photocarrier ensemble in the average field approximation, given by

$$L_d = \mu \tau \overline{F} \quad (3.3.2)$$

μ and τ being the drift mobility and the lifetime of the photocarriers, respectively. According to these equations and to the assumptions made, we can write

$$I_{ph} = I_{drift} = e\Phi_0 \int_0^{X_{sc}} \alpha \exp(-\alpha x) P(x, \overline{F}) \, dx \quad (3.3.3)$$

where Φ_0 is the photon flux corrected for the reflection at the electrolyte–film interface, assuming negligible reflections at the film–metal interface. By integration of Eq. (3.3.3) we get [133]

$$I_{\mathrm{ph}} = e\Phi_0 \frac{\alpha L_{\mathrm{d}}}{1+\alpha L_{\mathrm{d}}}\left[1 - \exp\left(-X_{\mathrm{sc}}\frac{1+\alpha L_{\mathrm{d}}}{L_{\mathrm{d}}}\right)\right] \quad (3.3.4a)$$

From Eq. (3.3.4a) follows a direct proportionality between the photocurrent and the optical absorption coefficient for $\alpha L_{\mathrm{d}} \gg 1$ and $\alpha X_{\mathrm{sc}} \ll 1$, as previously derived for crystalline materials [see Eq. (2.3.8)]. On the other hand, for $\alpha L_{\mathrm{d}} \ll 1$ a direct proportionality between I_{ph} and α is still assured by the fractional factor $\alpha L_{\mathrm{d}}/(1+\alpha L_{\mathrm{d}})$. According to these considerations, we can still assume as valid for amorphous SCs a direct proportionality between the photocurrent yield $I_{\mathrm{ph}}/e\Phi_0$ and the absorption coefficient α in the vicinity of the absorption edge. As in the case of crystalline materials, this allows us to replace α with the photocurrent yield in Eqs. (3.2.1) to (3.2.3), relating the absorption coefficient to the optical bandgap, and to derive the optical bandgap of amorphous semiconducting films from the photocurrent spectra (see Figs. 16, 17).

In order to test the ability of Eq. (3.3.4a) to reproduce the experimental findings reported in the literature for amorphous semiconductor films, we need expressions for the space-charge region and the average electric field in an a-SC. By assuming a constant DOS distribution within the mobility gap of the a-SC, in Section 3.1 we have shown that the potential distribution within the space-charge region is given by

$$\Phi(x) = \Delta\Phi_{\mathrm{sc}} \exp\left(-\frac{x}{x_0}\right) \quad (3.1.1)$$

where $x_0 = \varepsilon\varepsilon_0/e^2 N$ is the screening length of the a-SC, N is the DOS in $\mathrm{cm}^{-3}\,\mathrm{eV}^{-1}$, and $\Delta\Phi_{\mathrm{sc}} = U_{\mathrm{e}} - U_{\mathrm{fb}}$ is the total potential drop inside the a-SC. We will define the width of the space-charge region by using the condition $\Phi(X_{\mathrm{sc}}) = kT/e$, so that from Eq. (3.1.1) we derive

$$X_{\mathrm{sc}} = x_0 \ln\left(\frac{e\Delta\Phi_{\mathrm{sc}}}{kT}\right) \quad (3.3.5)$$

On the other hand, the average electric field can be calculated from Eq. (3.1.1) by integration over the space-charge region [111]:

$$\overline{F} = \frac{\Delta\Phi_{\mathrm{sc}}}{X_{\mathrm{sc}}}\left[1 - \exp\left(-\frac{X_{\mathrm{sc}}}{x_0}\right)\right] \quad (3.3.6)$$

In order to understand the role of different parameters in fitting the experimental photocurrent-vs-potential curves, we have performed numerical simulations based on Eqs. (3.3.4a) to (3.3.6). The most meaningful results are reported in Figures 28–30. From the theoretical curves shown in these figures it is seen that in the absence of initial recombination effects the photocharacteristics are always sublinear. As a general trend, for longer collection lengths Eq. (3.3.4a) gives a quadratic dependence of the photocurrent on $\Delta\Phi_{\mathrm{sc}}$, like that predicted by the Butler model. This finding confirms that linear I_{ph}^2-vs-U_{e} plots are not peculiar to crystalline materials, but can be

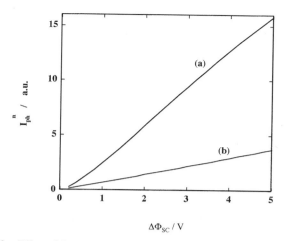

Fig. 28. Effect of the drift length on the theoretical dependence of the photocurrent on the band bending calculated from Eqs. (3.3.4a) to (3.3.6) in the case of unit efficiency of carrier generation ($\eta_g = 1$) with $x_0 = 40$ Å, $\alpha = 10^4\,\mathrm{cm}^{-1}$: (a) $\mu\tau = 8\times10^{-13}\,\mathrm{cm}^2\,\mathrm{V}^{-1}$, $n = 2$; (b) $\mu\tau = 2\times10^{-14}\,\mathrm{cm}^2\,\mathrm{V}^{-1}$, $n = 1.33$.

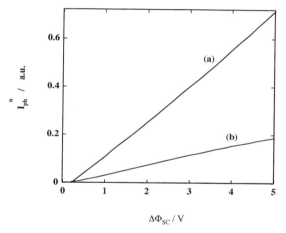

Fig. 29. Effect of the drift length on the theoretical dependence of the photocurrent on the band bending calculated from Eqs. (3.3.4a) to (3.3.6) in the case of unit efficiency of carrier generation ($\eta_g = 1$) with $x_0 = 135$ Å, $\alpha = 10^4\,\mathrm{cm}^{-1}$: (a) $\mu\tau = 2\times10^{-14}\,\mathrm{cm}^2\,\mathrm{V}^{-1}$, $n = 1.33$; (b) $\mu\tau = 10^{-11}\,\mathrm{cm}^2\,\mathrm{V}^{-1}$, $n = 2$.

observed also with a-SCs, with reasonable DOS distributions and collection lengths. We mention that sublinear behavior of the photocharacteristics at some wavelengths was observed for a-WO_3 anodic films [109]: in this case the same power law, $I_{\mathrm{ph}}^{4/3}$, was able to fit the experimental data in a large range of band bendings ($\Delta\Phi_{\mathrm{sc}} \cong 3$ V) and wavelengths ($270 \leq \lambda \leq 380$ nm), whilst at the shortest wavelength ($\lambda = 230$ nm) a linear dependence of I_{ph}^2 on the electrode potential was observed. These results are in agreement with our theoretical simulations in Figure 30, showing sublinear behavior with $n = 1.33$ for lower values of the absorption coefficient α, and n close to 2 for the highest values [curve (a)].

For other semiconducting anodic films (a-Nb_2O_5 and a-TiO_2) the I_{ph}-vs-U_{e} curves display linear or supralinear behavior at

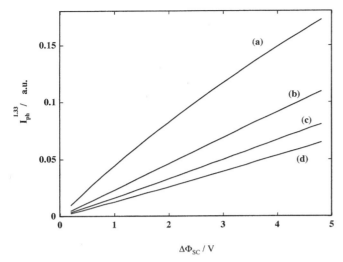

Fig. 30. Effect of the absorption coefficient on the theoretical dependence of the photocurrent on the band bending calculated from Eqs. (3.3.4a) to (3.3.6) in the case of unit efficiency of carrier generation ($\eta_g = 1$) with $x_0 = 40$ Å, $\mu\tau = 2 \times 10^{-13}$ cm^2 V^{-1}: (a) $\alpha = 8 \times 10^5$ cm^{-1}; (b) $\alpha = 2 \times 10^5$ cm^{-1}; (c) $\alpha = 8 \times 10^4$ cm^{-1}; (d) $\alpha = 1.5 \times 10^4$ cm^{-1}.

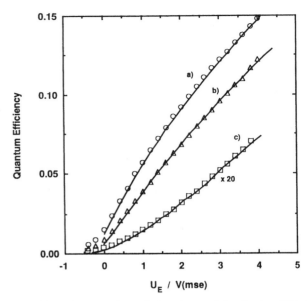

Fig. 31. Fitting of the experimental photocharacteristics for the a-TiO$_2$ film of Figure 18 according to Eq. (3.3.4b). Symbols are the experimental values. Lines are the theoretical curves calculated using the parameters $x_0 = 40$ Å, $\varepsilon = 15$, $\mu\tau = 2 \times 10^{-13}$ cm^2 V^{-1}. (a) $\lambda = 250$ nm, $\alpha = 8 \times 10^5$ cm^{-1}, $r_0 = 25$ Å; (b) $\lambda = 270$ nm, $\alpha = 6 \times 10^5$ cm^{-1}, $r_0 = 18$ Å; (c) $\lambda = 370$ nm, $\alpha = 1.5 \times 10^4$ cm^{-1}, $r_0 = 10$ Å. [*Electrochim. Acta* 38(1), 29–35 (1993)]; reprinted with permission from Elsevier Science.]

low photon energies, and sublinear behavior at higher photon energies. In order to fit the photocharacteristics recorded at different wavelengths, for these systems we need to introduce a variable efficiency of free carrier generation in Eq. (3.3.4a), which becomes

$$I_{ph} = e\Phi_0\eta_g\frac{\alpha L_d}{1+\alpha L_d}\left[1 - \exp\left(-X_{sc}\frac{1+\alpha L_d}{L_d}\right)\right] \quad (3.3.4b)$$

with η_g given by Eq. (3.2.4).

The results of the fitting performed by means of Eq. (3.3.4b) are reported in Figure 31, showing the influence of the photon energy on the shape of the photocharacteristics of the a-TiO$_2$–electrolyte junction. We stress that the theoretical curves were plotted against the band bending $\Delta\Phi_{sc}$, whilst in the experimental curves the photocurrent is reported as a function of the electrode potential. This fact must has taken into account in the fitting procedure, as reported in [111]. Small differences between the U_{fb}-values derived by the impedance study and those obtained from the fitting of the curves in Figure 31 could be traced to surface recombination effects and/or to changes in the Helmholtz potential drop, $\delta\Delta\Phi_H$, on going from dark to light conditions. In this last case strongly absorbed light should preferentially affect more the potential distribution close to the oxide–electrolyte interface [110, 111].

In the presence of multiple reflections the generation term in the integral of Eq. (3.3.3) must be modified to take into account the light fluxes reflected at the film–metal and film–electrolyte interfaces. It has be shown [134] that the corrections to the photocurrent-vs-potential curves are negligible for constant film thickness, whilst in the case of interference effects during the growth of the films (variable film thickness) remarkable effects can be observed (see Section 3.3.3).

3.3.2. Photocurrent Equations for the Insulating-Film–Electrolyte Junction

For passive films that are insulating (Ta$_2$O$_5$, HfO$_2$, ZrO$_2$, etc.) one may refer to a rigorous mathematical treatment given by Crandall [112] for deriving the photocurrent expression in thin film solar cells (p–i–n junctions) on the hypothesis of constant electric field and absence of trapping effects. According to this treatment, the photocurrent as a function of the film thickness d_f and the electric field F is given by

$$I_{ph} = e\Phi_0\alpha L_c\left[1 - \exp\left(-\frac{d_f}{L_c}\right)\right] \quad (3.3.7)$$

where $L_c = l_n + l_p = (\mu_n\tau_n + \mu_h\tau_h)F$ is the collection length, calculated as the sum of those pertaining to electrons (l_n) and holes (l_p).

However, an expression for the photocurrent as a function of the electrode potential can be easily derived from Eqs. (3.3.4), by taking into consideration that for insulating films, under otherwise identical conditions to those assumed previously, the electric field extends throughout the film thickness. In the hypothesis of a constant electric field, $F = \overline{F} = (U_e - U_{fb})/d_f$, the following equation is obtained from Eq. (3.3.4b):

$$I_{ph} = e\Phi_0\eta_g\frac{\alpha L_d}{1+\alpha L_d}\left[1 - \exp\left(-d_f\frac{1+\alpha L_d}{L_d}\right)\right] \quad (3.3.8)$$

We note that Eq. (3.3.8) is practically coincident with Crandall's expression (3.3.7) for the photocurrent in a thin film solar cell, as long as the condition $\alpha L_d \ll 1$ (usually valid for slightly absorbed light) is satisfied and the average parameter L_d for the ensemble of photogenerated electron–hole pairs is assumed coincident with the Crandall collection length L_c. The validity of this assumption follows from the fact that trapping effects are negligible, so that regardless of the microscopic mechanism of recombination, the numbers of electrons and holes that recombine must be (almost) equal. For strongly absorbed light ($\alpha L_d \gg 1$ and *a fortiori* $\alpha d_f \gg 1$) Eq. (3.3.8) gives the same saturation limit as the Crandall equation (3.3.7), but there are appreciable differences in the photocharacteristics before the saturation. In this last case the more rigorous Crandall equation will be used.

For values of film thickness and absorption coefficient obeying the relationship $\alpha d_f \ll 1$, uniform light absorption occurs within the film; in the presence of a reflecting film–metal interface, interference effects in the plots of photocurrent vs film thickness at constant electric field are expected, and this must be considered in the photocurrent expressions (see Section 3.3.3).

Under the hypothesis of unit efficiency of carrier generation, Eq. (3.3.7) predicts a linear or sublinear dependence of the photocurrent on the electrode potential at constant film thickness. Once again, in order to explain the change in the shape of the photocharacteristics with photon energy observed in thin a-Ta$_2$O$_5$ films (see Fig. 19), we need to introduce in Eq. (3.3.7) a variable efficiency of photocarrier generation, η_g, as in Eq. (3.3.8). Figure 32 displays the fitting of the photocharacteristics, recorded at two different wavelengths, performed by means of Eq. (3.3.7) modified with the factor η_g. The fitting parameters are reported in the figure. We stress that the extrapolation to zero photocurrent of the interpolating curves gives a positive band bending. This means that, analogous to the case of semiconducting films, the true flat-band potential may be slightly more cathodic than the zero photocurrent value extrapolated from the I_{ph}^n-vs-potential plots. This finding must be taken into account for a correct estimate of the flat-band potential of insulating films.

Finally we mention that in the presence of trapping effects the electric field can no longer be assumed constant throughout the whole insulating film. In this case the expressions of the photocharacteristics for the insulating-film–electrolyte junction can been derived in the frame of the Goodman–Rose theory [113] of double extraction of uniformly generated pairs, taking into account the effect of the generation efficiency [110]. In dimensionless form,

$$\frac{I_{ph}}{I_{ph0}} = (1+b)\frac{-b+\left[b^2+4\frac{1+b^2}{(1+b)^2}\frac{V}{V_0}\left(\frac{\eta_{g2}}{\eta_{g1}}-b\right)\right]^{1/2}}{2\left(\frac{\eta_{g2}}{\eta_{g1}}-b\right)} \quad (3.3.9)$$

where $b = l_n/l_p$ is the drift length ratio between electrons and holes, $V = \Delta\Phi_f$ is the total voltage drop across the insulating film, η_1 and η_2 are the generation efficiencies in the two regions of the insulator, and V_0 and I_{ph0} are the saturation voltage and photocurrent, respectively [110].

It is noteworthy that when $\eta_g = 1$ everywhere in the film, the Goodman–Rose model predicts:

(a) under uniform illumination (slightly absorbed light), a linear dependence of the photocurrent on the electrode potential for low potential drops, followed by a square root dependence in the high-field regime and for $l_n/l_p \gg 1$ or $l_n/l_p \ll 1$;

(b) a square root dependence of the photocurrent on the electrode potential under strong light excitation, where space-charge-limited currents can be expected.

These last effects can be neglected as long as the photogenerated charge injected is much less than the charge stored in the dielectric film [135].

3.3.3. Interference Effects in the Photocurrent-vs-Thickness Curves

In order to fit the interference effects on the photocurrent recorded at different wavelengths during the growth of photoconducting oxide films on valve metals, we must consider the variation of the oxide thickness d_{ox} with the formation potential V_f:

$$d_{ox} = d_0 + A V_f \quad (3.3.10)$$

where d_0 is the initial (native) film thickness and A (in nanometers per volt) is the anodizing ratio, i.e., the reciprocal of the anodizing electric field (assumed constant during the high-field film growth). The latter quantity is a known function of the growth rate [22, 23].

In a previous work [121] we have shown that a good fit to the experimental data can be obtained by using the transport equation derived by Many [136] for insulating photoconductors under blocking conditions. In such a treatment space-charge effects are neglected and position-independent lifetimes are assumed for both photocarriers. Moreover, the transport equations for the two photocarriers were solved independently, in contrast

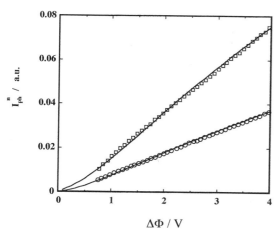

Fig. 32. Fitting of the experimental photocharacteristics for the Ta$_2$O$_5$ film of Figure 19 ($d_{ox} = 20$ nm) according to Eq. (3.3.7) multiplied by the efficiency of carrier generation. Symbols are the experimental values. Lines are the theoretical curves calculated using the parameters $\mu\tau = 10^{-13}$ cm^2 V^{-1}, $\varepsilon = 11$. $\bigcirc$, $\lambda = 230$ nm, $\alpha = 3.5 \times 10^5$ cm^{-1}, $r_0 = 20$ Å; $\square$, $\lambda = 300$ nm, $\alpha = 1.5 \times 10^4$ cm^{-1}, $r_0 = 15$ Å.

with the Crandall model. In this case the photocurrent measurements are performed during the growth of the anodic films, when large ionic fluxes are simultaneously flowing through the oxide films, thus the use of the model proposed by Many in the fitting procedure seemed to us more physically sound. However, a check of fitting with the Crandall equation (3.3.7) has been also performed in order to examine the differences between the two models.

According to Many [136], the transport equations for the single carriers are

$$J_i = eG_0(\Phi_0, \lambda, \alpha, \tilde{n}_j, d_{ox})l_i$$
$$\times \left\{ 1 - \frac{l_i}{d_{ox}}\left[1 - \exp\left(-\frac{d_{ox}}{l_i}\right)\right]\right\} \quad (3.3.11)$$

where i = n or p for electrons and holes, respectively, l_i are the electron and hole drift lengths, and the total photocurrent is given by $I_{ph} = J_n + J_p$. The interference effects are included in the volume excitation rate term G_0, which is a function of the wavelength λ, the absorption coefficient α, the film thickness d_{ox}, the incident photon flux Φ_0, and the complex refractive indexes $\tilde{n}_j$ of metal, oxide, and solution.

In a first application of Eq. (3.3.11), we used for G_0 the simple expression

$$G_0(\Phi_0, \lambda, \alpha, \tilde{n}_j, d_{ox})$$
$$= e\Phi_0\left[1 - R(\lambda, \alpha, \tilde{n}_j, d_{ox})\right]\frac{1 - \exp(-\alpha d_{ox})}{d_{ox}} \quad (3.3.12)$$

R being the total reflectivity of the metal–oxide–electrolyte junction [137]. An example of application of Eqs. (3.3.11) and (3.3.12) for interpolating the experimental photocurrent-vs-thickness curves is reported in Figure 33, which refers to Nb$_2$O$_5$ films during anodic growth at constant rate ($dV_f/dt = $ const.).

The use of Eq. (3.3.12) for the case of multiple light passes through the passive film underestimates the contribution to the measured photocurrent from internal reflection and was subsequently criticized by Sukamto et al. [134], who proposed a more appropriate expression for the generation rate in the presence of multiple internal reflections:

$$G_0(\Phi_0, \lambda, \alpha, \tilde{n}_j, d_{ox}) = \frac{1}{d_{ox}}\int_0^{d_{ox}} g'(x)\,dx \quad (3.3.13)$$

with

$$g'(x) = A_1^{(\infty)}\exp(-\alpha x) + A_2^{(\infty)}\exp(\alpha x)$$
$$+ A_3^{(\infty)}\cos\left(\frac{4\pi n_2}{\lambda}x\right)$$
$$+ A_4^{(\infty)}\sin\left(\frac{4\pi n_2}{\lambda}x\right) \quad (3.3.14)$$

Here n_2 is the oxide refractive index (real part), and the coefficients $A_i^{(\infty)}$ are functions of the complex refractive index of the media involved and of the complex reflection and transmission coefficients of the light at the phase boundaries [134].

Figure 34 displays a comparison between the generation terms in the case of Figure 33, calculated by means Eq. (3.3.12)

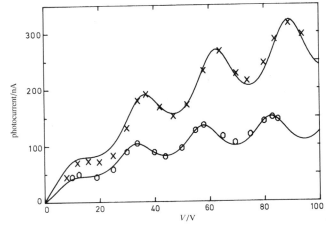

Fig. 33. Photocurrent intensity vs formation voltage for Nb$_2$O$_5$ films growing tensiodynamically at different growth rates and under irradiation ($\lambda = 340$ nm). Symbols are the experimental values. Solid lines are the theoretical fitted curves obtained with $\alpha = 8000$ cm^{-1}, $n_{met} = 2.56 - i2.28$, $n_{sol} = 1.55$, $n_{ox} = 2.7 - i0.0216$. $\times$: 100 mV s^{-1}; fitting parameters $A = 23.5$ Å V^{-1}, $b = 36$, $\mu_h\tau_h = 7 \times 10^{-14}$ cm^2 V^{-1}. $\bigcirc$: 20 mV s^{-1}; fitting parameters $A = 25.3$ Å V^{-1}, $b = 30$, $\mu_h\tau_h = 4.5 \times 10^{-14}$ cm^2 V^{-1}. [*J. Chem. Soc. Faraday Trans. 1* 85(10), 3309–3326 (1989); reprinted with permission from the Royal Society of Chemistry.]

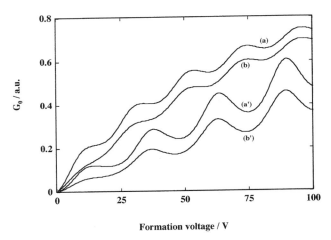

Fig. 34. Theoretical generation term G_0 as a function of film thickness for the sample of Figure 33, calculated by means of Eq. (3.3.12) (curves b and b′) and by means of Eqs. (3.3.13) and (3.3.14) (curves a and a′) at different wavelengths. a,b: $\lambda = 300$ nm; parameters $A = 23.5$ Å V^{-1}, $\alpha = 55,000$ cm^{-1}, $n_{met} = 2.64 - i2.42$, $n_{sol} = 2.3$, $n_{ox} = 3.0 - i0.1313$. a′,b′: $\lambda = 340$ nm; parameters as in Figure 33.

and by means of Eqs. (3.3.13) and (3.3.14). Noticeably, the positions of the maxima and minima are practically unchanged, being largely determined by the anodizing ratio. On the contrary, large differences (up to more than 40%) are evident in the generated carrier density for thin films ($d_{ox} \leq 20$ nm), as expected. With increasing photon energy and/or film thickness the differences between the two expressions progressively decline, by up to few percent for $\lambda = 300$ nm and $d_{ox} > 20$ nm.

Figure 35 shows the fitting of two curves of photocurrent vs formation potential for anodic Nb$_2$O$_5$ during growth under illumination with different wavelengths, by using Many's equa-

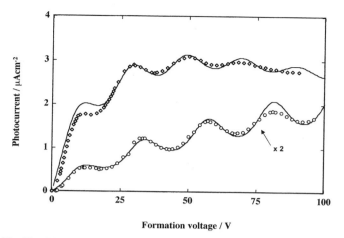

Fig. 35. Fitting of the curves of photocurrent vs formation voltage for Nb_2O_5 films growing tensiodynamically at different growth rates and under irradiation with different wavelengths. Symbols are the experimental values. Solid lines are the theoretical curves calculated by means of Eqs. (3.3.11), (3.3.13), and (3.3.14). $\Diamond$: $dV_f/dt = 50$ mV s^{-1}, $\lambda = 300$ nm; fitting parameters $A = 24$ Å V^{-1}, $\alpha = 55,000$ cm^{-1}, $n_{met} = 2.64 - i2.42$, $n_{sol} = 2.3$, $n_{ox} = 3.0 - i0.1313$, $b = 25$, $\mu_h\tau_h = 5 \times 10^{-14}$ cm^2 V^{-1}. $\bigcirc$: $dV_f/dt = 20$ mV s^{-1}, $\lambda = 340$ nm; fitting parameters $A = 25.9$ Å V^{-1}, $\alpha = 8000$ cm^{-1}, $n_{met} = 2.56 - i2.28$, $n_{sol} = 1.55$, $n_{ox} = 2.7 - i0.0216$, $b = 30$, $\mu_h\tau_h = 8 \times 10^{-13}$ cm^2 V^{-1}.

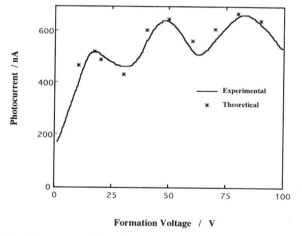

Fig. 36. Photocurrent vs formation voltage recorded during the tensiodynamic anodization (growth rate 200 mV s^{-1}) of an Al–Ti 50-at. % alloy in boric acid + borax (pH = 7.5) electrolyte under monocromatic irradiation ($\lambda = 310$ nm). Symbols represent the theoretical values calculated by means of Eqs. (3.3.7), (3.3.13), and (3.3.14). Fitting parameters: $\alpha = 1.4 \times 10^5$ cm^{-1}, $n_{met} = 0.8 - i3.2$, $n_{ox} = 2.5$, $\mu\tau = 10^{-11}$ cm^2 V^{-1}, $A = 17.7$ Å V^{-1}. [*Corrosion Sci.* 40(7), 1087–1108 (1998); reprinted with permission from Elsevier Science.]

tion (3.3.11) for the transport term and Sukamto et al.'s [134] equations (3.3.13) and (3.3.14) for the generation term. With respect to the previous fitting (see Figure 33 and [121]) we now observe that larger drift length values must be used for the photocarriers, whilst the anodizing ratio does not change appreciably. Moreover, more satisfactory agreement with the experimental points in Figure 35 is obtained at small to intermediate thicknesses, but not at the largest thicknesses, maybe owing to a roughening of the metal–oxide interface that hinders light reflection. This last finding deserves further investigation.

The fitting of the experimental curves performed using the Crandall equation (3.3.7) for calculating the transport term gives less satisfactory results for all the investigated anodic oxides on pure metals. In other cases, especially when the optical parameters of the oxide phase were not perfectly known, the simpler Crandall equation (3.3.7) was employed to fit the experimental interference curves, coupled to Eqs. (3.3.13) and (3.3.14) for the generation rate. An example, dealing with an anodic oxide growing on an Al–Ti alloy [124], is given in Figure 36.

Further interesting aspects of the use of multiple internal interference effects under both potentiodynamic and galvanostatic conditions for detecting the kinetics of growth of passive films on valve metals can be found in [120, 121].

4. QUANTITATIVE USE OF PCS FOR THE CHARACTERIZATION OF PASSIVE FILMS ON METALS AND ALLOYS

In the previous sections we have evidenced how a PCS study of a metal–passive-film–electrolyte junction can provide useful information on the physicochemical properties of the films (E_g, U_{fb}) as well as on the mechanisms of generation and transport of photocarriers in thin photoconducting films. Part of this information is preliminary for locating the energetic levels of the junction and then for understanding the ETR mechanisms in thin passive films (see [17] and references therein). From the point of view of corrosion studies, it is essential to attempt a more quantitative use of PCS for getting information also on the composition of passive films grown in different conditions. For this purpose, the only measured quantity that can be related to the passive film composition is the optical bandgap. Examples of determination of the optical gap for passive oxide film on pure metals from the photocurrent spectra are shown in Figures 37–42.

In Section 3 we mentioned, that according to Mott and Davis [90], the value of the mobility gap should be quite close to the bandgap of the crystalline counterpart, as long as the short-range order is the same in the two phases (crystalline and amorphous); the difference between the two values arises from the localization effects due to the absence of long-range order in the disordered material. In Table III the values of the mobility gap, E_g^m, for a number of investigated passive films are reported, together with the bandgap of the crystalline counterparts, E_g, and the difference, $\Delta E_{am} = E_g^m - E_g$. Some data in the table are relative to passive films grown anodically on valve-metal electrodes (from Zr to Mo), for which an amorphous or microcrystalline structure is generally accepted. For these films there is also a large consensus about the stoichiometry, which is practically coincident with that of the crystalline counterparts. We stress that for these films the value of ΔE_{am} is in very good agreement with the estimate of the effect of long-range disorder in amorphous materials (see Section 3). It is quite reasonable in such a case to assume the parameter ΔE_{am} as a rough measure of the localization of the electronic states near the mobility

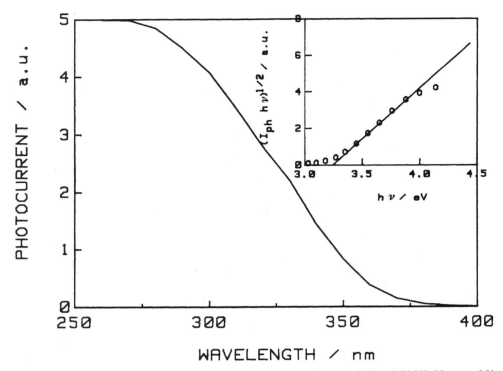

Fig. 37.   Photocurrent action spectrum for an anodic film grown on Ti metal at 2 V/s in 0.5 M H_2SO_4 up to 9 V (MSE). Electrode potential: +2 V (MSE). Inset: determination of the optical bandgap. [*Electrochim. Acta* 38(1), 29–35 (1993); reprinted with permission from Elsevier Science.]

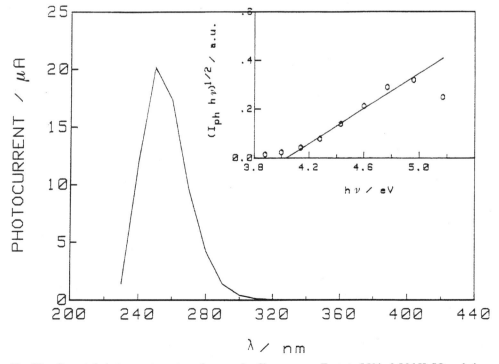

Fig. 38.   Corrected photocurrent spectrum for a passive film grown on Ta up to 5 V in 0.5 M H_2SO_4 solution. Electrode potential: +3 V (MSE). Inset: determination of the optical bandgap. [*Corrosion Sci.* 35(1–4), 801–808 (1993); reprinted with permission from Elsevier Science.]

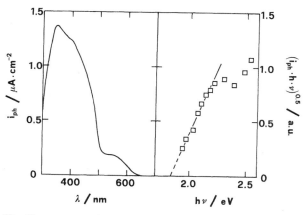

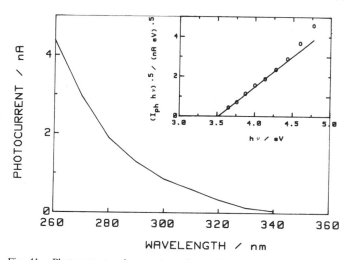

Fig. 39. Photocurrent action spectrum (left side) and determination of the optical bandgap (right side) for a passive film grown on Cu in 0.1 M Na_2SO_4 solution (initial pH 5) after 24 h of immersion at $U_{corr} = -0.087$ V (SCE). [*Electrochim. Acta* 30(3), 315–324 (1985); reprinted with permission from Elsevier Science.]

Fig. 41. Photocurrent action spectrum for an air-grown passive film on Cr metal soon after immersion in 0.6 M NaCl solution. Electrode potential: -0.9 V (MSE). Inset: determination of the optical bandgap assuming indirect optical transitions. [*J. Electrochem. Soc.* 137, 2411–2417 (1990); reprinted with permission from The Electrochemical Society, Inc.]

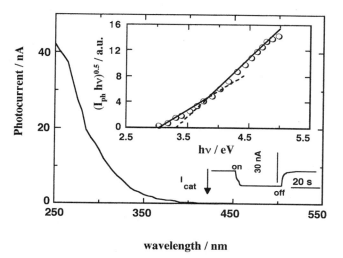

Fig. 40. Photocurrent action spectrum and total current in the dark and under illumination ($\lambda = 300$ nm) vs time for a passive layer grown potentiostatically on Ni in aerated 0.25 M Na_2HPO_4 solution. Electrode potential: -0.45 V (HgO). Inset: determination of the optical bandgap. [*Mater. Sci. Forum* 185–188, 435–446 (1995)]; reprinted with permission from Trans Tech Publications, Ltd.]

Fe_2O_3. Moreover, recent STM investigation performed on passive films grown on Ni single crystal in acidic solutions [144] suggests a bilayer structure with crystalline NiO underlying a hydrated surface layer. Such a result is in agreement with our PCS study on Ni [128], indicating an optical bandgap of the passive film very near to the bandgap value of crystalline NiO and a photocurrent tail at lower photon energies, attributed to a partially hydrated overlayer (see Fig. 40 and Section 4.3).

PCS studies performed in recent decades by various authors on corrosion layers grown on metal in different experimental conditions and aggressive environments have shown definitely the usefulness of this technique as a tool for getting indirect information on composition and crystallinity of the films, at least in the case of pure metals [14, 15, 74, 75, 145, 146]. In order to allow a more quantitative application of PCS for scrutinizing passive films on alloys and for inferring the film composition on the basis of bandgap measurements, we need to correlate in some way the bandgap with the composition of the films. In the next subsections we will present some correlations, proposed by us, which help to achieve that purpose.

4.1. Semiempirical Correlation between the Optical Bandgap of Crystalline Oxides and Their Composition

The theoretical prediction of the physical properties of inorganic solids and notably of the forbidden gap in insulating and semiconducting materials is a formidable task. Models have been proposed to explain the behavior of different classes of materials [147–151], but they are of limited usefulness in the case of corrosion layers, whose composition and morphology are very often complex and rather controversial. For this reason, semiempirical models and correlations between the bandgap of solids and the compositional parameters should be more helpful from a practical point of view.

edges. This interpretation is also in agreement with the results of a recent PCS investigation of passive films grown in different conditions on Ti metal [85], showing that the mobility gap values tend to the bandgap of the crystalline phases when TiO_2 films begin to crystallize.

In the same table we report also the measured optical gap values of anodic films on base metals (see Figs. 39–42 and [75, 128, 138–140]), for which ΔE_{am} is close to zero, with the notable exception of the air-formed films on Cr metal (see Fig. 41 and [139]. It is noteworthy that for the passive films grown on Fe metal, under conditions very similar to those used in our PCS measurements (see below), recent experimental results based on in situ STM [141], XANES [142], and X-ray scattering [143] have confirmed a microcrystalline structure close to that of

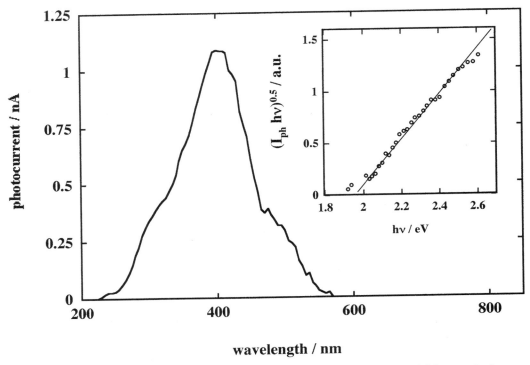

Fig. 42.   Photocurrent spectrum, corrected for the photon emission of the light source, recorded for a passive layer on Fe grown potentiostatically at +0.81 V (SHE) for 60 min in boric acid + borax electrolyte (pH 8.4). Electrode potential: +0.81 V (SHE). Inset: determination of the optical bandgap assuming indirect optical transitions.

On the basis of previous correlations between bandgap of inorganic solids and the single-bond energy [152–154], some years ago we proposed [76] a relationship between the bandgap of crystalline oxides, MO_y, and the average bond energy, D_{A-B}. By using for this last quantity the Pauling equation [155] for the general case of a polyatomic oxide molecule, MO_y, the following expression was derived for the bandgap of the crystalline oxide [76]:

$$E_g = 2\left\{E_I(X_M - X_O)^2 + \frac{1}{2y}\left[(D_{M-M} + yD_{O-O}) - R\right]\right\} \quad (4.1.1)$$

where, according to Phillips [148], E_I is the extra-ionic energy (orbitally dependent), assumed "to vary with hybridization configuration, i.e., with different atomic coordinations in different crystal structures." For ionic compounds, e.g., alkali halides, a value of $E_I = 1$ eV per atom pair is assumed, whilst a different value is expected for hybridized bonds involving d-orbitals. In Eq. (4.1.1), X_M and X_O are the electronegativities, on the Pauling scale, of metal and oxygen, respectively, y is the stoichiometric coefficient of the oxide MO_y, and D_{M-M} and D_{O-O} are the bond energies of diatomic molecules in the gas phase. R represents a repulsive contribution due to the difference between the lattice and bond energies [76].

As long as we can assume constant the quantity

$$\Xi = \frac{1}{2y}\left[(D_{M-M} + yD_{O-O}) - R\right] \quad (4.1.2)$$

Eq. (4.1.1) can be rewritten as

$$E_g = 2\left[E_I(X_M - X_O)^2 + \Xi\right] \quad (4.1.3)$$

Equation (4.1.3) allows us to correlate the optical bandgap of oxides with the square of the difference between the electronegativities of metal and oxygen (the "extra-ionic energy" of Pauling [155]). In order to test its validity as well as the constancy of the quantity Ξ, we have plotted in Figure 43 the bandgap values of crystalline oxides as a function of $(X_M - X_O)^2$. Two different straight lines fit the experimental data pertaining to sp-metal and d-metal oxides: this result is rationalized, in the frame of Eq. (4.1.3), by taking into account that different values for the parameter E_I (and thus different slopes of the fitted lines) are expected on going from sp-metal to d-metal oxides.

In comparison with previous correlations [152–154], Eq. (4.1.3) displays:

(a) two clearly separated interpolating lines with slopes of about 2.17 and 1.35 eV, respectively, for sp-metal and d-metal oxides, with a few exceptions (to be discussed);

(b) better agreement with the experimental data, as evidenced by the higher correlation coefficients for both interpolating lines;

(c) the ability to explain the experimental finding, pertaining to some oxides, of a decreasing bandgap at higher oxidation states of the metallic cation (Sn, Cu,

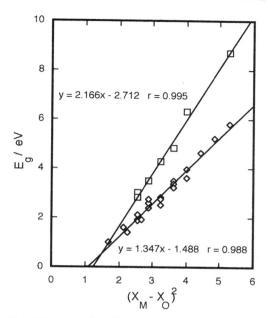

Fig. 43. Optical bandgap values E_g vs the square of the difference of Pauling electronegativities of crystalline oxides: □, sp-metal oxides; ◇, d-metal oxides. [*J. Phys. Chem. B* 101(14), 2519–2525, (1997)]; reprinted with permission from the American Chemical Society.]

Fe), in agreement with the expected change of the Pauling electronegativity parameter for the cation with its oxidation state.

The best-fitting correlations of Figure 43 are represented by the following equations:

$$E_g = 1.35(X_M - X_O)^2 - 1.49 \, \text{eV} \qquad (4.1.4a)$$

for d-metal oxides, and

$$E_g = 2.17(X_M - X_O)^2 - 2.71 \, \text{eV} \qquad (4.1.4b)$$

for sp-metal oxides. It is interesting to note that, from the fitted curve for d-metal oxides [Eq. (4.1.4a)], metallic behavior ($E_g \cong 0$) is expected for metals having Pauling electronegativity around 2.40 or more, in agreement with the common experience that oxides on noble metals (Au, Ir, Ru), having the highest electronegativity parameters, usually display metallic conductivity.

We mention that in Figure 43 the point for NiO lies neatly on the line for sp-metal oxides, and that the points for three non-transition-metal oxides (PbO, In₂O₃, Tl₂O₃) are better fitted as d-metal oxides. This originates an intriguing dividing line in the periodic table between d- and sp-metal oxides along the diagonal between Zn, In, Pb and Ga, Sn, Bi, with some of sp-metals (In, Tl, Pb) of higher atomic number showing d-like behavior in their average bond strength.

For the electronegativity values in Figure 43 we have used, for all calculations, but one, the electronegativities of the Pauling scale, integrated with the Gordy–Thomas values [156]. The exception is Tl(III) oxide, for which the value given by Allred [157] has been preferred. It is noteworthy that the

electronegativities of the elements calculated from the best-fitting straight lines of Figure 43 [Eqs. (4.1.4)] on the basis of the experimental bandgaps differ from the Pauling values by about 0.05, which is more or less the uncertainty in the values of electronegativity given by Pauling [155].

This observation suggests a very interesting application of Eqs. (4.1.4) in the case of metals giving cations with different oxidation numbers, for which rather different values of the electronegativity parameter are reported in the literature. It may be reasonable to derive the electronegativity values and the corresponding metal oxidation state in the corrosion layer from values of the E_g given by Eqs. (4.1.4). As an example of such a use of PCS, we consider the case of vanadium oxides, for which, according to different authors [155–158], X_M-values ranging between 1.35 and 1.9 are reported on going from +3 to +5 oxidation state. There has been recently reported for orthorhombic V₂O₅ films an optical bandgap of 2.3 eV [159], which inserted in Eq. (4.1.4a) gives a value $X_M = 1.824$ for V⁵⁺; the electronegativity of V⁵⁺; the electronegativity parameter derived by our procedure is in very good agreement with the value derived by using the bond energy scale by Haissinsky [158], but it is slightly lower than the value 1.9 based on the atomic radius and the nuclear screening formula suggested by Gordy and Thomas [156].

Another example deals with the corrosion layers grown on Fe metal. In this case the identification of the exact composition of the surface film is problematic, in that by varying the experimental conditions (solution pH, anions in solution, electrode potential, electrochemical history) different oxidation states of the metallic cation (+2 or +3), hydration degrees (to be discussed), and crystallographic structures can be encountered. Analogous situation happens for other metals (e.g., Cu and Ni). In these cases additional information, both theoretical (the thermodynamic predictions of phase stability) and experimental (coming from other in situ techniques— see Section 1), could be required. In Figure 42 is shown a photocurrent spectrum for a passive layer grown on Fe in borate buffer solution (pH 8.4), from which an indirect gap of 1.95 eV is derived. This film was formed potentiostatically at high anodic potential, under experimental conditions identical to those employed in previous studies carried out by X-ray scattering [143] and in situ AFM [160]. Those studies concluded that the structure of this film is very close to that of crystalline γ-Fe₂O₃. Notably, the measured bandgap value is almost coincident with that calculated from Eq. (4.1.4a) for crystalline Fe₂O₃ (1.96 eV), based on the reported value of $X_M = 1.90$ for Fe³⁺. Moreover, a bandgap of 1.90 eV has been quoted for Fe₂O₃ in a recent review on transition metal oxides [151].

Once the values of the Pauling electronegativity parameter are known for the different cationic metals in various oxidation states, Eqs. (4.1.4) can be employed for determining the composition of mixed oxides and passive films grown on metallic alloys by means of a PCS investigation (see next subsection).

4.2. Generalized Correlations for Mixed Oxides and Passive Films on Binary Alloys

For a quantitative application of the PCS technique in corrosion and passivity studies, it is very important to attempt to generalize the previous results to more complex systems like mixed oxides, hydroxides, and oxyhydroxides. With this aim, in the case of mixed oxides containing different cations, we have proposed [76] to use Eqs. (4.1.4) for the optical bandgap correlations, taking for X_M the cationic group average electronegativity $\overline{X}_c$ defined by using the arithmetic mean of those pertaining to the single cations:

$$\overline{X}_c = \frac{aX_A + bX_B}{a + b} \tag{4.2.1}$$

where a and b represent the stoichiometric coefficients of the mixed oxide $A_aB_bO_o$, and X_A and X_B are the electronegativities of the two metallic cations in the oxide.

A preliminary test of this hypothesis was performed by comparing the experimental values of bandgap, reported in the literature for some sp- and d-metal mixed oxides, with those calculated by means of Eqs. (4.1.4) and (4.2.1). Reasonably good agreement between experimental and calculated values was observed for mixed oxides containing only d-metal or only sp-metal cations [76]. More recent results, obtained both on mixed iron–titanium oxides (Fe_2O_3–TiO_2 anatase) of different composition [161, 162] and on passive films grown on magnetron-sputtered Mo–Ta metallic alloys [108], seem to support definitely the use of Eq. (4.1.4a) in connection with Eq. (4.2.1) for mixed d,d-metal oxides.

In Table V we report the data on the indirect optical gap given in [161] for different Fe_2O_3–TiO_2 mixed oxides, formed by metal–organic chemical vapor deposition (MOCVD) on Pt substrates, spanning the whole range of composition including pure phases. In the same table are reported also, for each composition, the values of the average cationic electronegativity $\overline{X}_c$, calculated according to Eq. (4.2.1), as well as the theoretical bandgap values for the mixed oxide, estimated according to Eq. (4.1.4a). In addition to the very good general

agreement between experimental and theoretical data in Table V, it is particularly interesting to observe that for the two samples with highest Ti content, for which the authors report an amorphous structure [161], the experimental optical gaps are slightly larger than the values estimated for the crystalline counterparts ($E_g^m = E_g + 0.15$ eV). This finding supports the suggestion, reported in Section 4.1, that amorphous phases display slightly larger (0.2–0.3 eV) optical gap values then their crystalline counterparts, provided that no change in the composition and short-range order occur in the two phases.

The results of Table V are encouraging for quantitative use of PCS in the case of passive films grown on metallic alloys. In this case we need to take into account also the possible amorphous structure of the films, which could affect in an unpredictable way the value of the optical bandgap. In a very recent study of the PCS characterization of thin passive films grown on magnetron-sputtered Mo–Ta alloys [108] we suggested modifying Eq. (4.1.4a) slightly by introducing the quantity ΔE_{am}, which takes account of the difference in optical bandgap values between amorphous passive film and crystalline phase. According to these considerations, for amorphous passive films grown on d,d-metallic alloys Eq. (4.1.4a) becomes

$$E_g^f - \Delta E_{am} = 1.35(\overline{X}_c - X_O)^2 - 1.49 \text{ eV} \tag{4.2.2}$$

where $\overline{X}_c$ is the average cationic electronegativity, given by Eq. (4.2.1), and the optical bandgap E_g^f of the passive film is corrected by ΔE_{am} before using the correlation valid for crystalline phases. For amorphous pure Ta_2O_5 and MoO_3 films slightly different ΔE_{am}-values (0.15 and about 0.2 eV, respectively) were estimated from our own photoelectrochemical data [108]. In order to take into account this variable contribution of the amorphous structure to ΔE_{am}, in the case of mixed oxides we assumed a change of ΔE_{am} with the composition of the films: a value of 0.2 eV was assumed for films grown on alloys with higher Mo content (>50 at. %), whilst a value of 0.15 eV was assumed at lower Mo content in the metallic alloys. Using such estimates for ΔE_{am} and on the basis of the experimental values of E_g^f, by means of Eq. (4.2.2) it was possible to derive the $\overline{X}_c$-value; from this, using the known values of the Pauling electronegativities of the two cations in Eq. (4.2.1), we have been able to plot the cationic fractions [$a/(a + b)$ and $b/(a + b)$] for passive films grown in different conditions as a function of the atomic ratio in the sputtered alloy (see Fig. 44). It is encouraging that our quantitative estimate of the passive films' composition gave a trend in good agreement with that inferred by Park et al. [163] based on a XPS analysis of the corrosion surface films. Moreover, for a corrosion film formed under identical experimental conditions the same composition was inferred by using the XPS study [163] and by using the PCS analysis based on Eq. (4.2.2) [108].

For mixed oxides involving both sp-metal and d-metal cations, different behavior was observed depending on the difference between the electronegativities of the two cations [76]. From the analysis of the experimental data pertaining to passive layers grown on sp,d-metal alloys, it was postulated in [76]

Table V. Average Cationic Electronegativity and Experimental [161] and Theoretical Optical Bandgap Values, Calculated According to Eqs. (4.2.1) and (4.1.4a), for Different Fe_2O_3–TiO_2 Mixed Films

Film composition	$\overline{X}_c$	E_g^{exp} (eV)	E_g^{theor} (eV)
Fe_2O_3	1.914	1.90	1.90
Fe–10 at. % Ti	1.875	2.09	2.07
Fe–20 at. % Ti	1.85	2.17	2.18
Fe–40 at. % Ti	1.80	2.35	2.40
Fe–50 at. % Ti	1.775	2.50	2.53
Fe–55 at. % Ti	1.7625	2.60	2.58
Fe–75 at. % Ti	1.7125	2.95	2.80
Fe–90 at. % Ti	1.675	3.15	3.00
TiO_2 (anatase)	1.634	3.20	3.20

that the correlation (4.1.4a) relative to d-metal oxides can be used in connection with Eq. (4.2.1) for the average cationic electronegativity, provided that the difference between the electronegativity values of the two cations is lower than about 0.5. It is still unclear what is the upper limit on the atomic fraction of sp-metal cation for observing such behavior; some preliminary results on passive films grown on Al–Ti (see Fig. 45 for the

photocurrent spectrum and the determination of the optical gap) and Al–W metallic alloys, having $\Delta X_c < 0.5$ [124, 164], seem to indicate that the correlation (4.1.4a) can be valid also for such mixed passive films when the cationic fraction of the d-metal in the oxide film is equal to or higher than 30%. For lower atomic fractions of the d-metal in the oxide film it seems reasonable to assume (4.1.4b) as a valid correlation, although experimental verification of this hypothesis is still lacking. When the difference between the electronegativities of the sp-metal and d-metal cations is equal to or greater than 0.5, both correlations (4.1.4) failed to give reasonable agreement with the experimental results [76]; it has been observed that the optical bandgap of such mixed oxides is often almost coincident with that of the pure d-metal oxide. However, very few experimental data concerning passive films grown on metallic alloys having such large electronegativity differences are at our disposal to date.

Obviously, in the case of mixed oxides containing more than two cationic species, the PCS technique is not able to give the composition of the film in the absence of further independent information about the content of some cationic species.

Finally we mention that the formation of pure unmixed oxides has been revealed by PCS spectroscopy in the case of corrosion layers grown on glassy metallic Zr–Cu alloys obtained by melt spinning [165]. A careful inspection of both the anodic and the cathodic photocurrent spectra and of the photocharacteristics recorded at different wavelengths (see Fig. 46)

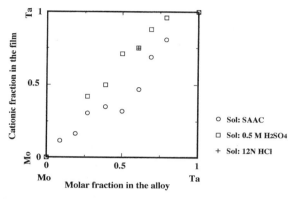

Fig. 44. Cationic fraction estimated by means of Eqs. (4.2.1) and (4.2.2) for passive films grown on different sputtered Mo–Ta alloys in various electrolytes (SAAC = 0.1 M sodium acetate in 2-wt % aqueous acetic acid), as a function of the metal alloy composition. [*J. Electrochem. Soc.* 147, 1366–1375 (2000); reprinted with permission from The Electrochemical Society, Inc.]

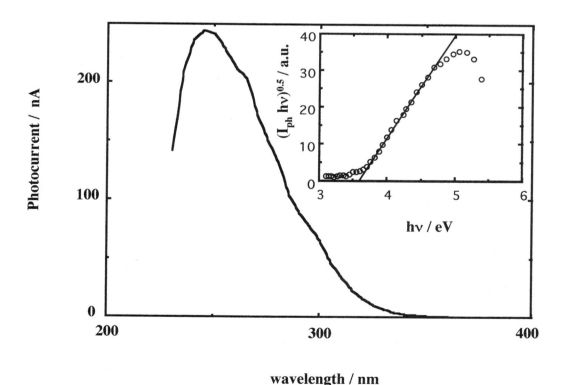

Fig. 45. Photocurrent action spectrum obtained by free immersion of an Al–Ti 50-at. % alloy in 0.1 M ammonium diphosphate electrolyte (pH 4.7) at the corrosion potential $U_{corr} = -0.71$ V (MSE). The inset shows the determination of the optical gap for the passive layer. [*Corrosion Sci.* 40(7), 1087–1108 (1998); reprinted with permission from Elsevier Science.]

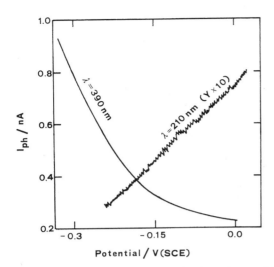

Fig. 46. Dependence of the photocurrent on the electrode potential at two different wavelengths for a passive layer formed on a glassy $Cu_{60}Zr_{40}$ alloy after 5 h of free corrosion in 0.1 M Na_2SO_4 solution.

has suggested the formation on the metallic surface of a passive film consisting of pure p-type semiconducting Cu_2O and insulating ZrO_2 oxides [165]. This finding confirms the ability of PCS to discriminate between the formation of mixed and pure oxide phases in the case of metallic alloys.

4.3. Correlations for Hydroxides and Oxyhydroxides

For further progress toward the possibility of using PCS in a quantitative way for the investigation of corrosion layers grown in different conditions, we need to correlate the optical bandgap of hydroxides and oxyhydroxides with their composition. The experimental data collected on a number of systems indicate that hydrated oxide films and hydroxides display lower optical bandgap values than the corresponding anhydrous oxides. This finding can be rationalized on the basis of the correlations between the bandgap of the films and the electronegativity of their constituents, illustrated in Sections 4.1 and 4.2. In the case of hydroxide phases we can postulate that the bandgap depends on the difference between the electronegativities of the metal cation, X_M, and the hydroxyl group, X_{OH}. The latter can be calculated as the arithmetic mean of those pertaining to oxygen (3.5) and hydrogen (2.2): in this case a value of 2.85 is obtained for X_{OH}, in accordance with other authors [166]. Alternatively, a value of 2.77 could be estimated for X_{OH} as the geometric mean of the electronegativities of oxygen and hydrogen.

By analogy with the procedure followed for mixed oxides, in the presence of a variable number of OH groups in the molecular unit we may define the average electronegativity of the anionic group, $\overline{X}_{an}$, in a generic oxyhydroxide having formula $MO_{y-m}OH_{2m}$ ($0 \leq m \leq y$), as the arithmetic mean of those relative to oxygen and hydroxyl anions:

$$\overline{X}_{an} = \frac{2m X_{OH} + (y - m) X_O}{y + m} \quad (4.3.1)$$

Alternatively, $\overline{X}_{an}$ can be estimated as the geometric mean of the electronegativity values of the O^{2-} and OH^- groups:

$$\overline{X}_{an}^g = \left(X_{OH}^{2m} X_O^{y-m} \right)^{1/(y+m)} \quad (4.3.2)$$

In the absence of more precise indications, in the following we will adopt Eq. (4.3.1) for a comparison with the experimental results. However, regardless of the final choice, from the previous definitions of $\overline{X}_{an}$ it follows that with increasing number of OH groups in the molecule the average anionic electronegativity decreases from the value of oxygen ($m = 0$) to that of the hydroxyl group ($m = y$).

In order to interpret some experimental data pertaining to passive layers on metals, for which the formation of hydroxides was inferred on the basis of both the PCS study and thermodynamic considerations, or was confirmed by surface analytical investigations (XPS and XANES), we assumed [76], in agreement with the previous findings on pure and mixed oxides, that also for oxyhydroxides and hydroxides the optical bandgap value is proportional to the square of the difference between the electronegativity of the metallic cation and the average anionic electronegativity, $(X_M - \overline{X}_{an})^2$. In the electrochemical literature there were a few, experimentally well-investigated systems that seemed to support our assumption. Further studies, carried out in our laboratory on selected systems, have now provided a number of experimental data sufficient to derive numerical correlations regarding the optical bandgap of hydroxide films on both sp- and d-metals.

4.3.1. sp-Metal Hydroxides (Al, Mg, Sn, Ni)

For pure (99.99%) aluminum electrodes anodized up to about 10 V in quasi-neutral aqueous solutions containing inorganic salts, an anodic photocurrent was recorded only when the Al surface was previously submitted to a mechanical treatment. In the case of electropolished surfaces no anodic photocurrent was observed, unless the anodization process occurred in organic-salt-containing solutions [125, 132]. In order to explain this finding we suggested, in the case of mechanically treated electrodes, the formation of hydrated phases having optical bandgap values E_g^{opt} lower than that estimated for the Al_2O_3 crystalline film (about 6.3 eV [167]). After anodization of aluminum surfaces scraped by a razor blade, a photocurrent spectrum giving $E_g^{opt} \cong 3.0$ eV was recorded, from which the formation of an $Al(OH)_3$ surface layer on the top of the passive film was inferred [132]. In fact, the presence of thin hydrated oxides on scraped Al surfaces after short immersion times in quasi-neutral aqueous solutions has been revealed by an XPS analysis [168].

Figure 47 shows the photocurrent spectrum of a corrosion layer formed on tin (purity: 99.999%) after anodizing in 1 M NaOH solution up to 1.4 V (HgO) at low scan rate (1 mV/s). An optical bandgap of 2.35 eV is determined by assuming indirect optical transitions (see inset of the figure). By taking into account that for anhydrous SnO_2 passive films grown in neutral and acidic solution under suitable conditions [169] we measured an E_g^{opt} equal to 3.50 eV (in agreement with results

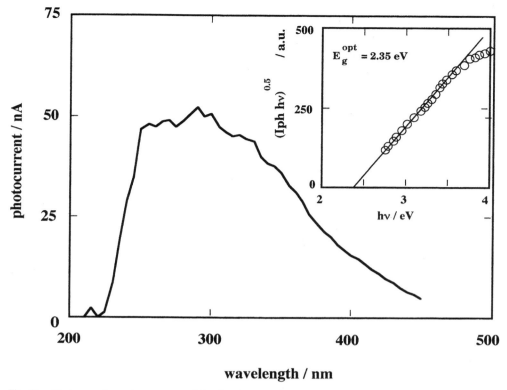

Fig. 47. Photocurrent spectrum, corrected for the photon emission of the light source, recorded in 1 M NaOH solution for a Sn electrode anodized at 10 mV/s up to +1.4 V (HgO). Inset: determination of the optical bandgap assuming indirect optical transitions.

reported in the literature for single-crystal SnO_2 [170]), we have concluded that the lower bandgap measured for the corrosion layer of Figure 47 formed in strongly alkaline solution must be attributed to the formation of a hydrated layer having a composition close to $Sn(OH)_4$.

The anodic photocurrent spectrum and the optical bandgap determination for a corrosion film grown on Mg electrode (purity: 99.9%) in 1 M NaOH solution are reported in Figure 48. For this metal no sharp influence of the metal surface treatment (scraping or chemical etching in acidic solution, after the initial polishing with metallurgical paper) on the E_g^{opt} values was observed. An optical bandgap of about 4.30 eV was obtained for the specimen of Figure 48, whilst a slightly lower value (around 4.15 eV) was obtained for chemicaly etched surfaces. In view of the possible effects of the disordered structure on the E_g^{opt}-values [see the discussion dealing with Eq. (4.2.2)], an average value of 4.25 eV will be assumed for the optical bandgap of passive films grown on Mg in strongly alkaline solutions. Since optical bandgap values ranging between 7.8 and 8.7 eV are reported in the literature for anhydrous MgO [171, 172], it is reasonable to attribute the measured E_g^{opt} values to hydrated oxides having composition close to $Mg(OH)_2$, in agreement also with the thermodynamic expectations based on the Pourbaix diagram [16].

In order to get the correlation for sp-metal hydroxides, in Table VI we report the optical bandgap measured for passive layers formed on Al, Mg, and Sn; the Pauling electronegativity

of the metal; and the square of the electronegativity difference $(X_M - X_{OH})$ between the different sp-metal cations and the OH group. From these data it is possible to derive the following empirical expression of E_g^{opt} as a function of the difference of electronegativity:

$$E_g^{opt} = 1.21(X_M - X_{OH})^2 + 0.90 \qquad (4.3.3)$$

In comparison with the analogous correlation obtained for sp-metal oxides [Eq. (4.1.4b)], the most relevant difference in Eq. (4.3.3) comes from the second term of the right side, suggesting that sp-metal hydroxides always present a finite optical bandgap, on the order of 1.70 eV for the most electronegative sp-metals.

It is worth noting that from Eq. (4.3.3) an optical bandgap of 2.23 eV is derived for $Ni(OH)_2$, in very good agreement both with the value extrapolated by us by assuming indirect optical transitions from the photocurrent spectra reported for passive films on Ni ($E_g^{opt} = 2.20$ eV [14]) and with the value reported in the literature for β-$Ni(OH)_2$ deposited cathodically ($E_g^{opt} = 2.30$ eV [173]). These findings confirm once more that in the proposed correlations Ni^{2+} oxide and hydroxide conform to the sp-metal behavior.

4.3.2. d-Metal Hydroxides (Cr, Zr, Y)

Upon immersing pure (99.99%) chromium metal in H_2SO_4 solution, the formation of a passive film having chemical com-

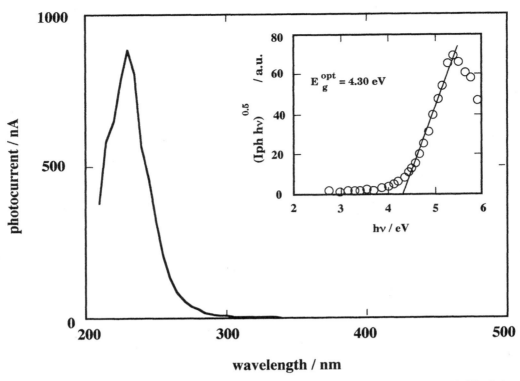

Fig. 48.   Photocurrent spectrum, corrected for the photon emission of the light source, recorded in 1 M NaOH solution for a Mg electrode chemically etched after anodization at 10 mV/s up to +0.8 V (HgO). Electrode potential: +0.4 V (HgO). Inset: determination of the optical bandgap assuming indirect optical transitions.

Table VI.   Experimental Optical Bandgap Values, Cation Electronegativities, and Squared Differences in Electronegativity between Metal and Hydroxyl Group for sp-Metal Hydroxides

Hydroxide	E_g^{opt} (eV)	X_M	$(X_M - X_{OH})^2$
$Mg(OH)_2$	4.25	1.2	2.7225
$Al(OH)_3$	3.0	1.5	1.8225
$Sn(OH)_4$	2.3	1.8	1.1025

position close to $Cr(OH)_3$ and a corresponding optical bandgap of 2.40 ± 0.05 eV was inferred by us some years ago [139, 140]. In the case of Cr and Fe–Cr alloys, the formation of a $Cr(OH)_3$ phase also in acidic solutions has been inferred by EXAFS, XANES, and XPS measurements [174–176], confirming our suggestion. Based on these findings, in the following we will attribute to the $Cr(OH)_3$ phase the E_g^{opt}-value 2.40 eV measured for passive films grown on chromium in sulfuric acid solution at negative electrode potentials.

For passive films grown on Zr metal, photocurrent spectra similar to that reported in Figure 49 have been detected in alkaline solution after anodization of specimens submitted to different treatments of the metallic surface [129]. The presence of two absorption thresholds in the photocurrent spectrum has been explained by us with the formation of a duplex ZrO_2–

$Zr(OH)_4$ film, with the thin external hydroxide surface layer having an optical bandgap of 2.75 ± 0.1 eV and the internal ZrO_2 oxide layer having a forbidden gap equal to 4.60 ± 0.1 eV. This last value is in agreement both with the literature data pertaining to the theoretical bandgap value (4.51 eV) of monoclinic ZrO_2 [177], and with the prediction based on Eq. (4.1.4a), which predicts a bandgap of about 4.45 eV for the crystalline anhydrous oxide. The small difference between the optical bandgap of the passive film and of the crystalline phase (≤ 0.2 eV) could be traced to the amorphous or nanocrystalline nature of the passive film on Zr, strongly depending on the anodizing process and the initial surface preparation [178]. On the other hand, the formation on anodized zirconium metal of an external hydroxide layer having a lower absorption threshold, suggested by us on the basis of a photoelectrochemical study [129], is in agreement with other ESCA and EIS investigations carried out on anodized Zr metal [178, 179].

Finally, in a recent investigation on the electrochemical behavior of Y metal [180], the photocurrent spectrum shown in Figure 50 was recorded for a passive film grown in borate buffer solution (pH 10) on a mechanically abraded metallic surface. To explain this finding, the presence of a $Y(OH)_3$ surface phase having an optical bandgap of 3.05 ± 0.05 eV has been suggested, considering that this phase is thermodynamically stable in aqueous solutions at high pH values [16] and that much larger bandgap values (5.40–5.80 eV) have been reported in the litera-

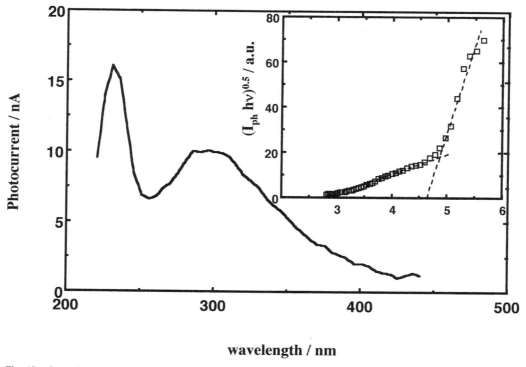

Fig. 49. Raw photocurrent spectrum (not corrected for photon efficiency) recorded for a Zr electrode treated mechanically after anodization in 0.1 M NaOH solution at 100 mV/s up to +5.0 V (HgO). Electrode potential: +3.0 V (HgO). Inset: determination of the optical bandgap assuming indirect optical transitions. [*Electrochim. Acta* 41(16), 2511–2522 (1996); reprinted with permission from Elsevier Science.]

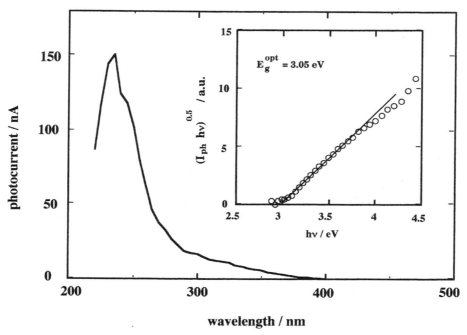

Fig. 50. Photocurrent spectrum, corrected for the photon emission of the light source, recorded in borate buffer solution (pH 10) for a Y electrode abraded mechanically after anodization at 200 mV/s up to +9 V (NHE). Electrode potential: +3.1 V (NHE). Inset: determination of the optical bandgap assuming indirect optical transitions.

Table VII.   Experimental Optical Bandgap Values, Cation Electronegativity, and Squared Difference of Electronegativity between Metal and Hydroxyl Group for d-Metal Hydroxides

Hydroxide	E_g^{opt} (eV)	X_M	$(X_M - X_{OH})^2$
Y(OH)$_3$	3.05	1.25	2.560
Zr(OH)$_4$	2.75	1.4	2.102
Cr(OH)$_3$	2.40	1.6	1.562

ture for Y_2O_3 [171, 181]. More details on structure of corrosion layers grown on yttrium metal can be found elsewhere [180].

In Table VII we report the optical bandgap values of the d-metal hydroxides investigated together with the Pauling electronegativity of the cations and the square of the difference of electronegativity ($X_M - X_{OH}$). From the data it is possible to derive also for hydroxide layers grown on d-metals an empirical correlation between E_g^{opt} and the electronegativity of the hydroxide constituents:

$$E_g^{opt} = 0.65(X_M - X_{OH})^2 + 1.38 \qquad (4.3.4)$$

As in the case of sp-metal hydroxides, on the basis of Eq. (4.3.4) an optical bandgap different than zero is expected even for hydroxides of transition metals having very high electronegativity parameter. Moreover, analogously to the previous correlations (4.1.4a) and (4.1.4b) for anhydrous oxides, the slope of the straight line fitting the correlation for d-metal hydroxides [Eq. (4.3.4)] is noticeably lower than that obtained for the sp-metal hydroxides [Eq. (4.3.3)].

Owing to the lack of information on the bandgap values of hydroxides, the correlations (4.3.3) and (4.3.4) between the optical bandgap and the difference of electronegativity of the hydroxide constituents are based on a restricted number of systems. Nevertheless, it is possible to make some general considerations on the behavior of the investigated systems which further support the validity of these correlations. By comparing the correlations (4.1.4a) and (4.3.4), relative to transition metals, it is found that larger bandgap values are expected for the hydroxide (or hydrated oxide) phase than for the corresponding anhydrous oxide, when the cation electronegativity is higher than 1.95. This finding can help to rationalize the experimental results reported for gold electrodes as well as for other noble metals. In the case of gold an optical bandgap of about 1.50 eV is expected for the hydroxide phase, in comparison with the zero bandgap value expected for the anhydrous oxide. Analogously, for platinum metal ($X_M = 2.1$) an increase of the optical gap is calculated from our correlations on going from the oxide ($E_g = 1.16$ eV) to the hydroxide phase (1.75 eV). Although the nature and the photoelectrochemical behavior of the oxide films grown on these noble metals are still under debate [126, 182–184], this finding could help to explain the complex behavior observed experimentally.

4.3.3. *Mixed Hydroxides and Oxyhydroxides*

By analogy with the case of mixed oxides, it is possible to conceive a straightforward extension to mixed hydroxides of the previous correlations between the optical bandgap of hydroxides and the difference in electronegativity between the metallic cation and the OH group, by using the concept of average cationic electronegativity. However, we are not aware up to now of experimental data pertaining to the optical bandgap values and the film composition of mixed hydroxides, although it is possible to imagine systems for accomplishing such a test. From a practical point of view, it would also be useful to relate the optical bandgap data to a variable hydration degree of the passive films and to find a relationship between data on the anhydrous phase and those on the hydroxide phase.

Careful PCS investigations carried out in the literature [132, 139, 140, 169, 180] have detected optical gap values intermediate between that for the pure anhydrous oxide and that for the hydroxide, for corrosion layers grown on both sp-metals and d-metals (Cr, Y, Al, and Sn). The present authors have attributed such intermediate optical bandgap values to the formation of oxyhydroxide phases of general formula $MO_{y-m}OH_{2m}$, having a variable hydration degree m. An example is shown in Figure 51, where the optical gap values of different Cr^{3+} oxyhydroxides are reported as a function of the density of the phases, which is related to their hydration degree [140, 185]. This interpretation, initially suggested for passive films grown on chromium metal in different electrolytes on the basis of a photoelectrochemical study, has since been used to interpret the PCS data relative to passive films grown on other metals too. In order to rationalize the behavior depicted in Figure 51, we proposed on a heuristic basis the following dependence of the optical gap of an oxyhydroxide phase and its hydration

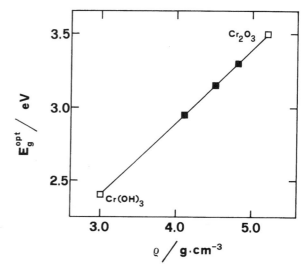

Fig. 51.   Dependence of the optical bandgap of chromium oxyhydroxides on their density. [*Corrosion Sci.* 31(7), 721–726 (1990); reprinted with permission from Elsevier Science.]

degree m:

$$E_g^{hyd} = \frac{E_g^{anh}}{1 + K_{ox}m} \qquad (4.3.5)$$

where E_g^{hyd} and E_g^{anh} are the optical gaps of the hydrated and of the anhydrous oxide, respectively, and K_{ox} is a constant for each system containing a specified cation, M^{z+}, which can be determined once the optical bandgap values for the oxide ($m = 0$) and for the corresponding hydroxide ($m = y$) are known. Equation (4.3.5) has been used for different systems, and it seems to address the problem of the influence of the hydration degree on the bandgap values of passive films and corrosion layers.

As for passive films on metallic alloys, we comment briefly on the data on the optical bandgap of passive films on stainless steels and the compositional data obtained by XPS analysis [186, 187]. In [186] it was shown that the films grown in sulfuric acid at sufficiently anodic electrode potentials (0.5 V SHE) on a single-crystal Fe–18Cr–13Ni alloy consist of an outer hydrated layer, containing a chromium hydroxide phase with a very small portion of $Ni(OH)_2$, and an internal layer of a mixed chromium-iron oxide. Although the relative composition is a function of the aging time, the data reported in the literature can be schematized by assuming that on such an alloy the external passive film is essentially a $Cr(OH)_3$ layer and the internal one is a mixed Cr–Fe oxide with Cr atomic percentage ranging between 62% and 83%. It is perhaps not surprising that bandgap values around 2.40 eV, very near to that quoted for $Cr(OH)_3$ in Table VII, have been reported in the literature [188] for austenitic stainless steels having composition similar to that investigated by Maurice et al. [186], whilst optical bandgaps ranging between 3.12 and 2.81 eV are predicted by Eqs. (4.2.1) and (4.1.4a) for the internal anhydrous layer. Considering that the external layer is also thicker, it is not surprising that the response from the internal layer can be missed by PCS. It could be of some value to investigate such a system under otherwise identical conditions, in order to exploit more deeply the ability of PCS to scrutinize such very complex systems.

In addition, if we take into account both the possible influence of the lattice disorder on the optical bandgaps of these phases and the possible formation of mixed oxyhydroxides having intermediate E_g^{opt}-values, it becomes evident that a PCS investigation alone might be not sufficient for the determination of the exact composition of passive films; additional information coming from other (possibly in situ) techniques, like EXAFS, XANES, and XRD, may be required. However, it is our opinion that also for these very complex systems a careful use of the PCS technique may provide very useful information on the composition of the passivating layers.

Acknowledgments

The authors express their gratitude to all coworkers, including undergraduate and Ph.D. students, who collaborated over the years in their research work. Special thanks to Dr. M. C. Romano, to G. Tuccio, and to Prof. A. Di Paola for their valuable collaboration. Financial assistance from Italian MURST and CNR (Roma), as well as the support of Becromal S.p.a. (Milano), is gratefully acknowledged.

REFERENCES

1. R. P. Frankental and J. Kruger, Eds., "Passivity of Metals," Proceedings of the 4th International Symposium on Passivity. Electrochemical Society, Pennington, 1978.
2. M. Froment, Ed., "Passivation of Metals and Semiconductors," Proceedings of the 5th International Symposium on Passivity. Elsevier, Oxford, 1983.
3. N. Sato and K. Hashimoto, Eds., "Passivation of Metals and Semiconductors," Proceedings of the 6th International Symposium on Passivity. Pergamon Press, Oxford, 1990.
4. B. R. MacDougall, R. S. Alwitt, and T. A. Ramanarayanan, Eds., "Oxide Films on Metals and Alloys," PV 92-22. Electrochemical Society, Pennington, 1992; K. Hebert and G. E. Thompson, Eds., "Oxide Films on Metals and Alloys," PV 94-25. Electrochemical Society, Pennington, 1994.
5. K. E. Heusler, Ed., "Passivation of Metals and Semiconductors," Proceedings of the 7th International Symposium on Passivity. Trans Tech. Publications, Zurich, 1995.
6. F. Mansfeld, A. Asphahani, H. Böhni, and R. Latanision, Eds., "H.H. Uhlig Memorial Symposium," PV 94-26. Electrochemical Society, Pennington, 1995.
7. P. Natishan, H. S. Isaacs, M. Janik-Czackor, V. A. Macagno, P. Marcus, and M. Seo, Eds., "Passivity and Its Breakdown," PV 97-26. Electrochemical Society, Pennington, 1998.
8. R. G. Kelly, G. S. Frankel, P. M. Natishan, and R. C. Newman, Eds., "Critical Factors in Localized Corrosion III," PV 98-17. Electrochemical Society, Pennington, 1999.
9. M. Seo, B. MacDougall, H. Takahashi, and R. G. Kelly, Eds., "Passivity and Localized Corrosion," PV-99-27. Electrochemical Society, Pennington, 1999.
10. J. Kruger, G. G. Long, D. R. Black, and M. Kuriyama, *J. Electroanal. Chem.* 180, 603 (1983).
11. D.C. Koningsberger and R. Prins, "X-ray Absorption: Principles, Applications, Techniques of EXAFS, SEXAFS and XANES." Wiley, New York, 1988.
12. A. J. Arvia, in "Spectroscopy and Diffraction Techniques in Interfacial Electrochemistry" (C. Gutierres and C. Melendres, Eds.). Kluver Academic, Boston, 1990.
13. "The Liquid/Solid Interface at High Resolution," Faraday Discussion no.94, The Royal Society of Chemistry, Lodon (1992).
14. U. Stimming, *Electrochim. Acta* 31, 415 (1986).
15. L. Peter, in "Comprehensive Chemical Kinetics" (R. G. Compton, Ed.), Vol. 29, p. 382. Elsevier Science, Oxford, 1989.
16. M. Pourbaix, "Atlas of Electrochemical Equilibria in Aqueous Solutions." Pergamon Press, Oxford, 1966.
17. W. Schmickler and J. W. Schultze, in "Modern Aspects of Electrochemistry" (J. O'M Bockris, B. E. Conway, and R. E. White, Eds.), Vol. 17, p. 357. Plenum Publishing Corporation, New York, 1986.
18. H. Gerischer, *Electrochim. Acta* 35, 1677 (1990).
19. H. Gerischer, *Corrosion Sci.* 29, 257 (1989).
20. H. Gerischer, *Corrosion Sci.* 31, 81 (1990).
21. J. W. Schultze and M. M. Lohrengel, *Electrochim. Acta* 45, 2499 (2000).
22. L. Young, "Anodic Oxide Films." Academic Press, London, 1961.
23. D. A. Vermilyea, in "Advances in Electrochemistry and Electrochemical Engineering" (P. Delahay, Ed.), Vol. 3, p. 211. Interscience Publishers, New York, 1963.

24. V. P. Parkhutik, J. M. Albella, and J. M. Martinez-Duarte, in "Modern Aspects of Electrochemistry" (B. E. Conway, J. O'M Bockris, and R. E. White, Eds.), Vol. 23, p. 315. Plenum Press, New York, 1992.

25. M. A. Butler, *J. Electrochem. Soc.* 131, 2185 (1984).

26. R. Peat, A. Riley, D. E. Williams, and L. M. Peter, *J. Electrochem Soc.* 136, 3352 (1989).

27. D. E. Williams, A. R. J. Kucernak, and R. Peat, in "Faraday Discussion no.94," p.149. The Royal Society of Chemistry, London, 1992.

28. H. Gerischer, in "Advances in Electrochemistry and Electrochemical Engineering" (P. Delahay, Ed), Vol. 1, p. 139. Interscience Publishers, New York, 1961.

29. V. A. Myamlin and Yu. V. Pleskov, "Electrochemistry of Semiconductors." Plenum Press, New York, 1967.

30. H. Gerischer, in "Physical Chemistry. An Advanced Treatise" (H. Eyring, D. Henderson, and W. Jost, Eds.), Vol. IXA, p. 463. Academic Press, New York, 1970.

31. "Faraday Discussion No.70," The Royal Society of Chemistry, London, 1980.

32. S. R. Morrison, "Electrochemistry at Semiconductor and Oxidized Metal Electrodes." Plenum Press, New York, 1980.

33. Yu. V. Pleskov and Yu. Ya. Gurevich, "Semiconductor Photoelectrochemistry." Consultants Bureau, New York, 1986.

34. A. Hamnett, in "Comprehensive Chemical Kinetics" (R. G. Compton, Ed.), Vol. 27, p. 61. Elsevier Science, Oxford, 1987.

35. J. O'M. Bockris and S. U. M. Khan, "Surface Electrochemistry." Plenum Press, New York, 1993.

36. H. L. F. von Helmholtz, *Ann. Phys. Chem.* 7, 337 (1879).

37. G. Gouy, *Ann. Chim. Phys.* 29, 145 (1903).

38. O. Stern, *Z. Elektrochem.* 30, 508 (1924).

39. D. C. Grahame, *Chem. Rev.* 47, 441 (1947).

40. M. Green, in "Modern Aspects of Electrochemistry" (B. E. Conway and J. O'M. Bockris, Eds.), Vol. 2, p. 343. Butterworths, London, 1959.

41. R. Parsons, in "Advances in Electrochemistry and Electrochemical Engineering" (P. Delahay, Ed.), Vol. 1, p. 1. Interscience, New York, 1961.

42. P. Delahay, "Double Layer and Electrode Kinetics." Interscience, New York, 1966.

43. R. Parsons, *Chem. Rev.* 90, 813 (1990), and references therein.

44. A. Wieckowski, Ed., "Interfacial Electrochemistry." Marcel Dekker, New York, 1999.

45. S. Trasatti, in "Comprehensive Treatise of Electrochemistry" (J. O'M. Bockris, B. E. Conway, and E. Yeager, Eds.), Vol. 1, p. 45. Plenum Press, New York, 1980.

46. E. H. Rodherick, "Metal–Semiconductor Contacts." Clarendon, Oxford, 1980.

47. J. F. Dewald, *Bell Syst. Tech. J.* 39, 615 (1960).

48. J. F. Dewald, *J. Phys. Chem. Solids* 14, 155 (1960).

49. G. Cooper, J. A. Turner, and A. J. Nozik, *J. Electrochem. Soc.* 129, 1973 (1982).

50. H. O. Finklea, *J. Electrochem. Soc.* 129, 2003 (1982).

51. H. Gerischer and R. McIntyre, *J. Chem. Phys.* 83, 1363 (1985).

52. D. S. Ginley and M. A. Butler, in "Semiconductor Electrodes" (H. O. Finklea, Ed.), p. 335. Elsevier Science, Oxford, 1988.

53. M. Tomkiewicz, *J. Electrochem. Soc.* 126, 1505 (1979).

54. W. P. Gomes and F. Cardon, in "Progress in Surface Science," Vol. 12, p. 155. Pergamon Press, Oxford, 1982.

55. W. H. Laflère, R. L. Van Meirhaeghe, F. Cardon, and W. P. Gomes, *Surface Sci.* 59, 401 (1976).

56. W. H. Laflère, R. L. Van Meirhaeghe, F. Cardon, and W. P. Gomes, *J. Appl. Phys. D* 13, 2135 (1980).

57. M. A. Butler, *J. Appl. Phys.* 48, 1914 (1977).

58. W. W. Gärtner, *Phys. Rev.* 116, 84 (1959).

59. S. R. Morrison, *J. Vac. Sci. Technol.* 15, 1417 (1978).

60. R. H. Wilson, *J. Appl. Phys.* 48, 4292 (1977).

61. H. Reiss, *J. Electrochem. Soc.* 125, 937 (1978).

62. J. Reichmann, *Appl. Phys. Lett.* 36, 574 (1980).

63. W. J. Albery, P. N. Bartlett, A. Hamnett, and M. P. Dare-Edwards, *J. Electrochem. Soc.* 128, 1492 (1981).

64. F. El Guibaly and K. Colbow, *J. Appl. Phys.* 53, 1737 (1982).

65. R. U. E't Lam and D. R. Franceschetti, *Mater. Res. Bull.* 17, 1081 (1982).

66. F. Di Quarto, A. Di Paola, S. Piazza, and C. Sunseri, *Solar Energy Mater.* 11, 419 (1985).

67. M. L. Meade, "Lock-In Amplifiers: Principles and Applications." Peter Peregrinus, London, 1983.

68. W. J. Albery and P. N. Bartlett, *J. Electrochem. Soc.* 129, 2254 (1982).

69. J. Li and L. M. Peter, *J. Electroanal. Chem.* 193, 27 (1985).

70. J. Li and L. M. Peter, *J. Electroanal. Chem.* 199, 1 (1986).

71. R. Peat and L. M. Peter, *J. Electroanal. Chem.* 209, 307 (1986).

72. L. M. Peter, *Chem. Rev.* 90, 753 (1990).

73. P. C. Searson, D. D. Macdonald, and L. M. Peter, *J. Electrochem. Soc.* 139, 2538 (1992).

74. F. Di Quarto, S. Piazza, and C. Sunseri, in "Current Topics in Electrochemistry" (J. C. Alexander, Ed.), Vol. 3, p. 357. Council of Scientific Research Integration, Trivarandum, 1994.

75. F. Di Quarto, S. Piazza, and C. Sunseri, *Mater. Sci. Forum* 192–194, 633 (1995).

76. F. Di Quarto, C. Sunseri, S. Piazza, and M. C. Romano, *J. Phys. Chem. B* 101, 2519 (1997).

77. E. Becquerel, *C.R. Hebl. Séances Acad. Sci.* 9, 561 (1839).

78. U. König and J. W. Schultze, in "Interfacial Electrochemistry" (A. Wieckowski, Ed.), p. 649. Marcel Dekker, New York, 1999.

79. D. Stutzle and K. E. Heusler, *Z. Phys. Chem. Neue Folge* 65, 201 (1969).

80. K. E. Heusler and M. Schulze, *Electrochim. Acta* 20, 237 (1975).

81. U. Stimming and J. W. Schultze, *Ber. Bunsenges. Phys. Chem.* 80, 1297 (1976).

82. B. Cahan and C. T. Chen, *J. Electrochem. Soc.* 129, 474 (1982).

83. F. Di Quarto, C. Sunseri, and S. Piazza, *Ber. Bunsenges. Phys. Chem.* 90, 549 (1986).

84. F. Di Quarto, S. Piazza, and C. Sunseri, *Ber. Bunsenges. Phys. Chem.* 91, 437 (1987).

85. S. Piazza, L. Calà, C. Sunseri, and F. Di Quarto, *Ber. Bunsenges. Phys. Chem.* 101, 932 (1997).

86. A. F. Joffe and A. R. Regel, *Prog. Semicond.* 4, 237 (1960).

87. N. F. Mott, *Contemp. Phys.* 10, 125 (1969).

88. M. H. Cohen, H. Fritzsche, and S. R. Ovshinsky, *Phys. Rev. Lett.* 22, 1065 (1969).

89. D. Adler, "Amorphous Semiconductors." CRS Press, Cleveland, 1971.

90. N. F. Mott and E. A. Davis, "Electronic Processes in Non-crystalline Materials," 2nd ed., Clarendon, Oxford, 1979.

91. J. Tauc, "Amorphous and Liquid Semiconductors." Plenum Press, London, 1974.

92. M. H. Brodsky, Ed., "Amorphous Semiconductors." Springer-Verlag, Berlin, 1979.

93. D. Adler, B. B. Schwartz, and M. C. Steele, Eds., "Physical Properties of Amorphous Materials." Plenum Press, New York, 1985.

94. F. Di Quarto, S. Piazza, and C. Sunseri, *Electrochim. Acta* 35, 99 (1990).

95. F. Di Quarto, V. O. Aimiuwu, S. Piazza, and C. Sunseri, *Electrochim. Acta* 36, 1817 (1991).

96. S. Piazza, C. Sunseri, and F. Di Quarto, *AIChE J.* 38, 219 (1992).

97. W. E. Spear, P. G. Le Comber, and A. J. Snell, *Philos. Mag. B* 38, 303 (1978).

98. R. A. Abram and P. J. Doherty, *Philos. Mag. B* 45, 167 (1982).

99. J. D. Cohen and D. V. Lang, *Phys. Rev. B* 25, 5321 (1982).

100. D. V. Lang, J. D. Cohen, and J. P. Harbison, *Phys. Rev. B* 25, 5285 (1982).

101. I. W. Archibald and R. A. Abram, *Philos. Mag. B* 48, 111 (1983).

102. I. W. Archibald and R. A. Abram, *Philos. Mag. B* 54, 421 (1986).

103. W. E. Spear and S. H. Baker, *Electrochim. Acta* 34, 1691 (1989).

104. C. Da Fonseca, M. G. Ferreira, and M. Da Cunha Belo, *Electrochim. Acta* 39, 2197 (1994).

105. See [90, p. 272–274].

106. S. Piazza, C. Sunseri, and F. Di Quarto, *J. Electroanal. Chem.* 293, 69 (1990).

107. G. D. Cody, in "Semiconductors and Semimetals" (J. I. Pankove, Ed.), Vol. 21, Part B, p. 11. Academic Press, London, 1984.

108. M. Santamaria, D. Huerta, S. Piazza, C. Sunseri, and F. Di Quarto, *J. Electrochem. Soc.* 147, 1366 (2000).

109. F. Di Quarto, G. Russo, C. Sunseri, and A. Di Paola, *J. Chem. Soc. Faraday Trans. I* 78, 3433 (1982).

110. F. Di Quarto, S. Piazza, R. D'Agostino, and C. Sunseri, *J. Electroanal. Chem.* 228, 119 (1987).

111. F. Di Quarto, S. Piazza, and C. Sunseri, *Electrochim. Acta* 38, 29 (1993).

112. R. Crandall, *J. Appl. Phys.* 54, 7176 (1983).

113. A. M. Goodman and A. Rose, *J. Appl. Phys.* 42, 2883 (1971).

114. See [90, p. 96].

115. F. Di Quarto, C. Gentile, S. Piazza, and C. Sunseri, *Corrosion Sci.* 35, 801 (1993).

116. L. Onsager, *Phys. Rev.* 54, 554 (1938).

117. D. M. Pai and R.C. Enck, *Phys. Rev. B* 11, 5163 (1975).

118. F. Di Quarto, A. Di Paola, and C. Sunseri, *J. Electrochem. Soc.* 127, 1016 (1980).

119. F. Di Quarto, S. Piazza, and C. Sunseri, *J. Electrochem. Soc.* 130, 1014 (1983).

120. F. Di Quarto, S. Piazza, and C. Sunseri, *Corrosion Sci.* 31, 267 (1990).

121. F. Di Quarto, S. Piazza, and C. Sunseri, *J. Chem. Soc. Faraday Trans. 1* 85, 3309 (1989).

122. F. Di Quarto, S. Piazza, R. D'Agostino, and C. Sunseri, *Electrochim. Acta* 34, 321 (1989).

123. K. Leitner, J. W. Schultze, and U. Stimming, *J. Electrochem. Soc.* 133, 1561 (1986).

124. S. Piazza, G. Lo Biundo, M. C. Romano, C. Sunseri, and F. Di Quarto, *Corrosion Sci.* 40, 1087 (1998).

125. F. Di Quarto, S. Piazza, A. Splendore, and C. Sunseri, in "Proceedings of the Symposium on Oxide Films on Metals and Alloys" (B. R. MacDougall, R. S. Alwitt, and T. A. Ramanarayanan, Eds.), Vol. 92-22, p. 311. Electrochemical Society, Pennington, 1992.

126. T. Watanabe and H. Gerischer, *J. Electroanal. Chem.* 122,73 (1981).

127. S. Piazza, A. Splendore, A. Di Paola, C. Sunseri, and F. Di Quarto, *J. Electrochem. Soc.* 140, 3146 (1993).

128. C. Sunseri, S. Piazza, and F. Di Quarto, *Mater. Sci. Forum* 185–188, 435 (1995).

129. F. Di Quarto, S. Piazza, C. Sunseri, M. Yang, and S.-M. Cai, *Electrochim. Acta* 41, 2511 (1996).

130. Yu. Ya. Gurevich, Yu. V. Pleskov, and Z. A. Rotenberg, "Photoelectrochemistry." Consultants Bureau, New York, 1980.

131. R. H. Fowler, *Phys. Rev.* 38, 45 (1931).

132. G. Tuccio, S. Piazza, C. Sunseri, and F. Di Quarto, *J. Electrochem. Soc.* 146, 493 (1999).

133. S. U. Khan and J. O'M. Bockris, *J. Appl. Phys.* 52, 7270 (1981).

134. J. P. H. Sukamto, W. H. Smyrl, C. S. McMillan, and M. R. Kozlowski, *J. Electrochem. Soc.* 139, 1033 (1992).

135. A. Doghmane and W. E. Spear, *Philos. Mag. B* 53, 463 (1986).

136. A. Many, *J. Phys. Chem. Solids* 26, 575 (1965).

137. O. S. Heavens, "Optical Properties of Thin Solid Films." Dover, New York, 1965.

138. F. Di Quarto, S. Piazza, and C. Sunseri, *Electrochim. Acta* 30, 315 (1985).

139. C. Sunseri, S. Piazza, and F. Di Quarto, *J. Electrochem. Soc.* 137, 2411 (1990).

140. F. Di Quarto, S. Piazza, and C. Sunseri, *Corrosion Sci.* 31, 721 (1990).

141. M. P. Ryan, R. C. Newman, and G. E. Thompson, *J. Electrochem. Soc.* 142, L177 (1995).

142. A. J. Davenport and M. Sansone, *J. Electrochem. Soc.* 142, 725 (1995).

143. M. F. Toney, A. J. Davenport, L. J. Oblonsky, M. P. Ryan, and C. M. Vitus, *Phys. Rev. Lett.* 79, 4282 (1997).

144. D. Zuili, V. Maurice, and P. Marcus, *J. Electrochem. Soc.* 147, 1393 (2000).

145. T. D. Burleigh, *Corrosion* 45, 464 (1989).

146. J. P. Sukamto, C. S. McMillan, and W. Smyrl, *Electrochim. Acta* 38, 15 (1993).

147. J. B. Goodenough, in "Progress in Solid State Chemistry" (H. Reiss, Ed.), Vol. 5, p. 145. Pergamon, Oxford, 1971.

148. J. C. Phillips, "Bonds and Bands in Semiconductors." Academic Press, New York, 1973.

149. J. Honig, in "Electrodes of Conductive Metallic Oxides" (S. Trasatti, Ed.), Part A, p. 2. Elsevier Science, Amsterdam, 1980.

150. J. A. Alonso and N. H. March, "Electrons in Metals and Alloys." Academic Press, New York, 1989.

151. P. A. Cox, "Transition Metal Oxides," Oxford Science Publications. Clarendon, Oxford, 1992.

152. P. Manca, *J. Phys. Chem. Solids* 2, 268 (1961).

153. A. K. Vijh, *J. Phys. Chem. Solids* 30, 1999 (1969).

154. A. K. Vijh, in "Oxides and Oxide Films," Vol. 2, p. 1. Marcel Dekker, New York, 1973.

155. L. Pauling, "The Nature of Chemical Bond," Chap. 3. Cornell University Press, Ithaca, NY, 1960.

156. W. Gordy and W. J. O. Thomas, *J. Phys. Chem.* 24, 439 (1956).

157. A. L. Allred, *J. Inorg. Nucl. Chem.* 17, 215 (1961).

158. M. Haissinsky, *J. Phys. Radium* 7, 7 (1946).

159. C. V. Ramana, O. M. Hussain, B. S. Naidu, C. Julien, and M. Balkanski, *Mater. Sci. Eng. B* 52, 32 (1998).

160. J. Li and D. J. Meier, *J. Electroanal. Chem.* 454, 53 (1998).

161. H. Kim, N. Hara, and K. Sugimoto, *J. Electrochem. Soc.* 146, 955 (1999).

162. F. Di Quarto, "The 1999 Joint International Meeting of the Electrochemical Society," Abstract 1937. Honolulu, HI, 1999.

163. P. Y. Park, E. Akiyama, A. Kawashima, K. Asami, and K. Hashimoto, *Corrosion Sci.* 37, 397 (1996).

164. G. Tuccio, F. Di Quarto, S. Piazza, C. Sunseri, G. E. Thompson, and P. Skeldon, "194th Electrochemical Society Meeting," Abstract 197. Boston, 1998.

165. S. Piazza, C. Sunseri, and F. Di Quarto, "Proceedings 9th European Congress on Corrosion," Abstract FU-140. Utrecht, 1989.

166. D. W. Smith, *J. Chem. Educ.* 67, 559 (1990).

167. W. Y. Ching and Y.-N. Xu, *J. Am. Ceram. Soc.* 77, 404 (1994).

168. W. C. Moshier, G. D. Davis, and J. S. Ahearn, *Corrosion Sci.* 27, 785 (1987).

169. M. C. Romano, S. Piazza, C. Sunseri, and F. Di Quarto, "Proceedings of the Symposium on Passivity and its Breakdown" (P. M. Natishan, H. S. Isaacs, M. Janik-Czachor, V. A. Macagno, P. Marcus, and M. Seo, Eds.), Vol. 97-26, p. 813. Electrochemical Society, Pennington, 1998.

170. M. S. Wrighton, D. L. Morse, A. B. Ellis, D. S. Ginley, and H. B. Abrahamson, *J. Am. Chem. Soc.* 98, 44 (1976).

171. E. D. Palik, Ed., "Handbook of Optical Constants of Solids II." Academic Press, New York, 1991.

172. R. L. Nelson and J. W. Hale, in "Faraday Discussion No.52," p. 77. Royal Society of Chemistry, London, 1980.

173. M. K. Carpenter and D. A. Corrigan, *J. Electrochem. Soc.* 136, 1022 (1989).

174. M. Kerkar, J. Robinson, and A. J. Forthy, in "Faraday Discussion No.89," p. 31. Royal Society of Chemistry, London, 1990.

175. L. J. Oblonsky, M. P. Ryan, and H. S. Isaacs, *J. Electrochem. Soc.* 145, 1922 (1998).

176. P. Marcus and V. Maurice, in "Interfacial Electrochemistry," (A. Wieckowski, Ed.), p. 541. Marcel Dekker, New York, 1999.

177. F. Zandiehnadem, R. A. Murray, and W. Y. Ching, *Physica B* 150, 19 (1988).

178. Y. Kurima, G. E. Thompson, K. Shimizu, and G. C. Wood, "Proceedings of the 7th International Symposium on Oxide Films on Metals and Alloys" (K. R. Hebert and G. E. Thompson, Eds.), Vol. 94-25, p. 256. Electrochemical Society, Pennington, 1994.

179. C. N. Panagopulos, *Thin Solid Films* 137, 135 (1986).

180. S. Piazza, S. Caramia, C. Sunseri, and F. Di Quarto, in "Proceedings of the Symposium on Passivity and Localized Corrosion" (M. Seo, B. MacDougall, H. Takahashi, and R. G. Kelly, Eds.), PV-99-27, p. 317. Electrochemical Society, Pennington, 1999.

181. K. G. Cho, D. Kumar, S. L. Jones, D. G. Lee, P. H. Holloway, and R. K. Singh, *J. Electrochem. Soc.* 145, 3456 (1998).

182. T. Watanabe and H. Gerischer, *J. Electroanal. Chem.* 117,185 (1981).

183. S. Gottesfeld, G. Maia, J. B. Floriano, G. Tremiliosi-Filho, E. A. Ticianelli, and R. Gonzales, *J. Electrochem. Soc.* 138, 3219 (1991).

184. A. J. Rudge, L. M. Peter, G. A. Hards, and R. J. Potter, *J. Electroanal. Chem.* 306, 253 (1996).

185. A. M. Sukhotin, Yu. P. Kostinov, and E. G. Kuz'mina, *Elektrokhimiya* 21, 1149 (1985).

186. V. Maurice, W. P. Yang, and P. Marcus, *J. Electrochem. Soc.* 145, 909 (1998).

187. L. Wegrelius, F. Falkenberg, and I. Olefjord, *J. Electrochem. Soc.* 146, 1397 (1999).

188. A. Di Paola, F. Di Quarto, and C. Sunseri, *Corrosion Sci.* 26, 935 (1986).

Chapter 9

ELECTRON ENERGY LOSS SPECTROSCOPY FOR SURFACE STUDY

Takashi Fujikawa

Graduate School for Science, Chiba University, Chiba 263-8522, Japan

Contents

1. INTRODUCTION

The characteristic energy losses of electron beams penetrating a thin film or scattered from a solid surface can give important information on the nature of the film or the surface. Electron energy loss spectroscopy (EELS) is carried out from about 1 eV to about 100 keV. The low-energy region is used primarily in surface studies where the investigation is focused on the energies of vibrational states associated with adsorbed molecules. The energy loss spectrum contains discrete peaks corresponding to the vibrational states of adsorbed molecules [1]. At higher energies, the dominant peak corresponds to a plasmon loss or losses [2]. The spectra are interpreted in terms of the complex dielectric function, similar to the optical method. Core excitation EELS offers local structural information similar to that provided by the X-ray absorption near-edge structures (XANES) and the extended X-ray absorption fine structures (EXAFS). These spectroscopies, however, are only accessible

Handbook of Thin Film Materials, edited by H.S. Nalwa
Volume 2: Characterization and Spectroscopy of Thin Films

ISBN 0-12-512910-6/$35.00

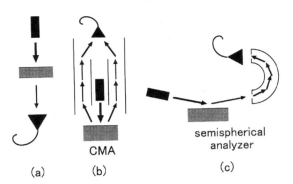

Fig. 1. The different experimental EELS systems for core-loss studies. (a) Transmission mode. (b) Backscattering mode. (c) Grazing incidence and small-takeoff-angle reflection mode.

at a synchrotron radiation facility. In contrast to this limited accessibility, extended energy-loss fine structure (EELFS) measurement can be easily performed in a laboratory system. This spectroscopy has become a valuable tool for local structural investigation of solid surfaces [3, 4].

EELFS was initially observed in an energy spectrum of transmission electron microscopy (TEM) at high energy (about 100 keV) [3]. In this case the observed spectra reflect bulk structures (see Fig. 1a). On the other hand, reflection EELS is sensitive to surface structures. We can classify reflection EELS into two classes (see Fig. 1). The first method counts back scattered electrons (b), and the second measures small-angle scattering electrons (c). Of course, the latter is more surface sensitive.

EELFS may be classified into EXELFS (extended energy-loss fine structure) and ELNES (energy-loss near-edge structure) in a manner similar to that of EXAFS and XANES in X-ray absorption fine structures (XAFS). EELS measures the momentum distribution of electrons generated by inelastic scattering of a monochromatic beam of electrons. When the incident electron energy ϵ_p exceeds the threshold for exciting an electron in the core level, the distribution of energies of inelastically reflected electrons will show structure at $\epsilon_p - \Delta$, where Δ is the excitation energy.

The energy loss Δ corresponds to the photon energy in X-ray absorption spectroscopy. If we can apply the dipole selection rule in EELFS analyses as used in XAFS analyses, we already have analytical tools for the interpretation of EELFS data. A simple theory in which the scattering wave functions of a probe electron are approximated by plane waves, that is, the Bethe theory, gives a simple conclusion that the dipole approximation works well under small-angle scattering conditions [5–8]. As is well known, electron–surface interaction is rather strong, so electron beam cannot penetrate deep into the bulk region. The Bethe approximation is too simple to study EELS spectra from solid surfaces.

To use EELFS as a practical tool of surface analyses, it is inevitably necessary to develop a reliable theory to correctly describe the inelastic scattering of electron beams for any scattering angle and any incident energy. The general theory of electron scattering from solids and their surfaces has been developed by Fujikawa and Hedin [9]. It describes important many-body effects in a physically transparent and acceptable expressions. Based on this general theory, an EELFS theory has been proposed, where important ingredients are damping wave functions of a probe electron and a secondary excited electron under the influence of the corresponding optical potentials [10–13]. In these papers they have shown that EELFS in the small-angle scattering reflection mode has remarkable features, such as surface sensitivity; it is a simple theoretical formula from which we can easily obtain useful local structural information.

Based on the above theoretical consideration, Usami and his co-workers have constructed an apparatus that satisfies the above theoretical requirement [14–16]. They have measured the core excitation EELFS spectra of atomic adsorbates (C, N, O) on single crystals and polycrystalline metals by reflection EELFS methods [17, 18].

In this article we review the EELFS theory, starting from the general scattering theory, and give several examples of how the EELFS technique can be applied to surface structural and electronic structural analyses.

2. GENERAL SCATTERING THEORY

2.1. Potential Scattering Theory

In EELS processes electron scattering plays a central role. This section thus discusses introductory general scattering theory. Based on the present basic theory, further discussion will be given in the later sections. For the later discussions, we consider here the simplest problem in quantum scattering theory, that is, the potential scattering theory [19–21].

2.1.1. Boundary Condition

The boundary condition plays an important role in distinguishing between incoming and outgoing waves. Let us assume that $\phi_{\mathbf{p}}^{\pm}(\mathbf{r})$ is a stationary state solution for a local potential $V(\mathbf{r})$ ($V(\mathbf{r}) \to 0$ as $r \to \infty$) that satisfies the Schrödinger equation

$$\left[-\frac{\hbar^2}{2m}\nabla^2 + V(\mathbf{r}) \right]\phi_{\mathbf{p}}(\mathbf{r}) = \epsilon_p \phi_{\mathbf{p}}(\mathbf{r}) \qquad (1)$$

From now on our discussion is focused on electron scattering, so that it is convenient to use atomic units, that is, $m = e = \hbar = 1$. We rewrite Eq. (1) as

$$(\epsilon_p - T_e)\phi_{\mathbf{p}}(\mathbf{r}) = V(\mathbf{r})\phi_{\mathbf{p}}(\mathbf{r}) \qquad (2)$$

where T_e is the kinetic energy operator of a particle. Far from the target the potential $V(\mathbf{r})$ disappears, and the solution $\phi_{\mathbf{p}}(\mathbf{r})$ is written as the superposition of the corresponding homogeneous solution, the plane wave $\phi_{\mathbf{p}}^0(\mathbf{r}) = (2\pi)^{-3/2}\exp(i\mathbf{p}\cdot\mathbf{r})$ and the spherical wave,

$$\phi_{\mathbf{p}}(\mathbf{r}) \to \phi_{\mathbf{p}}^0(\mathbf{r}) + \frac{1}{(2\pi)^{3/2}}\frac{\exp(\pm ipr)}{r}f^{\pm} \qquad (r \to \infty) \quad (3)$$

The quantity $f^{\pm}$ is called the scattering amplitude. It plays a central role in scattering theory, and we usually use the simplified notation f instead of f^+. To solve the differential equation (2) it is convenient to introduce Green's function g_0, which satisfies the equation

$$(\epsilon_p - T_e)g_0(\mathbf{r}, \mathbf{r}') = \delta(\mathbf{r} - \mathbf{r}') \tag{4}$$

By use of Green's function, the general solution of Eq. (2) satisfies the inhomogeneous integral equation

$$\phi_{\mathbf{p}}(\mathbf{r}) = \phi_{\mathbf{p}}^0(\mathbf{r}) + \int g_0(\mathbf{r}, \mathbf{r}')V(\mathbf{r}')\phi_{\mathbf{p}}(\mathbf{r}')\, d\mathbf{r}' \tag{5}$$

To incorporate the boundary conditions, we must choose the appropriate g_0, which is clearly a function of $\mathbf{r} - \mathbf{r}'$ only, so we put

$$g_0(\mathbf{r} - \mathbf{r}') = \int \frac{d\mathbf{q}}{(2\pi)^3}\hat{g}(\mathbf{q})\exp[i\mathbf{q}\cdot(\mathbf{r} - \mathbf{r}')] \tag{6}$$

Upon substituting Eq. (6) into Eq. (4), we have

$$\tfrac{1}{2}(p^2 - q^2)\hat{g}(\mathbf{q}) = 1 \tag{7}$$

The solution of Eq. (7) is not uniquely determined, and we should introduce boundary condition whose physical meanings are clarified in a wave packet treatment as discussed later [19]. Physical solutions of Eq. (7) are then

$$\hat{g}^{\pm}(\mathbf{q}) = \left[\tfrac{1}{2}(p^2 - q^2) \pm i\eta\right]^{-1} \qquad (\eta \to +0) \tag{8}$$

We obtain

$$g_0^{\pm}(\mathbf{r} - \mathbf{r}') = \int d\mathbf{q}\, \frac{\phi_{\mathbf{q}}^0(\mathbf{r})\phi_{\mathbf{q}}^0(\mathbf{r}')^*}{\epsilon_p - \epsilon_q \pm i\eta} \tag{9}$$

after we substitute Eq. (8) into Eq. (6). First we integrate over the orientation of $\mathbf{q}$, which yields

$$g_0^{\pm}(r) = \frac{1}{2\pi^2 ir}\int_{-\infty}^{\infty} dq\, \frac{q\exp(iqr)}{p^2 - q^2 \pm i\eta} \tag{10}$$

Since $r > 0$, the counter can be closed in the upper half-plane. The residue theorem then yields

$$g_0^{\pm}(r) = -\frac{1}{2\pi}\frac{\exp(\pm ipr)}{r} \tag{11}$$

This gives us the basic integral equation of scattering theory from Eq. (5),

$$\phi_{\mathbf{p}}^{\pm}(\mathbf{r}) = \phi_{\mathbf{p}}^0(\mathbf{r}) - \frac{1}{2\pi}\int \frac{\exp(\pm ip|\mathbf{r} - \mathbf{r}'|)}{|\mathbf{r} - \mathbf{r}'|}V(\mathbf{r}')\phi_{\mathbf{p}}^{\pm}(\mathbf{r}')\, d\mathbf{r}' \tag{12}$$

We now verify that Eq. (12) has the asymptotic form shown by Eq. (3). Let $r \to \infty$, and assume that V decreases rapidly, so that the condition $r \gg r'$ is satisfied everywhere. Then

$$p|\mathbf{r} - \mathbf{r}'| = pr\sqrt{1 + (r'/r)^2 - 2(\hat{\mathbf{r}}\cdot\mathbf{r}')/r} \approx pr - \mathbf{p}'\cdot\mathbf{r}' \tag{13}$$

where $\mathbf{p}' = p\hat{\mathbf{r}}$. Note that $|\mathbf{p}| = |\mathbf{p}'|$. Thus the wave function $\phi_{\mathbf{p}}^{\pm}$ has the desired asymptotic form,

$$\phi_{\mathbf{p}}^{\pm}(\mathbf{r}) \approx \phi_{\mathbf{p}}^0(\mathbf{r}) - \frac{1}{2\pi}\frac{\exp(\pm ipr)}{r}\int \exp(\mp i\mathbf{p}'\cdot\mathbf{r}')$$
$$\times V(\mathbf{r}')\phi_{\mathbf{p}}^{\pm}(\mathbf{r}')\, d\mathbf{r}' \tag{14}$$

The comparison between Eq. (14) and Eq. (3) gives the integral expression of $f^{\pm}$, and we give the explicit formula of $f(= f^+)$,

$$f(\mathbf{p}', \mathbf{p}) = -4\pi^2\langle\phi_{\mathbf{p}'}^0, V\phi_{\mathbf{p}}^+\rangle \tag{15}$$

In the same way $f^-(\mathbf{p}', \mathbf{p})$ is given by

$$f^-(\mathbf{p}', \mathbf{p}) = -4\pi^2\langle\phi_{-\mathbf{p}'}^0, V\phi_{\mathbf{p}}^-\rangle \tag{16}$$

We see that $\phi_{\mathbf{p}}^+$ is closely related to $\phi_{\mathbf{p}}^-$ as

$$\phi_{\mathbf{p}}^-(\mathbf{r}) = \phi_{-\mathbf{p}}^+(\mathbf{r})^* \tag{17}$$

We now turn to a detailed examination of the physical significance of the continuum solutions and their precise connection with collision phenomena. These phenomena are best understood from a time-dependent point of view [19]. Our task is then to find the solution of the Schrödinger equation,

$$i\hbar\frac{\partial\Phi(\mathbf{r}, t)}{\partial t} = h(\mathbf{r})\Phi(\mathbf{r}, t)$$

where $h(\mathbf{r}) = T_e + V(\mathbf{r})$. We shall show that the desired solution is simply

$$\Phi^{\pm}(\mathbf{r}, t) = \int d\mathbf{q}\, \phi_{\mathbf{q}}^{\pm}(\mathbf{r})\chi(\mathbf{q})\exp(-i\epsilon_q t) \tag{18}$$

where $\phi_{\mathbf{q}}^{\pm}$ (in atomic unit) is the solution of Eq. (12). The function $\chi(\mathbf{q})$ is chosen to make Φ have the appropriate properties. Let first study the effects of the plane wave part $\phi_{\mathbf{q}}^0$ on Φ. We shall assume that the spread Δq of χ is much smaller than p, $\Delta q/p \ll 1$. Then we obtain

$$q^2 = |\mathbf{p} + \Delta\mathbf{q}|^2 = p^2 + 2\mathbf{p}\cdot\Delta\mathbf{q} + \Delta q^2 \approx p^2 + 2\mathbf{p}\cdot\Delta\mathbf{q} \tag{19}$$

Now we can write the wave packet of the plane wave part of Eq. (12)

$\Phi_0(\mathbf{r}, t)$

$$= \int d\mathbf{q}\, \phi_{\mathbf{q}}^0(\mathbf{r})\chi(\mathbf{q})\exp(-i\epsilon_q t)$$
$$\approx \exp(-i\epsilon_p t)\int \frac{d\mathbf{q}}{(2\pi)^{3/2}}\exp(-i\mathbf{p}\cdot\Delta\mathbf{q}t)\chi(\mathbf{q})\exp(i\mathbf{q}\cdot\mathbf{r})$$
$$= \exp(i\epsilon_p t)\int \frac{d\mathbf{q}}{(2\pi)^{3/2}}\exp[i\mathbf{q}\cdot(\mathbf{r} - \mathbf{v}_p t)]\chi(\mathbf{q}) \tag{20}$$

where $\mathbf{v}_p = \mathbf{p}$ is the group velocity of the wave packet. By comparing the first and final expressions of Eq. (20), we can see that Φ_0 is simply represented by

$$\Phi_0(\mathbf{r}, t) \approx \exp(i\epsilon_p t)\Phi_0(\mathbf{r} - \mathbf{v}_p t, 0) \tag{21}$$

This demonstrates that the packet indeed moves with velocity $\mathbf{v}_p$ along the classical trajectory without distortion.

We now turn to the term $\Phi_1^\pm(\mathbf{r}, t)$ in $\Phi^\pm$ involving the spherical wave part of Eq. (12),

$$\Phi_1^\pm(\mathbf{r}, t) = -\frac{1}{2\pi} \int d\mathbf{q} \int d\mathbf{r}' \frac{\exp(\pm iq|\mathbf{r} - \mathbf{r}'|)}{|\mathbf{r} - \mathbf{r}'|}$$
$$\times V(\mathbf{r}')\phi_\mathbf{q}^\pm(\mathbf{r}') \exp(-i\epsilon_q t)\chi(\mathbf{q}) \qquad (22)$$

We shall refer to $\Phi_1^\pm$ as the *scattered packet*. The $\mathbf{q}$-integration in (22) will give a negligible contribution unless we are at the point of the stationary phase, which is determined by

$$\frac{\partial}{\partial \mathbf{q}}\left[u(\mathbf{q}) - \epsilon_q t \pm q|\mathbf{r} - \mathbf{r}'| + \lambda_\mathbf{q}^\pm(\mathbf{r}')\right]_\mathbf{p} = \mathbf{0}$$

where u is the phase of $\chi(\mathbf{q})$ and $\lambda_\mathbf{q}^\pm(\mathbf{r}')$ is the phase of $\phi_\mathbf{q}^\pm(\mathbf{r}')$. Because of the presence of $V(\mathbf{r}')$ in the integrand in Eq. (22), the variable $\mathbf{r}'$ is confined to the region specified by $r' < a$, where a is the range of the potential. In analyzing Eq. (22), we must therefore bear in mind that $r' < a$. Since $\Phi_1(\mathbf{r}, 0)$ is assumed to have its maximum at $-z_0\hat{\mathbf{p}}$ (z_0 is macroscopic), we require the stationary phase point of the integrand of (18) to be

$$\frac{\partial u}{\partial \mathbf{q}} = z_0\hat{\mathbf{p}} \qquad (23)$$

The stationary phase point of $\Phi_1^\pm$ is therefore determined from

$$\left(z_0 - v_p t \pm |\mathbf{r} - \mathbf{r}'|\right)\hat{\mathbf{p}} + \frac{\partial \lambda_\mathbf{q}^\pm(\mathbf{r}')}{\partial \mathbf{q}}\bigg|_\mathbf{p} = \mathbf{0} \qquad (24)$$

If $\mathbf{r}$ in the defining equation for $\phi_\mathbf{q}^\pm(\mathbf{r})$ is microscopic, there are simply no macroscopic lengths in Eq. (12), and all lengths characterizing $\phi_\mathbf{q}^\pm(\mathbf{r})$ for such values of $\mathbf{r}$ must themselves be microscopic. In particular, the length $|\partial \lambda_\mathbf{q}^\pm/\partial \mathbf{q}|$ should be microscopic.

At time t_0 the wave packet will actually hit the target. For $t \ll t_0$, $z_0 - v_p t + |\mathbf{r} - \mathbf{r}'|$ is a macroscopic length for all values of $\mathbf{r}$. Because $\frac{\partial \lambda}{\partial \mathbf{q}}$ is microscopic, Eq. (24) cannot be satisfied for such values of t throughout all space. Hence Φ_1^+ vanishes for $t \ll t_0$, and $\Phi^+(\mathbf{r}, t) \to \Phi_0(\mathbf{r}, t)$ for such early times. We have therefore shown that $\Phi^+(\mathbf{r}, t)$ does satisfy the required initial conditions. When t reaches the value t_0, $z_0 - v_p t \approx 0$, and then Eq. (24) can be satisfied for microscopic values of $\mathbf{r}$. This simply says that when $\Phi_1^+(\mathbf{r}, t)$ actually strikes the target, the scattered wave begins to form in the range of the potential, $r < a$. By the same token, when $t \gg t_0$, $z_0 - v_p t$ is a macroscopic negative value, and $|\mathbf{r} - \mathbf{r}'|$ ($r' < a$) must be macroscopic if Eq. (24) is to be satisfied. Thus for $t \gg t_0$, Φ_1^+ is nonvanishing in a shell of radius $v_p t - z_0 = v_p(t - t_0)$. In the same way, if we use the incoming wave in Eq. (24), we find that Φ_1^- vanishes for $t \gg t_0$, and $\Phi^-(\mathbf{r}, t) \to \Phi_0(\mathbf{r}, t)$ for such later times.

We now study the behavior of Φ_1^+ for $r \gg a$ and $t \gg t_0$. In this case $v_p t$ is macroscopic, and Eq. (24) is satisfied for macroscopic r. Since $r \gg r'$ everywhere,

$$|\mathbf{r} - \mathbf{r}'| \approx r - \hat{\mathbf{r}} \cdot \mathbf{r}' \qquad (25)$$

and we have the asymptotic form of Φ_1^+ from Eq. (22) by using the symbol defined by $\mathbf{p}' = p\hat{\mathbf{r}}$ at the infinite future,

$$\Phi_1^+(\mathbf{r}, t) \approx \frac{1}{r} \int \frac{d\mathbf{q}}{(2\pi)^{3/2}} \chi(\mathbf{q}) \exp[i(qr - \epsilon_q t)]f(\mathbf{p}', \mathbf{q}) \quad (26)$$

We shall now make the additional assumption that $f(\mathbf{p}', \mathbf{q})$ varies sufficiently slowly in the interval Δq to permit its replacement by $f(\mathbf{p}', \mathbf{p})$. As before, $\epsilon_q \approx -\epsilon_p + \mathbf{q} \cdot \mathbf{v}_p$, and qr is approximately replaced by $(\mathbf{q} \cdot \hat{\mathbf{p}})r$; then $\Phi_1^+(\mathbf{r}, t)$ becomes

$$\Phi_1^+(\mathbf{r}, t) \approx \frac{f(\mathbf{p}', \mathbf{p})}{r} \exp(i\epsilon_p t)\Phi_0(\hat{\mathbf{p}}r - \mathbf{v}_p t, 0) \qquad (t \to \infty)$$
$$(27)$$

In the same way we obtain the asymptotic behavior of Φ_1^- at the infinite past,

$$\Phi_1^-(\mathbf{r}, t) \approx \frac{f^-(\mathbf{p}', \mathbf{p})}{r} \exp(i\epsilon_p t)\Phi_0(-\hat{\mathbf{p}}r - \mathbf{v}_p t, 0)$$
$$(t \to -\infty) \qquad (28)$$

The propagation of the wave packets $\Phi^\pm$ is schematically shown in Figure 2. From the above discussion we see that we can select the proper boundary condition for each experimental method. For example, we have to use the boundary condition for photoelectrons because we measure their momenta at $t \gg 0$. In the case of EELS, the incident wave with momentum $\mathbf{p}$ is described by $\Phi_\mathbf{p}^+$, and the scattered wave with momentum $\mathbf{p}'$ is described by $\Phi_{\mathbf{p}'}^-$. For practical purposes we use the time-independent stationary states $\phi_\mathbf{p}^\pm$ instead of the wave packets $\Phi_\mathbf{p}^\pm$.

2.1.2. Partial Wave Expansion

When the potential $V(\mathbf{r})$ is spherically symmetric, it is convenient to use an angular momentum representation of the scattering functions $\phi_\mathbf{p}^\pm$. We can write the plane wave $\phi_\mathbf{p}^0$ as

$$\phi_\mathbf{p}^0 = \sqrt{2/\pi} \sum_L i^l j_l(pr)Y_L(\hat{\mathbf{r}})Y_L^*(\hat{\mathbf{p}}) \qquad (29)$$

The total wave function $\phi_\mathbf{p}^+$ can also be written as a superposition of partial waves,

$$\phi_\mathbf{p}^+ = \sqrt{2/\pi} \sum_L i^l R_l(pr)Y_L(\hat{\mathbf{r}})Y_L^*(\hat{\mathbf{p}}) \qquad (30)$$

From Eq. (5) R_l should satisfy the integral equation

$$R_l(pr) = j_l(pr) + \int g_p^{(l)}(r, r')V(r')R_l(pr')r'^2 \, dr' \quad (31)$$

where $g_p^{(l)}(r, r')$'s are expansion coefficients of the free propagator g_0 given by Eq. (9),

$$g_0(\mathbf{r} - \mathbf{r}') = \sum_L g_p^{(l)}(r, r')Y_L(\hat{\mathbf{r}})Y_L^*(\hat{\mathbf{r}}') \qquad (32)$$

Before the Collision

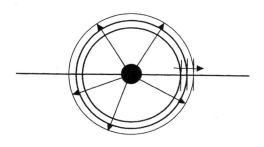

After the Collision

(a)

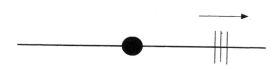

After the Collision

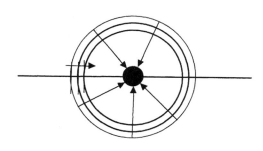

Before the Collision

(b)

Fig. 2.   The wave packets (a) Φ^+ and (b) Φ^- before and after they arrive at the scattering center.

This is written in integral representation and in terms of the spherical Hankel function h_l and the spherical Bessel function j_l,

$$g_p^{(l)}(r, r') = \frac{1}{\pi} \int_{-\infty}^{\infty} dq \, \frac{j_l(qr) j_l(qr') q^2}{p^2 - q^2 + i\eta}$$
$$= -2ip h_l(pr_>) j_l(pr_<) \quad (33)$$

where $r_> = \max(r, r')$ and $r_< = \min(r, r')$. When $\mathbf{r}$ is outside the potential region, the condition $r > r'$ is satisfied. We now

define the phase shifts $\delta_l(p)$ as

$$\exp[2i\delta_l(p)] = 1 - 4ip \int j_l(pr) V(r) R_l(pr) r^2 \, dr \quad (34)$$

The asymptotic form of the radial solution $R_l(pr)$ is written from Eq. (31),

$$R_l(pr) \approx \frac{i^{-l-1}}{2pr} \left\{ \exp[ipr + 2i\delta_l(p)] - (-1)^l \exp[-ipr] \right\}$$
$$= \frac{\exp[i\delta_l(p)]}{pr} \sin\left\{ pr - \frac{l\pi}{2} + \delta_l(p) \right\} \quad (35)$$

As $\exp(ipr)/r$ represents the outgoing spherical wave and $\exp(-ipr)/r$ the incoming spherical wave, we obtain a condition for the probability conservation,

$$\left| \exp(2i\delta_l) \right| = \left| (-1)^l \right| = 1$$

that is, the phase shifts δ_l should be real.

It is important to relate the phase shifts to the scattering amplitude defined by Eq. (3). For this purpose we calculate the asymptotic form of $\phi_{\mathbf{p}}^+ - \phi_{\mathbf{p}}^0$ at $r \gg a$ with the use of Eqs. (29), (30), and (35),

$$\phi_{\mathbf{p}}^+ - \phi_{\mathbf{p}}^0 \approx (2\pi)^{-3/2} \frac{\exp(ipr)}{r} f(\mathbf{p}', \mathbf{p})$$
$$= (2\pi)^{-3/2} \frac{\exp(ipr)}{r} 4\pi$$
$$\times \sum_L \frac{\exp(2i\delta_l) - 1}{2ip} Y_L(\hat{\mathbf{r}}) Y_L^*(\hat{\mathbf{p}}) \quad (36)$$

Therefore we can write the scattering amplitude in terms of the phase shifts,

$$f(\mathbf{p}', \mathbf{p}) = -4\pi \sum_L t_l(p) Y_L(\hat{\mathbf{p}}') Y_L^*(\hat{\mathbf{p}})$$
$$= -\sum_{l=0}^{\infty} (2l+1) t_l(p) P_l(\cos\theta) \quad (37)$$

$$t_l = -\frac{\exp(2i\delta_l) - 1}{2ip} \quad (38)$$

Here θ is the angle between $\hat{\mathbf{p}}$ and $\hat{\mathbf{p}}'$, and $t_l(p)$ is closely related to the T matrix discussed later.

We can demonstrate the optical theorem where the total scattering cross section $\sigma(p)$ is related to the forward scattering amplitude [5, 19–21],

$$\sigma(p) = \frac{4\pi}{p} \text{Im} \, f(\mathbf{p}, \mathbf{p}) \quad (39)$$

2.2. Formal Scattering Theory

2.2.1. T Matrix

We now follow Gottfried's treatment for the time-dependent approach to general scattering theory [19]. Let $|t\rangle$ be a solution of the complete Schrödinger equation,

$$\left(i\frac{\partial}{\partial t} - H \right) |t\rangle = 0 \quad (40)$$

The Hamiltonian $H = H_0 + V$ is assumed to be time independent. For early times $|t\rangle$ tends to $|t\rangle_0$, where

$$\left(i\frac{\partial}{\partial t} - H_0\right)|t\rangle_0 = 0 \tag{41}$$

Although $|t\rangle_0$ coincides with $|t\rangle$ for $-t \gg 0$, the former is not a solution of Eq. (40) once V comes into play.

We may now construct an integral equation that incorporates Eq. (40) and the initial condition. As usual, we write

$$\left(i\frac{\partial}{\partial t} - H_0\right)|t\rangle = V|t\rangle \tag{42}$$

and introduce a Green's function (Green's operator) that satisfies

$$\left(i\frac{\partial}{\partial t} - H_0\right)G(t - t') = \delta(t - t') \tag{43}$$

In particular, we use the retarded boundary condition, i.e., $G^{(+)}(t - t') = 0$ if $t < t'$. Then the solution is written as

$$G^{(+)}(t - t') = -i\theta(t - t')\exp\left[-iH_0(t - t')\right] \tag{44}$$

The desired integral equation is then

$$\left|t^{(+)}\right\rangle = |t\rangle_0 + \int_{-\infty}^{\infty} G^{(+)}(t - t')V\left|t'^{(+)}\right\rangle dt' \tag{45}$$

The notation $|t^{(+)}\rangle$ is intended to emphasize that the solution of Eq. (45) will tend to the free wave packet $|t\rangle_0$ as $t \to -\infty$.

We shall require these special states that are actually eigenstates of H. These states shall be denoted by $|a^{(+)}, t\rangle = |a^{(+)}\rangle \exp(-iE_a t)$. The symbol a incorporates all of the quantum numbers necessary for the specification of the state. The corresponding stationary state belonging to H_0 is written as $|a, t\rangle = |a\rangle \exp(-iE_a t)$. Because we are concerned with states belonging to the continuous rspectrum we have to assign the same energy eigenvalue E_a to both $|a\rangle$ and $|a^{(+)}\rangle$. Returning to Eq. (45), we have

$$\left|a^{(+)}\right\rangle = |a\rangle - i\int_{-\infty}^{0}\exp\left[i(H_0 - E_a)t\right]V\left|a^{(+)}\right\rangle dt \tag{46}$$

The singular integral is calculated by inserting the factor $e^{\eta t'}(\eta \to +0)$ into the integrand of Eq. (45). We thus can integrate Eq. (46) to obtain

$$\left|a^{(+)}\right\rangle = |a\rangle + \frac{\exp(\eta t)}{E_a - H_0 + i\eta}V\left|a^{(+)}\right\rangle \tag{47}$$

This equation is called the Lippmann–Schwinger equation. The infinitesimal η is important to ensure that the state tends to a free packet in the past. From Eq. (47) we have a formal solution for the full Hamiltonian $H = H_0 + V$, when $e^{\eta t}$ is replaced by 1,

$$\begin{aligned}\left|a^{(+)}\right\rangle &= |a\rangle + \frac{1}{E_a - H + i\eta}V|a\rangle \\ &= |a\rangle + G^+(E_a)V|a\rangle \\ &= \frac{i\eta}{E_a - H + i\eta}|a\rangle \end{aligned} \tag{48}$$

where $G^+(E) = 1/(E - H + i\eta)$. Let consider the transition $a \to b \ (a \neq b)$. Because of the orthogonality condition, we have, from Eq. (47),

$$\langle b|a^{(+)}\rangle = \frac{\exp(\eta t)}{E_a - E_b + i\eta}\langle b|V|a^{(+)}\rangle$$

From this we can calculate the transition rate $w_{a\to b}$, in the limit $\eta \to 0$,

$$\begin{aligned}w_{a\to b} &= \frac{d}{dt}\left|\langle b|a^{(+)}\rangle\right|^2 \to 2\pi\left|\langle b|V|a^{(+)}\rangle\right|^2\delta(E_a - E_b) \\ &= 2\pi\left|\langle b|T(E_a)|a\rangle\right|^2\delta(E_a - E_b) \end{aligned} \tag{49}$$

Now we introduce the T matrix, defined by

$$T(b, a) = \langle b|V|a^{(+)}\rangle = \langle b|T|a\rangle \tag{50}$$

Substituting Eq. (47) into Eq. (50), we obtain an equation for the T matrix for any state $|a\rangle$ and $|b\rangle$,

$$T(b, a) = \langle b|V|a\rangle + \langle b|VG_0^+(E_a)T|a\rangle \tag{51}$$

$$G_0^+(E) = (E - H_0 + i\eta)^{-1} \tag{52}$$

Here the operator identity

$$A^{-1} = B^{-1} + B^{-1}(B - A)A^{-1} = B^{-1} + A^{-1}(B - A)B^{-1}$$

is useful for showing that

$$\begin{aligned}G^+(E) &= G_0^+(E) + G_0^+(E)VG^+(E) \\ &= G_0^+(E) + G^+(E)VG_0^+(E) \end{aligned} \tag{53}$$

Equation (51) is valid for any eigenstates of H_0, $|a\rangle$ and $|b\rangle$, and we thus have an operator equation for T,

$$\begin{aligned}T(E) &= V + VG_0^+(E)T(E) \\ &= V + VG_0^+V + VG_0^+VG_0^+V + \cdots \end{aligned} \tag{54}$$

The full Green's function $G^+(E)$ is related to T by Eqs. (53) and (54) as

$$G^+(E) = G_0^+(E) + G_0^+(E)T(E)G_0^+(E) \tag{55}$$

In the first-order approximation, the transition rate is represented by

$$w_{a\to b} \approx 2\pi\left|\langle b|V|a\rangle\right|^2\delta(E_a - E_b) \tag{56}$$

which is known as Fermi's golden rule and is widely used.

We can show that

$$I_a = \sum_b w_{a\to b} = -2\,\mathrm{Im}\langle a|T(E_a)|a\rangle \tag{57}$$

This is also a kind of optical theorem (see Eq. (39)), and is used for the discussion of photoabsorption processes [22].

2.2.2. Gell-Mann, Goldbeger Theory

We now discuss the formal scattering theory, following the celebrated paper by Gell-Mann and Goldbeger [23] (see also [20, 21]).

We remove, in the usual way, the time dependence of the state vector associated with H_0 by a unitary transformation to the interaction representation. Let define a new state vector $\hat{\psi}$ in the interaction representation

$$\hat{\psi}(t) = \exp(iH_0 t)\psi(t) \tag{58}$$

The interaction representation reduces to the Schrödinger representation at time $t = 0$. The new state vector satisfies

$$i\frac{\partial \hat{\psi}(t)}{\partial t} = \hat{V}(t)\hat{\psi}(t) \tag{59}$$

where $\hat{V}(t) = \exp(iH_0 t)V\exp(-iH_0 t)$. We introduce the unitary time development operator $\hat{U}(t, t_0)$ such that

$$\hat{\psi}(t) = \hat{U}(t, t_0)\hat{\psi}(t_0) \tag{60}$$

for each solution $\hat{\psi}(t)$ of Eq. (59). $\hat{U}(t, t_0)$ has two obvious properties that should be noted:

$$\hat{U}(t, t) = 1, \qquad \hat{U}(t, t_0) = \hat{U}(t, t')\hat{U}(t', t_0) \tag{61}$$

It is clear that from Eqs. (58), (59), and (60) that

$$\hat{U}(t, t_0) = \exp(iH_0 t)\exp\{-iH(t - t_0)\}\exp(-iH_0 t_0) \tag{62}$$

$$i\frac{\partial \hat{U}(t, t_0)}{\partial t} = \hat{V}(t)\hat{U}(t, t_0) \tag{63}$$

Integrating both sides of Eq. (63) from t_0 to t, we have

$$\hat{U}(t, t_0) = 1 - i\int_{t_0}^{t} dt'\, \hat{V}(t')\hat{U}(t', t_0) \tag{64}$$

The formal solution of Eq. (64) by iteration can be written, with the aid of Dyson's time ordering operator $\hat{T}$, as

$$\hat{U}(t, t_0) = \hat{T}\exp\left(-i\int_{t_0}^{t} dt'\, \hat{V}(t')\right) \tag{65}$$

To allow t_0 to approach $-\infty$ in $\hat{U}(t, t_0)$, we take

$$\hat{U}(t, -\infty) = \lim_{\eta \to +0} \eta \int_{-\infty}^{0} dt'\, \exp(\eta t')\hat{U}(t, t') \tag{66}$$

and in an analogous way, we define

$$\hat{U}(\infty, t) = \lim_{\eta \to +0} \eta \int_{0}^{\infty} dt'\, \exp(-\eta t')\hat{U}(t', t) \tag{67}$$

The operator $\hat{U}(t, -\infty)$ carries $|a\rangle$ into $|a^{(+)}\rangle$ from Eq. (48),

$$\hat{U}(t, -\infty)|a\rangle = \hat{U}(t, 0)\hat{U}(0, -\infty)|a\rangle$$
$$= \hat{U}(t, 0)\frac{i\eta}{E_a - H + i\eta}|a\rangle$$
$$= \hat{U}(t, 0)|a^{(+)}\rangle \tag{68}$$

Similarly the operator $\hat{U}(\infty, 0)$ carries $|a\rangle$ into the eigenstates $|a^{(-)}\rangle$ of the total Hamiltonian H as an outgoing wave,

$$\hat{U}(0, \infty)|a\rangle = |a^{(-)}\rangle \tag{69}$$

It is well known that in the interaction representation Heisenberg's S matrix takes the form $\hat{U}(\infty, -\infty)$,

$$S = \hat{U}(\infty, -\infty) = \hat{U}(\infty, 0)\hat{U}(0, -\infty) \tag{70}$$

We thus have a matrix element with the aid of Eqs. (68) and (69),

$$S(a, b) = \langle a|\hat{U}(\infty, -\infty)|b\rangle = \langle a^{(-)}|b^{(+)}\rangle \tag{71}$$

This is also rewritten in terms of the T matrix, after we substitute Eq. (48) into (71),

$$S(a, b) = \delta_{a,b} - 2\pi i\delta(E_a - E_b)\langle a|T(E_a)|b\rangle \tag{72}$$

We can prove that the S matrix is unitary in a variety of ways [20, 21, 23],

$$SS^\dagger = S^\dagger S = 1 \tag{73}$$

2.2.3. Watson's Theorem

As an application of the previous subsection, we consider briefly a scattering process in which two potentials are acting. This process plays an important role in the discussion of resonance scattering later. We consider a Hamiltonian of the form

$$H = H_0 + U + V$$

We are interested in obtaining an expression for the exact T matrix associated with two potential $U + V$. Here we assume that $|\phi_a\rangle$, $|\chi_a\rangle$, and $|\psi_a\rangle$ are the eigenstates of the Hamiltonian, H_0, $H_0 + U$, and H,

$$H_0|\phi_a\rangle = E_a|\phi_a\rangle$$
$$(H_0 + U)|\chi_a\rangle = E_a|\chi_a\rangle \tag{74}$$
$$H|\psi_a\rangle = E_a|\psi_a\rangle$$

The state $|\chi_b^-\rangle$ is related to $|\phi_b^-\rangle$, and $|\psi_a^+\rangle$ to $|\phi_b^+\rangle$ from Eq. (47),

$$|\chi_b^-\rangle = |\phi_b^-\rangle + \frac{1}{E_b - H_0 - i\eta}U|\chi_b^-\rangle \tag{75}$$

$$|\psi_a^+\rangle = |\phi_a^+\rangle + \frac{1}{E_a - H_0 + i\eta}(U + V)|\psi_a^+\rangle \tag{76}$$

The transition matrix on the energy shell, $E_a = E_b$, is given with the use of Eqs. (75) and (76),

$$T(b, a) = \langle\phi_b^-|U + V|\psi_a^+\rangle$$
$$= \langle\chi_b^-|U + V|\psi_a^+\rangle$$
$$\quad - \langle\chi_b^-|U\frac{1}{E_a - H_0 + i\eta}(U + V)|\psi_a^+\rangle \tag{77}$$
$$= \langle\chi_b^-|U + V|\psi_a^+\rangle - \langle\chi_b^-|U|\psi_a^+ - \phi_a^+\rangle$$
$$= \langle\chi_b^-|V|\psi_a^+\rangle + \langle\chi_b^-|U|\phi_a^+\rangle$$

Similarly, we can show that $\langle\chi_b^-|U|\phi_a^+\rangle$ is rewritten as $\langle\phi_b^-|U|\chi_a^+\rangle$ from Eq. (48),

$$\langle\chi_b^-|U|\phi_a^+\rangle = \left\langle\left(1 + \frac{1}{E_b - H_0 - U - i\eta}U\right)\phi_b^-\bigg|U|\phi_a^+\right\rangle$$
$$= \langle\phi_b^-|U\left(1 + \frac{1}{E_b - H_0 - U + i\eta}U\right)|\phi_a^+\rangle$$
$$= \langle\phi_b^-|U|\chi_a^+\rangle$$

The final useful formula for T matrix elements is thus given for two-potential scattering, which is referred to as Watson's theorem [21],

$$T(b, a) = \langle \phi_b^- | U | \chi_a^+ \rangle + \langle \chi_b^- | V | \psi_a^+ \rangle \tag{78}$$

From this expression we can readily obtain a distorted wave approximation,

$$T(b, a) = \langle \phi_b^- | U | \chi_a^+ \rangle + \langle \chi_b^- | V | \chi_a^+ \rangle + \cdots \tag{79}$$

2.3. Site T Matrix Expansion

In this subsection we consider the electron scattering from many-atomic systems, where the scattering potential $V(\mathbf{r})$ is given as a sum of each atomic potential v_α.

2.3.1. Site T Matrix for Spherically Symmetric Potential

Here we consider the scattering from a localized spherically symmetric potential $v(r)$ at site 0. In the coordinate representation, from Eq. (54) the T matrix is written as

$$t(\mathbf{r}, \mathbf{r}') = v(r)\delta(\mathbf{r} - \mathbf{r}') + v(r)g_0(\mathbf{r} - \mathbf{r}')v(r')$$
$$+ v(r) \int g_0(\mathbf{r} - \mathbf{r}_1)v(r_1)g_0(\mathbf{r}_1 - \mathbf{r}') \, d\mathbf{r}_1 v(r') + \cdots \tag{80}$$

where g_0 is the free propagator shown by Eq. (32). We first note that

$$\delta(\mathbf{r} - \mathbf{r}') = \frac{\delta(r - r')}{r^2} \sum_L Y_L(\hat{\mathbf{r}})Y_L^*(\hat{\mathbf{r}}') \tag{81}$$

Substituting Eq. (32) into Eq. (80), we see that $t(\mathbf{r}, \mathbf{r}')$ is represented in terms of spherical harmonics,

$$t(\mathbf{r}, \mathbf{r}') = \sum_L t_l(r, r')Y_L(\hat{\mathbf{r}})Y_L^*(\hat{\mathbf{r}}') \tag{82}$$

where

$$t_l(r, r') = \frac{v(r)\delta(r - r')}{r^2} + v(r) \Big\{ g^{(l)}(r, r')$$
$$+ \int g^{(l)}(r, r_1)v(r_1)g^{(l)}(r_1, r')r_1^2 \, dr_1 + \cdots \Big\} v(r') \tag{?}$$

The scattering wave function $\phi_{\mathbf{p}}^+$ was given in terms of the plane wave $\phi_{\mathbf{p}}^0$ and the free propagator as in Eq. (12) and is written in terms of t,

$$\phi_{\mathbf{p}}^+ = \phi_{\mathbf{p}}^0 + g_0 v \phi_{\mathbf{p}}^+$$
$$= \phi_{\mathbf{p}}^0 + g_0(v + vg_0v + \cdots)\phi_{\mathbf{p}}^0$$
$$= \phi_{\mathbf{p}}^0 + g_0 t \phi_{\mathbf{p}}^0 \tag{83}$$

Substitution of Eqs. (32) and (82) into Eq. (83) yields the following formula for R_l defined in Eq. (30) in the case where $\mathbf{r}$ is outside the region of v,

$$R_l(pr) = j_l(pr) - iph_l(pr)t_l(p) \tag{84}$$

$$t_l(p) = 2 \int j_l(pr)j_l(pr')t_l(r, r')r^2r'^2 \, dr \, dr' \tag{85}$$

On the other hand, from Eqs. (31) and (34), we have for $R_l(pr)$ outside the potential region of v

$$R_l(pr) = j_l(pr) + h_l(pr)\frac{\exp(2i\delta_l) - 1}{2} \tag{86}$$

The comparison between Eq. (84) and Eq. (86) yields an important relation (see also Eq. (38)),

$$t_l(p) = -\frac{\exp(2i\delta_l) - 1}{2ip} \tag{87}$$

This shows that $t_l(p)$, defined by Eq. (85), is the same as the definition given in Eq. (37).

2.3.2. Radial Green's Functions for Spherically Symmetric Potential and Origin Shift Formula

We derive the radial Green's function for spherically symmetric potential $v(r)$ and several useful expressions to shift the origin around which angular momentum expansion is applied.

From Eq. (55), we have $g = g_0 + g_0 t g_0$, where t is expanded as Eq. (82) in angular momentum representation,

$$g(\mathbf{r}, \mathbf{r}') = \sum_L g_l(r, r')Y_L(\hat{\mathbf{r}})Y_L^*(\hat{\mathbf{r}}') \tag{88}$$

As g satisfies

$$(\epsilon_p - T_e - v)g(\mathbf{r}, \mathbf{r}') = \delta(\mathbf{r} - \mathbf{r}') \tag{89}$$

the radial part $g_l(r, r')$ should satisfy the radial equation

$$\left[\epsilon_p + \frac{1}{2r^2} \left(\frac{\partial}{\partial r} \left(r^2 \frac{\partial}{\partial r} \right) - l(l+1) \right) - v(r) \right] g_l(r, r') = \frac{\delta(r - r')}{r^2} \tag{90}$$

General properties of a one-dimensional Green's function like g_l should be used to obtain the explicit expression. It should be continuous, but its derivative should be discontinuous at $r = r'$, and then we have a condition on it by integrating both sides of Eq. (90) from $r' - \Delta$ to $r' + \Delta$ ($\Delta \to 0$),

$$\left. \frac{\partial g_l(r, r')}{\partial r} \right|_{r=r'-\Delta}^{r=r'+\Delta} = \frac{2}{r'^2} \tag{91}$$

Let us find a solution of Eq. (90) in the form

$$g_l(r, r') = a_l R_l(r_<) f_l(r_>) \tag{92}$$

Equation (90) implies that g_l is analytic everywhere except $r = r'$, so that R_l should be regular at $r = 0$ and satisfies the homogeneous equation corresponding to Eq. (90),

$$\left[\epsilon_p + \frac{1}{2r^2} \left(\frac{\partial}{\partial r} \left(r^2 \frac{\partial}{\partial r} \right) - l(l+1) \right) - v(r) \right] R_l(r) = 0 \tag{93}$$

Thus R_l has to be the same as that given by Eq. (84). Both f_l and R_l depend on p; f_l is just the spherical Hankel function h_l outside the potential region. If the potential is switched off, g_l is simply to be reduced to $g^{(l)}$, shown by Eq. (33).

Substituting Eq. (92) into (91), we have

$$a_l W(R_l, f_l)_r = \frac{2}{r^2} \tag{94}$$

where $W(F, G)$ is Wronskian defined by $W(F, G) = FG' - F'G$. Outside the potential region, R_l and f_l should have the following asymptotic formulas from Eq. (86),

$$R_l(r) \approx \exp(i\delta_l)\sin(pr - l\pi/2 + \delta_l)/pr \quad (95)$$

$$f_l(r) \approx i^{-l-1}\exp(ipr)/pr \quad (96)$$

When we calculate the Wronskian $W(R_l, f_l)_r$ far from the potential region $pr \gg 1$, we obtain the simple expression $W(R_l, f_l)_r \to i/pr^2$, so that we get $a_l = -2ip$. Finally we obtain the explicit formula of g_l as

$$g_l(r, r') = -2ipR_l(r_<)f_l(r_>) \quad (97)$$

Next we derive the formula to shift the origin around which angular momentum expansion is applied. From now on we use k instead of p. First we calculate the following integral representation for the free propagator (see Eq. (9)),

$$g_0(\mathbf{r} + \mathbf{R} - \mathbf{r}') = \frac{2}{(2\pi)^3}\int\frac{\exp[\mathbf{q}\cdot(\mathbf{r} + \mathbf{R} - \mathbf{r}')]}{k^2 - q^2 + i\eta}d\mathbf{q} \quad (98)$$

By use of the angular momentum representation of each plane wave shown by Eq. (29) we find that $g_0(\mathbf{r} + \mathbf{R} - \mathbf{r}')$ is also written as

$$g_0(\mathbf{r} + \mathbf{R} - \mathbf{r}')$$
$$= 16\sum_{L_1L_2L_3}i^{l_1+l_2-l_3}\int_0^\infty\frac{j_{l_1}(qR)j_{l_2}(qr)j_{l_3}(qr')q^2}{k^2 - q^2 + i\eta}dq$$
$$\times Y_{L_1}(\hat{\mathbf{R}})Y_{L_2}(\hat{\mathbf{r}})Y_{L_3}^*(\hat{\mathbf{r}}')G(L_1L_2|L_3) \quad (99)$$

Noting that $j_l(-r) = (-1)^l j_l(r)$, $j_l(r) = [h_l(r) + h_l^{(2)}(r)]/2$, and that the Gaunt integral $G(L_1L_2|L_3)$ defined by $\int Y_{L_1}Y_{L_2}Y_{L_3}^*d\hat{\mathbf{r}}$ does not vanish only for $l_1 + l_2 + l_3 = $ even, we can replace the integral $\int_0^\infty dq$ with $\frac{1}{2}\int_{-\infty}^\infty dq$ in Eq. (99). As we can assume that $R > r + r'$, the integral in Eq. (99) now becomes

$$\frac{1}{4}\int_{-\infty}^\infty dq\frac{[h_{l_1}(qR) + h_{l_1}^{(2)}(qR)]j_{l_2}(qr)j_{l_3}(qr')q^2}{k^2 - q^2 + i\eta}$$

The asymptotic forms of the first and second kinds of Hankel functions are

$$h_l(x) \to i^{-l-1}\exp(ix)/x, \qquad h_l^{(2)}(x) \to i^{l+1}\exp(-ix)/x \quad (100)$$

From these relations $h_{l_1}(qR)j_{l_2}(qr)j_{l_3}(qr')$ behaves like $\exp[iq(R \pm r \pm r')]$ on the very large semicircle in the upper half complex plane, and the counter-integration on the semicircle completely vanishes. The residue analyses give us the result of the integration,

$$\frac{1}{4}\int_{-\infty}^\infty dq\frac{h_{l_1}(qR)j_{l_2}(qr)j_{l_3}(qr')q^2}{k^2 - q^2 + i\eta}$$
$$= -\frac{\pi ik}{4}h_{l_1}(kR)j_{l_2}(kr)j_{l_3}(kr') \quad (101)$$

Noting that $h_l^{(2)}(-x) = (-1)^l h_l(x)$ and using the parity of spherical Bessel functions mentioned above, we find that the integral including the function $h_l^{(2)}$ is the same as Eq. (101). Then we obtain the result,

$$g_0(\mathbf{r} + \mathbf{R} - \mathbf{r}') = -8\pi ik\sum_{L_1L_2L_3}i^{l_1+l_2-l_3}h_{l_1}(kR)j_{l_2}(kr)j_{l_3}(kr')$$
$$\times Y_{L_1}(\hat{\mathbf{R}})Y_{L_2}(\hat{\mathbf{r}})Y_{L_3}^*(\hat{\mathbf{r}}')G(L_1L_2|L_3) \quad (102)$$

Substituting $\mathbf{r} = 0$ in Eq. (102) and using the relations $j_l(0)Y_L(\hat{\mathbf{r}}) = \delta_{L,00}/\sqrt{4\pi}$, $G(L_100|L_3) = \delta_{L_1,L_3}/\sqrt{4\pi}$, we obtain the formula shown by Eq. (33). We can see that $|\mathbf{R}+\mathbf{r}| > r'$ if $R > r + r'$. From Eq. (33) we thus have

$$g_0(\mathbf{r} + \mathbf{R} - \mathbf{r}')$$
$$= -2ik\sum_L h_l(k|\mathbf{r} + \mathbf{R}|)j_l(kr')Y_L(\widehat{\mathbf{r} + \mathbf{R}})Y_L^*(\hat{\mathbf{r}}') \quad (103)$$

Comparison of Eq. (102) with Eq. (103) yields an important relation,

$$h_l(k|\mathbf{r} + \mathbf{R}|)Y_L(\widehat{\mathbf{r} + \mathbf{R}})$$
$$= 4\pi\sum_{L_1L_2}i^{l_1+l_2-l}h_{l_1}(kR)j_{l_2}(kr)Y_{L_1}(\hat{\mathbf{R}})Y_{L_2}(\hat{\mathbf{r}})G(L_1L_2|L) \quad (104)$$

Now we define $G_{LL'}(k\mathbf{R})$ as

$$G_{LL'}(k\mathbf{R}) = -4\pi ik\sum_{L_1}i^{l_1}h_{l_1}(kR)Y_{L_1}(\hat{\mathbf{R}})G(L_1L|L') \quad (105)$$

then $g_0(\mathbf{r} + \mathbf{R} - \mathbf{r}')$ is written in terms of $G_{LL'}(k\mathbf{R})$,

$$g_0(\mathbf{r}+\mathbf{R}-\mathbf{r}') = 2\sum_{LL'}i^{l-l'}G_{LL'}(k\mathbf{R})j_l(kr)Y_L(\hat{\mathbf{r}})j_{l'}(kr')Y_{L'}^*(\hat{\mathbf{r}}') \quad (106)$$

From Eqs. (104) and (105) we have a useful relation for shifting the origin,

$$-ikh_l(k|\mathbf{r} + \mathbf{R}|)Y_L(\widehat{\mathbf{r} + \mathbf{R}}) = \sum_{L'}i^{l'-l}G_{L'L}(k\mathbf{R})j_{l'}(kr)Y_{L'}(\hat{\mathbf{r}}) \quad (107)$$

When we put $\mathbf{r} = 0$ in Eq. (98) and in Eq. (102), we obtain an integral representation,

$$-2ik\sum_L h_l(kR)j_l(kr)Y_L(\hat{\mathbf{R}})Y_L^*(\hat{\mathbf{r}})$$
$$= \frac{2}{(2\pi)^3}\int\frac{\exp[i\mathbf{q}\cdot(\mathbf{R} - \mathbf{r})]}{k^2 - q^2 + i\eta}d\mathbf{q} \quad (108)$$

We now multiply both sides of Eq. (108) by $j_{l'}(k'r)Y_{L'}(\hat{\mathbf{r}})$ and integrate over $\mathbf{r}$, and then we obtain

$$-2ikh_{l'}(kR)Y_{L'}(\hat{\mathbf{R}})\frac{\delta(k - k')}{k^2}$$
$$= \frac{8\pi i^{-l'}}{(2\pi)^3}\int\frac{\exp(i\mathbf{q}\cdot\mathbf{R})}{k^2 - q^2 + i\eta}\frac{\delta(k' - q)}{q^2}Y_{L'}(\hat{\mathbf{q}})d\mathbf{q} \quad (109)$$

where we have used the orthogonality relation,

$$\int j_l(kr)Y_L^*(\hat{\mathbf{r}})j_{l'}(k'r)Y_{L'}(\hat{\mathbf{r}})d\mathbf{r} = \frac{\pi}{2}\delta_{LL'}\delta(k - k')/k^2 \quad (110)$$

Again we integrate Eq. (109) over k', and we find that

$$-2ikh_l(kR)Y_L(\hat{\mathbf{R}}) = \frac{i^{-l}}{\pi^2} \int \frac{\exp(i\mathbf{q}\cdot\mathbf{R})}{k^2-q^2+i\eta} Y_L(\hat{\mathbf{q}})\,d\mathbf{q} \quad (111)$$

The substitution of Eq. (111) into Eq. (105) yields an integral representation of $G_{LL'}(k\mathbf{R})$,

$$G_{LL'}(k\mathbf{R}) = \frac{2}{\pi} \int \frac{\exp(i\mathbf{q}\cdot\mathbf{R})}{k^2-q^2+i\eta} Y_L^*(\hat{\mathbf{q}})Y_{L'}(\hat{\mathbf{q}})\,d\mathbf{q} \quad (112)$$

In the above derivation we have utilized the identity

$$\sum_{L_1} Y_{L_1}(\hat{\mathbf{q}})G(L_1L|L')$$
$$= \int d\hat{\mathbf{Q}} \sum_{L_1} Y_{L_1}(\hat{\mathbf{q}})Y_{L_1}^*(\hat{\mathbf{Q}})Y_L^*(\hat{\mathbf{Q}})Y_{L'}(\hat{\mathbf{Q}})$$
$$= \int d\hat{\mathbf{Q}}\delta(\hat{\mathbf{q}}-\hat{\mathbf{Q}})Y_L^*(\hat{\mathbf{Q}})Y_{L'}(\hat{\mathbf{Q}}) = Y_L^*(\hat{\mathbf{q}})Y_{L'}(\hat{\mathbf{q}}) \quad (113)$$

By use of the integral representation (112) and the parity of the spherical harmonics $Y_L(-\hat{\mathbf{r}}) = (-1)^l Y_L(\hat{\mathbf{r}})$, we have a symmetry relation of $G_{LL'}(k\mathbf{R})$,

$$G_{LL'}(-k\mathbf{R}) = (-1)^{l+l'}G_{LL'}(k\mathbf{R}) \quad (114)$$

Furthermore, we obtain another symmetric relation by use of the property $Y_L(\hat{\mathbf{r}}) = (-1)^m Y_{\bar{L}}^*(\hat{\mathbf{r}})$, where $\bar{L} = (l, -m)$,

$$G_{LL'}(k\mathbf{R}) = (-1)^{m+m'}G_{\bar{L}'\bar{L}}(k\mathbf{R}) \quad (115)$$

Combining Eq. (114) and Eq. (115), we obtain the following useful relation:

$$G_{LL'}(-k\mathbf{R}) = (-1)^{l+l'+m+m'}G_{\bar{L}'\bar{L}}(k\mathbf{R}) \quad (116)$$

These symmetric relations reduce the number of matrix elements to be calculated and save storage memory in XANES and ELNES analyses [24, 25, 144].

To reduce the computation time further, some recurrence relations between $G_{LL'}(k\mathbf{R})$'s are useful, which are discussed in [25]. The direct calculation of $G_{LL'}(k\mathbf{R})$ requires much computation time, but the simplest elements, such as $G_{00,L}(k\mathbf{R})$, are easily calculated from Eq. (105),

$$G_{00,L}(k\mathbf{R}) = -\sqrt{4\pi}c_l(z)Y_L(\hat{\mathbf{R}})\exp(ikR)/R$$

where $z = i/(2kR)$ and $c_l(z)$ is defined by the relation

$$h_l(kR) = i^{-l-1}\frac{\exp(ikR)c_l(z)}{kR} \quad (117)$$

To obtain other elements of $G_{LL'}(k\mathbf{R})$ conserving m and m', we apply the relation

$$\int Y_{10}(\hat{\mathbf{r}})Y_{L_3}(\hat{\mathbf{r}})Y_{L_1}(\hat{\mathbf{r}})Y_{L_2}^*(\hat{\mathbf{r}})\,d\hat{\mathbf{r}}$$
$$= \sum_{l_4} G(L_110|l_4)G(L_3l_4m_1|L_2)$$
$$= \sum_{l_4} G(L_210|l_4)G(L_3L_1|l_4) \quad (118)$$

In the above equation the sum over l_4 is restricted to $l_1 \pm 1$ and $l_2 \pm 1$, so that we have a simple relation between those Gaunt integrals. From Eq. (105) we can obtain the recurrence relation between $G_{LL'}$ conserving m and m' with the aid of Eq. (118),

$$a_{l+1,m}G_{l+1,m;L'} + a_{l,m}G_{l-1,m;L'}$$
$$= a_{l'+1,m'}G_{L;l'+1,m'} + a_{l',m'}G_{L;l'-1,m'} \quad (119)$$

where

$$a_{lm} = \sqrt{\frac{(l+m)(l-m)}{(2l+1)(2l-1)}} \quad (120)$$

For the special case where $l = 0$, we obtain the relation

$$a_{1m}G_{1m;L'} = (a_{l'+1,m'}G_{00;l'+1,m'} + a_{l'm'}G_{00;l'-1,m'})\delta_{m,0} \quad (121)$$

From Eq. (121), we can obtain the expressions for $G_{10;L'}$ in terms of $G_{00;l'\pm1.m'}$, and from Eq. (119) $G_{20;L'}$ in terms of $G_{10;l'\pm1.m'}$ and $G_{00;L'}$, and so on.

To obtain the matrix elements such as $G_{l,\pm1;L'}$ from $G_{l0;L'}$'s, we have to derive a recurrence relation to change m. We now use the relation between the Gaunt integrals,

$$\int Y_{1,\pm1}(\hat{\mathbf{r}})Y_{L_3}(\hat{\mathbf{r}})Y_{L_1}(\hat{\mathbf{r}})Y_{L_2}^*(\hat{\mathbf{r}})\,d\hat{\mathbf{r}}$$
$$= \sum_{l_4} G(L_11,\pm1|l_4)G(L_3l_4m_1\pm1|L_2)$$
$$= -\sum_{l_4} G(L_21,\mp1|l_4)G(L_3L_1|l_4) \quad (122)$$

From Eq. (105) we obtain the recurrence relations changing m and m' by ±1,

$$b_L^+G_{l+1,m+1;L'} + b_L^-G_{l-1,m+1;L'}$$
$$= -b_{l',-m'}^+G_{L;l'+1,m'-1} - b_{l',-m'}^-G_{L;l'-1,m'-1}$$
$$b_{l,-m}^+G_{l+1,m-1;L'} + b_{l,-m}^-G_{l-1,m-1;L'}$$
$$= -b_{L'}^+G_{L;l'+1,m'+1} - b_{L'}^-G_{L;l'-1,m'+1} \quad (123)$$

where

$$b_L^+ = \sqrt{\frac{(l+m+1)(l+m+2)}{2(2l+1)(2l+3)}}$$

$$b_L^- = -\sqrt{\frac{(l-m)(l-m-1)}{2(2l+1)(2l-1)}} \quad (124)$$

From these recurrence relations and symmetric relations we can calculate all of the matrix elements of $G_{LL'}$.

The Green's function g_A for the full potential v_A at site A is expanded in angular momentum representation as in Eq. (88). Now let us consider the situation where $\mathbf{r}$ is in the region at site $\mathbf{R}$, and $\mathbf{r}'$ is in the region A. In this case $|\mathbf{r}+\mathbf{R}| > r'$ is satisfied, and f_l is simply replaced by h_l. We can thus apply the origin shift formula (104) and have an origin shift formula of g_A,

$$
\begin{aligned}
g_A&(\mathbf{r} + \mathbf{R}, \mathbf{r}') \\
&= g_A(\mathbf{r}', \mathbf{r} + \mathbf{R}) \\
&= 2 \sum_{LL'} i^{l-l'} \exp(i\delta_{l'}^A) G_{LL'}(k\mathbf{R}) j_l(kr) Y_L(\hat{\mathbf{r}}) \tilde{R}_{l'}(r') Y_{L'}^*(\hat{\mathbf{r}}')
\end{aligned}
$$

$$(125)$$

where $R_l(r')$ is written as $R_l(r') = \exp(i\delta_l^A)\tilde{R}_l(r')$. The radial function $\tilde{R}$ is real for the real potential v_A (see Eq. (95)). Of course, Eq. (125) is reduced to Eq. (106) when the potential v_A is switched off. This formula is very important for the study of EELFS spectra as discussed later.

2.3.3. Site T Matrix Expansion for Nonoverlapping Potential

When the potential is given as a sum of nonoverlapping atomic potentials $V = \sum_\alpha v_\alpha$, we have an expression for the total T matrix expanded in terms of site T matrix from Eq. (54),

$$
\begin{aligned}
T &= V + Vg_0V + Vg_0Vg_0V + \cdots \\
&= \sum_\alpha v_\alpha + \sum_{\alpha\beta} v_\beta g_0 v_\alpha + \sum_{\alpha\beta\gamma} v_\gamma g_0 v_\beta g_0 v_\alpha + \cdots \\
&= \sum_\alpha (v_\alpha + v_\alpha g_0 v_\alpha + v_\alpha g_0 v_\alpha g_0 v_\alpha + \cdots) \\
&\quad + \sum_{\alpha \neq \beta} (v_\beta + v_\beta g_0 v_\beta + \cdots) g_0 (v_\alpha + v_\alpha g_0 v_\alpha + \cdots) + \cdots \\
&= \sum_\alpha t_\alpha + \sum_{\alpha \neq \beta} t_\beta g_0 t_\alpha + \sum_{\gamma \neq \beta \neq \alpha} t_\gamma g_0 t_\beta g_0 t_\alpha + \cdots
\end{aligned}
$$

$$(126)$$

Here we define the site T matrix at site α,

$$
\begin{aligned}
t_\alpha &= v_\alpha + v_\alpha g_0 v_\alpha + v_\alpha g_0 v_\alpha g_0 v_\alpha + \cdots \\
&= v_\alpha + v_\alpha g_0 t_\alpha
\end{aligned}
$$

$$(127)$$

By use of the site T matrix expansion of T, we have a useful expression for the Green's function g from Eq. (53),

$$
g = \sum_\alpha g_0 \left(1 + \sum_{\beta \neq \alpha} t_\beta + \sum_{\gamma \neq \beta \neq \alpha} t_\gamma g_0 t_\beta + \cdots \right) g_\alpha
$$

$$(128)$$

where $g_\alpha = g_0 + g_0 t_\alpha g_0$ is the full Green's function under the influence of the potential v_α at site α.

These formulas are extensively used in the EELFS analyses discussed in Section 5.

3. MANY-BODY ELECTRON SCATTERING THEORY

In this section we discuss some rather general formulas for electron scattering from solids [9], which describe the effects of losses. The results are given in terms of one-electron expressions involving dumped one-electron functions and optical potentials, which are important for explaining why EELS spectra are surface sensitive.

3.1. Basic Theory of Electron Scattering

At a high enough energy of the scattering electron, its identity with the electrons in the target may be neglected. We can then use a product space for the total state vector, with the scattering electron in one space and the electrons in the target in another. Exchange effects between scattered and target electrons are thus neglected. We use the following Hamiltonian:

$$H = T_e + H_s + V_{es} \tag{129}$$

T_e is the kinetic energy operator for the scattering electron. H_s is the many-body Hamiltonian for the target. V_{es} is the total Coulomb potential between the scattering electron and the target,

$$V_{es} = \sum_{\mathbf{k}_1 \mathbf{k}_2} \sum_{l_1 l_2} c_{\mathbf{k}_1}^\dagger c_{\mathbf{k}_2} \left[\langle \mathbf{k}_1 l_1 | v | \mathbf{k}_2 l_2 \rangle c_{l_1}^\dagger c_{l_2} + \langle \mathbf{k}_1 | V_{en} | \mathbf{k}_2 \rangle \right] \tag{130}$$

where V_{en} is the Coulomb interaction between the scattering electron and the positive nuclei, and $\langle \cdots | v | \cdots \rangle$ is the usual two-electron matrix element of the Coulomb potential $1/r_{12}$. The one-electron states of the scattering electron are denoted by $\mathbf{k}$, and those of electrons in the target by l. The indices $\mathbf{k}$ specify both momentum and spin. The indices l correspond to whatever quantum numbers are pertinent for the electrons in the target, such as crystal momentum, band index, and spin.

It is convenient to introduce some subsidiary quantities defined by

$$
\begin{aligned}
H &= H_0 + V \\
H_0 &= T_e + \langle \Phi_0 | V_{es} | \Phi_0 \rangle + H_s = h_0 + H_s \\
V &= V_{es} - \langle \Phi_0 | V_{es} | \Phi_0 \rangle \\
h_0 &= T_e + V_H
\end{aligned}
$$

$$
\begin{aligned}
V_H &= \langle \Phi_0 | V_{es} | \Phi_0 \rangle \\
&= \sum_{\mathbf{k}_1 \mathbf{k}_2} \sum_{l_1 l_2} c_{\mathbf{k}_1}^\dagger c_{\mathbf{k}_2} \langle \mathbf{k}_1 l_1 | v | \mathbf{k}_2 l_2 \rangle \langle \Phi_0 | c_{l_1}^\dagger c_{l_2} | \Phi_0 \rangle \\
&\quad + \langle \mathbf{k}_1 | V_{en} | \mathbf{k}_2 \rangle
\end{aligned}
$$

Equations (131), (132), (133), (134) respectively.

$$(131)$$
$$(132)$$
$$(133)$$
$$(134)$$

where $|\Phi_0\rangle$ is the ground state of H_s. The states with index $\mathbf{k}$ in Eqs. (130) and (134) are scattering states of h_0 with outgoing-wave boundary conditions

$$h_0 |\phi_{\mathbf{p}}^+\rangle = \epsilon_p |\phi_{\mathbf{p}}^+\rangle \tag{135}$$

and $\epsilon_p = p^2/2$. The transition matrix for scattering $0\mathbf{p} \rightarrow n\mathbf{p}'$ can now be written from Eq. (50) as

$$T(n\mathbf{p}', 0\mathbf{p}) = \langle n\mathbf{p}' | V_{es} | \Psi_{0\mathbf{p}}^+ \rangle \tag{136}$$

where $|n\mathbf{p}'\rangle$ is an eigenstate of $T_e + H_s$, $|n\mathbf{p}'\rangle = |\Phi_n\rangle |\phi_{\mathbf{p}'}^0\rangle$ (where $|\Phi_n\rangle$ is an eigenstate of H_s, $H_s |\Phi_n\rangle = E_n |\Phi_n\rangle$, and $|\phi_{\mathbf{p}'}^0\rangle$ is a plane-wave state) and $|\Psi_{0\mathbf{p}}^+\rangle$ is an eigenstate (outgoing-wave solution) of H. We write this latter state, using the Lippman–Schwinger equation (48), as

$$|\Psi_{0\mathbf{p}}^+\rangle = [1 + G(E)V] |\Phi_{0\mathbf{p}}^+\rangle \tag{137}$$

where $G(E) = (E - H + i\eta)^{-1}$, η is a positive infinitesimal, and $|\Phi_{0\mathbf{p}}^+\rangle$ is an eigenstate of H_0. $|\Phi_{0\mathbf{p}}^+\rangle$, like $|n\mathbf{p}'\rangle$, is a direct product $|\Phi_{0\mathbf{p}}^+\rangle = |\Phi_0\rangle|\phi_{\mathbf{p}}^+\rangle$, and $|\Phi_0\rangle(= |0\rangle)$ is the ground state of H_{s}. The energy E is the total energy, $E = E_0 + \epsilon_p = E_n + \epsilon_{p'}$.

We now have for the transition matrix,

$$
\begin{aligned}
T(n\mathbf{p}'; 0\mathbf{p}) &= \langle n\mathbf{p}'|V_{\mathrm{es}}[1 + G(E)V]|\Phi_{0\mathbf{p}}^+\rangle \\
&= \langle\phi_{\mathbf{p}'}^0|\langle n|V_{\mathrm{es}}[1 + G(E)V]|0\rangle|\phi_{\mathbf{p}}^+\rangle \quad (138)
\end{aligned}
$$

$T(n\mathbf{p}'; 0\mathbf{p})$ is thus given as a matrix element of a one-electron operator $\langle n|V_{\mathrm{es}}[1 + G(E)V]|0\rangle$. This operator in turn is a matrix element involving correlated target wave functions Φ_n and Φ_0.

To evaluate this complicated operator we make an expansion of $G(E)$ in terms of diagonal operators of the Van Hove type. Such an expansion can be done by projection operator techniques. We introduce the projection operators,

$$
P = |0\rangle\langle 0|, \qquad Q = 1 - P \quad (139)
$$

We then define a new unperturbed Hamiltonian and its corresponding perturbation,

$$
\tilde{H}_0 = H_0 + QVQ, \qquad \tilde{V} = PVQ + QVP \quad (140)
$$

Because $PVP = 0$ [cf. Eq. (132)], we have $H = H_0 + V = \tilde{H}_0 + \tilde{V}$. From our definitions in Eq. (140) it trivially follows that

$$
\begin{aligned}
\tilde{V}|0\rangle &= V|0\rangle \\
\tilde{H}_0|0\rangle &= H_0|0\rangle = (E_0 + h_0)|0\rangle \quad (141)
\end{aligned}
$$

The Green's function related to $\tilde{H}_0$,

$$
\tilde{G}_0 = (E - \tilde{H}_0 + i\eta)^{-1} \quad (142)
$$

has the properties

$$
\begin{aligned}
Q\tilde{G}_0 P &= P\tilde{G}_0 Q = 0, \\
P\tilde{G}_0 P &= PG_0 P = P(E - H_0 + i\eta)^{-1} \quad (143)
\end{aligned}
$$

as easily follows from the identities

$$
\tilde{G}_0 = G_0 + G_0 QV Q\tilde{G}_0 = G_0 + \tilde{G}_0 QV QG_0 \quad (144)
$$

We want to rewrite $(1 + GV)|0\rangle$ in Eq. (138), which by Eq. (141) equals $(1 + G\tilde{V})|0\rangle$. Expanding G in powers of $\tilde{V}$,

$$
G = \tilde{G}_0 + \tilde{G}_0\tilde{V}\tilde{G}_0 + \tilde{G}_0\tilde{V}\tilde{G}_0\tilde{V}\tilde{G}_0 + \tilde{G}_0\tilde{V}\tilde{G}_0\tilde{V}\tilde{G}_0\tilde{V}\tilde{G}_0 + \cdots \quad (145)
$$

and collecting even and odd powers of $\tilde{V}$ in the expansion of $1 + G\tilde{V}$, we have

$$
\begin{aligned}
1 + G\tilde{V} &= (1 + \tilde{G}_0\tilde{V})(1 - \tilde{G}_0\tilde{V}\tilde{G}_0\tilde{V})^{-1} \\
&= (1 + \tilde{G}_0\tilde{V})(E - \tilde{H}_0 - \tilde{V}\tilde{G}_0\tilde{V} + i\eta)^{-1} \\
&\quad \times (E - \tilde{H}_0 + i\eta) \quad (146)
\end{aligned}
$$

By Eqs. (141) and (143) we can replace $(E - \tilde{H}_0 + i\eta)|0\rangle$ with $(\epsilon_p - h_0 + i\eta)|0\rangle$ and $(E - \tilde{H}_0 - \tilde{V}\tilde{G}_0\tilde{V} + i\eta)|0\rangle$ with $[\epsilon_p - h_0 - \Sigma(E) + i\eta]|0\rangle$, where we define the non-hermitian optical potential for the state $|0\rangle$,

$$
\Sigma(E) = \langle 0|\tilde{V}\tilde{G}_0(E)\tilde{V}|0\rangle \quad (147)
$$

Collecting our results, we obtain

$$
\begin{aligned}
(1 + GV)|0\rangle|\phi_{\mathbf{p}}^+\rangle &= (1 + \tilde{G}_0\tilde{V})|0\rangle[\epsilon_p - h_0 - \Sigma(E) + i\eta]^{-1} \\
&\quad \times (\epsilon_p - h_0 + i\eta)|\phi_{\mathbf{p}}^+\rangle \quad (148)
\end{aligned}
$$

We can now define a one-electron function for a dumped outgoing wave by

$$
\begin{aligned}
|\psi_{\mathbf{p}}^+\rangle &= [\epsilon_p - h_0 - \Sigma(E) + i\eta]^{-1}(\epsilon_p - h_0 + i\eta)|\phi_{\mathbf{p}}^+\rangle \\
&= i\eta[\epsilon_p - h_0 - \Sigma(E) + i\eta]^{-1}|\phi_{\mathbf{p}}^+\rangle \\
&= \left(1 + [\epsilon_p - h_0 - \Sigma(E) + i\eta]^{-1}\Sigma(E)\right)|\phi_{\mathbf{p}}^+\rangle \quad (149)
\end{aligned}
$$

Clearly from Eq. (48) $|\psi_{\mathbf{p}}^+\rangle$ satisfies the equation

$$
[\epsilon_p - h_0 - \Sigma(E) + i\eta]|\psi_{\mathbf{p}}^+\rangle = 0 \quad (150)
$$

We now have a formally exact expression (within our basic approximations, stated at the beginning of this section) for the scattering amplitude

$$
T(n\mathbf{p}'; 0\mathbf{p}) = \langle\phi_{\mathbf{p}'}^0|\langle n|V_{\mathrm{es}}[1 + \tilde{G}_0(E)V]|0\rangle|\psi_{\mathbf{p}}^+\rangle \quad (151)
$$

This expression is valid for inelastic as well as elastic scattering. If we compare Eqs. (151) and (138), the only differences are that $|\phi_{\mathbf{p}}^+\rangle$ is replaced by a dumped wave $|\psi_{\mathbf{p}}^+\rangle$ and that G is replaced by $\tilde{G}_0$. We obtained our new expression for T by rearranging the perturbation expansion in V in such a way that the state $|0\rangle$ never appears as an intermediate state. Thus the effects of coherent scattering are summed exactly and taken into account by the dumped wave function $|\psi_{\mathbf{p}}^+\rangle$. The quantity Σ is an optical potential, which will be discussed in detail later. To the lowest approximation Σ is an imaginary constant inside the solid and zero outside; hence the solutions of Eq. (149) are dumped states.

We note that T is still given by a one-electron formula and that the one-electron operator $\langle n|V_{\mathrm{es}}(1 + \tilde{G}_0 V)|0\rangle$ is still defined from a matrix element involving correlated target wave functions.

3.2. Elastic Scattering

Before turning to inelastic scattering in the next subsection, we give a short discussion of elastic scattering, showing how our results in the previous subsection relate to the exact formula obtained by Bell and Squires [26]. From Eq. (151) we have

$$
T(0\mathbf{p}'; 0\mathbf{p}) = \langle\phi_{\mathbf{p}'}^0|\langle 0|V_{\mathrm{es}}(1 + \tilde{G}_0 V)|0\rangle|\psi_{\mathbf{p}}^+\rangle \quad (152)
$$

and from Eqs. (141) and (147),

$$
\begin{aligned}
\langle 0|V_{\mathrm{es}}(1 + \tilde{G}_0 V)|0\rangle &= \langle 0|V_{\mathrm{es}}|0\rangle + \langle 0|\tilde{V}\tilde{G}_0\tilde{V}|0\rangle \\
&= V_{\mathrm{H}} + \Sigma(E) \quad (153)
\end{aligned}
$$

Thus

$$
T(0\mathbf{p}'; 0\mathbf{p}) = \langle\phi_{\mathbf{p}'}^0|[V_{\mathrm{H}} + \Sigma(E)]|\psi_{\mathbf{p}}^+\rangle \quad (154)
$$

which is the same expression as obtained by Bell and Squires [26], except that $\Sigma(E)$ as defined in Eq. (147) lacks the first-order (unscreened) exchange term and the second-order exchange terms. It contains, however, the very important dynamical polarization (cf. next subsection). The lack of the exchange terms is clearly due to the fact that we have regarded the scattering electron as distinguishable from the electrons in the target. We will find later that we can easily recapture exchange (cf. Section 3.4).

From Watson's theorem for two potential scattering (see Section 2.2.3, particularly Eq. (78)), we can rewrite Eq. (154) as

$$T(0\mathbf{p}'; 0\mathbf{p}) = \langle \phi_{\mathbf{p}'}^0 | V_H | \phi_{\mathbf{p}}^+ \rangle + \langle \phi_{\mathbf{p}'}^- | \Sigma(E) | \psi_{\mathbf{p}}^+ \rangle \qquad (155)$$

3.3. Inelastic Scattering

In the last subsection we specialized our general expression for electron scattering to the case of elastic scattering. In this subsection we will take up the main topic, inelastic scattering.

The general expression for electron scattering is given in Eq. (151). $\tilde{G}_0$ given by Eq. (142) contains the interaction potential $V = V_{es} - \langle 0|V_{es}|0 \rangle$. We will do a perturbation expansion in V in such a way that all "coherent" contributions are eliminated. By a "coherent contribution" we mean a term in which two intermediate states are equal, or one intermediate state coincides with the initial 0 or the final n state of the target (which are different for inelastic scattering). Such an expansion can be made with projection–operator techniques.

We use a slightly different definition of H_0 and V; H_0 now includes the diagonal part of V_{es}, and V has zero diagonal elements, $\langle n|V|n \rangle = 0$,

$$H_0 = T_e + H_s + \sum_n P_n V_{es} P_n$$

$$V = V_{es} - \sum_n P_n V_{es} P_n$$

$$H = H_0 + V \qquad (156)$$

Q_{0n} projects away the states 0 and n,

$$Q_{0n} = 1 - P_0 - P_n \qquad (157)$$

where as before, $P_n = |n\rangle\langle n|$. We now put a prime on the quantities used in Sections 3.1 and 3.2; $H_0' = T_e + H_s + \langle 0|V_{es}|0 \rangle$ and $V' = V_{es} - \langle 0|V_{es}|0 \rangle$. $H = T_e + H_s + V_{es}$ is unchanged. The projection operators in (140) refer to eigenstates of H_s. The full space of eigenstates to H_0 is given by the product space of eigenstates of H_s and of $h_n = T_e + \langle n|V_{es}|n \rangle$, $\{|n\rangle, \phi_{\mathbf{p}}^+\}$. In Eq. (151) we have the term $\tilde{G}_0 Q = Q(E - H_0' - QV'Q + i\eta)^{-1}$. Here Q is the same as that in Eq. (139). Furthermore, $\tilde{G}_0 Q = G_{\bar{0}}$ since

$$H_0' + QV'Q = T_e + H_s + \langle 0|V_{es}|0 \rangle + Q(V_{es} - \langle 0|V_{es}|0 \rangle)Q$$
$$= T_e + H_s + P V_{es} P + Q V_{es} Q$$
$$= H_0 + QVQ \qquad (158)$$

where $G_{\bar{0}}$ is defined by ($Q = Q_0$)

$$G_{\bar{0}} = Q_0(E - H_0 - Q_0 V Q_0 + i\eta)^{-1} \qquad (159)$$

Equation (151) now becomes

$$T(n\mathbf{p}'; 0\mathbf{p}) = \langle \phi_{\mathbf{p}'}^0 | \langle n | V_{es} + V_{es} G_{\bar{0}}(E) V_{es} |0\rangle | \psi_{\mathbf{p}}^+ \rangle \qquad (n \neq 0)$$
$$(160)$$

First we will rewrite this equation so that $G_{\bar{0}}$ is replaced by $G_{\bar{0}n}$, where

$$G_{\bar{0}n} = Q_{0n} \tilde{G}_{0n} = Q_{0n}(E - H_0 - Q_{0n} V Q_{0n} + i\eta)^{-1} \qquad (161)$$

We insert $1 = P_n + Q_n$ to the right of $G_{\bar{0}}$ in Eq. (160). The contribution with P_n is $\langle n|V_{es} + V_{es} G_{\bar{0}} P_n V_{es}|0\rangle = (1 + \langle n|V_{es} G_{\bar{0}}|n\rangle)\langle n|V_{es}|0\rangle$. The remaining contribution is $\langle n|V_{es} G_{\bar{0}} Q_n V_{es}|0\rangle$. Since $Q_{0n} = Q_0 Q_n$ and $Q_0 = Q_{0n} + P_n$, we have

$$G_{\bar{0}} Q_n = G_{\bar{0}} Q_{0n}$$
$$= G_{\bar{0}n} + G_{\bar{0}}(Q_{0n} V P_n + P_n V Q_{0n}) G_{\bar{0}n}$$
$$= G_{\bar{0}n} + G_{\bar{0}} P_n V G_{\bar{0}n} = G_{\bar{0}n} + G_{\bar{0}} P_n V_{es} G_{\bar{0}n} \qquad (162)$$

which gives

$$\langle n|V_{es} G_{\bar{0}} Q_n V_{es}|0\rangle$$
$$= \langle n|V_{es}(G_{\bar{0}n} + G_{\bar{0}} P_n V_{es} G_{\bar{0}n}) V_{es}|0\rangle$$
$$= (1 + \langle n|V_{es} G_{\bar{0}}|n\rangle)\langle n|V_{es} G_{\bar{0}n} V_{es}|0\rangle \qquad (163)$$

Combining this with our first term of Eq. (160) gives

$$T(n\mathbf{p}'; 0\mathbf{p})$$
$$= \langle \phi_{\mathbf{p}'}^0 | (1 + \langle n|V_{es} G_{\bar{0}}|n\rangle)\langle n|V_{es} + V_{es} G_{\bar{0}n} V_{es}|0\rangle | \psi_{\mathbf{p}}^+ \rangle$$
$$(164)$$

The first step, replacing $G_{\bar{0}}$ with $G_{\bar{0}n}$, is now completed.

It remains to rewrite $\langle \phi_{\mathbf{p}'}^0 | (1 + \langle n|V_{es} G_{\bar{0}}|n\rangle)$ as a dumped wave function. For that purpose we use expansion in diagonal Green's functions developed by Hedin [27]. We note that $V = QVQ = V_{0n} + \tilde{V}_{0n}$, where $V_{0n} = Q_{0n} V Q_{0n}$, $\tilde{V}_{0n} = P_n V Q_{0n} + Q_{0n} V P_n$, and $G_{\bar{0}}|n\rangle = (P_n + Q_{0n}) G_{\bar{0}} P_n |n\rangle$. We now introduce an *unperturbed* Green's function,

$$\tilde{G}_{\bar{0}n} = (E - H_0 - V_{0n} + i\eta)^{-1} \qquad (165)$$

Expanding $G_{\bar{0}}$ in powers of $\tilde{V}_{0n}$, we obtain the result

$$P_n G_{\bar{0}} P_n = P_n (\tilde{G}_{\bar{0}n}^{-1} - \tilde{V}_{0n} \tilde{G}_{\bar{0}n} \tilde{V}_{0n})^{-1} P_n$$
$$= P_n [E - H_0 - \Sigma_{0n}(E)]^{-1} P_n \qquad (166)$$

where the *optical potential* $\Sigma_{0n}(E)$ is defined by

$$\Sigma_{0n}(E) = \langle n|V G_{\bar{0}n} V|n\rangle \qquad (167)$$

the properties of which are discussed later in detail. For convenience we introduce a diagonal Green's function $G_{0n}^d(E)$,

$$G_{0n}^d(E) = Q_0 P_n (E - H_0 - \Sigma_{0n}(E) + i\eta)^{-1} \qquad (168)$$

Equation (166) is thus written as $P_n G_{\bar{0}} P_n = P_n G_{0n}^d P_n$. In the same way we have the relation $Q_{0n} G_{\bar{0}} P_n = G_{\bar{0}n} V P_n G_{0n}^d P_n$.

With the use of this identity and Eq. (166), we can get an expansion in terms of the diagonal Green's function [9, 27],

$$G_{\bar{0}}|n\rangle = (1 + G_{\bar{0}\bar{n}}V)G_{0n}^{\mathrm{d}}|n\rangle$$

$$= \Big(G_{0n}^{\mathrm{d}} + \sum_k G_{0nk}^{\mathrm{d}} V G_{0n}^{\mathrm{d}}$$

$$+ \sum_{k \neq l} G_{0nkl}^{\mathrm{d}} V G_{0nk}^{\mathrm{d}} V G_{0n}^{\mathrm{d}} \cdots \Big)|n\rangle \quad (169)$$

Taking a particular term in this expansion, the same intermediate state can never appear more than once. The expansion describes "coherent" propagation, described by the diagonal operators, interrupted by inelastic scattering events. The optical potential provides a shift in energy and some damping in the coherent propagation.

From Eq. (169) we have

$$G_{\bar{0}}|n\rangle = (1 + G_{\bar{0}\bar{n}}V)Q_0(E - H_0 - \Sigma_{0n} + i\eta)^{-1}|n\rangle$$

$$= (1 + G_{\bar{0}\bar{n}}V_{\mathrm{es}})Q_0|n\rangle(\epsilon_{p'} - h_n - \Sigma_{0n} + i\eta)^{-1} \quad (170)$$

with $h_n = T_{\mathrm{e}} + \langle n|V_{\mathrm{es}}|n\rangle = T_{\mathrm{e}} + V_n$, and thus

$$1 + \langle n|V_{\mathrm{es}}G_{\bar{0}}|n\rangle$$

$$= 1 + \langle n|V_{\mathrm{es}}(1 + G_{\bar{0}\bar{n}}V_{\mathrm{es}})|n\rangle(\epsilon_{p'} - h_n - \Sigma_{0n} + i\eta)^{-1}$$

$$= 1 + (V_n + \Sigma_{0n})(\epsilon_{p'} - T_{\mathrm{e}} - V_n - \Sigma_{0n} + i\eta)^{-1} \quad (171)$$

We now have

$$\langle \phi_{\mathbf{p}'}^0|(1 + \langle n|V_{\mathrm{es}}G_{\bar{0}}|n\rangle)$$

$$= \langle \phi_{\mathbf{p}'}^0|[1 + (V_n + \Sigma_{0n})(\epsilon_{p'} - T_{\mathrm{e}} - V_n - \Sigma_{0n} + i\eta)^{-1}]$$

$$= \langle \psi_{\mathbf{p}'}^{0n-}| \quad (172)$$

The dumped wave function $\psi_{\mathbf{p}'}^{0n-}$ satisfies the equation

$$\langle \psi_{\mathbf{p}'}^{0n-}|(\epsilon_{p'} - T_{\mathrm{e}} - V_n - \Sigma_{0n} + i\eta) = 0 \quad (173)$$

Combining Eqs. (172) and (164), we have now derived the general expression for inelastic scattering amplitude,

$$T(n\mathbf{p}'; 0\mathbf{p}) = \langle \psi_{\mathbf{p}'}^{0n-}|\langle n|V_{\mathrm{es}} + V_{\mathrm{es}}G_{\bar{0}\bar{n}}V_{\mathrm{es}}|0\rangle|\psi_{\mathbf{p}}^+\rangle, \quad n \neq 0 \quad (174)$$

The Green's function $G_{\bar{0}\bar{n}}$ can be expanded in powers of V with the use of Eq. (169); to second order we have

$$T(n\mathbf{p}'; 0\mathbf{p}) = \langle \psi_{\mathbf{p}'}^{0n-}|\langle n|V_{\mathrm{es}} + \sum_l V_{\mathrm{es}}G_{0nl}^{\mathrm{d}}V_{\mathrm{es}} + \cdots |0\rangle|\psi_{\mathbf{p}}^+\rangle,$$

$$n \neq 0 \quad (175)$$

where G_{0nl}^{d} is diagonal with respect to the target Hamiltonian H_{s},

$$G_{0nl}^{\mathrm{d}} = Q_{0n}P_l(E - H_0 - \Sigma_{0nl} + i\eta)^{-1}$$

$$= Q_{0n}P_l(E - E_l - h_l - \Sigma_{0nl} + i\eta)^{-1} \quad (176)$$

where Σ_{0nl} is an excited-state optical potential and h_l is a Hamiltonian for a scattering electron,

$$\Sigma_{0nl} = \langle l|V_{\mathrm{es}}G_{\bar{0}\bar{n}\bar{l}}V_{\mathrm{es}}|l\rangle \quad (177)$$

All of this may look rather abstract and useless. However for extended excitations, as discussed later (Section 4.3), $\Sigma_{0nl} = \Sigma_l$. Furthermore, as discussed in Section 4.3, $\Sigma_l(E) = \Sigma_0(E - E_l + E_0)$ and $h_0 \approx h_l$. We then have

$$\psi_{\mathbf{p}'}^{0n-} \approx \psi_{\mathbf{p}'}^-$$

$$\Sigma_{0nl}(E) \approx \Sigma_0(E - E_l + E_0)$$

$$h_l \approx h_0 \quad (178)$$

Adopting Eq. (178), we can write Eq. (175),

$$T(n\mathbf{p}'; 0\mathbf{p}) = \langle \psi_{\mathbf{p}'}^-|\langle n|V_{\mathrm{es}}|0\rangle$$

$$+ \sum_{l \neq 0, n} \langle n|V_{\mathrm{es}}|l\rangle g_0(\epsilon_p + E_0 - E_l)\langle l|V_{\mathrm{es}}|0\rangle$$

$$+ \cdots |\psi_{\mathbf{p}}^+\rangle \quad (179)$$

where $g_0(\omega) = [\omega - h_0 - \Sigma_0(\omega + E_0) + i\eta]^{-1}$ is the one-electron damping Green's function.

In the case of very high energy excitation, the wave functions for the scattering electron $\psi_{\mathbf{p}'}^{0n-}$ and $\psi_{\mathbf{p}}^+$ can be taken as plane waves, and only the lowest order term in V_{es} is important. We can then Fourier transform the Coulomb interaction in V_{es} to obtain a well-known Bethe approximation [5],

$$T(n\mathbf{p}'; 0\mathbf{p}) = \langle n|\sum_j \exp(-i\Delta\mathbf{P} \cdot \mathbf{r}_j)|0\rangle/(\Delta P)^2 \quad (180)$$

where $\Delta\mathbf{P} = \mathbf{p}' - \mathbf{p}$ is the momentum transfer of the scattering electron. In the second quantized form, this is written as

$$T(n\mathbf{p}'; 0\mathbf{p}) = \int \exp(-i\Delta\mathbf{P} \cdot \mathbf{r})\langle n|\rho(\mathbf{r})|0\rangle \, d\mathbf{r}/(\Delta P)^2 \quad (181)$$

where $\rho(\mathbf{r}) = \psi^\dagger(\mathbf{r})\psi(\mathbf{r})$ is the density operator. Eventually one obtains the differential scattering cross section for the energy loss with $E = \epsilon_p - \epsilon_{p'}$,

$$\frac{d^2\sigma}{dE \, d\Omega} = \left[\frac{2}{(\Delta P)^2}\right]^2 \frac{p'}{p} S(\Delta\mathbf{P}, E) \quad (182)$$

where

$$S(\Delta\mathbf{P}, E) = \sum_n \int \exp[i\Delta\mathbf{P} \cdot (\mathbf{r}' - \mathbf{r})]\langle 0|\rho(\mathbf{r}')|n\rangle$$

$$\times \langle n|\rho(\mathbf{r})|0\rangle \, d\mathbf{r} \, d\mathbf{r}' \, \delta(E_n - E_0 - E)$$

$$= \int \exp[i\Delta\mathbf{P} \cdot (\mathbf{r}' - \mathbf{r})]\langle 0|\delta\rho(\mathbf{r}', t)\delta\rho(\mathbf{r})|0\rangle$$

$$\times \exp(iEt)\frac{dt}{2\pi} \, d\mathbf{r} \, d\mathbf{r}' \quad (183)$$

where $\delta\rho(\mathbf{r}) = \rho(\mathbf{r}) - \langle 0|\delta\rho(\mathbf{r})|0\rangle$ is the density fluctuation operator. The Heisenberg representation $\delta\rho(\mathbf{r}', t)$ is defined as usual as $\delta\rho(\mathbf{r}', t) = \exp(iHt)\delta\rho(\mathbf{r}')\exp(-iHt)$. The correlation function $\langle 0|\delta\rho(\mathbf{r}', t)\delta\rho(\mathbf{r})|0\rangle$ is just the reducible polarization propagator $i\Pi^>(\mathbf{r}'t, \mathbf{r})$. The dynamic scattering factor S is written in terms of the reducible polarization propagator. It is not easy to handle the reducible polarization Π, so we relate it to the irreducible polarization propagator P and the screening

Coulomb interaction W [28] with the notation $1 = (\mathbf{r}_1, \sigma_1, t_1)$,

$$\Pi(1, 2) = P(1, 2) + \int d3\, d4\, P(1, 3) W(3, 4) P(4, 2) \quad (184)$$

Note that we need only the greater part of the second term of the right-hand side of Eq. (184). For that purpose we can use a skeleton diagram expansion in the Keldysh formulation [28, 29]. Some useful properties of S for electron gas have been discussed in various excellent textbooks [2, 30] and are not repeated here. It is important to note, however, that the simple expression in Eq. (180) is valid only under very special conditions, where both $\psi_{\mathbf{p}'}^-$ and $\psi_{\mathbf{p}}^+$ are replaced by the plane waves $\phi_{\mathbf{p}'}^0$ and $\phi_{\mathbf{p}}^0$. This condition is satisfied when both the incident and the scattered electrons, that is, the probe electron, have high enough energy compared with the electron–target interaction energy. This problem is studied in Section 5 in detail.

3.4. Inclusion of the Direct Exchange Effects

The full Coulomb interaction between all of the electrons is

$$V_c = \frac{1}{2} \sum_{pqrs} \langle pq|rs \rangle c_p^\dagger c_q^\dagger c_s c_r \quad (185)$$

As in Section 3.1 we divide the one-electron states into one set $\{l\}$ to be used for the electrons in the solid, and another set $\{\mathbf{k}\}$ for the scattering electrons. The dividing line is taken at some energy ϵ_0, chosen so that the electrons in the solid can be well correlated. To be more specific, we calculate the ground state of the solid in Hartree–Fock theory and choose as $\{l\}$ the ground-state orbitals plus the virtual orbitals up to the energy ϵ_0, which may be, say, some 50 eV.

We now pick from the full Coulomb interaction V_s the part that describes the interaction between the electrons in the solid,

$$V_s = \frac{1}{2} \sum_{l_1 l_2 l_3 l_4} \langle l_1 l_2 | l_3 l_4 \rangle c_{l_1}^\dagger c_{l_2}^\dagger c_{l_4} c_{l_3} \quad (186)$$

and a part that is bilinear in both $\mathbf{k}$ and l,

$$V_{es} - V_{en} = \sum_{\mathbf{k}_1 \mathbf{k}_2 l_1 l_2} \langle \mathbf{k}_1 l_1 || \mathbf{k}_2 l_2 \rangle c_{\mathbf{k}_1}^\dagger c_{\mathbf{k}_2} c_{l_1}^\dagger c_{l_2} \quad (187)$$

Here $\langle || \rangle$ stands for the antisymmetrized matrix element

$$\langle pq || rs \rangle = \langle pq|v|rs \rangle - \langle pq|v|sr \rangle \quad (188)$$

What is left in V_c then are terms with one, three, or four $\mathbf{k}$ indices, plus terms of type $c_{\mathbf{k}_1}^\dagger c_{\mathbf{k}_2}^\dagger c_{l_1} c_{l_2}$ and their Hermitian conjugates. A term with one $\mathbf{k}$ index can destroy the scattering electron and couple to a resonant quasi bound state. We will discuss such processes in the next subsection. The terms in V_c that we have included with H_s and V_{es} are expected to be the most important, except close to the threshold or to a resonance.

Unlike in Section 3.1 we now fully account for the identity of the electrons. The full Hamiltonian is formally the same as before, but now V_{es} is defined by Eq. (187). The T matrix is

$$T(n\mathbf{p}'; 0\mathbf{p}) = \langle n\mathbf{p}' | V_{es} | \Psi_{0\mathbf{p}}^+ \rangle \quad (189)$$

Here $|n\mathbf{p}'\rangle = |n\rangle |\phi_{\mathbf{p}'}^0\rangle$ as defined in Section 3.1, and $|\Psi_{0\mathbf{p}}^+\rangle$ is an eigenstate of H. Again we write H as $H = H_0 + V$, with

$$H_0 = T_e + \sum_{\mathbf{k}_1, \mathbf{k}_2} V(\mathbf{k}_1, \mathbf{k}_2) c_{\mathbf{k}_1}^\dagger c_{\mathbf{k}_2} + H_s \quad (190)$$

and

$$\begin{aligned} V &= V_{es} - \sum_{\mathbf{k}_1, \mathbf{k}_2} V(\mathbf{k}_1, \mathbf{k}_2) c_{\mathbf{k}_1}^\dagger c_{\mathbf{k}_2} \\ &= \sum_{\mathbf{k}_1 \mathbf{k}_2 l_1 l_2} \langle \mathbf{k}_1 l_1 || \mathbf{k}_2 l_2 \rangle c_{\mathbf{k}_1}^\dagger c_{\mathbf{k}_2} \left(c_{l_1}^\dagger c_{l_2} - \langle 0|c_{l_1}^\dagger c_{l_2}|0\rangle \right) \quad (191) \end{aligned}$$

where

$$V(\mathbf{k}_1, \mathbf{k}_2) = \sum_{l_1 l_2} \langle \mathbf{k}_1 l_1 || \mathbf{k}_2 l_2 \rangle \langle 0|c_{l_1}^\dagger c_{l_2}|0\rangle + \langle \mathbf{k}_1 | V_{en} | \mathbf{k}_2 \rangle \quad (192)$$

For $|\Psi_{0\mathbf{p}}^+\rangle$ we now have

$$|\Psi_{0\mathbf{p}}^+\rangle = (1 + GV)|\Phi_{0\mathbf{p}}^+\rangle \quad (193)$$

with $|\Phi_{0\mathbf{p}}^+\rangle = c_{\mathbf{p}}^\dagger|0\rangle$ and thus

$$T(n\mathbf{p}'; 0\mathbf{p}) = \langle n\mathbf{p}' | V_{es}(1 + GV) | \Phi_{0\mathbf{p}}^+ \rangle \quad (194)$$

Here G as usual is given by $(E - H + i\eta)^{-1}$, and $c_{\mathbf{p}}^\dagger$ corresponds to an eigenstate of $T_e + \sum_{\mathbf{k}_1, \mathbf{k}_2} V(\mathbf{k}_1, \mathbf{k}_2)$ with outgoing-wave boundary conditions. Despite that we now allow for the identity of electrons, we can follow the development in Section 3.1 closely. This is due to the fact that V only contains the operators $c_{\mathbf{k}}$ as $c_{\mathbf{k}_1}^\dagger c_{\mathbf{k}_2}$. Thus when we expand G in powers of V, each term with its product of V operators will, when it operates on $|\Phi_{0\mathbf{p}}^+\rangle$, only create states of the type $|\Phi_{n\mathbf{p}}^+\rangle = c_{\mathbf{p}}^\dagger|n\rangle$, i.e., with only one scattering electron present. The expression for T thus becomes a sum of matrix products with the matrices labeled by $\mathbf{k}$ and n. These same matrix-element products, however, are also reproduced if we use a product space $|n\rangle |\phi_{\mathbf{p}}^+\rangle$. This shows that we can work with direct product states, despite the fact that we consider the electrons as identical. Hence we can take over all results from Sections 3.2–3.3, only by replacing the Coulomb matrix element $\langle \mathbf{k}_1 l_1 | v | \mathbf{k}_2 l_2 \rangle$ with the antisymmetrized matrix element $\langle \mathbf{k}_1 l_1 || \mathbf{k}_2 l_2 \rangle$.

For elastic scattering we obtain

$$T(0\mathbf{p}'; 0\mathbf{p}) = \langle 0\mathbf{p}' | V_{HF} + \Sigma | \Phi_{0\mathbf{p}}^+ \rangle \quad (195)$$

Here V_{HF} is the Hartree–Fock potential, and Σ is given by the same expression as in Section 3.1, except for the replacement with matrix elements $\langle \mathbf{k}_1 l_1 || \mathbf{k}_2 l_2 \rangle$. The explicit expression for inelastic scattering is the same as in Section 3.2, except for the replacement with matrix elements. As a very simple explicit example of the use of Eq. (174), we write down the expression for spin flip scattering of an electron from a hydrogen atom. To lowest order in V_{es} we have from Eqs. (155), (174), and (187),

$$\begin{aligned} &T(n\mathbf{p}'; 0\mathbf{p}) \\ &\approx \langle \phi_{\mathbf{p}'}^0 | \langle n | V_{es} | 0 \rangle | \phi_{\mathbf{p}}^0 \rangle = \langle \phi_{\mathbf{p}'}^0 | V_{en} | \phi_{\mathbf{p}}^0 \rangle \delta_{n,0} \\ &\quad + \sum_{\mathbf{k}_1 \mathbf{k}_2 l_1 l_2} \langle \mathbf{k}_1 l_1 || \mathbf{k}_2 l_2 \rangle \langle \phi_{\mathbf{p}'}^0 | c_{\mathbf{k}_1}^\dagger c_{\mathbf{k}_2} | \phi_{\mathbf{p}}^0 \rangle \langle n | c_{l_1}^\dagger c_{l_2} | 0 \rangle \quad (196) \end{aligned}$$

To simplify notation we only write out the spin symbols. Equation (196) then becomes

$$T(s'_e, s'_t; s_e, s_t) \approx \langle \phi^0_{\mathbf{p}'} | V_{en} | \phi^0_{\mathbf{p}} \rangle \delta_{n,0} + \langle s'_e, s'_t || s_e, s_t \rangle \quad (197)$$

where s'_t, (s_t) stands for the spin of the target hydrogen atom in its final (initial) state, etc. Writing out explicitly the direct and exchange contributions, we have

$$T(s'_e, s'_t; s_e, s_t) \approx \delta(s'_e, s_e)\delta(s'_t, s_t)V_D - \delta(s'_e, s_t)\delta(s'_t, s_e)V_{ex} \quad (198)$$

where $V_D = \langle \mathbf{p}', 1s|v|\mathbf{p}, 1s \rangle + \langle \mathbf{p}'|V_{en}|\mathbf{p} \rangle$ and $V_{ex} = \langle \mathbf{p}', 1s|v|1s, \mathbf{p} \rangle$. If the target electron has spin up, there are three possibilities:

$$T(\uparrow\uparrow, \uparrow\uparrow) = V_D - V_{ex}, \qquad T(\downarrow\uparrow, \downarrow\uparrow) = V_D,$$
$$T(\uparrow\downarrow, \downarrow\uparrow) = -V_{ex} \quad (199)$$

To obtain the total scattering, we choose the target spin up. For an unpolarized beam of scattering electrons, we then get the well-known expression for the total scattering cross section σ,

$$\sigma = \left(|V_D - V_{ex}|^2 + |V_D|^2 + |V_{ex}|^2\right)/2 \quad (200)$$

3.5. Processes that Do Not Conserve the Number of Scattering Electrons

So far we have discussed only Coulomb interactions that conserve the number of scattering electrons, that is, terms like $c^\dagger_{\mathbf{k}} c_{\mathbf{k}} c^\dagger_l c_{l'}$. In addition we have terms like $c^\dagger_{\mathbf{k}} c^\dagger_{\mathbf{k}'} c_l c_{l'}$, $c^\dagger_{l_1} c_{\mathbf{k}} c^\dagger_{l_2} c_{l_3}$, $c^\dagger_{\mathbf{k}} c_{\mathbf{k}} c^\dagger_{\mathbf{k}_1} c_l$, and $c^\dagger_{\mathbf{k}_1} c_{\mathbf{k}_2} c^\dagger_{\mathbf{k}_3} c_{\mathbf{k}_4}$ and their Hermitian conjugates. These terms, e.g., can create "secondaries," i.e., promote electrons in the solid to scattering electrons. Here we will only discuss effects when the scattering electron is virtually absorbed and we reach an $(N + 1)$-electron state of the target. This process is of particular interest when the $(N + 1)$-electron state is a resonant state. The Hamiltonian then has the form

$$H = H_0 + V + \delta V = T_e + H_s + V_{es} + \delta V \quad (201)$$

where δV is the contribution from terms with one $c_{\mathbf{k}}$ operator. We derive the transition matrix with the use of Watson's theorem, as shortly described in Section 2.2.3, taking V and δV as the two perturbations. This gives

$$T'(n\mathbf{p}'; 0\mathbf{p}) = T(n\mathbf{p}'; 0\mathbf{p}) + \langle \Psi^-_{n\mathbf{p}'} | \delta V | \Psi'^+_{0\mathbf{p}} \rangle \quad (202)$$

where T' is the new and T is the previous transition matrix (corresponding to $\delta V = 0$), and $|\Psi'^+_{0\mathbf{p}}\rangle$ is an eigenstate of H in Eq. (201). The Lippman–Schwinger equation (48) gives

$$T'(n\mathbf{p}'; 0\mathbf{p})$$
$$= T(n\mathbf{p}'; 0\mathbf{p}) + \langle \Psi^-_{n\mathbf{p}'} | \delta V(1 + G'\delta V) | \Psi^+_{0\mathbf{p}} \rangle$$
$$= T(n\mathbf{p}'; 0\mathbf{p})$$
$$+ \langle \Phi^-_{n\mathbf{p}'} | (1 + VG)\delta V(1 + G'\delta V)(1 + GV) | \Phi^+_{0\mathbf{p}} \rangle \quad (203)$$

Here G' is the Green's function related to H, G is the one related to $H_0 + V$, and $\Phi^-_{n\mathbf{p}'}$ and $\Phi^+_{0\mathbf{p}}$ are functions relating an

eigenstate of H_0. Since V does not change the number of scattering electrons, the state $(1 + GV)|\Phi^+_{0\mathbf{p}}\rangle$ has only one scattering electron, as does $\langle \Phi^-_{n\mathbf{p}'} | (1 + VG)$. Since δV also has only one scattering electron, the term linear in δV drops out, and we have

$$T'(n\mathbf{p}'; 0\mathbf{p})$$
$$= T(n\mathbf{p}'; 0\mathbf{p})$$
$$+ \langle \Phi^-_{n\mathbf{p}'} | (1 + VG)\delta V G'\delta V(1 + GV) | \Phi^+_{0\mathbf{p}} \rangle \quad (204)$$

If we neglect the inelastic scattering contributions from the $(1 + GV)$ operators and keep only their diagonal parts, Eq. (204) becomes

$$T'(n\mathbf{p}'; 0\mathbf{p}) = T(n\mathbf{p}'; 0\mathbf{p}) + \langle n|c'_{\mathbf{p}'}\delta V G'\delta V c'^\dagger_{\mathbf{p}}|0 \rangle \quad (205)$$

where $c'_{\mathbf{p}'}$ and $c'^\dagger_{\mathbf{p}}$ correspond to dumped one-electron states. If we consider an energy E close to that of a resonance state where only one term in the expansion of G' dominates, we have

$$G' \approx |R\rangle \langle R|(E - E_R + i\Gamma_R)^{-1} \quad (206)$$

where $|R\rangle$ is the resonance state, E_R is its energy, and Γ_R is its width. We then find, as expected, that Eq. (205) gives a Fano-type resonance expression. This type of resonance can occur for both elastic and inelastic scattering.

3.6. Fukutome–Low Scattering Theory

So far we have discussed electron scattering from many-electron systems, where scattering electrons are distinguishable from the electrons in the target. The above theory, however, provides us with a transparent physical interpretation of the EELS spectra in terms of the damping functions $\psi^{0n-}_{\mathbf{p}'}$ and $\psi^+_{\mathbf{p}}$, the associated optical potentials, and loss amplitudes like $\langle n|V_{es}|0\rangle$. In contrast to the above theory, the Fukutome–Low theory is not so convenient for practical purposes; however, it is useful for understanding the theoretical framework of many-body electron scatterings. We here follow the paper by Fukutome [31].

An electron scattering state satisfying the incoming $(-)$ or outgoing $(+)$ wave boundary condition is written, by the use of Eq. (48), in the form

$$|\Psi^{(\pm)}_{n\mathbf{p}}\rangle = \frac{\pm i\eta}{E - H \pm i\eta} c^{(\pm)\dagger}_{\mathbf{p}}|n\rangle \quad (207)$$

where $c^{(\pm)\dagger}_{\mathbf{p}}$ is the creation operator of the scattering electron in the state $\phi^\pm_{\mathbf{p}}$. The state satisfies the $(+)$ or $(-)$ boundary condition at $r \to \infty$, which does not determine $\phi^\pm_{\mathbf{p}}$. We determine $\phi^\pm_{\mathbf{p}}$ so as to satisfy

$$h_n\phi^\pm_{\mathbf{p}}(x) - \int v(\mathbf{r} - \mathbf{r}')\langle \psi^\dagger(x')\psi(x)\rangle_n \phi^\pm_{\mathbf{p}}(x')dx' = \epsilon_n\phi^\pm_{\mathbf{p}}(x) \quad (208)$$

where $\langle \rangle_n$ is the expectation value with respect to the target state $|n\rangle$ and $v(\mathbf{r}-\mathbf{r}')$ is the bare Coulomb potential. The one-electron Hamiltonian h_n is defined as $h_n = T_e + \langle n|V_{es}|n\rangle$ as before. Equation (208) should be solved with a $(+)$ or $(-)$ boundary condition. The approximation (208) is called the stationary target approximation.

Equation (207) is rewritten as

$$|\Psi_{n\mathbf{p}}^{(\pm)}\rangle = c_{\mathbf{p}}^{(\pm)\dagger}|n\rangle + \frac{1}{E - H \pm i\eta}V_{n\mathbf{p}}^{(\pm)}|n\rangle \tag{209}$$

where

$$V_{n\mathbf{p}}^{(\pm)} = \left[H, c_{\mathbf{p}}^{(\pm)\dagger}\right] - \epsilon_p c_{\mathbf{p}}^{(\pm)\dagger} \tag{210}$$

The state $\phi_{\mathbf{p}}^{(\pm)}$ depends on the target state $|n\rangle$ through the stationary target equation (208), so that $V^{(\pm)}$ (defined by Eq. (210)) depends on the state $|n\rangle$, which is also written as

$$\begin{aligned}
V_{n\mathbf{p}}^{(\pm)} &= \int v(\mathbf{r} - \mathbf{r}')\psi^\dagger(x)\{\rho(x') - \langle\rho(x')\rangle_n\}\phi_{\mathbf{p}}^{(\pm)}(x)\,dx\,dx' \\
&\quad + \int v(\mathbf{r} - \mathbf{r}')\psi^\dagger(x) \\
&\qquad \times \langle\psi^\dagger(x')\psi(x)\rangle_n\phi_{\mathbf{p}}^{(\pm)}(x')\,dx\,dx' \tag{211} \\
&= \sum_k \sum_{ll'} c_k^\dagger\langle kl\|\mathbf{p}l'\rangle\left(\tfrac{1}{2}c_l^\dagger c_{l'} - \langle c_l^\dagger c_{l'}\rangle_n\right) \tag{212}
\end{aligned}$$

The second expression of $V_{n\mathbf{p}}^{(\pm)}$ in Eq. (212) is obtained by Hedin et al. [32]. In $\langle c_l^\dagger c_{l'}\rangle_n$ one-electron levels l and l' should refer to the virtual occupied one-electron levels in the state $|n\rangle$. This is also the case for the one-electron state l and l' in the first term. When ϵ_p is large enough, ϵ_k should also be high so as to obtain a finite value of the integral $\langle kl\|\mathbf{p}l'\rangle$. In the sum $(1/2)\sum c_k^\dagger\langle kl\|\mathbf{p}l'\rangle c_l^\dagger c_{l'}$, either k or l can be $\mathbf{p}'$, and antisymmetric properties of the integral $\langle kl\|\mathbf{p}l'\rangle$ and $c_k^\dagger c_l^\dagger$ get rid of the factor 1/2. We thus have an approximation for the "potential" $V_{n\mathbf{p}}^{(\pm)}$,

$$V_{n\mathbf{p}}^{(\pm)} \approx \sum_{\mathbf{p}'}\sum_{ll'} c_{\mathbf{p}'}^\dagger\langle\mathbf{p}'l'\|\mathbf{p}l\rangle\left(c_{l'}^\dagger c_l - \langle c_{l'}^\dagger c_l\rangle_n\right) \tag{213}$$

Let us examine the scattering potential V defined by Eq. (191) on $c_{\mathbf{p}}^\dagger|n\rangle$. We find that

$$\begin{aligned}
Vc_{\mathbf{p}}^\dagger|n\rangle &= \sum_{\mathbf{k}\mathbf{k}'}\sum_{ll'}\langle\mathbf{k}'l'\|\mathbf{k}l\rangle\left(c_{l'}^\dagger c_l - \langle c_{l'}^\dagger c_l\rangle_n\right) \\
&\quad \times \left[c_{\mathbf{k}'}^\dagger\delta_{\mathbf{p},\mathbf{k}} - c_{\mathbf{k}'}^\dagger c_{\mathbf{p}}^\dagger c_{\mathbf{k}}\right]|n\rangle \tag{214}
\end{aligned}$$

For high energy scatterings, $c_{\mathbf{k}}|n\rangle = 0$ is satisfied. Thus we have

$$Vc_{\mathbf{p}}^\dagger|n\rangle \approx V_{n\mathbf{p}}^{(\pm)}|n\rangle \tag{215}$$

In this approximation, the scattering state $|\Psi_{n\mathbf{p}}^{(\pm)}\rangle$ is again given by Eq. (193) (0 can be replaced by n).

We now have the expected result. At high energies we can treat the scattering electron as a distinguishable electron interacting with the density fluctuations of the target system [32].

From the general expression (71), the S matrix of electron scattering is given by

$$S(n\mathbf{p}', 0\mathbf{p}) = \langle\Psi_{n\mathbf{p}'}^-|\Psi_{0\mathbf{p}}^+\rangle \tag{216}$$

The one-electron S matrix in the stationary target approximation for the state n is given by

$$s(\mathbf{p}', \mathbf{p})_n = \langle\phi_{\mathbf{p}'}^-|\phi_{\mathbf{p}}^+\rangle \tag{217}$$

where both $\phi_{\mathbf{p}'}^-$ and $\phi_{\mathbf{p}}^+$ are the solutions of Eq. (208) for the target state $|n\rangle$. The plus and minus operators are related to each other via $s(\mathbf{p}', \mathbf{p})_n$ as

$$c_{\mathbf{p}}^{(+)\dagger} = \sum_{\mathbf{k}} c_{\mathbf{k}}^{(-)\dagger}s(\mathbf{k}, \mathbf{p})_n, \qquad V_{n\mathbf{p}}^{(+)} = \sum_{\mathbf{k}} V_{n\mathbf{k}}^{(-)}s(\mathbf{k}, \mathbf{p})_n \tag{218}$$

Using the relation

$$\frac{1}{E - H + i\eta} = \frac{1}{E - H - i\eta} - 2\pi i\delta(E - H)$$

we see that the plus scattering state is related to the minus one as

$$|\Psi_{0\mathbf{p}}^+\rangle = \sum_{\mathbf{q}} s(\mathbf{q}, \mathbf{p})_0\left\{|\Psi_{0\mathbf{q}}^-\rangle - 2\pi i\delta(E_0 + \epsilon_p - H)V_{0\mathbf{q}}^{(-)}|0\rangle\right\} \tag{219}$$

Substitution of Eq. (219) into Eq. (216) yields the expression of the S matrix,

$$\begin{aligned}
S(n\mathbf{p}', 0\mathbf{p}) &= s(\mathbf{p}', \mathbf{p})_0\delta_{n0} - 2\pi i\delta(E_0 + \epsilon_p - E_n - \epsilon_{p'}) \\
&\quad \times \sum_{\mathbf{q}}\hat{T}(n\mathbf{p}', 0\mathbf{q})s(\mathbf{q}, \mathbf{p})_0 \tag{220}
\end{aligned}$$

where

$$\hat{T}(n\mathbf{p}', 0\mathbf{q}) = \langle\Psi_{n\mathbf{p}'}^-|V_{0\mathbf{q}}^{(-)}|0\rangle \tag{221}$$

We see that the amplitude $\hat{T}$ is related to the ordinary T matrix defined by Eq. (50) as

$$T(n\mathbf{p}', 0\mathbf{p}) = \sum_{\mathbf{q}}\hat{T}(n\mathbf{p}', 0\mathbf{q})s(\mathbf{q}, \mathbf{p})_0 \qquad (n \neq 0)$$

Let us derive an equation to determine the amplitude $\hat{T}$. Substituting Eqs. (209) and (210) into Eq. (221), we have

$$\begin{aligned}
\hat{T}(n\mathbf{k}, 0\mathbf{q}) &= \langle n|\{c_{\mathbf{k}}^{(-)}, V_{0\mathbf{q}}^{(-)}\}|0\rangle - \langle n|V_{0\mathbf{q}}^{(-)}c_{\mathbf{k}}^{(-)}|0\rangle \\
&\quad + \langle n|V_{n\mathbf{k}}^{(-)\dagger}\frac{1}{E_n + \epsilon_k - H + i\eta}V_{0\mathbf{q}}^{(-)}|0\rangle \tag{222}
\end{aligned}$$

The first term designated as $\hat{T}^0(n\mathbf{k}, 0\mathbf{q})$ is shown to be 0 if $n = 0$, and

$$\begin{aligned}
&\hat{T}^0(n\mathbf{k}, 0\mathbf{q}) \\
&= \int v(\mathbf{r} - \mathbf{r}')\phi_{\mathbf{k}}^{(-)}(x)^*\{\langle n|\rho(x')|0\rangle\phi_{\mathbf{q}}^{(-)}(x) \\
&\quad - \langle n|\psi^\dagger(x)\psi(x')|0\rangle\phi_{\mathbf{q}}^{(-)}(x')\}\,dx\,dx' \qquad (n \neq 0) \tag{223}
\end{aligned}$$

This amplitude describes the direct inelastic scattering amplitude, which vanishes for the elastic channel $n = 0$. This amplitude can be compared with the first term in the general inelastic scattering amplitude (174), however, no damping effect is taken into account in Eq. (223).

The second term of Eq. (222) is usually negligible. The intermediate states in the third term are restricted to $N + 1$-electron states. We now make an approximation to restrict the intermediate states only to incoming scattering states $|\Psi_{l\mathbf{p}}^-\rangle$'s and discrete bound states $|\Psi_a\rangle$'s of the electron + target system with $N+1$ electrons. Then Eq. (222) becomes a nonlinear equation for $\hat{T}$ as far as we neglect the second term,

$$\hat{T}(n\mathbf{k}, 0\mathbf{q}) = \hat{T}^0(n\mathbf{k}, 0\mathbf{q}) + \sum_a \frac{V(a, n\mathbf{k})^* V(a, 0\mathbf{q})}{E_n + \epsilon_k - E_a + i\eta}$$
$$+ \sum_{l\mathbf{p}} \frac{\hat{T}(l\mathbf{p}, n\mathbf{k})^* \hat{T}(l\mathbf{p}, 0\mathbf{q})}{E_n + \epsilon_k - E_l - \epsilon_p + i\eta} \quad (224)$$

where $V(a, n\mathbf{k}) = \langle \Psi_a | V_{n\mathbf{k}}^{(-)} | n \rangle$ and $H|\Psi_a\rangle = E_a|\Psi_a\rangle$. This equation is called the Low equation, derived by Fukutome [31] for electron scattering. The second term is important only when $\epsilon_k \approx E_a - E_n$; this term is closely related to the resonance scatterings. The third term describes the multiple scatterings. If we can solve the complicated nonlinear Low equation (224) for the amplitude $\hat{T}$, we could relate it to Eq. (174). Near resonance energy, Eq. (224) can be solved. Multiple plasmon loss in X-ray photoelectron spectroscopy (XPS) spectra was approximately calculated by use of Eq. (224) where the damping of the probe electrons was neglected [141, 142]. For EELS analyses the Fukutome–Low theory can be applied; however, it is convenient for the description of EELS from molecules and atoms [33].

4. OPTICAL POTENTIALS

We will discuss the properties of the optical potentials introduced in Sections 3.1–3.3. We first show their relation to the potential of Francis and Watson. We then compare the expressions, now including exchange, with the "GW" approximation. Finally we introduce the quasi-boson representation and derive explicit expressions for the potentials in terms of the coupling parameters. In particular, we derive the relation between the excited-state and ground-state potentials. In the last section we discuss the atomic optical potential in solids that are crucial in EELFS, XAFS, and X-ray photoelectron diffraction (XPD) analyses.

4.1. Relation to the Francis–Watson Optical Potential

We first give a short summary of the properties of the Francis and Watson potential [34], following the presentation by Dederichs [35]. An operator F is defined such that

$$\Psi_{0\mathbf{p}}^+ = F|0\rangle|\psi_{\mathbf{p}}^+\rangle \quad (225)$$

where the wave functions and Hamiltonian are defined in the same way as in Eqs. (129)–(134), except for $|\psi_{\mathbf{p}}^+\rangle$, the "coherent" one-electron wave function, which is defined by

$$|\psi_{\mathbf{p}}^+\rangle = \int \langle 0|\Psi_{0\mathbf{p}}^+\rangle d\tau = \langle 0|F|0\rangle|\psi_{\mathbf{p}}^+\rangle \quad (226)$$

with the integration running over the coordinates of the target only.

From the Lippman–Schwinger equation,

$$|\Psi_{0\mathbf{p}}^+\rangle = |0\rangle|\phi_{\mathbf{p}}^0\rangle + (E - H_s - T_e + i\eta)^{-1} V_{es}|\Psi_{0\mathbf{p}}^+\rangle \quad (227)$$

it readily follows that $|\psi_{\mathbf{p}}^+\rangle$ satisfies

$$|\psi_{\mathbf{p}}^+\rangle = |\phi_{\mathbf{p}}^0\rangle + (\epsilon_p - T_e + i\eta)^{-1} U|\psi_{\mathbf{p}}^+\rangle \quad (228)$$

where $U = \langle 0|V_{es}F|0\rangle$ and $\phi_{\mathbf{p}}^0$ is a plane wave. Furthermore, F satisfies the relation from Eqs. (225), (227), and (228) for the state $|0\rangle$,

$$F = 1 + (E - T_e - H_s + i\eta)^{-1}(V_{es}F - \langle 0|V_{es}F|0\rangle) \quad (229)$$

We will now show that $U = V_H + \Sigma$, and that thus the $|\psi_{\mathbf{p}}^+\rangle$ of Eq. (225) is identical with the damping function $|\psi_{\mathbf{p}}^+\rangle$ of Eq. (150). By some manipulations, noting that from Eq. (226) $\langle 0|F|0\rangle = 1$, Eq. (229) becomes

$$F = 1 + G_0(VF - \langle 0|VF|0\rangle) \quad (230)$$

where $G_0(E) = (E - H_0 + i\eta)^{-1}$ and $V = V_{es} - \langle 0|V_{es}|0\rangle$ (H_0 is defined by Eq. (131)).

Equation (230) gives (note that $\langle 0|V|0\rangle = 0$ and $\langle 0|VG_0(E)|0\rangle = 0$)

$$\delta U = U - V_H = \langle 0|VF|0\rangle$$
$$= \langle 0|V|0\rangle + \langle 0|VG_0(E)(VF - \langle 0|VF|0\rangle)|0\rangle$$
$$= \langle 0|VG_0(E)VF|0\rangle \quad (231)$$

We already see that if we approximate F with 1, δU equals Σ (cf. Eq. (147)) in a low-order approximation.

Since $P|0\rangle = |0\rangle$ and $PVP = 0$, we can rewrite δU as

$$\delta U = \langle 0|VG_0VF|0\rangle = \langle 0|PVQG_0QVFP|0\rangle \quad (232)$$

To evaluate $QVFP$ we insert $P + Q = 1$ before F,

$$QVFP = QVQFP + QVPFP = QVQFP + QVP$$

Multiplying Eq. (230) from the left by Q and from the right by P, we have

$$QFP = QG_0VFP = G_0QVFP = G_0QV(P + Q)FP$$
$$= G_0(QVP + QVQFP)$$

We can now solve for QFP to obtain $QFP = (1 - G_0QVQ)^{-1}G_0QVP$ and

$$QVFP = QV(1 - G_0QVQ)^{-1}G_0QVP + QVP$$
$$= Q[1 + V(1 - G_0QVQ)^{-1}G_0Q]VP \quad (233)$$

which finally gives the result

$$\delta U = \langle 0|PVQG_0QVFP|0\rangle$$
$$= \langle 0|PVQG_0Q[1 + V(1 - G_0QVQ)^{-1}G_0]V|0\rangle$$
$$= \langle 0|PVQ[G_0 + G_0QVQ(E - H_0$$
$$- QVQ + i\eta)^{-1}]V|0\rangle$$
$$= \langle 0|V(E - H_0 - QVQ + i\eta)^{-1}V|0\rangle \quad (234)$$

If we now compare with the expression for Σ in Eq. (147), we easily see that $\Sigma = \delta U$, which is what we set out to prove.

4.2. Comparison with the GW Approximation of the Optical Potential

Next we want to show the connection between our approximate $\Sigma = \delta U$ and the GW expression for the self-energy that is connected with the one-electron Green's function [36]. First we write our expression for Σ in Eq. (147) more explicitly, keeping the lowest-order term,

$$\Sigma(E) = \sum_{m \neq 0} \langle 0|V_{es}|m\rangle G_m^d(E)\langle m|V_{es}|0\rangle$$

$$= \sum_{m \neq 0} \langle 0|V_{es}|m\rangle$$

$$\times \left(E - E_m - h_0 - \Sigma(E - E_m + E_0) + i\eta\right)^{-1}$$

$$\times \langle m|V_{es}|0\rangle \qquad (235)$$

where we have used the approximation (178) for the optical potential. We insert the explicit expression for V_{es} and define coupling functions,

$$\langle m|V_{es}|0\rangle = \sum_{\mathbf{k},\mathbf{k}'} V^m(\mathbf{k}, \mathbf{k}')c_{\mathbf{k}}^\dagger c_{\mathbf{k}'}$$

$$V^m(\mathbf{k}, \mathbf{k}') = \sum_{ll'}\langle \mathbf{k}l|v|\mathbf{k}'l'\rangle\langle m|c_l^\dagger c_{l'}|0\rangle \qquad (236)$$

to obtain

$$\Sigma(E) = \sum_{m \neq 0}\sum_{\mathbf{kk'q}} \frac{V^m(\mathbf{q}, \mathbf{k})^* V^m(\mathbf{q}, \mathbf{k}')c_{\mathbf{k}}^\dagger c_{\mathbf{k}'}}{E - E_m - \epsilon_q - \langle\mathbf{q}|\Sigma|\mathbf{q}\rangle + i\eta} \qquad (237)$$

where in the denominator we have kept only the diagonal part of Σ. Substituting $E = E_0 + \epsilon_p$ and $\omega_m = E_m - E_0$, and using the subscript ca to indicate that our expression appears in a theory of scattering, we have

$$\Sigma_{sca}(E) = \sum_{m \neq 0}\sum_{\mathbf{kk'q}} \frac{V^m(\mathbf{q}, \mathbf{k})^* V^m(\mathbf{q}, \mathbf{k}')c_{\mathbf{k}}^\dagger c_{\mathbf{k}'}}{\epsilon_p - \omega_m - \epsilon_q - \langle\mathbf{q}|\Sigma|\mathbf{q}\rangle + i\eta} \qquad (238)$$

The GW approximation for the self-energy can be written as

$$\Sigma_{GW}(\mathbf{r}, \mathbf{r}'; \omega)$$

$$= \frac{i}{2\pi}\int \exp(i\omega'\eta)G(\mathbf{r}, \mathbf{r}'; \omega + \omega')W(\mathbf{r}, \mathbf{r}'; \omega')d\omega' \qquad (239)$$

with

$$G(\mathbf{r}, \mathbf{r}'; \omega) = \sum_s \frac{f_s(\mathbf{r})f_s(\mathbf{r}')^*}{\omega - \epsilon_s + i\eta} + \sum_t \frac{g_t(\mathbf{r})g_t(\mathbf{r}')^*}{\omega - \epsilon_t - i\eta}$$

$$f_s(\mathbf{r}) = \langle 0, N|\psi(\mathbf{r})|s, N+1\rangle$$

$$g_t(\mathbf{r}) = \langle t, N-1|\psi(\mathbf{r})|0, N\rangle$$

$$\epsilon_s = E(s, N+1) - E(0, N)$$

$$\epsilon_t = E(0, N) - E(t, N-1) \qquad (240)$$

$$W(\mathbf{r}, \mathbf{r}'; \omega) = \int \epsilon^{-1}(\mathbf{r}, \mathbf{r}_1; \omega)v(\mathbf{r}_1 - \mathbf{r}')d\mathbf{r}_1 \qquad (241)$$

$$\epsilon^{-1}(\mathbf{r}, \mathbf{r}'; \omega) = \delta(\mathbf{r} - \mathbf{r}') + \sum_{m \neq 0}\int d\mathbf{r}_1 v(\mathbf{r} - \mathbf{r}_1)$$

$$\times \left[\frac{\rho_m(\mathbf{r}_1)\rho_m(\mathbf{r}')^*}{\omega - \omega_m + i\eta} - \frac{\rho_m(\mathbf{r}_1)^*\rho_m(\mathbf{r}')}{\omega + \omega_m - i\eta}\right] \qquad (242)$$

Here ϵ^{-1} is the inverse dielectric response function, which is expressed in the spectral form containing matrix elements of the charge density $\rho(\mathbf{r})$, $\rho_m(\mathbf{r}) = \langle m|\rho(\mathbf{r})|0\rangle$, and excitation energy $\omega_m = E_m - E_0$. Introducing the fluctuation potential,

$$V^m(\mathbf{r}) = \int v(\mathbf{r} - \mathbf{r}')\rho_m(\mathbf{r}')d\mathbf{r}' \qquad (243)$$

we have

$$W(\mathbf{r}, \mathbf{r}'; \omega) = v(\mathbf{r} - \mathbf{r}')$$

$$+ \sum_{m \neq 0}\left[\frac{V^m(\mathbf{r})V^m(\mathbf{r}')^*}{\omega - \omega_m + i\eta} - \frac{V^m(\mathbf{r})^*V^m(\mathbf{r}')}{\omega + \omega_m - i\eta}\right] \qquad (244)$$

Since we have both G and W on their spectral forms, we can do the integration in Eq. (239) analytically:

$$\Sigma_{GW}(\mathbf{r}, \mathbf{r}'; \omega) = -v(\mathbf{r} - \mathbf{r}')\langle 0|\psi^\dagger(\mathbf{r}')\psi(\mathbf{r})|0\rangle$$

$$+ \sum_{m \neq 0, s} \frac{V^m(\mathbf{r})^*V^m(\mathbf{r}')f_s(\mathbf{r})f_s(\mathbf{r}')^*}{\omega - \omega_m - \epsilon_s + i\eta}$$

$$+ \sum_{m \neq 0, t} \frac{V^m(\mathbf{r})V^m(\mathbf{r}')^*g_t(\mathbf{r})g_t(\mathbf{r}')^*}{\omega + \omega_m - \epsilon_t - i\eta} \qquad (245)$$

The first term is the exchange potential, and it is included in $\langle 0|V_{es}|0\rangle$ when we use antisymmetrized Coulomb matrix elements as discussed in Section 3.4. It is thus only the last two terms in Σ_{GW} that should be compared with Σ_{sca}. Because of the large denominator, the third term can be neglected, and it is thus only the second term in Σ_{GW} that should be compared with Σ_{sca}. To do this we write this term, Σ_{pol}, expressed in creation and annihilation operators. For this purpose we apply a Hartree–Fock approximation for $\{f_s\}$, which are replaced by Hartree–Fock vacant orbitals, and these are approximated by scattering functions ($f_s \approx \phi_\mathbf{q}$),

$$\Sigma_{pol}(\omega) = \sum_{\mathbf{kk'}}\langle\mathbf{k}|\Sigma_{pol}(\omega)|\mathbf{k}'\rangle c_\mathbf{k}^\dagger c_{\mathbf{k}'}$$

$$= \sum_{\mathbf{kk'q}}\sum_{m \neq 0} \frac{V^m(\mathbf{q}, \mathbf{k})^* V^m(\mathbf{q}, \mathbf{k}')c_\mathbf{k}^\dagger c_{\mathbf{k}'}}{\omega - \omega_m - \epsilon_q + i\eta} \qquad (246)$$

We actually should have used antisymmetrized matrix elements not only in the first term of Σ_{GW}, but also in the second and third. This is of minor importance, however, because the exchange elements are much smaller than the direct elements, and hence the difference in a higher-order contribution like Σ_{pol} is relatively small. Except for the nonlinear correction in Eq. (238), Σ_{pol} is the same as Σ_{sca}.

4.3. Quasi-boson Representation for the Optical Potential

It is useful to work out an explicit expression for Σ_{sca} in the quasi-boson representation, because we can then discuss the physics in more detail, and in particular we can discuss the difference between ground-state and excited-state potentials.

Quasi-boson approximation has been used for the discussion of collective properties of electrons in metals [145]. This approximation is also widely used in nuclear many-body theory, and in some excellent textbooks (e.g., [146]).

As the Hamiltonian $H = H_0 + V$, we now take

$$
H_0 = \sum_m \omega_m a_m^\dagger a_m + \sum_{\mathbf{k}} \epsilon_k c_{\mathbf{k}}^\dagger c_{\mathbf{k}}
$$
$$
V = \sum_{\mathbf{k}\mathbf{k}'} \sum_m [V^m(\mathbf{k}, \mathbf{k}') a_m c_{\mathbf{k}}^\dagger c_{\mathbf{k}'} + V^m(\mathbf{k}, \mathbf{k}')^* a_m^\dagger c_{\mathbf{k}'}^\dagger c_{\mathbf{k}}]
$$
(247)

Here $c_{\mathbf{k}}^\dagger$ and ϵ_k correspond as before to the scattered electron, and a_m corresponds to excitations in the solid and obey boson commutation relations; the ω_m are excitation energies for the solid. When the ground-state optical potential operates on states with only one photoelectron present, it becomes

$$
\Sigma(E) = \langle 0|V(E - H_0 - QVQ)^{-1}V|0\rangle
$$
$$
= \sum_{jj'} \langle 0|V|j\rangle\langle j|(E - H_0 - QVQ)^{-1}|j'\rangle\langle j'|V|0\rangle
$$
$$
\approx \sum_{\mathbf{k}\mathbf{k}'\mathbf{q}} \sum_{m\neq 0} \frac{V^m(\mathbf{q}, \mathbf{k})^* V^m(\mathbf{q}, \mathbf{k}') c_{\mathbf{k}}^\dagger c_{\mathbf{k}'}}{\epsilon_p - \omega_m - \epsilon_q + i\eta}
$$
(248)

where the ground state $|0\rangle$ is for the boson system, and we have neglected the QVQ term. This is precisely the same equation as (246), and we thus have to identify the coupling coefficients $V^m(\mathbf{k}, \mathbf{k}')$ in Eq. (247) with those appearing in Eqs. (236), (243), and (246). Clearly also the ω_m are excitation energies for the solid.

We now consider an excited-state optical potential. Let us take $|j\rangle$ as a state with one boson present, $|j\rangle = a_m^\dagger|0\rangle$. We now get one contribution to Σ from $a_m(m = n)$ in V, which leads to an intermediate state with no boson, and one contribution from $a_m^\dagger(m = n)$, which leads to an intermediate state with two bosons of type n. For bosons that correspond to extended excitations the coupling coefficients are proportional to $1/\sqrt{\Omega}$, where Ω is the volume of the solid. We can thus neglect the two contributions with $m = n$, as in fact we can neglect any finite number of contributions. For contributions with $m \neq n$ we obtain exactly the same expression as in the ground-state case, except that we have to replace the constant ϵ_p with $\epsilon_p - \omega_n$. Similarly, if we have a number of bosons present in the state $|j\rangle$ with total energy $\omega_1 + \omega_2 + \omega_3 + \cdots$, we have to replace ϵ_p with $\epsilon_p - (\omega_1 + \omega_2 + \omega_3 + \cdots)$, and thus we have the relation for Σ stated in Eq. (178).

In our discussion we have neglected the QVQ term in the denominator. However, Eq. (178) still holds when QVQ is in-cluded. To see this we consider the expansion

$$
\Sigma_j(E) = \sum_{k\neq j} \langle j|V_{\text{es}} G_{jk}^{\text{d}} V_{\text{es}}|j\rangle
$$
$$
+ \sum_{l\neq k\neq j} \langle j|V_{\text{es}} G_{jkl}^{\text{d}} V_{\text{es}} G_{jk}^{\text{d}} V_{\text{es}}|j\rangle + \cdots
$$
(249)

Index l is different from k and j indices. We consider the case with extended excitations and solve Eq. (249) by iteration. In the first iteration all Σ in the diagonal G's on the right-hand side are set equal to zero. This gives first iterate $\Sigma_j(E)$, which satisfies $\Sigma_j(E) = \Sigma_0(E - E_j)$, where Σ_0 is the first-iterate ground-state optical potential. In the next order all propagators $G_{jk}^{\text{d}}(E)$ take the form $[E - E_k - h_0 - \Sigma_0(E - E_k)]^{-1}$. Also, in the second iterate we can absorb the higher intermediate energies in the expansion of $\Sigma_j(E)$ as compared with $\Sigma_0(E)$ by replacing E with $E - E_j$, and thus the relation $\Sigma_j(E) = \Sigma_0(E - E_j)$ still holds, as it will also in all higher-order iteration.

4.4. Atomic Optical Potential in Solids

In the case of EELFS, both inelastic scattering of the probe electrons and elastic scattering of secondary excited photoelectrons play an important role [10]. One is the optical potential on $\psi_{\mathbf{p}}^+$ for the ground state of the target Σ_0. The second one works on EXAFS electron function $\psi_{\mathbf{k}}^-$, and the third, on the scattered probe electron function $\psi_{\mathbf{p}'}^-$. These two optical potentials include the core-hole effects [10]. Our discussion will be focused on the calculation of the ground-state optical potential Σ_0 in this article.

It has long been recognized that elastic scattering of electrons is determined by the self-energy for the one-electron Green's function (or optical potential) [26, 36, 37]. More recently it has been shown that also in other spectroscopies like XPS [38], EXAFS [39], and EELFS [9, 10] the optical potential and its associated one-electron damping function play an important role.

Results for elastic electron-helium and neon scattering at 400 eV by Byron and Joachain [40, 40a], obtained with the use of a simplified local polarization, clearly show the important contribution of the polarization potential particularly for small-angle scattering. Fujikawa and Hedin have developed a theory for a practical, self-consistent, and nonlocal optical potential in a solid [41] and applied it to elastic electron-He [42], -Ne, and -Ar [43] scattering, where good agreement with the experimental results was obtained.

To calculate the elastic scattering from atoms in a solid, a muffin-tin-type potential is constructed, from which a set of phase shifts is obtained. The construction of an effective one-electron potential requires knowledge of the electronic charge density and a theory for calculating the exchange-correlation potential. Most potentials used so far are at best based on Hartree–Fock theory, using nonlocal exchange.

A simplified local exchange potential was proposed by Slater [44]. For electrons with a uniform density ρ in a large box, the

Hartree–Fock exchange contribution can be found analytically. Slater averaged this exchange contribution for electrons below the Fermi surface and assumed that the exchange potential in a solid could be approximated by a local potential whose value at point $\mathbf{r}$ was set equal to this electron gas averaged exchange taken at the solid-state density $\rho(\mathbf{r})$, which gives the Slater exchange potential,

$$V_{\text{ex}}(\mathbf{r}) \approx -(3/2)\big(3\rho(\mathbf{r})/\pi\big)^{1/3} \qquad (250)$$

This potential is widely used, not only for electrons below the Fermi level, but also for scattering problems, where it is basically incorrect. Thus for high energies exchange scattering can be neglected, whereas the Slater exchange potential still has an influence. For scattering problems the exchange potential before averaging is more motivated. This so-called Dirac–Hara potential is local and energy dependent and has been found to be successful for electron scattering from atoms and molecules [45]. Another variant is the X_α potential. Here the Slater potential is multiplied by an empirical parameter α, $V_{X_\alpha} = \alpha V_{\text{ex}}$, where α often is taken as 2/3.

These methods employ the Hartree–Fock approximation for the uniform electron gas. One expects that going beyond the Hartree–Fock approximation to the electron self-energy within a local density approximation should give an improved scattering potential. Following Hedin and Lundqvist [46], Lee and Beni applied such a potential to electron scattering from atoms in the intermediate energy region, using the plasmon pole approximation to the GW self-energy, and showed that this potential gives an excellent description of the EXAFS data for Br_2, $GeCl_4$, and Ge [47].

Several authors have compared these potentials for LEED [48–50] and XAFS calculations [47, 51–55]. In principle the Hedin–Lundqvist potential should give the best result among those methods described above. However, the results are not so clear-cut; sometimes it gives a poorer result than the X_α potential [48].

4.4.1. Basic Theory

The optical potential for the target state $|0\rangle$ is given to the lowest nontrivial order by a Van Hove-type expansion [9],

$$\Sigma_0(E) = \sum_{n(\neq 0)} \frac{\langle 0|V|n\rangle\langle n|V|0\rangle}{E - E_n - h - \Sigma_0(E - \omega_n) + i\eta} \qquad (251)$$

where $E = E_0 + \epsilon_k$, $\omega_n = E_n - E_0$, and V is the interaction potential between the scattering electron and the target,

$$V = \sum_{\mathbf{pq}} \left[\sum_{lm} c_l^\dagger c_m \langle \mathbf{p}l||\mathbf{q}m\rangle + \langle \mathbf{p}|V_{\text{en}}|\mathbf{q}\rangle \right] c_{\mathbf{p}}^\dagger c_{\mathbf{q}} \qquad (252)$$

Here V_{en} is the Coulomb interaction between the scattering electron and the positive nuclei in the target, and $\langle \mathbf{p}l||\mathbf{q}m\rangle$ is an antisymmetrized Coulomb matrix element [9]. The one-electron states of the scattering electron are denoted by $\mathbf{p}$ and $\mathbf{q}$, and those of electrons in the target by l and m. The

interaction V_{en} gives no contribution to the inelastic matrix elements $\langle 0|V|n\rangle$ ($n \neq 0$). When we consider core excitation processes as in XAFS and XPS spectra, the core hole effects are included only in the one-electron scattering Hamiltonian $h = T_e + \langle 0|V|0\rangle$, where T_e is the kinetic energy operator; we can handle the core hole optical potential in the same way as the ground-state optical potential, except for some minor differences [10, 11]. Furthermore, we have shown that the optical potential given by Eq. (251) is roughly equivalent to the self-energy in the GW approximation [9], so that we will develop a practical method of calculating the atomic optical potential in solids and surfaces based on the GW approximation.

Detailed discussions of the crystal potential were given long ago by Hedin [36, 37]. In [37, p. 129], the following expression for the self-energy (optical potential) is given:

$$\Sigma_0 = G^v W^v + V_{\text{ex}}^c + G^v W^v P^c W^v + \cdots \qquad (253)$$

Here $G^v W^v$ is the self-energy from the valence electrons, V_{ex}^c is the bare exchange, and $G^v W^v P^c W^v$ the screened polarization potential from the ion cores. The precise definitions of these quantities are given below.

It is common in, say, LEED calculations to approximate $\text{Im}\, G^v W^v$ by a spatially constant, energy-dependent function, $\text{Im}\, \Sigma_0(q, \epsilon(q))$, where $\Sigma_0(q, \omega)$ is the electron gas result. We do the same for the real part. For practical applications q is replaced by a density-dependent function $q(r)$ [56].

The next matter to consider is the effect of screening in the core-polarization term. For a free atom the screening is small, and we have to a good approximation $G^v v P^c v$, where v is the bare Coulomb potential. For long distances from the ion core it further reduces to the well-known local potential $-\alpha e^2/r^4$, where α is the dipole polarizability [36]. In a solid we expect for simple physical reasons that for long distances we should have a statically screened polarization potential, $G^v W^v(0) P^c W^v(0)$. Because an r^{-4} potential already is very weak, the additional screening should make it negligible outside the Wigner–Seitz cell of the ion under consideration. Inside the Wigner–Seitz cell, on the other hand, we do not expect much screening to take place because of the cost in kinetic energy of localizing the screening charge.

The full RPA polarization propagator is [37]

$$P(\mathbf{r}, \mathbf{r}'; \omega) = -\sum_k^{\text{unocc}} \sum_l^{\text{occ}} \frac{2(\epsilon_k - \epsilon_l)}{(\epsilon_k - \epsilon_l)^2 - \omega^2} f_{kl}(\mathbf{r}) f_{kl}^*(\mathbf{r}')$$

$$f_{kl}(\mathbf{r}) = \int \psi_k(x)\psi_l^*(x)\,d\xi \qquad (254)$$

Here $x = (\mathbf{r}, \xi)$ includes both space and spin variables. The sum over k runs over unoccupied electron states, and l runs over the occupied core and valence electron states. By splitting the summation over l into core and valence contributions, P can be written as a sum of core and valence parts,

$$P = P^c + P^v \qquad (255)$$

Similarly, we can split the summation over k in the expression for the one-electron Green's function,

$$G(x, x'; \omega) = \sum_k^{\text{occ+unocc}} \frac{\psi_k(x)\psi_k^*(x')}{\omega - \epsilon_k} \qquad (256)$$

to obtain

$$G = G^c + G^v \qquad (257)$$

The symbol $G^v v P^c v$ stands for a convolution in energy space [41], which can be done analytically, giving

$$[G^v v P^c v](x, x'; \omega)$$
$$= \sum_k^{\text{unocc}} \sum_l^{\text{core}} \sum_{k'}^{\text{valence}} \frac{v_{kl}(\mathbf{r})\psi_{k'}(x)\psi_{k'}^*(x')v_{kl}^*(\mathbf{r}')}{\omega - \omega_{kl} - \epsilon_{k'}} \qquad (258)$$

where $\omega_{kl} = \epsilon_k - \epsilon_l$, and $v_{kl}(\mathbf{r}) = \int v(\mathbf{r} - \mathbf{r}')\psi_k^*(\mathbf{r}')\psi_l(\mathbf{r}')d\mathbf{r}'$. The more tightly bound the core level l is, the smaller its contribution to $v_{kl}(\mathbf{r})$, due to the smaller overlap with the unoccupied function k. Thus the outermost core level will give the dominant contributions.

We replace ω_{kl} with a constant Δ, the average excitation energy. This approximation has been very successful in the free atom case [40]. We define a function $A(\mathbf{r}, \mathbf{r}')$,

$$A(\mathbf{r}, \mathbf{r}') = \sum_k^{\text{unocc}} \sum_l^{\text{core}} v_{kl}(\mathbf{r})v_{kl}^*(\mathbf{r}')$$
$$= \int v(\mathbf{r} - \mathbf{r}_1)v(\mathbf{r}' - \mathbf{r}_2)[\delta(x_1 - x_2) - \rho(x_1, x_2)]$$
$$\times \rho^c(x_2, x_1) dx_1 dx_2 \qquad (259)$$

where the last equality follows by closure, and ρ and ρ^c are the one-electron density matrices for all electrons, and for core electrons, respectively. With $\omega_{kl} = \Delta$ we then have

$$V^{\text{pol}}(x, x'; \omega) = [G^v v P^c v](x, x'; \omega)$$
$$= A(\mathbf{r}, \mathbf{r}')G^v(x, x'; \omega - \Delta) \qquad (260)$$

With the use of closure we avoid the summation over the unoccupied states. Still the density matrix $\rho(x_1, x_2)$ contains a sum over the occupied Bloch functions. We will take here a simplified approach and represent the sum over Bloch functions in each atomic cell α by a sum over localized functions $R_m^\alpha Y_{lm}$. This is well motivated for rare gas solids, but a more serious approximation is needed for, say, metals. In a more accurate treatment one could consider representing the Bloch functions with muffin-tin orbitals, but this would mean an integration in k-space, which would substantially increase the computational work.

The core hole optical potential Σ_{0*} on EXAFS electrons is also given by the product AG as in Eq. (260). The quasi-boson approximation provides us with the useful result that A is the same as the one in Σ_0, whereas $G(\omega - \Delta^*)$ includes the hole effects in the static and the optical potentials. Here Δ^* is the average excitation energy in the presence of the core hole. The core hole optical potential for the function $\psi_{\mathbf{p}'}^-$ is obtained by energy shift of Σ_{0*} [10].

If $\mathbf{r}$ is in cell α and the core function l is in a neighboring cell β, then $v_{kl}(\mathbf{r})$ is small and depends on the dipole matrix element between k and l. To lowest order the interatomic contribution to A from the first term in Eq. (259) is (taking the core functions as completely localized)

$$\frac{N_c^\beta}{|\mathbf{r} - \mathbf{R}_{\alpha\beta}||\mathbf{r}' - \mathbf{R}_{\alpha\beta}|} \qquad (261)$$

where N_c^β is the total number of core electrons in cell β, and $\mathbf{R}_{\alpha\beta} = \mathbf{R}_\alpha - \mathbf{R}_\beta$. This large contribution is exactly canceled by the second term in Eq. (259), since $\int \rho(x_1, x_2) \rho^c(x_2, x_1) dx_1 dx_2 = N_c^\beta$ if the integration over x_1 and x_2 is restricted to cell β. We will evaluate the interatomic contributions to A from a multipole expansion, using a bare potential $v(\mathbf{r} - \mathbf{r}_1)$. The results are taken as an upper limit to what we can expect to have with a properly screened potential W.

Byron and Joachain [40, 40a] simplified the expression for the core polarization potential in Eq. (260) and were able to obtain an explicit local approximation. They also approximated the exchange potential by taking the electron gas expression with the local electron density. These approximations for the optical potential gave quite good results for electron-atom scattering [40, 40a]. It is not easy to assess the accuracy of the many different approximations made by Byron and Joachain [40, 40a], and their approximations are also difficult to generalize to the solid-state case. Instead we prefer to actually evaluate the nonlocal expressions for both the exchange and polarization potentials from the ion cores.

4.4.2. Atomic Optical Potentials in Solids

When one studies electron spectroscopies such as LEED and XAFS, the elastic scattering from each atomic site is a crucial physical process. It is therefore important to separate the total effective one-electron potential V_{cryst} into atomic scattering potentials v_α centered at sites α. The v_α are usually spherically averaged to simplify the formulas. The one-electron Green's function can then be represented by a T-matrix expansion in terms of a homogeneous medium Green's function $g_0 = (\epsilon_p - T_e - \Delta + i\Gamma)^{-1}$. When $\mathbf{r} \in \alpha$ and $\mathbf{r}' \in \beta$, we have to lowest order

$$g(\mathbf{r}, \mathbf{r}') = g_0 + g_0 t_\alpha g_0 + g_0 t_\beta g_0 + \cdots \qquad (262)$$

where t_α is the T matrix for site α. The damping propagator g_0 is already small at the nearest neighbor distance. The energy-independent term A defined by Eq. (259) also decreases with the distance $|\mathbf{r} - \mathbf{r}'|$ (see Eq. (272)). Hence it is enough to consider only one-site contributions to $g(\mathbf{r}, \mathbf{r}')$ at higher energies, say $\epsilon_p \geq 100$ eV. For a spherically symmetric potential v_α, the Green's function g_α is expanded as

$$g_\alpha(\mathbf{r}, \mathbf{r}') = \sum_L g_l^\alpha(r, r')Y_L(\hat{r})Y_L^*(\hat{r}') \qquad (263)$$

$$g_l^\alpha(r, r') = -2i\bar{p}\exp(i\delta_l^\alpha)R_l^\alpha(\bar{p}r_<)f_l^\alpha(\bar{p}r_>)$$

where δ_l^α is the phase shift of the lth partial wave at site α, $\bar{p} = [2(\epsilon_p - \Delta + i\Gamma)]^{1/2}$, R_l^α and f_l^α are the regular and irregular solutions for the potential v_α, $r_> = \max(r, r')$, and $r_< = \min(r, r')$. In the large $\mathbf{r}$ limit they have the asymptotic forms

$$R_l^\alpha(\bar{p}r) \approx \sin(\bar{p}r - l\pi/2 + \delta_l^\alpha)/(\bar{p}r), \qquad f_l^\alpha(\bar{p}r) \approx h_l(\bar{p}r)$$
(264)

where h_l is the spherical Hankel function of lth order.

Next we investigate how to calculate the energy-independent term $A(\mathbf{r}, \mathbf{r}')$. In each atomic region α we average the core charge density to obtain a spherically symmetric function,

$$d_\alpha^c(r) = \frac{1}{4\pi} \int d\hat{\mathbf{r}} \rho^c(\mathbf{r})$$
(265)

From the first term of the large parenthesis in Eq. (259), we have an intraatomic contribution to $A(\mathbf{r}, \mathbf{r}')$,

$$I_\alpha(\mathbf{r}, \mathbf{r}') = \sum_L \left(\frac{4\pi}{2l+1}\right)^2 I_\alpha(r, r')_l Y_L(\hat{\mathbf{r}}) Y_L^*(\hat{\mathbf{r}}')$$
(266)

where both $\mathbf{r}$ and $\mathbf{r}'$ belong to atomic region α, and $\mathbf{r}_1 (= \mathbf{r}_2)$ is in region α. $I_\alpha(r, r')_l$ is

$$I_\alpha(r, r')_l = S_1^l + S_2^l + S_3^l$$
(267)

$$S_1^l = (rr')^{-l-1} \int_0^{r_<} d_\alpha^c(r_1) r_1^{2l+2} \, dr_1$$
(268)

$$S_2^l = \frac{r_<^l}{r_>^{l+1}} \int_{r_<}^{r_>} d_\alpha^c(r_1) r_1 \, dr_1$$
(269)

$$S_3^l = (rr')^l \int_{r_>}^{R_\alpha} d_\alpha^c(r_1) r_1^{-2l} \, dr_1$$
(270)

where R_α is the radius of the atomic region α.

In addition to this intraatomic contribution to the first term of Eq. (259) we have the interatomic term, where $\mathbf{r}$ and $\mathbf{r}'$ are still in region α, whereas $\mathbf{r}_1 (= \mathbf{r}_2)$ is in region β ($\beta \neq \alpha$). This term is given by

$$I_\alpha^\beta(\mathbf{r}, \mathbf{r}')$$
$$= \sum_{LL'} \left(\frac{4\pi}{2l+1}\right)^2 |F_{LL'}(\mathbf{R}_{\beta\alpha})|^2 (rr')^l Y_{L'}(\hat{\mathbf{r}}) Y_{L'}^*(\hat{\mathbf{r}}') \langle r^{2l} \rangle_\beta$$
$$= \sum_L F_L(r, r'; \mathbf{R}_{\beta\alpha}) Y_L(\hat{\mathbf{r}}) Y_L^*(\hat{\mathbf{r}}')$$
(271)

where in terms of Gaunt's integral $G(LL'|L'') \equiv \int Y_L(\hat{\mathbf{r}}) \times Y_{L'}(\hat{\mathbf{r}}) Y_{L''}^*(\hat{\mathbf{r}}) d\hat{\mathbf{r}}$,

$$F_{LL'}(\mathbf{R}_{\beta\alpha}) = \frac{4\pi(-1)^l(2l+2l'-1)!!}{(2l-1)!!(2l'+1)!!R_{\beta\alpha}^{l+l'+1}}$$
$$\times G(l+l', m-m', L'|L)$$
$$\times Y_{l+l', m'-m}(\hat{\mathbf{R}}_{\beta\alpha})$$
(272)

$$F_L(r, r'; \mathbf{R}_{\beta\alpha}) = \sum_{L'} \left(\frac{4\pi}{2l'+1}\right)^2 |F_{L'L}(\mathbf{R}_{\beta\alpha})|^2$$
$$\times (rr')^l \langle r^{2l} \rangle_\beta$$
(273)

$$\langle r^{2l} \rangle_\beta = \int_0^{R_\beta} d_\beta^c(r) r^{2l+2} \, dr$$
(274)
$$\mathbf{R}_{\beta\alpha} = \mathbf{R}_\beta - \mathbf{R}_\alpha$$

We note that $\sum_{m'} |F_{LL'}(\mathbf{R}_{\beta\alpha})|^2$ depends on m. To get rid of this difficulty, we use the spherically averaged value of $\sum_m |F_{LL'}(\mathbf{R}_{\beta\alpha})|^2$. Finally we have the averaged expression of F_L in Eq. (271) in terms of the Clebsch–Gordan coefficient $\langle l0l'0|l+l'0\rangle$, which is denoted F_l^0:

$$F_l^0(r, r'; R_{\beta\alpha}) = (rr')^l \sum_{l'} \left\{ \frac{4\pi(2l+2l'-1)!!}{(2l+1)!!(2l'+1)!!R_{\alpha\beta}^{l+l'+1}} \right\}^2$$
$$\times \langle r^{2l'} \rangle_\beta (2l'+1)\langle l'0l0|l+l'0\rangle^2$$
(275)

Because of the small extension of the core functions, the dominant contribution comes from $l' = 0$,

$$F_l^0(r, r'; R_{\beta\alpha}) = \frac{4\pi}{(2l+1)^2} \frac{(rr')^l}{R_{\alpha\beta}^{2l+2}} N_c^\beta$$

This is the same result as obtained from averaging over $\hat{\mathbf{R}}_{\alpha\beta}$ in Eq. (261).

The second term of the large parenthesis in Eq. (259) is more difficult to calculate in general. We note that

$$\rho(x_1, x_2)\rho^c(x_2, x_1) = \sum_i^{occ} \sum_j^{core} d_{ij}(x_1) d_{ij}^*(x_2)$$
(276)

where $d_{ij}(x) = \phi_i(x)\phi_j^*(x)$, and we assume that the core orbital j is localized on site α. We take both $\mathbf{r}_1$ and $\mathbf{r}_2$ to be in the same region α. For the occupied states we will not use Bloch functions but instead take a simplified approach and use localized functions $R_n^\alpha Y_{L_n}$. This is well motivated for rare gas solids, but a more serious approximation is needed for, say, metals.

By spherically averaging $d_{ij}(\mathbf{r})$ at each site α, we can get a simple representation for Eq. (276) as

$$\rho(\mathbf{r}_1, \mathbf{r}_2)\rho^c(\mathbf{r}_2, \mathbf{r}_1)$$
$$\approx \sum_m^{core} d_m^\alpha(r_1) d_m^\alpha(r_2) + \sum_m^{core} \sum_{n \neq m}^{occ} d_{mn}^\alpha(r_1) d_{mn}^\alpha(r_2)$$
$$\mathbf{r}_1, \mathbf{r}_2 \in \beta$$
(277)

where m and n stand for atomic states 2s, $2p_+$, $2p_0$, $2p_-$, and so on, in atom α that have the same angular quantum number $L = (l, m)$. The quantity $d_m^\alpha(= d_{mm}^\alpha)$ is the spherically averaged electron density of the mth atomic functions at site α, and d_{mn}^α is the cross charge from the mth and nth atomic functions on site α written in terms of the radial parts of the shell wave functions, $d_{mn}^\alpha(r) = R_m^\alpha(r) R_n^\alpha(r)^*$. As mentioned, R_m^α refers to a localized function, which for metals has a fractional occupation number. When we use this simple approximation in Eq. (259), the interatomic contribution to A can be written as

$$J_\alpha(r, r') = \sum_m J_m^\alpha(r) J_m^\alpha(r') + \sum_{m \neq n} J_{mn}^\alpha(r) J_{mn}^\alpha(r')$$
(278)

with

$$J_m^\alpha(r) = 4\pi \left\{ \frac{1}{r} \int_0^r r_1^2 d_m^\alpha(r_1)\, dr_1 + \int_r^\infty r_1 d_m^\alpha(r_1)\, dr_1 \right\} \quad (279)$$

We see that J_α has only a spherically symmetric contribution.

The interatomic contribution J_α^β is not difficult to evaluate as

$$J_\alpha^\beta(r, r') = \sum_m^{core} J_m^{\alpha\beta}(r) J_m^{\alpha\beta}(r') \quad (280)$$

where by noting that $r_\alpha = r$, $J_m^{\alpha\beta}(r)$ is given as

$$J_m^{\alpha\beta}(r) = \frac{1}{4\pi} \int d\hat{\mathbf{r}}_\alpha \frac{n_m^\beta}{|\mathbf{r}_\alpha - \mathbf{R}_{\beta\alpha}|} \quad (\alpha \neq \beta) \quad (281)$$

Here n_m^β is the number of electrons on site β, and $\mathbf{R}_{\beta\alpha} = \mathbf{R}_\beta - \mathbf{R}_\alpha$. If we evaluate the angular integral in Eq. (281), we have

$$J_m^{\alpha\beta}(r) = \frac{n_m^\beta}{R_{\beta\alpha}}, \qquad J_\alpha^\beta(r, r') = \frac{N_c^\beta}{R_{\beta\alpha}^2}$$

Cross terms such as d_{mn}^β give no contribution to $J_m^{\alpha\beta}$ because of the orthogonality between mth and nth shell functions. Therefore $A(\mathbf{r}, \mathbf{r}')$ can be written as

$$A(\mathbf{r}, \mathbf{r}') = \sum_L A(r, r')_l Y_L(\hat{r}) Y_L^*(\hat{r}') \quad (282)$$

where A is a sum of one- and two-center terms,

$$A(r, r')_l = A_\alpha(r, r')_l + \sum_{\beta \neq \alpha} A_\alpha^\beta(r, r')_l \quad (283)$$

$$A_\alpha(r, r')_0 = (4\pi)^2 \left[I_\alpha(r, r')_0 - 4\pi J_\alpha(r, r')_0 \right] \quad (284)$$

$$A_\alpha(r, r')_l = \left(\frac{4\pi}{2l+1} \right)^2 I_\alpha(r, r')_l \quad (l \geq 1) \quad (285)$$

$$A_\alpha^\beta(r, r')_l = \frac{(4\pi)^2}{3} \left(\frac{l+1}{2l+1} \right) \frac{\langle r^2 \rangle_\beta}{R_{\beta\alpha}^{2l+4}} (rr')^l + \cdots \quad (286)$$

We find a good convergence for the two-center sum, the second term in Eq. (283), when we include the surrounding atoms up to the third shell for the systems considered here.

The optical potential can be given by Eq. (287) after the spherical averaging of the potential over $\hat{r}$ and $\hat{r}'$ in the same atomic region,

$$V^{pol}(\mathbf{r}, \mathbf{r}'; E_0 + \epsilon_p) = \sum_L V_l^{pol}(r, r'; \epsilon_p) Y_L(\hat{r}) Y_L^*(\hat{r}') \quad (287)$$

where V_l^{pol} is expressed in terms of $A_{l'}$, $g_{l''}$, and Clebsh–Gordan coefficients,

$$V_l^{pol}(r, r'; \epsilon_p) = \frac{1}{4\pi} \sum_{l'l''} \frac{(2l'+1)(2l''+1)}{2l+1} \langle l'0l''0|l'+l''0 \rangle^2$$
$$\times A_{l'}(r, r') g_{l''}^\alpha(r, r'; \bar{p}) \quad (288)$$

Figure 3 shows the $A_l(r, r')$ ($l = 0, 1, 2$) for hydrogen atoms where r is fixed at 0.1, 0.5, 1.0, and 2.0 a.u. [41]. Both A_1

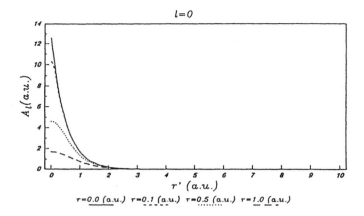

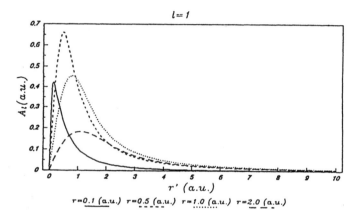

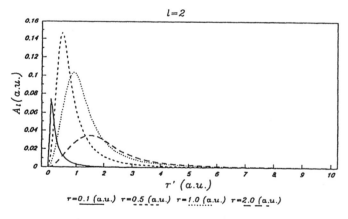

Fig. 3. $A_l(r, r')$ ($l = 0, 1, 2$) for a hydrogen atom. r is fixed at 0.1, 0.5, 1.0, and 2.0 a.u. [41].

and A_2 vanish at $r' = 0$, which can be easily verified from Eqs. (268)–(270), and have a maximum whose position shifts to larger r' with r. In contrast, A_0 is a smoothly and rapidly decreasing function of r and r'. We see that A_0 is much larger than A_1 and A_2 inside the atomic region. At small r, A_0 is dominant, and the relative importance decreases with l. At large $r (> R_\alpha)$, A_0 becomes small faster than $A_l (l \geq 1)$, as shown in those figures, and A_1 becomes the most important. Calculated results for Ne atom also show quite similar behavior.

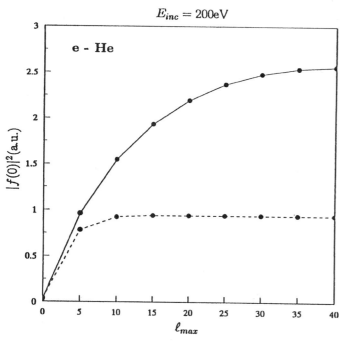

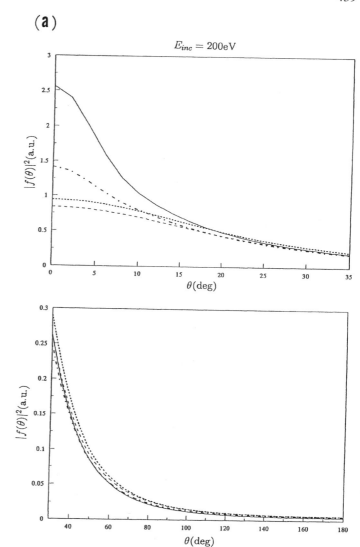

Fig. 4. Forward differential cross section in a.u.^{2} for electron elastic scattering from He at 200 eV (a) and 500 eV (b) [42]. The convergence is investigated in partial wave expansion; the solid line (dashed line) is the result for the HF + optical (HF only) potential with $\Delta = 20$ eV.

As demonstrated above it is enough only to include A_0 and A_1 in the expansion in Eq. (288). We thus obtained an explicit expression for V_l^{pol},

$$V_l^{\text{pol}}(r, r'; \epsilon_p)$$
$$= \frac{1}{4\pi} \Big[A_0(r, r') g_l(r, r'; \bar{p}) + A_1(r, r') \tilde{g}_l(r, r'; \bar{p}) \Big] \quad (289)$$

where $\tilde{g}_l$ is defined by

$$\tilde{g}_l(r, r'; \bar{p}) = \frac{l}{2l+1} g_{l-1}(r, r'; \bar{p}) + \frac{l+1}{2l+1} g_{l+1}(r, r'; \bar{p})$$
$$(290)$$

Each $g_{l''}$ includes the radial solution for the potential $\Sigma_{l''}(\epsilon_p)$ to be determined, so that our optical potential Σ_l can be solved self-consistently. Note that V_l^{pol} depends on g_l and g_{l+1}, so that we have to solve coupled self-consistent equations.

4.4.3. Results and Discussion

Figure 4 shows the forward differential cross section (in atomic units) for electron elastic scattering by He at 200 eV [42]. The convergence is investigated in partial wave expansion; the solid line (dashed line) is the result for the Hartree–Fock+the optical potential (only the Hartree–Fock potential). We see that good convergence is obtained for $l_{\max} = 10$ for the Hartree–Fock potential, whereas many partial waves are necessary to obtain good convergence for the Hartree–Fock + the optical potential discussed above, converged at $l_{\max} = 35$. Here $l_{\max}$

Fig. 5. Differential cross section in arbitrary units for electron elastic scattering by He at 200 eV (a) and 500 eV (b) [42]. The solid (dash-dotted) line shows the SCF (non-SCF) result for the HF + optical potential with $\Delta = 20$ eV. The dotted (dashed) line shows the result for the Hartree–Fock (Hartree) potential.

corresponds to the impact parameter in the classical collision theory, the long-range polarization part is mainly responsible for this large impact parameter.

Figure 5 shows the importance of the self-consistency in the optical potential calculations [42]. Nonself-consistent field (SCF) means the results for the first iteration calculation, including both the exchange and the optical potential when we use the Hartree solution as input data. In the small-angle region ($\theta < 30°$), SCF iteration, which is a nonlinear effect, plays an important role at 200 eV, and in the region $\theta < 15°$ at 500 eV. The results for the Hartree–Fock and Hartree solutions are shown for comparison. These two approximations give quite poor results in the small-angle scatterings.

Figure 6 also shows the forward differential cross section in a.u.^{2} for e-Ne (a) and -Ar(b) at 200 eV [43]. The con-

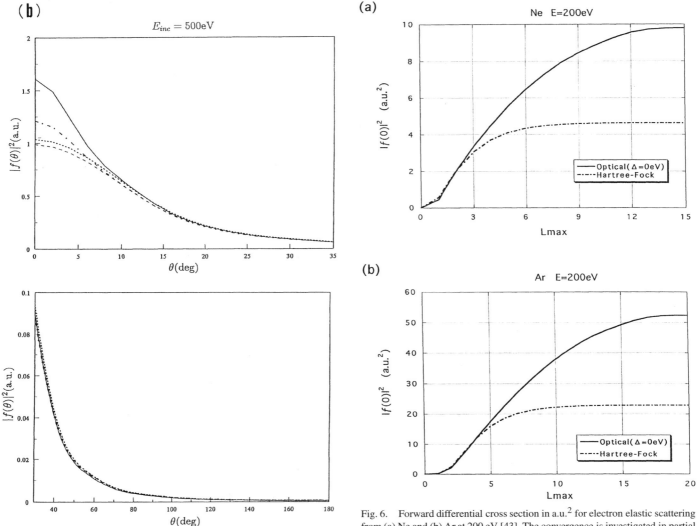

(b)

$E_{inc} = 500\text{eV}$

(a)

Ne E=200eV

(b)

Ar E=200eV

Fig. 5.   (Continued).

Fig. 6.   Forward differential cross section in a.u.^{2} for electron elastic scattering from (a) Ne and (b) Ar at 200 eV [43]. The convergence is investigated in partial wave expansion; the solid line (dashed line) is the result for the HF + optical (HF only) potential with $\Delta = 0$ eV.

vergence of the partial wave expansion is studied here. We see that good convergence is found for $l_{\max} = 3$ (Ne) and $l_{\max} = 8$ (Ar) for the Hartree–Fock potential, whereas more partial waves are necessary to get good convergence for the Hartree–Fock + the optical potential. The calculated result converges for $l_{\max} = 15$ (Ne) and $l_{\max} = 18$ (Ar). In comparison with the result for e-He scatterings, we find rather rapid convergence in the partial wave expansion for the e-Ne and -Ar scatterings.

Figure 7 shows the importance of the self-consistency in the optical potential calculations [43]. In the small-angle region $\theta < 5°$ the SCF iteration has some effect at 200 eV, whereas it is not important at 700 eV. The results for the Hartree–Fock and Hartree are shown for comparison; quite large differences are found at the small-angle scatterings.

Figure 8 shows the differential cross section in a.u.^{2} as a function of scattering angle for electron elastic scattering scattering by Ne at 200 and 400 eV for different values of Δ [43].

The solid line shows the result for the parameter $\Delta = 0$ eV, the dashed line the result for $\Delta = 20$ eV, and the dotted line the result for $\Delta = 40$ eV. Some experimental results and the result from the Hartree–Fock calculation are also shown for comparison. Only in the small-angle region ($\theta < 20°$) do the calculated results depend on the parameter Δ, though the dependence is weak. For $\Delta = 20$ eV, good agreement with experiment is obtained.

Figure 9 shows the calculated scattering cross sections from Ar at 200 and 400 eV for the same three Δ values [43]. For comparison, some available experimental results and the result from the Hartree–Fock calculation are also shown. Satisfactory agreement is found for both Figure 9a and Figure 9b; however, the Hartree–Fock calculation gives a poor result in the small-angle region, as observed in e-He and -Ne scatterings. The calculated results are not so sensitive to this parameter compared with the results for e-He and -Ne scatterings.

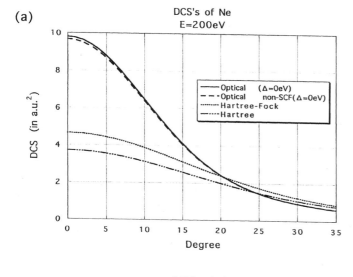

(a)

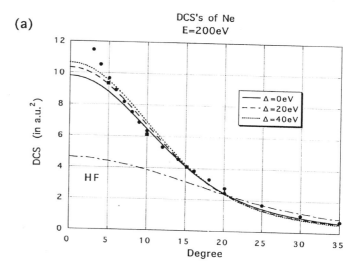

(a)

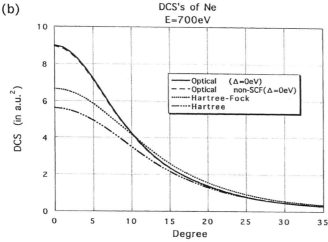

(b)

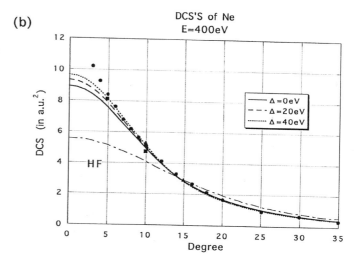

(b)

Fig. 7. Differential cross section (DCS) in a.u.^{2} as a function of scattering angle for electron elastic scattering from Ne at (a) 200 eV and (b) 700 eV [43]. The solid (dashed) line shows the SCF (non-SCF) result for the HF + optical potential with $\Delta = 0$ eV. The dotted (three-dot-dashed) line shows the result for the Hartree–Fock (Hartree) potential.

Fig. 8. Differential cross section (DCS) in a.u.^{2} as a function of scattering angle for electron elastic scattering from Ne at (a) 200 eV and (b) 400 eV [43]. The solid line shows the result for the parameter, $\Delta = 0$ eV; the dashed line, 20 eV; and the dotted line, 40 eV. Some experimental results are also shown for comparison. The calculated result for the Hartree–Fock approximation is also shown.

The Hedin–Lundqvist potential has been compared with the present theoretical approach [143]: the Hedin–Lundqvist potential overestimates the small-angle scattering intensity. For the valence electron optical potential we can safely use a local density approximation because the charge density changes fairly slowly. For these electrons plasmon-pole approximation can also be applied, whereas these two approximations present serious problems for core electrons. This is why the Hedin–Lundqvist potential for all electrons (including core electrons) does not give such good results.

5. THEORY OF DEEP CORE EXCITATION EELS

In this section we now apply the general electron scattering theory developed in the previous sections to deep core excitation EELS; in particular, EELFS is studied [10, 11]. In this case,

the three different damping wave functions are considered for a secondary excited electron from a deep core and a probe electron before and after the core excitation. Correspondingly, we have to calculate three different optical potentials, even if we apply the energy shift theorem for the optical potentials with different energy arguments. We apply site T matrix expansion for these wave functions to explicitly discuss the EELFS and the intensity oscillation due to the elastic scatterings of probe electrons.

5.1. Basic Formulas of Deep Core Excitation EELS

We consider the amplitude of inelastic scattering in the transition $0\mathbf{p} \rightarrow n\mathbf{p}'$, where the target is excited from the ground state $|0\rangle$ to an excited state $|n\rangle$ ($n \neq 0$) and the probe electron

(a)

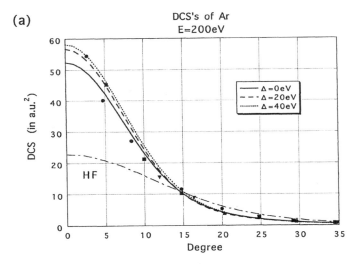

(b)

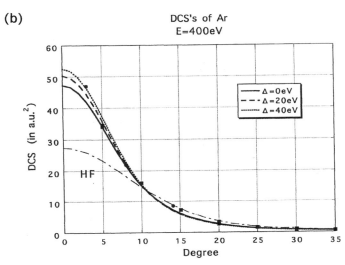

Fig. 9.   As in Figure 8, but for e-Ar scatterings [43].

analyze X-ray absorption spectra,

$$I(\mathbf{p}', \mathbf{p}) = -2 \operatorname{Im}\langle 0|U(\mathbf{p}', \mathbf{p})^{\dagger}(E - H + i\eta)^{-1}U(\mathbf{p}', \mathbf{p})|0\rangle$$
$$(\eta \to \infty) \qquad\qquad (294)$$

where $E = E_0 + \Delta E$ and H is the full many-body Hamiltonian. The difference between EELFS and XAFS only comes from the difference between $U(\mathbf{p}', \mathbf{p})$ and H_{ep} (electron–photon interaction Hamiltonian). Hedin has developed a very useful EXAFS expression, starting from Eq. (294) and using quasi-boson approximation [39]. An alternative approach has been developed by Fujikawa [57] with the use of many-body scattering theory. Here we follow the discussion in [28].

Above the core threshold $U(\mathbf{p}', \mathbf{p})$ can be written as

$$U(\mathbf{p}', \mathbf{p}) = \sum_{\mathbf{k}} \langle \mathbf{k}|u(\mathbf{p}', \mathbf{p})|c\rangle c_{\mathbf{k}}^{\dagger} b + \text{h.c.} \qquad (295)$$

$b(c_{\mathbf{k}}^{\dagger})$ is the annihilation (creation) operator of the core state ϕ_{c} (photoelectron state), and h.c. means the hermitian conjugate of the first term. To specify the problem we introduce the Hamiltonian widely used to study deep core processes [22],

$$H = H_{\mathrm{v}} + h + V + V_{\mathrm{c}}bb^{\dagger} + \epsilon_{\mathrm{c}}b^{\dagger}b \qquad (296)$$

Here H_{v} is the full many-electron Hamiltonian for valence electrons, V_{c} is the interaction between the core hole and valence electrons, and ϵ_{c} is the core electron energy. From the interaction operator V_{es} (see Eq. (130)) between the photoelectron and the other particles in the solid, the diagonal (elastic) interaction V_{d} and the off-diagonal (inelastic) interaction V can be constructed by use of the projection operator $P_n(= |n_{\mathrm{v}}^*\rangle\langle n_{\mathrm{v}}^*|)$,

$$V_{\mathrm{d}} = \sum_{n} P_n V_{\mathrm{es}} P_n, \qquad V = V_{\mathrm{es}} - V_{\mathrm{d}} \qquad (297)$$

The projection operator P_n is now defined for the hole state $|n_{\mathrm{v}}^*\rangle$ that satisfies the eigenvalue equation

$$H_{\mathrm{v}}^*|n_{\mathrm{v}}^*\rangle = E_n^{\mathrm{v*}}|n_{\mathrm{v}}^*\rangle, \qquad H_{\mathrm{v}}^* = H_{\mathrm{v}} + V_{\mathrm{c}}$$

The effective one-electron Hamiltonian h for a photoelectron is now given by

$$h = T_{\mathrm{e}} + V_{\mathrm{d}} \qquad (298)$$

Within the present approximation shown by Eq. (296), the initial state $|0\rangle$ is written by the product $|0\rangle = |0_{\mathrm{v}}\rangle|c\rangle$, where $|0_{\mathrm{v}}\rangle$ is the ground state of $H_{\mathrm{v}} : H_{\mathrm{v}}|0_{\mathrm{v}}\rangle = E_0^{\mathrm{v}}|0_{\mathrm{v}}\rangle$. The loss intensity is then

$$I(\mathbf{p}', \mathbf{p}) = -2 \operatorname{Im}\langle 0_{\mathrm{v}}|T^{\dagger}(E - H_{\mathrm{v}}^* - h - V + i\eta)^{-1}T|0_{\mathrm{v}}\rangle \quad (299)$$

where T is an operator that creates photoelectrons:

$$T = \sum_{\mathbf{k}} \langle \mathbf{k}|u|c\rangle c_{\mathbf{k}}^{\dagger} \qquad (300)$$

This operator, if used on a photoelectron vacuum state, is then replaced by $u|c\rangle$ by use of the approximate closure relation

$$\sum_{\mathbf{k}} |\mathbf{k}\rangle\langle \mathbf{k}| \approx 1$$

is scattered from $\mathbf{p}$ to $\mathbf{p}'$, which is given by Eq. (179) in Section 3.3. When we consider the high-energy excitation where $\epsilon_p, \epsilon_{p'} \gg V_{\mathrm{es}}$ is satisfied, we can neglect the second term of Eq. (179) and change the order of integration as in Eq. (180); then we have

$$T(n\mathbf{p}'; 0\mathbf{p}) \approx \langle n|U(\mathbf{p}', \mathbf{p})|0\rangle \qquad (291)$$
$$U(\mathbf{p}', \mathbf{p}) = \sum_{ll'} \langle l|u(\mathbf{p}', \mathbf{p})|l'\rangle c_l^{\dagger} c_{l'}, \quad \langle l|u(\mathbf{p}', \mathbf{p})|l'\rangle$$
$$= \langle \mathbf{p}'l\||\mathbf{p}l'\rangle \qquad (292)$$

The loss intensity $I(\mathbf{p}', \mathbf{p})$ from the ground state $|0\rangle$ is now given by

$$I(\mathbf{p}', \mathbf{p}) = 2\pi \sum_{n} |\langle n|U(\mathbf{p}', \mathbf{p})|0\rangle|^2 \delta(E_n - E_0 - \Delta E) \quad (293)$$

where $\Delta E = \epsilon_p - \epsilon_{p'}$ is the loss energy. One way to treat the loss intensity utilizes a standard expression frequently used to

This approximation can be good well above the threshold. By use of these simplifications, a compact formula first derived by Hedin is obtained [39] for XAFS and EELFS analyses,

$$I(\mathbf{p}', \mathbf{p}) = -2 \operatorname{Im} \langle c | u(\mathbf{p}', \mathbf{p})^* \hat{G}(E) u(\mathbf{p}', \mathbf{p}) | c \rangle$$
$$\hat{G}(E) = \langle 0_v | (E - H_v^* - h - V + i\eta)^{-1} | 0_v \rangle \quad (301)$$

By inserting the closure relation $\sum_n |n_v^*\rangle\langle n_v^*| = 1$, $\hat{G}$ can be written as

$$\hat{G}(E) = |S_0|^2 \langle 0_v^* | G(E) | 0_v^* \rangle$$
$$+ \sum_{n>0} \left(S_n^* S_0 \langle n_v^* | G(E) | 0_v^* \rangle + S_n S_0^* \langle 0_v^* | G(E) | n_v^* \rangle \right)$$
$$+ \sum_{n>0} |S_n|^2 \langle n_v^* | G(E) | n_v^* \rangle$$
$$+ \sum_{n \neq m > 0} S_m^* S_n \langle m_v^* | G(E) | n_v^* \rangle \quad (302)$$

where $G(E) = (E - H_v^* - h - V + i\eta)^{-1}$, and the intrinsic loss amplitude S_n is given by

$$S_n = \langle n_v^* | 0_v \rangle = \langle n_v^* | b | 0 \rangle$$

With the use of the technique discussed in Section 3.3, the diagonal part of $G(E)$ is simply written as

$$\langle n_v^* | G(E) | n_v^* \rangle$$
$$= g^c(\epsilon - \omega_n) = \left[\epsilon - \omega_n - h - \Sigma(E - \omega_n) + i\eta \right]^{-1}$$
$$(303)$$

where $\omega_n = E_n^{v*} - E_0^{v*}$ is the excitation energy of the target and $\epsilon = E - E_0^*$ is the kinetic energy of photoelectrons in the normal ionization process $|0\rangle|c\rangle \rightarrow |0_v^*\rangle$. The optical potential is responsible for the damping of photoelectrons during the propagation in solids.

The off-diagonal part of $G(E)$ is systematically calculated by use of the diagonal Green's function expansion as shown by Eq. (169), which yields

$$\langle 0_v^* | G(E) | n_v^* \rangle = g^c(\epsilon) \langle 0 | V | n \rangle g^c(\epsilon - \omega_n)$$
$$+ \sum_{j(\neq 0, n)} g^c(\epsilon) \langle 0 | V | j \rangle g^c(\epsilon - \omega_j)$$
$$\times \langle j | V | n \rangle g^c(\epsilon - \omega_n) + \cdots \quad (304)$$

Finally we obtain a useful formula for the study of EELFS and ELNES spectra,

$$I(\mathbf{p}', \mathbf{p}) = -2 \operatorname{Im} \langle c | u^* g^c(\epsilon) u | c \rangle |S_0|^2$$
$$- 2 \sum_{n>0} \operatorname{Im} \left[\langle c | u^* g^c(\epsilon - \omega_n) V_{n0} g^c(\epsilon) u | c \rangle S_n^* S_0 \right.$$
$$+ \langle c | u^* g^c(\epsilon) V_{0n} g^c(\epsilon - \omega_n) u | c \rangle S_n S_0^* \left. \right]$$
$$- 2 \sum_{n>0} \operatorname{Im} \langle c | u^* g^c(\epsilon - \omega_n) u | c \rangle |S_n|^2$$
$$- 2 \sum_{n \neq m > 0} \operatorname{Im} \left[\langle c | u^* g^c(\epsilon - \omega_n) \right.$$
$$\times V_{nm} g^c(\epsilon - \omega_m) u | c \rangle S_m S_n^* \left. \right] - \cdots \quad (305)$$

where $V_{nm} = \langle n_v^* | V_{es} | m_v^* \rangle$ is the extrinsic loss amplitude for the transition, $m_v^* \rightarrow n_v^*$. The first term of Eq. (305) describes the normal EELS intensity without extrinsic and intrinsic losses of the photoelectrons. This term is the most important. The second term describes the interference between the intrinsic and the extrinsic losses, the third term describes the intrinsic losses, and the last term describes the higher order terms, which includes both the intrinsic and the extrinsic losses at least in third order. Equation (305) is useful for practical purposes, because this formula can properly describe the damping of photoelectrons during propagation in solids.

5.2. Suppression of Loss Structures

EELFS spectra are usually analyzed without considering the interference term (the second term of Eq. (305)). In this approximation we can expect an abrupt jump of the EELS intensity of the additional excitation of outer electrons [59]. However, such structures have not been observed in EELFS and XAFS spectra.

In the case of XPS spectra the destructive interference between the intrinsic and the extrinsic losses has been well established from experimental results for plasmon losses [60, 61]. For excitation by low-energy photons, the loss spectra are featureless and are lost in the background; the two losses are canceled at the threshold because of the strong quantum interference. A similar situation can be expected for EELFS and XAFS: for XAFS the subtle cancellation of these losses has been explained based on many-body scattering theory [39, 57, 62].

To study this problem in detail, Eq. (305) is not convenient because the imaginary part of the optical potential $\operatorname{Im} \Sigma$ has some characteristic features around the loss threshold [37]. The effective one-electron operator $\hat{G}$ for photoelectrons in Eq. (302) can be written in terms of the core hole excitation operator X_c defined by

$$X_c = \sum_{n>0} |n_v^*\rangle (S_n/S_0) \langle 0_v^*| \quad (306)$$

as

$$\hat{G} = |S_0|^2 \langle 0_v^* | G(E) + X_c^\dagger G(E) + G(E) X_c + X_c^\dagger G(E) X_c | 0_v^* \rangle$$
$$(307)$$

By noting that

$$P X_c = X_c Q = 0$$
$$P V P = 0 (P = P_0 = |0_v^*\rangle\langle 0_v^*|, \; Q = 1 - P)$$

and

$$X_c P = Q X_c = X_c$$

the relation

$$PGP = PG_0P + PG_0VGVG_0P \quad (308)$$

is obtained, where $G_0(E) = (E - H_v^* - h + i\eta)^{-1}$. Finally a simple formula to describe the cancellation of loss structures is obtained,

$$\hat{G}(E) = |S_0|^2 \langle 0_v^* | G_0 + (G_0 V + X_c^\dagger) G (V G_0 + X_c) | 0_v^* \rangle \quad (309)$$

As demonstrated in several papers on XPS theory, the interference term $(V G_0 + X_c)|0_v^*\rangle \Delta|c\rangle$ is responsible for the suppression of the loss structures when the kinetic energy of a photoelectron is comparable to the loss excitation energy, $\epsilon_k \approx \omega_n$ [32, 38, 63, 63a, 64, 64a]. Therefore we can expect that the suppression of the loss structures at threshold is also found in EELFS and XAFS spectra. This result does not mean that unknown peaks, which cannot be analyzed within a one-electron approximation, can be safely interpreted in terms of shake effects as shake-up or shake-off processes.

From the above discussion, in the ELNES region ($\epsilon < 50$ eV), the loss intensity is approximated by

$$I(\mathbf{p}', \mathbf{p}) \approx -2 \operatorname{Im} \langle c | u^* \hat{g}^c(\epsilon) u | c \rangle \quad (310)$$

where $\hat{g}^c(\epsilon)$ is also a core hole one-electron Green's function without optical potential, that is, without damping. On the other hand, in the EXELFS region ($\epsilon > 50$ eV) the loss intensity is approximated by neglecting the interference terms in Eq. (305)

$$I(\mathbf{p}', \mathbf{p}) \approx -2 \operatorname{Im} \sum_n \langle c | u^* g^c(\epsilon - \omega_n) u | c \rangle |S_n|^2 \quad (311)$$

In the above equation g^c describes the damping.

5.3. Multiple Scattering Expansion of a Secondary Excited Electron

We first consider the multiple scatterings of photoelectrons, which cause oscillations in the EELS intensity compared with XAFS. In Section 5.4 we also consider the multiple scatterings of probe electrons.

5.3.1. EELFS Spectra

Now we separate the core hole optical potential Σ_{0*} into a real part $\operatorname{Re} \Sigma_{0*}$ and an imaginary part $-i\Gamma_{0*}$. For the very high energy case ($\epsilon_p > 200$ eV), Γ_{0*} can be constant; $\Gamma_{0*} = 0$ outside the solid and $\Gamma_{0*} > 0$ inside. We also separate the effective one-electron real potential $\langle 0* | V_{es} | 0* \rangle + \operatorname{Re} \Sigma_{0*}$ into each atomic scattering potential

$$\langle 0* | V_{es} | 0* \rangle + \operatorname{Re} \Sigma_{0*} = \sum_\alpha v_\alpha \quad (312)$$

where v_α is a one-electron effective potential centered on site α. It is Hermitian, nonlocal, and approximately spherically symmetric, and depends parametrically on energy. The core orbital ϕ_c is strongly localized at site A, and it is sufficient to pick up the following expansion of g^c,

$$g^c = g_A^c + \sum_{\alpha \neq A} g_A^c t_\alpha g_A^c + \sum_{\beta \neq \alpha \neq A} g_A^c t_\beta g_0 t_\alpha g_A^c + \cdots \quad (313)$$

where $t_\alpha = v_\alpha + v_\alpha g_0 t_\alpha$ is the site T matrix st α (Section 2.3.1, Eq. (127)), and the core Green's function at site A is given by

$$g_A^c = g_0 + g_0 t_A g_0 = (\epsilon - T_e - v_A + i\eta)^{-1} \quad (314)$$

The free propagator g^0 now includes the imaginary part of the potential $g^0(\epsilon) = (\epsilon - T_e + i\Gamma_{0*})^{-1}$, which can describe the damping of the electron propagation.

First we consider the EXELFS spectra. The direct loss intensity I^0 where a photoelectron suffers no elastic scattering is calculated by use of Eq. (310) or (311). We can neglect the energy difference between $g_A^c(\epsilon)$ and $g_A^c(\epsilon - \omega_n)$ in Eq. (311), because it contributes to the atomic excitation cross section which shows smooth variation in energy. We thus have the atomic loss intensity by noting that $\sum_n |S_n|^2 = 1$,

$$I^0(\mathbf{p}', \mathbf{p}) = -2 \operatorname{Im}[\langle c | u^* g_A^c(\epsilon) u | c \rangle] \quad (315)$$

which can be written in terms of partial waves $R_l(kr) Y_L(\hat{\mathbf{r}})$ at site A ($\epsilon = k^2/2$),

$$I^0(\mathbf{p}', \mathbf{p}) = 4k \sum_L |\langle R_l Y_L | u | c \rangle|^2 \quad (316)$$

The single scatterings loss intensity with regard to photoelectrons is given with the use of Eq. (311) and the second term of Eq. (313) as

$$I^1(\mathbf{p}', \mathbf{p})$$
$$= -8 \sum_n |S_n|^2 \operatorname{Im} \sum_{\alpha(\neq A)} \langle c | u^* g_A^c(\epsilon - \omega_n) t_\alpha g_A^c(\epsilon - \omega_n) u | c \rangle \quad (317)$$

In contrast to I^0 calculation, each channel gives EXAFS-like rapid oscillation, and we have to explicitly keep the energy difference in g_A^c. For the spherically symmetric potential at each site, I^1 is explicitly written in terms of phase shifts at A and α,

$$I^1(\mathbf{p}', \mathbf{p}) = -4 \sum_n |S_n|^2 \operatorname{Im} \Bigg\{ \sum_{L_1} \sum_{L_2} \sum_{L_3} i^{l_3 - l_1} \langle c | u^* | R_{l_3} Y_{L_3} \rangle$$
$$\times \exp[i(\delta_{l_3}^A + \delta_{l_1}^A)]$$
$$\times \sum_\alpha G_{L_3 L_2}(\kappa_n; -\mathbf{R}_\alpha) t_{l_2}^\alpha G_{L_2 L_1}(\kappa_n; \mathbf{R}_\alpha)$$
$$\times \langle R_{l_1} Y_{L_1} | u | c \rangle \Bigg\} \quad (318)$$

where κ_n is the principal value of $\sqrt{2(\epsilon - \omega_n + i\Gamma)}$ and $k_n = \sqrt{2(\epsilon - \omega_n)}$, $R_l = R_l(k_n r)$. The T matrix t_l^α is given in terms of phase shifts as Eq. (87). Here $G_{LL'}$ denotes the damping outgoing propagator in an angular momentum representation as shown in Eq. (105). Higher order scattering terms such as double scattering can easily be calculated in the same way; here the double scattering term is shown by

$$I^2(\mathbf{p}', \mathbf{p}) = -4 \sum_n |S_n|^2 \sum_{L_1 L_2} \sum_{L_3 L_4} \sum_{\alpha \neq \beta} \operatorname{Im} \Big\{ \langle c | u^* | R_{l_4} Y_{L_4} \rangle$$
$$\times \exp[i(\delta_{l_4}^A + \delta_{l_1}^A)] i^{l_4 - l_1}$$
$$\times G_{L_4 L_3}(\kappa_n; -\mathbf{R}_\beta) t_{l_3}^\beta G_{L_3 L_2}(\kappa_n; \mathbf{R}_\beta - \mathbf{R}_\alpha) t_{l_2}^\alpha$$
$$\times G_{L_2 L_1}(\kappa_n; \mathbf{R}_\alpha) \langle R_{l_1} Y_{L_1} | u | c \rangle \Big\} \quad (319)$$

In Eqs. (318) and (319) we have to include a lot of partial waves in the EXELFS region. The direct calculation of I^1 and

I^2 with the use of these formulas is not practical. In the next subsection we discuss a separable formula that is useful in the high-energy region $kR \gg 1$.

5.3.2. z Axis Propagator

The angular momentum representation of the propagator $G_{LL'}$ in Eq. (105) has characteristic features if **R** is on the z axis, as pointed out by Fritsche [65] and Rehr and Albers [66]. First we note that from Eq. (105),

$$g_{ll'}^m(k, R) \equiv G_{LL'}(kR\hat{z})$$
$$= -\sqrt{4\pi} k \sum_{l_1} i^{l_1+1} h_{l_1}(kR) \sqrt{2l_1 + 1} G(l_1 0 L | L') \quad (320)$$

We can show the symmetric relations

$$g_{ll'}^m = g_{ll'}^{-m} = g_{l'l}^m \quad (321)$$

and from these relations we can restrict our discussion to the case $l' \geq l \geq m \geq 0$. With the aid of the recurrence relations (119) and (123), we have

$$\frac{1}{kR} g_{ll'}^l(k, R) = i\sqrt{\frac{2l+2}{2l+3}} \frac{g_{l+1,l'}^{l+1}(k, R)}{\sqrt{(l'-l)(l'+l+1)}} \quad (322)$$

Rehr and Albers have derived a very useful separable formula starting from Eq. (320) [66]. The spherical Hankel function is written as the integral formula

$$h_l(z) = -i^{-l} \int_1^{i\infty} \exp(izt) P_l(t) \, dt$$

By analytical continuation of the interval $|t| \leq 1$ to all complex planes, we have from Eq. (320)

$$g_{ll'}^m(k, R) = 4\pi i k \int_1^{i\infty} \exp(ikRt) Y_{lm}^*(\theta, \phi) Y_{l'm}(\theta, \phi) \, dt$$
$$= 4\pi i k N_{lm} N_{l'm} J_{ll'}^m \quad (323)$$

$$J_{ll'}^m = \int_1^{i\infty} \exp(ikRt)(1 - t^2)^m \tilde{P}_{lm}(t) \tilde{P}_{l'm}(t) \, dt \quad (324)$$

where $t = \cos\theta$. As $m = m'$, the above integral does not depend on ϕ. The factor N_{lm} is given by

$$N_{lm} = (-1)^m \sqrt{\frac{(2l+1)(l-m)!}{4\pi(l+m)!}} \qquad (m \geq 0)$$

and $\tilde{P}_{lm}(t)$ is a polynomial defined by $\tilde{P}_{lm}(t) = d^m/dt^m P_l(t)$. The Cauchy theorem can change the line integral $\int_1^{i\infty}$ to $\int_1^{-\infty}$. Thus we rewrite Eq. (324) as a Laplace transform,

$$J_{ll'}^m = -\frac{\exp(i\rho)}{i\rho} I(z)$$

$$I(z) = \frac{1}{z} \int_0^\infty \exp(-x/z)(2x - x^2)^m$$
$$\times \tilde{P}_{lm}(1 - x) \tilde{P}_{l'm}(1 - x) \, dx \quad (325)$$

where $\rho = kR$ and $z = 1/i\rho$. The function $I(z)$ is considered as a Laplace transform of $a(x)b(x)$, where $a(x) =$

$(2x - x^2)^m \tilde{P}_{lm}(1 - x)$ and $b(x) = \tilde{P}_{l'm}(1 - x)$,

$$I(z) = \frac{1}{z} \int_0^\infty \exp(-x/z) a(x) b(x) \, dx \quad (326)$$

When we define the Laplace transform of a and b,

$$A(z) = \frac{1}{z} \int_0^\infty \exp(-x/z) a(x) \, dx$$
$$B(z) = \frac{1}{z} \int_0^\infty \exp(-x/z) b(x) \, dx \quad (327)$$

a and b are given by the inverse Laplace transform with the use of some appropriate closed contour Γ,

$$a(x) = -\frac{1}{2\pi i} \int_\Gamma \frac{dz}{z} A(z) \exp(x/z)$$
$$b(x) = -\frac{1}{2\pi i} \int_\Gamma \frac{dz}{z} B(z) \exp(x/z) \quad (328)$$

The substitution of Eq. (328) into Eq. (326) yields the integral representation

$$I(z) = \int_\Gamma \frac{dz_1}{2\pi i} \frac{dz_2}{2\pi i} \frac{A(z_1) B(z_2)}{z_1 z_2 - z_1 z - z_2 z}$$
$$= \int_\Gamma \frac{dz_1}{2\pi i} \frac{A(z_1) B(z + z^2/(z_1 - z))}{z_1 - z} \quad (329)$$

From Eq. (322) we can relate $g_{ll'}^l$ to $g_{0l'}^0$

$$g_{ll'}^l = z^l \sqrt{\frac{(2l+1)!!(l+l')!}{(2l)!!(l'-l)!}} g_{0l'}^0 \quad (330)$$

and $g_{0l'}^0$ is simply $-\exp(i\rho)\sqrt{2l'+1} c_{l'}(z)/R$, where the lth polynomial $c_l(z)$ is defined by Eq. (117). Noting that $\tilde{P}_{ll}(x) = (2l-1)!!$, we can relate it to $A(z)$ from Eq. (325),

$$g_{ll'}^l = -\frac{\exp(i\rho)}{\rho} \sqrt{\frac{(2l+1)(2l'+1)(l'-l)!}{(2l)!(l'+l)!}} (2l-1)!! k A(z)$$
$$(l' \geq l) \quad (331)$$

The comparison of these two different expressions (330) and (331) gives an explicit formula of $A(z)$ in terms of $c_{l'}$, after we interpret l as m, and l' as l,

$$A(z) = z^m c_l(z) \frac{(l+m)!}{(l-m)!} \quad (332)$$

We can calculate $B(z)$ directly from the definition (327):

$$B(z) = (-1)^m \sum_{r=m}^l P_{l,r} z^{r-m}$$
$$P_{l,r} = \frac{(-1)^r (l+r)!}{2^r r! (l-r)!} \quad (333)$$

We note that $c_l(z) = \sum_{r=0}^l P_{l,r} z^r$. Substituting Eqs. (332) and (333) into Eq. (329), we obtain a closed expression of $I(z)$ in terms of $c_l(z)$'s and their derivatives [66],

$$I(z) = (-1)^m \frac{(l+m)!}{(l-m)!} \sum_{r=0}^{\lambda} \frac{c_l^{(r)}(z)c_{l'}^{(m+r)}(z)}{r!(m+r)!} z^{m+2r} \quad (334)$$

where $\lambda = \min[l, l' - m]$. The z axis propagator $g_{ll'}^m$ is thus separable as

$$g_{ll'}^m = \frac{\exp(ikR)}{R} \sum_{r=0}^{\lambda} \tilde{\gamma}_{mr}^l(z) \gamma_{mr}^{l'}(z)$$

$$\tilde{\gamma}_{mr}^l(z) = \frac{2l+1}{N_{lm}} \frac{c_l^{(r)}(z)}{r!} z^r \quad (335)$$

$$\gamma_{mr}^{l'}(z) = N_{l'm}(-1)^{m+1} \frac{c_{l'}^{(m+r)}(z)}{(m+r)!} z^{m+r}$$

The integral representation (112) helps us to write $G_{LL'}$ in terms of the z axis propagator by use of the Euler rotation Ω and the rotation matrix D (Wigner D function), which rotates the z axis to the $\mathbf{R}$ direction,

$$G_{LL'}(k\mathbf{R}) = \sum_s D_{ms}^{(l)}(\Omega) D_{m's}^{(l')}(\Omega)^* g_{ll'}^s(k, R)$$

$$= \frac{\exp(ikR)}{R} \sum_{rs} \tilde{\Gamma}_{sr}^L(z) \Gamma_{sr}^{L'}(z)$$

$$\tilde{\Gamma}_{sr}^L(z) = D_{ms}^{(l)}(\Omega) \tilde{\gamma}_{|s|,r}^l(z)$$

$$\Gamma_{sr}^{L'}(z) = D_{m's}^{(l')}(\Omega)^* \gamma_{|s|,r}^{l'}(z) \qquad (r \geq 0) \quad (336)$$

Each of the (s, r) terms is in the order of $z^{|s|+2r}$.

As an example, we calculate the amplitude $G(\mathbf{R}_\gamma - \mathbf{R}_\beta)t^\beta \times G(\mathbf{R}_\beta - \mathbf{R}_\alpha)$ with the use of the separable formula (336),

$$\sum_{L'} G_{L_1 L'}(\mathbf{R}_\gamma - \mathbf{R}_\beta) t_{l'}^\beta G_{L'L}(\mathbf{R}_\beta - \mathbf{R}_\alpha)$$

$$= \frac{\exp[ik(R_{\gamma\beta} + R_{\beta\alpha})]}{R_{\gamma\beta} R_{\beta\alpha}}$$

$$\times \sum_{sr} \sum_{s'r'} \tilde{\Gamma}_{s'r'}^{L_1}(z_{\gamma\beta}) F_{s'r',sr}^\beta(z_{\gamma\beta}, z_{\beta\alpha}) \Gamma_{sr}^L(z_{\beta\alpha}) \quad (337)$$

where

$$F_{s'r',sr}^\beta(z_{\gamma\beta}, z_{\beta\alpha}) = \sum_{L'} \Gamma_{s'r'}^{L'}(z_{\gamma\beta}) t_{l'}^\beta \tilde{\Gamma}_{sr}^{L'}(z_{\beta\alpha}) \quad (338)$$

is the spherical wave scattering amplitude at site β in the $\alpha \rightarrow \beta \rightarrow \gamma$ scattering process. Here we define $R_{\alpha\beta} = |\mathbf{R}_\gamma - \mathbf{R}_\beta|$, $z_{\gamma\beta} = 1/(ikR_{\gamma\beta})$. The lowest order (the most important) term is $(s, r) = (0, 0)$, and the next order terms are $(s, r) = (\pm 1, 0)$ under the condition $kR \gg 1$. For example, $F_{00,00}^\beta(z_{\gamma\beta}, z_{\beta\alpha})$ is the most important and is quite similar to the plane wave scattering amplitude shown by Eq. (37),

$$F_{00,00}^\beta(z_{\gamma\beta}, z_{\beta\alpha}) = F(\gamma\beta\alpha; k)$$

$$= -\sum_l (2l+1) t_l^\beta(k) P_l(\cos\theta_{\gamma\beta\alpha})$$

$$\times c_l(z_{\gamma\beta}) c_l(z_{\beta\alpha}) \quad (339)$$

where $\theta_{\gamma\beta\alpha}$ is the scattering angle at site β in the scattering process. When $c_l(z)$ is replaced by 1, it is reduced to the ordinary plane wave scattering amplitude. Typically we find rapid convergence of the (s, r) expansion in the EXELFS region [66, 67]. The damping effect is introduced through κ_n.

For simplicity only I^1 and I^2 are explicitly shown here in the lowest order Rehr–Albers expansion. With these approximations we have

$$I^1 = 4 \sum_n |S_n|^2 \text{Im}\left\{ \sum_\alpha \frac{\exp(2i\kappa_n R_\alpha)}{R_\alpha^2} \right.$$

$$\times \sum_{L_1} \sum_{L_2} \langle c|u^\dagger|L_2\rangle \exp[i(\delta_{l_2}^A + \delta_{l_1}^A)]$$

$$\times Y_{L_2}^*(-\hat{\mathbf{R}}_\alpha) i^{l_2-l_1} c_{l_2}(z_\alpha) F(A\alpha A; \kappa_n) c_{l_1}(z_\alpha)$$

$$\left. \times Y_{L_1}(\hat{\mathbf{R}}_\alpha) \langle L_1|u|c\rangle \right\} \quad (340)$$

$$I^2 = 4 \sum_n |S_n|^2 \sum_{\alpha\neq\beta} \text{Im}\left\{ \frac{\exp[i\kappa_n(R_\beta + R_{\beta\alpha} + R_\alpha)]}{R_\beta R_{\beta\alpha} R_\alpha} \right.$$

$$\times \sum_{L_1 L_2} \langle c|u^\dagger|L_2\rangle \exp[i(\delta_{l_2}^A + \delta_{l_1}^A)] i^{l_2-l_1}$$

$$\times Y_{L_2}^*(-\hat{\mathbf{R}}_\beta) c_{l_2}(z_\beta) c_{l_1}(z_\alpha) F(A\beta\alpha; \kappa_n) F(\beta\alpha A; \kappa_n)$$

$$\left. \times Y_{L_1}(\hat{\mathbf{R}}_\alpha) \langle L_1|u|c\rangle \right\} \quad (341)$$

where $|L_2\rangle = |R_{l_2} Y_{L_2}\rangle$ and $|c\rangle = |\phi_c\rangle$.

Rehr and his coworkers have developed a standard computer code called FEFF for XAFS analyses based on Rehr–Albers expansion [147, 148]. They used efficient path-by-path multiple scattering calculations. In principle we can apply this code to EXELFS analyses; however, as discussed in Section 5.4, probe electrons suffer elastic scatterings in addition to core excitation loss, which raises a problem with regard to σ_0.

5.3.3. ELNES Spectra

Rehr–Albers expansion is very useful for the study of EXELFS spectra where excited EXAFS electrons have quite high energy ($kR \gg 1$). On the other hand, in the ELNES region ($kR \approx 1$), the convergence of the Rehr–Albers expansion shows slow convergence. In this energy region, the number of partial waves to be taken into account is also small. Scatterings from surrounding atoms are strong enough that the multiple scattering renormalization is crucial. From Eqs. (318) and (319), the full multiple scattering resummation ELNES formula is obtained in the closed form as a typical XANES formula [24, 25, 57],

$$I^\infty(\mathbf{p}', \mathbf{p}) = I^0(\mathbf{p}', \mathbf{p}) + I^1(\mathbf{p}', \mathbf{p}) + I^2(\mathbf{p}', \mathbf{p}) + \cdots$$

$$= -4 \text{Im} \sum_{LL_1} \{i^{l_1-l} (t_{l_1}^A)^{-1} \exp[i(\delta_{l_1}^A + \delta_l^A)]$$

$$\times \langle c|u^\dagger|L_1\rangle [(1-X)^{-1}]_{L_1L}^{AA}$$

$$\left. \times \langle L|u|c\rangle \right\} \quad (342)$$

where the matrix $X^{\alpha\beta}$ is defined by

$$X^{\alpha\beta}_{L_1 L_2} = t^{\alpha}_{l_1}(k) G_{L_1 L_2}(k\mathbf{R}_\alpha - k\mathbf{R}_\beta)(1 - \delta^{\alpha\beta}) \quad (343)$$

This matrix element describes a physical process where a photoelectron propagates from site β with angular momentum L_2 to site α and is scattered with angular momentum L_1. In the EXELFS region core electrons play an important role in backscattering, whereas outer electrons play an important role in the multiple scatterings in the ELNES region (see also Section 4.4). The T matrix reflects the electronic structure, whereas G reflects the geometric structure. Thus ELNES spectra can give us useful information on electronic and geometric structures around the excited atom.

As discussed in Section 5.2, the loss effects of ELNES electrons are not so important. It is thus sufficient to consider only the main channel.

The inverse matrix $(1 - X)^{-1}$ includes infinite-order full multiple scatterings inside a cluster. In some cases this direct calculation is formidable, because the dimension of the matrix X amounts to 1000 for a cluster composed of 40 atoms and for a partial wave sum up to g wave. Equation (342) shows that we need only a small part of the large matrix $(1 - X)^{-1}$ to calculate ELNES spectra. We now separate the atoms in the cluster into two groups, a near region including the core excited atom A (group 1) and a far region (group 2). For example, group 1 has only the atom A, or in the other cases it includes the atom A and its nearest neighbors. We can divide the matrix $1 - X$ according to the above partitioning as

$$1 - X = \begin{pmatrix} A_{11} & A_{12} \\ A_{21} & A_{22} \end{pmatrix}$$

Correspondingly we can write the matrix $(1 - X)^{-1}$ as

$$(1 - X)^{-1} = \begin{pmatrix} \bar{A}_{11} & \bar{A}_{12} \\ \bar{A}_{21} & \bar{A}_{22} \end{pmatrix}$$

The submatrix that we have to calculate is only $\bar{A}_{11}$, and it is written in terms of the submatrix A_{ij}'s,

$$\bar{A}_{11} = \left(A_{11} - A_{12} A_{22}^{-1} A_{21}\right)^{-1}$$

From the above relation we have an expression for the submatrix $\bar{A}_{11}$,

$$\left[(1 - X)^{-1}\right]^{AA} = \left(\left[(1 - X)^{11} - X^{12}\{(1 - X)^{-1}\}^{22} X^{21}\right]^{-1}\right)^{AA} \quad (344)$$

We can expand $[(1 - X)^{-1}]^{22}$ in the power series of X; we should note that the power series is already in the inverse $[\cdots]^{-1}$, so that the low order expansion still includes a infinite partial sum of the multiple scatterings [24, 25]. Numerical calculations demonstrated the efficiency; for example, the computation time is only about 5% of that for the full multiple scattering calculation for a cluster with 27 Fe atoms [25].

5.4. Multiple Scattering Expansion of a Probe Electron

5.4.1. Basic Formulas

Both Eqs. (310) and (311) can be used for all excitation operators u; for example, we can consider the X-ray absorption processes. Hereafter we will specify the operator u for high-energy electron impact excitation. The basic ingredients needed to calculate u are the damping wave functions $\psi^+_{\mathbf{p}}$ and $\psi^-_{\mathbf{p}'}$ for probe electrons in the solid; they are calculated by use of the site T matrix expansion as discussed in Section 2.3.3. From Eq. (126) both $\psi^+_{\mathbf{p}}$ and $\psi^-_{\mathbf{p}'}$ are expanded,

$$\psi^+_{\mathbf{p}} = \sum_{i=0} \psi^{(i)}_{\mathbf{p}}, \qquad \psi^-_{\mathbf{p}'} = \sum_{i=0} \psi^{(i)}_{\mathbf{p}'} \quad (345)$$

where $\psi^{(0)}_{\mathbf{p}}$ is the damping plane wave (PW), whose wave vector $\mathbf{p}$ is given by $\hat{\mathbf{p}}(p_1 + ip_2) = \hat{\mathbf{p}}\bar{p}$ in simple PW $\phi^0_{\mathbf{p}}$; $\bar{p}$ is the principal value of $\sqrt{2(\epsilon_p + i\Gamma)}$. $\psi^{(1)}_{\mathbf{p}}$ includes single elastic scattering of the incident electron, $\psi^{(2)}_{\mathbf{p}}$ double elastic scatterings, and so on. The lowest order Rehr–Albers expansion for $G_{LL'}$ can express $\psi^{(1)}_{\mathbf{p}}$ as

$$\psi^{(1)}_{\mathbf{p}}(\mathbf{r}) = \sum_{\alpha(\neq A)} \bar{\psi}^{(0)}_{A\alpha p}(\mathbf{r}) F(A\alpha p; \bar{p}) \exp\{i(\bar{p}R_\alpha + \bar{\mathbf{p}} \cdot \mathbf{R}_\alpha)\}/R_\alpha \quad (346)$$

where $F(A\alpha p; \bar{p})$ is also the scattering amplitude with spherical wave correction, however, it includes only linear terms of c_l in comparison with $F(\gamma\beta\alpha)$ given by Eq. (339),

$$F(A\alpha p; \bar{p}) = -\sum_l (2l + 1) c_l(z_\alpha) t^{\alpha}_l(\bar{p}) P_l(\cos\theta_{A\alpha p}) \quad (347)$$

where $\theta_{A\alpha p}$ is the angle between $\mathbf{p}$ and $-\mathbf{R}_\alpha$ and z_α is defined by $i/(\bar{p}R_\alpha)$. The damping PW with spherical wave correction $\bar{\psi}^{(0)}_{A\alpha p}$ propagates toward $-\mathbf{R}_\alpha$ with kinetic energy ϵ_p, which is represented by

$$\bar{\psi}^{(0)}_{A\alpha p}(\mathbf{r}) = \sqrt{\pi/2} \sum_L i^l j_l(\bar{p}r) c_l(z_\alpha) Y^*_L(\hat{\mathbf{r}}) Y_L(-\hat{\mathbf{R}}_\alpha) \quad (348)$$

Higher order terms, such as the double scattering term, are given in the same way by

$$\psi^{(2)}_{\mathbf{p}}(\mathbf{r}) = \sum_{\alpha \neq \alpha'(\neq A)} \bar{\psi}^{(0)}_{A\alpha' p}(\mathbf{r}) F(A\alpha'\alpha, \bar{p}) F(\alpha'\alpha p; \bar{p})$$
$$\times \exp\{i\bar{p}(R_{\alpha'} + R_{\alpha'\alpha}) + i\bar{\mathbf{p}} \cdot \mathbf{R}_\alpha\}/(R_\alpha R_{\alpha'\alpha}) \quad (349)$$

We can also explicitly write down the single scattering function after the loss $\psi^{(1)}_{\mathbf{p}'}$ and the double scattering function $\psi^{(2)}_{\mathbf{p}'}$ and so on in the same approximation,

$$\psi^{(1)}_{\mathbf{p}'}(\mathbf{r}) = \sum_{\beta(\neq A)} \psi^{(0)}_{p'\beta A}(\mathbf{r}) \exp\{-i(\bar{p}'R_\beta + \bar{\mathbf{p}}' \cdot \mathbf{R}_\beta)\}$$
$$\times F(p'\beta A; \bar{p}')^*/R_\beta \quad (350)$$

$$\psi_{\mathbf{p}'}^{(2)}(\mathbf{r}) = \sum_{\beta \neq \beta'(\neq A)} \psi_{p'\beta'\beta}^{(0)}(\mathbf{r})$$
$$\times \exp\{-i\,\bar{p}'(R_\beta + R_{\beta\beta'}) + i\bar{\mathbf{p}}' \cdot \mathbf{R}_{\beta'}\}$$
$$\times F(p'\beta'\beta; \bar{p}')^* F(\beta'\beta A; \bar{p}')^* / (R_\beta R_{\beta\beta'}) \quad (351)$$

where $\bar{\mathbf{p}}'$ is defined by $\bar{\mathbf{p}}' = \hat{\bar{\mathbf{p}}}'\bar{p}'$ and $\bar{p}'$ is the principal value of $\sqrt{2(\epsilon_{p'} - i\Gamma_f)}$. Substituting Eq. (345) into Eq. (292), we can express u in the form

$$u(\mathbf{p}'; \mathbf{p}) = \sum_{j=1}^{\infty} A(\mathbf{p}'; \mathbf{p}) \quad (352)$$

$$\langle l|A(\mathbf{p}'; \mathbf{p})_1|m\rangle = \langle \mathbf{P}'^{(0)}l|\mathbf{p}^{(0)}m\rangle$$
$$\langle l|A(\mathbf{p}'; \mathbf{p})_2|m\rangle = \langle \mathbf{P}'^{(1)}l|\mathbf{p}^{(0)}m\rangle$$
$$\langle l|A(\mathbf{p}'; \mathbf{p})_3|m\rangle = \langle \mathbf{P}'^{(0)}l|\mathbf{p}^{(1)}m\rangle$$
$$\langle l|A(\mathbf{p}'; \mathbf{p})_4|m\rangle = \langle \mathbf{P}'^{(2)}l|\mathbf{p}^{(0)}m\rangle \quad (353)$$
$$\langle l|A(\mathbf{p}'; \mathbf{p})_5|m\rangle = \langle \mathbf{P}'^{(0)}l|\mathbf{p}^{(2)}m\rangle$$
$$\langle l|A(\mathbf{p}'; \mathbf{p})_6|m\rangle = \langle \mathbf{P}'^{(1)}l|\mathbf{p}^{(1)}m\rangle$$

where $\mathbf{p}^{(2)}$ is the abbreviation of $\psi_{\mathbf{p}}^{(2)}$, and so on. The A_1 term describes the direct deep core excitation inelastic scattering, whereas A_2 and A_3 describe the inelastic scattering at site A accompanied by single elastic scatterings after (A_2) or before (A_3) the inelastic scattering. A_4, A_5, and A_6 describe the second-order processes with regard to elastic scattering of the probe electrons. In the above expressions we neglect the bare exchange effect, which becomes negligibly small in the high energy excitation. Schematic representation of these terms is given in Figure 10.

We now have the explicit formula for the excitation operator $A(\mathbf{p}'; \mathbf{p})_1$,

$$A(\mathbf{p}'; \mathbf{p})_1 = \frac{\exp(-i\,\Delta\mathbf{p}_1 \cdot \mathbf{r})}{2\pi^2(\Delta p_1)^2} \exp\{-(l_A p_2 + l'_A p'_2)\} \quad (354)$$

where $\Delta\mathbf{p}_1$ is the real part of the momentum transfer, $\Delta\mathbf{p}_1 = \mathbf{p}'_1 - \mathbf{p}_1$, and l_A and l'_A are the distance between site A and the solid surface measured in the directions $\mathbf{p}_1$ and $\mathbf{p}'_1$.

In Eqs. (345), (348), (349), and (350), if we replace $\tilde{\psi}^{(0)}$ with $\psi^{(0)}$ in the core region and use the same approximation to derive Eq. (353), we obtain the explicit formulas for A_2 and A_3,

$$A(\mathbf{p}'; \mathbf{p})_2$$
$$= \frac{1}{2\pi^2} \sum_{\beta(\neq A)} \frac{\exp\{i R_\beta(\bar{p}' - p'_1 \cos\theta_{p'\beta A} - i\,\Delta\mathbf{p}_1 \cdot \mathbf{R}_A)\}}{R_\beta|\mathbf{p}'_{\beta A} - \mathbf{p}'_1|^2}$$
$$\times F(p'\beta A; \bar{p}')$$
$$\times \exp\{-(l_A p_2 + l'_\beta p'_2) - i(\mathbf{p}'_{\beta A} - \mathbf{p}_1) \cdot (\mathbf{r} - \mathbf{R}_A)\} \quad (355)$$

$$A(\mathbf{p}'; \mathbf{p})_3$$
$$= \frac{1}{2\pi^2} \sum_{\alpha(\neq A)} \frac{\exp\{i R_\alpha(\bar{p} - p_1 \cos\theta_{A\alpha p} - i\,\Delta\mathbf{p}_1 \cdot \mathbf{R}_A)\}}{R_\alpha|\mathbf{p}_{A\alpha} - \mathbf{p}_1|^2}$$
$$\times F(A\alpha p; \bar{p})$$
$$\times \exp\{-(l_\alpha p_2 + l'_A p'_2) - i(\mathbf{p}'_1 - \mathbf{p}_{A\alpha}) \cdot (\mathbf{r} - \mathbf{R}_A)\}$$

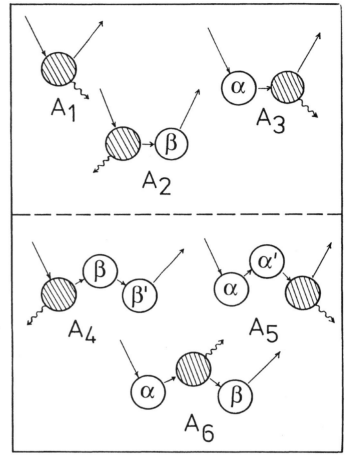

Fig. 10. The inelastic scattering amplitudes are schematically shown. The hatched (empty) circles stand for the atomic site at which the inelastic (elastic) scattering occurs. The arrows indicate the propagation of the electron waves. A_1 shows the direct inelastic scattering amplitude accompanied by no elastic scattering. $A_2(A_3)$ shows the one accompanied by a single elastic scattering after (before) the inelastic scattering. In the same way, A_4, A_5, and A_6 describe the second-order contributions; A_4, for example, shows the inelastic scattering amplitude with double elastic scattering after the deep core excitation.

where $\mathbf{p}'_{\beta A}$ and $\mathbf{p}_{A\alpha}$ are defined by $\mathbf{p}'_{\beta A} = \hat{\mathbf{R}}_\beta p'_1$ and $\mathbf{p}_{A\alpha} = -\hat{\mathbf{R}}_\alpha p_1$. The higher order terms such as A_4, A_5, etc., can easily be obtained in the same way.

By use of the expansion shown by Eq. (351), the intensity I^0 (see Eq. (316)) is represented by the sum of $\hat{A}_{ij}$'s, which are defined by

$$I^0 = \sum_{ij} \hat{A}_{ij}^0$$
$$\hat{A}_{ij}^0 = \frac{2k}{\pi} \sum_L \mathrm{Re}\{\langle \phi_c|A_i^\dagger|R_l Y_L\rangle\langle R_l Y_L|A_j|\phi_c\rangle\} \quad (356)$$

It is easy to show that $\hat{A}_{ij}^0 = \hat{A}_{ji}^0$. For simplicity we will consider the excitation from the core orbital $\phi_c(\mathbf{r}) = R_{l_c}(r)Y_{L_c}(\hat{\mathbf{r}})$ with angular momentum quantum number $L_c = (0, 0)$, that is, the excitation from K and L_1 edges and so on. In this simple

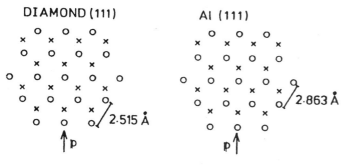

DIAMOND (111) **Al (111)**

2.515 Å

2.863 Å

↑p   p↑

o **1st layer**

× **2nd layer**

Fig. 11.   The surface structures of diamond (111) and Al (111) surfaces. The arrows indicate the directions of the incident electrons.

case $\hat{A}_{11}^0$ is given by

$$\hat{A}_{11}^0 = \hat{m}(0, k; \Delta\mathbf{p}_1, \Delta\mathbf{p}_1) \exp\{-2(l_A p_2 + l_A' p_2')\} \quad (357)$$

where $\hat{m}(0, k; \mathbf{p}, \mathbf{q})$ is defined by

$$\hat{m}(0, k; \mathbf{p}, \mathbf{q}) = \frac{32\pi k}{p^2 q^2} \sum_l (2l + 1)\rho(k, p)_l \rho(k, q)_l P_l(\cos\theta_{pq}) \quad (358)$$

In the above formula, θ_{pq} is the angle between $\mathbf{p}$ and $\mathbf{q}$, and $\rho(k, q)_l$ is the radial integral defined by

$$\rho(k, q)_l = \int R_l(kr) j_l(qr) R_0(r) r^2 \, dr \quad (359)$$

We can also obtain the explicit formulas for $\hat{A}_{12}$, $\hat{A}_{13}$, etc., which include the effects of elastic scattering of a probe electron. Here only those for $\hat{A}_{12}$ and $\hat{A}_{13}$ are shown:

$$\hat{A}_{21}^0 = 2 \sum_{\beta(\neq A)} \text{Re}\big[\exp\{-i R_\beta(\bar{p}' - p_1' \cos\theta_{p'\beta A})\}$$
$$\times F^*(p'\beta A; \bar{p}')\big]\hat{m}(l_c, k; \mathbf{p}_{\beta A}' - \mathbf{p}, \Delta\mathbf{p}_1)$$
$$\times \exp[-\{2l_A p_2 + (l_A' + l_\beta')p_2'\}]/R_\beta \quad (360)$$

$$\hat{A}_{31}^0 = 2 \sum_{\alpha(\neq A)} \text{Re}\big[\exp\{-i R_\alpha(\bar{p} - p_1 \cos\theta_{A\alpha p})\}$$
$$\times F^*(A\alpha p; \bar{p})\big]\hat{m}(l_c, k; \mathbf{p}' - \mathbf{p}_{A\alpha}, \Delta\mathbf{p}_1)$$
$$\times \exp[-\{2l_A' p_2' + (l_A + l_\alpha)p_2\}]/R_\alpha$$

The explicit formulas of $\hat{m}$ for $l_c = 1$ and 2 are given in [10]. In Eqs. (357) and (360), the damping effects of probe electrons are explicitly included as the exponential damping factors, which are important for the study of the surface sensitivity of EELFS spectra in a variety of modes (see Fig. 1).

5.4.2. Calculated Results

We discuss here the effects of elastic scatterings of probe electrons on EELS. First we discuss the grazing incidence and small takeoff angle EELS from diamond (111) and Al(111)

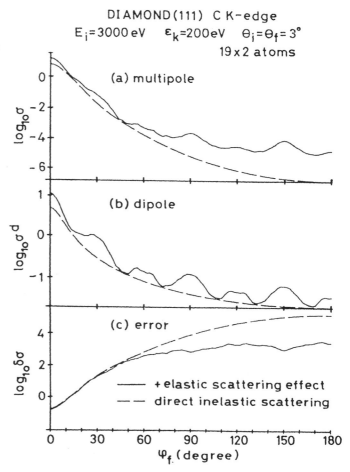

DIAMOND (111) C K-edge
$E_i = 3000\,\text{eV}$ $\epsilon_k = 200\,\text{eV}$ $\theta_i = \theta_f = 3°$
19×2 atoms

(a) multipole

(b) dipole

(c) error

—— + elastic scattering effect

- - - direct inelastic scattering

φ_f (degree)

Fig. 12.   The calculated carbon 1s EELS intensity from a diamond (111) surface as a function of ϕ_f [68]. In (b) we show the EELS intensity calculated in the dipole approximation, and in (a) we include the effects from the multipole excitation. In (c) the relative error of the dipole approximation is shown for the direct EELS and the EELS with single scatterings of probe electrons.

surfaces (see Fig. 1c). Their surface structures and the direction of incident electrons are shown in Figure 11. We use the cluster approximation to describe the processes of scattering from the surfaces. For small incident (θ_i) and takeoff angles (θ_f), $\theta_i = \theta_f = 3°$, the two-layer cluster model is sufficient; that is, this mode is very surface sensitive.

Figure 12 shows the calculated C 1s EELS intensity $\hat{A}_{11}^0 + 2(\hat{A}_{12}^0 + \hat{A}_{13}^0)$ from the diamond (111) surface as a function of ϕ_f under the conditions $\theta_i = \theta_f = 3°$, $\epsilon_p = 3$ keV, and $\epsilon_k = 200$ eV [68]. In Figure 12b the intensity of the dipole approximation, where we use only one term $l = 1$ in the sum (358), is shown, whereas in Figure 12a we include the effects from the multipole excitation processes in Eq. (358). In Figure 12c we show the relative error $\delta\sigma$, defined by

$$\delta\sigma = \big[I^0(\text{dipole}) - I^0(\text{multipole})\big]/I^0(\text{multipole})$$

The solid lines show the calculated results including the elastic scatterings of probe electrons $\hat{A}_{11}^0 + 2(\hat{A}_{12}^0 + \hat{A}_{13}^0)$, and the dashed lines show the results without the elastic scatterings $\hat{A}_{11}^0$

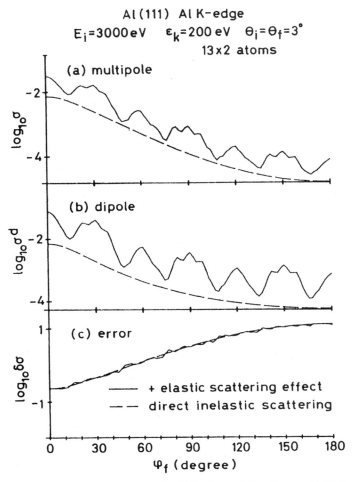

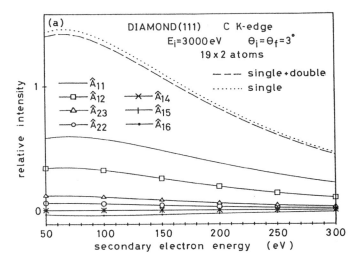

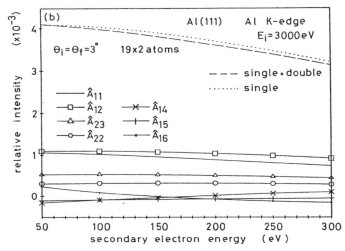

Fig. 13.   As in Figure 12, except for Al K-edge excitation from an Al (111) surface [68].

Fig. 14.   Contributions from $\hat{A}_{ij}^0$ terms as a function of ϵ_k for diamond (a) and Al (111) (b) surfaces. Multipole excitation processes are fully taken into account [12].

given by Eq. (356). The elastic scattering effects give some structures on the monotonic behavior to $\hat{A}_{11}^0$. However, their additional features for the dipole excitation are different from those for the multipole excitation. In the latter case, the elastic scattering has much influence at large ϕ_f because the dipole EELS intensity slowly decreases with ϕ_f in comparison with the multipole EELS intensity. The relative error $\delta\sigma$ is shown in Figure 12c. If we neglect the elastic scattering of probe electrons, the relative error $\delta\sigma$ increases monotonically with ϕ_f. On the other hand, it is much suppressed when the elastic scatterings are taken into account.

Next we consider the Al K-edge EELS from on Al(111) surface. The elastic scatterings are stronger in Al than in diamond, so we can expect more prominent effects in the EELS spectra from Al (111) surfaces. Figure 13 shows the same effects as Figure 12, except for Al [68]; the elastic scatterings of probe electrons give rise to large effects, as expected. The elastic scatterings produce similar structures in ϕ_f scan in both dipole and multipole excitation EELS, so that $\delta\sigma$ is nearly the same.

Next we discuss the energy scan mode of K-edge EELS from the diamond (111) and Al (111) surfaces [12]. Figure 14 repre-

sents the results for the forward scattering ($\theta_i = \theta_f = 3°$) mode (see Fig. 1c). Each contribution of $\hat{A}_{ij}^0$ terms is also shown. Figure 14a clearly shows that the main contribution arises from the direct inelastic scattering term $\hat{A}_{11}^0$. All of these terms $\hat{A}_{ij}^0$ are featureless functions of ϵ_k irrespective of their oscillating factor, say $R_\beta(\bar{p}' - p_1' \cos\theta_{p'\beta A})$ in $\hat{A}_{12}^0$ given by Eq. (360). Under this small-angle scattering condition, most of the scattering angles $\theta_{p'\beta A}$ contributing to $\hat{A}_{12}^0$ are small enough, and the above phase factor is not important. Therefore we cannot observe the plural oscillation in this geometrical setup. The three terms containing double scatterings, $\hat{A}_{14}^0$, $\hat{A}_{15}^0$, $\hat{A}_{16}^0$, are much smaller than the other terms because the elastic scatterings from carbon atoms are very weak. In comparison with the EELS from the diamond (111) surface, the elastic scatterings are more important for the EELS from the Al surface. In this case $\hat{A}_{12}^0$ is slightly larger than the direct loss intensity $\hat{A}_{11}^0$, but the three terms, $\hat{A}_{14}^0$, $\hat{A}_{15}^0$, $\hat{A}_{16}^0$, are small enough to be neglected. As in

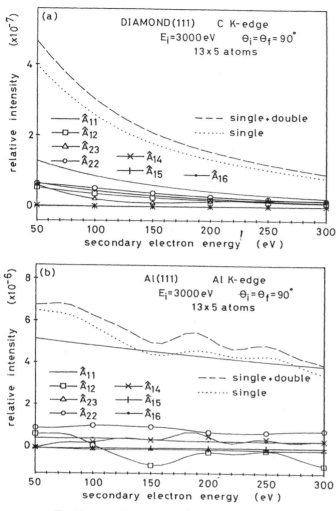

Fig. 15. As in Figure 14, but for the backscattering.

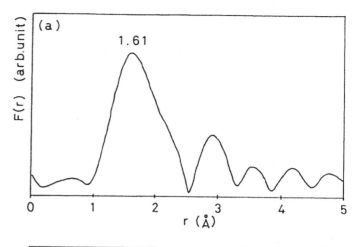

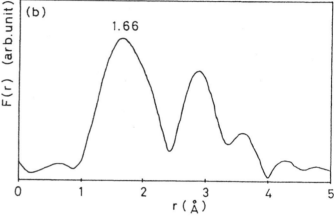

Fig. 16. Fourier transform of the spectra in Figure 15b after the subtraction of the monotonic part [12]. Only the single elastic scattering effects are included in (a), and the single and the double elastic scatterings are included in (b). The transform is performed in the range $50 < \epsilon_k < 300$ eV. The phase correction is applied in the transform.

the case of EELS from diamond, none of these terms show oscillatory behavior, for the reason given above.

Several workers have made use of a backscattering experimental setup (see Fig. 1b) [4, 69]. The EXELFS spectra observed in this mode have also been analyzed with the use of conventional EXAFS analyses; however, we have had no sound theoretical background for the analyses. Figure 15 shows the energy scan mode of (a) carbon and (b) aluminum K-edge EELS from diamond and Al (111) surfaces under backward reflection conditions ($\theta_i = \theta_f = 90°$), where ϵ_p is fixed at 3 keV but $\epsilon_{p'}$ is scanned [12]. We use five layers to obtain good convergence with regard to the cluster size. This setup is less surface sensitive than the mode in Figure 1c and discussed above. All of $\hat{A}_{ij}^0$'s are much smaller than the corresponding terms in Figure 14 because of the small atomic excitation matrix elements $\hat{m}$ in Eq. (358) in the large-angle scatterings. In Eq. (359) $\rho(k, \Delta p)_l$ should be small enough for large Δp in the spherical Bessel function $j_l(\Delta p)$. Although the large-angle scattering EELS are expected to give oscillation (see Eq. (360)), we do not find such oscillation for the C $1s$ EELS from diamond

(111) surfaces. This unexpected result is explained by the fact that the carbon atom is a very weak scatterer for fast electrons ($\epsilon_{p'} > 2.5$ keV). On the other hand, we find prominent oscillations in $\hat{A}_{12}^0$ and $\hat{A}_{14}^0$, and small oscillations in $\hat{A}_{22}^0$ and $\hat{A}_{23}^0$ for the backscattering EELS from Al (111) because the elastic scattering is not so weak.

In the EXELFS analyses, the Fourier transform is usually used to obtain the distance from an excited atom to surrounding atoms [4]. The EELFS oscillation is caused by the interference of the electron waves ejected from a deep core orbital, which is described by Eq. (318). However, as we have observed in Figure 15b, the interference between elastically scattered electron waves of probe electrons can show the other kind of oscillation, similar to XPD oscillation, particularly under the large-angle scattering condition [70]. We now apply the Fourier analyses to the EELS spectra shown in Figure 15b. If we misunderstand the oscillation as the EXELFS and apply the conventional Fourier transform technique to such an EELS spectrum, we may obtain incorrect information about the bond length. Figure 16a shows the result for the single scattering approximation, and

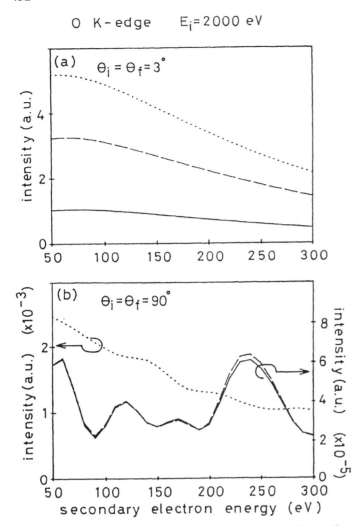

O K-edge $E_i = 2000$ eV

(a) $\theta_i = \theta_f = 3°$

(b) $\theta_i = \theta_f = 90°$

secondary electron energy (eV)

Fig. 17. As in Figure 14, but for p(2 × 2)O/Ni(001) (solid line), c(2 × 2)O/Ni(001) (dashed line), and Ni(001) (dotted line) surfaces in two detection modes: (a) $\theta_i = \theta_f = 3°$, (b) 90°. The EELS intensities are normalized to those in (a) [12].

Figure 16b shows the result for the single + double scattering approximation [12]. In Figure 16a we find a strong peak at 1.61 Å, and in Figure 16b we find a strong peak at 1.66 Å. If the oscillation is true EXELFS due to excited electrons from a deep core, we can obtain a true bond length of 2.01 Å. If the plural oscillation happens to give a realistic length, we could be led to an incorrect conclusion.

Calculated results for more complicated surfaces help to reveal the surface sensitivity of the grazing incidence and small-takeoff-angle EELS. Here we illustrate the calculated results for adsorbed systems p(2×2) and c(2×2) O/Ni(001) and a single crystalline NiO(001) surface [12]. Adsorbed oxygen atoms are on hollow sites with the shortest O-Ni distance (1.96 Å) for both systems. The calculated EELS spectra for probe electrons of 2 keV are shown in Figure 17, where (a) and (b) are spectra at $\theta_i = \theta_f = 3°$ and 90°. In the forward scattering shown in Figure 17a, the three spectra are smoothly decreasing func-

tions of ϵ_k. The EELS intensity reflects the density of oxygen atoms in the surfaces. Under this very small-angle scattering condition, the observed EELS is very sensitive to the first layer, and then EELS intensities for the adsorbed systems are not so small compared with that from the oxide surface. However, the EELS spectrum for the NiO surface shown in Figure 17b is quite different from those for c(2×2) and p(2×2) surfaces, and the spectra for the adsorbed systems are nearly the same. The EELS intensity for the oxide surface is about 100 times larger than those for the adsorbed surfaces. In the case of adsorbed systems, if the elastic scatterings from the surrounding oxygen atoms, which are in the first layer, mainly contribute to the EELS spectra, we would observe different behaviors between them because of different atomic arrangements around excited oxygen atoms. Therefore the elastic scattering not from O atoms but from Ni atoms in the substrate plays an important role in the backward reflection because Ni is a stronger scatterer than O. The relative atomic configurations between excited O and the nearby Ni atoms are the same in these adsorbed systems, and we find the same EELS spectra for the adsorbed systems. The plural oscillation is stronger than that found for Al surfaces because Ni is a much stronger scatterer than Al.

Other approaches to the elastic scatterings of probe electrons are also found in some of the literature; they are based on the long-range order theory in terms of Bloch waves [71–73]. Of course, this technique can be applied only to well-ordered surfaces.

Mila and Noguerra also proposed a short-range order theory that takes the single elastic scattering of fast electrons only after the core excitation [74]. Furthermore, they simplified the EELS formulas by averaging over 4π. Because of the presence of the surface, 4π averaging is impossible in practice and is an oversimplification.

5.4.3. EXELFS Formulas

The single elastic scattering intensity I^1 with regard to the photoelectron is also represented by the sum of each contribution of elastic scattering within the intrinsic approximation (see Section 5.2),

$$I^1(\epsilon) = \sum_n |S_n|^2 \sum_{ij} \hat{A}_{ij}^1(\epsilon - \omega_n) \qquad (361)$$

where each $\hat{A}_{ij}^1$ is given in the same way as defined by Eq. (356). Here we have to calculate $\hat{A}_{ij}^1$ for different channels of core hole states. For the excitation from a deep s orbital we obtain the explicit formula of $\hat{A}_{11}^1$ from Eqs. (339) and (340),

$$\hat{A}_{11}^1 = \frac{32\pi}{(\Delta p_1)^4} \sum_n |S_n|^2 \sum_{\gamma(\neq A)} \text{Im}[\exp(2i\kappa_n R_\gamma) F(A\gamma A; \kappa_n)$$
$$\times \Gamma_\gamma(\Delta\mathbf{p}_1; \kappa_n) \Gamma_\gamma(-\Delta\mathbf{p}_1; \kappa_n)]$$
$$\times \exp\{-2(l_A p_2 + l'_A p'_2)\}/R_\gamma^2 \qquad (362)$$

where $\Gamma_\gamma(\Delta \mathbf{p}_1; \kappa_n)$ is given by

$$\Gamma_\gamma(\Delta \mathbf{p}_1; \kappa_n) = \sum_l (2l + 1)i^{-l}P_l(\cos\theta_\gamma)\rho(k, \Delta p)_l$$
$$\times c_l(z_\gamma)\exp(i\delta_l^A) \qquad (363)$$

where θ_γ is the angle between the momentum transfer $\Delta\mathbf{p}$ and $\mathbf{R}_\gamma$. $\Gamma_\gamma(-\Delta\mathbf{p}_1; \kappa_n)$ can be obtained, if we replace $\cos\theta_\gamma$ with $-\cos\theta_\gamma$ in Eq. (363). In the very small Δp limit, we can show that the ratio $\hat{A}_{11}^1/\hat{A}_{11}^0$ is reduced to an ordinary EXAFS formula, which will be checked below. In this limit, $j_l(\Delta pr)$ is approximately given by $j_l(\Delta pr) \approx (\Delta pr)^l/(2l + 1)!!$, and $\rho(k, \Delta p)_0$ vanishes because of orthogonality between the s wave radial part $R_0(kr)$ and the core radial part $R_0^c(r)$. Then the main contribution arises from $\rho(k, \Delta p)_1$; Γ_γ is approximated by

$$\Gamma_\gamma(\Delta p; \kappa) \approx 3i^{-1}\cos\theta_\gamma\rho(k, \Delta p)_1 c_1(z_\gamma)\exp(i\delta_1^A) \quad (364)$$

After we substitute the approximate expression shown by Eq. (364) into Eq. (362), we obtain the ordinary EXAFS formula for the ratio $\hat{A}_{11}^1/\hat{A}_{11}^0$,

$$\hat{A}_{11}^1/\hat{A}_{11}^0$$
$$\approx -3\sum_n \frac{|S_n|^2}{\kappa_n}\sum_{\gamma(\neq A)}\cos^2\theta_\gamma\,\mathrm{Im}[\exp\{2i(\kappa_n R_\gamma + \delta_1^A)\}$$
$$\times F(A\gamma A; \kappa_n)c_1(z_\gamma)^2]/R_\gamma^2 \qquad (365)$$

The oscillating term Z_γ in EXELFS, which is given by the argument of Im in Eq. (362), is quite similar to the corresponding term $\tilde{Z}_\gamma$ in EXAFS, which is given by

$$\tilde{Z}_\gamma = -9\exp(2ikR_\gamma + 2i\delta_1^A)F(A\gamma A; \kappa_n)\rho(1)^2\cos^2\theta_\gamma$$
$$(366)$$

In the very small Δp limit, we have shown that Z_γ is reduced to $\tilde{Z}_\gamma$ because of the orthogonality between $R_0^c(r)$ and $R_0(kr)$. When they use high-energy ($\epsilon_p \geq 100$ keV) EELFS in transmission mode, this condition is satisfied, and they obtain the EELFS spectra equivalent to the corresponding XAFS spectra [3].

When we study EELFS spectra in the reflection mode, we should carefully investigate the energy and angular dependence of Z_γ [58]. Figure 18 shows the absolute value of the ratio of $(2l + 1)\rho(l)$ to $3\rho(1)$ as a function of a scattering angle for $l = 0$–6, where the fast probe electron (2400 eV) excites an oxygen $1s$ electron. The photoelectron kinetic energy ϵ_k is 50 eV in Figure 18a and 350 eV in Figure 18b; EELFS measurements are usually carried out for the energy range $\epsilon_k = 50$–350 eV. Because the relative contribution of the s-wave $|\rho(0)/3\rho(1)|$ at forward scattering is about 10% for $\epsilon_k = 50$ eV and about 5% for 350 eV, transition to the s-wave cannot be neglected, even in the forward scattering in these energy ranges. As the binding energy of an oxygen $1s$ orbital is 532 eV, Δp_1 is about 3–5 Å^{-1}, corresponding to $\epsilon_k = 50$–350 eV. The wave vector of X-ray plays the same role as Δp_1 in the X-ray absorption,

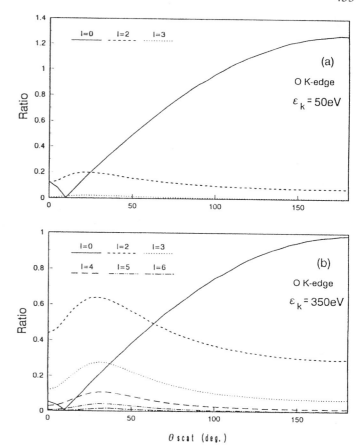

Fig. 18. The ratio $|(2l + 1)\rho(l)/3\rho(1)|$ as a function of the scattering angle for $\epsilon_k = 50$ eV (a) and 350 eV (b) [58].

whereas it is small (only 0.3–0.5 Å^{-1}) in the EXAFS energy region. In contrast to the forward scattering, the s-wave plays a more dominant role in the backscattering. A similar result has been obtained by Tomellini and Ascarelli [70]. Furthermore, we can see that $\rho(0)$ happens to vanish at a scattering angle of 10°. In the low-energy region ($\approx$50 eV), the excitation amplitude of the partial waves of $l = 2, 3, \ldots$ becomes negligibly small, and the amplitude of the s-wave disappears at this angle, so that the dipole approximation in the sense of the irreducible tensorial expansion works well (see Eq. (363)). Thus we can expect that the electron ELNES would be equivalent to the corresponding XANES under this condition. Here we have only studied $\hat{A}_{11}^1$; however, the $\hat{\Gamma}$ dependence is nearly the same in the multiple scattering ELNES formula (342) as in $\hat{A}_{11}^1$. An experimental proof for the above theoretical consideration is also shown later.

We are very familiar with the polarization-dependent anisotropy of EXAFS $\cos^2\theta_\gamma$ as shown by Eq. (365). This simple anisotropic behavior is due to the electric dipole transition approximation. In EELFS, however, several partial waves have nonnegligible contributions to Γ_γ as described above, so that the phase function of EELFS, $\arg Z_\gamma$, in Eq. (362) is different from that of EXAFS, $\arg \tilde{Z}_\gamma$. Furthermore, this phase func-

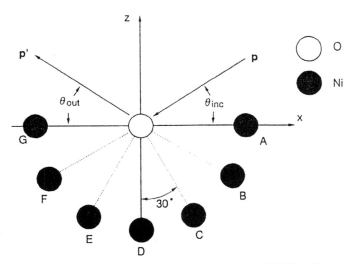

Fig. 19. A model system for the study of the anisotropy of EELFS. All atoms and **p** and **p**′ are in the same plane.

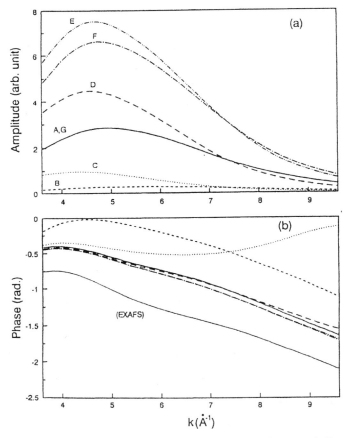

Fig. 20. $|Z_\alpha|$ and arg Z_α as a function of k for each scattering atom A–G, as shown in Figure 19; $\theta_i = \theta_f = 5°$ [58].

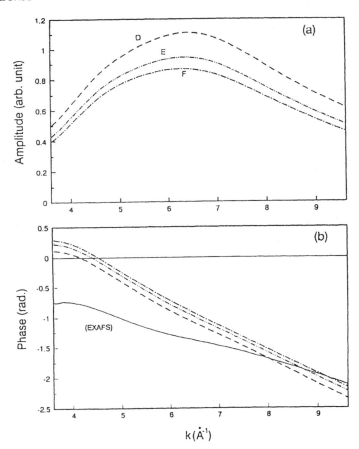

Fig. 21. As in Figure 20, except for the scattering angle; $\theta_i = \theta_f = 90°$ [58].

tion of EELFS depends on the position of surrounding atoms through $P_l(\cos\theta_\gamma)$, as shown in Eq. (363). This anisotropy seems more complicated than that of EXAFS.

Figure 19 shows the model O-Ni system used in this calculation. The calculation has been made for two different modes: the small-angle scattering ($\theta_{inc} = \theta_{out} = 5°$) and the backscattering ($\theta_{inc} = \theta_{out} = 90°$). The other parameters are $\epsilon_p = 2500$ eV and $\epsilon_k = 50$–350 eV. Figure 20 shows the amplitudes and phases of Z_γ as a function of the wave number k of the secondary (EXAFS) electron for the small-angle scattering [58]. As shown in Figure 20a, the amplitudes are sensitive to the position of the scattering atoms; the atoms E and F have large amplitudes, whereas B and C have small ones. From Figure 20b, we find that the important scatterers like E, F, A, G, and D show similar phase curves. This is a desirable result for EELFS analyses. We can expect that the experimental EELFS signal arises mainly from the atoms with large $|\Gamma_\gamma|$, and their phases are nearly the same, so that the single phase function is sufficient for the EELFS analyses, which is approximately calculated in the dipole selection rule (364).

Figure 21 shows the amplitudes and phases of Z_γ for the backscattering [58]. From the symmetry consideration, it is enough to show the results for three atoms (D, E, and F). The three amplitudes show quite small differences compared with those in Figure 20a. This is due to the fact that $|\rho(0)|$ is important in this large-angle scattering mode as described above; the s-wave propagates isotropically. The phase curves in Figure 21b are almost the same; however, they are clearly different from the EXAFS phase curve. Therefore we can conclude that EELFS in the backscattering mode gives rather isotropic infor-

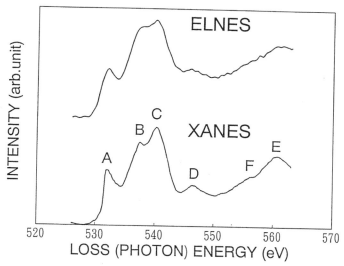

Fig. 22. Oxygen K-edge ELNES spectrum of a single-crystalline NiO(100) surface [18]. A XANES spectrum measured by Davoli et al. [149] is also shown. The ELNES was measured with $(E_p, \theta_{\mathrm{inc}}, \theta_{\mathrm{out}}) = (2800\ \mathrm{eV}, 4°, 6°)$. The probe current was 30 nA.

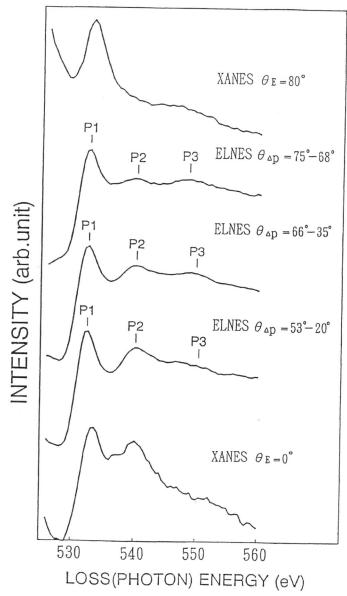

Fig. 23. Dependence of the ELNES spectrum on the direction of the momentum transfer $\Delta\mathbf{p}$ [18]. $\theta_{\Delta p}$ denotes the angle of $\Delta\mathbf{p}$ with respect to the surface plane. $(E_p, \theta_{\mathrm{inc}}, \theta_{\mathrm{out}})$ for the ELNES measurements were (2800 eV, 12°, 3°) for $\theta_{\Delta p} = 75°$ to 68°; (2800 eV, 4°, 6°) for $\theta_{\Delta p} = 66°$ to 35°, and (2000 eV, 3°, 3°) for $\theta_{\Delta p} = 53°$ to 20°. The probe current was 30 nA. The XANES spectra were measured by Norman et al. [150], and θ_E is the angle between the electric field of the X-ray and the surface plane.

mation about the surrounding atoms, and the phase function is quite different from that of EXAFS.

The reflection-ELNES spectra of a single-crystalline NiO (100) surface were measured, and the result is shown in Figure 22 [18]. The XANES spectra of NiO (100) obtained by Davoli et al. with the electron partial-yield method [149] are also shown in the same figure for comparison. The ELNES gives the same features $A-F$ as the XANES. This suggests that the ELNES spectrum originates predominantly from the dipole-allowed transition. Hayashi et al. measured O K-edge ELNES spectra of NiO for different $\Delta\mathbf{p}$ by varying the primary electron energy from 2000 to 2800 eV as well as the scattering angle from 6° to 15° [18]. The ELNES spectra obtained were quite similar to the corresponding XANES spectra. These results experimentally indicate that the reflection-ELNES spectra can be analyzed within the dipole selection rule when $\Delta\mathbf{p}$ is small.

The oxygen K-edge XANES spectra for $p(2\times2)$ and $c(2\times2)$ were reported to depend on the orientation of the X-ray electric field vector $\mathbf{E}$ [150]. The momentum transfer $\Delta\mathbf{p}$ in the reflection-EELFS measurement is expected to play a role corresponding to $\mathbf{E}$ in the XANES. The direction of $\Delta\mathbf{p}$ depends on an incident angle θ_{inc}, a take-off angle θ_{out}, the primary electron energy E_p, and the loss energy. Oxygen K-edge ELNES spectra for $p(2 \times 2)$ were measured as a function of $\Delta\mathbf{p}$ and are shown in Figure 23 [18]. Even though θ_{inc}, θ_{out}, and E_p are kept constant, the direction of $\Delta\mathbf{p}$ changes in some range with the change of the loss energy ΔE according to the relation $p' = \sqrt{2(E_p - \Delta E)}$, where p' is the momentum of scattered electrons. XANES spectra for Ni(100)-$p(2 \times 2)$-O, also shown in this figure, were measured by Norman et al. at grazing ($\mathbf{E}$ along a direction of 80° to the surface plane) and normal ($\mathbf{E}$ parallel to the surface plane) incidence [150]. The ELNES spectra

depend on the direction of $\Delta\mathbf{p}$, which is denoted by the angle from the surface plane $\theta_{\Delta p}$. As $\Delta\mathbf{p}$ changes its direction from the surface normal ($\theta_{\Delta p} = 75°$ to 68°) to the grazing direction ($\theta_{\Delta p} = 53°$ to 20°), feature $P2$ grows and feature $P3$ shifts to the higher energy side on the loss energy axis. Comparing the XANES for $\theta_E = 80°$ with the ELNES for $\theta_{\Delta p} = 75°$ to 68°, and the ELNES for $\theta_{\Delta p} = 53°$ to 20° with the XANES for $\theta_E = 0°$, we find a close correspondence between the ELNES and XANES.

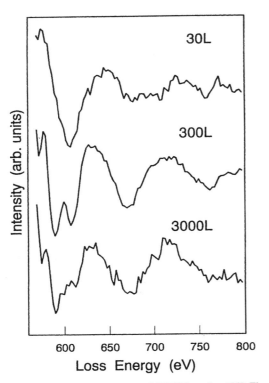

Fig. 24. Oxygen K-edge EXELFS for an O/Ni(100) surface [18]. The primary electron energy, the probe current, and the scattering angle are 2000 eV, 25 nA, and 10° for 30 L and 300 L, and 2500 eV, 50 nA, and 8° for 3000 L.

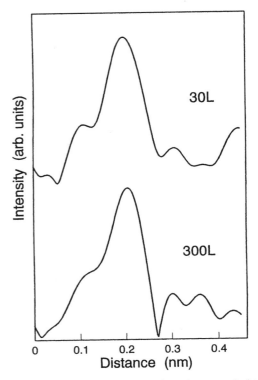

Fig. 25. Absolute value of the Fourier-transformed curves calculated for the 30 L and 300 L curves shown in Figure 24, with the use of the phase shift correction [18].

Figure 24 shows O K-edge EXELFS spectra for an O-adsorbed Ni (100) surface for 30 L, 300 L, and 3000 L oxygen exposures. The 3000 L O_2-dosed surface is identified as NiO by the ELNES spectrum. Using the fact that the Ni-O distance for a NiO single crystal is 2.08 Å, we obtained the phase shift correction by filtering the Fourier-transformed data of the EXELFS for the 3000 L dose. Figure 25 shows the absolute value of the Fourier-transformed data for the 30 L and 300 L EXELFS shown in Figure 24 after the phase shift correction. The nearest-neighbor interatomic distance of Ni-O is 1.97 Å for 30 L and 2.06 Å for 300 L. This interatomic distance for 30 L or $c(2 \times 2)$-O is close to the 1.95–1.98 Å estimated by SEXAFS [151, 152]. On the other hand, the interatomic distance for 300 L is nearly equal to that of the NiO single crystal, although the LEED pattern for 300 L exhibits a $c(2 \times 2)$-rich structure. This result is due to the fact that EXELFS is sensitive to the local structure and LEED to the long-range order.

Hitchcock and Tyliszczak have made an extensive review of the reflection-EELFS technique and results up to 1994 [153]. After this review article was published, several studies used reflection-EELFS [16, 18, 154].

6. THERMAL EFFECTS ON EELFS

Thermal and static disorder affect the amplitude and phase of XAFS and EELS. As discussed before, we have two different types of interference, EXAFS and plural XPD-like interference. The latter plays a minor role in the small-angle scattering modes shown in Figure 1a and c. In this case only XAFS thermal factors are important, and our discussion will be focused on these factors here. XPD thermal factors are discussed in [106, 114], which are also useful for the discussion on EXELFS thermal factors.

Theoretical aspects of temperature dependence in EXAFS were first discussed by Beni and Platzman within the framework of harmonic vibration for nuclei motion and plane wave approximation [75]. The EXAFS oscillation is given for K-shell excitation,

$$\chi(k) = -\sum_\alpha \frac{1}{kR_\alpha^2} \sin\{2kR_\alpha + \arg f_\alpha(\pi) + 2\delta_1^A\}$$
$$\times |f_\alpha(\pi)| e^{-2k^2\sigma^2} \qquad (367)$$

where $\mathbf{k}$ is the wave vector of the scattering electron and $\mathbf{R}_\alpha$ is the position of a surrounding atom α. The backscattering amplitude $f_\alpha(\pi)$ from an atom α and the p wave phase shift δ_1^A at the X-ray absorbing atom A are important factors in the description of the elastic scattering of photoelectrons.

The first experimental attempt to extract information about the Debye–Waller factor in EXAFS, σ in Eq. (367), was made by Greegor and Lytle [76], based on the above theory. The change in the Debye–Waller factor was studied from the slope $\ln|\chi_1/\chi_2|$ plots vs. k^2 and by use of least-squares fitting, which is now widely referred to as the *ratio method*. Here χ_1 and χ_2

are EXAFS functions at different temperatures T_1 and T_2. Following these pioneering works, several authors have developed more reliable theoretical methods for the study of temperature effects on EXAFS in the framework of harmonic vibration approximation. Sevillano et al. used a more realistic projected density of states [77] than did the correlated Debye model used by Beni and Platzman [75]. For more complicated samples such as AgI, Dalba et al. discussed the strength and limitations of Einstein and correlated Debye models in interpreting the experimental results [78, 78a]. They found that the experimental behavior is satisfactorily reproduced by the mixed model (Debye + Einstein).

All of the above approaches are devised for crystalline samples. To study EXAFS Debye–Waller factors of amorphous samples, Lottici [79] has applied the interesting relation between the projected phonon density of states $\rho_1(\omega)$ and the normalized phonon density of states $\rho(\omega)$; $\rho_1(\omega) = \omega^2 \rho(\omega) / \langle \omega^2 \rangle$, derived by Knapp et al. [80]. He has calculated the EXAFS Debye–Waller factor of amorphous arsenic with the aid of $\rho(\omega)$ reduced from Raman intensity and neutron inelastic scattering data. He has found that the experimental temperature dependence of the factor is well explained with the use of no fitting parameter.

For most samples the harmonic approximation can reasonably be applied to EXAFS analyses; however, anharmonic effects play some important role in some special samples such as CuBr [81], RbCl, and NaBr [82]. Stern et al. measured that XAFS of lead from 10 K through and above the melting transition at 600 K and found discontinuities of the first four cumulants on melting, except for the second-order cumulant [83, 83a]. The temperature dependence of SEXAFS of half a monolayer of adsorbed systems, O and N atoms on Ni (100) and Cu (100) surfaces, indicates an enhanced anharmonic contribution [84]. From this measurement it is concluded that the thermal expansion along the internuclear bond axis for O and N chemisorbed on Cu and Ni is 3 to 10 times larger than that in bulk systems. Tranquanda and Ingalls first applied cumulant expansion to analyze anharmonic effects on the EXAFS Debye–Waller factor, and estimated third- and fourth-order cumulant with the use of local distribution for nuclei [81]. These results are interesting; however, their analyses are completely based on classical approximation.

Beyond harmonic approximation, first-principle quantum statistical approaches have been developed by several authors. Frenkel and Rehr found a simple relation between the thermal expansion coefficient and the EXAFS third-order cumulant with the use of a correlated Einstein model and thermodynamic perturbation theory [85]. This discussion is based on a local vibration picture in which phonon dispersion relations are neglected. This model is successfully extended to be applied to more realistic systems by Yokoyama et al. [86]. Different perturbation approaches are developed based on temperature Green's function methods by Fujikawa and Miyanaga [87–89], which are discussed in Section 6.1.

The above perturbation approaches are closely related to the cumulant expansion; we can safely use these approaches

to relatively weak anharmonic systems. On the other hand, a real space approach proposed by Yokoyama et al. has been widely used to relate EXAFS Debye–Waller factors to interatomic potential, even for strongly anharmonic systems [90]. In these analyses direct information on the bare interatomic potential is obtained; however, the applicability is limited because of the classical approximation used there. The classical approximation can be applied at high temperatures, even though the anharmonicity is strong. Useful methods including quantum effects have been developed in real space representations with the use of path-integral approaches [91–95]. In comparison with the perturbation approaches, we can easily obtain close relation to the classical approach, and we can apply these approaches to strongly anharmonic systems, where the cumulant expansion does not work.

In addition to these anharmonic effects, spherical wave (SW) effects on Debye–Waller factors play an important role in the study of temperature effects in EXAFS spectra. Theoretical works demonstrate the peak shift in Fourier-transformed EXAFS spectra due to SW thermal factors [96–101]. A simple but rather accurate method for handling XAFS SW thermal factors due to atomic vibration was recently proposed to describe single and multiple scattering EXAFS temperature effects and XANES temperature effects where the SW effects are crucial [102, 103].

In this section we review the important theoretical problems in calculating EELFS and XAFS thermal factors, which are closely related to the experimental analyses of these spectra. We focus our attention on the perturbation theory for weak anharmonic systems and the path-integral theory for strongly anharmonic systems. We also review the recent progress in the theory in handling spherical wave effects in XAFS spectra [96–101, 140] and Franck–Condon factors in XAFS spectra [28, 104, 105].

6.1. Perturbation Approach to the EELFS and EXAFS Debye–Waller Factors

In this section, we discuss the analysis of the XAFS Debye–Waller factor in the framework of the perturbation approach. The advantage of the perturbation method is mathematically analytic, so it is easy to obtain the potential parameters from the experimental XAFS spectra. On the other hand, there is a disadvantage: the perturbation method breaks down for strongly anharmonic systems.

6.1.1. Basic Theory

At first we present the theoretical formulation for the XAFS Debye–Waller factor using the perturbation method developed by Fujikawa and Miyanaga [87]. In the single scattering plane wave theory of EXAFS, the oscillation part of the absorption coefficient χ for a photon frequency ω is given by Eq. (367), and the corresponding one in EELFS is given by Eq. (362).

To study the effects of the thermal motion of atoms on XAFS, we should evaluate the ensemble average of Eq. (367)

over a canonical ensemble. The procedure for averaging the thermal motion is first to replace $\mathbf{R}_\alpha$ with $\mathbf{R}_\alpha^0 + \mathbf{u}_\alpha - \mathbf{u}_A$ and retain all of the leading terms in $|\mathbf{u}_\alpha - \mathbf{u}_A|/R_\alpha^0$. Here $\mathbf{u}_\alpha$ and $\mathbf{u}_A$ are, respectively, the αth atom and X-ray absorbing atom displacement vectors, and $\mathbf{R}_\alpha^0$ is the position vector of the αth atom measured from the core excited atom A in equilibrium. In the case of EXAFS, an evaluation of the temperature dependence requires calculation of the average,

$$\langle \exp(2ik\Delta_\alpha) \rangle \equiv \mathrm{Tr}\big[e^{-\beta H} \exp(2ik\Delta_\alpha) \big] / \mathrm{Tr}\big(e^{-\beta H} \big)$$
$$\beta = 1/k_B T \tag{368}$$

where H is the full Hamiltonian of the nuclei vibration including anharmonic phonon effects, and Δ_α is defined as

$$\Delta_\alpha = \hat{\mathbf{R}}_\alpha^0 \cdot (\mathbf{u}_\alpha - \mathbf{u}_A) \tag{369}$$

If we include the damping effect and small correction factor in the order of $1/R_\alpha^2$, a minor change should be incorporated [81, 107]. This is quite easy, so that we only consider the simplest form shown by Eq. (369)

The most convenient way to evaluate such a thermal average is to apply the cumulant techniques as usually used in quantum statistical mechanics [108–111]. We can get the following expansion:

$$\langle \exp(2ik\Delta_\alpha) \rangle = \exp\left\{ \sum_{n=1}^{\infty} \frac{(2ik)^2}{n!} M_n \right\} \tag{370}$$

where M_n is the nth order cumulant, and the cumulants are explicitly related to low-order moments as

$$M_1 = \langle \Delta_\alpha \rangle$$
$$M_2 = \langle \Delta_\alpha^2 \rangle - \langle \Delta_\alpha \rangle^2$$
$$M_3 = \langle \Delta_\alpha^3 \rangle - 3\langle \Delta_\alpha^2 \rangle \langle \Delta_\alpha \rangle + 2\langle \Delta_\alpha \rangle^3$$
$$M_4 = \langle \Delta_\alpha^4 \rangle - 4\langle \Delta_\alpha^3 \rangle \langle \Delta_\alpha \rangle - 3\langle \Delta_\alpha^2 \rangle^2$$
$$+ 12\langle \Delta_\alpha^2 \rangle \langle \Delta_\alpha \rangle^2 - 6\langle \Delta_\alpha \rangle^4 \tag{371}$$

All of cumulants are real, so that the EXAFS formula should be read up to the fourth-order cumulant [81, 110],

$$\chi(k) = - \sum_\alpha \frac{1}{k(R_\alpha^0)^2} \mathrm{Im}\left[\exp\big(2ikR_\alpha^0 + 2i\delta_1^A\big) f_\alpha(\pi) \right.$$
$$\times \exp\left\{ 2ik\left(M_1 - \frac{2}{3}k^2 M_3 \right) \right\} \right]$$
$$\times \exp\left(-2k^2 M_2^2 + \frac{2}{3}k^4 M_4 \right) \tag{372}$$

6.1.2. Evaluation of Moments and Cumulants

To include anharmonic phonon effects by applying perturbation theory, we now write the vibrational total Hamiltonian H as the sum

$$H = H_0 + H_1 \tag{373}$$

where H_0 is the harmonic term and H_1 is the anharmonic term. For the present purposes it is convenient to express these operators in terms of $A_{\mathbf{k}j}$ and $B_{\mathbf{k}j}$, where they are related to the phonon annihilation (creation) operator $b_{\mathbf{k}j}$ ($b_{\mathbf{k}j}^\dagger$) for crystal momentum $\mathbf{k}$ and phonon branch j [112].

$$A_{\mathbf{k}j} = b_{\mathbf{k}j} + b_{-\mathbf{k}j}^\dagger$$
$$B_{\mathbf{k}j} = b_{\mathbf{k}j} - b_{-\mathbf{k}j}^\dagger \tag{374}$$

They satisfy the commutation relation

$$[A_{\mathbf{k}j}, A_{\mathbf{k}'j'}] = [B_{\mathbf{k}j}, B_{\mathbf{k}'j'}] = 0$$
$$[A_{\mathbf{k}j}, B_{\mathbf{k}'j'}] = -2\Delta(\mathbf{k} + \mathbf{k}')\delta_{jj'} \tag{375}$$

In writing these commutation relations we have introduced the function

$$\Delta(\mathbf{k}) = \frac{1}{N} \sum_l \exp[i\mathbf{k} \cdot \mathbf{x}(l)] \tag{376}$$

where $\mathbf{x}(l)$ is the lattice site of the l primitive unit cell, and N is the number of unit cells. It equals unity if $\mathbf{k}$ is a translational vector of the reciprocal lattice and vanishes otherwise. We can easily find that

$$A_{\mathbf{k}j} = A_{-\mathbf{k}j}^\dagger$$
$$B_{\mathbf{k}j} = -B_{-\mathbf{k}j}^\dagger \tag{377}$$

In terms of these operators we can show that for the q Cartesian component of the κth atom in the lth primitive unit cell, the atomic displacements and momenta become

$$u_q(l\kappa) = \sqrt{\frac{\hbar}{2NM_\kappa}} \sum_{\mathbf{k}j} \frac{e_q(\kappa|\mathbf{k}j)}{\sqrt{\omega_j(\mathbf{k})}} \exp\{i\mathbf{k} \cdot \mathbf{x}(l)\} A_{\mathbf{k}j} \tag{378}$$

$$p_q(l\kappa) = -i\sqrt{\frac{\hbar M_\kappa}{2N}} \sum_{\mathbf{k}j} \sqrt{\omega_j(\mathbf{k})} e_q(\kappa|\mathbf{k}j)$$
$$\times \exp\{i\mathbf{k} \cdot \mathbf{x}(l)\} B_{\mathbf{k}j} \tag{379}$$

where $\omega_j(\mathbf{k})$ is the phonon spectrum and $e_q(\kappa|\mathbf{k}j)$ is the corresponding eigenvectors of the dynamical matrix. These equations show that $A_{\mathbf{k}j}$ and $B_{\mathbf{k}j}$ are Fourier coefficients of the atomic displacements and momenta. In terms of these new operators, the harmonic Hamiltonian H_0 takes the form [112]

$$H_0 = \sum_{\mathbf{k}j} \frac{1}{4} \hbar\omega_j(\mathbf{k})\big[B_{\mathbf{k}j}^\dagger B_{\mathbf{k}j} + A_{\mathbf{k}j}^\dagger A_{\mathbf{k}j} \big] \tag{380}$$

To simplify the complicated anharmonic Hamiltonian we introduce abbreviations like $\lambda_1 = (\mathbf{k}_1, j_1)$ and $-\lambda = (-\mathbf{k}, j)$. Then the cubic and quartic anharmonic terms are shown as

$$H_1 = \sum_{\lambda_1, \lambda_2, \lambda_3} V(\lambda_1, \lambda_2, \lambda_3) A_{\lambda_1} A_{\lambda_2} A_{\lambda_3}$$
$$+ \sum_{\lambda_1, \ldots, \lambda_4} V(\lambda_1, \ldots, \lambda_4) A_{\lambda_1} \ldots A_{\lambda_4} \tag{381}$$

where $V(\lambda_1, \ldots, \lambda_n)$ can be represented by use of an nth-order atomic force constant $\Phi_{q_1, \ldots, q_n}$,

$$
\begin{aligned}
V(\lambda_1, &\ldots, \lambda_n) \\
&= \frac{1}{n!} (\hbar/2N)^{n/2} \frac{N\Delta(\mathbf{k}_1 + \cdots \mathbf{k}_n)}{\{\omega_{j_1}(\mathbf{k}_1) \cdots \omega_{j_n}(\mathbf{k}_n)\}^{1/2}} \\
&\quad \times \sum_{\kappa_1 q_1} \sum_{\kappa_2 q_2} \cdots \sum_{\kappa_n q_n} \Phi_{q_1 q_2 \cdots q_n}(0\kappa_1, l_1\kappa_2, \ldots, l_n\kappa_n) \\
&\quad \times \frac{e_{q_1}(\kappa_1|\lambda_1)}{\sqrt{M_{\kappa_1}}} \cdots \frac{e_{q_n}(\kappa_n|\lambda_n)}{\sqrt{M_{\kappa_n}}} \\
&\quad \times \exp[i\{\mathbf{k}_2 \cdot \mathbf{x}(l_2) + \cdots + \mathbf{k}_n \cdot \mathbf{x}(l_n)\}]
\end{aligned} \tag{382}
$$

Now we can represent the projected deviation Δ_α given by Eq. (369) in terms of A_λ as

$$
\Delta_\alpha = \sum_\lambda f_\lambda A_\lambda \tag{383}
$$

where f_λ is the abbreviation of $f_\lambda(\alpha A)$, which is given as

$$
\begin{aligned}
f_\lambda(\alpha A) &= \sqrt{\frac{\hbar}{2N\omega_j(\mathbf{k})}} \hat{\mathbf{R}}_\alpha \\
&\quad \times \left\{ \frac{\mathbf{e}(\alpha|\mathbf{k}j)}{\sqrt{M_\alpha}} \exp[i\mathbf{k}\cdot\mathbf{x}(l)] - \frac{\mathbf{e}(A|\mathbf{k}j)}{\sqrt{M_A}} \right\}
\end{aligned} \tag{384}
$$

We assume that atom α is the αth atom in the lth cell and atom A is the Ath atom in the 0th cell. We can easily show that

$$
f_\lambda = f_{-\lambda}^* \tag{385}
$$

With the use of these fundamental quantities we can calculate the Debye–Waller factor including anharmonic effects as shown below [87].

6.1.3. Second-Order Cumulant in the Harmonic Approximation

We show that the ordinary Debye–Waller factor is obtained if we neglect the anharmonic effect in the calculation of the second-order cumulant. Here we extensively apply a harmonic phonon Green's function for later discussion. In this approximation the first-order moment $\langle \Delta_\alpha \rangle_0$ disappears, where the subscript 0 denotes the average over the harmonic states,

$$
\langle X \rangle_0 = \mathrm{Tr}\left[\exp(-\beta H_0)X\right]/\mathrm{Tr}\left[(-\beta H_0)\right] \tag{386}
$$

In the average $\langle X \rangle$ the unperturbed Hamiltonian H_0 is replaced by the full Hamiltonian H. The second-order moment in the harmonic approximation is given from Eq. (383) by

$$
\langle \Delta_\alpha^2 \rangle_0 = \sum_{\lambda, \lambda'} f_\lambda f_{\lambda'} \langle A_\lambda A_{\lambda'} \rangle_0 \tag{387}
$$

The average value $\langle A_\lambda A_{\lambda'} \rangle_0$ is related to the harmonic phonon Green's function $G_{\lambda\lambda'}^0(\tau)$ as

$$
\langle A_\lambda A_{\lambda'} \rangle_0 = \langle \hat{A}_\lambda(0) \hat{A}_{\lambda'}(0) \rangle_0 = -G_{\lambda\lambda'}^0(0) \tag{388}
$$

$$
\hat{A}_\lambda(\tau) \equiv \exp(\tau H_0) A_\lambda \exp(-\tau H_0) \tag{389}
$$

The harmonic phonon Green's function $G_{\lambda\lambda'}^0(\tau)$ is defined and simply represented [30, 113] as

$$
\begin{aligned}
G_{\lambda\lambda'}^0(\tau) &= -\langle \hat{A}_\lambda(\tau) \hat{A}_{\lambda'}(0) \rangle \\
&= -\delta_{\lambda,-\lambda}\{ \langle n_\lambda + 1 \rangle \exp(-\hbar|\tau|\omega_\lambda) \\
&\quad + \langle n_\lambda \rangle \exp(\hbar|\tau|\omega_\lambda) \}
\end{aligned} \tag{390}
$$

where $\langle n_\lambda \rangle$ is the average phonon occupation number,

$$
\langle n_\lambda \rangle = \frac{1}{\exp(\hbar\omega_\lambda\beta) - 1} \tag{391}
$$

By use of the relations shown by Eqs. (387), (388), (390), and (391), we obtain the expression

$$
\begin{aligned}
\langle \Delta_\alpha^2 \rangle_0 &= \sum_\lambda |f_\lambda|^2 \coth(\hbar\omega_\lambda\beta/2) \\
&= \frac{\hbar}{2N} \sum_\lambda \frac{1}{\omega_\lambda} \left(\frac{|a_\alpha|^2}{M_\alpha} + \frac{|a_A|^2}{M_A} - 2\frac{1}{\sqrt{M_\alpha M_A}} \right. \\
&\quad \left. \times \mathrm{Re}\{a_\alpha a_A^* \exp(i\mathbf{k}\cdot\mathbf{x}(l))\} \right) \coth\left(\frac{\hbar\omega_\lambda\beta}{2} \right)
\end{aligned} \tag{392}
$$

$$
a_\alpha = \hat{\mathbf{R}}_\alpha^0 \cdot \mathbf{e}(\alpha|\lambda), \qquad a_A = \hat{\mathbf{R}}_A^0 \cdot \mathbf{e}(A|\lambda) \tag{393}
$$

which was first derived by Beni and Platzman [75]. This is alternative derivation of the EXAFS Debye–Waller factor in the harmonic approximation, which can be generalized to include anharmonic effects. In the special case of a monatomic Bravais lattice, $a_\alpha = a_A$, so that we obtain the well known result [75]

$$
\langle \Delta_\alpha^2 \rangle = \sigma_0^2 = \frac{\hbar}{NM} \sum_\mathbf{k} \frac{1 - \cos(\mathbf{k}\cdot\mathbf{R}_\alpha^0)}{\omega(\mathbf{k})} |a_\alpha|^2 \coth\left(\frac{\hbar\omega(\mathbf{k})\beta}{2} \right) \tag{394}
$$

In the harmonic approximation all other cumulants equally vanish because of Mermin's theorem [134].

6.1.4. First-Order Cumulant

Beyond the harmonic vibration approximation, the first-order moment $\langle \Delta_\alpha \rangle$ does not necessarily vanish and can be calculated with the aid of the many-body perturbation approach by the formula [30, 113]

$$
\langle \Delta_\alpha \rangle = \frac{\sum_\lambda f_\lambda \langle A_\lambda S(\beta) \rangle_0}{\langle S(\beta) \rangle_0} \tag{395}
$$

where $S(\beta)$ is defined by

$$
\begin{aligned}
S(\beta) &= \sum_{n=0}^\infty \frac{(-1)^n}{n!} \int_0^\beta d\tau_1 \cdots \\
&\quad \times \int_0^\beta d\tau_n T\left[\hat{H}_1(\tau_1) \cdots \hat{H}_1(\tau_n) \right]
\end{aligned} \tag{396}
$$

$$
\begin{aligned}
\langle A_\lambda S(\beta) \rangle_0 &= \langle A_\lambda \rangle_0 - \int_0^\beta d\tau \langle T[A_\lambda \hat{H}_1(\tau)] \rangle_0 \\
&\quad + \frac{1}{2} \int_0^\beta d\tau_1 \int_0^\beta d\tau_2 \langle T[A_\lambda \hat{H}_1(\tau_1) \hat{H}_1(\tau_2)] \rangle_0 \cdots
\end{aligned} \tag{397}
$$

The first term of the right-hand side vanishes as mentioned above. The second term has a finite contribution from the cubic anharmonicity, and so on. In the denominator of Eq. (395) it is represented as the 0th order of anharmonicity + first-order term + $\cdots$, so that we can write the formula for $\langle \Delta_\alpha \rangle$ up to the first-order anharmonicity as

$$\langle \Delta_\alpha \rangle_1 = -\sum_\lambda \sum_{\lambda_1 \lambda_2 \lambda_3} f_\lambda V(\lambda_1 \lambda_2 \lambda_3)$$

$$\times \int_0^\beta d\tau \langle T[A_\lambda \hat{A}_{\lambda_1}(\tau) \hat{A}_{\lambda_2}(\tau) \hat{A}_{\lambda_3}(\tau)] \rangle_0 \quad (398)$$

which can be factorized with the aid of Wick's theorem [30, 113],

$$\langle \Delta_\alpha \rangle_1 = -3 \sum_{\lambda \nu} V(-\lambda, \nu, -\nu) f_\lambda G_\nu^0(0) \int_0^\beta d\tau G_\lambda^0(-\tau) \quad (399)$$

Here the factor 3 arises from the symmetry of $V(\lambda_1, \lambda_2, \lambda_3)$, $V(\lambda_1, \lambda_2, \lambda_3) = V(\lambda_2, \lambda_1, \lambda_3) = V(\lambda_2, \lambda_3, \lambda_1) = \cdots$. Substitution of Eq. (390) into Eq. (399) yields the simple expression for $\langle \Delta_\alpha \rangle$,

$$\langle \Delta_\alpha \rangle_1 = -6 \sum_{\lambda \nu} V(-\lambda, \nu, -\nu) \coth(\beta \hbar \omega_\nu / 2) f_\lambda / \hbar \omega_\lambda \quad (400)$$

To directly calculate the right-hand side of Eq. (400), we should apply the symmetry of V,

$$V(\lambda_1, \lambda_2, \lambda_3) = V(-\lambda_1, -\lambda_2, -\lambda_3)^* \quad (401)$$

The periodicity of a crystal has the consequence that the Fourier-transformed anharmonic force constant vanishes unless the sum of the wave vectors in its argument equals a translation vector of the reciprocal lattice as shown by Eq. (382). From this restriction we find that $V(-\lambda, \nu, -\nu) \propto \Delta(\mathbf{k})$ for $\lambda = (\mathbf{k}, i)$ and $\nu = (\mathbf{q}, j)$. Now we can write $\langle \Delta \rangle_1$ in the form

$$\langle \Delta_\alpha \rangle_1 = -6 \sum_i \frac{f_{0i}}{\hbar \omega_i(0)} \sum_\nu V(\mathbf{0}i, \mathbf{q}j, -\mathbf{q}j) \coth(\beta \hbar \omega_\nu / 2) \quad (402)$$

From Eq. (401) we can see that $V(\mathbf{0}i, \mathbf{q}j, -\mathbf{q}j)$ should be real. Furthermore, we put $\mathbf{k} = \mathbf{0}$ in Eq. (384) and find that f_{0i} vanishes for acoustic phonon branches; therefore the branches to be considered in the sum over i in Eq. (402) are restricted to optical phonon branches. This result shows that the lowest contribution to the first-order cumulant $\langle \Delta_\alpha \rangle_1$ can be neglected for monatomic Bravais lattices like Cu, Si crystal, and so on. Furthermore, even for more complex crystals, which have a center of inversion at every ion site like NaCl, we have the relation

$$V(\mathbf{0}i, \mathbf{q}j, -\mathbf{q}j) = 0 \quad (403)$$

In this case $\langle \Delta_\alpha \rangle_1$ given by Eq. (402) equally vanishes.

6.1.5. Second-Order Cumulant Including Anharmonic Effects

If we include anharmonic effects on the calculation of $\langle \Delta^2 \rangle$, the average $\langle \cdots \rangle_0$ is just replaced by $\langle \cdots \rangle$ in Eq. (387), and we note

that instead of a harmonic phonon Green's function G^0, the full phonon Green's function G

$$G_{\lambda \lambda'}(0) = -\langle A_\lambda A_{\lambda'} \rangle \quad (404)$$

should be used. The first-order correction arises from quartic anharmonicity, so that we have the first-order correction term

$$G_{\lambda \lambda'}^1(0) = -12 \sum_\nu V(-\lambda, -\lambda', \nu, -\nu) G_\nu^0(0)$$

$$\times \int_0^\beta d\tau G_\lambda^0(-\tau) G_{\lambda'}^0(-\tau) \quad (405)$$

The integration is easily performed, and the first-order correction in the moment $\langle \Delta_\alpha^2 \rangle$ is given by

$$\langle \Delta_\alpha^2 \rangle_1 = 24 \sum_{\lambda \lambda'} \sum_\nu V(-\lambda, -\lambda', \nu, -\nu) f_\lambda f_{\lambda'} \coth\left(\frac{\beta \hbar \omega_\nu}{2}\right) / \hbar$$

$$\times \frac{1}{\omega_\lambda^2 - \omega_{\lambda'}^2} \{\omega_\lambda \langle 2n_{\lambda'} + 1 \rangle - \omega_{\lambda'} \langle 2n_\lambda + 1 \rangle\} \quad (406)$$

where the subscript 1 denotes the average including the first-order correction due to the quartic anharmonicity.

6.1.6. Third-Order Cumulant

Up to the first-order anharmonic correction the third-order cumulant M_3 is approximately represented by the use of the lowest order moments,

$$M_3 \approx \langle \Delta_\alpha^3 \rangle_1 - 3\langle \Delta_\alpha^2 \rangle_0 \langle \Delta_\alpha \rangle_1 \quad (407)$$

Both $\langle \Delta_\alpha \rangle_1$ and $\langle \Delta_\alpha^2 \rangle_0$ have been obtained as shown by Eqs. (402) and (392), so that we have to estimate the third-order moment $\langle \Delta_\alpha^3 \rangle_1$, which is written with the use of f_λ and A_λ:

$$\langle \Delta_\alpha^3 \rangle_1 = \sum_{\lambda_1 \lambda_2 \lambda_3} f_{\lambda_1} f_{\lambda_2} f_{\lambda_3} \langle A_{\lambda_1} A_{\lambda_2} A_{\lambda_3} \rangle_1 \quad (408)$$

By the same approximate procedure as used for the first order cumulant calculation, where the first-order of cubic anharmonicity is taken into account, we obtain the expression for that moment,

$$\langle \Delta_\alpha^3 \rangle_1 = -18 \sum_{\lambda_1 \lambda_2 \nu} |f_{\lambda_1}|^2 f_{\lambda_2} \frac{V(-\lambda_2, \nu, -\nu)}{\hbar \omega_{\lambda_2}}$$

$$\times \coth\left(\frac{\beta \hbar \omega_{\lambda_1}}{2}\right) \coth\left(\frac{\beta \hbar \omega_\nu}{2}\right)$$

$$- 6 \sum_{\lambda_1 \lambda_2 \lambda_3} f_{\lambda_1} f_{\lambda_2} f_{\lambda_3} V(\lambda_1 \lambda_2 \lambda_3)^* \langle n_{\lambda_1} \rangle \langle n_{\lambda_2} \rangle \langle n_{\lambda_3} \rangle$$

$$\times \int_0^\beta d\tau G_{\lambda_1}^0(\tau) G_{\lambda_2}^0(\tau) G_{\lambda_3}^0(\tau) \quad (409)$$

The first term of Eq. (409) is canceled by $3\langle \Delta_\alpha^2 \rangle_0 \langle \Delta_\alpha \rangle_1$ after we substitute Eqs. (402) and (392) into Eq. (409). The final expression of the third-order cumulant M_3 is thus simply given by [87]

$$M_3 = -\frac{12}{\hbar} \sum_{\lambda_1 \lambda_2 \lambda_3} f_{\lambda_1} f_{\lambda_2} f_{\lambda_3} V(\lambda_1 \lambda_2 \lambda_3)^* \langle n_{\lambda_1} \rangle \langle n_{\lambda_2} \rangle \langle n_{\lambda_3} \rangle$$
$$\times \left\{ \frac{\exp[\hbar\beta(\omega_{\lambda_1} + \omega_{\lambda_2} + \omega_{\lambda_3})] - 1}{\omega_{\lambda_1} + \omega_{\lambda_2} + \omega_{\lambda_3}} \right.$$
$$+ \frac{\exp[\hbar\beta(\omega_{\lambda_1} + \omega_{\lambda_2})] - \exp[\hbar\beta\omega_{\lambda_3}]}{\omega_{\lambda_1} + \omega_{\lambda_2} - \omega_{\lambda_3}}$$
$$+ \frac{\exp[\hbar\beta(\omega_{\lambda_1} + \omega_{\lambda_3})] - \exp[\hbar\beta\omega_{\lambda_2}]}{\omega_{\lambda_1} + \omega_{\lambda_3} - \omega_{\lambda_2}}$$
$$\left. + \frac{\exp[\hbar\beta(\omega_{\lambda_2} + \omega_{\lambda_3})] - \exp[\hbar\beta\omega_{\lambda_1}]}{\omega_{\lambda_2} + \omega_{\lambda_3} - \omega_{\lambda_1}} \right\} \quad (410)$$

6.1.7. Fourth-Order Cumulant

From Eq. (371) the fourth-order cumulant M_4 is approximately given by the following approximation up to the first-order anharmonic correction,

$$M_4 = \langle \Delta_\alpha^4 \rangle_0 - 3\langle \Delta_\alpha^2 \rangle_0^2 + \langle \Delta_\alpha^4 \rangle_1 - 6\langle \Delta_\alpha^2 \rangle_0 \langle \Delta_\alpha^2 \rangle_1 + \cdots \quad (411)$$

With the use of the simple algebra based on Wick's theorem [113, 114], we can show that

$$\langle \Delta_\alpha^4 \rangle_0 = 3\langle \Delta_\alpha^2 \rangle_0^2 \quad (412)$$

Therefore M_4 vanishes within the harmonic vibration approximation as expected. The first-order anharmonic vibration correction comes from the third- and fourth-order terms of the right-hand side of Eq. (411). Both $\langle \Delta_\alpha^2 \rangle_0$ and $\langle \Delta_\alpha^2 \rangle_1$ have already been represented in a closed form, as shown by Eqs. (392) and (406), so that the work we have to do is only to estimate $\langle \Delta_\alpha^4 \rangle_1$, which comes from quartic anharmonicity. The corresponding term of $\langle \Delta_\alpha^4 \rangle_1$ is given by

$$\langle \Delta_\alpha^4 \rangle_1 = \frac{72}{\hbar} \sum_{\lambda_1} |f_{\lambda_1}|^2 G_{\lambda_1}^0(0) \sum_{\nu_1 \nu_2 \nu_3} f_{-\nu_1} f_{-\nu_2}$$
$$\times V(\nu_1, \nu_2, \nu_3, -\nu_3) G_{\nu_3}^0(0)$$
$$\times \frac{2}{\omega_{\nu_1}^2 - \omega_{\nu_2}^2} \left[\omega_{\nu_1} \{ \langle 2n_{\nu_2} \rangle + 1 \} - \omega_{\nu_2} \{ \langle 2n_{\nu_1} \rangle + 1 \} \right]$$
$$+ 24 \sum_{\lambda_1 \cdots \lambda_4} f_{\lambda_1} \cdots f_{\lambda_4} V(\lambda_1 \cdots \lambda_4)^*$$
$$\times \int_0^\beta d\tau G_{\lambda_1}^0(\tau) \cdots G_{\lambda_4}^0(\tau) \quad (413)$$

We can see that the first term of the right hand side is exactly canceled by $-6\langle \Delta_\alpha^2 \rangle_0 \langle \Delta_\alpha^2 \rangle_2$ in M_4 with the use of Eqs. (392) and (406). The expression of M_4 is therefore given after the integral in the second term of Eq. (413) is evaluated [87]:

$$M_4 = \frac{48}{\hbar} \sum_{\lambda_1 \cdots \lambda_4} f_{\lambda_1} \cdots f_{\lambda_4} \langle n_{\lambda_1} \rangle \cdots \langle n_{\lambda_4} \rangle$$
$$\times \left\{ \frac{\exp[\hbar\beta(\omega_{\lambda_1} + \omega_{\lambda_2} + \omega_{\lambda_3} + \omega_{\lambda_4})] - 1}{\omega_{\lambda_1} + \omega_{\lambda_2} + \omega_{\lambda_3} + \omega_{\lambda_4}} \right.$$

$$+ \frac{\exp[\hbar\beta(\omega_{\lambda_1} + \omega_{\lambda_2} + \omega_{\lambda_3})] - \exp[\beta\hbar\omega_{\lambda_4}]}{\omega_{\lambda_1} + \omega_{\lambda_2} + \omega_{\lambda_3} - \omega_{\lambda_4}}$$
$$+ \cdots \frac{\exp[\hbar\beta(\omega_{\lambda_1} + \omega_{\lambda_2})] - \exp[\beta\hbar(\omega_{\lambda_3} + \omega_{\lambda_4})]}{\omega_{\lambda_1} + \omega_{\lambda_2} - \omega_{\lambda_3} - \omega_{\lambda_4}}$$
$$\left. + \cdots \right\} \quad (414)$$

From these equations we can directly calculate the cumulants when we have some information on $V(\lambda_1 \lambda_2 \lambda_3)$, $V(\lambda_1 \cdots \lambda_4)$, and phonon dispersion relation ω_λ.

6.1.8. Monatomic Chain Model

We now discuss the XAFS Debye–Waller factor for the simplest model of monatomic linear chains with nearest-neighbor interaction described by Morse potential [88, 89],

$$V(x) = D(\exp(-2\alpha x) - 2\exp(-\alpha x)) \quad (415)$$

In Eq. (415), x is the scaled relative deviation defined by $x = (l - a)/a$, where l is the instantaneous distance between the nth and $n + 1$th atoms and a is the equilibrium distance between these two atoms. We can expand $V(x)$ as

$$V(x) = -D + D\alpha^2 x^2 - D\alpha^3 x^3 + \frac{7}{12} D\alpha^4 x^4 + \cdots \quad (416)$$

The first term in Eq. (416) can be neglected because it just contributes to the energy shift. From now on we consider the weak anharmonicity, so that we keep cubic and quartic anharmonicity in Eq. (416).

First we study the site dependence of the harmonic Debye–Waller factor given by Eq. (394). With the use of an important parameter ω_m ($= \sqrt{8D\alpha^2/Ma^2}$), we can write the phonon spectrum as $\omega(k) = \omega_m |\sin(ka/2)|$. The temperature dependence of the harmonic Debye–Waller factor for the ρth shell $\langle \Delta_\rho^2 \rangle_0$ ($\rho = 1, 2, 3,$ and 4) is shown as a function of dimensionless temperature $\tilde{T}_m = k_B T / \hbar \omega_m$ in Figure 26 compared with the result for the nearest-neighbor classical model, which is a linear function of T [88]. We can see that all

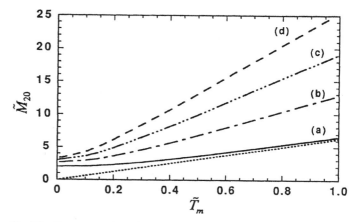

Fig. 26. Temperature dependence of the harmonic EXAFS Debye–Waller factor $\langle \Delta_\rho^2 \rangle_0$ ($\rho = 1$ (a), 2 (b), 3 (c), and 4 (d)). The dotted line is the result for the classical approximation [88].

curves approach the straight line for the classical approximation in the high-temperature region. On the other hand, the dimensionless second-order harmonic cumulant $\tilde{M}_\rho^0$ defined by $\langle \Delta_\rho^2 \rangle_0 / (\hbar a / \pi \alpha \sqrt{2DM})$ approaches corresponding constant values in the low-temperature limit. The predicted behaviors of $\tilde{M}_\rho^0$ are very similar to the experimental ones [78, 78a, 81]. We can see that the Debye–Waller factor for the atoms located at a greater distance from an X-ray-absorbing atom is larger than that for the atoms located at the shorter distance. It is to be noted that the present EXAFS Debye–Waller factor diverges as $\rho \to \infty$. In such a case this factor is reduced to the one without an atomic correlation. In other words, this factor is converted to the diffraction Debye–Waller factor in crystal, which diverges for a one-dimensional phonon system [115]. From this we can understand the site dependence of the second-order cumulant in the harmonic approximation. Experimentally this site dependence cannot be confirmed easily, because of the considerable multiple scattering contribution for long-distance shells. However, in [78, 78a, 81] the site dependence of the EXAFS Debye–Waller factors has been carefully studied for first and second shells, and the similar temperature and site dependence predicted here for CuBr [81] and AgI [78, 78a] was observed. At room temperature ($T = 300$ K) $\tilde{T}_m$ is estimated to be about 0.66 for Cu metals with the use of the Morse parameters proposed by Yokoyama et al. [90]: $D = 0.74$ eV, $\alpha = 4.531$, $a = 2.545$ Å. Our calculation gives a value of $\langle \Delta_1^2 \rangle_0$ of about 5.2×10^{-3} Å^2 in the quantum statistical approach, but 4.5×10^{-3} Å^2 in the classical approximation, which are comparable with the value 7.7×10^{-3} Å^2 given by Sevillano et al. [77].

Figure 27 shows the temperature dependence of the third-order cumulant M_3 of the first nearest neighbor for Cu metal by use of the Morse parameters of Yokoyama et al. [90]. The solid line is the result including the quantum effect and the dashed line is the one for the classical approximation [88]. At 300 K both are about 1.2×10^{-4} Å^3, which is quite close to the value reported by Yokoyama et al., 1.3×10^{-4} Å^2 at 295 K [90]. From this figure the classical approximation is very good above 200 K, which is contrast to the result for the harmonic second-order cumulant M_{20}, where the quantum correction is important, even at room temperature. Below 100 K, however, a remarkable difference is found between the two curves because of the zero point vibration.

6.1.9. Diatomic Chain Model

Next we discuss the EXAFS Debye–Waller factor for a diatomic chain model in which atoms A and B are arranged alternately with atomic masses M_A and M_B [88]. Figure 28 shows the temperature dependence of the harmonic second order cumulant for the nearest-neighbor correlation $\langle \Delta_1^2 \rangle_0$ and each contribution from acoustic and optical phonons for the diatomic chain model with $M_A = 4M_B$. For convenience we show the dimensionless second-order cumulant in the harmonic approximation defined by

$$\tilde{M}_{20} = \langle \Delta_1^2 \rangle_0 \bigg/ \left(\frac{\hbar}{\pi \sqrt{5\gamma\mu}} \right) \tag{417}$$

as a function of dimensionless temperature $\tilde{T}_d$ defined by

$$\tilde{T}_d = \frac{k_B T}{\hbar \omega_d}, \qquad \omega_d = \sqrt{\frac{\gamma}{2\mu}}$$

$$\gamma = \frac{8D\alpha^2}{a^2} \tag{418}$$

where μ is reduced mass, $\mu = M_A M_B / (M_A + M_B)$, and a is the equilibrium interatomic distance. We can separately evaluate the contribution from acoustic phonons (Fig. 28b) and and optical phonons (Fig. 28c). The total contribution is shown by Figure 28a compared with the corresponding classical result shown by Figure 28d. In this case acoustic and optical phonons give the same contributions within the classical approximation, so that the contribution from acoustic phonon is the same as that

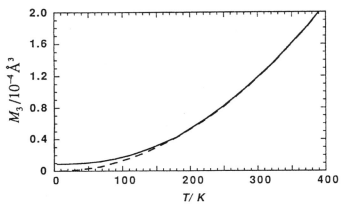

Fig. 27. Calculated temperature dependence of the third-order cumulant M_3 given by Eq. (410) from a first-shell atom for Cu metals [88]. The morse parameters used here are given by Yokoyama et al. [90]. The solid line is the result for quantum statistical theory, and the dashed line is the result for the classical approximation.

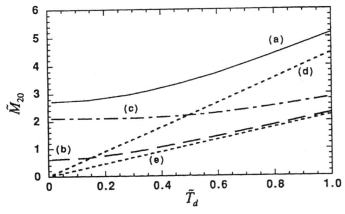

Fig. 28. Temperature dependence of the dimensionless second-order cumulant $\tilde{M}_{20}$ given by Eq. (417) for the diatomic model chains with the parameters $M_A = 4M_B$ [88]. The total contribution is shown by (a), from acoustic phonons by (b), and from optical phonons by (c). The total classical contribution is shown by (d), together with the classical contribution from acoustic (and optical) phonons by (e).

from optical phonons shown by Figure 28e [88]. In the low-temperature region optical phonons contribute significantly to the EXAFS Debye–Waller factor in comparison with the acoustic phonon. In contrast to the low-temperature region, their contributions are comparable in the high-temperature region. This result is interesting because optical phonons are not excited in the low-temperature region, whereas acoustic phonons are easily excited, even in the low-temperature region, because $\omega_{\text{acoustic}}(k) \ll \omega_{\text{optical}}(k)$ as $k \to 0$. At very low temperatures only a zero point vibration can contribute to the Debye–Waller factor. In the case of EXAFS the relative displacement is the measure of the Debye–Waller factor, and it is larger for optical phonons than that for acoustic phonons because atoms A and B in the same unit cell are displaced in the opposite direction for the optical branch, whereas they move in the same direction for the acoustic branch.

In contrast, all phonons can be excited in the high-temperature region, and we can expect the nearly same contribution from the two types of phonons. For this model phonon system we can prove that the two contributions are exactly the same in the high-temperature region [88]. If we extrapolate the curves b and c to the very high temperature region, they must coincide. As a whole we again find a result similar to those for monatomic chains; all curves behave as const. × T in the high-temperature region and T^0 in the low-temperature region. This figure shows that the acoustic phonon contribution is well-approximated by the classical approximation, whereas the optical phonon contribution includes a considerable amount of quantum effect, even in the high-temperature region.

In comparison with the results for the first shell, we find that the contribution from the acoustic mode is 4 times larger, whereas the contribution from the optical vibration is negligibly small for the second-shell Debye–Waller factor. We consider an A–A atomic correlation for the second-shell EXAFS, and the relative displacement of that pair is quite small, even for optical modes.

Figure 29 shows the dimensionless third-order cumulant $\tilde{M}_3$ for a first-shell atom in a diatomic chain, where $\tilde{M}_3$ is defined as $\tilde{M}_3 = M_3/(32\alpha\hbar^2/5a\pi^2\gamma\mu)$ [88]. It should be noted that there are two different types of cross terms, O-O-A and O-A-A (O = optical, A = acoustic). The result shows that the contribution from the O-O-A cross term is dominant in the third-order cumulant. The result shown in this figure can be considered as a model observed for M_3 in EXAFS spectra of $Zn(H_2O)_6^{+2}$ complex measured by Miyanaga et al. [116]. The ratio of atomic mass for nearest neighbor Zn–O pair is about 4. They found a close relation between the anharmonicity of the atomic vibration and the efficiency of the ligand exchange reaction rate. The observed temperature dependence of the third cumulant is well reproduced by the present calculation. In the high-temperature region we find that $\tilde{M}_3$ behaves like T^2, as expected from the classical theory.

Figure 30 shows the temperature dependence of the dimensionless fourth cumulant $\tilde{M}_4$ for the first shell defined by $\tilde{M}_4 = M_4/\{(56\alpha^2/a^2)(\hbar/\pi\sqrt{5\gamma\mu})^3\}$. In this case we classify five different contributions, O-O-O-O, O-O-O-A, O-O-A-A, O-A-A-A, and A-A-A-A. In this case O-O-A-A is the most important; however, it is difficult to explain the reason for this. At high temperature it behaves like T^3, as predicted by the classical approximation [87].

When we analyze the temperature dependence of the experimental Debye–Waller factor for complex systems, we use a Debye and/or Einstein approximation. It is important to check here the reliability of these approximations for the simple diatomic chain model. The Debye model and the Einstein model are applied to the acoustic mode and the optical mode, respectively, for the diatomic chain model. Figure 31 shows the temperature dependence of the scaled harmonic cumulant $\tilde{M}_{20}$ [89]. The Debye and Einstein approximations are compared with the *ab initio* result discussed above. For the harmonic second-order cumulant, both Debye and Einstein approximations reproduce the *ab initio* results well. On the other hand, the

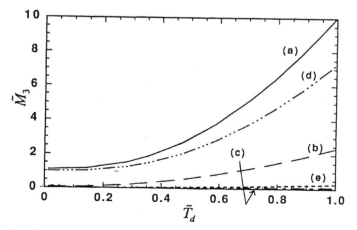

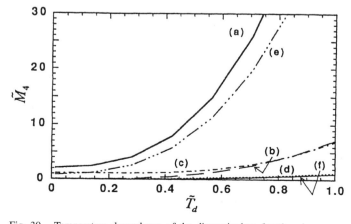

Fig. 29. Temperature dependence of the dimensionless third-order cumulant $\tilde{M}_3$ for a first-shell atom in a diatomic model chain [88]. (a) Total contribution, (b) the contribution from acoustic phonon, (c) that from optical phonon, (d) that from O-O-A cross term, (e) that from O-A-A cross term.

Fig. 30. Temperature dependence of the dimensionless fourth-order cumulant $\tilde{M}_4$ for a first-shell atom in a diatomic model chain. (a) Total contribution, (b) the contribution from acoustic phonon, (c) that from optical phonon, (d) that from O-O-O-A cross term, (e) that from O-O-A-A cross term, (f) that from O-A-A-A cross term.

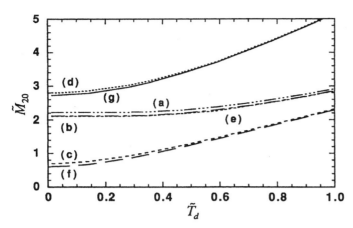

Fig. 31.   Temperature dependence of the harmonic contribution to the Debye–Waller factor in a diatomic chain with $M_A = 4M_B$ [89]. (a and b) The optical contribution with (a) the Einstein I model and (b) the Einstein I' model (dotted line). (c) The acoustic contribution with the Debye I model. (d) Total contribution from the Debye I and Einstein I' models. (e) *Ab initio* result of the optical mode (dash-dotted line). (f) *Ab initio* result of the acoustic mode. (g) *Ab initio* result of the total contribution.

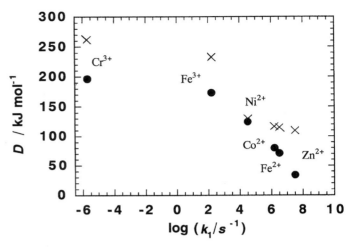

Fig. 32.   Relation between the dissociation energy D determined from EXAFS cumulant analyses and the water exchange reaction rate constant, $\log k_1$ [121]. The data points with crosses represent the theoretical results of Akesson et al. [122].

reproducibility of these approximations becomes worse for the third-order cumulant and more pronounced for the fourth-order cumulant [89]. Dalba et al. found that the experimental behavior of the Debye–Waller factor for AgI is well reproduced by the mixed model of Debye and Einstein approximations [78, 78a]. It has been found that the low-frequency optical modes have much more of an effect on the second-shell than the first-shell M_{20} [78, 78a].

So far we have discussed the Debye–Waller factor for a one-dimensional model in the framework of the perturbation method. It is possible to apply this method to two- and three-dimensional diatomic models [117]. The temperature dependence of the EXAFS Debye–Waller factors for the first and second nearest neighbor atoms in highly symmetric two- and three-dimensional crystals shows behavior quite similar to those of one-dimensional crystals. This is because the EXAFS Debye–Waller factors reflect projected relative deviation within the single scattering approximation. If we consider the multiple scattering EXAFS Debye–Waller factors, we can expect some observable difference between low-dimensional and three-dimensional systems, even for highly symmetric crystals.

Equation (372) shows that the different k dependence enables us to obtain direct information about each cumulant M_n. On the other hand, in the classical approximation we can closely relate the EXAFS thermal factor to the interatomic potential V, as shown by Eq. (428), where the effective interatomic potential V_L should be replaced by V. Yokoyama et al. first applied this relation to obtain information on V in Ag and Pd clusters [118] and in some crystals [86]. Dalba et al. studied the pair potential between Ag and I in β-AgI [119], where they observed strong anharmonicity. In solutions, the interatomic potential between metal ions and the surrounding solvent molecules is studied with the use of the temperature dependence of EXAFS for Br^- ions [120] and 3d metal ions

[121] in solutions. We show here an interesting example of the relation between the interatomic potential obtained from the EXAFS Debye–Waller factor and the chemical reaction rate constant [122]. In the classical approximation, the second-order harmonic cumulant $M_{20} (= \sigma^2)$ for a diatomic system with a Morse potential is proportional to the temperature T [87, 88] as

$$\sigma^2 \propto \frac{a^2}{D\alpha^2} k_B T \qquad (419)$$

On the other hand, the third-order cumulant M_3 for the same system is proportional to T^2 in the classical approximation as [87, 88]

$$M_3 \propto \frac{a^3}{D^2\alpha^3} (k_B T)^2 \qquad (420)$$

From the experimental measurements of the temperature dependence of the EXAFS cumulants, we can determine the coefficients of $\eta = \sigma^2/T$ and $\xi = M_3/T^2$ and relate them to the dissociation energy D in the Morse function in Eq. (415),

$$D = \kappa \frac{\eta^3}{\xi^2} \qquad (421)$$

where factor κ is a constant value calculated from the diatomic model [88]. The Arrhenius theory is applied to the water-exchange reaction in the solution,

$$k_1 = A \exp(-E_a/k_B T) \qquad (422)$$

where k_1 is the ligand water-exchange rate constant, E_a is the activation energy, and A is the frequency factor. Under the assumption that the water-exchange dissociative reaction proceeds predominantly, the Morse potential can be regarded as a model adiabatic potential for this reaction; thus D can be considered as E_a. In Figure 32 the dissociation energy D determined from the EXAFS cumulant analysis for 3d metal ions in aqueous solutions is plotted against the logarithm of the water-exchange rate constant k_1 [121]. The D value has a good correlation with the water-exchange rate constant, and it is interesting to note that the extrapolation to $D = 0$ comes close to the point of $\log k_1 \approx 10$ for the diffusion-limited reaction in water. For this system, the dissociation energy D was calculated by the *ab initio* SCF method by Akesson et al. [122] (cross points in Fig. 32). We can find that the D values obtained from EXAFS are comparable to that from the theoretical calculation. As demonstrated above, in the classical approximation, we can directly obtain the interatomic potential not only in solid states but also in solutions, surfaces, and catalysis. The cumulant analyses of the EXAFS Debye–Waller factors give us useful information about the dynamical properties of more complex systems.

6.2. Path-Integral Approaches to EELFS and EXAFS Debye–Waller Factors

Next we discuss the path-integral approach to XAFS Debye–Waller factors to go beyond the classical approximation. The path-integral approach is applicable to strong anharmonic systems, and numerical calculations are required to evaluate the Debye–Waller factor. Formulation of the XAFS Debye–Waller factor in the framework of the path-integral theory was first performed by Fujikawa et al. and applied to the double-well potential [91] and Morse potential [92, 93]. The direct comparison of the experimental XAFS data with the path-integral calculation was performed for Br_2 molecules and Kr and Ni crystals by Yokoyama [94]. He discussed the advantages and disadvantages of the path-integral approach in comparison with the perturbation method. All of these works demonstrate the power of path-integral approaches for the study of strongly anharmonic systems.

First, we introduce the theoretical basis of the path-integral approach to the XAFS Debye–Waller factor developed by Fujikawa et al. [91].

6.2.1. Basic Theory

Let consider diatomic systems in a reservoir at temperature T whose relative vibrational motion is described by the Hamiltonian

$$H = \frac{p^2}{2\mu} + V(q) \qquad (423)$$

where μ is the reduced mass and q is the instantaneous interatomic distance. When we deal with the statistical average of an operator A, we should calculate the trace

$$\langle A \rangle = \frac{1}{Z} \text{Tr}\langle A\rho \rangle \qquad (424)$$

where ρ is the density operator defined by $\rho = \exp(-\beta H)$, $\beta = 1/k_B T$, and Z is the partition function for the system. The trace can be calculated by applying Feynman's path integral techniques; however, instead of summing over all paths in just one step, one can classify the paths into two groups as proposed by Feynman [123]. One group consists of an average

(quasi-classical) path $\bar{q}$ given by

$$\bar{q} = \frac{1}{\beta} \int_0^\beta du\, q(u) \qquad (\hbar = 1) \qquad (425)$$

and the other group consists of quantum fluctuation around $\bar{q}$. The average path is the same as the classical path in the high-temperature limit ($\beta \to 0$). To use the nonperturbation method based on the path-integral technique, we approximate the instantaneous potential $V(q(u))$ by a trial potential quadratic in the fluctuation path [124, 125],

$$V \cong V_0(q, \bar{q}) = w(\bar{q}) + \frac{\mu\omega(\bar{q})^2}{2}(q - \bar{q})^2 \qquad (426)$$

Now the parameters w and ω are to be optimized, so that the trial reduced density well approximates the true reduced density [124]. A variational approach that gives the same result as the self-consistent approximation is also possible [124, 125]. Final expression for the average of a local operator A can be represented in terms of the probability density just as in classical statistical mechanics (from now on q is used instead of $\bar{q}$ for brevity),

$$\langle A \rangle = \int A(q)P(q)\, dq \qquad (427)$$

This expression, however, includes the important quantum effects, and the probability density is represented by

$$P(q) = \frac{1}{Z}\sqrt{\frac{\mu}{2\pi\beta}} \exp[-\beta V_L(q)] \qquad (428)$$

where the local effective potential $V_L(q)$ is defined by [124]

$$\exp[-\beta V_L(q)] = \int dq' \exp[-V_e(q + q')]\frac{1}{\sqrt{2\pi\alpha(q + q')}}$$
$$\times \exp[-q'^2/2\alpha(q + q')] \qquad (429)$$

Now we have used the relations

$$V_e(q) \equiv w(q) + \frac{1}{\beta}\ln\left(\frac{\sinh f}{f}\right)$$
$$f(q) = \beta\omega(q)/2$$
$$\alpha(q) = (\coth f - 1/f)/(2\mu\omega(q)) \qquad (430)$$

The local effective potential $V_e(q)$ is reduced to the bare potential $V(q)$ in the high-temperature limit. In the EXAFS analyses the operator A should be $\exp(2ik\Delta_\alpha)$, where k is the wave vector of an ejected photoelectron, and Δ_α is the projected relative displacement [75], which is simply given by $\Delta q = q - q_0$ (q_0 is the equilibrium interatomic distance) in one-dimensional cases [81]. So what we should calculate to study EXAFS thermal factors is the thermal average including the quantum fluctuation, given with the use of the probability density defined by Eq. (428),

$$\langle\exp(2ik\Delta q)\rangle = \frac{1}{Z}\sqrt{\frac{\mu}{2\pi\beta}}\int \exp(2ik\Delta q)\exp[-\beta V_L(q)]\, dq \qquad (431)$$

in the PW approximation. We now shift the origin for the potential V to $q_0 = 0$. This expression clearly shows that the widely used classical real space representation is reproduced with some modification including the quantum fluctuation effects; the original interatomic potential V should be replaced by the local effective potential V_L, which is temperature dependent and tends to be $V(q)$ at high temperatures from physical considerations.

What we should calculate to study EXAFS thermal factors by phonon effects is the thermal average including the quantum fluctuation, given with the use of the probability density defined by Eq. (428),

$$g(k) = \langle \exp(2ihq) \rangle = \int_{-\infty}^{\infty} \exp(2ikq) P(q) \, dq \quad (432)$$

When all of the integrals in the cumulants and the cumulant expansion converge, the thermal damping function $g(k)$ can be written as

$$g(k) = \exp\left\{ -2k^2 \langle q^2 \rangle_c + \tfrac{2}{3} k^4 \langle q^4 \rangle_c - \cdots \right\}$$
$$\times \exp\left[i \left\{ -2k \langle q \rangle_c + \tfrac{4}{3} k^3 \langle q^3 \rangle_c - \cdots \right\} \right] \quad (433)$$

In comparison with the results for the symmetric potentials, the odd orders of the cumulant expansion can contribute to the phase factor. This expansion is useless for the Morse potential in the high-temperature region, because each cumulant $\langle q_n \rangle_c$ diverges for this potential. In this case our choice is only a direct calculation of $g(k)$ in Eq. (432).

6.2.2. Application to Double-Well Potential

First we consider the double-well potential

$$V(q) = \epsilon \left(q^2 - \sigma^2 \right)^2 \quad (434)$$

which gives the absolute minimum $V(q)_{\min} = 0$ at $q = \pm \sigma$. The double-well potential can be useful in models used to study high-T_c super conductors [127] and perovskite ferroelectric materials [128]. An important parameter for describing quantum effects is $\bar{g}$, defined by $\omega_0 / \epsilon \sigma^4$; the weak (strong) quantum effects are expected when $\bar{g}$ is small (large). In the following application we shall use the unit $\sigma = \mu = 1$ for simplicity. With the use of the self-consistent method described in the previous subsection, we have the relations corresponding to Eq. (430):

$$V_e(q)/\epsilon = \left(q^2 - 1 \right)^2 - 3\alpha(q)^2 + \frac{1}{\beta} \ln\left(\frac{\sinh f(q)}{f(q)} \right) \quad (435)$$

$$f(q)^2 = \frac{\beta^2 \bar{g}^2}{8} \quad (436)$$

$$\alpha(q) = \frac{\beta \bar{g}^2}{32} \left(\coth f(q) - \frac{1}{f(q)} \right) \Big/ f(q) \quad (437)$$

The above coupled equation can be solved with the conventional numerical technique, as shown in the previous subsection [126].

Figure 33 shows the $\bar{g}$ dependence of the second-order cumulant $\langle q^2 \rangle_c$ as a function of temperature for $\bar{g} = 1$ (a), 5 (b), and 10 (c) [91]. For comparison the classical results are also shown by a dotted line. For small $\bar{g} (= 1)$, the difference between the classical and the quantum calculations is very small in the whole temperature region, whereas the large $\bar{g}$ values $(= 5$ and $10)$ give rise to large quantum effects, particularly at low temperature. All of the classical results give the same value of $\langle q^2 \rangle_c$ at $T = 0$ K; this is explained by the fact that a classical particle in the double-well potential is frozen at one of the bottoms of the potential well, at $q = \pm 1$. So we obtain the result $\langle q^2 \rangle_c = 1$ in the classical approximation at $T = 0$ K. Of course the classical approximation does not work at such low temperatures, and the tunneling effects play an important role through the potential barrier at $q = 0$. When $\bar{g} = 1$, both the classical and the quantum cumulants $\langle q^2 \rangle_c$ show nearly the same temperature dependence because of weak quantum effects for small $\bar{g}$. At low temperatures the second cumulants $\langle q^2 \rangle_c$ show quite different behaviors for the classical and the quantum approaches because the quantum effects cannot be neglected in the low-temperature region, $T < 2$. It should be noted that the quantum effect is not so large in comparison with those found in Figures 33b and c. On the other hand, the classical and the quantum $\langle q^2 \rangle_c$ show quite different behaviors at low temperature for $\bar{g} = 5$ and 10.

Figure 34 shows the fourth-order cumulant $\langle q^4 \rangle_c$ for the three different $\bar{g}$. The result obtained is similar to that observed in the discussion for Figure 29; the quantum effect is large for large $\bar{g}$ values because of the large tunnel probability, and the deviation from the classical approximation for $\langle q^4 \rangle_c$ is larger than that for $\langle q^2 \rangle_c$.

Figure 35a shows the comparison of EXAFS damping factors $g(k)$ defined by Eq. (432) for the classical and quantum probability density $P(q)$ shown in Figure 35b, and the cumulant expansion given by Eq. (433) up to the second- and fourth-order cumulants. The parameter $\bar{g}$ is set to be 5 and T to be 0.16. At this temperature the classical-like probability density $P(q)$ is obtained as shown in Figure 35b. The two damping functions are nearly the same for the classical and quantum probability density; however, this result is quite different from the result obtained with the use of the cumulant expansion. In particular we find that $g(k)$ obtained the without the use of the cumulant expansion is negative in the region $0.9 < k < 2.0$. As far as this expansion can be applied, $g(k)$ should be positive for any value of k because the argument of the exponential function is real. The negative $g(k)$ found in Figure 35a clearly indicates the inapplicability of the cumulant expansion, whereas the classical approximation is good.

At lower temperature $T = 0.02$ the slow damping of $g(k)$ is observed for the three quantum approaches in Figure 36a as a function of k as expected, though the classical probability density gives rise to a behavior completely different from those of the other three quantum approaches. In particular the cumulant expansion up to fourth order gives a satisfactory result. The two curves for the self-consistent approach and the cumulant approach up to fourth order cannot be distinguished in this figure.

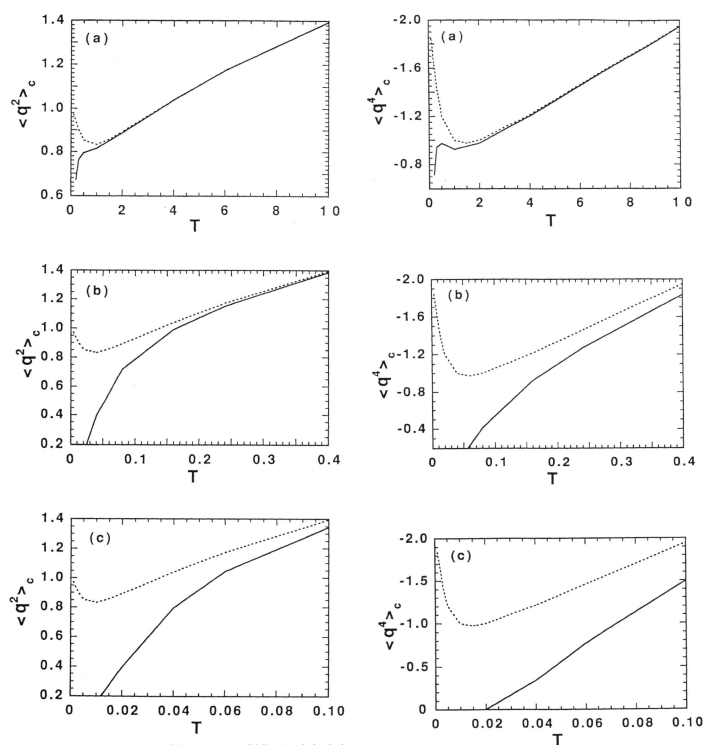

Fig. 33. Temperature dependence of the quantum (solid lines) and classical (dotted lines) second-order cumulants for the double-well potential at three different parameters, $\bar{g} = 1$ (a), 5 (b), and 10 (c) [91].

Fig. 34. Temperature dependence of the quantum (solid lines) and classical (dotted lines) fourth-order cumulants for the double-well potential at three different parameters, $\bar{g} = 1$ (a), 5 (b), and 10 (c) [91].

The very poor result for the classical approximation is due to the interesting behavior of $P(q)$ as shown in Figure 36b: the classical approximation gives two peaks at $q \approx \pm 1$, whereas the quantum approach gives only the single peak at $q \approx 0$.

In this case the quantum $g(k)$ is well approximated by the cumulant expansion up to fourth order; the cumulant expansion shows rapid convergence.

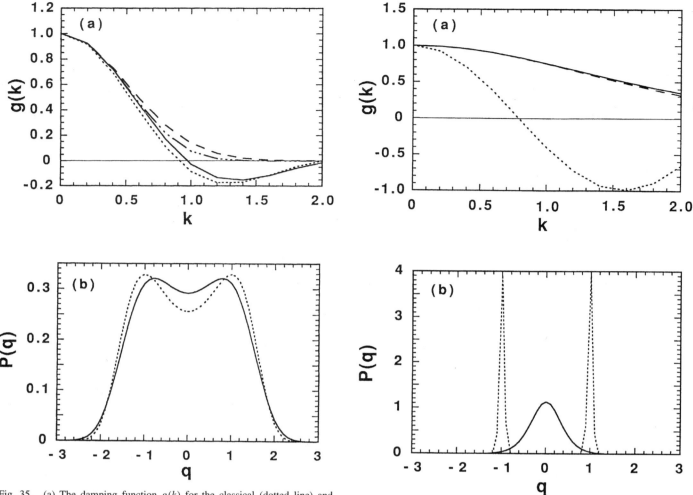

Fig. 35. (a) The damping function $g(k)$ for the classical (dotted line) and quantum (solid line) probability density, and also for the cumulant expansion up to second (dashed line) and fourth order (three-dots-dashed line), at $\bar{g} = 5$ and $T = 0.16$. (b) The probability density $P(q)$ for the classical (dotted line) and the quantum (solid line) methods [91].

Fig. 36. As in Figure 35, except that $T = 0.02$ [91].

It is interesting to see that even for the same double-well potential $g(k)$ shows quite different behaviors for different temperatures. At low temperatures the cumulant expansion can be applied, whereas the classical approximation breaks. On the other hand, the classical approximation can be applied, whereas the cumulant approximation breaks at high temperatures. The criteria for applying the classical approximation are given by the effective probability density $P(q)$. If two prominent peaks are found, the classical approximation can be safely applied.

To closely compare the calculated result with the corresponding experimental results we should specify the systems to be considered. As a model system the Cu–O diatomic system in $YBa_2Cu_3O_{7-\delta}$ is considered. Several experimental results indicate that the axial oxygen moves in a double-well potential. Mustre de Leon et al. [127] used an approximation; $\epsilon\sigma^4 = 200\, k_B$, $\sigma = 0.1$ Å. The Einstein temperature for this system is estimated as $\theta = 523$ K by Crozier et al. [129]. With the

use of these parameters we can calculate the damping function $g(k)$ for several temperatures, which are shown in Figure 37: $T = 300$ K (a), 80 K (b), and 30 K (c) [91]. For this system the calculated results show that neither the cumulant expansion nor the classical approximation works well at the three temperatures considered here. Mustre de Leon et al. [127] found phase inversion of the EXAFS envelope function; this result is consistent with the present calculations.

6.2.3. Application to Morse Potential

In this section the method described in Section 6.2.1 is applied to diatomic systems in a heat bath, which vibrate around an equilibrium distance q_0 in a Morse potential,

$$V(q) = D\big[\exp\{-2\Gamma(q - q_0)\} - 2\exp\{-\Gamma(q - q_0)\}\big] \quad (438)$$

where the parameter D describes the depth or the dissociation energy and Γ the curvature of the potential. The self-consistent approach described in the previous subsection gives the set of

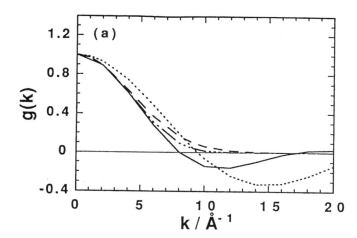

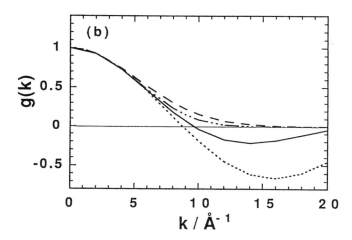

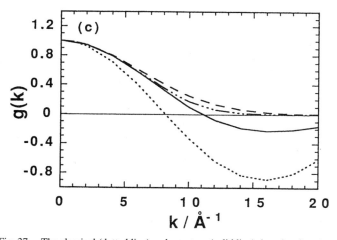

Fig. 37. The classical (dotted line) and quantum (solid line) damping function $g(k)$ for the Cu–O system with the double-well potential at (a) $T = 300$ K, (b) $T = 80$ K, and (c) $T = 30$ K [91]. The cumulant expansion for $g(k)$ are also shown up to the second order (dashed lines) and fourth order (three-dots-dashed lines). The parameters used for Cu–O in $YBa_2Cu_3O_{7-\delta}$ were proposed by Mustre de Leon et al. [127] and by Crozier et al. [129].

equations in terms of f_1 and f_2, which are defined by

$$f_1(q) = \exp\{-\Gamma(q - q_0) + \alpha\Gamma^2/2\} \qquad (439)$$

$$f_2(q) = \exp\{-2\Gamma(q - q_0) + 2\alpha\Gamma^2\} \qquad (440)$$

We have to solve the nonlinear equations given by

$$\mu\omega(q)^2 = 2D\Gamma^2[2f_2(q) - f_1(q)] \qquad (441)$$

$$w(q) = [1 - 2D\Gamma^2\alpha(q)]f_2(q)$$
$$\quad - [2 - D\Gamma^2\alpha(q)]f_1(q) \qquad (442)$$

where $\alpha(q)$ and $f(q)$ are given by Eq. (430). We can show that the moments $\langle q^n \rangle$ diverge in the classical approximation for the Morse potential. However, they converge because of the quantum fluctuation at low temperatures as shown below.

The Morse potential is asymmetric around the equilibrium position q_0, and the thermal factor defined by Eq. (438) for the EXAFS Debye–Waller factor should be complex.

Figure 38 shows the probability densities given by Eq. (428) compared with those in the classical approximation for this potential with $D = 1$ (except for the three-dotted line in (b), where $D = 5$), $\Gamma = 0.1$ Å^{-1}, $q_0 = 2.5$ Å [92, 93]. For example, a typical value of D is 0.8 eV; $T = 930$ K is obtained at a reduced temperature t ($= k_BT/D$) = 0.1. This value is obtained when we assume that the Einstein temperature $\theta = 523$ K for a Cu–O diatomic system [128]. At high temperatures ($t = 0.1$, Figure 38a), the quantum probability density is nearly the same as the classical one. At low temperatures ($t = 0.02$, Fig. 38b) the quantum probability distribution shifts its maximum position to 2.6 Å because of the tunneling effects, as observed for the double-well potential.

The EXAFS thermal factors defined by Eq. (432) can be written for the asymmetric potential, as they can for the Morse potential, by

$$g(k) = |g(k)| \exp\{ik\psi(k)\} \qquad (443)$$

Figure 39a shows the EXAFS thermal damping function and Figure 39b shows the phase $\psi(k)$ calculated with the use of the path-integral approach for three different temperatures [92]. Of course, the damping function decays rapidly at high temperatures, as shown in Figure 39a. On the other hand, we cannot easily understand the behaviors of the phase. They are rather sensitive to the temperature, as shown in Figure 39b. At low temperatures ($t = 0.02$), $\psi(k)$ is positive and has a broad peak at ~ 6 Å. As the temperature increases $\psi(k)$ is negative for large k (> 2 Å^{-1}). The latter behavior can be expected from the temperature dependence of $P(q)$. The probability density shifts to the large q side at low temperatures because of the tunneling effect, which gives rise to a positive first cumulant. Therefore $\psi(k)$ is an increasing linear function of k in the low energy region. The tails around the peak to the large q side give a positive third-order cumulant, which is a decreasing function as k^3 in the intermediate energy region. At high temperatures third-

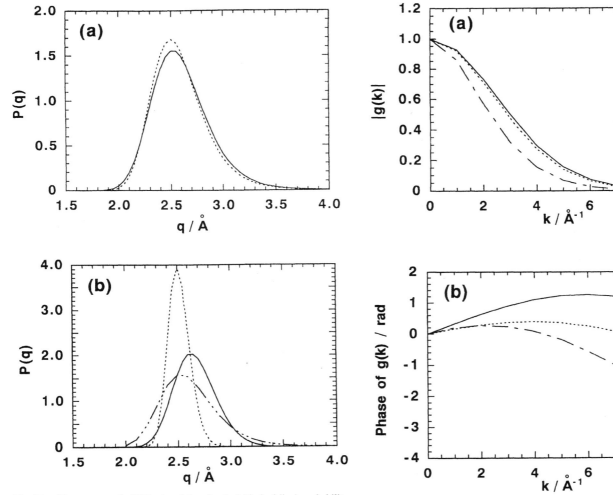

Fig. 38. The quantum (solid line) and the classical (dashed line) probability density functions for the Morse potential at high temperature ($t = 0.1$) (a) and at low temperature ($t = 0.02$) (b) for $D = 1$. Three dot-dashed line in (b) represents the quantum probability density for $D = 5$ [92, 93].

Fig. 39. (a) EXAFS thermal damping function $|g(k)|$ and (b) phase $\psi(k)$ as a function of k for $t = 0.02$ (solid line), $t = 0.05$ (dashed line), and $t = 0.1$ (dot-dashed line) [92].

order cumulant plays an important role, and deviation from the linear function of k is still found, even in the low-energy region. This discussion works as far as the cumulant expansion converges at low temperature, $t = 0.02$ and 0.05.

Figure 40 shows the damping function and phase function, at low temperatures ($t = 0.02$) in comparison with the cumulant results [92]. At low temperatures all of the cumulant expansions up to the fourth-order terms converge because of the weak anharmonicity. Both of the cumulant expansions up to the second- and fourth-order terms give satisfactory results; the latter result in particular is good enough, whereas the classical approximation gives a poor result. In Figure 40b, the first-order cumulant gives a linear function of k, which is a poor approximation for large k. If we take the third-order cumulant into account, the result is much improved. The classical approximation, on the other hand, is quite bad for the phase factor, too. The quantum damping function is much smaller than that in the classical approximation.

The path-integral effective potential method was applied to the calculation of the EXAFS cumulants up to third or fourth order for diatomic Br_2 and solid Kr and Ni by Yokoyama [94]. In the case of solid Kr the pair-potential approximation was employed to describe van der Waals interactions, whereas for Ni metal the embedded atom method that accounts for many-body interactions in metallic bonds was applied. The evaluated EXAFS cumulants were compared with experimental results, and excellent agreement was obtained. In particular, Yokoyama obtained good agreement between the calculation and the EXAFS cumulant for higher nearest-neighbor shells in Ni as well as for the first nearest one, by taking into account the many degrees of vibrational freedom and three-dimensional periodicity. He also pointed out the difficulty of the effective potential calculation for the third-order cumulant for Br_2; a strange decrease was found at temperatures lower than ~ 100 K.

The path-integral effective potential method discussed here is applicable to the study of the quantum correction for the

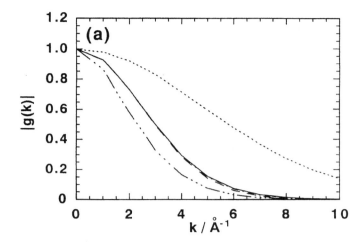

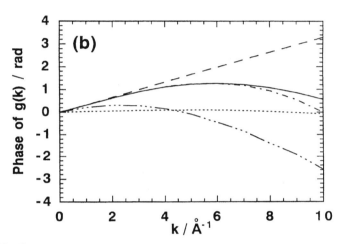

Fig. 40. (a) EXAFS thermal damping function $|g(k)|$ and (b) phase $\psi(k)$ as a function of k for $t = 0.02$ (solid line). The results for the cumulant expansion are also shown up to the second (first) order by the long dashed lines and up to the fourth (third) order by dash-dotted lines in (a) ((b)). The dash-dotted lines are nearly the same as the quantum results. Three-dot-dashed lines in (a) and (b) represent the quantum results for $D = 5$ [92].

molecular dissociative reaction rate [130, 131]. We can calculate the decrease in the effective dissociation energy D in the Morse potential due to the quantum tunneling effect, with the use of the path-integral method. The temperature dependence of the dissociative chemical reaction rate is discussed in connection with the temperature dependence of the effective dissociation energy [131]. Combining the EXAFS Debye–Waller factor measurements and the path-integral effective potential calculation provides us with a new tool for the study of the quantum effect on the chemical reaction rate, with the use of XAFS spectroscopy. The effective potential method can explain the deviation from the classical Arrhenius plot for dissociative reactions.

To study temperature effects in backscattering EXELFS, it is also important to investigate XPD thermal factors. Such factors are studied by Miyanaga et al. [155, 156] for strongly anharmonic systems with the use of the path-integral technique.

6.3. Spherical Wave Effects on EELFS and XAFS Debye–Waller Factors

Here we discuss SW effects on Debye–Waller factors to study the temperature effects in EELFS and ELNES and in EXAFS and XANES analyses.

6.3.1. Single Scattering SW Debye–Waller Factors

The single scattering K-edge EXAFS formula is given for randomly oriented samples [132],

$$\chi(k) = -k \sum_{\alpha} \text{Im}\left[\sum_{l} t_l^{\alpha}(k) \{ l h_{l-1}^2(kR_{\alpha}) \right.$$
$$\left. + (l+1)h_{l+1}^2(kR_{\alpha})\} \exp(2i\delta_1^A) \right] \quad (444)$$

In the plane wave approximation the spherical Hankel function $h_l(kR_{\alpha})$ is approximated by $i^{-l-1}\exp(ikR_{\alpha})/kR_{\alpha}$, which converts the above formula to the well-known EXAFS formula.

The procedure for averaging thermal motion is first to replace $\mathbf{R}_{\alpha}$ by $R_{\alpha}^0 + \Delta_{\alpha}$. Second, we apply Brouder's formula for the spherical Hankel function with small displacement [96],

$$\sqrt{2l+1}h_l(kR_{\alpha}^0 + k\Delta_{\alpha}) = \sum_{\lambda} \sqrt{2\lambda+1}h_{\lambda}(kR_{\alpha}^0)J_{\lambda l}(k\Delta_{\alpha})$$
$$(445)$$

where the matrix J is an orthogonal representation of a one-parameter Lie group, and Lie's theorem states that there exists a real antisymmetric matrix A such that

$$J(k\Delta_{\alpha}) = \exp(k\Delta_{\alpha}A) \quad (446)$$
$$A_{ll'} = i^{l-l'+1}\sqrt{(2l+1)(2l'+1)}\langle l0l'0|1\rangle^2/3 \quad (447)$$

We substitute Eq. (445) into Eq. (444) and take the average of it over a canonical ensemble. Then we have

$$\chi(k) = -k \sum_{\alpha} \text{Im}\left[\exp(2i\delta_1^A) \sum_{l} t_l^{\alpha} \sum_{\lambda\lambda'} \sqrt{(2\lambda+1)(2\lambda'+1)} \right.$$
$$\times h_{\lambda}(kR_{\alpha}^0)h_{\lambda'}(kR_{\alpha}^0)$$
$$\left. \times \langle [J(k\Delta_{\alpha})B^lJ(-k\Delta_{\alpha})]_{\lambda\lambda'} \rangle \right] \quad (448)$$

In the above equation we introduce the diagonal matrix B defined by

$$B_{\mu\mu}^l = \delta_{\mu\mu}(b_l^-\delta_{\mu,l-1} + b_l^+\delta_{\mu,l+1}), \qquad b_l^- = \frac{l}{2l-1},$$
$$b_l^+ = \frac{l+1}{2l+3} \quad (449)$$

The ensemble average is now represented by the symbol $\langle \rangle$. By applying the Campbell–Baker–Hausdorff formula, we obtain

$$\langle J(k\Delta_{\alpha})B^lJ(-k\Delta_{\alpha})\rangle$$
$$= B^l + k\langle\Delta_{\alpha}\rangle[A, B^l] + \frac{k^2}{2}\langle\Delta_{\alpha}^2\rangle[A, \bullet]^2B^l + \cdots \quad (450)$$

where $[A, \bullet]^2 B^l = [A, [A, B^l]]$, and so on. To relate the expression (450) to the widely used plane wave cumulant expression, we rewrite Eq. (448) as

$$\langle J(k\Delta_\alpha) B^l J(-k\Delta_\alpha)\rangle = \sum_{n=0}^{\infty} \frac{(2ik)^n}{n!}\langle\Delta_\alpha^n\rangle B^l + \sum_{n=1}^{\infty} \frac{k^n}{n!} Y_n^l \langle\Delta_\alpha^n\rangle \quad (451)$$

where we define Y_n^l as

$$Y_n^l = ([A, \bullet]^n - (2i)^n) B^l = \tilde{Y}_n B^l \quad (452)$$

$$\langle J(k\Delta_\alpha) B^l J(-k\Delta_\alpha)\rangle$$

$$= \exp\left\{\sum_{n=1}^{\infty} \frac{(2ik)^n}{n!}\langle\Delta_\alpha^n\rangle_c\right\} B^l$$

$$+ \left[\exp\left(\sum_{n=1}^{\infty} \frac{k^n}{n!}\langle\Delta_\alpha^n\rangle_c \tilde{Y}_n\right) - 1\right]$$

$$\times \exp\left\{\sum_{n=1}^{\infty} \frac{(2ik)^n}{n!}\langle\Delta_\alpha^n\rangle_c\right\} B^l \quad (453)$$

Substituting the first term of Eq. (453) into Eq. (448), we obtain the expression

$$\chi^p(k) = \sum_\alpha \frac{1}{k(R_\alpha^0)^2} \text{Im}\left[\exp(2ikR_\alpha^0 + 2i\delta_1^A)\tilde{f}_\alpha(\pi, z_\alpha)\right.$$

$$\left.\times \exp\left\{\sum_{n=1}^{\infty} \frac{(2ik)^n}{n!}\langle\Delta_\alpha^n\rangle_c\right\}\right] \quad (454)$$

which will be called the *thermal plane wave part*, where we have used relation (117) and

$$z_\alpha = 1/ikR_\alpha \quad (455)$$

The spherical wave backscattering amplitude $\tilde{f}_\alpha(\pi, z_\alpha)$ is used instead of the ordinary plane wave backscattering amplitude written in terms of $c_l(z_\alpha)$ defined by Eq. (455),

$$\tilde{f}_\alpha(\pi, z_\alpha) = -\sum_{l=0}(-1)^l t_l^\alpha\{lc_{l-1}(z_\alpha)^2 + (l+1)c_{l+1}(z_\alpha)^2\} \quad (456)$$

The factor $c_l(z_\alpha)$ describes the spherical wave effect of a photoelectron wave, which is the lth-order polynomial of z_α, and tends to 1 in the small limit of z_α, $\lim_{z_\alpha\to 0} c_l(z_\alpha) = 1$. Therefore $\chi^p(k)$ approaches the ordinary plane wave EXAFS formula when $z_\alpha \to 0$ ($R_\alpha \to 0$). The second term of Eq. (453) yields the term *thermal spherical wave part* χ^s; it gives a finite contribution only when the spherical wave correction is large. The explicit formula for χ^s is given in [100].

Figure 41 shows four different EXAFS for an Ag$_2$ cluster. χ_p is the ordinary PW EXAFS function with second- and third-order cumulants, and χ^p is the SW EXAFS function with *PW thermal factors* as defined in Eq. (454) [100]. χ^s is the one with *SW thermal factors*, as determined from the second term of Eq. (453), and χ_s is the full SW EXAFS function,

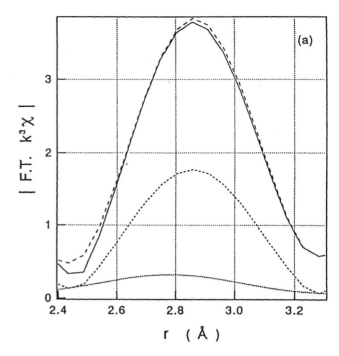

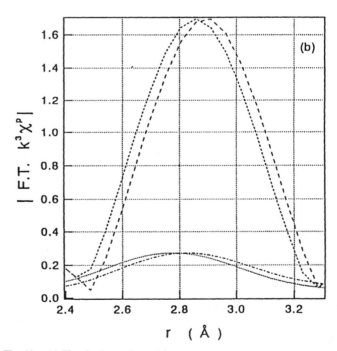

Fig. 41. (a) The absolute values of the Fourier-transformed EXAFS function for an Ag$_2$ cluster [100]. The short dashed lines show the Fourier-transformed EXAFS function at 30 K; the dotted lines show the function at 295 K. (b) As in (a), but for $\chi^p(k)$ at 30 K (short and long dashed lines). (c) As in (b), but for χ^s at 30 K with PW (short dashed lines) and SW phase correction (long dashed line).

$\chi_s = \chi^s + \chi^p$. The solid lines are the results for static approximation, where all cumulants are set to be 0. The dashed and dotted lines are for 30 K and 295 K. For this system χ^s/χ^p

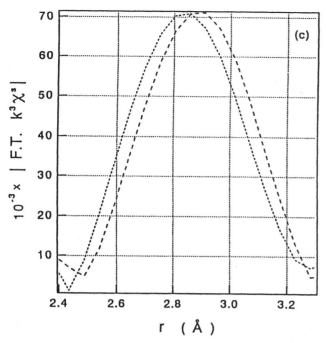

Fig. 41. (Continued.)

is about 0.05. It is interesting to note that χ^s at 295 K is larger than that at 30 K in the small k region, $k < 5.5$ a.u.$^{-1}$, whereas the former rapidly decreases with k because of the exponential Debye–Waller factor. It is also important to note that the phase factor of χ^s is quite different from that of χ^p. This phase difference gives rise to a peak shift if we apply Fourier transformation to the EXAFS function [96, 100, 101].

The approach described above is rather complicated and time consuming, even for the single scattering EXAFS calculations. It is difficult use beyond the single scattering approximation. Fujikawa et al. [102, 103] proposed a simple but quite accurate method for handling XAFS SW thermal factors, which is easily taken into account in multiple scattering EXAFS and XANES formulations. The details of the mathematics are quite different, however.

Instead of the exact relation (445), we use an approximate formula for the spherical Hankel function,

$$h_l(kR) \approx i^{-l-1} \frac{\exp(ikR)}{kR} \left\{ 1 + \frac{l(l+1)}{2k^2R^2} \right\}^{1/2} \exp\left\{ i \frac{l(l+1)}{2kR} \right\}$$

(457)

This approximation is satisfactory for large kR and l. With the use of the relations $R_\alpha = R_\alpha^0 + \Delta_\alpha$ and Eq. (445), we can calculate the single scattering EXAFS function including the SW thermal factor. We find a good agreement between the present approximate and the previous accurate calculations [102, 103].

6.3.2. SW XANES Debye–Waller Factors

In contrast to EXAFS, the temperature dependence in XANES spectra is usually small enough. In some special cases we

should take the thermal vibration effects into account for systems with large thermal fluctuation. For example, recent work on Br K-edge XANES measurement of bromonaphthalene in supercritical Xe solvent clearly shows the importance of thermal vibration, even in the XANES spectra [133]. To consider the thermal factors in XANES spectra we first consider the plane wave multiple scattering thermal factors within the harmonic approximation [102, 103]. An Nth-order multiple scattering factor is calculated from the thermal average,

$$\left\langle \exp\left(ik \sum_{n=1}^{N+1} \Delta_n \right) \right\rangle$$

(458)

where $\Delta_n = \hat{\mathbf{R}}_{n,n-1} \cdot (\mathbf{u}_n - \mathbf{u}_{n-1})$, and $\mathbf{R}_{n,n-1} = \mathbf{R}_n^0 - \mathbf{R}_{n-1}^0$. With the application of Mermin's theorem [134], the average of Eq. (454) is rewritten as

$$\exp\left[-\frac{k^2}{2} \left(\sum_{n=1}^{N+1} \langle \Delta_n^2 \rangle + 2 \sum_{n>m} \langle \Delta_n \Delta_m \rangle \right) \right]$$

(459)

We can expect that the second term, $2 \sum_{n>m} \langle \Delta_n \Delta_m \rangle$, is small because of the random phase cancellation for the complicated MS correlation in solids [102, 103]. In this approximation the plane-wave thermal MS factor can be given by

$$\exp\left[-\frac{k^2}{2} \left(\sum_{n=1}^{N+1} \langle \Delta_n^2 \rangle \right) \right]$$

(460)

As an example, we consider a double scattering path $A \to \alpha \to \beta \to A$. Here we use the expansion formula for the Green's function,

$$G_{LL'}(k\mathbf{R}^0 + k\mathbf{u}) = \sum_{L_1} G_{LL_1}(k\mathbf{R}^0) J_{L_1L}(k\mathbf{u})$$

(461)

$$J_{L_1L}(k\mathbf{u}) = \exp(ik\mathbf{u} \cdot \mathbf{M})_{L_1L}$$
$$= \int Y_{L_1}^*(\hat{\mathbf{k}}) \exp(i\mathbf{k} \cdot \mathbf{u}) Y_L(\hat{\mathbf{k}}) \, d\hat{\mathbf{k}}$$

(462)

where $\mathbf{M}$ is a vector matrix whose explicit formulas are shown in [135, 136]. Using Eq. (462), we have the thermal average of the double scattering term,

$$\sum_{\alpha\beta} \langle G_{LL_1}(-\mathbf{R}_\beta^0 + \mathbf{u}_{A\beta}) t_{l_1}^\beta G_{L_1L_2}(\mathbf{R}_{\beta\alpha}^0 + \mathbf{u}_{\beta\alpha}) t_{l_2}^\alpha$$
$$\times G_{L_2L}(\mathbf{R}_\alpha^0 + \mathbf{u}_{\alpha A}) \rangle$$
$$= \sum_{\alpha\beta} \sum_{L_3L_4L_5} \langle G_{LL_3}(-\mathbf{R}_\beta^0) J_{L_3L_1}(k\mathbf{u}_{A\beta}) t_{l_1}^\beta G_{L_1L_4}(\mathbf{R}_{\beta\alpha}^0)$$
$$\times J_{L_4L_2}(k\mathbf{u}_{\beta\alpha}) t_{l_2}^\alpha G_{L_2L_5}(\mathbf{R}_{\alpha A}^0) J_{L_5L}(k\mathbf{u}_{\alpha A}) \rangle$$

(463)

where $\mathbf{u}_{\beta\alpha} = \mathbf{u}_\beta - \mathbf{u}_\alpha$. By using the lowest order approximation for $J(k\mathbf{u})$ with respect to the SW correction,

$$J(k\mathbf{u}) = \exp(ik\mathbf{u} \cdot \mathbf{M})$$
$$= \exp(ik\mathbf{u} \cdot \hat{\mathbf{R}})\left[1 + ik\mathbf{u} \cdot (\mathbf{M} - \hat{\mathbf{R}}) + \cdots \right]$$
$$= \left[1 + (\mathbf{M} - \hat{\mathbf{R}}) \cdot \frac{\partial}{\partial \hat{\mathbf{R}}} + \cdots \right] \exp(ik\mathbf{u} \cdot \hat{\mathbf{R}})$$

(464)

we can rewrite the thermal average (463) with the aid of Eq. (464); its matrix form is given by

$$\sum_{\alpha\beta} G^{A\alpha}(T) X^{\alpha\beta}(T) X^{\beta A}(T) = \left[G(T) X(T) X(T) \right]^{AA} \quad (465)$$

where the temperature-dependent propagator $G(T)$ is defined by

$$G^{A\alpha}(T) = G\left(-\mathbf{R}_{\alpha}^{0}\right)\left[1 + k^{2}\left(\mathbf{M} + \hat{\mathbf{R}}_{\alpha}^{0}\right) \cdot U^{A\alpha}\hat{\mathbf{R}}_{\alpha}^{0}\right]$$
$$\times \exp\left(-\frac{k^{2}}{2}\hat{\mathbf{R}}_{\alpha}^{0} \cdot U^{A\alpha}\hat{\mathbf{R}}_{\alpha}^{0}\right) \quad (466)$$

and the related matrix $X(T)$ is defined with the use of the temperature-independent matrix X,

$$X^{\alpha\beta}(T) = X^{\alpha\beta}\left[1 - k^{2}\left(\mathbf{M} - \hat{\mathbf{R}}_{\alpha\beta}^{0}\right) \cdot U^{\alpha\beta}\hat{\mathbf{R}}_{\alpha\beta}^{0}\right]$$
$$\times \exp\left(-\frac{k^{2}}{2}\hat{\mathbf{R}}_{\alpha\beta}^{0} \cdot U^{\alpha\beta}\hat{\mathbf{R}}_{\alpha\beta}^{0}\right)$$
$$X_{LL'}^{\alpha\beta} = t_{l}^{\alpha} G_{LL'}(\mathbf{R}_{\alpha} - \mathbf{R}_{\beta})(1 - \delta_{\alpha\beta}) \quad (467)$$

The temperature-dependent tensor $U_{ij}^{\alpha\beta}$ is defined by the thermal average $U^{\alpha\beta} = \langle (\mathbf{u}_{\alpha} - \mathbf{u}_{\beta})_{i} \cdot (\mathbf{u}_{\alpha} - \mathbf{u}_{\beta})_{j} \rangle$. The approximation shown by Eqs. (465) and (467) enables us to renormalize the full multiple scattering series. We thus obtain the temperature-dependent XANES formula excited from a core orbital $\phi_{c}(\mathbf{r}) = R_{l_c}(r) Y_{L_c}(\hat{\mathbf{r}})$ at site A,

$$\sigma(T) = -\frac{8}{3} \text{Im}\left[\sum_{m_c L L'} i^{l-l'} e^{i(\delta_l^A + \delta_{l'}^A)} \rho_c(l) \rho_c(l') G(L_c 10|L) \right.$$
$$\left. \times G(L_c 10|L')(t^{-1})_{LL}^{AA}\left([1 - X(T)]^{-1}\right)_{LL'}^{AA} \right] \quad (468)$$

where $G(LL'|L'')$ is the Gaunt integral and $\rho_c(l)$ is the radial dipole integral between the radial part $R_{l_c}(r)$ and the lth partial wave of a photoelectron $R_l(kr)$. The phase shift of the lth partial wave is δ_l^A at the X-ray absorption site A. The above formula shows an interesting feature of the SW thermal factor, which can be calculated from the PW thermal factor.

For isotropic systems the thermal factor in Eqs. (466) and (467) is further simplified as

$$X^{\alpha\beta}(T) = X^{\alpha\beta}\left[1 - k^{2}\sigma_{\alpha\beta}^{2}\left(\mathbf{M} \cdot \hat{\mathbf{R}}_{\alpha\beta}^{0} - 1\right)\right] \exp\left(-k^{2}\sigma_{\alpha\beta}^{2}/2\right)$$
$$(469)$$

where $\sigma_{\alpha\beta}^{2} = \langle \Delta_{\alpha\beta}^{2} \rangle$.

As an example we study the temperature dependence of Br K-edge XANES spectra of bromonaphthalene solved in supercritical xenon, which are recently measured by Murata et al. [133]. A detailed analysis has been made by Hayakawa et al. [137]; here we give a short discussion of the result. Supercritical fluids are characterized by clustering, so that we expect several Xe atoms to surround the X-ray-absorbing Br atom. We consider the model shown in Figure 42; the six Xe atoms form a benzene-like structure because they should be closely packed, and the plane is assumed to be normal to the Br–C axis, although the Br atom expels the central Xe atom. The nearest

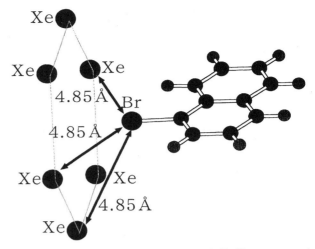

Fig. 42. A possible Xe–Br Nph interaction model. Six Xe atoms comprise a benzene-like structure whose plane is normal to the Br–C axis; the center of the plane and the Br–C axis are in a collinear arrangement [137].

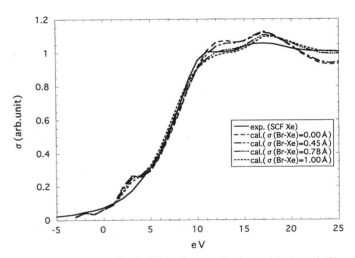

Fig. 43. Calculated Br K-edge XANES spectra for the models shown in Figure 42, with different $\sigma_{\text{Br-Xe}}$ values, 0.00–1.00 Å. The nearest Br–Xe distance is fixed at 4.85 Å [137].

Br–Xe distance is fixed at 4.85 Å, which is much larger than the sum of the van der Waals radii, 4.15 Å, because side hydrogen atoms prohibit a Xe–Br distance shorter than 4.85 Å. Figure 43 shows the calculated results for different $\sigma_{\text{Br-Xe}}$ values (0.00–1.00 Å) compared with the observed spectrum. The models with small $\sigma_{\text{Br-Xe}}$ (0.00–0.45 Å) overestimate the intensity at 11–12 eV and ~23 eV, whereas the large thermal fluctuations (0.78 and 1.00 Å) give quite good agreement, except at in the 16–17 eV energy region. This disagreement is not found for the expanded Xe coordination, where the Xe–Br distance is 5.00 Å. For this model, thermal fluctuation plays an important role.

In this subsection we show how the XAFS Debye–Waller factors can be used to obtain useful information about local force constants or local atomic vibrations around an X-ray-absorbing atom. The cumulant analyses are useful for the study of weak anharmonic systems, whereas the path-integral approaches are useful in the study of strong anharmonic systems. The former method is based on perturbation theory, where temperature Green's functions play an important role. The latter method can be closely compared with the classical approximation, which provides us with a transparent physical intuition about the interatomic pair potential.

In EELFS and XAFS spectra the core excitation process is crucial, so that we can expect some contribution from Franck–Condon effects in addition to the Debye–Waller effects. This problem has been discussed on the basis of a many-body scattering theory [138, 139] and the nonequilibrium Green's function theory [28]. It is shown that the side band effects play an essential role in the recovery of the XAFS formulas, which include only Debye–Waller factors. That is, we can see that the Franck–Condon factor is not important after delicate theoretical discussion.

7. CONCLUDING REMARKS

In this chapter some basic aspects of EELFS are discussed in detail, in particularly from a theoretical point of view. Three different experimental modes considered here have characteristic features. Among them, the small-angle reflection EELFS spectra have some advantages over backscattering EELFS for the purpose of the surface studies: high surface sensitivity and a sound theoretical background, for use as well-established XAFS analytical tools.

Some important basic problems have not been discussed in this article. One is the lifetime problem of deep cores, which has been discussed thoroughly in an excellent article by Almbladh and Hedin [22]. Other interesting problem is the use of non-muffin-tin potential for the EELFS analyses. For highly anisotropic systems with sparse packing, the muffin-tin approximation is poor. Some successful approaches are now found in the literature beyond the muffin-tin approximation [157–159].

Although we have encountered some difficult problems, we can take advantage of the feasibility of EELFS as a technique complementary to XAFS.

Acknowledgments

The author is grateful to Prof. Lars Hedin for the valuable discussion on the basic theory of EELS. He also thanks Prof. S. Usami for the long collaboration on the EELFS project supported by the TOSOH company, and Prof. J. J. Rehr and Prof. T. Miyanaga for the discussion on the temperature effects in XAFS. He is also grateful to Prof. A. P. Hitchcock, Dr. T. Yikegaki, S. Takatoh, K. Hatada, and members of the TOSOH research group for their valuable discussion of the various aspects of EELFS discussed in this article.

REFERENCES

1. R. F. Wills, A. A. Lucas, and G. D. Mahan, in "The Chemical Physics of Solid Surfaces and Heterogeneous Catalysis" (D. A. King and D. P. Woodruff, Eds.), Vol. 2, p. 59. Elsevier, New York, 1983.
2. P. Schattschneider, "Fundamentals of Inelastic Scattering." Springer-Verlag, Vienna/New York, 1986.
3. R. F. Egerton, in "Electron Energy-Loss Spectroscopy in the Electron Microscope," 2nd ed. Plenum, New York/London, 1996.
4. M. De Crescenzi and M. N. Piancastelli, "Electron Scattering and Related Spectroscopies." World Scientific, Singapore, 1996.
5. H. Bethe and R. Jackiw, in "Intermediate Quantum Mechanics," 3rd ed., Chap. 17. Benjamin, Elmsford, NY, 1986.
6. T. Fujikawa, S. Takatoh, and S. Usami, *Jpn. J. Appl. Phys.* 27, 348 (1988).
7. P. Aebi, M. Erbudak, F. Vanini, D. D. Vvedensky, and G. Kostorz, *Phys. Rev. B: Solid State* 41, 11760 (1990).
7a. P. Aebi, M. Erbudak, F. Vanini, D. D. Vvedensky, and G. Kostorz, *Phys. Rev. B: Solid State* 42, 5369 (1990).
8. B. Luo and J. Urban, *Surf. Sci.* 239, 235 (1990).
9. T. Fujikawa and L. Hedin, *Phys. Rev. B: Solid State* 40, 11507 (1989).
10. T. Fujikawa, *J. Phys. Soc. Jpn.* 60, 3904 (1991).
11. T. Fujikawa, *Surf. Sci.* 269/270, 55 (1992).
12. T. Yikegaki, N. Yiwata, T. Fujikawa, and S. Usami, *Jpn. J. Appl. Phys.* 29, 1362 (1990).
13. T. Fujikawa, T. Yikegaki, and S. Usami, *Catal. Lett.* 20, 149 (1993).
14. S. Usami, T. Fujikawa, K. Ota, T. Hayashi, and J. Tsukajima, *Catal. Lett.* 20, 159 (1993).
15. T. Yikegaki, H. Shibata, K. Takada, S. Takatoh, T. Fujikawa, and S. Usami, *Vacuum* 41, 352 (1990).
16. T. Hayashi, K. Araki, S. Takatoh, T. Enokijima, T. Yikegaki, T. Futami, Y. Kurihara, J. Tsukajima, K. Takamoto, T. Fujikawa, and S. Usami, *Appl. Phys. Lett.* 66, 25 (1995).
17. T. Hayashi, K. Araki, S. Takatoh, T. Enokijima, T. Yikegaki, J. Tsukajima, T. Fujikawa, and S. Usami, *Jpn. J. Appl. Phys.* 32, 182 (1993).
18. T. Hayashi, K. Araki, S. Takatoh, T. Enokijima, T. Yikegaki, T. Futami, Y. Kurihara, J. Tsukajima, K. Takamoto, T. Fujikawa, and S. Usami, *Jpn. J. Appl. Phys.* 34, 3255 (1995).
19. K. Gottfried, "Quantum Mechanics." Benjamin, Elmsford, NY, 1966.
20. R. G. Newton, "Scattering Theory of Waves and Particles," 2nd ed. Springer-Verlag, Berlin/New York, 1982.
21. C. J. Joachain, "Quantum Collision Theory." North-Holland, Amsterdam, 1975.
22. C.-O. Almbladh and L. Hedin, in "Handbook on Synchrotron Radiation" (E. E. Koch, Ed.), Vol. 1B, p. 607. North-Holland, Amsterdam, 1983.
23. M. Gell-Mann and M. L. Goldberger, *Phys. Rev.* 91, 398 (1953).
24. T. Fujikawa and N. Yiwata, *Surf. Sci.* 357/358, 60 (1996).
25. T. Fujikawa, R. Yanagisawa, N. Yiwata, and K. Ohtani, *J. Phys. Soc. Jpn.* 66, 257 (1997).
26. J. S. Bell and E. J. Squires, *Phys. Rev. Lett.* 3, 96 (1959).
27. L. Hedin, in "Recent Progress in Many-Body Theories" (A. J. Kallio, E. Pajanne, and R. F. Bishop, Eds.), Vol. 1, p. 307. Plenum, New York, 1988.
28. T. Fujikawa, *J. Phys. Soc. Jpn.* 68, 2444 (1999).
29. J. Rammer and H. Smith, *Rev. Mod. Phys.* 58, 323 (1986).
30. G. D. Mahan, "Many-Particle Physics," 2nd ed. Plenum, New York, 1990.
31. H. Fukutome, *Prog. Theor. Phys.* 67, 1776 (1982).
32. L. Hedin, J. Michials, and J. Inglesfield, *Phys. Rev. B: Solid State* 58, 15565 (1998).
33. T. Fujikawa, *J. Phys. Soc. Jpn.* 57, 306 (1988).
34. N. C. Francis and K. M. Watson, *Phys. Rev.* 92, 291 (1953).
35. P. H. Dederichs, "Solid State Physics" (H. Ehrenreich, F. Seitz, and D. Turnbull, Eds.), Vol. 27, p. 135. Academic Press, New York, 1972.
36. L. Hedin, *Phys. Rev. A: At., Mol., Opt. Phys.* 139, 796 (1965).

37. L. Hedin and S. Lundqvist, in "Solid State Physics" (H. Ehrenreich, F. Seitz, and D. Turnbull, Eds.), Vol. 23, p. 1. Academic Press, New York, 1969.

38. W. Bardyszewski and L. Hedin, *Phys. Scr.* 32, 439 (1985).

39. L. Hedin, *Physica B* 158, 344 (1989).

40. F. W. Byron and J. C. Joachain, *Phys. Rev. A: At., Mol., Opt. Phys.* 9, 2559 (1974).

40a. F. W. Byron and J. C. Joachain, *Phys. Rev. A: At., Mol., Opt. Phys.* 15, 128 (1977).

41. T. Fujikawa, A. Saito, and L. Hedin, *Jpn. J. Appl. Phys.* S32-2, 18 (1993).

42. T. Fujikawa, T. Yikegaki, and L. Hedin, *J. Phys. Soc. Jpn.* 64, 2351 (1995).

43. T. Fujikawa, K. Hatada, T. Yikegaki, and L. Hedin, *J. Electron Spectrosc. Relat. Phenom.* 88-91, 649 (1998).

44. J. C. Slater, *Phys. Rev.* 81, 385 (1951).

45. S. Hara, *J. Phys. Soc. Jpn.* 22, 710 (1967).

46. L. Hedin and B. I. Lundqvist, *J. Phys. C: Solid State Phys.* 4, 2064 (1971).

47. P. A. Lee and G. Beni, *Phys. Rev. B: Solid State* 15, 2862 (1977).

48. P. M. Echenique, *J. Phys. C: Solid State Phys.* 9, 3193 (1976).

49. P. M. Echenique and D. J. Titterington, *J. Phys. C: Solid State Phys.* 10, 625 (1977).

50. M. S. Woolfson, S. J. Gurman, and B. W. Holland, *Surf. Sci.* 117, 450 (1982).

51. D. Lu and J. J. Rehr, *Phys. Rev. B: Solid State* 37, 6126 (1988).

52. R. Gunnela, M. Benfatto, A. Marcelli, and C. R. Natoli, *Solid State Commun.* 76, 109 (1990).

53. J. Chaboy, *Solid State Commun.* 99, 877 (1996).

54. A. L. Ankudinov and J. J. Rehr, *J. Phys. (Paris)* 7C2, 121 (1997).

55. M. Roy and S. J. Gurman, *J. Synchrotron Radiat.* 6, 228 (1999).

56. L. J. Sham and W. Kohn, *Phys. Rev.* 145, 561 (1966).

57. T. Fujikawa, *J. Phys. Soc. Jpn.* 62, 2155 (1993).

58. T. Yikegaki, T. Hayashi, T. Enokijima, S. Takatoh, K. Araki, J. Tsukajima, and T. Fujikawa, *J. Appl. Phys.* S32, 73 (1993).

59. J. J. Rehr, E. A. Stern, R. L. Martin, and E. R. Davidson, *Phys. Rev. B: Solid State* 17, 560 (1978).

60. R. S. Williams, R. S. Wenen, G. Apai, J. Stöhr, D. A. Shirley, and S. P. Kowalczyk, *J. Electron Spectrosc. Relat. Phenom.* 12, 477 (1977).

61. R. Z. Bachrach and A. Bianconi, *Solid State Commun.* 42, 529 (1982).

62. J. J. Rehr, W. Bardyszewski, and L. Hedin, *J. Phys. (Paris)* 7 C2, 97 (1997).

63. J. E. Inglesfield, *Solid State Commun.* 40, 467 (1981).

63a. J. E. Inglesfield, *J. Phys. C: Solid State Phys.* C16, 403 (1983).

64. T. Fujikawa, *J. Phys. Soc. Jpn.* 55, 3244 (1986).

64a. T. Fujikawa, in "Core-Level Spectroscopy in Condensed Systems" (J. Kanamori and A. Kotani, Eds.), p. 213. Springer-Verlag, Berlin, 1988.

65. V. Fritsche, *J. Phys.: Condens. Matter* 2, 9735 (1990).

66. J. J. Rehr and R. C. Albers, *Phys. Rev. B: Solid State* 41, 8139 (1990).

67. Y. Chen, F. J. Garcia de Abajo, A. Chasse, R. X. Ynzunza, A. P. Kaduwela, M. A. Van Hove, and C. S. Fadley, *Phys. Rev. B: Solid State* 58, 13121 (1998).

68. T. Fujikawa, S. Takatoh, and S. Usami, *Jpn. J. Appl. Phys.* 28, 1683 (1989).

69. J. Derien, E. Chaint, M. De Crescenzi, and C. Noguerra, *Surf. Sci.* 189/190, 590 (1987).

70. M. Tomellini and P. Ascarelli, *Solid State Commun.* 72, 371 (1989).

71. K. Saldin and P. Rez, *Philos. Mag. B* 55, 481 (1987).

72. K. Saldin, *Philos. Mag. B* 56, 515 (1987).

73. K. Saldin, *Phys. Rev. Lett.* 60, 1197 (1988).

74. F. Mila and C. Noguerra, *J. Phys. C: Solid State Phys.* 20, 3863 (1987).

75. G. Beni and P. M. Platzman, *Phys. Rev. B: Solid State* 14, 1514 (1976).

76. R. B. Greegor and F. W. Lytle, *Phys. Rev. B: Solid State* 20, 4902 (1979).

77. E. Sevillano, H. Meuth, and J. J. Rehr, *Phys. Rev. B: Solid State* 20, 4908 (1979).

78. G. Dalba, P. Fornasini, F. Rocca, and S. Mobilio, *Phys. Rev. B: Solid State* 41, 9668 (1990).

78a. G. Dalba and P. Fornasini, *J. Synchrotron Radiat.* 4, 243 (1997).

79. P. P. Lottici, *Phys. Rev. B: Solid State* 35, 1236 (1987).

80. G. S. Knapp, H. K. Pan, and J. M. Tranquanda, *Phys. Rev. B: Solid State* 32, 2006 (1985).

81. J. M. Tranquanda and R. Ingalls, *Phys. Rev. B: Solid State* 28, 3520 (1983).

82. J. M. Tranquanda, in "EXAFS and Near Edge Structure III" (K. O. Hodgeson, B. Hedman, and J. E. Penner-Hahn, Eds.), p. 74. Springer-Verlag, Berlin, 1984.

83. E. D. Stern, P. Livins, and Z. Zhang, in "X-ray Absorption Fine Structure" (S. S. Hasnain, Ed.), p. 58. Ellis Horwood, 1991.

83a. E. D. Stern, P. Livins, and Z. Zhang, *Phys. Rev. B: Solid State* 43, 8850 (1991).

84. H. Rabus, D. Arvanitis, T. Lederer, and K. Baberscheke, "X-ray Absorption Fine Structure," Phys. Rev. B: Solid State 43, 193 (1991).

85. A. I. Frenkel and J. J. Rehr, *Phys. Rev. B: Solid State* 48, 585 (1993).

86. T. Yokoyama, K. Kobayashi, T. Ohta, and A. Ugawa, *Phys. Rev. B: Solid State* 53, 6111 (1996).

87. T. Fujikawa and T. Miyanaga, *J. Phys. Soc. Jpn.* 62, 4108 (1993).

88. T. Miyanaga and T. Fujikawa, *J. Phys. Soc. Jpn.* 63, 1036 (1994).

89. T. Miyanaga and T. Fujikawa, *J. Phys. Soc. Jpn.* 63, 3683 (1994).

90. T. Yokoyama, T. Satsukawa, and T. Ohta, *Jpn. J. Appl. Phys.* 28, 1905 (1985).

91. T. Fujikawa, T. Miyanaga, and T. Suzuki, *J. Phys. Soc. Jpn.* 66, 2897 (1997).

92. T. Miyanaga and T. Fujikawa, *J. Phys. Soc. Jpn.* 67, 2930 (1998).

93. T. Miyanaga and T. Fujikawa, *J. Synchrotron Radiat.* 6, 296 (1999).

94. T. Yokoyama, *Phys. Rev. B: Solid State* 57, 3423 (1998).

95. T. Yokoyama, *J. Synchrotron Radiat.* 6, 323 (1999).

96. C. Brouder, *J. Phys. C: Solid State Phys.* 21, 5075 (1988).

97. P. Rennert, *J. Phys.: Condens. Matter.* 4, 4315 (1992).

98. P. Rennert, *Jpn. J. Appl. Phys.* S32-2, 79 (1993).

99. O. Speder and P. Rennert, *Czech. J. Phys.* 43, 1015 (1993).

100. T. Fujikawa, M. Yimagawa, and T. Miyanaga, *J. Phys. Soc. Jpn.* 64, 2047 (1995).

101. M. Yimagawa and T. Fujikawa, *J. Electron Spectrosc. Relat. Phenom.* 79, 29 (1996).

102. T. Fujikawa, J. J. Rehr, Y. Wada, and S. Nagamatsu, *J. Phys. Soc. Jpn.* 68, 1259 (1999).

103. T. Fujikawa, J. J. Rehr, Y. Wada, and S. Nagamatsu, *J. Synchrotron Radiat.* 6, 317 (1999).

104. T. Fujikawa, *J. Phys. Soc. Jpn.* 65, 87 (1996).

105. T. Fujikawa, *J. Electron Spectrosc. Relat. Phenom.* 79, 25 (1996).

106. T. Yanagawa and T. Fujikawa, *J. Phys. Soc. Jpn.* 65, 1832 (1996).

107. T. Ishii, *J. Phys. Condens. Matter* 4, 8029 (1992).

108. R. Kubo, *J. Phys. Soc. Jpn.* 17, 1100 (1962).

109. E. A. Stern and S. Heald, "Handbook on Synchrotron Radiation" (E. E. Koch, Ed.), Vol. 1B, p. 955. North-Holland, Amsterdam, 1983.

110. G. Bunker, *Nucl. Instrum. Methods* 207, 437 (1983).

111. E. D. Crozier, J. J. Rehr, and R. Ingalls, "X-ray Absorption: Principles, Applications, Techniques of EXAFS, SEXAFS and XANES" (D. C. Koningsberger and R. Prins, Eds.), p. 373. Wiley, New York, 1988.

112. A. A. Maradudin, "Dynamical Properties of Solids" (G. K. Horton and A. A. Maradudin, Eds.), Vol. 1, p. 1. North-Holland, Amsterdam, 1974.

113. T. H. K. Barron and M. L. Klein, "Dynamical Properties of Solids" (G. K. Horton and A. A. Maradudin, Eds.), Vol. 1, p. 391. North-Holland, Amsterdam, 1974.

114. T. Fujikawa, K. Nakayama, and T. Yanagawa, *J. Electron Spectrosc., Relat. Phenom.* 88-91, 527 (1998).

115. N. M. Ashcroft and N. D. Mermin, "Solid State Physics." Holt, Rinehart & Winston, 1976.

116. T. Miyanaga, H. Sakane, and I. Watanabe, *Bull. Chem. Soc. Jpn.* 68, 819 (1995).

117. T. Miyanaga, H. Katsumata, T. Fujikawa, and T. Ohta, *J. Phys. IV* 7C2, 225 (1997).

118. T. Yokoyama and T. Ohta, *Jpn. J. Appl. Phys.* 29, 2052 (1990).

119. G. Dalba, P. Fornasini, and F. Rocca, *Phys. Rev. B: Solid State* 47, 8502 (1993).

120. Y. Sawa, T. Miyanaga, H. Tanida, and I. Watanabe, *J. Chem. Soc. Faraday Trans.* 91, 4389 (1995).

121. T. Miyanaga, H. Sakane, and I. Watanabe, *Phys. Chem. Chem. Phys.*, to appear.

122. R. Akesson, L. G. M. Pettersson, M. Sandstrom, and U. Wahlgren, *J. Am. Chem. Soc.* 116, 8705 (1994).

123. R. P. Feynman, "Statistical Mechanics." Benjamin, Elmsford, NY, 1972.

124. A. Cuccoli, R. Giachetti, V. Toggnetti, R. Vaia, and P. Verrucchi, *J. Phys. Condens. Matter* 7, 7891 (1995).

125. R. P. Feynman and H. Kleinert, *Phys. Rev. A: At., Mol., Opt. Phys.* 34, 5080 (1986).

126. R. Giachetti and V. Toggnetti, *Phys. Rev. B: Solid State* 33, 7647 (1986).

127. J. Mustre de Leon, S. D. Conradson, I. Batistic, A. R. Bishop, I. D. Raistrick, M. C. Aronson, and F. H. Garzon, *Phys. Rev. B: Solid State* 45, 2447 (1992).

128. R. E. Cohen, *Nature (London)* 356, 136 (1992).

129. E. D. Crozier, N. Alberding, K. R. Bauchspiess, A. J. Seary, and S. Gygax, *Phys. Rev. B: Solid State* 36, 8288 (1987).

130. J. D. Doll, *J. Chem. Phys.* 81, 3536 (1984).

131. T. Miyanaga and T. Fujikawa, *Chem. Phys. Lett.* 308, 78 (1999).

132. W. L. Schaich, *Phys. Rev. B: Solid State* 29, 6513 (1984).

133. T. Murata, K. Nakagawa, Y. Ohtsuki, I. Shimoyama, T. Mizutani, K. Kato, K. Hayakawa, and T. Fujikawa, *J. Synchrotron Radiat.* 5, 1004 (1998).

134. N. D. Mermin, *J. Math. Phys.* 7, 1038 (1966).

135. C. Brouder and J. Goulon, *Physica B* 158, 351 (1989).

136. V. Fritzshe, *J. Phys. Condens. Matter* 2, 9735 (1990).

137. K. Hayakawa, K. Kato, T. Fujikawa, T. Murata, and K. Nakagawa, *Jpn. J. Appl. Phys.* 38, 6423 (1999).

138. T. Fujikawa, *J. Phys. Soc. Jpn.* 65, 87 (1996).

139. T. Fujikawa, *J. Electron Spectrosc. Relat. Phenom.* 79, 25 (1996).

140. P. Rennert, *J. Phys. (Paris)* 7 C2, 147 (1997).

141. T. Fujikawa, *Z. Phys. B: Condens. Matter* 54, 215 (1984).

142. T. Fujikawa, *Z. Phys. B: Condens. Matter* 67, 111 (1987).

143. T. Fujikawa, K. Hatada, and L. Hedin, *Phys. Rev. B: Solid State* 62, 5387 (2000).

144. T. Fujikawa, R. Yanagisawa, and N. Yiwata, *J. Phys. (Paris)* 7 C2, 99 (1997).

145. J. Arponen and E. Pajanne, *Ann. Phys. (Leipzig)* 91, 450 (1976).

146. J. P. Blaizot and G. Ripka, "Quantum Theory of Finite Systems." MIT Press, Cambridge, MA, 1986.

147. J. J. Rehr, J. Mustre de Leon, S. I. Zabinsky, and R. C. Albers, *J. Am. Chem. Soc.* 113, 5135 (1991).

148. S. I. Zabinsky, J. J. Rehr, A. Ankudinov, R. C. Albers, and M. J. Eller, *Phys. Rev. B: Solid State* 52, 2995 (1995).

149. I. Davoli, A. Marcelli, A. Bianconi, M. Tomellini, and M. Fanfoni, *Phys. Rev. B: Solid State* 33, 2979 (1986).

150. D. Norman, J. Stöhr, R. Jaeger, P. J. Durham, and J. B. Pendry, *Phys. Rev. Lett.* 51, 2052 (1983).

151. J. Stöhr, R. Jaeger, and T. Kendelewicz, *Phys. Rev. Lett.* 49, 142 (1982).

152. L. Wenzel, D. Arvanitis, W. Daum, H. H. Rotermund, J. Stöhr, K. Baberschke, and H. Ibach, *Phys. Rev. B: Solid State* 36, 7689 (1987).

153. A. P. Hitchcock and T. Tyliszczak, *Surf. Rev. Lett.* 2, 43 (1995).

154. J. Tsukajima, K. Arai, S. Takatoh, T. Enokijima, T. Hayashi, T. Yikegaki, A. Kashiwagi, K. Tokunaga, T. Suzuki, T. Fujikawa, and S. Usami, *Thin Solid Films* 281/282, 318 (1996).

155. T. Miyanaga and T. Fujikawa, *J. Electron Spectrosc. Relat. Phenom.* 88-91, 523 (1998).

156. T. Miyanaga, T. Suzuki, and T. Fujikawa, *J. Synchrotron Radiat.* 7, 95 (2000).

157. C. R. Natoli, M. Benfatto, and S. Doniach, *Phys. Rev. A., At., Mol., Opt. Phys.* 34, 4682 (1986).

158. D. L. Foulis, F. Pettifer, and P. Sherwood, *Physica B* 208/209, 68 (1995).

158a. D. L. Foulis, F. Pettifer, and P. Sherwood, *Europhys. Lett.* 29, 647 (1995).

159. A. Gonis, "Green's Functions for Ordered and Disordered Systems." North-Holland, Amsterdam, 1992.

Chapter 10

THEORY OF LOW-ENERGY ELECTRON DIFFRACTION AND PHOTOELECTRON SPECTROSCOPY FROM ULTRA-THIN FILMS

Jürgen Henk*

Max-Planck-Institut für Mikrostrukturphysik, Halle/Saale, Germany

Contents

1. INTRODUCTION

If an extended condensed-matter system is reduced in its dimensionality, it can exhibit novel physical properties. These finite-size or quantum-size effects are fascinating from a fundamental point of view as being likewise important with regard to technological applications. For example, quantum wells and superlattices build up by semiconductors or magnetic metals are prominent examples for intensively investigated and applied systems [1, 2].

When reducing the dimension from three to two, one goes from the "bulk" system to an "ultra-thin" film. For the latter, the extension in one dimension is small compared to those in the two other dimensions. What, however, actually is meant by "ultra-thin" depends on the physical quantity under consideration. Because this chapter is on electron spectroscopies, ultra-thin films can be regarded as of a few nanometer thickness. The properties of ultra-thin films are obviously determined by the material of the "bulk" system. However, because films are (usually) realized by evaporation on a substrate, their properties can significantly be influenced by the film–substrate interface, especially for films a few angstroms thick.

Probably the most striking fact of films is that electrons can become confined to the film and thus show discrete energy levels; that is, the electron energies become quantized with respect to the direction normal to the film. Many properties of films can be attributed to these quantum-well states (QWSs). For example, the giant magnetoresistance (GMR) effect [3], that is, the change of the resistance of a stack of magnetic films (usually named multilayer or superlattice) due to the arrangement of the magnetic moments within each film, can be explained in terms of QWSs. The GMR is still of interest in fundamental science as well as it is applied in modern computer hard disks [4, 5].

The key to the understanding of the various physical phenomena in these systems lies in a detailed knowledge of both

---

*To my sons Kai-Hendrik and Jan-Malte.

Handbook of Thin Film Materials, edited by H.S. Nalwa
Volume 2: Characterization and Spectroscopy of Thin Films

ISBN 0-12-512910-6/$35.00

their electronic and geometric structures, which are intimately related. Two prominent experimental techniques that allow for detailed investigations of these properties are low-energy electron diffraction (LEED) and angle-resolved photoelectron spectroscopy (ARPES). Both have been very successfully applied to semi-infinite systems. However, applied to an ultra-thin film on a substrate, features in the spectra occur that are closely related to the particular electronic states within the film; that is, they are direct manifestations of QWSs in the spectra. Therefore, these signatures give access to the electronic and geometrical structures of the film–substrate system and can further be related to more fundamental quantities, for example, the thickness and the magnetization of the film.

Because the interpretation of experimentally obtained LEED and ARPES data is rather difficult without theoretical aid, there is need for a theoretical description of the electronic structure of the film–substrate system and, based on this, of the spectroscopies. Basic physical properties can be obtained by *ab initio* total-energy calculations. But these often do not provide a direct link to spectroscopically measured quantities, such as, for example, LEED intensities or photocurrents. Therefore, theoretical methods have to be developed that are based on *ab initio* results and yield spectroscopic quantities that can directly be compared to experimental data.

This chapter first introduces the geometrical and electronic description of ultra-thin films (Section 2). In Section 3, theories of LEED at various levels of sophistication together with some applications are presented. Subsequently, theories of ARPES are given (Section 4). A complete coverage of ARPES is clearly beyond the scope of this chapter. Therefore, it focuses on photoemission from valence states that are excited by light within the vacuum-ultraviolet (VUV) range of photon energies, the latter being well suited for the study of QWSs. Further, important experimental results are used to illustrate the theoretical findings. Of particular interest are films formed either by normal or by magnetic metals. A description of magnetic systems is best given by means of a relativistic theory, rather than incorporating the electron's spin into a nonrelativistic theory. In this way, exchange and spin-orbit interaction are treated on an equal footing (which is essential for understanding some of the effects). Therefore, multiple-scattering theory is introduced in its relativistic form. This requires some background knowledge of quantum and solid-state physics. For the reader interested in more details, references to review articles and books as well as to important original articles are given.

2. THIN FILMS AND QUANTIZED ELECTRONIC STATES

In this section, we introduce the geometry of films on a substrate (Section 2.1) and describe electronic states within a film (Section 2.2).

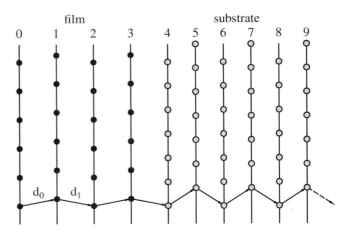

Fig. 1. Geometrical arrangement of an ultra-thin film grown on a substrate. The film consists of layers (vertical lines) $j = 0, \ldots, 3$ connected by translation vectors $\mathbf{d}_j$. The substrate is made up by all layers $j \geq 4$. Sites are represented by black (film) or gray (substrate) circles.

2.1. Films, Substrates, and Lattices

Before turning to the description of electrons confined to an ultra-thin film, one has to consider the geometry of the system under consideration. An ultra-thin film can be regarded as stack of layers, the basic object, grown on a substrate. Each layer consists of sites (circles in Fig. 1) that form a lattice.[1] In principle, these two-dimensional lattices may vary from layer to layer. This is schematically depicted in Figure 1: The distance between next-nearest sites in the film and the substrate layers differs. Especially at the surface (layer index $j = 0$), sites can be rearranged in such a way that the two-dimensional lattice at the surface differs from those of the other layers. For example, the (110) surface of Pt shows a reconstruction where each second row of Pt atoms is missing [the "missing-row" reconstruction of Pt(110)1 × 2 [6]].

Each layer j is connected to the next-nearest layer $j + 1$ by a translation vector $\mathbf{d}_j$. The latter may be different for each layer, in particular, at the surface where usually $\mathbf{d}_0$ can differ considerably from that of the other layers; that is, there is layer relaxation.

Geometrical rearrangements such as relaxation and reconstruction usually take place near the surface. Deep in the interior of the system, all layers are identical; that is, they show the same lattice and the same translation vector $\mathbf{d}$. We take this feature as the definition of the bulk system (i.e., an infinitely repeated arrangement of identical layers). Note that for some systems the repeated entity does not consist of a single layer but of a stack of layers. One can comprise this stack into a principal layer and use this as the basic quantity. For example, in Figure 1 a principal layer would consist of two substrate layers.

---

[1] A two-dimensional (three-dimensional) lattice is a set of vectors that can be expressed as the set of all integral linear combinations of two (three) basis vectors, not all along the same direction (in the same plane).

After considering the basic geometry of film and substrate, we can turn to the structure of the layers (i.e., the two-dimensional lattices).

2.1.1. Two-Dimensional Crystal Structures

Because the surface breaks the three-dimensional periodicity of the bulk, the 14 Bravais lattices [7] of the latter are no longer suitable for the description of semi-infinite systems. Instead, one is concerned with five types of two-dimensional nets, the translation vectors of which connect identical positions (see, e.g., [6, Chap. 3]).

Consider the unit cell[2] of the substrate spanned by the vectors $\mathbf{a}_1$ and $\mathbf{a}_2$ [8, 9],

$$\mathbf{a}_1 = a_{11}\mathbf{e}_1 + a_{12}\mathbf{e}_2 \tag{1a}$$
$$\mathbf{a}_2 = a_{21}\mathbf{e}_1 + a_{22}\mathbf{e}_2 \tag{1b}$$

where $\mathbf{e}_1$ and $\mathbf{e}_2$ are orthonormal vectors. The preceding equations establish the matrix A,

$$A = \begin{pmatrix} a_{11} & a_{12} \\ a_{21} & a_{22} \end{pmatrix} \tag{2}$$

It can easily be verified that the area of the unit cell F_A is given by the determinant of A, $\det(A)$.[3] Note that each translational vector reads $\mathbf{t} = m\mathbf{a}_1 + n\mathbf{a}_2$ with m and n integers, that is, $m, n \in \mathbb{Z}$. The set of vectors $\mathbf{t}$ defines the two-dimensional lattice. The five two-dimensional Bravais lattices are shown in Figure 2.

The unit cell of the film layers is spanned by the vectors $\mathbf{b}_1$ and $\mathbf{b}_2$, which, in analogy to Eq. (1), gives rise to a matrix B. Both sets of vectors are connected linearly by

$$\mathbf{b}_1 = n_{11}\mathbf{a}_1 + n_{12}\mathbf{a}_2 \tag{3a}$$
$$\mathbf{b}_2 = n_{21}\mathbf{a}_1 + n_{22}\mathbf{a}_2 \tag{3b}$$

which establishes the matrix N. The areas of the unit cells are related by $F_B = F_A \det(N)$.

All periodic lattice structures can be cast into the following three categories [9, 12]:

Simply related structures have all $n_{ij} \in \mathbb{Z}$. Thus, $\det(N)$ is also an integer. For example, commensurate adsorbate layers belong to this class.

Coincident structures have rational n_{ij}. Incommensurate adsorbed layers belong to this class.

Incoherent structures have irrational n_{ij}. In contrast to the other two cases, there is no common periodicity of substrate and film.

---

[2] A unit cell is a region that fills space without overlapping when translated through some subset of the vectors of a Bravais lattice [7].

[3] A general definition of the area which is valid for all spatial dimensions is given by the outer product $\mathbf{a}_1 \wedge \mathbf{a}_2$ of the corresponding geometric algebra [10, 11]. In two dimensions, it corresponds to the determinant, in three dimensions to the cross product $|\mathbf{a}_1 \times \mathbf{a}_2|$.

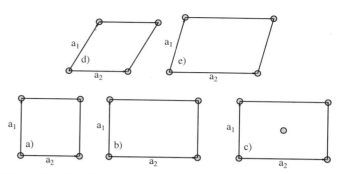

Fig. 2. Two-dimensional Bravais lattices: (a) square ($|\mathbf{a}_1| = |\mathbf{a}_2|$, $\alpha = 90°$), (b) primitive ($|\mathbf{a}_1| \neq |\mathbf{a}_2|$, $\alpha = 90°$), (c) centered rectangular ($|\mathbf{a}_1| \neq |\mathbf{a}_2|$, $\alpha = 90°$), (d) hexagonal ($|\mathbf{a}_1| = |\mathbf{a}_2|$, $\alpha = 60°$), and (e) oblique ($|\mathbf{a}_1| \neq |\mathbf{a}_2|$, α arbitrary), with $\alpha = \angle(\mathbf{a}_1, \mathbf{a}_2)$.

A lattice can be represented either in direct space (covariant representation, $\mathbf{a}_i$) or in reciprocal space (contravariant representation, $\mathbf{g}_i$). Both are related by

$$\mathbf{g}_i \cdot \mathbf{a}_j = \delta_{ij} \qquad i, j = 1, 2 \tag{4}$$

The vectors $\mathbf{g}_i$ from the basis of the reciprocal lattice. Introducing a matrix G by

$$\mathbf{g}_1 = \mathbf{e}_1 g_{11} + \mathbf{e}_2 g_{21} \tag{5a}$$
$$\mathbf{g}_2 = \mathbf{e}_1 g_{12} + \mathbf{e}_2 g_{22} \tag{5b}$$

gives, using Eq. (4), $G = A^{-1}/F_A$. Thus, the area of the unit cell of the reciprocal lattice is $F_G = 1/F_A$. Defining in analogy to Eq. (5) the reciprocal net of the film layers, that is, the matrix H, and the relating matrix M, the latter is given by $M = N^{-1}$.

As we have seen, the space can be filled with (conventional) unit cells. One can also fill the space by cells with the full symmetry of the Bravais lattice, for example, with Wigner–Seitz cells. The latter are defined as regions in space about a lattice point that are closer to that particular point than to any other lattice point. The Wigner–Seitz cell of the reciprocal lattice is called the (first) Brillouin zone [7].

2.2. Quantum-Well States

In 1928, Bloch showed that the energy spectrum of an electron in a solid consists of continuous stripes of allowed energies that are separated by regions of forbidden energies [13]. These stripes are called energy bands; correspondingly, the forbidden intervals are named bandgaps. The energy E of an electron within a band is continuous with respect to the momentum $\mathbf{k}$, which gives rise to the concept of band structure, $E(\mathbf{k})$. Undoubtedly, this concept is one of the most successful in condensed-matter physics. In the following, we discuss the energy spectrum of electrons confined to a film.

Because the geometrical structure and the electronic structure are closely related, several descriptions of electrons confined to a film will be presented (for an introduction, see [14]). In Section 2.2.1, the simple free-electron model will be introduced. The envelope picture presented in Section 2.2.2 can be regarded as a refinement of the preceding. A tight-binding description is given in Section 2.2.3.

2.2.1. Free-Electron Description of Quantum-Well States

As a preliminary, we recall the well-known basics of the electronic structure of an electron in a potential with rectangular shape, the so-called rectangular well (see, e.g., [15]). The well is assumed to be infinitely extended in the xy plane and to extend from $-a/2$ to $a/2$ along the z axis. Thus, it can be regarded as a simple representation of a film. According to the model of a quantum well, the film is structureless in the xy plane, and, thus, the potential function $V(\mathbf{r})$ depends exclusively on z,

$$V(\mathbf{r}) = \begin{cases} U & z < -a/2 \\ 0 & -a/2 \leq z \leq a/2 \\ U & a/2 < z \end{cases} \quad (6)$$

The three different regions along the z axis are denoted I, II, and III (cf. Fig. 3). For convenience, we introduce the in-plane vector $\boldsymbol{\varrho} = (x, y)$. Therefore, $\mathbf{r} = (\boldsymbol{\varrho}, z)$. Because the potential is symmetric with respect to the plane $z = 0$, that is, $V(\boldsymbol{\varrho}, z) = V(\boldsymbol{\varrho}, -z)$, the electronic wave functions $\Psi(\boldsymbol{\varrho}, z)$ have to have either even (+) or odd (−) parity: $\Psi(\boldsymbol{\varrho}, z) = \pm\Psi(\boldsymbol{\varrho}, -z)$. Further, due to the translational invariance of the potential in the xy plane, that is, $V(\boldsymbol{\varrho}, z) = V(\boldsymbol{\varrho}', z)$ for all $\boldsymbol{\varrho}$ and $\boldsymbol{\varrho}'$, the wave functions obey Floquet's theorem, $\Psi(\boldsymbol{\varrho} + \boldsymbol{\varrho}', z) = \lambda(\boldsymbol{\varrho}')\Psi(\boldsymbol{\varrho}, z)$. Obviously, $\lambda(\mathbf{0}) = 1$ and $\lambda(\boldsymbol{\varrho} + \boldsymbol{\varrho}') = \lambda(\boldsymbol{\varrho})\lambda(\boldsymbol{\varrho}')$, which is the functional equation of the exponential function. Because $|\Psi(\mathbf{r})|^2$ is the probability of finding an electron at $\mathbf{r}$, the wave functions have to be square integrable ($\Psi \in \mathcal{L}_2$), that is, $\int |\Psi(\mathbf{r})|^2 dr^3 < \infty$. With a normalized Ψ (i.e., $\|\Psi\| = 1$), it follows that $|\lambda(\boldsymbol{\varrho})| = 1$ for all $\boldsymbol{\varrho}$. Therefore, one can write $\lambda(\boldsymbol{\varrho})$ as $\lambda(\boldsymbol{\varrho}) = \exp(i\mathbf{k}_\| \cdot \boldsymbol{\varrho})$. In conclusion, the wave functions can be written as

$$\Psi(\boldsymbol{\varrho}, z) = \exp(i\mathbf{k}_\| \cdot \boldsymbol{\varrho})\Psi(0, z) \quad (7)$$

which (i) establishes a classification of the wave functions with respect to $\mathbf{k}_\|$ and (ii) reduces the initially three-dimensional problem to a one-dimensional problem [which has to be solved for $\Psi(0, z)$; the in-plane wave vector $\mathbf{k}_\|$ acts as a parameter in reciprocal space].

We now solve the time-independent Schrödinger equation[4] for the potential of Eq. (6),

$$\left(-\frac{1}{2}\nabla^2 + V(\mathbf{r})\right)\Psi(\mathbf{r}) = E\Psi(\mathbf{r}) \quad (8)$$

which reduces, using Eq. (7), to the one-dimensional problem[5]

$$\left(-\frac{1}{2}\partial_z^2 + V(z)\right)\Psi(0, z) = \left(E - \frac{1}{2}\mathbf{k}_\|^2\right)\Psi(0, z) \quad (9)$$

In region II, the potential vanishes and one is left with

$$-\frac{1}{2}\partial_z^2\Psi^{(II)}(0, z) = \left(E - \frac{1}{2}\mathbf{k}_\|^2\right)\Psi^{(II)}(0, z) \qquad \mathbf{r} \in II \quad (10)$$

which is immediately solved by plane waves,

$$\Psi^{(II)}(\mathbf{r}) = A^{(II)}\exp(i\mathbf{k}_\| \cdot \boldsymbol{\varrho})\exp(ik_\perp^{(II)}z) \quad (11)$$

The z component of the wave vector, $k_\perp^{(II)}$, is related to the energy by

$$k_\perp^{(II)} = \sqrt{2E - \mathbf{k}_\|^2} \quad (12)$$

In region I, the Schrödinger equation reads

$$-\frac{1}{2}\partial_z^2\Psi^{(I)}(0, z) = \left(E - \frac{1}{2}\mathbf{k}_\|^2 - U\right)\Psi^{(I)}(0, z) \qquad \mathbf{r} \in I \quad (13)$$

and, using the preceding results, Eqs. (11) and (12),

$$\Psi^{(I)}(\mathbf{r}) = A^{(I)}\exp(i\mathbf{k}_\| \cdot \boldsymbol{\varrho})\exp(ik_\perp^{(I)}z) \quad (14)$$

with

$$k_\perp^{(I)} = \sqrt{2(E - U) - \mathbf{k}_\|^2} \quad (15)$$

Note that $k_\perp^{(I)}$ can be either real [$2(E - U) \geq \mathbf{k}_\|^2$] or imaginary [$2(E - U) < \mathbf{k}_\|^2$]. For region III, one obtains the same results as for region I.

The next step is to construct the wave function of the entire system. Therefore, one has to match both the wave function and its derivative at the region boundaries, $z = \pm a/2$, for a given energy E and $\mathbf{k}_\|$. Because we are interested in bound states, that is, those confined to the well, we choose $U > 0$ and assume $0 < 2E - \mathbf{k}_\|^2 < 2U$.[6] In this case, $k_\perp^{(I)}$ and $k_\perp^{(II)}$ are imaginary and real, respectively. Setting $\kappa^{(I)} = ik_\perp^{(I)} > 0$, the ansatz for the wave function reads, using Eqs. (11) and (14),

$$\Psi(0, z) = \begin{cases} A^{(I)}\exp(\kappa^{(I)}z) & \mathbf{r} \in I \\ A^{(II)}\left(\exp(ik_\perp^{(II)}z) \pm \exp(-ik_\perp^{(II)}z)\right) & \mathbf{r} \in II \\ \pm A^{(I)}\exp(-\kappa^{(I)}z) & \mathbf{r} \in III \end{cases} \quad (16)$$

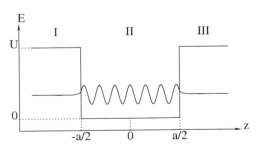

Fig. 3. Quantum well of rectangular shape. The potential $V(z)$ is given as in Eq. (6) with $U > 0$. In the three regions I, II, and III, representatives of the wave functions with energy $0 < E < U$ are shown schematically in addition.

---

[4]We use atomic units throughout (i.e., $e = \hbar = m = 1$). Lengths are in bohrs ($1a_0 = 0.529177$ Å), energies in hartrees ($1H = 2$ Ry, 1 Ry = 13.6058 eV). The speed of light is given by the reciprocal fine-structure constant, $c = 137.036$.

[5]The notation ∂_z (∂_z^2) is short for d/dz (d^2/dz^2).

[6]There is at least one bound state in this case [16].

where the upper signs are for even parity, the lower ones for odd parity. Note that in region II the two linearly independent solutions $\exp(\pm i k_\perp^{(II)} z)$ were superposed. In regions I and III, one would have had the two solutions $\exp(\pm \kappa^{(I)} z)$ but only one of them is square integrable in the respective region. The matching conditions at $z = a/2$ yield

$$\pm A^{(I)} \exp(-\kappa^{(I)} a/2)$$
$$= A^{(II)} \big(\exp(i k_\perp^{(II)} a/2) \pm \exp(-i k_\perp^{(II)} a/2) \big) \quad \text{(17a)}$$

$$\mp A^{(I)} \kappa^{(I)} \exp(-\kappa^{(I)} a/2)$$
$$= i A^{(II)} k_\perp^{(II)} \big(\exp(i k_\perp^{(II)} a/2) \mp \exp(-i k_\perp^{(II)} a/2) \big) \quad \text{(17b)}$$

(The conditions for $z = -a/2$ give no additional equations.) Because the coefficients $A^{(I)}$ and $A^{(II)}$ remain arbitrary until Ψ is normalized, the two conditions are equivalent to the requirement that the logarithmic derivative of Ψ, $(\partial_z \Psi(0, z))/\Psi(0, z)$, is continuous at $z = a/2$. Using Euler's formulas, one eventually obtains transcendental equations in the wave numbers,

$$\kappa^{(I)} \cos(k_\perp^{(II)} a/2) = k_\perp^{(II)} \sin(k_\perp^{(II)} a/2) \quad \text{even parity} \quad \text{(18a)}$$

$$-\kappa^{(I)} \sin(k_\perp^{(II)} a/2) = k_\perp^{(II)} \cos(k_\perp^{(II)} a/2) \quad \text{odd parity} \quad \text{(18b)}$$

which can be solved numerically.

At this point, it it illustrative to consider a well with infinitely high barriers. The limit $U \to \infty$ implies $\kappa^{(I)} \to 0$, and, therefore, the wave function vanishes in regions I and III. Thus, there is no requirement for the z derivative of $\Psi(0, z)$, and the remaining matching condition $\Psi^{(II)}(0, \pm a/2) = 0$ immediately gives

$$k_{\perp n}^{(II)} = \frac{\pi}{a} \begin{cases} (2n+1) & n \in \mathbb{N}_0 & \text{even parity} \\ (2n) & n \in \mathbb{N} & \text{odd parity} \end{cases} \quad \text{(19)}$$

which can be combined to

$$k_{\perp m}^{(II)} = \frac{m\pi}{a} \qquad m \in \mathbb{N} \quad \text{(20)}$$

Therefore, the energy levels E_m are given by

$$E_m = \frac{1}{2} \left[\left(\frac{m\pi}{a} \right)^2 + \mathbf{k}_\parallel^2 \right] \qquad m \in \mathbb{N} \quad \text{(21)}$$

The lowest energy level is for an even-parity state, and levels of even and odd states alternate. Because the Hamiltonian is invariant under time reversal,[7] both Ψ and Ψ^* are solutions of the Schrödinger equation (8) to the same energy.[8] This is expressed by the fact that both $\mathbf{k}$ and $-\mathbf{k}$ lead to the same energy, or, more specifically, $E_m = E_{-m}$. Therefore, it is sufficient to consider only positive wave numbers $k_{\perp m}^{(II)}$.

In comparison to the case of free electrons, that is, $V(\mathbf{r}) = 0$ for all $\mathbf{r}$, with energy band $E = \mathbf{k}^2/2$, the energy of the electrons confined to the well is quantized. In other words: the spectrum E_m is discrete, as is shown in Figure 4. The electronic

states Ψ_m are called quantum-well states (QWSs) and have even (odd) parity for even (odd) m.

If one identifies the width a of the quantum well with a certain number N of layers with thickness d, $a = Nd$, the energy levels are given by

$$E_m(N) = \frac{1}{2} \left[\left(\frac{m\pi}{Nd} \right)^2 + \mathbf{k}_\parallel^2 \right] \qquad m \in \mathbb{N} \quad \text{(22)}$$

These are shown in Figure 5. Connecting energy levels with identical quantum number m gives the typical $1/N^2$ behav-

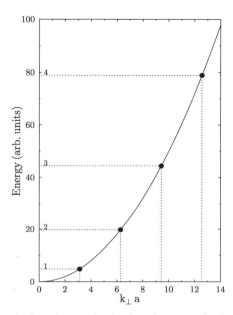

Fig. 4. Quantization of energy levels of an electron confined to a quantum well with infinitely high barriers ($U = \infty$). The solid line is the parabolic free-electron band, $E = k_\perp^2/2$. Filled circles represent the discrete spectrum of the confined electron. The allowed wave numbers $k_{\perp m} = m\pi/a$ are equidistant, cf. Eq. (20), and marked by dotted lines. The quantum number m is given at the left. The width of the well is a.

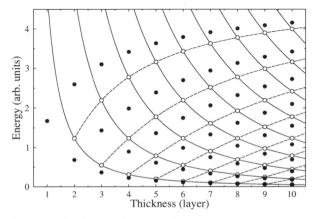

Fig. 5. Energy levels of an electron confined to a quantum well versus well thickness a. The latter is given in layers, $a = Nd$, d being the interlayer distance. Empty circles indicate energy levels for a well with infinitely high barriers. Solid lines connect levels with the same quantum number m, cf. $E_m(N)$ in Eq. (22). Dashed lines, however, connect levels with identical $m - N$. Filled circles indicate energy levels for a well with finite barriers, $U = 4.5$ ($\beta = 1.5$).

---

[7] The nonrelativistic time-reversal operator is complex conjugation, K_0. Its relativistic counterpart reads $-i\sigma_y K_0$, σ_y being a Pauli matrix.

[8] This is called Kramer's degeneracy [17].

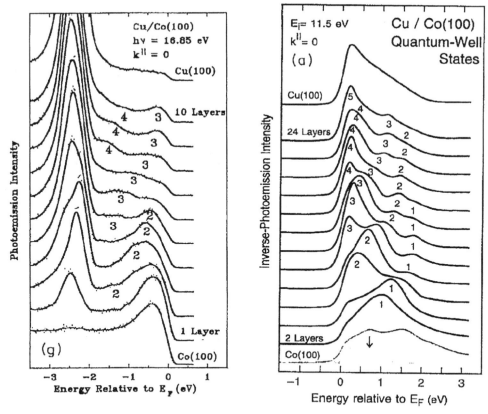

Fig. 6. Experimental photoelectron spectroscopy from quantum-well states of Cu films on fcc Co(100) for $\mathbf{k}_\parallel = 0$. Left: photoemission for 16.85 eV photon energy. The lowest spectrum shows the intensity for the uncovered substrate [fcc Co(100)], the uppermost that of semi-infinite Cu(100). Spectra for Cu film thicknesses from 1 layer up to 10 layers are shown in between. Intensity maxima that are related to quantum-well states are labeled by numbers. Right: inverse photoemission with 11.5 eV initial energy. The same quantum-well states as in the left panel are detected but above the Fermi energy E_F (cf. the numbers associated with the intensity maxima). Reprinted with permission from J. E. Ortega, F. J. Himpsel, G. J. Mankey, and R. F. Willis, *Phys. Rev. B* 47, 1540 (1993). Copyright 1993, by the American Physical Society.

ior (solid lines in Fig. 5). If levels with identical $m - N$ are connected (dashed lines in Fig. 5), one obtains a more tight-binding-like visualization, as we will see later.

We now return to the well with finite barriers. Defining $\xi = k_\perp^{(\mathrm{II})} a/2$, $\beta = a\sqrt{U/2}$, and $\eta = \cot\xi$, the matching conditions Eq. (18) can be written as

$$\eta = \frac{\xi}{\sqrt{\beta^2 - \xi^2}} \qquad \text{even parity} \qquad (23a)$$

$$-\eta = \frac{\sqrt{\beta^2 - \xi^2}}{\xi} \qquad \text{odd parity} \qquad (23b)$$

From this, one can extract that (i) all bound states are nondegenerate, (ii) even and odd solutions alternate with increasing energy, and (iii) the number of bound states is finite and equals $M + 1$ if $M\pi < 2\beta \leq (M + 1)\pi$. The first two statements have already been observed for the case of infinitely high barriers ($U \to \infty$, $\beta \to \infty$). Note that the well is sufficiently characterized by the dimensionless parameter β.

The energy levels for a well with finite β (finite height) are shown in Figure 5 as filled circles. For the parameters chosen,

the number of states is equal to the number of layers. The global behavior in dependence on the film thickness is close to that for a well with $\beta = \infty$.

The preceding dispersion relations $E_m(N)$ for the QWSs can be experimentally determined by ARPES and angle-resolved inverse photoelectron spectroscopy (ARIPES); see Section 4. As a prototypical example, we address Cu films grown on face-centered cubic (fcc) Co(100). The photoelectron spectra taken for emission normal to the surface (i.e., $\mathbf{k}_\parallel = 0$) are shown in Figure 6. The intensity maxima related to the QWSs are labelled by the quantum numbers m (ARPES has access to the initial states below the Fermi energy E_F, whereas ARIPES probes states above E_F). The energy dispersion $E_m(N)$ as obtained from the spectra is shown in Figure 7. The dependence of the energy position on film thickness follows the theoretically determined one in very good agreement (lines in Fig. 7). These lines correspond to those for $m - N$ as obtained for the simple "rectangular-well" model (dashed lines in Fig. 5). Note that the bulk-band structure of Cu, in particular, the sp valence band, plays the same role as the free-electron band in the preceding quantum-well model (cf. Figs. 4 and 5).

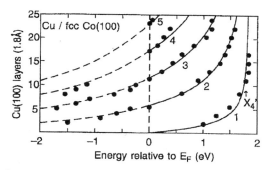

Fig. 7. Quantum-well states of Cu films on fcc Co(100) as obtained by angle-resolved photoelectron spectroscopy for $\mathbf{k}_\parallel = 0$. Shown are energy positions of quantum-well states versus film thickness—as obtained from the spectra in Figure 6 (filled circles). Lines are theoretically predicted values using the bulk-band structure of Cu along the $\Gamma-\Delta-X$ direction (cf. also Fig. 5). Reprinted with permission from J. E. Ortega, F. J. Himpsel, G. J. Mankey, and R. F. Willis, *Phys. Rev. B* 47, 1540 (1993). Copyright 1993, by the American Physical Society.

In the basic theory discussed previously, we assumed a symmetric quantum well, $V(\varrho, z) = V(\varrho, -z)$, thus ignoring any effects of the substrate that would affect the potential of the film at the film–substrate interface. It is straightforward to enhance the model to a different barrier height at the film–substrate boundary with respect to that at the vacuum–film boundary. Further, one can introduce additional potential steps in order to mimic polarization effects due to the substrate.

2.2.2. Envelope Picture of Quantum-Well States

In the preceding section, we employed a description of quantum-well states based on plane waves; for example, we assumed free electrons in the three regions of the system and neglected any "internal structure" of the potential in the film region. Now we take into account the crystal potential in an approximation, which was originally introduced for semiconductor superlattices [18, 19].

We start with the Schrödinger equation (8) for the infinitely extended system, the so-called bulk system, where $V(\mathbf{r})$ is now the potential of the bulk. The latter can be written as a sum over site, potentials $V(\mathbf{r}) = \sum_i V(\mathbf{r} - \mathbf{R}_i)$, where the vectors $\mathbf{R}_i$ form a lattice; that is they can be expressed in terms of basic translation vectors $\mathbf{a}_j$, $j = 1, 2, 3$: $\mathbf{R}_i = \sum_{j=1}^{3} n_{ij}\mathbf{a}_j$, $n_{ij} \in \mathbb{Z}$ (cf. Section 2.1.1). Therefore, the crystal potential is translationally invariant with respect to lattice vectors $\mathbf{R}$, $V(\mathbf{r}) = V(\mathbf{r}+\mathbf{R})$. Following the argumentation in Section 2.2.1 [Floquet's theorem; cf. also Eq. (7)], the wave functions fulfill the Bloch condition

$$\Psi(\mathbf{r} + \mathbf{R}; \mathbf{k}) = \exp(i\mathbf{k} \cdot \mathbf{R})\Psi(\mathbf{r}; \mathbf{k}) \qquad (24)$$

Because the wave functions have to be square integrable, the wave vector $\mathbf{k}$ has to be real. These wave functions are called Bloch states and can be classified with respect to $\mathbf{k}$. Their energy eigenvalues, the bulk-band structure, are denoted $E(\mathbf{k})$.

The electrons have to be confined to the film, which is again considered as infinitely extended in the xy plane. We therefore introduce two boundaries, the vacuum–film interface (i) and the film–substrate interface (s). The interfaces are assumed only to change the boundary conditions with respect to the bulk case but to leave the potential in the film region unaltered; that is, effects due to the substrate or the vacuum region are ignored. Therefore, the effect of the interfaces on the wave functions can be described by reflectivities R_s and R_i as well as phase shifts ϕ_i and ϕ_s ($R_s, R_i, \phi_s, \phi_i \in \mathbb{R}$). For $\mathbf{k}_\parallel = 0$, fixed ϱ, and a film of thickness Nd (d being the interlayer distance), a 'round trip' of a Bloch wave in the film region yields a total interference factor of

$$I = R_s R_i \exp\left[i(2k_\perp Nd + \phi_s + \phi_i)\right] \qquad (25)$$

Assuming complete reflection at the interfaces, $R_s R_i = 1$, constructive interference requires that the exponential factor in the preceding equation be equal unity. Or, equivalently,

$$2k_\perp Nd + \phi_s + \phi_i = 2n\pi \qquad n \in \mathbb{Z} \qquad (26)$$

Introducing $\phi = \phi_s + \phi_i$, one arrives at the condition

$$k_\perp = \frac{2n\pi - \phi}{2Nd} \qquad n \in \mathbb{Z} \qquad (27)$$

In conclusion, the boundary conditions at the interface restrict the allowed values of $k_\perp$ to those compatible with the "round-trip" criterion. The energies of the QWSs are therefore given by $E(k_\perp)$ with $k_\perp$ from Eq. (27). For a rectangular well with infinitely high barriers, the phase shift ϕ is either 0 (odd parity, $\phi_s = \phi_i = 0$) or 2π (even parity, $\phi_s = \phi_i = \pi$), which immediately yields Eq. (22).

The preceding $k_\perp$ quantization allows for an accurate determination of the bulk-band structure $E(k_\perp)$ by ARPES [20–22]. For fixed $\mathbf{k}_\parallel$, one measures photoemission intensities for films with different numbers of layers N and thus determines $E(k_\perp)$. For semi-infinite systems, this method does not work because there is no restriction for $k_\perp$; that is, it is not conserved in the photoemission process and therefore remains unknown.

Now consider a Bloch state with energy $E(k_\perp)$ in the range of a bulk band. Its wave function can be written as

$$\Psi(z; k_\perp) = \exp(ik_\perp z)u(z) \qquad (28)$$

where $u(z)$ is periodic with the interlayer distance, $u(z + d) = u(d)$, which follows immediately, from Eq. (24). We now assume that $k_\perp$ takes a value $k_\perp^{edge}$ close to a band edge; for example, $k_\perp^{edge} = 0$ or π/d. A wave function at another energy than the bulk-band edge but within the bulk-band range, and therefore $k_\perp \neq k_\perp^{edge}$, can be approximated by

$$\Psi(z; k_\perp) \sim F(z)\Psi\left(z; k_\perp^{edge}\right) \qquad (29)$$

where $F(z)$ is slowly varying: The wave function Ψ is given by $\Psi(k_\perp^{edge})$ but modulated by the envelope $F(z)$. According to the previous consideration on interference within a film, the envelope is given by $\exp(ik_\perp^{env}z)$. Thus, the total wave number $k_\perp$ reads $k_\perp = k_\perp^{env} + k_\perp^{edge}$. Note that, a Bloch state with $k_\perp^{edge}$ can occur as QWS if it fulfills the boundary conditions, that is, $k_\perp^{edge} = k_\perp$. In this case, $F(z) = 1$. Further, the envelope accounts for the correct boundary condition at the interfaces and obviously, according to Eq. (27), $k_\perp^{\text{env}}$ depends on the phase shift ϕ. If $k_\perp^{\text{edge}}$ is given by π/d, we have

$$k_\perp^{\text{env}} = \frac{2(n - N)\pi - \phi}{2Nd} \qquad n \in \mathbb{Z} \qquad (30)$$

It is worth noting again that the microscopic details at the interfaces are "smeared out" in the preceding envelope approximation. Therefore, it should be applied only in cases of thick films in which there is a bulklike potential of considerable spatial extent. In very thin films, for example, those in which the interfaces are very close, the previous approximation becomes questionable.

2.2.3. Tight-Binding Description of Quantum-Well States: Rare-Gas Films on a Metallic Substrate

We now address a description of quantum-well states based on the envelope theory. This model has been introduced by Grüne and co-workers in order to describe photoemission experiments of rare-gas films on metallic substrates [22]. Due to the inert behavior of the rare gases, bulk- as well as surface-band structures can be described very well within tight-binding theory. For example, the Xe $5p$ states are split due to spin-orbit coupling (SOC) into $5p_{j=1/2}$ and $5p_{j=3/2}$ states, j denoting the total angular momentum. In layered structures (as well as in the bulk of fcc-Xe), the $5p_{j=3/2}$ states are further split due to lateral interactions, for example, the overlap of orbitals located at different Xe sites [23–25]. Tight-binding models have proven to reproduce the experimentally determined energy dispersions very well [26, 27].

The main problem in describing the energy dispersion of Xe states on a metal substrate measured by means of ARPES is to account for the change in the Xe binding energy in dependence on the amount of adsorbed Xe. In particular, the binding energies of states located at the Xe–metal interface are substantially smaller than those of states in the other Xe layers. This behavior can be attributed either to a change in the work function,[9] which is due to the adsorption of Xe, or to the image force acting on the hole, which has been created in the photoemission process (cf. Section 4). The first effect is a so-called initial-state effect because it is present in the ground state of the system. The second is a final-state effect because it occurs in the excited system, that is, in the state with the photoelectron missing. To account for these effects, the simple rectangular well has to be extended as shown in Figure 8. Instead of three regions (I, II, and III), we now have four (A, B, C, and D). In the initial-state model, the change of the work function is V_B, which is known from experiment. The Fermi energy E_F is fixed by the substrate. The binding energies are $E = V_D + V_B - \Phi_M - E_F$ and $E = V_D + \Phi_M - E_F$ in the initial-state and the final-state model, respectively.

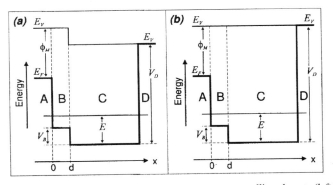

Fig. 8. Quantum-well models for rare-gas films on a metallic substrate (left, initial-state model; right, final-state model). E is the electron energy and Φ_M the work function of the substrate (no film present). The four regions A, B, C, and D are the substrate, the rare-gas layer next to the substrate, the remaining rare-gas layers, and the vacuum, respectively. Reprinted with permission from M. Grüne, T. Pelzer, K. Wandelt, and I. T. Steinberger, *J. Electron Spectrosc. Relat. Phenom.* 98–99, 121 (1999). Copyright 1999, by Elsevier Science.

Due to the confinement of the electrons to the Xe film, one assumes exponentially decaying plane waves in regions A (substrate) and D (vacuum) with constants κ_A and κ_D [cf. Eq. (16)]. The latter are given by

$$\kappa_A = \sqrt{2(V_D + V_B - \Phi_M - E)} \qquad (31a)$$
$$\kappa_D = \sqrt{2(V_D - E)} \qquad (31b)$$

in the initial-state model, which we focus on in the following discussion. In region C, the Xe film, one writes the wave function according to the envelope theory as

$$\Psi_C(z) = C \sin(k_C z + \delta_C) u(z) \qquad (32)$$

where $\sin(k_C z + \delta_C)$ is the envelope and $u(z)$ is periodic with respect to the interlayer spacing d: $u(z) = u(z + d)$. One further assumes that u is symmetric with respect to the Xe layers, $\partial_z u(z) = 0$ at $z = nd$, $n = 0, \ldots, N$, where N is the number of Xe layers. In region B, the wave function takes the same form as in Eq. (32) but with labels B. Note that $u(z)$ need not be explicitly specified. The band structure $E(k)$ of bulk Xe is approximated by

$$E(k_B) = -\gamma_B \left[1 + \cos(k_B d)\right] + V_B - 2(\gamma_B - \gamma_C) \qquad (33a)$$
$$E(k_C) = -\gamma_C \left[1 + \cos(k_C d)\right] \qquad (33b)$$

that is, one uses the energy dispersion of a simple tight-binding model. Note that γ_B and γ_C are allowed to differ. As in the case of a rectangular quantum well, the dispersion relation is obtained via matching the wave functions at the interfaces. This yields the transcendental equation

$$k_C \tan\left[k_B d + \arctan(k_B/\kappa_A)\right]$$
$$= -k_B \tan\left[(N - 1)k_C d + \arctan(k_C/\kappa_D)\right] \qquad (34)$$

which is used to determine the energies of the QWSs. After eliminating κ_A, κ_D, and k_B, one eventually obtains an equation in k_C with adjustable parameters γ_B, γ_C, and V_D that have to be determined by comparing the experimental QWS energies with

---

[9]The work function is the energy difference between the vacuum level and the Fermi level.

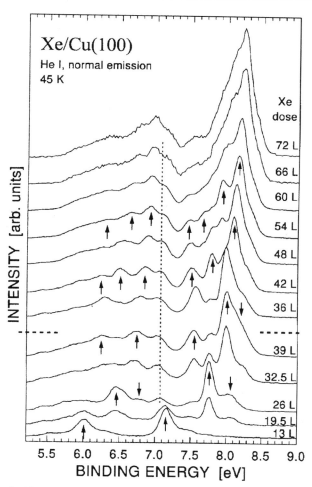

Fig. 9. Experimental photoelectron spectroscopy from Xe films on Cu(100) in normal emission ($\mathbf{k}_{\parallel} = 0$) and unpolarized light with 21.22 eV photon energy He(I). The Xe dose in langmuirs (L) is denoted on the right of each spectrum. Arrows indicate intensity maxima attributed to spin-orbit split $5p_{j=1/2}$ and $5p_{j=3/2}$ quantum-well states. Reprinted with permission from M. Grüne, T. Pelzer, K. Wandelt, and I. T. Steinberger, *J. Electron Spectrosc. Relat. Phenom.* 98–99, 121 (1999). Copyright 1999, by Elsevier Science.

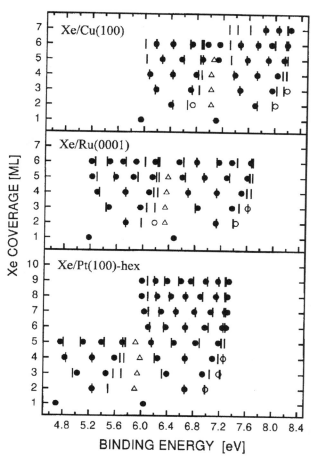

Fig. 10. Energies of quantum-well states of Xe films on metallic substrates [from bottom to top, Xe/Pt(100), Xe/Ru(0001), and Xe/Cu(100)]. Circles and triangles indicate experimentally determined values [for Xe/Cu(100)], cf. Fig. 9]. Vertical lines mark values obtained from the theory of Section 2.2.3. Reprinted with permission from M. Grüne, T. Pelzer, K. Wandelt, and I. T. Steinberger, *J. Electron Spectrosc. Relat. Phenom.* 98–99, 121 (1999). Copyright 1999, by Elsevier Science.

that obtained by theory for all film thicknesses. Note that the adjustable parameters do not depend on the number of layers N.

The experimental photoelectron spectra for Xe films on Cu(100) are shown in Figure 9. For very low Xe coverage (cf. the lowest spectrum), one observes two Xe-derived maxima with binding energies that differ considerably from those at higher coverages. This can be attributed to the different potentials in regions B and C. At higher Xe coverages, the number of QWSs increases, as can be seen in Figure 10. For all three substrates, the pattern of energy positions is similar except for the specific binding energies. Each pattern can be divided into two groups, one for the $5p_{j=1/2}$ states, the other for the $5p_{j=3/2}$ states. This separation accomplishes the fitting procedure because each group can be fitted separately. For film thicknesses $N \geq 2$, each pattern is almost symmetric in energy, which can be attributed to the underlying tight-binding band-structure of Xe, Eq. (33) (see also Section 4.2.4). However, distinct deviations from the symmetry occur: Open circles

in Figure 10 denote so-called extra peaks. For a film N layers thick, $N - 1$ energy positions follow closely that of a quantum well with infinitely high barriers and tight-binding band structure [cf. Eq. (27)]. The behavior of the Nth value can also be attributed to the potential difference between regions B and C. In conclusion, the sophisticated quantum-well model presented before is able to reproduce and to explain most, if not all, of the features found in experiment. Finally, it should be noted that initial-state and final-state models yield almost identical QWS dispersions. Therefore, one cannot judge from the preceding analysis whether the effects appear in the ground state or in the final state.

3. LOW-ENERGY ELECTRON DIFFRACTION

3.1. Introduction and History

Scattering of electrons from solid surfaces is one of the paradigms of quantum physics. The pioneering experiments were performed by Davisson and Germer [28, 29] on a Ni single crystal with (111) orientation and confirmed de Broglie's concept of the wave nature of particles [30–32], a concept at the very heart of quantum mechanics (wave-particle dualism). Already in these early works they recognized the potential of low-energy electron diffraction (LEED) as a tool for the determination of surface structures [33, 34] and applied it to gas-adsorbate layers on Ni(111) [35]. This success was only possible due to two important properties of LEED: surface sensitivity and interference.

The schematical setup of a LEED experiment is shown in Figure 11. A monoenergetic beam of electrons with kinetic energy E impinges on the sample. The reflected electron beams are detected and analyzed with respect to their direction and energy. Usually, one detects only elastically reflected electrons (for which the energy is conserved) and uses incidence normal to the surface. Therefore, set of LEED spectra—or $I(E)$ curves—represents the current I of each reflected beam versus the initial energy E. Note that the reflected intensities are roughly as large as 1/1000 of the incoming intensity.

Surface Sensitivity. In 1928, Davisson and Germer observed an attenuation of the electron-beam intensity with sample thickness [36]. Electrons in a LEED experiment have a typical kinetic energy in the range of 20 to 500 eV.[10] Due to the interaction of the incoming electron with the electrons in the sample, the former penetrates into the solid only a few angstroms. Typical penetration lengths taken from the "universal curve" (see, e.g., Fig. 12) range from 5 to 10 Å [38, 39]. Therefore, LEED spectra usually carry less information about the geometrical structure of the volume of the solid, that is, the bulk, than of the solid's surface region.

Interference. De Broglie showed that a particle with momentum $\mathbf{p}$ can be associated with a wave with wavelength $\lambda = 2\pi/p$ ($p = |\mathbf{p}|$). For example, an electron in vacuum can be described by a plane wave

$$\Psi(\mathbf{r}, t) = \exp\left[i(\mathbf{k} \cdot \mathbf{r} - \omega t)\right] \tag{35}$$

with wave number $k = 2\pi/\lambda$ and energy $E = \omega = k^2/2$. De Broglie's picture of electrons as waves and the interpretation of the Davisson–Germer LEED experiments lead to the question: *Are electrons waves?* [40]. Comparison was made to X-ray scattering in view of the determination of structural information and Davisson came to the conclusion that if X-rays are waves then electrons are, too. However, he admitted that the picture of electrons as particles is better suited for the explanation of the Compton effect or the photoelectric effect (cf. Section 4 on angle-resolved photoelectron spectroscopy).

In the wave picture of electrons, the LEED experiment can be regarded as follows. An incoming plane wave, the incident beam, is scattered at each site and the outgoing plane waves, the outgoing beams, are measured. Both amplitude and phase of each outgoing wave are determined by the scattering properties and the position of each scatterer. For example, a change in the position of a scatterer will change the wave pattern in the solid and, therefore, will affect both amplitudes and phases of the outgoing waves. Because the LEED current of a beam is given by the wave amplitude, it carries information on both positions and scattering properties of the sites. This mechanism can be used, for instance, to obtain images of the geometrical structure in configuration space by LEED holography [41].

Although LEED is sensitive to the outermost region of the sample, it is capable of detecting fingerprints of the electronic states of the film–substrate system. As mentioned previously,

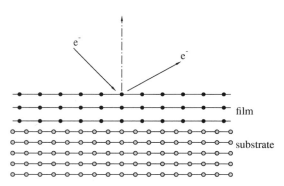

Fig. 11. Scheme of the LEED setup. An incoming beam of electrons e^- is elastically scattered by the solid. The latter is considered as a compound of the substrate (gray circles) and a thin film (black circles). A reflected electron beam is detected. The dashed–dotted arrow represents the surface normal.

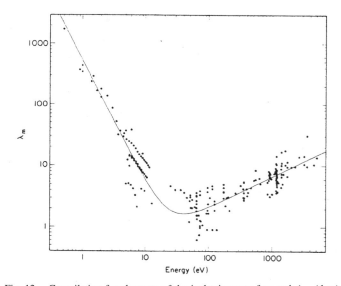

Fig. 12. Compilation for elements of the inelastic mean free path λ_m (dots) in monolayers as a function of energy above the Fermi level. This "universal curve" is almost independent of the solid, for example, surface orientation or elemental composition. The solid line serves as a guide to the eye. For details, see [37]. Reprinted with permission from M. P. Seah and W. A. Dench, *Surf. Interface Anal.* 1, 2 (1979). Copyright 1979, by John Wiley & Sons.

---

[10]Experiments with energies below this range are called very low energy electron diffraction (VLEED), those with higher energies medium-energy electron diffraction (MEED). At even higher energies, one uses gracing electron incidence and emission to obtain surface sensitivity, that is, reflecting high-energy electron diffraction (RHEED).

the LEED intensities depend on the electronic structure of the sample above the vacuum level and, therefore, contain information of the electronic structure of the entire film. In particular, quantized states that are confined to the film have a pronounced effect on the LEED spectra.

It took considerable time to develop theories that include multiple scattering of the LEED electron [42–44], which obviously is necessary for a proper description of LEED spectra. Textbooks that introduce to the field and present computer codes for the calculation of $I(E)$ spectra were written by Pendry [45] as well as van Hove et al. [46]. Van Hove and Tong also provide review articles [47, 48].

Additional information can be obtained if one uses a spin-polarized beam of incoming electrons, that is, spin-polarized low-energy electron diffraction (SPLEED), and uses a spin-sensitive detector, for example a Mott detector or a SPLEED detector. Interestingly, the latter exploits the LEED mechanism itself for a spin resolution in the experiment. Pioneering works were made by Feder [49–51] on the theoretical and by Kirschner [52] on the experimental side.

3.2. Theories

In the following, a theoretical description of LEED from semi-infinite solids covered by thin films (cf. Fig. 11) is presented. The kinematical theory focuses mainly on interference effects (Section 3.2.1), whereas in the pseudopotential theory a close connection of LEED intensities to the electronic structure is established (Section 3.2.2). The multiple-scattering theory gives an introduction to state-of-the-art calculations of LEED from layered systems (Section 3.2.3).

3.2.1. Kinematical Theory

According to de Broglie's wave picture of electrons, we consider as an introduction elastic scattering of a plane wave by a one-dimensional periodic structure (cf. Fig. 13). Constructive interference occurs if the phase difference between outgoing beams scattered from neighboring sites is an integer multiple of the wavelength λ,

$$a(\sin\varphi - \sin\varphi_0) = h\lambda \qquad h \in \mathbb{Z} \tag{36}$$

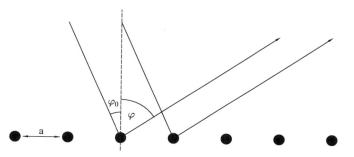

Fig. 13. Scattering from a one-dimensional periodic structure with lattice constant a. The incoming plane wave impinges with a polar angle of φ_0 relative to the normal axis (dashed line). The diffracted wave is outgoing with an angle φ.

This is the so-called Laue condition, and h is the order of diffraction. Note that the hth order of diffraction corresponds to the nhth order of the same periodic structure but with lattice constant na.

The preceding consideration can easily be applied to two-dimensional periodic structures, that is, a two-dimensional lattice with basis vectors $\mathbf{a}_1$ and $\mathbf{a}_2$; cf. Section 2.1.1. The Laue condition then reads

$$\sin\varphi - \sin\varphi_0 = \lambda/d_{hk} \tag{37}$$

with $d_{hk} = |h\mathbf{a}_1 + k\mathbf{a}_2|$ denoting the length of a vector of the direct (covariant) lattice. In other words, the diffraction pattern, that is, the set of angles φ and φ_0 for which there is constructive interference, yields the geometry of the direct lattice. h and k are referred to as the Miller indices. The Laue conditions can further be written as

$$(\mathbf{s} - \mathbf{s}_0) \cdot \mathbf{a}_i = h_i\lambda \qquad i = 1, 2, \ h_i \in \mathbb{Z} \tag{38}$$

where the vectors $\mathbf{s}$ and $\mathbf{s}_0$ specify the directions of the incoming beam and the diffracted beams, respectively. Expanding $\Delta\mathbf{s} = \mathbf{s} - \mathbf{s}_0$ in the basis of the reciprocal lattice, $\Delta\mathbf{s} = \mathbf{g}_1\zeta_1 + \mathbf{g}_2\zeta_2$, one sees immediately that $\Delta\mathbf{s} = (\mathbf{g}_1h_1 + \mathbf{g}_2h_2)\lambda$. Or, in other words, the diffraction pattern is directly a representation of the reciprocal (contravariant) lattice. Because the Laue conditions pick out discrete directions of reflection, each of which is associated with a pair of Miller indices (h, k), one usually calls the reflected wave functions LEED beams and indicates them by (h, k). The results obtained so far can be cast into the sketch of a LEED experiment as shown in Figure 14.

Up to now, only the diffraction pattern has been considered; intensities of the diffracted beams have been ignored. A first approach for calculating intensities is the kinematical theory,

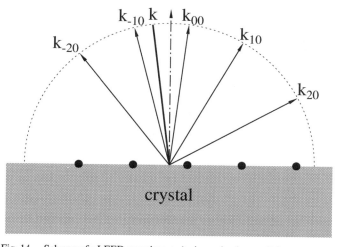

Fig. 14. Scheme of a LEED experiment. An incoming beam of electrons specified by its wave vector $\mathbf{k}$ (incoming arrow) is scattered by the solid (gray area). The elastically reflected electron beams are indicated by their Miller indices h and k of the surface-reciprocal lattice vectors, their wave vector denoted as $\mathbf{k}_{hk}$ (outgoing arrows). Filled circles represent atoms at the surface. The surface normal is given by the dashed–dotted arrow. Because the electrons are reflected elastically, the lengths of the individual wave vectors are identical (cf. the dotted semicircle).

which assumes that the interaction of the LEED electron with the solid is weak. Therefore, only single-scattering events have to be considered; multiple scattering is neglected. This approximation gives reasonable results for scattering of X-rays and fast electrons (e.g., for Compton profiles), but worse results for slow electrons (e.g., for LEED or VLEED). However, the basic idea can be regarded as the starting point for a dynamical theory, that is, a theory that takes into account multiple scattering.

Consider a LEED experiment from a semi-infinite solid that consists of a substrate covered by a thin film (cf. Fig. 11). The incoming LEED beam with incidence direction $\mathbf{s}_0$ is represented by the plane wave $\Psi_{\mathrm{inc}} = \Psi_0 \exp(i\mathbf{k}_0 \cdot \mathbf{r})$, where the momentum $\mathbf{k}_0 = 2\pi\mathbf{s}_0/\lambda$. An outgoing wave scattered from site $\mathbf{R}'$ and detected at $\mathbf{R}$ is given by

$$\Psi_{\mathbf{R}'} = \Psi_0 \underbrace{\frac{\exp(i\mathbf{k} \cdot \mathbf{R})}{R}}_{1} \underbrace{f(\mathbf{k}, \mathbf{k}_0; \mathbf{R}')}_{2} \underbrace{\exp\left[i(\mathbf{k} - \mathbf{k}_0) \cdot \mathbf{R}'\right]}_{3} \quad (39)$$

The first term [denoted 1 in Eq. (39)] is a spherical wave; the second is the atomic structure factor and describes scattering of plane waves at site $\mathbf{R}'$. In a dynamical theory, it depends on both directions $\mathbf{k}$ and $\mathbf{k}_0$, whereas in the kinematical theory it is assumed to depend only on $\mathbf{k} - \mathbf{k}_0$: $f(\mathbf{k}, \mathbf{k}_0; \mathbf{R}') = f(\mathbf{k} - \mathbf{k}_0; \mathbf{R}')$. The third term takes care of the phase difference relative to the origin of the coordinate system. Note that the detector is assumed to be positioned at a large distance away from the sample. Therefore, the spherical waves outgoing from each scatterer can be replaced in good approximation by plane waves at the detector position.

Because only single scattering is considered, the total outgoing wave function is the sum of those outgoing from the substrate and those from the film, $\Psi_{\mathrm{out}} = \Psi_{\mathrm{out}}^{\mathrm{sub}} + \Psi_{\mathrm{out}}^{\mathrm{film}}$. We first consider the substrate, the lattice of which is defined by the basis vectors $\mathbf{a}_1$, $\mathbf{a}_2$, and $\mathbf{a}_3$. The first two are parallel to the surface ($a_{1z} = a_{2z} = 0$), and the latter points from one layer to the neighboring one ($a_{3z} \neq 0$; $\mathbf{a}_3$ corresponds to the interlayer vector $\mathbf{d}$ introduced previously). Further, the structure factor is assumed as identical for all sites: $f^{\mathrm{sub}}(\mathbf{k} - \mathbf{k}_0; \mathbf{R}') = f^{\mathrm{sub}}(\mathbf{k} - \mathbf{k}_0)$. The outgoing wave then reads

$$\Psi_{\mathrm{out}}^{\mathrm{sub}} \sim f^{\mathrm{sub}}(\mathbf{k} - \mathbf{k}_0) \sum_{n_1 n_2 n_3} \exp\left[i(\mathbf{k} - \mathbf{k}_0) \cdot (n_1\mathbf{a}_1 + n_2\mathbf{a}_2 + n_3\mathbf{a}_3)\right]$$

$$(40)$$

and is the product of a single-site structure factor (f^{sub}) and a geometrical part that depends only on the lattice structure. This feature—the separation of scattering properties and the geometrical arrangement of the scatterers—is also present in the Korringa–Kohn–Rostoker (KKR) multiple-scattering approach presented in Section 3.2.3. With the definitions

$$S_i^{\mathrm{sub}}(\mathbf{k} - \mathbf{k}_0) = \sum_{n_i} \exp\left[in_i(\mathbf{k} - \mathbf{k}_0) \cdot \mathbf{a}_i\right] \qquad i = 1, 2, 3 \quad (41)$$

the intensity I^{sub} at the detector position $\mathbf{R}$ solely from the substrate reads

$$I^{\mathrm{sub}} \sim |\Psi_{\mathrm{out}}^{\mathrm{sub}}|^2 \sim \left|f^{\mathrm{sub}}(\mathbf{k} - \mathbf{k}_0)\right|^2 \prod_{i=1}^{3} \left|S_i^{\mathrm{sub}}(\mathbf{k} - \mathbf{k}_0)\right|^2 \quad (42)$$

One now has to consider the sums S_i^{sub}, $i = 1, 2, 3$. As Davisson and Germer observed [36], the electron beams are attenuated in a direction normal to the surface. Thus, an empirical attenuation factor μ is introduced, which takes care of the surface sensitivity of the LEED experiment (see Fig 12). Then S_3^{sub} reads

$$S_3^{\mathrm{sub}}(\mathbf{k} - \mathbf{k}_0) = \sum_{j=0}^{\infty} \exp\left[ij(\mathbf{k} - \mathbf{k}_0) \cdot \mathbf{a}_3 - j\mu\right] \quad (43)$$

Summing up the geometrical series yields

$$\left|S_3^{\mathrm{sub}}(\mathbf{k} - \mathbf{k}_0)\right|^2 = \left\{1 - 2\exp(-\mu)\cos\left[(\mathbf{k} - \mathbf{k}_0) \cdot \mathbf{a}_3\right] + \exp(-2\mu)\right\}^{-1} \quad (44)$$

which is maximal if $\cos[(\mathbf{k} - \mathbf{k}_0) \cdot \mathbf{a}_3] = 1$ or, equivalently, if $(\mathbf{k} - \mathbf{k}_0) \cdot \mathbf{a}_3 = 2\pi h_3$ with integer h_3. This establishes the third Laue condition; cf. Eq. (38). Analogously, the sums S_1 and S_2 give the two Laue conditions for planar diffraction.

The effect of the attenuation on the third Laue condition is shown in Figure 15a, where $|S_3(\mathbf{k} - \mathbf{k}_0)|^2$ is presented for several values of μ. An increase of the latter reduces the ratio of the maxima at $2\pi h_3$ to the minima at πh_3. In other words, strong attenuation weakens the third Laue condition.

Now we consider the effect of the film on the electron diffraction. For a film with N_3 layers, the outgoing wave func-

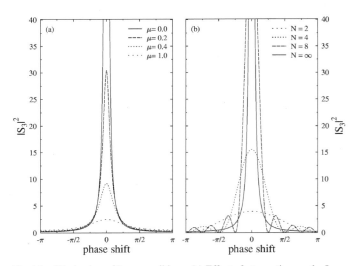

Fig. 15. Weakening of Laue conditions. (a) Effect of attenuation on the Laue condition. The lattice factor $|S_3|^2$, cf. Eq. (44), is shown for attenuation factors $\mu = 0.0, 0.2, 0.4$, and 1.0 versus the phase shift $(\mathbf{k} - \mathbf{k}_0) \cdot \mathbf{a}_3$. The lattice factor is maximal at multiples of 2π, in accordance with the third Laue condition. An increase in μ leads to weakening of the latter. (b) Effect of a finite number of layers on the Laue condition. The lattice factor $|S_3|^2$, cf. Eq. (47), is shown for $N = 2, 4, 8$, and ∞ layers versus the phase shift $(\mathbf{k} - \mathbf{k}_0) \cdot \mathbf{b}_3$. The lattice factor is maximal at multiples of 2π, in accordance with the third Laue condition but additional maxima occur for finite N. A decrease in N leads to weakening of the Laue condition. The attenuation factor chosen is $\mu = 0.01$.

tion is

$$\Psi_{\text{out}}^{\text{film}} \sim f^{\text{film}}(\mathbf{k} - \mathbf{k}_0) \prod_{i=1}^{3} S_i^{\text{film}} \quad (45)$$

with the definition

$$S_i^{\text{film}}(\mathbf{k} - \mathbf{k}_0) = \sum_{n_i} \exp\left[in_i(\mathbf{k} - \mathbf{k}_0) \cdot \mathbf{b}_i\right] \quad i = 1, 2, 3 \quad (46)$$

The lattice factors S_1^{film} and S_2^{film} establish the first two Laue conditions for the film structure, $(\mathbf{k} - \mathbf{k}_0) \cdot \mathbf{b}_i = 2\pi h_i$, h_i integer. In S_3^{film}, one has to sum over the finite number N_3 of film layers, which yields

$$\left|S_3^{\text{film}}(\mathbf{k} - \mathbf{k}_0)\right|^2$$
$$= \frac{1 - 2\exp(-\mu N_3)\cos\left[N_3(\mathbf{k} - \mathbf{k}_0) \cdot \mathbf{b}_3\right] + \exp(-2\mu N_3)}{1 - 2\exp(-\mu)\cos\left[(\mathbf{k} - \mathbf{k}_0) \cdot \mathbf{b}_3\right] + \exp(-2\mu)}$$
$$(47)$$

Obviously, for $N_3 = 1$ there is no third Laue condition $[|S_3^{\text{film}}(\mathbf{k} - \mathbf{k}_0)|^2 = 1]$ because there is no interference in the direction normal to the film. The effect of finite N_3 on $|S_3|^2$ is shown in Figure 15b. A large number of layers leads to sharp global maxima in the lattice factor (compare the cases $N_3 = 8$ and $N_3 = \infty$); however, $N_3 - 2$ additional maxima occur in $[0, 2\pi]$. A decrease of N_3 weakens the third Laue condition considerably (cf. the case $N_3 = 2$). In the LEED $I(E)$ spectra, this would lead to broad maxima of the "kinematical peaks".

Finally, consider the total intensity I^{tot} of the entire film–substrate system. Because the outgoing wave function is the sum of that arising from the substrate and that of the film, $\Psi_{\text{out}} = \Psi_{\text{out}}^{\text{sub}} + \Psi_{\text{out}}^{\text{film}}$, the intensity reads

$$I^{\text{tot}} \sim |\Psi_{\text{out}}^{\text{sub}} + \Psi_{\text{out}}^{\text{film}}|^2 \quad (48a)$$
$$\sim I^{\text{sub}} + I^{\text{film}} + 2\,\text{Re}\left[(\Psi_{\text{out}}^{\text{sub}})^* \Psi_{\text{out}}^{\text{film}}\right] \quad (48b)$$

The third term is due to interference of the electron's wave between the film and the substrate structure. Its effect on the LEED intensity is shown in Figure 16 for a film with $N_3 = 4$ layers and a simply related structure ($b_3 = 3a_3$; cf. panel a). Obviously, the total LEED intensity is not the sum of the film and substrate intensity, as is evident from the maxima at multiples of 2π. If the structures are incoherently related (panel b), the LEED intensity becomes irregular, in particular, that of the substrate-related maxima. In a polar plot (shown as insets in Fig. 16), a simply related structure shows n clubs if $\mathbf{b}_i = n\mathbf{a}_i$, which, for the case shown here, leads to three clubs. If the film and the substrate structure are incoherently related, that is, the ratio n is irrational, the pattern becomes dense in the polar plot. For $n = 2\sqrt{3} \sim 3.4641$, the substrate-related maxima at $2\pi n$ dominate (cf. the club aligned along the x axis in the upper right inset of Fig. 16).

The periodicity at the surface or in the film must not be identical to that of the substrate (cf. Section 2.1.1). But because the periodicity of the latter can easily be recognized in the LEED pattern, it is convenient to use it as the reference [cf. the matrix G in Eq. (5)]. A spot in the surface reciprocal net, which

has integer indices n_1 and n_2, can be expressed in the substrate reciprocal net via $(m_1, m_2)G = (n_1, n_2)H$, which yields the indices $(m_1, m_2) = (n_1, n_2)M$.

If there are N scatterers in the unit cell located at τ_j, $j = 1, \ldots, N$, the kinematical structure factor is given by

$$F(\mathbf{k} - \mathbf{k}_0; \mathbf{R}) = \sum_{j=1}^{N} f_j(\mathbf{k} - \mathbf{k}_0; \mathbf{R}) \exp\left[i(\mathbf{k} - \mathbf{k}_0) \cdot \tau_j\right] \quad (49)$$

Again, multiple scattering has been neglected, in contrast to a dynamical theory. Therefore, in order to treat several scatterers in the unit cell, one has to replace $f(\mathbf{k} - \mathbf{k}_0; \mathbf{R})$ by $F(\mathbf{k} - \mathbf{k}_0; \mathbf{R})$.

In the kinematical theory, the electronic structure of the sample enters via the atomic form factors, that is, more or less indirectly because these do not provide access to details of the electronic structure of both substrate and sample [e.g., band structure $E(\mathbf{k})$ and density of states (DOS)]. To illustrate the close connection between LEED intensities and electronic structure, we introduce a LEED theory based on empirical pseudopotentials.

3.2.2. Pseudopotential Description of Low-Energy Electron Diffraction

The following approach to LEED calculations is based on the considerations that (i) the scattering of the electrons inside the solid can be calculated with well-established band structure methods; (ii) the LEED state inside the sample is expanded into Bloch waves of the infinite solid, taking into account the correct boundary conditions in both the vacuum and the bulk; and (iii) lifetime effects can easily be incorporated via complex energies [53]. Employing the band structure of the solid can help to interpret LEED spectra or—the other way round—to determine the band structure by adjusting the potential parameters used in the calculations in such way that experimental and theoretical LEED, VLEED, or target-current spectra[11] come as close as possible. The latter band structure mapping has been proven to give good results for both metals and layered semiconductors [54] (for a short review, see [55]).

One approach for calculating LEED intensities is based on empirical local pseudopotentials. If the scattering of the electrons is weak, as can be expected because the band structure at typical LEED energies is more or less free-electron-like (cf. also the preceding kinematical treatment), the Coulomb part of the atomic potentials can be neglected, because it is screened by the core-level electrons. In the bulk of the solid, the potential is expanded in a Fourier series with a few reciprocal lattice vectors,

$$V(\mathbf{r}) = \sum_{\mathbf{G}} V_{\mathbf{G}} \exp(-i\mathbf{G} \cdot \mathbf{r}) \quad (50)$$

---

[11] Target-current spectroscopy (TCS) measures the current flowing through the sample that is due to the incident electron beam. It can therefore be regarded as complementary to LEED because in TCS the *transmitted* current is measured, whereas in LEED it is the *reflected* current.

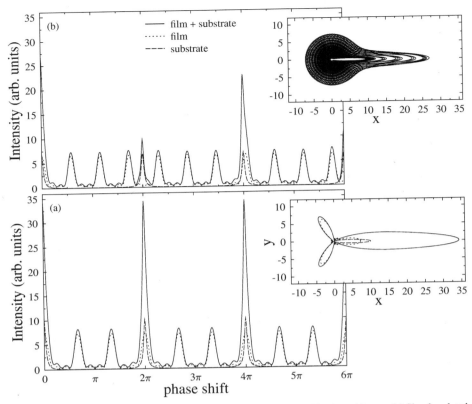

Fig. 16.   Interference effect on the LEED intensity from a thin film with $N_3 = 4$ layers. (a) Simply related structure ($b_3 = 3a_3$) and (b) incoherent structure ($b_3 = 2\sqrt{3}a_3$). Shown are the total intensity (I^{tot}, solid line), the film intensity (I^{film}, dotted line), and the substrate intensity (I^{sub}, dashed line) versus the phase factor. Atomic form factors f^{film} and f^{sub} were chosen as identical. The attenuation factor is $\mu = 0.1$. The insets on the right show the same data as in (a) and (b) but as polar plots and within an extended range of the phase shift.

where $\mathbf{G}$ is a vector of the bulk reciprocal lattice. For a given wave vector $\mathbf{k}$, the wave function is also expanded in a Fourier series,

$$\Phi(\mathbf{r}; \mathbf{k}) = \sum_{\mathbf{G}} \alpha_{\mathbf{G}}(\mathbf{k}) \exp\left[i(\mathbf{k} + \mathbf{G}) \cdot \mathbf{r}\right] \quad (51)$$

Inserting $\Phi(\mathbf{r}; \mathbf{k})$ and $V(\mathbf{r})$ into the Schrödinger equation, one arrives at the secular equation

$$\sum_{\mathbf{G}} \left\{ \left[(\mathbf{k} + \mathbf{G})^2 - 2E\right]\delta_{\mathbf{G}\mathbf{G}'} + 2V_{\mathbf{G} - \mathbf{G}'} \right\} \alpha_{\mathbf{G}}(\mathbf{k}) = 0 \quad \forall \mathbf{G}' \quad (52)$$

which can be solved numerically using standard eigenproblem routines. The preceding equation is not very well suited for LEED calculations, because in LEED one wants to know the bulk-band structure for given energy E and $\mathbf{k}_{\parallel}$. In other words, we do not want to compute $E(\mathbf{k}_{\parallel}, k_{\perp})$ but $k_{\perp}(E, \mathbf{k}_{\parallel})$. Rewriting Eq. (52) as

$$\left[(k_{\perp} + G'_{\perp})^2 - 2E\right]\alpha_{\mathbf{G}'}$$
$$+ \sum_{\mathbf{G}} \left[2V_{\mathbf{G} - \mathbf{G}'} + (\mathbf{k}_{\parallel} + \mathbf{G}_{\parallel})^2\delta_{\mathbf{G}\mathbf{G}'}\right]\alpha_{\mathbf{G}'} = 0 \quad \forall \mathbf{G}' \quad (53)$$

and defining

$$A_{\mathbf{G}\mathbf{G}'} = \left(G_{\perp} - \sqrt{2E}\right)\delta_{\mathbf{G}\mathbf{G}'} \quad (54a)$$

$$B_{\mathbf{G}\mathbf{G}'} = \left(G_{\perp} + \sqrt{2E}\right)\delta_{\mathbf{G}\mathbf{G}'} \quad (54b)$$

$$C_{\mathbf{G}\mathbf{G}'} = \left(\mathbf{k}_{\parallel} + \mathbf{G}_{\parallel}\right)^2\delta_{\mathbf{G}\mathbf{G}'} + 2V_{\mathbf{G} - \mathbf{G}'} \quad (54c)$$

one arrives at the matrix equation

$$\begin{pmatrix} -A & -C \\ 1 & -B \end{pmatrix} \begin{pmatrix} \alpha \\ \beta \end{pmatrix} = k_{\perp} \begin{pmatrix} \alpha \\ \beta \end{pmatrix} \quad (55)$$

Instead of an eigenvalue problem of a hermitian matrix, one eventually is dealing with one that is twice the size and non-hermitian. Therefore, the number of eigenfunctions and eigenvalues is twice the number of vectors $\mathbf{G}$ taken into account.

We now have to consider the boundary conditions. In the bulk, each eigenfunction fulfills Floquet's theorem,

$$\Psi^{(j)}(\mathbf{r} + \mathbf{R}) = \lambda^{(j)}(\mathbf{R})\Psi^{(j)}(\mathbf{r}) \quad \forall \mathbf{R} \quad (56)$$

where $\mathbf{R}$ is a translation vector of the bulk. Because we are interested in the band structure for given $\mathbf{k}_{\parallel}$, the proportionality factor $\lambda^{(j)}$ can conveniently be written as $\lambda^{(j)}(\mathbf{R}) = \exp(i(\mathbf{k}_{\parallel} \cdot \mathbf{R}_{\parallel} + k_{\perp}^{(j)}R_{\perp}))$ (cf. Section 2.2.1). And because the eigenvalue problem is no longer hermitian, $k_{\perp}$ can be complex even for real energies E. This feature gave rise to Heine's concept of the complex band structure [56].

According to Chang [57], the bands associated with $k_\perp^{(j)}$ can be cast into the following categories:

Real bands correspond to the conventional band structure and have $\text{Im}\, k_\perp^{(j)} = 0$. Thus, $|\lambda^{(j)}| = 1$ and the wave functions $\Psi^{(j)}$ are the Bloch states [13].

Imaginary bands of the first kind have $\text{Re}\, k_\perp^{(j)} = 0$ and $\text{Im}\, k_\perp^{(j)} \neq 0$.

Imaginary bands of the second kind have $\text{Re}\, k_\perp^{(j)} = k_{\perp\max}$ and $\text{Im}\, k_\perp^{(j)} \neq 0$.

Complex bands have $\text{Re}\, k_\perp^{(j)} \neq 0$, $\text{Re}\, k_\perp^{(j)} \neq k_{\perp\max}$, and $\text{Im}\, k_\perp^{(j)} \neq 0$.

Here, $k_{\perp\max}$ refers to the boundary of the (first) Brillouin zone in the z direction. In the infinite solid, all wave functions $\Psi^{(j)}$ have to be square integrable ($\Psi^{(j)} \in \mathcal{L}_2$). That means only Bloch states are proper solutions of the Schrödinger equation. Due to the surface, the integration range can be taken as one half-space, say $z > 0$, and the set of proper wave functions can be extended to those the amplitude of which decays in direction toward the bulk. Thus, all imaginary and complex bands with $\text{Im}\, k_\perp^{(j)} < 0$ have to be considered in addition to the real bands. The wave functions associated with these bands are usually called evanescent states.

The next step is to construct the wave function of the LEED electron, the LEED state. In the vacuum, which is characterized by the fact that the potential is, the incoming beam is represented by an incoming plane wave with surface-parallel wave vector

$$\mathbf{k}_\| = \sqrt{2E}\,\sin\vartheta_e \begin{pmatrix} \cos\varphi_e \\ \sin\varphi_e \end{pmatrix} \qquad (57)$$

where E is the kinetic energy of the incident electron. ϑ_e and φ_e are the polar and the azimuthal angles of electron incidence, respectively. The wave function in the vacuum thus reads

$$\Phi(\mathbf{k}_\|; E) = \underbrace{\exp[i(\mathbf{k}_\| \cdot \boldsymbol{\varrho} + \kappa_0 z)]}_{\text{incoming}}$$
$$+ \underbrace{\sum_{\mathbf{g}} \varphi_{\mathbf{g}} \exp\{i[(\mathbf{k}_\| + \mathbf{g}) \cdot \boldsymbol{\varrho} - \kappa_{\mathbf{g}} z]\}}_{\text{outgoing}} \qquad (58)$$

Here, we have again decomposed the spatial vector $\mathbf{r}$ into a surface-parallel component $\boldsymbol{\varrho} = (x, y)$ and a perpendicular component z. The perpendicular component of the wave vector is given by

$$\kappa_{\mathbf{g}} = \sqrt{2E - (\mathbf{k}_\| + \mathbf{g})^2} \qquad (59)$$

Note that for beams that cannot escape from the solid into the vacuum $\kappa_{\mathbf{g}}$ is imaginary and their wave functions are damped in direction toward the vacuum. The intensities of the reflected beams, which are indicated by surface-reciprocal lattice

vectors $\mathbf{g}$, are proportional to $|\varphi_{\mathbf{g}}|^2$. Inside the solid, proper solutions of the half-space problem can be expanded into Bloch and evanescent states,

$$\Psi(\mathbf{k}_\|; E) = \sum_j t^{(j)} \Psi^{(j)}(\mathbf{k}_\|, E; k_\perp^{(j)}) \qquad (60)$$

To determine the coefficients $\varphi_{\mathbf{g}}$ and $t^{(j)}$, one requires that at a certain coordinate z_0 both the wave function and its derivative with respect to z are continuous. In matrix form, this requirement can be written as a relation between the coefficients of the bulk states ($t^{(j)}$) and those of the incoming ($\varphi_{\mathbf{g}}^+$) and outgoing ($\varphi_{\mathbf{g}}^-$) LEED beams,

$$B\mathbf{t} = A^+\varphi^+ + A^-\varphi^- \qquad (61a)$$
$$B'\mathbf{t} = A^{+'}\varphi^+ + A^{-'}\varphi^- \qquad (61b)$$

with

$$B_{\mathbf{g}j} = \sum_{\mathbf{G}} \alpha_{\mathbf{G}}^{(j)} \exp[i(k_\perp^{(j)} + G_\perp)z_0]\delta_{\mathbf{g}\mathbf{G}_\|} \qquad (62a)$$

$$A_{\mathbf{g}\mathbf{g}'}^{\pm} = \exp(\pm i\kappa_{\mathbf{g}} z_0)\delta_{\mathbf{g}\mathbf{g}'} \qquad (62b)$$

and analogous expressions for the matrices of the z derivatives, the latter being indicated by a prime ($A^{\pm'}, B'$). The incoming amplitudes φ^+ are represented by the vector that contains a 1 at the row of the $(0, 0)$ beam and 0s otherwise. After some manipulation, one arrives at relations between the incoming amplitude φ^+ and the coefficients $\mathbf{t}$ of the solid's eigenfunctions as well as coefficients φ^- of the outgoing beams,

$$\mathbf{t} = [(B' - A'(A^-)^{-1}B)]^{-1}$$
$$\times [A^{+'} - A^{-'}(A^-)^{-1}A^+]\varphi^+ \qquad (63a)$$

$$\varphi^- = [(A^{-'} - B'(B)^{-1}A^-)]^{-1}$$
$$\times [A^{+'} - B'(B)^{-1}A^+]\varphi^+ \qquad (63b)$$

Note that one equation can be obtained by the other by simultaneously replacing A^- with B as well as $A^{-'}$ with B'.

With regard to the symmetry of the setup, not all of the wave functions $\Psi^{(j)}$ can couple to the LEED beams. Because the incident plane wave is totally symmetric at the detector [58], it belongs to the trivial representation of the "small group" of $\mathbf{k}_\|$ [59]. Thus, only those coefficients $t^{(j)}$ of wave functions $\Psi^{(j)}$ that belong to the same representation are nonzero.[12] This can easily be seen by the matching procedure which involves integration over the surface plane $z = z_0$.

The relation of the coefficients $t^{(j)}$ and the amplitudes of the outgoing beams $\varphi_{\mathbf{g}}$ allows a first interpretation of LEED $I(E)$ spectra. For example, if there is a bandgap at the energy of the incident beam, the latter cannot couple to Bloch states but to evanescent states inside the solid and thus has to be strongly reflected. This leads to a maximum in the $I(E)$ spectrum. In turn, if the incoming beam can couple very well to Bloch states,

---

[12] For example, if the incident beam impinges in a mirror plane of the solid, its plane wave is even under the associated reflection. Therefore, only wave functions $\Psi^{(j)}$ that are also even have nonvanishing expansion coefficients $t^{(j)}$.

the current is propagating toward the interior of the solid and the reflected intensity drops. This establishes a close connection between LEED intensities and electronic structure.

This method of wave function matching was first applied to LEED by Pendry [60–62]. It can also be applied for the determination of the band structure above the vacuum level, as has been demonstrated for Cu and semiconductors by Strocov and co-workers [63, 64]. Because the LEED wave functions appear also in photoemission as the final state (see Section 4), this method can be used in photoemission calculations [65–67]. On the mathematical problems related to this technique, we refer the reader to [68, 69].

The advantages of this method include the following: (i) In a LEED calculation, one is looking for the intensity for a given primary energy E and wave vector $\mathbf{k}_\parallel$ of the incoming beam. Due to the broken translational symmetry perpendicular to the surface, $k_\perp$ is not a "good quantum number;" that is, it is not conserved. (ii) Inelastic effects can be taken into account by an "optical potential," the imaginary part of which leads to broadening of the maxima in the $I(E)$ spectra (cf. Fig. 15 and Section 3.3).

However, problems occur in cases of ultra-thin films. Due to the confinement of the electronic states, the energy levels can become quantized (quantum-well states) and the description of LEED in terms of the bulk-band structure may not be appropriate. In the next section, we will introduce an approach that is by far better suited to the description of ultra-thin films and their electronic properties.

3.2.3. Dynamical Theory

In contrast to a kinematical theory (see Section 3.2.1), a dynamical theory considers multiple scattering of electrons inside the solid. Here, we can only give a brief survey and refer for a comprehensive description to textbooks by Mertig et al. [70], Weinberger [71], and Gonis [72].

The main task of a multiple-scattering theory for LEED is to determine the scattering properties of the whole semi-infinite solid. This is achieved by consecutive calculations of the scattering properties of a single site (sometimes loosely denoted as an "atom"), a single layer, stacks of layers, and eventually the entire solid. This step-by-step procedure gives a great flexibility concerning the actual arrangement of scatterers. Connected with these steps is a change of the basis in which the calculations are performed. For example, scattering from a single site is conveniently formulated in an angular-momentum basis that gives rise to so-called partial waves. Or scattering from layers is conveniently formulated in a plane wave basis.

The multiple-scattering theory as formulated in the following can be traced back to the original work of Korringa [73] as well as that of Kohn and Rostoker [74] who formulated it for three-dimensional systems. Because for surface problems a formulation in terms of layers is often more appropriate, the method that uses layers as an essential object is known as layer-KKR (see, e.g., [75]), where KKR stands for the initials of the inventors, Korringa, Kohn, and Rostoker.

Angular Momenta. Before turning to the multiple-scattering theory, a brief review of the basic properties of the angular-momentum operator $\mathbf{l} = (l_x, l_y, l_z)$ and the spin operator $\mathbf{s} = (s_x, s_y, s_z)$ is given. With the "ladder operators" $l_\pm = l_x \pm il_y$, $s_\pm = s_x \pm is_y$, and the total angular momentum $\mathbf{j} = \mathbf{l} + \mathbf{s}$, one has $2\mathbf{l} \cdot \mathbf{s} = 2l_z s_z + l_+ s_- + l_- s_+$ and the commutational rules $[j_z, \mathbf{s} \cdot \mathbf{l}] = 0$, $[\mathbf{j}^2, \mathbf{s} \cdot \mathbf{l}] = 0$, and $[s_z, \mathbf{s} \cdot \mathbf{l}] = -[l_z, \mathbf{s} \cdot \mathbf{l}]$.

Spherical harmonics are eigenfunctions of $\mathbf{l}^2$ and l_z, $\mathbf{l}^2 Y_l^m = l(l+1)Y_l^m$ and $l_z Y_l^m = mY_l^m$. They obey the relation $(Y_l^m)^* = (-1)^m Y_l^{-m}$. Further, one has

$$l_z s_z Y_l^{\mu-\tau} \chi^\tau = (\mu - \tau)\tau Y_l^{\mu-\tau} \chi^\tau \tag{64a}$$

$$l_+ s_- Y_l^{\mu-\tau} \chi^\tau = \sqrt{l(l+1) - \left(\mu + \frac{1}{2}\right)\left(\mu + \frac{3}{2}\right)}$$
$$\times Y_l^{\mu+1/2} \chi^- \delta_{\tau+} \tag{64b}$$

$$l_- s_+ Y_l^{\mu-\tau} \chi^\tau = \sqrt{l(l+1) - \left(\mu + \frac{1}{2}\right)\left(\mu - \frac{1}{2}\right)}$$
$$\times Y_l^{\mu-1/2} \chi^+ \delta_{\tau-} \tag{64c}$$

with the Pauli spinors

$$\chi^+ = \begin{pmatrix} 1 \\ 0 \end{pmatrix} \tag{65a}$$

$$\chi^- = \begin{pmatrix} 0 \\ 1 \end{pmatrix} \tag{65b}$$

The latter are quantized with respect to the z direction; that is, they obey $\sigma_z \chi^\tau = \tau \chi^\tau$, $\tau = \pm$. μ is half-integer and runs from $-l - 1/2$ to $l + 1/2$. The Pauli matrices σ_i, $i = x, y, z$, read

$$\sigma_x = \begin{pmatrix} 0 & 1 \\ 1 & 0 \end{pmatrix} \tag{66a}$$

$$\sigma_y = \begin{pmatrix} 0 & -i \\ i & 0 \end{pmatrix} \tag{66b}$$

$$\sigma_z = \begin{pmatrix} 1 & 0 \\ 0 & -1 \end{pmatrix} \tag{66c}$$

The relativistic "companions" of the spherical harmonics are obtained by coupling $\mathbf{l}$ and $\sigma = 2\mathbf{s}$ and are given by [76]

$$\chi_\kappa^\mu = \sum_\tau C\left(l\frac{1}{2}j; \mu - \tau, \tau\right) Y_l^{\mu-\tau} \chi^\tau \tag{67}$$

They are eigenfunctions of $\sigma \cdot \mathbf{l} + 1$ with eigenvalues $\kappa = (j + 1/2)^2 - l(l+1)$. The coefficients C are the well-known Clebsch–Gordan coefficients,

$$C\left(l\frac{1}{2}j; \mu - \tau, \tau\right)$$

$$= \frac{1}{\sqrt{2l+1}} \begin{cases} & \tau = + & \tau = - \\ & \sqrt{l + \mu + \frac{1}{2}} & \sqrt{l - \mu + \frac{1}{2}} \\ j = l + \frac{1}{2} & & \\ & -\sqrt{l - \mu + \frac{1}{2}} & \sqrt{l + \mu + \frac{1}{2}} \\ j = l - \frac{1}{2} & & \end{cases} \tag{68}$$

Table I.   Relationship between Relativistic Quantum Numbers κ and Nonrelativistic Ones l and j

κ	-5	-4	-3	-2	-1	$+1$	$+2$	$+3$	$+4$
l	4	3	2	1	0	1	2	3	4
$\bar{l}$	5	4	3	2	1	0	1	2	3
j	$\frac{9}{2}$	$\frac{7}{2}$	$\frac{5}{2}$	$\frac{3}{2}$	$\frac{1}{2}$	$\frac{1}{2}$	$\frac{3}{2}$	$\frac{5}{2}$	$\frac{7}{2}$

Note: Positive values of κ correspond to $j = l - (1/2)$, negative to $j = l + (1/2)$.

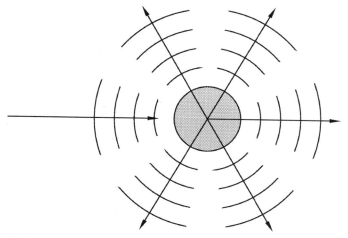

Fig. 17.   Schematic view of scattering by a single potential (gray circle). The incoming and the outgoing partial waves are visualized by arrows and arcs.

For $j = l + (1/2)$ ($j = l - (1/2)$), one has $\kappa = -l - 1$ ($\kappa = l$); cf. Table I. With $r\sigma_r = \sum_{i=x,y,z} r_i \sigma_i$, it is $\sigma_r \chi_\kappa^\mu = -\chi_{-\kappa}^\mu$. Where possible, angular-momentum quantum numbers are combined to give a compound quantum number, in the non-relativistic case $L = (l, m)$, in the relativistic case $\Lambda = (\kappa, \mu)$. Finally, we define $S_\kappa = \kappa / |\kappa|$, $\bar{l} = l - S_\kappa$, and $\bar{\Lambda} = (-\kappa, \mu)$.

Scattering by a Single Site.   The first step in a multiple-scattering calculation is to determine the scattering properties of a single site. The basic idea is to expand both the incoming wave function and the scattered wave function with respect to angular momentum. For spherical potentials, the scattering problem is then solved for each angular momentum separately, giving rise to partial waves (cf. Fig. 17).

We assume that the potential inside the solid V_{solid} can be written as a sum over site-dependent potentials V_i,

$$V_{\text{solid}}(\mathbf{r}) = \sum_i V_i(\mathbf{r} - \mathbf{R}_i) \qquad (69)$$

where the sum is over all sites. Usually, the site potentials are considered to be muffin-tin shape; that is, they are nonzero inside a sphere, spherically symmetric,

$$V_i(\mathbf{r}_i) = \begin{cases} V_i(r) & r \leq r_{\text{mt}i} \\ 0 & \text{otherwise} \end{cases} \qquad (70)$$

and the spheres of different sites do not overlap. Here, r_{mt} is the muffin-tin radius. Further, we assume that $\lim_{r \to 0} r^2 V(r) = 0$

[71]. Of course, generalizations to nonspherical as well as space-filling potentials have been developed (see, e.g., [77, 78]). In particular, the effect of nonspherical potentials on SPLEED intensities has been investigated by Krewer and Feder [79, 80].

Now the calculation of scattering phase shifts and of the single-site scattering matrix, which is a central quantity in multiple-scattering theory, is presented for the nonrelativistic case; that is, the radial Schrödinger equation is solved. For a free particle, the potential function vanishes in the whole space, $V(\mathbf{r}) = 0$. Because the potential is obviously spherically symmetric, the angular momentum l is conserved, l being a "good" quantum number. Therefore, the solutions of the radial Schrödinger equation can be characterized and indexed by l. For a particular value of l, the radial Schrödinger equation reads

$$\left(\partial_r^2 - \frac{l(l+1)}{r^2} + p^2 \right) P_l(r) = 0 \qquad (71)$$

where $p^2 = 2E$, $E > 0$. P_l is connected to the radial solution R_l by $P_l = r R_l$. Thus, the differential equation for R_l reads

$$\left(\partial_r^2 + \frac{2}{r} \partial_r + p^2 - \frac{l(l+1)}{r^2} \right) R_l(r) = 0 \qquad (72)$$

The solutions $R_l(r)$ can be characterized as follows:

Regular solutions are finite in the limit $r \to 0$. For example, R_l, being proportional to a spherical Bessel function $j_l(pr)$ (discucced later), is regular. In other words, for $r \to 0$, R_l behaves like r^l and P_l correspondingly like r^{l+1}.

Irregular solutions diverge for $r \to 0$. R_l, which is proportional to a spherical Neumann function $n_l(pr)$ (discussed later), is irregular; that is, for $r \to 0$, R_l behaves like r^{-l-1} and P_l like r^{-l}.

Incoming and outgoing solutions. For large r, the "centrifugal" term $l(l+1)/r^2$ vanishes. Therefore, solutions should behave asymptotically as plane waves $\exp(\pm ipr)$. By linear combination of regular and irregular solutions, the required asymptotical behavior can be obtained. Spherical Hankel functions (discussed later) are these solutions. The incoming (outgoing) solution is given by zh_l^+ (zh_l^-), $z = pr$. Note that both Hankel functions are irregular.

For negative energy, $E < 0$, p is imaginary. Using the substitution $\tilde{p} = ip$, $\tilde{p} > 0$, the radial Schrödinger equation can be transformed into that for $E > 0$, Eq. (71) or Eq. (72), respectively. The solutions that can be obtained from the preceding but with argument $\tilde{p}r$ are the modified Bessel (i_l), Neumann (m_l), and Hankel functions (k_l). The latter obey the relations $i_l(pr) = (-i)^l j_l(ipr)$, $m_l(pr) = (-i)^{l+1} n_l(ipr)$, and $k_l^+(pr) = (-i)^{-l} h_l^+(ipr)$, respectively [81].

For a nonvanishing potential $V(r)$ of muffin-tin shape, it is convenient to match the solutions inside the sphere to those outside the sphere (free space) at the muffin-tin radius r_{mt}; that is, one requires that both the wave function and its r derivative should be continuous. For $E > 0$ this yields, at r_{mt},

$$\sum_L R_l(r) Y_L(\widehat{\mathbf{r}}) = \sum_L [A_l j_l(pr) + B_l n_l(pr)] Y_L(\widehat{\mathbf{r}}) \quad (73a)$$

$$\sum_L \partial_r R_l(r) Y_L(\widehat{\mathbf{r}}) = \sum_L [A_l \partial_r j_l(r) + B_l \partial_r n_l(pr)] Y_L(\widehat{\mathbf{r}}) \ (73b)$$

with regular R_l and $\widehat{\mathbf{r}} = \mathbf{r}/r$. The unknown coefficients A_l and B_l are traditionally chosen as $\cos \delta_l$ and $-\sin \delta_l$. Thus, one has, using the orthonormality of spherical harmonics,

$$R_l(r) = \cos \delta_l j_l(pr) - \sin \delta_l n_l(pr) \quad (74a)$$

$$\partial_r R_l(r) = \cos \delta_l \partial_r j_l(pr) - \sin \delta_l \partial_r n_l(pr) \quad (74b)$$

After some manipulations, one arrives at

$$\tan \delta_l = \frac{(\partial_r R_l) j_l - R_l(\partial_r j_l)}{(\partial_r R_l) n_l - R_l(\partial_r n_l)} \quad (75)$$

δ_l is called the scattering phase shift of angular momentum l.[13] With the Wronskian $W(f, g) = f \partial_r g - g \partial_r f$, the preceding equation can be compactly written as $\tan \delta_l = W(j_l, R_l)/W(n_l, R_l)$. Using incoming Hankel functions instead of Neumann functions, that is,

$$R_l(r) = (j_l(pr) - i p t_l h_l^+(pr)) \exp(-i\delta_l) \quad (76)$$

one obtains the relation between the scattering phase shifts and the (l-diagonal) single-site scattering matrix t,

$$t_l = -\frac{1}{p} \sin \delta_l \exp(i \delta_l) \quad (77)$$

from which follows the "optical theorem" $\mathrm{Im}(t_l) = -p \, t_l \, t_l^*$. The real and imaginary parts of the t matrix fulfill $\mathrm{Re}(t_l) = -(\sin 2\delta_l)/(2p)$ and $\mathrm{Im}(t_l) = -(\sin^2 \delta_l)/p$. Instead of the t matrix, sometimes the scattering amplitude $f_l = -p t_l$ or the reactance $K_l = -(\tan \delta_l)/p$ is used.

Scattering by a Single Site: The Relativistic Case. The Dirac equation for a single site with an effective magnetic field $\mathbf{B}(\mathbf{r})$ included reads

$$[c \boldsymbol{\alpha} \cdot \mathbf{p} + \beta m c^2 + v(\mathbf{r}) + \beta \boldsymbol{\sigma} \cdot \mathbf{B}(\mathbf{r})] \Psi(\mathbf{r}) = E \Psi(\mathbf{r}) \quad (78)$$

with

$$\alpha = \begin{pmatrix} 0 & \sigma \\ \sigma & 0 \end{pmatrix} \quad (79a)$$

$$\beta = \begin{pmatrix} 1 & 0 \\ 0 & -1 \end{pmatrix} \quad (79b)$$

---

[13] The required derivatives of the spherical Bessel and Neumann functions can be calculated via the relations

$$\left(\frac{1}{z}\partial_z\right)^m [z^{n+1} f_n] = z^{n-m+1} f_{n-m}$$

$$\left(\frac{1}{z}\partial_z\right)^m [z^{-n} f_n] = (-1)^m z^{-n-m} f_{n+m}$$

for $m = 1$. f_n is one of j_l, n_l, or $h_l^{\pm}$ [81].

Here, α and β are 4×4 matrices and Ψ is a Dirac spinor (4-spinor). The potential matrix $V(\mathbf{r})$ is defined by

$$V(\mathbf{r}) = \begin{pmatrix} v(\mathbf{r}) + \sigma \cdot \mathbf{B}(\mathbf{r}) & 0 \\ 0 & v(\mathbf{r}) - \sigma \cdot \mathbf{B}(\mathbf{r}) \end{pmatrix} \quad (80)$$

The solution of the Dirac equation for general potentials has been addressed by Tamura [82]. However, we restrict ourselves again to muffin-tin potentials. Thus, the direction of the effective magnetic field can be chosen conveniently along the z axis, that is, $v(\mathbf{r}) = v(r)$ and $\mathbf{B}(\mathbf{r}) = B(r)\mathbf{e}_z$. Because the effective magnetic field is coupled only to the spin, one can introduce spin-dependent potentials, $v_{\pm}(r) = v(r) \pm B(r)$. The potential matrix then reads

$$V(\mathbf{r}) = \begin{pmatrix} v_+(r) & 0 \\ 0 & v_-(r) \end{pmatrix} \quad (81)$$

For the wave function, we make the ansatz

$$\langle \mathbf{r} | \Psi \rangle = \frac{1}{r} \sum_\Lambda \begin{pmatrix} f_\Lambda(r) \langle \widehat{\mathbf{r}} | \chi_\Lambda \rangle \\ i g_\Lambda(r) \langle \widehat{\mathbf{r}} | \chi_{\overline{\Lambda}} \rangle \end{pmatrix} = \sum_\Lambda \begin{pmatrix} \Psi_\Lambda(r) \langle \widehat{\mathbf{r}} | \chi_\Lambda \rangle \\ i \Phi_\Lambda(r) \langle \widehat{\mathbf{r}} | \chi_{\overline{\Lambda}} \rangle \end{pmatrix} \quad (82)$$

that is, $\Psi_\Lambda(r) = f_\Lambda(r)/r$ and $\Phi_\Lambda(r) = g_\Lambda(r)/r$. The index Λ combines the relativistic angular-momentum quantum numbers κ and μ, $\Lambda = (\kappa, \mu)$ and $\overline{\Lambda} = (-\kappa, \mu)$ (discussed previously). Inserting the preceding ansatz into the Dirac equation yields a set of coupled equations for f_Λ and g_Λ,

$$c\partial_r f_\Lambda = -c\frac{\kappa}{r} f_\Lambda + (E + c^2 - v)g_\Lambda + B \sum_{\Lambda'} \langle \chi_{\overline{\Lambda}} | \sigma_z | \chi_{\overline{\Lambda'}} \rangle g_{\Lambda'} \quad (83a)$$

$$c\partial_r g_\Lambda = c\frac{\kappa}{r} g_\Lambda - (E - c^2 - v)f_\Lambda + B \sum_{\Lambda'} \langle \chi_\Lambda | \sigma_z | \chi_{\Lambda'} \rangle f_{\Lambda'} \quad (83b)$$

Note that in the nonmagnetic case ($B = 0$), one has

$$c\partial_r f_\Lambda = -c\frac{\kappa}{r} f_\Lambda + (E + c^2 - v)g_\Lambda \quad (84a)$$

$$c\partial_r g_\Lambda = c\frac{\kappa}{r} g_\Lambda - (E - c^2 - v)f_\Lambda \quad (84b)$$

that is, the solutions f_Λ and g_Λ are independent of the magnetic quantum number μ.

The matrix elements of σ_z can easily be obtained from the definition of χ_Λ. From the restriction $l' = l$, one has two cases, $\kappa' = \kappa$ and $\kappa' = -\kappa - 1$, which give

$$\langle \chi_\Lambda | \sigma_z | \chi_{\Lambda'} \rangle = \begin{cases} -\dfrac{2\mu}{2\kappa + 1} & \text{for } \kappa' = \kappa \\ \sqrt{1 - \left(\dfrac{2\mu}{2\kappa + 1}\right)^2} & \text{for } \kappa' = -\kappa - 1 \\ 0 & \text{otherwise} \end{cases} \quad (85)$$

Inserting Eq. (85) into Eq. (83), one finds terms that couple angular momenta l and $l+2$. As Ackermann has shown [83], these can be neglected due to the missing singularity of $B(r)$ at the origin. Thus, only partial waves with total angular momentum

$j = l + (1/2)$ and $j = l - (1/2)$ are coupled. Eventually, one arrives at a system of four coupled differential equations of first order,

$$c\partial_r f_{\kappa\mu} = -c\frac{\kappa}{r} f_{\kappa\mu}$$
$$+ \left(E + c^2 - v + B\frac{2\mu}{2\kappa - 1} \right) g_{\kappa\mu} \quad (86a)$$

$$c\partial_r g_{\kappa\mu} = c\frac{\kappa}{r} g_{\kappa\mu} - \left(E - c^2 - v - B\frac{2\mu}{2\kappa + 1} \right) f_{\kappa\mu}$$
$$- B\sqrt{1 - \left(\frac{2\mu}{2\kappa - 1}\right)^2} f_{-\kappa-1,\mu} \quad (86b)$$

$$c\partial_r f_{-\kappa-1,\mu} = -c\frac{\kappa + 1}{r} f_{-\kappa-1,\mu}$$
$$+ \left(E + c^2 - v - B\frac{2\mu}{2\kappa + 3} \right) g_{-\kappa-1,\mu} \quad (86c)$$

$$c\partial_r g_{-\kappa-1,\mu} = -c\frac{\kappa + 1}{r} g_{-\kappa-1,\mu}$$
$$- \left(E - c^2 - v + B\frac{2\mu}{2\kappa + 1} \right) f_{-\kappa-1,\mu}$$
$$- B\sqrt{1 - \left(\frac{2\mu}{2\kappa + 1}\right)^2} f_{\kappa,\mu} \quad (86d)$$

As in the nonrelativistic case, two types of solutions can be distinguished due to their behavior in the vicinity of the origin, that is, regular and irregular solutions. Regular and irregular partial waves can be written as

$$\langle \mathbf{r} | \Psi_\Lambda^x \rangle = \sum_{\Lambda'} \begin{pmatrix} \Psi_{\Lambda'\Lambda}^x(r) \langle \widehat{\mathbf{r}} | \chi_{\Lambda'} \rangle \\ i\Phi_{\Lambda'\Lambda}^x(r) \langle \widehat{\mathbf{r}} | \chi_{\overline{\Lambda'}} \rangle \end{pmatrix} \quad x = \text{reg, irr} \quad (87)$$

and show the asymptotics

$$\langle \mathbf{r} | \Psi_\Lambda^{\text{reg}} \rangle$$
$$\rightarrow \sum_{\Lambda'} \begin{pmatrix} [j_l(kr)\delta_{\Lambda'\Lambda} + h_{l'}^+(kr) t_{\Lambda'\Lambda}] \langle \widehat{\mathbf{r}} | \chi_{\Lambda'} \rangle \\ i S_{\kappa'} \dfrac{ck}{E + c^2} [j_{\bar{l}}(kr)\delta_{\Lambda'\Lambda} + h_{\bar{l}'}^+(kr) t_{\Lambda'\Lambda}] \langle \widehat{\mathbf{r}} | \chi_{\overline{\Lambda'}} \rangle \end{pmatrix}$$
$$(88a)$$

$$\langle \mathbf{r} | \Psi_\Lambda^{\text{irr}} \rangle \rightarrow \sum_{\Lambda'} \begin{pmatrix} h_{l'}^+(kr)\delta_{\Lambda'\Lambda} \langle \widehat{\mathbf{r}} | \chi_{\Lambda'} \rangle \\ i S_{\kappa'} \dfrac{ck}{E + c^2} h_{\bar{l}'}^+(kr)\delta_{\Lambda'\Lambda} \langle \widehat{\mathbf{r}} | \chi_{\overline{\Lambda'}} \rangle \end{pmatrix} \quad (88b)$$

for $r \rightarrow \infty$. The Wronskian is given by

$$\{\langle \Psi_\Lambda^{\text{reg}} |, |\Psi_{\Lambda'}^{\text{irr}} \rangle\} = \sum_{\Lambda''} cr^2 \big(\Psi_{\Lambda''\Lambda}^{\text{reg}}(r) \Phi_{\Lambda''\Lambda'}^{\text{irr}}(r) -$$
$$\Phi_{\Lambda''\Lambda}^{\text{reg}}(r) \Psi_{\Lambda''\Lambda'}^{\text{irr}}(r) \big) \quad (89a)$$

$$= \frac{ic^2}{k(E + c^2)} \delta_{\Lambda\Lambda'} \quad (89b)$$

which is independent of r.

Now one can calculate the single-site t matrix. The incoming partial wave is given by

$$\langle \mathbf{r} | J_\Lambda \rangle = \begin{pmatrix} j_l(kr) \langle \widehat{\mathbf{r}} | \chi_\Lambda \rangle \\ i S_\kappa \dfrac{ck}{E + c^2} j_{\bar{l}}(kr) \langle \widehat{\mathbf{r}} | \chi_{\overline{\Lambda}} \rangle \end{pmatrix} \quad (90)$$

With

$$\langle \mathbf{r} | H_\Lambda^{(i)} \rangle = \begin{pmatrix} h_l^i(kr) \langle \widehat{\mathbf{r}} | \chi_\Lambda \rangle \\ i S_\kappa \dfrac{ck}{E + c^2} h_{\bar{l}}^i(kr) \langle \widehat{\mathbf{r}} | \chi_{\overline{\Lambda}} \rangle \end{pmatrix} \quad i = \pm \quad (91)$$

the total wave function is given by

$$\langle \mathbf{r} | \Psi \rangle = \sum_\Lambda \big(A_\Lambda \langle \mathbf{r} | J_\Lambda \rangle + B_\Lambda \langle \mathbf{r} | H_\Lambda^{(+)} \rangle \big) \quad (92)$$

The coefficients A_Λ (incoming) and B_Λ (outgoing) are connected by the single-site t matrix,

$$B_\Lambda = \sum_{\Lambda'} t_{\Lambda\Lambda'} A_{\Lambda'} \quad (93)$$

and can be obtained either by wave function matching [71] or by exploiting the Wronskians [82]. Due to the potential considered here, only those elements of t that belong to partial waves coupled by the radial Dirac equation are nonzero; that is, $t_{\Lambda\Lambda'} = 0$ if $\kappa' \notin \{\kappa, -\kappa - 1\}$ or $\mu' \neq \mu$. In the nonmagnetic case, the t matrix is diagonal in κ.

Many effects in electron spectroscopies of magnetic systems rely on the simultaneous presence of magnetization and spin-orbit coupling (SOC). For some purposes (e.g., testing and model calculations), it is desirable to vary the strengths of these. In nonrelativistic theories that include SOC as a perturbation, its strength can easily be changed by scaling the respective coupling constant. In relativistic theories, one usually sets the speed of light c to a rather large value, with the drawback that all relativistic effects (mass term, Darwin term) are changed, too. However, based on the scalar-relativistic approximation [84–86], one can derive an equation that interpolates between the fully relativistic (Dirac) and the scalar-relativistic Schrödinger equation [87–89].

Scattering by a Single Layer. After having solved the single-site problem, for example, having obtained the single-site t matrix, we now have to calculate the scattering properties of a single layer, the essential object in layer-KKR. In the following, only the case of one site per layer unit cell will be addressed. For each beam that is characterized by the reciprocal lattice vector $\mathbf{g}$, define the wave vector $\mathbf{k}_\mathbf{g}^\pm$ by

$$\mathbf{k}_\mathbf{g}^\pm = \begin{pmatrix} \mathbf{k}_\| + \mathbf{g} \\ \pm\sqrt{k^2 - (\mathbf{k}_\| + \mathbf{g})^2} \end{pmatrix} \quad (94)$$

with $c^2 k^2 = E^2 - c^4$. The $+ (-)$ sign refers to plane waves propagating or decaying in the $+z (-z)$ direction. The wave fields incident on (Ψ_{inc}) and outgoing from (Ψ_{out}) the layer can

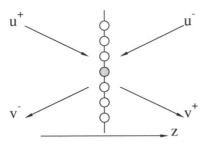

Fig. 18. Schematic view of scattering by a layer, that is, a two-dimensional periodic arrangement of scatterers (circles). The reference scatterer is represented as a gray circle. Incoming (outgoing) beams are labeled $u^\pm$ ($v^\pm$) with respect to the propagation direction ($\pm z$).

be written as

$$\Psi_{\text{inc}}(\mathbf{r}) = \sum_{\mathbf{g}\tau} \left[u_{\mathbf{g}\tau}^+ \exp(i\mathbf{k}_{\mathbf{g}}^+ \cdot \mathbf{r}) + u_{\mathbf{g}\tau}^- \exp(i\mathbf{k}_{\mathbf{g}}^- \cdot \mathbf{r}) \right] \chi^\tau \quad (95a)$$

$$\Psi_{\text{out}}(\mathbf{r}) = \sum_{\mathbf{g}\tau} \left[v_{\mathbf{g}\tau}^+ \exp(i\mathbf{k}_{\mathbf{g}}^+ \cdot \mathbf{r}) + v_{\mathbf{g}\tau}^- \exp(i\mathbf{k}_{\mathbf{g}}^- \cdot \mathbf{r}) \right] \chi^\tau \quad (95b)$$

Arranging the coefficients $u_{\mathbf{g}\tau}^\pm$ and $v_{\mathbf{g}\tau}^\pm$ into column vectors, the connection between these is defined in terms of the scattering matrix M of the layer (see Fig. 18),

$$\begin{pmatrix} \mathbf{v}^+ \\ \mathbf{v}^- \end{pmatrix} = \begin{pmatrix} M^{++} & M^{+-} \\ M^{-+} & M^{--} \end{pmatrix} \begin{pmatrix} \mathbf{u}^+ \\ \mathbf{u}^- \end{pmatrix} \quad (96)$$

Thus, the M matrix corresponds to the single-site t matrix but for an entire layer; cf. Eq. (93).

To calculate the M matrix, one expands the incoming wave into spherical waves. Using

$$\exp(i\mathbf{k} \cdot \mathbf{r})\chi^\tau = \sum_\Lambda a_{\Lambda\tau}(\widehat{\mathbf{k}}) j_l(kr) \chi_\Lambda(\widehat{\mathbf{r}}) \qquad \tau = \pm \quad (97)$$

with coefficients [cf. Eq. (67)]

$$a_{\Lambda\tau}(\widehat{\mathbf{k}}) = 4\pi i^l C\left(l \frac{1}{2} j; \mu - \tau, \tau \right) \left(Y_l^{\mu-\tau}(\widehat{\mathbf{k}}) \right)^* \qquad \tau = \pm \quad (98)$$

one obtains for the wave field

$$\sum_{\mathbf{g}} u_{\mathbf{g}\tau}^\pm \exp(i\mathbf{k}_{\mathbf{g}}^\pm \cdot \mathbf{r})\chi^\tau = \sum_\Lambda A_{\Lambda\tau}^{\pm(0)} j_l(kr)\chi_\Lambda(\widehat{\mathbf{r}}) \qquad \tau = \pm \quad (99)$$

with $A_{\Lambda\tau}^{\pm(0)} = \sum_{\mathbf{g}} a_{\Lambda\tau}(\widehat{\mathbf{k}}_{\mathbf{g}}^\pm) u_{\mathbf{g}\tau}^\pm$ and $k = \sqrt{2E + E^2/c^2}$. The incoming wave is multiply scattered at each site of the layer, which leads to the wave field $\sum_\Lambda A_{\Lambda\tau} j_l(kr)\chi_\Lambda(\widehat{\mathbf{r}})$ incident at the reference atom. The wave field incident at site $\mathbf{R}_j$ is given by

$$\sum_\Lambda A_{\Lambda\tau} j_l(kr_j)\chi_\Lambda(\widehat{\mathbf{r}}_j) \exp(i\mathbf{k}_\parallel \cdot \mathbf{R}_j) \quad (100)$$

with $\mathbf{r}_j = \mathbf{r} - \mathbf{R}_j$. The outgoing wave field from this site is

$$\sum_\Lambda B_{\Lambda\tau} h_l^+(kr_j)\chi_\Lambda(\widehat{\mathbf{r}}_j) \exp(i\mathbf{k}_\parallel \cdot \mathbf{R}_j) \quad (101)$$

and the coefficients $B_{\Lambda'\tau}$ and $A_{\Lambda\tau}$ are connected by the t matrix, Eq. (93). The total incident wave field at the reference atom can be separated into a direct part, which stems from the non-layer region (incident onto the layer), and a layer part, which is due to multiple scattering within the layer. The latter reads

$$\sum_\Lambda A_{\Lambda\tau}^{\text{layer}} j_l(kr)\chi_\Lambda(\widehat{\mathbf{r}})$$

$$= \sum_j{}' \exp(i\mathbf{k}_\parallel \cdot \mathbf{R}_j) \sum_{\Lambda'} B_{\Lambda'\tau} h_{l'}^+(kr_j)\chi_{\Lambda'}(\widehat{\mathbf{r}}_j) \quad (102)$$

where the first sum (indicated by a prime) is over all sites within the layer except the reference atom. The coefficients $A_{\Lambda\tau}^{\text{layer}}$ can be obtained with the help of the layer structure constant $G_{\Lambda\Lambda'}(-\mathbf{R}_j)$ (for a detailed discussion, see [42, 72]), which obeys

$$h_l^+(kr_j)\chi_\Lambda(\widehat{\mathbf{r}}_j) = \sum_{\Lambda'} G_{\Lambda\Lambda'}(-\mathbf{R}_j) j_{l'}(kr_j)\chi_{\Lambda'}(\widehat{\mathbf{r}}_j) \quad (103)$$

The relativistic structure constant is related to the nonrelativistic one by Clebsch–Gordan coefficients,

$$G_{\Lambda\Lambda'}(-\mathbf{R}_j) = \sum_\tau C\left(l \frac{1}{2} j; \mu - \tau, \tau \right) G_{l,\mu-\tau;l',\mu'-\tau}(-\mathbf{R}_j)$$

$$\times C\left(l' \frac{1}{2} j'; \mu' - \tau, \tau \right) \quad (104)$$

Thus, one has

$$A_{\Lambda\tau}^{\text{layer}} = \sum_j{}' \exp(i\mathbf{k}_\parallel \cdot \mathbf{R}_j) \sum_{\Lambda'} B_{\Lambda'\tau} G_{\Lambda'\Lambda}(-\mathbf{R}_j) \quad (105)$$

It is convenient to introduce a "multiple-scattering matrix" X by $A_{\Lambda\tau}^{\text{layer}} = \sum_{\Lambda'} A_{\Lambda'\tau} X_{\Lambda'\Lambda}$, which is easily calculated as

$$X_{\Lambda''\Lambda} = \sum_{\Lambda'} t_{\Lambda''\Lambda'} \sum_j{}' \exp(i\mathbf{k}_\parallel \cdot \mathbf{R}_j) G_{\Lambda'\Lambda}(-\mathbf{R}_j) \quad (106)$$

The total incoming wave field at the reference atom is then given by

$$A_{\Lambda\tau} = A_{\Lambda\tau}^{+(0)} + A_{\Lambda\tau}^{-(0)} + A_{\Lambda\tau}^{\text{layer}} = A_{\Lambda\tau}^{+(0)} + A_{\Lambda\tau}^{-(0)} + \sum_{\Lambda'} A_{\Lambda'\tau} X_{\Lambda'\Lambda} \quad (107)$$

or, in matrix notation, $A = A^{(0)} + AX$, from which one obtains $A = A^{(0)}(1 - X)^{-1}$. Eventually, the blocks of the M matrix can be written formally as

$$M^{s,s'} = \delta_{s,s'} + a^s t (1 - X)^{-1} b^{s'} \qquad s, s' = \pm \quad (108)$$

where $a_{\Lambda\tau}(\widehat{\mathbf{k}}_{\mathbf{g}}^\pm)$ transforms from angular-momentum into plane wave representation [cf. Eq. (98)], $b_{\Lambda\tau}(\widehat{\mathbf{k}}_{\mathbf{g}}^\pm)$ vice versa.

Scattering by a Double Layer. To calculate the scattering properties of an arbitrary stack of layers, one starts with the M matrix of a double layer, that is, a stack of two layers. By consecutive application of the following computational scheme,

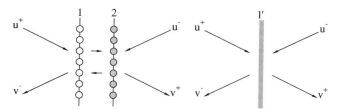

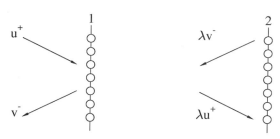

Fig. 19. Schematic view of scattering by a double layer, that is, a stack of two layers (1 and 2, left). The scattering properties can be cast into an effective scatterer (1', right).

Fig. 20. Bloch waves in multiple-scattering theory. For identical layers (1 and 2), plane waves on the right ($\lambda\mathbf{u}^+$, $\lambda\mathbf{v}^-$) are multiples of those on the left ($\mathbf{u}^+$, $\mathbf{v}^-$) due to Floquet's theorem.

one eventually obtains the M matrix of the stack. The "layer-doubling" algorithm for computation of the bulk reflection matrix (discussed later) is also based on this method.

One starts with a stack consisting of two layers, labeled 1 and 2, which need not be identical; cf. Figure 19. The result is the M matrix of this stack, labeled 1',

$$\begin{pmatrix} \mathbf{v}^+ \\ \mathbf{v}^- \end{pmatrix} = \begin{pmatrix} M_{1'}^{++} & M_{1'}^{+-} \\ M_{1'}^{-+} & M_{1'}^{--} \end{pmatrix} \begin{pmatrix} \mathbf{u}^+ \\ \mathbf{u}^- \end{pmatrix} \quad (109)$$

Summing up all multiple-scattering events that are due to reflection at each layer, one obtains for the amplitudes of the plane waves

$$\begin{aligned} \mathbf{v}^+ &= M_2^{++} P^+ (1 - M_1^{+-} P^- M_2^{-+} P^+)^{-1} M_1^{++} \mathbf{u}^+ \\ &\quad + \big[M_2^{+-} + M_2^{++} P^+ M_1^{+-} P^- \\ &\quad \times (1 - M_1^{-+} P^+ M_1^{+-} P^-)^{-1} M_2^{--} \big] \mathbf{u}^- \quad (110a) \end{aligned}$$

$$\begin{aligned} \mathbf{v}^- &= \big[M_1^{-+} + M_1^{--} P^- M_2^{-+} P^+ \\ &\quad \times (1 - M_1^{+-} P^- M_2^{-+} P^+)^{-1} M_1^{++} \big] \mathbf{u}^+ \\ &\quad + M_1^{--} P^- (1 - M_2^{-+} P^+ M_1^{+-} P^-)^{-1} M_2^{--} \mathbf{u}^- \quad (110b) \end{aligned}$$

where we have used $1 + x + x^2 + x^3 + \cdots = (1-x)^{-1}$. For convenience, matrices $N^{\pm\pm}$ are introduced, that is, $M^{\pm\pm}$ matrices enhanced by plane wave propagators $P^\pm$, $N^{++} = P^+ M^{++}$, $N^{+-} = P^+ M^{+-} P^-$, $N^{-+} = M^{-+}$, and $N^{--} = M^{--} P^-$, which yields

$$N_{1'}^{++} = N_2^{++} (1 - N_1^{+-} N_2^{-+})^{-1} N_1^{++} \quad (111a)$$

$$N_{1'}^{+-} = N_2^{+-} + N_2^{++} N_1^{+-} (1 - N_2^{-+} N_1^{+-})^{-1} N_2^{--} \quad (111b)$$

$$N_{1'}^{-+} = N_1^{-+} + N_1^{--} N_2^{-+} (1 - N_1^{+-} N_2^{-+})^{-1} N_1^{++} \quad (111c)$$

$$N_{1'}^{--} = N_1^{--} (1 - N_2^{-+} N_1^{+-})^{-1} N_2^{--} \quad (111d)$$

The elements of the diagonal matrices $P^\pm$ are defined by

$$P_{\mathbf{g}\tau, \mathbf{g}'\tau'}^\pm = \exp(i\mathbf{k}_{\mathbf{g}}^\pm \cdot \mathbf{d}) \delta_{\mathbf{g}\mathbf{g}'} \delta_{\tau\tau'} \quad (112)$$

where $\mathbf{d}$ is the translation vector from layer 1 to layer 2 (cf. Fig. 1). Note the relation between $N_{1'}^{++}$ and $N_{1'}^{--}$ as well as that between $N_{1'}^{+-}$ and $N_{1'}^{-+}$: By interchanging $+ \leftrightarrow -$ and $1 \leftrightarrow 2$ in one expression, one obtains the other.

The preceding procedure can be used to calculate iteratively the bulk reflection matrix R_{bulk}^{-+}, known as the "layer-doubling" method. The bulk is defined as an infinitely repeated arrangement of identical layers or stacks of layers (principal layers).

One first calculates the M matrix of a double layer, subsequently the M matrix of a doubled double layer, which yields M of a stack of four layers, and so on. After n iterations, the M matrix of 2^n layers is obtained. The bulk reflection matrix is eventually given by M^{-+} of the 2^n-layer stack. The "layer-doubling" procedure is repeated until (i) the change of the bulk reflection matrix of the 2^{n-1} stack and the 2^n stack is small enough to be regarded as negligible or (ii) the incoming waves $\mathbf{u}^+$ are absorbed within the stack; that is, M^{++} is close enough to 0. In practical LEED calculations, which take into account the mean free path via an absorptive optical potential, three or four iterations are regarded as sufficient.

Bloch Wave Method. The M matrix of a layer or a stack of layers can be used to compute the bulk-band structure $k_\perp(E, \mathbf{k}_\parallel)$. Consider identical layers 1 and 2, which are connected by a vector $\mathbf{d}$. On the left-hand side of layer 1, we have incoming and outgoing waves $\mathbf{u}^+$ and $\mathbf{v}^-$. Due to Floquet's theorem, the plane waves on the left-hand side of layer 2 are these plane waves but multiplied by a factor (λ in Fig. 20). Thus, the outgoing waves $\lambda\mathbf{u}^+$ and $\mathbf{v}^-$ are related to the incoming waves $\mathbf{u}^+$ and $\lambda\mathbf{v}^-$ by

$$\lambda\mathbf{u}^+ = N^{++}\mathbf{u}^+ + \lambda N^{+-}\mathbf{v}^- \quad (113a)$$

$$\mathbf{v}^- = N^{-+}\mathbf{u}^+ + \lambda N^{--}\mathbf{v}^- \quad (113b)$$

In matrix notation, the eigenvectors and eigenvalues can be obtained from the generalized eigenproblem

$$\begin{pmatrix} N^{++} & 0 \\ -N^{-+} & 1 \end{pmatrix} \begin{pmatrix} \mathbf{u}^+ \\ \mathbf{v}^- \end{pmatrix} = \lambda \begin{pmatrix} 1 & -N^{+-} \\ 0 & N^{--} \end{pmatrix} \begin{pmatrix} \mathbf{u}^+ \\ \mathbf{v}^- \end{pmatrix} \quad (114)$$

which can be solved by standard numerical program packages. However, by some algebra the preceding equation can be transformed into standard form, $Q\mathbf{c}_n = \lambda_n \mathbf{c}_n$, with the blocks of the matrix Q given by

$$Q^{++} = N^{++} - N^{+-}(N^{--})^{-1} N^{-+} \quad (115a)$$

$$Q^{+-} = N^{+-}(N^{--})^{-1} \quad (115b)$$

$$Q^{-+} = -(N^{--})^{-1} N^{-+} \quad (115c)$$

$$Q^{--} = (N^{--})^{-1} \quad (115d)$$

Here, the eigenvalue λ_n is in general complex and $\mathbf{c}_n$ is a $4N_{\mathbf{g}}$ vector, $N_{\mathbf{g}}$ being the number of reciprocal lattice vectors $\mathbf{g}$ taken into account. The upper (lower) $2N_{\mathbf{g}}$ components of $\mathbf{c}_n$ describe waves propagating or decaying in the $+z$ direction ($-z$ direction). The wave vector $\mathbf{k}$ of the Bloch waves can be decomposed into components parallel, $\mathbf{k}_\parallel$, and perpendicular, $k_\perp$, to the layer. Thus, the eigenvalue can be written as $\lambda_n = \exp(i\mathbf{k}_n \cdot \mathbf{d})$ from which $k_{n,\perp}$ is obtained as

$$k_{n,\perp} = \frac{-i}{d_\perp}(\ln \lambda_n - i\mathbf{k}_\parallel \cdot \mathbf{d}_\parallel) \tag{116}$$

In conclusion, the bulk-band structure has been computed as $k_\perp(E, \mathbf{k}_\parallel)$.

Eigenfunctions with $|\lambda_n| = 1$ belong to the real band structure; that is, they fulfill the Bloch condition. But even if the energy is real, the set of eigenvalues $\{\lambda_n\}$ consists of values with modulus greater than or less than 1. In particular, the norms of the corresponding wave functions increase or decrease when propagating across a layer. It is clear that these solutions cannot be normalized in the bulk because only square-integrable functions ($\mathcal{L}_2$ functions) belong to the Hilbert space. If, however, a surface is present the normalization has to be carried out only in the half-space $z > 0$. Thus, in addition to Bloch states solutions with $|\lambda_n| < 1$ are allowed. The latter are the evanescent states, that is, those states with decreasing amplitude when propagating into the interior of the semi-infinite solid.

Once the eigenfunctions and eigenvalues of the Q matrix are known, they can be used to calculate the reflection matrix of the bulk. The $4N_\mathbf{g}$ eigenvectors can be classified into those that (i) decay exponentially in the z direction, (ii) decay in the $-z$ direction, (iii) have positive probability current perpendicular to the layers, and (iv) have negative probability current perpendicular to the layers. The eigenvectors can be arranged such that those $2N_\mathbf{g}$ of them that decay in the $+z$ direction and the current of which is positive comprise the first $2N_\mathbf{g}$ rows of the eigenvector matrix V. The relation between plane waves and eigenvectors then reads

$$\begin{pmatrix} \mathbf{c}^+ \\ \mathbf{c}^- \end{pmatrix} = \begin{pmatrix} V^{++} & V^{+-} \\ V^{-+} & V^{--} \end{pmatrix} \begin{pmatrix} 0 \\ \mathbf{u}^- \end{pmatrix} \tag{117}$$

Here, $\mathbf{u}^+$ equals 0 because the outgoing waves in the $+z$ direction of an infinitely thick slab vanish. In practical calculation, this is assured by adding a small imaginary constant to the energy or to the potential in order to simulate inelastic scattering of the electrons [53]. In this way, there are, strictly speaking, no Bloch states, only evanescent states that decay in either the $+z$ or the $-z$ direction. Eliminating $\mathbf{u}^-$ in the preceding equation yields the bulk reflection matrix R_{bulk}^{-+},

$$\mathbf{c}^- = V^{--}(V^{-+})^{-1}\mathbf{c}^+ = R_{\text{bulk}}^{-+}\mathbf{c}^+ \tag{118}$$

Scattering from the Surface Barrier. At this point, one has to consider the surface region of the solid. In multiple-scattering theory, the surface barrier is treated as an additional layer that enters via its M matrix. For a step barrier, which appears to be sufficient in calculations for typical LEED energies, the latter can easily be calculated as

$$M_{\mathbf{g}\tau,\mathbf{g}'\tau'}^{++} = 2k_{\perp\mathbf{g}}/(k_{\perp\mathbf{g}} + k_{\perp\mathbf{g}}^{\text{in}})\delta_{\mathbf{g}\mathbf{g}'}\delta_{\tau\tau'} \tag{119a}$$

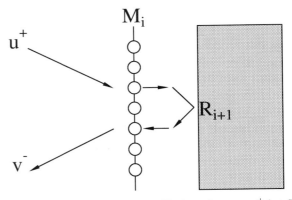

Fig. 21. Reflection of a stack of layers. The incoming wave $\mathbf{u}^+$ is reflected into $\mathbf{v}^-$ by layer i (which is described by its scattering matrices $M_i^{\pm\pm}$) and the stack of layers with index greater than i (which is described by the reflection matrix R_{i+1}^{-+}).

$$M_{\mathbf{g}\tau,\mathbf{g}'\tau'}^{+-} = (k_{\perp\mathbf{g}}^{\text{in}} - k_{\perp\mathbf{g}})/(k_{\perp\mathbf{g}} + k_{\perp\mathbf{g}}^{\text{in}})\delta_{\mathbf{g}\mathbf{g}'}\delta_{\tau\tau'} \tag{119b}$$

$$M_{\mathbf{g}\tau,\mathbf{g}'\tau'}^{-+} = (k_{\perp\mathbf{g}} - k_{\perp\mathbf{g}}^{\text{in}})/(k_{\perp\mathbf{g}} + k_{\perp\mathbf{g}}^{\text{in}})\delta_{\mathbf{g}\mathbf{g}'}\delta_{\tau\tau'} \tag{119c}$$

$$M_{\mathbf{g}\tau,\mathbf{g}'\tau'}^{--} = 2k_{\perp\mathbf{g}}^{\text{in}}/(k_{\perp\mathbf{g}} + k_{\perp\mathbf{g}}^{\text{in}})\delta_{\mathbf{g}\mathbf{g}'}\delta_{\tau\tau'} \tag{119d}$$

with

$$k_{\perp\mathbf{g}} = \sqrt{2E - (\mathbf{k}_\parallel + \mathbf{g})^2} \tag{120a}$$

$$k_{\perp\mathbf{g}}^{\text{in}} = \sqrt{2(E + V_0) - (\mathbf{k}_\parallel + \mathbf{g})^2} \tag{120b}$$

Here, V_0 is the inner potential, that is, the energy shift of the muffin-tin potentials inside the solid relative to the vacuum level, which is chosen as 0 eV for convenience.

In VLEED, however, a barrier with image-potential asymptotics, that is, $V(z) \approx 1/(4z)$, is better suited [90]. The JJJ barrier, which interpolates smoothly between the image potential in the vacuum and the constant inner potential V_0 in the solid, was introduced by Jones, Jennings, and Jepsen [91] (thus the name JJJ) and improved by Tamura and Feder [92]. Besides simple integration of the Schrödinger equation in the surface region with standard numerical methods, highly sophisticated, methods have been proposed that take into account the corrugation, that is, the variation of the potential parallel to the surface [67, 68, 93, 94].

Calculation of the Total Reflection Matrix. To compute the reflected intensities $I_\mathbf{g}(E)$, the reflection matrix of the semi-infinite system has to be calculated. If the scattering matrices of all layers and the bulk reflection matrix have been determined, the total reflection matrix R_{tot} can be calculated step by step. Consider the scattering matrices $M_i^{\pm\pm}$ of layer i and the reflection matrix R_{i+1}^{-+} of the stack of layers with index greater than i (cf. Fig. 21). Then the reflection matrix R_i^{-+} of the stack comprising all layers with index greater than or equal to i is given by

$$R_i^{-+} = M_i^{-+} + M_i^{--}P^-R_{i+1}^{-+}(1 - P^+M_i^{+-}P^-R_{i+1}^{-+})^{-1}$$
$$\times P^+M_i^{++} \tag{121}$$

The first term is solely due to reflection at layer i, whereas the second term takes into account multiple scattering between layer i and the stack of layers on the r.h.s. of layer i.

Starting with the bulk reflection matrix R_{bulk}^{-+} and the scattering matrices $M_n^{\pm\pm}$ of the first non-bulk-like layer n, one calculates the reflection matrix R_{n-1}^{-+} of this stack according to the preceding scheme. For example, from R_{n-1}^{-+} and $M_{n-2}^{\pm\pm}$, one computes R_{n-2}^{-+}, and so forth. Eventually, one obtains the reflection matrix R^{tot} of the entire semi-infinite system.

Spin-Polarized Low-Energy Electron Diffraction. Now we have gathered all ingredients for calculating SPLEED. After computing the single-site t matrices of the scatterers, one computes the M matrices of each layer (including the surface barrier). From these, one calculates the bulk reflection matrix, from which, in turn, the reflection matrix R^{tot} of the entire semi-infinite system can be computed. This step-by-step procedure allows for a great flexibility with respect to the geometrical setup of the solid. Therefore, it is not only suited to semi-infinite systems that are built by a single type of layer, but can handle any arrangement of layers, in particular, ultra-thin films on a substrate [95].

In SPLEED, one uses a spin-polarized incoming electron beam with spin polarization $\mathbf{P}^{\text{in}}$ and measures intensity $I_{\mathbf{g}}$ and spin polarization $P_{\mathbf{g}}^{\text{out}}$ of the reflected beams. The density matrix ϱ^{in} of the incoming electron beam is related to the spin polarization $\mathbf{P}^{\text{in}}$ by $\varrho^{\text{in}} = (1 + \mathbf{P}^{\text{in}} \cdot \sigma)/2$. The density matrix $\varrho_{\mathbf{g}}^{\text{out}}$ of an outgoing beam is therefore given by $\varrho_{\mathbf{g}}^{\text{out}} = \varrho_{\mathbf{g}} \varrho^{\text{in}} \varrho_{\mathbf{g}}^{\dagger}$ [96, 97]. $\varrho_{\mathbf{g}}$ is given by $\varrho_{\mathbf{g}\tau\tau'} = R_{\mathbf{g}\tau, 0\tau'}^{\text{tot}} (E_{\perp\mathbf{g}}/E_{\perp 0})^{1/4}$ with the "effective" energies $E_{\perp\mathbf{g}} = 2E - (k_{\parallel} + \mathbf{g})^2$. Intensity and spin polarization of the outgoing LEED beams are obtained from $\varrho_{\mathbf{g}}^{\text{out}}$ by $I_{\mathbf{g}} = \text{tr}(\varrho_{\mathbf{g}}^{\text{out}})$ and $\mathbf{P}_{\mathbf{g}}^{\text{out}} = \text{tr}(\sigma \varrho_{\mathbf{g}}^{\text{out}})/I_{\mathbf{g}}$.

In a typical SPLEED experiment from ferromagnetic systems, one chooses the spin polarization of the incoming beam $\mathbf{P}^{\text{in}}$ either parallel or antiparallel to some direction $\hat{\mathbf{p}}$. The sample magnetization $\mathbf{M}$ can be aligned, for example, via an external magnetic field, parallel or antiparallel to some other direction $\hat{\mathbf{m}}$. Therefore, one can detect for each outgoing beam four intensities $I_{\mathbf{g}}^{\pm\pm}$ where the first (second) superscript refers to $\mathbf{P}^{\text{in}}$ ($\mathbf{M}$). With $I_{\mathbf{g}}$ being the sum of these four intensities, one defines the asymmetries

$$A_{\mathbf{g}}^{\text{so}} = (I_{\mathbf{g}}^{++} + I_{\mathbf{g}}^{+-} - I_{\mathbf{g}}^{-+} - I_{\mathbf{g}}^{--})/I_{\mathbf{g}} \quad (122a)$$

$$A_{\mathbf{g}}^{\text{ex}} = (I_{\mathbf{g}}^{++} - I_{\mathbf{g}}^{+-} - I_{\mathbf{g}}^{-+} + I_{\mathbf{g}}^{--})/I_{\mathbf{g}} \quad (122b)$$

$$A_{\mathbf{g}}^{\text{un}} = (I_{\mathbf{g}}^{++} - I_{\mathbf{g}}^{+-} + I_{\mathbf{g}}^{-+} - I_{\mathbf{g}}^{--})/I_{\mathbf{g}} \quad (122c)$$

where the labels "so," "ex," and "un" are short for spin-orbit coupling, exchange, and unpolarized, respectively. In A^{so}, effects due to the magnetization cancel and thus leave SOC as the main source of the intensity difference. Analogously, SOC effects cancel in A^{ex} but effects due to exchange are kept. A^{un} is the asymmetry of an unpolarized incoming beam due to magnetization reversal (cf. [97]).

As mentioned in Section 3.1, LEED provides information on the geometrical arrangement of the scattering sites. Thus, by comparing experimental with theoretical $I(E)$ spectra, one can determine the geometric structure of a sample by a systematical search; for example, by variation of structural parameters (the positions of the scatterers) and nonstructural parameters (the mean free path), one tries to minimize an objective function, which gives a measure of the agreement between experiment and theory (see, e.g., [6, p. 38]). These objective functions are called reliability factors, or for short R factors, and take into account energy positions and width of intensity maxima [98, 99]. Further, via SPLEED analysis, one is able to determine nongeometrical quantities. For example, the enhancement of the magnetic moment at the (110) surface of Fe has been found by comparing experimental and theoretical $I(E)$ spectra, treating the surface magnetic moment as a fit parameter [100] (for an *ab initio* calculation of magnetic properties of surfaces, see [101] and references therein).

Temperature effects occur prominently as thermal vibrations of the scattering sites that reduce the reflected intensities and increase the background. In LEED, this is approximately taken into account by multiplying the elements of the single-site t matrix by effective Debye–Waller factors [7, 12], thus ignoring the correlation of the motion (e.g., phonons), which might appear important because of multiple scattering (see, e.g., [102]).

3.3. Application: Quantum-Well Resonances in Ferromagnetic Co Films

As a typical example of the manifestation of electron scattering within an ultra-thin film, we address quantum-well resonances (QWRs) (see Fig. 22). As introduced in Section 2, QWSs are quantized electronic states that are confined to an ultra-thin film. Strict confinement on the substrate side of the film is possible if there is a gap in the substrate band structure. On the vacuum side, confinement is achieved by the surface barrier (cf. "QWS" in Fig. 22). At energies above the vacuum level, there is no total reflection at the surface barrier and, therefore, the electronic states in the film are not strictly confined. In other words, they are resonant with the free-electron states in the vacuum (cf. "QWR" in Fig. 22). In a ferromagnetic film on a nonmagnetic substrate, QWRs are characterized as exchange-split pairs of peaks in the $\mathbf{k}_{\parallel}$-resolved and layer-projected density of states (DOS), that is, the Bloch spectral function. Therefore, at the energy of a QWR, LEED electrons incident from the vacuum side have highest probability of entering the film and, consequently, a minimal reflection coefficient; that is, QWRs show up as minima in the SPLEED spectra.

Similar connections exist between the electronic structure of semi-infinite systems and other electron spectroscopies. It is well known that the LEED reflectivity is maximal for energies in bulk-band gaps (see Section 3.2.2) and tends to have minima at energies near band edges [45]. These minima correspond to maxima in the fine structure of secondary electron emission spectra [103]. Likewise, band edges are mirrored by maxima in the target current absorbed by the crystal [104] because the target current is complementary to the LEED reflectivity. Therefore, it could also be used to study QWRs.

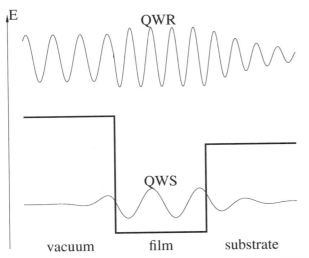

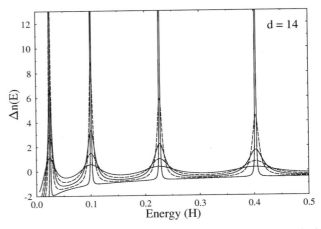

Fig. 22.   Quantum-well resonance (QWR) and quantum-well state (QWS) in an ultra-thin film. The film is depicted as a quantum well (thick solid lines; the abscissa is the spatial extent normal to the sample surface, the ordinate is the energy E). The wave functions of a QWS and a QWR are schematically given by thin solid lines.

Fig. 23.   Change of the density of states $\Delta n(E)$ due to confinement of a free electron to a rectangular quantum well, cf. Eq. (124), versus energy for confinement strengths $R = 0.2$, 0.4, 0.6, 0.8, and 0.99 (alternating solid and dashed lines). For a well thickness of $d = 14$, the maxima correspond to quantum-well states or resonances ($R < 1$).

The change of the DOS due to QWSs or QWRs can easily be computed for a rectangular quantum well [105]. A plane wave with $\mathbf{k}_\parallel = 0$ is reflected at the surface-side and the bulk-side boundary with reflection coefficients $R_s \exp(i\phi_s)$ and $R_i \exp(i\phi_i)$, respectively (R_s, $R_i \in \mathbb{R}$). The accumulated phase along a round trip, that is, reflection at a boundary, propagation, reflection at the other boundary, and once again propagation, is $\Delta\phi = 2kd + \phi_s + \phi_i$, with k being the wave number of the electron and d the thickness of the well. The change of the DOS $\Delta n(E)$ should be proportional to $\cos(\Delta\phi)$, the strength of the confinement $R = R_i R_s$, the thickness of the well d, and the unit DOS $2\partial_E k(E)/\pi$. For j round trips, one has to replace R by R^j and $\Delta\phi$ by $j\Delta\phi$. Thus, $\Delta n(E)$ is given by

$$\Delta n(E) = \frac{2d}{\pi}\big(\partial_E k(E)\big)\sum_{j=1}^{\infty} R^j \cos(j\,\Delta\phi) \qquad (123)$$

which, after some manipulation, can be written as

$$\Delta n(E) = \frac{2d}{\pi}\frac{R[\cos(\Delta\phi) - R]}{1 - 2R\cos(\Delta\phi) + R^2}\partial_E k(E) \qquad (124)$$

The important factor is the second one, which reflects the interference of the electron. The last term takes into account the electronic structure of the infinitely extended ("bulk") system. In Figure 23, $\Delta n(E)$ is shown for a free electron; that is, $\partial_E k(E) = 1/k(E)$. If the phase shift is a multiple of 2π, constructive interference leads to maxima in $\Delta n(E)$. These maxima are very sharp (quasi δ peaks) if R is close to 1 (cf. $R = 0.99$ in Fig. 23). If R is considerably small, as is expected for a QWR, $\Delta n(E)$ shows smooth oscillations that become broader with increasing energy. This behavior can be attributed to $\partial_E k(E)$.

As a prototypical system, we now address SPLEED from Co films on W(110) (for details, see [106]). Its layer-by-layer

growth [107–110] allows for a study of the evolution of QWRs with increasing number n of Co layers. Note that initially a pseudomorphic body-centered cubic (bcc) (110) monolayer is formed, which upon further Co deposition becomes a monolayer with approximate hexagonally closed packed (hcp) structure. Subsequent layers grow in the form of hcp Co(0001). It is further important that the magnetization of the Co film is parallel to the surface for all film thicknesses ($\mathbf{M} \parallel [1\bar{1}00]$ of the hcp Co film).

Experimental. Although this chapter is mainly concerned with theory, some words on the experimental aspects are in order. To estimate the film thickness during the growth process, one monitors the Co evaporation rate, for example, by a quartz crystal oscillator. This procedure gives an accuracy of better than 2% in thickness calibration. Spectra were measured by means of a spin-polarized low-energy electron microscope (SPLEEM), which, for normal incidence ($\mathbf{k}_\parallel = 0$), allows for the observation of the specular reflected beam ($\mathbf{g} = 0$) from single ferromagnetic domains [111]. The illumination system produces an electron beam with spin polarization $\mathbf{P}^{\mathrm{in}}$ perpendicular to the direction $\mathbf{k}$ of the beam, that is, parallel to the surface. Thus, in-plane magnetization $\mathbf{M}$ can be studied. The degree of polarization P^{in} was about 20%. Spectra were taken in remanence because even weak magnetic fields would deflect the low-energy electrons because of the Lorentz force.

Specular SPLEED intensities I_+ and I_- were recorded for $\mathbf{P}^{\mathrm{in}}$ parallel and antiparallel to $\mathbf{M}$. As mentioned previously, minima in I_+ (I_-) directly reflect majority (minority) spin QWRs. The asymmetry $A^{\mathrm{ex}} = (I_+ - I_-)/(I_+ + I_-)$ is identical to the exchange-induced asymmetry A^{ex} [cf. Eq. (122)], because in this highly symmetric setup effects due to SOC do not show up in an asymmetry, although they are present in nature.

Computational and Model Assumptions. Numerical calculations were performed within the framework of spin-polarized relativistic layer-KKR (SPRLKKR) theory for ferromagnetic systems consisting of arbitrary combinations of commensurate atomic monolayers, as briefly described in Section 3.2.3. The spin- and layer-resolved Bloch spectral function was obtained as the imaginary part of the Green function (see Section 4.2.3).

Because hcp Co films are incommensurate with the W(110) substrate, only the specular beam ($\mathbf{g} = 0$) is common to the beam sets of hcp Co(0001) and bcc W(110) [110]. However, the previous layer-KKR (LKKR) approach is applicable only to commensurate systems. Therefore, either the substrate plus pseudomorphic Co layers or a standalone hcp Co film, that is, a Co film without substrate, can be treated. The latter should be a reasonable approximation in the case of thicker films, but the influence of the substrate can certainly not be ignored in the case of a few monolayers (MLs) because of the mean free path (cf. Fig. 12). The effect of the substrate can be included in the following approximation: One calculates the total reflection matrix R^{tot} from semi-infinite W(110) without surface barrier. From this R^{tot}, the specular reflection coefficient R^{tot}_{00} is extracted and taken as the only nonvanishing element of the reflection matrix at the substrate side. To simulate lifetime effects, the imaginary part V_{i} of the optical potential was taken as energy dependent. Because a surface barrier of step shape is not appropriate for VLEED, a smooth JJJ form [91] was chosen.

Theoretical and Experimental Results. Numerical results of QWRs and their manifestation in SPLEED are presented in Figure 24 for 4-ML Co on W(110). First, the bulk-band structure above the vacuum level of hcp Co(0001) along the line Γ–A is addressed. Although in the presence of SOC bands of ferromagnets should be characterized with respect to magnetic double groups (cf. [112]), a classification in terms of nonmagnetic single groups and majority/minority spin remains useful in large parts of the band structure, in particular, those in parts where there are no bandgaps induced by SOC. Between 2.0 and 17.5 eV above the vacuum level (0 eV), there is a rather steep spin-split band that belongs to the Λ^1 representation, separated by a narrow gap at 17.5 eV from a flat band of the same representation. Further, there are fairly flat bands of dominant Λ^3 representation above 21 eV. The spin polarization expectation value for the majority (minority) bands departs from $+1$ (-1) only by a few hundredths, which indicates a negligible influence of SOC.

For each of the four Co monolayers, the Bloch spectral function for $\mathbf{k}_{\parallel} = 0$ (Fig. 24b) is seen to have four widely spaced pairs of majority and minority spin peaks between 2 and 17 eV, and four narrowly spaced pairs between 17 and 20 eV (the pair around 16.5 eV looks more like a shoulder because it is close to the strong peak at 17.5 eV). These two sets of Λ^1 QWRs are related to the steep and to the flat Λ^1 band, respectively. In Figure 24b, the first three pairs of QWRs are indicated by arrows, the $k_{\perp}$ values of which are shown in panel a. For the majority Λ^1 band, one finds $k_{\perp}$ values of 0.40 $2\pi/c$, 0.77 $2\pi/c$, and 0.76 $2\pi/c$, whereas for the minority band one finds 0.40 $2\pi/c$,

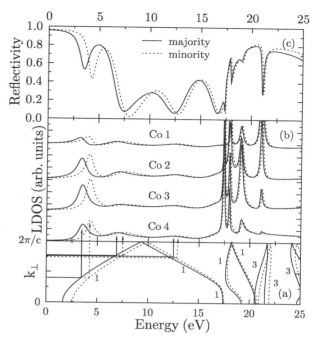

Fig. 24. Ferromagnetic hcp Co. (a) Bulk-band structure along [0001] with bands of mainly majority (solid lines) and minority spin (dashed lines) and dominant spatial representations Λ^1 and Λ^3 indicated by numbers 1 and 3. (b) Λ^1 contributions to the layer- and spin-resolved density of states (LDOS) for $\mathbf{k}_{\parallel} = 0$ of a 4-ML film of Co on W(110); solid and dashed lines indicate majority and minority spin, respectively; the first ML (Co 1) is at the surface, the fourth (Co 4) adjacent to the substrate. (c) SPLEED from 4-ML Co on W(110): spin-dependent intensity versus energy curves I_+ (solid) and I_- (dashed) of the specular beam for normal incidence, calculated with the same lifetime broadening $V_{\text{i}} = 0.05$ eV as in (b). The quantum-well resonances are indicated by arrows in (b); see text. The vacuum level is 0 eV. Reprinted with permission from T. Scheunemann, R. Feder, J. Henk, E. Bauer, T. Duden, H. Pinkvos, H. Poppa, and K. Wurm, *Solid State Commun.* 104, 787 (1999). Copyright 1997, by Elsevier Science.

0.77 $2\pi/c$, and 0.74 $2\pi/c$ (note that all bands are back-folded at the Brillouin zone boundary $k_{\perp} = 2\pi/c$). According to the theories presented in Section 2.2, these values nicely coincide with those obtained from Eq. (20) with $a = 5c/4$. Because the substrate is nonmagnetic, the reflection properties at the Co/W interface are identical for spin-up and spin-down electrons. Therefore, one would expect the same $k_{\perp}$ for majority and minority QWRs, as can be observed from Figure 24. The small differences in $k_{\perp}$ can be attributed to the different energies of majority, and minority QWRs, which are due to the spin-dependent potential within the Co film; cf. Eq. (81).

Decomposition of the Bloch spectral function according to angular momenta reveals that the first set of resonances, that is, those QWRs below 17.5 eV, has comparable s, p, d, and higher contributions, whereas in the second set the d and higher contributions far outweigh the s and p parts. The exchange splitting is largest for the pair around 3 eV and decreases with increasing energy. The difference between the local density of states (LDOS) of the first layer and that of the fourth is due to the fact that the film is bounded by vacuum on one side and by W(110) on the other. Because of their Λ^1 representation, the

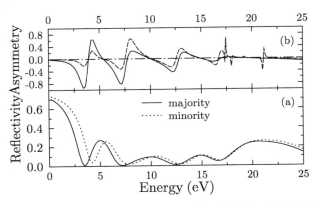

Fig. 25. Ferromagnetic hcp Co: (a) Same as Figure 24c, but calculated with the larger energy-dependent lifetime broadening V_i. (b) SPLEED asymmetry $A = (I_+ - I_-)/(I_+ + I_-)$ obtained from the intensities $I_\pm$ in panels (c) in Figure 24 (dashed line) and (a) (solid line). Reprinted with permission from T. Scheunemann, R. Feder, J. Henk, E. Bauer, T. Duden, H. Pinkvos, H. Poppa, and K. Wurm, *Solid State Commun.* 104, 787 (1997). Copyright 1997, by Elsevier Science.

resonances can couple to scattering solutions outside the film and thus become observable by SPLEED, as mentioned previously. In contrast, the two sets of four LDOS peaks associated with the Λ^3 bands above 20 eV correspond to QWRs that cannot be accessed from the vacuum side for $\mathbf{k}_\parallel = 0$.

We now proceed to specular-beam SPLEED spectra I_+ (majority) and I_- (minority) for normal incidence. The incoming electrons are fully polarized parallel and antiparallel to the majority spin direction of the Co film. Figure 24c shows SPLEED spectra that were calculated with the same lifetime broadening as the Bloch spectral function; for example, the imaginary part of the optical potential was chosen as $V_i = 0.05$ eV. The most important features of the I_+/I_- pair are broad valleys between 2 and 17 eV and four pairs of narrow dips between 17 and 22 eV. All these minima occur exactly at the energies of QWRs, which establishes the one-to-one correspondence between SPLEED minima and QWRs.

To come closer to experiment, the lifetime broadening V_i was increased, as shown in Figure 25. The larger energy-dependent V_i, which is on the order of 2 eV, leads to broader maxima in the $I_\pm(E)$ spectra, as has been demonstrated by means of a model calculation (Fig. 15). For example, the sharp minima between 17 and 20 eV are completely smeared out. The fingerprints of the QWRs between 2 and 17 eV are thus preserved, whereas those of the resonances above 17 eV are obscured by lifetime effects. Compared to Figure 24c, the first two valleys are only slightly broader, but significantly deeper. The latter is plausible because, roughly speaking, "more electrons disappear" in inelastic channels inside the solid and are consequently missing in the elastically reflected channel. In the exchange-induced asymmetry $A^{\mathrm{ex}} = (I_+ - I_-)/(I_+ + I_-)$, shown in panel b, the exchange-split QWRs manifest themselves as pronounced $-/+$ features.

In Figure 26, experimental and calculated SPLEED spectra Co films on W(110) with 0 ML to 8 ML thickness are shown. Note that the spectra for uncovered W(110) are not

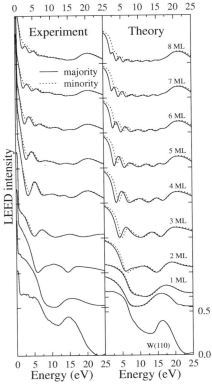

Fig. 26. Spin-polarized low-energy electron diffraction from clean W(110) and from n-ML ferromagnetic Co on W(110) with $n = 1, \ldots, 8$ as indicated: spin-dependent intensity versus energy curves I_+ (solid) and I_- (dashed) of the specular beam for normal incidence. Experiment (left panel) and theory (right panel). For 1-ML Co, the lower pair of curves was obtained for a pseudomorphic overlayer with Co atoms in bulklike W positions, the upper pair for an incommensurate overlayer with the lateral geometry of hcp Co(0001). The tick mark below each curve indicates its zero line. The intensities are normalized to the primary beam intensity, with the distance between 0 and 0.5 at the bottom of the right panel giving the scale. Reprinted with permission from T. Scheunemann, R. Feder, J. Henk, E. Bauer, T. Duden, H. Pinkvos, H. Poppa, and K. Wurm, *Solid State Commun.* 104, 787 (1997). Copyright 1997, by Elsevier Science.

spin dependent because of the high symmetry of the normal-incidence geometry and the absence of ferromagnetism. For 1-ML Co on W(110), the measured spectrum is, in contrast to the calculated one, still spin independent. This may be explained by a Curie temperature T_C below room temperature, at which the data were taken. For example, a 2-ML Co film on Cu(001) shows T_C of about 320 K compared to a bulk value of 1388 K [113]. Another reason might be perpendicular magnetization, which cannot be detected by SPLEED with incoming electrons polarized parallel to the surface. The similarity of the measured 1-ML spectrum to the spectrum for uncovered W and its agreement with the 1-ML spectrum calculated for Co in W sites indicates pseudomorphic growth of the first Co monolayer. In the theoretical spectrum for an incommensurate monolayer with the lateral geometry of hcp Co(0001), the second peak is shifted upward by about 2 eV. The rather broad minimum between 7.5 and 12.5 eV is associated with a QWR.

With increasing number n of Co monolayers, the calculated spectra in Figure 26 show a systematic evolution of the minima, that is, the signatures of the QWRs, the number of the latter being equal to n. Comparing with experiment, one notices good agreement for 2 and 4 ML. For 3 ML, the experimental spectrum duly exhibits three minima but the third one is higher in energy and so is the subsequent peak. On the grounds of the evolution of the calculated spectra with thickness n, the experimental data for 3 ML indicate a surface morphology other than three hcp monolayers. From 5 ML upward, the experimental spectra differ from theory by having only $n - 1$ minima instead of n. It is, however, striking that the experimental curves for 6, 7, and 8 ML are in good agreement with the calculated ones for 5, 6, and 7 ML, respectively.

The preceding theoretical and experimental results show that the electronic structure of ultra-thin films at energies above the vacuum level can comprise pairs of exchange-split QWRs. The calculated as well as the experimental SPLEED spectra have minima, which exactly coincide in energy with maxima in the Bloch spectral function of the same spin and the same spatial representation.

4. PHOTOELECTRON SPECTROSCOPY

4.1. Introduction and History

The first publications on the photoelectric effect appeared in the 19th century (for detailed coverage of the history of photoelectron spectroscopy (PES), see [114]). In 1887, Hertz discovered that a discharge between metal plates (electrodes) occurs at a lesser voltage than usual if the cathodes are illuminated by ultraviolet light [115]. One year later, Hallwachs found that a negatively charged metal plate connected to an electroscope becomes discharged if illuminated. A positively charged metal plate, however, remains charged [116]. These findings led to intensive experimental work, for example, by Elster and Geitel [117, 118] as well as Lenard [119]. The latter observed that irradiated metals send electrons into their surroundings. The velocity of the photoelectrons, however, does not depend on the intensity of the incoming radiation, but the number of electrons increases with light intensity. These observations could not be explained within classical theories. In 1905, Einstein explained, following Planck's ideas, the photoelectrical effect by quanta of the electromagnetic radiation, that is, by photons [120]. For this he achieved the Nobel Prize in 1919.

Because of the comparably large experimental effort, it took considerable time to use the photoelectric effect as a spectroscopic method. Siegbahn used X-rays to determine the electronic structure of the inner shells of atoms [121], a method first termed electron spectroscopy for chemical analysis (ESCA) and nowadays better known as X-ray photoelectron spectroscopy (XPS). Using light in the ultraviolet regime, which can be produced by rare-gas discharge lamps, instead of X-rays and using angle-resolved detection of the photoelectrons instead of angle-integrated detection allowed for the determination of the valence-band structure of solids [122]. As in LEED, surface sensitivity plays an important role: The detected photoelectrons

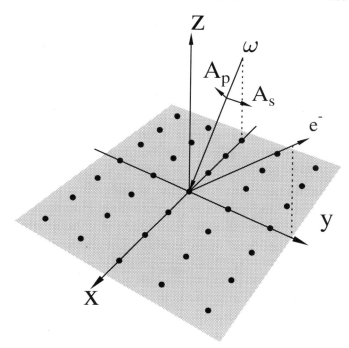

Fig. 27. Setup of photoemission. Light with photon energy ω impinges on the solid surface (gray area, with dots representing surface atoms). Its polarization components are denoted A_s and A_p, the photoelectrons as e^-.

have their origin in the first few layers of the solid and, therefore, the spectra contain contributions from both the surface and the bulk. A considerable impact on the development of angle-resolved photoelectron spectroscopy (ARPES) or, as commonly but less accurately named, photoemission was obtained by synchrotron radiation facilities [123–125]. These allow the use of photon energies from the vacuum-ultraviolet (VUV) to the X-ray regime. Further, they provide both circularly and linearly polarized light, which can be used to determine the symmetry of electronic states via dipole selection rules.

Figure 27 shows a rather typical setup of a photoemission experiment [6, p. 73]. The incoming light is characterized by the incidence direction, the photon energy ω, and its polarization. The latter is described by the components of the electrical-field vector $\mathbf{A}$ with components $\mathbf{A}_p$ and $\mathbf{A}_s$ lying parallel and perpendicular[14] to the plane of incidence, which is spanned by the surface normal and the incidence direction (xz plane in Fig. 27), respectively. A list of commonly used light polarizations is given in Table II. The electrons (e^- in Fig. 27) are detected with respect to their kinetic energy, their outgoing direction (angle-resolved photoemission), and their spin (spin-resolved photoemission).

In an energy diagram (Fig. 28), photoemission can be regarded as a three-step process [126]. In the initial stage, the electronic states are occupied up to the Fermi level E_F. The incoming photon with energy ω excites one electron into an unoccupied state with energy larger than E_F. The latter then propagates toward the surface. If the electron energy is larger

---

[14]Historically, s stands for German *senkrecht*, that is, perpendicular.

Table II.   Light Polarization in Photoemission

Polarization	A_s	A_p	Remark
s	1	0	Linear
p	0	1	Linear
σ_+	1	i	Right-handed circular
σ_-	1	$-i$	Left-handed circular

Note: The components A_s and A_p of the electrical-field vector are given for both linearly and circularly polarized light. Unpolarized light is an incoherent superposition of either s- and p-polarized or left- and right-handed circularly polarized light.

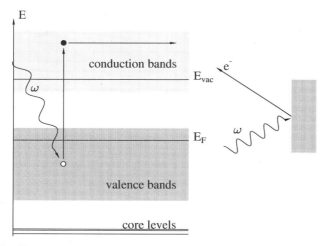

Fig. 28.   Energy scheme of photoemission. Right panel: Incident light with photon energy ω, represented by the wavy line, impinges on the solid surface (gray area, right) and outgoing electrons e^- leave the solid. Left panel: An electron (filled circle) is excited into a state above the vacuum level E_{vac}, leaving behind a hole (empty circle) below the Fermi energy E_F. The electron, which stems either from the valence-band (gray area) or from the core-level (horizontal lines) regime, leaves the solid and is subsequently detected.

than the vacuum level E_{vac}, the photoelectron can leave the solid and propagate toward the detector, leaving behind the solid with one hole. The energy difference between E_{vac} and E_F is the work function Φ, typically about 5 eV.

There are three main modes in photoemission: (i) The most commonly used is the energy distribution curve (EDC) mode in which the photon energy is kept constant. Therefore, variation of the kinetic energy of the photoelectrons corresponds to a variation in the initial-state energy (cf. Fig. 29). (ii) In constant initial-state (CIS) mode, both the photon energy and the kinetic energy are varied in such a way that the energy of the initial state remains constant. (iii) In constant final-state (CFS) mode, the kinetic energy is kept fixed and the photon energy is varied. This leads to a variation in the initial-state energy. Because CIS and CFS require variable photon energy, they are not possible with laboratory rare-gas discharge lamps, which provide only fixed photon energies, for example, $\omega = 21.22$ eV from He(I) discharge.

A simple theoretical description of the process uses Fermi's golden rule. The transition probability w_{fi} between the initial state $|\Psi_i\rangle$ with energy E_i and the final state $|\Phi(\mathbf{k}_\parallel, E_f)\rangle$ with

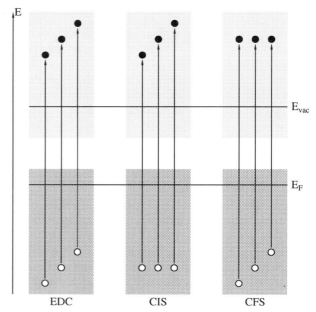

Fig. 29.   Energy modes of photoemission. In energy distribution curve (EDC) mode (left), the kinetic energy of the photoelectrons (black circles) is varied while keeping the photonenergy constant. In constant initial-state (CIS) mode (middle), the energy of the initial state (hole, white circles) is kept constant while the photon energy is varied, which, in turn, leads to variable kinetic energies. In constant final-state (CFS) mode (right), the kinetic energy is kept fixed while the photon energy is varied. The Fermi energy is E_F, the vacuum level E_{vac}.

energy $E_f = E_i + \omega$ and surface-parallel momentum $\mathbf{k}_\parallel$ is given by

$$w_{fi} = \left| \langle \Phi(\mathbf{k}_\parallel, E_f) | D | \Psi_i \rangle \right|^2 \delta(E_f - E_i + \omega) \qquad (125)$$

The transition is mediated by the dipole operator D, which can be approximated as $\mathbf{A} \cdot \mathbf{p}$, with $\mathbf{p}$ denoting the momentum operator and $\mathbf{A}$ the vector potential of the incident light. The photocurrent j for kinetic energy E_{kin} and $\mathbf{k}_\parallel$ of the photoelectrons is then given by

$$j \sim \sqrt{E_{\text{kin}}} \sum_i w_{fi} \qquad (126)$$

The electron detection angles ϑ_e and φ_e as well as the kinetic energy determine the surface-parallel momentum $\mathbf{k}_\parallel$,

$$\mathbf{k}_\parallel = \sqrt{2E_{\text{kin}}} \begin{pmatrix} \cos \varphi_e \\ \sin \varphi_e \end{pmatrix} \sin \vartheta_e \qquad (127)$$

One of the most successful methods for the analysis of electronic states in the valence regime is angle-resolved photoemission using VUV light (for reviews on ARPES, see [114, 127–129]). For fixed photoelectron detection angles, the experimentally obtained intensity maxima disperse in binding energy with photon energy. The interpretation of this behavior is usually based on the very popular and successful direct-transition approximation, which relates the energy position of the intensity maxima to the bulk-band structure $E(\mathbf{k})$. This relies on the assumption that the wave vector $\mathbf{k}$ is conserved in the excitation process. A survey of the theoretical aspects is given next.

4.2. Theory

4.2.1. Historical Sketch

First, a few milestones of photoemission theory will be sketched, without claim of completeness. In 1964, Adawi showed within scattering theory that the final state $|\Phi\rangle$ can be described by an incoming plane wave [130]. Mahan addressed the angular distribution of the photoelectrons and pointed out that the results obtained via Green functions and via quadratic response are equivalent [131, 132]. Regarding surface sensitivity, model calculations by Schaich and Ashcroft yielded that the photoeffect at the surface and in the bulk can be of the same order of magnitude [133]. Langreth determined the scattering of the photoelectrons versus several parameters, in particular, the escape depth [134]. Because of the similarity of some aspects of LEED and photoemission, in particular, the state of the photoelectron, several works based on LEED theory were published. Especially, multiple scattering of the outgoing electrons was treated by Pendry [135]. Taking into account all ingredients of photoemission theory, for example, band structures, transition-matrix elements, scattering, and surface effects, Pendry developed a dynamical theory that eventually led to the PEOVER computer program [136, 137].

Relativistic effects were considered by Ackermann and Feder [83] as well as by Borstel's group [138–140]. The latter addressed, in particular, optical orientation, that is, the alignment of the photoelectron's spin parallel or antiparallel to the helicity of the incoming circularly polarized light [141]. In a series of publications, Feder's group predicted that the photoelectron shows a spin polarization even in the case of linearly polarized light [142–144] (see also [145]). All their predictions were fully confirmed experimentally [146–148].

In addition, multiple-scattering theories for ferromagnets [149] found entrance into the photoemission theories (for an overview, see [51]). Relativistic theories for spin-polarized systems, in particular, spin-polarized relativistic layer-KKR (SPRLKKR), proved to be successful in the explanation of magnetic dichroism in photoemission, that is, the change of the photocurrent due to reversal of the magnetization direction by an otherwise fixed setup [150]. This allowed for a detailed understanding of the electronic structure of ferromagnets, in particular, the delicate interplay of spin-orbit coupling (SOC) and exchange [151–153]. Further, multiple-scattering theory allows for the treatment of disorder, for example, substitutional alloys [154]. This can also be used to describe photoemission from ferromagnets at elevated temperatures [155].

In the following, we portray the photoemission formalism as derived by Feibelman and Eastman [156], which can be seen as the basis of most, if not all, applied photoemission theories today.

4.2.2. Formalism of Photoemission

The nonrelativistic theory of Feibelman and Eastman [156] is based on the work of Caroli et al. [157] who applied Keldysh's Green function formalism for nonequilibrium systems. Consider a semiinfinite solid, the electrons of which are assumed independent. Monochromatic light with vector potential $\mathbf{A}(\mathbf{r}, t) = \mathbf{A}_0(\mathbf{r}) \cos(\omega t)$ impinges on this solid. The Hamiltonian of the entire system is then given by

$$H(t) = \frac{1}{2}\left(\mathbf{p} + \frac{1}{c}\mathbf{A}(\mathbf{r}, t)\right)^2 + V(\mathbf{r}) \tag{128}$$

where $V(\mathbf{r})$ is the crystal potential. We separate $H(t)$ into the time-independent reference Hamiltonian H_0 and the time-dependent perturbation $H_1(t)$, $H(t) = H_0 + H_1(t)$ with

$$H_0 = \frac{\mathbf{p}^2}{2} + V(\mathbf{r}) \tag{129a}$$

$$H_1(t) = \frac{1}{2c}\left[\mathbf{p} \cdot \mathbf{A}(\mathbf{r}, t) + \mathbf{A}(\mathbf{r}, t) \cdot \mathbf{p}\right] + \frac{\mathbf{A}(\mathbf{r}, t)^2}{2c^2} \tag{129b}$$

The eigenstates of H_0 form a complete set and obey $H_0|n\rangle = E_n|n\rangle$. Our task is to calculate the current density $\mathbf{j}$, which is due to a state $|\Psi\rangle$ at the detector position $\mathbf{R}$,

$$\langle\Psi|\mathbf{j}(\mathbf{R})|\Psi\rangle = \frac{i}{2}\left(\Psi^*\nabla\Psi - \Psi\nabla\Psi^*\right)_{\mathbf{R}} \tag{130}$$

where the expectation value is evaluated at $\mathbf{R}$. Following time-dependent perturbation theory, $H_1(t)$ is adiabatically switched on. For the time development of an eigenstate $|n\rangle$, one has in first order

$$|\Psi_{nI}(t)\rangle = |n\rangle - i\int_{-\infty}^{t} H_{1I}(t')|n\rangle\, dt' \tag{131}$$

where the subscript I denotes the interaction picture. The contribution of state $|n\rangle$ to the current density,

$$\langle\mathbf{j}_n(\mathbf{R})\rangle = \langle\Psi_{nI}(t)|\mathbf{j}_I(\mathbf{R}, t)|\Psi_{nI}(t)\rangle \tag{132}$$

is straightforward with the following considerations: (i) The diamagnetic term of the dipole operator, $\mathbf{A}^2/(2c^2)$, is neglected. (ii) Terms in which the current density operator acts on bound states give no contribution to $\mathbf{j}(\mathbf{R})$ because these electrons cannot leave the solid. (iii) Translating the dipole operator from the interaction picture into the Schrödinger picture yields exactly one term, which corresponds to the excitation of $|n\rangle$ into a state with energy $E_n + \omega$. In summary, one arrives at

$$\langle\mathbf{j}_n(\mathbf{R})\rangle = \langle n|O G^a(E_n + \omega)\mathbf{j}(\mathbf{R})G^r(E_n + \omega)O|n\rangle \tag{133}$$

with $O = (\mathbf{p} \cdot \mathbf{A}_0 + \mathbf{A}_0 \cdot \mathbf{p})/2c$ and advanced (G^a) as well as retarded (G^r) Green functions

$$G^a(E) = \sum_m \frac{|m\rangle\langle m|}{E - E_m - i\eta} \tag{134a}$$

$$G^r(E) = \sum_m \frac{|m\rangle\langle m|}{E - E_m + i\eta} \tag{134b}$$

($\eta > 0$). Inserting Eq. (130) into Eq. (133) and introducing the nonlocal spectral density

$$G^+(E) = 2\pi i \sum_n |n\rangle\langle n|\delta(E - E_n) \tag{135}$$

one eventually obtains

$$R^2\langle j(\mathbf{R})\rangle = -\frac{K}{32\pi^2 c^2}\langle\Phi^*(E+\omega)|\Delta\,\mathrm{Im}\,G^+(E)\Delta^\dagger|\Phi^*(E+\omega)\rangle \tag{136}$$

$K = \sqrt{2(E+\omega)}$ is the momentum of the photoelectron. The time-reversed final state $|\Phi\rangle$ fulfills

$$\Phi(\mathbf{r}, E+\omega) = \exp(i\mathbf{K}\cdot\mathbf{r}) + \int dr'\, G^r(\mathbf{r},\mathbf{r}'; E+\omega)V(\mathbf{r}')$$
$$\times \exp(i\mathbf{K}\cdot\mathbf{r}') \tag{137}$$

which establishes the connection of photoemission with LEED. First, $|\Phi\rangle$ is a superposition of an incoming plane wave, the first term in Eq. (137), and outgoing waves. The latter are represented by the integral in Eq. (137), which gives nonzero contributions only inside the crystal, in particular, where $V(\mathbf{r}') \neq 0$. The retarded Green function propagates electrons from the interior of the solid ($\mathbf{r}'$) to the vacuum region ($\mathbf{r}$). Therefore, the integral can be regarded as giving rise to reflected beams in vacuum. In short, $|\Phi\rangle$ is a state suitable for the description of a LEED experiment. $|\Phi^*\rangle$ is known as a time-reversed LEED state.

As a last step, we observe that the final state can be written as $|\Phi^*\rangle = G^r(E+\omega)|\Phi_0^*\rangle$ where $|\Phi_0^*\rangle$ is the plane wave at the detector position. The expression for the photocurrent then eventually reads

$$j \sim -\langle\Phi_0^*|G^a(E+\omega)\Delta\,\mathrm{Im}\,G^+(E)\Delta^\dagger G^r(E+\omega)|\Phi_0^*\rangle \tag{138}$$

Equation (138) can be represented by the Feynman diagram shown in Figure 30. Its interpretation is straightforward if Eq. (138) is read from the right. First, the photoelectron state $|\Phi_0^*\rangle$ with energy $E+\omega$ is propagated by the retarded Green function G^r from the detector to the interior of the solid. Subsequently, the dipole operator $\Delta^\dagger$ mediates a deexcitation to initial states with energy E, which are described by the nonlocal density of states, $\mathrm{Im}\,G^+$. These are excited into the outgoing photoelectron state $\langle\Phi_0^*|G^a$ by the dipole operator Δ. The diagram in Figure 30 is that of lowest order. Higher order diagrams include, for example, the (screened) Coulomb interaction between the final and the initial states, that is, scattering between the photoelectron and the remaining hole. These terms are, for instance, essential for the description of the resonant behavior of photoemission intensities from Pd [158]. Usually, one neglects higher order terms; that is, one assumes the sudden approximation [159–161].

Applying the Dirac identity,

$$\lim_{\eta\to 0^+}\frac{1}{x\pm i\eta} = \mathcal{P}\left(\frac{1}{x}\right)\mp i\delta(x) \tag{139}$$

which establishes for any state $|\Xi\rangle$ the relation [cf. Eq. (134); $\mathcal{P}$ stands for principal value]

$$\mathrm{Im}\langle\Xi|G(E+i\eta)|\Xi\rangle = -\pi\sum_m|\langle\Xi|m\rangle|^2\delta(E-E_m) \tag{140}$$

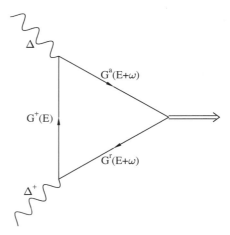

Fig. 30. Feynman diagram of photoemission according to Eq. (138). The double line represents the detected state $|\Phi_0^*\rangle$ with energy $E+\omega$ and surface-parallel momentum $\mathbf{k}_\parallel$. Green functions G^a, G^r, and G^+ are represented by arrow-decorated lines, photons by wavy lines.

the photocurrent can be expressed in golden-rule form,

$$j \sim \sum_m|\langle\Phi^*(E+\omega)|\Delta|m\rangle|^2\delta(E+\omega-E_m) \tag{141}$$

The main difference between the golden-rule and the Green function expression, Eq. (138), is that the former holds only for real energies, whereas the latter can also be applied for complex energies. In other words, the golden rule is only valid for infinite lifetimes, whereas the Green function takes into account finite lifetimes of both photoelectrons and holes.

The description of electron scattering in a many-body theory leads to quasi-particle states and to the self-energy Σ, sometimes denoted optical potential [162]. In the lowest order approximation, the self-energy is local and homogeneous. Its real part gives rise to shifts of the quasi-particle energies. For example, fundamental bandgaps obtained from density-functional calculations for zinc-blende semiconductors are too small with respect to the experimental values; inclusion of Σ in the GW approximation [163] increases the bandgaps considerably. Another example is Ni where the experimental energy shift between spin-split bands is about 0.3 eV, in comparison to 0.6 eV from density-functional calculations [164]. The imaginary part of Σ accounts for the lifetime of the quasi-particles; that is, an increase of $\mathrm{Im}\,\Sigma$ leads to broader photoemission spectra [165, 166]. This can be attributed to the spectral function $A(E) = -\mathrm{Im}\,\mathrm{Tr}\,G(\mathbf{r},\mathbf{r}; E)/\pi$,

$$A(E) = \sum_m\begin{cases}\delta(E-E_m)\\ \quad\text{real energies}\\ \dfrac{\Gamma}{\pi}\dfrac{1}{(E-E_m)^2+\Gamma^2}\\ \quad\text{complex energies}\\ \dfrac{\mathrm{Im}\,\Sigma(E_m)}{(E-E_m-\mathrm{Re}\,\Sigma(E_m))^2+(\mathrm{Im}\,\Sigma(E_m))^2}\\ \quad\text{general case}\end{cases} \tag{142}$$

4.2.3. Formulation within Multiple-Scattering Theory

As we have seen in Section 3, multiple-scattering theory provides an excellent description of the LEED process. In the previous, we have further established a close connection between LEED and photoemission; in particular, the final state in photoemission is a time-reversed LEED state. Guided by these findings, photoemission should also nicely be described in terms of multiple-scattering theory. Instead of sketching Pendry's formulation [136, 167], which provides a very fast algorithm (in terms of computational time) but a rather technical theoretical description, we present a formulation in terms of Green functions. Of course, both methods give identical results.

In the formulation of the LKKR method for LEED, we have obtained the scattering properties of the entire semi-infinite system by consecutively treating the scattering of smaller entities: the single-site t matrix for scattering from atoms, the M and Q matrices for layers, and from these eventually the reflection matrix of the semi-infinite solid. This idea appears again in the calculation of the Green function. We shall start with the Green function of free space, then treat an empty layer embedded in the otherwise occupied system, and eventually treat a full layer.

Before turning to the investigation of the Green function, we sketch the very basis of multiple-scattering theory, the Lippmann–Schwinger equation and the Dyson equation.

Lippmann–Schwinger and Dyson Equations. Consider an eigenstate $|\Phi^0\rangle$ of a reference Hamiltonian H^0, $H^0|\Phi^0\rangle = E|\Phi^0\rangle$. The associated Green function G^0 fulfills $(z - H^0) \times G^0 = 1$, with $z = E + i\eta$, $\eta = 0^+$. The eigenfunction $|\Phi\rangle$ of the Hamiltonian $H = H^0 + V$, V being the perturbation, fulfills $H|\Phi\rangle = E|\Phi\rangle$ and can be obtained via the Lippmann–Schwinger equation

$$|\Phi\rangle = |\Phi^0\rangle + G^0 V |\Phi\rangle \qquad (143)$$

Solving for $|\Phi\rangle$ yields formally $|\Phi\rangle = (1 - G^0 V)^{-1}|\Phi^0\rangle$. Introducing the transition operator $T = V(1 - G^0 V)^{-1}$ gives $V|\Phi\rangle = T|\Phi^0\rangle$, and $|\Phi\rangle = (1 + G^0 T)|\Phi^0\rangle$.

The Dyson equation for the Green function G with $(z - H) \times G = 1$ can be obtained from $(z - H^0)G = 1 + VG$,

$$G = G^0 + G^0 V G = G^0 + G V G^0 \qquad (144)$$

Or, in terms of T, $G = G^0 + G^0 T G^0$, which immediately gives $VG = TG^0$. This result can also be used for the wave function $|\Phi\rangle$, $|\Phi\rangle = |\Phi^0\rangle + GV|\Phi^0\rangle$.

Free-Space Solutions. We recall briefly basic properties of the solutions of the free-space Dirac equation for a given complex energy E. Because in this case the Hamiltonian is no longer hermitian, one has to deal with left-hand side (superscript L) and right-hand side (superscript R) wave functions [168]. The former obey $\langle \Psi^L | H = E \langle \Psi^L |$, the latter $H|\Psi^R\rangle = E|\Psi^R\rangle$. In general, the l.h.s. solutions are not the hermitian conjugate of the r.h.s. solutions. In plane wave representation, the free-space

solutions $[V(\mathbf{r}) = 0]$ read [76]

$$\langle \mathbf{r} | f_{\mathbf{k}\tau}^R \rangle = \sqrt{\frac{E + c^2}{2c^2}} \begin{pmatrix} \chi^\tau \\ \dfrac{c\sigma \cdot \mathbf{k}}{E + c^2} \chi^\tau \end{pmatrix} \exp(i\mathbf{k} \cdot \mathbf{r}) \qquad (145)$$

and

$$\langle f_{\mathbf{k}\tau}^L | \mathbf{r} \rangle = \sqrt{\frac{E + c^2}{2c^2}} \left((\chi^\tau)^T, (\chi^\tau)^T \dfrac{c\sigma \cdot \mathbf{k}}{E + c^2} \right) \exp(-i\mathbf{k} \cdot \mathbf{r}) \qquad (146)$$

In angular-momentum representation, they are given by

$$\langle \mathbf{r} | z_\Lambda^R \rangle = \sqrt{\frac{E + c^2}{2c^2}} \begin{pmatrix} z_l(kr)\langle \hat{\mathbf{r}} | \chi_\Lambda \rangle \\ ick S_\kappa \\ \dfrac{ick S_\kappa}{E + c^2} z_{\bar{l}}(kr)\langle \hat{\mathbf{r}} | \chi_{\bar{\Lambda}} \rangle \end{pmatrix} \qquad (147)$$

and

$$\langle z_\Lambda^L | \mathbf{r} \rangle = \sqrt{\frac{E + c^2}{2c^2}} \left(z_l(kr)\langle \chi_\Lambda | \hat{\mathbf{r}} \rangle, \dfrac{-ick S_\kappa}{E + c^2} z_{\bar{l}}(kr)\langle \chi_{\bar{\Lambda}} | \hat{\mathbf{r}} \rangle \right) \qquad (148)$$

with $z = j$ for regular or $z = h$ for irregular solutions. The wave number k is $\sqrt{E^2 - c^4}/c$. Each representation can be transformed into the other by

$$\langle \mathbf{r} | f_{\mathbf{k}\tau}^R \rangle = \sum_\Lambda \langle \mathbf{r} | j_\Lambda^R \rangle a_{\Lambda\tau}(\hat{\mathbf{k}}) \qquad (149)$$

with $a_{\Lambda\tau}(\hat{\mathbf{k}})$ from Eq. (98).

Free-Electron Green Function. The retarded Green function of free space obeys the Dirac equation

$$(E - H)G_0^+(\mathbf{r}, \mathbf{r}'; E) = \begin{pmatrix} 1 & 0 \\ 0 & 1 \end{pmatrix} \delta(\mathbf{r} - \mathbf{r}') \otimes \delta_{\text{spin}} \qquad (150)$$

with the Hamiltonian $H = c\alpha \cdot \mathbf{p} + c^2 \beta$ and $V(\mathbf{r}) = 0$. δ_{spin} denotes the Kronecker δ in spin space, $\delta_{\text{spin}} = \sum_{\tau=\pm} \chi^\tau(\chi^\tau)^T$. For given energy E and wave vector $\mathbf{k}_\parallel$, G_0^+ is given in plane wave representation by

$$G_0^+(\mathbf{r}, \mathbf{r}') = \frac{1}{i F_A} \sum_{\mathbf{g}\tau} \frac{1}{k_{\mathbf{g}\perp}} \begin{cases} \langle \mathbf{r} | f_{\mathbf{k}_\mathbf{g}^+\tau}^R \rangle\langle f_{\mathbf{k}_\mathbf{g}^+\tau}^L | \mathbf{r}' \rangle & z > z' \\ \langle \mathbf{r} | f_{\mathbf{k}_\mathbf{g}^-\tau}^R \rangle\langle f_{\mathbf{k}_\mathbf{g}^-\tau}^L | \mathbf{r}' \rangle & z < z' \end{cases} \qquad (151)$$

$\mathbf{k}_\mathbf{g}^\pm$ is taken from Eq. (94) and the $+ (-)$ sign refers to the case $z > z'$ $(z < z')$. F_A is the area of the two-dimensional layer unit cell. In angular-momentum representation, $G_0^+(\mathbf{r}, \mathbf{r}')$ reads

$$G_0^+(\mathbf{r}, \mathbf{r}') = -ik \sum_\Lambda \begin{cases} \langle \mathbf{r} | j_\Lambda^R \rangle\langle h_\Lambda^L | \mathbf{r}' \rangle & r < r' \\ \langle \mathbf{r} | h_\Lambda^R \rangle\langle j_\Lambda^L | \mathbf{r}' \rangle & r > r' \end{cases} \qquad (152)$$

With $r_>$ $(r_<)$ being the larger (smaller) of r and r', the Green function can be written in a more compact form as

$$G_0^+(\mathbf{r}, \mathbf{r}') = -ik \sum_\Lambda \langle \mathbf{r}_< | j_\Lambda^R \rangle\langle h_\Lambda^L | \mathbf{r}_>' \rangle \qquad (153)$$

keeping in mind the two cases in Eq. (152).

Lippmann–Schwinger Equation and Single-Site Green Function. The regular solutions $|J_\Lambda^R\rangle$ fulfill the Lippmann–Schwinger equation $|J^R\rangle = |j^R\rangle + G_0^+ V|J^R\rangle$, or explicitly for the radial part,

$$\langle r|J_\Lambda^R\rangle = \langle r|j_\Lambda^R\rangle - ik\langle r|h_\Lambda^R\rangle \int_0^r dr'\, r'^2 \langle j_\Lambda^L|r'\rangle V(r')\langle r'|J_\Lambda^R\rangle$$
$$- ik\langle r|j_\Lambda^R\rangle \int_r^\infty dr'\, r'^2 \langle h_\Lambda^L|r'\rangle V(r')\langle r'|J_\Lambda^R\rangle \quad (154)$$

The irregular solutions obey $|H^R\rangle = |h^R\rangle + G_0^+ V|H^R\rangle$ and

$$\int_0^\infty dr\, r^2 \langle h^L|r\rangle V(r)\langle r|H^R\rangle = 0 \quad (155)$$

where $|H^R\rangle$ and $|h^R\rangle$ are either of the first $(+)$ or of the second $(-)$ kind [81]. Analogously to the spherical Bessel and Hankel functions, the regular and irregular solutions are related by $|J^R\rangle = (|H^{R(+)}\rangle + |H^{R(-)}\rangle)/2$.

To determine the single-site Green function $G^+(\mathbf{r}, \mathbf{r}')$, one starts from the expression for G^+ in terms of eigenfunctions of the Hamiltonian,

$$G^+(\mathbf{r}, \mathbf{r}') = \int \frac{dk^3}{(2\pi)^3} \frac{\langle\mathbf{r}|\Psi(\mathbf{k}, E)\rangle\langle\Psi(\mathbf{k}, E)|\mathbf{r}'\rangle}{E + i\eta - k^2} \quad (156)$$

and obtains with the regular solutions after integration over angles in reciprocal space

$$G^+(\mathbf{r}, \mathbf{r}') = \frac{2}{\pi} \sum_\Lambda \int_0^\infty dk\, \frac{\langle\mathbf{r}|J_\Lambda^R(k, E)\rangle\langle J_\Lambda^L(k, E)|\mathbf{r}'\rangle}{E + i\eta - k^2} \quad (157)$$

Because the integrand is even in k, the integration can be extended to the interval $[-\infty, +\infty]$, allowing the integral to be treated as a contour integral. Defining $\alpha = p + i\eta/(2p)$, $p = \sqrt{E}$, and noticing that $\eta = 0^+$, one has $\alpha^2 = E + i\eta$. Thus, the preceding integral contains the factor

$$\frac{k^2}{\alpha^2 + k^2} = -\frac{1}{2}\left(\frac{2k}{k-\alpha} + \underbrace{\frac{k}{k+\alpha} - \frac{k}{k-\alpha}}\right) \quad (158)$$

(k^2 in the numerator on the l.h.s. is due to the path element). The underbraced term is odd in k and thus gives no contribution to the integral. Therefore, one is left with

$$G^+(\mathbf{r}, \mathbf{r}') = \frac{1}{\pi} \sum_\Lambda \int_{-\infty}^\infty dk\, \frac{\langle\mathbf{r}|J_\Lambda^R(k, E)\rangle k \langle J_\Lambda^L(k, E)|\mathbf{r}'\rangle}{k - \alpha} \quad (159)$$

Replacing $\langle J^L(k, E)|\mathbf{r}'\rangle$ by the sum of $\langle H^{L(+)}(k, E)|\mathbf{r}'\rangle$ and $\langle H^{L(-)}(k, E)|\mathbf{r}'\rangle$, the integral can be performed. Taking the limit $\lim_{\eta\to 0^+}$ eventually yields

$$G^+(\mathbf{r}, \mathbf{r}') = -ik \sum_\Lambda \langle\mathbf{r}_<|J_\Lambda^R(k, E)\rangle\langle H_\Lambda^{L(+)}(k, E)|\mathbf{r}_>\rangle \quad (160)$$

Alternative representations of the single-site Green function follow immediately from operator equations for G. For example, $G = G^0 + G^0 T G^0$ yields [72]

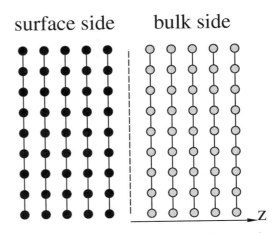

Fig. 31. Arrangement of layers for the calculation of the empty-layer Green function. An empty layer (dashed line) is embedded between stacks of layers (left, surface side, black circles; right, bulk side, gray circles).

$$G^+(\mathbf{r}, \mathbf{r}') = -ik \sum_\Lambda \langle\mathbf{r}_<|j_\Lambda^R(k, E)\rangle\langle h_\Lambda^{L(+)}(k, E)|\mathbf{r}_>\rangle$$
$$- k^2 \sum_{\Lambda\Lambda'} \langle\mathbf{r}_<|h_\Lambda^R(k, E)\rangle t_{\Lambda\Lambda'}$$
$$\times \langle h_{\Lambda'}^{L(+)}(k, E)|\mathbf{r}_>\rangle \quad (161)$$

Empty-Layer Green Function. Now consider a solid buildup by layers from which one layer of scatterers is removed (cf. Fig. 31). We will denote this layer with index i as "empty". For semi-infinite solids, one side belongs to the surface region, the other to the bulk region. Note that the following method can treat any kind of embedded layers, the embedding regions may consist of vacuum, films, or semiinfinite solids, respectively. For the Green function of the empty layer, one makes the ansatz

$$G_{EL}^+(\mathbf{r}_i, \mathbf{r}_i') = G_0^+(\mathbf{r}_i, \mathbf{r}_i') + \sum_{\Lambda\Lambda'} \langle\mathbf{r}_i|j_\Lambda^R\rangle D_{\Lambda\Lambda'}^{ii} \langle j_\Lambda^L|\mathbf{r}_i'\rangle \quad (162)$$

with arguments E and $\mathbf{k}_\parallel$ dropped. The coordinates are taken with respect to layer i, $\mathbf{r}_i = \mathbf{r} - \mathbf{R}_i$. G_0^+ is the retarded Green function of free space, Eq. (152). The coefficients $D_{\Lambda\Lambda'}^{ii}$ have to be determined by the boundary conditions: the Bloch condition parallel to the layers,

$$G_{EL}^+(\mathbf{r}_i + \mathbf{R}, \mathbf{r}_i') = \exp(i\mathbf{k}_\parallel \cdot \mathbf{R})\, G_{EL}^+(\mathbf{r}_i, \mathbf{r}_i') \quad (163)$$

($\mathbf{R}$ is a vector of the layer lattice) and correct reflection at both the surface and the bulk side. Thus, $D_{\Lambda\Lambda'}^{ii}$ is decomposed into $D_{\Lambda\Lambda'}^{ii} = A_{\Lambda\Lambda'}^{ii} + B_{\Lambda\Lambda'}^{ii}$. $A_{\Lambda\Lambda'}^{ii}$ is the structure constant of layer i, which obeys

$$A_{\Lambda\Lambda'}^{ii} = -ik\frac{E + c^2}{c^2} \sum_{\mathbf{R}\neq 0} G_{\Lambda\Lambda'}^{ii}(-\mathbf{R}) \exp(i\mathbf{k}_\parallel \cdot \mathbf{R}) \quad (164)$$

Further, one has $h_l^+(k|\mathbf{r} - \mathbf{R}|)\langle\widehat{\mathbf{r} - \mathbf{R}}|\chi_\Lambda\rangle = \sum_{\Lambda'} G_{\Lambda\Lambda'}^{ii}(-\mathbf{R}) \times j_{l'}(kr)\langle\widehat{\mathbf{r}}|\chi_{\Lambda'}\rangle$.

To determine the coefficients $B^{ii}_{\Lambda\Lambda'}$, it is convenient to switch from angular-momentum $(\kappa\mu)$ to plane wave representation $(\mathbf{g}\tau)$, which yields

$$B^{ii}_{\Lambda\Lambda'} = \sum_{s,s'=\pm} \sum_{\mathbf{g}\tau} \sum_{\mathbf{g}'\tau'} a_{\Lambda\tau}(\mathbf{k}^s_\mathbf{g}) W^{ss'}_{\mathbf{g}\tau,\mathbf{g}'\tau'} \left(a_{\Lambda'\tau'}(\mathbf{k}^{s'}_{\mathbf{g}'})\right)^* \quad (165)$$

Thus, the matrix B^{ii} is completely determined by the matrices $W^{\pm\pm}$. Considering the reflection at the boundaries of the empty layer, one obtains the following matrix equation:

$$\begin{pmatrix} -R^{+-} & 1 \\ 1 & -R^{-+} \end{pmatrix} \begin{pmatrix} W^{++} & W^{+-} \\ W^{-+} & W^{--} \end{pmatrix}$$
$$= \frac{1}{iA} \begin{pmatrix} R^{-+} & 0 \\ 0 & R^{+-} \end{pmatrix} \begin{pmatrix} K & 0 \\ 0 & K \end{pmatrix} \quad (166)$$

R^{+-} and R^{-+} are the reflection matrices at the surface side and the bulk side, respectively, which can easily be computed from the scattering matrices of the layers forming the surface and the bulk side stacks. Further, $K_{\mathbf{g}\tau,\mathbf{g}'\tau'} = \delta_{\mathbf{g}\tau,\mathbf{g}'\tau'}/k_{\mathbf{g}\perp}$. Eventually, one obtains

$$W^{++} = R^{+-}W^{-+} \quad (167a)$$

$$W^{+-} = \frac{1}{iF_A}(1 - R^{+-}R^{-+})^{-1}R^{+-}K \quad (167b)$$

$$W^{-+} = \frac{1}{iF_A}R^{-+}(1 - R^{+-}R^{-+})^{-1}K \quad (167c)$$

$$W^{--} = R^{-+}W^{+-} \quad (167d)$$

Note that for a standalone layer, that is, a layer without vacuum embedding on both sides, $R^{+-} = R^{-+} = 0$, which leads to $W^{\pm\pm} = 0$, and the matrix D is given by the structure constant A alone.

It is illustrative to write the empty-layer Green function in terms of outgoing and incoming plane waves. With the definitions

$$\langle v^\pm_{\mathbf{g}\tau}|\mathbf{r}'_i\rangle = \frac{1}{iF_A k_\perp}\langle f^L_{\mathbf{k}^\pm_\mathbf{g}\tau}|\mathbf{r}'_i\rangle$$
$$+ \sum_{s'=\pm}\sum_{\mathbf{g}'\tau'} W^{\pm s'}_{\mathbf{g}\tau,\mathbf{g}'\tau'}\langle f^L_{\mathbf{k}^{s'}_{\mathbf{g}'}\tau'}|\mathbf{r}'_i\rangle \quad (168a)$$

$$\langle \tilde{v}^\pm_{\mathbf{g}\tau}|\mathbf{r}'_i\rangle = \sum_{s'=\pm}\sum_{\mathbf{g}'\tau'} W^{\pm s'}_{\mathbf{g}\tau,\mathbf{g}'\tau'}\langle f^L_{\mathbf{k}^{s'}_{\mathbf{g}'}\tau'}|\mathbf{r}'_i\rangle \quad (168b)$$

one has in matrix form

$$\begin{pmatrix} \mathbf{v}^+ \\ \mathbf{v}^- \end{pmatrix} = \begin{pmatrix} \dfrac{1}{iF_A}K + W^{++} & W^{+-} \\ W^{-+} & \dfrac{1}{iF_A}K + W^{--} \end{pmatrix}\begin{pmatrix} \mathbf{f}^{L+} \\ \mathbf{f}^{L-} \end{pmatrix} \quad (169a)$$

$$\begin{pmatrix} \tilde{\mathbf{v}}^+ \\ \tilde{\mathbf{v}}^- \end{pmatrix} = \begin{pmatrix} W^{++} & W^{+-} \\ W^{-+} & W^{--} \end{pmatrix}\begin{pmatrix} \mathbf{f}^{L+} \\ \mathbf{f}^{L-} \end{pmatrix} \quad (169b)$$

and the empty-layer Green function is given by

$$G^+_{EL}(\mathbf{r}_i,\mathbf{r}'_i) = \sum_{\mathbf{g}\tau}\langle\mathbf{r}_i|f^R_{\mathbf{k}^\pm_\mathbf{g}\tau}\rangle\langle v^\pm_{\mathbf{g}\tau}|\mathbf{r}'_i\rangle + \sum_{\mathbf{g}\tau}\langle\mathbf{r}_i|f^R_{\mathbf{k}^\mp_\mathbf{g}\tau}\rangle\langle\tilde{v}^\mp_{\mathbf{g}\tau}|\mathbf{r}'_i\rangle$$
$$(170)$$

where the upper (lower) signs are for the case $z > z'$ $(z < z')$. According to this decomposition, the Green function can be interpreted as a propagator: From $\mathbf{r}'_i$ there are outgoing plane waves into the $+z$ direction ($\mathbf{v}^+$ and $\tilde{\mathbf{v}}^+$) and into the $-z$ direction ($\mathbf{v}^-$ and $\tilde{\mathbf{v}}^-$) that are collected at $\mathbf{r}_i$ via $\mathbf{f}^R$. This interpretation is quite helpful in obtaining the interlayer part of the Green function.

Full-Layer Green Function. For the layer-diagonal Green function for the full layer i, we make the ansatz[15]

$$G(\mathbf{r}_i,\mathbf{r}'_i) = G^+(\mathbf{r}_i,\mathbf{r}'_i) + \sum_{\Lambda\Lambda'}\langle\mathbf{r}_i|J^{iR}_\Lambda\rangle U^{ii}_{\Lambda\Lambda'}\langle J^{iL}_{\Lambda'}|\mathbf{r}'_i\rangle \quad (171)$$

with the single-site Green function G^+ for a site at layer i, Eq. (160). The regular solutions of the single-site problem fulfill the Lippmann–Schwinger equation for the site potential $V_i(\mathbf{r})$,

$$\langle\mathbf{r}_i|J^{iR}_\Lambda\rangle = \langle\mathbf{r}_i|j^R_\Lambda\rangle + \int_{\Omega_i} G^+_0(\mathbf{r}_i,\mathbf{r}'_i)V_i(\mathbf{r}'_i)\langle\mathbf{r}'_i|J^{iR}_\Lambda\rangle dr'^3_i \quad (172)$$

Thus, the single-site t matrix is given by

$$t^{iR}_{\Lambda\Lambda'} = -ik\int_{\Omega_i}\langle j^L_\Lambda|\mathbf{r}_i\rangle V_i(\mathbf{r}_i)\langle\mathbf{r}_i|J^{iR}_{\Lambda'}\rangle dr^3_i \quad (173)$$

Further, $t^{iL}_{\Lambda'\Lambda} = t^{iR}_{\Lambda\Lambda'}$. The single-site solutions show the asymptotics as in Eq. (88). The coefficients $U^{ii}_{\Lambda\Lambda'}$ are determined by the Dyson equation for G,

$$G(\mathbf{r},\mathbf{r}') = G_{EL}(\mathbf{r},\mathbf{r}') + \int_\Omega G_{EL}(\mathbf{r},\mathbf{r}'')V(\mathbf{r}'')G(\mathbf{r}'',\mathbf{r}')dr''^3$$
$$(174)$$

Using the asymptotics for the single-site solutions, one obtains after some manipulation an equation for $U^{ii}_{\Lambda\Lambda'}$, which in matrix form reads

$$U^{ii} = \left(1 - \frac{i}{k}D^{ii}t^{iR}\right)^{-1}D^{ii} \quad (175)$$

where the indices of U^{ii}, D^{ii}, and t^{iR} run over all Λ.

Now we turn to the calculation of the non-layer-diagonal parts of the Green function, that is, the matrices U^{ij} for $i \neq j$. As we have seen, the empty-layer Green function can be interpreted in terms of outgoing and incoming plane waves, Eq. (170). Now the incoming waves do not belong to the empty layer (with index j) but to the full layer (with index i); see Figure 32. The first task is to find the transfer matrix from layer j to layer i. Therefore, the reflection matrix R^{+-}_{surf} at the surface side, the scattering matrix M_{slab} of the layers sandwiched between layers j and i, and the reflection matrix R^{-+}_{bulk} at the bulk side have to be computed according to the methods presented in Section 3. Then the matrix U^{ij} can be computed according to the following scheme:

(i) The wave fields outgoing from layer j in the $\pm z$ direction are given by $\mathbf{v}^\pm_j$ in the plane wave representation, Eq. (169). These have to be

---

[15] Note that the ansatz shows the same structure as in Eq. (162).

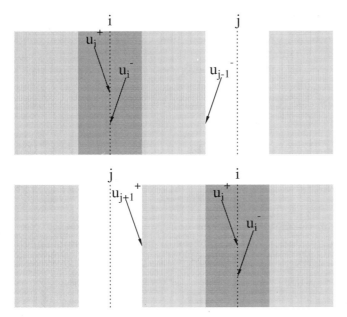

Fig. 32. Arrangement in the calculation of the interlayer part of the empty-layer Green function in the cases $i > j$ (top) and $i < j$ (bottom). The empty layer is labeled j (white), the full layer i (dark gray). Left (right) the surface (bulk) side is displayed. The stack is shown in between layers j and i. Arrows represent outgoing ($\mathbf{u}_{j+1}^+$, $\mathbf{u}_{j-1}^-$) and incoming ($\mathbf{u}_i^{\pm}$) wave functions.

propagated to the boundaries of the layer stack and give rise to $\mathbf{u}_{j+1}^+ = P^+ \mathbf{v}_j^+$ and $\mathbf{u}_{j-1}^- = P^- \mathbf{v}_j^-$.

(ii) The wave field impinging on layer i, which is due to the wave field outgoing from layer j, has to be calculated. For the case $i > j$, that is, the empty layer on the surface side (cf. lower part of Fig. 32), one has

$$\mathbf{u}_i^+ = (1 - M_{\text{slab}}^{+-} R_{\text{bulk}}^{-+})^{-1} M_{\text{slab}}^{++} \mathbf{u}_{j+1}^+ \quad (176a)$$

$$\mathbf{u}_i^- = R_{\text{bulk}}^{-+}(1 - M_{\text{slab}}^{+-} R_{\text{bulk}}^{-+})^{-1} M_{\text{slab}}^{++} \mathbf{u}_{j+1}^+$$
$$= R_{\text{bulk}}^{-+} \mathbf{u}_i^+. \quad (176b)$$

For the case $i < j$, that is, the empty layer on the bulk side (cf. upper part of Fig. 32), analogously

$$\mathbf{u}_i^- = (1 - M_{\text{slab}}^{-+} R_{\text{surf}}^{+-})^{-1} M_{\text{slab}}^{--} \mathbf{u}_{j-1}^- \quad (177a)$$

$$\mathbf{u}_i^+ = R_{\text{surf}}^{+-}(1 - M_{\text{slab}}^{-+} R_{\text{surf}}^{+-})^{-1} M_{\text{slab}}^{--} \mathbf{u}_{j-1}^-$$
$$= R_{\text{surf}}^{+-} \mathbf{u}_i^+. \quad (177b)$$

(iii) To express the incoming wave field in terms of regular solutions of layer i, one has to multiply $\mathbf{u}_i^{\pm}$ by $(1 - i A^{ii} t^i / k)^{-1}$, where A^{ii} and t^i are the structure constant and the single-site t matrix of layer i, respectively.

(iv) To obtain the "full layer" j, one has to solve the Dyson equation. Applying the similar considerations as for the layer-diagonal part, $\mathbf{v}_j^{\pm}$ has to be multiplied by $-i(1 + i t^j U^{jj} / k)$. In summary, the Green

function is given by

$$G(\mathbf{r}_i, \mathbf{r}_j') = G^+(\mathbf{r}_i, \mathbf{r}_i')\delta_{ij} + \sum_{\Lambda \Lambda'} \langle \mathbf{r}_i | J_\Lambda^{iR} \rangle U_{\Lambda \Lambda'}^{ij} \langle J_{\Lambda'}^{jL} | \mathbf{r}_j' \rangle \quad (178)$$

Scattering-Path Operator and Scattering Solutions. Certain equations of multiple-scattering theory become nicer when they are not formulated in terms of regular and irregular solutions of the single-site problem but in terms of scattering solutions; for example, the matrices U^{mn} are replaced by the scattering-path operators (SPOs) τ^{mn}. The latter are the on-the-energy-shell matrices of the transition operator T. Because the derivation of τ^{mn} can be found in several textbooks (see, e.g., [70–72]), we just recall some basic properties.

The SPO τ^{mn} from site n to site m obeys $\tau^{mn} = t^{mR}\delta_{mn} + t^{mR} \sum_{k \neq m} G^{mk} \tau^{kn}$, where G^{mk} is the structure constant. The matrices U and τ are closely related,

$$U^{mn} = (t^{mR})^{-1}(\tau^{mn} - t^{mR}\delta_{mn})(t^{nL})^{-1} \quad (179)$$

Consider, for example, the Green function that propagates an electron from a site $\mathbf{R}_n$ to a site $\mathbf{R}_m$, $m \neq n$. Taking $\mathbf{r}_n$ and $\mathbf{r}_m$ outside the muffin-tin sphere and using the asymptotic behavior of $|J^{nR}\rangle$ and $\langle J^{mL}|$, cf. Eq. (88), one arrives at

$$G(\mathbf{r}_n, \mathbf{r}_m) = \langle \mathbf{r}_n | j^R (t^{mR})^{-1} + h^R \rangle \tau^{mn} \langle j^L (t^{nL})^{-1} + h^L | \mathbf{r}_m \rangle \quad (180)$$

The preceding equation establishes the regular scattering solutions [169]. Within the muffin-tin sphere, they are given by

$$|Z_\Lambda^R \rangle = \sum_{\Lambda'} |J_{\Lambda'}^R\rangle (t^R)_{\Lambda' \Lambda}^{-1} \quad (181a)$$

$$\langle Z_\Lambda^L | = \sum_{\Lambda'} (t^L)_{\Lambda \Lambda'}^{-1} \langle J_{\Lambda'}^L | \quad (181b)$$

Analogously, the Green function at site $\mathbf{R}_n$ can be written as

$$G(\mathbf{r}_n, \mathbf{r}_n') = \langle \mathbf{r}_n | Z^{nR} \rangle \tau^{nn} \langle Z^{nL} | \mathbf{r}_n' \rangle - ik \langle \mathbf{r}_n | Z^{nR} \rangle \langle \tilde{J}^{nL} | \mathbf{r}_n' \rangle \quad (182)$$

where $|\tilde{J}\rangle$ is an irregular scattering solution. The main properties of the single-site solutions are given in Table III.

Screened-KKR Methods. The Green function G of a system can be obtained from the Dyson equation with respect to the free-space Green function G_0, $G = G_0 + G_0 t G$, where t is the single-scattering matrix. The SPO τ is implicitly defined by $G = G_0 + G_0 \tau G_0$, which leads to $\tau = (t^{-1} - G_0)^{-1}$ or $G = t^{-1} \tau t^{-1} - t^{-1}$.

The preceding formalism is based on free space as a reference system but can, in fact, rely on any other reference system [170]. The Green function G_r of the reference system is given by $G_r = G_0(1 - t_r G_0)^{-1}$. Now G can be expressed in terms of G_r as $G = G_r(1 - \Delta t\, G_r)^{-1}$ with $\Delta t = t - t_r$. For the SPO, one obtains $\tau_\Delta = [(\Delta t)^{-1} - G_r]^{-1}$. Now, one can ask for a reference system that (i) can easily be computed and (ii) allows for a rapid computation of the Green function G by means of $G = (\Delta t)^{-1} \tau_\Delta (\Delta t)^{-1} - (\Delta t)^{-1}$. One answer is the screened-KKR method, that is, the transformation of the KKR equations into a tight-binding (TB) form.[16] In TB calculations, one exploits that the interaction integrals, which describe the hopping of the electron from one site to another, decay rapidly in configuration space, so that only a few nearest neighbor shells have to be taken into account. This approximation allows for the use of a variety of computational methods to obtain the Green function. Of practical interest is the renormalization scheme or, in the case of disordered systems, the recursion method [171].

Within the screened-KKR method, one uses a reference system that allows a fast computation of τ_Δ. Because Δt is site diagonal, a perfectly suited G_r should also be site diagonal. This, however, cannot be the case in real solid systems. Thus, one tries to construct a reference Green function, that decays as fast as possible in, configuration space. One way is to follow Anderson who invented the screening method and to exploit the scaling properties of the preceding equations in order to construct a "most screened" set of basis functions [172]. Another, more practical way builds the reference system by repulsive ($V_0 > 0$) well-shaped and spherically symmetric potentials of muffin-tin form [173],

$$V(\mathbf{r}) = \begin{cases} V_0 & r < r_{\mathrm{mt}} \\ 0 & \text{otherwise} \end{cases} \qquad (183)$$

the t matrix of which can easily be calculated analytically. The screening, that is, the exponential decay of the SPO in real space, can be tuned by the height of the potential wells. Further, in order to compute the reference Green function, one can employ the translational properties of the reference system, in particular, the Bloch property.

Photoemission Final State. The final state in photoemission, the time-reversed LEED state, can be computed either directly via the reflection and transmission matrices of the layers [136] or within the Green function formalism presented here. The LEED state $\Phi_\tau^{\mathrm{LEED}}(\mathbf{r}; E, \mathbf{k})$ at energy E, wave vector $\mathbf{k}$, and spin τ fulfills the Lippmann–Schwinger equation

$$\Phi_\tau^{\mathrm{LEED}}(\mathbf{r}; E, \mathbf{k}) = \langle \mathbf{r}|f_{\mathbf{k}\tau}^R\rangle + \int d^3r' G(\mathbf{r}, \mathbf{r}'; E) V(\mathbf{r}')\langle \mathbf{r}'|f_{\mathbf{k}\tau}^R\rangle \qquad (184)$$

that is, the relativistic version of Eq. (137), with the free-space solution $|f_{\mathbf{k}\tau}^R\rangle$ and the potential in the solid V. Instead of

using the plane wave representation, we turn again to angular-momentum representation. With Eqs. (147) and (149) for regular solutions ($z = j$), the LEED state at site $\mathbf{R}_n$ is given by

$$\Phi_\tau^{\mathrm{LEED}}(\mathbf{r}_n; E, \mathbf{k})$$
$$= \exp(i\mathbf{k}\cdot\mathbf{R}_n)\sum_\Lambda \langle \mathbf{r}_n|j_\Lambda^R\rangle a_{\Lambda\tau}(\widehat{\mathbf{k}})$$
$$+ \sum_m \exp(i\mathbf{k}\cdot\mathbf{R}_m)\sum_\Lambda \int_{\Omega_m} d^3r'_m\, G(\mathbf{r}_n, \mathbf{r}'_m; E)$$
$$\times V(\mathbf{r}'_m)\langle \mathbf{r}'_m|j_\Lambda^R\rangle a_{\Lambda\tau}(\widehat{\mathbf{k}}) \qquad (185)$$

with G from Eq. (171). Inserting the interlayer contribution to the Green function, one arrives at a term

$$\sum_m \exp(i\mathbf{k}\cdot\mathbf{R}_m)\sum_{\Lambda\Lambda'\Lambda''}\langle \mathbf{r}_n|J_{\Lambda'}^R\rangle U_{\Lambda'\Lambda''}^{nm}$$
$$\times \underbrace{\int_{\Omega_m} d^3r'_m \langle J_{\Lambda''}^L|\mathbf{r}'_m\rangle V(\mathbf{r}'_m)\langle \mathbf{r}'_m|j_\Lambda^R\rangle}_{=t_{\Lambda''\Lambda}^L/(-ik)} a_{\Lambda\tau}(\widehat{\mathbf{k}})$$
$$= \frac{i}{k}\sum_m \exp(i\mathbf{k}\cdot\mathbf{R}_m)\sum_{\Lambda\Lambda'\Lambda''}\langle \mathbf{r}_n|J_{\Lambda'}^R\rangle U_{\Lambda'\Lambda''}^{nm} t_{\Lambda''\Lambda}^L a_{\Lambda\tau}(\widehat{\mathbf{k}}) \qquad (186)$$

For the evaluation of the remaining single-site terms, one exploits the Lippmann–Schwinger equation (172) for the regular solutions. Thus, these terms reduce to $|J_\Lambda^{nR}\rangle$. In summary, the LEED state at site $\mathbf{R}_n$ is given by

$$\Phi_\tau^{\mathrm{LEED}}(\mathbf{r}_n; E, \mathbf{k})$$
$$= \exp(i\mathbf{k}\cdot\mathbf{R}_n)\sum_\Lambda\langle \mathbf{r}_n|J_\Lambda^R\rangle a_{\Lambda\tau}(\widehat{\mathbf{k}}) + \frac{i}{k}\sum_m \exp(i\mathbf{k}\cdot\mathbf{R}_m)$$
$$\times \sum_{\Lambda\Lambda'\Lambda''}\langle \mathbf{r}_n|J_{\Lambda'}^R\rangle U_{\Lambda'\Lambda''}^{nm} t_{\Lambda''\Lambda}^L a_{\Lambda\tau}(\widehat{\mathbf{k}}) \qquad (187)$$

A slightly more compact form is obtained using the SPO,

$$\Phi_\tau^{\mathrm{LEED}}(\mathbf{r}_n; E, \mathbf{k}) = \frac{i}{k}\sum_m \exp(i\mathbf{k}\cdot\mathbf{R}_m)\sum_{\Lambda\Lambda'}\langle \mathbf{r}_n|Z_\Lambda^R\rangle \tau_{\Lambda\Lambda'}^{nm} a_{\Lambda\tau}(\widehat{\mathbf{k}}) \qquad (188)$$

It is worth mentioning that a construction of the final state in terms of Bloch states has been given by Bross [174] (see also [69]).

Transition-Matrix Elements and Photocurrent. A photoemission theory is incomplete as long as the transition-matrix elements are not taken into account. The interaction of an electron with incoming monochromatic light of frequency ω and with wave vector $\mathbf{q}$ is relativistically described by the Hamiltonian

$$H'(\mathbf{r}, t) = \alpha \cdot \mathbf{A}(\mathbf{r}, t) = \alpha \cdot \mathbf{A}_0 \exp(i\mathbf{q}\cdot\mathbf{r} - \omega t) \qquad (189)$$

Upper and lower components of the Dirac spinors are coupled via α, Eq. (79). Usually, one decomposes $\mathbf{A} = (A_x, A_y, A_z)$ into (A_+, A_-, A_z) with $A_\pm = A_x \pm i A_y$, that is, into contribution from left- and right-handed circularly polarized light as well as linearly polarized light.

---

[16]Therefore, screened methods are sometimes loosely denoted as TB methods (TB-KKR, TB-LMTO, where LMTO is short for linearized muffin-tin orbital).

The transition-matrix elements between the Dirac spinors at a particular site are integrals over the muffin-tin sphere. Integration over angles, that is, matrix elements of $\alpha \cdot \mathbf{A}$ of the central-field spinors $|\chi_\Lambda\rangle$, yield the well-known "atomic" selection rules [175]: The angular momentum l has to differ by 1, $\Delta l = \pm 1$, and its z projection m is conserved for linearly polarized light, $\Delta m = 0$, or for circularly polarized light it is changed by 1, $\Delta m = \pm 1$. The integration over the radial part has to be computed numerically. The Green function of the initial state consists of both regular and irregular solutions, the time-reversed LEED state of regular solutions only. Therefore, one has to consider two general types of matrix elements. For those between regular solutions, one has

$$M^{(1)}_{i\Lambda'\Lambda} = \int_0^{R_{\mathrm{mt}}} \langle J_{\Lambda'}|r\rangle \alpha_i A_i \langle r|J_\Lambda\rangle dr \qquad i = \pm, z \quad (190)$$

The single-site term of the Green function gives rise to a double integral, $M^{(2)}_{i\Lambda'\Lambda}$, due to the selection of regular and irregular wave functions with respect to r; see Eq. (160). In the preceding definition, we have suppressed the indices L and R, which are due to the occurrence of left and right wave functions.

Eventually, the spin-density matrix ρ of the photoelectron is given by [cf. Eq. (138)]

$$\rho_{\tau\tau'} \sim \langle \Phi_{0\tau}|G^a(E+\omega)\Delta \operatorname{Im} G^+(E)\Delta^\dagger G^r(E+\omega)|\Phi_{0\tau'}\rangle$$
$$\tau, \tau' = \pm \qquad (191)$$

which, besides the computation of the photocurrent $I = \operatorname{tr}(\rho)$, allows for computation of the spin polarization of the photoelectrons $\mathbf{P} = \operatorname{tr}(\sigma\rho)/\operatorname{tr}(\rho)$. Note that due to SOC the spin polarization can be nonzero even from nonmagnetic solids (see, e.g., [150]).

Because the final state is a time-reversed LEED state, it shows the same symmetry properties as a LEED state; cf. Section 3.2.2. This, together with knowledge of the light polarization, allows for a detailed group-theoretical analysis of photoemission from (ferromagnetic) surfaces [58, 176]. In particular, it reveals the initial states for which dipole transitions are allowed or forbidden (see also [97]).

A popular approximation for analyzing experimental spectra is the direct-transition model. If the initial states are taken as Bloch states, for example, by neglecting the surface of the sample, the normal component $k_\perp$ of the wave vector is conserved in the transition process. This allows for a mapping of the intensity maxima to the band structure $k_\perp(E, \mathbf{k}_\parallel)$.

Before turning to applications of the multiple-scattering theory of photoemission, basic properties of photoemission from ultra-thin films should be addressed.

4.2.4. Simple Theory of Photoemission from Ultra-Thin Films

In recent experimental work on photoemission from films with thicknesses of a few layers, the photoemission intensity maxima show dispersion, as in the bulk case, and can be well described within the direct-transition model using bulk-band structures.

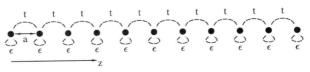

Fig. 33. Linear chain along the z axis with $n = 11$ equidistant sites with intersite distance a. The tight-binding parameters, cf. Eq. (192), are the on-site energy ϵ and the next nearest neighbor hopping energy t, visualized by dashed lines.

This led to the conclusion that even ultra-thin films "show a (bulk) band structure" [177, 178]. This behavior can be understood within a simple theory of photoemission from linear chains [179], which will be presented in the following (for an approach based on Green functions, see [180]).

There are two limits in which valence electrons in a film can be described (see Section 2 and, e.g., [21, 181–184]): (i) In a plane wave representation, free electrons can be confined to a quantum well. (ii) In a tight-binding description, electrons are allowed to hop only within a finite number of layers. The prototypical realization of the latter model is rare-gas layers on a (metal) substrate (cf. Section 2.2.3).

The film is represented by a linear chain oriented along the z axis, that is, perpendicular to the surface, with n equidistant sites i, $i = 1, \ldots, n$, with one orbital $|\Phi_i\rangle$ per site. The latter is located at $ia\mathbf{e}_z$, a denoting the intersite distance (Fig. 33) [185–188]. Further, the overlap between the normalized orbitals located on different sites is assumed to be 0, $\langle \Phi_i|\Phi_j\rangle = \delta_{ij}$. Neglecting the substrate completely, the elements of the Hamiltonian matrix $H^{(n)}$ read

$$H^{(n)}_{ij} = \epsilon\delta_{ij} + t\delta_{|i-j|,1} \qquad i, j = 1, \ldots, n \quad (192)$$

with on-site energies $\epsilon = \langle \Phi_i|H^{(n)}|\Phi_i\rangle$ and next nearest neighbor hopping energies $t = \langle \Phi_i | H^{(n)} | \Phi_{i+1}\rangle$. The eigenvalues of $H^{(n)}$ can be written as

$$\lambda^{(n)}_i = \epsilon + 2t\cos\bigl(k^{(n)}_i a\bigr) \qquad i = 1, \ldots, n \quad (193)$$

with $k^{(n)}_i = \pi i/[a(n+1)]$. In the case of a single site where there is obviously no hopping, $\lambda^{(1)}_1 = \epsilon$. For an infinite chain (i.e., in the limit $n \to \infty$), $k^{(n)}_i a$ is dense in $[0, \pi]$ and, thus, the eigenvalues represent the bulk-band structure $E(k) = \epsilon + 2t\cos(ka)$; cf. Figure 34. An eigenfunction $|\Psi^{(n)}_i\rangle$ of $H^{(n)}$ with energy $\lambda^{(n)}_i$ can be written as

$$|\Psi^{(n)}_i\rangle = \sum_{j=1}^n c^{(n)}_{ij}|\Phi_j\rangle \qquad i = 1, \ldots, n \quad (194)$$

the coefficients $c^{(n)}_{ij}$ of which can be calculated iteratively by

$$c^{(n)}_{ij} = 2\cos\bigl(k^{(n)}_i a\bigr)c^{(n)}_{i,j-1} - c^{(n)}_{i,j-2} \qquad j = 2, \ldots, n \quad (195)$$

with $c^{(n)}_{i0} = 0$ and $c^{(n)}_{i1} = 1$. The additional relation $2\cos(k^{(n)}_i a)c^{(n)}_{i,n} = c^{(n)}_{i,n-1}$ ensures that $k^{(n)}_i$ has to be chosen properly. It can easily be shown that $c^{(n)}_{ij} \sim \sin(\pi i j/a(n+1))$. Strictly speaking, $k^{(n)}_i$ is not a wave number as it is in the case

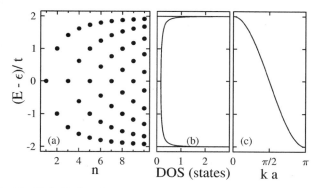

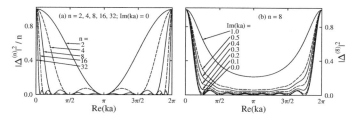

Fig. 35. The "k-conservation" function $|\Delta^{(n)}|^2$, as defined in Eq. (198). (a) Dependence on the number of sites n in the linear chain. n ranges from 2 to 32, as indicated, Im$(ka) = 0$. (b) Dependence on Im(ka). Im(ka) ranges from 0.0 to 1.0, $n = 8$.

Fig. 34. Tight-binding electronic structures of linear chains. The tight-binding parameters are ϵ and t; cf. Eq. (192). (a) Eigenenergies $\lambda_i^{(n)}$ [dots; cf. Eq. (193)] of chains with $n = 1, \ldots, 10$ sites. (b) Density of states (DOS) of the infinite chain. (c) Band structure $E(k) = \epsilon + 2t \cos(ka)$ of the infinite chain. Reprinted with permission from J. Henk and B. Johansson, *J. Electron Spectrosc. Relat. Phenom.* 105, 187 (1999). Copyright 1999, by Elsevier Science.

of Bloch states because there is no translational symmetry and therefore no periodicity. Due to the inversion symmetry, the eigenstates $\Psi_i^{(n)}$ show the expected even–odd alternation [cf. Eq. (19)], and the number of nodes in the wave function increases with $|k_i^{(n)}|$.

The photoelectron state $|\Psi_f\rangle$ can crudely be approximated by a single plane wave, $\langle \mathbf{r}|\Psi_f\rangle = \exp(i\mathbf{k}_f \cdot \mathbf{r})$, as is often done in the interpretation of experimental data. This way, quantum-size effects in the upper band structure that show up, for example, in LEED (cf. Section 3.3) are ignored. The wave vector $\mathbf{k}_f$ is determined by both the position of the detector and the energy of the photoelectron, $E_f \sim k_f^2$. The photocurrent $I_i^{(n)}$ at photon energy ω from the initial state $|\Psi_i^{(n)}\rangle$ is given by Fermi's golden rule, Eq. (141). Inserting the previous expressions for the wave functions and defining the Fourier-transformed atomic wave function $F(\mathbf{k})$ by $F(\mathbf{k}) = \int \Phi^*(\mathbf{r}) \exp(i\mathbf{k}\cdot\mathbf{r}) d^3r$, one obtains eventually

$$I_i^{(n)} \propto |\mathbf{E}\cdot\mathbf{k}_f|^2 |F(\mathbf{k}_f)|^2 \left|\Delta_i^{(n)}(k_{f\perp})\right|^2 \delta\left(E_f - \omega - \lambda_i^{(n)}\right) \quad (196)$$

The function $\Delta_i^{(n)}(k)$, which is defined by

$$\Delta_i^{(n)}(k) = \sum_{j=1}^{n} c_{ij}^{(n)} \exp(ikja) \quad (197)$$

determines considerably the dependence of the photocurrent on the photon energy and, thus, should be discussed in more detail. Obviously, $\Delta_i^{(n)}$ is periodic with period $2\pi/a$. In the case of a single site, $n = 1$, $|\Delta^{(1)}(k)| = 1$ and the photon energy dependence of the photocurrent is determined solely by $F(\mathbf{k}_f)$. In the case of an infinite chain, $n \to \infty$, strict wave vector conservation, $\Delta^{(\infty)}(k_{f\perp}) = \delta(k_{f\perp} - k)$, is obtained; that is, the direct-transition model is recovered.

Setting all $c_{ij}^{(n)} = 1$ leads to a geometrical series for $\Delta^{(n)}$,

$$\Delta^{(n)}(k) = \begin{cases} n & \text{for } k = 0 \\ (q^n - 1)/(q - 1) & \text{otherwise} \end{cases} \quad q = \exp(ika) \quad (198)$$

Obviously, $\Delta^{(n)}$ shows $n - 1$ zeros in $[0, 2\pi]$ at $ka = 2\pi i/n$ with $i = 1, \ldots, n - 1$. Its absolute value increases with n in the vicinity of $k = 0$, whereas it decreases in the interior of the interval $[0, 2\pi]$. In short, $\Delta^{(n)}$ is an approximation of Dirac's δ function [189] (cf. Fig. 35). The main photoemission intensity comes from the region around $k = 0$ (i.e., $k_{f\perp} = k_i^{(n)}$), but additional intensity maxima, which are due to the oscillatory behavior of $\Delta^{(n)}$, should occur.

So far, we have considered only the case of infinite lifetime of the photoelectron. Introducing a finite lifetime leads to a complex wave number [53], which results in an additional weakening of the k conservation, as is also shown in Figure 35. $\Delta^{(n)}$ decreases rapidly around $k = 0$ with increasing Im(ka) (as is evident from the geometrical series), but the oscillatory behavior is still visible, except for very strong damping, for example, Im$(ka) = 0.5$ in Figure 35b.

Photoemission from chains with length of 5 and 10 sites is compared in Figure 36. The intensities were obtained from Eq. (196) with $|\mathbf{E}\cdot\mathbf{k}_f|^2|F(\mathbf{k}_f)|^2$ set to 1, but $\Delta_i^{(n)}$ calculated with the coefficients $c_{ij}^{(n)}$ obtained from Eq. (195). At the bottom of each box, the initial-state band structure $E(k_{f\perp})$ is shown (note that $k_{f\perp}$ is related to the photon energy ω by $\mathbf{k}_f^2 \sim \lambda_i^{(n)} + \omega$). The individual photoemission intensities show main maxima at $k_i^{(n)}$, that is, $E(k_{f\perp}) = \lambda_i^{(n)}$. In other words, one obtains approximate $k_\perp$ conservation. However, the intensities show oscillatory behavior (cf. Fig. 35). Further, the main maxima for $n = 5$ (Fig. 36b) are much broader than those for $n = 10$ (Fig. 36a) due to the weakening of the k conservation for shorter chains.

The finite photoelectron lifetime can be modeled using complex energies [53], which leads to complex $k_{f\perp}$. Its effect is addressed for a 10-site chain in Figure 37. For a rather large lifetime [Im$(k_{f\perp}a) = 0.2$, in Fig. 37b], there are still oscillations with $k_{f\perp}$ in the photoemission intensities from the individual initial states. These become smeared out for decreasing lifetime [e.g., Im$(k_{f\perp}a) = 0.5$ in Fig. 37a]. However, the intensities follow the bulk-band structure in both cases. At a fixed photon energy or a fixed Re$(k_{f\perp})$, for example, Re$(k_{f\perp}) = 0$, the EDC becomes broader with increasing Im$(k_{f\perp})$, which is due to the smearing out of the individual maxima and not to the uncertainty in $k_{f\perp}$.

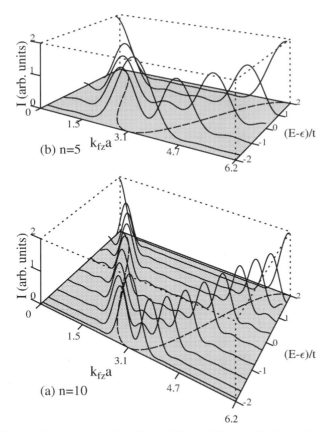

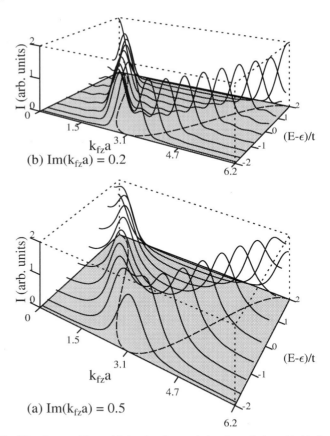

Fig. 36. Photoemission from linear chains with lengths 10 (a) and 5 sites (b), respectively. The intensity I is shown for each initial state at energy $\lambda_i^{(n)}$ [cf. Eq. (193)] for final-state wave numbers $k_{f\perp}$ ranging from 0 to $2\pi/a$ and $\mathrm{Im}(k_{f\perp}a) = 0$. The initial-state band structure $E(k) = \epsilon + 2t\cos(ka)$ (dashed) is shown at the bottom of each box. Intensities are scaled to the same maximum in each box. Reprinted with permission from J. Henk and B. Johansson, *J. Electron Spectrosc. Relat. Phenom.* 105, 187 (1999). Copyright 1999, by Elsevier Science.

Fig. 37. Same as Figure 36, but for photoemission from chains with 10 sites for $\mathrm{Im}(k_{f\perp}a) = 0.5$ (a) and $\mathrm{Im}(k_{f\perp}a) = 0.2$ (b), respectively. Reprinted with permission from J. Henk and B. Johansson, *J. Electron Spectrosc. Relat. Phenom.* 105, 187 (1999). Copyright 1999, by Elsevier Science.

In summary, photoemission from ultra-short chains shows the following properties: (i) The confinement of the valence electrons to the chain leads to a weakening of the wave number conservation: the shorter the chain, the broader the photoemission maxima in $k_{f\perp}$. (ii) Besides the periodicity with $2\pi/a$, individual photoemission intensities show oscillations with $k_{f\perp}$, the number of which is proportional to the chain length. These oscillations become smaller in intensity with decreasing photoelectron lifetime [increasing $\mathrm{Im}(k_{f\perp})$]. (iii) Even for very small lengths, the main maxima in the photoemission intensity follow the initial-state bulk-band structure, despite the fact that the initial-state energies are discrete (quantum-well states).

4.3. Applications

In the following, we focus on theoretical photoemission results for metallic films on metal substrates, which were obtained by multiple-scattering methods. Further, representative experimental data that show fingerprints of QWSs are presented.

4.3.1. Ultra-Thin Cu Films on fcc Co(100)

Hansen et al. performed photoemission experiments for ultrathin Cu films on fcc Co with an identical number of Cu layers but different crystallographic orientation of the substrate [190]. For 14 layers of Cu on fcc Co(111), they found a bulklike dispersion in the Cu sp states but three quantized states with fixed energy for fcc Co(100) and fcc Co(110) substrates (see Fig. 38). These findings were explained by the bulk-band structure of Co: Only in the latter two cases do bandgaps lead to confinement of the valence electrons to the Cu films and thus to QWSs. Further, it was observed that for the (100) and (110) films the photoemission intensities from the QWSs behave similarly to those of semi-infinite Cu(100) or Cu(110), respectively, which can also be understood by means of photoemission from linear chains (Section 4.2.4). A closer look at the intensity variations, however, gives hints that the maxima show more structure in their dependence on both the binding energy and the photon energy.

Ultra-thin Cu films on fcc Co(001) lend themselves support as prototypical systems because of extensive experimental and theoretical work. However, experimental [183, 191–194] and theoretical [195] investigations dealing with Cu/Co have focused mainly on the properties of QWSs as a function of the film thickness (e.g., binding energy and spin polarization). Usu-

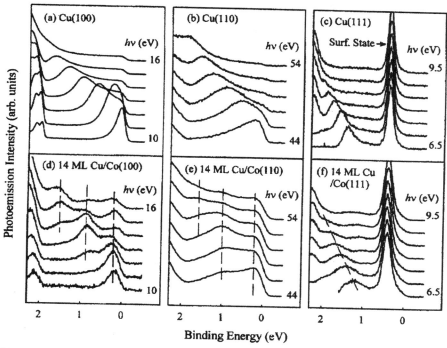

Fig. 38. Experimental normal-emission photoelectron spectra for single-crystal Cu surfaces and for Cu films 14 layers in thickness on Co substrates for (100) (left), (110) (middle), and (111) (right) orientation. Photon energies are evenly spaced between the lower and upper bounds indicated in each graph. The vertical dashed lines for Cu/Co(100) and Cu/Co(110) indicate the position of intensity maxima related to quantum-well states, while the dashed line for Cu/Co(111) indicates the dispersion of a peak related to emission from a Cu bulk band. Reprinted with permission from E. D. Hansen, T. Miller, and T.-C. Chiang, *J. Phys.: Condens. Matter* 9, L435 (1997). Copyright 1997, by the Institute of Physics.

ally, such analyses were performed at a fixed photon energy. In the following, we focus on a few film thicknesses but extend the analysis to variable photon energy in order to work out the manifestation of quantum-size effects in photoemission. Photoemission from Cu/Co(001) is analyzed by means of calculations within the one-step model of photoemission based on multiple-scattering theory (SPRLKKR), as presented in Section 4.2.3.

Quantum-Well States in Cu Films on fcc Co(001). Typical photoemission spectra from Cu/Co(001) for various thicknesses of the Cu films are shown in Figure 39. The intensity maxima are labeled by numbers that refer to the QWSs in the film (cf. Fig. 7). With increasing thickness, the maxima disperse to higher energies. At a fixed binding energy, for example, at the Fermi level E_F, the intensity is higher if a QWS crosses this energy than if there is no QWS at that particular energy. This gives rise to oscillations in the photoemission intensity at a fixed binding energy, as shown in the top panel of Figure 39 (and discussed later).

Before turning to the photoemission results, the electronic structure of Cu films on fcc Co(001) at $\overline{\Gamma}$, which is relevant for normal emission, $\mathbf{k}_\parallel = 0$, is briefly analyzed. The perpendicular component $k_\perp$ of the wave vector takes values from the direction $\Gamma - \Delta - X$ in the bulk Brillouin zone. The Cu *sp* band belongs to the double-group representation Δ_6; the related

wave functions show a prominent Δ^1 single-group (spatial) contribution (for applications of group theory in solid-state physics, see [59, 196]). To confine these electrons completely within the Cu film, the Co substrate has to have a gap in the Δ^1 bands. This is the case for minority electrons below -0.65 eV (light-gray area in Fig. 40), for majority electrons below -2.09 eV (dark-gray area in Fig. 40).

To distinguish among surface states, interface states, and QWSs, one calculates the layer-resolved Bloch spectral function (LDOS) for the whole Cu film and the subsequent Co layers. Surface and interface states, the energies of which may also lie in a bandgap of Co, are localized at the respective boundary (e.g., vacuum/Cu or Cu/Co). This means that the corresponding maxima in the LDOS decrease with distance from the boundary. Quantum-well states, however, show maxima in the whole Cu film but decreasing maxima in the Co substrate. The latter can be attributed to the gap in the bulk-band structure of Co because the QWSs cannot couple to Bloch states but to evanescent states in the Co substrate. Further, the energetic position of surface and interface states is expected not to depend significantly on the number of Cu layers, whereas QWSs should show the typical dispersion with film thickness (see Section 2 and Fig. 39).

The Bloch spectral function for a 14-ML Cu film shows two sharp maxima of minority spin character with energies -1.52 and -0.80 eV, respectively, which are denoted as QWSs *A*

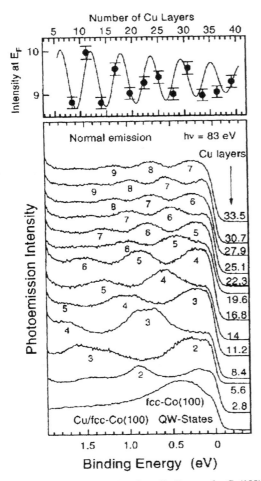

Fig. 39. Experimental photoemission from Cu films on fcc Co(100) for $\mathbf{k}_\parallel = 0$ and 83 eV photon energy. The lower panel shows intensity versus Cu film thickness (as indicated on the right of each spectrum). Intensity maxima related to quantum-well states are labeled by numbers (cf. also Fig. 7). The upper panel depicts intensity modulation at the Fermi level (0 eV) versus film thickness. Reprinted with permission from P. Segovia, E. G. Michel, and J. E. Ortega, *Phys. Rev. Lett.* 77, 3455 (1996). Copyright 1996, by the American Physical Society.

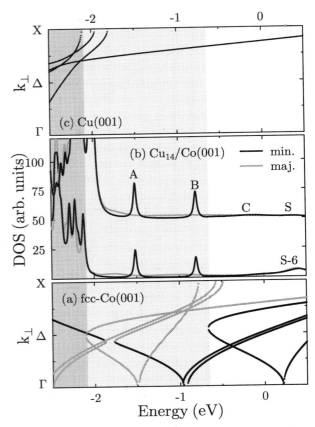

Fig. 40. Spin-resolved relativistic electronic structure of 14-ML Cu on fcc Co(001) for $\overline{\Gamma}$ ($\mathbf{k}_\parallel = 0$, $\Gamma-\Delta-X$ in the bulk Brillouin zone). (a) Band structure of fcc Co(001) along $\Gamma-\Delta-X$. The sliding gray scale of the bands indicates dominant majority (minority) spin orientation with black (light gray). (b) Density of states of 14-ML Cu on fcc Co(001) for the outermost (S) and a central ($S - 6$) layer with black (light gray) lines indicating minority (majority) spin orientation. (c) Same as (a), but for Cu(001). Gray areas indicate gaps in the Co band structure: dark gray for both majority and minority electrons, light gray for minority electrons with prominent Δ^1 spatial symmetry. The latter leads to confinement of minority electrons in the Cu film; see maxima A and B in panel (b). For C, see text. The Fermi energy is at 0 eV. Reprinted with permission from J. Henk and B. Johansson, *J. Electron Spectrosc. Relat. Phenom.* 105, 187 (1999). Copyright 1999, by Elsevier Science.

and B in Figure 40b. The latter agree reasonably well with those obtained experimentally by Hansen and coworkers [190], who found QWSs at -1.5 and -0.9 eV (see also Fig. 39 and [191, 192]). At energies larger than -0.65 eV, the Bloch spectral function shows weak maxima, which may also be associated with QWSs but lack the complete confinement due to the weak reflection at the Cu/Co interface at these energies [197]. There is no even–odd alternation of the QWSs, as found in the simple tight-binding model (Section 4.2.4), due to the lack of inversion symmetry in the Cu film.

The reflectivity at the Cu/Co boundary, as obtained theoretically by Dederichs and co-workers [197], is shown in Figure 41. In the lowest panel, the reflectivity of the Bloch state, which is associated with the Cu sp valence band, is shown for incidence on a Co film with 20 ML thickness. At energies less than -0.6 eV, there is almost complete reflection in the minority spin channel, in accordance with the band structure shown

in Figure 40. Above -0.6 eV, there are QWSs in the Co film, which reduce the reflectivity; cf. the pronounced minima in the reflectivity. This picture corresponds nicely to that of the quantum-well resonances in LEED (see Section 3.3): Here, the incoming wave is the Cu Bloch state, whereas in LEED it is the electron beam.

Manifestation of Quantum-Size Effects in Photoemission. As a prototypical example, normal photoemission ($\mathbf{k}_\parallel = 0$) with p-polarized light that impinges with a polar angle of 45° onto the surface is discussed in detail. In Figure 42, photoemission from semi-infinite Cu(001) is compared to that of 14-ML Cu on fcc Co(001). For the former (Fig. 42a), the intensity at energies below -2 eV stems from the d-band regime. The maximum that disperses from the Fermi energy at 10 eV photon energy down to -2 eV at 17 eV photon energy is due to emission

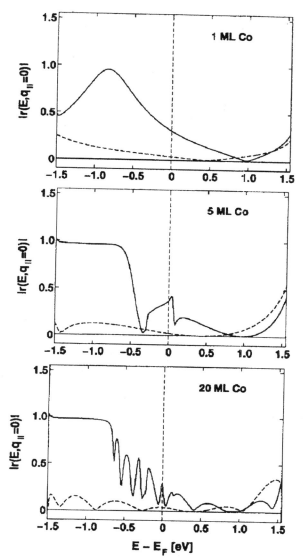

Fig. 41. Reflectivity r of a Cu Bloch wave of Δ_1 representation (along the $\Gamma-\Delta-X$ direction, $\mathbf{k}_\parallel = 0$) at the Cu/Co interface versus Co film thickness [from top to bottom, 1-ML, 5-ML, and 20-ML Co on Cu(001)]. Note that distinct minima occur which can directly be attributed to quantum-well states in the Co film (cf. [198]). Solid lines, minority spin; dashed lines, majority spin. Reprinted with permission from P. H. Dederichs, K. Wildberger, and R. Zeller, *Physica B* 237–238, 239 (1997). Copyright 1997, by Elsevier Science.

from the Cu sp band (cf. Fig. 40c). The direct-transition model can be used to explain the widths of these maxima: The sp band and the final-state band are almost parallel in the band structure and, thus, there is a certain energy range where the difference in the respective $k_\perp$ is rather small [166]. The slightly weakened $k_\perp$ conservation results therefore in a broad maximum. For the 14-ML Cu film on fcc Co(001) (Fig. 42b), the energies of the QWSs lead to narrow maxima [199]. The two sharp peaks, A and B, correspond to those found in the LDOS (Fig. 40b). The intensity distribution of structure C, however, agrees with that found for Cu(001), which can be also explained by the LDOS: Near E_F there are no strictly confined electronic

states in the Cu film because the reflection at the Cu/Co interface is small. This qualitative difference between A and B on the one hand and C on the other is further established in the photoelectron spin polarization. A and B show strong minority polarization ($P \approx -0.75$), whereas C is weakly spin polarized ($P \approx -0.05$), as expected from the LDOS.

The intensity variation with photon energy of maxima A and B is similar to that found for semi-infinite Cu(001) at the respective binding energies, a finding that confirms nicely both the simple theory presented in Section 4.2.4 and the experiment. At this point, quantum-size effects seem to occur only in the widths of those intensity maxima that are associated with QWSs [199]. This feature should be observable with high-resolution photoemission techniques [200]. However, hints about this behavior may be seen, for example, in the work by Hansen et al. (Fig. 38).

Further pronounced manifestations of quantum-size effects in photoemission are intensity oscillations with photon energy. These can be observed in the CIS mode of photoemission (Fig. 29): The initial-state energy is chosen as that of a QWS and the photon energy is varied while keeping $\mathbf{k}_\parallel$ fixed. The results for semi-infinite Cu(001) and 14 ML on fcc Co(001) are shown in Figure 43 where the initial-state energies were chosen as those of QWSs A, B, and C.

For semi-infinite Cu(001), one observes for each initial-state energy a dominating maximum and a few smaller maxima and shoulders (Fig. 43a). The former directly reflects the $k_\perp$ conservation; the latter can be explained by the final-state band structure. Further, because the wave function of the initial state does not change rapidly with energy, as is evident from the band structure, the three CIS spectra show almost the same shape, which appears only shifted in photon energy (see the inset in Fig. 43a). In other words, the CIS spectral shapes are governed by the final states. The fine structure for the energy of state C is slightly more pronounced when compared to those for A and B because of the larger photoelectron lifetime, which decreases with kinetic energy. The most important observation, however, is the absence of significant oscillations with photon energy.

For the 14-ML film, one finds similar behavior regarding the overall CIS intensity distribution (Fig. 43b). In particular, the relative heights of the main maxima for A, B, and C are close to their counterparts of semi-infinite Cu(001). The main differences are distinct intensity oscillations, which become clearly visible in the insets showing the logarithm of the intensities. In particular, maxima A and B show almost the same oscillation period, which is indicated by vertical lines in the inset of Figure 43b. The period for maximum C, however, differs significantly from those of A and B. Further, the spectral shapes of A and B are nearly identical and again differ from that of C; in particular, the double-peak structure near the maximum intensity occurs for both A and B but is missing for C. This double-peak structure is clearly due to the quantum-size induced oscillations of the CIS intensity. These findings show directly the different confinement strengths of the QWSs: strict confinement for A and B, less confinement for C.

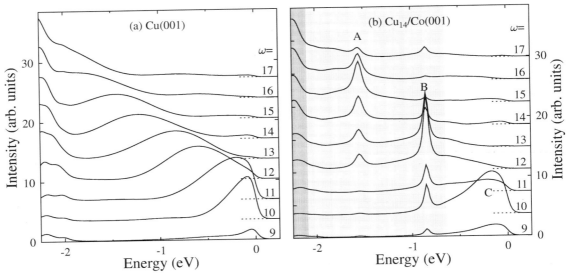

Fig. 42. Photoemission for $\mathbf{k}_\parallel = 0$ with p-polarized light incident at 45° off-normal from Cu(001) (a) and 14-ML Cu on fcc Co(001) (b). The photon energy ω ranges from 9 eV (bottom spectra) to 17 eV (top spectra), as indicated on the right. Gray areas in (b) indicate gaps in the Co band structure as in Figure 40. A, B, and C refer to quantum-well states (see text and Fig. 40). The Fermi energy is at 0 eV. Reprinted with permission from J. Henk and B. Johansson, *J. Electron Spectrosc. Relat. Phenom.* 105, 187 (1999). Copyright 1999, by Elsevier Science.

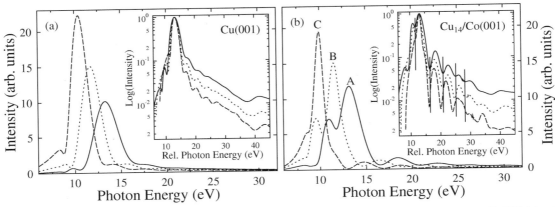

Fig. 43. Constant initial-state photoemission for $\mathbf{k}_\parallel = 0$ with p-polarized light incident at 45° off-normal from Cu(001) (a) and 14-ML Cu on fcc Co(001) (b). The initial-state energies are chosen as those of quantum-well states A (solid lines), B (dotted lines), and C (dashed lines); see text as well as Figures 40 and 42. Insets show the logarithms of the intensities, which are normalized to 1 and shifted in energy such that the maximum intensity is at 13 eV (relative photon energy). Vertical lines in the inset of (b) indicate intensity minima of state A. Reprinted with permission from J. Henk and B. Johansson, *J. Electron Spectrosc. Relat. Phenom.* 105, 187 (1999). Copyright 1999, by Elsevier Science.

Finally, the dependence of the oscillatory behavior on film thickness is addressed. According to Section 4.2.4, the oscillation period should decrease with increasing film thickness. For quantum-well state A of the 14-ML film and the corresponding states for films with thicknesses of 9 ML, 19 ML, 24 ML, and 29 ML, the energy positions are almost identical. This means that hole and photoelectron lifetimes are also almost identical, and the main differences in the CIS spectra can unambiguously be attributed to the difference in film thickness. Constant initial-state spectra for the various film thicknesses are shown in Figure 44. Both the width of the main maximum and the oscillation period decrease with film thickness. Further, the intensity at higher photon energies decreases with film thick-

ness, which is also evident from Section 4.2.4, particularly from Figure 35. The double-peak structure can clearly be attributed to the quantum-size induced oscillations: For the Cu film, the main intensity maximum is broadened with respect to the semi-infinite case due to the weakening of the $k_\perp$ conservation, as is evident from Figure 44. This maximum is "divided" in two due to the intensity oscillations (cf. the dashed–dotted guideline in Fig. 44). With increasing film thickness, the double-peak structure disappears.

In summary, quantum-size effects in photoemission from ultra-thin films manifest themselves in the following features: (i) Strict confinement of valence electrons to the film leads to a weakening of the $k_\perp$ conservation: the thinner the film,

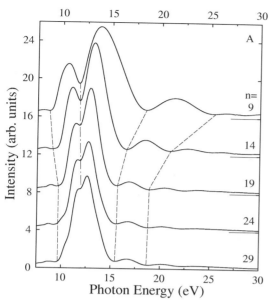

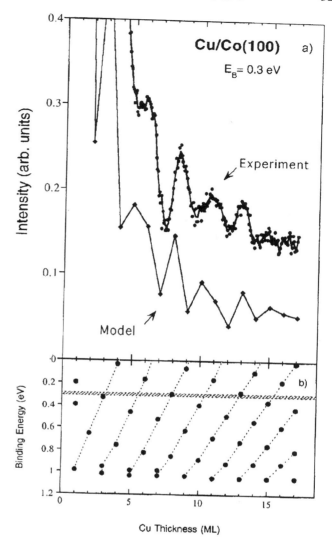

Fig. 44. Constant initial-state photoemission for $\mathbf{k}_\parallel = 0$ with p-polarized light incident at 45° off-normal from Cu films on fcc Co(001). The initial-state energies are chosen as those of quantum-well state A for selected film thicknesses n from 9 ML (top spectrum) to 29 ML (bottom spectrum), as indicated on the right. Dashed and dashed–dotted lines visualize the behavior of oscillations. Short horizontal lines represent zero intensity for each respective spectrum. Reprinted with permission from J. Henk and B. Johansson, *J. Electron Spectrosc. Relal. Phenom.* 105, 187 (1999). Copyright 1999, by Elsevier Science.

the broader the photoemission maxima. (ii) Photoemission intensities from individual QWSs show oscillations with photon energy, the period of which decreases with film thickness. (iii) Even for films only a few layers thick, the main maxima in the photoemission intensity follow the initial-state bulk-band structure, despite the fact that the initial-state energies are discrete.

Oscillations of Photoemission Intensity with Film Thickness. Another example of investigating QWSs at a fixed binding energy is illustrated in Figure 45. Kläsges et al. recorded experimentally the photocurrent at 0.3 eV binding energy and fixed photon energy ($\omega = 77$ eV) of Cu films on Co(100) for a variety of film thicknesses (1 ML–17 ML) [194]. Besides a global decrease of the intensity with film thickness, they found significant oscillations in the current, which, of course, can be attributed to QWSs. The period of the maxima was determined as 2.3 ML ± 0.1 ML.

Results of electronic-structure calculations of Cu films on Co(100) are shown in Figure 45b. Dederich's group calculated self-consistently the Bloch spectral function $A_B(\mathbf{k}_\parallel, E) = -\text{Im}\, G(\mathbf{k}_\parallel; E)/\pi$. Each maximum in A_B in the sp-band range indicates a QWS (filled circles in Fig. 45b). Again, one finds the familiar dispersion with film thickness, as discussed in Section 2.2. With increasing film thickness, the QWSs disperse in energy toward the Fermi level, as visualized, by the dashed lines. The latter cross the binding energy of 0.3 eV with a pe-

Fig. 45. Photoemission intensity from Cu films on Co(100) at 0.3 eV binding energy versus Cu film thickness. The emission angle chosen is 12° ($\mathbf{k}_\parallel = 0.94$ Å^{-1}). (a) Shows experimental data (dots) and results of a model calculation (solid line). Binding energies of quantum-well states are shown in (b). Dashed lines serve as a guide to the eye. The hatched area indicates the binding energy as chosen in the photoemission experiment. Reprinted with permission from R. Kläsges, D. Schmitz, C. Carbone, W. Eberhardt, P. Lang, R. Zeller, and P. H. Dederichs, *Phys. Rev. B.* 57, R696 (1998). Copyright 1998, by the American Physical Society.

riod of 2.4 ML, which corresponds nicely to the experimentally obtained value. Because the photoemission intensity shows a maximum at the energy position of a QWS, the dispersion of the QWSs can be translated into a dispersion of the photoemission maxima. As shown in Figure 45a, a model photoemission calculation reproduces all general features found in the experimental results, in particular, the global decay and the oscillations. The model assumes that each QWS contributes to the photocurrent with a finite peak width corresponding to the experimental energy resolution. Further, this intensity is expected to be proportional to the inverse of the film thickness. The background intensity due to the Co substrate is approximated as decaying exponentially with Cu thickness, in accordance with the mean free path of the photoelectron (Fig. 12). This causes the global decay of the intensity. Missing are, however, interference effects in both the initial and the final state of the photoemission process. These can, for example, lead to significant changes in the photocurrent that call into question the one-to-one correspondence between maxima in the Bloch spectral function and the photocurrent maxima. These effects have been observed, for example, for Co/Cu(001) [201] and Au/Ag(111) [202]: Due to destructive interference in the final state at a particular kinetic energy, only one of two QWSs of the 2-ML films has been observed in both experiment and theory although the layer- and symmetry-resolved Bloch spectral function of both states shows maxima of comparable height. The effect of interference is discussed in more detail in the following Section.

4.3.2. Quantum-Well States and Interference: Ag on Fe(001)

Particle wave duality is one of the fundamental features of quantum mechanics. We now investigate how photoemission from a thin film establishes an almost perfect analogy between a standing electromagnetic wave caught between two mirrors and an electron confined to a thin film.

Consider an electron confined to a quantum well, the latter, for example, being realized by a thin film (Fig. 46). The electron is reflected at both the surface side (s) and the substrate side (interface, i) of the film with reflection coefficients $R_s = |R_s| \exp(i\phi_s)$ and $R_i = |R_i| \exp(i\phi_i)$, respectively. The propagation from one side to the other is taken into account via the phase factors P^+ and P^-. For $\mathbf{k}_\parallel = 0$, these read $P^\pm = \exp(ik_\perp Nd)$, where N is the number of layers of the film, d is the interlayer distance, and $k_\perp$ is the wave number of the electron. The Bohr–Sommerfeld quantization rule, which is well known from the theory of atomic spectra, then reads $2k_\perp Nd + \phi_s + \phi_i = 2n\pi$, $n \in \mathbb{Z}$. In other words, constructive interference occurs if the accumulated phase shift is an even multiple of π. This relation is known as the round-trip criterion

(see Sections 2.2.2 and 3.3) and holds for perfectly reflecting boundaries ($|R_s| = |R_i| = 1$).

The more general case can be discussed in terms of an interference factor I. For each round trip, the wave function of the electron is changed by the factor $P^+ R_s P^- R_i$, with $P^\pm = \exp(i\mathbf{k}^\pm \cdot \mathbf{d})$. For $\mathbf{k}_\parallel = 0$, the interference factor I then becomes

$$
I = \sum_{j=0}^{\infty} (P^+ R_s P^- R_i)^j
$$
$$
= \left(1 - R \exp\left(i(\phi + 2k_\perp Nd)\right) \exp\left(-Nd/\lambda\right)\right)^{-1} \quad (199)
$$

with the definitions $R = |R_s R_i|$ and $\phi = \phi_s + \phi_i$. Note that the mean free path λ is taken into account. The modulus of the interference factor is given by

$$
|I|^2 = \left(1 - R \exp\left(-\frac{Nd}{\lambda}\right)\right)^{-2}
$$
$$
\times \left[1 + \left(\frac{2f}{\pi}\right)^2 \sin^2\left(k_\perp Nd + \frac{\phi}{2}\right)\right]^{-1} \quad (200)
$$

with

$$
f = \frac{\pi \sqrt{R} \exp\left(-Nd/2\lambda\right)}{1 - R \exp\left(-Nd/\lambda\right)} \quad (201)
$$

f is the finesse (i.e., the ratio of peak separation and peak width) of a Fabry–Pérot interferometer with an absorptive medium ($\lambda < \infty$). Such a device was invented by Fabry and Pérot in 1899 [203]. Equation (200) establishes the close analogy between interference of electromagnetic waves and of electrons (cf., e.g., [204]).

The first factor in Eq. (200), $(1 - R \exp(-Nd/\lambda))^{-2}$, depends on both the mean free path λ and the reflectivity R, quantities that are expected to depend rather smoothly on energy. The same holds for the finesse f and the phase shift ϕ. Therefore, the modulation of the interference factor can be mainly attributed to the wave number $k_\perp$. Assuming the first factor, the finesse, and the phase shift as energy independent, the interference factor becomes approximately

$$
|I|^2 \approx \left[1 + \left(\frac{2f}{\pi}\right)^2 \sin^2\left(k_\perp(E)Nd + \frac{\phi}{2}\right)\right]^{-1} \quad (202)
$$

which is shown in Figure 47. Maxima in $|I|^2$ occur if $\sin^2(k_\perp(E)Nd + \phi/2) = 0$ or, equivalently, if $k_\perp(E) = (n\pi - \phi/2)/(Nd)$, $n \in \mathbb{Z}$.

The relevant quantities that determine the interference can be cast into two categories. (i) The wave number $k_\perp(E)$ and the mean free path $\lambda(E)$ depend on the band structure of the film material. Because the electrons can be described as quasi-particles, the band structure $k_\perp(E, \mathbf{k}_\parallel)$ is, in general, complex (cf. Section 3.2.2). The imaginary part of $k_\perp$ is related to the mean free path by the group velocity $v_\perp = \partial_{k_\perp} E(k_\perp)$, $\lambda = v_\perp / \operatorname{Im} k_\perp(E)$. (ii) The reflectivity R and the phase shift ϕ depend on the quantum-well boundaries, in particular, on the reflection properties at the film–substrate interface. For thick films, R and ϕ can be regarded as independent of the film thickness Nd. Therefore, one can expect to determine them by

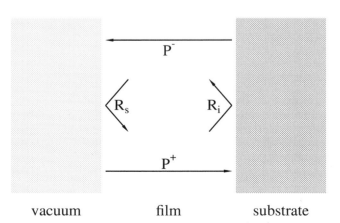

Fig. 46. Electron confined to a quantum well. Arrows $P^\pm$ denote propagation between the interfaces (vacuum–film and film–substrate). R_i and R_s are the reflectivities at the film boundaries.

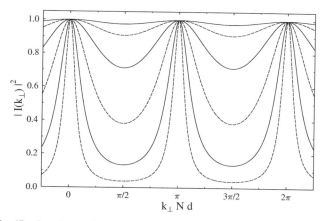

Fig. 47. Interference factor $I(k_\perp)$ for a Fabry–Pérot-type quantum well as a function of wave number $k_\perp$ for finesses 0.5, 1.0, 2.0, 4.0, 8.0, and 16.0 (from top to bottom, alternating solid and dashed lines). N and d are the number of layers and the interlayer distance of the film, respectively. The phase shift ϕ is chosen as 0; cf. Eq. (202).

ARPES from films with different thicknesses Nd. For very thin films, however, the interfaces (surface side and substrate side) cannot be separated and R and ϕ should differ considerably from their values for thicker films. This becomes evident by comparing reflectivities for various film thicknesses, as, for example, shown in Figure 41. The peak positions depend on $k_\perp$ and ϕ, the peak widths on R and λ.

If the film were infinitely thick, the photocurrent J_s would be expressed in the form of Fermi's golden rule, Eq. (141). However, for a finite thickness, the initial state $|i\rangle$ can be seen as modulated by the interference factor I, Eq. (200), and thus is given by $I_i|i\rangle$ [205]. Therefore, the photocurrent from the film with finite thickness reads

$$J_{qw} \sim \sum_i |I_i|^2 |\langle f|\mathbf{A} \cdot \mathbf{p}|i\rangle|^2 \delta(E_f - \omega - E_i) \quad (203)$$

The task to determine the interference-determining quantities $k_\perp$, ϕ, R, and λ might be complicated by several facts. (i) The growth of the film material on the substrate should be in the layer-by-layer mode, which leads to well-defined film boundaries and minimizes film imperfections. (ii) The electronic properties of the film and the substrate should "match". In other words, they should allow for QWSs; for example, there has to be a gap in the band structure of the substrate. (iii) In general, several initial states $|i\rangle$ contribute to the total photocurrent in the considered energy range and, thus, one has to deal with a set of parameters for each initial state. Fortunately, there are systems in which only a single initial state is present in the considered energy range.

Paggel et al. reported on an experimental investigation of the interference properties of Ag films on Fe(001) [206], another prototypical system. In the considered initial-energy range (-2 eV up to 0 eV) and the chosen photon energy, the spectrum for semi-infinite Ag looks almost structureless. Therefore, the intensity modulations due to the Ag films can be easily identified (cf. Fig. 48). Further, Ag films grow in a layer-by-layer

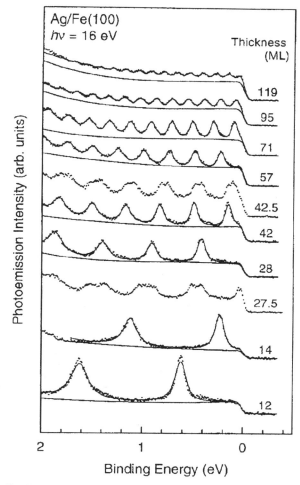

Fig. 48. Photoemission spectra for Ag films on Fe(001) for various film thicknesses in monolayers (ML), as indicated on the right of each spectrum. The experimental data (given by dots) were recorded for normal emission ($\mathbf{k}_\parallel = 0$) and 16 eV photon energy. Solid lines correspond to the fitted interference spectra and the background. Reprinted with permission from J. J. Paggel, T. Miller, and T.-C. Chiang, Science 283, 1709 (1999). Copyright 1999, by the American Association for the Advancement of Science.

mode on Fe(001), which allows for a very accurate thickness calibration via the observed intensity modulation. As an example, the spectrum for a film with thickness 27.5 ML shows simultaneously the peak structures of films with 27 and 28 ML. The modulated intensity appears as a superposition of those of the latter films (this holds, in addition, for the 42.5-ML film).

To determine the reflectivity R, the phase shift ϕ, and the mean free path λ, one has to know the initial-state band structure of Ag. The latter can be obtained either by the photoemission experiment itself, for instance, via band mapping using various photon energies or by a band structure calculation. The relevant band in the considered energy range is the sp valence band, which is roughly a free-electron parabola; see Figure 49a (cf., e.g., the band structure of Cu in Figure 40, which shows an sp band, too). Applying a fitting procedure, Paggel et al. obtained $R(E)$, $\phi(E)$, and $\lambda(E)$, which are assumed to be independent of the film thickness. The resulting

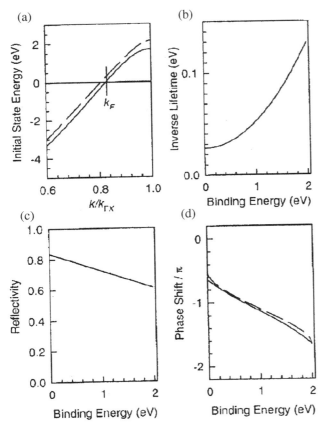

Fig. 49. Electronic structure of Ag films on Fe(100). Panel a shows the Ag band structure near the Fermi energy (i.e., the *sp* band range). The solid line corresponds to the experimentally determined dispersion of the *sp* band, the dashed line to one theoretically obtained. Panel b depicts the inverse lifetime as obtained from the experimental data. Panels c and d show the reflection coefficients and the phase shift at the Ag/Fe interface. Reprinted with permission from J. J. Paggel, T. Miller, and T.-C. Chiang, *Science* 283, 1709 (1999). Copyright 1999, by the American Association for the Advancement of Science.

theoretical modulated intensities are shown in Figure 48 and match almost perfectly the experimental ones for all film thicknesses. Note that the peak shapes in the experimental spectra agree very well with that of the interference factor I shown in Figure 47. The determined reflectivity R was less than unity ($R \in [0.6, 0.85]$, Fig. 49c), indicating that the electrons are not strictly confined to the Ag film. This can be attributed to the fact that the bandgap in the Fe substrate is not an absolute one but a hybridization bandgap. In the former case, there are no states in the bandgap energy range, whereas in the latter, there are states with different spatial symmetry, which, however, are mixed by SOC. Further, the small, but not negligible lattice mismatch between Ag and Fe leads to nonspecular reflection at the boundaries. The inverse lifetime depends quadratically on the binding energy (Fig. 49b and [207]), as in Fermi liquid theory.

Acknowledgments

It is a great pleasure to thank those colleagues who have made the present chapter possible. Unfortunately, it is nearly impossi-

ble to acknowledge them all. Therefore, I want to mention only four to whom I am especially grateful: Roland Feder, Samed Halilov, Thomas Scheunemann, and Eiichi Tamura.

REFERENCES

1. L. Bennett and R. Watson, Eds., "Magnetic Multilayers." World Scientific, Singapore, 1993.
2. J. Bland and B. Heinrich, "Ultrathin Magnetic Structures." Springer-Verlag, Berlin, 1994.
3. A. Barthélémy, A. Fert, and F. Petroff, *In* "Handbook of Magnetic Materials" (K. H. J. Buschow, Ed.), Vol. 12, p. 1. Elsevier, Amsterdam, 1999.
4. J. M. Daughton, *J. Magn. Magn. Mater.* 192, 334 (1999).
5. U. Hartmann, Ed., "Magnetic Multilayers and Giant Magnetoresistance: Fundamentals and Industrial Applications," Springer Series in Surface Sciences, Vol. 37. Springer-Verlag, Berlin, 1999.
6. A. Zangwill, "Physics at Surfaces." Cambridge Univ. Press, Cambridge, UK, 1988.
7. N. Ashcroft and N. Mermin, "Solid State Physics." Holt–Saunders, London, 1976.
8. L. Brillouin, "Wave Propagation in Periodic Structures." McGraw–Hill, New York, 1946.
9. R. L. Park and H. H. Madden, *Surf. Sci.* 11, 188 (1968).
10. S. Gull, A. Lasenby, and C. Doran, *Found. Phys.* 23, 1175 (1993).
11. T. G. Vold, *Am. J. Phys.* 61, 491 (1993).
12. G. Ertl and J. Küppers, "Low Energy Electrons and Surface Chemistry," Chap. 9, p. 201. VCH, Weinheim, 1985.
13. F. Bloch, *Z. Phys.* 52, 555 (1928).
14. P. Harrison, "Quantum Wells, Wires and Dots." Wiley, Chichester, 2000.
15. E. Merzbacher, "Quantum Mechanics," 2nd ed. Wiley, New York, 1970.
16. K. R. Brownstein, *Am. J. Phys.* 68, 160 (2000).
17. H. A. Kramers, *Koniklije Akademie van Wetenschapen* 33, 959 (1930).
18. G. Bastard, *Phys. Rev. B* 24, 5693 (1981).
19. G. Bastard, *Phys. Rev. B* 25, 7584 (1982).
20. P. D. Loly and J. B. Pendry, *J. Phys. C: Solid State Phys.* 16, 423 (1983).
21. J. E. Ortega, F. J. Himpsel, G. J. Mankey, and R. F. Willis, *Phys. Rev. B* 47, 1540 (1993).
22. M. Grüne, T. Pelzer, K. Wandelt, and I. T. Steinberger, *J. Electron Spectrosc. Relat. Phenom.* 98–99, 121 (1999).
23. K. Horn, M. Scheffler, and A. M. Bradshaw, *Phys. Rev. Lett.* 41, 822 (1978).
24. M. Scheffler, K. Horn, A. M. Bradshaw, and K. Kambe, *Surf. Sci.* 80, 69 (1979).
25. K. Kambe, *Surf. Sci.* 105, 95 (1981).
26. P. Trischberger, H. Dröge, S. Gokhale, J. Henk, H.-P. Steinrück, W. Widdra, and D. Menzel, *Surf. Sci.* 377–379, 155 (1996).
27. W. Widdra, P. Trischberger, and J. Henk, *Phys. Rev. B* 60, R5161 (1999).
28. C. Davisson and L. H. Germer, *Nature* 119, 558 (1927).
29. C. J. Davisson and L. H. Germer, *Phys. Rev.* 30, 705 (1927).
30. L. de Broglie, *Comte Rendu* 179, 676 (1924).
31. L. de Broglie, *Comte Rendu* 179, 1039 (1924).
32. L. de Broglie, *Comte Rendu* 180, 498 (1925).
33. J. F. van Veen and M. A. van Hove, Eds., "The Structure of Surfaces II," Springer Series in Surface Sciences, Vol. 11. Springer-Verlag, Berlin, 1988.
34. S. Y. Tong, M. A. van Hove, K. Takayanagi, and X. D. Xie, Eds., "The Structure of Surfaces III," Springer Series in Surface Sciences, Vol. 24. Springer-Verlag, Berlin, 1991.
35. C. Davisson and L. H. Germer, *Phys. Rev.* 31, 307 (1928).
36. C. Davisson and L. H. Germer, *Phys. Rev.* 31, 155 (1928).
37. M. P. Seah and W. A. Dench, *Surf. Interface Anal.* 1, 2 (1979).
38. Z.-J. Ding and R. Shimizu, *Surf. Sci.* 222, 313 (1989).
39. J. Rundgren, *Phys. Rev. B* 59, 5106 (1999).
40. C. J. Davisson, *J. Franklin Inst.* 205, 597 (1928).

41. K. Heinz, U. Starke, and J. Bernhardt, *Prog. Surf. Sci.* 64, 163 (2000).
42. K. Kambe, *Z. Naturforsch., A: Phys. Sci.* 22, 322 (1967).
43. K. Kambe, *Z. Naturforsch., A: Phys. Sci.* 22, 422 (1967).
44. K. Kambe, *Z. Naturforsch., A: Phys. Sci.* 22, 1280 (1967).
45. J. B. Pendry, "Low Energy Electron Diffraction." Academic Press, London, 1974.
46. M. A. van Hove, W. H. Weinberg, and C. M. Chan, "Low-Energy Electron Diffraction." Springer-Verlag, Berlin, 1986.
47. S. Y. Tong, *In* "Progress in Surface Science" (S. G. Davisson, Ed.), Vol. 7, p. 1. Pergamon, London, 1975.
48. M. A. van Hove and S. Y. Tong, "Surface Crystallography by LEED: Theory, Computation and Structural Results," Springer Series in Chemical Physics, Vol. 2. Springer-Verlag, Berlin, 1979.
49. R. Feder, *Solid State Commun.* 31, 821 (1979).
50. R. Feder and J. Kirschner, *Surf. Sci.* 103, 75 (1981).
51. R. Feder, *In* "Polarized Electrons in Surface Physics" (R. Feder, Ed.). Advanced Series in Surface Science, Chap. 4, p. 125. World Scientific, Singapore, 1985.
52. J. Kirschner and R. Feder, *Phys. Rev. Lett.* 42, 1008 (1979).
53. J. C. Slater, *Phys. Rev.* 51, 840 (1937).
54. V. N. Strocov, R. Claessen, G. Nicolay, S. Hüfner, A. Kimura, A. Harasawa, S. Shin, A. Kakizaki, P. O. Nilsson, H. I. Starnberg, and P. Blaha, *Phys. Rev. Lett.* 81, 4943 (1998).
55. I. Bartoš, *Prog. Surf. Sci.* 59, 197 (1998).
56. V. Heine, *Proc. Phys. Soc.* 81, 300 (1963).
57. Y.-C. Chang, *Phys. Rev. B* 25, 605 (1982).
58. J. Hermanson, *Solid State Commun.* 22, 9 (1977).
59. T. Inui, Y. Tanabe, and Y. Onodera, "Group Theory and Its Applications in Physics," 1st ed., Springer Series in Solid State Sciences, Vol. 78. Springer-Verlag, Berlin, 1990.
60. J. B. Pendry, *J. Phys. C: Solid State Phys.* 2, 1215 (1969).
61. J. B. Pendry, *J. Phys. C: Solid State Phys.* 2, 2273 (1969).
62. J. B. Pendry, *J. Phys. C: Solid State Phys.* 2, 2283 (1969).
63. V. N. Strocov, H. I. Starnberg, and P. O. Nilsson, *J. Phys.: Condens. Matter* 8, 7539 (1996).
64. V. N. Strocov, H. I. Starnberg, and P. O. Nilsson, *J. Phys.: Condens. Matter* 8, 7549 (1996).
65. J. Henk, W. Schattke, H.-P. Barnscheidt, C. Janowitz, R. Manzke, and M. Skibowski, *Phys. Rev. B* 39, 13286 (1989).
66. J. Henk, J.-V. Peetz, and W. Schattke, *In* "20th International Conference on the Physics of Semiconductors" (E. M. Anastassakis and J. D. Joannopoulos, Eds.), Vol. 1, pp. 175–178. World Scientific, Singapore, 1990.
67. J. Henk, W. Schattke, H. Carstensen, R. Manzke, and M. Skibowski, *Phys. Rev. B* 47, 2251 (1993).
68. S. Lorenz, C. Solterbeck, W. Schattke, J. Burmeister, and W. Hackbusch, *Phys. Rev. B* 55, R13432 (1997).
69. W. Schattke, *Prog. Surf. Sci.* 64, 89 (2000).
70. I. Mertig, E. Mrosan, and P. Ziesche, "Multiple Scattering Theory of Point Defects in Metals: Electronic Properties," Teubner-Texte zur Physik, Vol. 11. Teubner, Leipzig, 1987.
71. P. Weinberger, "Electron Scattering Theory of Ordered and Disordered Matter," Clarendon, Oxford, 1990.
72. A. Gonis, "Green Functions for Ordered and Disordered Systems," Studies in Mathematical Physics, Vol. 4. North-Holland, Amsterdam, 1992.
73. J. Korringa, *Physica* 13, 392 (1947).
74. W. Kohn and N. Rostoker, *Phys. Rev.* 94, 1111 (1954).
75. J. M. MacLaren, S. Crampin, D. D. Vvedensky, and J. B. Pendry, *Phys. Rev. B* 40, 12164 (1989).
76. E. M. Rose, "Relativistic Electron Theory." Wiley, New York, 1961.
77. S. Bei der Kellen and A. J. Freeman, *Phys. Rev. B* 54, 11187 (1996).
78. A. Gonis, P. Turchi, J. Kudrnovský, V. Drchal, and I. Turek, *J. Phys.: Condens. Matter* 8, 7869 (1996).
79. J. W. Krewer, "Beugung spinpolarisierter langsamer Elektronen (SPLEED) mit nicht-sphärischen Potentialen," Ph.D. Dissertation, Universität Duisburg, 1990.
80. J. W. Krewer and R. Feder, *Physica B* 172, 135 (1991).
81. M. Abramowitz and I. Stegun, Eds., "Handbook of Mathematical Functions." Dover, New York, 1965.
82. E. Tamura, *Phys. Rev. B* 45, 3271 (1992).
83. B. Ackermann, "Relativistische Theorie der Photoemission und Streuung langsamer Elektronen von ferromagnetischen Oberflächen," Ph.D. Thesis, Universität Duisburg, 1985.
84. D. D. Koelling and B. N. Harmon, *J. Phys. C: Solid State Phys.* 10, 3107 (1977).
85. H. Gollisch and L. Fritsche, *Phys. Status Solidi B* 86, 156 (1978).
86. T. Takeda, *J. Phys. F: Met. Phys.* 9, 815 (1979).
87. E. Tamura, private communication.
88. H. Ebert, H. Freyer, A. Vernes, and G. Y. Guo, *Phys. Rev. B* 53, 7721 (1996).
89. H. Ebert, H. Freyer, and M. Deng, *Phys. Rev. B* 56, 9454 (1997).
90. L. A. Mac Coll, *Phys. Rev.* 56, 699 (1939).
91. R. Jones, P. Jennings, and O. Jepsen, *Phys. Rev. B* 29, 6474 (1984).
92. E. Tamura and R. Feder, *Z. Phys. B: Condens. Matter* 81, 425 (1990).
93. C. S. Lent and D. J. Kirkner, *J. Appl. Phys.* 67, 6353 (1990).
94. Y. Joly, *Phys. Rev. Lett.* 68, 950 (1992).
95. R. Feder, *J. Phys. C: Solid State Phys.* 14, 2049 (1981).
96. J. Kessler, "Polarized Electrons," 2nd ed., Springer Series on Atoms and Plasmas, Vol. 1. Springer-Verlag, Berlin, 1985.
97. R. Feder, Ed., "Polarized electrons in surface physics," Advanced Series in Surface Science. World Scientific, Singapore, 1985.
98. E. Zanazzi and F. Jona, *Surf. Sci.* 62, 61 (1977).
99. J. B. Pendry, *J. Phys. C: Solid State Phys.* 13, 937 (1980).
100. E. Tamura, R. Feder, G. Waller, and U. Gradmann, *Phys. Status Solidi B* 157, 627 (1990).
101. O. Hjortstam, J. Trygg, J. M. Wills, B. Johansson, and O. Eriksson, *Phys. Rev. B* 53, 9204 (1996).
102. I. Delgadillo, H. Gollisch, and R. Feder, *Phys. Rev. B* 50, 15808 (1994).
103. R. Feder, B. Awe, and E. Tamura, *Surf. Sci.* 157, 183 (1985).
104. E. Tamura, R. Feder, J. Krewer, R. E. Kirby, E. Kisker, E. L. Garwin, and F. K. King, *Solid State Commun.* 55, 543 (1985).
105. P. Bruno, *J. Phys.: Condens. Matter* 11, 9403 (1999).
106. T. Scheunemann, R. Feder, J. Henk, E. Bauer, T. Duden, H. Pinkvos, H. Poppa, and K. Wurm, *Solid State Commun.* 104, 787 (1997).
107. B. Johnson, P. Berlowitz, D. Goodman, and C. Bartholomew, *Surf. Sci.* 217, 13 (1989).
108. J. G. Ociepa, P. J. Schultz, K. Griffiths, and P. R. Norton, *Surf. Sci.* 225, 281 (1990).
109. M. Tikhov and E. Bauer, *Surf. Sci.* 232, 73 (1990).
110. H. Knoppe and E. Bauer, *Phys. Rev. B* 48, 1794 (1993).
111. E. Bauer, *Rep. Prog. Phys.* 57, 895 (1994).
112. L. M. Falicov and J. Ruvalds, *Phys. Rev.* 172, 498 (1968).
113. C. M. Schneider, P. Bressler, P. Schuster, J. Kirschner, J. J. de Miguel, and R. Miranda, *Phys. Rev. Lett.* 64, 1059 (1990).
114. H. Bonzel and C. Kleint, *Prog. Surf. Sci.* 49, 107 (1995).
115. H. Hertz, *Ann. Phys. Chem. Neue Folge* 31, 983 (1887).
116. W. Hallwachs, *Ann. Phys. Chem. Neue Folge* 33, 301 (1888).
117. J. Elster and H. Geitel, *Ann. Phys. Chem. Neue Folge* 38, 40 (1889).
118. J. Elster and H. Geitel, *Ann. Phys. Chem. Neue Folge* 38, 497 (1889).
119. P. Lenard, *Ann. Phys.* 8, 149 (1902).
120. A. Einstein, *Ann. Phys.* 17, 132 (1905).
121. K. Siegbahn, C. Nordling, A. Fahlman, R. Nordberg, K. Hamrin, J. Hedman, G. Johanson, T. Bergmark, S.-E. Karlsson, I. Lindgren, and B. Lindberg, "ESCA–Atomic, Molecular and Solid State Structure Studied by Means of Electron Spectroscopy." Almqvist & Wiksell, Uppsala, 1967.
122. E. O. Kane, *Phys. Rev. Lett.* 12, 97 (1964).
123. M. Campagna and R. Rosei, Eds., "Photoemission and Absorption Spectroscopy of Solids and Interfaces with Synchrotron Radiation." North-Holland, Amsterdam, 1990.
124. R. Z. Bachrach, Ed., "Technique," Synchrotron Radiation Research: Advances in Surface and Interface Science, Vol. 1. Plenum, New York, 1992.
125. R. Z. Bachrach, Ed., "Issues and Technology," Synchrotron Radiation Research: Advances in Surface and Interface Science, Vol. 2. Plenum, New York, 1992.

126. C. Berglund and W. Spicer, *Phys. Rev.* 136, A1030 (1964).

127. M. Cardona and L. Ley, Eds., "Photoemission in Solids I," Topics in Applied Physics, Vol. 26. Springer-Verlag, Berlin, 1978.

128. A. Liebsch, *In* "Photoemission and the Electronic Properties of Surfaces" (B. Feuerbacher, B. Fitton, and R. F. Willis, Eds.), p. 167. Wiley, Chichester, 1978.

129. S. V. Kevan, Ed., "Angle-Resolved Photoemission: Theory and Current Applications." Elsevier, Amsterdam, 1992.

130. I. Adawi, *Phys. Rev. A* 134, 788 (1964).

131. G. D. Mahan, *Phys. Rev. B* 2, 4334 (1970).

132. G. D. Mahan, *Phys. Rev. Lett.* 24, 1068 (1970).

133. W. L. Schaich and N. W. Ashcroft, *Phys. Rev. B* 3, 2452 (1971).

134. D. C. Langreth, *Phys. Rev. B* 3, 3120 (1971).

135. J. B. Pendry, *J. Phys. C: Solid State Phys.* 8, 2431 (1975).

136. J. B. Pendry, *Surf. Sci.* 57, 679 (1976).

137. J. F. L. Hopkinson, J. B. Pendry, and D. J. Titterington, *Comput. Phys. Commun.* 19, 69 (1980).

138. G. Thörner and G. Borstel, *Phys. Status Solidi B* 126, 617 (1984).

139. J. Braun, G. Thörner, and G. Borstel, *Phys. Status Solidi B* 130, 643 (1985).

140. J. Braun, G. Thörner, and G. Borstel, *Phys. Status Solidi B* 144, 609 (1987).

141. M. Wöhlecke and G. Borstel, *In* "Optical Orientation" (F. Meier and B. P. Zakharchenya, Eds.), North-Holland, Amsterdam, 1984.

142. E. Tamura, W. Piepke, and R. Feder, *Phys. Rev. Lett.* 59, 934 (1987).

143. E. Tamura and R. Feder, *Europhys. Lett.* 16, 695 (1991).

144. J. Henk and R. Feder, *Europhys. Lett.* 28, 609 (1994).

145. B. Ginatempo, P. J. Durham, B. L. Gyorffy, and W. M. Temmerman, *Phys. Rev. Lett.* 54, 1581 (1985).

146. B. Schmiedeskamp, B. Vogt, and U. Heinzmann, *Phys. Rev. Lett.* 60, 651 (1988).

147. B. Schmiedeskamp, N. Irmer, R. David, and U. Heinzmann, *Appl. Phys. A* 53, 418 (1991).

148. N. Irmer, F. Frentzen, S.-W. Yu, B. Schmiedeskamp, and U. Heinzmann, *J. Electron Spectrosc. Relat. Phenom.* 78, 321 (1996).

149. R. Feder, F. Rosicky, and B. Ackermann, *Z. Phys. B: Condens. Matter* 52, 31 (1983).

150. R. Feder and J. Henk, *In* "Spin-Orbit Influenced Spectroscopies of Magnetic Solids" (H. Ebert and G. Schütz, Eds.), Lecture Notes in Physics, Vol. 466, p. 85. Springer-Verlag, Berlin, 1996.

151. W. Kuch, A. Dittschar, K. Meinel, M. Zharnikov, C. Schneider, J. Kirschner, J. Henk, and R. Feder, *Phys. Rev. B* 53, 11621 (1996).

152. A. Fanelsa, E. Kisker, J. Henk, and R. Feder, *Phys. Rev. B* 54, 2922 (1996).

153. A. Rampe, G. Güntherodt, D. Hartmann, J. Henk, T. Scheunemann, and R. Feder, *Phys. Rev. B* 57, 14370 (1998).

154. P. J. Durham, *J. Phys. F: Met. Phys.* 11, 2475 (1981).

155. P. J. Durham, J. Staunton, and B. L. Gyorffy, *J. Magn. Magn. Mater.* 45, 38 (1984).

156. P. Feibelman and D. Eastman, *Phys. Rev. B* 10, 4932 (1974).

157. C. Caroli, D. Lederer-Rozenblatt, B. Roulet, and D. Saint-James, *Phys. Rev. B* 8, 4552 (1973).

158. H. Gollisch, D. Meinert, E. Tamura, and R. Feder, *Solid State Commun.* 82, 197 (1992).

159. W. Schattke, *Prog. Surf. Sci.* 54, 211 (1997).

160. L. Hedin, J. Michiels, and J. Inglesfield, *Phys. Rev. B* 58, 15565 (1998).

161. L. Hedin, *J. Phys.: Condens. Matter* 11, R489 (1999).

162. T. Fujikawa and L. Hedin, *Phys. Rev. B* 40, 11507 (1989).

163. S. Lundqvist and N. H. March, Eds., "Theory of the Inhomogenous Electron Gas." Plenum, New York, 1983.

164. A. Liebsch, *Phys. Rev. Lett.* 43, 1431 (1979).

165. A. Goldmann, R. Matzdorf, and F. Theilmann, *Surf. Sci.* 414, L932 (1998).

166. R. Matzdorf, *Surf. Sci. Rep.* 30, 154 (1998).

167. J. Braun, *Rep. Prog. Phys.* 59, 1267 (1996).

168. P. M. Morse and H. Feshbach, "Methods of Theoretical Physics," Vol. 1. McGraw–Hill, New York, 1953.

169. J. S. Faulkner and G. M. Stocks, *Phys. Rev. B* 21, 3222 (1980).

170. P. Braspenning and A. Lodder, *Phys. Rev. B* 49, 10222 (1994).

171. G. Grosso, S. Moroni, and G. P. Parravicini, *Phys. Scr., T* 25, 316 (1989).

172. L. Szunyogh, B. Újfalussy, and P. Weinberger, *Phys. Rev. B* 51, 9552 (1995).

173. K. Wildberger, R. Zeller, and P. H. Dederichs, *Phys. Rev. B* 55, 10074 (1997).

174. H. Bross, *Z. Phys. B: Condens. Matter* 28, 173 (1977).

175. J. Henk, A. M. N. Niklasson, and B. Johansson, *Phys. Rev. B* 59, 13986 (1999).

176. J. Henk, T. Scheunemann, S. Halilov, and R. Feder, *J. Phys.: Condens. Matter* 8, 47 (1996).

177. C. M. Schneider, J. J. de Miguel, P. Bressler, P. Schuster, R. Miranda, and J. Kirschner, *J. Electron Spectrosc. Relat. Phenom.* 51, 263 (1990).

178. W. Kuch, A. Dittschar, M. Salvietti, M.-T. Lin, M. Zharnikov, C. M. Schneider, J. Camarero, J. J. de Miguel, R. Miranda, and J. Kirschner, *Phys. Rev. B* 57, 5340 (1998).

179. J. Henk and B. Johansson, *J. Electron Spectrosc. Relat. Phenom.* 105, 187 (1999).

180. A. Beckmann, *Surf. Sci.* 349, L95 (1996).

181. R. Paniago, R. Matzdorf, G. Meister, and A. Goldmann, *Surf. Sci.* 325, 336 (1995).

182. R. Schmitz-Hübsch, K. Oster, J. Radnik, and K. Wandelt, *Phys. Rev. Lett.* 74, 2995 (1995).

183. P. Segovia, E. G. Michel, and J. E. Ortega, *Phys. Rev. Lett.* 77, 3455 (1996).

184. F. G. Curti, A. Danese, and R. A. Bartynski, *Phys. Rev. Lett.* 80, 2213 (1998).

185. H. Hoekstra, *Surf. Sci.* 205, 523 (1988).

186. J. Henk and W. Schattke, *Comput. Phys. Commun.* 77, 69 (1993).

187. S. V. Halilov, J. Henk, T. Scheunemann, and R. Feder, *Surf. Sci.* 343, 148 (1995).

188. J. Heinrichs, *J. Phys.: Condens. Matter* 12, 5565 (2000).

189. W.-H. Steeb, "Hilbert Spaces, Generalized Functions and Quantum Mechanics." B.I. Wissenschaftsverlag, Mannheim, 1991.

190. E. D. Hansen, T. Miller, and T.-C. Chiang, *J. Phys.: Condens. Matter* 9, L435 (1997).

191. K. Garrison, Y. Chang, and P. Johnson, *Phys. Rev. Lett.* 71, 2801 (1993).

192. C. Carbone, E. Vescovo, O. Rader, W. Gudat, and W. Eberhardt, *Phys. Rev. Lett.* 71, 2805 (1993).

193. C. Carbone, E. Vescovo, R. Kläsges, D. Sarma, and W. Eberhardt, *Solid State Commun.* 100, 749 (1996).

194. R. Kläsges, D. Schmitz, C. Carbone, W. Eberhardt, P. Lang, R. Zeller, and P. H. Dederichs, *Phys. Rev. B* 57, R696 (1998).

195. P. van Gelderen, S. Crampin, and J. Inglesfield, *Phys. Rev. B* 53, 9115 (1996).

196. C. Bradley and A. Cracknell, "The Mathematical Theory of Symmetry in Solids." Clarendon, Oxford, 1972.

197. P. H. Dederichs, K. Wildberger, and R. Zeller, *Physica B* 237–238, 239 (1997).

198. P. Bruno, *Phys. Rev. B* 52, 411 (1995).

199. J. J. Paggel, T. Miller, and T.-C. Chiang, *Phys. Rev. Lett.* 81, 5632 (1998).

200. R. Matzdorf, A. Gerlach, R. Hennig, G. Lauff, and A. Goldmann, *J. Electron Spectrosc. Relat. Phenom.* 94, 279 (1998).

201. D. Reiser, J. Henk, H. Gollisch, and R. Feder, *Solid State Commun.* 93, 231 (1995).

202. F. Frentzen, J. Henk, N. Irmer, R. David, B. Schmiedeskamp, U. Heinzmann, and R. Feder, *Z. Phys. B: Condens. Matter* 100, 575 (1996).

203. C. Fabry and A. Pérot, *Ann. Chim. Phys.* 19, 115 (1899).

204. M. Born and E. Wolf, "Principles of Optics: Electromagnetic Theory of Propagation, Interference and Diffraction of Light," 3rd ed. Pergamon, Oxford, 1965.

205. P. Voisin, G. Bastard, and M. Voos, *Phys. Rev. B* 29, 935 (1984).

206. J. J. Paggel, T. Miller, and T.-C. Chiang, *Science* 283, 1709 (1999).

207. A. Beckmann, *Surf. Sci.* 326, 335 (1995).

Chapter 11

IN SITU SYNCHROTRON STRUCTURAL STUDIES OF THE GROWTH OF OXIDES AND METALS

A. Barbier, C. Mocuta, G. Renaud

CEA/Grenoble, Département de Recherche Fondamentale sur la Matière Condensée
SP2M/IRS, 38054 Grenoble Cedex 9, France

Contents

1. INTRODUCTION

When dealing with thin film growth, a wealth of open questions must be answered for each materials combination. Among the major questions that immediately arise, we may cite the following: What is the growth mode and what is the resulting morphology of the layer? Is the growth epitaxial and what are the epitaxial relationships? What is the crystalline quality of the layer? What are the structural defects, which are known to have important effects on the properties of devices? What is the residual strain in the grown film and what are the mechanisms allowing the relaxation of this strain? Where do the adsorbate atoms rest on the surface? What kind of atoms (surfactants) or gases might be added to modify the growth mode, to render it more 2D (layer by layer) or better controlled 3D? How can we quantify the growth laws and especially the 3D cluster

Handbook of Thin Film Materials, edited by H.S. Nalwa
Volume 2: Characterization and Spectroscopy of Thin Films
Copyright © 2002 by Academic Press
All rights of reproduction in any form reserved.

growth and how do we control it to eventually obtain controlled self-organization? How can the physical properties of the heterostructures like reactivity, magnetism, electron transport, or light emission, for example, be connected to their crystalline quality? In this respect, is the morphology the dominant parameter or does the structure eventually influence the overall properties of the overlayer? How can we access buried interfaces; are they ordered or not? What is the role of the crystalline quality and cleanliness of the substrates?

The situation is even more complicated when we consider that all of these questions should be addressed throughout the growth. As a matter of fact, during growth many parameters may change. This affect not only the incoming atoms. Indeed, except for 2D growth, during the elaboration of the layer the atoms impinge on an evolving surface, for instance from an initial flat and clean substrate to an assembly of clusters or, finally, a continuous overlayer. All of these situations are very different, and the relative importance of each energetic term defining the behavior of the incoming atoms may change throughout the growth process. Thus it is only once we are able to answer most of the questions underlying growth, for each thickness, throughout the growth, that we may understand the physical processes that underlie the formation of a given interface. The description of the film, as a whole, after growth, can lead to erroneous interpretations if we do not know what happened to the film throughout its growth. Within this framework, the investigation of the structure of the film, *in situ*, during the growth itself is mandatory for an in-depth understanding and identification of the pertinent parameters. Such a fine-tuned description should then, in turn, allow theoreticians to evaluate the importance of each parameter and to elaborate improved models and potentials to describe thin film growth more adequately.

Surface and interface techniques and, in particular, synchrotron light provide a wealth of *in situ* and *ex situ* methods, allowing the investigation of almost any type of materials or physical properties. However, many of them will not be able to address all of the questions concerning the growth. Near-field techniques, although they are most helpful in providing an understanding surface structures, can provide only a top view of a surface, with or without a deposit on it. The information depth is thus extremely limited. The morphology of an overlayer can be more easily accessed, although the image will result from the convolution of the tip shape and the actual morphology that could lead to artifacts. Electrons, ions, or atom diffraction experiments undergo similar limitations, because their penetration depth is limited to a very few atomic planes, thus giving access only to the structure of the top layers. Spectroscopies are another important class of techniques. They often rely on charged particles, especially electrons. The electron mean free path in the material, which ranges roughly from 0.5 to 5 nm, thus limits the information depth. Moreover, because the inner shells of the atoms are probed and their energies are discrete, one does not generally have a choice of information depth. With such techniques the growth of a few layers is generally well described, although chemical reactions or bond formation of the atoms may transform the spectra and render a quantitative interpretation difficult or impossible. Finally, electron microscopy is a powerful technique that fully enables the investigation of buried interfaces but in a destructive way, preventing any investigation of growth modes.

On the one hand, increasing the information depth requires the use of particles that interact only weakly with matter. On the other hand, extracting useful information in a realistic counting time, for a number of atoms as limited as a tenth of a monolayer, will require very high fluxes to compensate for the weak interaction. This apparent contradiction has been solved with the advent of third-generation hard X-ray synchrotron sources. Of course, the number of possible techniques is reduced, but in this field, synchrotron light, especially through diffraction, gives insights into the intimate mechanisms *in situ* during growth. Such an approach allows an understanding of each step of the formation of an interface and avoids speculation from the observation of the final state. Because the information depth is as large as from submonolayer deposits up to micron-thick layers for hard X-rays, and because there are no limitations in experimental environment, the possible investigations represent wide and comparatively virgin fields. Indeed, X-ray experiments can be performed at almost all pressures, from ultrahigh vacuum to high-pressure conditions, with reactive gases or not, as well as at all temperatures, from a few Kelvin to thousands of degrees.

When dealing with surfaces and X-ray diffraction one may observe that we have access to the Fourier transform of the object under investigation (or, more exactly, its autocorrelation function). Of course, the phase information is lost and only the intensity can be measured, but it remains that each feature in direct real space has a counterpart in the reciprocal Fourier space. Intrinsically the information exists somewhere in reciprocal space. Specific techniques and equipment were designed to allow a quantitative record of the intensity in the regions of reciprocal space that are of interest with respect to surface and growth studies. Choosing the pertinent measurement and acquisition conditions is an important task in this respect, but the quality of the resulting data is noteworthy because they are quantitative and allow the determinication of exact parameters with their error bars. Moreover, X-ray diffraction as we use it in this chapter is kinematical; the experiments will thus not be hampered by multiple scattering effects or transformations of the electronic structure.

In Section 2 we describe the basics of two methods that we have used and developed for our investigations: grazing incidence X-ray diffraction (GIXD) and grazing incidence small-angle X-ray scattering (GISAXS). These techniques make it possible to quantitatively tackle all of the open questions, although not all of them will be answered for all systems. Through the examples, none of the questions enumerated above will be ignored. In some situations, complementary techniques must be used, after the growth, and in general the knowledge acquired during the growth plus the investigation after the growth will make it possible to draw a coherent and complete picture of the whole growth process.

To illustrate these powerful synchrotron-based *in situ* techniques, we give examples of interfaces that could not be investigated by other tools. In all cases, at least one component, substrate, film, or even both, will be an insulator, thus preventing the investigation, in an extended thickness range, from other techniques. This applies, in particular, to the investigation of buried interfaces *in situ*, during growth, and to insulating surfaces. The net result is a comparatively poor understanding of the physics and overall properties of these materials. However, these materials are neither rare nor unimportant. As a matter of fact, most crystallized matter is insulating or poorly conducting, and the way in which minerals interact with the atmosphere or water determines erosion and aging mechanisms that are of growing interest, for instance, with respect to waste storage. Thus, surfaces or interfaces of ceramic materials are very common in nature. Understanding their properties is a very important issue if one wishes to understand the formation processes and the reactivity of these materials. Ceramic surfaces are increasingly studied both theoretically and experimentally, because they are also involved in many rapidly innovating industrial sectors. Until very recently they were mainly used as supports for catalysis or thin film growth. Recent advances in their synthesis have extended the range of potential new applications. Indeed, ceramics as well as semiconductors or metals provide a large diversity of intrinsic properties: they might exhibit different gap widths, and some of them are ferrimagnetic, ferromagnetic, or antiferromagnetic. Unfortunately, because of their intrinsic insulating properties, only very little is known about the crystallographic structure of oxide surfaces and of the interfaces they may form with other materials. The present chapter deals with these surfaces and growth on them.

In Section 3 we examine in detail two single-crystal oxide surfaces that are routinely used in metal growth and semiconductor and superconductor thin film growth: MgO(001) and α-Al$_2$O$_3$(0001). We show that they can be obtained with an excellent and controlled crystalline quality. Examination of the growth of metals on MgO(001), reported in Section 4, was undertaken with respect to the lattice parameter mismatch and the adhesion energy, both parameters increase from Ag to Pd and finally to Ni.

Polar NiO(111) and CoO(111) surfaces exhibit attractive physical properties and thus are particularly interesting. They are antiferromagnets, and only a few years ago interfacial magnetic ordering effects on NiO-based films were observed, including evidence of the interlayer exchange interaction and antiferromagnetic ordering along the (111) planes in superlattices. There are also hints of an enhanced reactivity of NiO(111) films. In contrast to MgO or α-Al$_2$O$_3$(0001), NiO and CoO play an active role in the heterostructures built on them. They are substrates, but they are also responsible for the magnetic exchange coupling in spin valve devices. Despite these interesting properties, single-crystal surface studies were only undertaken in very recent years. The preparation conditions for NiO(111) and CoO(111) single crystals are be discussed in Section 3. Whereas the NiO(111) single crystals can be produced in a quality similar to that of the other oxide substrates, CoO(111) remains stabilized by a nonstoichiometric surface layer. The growth of exchange-coupled ferromagnetic metallic layers (Co

and Ni$_{80}$Fe$_{20}$) on single-crystalline NiO(111) is the subject of Section 5. The formation of the ferromagnetic film is characterized from the very beginning, as well as a reactive interface formation in the NiFe/NiO case. For this last system, the use of other techniques (atomic force microscopy, transmission electron microscopy) was necessary to achieve a complete picture of the interface. Finally, the growth of NiO(111) itself as epitaxial films on other materials like α-Al$_2$O$_3$(0001) and Au(111) is discussed in Section 6.

2. EXPERIMENTAL TECHNIQUES

2.1. Introduction: Particularities of Metal/Oxide Studies

Although many oxides are commonly used as substrates for thin film growth, their surface structures and properties are still not very well described, in contrast to metal or semiconductor surfaces. The control of the crystalline bulk structure is not the limiting issue, inasmuch as bulk Al$_2$O$_3$ or MgO may be grown in large quantities with a quality closer to that of Si than to those of many metals. From another point of view, although metals often have poorer bulk crystalline quality, because of their generally small fusion temperature, the polishing-induced polycrystalline texture can easily be removed by gentle annealing at moderate temperatures (below 1000 K). In contrast, many oxides have high fusion temperatures, and the atomic mobility in the surface region during gentle annealing is comparatively very small. The adequate preparation conditions for oxides are specific and require high temperatures, which imposes a design of specific furnaces that is often not available in standard ultrahigh vacuum (UHV) preparation chambers. Second, oxides are generally of complex structure, inasmuch as they are made of at least two elements, making their investigation more difficult than that of pure elements. However, the most stringent limitation comes from their insulating character. Indeed, many oxides are good insulators, and because most of the common surface science techniques are based on the backscattering of charged particles (electron spectroscopy and diffraction, tunnel microscopy), the charge buildup severely limits their use. For clean surfaces, the use of float guns or small densities of incident particles allows for some investigations, but as soon as metallic particles are deposited, the charge buildup becomes so large and uncontrollable that no clear conclusion can be reached. The situation is better when the incident particles have no charge, as is the case for photons. Some perturbations still occur if the analyzed scattered particles possess a charge, like electrons in X-ray photoelectron spectroscopy (XPS). Thus it is clear that the methods best-suited for the investigation of oxide surfaces have to be based on neutral incident and neutral scattered particles. Of course, these methods also have their drawbacks and limitations. In particular, the generally small interaction cross sections of matter with X-rays require high incidence fluxes on the samples. The use of synchrotron light then becomes a necessity. Moreover, even with high fluxes, geometrical constraints and sample requirements (see Section 3.1) will apply, and, in particular, grazing incidence geometry will be used to enhance the surface signal with respect to the bulk contribution.

We describe in the next section some grazing incidence X-ray methods. They are techniques of choice for the investigation of any type of surface (metals, semiconductors, or insulators), but are among the very unusual methods that apply as well to insulating oxide surfaces and interfaces. We describe the diffraction and diffusion of X-rays by a surface and focus on practical considerations, limitations, and orders of magnitude. For a more comprehensive description of the well-known and well-documented standard diffraction and crystallography, the reader may refer to standard textbooks and reviews [1–10].

2.2. Refraction from Surfaces

When light crosses a flat interface between two different materials, some part of the beam is reflected and some is transmitted through the interface. Depending on which material is the optically denser medium, the propagation direction in the second medium will be closer to or further from the normal of the interface. This very general phenomenon is called refraction, and it occurs for all wavelengths and all types of interfaces, crystalline or not. Mathematically, the behavior is well understood by introducing the index of refraction. For visible light, air has the smaller index of refraction, and total reflection is known to occur within materials: an evanescent wave that is damped exponentially in the material appears instead of emergent light on the other side of the interface, under internal incidence conditions given by Snell's law.

In the case of a sharp interface between the vacuum and a surface of a material of wavelength-dependent index of refraction n, the geometry that applies is given in Figure 1. An incident beam made of a linearly polarized plane wave impinges on a surface at an angle α_i, with an amplitude $\mathbf{E}_i$, and along a wave vector $\mathbf{k}_i$; the reflected beam leaves at an angle α_f, at amplitude $\mathbf{E}_f$, and along a wave vector $\mathbf{k}_f$; and the transmitted beam makes an angle α_t with the surface and has an amplitude $\mathbf{E}_t$ and a wave vector $\mathbf{k}_t$. Applying Snell's law gives

$$\cos(\alpha_t) \cdot n = \cos(\alpha_i) \quad \text{and} \quad \alpha_f = \alpha_i \qquad (1)$$

As long as $n > 1$ total reflection cannot occur when light travels from the vacuum to the material, even if $\alpha_i = 0$. Fortunately, unlike visible light, when hard X-rays (i.e., the energy is above several keV) are considered, the index of refraction is generally less than unity and can be written as

$$n = 1 - \delta - i\beta \qquad (2)$$

with

$$\delta = \frac{\lambda^2 \cdot e^2 \cdot N_A \cdot \rho}{2\pi mc^2} \cdot \frac{\sum_{j \in \text{Cell}} (Z_j - f_j')}{\sum_{j \in \text{Cell}} A_j} \qquad (2a)$$

and

$$\beta = \frac{\lambda^2 \cdot e^2 \cdot N_A \cdot \rho}{2\pi mc^2} \cdot \frac{\sum_{j \in \text{Cell}} f_j''}{\sum_{j \in \text{Cell}} A_j} = \frac{\lambda\mu}{4\pi} \qquad (2b)$$

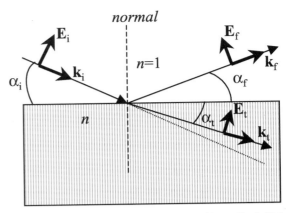

Fig. 1.   Refraction and reflection of a plane wave with amplitude $\mathbf{E}_i$ incident upon the interface between a vacuum and a material of index n.

where the summation is over all atomic species j present in the unit cell; N_A, $(Z_j - f_j')$, f_j'', A_j, ρ, μ, and λ are, respectively, Avogadro's number, the scattering factor, the anomalous dispersion factor, the atomic weight of species j, the density, the photoelectric absorption coefficient, and the wavelength.

Thus in the case of hard X-rays, total external reflection occurs on the vacuum side (although for very grazing angles, because δ and β are, respectively, in the 10^{-5} and 10^{-6} range), leaving a critical angle for total external reflection $\alpha_c \approx \sqrt{2 \cdot \delta}$ in the 0.1–0.6° range. It is important to note that the value of $n < 1$ is the key that allows surface investigations with hard X-rays; in turn, it also fully defines the geometry of the experiments. When $\alpha_i < \alpha_c$, the component of the transmitted wavevector normal to the surface becomes imaginary and the refracted wave is exponentially damped as a function of the distance below the surface and is an evanescent wave traveling parallel to the surface. The $1/e$ depth of penetration for the intensity of the X-rays becomes

$$\Lambda = \frac{\lambda}{4 \cdot \pi \cdot \text{Im}(\sqrt{\alpha_i^2 - \alpha_c^2 - 2i\beta})} \qquad (3)$$

The reflection and the transmission coefficients of the surface are critically dependent on α_i, and their variations are given by Fresnel's formulae,

$$R(\alpha_i) = \frac{I_f}{I_i} = \left| \frac{\sin\alpha_i - \sqrt{n^2 - \cos^2\alpha_i}}{\sin\alpha_i + \sqrt{n^2 - \cos^2\alpha_i}} \right|^2$$

$$T(\alpha_i) = \frac{I_t}{I_i} = \left| \frac{2\sin\alpha_i}{\sin\alpha_i + \sqrt{n^2 - \cos^2\alpha_i}} \right|^2 \qquad (4)$$

Let us now discuss some of the important features of these formulas. Figure 2 reproduces the calculated Λ and T for some typical situations. For $\alpha < \alpha_c$, $R = 1$ and total external reflection occurs, and, as expected, the penetration depth is a minimum and is in the nanometer range, highlighting the great surface sensitivity that can be achieved. The departure from unity of n depends linearly on δ, showing that α_c increases with the density of the material; light elements like MgO and Al_2O_3 thus have particularly small critical angles. Moreover, through

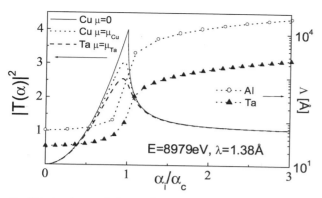

Fig. 2. Calculated transmission coefficients (left ordinate) as a function of the reduced grazing incidence angle, α_i/α_c, for a nonabsorbing Cu surface (—), a real Cu surface (···), and a real Ta surface (---). Penetration depths (right ordinate) versus α_i/α_c for real Al (o) and Ta (▲) surfaces.

their small absorption coefficients they also have small β values, and thus larger Λ, making grazing incidence conditions even more desirable for such surfaces (to limit Λ). From another point of view, the reflectivity falls off rapidly as α_i^{-4} when $\alpha_i \gg \alpha_c$, enabling tunable depth analysis and in turn the investigation of buried interfaces under up to micrometer-thick capping layers; it also allows growth monitoring up to fairly thick deposits. For a crystalline surface the diffracted intensity will be proportional to the transmission coefficient because it is related to the electrical field strength at the dielectric boundary, i.e., at the surface. Although the wave does not propagate below α_c, the signal will be enhanced by a factor of 4 for a nonabsorbing material when $\alpha_i = \alpha_c$. Because of time microreversibility, the diffracted beam experiences exactly the same refraction effects, and a second enhancing factor can be obtained when the exit angle from the surface is equal to α_c. A comprehensive discussion of the refraction effects of the outgoing beam can be found in [1]. This advantage exists for all elements, although it is damped by increasing absorption, i.e., for heavier elements (Fig. 2). However, working at grazing incidence has another very useful feature: it allows a drastic reduction of the background. As a matter of fact, except for the fluorescence contribution, most of the background originates from the bulk (thermal diffuse scattering, point defect scattering, etc.) and can be removed or heavily reduced by limiting the incidence and/or the emergence angle to α_c. The only other less efficient way of overcoming the bulk background, at least the thermal diffuse scattering, is to cool down the sample to very low temperatures, thus reducing the thermal agitation of the atoms.

For light elements, reducing the penetration depth and the background and the necessity to enhance the weak scattering often imposes the requirement of working at the critical angle for total external reflection. In such conditions great care must be taken to keep the incident angle strictly constant throughout the data collection. Large intensity variations are obtained for very small variations of the incidence angle, because $\alpha_i = \alpha_c$ corresponds to the maximum of the $T(\alpha_i)$ function. Working at $2\alpha_c$ or $3\alpha_c$ allows for more comfortable measurement conditions, if the background remains acceptable. In any event,

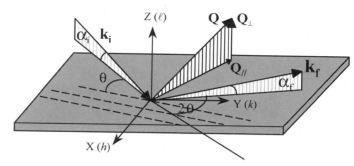

Fig. 3. Grazing incidence X-ray diffraction geometry. See text for the definitions of the notations.

grazing incidence often remains the mandatory condition in investigations of surfaces with hard X-rays. This principle is used for grazing incidence X-ray diffraction (GIXD) as well as for grazing incidence small-angle X-ray scattering (GISAXS), which are the two main methods that we have used to investigate metal/oxide interfaces.

2.3. Grazing Incidence X-Ray Diffraction

2.3.1. Surface Diffraction

Considering the remarks of the previous section, the general geometry of a GIXD experiment will be as depicted in Figure 3. An incident beam with wavevector $\mathbf{k}_i$ falls on a surface under an incidence angle α_i that is kept close to α_c. If the material is a single crystal, the reflected beams will organize, for given incidence conditions with respect to the atomic planes, along well-defined directions with a wavevector $\mathbf{k}_f$ and an exit angle α_f: They are diffracted by the well organized array of atoms. It is convenient to define the momentum transfer $\mathbf{Q} = \mathbf{k}_f - \mathbf{k}_i$ and to decompose it into components parallel ($\mathbf{Q}_\parallel$) and perpendicular ($\mathbf{Q}_\perp$) to the surface.

For hard X-rays and small objects the kinematical approximation of single scattering is valid [6]. The intensity, $I(\mathbf{Q})$, elastically scattered in a direction defined by the momentum transfer $\mathbf{Q}$ is proportional to the square modulus of the coherent addition of the amplitudes scattered by all electrons in the diffracting object. The field seen at large distance R from a scattering electron of charge e and mass m at $\mathbf{r}'$ is given by the well-known Thomson formulae. Within the Born approximation a single atom at $\mathbf{r}$ diffracts an amplitude obtained by integrating over its electronic distribution function $\rho(\mathbf{r}')$ about $\mathbf{r}$. Defining the atomic form factor, $f(\mathbf{Q})$, as the Fourier transform of $\rho(\mathbf{r}')$, a unit cell of a crystal with N atoms diffracts an amplitude

$$A_{\text{Cell}} = \sqrt{P}\, A_0 \, \frac{e^2}{mc^2 R} \sum_{j=1}^{N} \left(f_j(\mathbf{Q}) \cdot e^{i\mathbf{Q}\cdot\mathbf{r}_j} \cdot e^{-M_j} \right)$$

$$= \sqrt{P}\, A_0 \, \frac{e^2}{mc^2 R} \cdot F(\mathbf{Q}) \tag{5}$$

which defines the structure factor, $F(\mathbf{Q})$, of the unit cell, and where P, A_0^2, and e^{-M_j} are, respectively, the polarization factor, the incident intensity in photons per unit area per second,

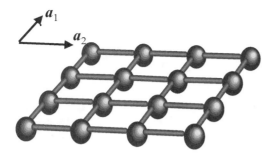

Fig. 4.   Free standing surface layer of atoms.

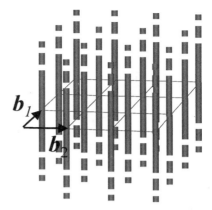

Fig. 5.   Reciprocal space of Figure 4.

and the Debye–Waller factor [6, 11] for species j. M may be written as $B \sin^2 \theta / \lambda^2$, where $B = 8\pi^2 \langle u_x^2 \rangle$ is called the temperature factor and $\langle u_x^2 \rangle$ is the mean square component of vibration of the atom along the direction of momentum transfer. The thermal motion can yield to a single B value for all atoms or individual B's for each atom or even components of a tensor describing the anisotropic thermal motion. For example, it has been shown that the thermal motion at the surface can be enhanced compared with its bulk [12] counterpart, although this is not necessarily the case. The polarization factor, P, describes the dependence of $\mathbf{E}_i$ on the polarization of the incoming wave. Because the direction of the electrical field determines the direction of the electron motion that will radiate the wave, the angle 2θ between the incident and the exit beams will modulate the observed intensity. When $\mathbf{E}_i$ is normal to the scattering plane (the plane spanned by $\mathbf{k}_i$ and $\mathbf{k}_f$) P is unity, and when $\mathbf{E}_i$ is in the scattering plane $P = \cos^2 2\theta$.

If we now consider a two-dimensional surface built with N_1 and N_2 atoms along the $\mathbf{a}_1$ and $\mathbf{a}_2$ directions as shown in Figure 4, we will obtain the scattered intensity by summing the amplitudes over the $N_1 \times N_2$ unit cells. It is convenient to express $\mathbf{Q} = \mathbf{q}_1 + \mathbf{q}_2 + \mathbf{q}_3$ in the reciprocal space basis as $\mathbf{Q} = h\mathbf{b}_1 + k\mathbf{b}_2 + \ell\mathbf{b}_3$ with the reciprocal $\mathbf{b}_i$ vectors related to the direct vectors by $\mathbf{b}_i = 2\pi(\mathbf{a}_j \times \mathbf{a}_k)/(\mathbf{a}_i, \mathbf{a}_j, \mathbf{a}_k)$, where $\mathbf{a}_3$ can be any vector perpendicular to the $(\mathbf{a}_1, \mathbf{a}_2)$ plane. As a convention in surface diffraction, ℓ (resp. Z) is always chosen perpendicular to the surface, and h and k (resp. X and Y) span the surface plane (Fig. 3). Defining also the $S_N(x) = \sin^2(Nx/2)/\sin^2(x/2)$ function, the intensity of the total scattered signal from the surface is

$$I_S^{2D}(\mathbf{Q}) = PA_0^2 \frac{e^4}{m^2c^4R^2} |F(\mathbf{Q})|^2 S_{N_1}(\mathbf{Q} \cdot \mathbf{a}_1) \cdot S_{N_2}(\mathbf{Q} \cdot \mathbf{a}_2) \quad (6)$$

For large N_1 and N_2 values, $I_S^{2D}(\mathbf{Q})$ yields significant intensities only when both Laue conditions are fulfilled simultaneously: $\mathbf{Q} \cdot \mathbf{a}_1 = 2\pi h$ and $\mathbf{Q} \cdot \mathbf{a}_2 = 2\pi k$, where h and k are integers, defining a two-dimensional reciprocal lattice. Because the intensity is independent of $\mathbf{q}_3$, the scattering is diffuse in the direction perpendicular to the surface and the reciprocal space is made of continuous rods as shown in Figure 5. The intensity in the diffraction rods reduces to

$$I_{hk}^{2D} = PA_0^2 \frac{e^4}{m^2c^4R^2} |F_{hk}|^2 N_1^2 N_2^2 \quad (7)$$

The expression of I_{hk}^{2D} is valid as long as only exactly one layer diffracts. This is the case when the surface layer has a periodicity different from that of the bulk, i.e., in the case of surface reconstruction or if a 2D film with a mismatched lattice parameter is deposited on a substrate. Both situations will be illustrated by reconstruction analysis of NiO(111) and α-Al$_2$O$_3$(0001) (see Section 3) and by the structural description of a NiO(111) layer deposited on Au(111) (see Section 6). As soon as there are several layers in the perpendicular direction, the diffraction rods become modulated because the periodicity in the third direction must be taken into account and the intensity will take the usual form,

$$I_{hk\ell}^{3D} = PA_0^2 \frac{e^4}{m^2c^4R^2} |F_{hk\ell}|^2$$
$$\times S_{N_1}(\mathbf{Q} \cdot \mathbf{a}_1) \cdot S_{N_2}(\mathbf{Q} \cdot \mathbf{a}_2) \cdot S_{N_3}(\mathbf{Q} \cdot \mathbf{a}_3) \quad (8)$$

For small N_3 values, the rods will modulate along $\mathbf{Q}_\perp$. In contrast, for large N_3 values, the intensity will concentrate along a discrete array in the third direction too, leading to a third Laue condition: $\mathbf{Q} \cdot \mathbf{a}_3 = 2\pi\ell$, where ℓ is an integer. This situation corresponds to the classical 3D diffraction, where Bragg peaks correspond to the possible h, k, and ℓ's.

An intermediate situation is obtained when a physical surface is considered (i.e., the truncation of a bulk material, that is only semi-infinite), and Eq. (8) will no longer apply because it assumes infinite extension of the diffracting object in all three directions. Taking into account the summation from $-\infty$ to 0 along $\mathbf{a}_3$ of $N_1 \times N_2$ surface unit cells, the scattered intensity for a perfectly sharp surface becomes

$$I_{hk\ell}^{CTR} = PA_0^2 \frac{e^4}{m^2c^4R^2} |F_{hk\ell}|^2 S_{N_1}(\mathbf{Q} \cdot \mathbf{a}_1) \cdot S_{N_2}(\mathbf{Q} \cdot \mathbf{a}_2) \cdot I_\ell^{CTR} \quad (9)$$

with

$$I_\ell^{CTR} = \frac{1}{2\sin^2(\mathbf{Q} \cdot \mathbf{a}_3/2)} \quad (10)$$

The intensity variation along a crystal truncation rod (CTR) now contains Bragg peaks for integer values of h, k and ℓ, not excluded by the 3D extinction rules that are still valid, and diffuse scattering in between, as shown in Figure 6. Importantly,

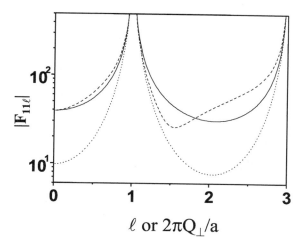

Fig. 6. Evolution of the calculated intensity along the $\langle 11\ell \rangle$ rod of a Ag(001) surface (face-centered cubic structure) as a function of the perpendicular momentum transfer $Q_\perp$ (or Q_z or ℓ) for a perfectly sharp surface (——), an outside oriented relaxation of $+5\%$ of the last Ag layer (– – –) and a surface with a 4-Å roughness (β^n model with $\beta = 0.6$) and no relaxation ($\cdots$).

this means that, insofar as the two first Laue conditions are fulfilled, ℓ must be considered as a continuous variable for surface diffraction. Even half-way from the Bragg peaks some intensity remains, because the last term in Eq. (9) is then $1/2$ unity; this remaining intensity is comparable to the intensity of a single layer, $I_S^{2D}(\mathbf{Q})$. Real surfaces have roughness; i.e., the probability distribution of the atoms is less sharp than a simple step. In that case the intensity will concentrate more near the Bragg peaks and decrease in the zone center. To take this effect into account, a functional form used to fit the data has been derived [13] within the β^n model for simple unit cells, and the adequate CTR term becomes

$$I_\ell^{CTR} = \frac{(1-\beta)^2}{1+\beta^2 - 2\beta \cos(\mathbf{Q} \cdot \mathbf{a}_3)} \cdot \frac{1}{\sin^2(\mathbf{Q} \cdot \mathbf{a}_3/2)} \quad (11)$$

with $0 < \beta < 1$ representing the roughness and where $\beta = 0$ (resp. 1) corresponds to a perfectly flat (resp. infinitely rough) surface. This model gives generally good results, although it can give significant occupancies for planes fairly far away from the average surface. The sensitivity of the CTRs to roughness is well illustrated in Figure 6, where the decrease in intensity in the zone center can easily represent an order of magnitude for roughness well below a nanometer. Alternative models, in which the roughness is treated like an additional Debye–Waller factor, were also proposed [14].

A last important feature of the CTRs is their great sensitivity to surface relaxation. Relatively small variations in the last layer spacing are able to produce measurable asymmetries in the shape of the intensity variation along CTRs. If, for example, the last interplanar distance is b instead of a_3, the intensity along the CTR will obey

$$I_\ell^{CTR} = \left| \frac{e^{i\pi\ell}}{2i \sin(\pi\ell)} + e^{i2\pi\ell b/a_3} \right|^2 \quad (12)$$

The result of a b/a_3 ratio of 1.05 in the case of a Ag(001) surface is illustrated in Figure 6. The effect increases with increasing perpendicular momentum transfer. In practice, the geometry of the diffractometer and the wavelength will limit the maximum reachable ℓ value.

We have described the main features that may occur in the reciprocal space when surfaces are considered. Let us now concentrate on the way in which these intensities can be efficiently measured and the underlying practical limitations.

2.3.2. Practical Considerations

In practice, the reciprocal space is a space of directions; with respect to the scattering geometry from Figure 3, at least four settable angles are needed: one for the incidence angle, a rotation of the sample around its normal to bring the atomic planes into diffraction conditions, and two degrees of freedom to position the detector arm to reach all $\mathbf{Q}_\parallel$ and $\mathbf{Q}_\perp$ positions in space. Because an extra angular degree of freedom is available (three angles are enough to define any direction of $\mathbf{Q}$), one condition can be imposed. For standard four-circle diffractometers, the most popular working conditions in surface diffraction are incidence fixed, emergence fixed, or incidence equals emergence. Including one of these conditions, all (h, k, ℓ) positions in reciprocal space are connected in a unique manner to an angular setting. It is convenient to add two additional cradles below the sample to align the optical surface.

Prevention of contamination during the study of surfaces or interfaces requires UHV conditions. From the expression of $I_S^{2D}(\mathbf{Q})$ it appears that the diffracted intensity is proportional to the incident flux, and, through $F(\mathbf{Q})$, it is proportional to the square of the atomic number Z. A comparison of (8) and (6) shows that bulk scattering is roughly six orders of magnitude larger than surface scattering. Moreover, if A is the area of the unit cell, the intensity falls off as A^2, which strongly reduces the diffracted intensity for large unit cells occurring for reconstructed surfaces; the increased number of atoms only partially compensates for this effect. Thus, although surface diffraction experiments can be carried out with a laboratory rotating anode for very dense materials with high Z, they becomes almost impossible for lighter elements and large reconstructions that will require high flux synchrotron radiation sources like the ESRF [15] (European Synchrotron Radiation Facility, Grenoble, France). When deposited layers of only a few monolayers are considered, the flux condition becomes even more stringent. Importantly, the synchrotron beam is strongly polarized in the horizontal plane. For crystallography, vertical sample geometry is then preferred because the polarization factor will remain unity for all in-plane reflections, whereas in the horizontal sample orientation the scattering at 90° would vanish. Considering all of the previous remarks, several UHV diffractometers, all dedicated to *in situ* surface diffraction, were recently built on various synchrotrons; they all have about the same overall characteristics [16–23]. The GIXD experiments of our group were mainly performed with the setups located at the ESRF [15] on beamlines ID03 [21], ID32 [22], and BM32-SUV [17]. The last

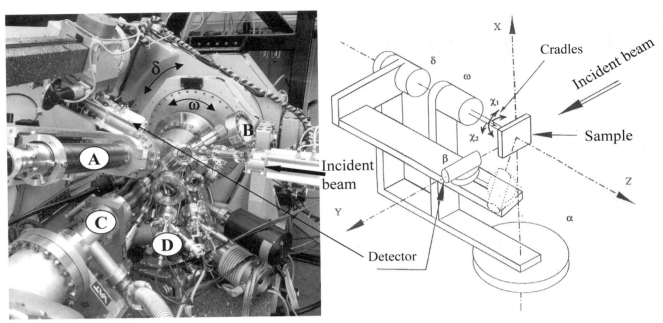

Fig. 7. (Left) Photograph of the BM32 4-circle diffractometer. The UHV chamber, with a base pressure of 2×10^{-11} mbar, is equipped with a small input and a large exit Be window that is transparent to high-energy X-rays. (A) An Auger analyzer. (B) A reflection high-energy electron diffraction (RHEED) facility. (C) A fast transfer system. (D) Several evaporation sources. (Right) Geometrical principle of the diffractometer and definition of the angles.

setup is installed on a bending magnet beamline and is represented in Figure 7.

Alternatively, the bulk truncation can be understood as the multiplication of an infinite lattice by a step function. In reciprocal space, this leads to the convolution of the reciprocal space of the infinite 3D crystal with the Fourier transform of a step. The result is a smearing of the intensity of each Bragg point in the direction perpendicular to the surface. Note that in the case of a crystalline surface and where optical surfaces are not identical (because of miscuts or vicinal surfaces), the smearing is still perpendicular to the optical surface and not to the lattice, leading to partial CTRs emerging from each Bragg peak without continuity along ℓ. The intensity along the CTR is not modified by a miscut, but the correct (h, k) position, for a given ℓ, at which the rocking scan must be performed will be affected and must be refined experimentally [24]. A convenient practical criterion is to remember that the derived roughness should not depend on which rod is measured, nor should it depend on whether the rod is above or below the Bragg position [25]. Inadequate measurements when a miscut is present lead to additional asymmetry and thus to overestimated roughness and to erroneous relaxation values. The measurements must thus be carried out with great care; the observation of a diffracted beam is only quantitatively significant if all angles are perfectly defined and if the (h, k, ℓ) position has been adequately refined.

2.3.3. Data Collection, Integrated Intensities, and Corrections

In contrast to bulk diffraction, the intensity cannot be recorded simply by placing a detector with a wide enough aperture at the right (h, k, ℓ) positions, because surface diffraction is diffuse, at least in the direction perpendicular to the surface. Peak broadening may also occur in various well-defined directions, depending on its physical origin. Some of the possible features are reported in Figure 8. When the diffracting object has a finite size D (for example, an island) it means that the number of planes N_1 and/or N_2 in $I_S^{2D}(\mathbf{Q})$ will be small and the Laue condition will be partially relaxed: the intensity is no longer strictly peaked at integers (h and k). The finite domain size leads to a constant broadening with respect to $\mathbf{Q}_\parallel$ (Fig. 8a) and is related to the angular width $\Delta\omega$ by

$$D = \frac{2\pi}{Q_\parallel \cdot \Delta\omega} \tag{13}$$

Relative disorientation between individual grains will lead to a constant angular broadening (in $\mathbf{Q}_\parallel$) because each grain has its own reciprocal space with some angular deviation from the average reciprocal space (Fig. 8b). Finally, when a parameter distribution is present, reciprocal spaces with different basis vectors will superimpose about the average lattice, and the diffraction features will accordingly broaden (Fig. 8c and d). Measuring the angular and radial in-plane widths of several orders of diffraction (i.e., as a function of $Q_\parallel$) allows decorrelation of at least the two first effects. This approach is used and discussed in Section 5.2.2.

Because all of these lineshapes may coexist at any point in the reciprocal lattice, the pertinent intensity is the integrated intensity that is obtained by scanning one or several angles through the investigated position. To put it simply, it allows collection of all of the intensity that should ideally be present at a given point in the reciprocal space. In practice this corresponds

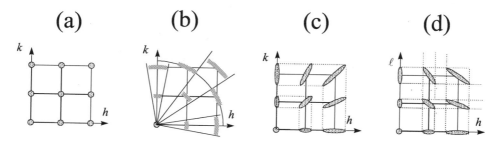

Fig. 8. Schematic drawing of the major sources of peak broadening. (a) Finite domain size. (b) Mosaic spread. (c) In-plane parameter distribution. (d) In-plane and out-of-plane parameter distribution.

to scanning one direction while integrating along the perpendicular direction with sufficiently opened slits, i.e., an adapted resolution function. This strategy applies well for in-plane measurements and works in principle for measuring CTRs through a scan along ℓ, but other broadening perpendicular to the CTR is often so great that not all intensity reaches the detector. Moreover, scanning along a CTR implies a perfect stability of all angles of the diffractometer. Successive rocking scans with a constant angular speed, Ω, around the surface normal (ω scans with the SUV setup in Fig. 7) for discrete ℓ values are thus the usual and preferred measurement strategy, although the slice of integrated intensity now depends on the aperture of the slits before the detector. The measured intensity thus depends on geometrical settings; fortunately, in grazing incidence $\mathbf{Q}_\perp$ is also perpendicular to the surface plane, and the integrated momentum width $\Delta Q_\perp$ is nearly constant for small ℓ's and is simply related to the exit slit size L by

$$\Delta Q_\perp = \frac{2\pi L}{\lambda R} \cos\beta \qquad (14)$$

Because the integration is performed over an angular unit volume ($d\alpha\, d\beta\, d\delta$) and not a momentum unit volume ($dq_1\, dq_2\, dq_3$), the measured intensity will depend on the relationship between the reciprocal space coordinate and the angular coordinate. The corresponding factor that divides the intensity is called the Lorentz factor, $\mathcal{L}$. In particular, this means that the intensity will depend on the type of scan that is performed and that each type of scan will produce a different Lorentz factor. Finally, the measured integrated intensity will also sit on a background $\mathcal{B}$ and will be proportional to the active area of the sample defined by the slits before and after the sample because they define the number of atoms ($N_1 \times N_2$) that contribute to the signal. Finally, if ω is the rocking angle, the integrated intensity that is really measured during a rocking scan is

$$I_{hk\ell}^{\mathrm{Mes}} = \int_\omega \left\{ \mathcal{B} + \frac{PA_0^2}{\mathcal{L}\Omega} \frac{e^4}{m^2 c^4 R^2} |F_{hk\ell}|^2 \right.$$
$$\left. \times N_1 \cdot N_2 \cdot \int_{\ell-\Delta Q_\perp/4\pi}^{\ell+\Delta Q_\perp/4\pi} I_u^{\mathrm{CTR}}\, du \right\} d\omega \quad (15)$$

The measured intensity in itself is not of direct use. The important quantity is $|F_{hk\ell}|$, although the phase is definitively lost. Deducing a quantity proportional to the structure factor from a

measurement requires a correction of the integrated intensity for the factors that are introduced by the experiment and the geometry. The background subtraction (generally a linear regression) requires large enough scans to reach points where no scattering contributes. Generally, a beam monitor is mounted before the sample to normalize the intensity with respect to the incident photon flux and in turn permits rescaling of scans that were not performed at the same speed. Another correction comes from the integration in (15); because $\Delta Q_\perp$ varies slowly and because the slope of the rod along ℓ may vary much faster, the actual ℓ corresponding to a measurement is shifted when the slope is large. Finally, the Lorentz, area, and polarization factors depend on the diffractometer geometry and on the beamline characteristics that include the degree of linear/elliptical/circular polarization and must be calculated for each setup. The reader may refer to the reference adapted to a given setup for these corrections [26–30].

Once the structure factors have been extracted from the measured signal, they have to be averaged with their symmetry-related equivalents to find the right symmetry of the signal, and complementarily, to deduce the systematic error, which generally is close to 10% for GIXD experiments.

In the absence of a model and because the phase information is missing, the Patterson map analysis is a convenient tool to test at least in-plane projected structures for reconstructions. The Fourier transform of the structure factor moduli does not give the electron density map in the unit cell, $\rho(\mathbf{r})$, but the density–density correlation function (or autocorrelation function). Its planar section in the direct space is referred to as the Patterson map. It can be written as

$$P(\mathbf{r}) = \sum_{hk\ell} |F_{hk\ell}|^2 \cdot e^{-i\mathbf{Q}\cdot\mathbf{r}} = \int \rho(\mathbf{r})\rho(\mathbf{r}+\mathbf{r}')\, d^3\mathbf{r}'$$
$$= \langle \rho(\mathbf{r})\rho(0) \rangle \qquad (16)$$

For in-plane measurements (i.e., $\ell = 0$), because of Friedel's law ($|F_{h,k,0}| = |F_{-h,-k,0}|$), the Patterson function is real and reduces to

$$P(\mathbf{r}) = 2\sum_{hk} |F_{hk}|^2 \cos\big(2\pi(hx + ky)\big) \qquad (17)$$

where x and y are the coordinates within the unit cell.

To test the agreement between a model structure and the experimental data, two criteria can be used. The first is the chi-squared, χ^2, approach. For N measured diffraction peaks, p parameters in the model, and an experimental uncertainty $\sigma_{hk\ell}^2$ for an (h, k, ℓ) peak, one can write

$$\chi^2 = \frac{1}{N-p} \sum_{hkl} \left(\frac{|F_{hk\ell}^{\text{exp}}| - |F_{hk\ell}^{\text{calc}}|}{\sigma_{hk\ell}} \right)^2 \tag{18}$$

A good agreement is obtained when χ^2 is close to 1, and no new parameter should then be introduced in the model. The second criterion is the reliability factor R, which is given by

$$R = \frac{\sum_{hk\ell} ||F_{hk\ell}^{\text{exp}}| - |F_{hk\ell}^{\text{calc}}||}{\sum_{hk\ell} |F_{hk\ell}^{\text{exp}}|} \tag{19}$$

when R approaches $(1/N) \sum_{hk\ell} (\sigma_{hk\ell}/|F_{hk\ell}^{\text{exp}}|)$ the agreement is good. Both criteria are helpful in discriminating between different possible models.

2.3.4. Application to Structure Determination and Growth Mode Studies

When a reconstructed surface is investigated, the accuracy of the final model will depend on the type of data that were measured. To obtain in-plane diffraction peaks for a complete structural determination, reconstruction rods and CTRs are needed. Peaks in the Patterson map allow identification of interatomic vectors and thus elaboration of projected possible 2D models. Once a convenient model is found, the perpendicular atomic positions z can be introduced, and the modulations along reconstruction rods can be reproduced, hence resolving the structure of the atomic planes contributing to the reconstruction and those that do not belong to the bulk. Finally, taking into account CTRs should allow determination of the registry of the reconstructed cell with respect to the underlying bulk. Only models that are able to simultaneously reproduce all three features (Patterson map, rods, and CTR) are acceptable. It is exactly this approach that has been used for the determination of the structure of p(2 × 2)-NiO(111) reconstructed surfaces. The resulting model, however, may not be unique and remains strongly dependent on the number of reflections that are measured. As a general rule, the level of confidence in a model increases with the number of measured reflections. If only in-plane data and rods are available, only the internal structure of the reconstructed unit mesh can be obtained. This was the case for p(2 × 2)-NiO(111)/Au(111) layers because of the small lattice mismatch between Au and NiO. Finally, if the scattering material is too light, only the in-plane reflections may be accessible, as is the case for the α-Al$_2$O$_3$(0001) reconstructions. For these studies it is of practical use to note that as a general rule, the h, k, and ℓ indexes are always expressed in reciprocal lattice units (r.l.u.).

When an epitaxial thin film is deposited on a substrate surface, all of the features in reciprocal space that have previously been discussed may exist for the epilayer, plus interferences between the film and the substrate. The deposited film will exhibit its own reciprocal lattice, which will superimpose on the reciprocal space of the substrate. Because the penetration depth of X-rays is large, attenuation of the substrate features will only occur for fairly thick overlayers (several nanometers). If the film is fully incoherent, with a parameter different from the substrate, both reciprocal spaces will be fully resolved and can be studied separately by Fourier filtering. The peak positions during growth give a direct access to the strain relaxation of the layer and the widths of the peaks to the quality of the growing film. Such investigations were performed for the growth of Co and Ni$_{80}$Fe$_{20}$ on NiO(111) (see Section 5) and for the native Co$_3$O$_4$ overlayer on CoO(111) (see Section 3.5). It may happen that the overlayer adopts different possible variants because the symmetries of the film and the substrate do not match. Each variant will exhibit its own reciprocal space, and the relative intensities between the peaks allow quantification of the structural composition of the film. The Ni(110) epitaxy on MgO(001) (see Section 4.3) corresponds to such a situation. Because GIXD allows investigation of the reciprocal space in 3D, it is also easily able to discriminate and to quantify different out-of-plane stacking with the same in-plane structure, like twins and stacking faults in the face-centered cubic (FCC) structures observed for Co and Ni$_{80}$Fe$_{20}$ on NiO(111).

Although pseudomorphic growth is quite rare (exact parameter match), it happens often that at least some of the epilayer atoms occupy positions continuing the staking of the substrate. In such a case these atoms are correlated via the substrate, and they will diffract at the same $\mathbf{Q}_\parallel$ positions as the substrate. But because they also belong in some way to the substrate stacking, it is not the intensities that will add at these $\mathbf{Q}_\parallel$ positions, but the amplitudes leading to constructive or destructive interferences. In practice the substrate CTRs will be modulated and become asymmetric, basically for exactly the same reason that relaxations make them asymmetric. The effect is particularly marked when the substrate is the lighter scatterer. The detailed investigation of the modulations along the CTRs during growth allows the extraction of the registry of the epitaxial atoms. This has been done for Ag, Pd, and Ni films on MgO(001) (see Section 4).

Finally, not only arrangements of atoms may diffract, but also well-organized superstructures of such defects as, for example, dislocations, that may organize in networks to accommodate the lattice mismatch between two materials. Such a dislocation array will produce in-plane satellites visible between the substrate and epilayer peaks. The dislocation networks between Ag/MgO(001) and Pd/MgO(001) were investigated in detail (see Sections 4.1.3 and 4.2.2).

Although GIXD is a very powerful method, it has a very poor chemical resolution, insofar as the atomic species have close atomic numbers, Z. In the case of Ni$_{80}$Fe$_{20}$ on NiO(111), the interface can be rendered reactive and diffusion occurs, but all metallic atoms have close Z numbers. For this situation we have used energy-filtered electron transmission microscopy (EF-TEM) and high-resolution TEM as a complementary method to determine the nature of the diffuse interface (see Section 5.3).

2.4. Grazing Incidence Small-Angle X-Ray Scattering

Determining the morphology of islands during their growth on a substrate is a very important step in the fabrication control of nanometer-sized objects (nano-objects). For this reason, a new method, called grazing incidence small-angle X-ray scattering (GISAXS), has been developed in the last decade [31–33] and very recently applied *in situ*, in UHV, in real time during growth [34, 35]. We briefly describe below some useful characteristics of this new method.

We have seen up to now that the periodicity at the atomic level, i.e., with typical periods of a few 0.1 nm, can be characterized by measurements of the scattered intensity in reciprocal space far from the origin. If objects of much larger dimensions, typically between a few nanometers and several tens of nanometers, are present in the sample, additional scattering will be found close to the origin of the reciprocal space. This scattering contains information on the density, the shape, and the organization of these large objects. Its use in the measurement and analysis for bulk samples is the basis of a very well-known and old method called small-angle X-ray scattering (SAXS) [36], for which measurements are usually performed in transmission.

This method has recently been extended to analyze the morphology of nanometer-scale particles deposited on or embedded below the surface of a sample, by combination of the SAXS technique with grazingincidence conditions, thus making it surface sensitive [31]. This new technique of GISAXS is performed under conditions close to total external reflection conditions. It was first developed to investigate *ex situ* the morphology of aggregates deposited on a substrate. One of the most exciting possibilities of GISAXS is the *in situ* investigation, in UHV, of the evolution of the morphology of deposits growing in 3D on a substrate, when the use of imaging techniques is difficult. This is the case of many metal/oxide interfaces, because the atomic force microscope (AFM) tip is big and often moves the metal clusters, and because a scanning tunneling microscope (STM) cannot be used on an insulating substrate. Examples of such *in situ* investigations [34, 35] are given in this chapter.

The experimental geometry of GISAXS is schematically represented in Figure 9. The incident beam impinges on the sample under grazing incidence close to α_c, and a 2D detector is placed behind, thus recording the intensity in the $(\mathbf{Q}_\parallel, \mathbf{Q}_\perp)$ plane of the reciprocal space. This scattering contains information on the islands' shape, height, and lateral size, as well on the organization of the islands with respect to each other. Because this scattering is very small, the direct, transmitted, and reflected beams are completely stopped by a beam stop before the detector, to avoid saturation.

To get accurate morphology characteristics of the islands, it is extremely important to carry out a precise, quantitative GISAXS analysis. Because the investigated distances are large compared with interatomic distances, the particles are generally treated as a continuum.

For a dense system of N_p identical and isotropic particles, the scattered intensity $I(\mathbf{Q})$ can be expressed as

$$I(\mathbf{Q}) = A \cdot N_p \cdot P(\mathbf{Q}) \cdot S(\mathbf{Q}) \cdot T(\alpha_f) \qquad (20)$$

where A is a constant, $P(\mathbf{Q})$ is the form factor of one particle, $S(\mathbf{Q})$ is the interference function, and $T(\alpha_f)$ is the transmission factor that reflects the effect of refraction of the scattered wave.

The form factor $P(\mathbf{Q})$ is the square of the amplitude $F(\mathbf{Q})$ scattered by a single island of volume V:

$$F(\mathbf{Q}) = \int_V \rho(\mathbf{r}) e^{-i\mathbf{Q}\cdot\mathbf{r}} \, dV \qquad (21)$$

where ρ is the electronic density within the island.

The interference function $S(\mathbf{Q})$ enters the expression only if the particles are correlated. $S(\mathbf{Q})$ is related to the particle–particle pair correlation function $g(\mathbf{r})$ by

$$S(\mathbf{Q}) = 1 + \rho_S \int \big(g(\mathbf{r}) - 1\big) e^{-i\mathbf{Q}\mathbf{r}} \, d\mathbf{r} \qquad (22)$$

where ρ_S is the surface density of the islands. This statistical $g(\mathbf{r})$ function thus tells us how the particles are distributed with respect to each other.

In general, the analysis is complicated by the fact that the different islands are not all identical. Assuming a constant shape, such as truncated pyramids, the islands generally exhibit height and lateral size distributions. The height and lateral size of a given island are generally correlated, and hence these distributions are not independent. To a first approximation, however, the mathematical treatment can be simplified by assuming independence. According to microscopy studies, the dimensional parameters of the islands generally follow a log-normal distribution. For the lateral size R (respectively, the height H), the maximum is denoted as μ_R (resp. μ_H), and the distribution parameter is σ_R (resp. σ_H):

$$
\begin{aligned}
I = A N_p \int_0^\infty \int_0^\infty & P(\mathbf{Q}, R, H) \times S(\mathbf{Q}) \\
& \times \frac{e^{(-1/2) \times ((\log R - \log \mu_R)/\log \sigma_R)^2}}{\sqrt{2 \times \pi} \times \log \sigma_R} \\
& \times \frac{e^{(-1/2) \times ((\log H - \log \mu_H)/\log \sigma_H)^2}}{\sqrt{2 \times \pi} \times \log \sigma_H} \, dR \, dH
\end{aligned} \qquad (23)
$$

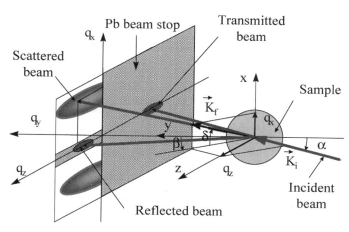

Fig. 9. Typical geometry for a GISAXS experiment.

In practice, the form factor $P(\mathbf{Q}, R, H)$ can be analytically calculated for most simple shapes usually assumed by islands growing on a surface, such as truncated cylinders, ellipsoids, or pyramids. The resulting intensities have characteristic profiles with a series of well-defined zeros (minima when dealing with experimental data). The position of the minima as well as the profile of the intensity unambiguously allows determination of the shape, as well as estimation of the average height and lateral size. If the particles are correlated, the main resulting feature is an interference peak as a function of $Q_{\parallel}$, whose position Q_p directly yields a rough estimation of the average center-to-center interparticle distances D, according to $D = 2\pi/Q_p$ [37–40]. However, this is a crude approximation that yields a systematic error. Finally, the distributions of dimensional parameters yield large variations in the intensity in the minima, from which the distribution parameters σ can be estimated.

Going further (i.e., refining the average dimensional parameters, the inter-island distance D, and widths of the distributions) requires fitting the intensity with simple approximate analytical formula and, finally, simulations by performing the whole Fourier transform and integration over the distributions, which is computer-time consuming. In the end, it is possible to wholly reproduce the 2D pictures.

Note that the above treatment implicitly used the kinematical (i.e., Born) approximation. It can be shown that this is valid when the substrate reflectivity is not too high, i.e., for light substrates as compared with the deposit [41]. If this is not the case, a more complicated treatment must be performed within the distorted wave born approximation (DWBA) [42].

3. PREPARATION AND STRUCTURE OF CLEAN SURFACES

3.1. Specific Considerations

Only a few oxide surfaces have been quantitatively investigated by GIXD [4]: mainly the sapphire α-Al$_2$O$_3$(0001) surface, the MgO(001) surface, and the rutile TiO$_2$(110) surface, and more recently the NiO(111), CoO(111), and ZnO(0001) surfaces. In all cases, specific experimental considerations apply, particularly for the measurements of the CTRs, to determine the atomic relaxation of the surface. The two first systems are very light scatterers ($\langle Z_{MgO}\rangle = \langle Z_{Al_2O_3}\rangle = 10$). The possible intensity that can be collected depends on the material and is given by Eqs. (6), (8), or (15), depending on the type of scattering, and cannot be increased. Therefore, to get measurable CTRs, the intensity needs to be concentrated over a very narrow angular range along these CTRs. This requires two conditions: first a very good starting single crystal, with a very small mosaic spread, and, second, a very flat surface on the length scale of the coherence length of the X-ray beam, which is typically 1 μm. These conditions are not trivial, because commercially available sapphire or MgO single-crystal surfaces have rocking curve widths on the order of a few 0.01° and a typical r.m.s. roughness of 1 nm. For α-Al$_2$O$_3$ and MgO(001), both the bulk crystalline quality and the surface flatness can be improved

by annealing in air or under partial oxygen pressure at high temperature ($\sim$1500°C). The rocking curve full width at half-maximum then decreases from $\sim$0.03° down to 0.0025° in the case of sapphire, and from $\sim$0.01° down to 0.001° in the case of MgO, thus yielding a 10-fold enhancement of the peak intensity along the CTRs, while leading to a negligible r.m.s. roughness of less than 0.05 nm in the case of α-Al$_2$O$_3$(0001) and less than 0.25 nm in the case of MgO(001). These r.m.s. roughness values were deduced from fits of the measured CTRs in both cases and confirmed by AFM measurements. However, in both cases, the high-temperature anneal has the drawback of enabling surface segregation of bulk impurities. This phenomenon can be minimized in the case of sapphire by limiting the duration of the annealing to a few hours, while Ca segregates on the MgO(001) surface. In the case of TiO$_2$, the experiments seem to be less stringent, because titanium is a heavier scatterer, and very good single crystals with low surface roughness are readily available from the suppliers. As soon as the atomic species become heavier, as in the case of the NiO(111) surface ($\langle Z_{NiO}\rangle = 18$), the above constraints could be expected to relax dramatically. The intensity scattered by the surface is larger, and the bulk background decreases because the X-ray absorption is larger. However, NiO(111) and CoO(111) crystals were practically ignored in the past, and commercial wafers have mosaic spreads as large as 1.7°. In the NiO(111) case the crystalline quality can be improved through the same method used for MgO(001) and α-Al$_2$O$_3$(0001): high-temperature annealing. Figure 10a shows the progressive improvement of the rocking curve of an in-plane NiO(111) Bragg peak with increased annealing temperature. GIXD has, indeed, the nice ability to allow investigations at any temperature. Direct annealing reduces the mosaicity to 0.1° before the crystal melts. Further improvements need better starting crystals (i.e., crystals annealed before cutting and polishing). However, quantitative CTR measurements can be performed on samples with a 0.5° mosaic spread. Such crystals are still not adequate for investigation of the p(2×2) reconstruction, because a unit cell that is four times larger reduces the intensity by an order of magnitude, and the surface layers of the reconstructions are incomplete, reducing the $\langle Z_{NiO}^{Surf}\rangle$ down to 13 (for the Ni-terminated octopolar reconstruction, see Section 3.4). The ultimate quality for NiO(111) that we could achieve was a 0.02° mosaic spread and surface morphologies like that of the AFM image reported in Figure 10b and is sufficient for investigating the reconstruction. The annealing procedure works fine when the oxide has only one stable stoichiometry; in the case of CoO(111) it failed and the crystals could not be improved.

Another difficulty is the large noise due to the bulk of the sample. Indeed, if we wish to atlain a high accuracy for the atomic coordinates, it is necessary to measure the CTRs over an extended range of $Q_{\perp}$, which requires an X-ray beam of large enough energy. The energy was fixed at 23 keV for the α-Al$_2$O$_3$(0001) surface, and 18 keV for the MgO(001) and NiO(111) surfaces. At this energy, as soon as the incident angle is larger or equal to the critical angle for total external reflection, a large background scattering is present, presumably arising from bulk point defects, that overcomes the surface scat-

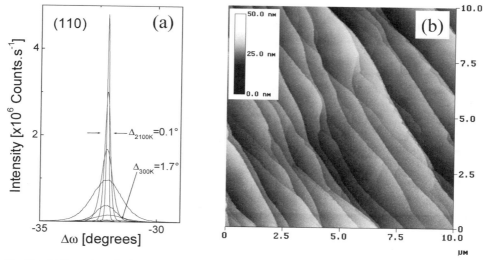

Fig. 10. (a) Scans through a Bragg peak of NiO(111) that lies in the surface plane during *in situ* recrystallization of a NiO(111) sample from room temperature up to 2100 K. (b) AFM image of a NiO(111) single crystal after optimal preparation. The terraces are several μm^2 wide, and the mosaic spread is about 0.02°. Reprinted with permission from [91], © 1997, Elsevier Science.

tering. Consequently, the experiment must be performed under very stringent conditions, with the incident angle kept below the critical angle for total external reflection for MgO(001) and α-Al_2O_3(0001) and close to it for NiO(111). Because in this region the amplitude of the wavefield varies very rapidly with the incident angle, the diffractometer as well as the sample alignment have to be of the topmost quality. This, of course, also requires that the sample surface be very flat and well defined.

3.2. MgO(001)

The MgO(001) surface has been the object of numerous studies because it is widely used as a substrate for the epitaxial growth of metals [43–46] and as a model support for finely dispersed catalytic particles. It is also used as a substrate to grow high-temperature superconductors because of its close lattice match to $YBa_2Cu_3O_{7-x}$ and its small chemical reactivity. Because the detailed structure and morphology of the surface may play a significant role in the overlayer properties, well-defined procedures for preparing MgO surfaces of very high quality are important in a number of areas of surface physics and material science.

A precise knowledge of the MgO(001) surface atomic structure is also important because this surface is often chosen as a model system for testing calculations on ionic oxides [47]. In particular, the top plane relaxation and the differential relaxation between anions and cations (rumpling) have been extensively studied, both theoretically [48–55] and experimentally [56–65]. These relaxations were always found to be extremely small on this surface. As shown below, when it is performed on a surface of very high quality, GIXD is well adapted for studying such small deviations from the ideally truncated surface.

The procedure for preparing surfaces of high quality necessarily involves a first step of annealing at high temperature

(1500–1600°C). However, this annealing also results in a strong surface segregation of bulk impurities, mainly calcium.

AFM and fluorescence scanning electron microscopy (SEM) showed that the annealed MgO(001) surface was composed of very large terraces, several 100 nm wide and running over distances of microns, separated either by monolayer high steps, or by steps that were several nanometers high. In addition, large, nearly equidistant "droplets" of hemispherical shape with diameters varying from 0.1 μm to 10 μm were found on these terraces. The droplets were found to have a rich chemical composition: Mg, P, Ca, Si, C, O and V, with ratios of $\sim 100:60:30:15:10:3:1$, respectively. In between, the terraces are atomically flat. Outside the droplets, on the flat surface, only Ca, in a quantity of about one monolayer, was detected.

Low-energy electron diffraction (LEED), Auger electron spectroscopy (AES), and GIXD were performed in UHV. In addition to the main diffraction spots, which were very sharp, half-order diffuse spots were present, corresponding to a $(\sqrt{2} \times \sqrt{2})$ R45° surface reconstruction. The atomic structure of this reconstruction was investigated by quantitative measurements and analysis of the MgO(001) CTRs [43].

On the MgO(001) surface, there are two nonequivalent CTRs: (i) "strong" ones with h and k even, the intensity of which is proportional to the square of the sum of the atomic form factors of O and Mg, and (ii) "weak" ones, with h and k odd, whose intensity is proportional to the square of the difference of the form factors. When h and k have different parities, the CTRs are forbidden by symmetry.

In view of GIXD studies of the clean surface and of growing metal/MgO interfaces, a procedure was developed to remove the Ca surface contamination while keeping the MgO surface as perfect as possible. For this purpose, the surface was etched by Ar^+ bombardment at 1550°C, which is a temperature high enough to allow the surface to reorder faster than it disorders, and to keep its smoothness. After this treatment, oxygen or magnesium vacancies could be expected on the surface. To restore the surface stoichiometry, a procedure suggested by several groups [62, 63, 66] was followed: the sample was annealed for 15 min at 700°C at a partial oxygen pressure of 10^{-4} mbar. The surface cleanliness was checked with the use of AES, and no remaining impurities were found, to the level of 1% of a monolayer. After this preparation, the samples were never exposed to air, to avoid the well-known attack of the surface by water vapor [67].

The GIXD measurements on this clean surface were aimed at the determination of the roughness, the relaxation ρ, and the rumpling ε, defined according to $\rho = (1/2)(\varepsilon_1 + \varepsilon_2)$ and $\varepsilon = \varepsilon_1 - \varepsilon_2$, where ε_1 and ε_2 are, respectively, the fractional displacements of the surface anions and cations, expressed as a percentage of the bulk interplanar distance (2.106 Å) perpendicular to the surface. The "strong" (20ℓ) CTR was mostly used to determine the r.m.s. roughness and the "weak" (11ℓ) one to determine the surface relaxation.

On the surfaces prepared as described above, both CTRs, measured by rocking the sample, were everywhere above background, with a very small width (~0.01°), always resolution limited, whatever the experimental resolution, which was a confirmation of the high crystalline quality and a first indication of a small surface roughness. The Lorentzian shape (FWHM 0.01°) at the in-plane anti-Bragg location $(1, 1, 0.05)$ indicates an exponentially decaying height–height correlation function with a terrace length of ~600 nm.

Figure 11 shows the (20ℓ) and (11ℓ) CTRs for clean MgO(001). Figure 11 illustrates the sensitivity of the (11ℓ) CTR to rumpling and relaxation, which is obtained only if the roughness is small enough, and when the CTR is measured over an extended range. The two CTRs were simultaneously fitted with four parameters: an overall scale factor, the relaxation and rumpling in the top plane, and the r.m.s. roughness. The data could not be fitted by restricting the step heights to multiple values of the MgO lattice parameter, i.e., to an even number of atomic planes, which introduces a clear maximum between Bragg peaks. All step height possibilities had to be introduced, which indicates that most steps are presumably only one atomic plane high (i.e., 2.1 Å). The Debye–Waller factor was fixed at its bulk value of 0.3 Å^2 [68] for all ions. The normalized chi-squared agreement factor of 1.1 ± 0.1 was very close to the ideal value of 1, which shows that no new parameter, such as atomic relaxations of deeper atoms, could be added.

All substrates prepared according to our new procedure yielded the same roughness value of 2.4 ± 0.1 Å, which is also the value determined on the Ca-segregated surfaces. One could suggest that further decrease of the roughness would be achieved by stopping the ion sputtering before starting to lower the annealing temperature. This is not obvious, because the time for annealing without bombardment is limited by the inevitable new segregation of impurities from the bulk.

Because the relaxation and rumpling are both very small, slightly different values were found on the different substrates.

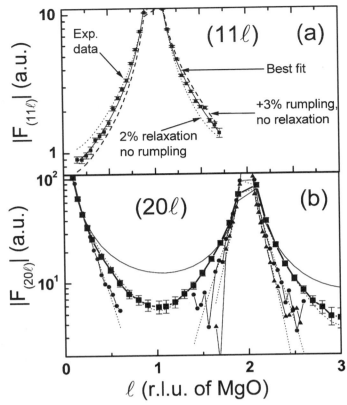

Fig. 11.   Modulus of the structure factor of the $\langle 11\ell \rangle$ (a) and $\langle 20\ell \rangle$ (b) CTRs of the clean MgO(001) surface, as a function of the perpendicular momentum transfer ℓ, in reciprocal lattice units of MgO, after 20 min of Ar$^+$ ion bombardment at 1500°C (■ with error bars). For the $\langle 11\ell \rangle$ CTR, the continuous line is the best result of a simultaneous fit of the $\langle 20\ell \rangle$ and $\langle 11\ell \rangle$ data. Calculated curves without rumpling and with a 2% relaxation (short dashed line) and with 3% rumpling and no relaxation (long dashed line) illustrate the sensitivity of the $\langle 11\ell \rangle$ CTR to rumpling and relaxation. The measured $\langle 20\ell \rangle$ CTR is represented for different surface states: after 20 min of Ar$^+$ ion bombardment at 1500°C (■ linked with a thick line) and after 30 min (●) and 2 h (▲) of Ar$^+$ ion bombardment at 900°C. Dotted lines correspond to the best fits, which yield, respectively, r.m.s. roughness values of (a) 2.4 Å, (b) 4 Å, and (c) 6 Å. The $\langle 20\ell \rangle$ CTR calculated for a perfectly flat surface is also shown (thin continuous line) for comparison. The rough surfaces are obviously not suitable for a quantitative study. Reprinted with permission from [43], © 1998, Elsevier Science.

The average values of $\rho = (-0.56 \pm 0.35)\%$ and $\varepsilon = (1.07 \pm 0.5)\%$ are thus given, with the error bar estimated from the uncertainties of each fit, and from the different values obtained.

In the original paper [43], these values are compared with previous experimental and theoretical determinations. Thanks to the high substrate quality and the extended measurement range, the error bars are significantly smaller in the present study. Many early shell model calculations and several experiments (reflection high-energy electron diffraction (RHEED) [68], He diffraction [61], ICISS [59], and SEELFS [62]) yielded much too large rumpling values. In most cases, this can be attributed to an inadequate substrate preparation, i.e., exposure to air before introduction in the UHV chamber. Most other theoretical or experimental results are close to the present ones, especially the latest one obtained by

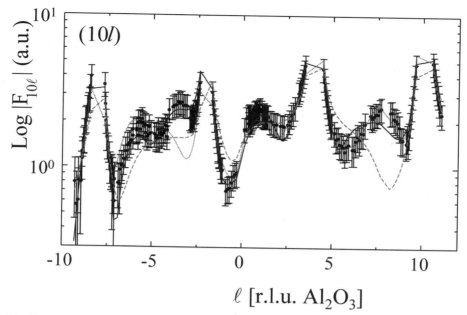

Fig. 12. $\langle 10\ell \rangle$ CTR or the α-Al$_2$O$_3$(0001)-(1 × 1) surface. Experimental (solid circles with error bars) and best-fit models for each possible termination: single Al layer (thick solid line), double Al layer (dashed line), and oxygen-terminated surfaces (dotted line). The logarithm of the structure factor is reported as a function of the out-of-plane momentum transfer in reciprocal lattice units of Al$_2$O$_3$. Reprinted with permission from [69], © 1997, Elsevier Science.

medium energy ion scattering [65], which yields similar values of the relaxation and rumpling with larger error bars. The precise values reported in the present work might help further refinement of the theoretical calculations.

In summary, a new procedure has been developed to prepare MgO(001) surfaces of very high quality. These surfaces are ideally suited to performing GIXD measurements. This offers the opportunity to investigate the atomic structure and morphology of metal/MgO interfaces by this technique, during *in situ* deposition in UHV by molecular beam epitaxy, from the very early stages of submonolayer deposition up to fairly thick metallic layers. Such measurements are presented in Section 4.

3.3. α-Al$_2$O$_3$(0001)-(1 × 1) and Its Reconstructions

3.3.1. *Termination and Relaxation of the Unreconstructed Surface*

The (0001) surface of sapphire (α-alumina, corundum) is one of the most widely used substrates for the growth of metal, semiconductor, or high-temperature superconductor thin films [69, 70]. It is also used as a substrate in silicon-on-sapphire (SOS) technology. Moreover, its initial state is known to play a role on the overlayer properties [71].

Despite many theoretical calculations of the α-Al$_2$O$_3$(0001) surface structure and relaxation [72–77], the nature (Al or O) of the terminating plane of the unreconstructed surface is still an open topic because of the lack of experimental results. The single Al-terminated surface is favored by electrostatic considerations as well as surface energy calculations [72]. Indeed, for an Al termination no dipole moment is left across the surface, and only the longer, and thus the weaker, anion–

cation bonds are broken. However, the (1 × 1) structure is also experimentally observed on alumina surfaces heated in an oxygen-rich atmosphere and could thus be suspected to be oxygen terminated. With regard to the Al-terminated surface, large relaxations have been predicted by pair-potential calculations [73]. More recently, *ab initio* calculations, by the density functional theory combined with pseudopotential techniques [74, 75], predicted very large relaxation of the last atomic plane (−87%), whereas Hartree–Fock calculations [76] yielded smaller, although still sizable, relaxation (−40%). A tight binding, total-energy method [77] also predicted large out-of-plane relaxations of the Al planes and in-plane displacements of the oxygen atoms.

The aim of the GIXD study was to determine experimentally, for the first time, the nature of the terminating plane and the relaxations of the first few atomic planes below the surface by quantitative measurement and analysis of the CTRs. The experiments were performed on a first sample with the LURE W21 beamline and diffractometer [22] and on a second sample with the ESRF [15] ID03 surface diffraction setup. Both samples were first annealed in air for 3 h at 1500°C, resulting in a surface with wide, atomically flat terraces, and a good near-surface crystalline quality. The two data sets yielded exactly the same conclusions. During the second, more precise experiment, eight CTRs were measured over an extended range of perpendicular momentum transfer from 0 up to 7.2 Å^{-1}, and corresponding to 873 nonequivalent reflections. The (10ℓ) CTR is reported in Figure 12.

Bulk sapphire has rhombohedral symmetry, which is usually treated as hexagonal (space group $R\bar{3}c$), with 30 atoms (six Al$_2$O$_3$ units) per primitive unit cell. The lattice parameters ($a = b = 4.7570$ Å, $c = 12.9877$ Å) and the internal coordinates ($x = 0.3063$, $z = 0.3522$) are taken from [78]. The bulk unit cell consists of an alternated stacking, along the c-axis, of two Al^{3+} planes (12 in the unit cell) with one atom per plane, and one oxygen plane (six in the unit cell) with three O^{2-} ions arranged with a threefold symmetry. From a crystallographic point of view, no two planes among the 18 of the unit cell are equivalent, under any translation. For each possible chemical termination (oxygen, single Al, or double Al) there are thus six crystallographically nonequivalent terminations. Fortunately, only two of these yield nonequivalent diffraction patterns. The (1×1) surface has $p3$ symmetry, with the threefold axis lying on the Al atoms. Hence, only their coordinate along the c-axis can vary, whereas oxygen can also move parallel to the surface plane as long as the atoms keep their threefold symmetry around the Al sites. Because of this symmetry, they must also remain coplanar. The surface model included displacements of three Al_2O_3 units plus those of the terminating planes (15, 16, or 17 parameters, depending on the termination). Because of the extremely high sensitivity of CTRs to atomic positions, deeper atomic displacements can enhance fit quality, even with very small values, but they are not significant for the structure. Al^{3+} and O^{2-} atomic scattering factors were used for all atoms, although theoretical arguments [76] as well as an experimental AES study [79] suggest that surface atoms may have a different charge. This should be of little importance, especially at large momentum transfer. Debye–Waller parameters were taken to be isotropic and were set to the bulk values [78] for all atoms.

Least-squares fitting and visual inspection (Fig. 12) clearly allowed the oxygen and double Al terminations to be ruled out. The surface is thus terminated with a single Al layer, which yielded $\chi^2 = 0.92$. The surface structure is schematically shown in Figure 13. The top two planes undergo large displacements from their bulk positions (0.34 Å and 0.23 Å, respectively). The top Al layer moves down toward the bulk, so that the interplanar spacing with the nearest-neighbor oxygen layer is reduced by 51%. The underlying oxygen atoms shift mainly parallel to the surface plane and are repelled from the first layer Al sites, moving almost radially toward the second layer. This is an almost bond-length conservative motion: the bond length is only 4.5% ($\pm 2.5\%$) shorter than the bulk nearest-neighbor value. Displacements below the first two planes are smaller: relaxations are $+16\%$, -29%, and $+20\%$ for the next three interplanar spacings.

These results were compared with theoretical calculations. The single Al termination was always the predicted one, because it is the only termination that is "autocompensated" (both charge neutral and chemically stable) and with no electric dipole moment across the surface. Regarding the surface relaxations, all theoretical studies agreed with the large negative relaxation of the top Al layer. Godin and LaFemina [77] suggested a rehybridization of the top Al atoms from the fourfold coordinated arrangement (which may be thought of as a sp^3 configuration) to a nearly perfect sp^2 configuration. More generally, all authors agree on the stronger bonding between Al and

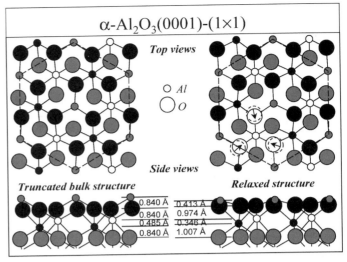

Fig. 13. Illustration of the sapphire (0001) surface: truncated bulk and best-fit model obtained from CTR measurements. The two top planes are strongly shifted with respect to their bulk positions in a nearly bond length conservative displacement. Reprinted with permission from [69], © 1997, Elsevier Science.

oxygen atoms near the surface, which is the general trend of the GIXD results; bond lengths are shortened by 4.5% to 6.1% in the first four planes. The results are also in very good agreement with very recent theoretical [80] and experimental [81] results.

To the author's knowledge, this was the first experimental determination of the relaxations of an oxide surface by GIXD and the first experimental determination of the relaxation and termination of the α-Al_2O_3(0001) surface. Another determination was made performed very recently by combining time-of-flight scattering and recoiling spectrometry with LEED and classical ion trajectory simulations [81], with essentially the same results.

3.3.2. Projected Atomic Structure of the α-Al_2O_3(0001)($\sqrt{31} \times \sqrt{31}$)R ± 9° Reconstruction

When the α-Al_2O_3(0001) surface is heated to high temperatures in UHV, several reconstructions appear: ($\sqrt{3} \times \sqrt{3}$)R30° around 1100°C, ($2\sqrt{3} \times 2\sqrt{3}$)R30° around 1150°C, ($3\sqrt{3} \times 3\sqrt{3}$)R30° around 1250°C, and finally ($\sqrt{31} \times \sqrt{31}$)-R±9° around 1350°C [82]. Although their electronic structure and symmetry were well characterized [83–85], their atomic structure remained essentially unknown. The ($\sqrt{31} \times \sqrt{31}$)-R±9° reconstruction is of particular interest because it has been reported to help epitaxy and enhance adhesion in some cases [71] and because it is unusually stable, even after air exposure. A structural model for this reconstruction had been proposed three decades ago [85]. The LEED pattern was interpreted as the superposition of two reciprocal lattices: that of the hexagonal substrate and that of a nearly cubic overlayer with composition Al_2O or AlO, plus the interference pattern because of double diffraction. However, this model remained controversial. In particular, this interpretation did not include a supercell formation with atomic relaxations. The aim of the GIXD study was to analyze the α-Al_2O_3(0001)($\sqrt{31} \times \sqrt{31}$)R±9° recon-

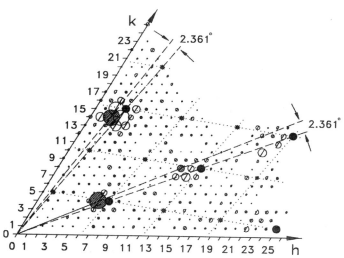

Fig. 14. Experimental diffraction pattern, indexed in the reciprocal space of the reconstructed unit cell (in 1/6 of the $\ell = 0.12$ reciprocal plane). The radii of the right-hand halves of the open circles are proportional to the experimental structure factors, and the left-hand open circles are calculated from the best model ($\chi^2 = 1.2$). Black disks represent bulk allowed and CTR reflections. The bulk unit mesh is superposed as dotted lines. The three main diffraction peaks of the reconstruction, corresponding to the "parent" phase, are hatched. Reprinted with permission from [82], © 1994, American Physical Society.

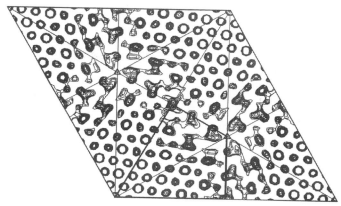

Fig. 15. Experimental Patterson map of the α-Al$_2$O$_3$(0001)-($\sqrt{31} \times \sqrt{31}$)R9° reconstruction in the whole reconstructed unit cell. Lines are only guides to locate the three-fold axes and the centered two-fold axis of the Patterson p6 symmetry. Bold lines delimit the asymmetric unit cell of the Patterson map. Reprinted with permission from [82], © 1994, American Physical Society.

struction to get unambiguous answers concerning the presence of a supercell, and ultimately to determine its atomic structure. Clearly, in the case of a reconstruction or of a thin layer, multiple scattering of X-rays is completely negligible. Hence, the GIXD pattern can be fully interpreted with the use of the kinematic theory of diffraction, where only the single scattering events are taken into account.

The α-Al$_2$O$_3$(0001) single crystals were first annealed in air at 1500°C for 3 h, and next heated to ~1350°C for ~20 min in UHV to obtain the ($\sqrt{31} \times \sqrt{31}$)R±9° reconstruction. Measurements were made with the LURE W21 beamline and diffractometer [22]. A large number, 366 (of which 267 were nonequivalent), of in-plane reflections arising from the reconstruction were measured. All peaks were exactly centered at the expected positions to within 0.001° of azimuthal rotation, which showed that the surface reconstruction is perfectly commensurate with the underlying bulk lattice. Their width and Lorentzian shape indicated an exponential decay in correlations with a decay length of ~50 nm. Several reconstruction diffraction rods were also measured. The absence of symmetry of the rod intensity with respect to $\ell = 0$ showed that the reconstruction has the minimal hexagonal symmetry p3.

The experimental diffraction pattern (Fig. 14) has sixfold symmetry. Measurable intensity was found at all reciprocal lattice points of the reconstructed unit cell, even far from bulk Bragg peaks. Because X-ray scattering by surfaces is in essence kinematical, this result contradicts previous interpretations of the LEED pattern [85] in terms of multiple electron scattering due to the coincidence of lattice sites between a rearranged surface layer with a small unit cell and the hexagonal substrate. In that case, X-ray diffraction peaks other than bulk would be found only at the reciprocal lattice points of the surface and bulk

unit cells. The X-ray diffraction intensity distribution proved that there indeed is a genuine ($\sqrt{31} \times \sqrt{31}$)R±9° supercell formation with atomic relaxations.

The diffraction pattern was shown to be qualitatively very similar to that predicted [86, 87] in the case of rotational epitaxy of an hexagonal overlayer, which is expanded and rotated with respect to an ideal overlayer R in perfect registry. The main peaks (hatched in Fig. 14) correspond to the first-order approximation (called the "parent" phase) of the adsorbed structure. Their locations yield the expansion, 10.62%, and rotation, 2.361°, applied to the R phase to obtain this rigid hexagonal "parent" phase. The other diffraction peaks are satellites corresponding to the static distortions of this parent phase and possibly to additional disorder.

Figure 15 shows the experimental pair-correlation (Patterson) function. Most Patterson peaks have a nearly perfect hexagonal arrangement. The positions of these peaks can be directly constructed by a rotation of ~1.4° followed by a small expansion of the projected atomic positions of an FCC(111) stacking on top of the oxygen HCP(0001) stacking of the underlying bulk lattice. Many possible models were tested before the final one was proposed. The reconstructed structure, schematically shown in Figure 16, was interpreted as a tiling of domains bearing a close resemblance to that of two metal Al(111) planes, separated by a hexagonal network of domain walls. In the middle of domains, the overlayers are well ordered, with a lattice parameter very close to that of metallic Al (expansion of 4% with respect to the registered state) and a small rotation (~1.4° with respect to the R state) with the epitaxial relationships (111)Al∥(0001)Al$_2$O$_3$ and [$\bar{1}$10]Al∥(R1.4°)[11$\bar{2}$0]Al$_2$O$_3$. In the domain walls, large expansion and rotation and even loss of honeycomb network topology were found. The observed structure was interpreted as due to rotational epitaxy with nonlinear distortions.

The study described in the previous paragraph (Section 3.3.1) showed that the unreconstructed α-Al$_2$O$_3$(0001) surface is terminated by an Al layer with 1/3 compact packing. Hence, starting from this surface, removing the two last O planes would

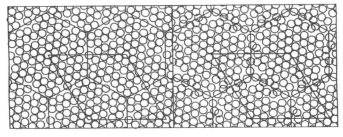

Fig. 16. Several domains of the projected atomic structure of the α-Al$_2$O$_3$ (0001)-($\sqrt{31} \times \sqrt{31}$)R9° reconstruction, where the unit cells as well as domain walls are drawn. The two constituting Al planes are shown separately, with evidence of one being much better ordered than the other. Numerical relaxation has shown that the ordered layer could be associated with the second layer, and the more disordered one with the layer adjacent to the substrate. Reprinted with permission from [82], © 1994, American Physical Society.

leave five Al layers with 1/3 compact packing occupancy at the surface, which is the observed 5/3 filling ratio. It was then suggested that the reconstruction is obtained after evaporation of the two upper oxygen layers of the unreconstructed surface. The physical origin of the ~4% expansion in the domains is clear, because the overlayer is very close to bulk Al and registry. A minimum-energy numerical simulation of the two Al planes, interacting with each other and with the substrate via a Lennard–Jones potential, was performed. The observed atomic structure was shown to be consistent with previous studies of the ($\sqrt{31} \times \sqrt{31}$)R±9° reconstruction [83–85, 88], which yielded an Al enrichment, intermediate oxidation states of surface aluminum atoms, and a reduced surface band gap. It is also consistent with the observation [85, 89] of a ($\sqrt{31} \times \sqrt{31}$)R±9° reconstruction during the first stage of Al deposition (between 0.4 and 2.5 Al(111) monolayer coverage) on an α-Al$_2$O$_3$(0001) surface with (1 × 1) structure, followed by Al(111) domain growth for larger coverage. Thus, a fundamental question was opened concerning the process and dynamics of this reconstruction formation by different routes: reduction or Al deposition. This motivated a structural study of the intermediate reconstructions briefly reported in the next paragraph.

3.3.3. GIXD Studies of the $(2P\sqrt{3} \times 2\sqrt{3})R30°$ and $(3\sqrt{3} \times 3\sqrt{3})R30°$ Reconstruction

The $(2\sqrt{3} \times 2\sqrt{3})$R30° and $(3\sqrt{3} \times 3\sqrt{3})$R30° reconstructions were recently investigated by GIXD at the ESRF [15], on the ID03 and BM32 beamlines, under experimental conditions very similar to those of the ($\sqrt{31} \times \sqrt{31}$)R±9° study [90]. Both reconstructions can be prepared in a very well-defined state, with large domain sizes, and very similar data were obtained in the two cases. Figure 17, which shows in-plane radial scans along the [h00] direction for $(2\sqrt{3} \times 2\sqrt{3})R30°$ and $(3\sqrt{3} \times 3\sqrt{3})$-R30° reconstructions, illustrates this high structural quality. The experimental Patterson map of the $(3\sqrt{3} \times 3\sqrt{3})R30°$ reconstruction is shown in Figure 18. Out-of-plane rod measurements showed that, as in the ($\sqrt{31} \times \sqrt{31}$)R±9° case, the thickness of the reconstruction is limited to one or two atomic planes.

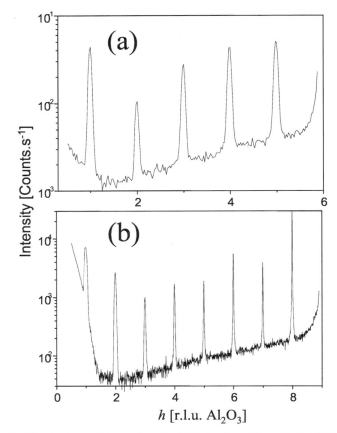

Fig. 17. In-plane radial scan along the ⟨100⟩ direction of the α-Al$_2$O$_3$(0001)-$(2\sqrt{3} \times 2\sqrt{3})$R30° (a) and $(3\sqrt{3} \times 3\sqrt{3})$R30° (b) reconstructions. h is in reciprocal lattice units of the reconstructed unit cells. Note the very intense and narrow reconstruction peaks, indicating very large domain sizes of several microns.

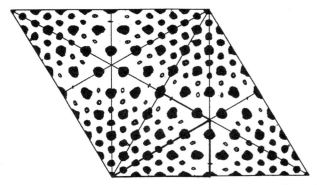

Fig. 18. Experimental Patterson map of the α-Al$_2$O$_3$(0001)-$(3\sqrt{3} \times 3\sqrt{3})$-R30° reconstruction in the whole reconstructed unit cell. Bold lines delimit the asymmetric unit cell of the Patterson map. Reprinted with permission from [4], © 1998, Elsevier Science.

Although the analysis has not yet been performed, a qualitative comparison with the diffraction pattern and Patterson map of the ($\sqrt{31} \times \sqrt{31}$)R±9° reconstruction shows that the structures are likely to have the same origin, except that no rotation is involved in the case of the $(3\sqrt{3} \times 3\sqrt{3})$R30° reconstruction. More precisely, the structure of the $(3\sqrt{3} \times 3\sqrt{3})$R30° reconstruction is likely to consist of an overlayer with hexagonal symmetry, made of one or several planes close to compact planes, and with a lattice parameter slightly larger ($\sim$12%) than that of sapphire, such that, along the [110] directions of sapphire, the two lattices coincide every 9 (110) d-spacing of sapphire, and 8 (110) d-spacing of the overlayer, yielding a $(3\sqrt{3} \times 3\sqrt{3})$R30° superlattice unit cell of $\sim$25-Å periodicity. Such a rigid overlayer would yield only the strongest peaks of the diffraction pattern. All of the other, weaker peaks would correspond to harmonics in Fourier decomposition, arising from small displacements of the atomic positions with respect to the average "parent" rigid lattice. A quantitative analysis with modeling of the compact overlayer is required to get a more detailed picture of the structure.

3.4. NiO(111)

The structures of electrostatically polar (111) surfaces of the rock-salt oxides (NiO, CoO, MnO, and MgO) have been the focus of many studies in very recent years and were long considered a mystery in surface science because they are difficult to investigate both experimentally and theoretically [91–98]. Because the bulk structure has alternating cationic and anionic sheets along the [111] direction, the simple truncated surfaces have a divergent electrostatic energy, in theory making them highly unstable [99]. Thus, the polar rock-salt surfaces were long believed to be unstable, according to Tasker [100] and to early experimental evidence of (100) faceting on MgO(111) [101]. However, their technological importance is growing because their electrostatic specificity provides properties that could be particularly interesting, as in catalysis, for example [102]. Furthermore, the (111) plane of NiO is a highly interesting place to perform exchange coupling of ferromagnetic films [103] in the newest giant magnetoresistive sensors [104].

Wolf recently predicted that such surfaces may be stabilized by a particular p(2 × 2) "octopolar" reconstruction, which cancels the divergence of the electric field in the crystal [105]. A schematic representation of this nonstoichiometric surface structure, which organizes in a p(2 × 2) surface reconstruction, is shown in Figure 19. Indeed, NiO(111) surfaces were known to naturally exist as facets on small NiO single crystals [106] and as thin films with a p(2 × 2) structure on gold and nickel substrates [107, 108]. Early experiments on NiO(111) also showed complex reconstructions attributed to Si segregation [109]. Thus the real structure of polar oxide surfaces and the importance of the electrostatic criterion were really puzzling, and the possibility of preparing NiO(111) surfaces of high crystalline quality and known structure could open new possibilities for theorists and experimentalists in the fields of highly correlated materials [110], magnetism [111–113], and catalysis [102]. The aim of our GIXD experiments was thus to establish the optimal preparation conditions, if they exist, of NiO(111), to investigate its stability and reconstruction ability, and finally to use it as a substrate for the growth of ferromagnetic and well-controlled thin films. The situation rapidly appeared to be much more complex than for MgO(001) or

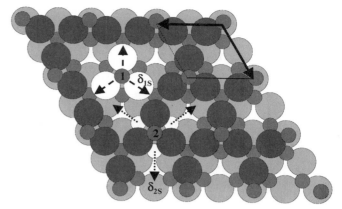

Fig. 19. Schematic drawing of the Ni-terminated octopolar reconstruction of NiO(111). Small circles stand for Ni atoms and the large ones for O atoms. The first and second planes are, respectively, 75% and 25% vacant. The thick arrows indicate the basis vectors of the p(2 × 2) surface lattice mesh. The symmetry-related radial relaxations δ_{1S} and δ_{2S} are respectively represented around apex atoms labeled 1 and 2 by dashed and dotted arrows and apply to the second oxygen and the third nickel layers. The corresponding atoms are whited out.

α-Al$_2$O$_3$ because of a different surface chemistry, likely driven by the electrostatic properties of the surface.

To elucidate this puzzling topic of the polar NiO(111) surface, the first GIXD experiments [91] were carried out on the SUV-BM32 setup [17] at ESRF [15]. The beam energy was set at 18 keV to avoid fluorescence background from the Ni K-edge. The beamline optics were doubly focused. The beam size at the sample was 0.3 mm(H) × 1 mm (V). The incidence angle was set at the critical angle for total external reflection (0.17°). The NiO(111) sample (purity 99.9%) was provided, aligned to better than 0.1°, and polished by Crystal GmBH (Berlin). The surface basis vectors describe the triangular lattice that is appropriate for (111) surfaces, they are related to the bulk ones by $\mathbf{a}_s = [\bar{1}10]_{Cube}/2$, $\mathbf{b}_s = [0\bar{1}1]_{Cube}/2$, and $\mathbf{c}_s = [111]_{Cube}$. The h and k indexes are chosen to describe the in-plane momentum transfer, $Q_{\parallel}$, and ℓ of the perpendicular one, $Q_{\perp}$. All three indexes are expressed in reciprocal lattice units (r.l.u.) of NiO(111).

To check the stability of the bulk lattice against decomposition and to follow the improvement of crystal quality, in-plane rocking scans of the (110) Bragg peak were performed on another sample during *in situ* annealing at increasing temperatures until melting (2200 K). Figure 10a shows that the near-surface crystalline quality improves drastically during the temperature treatment: the initial mosaic spread of about 2° falls to 0.1° without a change in the structure because the integrated intensity remains constant. Other crystals were airannealed up to 1800 K, but the surface morphology, checked by AFM, showed many steps, hexagonal shaped holes, and defects. Only below 1300 K did the direct annealing of the crystal show unchanged surface morphologies. The crystal used in the first study was thus first annealed in air at 1273 K for 3 h and introduced in the chamber just after cooling. The mosaic spread of the sample measured on in-plane and out-of-plane Bragg reflections was 0.3°–0.4°. The detector slits were adapted to integrate the intensity over this angular width.

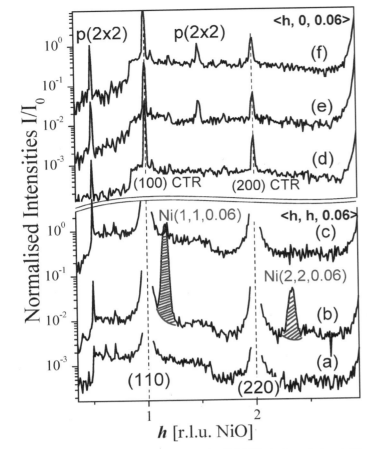

Fig. 20. In-plane measurements along the $\langle h\,h\,0.06 \rangle$ (a, b, and c) and the $\langle h\,0\,0.06 \rangle$ (d, e, and f) directions of the NiO(111) surface after air annealing (a and d), UHV annealing (b and e), and *in situ* postoxidation (c and f) under 2.5×10^{-6} mbar O_2 for 45 min at 860 K. Note that the strong CTRs of NiO are present in the $\langle h\,0\,0.06 \rangle$ direction, showing that the topmost surface planes of NiO(111) are very well defined. Ni appears in the form of islands during decomposition because only Bragg peaks and no CTRs are observed. The curves were shifted for the sake of clarity, but the relative intensities are comparable. The ordinate scale is logarithmic. Reprinted with permission from [91], © 1997, Elsevier Science.

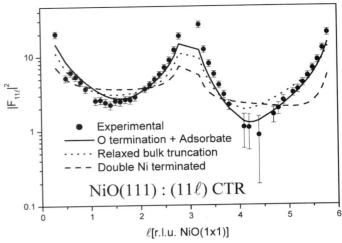

Fig. 21. $\langle 11\ell \rangle$ CTR of NiO(111) for an air-annealed crystal with a mosaic spread of 0.4° compared with a simple bulk truncation with relaxation ($\cdots$), an O-terminated surface with C contamination (—), and a metallic surface covered by two layers of pure Ni (−−−). Reprinted with permission from [92], © 1998, American Physical Society.

The surface was next annealed at 860 K for 30 min resulting in desorption of the typical air contaminants. This leads to a slight decrease in the CTR intensity, indicating a limited surface roughening. The in-plane measurements (Fig. 20b and e) reveal two new and strong features: extra peaks at each half-integer value corresponding to the p(2×2) reconstruction, and Bragg peaks at the exact positions expected for epitaxial relaxed FCC Ni(111) ($h = 1.17$ and 2.24 in Fig. 20b). Out-of-plane scans confirm the 3D character of the metallic Ni. The lack of asymmetry or interference along the NiO CTRs indicates that no significant amount of Ni is pseudomorphic. Direct annealing of the surface thus leads to reduction through a loss of O atoms. In regions not covered by Ni, the surface exhibits the p(2×2) reconstruction, and no macroscopic faceting was observed.

The Ni clusters were then removed by heating the surface at a partial pressure of oxygen of 2.5×10^{-6} mbar at 860 K for 45 min. Only the p(2×2) reconstruction remains (Fig. 20c and f). Again, no macroscopic facetting was observed.

A quantitative description of the surface structure requires better crystals with a higher signal-to-noise ratio, but the previously described encouraging first results showed, at least, that the NiO(111) surface is quite stable, although not that easy to manipulate. It was thus worth improving the crystal preparation. Improved crystals showed that in fact the (1×1) assumption was erroneous because the p(2×2) reconstruction structure that applies after air-annealing has small structure factors, which were below the bulk background, with samples having mosaic spreads of 0.4°.

Optimal crystals can be produced with the following procedure. High-quality surfaces with good morphology were obtained by annealing the NiO boule at 1850 K for 24 h in air, then cutting, polishing, and re-annealing at 1300 K for 3 h, which yields a flat, shiny surface. A mosaicity of 0.054° and a typical domain size of 1800 Å were obtained. Such crystals have high crystalline quality up to the surface and are well suited

Without outgassing, strong crystal truncation rods (Fig. 20 and [91]) were measured perpendicular to the surface up to the highest accessible $Q_\perp$ values ($\ell = 6$). At the out-of-phase conditions (between Bragg reflections) the rods were still intense and sharp, indicating a surface of very slight roughness. No oscillations were observed by X-ray specular reflectivity measurements, only an inflection at $\sim 1.5°$, which could correspond to a very thin (less than 10 Å) adsorbed layer on top of the surface. In-plane measurements (Fig. 20a and d) along the high-symmetry directions showed only a weak first-order p(2×2) reconstruction peak, leading to the idea that the surface structure could be mainly (1×1) at this stage. No contribution from ordered adsorbates, metallic Ni, or (100) facets could be detected. A first attempt to interpret the CTRs with a (1×1) surface cell showed that a simple truncation of the bulk or a metallic Ni-covered surface was not the actual structure. At least surface contamination has to be introduced to reproduce the shape of the CTRs, as shown, for example, in Figure 21 [92].

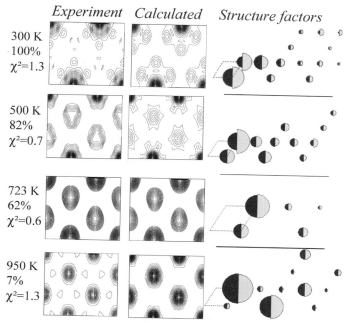

	Experiment	Calculated	Structure factors
300 K 100% $\chi^2=1.3$			
500 K 82% $\chi^2=0.7$			
723 K 62% $\chi^2=0.6$			
950 K 7% $\chi^2=1.3$			

Fig. 22. Structural evolution of p(2 × 2) reconstruction of NiO(111) with respect to the temperature, under 1×10^{-5} mbar O_2, from room temperature (top) to 950 K (bottom), and comparison with a model of a combination of octopolar and spinel-like reconstructions. First column: temperature, ratio of octopolar reconstruction in the model and χ^2 for the in-plane data. Second and third columns: experimental and calculated Patterson maps. Fourth column: experimental (right half-circles) and calculated (left half-circles) structure factors for NiO(111)-p(2 × 2) measured at the four different temperatures.

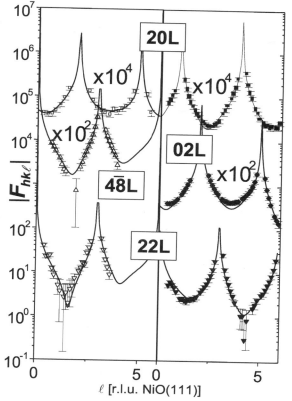

Fig. 23. NiO(111)-p(2 × 2) crystal truncation rods measured for a non-UHV-annealed sample (right) and a sample annealed at 950 K under oxygen (left). The ⟨20ℓ⟩ and ⟨22ℓ⟩ rods were measured for both situations. The straight lines represent the best fits for the octopolar model (right) and the spinel-like model (left). Reprinted with permission from [95], © 2000, American Physical Society.

for detailed GIXD investigations. In the vacuum chamber, such surfaces were immediately p(2 × 2) reconstructed with some residual C and sometimes Ca. GIXD data were recorded after such an *ex situ* sample preparation because further treatments proved ineffective. Annealing under up to 10^{-4} mbar O_2 at 700 K removes the C contamination but drastically transforms the internal structure of the reconstruction. An O_2 sputtering at 2 keV and an anneal in air at 1000 K removes the Ca contamination but leaves a large surface mosaicity, making a complete analysis also impossible. Nonetheless, because the first orders of the reconstruction retain the same relative intensities, the Ca is not responsible for the reconstruction or for its internal structure. Thus, our *ex situ* preparation has given the best surface.

For single crystals prepared according to our optimized preparation method, the p(2 × 2) reconstruction obtained, after air annealing, can be measured quantitatively. However, the intensity remains weak and the measurements were carried out on the ID03 undulator beamline at ESRF [15] at a photon energy of 18 keV and an incidence angle of 0.17° [94]. Because the single crystals are always reconstructed, it is more convenient to use the p(2 × 2) lattice to index the reciprocal space (i.e., the crystallographic basis vectors for the surface unit cell are related, for the rest of this chapter to the bulk basis by $\mathbf{a}_{surf} = [\bar{1}10]_{Cube}$, $\mathbf{b}_{surf} = [0\bar{1}1]_{Cube}$, and $\mathbf{c}_{surf} = [111]_{Cube}$). For the as-prepared NiO(111) single crystal, the in-plane and out-of-plane structure factors are reported in Figures 22 (top) and 23 (right). The in-plane scattering of the p(2 × 2) patterns

was measured quantitatively by rocking scans at all accessible positions belonging to the reconstruction. A total of 33 nonzero peaks were measured at $\ell = 0.1$. The symmetry of the diffraction pattern is $P6mm$, leaving 14 nonequivalent peaks with a systematic uncertainty of 12%.

The measured and calculated in-plane scattering can be directly compared, using the octopolar model predicted by Wolf [105] with both Ni- and O-termination. Reproducing the crystal truncation rods (CTRs) restricted the solution to the Ni termination uniquely, plus the following symmetry-related atomic relaxations in each atomic layer p ($p = 0$ for the apex layer of atoms). Note that three of the four atoms in layer p of the unit cell have symmetry-related vertical displacements ζ_{pS}, whereas the independent atom has a vertical displacement ζ_p. Likewise, δ_{pS} defines a radial displacement of symmetry-related atoms away from the in-plane position of the apex atom (Fig. 19). The ρ_{pS} terms define the possible rotational displacements. Detailed fitting reveals that all δ_{pS} and ρ_{pS} terms are negligible, except $\delta_1 = 0.117 \pm 0.015$ Å, which indicates a dilation of the threefold hollow site where the apex atom rests. The least-squares refinement converges for a 0.2 Å r.m.s. roughness, $\zeta_0 = 0.06 \pm 0.02$ Å, $\zeta_{1S} = -0.17 \pm 0.02$ Å, $\zeta_2 = 0.13 \pm 0.01$ Å, $\zeta_{2S} = -0.029 \pm 0.007$ Å, $\zeta_3 = 0.23 \pm 0.12$ Å, $\zeta_{3S} = -0.09 \pm 0.02$ Å, $\zeta_4 = 0.02 \pm 0.01$ Å,

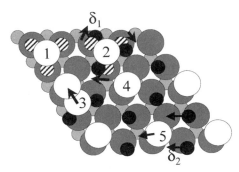

Fig. 24. Structural transformation steps needed to obtain the spinel configuration starting from the octopolar reconstruction. Large (resp. small) circles stand for O (resp. Ni) atoms. (1) O-terminated octopolar reconstruction. (2) Rotation and centrifuge motion of the Ni atoms. (3 and 4) Translation of an O atom on top of a Ni atom. (5) Global $\langle 010 \rangle/3$ shift of the reconstructed layer with respect to the bulk. The final spinel-like configuration is reported in Figure 26. Reprinted with permission from [95], © 2000, American Physical Society.

and $\zeta_{4S} = -0.005 \pm 0.003$ Å. For the 138 measured structure factors a global χ^2 of 1.5 is obtained with nine structural parameters, the roughness, and a scale factor. The agreement is good, and further relaxations do not significantly improve the fit. The relaxations extend deeply into the crystal. The smallest bond length in the refined model is 1.9 Å (i.e., a 10% contraction) and is located between the last complete layer ($p = 2$) and the 25% vacant layer of the octopole ($p = 1$). The same procedure with an O-terminated surface did not converge, and the best χ^2 was 30. In conclusion, the theoretically predicted octopolar reconstruction effectively applies to NiO(111) to overcome the divergence of the electrical field. The observed relaxations remain very limited with respect to the ideal structure.

Let us now concentrate on the structure obtained during annealing under partial oxygen pressure to avoid the formation of Ni clusters. This was achieved by using a partial oxygen pressure of 1×10^{-5} mbar during annealing. The in-plane scattering has been measured at several temperatures and is reported in Figure 22. The continuous evolution of the structure factors is obvious, and one may suggest that either the transformation is continuous or two different structures combine to yield the observed scattering. In any case, it is necessary to first describe the final structure, which is expected to be close to the final transformation state or the second surface structure. Fortunately, the high-temperature reconstruction yields much larger intensities than the octopolar reconstruction, and several reconstruction rods could be measured in this case. The out-of-plane CTRs are reported in Figure 23 and still resemble the octopolar reconstruction, showing that the number of planes involved remains close to that of the octopolar reconstruction. Thus we have taken this structure as a starting point to develop a model able to reproduce the observed structure factors (Fig. 24 (1)). Already in early interpretations it was obvious that a centrifuge rotation, δ_1, which consists of a displacement along the three equivalent $\langle 100 \rangle$ directions (arrows in Fig. 24 (2)), was mandatory to describe the structure after annealing under oxygen in UHV [114]. Further improvements of the fit could be obtained by using the O-terminated surface only (as depicted in Fig. 24).

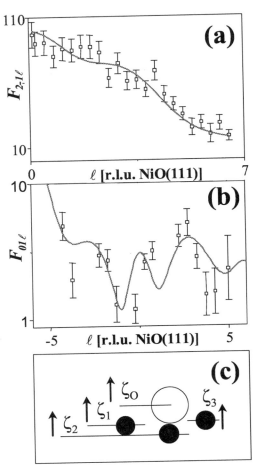

Fig. 25. Out-of-plane configuration for NiO(111) single-crystal substrates annealed at high temperature (1000 K) under partial O_2 pressure. The straight lines are calculated for the spinel configuration. (a) $\langle 2\bar{1}\ell \rangle$ reconstruction rod. (b) $\langle 01\ell \rangle$ reconstruction rod. (c) Structural configuration for the best fit in the spinel configuration. Large (resp. small) circles stand for O (resp. Ni) atoms. Reprinted with permission from [95], © 2000, American Physical Society.

The next determining step is not intuitive, because it consists of the positioning of the apex atom on top of one particular Ni atom, as shown by the arrow in Figure 24 (3); the resulting position corresponds to Figure 24 (4). One could argue that this might lead to nonphysical perpendicular distances. In fact, the fits of the reconstruction rods (Fig. 25a and b), which are highly sensitive to the out-of-plane stacking in the reconstructed unit cell, fortunately converge then only for a very realistic stacking (Fig. 25c). The three-symmetry nonequivalent Ni atoms of the second layer adopt a configuration in which the interatomic distances are close to the bulk values, but one of the atoms does not move, another moves upward, and a third moves downward. At the same time, the O atom moves away from the surface at the right position to obtain the expected Ni—O bond length. At that point, the proposed structure is not able to reproduce the CTRs at all. This becomes possible only after a global shift, δ_2 along the $\langle 010 \rangle$ direction, of the reconstruction with respect to the bulk, as indicated in Figure 24. The best fit of the reconstruction rods is obtained with δ_2 exactly equal to two-thirds of the reconstruction lattice parameter, as shown in Figure 26 (left-

NiO(111)-spinel like

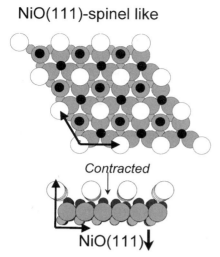

Contracted

NiO(111)↓

Fig. 26. Final model derived from GIXD data of the structure of high-temperature NiO(111) crystals annealed in UHV under partial oxygen pressure. Large (resp. small) circles stand for O (resp. Ni) atoms.

hand side). The final positioning allows the upward moving Ni atom to sit directly on top of an O atom, whereas the downward moving Ni atom rests at a three-fold site that offers the necessary space. The registry of the reconstruction finally seems very realistic, although the structure is now very different from the initial octopole. The best fit yields a global χ^2 of 1.85 over 119 structure factors with $\delta_1 = 0.84 \pm 0.05$ Å, $\delta_2 = 1.95 \pm 0.09$ Å, $\zeta_1 = +0 \pm 0.02$ Å, $\zeta_3 = +0.24 \pm 0.02$ Å, $\zeta_2 = -0.29 \pm 0.03$ Å, and $\zeta_O = +0.76 \pm 0.05$ Å. The χ^2 over the 11 nonequivalent in-plane structure factors is 2.

Combining this O-terminated structural model with the Ni-terminated octopolar reconstruction obtained after the room temperature annealing allows easy reproduction of the scattering observed at the intermediate temperature. The best fits for 300 K, 503 K, 723 K, and 950 K were obtained with 100%, 82%, 62%, and 7% of octopolar reconstruction and in-plane χ^2's of 1.3, 0.7, 0.6, and 1.3, respectively. The quality of the fits is very satisfactory, as can be seen in Figures 22 and 23. Note that for the 950 K situation, the in-plane χ^2 is strongly reduced by the addition of 7% octopolar reconstruction, showing that the transformation is still not fulfilled. When this combination is used to fit the 950 K data, the global χ^2 reduces from 2 to 1.7 over the 119 structure factors. The two proposed structures are thus sufficient to completely reproduce the data at all temperatures.

Although the Ni_3O_4 spinel bulk phase does not exist, the annealed crystal adopts the configuration we would expect from such a surface over the four last planes. The nonextension of Ni_3O_4 toward the bulk is likely to be the origin of the protective effect of the spinel-like surface layer. It is interesting to note that for this configuration the reconstruction unit mesh contains one atom of oxygen in the top layer and three atoms of Ni below; the crystal is thus reduced compared with the octopolar reconstruction, and the divergence of the electrostatic potential is avoided as well. The Patterson maps observed for the Ni_3O_4

spinel-like surface structure are identical to that of a Co_3O_4 surface spinel layer (see below), strongly supporting this structural model. Interestingly, the transformation is fully reversible by air exposure, switching the crystal back to its oxidized (octopolar reconstructed) state. Moreover, once the Ni_3O_4 spinel-like surface layer is formed, no decomposition with the appearance of Ni clusters could be observed, showing that the surface chemistry has strongly changed.

Dosing the surfaces with up to 10^6 L at a partial H_2O pressure of 10^{-6} mbar in the chamber tested the eventual role of hydroxyl groups. No effect could be observed. The same conclusion applies to the adsorption of more oxidizing gazes like NO and NO_2 under similar conditions. They were found to be unable to transform the spinel configuration back to the octopolar one. The stability of the reconstruction is attributed to the high quality of the NiO(111) surfaces, which contain only a very small density of defects, which are believed to be mandatory for the initiation of reaction with these gases.

Apart from the fact that the NiO(111) surface is always p(2 × 2) reconstructed to cancel the diverging electrical field because of the polarity, it behaves like most oxides: it is almost insensitive to contamination or air exposure, it may decompose through UHV annealing, and it may exhibit a surface layer that prevents decomposition. The structural models, derived from GIXD measurements, for the p(2×2) reconstructions of NiO(111) are fully coherent with these properties. The p(2 × 2) reconstructions of NiO(111) are based on two states: the Ni-terminated octopolar reconstruction, which corresponds to the oxidized surface state, and the O-terminated $Ni_3O_4(111)$-like spinel configuration, which stands for the reduced surface state and is more stable against decomposition.

Fortunately, NiO(111) can be obtained and prepared in both pure extreme states (fully reduced and fully oxidized) in common UHV preparation conditions. This may not necessarily be the case for other polar surfaces that could perhaps exhibit combinations of structures or phases (like CoO), making their analysis even more difficult, although these difficulties are not directly related to the polarity problem. From our studies on NiO(111) it clearly appears that the reconstruction scheme proposed to stabilize polar surfaces is correct. Because we have established the optimal conditions for the preparation of very high quality NiO(111) substrates and understood the chemistry of these surfaces, we can use them as substrate for metal growth, as will be detailed in Sections 4.2 and 4.3.

3.5. CoO(111)

CoO, like MgO and NiO, has a very simple cubic rock-salt structure, in which pure O and pure Co(111) planes alternate along the cubic ⟨111⟩ direction [115]. Exactly as for the NiO(111) case, the (111) surface is polar, and an ideally cut surface should be (theoretically) unstable or p(2 × 2) reconstructed, at least if these assumptions are general rules. CoO(111) could be expected *a priori* to have a behavior very similar to that of NiO. Indeed, the two share the same structure and have nearly the same lattice parameter. In addition, in CoO, as well as in

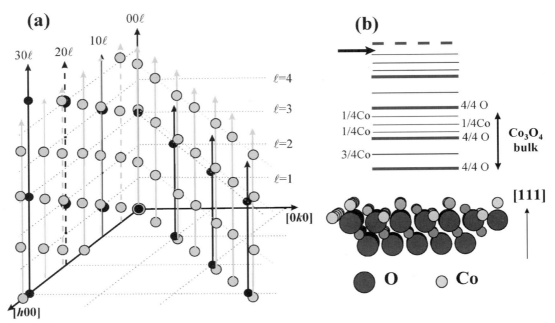

Fig. 27.   (a) The reciprocal space of CoO(111) (black circles) and Co₃O₄(111) (gray circles) in CoO reciprocal lattice units (r.l.u.) indexed in a triangular unit cell. Bragg peaks in the ⟨h0ℓ⟩ and ⟨hhℓ⟩ planes are shown, for both CoO(111) and Co₃O₄(111). Straight thick lines (respectively, dashed and thin lines) represent CTRs with (resp. without) in-plane Bragg peaks. (b) Schematic representation of the structure of Co₃O₄(111). The horizontal arrow points to the position of the truncation used in simulating data. Reprinted with permission from [115], © 2000, Elsevier Science.

NiO, the atoms in the (111) pure Co (Ni) planes are spin uncompensated [116, 117]. The antiferromagnetic ordering is along the ⟨111⟩ direction, exchange-moderated by the pure oxygen planes. The Néel temperature of CoO (T_N = 292 K) is lower than that for NiO (520 K), but with a higher unidirectional anisotropy. Mixed Ni and Co oxide or multilayers combine almost linearly the advantages of each of the simple antiferromagnetic oxides: high anisotropy and high working temperature [118–123]. Because both oxides are highly resistant to corrosion they are both good candidates for the building of giant magnetoresistive read heads [118, 124–128].

The aim of the GIXD study was thus to establish preparation conditions and to make a structural quantitative investigation of the CoO(111) single crystal surface in a way similar to that of the NiO(111) study. All of the measurements on CoO(111) were performed on the BM32-SUV setup [17] at the ESRF [15]. The samples were illuminated by a well-focused 18-keV X-ray beam (350 μm (horizontal) × 500 μm (vertical) FWHM) under an incidence angle α on the order of the critical angle for total external reflection of CoO (α_c = 0.16° at 18 keV). When information from a larger depth was required, the incidence angle was increased to higher values, up to 0.6°. Several samples were investigated in the UHV chamber.

Polished single CoO(111) crystals of purity 99.9% were provided by Crystal GmBH (Berlin). The samples were cut and aligned to better than 0.1° before polishing. A triangular unit cell was chosen to describe the three-fold symmetry of the (111) surface plane. The unit vectors are related to the cubic ones by $\mathbf{a}_s$ = [$\bar{1}$10]$_{Cube}$/2, $\mathbf{b}_s$ = [0$\bar{1}$1]$_{Cube}$/2, $\mathbf{c}_s$ = [111], with

$a_s = b_s$ = 3.002 Å; c_s = 7.3533 Å; α = 90°, β = 90°, and γ = 120°. The h and k indexes describe the in-plane momentum transfer and the ℓ index, the perpendicular one, expressed in reciprocal lattice units (r.l.u.). The ⟨h00⟩ and ⟨0k0⟩ in-plane directions are defined in such a way that a Bragg peak is found at the (1, 0, 1) position in the reciprocal space.

Very surprisingly, the air annealing, which allowed efficient recrystallization of NiO(111), completely failed for CoO(111). After annealing at 1300 K the crystals lost their surface polish and the stoichiometry was severely changed. Fortunately, directly after polishing, the CoO(111) single-crystal surfaces exhibit acceptable (<1°) and much better surface mosaic spreads than NiO(111), thus allowing for a direct investigation by GIXD. The presence of contaminants was checked by AES. Carbon was the unique contaminant and an oxygen annealing at 550 K for 10 min under $p(O_2)$ = 10^{-5} mbar proved to be effective at cleaning the surface.

The stacking of the Co and O atoms in the unit cell gives rise to the reciprocal space shown in Figure 27 (black circles). Several directions in the reciprocal space are particularly interesting. Figure 28 presents a scan along the in-plane ⟨h 0 0⟩ direction; obviously many more peaks than expected appear. Positions corresponding to the expected CoO(111) structure (integer h indexes) are marked by arrows. As can be seen, by comparison with Figure 27, scattering at the in-plane (100) and (200) positions corresponds to the intersection of (10ℓ) and (20ℓ) CTRs of CoO(111) with the ℓ = 0 plane and arises from the truncation of the surface. These features look like the ones observed for NiO(111) and correspond well to the rock-

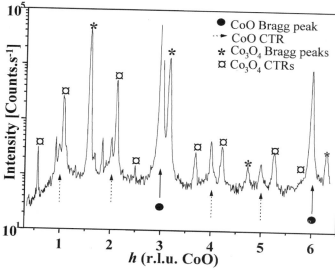

Fig. 28. In-plane scan along the $\langle h00 \rangle$ direction. The positions of scattered signals from different structures are indicated. The most intense Bragg peak of CoO (at $h = 3$) was avoided because of its large intensity. CoO CTRs appear at integer positions of h (arrows). Peaks marked by * and o correspond to the spinel structure and the filled circles to the $(3, 0, 0)$ and $(6, 0, 0)$ CoO Bragg peaks. Reprinted with permission from [115], © 2000, Elsevier Science.

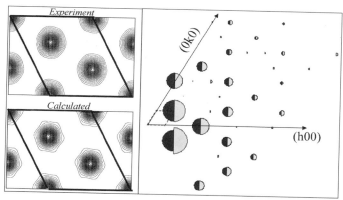

Fig. 29. Comparison of the structure factors deduced from the extra peaks in the CoO(111) scattering with a model based on metallic terminated $Co_3O_4(111)$ in a spinel structure. Only interatomic distances and angles characteristic of the Co_3O_4 structure appear here because no scattering from the CoO(111) lattice can exist at these locations. (Left top) Self-correlation electron density map (Patterson map) for the in-plane reflections. (Left bottom) Calculated self-correlation map for a Co-terminated $Co_3O_4(111)$ surface. (Right) Experimental (right half-circles) and calculated (left half-circles) structure factors. For the unit cells in the maps (thick lines) the edge of the cell is 5.716 Å and the real angle is 120°.

salt structure of CoO(111). But extra peaks were systematically found (Fig. 28) at positions corresponding to another crystallographic structure that has in-plane parameters $a_1 = b_1 = 5.716$ Å and an out-of-plane parameter $c_1 = 9.9$ Å, almost twice those of CoO(111). The small width in h and k shows that the in-plane lattice parameter is very well defined. The surface mosaic spread of this structure is on the order of the CoO one, $\sim 0.8°$. However, it is not a surface reconstruction, inasmuch of as the out-of-plane Bragg peaks of this structure are peaked and correspond to an estimated thickness of about 50 Å.

Integrated intensities of all of the extra in-plane Bragg peaks (o in Fig. 28) were quantitatively measured at $\ell = 0.1$. The experimental data set had a $P6mm$ symmetry, indicating an overlayer of cubic lattice, with a cubic lattice parameter of 8.083 Å, which corresponds to that of the bulk spinel phase of cobalt oxide: $Co_3O_4(111)$ (8.085 Å). A set of 34 nonequivalent reflections was deduced from a total of 150 measured peaks (systematic error 17%). The experimental diffraction pattern could indeed be well simulated on the basis of the spinel $Co_3O_4(111)$ unit cell with the epitaxial relationship $Co_3O_4(111) \| CoO(111)$ and $Co_3O_4 \langle 100 \rangle \| CoO \langle 100 \rangle$, provided that the metallic surface termination shown in Figure 27b is used in the model. Other terminating planes of the spinel structure were not able to reproduce the experimental data. A comparison between the experimental and calculated Patterson maps is drawn in Figure 29 ($\chi^2 = 1.17$ and $R = 0.176$). The agreement is very good. Interestingly, the extra peaks used here would correspond in the $p(2 \times 2)$ NiO(111) case to the in-plane reconstruction peaks, and in fact the Patterson maps of the reduced NiO(111) surface and the one shown in Figure 29 for CoO(111) are very close.

We thus conclude that the CoO(111) polished surface is stabilized by the presence of a thick spinel $Co_3O_4(111)$ surface layer. The CTR scattering at the (100) and (200) positions is much weaker than for the NiO(111) surface, which seems to indicate that the interface between the $Co_3O_4(111)$ layer and the CoO(111) crystal is rough or diffuse (i.e., an oxygen gradient).

For use as a pinning antiferromagnetic layer and for an understanding of the exchange coupling phenomenon, CoO(111) surfaces with a 1:1 Co:O atom ratio are mandatory. The presence of a $Co_3O_4(111)$ layer is expected to modify the exchange coupling. Indeed, bulk $Co_3O_4(111)$ is known to be ferrimagnetic. However, studies of magnetic exchange phenomena at ferrimagnetic/antiferromagnetic or ferromagnetic/ferrimagnetic interfaces [129–131] were performed. Moreover, the magnetic properties of the spinel thin layer may be different from the bulk ones, rendering the situation even more complex. An antiferromagnetic substrate should thus not exhibit different stoichiometries near the surface. Because air annealing turned out to be unable to restore good surfaces, we have checked the possibility of removing the Co_3O_4 layer and improving the crystallographic quality of the CoO surface by the usual procedures in surface science: annealing in UHV or under O_2 and Ar^+ etching.

Annealing in UHV leads to the formation of small metallic Co clusters on the surface, which is similar to the NiO case [91]. The X-ray scattered signal arising from (and so characteristic of) the Co_3O_4 layer was found to be approximately constant during the whole annealing process, at 500 or 800 K (Fig. 30). While metallic Co is formed, the quantity of the spinel remains roughly constant. Distinguishing between the formation of metallic Co in the spinel or in the CoO is not possible from these data. A post-annealing at 800 K and 10^{-5} mbar O_2 totally oxidized the Co clusters in a few minutes, but the Co_3O_4 layer

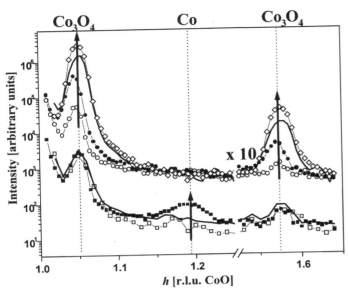

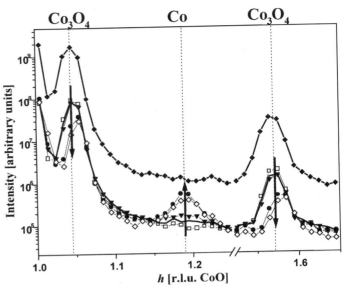

Fig. 30. In-plane scans along ⟨hh0⟩ during UHV and oxygen annealing (open squares: as-polished sample; black line: UHV annealed at 500 K; filled squares: UHV annealed at 800 K; open circles: oxygen annealed at 800 K, 10 min; filled circles: 800 K, O_2, 60 min; gray line: 800 K, O_2, 10 h; open diamonds: 800 K, O_2, 12 h). Quick annealing in oxygen (less than 10 min) oxidizes the metallic Co formed during the UHV annealing but increases the amount of Co_3O_4. The arrows indicate the evolution of the intensity between successive measurements. Reprinted with permission from [115], © 2000, Elsevier Science.

Fig. 31. In-plane ⟨hh0⟩ scans during Ar$^+$ etching (bottom curves) showing the onset of metallic Co during sputtering and the decrease in the spinel thickness (open squares: reference, as polished sample; line: after 10 min of Ar$^+$ bombardment; filled triangle: 30 min of Ar$^+$ bombardment; filled circles: after 90 min of Ar$^+$ bombardment; open diamonds: 110 min of Ar$^+$ bombardment). Annealing under O_2 restores the Co_3O_4 layer (top curve, filled diamonds: 10 min of O_2 annealing). The arrows indicate the evolution of the intensity between successive measurements. Reprinted with permission from [115], © 2000, Elsevier Science.

is still present. Moreover, its scattered intensity increases with increasing oxidation time, as can be seen in Figure 30 from the Co_3O_4 peak, showing a development of the spinel. UHV annealing thus proved ineffective in removing the spinel layer.

Another polished sample (covered by a ~50-Å-thick Co_3O_4 layer) was Ar$^+$ bombarded, first at room temperature and then at 800 K ($U = 800$ V, $p(Ar^+) = 2.6 \times 10^{-5}$ mbar, drain current ~10 μA). At high temperatures, Co in-plane Bragg peaks are observed, showing again the formation of metallic Co, presumably in the form of flat and extended islands, inasmuch as scattering is found in the surface plane, at intersections with positions of the Co CTRs, far from Co Bragg peaks. Meanwhile, the scattering from the Co_3O_4 layer decreases (Fig. 31), proving definitively that the Co_3O_4 layer sits on top of the CoO(111) crystal. In other words, we deal with a top spinel layer, and not with a buried interface. The crystallographic quality of the metallic Co is poor, with a measured mosaic spread of 2.6°. Scans along Co rods, perpendicular to the surface, make it possible to distinguish the different stacking in the Co islands, as these scans pass through Bragg peaks of FCC, twinned-FCC, and HCP Co. Co is mainly of FCC stacking (continuing that of CoO(111)) with small quantities of twinned FCC (rotated by 60° with respect to the FCC) and HCP. This result is similar to what was obtained for the growth of Co on the polar NiO (111) surface [132], indicating the strong influence of polar oxide surfaces on metallic films.

The Ar$^+$ bombardment thus reduces the thickness of the Co_3O_4 layer, but with formation of metallic Co islands. Regardless of the sputtering time, it was not possible to completely

remove the spinel surface layer. With annealing in a partial pressure of oxygen, the Co clusters rapidly oxidized and transformed again in the spinel oxide. All of these observations show that Co_3O_4(111) is likely to have a much lower surface energy than CoO(111) or any possible stabilizing reconstruction. This behavior probably explains why Co crystals become almost impossible to clean once they have been oxidized [133].

In summary, whatever the preparation, the polished (111) surface of CoO(111) single crystals is covered by an (111) oriented epitaxial Co_3O_4 layer (epitaxial relationship Co_3O_4(111) ∥CoO(111) and Co_3O_4⟨100⟩∥CoO⟨100⟩), which can be prepared free of any contaminant. This may be the major reason why $Ni_{1-x}Co_xO$ compounds exhibit spinel structures when $x > 0.6$ [134]. Air annealing of the sample to improve the crystallographic quality of the surface does not reestablish the Co : O 1 : 1 surface stoichiometry but destroys the sample. UHV annealing or Ar$^+$ bombardment reduces the thickness of the Co_3O_4 layer, but induces the formation of flat metallic Co(111) islands, which are reoxidized upon annealing in oxygen. With oxygen annealing, the signal coming from metallic Co clusters vanishes. Longer oxygen annealing yields the transformation of CoO into Co_3O_4. This shows a better stability of the spinel structure.

Unlike NiO(111), which is stabilized by a p(2 × 2) reconstruction, the CoO(111) surface is stabilized by a Co_3O_4 spinel structure, in which the problem of the diverging potential is also avoided. This raises questions about some of the results concerning ferromagnetic metals/CoO exchange-coupled interfaces, because the bulk Co_3O_4 is ferrimagnetic. If the Co_3O_4

layer were always present, it would modify magnetic coupling (however, exchange-coupled interfaces were found with the use of ferrimagnetic/antiferromagnetic interfaces). A similar situation appears for the NiFe/NiO(111) single-crystal interface (Section 5.3). In defined conditions, a controlled spinellike interface can be produced, and the magnetic measurement data support a modification of the permalloy-based spin-valve sensor characteristics when the spinel interface is formed. Thorough knowledge of the surface stoichiometry and structure is thus a very important issue in the framework of magnetic coupling. Although magnetic coupling with the spinel cannot be excluded, the knowledge of the structure of the crystal on which the deposit is made is of great importance to an understanding of the magnetic properties.

It is possible that preparation techniques like RF sputtering can yield an entirely stoichiometric CoO film. However, the poor crystallographic quality of such films makes a quantitative characterization very difficult or even impossible. The establishment of the correlation between the structure and the magnetic properties in exchanged-coupled systems of this type thus remains experimentally difficult.

4. MODEL METAL/OXIDE SYSTEMS

Studying thin films by performing *in situ* GIXD measurements during metal deposition, from the very early stages up to fairly thick films, allows one to address many fundamental questions in surface science, some of which are listed here.

First, is there epitaxy, and if yes, what are the orientational relationships? What is the structural quality of the growing film? What is the registry of the metal with respect to the substrate, i.e., the adsorption site and the interfacial distance between the last oxide plane and the first metal plane? For instance, in the case of a metal/MgO(001) interface, is the metal on top of the O ions or the Mg ions of the last MgO(001) plane or in between, above the octahedral site?

How does this interfacial distance evolve with the thickness of the metal film? What are the growth mode and the morphology of the growing film? How is the accommodation of the lattice parameter misfit performed? When does the transition from elastic to plastic relaxation happen? Which defects are involved in the process of plastic relaxation; stacking faults or interfacial dislocations? What is the interplay between these structural relaxation processes and the morphology? Finally, can we improve the structure and morphology of thick films by UHV annealing? Can annealing studies provide kinetic information?

When the film thickness is larger than a few hundred angstroms, the top of the metal film is generally close to being fully relaxed to its bulk lattice parameter. From the point of view of elasticity, the relaxation of an epitaxial film on a substrate is governed by the lattice parameter misfit f. In the case of parallel epitaxy, like that of Ag/MgO(001), f is defined by $f = (a_f - a_s)/a_s$, where a_f and a_s are, respectively, the film and substrate bulk lattice parameters. For small values

of f ($<10\%$), the lattice parameter relaxation is generally realized by localized, ordered misfit dislocations [135–137], in which case the interface is said to be "semi-coherent." When f is larger, the density of misfit dislocations becomes so large that they cannot remain localized and organized, and the proportion of the interface in "poor epitaxy" increases. In that case, the interface is said to be "incoherent." There are numerous exceptions, however. For instance, the Pd/η-Al$_2$O$_3$(111) interface is incoherent with $f = 2.7\%$ [138] as are the Au/MgO(001) interface ($f = 3\%$) [139–141], the Cu/Al$_2$O$_3$ interface ($f = 10\%$) [142], and the Ag/CdO interface [143], whereas the Au/ZrO$_2$(111) interface is semicoherent, with a very large misfit of 22% [138]. The technique of choice for investigating interfacial dislocations is high-resolution transmission electron microscopy (HR-TEM) [138]. However, HR-TEM does not always provide all of the required information concerning dislocations and may in some cases lead to erroneous conclusions. In these cases, it may be useful to resort to GIXD, as will be shown in the few examples below.

Three examples will be given of *in situ* studies during growth for the three FCC metals Ag, Pd, and Ni, with cube-on-cube epitaxy on MgO(001) (with increasing misfits of 3%, 7.6%, and 16.4%, respectively). The Ag/MgO(001) system has been chosen by most theoreticians as a model because it is one of the simplest metal/oxide interfaces: the MgO(001) surface is nonpolar, epitaxial relationships are particularly simple (square/square) with a small lattice misfit (3%), and chemical and charge transfer contributions to bonding are negligible. Many studies have been devoted to the Pd/MgO(001) interface because it is a model catalyst. The Ni/MgO(001) interfaces are also of particular interest because they are simple transition metal/oxide interfaces in which the metal is ferromagnetic.

The analysis of interfacial dislocation networks by GIXD will be described in the cases of the Ag/MgO(001) and Pd/MgO(001) interfaces.

4.1. Ag/MgO(001)

4.1.1. In Situ Studies of the First Stages of Formation of the Ag/MgO(001) Interface

In this model metal/oxide system, all theoretical calculations minimize the interfacial energy with respect to two structural parameters: the silver adsorption site (on top of O atoms, on top of Mg atoms, or in between, above the octahedral sites of the substrate), and the interfacial distance between the MgO(001) surface and the first Ag(001) plane [44, 144–146]. However, although crucially needed by theoreticians to evaluate and refine their models [46], no accurate experimental determination of these parameters was available at the time the GIXD study was performed. The aim of the GIXD study was to determine these parameters and their evolution with the Ag thickness and to analyze the growth mode.

The scattered X-ray intensity measured during radial in-plane scans along $(h, h, \ell = 0.1)$ and $(h, 0, \ell = 0.1)$ is shown in Figure 32 around $h = 2$, for deposited thickness θ ranging from 0 to 72 Ag monolayers (ML). These scans show

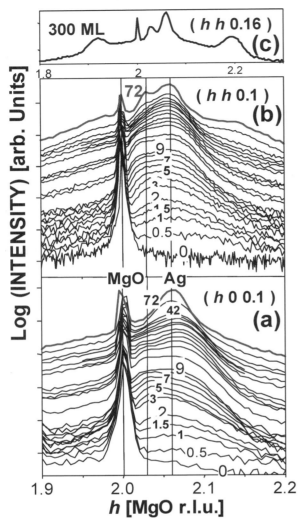

Fig. 32.   Radial scans at $\ell = 0.1$ along the $\langle h\,0\,0.1\rangle$ (a) and $\langle h\,h\,0.1\rangle$ (b) directions, as a function of the amount of deposited Ag (0, 0.5, 1, 1.5, 2, 3, 4, 5, 6, 7, 8, 9, 10, 11, 13, 15, 17, 19, 22, 25, 28, 32, 36, 42, and 72 monolayers (ML)), which is indicated above the corresponding curves. The curves corresponding to the different deposits have been shifted vertically for clarity. Vertical lines indicate the $h = 2$, $h = 2.03$, and $h = 2.06$ positions. The top figure (c) is a scan at $\ell = 0.16$ along the $\langle h\,h\,0.16\rangle$ direction, performed at 300 ML. Reprinted with permission from [145], © 1997, Elsevier Science.

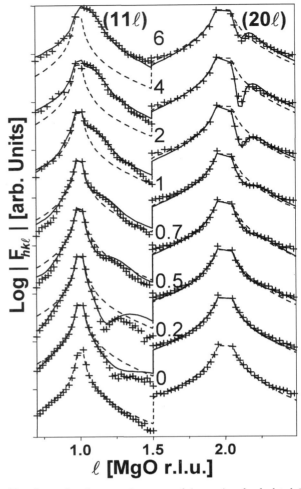

Fig. 33.   Comparison between the measured (crosses) and calculated (solid lines) CTRs for various Ag thicknesses. The dashed lines correspond to the clean MgO. The $\langle 11\ell\rangle$ and $\langle 20\ell\rangle$ CTRs have been represented on the same ℓ-scale, although they are at different h, k values. The curves corresponding to the different amounts of deposited Ag are shifted vertically for clarity. Reprinted with permission from [4], © 1998, Elsevier Science.

the presence of relaxed Ag(001) in cube/cube epitaxy with the MgO(001) substrate from the very beginning of Ag deposition, as revealed by the broad peak in both directions, approximately centered at the expected position ($h = 2.062$) for bulk Bragg peaks of Ag in cube on cube epitaxy. The integrated intensity of this relaxed Ag component increases linearly with θ, showing that, at all deposits, most of the Ag is incorporated into this relaxed part. However, at the early stages of Ag deposition, another significant effect is present: between 0 and 2 ML, a significant decay of the MgO $(2, 2, \ell = 0.1)$ intensity occurs. This decrease originates from a destructive interference between the waves scattered by the Ag layer and the substrate on the $(2, 2, 0.1)$ MgO CTR. This implies that at least part of the deposited Ag is initially perfectly on site, i.e., exactly lo-

cated above atoms of the substrate. Rocking scans of the MgO $(2, 2, 0.1)$ peak are resolution limited (0.003°) whether or not Ag is present and correspond to a correlation length of the registered Ag of at least 2000 nm. This indicates that the epitaxial site is perfectly well defined: the Ag atoms responsible for this destructive interference effect are correlated via the substrate. This registered part can be selected in reciprocal space, because it yields rods that are located at exactly the same integer (h, k) values as the MgO CTRs. The interference between the waves scattered by the MgO substrate and by the registered Ag film yields modulations of the intensity along the bulk CTR directions, which can be analyzed to determine the structural parameters of interest: site of epitaxy and interfacial distance.

For this purpose, the (11ℓ) and (20ℓ) rods were measured as a function of θ (Fig. 33). The sign of the interference (destructive on the high ℓ side of both the (111) and (202) Bragg peaks of MgO, at least for very small amounts deposited, because for large amounts, the MgO(11ℓ) CTR is rapidly overcome by the

Ag CTR) at the first stages of deposition unambiguously allows the assignment of the epitaxial site: the Ag atoms sit on top of oxygen atoms of the substrate. The location of the Ag intensity in a small ℓ range on the high ℓ side of the MgO Bragg peaks indicates that the surface of the registered part is rougher than the substrate's surface.

Quantitative analysis of the Ag/MgO CTRs was carried out with four parameters: (i) the total occupancy of registered Ag (i.e., the amount of Ag ML that is perfectly on site); (ii) the interfacial distance; (iii) the average out-of-plane distance between registered Ag; and (iv) the additional roughness of the registered Ag film with respect to the substrate. These parameters are reported in Figure 34 as a function of deposited thickness. The main results are that only a small fraction of the deposited Ag, amounting to 10% until $\theta \approx 5$ ML is perfectly on site, and that the thickness of this registered part is always larger than the equivalent thickness deposited, which shows that the growth is three-dimensional from the very beginning. Despite the fact that most of the Ag deposited is relaxed, the selection of the registered fraction on CTRs allowed a determination of the parameters of interest. The interplanar distance in Ag is very close to that in bulk silver, which is consistent with registered Ag surrounded by relaxed Ag. The interfacial distance is found to increase at the beginning of deposition and stabilize around an average value of 2.5 Å. In the first stages of growth, the smaller interfacial distance may originate from the different local environment of Ag atoms at the interface. Indeed, all theoretical studies indicate that there is little charge transfer between Ag and O, when a bulk Ag crystal above the MgO(001) surface is considered, but this may not be the case for isolated atoms or very thin films on the surface. These results were next compared with the theoretical models of the Ag/MgO(001) interface. As far as the site of epitaxy is concerned, the image interaction model predicts that the Ag atoms are above the octahedral sites of the MgO(001) surface, but this translational state was shown to result from the hard-core repulsion used [147]. The most recent *ab initio* calculations [148, 149] show that the energetically favored configuration is for Ag on top of oxygen atoms of the substrate. The GIXD results experimentally demonstrate this latter conclusion for the first time. They were later confirmed by an EXAFS investigation [150]. Regarding the interfacial distance, the experimental steady-state value is very close to the most recent *ab initio* calculations: 2.49 Å [149, 151]. This, along with the adsorption site, shows that recent *ab initio* calculations give a good description of the Ag/MgO(001) interface.

This quantitative analysis thus demonstrated for the first time the possibility of characterizing by GIXD a small fraction of the Ag film (those Ag atoms that are perfectly on top of substrate sites) and of precisely determining the epitaxial site and interfacial distance and their evolution from the very early stages of deposition to thick deposits. The next question was, what can we learn from GIXD about the morphology and the structure of the major part of the Ag film?

As far as the morphology is concerned, because the adhesion energy of Ag on Ag is larger than that of Ag on MgO (1.36 eV and 0.45 eV, respectively [152]), Ag is expected to grow in

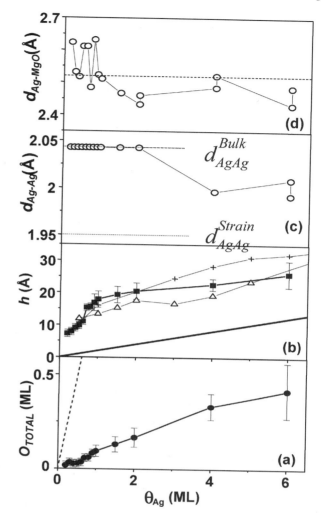

Fig. 34. Evolution with the amount θ of deposited Ag (in equivalent ML) of (a) the total amount of "on-site" Ag expressed in number of ML (solid circles with error bars), compared with the total amount deposited (dashed line); and (b) the "on-site" Ag thickness (solid squares). This is compared with the average island height deduced from two independent GISAXS measurements with the 1D detector (crosses and open triangles). The total equivalent thickness of Ag deposited is also represented (solid line). (c) The interplane distance d_{AgAg} in Ag, perpendicular to the surface, compared with the distances d_{AgAg}^{Bulk} expected for bulk Ag, and d_{AgAg}^{Strain} calculated according to isotropic elasticity for Ag strained in-plane to the MgO lattice parameter (dashed lines). (d) Interfacial distance d_{Ag-MgO} deduced from the fits of the CTRs (open circles) and average interfacial distance (dashed line).

the form of islands. The growth morphology was actually probed by performing GISAXS measurements as a function of θ (Fig. 35). A measurable peak appears at $\theta = 0.5$ ML and then increases in intensity with θ, while becoming narrower and moving toward the origin of reciprocal space. This small angle pattern is typical of small correlated islands that nucleate, grow and coalesce as the film gets thicker. The deposited Ag therefore rapidly evolves toward the classical island growth regime (Volmer–Weber). The average distance between islands can easily be deduced from the location of the intensity maximum, and the average island diameter can also be deduced

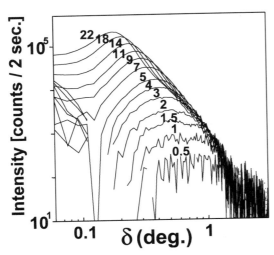

Fig. 35. Intensity scattered at small angles, as a function of the in-plane scattering angle δ, for different amounts of deposited Ag. The amount of Ag in ML is indicated above the corresponding curve. The intensity measured on the clean MgO was subtracted.

by several means, assuming given shapes of the islands, for instance, hemispherical in the present case.

If GISAXS can provide useful information on the morphology, GIXD (performed at wide angles) can provide very detailed information on the structure of the relaxed part of the Ag film. For that purpose, rocking scans were systematically performed on the relaxed Ag Bragg peaks as a function of θ. They were all found to be of Lorentzian lineshape, perfectly centered on the $\langle 110 \rangle$ direction, which shows that there is no rotation between the relaxed structure and the substrate. The lineshape corresponds to exponentially decaying correlation functions with small ($\sim$100 Å) correlation lengths increasing with θ. The corresponding "domain size" and its evolution with θ were found to be in good agreement with the island size deduced from GISAXS because of the excellent substrate and epilayer qualities (see Section 5.2 for an example in which domain size and GISAXS island sizes do not match).

Further information on the relaxation process of the lattice parameter was gained by looking in more detail at the intensity distribution along the radial scans of Figure 33. Apart from a broadening induced by the finite island size, the lineshape mainly reveals the distribution of lattice parameters in the silver film. Between 0 and 4 ML, the scattering is composed of only one component, the center of which progressively shifts from $h = 2.06$, corresponding to Ag fully relaxed to its bulk lattice parameter, toward an intermediate value, $\sim$2.03 for $\theta \approx 4$ ML. Above $\sim$4 ML, in both directions, the scattering progressively splits into two components centered respectively around $h = 2.03$ and $h = 2.06$, and whose exact positions evolve with θ. Whereas these two components remain up to large thickness along the $(h\,h\,0.1)$ direction, the intermediate component, around $h = 2.03$, progressively disappears along the $(h\,0\,0.1)$ direction, for thicknesses larger than 20 ML. We will see in Section 4.13 that, for thick enough films, the satellite around $(2.03\,2.03\,\ell)$ arises from the formation of a well-ordered

network of interfacial dislocations releasing the lattice parameter misfit between Ag and MgO. This ordered network does not yield any satellite around $(2.03\,0\,\ell)$, because this is the location of an extinction for this structure. Thus, at intermediate thickness, between $\theta \approx 4$ and 20 ML, the scattering around $h = 2.03$ in both directions does not arise from a new interfacial supercell, but rather is due to an inhomogeneous distribution of lattice parameters within the Ag islands. The observed evolution was analyzed as follows. At the very beginning (below 1 ML), the relaxed Ag fraction is made of very small islands of fully relaxed Ag, with the lattice parameter of bulk Ag. At this stage, the width in radial scans is completely dominated by the finite size effect. As shown by GISAXS and by the decreasing widths of Ag scattering, both radially and transversely, the islands next become larger. At the same time, the Ag becomes more strained by the MgO substrate, with an average in-plane lattice parameter intermediate between that of MgO and Ag for $\theta \approx 4$ ML. This is likely connected to an increased interfacial area over Ag volume ratio, i.e., to a decrease in the aspect ratio (height/width) of the islands. Up to 4 or 5 ML, radial scans are composed of only one contribution, because the strain in Ag is homogeneous. In other words, there is continuity between net planes in the MgO substrate and in the Ag islands: the Ag islands are said to be coherent with the substrate. The elastic strain energy stored in the islands increases as the islands grow in size, up to a point, around 4–5 ML, where it becomes energetically more favorable for the islands to release part of the strain by introducing a defect, such as a stacking fault or an interfacial dislocation. At this stage, the coherency with the substrate is lost, and the misfit relaxation is said to be plastic, as opposed to elastic when the islands were still coherent. Therefore, the strain within the islands becomes inhomogeneous, leading to the two components observed along both radial directions. The component around $h \approx 2.06$ is due to fully relaxed Ag, in which the strain is indeed homogeneous, whereas the component around $h \approx 2.03$ arises from the regions surrounding the cores of the interfacial dislocations, where the strain is strongly inhomogeneous. A detailed study was performed to determine whether these structural defects were appearing in the center of islands or near their edges. The observed intensity distribution in the two directions could only be explained by locating the net plane discontinuity at the edges of the islands. This is the first experimental evidence that misfit dislocations nucleate at the edges of islands during the growth of this kind of metal/oxide system.

The intensity of the shoulder around $h \approx 2.03$ decreases above 20–30 ML, until complete disappearance above 50 ML. This can be related to the beginning of the coalescence of the Ag islands around 20 ML, as shown, for instance, by the increase in the critical angle for total external reflection, which is due to an increase in the average density of the Ag layer. As islands coalesce, the interfacial misfit dislocations can reorganize into the energetically favored ordered interfacial network, resulting in an extinction of the intensity around $(2.03\,0\,\ell)$ and a well-defined satellite around $(2.03\,2.03\,\ell)$.

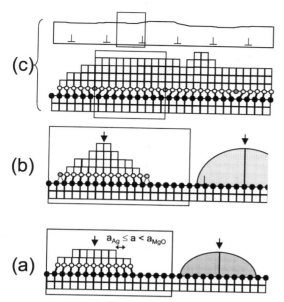

Fig. 36. Schematic representation of the morphology and structure during the first stages of growth of Ag on MgO(001) at room temperature, as a function of the amount of Ag deposited, θ. A side view of the atomic positions within these islands is depicted. For all deposited amounts below 30 ML, the deposit consists of Ag islands with a height-to-width ratio of $\sim 0.37 \pm 0.05$. The growth is decomposed into three stages: (a) For $0 < \theta \leq 4$–6 ML, the Ag islands are coherent with the MgO. Their lateral size is smaller than 90 Å. Their in-plane lattice parameter is equal to that of bulk Ag at 0.5 ML, and then becomes intermediate between that of bulk Ag and that of MgO between 0.5 and 4–6 ML. (b) Around 4–6 ML, on the average, the islands reach a critical size (~ 90 Å) above which disordered misfit dislocations are introduced near their edges. (c) Above 30 ML, the film becomes continuous, and the dislocations reorder to form a square network. On all figures, the arrows locate the presence of a column in Ag that is exactly "on site." The supercell used to calculate the CTRs is shown schematically.

In summary, this GIXD study provides much information concerning the growth and relaxation at the Ag/MgO interface, as summarized in Figure 36. The growth of Ag on MgO(001) was shown to be always of the 3D Volmer–Weber type. Between 0 and 4 ML, coherent Ag islands form and grow laterally, with decreasing aspect ratio. Ag is fully relaxed on the edges at the very beginning and becomes increasingly strained to the MgO in-plane lattice parameter as the islands grow laterally. Around 4 ML, the islands reach a lateral critical size of ~ 100 Å for which interfacial defects such as misfit dislocations and stacking faults are naturally introduced at the edges. The coalescence happens around 10–30 ML. Above 30 ML, the film is continuous, and interfacial dislocations rearrange on a regular network. The fraction of Ag that is on site is made of columns located at the center of islands before coalescence and between dislocation lines after.

4.1.2. Surfactant-Assisted Growth of Ag on MgO(001)

Finding new ways to improve the adhesion of a metal on a ceramic surface is of great importance for many applications. The growth of Ag on MgO(001) is 3D because Ag does not

thermodynamically "wet" MgO(001), but also because of the lattice misfit. The introduction of adequate surfactants to promote 2D with respect to 3D growth has been found to be very efficient in numerous systems, in both homoepitaxy [153–155] and heteroepitaxy [153, 154]. In particular, antimony (Sb) has been shown (actually by GIXD) to be a surfactant during the homoepitaxial growth of Ag, of both (111) [153] and (100) [154] orientations. For the Ag/MgO system, Sb could be expected to induce several kinetic effects promoting a 2D growth, like an increase of the nucleation density, the concentration of lattice accommodation in Sb-rich regions, the earlier formation of the misfit dislocation network, as well as the earlier appearance of the coalescence or the suppression of stacking faults due to the dendritic shape of islands inducing connections between islands, as has been observed, for instance, in the case of the homoepitaxial growth of Cu(111) [155].

A GIXD study was thus performed to investigate the effect of Sb as a surfactant during the heteroepitaxial growth of Ag(001) on top of MgO(001). The same measurements as above were performed under two conditions for Sb deposition: after deposition of 0.2 ML of Sb on the bare substrate on the one hand, and after deposition of ~ 1 ML of Ag before deposition of 0.2 ML of Sb on the other hand, followed by Ag growth. Whatever the growth conditions, only extremely small differences were observed in measurements made with and without Sb. The conclusion of this study is that Sb does not modify the structure, the morphology, or the kinetics of the growth of the Ag/MgO(001) interface. Complementary AES experiments were performed which showed that Sb indeed does not wet MgO(001), at least for alternated deposition. Nevertheless, the GIXD technique could be useful for testing other ideas to modify growth, such as the association of a wetting tensioactive element like Fe with a surfactant, or growth under partial oxygen (or CO) pressure, which is found to improve wetting in some cases.

4.1.3. Thick Ag Films on MgO(001)

For the Ag/MgO(001) interface, the misfit is $f = -2.98\%$ [156, 157]. Hence, a semicoherent interface would be expected. Ordered, localized interfacial misfit dislocations had actually been observed by HR-TEM [152]. The HR-TEM conclusion was that the dislocation lines are oriented along the $\langle 100 \rangle$ directions, with a $1/2\langle 100 \rangle$ Burgers vector. This could only be explained by the coexistence of two possible epitaxial sites for silver in the regions of "good match" between the Ag film and the MgO substrate: regions where the silver atoms sit above oxygen ions of the last substrate plane, and regions where they sit above magnesium ions. This conclusion was very surprising, because all theoretical calculations of the epitaxial site performed so far arrived at only one kind of epitaxial site, above O ions, which was also our experimental conclusion from GIXD [44]. These discrepancies motivated a GIXD investigation of the interfacial dislocation network.

According to the different possible epitaxial sites, different coincident site lattices (CSLs) may be considered. In "good

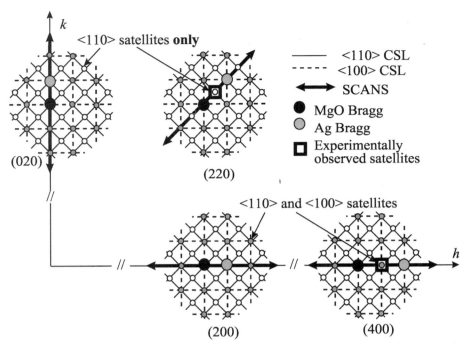

Fig. 37. Schematic representation of the $(hk0)$ interfacial plane of the reciprocal lattice of the Ag/MgO(001) interface with an interfacial network of misfit dislocations. The MgO and Ag Bragg peaks are respectively represented by large black and gray disks. The reciprocal lattices of the two possible interfacial misfit dislocation networks are also shown, as grids, with continuous lines for the ⟨110⟩ CSL and a dashed line for the ⟨100⟩ CSL. The locations of satellites from the interfacial network are represented as gray disks for the satellites that are common to the two CSL, and as open circles for those satellites that pertain only to the ⟨110⟩ CSL. The experimental radial scans performed on the different samples are also indicated. A scan along the ⟨110⟩ reciprocal direction, between the MgO and Ag Bragg peaks, should make it possible to distinguish unambiguously between the two possible network orientations.

match" regions, because of the symmetries of the MgO(001) plane, the Ag atoms may sit either above oxygen ions of the substrate, or above magnesium ions, or in between, above the octahedral sites, with two possible variants [152]. If they sit above only one of the possible epitaxial sites, a square CSL of 97 Å periodicity oriented along ⟨110⟩ directions is obtained. If there are two equivalent epitaxial sites, for instance, O and Mg, or two variants of the octahedral site, then a square network oriented along ⟨100⟩ directions is obtained, of 69 Å periodicity, $\sqrt{2}$ smaller than for the ⟨110⟩ CSL. As illustrated in Figure 37, it is possible to distinguish between the ⟨100⟩ and ⟨110⟩ dislocation networks by performing X-ray scattering along the ⟨h00⟩ and ⟨hh0⟩ directions of the reciprocal space. Indeed, along the ⟨hh0⟩ direction, the satellite periodicity is double in the case of a ⟨110⟩ CSL with respect to the ⟨100⟩ case.

Many different samples were studied, with different Ag thicknesses ranging from 5 to 150 nm, different substrate surface preparations, and different miscuts of the MgO(001) substrates ranging from 0° to 3°. In all cases, as shown in Figure 38 for a miscut substrate and in Figure 39 for a flat substrate, a satellite was found between the MgO(220) and Ag(220) Bragg peaks, which unambiguously demonstrates that the dislocation network is of ⟨110⟩ orientation and is sufficiently ordered to yield at least a first-order satellite diffracted by this network.

In addition to the orientation, many other features were deduced from these measurements. On all substrates with a significant miscut, a large background of diffuse scattering was found in the region of the Bragg peaks. It takes the form of two shoulders, symmetric with respect to the Ag peak, with a significant diffuse scattering between. As shown in Figure 40, on large radial scans taken at different values of the perpendicular momentum transfer ℓ (in r.l.u. of MgO), the separation between these two peaks increases with ℓ, in such a way that they are aligned on rods along the ⟨$\bar{1}\bar{1}1$⟩ and ⟨111⟩ directions emanating from the Ag Bragg peak located at (2.061 2.061 0.04). These rods were shown to originate from stacking faults along (111) planes [158]. Another interesting feature seen in Figure 40 is the peak measured around (2.404 2.404 0.35), which was shown to arise from twin formations, corresponding to two crystals of reverse FCC stacking, with a mirror plane at the fault location. Twins no longer produce rods, but produce additional peaks at (220) + 1/3(111) and (220) + 2/3 ($\bar{1}\bar{1}1$), in reciprocal lattice units of Ag.

Hence, these data showed that, in addition to the interfacial dislocation network, growth faults are present within the Ag thin film, mainly stacking faults and twins along (111) planes. These faults are likely to occur during coalescence of neighboring islands of different stacking. The density of stacking faults

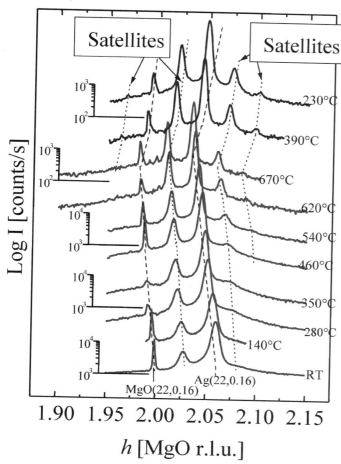

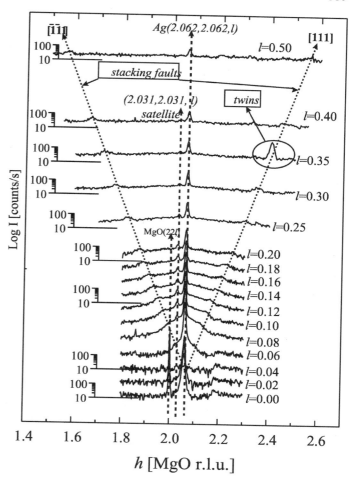

Fig. 38. Radial scans along the $\langle h\,h\,0.1\rangle$ direction around the (220) Bragg peak at different temperatures during heating, followed by cooling to room temperature, for a 1500-Å-thick Ag film grown by MBE on an MgO(001) substrate with a 2° miscut.

Fig. 40. Radial scans $\langle h\,h\,\ell\rangle$ around $h = 2$ on a 1500-Å-thick Ag film with a 2.5° miscut. A vertical translation, proportional to the ℓ coordinate, has been introduced between the different scans. In addition to the MgO and the Ag CTRs and to the dislocation satellite rod, there are additional rods of scattering, oriented along the $\langle 111\rangle$ directions, crossing the relaxed Ag peak around $\ell \approx 0.04$. These rods are due to stacking faults in the silver film. The peak around $h = 2.404$ and $\ell = 0.35$ arises from twinned Ag domains. The shift in ℓ (0.04) of the origin of the stacking fault rods is due to refraction in the Ag film.

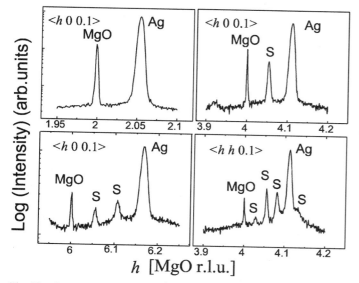

Fig. 39. Radial scans on a 1500-Å Ag film on a MgO(001) substrate with a very small residual miscut (<0.1°), after high-temperature annealing up to 770°C. The diffraction satellites from the interfacial network of misfit dislocations are labeled S.

was found to increase with the step density of the substrate, i.e., its miscut.

To get further information on the interfacial dislocation network, measurements of other diffraction satellites were required. This implies samples with a better ordered dislocation network yielding more satellites of larger intensity and background diffuse scattering below these satellites that is as small as possible. To meet these requirements, the samples were annealed at different temperatures. Because the MgO thermal expansion parameter (13.5×10^{-6} K^{-1}) is smaller than the Ag thermal expansion parameter (19.0×10^{-6} K^{-1}) [159], the lattice parameter misfit f decreases with increasing temperature. Hence, at equilibrium, the period of the dislocation network is expected to increase with temperature, resulting in a larger separation between dislocations. Figure 38 illustrates the evolution of the intensity along radial scans during increasing and next decreasing annealing temperatures. During heating, all peaks shift toward smaller h values because of thermal expansion. Several features were deduced from the observed evolution. First, upon annealing, the background due to the stacking faults decreases and disappears completely around 300°C. Second, the main satellite always remains exactly centered between the MgO and Ag peaks, whatever the temperature. This indicates that, whatever the temperature, the period of the dislocation network changes for it to be exactly on the CSL. Third, annealing clearly induces a recrystallization of the Ag thin film, as revealed by a narrowing of the Ag peak in both the radial and transverse directions. Transverse measurements of different orders of diffraction by the Ag film revealed that the crystalline quality is limited by a finite mosaic spread, which decreases from ~0.25° down to ~0.1° after annealing. At the same time the dislocation network strongly reorders, as revealed by the much larger intensity of the satellites and by the appearance of additional satellites, both on the right of the Ag peak and on the left of the MgO peak. Around 350°C, the second satellite on the right of the Ag peak becomes clearly visible, and two other satellites appear around 620°C.

The above results demonstrate that the annealed sample is, as expected, the best suited for quantitative measurements of the dislocation satellites. For this purpose, all measurable dislocation satellites were recorded. An incident angle of $\alpha = 2\alpha_c$ and an ℓ-value of 0.1 were selected because they yielded the optimal satellite intensity over background ratio. Figure 39 shows the radial measurements, which all again confirm the $\langle 110 \rangle$ orientation of the dislocation network. All in-plane peaks had a transverse FWHM of $0.1° \pm 0.02°$, thus limited by the Ag film mosaic spread. They were integrated both radially and transversely and corrected for the active area, monitor normalization, polarization, and appropriate Lorentz corrections. One important goal of this study was to quantitatively compare the experimental results with calculated intensities, to test the ability of the available models to calculate the main characteristics of the dislocation network, as probed by X-ray diffraction. An analytical model [160, 161], developed in the framework of linear elasticity in a continuous and isotropic medium, was used to compute the complete 3D displacement field (that is, all atomic positions) within a Ag(001) film of finite thickness in semicoherent epitaxy with MgO(001). The structure factors of all main satellites were next calculated and fitted to the data with only three parameters: a scale factor common to all satellite peaks, a different scale factor for Ag Bragg peaks to take into account the finite thickness simulated, and a Debye–Waller factor taken identical for all atoms. The χ^2 factor was found to be very sensitive to the ratio $k = \mu_{Ag}/\mu_{MgO}$ of the respective shear moduli of the film and the substrate. The best agreement was obtained for $k = 0.28$, which yields $\chi^2 = 1.5$, close to the ideal value of 1.

This simulation showed that the substrate is effectively deformed upon epitaxy of the Ag film, and this deformation significantly contributes to the experimental data. The best agreement was found for $k = 0.28$, which is the value deduced from the bulk shear moduli. This clearly demonstrates

that the theory of elasticity is well adapted to calculation of the deformation field in both the substrate and the film, despite the hypothesis of elastic isotropy of both media, and is well adapted to calculation of the properties of the interface, without invoking a weaker "effective interfacial modulus." These results validate the choice of this quantitative model and show that a numerical simulation that would be required to better describe the detailed structure of the very core of the dislocation is not necessary to reproduce these data. This actually implies that for this ℓ value, the scattered intensity is mainly sensitive to atomic displacement far from the dislocation cores, rather than to the dislocation core itself. Further measurements of the dislocation satellites as a function of the perpendicular coordinate ℓ would be required to determine if, for large ℓ values, X-ray scattering could be sensitive to the atomic structure of the dislocation cores.

Additional information was gained by analyzing the residual deformation in Ag, which was found to depend upon the thickness of the film, the density of steps on the substrate, and the state (as-grown or annealed) of the samples. These observations were discussed in detail. In particular, the residual deformations measured on the different samples before and after annealing were compared with the calculated deformation as a function of the film thickness. Except for the as-grown samples on the flat substrate, the experimental residual deformations were always larger than the calculated residual deformations. For thin films, very large and anisotropic residual deformation were found, which were explained by the fact that the 100-Å- and 200-Å-thick films have not yet fully coalesced. Because for 3D islands, the deformation, and thus the strain, is imposed only on the atomic plane that is in contact with the substrate, the relaxation may partly proceed on the edges. The observed nonbiaxial deformation was then related to the elastic anisotropy of Ag. The decreasing difference between the residual deformations along the two kinds of directions with increasing thickness was connected to the fact that, when the film gets thicker, a smaller fraction is in the form of islands, and hence the deformation tends to be more biaxial.

For thick, continuous films, during increasing annealing temperature, the difference in lattice parameters deduced from the MgO(2 2 0.1) and Ag(2 2 0.1) peak positions was found to behave as expected according to the respective thermal expansion coefficients of Ag and MgO. In contrast, whatever the final temperature reached, the difference in lattice parameters back to room temperature was smaller after high-temperature annealing than before. On miscut samples, a room temperature misfit value of $2.8 \pm 0.02\%$ was found after annealing, compared with 2.98% before annealing. The corresponding difference in lattice parameter amounts to $2.82 \pm 0.05\%$, i.e., a residual deformation of $+0.16\%$ in silver, which is larger by far than the theoretical elastic deformation for 1500-Å-thick films. On the flat sample, the residual deformation found after high-temperature annealing, $0.31 \pm 0.05\%$, was even larger.

These observations were discussed in detail and were shown to prove the existence of an energetic barrier to the nucleation of misfit dislocations. A brief summary of the discussion is given hereafter. If we suppose that all of the relaxation is due to the misfit edge dislocations, the dislocation density is directly related to the misfit parameter f. At 1100 K, a state is reached where $f \approx 2.5\%$, which corresponds to 17% less dislocations than at room temperature. Because the misfit measured after cooling is larger than 2.5%, there must have been nucleation of new dislocations during cooling. Because the gliding plane is the interface plane, this introduction must proceed at the edges of islands or of the silver film. However, because the substrate steps pin the dislocations, the larger the step density, the smaller the number of dislocations that are eliminated at high temperature. The excess dislocations must stay on their terrace and should not stay pinned by the steps, because in that case, the larger the step density, the larger the residual deformation should be, which is contrary to the experimental observation. Hence, the larger the step density, the smaller the number of dislocations that have to be reintroduced during cooling, because steps are a reservoir of dislocations. Finally, it is found that the larger the number of dislocations that have to be reintroduced during cooling, the larger the residual deformation, which can only be explained by the existence of a barrier to the nucleation of new dislocations.

This study also provided important information concerning the initial formation of interfacial dislocations during growth. Indeed, on all 1500-Å-thick as-grown samples, the Ag film was fully relaxed. This implies that the introduction of dislocations during growth is a progressive process, probably linked with the coalescence of islands, rather than a simple nucleation followed by gliding, because in the latter case, a significant residual deformation should be observed. This deduction is in good agreement with the model for the introduction of misfit dislocations that was deduced from GIXD measurements during the *in situ* growth.

4.1.4. Shape Evolution of the Islands during Growth of Ag on MgO(001)

GIXD and GISAXS are well suited for characterising the structure and morphology of metal/oxide interfaces during their growth. However, the growth mode can only be extracted through accurate modeling. The Ag/MgO(001) system is particularly interesting to investigate with these techniques, because most other techniques fail [162]. The very small adhesion energy prevents the use of AFM because the islands are likely to be displaced or modified by the AFM tip. To achieve a more complete quantitative description of the growth morphology, we have used the recently proposed formalism of the generalized model for interface description (GMID) [163]. We have also included in the interpretation previous XPS data, to draw a picture of the growth process that is as accurate as possible. The GMID modelling is well suited to describing the growth of islands with realistic truncated pyramid shapes, atom per atom. It can be very efficiently combined with the GISAXS data previously presented to extract the actual evolution of the morphology during growth.

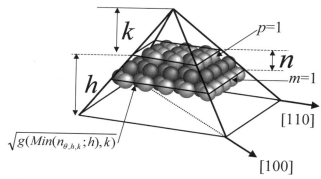

Fig. 41. Schematic representation of the (110) oriented islands that were taken into account by the GMID to describe the growth of Ag/MgO(001).

The GMID model contains in a unified formalism all classical growth modes and is able to generate all possible morphologies within a given symmetry. Here we focus only on the derivation of the quantities measured by GISAXS. The island shape consists of truncated four-fold pyramids exhibiting (111) facets with an index p counting the atomic layers from the top (Fig. 41). For a given coverage θ (expressed in monolayers) the growth morphology is characterized by (i) a particular value h of the index p for which the lateral growth stops, either because the islands touch each other or because island repulsion prevents the complete wetting of the surface (in which case we introduce a fraction ε of the surface that is never covered; h determines the density of islands on the surface); (ii) the number of layers k that are truncated from the top of the pyramid; (iii) the number of layers $n_{\theta,h,k}$, effectively present in the island and that are built below the k missing layers; and (iv) the number of atoms $q_{\theta,h,k}$ that remain after the completion of a situation with $n_{\theta,h,k}$ layers and that are insufficient to proceed to islands with $n_{\theta,h,k} + 1$ layers. The $q_{\theta,h,k}$ atoms are statistically spread over the islands according to the number of atoms each layer must receive to progress to the next island size.

For a given island shape it is assumed that there exists a generating function $g(p)$, which returns the number of atoms on level $p > 0$. Figure 42 gives an example of the construction of (100) oriented truncated pyramids. The amount $g(p)$ is the number of atoms at level p; thus $g(p_{\max})/g(h)$ is the coverage of the surface. The sum over $g(p)$ represents the number of atoms in an island, and $\sum g(p)/g(h)$ is the number of monolayers that have been deposited, Θ.

Θ is connected to the g function by

$$\Theta = (1 - \varepsilon)\left\{ \left(\sum_{p=1}^{\mathrm{Min}(h;n)} \frac{g(p)}{g(h)} \right) + \frac{q}{g(h)} + \left[\mathrm{Max}(n; h) - h \right] \right\}$$

(24)

Because $n(\Theta)$ and $q(\Theta)$ define the morphology, this equation must be reversed. The analytical solution can be written if we define $s(n) = \sum_p g(p)$ and $\theta = \Theta/(1 - \varepsilon)$ and if n_s, the root of $n_s = s^{-1}(\theta \cdot g(h))$, can be found. Then, $n(\theta)$ and $q(\theta)$

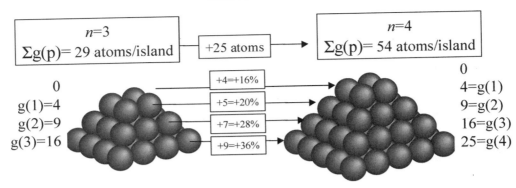

Fig. 42. Example of the construction of ⟨100⟩ oriented truncated pyramids within the GMID formalism, $g(p + k) = (p + k)^2$. Outer left and right columns indicate the number of atoms in each layer. To proceed from three layers to four layers, the island has to receive 25 additional atoms. The central column indicates the percentages of atoms that each layer has to receive to reach the $n = 4$ situation. When fewer than 25 atoms are added, they will be put on each layer according to the percentage of atoms they have to receive, and n will stay at 3 until 25 atoms are added. The number of layers depends on the generating function and must be recalculated for each thickness. In this example the first layer is truncated.

take an analytical form,

$$n(\theta) = \mathrm{Int}(n_s)\big[1 - H\big(\mathrm{Int}(n_s) - h\big)\big]$$
$$+ \left[\mathrm{Int}\left\{\theta - \frac{s(h)}{g(h)}\right\} + h\right] \cdot H\big[\mathrm{Int}(n_s) - h\big] \quad (25a)$$

$$q(\theta) = g(h) \cdot \mathrm{Frac}\left[\theta - \frac{s[\mathrm{Min}(\mathrm{Int}(n_s; h))]}{g(h)}\right] \quad (25b)$$

In the Ag/MgO(001) case, truncated pyramids with ⟨110⟩ orientation apply well, the corresponding generating function is $g(p, k) = 2 \cdot (p + k)^2$, and all of the equations of the GMID can be solved analytically. The sum function is

$$s(n, k) = \sum_{p=1}^{n} g(p, k)$$
$$= 2k^2n + 2kn^2 + 2kn + \frac{2n^3}{3} + n^2 + \frac{n}{3} \quad (26)$$

and

$$n_s = s^{-1}\big(\theta \cdot g(h)\big) = \mathrm{Int}\left(\Re\left\{\Delta + \frac{1}{12\Delta} - k - \frac{1}{2}\right\}\right) \quad (27)$$

with

$$\Delta = \left[\frac{k^3}{2} + \frac{3k^2}{4} + \frac{k}{4} + \frac{3}{2}\theta(h + k)^2\right.$$
$$+ \frac{1}{72}\left[-3 + 18^2\{36\theta^2h^4 + k(12\theta(h^2 + 12\theta h^3)\right.$$
$$+ k(1 + 12\theta(3h^2 + 2h + 18\theta h^2$$
$$+ k(6(1 + 2\theta(1 + 2(h^2 + 3h + 2\theta h))$$
$$+ k(13 + 12\theta(4h + 3(1 + \theta)$$
$$\left.\left.+ k(12(1 + 2\theta) + 4k)))))))\}\right]^{1/2}\right]^{1/3} \quad (28)$$

thus linking the coverage in a unique way to the morphology. From SEM it is known that, even for very thick deposits, about

10% of the surface will never be covered; thus $\varepsilon = 0.1$ in the present case.

The GMID returns $n_{\theta,h,k}$ and the way in which the remaining atoms have to be put on the pyramids. The lattice mesh of Ag is $a_{Ag} = 4.17$ Å. Let us now derive the quantities of interest in the present study. The height of a pyramid, H, is the number of layers times the height of one layer: $H(\theta, h, k) = n_{\theta,h,k} \cdot a_{Ag}/2$. Because the basis layer of the island contains $g(\mathrm{Min}(n_{\theta,h,k}; h), k)$ atoms (see Fig. 42) the lateral size, S, of the island along the ⟨100⟩ direction is given by $S(\theta, h, k) = \sqrt{g(\mathrm{Min}(n_{\theta,h,k}; h), k)} \cdot a_{Ag}/\sqrt{2}$. The interisland distance, D, along the ⟨100⟩ measurement direction is $D(\theta, h, k) = \sqrt{g(h, k)} \cdot a_{Ag}/\sqrt{1 - \varepsilon}$. These are exactly the quantities one can extract from the GISAXS data. The formulation allowing estimation of the XPS intensity is complex and has been described in detail in [163].

Varying h and k permits the exploration of a huge number of possible morphologies. Intuitively, calculating the different quantities for which data are available and drawing the confidence ratio, R, with respect to h and k should allow a determination of the actual morphology of the Ag islands. As can be seen in Figure 43, this approach leads to incoherent results. Each data set providing different best couples (h, k) and the corresponding fits remain extremely poor. Sometimes, as for the interisland distance, the minimum is not defined at all and belongs to a one-dimensional space of solutions. Thus there is no defined shape of static islands that allows reproduction of all of the data. The islands must evolve in some way during the growth. This analysis can also be quantitatively performed with the GISAXS data and the GMID formalism.

It can be shown that k is connected to the experimental GISAXS data by $k = (S - 2H)/a_{Ag}$. The linear law $k(\theta) = (2.36 \pm 0.60) + (0.83 \pm 0.06)\theta$ reproduces the behavior well (Fig. 44d). From k and D one can extract h: $h = -k + D/(a_{Ag} \cdot \sqrt{2/(1 - \varepsilon)})$. Here again a linear law $h(\theta) = (10.96 \pm 0.40) + (0.96 \pm 0.04)\theta$ applies (Fig. 44d). At this stage the growth is completely characterized. Using these laws

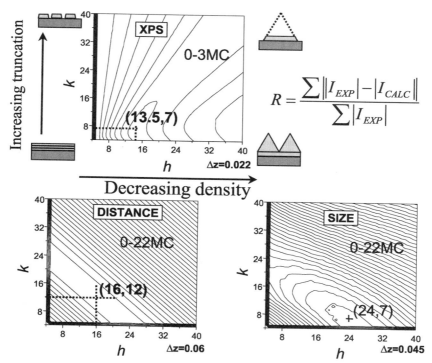

$$R = \frac{\sum \left\| I_{EXP} \right| - \left| I_{CALC} \right\|}{\sum \left| I_{EXP} \right|}$$

Fig. 43. Confidence ratio, R, maps for constant h and k drawn for XPS data (top) and the GISAXS interisland distance (bottom left) and island size (bottom right). The Δz's indicate the interval size between levels in the map. The best h and k are indicated in each map.

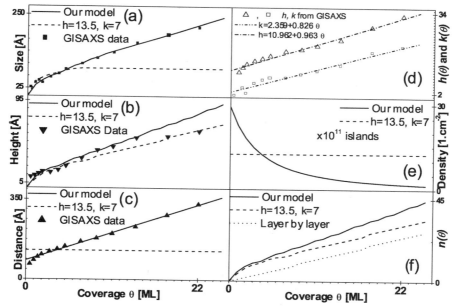

Fig. 44. (a, b, and c) Experimental size (■), height (▼), and interisland distance (▲) compared with the evolutive shape growth model (straight) and to the best ($h = 13.5$, $k = 7$) couple deduced from the XPS data alone (dashed). (d) h (□) and k (△) deduced from the GISAXS data and the best linear laws. (e) Calculated island density for the evolving shape model (straight) and for ($h = 13.5$, $k = 7$) (dashed). (f) Number of layers in the islands for the evolving shape growth (straight), the best XPS (h, k) (dashed) and a layer-by-layer growth (dotted).

and the GMID formalism, one can now calculate any quantity of interest and evaluate experimental measurements. The agreement (XPS and GISAXS; Fig. 44a–c) is always very good. Note that there is no fitting parameter, because everything is directly calculated from the growth laws. One can also extract quantities that describe the growth more intuitively. The island density per

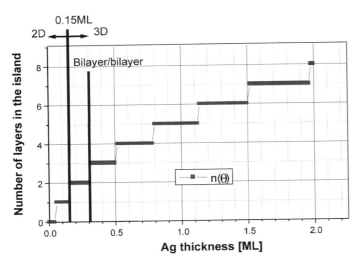

Fig. 45. Detailed evolution of the number of layers in a Ag island during growth, from the quantitative growth laws.

square centimeter is given by $n_{Ag} = D^2/2 = (7.864 \times 10^{-8}\theta + 5.855 \times 10^{-7})^{-2}$ (Fig. 44e); the number of layers $n_{\theta,h,k}$ present in the islands with respect to θ is reported in Figure 44f.

The characterization of the growth thus remains a difficult task because for a given data set it is possible to extract an (h, k) couple that is unable to reproduce other measurements. Moreover, the different measurements are more or less sensitive to the morphology: XPS data are only poorly able to separate the different possibilities obtained from the model (compare the $h = 13.5$, $k = 7$ case with the final model), whereas other data sets, like the particle size, are highly sensitive to variations. These observations show that care must be taken when growth modes are described. The ability of a model to reproduce a given data set is obviously not enough to conclude.

The present analysis shows that the growth of Ag on MgO(100) is, in essence, of an evolving nature. Snapshots at different thicknesses will not be similar in particle size, density, and shape. But the growth laws show that the overall shape is always the same (constant aspect ratio of 0.36 for $\theta > 0.2$ monolayers), because h is directly connected to k by $h = 8.23 + 1.16k$. On the other hand, the density is only governed by the thickness, and its behavior is compatible with a nucleation, growth, and coalescence behavior.

To compare our present interpretation with previous ones, we have to extract the behavior in the submonolayer regime (Fig. 45). One can show that from 0 to 0.15 ML only 1-ML-thick islands form; then the second layer starts its growth up to 0.3 ML, where the third layer begins its growth, and so on. Thus the interpretations of a pseudo–Stranski–Krastanov growth up to 0.1 ML or a bilayer-by-bilayer growth up to 0.4 ML are consistent with the actual growth mode, but they are inaccurate at higher coverage [164]. A description of the growth of islands with variable shapes and densities was not possible in simpler approaches because the shape is then generally fixed *a priori*.

Finally, the model presented here, combined with GISAXS data and GIXD, has shown its ability to reproduce all of the available data within an evolving shape behavior of the Ag is-

lands during their growth. Indeed, the present description is fully coherent with the Ag diffraction peak width evolution during growth. For the Ag/MgO(001) interface we have thus extracted precise and quantitative growth laws for growth at room temperature.

4.2. Pd/MgO(001)

4.2.1. In Situ Studies of the First Stages of Formation of the Pd/MgO(001) Interface

Unlike the Ag/MgO(001) interface, the Pd/MgO(001) interface had been the subject of a large number of studies [165–178], mainly to investigate the kinetics of nucleation in the submonolayer regime between 400 K and 800 K. Whatever the temperature, the growth was found to be 3D (Volmer–Weber), with nucleation, growth, and coalescence of clusters. These clusters are single-crystal particles, fully relaxed and in cube/cube epitaxy, excepted for the first layer in contact with the MgO(001), which, according to HR-TEM results [174], would be perfectly accommodated. No twins could be detected by HR-TEM [174, 176], and a SEELFS study performed at the Pd $N_{2,3}$ edge [179] concluded that the Pd atoms adsorb on top of Mg ions. In addition, an HR-TEM study of the Pd/MgO(001) interface formed by internal oxidation [176] concluded that the 7.6% misfit is accommodated by a network of interfacial dislocations of ⟨110⟩ orientation and $1/2$⟨110⟩ Burgers vector, in agreement with the O-lattice theory.

The GIXD experiments were aimed at studying the growth morphology at room temperature, the epitaxial site, and interfacial distance, as well as characterizing the interfacial dislocation network. They were performed during the *in situ* growth at room temperature of Pd on MgO(001) substrates of high quality, prepared according to the procedure described in Section 3.1 The ID32 ESRF [15] beamline and the W21 surface diffractometer setup were used.

The (20ℓ) and (31ℓ) CTRs measured on the clean MgO(001) surface and after the room temperature deposition of $\theta = 1$ ML of Pd are shown in Figure 46. The large modification induced by deposition shows that, as in the Ag/MgO case, a significant fraction of the Pd deposited is in perfect registry (i.e., on site or pseudomorphic), and the sign of the interference makes it possible to rule out the octahedral epitaxial site. A quantitative fit of the integrated and corrected CTR intensities was performed, yielding with high accuracy the amount of Pd in registry ($\theta = 0.5$ ML), the additional roughness of the pseudomorphic fraction (2.7 Å (r.m.s.)), and an interfacial distance of 2.216 ± 0.02 Å $= 1.05 \times d_{(002)}^{MgO}$. This (20ℓ) CTR is not sensitive to the difference between O and Mg epitaxial sites, unlike the (11ℓ) or (31ℓ) rods, which are very sensitive to the actual site. This is illustrated in Figure 46b and c, which show the experimental and calculated (31ℓ) CTRs for 1 ML of Pd deposited, superimposed on the CTRs of the bare substrate. The qualitative comparison makes it possible to conclude that Pd is above the O ions and not the Mg ions. This is in contradiction to a previous SEELFS investigation, but both the epitaxial site and interfacial

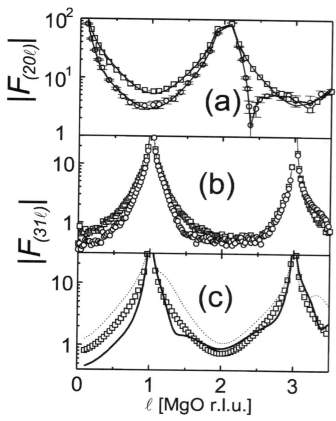

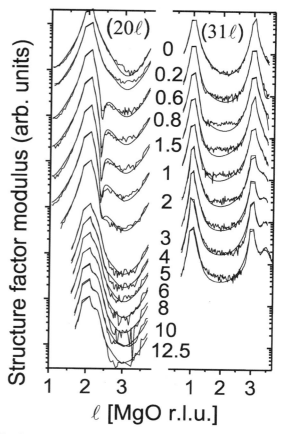

Fig. 46. Modulus of the structure factor of the $\langle 20\ell \rangle$ and $\langle 31\ell \rangle$ CTRs for the bare substrate and for 1 ML of Pd deposited. (a) $\langle 20\ell \rangle$ CTR, obtained by integration and correction of rocking scan measurements, for the bare MgO(001) substrate (open squares with the dashed line showing the best fit) and for $\theta = 1$ ML of Pd deposited (open circles with error bars; the continuous line is the best fit). The best fits yielded the following parameter values: $d_{Pd\text{-}MgO} = 2.22 \pm 0.02$ Å, $O_{\text{on-site}} = 0.5 \pm 0.1$ ML, $\sigma_{Pd} = 2.7 \pm 0.3$ Å, and $d_{Pd\text{-}Pd} = 1.86 \pm 0.03$ Å. (b) ℓ-scan measurements of the $\langle 31\ell \rangle$ CTR for the bare substrate (open squares) and after deposition of $\theta = 1$ ML (open circles). (c) Calculated $\langle 31\ell \rangle$ CTR, for the bare substrate (open squares) and for 1 ML of Pd deposited, with Pd above either Mg ions (dashed line) or O ions (thick continuous line).

Fig. 47. Comparison of the measured (rough line) and calculated (smooth line) $\langle 20\ell \rangle$ and $\langle 31\ell \rangle$ CTRs during the room temperature growth of Pd on MgO(001). The modulus of the structure factor is reported as a function of the out-of-plane momentum transfer. The two CTRs have been simultaneously fitted over a large range of out-of plane momentum transfer. The amount θ (in ML) of Pd deposited is indicated in the figure. The curves were vertically shifted for clarity.

distance are in very good agreement with a recent theoretical calculation [180] yielding 2.18 Å. The evolution of the CTRs with θ is very similar to that in the Ag/MgO(001) case: above $\theta \approx 1$ ML, only a small fraction of the Pd deposited remains on site, most of the film being relaxed. If the epitaxial site is identical to the case of the Ag/MgO interface, the interfacial distance is much smaller (2.216 Å as compared with 2.45 Å in the case of the Ag/MgO interface), likely because of the much stronger bonding in the case of a transition metal like Pd, as compared with a noble one like Ag.

For all deposits between $\theta = 0$ and 12.5 ML, the $\langle 20\ell \rangle$ and $\langle 31\ell \rangle$ CTRs measured in ℓ-scans were simultaneously fitted over the ranges $\ell = 1$ to 3.7 and 0.5 to 3.7, respectively, assuming the oxygen site. The best fits of the experimental data are reported in Figure 47, and the corresponding parameters, in Figure 48. For all deposits, the agreement is good, which shows that the chosen model is adequate. For $\theta > 0.8$ ML, all four fitting parameters are well decorrelated and can be fitted independently. For $\theta = 0.58$ and 0.2 ML, $d_{Pd\text{-}MgO}$ and $d_{Pd\text{-}Pd}$ were found to be correlated.

$O_{\text{on-site}}$ (Fig. 48a) and σ_{Pd} (Fig. 48b) are found to first increase quickly with θ, and then slowly reach asymptotes around $\sim$1.3 ML and 6 Å, respectively. The r.m.s. roughness reaches 4.7 Å and 6.2 Å for $\theta = 5$ and 12.5 ML, respectively (i.e., for equivalent deposited thicknesses of 9.72 Å and 24.3 Å, respectively). It thus always remains much smaller than the equivalent deposited thickness. $d_{Pd\text{-}Pd}$ (Fig. 48c) decreases from 1.895 Å for $\theta = 0.2$ ML to 1.79 Å for $\theta = 4$ ML, and next stays nearly constant, with only a very slight decrease down to 1.785 Å for $\theta = 12.5$ ML. Finally, $d_{Pd\text{-}O}$ (Fig. 48d) shows a peculiar behavior: it first decreases from $\sim$2.23 Å at $\theta = 0.5$ ML to 2.15 Å for $\theta = 4$ ML, and then increases to reach a steady-state value of 2.22 $\pm$0.02 Å above 10 ML. Note that all of these $d_{Pd\text{-}O}$ values are very close to each other. A remarkable feature is that, although they were fitted independently, the same parameters are obtained for $\theta = 1$ ML after fitting of the CTRs measured in rocking scans and in ℓ-scans, although the fits are performed over a very large range of ℓ values. This demonstrates the adequacy of our intensity corrections for ℓ-scans.

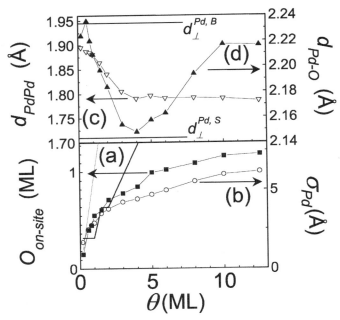

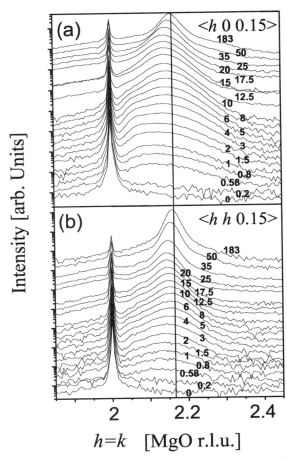

Fig. 48. (a) Evolution with the amount θ of deposited Pd of the four parameters used for fitting the CTRs. (a) The total amount of "on-site" Pd expressed as the number of ML (uncertainty estimated to ±0.1 ML; lower curve, solid squares, left scale). The dashed line is the equivalent amount of deposited θ. (b) The r.m.s. roughness of the "on-site" Pd (uncertainty estimated to ±0.3 Å; lower curve, open circles, right scale). The continuous line is the equivalent height of the Pd deposited. (c) The (002) interplane distance in Pd, d_{PdPd} (uncertainty estimated to ±0.03 Å; upper curve, open triangles, left scale). Horizontal lines indicate the bulk interplane distance $d_{\perp}^{Pd,B}$ and the fully strained interplane distance $d_{\perp}^{Pd,S}$ calculated in the framework of the linear elasticity theory. (d) The interfacial distance d_{Pd-O} (uncertainty estimated to ±0.02 Å; upper curve, solid triangles, right scale).

Fig. 49. Radial scans during the room temperature growth of Pd on MgO(001), along the $\langle h\,0\,0.15 \rangle$ (a) and $\langle h\,h\,0.15 \rangle$ (b) directions, as a function of the amount θ of deposited Pd. The different amounts (0, 0.2, 0.58, 0.8, 1, 2, 3, 4, 5, 6, 8, 10, 12.5, 15, 17.5, 20, 25, 35, 50, and 183 ML) are indicated near the corresponding curves. The successive scans were vertically shifted (multiplication by a factor of 2) for clarity. A vertical line indicates the position, $h = 2.1645$, expected for bulk Pd. All of the scans shown here were performed with the same fixed incident angle of 0.07°.

The misfit relaxation process was investigated by performing in-plane radial and transverse scans during growth. Figure 49 shows the evolution of the intensity along the $\langle h\,h\,0.15 \rangle$ and $\langle h\,0\,0.15 \rangle$ directions crossing the MgO CTR (at $h = 2$) and the Pd rod, expected at $h = 2.165$ for fully relaxed Pd. In contrast to the Ag/MgO measurements reported above, the incident angle was kept constant, smaller than the critical angle for total external reflection for both materials. Consequently, only the structure of the top ~30 Å is probed, and, for thicker deposits, the measurements become insensitive to the interfacial structure. Despite the much larger misfit, the trends are very similar to those in the Ag/MgO case. Relaxed Pd is found from the very beginning of deposition. At 1 ML of Pd deposited, according to the above result on the registered fraction, half is on site, and half is already relaxed. Although GISAXS was not performed, the growth is almost certainly 3D, inasmuch as this is the conclusion of all previous studies, and is completely consistent with the X-ray data. Above $\theta \approx 1$ ML, the similar evolutions along the two azimuths are very close to that of the Ag/MgO growth, despite a lattice parameter misfit that is more than twice as large. The differences with the Ag/MgO case are subtle. Unlike Ag, Pd is never fully relaxed to its bulk lattice parameter, and its average lattice parameter decreases continuously with increasing thickness. Thus it is likely that in the Pd/MgO case,

the islands are flatter at the beginning than for Ag/MgO, resulting in more strained islands. This deduction is in agreement with HR-TEM results, obtained for higher substrate temperatures. Another difference is that the splitting of both the (220) and (200) Pd peaks into two contributions, which corresponds to the introduction of misfit dislocations at the edges of the islands, happens earlier, around $\theta \approx 1.5$ ML. In addition, although special scans were performed to detect the appearance of growth faults, no stacking faults or twins could be detected during the growth of Pd. Finally, the lattice parameter misfit in thick enough films, as in the Ag/MgO case, is relaxed by an interfacial network of misfit dislocations. This is indicated by the satellite between the MgO and Pd peaks along the $\langle h\,h\,0.15 \rangle$ direction, which is observed when the incident angle is increased above the critical angle for total external reflection, thus making it possible to probe the whole Pd film.

Additional conclusions can be drawn from Figure 50, in which the evolution of some parameters is reported as a function of the amount of Pd deposited. The critical angle for total

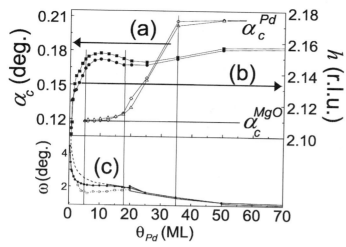

Fig. 50. Evolution, with the amount θ of Pd deposited, of (a) the critical angle for total external reflection, as measured by locating the maximum of the intensity of α-scans performed on the (220) (open diamonds) and (200) (open triangles) Pd Bragg peaks (upper curve, left scale); (b) the mean h position of the (220) (black squares) and (200) (black disks) Pd Bragg peaks in radial scans, as deduced by Lorentzian least-square fits of the peaks (upper curve, right scale); and (c) FWHM $\omega(Q)$ of the (200) (dashed line), (220) (continuous line with filled circles) and (400) (dotted line with open squares) Pd Bragg peaks measured in rocking scans (lower curve, left scale).

external reflection, α_c, measured on the Pd Bragg peaks in both azimuths, is found to be constant below $\theta = 18$ ML and equal to the critical angle of the MgO substrate. Above 18 ML, it increases linearly until it reaches the critical angle for bulk Pd, for $\theta = 35$ ML, and stays constant at this value. This evolution can be interpreted as follows: below 18 ML, all of the deposit is in the form of small islands, the incident X-ray beam is in transmission through these islands, and thus no increase in α_c is found. At 18 ML, the coalescence starts, resulting in some wide islands with a fraction of the surface flat, on which the X-ray beam is totally reflected. During the coalescence process, two effects result in an observed increase in α_c. The first is that more and more islands widen and have a fraction of their surface that is flat and thus externally reflect X-rays. The second is an increase in the film average density toward that of bulk Pd. Above $\theta \approx 35$ ML, the film is continuous. The evolution of α_c thus clearly allows location of the coalescence between 18 and 35 ML. This deduction is corroborated by the evolution of the angular width of the Pd(200), Pd(220), and Pd(400) peaks as a function of θ, also reported in Figure 50. Whereas below $\theta = 18$ ML, these widths are inversely proportional to the modulus of the momentum transfer, which shows that they are dominated by the finite size of the Pd islands, they become equal and independent of it above 18 ML, which shows that they are then dominated by the mosaic spread of the Pd film. The positions of the (220) and (200) Pd Bragg peaks are also reported as a function of θ in Figure 50. They clearly increase very quickly between 0 and 5 ML deposited and then continue to increase only very slowly. This indicates that the plastic relaxation process is mostly complete at 5 ML; nearly all necessary interfacial dislocations are introduced earlier.

In summary, this *in situ* GIXD study of the Pd/MgO(001) interface confirmed the possibility of unambiguously determining the epitaxial site, above O ions, and interfacial distance, to locate the coalescence and the onset of misfit dislocations.

A previous GIXD study on this system deserves to be mentioned here [178]. Films of different thicknesses between 5 and 400 Å were grown in a separate molecular beam epitaxy (MBE) chamber by electron beam evaporation of Pd at a rate of 1 Å/s (100 times faster than the rate of the study described above) onto MgO(001) substrates held at 600°C. They were characterized *ex situ* with the use of a standard laboratory X-ray source. Despite the poor experimental resolution, the average lattice parameter for the (001) epitaxy was deduced as a function of the deposited thickness. Contrary to our work, no dislocation network was detected, and part of the deposited Pd film was found to have a different epitaxial relationship: Pd(111)∥MgO(001), with four variants. This is likely to be induced by defects or contamination of the MgO surface.

4.2.2. Thick Pd Films on MgO(001)

It has been shown in Sections 4.1.1 and 4.2.1 that the growth and relaxation processes at the Ag/MgO(001) and Pd/MgO(001) interfaces are very much alike. This may be connected with the fact that, despite the much larger lattice parameter misfit ($f = -7.6\%$) for Pd (compared with -2.98% for Ag), the bonding of Pd with MgO is stronger. As in the Ag/MgO case, an interfacial misfit dislocation network was detected by GIXD, which is in agreement with a previous HR-TEM observation of the Pd/MgO interface obtained by internal oxidation [176]. In both systems, the metal lattice parameter is smaller than that of the oxide. However, if the thermal expansion coefficient of Ag is larger than that of MgO, which results in a smaller misfit and thus a smaller dislocation density at high temperature, the opposite is true for Pd. The Pd thermal expansion coefficient (11×10^{-6} K^{-1}) is smaller than that of MgO (13.5×10^{-6} K^{-1}), and hence the misfit increases with temperature, corresponding to a larger density of misfit dislocations. To test the hypothesis of an energetic barrier to the nucleation of dislocations, it was natural to perform on this system an experiment similar to the one described in the case of Ag.

The 356-Å-thick Pd film deposited *in situ* on the high-quality MgO(001) surface already presented an interfacial dislocation network, which was fairly disordered, as shown by the small intensity and large width of the main dislocation satellite. This film was thus annealed at increasing temperatures. As in the Ag/MgO case, this resulted in a dramatic recrystallization of the Pd film and reordering of the dislocation network, as shown in Figure 51. Before annealing, only a small and broad satellite was present between the MgO and Pd peaks, and the Pd peak was very large in the radial direction, revealing a large distribution of lattice parameters in the Pd film. Rocking scans on different Pd peaks also showed an in-plane mosaic spread as large as 1°. Annealing resulted in a strong increase and narrowing of the Pd peak along the radial direction, revealing a much narrower distribution of lattice parameters. A dramatic decrease in

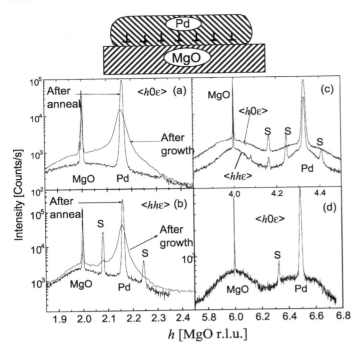

Fig. 51. Logarithm of the intensity measured during a radial scan along the $\langle h\,k\,\varepsilon = 0.15 \rangle$ directions around $h = 2$, 4, and 6 for the as-grown 183-ML-thick Pd(001) film on the MgO(001) substrate, and after 30 min of annealing around 750°C. The peaks at $h = 2$, 4, and 6 correspond to the MgO CTRs, and the peak at multiples of $h = 2.165$ are the Bragg peaks of the Pd film. The peaks marked S are satellites of diffraction by the interfacial network of misfit dislocations. This network and the Pd film are clearly much better ordered after this high-temperature annealing.

the mosaic spread, from 1° to 0.13°, was also observed. Annealing also resulted (Fig. 51) in a strong increase and narrowing of the main satellite, and in the appearance of a very well-defined and narrow second satellite to the right of the Pd Bragg peak. This phenomenon of film recrystallization together with reordering of the dislocation network is thus probably general.

Interestingly, and contrary to the Ag/MgO(001) case, the Pd film was not fully relaxed after growth (with a peak position of $h = 2.157$), wereas annealing led to full relaxation ($h = 2.163$). Hence, in that case, the equilibrium density of dislocations is not obtained during the growth, but because the misfit increases during annealing, new dislocations are introduced at the edges during heating. Although most of these new dislocations were eliminated during cooling, some of them remained for the interface to reach its equilibrium configuration at room temperature.

4.3. Ni/MgO(001)

The Ni/MgO(001) large lattice parameter mismatch (16.4%) addresses the general problem of strained growth for which the build-up stress is not expected to be simply released by misfit dislocations [181]. Recent theoretical investigations showed that Ni is one of the rare metals that is expected to strongly interact with MgO(001) [182, 183]: a large adhesion energy (0.62 eV/atom) and strong bonding (1.24 eV) are predicted

when the epitaxial site is O. All previous experimental works [141, 184–194] agree that the growth of Ni on MgO(001) at room temperature is polycrystalline, but from their results, we readily conclude that growth is strongly dependent on the deposition conditions. Cube-on-cube (CC) epitaxy combined with a dislocation network was reported for sputtered films prepared at 580 K and studied by HR-TEM [187]. Pure CC epitaxy followed by Ni(110) growth was reported for MBE films prepared between 700 K and 900 K [186]. Interdiffusion and Ni(111) growth were observed at high temperature and high pressure [141].

The Ni/MgO(001) system also has interesting magnetic properties because Ni is a ferromagnetic element. An early study [188] showed that substrate defects and film strain have large effects on the magnetic domain structure. Several studies [189–194] used the magnetism of the Ni atoms to indirectly determine the residual strain in the Ni film.

Because all previous investigations of the Ni/MgO(001) system were performed on films prepared at temperatures above 600 K, issues about intrinsic strain at room temperature can only be extrapolated [189]. Moreover, none of these previous contributions described the structural and morphological evolution of the film during growth. Thus it was interesting to use the GIXD approach to determine the intrinsic growth mode at room temperature, the adsorption site, and the relaxation processes for this interface with large lattice mismatch and high bonding energy.

Ni was evaporated with the use of an electron-bombarded Ni rod of purity 99.99% on our high-quality MgO(001) substrates. The growth of the Ni film was investigated for coverages, θ, between 0 and 125 ML, i.e., from 0 to 25.4 nm. Deposition and measurements were alternated: our film thicknesses thus represent cumulative quantities. The GIXD experiments were performed on the BM32-SUV station [17] at ESRF [15]. The beam energy was set at 18 keV to avoid the fluorescence background of the Ni K-edge and to access an extended range of momentum transfer. To obtain a good signal-to-background ratio during the initial growth measurements, the incidence angle was set at $\alpha = 0.09°$, which is smaller than the critical angle for total external reflection of MgO (0.122° at a beam energy of 18 keV). The Miller indexes are relative to the reciprocal lattice units (r.l.u.) of MgO(001). The h and k indexes describe the in-plane reciprocal space, whereas the ℓ index is the out-of-plane component.

To determine the epitaxial relationships, in-plane radial scans were performed in situ for increasing coverages in the $\langle h00 \rangle$ (Fig. 52) and in the $\langle hh0 \rangle$ directions around $h = 2$. Below 1 ML, no signal attributable to Ni was found in these regions of the reciprocal space. For higher coverages, well-defined peaks appear where expected for relaxed Ni(001) in a CC epitaxial relationship (CC-Ni(001)) with respect to MgO. The intensity of these peaks increases for each new deposit and the peaks progressively shift toward the expected positions for fully relaxed Ni ($h = 2.39$). However, CC-Ni(001) at $\theta = 125$ ML is still not completely relaxed.

Besides these dominant contributions, other features successively appear with increasing thickness. These additional peaks arise from a Ni(110)‖MgO(001) epitaxial relationship with four different in-plane orientations, which we call A-, B-, C- and D-Ni(110), as shown in Figure 53. The epitaxial relationships are

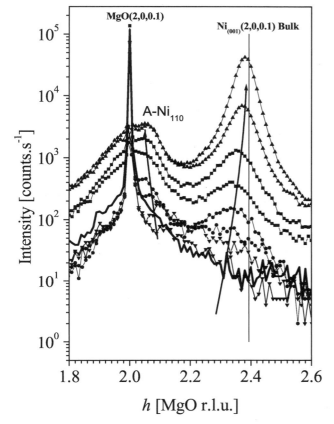

Fig. 52. Evolution of the in-plane $\langle h\,00.1\rangle$ scan, around $h = 2$, during the growth of Ni on MgO(001). The arrows indicate the evolution of the peak positions. The incidence angle was progressively increased from 0.09° to 0.2° to compensate for the refraction of the Ni layer. Only selected scans are reported for clarity. From bottom to top, 0, 1, 2, 6, 12.5, 17.5, 27.5, 50, 125 ML of Ni.

reported in Table I and are enough to reproduce all observed in-plane peaks [181].

Out-of-plane scans along the (20ℓ) and the (11ℓ) (Fig. 54) CTRs were measured at all coverages. Although no signal from the first monolayer was detectable on the in-plane scans, the (20ℓ) CTR shows a strong intensity decrease of half an order of magnitude between 0 and 1 ML. A similar intensity decrease is observed between 1 and 125 ML. The (11ℓ) CTR shows the same kind of behavior for $\ell < 1$. These interferences arise, as they do for Ag/MgO(001) and Pd/MgO(001), because large fraction of the Ni atoms sit on top of the substrate sites in the submonolayer regime. The modeling again indicates the O adsorption site.

The (11ℓ) CTR (Fig. 54) exhibits other peculiar features. Up to a few monolayers a broad shoulder develops on the right-hand side of the (111) Bragg peak of MgO. At 10 ML it clearly decomposes into two contributions, the first at $\ell = 1.52$ and the second at $\ell = 1.69$. The peak at $\ell = 1.69$ increases up to 12 ML, and then decreases, whereas the peak at $\ell = 1.52$ narrows and intensifies continuously. At 125 ML only a well-defined peak at $\ell = 1.52$ remains. Two reflections originating

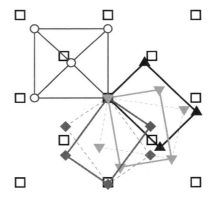

Fig. 53. Schematic representation, in real space, of the epitaxial relationships observed during the growth of Ni on MgO(001). □, MgO(001); ○, Ni(001); ♦, A-Ni(110); ▲, B-Ni(110); ▼, C-Ni(110); ◆, D-Ni(110). The possible variants are not drawn for the sake of clarity.

Table I. Epitaxial Relationships for the Ni(110) Stacking[a]

	$(110)\text{Ni}\,\|\,(001)\text{MgO}$					
	Epilayer basis: $([001]\text{Ni};\,[1\bar{1}0]\text{Ni})$ $\gamma = ([001]\text{Ni}\,\hat{}\,[100]\text{MgO})$				$([1\bar{1}0]\text{Ni};\,[001]\text{Ni})$ $\gamma = ([1\bar{1}0]\text{Ni}\,\hat{}\,[100]\text{MgO})$	
$\gamma + n \cdot 90°$	Type	Epitaxial relationship		$\gamma + n \cdot 90°$	Type	Epitaxial relationship
−35.264°	A	$[\bar{1}12]\text{Ni}\,\|\,[100]\text{MgO}$		+35.264°	A	$[1\bar{1}2]\text{Ni}\,\|\,[010]\text{MgO}$
45°	B	$[001]\text{Ni}\,\|\,[110]\text{MgO}$		45°	B	$[1\bar{1}0]\text{Ni}\,\|\,[110]\text{MgO}$
9.736°	C	$[\bar{1}12]\text{Ni}\,\|\,[110]\text{MgO}$		−9.736°	C	$[1\bar{1}2]\text{Ni}\,\|\,[110]\text{MgO}$
0°	D	$[001]\text{Ni}\,\|\,[100]\text{MgO}$		0°	D	$[1\bar{1}0]\text{Ni}\,\|\,[100]\text{MgO}$

[a] The angle γ defines the relative in-plane orientation between a particular Ni(110) mesh and the [100]MgO(001) direction. The angles are given modulo 90° because of the fourfold symmetry of the substrate. The type of stacking refers to the observed peaks: A occurs between 0 and 10 ML, B between 10 and 50 ML, C after 50 ML, and D is always extremely weak.

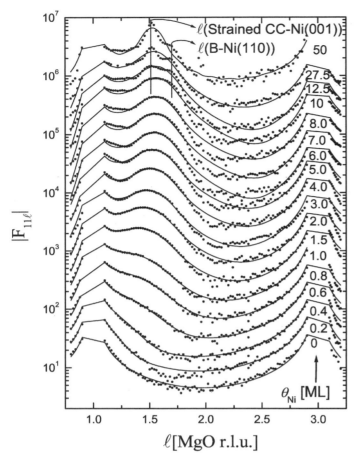

Fig. 54. ⟨11ℓ⟩ CTR versus Ni coverage. From bottom to top, 0, 0.2, 0.4, 0.6, 0.8, 1.0, 1.5, 2.0, 3.0, 4.0, 5.0, 6.0, 7.0, 8.0, 10.0, 12.5, 27.5, 50 ML. Comparison is made for each thickness between experimental data (·) and the best fit (—) obtained with an oxygen epitaxial site and the four parameters reported in Figure 55 for the Ni fraction that sits "on site."

from B-Ni(110) (see Table I) appear at $\ell = 1.69$, but with $h = 0.8$ and 1.2. For small-domain Ni(110), these two Bragg peaks are broad in the h and k directions, yielding the contribution at $\ell = 1.69$. Below 27 ML the CC-Ni film is clearly dominated by a finite domain size effect (≈ 30 Å), most probably due to the finite size of crystallites, and above by a fairly large mosaic spread ($\approx 4°$). The peak at $\ell = 1.52$ in the (11ℓ) CTR corresponds to pseudomorphic Ni with its in-plane lattice parameter, $a_\parallel$, strained to that of MgO ($\Delta a_\parallel/a_\parallel = 16.4\%$). From the elastic theory, the out-of-plane parameter, $a_\perp$, is given by $\Delta a_\perp$, with $\Delta a_\perp/a_\perp = -2 \cdot C_{12} \cdot \Delta a_\parallel/C_{11} \cdot a_\parallel = -20\%$, where C_{11} and C_{12} are the elastic constants of Ni (2.46×10^{11} Pa and 1.5×10^{11} Pa, respectively) [195]. Converted into the MgO r.l.u. this parameter variation gives the $\ell \approx 1.5$ contribution. With this assumption of pseudomorphic or "on site" Ni atoms, the (11ℓ) CTR was quantitatively fitted (Fig. 54). In addition to the epitaxial site, four other parameters were taken into account: the distance between the last MgO plane and the first Ni plane, $d_{\text{Interface}}$; the distance between Ni planes parallel to the surface plane, $d_{\text{Ni-Ni}}$; the total occupancy or amount of the "on

site" Ni fraction, and, finally, the additional roughness of the Ni film with respect to the substrate. In agreement with the theoretical calculations [182], the sign of the interference along the (11ℓ) CTR unambiguously showed that the Ni atoms sit above the oxygen atoms of the last MgO plane. Figure 54 shows that, within this model, the agreement between the experimental and calculated structure factors along the (11ℓ) CTR is good up to 4 ML, fair up to 12 ML; and unsatisfactory above this because of the contribution from the Ni(110) phase at $\ell = 1.69$. The values of the four previously described parameters for all thicknesses are reported in Figure 55. The patterned area corresponds to the values obtained for thicknesses above 12 ML, because the fits are poor in this region these values must be taken with care. In the very early stages of growth the distance between two successive Ni layers is close to that of bulk Ni (most likely isolated atoms or small clusters); it decreases rapidly with increasing coverage down to a value close to that expected from the elasticity theory. The fraction of the Ni layer that sits "on site" increases slowly at first and then more firmly, with the coverage, and reaches about 1 ML for a total coverage of 10 ML; afterward it saturates. The amount of "on site" Ni always remains well below the coverage; indeed, if all of the Ni were pseudomorphic, the experimental points would be situated on the straight line in Figure 55b. The interfacial distance is found to be ≈ 1.82 Å, close to the bulk Ni interplane distance. In the validity domain of the fits (i.e., below 12 ML), this parameter remains almost constant. The additional roughness of the Ni layer passes through a minimum at 0.6 ML and then essentially increases with respect to the coverage (Fig. 55d).

From our experiments, we can deduce the growth mode sketched in Figure 56. For the very early stages of growth (<0.6 ML), nucleation of crystallites occurs, with a composition of roughly one-third on-site Ni and two-thirds partially relaxed Ni. At about 1 ML the accumulated strain becomes too large; most probably the elastic energy becomes larger than the strong Ni–MgO interaction, and CC-Ni(001) and A-Ni(110) grow to 10 ML (Fig. 52). A-Ni(110) only partially relaxes the strain. Around 10 ML all non-CC-Ni(001) strained Ni phases (i.e., on site Ni and A-Ni(110)) stop their growth. This behavior could indicate the start of a coalescence process in the film (Fig. 56c). Finally, above 10 ML, the other Ni(110) in plane orientations successively start their growth. The convergence of the rocking scan widths around 27 ML may indicate the end of the coalescence. The interfacial distance (Fig. 55c) remains almost insensitive to the coverage in the validity domain of our analysis. The average distance on top of O sites is found to be 1.82 ± 0.02 Å, which is close to the value of 1.87 Å deduced from recent *ab initio* calculations [182, 183]. To summarize, below 10 ML the interface dominates the growth of the clusters, and after the coalescence the Ni surface energy becomes the dominant term.

No satellites corresponding to dislocations could be detected after the room temperature growth. From a TEM point of view, the whole film would correspond to a polycrystalline layer, although it can be shown from elastic theory arguments that the different Ni(001) and Ni(110) stackings are fully able to

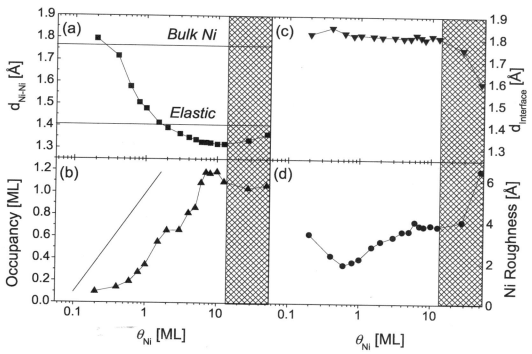

Fig. 55. Parameters used for the fit of the ⟨11ℓ⟩ CTR versus Ni coverage. (a) Distance between two successive Ni planes parallel to the surface. (b) Amount (or occupancy) of Ni that sits on substrate sites. (c) Interfacial distance and (d) roughness of the Ni film. The patterned areas correspond to thickness regions in which the fits are not reliable because some effects (interference with the (110) Ni) were not taken into account in the model.

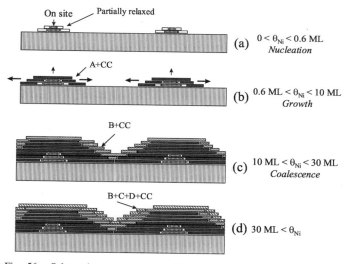

Fig. 56. Schematic representation of the successive stages of the Ni on MgO(001) growth with respect to the coverage.

relax the strain in the Ni layer and thus compete with dislocations [181]. Note that the relaxation is anisotropic with such a process and matches well the in-plane uniaxial magnetic anisotropy previously reported [192, 193]. The Ni film can be made single crystalline, with a good quality (mosaicity <0.7°), in CC epitaxy after annealing at 1250 K. The interface remains, within this treatment, chemically sharp. However, even for the annealed film no dislocation network peak could be detected,

in contrast to the previous TEM report [187]. If such a network were to exist, we should have observed the corresponding satellite peak between the MgO and the Ni Bragg reflections, as in the Ag and Pg/MgO(001) cases. We have seen that the relaxation process is dominated by the different Ni orientations; however, after annealing none of them remain. Thus either the dislocation network does not exist, at least in our preparation conditions, or it remains very disordered and does not give rise to any observable reflection because of the large lattice mismatch.

In summary, the GIXD study has shown that, at room temperature, Ni grows in a rather complex 3D way on MgO(001). The growth is dominated by the more stable CC epitaxial relationship together with different Ni(110)/MgO(001) orientations, which favor anisotropic strain relaxation in the Ni(001) layer. The epitaxial site is oxygen, which is confirmed to be the preferred absorption site on MgO(001). The coalescence of the film occurs between 10 and 27 ML. No ordered misfit dislocation network was found, even for a recrystallized layer.

4.4. Comparison between the Different Metal/MgO(001) Interfaces

The above studies have shown that, for thick films that have already coalesced, the lattice parameter misfit at the Ag/MgO(001) and Pd/MgO(001) interfaces is relaxed by a regular network of dislocations. This network can be characterized in detail by GIXD, which can provide the orientation of the dislocation lines, as well as their Burger's vectors. In both cases, annealing strongly improves the crystalline quality of the film and the ordering of the dislocation network. The mosaic spread decreases and the domain size increases. In the Ni/MgO(001) case, the misfit is too large for the relaxation to proceed via misfit dislocations alone, and clusters containing other orientations are preferred. However, annealing also leads to a drastic recrystallization and to complete disappearance of the Ni with (110) orientation; only CC Ni(001)‖MgO(001) is left, which demonstrates that, despite the large misfit, it is the most stable orientation relationship. Misfit dislocations are likely to be present at the interface to relax the misfit but are probably not ordered, because they were not detected by GIXD.

5. EXCHANGE-COUPLED SYSTEMS

5.1. Specific Considerations: Magnetism Versus Metal/Oxide

Among metal/oxide interfaces, the ones including a ferromagnetic metal on an antiferromagnetic oxide (F/AF) substrate are of special interest. Because of the very peculiar and puzzling magnetic exchange coupling that occurs at this interface [196, 197], they are involved in the fabrication of spin valve sensors, which can be used as magnetic read heads in computer hard drives, position sensors, and magnetic random access memory (MRAM) elements. The phenomenon that allows the devices to function is the magnetic exchange coupling itself, which occurs at these interfaces: the ferromagnetic layer is magnetically pinned (the magnetic domains of the ferromagnet and of the antiferromagnet are coupled) by the antiferromagnetic layer, and the net result is a unidirectional magnetic anisotropy. This is characterized by a displacement of the hysteresis loop [196–198] along the anisotropy direction and/or an increase in the coercive field [121]. To explain the behavior of this phenomenon, interface properties such as roughness, diffusion, etc., are generally suspected, but many contradictory results were reported. An in-depth determination of the atomic structure of such interfaces seems thus to be needed to really understand their physical properties. For this purpose, precise experimental data are required, and GIXD is a technique of choice, because it is insensitive to charge build-up.

A spin valve sensor consists of at least four layers: a ferromagnetic layer deposited on a nonmagnetic layer deposited on a second ferromagnetic layer deposited on an antiferromagnetic layer or substrate (F2/NM/F1/AF sandwich, Fig. 57). The nonmagnetic layer has to be thick enough to avoid magnetic coupling between the ferromagnetic layers. Then, if it is a soft layer, the magnetization of the top ferromagnetic layer is free to rotate with an external field, whereas the magnetization of the second ferromagnetic layer remains pinned by the substrate and will only change its magnetization in very large external fields, if it is the hard magnetic layer. In other words, the magnetic configuration of the two ferromagnetic layers can be switched from parallel to antiparallel [199, 200] (Fig. 57). The resistance of the complete structure will be, respectively, small and large

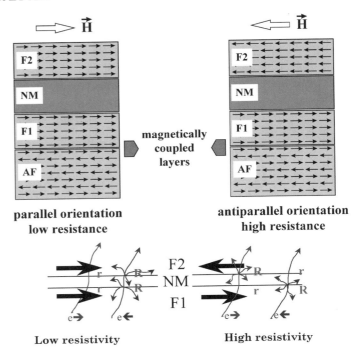

Fig. 57. Schematic representation of a spin-valve sandwich and of its functioning principle. Polarized conduction electrons from one metallic ferromagnetic layer are differentially scattered, depending on the orientation of the second ferromagnetic layer. Small external magnetic fields of a few Oe can switch the system from one situation to another.

because the scattering of the electrons at the interfaces is spin dependent. This resistance variation is the output signal when the magnetically stored information is read, which consists of magnetic domains written on a media. The soft sensing layer orients its magnetization along the stray field of the written information bit.

Multilayer spin valves are good candidates for magnetoresistive read heads because of their high sensitivity and high output signal [199, 201]. To suppress the noise arising from magnetic domain wall motion in the F layers, various AF films, such as FeMn [202–204], NiMn [205], NiO [206–208], α-Fe_2O_3 [203, 209], and CoO [210], were proposed as biasing layers. Indeed, the magnetic exchange coupling phenomenon has been observed for many systems: NiFe/FeMn [201, 203, 204, 211–213], NiFe/NiO [121, 214–218], NiFe/NiCoO [120–122], Fe_3O_4/NiO [129–131], NiFe/α-Fe_2O_3 [203], NiFe/CoO [210, 219, 220], Co/NiO [128, 206, 221, 222], and NiFe/NiMn [205, 223].

A wide range of materials and geometries were used for sensor fabrication. An overview reviewing the advantages and drawbacks for different materials choices and various multilayer layouts in spin-valve structures can be found in [200]. Successful designs must fulfill several stringent conditions; they must be (i) strongly corrosion resistant; (ii) easy to prepare, with reproducible and controllable properties; and (iii) stable over many cycles of use. It has recently been shown [121, 214–218] that interfaces in which the AF is an oxide are promising candidates as replacements for classical devices.

Although such interfaces are already in use at an industrial level to produce spin-valve giant magneto-resistive (GMR) read heads [124, 127, 224–227], the magnetic exchange coupling has puzzled materials scientists for more than 50 years.

The magnetic exchange coupling phenomenon is generally considered an interface effect [211], but some authors are questioning this, too [228]. The correlations between the magnetic properties of such interfaces and some of the structural characteristics were addressed in many studies: e.g., roughness [215, 216, 229–232], thickness of the ferromagnetic layer [211, 214] or of the antiferromagnetic layer [233, 234], crystal texture [216, 222, 223, 230, 232–235], structural stress in the substrate [217, 218], etching of the surface before deposition [229, 230], the dependence of the exchange bias on the magnetic field applied during the cooling through the Néel temperature of the substrate [236] or on the preparation conditions [204, 213], and its dependence on temperature [121, 201, 203]. One of the most mysterious reported facts certainly is a different behavior of the growth order: the properties appeared to be very different if the ferromagnetic layer is deposited on the AF or the reverse [210, 237].

Generally, for F/AF interfaces, the exchange coupling phenomenon was found to be manifested either by an increase in the coercivity with (almost) no exchange field, or a relatively high exchange field (several hundred of Oe), leaving the coercivity of the metallic layer unmodified. The rougher interfaces are found to have a higher coercivity and small exchange coupling [222], although some sputtered Co/NiO layers show only an increase in the coercivity, even for relatively smooth interfaces [238]. A finer grain size of the NiO film should enhance the exchange field [239, 240]. However, it is accepted that improved crystallization of the antiferromagnetic substrate should enhance the exchange coupling by providing a higher level of antiferromagnetic order [222, 239, 241], but correlation can exist (e.g., a smoother surface is related to smaller grain size [239]). As one can easily see from the open literature, there are still ambiguities between experimental results and controversies over theoretical models of the controlling factor(s) for the exchange field of the F/AF interfaces. A recent review on the exchange coupling phenomenon and on the open questions on this field can be found in [242].

The preparation techniques used in the previously cited studies (rf or dc magnetron sputtering, ion beam sputtering, vapor deposition, electron beam deposition, or even MBE; see also the reviews [242, 243] and references therein) yield polycrystalline and/or textured surfaces, which are not suitable for a complete and detailed structural characterization. In this case, many different parameters are intrinsically coupled, making the study of the influence of only one (or precisely one) very difficult.

In the early stages of spin-valve head development, AF metallic alloy materials such as FeMn were mostly used as exchange pinning layers [201, 203, 204, 211–213]. However, the poor corrosion resistance [202] of these alloys leads to serious problems for read head manufacture. To overcome the corrosion resistance problem, nonconducting transition metal oxide materials and, in particular, NiO emerged as potential choices for

the pinning material [121]. From a theoretical point of view, the oxide-based interfaces are also very interesting, because models have still to be developed and tested to describe these interfaces [223, 244, 245].

To the best of the authors' knowledge, a unified picture of the precise and individual role of all of the possible parameters on the exchange coupling phenomenon has not yet emerged from the literature. The microscopic origin of the exchange anisotropy effect remains a subject of speculation. Surprisingly, some results claim that the texture of the NiO, either for the spin compensated or uncompensated planes, does not seem to make a difference for the exchange coupling, contrary to what is expected from theory [215, 216, 235, 246, 247]. Contradictorily, some studies indicated the existence of an increased coupling when the uncompensated (111) surface of NiO is used [237], and others a decrease for this plane [232]. In any case, the most used AF layers in recent years were the spin-uncompensated (111) polar surfaces of transition metal oxides, like NiO, CoO, or NiCoO.

The present studies were intended to overcome all of these problems by with the use of single crystalline NiO(111) substrates and MBE growth techniques to provide a clear picture of the structure of some chosen situations. NiO(111) is a good prototypical material because GMR heads can be built on it and because we were able to produce high-quality single crystalline surfaces (see Section 3.4). In practice it is a good pinning layer for spin-valve devices. NiO is highly resistant to corrosion and has a relatively high Néel temperature ($T_N^{NiO} = 523$ K), which is adequate for applications (high blocking temperatures of the NiO based spin valves). Moreover, on NiO-based films, interfacial ordering effects were observed, as were interlayer exchange interactions [248] and AF ordering along the (111) planes in superlattices [129].

Two interfaces, Co/NiO(111) and $Ni_{81}Fe_{19}$/NiO(111), were investigated in detail. Different samples, prepared at different deposition temperatures, were studied in each case, for thicknesses in the 0–20-nm range, to determine the variation in crystalline quality with respect to the growth temperature. To get a complete picture of these films, different techniques were used in addition to GIXD: in situ GISAXS, ex situ AFM, and magnetic measurements (XMCD, vibrating sample magnetometry (VSM), and the magneto-optical Kerr effect (MOKE)). The magnetic features show clearly that the Co and $Ni_{81}Fe_{19}$ films are magnetically coupled with the single crystalline antiferromagnetic substrate, leaving them as very hard pinned magnetic layers. In the $Ni_{81}Fe_{19}$ (Permalloy, Py) case, complementary TEM measurements were performed. Although the present studies do not yet solve the question of the origin of the magnetic exchange coupling, they give, at least, a quantitative description of the structure of the interfaces.

The GIXD experiments on these interfaces were all performed on the SUV-BM32 station [17] at the ESRF [15]. The energy of the X-rays was always set at 18 keV, which is very far from any absorption edge (Co, O, or Ni) and which allows large momentum transfers. The incidence angle was set at or below the critical angle for total external reflection of the NiO(111) surface (0.17° at 18 keV), which gives the best signal-to-noise ratio for the signal coming from the surface.

The NiO(111) reciprocal space was always indexed with the use of a triangular surface unit cell. The basis vectors are related to the FCC cubic vectors by $\mathbf{a}_s = [\bar{1}10]_{Cube}/2$, $\mathbf{b}_s = [0\bar{1}1]_{Cube}/2$, $\mathbf{c}_s = [111]_{Cube}$, with $a_s = b_s = 2.949$ Å; $c_s = 7.235$ Å; $\alpha = 90°$, $\beta = 90°$, and $\gamma = 120°$. The substrate is always oriented such that a Bragg peak is observed at the $(1, 0, 1)$ position. The h and k indexes describe the in-plane momentum transfer, $Q_\parallel$, and the ℓ index describes the perpendicular momentum transfer, $Q_\perp$, expressed in reciprocal lattice units (r.l.u.). Co and Py both have lattice mismatches of 18% with respect to NiO. It is convenient to use NiO, Co, or Py unit cells to describe the reciprocal space units. Hence, we will use the indexes NiO, Co, and Py when we deal with the r.l.u. of NiO, Co, and Py, respectively, with 1 Co (Py) r.l.u. = 1.18 NiO r.l.u.

Working at a fixed incidence angle in total external reflection ($\alpha_{incident} = \alpha_c = 0.17°$) renders the region of the reciprocal space characterized by $\ell = 0$ (the in-plane scattering) inaccessible. Thus all reported in-plane scans ($\ell = 0$) correspond to directions very close to the surface plane, with a very small nonzero ℓ out-of-plane momentum transfer (in vacuum). Typical values used during all of our measurements are $\ell \approx 0.05$ r.l.u. of NiO (which corresponds to an exit angle set at α_c, too). All of the NiO(111) substrates were p(2×2) reconstructed and were annealed under 10^{-5} mbar O_2 to stabilize the spinel configuration, which does not decompose (see Section 2.4). The intensities reported in these studies were always quantitatively measured and corrected by adequate factors.

5.2. Co/NiO(111)

5.2.1. Growth Versus Temperature

Here we intend to give a detailed study of the Co/NiO(111) interface during its formation, starting from the very early stages of deposition up to a device-related thickness (nominal thickness, $\sim$20 nm Co) with different preparation conditions. Because the structure of the overlayer is generally dominated by the growth temperature, we have studied several deposition temperatures of the substrate (T_S). To address the wealth of opened questions about surface roughness, defects, and surfactant-assisted growth, the substrates were prepared accordingly. The different samples that were studied are summarized in Table II.

All of the NiO(111) substrates were spinel-p(2×2) reconstructed, but their crystalline quality varied along the period of the Co/NiO(111) study: the first samples had a weak reconstruction signal with a mosaic spread in the range of 0.4°–0.7°, and the last ones had excellent quality, 0.03° mosaic spread. We will see later that the controlled quality of the substrates allows highlighting of some well-known limits of the GIXD line-shape analysis.

Co was evaporated with the use of an electron-bombarded rod of 99.99% purity or high-temperature effusion cells. The

Table II.   Preparation Conditions of the GIXD-Investigated Co/NiO(111) Samples[a]

Sample	T_S (K)	Preparation conditions
1	300	Deposit of 20-nm Co in UHV
2	573	Deposit of 20-nm Co in UHV + surfactant effect of 10^{-8} mbar O_2
3	703	Deposit of 20-nm Co in UHV
4	773	Deposit of 20-nm Co in UHV
5	723	Deposit of 20-nm Co in UHV + O_2 etching before deposition

[a]In some of the situations, several substrates were used, with different preparation conditions of the NiO substrate or different final Co thickness.

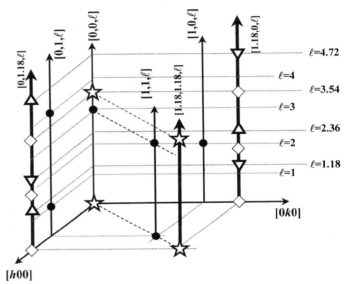

Fig. 58.   Reciprocal space of the Co/NiO interface. The surface plane is spanned by $\langle h00 \rangle$ and $\rangle 0k0 \rangle$. The Bragg peaks corresponding to the different stacking are represented by ● for NiO(111); △, ▽, and ◇ for the FCC, twinned FCC, and HCP Co, respectively. Open star, common peaks for all stacking (FCC, twinned FCC, and HCP). The solid lines perpendicular to the surface plane stand for the CTRs (the thin lines for the NiO CTRs, the thick lines for the Co CTRs).

deposition rate was calibrated with a quartz microbalance. Typical deposition rates on the order of 1–2 Å/min were used ($\sim$1275°C for the effusion cell). At each layer thickness, θ_{Co}, in-plane and out-of-plane scans were performed. The given thickness must thus be understood as cumulative quantities. After deposition a protective Ag or Au film of $\sim$2-nm thickness was applied for *ex situ* investigations.

The reciprocal lattice that can be expected from the 18% lattice mismatched Co/NiO(111) system is represented in Figure 58. It contains several possible forms of Co stacking: HCP, FCC (the stacking that continues the NiO stacking), and twinned FCC-Co (rotated by 60°). The in-plane directions are sensitive, to the total deposited epitaxial quantity of Co (common peaks of all crystalline structures), to the surface state (the presence of intersection of CTRs with the surface plane charac-

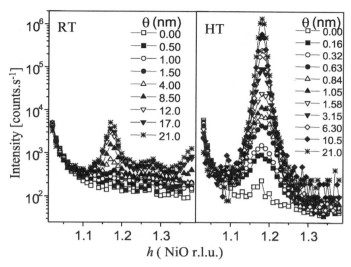

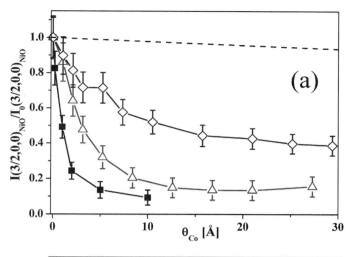

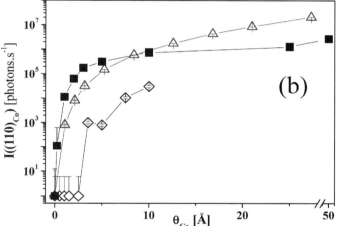

Fig. 59. Evolution of $\langle hh0 \rangle$ scans during the growth of Co on NiO(111) as a function of the thickness, $0 < \theta_{Co} < 21$ nm, for deposition at 300 K (left) and 773 K (right).

Fig. 60. Integrated intensities of the $(3/2, 0, 0)_{NiO}$ reconstruction peak (a) and of the $(1, 1, 0)_{Co}$ Bragg peak (b) with respect to the deposited Co thickness, θ_{Co}. The evolution for three substrate temperatures is shown: $T_S = 600$ K (■), $T_S = 300$ K (◇), and $T_S = 800$ K (△). The intensity of the reconstruction peaks is normalized with respect to the intensity measured for the uncovered surface, $I_0(3/2, 0, 0)$. The mosaic spreads were ∼0.4° for samples prepared at 300 K and 800 K and ∼0.03° for the one prepared at 600 K. The dashed line in (a) represents the attenuation of the surface signal due to X-ray absorption for a 2D Co film (absorption length in Co at 18 keV = 31.05 μm).

terizes flat Co surfaces), or to the coverage through the damping of the p(2 × 2) reconstruction peaks. Out-of-plane scans allow differentiation of the different ways of Co stacking. Indeed, the triangular lattice system is 6-fold in the plane, and 3-fold out of the plane, leaving for a FCC lattice three types of CTRs [13, 24, 249]: $\langle 00\ell \rangle$-like, with Bragg peaks at $\ell = 3 \cdot \kappa$; $\langle 10\ell \rangle$-like, with Bragg peaks at $\ell = 3 \cdot \kappa + 1$; and $\langle 20\ell \rangle$-like, with Bragg peaks at $\ell = 3 \cdot \kappa + 2$, κ is an integer. Introducing the twinned lattice (ACBACB... stacking), which is rotated by 60° with respect to the reference lattice (ABCABC... stacking) makes the whole reciprocal space 6-fold (if the amounts of twinned and direct lattices are identical) by superposition of the two lattices. Whereas FCC Co shows a Bragg peak at $(1, 0, 1)_{Co}$, the Bragg peak on the $\langle 10\ell \rangle_{Co}$ rod of the twinned FCC-Co appears at $\ell_{Co} = 2$. The actual $\langle 10\ell \rangle_{Co}$ rod is thus the superposition of the $\langle 10\ell \rangle_{Co}$ and $\langle 20\ell \rangle_{Co}$ rods of untwinned lattices. For example, along a $(1, 0, \ell)_{Co} = (1.18, 0, \ell)_{NiO}$ rod, the Bragg peaks originating from HCP, FCC, and twinned FCC stacking are separated, as shown in Figure 58.

The $(1.18, 1.18, 0)_{NiO} \equiv (1, 1, 0)_{Co}$ in-plane Bragg peak of Co is common to all stacking, and its intensity allows for the estimation of the fraction of Co in epitaxy. Surprisingly, for the room temperature deposit, the signal on the $(1, 1, 0)_{Co}$ peak is found to appear only for $\theta_{Co} > 1$ nm, whereas for the deposits at higher temperatures (HT, $T_S > 600$ K), it appears already in the very early stages of growth (Fig. 59, right). When θ_{Co} increases, the signal increases, but for room temperature deposition it always remains very weak (about 1/500 of the 773 K signal). The room temperature deposit is thus poorly ordered, whereas the HT deposit is epitaxic and well ordered.

Metallization of the NiO(111) surface is an alternative route to stabilizing the diverging electrostatic potential. The deposition of Co could be expected to kill the reconstruction, and hence the intensity scattered by the reconstruction can be used to monitor the surface coverage with respect to the deposited

thickness. The integrated intensity of the $(3/2, 0, 0)$ peak, which is the most intense in the spinel-configured p(2 × 2) reconstruction, has been used in several situations to monitor the coverage. Its evolution as well as that of the $(1, 1, 0)_{Co}$ peak are reported in Figure 60 with respect to θ_{Co}. The NiO surface reconstruction vanishes as expected because of the Co metallization. However, large differences in the behavior of the reconstruction signal and of the ordered Co signal are observed with respect to the growth temperature. For a 2-nm deposit the damping is a factor of 2.3 for room temperature and 7.6 for HT (773 K) deposition, corresponding to 53% and 87% surface coverage, respectively (Fig. 60). The $(1, 1, 0)_{Co}$ integrated intensity has a coherent evolution with the $\langle hh0 \rangle$ scans: undetectable signal below 1 nm for the room temperature deposit, even if the p(2 × 2) signal decreases, meaning that the metal is

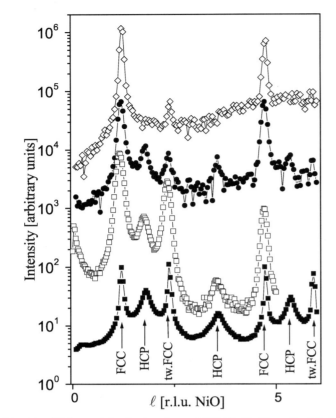

Fig. 61. ⟨10ℓ⟩$_{Co}$ rod for Co deposits on NiO(111). ■, 20-nm Co, RT deposited and annealed (973 K, 45 min). □, 2.5-nm Co deposited at 600 K. ●, 20-nm Co deposited at 730 K. ◇, 20-nm Co deposited at 800 K. For the sake of clarity, a vertical shift between scans was applied. Note the progressive selection of the FCC structure with increasing temperature.

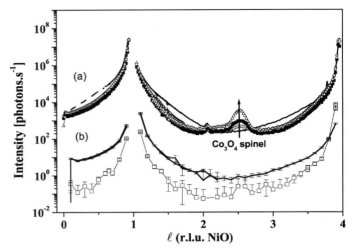

Fig. 62. Evolution of the ⟨10ℓ⟩ NiO CTR with respect to the Co thickness for a Co/NiO(111) sample grown at 600 K with (a) a residual partial O$_2$ pressure of 10^{-8} mbar and (b) in UHV (the intensity was divided by a factor of 100 for the sake of clarity). The straight lines stand for the substrate CTR before deposition. The thicknesses for the other curves are 2 nm (□), 0.5 nm (○), 1 nm (■), and 2.5 nm (△). The Co$_3$O$_4$(111) spinel peak only appears when the growth is performed under partial oxygen pressure.

effectively deposited, and significant signal from the submonolayer coverage for the HT deposit. We can thus conclude, with great confidence, that the growth of Co on NiO(111) is always 3D for any T_S between 300 K and 800 K. Moreover, vertical growth of the islands seems to be preferred for increasing θ_{Co}, because the surface coverage saturates asymptotically (Fig. 60) for all temperatures.

Let us now discuss the out-of-plane scattering. A scan along the ⟨10ℓ⟩$_{Co}$ rod makes it possible to distinguish between the different ways of Co stacking because it successively crosses Bragg peaks of HCP, FCC, and twinned FCC stacking (Fig. 58). Here, too, very different behaviors are observed with respect to the growth temperature. Significant scattering along the ⟨10ℓ⟩$_{Co}$ rod, for a 20-nm film, is only observed after annealing at high temperature, for example, for 45 min at 973 K. Then, peaks from FCC, twinned FCC, and HCP Co appeared (Fig. 61). The integrated and corrected intensities reveal equal proportions of all three ways of stacking, leading to the idea of the recrystallization of almost amorphous initial Co islands. For HT deposits, the intensity of the out-of-plane Co peaks can be directly recorded from the very first stages of the growth ($\theta_{Co} < 1$ ML). The out-of-plane signals, in the submonolayer regime, strongly support the 3D growth of Co. Interestingly,

the FCC stacking is strongly dominant, but residual contributions of all other ways of stacking are present for the deposit at 703 K. At 773 K, no HCP Co is found, although it is the most stable bulk structure of Co (Fig. 61). The total out-of-plane intensity of the annealed room temperature deposit is still 7.5 times smaller than the FCC peak intensities for HT deposits, showing that only about 40% of the Co islands recrystallized during the annealing.

All of the samples prepared in the 600–730 K temperature range contain a combination of FCC, twinned FCC, and HCP stacking (Fig. 61). Stacking other than FCC appears more and more firmly with decreasing temperature, and, below 600 K, faulted structures, like FCC rotated by 30°, start to appear. The ultimate limits certainly are the room temperature case, where the islands are strongly polycrystalline, and the 800 K growth, in which the Co islands are almost pure FCC single crystals.

Unlike Ag, Pd, and Ni/MgO(001), the CTRs do not provide here the expected information for the epitaxial site. The in-plane scans already indicate that the Co is fully relaxed from the beginning. Quantitative measurements of the NiO CTRs have only shown a progressive attenuation during growth, which is more marked, however, during the early stages of growth, mainly because of the derelaxation of the substrate; the asymmetric shape due to the perpendicular relaxation is lost (Fig. 62). This might also suggest a small roughening of the Co/NiO interface at the beginning and maybe a tiny reactive behavior. In any case, Co clearly adopts a well-defined epitaxial relationship with respect to NiO(111), but the epitaxial site obviously is not well defined.

Part of the answer to the question of interface reactivity is provided by the study of a Co/NiO(111) interface prepared at 600 K under an O$_2$ pressure of 10^{-8} mbar. The aim was to

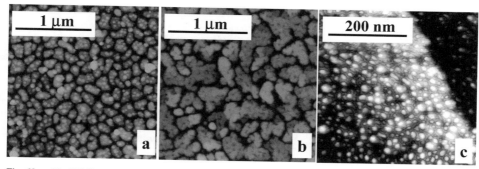

Fig. 63. Air-AFM images taken *ex situ* for different Co/NiO(111) samples. (a) 20-nm Co/NiO(111) (substrate mosaicity 0.4°, T_S = 800 K, 1.8-nm Ag capping). The Co is 3D, with typical island sizes of 100–200 nm. The small islands grown on the Co islands arise from the Ag capping layer. (b) 20-nm Co/NiO(111) (substrate mosaicity ∼0.05°, T_S = 800 K, 2-nm Au capping). Typical sizes for Co islands: 200–400 nm. (c) 20-Å Co/NiO(111) (substrate mosaicity ∼0.05°, T_S = 800 K, 1.5-nm Au capping). The average size of the islands (∼10–20 nm) is in good agreement with the size deduced from the GIXD domain size as well as from GISAXS. The height scale is 50 nm for (a) and (b) and 10 nm for (c).

detect an eventual "surfactant-like" effect associated with the increased GMR measured on spin valves made under such conditions [250]. Surfactants, like Au, Pb, or In [251, 252], are suspected to improve the interface roughness or to introduce layers that have higher spin-dependent scattering capabilities. In fact, the growth at 600 K under a 10^{-8} mbar O_2 pressure looks very similar to that observed in UHV at 600 K; again, mainly FCC Co is found, with a small amount of twinned FCC and HCP. To the difference, diffraction peaks between NiO Bragg peaks along the CTRs clearly appear. Their presence at all half integer index values, out of the surface plane and superposed on the p(2 × 2) reconstruction peaks in the surface plane, indicates a 3D structure. It continues to form during the whole growth process (arrows in Fig. 62). The lattice parameter determined from the position of the extra peaks, as well as the results obtained for the CoO(111) single crystals (see Section 3.5), strongly supports the formation of a $Co_3O_4(111)$ spinel phase, which is relaxed and in CC epitaxy with respect to the substrate. The strong increase in GMR reported for spin valves built under these conditions (up to 24.8% GMR [250]) are thus likely to be related to the formation of cobalt oxide rather than to a surfactant effect.

To summarize the growth mode study, the NiO(111) substrate temperature strongly influences the Co islands' crystalline structure, which varies from fairly polycrystalline to pure FCC single crystalline. The presence of oxygen favors the formation of cobalt oxides—partial oxidation occurred at the interface with NiO(111) in UHV, and strong oxidation occurred when a partial O_2 pressure was used during growth.

5.2.2. Morphology

The crystalline quality of the substrate seemed to have a major influence on the capacity of Co to wet the surface. Indeed, the sample grown at 600 K in Figure 60 had a mosaic spread of 0.03° and Co wets the surface quite well, whereas the other two samples had poorer wetting capabilities but also larger mosaic

spreads of about 0.4–0.7°. Thus it was interesting to characterize the morphology of the islands in more detail with respect to the substrate temperature, growth conditions, and substrate quality. GIXD provides access to the size of clusters in an indirect way through the domain size of the diffracting object. The mosaic spread and domain size are extracted from the evolution of the in-plane peak widths, $w(Q_\parallel)$, with respect to $Q_\parallel$, with $w^2(Q_\parallel) = ([2\pi/D]/Q_\parallel)^2 + (\Delta\omega)^2$, where D is the domain size and $\Delta\omega$ is the mosaic spread. A simple linear regression of the experimental $w^2 = f(1/Q_\parallel^2)$ graph gives the values of D and $\Delta\omega$. Comparison of the morphology with more direct methods, like AFM and GISAXS, will highlight the limitations of the GIXD approach and the role of the substrate.

For many samples the morphology was investigated after growth, *ex situ*, by AFM after the GIXD investigations. Large discrepancies were obtained between the AFM measurements and the deduced GIXD domain sizes. AFM measurements on a 18-Å Ag/200-Å Co/NiO(111) sample (NiO mosaicity ∼0.4°), prepared at 773 K, show that the Co film is not continuous, even for a 20-nm deposit (Fig. 63a). Moreover, it is found that Ag does not wet enough Co to efficiently cap the Co islands; thus Au has been chosen in the next samples. Figure 63b shows an AFM image for 20-nm Co/NiO(111) capped with Au for a high-quality substrate. The Co islands are much larger and might be percolated. The percolation can be determined more easily by a simple electrical measurement; sample a (resp. b) in Figure 63 is still (is no longer) insulating when the surface resistance is measured at a tip distance of about 10 mm. Figure 63c shows an AFM image taken for a 2-nm-thick Co film, in the same condition, confirming the onset of 3D growth since the very beginning of the deposition.

If we now try to compare the GIXD diffracting domain size with the AFM measurement, it appears that the apparent domain size depends on both the quality of the substrate (the size of the terraces) and the crystalline quality of the Co islands. From Figure 64 it is clear that for room temperature deposition, either as deposited or annealed, the diffracting domain size remains much too small to explain the p(2×2) signal decrease. On

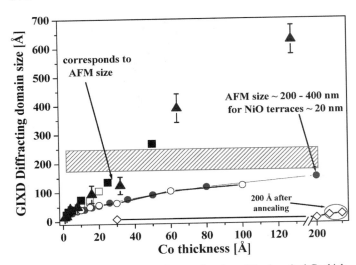

Fig. 64. GIXD-deduced domain size as a function of the deposited Co thickness and of the substrate crystalline quality: NiO(111) with 0.4° mosaicity with $T_S = 730$ K (○) and $T_S = 800$ K (●) and NiO(111) with 0.05° mosaicity with $T_S = 750$ K (▲), $T_S = 600$ K (□, ■) and $T_S = 300$ K (◇). The patterned area represents the terrace size for NiO with 0.4° mosaicity (180–250 Å).

Table III. Characteristics of the Growths of Co on NiO(111) Performed During GISAXS Measurements

Sample	NiO mosaicity (degrees)	T_S (K)	Thickness range (nm)
1	0.05°	480	0–3
2	0.05°	600	0–3
3	0.05°	800	0–2
4	0.5°	700	0–3

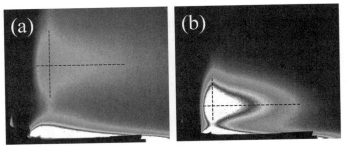

Fig. 65. Upper part of the GISAXS signal recorded on a CCD detector for (a) 0.3 nm and (b) 1.5 nm of Co deposited on NiO(111). The beam energy was 10 keV; the vertical is parallel to the surface, and the horizontal is perpendicular to it. The dashed lines are guides for the eye, to indicate the changes in shape and position of the GISAXS signal.

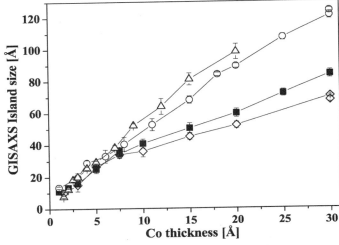

Fig. 66. Average Co island diameter from GISAXS data with respect to the growth temperature and the substrate crystalline quality, for NiO(111) with 0.5° mosaicity (■, $T_S = 700$ K) and 0.05° (△, $T_S = 800$ K; ○, $T_S = 600$ K; ◇, $T_S = 480$ K).

the other hand, the substrate effect is obvious if we compare the diffracting domain sizes for a deposition at 750 K on substrates with 0.4° and 0.05° mosaicity. The first saturates asymptotically around 18 nm, whereas the size continues to increase with thickness for the second. Moreover, the island sizes observed by AFM coincide perfectly with the GIXD diffracting domain size for the substrates with 0.05° mosaicity and completely diverge for the samples with 0.4° mosaicity. To complete the morphology study we have investigated growth by GISAXS (see Section 2.4), which is sensitive to the island size without the instrumental effect due to the AFM tip. The growths that were investigated by GISAXS are summarized in Table III.

From a practical point of view, the Co/NiO(111) system is not easy to study by GISAXS because the interface signal is roughly proportional to the density difference between the materials, and here the densities are close ($\rho_{Co} = 8.9$ g/cm^3 and $\rho_{NiO} = 6.67$ g/cm^3), thus limiting the signal. However, an acceptable signal could be recorded, even with a 2D detector, as can be seen in Figure 65. The experiments were performed on the ID32 beamline at ESRF [15] with the W21 diffractometer [22] and a special setup. Although with horizontal and vertical beam focusing, the total exposure time needed to obtain images like the ones reported in Figure 65 is several minutes,

indicating a very small signal. Thus, the scattering in the present case is more likely related to the Co/vacuum interface and corresponds, in a crude approximation, to the macro-roughness of the islands plus bulk rather than to the difference between the materials. However, the signal-to-noise ratio is good, and the evolution of the morphology can easily be monitored during the growth, as one can see by comparing Figure 65a with Figure 65b. It could be verified that the AFM island size and the GISAXS island sizes are always close, but GISAXS has the advantage of providing statistical average values. The *in situ* investigation of the growth through GISAXS thus gives direct access, in real time, to the morphology of the growing islands in conditions similar to those of GIXD.

The observed GISAXS island sizes are reported in Figure 66. Obviously, when the mosaicity of the substrate is very good (0.05°) and the crystalline quality of the Co is good, this size increases linearly with θ_{Co}. This evolution certainly stands for the real intrinsic growth mode of Co islands on NiO(111). Moreover, these are also the situations where the AFM and GIXD domain sizes are coherent in Figure 64. Now if either the crystalline quality of the Co is poor (growth at 480 K) or the mosaicity of the substrate is large (0.5° and $T_S > 600$ K) the island size ceases growing linearly (Fig. 66). A last key remark is to understand that 0.4° mosaicity represents terraces of about 20 nm on the NiO(111) surface. We can now draw a coherent picture of the morphology and of the GIXD limits. For excellent substrate mosaicities and T_S large enough to ensure good epitaxial growth ($T_S > 600$ K), the GIXD domain size, the GISAXS island size, and the AFM average island size are fully coherent. When the substrate has a limited mosaic spread, and thus small terraces and many defects, these defects propagate into the Co islands, and the GIXD domain size will always be limited by the initial mosaicity of the substrate, giving in turn the asymptotic behaviors reported in Figure 64. In this case the GIXD domain size is no longer a good measurement of the island size. On the other hand, the presence of the defects modifies the intrinsic growth mode; the defects are likely to pin the islands, thus preventing their growth and explaining the deviation from linear size increase in Figure 66. The last situation occurs when the crystalline quality within the Co island is so poor that the islands have defects breaking the coherence of the X-rays on a shorter length scale than the terrace size of the substrate. Then this last phenomenon will limit the observed GIXD domain size (room temperature growth in Fig. 64) and limit the growth of the islands themselves (Fig. 66, growth at 480 K).

To summarize, we have fully characterized the growth of the Co islands at different temperatures and on substrates of different crystalline quality. We have also shown that one and only one of the following lengths may limit the GIXD domain size: the distance between defects in the Co islands, the terrace size on the substrate, or, finally, the size of the islands. The GIXD domain size is thus only a good measurement of the island size when the well-crystallized islands have small extensions compared with the terrace size of the substrate.

5.2.3. Magnetic Properties

We have shown the possibility of building Co/NiO(111) interfaces of controlled crystalline quality. The next question is how the magnetic exchange coupling depends on this crystalline quality, especially in view of the divergent interpretations of magnetic exchange coupling. In our case the NiO grain size can be considered infinite compared with sputtered NiO films. Were the magnetic exchange coupling, for example, to be born at the grain boundaries, we would not observe this phenomenon on single crystal interfaces [132]. After the growth we measured the magnetic behavior of our Co/NiO(111) sample by MOKE and VSM.

The MOKE measurements for room temperature grown and annealed and high-temperature (HT = 700 K) samples are reported and compared in Figure 67a. The annealed room temperature deposited film has a large and distorted hysteresis loop, which is characteristic for polycrystalline films, with a coercive field $H_c = 200$ Oe. The HT sample shows a much squarer hysteresis loop, with $H_c = 410$ Oe. The factor of 3 in the saturation signal obtained at the external applied field of 600–700 Oe

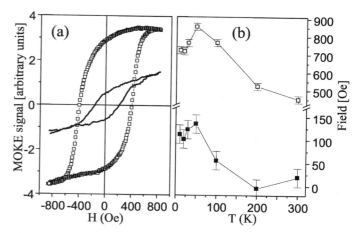

Fig. 67. Magnetic properties of 20-nm Co/NiO(111) films. (a) RT measured MOKE hysteresis loops for a RT deposited and annealed sample (45 min at 970 K) (—) and a HT (703 K) deposit (□). (b) VSM measurement of the evolution of the coercive (H_c, □) and exchange (H_{xc}, ■) fields with respect to the temperature.

can be well explained by the poorer reflectivity of the rougher RT sample. Obviously, the well-crystallized sample shows a much squarer hysteresis loop and acts as a harder (magnetically pinned) layer. To fully understand the exchange coupling, the magnetic properties of the 703 K deposit have been investigated by VSM with respect to the temperature (Fig. 67b). H_c increases with decreasing temperature, and below 150 K the exchange field (unidirectional shift of the hysteresis loop) appears, and it exhibits a behavior similar to that of H_c, with a maximum of 135 Oe at 50 K. These features undoubtedly confirm the onset of classical exchange coupling between the Co islands and the antiferromagnetic NiO(111) substrate at low temperatures.

As a general rule the samples prepared at least at 600 K exhibit features similar to those of the HT sample: large coercive fields at room temperature and increasing exchange fields below 150 K. Measurements for different Co thicknesses, in the 14–23-nm range, for samples prepared at 800 K showed the expected $H_c = A/\theta_{Co}$ law (where A is a constant) and no hysteresis loop shift at room temperature.

If we now consider the coercive field values in more detail, we observe that the H_c obtained for single-crystal substrates for $\theta_{Co} = 20$ nm is comparable to those obtained for spin valves prepared by sputtering [221, 222, 238]. Knowing that the thickness of the pinned layer in a Co-based spin valve is typically 3 nm, and considering the $1/\theta_{Co}$ dependence of H_c, the corresponding H_c for a 3-nm deposit on a single crystal would be as huge as 1200 Oe. Moreover, our interfacial magnetic energies ($J = M_s \theta_{Co} H_c$, where M_s is the spontaneous magnetization of the Co layer, 1420 emu/cm^3 for the bulk Co) are nearly an order of magnitude larger than those obtained for sputtered systems ($J = 1.27$ erg/cm^2 compared with $J = 0.11$ erg/cm^2 in Co-based spin valves [238]).

The coercive fields observed here are huge (at least an order of magnitude larger) compared with what could be expected from an equivalent freestanding layer. This phenomenon is thus connected to the magnetic exchange coupling as much as the hysteresis loop shift is. Obviously, in the single-crystal substrate case the increase in H_c is augmented and the H_{xc} field is reduced, leading to the idea that H_{xc} may be related to defects such as grain boundaries, whereas the large H_c is more intrinsically connected to the interface between a ferromagnet and an antiferromagnet.

To summarize the Co/NiO(111) study, we confirmed experimentally that the p(2 × 2) reconstruction of the NiO(111) single-crystal surface vanishes through metallization, and we have shown that the structural quality of the Co islands can be tuned from polycrystalline to perfectly single crystalline and that the related magnetic properties are strongly correlated with the structure. Moreover, the interface is strongly exchange-coupled for a single-crystal substrate, but it is the coercivity increase that is the basic phenomenon rather that the hysteresis loop shift. Because the structure is tunable, the magnetic properties are also tunable. However, the 3D growth is not favorable in this system inasmuch as it shows that by itself Co does not wet NiO(111) (i.e., a Co layer deposited by sputtering is far from thermodynamic equilibrium); thus devices based on this interface may always remain fragile and may have unforeseeable aging properties. This system also allowed verification of the limits of the finite domain size interpretation of the GIXD peak lineshape analysis, which only gives the distance between two defects that break the coherence of the X-ray scattering.

5.3. Ni$_{80}$Fe$_{20}$/NiO(111)

To improve our understanding of the magnetic exchange coupling phenomenon and related devices (spin valves), one needs pinned ferromagnetic layers that are continuous. An assembly of clusters like the ones observed during the growth of Co on NiO(111) are not adequate because the electron flow could avoid crossing the pinned interface. For this reason we have undertaken the study of other ferromagnetic layers based on the ferromagnetic transition metals (Fe, Ni, and Co). A similar approach to the Co case was used: *in situ* X-ray investigations, followed by *ex situ* magnetic measurements. The growth of permalloy (Py = Ni$_{80}$Fe$_{20}$) is a particularly interesting case, and its growth is described in this section. We focus on the growth characteristics and interface formation, which, in contrast to all of the other cases, can be rendered reactive. In addition to the GIXD study; the samples were characterized by HR-TEM and energy-filtered (EF)-TEM, which allows access to the concentration profile across the interface.

For these studies we always used high-quality NiO(111) single crystals, with mosaic spreads below 0.05°. The substrates were always prepared to exhibit the p(2 × 2) reconstruction spinel configuration, which stabilizes the surface against decomposition (see Section 2.4). The growth was studied by GIXD, and the surface unit cell was indexed as described previously for the Co/NiO(111) interface. The Py film has been obtained by codeposition with the use of two independent remote-controlled sources with electron-bombarded rods

(Ni and Fe, 99.99% purity). Each source was calibrated with a quartz microbalance, and the composition was checked after growth by chemical dissolution in HNO$_3$, which dissolves only the metals and not NiO, and dosing the Ni and Fe in solution. The typical deposition rate was 0.58 Å Py/min. During the growth, different thicknesses of Py (θ_{Py}) were investigated *in situ*, at each θ_{Py}, in-plane and out-of-plane GIXD measurements were carried out. Then, a new deposit of Py was performed. Thus, the given thickness must be understood as cumulative quantities. The deposition was continued until a total nominal thickness of the Py film of 20 nm was achieved. Different samples were prepared in the 600–650 K temperature range. Some samples were magnetically coupled *in situ* by an anneal at 800 K for 20 min and left to cool to room temperature under an external static field of about 500 Oe to generate the magnetic unidirectional exchange anisotropy and to recrystallize the samples. The annealing temperature is chosen to be between the Curie temperature of the ferromagnetic layer and the Néel temperature of the antiferromagnet, allowing for the magnetic saturation of the ferromagnet while the antiferromagnet is not ordered. During cooling down, in such a configuration, the unidirectional magnetic coupling occurs in the direction of the initial external magnetic field. After the UHV annealing, the film was capped by a protective 20-Å-thick Ag or Au capping layer.

5.3.1. Structure and Growth Mode Versus Temperature

Like Co/NiO(111), the Py/NiO(111) interface is 18% lattice mismatched. As can be seen in Figure 68, which presents a HR-TEM image of a room temperature Py deposit, room temperature and low-temperature deposits are fairly polycrystalline. Several grains can easily be identified on the image. Note the high quality of the NiO(111) single-crystal substrate, which shows no defect over the entire image. Moreover, the interface is chemically perfectly sharp and flat. This is quite similar to the Co/NiO(111) interface, with the exception that Py films of 10 nm are almost 2D and wet the NiO(111) substrate very well. Such situations are not well adapted for GIXD investigations, and the film quality is similar to that obtained by sputtering. Because we want to produce single crystalline interfaces, we have explored higher temperature growths.

We now describe the structure of the Py film when the growth takes place at 650 K. Because Py cannot adopt HCP stacking, the reciprocal space is identical to the Co/NiO(111) case without the HCP peaks (Fig. 58). The investigation of the same reciprocal space regions permits an understanding the structure of the Py layer during its growth. Figure 69 shows the evolution of the (110)$_{Py}$ peak with respect to the Py thickness though a $\langle hh0 \rangle$ scan with h close to 1. This Bragg peak contains all of the scattered intensity from the FCC and twinned FCC stacking (Fig. 58). The peak continuously intensifies from the very beginning of the deposition, showing that Py grows epitaxially, with a CC epitaxial relationship, from the early stages of growth. The (110)$_{Py}$ peak immediately appears at $h = 1.18$; similarly to Co, the Py film is thus always fully

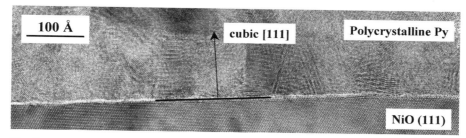

Fig. 68. HR-TEM image of a polycrystalline Py film grown at RT on NiO(111). On the HR-TEM image, several Py grains with different orientations with respect to the substrate can be seen.

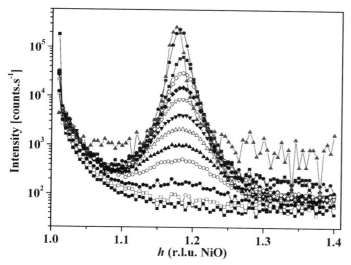

Fig. 69. In-plane scans along $\langle hh0 \rangle$ for different thicknesses of deposited Py between 0 and 20 nm. The thickness are, from bottom to top, 0, 2, 3, 6, 8, 12, 16, 24, 34, 50, 80, 120, and 200 Å (the last situation is the annealed film). The position of the NiO (110) Bragg peak (at $h = 1$, in the left part of the figure) has not been recorded to avoid the very high intensity of the NiO Bragg peak. The higher background for the last scans (large θ_{Py}; top curves) originates from the absorber used to avoid the saturation of the signal in the detector when it passes through the Py peak.

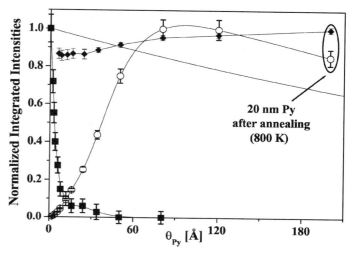

Fig. 70. Characteristic signals of growth with respect to the deposited Py thickness. ■, Evolution of the normalized $(3/2, 0, 0)_{NiO}$ intensity sensitive to the uncovered surface fraction. ○, Normalized $(1, 1, 0)_{Py}$ Bragg peak intensity sensitive to the ordered fraction of the film. ◆, FCC fraction of the film and (—) evolution of the $p(2 \times 2)$ signal that could be expected from the X-ray damping alone (absorption length in Py at 18 keV = 30.56 μm for an incidence angle of 0.17°). The absorption in the Py layer cannot explain the attenuation and disappearence of the NiO reconstruction signal.

relaxed. Scans along other high-symmetry directions ($\langle h00 \rangle$, $\langle h\overline{2h}0 \rangle$, $\langle 0k0 \rangle$, ...) confirmed our conclusions, and no other epitaxial relationships were found.

The NiO(111)-p(2×2) reconstruction signal has been used again to monitor the progressive surface coverage, and again the metallization effectively kills the reconstruction (Fig. 70) and heals the polarity problem. At $\theta_{Py} = 3$ nm, the uncovered surface signal (~3%) comes close to the noise. Importantly, the reconstruction signal decreases much faster than the attenuation of the beam and finally vanishes completely around the 4-nm Py deposit (Fig. 70), highlighting the good surface coverage and wetting of Py. Simultaneously, the evolution of the intensity of the $(1, 1, 0)_{Py}$ Py peak, characteristic of the ordered part of the metallic film, has been measured quantitatively. Its behavior is reported in Figure 70. It increases from the beginning until the thickness of the Py film limits this increase because of refraction around 8 nm; above this thickness it is always the same amount of Py that contributes to the GIXD signal.

To separate the different ways of stacking in the film, out-of-plane scans along the $\langle 10\ell \rangle_{Py}$ direction, which passes successively through FCC and twinned FCC Bragg peaks (Fig. 58), were performed. Several such scans are shown in Figure 71 for different Py coverages. The existence of narrow peaks in the ℓ direction supports a 3D growth mode. In particular, in the submonolayer regime, three layers are needed to distinguish between the two ways of stacking (ABCABC... and ACBACB...), and a 2D layer would yield a rod of constant intensity along ℓ (see Section 2). This shows that already for small quantities of deposited Py, islands form that have at least three atomic layers. Interestingly, the FCC part, which corresponds to the NiO(111) stacking, is dominant for all thicknesses. To accurately quantify this observation, which could also originate from different crystalline qualities, rocking scans at the exact locations of the $(1, 0, 1)_{Py}$ and $(1, 0, 2)_{Py}$ peaks were performed on both the FCC and twinned FCC peaks to get the integrated intensities. The widths along ℓ were also taken into account to refine the integration. After correction the con-

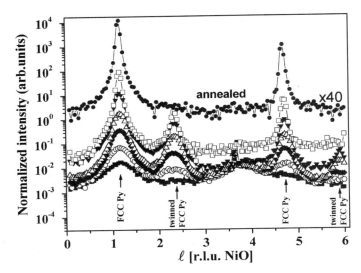

Fig. 71. Out-of-plane scans along the $\langle 1, 0, \ell \rangle_{Py}$ rod for different thicknesses of the Py film, θ_{Py}. From bottom to top, the Py thickness are 3, 6, 12, 24, 50, 120, and 200 Å. The last situation is the annealed film and was shifted on the graph (×40) for the sake of clarity.

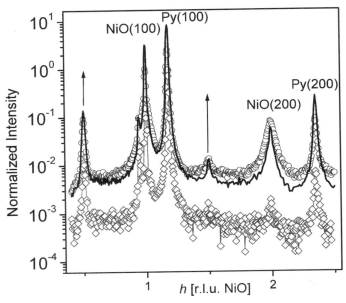

Fig. 72. $\langle h00 \rangle$ scan for a 20-nm Py/NiO(111) film deposited at 650 K and annealed at 800 K for 30 min at different incidence angles. $\Diamond$, $\alpha_i = 0.05°$; —, $\alpha_i = \alpha_c = 0.17°$; $\bigcirc$, $\alpha_i = 0.35°$. Arrows indicate the peaks that grow with the incidence angle.

centration η_{FCC} of the FCC part in the film can be deduced through

$$\eta_{FCC} = \frac{\Delta\ell((101)_{Py}) \cdot I((101)_{Py})}{\Delta\ell((101)_{Py}) \cdot I((101)_{Py}) + \Delta\ell((102)_{Py}) \cdot I((102)_{Py})} \tag{29}$$

The evolution of the fraction η_{FCC} with respect to the thickness is reported in Figure 70. The FCC stacking is always dominant (at least 80%), and at large θ_{Py}, the film is mostly in FCC stacking (95%). The larger error bars at low coverage are due to the smaller signal-to-noise ratio at the beginning of the growth. The residual twinned FCC stacking is here likely to occur from stacking faults rather than from an intrinsic growth mode in which FCC and twinned FCC stacking should be present in a 1 : 1 ratio.

This general behavior of the Py/NiO(111) interface is similar to that observed for Co/NiO(111): polycrystalline layers at low temperature and single crystalline layers at high deposition temperatures. As for Co/NiO(111), the FCC stacking, which continues the NiO(111) stacking, is selected during growth when enough atomic mobility is available (i.e., at high enough temperature). Moreover, this happens for fairly thick layers, indicating the strong influence of the polar surface, probably through the charge image term, on the crystalline structure of the overlayer.

There are, however, some important differences between the two systems. Although the initial growth mode is 3D in both cases, the Py film progressively covers the surface and coalescence occurs at about 30–40 Å. Because metal/oxide systems are likely to adopt 3D growth modes, this behavior deserved further investigation. Annealing and higher temperature growth will permit identification of the corresponding driving force.

5.3.2. Interface Compound

After the deposition of the 20 nm at 650 K, the sample was annealed at 800 K for 30 min to improve the crystalline quality of the Py film. The crystallographic structure of the sample was then reinvestigated after cooling to room temperature under a magnetic field to exchange couple the sample and check the modifications induced by annealing.

The annealing effectively completely washed out the twinned FCC Py: the characteristic peak at the $(1, 0, 2)_{Py}$ position disappeared (Figs. 70 and 71). Although the annealing recrystallized the Py layer and improved its mosaic spread (and thus showed larger signals in radial scans), a significant loss in the integrated intensity of the $(1, 1, 0)_{Py}$ and $(1, 0, 2)_{Py}$ peaks is observed (last point in Fig. 70). Surprisingly, in-plane scans along the $\langle h00 \rangle$, $\langle 0k0 \rangle$, and $\langle hh0 \rangle$ directions, for instance, show the presence of new features at almost half integer indexes (expressed in NiO r.l.u.). Their widths are much larger (5–7°) than those of the p(2 × 2) reconstruction (0.1°), indicating a different origin and not a reorganization of the reconstruction or of the film, which could lead to Py islands and uncovered NiO(111). Moreover, out-of-plane scans also show peaks at almost half-integer ℓ values, showing that this new structure is bulk-like: it has a fairly large thickness. Some $\langle h00 \rangle$ scans were performed at different incidence angles, ranging from grazing (0.05°) to large enough to cross the Py layer and the interface (0.35°) (Fig. 72). They allow the exclusion of a surface type structure. Scans at very grazing incidence do not show at all the peak located at $(3/2, 0, 0)$, and only a small signal from the peak at $(1/2, 0, 0)$ is observed. These features all increase faster than the signal-to-noise ratio when the incidence angle is increased. This unambiguously shows that this new phase is located at

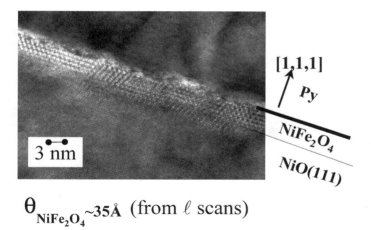

$$\theta_{\text{NiFe}_2\text{O}_4} \sim 35\text{Å} \text{ (from } \ell \text{ scans)}$$

Fig. 73. HR-TEM image of the interface of a 20-nm Py layer deposited on NiO(111) at 650 K and annealed at 800 K for 30 min.

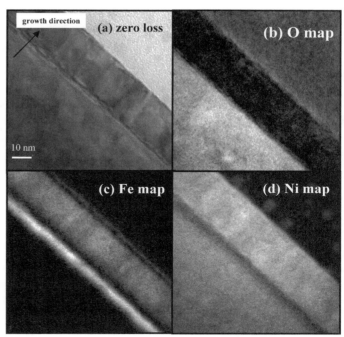

Fig. 74. EF-TEM images of the sample corresponding to Figure 73. The zero-loss image (a) shows the resolution and the overall morphology. (b), (c), and (d) are the maps for which the white zones in the image correspond to O, Fe, and Ni, respectively.

the interface between Py and NiO(111). Moreover, it is interesting to note that the Py thick film exhibits CTRs, indicating a flat top surface. Basing the interpretation on the lattice parameter alone indicates that a Fe_2NiO_4 phase is likely to occur. In any case, the interface obviously becomes reactive during annealing, and a diffuse structure with poor crystalline quality (mosaicity about 5–7°) appears at the interface. A quantitative characterization was not possible. Moreover, differentiating Ni and Fe positions and concentrations in the interfacial compound is not at all easy with GIXD, because all of these atoms have very close scattering cross sections.

Understanding this interface in more detail required additional HR-TEM and EF-TEM investigations. Transverse cross sections of the interface were examined, with a 3010 Jeol microscope working at 300 keV with a LaB_6 filament and equipped with a Gatan image filter [253]. The results confirm the data taken with X-rays, but also show some new features not accessible by GIXD.

The chemical analysis was performed to get concentration profiles of the three elements (Ni, Fe, and O) along the growth direction of the film. The so-called three-windows technique was used. It consists of taking images of the interface at precise emerging electron energies (in a band $\Delta E = 20$ eV) corresponding to characteristic absorption edges (ionization levels) of different elements (in our case, Ni L_{23}, Fe L_{23}, and O K edges, at energies O K = 532 eV, Ni L_{23} = 855 eV, and Fe L_{23} = 708 eV). For each element, two more images are recorded, at energies below the edge. They are used to deduce the background signal. Once the background is subtracted, the images obtained correspond to the chemical distribution of each of the elements. A semiquantitative analysis has been carried out, using the NiO substrate and the Py layer far from the diffuse interface as references for the chemical composition. Elemental distribution maps of Fe, Ni, and O with a lateral resolution on the order of 1 nm and chemical concentration profiles across the interface were obtained.

The interfacial diffuse layer, with a double unit cell with respect to NiO identified by GIXD, has effectively been observed on HR-TEM images (Fig. 73). The different layers are clearly separated and are easy to identify because the different structures yield different contrast. The poor crystalline quality of the diffuse region appears in TEM as well as in GIXD, and the images are compatible with a diffusion process. Note that NiO and the Py film are of excellent crystalline quality without any detectable defects. The lattice parameter of the inverse spinel structure of Fe_2NiO_4 (trevorit) is about twice as large as that of NiO, which leads to positions of new Bragg peaks at about half the distance between NiO peaks. Diffraction TEM images taken in the interface region confirm that the lattice parameter observed in GIXD corresponds only to the diffuse zone.

A detailed observation of the HR-TEM images reveals that the Py/spinel interface becomes wavy and remains quite smooth, whereas the spinel/NiO interface is much rougher. This leads to the idea of a diffusion process at the interface into the NiO(111) substrate. In quantifying this result more precisely, EF-TEM images were most helpful (Fig. 74). As concluded from the GIXD coverage estimation (Fig. 70) and the occurrence of strong Py CTRs (Fig. 72), the film is smooth and 2D (Fig. 74a). For oxygen the interface is sharp, showing that there is an oxide region and a metallic region without oxygen (Fig. 74b, O K edge). The iron (Fig. 74c, Fe L_{23} edge) and nickel (Fig. 74d, Ni L_{23} edge) are more interesting. Between the NiO substrate and the Py film, an Fe-rich (resp. Ni-poor) region alternates with an Fe-poor (resp. Ni-rich) region. However, these images are only qualitative.

A semiquantitative approach is possible here with the use of the NiO and $Ni_{80}Fe_{20}$ known compositions [253] and a refer-

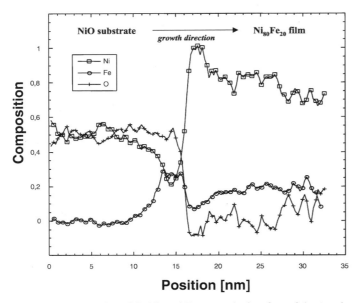

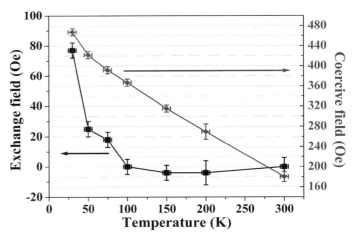

Fig. 75. Concentration of O, Ni, and Fe across the interface of the sample corresponding to Figure 73. The profiles were obtained by integrating across the interface slides of energy-filtered electron microscopy images.

Fig. 76. VSM-measured coercive field and magnetic exchange coupling for a 10-nm Py/NiO(111) sample grown at 630 K with respect to the measurement temperature.

ence trevorit (Fe_2NiO_4) sample, which permits determination of the scattering cross section for each element in each compound. The resulting composition profile across the interface is reported in Figure 75. As a matter of fact, the interfacial spinel compound does not correspond to stoichiometric Fe_2NiO_4, but it lacks in Fe. The interfacial layer is thus rather of the $Ni_\xi Fe_{3-\xi}O_4$ composition, where ξ varies continuously across the interface, between 1 and 2. Indeed, the lattice parameter of such a compound is not expected to significantly vary for $1 < \xi < 2$, making the investigation of the composition change through the interface nonobservable for GIXD. The diffusion profile of Fe into the NiO(111) substrate across the interface is responsible for the large composition range. Moreover, the interface is strongly wavy (Fig. 73), and because the interface morphology may also extend along the other directions, the chemical images and profiles are certainly blurred because of the integration through the sample. However, the HR-TEM estimated thickness for the interface compound is in the 2.5–3 nm range and is in good agreement with the thickness deduced by GIXD from the ℓ-widths of the spinel Bragg peaks (3.6 nm). The Ni-rich region was undetectable by GIXD because Ni and Py have close parameters.

Interestingly, HR-TEM images indicated a network of dislocations 1.5 nm apart in the Py film along the $[11\bar{2}]_{Cube}$ direction, which is close to the 1.77-nm value expected from the triangular coincidence network for Ni (or Py) on NiO(111). This network fully explains the full relaxation of the Py film. In contrast, no network could be detected in the spinel compound, showing that the spinel/NiO interface is coherent. In this latter case the relaxation is likely to occur through the wave-shaped plastic deformation of the interface.

One might now suspect the initial interface, built at 650 K, to already be diffuse. During the deposition process, as well as at

the end of the growth, plane scans, even at large incident angles ($\sim 2\alpha_c$), show no presence of additional peaks. From GIXD we concluded that diffusion starts to occur around ~ 630–650 K. However, to observe a signal coming from the spinel structure, it must be of a minimal crystalline quality and thickness. HR-TEM and EF-TEM would place the onset of diffusion at slightly smaller temperatures, because around 600 K a tiny contrast is observable on the Py/NiO interface, extending over a very few atomic planes.

In the present section we have shown that annealing a Py layer at temperatures above 650 K leads to an interfacial compound. For its quantification HR-TEM and EF-TEM proved to be excellent complementary techniques to GIXD.

5.3.3. Magnetic Properties

We have seen that the Py/NiO(111) system can be prepared with a controllable crystalline quality and that flat and continuous 2D films can be obtained. Py is a soft magnetic material and is only used as a soft sensing layer in sputtered spin valves. What about the magnetic properties of Py on single crystalline NiO(111)? Is it exchange coupled and is it usable? To answer these questions we have performed VSM measurements for Py films deposited in the 600–650 K temperature range.

The VSM hysteresis loops were measured from 300 K down to 15 K. Figure 76 shows the evolution of the coercivity and of the exchange field with respect to the temperature. The general behavior is very similar to that observed in the Co/NiO(111) case. The coercive field increases nearly linearly with decreasing temperature, and the magnetic exchange field appears below 100 K, proving that the Py layer undergoes unidirectional magnetic exchange coupling. Interestingly, the coercive field for a 10-nm Py film is already huge at room temperature (180 Oe), nearly two orders of magnitude larger that the coercive field expected for a sputtered permalloy film. Thus the Py layer is pinned by the NiO substrate and acts as a hard magnetic layer. Moreover, it has been shown for sputtered Py/NiO(111) films that the room temperature exchange field decreases with increasing NiO grain size [254], again highlighting the idea that the large increase in coercivity is a phenomenon that is more intrinsic to the exchange coupling than the exchange field, which is related to the grain boundaries in the antiferromagnet. It is thus likely that the uncompensated spins at the grain boundaries have a major role in the hysteresis loop unidirectional shift. Large coercive fields and exchange fields appearing only at low temperature seem to be characteristic for single crystalline ferromagnetic/antiferromagnetic interfaces.

As in the case of the Co/NiO interface, the coupling energy J was calculated ($J = M_s \theta_{Py} H_c$, where M_s is the spontaneous magnetization of the Py layer, 800 emu/cm^3 for the bulk Py). The value obtained for J is as large as $\sim$0.12 erg/cm^2; it is in fact fully comparable to the coupling energy of a sputtered Co/NiO interface (0.11 erg/cm^2 [238]). Hence, from a magnetic point of view, a single crystalline Py/NiO(111) interface is worth a sputtered Co/NiO(111) interface, and, thus, our Py layers can be used as pinned hard magnetic layers in fully epitaxial spin valves [255].

To summarize, growth conditions able to provide single crystalline 2D Py layers on NiO(111) were determined. The p(2 $\times$ 2) reconstruction of NiO(111) is confirmed to vanish through metallization. Annealing above 650 K leads to a spinel interfacial compound because of the diffusion of Fe into the NiO(111) substrate. The full characterization of the diffuse interface was only possible through the combination of GIXD, HR-TEM, and EF-TEM. The magnetic properties of the single crystalline Py/NiO(111) interface are surprising and allow the use of Py as a hard magnetic layer.

6. GROWTH OF NICKEL OXIDE

6.1. NiO(111)/α-Al$_2$O$_3$(0001)

In Section 5 we have seen that ferromagnetic films deposited on single crystalline NiO(111) have interesting magnetic properties. In particular, giant coercive fields can be obtained through exchange coupling, allowing for the use of fairly large ferromagnetic film thickness. Such a situation is very favorable with respect to device elaboration, because the final spin valve would be less sensitive to surface defects and pinholes. Because sputtered NiO(111) films were reported on α-Al$_2$O$_3$(0001) [256–258], it was interesting to investigate the MBE growth of NiO on this surface to determine the best growth conditions (if they exist) for obtaining single crystalline NiO(111) films. Moreover, for sputtered NiO(111) films, the influence of the crystalline quality on the magnetic properties for NiO films was investigated in previous studies [215, 216, 232, 234, 235, 239, 259, 260]. However, the crystallographic quantitative analysis of such films is quite difficult. In contrast, the use of MBE prepared films allows for a quantitative characterization because of their better crystallographic quality. We have thus undertaken the structural study of MBE-grown NiO(111) films on α-Al$_2$O$_3$(0001) with respect to the growth temperature (320–700°C) and to the thickness (29–200 nm).

Several samples were prepared, at four different deposition temperatures: 620, 700, 800, and 1000 K. NiO films were obtained by Ni evaporation in an oxygen atmosphere, by electron bombarding a metallic Ni rod of 99.99% purity. During Ni evaporation, the oxygen pressure was kept constant at 2×10^{-4} Pa, as was the deposition flux of the Ni source (0.2 nm/min, calibrated with a quartz microbalance). A thickness of 200 nm was achieved for all temperatures. Three additional samples were prepared with different thicknesses: 30 nm and 150 nm at 700 K and 170 nm at 1000 K. The thicknesses were checked by X-ray reflectivity. The reflectivity signal rapidly damps, showing that the prepared NiO surfaces are rough.

Polished and cleaned α-Al$_2$O$_3$(0001) substrates were used. All of our substrates were annealed at 1170 K under an oxygen pressure of $p(O_2) = 5 \times 10^{-3}$ Pa for 10 min and then checked by LEED and by X-rays: well-crystallized samples were obtained, with a near-surface mosaic spread of 0.004°–0.009° (for more details, see the preparation of sapphire substrates, Section 3.3).

LEED patterns were recorded for each sample at the end of the deposition, whereas the morphology was investigated by tapping-mode air-AFM. The crystallographic structure of the samples was then investigated by *ex situ* GIXD. The measurements were performed with a four-circle GMT diffractometer on the BM32 beamline at ESRF [15]. To reduce the absorption of X-rays by air, the samples were mounted in a small vacuum chamber (pressure in the 10^{-3} Pa range), basically consisting of a beryllium cylinder. In GIXD, the samples were illuminated by a well-focused (size at sample: 350 μm horizontal $\times$ 500 μm vertical FWHM) 18-keV X-ray beam under an incidence angle on the order of the critical angle for total external reflection of the sample. The penetration depth can easily be varied from a few nanometers up to tens of nanometers by increasing the incidence angle. An incidence angle of 0.3°, larger than the critical angles for total external reflection α_c of NiO (0.17°) and α-Al$_2$O$_3$ (0.13°), was used, to get the signal from the whole NiO film, even for thicknesses of tens of nanometers. At large incidence angles the α-Al$_2$O$_3$ substrate signal was also detected.

The reciprocal in-plane unit cells of α-Al$_2$O$_3$(0001) and relaxed NiO(111) are shown in Figure 77. The intersections of different CTRs with the surface plane are also indicated. The in-plane Bragg peaks unambiguously permit to identification of the presence of different in-plane orientational relationships between NiO and α-Al$_2$O$_3$(0001).

LEED patterns were taken *in situ* before and after oxygen annealing of the sapphire substrate, as well as once the targeted thickness of the NiO film was achieved. The typical patterns before and after the deposition of a film at 1000 K are shown in Figure 78 for the same sample. The reciprocal unit vectors of the substrate and of the NiO film are rotated by 30°. Moreover, the NiO film exhibits a six-fold symmetry. Compared with the possible epitaxial relationships (Fig. 77), the LEED patterns correspond to a FCC plus twinned FCC NiO(111) film on α-Al$_2$O$_3$(0001), with the NiO(111) in-plane cell rotated by

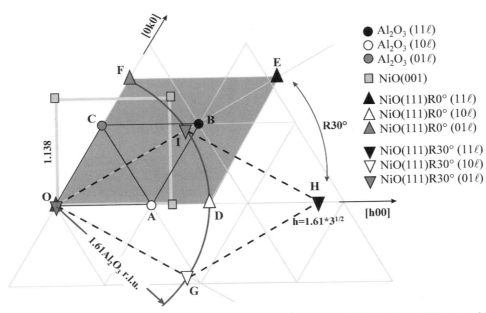

Fig. 77. In-plane view of the reciprocal space of the NiO/Al$_2$O$_3$(0001) interface. Different epitaxial relationships are shown. Represented reciprocal unit cells: Al$_2$O$_3$(0001) (OABC circles, continuous line); a possible orientation of NiO(001) (gray squares); NiO(111)R0° (ODEF, grey hexagonal cell with (111)$_{NiO}$∥(0001)$_{sapphire}$ $\langle h00\rangle_{NiO}$∥$\langle h000\rangle_{sapphire}$); NiO(111)R30° (OGHI, dashed line, (111)$_{NiO}$∥(0001)$_{sapphire}$ $\langle hh0\rangle_{NiO}$∥$\langle h000\rangle_{sapphire}$. For the sake of clarity, twinned orientations are not shown. They can easily be obtained by rotating the reciprocal unit cells of NiO by 60°. Symbols represent different types of NiO CTRs (in NiO r.l.u.): $\langle 11\ell\rangle$-like rods (filled symbols), with Bragg peaks at ℓ positions that can be expressed as $\ell = 3 \times \kappa$; $\langle 10\ell\rangle$-like rods (open symbols), with Bragg peaks at $\ell = 3 \times \kappa + 1$; and $\langle 01\ell\rangle$-like rods (gray symbols), with Bragg peaks at $\ell = 3 \times \kappa + 2$.

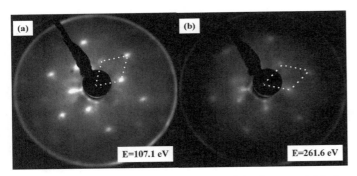

Fig. 78. LEED patterns for (a) the oxygen-annealed substrate and (b) after NiO deposition at 700°C. The projections in the surface plane of the reciprocal unit cells are also shown. Peaks appear at different energies after NiO deposition.

30° with respect to the sapphire substrate. The FCC variant will be called R30° and the twinned FCC variant (i.e., rotated by 90° with respect to Al$_2$O$_3$), R90°; the respective epitaxial relationships are NiO(111)∥α-Al$_2$O$_3$(0001), $\langle 110\rangle_{NiO}$∥$\langle 1000\rangle_{sapphire}$ and NiO(111)∥α-Al$_2$O$_3$(0001), $\langle 110\rangle_{NiO}$∥$\langle \bar{2}100\rangle_{sapphire}$.

AFM images were taken for all samples to determine the morphology of the NiO films. Some of the investigated situations are shown in Figure 79a–f. For all of our samples, a background layer made of triangular pyramids, which have different dimensions, depending on the preparation conditions, is found. The roughness of the NiO films varies with the thickness and deposition temperature. The smallest roughness is found

for the thinner NiO layer deposited at 1000 K, probably because the NiO islands are just about to be formed. The 200-nm thickness samples are very rough; the height of the NiO islands can be several tens of nanometers. The general trend is an increase in the roughness of the film with temperatures above 700 K. For the same thickness of NiO (200 nm) prepared at different temperatures, the sample prepared at 700 K has the smallest roughness. Figure 79a–e shows the morphology for two situations: the 200-nm NiO deposit at 620 K and 700 K. The images obtained for the other samples are quite similar. The shape and dimensions of the islands may change, as seen by AFM. Note that the color scale used in Figure 79c–f (i.e., where the height scale is not reported) does not represent the real measured heights but the variation in the amplitude of the vibrating AFM tip. In this case, the contrast is enhanced for small details in the morphology. Several features are visible. In images similar to Figure 79d the angle of the facets with respect to the surface plane can easily be measured. The quantitative measurement was made on real height images; these facets are found to form an angle of 54° ± 4° with respect to the mean surface plane. In addition, the angles between the corners of the triangular facets were found to be close to 120° ± 10° (threefold symmetry). Indeed, the angle between the (111) and (100) planes is 54.7°. The pyramid has a threefold symmetry axis perpendicular to the (111) plane; a top view shows angles between corners of facets of 120°. The pyramids thus expose (100) facets of NiO(111). (110) facets (expected angle of 35.26°) were not found.

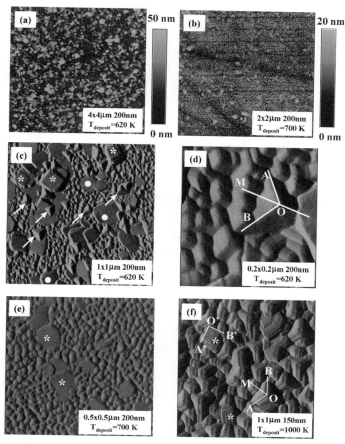

Fig. 79. Tapping-mode air images for different samples. The thickness of the NiO film and preparation temperature (T_{deposit}) are shown on each image, as is the scanned size. (a) and (b) real heights grayscale maps. (c) (100) faceted pyramids (FCC and twinned FCC, marked by arrows; the arrow on the right side of the figure shows two neighboring islands rotated with respect to each other by 60°), square (100) facets (*), and square corner-truncated (o) islands. (d) The angle measured along the OM direction is 54° ± 4° and $\widehat{AOB}$ = 120° ± 10°. (e) Hexagonal islands (*) on the triangular pyramid background layer. (f) Triangular pyramids (FCC and twinned FCC) with (100) facets tilted by 54° ± 4° (e.g., along the OM direction) and square (100) facets with $\underline{A'\widehat{O}B'}$ = 90° ± 7°.

The shape of the islands strongly depends on the preparation conditions. We first notice flat hexagonal-like shapes for the 700 K prepared samples, which can be assigned to the (111) NiO surface plane. AFM images show a tilt of about 12° ± 5° between the surface of these hexagons and the basal surface plane, along directions having again a threefold symmetry. Square islands, probably (100) facets, are also present for extreme preparation conditions (200 nm at 640 K and samples at 1000 K). In this case, the in-plane orientation cannot be addressed by AFM measurements. Some of these (100) facets have corners truncated by (111) facets (Fig. 79c and f).

Concerning the majority NiO structure, NiO(111), the presence of twins can be identified even in AFM images. Indeed, the triangular (100) faceted NiO islands, having a (111)$_{\text{NiO}}$ basal plane, will show different orientations of facets if twins are present. For example, in Figure 79c, two types of islands, with

facets rotated by 60° with respect to each other, can be seen. On the right-hand side of the image, an arrow shows two neighboring triangular islands with their bases rotated by 60°. The morphological analysis thus indicates the presence of twinned NiO(111) on the prepared layers.

If the AFM images allow the determination of the morphology of the NiO film and some crystallographic features, the epitaxial relationships mentioned in Figure 77 cannot be distinguished. This is due to the lack of information in AFM about the crystallographic orientation in the α-Al$_2$O$_3$(0001) substrate with respect to the recorded image. Indeed, atomic resolution is not possible, and the substrate is completely covered by the NiO film. Moreover, only a very small surface fraction is investigated in AFM. To fully characterize the NiO films, X-ray diffraction was used to probe their crystallographic structure. The α-Al$_2$O$_3$(0001) triangular unit cell parameters were $a = b = 4.754$ Å, $c = 12.99$ Å, $\alpha = \beta = 90°$, and $\gamma = 120°$ (see Section 3.3). Because the LEED patterns support the (111) epitaxy of NiO (Fig. 78), we describe NiO by the usual triangular unit cell, with $a_{\text{cube}} = 4.177$ Å, so the unit cell vectors are $a = b = 2.95$ Å and $c = 7.235$ Å, with $\alpha = \beta = 90°$ and $\gamma = 120°$. The in-plane cell parameter misfit is thus $(4.754 - 2.95)/4.754 = 37.9\%$.

Figure 77 illustrates how the different possible crystallographic structures and orientations of the NiO layer can be distinguished. Along a CTR of NiO(111), the Bragg peaks lay out of the plane of the surface, and they are well separated for the FCC and twinned FCC structures. The stacking perpendicular to the surface can be extracted from the specular $\langle 00\ell \rangle$ rod. The characteristic interplane distances in the samples are determined from the peak positions. Particular expected out-of-plane distances are 7.235 Å (c axis of the NiO(111)‖Al$_2$O$_3$(0001) FCC and twinned FCC stacking, giving peaks at $\ell = 5.4 \times \kappa$, expressed in Al$_2$O$_3$ reciprocal lattice units (r.l.u.), where κ is an integer) and 4.177 Å (c axis of NiO(100)‖Al$_2$O$_3$(0001), giving FCC characteristic peaks at $\ell = 2 \times 3.1 \times \kappa$ Al$_2$O$_3$ r.l.u., where κ is an integer). However, the in-plane orientation of the (111) or (100) NiO plane with respect to the substrate cannot be assessed from $\langle 00\ell \rangle$ measurements. For that purpose, large in-plane rocking scans, with constant $Q_\parallel$, for example, along the circular segment EH in Figure 77 and radial scans along $\langle h00 \rangle$ or $\langle hh0 \rangle$ directions, are necessary. The two sets of measurements were performed for all of our samples. Both NiO(100) and NiO(111) stacking was observed along the specular $\langle 00\ell \rangle$ rod. In the following, if not otherwise specified, α-Al$_2$O$_3$ r.l.u. are used.

In all cases, along the $\langle h00 \rangle$ direction of sapphire, Bragg peaks were found at $h = 1.61\sqrt{3} = 2.78$ r.l.u., indicating an in-plane NiO unit cell rotated by 30° with respect to that of the sapphire. This confirms not only that the LEED result is due to the surface layers, but that the structure is characteristic of the whole film. The NiO rod (at $2.78/3 = 0.93$ r.l.u., $\langle 0.93\ 0.93\ \ell \rangle$ at point I in Fig. 77) crosses Bragg peaks, which are distinct for FFC ($\ell = (3 \times \kappa + 1) \times 1.8$) and twinned FCC stacking ($\ell = (3 \times \kappa + 2) \times 1.8$, where κ is an integer; Fig. 80b). Rocking scans at FCC and twinned FCC positions permit an estimation of the

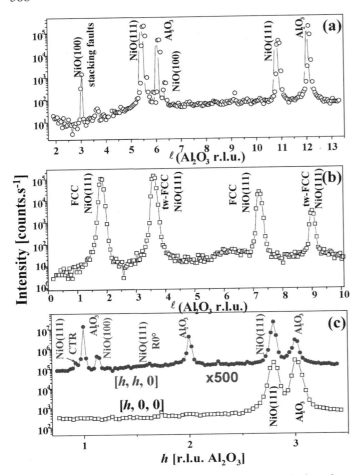

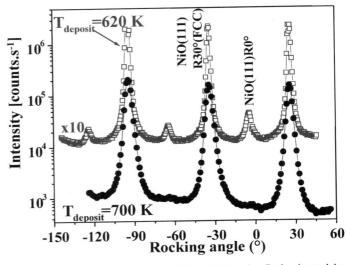

Fig. 81. Rocking scans at $Q_\parallel = 2.78$ α-Al_2O_3 r.l.u. Peaks denoted by NiO(111)R30° and NiO(111)R0° correspond to H and E points in Figure 77.

Fig. 80. X-ray intensities measured along different directions in the reciprocal space. (a) $\langle 0, 0, \ell \rangle$ $T_{deposit} = 320°C$. (b) $\langle 0.93, 0.93, \ell \rangle$ $T_{deposit} = 700°C$. (c) $\langle h, 0, 0 \rangle$ and $\langle h, h, 0 \rangle$ directions; $T_{deposit} = 700°C$. These scans were selected from different samples to highlight the possible NiO crystallographic structures with respect to the sapphire. Positions where scattering is expected are indicated.

$1.61\sqrt{3} = 0.93$ r.l.u. ("I" point in Fig. 77: it is the intersection of a NiO(111)R30° CTR with the surface plane), showing the presence of a (111) surface of FCC NiO(111). It confirms the existence of some relatively flat portions of a (111) surface for the deposited NiO.

Figure 80 shows examples of measured intensities along different directions in the reciprocal space. The positions where scattering is expected for the different structures are shown. The quality of different NiO variants of the film can be addressed by analyzing the peak widths at different in-plane $Q_\parallel$ positions to deduce the mosaicity and the diffracting domain size. The NiO(111) fraction (either FCC or twinned FCC) has a similar quality for all of the films: a mosaic spread between 0.8 and 1.7° and a diffracting domain size generally larger than 10 nm. The NiO(100) variant, when present, is of very poor crystallographic quality; the intensity of the signal is spread out over several degrees (5–8°). We did not find any preferential in-plane orientation of NiO(100) with respect to the sapphire unit cell. It is likely that the observed signal on the specular rod (yielding to characteristic distances in NiO(100)) comes, in fact, from (100) stacking faults in NiO(111) or from very small NiO(100) crystallites embedded in NiO(111).

Let us now discuss these results with respect to other NiO films. Sputtered NiO(111) films have already been used as pinning AF layers to elaborate spin-valve read heads [127, 202, 224]. For industrial applications, this preparation method is faster and cheaper than MBE, and the films obtained are much smoother. Relatively high deposition rates lead to sandwiches of several well-defined layers without pinholes between them, permitting the elaboration of functioning devices. However, this method also has some drawbacks; the NiO target crystallizes easily and may become rapidly unusable. It did not permit a complete understanding of the structure of the interfaces during growth or the role of each parameter (roughness, texture, diffusion at interface, etc.). Indeed, some of the reported results contradict each other. Generally, the exchange coupling is

quantity of both ways of stacking in the film. For all of our prepared films, the FCC and twinned FCC NiO(111) structures were found to be present in about the same proportion, with a small preference for the twin orientation (NiO(111) R90°). This result contrasts strongly with the Co and Py growths on NiO(111) (Sections 5.2 and 5.3). Indeed, in the present case the substrate is six-fold and the epilayer has three-fold symmetry. A resulting 1 : 1 nucleation rate of the two variants is thus not surprising.

Large in-plane rocking scans at $Q_\parallel = 1.61\sqrt{3} = 2.78$ r.l.u. indicate the presence of unrotated NiO(111)R0°. Corresponding twins of this unrotated structure are also present (R60° in Fig. 81). Integrated intensities show that less than 0.9% of the NiO layer adopts this unrotated structure for samples prepared in extreme conditions (620 K and 1000 K deposits). This is also confirmed by in-plane scans along the $\langle hh0 \rangle$ direction, where the NiO(111)R0° peak is expected at the (1.61, 1.61, 0) position in reciprocal space. Moreover, scanning along the $\langle hh0 \rangle$ in-plane direction, a small peak is present at $h = k =$

believed to be an interface effect [211], although some studies report different assumptions [228]. Some authors claim, in contradiction to existing theories [215, 216, 235, 246, 247], that the NiO texture, for the spin-compensated as well as for the uncompensated planes, does not really influence the exchange coupling. On the other hand, other studies report increases or decreases in the coupling when the uncompensated (111) surface of the NiO is used [232, 237]. Except for the better crystalline quality and the identification of all stacking, our results are comparable to similar studies of sputtered NiO films or NiO-CoO multilayers elaborated on sapphire [257, 258], where the presence of twins was explained as being induced by steps on the surface of the substrate. This structural behavior is thus likely to be intrinsic to the interface.

MBE-grown films represent an alternative approach in which films of good crystallographic quality can be expected at smaller deposition rates. One can also expect to better control each parameter. Unfortunately, our NiO films are much too rough, whatever the temperature and thickness growth conditions, to be used as antiferromagnetic substrates in a spin-valve device. Their magnetic properties for sensors are thus not accessible. In the range of explored parameters, the 700 K prepared film looks to be the smoothest.

Our GIXD data show that the growth of NiO on sapphire is epitaxial within a twinned FCC scheme and (100) stacking faults or inclusion of very small (100) crystallites. The NiO(111) unit cells are rotated by 30°, for the FCC and twinned FCC stacking with respect to the α-Al_2O_3(0001) unit cells. The low-temperature growth shows the additional presence of small quantities of NiO(111)R0°. The unit cell parameter evolution does not indicate strain in the NiO films, and only stoichiometric NiO was found, highlighting the high stability of this structure for nickel oxide. Quantitative measurements permitted estimation of the fraction of each structure as well as the corresponding crystallographic quality.

The complementary use of LEED patterns, AFM images, and X-ray diffraction showed that the crystallographic quality of MBE-grown films is comparable to that of sputtered films. However, our films are rougher in the whole range of the parameters we have investigated. Indeed, the small deposition rate in our case (a few angstroms per minute) favors the formation of islands with complex morphologies during growth. Conditions in which the NiO film is perfectly 2D were not found.

We do not expect a different magnetic behavior from our films with respect to sputtered ones (inasmuch as the qualities look comparable, except for the flatness of the surface), and because spin-valve elaboration needs flat interfaces it does not seem straightforward at all to build epitaxial spin valves on α-Al_2O_3(0001). The elaboration of single crystalline spin valves is a challenge that seems difficult to tackle with NiO films on sapphire.

The present results support a growth mode in which at the beginning, a layer made of micropyramids is formed. Depending on the growth conditions, these pyramids grow more or less homogeneously. Because the NiO(111) surface is polar and relatively stable (see Section 3.4), its surface energy remains very

large compared with that of NiO(100). For MBE growth of NiO films, this fact may prevent the elaboration of single crystalline NiO(111) epitaxial surfaces because of the onset of NiO(100) facets at small growth rates.

6.2. NiO(111)/Au(111)

The NiO(111) single-crystal surface is a good insulator at room temperature, and its conductivity drops sharply when the temperature decreases [261], preventing any investigation with electron-based techniques because of charge build-up. As a matter of fact, NiO(111) thin films, which are more adapted to many techniques, were investigated well before the single crystals because of the very intriguing predictions [98, 105] stating that this polar surface could be stable through the Ni-terminated [98] p(2 × 2) octopolar reconstruction [105]. A p(2 × 2) LEED pattern was reported for films grown at 550 K on Au(111) [94, 107] and for films grown at room temperature on Ni(111) [108]. On NiO/Ni(111) films, the pattern disappeared when water was introduced into the chamber because of hydroxyl adsorption and was restored upon annealing. However, this result contradicts a previous LEED work on the oxidation of Ni(111), where only c(2 × 2) and (7 × 7) reconstructions were reported [262]. On the other hand, only the reconstruction pattern has been reported, and not the structure of the reconstruction itself. Thus it was interesting to know if these films, grown in out-of-equilibrium conditions, adopt the oxidized or reduced NiO(111)-p(2 × 2) surface reconstruction to achieve stabilization. Because the reported LEED patterns for NiO/Ni(111) showed poor order and NiO/Au(111) showed excellent contrast, we have chosen to investigate the NiO/Au(111) interface by GIXD [94], which has proved to work quite well for single crystals (see Section 3.4).

The GIXD experiments were performed on the ID03 surface diffraction beamline at the ESRF [15] in ultra-high-vacuum conditions (10^{-10} mbar). The thin NiO film was measured at 17 keV and a 0.9° incidence angle because the bulk scattering from the Au(111) substrate was weak in the regions of interest. The crystallographic basis vectors for the surface unit cell describe the triangular lattice of the reconstruction. They are related to the bulk basis by $a_{surf} = [\bar{1}10]_{Cube}$, $b_{surf} = [0\bar{1}1]_{Cube}$, and $c_{surf} = [111]_{Cube}$. The h and k indexes describe the in-plane momentum transfer (in reciprocal lattice units (r.l.u.) of the NiO(111) reconstruction), and ℓ, the perpendicular momentum transfer.

The NiO(111) thin film was prepared *in situ* on Au(111) [107, 108]. The surface used for the growth had a mosaicity of 0.052° and a domain size of 2600 Å exhibiting the herringbone reconstruction [263]. During deposition the substrate was held at 615 K, which was found to be the optimal growth temperature, and the Ni was evaporated from an electron-bombarded Ni rod in a 2×10^{-5} mbar partial pressure of O_2. A perfect 2D growth has been observed (by reflection high-energy electron diffraction (RHEED) and GIXD) from three to eight monolayers before 3D crystallites formed. For the quantitative analysis we have chosen an intermediate thickness of 5 ML.

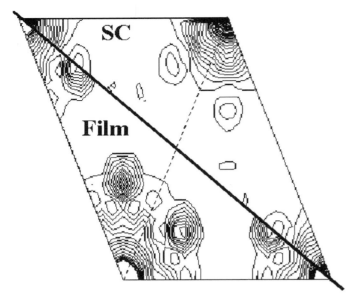

Fig. 82. Comparison of the Patterson maps of a 5-ML NiO(111) film on Au(111) and of the single crystal (SC).

The NiO(111) thin film was in good epitaxy, of surprisingly good crystalline quality, completely relaxed and p(2 × 2) reconstructed with 0.106° mosaicity. It had a domain size of 550 Å, was intense throughout the accessible region of reciprocal space, and had a perfect Lorentzian shape. This behavior is very different from the metal growth we discussed in Sections 4 and 5, which exhibited large mosaicities (several degrees) in the early stages of growth.

The in-plane scattering of the p(2 × 2) pattern was measured quantitatively by rocking scans at all accessible positions belonging to the reconstruction in reciprocal space. Fifty-nine nonzero in-plane peaks were measured at $\ell = 0.3$, and the diffraction pattern has $P3m1$ symmetry, leaving 32 nonequivalent peaks with a systematic error level of 11%. The largely different symmetries between the $P6mm$ single crystal and the thin film cannot be explained by the small ℓ difference, but indicate rather a different p(2 × 2) structure. In fact, both symmetries should be $P6mm$ in the strict $\ell = 0$ surface plane, but the thin film deviates more firmly from this in-plane symmetry than the single crystal does. A direct comparison of the Patterson maps for the thin film and the single crystal is reported in Figure 82, showing that the structures are unlike. Moreover, 13 diffraction rods with 322 nonzero structure factors were measured, and their periodicity along ℓ clearly indicates a reconstruction that is three atomic layers thick, further evidence that the thin film structure cannot match any of the known single crystal reconstructions. As intuitively expected from these simple observations, the octopolar reconstruction cannot reproduce the in-plane data, regardless of the relaxations or the termination ($\chi^2 > 8$).

A coherent juxtaposition of half Ni- and half O-terminated domains separated by single steps yields the experimental Patterson map (provided that the bases of the two octopoles have

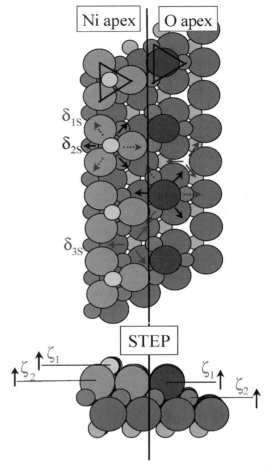

Fig. 83. Top and side views of two possible octopolar reconstructions with Ni- or O-terminated terraces (left and right, respectively), separated by a single step (solid line). The octopoles, on both sides of the step, have identical orientations, as shown by the triangles at the top of the figure. Large circles are oxygen atoms, and small circles are nickel. For either termination, the top two layers are 75% and 25% vacant compared with the bulk lattice. The possible symmetry-compatible relaxations, δ and ζ, are shown as arrows indicating a positive relaxation.

the same orientation). The structure was refined, as for the single crystal, with the use of the same relaxations plus a domain fraction, the roughness, and a scale factor. The structural model is shown in Figure 83. The symmetry-related atomic relaxations in each atomic layer are indexed with respect to p, where $p = 0$ for the apex layer of atoms. Three of the four atoms in layer p of the unit cell have symmetry-related vertical displacements ζ_{pS}, whereas the independent atom has vertical displacement ζ_p. Likewise, δ_{pS} defines a radial displacement of symmetry-related atoms away from the in-plane position of the apex atom. Because this model contains a large number of atoms, the convergence of the fitting procedure was difficult, mainly because of local minimums. We have thus preferred an approach with identical relaxations in the two domains, so that ζ_0 is the perpendicular relaxation of both apex atoms, and so on (Fig. 83). The best stable solution remained essentially unchanged when we allowed independent relaxations in

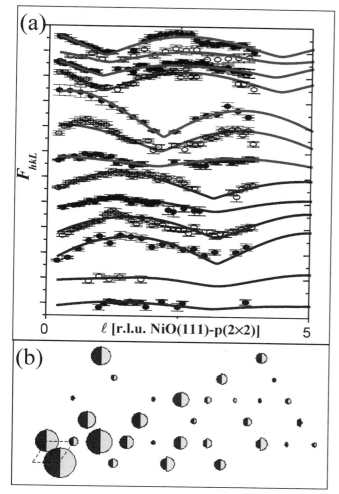

Fig. 84. Comparison between the experimental and calculated structure factors for a model with Ni- and O-terminated octopoles separated by a single step. (a) Diffraction rods: from bottom to top, 90ℓ, 71ℓ, $5\bar{1}\ell$, 41ℓ, 33ℓ, 30ℓ, 70ℓ, 32ℓ, 10ℓ, 31ℓ, 50ℓ, $7\bar{1}\ell$, and $4\bar{1}\ell$. ○ and ●, experimental data points. —, calculated structure factors. (b) In-plane structure factors. Right, experimental; left, calculated.

gle crystal exhibits only Ni-terminated double steps, and the thin film can adopt both terminations and therefore single steps as well. We note that the thin film was grown under nonequilibrium conditions. At 615 K, Ni has a limited mobility on NiO, inasmuch as Ni clusters appear only at higher temperatures on NiO [91, 93]. The atomic flux ratio of O : Ni was roughly 1000 : 1 on the surface. For these reasons, the growing surface necessarily passes through nonequilibrium structures to fill the incomplete lower layers. Indeed, a perfect Ni-terminated octopolar domain could be transformed into O termination via incorporation of a half-monolayer of nickel and a half-monolayer of oxygen atoms. We propose that the observed O-terminated regions are relatively stable metastable domains that are achieved during the special conditions of growth.

Because Au does not easily oxidize, we can assume that the interfacial layer is Ni. Ni- and O-terminated octopolar structures are electrostatically equivalent. For the thin film, each Au(111) step will place unlike layers at the same height, causing an accumulated electrostatic energy that may be responsible for the sudden change from two- to three-dimensional growth after 8 ML.

An alternative approach to the trial-and-error method for which the structure is guessed and the corresponding structure factors are calculated is the so-called direct methods. Information about the phases of the structure factors that is lost in a diffraction experiment is needed to restore the charge density. In the last few years, the direct methods approach has been adapted to 2D X-ray diffraction [264–266]. This method solves the phase problem by exploiting probability relationships between the amplitudes and the phases of the diffracted beams to determine a plausible initial solution. The method used for the $p(2 \times 2)$ NiO (111) structure (with the GIXD data under discussion) involved a minimum relative entropy algorithm combined with a genetic algorithm for global optimization [267, 268]. The algorithm searches for the set of phases with the lowest figures of merit. The corresponding solutions are used to create electron density maps that obey the imposed symmetry. In the final step, the atomic positions were refined by the least-squares method.

This approach cannot handle several structural domains but reaches an acceptable solution with a similar χ^2 of 1.37 over all of the data [269]. The corresponding structural model is drawn in Figure 85. It is essentially made of a particular combination of two embedded octopoles. The apex atom of the underlying O-terminated octopole is included in the basal plane of a more external Ni-terminated octopole (they are thus naturally oriented the same way). The net result is a 75% Ni vacant top plane, a full O plane, a 25% Ni vacant plane, and then bulk NiO(111). The relaxations are small. Thus the basic units needed to construct the two models are the same and the agreements are close. Here we hit another limitation of GIXD: the possible nonuniqueness of the model, even with good χ^2 values, due mainly to the fact that the phase information is lost. The direct methods model, however, exhibits 25% vacancies in the third layer, which does not seem easy to understand within

the two domains. Only three relaxations were nonnegligible. The best agreement was obtained for $\delta_{1S} = 0.096 \pm 0.008$ Å, $\delta_{2S} = 0.078 \pm 0.007$ Å, and $\zeta_0 = -0.103 \pm 0.028$ Å, with 50% Ni- and 50% O-terminated domains, a 1-Å r.m.s. roughness, and a global χ^2 of 1.4, close to the ideal value of 1. This solution reproduced all of the data well, as can be seen in Figure 84. The agreement is very good up to large momentum transfers, definitively supporting this two-variant octopolar reconstruction. Note that, in contrast to double steps, the single step belongs to the model itself and thus will not influence the coherence length (550 Å), i.e., the single steps may be located anywhere and do not necessarily form large continuous step edges. This might explain why STM investigations, although for thinner layers, have mainly reported double steps, because they really define the step edges [107].

Here again, Wolf's octopolar reconstruction is the basis needed for the interpretation of the data. However, the sin-

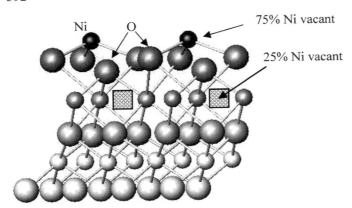

Ni   O   75% Ni vacant

25% Ni vacant

Fig. 85. Structural model for a 5-ML NiO(111) film on Au(111) derived from the direct methods approach. The patterned squares highlight the vacant Ni atoms in the third layer.

a growth process, and it is hardly of any help to explain the onset of 3D growth after 8 ML. On the other hand, the growth was performed at 615 K, and limited atomic mobility exists. Additional experiments would be necessary to unambiguously differentiate between the two possible models. In particular, the direct methods model would lead to ferromagnetic oxygen because of the vacancies. Unfortunately, a dichroism experiment seems very difficult to carry out on the O K-edge because of the tiny cross section, and the total expected magnetization is well below the detectable level of the most sensitive magnetometers. Thus we should conclude that unlike the NiO(111) single crystal, which exhibits a clear Ni-terminated octopolar reconstruction after air annealing, the NiO(111) thin film has another $p(2 \times 2)$ surface reconstruction that combines in one way or another the two octopolar terminations with small relaxation.

The stability of this surface against hydroxylation was tested by extensively dosing the surface with up to 10^6 Langmuir of H_2O at a partial pressure of 3×10^{-5} mbar at room temperature. Similarly to the single crystal case, no structural effect could be detected. This stands in contrast to the NiO(111) films of poor structural quality made by oxidation of Ni(111) [108, 270], which were found to be unstable against hydroxylation. We thus suggest that the reactivity of NiO(111) against water takes place mainly at defects (as for MgO [271]), which are almost absent on our surfaces and thin films.

Our main conclusion is that NiO(111) is terminated and stabilized by a $p(2 \times 2)$ reconstruction, but the internal configuration may change a great deal from one experimental condition to another. The electrostatic criterion is always fulfilled, however, and because the electrostatic driving force is strong enough to stabilize the surface reconstruction, even in air or water (only metallization seems to be more stable), it is clear that this criterion is very important when polar surfaces are considered. Interestingly, the octopolar reconstruction prediction and the observation of the $p(2 \times 2)$ surface cell were not enough to understand the polar NiO(111) surface. Note also that oxide thin films may not exhibit the same surface structure as their bulk counterpart. The studies carried out on thin oxide films may thus only be carefully extrapolated to the bulk surface.

7. CONCLUSIONS

Let us now re-examine the questions raised in the Introduction concerning the structure and morphology of growing interfaces. For all systems we have determined the growth modes, the crystalline quality of the epilayer, and the epitaxial relationships and quantitatively discussed the morphology with respect to the deposited thickness. For exchange-coupled interfaces we have seen that the structure plays a determining role in the overall properties of the heterostructure. Several relaxation mechanisms of the interfacial strain due to lattice mismatch were observed: well-ordered dislocation networks at the buried Ag and Pd on MgO(001) interfaces; several epitaxial relationships and variants that cancel the strain in Ni on MgO(001), and plastic deformation of the interface with growth of an interfacial compound in Py/NiO(111). For the Ag, Pd, and Ni on MgO(001), we have identified the O epitaxial site. The use of surfactants was also questioned, although the extrapolation from homoepitaxy to heteroepitaxy must be taken with care and some reactions like O with Co may only mimic a surfactant effect. The crystalline quality of the NiO(111) substrates on the nucleation of Co clusters was quantitatively investigated.

In the present review we have thus discussed a number of typical situations that may arise during growth, on substrates that are insulating or not. We have shown that in situ synchrotron radiation investigations are often able to draw clear and quantitative answers, for a very wide range of situations, ranging from perfectly sharp interfaces (metal/MgO(001)) to highly reactive ones (Py/NiO(111)) and from single crystalline 2D epitaxial growth (NiO(111)/Au(111)) to fairly polycrystalline islands (Co/NiO(111)). Thus these studies show that grazing incidence X-ray diffraction is not limited to the well-established description of surface reconstructions and relaxation, although this works nicely for oxides too, as seen for MgO(001), NiO(111), or α-Al$_2$O$_3$(0001). Describing the growth of metals or oxides on metals or oxides, especially when coupled to GISAXS, AFM, or TEM, can be described with a high level of detail. Thus although the number of techniques is reduced for insulating materials, the now available in situ synchrotron radiation techniques are really powerful and allow investigation of this class of materials with a high level of confidence and accuracy.

ACKNOWLEDGMENTS

Parts of this report are based on the work of former Ph.D. students of the group and, in particular, P. Guénard and O. Robach. Y. Samson and P. Bayle-Guillemaud, respectively, are gratefully acknowledged for their excellent AFM and TEM images and active participation in illuminating the dark zones of the GIXD data. The SP2M/NM laboratory is acknowledged for giving us the opportunity to use much of their magnetic characterization equipment. M. Noblet and O. Ulrich are acknowledged for efficient technical assistance during the experiments on BM32-SUV. The staffs of beamlines BM32, ID32, and ID03 are acknowledged for help and assistance during the experiments. E. Vlieg is acknowledged for providing the ROD software that was largely used in this work. We are also pleased to thank many experimentalists who participated in one or several of the experiments reported here: J. Jupille, M. Gautier-Soyer, A. Stierle, K. Peters, H. Kuhlenbeck, B. Richter, and O. Robach. It is difficult to cite here all of the scientists with whom we had illuminating discussions or fruitful collaborations, in particular we wish to thank J. Jupille, C. Noguera, H.-J. Freund, M. Gautier-Soyer, P. Humbert, C. A. Ventrice, Jr., L. D. Marks, and B. Dieny. Finally we are indebted to Prof. H. S. Nalwa for inviting us to write this chapter.

REFERENCES

1. R. Feidenhansl, *Surf. Sci. Rep.* 10, 105 (1989).
2. A. Guinier, "X-Ray Diffraction in Crystals, Imperfect Crystals, and Amorphous Bodies." Dover, New York, 1994.
3. R. W. James, "The Optical Principles of the Diffraction of X-Rays." Cornell Univ. Press, Ithaca, NY, 1965.
4. G. Renaud, *Surf. Sci. Rep.* 32, 1 (1998).
5. I. K. Robinson and D. J. Tweet, *Rep. Prog. Phys.* 55, 599 (1992).
6. B. E. Waren, "X-Ray Diffraction." Addison-Wesley, Reading, MA, 1969.
7. I. K. Robinson, "Surface Crystallography," Vol. 3, pp. 221–266. Elsevier, Amsterdam/New York, 1991.
8. M. M. Woolfson, "An Introduction to X-Ray Crystallography." Cambridge Univ. Press, Cambridge, UK, 1970.
9. W. Borchardt-Ott, "Crystallography," 2nd ed. Springer-Verlag, Berlin, 1995.
10. T. P. Russel, "Small-Angle Scattering," Vol. 3, pp. 379–470. Elsevier, Amsterdam/New York, 1991.
11. P. Debye, *Ann. Phys. (Leipzig)* 43, 49 (1914).
12. I. K. Robinson, *Phys. Rev. Lett.* 50, 1145 (1983).
13. I. K. Robinson, *Phys. Rev. B: Solid State* 33, 3830 (1986).
14. T. Harada, M. Asano, and Y. Mizumati, *J. Cryst. Growth* 116, 243 (1992).
15. ESRF, avaible at http://www.esrf.fr.
16. A. Akimoto, J. Mizuki, I. Hirosawa, and J. Matsui, *Rev. Sci. Instrum.* 60, 2362 (1989).
17. R. Baudoing-Savois, G. Renaud, M. De Santis, A. Barbier, O. Robach, P. Taunier, P. Jeantet, O. Ulrich, J. P. Roux, M. C. Saint-Lager, A. Barski, O. Geaymond, G. Berard, P. Dolle, M. Noblet, and A. Mougin, *Nucl. Instrum. Methods Phys. Res., Sect. B* 149, 213 (1999).
18. S. Brennan and P. Eisenberger, *Nucl. Instrum. Methods* 222, 164 (1984).
19. P. Claverie, J. Massies, R. Pinchaux, M. Sauvage-Simkin, J. Frouin, J. Bonnet, and N. Jedrecy, *Rev. Sci. Instrum.* 60, 2369 (1989).
20. P. H. Fuoss and I. K. Robinson, *Nucl. Instrum. Methods* 222, 171 (1984).
21. S. Ferrer and F. Comin, *Rev. Sci. Instrum.* 66, 1674 (1995).
22. G. Renaud, B. Villette, and P. Guénard, *Nucl. Instrum. Methods Phys. Res., Sect. B* 95, 422 (1995).
23. E. Vlieg, Van't Ent, A. P. de Jongh, H. Neerings, and J. F. Van der Veen, *Nucl. Instrum. Methods Phys. Res., Sect. A* 262, 522 (1987).
24. A. Munkholm and S. Brennan, *J. Appl. Crystallogr.* 32, 143 (1999).
25. A. Munkholm, S. Brennan, and E. C. Carr, *J. Appl. Phys.* 82, 2944 (1997).
26. C. Schamper, H. L. Meyerheim, and W. Moritz, *J. Appl. Crystallogr.* 26, 687 (1993).
27. M. F. Toney and D. G. Wiesler, *Acta Crystallogr., Sect. A* 49, 624 (1993).
28. E. Vlieg, *J. Appl. Crystalogr.* 30, 532 (1997).
29. O. Robach, Y. Garreau, K. Aïd, and M. B. Véron-Jolliot, *J. Appl. Cryst.* 33, 1006 (2000).
30. E. Vlieg, *J. Appl. Crystallogr.* 31, 198 (1998).
31. J. R. Levine, J. B. Cohen, Y. W. Chung, and P. Georgopoulos, *J. Appl. Crystallogr.* 22, 528 (1989).
32. A. Naudon and D. Thiaudière, *J. Appl. Crystallogr.* 30, 822 (1997).
33. D. Babonneau, A. Naudon, D. Thiaudière, and S. Lequien, *J. Appl. Crystallogr.* 32, 226 (1999).
34. G. Renaud, M. Noblet, A. Barbier, C. Revenant, O. Ulrich, Y. Borensztein, R. Lazzari, J. Jupille, and C. Henry, to be published.
35. G. Renaud, M. Noblet, A. Barbier, J. P. Deville, O. Fruchart, and F. Scheurer, unpublished results (2000).
36. G. Porod, "Small Angle X-ray Scattering," p. 37. Academic Press, San Diego, 1982.
37. J. R. Levine, J. B. Cohen, and Y. W. Chung, *Surf. Sci.* 248, 215 (1991).
38. A. Naudon, D. Babonneau, F. Petroff, and A. Vaurès, *Thin Solid Films* 319, 81 (1998).
39. M. Schmidbauer, T. Wiebach, H. Raidt, H. Hanke, R. Köhler, and H. Wawra, *J. Phys. D: Appl. Phys.* 32, A230 (1999).
40. D. Babonneau, J. Briatico, F. Petroff, T. Cabioc'h, and A. Naudon, *J. Appl. Phys.* 87, 3432 (2000).
41. H. You, K. G. Huang, and R. T. Kampwirth, *Physica B* 221, 77 (1996).
42. M. Rauscher, R. Paniago, H. Meyzger, Z. Kovats, J. Domke, J. Peisl, H. D. Pfannes, J. Schulze, and I. Eisele, *J. Appl. Phys.* 86, 6763 (1999).
43. O. Robach, G. Renaud, and A. Barbier, *Surf. Sci.* 401, 227 (1998).
44. P. Guénard, G. Renaud, and B. Villette, *Physica B* 221, 205 (1996).
45. Y. C. Lee, P. Tong, and P. A. Montano, *Surf. Sci.* 181, 559 (1987).
46. C. Li, R. Wu, A. J. Freeman, and C. L. Fu, *Phys. Rev. B: Solid State* 48, 8317 (1993).
47. G. Bordier and C. Noguéra, *Phys. Rev. B: Solid State* 44, 6361 (1991).
48. M. Causa, R. Dovesi, C. Pisani, and C. Roetti, *Surf. Sci.* 175, 551 (1986).
49. F. W. de Wette, W. Kress, and U. Schröder, *Phys. Rev. B: Solid State* 32, 4143 (1985).
50. J. Goniakowski and C. Noguera, *Surf. Sci.* 323, 129 (1995).
51. J. P. LaFemina and C. B. Duke, *J. Vac. Sci. Technol. A* 9, 1847 (1991).
52. G. V. Lewis and C. R. A. Catlow, *J. Phys. C: Solid State Phys.* 18, 1149 (1985).
53. A. J. Martin and H. Bilz, *Phys. Rev. B: Solid State* 19, 6593 (1979).
54. G. Pacchioni, T. Mierva, and P. S. Bagus, *Surf. Sci.* 275, 450 (1992).
55. P. W. Tasker and D. M. Duffy, *Surf. Sci.* 137, 91 (1984).
56. D. L. Blanchard, D. L. Lessor, J. P. LaFemina, D. R. Baer, W. K. Ford, and T. Guo, *J. Vac. Sci. Technol., A* 9, 1814 (1991).
57. C. G. Kinniburgh, *J. Phys. C: Solid State Phys.* 9, 2695 (1976).
58. P. A. Maksym, *Surf. Sci.* 149, 157 (1985).
59. H. Nakamatsu, A. Sudo, and S. Kawai, *Surf. Sci.* 194, 265 (1988).
60. M. Prutton, J. A. Walker, M. R. Welton-Cook, R. C. Felton, and Ramsey, *Surf. Sci.* 89, 95 (1979).
61. K. H. Rieder, *Surf. Sci.* 188, 57 (1982).
62. A. Santoni, D. B. Tran Thoai, and J. Urban, *Solid State Commun.* 68, 1039 (1988).
63. T. Urano and T. Kanaji, *Surf. Sci.* 134, 109 (1983).
64. M. R. Welton-Cook and W. Berndt, *J. Phys. C: Solid State Phys.* 15, 5691 (1982).
65. J. B. Zhou, H. C. Lu, T. Gustafsson, and P. Häberle, *Surf. Sci.* 302, 350 (1994).
66. V. E. Henrich, G. Dresselhaus, and H. J. Zeiger, *Phys. Rev. B: Solid State* 22, 4764 (1980).
67. C. Duriez, C. Chapon, C. R. Henry, and J. M. Rickard, *Surf. Sci.* 230, 123 (1990).
68. T. Gotoh, S. Murakami, K. Kinosita, and Y. Murata, *J. Phys. Soc. Jpn.* 50, 2063 (1981).
69. P. Guénard, G. Renaud, A. Barbier, and M. Gautier-Soyer, *Surf. Rev. Lett.* 5, 321 (1998).
70. P. Guénard, G. Renaud, A. Barbier, and M. Gautier-Soyer, *Mater. Res. Soc. Symp. Proc.* 437, 15 (1996).
71. M. Gautier, J. P. Duraud, and L. Pham Van, *Surf. Sci. Lett.* 249, L327 (1991).

72. J. Guo, D. E. Ellis, and D. J. Lam, *Phys. Rev. B: Solid State* 45, 13647 (1992).

73. W. C. Mackrodt, R. J. Davey, S. N. Black, and R. Docherty, *J. Cryst. Growth* 80, 441 (1987).

74. I. Manassidis and M. J. Gillan, *Surf. Sci. Lett.* 285, L517 (1993).

75. I. Manassidis and M. J. Gillan, *J. Am. Ceram. Soc.* 77, 335 (1994).

76. M. Causà, R. Dovesi, C. Pisani, and C. Roetti, *Surf. Sci.* 215, 259 (1989).

77. T. J. Godin and J. P. LaFemina, *Phys. Rev. B: Solid State* 49, 7691 (1994).

78. A. Kirfel and K. Eichhorn, *Acta Crystallogr., Sect. A* 46, 271 (1990).

79. G. C. Ndubuisi, J. Liu, and J. M. Cowley, *Micross. Res. Tech.* 20, 439 (1992).

80. V. E. Puchin, J. D. Gale, A. L. Shluger, E. A. Kotomin, J. Günster, M. Brause, and V. Kempter, *Surf. Sci.* 370, 190 (1997).

81. J. Ahn, and J. W. Rabalais, *Surf. Sci.* 388, 121 (1997).

82. G. Renaud, B. Villette, I. Vilfan, and A. Bourret, *Phys. Rev. Lett.* 73, 1825 (1994).

83. M. Gautier, J. P. Duraud, L. Pham Van, and M. J. Guittet, *Surf. Sci.* 250, 71 (1991).

84. S. Baik, D. E. Fowler, J. M. Blakeley, and R. Raj, *J. Am. Ceram. Soc.* 68, 281 (1985).

85. T. M. French and G. A. Somorjai, *J. Phys. Chem.* 74, 12 (1970).

86. H. Shiba, *J. Phys. Soc. Jpn.* 48, 211 (1980).

87. A. D. Novaco and J. P. McTague, *Phys. Rev. Lett.* 38, 1286 (1977).

88. E. Gillet and B. Ealet, *Surf. Sci.* 273, 427 (1992).

89. M. Vermeersch, R. Sporken, Ph. Lambin, and R. Caudano, *Surf. Sci.* 235, 5 (1990).

90. G. Renaud, M. Gautier-Soyer, and A. Barbier, unpublished results (2000).

91. A. Barbier and G. Renaud, *Surf. Sci. Lett.* 392, L15-20 (1997).

92. A. Barbier, G. Renaud, and A. Stierle, *Surf. Sci.* 402–404, 757 (1998).

93. A. Barbier, G. Renaud, C. Mocuta, and A. Stierle, *Surf. Sci.* 433–435, 761 (1999).

94. A. Barbier, C. Mocuta, H. Kuhlenbeck, K. F. Peters, B. Richter, and G. Renaud, *Phys. Rev. Lett.* 84, 2897 (2000).

95. A. Barbier, C. Mocuta, and G. Renaud, *Phys. Rev. B* 62, 16056 (2000).

96. J. J. M. Pothuizen, O. Cohen, and G. A. Sawatzky, *Mater. Res. Soc. Symp. Proc.* 401, 501 (1996).

97. V. I. Anisimov, F. Aryasetiawan, and A. I. Lichtenstein, *J. Phys.: Condens. Matter* 9, 767 (1997).

98. P. M. Oliver, G. W. Watson, and S. C. Parker, *Phys. Rev. B: Solid State* 52, 5323 (1995).

99. C. Noguéra, "Physics, and Chemistry at Oxide Surfaces." Cambridge Univ. Press, Cambridge, UK, 1996.

100. P. W. Tasker, *J. Phys. C: Solid State Phys.* 12, 4977 (1979).

101. V. E. Henrich, *Surf. Sci.* 57, 385 (1976).

102. D. Cappus, C. Xu, D. Ehrlich, B. Dillmann, C. A. Ventrice, Jr., K. Al-Shamery, H. Kuhlenbeck, and H. J. Freund, *Chem. Phys.* 177, 533 (1993).

103. W. H. Meiklejohn, *J. Appl. Phys.* 33, 1328 (1962).

104. S. Soeya, S. Nakamura, T. Imagawa, and S. Narishige, *J. Appl. Phys.* 77, 5838 (1995).

105. D. Wolf, *Phys. Rev. Lett.* 68, 3315 (1992).

106. J. M. Cowley, *Surf. Sci.* 114, 587 (1982).

107. C. A. Ventrice, Jr., Th. Bertrams, H. Hannemann, A. Brodde, and H. Neddermeyer, *Phys. Rev. B: Solid State* 49, 5773 (1994).

108. F. Rohr, K. Wirth, J. Libuda, D. Cappus, M. Bäumer, and H. J. Freund, *Surf. Sci. Lett.* 315, L977 (1994).

109. N. Floquet and L. C. Dufour, *Surf. Sci.* 126, 543 (1983).

110. M. Finazzi, N. B. Brookes, and F. M. F. de Groot, *Phys. Rev. B: Solid State* 59, 9933 (1999).

111. J. P. Hill, C. C. Kao, and D. F. McMorrow, *Phys. Rev. B: Solid State* 55, R8662 (1997).

112. G. Ju, A. V. Nurmikko, R. F. C. Farrow, R. F. Marks, M. J. Carey, and B. A. Gurney, *Phys. Rev. B: Solid State* 58, R11857 (1998).

113. V. Fernandez, C. Vettier, F. de Bergevin, C. Giles, and W. Neubeck, *Phys. Rev. B: Solid State* 57, 7870 (1998).

114. A. Barbier, G. Renaud, C. Mocuta, and A. Stierle, *Adv. Sci. Technol.* 19, 149 (1999).

115. C. Mocuta, A. Barbier, and G. Renaud, *Appl. Surf. Sci.* 162–163, 56 (2000).

116. T. Shishidou and T. Jo, *J. Phys. Soc. Jpn.* 67, 2637 (1998).

117. W. L. Roth, *J. Appl. Phys.* 31, 2000 (1960).

118. T. R. McGuire and T. S. Plaskett, *J. Appl. Phys.* 75, 6537 (1994).

119. R. P. Michel, A. Chaiken, and C. T. Wang, *J. Appl. Phys.* 81, 5374 (1997).

120. M. Tan, H.-C. Tong, S.-I. Tan, and R. Rottmayer, *J. Appl. Phys.* 79, 5012 (1996).

121. M. J. Carey and A. E. Berkowitz, *Appl. Phys. Lett.* 60, 3060 (1992).

122. M. J. Carey and A. E. Berkowitz, *J. Appl. Phys.* 73, 6892 (1993).

123. M. J. Carey, A. E. Berkowitz, J. A. Borchers, and R. W. Erwin, *Phys. Rev. B: Solid State* 47, 9952 (1993).

124. H. Yamane and M. Kobayashi, *J. Appl. Phys.* 83, 4862 (1998).

125. F. Voges, H. De Gronckel, C. Osthover, R. Schreiber, and P. Grunberg, *J. Magn. Magn. Mater.* 190, 183 (1998).

126. J. B. Restorff, M. Wun-Fogle, and S. F. Cheng, *J. Appl. Phys.* 81, 5218 (1997).

127. K. Nakamoto, Y. Kawato, Y. Suzuki, Y. Hamakawa, T. Kawabe, K. Fujimoto, M. Fuyama, and Y. Sugita, *IEEE Trans. Magn.* 32, 3374 (1996).

128. W. F. Egelhoff, Jr., T. Ha, R. D. K. Misra, Y. Kadmon, J. Nir, C. J. Powel, M. D. Stiles, R. D. McMichael, C. L. Lin, J. M. Sivertsen, J. H. Judy, K. Takano, A. E. Berkowitz, T. C. Anthony, and J. A. Brug, *J. Appl. Phys.* 78, 273 (1995).

129. D. M. Lind, S. P. Tay, S. D. Berry, J. A. Borchers, and R. W. Erwin, *J. Appl. Phys.* 73, 6886 (1993).

130. A. R. Ball, A. J. G. Leenaers, P. J. Van der Zaag, K. A. Shaw, B. Singer, D. M. Lind, H. Frederizike, and M. T. Rekveldt, *Appl. Phys. Lett.* 69, 1489 (1996).

131. P. Stoyanov, A. Gottschalk, and D. M. Lind, *J. Appl. Phys.* 81, 5010 (1997).

132. C. Mocuta, A. Barbier, G. Renaud, and B. Dieny, *Thin Solid Films* 336, 160 (1998).

133. A. Barbier, P. Ohresser, V. Da Costa, B. Carrière, and J.-P. Deville, *Surf. Sci.* 405, 298 (1998).

134. N. W. Nydegger, G. Couderc, and M. A. Langell, *Appl. Surf. Sci.* 147, 58 (1999).

135. J. H. Van der Merwe, *Philos. Mag. A* 45, 159 (1982).

136. J. H. Van der Merwe, *Philos. Mag. A* 45, 127 (1982).

137. J. H. Van der Merwe, *Philos. Mag. A* 45, 145 (1982).

138. T. Epicier and C. Esnouf, *J. Phys. III* 4, 1811 (1994).

139. R. H. Hoel, *Surf. Sci.* 169, 317 (1986).

140. R. H. Hoel, H. U. Habermeier, and M. Rühle, *J. Phys.* 46, C4 (1989).

141. R. H. Hoel, J. M. Penisson, and H. U. Habermeier, *J. Phys.* 51, C1 (1990).

142. F. Ernst, P. Pirouz, and A. H. Heuer, *Philos. Mag. A* 63, 259 (1991).

143. G. Necker and W. Mader, *Philos. Mag. Lett.* 58, 205 (1988).

144. G. Renaud, O. Robach, and A. Barbier, *Faraday Discuss.* 114, 157 (1999).

145. O. Robach, G. Renaud, A. Barbier, and P. Guénard, *Surf. Rev. Lett.* 5, 359 (1998).

146. O. Robach, G. Renaud, and A. Barbier, *Phys. Rev. B: Solid State* 60, 5858 (1999).

147. U. Schonberger, O. K. Andersen, and M. Methfessel, *Acta Metall. Mater.* 40, S1 (1992).

148. P. Blöchl, "Metal-Ceramic Interfaces." Pergamon, Oxford, 1990.

149. J. R. Smith, T. Hong, and D. J. Srolovitz, *Phys. Rev. Lett.* 72, 4021 (1994).

150. A. M. Flanck, R. Delaunay, P. Lagarde, M. Pompa, and J. Jupille, *Condens. Matter* 53, 1737 (1996).

151. L. Spiess, *Surf. Rev. Lett.* 3, 1365 (1997).

152. A. Trampert, F. Ernst, C. P. Flynn, H. F. Fischmeister, and M. Rühle, *Acta Metall. Mater.* 40, S227 (1992).

153. H. A. Van der Vegt, H. M. Van Pinxteren, M. Lohmeier, E. Vlieg, and J. M. C. Thornton, *Phys. Rev. Lett.* 68, 3335 (1992).

154. H. A. Van der Vegt, W. J. Huisman, P. B. Howes, and E. Vlieg, *Surf. Sci.* 330, 101 (1995).

155. H. A. Van der Vegt, J. Alvarez, X. Torrelles, S. Ferrer, and E. Vlieg, *Phys. Rev. B: Solid State* 52, 17443 (1995).

156. G. Renaud, P. Guénard, and A. Barbier, *Phys. Rev. B: Solid State* 58, 7310 (1998).
157. P. Guénard, G. Renaud, B. Vilette, M. H. Yang, and C. P. Flynn, *Scripta Metall. Mater.* 31, 1221 (1994).
158. M. S. Patterson, *J. Appl. Phys.* 8, 805 (1952).
159. "Handbook of Thermophysical Properties of Solid Materials." Macmillan, New York, 1961.
160. R. Bonnet and J. L. Verger-Gaugry, *Philos. Mag. A* 66, 849 (1992).
161. R. Bonnet, *Philos. Mag. A* 43, 1165 (1981).
162. A. Barbier, G. Renaud, and J. Jupille, *Surf. Sci.* 454–456, 979 (2000).
163. A. Barbier, *Surf. Sci.* 406, 69 (1998).
164. F. Didier and J. Jupille, *Surf. Sci.* 307–309, 587 (1994).
165. G. Renaud and A. Barbier, *Surf. Sci.* 433–435, 142 (1999).
166. G. Renaud, A. Barbier, and O. Robach, *Phys. Rev. B: Solid State* 60, 5872 (1999).
167. G. Renaud and A. Barbier, *Appl. Surf. Sci.* 142, 14 (1999).
168. C. Goyhenex, C. R. Henry, and J. Urban, *Philos. Mag. A* 69, 1073 (1994).
169. C. Goyhenex, M. Croci, C. Claeys, and C. R. Henry, *Surf. Sci.* 353, 475 (1996).
170. C. Goyhenex, M. Meunier, and C. R. Henry, *Surf. Sci.* 350, 103 (1996).
171. C. R. Henry, C. Chapon, C. Duriez, and S. Giorgio, *Surf. Sci.* 253, 177 (1991).
172. C. R. Henry, M. Meunier, and S. Morel, *J. Cryst. Growth* 129, 416 (1993).
173. C. R. Henry, *Surf. Sci. Rep.* 31, 231 (1998).
174. S. G. Giorgio, C. Chapon, C. R. Henry, and G. Nihoul, *Philos. Mag. B* 67, 773 (1993).
175. S. Bartuschat and J. Urban, *Philos. Mag. A* 76, 783 (1997).
176. P. Lu and F. Cosandey, *Acta Metall. Mater.* 40, S259 (1992).
177. K. Heinemann, T. Osaka, H. Poppa, and M. Avalos-Borja, *J. Catal.* 83, 61 (1983).
178. H. Fornander, J. Birch, L. Hultman, L. G. Petersson, and J. E. Sundgren, *Appl. Phys. Lett.* 68, 2636 (1996).
179. C. Goyhenex and C. R. Henry, *J. Electron Spectrosc. Relat. Phenom.* 61, 65 (1992).
180. J. Goniakovski, *Phys. Rev. B: Solid State* 57, 1 (1998).
181. A. Barbier, G. Renaud, and O. Robach, *J. Appl. Phys.* 84, 4259 (1998).
182. G. Pacchioni and N. Rösch, *J. Chem. Phys.* 104, 7329 (1996).
183. N. Rösch and G. Pacchioni, "Chemisorption and Reactivity on Supported Clusters, and Thin Films." Kluwer Academic, Dordrecht, 1997.
184. H. Bialas and K. Heneka, *Vacuum* 45, 79 (1994).
185. F. Reniers, M. P. Delplancke, A. Asskali, V. Rooryck, and O. Van Sinay, *Appl. Surf. Sci.* 92, 35 (1996).
186. G. Raatz and J. Woltersdorf, *Phys. Status Solidi A* 113, 131 (1989).
187. H. Nakai, H. Qiu, M. Adamik, G. Sáfran, P. B. Barna, and M. Hashimoto, *Thin Solid Films* 263, 159 (1995).
188. H. Sato, R. S. Toth, and R. W. Astrue, *J. Appl. Phys.* 33, 1113 (1962).
189. A. A. Hussain, *J. Phys.: Condens. Matter* 1, 9833 (1989).
190. H. Bialas and Li-Shing Li, *Phys. Status Solidi A* 42, 125 (1977).
191. N. I. Kiselev, Yu. l. Man'kov, and V. G. Pyn'ko, *Sov. Phys. Solid State* 31, 685 (1989).
192. H. Maruyama, H. Qiu, H. Nakai, and M. Hashimoto, *J. Vac. Sci. Technol., A* 13, 2157 (1995).
193. H. Qiu, A. Kosuge, H. Maruyama, M. Adamik, G. Safran, P. B. Barna, and M. Hashimoto, *Thin Solid Films* 241, 9 (1994).
194. M. R. Fitzsimmons, G. S. Smith, R. Pynn, M. A. Nastasi, and E. Burkel, *Physica B* 198, 169 (1994).
195. G. Simmons and H. Wang, "Single Crystal Elastic Constants and Calculated Aggregated Properties." MIT Press, Cambridge, MA, 1971.
196. W. H. Meiklejohn and C. P. Bean, *Phys. Rev.* 105, 904 (1957).
197. W. H. Meiklejohn and C. P. Bean, *Phys. Rev.* 102, 1413 (1956).
198. W. H. Meiklejohn, *J. Appl. Phys.* 33, 1328 (1962).
199. B. Dieny, V. S. Speriosu, S. S. P. Parkin, B. A. Gurne, D. R. Wilhoit, and D. Mauri, *Phys. Rev. B: Solid State* 43, 1297 (1991).
200. J. C. S. Kools, S. Lardoux, and F. Roozeboom, *Appl. Phys. Lett.* 72, 611 (1998).
201. C. Tsang and Kenneth Lee, *J. Appl. Phys.* 53, 2605 (1982).
202. S. L. Burkett, S. Kora, J. L. Bresowar, J. C. Lusth, B. H. Pirkle, and M. R. Parker, *J. Appl. Phys.* 81, 4912 (1997).
203. W. C. Cain, W. H. Meiklejohn, and M. H. Kryder, *J. Appl. Phys.* 61, 4170 (1987).
204. G. Choe and S. Gupta, *Appl. Phys. Lett.* 70, 1766 (1997).
205. T. Lin, D. Mauri, N. Staud, and C. Hwang, *Appl. Phys. Lett.* 65, 1183 (1994).
206. Harsh Deep Chopra, B. J. Hockey, P. J. Chen, W. F. Egelhoff, Jr., M. Wutting, and S. Z. Hua, *Phys. Rev. B: Solid State* 55, 8390 (1997).
207. H. J. M. Swatgen, G. J. Strijkers, P. J. H. Bloemen, M. M. H. Willekens, and W. J. M. De Jonge, *Phys. Rev. B: Solid State* 53, 9108 (1996).
208. Z.-H. Lu, T. Li, J.-J. Qiu, K. Xun, H.-L. Shen, Z.-Y. Li, and D.-F. Shen, *Chin. Phys. Lett.* 16, 65 (1999).
209. Y. Kawawake, Y. Sugita, M. Satomi, and H. Sakakima, *J. Appl. Phys.* 85, 5024 (1999).
210. T. Ambrose, K. Leifer, K. J. Hemker, and C. L. Chien, *J. Appl. Phys.* 81, 5007 (1997).
211. C. Tsang, N. Heiman, and K. Lee, *J. Appl. Phys.* 52, 2471 (1981).
212. R. Jungblut, R. Coehoorn, M. T. Johnson, J. aan de Stegge, and A. Reinders, *J. Appl. Phys.* 75, 6659 (1994).
213. O. Allegranza and M.-M. Chen, *J. Appl. Phys.* 73, 6218 (1993).
214. J. Ding and J.-G. Zhu, *J. Appl. Phys.* 79, 5892 (1996).
215. D.-H. Han, J.-G. Zhu, and J. H. Judy, *J. Appl. Phys.* 81, 4996 (1997).
216. D.-H. Han, J.-G. Zhu, J. H. Judy, and J. M. Sivertsen, *J. Appl. Phys.* 81, 340 (1997).
217. D.-H. Han, J.-G. Zhu, J. H. Judy, and J. M. Sivertsen, *Appl. Phys. Lett.* 70, 664 (1997).
218. D.-H. Han, J.-G. Zhu, J. H. Judy, and J. M. Sivertsen, *J. Appl. Phys.* 81, 4519 (1997).
219. T. J. Moran, J. M. Gallego, and I. K. Schuller, *J. Appl. Phys.* 78, 1887 (1995).
220. T. J. Moran and I. K. Schuller, *J. Appl. Phys.* 79, 5109 (1996).
221. H. D. Chopra, B. J. Hockey, P. J. Chen, R. D. McMichael, and W. F. Egelhoff, Jr., *J. Appl. Phys.* 81, 4017 (1997).
222. S. F. Cheng, J. P. Teter, P. Lubitz, M. M. Miller, L. Hoines, J. J. Krebs, D. M. Schaefer, and G. A. Prinz, *J. Appl. Phys.* 79, 6234 (1996).
223. T. Lin, C. Tsang, R. E. Fontana, and J. K. Howard, *IEEE Trans. Magn.* 31, 2585 (1995).
224. Y. Hamakawa, H. Hoshiya, T. Kawabe, Y. Suzuki, R. Arai, K. Nakamoto, M. Fuyama, and K. Sugita, *IEEE Trans. Magn.* 32, 149 (1996).
225. READRITE, available at http://www. readrite. com.
226. IBM, available at http://www.research.ibm.com/storage/oem/tech/eraheads.htm.
227. SEAGATE, available at http://www.seagate.com.
228. N. J. Gökemeijer, T. Ambrose, C. L. Chien, N. Wang, and K. K. Fung, *J. Appl. Phys.* 81, 4999 (1997).
229. D. G. Hwang, S. S. Lee, and C. M. Park, *Appl. Phys. Lett.* 72, 2162 (1998).
230. D. G. Hwang, C. M. Park, and S. S. Lee, *J. Magn. Magn. Mater.* 186, 265 (1998).
231. C.-M. Park, C.-I. Min, and C. H. Shin, *IEEE Trans. Magn.* 32, 3422 (1996).
232. J. X. Shen and M. T. Kief, *J. Appl. Phys.* 79, 5008 (1996).
233. C.-H. Lai, H. Matsuyama, R. White, T. C. Anthony, and G. G. Bush, *J. Appl. Phys.* 79, 6389 (1996).
234. C.-H. Lai, T. J. Regan, R. L. White, and T. C. Anthony, *J. Appl. Phys.* 81, 3989 (1997).
235. S.-S. Lee, D.-G. Hwang, C. M. Park, K. A. Lee, and J. R. Rhee, *J. Appl. Phys.* 81, 5298 (1997).
236. J. Nogues, D. Lederman, T. J. Moran, and I. K. Schuller, *Phys. Rev. Lett.* 76, 4624 (1996).
237. O. Kitakami, H. Takashima, and Y. Shimada, *J. Magn. Magn. Mater.* 164, 43 (1996).
238. C. Cowache, B. Dieny, S. Auffret, M. Cartier, R. H. Taylor, R. O'Barr, and S. S. Yamamoto, *IEEE Trans. Magn.* 34, 843 (1998).
239. M.-H. Lee, S. Lee, and K. Sin, *Thin Solid Films* 320, 298 (1998).

240. S. Soeya, M. Fuyama, S. Tadokoro, and T. Imagawa, *J. Appl. Phys.* 79, 1604 (1996).

241. C.-M. Park, K.-I. Min, and K. H. Shin, *J. Appl. Phys.* 79, 6228 (1996).

242. J. Nogues and I. K. Schuller, *J. Magn. Magn. Mater.* 192, 203 (1999).

243. J. C. S. Kools, *IEEE Trans. Magn.* 32, 3165 (1996).

244. B. A. Everitt, D. Wang, and J. M. Daughton, *IEEE Trans. Magn.* 32, 4657 (1996).

245. K. Matsuyama, H. Asada, S. Ikeda, K. Umezu, and K. Tauiguchi, *IEEE Trans. Magn.* 32, 4612 (1996).

246. G. Chern, D. S. Lee, H. C. Chang, and T.-H. Wu, *J. Magn. Magn. Mater.* 193, 497 (1999).

247. K. Takano, R. H. Kodama, A. E. Berkowitz, W. Cao, and G. Thomas, *Phys. Rev. Lett.* 79, 1130 (1997).

248. S. Soeya, S. Tadokoro, T. Imagawa, and M. Fuyama, *J. Appl. Phys.* 77, 5838 (1995).

249. S. R. Andrews and R. A. Cowley, *J. Phys. C: Solid State Phys.* 18, 6427 (1985).

250. W. F. Egelhoff, Jr., P. J. Chen, C. J. Powell, M. D. Stiles, R. D. McMichael, J. H. Judy, K. Takano, and A. E. Berkowitz, *J. Appl. Phys.* 82, 6142 (1997).

251. W. F. Egelhoff, Jr., P. J. Chen, C. J. Powell, M. D. Stiles, and R. D. McMichael, *J. Appl. Phys.* 79, 2491 (1996).

252. W. F. Egelhoff, Jr., P. J. Chen, C. J. Powell, M. D. Stiles, R. D. McMichael, C. L. Lin, J. M. Sivertsen, J. H. Judy, K. Takano, and A. E. Berkowitz, *J. Appl. Phys.* 80, 5183 (1996).

253. P. Bayle-Guillemaud, A. Barbier, and C. Mocuta, *Ultramicroscopy* 88, 99 (2001).

254. C.-H. Lai, T. C. Anthony, E. Iwamura, and R. L. White, *IEEE Trans. Magn.* 32, 3419 (1996).

255. C. Mocuta, A. Barbier, S. Lafaye, P. Bayle-Guillemaud, *Mater. Res. Soc. Symp. Proc.*, in press.

256. C. Mocuta, A. Barbier, G. Renaud, Y. Samson, and M. Noblet, *J. Magn. Magn. Mater.* 211, 283 (2000).

257. M. J. Carey, F. E. Spada, A. E. Berkowitz, W. Cao, and G. Thomas, *J. Mater. Res.* 6, 2680 (1991).

258. W. Cao, G. Thomas, M. J. Carey, and A. E. Berkowitz, *Scripta Metal. Mater.* 25, 2633 (1991).

259. C.-H. Lai, H. Matsuyama, and R. L. White, *IEEE Trans. Magn.* 31, 2609 (1995).

260. R. P. Michel, A. Chaiken, C. T. Wang, and L. E. Johnson, *Phys. Rev. B: Solid State* 58, 8566 (1998).

261. J. E. Keem, J. M. Honig, and L. L. Van Zandt, *Philos. Mag. B* 37, 537 (1978).

262. W. D. Wang, N. J. Wu, and P. A. Thiel, *Chem. Phys.* 92, 2025 (1990).

263. A. R. Sandy, S. G. J. Mochrie, D. M. Zehner, K. G. Huang, and D. Gibbs, *Phys. Rev. B: Solid State* 43, 4667 (1991).

264. L. D. Marks, *Phys. Rev. B: Solid State* 60, 2771 (1999).

265. L. D. Marks, E. Bengu, C. Collazo-Davila, D. Grozea, E. Landree, C. Leslie, and W. Sinkler, *Surf. Rev. Lett.* 5, 1087 (1998).

266. L. D. Marks, W. Sinkler, and E. Landree, *Acta Crystallogr.* 55, 601 (1999).

267. C. Collazo-Davila, D. Grozea, L. D. Marks, R. Feidenhans'l, M. Nielsen, L. Seehofer, L. Lottermoser, G. Falkenberg, R. L. Johnson, M. Gothelid, and U. Karlsson, *Surf. Sci.* 418, 395 (1998).

268. L. D. Marks, D. Grozea, R. Feidenhans'l, M. Nielsen, and R. L. Johnson, *Surf. Rev. Lett.* 5, 459 (1998).

269. N. Erdman, O. Warschkow, A. Barbier, D. E. Ellis, and L. D. Marks, submitted for publication.

270. N. Kitakatsu, V. Maurice, and P. Marcus, *Surf. Sci.* 411, 215 (1998).

271. P. Liu, T. Kendelewicz, and G. E. Brown, Jr., *Surf. Sci.* 412, 315 (1998).

Chapter 12

OPERATOR FORMALISM IN POLARIZATION-NONLINEAR OPTICS AND SPECTROSCOPY OF POLARIZATION-INHOMOGENEOUS MEDIA

I. I. Gancheryonok, A. V. Lavrinenko

Department of Physics, Belarusian State University, F. Skariny av. 4, Minsk 2220080, Belarus

Contents

Handbook of Thin Film Materials, edited by H.S. Nalwa
Volume 2: Characterization and Spectroscopy of Thin Films
Copyright © 2002 by Academic Press
All rights of reproduction in any form reserved.

1. FUNDAMENTALS OF THE THEORY OF INERACTION OF VECTOR LIGHT FIELDS WITH NONLINEAR MEDIA

In the general frame work the description of process of propagation of electromagnetic radiation in a medium is reduced to seeking the self-consistent solution of equations describing the action of radiation on a substance, and the influence of the latter on the radiation field. For complete correspondence of the theory and experiment it is necessary to undertake a quantum description of the medium as well as of the electromagnetic field. Indeed, in some only quantum effects can be exhibited: photon antigrouping, generation of squeezed states, etc. However, for a wide range of phenomena good consistency between theory and experiment is achieved with the use of a semiclassical description, i.e., description of the medium by the quantum equations, and of the radiation field by the classical Maxwell's equations. Further simplifications can be made under special conditions on the properties of the medium and the radiation.

Let us consider nonmagnetic electrically neutral media, which include the majority substances that are isotropic in the linear approximation. Then the character of light–matter interaction will be characterized first of all by the electric field E, and instead of Maxwell's equations it is convenient to use the wave equations [1]

$$(\nabla \times (\nabla \times E^+)) + \mu_0 \varepsilon_0 \tilde{\varepsilon} \frac{\partial^2 E^+}{\partial t^2} = -\mu_0 \frac{\partial^2 P^+}{\partial t^2} \quad (1.1a)$$

$$\nabla \cdot \varepsilon_0 \tilde{\varepsilon} E^+ = -\nabla \cdot P^+ \quad (1.1b)$$

where E^+ is the positive-frequency amplitude of the electric field $E = E^+ + E^-$, and $E^- = (E^+)^*$; $\tilde{\varepsilon}$ is the complex dielectric permittivity, defining the linear polarization of the medium $P_{\mathcal{L}}^+ = \varepsilon_0(\tilde{\varepsilon} - 1)E^+$; P^+ is the positive-frequency part of the nonlinear polarization of the medium; ε_0 and μ_0 are the electric and magnetic vacuum permittivities, respectively; and an asterisk denotes the complex conjugate.

In resonant media it is convenient to consider P^+ as the vector of polarization response associated with the resonant transitions [1]. The nonlinearity of this vector is considerable even for rather low radiation fields. Therefore dividing it into linear and nonlinear parts is frequently not meaningful. Then one can consider $\tilde{\varepsilon}$ as a permittivity determined by all nonresonant transitions of the medium. The polarization associated with these transitions remains linear even at very high radiation electric field; therefore, as a rule, $\tilde{\varepsilon}$ can be considered independent of E.

It should be taken into account that, because their effects are small, the optical nonlinearities have been observed only under the action of radiation of high spectral density [2], which is produced as a rule by lasers. Then the radiation field is conveniently presented as

$$E^+(r, t) = \sum_j E_j^+ \exp(-i\omega_j t + i k_j \cdot r) \quad (1.2)$$

where E_j^+ are complex amplitudes depending weakly [in comparison with $\exp(-i\omega_j t + i k_j \cdot r)$] on coordinates r and time t; j is the index numbering the waves distinguished by their frequency and direction of propagation; and k_j and ω_j are the wave vector and frequency of the jth wave, respectively.

The representation (1.2), in principle, can be used for any radiation field; however, it is the most convenient for laser radiation, owing to the restricted number of longitudinal modes. The decomposition (1.2) is also useful when it is known that the light beams are statistically independent and that their independence is not broken by interaction with the medium.

Let us restrict ourselves to the case of quasi-homogeneous and quasi-stationary radiation and present these requirements in the following way:

$$\tau\gamma \gg 1, \qquad |\Delta E^+| \ll k\left|\frac{\partial E^+}{\partial r}\right| \quad (1.3)$$

where τ is the characteristic time of changes of the amplitude E^+ and γ^{-1} is the longest relaxation time of resonance excitation.

Now the wave equations (1.1a), (1.1b) can be reduced to the following ones:

$$(k_j \cdot \nabla)E_j^+ + \frac{k_j}{v_j}\frac{\partial E_j^+}{\partial t} - \frac{i}{2}\left(\tilde{\varepsilon}\frac{\omega_j^2}{c^2} - k_j^2\right)E_j^2 = i\mu_0\omega_j^2 P_j^+ \quad (1.4)$$

where v_j is the group velocity of the jth wave, and P_j^+ is the Fourier component of P^+.

In the obtained system of equations the response of the medium to the action of a polarized electromagnetic field is determined by the polarization of the medium, which can be calculated with the help of quantum mechanical density matrix equations [3, 4]. This method allows one to take into account the interaction of the medium with an environment in an average sense, and treat the processes of relaxation phenomenologically, without in most cases restricting the completeness and correctness of description. On the other hand, the resonant character of interaction permits us [5] to treat only those states that interact strongly enough with radiation, referring interaction with remaining ones, as previously specified, to the linear part of the polarization of the vector response. Such as approach [4, 5] essentially simplifies the equations for the elements of the density matrix, which can be presented as

$$i\hbar\rho_{ij} = [\hat{V}\hat{\rho}]_{ij} + \rho_{ij}(-i\hbar\gamma_{ij} + \hbar\omega_{ij}), \qquad i \neq j$$

$$i\hbar\rho_{ij} = [\hat{V}\hat{\rho}]_{jj} + i\hbar\sum_k(\rho_{kk}\rho_{jj}^0 - \rho_{jj}\rho_{kk}^0)d_{jk} \quad (1.5)$$

where $\hbar\omega_{ij} = E_i - E_j$, E_α is the energy of an eigenstate of the quantum system, γ_{ij}^{-1} and $\gamma_{jj}^{-1} = (\sum_k d_{jk})^{-1}$ are the relaxation times for nondiagonal (cross-relaxation) and diagonal (longitudinal) elements of the density matrix, and $\rho_{\alpha\alpha}^0$ are the equilibrium values of the elements $\rho_{\alpha\alpha}$.

Solving the system (1.5), the vector P in a nonmagnetic medium can be found the relation

$$P = N\langle\text{Tr}(\hat{\mu}\hat{\rho})\rangle - \nabla N\langle\text{Tr}(\hat{q}\hat{\rho})\rangle \quad (1.6)$$

where N is the concentration of atoms (molecules), $\hat{\mu}$ is the operator of the electrical dipole moment, $\hat{q}$ is the operator of the electrical quadrupole moment, and the angular brackets denote the ensemble average (spatial average).

The second term in (1.6) is essential, for example, in nonlinear polarization spectroscopy (NPS) of atomic systems with dipole-forbidden transitions [4]. In solutions for complex molecules even symmetry-forbidden, transitions have electronic–vibrational nature and are well treated in the dipole approximation [6]. Thus this approximation can be used in our further calculations. It should be noted, however, that the electrical quadrupole moment of the medium, and also its magnetic dipole moment (which is comparable in magnitude), can be taken into account on the basis of these results [7].

Now let us specify the form of the interaction operator. In the dipole approximation it can be written as

$$\hat{V} = \sum_k \hat{\boldsymbol{\mu}}_k \cdot \boldsymbol{E}_k \tag{1.7}$$

The realization of the requirement of quasistationarity allows one to solve the system (1.5) in analytical form and, hence, to find explicitly the functional form $\boldsymbol{P}(\boldsymbol{E})$. However, there are difficulties to this approach in practice, and more often approximate methods are used and give good enough results [4, 8]. That is because the field strength of laser radiation, as a rule, is much less than strengths of the internal fields that determine the stability of atoms and molecules. In such an approach the functional relationship between $\boldsymbol{P}$ and $\boldsymbol{E}$ can be presented as a series [2]

$$\begin{aligned} P_\alpha = \varepsilon_0 \big(&K_1 \chi^{(1)}_{\alpha\beta} E_\beta + K_2 \chi^{(2)}_{\alpha\beta\gamma} E_\beta E_\gamma \\ &+ K_3 \chi^{(3)}_{\alpha\beta\gamma\delta} E_\beta E_\gamma E_\delta + \cdots \big) \end{aligned} \tag{1.8}$$

where $\hat{\chi}^{(i)}$ and K_i ($i = 1, 2, 3, \ldots$) are the susceptibility tensors of the medium of the ith order and the numerical coefficients, respectively; repeated indices imply summation.

The first term in (1.8), obviously, can be omitted, as it features linear processes, which were taken into account in the statement of the initial system of Eqs. (1.1a) and (1.1b). The second term is equal to zero in isotropic (centrosymmetric) media [2]. Therefore the basic role will be played by the third term (cubic in the electric field). According to [8], "it is possible to state unequivocally that the susceptibility $\hat{\chi}^{(3)}$ constitutes one of the major optical characteristics of gases, liquids, and solids." It is necessary to note that by using the cubic approximation, we restrict the intensities of the light beams interacting with the medium. However, as experiment [8] show, even in the case of fairly powerful laser fields, the $\chi^{(3)}$-approach gives good qualitative agreement with the experimental data. The use of the various model representations [4, 9] or approximations further permits one, if not to solve a problem completely, then to give qualitative indications of probable regularities.

Thus, summarizing, we can conclude that the interaction of the vector light field with a nonlinear medium can be described with the help of the Eqs. (1.5) and (1.6). For examination of light-induced anisotropy (photoanisotropy) of the medium we will apply the methods of the optics of anisotropic media [10–12].

2. TENSOR-OPERATOR APPROACH TO THE DESCRIPTION OF PHOTOANISOTROPIC MEDIA

At present, for the description of the vector field and the effects of its interaction with the medium, a number of methods [11–15] have been successfully applied. Some of them serve well in the simple cases; others should be applied for the solution of complex problems. It is necessary only that the chosen methods should ensure complete enough description of the evolution of the vector field and sufficient simplicity of the calculations.

The description of photoinduced anisotropy in initially isotropic media is quite complicated and not always possible analytically [11, 12]. In this connection the methods of classical crystal optics have been successfully applied to the solution of some problems of nonlinear optics (see, e.g., [16, 17]), and also of problems that have traditionally been treated with classical linear optics [12], though rigorously speaking they are nonlinear ones [18]. Moreover, the methods of classical crystal optics can be used when the equations describing the nonlinear interaction are linearized in an examined field. In the case of a medium with light-induced anisotropy (MLIA) such possibility is realized in an approximation of a given field that is widely used in nonlinear spectroscopy [8, 18]. It is supposed that the intensity of wave fields is much larger than that of other fields, so that under certain conditions the influence of the latter fields on the former ones can be neglected. Moreover, the change of some parameters (in particular, polarization) of the strong field is so small that one may consider the behavior of the weak field in the presence of a given strong (polarization) field taking into account the nonlinearity of the interaction. Such an approach (whose validity follows from the existence of sustainable polarization states of radiation in resonant media [4] and experimentally observed phenomenon of *forestalling* change of polarization of the weak radiation [19]) is used in the present chapter. In some cases the change of intensity of the exciting (pumping) field caused by absorption (or amplification) in resonant media will be taken into account too.

In the presence of two fields (a given strong field and a weak probe field), the part of nonlinear polarization (1.8) that is cubic in the field can be represented by a sum of terms, the most essential of which (linear in the probe field)) has the form

$$(P_1)_\alpha = \varepsilon_0 K_3 \chi^{(3)}_{\alpha\beta\gamma\delta} (E_0)_\beta (E_0^*)_\gamma (E_1)_\delta \tag{2.1}$$

where the coefficients 0 and 1 herein and after distinguish the strong and probe fields, respectively, and $\chi^{(3)}_{\alpha\beta\gamma\delta}$ are the elements of the fourth-rank cubic (third-order) susceptibility tensor, of which only three are independent, with [8]

$$\chi_{1111} = \chi_{1122} + \chi_{1212} + \chi_{1221} \quad \text{or} \quad \chi_{11} = \chi_{12} + \chi_{nn} + \chi_{n3} \tag{2.2}$$

The frequency arguments of all components are identical, and we will suppress the frequency dependence of χ_{ijkl}, K_3, and the electric fields.

It is necessary to underline that in the case of arbitrary directions of propagation of strong and weak waves, the evolution of the latter will be determined not by the vector $\boldsymbol{P}$ as such, but by its transverse component, which with the use of (2.1) and (2.2) can be written as follows [20]:

$$P_1^+ = \varepsilon_0 K_3 \chi_{1221} |E_0^+|^2 \hat{S}_\perp E_1^+ \tag{2.3}$$

where

$$\hat{S}_\perp = \hat{A}\hat{S}\hat{A} = \left(-n_1^{\times 2}\right)\hat{S}\left(-n_1^{\times 2}\right) \tag{2.4}$$

Here $n_i^\times$ is the antisymmetric second-rank tensor dual to the unit-norm real vector n representing the direction of propagation of the ith wave (wave normal), and we introduced a photoanisotropy tensor $\hat{S}$ (LIA tensor)

$$\hat{S} = \hat{I} + C_1\varepsilon_0 \otimes \varepsilon_0^* + C_2\varepsilon_0^* \otimes \varepsilon_0 \tag{2.5}$$

$$C_1 = \frac{\chi_{1122}}{\chi_{1221}}, \qquad C_2 = \frac{\chi_{1212}}{\chi_{1221}} \tag{2.6}$$

Here $\hat{I}$ is the unit tensor; ε_0 is the unit-norm polarization vector of the strong wave ($E_i = E_i\varepsilon_i$), and the sign $\otimes$ denotes the direct (dyadic) product of two vectors.

Now we can rewrite (2.4) in the following way:

$$\hat{S}_\perp = \hat{A} + C_1 a \otimes a^* + C_2 a^* \otimes a \tag{2.7}$$

where

$$a = (n_1 \times \varepsilon_0) \times n_1 = \hat{A}\varepsilon_0 \tag{2.8}$$

and $a \times b$ is the vector product of two vectors a and b.

For simplification of the further calculations we present a as an elliptic vector in the following manner:

$$a = |a|\left(1 + \eta^2\right)^{-1/2}(e_1 + i\eta e_2) \tag{2.9}$$

where e_1 and e_2 are the unit-norm vectors along the long and short axes, respectively, of the polarization ellipse; η is the ellipticity, defined as the ratio of lengths of these axes (short to long); and in order to avoid the terminological tangle in the modern periodic literature, here and further we shall use the notation of the monograph [14]: $|a|^2 = 1 - |n_1 \cdot \varepsilon_0|^2$.

Now the eigenvectors of the tensor $\hat{S}_\perp$ can be presented [20] as

$$E_\pm = e_1 + ik_\pm e_2 \tag{2.10}$$

where

$$\begin{aligned} k_\pm = &-\frac{(C_1 + C_2)(1 - \eta^2)}{2\eta(C_1 - C_2)} \\ &\pm \frac{[(C_1 + C_2)^2(1 + \eta^2)^2 - 16\eta^2 C_1 C_2]^{1/2}}{2\eta(C_1 - C_2)} \end{aligned} \tag{2.11}$$

For the eigenvalues of $\hat{S}_\perp$

$$\lambda_\pm = \frac{1}{2}\left[\operatorname{Tr}\hat{S}_\perp \pm \left(2\operatorname{Tr}\hat{S}_\perp^2 - \operatorname{Tr}^2\hat{S}_\perp\right)^{1/2}\right] \tag{2.12}$$

we obtain

$$\begin{aligned} \lambda_\pm = &1 + \frac{1}{2}(C_1 + C_2)|a|^2 \\ &\pm \left[\frac{1}{4}(C_1 + C_2)^2|a|^4 + C_1 C_2(a \times a^*)^2\right]^{1/2} \end{aligned} \tag{2.13}$$

In the frame defined by the unit vectors e_1 and e_2, the tensor $\hat{S}_\perp$ can be rewritten in the matrix form

$$\hat{S}_\perp = \begin{bmatrix} 1 + |a|^2\frac{C_1+C_2}{1+\eta^2} & -i\eta|a|^2\frac{C_1-C_2}{1+\eta^2} \\ i\eta|a|^2\frac{C_1-C_2}{1+\eta^2} & 1 + \eta^2|a|^2\frac{C_1+C_2}{1+\eta^2} \end{bmatrix} \tag{2.14}$$

or in terms of ellipticity angle ($\eta = \tan\varepsilon$), [21]

$$\hat{S}_\perp = \begin{bmatrix} 1 + (C_1 + C_2)|a|^2\cos^2\varepsilon & -\frac{i}{2}(C_1 - C_2)|a|^2\sin 2\varepsilon \\ \frac{i}{2}(C_1 - C_2)|a|^2\sin 2\varepsilon & 1 + (C_1 + C_2)|a|^2\sin^2\varepsilon \end{bmatrix} \tag{2.15}$$

From (2.14), (2.15) it is easy to see that in case $C_1 = C_2 = C$ and $\eta = 1$ we have degeneracy ($\lambda_+ = \lambda_- = 1 + |a|^2 C$); thus the tensor $\hat{S}_\perp$ becomes the usual λ-matrix. If C_1 also C_2 are real, $\hat{S}_\perp$ is Hermitian; hence $\lambda_\pm^* = \lambda_\pm$, and the eigenvectors corresponding to different eigenvalues are orthogonal in the sense that their scalar product [12]

$$E_+ \cdot E_-^* = 0 \tag{2.16}$$

Let us note one case where (2.16) is fulfilled. According to [22], it takes place when

$$\hat{S}_\perp^* \hat{S}_\perp = \hat{S}_\perp \hat{S}_\perp^* \tag{2.17}$$

i.e., when the photoanisotropy tensor is normal. To this fact we may add of [23, 24], where it is noted that the orthogonality of eigenvectors of the tensor $\hat{S}_\perp$ also follows from the condition

$$\operatorname{Im}\left(\frac{C_1 - C_2}{C_2 + C_2}\right) = 0 \tag{2.18}$$

For our further consideration the following decomposition (Silvester's interpolation formula) will be useful:

$$\hat{S} = \sum_{k=1}^{k'} \lambda_k \hat{G}_k \tag{2.19}$$

where $\hat{G}_k$ are the projection operators

$$\hat{G}_k\hat{G}_m = \delta_{km}G_m, \qquad \sum_{k=1}^{k'}\hat{G}_k = \hat{I} \tag{2.20}$$

which are determined as follows:

$$\hat{G}_k = \prod_{\substack{1 \le m \le k' \\ m \ne k}} \frac{\hat{S} - \lambda_m\hat{I}}{\lambda_k - \lambda_m} \tag{2.21}$$

These operators project any vector on the directions of the respective eigenvectors and in our case can be written as

$$\begin{aligned} \hat{G}_\pm = \pm\Bigg(&C_1 a \otimes a^* + C_2 a^* \otimes a \\ &- \hat{A}\left[\frac{1}{2}(C_1 + C_2)|a|^2 \mp \left[\frac{1}{4}(C_1 + C_2)^2|a|^4\right.\right. \\ &\left.\left.+ C_1 C_2(a \times a^*)^2\right]^{1/2}\right]\Bigg) \\ &\times \left[(C_1 + C_2)^2|a|^4 + 4C_1 C_2(a \times a^*)^2\right]^{-1/2} \end{aligned} \tag{2.22}$$

Now we pay attention to the problem of probe wave propagation along an axis of the chosen coordinate system. Suppose that all nonresonant processes affecting the change of the field E_j^+ are taken into account through the amplitude coefficient of linear losses, σ_j. Then the Eq. (1.4) for the probe wave can be reduced to the form [20]

$$\frac{dE_1^+}{d\xi} + \frac{\sigma_1}{2}E_1^+ = i\frac{3}{4}\frac{\sqrt{\varepsilon_0\mu_0\omega_1}}{n_1}\chi_{43}|E_0^+|^2\hat{S}_\perp E_1^+ = a\hat{S}_\perp E_1^+\gamma$$
$$n_1 = \sqrt{\operatorname{Re}\bar{\varepsilon}}$$
$$(2.23)$$

where, under the assumption of constancy of the vectors $\boldsymbol{\varepsilon}_0$ and $\boldsymbol{n}_1$, $\hat{S}_\perp$ represents a constant tensor of second rank, and Eq. (2.23) depends linearly on E_1^+. Introducing a factor

$$A(\xi) = a\xi \qquad (2.24)$$

the solution of the Eq. (2.23) (in the prescribed strong-field approximation) can be relatively easily constructed on the basis of a matrix representation of solution of the set of differential equations [20]

$$E_1^+(\xi) = \exp(-\sigma_1\xi)\big\{\hat{G}_+ \exp\big[A(\xi)\lambda_+\big] + \hat{G}_- \exp\big[A(\xi)\lambda_-\big]\big\}E_1^+(0) \qquad (2.25)$$

where $E_1^+(0) = E_{10}^+$ is the vector amplitude of the probe wave on the forward boundary of the nonlinear photoanisotropic medium.

Thus, the expression (2.25) represents the general solution for the probe field.

For completeness of consideration and with the purpose of comparing properties of photoanisotropic media and crystals, instead of using the susceptibility tensor formalism we can describe the dielectric response of the nonlinear medium by means of the dielectric permittivity tensor $\hat{\varepsilon}$. In the X–Y–Z coordinate system, where the Z-axis is chosen in the direction of $\boldsymbol{n}_0$, the matrix representation of $\hat{S}$ (2.5) is given as [21]

$$\hat{S} = \begin{bmatrix} 1 + (C_1 + C_2)\cos^2\varepsilon_{\text{pmp}} & -\frac{i}{2}(C_1 - C_2)\sin 2\varepsilon_{\text{pmp}} & 0 \\ \frac{i}{2}(C_1 - C_2)\sin 2\varepsilon_{\text{pmp}} & 1 + (C_1 + C_2)\sin^2\varepsilon_{\text{pmp}} & 0 \\ 0 & 0 & 1 \end{bmatrix}$$
$$= [S_{ij}] \qquad (2.26)$$

where ε_{pmp} is the ellipticity angle of the vector $\boldsymbol{\varepsilon}_0$.

Now, taking into account that for an isotropic substance the induced polarization P in tensor notation as

$$P_i = \varepsilon_0 \chi_{ij} E_j \qquad (2.27)$$

and in our case of initially isotropic media

$$P_i = P_i^{\text{L}} + P_i^{\text{NL}} = \varepsilon_0\big[\chi_0 E_i + \chi_{ij}^{\text{NL}} E_j\big] \qquad (2.28)$$

we can write the electric susceptibility χ_{ij} as

$$\chi_{ij} = \chi_0\delta_{ij} + K_3|E_0^+|^2\chi_{1221}S_{ij} = \chi_0\delta_{ij} + \chi_1 S_{ij} \qquad (2.29)$$

Here χ_0 is the linear scalar susceptibility; δ_{ij} are the components of $\hat{I}$.

The further procedure of finding $\hat{\varepsilon}$ is simple and leads to the result

$$\hat{\varepsilon} = \varepsilon_0\big[(1+\chi_0)\hat{I} + \chi_1\hat{S}\big] = g_1\hat{I} + g_2\boldsymbol{\varepsilon}_0\otimes\boldsymbol{\varepsilon}_0^* + g_3\boldsymbol{\varepsilon}_0^*\otimes\boldsymbol{\varepsilon}_0 \quad (2.30)$$

From (2.30) it follows that $\hat{\varepsilon}$ is an asymmetrical and non-Hermitian tensor. The matrix form of (2.30) is given as

$$\hat{\varepsilon} = \varepsilon_0 \begin{bmatrix} 1 + \chi_0 + \chi_1 S_{11} & \chi_1 S_{12} & 0 \\ \chi_1 S_{21} & 1 + \chi_0 + \chi_1 S_{22} & 0 \\ 0 & 0 & 1 + \chi_0 + \chi_1 S_{34} \end{bmatrix}$$
$$= \begin{bmatrix} \varepsilon_{11} & \varepsilon_{12} & 0 \\ \varepsilon_{21} & \varepsilon_{22} & 0 \\ 0 & 0 & \varepsilon_{33} \end{bmatrix} \qquad (2.31)$$

The tensor $\hat{\varepsilon}$ in the representation (2.30) or (2.31) is formally equivalent, in general, to the dielectric permittivity tensor of a biaxial dichroic crystal with gyrotropy. In some cases $\operatorname{Im}\varepsilon_{ij} = 0$, which means that the dichroism disappears, and the off-diagonal elements are connected by the relationship

$$\varepsilon_{ij} = ib = \varepsilon_{ij}^*, \qquad b \in \operatorname{Re} \qquad (2.32)$$

Then MLIA models an optically active, biaxial crystal with the gyration vector

$$\boldsymbol{G} = \frac{1}{2}\chi_1'(C_2 - C_1)\sin 2\varepsilon_{\text{pmp}}\,\boldsymbol{n}_0, \qquad \chi_1' \in \operatorname{Re} \quad (2.33)$$

Thus, $\boldsymbol{G}$ is parallel to $\boldsymbol{n}_1$, and hence the direction of rotation bears a fixed relation to the direction of propagation of a pump wave. The eigenpolarizations are preserved on reflection, so the net rotation is doubled, as in the case of Faraday rotation. This means [13] that light-induced gyrotropy (LIG) is a nonreciprocal effect. It is interesting to note that the direction of $\boldsymbol{G}$ depends on the handedness of the strong-field polarization ellipse [sign(ε)] and the properties of the MLIA (sign[$\chi_1'(C_2 - C_1)$]).

It is seen from Eq. (2.33) that in the case of a linearly polarized pump ($\varepsilon_{\text{pmp}} = 0$) and Kleinman symmetry ($C_1 = C_2 = 1$), which holds in many rigid dye solutions, the gyrotropy vanishes. In the case of circularly polarized pumping ($\varepsilon_{\text{pmp}} = \pm\pi/4$) and $C_1 = C_2$, MLIA is isotropic in the direction of $\boldsymbol{n}_0$, but anisotropy remains for other directions. Indeed, using the representation of the unit tensor on an axial basis, i.e., on the basis of the circular polarization vectors $\boldsymbol{\varepsilon}_0$, $\boldsymbol{\varepsilon}_0^*$ and the linear vector $\boldsymbol{n}_0$, we have

$$\hat{I} = \boldsymbol{n}_0\otimes\boldsymbol{n}_0 + \boldsymbol{\varepsilon}_0\otimes\boldsymbol{\varepsilon}_0^* + \boldsymbol{\varepsilon}_0^*\otimes\boldsymbol{\varepsilon}_0 \qquad (2.34)$$

and we can rewrite (2.30) as

$$\hat{\varepsilon} = (g_1 + g_2)\hat{I} - g\boldsymbol{n}_0\otimes\boldsymbol{n}_0 \qquad (2.35)$$

Here $g = g_2 = g_3$. The permittivity tensor given by (2.35) characterizes a unixial, generally dichroic crystal with optical axis $\boldsymbol{n}_0$. This is connected with the fact that a specific direction $\boldsymbol{n}_0$ remains in the media due to the transverse character of the electromagnetic waves.

Now let us estimate the magnitude of the LIG effect, with we will describe in terms of the rotational power, i.e., the rotation of the plane of polarization of a probe wave (wave length

Table I.   Estimated Light-Induced Rotary Power in a Variety of Substances

Phase	Substance	Ref.	λ_2 (nm)	Rotatory power (deg/mm)
Gas	Na vapor	[25]	589.1	1–2
Liquid	Ethanol solution of R6G dye	[26]	1064	15–20
Crystal	Silicon	[27]	1064	30
	α-Quartz[a]	[21]	589.3	27.7

[a] From the natural optical activity of right-handed α-quartz.

λ_1) per unit path length in the medium. Table I shows the corresponding values for different types of initially isotropic media. Comparing them with the value for α-quartz, we find that the LIG effect in condensed matter can be comparable to or even larger than natural optical activity.

Now let us apply Fedorov's formalism for the description of MLIA. Let us rewrite (2.30), defining the symmetric and antisymmetric parts of the tensor $\hat{\varepsilon}$:

$$
\begin{aligned}
\hat{\varepsilon} &= \left[\left(g_1 + \frac{g_2 + g_3}{2} \right) \hat{I} + \frac{g_2 + g_3}{2} \cos 2\varepsilon_H (\hat{\tau}_x - \hat{\tau}_y) \right]_S \\
&\quad + \left[i \frac{\sin 2\varepsilon_H}{2} (g_2 - g_3) e_z^\times \right]_{aS} \\
&= \hat{\varepsilon}_S + \hat{\varepsilon}_{aS} = \hat{\varepsilon}_S + iG^\times
\end{aligned}
\tag{2.36}
$$

where $\hat{\tau}_i$ denotes the projective dyadic $e_i \otimes e_i$. Further, for the tensor $\hat{\varepsilon}_S$, simple analysis shows that the eigenvectors and relevant eigenvalues are given by

$$
\begin{aligned}
e_1 &= e_x, & \lambda_1 &= g_1 + \frac{g_2 + g_3}{2}(1 + \cos 2\varepsilon_H) \\
e_2 &= e_y, & \lambda_2 &= g_1 + \frac{g_2 + g_3}{2}(1 - \cos 2\varepsilon_H) \\
e_3 &= e_z, & \lambda_3 &= g_1
\end{aligned}
\tag{2.37}
$$

We proceed now to finding the optical axes of the medium described by the tensor (2.36), defining them as real vectors n obeying [12] [according to (2.30)]

$$
n \times \varepsilon_0 = 0 \quad \text{and} \quad n \times \varepsilon_0^* = 0
\tag{2.38}
$$

The solutions of the Eqs. (2.38) are

$$
\begin{aligned}
n'_\pm &= \frac{\sqrt{\varepsilon_0^{*2}} \varepsilon_0 + \sqrt{\varepsilon_0^2} \varepsilon_0^* \pm i(\varepsilon_0 \times \varepsilon_0^*)}{|\varepsilon_0|^2 + |\varepsilon_0^2|} \\
&= \sqrt{1 - \tan^2 \varepsilon_{\text{pmp}}}\, e_x \pm \tan \varepsilon_{\text{pmp}}\, e_z
\end{aligned}
\tag{2.39}
$$

$$
\begin{aligned}
n''_\pm &= \frac{\sqrt{\varepsilon_0^{*2}} \varepsilon_0^* + \sqrt{\varepsilon_0^{*2}} \varepsilon_0 \pm i(\varepsilon_0^* \times \varepsilon_0)}{|\varepsilon_0^*|^2 + |\varepsilon_0^{*2}|} \\
&= \sqrt{1 - \tan^2 \varepsilon_{\text{pmp}}}\, e_x \mp \tan \varepsilon_{\text{pmp}}\, e_z
\end{aligned}
\tag{2.40}
$$

Thus, we in fact have four circular optical axes, of which, however, only two are distinct. Hence, media with LIA simu-

late, in general, properties of absorbing crystals of the lowest possible symmetry.

In conclusion, for completeness of our consideration we present here the external, overall Jones matrix $\hat{T}(\xi)$ of an optical device in the form of a MLIA of thickness ξ. Using the result of Azzam and Bashara [14], we can write the following relationship between $\hat{T}(\xi)$ and the differential propagation Jones matrix $\hat{N}$:

$$
\hat{N} = \frac{d\hat{T}(\xi)}{d\xi} \hat{T}^{-1}(\xi)
\tag{2.41}
$$

An expression for $\hat{N}(\xi)$ is given by the first equation of the system (2.23):

$$
\frac{dE_1^+}{d\xi} = \left(a\hat{S}_\perp - \frac{\sigma_1}{2}\hat{I} \right) E_1^+ = \hat{N} E_1^+
\tag{2.42}
$$

From (2.42) we can obtain, in the approximation of negligible pump and probe depletion, the matrix $\hat{T}(\xi)$ in the form of a matrix exponential,

$$
\hat{T}(\xi) = \exp(\hat{N}\xi) = \exp\left[\hat{H}(\xi)\right]
\tag{2.43}
$$

However, for collinear light beams the last assumption can be omitted and $\hat{T}(\xi)$ is still given in the same eq. (2.43), with

$$
\hat{H}(\xi) = \frac{2\pi i k_1 K_3 \varepsilon_0}{\sigma_0} \chi_{1221} |E_0^+|^2 \left[1 - \exp(-\sigma_0\xi) \right] \hat{S} - \frac{1}{2}\sigma_1 \xi
\tag{2.44}
$$

where σ_0 and σ_1 are the linear absorption coefficients, and E_0 is the input vector amplitude of the strong field.

Furthermore, we note that the matrix $\hat{T}(\xi)$ in Eq. (2.43) has the same eigenvectors as $\hat{H}$, with corresponding eigenvalues $t_i(\xi)$ given by $t_i(\xi) = \exp[h_i(\xi)]$. Here, $h_i(\xi)$ are the eigenvalues of $\hat{H}(\xi)$.

Thus, our classification of anisotropy properties is valid for the corresponding optical devices.

3. FEDOROV'S LIGHT BEAM TENSOR FORMALISM FOR DESCRIPTION OF VECTOR FIELD POLARIZATION

From the scheme of this chapter it will be clear that after describing the formalism of light–matter interactions (light-induced anisotropy) in Section 1 and the tensor theory for the analysis of medium properties in Section 2, we can now develop a coordinate-free technique for leading with partially polarized light. There are several reasons to include such a formalism here. The first one is to consider the most general case of light polarization through a direct coordinate-free description. The second one is to introduce a new degree of freedom into spectroscopic studies.

Also, we should take into account that every interaction of polarized light with refracting interfaces will change the polarization ratio, and thus we must apply some formalism that can deal with partially polarized or totally nonpolarized light. It is well known that partially polarized light can be described by the Stokes parameter formalism. The four Stokes parameters S_1, S_2, S_3, S_4 compose a four-component column, which, however, does not possess the properties of a vector—for example, it does not transform as a vector under coordinate system rotations. Instead we call it a vector parameter, or simply the set of Stokes parameters. They are defined by

$$S_1 = |E_1|^2 + |E_2|^2, \qquad S_2 = |E_1|^2 - |E_2|^2$$
$$S_3 = 2\,\text{Re}(E_1 E_2^*), \qquad S_4 = -2\,\text{Im}(E_1 E_2^*) \qquad (3.1)$$

where $E_1 = \boldsymbol{e}_1 \cdot \boldsymbol{E}$ and $E_2 = \boldsymbol{e}_2 \cdot \boldsymbol{E}$ are projections of the complex electric field vector $\boldsymbol{E}$ onto any two mutually perpendicular directions in the phase plane of the plane wave along the unit vectors $\boldsymbol{e}_{1,2}$. If we have a light beam that is a superposition of incoherent light waves with given wave normal $\boldsymbol{n}$, then instead of (3.1) we have

$$S_1^{(j)} = \left|E_1^{(j)}\right|^2 + \left|E_2^{(j)}\right|^2, \qquad S_2^{(j)} = \left|E_1^{(j)}\right|^2 - \left|E_2^{(j)}\right|^2$$
$$S_3^{(j)} = 2\,\text{Re}\left(E_1^{(j)} E_2^{(j)}\right), \qquad S_4 = -2\,\text{Im}\left(E_1^{(j)} E_2^{(j)*}\right) \qquad (3.2)$$

where $S_k = \sum_j S_k^{(j)}$, $k = 1, 2, 3, 4$. Between the Stokes parameters we have the relation $S_1^2 \geq S_2^2 + S_3^2 + S_4^2$, where equality takes place only for a totally polarized beam. Therefore, the positive normalized number $p = \sqrt{S_2^2 + S_3^2 + S_4^2}/S_1 \leq 1$ is the polarization ratio (degree of polarization) of the beam.

A difficulty connected with the use of Stokes parameters is that they are specific to a particular system of coordinates. On changing the system (as in the analysis of a reflection problem), one has to change the phase plane of the whole beam and recalculate the Stokes parameters accordingly.

Another way of describing partially polarized waves, which is free of these disadvantages, was introduced in [28] by Fedorov; see also the comprehensive book [12]. He proved that the tensor construction $\hat{\Phi} = \boldsymbol{E} \otimes \boldsymbol{E}^*$, called a light beam tensor (LBT), possesses all necessary properties of a partially polarized wave and thus can be used for a coordinate-free polarization description. For the incoherent bundle of waves with the same speed and direction of propagation the LBT is easily obtained [12] as

$$\hat{\Phi} = \sum_j \boldsymbol{E}^{(j)} \otimes \boldsymbol{E}^{(j)*} \qquad (3.3)$$

The general properties of the LBT follow directly from the defining expression (3.3). Namely, this is a Hermitian tensor with one zero eigenvalue corresponding to the eigenvector $\boldsymbol{n}$ ($\hat{\Phi}\boldsymbol{n} = \boldsymbol{n}\hat{\Phi} = 0$), and two others that are both real and positive.

Now let us consider, following [12], the connection between the LBT and the Stokes parameters. Consider an arbitrary circular vector $\boldsymbol{e} = (\boldsymbol{e}_1 + i\boldsymbol{e}_2)/\sqrt{2}$, orthogonal to the unit wave normal $\boldsymbol{n}$ ($\boldsymbol{e} \cdot \boldsymbol{n} = 0$). We can organize three invariants using the tensor $\hat{\Phi}$ and two vectors: $\boldsymbol{e}$ and its complex conjugate $\boldsymbol{e}^*$. They are

$$I_1 = I_1^* = \boldsymbol{e}^* \Phi \boldsymbol{e}, \quad I_2 = I_2^* = \boldsymbol{e}\Phi\boldsymbol{e}^*, \quad I_3 = \boldsymbol{e}\Phi\boldsymbol{e} \qquad (3.4)$$

Taking into account determination of the LBT (3.3) and the definitions $E_1^{(j)} = \boldsymbol{e}_1 \boldsymbol{E}^{(j)}$ and $E_2^{(j)} = \boldsymbol{e}_2 \boldsymbol{E}^{(j)}$, the invariants

(3.4) can be transformed to the expressions

$$I_1 = \frac{1}{2} \sum_j \left\{ \left(|E_1^{(j)}|^2 + |E_2^{(j)}|^2 \right) - i\left(E_1^{(j)*} E_2^{(j)} - E_1^{(j)} E_2^{(j)*} \right) \right\}$$

$$I_2 = \frac{1}{2} \sum_j \left\{ \left(|E_1^{(j)}|^2 + |E_2^{(j)}|^2 \right) + i\left(E_1^{(j)*} E_2^{(j)} - E_1^{(j)} E_2^{(j)*} \right) \right\}$$

$$I_3 = \frac{1}{2} \sum_j \left\{ \left(|E_1^{(j)}|^2 - |E_2^{(j)}|^2 \right) + i\left(E_1^{(j)*} E_2^{(j)} + E_1^{(j)} E_2^{(j)*} \right) \right\}$$

which immediately give us the relations

$$S_1 = I_1 + I_2 = \boldsymbol{e}^* \Phi \boldsymbol{e} + \boldsymbol{e}\Phi\boldsymbol{e}^*$$
$$S_2 = 2\,\text{Re}\,I_3 = 2\,\text{Re}(\boldsymbol{e}\Phi\boldsymbol{e})$$
$$S_3 = 2\,\text{Im}\,I_3 = 2\,\text{Im}(\boldsymbol{e}\Phi\boldsymbol{e}) \qquad (3.5)$$
$$S_4 = I_1 - I_2 = \boldsymbol{e}^* \hat{\Phi} \boldsymbol{e} - \boldsymbol{e}\hat{\Phi}\boldsymbol{e}^*$$

These formulas reflect the very important fact that the Stokes parameters belong to a certain system of coordinates, while the LBT is free of such connection. Moreover, as we shall see a bit later, Stokes parameters cannot be expressed solely in terms of invariants of the tensor $\hat{\Phi}$; otherwise they also would be coordinates-free. Further relations can be found with more direct invariants of the LBT. The trace of the LBT is $\hat{\Phi}_t = \sum_j \boldsymbol{E}^{(j)} \cdot \boldsymbol{E}^{(j)*} = \sum_j I_j = I$, which means that it is equal to the beam intensity (as a sum of intensities of incoherent waves in the beam). Thus, it is evident that $S_1 = \hat{\Phi}_t$ [12].

Let us consider totally nonpolarized light. We now have only one distinguished direction in the three-dimensional space — the direction along the wave normal unit vector $\boldsymbol{n}$. All the directions in the phase plane orthogonal to the wave normal are uniformly occupied by the electric fields of the waves in the beam. From the point of view of symmetry this means transverse isotropy of the bundle (one axis of infinite symmetry). We can have only one general of tensor conserving such symmetry [12]: $\hat{\Phi} = a + b\boldsymbol{n} \otimes \boldsymbol{n}$, where a and b are some scalars. But we know that $\boldsymbol{n}$ is the eigenvector of the tensor $\hat{\Phi}$ with zero eigenvalue, which means that $a + b = 0$ and $\hat{\Phi}_t = 2a = I$. Finally we obtain

$$\hat{\Phi} = \frac{1}{2} I(1 - \boldsymbol{n} \otimes \boldsymbol{n}) = -\frac{1}{2} I(\boldsymbol{n}^\times)^2 \qquad (3.6)$$

Squaring this equality, we get $\hat{\Phi}^2 = \frac{1}{4}(\hat{\Phi}_t)^2(\boldsymbol{n}^\times)^4 = -\frac{1}{4}(\hat{\Phi}_t)^2 \times (\boldsymbol{n}^\times)^2 = \frac{1}{2}(\hat{\Phi}_t\hat{\Phi})$, where we have taken into account (3.6) and the equality $(\boldsymbol{n}^\times)^3 = -\boldsymbol{n}^\times$. The trace of above relation leads us to the formula

$$(\hat{\Phi}^2)_t = \frac{1}{2}(\hat{\Phi}_t)^2 \qquad (3.7)$$

This formula can be considered as the condition for total depolarization of light; that is, if and only if it is satisfied, the beam is totally depolarized.

Another important particular case is connected with one polarized wave in a beam or with a beam consisting of coherent partial waves that give, after superposition, some definite polarization state that is unchanging in homogeneous media. The LBT id $\hat{\Phi} = E \otimes E^*$, and one can use the invariants of the tensor $\hat{\Phi}$. First of all we note that

$$(\hat{\Phi}^2)_t = (\hat{\Phi}_t)^2 \qquad (3.8)$$

which can be accepted as the principal condition for polarized light. Taking into account that the tensor $\hat{\Phi}$ is a dyadic (the direct Kronecker product of two vectors) and some of its invariants are equal to zero, we can write down the following:

$$I \equiv \hat{\Phi}_t = E \cdot E^*, \qquad |\hat{\Phi}| = 0, \qquad \bar{\bar{\Phi}} = 0,$$
$$K \equiv (\hat{\Phi}^2)_t = (E \cdot E^*)^2, \qquad L \equiv (\hat{\Phi}\hat{\Phi}^*)_t = |E^2|^2 \quad (3.9)$$
$$M \equiv i(n^\times \cdot \hat{\Phi})_t = in[E \cdot E^*] = S_4$$

These invariants give us the full description of the polarization state [12]. Thus linear polarization occurs under the condition $M = 0$. Moreover, we can find the direction of the electric vector of a wave by choosing any arbitrary vector x noncollinear with the wave normal n, then $e_0 = E/|E| = \hat{\Phi}x/\sqrt{(\hat{\Phi}x)^2}$. Circular polarization occurs when $L = 0$ and hence $M = \pm|E|^2 = \pm I$, where $+ (-)$ corresponds to right (left) circular polarization. In general $M \neq 0$, $L \neq 0$ for elliptic polarization, but the semiaxes of the ellipse of polarization are determined as $a_2 = \frac{1}{2}(I + \sqrt{L})$, $b_2 = \frac{1}{2}(I - \sqrt{L})$, and the right (left) direction of vector rotation corresponds to a positive (negative) sign of the invariant M.

Though the invariants K, L, M were introduced for polarized light, we can use them also for a partially polarized light beam. In this case only their expressions in terms of the LBT Φ are valid, but still they can connect the LBT with the Stokes parameters. These connections are

$$I = S_1, \qquad K = \frac{1}{2}\left(S_1^2 + S_2^2 + S_3^2 + S_4^2\right)$$
$$L = \frac{1}{2}\left(S_1^2 + S_2^2 + S_3^2 - S_4^2\right), \qquad M = S_4 \qquad (3.10)$$

In spite of the fact that the number of invariants is equal to the number of Stokes parameters, it is impossible to express all Stokes parameters directly in terms of I, K, L, M. On the one hand we have already mentioned that if that were possible, it would mean that the Stokes parameters had the property of coordinate independence, which is impossible due to their definition. On the other hand the parameters are not independent of each other, but are connected by relations easily derived from (3.10):

$$2K - M^2 = 2L + M^2 = K + L \qquad (3.11)$$

Let us consider the most general case of partially polarized light. The light can be presented as a superposition of polarized and nonpolarized parts of the beam. That immediately yields the inequality

$$\frac{1}{2}(\hat{\Phi}_t)^2 \leq (\hat{\Phi}^2)_t \leq (\hat{\Phi}_t)^2, \qquad \text{or} \quad \frac{1}{2}I^2 \leq K \leq I^2 \quad (3.12)$$

with equalities achieved only in the two limiting cases of totally polarized light (right equality sign) or nonpolarized light (left equality sign).

It is very useful to represent $\hat{\Phi}$ as a superposition of polarized ($\hat{\Phi}^p$) and nonpolarized ($\hat{\Phi}^n$) parts [12]:

$$\hat{\Phi} = \hat{\Phi}^p + \hat{\Phi}^n \qquad (3.13)$$

Such a decomposition has been proved to be unique [12]. For these parts we have

$$\hat{\Phi}^p = I_p e_0 \otimes e_0^*, \qquad \hat{\Phi}^n = \frac{1}{2}I_n(\hat{1} - n \otimes n) = -\frac{1}{2}I_n(n^\times)^2 \qquad (3.14)$$

where I_p and I_n are the intensities of the polarized and nonpolarized components of the light beam, respectively. Evidently the degree of polarization of a beam, p, can be obtained from these quantities in the following manner [12]:

$$p = \frac{I_p}{I} = \frac{\sqrt{2K - I^2}}{I}, \qquad I = I_p + I_n \qquad (3.15)$$

Using the invariants I and K permits us to obtain with (3.15) a coordinate-free equation for the degree of polarization, independent of the preliminary splitting into polarized and nonpolarized parts. A polarized beam has $p = 1$, and a nonpolarized beam has $p = 0$. The vector e_0 can be found with a linear polarizer. It corresponds to the orientation of the polarizers axis such that its transmittance attains its maximum. Furthermore, I_n is the minimal energy transmitted through the polarizer.

Now we will give a wide prospect for applications of the LBT formalism. Suppose we know the initial LBT $\hat{\Phi}_0$. The beam is incident on some substance with a plane interface. Neglecting nonlinear light–matter interactions, the electric field of a light wave undergoes linear transformations. For example, we consider transmission of light through the air–substance interface. We can interpret the resulting transformation as the action of a linear operator (Fresnel's transmission operator in our case) on the initial electric field vector E_0: $E^T = \hat{T}E_0$. Inasmuch as this linear operation is the same for all partial waves in a beam (the transmission operator is determined by the dielectric and magnetic properties of the medium and a vector of refraction of incident waves), it can be applied directly to the LBT as follows [12]:

$$\hat{\Phi}^t = \sum_j E^{(j)T} \otimes E^{(j)T*} = \sum_j \hat{T}E_0^{(j)} \otimes (\hat{T}E_0^{(j)})^*$$
$$= \sum_j \hat{T}E_0^{(j)} \otimes E_0^{(j)*}\hat{T}^+ = T\hat{\Phi}_0\hat{T}^+ \qquad (3.16)$$

where $\hat{T}^+ = \tilde{\hat{T}}^*$ denotes the Hermitian conjugation of the tensor $\hat{T}$.

In the important particular case of interaction of the beam with an ideal polarizer, we can find the intensity of the beam passed through the device with the help of Eq. (3.9): $I^T = (\hat{\Phi}^T)_t = (\hat{T}\hat{\Phi}_0\hat{T}^+)_t$. Now we take into account that the transmission operator for an ideal polarizer can be written in the form

$$\hat{T} \equiv \hat{P} = \boldsymbol{u}_P \otimes \boldsymbol{u}_p^* \qquad (3.17)$$

where unit-norm vector $\boldsymbol{u}_P$ is the vector of the polarizer ($\boldsymbol{u}_P^* \boldsymbol{u}_P = 1$). We obtain

$$\begin{aligned}
I^T &= (\hat{P} \hat{\Phi}_0 \hat{P}^+)_t = \left(\boldsymbol{u}_P \otimes \boldsymbol{u}_p^* \hat{\Phi}_0 \boldsymbol{u}_P \otimes \boldsymbol{u}_p^* \right)_t \\
&= (\boldsymbol{u}_P^* \boldsymbol{u}_P)(\boldsymbol{u}_p^* \hat{\Phi}_0 \boldsymbol{u}_P) = \boldsymbol{u}_p^* \hat{\Phi}_0 \boldsymbol{u}_P \qquad (3.18)
\end{aligned}$$

The same procedure must be repeated to obtain the signal intensity detected after the analyzer, whose transmission operator is

$$\hat{T} \equiv \hat{A} = \boldsymbol{u}_A \otimes \boldsymbol{u}_A^* \qquad (3.19)$$

Let us finally consider a system consisting of a polarizer, anisotropic medium, and analyzer. Knowing the transmission operator of each component of the system allows us to write down step by step the resulting expression for the detected signal, namely

$$\begin{aligned}
I &= (\hat{\Phi}^A)_t = (\hat{A} \hat{\Phi}^T \hat{A}^+)_t = (\hat{A} \hat{T} \hat{\Phi}^P \hat{T}^+ \hat{A}^+)_t \\
&= (\hat{A} \hat{T} \hat{P} \hat{\Phi}_0 \hat{P}^+ \hat{T}^+ \hat{A}^+)_t = (\boldsymbol{u}_A^* \hat{T} \boldsymbol{u}_P)(\boldsymbol{u}_p^* \hat{\Phi}_0 \boldsymbol{u}_P)(\boldsymbol{u}_P^* \hat{T}^+ \boldsymbol{u}_A) \\
&= \left| (\boldsymbol{u}_A^* \hat{T} \boldsymbol{u}_P) \right|^2 (\boldsymbol{u}_p^* \hat{\Phi}_0 \boldsymbol{u}_P) \qquad (3.20)
\end{aligned}$$

This formula once more emphasizes the advantages of direct methods in the optics and spectroscopy of anisotropic media.

We remark that because $\boldsymbol{n} \cdot \boldsymbol{E}^{(j)} \neq 0$, the phase normal vector $\boldsymbol{n}$ will no longer be an eigenvector of the LBT in anisotropic media, and most of the mentioned consequences for the LBT and its connection with the Stokes parameters will no longer hold.

It was shown in [29] that one can use another definition of the LBT when the magnetic field vector is expanded in partial waves:

$$\hat{\Phi}^H = \sum_j \boldsymbol{H}^{(j)} \otimes \boldsymbol{H}^{(j)*} \qquad (3.21)$$

Such a presentation of the LBT is preferable in the theory of light beam interactions with anisotropic media. This is because for the majority of media studied in spectroscopy we have magnetic permeability equal to 1, which means that though the electric field vector does not lie in the phase plane of the beam, the magnetic field does ($\boldsymbol{n} \cdot \boldsymbol{H}^{(j)} = \boldsymbol{n} \cdot \boldsymbol{H}^{(j)*} = 0$), so all the properties of the LBT are preserved.

4. PROPAGATION OF POLARIZED RADIATION IN AN ANISOTROPIC MEDIUM: EVOLUTION OF PROBE WAVE INTENSITY

The importance of analysis of the propagation of a light wave in an anisotropic medium was recognized long before the invention of lasers. With that development, however, the magnitude of effects accompanying propagation of light beams with high intensity in nonlinear media was considerably increased. Multi beam interaction schemes came to be applied in experimental studies, where beams propagated in media whose properties

were changed by other beams. In the present chapter we will consider different versions of nonlinear spectroscopy of media with LIA within the pump–probe scheme. We will start from the description of evolution of the probe beam intensity — in particular, propagation in a photoanisotropic medium.

Let us define the intensity as

$$I_1 = (\boldsymbol{E}_1 \otimes \boldsymbol{E}_1^\dagger) \qquad (4.1)$$

where the superscript $\dagger$ means Hermitian conjugation. Taking into account Eq. (2.30), we have [30]

$$\begin{aligned}
I_1 = \boldsymbol{E}_{10}^* \big\{ &\hat{G}_+^\dagger \hat{G}_+ \exp \left[2\xi \, \text{Re}(a\lambda_+) \right] \\
&+ \hat{G}_-^\dagger \hat{G}_- \exp \left[2\xi \, \text{Re}(a\lambda_-) \right] \\
&+ \hat{G}_+^\dagger \hat{G}_- \exp \left[(a^* \lambda_+^* + a\lambda_-)\xi \right] \\
&+ \hat{G}_-^\dagger \hat{G}_+ \exp \left[(a^* \lambda_-^* + a\lambda_+)\xi \right] \big\} \boldsymbol{E}_{10}
\end{aligned} \qquad (4.2)$$

This expression is rather complicated and in the general case leads to cumbersome analytical expressions. In the case of Hermiticity of $\hat{S}_\perp$ it is greatly simplified and can be reduced to the form

$$I_1 = \boldsymbol{E}_{10}^* \big\{ \hat{G}_+ \exp[2\lambda_+ \xi \, \text{Re} \, a] + \hat{G}_- \exp[2\lambda_- \xi \, \text{Re} \, a] \big\} \boldsymbol{E}_{10} \qquad (4.3)$$

Moreover, supposing weak enough excitation and expanding the exponents in a series up to the linear terms in $\text{Re} \, a$, we obtain

$$I_1 = I_{10} + 2\xi \, \text{Re} \, a \big\{ \lambda_+ \boldsymbol{E}_{10}^* \hat{G}_+ \boldsymbol{E}_{10} + \lambda_- \boldsymbol{E}_{10}^* \hat{G}_- \boldsymbol{E}_{10} \big\} \qquad (4.4)$$

where $I_{10} = (\boldsymbol{E}_{10} \otimes \boldsymbol{E}_{10}^\dagger)_t$.

We consider now the most frequent particular polarization geometry in experiment, when

(i) a probe wave is linearly polarized at angle $\theta_{10} = 45°$ to the major axis of the polarization ellipse of radiation in collinear pumping, or

(ii) it is circularly polarized.

Such conditions allow us to take into account the attenuation of exciting light owing to absorption and provide us in case (i) with following expression for I_1:

$$\begin{aligned}
I_1 &= \exp(-\sigma_1 \xi) I_{10} \big\{ 1 + (\lambda_+ + \lambda_-) f(\xi) \, \text{Re} \, a \big\}, \\
f(\xi) &= \frac{1 - \exp(-\sigma_0 \xi)}{\sigma_0}
\end{aligned} \qquad (4.5)$$

We conclude that in this case the intensity I_1 does not depend on the ellipticity of radiation of the pump wave. On the other hand, the relation obtained can be useful for investigating the dispersion $\text{Im} \, \chi_{43}$.

In case (ii) the required intensity can be written as

$$\begin{aligned}
I_1 = \exp(-\sigma_1 z) \\
\times \big\{ 1 + [2 + C_1 + C_2 + (C_1 - C_2) \sin 2\varepsilon_{\text{pmp}}] f(z) \, \text{Re} \, a \big\}
\end{aligned} \qquad (4.6)$$

This expression relates I_1 to ε_{pmp} and reduces to (4.5) for linearly polarized pumping or $C_1 = C_2$.

Yet another method of calculating I_1 is given in our paper [30].

5. SATURATION SPECTROSCOPY

In this section we extend results obtained earlier [31] to the case of arbitrary polarization of interacting waves. The observed signal in nonlinear spectroscopy (NS) can be expressed in the following Hermitian quadratic form [30]:

$$\Delta I_{\mathrm{NS}} = \mathbf{u}^*(\mathbf{E}_1 \otimes \mathbf{E}_1^*)\mathbf{u} \tag{5.1}$$

where $\mathbf{u}$ is a unit vector in the phase plane of a probe wave ($\mathbf{u} \cdot \mathbf{n}_1 = 0$). Here $\mathbf{u}$ can be considered representing an ideal general elliptical polarizer ($\mathbf{u}_P$) or analyzer ($\mathbf{u}_A$). In saturation spectroscopy we have ($\mathbf{u}_A^* \cdot \mathbf{u}_P = 1$) ($\mathbf{u}_P = \varepsilon_{10}$), so that within the framework of linearity in the parameter a, from (2.25) and (5.1) we obtain

$$\Delta I_{\mathrm{SS}} = I_{10}\{1 + 2\xi \operatorname{Re}[a\mathbf{u}_A^*\hat{S}\mathbf{u}_P]\} \tag{5.2}$$

$$\mathbf{u}_A^*\hat{S}\mathbf{u}_P = \mathbf{u}_A\hat{S}_\perp\mathbf{u}_P = 1 + C_1|\mathbf{u}_P\varepsilon_0^*|^2 + C_2|\mathbf{u}_P\varepsilon_0|^2 \tag{5.3}$$

The formulas (5.2) and (5.3) demonstrate that ΔI_{SS} strongly depends on the polarization of a probe beam as determined by a polarizer ($\mathbf{u}_P$) and the polarization state of the pump radiation. For parallel pump and probe waves and Hermitian $\hat{S}_0$ the Eq. (5.2) can be rewritten as

$$\Delta I_{\mathrm{SS}} = I_{10}\exp(-\sigma_1\xi)\{1 + 2f(\xi)\mathbf{u}_A^*\hat{S}\mathbf{u}_P\operatorname{Re}a\} \tag{5.4}$$

Moreover, under the same polarization conditions (5.4) reduces to (4.5) and (4.6).

6. NONLINEAR POLARIZATION SPECTROSCOPY

We begin this section with some terminological explanations. There are several often encountered terms for the version of nonlinear spectroscopy considered: here polarization spectroscopy, polarization modulation spectroscopy (Akhmanov and Koroteev [8]), and nonlinear polarization spectroscopy. Herein we use the last mentioned.

In NPS we have $\mathbf{u}_A^* \cdot \mathbf{u}_P = 0$. Then for the detected signal value, up to a real scalar factor, we have

$$\Delta I_{\mathrm{NPS}} \sim \left|C_1(\mathbf{u}_A^* \cdot \varepsilon_0)(\mathbf{u}_P \cdot \varepsilon_0^*) + C_2(\mathbf{u}_A \cdot \varepsilon_0^*)(\mathbf{u}_P \cdot \varepsilon_0)\right|^2 \tag{6.1}$$

The expression (5.4) allows us to analyze the polarization dependence of a NPS signal in the general case of elliptically polarized interacting waves. Usually in NPS one uses an elliptically polarized pump wave and a probe beam linearly polarized at an angle $\theta_{10} = \pi/4$ to the semimajor axis of the pump-wave ellipse [32]. Nevertheless, in Section 8.5 it will be shown that the selection $\theta_{10} = \pi/4$ is not always optimal. Hence it is interesting to obtain the expression for ΔI_{NPS} with arbitrary θ_{10}. Let us assume the X-axis of the Cartesian coordinate system to be along the major axis of the pump-wave ellipse. Then the polarization dependence ΔI_{NPS} can be described by the following function:

$$F = \cos^2 2\varepsilon_{\mathrm{pmp}}\sin^2 2\theta_{10} + |q|^2\sin^2 2\varepsilon_{\mathrm{pmp}} - \operatorname{Im}q\sin A\varepsilon_{\mathrm{pmp}}\sin 2\theta_{10} \tag{6.2}$$

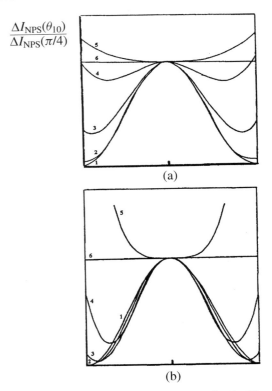

$$\frac{\Delta I_{\mathrm{NPS}}(\theta_{10})}{\Delta I_{\mathrm{NPS}}(\pi/4)}$$

(a)

(b)

Fig. 1. Dependence of registered NPS signal on the azimuth of linear polarization of the probe wave for: (a) benzene ($\operatorname{Re}q = -0.75$, $\operatorname{Im}q = 0.40$); (b) CS$_2$ ($\operatorname{Re}q = -0.16$, $\operatorname{Im}q = 0.32$). The angle of the ellipticity of the pumping wave is taken as parameter: $\varepsilon_{\mathrm{pmp}} = 0$ (curve 1); $\pi/20$ (2); $\pi/10$ (3); $3\pi/20$ (4); $\pi/5$ (5); $\pi/4$ (6).

It is seen from (6.2) that the function F is nonmonotonic in θ_{10}. Minimum values of F (and therefore of ΔI_{NPS}) are reached at values of the probe wave azimuth θ_{10} determined by

$$\theta_{10} = \theta_{10\mathrm{min}} = \begin{cases} \frac{1}{2}\arcsin(\operatorname{Im}q\tan 2\varepsilon_{\mathrm{pmp}}) \\ \pi/2 - \frac{1}{2}\arcsin(\operatorname{Im}q\tan 2\varepsilon_{\mathrm{pmp}}) \end{cases} \tag{6.3}$$

The corresponding values of a NPS signal are equal to

$$\frac{\Delta I_{\mathrm{NPS}}^{\min}}{\Delta \hat{I}} = (\operatorname{Re}q\sin 2\varepsilon_H)^2 \tag{6.4}$$

where $\Delta I_{\mathrm{NPS}} = \Delta\hat{I}$ when $\varepsilon_0 = \varepsilon_0^* = \mathbf{u}_A = \mathbf{u}_P$. It is interesting to remark that from (6.3) for $\varepsilon_{\mathrm{pmp}} = \pi/8$ we have

$$\sin 2\theta_{10\mathrm{min}} = \operatorname{Im}q \tag{6.5}$$

The graphs of polarization dependences for benzene (molecular oscillation 992 cm^{-1}) and CS$_2$ (Roman line 655.7 cm^{-1}) are shown in Figure 1a and b, in comparison with the above ΔI_{NPS} analysis. The values for the imaginary and real parts of the parameter q for benzene are taken from the paper [33], and for CS$_2$ are calculated on the basis of experimental data [34]. The ellipticity of the pump wave was taken as parameter. It is important to emphasize the following:

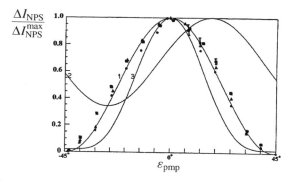

Fig. 2. Computational dependences $\Delta I_{NPS}/\Delta I_{NPS}^{max}\varepsilon_{pmp}$ when $\theta_{10} = \pi/4$, curve 1, $q = 0$; curve 2, $\mathrm{Re}\,q = -0.75$, $\mathrm{Im}\,q = 0.40$; curve 3, from [32]. Circles, triangles, and small squares show experimental values for MB, CV, and RB, respectively. The bars indicate typical errors of measurements.

- the angle $|\theta_{10}| = \pi/4$ is not optimum in the sense of obtaining maximum signal when the polarization of the pump wave is close to circular;
- the shape of the plot of ΔI_{NPS} essentially depends on $\mathrm{sign}(\mathrm{Im}\,q)$ (see Figure 1a and b, which differ only in the sign of $\mathrm{Im}\,q$).

We proceed now to the analysis of the cases $\theta_{10} = \pm\pi/4$. Then (6.2) can rewritten as follows:

$$F = \cos^2 2\varepsilon_{pmp} + |q|^2 \mp \mathrm{Im}\,q \sin 4\varepsilon_{pmp} \quad (6.6)$$

This expression differs significantly from Eq. (15) of [32], in which the authors derive another expression within the same approach. From (6.6) it is easy to see that the behavior of ΔI_{NPS} while scanning the elliptically of pumping is necessarily non-monotonic. Its extreme values are reached at

$$\mathrm{Im}\,q > 0: \quad \varepsilon_{pmp}^{max} = -\frac{1}{4}\arctan\frac{2\,\mathrm{Im}\,q}{1 - |q|^2} \quad \varepsilon_{pmp}^{min} = \varepsilon_{pmp}^{max} + \frac{\pi}{4}$$

$$\mathrm{Im}\,q < 0: \quad \varepsilon_{pmp}^{max} = \frac{1}{4}\arctan\frac{2\,\mathrm{Im}\,q}{1 - |q|^2} \quad \varepsilon_{pmp}^{min} = \varepsilon_{pmp}^{max} - \frac{\pi}{4}$$

$$(6.7)$$

Therefore, in the case of Hermitian $\hat{S}$ the maximum signal will be observed with linearly polarized pumping, and the minimum when the exciting radiation is circularly polarized.

It is important to point out that the measurement of $\varepsilon_{pmp}^{max(min)}$ allows one to determine the real and imaginary parts of q, which are spectroscopically informative quantities [21]. Another important feature of such measurements is that the measured parameters are independent of the pump power.

In Figure 2 the results of our experiment [19] on aqueous solutions of three dyes [crystal violet (CV), methyline blue (MB), and rhodamine-B (RB)] are shown, which are in good accord with the theoretical dependence (6.6) (curve 1) and are in obvious contrast with the theoretical (curve 3) and experimental data on an aqueous solution of a dye (malachite green (MG)) from [32]. Curve 2 corresponds to characteristic parameters of the 992 cm^{-1} line of benzole, so that $\varepsilon_{pmp}^{max} \cong 18°$ and $\varepsilon_{pmp}^{min} \cong 27°$. It is necessary to point out that curve 1 is constructed on the assumption $q = 0$, which (taking into consideration symmetry

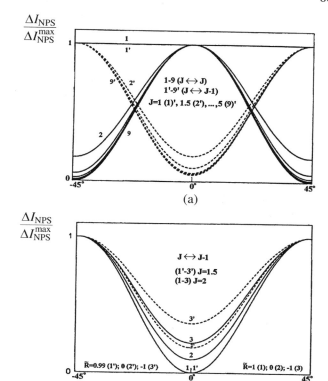

Fig. 3. Theoretical dependences of the normalized value of the NPS signal on the angle of ellipticity of the pump wave for ensembles of two-level particles. See text.

of tensor index permutations) leads to equality of three components of the cubic susceptibility tensor: $\chi_{12} = \chi_{44} = \chi_{43}$. Thus, there is strong evidence of Kleinman symmetry of the resonant tensor $\hat{\chi}^{(3)}$ in aqueous solutions of the dyes CV, MB, RB. The small values for $\mathrm{Re}\,q$ and $\mathrm{Im}\,q$ obtained in [32] hardly can be a basis for inferring violation of that symmetry in aqueous solutions of MG (and other dyes as reported in [32]). Moreover, the errors found in the fitting of theoretical to experimental values $\mathrm{Re}\,q$ and $\mathrm{Im}\,q$ are not indicated.

We now address atomic systems. The essential dependencies are displayed in Figure 3a, where effects of interatomic collisions and radiation trapping in an ensemble of two-level particles are neglected, and the decay rates for the levels are supposed to be approximately equal. The curves in Figure 3b are free from the limitation imposed by the last supposition, and parameter $\tilde{R}$ is determined as

$$\tilde{R} = \frac{\Delta J(\gamma_b - \gamma_a)}{|\Delta J|(\gamma_a + \gamma_b)} \quad (6.8)$$

It is easy to see from Figure 3 that the shape of the curves of ΔI_{NPS} is characterized by strong J-dependence for small J, which can be utilized for identification of unknown transitions as well as in the analysis of three-level cascade systems [30]. In [30] the corresponding analysis was made for the following models: (i) $J = 0 \rightarrow J = 1 \rightarrow J = 0$; (ii) neon

atom [$1s_4(J = 1) \rightarrow 2p_4(J = 2) \rightarrow 3s_2(J = 1)$], and
(iii) $J - 1 \rightarrow J \rightarrow J + 1$ or $J + 1 \rightarrow J \rightarrow J - 1$. The
values of q are 1, 15/7, and -5 respectively. For model (i)
$\Delta I_{NPS}(\varepsilon_{pmp}) = $ const was obtained, as well as for the following cascade systems: $J + 1 \rightarrow J \rightarrow J$ and $J \rightarrow J \rightarrow J + 1$
for $J = 3$.

We now give explicit forms of the function F for other polarization conditions. If the probe wave is circularly polarized,
then

$$F = \cos^2 2\varepsilon_{pmp} \qquad (6.9)$$

i.e., in media with Kleinman symmetry of the $\hat{\chi}^{(3)}$ components,
the expressions (6.2) and (6.9) coincide. If the probe wave is
elliptically polarized with the semimajor axis of the ellipse of
polarization along the X-axis, and the strong wave has linear
polarization, then

$$F \neq f(\varepsilon_{pmp}), \qquad e_0 \cdot e_x = \frac{\sqrt{2}}{2}$$
$$F = \sin^2 2\varepsilon_{PO}, \qquad e_0 \parallel e_x \quad \text{or} \quad e_0 \parallel e_y \qquad (6.10)$$

Here ε_{PO} is the angle of ellipticity of ε_{10}. Let us remark that in
the previously examined cases with circularly and elliptically
polarized probe radiation, the use of circular and elliptical analyzers, respectively, was assumed.

Finally, we take a look at the dependence of a NPS signal
on ξ in collinear geometry of the interacting waves. The dependence can be described by the following function:

$$F_\xi = \exp(-\sigma_1 \xi)\left[1 - \exp(-\sigma_0 \xi)\right]^2 \qquad (6.11)$$

From (6.11) it follows that the maximum NPS signal will be
observed at

$$\xi_{max} = \frac{\ln[(2\sigma_0 + \sigma_1)/\sigma_1]}{\sigma_0} \qquad (6.12)$$

In NPS the frequency detuning of interacting waves as a rule
is small, so that the corresponding optimum optical density can
be estimated as $D_{opt} = z_{max}\sigma_0 \cong \ln 3 \cong 1.1$. In [32] D_{opt} was
found to be 0.48. In a series of experimental works on NPS of
dye solutions [35, 19, 36] the samples under study had D a little
higher than D_{opt} from [32].

6.1. Wave Operator Formalism

Let us consider in detail the application of Fresnel's wave operator formalism to the problem of nonlinear polarization spectroscopy. The principles of this formalism have been given in a
number of papers (see, for example, [37–39] and the literature
cited therein). A surface impedance operator $\hat{\gamma}$ (this is a linear
operator of matrix form, so it may be considered as a tensor)
is one of the fundamental concepts of this theory. The operator
$\hat{\gamma}$ generalizes a scalar surface impedance that has been widely
known in optics and radio-wave theory for many years [40–42].
Another important concept is a normal refraction operator $\hat{N}$,
which generalizes the refractive index operator [43, 44]. This
operator (or tensor, as we showed earlier) describes the space

evolution of the field vector amplitudes of an electromagnetic
wave during propagation in anisotropic and gyrotropic media.
The surface impedance and normal refraction operators are very
useful in the theory of electromagnetic or elastic wave propagation in stratified anisotropic media [45, 46]. The boundary value
problem in such media can be rigorously formulated in terms
of the operators $\hat{\gamma}$ and $\hat{N}$, which depend on the characteristics
of the incident waves and the properties of the corresponding
media. Thus, the operator method of solving various boundary
value problem, including reflection and transmission of waves
at a single plane interface, may be applied to calculations where
high precision is needed.

Let the first medium, from which the plane harmonic wave
(of frequency ω)

$$E(r, t) = E_i \exp\left[i(km \cdot r - \omega t)\right] \qquad (6.13)$$

is obliquely incident onto the second one, be isotropic, and the
other medium be anisotropic and described by the permittivity
tensor $\hat{\varepsilon}$ and permeability tensor $\hat{\mu}$. Here, E_i is the complex
vector amplitude of the electric field strength, $k = \omega/c$, c is the
velocity of the light in vacuum, $m = b + m_n q$ is the refraction
vector [37] with tangential component b and normal component
$m_n q$, and q is the unit vector normal to the interface. The refraction vector m is connected with the wave normal n through an
index of refraction n:

$$m = nn \qquad (6.14)$$

It is convenient, especially in nonmagnetic media, to use the
vector of magnetic field strength H and its tangential component H_t (relative to the interface) [37–39]. It is well known that
for the plane waves

$$H = m \times E = m^\times E \qquad (6.15)$$

where $m \times E$ denotes the vector product of m and E, and $m^\times$ is
the antisymmetric second-rank tensor dual to the vector m [12].

The projection operator $\hat{G}$ is applied to project vectors on
the tangential plane:

$$H_t = \hat{G}H \qquad (6.16)$$

With the help of the dyadic $q \otimes q$ and unit tensor $\hat{I}$ one can
easily find that $\hat{G} = \hat{I} - q \times q$.

We follow the intrinsic notation [12], meaning in (6.16) con-
traction of tensor $\hat{G}$ with vector H: $(H_t)_i = \sum_{j=1}^{3} G_{ij} H_j$. The
direct manipulation with tensors and vectors as in (6.16) sim-
plifies the final expressions and provides results of great generality, eliminating the use of any coordinate system. Moreover,
the results obtained are suitable for computer use.

One can find the complex vector amplitudes of the reflected
waves (H_t^r) and transmitted waves (H_t^d) as shown in [37–39]:

$$H_t^r = \hat{r}H_t, \qquad H_t^d = \hat{d}H_t, \qquad \hat{r} + \hat{G} = \hat{d} \qquad (6.17)$$

where $\hat{r}$ and $\hat{d}$ are Fresnel's reflection and transmission operators, respectively. These operators are expressed by means
of the surface impedance operators for incident ($\hat{\gamma}^i$), reflected

$(\hat{\gamma}^r)$, and transmitted $(\hat{\gamma}^d)$ radiation as follows:

$$\hat{r} = (\hat{\gamma}^d - \hat{\gamma}^r)^- (\hat{\gamma}^i - \hat{\gamma}^d), \qquad \hat{d} = (\hat{\gamma}^d - \hat{\gamma}^r)^- (\hat{\gamma}^i - \hat{\gamma}^r) \quad (6.18)$$

where $(\hat{\gamma}^d - \hat{\gamma}^r)^-$ is a pseudo-inverse operator ("pseudo" because Fresnel's operators $\hat{r}$ and $\hat{d}$ are planar tensors, acting in the two-dimensional subspace, of the plane interface, so the ordinary inverse for these operators is not defined). The pseudo-inverse operator $\hat{\alpha}_t^-$ can be found through the algorithm presented in [37–39]:

$$\hat{\alpha}_t^- = \frac{\hat{\alpha}_t \hat{G} - \hat{\alpha}}{\bar{\hat{\alpha}}_t} \quad (6.19)$$

where $\bar{\hat{\alpha}}_t$ denotes the trace of the adjoint tensor [12]. The surface impedance tensor $\hat{\gamma}$ is introduced as a linear operator transforming the vector $\boldsymbol{H}_t$ into the vector $\boldsymbol{q} \times \boldsymbol{E}$:

$$\boldsymbol{q} \times \boldsymbol{E} = \hat{\gamma} \hat{\boldsymbol{H}}_t \quad (6.20)$$

These vectors lie at the plane interface. It has been shown [38, 39] that the surface impedance satisfies the Riccati tensor equation

$$\hat{\gamma} \hat{B} \hat{\gamma} + \hat{\gamma} \hat{A} - \hat{D} \hat{\gamma} - \hat{C} = 0 \quad (6.21)$$

where the tensor coefficients are

$$\begin{aligned} \hat{A} &= \frac{1}{\varepsilon_q} \boldsymbol{q}^\times \hat{\varepsilon} \boldsymbol{q} \otimes \boldsymbol{a} - \frac{1}{\mu_q} \boldsymbol{b} \otimes \boldsymbol{q} \hat{\mu} \hat{G} \\ \hat{B} &= \frac{1}{\varepsilon_q} \hat{G} \bar{\tilde{\varepsilon}} \hat{G} - \frac{1}{\mu_q} \boldsymbol{b} \otimes \boldsymbol{b} \\ \hat{C} &= -\frac{1}{\varepsilon_q} \boldsymbol{a} \otimes \boldsymbol{a} - \frac{1}{\mu_q} \boldsymbol{q}^\times \bar{\tilde{\mu}} \boldsymbol{q}^\times \\ \hat{D} &= -\frac{1}{\varepsilon_q} \boldsymbol{a} \otimes \boldsymbol{q} \hat{\varepsilon} \boldsymbol{q}^\times - \frac{1}{\mu_q} \hat{G} \hat{\mu} \boldsymbol{q} \otimes \boldsymbol{b} \\ \varepsilon_q &= \boldsymbol{q} \hat{\varepsilon} \boldsymbol{q}, \qquad \mu_q = \boldsymbol{q} \hat{\mu} \boldsymbol{q} \end{aligned} \quad (6.22)$$

in which a tilde denotes the transposed operator.

For a harmonic wave (6.13) in an isotropic medium the surface impedance is given by [39]

$$\hat{\gamma}^i = -\hat{\gamma}^r = \frac{\mu_i \hat{G} - \frac{b^2}{\varepsilon_i} \hat{\tau}_a}{m_n}$$

where the dyadic $\hat{\tau}_a = \boldsymbol{a}_0 \otimes \boldsymbol{a}_0$ is a projective operator in the direction orthogonal to the plane of incidence, $\boldsymbol{a}_0 = \boldsymbol{b}_0 \times \boldsymbol{q}$, $b = |\boldsymbol{b}|$, and $\boldsymbol{b}_0 = \boldsymbol{b}/b$. In an anisotropic medium two partial waves are excited [47]. They have different polarizations and refraction vectors $\boldsymbol{m}_\mp = \boldsymbol{b} + m_n^\mp \boldsymbol{q}$. Therefore, an operator $\hat{\gamma}$ (6.20) connects the vectors $\boldsymbol{H}_t$ and $\boldsymbol{q} \times \boldsymbol{E}$, which characterize a superposition of harmonic fields of both refracted waves. In the particular case when the crystal is nonmagnetic ($\hat{\mu} = \hat{G}$) and is cut in such a way that $\boldsymbol{q}$ is an eigenvector of the $\hat{\varepsilon}$ ($\boldsymbol{q}\hat{\varepsilon} = \hat{\varepsilon}\boldsymbol{q} = \varepsilon_q \boldsymbol{q}$), the tensor $\hat{\gamma}_d$ is expressed in the following form:

$$\begin{aligned} \hat{\gamma}_d = (m_n^+ + m_n^-)^{-1} \Big[&\hat{G} - m_n^+ m_n^- \frac{\boldsymbol{q}^\times \hat{\varepsilon}^{-1} \boldsymbol{q}^\times}{1 - \boldsymbol{a}\hat{\varepsilon}^{-1}\boldsymbol{a}} \\ &- \Big(\frac{1}{\varepsilon_q} + \frac{\varepsilon_q m_n^+ m_n^-}{|\hat{\varepsilon}|} - \boldsymbol{a}\bar{\hat{\varepsilon}}\boldsymbol{a} \Big) \hat{\tau}_a \Big] \end{aligned} \quad (6.23)$$

where $|\hat{\varepsilon}|$ is the determinant of $\hat{\varepsilon}$. The total field of refracted waves,

$$\boldsymbol{H}_t^d(\boldsymbol{r}, t) = \boldsymbol{H}_t^d(0) \exp\big[i(k\boldsymbol{b} \cdot \boldsymbol{r} - \omega t)\big] \exp(ikz\hat{N}) \quad (6.24)$$

contains an exponential operator

$$\exp\big(ikz\hat{N}\big)^k = \sum_{k=0}^\infty \big[(ikz)^k/k!\big]\hat{N}^k$$

where $z = \boldsymbol{q} \cdot \boldsymbol{r}$, and

$$\hat{N} = \hat{A} + \hat{B}\hat{\gamma} \quad (6.25)$$

The eigenvalues and eigenvectors of the normal refraction operator (6.25) yield the normal components $m_n^\pm$ of the refraction vector and the polarization states of harmonic partial refracted waves. The $m_n^\pm$ are the solutions of the quartic algebraic equation

$$m_n^4 - \hat{P}_t m_n^3 + (\bar{\hat{P}}_t - \hat{Q}_t)m_n^2 + (\hat{P}_t \hat{Q}_t - (\hat{P}\hat{Q})_t)m_n + \bar{\hat{Q}}_t = 0 \quad (6.26)$$

or one can find them by solving the equation of normals [12]

$$(\boldsymbol{m}\overline{\hat{\varepsilon}^{-1}\boldsymbol{m}})(\boldsymbol{m}\overline{\hat{\mu}^{-1}\boldsymbol{m}}) + (\hat{\varepsilon}^{-1}\boldsymbol{m}^\times \hat{\mu}^{-1}\boldsymbol{m}^\times)_t + 1 = 0 \quad (6.27)$$

The tensor coefficients $\hat{P}$ and $\hat{Q}$ in (6.25) are derived from $\hat{A}$, $\hat{B}$, $\hat{C}$, $\hat{D}$:

$$\hat{P} = \hat{A} + \hat{B}\hat{D}\hat{B}^-, \qquad \hat{Q} = \hat{B}(\hat{C} - \hat{D}\hat{B}^-\hat{A}) \quad (6.28)$$

Some complicated cases with degenerate tensor $\hat{N}$, where refracted waves with linear, quadratic, and cubic dependence on coordinates appear, are considered in detail in the papers [47, 48].

6.2. Reflection from Media with Light-Induced Anisotropy

Let us consider an initially isotropic nonlinear medium in which uniaxial anisotropy is induced by a powerful plane monochromatic pump wave normally incident from vacuum. If the pump wave is linearly polarized, then the dielectric permittivity tensor $\hat{\varepsilon}$ may be written in the form [49]

$$\hat{\varepsilon} = \varepsilon_0\big[(1 + \chi_0 + \chi_1)\boldsymbol{I} + \chi_1 C \boldsymbol{c} \otimes \boldsymbol{c}\big] \quad (6.29)$$

where ε_0 is the dielectric constant, χ_0 is the linear scalar susceptibility, $\chi_1 = \chi_{1221}$, $C = (\chi_{1122} + \chi_{1212})/\chi_1$, χ_{ijkl} are the components of the fourth-rank tensor of the third-order nonlinear susceptibility, and $\boldsymbol{c}$ is the unit vector along the induced optical axis ($\boldsymbol{c}$ is also the polarization vector of the pump field). Since the incident pump wave normal is perpendicular to the plane interface, the optical axis of the MLIA is in the plane interface. Therefore, one can write

$$\boldsymbol{c} = \cos\varphi\, \boldsymbol{a}_0 + \sin\varphi\, \boldsymbol{b}_0 \quad (6.30)$$

where φ is the azimuth measured from the direction of $\boldsymbol{a}_0$, which is perpendicular to the incidence plane. We point out that the substitution of $\boldsymbol{c}$ (6.30) into (6.29) yields

$$\hat{\varepsilon}\boldsymbol{q} = \boldsymbol{q}\hat{\varepsilon} = \varepsilon_q\boldsymbol{q} \quad (6.31)$$

i.e., q is the eigenvector of the dielectric permittivity tensor $\hat{\varepsilon}$ (6.29) with eigenvalue,

$$\varepsilon_q = \varepsilon_0(1 + \chi_0 + \chi_1) \qquad (6.32)$$

Therefore, we may use the exact analytical expression (6.23) for the surface impedance tensor, substituting in it the following formula for the permittivity tensor $\hat{\varepsilon}$ (6.29):

$$\hat{\varepsilon} = \varepsilon_q + \varepsilon_1\hat{\tau}_c \qquad (6.33)$$

where $\varepsilon_1 = \varepsilon_0\chi_1 C$, $\hat{\tau}_c$ denotes the projective dyadic operator $c \otimes c$, and we imply that the scalar ε_q is multiplied by a unit tensor $\hat{G}$, which from now we shall drop. A simple but cumbersome calculation gives us an expression for the impedance tensor of the MLIA:

$$\hat{\gamma}^d = \frac{\hat{X}}{\varepsilon_q(m_n^+ + m_n^-)} + \frac{m_n^+ m_n^-}{m_n^+ + m_n^-}\frac{\hat{X} + \varepsilon_1\hat{\tau}_c}{F} \qquad (6.34)$$

where the tensor $\hat{X}$ is

$$\hat{X} = \varepsilon_q\hat{\tau}_b + m_n^2\hat{\tau}_a, \qquad \hat{\tau}_b = \boldsymbol{b}_0 \otimes \boldsymbol{b}_0 \qquad (6.35)$$

and by F we denote

$$F = \varepsilon_q(\varepsilon_q + \varepsilon_1) - b^2(\varepsilon_q + \varepsilon_1\sin^2\varphi) \qquad$$

The eigenvalues $m_n^\pm$ of the normal refraction tensor are derived from (6.26): $m_n^- = \sqrt{\varepsilon_q}\cos\theta$ for the ordinary wave, and $m_n^\pm = \{\varepsilon_q - b^2 + \varepsilon_1[1 - (b^2\sin\varphi)/\varepsilon_q]\}^{1/2}$ for the extraordinary wave. The angle of refraction of the ordinary wave, θ_0, is simply derived from Snell's law:

$$\frac{\sin\theta_i}{\sin\theta_0} = \sqrt{\varepsilon_q} \qquad (6.36)$$

Further transformations are connected with the small magnitude of light-induced anisotropy. It means that $|\varepsilon_1| \ll |\varepsilon_q|$ and we may consider ε_1 as a small parameter for expansion of m_n^+ in a series. Neglecting terms higher than linear in ε_1, we obtain

$$m_n^+ \approx m_n^-\left[1 + \frac{\varepsilon_1(1 - \sin^2\varphi\sin^2\theta_0)}{2(m_n^-)^2}\right] \qquad (6.37)$$

Substitution of (6.37) into (6.34) gives the tensor of surface impedances in the form

$$\hat{\gamma}^d = \left[\frac{1}{m_n^- + \Gamma_b}\right]\hat{\tau}_b + \left[\frac{m_n^-}{\varepsilon_q + \Gamma_a}\right]\hat{\tau}_a + \lambda(\hat{\tau}_b - \hat{\tau}_a)\boldsymbol{q}^\times \qquad (6.38)$$

Here we introduce the following notation:

$$\Gamma_a = -\frac{\varepsilon_1\sin^2\varphi\cos^2\theta_i}{2\varepsilon_q m_n^-}$$

$$\Gamma_b = \frac{\varepsilon_1}{2\varepsilon_q(m_n^-)^3}\left((m_n^-)^2\sin^2\varphi + \varepsilon_q b^2\sin^2\varphi - \varepsilon_q\right) \qquad (6.39)$$

$$\lambda = \frac{\varepsilon_1\sin 2\varphi}{4\varepsilon_q m_n^-}$$

It is clear that $m_n^-/\varepsilon_q \gg \Gamma_a$, and $1/m_n^- \gg \Gamma_b$, Γ_a, Γ_b, λ; thus Γ_a, Γ_b, λ are quantities of first order in the small parameter

ε_1, and they give small additions to the expression for the tensor $\hat{\gamma}$ in the case of an isotropic dielectric:

$$\hat{\gamma}^d = \frac{1}{m_n^-}\hat{\tau}_b + \frac{m_n^-}{\varepsilon_q}\hat{\tau}_a \qquad (6.40)$$

This expression follows from (6.34) for the case of an isotropic medium. In vacuum or, with high accuracy, in air ($\hat{\varepsilon}_i = \hat{G}$, $\hat{\mu}_i = \hat{G}$, $n = 1$), the refraction vector $\boldsymbol{m}$ and wave normal $\boldsymbol{n}$ coincide. Let us rewrite the relation (6.38) on that assumption ($m_n = \cos\theta_i$, $b = \sin\theta_1$):

$$\hat{\gamma} = -\hat{\gamma}^r = m_n\hat{\tau}_a + \frac{1}{m_n}\hat{\tau}_b \qquad (6.41)$$

Now we are ready to derive Fresnel's reflection and transmission operators. Using the approximate formulae (6.38) and (6.41), it is easy to obtain from (6.18) tensors of reflection and transmission, generalizing the usual Fresnel coefficients. One has

$$\hat{d} = \hat{d}_0 + \hat{d}_1, \qquad \hat{r} = \hat{r}_0 + \hat{r}_1 \qquad (6.42)$$

where the norms of the tensors $\hat{d}_0$ and $\hat{r}_0$ are much greater than those of the tensors $\hat{d}_1$ and $\hat{r}_1$: $\|\hat{d}_1\| \ll \|\hat{d}_0\|$, $\|\hat{r}_1\| \ll \|\hat{r}_0\|$. The zero-order approximations for $\hat{d}_0$ and $\hat{r}_0$ can be written in the following form:

$$\hat{d}_0 = a_{0a}\hat{\tau}_a + d_{0b}\hat{\tau}_b = \frac{2m_n}{V}\hat{\tau}_a + \frac{2}{m_n\Lambda}\hat{\tau}_b$$
$$= \frac{2\sqrt{\varepsilon_q}\cos\theta_i}{\sqrt{\varepsilon_q}\cos\theta_i + \cos\theta_0}\hat{\tau}_a + \frac{2\sqrt{\varepsilon_q}\cos\theta_0}{\cos\theta_i + \sqrt{\varepsilon_q}\cos\theta_0}\hat{\tau}_b \qquad (6.43)$$

$$\hat{r}_0 = r_{0a}\hat{\tau}_a + r_{0b}\hat{\tau}_b = \left(\frac{2m_n}{V} - 1\right)\hat{\tau}_a + \left(\frac{2}{m_n\Lambda} - 1\right)\hat{\tau}_b$$
$$= \frac{\sqrt{\varepsilon_q}\cos\theta_i - \cos\theta_0}{\sqrt{\varepsilon_q}\cos\theta_i + \cos\theta_0}\hat{\tau}_a + \frac{\sqrt{\varepsilon_q}\cos\theta_0 - \cos\theta_i}{\sqrt{\varepsilon_q}\cos\theta_0 + \cos\theta_i}\hat{\tau}_b \qquad (6.44)$$

$$\Lambda = \frac{1}{m_n} + \frac{1}{m_n^-}, \qquad V = m_n + \frac{m_n^-}{\varepsilon_q} \qquad (6.45)$$

Direct comparison shows that the tensors $\hat{r}_0$ and $\hat{d}_0$ have two eigenvectors $\boldsymbol{a}_0$, $\boldsymbol{b}_0$ and two nonzero eigenvalues—the coefficients of the corresponding dyadic projections. These eigenvalues exactly coincide with Fresnel's reflection and transmission coefficients (see, for example, [50]). Thus, the operators $\hat{d}_0$ (6.43) and $\hat{r}_0$ (6.44) give us appropriate coefficients for reflection and transmission of TE or TM plane monochromatic waves at the isotropic media interface. One can obtain by direct calculation the equality

$$\hat{r}_0 + \hat{G} = \hat{d}_0 \qquad (6.46)$$

which means that the tangential components of the magnetic field strength vector are continuous across the boundaries in the main approximation. Taking into account the second equality in (6.17), we can write

$$\hat{r}_1 = \hat{d}_1 \qquad (6.47)$$

so it is sufficient to find the first-order approximation for the reflection operator. Further calculation yields the formula

$$\hat{r}_1 = \frac{2}{\Lambda V}\left[-\frac{m_n \Lambda}{V}\Gamma_a \hat{\tau}_a - \frac{V}{m_n \Lambda}\Gamma_b \hat{\tau}_b + \lambda\left(\frac{\hat{\tau}_a}{m_n} - m_n \hat{\tau}_b\right)\boldsymbol{q}^{\times}\right]$$

(6.48)

With the help of (6.39) and (6.45) this expression is reduced to the following form:

$$\hat{r}_1 = \frac{\varepsilon_1 m_n \sin^2\varphi \cos^2\theta_i}{m_n^-(\varepsilon_q m_n + m_n^-)^2}\hat{\tau}_a$$
$$- \frac{\varepsilon_1 m_n (\sin^2\varphi(\cos\theta_0 + \sin^2\theta_i) - 1)}{m_n^-(m_n^- + m_n)^2}\hat{\tau}_b$$
$$+ \frac{\varepsilon_1 m_n \sin 2\varphi \boldsymbol{q}^{\times}[(1/m_n)\hat{\tau}_b - m_n \hat{\tau}_a]}{2(m_n^- + m_n)(\varepsilon_q m_n + m_n^-)}$$

(6.49)

or with appropriate notation

$$\hat{r}_1 = r_{1a}\hat{\tau}_a + r_{1b}\hat{\tau}_b + \hat{r}_{1ab}$$

(6.50)

where r_{1a}, r_{1b} are scalar coefficients and $\hat{r}_{1ab}$ is a second-rank tensor. The initial tensors of reflection and transmissions $\hat{r}$ and $\hat{d}$ can be obtained by combining Eq. (6.42)–(6.45), (6.47), and (6.50):

$$\hat{r} = r_a\hat{\tau}_a + r_b\hat{\tau}_b + \hat{r}_1, \qquad \hat{d} = d_a\hat{\tau}_a + d_b\hat{\tau}_b + \hat{r}_1$$
$$r_a = r_{0a} + r_{1a}, \qquad r_b = r_{0b} + r_{1b},$$
$$d_a = d_{0a} + r_{1a}, \qquad d_b = d_{0b} + r_{1b}$$

(6.51)

Our subsequent scheme includes the following procedures:

1. Deduce the vector $\boldsymbol{H}_i$ (6.15) from the given vector amplitude $\boldsymbol{E}$ of the incident wave and the refraction vectors $\boldsymbol{m}_i$.
2. Project the vector amplitude $\boldsymbol{H}_i$ on the plane interface of the two media according to (6.16), depending on the vector $\boldsymbol{q}$.
3. Calculate the tangential component $\boldsymbol{H}_{rt}$ of the reflected wave (or transmitted wave if that is what is needed) with the help of (6.50), (6.51).
4. Recover the electric field strength vector of reflected wave $\boldsymbol{E}_r$ from the vector $\boldsymbol{H}_{rt}$ using the operator $\hat{v}$ [39]:

$$\boldsymbol{E} = \hat{v}\boldsymbol{H}_t, \qquad \hat{v} = -\boldsymbol{q}^{\times}\hat{\gamma} + \frac{1}{\varepsilon_q}\boldsymbol{q}\otimes(\boldsymbol{a} + \boldsymbol{q}\hat{\varepsilon}\boldsymbol{q}^{\times}\hat{\gamma})$$

(6.52)

Taking into account that the first medium is isotropic and $\hat{\gamma}_r = -\hat{\gamma}$ (6.41), we may simplify (6.52):

$$\boldsymbol{E}_r = \hat{v}_r\boldsymbol{H}_{rt}, \qquad \hat{v}_r = m_n\boldsymbol{q}^{\times}\hat{\tau}_a + \frac{1}{m_n}\boldsymbol{q}^{\times}\hat{\tau}_b + \boldsymbol{q}\otimes\boldsymbol{a} \quad (6.53)$$

Finally, we obtain the desired expression connecting the electric fields of the incident and reflected waves by combing Eq. (6.15)–(6.17), (6.53):

$$\boldsymbol{E}_r = \hat{R}_E\boldsymbol{E}_i$$

(6.54)

where the amplitude reflection operator for the electric field is defined as

$$\hat{R}_E = \hat{v}_r\hat{r}\boldsymbol{m}_i^{\times}$$

(6.55)

Substituting the relations for $\hat{v}_r$ (6.53), for $\hat{r}$ (6.51), and for the tensor $\boldsymbol{m}_i^{\times}$ dual to the refraction vector of the incident wave ($\boldsymbol{m}_i^{\times} = \boldsymbol{b}^{\times} + m_n\boldsymbol{q}^{\times}$) into (6.55) gives us the final formula:

$$\hat{R}_E = m_n r_a b(-\boldsymbol{a}_0^{\times})(\hat{\tau}_q + \hat{\tau}_b) - m_n^2 r_a\hat{\tau}_b + r_a b^2\hat{\tau}_q - r_b\hat{\tau}_a$$
$$- \sigma m_n\boldsymbol{q}^{\times}(\hat{\tau}_a - \hat{\tau}_b) + \lambda\boldsymbol{b}^{\times}(\hat{\tau}_a + \hat{\tau}_q)$$

(6.56)

where

$$\sigma = \frac{\varepsilon_1 m_n \sin 2\varphi}{2(m_n + m_n^-)(\varepsilon_q m_n + m_n^-)}$$

(6.57)

This expression is in a very useful form for applied calculations, because all quantities are explicitly connected with the natural basis of the boundary value problem, $(\boldsymbol{a}_0, \boldsymbol{b}_0, \boldsymbol{q})$.

6.3. Reflection Configuration for a Nonlinear Polarization Spectroscopy Detection Scheme

Now, we apply the obtained results to the reflection–transmission problem in the scheme of nonlinear polarization spectroscopy. A powerful normally incident pump wave induces anisotropy in a nonlinear medium. This anisotropy is taken properly into account through the complex dielectric permittivity tensor $\hat{\varepsilon}$ (6.29). Thus, we use the covariant expression (6.56) to derive the vector amplitude of the reflected probe wave. Let an obliquely incident plane wave $\boldsymbol{E}_0$ pass through a linear polarizer described by the unit vector $\boldsymbol{u}_p$. So we have the incident (on the MLIA) wave in the form $\boldsymbol{E}_i = \hat{P}\boldsymbol{E}_0$, where the dyadic $\hat{P} = \boldsymbol{u}_p\otimes\boldsymbol{u}_p$ is related to the polarizer action. If the reflected beam

$$\boldsymbol{E}_r = \hat{R}_E\hat{P}\boldsymbol{E}_0$$

(6.58)

is blocked by a crossed analyzer described by the dyadic operator $A = \boldsymbol{u}_A\otimes\boldsymbol{u}_A$, where $\boldsymbol{u}_A$ is parallel to the analyzer axis, we may write the detected field as

$$\boldsymbol{E} = \hat{A}\boldsymbol{E}_r = \hat{A}\hat{R}_E\hat{P}\boldsymbol{E}_0$$

(6.59)

For definiteness let us take one of the characteristic polarizations of the incident wave, for example, a TE wave of unit intensity. The polarizer transmits this wave without losses, which means that

$$\boldsymbol{E}_0 = \boldsymbol{a}_0, \qquad \boldsymbol{u}_p = \boldsymbol{a}_0, \qquad \boldsymbol{E}_i = \hat{P}\boldsymbol{E}_0 = (\boldsymbol{u}_p\cdot\boldsymbol{a}_0)\boldsymbol{u}_p = \boldsymbol{a}_0$$

Then from (6.54)–(6.56) it follows that

$$\boldsymbol{E}_r = \hat{R}_E\boldsymbol{a}_0 = (-r_b\hat{\tau}_a - \sigma m_n\boldsymbol{q}^{\times}\hat{\tau}_a + \sigma\boldsymbol{b}^{\times}\hat{\tau}_a)\boldsymbol{a}_0$$
$$= -r_b\boldsymbol{a}_0 - \sigma m_n(\boldsymbol{q}\times\boldsymbol{a}_0) + \sigma(\boldsymbol{b}\times\boldsymbol{a}_0)$$
$$= -r_b\boldsymbol{a}_0 - \sigma(m_n\boldsymbol{b}_0 + b\boldsymbol{q})$$

(6.60)

Here, we have taken into consideration that $\hat{\tau}_b\boldsymbol{a}_0 = \hat{\tau}_q\boldsymbol{a}_0 = 0$, $\hat{\tau}_a\boldsymbol{a}_0 = \boldsymbol{a}_0$, $\boldsymbol{q}\times\boldsymbol{a}_0 = \boldsymbol{b}_0$, $\boldsymbol{b}_0\times\boldsymbol{a}_0 = -\boldsymbol{q}$. The projective dyadic of the crossed analyzer has an axis

$$\boldsymbol{u}_A = -\cos\theta_i\boldsymbol{b} - \sin\theta_i\boldsymbol{q} = -m_n\boldsymbol{b}_0 - b\boldsymbol{q}$$

(6.61)

Therefore, the resulting electric field behind the analyzer is deduced from (6.59), (6.61) to be

$$\boldsymbol{E}_{\text{ex}} = \hat{A}\boldsymbol{E}_r = (\boldsymbol{u}_A \cdot \boldsymbol{E}_r)\boldsymbol{u}_A = (\sigma m_n^2 + \sigma b^2)\boldsymbol{u}_A = \sigma \boldsymbol{u}_A \quad (6.62)$$

Finally, for the detected signal in our approximations we have

$$I_{\text{NPS}} = |\sigma|^2 = \left| \frac{\varepsilon_1}{2(m_n^- + m_n)(\varepsilon_q m_n + m_n^-)} \right|^2 m_n^2 \sin^2 2\varphi \tag{6.63}$$

Here we have taken into account that ε_1 and m_n^- may be complex quantities due to absorbing properties of nonlinear medium.

Thus, we have verified theoretically the possibility in principle of using the reflection scheme in nonlinear polarization spectroscopy as proposed for the first time in [51]. On the other hand, the general theoretical expressions obtained seem to be very useful in the light of experimental research on nonlinear selective reflection [52, 53].

6.4. Noncollinear Geometry in Polarization-Sensitive Spectroscopy

The geometry of interacting waves is basic in nonlinear spectroscopy. Nevertheless, only a rather small amount of theoretical and experimental work has been dedicated to detailed study of that problem. Examples are the works of Saican [54] and Levenson's group [55], and the report [20]. In [54] the author limited himself to consideration of the scheme of backward four wave mixing. The authors of [55] offered a new experimental scheme ensuring full collineary of the interacting pump and probe beams in NPS. In [20] the geometry of a possible experiment with a cell of cylindrical form was described, the diameter of which substantially exceeds the cross-sectional dimensions of the light beams, for elimination of polarization changes of the beams during oblique passage through walls of the cell (see also Section 10). Another scheme of polarization-sensitive spectroscopy [31] has, however, continued in common use, where probing and exciting beams are incident on a cell of standard form (right parallelepiped), containing the investigated substance, at a small angle $\arccos(\boldsymbol{n}_0 \cdot \boldsymbol{n}_1)$, thus providing their spatial separation. The angle can be from several degrees [56] down to several milliradians [57]. Evidently, even at such small angles, the polarization state of radiation (pumping in the standard version of NPS, or probing in the version of NPS for observation of out-of-Doppler dichroism and birefringence induced by laser radiation [58]) in the investigated medium with unperturbed refractive index n_m will in general differ from the initial state in a medium with refractive index $\tilde{n}$ after passing through a plane wall of the cell made of a substance with refractive index n_{CW}. So the interaction that actually takes place is, for example, not with the circularly polarized radiation of the pump beam, but with an elliptically polarized wave. Note in this connection that the elliptical polarization, as a rule, is not constant and tends to shift during propagation in a nonlinear medium to a steady polarization state, which can be circular or linear [4]. On the other hand, even when the mentioned

polarization changes are small, they can have rather large significance in polarization-sensitive spectroscopy (especially for optically thin media), particularly in the quantitative interpretation of polarization spectra containing hyperfine structure of molecular lines [31]. Thus the investigations proposed in the following are realistic.

Let the exciting radiation have circular polarization in an isotropic medium surrounding the cell. Then, by applying Fresnel's operator formalism (6.42)–(6.45), it is possible to show that for $\arccos(\boldsymbol{n}_0 \cdot \boldsymbol{n}_1) \ll 1$ [39] in the phase plane of a probe wave, the pump beam will have a polarization ellipse with ellipticity

$$\eta = 1 - \arccos^2(\boldsymbol{n}_0 \cdot \boldsymbol{n}_1)\left[\frac{\tilde{n}^2}{n_{\text{CK}}n_{\text{C}}} + \frac{\tilde{n}}{n_{\text{CK}}} + \frac{\tilde{n}^2}{n_{\text{CK}}^2} - \frac{1}{2} \right] = 1 - \tilde{k} \tag{6.64}$$

and azimuth $\arccos(\boldsymbol{n}_0 \cdot \boldsymbol{n}_1) - \pi/2$ relative to the vector $\boldsymbol{u}_P$ specifying the orientation of a linear polarizer crossed with a linear analyzer $\boldsymbol{u}_A$.

One can also solve the inverse problem of determining the polarization state of a pump wave incident on a cell filled with the investigated substance, such that in the cell the projection of the polarization onto a phase plane of a probe wave gives a polarization ellipse that degenerates to a circle. From a calculation of least order in $\arccos(\boldsymbol{n}_0 \cdot \boldsymbol{n}_1)$ for an initial ellipticity of exciting radiation with the semimajor axis of the ellipse of polarization lying in the plane of incidence, one obtains [59] $\eta' = \eta$.

We would like to emphasize that despite the small value of η', the modern capabilities of the control and measurement of polarization of light (see, for example, [60]) make the solution of the given inverse problem quite practical.

7. SPECTROSCOPY OF OPTICAL MIXING

7.1. Linearly and Circularly Polarized Pump and Probe Waves

Let us explain the version of nonlinear spectroscopy mentioned in the heading of this section to avoid terminological misunderstanding. In this section we follow the terminology (SOM) of Akhmanov and Koroteev [8], though in a number of works this version of nonlinear spectroscopy is called optically heterodyned polarization interferometry [61].

As was shown in the previous section, NPS is a very sensitive experimental technique and has a number of advantages over saturation spectroscopy. In the standard version of NPS the intensity of weak radiation measured by a detector, after passage an analyzer crossed to its initial polarization, consists of the intensity of the desired signal plus the intensity of a noncoherent background. Contrariwise, in the heterodyne scheme a coherent component of a signal interferes with a coherent light wave of the same frequency. This is accomplished either by detuning the analyzer from the perfectly crossed position by a small angle [31], or by appropriate polarization of the probe radiation, obtained by placing a phase retarder before the investigated medium [62]. The pump wave in both cases, as a rule, is reshaped by linear or circular polarization.

To begin, we conduct a calculation of the registered intensivity of the SOM signal. On the basis of the relation (5.1), keeping the terms that are quadratic in the parameter $A = A(\xi)$ [see (2.25)], for the case $n_0 \parallel n_1$ it is easy to obtain [63]

$$\Delta I_{NS} = \big| I_{10} \exp(-\sigma_1 \xi)\big| (u_A^* \cdot u_P) + A(u_A^* \hat{S} u_P)$$
$$+ \frac{1}{2} A^2 \big[(\lambda_+ + \lambda_-) u_A^* \hat{S} u_P - \lambda_+ \lambda_- (u_A^* \cdot u_P) \big]\big|^2$$
$$= \bar{I} \big[|(u_A^* u_P)|^2 \big(1 - \mathrm{Re}(\lambda_+ \lambda_- A^2) \big)$$
$$+ \mathrm{Re}\big\{ \big((u_A \cdot u_P^*) u_A^* \hat{S} u_P \big) (2A + (\lambda_+ + \lambda_-) A^2) \big\}$$
$$+ |A u_A^* \hat{S} u_P|^2 \big] \qquad (7.1)$$

where $\bar{I}$ is the intensity of the probe wave passed through the unexcited sample. Let us define the following notation:

$$\lambda_+ + \lambda_- = \lambda_\Sigma = 2 + C_1 + C_2,$$
$$\lambda_+ \lambda_- = \lambda_\Pi = 1 + C_1 + C_2 + C_1 C_2 \sin^2 2\varepsilon_{pmp} \qquad (7.2)$$

Notice that λ_Σ is an invariant in relation to the polarization state of the pump wave.

The expressions (7.1) and (7.2) allow us to calculate a SOM signal in the most general case, where the interacting waves have arbitrary elliptical polarizations and linear absorbtion takes place in a medium with light-induced anisotropy. Moreover, in the approximation of linearity in the emitted pump power, the optimum length of a sample for observation of a maximum SOM signal can be obtained:

$$\xi_{opt} = \frac{1}{\sigma_0} \ln \left[\frac{\sigma_0 + \sigma_1}{\sigma_1} \right] \qquad (7.3)$$

so that, neglecting the dispersion of the linear absorption coefficient for small Δ_ω, we have $\xi_{opt} \cong \ln 2$. Thus, optically thinner samples should be used in the SOM technique than in nonlinear spectroscopic ellipsometry and NPS, where $\xi_{opt} \cong \ln 3$.

Hereinafter we limit our consideration to the most common cases in practice: (i) the tensor $\hat{S}$ is Hermitian; (ii) the weak wave has circular polarization or linear polarization with arbitrary azimuth θ_{10} in relation to an axis X; (iii) the pump wave is circularly polarized or linearly polarized with zero azimuth; or (iv) u_A is a real vector (i.e., light passed through the investigated medium is registered with a linear analyzer, oriented at an angle δ, as shown in Fig. 4).

7.1.1. Pump and Probe Waves with Linear Polarization

The corresponding polarization conditions are shown in Figure 4a. The expression (7.1) can be rewritten as

$$\Delta I_{1l}^{0l} = \bar{I}_1 \Big[\cos^2 \phi \big(1 + 2 \mathrm{Re}\, A + (\lambda_\Sigma - \lambda_\Pi) \mathrm{Re}\, A^2 + |A|^2 \big)$$
$$+ 2 \cos \phi \cos \theta_{10} \cos \delta (C_1 + C_2)$$
$$\times \left(\mathrm{Re}\, A + \frac{1}{2} \lambda_\Sigma \mathrm{Re}\, A^2 + |A|^2 \right)$$
$$+ \cos^2 \theta_{10} \cos^2 \delta |A|^2 (C_1 - C_2)^2 \Big] \qquad (7.4)$$

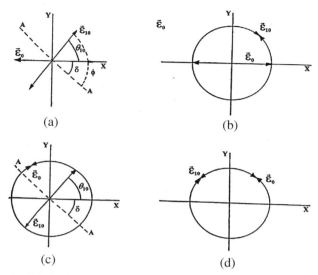

Fig. 4. Polarization conditions for the pump and probe waves. AA is the transmission axis of the linear analyzer.

Hereinafter the indices l and c denote linearly and circularly polarized light, and $\varphi = \theta_{10} - \delta$.

The expression (7.4) is more general than the one obtained earlier [31]. For example, it contains terms linear in the parameter $\mathrm{Im}\, A$ not only for $\theta_{10} = \pi/4$, as in [64], but also for an arbitrary initial azimuth of the probe wave. Moreover, we notice that the quadratic terms describing, in general, contributions of both light-induced birefringence (LIB) and light-induced dichroism (LID) can distort the waveform of pure LID signal, which the authors of [58, 64] have proposed to measure under special polarization conditions.

We proceed now to the estimation of quantities that can be measured experimentally. According to the notation in [64], we define the pure linear LID as

$$LD = \Delta I(\delta = 0) - \Delta I(\delta = \pi/2) \qquad (7.5)$$

Then, after straightforward calculations, from (7.4) we have

$$LD = \bar{I}_1 \Big[\cos 2\theta_{10} \big(1 + 2 \mathrm{Re}\, A + (\lambda_\Sigma - \lambda_\Pi) \mathrm{Re}\, A^2 + |A|^2 \big)$$
$$+ 2 \cos^2 \theta_{10} (C_1 + C_2) \mathrm{Re}\, A (1 + \lambda_\Sigma \mathrm{Re}\, A) \Big] \qquad (7.6)$$

When $\theta_{10} = \pi/4$, the background is suppressed, and according to the results of [64], only the LID signal can be registered. However, Eq. (7.6) shows that it is necessary to take into consideration the contribution of the quadratic term, which is determined entirely by the imaginary part of the effective cubic susceptibility.

Although the generality is lost, it is worthwhile to limit further consideration to the following three cases:

- The first is the case of $\theta_{10} = \pi/4$ and $\delta = -\pi/4 + \beta$. Supposing $|\beta|, |A| \ll 1$, from (7.4) we obtain

$$\Delta I_{1l}^{0l} = \bar{I}_1 \Big[\beta^2 + \beta(C_1 + C_2) \mathrm{Re}\, A + \frac{1}{4}(C_1 + C_2)^2 |A|^2 \Big] \qquad (7.7)$$

This expression is basic for the well-known SOM technique with adjustable analyzer.

- The second case of spectroscopic interest occurs then the azimuth of the probe wave is equal to the *magic angle* ($\tan^2 \theta_{10} = 2$) [65] and the analyzer is oriented along ε_{10}. In this case the expression for signal intensity takes the following form:

$$
\begin{aligned}
\Delta I_{\tan^2 \theta_{10}=2} \\
= \bar{I}_1 \Bigg[1 + 2 \left\{ 1 + \frac{1}{3}(C_1 + C_2) \right\} \operatorname{Re} A \\
+ \left\{ \lambda_\Sigma - \lambda_\Pi + \frac{1}{3}(C_1 + C_2)\lambda_\Sigma \right\} \operatorname{Re} A^2 \\
+ |A|^2 \left\{ 1 + \frac{2}{3}(C_1 + C_2) + \frac{1}{9}(C_1 + C_2)^2 \right\} \Bigg]
\end{aligned}
\tag{7.8}
$$

or, under Kleiman's conditions,

$$
\begin{aligned}
\Delta I_{\tan^2 \theta_{10}=2} = \bar{I}_1 \Bigg[1 + \frac{10}{3} \operatorname{Re} A \\
+ \left(\frac{5}{3} \lambda_\Sigma - \lambda_\Pi \right) \operatorname{Re} A^2 + \frac{25}{9} |A|^2 \Bigg]
\end{aligned}
\tag{7.9}
$$

- Suppose the probe radiation is linearly polarized with $\theta_{10} = \pi/4$, and the angular position of the analyzer is determined by the *mystical angle* ($\tan \delta = 2$) [65, 66]. Taking into consideration that in this case $\cos^2 \varphi = 9/10$, and supposing $C_1 = C_2 = 1$, we obtain the intensity on the detector as

$$
\begin{aligned}
\Delta I_{\tan \theta = 2} = \bar{I}_1 \Bigg[\frac{9}{10} + 3 \operatorname{Re} A \\
+ \left(\frac{3}{2} \lambda_\Sigma - \frac{9}{10} \lambda_\Pi \right) \operatorname{Re} A^2 + \frac{5}{2} |A|^2 \Bigg]
\end{aligned}
\tag{7.10}
$$

7.1.2. Linear Polarization of the Pump Wave and Circular Polarization of the Probe Wave

In this case (see Fig. 4b) the SOM signal is determined as

$$
\begin{aligned}
\Delta I_{1c}^{0l} = \frac{1}{2} \bar{I}_1 \Bigg[1 - \lambda_\Pi \operatorname{Re} A^2 \\
+ \left\{ 1 + \cos^2 \delta (C_1 + C_2) \right\} (2 \operatorname{Re} A + \lambda_\Sigma \operatorname{Re} A^2) \\
\pm (C_1 + C_2) \sin 2\delta \left(\operatorname{Im} A + \frac{1}{2} \lambda_\Sigma \operatorname{Im} A^2 \right) \\
+ |A|^2 \left\{ 1 + 2\cos^2 \delta (C_1 + C_2) + \cos^2 \delta (C_1 + C_2)^2 \right\} \Bigg]
\end{aligned}
\tag{7.11}
$$

Here the signs $+$ and $-$ correspond to right- and left-handed circular polarizations of the probe wave.

A more general result than in [31, 64] follows from (7.11):

$$
\Delta I_{1c}^{0l}(\delta) - \Delta I_{1c}^{0l}(-\delta) = \pm \bar{I}_1 (C_1 + C_2) \sin 2\delta \operatorname{Im} A (1 + \lambda_\Sigma \operatorname{Re} A)
\tag{7.12}
$$

Thus, the signal of linear LIB is subjected to influence by LID within the framework of the adopted assumptions and reaches its maximum value at $\delta = \pi/4$. Similarly the quantity

$$
\Delta I_{1c}^{0l}(\delta = 0) - \Delta I_{1c}^{0l}(\delta = \pi/2) = \bar{I}_1 (C_1 + C_2) \operatorname{Re}(1 + \lambda_\Sigma \operatorname{Re} A)
\tag{7.13}
$$

contains a quadratic term, which induces deformation of a contour of the LID signal. Notice also the feasibility of a linear LIB technique, using the sense of the rotation of the probe wave. From (7.11) it is easy to see that the difference in SOM signals for right- and left-hand circular polarizations of a probe wave leads to (7.12).

7.1.3. Circularly Polarized Pump Wave and Linearly Polarized Probe Wave

In this case (see Fig. 4c) the SOM signal can be computed by the formula

$$
\begin{aligned}
\Delta I_{1l}^{0c} = \bar{I}_1 \Bigg[\cos^2 \phi \bigg\{ 1 - \lambda_\Pi \operatorname{Re} A^2 \\
+ \left(1 + \frac{1}{2}(C_1 + C_2) \right) \left(2 \operatorname{Re} A + \frac{1}{2} \lambda_\Sigma \operatorname{Re} A^2 \right) \\
+ |A|^2 \left(1 + \frac{1}{2}(C_1 + C_2) \right)^2 \bigg\} \\
\pm \frac{1}{2}(C_1 - C_2) \sin 2\phi \left(\operatorname{Im} A + \frac{1}{2} \lambda_\Sigma \operatorname{Im} A^2 \right) \\
+ \frac{1}{4} |A|^2 (C_1 - C_2)^2 \sin^2 \phi \Bigg]
\end{aligned}
\tag{7.14}
$$

Here by the signs $+$ and $-$ differnt directions of handedness of the pump polarizations are distinguished.

It is easy to see from (7.14) that the difference

$$
\Delta I_{1l}^{0c}(\phi) - \Delta I_{1l}^{0c}(-\phi) = \pm \bar{I}_1 (C_1 - C_2) \sin 2\phi \operatorname{Im} A (1 + \lambda_\Sigma \operatorname{Re} A)
\tag{7.15}
$$

contains the information about circular LIB. As before, however, for observation of the pure LIB signal, limitations on the intensity of exciting light are demanded. The maximum signal amplitude is expected for $\varphi = \pi/4$. In that circumstance, the ratio of maximum intensities from Eqs. (7.15) and (7.12) is equal to $(C_1 - C_2)/(C_1 + C_2) = -q$. The informativeness of the parameter q will be shown in the following chapters.

We proceed now to the analysis of the standard SOM scheme, when $\phi = \pi/2 - \beta$ and also the absolute values β and A are small. The required formula for detected intensity in this case looks like

$$
\Delta I_{1l}^{0c} = \bar{I}_1 \left[\beta^2 \pm \beta (C_1 - C_2) \operatorname{Im} A + \frac{1}{4}(C_1 - C_2)^2 |A|^2 \right]
\tag{7.16}
$$

Let us rewrite Eqs. (7.7) and (7.16) in more compact form, taking advantage of the notation $\Delta\lambda_L^2 = -(C_1 + C_2)$ and $\Delta\lambda_c = C_2 - C_1$:

$$\Delta I_{1l}^{0l} = \bar{I}_1\left[\beta^2 - \beta\Delta\lambda_L \, \mathrm{Re}\, A + \frac{1}{4}\Delta\lambda_L^2 |A|^2\right] \qquad (7.17)$$

$$\Delta I_{1l}^{0c} = \bar{I}_1\left[\beta^2 \mp \beta\Delta\lambda_C \, \mathrm{Im}\, A + \frac{1}{4}\Delta\lambda_C^2 |A|^2\right] \qquad (7.18)$$

The Eqs. (7.17) and (7.18) suggest two independent possible methods of measuring the effective nondegenerate cubic susceptibility:

1. *Rotation of the analyzer*. In this case there is a position $AA(\beta_{\min})$ where the detected intensity is minimal:

$$\beta_{\min}^l = \frac{1}{2}\Delta\lambda_L \, \mathrm{Re}\, A, \qquad \beta_{\min}^c = \pm\frac{1}{2}\Delta\lambda_C \, \mathrm{Im}\, A \tag{7.19}$$

$$\Delta I_{\min(\beta)}^l = \frac{1}{4}\bar{I}_1\Delta\lambda_L^2 (\mathrm{Im}\, A)^2$$
$$\Delta I_{\min(\beta)}^c = \frac{1}{4}\bar{I}_1\Delta\lambda_C^2 (\mathrm{Re}\, A)^2 \tag{7.20}$$

From measurements of $\beta_{\min}$ and $\Delta I_{\min}$ with linearly or circularly polarized pumping, information about the dispersion of the imaginary and real parts of the effective $\hat\chi^{(3)}$ can be obtained. An experimentally obtained parabolic dependence of the SOM signal on the angular position of the analyzer [67] will verify the validity of this philosophy of measurement. The physical reason for such nonmonotonic behavior can be easily understood from Figure 5. Owing to light-induced anisotropy, there is a certain deformation of the initial linear polarization of the probe radiation: it becomes elliptically polarized with rotated axes of the ellipse. Then, adjusting the analyzer, we can find a position where the axis of the analyzer becomes perpendicular to the semimajor axis of the polarization ellipse of the pump radiation. In this case the detected signal is determined only by the induced ellipticity and therefore corresponds to its minimum value. At the same time $\beta_{\min}$ determines the rotation angle of the plane of polarization.

2. *Varying the pump wave intensity*. In this case we have

$$\frac{\Delta I_{\min(I_0)}^{l(c)}}{\bar{I}} = \frac{\mathrm{Im}^2 A(\mathrm{Re}^2 A)}{|A|^2} = \frac{\mathrm{Re}^2\chi_{43}(\mathrm{Im}^2\chi_{43})}{|\chi_{43}|^2}$$
$$= \cos^2\alpha(\sin^2\alpha) \tag{7.21}$$

where $\bar{I} = \bar{I}_1\beta^2$ is the intensity of the probe beam at the detector in the absebce of pumping and $\tan\alpha = \mathrm{Im}\,\chi_{43}/\,\mathrm{Re}\,\chi_{43}$.

The result (7.20) opens a new possibility for direct measurement of the phase of the components of the complex tensor $\hat\chi^{(3)}$. We emphasize the apparent simplicity of the method (in com-

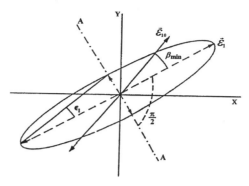

Fig. 5. Input (ε_{10}) and output (ε_1) polarizations of the probe wave and the angular position of the analyzer for observation of the minimal SOM signal due to light-induced ellipticity ($\tan\varepsilon_1$). $\beta_{\min}$ coincides with the light-induced rotation angle of the polarization plane of probe radiation.

parison, for example, with the technique of polarization beats [34]) and its applicability to a wide variety of media.

7.1.4. Circularly Polarized Pump and Probe Waves

In this case probe radiation passes through the excited sample without changing the state of polarization. Then, supposing for definiteness that the circular polarization of the pump radiation is right-handed, we have

$$\Delta I_{1c+(-)}^{0c} = \frac{1}{2}\bar{I}_1\Big[1 - \lambda_\Pi\,\mathrm{Re}\, A^2$$
$$+ \{1 + C_1(C_2)\}\{2\,\mathrm{Re}\, A + \lambda_\Sigma\,\mathrm{Re}\, A^2\}$$
$$+ |A|^2\{1 + C_1(C_2)\}^2\Big] \tag{7.22}$$

It is important to note that the difference of two SOM signals,

$$\Delta_{1c+}^{0c} - \Delta_{1c-}^{0c} = \bar{I}_1(C_1 - C_2)\,\mathrm{Re}\, A(1 + \lambda_\Sigma\,\mathrm{Re}\, A) \tag{7.23}$$

gives the value of the circular LID signal if the experimental conditions allow one to neglect the terms quadratic in the pump beam intensity. Again we point out the rich spectrum of possible applications of the given technique for measurement of both the phase $\chi^{(3)}$ and the parameter q [68].

7.2. Elliptically Polarized Interacting Waves

We give here a concise presentation of the outcomes obtained by us in [61, 69, 70].

7.2.1. Elliptically Polarized Probe Wave and Linearly Polarized Pump Beam

Let us consider first the case when a retarder with a small enough birefringence reshapes the elliptical polarization of the probe radiation. This case includes the situation when the initial linear polarization of the probe beam is deformed by a spurious birefringence in the walls of the cell or in some other optical element. It is possible to show by calculation that in this case the detected signal can be written as [61]

$$\Delta I = \bar{I}_1 \left\{ \alpha_P^2 + \alpha_A^2 + \beta^2 + b^2 \cos^2 2\gamma + \beta (C_1 + C_2) \, \mathrm{Re} \, A \right.$$
$$\left. + b \cos 2\gamma (C_1 + C_2) \, \mathrm{Im} \, A + \frac{1}{4} (C_1 + C_2)^2 |A|^2 \right\} \tag{7.24}$$

where the parameter b determines the extent of the linear birefringence, and γ the angular position of the retarder; the parameters α_i characterize the nonideality of both polarized and analyzer.

The analysis of Eq. (7.24) shows that it is possible by tuning the analyzer to provide separate observation of LIB and LID (experimental confirmation of this technique is given in Section 6). Moreover, when $\beta = 0$ we have

$$b_{\min} = \frac{\eta_1}{\cos 2\gamma}, \qquad (\cos 2\gamma)_{\min} = \frac{\eta_1}{b},$$
$$\frac{\Delta I_{\min}}{\bar{I}_1} = \Delta \theta_1^2 + \alpha_P^2 + \alpha_A^2 \tag{7.25}$$

where η_1 and $\Delta \theta_1$ are the ellipticity of the probe beam's polarization and the rotation angle its plane of polarization.

The Eqs. (7.25) offer a means of measurement of polarization changes of probe radiation and can be rather useful for the technique of nonlinear spectroscopic ellipsometry introduced in Section 8.5.

An algorithm for $\hat{\chi}^{(3)}$, closely following the preceding discussion, is described in our work [69], where all necessary expressions are adduced. The essence of it consists in variation of the ellipticity of the probe beam radiation with fixed azimuth.

7.2.2. Elliptically Polarized Pump Beam and Linearly Polarized Probe Wave

The main results are given in [59, 71], where the most general expression for the detected signal with allowance for possible depolarization of probe radiation is obtained. As in the NPS scheme, the measuring technique for spectroscopic parameters is based on the variation of the ellipticity of the pump radiation. The signal-to-noise ratio is analyzed, and the optimal value for the parameter β is found. Returning to the subject of the Section 6.4, on the basis of the results obtained there one can conclude that the influence of noncollinearity of the interacting waves on the form of the profile of the resonance in NPS can be made negligible by maling the angle between vectors $\boldsymbol{n}_0$ and $\boldsymbol{n}_1$ small enough.

8. PRINCIPLE OF NONLINEAR SPECTROSCOPIC ELLIPSOMETRY

In the present section we proceed from consideration of general arrangements in polarization-sensitive spectroscopy to presentation of a new approach to nonlinear laser spectroscopy that has been developed intensively in recent years—the NSE method.

Classical ellipsometry, as a rule, is used as an optical method for investigating the properties of an interface between two media or of a third medium (film) located between them, and for observing phenomena in such conditions [14]. The change of polarization is the basis of all such methods. This change can take place when the light beam is reflected from an interface or when it passes through an investigated anisotropic medium. As we have seen, such anisotropy can be induced in an originally isotropic medium owing to nonlinear interaction with powerfull enough electromagnetic radiation. Then the state of polarization of a weak (probe) light beam (which, pursuant to the principles of classical ellipsometry, must not appreciably influence the medium under investigation and therefore must have small enough intensity) can change while passing through the medium or reflecting from its surface. On the other hand, as a result of nonlinear interaction of light waves in the investigated sample, the appearance of a new spectral component in the secondary radiation is possible. The state of polarization of this component depends on properties of the medium. This allows us to identify optical methods of investigating (a) spectroscopic properties of a medium through analysis of its polarization properties and (b) dispersion of polarization parameters of secondary radiation from media with nonlinear photoanisotropy as varieties of *nonlinear spectroscopic ellipsometry of polarization-nonuniform media* [72]. We note that the technique of coherent ellipsometry in Raman scattering (RS) due to Akhmanov and Koroteev [8] is rather close in philosophy to the NSE.

8.1. Polarization Changes of the Probe Wave during Propagation in a Medium with Light-Induced Anisotropy

8.1.1. General Formalism

For the description of propagation of polarized light in a photoanisotropic medium, in this subsection we take advantage of the formalism of the differential matrix of propagation, $\hat{N}$, due to Azzam and Bashara (see Section 2). With this aim, we limit ourselves to the case of collinear interacting beams propagating along the Z-axis, so that Eq, (2.41) can be rewritten in the following way:

$$\frac{\mathrm{d} \boldsymbol{E}_1}{\mathrm{d} z} = \left(a \hat{S} - \frac{\sigma_1}{2} \hat{I} \right) \boldsymbol{E}_1 = \hat{N} \boldsymbol{E}_1 \tag{8.1}$$

The authors of [14] showed that when the changes of vector amplitude of a light wave are described by a differential equation in the form (8.1), the behavior of the state of polarization can be treated independently. The corresponding equation is

$$\frac{\mathrm{d} \chi_1}{\mathrm{d} z} = -N_{12} \chi_1^2 + (N_{22} - N_{11}) \chi_1 + N_{21} \tag{8.2}$$

where N_{ij} are matrix elements, $\chi_1 = E_{1v}/E_{1u}$ is a complex variable determining the polarization state of the probe wave, and E_{1u} and E_{iv} are the components of the vector $\boldsymbol{E}_1$ along the basis vectors $\boldsymbol{u}$ and $\boldsymbol{v}$.

Taking into account the explicit form of N_{ij} given by

$$N_{ij} = aS_{ij} - \frac{1}{2}\sigma_1\delta_{ij} \tag{8.3}$$

the Eq. (8.2) is transformed as follows:

$$\frac{d\chi_1}{dz} = -aS_{12}\chi_1^2 + a(S_{22} - S_{11})\chi_1 + aS_{21} \tag{8.4}$$

This demonstrates that the evolution of the probe wave polarization state does not depend on its amplitude changes caused by linear absorption in a resonant medium. In our case the matrix $\hat{N}$ is a function of the parameter z due to the linear attenuation of the pump wave, so that (8.4) is a general Riccati equation. The right side of Eq. (8.4) can be substantially simplified in the basis of the eigenvectors of the tensor $\hat{S}$:

$$\frac{d\chi_1}{dz} = a(\lambda_- \lambda_+)\chi_1 = a\delta\lambda\chi_1 \tag{8.5}$$

where λ_+ and λ_- are the diagonal members (eigenvalues) of the matrix $\hat{S}$ (see Section 2). The Eq. (8.5) is a homogeneous differential equation of the first order, the solution of which can be written

$$\chi_1(z) = \chi_{10} \exp[\Delta\lambda A(z)] \tag{8.6}$$

with

$$A(z) = i\frac{3}{4}\frac{\omega_1}{cn_1}\chi_{43}\int_0^z |E_0(x)|^2 dx \tag{8.7}$$

Here χ_{10} determines the initial state of polarization of the probe wave.

Now the behavior of the parameters of the polarization ellipse (azimuth θ_1 and ellipticity η_1) can be easily obtained from the function $\chi_1(z)$. When the basis vectors u and v are chosen along the coordinate unit vectors e_x and e_y, the following relations hold:

$$\tan 2\theta_1 = \frac{2\,\mathrm{Re}\,\chi_1}{1 - |\chi_1|^2} = a_1 \tag{8.8}$$

$$\sin 2\varepsilon_1 = \frac{2\,\mathrm{Im}\,\chi_1}{1 + |\chi_1|^2} = b_1 \tag{8.9}$$

where $\eta_1 = \tan\varepsilon_1$. In the basis of circular vectors (e^+, e^-) we have

$$\theta_1 = \frac{1}{2}\arg\chi_1 \tag{8.10}$$

$$\eta_1 = \frac{|\chi_1| - 1}{|\chi_1| + 1} \tag{8.11}$$

In an arbitrary basis it is possible to take advantage of the bilinear transformation $\chi_{1uv} \to \chi_{1xy}$ with

$$\chi_{1xy} = \frac{f_{22}\chi_{1uv} + f_{21}}{f_{12}\chi_{1uv} + f_{11}} \tag{8.12}$$

where f_{ij} are the elements of the matrix $\hat{f}$ diagonalizing $\hat{S}$:

$$\hat{f}^{-1}\hat{S}\hat{f} = \begin{pmatrix} \lambda_1 & 0 \\ 0 & \lambda_2 \end{pmatrix} \tag{8.13}$$

8.1.2. The Case of a Linearly Polarized Pump Wave

Here it is convenient to select a Cartesian basis, in which the matrix $\hat{S}$ becomes diagonal. Let us take the polarization of the pump wave along the X, axis. Further calculation is possible in the most general case of an elliptically polarized probe wave, but the final expressions for θ_1 and η_1 are rather awkward and do not lend themselves to analytical treatment. Therefore we limit our consideration here to the case most widely used in experiment, namely linearly and circularly polarized probes. We note in passing that the case of an elliptically polarized probe wave with semimajor axis along e_x or e_y is analyzed in our paper [72].

8.1.2.1. Linearly Polarized Probe Beam ($\chi_{10} = \tan\theta_{10}$)

In this case the induced rotation of the plane of polarization ($\Delta\theta_1 = \theta_{10} - \theta_1$) and the ellipticity of the probe are determined by the expressions [68]

$$\Delta\theta_1 = \frac{1}{2}\arctan\left(\frac{\tan 2\theta_{10} - a_1}{1 + a_1\tan 2\theta_{10}}\right) \tag{8.14}$$

$$\eta_1 = \tan\frac{1}{2}\arcsin b_1 \tag{8.15}$$

It is convenient to use numerical analysis for these quantities, owing to the rather lengthy expressions for $\Delta\theta_1$ and η_1 that are obtained when all parameters, are substituted in (8.14) and (8.15). The azimuthal dependences of $\Delta\theta_1$ and η_1 for different intensity levels of pumping are shown in Figures 6 and 7. Both figures demonstrate that the optimum input azimuth of polarization of the probe wave for observation of the maximum polarization change differs from the value $\pi/4$ given in the literature [65], and this difference increreases with increase of emission power of pumping. Moreover, the values of θ_{10} at which $\Delta\theta_1$ and η_1 are maximal do not coincide, i.e., $\theta_{\Delta\theta_1} \neq \theta_\eta$. As follows from (8.14) and (8.15) with attention to (8.6)–(8.9), the light-induced linear dichroism $\mathrm{Im}(\chi_{1122} + \chi_{1212}) = \mathrm{Im}\,\chi_{\mathrm{eff}}^{(3)}$ and the birefringence both give contributions to the rotation of the plane of polarization and participate in the genesis of elliptical polarization. However, under weak enough excitation the real and

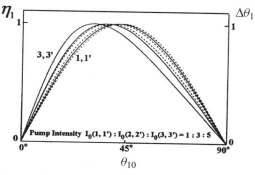

Fig. 6. Theoretical dependences of the normalized light-induced ellipticity (solid curves) and turning angle of the polarization plane (dashed curves) of the probe wave on its input azimuth θ_{10}. The intensity of pumping is taken as parameter, and the imaginary and real part of $\chi_{\mathrm{eff}}^{(3)}$ are assumed to be equal.

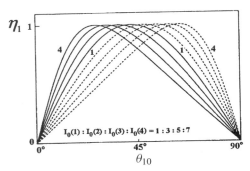

Fig. 7.   Azimuthal dependence of the light-induced ellipticity for different values of the intensity of pumping: solid curves, positive $\mathrm{Im}\,\chi_{\mathrm{eff}}^{(3)}$; dashed curves, negative $\mathrm{Im}\,\chi_{\mathrm{eff}}^{(3)}$; $|\,\mathrm{Im}\,\chi_{\mathrm{eff}}^{(3)}/\,\mathrm{Re}\,\chi_{\mathrm{eff}}^{(3)}|\cong 1$.

imaginary parts can be measured separately [72], since in this case $\Delta\theta_1$ is determined by $\mathrm{Im}\,\chi_{\mathrm{eff}}^{(3)}$, and η_1 by $\mathrm{Re}\,\chi_{\mathrm{eff}}$.

Another method separation follows at once from the non-monotonic behavior of η_1. Rather simple calculations result in the following:

$$\tan\theta_\eta = \pm\exp\left(\Delta\lambda\,\mathrm{Re}\,A(z)\right) \qquad (8.16)$$

$$\eta_{1\max} = \pm\tan\left(\tfrac{1}{2}\Delta\lambda\,\mathrm{Im}\,A(z)\right) \qquad (8.17)$$

$$(\Delta\theta_1)_{\theta_{10}=\theta_\eta} = \mp\arctan\left[\sinh(\Delta\lambda\,\mathrm{Re}\,A(z))\right] \qquad (8.18)$$

Thus, measuring θ_η and $\eta_1\max$ we have a capability of separate measurement of the real and imaginary parts of the nonlinear susceptibility. Moreover, the expression (8.18) enables us to check our measurements of $\mathrm{Im}\,\hat\chi_{\mathrm{eff}}^{(3)}$. Evidently, one may expect similar results from measurements of $\theta_{\Delta\theta_1}$ as well as from the corresponding polarization changes of the probe wave. In this case, however, it is necessary to solve an algebraic equation of sixth degree.

8.1.2.2.   Circularly Polarized Probe Beam

Here $\chi_{10} = \pm i$, where the sign $+$ corresponds to right-handed and $-$ to left-handed circular polarization of the probe wave, which in both cases is transformed to elliptical. Note that the effect of polarization deformation is weak and appears only in the quadratic approximation of the intensity of the pump wave. Then

$$\tan 2\theta_1 = \mp\frac{\mathrm{Re}\,\chi_{43}}{\mathrm{Im}\,\chi_{43}} \qquad (8.19)$$

$$\sin 2\varepsilon = \pm\left[1 - \left(\Delta\lambda\,\mathrm{Re}\,A(z)\right)^2 - \tfrac{1}{2}\Delta\lambda^2\left|A(z)\right|^2\right] \qquad (8.20)$$

Here and hereinafter, if it is not mentioned specially, we consider $\hat S$ as a Hermitian matrix. Thus, having measured the effect of the azimuth of the ellipse of polarization of the probe beam on the output of the investigated medium, we can determine the phase of the nondegenerate third-order complex susceptibility. This method is considered preferable due to the sharp response of its ellipsometric measurements in comparison with other recently suggested techniques [73].

8.1.3.   The Case of a Circularly Polarized Pump Wave

Selecting as basis the circular unit vectors and supposing the probe beam linearly polarized with azimuth θ_{10}, from (8.6), (8.10), and (8.11) we have

$$\theta_1 = \theta_{10} + \tfrac{1}{2}\Delta\lambda\,\mathrm{Im}\,A(z) \qquad (8.21)$$

$$\eta_1 = \tanh\left(\tfrac{1}{2}\Delta\lambda\,\mathrm{Re}\,A(z)\right) \qquad (8.22)$$

Thus, NSE with circularly polarized pump beam and linearly polarized probe beam also permit us to make independent measurements of the spectral behavior of the imaginary and real parts of the effective cubic susceptibility, which is this case are determined through the difference of certain components of the tensor $\hat\chi^{(3)}$. It is remarkable that, in contrast with the case of Section 8.1.2, under circular polarization ox exciting radiation the rotation of the polarization plane of the probe wave is caused by the (circular) birefringence, but the induced ellipticity is determined by the induced dichroism. To conclude this sub-subsection, we first note that results similar to our were obtained by the authors of [74, 75]. However they applied a rather general formalism within which, as they themselves remarked, it is impossible to interpret actual experimental results. We observe that the Eqs. (8.21) and (8.22), unlike the similar equations for the case of linearly polarized pumping [72], do not contain any periodic functions.

8.1.4.   The Case of an Elliptically Polarized Pump Wave

This case is treated in detail in our work [72]. Here it is important only to note the following: The bilinear transformation (8.12) becomes

$$\chi_{1xy} = \frac{\cos\Psi\,\chi_{10uv}\exp(\Delta\lambda\,A(z)) - \sin\Psi}{-\sin\Psi\,\chi_{10uv}\exp(\Delta\lambda\,A(z)) + \cos\Psi} \qquad (8.23)$$

where

$$\chi_{10uv} = \frac{\dfrac{\sin(\psi-\theta_{10})}{\cos(\psi-\theta_{10})} + i\tan\varepsilon_{10}\dfrac{\cos(\psi+\theta_{10})}{\cos(\psi-\theta_{10})}}{1 - i\tan\varepsilon_{10}\tan\Psi - \theta_{10}} \qquad (8.24)$$

$$\sin 2\Psi = i\frac{C_2 - C_1}{C_1 + C_2}\tan 2\varepsilon = iq\tan 2\varepsilon = iK \qquad (8.25)$$

The derivation of (8.25) will be presented in Section 9. Further, despite some loss of generality, we proceed to consideration of a concrete case realized in an experiment by Levinson's group [32], where the probe radiation is linearly polarized with azimuth $\theta_{10} = \pi/4$. Then $\chi_{10uv} = 1$ for the parameters $\Delta\theta_1$, ε_1 we have

$$\cot\Delta\theta_1 = \frac{2\exp(\Delta\lambda\,\mathrm{Re}\,A(z))\cos(\Delta\lambda\,\mathrm{Im}\,A(z))}{1 - \exp(2\Delta\lambda\,\mathrm{Im}\,A(z))} \qquad (8.26)$$

$$\sin 2\varepsilon_1 = \frac{2\exp(\Delta\lambda\,\mathrm{Re}\,A(z))\sin(\Delta\lambda\,\mathrm{Im}\,A(z)) - K}{1 + \exp(2\Delta\lambda\,\mathrm{Re}\,A(z))} \qquad (8.27)$$

where we have supposed that $|\Psi| \ll 1$.

The expressions (8.26) and (8.27) are nearly identical to those that were obtained for linearly polarized pumping [72] [the difference consists only in an additional term in the numerator of (8.27)]. The comparison favors the use of exciting radiation with linear polarization in NSE, as being more stable in resonant media [4].

9. NUMERICAL EVALUATION OF AN EFFECTIVE NONLINEAR SUSCEPTIBILITY IN THE FRAMEWORK OF NSE

In the beginning we consider the case when biharmonic pumping and probe radiation interact with the same electronic transition of the resonant medium. As far as we know, till now only two works have been dedicated to the experimental study of polarization changes of the probe beam under these conditions. We analyze the data of [76] and [77] on 3-amino-N-methylphthalimide in glycerol and on cryptocianine in ethanol, respectively. In the given field approximation and the condition of orthogonal excitattion (also we suppose $C_1 = C_2 = 1$), expressions for the real and imaginary parts of χ_{1221} are collected in Table II. The authors of [76] examined the optical anisotropy in solutions of polyatomic molecules in an excited state. They did not observe any phase anisotropy resulting in induced elliptical polarization of a probe beam in accordance with the analysis conducted above. This means that Re χ_{1221} is negligibly small in these conditions. In contrast to this, the authors of [77] observed only induced birefringence in alcohol solution of cryptocyanine, having selected in an appropriate way the wavelength and the power of the probe beam.

Also in Table II we adduce our estimations for the effective $\hat{\chi}^{(3)}$ in two-photon [78] and double [79] resonances in liquid and gaseous media. In the former case the absorption (and therefore the dichroism) is essentially suppresed, and the induced gyrotropy leads to simple rotation of the plane of polarization of the probe radiation. In the latter case the estima-

tions were made for the cascade transition ($J = 0 \rightarrow J = 1 \rightarrow J = 0$) in helium. Analysis of the data in the table indicates extremely large values of Im χ_{1221} for 3-amino-N-methylphthalimide in glycerol. These values are comparable with the published data for semiconductors [80] and benzporphyrins [72] and exceed by 4–5 orders of maqnitude the corresponding values for benzene and carbon disulfide (the latter material being traditionally used for shutters and correlators). Large values of $\chi_{\rm eff}^{(3)}$ in gaseous media, despite the rather small density of particles interacting with the radiation, can be explained by the sharply resonant conditions of nonlinear interaction in such systems.

10. THE CONCEPT OF NORMAL WAVES IN PHOTOANISOTROPIC MEDIA

It is easy to see that the symetrical part of the tensor $\hat{S}_\perp$ (2.7) is a diagonal matrix. Then the diagonalizing matrix $\hat{D}$ (for $\hat{S}_\perp$) can be represented as the matrix of a shift tranformation by a complex angle $\Psi = \alpha + i\beta$ [21]:

$$\hat{D} = \begin{pmatrix} \cos \Psi & \sin \Psi \\ \sin \Psi & \cos \Psi \end{pmatrix} \qquad (10.1)$$

It is necessary to note here that, in general, a transformation in the form (10.1) is not a unitary matrix. $\hat{D}$ coincides with a unitary matrix (up to a constant factor) only under the condition $\alpha = 0$. Then

$$\hat{D}^+ = \hat{D}^{-1} \cosh 2\beta \qquad (10.2)$$

Now we require that the shift transformation by the angle Ψ should diagonalize the matrix $\hat{S}_\perp$. For this purpose the angle Ψ should satisfy the requirements (8.25). Then, the column vectors of $\hat{D}^{-1}$ are eigenvectors of the tensor $\hat{S}_\perp$ in the chosen

Table II.   Numerical Estimates of the Cubic Susceptibility Obtained by NSE[a]

Substance	λ_0 (nm)	λ_1 (nm)	Im $\chi_{43}(\omega_1; \omega_0, \omega_1, -\omega_0)$ (esu)	Re $\chi_{43}(\omega_1; \omega_0, \omega_1, -\omega_0)$ (esu)
3-amino-N-methyl-phthalimide in glycerol	347	490	-3×10^{-10}	?
		500	-3×10^{-10}	?
		510	-1×10^{-9}	?
Cryptocyanine in ethanol	694	900	?	1×10^{-13}
			Im$(\chi_{44} - \chi_{12})$ (esu)	Re$(\chi_{44} - \chi_{12})$ (esu)
RGG in Ethanol	1064	1064	?	1×10^{-11}
Na	615.9	589.1	?	2×10^{-2}
Ne [74]	609.4	633.0	10^{-7}–10^{-8}	10^{-7}–10^{-8}

[a] A question mark means that the value must be much less than estimated, because the type of anisotropy was not found experimentally.

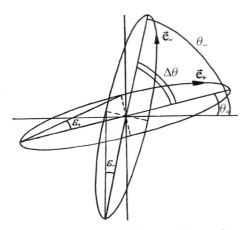

Fig. 8. Polarization characteristics of the normal waves of a medium with light-induced anisotropy in the general case.

basis:

$$\hat{D}^{-1}(\Psi) = (\cos 2\Psi)^{-1} \begin{pmatrix} \cos \Psi & -\sin \Psi \\ -\sin \Psi & \cos \Psi \end{pmatrix}$$

$$= (\cos 2\Psi)^{-1} \hat{D}(-\Psi) \qquad (10.3)$$

To define the polarization state of these vectors it is convenient to employ the formalism of complex trigonometric functions proposed by Fedorov [11]. If θ and η are the azimuth and ellipticity of the ellipse of polarization, then a polarization variable $\chi = E_u/E_v$ (where v and u are the basis states of polarization) can be represented in the form

$$\chi = \tan(\theta + i\gamma) \qquad (10.4)$$

where $\tanh \gamma = \eta$.

In our case, from (10.4) we have two normal waves with the following values of the polarization variable:

$$\chi_+ = -\tan \Psi, \qquad \chi_- = \tan \left(\frac{\pi}{2} + \Psi \right) \qquad (10.5)$$

$$\chi_+ \chi_- = 1$$

Thus, the normal waves, in general, have an elliptical polarization with nonorthogonal directions of major axes and equal axial ratios, but traced in opposite directions; therefore, they are not mutually orthogonal. The corresponding vectors are represented qualitatively in Figure 8. We note that the angle between the semimajor axes of the ellipses of polarization ($\Delta\theta$) is equal to $\pi/2 + 2\alpha$, and the corresponding values of ellipticities satisfy the equality $\eta_+ = -\eta_- = -\tan \beta (\tan \varepsilon_\pm = \eta_\pm)$. Moreover, $\Delta\theta$ and $\eta_\pm$ can be determined from the set of equations

$$\frac{\text{Im}^2 K}{\cosh^2 2\beta} + \frac{\text{Re}^2 K}{\sinh^2 2\beta} = 1$$

$$\frac{\text{Im}^2 K}{\sin^2 2\alpha} - \frac{\text{Re}^2 K}{\cos^2 2\alpha} = 1 \qquad (10.6)$$

Here it is necessary to point out that the mentioned polarization parameters of normal waves ($\Delta\theta$ and $\eta_\pm$) do not depend on

the intensity of the pump wave and are in crucial manner determined by its polarization. Thus it is natural to expect noticeable increase of accuracy of measurements in comparison with the standart scheme of NPS with a completely blocked analyzer, in which the detected signal is, in general, a nonlinear function of L_0. As a consequence, the signal undergoes strong fluctuations even when the pump intensity has rather weak instabilities. On the other hand, the polarization parameters of the exciting field can be fixed rather rigidly.

Furthermore, although some generality is lost, it is worthwhile to consider a set of particular cases that are common in practice.

10.1. Real K

We begin with the case when K is a real parameter. Then the system (10.6) is reduced to the following:

$$\begin{aligned} \theta_+ &= \alpha = 0, & \sinh 2\beta &= K \\ \theta_- &= \frac{\pi}{2}, & \sinh 2\beta &= -K \end{aligned} \qquad (10.7a)$$

or

$$\begin{aligned} \theta_+ &= 0, & \eta_+ &= \frac{1}{K} - \sqrt{1 + \frac{1}{K^2}} \\ \theta_- &= \frac{\pi}{2}, & \eta_- &= -\eta_+ \end{aligned} \qquad (10.7b)$$

and

$$\Psi_+ \Psi_-^* = -1 \qquad (10.8)$$

The last relationship represents the condition of orthogonality of the polarization states.

Thus when K is real, the normal waves have, in general, elliptical polarization with ellipticities equal in value but traced in opposite directions, and orthogonal semimajor axes. The value $|\eta_\pm|$ depends on the ratio q and the ellipticity of the vector a. Because the expression for q in terms of the spectroscopic parameters of the medium is determined by the analysis of a certain mechanism of interaction, the method of normal waves permits, for example, effective spectroscopic investigations of simple quantum systems, as well as the identification of quantum transitions in such systems. So we evaluate [68], the q-parameters for inhomogeneously widened two-level atoms with relaxational constants γ_a and γ_b:

$$q = 5 \left[(2J - 1)(2J + 3) \right]^{-1}$$
$$\text{for transitions} \quad \Delta J = 0 \quad (J \leftrightarrow J) \qquad (10.9)$$

and

$$q = \frac{10J^2 - 5\tilde{R}J - 5}{2J^2 - 5\tilde{R}J + 3} \quad \text{for} \quad \Delta J = \pm 1 \quad (J \leftrightarrow J - 1) \qquad (10.10)$$

It is relevant to note that for transitions $J = 1 \leftrightarrow J = 0$ and $1 \leftrightarrow 1$ we have $q \equiv 1$ and therefore the polarization of one of normal waves coincides with that of the pump wave, and the other one is polarized orthogonally. When a medium with light-induced anisotropy is an ensemble of atomic systems with resonant transition $1/2 \leftrightarrow 1/2$, the normal waves have circular polarization irrespective of the polarization of the pump wave. However, in observations of the radiation of actual atomic media in experimental conditions the relaxation processes manifested through the polarization moments of higher order can be exhibited. In particular, $q = \gamma_{al}/\gamma_{or}$ for the transition $1 \rightarrow 0$, where γ_{or} and γ_{al} are the relaxation constants of orientation and arrangement to the top level. Moreover, it is possible to estimate the ratio γ_{al}/γ_{or} on the basis q-measurements, using limitations on the relaxation constant of the polarization moments [81] [in our case $\min[(\gamma_{al} - \gamma_b)/(\gamma_{or} - \gamma_b)] = 0.6000$]. The method of normal waves can also be successfully applied for measurement of invariant of the tensor Raman (RT). Using results in [82], it is easy to show that q in this case looks like

$$q = \frac{G^A + G^S - 2G^0}{G^A + \frac{1}{5}G^S + 2G^0} \qquad (10.11)$$

where G^0, G^S, and G^A are invariants of the isotropic, anisotropic, and antisymetric parts of the RT tensor. On the other hand, if the RT is symmetrical and the nonresonant background is small enough, the expression for q reduces to

$$q = \frac{3\rho - 1}{1 - \rho} \qquad (10.12)$$

Here ρ is the depolarization ratio.

Other capabilities of the normal wave method (for investigations in nonlinear spectroscopy) are reviewed in our paper [21]. It is necessary, however, to note that in a number of cases (as, for example, in the case of forced backward Rayleigh scattering) the absolute value $|\eta_\pm|$ and depends only on the ellipticity of the strong wave, so that the method of normal waves becomes un informative from the spectroscopic point of view.

10.2. Small K or Nearly Linear Polarization

The second case having practical interest occurs when K (and therefore q) is small enough (for example, for many dye solutions with known linearity of the electrical dipole moment of the π–π^* transition; the value of $\mathrm{Re}\, q$ and $\mathrm{Im}\, q$ for aqueous solutions of malachite green measured by Levenson's group [32] are according 0.14 and 0.03) or the polarization of the vector $\boldsymbol{a}$ is close to linear. Then it is possible to write

$$\sin 2\Psi \cong 2\Psi = iK \qquad (10.13)$$

and the set of Eqs. (10.6) is transformed to

$$\begin{array}{ll}
\theta_+ = \frac{1}{2}\,\mathrm{Im}\, q \tan 2\varepsilon, & \eta_+ = -\frac{1}{2}\,\mathrm{Re}\, q \tan 2\varepsilon \\[2mm]
\theta_+ = \frac{\pi}{2} - \frac{1}{2}\,\mathrm{Im}\, q \tan 2\varepsilon, & \eta_- = -\eta_+
\end{array} \qquad (10.14)$$

The difference in the normal mode azimuths $\Delta\theta$ is accordingly equal to

$$\Delta\theta = \theta_- - \theta_+ = \frac{\pi}{2} - \mathrm{Im}\, q \tan 2\varepsilon \qquad (10.15)$$

Now, using Eqs. (10.14) and (10.15), within the framework of the method of normal waves it is possible to obtain a greater variety of spectroscopic information. In particularly, in the vinicity of an isolated Raman mode with Lorentzian line shape, in the notation of [33] we have

$$\begin{aligned}
\mathrm{Im}\, q &= \frac{(1 - 3\rho)\Delta}{\Gamma} \\[2mm]
\mathrm{Re}\, q &= \frac{(\rho - 1)\,\mathrm{Im}^2 q}{1 - 3\rho}
\end{aligned} \qquad (10.16)$$

where $\Delta = N(\alpha_{11}^R)/48hc\chi_{NR}$, in which α_{11}^R is the diagonal RT tensor element. If Γ is the spectral half-width of the RT line, it is supposed that $|(1 - \rho)\Delta/\Gamma| \ll 1$, and that the frequency detuning of interacting waves is close to the Raman frequency.

From (10.14) it is easy to see that the measuring procedure can be greatly simplified, as the normal waves are practically linearly polarized, and only $\Delta\theta$ need be measured.

If the conditions of applicability of (10.14) and (10.15) are not met — as takes place, for example, for the Raman benzene mode (992 cm^{-1}) [33], the expressions for $\mathrm{Im}\, q$ and $\mathrm{Re}\, q$ become

$$\begin{aligned}
\mathrm{Im}\, q &= \frac{\frac{(1-3\rho)\Delta}{\Gamma}}{1 + \left[\frac{(\rho-1)\Delta}{\Gamma}\right]^2} \\[3mm]
\mathrm{Re}\, q &= \frac{\frac{(1-3\rho)(\rho-1)\Delta^2}{\Gamma}}{1 + \left[\frac{(\rho-1)\Delta}{\Gamma}\right]^2}
\end{aligned} \qquad (10.17)$$

Furthermore, using numerical evaluations obtained from [33] [$\rho = 0.02$ and $(1 - \rho)\Delta/\Gamma = 1.8$], we obtain $\mathrm{Re}\, q = -0.75$ and $\mathrm{Im}\, q = 0.40$. Selecting an ellipticity of collinear pumping $|\eta| = 0.10$–0.15 (to satisfy the condition of small K—the approximation of this case, as estimates of the polarization parameters of normal modes the values $|\pi/2 - \Delta\theta| = 2$–$3°$ and $|\eta_\pm| = (4$–$6) \times 10^{-2}$ can be considered. Taking into account that the present ellipsometric technique have reached a very high level of sensitivity, we conclude that these quantities are quite possible to measure. We note here also that for natural crystals the ellipticity can be considerably smaller [83].

10.3. Non-Collinear Elliptically Polarized Pump Wave

We address now ourselves to the expression (8.25), which is rather close in form to the relationships obtained in [23, 24] for parallel light fluxes, where instead of the ellipticity $\eta = \tan\varepsilon$ the proper ellipticity of the pump wave, $\eta_0 = \tan\varepsilon_{pmp}$, is considered. This resemblance is quite understandable, because during the projection the structure of $\hat{S}$ basically coincides with that of $\hat{S}_\perp$. Actually this entire operation is reduced to the projection of the vector $\boldsymbol{\varepsilon}_0$ on the plane orthogonal to $\boldsymbol{n}_1$. However, in general, the vectors $\boldsymbol{a}_1$ and $\boldsymbol{a}_2$ are not straightforward projections of the directions of the axes of the ellipse of polarization of the strong field; therefore we fail to express the connection between the parameters η and η_0 in a simple form.

At the same time, for a number of special cases that connection has a simple form. For example, if $\boldsymbol{n}_1$ lies in the plane containing n_0 and one of axes of the ellipse ε_0, then [84]

$$\eta = \eta_0(n_1 \cdot n_0)^{\pm 1} = \eta_0(\cos \Phi)^{\pm 1} \qquad (10.18)$$

where the sign $+$ $(-)$ signifies that the plane is determined by the major (minor) axis of the ellipse. Let us note three specific cases.

10.3.1.

When the strong field is linearly polarized ($\varepsilon_0 \cdot \varepsilon_0^* = 0$), the vector a is real ($\eta = 0$). The normal waves also have linear polarization, the directions of which are determined by the vectors a and $n_1 \times a$. The proof of this conclusion can be found in the results of [85, 78], where ethanol solutions of 3-metylcyanin perchlorate and rhodamine 6G were investigated, respectively.

10.3.2.

The orthogonal version of pumping is of peculiar interest, since it is widely used in the experimental schemes of liquid amplifiers and generators. It this case $\tan \varepsilon = 0$ irrespective of the pump wave polarization. The normal waves are polarized linearly, so that $e_+ = n_0$ and $e_- = n_1 \times n_0$. Further, using the general form of representation of vector a in the basis of the eigenvectors of the tensor $\hat{S}_\perp$,

$$a = (n_0 \cdot a)n_0 + \left[(n_1 \times n_0) \cdot a\right](n_1 \times n_0) \qquad (10.19)$$

Eq. (2.7) can be rewritten as follows:

$$\hat{S}_\perp = \hat{A} + C_1 + C_2 \left|(n_1 \times n_0) \cdot \varepsilon_0\right|^2 (n_1 \times n_0) \otimes (n_1 \times n_0) \qquad (10.20)$$

The relation (10.20) demonstrates that in this case, a medium with light-induced anisotropy models the properties of a uniaxial absorbing crystal.

For an experimental check of this conclusion, experiments with the spectrometer described in Section 2 were conducted as follows: as the medium with light-induced anisotropy was used a dye solution (cryptocyanine in ethanol, $N = 2 \cdot 10 \, \frac{Mol}{l}$) in a cell with an effective length 1 cm, arranged between a crossed polarizer and analyzer (degree of crossing 0.98–0.99), excited by the radiation of a ruby laser ($I_0 \cong 50 \, \text{MW/cm}^2$, $\tau_{imp} \cong 30$ ns); the probe beam was produced by a dye laser pumped by the radiation of the same ruby laser. The wavelength and the power of the probe beam were selected so that together with induced dichroism (ar parallel linear polarizations of the two waves, the signal amplification reached the value 2), birefringence also occurred (the detuning of the wavelength of the liquid laser from the absorption band maximum was 100–110 nm). The radiation of the ruby laser was in the absorption band of the dye, so that the effects of the induced anisotropy were resonant and the contributions of other mechanisms (for example, the optical Kerr effect and orientational nonlinearity) were negligibly small [77].

In the experiment the ratio of the photocell signals to the output A of the analyzer was measured with and without excitation of the dye solution at different orientations of the polarization vectors of the probe wave (ε_{10}) and the strong wave (ε_0). For linear polarization of both waves, ε_0 was usually fixed orthogonal to the plane of the wave vectors, and the orientation of ε_{10} was changed. A maximum increase of the signal (with correction for amplification) by a factor of 5–6 was observed when ε_0 and ε_{10} where at an angle $\pi/4$, ant the increase completely vanished ar parallel and orthogonal orientations. The increase of the signal was also absent when the polarization state of the strong wave was changed to circular with the help of retarders. Likewise there was no signal increase at any orientation of ε_{10} in the case when, with linear polarization of the strong field, the vector ε_0 coincided with n_1. This is quite understandable, because in view of (10.20) we get propagation of the probe wave along the optical axis.

Now we return to (8.25). The value of q is determined by the nature of the interaction of light waves with the nonlinear medium and can be calculated on the basis of model representations [23, 24, 86]. For some particular cases the results of such calculations are given in [21, 68].

Special interest attaches to the value $q = 0$, which can take place, for example, for composite molecules [87]. In fact the relationships (2.11) and (2.19) were obtained on the supposition that $C_1 \neq C_2$. However, the formal substitution $q = 0$ in (8.25) gives the same result as the straightforward calculation— the normal waves are linearly polarized along a_1 and a_2 irrespective of the strong field polarization.

Besides the direct problem, the inverse one can be also formulated—on the basis of the known structure of normal waves, determine the ratios of the nonlinear susceptibility tensor components. For its solution the authors of [23, 24] offered a method based on the results of investigation of the case of parallel luminous fluxes. For this case $\eta = \eta_0$, and if q is real, K is real too (see (8.25)) and determines the ellipticity of the normal waves. Now at given polarization of the strong field it is necessary to find a wave whose polarization remains invariable during passage through the cell with the substance under investigation, and them it is easy to find q with the help of (8.25). The problem is greatly simplified by the condition that in the considered case the directions of the axes of the strong field polarization ellipses and the normal waves coincide and can be fixed beforehand.

However, the practical implementation of the given method raises a number of problems, the most crucial of which is connected with the fulfillment of the polarization conditions for the strong field approach. In standard NPS schemes changes of polarization parameters of the strong field are usually negligible [31] in a thin enough layer of examined matter, though there is a substantial in polarization of the probe beam. In the present case this neglect, is not acceptable, as the method of normal waves is based on the minimization of the polarization effects of the probe beam. Therefore, even a small change of polarization of the strong field can result in large errors. As shown earlier, the nonuniformity of anisotropy, arising owing to absorption, for the case of parallel waves does not play any role. A decrease of the cell wall thickness or the intensity of fields does not lead to an appreciable fall in the light-induced anisotropy effects and therefore does not solve the problem.

A natural solution would be the use of a strong field with a polarization state that would remain stable despite self-effects, i.e. we arrive at the same problem of the normal waves, but also for the strong field. Thus, once again, the problem reduces to solving the eigenvector equation

$$\hat{S}\boldsymbol{\varepsilon}_0 = \lambda \boldsymbol{\varepsilon}_0 \qquad (10.21)$$

the notrivial solutions of which give either linear or circular polarization. But none of these solutions is suitable for the considered method.

So it is impossible to eliminate the disadvantages of the normal wave method [23, 24] due to the self-action of the strong field. Therefore, we suggest another method of spectroscopic application of normal waves. The induced anisotropy in the investigated medium is produced by counterpropagated beams of circular polarization, which allows us not only to preserve the polarization of the strong field throughout the interaction volume, but also to reduce substantially the inhomogeneity due to absorption. The probe flux is directed at a large angle, $\arccos(\boldsymbol{n}_1 \cdot \boldsymbol{n}_0)$, and the change of its polarization preserves the directions of the axes of the ellipse—one axis is in the plane of the vectors $\boldsymbol{n}_1$ and $\boldsymbol{n}_0$; the second one is orthogonal to it. The relation (8.25) in this case takes the form

$$\sin 2\Psi = iq \frac{\cos(\boldsymbol{n}_1 \cdot \boldsymbol{n}_0)}{\sin^2(\boldsymbol{n}_1 \cdot \boldsymbol{n}_0)} \qquad (10.22)$$

In the proposed experimental scheme, the study of the normal waves can be conducted in two ways: either the polarization of the probe beam is selected for fixed angles $\arccos(\boldsymbol{n}_1 \cdot \boldsymbol{n}_0)$, or at fixed polarization of the probe beam the angle of convergence $\arccos(\boldsymbol{n}_1 \cdot \boldsymbol{n}_0)$ is varied. For elimination of errors connected with the change of polarization of radiation during oblique transmission through the cell walls, it is expedient to use a cylindrical cell with diameter much greater then the diameters of the luminous fluxes.

We return now to Eq. (2.23). In the strong field approximation its solution can be written as (2.25). As the operators $\hat{G}_\pm$ project any vector on the directions of the eigenvectors $\boldsymbol{E}_\pm = E_\pm \boldsymbol{e}_\pm$, Eq. (2.25) can be viewed as a decomposition of the probe wave amplitude into normal waves. Then the induced birefringence and dichroism can be viewed as consequences of the difference in phase velocities and absorption coefficients of normal waves. These effects disappear either when $\boldsymbol{E}_{10}$ coincides with $\boldsymbol{E}_\pm$, or when the eigenvalues $\lambda_\pm$ become equal. The latter is possible only in the case $|\eta| = 1$ and $C_1 = C_2$, but then, in the direction so determined (η depends on $\boldsymbol{n}_1$), no anisotropy is induced at all.

The relation (2.25) is also applicable to an amplifying (inverted) medium. Owing to the inequality of the eigenvalues at larhe enough ξ, one of the exponents begins to dominate, and the polarization of the amplified wave tends to that of the corresponding normal wave. A similar situation arises in generating radiation in gas lasers [88], where the anisotropy is determined by the generated wave. Therefore we can conclude that the approach developed will be useful for describing the operation of

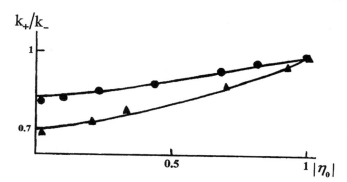

Fig. 9. Dependence of the induced dichroism of amplification, k_+/k_-, on the modulus of ellipticity of longitudinal pumping radiation. Circlets and triangles: results obtained in [89] for dye alcohol solutions of rhodamine 6G and TS-19, respectively.

those laser systems where the pumping is effected by optical radiation.

It is first of all in liquid lasers based on organic dye solutions that induced dichroism is essential for amplification [88, 89]. For its description we need the gain factors k_+ and k_-, which correspon to two eigenpolarizations of the probe (generated) wave in a Fabry–Perot resonator:

$$k_\pm \sim \text{Im} \left[\chi_\Lambda + \boldsymbol{e}_\pm^* \hat{\chi}^{(3)} \boldsymbol{e}_\pm \right]$$
$$\chi_\Lambda = \tilde{\varepsilon} - 1, \qquad \hat{\chi}^{(3)} = \chi_{12} |\boldsymbol{E}_0|^2 \hat{S}_\perp \qquad (10.23)$$

Now taking into consideration (2.7), (2.8) for media with Kleinman symmetry of the $\hat{\chi}^{(3)}$ components, we have

$$\frac{k_+}{k_-} = \frac{(1 - |\boldsymbol{n} \cdot \boldsymbol{\varepsilon}_0|^2)(1 + 3\eta^2) + d(1 + \eta^2)}{(1 - |\boldsymbol{n} \cdot \boldsymbol{\varepsilon}_0|^2)(3 + \eta^2) + d(1 + \eta^2)} \qquad (10.24)$$

where $d = \text{Im} \chi_\Lambda / \text{Im} \boldsymbol{e}_+^* \hat{\chi}^{(3)} \boldsymbol{e}_+$ and $\boldsymbol{n}$ is the unit vector along the axis of the resonator.

From (10.24) it follows that the dichroism of amplification fails only in two cases:

- The vector $\boldsymbol{\varepsilon}_0$ is real and oriented along the axis of the resonator.
- $\eta = \pm 1$.

Experimental research on the induced dichroism of amplification in alcohol dye solution of rhodamine 6G and TS-19 in the longitudinal scheme of pumping was conducted by the authors of [90]. Fitting of the experimental dependence with the theoretical dependence (10.24) (see Fig. 9), by the least squares method, gives the best correlation with $d = 8.9 \pm 1.9$ (at a level of pumping $I_0 = 360\,\text{MW/cm}^2$) and $d = 3.6 \pm 0.8$ (at $I_0 = 50\,\text{MW/cm}^2$) for rhodamine 6G and TS-19, respectively.

10.3.3.

In collinear geometry of interacting waves, $\boldsymbol{a} \equiv \boldsymbol{\varepsilon}_0$ and the ellipticity η has a simple physical sense, since $\eta \equiv \eta_0$. Then it is easy to deduce from the expression for $\hat{S}$ that for circular polarization of the exciting radiation the normal waves are also circularly polarized and differ only in their handedness. If the pumping wave is polarized linearly, then

$$\hat{S} = \hat{I} + (C_1 + C_2)\boldsymbol{\varepsilon}_0 \otimes \boldsymbol{\varepsilon}_0 \qquad (10.25)$$

which corresponds to a uniaxial (transverse, isotropic) tensor. The optical axis of the corresponding anisotropic medium is parallel to $\boldsymbol{\varepsilon}_0$, and the linear polarizations of the normal waves $e_\pm$ are directed along and perpendicular to the optical axis. The obtained results are in good agreement with theoretical predictions of Kaplan [91], concerning the existence of four nonlinear eigenfunctions (normal waves) for two sufficiently intense beams with linear or circular polarization propagating in opposite directions in an originally isotropic medium.

11. METHOD OF COMBINATION WAVES IN NSE

11.1. Codirected Pump and Probe Waves

The standard NSE technique involves the analysis of interaction of two polarized luminous fluxes that differ greatly in intensity. The possibility of a wave at a combination frequency either is completely eliminated in a scheme of counterpropagating waves, or is neglected in an approximation that is linear in the intensity of the strong field. However, in the polarization version of CARS, which is rather close in spirit to NSE, it is shown [8] that the use of combination waves allows one to increase the resolution and to expand the capabilities of polarization methods. Therefore, examination of the information that can be obtained by the combination wave investigations in the framework of NSE is of great interest.

In the approximation of weak saturation, that problem can be solved within the framework of the standard technique [4], where the influence of weak fields on strong ones is neglected. To simplify the calculations, the difference of field frequencies is assumed to be small, and the dispersion of the medium is assumed to be weak. Supposing the propagation directions of the luminous fluxes to be identical (and to coincide with the Z-axis of the laboratory coordinate system), which corresponds to the peak efficiency of wave excitation at the combined frequency [92], the system of reduced equations for probe and combination waves can be conveniently presented in the matrix form [93]

$$\frac{d}{dz}\begin{bmatrix} E_1 \\ E_2^* \end{bmatrix} + \frac{1}{2}\hat{\Sigma}\begin{bmatrix} E_1 \\ E_2^* \end{bmatrix} = \hat{S}\begin{bmatrix} E_1 \\ E_2^* \end{bmatrix} = \hat{S}\hat{E} \qquad (11.1)$$

where E_1 and E_2 are the vectorial amplitudes of probe and combined on waves, respectively, $\hat{\Sigma}$ is the tensor of linear losses, and the matrix S is constructed from the tensors of nonlinear susceptibility $\hat{\chi}_k(\omega_k)$ and parametric coupling $\hat{\beta}_{kk'}$ as follows:

$$S = \begin{bmatrix} g_1\hat{\chi}_1 & g_1\hat{\beta}_{12} \\ g_2^*\hat{\beta}_{21}^* & g_2^*\hat{\chi}_2^* \end{bmatrix} \qquad (11.2)$$

where $g_k = i\omega_k/cn$.

Insofar as the tensors $\hat{\chi}_k$ and $\hat{\beta}_{kk'}$ are quadratic in the strong field [4], the solution of Eq. (11.1) will depend strongly on the

character of the field evolution. As we saw earlier, irrespective of the nature of the intensity change $I_0(z) = I_0 f(z)$, the polarization of the strong field is preserved. Then matrix S can be reduced to

$$S = f(z)\begin{bmatrix} g_1 b_1 \hat{S} & g_1 b_{12}\hat{S}' \\ g_2^* b_{21}^*\hat{S}'^* & g_2^* b_2^*\hat{S}^* \end{bmatrix} = f(z)\tilde{S} \qquad (11.3)$$

The tensor $\hat{S}$ (2.5) and $\hat{S}'$ are distinct here, and the parameters b_k, $b_{kk'}$, C_k (complex, in general) are determined by the nature of the nonlinear medium and the character of the interaction, and $\hat{S}' = \hat{I}(\vec{\varepsilon}_0 \cdot \vec{\varepsilon}_0) + C_3\vec{\varepsilon}_0 \otimes \vec{\varepsilon}_0$.

Now the solution of Eq. (11.1) can be written

$$\hat{E}(z) = \exp\left(-\frac{1}{2}\hat{\Sigma}z\right)\exp\left[\kappa(z)\tilde{S}\right]\hat{E}(0) \qquad (11.4)$$

where $\kappa(z) = \int_0^z f(x)\,dx$. Using further the approximation of small saturation and the boundary conditions $E_{20} = 0$, the solution for the vectors of amplitudes becomes [93]

$$E_1(z) = \exp\left(-\frac{1}{2}\sigma z\right)\left[1 + u(z)\hat{S}\right]E_{10} \qquad (11.5)$$

$$E_2(z) = \exp\left(-\frac{1}{2}\sigma z\right)v(z)\hat{S}'E_{10}^* \qquad (11.6)$$

where $u(z) = g_1 b_1 \kappa(z)$, $v(z) = g_2 b_{21}\kappa(z)$.

The obtained solution demonstrates that in the chosen approximation the relation $|E_1^+(z)| \gg |E_2^+(z)|$ is valid, and the influence of the combination wave on the probe wave can be neglected. This means that for the probe wave all those conclusions are valid that were obtained in NSE spectroscopy disregarding the combination waves. On the other hand, (11.6) implies that the combination wave amplitude and, therefore, the intensity as a function of length of the cell has a maximum. So in the approximation of exponential damping of the strong field, $z_{max} = (\ln 3)/\sigma$, and in the nonresonant case z_{max} is on the order of centimeters. In the case of resonanse its value can be much less. It is clear that this condition should be taken into account in order to obtain the optimum signal.

Let us proceed now to the analysis of the polarization properties of the combination wave. The vector $E_2^+(z)$ is decomposed into two orthogonal unit vectors e_1 and e_2, and a polarization parameter is introduced [4],

$$\alpha_2 = -i\frac{E_2 \cdot e_2}{E_2 \cdot e_1} \qquad (11.7)$$

Then, considering (11.6), we have

$$\alpha_2 = -i\frac{e_2\hat{S}'E_{10}^*}{e_1\hat{S}'E_{10}^*} \qquad (11.8)$$

i.e., within the framework of the adopted assumptions the polarization of the combination wave is constant and does not depend on the intensity of the strong field. Thus, as in the polarization version of CARS, the use of combination waves in NSE allows one to eliminate a sourse of strong noise, namely, the intensity fluctuations of the strong field.

Table III.   The Polarization Parameter of the Combined Wave for Some Particular Cases

Case	Parameter
(i)	$\alpha_2^\kappa = 0$
(ii)	$-\alpha_2^\kappa = -i\frac{\tan\varphi}{1+C_3}$
(iii)	$\alpha_2^\vartheta = \frac{1}{1+C_3}$

We now analyze (11.6) for the three most frequent cases in experiment:

(i) A weak field with initial linear polarization in a circularly polarized strong field.

(ii) Both waves linearly polarized ($\boldsymbol{\varepsilon}_0 \cdot \boldsymbol{\varepsilon}_{10} = \cos\varphi$).

(iii) Linear polarization of a strong field and initial circular polarization of a weak field.

It is convenient to carry out the calculations of the polarization parameter in a circular basis (α_2^κ) for the first case, and in a Cartesian one (α_2^ϑ) for the remaining two. The results of the calculations are collected in Table III.

Thus, in case (i) the combination wave appearing in the medium coincides in polarization with the pump wave (see also [4]). For case (ii) the wave with amplitude E_2 for the medium with real C_3 is polarized linearly at an angle $\varphi' = \arctan[(\tan\varphi)/(1 + C_3)]$ to the vector $\boldsymbol{\varepsilon}_0$. Then from the polarization measurements it is possible to determine directly the parameter C_3, i.e. to extract the spectroscopic information. For case (iii) the combination wave has an elliptical polarization, and the degree of ellipticity is $\eta_2 = \alpha_2^\vartheta$.

Let us pay special attention to the case of circularly polarized pumping, when the tensor of parametric coupling becomes an ordinary projective matrix. In that case the polarization state of the combination wave always coincides with the polarization state of the pump wave, except in the situation with an identically polarized probe wave. Then no combination wave should be generated in an initially isotropic medium, and no third harmonics either [96]. These peculiarities are determined by the angular momentum conservation law for photons participating in parametric processes.

We proceed now to the consideration of a particular experimental situation realized by the authors of [94] within the framework of polarization condition (iii). The three-wave mixing was realized in mercury vapor under the conditions of the $6s\,{}^1S_0-7s\,{}^1S_0$ two-photon resonance. It was found that the combination wave replicates the polarization of the probe beam if the latter is polarized parallel or perpendicular to the linear polarization of the pump radiation. This result is in good agreement with Eq. (11.5). On the other hand, in the case of circular polarization of the probe radiation, the authors of [94] came to the conclusion that the light at the combination frequency also is polarized circularly but with an opposite handedness. It follows from our analysis that this situation can occur only if $C_3 = 0$. When two frequencies ω_λ and ω_μ satisfy the condition for a two-photon resonance between states A and B, i.e., $\omega_\lambda + \omega_\mu = \omega_B - \omega_A$, one can discover that $\chi_{\sigma\lambda\mu\nu}(\omega_\sigma; \omega_\lambda, \omega_\lambda, \omega_\lambda)$ consists of terms proportional to $\boldsymbol{\mu}_{BC}^\lambda \cdot \boldsymbol{\mu}_{CA}^\mu$ or $\boldsymbol{\mu}_{BC}^\mu \cdot \boldsymbol{\mu}_{CA}^\lambda$, where C is the intermediate state and $\boldsymbol{\mu}^\lambda$ is the component of the dipole moment parallel to the electric vector of the λ-wave. Thus we have that $\chi_{1122}(\omega_\sigma; \omega_\lambda, \omega_\mu, \omega_\nu) = 0$ when states A and B have $J = 0$. Hence, if we neglect the influence of nonresonant processes, the conclusion reached in [94] for circularly polarized probe waves in the conducted experiment is valid. However, the published data do not provide clear experimental verification for it. From our point of view, additional experiments with different polarizations of the probe wave should be conducted to explore polarization features of three-wave mixtures in mercury vapor. Moreover, as follows from Table III, for example, for distinct linear polarizations of the probe and pump waves the polarization state of the combination wave depends on the angle φ. Thus, it is rather difficult to agree in general with the statement of the authors [97] that "the generated radiation does not remember polarization of pumping radiation" in the context of the general approach developed by us here.

It is necessary to point out that the mentioned anisotropic properties of the tensor of parametric coupling hold also in a region of strong linearly or circularly polarized pump radiation [4], when the conditions of applicability for Eqs. (11.4) and (11.5) are not met.

11.2. Noncollinear Geometry of Interacting Waves

As a rule, the actual experimental scheme of investigations of multibeam interaction in nonlinear media is to some extent non-collinear. On the other hand, noncollinear pumping geometry is an indispensable condition for a number of experimental methods. So, for example, if we the change of polarization of a probe beam (the initial polarization of which does not coincide with any of the eigenpolarizations of the anisotropic medium) propagating obliquely to the strong beam and inducing an anisotropy in the medium, with maintenance their spatial overlap, then we can determine the distribution of the local density of matter. Therefore there is an urgent need for a corresponding theoretical analysis.

We begin our consideration with the nonlinear polarizability of the medium at the combination frequency. In the notation of Akhmanov and Koroteev and taking into account (2.2) we have

$$P(\omega_2) = 3\chi_{43}\left[(E_0 \cdot E_0)E_1^* + \frac{2\chi_{12}}{\chi_{43}}(E_0 \cdot E_1^*)E_0\right] \quad (11.9)$$

or

$$P(\omega_2) = 3\chi_{43}E_0^2\hat{S}'E_1^* \quad (11.10)$$

Further, to calculate the emission power detected in the experiment, it is necessary to take into account the propagation of secondary radiation in the medium. For this purpose we take advantage of the Cartesian coordinate system shown in Figure 10, where the axes Z and Z' are chosen along the vectors k_0 and $n_2 = (2k_0 - k_1)/|2k_0 - k_1|$ respectively, and the axes Y and Y' are perpendicular to k_0 and k_1. Then the transport equation for

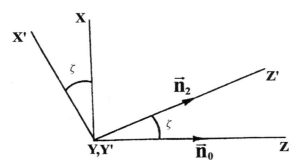

Fig. 10. Coordinate system used for the analysis of noncollinear geometry of interacting waves. All notation is explained in the text.

the slowly varying vector amplitude of the combination wave can be written in the form [95]

$$ik_2 \frac{dE_2}{dz'} = \frac{2\pi\omega_2^2}{c^2} P_\perp^{(3)}(\omega_2) \exp(i\,\Delta k\,z') \qquad (11.11)$$

where $k_2 = |\mathbf{k}_2|$, the subscript $\perp$ means the component of the vector perpendicular to the Z-axis Z', and

$$\Delta k = |2k_0 - k_1| - k_2$$

Having executed the rotation about the Y-axis given by

$$\hat{R}_y(\zeta) = \begin{pmatrix} \cos\zeta & 0 & -\sin\zeta \\ 0 & 1 & 0 \\ \sin\zeta & 0 & \cos\zeta \end{pmatrix} \qquad (11.12)$$

we have

$$\hat{R}_y(\zeta)\mathbf{P}(\omega_2) = 3\chi_{1221}E_0^2\,\hat{R}_y(\zeta)\hat{S}'\hat{R}_y(-\zeta)\mathbf{E}_1^*$$
$$= 3\chi_{1221}E_0^2\big[\hat{I}(\boldsymbol{\varepsilon}_0\cdot\boldsymbol{\varepsilon}_0) + C_3\mathbf{a}\otimes\mathbf{a}\big]\hat{R}_y(\zeta)\mathbf{E}_1^* \qquad (11.13)$$

where $\mathbf{a} = \hat{R}_y(\zeta)\boldsymbol{\varepsilon}_0$.

Now, using the decomposition $\boldsymbol{\varepsilon}_0 = \varepsilon_{0x}\mathbf{e}_x + \varepsilon_{0y}\mathbf{e}_y$, we can write $\mathbf{P}_\perp^{(3)}(\omega_2)$ as

$$\tilde{\mathbf{P}}_\perp^{(3)}(\omega_2) = 3\chi_{43}E_0^2\hat{S}'_\perp\tilde{\mathbf{E}}_1^* \qquad (11.14)$$

where

$$\hat{S}'_\perp = \begin{pmatrix} \boldsymbol{\varepsilon}_0\cdot\boldsymbol{\varepsilon}_0 + C_3\varepsilon_{0x}^2\cos^2\zeta & C_3\varepsilon_{0x}\varepsilon_{0y}\cos\zeta & C_3\varepsilon_{0x}\sin\zeta\cos\zeta \\ C_3\varepsilon_{0x}\varepsilon_{0y}\cos\zeta & \boldsymbol{\varepsilon}_0\cdot\boldsymbol{\varepsilon}_0 + C_3\varepsilon_{0y}^2 & C_3\varepsilon_{0x}\varepsilon_{0y}\sin\zeta \\ 0 & 0 & 0 \end{pmatrix} \qquad (11.15)$$

where the tilde means that the $X'Y'Z'$ coordinate system is to be used as the basis. When $\chi_{1221} = 0$, the term $\boldsymbol{\varepsilon}_0\cdot\boldsymbol{\varepsilon}_0$ should be eliminated from diagonal elements of (11.15).

On the assumption that the medium is illuminated by homogeneous plane waves, the integration of Eq. (11.11) results in the following expression for the amplitude of the combination output wave exiting from the medium:

$$\tilde{\mathbf{E}}_2 \sim \tilde{\hat{S}}'_\perp \tilde{\mathbf{E}}_{20}^* \qquad (11.16)$$

where we have omitted a scalar factor, which is inessential in the polarization analysis. The analytical representation of this

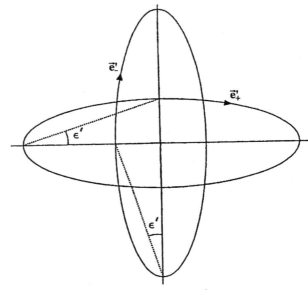

Fig. 11. Portrait of the eigenvectors of tensor $\tilde{\hat{S}}'_\perp$. The angle of ellipticity, ε', is the same for both vectors.

factor can be found in [96], where the effect of linear absorption of interacting beams is also taken into account.

Equation (11.16) looks like a linear tensor transformation of the vector $\tilde{\mathbf{E}}_{20}^-$ and formally coincides with the law of interaction of an incident wave with an optical system in classical ellipsometry [14]. (It is also interesting that within the framework of a model of the coherent photovoltaic effect, the field dependence of the current in the medium can be described on the basis of the same type of equation as (11.16) [97].) Moreover, as the eigenvectors of the tensor $\tilde{\hat{S}}'_\perp$ are entirely determined by the polarization of the pump wave and therefore are known a priori, the methods of classical ellipsometry allow us to conduct ellipsometric measurement of the ratio of the eigenvalues of $\tilde{\hat{S}}'_\perp$. As we shall show, that ratio depends on a number of spectroscopic parameters of the nonlinear medium, which can be determined with the high accuracy characteristic of classical ellipsometry.

To find the eigenvectors and eigenvalues of $\tilde{\hat{S}}'_\perp$, we apply the formalism developed by us in Section 10. Then for nonzero eigenvalues ($\lambda_\pm$) and corresponding values $\chi_\pm$ for the eigenvectors of the matrix (11.15) we have

$$\lambda_+ = \boldsymbol{\varepsilon}_0\cdot\boldsymbol{\varepsilon}_0 + C_3(\varepsilon_{0x}^2\cos^2\zeta + \varepsilon_{0y}^2), \qquad \lambda_- = \boldsymbol{\varepsilon}_0\cdot\boldsymbol{\varepsilon}_0$$

$$\chi_+ = \frac{\mathbf{e}_+\cdot\mathbf{e}'_y}{\mathbf{e}_+\cdot\mathbf{e}'_x} = \frac{\varepsilon_{0y}}{\varepsilon_{0x}\cos\zeta}, \qquad (11.17)$$

$$\chi_- = \frac{\mathbf{e}\cdot\mathbf{e}'_y}{\mathbf{e}\cdot\mathbf{e}'_x} = -\frac{\varepsilon_{0x}\cos\zeta}{\varepsilon_{0y}}$$

Thus, $\chi_+\chi_- = -1$, whence it follows that the eigenvectors of $\tilde{\hat{S}}'_\perp$ in the case of an elliptical vector $\boldsymbol{\varepsilon}_0$ are determined as (see Fig. 11)

$$\mathbf{e}'_+ = \boldsymbol{\varepsilon}_0, \qquad \mathbf{e}'_+\cdot\mathbf{e}'_- = 0 \qquad (11.18)$$

From Eq. (11.16) it follows that when the complex conjugate polarization state (the polarization state with opposite handedness) of the probe radiation entering the medium coincides with the polarization of one of the eigenvectors, the combination wave replicates this polarization.

In our work [96] this analysis is extended to the case when $\omega_2 = 2\omega_0 + \omega_1$. In this subsection, for definiteness, the process of frequency mixing of type $2\omega_0 - \omega_1$ will be called *case 1*, and the process of type $2\omega_0 + \omega_1$ will be called *case 2*. In case 2, replication of the polarization of the probe wave in the combination wave happens when the initial polarization of the probe (instead of its conjugate as in case 1) coincides with one of eigenpolarizations of $\tilde{S}'_\perp$.

We proceed now to consideration of two limiting cases [95, 98] (i) $\chi_{43} = 0$ and (ii) $C_3 = 0$.

  (i) Three-wave mixing is forbidden if the initial polarization of the probe wave coincides with e'_-. As the eigenvectors are perpendicular to the Z'-axis, conditions for that are fulfilled only with linear polarization of the probe beam along the Y-axis. Then the Y-polarized probe wave leads to the generation of a Y-polarized combination wave when the exciting radiation is polarized along the X- or the Y-axis.

  (ii) Three-wave mixing is forbidden for a circularly polarized pump wave. The tensor $\tilde{S}'_\perp$ becomes isotropic, and the polarization of the combination wave is conjugate (case 1) or identical (case 2) to the polarization $(n_2 \times E_{10}) \times n_2$.

11.3. Informativeness of a Variant NSE Based on the Measurement of the Ratio of Eigenvalues of the Tensor of Parametric Coupling

Examples that illustrate the informativeness of the ratio λ_+/λ_- are:

  (1) active Raman spectroscopy with degenerate frequencies [8];
  (2) generation of a combined on wave in low-viscosity solutions of complex molecules under the conditions of one-quantum electronic resonance [99];
  (3) four-photon spectroscopy inside an absorption line, when apart from four-photon process on the resonant electronic nonlinearity it is necessary to take into consideration Rayleigh type coherent scattering on the standing (or running) diffraction grating of refractive index induced in the medium by the beats of waves irradiating it [90];
  (4) three-wave mixing in two-level quantum systems; and
  (5) mixing in p-type Ge and GaAs chips [100, 101].

Further, we restrict ourselves to consideration of quasi-collinear geometry of interacting waves, thereby greatly simplifying the final result. We now discuss these examples in detail:

  (1) In this case, for the measured quantity we have

$$\frac{\lambda_+}{\lambda_-} = \frac{g_0^2 + \frac{4}{45}g_a^2}{\frac{1}{15}g_a^2 - \frac{1}{6}g_{as}^2} \qquad (11.19)$$

where g_0, g_a, and g_{as} are invariants of the molecular RS tensor, describing its isotropic, anisotropic, and antisymmetric parts. Thus, in this case, with two known invariants of the RS tensor, on the basis of the measured value of λ_+/λ_- the third invariant can be determined.

  (2) In the approximation of isotropic rotational diffusion we have [100]

$$\frac{\lambda_+}{\lambda_-} = \frac{\frac{4}{3}\left[\frac{1}{3}\tilde{\gamma} + \frac{A+\frac{1}{2}}{5(\tilde{\gamma}+\gamma_r)}\right]}{\left[\frac{A-1}{3(\tilde{\gamma}+\frac{1}{3}\gamma_r)+\frac{A+\frac{1}{3}}{5(\tilde{\gamma}+\gamma_r)}}\right]}, \qquad \tilde{\gamma} = \gamma_1 + i(\omega_1 - \omega_0)$$

$$(11.20)$$

Here γ_1 and γ_r are the rates of longitudinal and rotational relaxation, respectively, and the parameter A ($0 \leq A \leq 1$) characterizes the anisotropy of electrical dipole moment of the resonant transition: $A = 0$ for a circular dipole, $A = 1$ for a linear dipole. The value of γ_r is the same in the ground and excited states. In the case of coincident or close enough frequencies of interacting waves ($|\Delta\omega| \ll \gamma_1, \gamma_r$), for the majority of solutions of composite molecules ($A \cong 1$) the elliptically measured value of λ_+/λ_- allow us to determinate the ratio γ_r/γ_1. Also, it may be possible to estimate the parameter A and thereby the anisotropy of the electrical dipole moment for a resonant transition with known relaxation parameters of the molecules in solution and with frequency detuning of the pump and probe waves.

  (3) Under the given conditions, the ellipsometric measurement of $\lambda_+/\lambda_- = 3 + 2\chi_R/\chi_E$ allows one to determine the ratio of the coherent Rayleigh (χ_R) to the electronic (χ_E) contribution to the cubic susceptibility.

  (4) Let us distinguish here ensembles of nondegenerate two-level quantum systems, when

$$\frac{\lambda_+}{\lambda_-} = \frac{A+2}{2A-1} = \begin{cases} 3 & \text{(linear dipole model)} \\ -2 & \text{(circular dipole model)} \end{cases}$$

$$(11.21)$$

and two-level systems with degenerate magnetic quantum number, when

$$\frac{\lambda_+}{\lambda_-} = \begin{cases} \dfrac{3J^2 + 3J - 1}{(J-1)(J+2)}, & J \leftrightarrow J \\ -\dfrac{4J^2 + 5\tilde{R}J + 1}{2(J+1)(J-1)}, & J \leftrightarrow J-1 \end{cases}$$

$$(11.22)$$

When $\gamma_a \cong \gamma_b$, (11.22) can be rewritten in the compact form

$$\frac{\lambda_+}{\lambda_-} = \frac{1}{3}\frac{2W(0) + 3W(1) + W(2)}{W(2) - W(1)} \qquad (11.23)$$

where $W(K)$ is the square of the Rack factor $W(11J_aJ_b; KJ_a)$. It is interesting to note that in the case of one-photon resonance for alkali atoms we have $C_3 = -2$ for the D_1-line and $C_3 = -2(4\bar{R} + 1)/(\bar{R} + 1)$ for the D_2-line, where $\bar{R}$ is the decay rate ratio for $3\,{}^2P_{3/2}$ and $3\,{}^2S_{1/2}$ conditions. Nakayama [102] has shown that such information can hardly be obtained within the framework of standard saturation spectroscopy.

(5) In this case the ratio of eigenvalues demonstrates that for the description of the light-induced anisotropy effects in such systems the circular dipole model can be rather useful, since $\lambda_+/\lambda_- = -2$.

12. NONLINEAR OPTICAL ELLIPSOMETER

The development of photometric ellipsometry during recent years has necessarily involved the use of nonlinear optical phenomena, inasmuch as their ocurrence significally depends on the polarization of radiation disturbing the nonlinear medium. As an example of direct application of such effects we may consider [103] the analysis of polarization of laser radiation, based on investigations of the efficiently of second-harmonic generation. This approach provides sensitivity 2–3 times higher than when using a commercial Glan prism; however, it is suitable, generally speaking, only for the analysis of intense linearly polarized radiation. Moreover, the necessity of using a special type of nonlinear crystals sharply narrows the spectral band within which the analysis of polarization can be conducted. Here we consider the possibility of using the process of parametric generation for the analysis of the polarization state of a luminous flux.

For calculation we take advantage of the expression (11.5), obtained for the amplitude of secondary radiation at the combination frequency $2\omega_0 - \omega_1$. It is easy to show that the polarization dependence of the intensity of this radiation will be determined by the following function:

$$\Xi = |\varepsilon_0^2|^2 + 2\,\mathrm{Re}\left\{C_3(\varepsilon_0^2)^*(\varepsilon_0 \cdot \varepsilon_{10})(\varepsilon_0 \cdot \varepsilon_{10}^*)\right\} + |C_3|^2|\varepsilon_0 \cdot \varepsilon_{10}^*|^2 \tag{12.1}$$

This expression demonstrates that it is possible to investigate the polarization of one of the input waves for given polarization of the other one by observing the efficiency of generation of the combination wave. Moreover, from the same expression it follows that a simpler situation takes place when the polarization of the strong field is given. When the latter has a linear polarization, (12.1) is transformed to

$$\Xi = 1 + \left(2\,\mathrm{Re}\,C_3 + |C_3|^2\right)\frac{\cos^2\theta_{10} + \eta_1^2\sin^2\theta_{10}}{1 + \eta_1^2} \tag{12.2}$$

where η_{10} is the degree of ellipticity of the probe wave, and θ_{10} is the angle between the vector ε_0 and the major semiaxis of an ellipse of polarization ε_{10}. Now it is enough to make measurements at three independent orientations of ε_0 to determine the parameters θ_{10}, η_{10}, i.e. to describe the polarization of the probe wave completely.

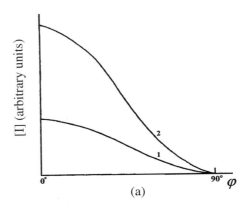

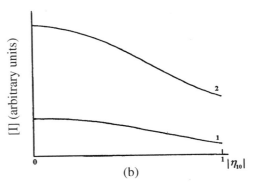

Fig. 12. Polarization dependences of intensity of the combination wave: (a) the interacting waves are polarized linearly with angle φ between polarization vectors; (b) the weak beam is polarized elliptically (ellipticity η_{10}) with the semimajor axis along the polarization vector of the pump wave.

The relation (12.2) for two particular cases is presented graphically in Fig. 12a and b for illustration:

(a) the probe wave has linear polarization ($\eta_{10} = 0$), and the angle φ between ε_0 and ε_{10} is varied;

(b) the weak beam is polarized elliptically, the major axis of the ellipse is parallel to ε_0 ($\theta_{10} = 0$), and the modulus of ellipticity $|\eta_{10}|$ is varied.

Curves 1 and 2 in both figures are constructed for media medeled by ensembles of chaotically oriented linear ($C_3 = 2$) and circular dipoles ($C_3 = -3$), respectively [4]. The transitions in actual media can be represented by the superposition of linear and circular dipoles, so that the corresponding curve passes between curbes 1 and 2.

When the polarization of a weak wave is given, the situation turns out to be more complex, though some simplifications can be achieved with the use of media for which the parameter C_3 is real. This requirement is satisfied, for example, for rigid assemblies of the majority of complex molecules. Equation (12.1) then is reduced to the form

$$\Xi = \left[\frac{1 - \eta_0^2}{1 + \eta_0^2}\right]^2 + 2C_3\left[\frac{1 - \eta_0^2}{1 + \eta_0^2}\right]\frac{\cos^2\theta_0 - \eta_0^2\sin^2\theta_0}{1 + \eta_0^2}$$
$$+ C_3^2\frac{\cos^2\theta_0 + \eta_0^2\sin^2\theta_0}{1 + \eta_0^2} \tag{12.3}$$

where η_0 and θ_0 have the same sense as in the previous case, but now for the strong wave. The rest of the procedure is similar to that previously reviewed; however, as can be seen from (12.3), the calculation of the polarization parameters is more cumbersome.

The analysis conducted in the present subsection demonstrates that the technique considered allows one to study an arbitrary polarization of radiation over a broad range of intensities. Moreover, in the suggested design of an ellipsometer with the use of circulating dye, such disadvantages of the usual nonlinear-optical polarimeter [103] as the inhomogeneity of the generating crystal and thermal effects can be avoided.

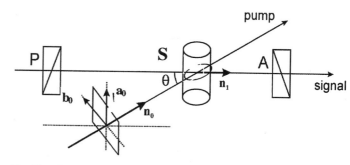

Fig. 13. Geometry of the problem and axis system for theoretical analysis. P and A denote the polarizer and the analyzer, respectively. S is the sample. Other symbols are defined in the text.

13. NONLINEAR LIGHT-INDUCED ANISOTROPY OF AN ISOTROPIC MEDIUM WITH PARTIALLY POLARIZED LIGHT

We work in an approximation of second order in the pump field and first order in the probe field. Cumbersome but straightforward calculation give the following result for the LIA tensor (2.5) [104]:

$$\hat{S} = \left(1 + \frac{C}{2}\right)\hat{I}$$
$$+ \frac{1}{2}\left[- C\hat{\tau}_n + pC\cos 2\eta\left(\hat{\tau}_a - \hat{\tau}_b\right)\left(\cos 2\hat{I} - \sin 2\varphi\, \boldsymbol{n}_0^\times\right) \right.$$
$$\left. + ipS\sin 2\eta\, \boldsymbol{n}_0^\times\right] \qquad (13.1)$$

where $C = C_1 + C_2$, $S = C_1 - C_2$, p is the degree of polarization (i.e. the ratio of the intensity of the polarized component to the total intensity I_0 of the light, $0 \le p \le$), $\hat{\tau}_a = \boldsymbol{a} \otimes \boldsymbol{a}$, $p = 0$ for nonpolarized and $p = 1$ for fully polarized light, $\hat{\tau}_n = \boldsymbol{n}_0 \otimes \boldsymbol{n}_0$, $\hat{\tau}_n = \boldsymbol{a}_0$, $\hat{\tau}_b = \boldsymbol{b}_0 \otimes \boldsymbol{b}_0$, $\hat{I} = \hat{1} - \hat{\tau}_n$, and $\boldsymbol{n}_0^\times$ is the second-rank antisymmetric tensor dual to the vector $\boldsymbol{n}_0$ [107, 12, 39]. The contraction of the tensor $\boldsymbol{n}_0^\times$ with any vector $\boldsymbol{v}$ means their vector prouct: $\boldsymbol{n}_0^\times \boldsymbol{v} = \boldsymbol{n}_0 \times \boldsymbol{v}$. The angle parameters characterizing the polarization of the pump wave are the azimuth angle φ, measured from the vector $\boldsymbol{a}_0$, and the ellipticity angle η. The vector $\boldsymbol{n}_0$ is the unit wave vector of the pump beam, and the triple of vectors $\boldsymbol{a}_0, \boldsymbol{b}_0, \boldsymbol{n}_0$ sets the orthonormal basis.

The standard geometry of the two-wave noncollinear spectroscopic experiment is shown in Figure 13. The orthonormal basis $\boldsymbol{a}_0, \boldsymbol{b}_0, \boldsymbol{n}_0$ is associated with the unit wave normal of the pump wave and its phase plane. The polarized and analyzer are depicted by capital letters P and A; the cylindrical cell containing the medium under investigation, by S. The vector $\boldsymbol{a}_0$ is perpendicular to the plane formed by unit wave normals $\boldsymbol{n}_0$ and $\boldsymbol{n}_1$ of the two waves interacting with the medium.

Now we project the LIA tensor $\hat{S}$ onto the phase plane of the probe wave. This can be done with the help of projection operators $\hat{I}_1 = \hat{1} - \hat{\tau} = \hat{1} - \boldsymbol{n}_1 \otimes \boldsymbol{n}_1$, where $\boldsymbol{n}_1 = \cos\theta \boldsymbol{n}_0 + \sin\theta \boldsymbol{b}_0$, that is,

$$\hat{S}_\perp = \hat{I}_1 \hat{S} \hat{I}_1$$

where the direct designation convention of operator algebra is used: two contractions on corresponding indices of adjacent tensors are implied. The result is

$$\hat{S}_\perp = \left(1 + \frac{C}{2}\right)\hat{I}_1$$
$$+ \frac{1}{2}\left\{ - C\sin^2\theta\left[\sin^2\theta\hat{\tau}_n + \cos^2\theta\hat{\tau}_b \right.\right.$$
$$\left. - \sin\theta\cos\theta(\boldsymbol{b}_0 \otimes \boldsymbol{n}_0 + \boldsymbol{n}_0 \otimes \boldsymbol{b}_0)\right]$$
$$+ pC\cos 2\eta\left[\cos 2\varphi\left(\hat{\tau}_a - \cos^4\theta\hat{\tau}_b - \sin^2\theta\cos^2\theta\hat{\tau}_n \right.\right.$$
$$\left. + \sin\theta\cos^3\theta(\boldsymbol{b}_0 \otimes \boldsymbol{n}_0 + \boldsymbol{n}_0 \otimes \boldsymbol{b}_0)\right) \qquad (13.2)$$
$$- \sin 2\varphi\left(\sin\theta\cos\theta(\boldsymbol{n}_0 \otimes \boldsymbol{a}_0 + \boldsymbol{a}_0 \otimes \boldsymbol{n}_0) \right.$$
$$\left.\left. - \cos^2\theta(\boldsymbol{a}_0 \otimes \boldsymbol{b}_0 + \boldsymbol{b}_0 \otimes \boldsymbol{a}_0))\right]$$
$$+ ipS\sin 2\eta\left(\boldsymbol{n}_0^\times - \sin\theta\cos\theta\right.$$
$$\left.\left. \times (\boldsymbol{n}_0 \otimes \boldsymbol{a}_0 - \boldsymbol{a}_0 \otimes \boldsymbol{n}_0 + \sin^2\theta(\boldsymbol{a}_0 \otimes \boldsymbol{b}_0 - \boldsymbol{b}_0 \otimes \boldsymbol{a}_0)))\right\}$$

This expression can be drastically simplified if we use the basis $\boldsymbol{a}_0, \boldsymbol{b}_1, \boldsymbol{n}_1$ instead of the initial basis $\boldsymbol{a}_0, \boldsymbol{b}_0, \boldsymbol{n}_0$, where $\boldsymbol{b}_1 = \boldsymbol{n}_1 \times \boldsymbol{a}_0$ is the vector product of $\boldsymbol{n}_1$ and $\boldsymbol{a}_0$. We get

$$\hat{S}_\perp = \left(1 + \frac{C}{2}\right)\hat{I}_1 + \frac{1}{2}\left\{ - C\sin^2\theta\hat{\tau}_{b1}\right.$$
$$+ pC\cos 2\eta\left[\cos 2\varphi\left(\hat{\tau}_a - \cos^2\theta\hat{\tau}_{b1}\right)\right.$$
$$\left. - \sin 2\varphi\cos\theta\left(\hat{\tau}_a - \hat{\tau}_{b1}\right)\boldsymbol{n}_1^\times\right]$$
$$\left. + ipS\sin 2\eta\cos\theta\boldsymbol{n}_1^\times\right\} \qquad (13.3)$$

Here we use the new projective operator $\hat{\tau}_{b1} = \boldsymbol{b}_1 \otimes \boldsymbol{b}_1$. Equation (13.3) is the starting point of further investigations. As far as we know, such an expression for the LIA tensor has been derived here for the first time. In spite of its rather bulky appearance, (13.3) is comprehensive and ready for direct application to the analysis of observed signals. The tensor $\hat{\kappa}_\perp$ is a full tensor in the two-dimensional subspace, or it can be considered as a tensor in three dimensions with one eigenvalue, corresponding to the eigenvector $\boldsymbol{n}_1$, equal to zero. Here θ is the angle of noncollinearity (see Fig. 13) — the angle between the interacting beams — and I_0 is the intensity of the pump wave. The explicit dependences on the polarization parameters η and φ of the pumping radiation as well as on the angle θ, the degree of polarization p, and the combinations C and S of nonlinear susceptibility components make this formula a universal tool for the theoretical analysis of various spectroscopic schemes.

13.1. Saturation Spectroscopy

The signal in SS is given by the following expression [30]:

$$\Delta I_{SS} = I_{10}\{1 + 2z\,\text{Re}\,[Au_A^*\hat{S}_\perp u_P]\} \tag{13.4}$$

where z is the distance in the medium with LIA along the propagation direction of the probe wave, I_{10} is the input intensity of the probe wave, $A = i\frac{3}{4}(\omega/cn)\chi_{1221}$, ω and n are the frequency of the probe wave and the linear refractive index of the medium at this frequency, c is the speed of light in vacuum, and u_A and u_P are the unit polarization vectors of the analyzer and polarizer. We can assume them to be in general elliptic complex vectors [12, 105], but for simplification of the further analysis we will consider these vectors as real. That is equivalent to saying that only linear polarization of the probe wave will be considered. The expression (13.4) is obtained in the approximation of linearity in A.

Introducing α as the angle of orientation of the polarizer P relative to the vector a_0, we have the following formula for the vector of the polarizer:

$$u_P = \cos\alpha\, a_0 + \sin\alpha\, b_1 \tag{13.5}$$

Substitution of (13.5) and the equality $u_A = u_P$ into (13.4) finally, after some manipulations, gives the contraction

$$u_A\hat{S}_\perp u_P = 1 + \frac{C}{2}(1 - \sin^2\theta\sin^2\alpha)$$
$$+ \frac{pC}{2}\cos 2\eta\big[\cos 2\varphi(\cos 2\alpha + \sin^2\alpha\sin^2\theta)$$
$$+ \sin 2\varphi\sin 2\alpha\cos\theta\big] \tag{13.6}$$

This result allow us to write down the signal intensity (13.4) in the form

$$\Delta I_{SS} = I_{10}\bigg\{1 + 2z\,\text{Re}\,\bigg[A\bigg\{1 + \frac{C}{2}(1 - \sin^2\theta\sin^2\alpha)$$
$$+ \frac{pC}{2}\cos 2\eta\big[\cos 2\varphi(\cos 2\alpha + \sin^2\alpha\sin^2\theta)$$
$$+ \sin 2\varphi\sin 2\alpha\cos\theta\big]\bigg\}\bigg]\bigg\} \tag{13.7}$$

If the medium possesses Kleinman symmetry ($C_1 = C_2 = 1$, $C = 2$), then from (13.7) follows

$$\Delta I_{SS} = I_{10}\big\{1 + 2z\,\text{Re}\,A\big\{1 + I_0(1 - \sin^2\theta\sin^2\alpha)$$
$$+ pI_0\cos 2\eta\big[\cos 2\varphi(\cos 2\alpha + \sin^2\alpha\sin^2\theta)$$
$$+ \sin 2\varphi\sin 2\alpha\cos\theta\big]\big\}\big\} \tag{13.8}$$

It means that Eq. (13.8) can be used for study of the imaginary part of the nondegenerate third-order susceptibility (Re A $\propto$ Im χ_{1221}). The signal intensity for nonpolarized pumping $p =$

0 will be the same as for circularly polarized pumping ($p = 1$, $\eta = \pi/4$):

$$\Delta I_{SS} = I_{10}\{1 + z\,\text{Re}\,(AC[2 + (1 - \sin^2\theta\sin^2\alpha)])\}$$

In the particular case of polarized orientation exactly along the vector a_0 ($\alpha = 0$), the general expression (13.7) is reduced to

$$\Delta I_{SS} = I_{10}\bigg\{1 + 2z\,\text{Re}\,\bigg[A\bigg\{1 + \frac{C}{2} + \frac{pC}{2}\cos 2\eta\cos 2\varphi\bigg\}\bigg]\bigg\}$$

and the dependences on the polarization parameters of the pump wave appear only in one product of cosine functions of double angles.

13.2. Nonlinear Polarization Spectroscopy

In the NPS scheme the probe beam is blocked by the precisely crossed analyzer in order to observe only the forward scattering contributions. The detected signal is given by the Hermitian quadratic form [12, 30]

$$\Delta I_{NS} = u_A^*(E_1 \otimes E_1^*)u_P$$

which can be reduced by expanding the vector amplitude of the output probe beam E_1 in a series in A with terms up to the second order. We obtain

$$\Delta I_{NS} = |A|^2 z^2 I_{10}|u_A^*\hat{S}_\perp u_P|^2 \tag{13.9}$$

For crossed polarizers we have

$$u_P = \cos\alpha\, a_0 + \sin\alpha\, b_1, \qquad u_A = -\sin\alpha\, a_0 + \cos\alpha\, b_1 \tag{13.10}$$

and after substitution of (13.3) and (13.10) in (13.9) and some straightforward manipulations, we get

$$\Delta I_{NS} = |A|^2 z^2 I_{10}\frac{1}{4}\bigg| - C\frac{\sin 2\alpha}{2}\sin^2\theta$$
$$+ pC\cos 2\eta\bigg(-\frac{\sin 2\alpha\cos 2\varphi}{2}(1 + \cos^2\theta)$$
$$+ \cos\theta\sin 2\varphi\cos 2\alpha\bigg) + ipS\sin 2\eta\cos\theta\bigg|^2 \tag{13.11}$$

If the pump beam is totally unpolarized (the degree of polarization $p = 0$), then (13.11) is considerably reduced:

$$\Delta I_{NPS} = \frac{K}{\sin^2\theta}\bigg[-\frac{C}{2}\sin 2\alpha\sin^2\theta\bigg]^2 \tag{13.12}$$

The NPS signal grows with increasing crossing angle θ, and it vanishes for certain positions or the polarizer, e.g., when $\alpha = 0, \pm\pi/2, \pm\pi, \ldots$. Contrariwise, for fully polarized pumping simplifications are minimal and one should make further assumptions. In general we have

$$\Delta I_{NS} = |A|^2 z^2 I_{10}\frac{1}{4}\bigg| - C\frac{\sin 2\alpha}{2}\sin^2\theta$$
$$+ C\cos 2\eta\bigg(-\frac{\sin 2\alpha\cos 2\varphi}{2}(1 + \cos^2\theta)$$
$$+ \cos\theta\sin 2\varphi\cos 2\alpha\bigg) + iS\sin 2\eta\cos\theta\bigg|^2 \tag{13.13}$$

First of all we will analyze the case of linear polarized pumping ($\eta = 0$). Then $\cos 2\eta = 1$ and $\sin 2\eta = 0$, so after some transformations we obtain from (13.13)

$$\Delta I_{NS} \propto |C|^2 \Bigg(\cot\theta \sin 2\varphi \cos 2\alpha$$

$$- \sin 2\alpha \frac{\cos^2\varphi - \sin^2\varphi \cos^2\theta}{\sin\theta} \Bigg)^2 \quad (13.14)$$

which gives the dependence of the NPS signal on the azimuth φ, the crossing-beam angle θ, and the angle α of the polarizer orientation. This dependence is still rather complicated.

If $\varphi = 0$ (the most general case in practice), then [106–108]

$$I \propto \left(\frac{\sin 2\alpha}{\sin\theta} \right)^2 = \sin^2 2\alpha (1 + \cot^2\theta)$$

If the polarization is oriented so that $\alpha = 0, \pm\pi/2, \pm\pi$, the signal will be suppressed. This is because the polarization vector of the probe wave becomes an eigenvector of the LIA tensor. The probe beam polarization is not changed during the propagation, and hence it is completely blocked by the analyzer. If the polarization orientation of the pump wave is orthogonal to the vector $\boldsymbol{a}_0$ ($\varphi = \pm\pi/2$), we then obtain

$$\Delta I_{NS} \propto \cot^2\theta \sin^2\alpha$$

In the particular case of circularly polarized pumping ($\eta = \pm\pi/4$), Eq. (13.13) can be reduced to the form

$$\Delta I_{NS} \propto \frac{C^2}{4} \sin^2 2\alpha \sin^2\theta + S^2 \cot^2\theta \quad (13.15)$$

As one can see, an exact $\cot\theta$ dependence follows if the angle $\alpha = 0, \pm\pi/2, \pm\pi$. Unfortunately for a direct comparison with experimental data, in [109, 110], where such a dependence was measured in the experiment, only the polarization states of the interacting beams (linear polarization for the probe beam, and circular polarization for pumping) are given, without a clear description of the experimental geometry. Therefore, it is impossible to provide more detailed comparison of our theory with the experimental results in this case. It is important to note the difference between the dependences for unpolarized (13.12) and circularly polarized (13.15) pumping. This difference is considerable for small angles θ and becomes negligible for $\theta \to \pi/2$; in addition, it depends on the coefficients C and S in a different manner. However, under the Kleinman symmetry conditions (when $S = 0$), both polarization states of the pump beam (totally unpolarized and circularly polarized) cause the same dependence of the NPS signal on the crossing angle, which substantially differs from the actually observed $\cot^2\theta$ dependence.

In order to show the influence of anisotropy of a resonant transition dipole moment for the corresponding models of media with LIA [111], we find the intensities of the NPS signal for two cases: the linear dipole moment [$C_1 = C_2 = 1 \Rightarrow C = 2, S = 0$; Eq. (13.16a)] and the circular one [$C_1 = 1, C_2 = -2/3 \Rightarrow C = 1/3, S = 5/3$;

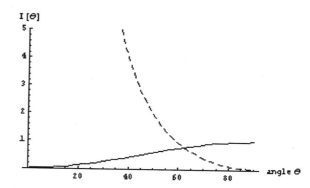

Fig. 14. Intensity of the signal in arbitrary units for a circularly polarized pump beam versus crossing angle of the beams in degrees, $\alpha = \pi/4$. The solid line is for a medium with LIA as an ensemble ol linear dipoles, and the dashed line for circular dipoles.

Eq. (13.16b)]:

$$\Delta I_{NS} \propto \sin^2\theta \quad (13.16a)$$

$$\Delta I_{NS} \propto \sin^2 2\alpha \sin^2\theta + \frac{25}{9} \cot^2\theta \quad (13.16b)$$

The dependences are distinctly different. They are plotted together in Figure 14.

The dependence of the detected signal on the crossing angle permits us to suggest a measurement technique where systematic variation of the beam crossing angle is a sensitive tool for clarification of the anisotropic properties of the resonance transition dipole moment. Moreover, there is a beam crossing angle θ_m ($7\sin^2\theta_m = 20\cot^2\theta_m$, $\theta_m = 62.3°$) for which the influence of anisotropy of the resonance transition dipole moment disappears. This arrangement can be considered as a kind of the beam-crossing magic angle.

13.3. Spectroscopy of Optical Mixing

So far much work has been done on various aspects of SOM. However, the polarization dependence of the SOM signal was studied only in the paper [63] and only for collinear interacting waves. Therefore, here we are going to give a full polarization description of the measured signal in the scheme of a SOM experiment. As in Sections 13.1–13.2, we use plane wave approximations for the beams and neglect self-acting effects of the fields in the nonlinear medium. Moreover, we assume that absorption of the probe beam is negligible.

We adopt the expression for the intensity of the detected signal within the SOM scheme from [63]:

$$\Delta I_{SOM} = I_0 \Bigg| \boldsymbol{u}_A^* \cdot \boldsymbol{u}_P + zA(\boldsymbol{u}^* \hat{\kappa}_\perp \boldsymbol{u}_P)$$

$$+ \frac{1}{2}z^2 A^2 \big[(\lambda_+ + \lambda_-)\boldsymbol{u}_A^* \hat{\kappa}_\perp \boldsymbol{u}_P - \lambda_+ \lambda_- \boldsymbol{u}_A^* \cdot \boldsymbol{u}_P \big] \Bigg|^2 \quad (13.17)$$

Here $\lambda_\pm$ are the eigenvalues of the LIA tensor $\hat{\kappa}_\perp$ (13.3), which can be easily obtained analytically from (13.3). Some tensorial algebra calculations lead to the needed combinations

$$\lambda_+ + \lambda_- = \lambda_\Sigma = 2 + C + \frac{C}{2}\sin^2\theta[p\cos 2\eta\cos 2\varphi - 1]$$

$$\lambda_+\lambda_- = \lambda_\Pi = \left(1 + \frac{C}{2} + \frac{pC}{2}\cos 2\eta\cos 2\varphi\right)$$
$$\times \left(1 + \frac{CI_0}{2}\cos^2\theta - \frac{pC}{2}\cos 2\eta\cos 2\varphi\cos^2\theta\right)$$
$$- \frac{p^2}{4}\cos^2\theta\left(C^2\cos^2 2\eta\sin^2 2\varphi + S^2\sin^2 2\eta\right)$$
$$(13.18)$$

The arrangement of the polarizer and analyzer is the following:

$$\boldsymbol{u}_P = \cos\alpha\,\boldsymbol{a}_0 + \sin\alpha\,\boldsymbol{b}_1, \qquad \boldsymbol{u}_A = -\sin\alpha - \delta\boldsymbol{a}_0 + \cos\alpha - \delta\boldsymbol{b}_1$$
$$(13.19)$$

where the angle δ is the small angle of analyzer detuning from the crossed position. Substitution of (13.19) in (13.17) yields, after cumbersome calculations, two basic expressions:

$$\boldsymbol{u}_A\hat{\kappa}_\perp\boldsymbol{u}_P = \left(1 + \frac{C}{2}\right)\sin\delta$$
$$+ \frac{1}{2}\big[pC\cos 2\eta\sin 2\varphi\cos\theta\cos(2\alpha - \delta)$$
$$- pC\cos 2\eta\cos 2\varphi\sin(\alpha - \delta)\cos\alpha$$
$$- C\sin^2\theta\cos(\alpha - \delta)\sin\alpha \qquad (13.20)$$
$$- pC\cos 2\eta\cos 2\varphi\cos^2\theta\sin\alpha\cos(\alpha - \delta)$$
$$+ ipS\sin 2\eta\cos\theta\cos\delta\big]$$

$$\boldsymbol{u}_A\cdot\boldsymbol{P}_P = \sin\delta$$

However, if one collects all parts of the expression (13.17) together, the result will be unreadable due to its huge appearance. From now we shall make some assumptions and treat different cases of the pump polarization.

13.4. Linear Pumping

In this case $\sin 2\eta = 0$, $\cos 2\eta = 0$, and some terms vanish. Without loss of generality we assume that the linearly polarized pump wave has a certain orientation, namely $\varphi = 0$. The signal intensity is [111]

$$\Delta I_L = I_{10}\bigg|\sin\delta + zA\bigg\{\left(1 + \frac{C}{2}\right)\sin\delta$$
$$+ \frac{1}{2}\big[-pC\sin(\alpha - \delta)\cos\alpha - C\sin^2\theta\cos(\alpha - \delta)\sin\alpha$$
$$- pC\cos^2\theta\sin\alpha\cos(\alpha - \delta)\big]\bigg\}$$
$$+ \frac{A^2z^2}{2}\bigg[\left(2 + C + \frac{C}{2}\sin^2\theta(p - 1)\right)$$
$$\times \bigg\{\left(1 + \frac{C}{2}\right)\sin\delta$$
$$+ \frac{1}{2}\big[-pC\sin(\alpha - \delta)\cos\alpha - C\sin^2\theta\cos(\alpha - \delta)\sin\alpha$$
$$- pC\cos^2\theta\sin\alpha\cos(\alpha - \delta)\big]\bigg\}\bigg]$$

$$- \frac{A^2z^2}{2}\sin\delta\bigg[\left(1 + \frac{C}{2} + p\frac{C}{2}\right)$$
$$\times \left(1 + \frac{C}{2}\cos^2\theta - p\frac{C}{2}\cos^2\theta\right)\bigg]\bigg|^2$$

The next step is connected with neglecting terms of order higher than the second. We take into account that

$$|\delta|, \ |A| \ll 1 \qquad (13.21)$$

so we are able to retain only terms like δ^2, δA, A^2. With the help of simple mathematical calculations the quantity of interest can be found as

$$\Delta I_L = I_{10}\bigg|\delta - zA\frac{C\sin 2\alpha}{4}\big[\sin^2\theta + p\cos^2\theta + p\big]\bigg|^2 \quad (13.22)$$

Nonpolarized pumping gives (in the case of real C)

$$\Delta I_N = I_{10}\bigg|\delta - zA\frac{C\sin 2\alpha}{4}\sin^2\theta\bigg|^2$$
$$= I_{10}\bigg(\delta^2 - \delta z\frac{C\sin 2\alpha}{2}\sin^2\theta\,\mathrm{Re}\,A$$
$$+ z^2|A|^2\frac{C^2\sin^2 2\alpha}{16}\sin^4\theta\bigg) \qquad (13.23)$$

In contrast, the polarized pumping leads to a formula with pure $\sin^{-2}\theta$ dependence:

$$\Delta I_L = I_{10}\bigg|\delta - zA\frac{C\sin 2\alpha}{2}\bigg|^2$$
$$= I_{10}\bigg(\delta^2 - \delta z\,\mathrm{Re}\,AC\sin 2\alpha + z^2|A|^2\frac{C^2\sin^2 2\alpha}{4}\bigg)$$
$$(13.24)$$

13.5. Circular Pumping

In this case we have $\eta = \pm\pi/4$ and $\sin 2\eta = \pm 1$, depending on the direction of polarization vector rotation. Substituting of the formulas (13.18), (13.20) into (13.17) yields [111]

$$\Delta I_C = I_{10}\bigg|\sin\delta + zA\bigg\{\left(1 + \frac{C}{2}\right)\sin\delta$$
$$+ \frac{1}{2}\big[-C\sin^2\theta\cos(\alpha - \delta)\sin\alpha \pm ipS\cos\theta\cos\delta\big]\bigg\}$$
$$\times \frac{A^2z^2}{2}\bigg[\left(2 + CI_0 - \frac{CI_0}{2}\sin^2\theta\right)$$
$$\times \bigg\{\left(1 + \frac{C}{2}\right)\sin\delta$$
$$+ \frac{1}{2}\big[-C\sin^2\theta\cos(\alpha - \delta)\sin\alpha \pm ipS\cos\delta\cos\delta\big]\bigg\}\bigg]$$
$$- \frac{A^2z^2}{2}\sin\delta\bigg[\left(1 + \frac{C}{2}\right)\left(1 + \frac{C}{2}\cos^2\delta\right)$$
$$- \frac{p^2\cos^2\theta}{4}S^2\bigg]\bigg|$$
$$(13.25)$$

The extraction of the terms not higher than second order in the small parameters $|\delta|$ and $|A|$ gives

$$\Delta I_C = I_{10}\left|\delta + \frac{zA}{2}\left[-C\sin^2\theta\frac{\cos 2\alpha}{2}\pm ipS\cos\theta\right]\right|^2 \quad (13.26)$$

For nonpolarized pumping we obtain once more the expression (13.22). It is rather interesting to compare the resulting signal under such pumping with the case of a circularly polarized pump wave. Let us put $p = 1$ into the formula (13.26):

$$\Delta I_C = I_{10}\left|\delta + \frac{zA}{2}\left[-C\sin^2\theta\frac{\cos 2\alpha}{2}\pm iS\cos\theta\right]\right|^2 \quad (13.27)$$

This expression generalizes results of the paper [114]. It reduces to the expression from that work on setting $\theta = 0$. The principal defference from the mentioned paper is the presence of the term $C\sin^2\theta(\cos 2\alpha)/2$, and hence it is logical to put $\cos 2\alpha = \pm 1$ for the analysis. Considering the parameters C and S as real, we derive from Eq. (13.27)

$$\Delta I_C = I_{10}\left[\delta^2 \mp \frac{zC\sin^2\theta}{2}\delta\,\mathrm{Re}\,A \pm zS\cos\theta\delta\,\mathrm{Im}\,A \right.$$
$$\left. + \frac{z^2}{4}\left(\frac{C^2}{4}\sin^4\theta + S^2\cos^2\theta\right)|A|^2\right] \quad (13.28)$$

14. METHODS FOR MEASURING NONLINEAR SUSCEPTIBILITY

Following the consideration of similar dependences in [63], we can treat three independent possible methods of measuring nondegenerate components of third-order nonlinear susceptibility.

1. The first would be based on the adjustment of the analyzer; the angle δ is examined as the independent variable. One can see parabolic dependence on δ in (13.23), (13.24), (13.27). Therefore, a position of the analyzer exists where the detected signal reaches a minimum. Straightforward calculations give

$$\delta_{\min}^C = \mp\frac{z}{2}\left(S\cos\theta\,\mathrm{Im}\,A - \frac{C\sin^2\theta}{2}\mathrm{Re}\,A\right)$$
$$\delta_{\min}^N = \frac{zC\sin 2\alpha\sin^2\theta\,\mathrm{Re}\,A}{2} \qquad (14.1)$$
$$\delta_{\min}^L = \frac{zC\sin 2\alpha\,\mathrm{Re}\,A}{2}$$

for the values of the angle variables, and

$$\Delta I_{\min}^C = \frac{z^2}{4}\left(\frac{C\sin^2\theta}{2}\mathrm{Im}\,A + S\cos\theta\,\mathrm{Re}\,A\right)^2 I_{10}$$
$$\Delta I_{\min}^N = \frac{z^2C^2\sin^2 2\alpha\sin^4\theta(\mathrm{Im}\,A)^2}{16}I_{10} \qquad (14.2)$$
$$\Delta I_{\min}^L = \frac{z^2C^2\sin^2 2\alpha(\mathrm{Re}\,A)^2}{16}I_{10}$$

for the minimal intensities of the detected signal. Superscripts C, N, L denote circularly polarized,

nonpolarized, and linearly polarized pump wave. From the measurements of $\delta_{\min}$ and $\Delta I_{\min}$ with linearly and circularly polarized pump, we can obtain information about the real and imaginary parts of the third-order susceptibilities. The reason for the existence of such a minimum is explained in [63].

2. The second method would involve the analysis of the minimum SOM signal with varying pump beam intensity. The analysis of the corresponding functions for monotonic behavior gives exactly the same expression as in [63], except for the case of circularly polarized pumping, when we have

$$\frac{\Delta I_{\min}^C}{\delta^2} = \frac{\left(\frac{C\sin^2\theta}{2}\mathrm{Im}\,A + S\cos\theta\,\mathrm{Re}\,A\right)^2}{|A|^2\left(\frac{C^2\sin^4\theta}{4} + S^2\cos^2\theta\right)}I_{10} \quad (14.3)$$

This result, as well as formulas for the linearly polarized pump wave, show a new possibility for direct measurement of the phase of nonlinear susceptibility components.

3. The third method is connected with the nonmonotonicity of the crossing angle θ in Eqs. (13.23), (13.24), (13.28). In Eq. (13.23) one can see the parabolic dependence on $\sin\theta$. It is easy to find the minimum of the detected signal intensity

$$\Delta I_{\min}^N = \delta^2\frac{(\mathrm{Im}\,A)^2}{|A|^2}I_{10} \qquad (14.4)$$

when

$$\sin\theta_{\min} = \frac{4\delta\,\mathrm{Re}\,A}{lC\sin 2\alpha|A|^2}$$

Equation (13.24) has quadratic dependence on $1/\sin\theta$, so similar derivations give us the result that the minimal signal will be expressed by (14.4), and will occur at

$$\frac{1}{\sin\theta_{\min}} = \frac{2\delta\,\mathrm{Re}\,A}{lC\sin 2\alpha|A|^2}$$

The most complicated θ-dependence is in Eq. (13.28). We treat it in two particular cases of dipole anisotropy: the linear dipole moment ($C_1 = C_2 = 1 \Rightarrow C = 1$, $S = 0$) and the circular one ($C_1 = 1$, $C_2 = -2/3 \Rightarrow C = 1/3$, $S = 5.3$). The data allow us to obtain formulas written in a corresponding manner:

$$\Delta I_{Cl} = I_{10}\left[\delta^2 \mp l\sin\theta\delta\,\mathrm{Re}\,A + \frac{l^2\sin^2\theta|A|^2}{4}\right]$$

$$\Delta I_{Cc} = \left[\delta^2 \mp l\delta\left(\frac{\sin\theta}{6}\mathrm{Re}\,A - \frac{5}{3}\cot\theta\,\mathrm{Im}\,A\right)\right.$$
$$\left. + \frac{l^2}{36}\left(\frac{\sin^2\theta}{4} + 25\cot^2\theta\right)|A|^2\right] \quad (14.5)$$

The first equation in (14.5) reflects quadratic dependence on $\sin\theta$, giving us the expressions (14.4) for minimal intensity of the signal with

$$\sin \theta_{\min} = \pm \frac{\delta \operatorname{Re} A}{l I_0 |A|^2}$$

The second equation in (14.5) is not so straightforward to analyze. But plotting the θ-dependence of this formula with some model values of the parameters proves its monotonic behavior. Thus, there are no extreme values of the angle θ.

These techniques also can be used for phase analysis of the susceptibility components in the same manner as in methods. The advantage of this method is that tuning the crossing angle θ is sometimes easier than varying the light intensity.

Acknowledgments

I. I. G. thanks the INTAS-Belarus-1997 project for partial support of the research presented and Rector of the Republican Institute of Higher Education Prof. P. Brigadin for his attention and support. We would like to express our special thanks to the Editors for their great job.

REFERENCES

1. P. A. Apanasevich and A. A. Afanasiev, "Parametric Interaction of Light Waves in Resonant Media," Preprint, IF AN BSSR, Minsk, 1972.
2. N. Blombergen, "Nonlinear Optics," p. 424. Mir, Moscow, 1966.
3. V. M. Petnikova, S. A. Pleshanov, and V. V. Shuvalov, *Herald Moscow State Univ.* 6, 24 (1984).
4. P. A. Apanasevich, "Foundations of the Theory of Interaction of Light with Media," p. 495. Nauka i Tekhnika, Minsk, 1977.
5. R. Pantel and G. Pythoph, "Foundations of Quantum Electronics," p. 384. Mir, Moscow, 1972.
6. Y. B. Band and R. Balvi, *Phys. Rev. A* 36, 3203 (1987). Y. B. Band, *Phys. Rev. A* 34, 326 (1986).
7. N. S. Onischenko, *Herald Belaruss. State Univ.* 9 (1985).
8. S. A. Akhmanov and N. I. Koroteev, "Methods of Nonlinear Optics in the Spectrosscopy of Light Scattering," p. 544. Nauka, Moscow, 1981.
9. Sh. D. Kakichashvili and Ya. A. Shvaitser, *J. Appl. Spectrosc.* 6, 1022 (1985).
10. L. M. Barkovskii, G. N. Borzdov, and F. I. Fedorov, "The Wave Operators in Optics," Preprint 304, p. 46. IF AN BSSR, Minsk, 1983.
11. F. I. Fedorov, "Optics of Anisotropic Media," p. 351. IF AN BSSR, Minsk, 1958.
12. F. I. Fedorov, "Theory of Gyrotropy," p. 456. Nauka i Tekhnika, Minsk, 1976.
13. V. N. Severikov, "Computational Methods of Polarization of Natural Types of Laser Resonator Oscillations," Preprint 165, p. 33. IF AN BSSR, Minsk, 1978.
14. R. Azzam and N. Bashara, "Ellipsometry and Polarized Light," North-Holland, Amsterdam, 1987.
15. M. A. Kashan, *Opt.* 66, 205 (1984).
16. L. I. Burov, Fam Vy Tkhin', and A. M. Sarzhevskii, *Opt. Spectrosc.* 46, 945 (1979).
17. K. I. Rudik and O. I. Yaroshenko, *Quantum Electron.* 584 (1981) (in Russian).
18. N. Bloembergen, Ed., "Nonlinear Spectroscopy," p. 586. Mir, Moscow, 1979.
19. I. I. Gancheryonok and V. A. Gaisyonok, *GTF* 19, 63 (1993) (in Russian).
20. L. I. Burov and I. I. Gancheryonok, *Opt. Spectrosc.* 60, 567 (1986) (in Russian).
21. I. I. Gancheryonok, *Jpn. J. Appl. Phys.* 31, 3862 (1992).
22. G. Korn and A. Korn, "The Manual on Mathematica" (in Russian), p. 436. Nauka, Moscow, 1973.
23. A. Tumaikin, W. Van Haeringen, and D. Lenstra, *Physica C* 114, 251 (1982).
24. A. M. Tumaikin and V. I. Turtsmanovich, *Radiophysics* 28, 51 (1985) (in Russian).
25. P. F. Liao and G. C. Blorklund, *Phys. Rev. Lett.* 36, 584 (1976).
26. V. M. Arutyanyan, S. A. Agadzhanyan, A. Zh. Muradyan, A. A. Oganyan, and T. A. Papazyan, *Opt. Spectrosc.* 58, 275 (1985) (in Russian).
27. A. A. Polyakov, V. N. Trukhin, and I. D. Yaroshetskii, *Sov. Tech. Phys. Lett.* 8, 439 (1982).
28. F. I. Fedorov, *J. Appl. Spectrosc.* 2, 523 (1965) (in Russian).
29. L. M. Barkovskii and F. I. Fedorov, *Opt. Spectrosc.* 34, 1193 (1973) (in Russian).
30. I. I. Gancheryonok, *Rev. Laser Eng.* 20, 813 (1992).
31. V. Demtreder, "Laser Spectrposcopy" (in Russian). Nauka, Moscow, 1985. A. L. Schawlow, *Rev. Mod. Phys.* 54, 697 (1982).
32. J. J. Song, J. H. Lee, and M. D. Levenson, *Phys. Rev. A* 17, 1439 (1978).
33. M. D. Levenson and J. J. Song, *J. Opt. Soc. Am.* 66, 641 (1976).
34. H. Ma, A. S. L. Gomes, and C. B. Araújo, *Opt. Lett.* 17, 1052 (1992).
35. A. Marcano, L. Marquez, L. Aranguren, and M. Salazar, *J. Opt. Soc. Am. B* 7, 2145 (1990). A. Marcano, L. Aranguren, and J. L. Paz, in "Nonlinear Phenomena in Fluids, Solids and Other Complex Systems" (P. Cordero and B. Nachtergaele, Eds.), p. 405. Elsevier Science, Amsterdam, 1991.
36. S. Saikan and J. Sei, *J. Chem. Phys.* 79, 4146 (1983).
37. L. M. Barkovskii and G. N. Borzdov, *Opt. Spectrosc.* 39, 150 (1975) (in Russian).
38. G. N. Borzdov, L. M. Barkovskii, and V. I. Lavrukovich, *J. Appl. Spectrosc.* 25, 529 (1976) (in Russian).
39. L. M. Barkovskii, G. N. Borzdov, and A. V. Lavrinenko, *J. Phys. A Math. Gen.* 20, 1095 (1987).
40. K. G. Budden, "Radio Waves in the Ionosphere," Cambridge University Press, Cambridge, 1961.
41. P. M. Morse and H. Feshbach, "Methods of Theoretical Physics," McGraw-Hill, New York, 1953.
42. L. M. Brekhovskikh, "Waves in Layered Media" (in Russian). Nauka, Moscow, 1973.
43. L. M. Barkovskii, *Sov. Phys. Crystallogr.* 21, 445 (1976) (in Russian).
44. L. M. Barkovskii, *J. Appl. Spectrosc.* 30, 115 (1979) (in Russian).
45. L. M. Barkovskii, G. N. Borzdov, and A. V. Lavrinenko, *Dokl. Akad. Nauk BSSR* 32, 424 (1987) (in Russian).
46. L. M. Barkovskii, G. N. Borzdov, and A. V. Lavrinenko, *Sov. J. Acoust.* 33, 798 (1987) (in Russian).
47. G. N. Borzdov, *J. Mod. Opt.* 37, 281 (1990).
48. G. N. Borzdov, *Sov. J. Crystallogr.* 35, 535 (1990) (in Russian).
49. I. I. Gancheryonok and A. V. Lavrinenko, *Opt. Appl.* 25, 93 (1995).
50. M. Born and E. Wolf, "Principes of Optics," Pergamon, Oxford, 1968.
51. N. K. Rumyantseva, V. S. Smirnov, and A. M. Tumaikin, *Opt. Spectrosc.* 46, 76 (1979) (in Russian).
52. O. A. Rabi, A. Amy-Klein, and M. Ducloy, "Technical Digest of 4th European Quantum-Electronics Conference" (P. De Natale, R. Meucci, and S. Pelli, Eds.), Vol. 1, p. 907. Firenze, Italy, 1993.
53. Yu. P. Svirko, N. I. Zheludev, "Polarization of Light in Nonlinear Optics," John Wiley & Sons, Chister, 1998. 231 p.
54. S. Saikan, *J. Phys. Soc. Jpn.* 50, 230 (1981).
55. M. A. Scarparo, J. J. Song, and J.H. Lee, *Appl. Phys. Lett.* 35, 490 (1979).
56. K. Kubota, *J. Phys. Soc. Jpn.* 29, 986 (1970).
57. J. B. Kim, J. M. Lee, and J. S. Kim, *Opt. Lett* 16, 511 (1991).
58. J.-C. Keller and C, Delsart, *Opt. Commun.* 20, 147 (1977).
59. I. I. Gancheryonok, A. V. Lavrinenko, and V. A. Gaisyonok, *Pisma Zh. Tekhn. Fiz.* 20, 53 (1994) (in Russian).
60. A. S. Chirkin and V. I. Emel'yanov, Eds., "Nonlinear Optical Processes in Solids," *Proc. SPIE* 1841 (1991).
61. I. I. Gancheryonok, *Sci. Int.* 4, 307 (1992).

62. A Owyoung, *IEEE J. Quant. Electron.* QE-14, 192 (1978).

63. I. I. Gancheryonok, Y. Kanematsu, and T. Kushida, *J. Phys. Soc. Jpn.* 62, 1964 (1993).

64. C. Delsart and J.-C. Keller, *J. Appl. Phys.* 49, 3362 (1978).

65. H. E. Lessing and J. Von Jena, "Laser Handbook," p. 793. North-Holland, Amsterdam, 1986.

66. D. S. Alavi, R. S. Hartman, and D. H. Waldeck, *J. Chem. Phys.* 92, 4055 (1990).

67. N. Preffer, F. Charra, and J. M. Nunzi, *Opt. Lett.* 16, 1987 (1991).

68. I. I. Gancheryonok, S. Saikan, and T. Kushida, *Thin Solid Films* 234, 380 (1993).

69. V. A. Gaisyonok, I. I. Gancheryonok, A. P. Klischenko, and T. Kushida, "Technical Digest of the Fourth European Quantum Electronics Conference," Vol. 1, p. 258. GNEQP, INO, LENS, Firenze, Italy, 1993.

70. I. I. Gancheryonok, I. V. Gaisyonok, P. G. Zhavrid, and V. A. Gaisyonok, *Proc. SPIE* 2340, 60 (1994).

71. I. I. Gancheryonok and V. A. Gaisyonok, *Opt. Spectrosc.* 75, 1296 (1993) (in Russian).

72. I. I. Gancheryonok, *Rev. Laser Eng.* 20, 502 (1992).

73. H. Ma, L. H. Acioli, A. S. L. Gomes, and C.B. Araújo, *Opt. Lett.* 16, 630 (1991).

74. B. Stahlberg, K.-A. Suominen, P. Jungner, and T. Fellman, *J. Opt. Soc. Am. B* 8, 2020 (1991).

75. K.-A. Suominen, S. Stenholm, and B. Stahlberd, *J. Opt. Soc. Am. B* 8, 1899 (1991).

76. K. I. Rudik, L. G. Pikulik, and V. A. Chernyavskii, *J. Appl. Spectrosc.* 45, 283 (1986) (in Russian).

77. A. M. Bonch-Bruevich, T. K. Rasumova, and I. O. Starobogatov, *Opt. Spectrosc.* 44, 957 (1978) (in Russian).

78. V. M. Arutyunyan, S. A. Agadzhanyan, and A. Zh. Muradyan, *Opt. Spectrosc.* 58, 459 (1985) (in Rusian).

79. F. Liao and G. C. Bjorklund, *Phys. Rev. A* 15, 2009 (1977).

80. N. N. Zheludev, Yu. P. Svirko, "Nonlinear Polarization Spectroscopy of Semiconductors," Moscow: VINITI, 1990. – P. 82–182. (in Russian).

81. M. P. Auzinsh, *Chem. Phys. Lett.* 198, 305 (1992).

82. A. Yariv and P. Yeh, "Optical Waves in Crystals," Wiley-Interscience, New York, 1984.

83. S. Saikan, *Jpn. J. Appl. Phys.* 23, L718 (1984).

84. L. I. Burov and I. I. Gancheryonok, *Opt. Spectrosc.* 60, 567 (1986) (in Russian).

85. A. I. Kurasbediani and V. V. Mumladze, "Optoelektr. Kvant. Elekt. Appl. Opt." (in Russian), p. 122. Metsniereba, Tbilisi, 1980.

86. L. I. Burov and I. I. Gancheryonok, *Herald BGY Ser. 1*, 5 (1985) (in Russian).

87. L. I. Burov and I. I. Gancheryonok, *Herald BGY Ser. 1, 61* (1984) (in Russian).

88. V. A. Pilipovich and A. A. Kavalyov, p. 173. Nauka ir Tekhnika, Minsk, 1980 (in Russian).

89. A. P. Voitovich and V. N. Severikov, "Lasers with Anisotropic Resonators" (in Russian). Nauka i Tekhnika, Minsk, 1988.

90. V. P. Novikov and M. A. Novikov, *J. Appl. Spectr.* 31, 894 (1979) (in Russian).

91. A. E. Kaplan, *Opt. Lett.* 8, 560 (1983).

92. S. G. Zeiger, "Theoretical Principles of Laser Spectroscopy of Saturation" (in Russian). LGY, Leningrad, 1979.

93. L. I. Burov and I. I. Gancheryonok, *J. Appl. Spectrosc.* 44, 328 (1986) (in Russian).

94. T. Tsukiyama, M. Tsukakoshi, and T. Kasuya, *Opt. Commun.* 81, 327 (1991).

95. I. I. Gancheryonok and T. Kushida, *J. Phys. Soc. Jpn.* 62, 3071 (1993).

96. T. Yajima and H. Souma, *Phys. Rev. A* 17, 309 (1978).

97. M. V. Entin, *Fiz. Tekhn. Poluprovodnikov* 23, 1066 (1989) (in Russian).

98. I. I. Gancheryonok, "Nonlinear Photoanisotropy of Isotropic Resonance Media," Minsk: Belarusian State Univ. Press, 2000. – 209 p.

99. I. I. Gancheryonok, "Nonlinear Photoanisotropy of the Solution of Complex Molecules" (in Russian), Dissertation, Cand. Phys.-Math. Sci. 01.04.05. Belarusian State Univ. Minsk, 1987.

100. I. I. Gancheryonok and V. A. Gaisyonok, *Pisma Zh. Tekhn. Fiz.* 19, 1 (1993) (in Russian).

101. E. L. Ivchenko, *Fiz. Tverdogo Tela* 28, 3660 (1986) (in Russian).

102. S. Nakayama, *Rev. Laser Eng.* 14, 129 (1986).

103. S. M. Saltiel and P. D. Yankov, Thesis, p. 79. Moscow University, 1985.

104. A. V. Lavrinenko, I. I. Gancheryonok, and T. Dreier, *J. Opt. Soc. Am. B,* 18, 225 (2001).

105. F. I. Fedorov, "Theory of Elastic Waves in Crystals," Plenum Press, New York, 1968.

106. A. V. Lavrinenko, I. I. Gancheryonok, and D. N. Chigrin, *Proc. SPIE* 3580, 2 (1998).

107. A. V. Lavrinenko and I. I. Gancheryonok, *Tech. Phys. Lett.* 25, 100 (1999).

108. A. V. Lavrinenko and I. I. Gancheryonok, *Opt. Spectrosc.* 86, 889 (1999).

109. K. Nyholm, R. Fritzon, and M. Alden, *Opt. Lett.* 18, 1672 (1993).

110. G. Zizak, J. Lanauze, and J. D. Winefordner, *Appl. Opt.* 25, 3242 (1986).

111. A. V. Lavrinenko, I. I. Gancheryonok, and T. Dreier, *J. Opt. B: Quantum. Semiclass.* 3, S202 (2001).

Chapter 13

SECONDARY ION MASS SPECTROMETRY AND ITS APPLICATION TO THIN FILM CHARACTERIZATION

Elias Chatzitheodoridis, George Kiriakidis
IESL/FORTH, Crete, Greece

Ian Lyon
Manchester University, England

Contents

Handbook of Thin Film Materials, edited by H.S. Nalwa
Volume 2: Characterization and Spectroscopy of Thin Films

ISBN 0-12-512910-6/$35.00

1. INTRODUCTION

Secondary ion mass spectrometry (SIMS) is a technique that uses ion beams to bombard the surfaces of sample materials and extract secondary ions for analysis. The secondary ions are focused into a mass spectrometer, analyzed by mass, and measured. The idea of producing secondary ions by bombarding the surfaces of solids with primary ions dates back to the middle 1930s with the work of Arnot and Milligan [1]. The first commercial SIMS instrument was produced in 1961 [2] and has continued to evolve to the present day.

SIMS is a technique that can provide a variety of information and which has certain advantages over other analytical techniques. Elemental and isotopic information can be obtained along with compositional or isotopic ion imaging, and when it is combined with depth profiling one can acquire compositional maps in two or three dimensions. Sensitivity is high because of the absence of inherent background and results in low detection limits, down to the ppb (parts per billion) range. The analytical dynamic range covers the whole periodic table, starting from hydrogen. Depth resolution can be made extremely small, down to a few tenths of nanometer, by varying the ion beam energy. Spatial resolution is also high (in the submicron range), depending on the ion source used, and sample preparation of solid samples is minimal.

Because of these characteristics, SIMS has been extensively used in the research of thin films. The growth process of thin films requires compositional precision and purity and must be optimized to enhance their physical and chemical properties. Doping and impurity contents are also important chemical information, which can be acquired with SIMS. For example, with depth profiling the efficiency of the doping process is tested, and detailed studies of the interaction of multilayer films at their interfaces or with the substrate can be performed. Although not extensively used, crystallographic information can also be obtained. Furthermore, the compositional distribution in three dimensions such as gradient layers (gradient change of composition or phase with depth), multilayer films (abrupt changes of chemistry with depth), and phase separation of chemically complicated films, are most commonly studied. Finally, thin films generally go through processing that affects their chemistry and structure. SIMS can study the effect of the annealing process on the distribution of certain elements or reactions at the interfaces of different materials. In SIMS, ions are focused into a fine beam, and, thus, *in situ* analysis of very small volumes of material is possible.

Instrumental advances have made this technique a very important tool for elemental, isotopic and molecular analysis of any kind of solid materials and their surfaces. Variations of the technique exist, depending on the instrumentation used, the kind of the bombarding species, and the mode of analysis. Different sputtering ions are used for different purposes (Cs^+, O^-, Ar^+, and Ga^+ are used in the main), and different spectrometers may be used to mass disperse the ions. Each of these techniques has strengths and applicability.

Time-of-flight mass spectrometry (ToF-SIMS) takes advantage of the difference in traveling times of the ions from the sample to the collector as a consequence of their difference in mass. It is more appropriate for large molecules, and it is used mainly for the analysis of organic materials. With the improvement of fast counting electronics, ToF-SIMS also gives the possibility of fast analysis of the whole spectrum of masses at once in simple instrument geometries.

Quad-SIMS uses a quadrupole mass analyzer and is appropriate for relatively low cost, robust applications.

Magnetic sector mass dispersion can achieve very high mass resolution and is therefore best suited to the accurate measurement of isotope ratios.

Considering the mode of operation, static SIMS is generally used for the characterization of surfaces at very low sputtering rates (removing only one monolayer in several hours), and dynamic SIMS is used for bulk analysis or depth profiling (up to a few microns per hour). All of these different mass spectrometer types and modes of operation will be considered in Section 4.

There are some practical limitations in SIMS. Depth profiling is limited by recoil in the sputtering process, resulting in compositional mixing and a so-called thickness broadening of the film. Matrix effects are common, resulting from changes in ion yields of species, depending upon variations in sample chemistry. Ion yields may also vary with crystallographic orientation.

Thus, to be quantitative, the technique requires extensive calibration and study of a wide range of materials as well as the use of complementary techniques. Like SIMS, some of the complementary techniques are destructive, but some are not. Destructive techniques are *Auger-electron emission spectrometry* (AES), *photoelectron spectrometry* (XPS), and *glow discharge optical emission spectrometry* (GDOES), which use an ion beam to sputter off surface material and analyze it. From these methods, AES can provide high-resolution depth profiles (lateral resolution values in the range of a few nanometers) and chemical information. XPS can provide chemical bonding information for organic and inorganic materials. GDOES is a quick method with low vacuum requirements, and sputtering rates of up to 100 nm/s. Nondestructive methods for depth profiling of films include *total reflection X-ray fluores-*

cence (TXRF); *Rutherford backscattering* (RBS), which uses He backscattered atoms and has a resolution of 2–20 nm and can provide quantitative composition measurements and thickness measurements, and *analytical transmission electron microscopy* (ATEM), which has a high depth resolution at a few nanometers but requires special sample preparation (very thin films).

This work introduces the SIMS technique and its application in thin films research. Following this introduction, Sections 2–8 cover the fundamentals of the technique, instrumentation, and the methodology of analysis and characterization, and Section 9 gives some examples of the problems that SIMS has been used to solve and the materials analyzed.

In more detail, Section 2 describes the sputtering process and consequent effects, such as recoil mixing, emission processes, and ionization. It refers briefly to available models that simulate the above processes and introduces some terminology. It concludes with a description of the damage that the sputtering process causes to the sample and the analytical significance of that damage. A discussion of difficulties characteristic of the technique, such as matrix and fractionation effects, and preferential sputtering during the initial steps of sputtering, are discussed in Section 3. Section 4 describes the overall design and the specific components of a SIMS instrument. The ways in which a SIMS instrument can be operated are introduced briefly. Some parameters that are used to characterize and compare the current instruments are explained in Section 5.

Quantification is discussed in some detail in Section 6. Theoretically, SIMS has a capability for standardless analysis: this means that the absolute composition of an unknown sample can be measured without reference to a standard material. Here it is explained why this has not yet been realized. A short reference is also made to several models that tried to predict the composition of unknown samples from sputtering yields and other parameters.

Sections 6 and 7 also provide practical procedures for the quantification of thin films, which are offered with the warning that they must be used with care because they do not apply to all materials or all analytical conditions.

In Section 7, depth profiling is introduced. This is an important methodology that has found many applications in thin film analysis. When material is sputtered away from the sample, the surface is dynamically changed and material is removed. Craters are formed that become deeper with bombardment time. This introduces a third dimension to SIMS analysis: chemical information with depth.

Section 8 introduces ion imaging, a result of either the scanning capabilities of SIMS instruments or of specialized ion optics. Chemical mapping is a combination of ion imaging with quantitative analysis, in which the absolute composition of an area of the surface of the sample may be determined. When depth profiling is included, then three-dimensional maps are acquired.

Section 9 reviews the applications of SIMS to thin films. The material here is organized into smaller paragraphs, each describing a certain application or material. This study is not completely bibliographically exhaustive, and there may be some omissions. However, an effort has been made to give a good understanding of the SIMS technique and its possible applications to thin film characterization.

2. FUNDAMENTALS OF THE SIMS TECHNIQUE

2.1. Basic Characteristics

The SIMS technique is capable of high-sensitivity microvolume elemental and isotopic *in situ* analysis, surface analysis, depth profiling, and element/ion imaging capabilities in three dimensions. For these reasons the technique has been further developed, and it is now one of the most significant tools for the characterization of various materials. However, SIMS has been mainly used as a qualitative or semiquantitative method. This is due to the many factors that occur during the analysis and give rise to a degree of irreproducibility in the measurements. Only recently it has been possible to use this method as an analytical technique capable of quantitative analysis. Many instrumental improvements have caused this to come about, as has the development of appropriate methodologies. Yet, there are still difficulties, and SIMS can be classified as a routine quantitative analytical technique only in the cases where methodologies for a certain group of materials have been fully developed. This means that there is not a small set of unified reference materials (standards) that can be used for all of the unknown materials.

Because reference materials could not be applied in a simple way, several researchers have considered the possibility of quantitative SIMS analysis without the use of reference materials. Systematic research and a considerable amount of physical modeling and simulation have shown that a full understanding of the physical processes involved has still not been achieved and will not be for some time. This process will be more fully considered in Section 6.

For some analytical methods only one or a small set of standards is required. As mentioned earlier, this is not true for SIMS, because it is a matrix-dependent method, which means that the composition of the standard must match the composition of the unknown material. Every sample must be treated as a unique case, or systematic work must have been done previously on the dependence of measurements on the specific matrix. Such work has yielded calibration curves or tables showing relative sensitivity factors (RSFs) for different sputtered species from a given matrix. Data and procedures for the quantification of elements in a variety of matrices such as semiconductors [3–5] and diamond [6], organic matrices [7], glasses [8], trace elements in silicates [9] and multielement glasses [10], and rare earth elements (REEs) in phosphates [11] and metals [12, 13] can be found in the available bibliography.

Two sets of basic effects can be considered problematic during SIMS quantitative analysis. First, effects that are related to sputtering and ionization processes (ionization effects) and second, instrumental effects, which occur after the ions have been created and during their path through the transmission optics into the spectrometer and onto the collectors.

The second group, instrumental effects, is easier to estimate with the use of standards, because they depend on instrumental factors such as the optical transmission of the instrument and the detector efficiency. These can be combined into one factor called the instrumental transmission factor (f), which can be expressed as their product, or as the ratio of the sputtered ions that enter the mass spectrometer and reach the collectors divided by the total number of ions created during the sputtering process (we will discuss these factors in more detail later). Consequently, the most important problems left are ionization effects because they vary between different samples in an unpredictable manner. Among these effects are:

(1) matrix effects, which are related to the chemistry and the structure of the samples;
(2) type of primary ions and the type of secondary ions created after bombardment;
(3) the orientation of primary beam relative to the possible crystallographic axes of the sample and its atomic structure;
(4) the angle of incidence of the primary beam and the angle of the secondary emission of the ions relative to the surface of the sample;
(5) the presence of reactive elements or other contamination on the sample surface, which will change the matrix composition and hence will complicate the matrix effects;
(6) the energy or velocity of the secondary ions; and
(7) the ionization level of the secondary ions. There is no unified way to treat and measure the ionization effects because they can change a great deal with small changes in the sample's conditions.

2.2. Bombardment, Recoil, and the Sputtering Processes

At high energy, a primary ion colliding with a surface transfers its energy to atoms in the sample. If the energy that is transferred by the primary ions is greater than the binding energy of the atoms of the sample, then the created recoil (displaced and moving) atoms can collide with other atoms in the sample and, through binary energy transfer, initiate a series of cascades. If these misplaced atoms are located close to or at the surface of the sample and have sufficient energy to break their bonds with the neighboring atoms, they may leave the surface of the sample as sputtered secondary atoms. The collisions between primary atoms and atoms in the sample could be elastic, if energy and momentum are conserved, or inelastic when energy is lost in electronic excitation. Under common sputtering conditions, the estimated lifetime and dimension of a cascade are on the order of 10^{-11} to 10^{-12} s and 100 Å, respectively [14].

The primary ions can be implanted in the sample or resputtered immediately or later. The depth of the implanted primary atoms depends on the bombarding energy and can be simulated with a Gaussian curve for amorphous materials. In crystalline materials, though, the concentration profile differs from the Gaussian, and higher depths can be achieved in the channeling

directions. During ion bombardment the implantation profile builds up until a steady state is reached.

Primary recoil atoms can create secondary recoil atoms by a series of binary collisions. This can be maintained if the initial energy of the bombarding species is very high and most of it is transferred to the recoil atoms. At lower energies, or when recoil atoms have lost most of their energy, the energy is transferred to lattice vibrations. Recoil mixing can be produced by the displacement of target atoms to greater depths. This is a very important effect in depth profiling analysis because it reduces the depth resolution of the profile. There is a family of computer codes simulating such effects, collectively known as TRIM-type codes. The T-DYN (TRIM-DYNAMIC) code has been developed [15] specifically for the case of recoil mixing and film broadening during SIMS bombardment. The code can successfully simulate sputtering and recoiling of atomic species and the broadening of very thin metal films (marker broadening) in Si and Ge amorphous matrices (Fig. 1) when no strong chemical effects are involved in the process. Simulation and experiments agree well and show that recoiling creates profiles with a highly skewed Gaussian curve, extended to the deeper parts of the sample.

Some of the recoil target atoms are backscattered and have enough energy to escape from the sample. These atoms are mainly neutral, but a very small proportion of them (generally much less than 1%) are ions [16]. These ions could be positively or negatively, singly or multiply charged. In addition to atomic species, molecular species and heterogeneous or homogeneous clusters of atoms can also be emitted. Emission of electrons by Auger processes or by energy transfer also takes place, and electromagnetic radiation is possible, from infrared, visible, and ultraviolet to the X-ray region. Singly charged mono-atomic ions usually dominate the sputtered species, and

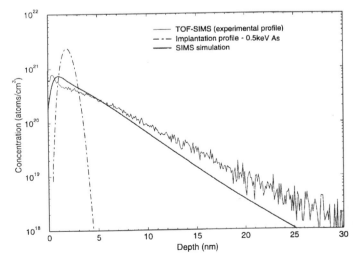

Fig. 1. A shallow implant of 0.5 keV As^+ originally restricted in the first 5 nm (broken-dotted line) is broadened to about 20 nm during SIMS analysis (noisy line). TRIM-DYN simulations can approximate the effect (heavy-dark line). From [15] with permission from Elsevier Science.

whether a species will be sputtered as a positive or negative ion is controlled by its electron affinity [16].

For SIMS, only the ionic species are used because only charged particles can be accelerated and analyzed by mass spectrometers. Several models have been proposed to explain the sputtering and ionization processes, such as Sigmund's collision cascade model, the bond-breaking model, the surface excitation model, Scroeer's perturbation model and variations, which are discussed in some detail in [14, 17–19]. There is not a single model that precisely describes all of the physical processes occurring during ion bombardment and can apply in all kinds of materials. This is the major weakness of the SIMS technique.

2.3. Energy and Angular Distribution of Sputtered Ions and Ion Yields

Sputtered secondary ions leave the surface of the sample with a wide spread of energies. The distribution curve of ion energies peaks at very low values, typically at about 10 eV. The distribution curve is very steep from the low energy side starting from zero energy and slowly decreases on the high-energy side (Fig. 2). The energies can extend to more than 1000 eV, depending on the primary beam energy and the binding energy of the target atoms. A knee on the curve is observed between 10 and 100 eV, which is possibly an indication of more than a single mechanism of ion formation [16, 20]. Further systematic observations suggest that both the position and the width of the peak depend on the element or the isotope and its chemical characteristics, whereas the magnitude and the shape of the peak seem to be a function of various factors, the most important being the chemistry and the chemical environment at the point of bombardment. The width of the energy distribution curve is also a function of the chemical bond that the element forms with

oxygen (the stronger the bond with oxygen, the broader the energy distribution curve of the element). A systematic pattern of the magnitude of the peak is formed when it is plotted against the atomic number of the elements present in the same matrix [10, 16].

The lower the probability that an element will escape as an ion from a given sample, the higher is the proportion of high-energy ions in the energy distribution [20]. Moreover, different isotopes have slight differences in their energy distribution curves, and it has been observed that there is a variable light to heavy (L/H) isotope enrichment over the energy spread, and the plateau formed in the L/H curve moves to higher energies with increasing mass [13]. It has also been observed that the negative ions have a sharper drop-off at their high-energy part of the curve compared to the peak, relative to the positive ions. Molecular ions show the same effects, but their peak widths are much narrower relative to their mono-atomic ions [16]. This is used as a practical method for resolving mono-atomic isotopic species from their molecular interferences with an energy windowing technique [16]. Generally the yield of diatomic ions is less than 1% compared with the number of mono-atomic ions although for some elements such as cobalt or tungsten, this could go up to 7% [16]. This observation is generally applicable to metals but silicates behave differently [10].

Secondary ions leave the surface of the sample at a wide range of angles. The sputtered flux of the secondary ions can be approximated with a cosine angular distribution [14]. In general the angular distribution depends on the charge and energy of the sputtered species, the angle of incidence of the primary beam, and the crystallography of the target sample. In polycrystalline samples, the high energy and multiply charged ions show flat distributions relative to their lower energy and singly charged counterparts. The mono-crystalline targets show a significantly different dependence that is mostly related to their crystallography and due to channeling of the primary ions along lattice directions or other effects [14]. Crystallographic dependencies can be lost by increasing the partial oxygen pressure or by selectively analyzing higher energy ions.

The bombarding species and its energy play an important role in the enhancement of sputtering yields. There is a correlation between the sputtering atomic yield and primary ion energy, and a plateau is observed at high primary ion energies. Chemical sputtering is the enhancement of the ionic yield of the target species, with reactive elements as a projectile. For example, Cs^+ bombardment increases the negative ion yield of a variety of elements such as hydrogen, oxygen, noble metals, and other electronegative elements [21–23]. O^+ and O^- bombardment, on the other hand, increase the positive ion yields [24].

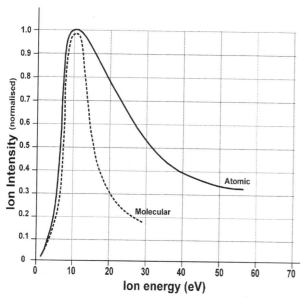

Fig. 2. Schematic demonstrating the relative energy distributions of secondary mono-atomic (wide; solid line) and molecular (narrow; broken line) ions.

2.4. Topographic and Compositional Damage During Bombardment

During ion bombardment the target surface is under continuous change because of destructive sputtering and the extraction of material. Apart from the expected craters (Fig. 3) [25] or

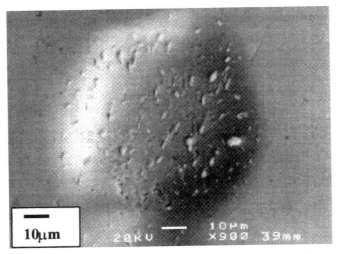

Fig. 3. SEM image showing a crater, that was formed by bombarding the surface of a magnetite crystal with a Cs$^+$ primary ion beam. From [25].

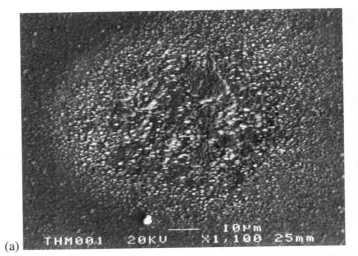

(a)

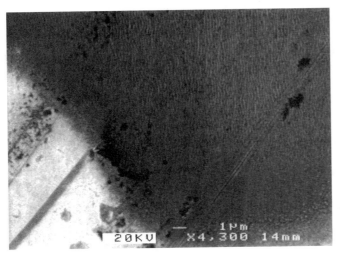

Fig. 4. SEM image of the waving surface topography, that was developed after sputtering of the surface of a silicate glass material (NBS 610 glass) with a Cs$^+$ ion beam. From [25].

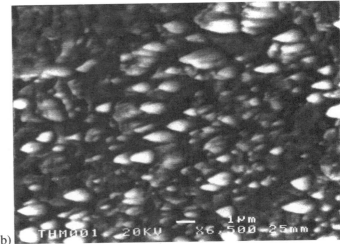

(b)

Fig. 5. (a) Surface topography created by Cs$^+$ bombardment on a pure copper metal (SEM image of a "burn mark" on a copper surface). Many cones are formed, the magnification of which is given in (b). The size of many of these cones at the main axis (vertical to the sample's surface) is more than 1 μm. From [25].

the wavy topography of the surface (Fig. 4) [25], there are additional changes in the topography of the surface [26–28] and in the composition. These topographical effects include crystal defects, such as dislocations, clusters and impurities within the lattice, amorphization, boundaries of crystals; impurities on the surface, inclusions, and impurities of the primary beam. All of these can lead to differential sputtering along the bombarded surface, resulting in the creation of distinct topographical features in both crystal and amorphous materials.

During bombardment, etch pits and pyramids are common in crystals [29], and they generally develop along the crystallographic surfaces of the material. Cones and pyramids, with distinct facets and edges, are observed to develop, mainly with an axis parallel to the primary beam direction (Fig. 5); differential sputtering of the different element constituents could also be responsible. However, they are also observed even in very pure materials. Some of these forms have been studied and simu-

lated with the use of erosion theory (see [14, 30] for a complete discussion and references on these topographic features). These topographic features are undesirable because they significantly reduce the depth resolution during depth profiling. Their size can be more than 1 μm. They can also produce apparent compositional changes if the sputtering rate changes.

A number of methods have been investigated that are aimed at eliminating these effects, including rotation of the sample [31–36] in combination with low-energy primary ion beams or with the use of two primary beams with different incident angles [37]. It has also been observed that bombardment with reactive primary ions, such as cesium or oxygen, can virtually eliminate these effects [14, 21, 24].

Compositionally induced effects are of similar importance to the topographically induced effects due to ion bombardment. These include: primary ion implantation, diffusion, preferential sputtering, and the already discussed recoil mixing. Cascade mixing (displacement of target atoms due to collision by recoil atoms) and recoil mixing are similar processes, and they have a primary importance in depth profile analysis because they reduce the depth resolution. Both are responsible for the broadening effect (marker broadening). It has been observed that the broadening effect depends on the velocity spectrum and the spatial distribution of the recoils and increases with increasing energy of the projectile. It also depends on the physical properties of the target material. Minimization of such effects can be achieved by using heavy projectiles at oblique angles. A detailed description and references on these effects can be found in [14], which further concludes that primary recoil implantation has the same importance as cascade mixing. The only difference is that mixing dominates at higher depths and has significant spatial dimensions. Recoil implantation can be preferential to mass [38].

3. FRACTIONATION EFFECTS

Isotopes or even elements are affected to a degree different from that of physical or chemical processes, because of their difference in mass. As a result, an isotope of an element of a system can be enriched compared with another if this system undergoes a transformation from one state to another. The enrichment is directly analogous to the reverse of the mass difference. This section refers to this type of fractionation effect, which, however, is caused by the sputtering and ionization process itself.

3.1. Energy-Related Mass Fractionation

There is a mass fractionation that is related to the energy of the ions [13, 20, 39]. This can be demonstrated with only three isotope plots [13]. By changing the voltage offset of the sample or by physically moving an energy window, which selects ions after electrostatic dispersion, it is possible to collect and measure secondary ions of different energies. It has been observed that the ratios measured form a straight line, which also goes through the accepted value, suggesting a true mass-dependent fractionation.

This effect introduces another complication. Spatial energy inhomogeneities of the secondary ions exist over the crater created by the primary ion beam. Consequently, with the collection of different parts of the sputtered cloud of the secondary ions different ratios are produced as well. These ratios occur along the same fractionation line given by changing the energy of the analyzed ion. In general, higher proportions of ions with higher energies are produced at the deepest parts of the crater when the primary beam is oblique to the sample surface, or again near the center of the crater when the angle of incidence is vertical to the sample surface. The ratios determined by collection of secondary ions from different parts of the cloud have been found to vary by more than 4%/amu for element B [13], suggesting that a careful tuning and collection of the secondary optics is necessary.

In conclusion, this energy-related mass fractionation causes variations in isotopic ratios when secondary ions of different energies are measured. The energy distribution curves differ between the different isotopes of an element resulting in different L/H ratios, where L and H represent the light and heavy isotopes of a single element. These observations have been made for pure metals and during steady-state conditions [13], avoiding the other effect of L/H enrichment observed at the initial stages of sputtering. It is also suggested that these fractionation effects are mostly related to ionization rather than to the sputtering processes.

3.2. Preferential Sputtering at the Initial Stages of Sputtering

Most research work, which involves depth profiling, tends either to ignore the very first data obtained during profiling or at least to study the related effect or to define its duration and consequently its depth. This effect is known as preferential sputtering, a preferential enrichment of the light isotope compared with the heavy one in the sputtered material. Effects such as preferential sputtering changing to steady-state conditions after some time are observed not only in isotopes, but are commonly observed in measured element ratios as well. Because they strongly correlate with depth or primary ion energy, they could possibly be used during depth profiling to estimate sputtering rates [40].

The light-to-heavy (L/H) isotopic enrichment at the initial stages of sputtering is a very interesting observation. The effect decreases with sputtering time until a steady-state value is reached [13, 41, 42]. The steady-state isotope ratio, however, is different from the true value of the bulk sample. The direction of change of this L/H enrichment, which is present until a steady-state condition is attained, is always the same, toward the real value (always decreasing). The L/H enrichment at zero-fluency, just when the first isotopes are sputtered, is not constant with species. It varies for different ion pairs and is not simply related to the mass difference of the isotopes. One constant, however, is that the magnitude of the enrichment of a specific isotope pair relative to its ratio at steady state seems to be the same for different materials. It is not related to charging problems [41]. Finally, the steady-state value is reached at depths between 150 and 250 Å [41].

The processes that first create the L/H enrichment and then the steady state are not well understood. It has been suggested [41] that the ionization probability is approximately constant during the sputtering process, from the zero-flux state to the steady state, and that the L/H effect is due to the differences in sputtering efficiency of different isotopes. The steady-state ion composition would then be simply related to the bulk sample composition but not equal to it, because of instrumental effects. Some support for this idea is shown by the observation that the relative ionization probability of two ions depends linearly on the inverse of the velocity with which the ions leave the surface of the sample [43]. According to this observation, shifts of more than 30 eV in the energy of the secondary ions are required to make the observed large fraction- ation. Consequently, this is an effect caused by the sputtering process [41]. The opposing view [13] that the fractionation arises from variable ionization has support, however, in that if fractionation were due to the sputtering process alone, then the light isotope would be preferentially sputtered and the composi- tion of the sample would be continuously altered. However, this is not the case, and therefore ionization effects probably have to be considered. Furthermore, a strong correlation between heavy isotope depletion and the inverse velocity of the secondary ions has been observed [20]. This leads to the conclusion that ion- ization processes are the main parameter for L/H enrichment. Extensive studies of this effect [38] suggest that preferential sputtering of the lighter isotope over the heavier isotope actu- ally changes the composition of the surface of the sample. The depth of the change is about the same with the primary beam penetration, and a steady state is reached after a layer of that depth has been sputtered away. During steady state the sputtered atoms have the same composition as the bulk composition of the sample because equilibrium is reached and further enrich- ment cannot occur. It has also been observed [38] that for low energies there is a strong correlation between the preferential sputtering effect and the projectile mass and energy, which is lost at high energies (primary beam energy more than 20 keV). Moreover, experiments on Ta_2O_5 have shown that highly graz- ing bombardment, at angles of about 80°, reduces preferential sputtering [44].

3.3. Matrix Effects

SIMS analysis is very sensitive to the chemical composition of the target (matrix effect). This is the effect that matrix composi- tion has on the yield of the extracted ions; it significantly affects the quantification of thin-film depth profiles or bulk composi- tional analyses. The ion yield is enhanced by the presence of reactive elements within the matrix or on the surface of the sample as a residual gas. For example, increased quantity of oxygen in the matrix or on the surface of the sample enhances the positive secondary ion yield of the elements by different amounts, depending on the oxygen affinity of each element [24, 45, 46]. In the same way Cs^+ enhances negative ions [46]. For this reason, oxygen and cesium primary beams are used, and their presence enhances the positive or the negative ion yields, respectively. With higher yields, variability is reduced but not eliminated. Other parameters, such as the substrate sticking coefficient, recoil implantation efficiencies, and the sputtering yield of the substrate, may cause variations in yield if they dif- fer from matrix to matrix. Systematics of relative secondary ion yields of negative ions produced with Cs^+ and of positive ions produced with O^- plotted against the atomic number show that elements with high ionization potentials have higher yields when they are sputtered with Cs^+, whereas the opposite is ob- served with O^- [22].

Systematic and reproducible fractionation due to matrix ef- fects has been observed in pure metals [13] and in silicate minerals [9]. It has been shown that for silicates there is a linear relationship between the ionization yield and concentration that could be used for quantitative analysis of the silicates.

3.4. Other Fractionation Effects

Other factors have been reported in the past to affect either quantification or isotope ratio measurements. There is evidence that the angle of collection of the secondary ions plays an important role in sputtering fractionation. It has been demon- strated that isotopically light material is sputtered back along the primary beam's direction, whereas heavier material is sput- tered along more oblique angles [42]. However, such effects have not been observed for metals, and they are probably mini- mized when high-energy ion beams are used [20].

Regarding the energy of the primary ions, it has been ob- served that both ion yield and isotopic fractionation increase with decreasing primary beam energy [13, 39], whereas the type of bombarding species affects the fractionation factor: the higher the mass of the bombarding species, the higher the frac- tionation. In the same way, the fractionation factor depends on the type of the emitted secondary species and the matrix from which it is emitted [13]. Furthermore, the ionization level of the secondary ions and the fluency of the primary ions are also factors that affect fractionation. Temperature could also affect ion yields in crystals, and a correlation has been found with the channeling directions of crystals [47]. Charging problems within the sputtered area can also shift the position of the en- ergy distribution curve variably and randomly in front of the energy window, introducing unpredictable results. Thus, selec- tive suppression or extraction of the sputtered species is used in a well-controlled manner [48].

Fragmentation and recombination processes, which form ei- ther molecular ions from their atomic counterparts or atomic ions from molecular ions, influence fractionation [13]. It is important to mention, however, that such effects can be used positively, as in the case of MCs^+ ions, where M is any neu- tral secondary atom of the sample produced during sputtering with a Cs^+ ion beam. In such a case it has been observed that matrix effects are minimized [49–55] or eliminated under cer- tain conditions [37]. There are also cases in which this is not clearly demonstrated [37] and quantification can still be matrix dependent [56].

4. INSTRUMENTATION

4.1. Basics of a SIMS Instrument

A SIMS instrument is basically a sensitive mass spectrometer. The major difference from the conventional mass spectrometer is in the process of ion formation. The ion beam sources (ion guns) ionize the bombarding species, and specialized electro- static devices (primary ion optical components) are responsible for focusing them into a fine beam. The beam can be kept steady or scanned over the sample's surface and sputters secondary ions.

Similar ion optical components are also used to extract and focus the produced secondary ions onto the entrance slit of the mass analyzer (secondary ion optics). The energy and the angular distribution of the secondary ions are very wide. The electrostatic components are thus designed to minimize the distribution of the secondary ions and match the acceptance requirements of the analyzer for higher transmission (lower loss of secondary ions that collide with the walls of the instrument). Part of the primary ion optics, the secondary ion optics and the sample reside in the source housing. This is separated from the mass spectrometer by a fine slit (the source slit). The two parts of the system can be considered independent and are under ultra-high vacuum. The primary ion gun (e.g., a gas source duoplasmatron) may output a gas load into the vacuum system, which must be removed by a differentially pumped chamber.

Only a microvolume of the sample is analyzed, and ultra-high vacuum conditions are essential for preventing contamination by residual gases residing in the instrument. This improves its analytical sensitivity. Depending on the type of mass analyzer (magnetic, time-of-flight, quadrupole), different arrangements can be employed, some utilizing multicollection facilities. Computerized control and power supply systems are also necessary to control and supply every element independently and to process the rapidly acquired data, especially during ion imaging.

Apart from the secondary ions, other radiation and particles are created, such as electrons. A scintillator can be used to convert these electrons into photon currents and the SIMS instrument can operate as a low-resolution secondary electron microscope (SEM).

In this section we describe most of the elements that make up a SIMS instrument (also called an ion probe, Fig. 6). Instrumental problems, such as charging of the surface of insulating samples during bombardment, are also discussed.

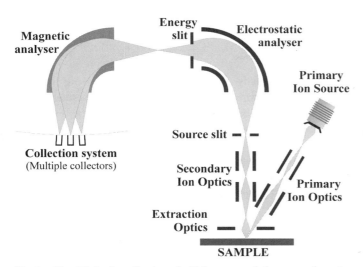

Fig. 6. Simplified schematic view of a high mass-resolution magnetic sector SIMS instrument with multiple collectors. The primary, extraction and secondary optics system is more complicated than shown here.

4.2. Theoretical Background

The source housing of a SIMS instrument comprises elements of both the primary and secondary ion optics. These elements include cylindrical (condenser) or planar lenses, which when an electrostatic field is applied to them can focus the primary ion species to a fine spot on the sample, or focus and transport the secondary ions into the entrance slit of the mass spectrometer. Their operation is similar to that of visible light optical lenses. The movement of ionic species through these elements is constrained by Liouville's theorem. This theorem states that during the motion of an ideal system of ions (an ideal system is one in which the ions do not interact between each other) the local density of the ions in position and momentum space (phase space) stays constant [14, 57]. A consequence of this is that a system of electrostatic lenses carries out geometrical transformations of the extracted ions, which can be expressed in a simplified form with the Helmholtz–Lagrange invariant in Gaussian optics,

$$a_1 x_1 V_1^{1/2} = a_2 x_2 V_2^{1/2}$$

Here a is the angular and x the spatial spread of the beam in a particular direction, and V is the voltage or energy of the ions at two different positions 1 and 2 along their path. This is generally useful during the design of the instrument for calculating the minimum spot size required to achieve full transmission through the entrance slit to the mass spectrometer.

In a real system, however, the optical elements of the instrument or even the forces between the moving ions themselves (known as the space charge resulting from their mutual charge repulsion) introduce various aberrations. These aberrations affect the diameter and shape of the primary and secondary beams. Serious aberrations of this kind are the on-axis chromatic aberrations (energy spread of the accepted ions) and spherical aberrations (horizontal and angular divergence). A direct application of this is the calculation of the size of the ion image formed at the collectors in comparison with the image at the source slit. Careful design of the optical system of the instrument is intended to minimize aberrations. In addition specific optical elements and slits, such as the electrostatic analyzers, the alpha slit (a slit located just before the source slit at the entrance of the mass spectrometer), and the energy window are used to reduce angular, space, and energy spread with a consequent loss of transmission. The slits are used to impose geometrical constraints on the moving ions, and only those that can go through the opening proceed further into the analyzers, while the rest collide with the walls of the slits.

4.3. Primary Ion Beams and Ion Guns

A variety of bombarding species can be produced when different types of ion sources are used. Bombarding species such as noble gases, chemically active gases, and metals have been tested [58, 22–24, 45]. The bombarding species is generally in a charged state (as in SIMS), but neutral atoms can be also used to avoid sample charging (in secondary neutral mass spectrometry, SNMS), especially when insulators have to be analyzed. Chemically active elements, such as oxygen or cesium, are most commonly used because they greatly enhance the secondary yields of positive and negative ions, respectively [21, 24, 45]. Primary ion sources such as duoplasmatrons have been used for gases and surface ionization, and liquid metal sources have been used for metallic primary beams. In all types of primary sources the primary optics system is basically the same.

The duoplasmatron is a device that uses a low-pressure discharge to ionize elements in the gaseous phase. It may be used to generate species such as Ar^+, O_2^+, O^-, N_2^+, CN^-, and F^- [58]. Duoplasmatrons are a standard primary ion source in SIMS instruments, capable of creating high primary beam currents of up to several microamperesQQ and (after suitable collimation and loss of ion current) very small spot sizes, generally less than 1 μm for positive primaries. Primary beam purification is essential with gas discharge plasma sources because of the variety of chemical species produced, and a magnetic sector field or a Wien filter can be used for this purpose. These filters can also be used to select different bombarding species when multielement gases are used, such as a mixture of argon and oxygen in the same bottle.

Surface ionization sources are based on the thermal ionization of neutral atoms evaporated from a heated surface. The most common bombarding species is cesium (Cs^+). The low energy spread and the high intrinsic brightness of that type of ion gun are an advantage over duoplasmatrons, giving them the capacity to form smaller and chemically pure ion beams [14]. However, the toxicity of Cs makes necessary some safety precautions when it is operated, for example, when it is warmed up or shut down and during refill. The main component of the Cs gun is the reservoir, which is heated to about 250°C to vaporize the Cs metal. The vapor is then ionized by passing it through a porous tungsten plug (or tungsten frit), which is heated to about 1100°C. Then, ions of the mono-isotopic $^{133}Cs^+$ are accelerated to high voltage (usually about 10 kV).

The above sources are the most common sources, and many instruments incorporate both in different geometrical arrangements, sharing the same primary column. Special electrostatic optical devices are at the entrance of the primary column to select either ion source (beam combiner). A beam combiner is a pair of curved plates and deflects either of the beams onto an aperture. The beam is focused by a set of electrostatic lenses (condensers). The deflectors correct the path of the ion beam by aligning it, but when they are located close to the sample they are used to scan the beam on the surface of the sample. A beam aperture is placed at a conjugate point to the sample and is used to define a round and uniformly illuminated spot (Koehler illumination). Astigmatism of the beam is corrected by a set of eight pin-shaped stigmators positioned close to the final lens (objective lens). A pair of horizontal deflector plates (dog leg) deflects the beam parallel to its axis, preventing neutrals from hitting the sample. To achieve optimum primary beam transmission (highest current on the sample) with the smallest spot size, all of the voltages applied to the electrostatic optical ele-

ments are carefully adjusted in a recursive operation. Imaging of samples with a specific geometry, such as microscopy metal grids, with the use of secondary electrons or ions can also help in the adjustment of the optimum focusing conditions.

4.4. The Sample System and Secondary Ion Optics

An advantage of SIMS is the very easy sample handling requirements. Generally the samples are not treated, and raw materials can be analyzed. However, the sample surfaces are preferably flat to reduce any distortions of the high extraction field that is applied to them. The surfaces of insulating samples must be made conducting, to minimize charging effects produced by the charged bombarding species.

Because SIMS is a surface analytical technique, contamination of the sample's surface must be minimized. It is, however, almost impossible to avoid contamination, even during sputtering, because atoms from the residual gases of the surrounding environment still arrive on the surface of the sample, even at the location under bombardment. Consequently, an ultra-high vacuum (UHV) is desirable in the source housing, the part of the instrument where the sample is positioned for analysis. Precleaning of the surfaces under vacuum conditions or by low-density ion bombardment with an inert element can be used to minimize surface contaminants. Storage housings located just before the source housing have carousels to hold a number of samples; these act as a clean environment where the samples can be stored before analysis and as an air lock between the ultra-high vacuum of the source housing and the atmosphere. A sample manipulator probe is used in most modern instruments to collect the sample from the carousel and deposit it on the holder in the source housing.

Ultra-high-vacuum conditions are equally important in the mass analyzer itself, as this reduces any effects produced by the collision of secondary ions with residual gas molecules and atoms, which scatter the ion beam and consequently reduce the mass resolution. The vacuum in the mass analyzer is usually in the range of 10^{-9} to 10^{-11} mbar.

4.5. Mass Analyzer

It has been already mentioned that mass analysis is most commonly performed with a magnetic field generated by an electromagnet or by time-of-flight measurements. The magnetic analyzer discriminates against the mass of the ions when the centrifugal force, mu^2/R, is balanced by the Lorentz force, euB. For a given magnetic field (B) and for singly charged (e) mono-energetic (same velocity u) ions, ions of different mass (m) are deflected, each at a different radius (R). Consequently, for a given radius only one mass is collected, according to the equation

$$R = \frac{1}{B}\sqrt{\frac{2\,mV}{e}}.$$

The magnetic analyzer also has focusing capabilities, and the ions of a single mass focus onto a single point, where they are measured by the collector. The minimum isotope separation can be calculated when the geometrical characteristics of the magnet are known. With the positioning of more collectors at these well-defined locations, simultaneous collection of more than one isotope can be achieved (multicollection). The large energy spread of secondary ions creates a serious problem for achieving high mass resolution that is overcome by the use of double focusing. Ions focused on the source slit of the mass spectrometer are dispersed in energy, so that isotopes of a given mass enter the magnetic field at different positions according to their energy, and the momentum dispersion of the magnetic field is arranged geometrically such that ions of the same mass are brought to a focus at the same point irrespective of their initial energy.

Time-of-flight instruments are based on the difference in arrival time of the ions on a single collector. A pulse of the primary ion beams produces a number of secondary ions, which are accelerated to have the same kinetic energy. Light ions arrive at the collector earlier than heavier ions. The collector receives bunches of ions, each having ions of the same mass, and, if it is fast enough, it can discriminate between the different groups of ions of increasing mass. Different time intervals correspond to ions of different mass.

Time-of-flight (ToF) instruments have the advantage of collecting ions of all masses during a small time interval. Repeated pulses reduce the counting errors and improve the statistics of the measurement. Fast collectors and counting electronics improve the mass discrimination of ToF systems. However, they are based on one collector, which has to be able to measure low- and high-intensity signals. Generally, the dynamic range of a single detector is not high enough, and this poses a limitation to ToF instruments. ToF instruments are also very capable in analyzing heavy ions, generally ion blocks of atoms common to the analysis of organic materials.

4.6. Detection System

Single-collector and multicollector SIMS instruments are available in the market. In a magnetic analyzer more than one detector (or collector) can be used, and these are positioned in the focal plane of the magnet. Multicollection is desirable for SIMS with a magnetic analyzer because

(a) It can significantly reduce the time of analysis and the sample consumption. This improves the depth resolution of the analysis.

(b) Fluctuations of the magnitude of the secondary ion signal, caused mainly by primary beam fluctuations or sample charging, affect all measured ionic species equally and to a first order do not affect the elemental or isotopic ratio.

(c) The dynamic range of the mass spectrometer is greatly increased with the use of different types of collectors.

This last feature is important for isotopic analyses where one or more isotopes of interest have very low abundance relative to others and a single detector has an insufficient dynamic range.

This also applies to elemental analysis when trace elements, for example, must be compared with an element of the main matrix, a common aspect of quantitative characterization of thin films. However, a drawback of multicollection is that it is impossible to measure masses that are very far from each other because the focusing plane of the magnet is generally only several mass units wide.

In multicollection systems, the different detectors have different sensitivities, and absolute ratios therefore cannot be easily acquired. In this case frequent calibration must be performed using standards. In electron multiplier detectors a further problem is dead time, introduced if the beam intensity is high. Dead time is defined as the time required by a detector to be free from the previous signal. In more detail, when an ion hits the surface of an electron multiplier and deposits its charge, more electrons are created and produce an avalanche of electrons. The dynode material of the multiplier then requires a time to recharge, governed by the resistance and capacitance of the system. This is typically a few tens of nanoseconds. If a second ion enter during this recharge time, it is unlikely to generate a sufficiently large electron avalanche to be detected. Faster collectors, more sophisticated counting systems, or statistics can improve or help to estimate the real value (although statistical methods should be avoided).

4.7. Electron Gun and Charge Compensation

The use of ionized beams in SIMS results in the deposition of charge locally on the surface of the sample. This is not a problem with most conductive materials, because the charge can be dissipated from the bombarded area, but it is a problem for insulating samples such as most oxides. Systematic or random errors can be introduced into the measurements because of charging due to fractionation of isotopic or elemental ratios, since they are energy dependent.

The charging effect is greater when positive primary beams (such as Cs^+) are used in contrast to the negative beams. This is probably due to the fact that with negative beams the excess electrons can move very easily in the surrounding sample, especially when the surface of the sample is coated with a conductive material. However, free electrons are not always available to compensate for positive charge when positive primary beams are used.

Charging effects can be observed as changes in the initial energy of the secondary ions. This leads to a mismatch with the acceptance of the secondary ion optics, and slowly the absolute ion yield is reduced. In this case, the loss of the secondary beam can be generally corrected by slightly offsetting the voltage of the sample; the secondary ions retrieve their initial energy and correctly follow the path through the electrostatic analyzer (ESA) and the energy window. In severe charging, either the secondary ions are completely suppressed or the primary ion beam is deflected to another area and the secondary ion signal is lost. Then, charging must be compensated for by other means.

Several methods have been proposed to compensate for charging problems, such as the use of very thin samples, low primary beam currents, metal grids, or apertures to surround the analyzed area; increasing the conductivity of the sample by thermal heating, bombarding the sample with metals; and others [59]. However, all of these methods severely affect the structure of the thin films or the result, especially during quantitative analysis or imaging. The use of primary beams of neutral species, such as in SNMS, reduces charging effects.

The most common way to compensate for charging effects, currently found on all SIMS instruments, is the use of electron flood guns [60, 61], in most cases combined with a conducting coating of the surface such as carbon, gold, or aluminum. Depending on the mode of analysis (positive or negative), electron flood guns must be capable of a large dynamic range of energies (typically up to 10 keV). When operation is in negative mode, the sample is offset at a very high negative voltage. This decelerates the coming electrons, and an electron nebula is formed in front of the sample at optimized conditions of the electron gun energy. Self-compensation is achieved, depending on the sample's charging condition [61]. For analyses in the positive mode the electrons are accelerated toward the sample, and the electron gun can be set to lower energies. In either case, it might be difficult to set both the primary ion beam and the electron beam to focus at the same point on the sample, while at the same time it is impossible to visualize the beams. Generally, experience has to be gained with samples for which visualization is possible (such as fluorescent surfaces or carbon tabs [61]) under the usual analytical conditions. Then, when the desired samples are analyzed, the electron beam energy can be set so that the secondary ion current maximizes and stabilizes.

4.8. Control System and Other Units

Apart from the systems described above, a SIMS instrument also includes more devices that are useful for performing additional tasks. A control unit is generally based on a computer with the appropriate interfaces to acquisition and power supplies when digital control is required. A scintillator and a photo-multiplier with a control unit are present to acquire secondary electron images from the scattered electrons produced during ion beam bombardment. A microscope or optical camera and lighting devices can provide an optical image of the analyzed sample.

4.9. Modes of Operation

The versatility of the SIMS analytical technique is due to the many modes of operation available. The analysis can be restricted to either the very surface of the sample (static SIMS), slowly analyzing monolayers of the sample's surface with minimum destruction, or rapidly removing monolayers from the surface of the sample (dynamic SIMS). With dynamic SIMS the surface of the analyzed sample changes dynamically with time, and the physical, chemical, and structural nature of the surface is altered or destroyed. This destructive process and the fast removal of material accomplished with dynamic SIMS is not only useful for bulk analysis of samples, but also makes depth profiling possible, providing a third dimension for analyses. In imaging mode the primary beam can be scanned over a certain area of the sample, and chemical analysis can be made or structural data can be acquired in the two horizontal dimensions. Ion imaging combined with depth profiling gives true three-dimensional information for the sample. This information can also be acquired for a whole range of masses, working in positive or negative mode.

Depth profiling is useful for thin film characterization, providing information on the chemical structure of the films in the third dimension, the interactions and reactions at the interface, and the depth of films at high resolution.

SIMS analysis can be made semiquantitative when standards, calibration curves, or physical models are used, although quantitative information can only be acquired under fully controlled operating conditions with the use of well-defined standards, sometimes assisted by good physical models. The modes of operation and quantitative analysis with SIMS will be discussed in more detail in the following sections.

SIMS is also well known as a surface characterization technique. During static SIMS, a single ion of the primary beam affects only a certain area of the sample. If the density of the bombarding beam (primary ion fluency) is very small, generally equal to or less than 1 nA/cm^2, the probability of a bombarding ion hitting an already damaged area is very small. Under such low fluencies the lifetime of a monolayer can be several hours, and the very low secondary ion yields make high-sensitivity and high-transmission instruments essential. Static SIMS is a very sensitive technique used in the study of the elemental composition and the chemical structure of the monolayers of the sample surfaces. Some applications are monolayer analysis, surface reaction studies, catalyst characterization, polymers, oxidation of surfaces, and the detection, identification, and structural analysis of nonvolatile organic molecules. Static SIMS is generally done with a time-of-flight mass spectrometer, because of its high transmission and the capability of multiple species analysis.

5. FACTORS CHARACTERIZING A SIMS INSTRUMENT

5.1. Instrumental Transmission and Detection Efficiency

Instrumental limitations result in elemental and isotopic fractionation during analyses as an effect of the differences in mass and energy of the secondary ions. Loss of transmission may occur in the ion optical elements because of second-order aberrations and on slits. Differences in the efficiency of the different collectors, when multicollection systems are used, demand that appropriate calibration procedures be adopted. Detector efficiency usually depends of on momentum, resulting in mass-dependent fractionation. Finally, pressure plays an important role in the degradation of the ion beams due to the collisions of the ions with residual gases in the spectrometer, resulting in an increase in the background levels. The instrumental fractionation is mass dependent, and there is evidence that it follows an exponential power law [14].

The overall transmission of a SIMS instrument is defined as the ratio of the secondary ion current as measured on the collectors divided by the total ion current produced by the sputtered secondary ions at the source. For an isotope A of mass M with a fractional isotope abundance α_M and concentration in the sample of $c(A)$ analyzed with a primary beam of current I_p and giving an ion yield during sputtering of $Y^+(A)$, the instrumental transmission factor f^+ is estimated by the formula that follows and varies considerably between different instruments [14]:

$$f^+(A, M) = \frac{I_s^+(A, M)}{I_p \cdot Y^+(A) \cdot \alpha_M \cdot c(A)}$$

This factor takes into account the efficiency of the transmission by the extraction optics into the mass analyzer, the transmission through the mass analyzer, and the detection efficiency. It should be the same for similar instruments under the same conditions. However, variations are possible because of the differences in the initial energy and angle of emission, as well as the mass of the measured secondary ion.

5.2. Mass Resolving Power, Mass Resolution, and Peak Interference

During isotopic or elemental analyses two different species may have only slightly different masses and as a consequence appear very close in the mass spectrum. The contribution of ions from the interfering mass to the mass of interest can introduce variable systematic errors. Various methods have already been developed, generally with the use of computer program, to analyze spectra with peaks acquired at low mass resolution and to predict the contribution of the interfering masses [62, 63]. However, working at high mass resolution, especially when working with chemically complex materials, is a better way to overcome such problems [64, 65].

Two definitions are generally used to express the capability of a SIMS instrument to resolve between two interfering peaks (Fig. 7). Mass resolving power (RP) is defined as $RP = M/\Delta M = M/(B - A)$, where M is the mass for which the peak scan has been acquired and $\Delta M = (B - A)$ is the mass difference of the two sides of the peak at 10% height. Mass resolution (MR), on the other hand, is defined as $MR = M/\Delta M = M/(C - A)$, where M is again the mass of the peak and $\Delta M = (C - A)$ is the mass difference between the one side of the peak at 10% height (point A) and the mass at which the peak flat starts (point C). Both definitions give dimensionless numbers, and both can be applied even when only a single peak is available. From the definitions it is straightforward that mass resolution gives higher values than mass resolving power. Furthermore, the RP values depend largely on the width of the detector slit, whereas the MR values on the width of the secondary ion beams currying the signal; however, we should

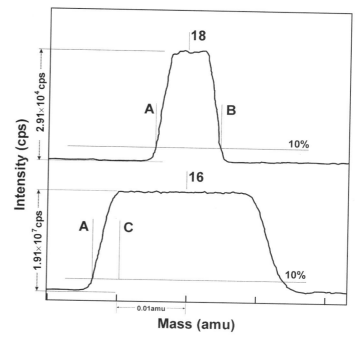

Fig. 7. Peak shape with Gaussian sides of real mass scans (isotopes 16 and 18 of oxygen [25]) and the definitions of mass resolving power and mass resolution (see text for more details). The ratio $^{18}O/^{16}O$ is fractionated compared with the natural abundance ratio, mainly by instrumental factors.

always take care that the beam width is smaller than the collector slit.

5.3. Abundance Sensitivity

The abundance sensitivity is a measure of the ability of a mass spectrometer to measure a low-abundance peak in close proximity to a high-abundance one. It is defined as the ratio of the intensity of the background in the wings of a peak measured on a mass unit away from that peak to the intensity of the peak itself. This is also a measure of the smallest ratio that can be measured between two peaks with very close masses but very different intensities.

Abundance sensitivity is a dimensionless number. It is also different at different masses and at different resolving power; consequently, a standard reference mass and resolving power have been defined to be used as the absolute reference between different instruments. A measurement of the intensity ratio I_{237}/I_{238} with a mass resolving power of 400 gives a standardized measure of the abundance sensitivity achieved by the instrument. It is actually a measurement of the contribution of the very strong peak ^{238}U to the background, which is measured at mass 237 [14].

The background intensity in the wings of a peak arises principally from the scattering of ions by residual gas in the mass spectrometer. The abundance sensitivity is therefore dependent on the pressure in the analyzer and the source. The lower the pressure of the system, the less the scattering of the ions there is and the higher the abundance sensitivity is. In more detail, as the ions travel from the source through the analyzer, they collide with residual gases present in the analyzer. These collisions could be elastic or inelastic and generally increase the background around the peak. Elastic collisions change the direction of the ions and contribute equally to the background at both the lower and higher mass sides of the peak. Inelastic collisions can change both the direction and the energy of the ions. When the energy of the ions is changed it is generally reduced so that inelastic collisions contribute more to the background of the low mass side of the peak. Because inelastic collisions are more common than elastic, the abundance sensitivity measured at the high mass side of the major peak is generally better than that at the low mass side. The abundance sensitivity of a SIMS instrument can be increased by

(1) achieving better ultra-high-vacuum conditions in both the analyzer and the source—lower pressure means fewer residual gas atoms and molecules and consequently fewer collisions with less scattering;
(2) using a second ESA for a further energy discrimination step; and
(3) increasing the resolving power of the analyzer (by increasing the dispersion length of the magnetic analyzer), which also increases the distance between the peaks and reduces their widths.

The abundance sensitivity is a crucial parameter that determines the background level in the neighborhood of intense peaks and hence detection levels for rare elements or isotopes. At optimum conditions the detection limit can be very low. It has been demonstrated [66] that Te atoms doped in GaAs at an atomic density of about 10^{12} atoms/cm^3 (roughly 0.23 ppb) can be detected at short acquisition times (about 100 s). Favorable conditions are high secondary ion yields (enhance with electronegative or electro-positive ion beams, such as O_2^- or Cs^+) and no mass interferences. The high ionization probability of Te (on the order of 50–70%) is already known ([3], where high ionization probability is indicated by the low relative sensitivity factor (*RSF*) values; see the next section for a definition of *RSF*). An alternative to high mass resolution for resolving molecular interferences is energy filtering. This technique relies upon the molecular spectrum decreasing far more rapidly with increasing secondary ion energy than for atomic ions. At very high secondary ion energies (> 100 eV) the molecular contribution at a particular mass can often be negligible.

The abundance sensitivity must be optimized for the characterization of thin films and other modern materials. Size minimization and high integration of today's devices make them more prone to contamination and defects or, device operation can be better optimized with, for example, precise definition of doping quantities. Consequently, it is important to know how to set the instrument for high abundance sensitivity and to be able to estimate it.

6. QUANTIFICATION

6.1. Theoretical Background

SIMS could be capable of quantitative analysis without the use of standards if there were available a full understanding of all of the details of the sputtering mechanism and how this varies between different chemical matrices. However, because this situation has not yet been realized, semiquantitative methods are used that have fewer problems, as the absolute value for the concentration of a species in the sample is not required. The major factor that arises is due to abundance partitioning between ions, neutral atoms and molecules, as these species are sputtered from the sample surface. Generally $\ll 1\%$ of the sputtered species are ions, and changes in sputtering conditions and the matrix can therefore have dramatic effects on the ionization efficiency (and hence, ion yield). Further fractionation factors are introduced by instrumental fractionation. Although these factors are not simply related to the masses of the species they are easier to quantify when combined into a single parameter. We have already discussed in the previous paragraphs the effects introduced in SIMS, which are very difficult to control. Taking into consideration these problems, we now see that standardless quantitative analysis is very difficult to achieve. The easiest way to achieve this kind of analysis is either to create a physical model, which precisely explains the ion yields of elements and isotopes under all of the different conditions and from all of the different matrices, or to make a very systematic study of the behavior of the different species in the same range of samples and conditions. The second approach is probably impossible because of the large number of observations needed. It can be used, however, to make observations that can contribute to the development of a single physical model of the prediction of ion yields from different species sputtered from any material. Currently, no such model exists, and the only solution for quantitative analysis is the use of standards, which is also problematical, because SIMS is very sensitive to matrix effects. Nevertheless, many successful attempts have been made in both directions, suggesting that it is possible to achieve reasonable quantitative analysis with SIMS.

Quantitative information for the sample can be acquired by the measurement of the secondary ion current $I_s^+(A, M)$ of an isotope A with mass M, which is linked with concentration according to the following formula:

$$I_s^+(A, M) = I_p \cdot Y_{total}(A) \cdot a^+ \cdot c(A) \cdot a_M \cdot f^+(A, M)$$

Here, I_p is the primary ion current, $Y_{total}(A)$ is the total sputtering yield of the isotope A, a^+ is the ionization probability, $c(A)$ is the concentration of the isotope A in the sample, a_M is the fractional isotope abundance of A, and $f^+(A, M)$ is the instrumental fractionation factor [14]. Some of the above parameters can be measured; others can be estimated with the use of standards, but still, some are very sensitive to matrix effects, making general-purpose quantitative analysis procedures impossible to establish.

6.2. Computational Models for Quantification

An approach taken by several researchers in the past was the creation of physical models that will precisely predict the concentration of any element or isotope if information for the instrument's operating parameters or information about the sample and the SIMS spectrum are given. All include some physical factors in their calculations, such as the binding energy during the sputtering process, the work function during the ion emission process, or the influence of the ionization energy on the ion yield. However, some of these models rely on the use of internal standards. Such samples can be developed from well-known samples that have been analyzed with other quantitative methods or have been created by mixing reservoirs of known composition and concentration or, again, by implanting known quantities of isotopes in the matrix of the sample. The results generally show uncertainties between 10% and 20%. It is beyond the scope of this study to give a detailed description of these models. An extensive review of all of these methods can be found in [14], and a comprehensive collection of them can be found in [57, 67, 68]. However, a very brief description of some of these models follows.

The kinetic model [69–71] is based on collision cascades during the bombardment, which release electrons and neutral species. The neutral species are later de-excited via Auger processes to ions. The secondary ions, which are taken into account in this model, are those with kinetic energies higher than 30 eV. There are no fitting parameters needed, and the system works very well for pure metals, which are bombarded with noble gas ions. The auto-ionization model [72–74] assumes an inner shell excitation into an auto-ionizing state. A relaxation follows via an Auger process when the sample is bombarded with inert gas ions. Deviations of the analytical results are within a factor of 2. The surface effect models comprise a group of models [75–77] that are based on the idea that all of the processes happen very close to or on the surface of the sample. When a particle is ejected it changes its electronic structure, and the probability that this ion will escape to a certain distance from the surface is then a measure used to calculate the ion yield. Only semiquantitative analyses can be obtained with this model, in as much as concentrations can only be defined to within a factor of 2. The thermodynamic models are also a group of models that comprise the local thermal equilibrium model (LTE) [78, 79], and the simplified two-fitting parameters LTE model (SLTE) [80, 81]. According to these models, plasma is created locally on the sample when it is bombarded, which is in complete thermodynamic equilibrium. It is assumed that the created plasma has the same composition as the sample and that neutralization processes are not taken into account. Temperature is introduced as a factor to correct for matrix effects. The thermodynamic models are internal standard methods because they take into account the measurements of at least two elements with known concentrations in the same matrix. Computer algorithms have also been developed by iterative methods. Such algorithms are CARISMA [82] and QUASIE [81]. These give similar results with an accuracy of about 20%, classifying them as semiquantitative methods [14]. QUASIE uses the collision cascade theory, which suggests that sputtering and ionization arise because of collision cascades initiated during bombardment. Again, a temperature parameter must be introduced in the algorithm, which is calculated from the kinetic energy of some reference ions and, consequently, depends on them as well as on the matrix and the oxygen content.

When isotopic analyses are required, in principle they should be easier to perform because we are only interested in isotopic ratios. However, most of the already discussed difficulties are still present. Matrix effects, preferential sputtering, and, to some extent, recoil mixing result in mass-dependent fractionation. However, the most common differences are related to the ionization probabilities and the energy differences, which are related to the mass differences in the isotopes, also seen as differences of the energy distribution curves [83]. This could be minimized by collecting the whole range of energies during isotopic measurements (wide energy windows), something that it is instrumentally difficult because of the limited range of acceptance of the analyzer. As a consequence, other methods should be used for the correction of the instrumental fractionation based on either external or internal calibrations with standards [84]. With the external calibration method the measured isotopic composition of the unknown sample has to be compared with the known isotopic composition of a standard that must have the same or similar chemical composition to avoid matrix effects, and the measurements must be made under conditions that are as close as possible to being identical. For elements with only two stable isotopes (such as H, Li, B, C, and N) external calibration has to be applied, but for elements with more than one isotope (such as O, Si, etc.) both internal and external calibration can be applied. Internal calibration is based on the use of a third isotope to calibrate for any fractionation that causes the results to deviate from the expected mass-dependent fractionation. It has the advantage that precise results are acquired because the measurement and the calibration are done simultaneously and consequently under the same instrumental and physical conditions. However, to use this method we must be sure that no other nonlinear fractionation effects, due to mechanisms such as radioactive decay, nucleosynthetic processes, and photochemical reactions, have changed the composition of any of the reference isotopes. Possible mass interferences can be corrected instrumentally by energy windowing, peak stripping, or much better by completely resolving the interfering peaks by mass with high-resolving-power mass analyzers.

Isotopic analysis is only used in special circumstances in thin film characterization or in the characterization of related technological materials. However, there are some cases where thin films are doped with materials of certain isotopic composition and isotopes are then studied to investigate chemical or physical effects [85, 86].

6.3. Practical Methods of Quantification

6.3.1. Calibration Curves

Quantification with calibration curves is based on a large set of standard materials of similar matrix composition to reduce matrix effects, which are very well characterized by other quantitative methods. These materials are selected so that they either have a specific element at different concentrations or they are doped later with the same element but at variable quantities. The SIMS instrument is then kept within reproducible conditions during the analyses of these standard materials and all fractionation factors are combined into one. The results form a calibration curve (Fig. 8), when they are plotted as the concentration of the element of interest against the ratio of its ion current in the unknown sample to the ion current in the reference material. The bulk matrix composition of the unknown sample must again be similar to that of the standard, and the concentration of the analyzed ion must be within the range of concentrations covered by the calibration curve used.

The calibration curves are generally continuous and increase monotonically with concentration. For low concentrations of the element of interest (generally below 1%) they can be considered as linear. An example of the development and uses of these curves (also called working curves) is given by [11] for the quantitative measurement of REE in phosphates; the precision of the method is better than 10%. A compilation of quantitative studies with the use of calibration curves [57] concludes that the precision of the analyses ranges between ±15% and a factor of 2, and results with less spread are acquired when the sample is saturated with oxygen.

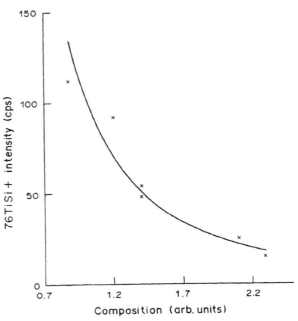

Fig. 9. A curve relating the intensity of ^{76}TiSi$^+$ secondary ions to composition of the TiSi$_x$ thin films (x ranges between 0.88 and 2.30). From [87] with permission from Elsevier Science.

This last characteristic can also be used to find out which are the matrix properties of the sample: the higher the oxygen concentration at the area of the analysis, the lower the matrix effects and the more numerous the different cluster oxide ions produced. Cluster oxide ions can be used by the method of fingerprinting spectra to calculate calibration curves, which do not depend on the matrix properties of the sample. Quantitative results of fingerprinting methods have a standard deviation of 10%, when the standard and the sample match in composition and the instrumental parameters are the same [57]. Another advantage is that we do not have to apply this method to matrices with high oxygen concentration, such as some glasses, because matrix effects are self-compensated, and the simple calibration curves can be used.

An example of the use of calibration curves in thin film characterization for technological applications is in the metallization of VLSI chips very large scale integrated with the use of titanium silicide films [87]. The resistivity of these films is a function of their composition phase. With SIMS the secondary ion yields of ^{76}TiSi$^+$ are acquired, which with the use of the calibration curve shown in Figure 9 can be translated into the exact composition phase of the TiSi$_x$ films. From this composition, the resistivity of the films can be directly deduced. The provided calibration curves can only be used for quantification purposes when the Si atomic ratio ranges between 0.88 and 3.20 in the films.

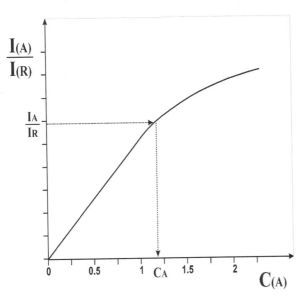

Fig. 8. A schematic representation of a calibration curve: I_A is the measured signal of ion A in the unknown sample by reference to the signal I_R (ratio I_A/I_R) of the same element in a known sample. The concentration c_A of A in the unknown sample is acquired from the curve. The concentration is expressed as a percentage. Generally for very low values ($< 1\%$) these curves are linear.

6.3.2. Relative Sensitivity Factors

A more modern method of quantification by SIMS uses (RSFs). These are values with the dimension of number density (atoms/cm^3). Multiplying the RSF value by the relative ion yield of an element should result in the concentration of this element in the unknown sample. They have to be free of matrix or instrumental effects, and they should be transferable through different instruments and laboratories.

More specifically, based on the observation that impurities and matrix elements are affected in a similar way by matrix compositional changes, their ratio can be used to correct measurements for matrix-effect free analysis [14, 88]. Consequently, for one isotope of an element A present in a matrix and calibrated to one isotope of the element M of the matrix, the RSF values are calculated according to the formula [3]

$$RSF_{A,M} = n_A \left(\frac{I_A/\gamma_A}{I_M/\gamma_M} \right)^{-1}$$

In this formula I_A and I_M are the intensities of the secondary ions of the selected isotopes of the elements A and M, γ_A and γ_M are the abundances of the isotopes of these elements, and n_A is the number density of the elements expressed in units of atoms/cm^3 (for example, in GaAs, n_{Ga} and n_{As} are 2.2×10^{22} atoms/cm^3).

Individual values or tables compiled for a large number of elements and for matrices of some technological interest have been published. Research work is continuing with the aim of covering more elements or matrix compositions or creating a better definition of these values. This work will not attempt to compile such values. The reader is advised either to refer to the publications cited here or, better, to search for updated values in more recent bibliographies. Corrections to the formalism are already present (such as the scaled sensitivity ratios, SSRs [88]), and some might have the potential to change the current views. However, in this work only the general trends of the data will be presented in a graphical format.

Two round-robin studies have been performed in the past few years that involve a number of laboratories around the world, different instrumentation, and different analytical conditions [3]. The RSF values were studied for GaAs matrices highly doped by several elements. Single-element doped crystals were prepared that contained Si, Cr, Zn, In, and Te, and multielement crystals were doped with Mn, Fe, and Cu. All samples, plus a blank one, contained B trapped from B_2O_3 during the growth of the crystal. Within this work the RSF values of the above elements are given for positive ions produced with O_2^+ ion beams and for negative ions and MCs^+ molecular ions (where M is the measured element) produced by Cs$^+$ ion beams. Variations of the results ranging within $\pm 50\%$ are observed between the different instruments, probably caused by the difference in the angle of incidence of the primary beam. The impact angle significantly affects ionization probabilities, and these variations seem to be higher for elements with high RSF values (less sensitive elements). In similar studies [4, 5], semiconductors and silicon oxides were doped with a large number of elements. RSF values of more than 73 elements implanted in 23 matrices made of Si, Ge, Ga, As, Al, Ti, and their combinations are given in [4] for positive and negative ions.

RSF systematics are also available for SiO$_2$, semiconductors, and diamond [5, 6]. In more detail, RSF values are given

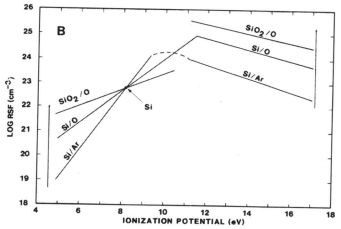

Fig. 10.   Trend lines of RSF values for elements analyzed by SIMS from Si or SiO$_2$ materials sputtered with O or Ar primary ions. From [5] with permission from the author and the American Institute of Physics.

for 25 elements implanted in matrices of unoxygenated Si (with Ar bombardment), partially oxygenated Si (oxygen bombardment), and fully oxygenated Si (SiO$_2$) [5]. The results are compensated for charging effects with the use of electron beams and Au coating. The normalizations used to acquire the RSF factors are relative to ^{28}Si$^+$ for pure silicon and for partially oxidized Si and relative to ^{30}Si$^+$ for SiO$_2$. Two trends have been observed (Fig. 10), a positive trend for ionization potentials between 4 and 10.5 eV, and a slightly negative trend for potentials over 10.5 eV. The first trend was already known as the "universal trend" [89, 90]. Within this trend range a number of lines with different inclinations are observed, which depend on the oxygen content (steeper lines correspond to pure Si). When these are calculated as effective temperature, they agree with the estimations provided by the "local thermal equilibrium model" [90]. This set of lines goes through one point, which by definition must be the ionization potential of Si. Elements with ionization potentials over 10.5 eV are halogens, O, C, and H and show different mechanisms of origin; the mechanism is possibly the increased competition from elements more electronegative than oxygen in oxygenated matrices. In conclusion:

(a) when the content of the surface oxygen is increased the Si$^+$ ion yield from Si increases as well. This causes rotation relative to the Si ionization potential, and this is explained as an increase in the higher ionization potential element yields compared with the lower ionization potential yields.

(b) Also, elements with ionization potentials lower than 4 eV (such as K, Rb, Cs) show 100% ionization. Because of this, only the Si$^+$ yield increases with increasing oxygen content, and, consequently, the RSF factors decrease.

(c) Competition of the highly electromagnetic elements with oxygen causes the observed effects; here we take into consideration the bonds that must break to form positively charged ions (see also [91]).

(d) A general conclusion is that the *RSF* values or the ion yields vary significantly even for a single matrix, depending upon the oxygen content of the surface from which they are emitted.

In the case of organic matrices [7] *RSF* values have been defined for quantification purposes and depth profiling on polymers poly(methyl methacrylat) (PMMA, Kapton and epoxy), which are implanted with more than 40 elements (Li, B, Na, 30Si, Fe, Ge, In, Cl, Be, S, Mg, Al, K, Ca, Cr, Ag, etc.). Measurements were made with Cameca IMS 3f or 4f instruments, with the use of 8 keV O_2^+ primary beams to extract positive ions and 14.5 Cs^+ for negative ions. The analyzed surfaces were coated with 50 nm Au, and charge compensation was performed with a flooding electron gun while depth profiles were measured with an Alphastep profilometer. It is observed that the patterns for both positive and negative ions versus ionization potential or electron affinity are like those for semiconductors Si and GaAs [4], for insulators [92], and for metals [12]. In Figure 11 a graph is given with a cumulative plot of *RSF* values for all studied elements versus their ionization potential or electron affinity. In this graph several branches are observed:

(a) the branch with electron affinity greater than 1.9 eV, which gives a constant $RSF = 3 \times 10^{21}$ cm^{-2};

(b) a line with a slope of 0.78 decades/eV for positive ions, which is similar to those of semiconductors and metals; and

(c) two lines with slopes of 0.32 decades/eV.

All of the above information shows that these *RSF* values can be applied for organics and polymers with a carbon density of

8×10^{21} cm^{-3}. Uncertainty values are large and within a factor of 3 for both negative and positive ions, probably because of the difficulties posed in the analysis of organics (crater depth measurement, volatility, change in structure or density, etc.). An effort was made to suppress molecular interferences with the use of an energy window offset by 75 V.

RSF values are also available for a large number of elements in $TiSi_2$, TiN, and TiW matrices [93] and in HgCdTe and CdTe matrices [94]. Furthermore, investigations on leached layers of glasses and their bulk glass show that they can be quantified with the use of similar *RSF* values, within error [8].

It has been experimentally observed that when a sample is bombarded with Cs^+, ionic species of the type MCs^+ are formed (*M* is any neutral elemental species of the sample) which can effectively be used for better quantitative results. According to some evidence [49, 50], these molecular ions are less prone to matrix effects and are used for quantitative depth profiling [53], ion imaging [49, 95], or film interface studies [96, 97]. Sputtered neutral atoms are probably less influenced by matrix composition, and with the available Cs^+ from the primary ion beam they combine to form MCs^+ molecular ions, also free of matrix effects. However, this observation is not always supported by all experimental results [37, 56], or mass interference limits the applicability of the method [52]. *RSF* values have still to be estimated based on a reference material, but they are largely matrix independent. The use of *RSF* values together with local atomic densities for each analyzed point can assist in the correct estimation of ion yields and assign relative depth scales to the analyzed sample volumes [49], especially when depth profiling is required. Local atomic densities can be determined in several ways or can be estimated from the respective pure elements.

The advantages of the use of MCs^+ ions are experimentally demonstrated on SiO_2 and Si films [50]. The intensity ratio of $SiCs^+$ is identical for SiO_2 and Si, when a 20% reduction of erosion rate in the oxide is considered, whereas in the case of the Si^+ ions it is at least two orders of magnitude higher. Moreover, the ratio of $SiCs^+:OCs^+$ in the oxide is actually 1 : 2, suggesting identical ionization probabilities. This is not only observed for samples with a variable concentration, but it is additionally demonstrated on depth profiles of multilayered Ni and Cr samples and on a-Si:H films grown on Si substrates. In the former case depth resolution is improved, and in the latter, Ar atoms trapped during the growth of the film could also be monitored.

6.4. Other Tools and Methods of Quantification

Secondary neutral mass spectrometry (SNMS) was first introduced in the early 1980s [98, 99]. Since then, it has been used mainly as a technique for comparison to SIMS. It has also been used in thin film research, and several studies have been made with the use of SNMS, such as depth profiling and characterization of multilayer structures [100], solar cells [18, 101], superconducting films and heterostructures [102], and providing *RSF* values for quantification or for the quantification of oxynitride

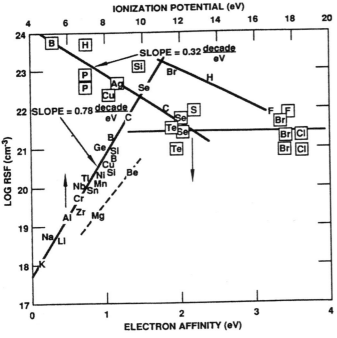

Fig. 11. SIMS *RSF* systematics plot for elements doped in polymers. From [7] with permission from the author and the American Institute of Physics.

films for ultra-large-scale integrated (ULSI) applications [103]. Because SNMS is based on the SIMS technique, several comparisons of these two techniques have been undertaken [52, 103], mainly because of the observed matrix-free [18, 101] analytical capabilities of laser-ionization SNMS [104–106] or e-beam ionization SNMS [107]. This makes the technique capable of far more accurate quantification [108, 109]. *RSF* values appear to be more stable [56, 107] when, at the same time, SNMS performs better with insulators, produces less topography, and can be sensitive in the ppm (part per million) range. Finally, although SIMS is still more sensitive than SNMS [18, 101, 110], the latter could assist the former at the very first steps of sputtering, before steady-state conditions are reached [110].

Despite the quantification limitations of both SIMS and SNMS [111] the advantages of other techniques are apparent and instrumental advances and computer-assisted interpretation methods [112] are expected to bring improvements. The latest SNMS instruments are now equipped with high-frequency mode (HFM) to resolve initial charging problems caused by secondary ions. Such systems have been efficiently used in diffusion studies between films and substrates [113] or in quantitative analysis of nonconducting silicate glasses and various oxide coatings [114]. HFM-SNMS uses a high-frequency plasma (10 kHz to 1 MHz) to compensate for charging. Quantification is performed with the use of *RSF* values, although variations within oxides and silicate glasses are high, and the same values cannot be applied to different systems [114]. The variations are of a factor of 3 and depend on the sample composition. However, HFM offers some advantages by increasing the stability within samples of a limited composition. In [114] *RSF* values are also available for oxides and glasses, as well as for calculation procedures.

Another quantification method, which can be characterized as process-specific, is the encapsulation-SIMS technique [115]. It is applied to oxide layers of n-type (100) Si wafers with the purpose of quantifying the oxygen content on the surface of silicon wafers. Investigations of encapsulation film erosion rates and film thickness are first performed on amorphous Si (α-Si) and poly-Si films grown on silicon wafers as an encapsulation layer. Ion cascade mixing causes peak broadening of both the native surface oxide layer and the interface layer (between encapsulation film and the wafer). Experiments have shown good separation between the sample surface and the interface, and reproducibility was increased for higher primary beam energies (also proved theoretically, see references in [115]). The study of the oxygen enhancement effects showed that over an oxide thickness of 0.66 nm the oxygen density is higher than expected by the thickness of the oxide layer and behaves nonlinearly. As shown with SIMS, this means that the self-enhancement of the oxygen intensity cause a nonlinearity when related to the oxide thickness and compared, for example, with Si. The mechanism is not well understood, but for quantification reasons, the oxygen content of native oxides of silicon wafers can be measured with a good accuracy when the interface between the encapsulation film and the silicon substrate is 8×10^{14} atoms/cm^2.

7. DEPTH PROFILING

7.1. General

Depth profiling is one of the most common characterization techniques of SIMS, especially when thin films are considered. The destructive operation of bombardment by ion beams allows the acquisition of compositional data with depth. Generally thin films involve a change of composition between individual layers or at interfaces, and, thus, compositional data can also provide measurement of the thickness of the films. A depth profile is a series of ion intensity measurements versus time. Further quantification is required to convert the ion intensity to actual concentration and the time into thickness. When I_A is the secondary ion intensity of an isotope of an element A, t is time, C_A is the concentration of A, d is depth, and f and g are conversion functions, then the conversion of the intensity–time relationship to concentration–depth can be simply expressed by

$$I_A = f(t) \quad \Rightarrow \quad C_A = g(d)$$

The first equation provides only qualitative information, and this is exactly what depth profiling analysis gives. The second equation results from the first after calibration and should give precise quantitative and structural information. Time versus depth and secondary ion intensity versus concentration are usually calibrated independently. This is not a problem when a single and relatively homogeneous material is analyzed, but common parameters are related in a complicated way, especially in the characterization of thin films, when at least two materials and one interface are present.

To convert ion intensity to concentration one has to consider what has been discussed already in the previous section about quantification. Converting time to thickness has also its own complications. Stable primary ion beam conditions (fluency and ion current: number of ions per unit area and per unit time, respectively) and stable conditions for secondary ion extraction and measurement can remove only instrumental effects. Additional complications are introduced that are intrinsic to the sputtering process, such as the variation in the sputtering rate due to matrix effects or recoil mixing of the atoms and roughness growth during sputtering. All reduce the precision of estimating the thickness of the films, and measurements need to be calibrated. A first and most commonly used approach for depth calibration is the estimation of etching rates in thick samples with composition similar to analyzed compositions; however, this cannot be used for very thin films.

For depth profiling, SIMS is used in the dynamic mode. Dynamic mode involves the fast and continuous transformation of the sample surface from an equilibrium condition to another as a result of the strong interaction of the primary beam atomic species with the atoms of the sample through a transfer of high kinetic energies. Apart from the removal of sample material in the form of atoms, clusters, and molecular species (a small portion of which are ionized), other processes, which soon come into equilibrium, also take place. Some of these processes been have already mentioned, but they are briefly summarized here:

(1) preferential sputtering and fractionation isotope effects,
(2) amorphization of the sample (loss of the crystal structure),
(3) primary particle implantation and mixing with the sample atomic species,
(4) recoil mixing (mixing of atoms from the surface of the sample with atoms deeper in the sample),
(5) changes in the bonding state of the surface atoms and creation of new complexes,
(6) changes in the element concentration of the sample due to the implantation of the primary beam species, and
(7) chemically driven segregation.

These effects add complexity to dynamic SIMS and may be enhanced or reduced under defined conditions; for example, sputtering with high-energy primary beams or with the use of reactive primary species, such as Cs^+ or O_2^-. The erosion rate of the sample surface can be very high, as much as several monolayers a second, and the surface of the sample continuously changes as new layers are exposed.

In depth profiling analysis high depth resolution is required. Although sputtering occurs within the three or four upper atomic layers of the sample, some artefacts increase this depth variably. Artefacts that can cause the depth resolution of the profiles to deteriovate sharply are the implantation of primary atoms in the analyzed sample, compositional changes, recoil mixing and the broadening effect of layers and interfaces, ion-induced topography and initial surface roughness, statistical precision expressed as the minimum volume that can be analyzed, the number of analyzed elements, non-uniformity of the primary beam and crater edge effects, and changes in focusing conditions due to the increase in the crater depth or charging of the insulating samples. To minimize some of these artefacts, some instrumental parameters have to be carefully controlled, such as beam uniformity, charging effects, and contamination from residual gases, or analytical methodologies and techniques must be applied, such as careful selection of the primary ion energy and fluency, sample rotation to reduce roughness, rastering of the primary ion beam, gated acquisition of measurements, and the use of chemically reactive primary species. All of these parameters can improve depth profiles.

Depth profiling is used extensively in semiconductor, thin film and diffusion studies for structural and chemical characterization of gradient layers and alternating layers in multilayer systems and to gain an understanding of interface structures [14, 116–118]. For these reasons, we will look at this method in more detail. In the following paragraphs some basic relationships that describe depth profiling with SIMS are given, and quantification methods for depth profiling are discussed. Finally, methods of calibrating depth profiles as well as improving depth resolution will complete this section.

7.2. Some Fundamental Relationships

During depth profiling, the primary ion beam is optimized and stabilized at the desired sputtering rate, then the sample is moved to a new area, and immediately acquisition starts at intervals and the elapsed time is monitored. The result is a table of values relating measured ion current (I_A) to time (t) according to the relationship $I_A = f(t)$. The ion current is actually measured and integrated for each time interval (Δt), and finally it is converted to number of ions ($N_A = I_A/e$, where e is the unit charge). This number differs from the number of ions sputtered because of losses in secondary ion optics and dead time in the detector. These factors are collectively expressed by the factor f_1.

Because measurements are taken at time intervals (Δt), that correspond to a depth (Δz), and because the primary ion beam has a certain average area (A_{avg}), a certain volume, $\Delta V = A_{avg}\Delta z$, is removed from the sample. This volume initially contains a number of atoms, N_0, of the analyzed element. Because with SIMS only isotopes are measured, a further correction of the above value has to be made so that $N_A = a_A N_0$, where a_A is a factor for correcting the isotopic abundance to the elemental abundance.

From the discussion in previous sections we already know that not all of these atoms are sputtered and converted to free ions. From the available atoms in a single volume unit (ΔV), some recoil deeper into the sample after collision with the primary ions or other recoiling atoms of the sample and remain, some are ejected as neutrals, some form random molecular complexes with other atoms, and finally, only about 1% of the contained atoms are converted to free ions and are collected by the secondary optics. Furthermore, other artefacts and phase chemical variations in the successive volumes (all expressed in a factor f_2) further affect the estimation of the number of produced ions. The expression $Y_A = f_2 N_0$ is a definition of the sputtered ion yield of a specific isotope of an element A.

Assembling all of the above in a general formula, the real content of ions in the analyzed volume can be expressed by the equation

$$N_A = \tau_d f_1 f_2 a_A \Delta V = \tau_d f_1 f_2 a_A A_{avg} \Delta z$$

This equation is true only for a specific time interval (that is, the volume analyzed at this time) and for a specific element/isotope; it does not express the whole profile, but its individual analyzed volumes. Considering that c is the concentration of the specific element/isotope in the analyzed volume, it can be rewritten as

$$N_0/A_{avg}\Delta z = \tau_d f_1 f_2 a_A = f_0 \quad \Rightarrow \quad c = f_0 = f(t)$$

The above is true because the Δz factor, which is combined in the overall factor f_0, is a function of time. A consequence of the previous relationship ($c = f(t)$), the relationship $c = f(z)$ is true, and, as a result, a more precise definition of the calibration curve $z = f(t)$ is required, where f expresses different functions. The relationship $z = f(t)$ can be approximated as linear only for totally homogeneous samples, and considerable care should be taken.

For a constant beam size and profile, the minimum Δz value achieved is the depth resolution of the specific profile. The minimum Δz that can be achieved by the specific instrument is the lateral depth resolution of the instrument. This value also depends on the abundance of the measured element in the sample because errors from counting statistics put a lower limit on the minimum analyzed volume size.

7.3. Quantitative Depth Profiling

The previous subsection was an attempt to explain with simplified formulation what depth profiling is. In this subsection, quantification will be briefly discussed, supported by more detailed formulation. This is only one task in interpreting SIMS depth profiles; the other is depth calibration, and it is discussed in the next subsection.

Although quantitative depth profiling could be approached in a way similar to that of the bulk quantification discussed in an earlier paragraph, it is different in the respect that small volumes of material have to be analyzed in a uniform way as the depth increases. However, this is not generally the case. For example, preferential sputtering (discussed in Section 3.2), produces variations during profiling. This phenomenon, together with the differences in the surface and bulk chemistry of the sample, affects the first values acquired during profiling and introduces difficulties into the thin film characterization [119]. Furthermore, the implantation of ions from the primary beam changes the chemistry, and only after equilibrium is achieved do the steady-state values reflect the true composition of the sample.

If we assume a sample with only two components A and B, preferential sputtering is defined as follows. If these components are present in equal abundance in the sample ($C_A = C_B$) the ion yields are not equal ($Y_A \neq Y_B$) because one atom is sputtered preferentially over the other. Moreover, the ratio Y_A/Y_B changes with depth until a steady state is reached. Furthermore, all ratios are different for different sample compositions, a property affected by the matrix effect (Section 3.3). These effects are very significant when thin films are analyzed and make quantitative measurements of thin films very difficult. Different methods have been investigated to minimize or at least to quantify such effects.

As with bulk quantification, MCs^+ ion signals were used for depth profiling to minimize matrix effects [120]. Quantitative values can be acquired from the formula $I_{MCs^+} = I_p c_M S_{MCs^+}$ [55, 120], where I denotes the primary ion current, Y is the total sputtering yield, c is the concentration, and S is the sensitivity factor of MCs^+ (the number of detected MCs^+ ions per sputtered M atom, which includes the instrumental transmission and the probability that a MCs^+ ion will be formed). Again this formula contains a sensitivity factor, and as in bulk quantification procedures, the RSF values must first be estimated and used. In a multielement sample the concentration of any element A can be estimated if the relative sensitivity factors are known, according to the formula $c_A = (I_{ACs}/S_{ACs})/\sum_{M=1}^{n}(I_{MCs}/S_{MCs})$, for all elements $M = 1, 2, 3, \ldots, n$. The sum of the previous equation is proportional to the total ion yield and can be monitored. If the atomic densities can be estimated, then the time- or depth-dependent erosion

rate can be calculated, as can the depth scale of a depth profile. At each time t the total eroded depth is given by the formula $z(t) = \int_0^t J_p[Y(t')/n(t')]\,dt'$, where J_p is the primary current density and $n(t')$ is the time-dependent atomic density of the sample [120]. For examples of how these corrections are applied to real depth profiles, the reader is referred to [120].

Efforts to reaching absolute values can be found in the literature, where parameters such as the type of primary ion species and energy or the type of secondary ion are optimized. For example, it is demonstrated that at certain bombarding energies matrix effects are reduced to the level where absolute values are approached [37]. These experiments were performed to assist the quantification of Si doping in SiGe HBT thin-layer heterostructures. A comparison was made between two primary beams (O_2 and Cs^+), and MCs^+ ions were detected to reduce matrix effects. Results were also compared with RBS measurements. The SiGe layers were produced with a variety of compositions ranging from 4% to 23.5% Ge, and the rest was Si. Different impact energies were also used, ranging from 3.6 to 7.5 keV for the Cs^+ primary beam, and from 3 to 10.5 keV for O_2^+. Gated collection was utilized to reduce crater-wall effects. In this systematic study [37], it was observed that O_2^+ bombardment generally shows more enriched samples, whereas Cs^+ produces slightly poorer samples. Profiles of Si, B, and Ge are shown in Figure 12 as an example of quantitative depth profiling under optimum conditions, which suppress matrix effects. Also in this work, acquisition of $GeCs^+$ or $SiCs^+$ ions is matrix-dependent, in contrast to previous discussion and other research work [50, 51].

Efforts were also made to define RSF values for quantitative depth profiling of silicon and aluminum oxynitride films [109]. A profile of a 40-μm-thick film is shown in Figure 13, calibrated for depth with the use of a profilometer. Also in this

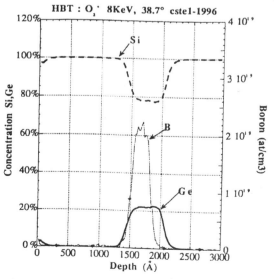

Fig. 12. Quantitative SIMS depth profiles of Ge, B, and Si profiled in a SiGe HBT heterostructure with a very thin boron-doped layer. From [37] with permission from Elsevier Science.

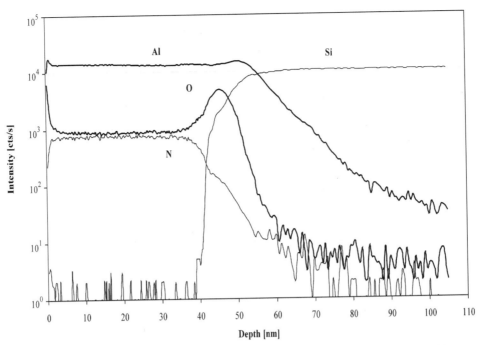

Fig. 13.   Depth-calibrated SIMS profile of a 40-nm Al-oxynitride film. The oxygen profile has a stable composition up to the 40 μm but then it increases at the interface with the substrate, which does not contain oxygen.
From [109] with permission from the author and Springer-Verlag.

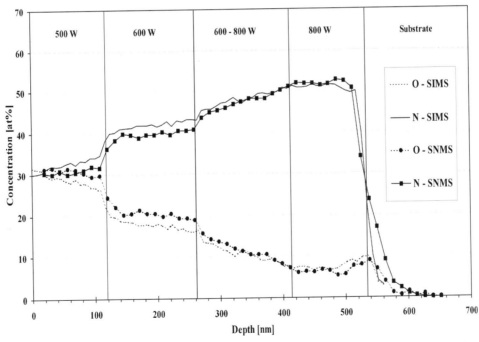

Fig. 14.   Comparison of quantitative depth profiles measured by SIMS and rf-SNMS. Steps in concentration
demonstrate variations in the sputtering power during film deposition (in W, top of the figure). SIMS and SNMS
follow these variations in the same way. From [109] with permission from the author and Springer-Verlag.

work, MCs^+ molecular ions were analyzed to reduce matrix
effects, and it was found that SIMS and rf-SNMS depth profiles
of oxygen and nitrogen compare relatively well for multilayer
films of varying composition (Fig. 14) after the application of

the same *RSF* values. Only oxygen is observed to be problematic at low concentration, exhibiting an error of 5 wt%, in
contrast to a 0.6 wt% of nitrogen. *RSF* values have been calculated in correlation with quantitative measurements acquired

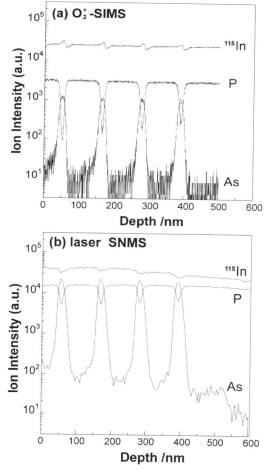

Fig. 15. Two depth profiles of In, P, and As measured from an In-GaAs(12 nm)–InP multilayer sample with the use of (a) O_2^+-SIMS and (b) laser ionization SNMS instruments. The SNMS profiles reflect true profiles, and the SIMS profile of In shows some peaks that are probably attributed to matrix effects. From [106] with permission from the author and Elsevier Science.

by the EPMA (electron probe microanalysis) technique. SIMS and laser-ionization SNMS depth profiles are also compared in Figure 15. The In profile in SIMS shows some matrix effects, whereas the SNMS profile reflects the real profiles better [106].

7.4. Depth Calibration Methods

Depth calibration is the conversion of sputtering time into depth. There is no easy way to make this calibration, except possibly for very homogeneous samples with stable sensitivity factors during profiling and under stable primary ion beams. Then the sputtering rate (designated by the symbol $\dot{z}$) is assumed to be constant ([14]; see also [117] for a simple discussion). In this case, the depth of the formed crater is measured with a stylus profilometer and then divided by the acquired number of time steps. This is the most common calibration method, especially for thick films, and, combined with beam scanning and gating of a smaller area, results in a better estimation of the depth and the sputtering rate because of

flat craters and improved depth resolution. A similar but more precise method is to create on a polished surface a number of craters with different depths, which are then measured with a stylus profilometer, and the erosion rates are calculated for each crater [8]. Inhomogeneities with depth can be estimated, as can erosion rates for each material of a multilayer structure when a large number of craters are acquired [8]. Several other methods based on stylus profilometer measurements have been proposed to estimate the depth scales and generally require sample preparation techniques, which are generally avoided in the literature because of sample contamination.

However, despite these best efforts, calibrations are still not totally accurate, especially when interfaces of very different materials are approached or during sputtering through them. For specific cases such as multilayers of very thin films with industrial interest, reference standards have been thoroughly characterized and can be analyzed before the unknown sample and compared. Such standards have been proposed in the literature, for example, GaAs/AlAs superlattice standards [121] or amorphous delta thin films of Ta_2O_5/SiO_2 multilayers and polycrystalline Ni/Cr multilayers [122]. Characterization with SIMS and AES has shown that Ar^+ primary ion beams induce very little roughness with little deterioration of the depth resolution, in contrast to O_2^+ ion beams. Lower kinetic energy ion beams and glancing (nonvertical) incident angles show improvement in the depth resolution. A round-robin study has been initiated to establish the material as an ISO/TC201 standard.

Similarly, amorphous multilayered delta thin films of Ta_2O_5/SiO_2 and polycrystalline Ni/Cr multilayers both grown on Si substrates were proposed as depth profile standards for SIMS calibrated depth profiling [122]. The standard comprises seven Ta_2O_5 18-nm-thick layers (chosen because of their smooth surface topography during sputtering) separated by 1-nm SiO_2 delta layers. Generally, ion beam mixing, surface segregation, preferential sputtering, and surface topography reduce depth profiling resolution. Optimizing the sputtering parameters, such as ion species, kinetic energy, and angle of incidence, may improve the analysis. The proposed depth profiling reference material is adequate because of the very low topography during bombardment. Various SIMS depth profiles made by different primary ion beams do not show significant deterioration of the depth resolution under commonly used SIMS conditions.

Calibration standards can be produced *in situ* by oxygen implantation in doped matrices where the doping element has an ionization efficiency that is very different from that of the matrix. Accordingly, a B-doped Si reference sample is proposed [123], as is a method for quantitatively estimating the matrix effects caused by the different degree of ionization enhancement of B and Si due to the presence of oxygen. The dependence of the B^+ and Si^+ relative sensitivity factors were studied *in situ* by first implanting oxygen with 1.9-keV O_2^+ normal incidence ion beams and then analyzing by SIMS.

Modeling can also be used to estimates depth from the $I = f(t)$ relationship but requires a great deal of *a priori* information for the conditions of the analysis, which are later approximated with computer calculations. In [124] a similar approach is described that involves an initial step of a simplified model of the $I = f(t)$ relationship and requires only a little *a priori* information on the $C = f(t)$ relationship. The parameters given to the computer program are some information on the sputtering conditions and values based on simple assumptions on the origin, transmission, and collection of the ion signals. After the first rough simulation, the data parameters are iteratively changed in an attempt to approach the shape of the measured profile. In more difficult situations further experiments are required to resolve the different artefacts. For more precise results, better estimation of the sputtering rate can be given to the computer program with the wedge crater technique [125]. Wedge crater sputtering is an effective method for the determination of the sputter rate and the depth function of complex layer systems. The wedge crater is prepared by mechanical methods (e.g., computer-controlled saw tooth or polishing) as a beveled (angled bottom) sample cutting through all films. After sputtering a profilometer is used to measure the difference in angle of each film and to determine the difference in sputtering rates of the materials/layers. To avoid mechanical treatment, computer-controlled primary ion beam deflection can also form a wedge crater [125]. A computer scans a finely focused ion beam on the surface of the sample and at the same time varies the intensity linearly between a minimum and a maximum value on one of the horizontal axes. To create the crater area, the ion beam is also scanned in the vertical direction but without varying the intensity. Consequently, for a single phase sample the bottom of the crater is inclined at one angle only, whereas for samples with more than one phase of different etching rates a number of different inclinations are observed. More that one profile is referenced to two crater marks on the sample and are averaged with the use of a stylus profilometer. The intensity profiles are then corrected from the etch rate estimation, which has been calculated in a simple way from the wedge crated profiles, the measured inclinations, and information on the variation of atomic densities in the analyzed volume. This method is easy to apply, but not all commercial SIMS instruments have the capability to vary the primary beam intensity in a controlled way and it is generally not used in the literature.

Finally, accurate depth measurement can be obtained by interferometric methods with a laser beam [126] or with ellipsometry. The accuracy is very high when SIMS instrumentation is adapted to include devices capable of measuring depth *in situ* during profiling.

7.5. High Depth Resolution

7.5.1. *Factors that Affect Depth Resolution*

Minimization and high-level integration of semiconductor devices result in optimized operation, reduced power consumption, and increased or even novel functionality when quantum levels are reached. Their characterization requires techniques capable of analyzing devices of small volume. SIMS instrumentation and analytical methods can be optimized for such

conditions up to certain limitations, mainly intrinsic to the process of sputtering. The question posed here is: What is the minimum volume that can be analyzed by SIMS?

To decrease the volume of analysis one must decrease the two horizontal dimensions of the sputtered area by improving the spot size of the primary ion beam. This involves techniques of producing electrostatic optics that focus the ion beams appropriately. However, this is not so important for most thin film applications. What is important to discuss here is how to improve depth resolution. This becomes even more significant for some applications, such as very thin films. One example is silicon oxynitride (ONO) films used for memory devices. It has been demonstrated [127] that SIMS is sensitive enough to analyze these devices with adequate depth resolution. Of similar importance to thin or ultrathin films is the study of the interfaces between different films or films and substrates. For example, the interfaces of Superconductor-Normal metal-Superconductor (SNS) or Josephson junctions of superconducting structures [128] or very thin superconducting multilayers based on the proximity effect [129] should either remain abrupt after annealing or be unaffected by roughening and chemical reactions. A study of the effects that cause the depth resolution of the above materials to deteriorate could assist the search for ways to improve this kind of analysis and provide more detailed information of the phenomena occurring at the interface during annealing.

Before we can proceed to further applications and solutions, one has to find the factors that affect depth resolution. One of the main factors affecting depth resolution is initial topography, which changes dynamically during sputtering and develops into a secondary topography (roughening), which can generally extend to a size greater than that of the actual thickness of the film. Secondary topography is severely enhanced by compositional differences within the bombarded area, crystal structure or other structural discontinuities. The film thickness broadens with increasing depth [118, 130] because interfaces operate as "magnifying glasses" for roughening [31]. Topography increases dynamically during sputtering, producing rippled surfaces, and starts forming early during profiling, generally at depths of 8–12 nm [129, 130]. Roughening can be decreased by normal incidence [128], low-energy ion beams at vacuum conditions [130], and sample rotation (see Section 7.5.2). Crater edge effects reduce depth resolution through mixing of material from the sides of the crater with material from the bottom. A common technique for overcoming this problem is to scan the primary beam in a larger area and to collect data in a smaller area inside the first (gating the signal). Similarly, when imaging ion probes are used with a wide, nonscanning beam, apertures can collect signal only from a smaller central area within the one bombarded by the primary beam.

The highest resolutions reported in the literature achieved with SIMS are in the range of 4 nm FWHM [131–133], 1.7 nm FWHM [134] for Ge and Si structures, or even 0.62 nm at the leading edge and 1.30 nm at the trailing edge of InGaAs/GaAs multiple quantum-well structures [118]. Such conditions are achieved with the use of low primary ion beams [129, 130] (although the opposite is observed in [135]), grazing incidence [35, 129, 136] and vacuum conditions ([130], nonchemical sputtering).

At low energy primary ion beam sputtering rates are low, and the detection limit or the dynamic range of the analyses can be increased by increasing the yield of the secondary ions. Without deteriorating the depth resolution too much, it has been demonstrated that chemical sputtering can be applied effectively [136]. This is demonstrated by sputtering Ge and B secondary ions from Si matrices with low-energy O_2^+ primary ion beams (700 eV) and at variable angles of incidence (about 50 to 75°) on a magnetic sector SIMS instrument. Under these conditions, the depth resolution for the Ge δ-layer was 1.6 nm FWHM and that for was B 2.4 nm FWHM, while keeping the erosion rates within 0.5 and 1.0 s^{-1} μA^{-1}. Also at 0.7 keV (at 71°) the sputtering rate was 1.9 atoms/ion, and the detection limit was in the range of 10^{15} atoms/cm^3. Even lower primary ion beam energies (250 eV) with normal incident O_2^+ have been used by successfully resolving B delta doping spikes with a spacing of 2 nm [137]. In some cases [35], the use of an O^+ primary ion beam show a preferential sputtering effect, highly enhancing at the In profiles of selenium-doped multilayer films of In_xGa_yAs/InP (InP thickness ranges from 50 nm to substrate thickness, whereas all In_xGa_yAs films are 12 nm thick). This is observed as high secondary ion intensities when the beam enters the In_xGa_yAs and very low intensities when it leaves the film. The effect disappears with Ar^+ bombarding, but small peaks are present, and for SNMS this part of the peak shows "dips," an effect that is probably not attributed to preferential sputtering, because the two techniques use the same mechanism for sputtering, but to in-depth changes or topography based on spikes (see Section 2.4). Very low energy bombardment with oxygen has also been used in [138] to resolve very thin films (in the range of 2–3 nm) of pure Ir and its silicides ($IrSi_1$–$IrSi_{1.75}$) grown on a Si substrate. The SIMS technique was used to characterize these films, in the attempt to solve problems like acquisition of quantitative information. It is suggested that very low energies are necessary to analyze such thin films and that mixing effects extend the film by half of its original thickness. Also, during quantification, the value measured just before the primary beam goes through the thin layer can be considered as an estimation of the absolute concentration.

The acquired depth resolution also depends on the analyzed species. In [136], for example, the Ge δ-layers are better resolved than the B δ-layers. This is probably an effect of a higher diffusion of B compared with Ge, which causes a redistribution of B during bombardment and increases the effective thickness of the layer [136]. During depth profiling, this is observed as a difference between the leading and the decay (trailing) lengths of the signal intensity of the film's analyzed component. The leading edge is always observed to be smaller than the decay length. The distortion of the shape of the depth profile is a common observation in the literature [128, 136], additionally accompanied by an actual shift of the position of the layer deeper into the sample, at depths that can extend as much as 4 nm under oxygen flooding. Under oxygen flooding the effect is more pronounced compared with profiles taken under ultra-high vacuum [130]. The shift of thin, δ-layers closer to the sample has been studied in detail in [135] and is attributed to mixing phenomena and changes in the sputtering rate in multilayer films. Monte Carlo simulations are in agreement with the experiment, except in the case of different bombarding ions. Experiment has shown that heavier primary ions shift the layer position less than lighter ions, an effect that probably occurs only at the first steps of the sputtering process and not during steady state. Moreover, molecular Si_2^+ signals provide better quantitative information than Si^+. Studies of Co secondary ions from Co-doped YBCO films grown on $LaAlO_3$ by laser ablation [128, 139] show different leading and trailing profile edges for different bombarding species (e.g., oxygen and xenon), and trailing edges become significantly long at incident angles larger than 45°, which is similar to the incident angles of both primary ion beams. It is evident that leading and trailing edges play an important role in depth resolution that is comparable to their rate in topography [128]. This is demonstrated in Figures 16 and 17. Figure 16 shows the YBCO/Co-YBCO/YBCO junction grown on the $LaAlO_3$ substrate. The $^{59}Co^+$ signal defines the intermediate, doped layer (Co-YBCO) with a Co/Cu ratio of about 3%. During high-resolution depth profiling, changes to the originally estimated thickness of the layers were observed. Careful studies with primary beams at different incident angles have shown that this is

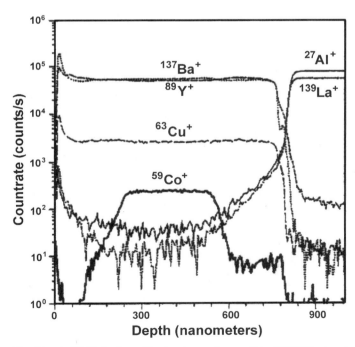

Fig. 16. A SIMS depth profile showing an intermediate YBCO/Co-doped YBCO/YBCO multilayer structure deposited on $LaAlO_3$. The observed larger trailing edge relative to the leading edge of the Co^+, which extents up to the interface with the substrate, suggests a SIMS artefact rather than an interdiffusion of Co to the neighboring layers. From [139] with permission from Elsevier Science.

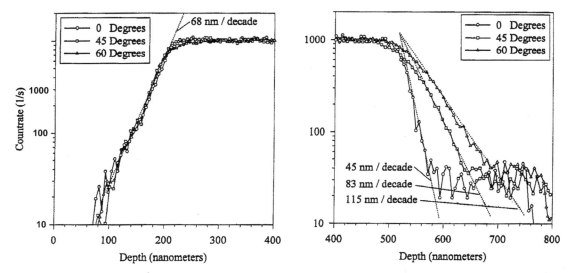

Fig. 17.   Two plots of the Co⁺ depth profiles showing the leading (left) and the trailing (right) edge at different angles of primary beam incidence. Normal incidence reduces effects, such as development of topography, that decrease the depth resolution. From [139] with permission from Elsevier Science.

accounted for not by a possible interdiffusion of Co from the middle layer to the others but mainly by the development of topography in the crater and recoil mixing. This is also demonstrated by AFM imaging [128]. Incidence angles less than 30° are suggested to reduce this effect. As already mentioned, the above has been verified with two different primary ion beams, Xe^+ and O_2^+ [128], that are easily observed at the trailing edge of the Co profile (Fig. 17). The leading edge is observed to be similar for each profile but with different slopes for the two primary ion beams.

Normal incidence bombardment can reverse depth resolution dependencies on the analyzed species [136]. Referring again to the example of Ge and B, it is possible that at grazing angles oxygen incorporation in the sample is less and segregation of Ge negligible, resulting in better depth resolution for Ge rather than B. However, at normal incidence the oxygen concentration on the surface is higher, causing oxidation and migration of Ge, which degrades the resolution. Consequently it is not clear what the best conditions are for improving depth resolution. Contrary evidence suggests that this can be improved at normal incidence, which can be extended up to 32° without significant deterioration [130], again, however, in vacuum conditions and for low primary ion energies.

Grazing incidence is supported by several researchers, but, in addition to topographic roughening, an increase in matrix effects is seen. It has been suggested [35, 105] that SNMS assisted by laser ionization instead of SIMS will yield high depth-resolution profiles under almost matrix-free conditions: grazing incidence at angles as great as 77° have shown the best resolution without degradation with depth. This is in contrast to the results of [128], where trailing edges become significantly long at incident angles larger than 45° and should degrade resolution. Also, according to [128], variable incidence angle varies sputtering rates considerably.

Summarizing the sometimes contradictory evidence above, one can conclude that low-energy bombardment is probably the preferred way to achieve high depth resolution [33] despite drawbacks such as the inhomogeneous erosion and the uneven and inclined crater bottoms [136]. It is not clear, however, how the type of the bombarding species and the angle of incidence affect the depth resolution. Matrix effects and primary ion density may cause these variations for different samples. When high resolution is required, it is suggested that different parameters are tested before analysis, and the procedure is optimized for the available instrumentation, sample and primary and secondary ion species. Additional techniques for improving depth resolution, such as sample rotation and deconvolution of depth profiles, are discussed in the following paragraphs.

7.5.2. The Advantages of Sample Rotation in Depth Resolution

A method of improving SIMS depth profiling resolution, especially at interfaces, is sample rotation. The method has been investigated in some detail [31–33, 35, 36] demonstrating that constant secondary ion yield and depth-independent depth resolution are achieved because sample rotation prevents ion-beam-induced roughness and reduces the effect of the inhomogeneity of low-energy ion beams. Rotation speeds do not generally exceed a few revolutions per minute.

Sample rotation is not as effective for all samples and their interfaces. For example, sample rotation is not necessary at semiconductor interfaces or at metal-to-metal interfaces, because resolution is not greatly affected by interface changes in semiconductors or between different metals [36]. On the contrary, it is highly effective on metallized semiconductors, which give highly broadened interfaces during depth profiling [36].

Furthermore, it is demonstrated that sample rotation improves resolution because it minimizes surface roughness.

For very high depth resolution however and for ultrathin films of semiconductor heterostructures, it has been shown that sample rotation can improve resolution by a factor of about 2 [31]. More specifically, sample rotation has been compared with stationary sputtering on heterostructures of ZnTe/GaAs,

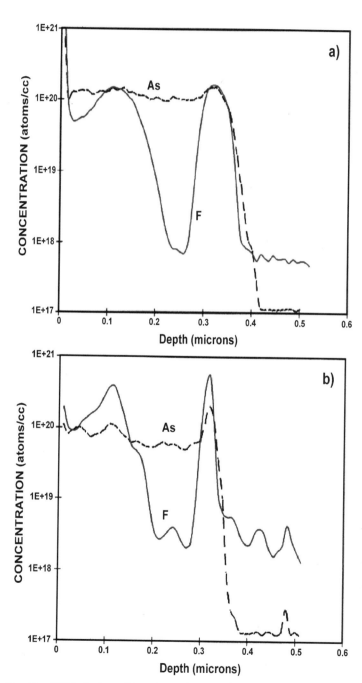

Fig. 18. SIMS depth profiles of a As$^+$- and F$^+$-doped polysilicon film demonstrating the advantages of sample rotation. (a) The depth profile of a static sample. (b) Similar depth profile with a sample rotation of 3 rpm. From [34] with permission from the Institute of Physics Publishing.

CdZnTe/ZnTe/GaAs, and their interfaces grown by MBE on GaAs substrates. High-resolution SIMS depth profiling has demonstrated that ZnTe/GaAs interfaces can be measured with a resolution of 23.9 nm for stationary samples, and a resolution of 14.8 nm has been achieved with sample rotation. The same result is observed for (CdZnTe/ZnTe)/GaAs, where stationary samples can be analyzed with a resolution of 24.6 nm, and rotated samples can be analyzed with a resolution of 16.9 nm. These results are similar to those obtained for other materials, such as InAlAs/InP [140]. Depth resolution during sample rotation can vary for different matrices and for different primary ion beams. It has been demonstrated by SEM [31] that, for stationary samples depth resolution increases at the interface and improves slowly after it. Rotation does not show the same effect. The experiments are supported by computer simulations, which suggest a relocation (lateral flow) of surface atoms on the order of a few nanometers (2–5 nm) at a near-surface layer of 2 nm thickness (for doses of 10^{16} ions/cm^2). A qualitative explanation is suggested that describes an up-lifting and a caving-in of atoms before the interface is reached [31]. In directional sputtering at an angle, which is the most common condition, this defect moves on the surface and increases in size. Sample rotation reduces these effects by averaging them.

Induced topography during sputtering is the reason for depth resolution reductions during sputtering of polysilicon films grown on silicon substrates and doped by As$^+$ and F$^+$ [34]. The effects of amorphous silicon films are much smaller, however, suggesting crystallinity as a major effect in roughness formation and growth. Roughness measurements and topography imaging have been performed before and after SIMS analysis with an AFM. The differences observed between static and rotating profiles are depicted in Figure 18. In these profiles a mostly uniform distribution of as is observed, whereas F is concentrated at the middle of the film, as a component deposited during implantation and close to the interface with the substrate (at about 0.32 μm), because of its migration during annealing. It is clear that more detail is achieved for the profile acquired with sample rotation. Topography images of stationary samples show the ripple formation demonstrated earlier in Figure 4.

7.5.3. Deconvolution of Depth Profiles

Growth technologies are quite capable of producing very fine films, but the available characterization methods, including SIMS, are not capable of resolving and characterizing them, despite efforts to improve depth resolution. SIMS, for example, is not capable of resolving films less than several nanometers in depth or of acquiring clear images of very sharp interfaces, especially under conditions of routine analysis. Instrumental improvements, mainly in the fabrication of specialized primary ion guns of very low energies, could provide some solutions. However, at present, computational procedures are used to model the sputtering process to try to extract high-quality depth profiles. Such computational procedures are restricted, however, because they are based on many simplifications and work only for certain sample models. Such a method is presented in detail [141–143] for δ-layers of boron doped in silicon under the assumption that there is no degradation in the resolution with depth. According to these works, δ-films separated by a distance of > 7 nm (optimal conditions) can be resolved, and a distance of > 9 nm can be resolved under routine conditions. However, estimation of the confidence level is an important factor for reliable results, and more work is still required in this direction.

Such methods generally reconstruct the original distribution, given the measured distribution and some parameters of the analytical conditions. In this way it is claimed that TRIM-DYNAMIC codes, which simulate layer broadening, could be reversed and successfully reconstruct the original distribution of metal layers in Si and Ge matrices [15]. Similarly, SIMS profiles of quantum wells and superlattice of III–V semiconductors have been modeled by convolving the true profiles with an analytical response function [118].

8. ION IMAGING WITH SIMS

High-speed acquisition of SIMS data by computerized systems is not only providing better and faster analysis, but also the possibility of storing and reproducing the acquired information. Consequently, a computer can be used to scan the primary beam on the surface of the sample, scan the secondary optics accordingly to secure stable transmission, switch between the masses at the mass analyzer, and store the acquired data. Scanning in two dimensions for one element is very fast and can be visualized in realtime on a computer screen. Stored data for multiple elements or isotopes can be retrieved and visualized in two or three dimensions. However, correct representation of the data fully depends upon the correct interpretation of the collected single analyses with the methods discussed in the previous sections for quantification and depth profiling. Furthermore, initial topography further affects overall resolution of the imaged area by differential sputtering. It is suggested that correlations between SIMS images and topography (obtained, for example, with an AFM instrument) should be studied to assist in the correct visualization of two- or three-dimensional SIMS data [144]. It has also been shown that, when instrumental factors are not involved, image correlation spectroscopy (ICS) or image cross-correlation spectroscopy (ICCS) can successfully confirm spatial relationships by comparing two-dimensional SIMS images acquired during depth profiling [145]. This is a mathematical approach and it is useful mainly when features close to the size of the resolution are considered.

Three-dimensional imaging, free of matrix effects due to the use of $M\text{Cs}^+$ ions and correction with RSF values, is demonstrated in [49]. However, a true three-dimensional reconstruction, suggested in [144], has to use information related to the original topography. Finally, correlations between images of different elements in two or three dimensions, line scans, and depth profiles of the analyzed volume are easy to plot by retrieving them for the existing three-dimensional data.

9. APPLICATIONS TO THIN FILM CHARACTERIZATION

9.1. Introduction

The high sensitivity of the SIMS technique is an advantage for thin film characterization. Thin films are low-volume materials with properties that are sometimes different from these of their bulkier counterparts. It is obvious, then, why precise control of their chemistry and their structural characteristics is so important.

SIMS has been extensively used for the characterization of thin films. However, recognizing the difficulties of acquiring quantitative results, it is mainly treated as a qualitative or semiquantitative technique, always supported by other physical techniques for analysis and characterization. In these cases, one will find only basic information on how SIMS has been applied in the specific case, such as that of the primary ion beam species, its energy, possibly the type of charge compensation used when insulating samples are analyzed, and in the mode in which it was operated, something that sometimes is straightforward from the application. This may give the reader the impression of a routine technique. In many cases, though, a detailed methodology of the analyses is given and the problems are discussed, and some are treated in a systematic way in the search for possible relations. These research works treat the technique with great care and support their findings with results from other techniques. Consequently, all researchers still treat the technique consciously and the reader is advised also to do so. Still, one must stress once again the following advantages, also summarized in [146]:

(a) SIMS can detect all of the elements of the periodic table, including hydrogen, and their molecular combinations;
(b) SIMS has the lowest detection limit of all analytical techniques;
(c) because it is a mass spectrometric technique, only isotopes are analyzed with applications in diffusion studies;
(d) static and Tof-SIMS can give information on one monolayer and they are excellent tools for surface analysis;
(e) ToF-SIMS is also a good tool for organic materials and can provide information on their chemical structure; and
(f) SIMS requires minimal or no sample preparation.

SIMS instruments can be used

(a) for dynamic analysis for bulk composition of very small structures;
(b) in static mode to analyze surfaces and monolayers;
(c) for profile analysis of one dimension, either in depth or across a line on the surface,
(d) for two-dimensional analysis of an area of the surface; and

(e) for three-dimensional analysis by combining the previous two, giving structural information (element distribution, interfaces, dissolution reactions, etc.).

SIMS has also been used on devices that result from thin films. Many of these applications, however, are scattered work, related only to individual materials, and only in a few cases is more systematic work on certain areas present. There is also a lack of information on specific applications, because of the industrial interest of the materials (e.g., the semiconductor industry). Consequently, the following work has to follow this structure. Moreover, the list of applications and examples is not exhaustive or even restrictive. The purpose of this work is to give a basic idea of the advantages and the problems encountered when the technique is used and the areas in which it has been used, supported by some examples. It is expected that the user will be able to apply the technique in new areas and in a correct way, taking the maximum amount of information that it can provide.

Thematically, SIMS has been used in materials science, surface science, and technology. In materials science, SIMS has been used to chemically and structurally characterize films, to optimize their growth procedures, and to control the growth environment, possible impurities, and sources of contamination. Types of layered materials that are more systematically studied are superconductors and, more specifically, YBCO films, to which a long paragraph is devoted here. New materials, such as synthetic diamonds or extra hard materials, are also very important, and their properties depend on their detailed chemistry. Other applications include the optimization of film alloys to produce electrodes and studies of the homogeneity of chemical coatings and some properties of polymers such as wettability or chemical changes and contamination of their surfaces. Surface characteristics of bulk materials are enhanced with thin films, and improvements in corrosion resistance or tribological properties are useful in aerospace technology or the production of biocompatible materials for artificial implants. SIMS has been also used to enhance the surface properties of substrates to improve the adhesion of films. Finally, some post-processing, such as annealing or doping, is common, and SIMS has been extensively used to study or control the effect on films.

Very important technological materials are the materials used to fabricate microelectronic and optoelectronic devices. These devices are generally multiply layered structures, and for completeness they are treated here as thin films. One would encounter applications such as radiation filtering and detection in the fabrication of devices such as thermistors, capacitors, and memories. SIMS was also used to improve insulators, optimize VLSI metallization, fabricate advanced buffer layers, or investigate new materials such as semiconducting diamonds or SiGe HBT heterostructures. New technologies demand high integration of high-quality devices and SIMS has found applications in high-density recording media, fabrication of LCD displays, the improvement of solar cells, quality testing of lithographic patterning, and, more recently, microelectro-mechanical systems (MEMS) assembly and packaging. Environmental aspects include applications to the search for thin-film gas sensors or the catalysis of automotive exhaustion.

9.2. Characterization and Optimization of Materials and Processes by SIMS

New and advanced materials, including thin films, should be designed and optimized with respect to their function, and with the purpose of finding technological applications. SIMS, because of its high sensitivity and high depth resolution, can assist in the chemical and structural optimization of thin films through the study of the films themselves, their method of growth, and the growth environment, and, finally, their stability during further processing (e.g., doping or annealing) or further testing. Briefly, by studying the surfaces of the substrates one can optimize them to improve growth, adhesion, and mechanical, thermal, or chemical stability at the interfaces. Compositional studies of films can provide detailed information on traces, impurities, and major chemistry with direct applications to the properties and the quality of the films. Further processing of the films, such as thermal treatment, might also result in self-organization or phase separation effects, which can be useful in the fabrication of quantum wells, etch stop layers or diffusion barriers. Furthermore, it is very common that thin films are used to alter or improve the surfaces of substrates. For example, oxide thin films grown on glasses improve their chemical and physical properties while they maintain transparency. Despite the sensitivity and the advantages over similar analytical techniques, SIMS must be supported by other techniques [147], such as SNMS, RBS, AES, and other spectroscopic techniques.

9.2.1. Control of Growth and the Growth Environment

An example where very good control of the initial steps of growth is expected to enhance the properties of grown films is in high-density recording (magnetic) media. A method of ultraclean (UC) sputtering of NiP/Al substrates was studied, in which Cr, CoNiCr, and CoCrTa films were deposited by DC magnetron sputtering [148]. Ultraclean conditions are achieved by dry etching with Ar just before film deposition. Magnetic measurements have shown that the coercive force H_c (the resistance of a magnetic material to changes in magnetization, measured as the field intensity necessary to demagnetize it when it is fully magnetized) increases with etching time. Systematic studies of ultracleaned and untreated samples made by SIMS (Fig. 19) have shown that this effect is related to oxygen that is absorbed on the substrate surface. The oxygen concentration is significantly reduced, even with slight cleaning, and results in an increase in H_c and the deterioration of magnetic properties. The readsorption of oxygen and other gas impurities on the surface does not reverse the effect.

During growth, impurities have to be minimized to acquire good materials with optimized properties. Because of its high sensitivity, SIMS is a good tool for studying impurities, and this will be demonstrated here with an example, the synthesis of n-type semiconducting diamond [149]. P-doped diamond

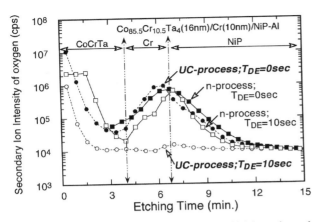

Fig. 19. Profiles showing how ultraclean sputtering, which is used as a cleaning procedure on substrates, can affect the oxygen content of surfaces. From [148] with permission from the author and Elsevier Science.

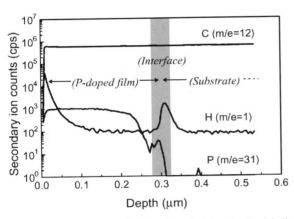

Fig. 20. A positive-ion SIMS depth profile of P-doped epitaxial diamond thin film (300-nm thickness) on a nondoped, synthetic diamond substrate. The higher content of hydrogen at 0.3 μm indicates that the interface with the substrate (shaded area) has been reached. The hydrogen content is higher because of contamination of the mechanically treated substrate surface. From [149] with permission from the author and the American Institute of Physics.

films were synthesized at high and low temperatures with vapor growth methods on diamond substrates, which were synthetic and mechanically polished. Because growth is performed from the vapor phase with PH_3/CH_4 precursors, the level of incorporated hydrogen is expected to be high, reducing the conduction properties of the diamond film. Optimized growth with the appropriate PH_3/CH_4 ratio and under substrate heating at very high temperatures has reduced the hydrogen content in the films [149]. The films have been characterized by SIMS depth profiling of the P and H elements (Fig. 20). Positive ions of the above elements were extracted with an O_2^+ primary beam. Quantification of the 1000 ppm P in the diamond films was estimated from *RSF* values, measured at an absolute concentration of 2.5×10^{19} cm^{-3}. Structurally, the depth profiles show an almost constant P composition, implying uniform implantation. SIMS revealed that hydrogen was high only at the interface with the substrate, which was possibly caused by trapping of H at the imperfections of the substrate, which were numerous

because of mechanical polishing. The film itself had reduced H content.

The effect of the growth technique and of the growth environment is demonstrated in the growth of Si_3N_4 films [150]. These films are chemically inert and thermally and mechanically stable, and they are hard and have good dielectric properties. Microelectronics and other technologies take advantage of these properties to fabricate oxidation masks, gate dielectrics, insulating layers, and passivation layers. When these films are grown by chemical vapor deposition (CVD), hydrogen is trapped in the films and the physical properties of the films deteriorate. Hydrogen is contained both in the SiH_4 target used by CVD and in the NH_3 atmosphere. Rf-magnetron sputtering, however, uses a Si target in a reactive nitrogen–argon atmosphere (N_2–Ar) and the hydrogen or oxygen involved comes from residual moisture in the environment. SIMS has helped to find the relation between bias voltage and hydrogen and oxygen content in the films, resulting in the observation that with increased voltage bias during sputtering deposition the hydrogen or oxygen content decreases to concentrations lower than 4×10^{20} atoms/cm^3 [150].

9.2.2. Optimizing Surface Properties and Adhesion with Thin Films

In many technological applications thin films are used to enhance the mechanical, chemical, electrical, optical, and other properties of substrate surfaces. However, to maintain these properties, the thin film should stably adhere to the surface of the substrate. Possible influences on the adhesion of thin films to substrates are the structural, topographical, and chemical properties of the substrate itself. SIMS can be used to study chemical properties, and it is known that generally gases or hydrocarbons form contamination layers that reduce adhesion.

SIMS, for example, was used to study the contamination that affects the adhesion of carbon, aluminum, chromium, and tungsten films deposited on steel and polished TA6V titanium alloy substrates. Oxygen and hydroxide radicals (O, OH) were detected, and it was demonstrated that sputter cleaning, with, for example, an argon ion beam, can reduce these effects [151]. Moisture diffused at the interfaces at room temperature is in a similar way the reason for reduced cohesion between TiN films and SiO_2, and sequences of ion images taken with SIMS provided the opportunity to visualize and study the effect in detail [152].

Adherence of a film can also be improved when the chemistry between the substrate surface and the film does not change abruptly. Implantation of some elements in the substrate may create the appropriate conditions to improve adherence. C and N elements have been implanted in Ti alloys [153]. Ti alloys are generally used in the aerospace technology and for biomedical purposes, and although they have many good properties, such as a low density, high corrosion resistance, and good mechanical properties, their tribological characteristics, such as friction and their wear resistance, are very weak. Thin films of nitride, carbide, and carbonitride can also improve these qualities. After implantation, SIMS depth profiling and RBS have demonstrated [153] that the independent implantation of C and N has reached the same depth, ensuring homogeneity, but the outermost Ti layer was partially oxidized.

A material commonly used to cover the surfaces of stainless steel is a film of tungsten carbide (WC). Tungsten carbide is a hard material (2200 HV) with a high melting point (about 2800°C) and a high thermal conductivity (1.2 J/(cm s K)) and, after diamond, has the highest modulus of elasticity (highly elastic), which is well above 700 GN/m^2. To further improve its hardness, it is doped with Ni. Estimations of the thickness of the grown films and compositional depth profiles have been acquired by SIMS [154], demonstrating that W, C, and Ni are uniform with depth. W and C are stoichiometric, allowing only for a slightly lower C signal, which is probably due to its lower ionization efficiency. SIMS has also revealed Co impurities, which probably have migrated from the stainless steel substrate, of which it is a component.

9.2.3. Post-Processing of Materials and the Resulting Effects

$Si_{1-x-y}Ge_xC_y$ alloy is a material that is applied in thin films on different substrates and can produce transistors and infrared detectors or high electron mobility devices. In the first case heterojunction bipolar transistors (HBT) with cut-off frequencies exceeding 100 GHz can be produced [155] by introducing carbon into the SiGe thin film, which is grown on Si. $Si_{1-x-y}Ge_xC_y$ alloy has a better lattice match with Si than with SiGe. Studies of SIMS and other methods have revealed a carbon self-organization into δ-films [155]. These fine films might have technological applications such as quantum wells, etch stops, or diffusion barriers. However, growing $Si_{1-x-y}Ge_xC_y$ alloy films is difficult with molecular beam epitaxy (MBE) or RTCVD (rapid thermal CVD) methods because of the low solubility of C in Si and the tendency for SiC to preferentially precipitate [156]. Films that were grown by MBE and further treated by high-temperature (950°C) rapid thermal oxidation (RTO) have been studied by SIMS and have shown that this treatment outgasses carbon and reduces the C content, especially close to the surface, but also deeper at the interface with the SiO$_2$ [156]. Furthermore, this introduces compressive stress. SIMS has also revealed a slight increase in carbon concentration close to the interface with the Si substrate [157] when these films are grown at about 450–560°C.

Thin films can be also produced by post-processing of another film by the induction of phase separation. Such processing, for example, has technological advantages because at the same time it separates films of Ti and Co alloy in a twin layer of a silicide (e.g., CoSi$_2$), which is a diffusion barrier of the Si from the substrate, and a surface transitional contact layer (TCL) film [158]. The first is functioning as a diffusion barrier layer (DBL) and the second as a TCL. Phase separation occurs in a N atmosphere; the gas reacts with the substrate to form the two layers. In the above study, SIMS was used mainly because of its advantages over AES, such as the higher sensitivity and

the capability to analyze molecular ions (CoTi$^+$, CoSi$^+$, and TiSi$^+$ when an O$_2^+$ primary beam is used, or SiN$^-$, TiN$^-$ with a Cs$^+$ primary ion beam). Furthermore, qualitative SIMS depth profiles were acquired to understand the interaction between Si and the thin films of Ti-Co and Ti-Co-N during thermal annealing. It has been observed that reaction of TiCo thin films with Si results in the formation of ternary silicide compounds CoTiSi instead of CoSi$_2$. With the introduction of N, a phase separation in Ti-Co-N thin films into two films occurs, resulting into a bottom CoSi$_2$ film and an upper TiN film. An excess of N will form Si$_3$N$_4$ at the top of the surface. SIMS, also supported by AES and XRD (X-ray diffraction), has suggested a number of possible reactions during annealing at 850°C, which lead to the thin film separation.

Migration of elements during post-processing is also demonstrated in the following two cases. Annealing is a common post-processing technique for thin films. It is used mainly to remove stress formed in the thin films during growth, to convert an amorphous film into a polycrystalline, to outgas volatiles that are trapped in the films during growth, etc. However, it can also initiate other chemical effects, and SIMS is capable of revealing these changes. For example, annealing has mobilized Si to diffuse from the substrate to the surface of the Fe film, which was deposited on Si [159]. It is common that transition metal thin films (Fe, Co, Ni, and others) deposited on Si substrates form various silicides after annealing. The effect of annealing on the element migration is graphically demonstrated in Figure 21. Diamond-like or carbon-based composite films are promising divertor materials for the new generation fusion energy reactors because of their excellent thermal properties. Trapping of ^{4}He$^+$ or H$^+$ is an important safety parameter for their use in such applications. SIMS depth profiling has demonstrated that migration of H starts at 700°C, and, finally, H is broadened to the whole thickness of the film [160]. Prior bombardment with He did not change the above behavior.

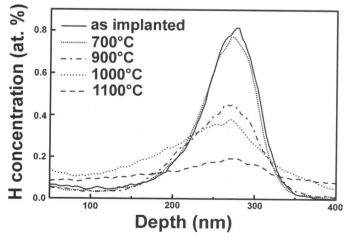

Fig. 21. The effect of annealing at different temperatures on hydrogen concentration profiles in carbon-based composite films, obtained by SIMS. From [160] with permission from the author and the American Institute of Physics.

9.2.4. Investigating Physical Properties of Materials

Corrosion effects can be enhanced by diffusion of elements into the materials in question. However, barrier layers can improve resistance to corrosion, and in harsh environments they are sometimes unintentionally grown, however, enhancing the properties of the used materials. The study of these materials in such environments is useful, and because these layers are generally thin and are present on the surfaces of the materials, they are also mentioned here. As a first example, oxide layers can be formed by oxidation as a result of oxygen diffusion, and with the intention of studying corrosion effects of reactor water on Zircaloy fuel rod claddings in pressurized water reactors, SIMS depth profiling and ion imaging were used to analyze these oxide layers [161]. In these reactors pH and reactivity are controlled by adding LiOH and H_3BO_3 to water. Li and B were found to diffuse in the $< 10~\mu m$ thin oxide layer, which is formed on the fuel rod claddings. It was also observed that Li concentrates closer to the surface of the oxide (in touch with the water), leaving a dense oxide layer closer to the metal/oxide interface to control corrosion, and because it was present at a low concentration, it could not initiate corrosion reactions.

Corrosion resistance of titanium alloys to an NaCl environment was also increased by implanting in them a specific quantity of nitrogen [162]. SIMS depth profiling helped to optimize the above material. In the same way, with the incorporation of yttrium into Fe-Ni-Cr alloys, oxide scales are formed at the surface of the alloys, protecting them from oxidation at high temperatures [163]. In this study, SIMS was capable of seeing the depth distribution of yttrium and of demonstrating the presence of an Y_2O_3 oxide 35 nm from the surface of the alloy, preventing further penetration of oxygen deeper into the sample. Similar studies have been performed for alumina or chromia alloys, but with less effect from yttrium [164]. Aluminum alloys can also be protected from corrosion when Ni is incorporated into the outer hydrated aluminum oxide; SIMS was used to study the properties of these films [165]. Corrosive-resistant Fe films could also be deposited at room temperature on Si wafers, and SIMS studies have shown a high-purity layer [166]. Corrosion preventions mechanical contact, such as between heads and hard disks, is increased by closer contact and the demand for higher storage densities. Among many techniques, ToF-SIMS was the most sensitive to cobalt migration toward the thin protective layer of the hard disk surface [167]. The effect is more pronounced for thinner protective layers and at normal humidity and temperature conditions of hard disk operation.

9.3. Superconducting Materials

The perfect conducting properties of the high-T_c superconducting films increase their technological potential as ideal chip and device interconnects. However, compatibility with other materials, such as doping materials or substrate materials, must first be achieved for optimized performance. Furthermore, growth procedures and environments must be improved. SIMS has helped researchers to study several aspects of YBCO films, such as their structural or chemical composition, the possible contamination from sources mainly related to the growth procedures, the migration and interdiffusion of doping elements between the layers or of elements from the substrate, optimization of their chemistry after doping with the aim of improving their electrical properties at even higher temperatures or their magneto-resistance properties. In the following subsection, a list of characteristic cases is presented to demonstrate the amount of information that can be acquired from the YBCO films during SIMS characterization.

9.3.1. Annealing Effects on Doped Elements and on the Interfaces of YBCO Heterostructures

When YBCO heterostructures are doped with different materials they acquire their superconducting properties. A usual step after doping is annealing, which, however, has an effect on the doping materials: they tend to migrate from the doped to the nondoped layers. In a similar way, elements from the substrate can also migrate to the neighboring layers. Both effects cause the deterioration of the efficiency of the structures and the choice of doping elements or of the substrates and the annealing conditions can be optimized and standardized by iterating between chemical characterization and characterizations of the electrical or other properties of the films. Because YBCO films are multilayered thin films, chemical characterization can easily be performed with depth profiling SIMS or SNMS, generally assisted by other techniques.

For example, SNMS has been used in comparison with AES techniques to show that when YBCO films are doped with rare earths there is no migration observed at annealing temperatures of 900°C [102]. SNMS was the technique chosen here to reduce the matrix effects introduced by SIMS. Y and Pr rare earths were the doping elements, and CVD techniques ware used to grow the heterostructures of $YBa_2Cu_3O_{7-\delta}$, $PrBa_2Cu_3O_{7-\delta}$, and $(YBa_2Cu_3O_{7-\delta}/PrBa_2Cu_3O_{7-\delta})_n$ films. Depth profiling of both annealed and as-deposited films has additionally demonstrated that CVD growth does not cause the deterioration of the interface quality of the heterostructures. This is depicted in Figure 22, where the depth profiles of the doping elements (Y and Pr) are compared with primary film elements (Cu and Ba) and the substrate (Ti and Sr). In this figure, two main YBCO layers can be distinguished (one doped with Y and a second with Pr), as can their interface at about 0.14 μm. The SNMS measurements have been performed for comparison on both powder and single crystals of these materials, and the acquired *RSF* values were identical. These values were used for the quantitative depth-profile characterization of the films and are available in [102].

From the previous sections it can be seen that Y and Pr are doping elements that produce films with stable chemistry and good electrical properties. Another doping element, Ag, can also increase the overall critical-current density (J_c) while at the same time giving to the films better mechanical characteristics. SIMS depth profiling, however, has shown that it is

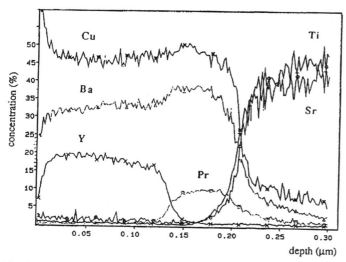

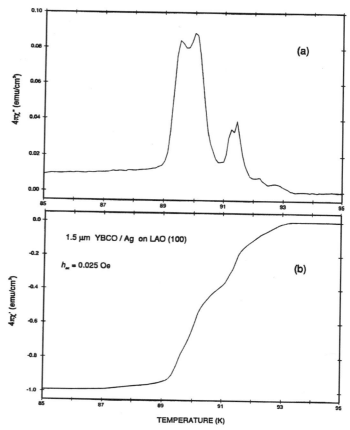

Fig. 22. An SNMS depth profile of $YBa_2Cu_3O_{7-\delta}$/ $PrBa_2Cu_3O_{7-\delta}$ heterostructures. Ti and Sr belong to the substrate, Cu and Ba are at high concentrations throughout the film, and Y is in the first layer and Pr is in the second. One interface of the heterostructures is seen here at about 0.14 μm. From [102] with permission from the author and Springer-Verlag.

highly mobile [168], a feature that makes Ag a less than useful doping element. More precisely, Ag-doped YBCO targets were used to grow thick YBCO films (> 2 μm) on $LaAlO_3$ substrates by pulsed laser deposition (PLD) with the purpose of producing coat conductors with increased current-carrying capabilities [168], which are useful for high-current applications in electrical engineering and power distribution. These films were characterized by X-ray fluorescence (XRF) and SIMS depth profiling. XRF was used for quantification purposes; it was estimated that Ag was contained in concentrations of 1 at.% over the entire surface of the samples, which had $T_c > 70$ K. SIMS depth profiling of several films demonstrated that this concentration decreases with depth (Fig. 23). It is possible that the observed nonuniformity of Ag with depth can explain the

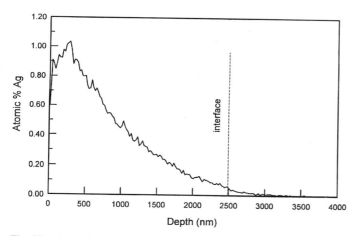

Fig. 23. A simple SIMS depth profile of Ag in a YBCO film, demonstrating how concentration of a doped element drops with depth. From [168] with permission from Elsevier Science.

Fig. 24. Imaginary (a) and real (b) part of ac susceptibility curves of Ag-doped YBCO films. The values of the imaginary part (a) show a second peak probably attributed to a second high-T_c component (at about 93 K) for the same film. SIMS has shown a migration of Ag to the surface layers, which possibly produces this effect. From [168] with permission from Elsevier Science.

multiple high-T_c components of some thick films (Fig. 24). The high mobility and surface energy of Ag forces it to migrate to the surface of the films, a phenomenon enhanced under substrate heating. Furthermore, Ag has a low solubility in YBCO, and it concentrates at the boundary grains, which are larger at the surface of the film [168].

9.3.2. Doping by Ion Implantation

Doped YBCO films are not only produced from predoped material sources, but also by ion implantation methods. However, ion implantation can severely affect the internal structure of the films, by the creation of nonuniform, Gaussian-shaped deposition profiles, crystal damage, or both. A variety of elements can be implanted at a variety of energies; each case must be studied individually. The first example is the case of YBCO films grown on a $LaAlO_3$ substrate and implanted with protons (H^+), oxygen (O), and gold (Au) at ion energies ranging between keV and MeV [169]. The purpose was to create thin film superconductor devices that would be used as magnetic thin film devices employing colossal magneto-resistance (CMR). TEM (tunneling electron microscopy), XTEM (cross-sectional

TEM), SEM, XRD, and SIMS were used together to study the structural changes. SIMS depth profiling in particular was used to study the distribution of the implanted ions. The profiles were calibrated for depth by measuring the sputtering time and calculated against the depth of the formed craters, which was measured with a stylus profilometer. As anticipated, depth profiles measured by SIMS of low-energy (50 keV) implantation of H^+ in a heavy matrix (such as the YBCO film) produced a Gaussian profile with a changing slope at the leading edge. Comparison of SIMS profiles for other implants, such as oxygen and gold, with TRIM simulations (see Section 2.2.) have shown that Gaussian profiles become more symmetrical and compare well with each other. This is attributed to reductions in the original crystallinity with increasing atomic mass of the bombarding species (TRIM simulations are more precise for amorphous materials).

Within the same work [169] the proton mass transport in YBCO films was obtained by SIMS depth profiling, through the study of H^- and OH^-. The Gaussian profile is also skewed, with a change of slope at the leading edge, reflecting pre-existing protons, which have diffused into the film before implantation. They give rise to a surface layer with a higher effective stopping power for the deuterons, and during implantation these protons are well mixed into a layer with uniform concentration, giving the step at mass 17 signal (probably $^{16}OH^-$). Diffusion of O [169] and deuterium (D or ^2H) [170] into the YBCO film (Fig. 25a) seems to occur easily inside or outside of the film and redistributes. Especially during annealing, protons diffuse without using OH as a carrier, whereas O diffuses in the film, keeping the initial concentration. D, however, is bonded to oxygen and OH shows a declining "diffusion-controlled" distribution to a similar depth of 150 nm within the as-received YBCO film [170]. Heavier atoms, such as Au implants, show no redistribution at temperatures lower than or equal to 650°C (Fig. 25b, [169]), although surface crystallization started. However, redistribution occurs at temperatures between 650°C and 700°C. Similar temperatures for ^2H and ^{18}O are about 175°C and 250–300°C, respectively.

Finally, in [169] a good example is provided that demonstrates the difficulties introduced during the interpretation of the acquired SIMS data. The problem is mass interference, observed as a mass-18 signal higher than expected from the natural abundance of ^{18}O. This is attributed to protons (^2H) combined with ^{16}O to form $^{18}(^{16}O_2H)$.

9.3.3. Contamination Sources

Contamination is an important parameter, significantly affecting the quality of superconducting properties. It is mainly related to growth procedures, the growth environment, and contamination or reaction phases originating from the substrate. Common contamination coming from the environment was found in YBCO films that are deposited by codeposition sputtering of four ion beams and three targets (Cu_2O, Y_2O_3, and $BaCO_3$) on $SrTiO_3$ substrates [171]. The fourth ion beam was supplying low-energy oxygen. Static SIMS was used, together

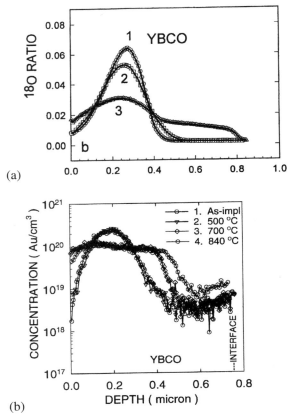

(a)

(b)

Fig. 25. Redistribution and peak widening of oxygen in a YBCO film at different annealing temperatures. (a) Three SIMS profiles show ^{18}O in YBCO before annealing (curve 1), after annealing at 300°C for 1 h (curve 2), and after rapid thermal annealing at 450°C for 2 min (curve 3). (b) The same effect is also observed for Au, but at higher temperatures. From [169] with permission from the author and Elsevier Science.

with AES and ESCA-XPS, to characterize the top monolayers of the surfaces of the films and compare them with monocrystal reference materials of the same composition. In the monocrystal reference materials traces of Al were found to be coming from the alumina crucible that was used during preparation. In contrast, the sputter codeposited samples have shown traces of residual gases, such as C, CO, CO_2, and H. Also in the mass spectra, CO was interfering with Si, which came from bulk diffusion during sputter deposition at 650°. AES and SIMS have shown identical results for the two samples.

Metal-organic chemical vapor deposition (MOCVD) is a well-established technique in the production of III–VI semiconductor optoelectronic devices (e.g., UHB-LEDs, ultra-high-brightness LEDs) or solar cells. Such a high-mass production technique could also be used in the production of cost-effective YBCO superconducting films [172, 173]. During the deposition procedures of the YBCO film, fluorinated precursors are used, which result in a quantity of fluorine remaining in the film [173, 174], which has to be removed by annealing [172] or reduced by the introduction of water into the flow of oxygen gas during growth [173]. SIMS demonstrated that annealing for 1 h in 10 mbar nitrogen at 800°C can reduce the original fluorine concentration from 250 ppm to about 50 ppm by evaporation [172].

The effect of a number of substrates, such as Al_2O_3, MgO, $SrTiO_3$, $ZrO_2-Y_2O_3$ (YSZ), and BaF_2, on the superconducting transition temperature (T_c) on YBCO films has been studied. Superconducting films can be deposited by ion-beam sputtering and e-beam evaporation [175], or even by pulsed injection CVD from an organometallic vapor [176]. SIMS depth profiling alone or in combination with other characterization methods has resolved several interface materials and the interdiffusion of elements that affect the T_c properties of the superconducting film. MgO substrates show little interaction between the phases, but T_c was again reduced. YSZ substrates, although they interact to form $BaZrO_3$ perovskite and although Zr was diffused in the film, have a small effect on the T_c properties [175]. However, an intermediate layer of CeO_2 can stop the Zr diffusion and improve epitaxy and the other electrical properties of the films [176].

Compatibility with the substrates and the doping materials and contamination sources are not the only parameters that should be monitored. Superconducting materials like YBCO are applied in multilayer devices and circuit structures, thus they have to be patterned and a lithographical step will be involved. Consequently, not only compatible insulating materials, but also an appropriate patterning process and etching materials have to be found. A first study [177] has shown that the $SrTiO_3$ (STO) can be considered a promising insulating layer in YBCO/STO/YBCO crossover structures. It was used with SIMS to study the interfaces and optimize the compatibility with the material photolithographic techniques and wet-etching process.

9.3.4. Proximity Effect and Other Superconductors

A type of periodically layered superconducting film based on Nb/Pb and Nb/Cu multilayers has properties that depend on the proximity effect, and consequently the quality of the interface and the presence of possible transition layers are also important. Generally, the Nb layers are about 20 nm thick, whereas the metallic layers range from less than 2 nm to 17 nm. The structures are deposited on Si substrates with magnetically enhanced sputtering methods. Very high-resolution SIMS depth profiling has been used effectively together with X-ray specular reflectivity (XSR) to study the above structures [129]. In this study, high depth resolution is achieved by very low intensity, highly grazing primary oxygen and cesium ion beams (< 2 keV and 5.5 keV, respectively) and has revealed all of the periods of the multilayer structure. SIMS has shown that residual oxygen from the growth chamber is trapped preferentially at the interface between Pb and Nb layers and with Nb (NbO), mainly because of its high reactivity. Preliminary experiments with the same multilayer films but grown in UHV by MBE do not show NbO formation. Furthermore, Nb matrix effects due to oxygen are not present in the SIMS profiles of the MBE-grown structures, whereas they are observed as high yields of Nb where Nb oxide was deposited in sputter-deposited layers.

Thin film superconducting oxide (SCO) materials are another type of interesting superconducting materials that can now be produced with a variety of methods as thin films. SIMS has been used to find compositional differences between the thin-film SCOs and the bulk SCOs that are produced by conventional methods [178].

9.4. Applications in Electronics, Optoelectronics, and Light Sensors

SIMS is a very common characterization technique in the semiconductor industry, mainly because of the very low detection limits compared with other techniques. It is possible to attain a detection limit as low as 2 ppb (= 10^{14} atoms/cm^3). This is achievable because of the larger area of analysis. However, for semiconductor VLSI, or UVLSI applications, the feature sizes are a challenge for the technique. VLSI, for example, requires a lateral resolution of 0.2 μm for doping elements of 10^{14} atoms/cm^3 concentration [179], which results in a detection limit of 200 ppm. Thus, higher currents at finer primary ion beams have to be adopted. A solution can be created by applying liquid metal ion beams that can be focused into submicron scales (generally less than 0.2 μm). To improve the secondary signal analyzed, a ToF instrument can also be used, which has high transmission, and post-ionization methods using lasers are also possible [179].

Because semiconductor devices are patterned structures, they may not be considered as thin films by the strict definition. However, a selection of examples is presented here, such as when special layers or interfaces are studied. Such materials can be, for example, silane agents (organo-silicates), of which only one monomolecular layer is enough to improve surface properties such as adhesion; when they are deposited in more molecular layers they are used for coupling. In both cases ToF-SIMS was used to study the quality and the effect of the layer on the substrates (see [180] and references therein). The areas in which SIMS can assist semiconductor research are mainly diffusion, migration, and segregation of elements between the different layers and the substrates, as well as chemical purity, doping concentration, and interface chemistry and structure. SIMS ion imaging has also been used to check the selectivity of deposition during lithographic patterning processes. Following is a selection of examples demonstrating the above studies for different devices.

9.4.1. Light Emission and Detection Devices

In optoelectronics, MBE-grown, the fabrication of ZnSe-based (II–VI) heterostructures have recently found many applications in devices emitting in the blue-green region of the optical spectrum. To achieve the above emission properties, layers containing Cd are deposited to generate the active quantum well regions. However, interdiffusion of these elements between the layers but also with the substrates (generally GaAs) reduces the quality of the devices [181]. Moreover, the diffusion of Ga into the ZnSe layer produces complex chemistry, which generates two more photoluminescence (PL) peaks at 2.0 and 2.3 eV. SIMS studies have demonstrated sharp compositional changes in the different layers before annealing. However, annealing induces interdiffusion of Ga from the substrate. Moreover, Cd from the Cd-containing layers broadens into the other layers. According to experiments and confirmation by SIMS, it has been demonstrated that thicker films do not reduce the effect. On the contrary, they give enough space between Cd and Ga from the substrate that they do not reach each other and improve the thermal stability of the device.

High-performance optoelectronic devices such as laser diodes emitting blue/UV light under room temperature conditions and continuous-wave operations with long lifetimes (> 10,000 h) [182] can be based on III-nitride semiconductors (e.g., GaN) [183]. Immediate applications of these structures are in the next generation digital video disk (DVD) players. GaN was deposited on sapphire substrates by MOCVD, and a thin (200-nm) layer (cap layer) of dielectric SiO_2 was deposited on top by electron beam evaporation [182, 183]. Previous results have shown that oxygen easily enters GaN and creates shallow donors equal in number to the concentration of the free electrons [182]. SIMS depth profiles (Fig. 26) were used to show that oxygen content in SiO_2-capped GaN has increased by an order of magnitude [182]. The reason for the degradation is oxygen impurities coming from the SiO_2 cap and not the Si, because when Si_xN_y is used instead of SiO_2 no significant degradation of the PL intensity is observed. Measurements have shown that in addition to the PL performance degradation of the devices, electrical effects also occur with increases in the electron density and mobility of GaN epilayers. Oxygen from SiO_2 seems to be the possible cause of this, and rapid thermal processing (RTP) under different annealing temperatures can recover the effect and improve PL.

A common impurity in GaN films is carbon. Taking advantage of the high sensitivity of SIMS, nondoped 5.5-μm-thick films were characterized for carbon impurities [183]. However, $^{13}C^-$ ions are difficult to ionize, but the detection of CN^- ions has improved the ion yield and, consequently, the sensitivity of the analysis by at least two orders of magnitude. It is expected that carbon detection can be improved in other films as well, such as Si_3N_4 or TiN, when measurement of CN^- ions is utilized.

Lasing materials can also be based on highly doped $Al_xGa_{1-x}As$/GaAs heterostructures, which are grown by MBE. These structures can not only lase but can operate as waveguides as well. Such structures have been investigated by SIMS depth profiling [184] and were calibrated with *RSF* factors. Depth profiles have demonstrated that Be outdiffuses significantly and segregates at growth directions during the growth of films. Although the high quantities of Al can hold higher concentrations of dopant Be, the Be diffuses to GaAs layers, deteriorating the quality of the structures. Upper doping limits must be followed to reduce redistribution.

Far-IR light can be detected by thin films of Ir and its silicides ($IrSi_1$–$IrSi_{1.75}$) grown on a Si substrate. The highest efficiency of the detector is achieved when the films are very thin, in the range of 2–3 nm, and the materials are of high purity. SIMS is capable of characterizing both of the above aspects. In [138] it is suggested that very low primary energies must be used to resolve the films, which will still be extended in thickness by half of their original thickness because of effects intrinsic to the SIMS techniques, such as recoil mixing. Quantification investigations, calibration curves, and comparisons with thicker samples suggest that the measurement acquired just before going through the film can be safely taken as the correct concentration. IR and far-IR detection can also be achieved by S- and Si-doped GaSb films. These films can also have other optical applications or can be used to fabricate laser diodes operating in the 0.8–2.5-μm range. When GaSb is doped with S or Si, it changes from a p-type to an n-type conductivity. These elements are used for doping [185] to avoid amorphization by the use of heavier ions (such as Te and Se), which later requires higher temperatures to anneal the low-melting-point GaSb (about 710°C). SIMS depth profiling of Si has shown that there is no redistribution, even at annealing temperatures as high as 600°C.

Light-emitting materials have also been fabricated by doping of a film of SiO_2 with S and N [186]. SIMS could show good N distributions inside the films, which were not affected by annealing.

9.4.2. Filter Coatings and Waveguides

Metal oxide multilayers are widely used as filter coatings in optical technology and industry. The chemical purity of these films results in better quality of the optical systems. In [100] the interfaces of SiO_2-TiO_2 and SiO_2-ZrO_2 multilayer combinations have been studied by SIMS and SNMS. Enhancement of the Zr^+ and Ti^+ ion signals has been observed, and this is attributed to the depletion of oxygen at the interfaces with the SiO_2 layer, which is a result of the preferential sputtering of

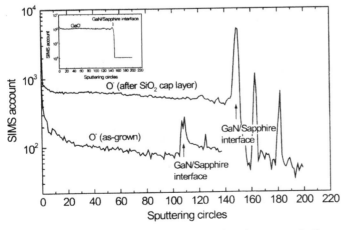

Fig. 26. A SIMS depth profile measurement showing a low-oxygen depth profile in as-grown GaN films and higher-oxygen GaN films capped with a SiO_2 film. From [182] with permission from the author and the American Institute of Physics.

oxygen compared with the two metals. This effect is not observed for SiO$_2$.

High-quality optical waveguides or optical coatings can be grown epitaxially from dielectric IIA fluorides (CaF$_2$, BaF$_2$, SrF$_2$, and their mixtures). Further applications of the same materials are found in microelectronics as insulators, gate dielectrics in field-effect transistors and lattice matching buffer layers, etc. They are promising for optoelectronic devices. The cubic fluorite structure is optically isotropic and has excellent transmission properties for wavelengths in the range of 0.3–5 μm. Contamination with carbon or oxygen during the deposition deteriorates these properties and has to be investigated. SIMS profiling has been used [187] to study epitaxially grown BaF$_2$ films. It is suggested that C and O are observed only at background levels, and the quality of the films is very high.

Optical waveguides can be fabricated by depositing and diffusing Ti into lithographically patterned LiNbO$_3$ substrates. It seems that the relationship between composition (acquired with SIMS depth profiling) and optical properties, such as the refractive index, is related to the difference in polarizability between titanium and lithium ions, but the strain induced by Ti in the crystal plays a minor effect [188].

9.4.3. Other Microelectronic and Electronic Materials

In microelectronic devices, the increased functionality at higher speeds with reduced energy consumption leads to ultra-large-scale integrated (ULSI) devices with ever smaller feature sizes. A primary requirement is technological developments not only in lithographic technologies, but also in materials science and growth techniques. Most structures are grown epitaxially into thin-layered structures (or, in other words, thin films), which are then patterned with small features by lithography. The application of SIMS in microelectronics has been already reviewed [180], with examples of studies of surface contamination, which originates from the different etching, lithographical, and metallization processes; of induced surface modifications such as silane growth for adhesion enhancement or protection against corrosion; and of the inspection of doping processes or growth control. Similar applications and examples selected from the recent literature are briefly reported here as well.

A way to dope silicon thin films is through the use of a gas mixture of silane and biborane (B$_2$H$_6$) or phosphine (PH$_3$) during their growth, which can be made by different CVD methods. It is useful to know which ratio of the above mixtures must be used to fabricate films with a defined concentration of the doping element. Such a study has been performed with the use of SIMS depth profiling [189], but a straightforward relationship of the precursor gas ratios with the final concentration of dopants in the solid thin film could not be defined. Depth profiles show a completely different behavior of the B and P dopants in the silicon film.

In semiconductor films, as with other thin films, element diffusion, migration, and segregation between layers and at interfaces can occur with or without further processing, such as

annealing. Gallium arsenide (GaAs) is a widely used semiconductor generally doped with Zn and Be, which tend to diffuse. More specifically, SIMS depth profiling has shown that there is significant interdiffusion of P and Sb between strained GaAsP/GaAs and GaAsSn/GaAs superlattices [190]. In both cases the interdiffusion was enhanced during annealing at increasing arsenic vapor pressures. Moreover, the interdiffusion coefficients were estimated at a variety of annealing temperatures and time periods, until a complete intermixing was reached.

Anisotropic diffusion of Al from the middle to the other layers of sandwiched structures of ZnSTe/ZnSTe:Al/ZnSTe is observed to occur, which can probably be attributed to channeling effects [191]. The studied structures were grown on GaAs substrates of different crystallographic orientations by MBE; the Al-doped layer was grown in a one-step process. SIMS depth profiling was performed with the use of two different instruments, and the measurements were calibrated by *RSF* functions. These measurements have shown that Al is stable at growth temperatures of up to 300°C. Annealing at 450°C and 550°C, however, initiates diffusion of Al to the neighboring barren layers. It was also observed that crystallographic orientation has an effect on the relative ratio of matrix atoms to Al during growth.

To understand the effect of epitaxial silicide layers such as CoSi$_2$ on the diffusion of B and Sb in the underlying Si, SIMS has been utilized to profile the two elements. It has been observed that Sb diffusion is enhanced, whereas B diffusion is highly reduced by the epitaxial silicide layer [192]. Consequently, stable nanoelectronic structures based on the LOCOSI process (local oxidation of silicide) can be produced (thin Si layers within CoSi$_2$ layers).

Copper germanides are now used for microelectronic interconnects because of their improved electrical characteristics and reduced oxidation effects, replacing Cu silicides. Cu germanides can be fabricated by diffusion of polycrystalline Cu thin films (poly-Cu) that are deposited on amorphous Ge (a-Ge) films followed by annealing [193]. Among other techniques, SIMS has been used for depth profiling and revealed that at annealing temperatures higher than about 430°C, Si from nonoxidized wafers diffuses through Ge, causing a change in resistivity of the multilayer structures. At lower annealing temperatures a stable Cu$_3$Ge thin film with fixed stoichiometry and abrupt interfaces is created.

Segregation and diffusion of P from *in situ* doped Si$_{1-x}$Ge$_x$ epitaxial films on Si at 750–850°C have been demonstrated with the use of SIMS depth profiling [194]. Depth profiles show that P diffuses and segregates in Si. Furthermore, segregation concentrations are higher in films with a high Ge/Si ratio.

Impurities affect the properties of semiconductor devices, as in the study of F impurities in Si-doped InAlAs/InGaAs heterojunction FETs (field-effect transistor), which are related to their electrical characteristics. Structures with and without SiN passivation layers were studied [195]. SIMS has shown that there is no significant diffusion of F impurities in the structure after removal of the passivation layer with a HF solution. However, in the case of SiN passivated structures, F accumulates in the Si-doped n-type InAlAs layer of the structure, affecting the electrical characteristics.

Impurities are also introduced during growth of the film microelectronic structures. The thinner the film and the smaller the structure, the higher is the effect of the impurities on the properties of the film. Gate dielectrics are an example that follows this trend for minimization. Silicon oxynitrides can replace silicon dioxide because they perform better and can be thinner [196]. Moreover, oxynitrides are grown in a N_2O ambient, and hydrogen contaminant is reduced because NH_3 is now avoided. SIMS depth profiling studies [196] have shown that nitrogen accumulates in high amounts at the oxynitridesilicon interface and has also been used to model the growth process.

Patterning with lithographic techniques is an important step in the fabrication of microelectronic devices. It is used in conjunction with deposition techniques to fabricate complicated layered structures. Deposition has to be highly selective; otherwise the operation and the efficiency of the device may be affected. An example, where a soft lithographic with polymer protection patterning in conjunction with additive CVD metallization is used, is given in [197], where micron-scale electronic Schottky diode devices are fabricated, based on Pt/Pt-silicide/Si-substrate arrangements. The involved steps are:

(a) patterning of a 30×10 array of micron-scale diode features on a silicon substrate with the use of a polymeric film, which was prepared by micro-moulding in capillaries (MIMIC), a soft lithographic patterning technique;
(b) etching that removed the oxide from the substrate surface; and
(c) metallization by selective platinum CVD that was used to form rectifying contacts with the substrate.

The polymeric film successfully served both as an oxide-etch resist before metallization and as a deposition-inhibiting surface for the selective deposition of platinum. SIMS ion imaging of Pt and C was used to confirm the selectivity of deposition. Pt represents the metallized areas and C the polymer areas [197].

Abrupt surfaces and interfaces are required in most electronic devices, and MBE or MOCVD techniques are generally used. These techniques are also used for the fabrication of heterostructures, superlattices, and quantum wells of the III–V semiconductors. In [118], GaAs/InGaAs/GaAs and Al-GaAs/InGaAs/AlGaAs quantum-well (QW) structures and In-GaAs/AlInAs superlattice (SL) structures were studied with SIMS using O_2^+ and Cs^+ ion beams and by monitoring the MCs^+ ions. In this work, depth resolution was a critical characterization parameter, and techniques such as reduced bombardment energy and initial good surface topography were adopted to improve depth resolution. Consequently, depth profiles of the QW structure managed to resolve In peaks after small differences in erosion speed were taken into account. It also demonstrated that the depth resolution can be preserved with depth, keeping the interfaces abrupt, and only a small shoulder is observed, corresponding to slower increase in signal at

the front edge, which can be also attributed to memory effects during growth. In [118], the highest resolution is also achieved using SIMS depth profiling on InGaAs/GaAs multiple-QW, demonstrating a 0.62-nm resolution at the leading edge and a 1.30-nm resolution at the trailing edge. The SIMS profiles have been modeled by convolving the true profiles with an analytical response function and calibrated using measurements from TEM. Effects such as blurring with increased depth, especially after 400 nm, which degrades the depth resolution, have also been observed. Topography roughness (ripples, cones, pyramids, and terraces) is also introduced and is related to impact angle and bombardment energy (down to 2 kV). Diffusion and segregation at the interfaces were also studied using SIMS, and a combination of high depth resolution and high sensitivity is suggested as the best tool to resolve them: segregation of In occurs at the front edge between the AlGaAs and InGaAs interfaces. Finally, MCs^+ molecular ions were used to minimize the matrix effects, and, together with the use of reference materials and RSF functions, concentration accuracies better than 2% were achieved.

9.5. Other Technological Materials

In this paragraph, thin film materials of special technological interest are described. This interest relies either on their mechanical and chemical properties or a combination of these and electrical properties. Consequently, hard materials such as diamonds and diamond-like and carbon nitride films are presented together. Because such materials are used as tools for mechanical processing of other materials, SIMS was used to study their interaction, a process that is mainly chemical diffusion of elements due to the high temperatures involved during cutting. SIMS also assisted in defining the exact chemistry of new materials of that kind.

Thin-film transistor displays, solar cells, transparent conducting films, and gas sensors are treated mainly as devices of technological interest, and therefore they are presented here. SIMS has been used to study parameters that will optimize the operation or efficiency of these devices. Perovskites are also mentioned here because of their multiple applications in technology, which are not only electronic or optical but mechanical as well. The section concludes with examples from organic films and processes in microsystems and chemical technology that involve films and interfaces.

9.5.1. Diamond and Other Hard Films

Diamond-like carbon films (DLC) or carbon-based composite films are promising divertor materials for the new generation of fusion energy reactors because of their excellent thermal properties. However, it is important to know the processes of hydrogen implantation, trapping, and transport in these films as well as possible contamination. These studies are performed by SIMS, among other techniques [160].

Diamond film materials are used for coatings of cutting tools because of their hardness. However, diamond is soluble in ferrous metals, and it cannot be used to cut them. The ternary boron–carbon–nitrogen system can produce super hard phases when it is deposited in films by microwave-CVD from a $B(N(CH_3)_2)_3$ precursor [198]. Their characterization by SIMS threw up new problems that required solution. These included the charging of the BN phases, which was eliminated by computer-controlled sample potential changes, and the low secondary ionization efficiencies of N, which required the analysis of molecular N-bearing species. The most significant problem was the matrix effect; quantification is not possible, even with the use of standards, and other techniques must be used. Other SIMS depth profiling studies [199] have shown that at high temperatures during cutting there is also an exchange of elements through diffusion processes between cubic boron nitride (cBN) tools and compacted graphite iron (CGI) materials, which reach a depth of about 20 μm.

Similar diamond films have been grown by CVD deposition from a $B(C_2H_5)_3$ gas precursor on a pretreated silicon substrate [200]. On these samples, the B/C ratio has been investigated with SIMS one- and three-dimensional depth profiling and ion imaging. It has been observed that B is five times higher in concentration when films are grown along {111} rather than {100}. Furthermore, the {111} samples have shown contamination from Al and Na traces. Cr traces are homogeneously distributed in the diamond films.

Hard materials with a variety of applications in optics, electronics, and coating are the oxynitride films of silicon and aluminum. A combination of technologies, including SIMS and rf-SNMS, has been used to study these film materials [109] and optimize their composition to achieve better film properties. SIMS has proved to be useful for low-concentration measurements, albeit with some disadvantages due to charging and extensive calculations for quantification. The rf-SNMS method is easier than SIMS at higher concentrations (e.g., for oxygen contents higher than 10 wt%) and can share the same *RSF* values with SIMS.

Carbon nitride (C_3N_4) films have been studied because of their mechanical properties; they are very hard and, in some cases, harder than diamond [201]. The films can be grown with reactive magnetron sputtering from a high-purity graphite target in nitrogen plasma. Understanding the chemical composition and bonding in these materials is of great interest; ToF-SIMS with a Cs^+ primary ion beam was used to sputter molecular fragments such as CN, C_2N, CN_2, C_3N, C_3N_2, and C_4N_3. These fragments and their relative intensities have been compared with the characteristic spectra of other diamond, diamond-like, or graphite materials, and there are no similarities, indicating that the film is a highly N-rich material [201].

9.5.2. Thin Film Transistor Display Materials

Technological developments of thin film transistor (TFT) displays are based on achieving high purity in the growth materials on large areas of appropriate substrates and optimization of the electrical properties (and in effect the optical properties) by controlling the impurity content and distribution. A commonly used material is silicon, either amorphous or polycrystalline (polysilicon). The following application examples show that SIMS depth profiling can successfully be used to investigate and optimize the above devices.

Thin films of amorphous silicon (a-Si:H) for TFT display applications can be grown over large areas by plasma CVD (PCVD) and other techniques. Then these are doped either by rf-ion sources [202] or by laser [203], where an KrF excimer laser at 248 nm is used to initiate doping due to good absorption by amorphous silicon and easy surface melting and crystallization. Common doping elements are hydrogen, phosphorous, and boron, which are used to create the required n+ or p+ junctions. SIMS depth profiling was used to analyze the doped films and to acquire information on the distribution of the doped elements. The results were used to compare experimental data with Monte Carlo simulations [202] and to optimize the number of laser pulses for more efficient doping [203]. This comparison is always useful because the doping effect can be calculated and the properties can be predicted. Furthermore, it was observed that peak positions (highest concentrations) of the phosphorus ($^{31}P^+$) profiles are observed at 6 nm and 25 nm, when doping occurs with energies of 6 kV and 30 kV, respectively, and 50 nm and 150 nm for hydrogen [202]. It was expected that hydrogen would have broader profiles relative to phosphorus because of its high mobility. However, both profiles are shallower relative to the calculated profiles, possibly because of etching, the last step in the process, which removed some surface material.

The growth of polysilicon thin films was optimized in [204] for lower temperatures with a new PCVD technique, at 450°C. This makes the growth process possible on ordinary glass substrates, which cannot tolerate high processing temperatures (> 600°C). However, impurities of oxygen in the grown films are demonstrated by qualitative SIMS. These impurities are probably introduced from the conventional vacuum chamber used for the growth process [204]. Further work has demonstrated that an *in situ* cleaning procedure, using fluorine or fluorine together with hydrogen in the plasma, eliminates the oxygen impurities in the films. From such reactants amorphous silicon has been developed elsewhere. SIMS helped to relate fluorine content with crystal size [205] and to investigate atomic-scale smooth interfaces and growth from a highly crystalline seed [206]. The way in which impurities and their distribution can affect the electrical and, as a consequence, the optical properties of the thin film diode liquid-crystal displays (TFD-LCD) is demonstrated in [207]. TFD-LCDs show an after-image effect, which can be reduced to levels that are not visible to the human eye when the current–voltage relation is improved. SIMS was then used and proved capable of revealing these impurity/compositional asymmetries, which cause the asymmetries between current and voltage during the Schottky effect conductivity of the TFD displays. Impurity atom species, such as C and P, are incorporated between the Ti and Ta electrodes during the deposition of the TaO_2 insulating films from the electrolyte solutions. From a variety of such anodizing solutions, only ammonium borate has shown that B atoms remain at the very surface of the film and reduce the after-image effect.

9.5.3. Solar Cells, Quantum Wells, and Conducting, Transparent Films

Solar cell structures can be fabricated from different materials and their structured combinations, such as QW $In_{1-x}Ga_xAs_y$ P_{1-y}/InP heterostructures [208], polycrystalline CdTe-CdS heterojunctions [19, 209], alternating layers of semiconducting and metallic materials [101], and hydrogenated μ-crystalline silicon (μc-Si:H, low band gap) film on an amorphous silicon (a-Si:H, high band gap) semiconductor [210]. To complete the devices they must be covered on one side with a metallic contact layer, such as Ag, and on the other side, to allow light go through, with a transparent film of low resistivity, such ITO (indium-titanium oxide), or an Al-doped ZnO, or an F-doped SnO_2 layer [101, 211]. The structures are fine and complicated, and impurities or structural quality might affect their efficiency. Once again, SIMS can provide a considerable amount of information on the structural and compositional status of these materials and has been used to help improve solar cell devices. The following are some selective examples from the literature explaining the above characterization procedures and the problems encountered during SIMS analysis.

Polycrystalline CdTe-CdS heterojunctions have been investigated [209] as new materials, that can be used for the production of low-cost solar cells. However, there is a problem: the interdiffusion of S between the films that form a layer of CdS_xTe_{1-x}, which reduces the efficiency of the cell by increasing light absorption. Depth profiling with nuclear reaction analysis (NRA) and SIMS has provided insight into the S diffusion in a CdTe monocrystal and in CdTe/CdS films, which is produced by thermal evaporation and during annealing. Both methods suggest that diffusion of S from CdS thin films to CdTe thin films is less than the diffusion occurring by thermal diffusion into the monocrystal of CdTe, and this is attributed to initial, stable chemical bonding in CdS thin films. SIMS profiles, despite the difficulties introduced by sample charging effects, have generally given more accurate diffusion coefficients and revealed a Si component in the films, which is a constituent during the sample (gas diffused) preparation phase.

CdTe thin films have also been successfully grown by MBE on InSb substrates [19] because of effective lattice matching. However, the quality of CdTe films is affected by the substrate growth temperature, the preparation of its surface before growth, and other growth parameters. Moreover, in diffuses from the substrate to the film. SIMS was used in this work to study depth profiles of the involved elements. A matrix effect was observed as a sudden increase in Cd^+ at the interface while all of the other elements changed abruptly. Monitoring of $InTe^+$, $InTe_2^+$ and $In_2Te_2^+$ signals suggests that the In_2Te_3 phase is probably present, because the above ions are possibly secondary ions that are produced from this phase; however, matrix effects cannot be ignored. Furthermore, for samples grown at high temperatures In diffuses (following Fick's second law) from the substrate and concentrates on the surface of CdTe. Only some Te diffuses into InSb.

Another type of solar cell is fabricated when several layers of C- or H-doped amorphous Si with different thicknesses are altered with metallic layers [101]. Such a p–i–n diode structure is completed by a metallic back layer, generally Ag, and a transparent but conducting front oxide film (TCO), such as an Al-doped ZnO, an F-doped SnO_2 layer or an ITO. The whole structure is supported by a glass substrate. It is important to know whether elements from the TCO layers contaminate the p-layers by diffusion. SIMS depth profiles have shown that there is no Zn contamination of the p-layers [17, 101], but SNMS studies [18] could not clearly resolve the differences between SnO_2 and ZnO. Furthermore, SIMS profiles of ^2D (deuterium) have shown that at different temperatures and for different plasma exposure times, only limited surface diffusion occurs, and, consequently, it is expected that the optical properties of ZnO will not be affected by diffusion of H from the p-doped layers [101]. At interfaces such as ZnO/SnO_2, SNMS profiles have shown a plasma pressure-dependent diffusion of Sn into ZnO, during ZnO sputter deposition. Finally, SNMS and high-mass-resolution SIMS depth profiles could not resolve possible P diffusion from the n-layer (P-doped a-Si:H) into the i-layer (a-Si:H). Technically, the high resolution was achieved by analyzing model systems comprising of simple parts of the complete structure [101]. The initial surface roughness of real TCO systems reduces depth resolution, and better studies were conducted with models with smoother surfaces. It is once again observed that SNMS profiles provide better results compared with SIMS because of the reduced matrix effects at the interfaces.

A hydrogenated μ-crystalline silicon (μc-Si:H, low band gap) film on an amorphous silicon (a-Si:H, high-band gap) semiconductor can produce efficient solar cells [212]. Such structures are easily oxidized when exposed to atmospheric conditions, which results in increasing dark conductivity (σ_{dark}) with time up to a saturation value, and porosity increases surface contamination from elements such as O, N, and C. SIMS has been used to investigate these effects ([ref:210] and references therein).

QW $In_{1-x}Ga_xAs_yP_{1-y}$/InP heterostructures have found applications in ultrafast lasers, electro-optic modulators, infrared detectors, and solar cells. Multiple quantum wells (MQWs) of such heterostructures have been grown by low-pressure metallographic CVD (MOCVD) [208]. One of the samples, the SIMS profile of which is depicted in Figure 27, had the following structure: substrate of p-type InP(001), 500 nm Zn-doped (p-doped) InP, 50 nm of nonintentionally doped InP, 16 layers of GaSP/InP forming a total 310 nm layer, 50 nm of InP and a layer of Si-doped (n-doped) InP emitter. This sample had gold contacts, and, a similar one had no contacts. SIMS depth profiling was used for both samples. It has been demonstrated that gold diffuses into the sample layers, resulting in electric field inconsistencies between the samples (not seen in the figure). Also, both SIMS (Fig. 27) and Franz–Keldysh oscillations (FKOs) suggest interdiffusion of Zn from the p-region to the i-region of the sample (16 layers).

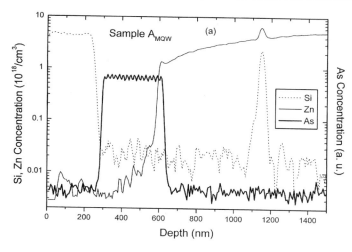

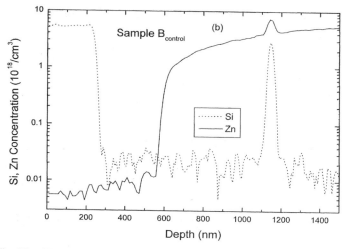

Fig. 27. SIMS profiles for (a) sample A_{MQW} and (b) sample $B_{CONTROL}$. The solid line is element Zn, the dotted line is Si, and the dark solid line is As (see text for more details). From [208] with permission from the author and the American Institute of Physics.

High-efficiency solar devices for selective solar thermal collection can be fabricated by depositing TiO_xN_y thin films on Si or Cu substrates [97]. Efficiency is reduced at operating temperatures higher than 200°C, when elements of the substrate migrate to the surface of the film and degrade its optical properties. In the case of Cu substrates, for example, SIMS depth profiles have shown that at about 350°C copper migrates to the free surface of the film and forms an oxide, which is also covered by a nitrogen-rich layer.

ITO coatings are materials that are applied to LCDs, solar cells, and photodetectors because of their visible light transparency and low resistivity. The substrate glasses usually contain Na, and during deposition or annealing, Na diffuses into the ITO films, reducing their optical properties. SIMS profiles have demonstrated that several nanometers sol-gel deposited TiO_2–SiO_2 barrier layers (T) prevent the diffusion of Na into the ITO films and that they are more efficient than SiO_2 layers alone [211].

9.5.4. Gas Sensors

Different combinations of materials, especially in the form of thin films, are used because of their selective sensitivity to different gases. The conductivity behavior of these materials, when in contact with these gases, is used to fabricate devices that can sense the presence of gases. The demand for sensitive gas sensing is increasing all the time, especially when one considers the consequences of modern pollution for the environment. Easy and quick sensing of a variety of gas pollutants is required in small and efficient devices for continuous monitoring. Such gases are the group of nitrogen oxides (NO and NO_2), which are significant pollutants. Thin-film sensors of tungsten trioxide (WO_3) deposited by sputtering on glass substrates are sensitive to NO_x [213]. SIMS depth profiling has been used to study the composition of films and their structural characteristics as element distribution with depth. The purpose of this has been to achieve a homogeneous and constant distribution of the elements in the film and to ensure high quality and good properties. During these studies, the matrix effect of SIMS was again observed at the transition from the film to the substrate as a peak shape that recovers soon.

NO_2 (but also CO_2, SO_2, Cl_2, CO, NH_3, and H_2) can be sensed by Al-doped and Al-coated NiO layers [214]. In [214] structural and chemical properties have been studied by a variety of techniques. SIMS was used to study the implantation profile of Al-doped NiO, and it was observed that there is an improvement of the gas sensor characteristics, such as sensitivity, selectivity, and signal quality. It was additionally observed that doped devices show better characteristics than the coated ones.

Another type of material used to sense CO, H_2, and Cl_2 gases is tin oxide (SnO_2). Tin oxide can be deposited as thin films by rf sputtering [215] on a silicon substrate, in polycrystalline form or with atomic layer epitaxy (ALE) [119]. To enhance its gas-sensing properties, this film can be doped either with Cu by dc sputtering with an additional surface activation by Pt doping [215] or with Sn to form an n-type semiconductor [119]. Doping with Sb in particular has not only increased conductivity but also improved the structural properties. In the first case the film was doped with different quantities of Cu to optimize the device in combination with SIMS, TEM, and XPS studies. SIMS and XPS have shown a constant concentration of Cu throughout the film; however, Pt is only observed as a very thin film on the surface of the SnO_2. Consequently, any electrical properties inside the film are independent of Pt but only on Cu. A small concentration of Cu enhances the sensing properties of SnO_2 and can control the selectivity for gases. In the second case [119], comparison of SIMS, XRF, and proton-induced X-ray emission (PIXE) compositional analyses of well-defined samples shows that SIMS is more sensitive at low concentrations, and it compares well with the other techniques at higher concentrations. However, uncertainty of measurements ranges between 8% and 22%. Depth profiles of SIMS can show that the growth of Sb inside SnO_2 can be defined precisely down to very fine layers with the pulsing-ratio method of ALE (variable number of pulses constantly altered for the two starting materials,

SnCl$_4$ and SbCl$_5$, during deposition, e.g., alternating 1 : 150 pulses, 1 : 600 pulses, etc.). SIMS required calibration samples, which were made by ion implantation of matrix-matched substrates. Another complication with the SIMS measurements was the difficulty of seeing the surface layer due to the first phase of sputtering limitations (preferential sputtering) or increased charging.

Sensor base materials used to detect reduced gases can be fabricated from n-type semiconducting polycrystalline Ga$_2$O$_3$ thin films by doping with SnO$_2$ [216]. To acquire the n-type character, the films are doped with Ti^{4+} or Zr^{4+}. The gas sensitivity can be improved by increasing the conductivity, and, consequently, Sn^{4+} was tested as a doping material. SIMS was used to profile 2-μm-thick films. After annealing, these films show a homogeneous distribution of SnO$_2$ throughout their thickness. Doping with Sn^{4+} has increased the conductivity of the films, with direct applications to chip size and energy consumption.

Thin-film metal halides, such as CuBr, are promising alternatives to metal oxides for gas sensing at low temperatures, and their easy preparation by dc sputtering is an advantage [217]. However, the nonohmic behavior of Au/CuBr/Au and Au/CuBr/Cu structures cannot be explained, and the authors recognize the significance of the SIMS technique and propose compositional measurements with SIMS depth profiling.

9.5.5. Perovskites

The significance of perovskites as technological materials is due to their multitude of applications. Sensors, actuators for micromotors and micropumps, memories, high-value capacitors, thermistors, infrared detectors, SAW delay lines, optical switches, field-effect transistors, high-frequency transducers, and buffer layers for superconducting films are some of the applications. Some common compositions are Pb/Zr/Ti (PZT), Pb/La/Zr/Ti (PLZT), and Ba/Sr/Ti oxides. Properties that make these materials important are their high ferroelectric and the piezoelectric properties. Perovskites can be deposited as thin or thick films by a variety of techniques. Following are some examples of how SIMS has been applied to improve these materials.

Thick films of PZT thin films (Pb(Zr,Ti)O$_3$) are difficult to preserve after annealing because of decreased mechanical stability [218]. It has been found, however, that increased growth pressure (for example, during sputtering) improves the stability of the films and their ferroelectric properties. With the use of SIMS analysis the lead, titanium and zirconium profiles have been measured before annealing. Despite sample charging problems due to the primary beam (O^{2-}), several profiles at thin films deposited under different conditions have shown that the content of Pb increases during growth and that there is a reverse relationship with the sputtering pressure. Consequently, higher growth pressures result in lower Pb content in the film, also increasing its mechanical stability due to reduced gradients with depth.

The above materials were also grown on Si(100), Au on Si(100), and TiN coated Si(100) substrates [219]. In this case, the SIMS technique was used, proving that diffusion between the TiN-coated Si(100) substrate and the film was less than that observed for the other combinations. Although all studied samples show piezoelectric properties without poling, the effective deposition of PZT on TiN-coated silicon substrates is reported.

The enhanced properties of perovskites are sensitive to composition, particularly to high Si content, which generally diffuses from the Si substrates [220]. Increased Si content affects the resistivity in electrical stress of the films, and intermediate materials (buffer layers) must be used to reduce this effect. PZT (PbZr$_x$Ti$_{1-x}$O$_3$), PLZT ((Pb$_{1-x}$La$_x$)(Zr$_{1-y}$Ti$_y$)$_{1-x/4}$O$_3$), BaTiO$_3$, and Bi$_4$Ti$_3$O$_{12}$ thin films have been deposited on a variety of substrates, and the effects of a direct contact or of the presence of Pt/Ti/SiO$_2$ and SrTiO$_3$ buffer layers have been studied. Positive and negative SIMS-depth profiles have shown that films of 0.1 μm of SrTiO$_3$ can effectively prevent Si diffusion in perovskite films.

In display applications, such as thin-film electroluminescent displays (TFELs), highly transparent films of BaTiO$_3$ and SrTiO$_3$ perovskites and their solid solution, (BaSr)TiO$_3$, are grown on ITO films by CVD, and they play an insulating role. The ITO films are grown on glass substrates by rf-magnetron sputtering. As seen previously, the effect of the substrates on the properties of the perovskite films is significant and SIMS was used to determine it as a function of deposition temperature [221]. Depth compositional profiles of (BaSr)TiO$_2$ films at their interface with ITO have been made to check for possible interdiffusion at structures grown at 350°C and 550°C. Some technical problems, such as charging, occurred, especially when researchers tried to pass a beam through the glass substrate. The two profiles have shown some interdiffusion phenomena between the two interfaces (perovskite/ITO and ITO/glass), which are strongly expressed at the higher temperature. It is suggested that a buffer layer has to be developed to prevent diffusion.

For devices such as thermistors, high-k capacitors, field-effect transistor nonvolatile memories and high-frequency transducers, a high dielectric constant is required. A high dielectric constant improves the storage capacity and at the same time reduces the price of the driving electronics. Perovskites of the composition (Ba$_x$Sr$_{1-x}$)TiO$_3$ combine the high dielectric constant of Ba in the crystal and the structural stability of SrTiO$_3$, properties that are significant to the production of large TFEL display panels [222]. For this study, such thin films were deposited on ITO layers, with buffer layers of 50–60 nm Y$_2$O$_3$ and Si$_3$N$_4$. SIMS depth profiles (Fig. 28) of buffered and nonbuffered structures have shown that Y$_2$O$_3$ acts as a better buffer for Sr (faster drop of the profile curve). Moreover, Si$_3$N$_4$ acts better with oxygen, which tends to diffuse into the substrate, leaving behind vacancies, which might be one of the reasons for the previously observed increase in the dissipation factor. The new structures show a reduction of the dielectric constant, but this is compensated for by the uniformity of structural and electrical properties throughout the film.

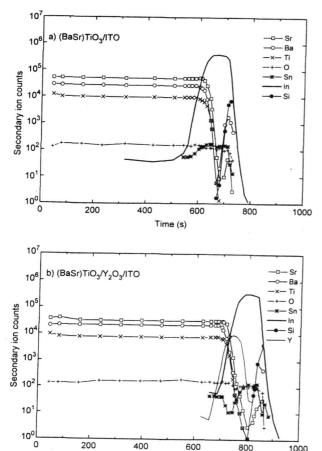

Fig. 28. The effect of buffering in producing high-quality electronic structures: SIMS depth profiles of compositional changes around the interface layers of nonbuffered, Y_2O_3 buffered and Si_3N_4-buffered $(BaSr)TiO_3$ thin films (440 nm in thickness). From [222] with permission from the author and the American Institute of Physics.

$SrTiO_3$ can also be doped with Ta. A film of 50-nm Ta_2O_5 is deposited on a $SrTiO_3$ film or sandwiched between two of these films, treated at 1100°C for 10–40 h and characterized by conductivity measurements and SIMS and Auger depth profiling [223]. The last two were used to study interdiffusion of the two layers after the thermal treatment. Experiments have shown that thermal diffusion is completed with treatment at 1100°C for 40 h.

The origin of oxygen in formed $PbTiO_3$ (PT) perovskite films was investigated with the use of oxygen isotopes [224, 225]. At the same time, improvements to the deposition method were made, a method based on pulsed laser deposition (PLD). Parameters such as the cooling step of the film in an oxygen atmosphere after deposition, the laser repetition rate, and the gas phase reactions have been optimized. The studies showed that during the cooling, between the laser pulses, about 15% of total oxygen content of the film originates from ambient gas, which is possibly attributable to surface exchange and vacancy diffusion.

9.5.6. Other Technological Materials

Contamination from detergents, airborne particles, or volatile organic compounds is introduced into microelectronic components during handling, assembly or treatment. Furthermore, it is common that polymer surfaces are reconstructed [226]. Such effects can be efficiently analyzed or even visualized by ion imaging, with static SIMS [227] or ToF-SIMS. It is demonstrated [228, 229] that ToF-SIMS is especially useful in the study of polymers and other organic materials, because information on the chemical structure can be acquired. Quantitative studies of organic materials can also be made just by using relative peak intensities only when the appropriate peaks from the ToF spectra are chosen [230], and the oligomer fragments of a monomer can be recognized in the ToF spectra, a significant aspect of the study of segregation problems [231]. This was demonstrated for the PEG part of a PEGMA monomer (poly(ethylene glycol)methacrylate) [230]. The damage to polymers such as PVC and PMMA during ion bombarding is considered important. In [232] a study of the change in ToF-SIMS positive and negative ion spectra is performed as a function of ion dose. Finally, SIMS help to reveal the phase decomposition undergone by some polymers, observed as layered structures with well-defined interfaces [226]. Instrumental aspects of how to effectively analyze organic and biological materials are described in [233], showing in particular that SF_5^+ primary ions can enhance secondary ion yields of characteristic molecular ions during operation at very high resolution. Both are demonstrated for a benzo(ghi)perylene and a cocaine ionic molecule by ion imaging. Further advantages of SF_5^+ sputtering are that sputtering rates from a glutamate film were about 37 times faster with primary ions compared with Ar^+ bombardment and that penetration and consequent damage are probably reduced with SF_5^+ molecular primary ions. Cleaning of surfaces of organics from common contamination by SF_5^+ bombardment is demonstrated with PMMA [233].

In microsystems and photonics, assembly and packaging are commonly done by flip-chip solder bonding. To solder bond two surfaces of semiconductors an initial metallic layer is required, which is generally a Pt film, because it is free from oxides. SnPb60/40 solder balls are then used to connect two Pt-metallized surfaces. SIMS was used to investigate Pt diffusion in SbPb60/40 soldering, and the sputtering rate was calibrated by sputtering from the back side which was actually an etchable InP wafer substrate with a 200-nm InGaAs layer as an etch-stop layer followed by three pairs of 10-nm InP and 10-nm InGaAs layers [234]. From run to run of SIMS depth profiling (Fig. 29), the sputtering rates could be compared and the depth resolution was estimated. Ti/Pt metallization is shown in a SIMS peak that is initially high because of matrix effects, just after penetration of the calibration layers, but then reaches equilibrium. Signals drop again when the Pt is shown to be intermixed with Sn and just after the increase in the Pb signal. The Pt dissolution into the solder follows the parabolic diffusion law. The consumption of Pt during soldering shows good temperature and time values, suggesting a stable top surface metallization layer for flip-chip (FC) bonding applications.

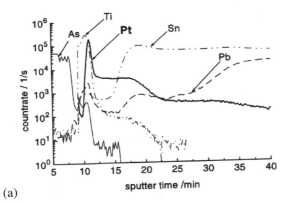

(a)

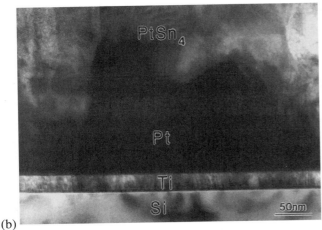

(b)

Fig. 29. (a) A SIMS depth profile from the back side of heat-treated samples: An InP wafer was etched away, and as a first layer InGaAs (200 nm) was followed by three quantum wells of InP(10 nm)/InGaAs(10 nm) with accurate thickness, to be used for an estimation of sputtering rates. Then a Ti(20 nm)/Pt(300 nm) metallization was sputtered, and then a SnPb60/40 foil of about 100 μm was placed on the InAs substrate which with heat treatment wetted the Pt metallized areas of the structure. A matrix effect is apparent in the profile, that is the high increase of Ti ions also increases the Pt ion yields, which later drop to equilibrium. (b) For comparison with the SIMS profiles, the TEM image of PtSn$_4$, which is the intermetallic phase formed at the interface between SnPb60/40 and Pt. EDX (energy dispersive X-ray) and X-ray goniometry have confirmed this material. From [234] with permission from the author and Elsevier Science.

In chemistry silicon is used to fabricate thin-film microelectrode arrays. These have higher efficiency relative to conventional macro-electrodes because of their improved diffusion and increased current density. ToF-SIMS reveals traces of Al and Ag impurities on the Pt microelectrode surfaces [235]. Al is probably introduced by an adhesive layer used before Pt deposition, which can be cleaned with dichromate-sulfuric acid. However, Cr is then introduced. Organic contaminants have not been detected.

REFERENCES

1. F. L. Arnot and J. C. Milligan, *Proc. Roy Soc. London Ser. A* 156, 538 (1937).
2. H. Liebl and R. F. K. Herzog, *J. Appl. Phys.* 34, 2893 (1963).
3. Y. Homma, F. Tohjou, A. Masamoto, M. Shibata, H. Shichi, Y. Yoshioka, T. Adachi, T. Akai, Y. Gao, M. Hirano, T. Hirano, A. Ihara, T. Kamejima, H. Koyama, M. Maier, S. Matsumoto, H. Matsunaga, T. Nakamura, T. Obata, K. Okuno, S. Sadayama, K. Sasa, K. Sasakawa, Y. Shimanuki, S. Suzuki, D. E. Sykes, I. Tachikawa, H. Takase, T. Tanigaki, M. Tomita, H. Tosho, and S. Kurosawa, *Surf. Interface Anal.* 26, 144 (1998).
4. R. G. Wilson and S. W. Novak, *J. Appl. Phys.* 69, 466 (1991).
5. S. W. Novak and R. G. Wilson, *J. Appl. Phys.* 69, 463 (1991).
6. R. G. Wilson, *Int. J. Mass Spectrom. Ion Processes* 143, 43 (1995).
7. R. G. Wilson, G. E. Lux, and C. L. Kirschbaum, *J. Appl. Phys.* 73, 2524 (1993).
8. A. A. Salem, G. Stingeder, M. Grasserbauer, M. Shreiner, K. H. Giessler, and F. Rauch, *J. Mater. Sci.* 7, 373 (1996).
9. A. Havette and G. Slodzian, *J. Phys. Lett.* 41, L247 (1980).
10. R. W. Hinton, *Chem. Geol.* 83, 11 (1990).
11. E. Zinner and C. Crozaz, *Int. J. Mass Spectrum. Ion Processes* 69, 17 (1986).
12. R. G. Wilson, F. A. Stevie, and P. H. Kahora, *Secondary Ion Mass Spectrom.* 487 (1992).
13. N. Shimizu and S. R. Hart, *Annu. Rev. Earth Planet. Sci.* 10, 483 (1982).
14. A. Benninghoven, F. G. Rüdenauer, and H. W. Werner, in "Chemical Analysis" (P. J. Eking and J. D. Winefordner, Eds.), Vol. 86. Wiley, New York, 1987.
15. J. Biersack, *Nucl. Instrum. Methods Phys. Res. Sect. B* 153, 398 (1999).
16. Z. Jurela, *Radiat. Effects* 19, 175 (1973).
17. A. Eicke and B. Bilger, *Surf. Interface Anal.* 12, 344 (1988).
18. H. C. Weller, R. H. Mauch, and G. H. Bauer, *Sol. Energy Mater. Sol. Cells* 27, 217 (1992).
19. A. T. S. Weet, Z. C. Feng, H. H. Hng, K. L. Tan, R. F. C. Farrow, and W. J. Choyke, *J. Phys.: Condens. Matter* 7, 4359 (1995).
20. J. C. Lorin, A. Havette, and G. Slodzian, *Secondary Ion Mass Spectrom.* 3, 140 (1982).
21. M. Bernheim and G. Slodzian, *Secondary Ion Mass Spectrom.* 3, 151 (1982).
22. H. A. Storms, K. F. Brown, and J. D. Stein, *Anal. Chem.* 49, 2023 (1997).
23. S. Y. Tang, E. W. Rothe, and G. P. Reck, *Int. J. Mass Spectrom. Ion Phys.* 14, 79 (1974).
24. C. A. Andersen, *Int. J. Mass Spectrom. Ion Phys.* 2, 61 (1969).
25. E. Chatzitheodoridis, The Development of a Multi-Collector Ion Microprobe for the Study of Oxygen Isotopes in Nature. Ph.D. Thesis, Manchester Univ., U.K., 1994.
26. G. Carter, M. J. Nobes, G. W. Lewis, J. L. Whitton, and G. Kiriakidis, *Vacuum* 34, 167 (1984).
27. G. W. Lewis, G. Kiriakidis, G. Carter, and M. J. Nobes, *Surf. Interface Anal.* 4, 141 (1982).
28. G. Carter, M. J. Nobes, I. V. Katardjiev, J. L. Whitton, and G. Kiriakidis, *Nucl. Instrum. Methods Phys. Res. Sect. B* 18, 529 (1987).
29. J. L. Whitton, G. Kiriakidis, G. Carter, G. W. Lewis, and M. J. Nobes, *Nucl. Instrum. Methods Phys. Res. Sect. B* 230, 640 (1984).
30. R. Smith and J. M. Walls, in "Methods of Surface Analysis, Techniques and Applications" (J. M. Walls, Ed.). Cambridge Univ. Press, Cambridge, U.K., 1989.
31. P. Konarski and M. Hautala, *Vacuum* 47, 1111 (1996).
32. H. C. Eun, *Thin Solid Films* 220, 197 (1992).
33. A. S. Hofmann, *Rep. Prog. Phys.* 61, 827 (1998).
34. S. J. Guilfoyle, A. Chew, N. E. Moiseiwitsch, D. E. Sykes, and M. Petty, *J. Phys.: Condens. Matter* 10, 1699 (1998).
35. Y. Higashi, T. Maruo, and Y. Homma, *Surf. Interface Anal.* 26, 220 (1998).
36. D. E. Sykes, *Surf. Interface Anal.* 28, 49 (1999).
37. G. Prudon, B. Gautier, J. C. Dupuy, C. Dubois, M. Bonneau, J. Delmas, J. P. Vallard, G. Bremond, and R. Brenier, *Thin Solid Films* 294, 54 (1997).
38. E. Taglauer, *Appl. Surf. Sci.* 13, 80 (1982).
39. A. J. T. Jull, *Int. J. Mass Spectrom. Ion Phys.* 41, 135 (1982).
40. E. Zinner, in "Proceedings of the 13th Annual Conference of the Microbeam Analysis Society," 1978.
41. H. Gnaser and I. D. Hutcheon, *Surf. Sci.* 195, 499 (1988).

42. W. A. Russell , D. A. Papanastassiou, and T. A. Tombrello, *Radiat. Effects* 52, 41 (1980).

43. H. Gnaser and I. D. Hutcheon, *Phys. Rev. B: Condensed Matter* 35, 877 (1987).

44. H. J. Kang, J. H. Kim, J. C. Lee, and D. W. Moon, *Radiat. Eff. Defects Solids* 142, 369 (1997).

45. C. A. Andersen, *Int. J. Mass Spectrom. Ion Phys.* 3, 413 (1970).

46. V. R. Deline, W. Katz, C. A. Evans, Jr., and P. Williams, *Appl. Phys. Lett.* 33, 832 (1978).

47. R. Laurent and C. Slodzian, *Radiat. Effects* 19, 181 (1973).

48. G. Kiriakidis, An Investigation of Materials by Secondary Ion Emission (SIMS), Ph.D. Thesis, Univ. of Salford, U.K. (1979).

49. H. Gnaser, *Fresenius J. Anal. Chem.* 358, 171 (1997).

50. H. Gnaser and H. Oechsner, *Fresenius J. Anal. Chem.* 341, 54 (1991).

51. Y. Gao, *J. Appl. Phys.* 64, 3760 (1988).

52. H. Gnaser, W. Bock, H. Oechsner, and T. K. Bhattacharayya, *Appl. Surf. Sci.* 70–71, 44 (1993).

53. P. Willich and U. Wischmann, *Mikrochim. Acta* Suppl. 15, 141 (1998).

54. H. Gnaser and H. Oechsner, *J. Appl. Phys.* 64, 257 (1994).

55. H. Gnaser, *J. Vac. Sci. Technol. A* 12, 452 (1994).

56. U. Breuer, H. Holzbrecher, M. Gastel, J. S. Becker, and H.-J. Dietze, *Fresenius' J. Anal. Chem.* 353, 372 (1995).

57. H. W. Werner, *Surf. Interface Anal.* 2, 56 (1980).

58. R. L. Hervig, R. M. Thomas, and P. Williams, in "New Frontiers in Stable Isotopic Research: Laser Probes, Ion Probes, and Small-Sample Analysis" (W. C. Shanks III, and R. E. Criss, Eds.), *U.S. Geol. Surv. Bull.* 1890, 137 (1988).

59. H. W. Werner and A. E. Morgan, *J. Appl. Phys.* 47, 1232 (1976).

60. R. L. Hervig, P. Williams, R. M. Thomas, S. N. Schauer, and L. M. Steele, *Int. J. Mass Spectrom. Ion Processes* 120, 45 (1992).

61. N. Reger, F. J. Stadermann, and H. M. Ortner, *Fresenius' J. Anal. Chem.* 358, 143 (1997).

62. W. Steiger and F. G. Rüdenauer, *Vacuum* 25, 409 (1975).

63. W. Steiger, F. G. Rüdenauer, J. Antal, and S. Kugler, *Vacuum* 33, 321 (1983).

64. F. Degrève, R. Figareet, and P. Laty, *Int. J. Mass Spectrom. Ion Phys.* 29, 351 (1979).

65. D. K. Bakale, B. N. Colby, and C. A. Evans Jr., *Anal. Chem.* 47, 1532 (1975).

66. H. Gnaser, *Surf. Interface Anal.* 25, 737 (1997).

67. P. Williams, *Appl. Surf. Sci.* 13, 241 (1982).

68. P. Williams, *Surf. Sci.* 90, 588 (1979).

69. P. Joyes, *J. Phys.: Condens. Matter* 29, 774 (1968).

70. P. Joyes, *J. Phys.: Condens. Matter* 30, 243 (1969).

71. P. Joyes, *J. Phys.: Condens. Matter* 30, 483 (1969).

72. G. Blaise and G. Slodzian, *J. Phys.: Condens. Matter* 31, 93 (1970).

73. G. Blaise and G. Slodzian, *J. Phys.: Condens. Matter* 35, 237 (1974).

74. G. Blaise and G. Slodzian, *J. Phys.: Condens. Matter* 35, 243 (1974).

75. Z. Sroubek, *Surf. Sci.* 44, 47 (1974).

76. J. M. Schroeer, T. N. Rhodin, and R. C. Bradley, *Surf. Sci.* 34, 571 (1973).

77. W. H. Gries, *Int. J. Mass Spectrom. Ion Phys.* 17, 77 (1975).

78. C. A. Andersen and J. R. Hinthorne, *Science* 175, 853 (1972).

79. C. A. Andersen and J. R. Hinthorne, *Anal. Chem.* 45, 1421 (1973).

80. D. S. Simons, J. E. Baker, and C. A. Evans, Jr., *Anal. Chem.* 48, 1341 (1976).

81. F. G. Ruedenauer and W. Steiger, *Vacuum* 26, 537 (1976).

82. C. A. Andersen and J. R. Hinthorne, in "Proceedings XL VIII 253," Carnegie-Mellon University, Pittsburgh, PA, 1972, p. 39A/4.

83. C. Slodzian, J. C. Lorin, and A. Havette, *J. Physique Lett.* 41, L555 (1980).

84. E. Zinner, in "New Frontiers in Stable Isotope Research: Laser Probes, Ion Probes, and Small-Sample Analysis" (W. C. Shanks III and R. E. Criss, Eds.). *U.S. Geol. Surv. Bull.* 1890, 145 (1988).

85. T. Kawada, K. Masuda, J. Suzuki, A. Kaimai, K. Kawamura, Y. Nigara, J. Mizusaki, H. Yugami, H. Arashi, N. Sakai, and H. Yokokawa, *Solid State Ionics* 21, 271 (1999).

86. R. J. Rosenberg, R. Zilliacus, E. L. Lakomaa, A. Rautiainen, and A. Maekelae, *Fresenius' J. Anal. Chem.* 354, 6 (1996).

87. S. L. Antonov, V. I. Belevsky, I. V. Gusev, A. A. Orlikovsky, K. A. Valiev, and A. G. Vasiliev, *Vacuum* 43, 635 (1992).

88. K. Wittmaack, *Secondary Ion Mass Spectrom.* 10, 657 (1996).

89. V. R. Deline, C. A. Evans-Jr, and P. Williams, *Appl. Phys. Lett.* 33, 578 (1978).

90. C. A. Andersen and J. R. Hinthorne, *Science* 175, 853 (1972).

91. P. Williams, *Secondary Ion Mass Spectrom.* 7, 15 (1990).

92. R. G. Wilson, F. A. Stevie, and C. W. Magee, "Secondary Ion Mass Spectrometry: A Practical Handbook for Depth Profiling and Bulk Impurity Analysis." Wiley, New York, 1989.

93. G. E. Lux, F. E. Stevie, P. M. Kahora, R. G. Wilson, and G. W. Cochran, *J. Vac. Sci. Technol., A* 11, 2373 (1993).

94. R. G. Wilson, *J. Appl. Phys.* 63, 5121 (1988).

95. H. Gnaser, *Surf. Interface Anal.* 24, 483 (1996).

96. R. Yue, Y. Wang, Y. Wang, and C. Chen, *Surf. Interface Anal.* 27, 98 (1999).

97. J. B. Metson and K. E. Prince, *Surf. Interface Anal.* 28, 159 (1999).

98. H. Oechsner, *Secondary Ion Mass Spectrom.* 3, 106 (1982).

99. K. H. Müller and H. Oechsner, *Mikrochim. Acta.* 10 (Suppl.), 51 (1983).

100. T. Albers, M. Neumann, D. Lipinsky, and A. Benninghoven, *Appl. Surf. Sci.* 70–71, 49 (1993).

101. M. Gastel, U. Breuer, H. Holzbrecher, J. S. Becker, H. J. Dietze, and H. Wagner, *Fresenius' J. Anal. Chem.* 358, 207 (1997).

102. C. Dubourdieu, N. Didier, O. Thomas, J. P. Sénateur, N. Valignat, Y. Rebane, T. Kouznetsova, A. Gaskov, J. Hartmann and B. Stritzker, *Fresenius' J. Anal. Chem.* 357, 1061 (1997).

103. M. Bersani, M. Fedrizzi, M. Sbetti, and M. Anderle, in "Characterization and Metrology for ULSI Technology: 1998 International Conference" (D. G. Seiler, A. C. Diebold, T. J. Shaffner, R. McDonald, W. M. Bullis, P. J. Smith, and E. M. Secula, Eds.). American Institute of Physics, New York, 1998, p. 892.

104. T. Maruo, Y. Higashi, T. Tanaka, and Y. Homma, *J. Vac. Sci. Technol. A* 11, 2614 (1993).

105. Y. Higashi, T. Maruo, Y. Homma, J. Kodate, and M. Miyake, *Appl. Phys. Lett.* 64, 2391 (1994).

106. Y. Higashi, *Spectrochim. Acta, Part B* 54, 109 (1999).

107. A. R. Bayly, J. Wolstenholme, and C. R. Petts, *Surf. Interface Anal.* 21, 414 (1994).

108. A. Wucher, *Mater. Sci. Forum* 287–288, 61 (1998).

109. S. Dreer, *Fresenius' J. Anal. Chem.* 365, 85 (1999).

110. U. Breuer, H. Holzbrecber, M. Gastel, J. S. Becker, and H. J. Dietze, *Fresenius' J. Anal. Chem.* 358, 47 (1997).

111. R. Bacon, S. Crain, L. Van Vaeck, and G. Williams, *J. Anal. At. Spectrom.* 13, 171R (1998).

112. A. G. Fitzgerald, H. L. L. Watton, B. E. Storey, J. S. Colligon, and H. Kheyrandish, *Surf. Interface Anal.* 22, 69 (1994).

113. H. Paulus, S. Kornely, J. Giber, and K. H. Muller, *Mikrochim. Acta* 125, 223 (1997).

114. R. Schmitz, G. H. Frischat, H. Paulus, and K. H. Müller, *Fresenius' J. Anal. Chem.* 358, 42 (1997).

115. H. Yamazaki and M. Takahashi, *Surf. Interface Anal.* 25, 937 (1997).

116. E. Zinner, *Scanning* 3, 57 (1980).

117. K. Wetzig, S. Baunack, V. Hoffmann, S. Oswald, and F. Prässler, *Fresenius' J. Anal. Chem.* 358, 25 (1997).

118. C. Gerardi, *Surf. Interface Anal.* 25, 397 (1997).

119. S. Lehto, R. Lappalainen, H. Viirola, and L. Niinistö, *Fresenius' J. Anal. Chem.* 355, 129 (1996).

120. H. Gnaser, *Secondary Ion Mass Spectrom.* 10, 335 (1996).

121. K. Yoshihara, D. W. Moon, D. Fujita, K. J. Kim, and K. Kajiwara, *Surf. Interface Anal.* 20, 1061 (1993).

122. K. J. Kim and D. W. Moon, *Surf. Interface Anal.* 26, 9 (1998).

123. K. Wittmaack, *Surf. Interface Anal.* 26, 290 (1998).

124. S. Oswald, V. Hoffmann, and G. Ehrlich, *Spectrochim. Acta, Part B* 49, 1123 (1994).

125. R. Voigtmann and W. Moldenhauer, *Surf. Interface Anal.* 13, 167 (1988).

126. J. E. Kempf and H. H. Wagner, in "Thin Film and Depth Profile Analysis" (H. Oechsner, Ed.). Springer-Verlag, Berlin/New York, 1984.

127. J. F. Moulder, S. R. Bryan, and U. Roll, *Fresenius' J. Anal. Chem.* 365, 83 (1999).

128. N. J. Montgomery, J. L. MacManus-Driscoll, D. S. McPhail, R. J. Chater, B. Moeckly and K. Char, *Thin Solid Films* 317, 237 (1998).

129. C. Gerardi, M. A. Tagliente, A. Del-Vecchio, L. Tapfer, C. Coccorese, C. Attanasio, L. V. Mercaldo, L. Maritato, J. M. Slaughter, and C. M. Falco, *J. Appl. Phys.* 87, 717 (2000).

130. K. Wittmaack, S. B. Patel, and S. F. Corcoran, in "Characterization and Metrology for ULSI Technology: 1998 International Conference" (D. G. Seiler, A. C. Diebold, T. J. Shaffner, R. McDonald, W. M. Bullis, P. J. Smith, and E. M. Secula, Eds.). American Institute of Physics, New York, CP449, 1998, p. 791.

131. P. C. Zalm, J. Vriezema, D. J. Gravesteijn, G. F. A. van de Walle, and W. B. de Boer, *Surf. Interface Anal.* 17, 556 (1991).

132. W. Vandervorst and T. Clarysee, *J. Vac. Sci. Technol., B* 10, 307 (1992).

133. P. C. Zalm, *Secondary Ion Mass Spectrom.* 10, 73 (1996).

134. J. B. Clegg, N. S. Smith, M. G. Dowsett, M. J. J. Theunissen, and W. B. de Boer, *J. Vac. Sci. Technol., A* 14, 2645 (1996).

135. H. J. Kang, W. S. Kim, D. W. Moon, H. Y. Lee, S. T. Kang, and R. Shimizu, *Nucl. Instrum. Methods Phys. Res., Sect. B* 153, 429 (1999).

136. Z.-X. Jiang, P. F. A. Alkemade, E. Algra, and S. Radelaar, *Surf. Interface Anal.* 25, 285 (1997).

137. M. G. Dowsett and D. P. Chu, *J. Vac. Sci. Technol., B* 16, 377 (1998).

138. J. M. Blanco, J. J. Serrano, J. Jiménez-Leube, T-R-M. Aguilar and R. Gwilliam, *Nucl. Instrum. Phys. Res., Sect. B* 113, 530 (1996).

139. N. J. Montgomery, J. L. MacManus-Driscoll, D. S. McPhail, B. Moeckly and K. Char, *J. Alloys Compounds* 251, 355 (1997).

140. P. Konarski, M. A. Herman, A. V. Kozhukhov, and V. I. Obodnikov, *Thin Solid Films* 267, 114 (1995).

141. B. Gautier, G. Prudon, and J. C. Dupuy, *Surf. Interface Anal.* 26, 974 (1998).

142. B. Gautie, J. C. Dupuy, B. Semmache, and G. Prudon, *Nucl. Instrum. Methods Phys. Res., Sect. B* 142, 361 (1998).

143. B. Gautier, J. C. Dupuy, R. Prost, and G. Prudon, *Surf. Interface Anal.* 25, 464 (1997).

144. M. L. Wagter, A. H. Clarke, K. F. Taylor, P. A. W. van der Heide, and N. S. McIntyre, *Surf. Interface Anal.* 25, 788 (1997).

145. M. Srivastava, N. O. Petersen, G. R. Mount, D. M. Kingston, and N. S. McIntyre, *Surf. Interface Anal.* 26, 188 (1998).

146. H. W. Werner, *Mater. Sci. Engin.* 42, 1 (1980).

147. K. Bange, *Fresenius' J. Anal. Chem.* 353, 240 (1995).

148. S. Kawakita, T. Sakurai, A. Kikuchi, T. Shimatsu, and M. Takahashi, *J. Magn. Magn. Mater.* 155, 172 (1996).

149. S. Koizumi, M. Kamo, Y. Sato, H. Ozaki, and T. Inuzuka, *Appl. Phys. Lett.* 71, 1065 (1997).

150. J. H. Kim and K. W. Chung, *J. Appl. Phys.* 83, 5831 (1998).

151. K. Koski, J. Hölsä, J. Ernoult, and A. Rouzaud, *Surf. Coat. Techn.* 80, 195 (1996).

152. G. Xu, M.-Y. He and D. R. Clarke, *Acta Materialia* 47, 4131 (1999).

153. M. Guemmaz, A. Mosser, J. J. Grob, J. C. Sens, and R. Stuck, *Surf. Coat. Techn.* 80, 53 (1996).

154. G. Zambrano, J. Prieto, F. Perez, C. Rincon, H. Galindo, L. Cota-Araiza, J. Esteve, and E. Martinez, *Surf. Coat. Techn.* 108–109, 323 (1998).

155. C. Guedj, X. Portier, A. Hairie, D. Bouchier, G. Calvarin, B. Piriou, B. Gautier, and J. C. Dupuy, *J. Appl. Phys.* 83, 5251 (1998).

156. W. K. Choi, J. H. Chen, L. K. Bera, W. Feng, K. L. Pey, J. Mi, C. Y. Yang, A. Ramam, S. J. Chua, J. S. Pan, A. T. S. Wee, and R. Liu, *J. Appl. Phys.* 87, 192 (2000).

157. S. Hearne, N. Herbots, J. Xiang, P. Ye, and H. Jacobsson, *Nucl. Instrum. Methods Phys. Res., Sect. B* 118, 88 (1996).

158. D. G. Gromov, A. I. Mochalov, V. P. Pugachevich, E. P. Kirilenko, and A. Yu. Trifonov, *Appl. Phys. A* 64, 517 (1997).

159. S. Banerjee, G. Raghavan, and M. K. Sanyal, *J. Appl. Phys.* 85, 7135 (1999).

160. E. Vainonen, J. Likonen, T. Ahlgren, P. Haussalo, J. Keinonen, and C. H. Wu, *J. Appl. Phys.* 82, 3791 (1997).

161. O. Gebhardt, *Fresenius' J. Anal. Chem.* 365, 117 (1999).

162. D. Krupa, J. Baszkiewicz, E. Jezierska, J. Mizera, T. Wierzchon, A. Barcz and R. Fillit, *Surf. Coat. Technol.* 111, 86 (1999).

163. E. Caudron, H. Buscail, Y. P. Jacob, C. Josse-Courty, F. Rabaste, and M. F. Stroosnijder, *Nucl. Instrum. Methods Phys. Res., Sect. B* 155, 91 (1999).

164. A. M. Huntz, S. C. Tsai, J. Balmain, K. Messaoudi, B. Lesage, and C. Dolin, *Mater. Sci. Forum* 251–254, 313 (1997).

165. A. N. Rider, D. R. Arnott, A. R. Wilson, I. Danilidis, and P. J. K. Paterson, *Surf. Interface Anal.* 24, 293 (1996).

166. K. Miyake and K. Ohashi, *Mater. Chem. Phys.* 54, 321 (1998).

167. C. Gao, *Mater. Res. Innovations* 1, 238 (1998).

168. C. Varanasi, R. Biggers, I. Maartense, T. L. Peterson, J. Solomon, E. K. Moser, D. Dempsey, J. Busbee, D. Liptak, G. Kozlowski, R. Nekkanti, and C. E. Oberly, *Physica C* 297, 262 (1998).

169. J. A. Kilner and Y. Li, *Nucl. Instrum. Methods Phys. Res., Sect. B* 139, 108 (1998).

170. Y. Li, J. A. Kilner, T. J. Tate, M. J. Lee, Y. H. Li, and P. G. Quincey, *Nucl. Instrum. Methods Phys. Res., Sect. B* 99, 627 (1995).

171. A. Gauzzi, H. J. Mathieu, J. H. James, and B. Kellett, *Vacuum* 41, 870 (1990).

172. B. Schulte, B. C. Richards, and S. L. Cook, *J. Alloys Compd.* 251, 360 (1997).

173. Y. Ito, Y. Yoshida, Y. Mizushima, I. Hirabayashi, H. Nagai, and Y. Takai, *Jpn. J. Appl. Phys., Part 2* 35, L825 (1996).

174. B. Pignataro, S. Panebianco, C. Consalvo, and A. Licciardello, *Surf. Interface Anal.* 27, 396 (1999).

175. J. J. Cuomo, M. F. Chisholm, D. S. Yee, D. J. Mikalsen, P. B. Madakson, R. A. Roy, E. Giess, and G. Scilla, *AIP Conf. Proc.* 165, 141 (1988).

176. A. Abrutis, V. Plaušinaitienė, A. Teišerskis, V. Kubilius, J.-P. Senateur, and F. Weiss, *Chem. Vapor Deposition* 5, 171 (1999).

177. J. A. Beall, M. W. Cromar, T. E. Harvey, M. E. Johansson, R. H. Ono, C. D. Reintsema, D. A. Rudman, S. E. Asher, A. J. Nelson, and A. B. Switzerlander, *IEEE Trans. Magn.* 27, 1596 (1991).

178. D. E. Ramaker, S. M. Hues, and F. L. Hutson, in "Pittsburgh Conference and Exposition on Analytical Chemistry and Applied Spectroscopy (Abstracts)," Pittsburgh, PA, Technical Paper 1234, 1988.

179. H. W. Werner and A. Torrisi, *Fresenius' J. Anal. Chem.* 337, 594 (1990).

180. H. W. Werner and H. van der Wel, *Anal. Methods Instrum.* 2, 111 (1995).

181. R. C. Tu, Y. K. Su, Y. S. Huang, and S. T. Chou, *J. Appl. Phys.* 84, 6017 (1998).

182. X. C. Wang, S. J. Xu, S. J. Chua, K. Li, X. H. Zhang, Z. H. Zhang, K. B. Chong and X. Zhang, *Appl. Phys. Lett.* 74, 818 (1999).

183. C. Takakuwa-Hongo and M. Tomita, *Surf. Interface Anal.* 25, 966 (1997).

184. A. Gaymann, M. Maier, and K. Köhler, *J. Appl. Phys.* 86, 4312 (1999).

185. M. V. Rao, A. K. Berry, T. Q. Do, M. C. Ridgway, P. H. Chi, and J. Waterman, *J. Appl. Phys.* 86, 6068 (1999).

186. J. Zhao, D. S. Mao, Z. X. Lin, X. Z. Ding, B. Y. Jiang, Y. H. Yu, X. H. Liu, and G. Q. Yang, *Nucl. Instrum. Methods Phys. Res., Sect. B* 149, 325 (1999).

187. H. Sato and S. Sugawara, *Jpn. J. Appl. Phys. Part 2* 32, L799 (1993).

188. F. Caccavale, A. Morbiato, M. Natali, C. Sada, and F. Segato, *J. Appl. Phys.* 87, 1007 (2000).

189. S. K. Wong, N. Du, P. K. John, and B. Y. Tong, *J. Non-Cryst. Solids* 110, 179 (1989).

190. U. Egger, M. Schultz, P. Werner, O. Breitenstein, T. Y. Tan, U. Gösele, R. Franzheld, M. Uematsu, and H. Ito, *J. Appl. Phys.* 81, 6056 (1997).

191. Z. H. Ma, T. Smith, and I. K. Sou, *J. Appl. Phys.* 86, 2505 (1999).

192. A. K. Tyagi, U. Breuer, H. Holzbrecher, J. S. Becker, H.-J. Dietze, L. Kappius, H. L. Bay, and S. Mantl, *Fresenius' J. Anal. Chem.* 365, 282 (1999).

193. Z. Wang, G. Ramanath, L. H. Allen, A. Rockett, J. P. Doyle, and B. G. Svensson, *J. Appl. Phys.* 82, 3281 (1997).

194. S. Kobayashi, M. Iizuka, T. Aoki, N. Mikoshiba, M. Sakuraba, T. Matsuura, and J. Murota, *J. Appl. Phys.* 86, 5480 (1999).

195. A. Wakejima, K. Onda, Y. Ando, A. Fujihara, E. Mizuki, T. Nakayama, H. Miyamoto, and M. Kuzuhara, *J. Appl. Phys.* 81, 1311 (1997).

196. S. Singhvi and C. G. Takoudis, *J. Appl. Phys.* 82, 442 (1997).

197. M. K. Erhardt and R. G. Nuzzo, *Langmuir* 15, 2188 (1999).

198. M. Griesser, H. Hutter, M. Grasserbauer, W. Kalss, R. Haubner, and B. Lux, *Fresenius' J. Anal. Chem.* 358, 293 (1997).

199. M. Gastel, U. Reuter, H. Schulz, and H. M. Ortner, *Fresenius' J. Anal. Chem.* 365, 142 (1999).

200. T. Kolber, K. Piplits, R. Haubner, and H. Hutter, *Fresenius' J. Anal. Chem.* 365, 636 (1999).

201. S. Lopez, H. M. Dunlop, M. Benmalek, G. Tourillon, M.-S. Wong, and W. D. Sproul, *Surf. Interface Anal.* 25, 827 (1997).

202. A. Yoshida, M. Kitagawa, and T. Hirao, *Jpn. J. Appl. Phys. Part 1* 32, 2147 (1993).

203. E. A. Al-Nuaimy, J. M. Marshall, and S. Muhl, *J. Non-Cryst. Solids* 227–230, 949 (1998).

204. T. Nagahara, K. Fujimoto, N. Kohno, Y. Kashiwagi, and H. Kakinoki, *Jpn. J. Appl. Phys. Part 1* 31, 4555 (1992).

205. L. J. Quinn, B. Lee, P. T. Baine, S. J. N. Mitchell, B. M. Armstrong, and H. S. Gamble, *Thin Solid Films* 296, 7 (1997).

206. T. Akasaka, D. He, Y. Miyamoto, N. Kitazawa, and R. Shimizu, *Thin Solid Films* 296, 2 (1997).

207. T. Hirai, K. Miyake, T. Nakamura, S. Kamagami, and H. Morita, *Jpn. J. Appl. Phys., Part 1* 31, 4582 (1992).

208. A. Jaeger, W. D. Sun, F. H. Pollak, C. L. Reynolds Jr., M. Geva, D. V. Stampone, M. W. Focht, O. Y. Raisky, W. B. Wang, and R. R. Alfano, *J. Appl. Phys.* 85, 1921 (1999).

209. D. W. Lane, G. J. Conibeer, S. Romani, M. J. F. Healy, and K. D. Rogers, *Nucl. Instrum. Methods Phys. Res., Sect. B* 136–138, 225 (1998).

210. J. Meier, S. Dubail, J. Cuperus, U. Kroll, R. Platz, P. Torres, J. A. Anna-Selvan, P. Pernet, N. Beck, N. Pellaton-Vaucher, Ch. Hof, D. Fischer, H. Keppner, and A. Shah, *J. Non-Cryst. Solids* 227–230, 1250 (1998).

211. R.-Y. Tsai, F.-C. Ho, and M.-Y. Hua, *Opt. Engin.* 36, 2335 (1997).

212. G. Kiriakidis, M. Marder, and Z. Hatzopoulos, *Sol. Energ. Mater.* 17, 25 (1988).

213. M. Penza, M. A. Tagliente, L. Mirenghi, C. Gerardi, C. Martucci, and G. Cassano, *Sens. Actuators, B* 50, 9 (1998).

214. M. Bögner, A. Fuchs, K. Scharnagl, R. Winter, T. Doll, and I. Eisele, *Sens. Actuators, B* 47, 145 (1998).

215. A. Galdikas, V. Jasutis, S. Kačiulis, G. Mattogno, A. Mironas, V. Olevano, D. Senuliene, and A. Setkus, *Sens. Actuators, B* 43, 140 (1997).

216. J. Frank, M. Fleischer, H. Meixner, and A. Feltz, *Sens. Actuators, B* 49, 110 (1998).

217. J. L. Seguin, M. Bendahan, P. Lauque, C. Jacolin, M. Pasquinelli, and P. Knauth, *Sens. Actuators, A* 74, 237 (1999).

218. E. Defaÿ, B. Semmache, C. Dubois, M. LeBerre, and D. Barbier, *Sens. Actuators, A* 74, 77 (1999).

219. P. Verardi, M. Dinescu, F. Craciun, R. Dinu, V. Sandu, L. Tapfer, and A. Cappello, *Sens. Actuators, A* 74, 41 (1999).

220. Y. C. Ling, J. P. Wang, M. H. Yeh, K. S. Liu, and I. N. Lin, *Appl. Phys. Lett.* 66, 156 (1995).

221. T. S. Kim, M. H. Oh, and C. H. Kim, *Jpn. J. Appl. Phys. Part 1* 32, 2837 (1993).

222. T. S. Kim, C. H. Kim, and M. H. Oh, *J. Appl. Phys.* 76, 4316 (1994).

223. E. B. Varhegyi, S. Jonda, I. V. Perczel, and H. Meixner, *Sens. Actuators, B* 47, 164 (1998).

224. N. Chaoui, E. Millon, J. F. Muller, P. Ecker, W. Bieck, and H. N. Migeon, *Mater. Chem. Phys.* 59, 114 (1999).

225. N. Chaoui, E. Millon, J. F. Muller, P. Ecker, W. Bieck, and H. N. Migeon, *Appl. Surf. Sci.* 138–139, 256 (1999).

226. J. Rysz, H. Ermer, A. Budkowski, M. Lekka, A. Bernasik, S. Wróbel, R. Brenn, J. Lekki, and J. Jedliski, *Vacuum* 54, 303 (1999).

227. L. Van Vaeck, A. Adriaens, and R. Gijbels, *Mass Spectrom. Rev.* 18, 1 (1999).

228. B. A. Keller and P. Hug, *Anal. Chim. Acta* 393, 201 (1999).

229. D. Pleul, H. Simon, and J. Jacobasch, *Fresenius' J. Anal. Chem.* 357, 684 (1997).

230. D. Briggs and M. C. Davies, *Surf. Interface Anal.* 25, 725 (1997).

231. S. D. Hanton, P. A. Cornelio Clark, and K. G. Owens, *J. Am. Soc. Mass Spectrom.* 10, 104 (1999).

232. D. Briggs and I. W. Fletcher, *Surf. Interface Anal.* 25, 167 (1997).

233. G. Gillen and S. Roberson, *Rapid Commun. Mass Spectrom.* 12, 1303 (1998).

234. J. F. Kuhmann, C.-H. Chiang, P. Harde, F. Reier, W. Oesterle, I. Urban, and A. Klein, *Mater. Sci. Engin., A* 242, 22 (1998).

235. M. Wittkampf, K. Cammann, M. Amrein, and R. Reichelt, *Sens. Actuators, B* 40, 79 (1997).

Chapter 14

A SOLID-STATE APPROACH TO LANGMUIR MONOLAYERS, THEIR PHASES, PHASE TRANSITIONS, AND DESIGN

Craig J. Eckhardt

Department of Chemistry, Center for Materials Research and Analysis, University of Nebraska–Lincoln, Lincoln, Nebraska, USA

Tadeusz Luty

Institute of Physical and Theoretical Chemistry, Technical University of Wrocław, Wrocław, Poland

Contents

Handbook of Thin Film Materials, edited by H.S. Nalwa
Volume 2: Characterization and Spectroscopy of Thin Films
Copyright © 2002 by Academic Press

ISBN 0-12-512910-6/$35.00

1. INTRODUCTION

1.1. General Characterization of Phases

Langmuir monolayers at the air–water interface exhibit very rich thermodynamic behavior [1]. They form a variety of phases with different degrees of translational and orientational disorder, from gas-like to solid-like phases. Although these phenomena have been known for a long time, recent progress in experimental techniques requires an understanding of the microscopic origins that lead to such fascinating thermodynamic processes. Phenomenological Landau-type theory [2] does not address this problem and treats the system only from the point of view of global symmetry. The richness of phases indicates that Langmuir monolayers are frustrated systems where local and global structures compete, as do their respective equilibria. Frustration arises because the cross-sectional area of the head groups is different, in general, from that of the attached alkane chain, making it impossible to fill space without introducing some strain into the layer configuration or into the molecule itself. This factor plays an even more important, if not essential, role for amphiphiles with large molecular groups attached to the tails. There are ongoing experimental, theoretical, and numerical attempts to gain knowledge of the ordering and molecular nature of the different phases and the transitions between them. In this chapter, we focus on the molecular origin, symmetry, and nature of intermolecular forces that determine translational and orientational order as well as fluctuations and instabilities in the Langmuir monolayers [3–7]. We will show the value of solid-state concepts in understanding these systems.

The detailed determination [8–10] of the phase behavior of these systems may be attributed to the advent of new techniques for their study: synchrotron X-ray diffraction [11, 12], atomic force microscopy [13], second harmonic generation [14], and Brewster angle [15] and polarized fluorescence [8] microscopies. (For a review of structural studies of ordered monolayers using atomic force microscopy, see [13].) These methods have significantly augmented the more traditional surface pressure (π)–area (A) isotherms [16, 17]. In particular, recent X-ray diffraction experiments have shown that Langmuir monolayer phases exhibit a variety of structures [10, 18–20]. (For a review of 2-D crystallography of Langmuir monolayers, see [12].)

The current view is that the phase behavior of a monolayer displays, in addition to gas-like and low-density liquid-like phases, mesomorphic and solid states and that the subtle and almost continuous changes between these phases even admit amorphous states. This is based on monolayer phase behavior of "classical" fatty acid amphiphiles that are taken to reveal the most subtleties in Langmuir film phase behavior [1]. Current theoretical approaches almost exclusively emphasize similarities of the monolayers to liquid-crystalline phases [21] rather than to crystalline properties and give significantly less consideration to the early observation [22] of great similarities to three-dimensional (3-D) crystalline phases. Within this framework it has been suggested that mesophases of Langmuir monolayers, observed in high-temperature regimes, are hexatic phases that display long-range orientational (algebraically decaying) and short-range translational (exponentially decaying) order. Crystalline phases that have algebraically decaying translational order are indeed observed at low temperatures.

Although there are many factors that may lead to detailed structure in the phase diagram arising from differences in translational and orientational order, predominant are the chemical nature, shape, and flexibility of the amphiphiles. This is exemplified by the striking difference in crystallinity at room temperature between amphiphiles comprising alkane chains and those that are perfluorinated [23, 24]. This has been attributed to the higher rigidity and interchain van der Waals interaction of the latter [25]. Moreover, for the homologous series of n-alkanoic acid molecules, the extent of two-dimensional (2-D) crystalline order is larger for systems with more attractive lattice potentials [25]. Thus, longer molecules favor crystallinity. Langmuir monolayers are, therefore, unusual systems; they combine features of 2-D and 3-D systems in a rather complicated way. The quasi-two-dimensionality may be attributed to the molecular tails that, by the very nature of their orientational flexibility, couple the 2-D system of heads with the 3-D system of tails. This combination makes Langmuir monolayers very exciting systems for the study of a variety of phase behaviors.

Grazing incidence X-ray diffraction experiments [18] have shown the monolayers to be crystalline in both compressed and uncompressed states. These crystalline films, essentially 2-D "powders," are weak X-ray scatterers [18, 26], and making a distinction between the "powder" and mesophases is neither easy nor definite. As noted by Peterson and Kenn [27], the mosaic structure of highly ordered mesophases is essentially indistinguishable from a polycrystalline texture, and the latter is the one that occurs commonly in alkanes. So, as noted in reference to the X-ray studies, the single peak of the "powder" sample has been interpreted as the triply degenerate peak of a hexatic [28]. But in fact, hexatic order has not been directly observed for Langmuir monolayers, and there is significant doubt that mesophases can be rigorously treated as hexatics [29]. Recent elegant studies of the shear elasticity of monolayers [30, 31] have confirmed that these phases are more crystalline in nature than hexatic. Furthermore, from a fundamental point of view, the intrinsic head–tail asymmetry of Langmuir monolayer molecules makes the system quasi-two-dimensional rather than strictly 2-D, the only dimensionality for which hexatics are defined. Hexatic structure obtains as a result of exponential decay of the displacement–displacement correlation function in 2-D systems composed of objects without rotational degrees of freedom. Molecular monolayers are quite different! The molecules interact with the aqueous subphase by hydrogen bonding of the head groups and the molecular tails are orientationally flexible. As shown by X-ray diffraction studies, the ordering of head groups on the water surface differs from that of the molecular tails in the air (viz. geometrical frustration mentioned above) [18–20].

Recent Monte Carlo simulations for a system of surfactant molecules grafted at interfaces suggest that mesophases of Langmuir monolayers, apparently intermediate between crystalline and liquid phases, are characterized by frozen crystalline head groups and fluidized tails of the molecules. Crystalline phases are characterized by ordering of both head groups and tails, whereas in liquid phases both head groups and molecular tails are fluidized. The melting of a crystalline Langmuir monolayer may go through mesophases considered to be a mixture of "clusters," i.e., domains of mesoscopic dimension [7]. The size of the domains depends on the length of the molecular tails that determine the direct coupling for the orientational ordering [7]. The shorter the tails and the smaller the clusters, the more easily they are fluidized, and the system approximates a 2-D net. Thus, if any Langmuir monolayer structures were hexatic-like, they would most likely be those formed by short tail amphiphiles. This conclusion has been treated more extensively by Kats and Lajzerowicz, who illustrate this finding with a schematic phase diagram [32].

Langmuir films are formed by the deposition of amphiphiles on a fluid subphase which we take to be pure water. The amphiphiles, molecules comprised of a polar or ionic "head" group and a hydrophobic "tail" terminating structure, are initially dissolved in a volatile solvent of minimal solubility in water. Upon evaporation of the solvent the amphiphiles, here taken to be alkanoic acids or alcohols, are singly and randomly distributed on the surface of the surface of the water with intermolecular separations much larger than their molecular size. Even under these conditions, aggregation may occur. However, at this point of no applied surface pressure, the molecules behave as a 2D ideal gas.

The surface pressure is monitored by a so-called Wilhelmy plate, which may be a known size of filter paper or a platinum foil, attached to a sensitive balance. The film on the surface exerts an effective force on the plate that can be related to the difference between the surface tension of the water and the monolayer film and is commonly referred to as the surface pressure of the film.

By closing barriers, usually made of Teflon, that enclose the film, the amphiphile intermolecular distances are decreased and liquid-like behavior is observed. This is reflected in the surface isotherm, the plot of surface pressure vs. area/molecule, by a rise in the surface pressure and decrease of the intermolecular distances to the order of molecular dimensions. Subsequent compression forces the amphiphiles to pack more closely together such that the "tails" are, on average, tilted to some angle from the normal to the surface with intermolecular distances small compared to the molecular "tail" cross section diameter. At this point, the film may be regarded as a 2D solid.

From a rigorous geometrical view, of course, the film is a 3D system since the molecule has finite extension of its "tail" out of the surface. However, from the physical standpoint, especially if we want to consider the film as a solid, another definition based on symmetry used, as long as only a monolayer is studied, no translation symmetry of the motif exist in the direction perpendicular to the surface thereby rendering the system physically 2D.

With sufficiently high applied surface pressure, the amphiphiles achieve a vertical arrangement thereby reaching the closest packing possible. Upon application of additional pressure, the film collapses and irreversibly forms a bilayer leading to a constant surface pressure but decreasing molecular area with decreasing distance between the moveable barriers.

By repeating this measurement at different temperatures a phase diagram can be obtained that exhibits many phases. The various solid phases are distinguished by the amphiphiles' degree of tilt and the directions in which the tilting occurs. It is this rich phase behavior that has occupied the interest of many experimentalists and theorists.

1.2. Representative Phase Diagrams

A generalized phase diagram for n-alkanoic acids is presented in Figure 1. It represents a rather inclusive compilation [33] based on a variety of measurements. Some phases, denoted by italics, require further confirmation but are included because most are consistent with the theory that will be developed here [3, 4]. The structural information for the depicted phases is displayed in Table I. This phase diagram can be further partitioned according to the method of measurement employed in the determination of each phase. Phases LS, L_2, L_2', L_2'', and CS are found by surface isotherm studies. All of these phases and two others, Ov and S, have been observed by optical methods such as Brewster angle microscopy and polarized fluorescence microscopy [33]. In addition, the three phases L_1', L_{2d}, and L_{2h} within the L_2 phase have been observed by X-ray scattering, as have the phases S' and L_2^* within the L_2' phase.

To characterize the phases shown by the phase diagram, and for a clearer distinction of theoretical models relevant to the

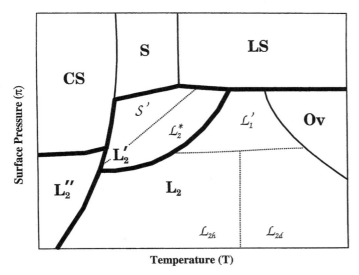

Fig. 1. Schematic phase diagram for *n*-alkanoic acids. The heavy phase equilibrium lines represent determinations by X-ray, optical, and surface isotherm measurements. The light phase equilibrium lines have been located by X-ray and optical methods. The broken phase equilibrium lines have been found only by X-ray measurement. Italic letters denote those phases that require further confirmation but are consistent with the proposed theoretical development.

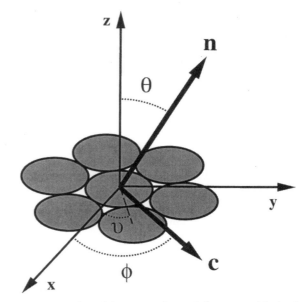

Fig. 2. Representation of the axes, angles, and directors used in the description of the orientation of rigid rod model *n*-alkanoic acid molecules. x, y, and z define the Cartesian coordinate system with polar angles θ and ϕ. υ is the angle formed by a crystallographic hexagonal "bond" axis with the x axis.

Table I. Characterization of *n*-Alkanoic Langmuir Monolayer Phases of Figure 1[e]

Phase	Lattice	Backbone order (related to T scale)	Tilt order (related to π scale)
LS	Hexagonal	Rotational disorder	Vertical
Ov	Rect/hex[a]	Rotational disorder	NNN
L_{2d}	Rect/hex[a]	Rotational disorder	NN
L_1'	Rect/hex[a]	Rotational disorder	NN^NNN
S	Rectangular	Static disorder[b]	Vertical
L_2^*	Rectangular	Parallel order[c]	NNN
L_{2h}	Rectangular	Parallel order[a]	NN
CS	Rect/oblique[d]	Herringbone order	Vertical
S'	Rect/oblique[d]	Herringbone order	NNN
L_2''	Rect/oblique[d]	Herringbone order	NN

[a] Molecules form a hexagonal, close-packed lattice in the plane perpendicular to their long axis.

[b] Molecular backbones are in two equivalent orientations in every lattice site.

[c] Molecular backbones are ordered parallel to each other.

[d] It is not certain if the herringbone order of molecular backbones is realized on a rectangular or oblique lattice.

[e] Italics indicate unconfirmed phases that do not appear in the phase diagram.

ensuing discussion, introduction of angles and directors specifying the orientation of molecules in the monolayer is desirable. These are defined in Figure 2, where the molecular chain is modeled as a rigid rod. In addition to two polar angles (θ, ϕ), there is the angle υ formed by one crystallographic hexagonal axis (a so-called bond) with the x axis of the coordinate system. The latter has been introduced to characterize hexatic order [34,

35], and it may be neglected in characterizing the aforementioned phases, i.e., by assuming hexatic rigidity, $\upsilon = 0$.

In the LS phase the molecules are, on average, not tilted ($\langle\theta\rangle = 0$), and indirect evidence suggests that the molecules are almost free rotors [18]. Those mesophases differing in the orientation of the **c** director (see Fig. 2) have $\langle\theta\rangle \neq 0$ and are characterized by the azimuthal angle ϕ. In the L_1' phase [10], $\langle\phi\rangle = \pm\alpha + 2\pi/6$, where α is the angle intermediate between 0 (molecules tilted in the direction toward nearest neighbors (NN), e.g., L_{2h} and L_{2d} phases) and $\pi/6$ (molecules tilted toward next nearest neighbors (NNN), e.g., Ov and L_2^* phases). The distinction between the L_2' and L_2^* phases is based on the tilt azimuth order and, for subphases within the L_2' and L_2 phases, is based on the order of the molecular backbones (see Table I). A coupling between the tilt azimuth order and the distortion [36] observed for these phases is clear evidence for translational–rotational coupling. Although not conclusive, transitions between LS and L_2 phases are believed to be continuous [1].

In the low-temperature regime, phases possess a higher degree of crystallinity. The S phase contains vertical molecules, thought to be orientationally disordered, on a distorted hexagonal lattice, which is, more precisely, a centered rectangular net [18]. The CS phase shows the long-range translational and orientational order of the spines of the molecules [9], but the pattern of the molecular backbones is not uniquely determined. It is believed that in the low-temperature regime the molecules are packed in a herringbone pattern analogous to such structures in 3-D solids [29]. Thus, lowering the temperature has the effect of reducing the symmetry of the high-pressure phase from hexagonal (LS) to centered rectangular (S) and, finally, to the possibly oblique superstructure (CS). This symmetry reduction is, no doubt, also related to translational–rotational coupling because the molecules become hindered rotors with decreasing temperature. All three phases, CS, S, and LS, may be compared with the 3-D crystalline structures of alkanes: herringbone crystal, distorted rotator without long-range herringbone order, and hexagonal rotator phases, respectively [29]. The order–disorder transition from the S to the CS phase is completely analogous to orientational transitions in molecular solids. It has also been shown that tilted phases can be associated with the same three categories as 3-D phases regarding the distortion and herringbone order that characterize the untilted phases [29].

The phase diagram presented in Figure 1 can be viewed as resulting from two tendencies in the thermodynamic behavior of the system. First, with decreasing temperature, there is an increasing order of molecular backbones, and the system passes from a rotationally disordered state to herringbone order through a centered rectangular lattice. Second, the effect of increasing the 2-D pressure is to suppress tilting of molecules, and the system evolves from NN tilting to the vertical structure by passing through a state with NNN tilting. Both types of orientational ordering are coupled to distortion of the hexagonal net. The two tendencies, in conjunction with the principle of continuity, provide a conceptual basis for an understanding of the formation of all phases. In this chapter we develop a general theory of the n-alkanoic acid Langmuir film phase diagram based on an intermolecular interaction model [3–5]. To illustrate the above-mentioned two tendencies in the diagram, we consider in detail two transitions: the swiveling transition, illustrating the mechanism of tilt change [6], and the orientational LS – S transition, to illustrate the orientational ordering within untilted phases [7].

Molecular dynamics [37–45] and Monte Carlo [46–49] simulations performed for Langmuir monolayers have involved a variety of approximations, ranging from conceptualization of the monolayer as a system of rods grafted onto a surface, through the inclusion of different interactions such as chain–chain and chain–surface, to different modelings of intermolecular potentials ranging from the idealized to the realistic. Most of these numerical experiments have concentrated on hexatic phases, with results supportive of the collective tilt measurement of experiment but with less attention being paid to the hexagonal lattice distortion. It remains for studies to focus on the microscopic aspects of the stabilities, structures, and transitions of Langmuir monolayer phases.

Computer simulations using realistic potentials offer much more accurate descriptions of monolayer ordering for specific systems than do microscopic and mean-field theories, but they cannot yield powerful analytical solutions available from the latter and thus fail to provide the broad conceptualizations and trends desired for understanding the phase behavior of Langmuir monolayers. Moreover, such theories are essential for better interpretation of both physical and computer experiments.

In an extensive study comparing intermolecular forces of Langmuir monolayers of perhydro and perfluoro amphiphiles, Cai and Rice [50, 51] developed a molecular theory based on density functional formalism to describe transitions between untilted (LS) and tilted mesophases. Numerical calculations [51, 52] using realistic Lennard–Jones potentials for reasonably strong molecule–surface interactions have shown that a mesophase with vertical molecules is the most stable. Thus, a delicate balance between the potential and chain–chain interaction dictates the tilting characteristics for the mesophases. The nature of the transitions depends critically on the intermolecular interactions. For perfluorinated amphiphiles, only a first-order phase transition between the ordered and disordered dilute phases is found, with no evidence of a continuous tilt transition. However, a continuous tilt transition is commonly found for perhydro systems. The crystallinity of Langmuir films of molecules without polar groups [53] may be attributed to the assumed increased rigidity of perfluoro chains over that of perhydro chains.

The model of grafted rods, where molecules are approximated by rigid, rodlike particles attached to a planar, impenetrable surface, has been most extensively studied [54–59]. The discrete version of the model, based on a spin-1 Ising-variable Hamiltonian, has been solved by mean-field [54] and renormalization-group [55] methods. The results deal with competing roles of interparticle and particle–surface interactions but are limited to transitions between LS and isotropic liquid phases. This model, augmented by continuous orientational variables, has been proposed by Somoza and Desai [59]. The advantage of the spin-1 model is that it is able to mimic two successive phase transitions, whereas all phenomenological models based on Landau theory introduce "coupling of phase transitions" [2, 60]. Phases are characterized by assumed order parameters, without a necessary relation to microscopic properties of the system.

1.3. Order Parameters

Consideration of the problem of order parameters used in the theoretical studies is crucial, and it is useful, at this point, to emphasize their importance to existing theories and to the one subsequently developed here. Following the theory of phase transitions, phases are conveniently characterized by order parameters and their symmetries, but these parameters have to be related to intermolecular interactions in the system. Let us represent a density of a system at any point, $\rho(\mathbf{x})$, as a sum of spatially and temporally averaged densities, ρ_0, and its fluctuation, $\delta\rho(\mathbf{x})$,

$$\rho(\mathbf{x}) = \rho_0 + \delta\rho(\mathbf{x}) \qquad (1.1)$$

A new phase is characterized by a nonzero, thermally averaged fluctuation, $\langle\delta\rho(\mathbf{x})\rangle \neq 0$, that can be treated as a generalized order parameter. To see how the density fluctuation is related to the degrees of freedom of a system, we consider the density at point $\mathbf{x}$ at a global equilibrium,

$$\rho(\mathbf{x}) = \rho_0\left[\exp(-\beta E(\mathbf{x}))/V^{-1}\int dV \exp(-\beta E(\mathbf{x}))\right] \quad (1.2)$$

where $\beta = 1/kT$ (k is Boltzmann's constant, T is temperature), V is the volume of the system, and $E(\mathbf{x})$ is the interaction energy of a molecule at $\mathbf{x}$ with the surrounding field formed by the rest of the system. The energy will be written in terms of variables $\{Y_i\}$ and fields, $\{F_i\}$,

$$E(\mathbf{x}) = -Y_i(\mathbf{x})F_i(\mathbf{x}) \tag{1.3}$$

The field, $\mathbf{F(x)}$, is a sum of macroscopic, uniform fields and a local field due to fluctuations in the degrees of freedom of the surrounding molecules,

$$\mathbf{F(x)} = \mathbf{F} + \delta\mathbf{F(x)} \tag{1.4}$$

where

$$\delta F_i(\mathbf{x}) = \sum_{x \neq x'} K_{ij}(\mathbf{x}, \mathbf{x}')y_j(\mathbf{x}') \tag{1.5}$$

$K_{ij}(\mathbf{x}, \mathbf{x}')$ is the correlation function for the fluctuating fields,

$$K_{ij}(\mathbf{x}, \mathbf{x}') = \beta\langle\delta F_i(\mathbf{x})\delta F_j(\mathbf{x}')\rangle \tag{1.6}$$

The density of the molecular degrees of freedom at site $\mathbf{x}'$ is

$$y_j(\mathbf{x}') = \delta\rho(\mathbf{x}')Y_j(\mathbf{x}') \tag{1.7}$$

The interaction energy is assumed to be smaller than the thermal energy. This requirement can be satisfied by taking the external field, $\mathbf{F}$, to be sufficiently small (see Eq. (1.3)). With this approximation and the global equilibrium condition (Eq. (1.2)), the density fluctuation becomes

$$\delta\rho(\mathbf{x}) = \beta\rho_0 Y_i(\mathbf{x})F_i(\mathbf{x}) \tag{1.8}$$

This relation is valid for a field-free system and, in the case of Langmuir monolayers, the global equilibrium will appear after the system relaxes to such a state. This process is known to be rather slow and thus can be taken as the reference state.

In the situation where an experiment on Langmuir monolayers is performed without allowing the system to relax to the global equilibrium, the condition for local equilibrium must be employed. Thus for the density fluctuation,

$$\delta\rho(\mathbf{x}) = -\beta\rho_0 \left[Y_i(\mathbf{x}) - \langle Y_i(\mathbf{x})\rangle\right]F_i(\mathbf{x}) \tag{1.9}$$

where $\langle Y_i(\mathbf{x})\rangle$ is the thermal average of the variable within the local system at $\mathbf{x}$. This equation demonstrates that, when a system is in local thermodynamic equilibrium (e.g., within a mesoscopic domain), the density fluctuation of the local degrees of freedom is determined only by the local fluctuation in these variables. For further discussion, global equilibrium is assumed, although the same treatment can be executed for the case of local equilibrium.

Now, it becomes evident that the density fluctuation is related to intermolecular interactions and the molecular degrees of freedom. Thus, the average $\langle\delta\rho(\mathbf{x})\rangle$ can be expressed as

$$\langle\delta\rho(\mathbf{x})\rangle = \sum_i a_i\langle Y_i(\mathbf{x})\rangle^2 \tag{1.10}$$

$\langle Y_i(\mathbf{x})\rangle$ and $\langle\sum_x Y_i(\mathbf{x})\rangle$ act as the local and global order parameters, respectively. Thus, the order parameters are just averages of fluctuations in the variables of a system. For the crystalline phases of Langmuir monolayers, these are the translational and orientational degrees of freedom of the amphiphilic molecules,

whereas for mesophases the order parameter may be the amplitude, $\rho_\mathbf{q}$, of the density fluctuation wave,

$$\delta\rho(\mathbf{x}) = N^{-1}\sum_\mathbf{q}\rho_\mathbf{q}\exp(i\mathbf{qx}) \tag{1.11}$$

characterized by the wave vector, $\mathbf{q}$. This amplitude is called the "weak crystallization" order parameter [2, 32]. For liquid-like and gas-like phases, it is just an isotropic density fluctuation, $\delta\rho = \rho - \rho_c$, where ρ_c stands for the critical density of the system. With these order parameters, we can describe a sequence of Langmuir monolayer phases as a continuous process of ordering, going from a gas-like phase ($\delta\rho < 0$), through a liquid-like phase ($\delta\rho > 0$), then mesophases, and finally crystalline phases ($\rho_\mathbf{q} \neq 0$, $\langle Y_i(\mathbf{q})\rangle$, $\langle Y_i\rangle$, $\langle Y_i(\mathbf{x})\rangle$).

Despite a large number of models and theories, orientational order in monolayers is usually studied by imposing azimuthal symmetry and examining the nematic order parameter, which, following the standard formulation [61], has the form

$$Q_{\alpha\beta} = Q(T)(\mathbf{n}_\alpha\mathbf{n}_\beta - (1/3)\delta_{\alpha\beta}) \tag{1.12}$$

where the $\mathbf{n}$ are unit vectors linked to the molecule, the subscripts refer to the laboratory frame, and δ is the Kroenecker symbol. In the context of Langmuir monolayers, the Q_{33} component ($\langle 3\cos^2\theta - 1\rangle$) is not the symmetry-breaking parameter, and the 2-D version of Eq. (1.12) has been used [2, 62–64]. We show that this limitation has no justification in microscopic considerations. Models that have discrete orientational degrees of freedom [54, 55] have used the nematic order parameter and have sought the so-called biaxial order associated with 2-D nematics, where the tilt order of molecules was neglected. Thus, Somoza and Desai [59] focused on tilt order but neglected biaxial order. They introduced the tilt order parameter, $\eta = \langle\sin\theta\cos\phi\rangle$. This arises naturally from using spherical harmonics for the description of a molecular orientational probability distribution function [59]. We, too, shall develop a description in terms of spherical harmonics and show their convenience and power for both microscopic theory and the free-energy expansion [3, 4].

Kaganer and co-workers [2, 62–64] have advocated a phenomenological Landau free-energy expansion with the use of two orientational order parameters, η to characterize tilt and a 2-D version of $Q_{\alpha\beta}$ to characterize molecular backbone orientation, and two other parameters to characterize the density wave and, separately, the herringbone ordering, associated with so-called weak crystallization order parameters. Coupled to each other, and based on purely phenomenological arguments, these parameters were used to characterize phases shown by the generic phase diagram [33]. The phenomenological description of Kaganer et al. has no relation to and does not attempt to connect the theory with microscopic origins of the stabilities or instabilities of Langmuir film phases. Furthermore, a recent review article by Kaganer et al. [2] incorrectly characterizes the microscopic molecular theories as being "severely restricted by the use of lattice models, i.e., the mass centers of the molecules are assumed to be fixed on a hexagonal lattice and the transition occurs between orientationally disordered and ordered states in the translationally ordered system." To the contrary, the essential point of the microscopic molecular model is that it does not place the amphiphile's centers of mass on a hexagonal lattice, but rather the head groups of the amphiphilic molecules form the hexagonal lattice and their tails possess orientational freedom [3–5]. In this context, we demonstrate that spherical harmonics offer a natural and logical set of variables to define vector and higher-rank tensor order parameters for Langmuir monolayer phases. In another conclusion, Kaganer et al. state that in Langmuir monolayers "translational and orientational ordering occur simultaneously, which demands off-lattice models for herringbone ordering" [2]. Antithetically, we show that Fourier transforms of the spherical harmonics may be exploited to describe translational ordering of the crystalline phases instead of the assumed separate weak crystallization order parameters. The translational and orientational ordering, their coupling and molecular origin, are very important to Langmuir monolayers, and, for this reason, a microscopic theory has to be developed [3, 4].

Numerical calculations have shown that tilting/nontilting transitions follow from competition between chain–chain and chain–surface interaction. Strong chain–surface interaction is needed to impair collective tilting. In the microscopic theory, which will be presented here, this is consistently explained by competition between rotational–rotational and rotational–translational coupling. On the other hand, the phenomenological description based on Landau theory "is not sensitive to the difference between a hexatic and a 2-D hexagonal crystal," as noted by Kaganer et al. [2], and thus, the important aspect of simultaneous translational and orientational ordering is missing in the studies using a phenomenological free-energy expansion based on the global symmetry of a system.

This contribution demonstrates and stresses the microscopic theory of Langmuir monolayer phases and their transitions by viewing them as essentially solid-state phenomena and describing them by exploiting their full symmetry. This leads to a microscopic theory that provides a consistent framework for understanding Langmuir monolayers and their phase behavior [3–7]. A series of steps is used to achieve this. First, we consistently use surface harmonics to describe fluctuations in molecular orientations. This produces a quite compact treatment of the rotational degrees of freedom, allows description of orientational fluctuations of any symmetry or magnitude, and gives a consistent and logical set of order parameters for the phase transitions. Second, we explicitly treat translational–rotational coupling, calculate effective rotational interactions, and analyze possible orientational instabilities from generalized rotational susceptibilities. Third, we calculate thermoelastic properties and find the free energy of the system in terms of the orientational order parameters that are average values of the surface harmonics [3]. In the next step, we discuss elastic aspects of the system by expressing the free energy alternatively in terms of strain tensor components and by exploiting the concept of elastic dipoles [4, 6]. This permits discussion of important macroscopic and mesoscopic aspects of the ferroelasticity of the Langmuir monolayers [7]. Finally, we describe

how the microscopic theory can be approximated and mapped onto a three-state spin-1 lattice gas model.

Contrary to most current theories of Langmuir monolayer phases that are formulated from a liquid-state viewpoint, this approach may be identified with well-established solid-state methods. We exploit physical concepts employed in describing translational–rotational coupling in molecular crystals with orientational disorder [65–70]. This approach may, on first consideration, appear to contradict the idea of making comparisons to mesophases. However, recognizing that all Langmuir monolayer phase transitions are related to orientational fluctuations, a powerful method for describing these must first be identified. It is this that generates the similarity to orientational disorder in solids. When we approach the hexatic phases and their transformations from this point of view, we concentrate on similarities between the mesophases and the 2-D solid state rather than on their differences.

We begin by defining the n-alkanoic Langmuir monolayer as a system of close-packed rigid rods of global hexagonal symmetry. The potential energy is then partitioned explicitly for single-molecule orientational potential and intermolecular couplings: rotational–rotational, translational–rotational, and translational–translational. Every part of the potential is derived for hexagonal symmetry, and the five lowest surface harmonics are used to describe the orientational fluctuations. The coupling matrices are expressed as Fourier transforms, thus forming a dynamical matrix for the system. With the expression of the translational–translational part in terms of elastic constants, the effective orientational potential is found. With this potential we calculate the rotational susceptibility matrix and analyze possible orientational instabilities related to different orientational fluctuations. Initially, the instabilities for not breaking translational symmetry are discussed and expected symmetry changes are predicted. Subsequently, we consider orientational instabilities with translational symmetry breaking and conclude with phase transitions to a superlattice with herringbone ordering. We show how the predicted orientational instabilities may drive phase transitions between the different phases found in the experimental phase diagram. Next we calculate the thermoelastic properties of the monolayer system and show how the structural (elastic) instabilities are driven by orientational ones. We then employ the concept of the elastic dipole and discuss the ferroelasticity of Langmuir monolayers as well as the swiveling transition.

2. MOLECULAR MODEL BUILDING

We now derive a model [3] that accounts for several features of the Langmuir monolayers relevant to the phase diagram of the system. The following concepts are employed in the model-building procedures:

i. The phase transitions result from instabilities that follow from a competition between intermolecular interactions. The competing interactions are identified and, in particular, focus is placed on direct and indirect (lattice-mediated) rotational interactions.

ii. Rotational degrees of freedom are important variables in the system and may be represented in terms of spherical harmonics.

iii. The sites of the 2-D lattice are noncentrosymmetric by the very nature of the Langmuir monolayer, and various translational–rotational couplings appear. These couplings are taken as the driving force for orientational and structural instabilities in the system.

The Langmuir monolayer model system is defined as a hexagonal planar lattice with the amphiphiles' tails perpendicular to the net. This clearly removes the molecular centers of mass from the hexagonal planar lattice. The amphiphiles' head groups are attached to the impenetrable lattice at the sites. To keep the model reasonably general, the detailed structure and chemical nature of the molecules are neglected. In principle, the cross section of a rigid-rod molecule mimics the structure and chemical nature of a particular molecule reasonably well [62].

The closest packed structure of such a Langmuir monolayer has C_{6v} point group symmetry that reflects the local environment around a molecular site. This is the symmetry of the so-called hexatic (LS) phase (see the Appendix for details), with one molecule in a planar primitive unit cell of the 2-D space group $p6m$ [71, 72]. The crystalline model of the LS phase assumes hexatic rigidity (the "bond" angle $\nu = 0$, Fig. 2), but it should be kept in mind that it is only a reference structure, a network.

The crystallinity of a phase depends on the correlation length that is determined by intermolecular couplings and temperature. At nonzero temperatures, fluctuations destroy the long-range order, and regions of uniform crystallinity will extend only over finite correlation lengths, L_x and L_y, which are, in general, different and vary from tenths to tens of thousands of times the size of the hexagonal lattice constant [18]. The correlation lengths can be used to define a discrete set of wave vectors, $\mathbf{q}$, in reciprocal space. The final size of the crystallinity of the LS phase, which serves as the reference structure, will then give a particular meaning to the special points of the Brillouin zone defined for the lattice (see the Appendix).

The position of the kth molecule on the 2-D lattice is denoted by $\mathbf{X}(k)$. This is where the head group of a molecule is grafted onto the impenetrable subphase lattice. Momentary displacements, $\mathbf{u}(k)$, are described relative to $\mathbf{R}(k)$, a well-defined equilibrium position,

$$\mathbf{X}(k) = \mathbf{R}(k) + \mathbf{u}(k) \tag{2.1}$$

The orientation of a rigid-rod molecule is given by $\Omega(k)$, which contains the polar angles, θ and ϕ (see Fig. 2).

The potential energy of the system is

$$V = \frac{1}{2} \sum_k \sum_{k'} V(kk') \tag{2.2}$$

which is approximated by the molecule–molecule pair potential,

$$V(kk') = V\left[\mathbf{X}(k), \mathbf{X}(k'); \Omega(k), \Omega(k')\right] \tag{2.3}$$

This potential can be modeled by any kind of (semi)empirical atom–atom potential, where details of the molecules are taken into account, or by nonrealistic potentials, where molecules are represented by some geometrical objects, as is often done in computer experiments.

We expand the potential in (2.2) to second order in translational displacements. The harmonic approximation for the displacements is incorporated into the potential,

$$V = V^R + V^{TR} + V^T \tag{2.4}$$

which is written as the sum of a purely rotational part, V^R, a translational–rotational part, V^{TR} (first-order in the displacement, $\mathbf{u}$), and a translational part, V^T, as the second-order term. These three contributions to the total energy may now be derived.

2.1. Rotational Potential

The rotational potential corresponds to the zeroth-order term in the expansion in (2.4) with respect to translational displacements, $\mathbf{u}$. Therefore, for the rotational potential we have

$$V^R = \frac{1}{2} \sum_k \sum_{k'} V\left[\mathbf{R}(k) - \mathbf{R}(k'); \Omega(k), \Omega(k')\right] \tag{2.5}$$

which describes the interaction of molecules in orientations $\Omega(k)$ and $\Omega(k')$, which are at equilibrium positions on the lattice. The potential is further decomposed into the single-molecule orienting potential, V_0^R (the term with $k = k'$ in the summation, Eq. (2.5)), and the rotational interaction,

$$V^R = V_0^R + \frac{1}{2} \sum_{k \neq k'} V\left[\Omega(k), \Omega(k')\right] \tag{2.6}$$

where

$$V_0^R = \sum_k V_0^R\left[\Omega(k)\right] \tag{2.7}$$

The single-molecule orientational potential is the sum of constituent orienting potentials experienced by a molecule at site k when all surrounding molecules are kept in their equilibrium positions, $\mathbf{R}(k')$, and with orientations $\Omega(k') = (\theta = 0, \phi = 0)$. This potential contains a contribution from the subphase of the monolayer. The single-molecule potential (Eq. (2.7)) possesses full hexagonal symmetry (C_{6v}).

The most convenient way to specify the potential, V_0^R, is to expand it in terms of spherical harmonics [73] that transform according to the totally symmetric representation of the C_{6v} point group. In principle, one has to expand the single-molecule orientational potential in terms of symmetry-adapted rotator functions. (For the theory of symmetry-adapted rotator functions that were originally introduced for solid methane [74], see [75–77].) The symmetry-adapted rotator functions transform according to the irreducible representations of the product of the site group and point group of a molecule [76]. This becomes important for nonlinear molecules when one wants to take into account, explicitly, the symmetry of the molecule. It has been shown [76] that for the case of a linear rod molecule ($C_{\infty v}$ symmetry), the symmetry-adapted functions depend on the spherical coordinates of the long molecular axis and, therefore, are just surface harmonics [73] relative to the site group.

For molecules in a Langmuir monolayer, the molecular axis vector spans only the upper half space ($z > 0$) above the surface. As pointed out by Somoza and Desai [59], expansion of the rotational potential with the full set of spherical harmonics, $Y_{l,m}(\theta, \phi)$, leads to a problem of redundancy or overcompleteness. Thus, following their suggestion, we restrict the set of spherical harmonics in problems of Langmuir monolayers to those for which

$$\left(\frac{\partial Y_{l,m}(\theta, \phi)}{\partial \theta}\right)_{\theta=\frac{\pi}{2}} = 0 \qquad (2.8)$$

i.e., limited to harmonics with $l + m =$ even.

For the site symmetry, C_{6v}, the single-molecule orientational potential, is written as

$$V_0^R(\Omega) = \alpha_0 + \sum_{l=2,4} \alpha_l Y_{l,0}(\Omega) + \sum_{l=6,8} \beta_l Y_l^{6c}(\Omega) + \cdots \quad (2.9)$$

where $Y_l^{6c} = (Y_{l,6} + Y_{l,-6})/\sqrt{2}$ and α_l, β_l, are the coefficients of the expansion. More explicitly, the orientational potential (2.9) is written in the form

$$V_0^R(\Omega) = a_0 + \sum_{n=1...} a_n \sin^{2n}\theta$$
$$+ \left(\sum_{n=1...} b_n \sin^{4-2n}\theta\right)\cos 6\theta + \cdots \quad (2.10)$$

The orientational probability distribution, $P(\Omega)$, for a molecule at the C_{6v} symmetry site is

$$P(\Omega) = Z_0^{-1}\exp[-\beta V_0^R(\Omega)] \qquad (2.11)$$

and the single-molecule rotational partition function is

$$Z_0 = \int d\Omega \exp[-\beta V_0^R(\Omega)] \qquad (2.12)$$

For a strong orientational potential (Eq. (2.9)), molecules are localized in quasidiscrete states (pocket states [78]) that often are further approximated by discrete states. The spin-1 model Hamiltonian considered for the Langmuir monolayer [54, 55] corresponds to such an approximation. By keeping the orientational potential expressed in terms of spherical harmonics, we allow a continuous change in molecular orientation.

The rotational–rotational coupling term, the second in Eq. (2.6), describes direct coupling between orientational fluctuations of different molecules ($k \neq k'$). Expressing the fluctuations in terms of surface harmonics [73], we write the interaction in the form

$$V^{RR} = \frac{1}{2}\sum_{k \neq k'}\sum Y_\alpha(k)J_{\alpha\beta}(kk')Y_\beta(k') \qquad (2.13)$$

where summation over repeated indices is assumed. The vector, $\mathbf{Y}(k)\{Y_\alpha[\Omega(k)]\}$, represents a set of surface harmonics that describe orientational fluctuations of the kth molecule, $J(kk')$, with elements, $J_{\alpha\beta}(kk')$, that represent the matrix of rotational–rotational coupling constants that couple the αth harmonic of the kth molecule to the βth harmonic of the k'th molecule. Equation (2.13) is a condensed notation for a more general form of the interaction potential between molecules [78, 79]. The coupling constants, $J_{\alpha\beta}(k, k')$, can be calculated from an assumed intermolecular potential between two molecules, k and k', and, with the use of the orthogonality properties of the surface harmonics,

$$J_{\alpha\beta}(kk') = \int d\Omega(k)$$
$$\times \int d\Omega(k')V[\mathbf{R}(k) - \mathbf{R}(k'); \Omega(k), \Omega(k')]$$
$$\times Y_\alpha[\Omega(k)]Y_\beta[\Omega(k')] \qquad (2.14)$$

The coefficients $J_{\alpha\beta}(k, k')$ consist of contributions from electrostatic, induction, dispersion, and repulsion interactions. For long molecules it is convenient to partition the potential into attractive and repulsive parts and relate the latter to the excluded area [59], $A[\Omega(k)\Omega(k')]$ (the area excluded by molecule k in orientation $\Omega(k)$ as seen by another molecule k' in orientation $\Omega(k')$). Consequently, the rotational–rotational coupling constants will be a sum of the corresponding contributions.

At this point we specify the set of surface harmonics describing the rigid rods' orientational fluctuations. Besause there is no symmetry constraint imposed on the set, the number of these variables depends on how much rotational freedom one expects for the molecules, i.e., on the strength of the single-molecule orientational potential. For Langmuir monolayers we do not expect a very strong single-molecule orientational potential, and, to keep the formalism reasonably clear, we limit consideration to surface harmonics with $l = 2$. Following the constraint (Eq. (2.8)), we take the following surface harmonics to represent variables of the orientational fluctuations of molecules,

$$(E_1) \qquad Y_1 = \frac{(Y_{1,1} + Y_{1,-1})}{\sqrt{2}} = cx \qquad (2.15)$$

$$(E_1) \qquad Y_2 = -i\frac{(Y_{1,1} - Y_{1,-1})}{\sqrt{2}} = cy \qquad (2.16)$$

$$(A_1) \qquad Y_3 = Y_{2,0} = c'(2z^2 - x^2 - y^2) \qquad (2.17)$$

$$(E_2) \qquad Y_4 = \frac{(Y_{2,2} + Y_{2,-2})}{\sqrt{2}} = c''(x^2 - y^2) \qquad (2.18)$$

$$(E_2) \qquad Y_5 = \frac{(Y_{2,2} - Y_{2,-2})}{\sqrt{2}} = c''xy \qquad (2.19)$$

where $c = (3/4\pi)^{1/2}$, $c' = (5/16\pi)^{1/2}$ and $c'' = (15/16\pi)^{1/2}$. Irreducible representations of the C_{6v} point group are given, and the Cartesian coordinates are $x = \sin\theta\cos\phi$, $y = \sin\theta\sin\phi$, and $z = \cos\theta$. This set of surface harmonics is representative of the problem and contains the lowest surface harmonics important to the description of different symmetries of the orientational fluctuations.

The functions Y_1 and Y_2 belong to the doubly degenerate E_1 irreducible representation of the C_{6v} point group and transform as the x and y components of a vector. The function Y_3 belongs to the totally symmetric representation A_1 and transforms as the symmetric components of a second-rank tensor. Finally, the functions Y_4 and Y_5 belong to the doubly degenerate E_2 representation and transform as the nonsymmetric (deviatoric) part of a second-rank tensor. The symmetry properties of the harmonics are important because they permit relation of every observable property, vectorial or tensorial, to statistical averages of the corresponding surface harmonics. We shall refer to this when discussing order parameters. Note that Y_1 and Y_2 are also used to represent the angular distribution of atomic orbitals, p_x and p_y, respectively, and Y_3, Y_4, and Y_5 are similarly associated with the d_{z^2}, $d_{x^2-y^2}$, and d_{xy} orbitals, respectively. This analogy will be helpful as the discussion proceeds.

With the set of surface harmonics (Eqs. (2.15)–(2.19)) we shall be able to define orientational order parameters of A_1 (non-symmetry-breaking) and E_1 and E_2 (symmetry-breaking) symmetries. Still, for a more complete set of surface harmonics, and the order parameters corresponding to them, the surface harmonics $Y_6 = (Y_{3,3} + Y_{3,-3})/\sqrt{2}$ of B_1 symmetry, $Y_7 = -i(Y_{3,3} - Y_{3,-3})/\sqrt{2}$ of B_2 symmetry, and $Y_8 = -i(Y_{6,6} - Y_{6,-6})/\sqrt{2}$ of A_2 symmetry need to be included. The surface harmonic Y_8 is a natural choice for the chiral order parameter. Here, we limit ourselves to the set given by Eqs. (2.15)–(2.19).

We introduce the Fourier transforms,

$$Y_\alpha(\mathbf{q}) = N^{-\frac{1}{2}} \sum_k Y_\alpha(k) \exp[i\mathbf{q}\mathbf{R}(k)] \qquad (2.20)$$

and

$$J_{\alpha\beta}(\mathbf{q}) = \sum_{k'} J_{\alpha\beta}(kk') \exp[i\mathbf{q}(\mathbf{R}(k) - \mathbf{R}(k'))] \qquad (2.21)$$

where $\mathbf{q}$ is the wave vector for the simple hexagonal planar lattice. There is a finite set of wave vectors determined by the extent of crystallinity of the reference phase. For such a set, the Fourier transform in Eq. (2.20) is equivalent to the "weak crystallization" parameter [2, 62–64]. The direct rotational–rotational part of the energy in reciprocal space is

$$V^{RR} = \frac{1}{2} \sum_\mathbf{q} Y_\alpha(\mathbf{q}) J_{\alpha\beta}(\mathbf{q}) Y_\beta(-\mathbf{q}) \qquad (2.22)$$

The matrix, $\mathbf{J}(\mathbf{q})$ can be calculated for six nearest neighbors by symmetry arguments (see the Appendix), and the elements of the matrix are specified in the Appendix. For $\mathbf{q} \Rightarrow 0$ the rotational–rotational matrix is diagonal, as required by C_{6v} point-group symmetry.

2.2. Translational–Rotational Coupling

The translational–rotational coupling term, V^{TR}, is the first term in the expansion of the total potential with respect to translational displacement, $\mathbf{u}$ (Eq. (2.4)). It is

$$V^{TR} = \sum_k \sum_{k'} V_i'[\mathbf{R}(k) - \mathbf{R}(k'); \Omega(k)] u_i(k') \qquad (2.23)$$

where V_i stands for the ith component of a force acting between the k and k' molecules at the equilibrium distance, $\mathbf{R}(k) - \mathbf{R}(k')$, when molecule k' is in its equilibrium orientation, $\Omega(k')$, and molecule k is in the orientation $\Omega(k)$. The force is calculated as

$$V_i'(kk'; \Omega(k)) = \left(\frac{\partial V[kk'; \Omega(k), \Omega(k')]}{\partial u_i(k')}\right)_{\Omega(k')=0} \qquad (2.24)$$

and represents an angular distribution. Therefore, we express the force in terms of orientational fluctuations of the kth molecule, e.g., in terms of the surface harmonics [65],

$$V_i'[kk'; \Omega(k)] = \sum_\alpha V_{i\alpha}(kk') Y_\alpha(k) \qquad (2.25)$$

The coupling constant $V_{i\alpha}(kk')$ couples the k'th molecule being displaced by the u_ith component of the displacement vector with the kth molecule with the orientational fluctuation Y_α. This constant is evaluated at the equilibrium distance between the molecules and is given by the equation

$$V_{i\alpha}(kk') = \int d\Omega \, V_i'[kk'; \Omega(k)] Y_\alpha(k) \qquad (2.26)$$

With the introduction of Fourier transforms,

$$u_i(\mathbf{q}) = N^{-\frac{1}{2}} \sum_k u_i(k) \exp[i\mathbf{q}\mathbf{R}(k)] \qquad (2.27)$$

and considering the translational–rotational coupling matrix,

$$V_{i\alpha}(\mathbf{q}) = \sum_{k'} V_{i\alpha}(kk') \exp[i\mathbf{q}(\mathbf{R}(k) - \mathbf{R}(k'))] \qquad (2.28)$$

the translational–rotational part of the energy becomes

$$V^{TR} = \sum_q u_i(\mathbf{q}) V_{i\alpha}(\mathbf{q}) Y_\alpha(-\mathbf{q}) \qquad (2.29)$$

Heeding the symmetry of the system (see Appendix), taking into account the translational–rotational coupling between nearest neighbors, and denoting the constants between molecules located at $(0, 0)$ and $(a, 0)$ as $V_{i\alpha}$, we construct the $\mathbf{V}(\mathbf{q})$ matrix (Eq. (2.30)),

$$V(\mathbf{q}) = \begin{pmatrix} 2v_{11}f_1(\mathbf{q}) & 0 & 2v_{13}f_2(\mathbf{q}) \\ 0 & 2v_{11}f_1(\mathbf{q}) & 2\sqrt{3}v_{13}f_3(\mathbf{q}) \\ 2v_{14}f_2(\mathbf{q}) & 2\sqrt{3}v_{14}f_3(\mathbf{q}) \\ -2\sqrt{3}v_{14}f_3(\mathbf{q}) & 2v_{14}f_2(\mathbf{q}) \end{pmatrix} \qquad (2.30)$$

where $f_1(q) = \cos 2\alpha + 2\cos\alpha\cos\beta - 3$, $f_2(q) = i(\sin 2\alpha + \sin\alpha\cos\beta)$, and $f_3(q) = i(\sin\alpha\sin\beta)$, where $\alpha = (1/2)q_x a$ and $\beta = (\sqrt{3}/2)q_y a$. a is the hexagonal lattice constant, and q_x and q_y are components of the wave vector in the orthogonal axis system (see the Appendix). In the limit $\mathbf{q} \Rightarrow 0$, there is no translational–rotational coupling for harmonics Y_1 and Y_2. This is due to the translational invariance of the system potential, which requires $\sum V_{i\alpha}(kk') = 0$ for every $(i\alpha)$ component [80]. At $\mathbf{q} \Rightarrow 0$, the system has C_{6v} symmetry, and the orientational fluctuations of type E_1 (Y_1 and Y_2 harmonics) can couple bilinearly to a displacement vector, $\mathbf{u}$, which transforms also as the E_1 representation. This coupling, a force, is compensated for when the system is at equilibrium.

At $\mathbf{q} \neq 0$, the system has lower symmetry than C_{6v}, and fluctuations of type E_2 and A_1 can couple bilinearly to the molecular displacements. In the limit $\mathbf{q} \Rightarrow 0$, the translational–rotational coupling matrix is approximated as

$$V(\mathbf{q} = 0) = i3a \begin{pmatrix} 0 & 0 & v_{13}q_x & v_{14}q_x & v_{14}q_y \\ 0 & 0 & v_{13}q_y & -v_{14}q_y & v_{14}q_x. \end{pmatrix} \quad (2.31)$$

Better insight into the physical meaning of the translational–rotational matrix in the limit $\mathbf{q} \Rightarrow 0$ can be gained from elasticity theory and the elastic dipole concept.

2.3. Translational Potential

The purely translational part of the system potential is conveniently expressed in terms of a translational–translational dynamical matrix, $M(\mathbf{q})$,

$$V^T = \frac{1}{2} \sum_q u_i(\mathbf{q}) M_{ij}(\mathbf{q}) u_j(-\mathbf{q}) \quad (2.32)$$

For the 2-D system, the dynamical matrix is 2×2 and consists of Fourier transform elements of translational–translational force constants (second derivatives of the energy with respect to displacements) between molecules. Explicitly, the matrix elements are

$$M_{11}(\mathbf{q}) = 2M_{11}(\cos 2\alpha - 1) + (M_{11} + 3M_{22})$$
$$\times (\cos\alpha \cos\beta - 1) \quad (2.33)$$
$$M_{22}(\mathbf{q}) = 2M_{22}(\cos 2\alpha - 1) + (M_{22} + 3M_{11})$$
$$\times (\cos\alpha \cos\beta - 1) \quad (2.34)$$
$$M_{12}(\mathbf{q}) = \sqrt{3}(M_{22} - M_{11}) \sin\alpha \sin\beta \quad (2.35)$$

where the M_{ii} are force constants between nearest molecules located at $(0, 0)$ and $(a, 0)$. When there are only central forces acting between the molecules, there is an interrelation between the M_{11} and M_{22} components and the dynamical matrix can be expressed in terms of only one parameter.

The dynamical matrix, $M(\mathbf{q})$, can be conveniently expressed in terms of elastic constants in the limit $\mathbf{q} \Rightarrow 0$. For the hexagonal lattice, the simplest form is

$$M(\mathbf{q}) = a^2 m^{-1} C_{11}^0 \begin{pmatrix} q_x^2 + \frac{1}{6}q_y^2 & \frac{2}{3}q_x q_y \\ \frac{2}{3}q_x q_y & \frac{1}{6}q_x^2 + q_y^2 \end{pmatrix} \quad (2.36)$$

where m is a molecular mass and C_{11}^0 is the bare elastic constant. We have assumed central forces between the molecules and that the Cauchy relation ($C_{66}^0 = C_{12}^0$) is obeyed.

Collecting all contributions to the potential energy, we write the total Hamiltonian for the system,

$$H = K^R + K^T + V_0^R + V^{RR} + V^{TR} + V^T \quad (2.37)$$

K^R and K^T represent the kinetic energy of the rotational and translational degrees of freedom, respectively. For linear rigid rotors,

$$K^R = \sum_k \frac{L^2(k)}{2I} \quad (2.38)$$

where $L^2 = (p_\theta^2 + p_\phi^2/\sin^2\theta)$ and I is the moment of inertia. We also have

$$K^T = \sum_k \frac{p^2(k)}{2m} \quad (2.39)$$

where p is the momentum operator and m is the molecular mass.

In the total Hamiltonian, V_0^R is described by Eqs. (2.9) and (2.10), and the other terms are written in the compact form,

$$V^{RR} = \frac{1}{2} \sum_q \mathbf{Y}(\mathbf{q})\mathbf{J}(\mathbf{q})\mathbf{Y}(-\mathbf{q}) \quad (2.40)$$

$$V^{TR} = \sum_q \mathbf{u}(\mathbf{q})\mathbf{V}(\mathbf{q})\mathbf{Y}(-\mathbf{q}) \quad (2.41)$$

$$V^T = \frac{1}{2} \sum_q \mathbf{u}(\mathbf{q})\mathbf{M}(\mathbf{q})\mathbf{u}(-\mathbf{q}) \quad (2.42)$$

The matrices of coupling constants form a 7×7 dynamical matrix for the system.

3. THERMODYNAMICS

First, we derive the effective orientational potential and susceptibilities. Next, we discuss possible orientational instabilities with and without translational symmetry breaking. Finally, we will analyze thermoelastic properties and derive the free energy for the system.

3.1. Effective Orientational Potential

The general structure of the Hamiltonian, as given by Eqs. (2.37)–(2.42), is similar to that used in describing spin–phonon coupling [81]. We follow this formalism and decouple the translations and rotations. The $\mathbf{u}(\mathbf{q})$ displacement is divided into two parts,

$$\mathbf{u}(\mathbf{q}) = \mathbf{u}^{el}(\mathbf{q}) + \mathbf{w}(\mathbf{q}) \quad (3.1)$$

where $\mathbf{u}^{el}(\mathbf{q})$ denotes the elastic lattice displacement for a given orientational configuration $\{\mathbf{Y}\}$ and $\mathbf{w}(\mathbf{q})$ is the vibrational part about the equilibrium position with the thermal average taken with the total Hamiltonian, $\langle \mathbf{w}(\mathbf{q}) \rangle = 0$. Substituting Eq. (3.1) into Eqs. (2.41) and (2.42) and minimizing the Hamiltonian with respect to $\mathbf{u}^{el}(\mathbf{q})$, we find

$$\mathbf{u}^{el}(\mathbf{q}) = -\mathbf{M}^{-1}(\mathbf{q})\mathbf{V}(-\mathbf{q})\mathbf{Y}(\mathbf{q}) \quad (3.2)$$

Substituting this back into the Hamiltonian, we obtain

$$H = H^{\mathrm{T}} + H^{\mathrm{R}} \tag{3.3}$$

with

$$H^{\mathrm{T}} = K^{\mathrm{T}} + \frac{1}{2} \sum_q \mathbf{w}(\mathbf{q}) \mathbf{M}(\mathbf{q}) \mathbf{w}(-\mathbf{q}) \tag{3.4}$$

$$H^{\mathrm{R}} = K^{\mathrm{R}} + V^{\mathrm{R}} + V^{\mathrm{RR}} - \frac{1}{2} \sum_q \mathbf{Y}(\mathbf{q}) \mathbf{L}(\mathbf{q}) \mathbf{Y}(-\mathbf{q}) \tag{3.5}$$

where

$$\mathbf{L}(\mathbf{q}) = \mathbf{V}(\mathbf{q}) \mathbf{M}^{-1}(\mathbf{q}) \mathbf{V}(-\mathbf{q}) \tag{3.6}$$

This matrix, $\mathbf{L}(\mathbf{q})$, gives indirect, lattice-mediated coupling between orientational fluctuations of the molecules.

One can understand this interaction in the following way. When the molecular orientation fluctuates, it produces a perturbation, which is transferred to other molecules through the translational–rotational coupling mechanism. Then, the surrounding molecules react and contribute to the orientational potential experienced by a reference molecule. The indirect rotational–rotational coupling contributes to the single-molecule orientational potential with the last term of the equation,

$$V^{\mathrm{R}} = V_0^{\mathrm{R}} - Y_\alpha L_{\alpha\beta}^{\mathrm{S}} Y_\beta \tag{3.7}$$

which gives the effective single-molecule orienting potential. The self-energy term is determined by the matrix,

$$L^{\mathrm{S}} = N^{-1} \sum_q L(\mathbf{q}) \tag{3.8}$$

In the mesoscopic limit ($\mathbf{q} \Rightarrow 0$) only orientational fluctuations of E_2 and A_1 symmetries are coupled by the indirect interaction. However, the matrix $\mathbf{L}(\mathbf{q} \Rightarrow 0)$ is not well defined because, as seen from Eq. (2.36), $\mathbf{M}(\mathbf{q}) \propto \mathbf{q}^2$ and $V(\mathbf{q}) \propto q$ (see Eq. (2.31)). This important point has been discussed in the context of translational–rotational coupling in alkali cyanide crystals [66] and is analogous to dipolar interactions [82]. The value of the matrix $\mathbf{L}(\mathbf{q} \Rightarrow 0)$ depends upon the direction from which the $\mathbf{q} \Rightarrow 0$ (Γ point) is approached. It thus depends on the shape of the system and is related to the domain problem and should not be confused with textures.

We write the effective rotational Hamiltonian,

$$H^{\mathrm{R}} = K^{\mathrm{R}} + V^{\mathrm{R}} - \frac{1}{2} \sum_q \mathbf{Y}(\mathbf{q})$$
$$\times \left[L(\mathbf{q}) - J(\mathbf{q}) - L^{\mathrm{S}} \right] \mathbf{Y}(-\mathbf{q}) \tag{3.9}$$

The subtraction of the self-energy term, $\mathbf{L}^s$, recognizes that the indirect interaction between a molecule and itself cannot contribute to the ordering of the molecules. Using a well-known procedure of statistical mechanics, we calculate the rotational susceptibility matrix, $\mathbf{X}(\mathbf{q})$, with elements defined as

$$X_{\alpha\beta}(\mathbf{q}) = \beta \left[\langle Y_\alpha(\mathbf{q}) Y_\beta(-\mathbf{q}) \rangle - \langle Y_\alpha(\mathbf{q}) \rangle \langle Y_\beta(-\mathbf{q}) \rangle \right] \tag{3.10}$$

The statistical averages are taken with the total rotational Hamiltonian,

$$\langle Y_\alpha(\mathbf{q}) \rangle = Z^{-1} \int d\Omega \, Y_\alpha(\mathbf{q}) \exp[-\beta H^{\mathrm{R}}] \tag{3.11}$$

where

$$Z = \int d\Omega \, \exp[-\beta H^{\mathrm{R}}] \tag{3.12}$$

Equation (3.11) defines order parameters corresponding to particular surface harmonics and their symmetries. The mean-field generalized susceptibility is given by [67]

$$X(\mathbf{q}) = X^0 \left[I - L'(\mathbf{q}) X^0 \right]^{-1} \tag{3.13}$$

where I is the diagonal unit matrix,

$$L'(\mathbf{q}) \equiv L(\mathbf{q}) - J(\mathbf{q}) - L^{\mathrm{S}}(\mathbf{q}) \tag{3.14}$$

and X^0 is the single-molecule rotational susceptibility calculated with the effective orientational potential, V^{R} (Eq. (3.7)). The matrix X^0 is diagonal by symmetry and is

$$X^0 = \beta \left[\chi_1, \chi_1, \chi_3, \chi_4, \chi_4 \right] \tag{3.15}$$

where $\chi_\alpha \equiv \langle Y_\alpha^2 \rangle$ is a weak function of temperature.

3.2. Orientational Instabilities

We now seek the orientational instabilities of the system. They can be identified from the condition $\mathbf{X}(\mathbf{q}) \Rightarrow \infty$, e.g.,

$$\det \left| (X^0)^{-1} - L'(\mathbf{q}) \right| = 0 \tag{3.16}$$

So, we have to find eigenvalues of the matrix. In principle, we should diagonalize the matrix for every $\mathbf{q}$ vector. This is possible and important when one considers a particular system for which the matrix elements are given numerically. Here, we analyze the instabilities in the model system for which only the symmetry is defined, and, therefore, we are constrained to those points in $\mathbf{q}$-space for which instabilities might be relevant to phase transitions in Langmuir monolayers. We shall consider the center of the Brillouin zone, the Γ-point ($\mathbf{q} \Rightarrow 0$), the Σ direction ($\mathbf{q} = (0, q_y)$), and the $\mathbf{M}$-point ($\mathbf{q} = (0, 2\pi/a\sqrt{3})$) [72]. (See the Appendix for details.)

First, we consider the orientational instabilities, which might drive phase transitions without translational symmetry breaking. These transitions seem to be most often found in the Langmuir monolayer systems in the high-temperature regime. We calculate $\mathbf{L}(\mathbf{q} \Rightarrow 0)$, and from Eqs. (3.6), (2.36), and (2.31) we find

$$L(q_x \Rightarrow 0) = \delta \begin{pmatrix} v_{13}^2 & v_{13}v_{14} & 0 \\ v_{13}v_{14} & v_{14} & 0 \\ 0 & 0 & 6v_{14}^2 \end{pmatrix} \tag{3.17}$$

where $\delta = 3m(C_{11}^0)^{-1}$. For the orthogonal direction, the matrix is

$$L(q_y \Rightarrow 0) = \delta \begin{pmatrix} v_{13}^2 & -v_{13}v_{14} & 0 \\ -v_{13}v_{14} & v_{14} & 0 \\ 0 & 0 & 6v_{14}^2 \end{pmatrix} \tag{3.18}$$

Introducing these matrices into Eq. (3.14) and calculating the inverse of the orientational susceptibility, $\mathbf{X}^{-1}(\mathbf{q} \Rightarrow 0)$, we find that the matrix is diagonal in Y_1, Y_2 (E_1 symmetry), and Y_5 (E_2 symmetry) and well defined for $\mathbf{q} \Rightarrow 0$. There is a 2×2 matrix on the diagonal that corresponds to a coupling (i.e., a hybridization) of Y_3 (A_1 symmetry) and Y_4 (E_2 symmetry) surface harmonics, and this matrix depends on the direction from which the $\mathbf{q} \Rightarrow 0$ point is approached (see Eqs. (3.17) and (3.18)). Therefore, there are three possibilities for which our model of the Langmuir monolayer will become unstable without translational symmetry breaking: (i) orientational fluctuations of E_1 symmetry, (ii) orientational fluctuations of E_2 symmetry, and (iii) orientational fluctuations of hybridized A_1 and E_2 symmetries.

For the E_1 symmetry fluctuations, the inverse of the orientational susceptibility is

$$X^{-1}(q \Rightarrow 0; E_1)$$
$$= \begin{pmatrix} (\beta\chi_1)^{-1} + 3(J_{11} + J_{22}) & 0 \\ 0 & (\beta\chi_1)^{-1} + 3(J_{11} + J_{22}) \end{pmatrix}$$
(3.19)

From Eq. (3.16) we find that the instability takes place at the temperature, $T_0(E_1)$,

$$T_0(E_1) = -k_B^{-1}\chi_1(T_0)3(J_{11} + J_{22}) \quad (3.20)$$

and that it is a result of competition between the single-molecule orientational potential that determines $\chi_1(T_0)$ and direct rotational–rotational interactions between molecules. This instability can happen only when the system gains enough energy by molecules tilting collectively (($J_{11} + J_{22}) < 0$). Thus, it represents only the kind of orientational-dependent intermolecular potential, whether it arises from short- or long-range interactions, that gives rise to a dipolar-like interaction (Y_1 and Y_2 represent the dipolar angular distributions) that will cause the instability. Moreover, the gain of energy must be large enough to overcome the influence of the single-particle orienting potential, which tends to keep molecules in a perpendicular orientation ($\chi_1 > 0$). *This is a general result that is independent of the assumption of the nature of the intermolecular potential.*

At $T_0(E_1)$ the system becomes simultaneously unstable against fluctuations described by both Y_1 and Y_2. We define the order parameter for the transition driven by this instability as

$$\eta(E_1) = a_1\langle Y_1\rangle + a_2\langle Y_2\rangle \quad (3.21)$$

i.e., a linear combination of parameters $\langle Y_1\rangle$ and $\langle Y_2\rangle$. For the general case, $a_1, a_2 \neq 0$, $\eta(E_1)$ measures the collective tilt of molecules in an arbitrary direction as seen from the more explicit form of the order parameter: $\eta(E_1) = c\langle\sin\theta\rangle(a_1\langle\cos\phi\rangle + a_2\langle\sin\phi\rangle)$; see Eqs. (2.15) and (2.16). Because the instability (Eq. (3.20)) is due to competition between the single-molecule orientational potential and direct rotational–rotational interaction, translational degrees of freedom do not contribute and there is no lattice deformation involved. The phase that would result from this instability will contain tilted molecules on the hexagonal, undeformed lattice.

When a_1 and $a_2 \neq 0$, the symmetry change at the transition is $C_{6v} \Rightarrow C_1$, with the resulting phase reminiscent of L'_1 on the phase diagram (Fig. 1). For $a_1 = 1, a_2 = 0$, and $\eta(E_1) = \langle Y_1\rangle$, molecules in the corresponding phase are tilted toward NN in an angular distribution analogous to a p_x orbital. For $a_1 = 0$, $a_2 = 1$, and $\eta(E_1) = \langle Y_2\rangle$, molecules are tilted toward NNN. The symmetry change in both cases is $C_{6v} \Rightarrow C_s$, with the phase suggestive of $L_{2d}(\langle Y_1\rangle)$ and Ov ($\langle Y_2\rangle$) in the schematic phase diagram (Fig. 1). This designation of phases is not final, because the observed tilt phases are all on a deformed lattice and a coupling with a lattice deformation will make the assignment more specific.

Phenomenologically, one can understand a coupling of E_1-type tilting with a deformation as a breaking of the hexagonal symmetry. The susceptibility matrix (Eq. (3.19)) then contains off-diagonal terms proportional to the symmetry-breaking deformation e_{xy} and ($e_{xx} - e_{yy}$) components of the strain tensor. Consequently, the degeneracy of the tilting instabilities of E_1 type is lifted, and discrimination between tilts toward NN and NNN is obtained. The deformation of the lattice will determine which of the instabilities will occur first; this can be analyzed from the 2×2 susceptibility matrix. The conclusion is that the elasticity of the system will dictate which of the instabilities will take place. Therefore, the interaction of the molecules with the subphase can be important.

Next, we discuss the instability of our Langmuir monolayer model system with respect to the orientational fluctuation of E_2 symmetry without translational symmetry breaking. This instability is expected when $(X(q \Rightarrow 0))_{55}^{-1} = 0$, and its temperature is calculated as

$$T_{ij}(E_2) = k_B^{-1}\chi_4(T_0)\left[6\delta V_{14}^2 - L_{44} - 3(J_{44} + J_{55})\right] \quad (3.22)$$

This instability arises from a competition between the single-molecule orientational potential, which determines χ_4, and the effective rotational–rotational interaction. The indirect, lattice-mediated coupling between orientational fluctuations, Y_5, helps the system to gain energy when molecules are collectively reoriented in the pattern described by the surface harmonic, Y_5. The energy gain measured by the first term in the square brackets of Eq. (3.22) is inversely proportional to the system elasticity (C_{11}^0) and critically depends on the strength of the translational–rotational coupling. The indirect rotational–rotational interaction should be very important in Langmuir monolayer systems and will be responsible for the orientational instabilities. For systems where direct interactions are described by quadrupolar-like interactions, more specifically by the xy component of a quadrupole, ($J_{44} + J_{55}) < 0$, the indirect interaction increases the temperature of the instability.

If an instability of the E_2 type takes place, the new phase is characterized by the order parameter,

$$\eta(E_2) = \langle Y_5\rangle \quad (3.23)$$

and the symmetry change expected at the transition is $C_{6v} \Rightarrow C_{2v}$. In this case, molecules are tilted collectively at every site with equal probability in the four directions, [1, 1], [1, −1], [−1, 1], and [−1, −1] (in the orthogonal axial system), and with the hexagonal lattice deformed according to the xy strain component. The deformation of the hexagonal lattice is involved in the E_2-type instability because the lattice-mediated indirect interaction contributes to the instability (Eq. (3.22)). For symmetry reasons, such a phase can be associated with the S region of the phase diagram (Fig. 1). The pattern of orientational disorder assumed for phase S, due to two mutually orthogonal orientations of a molecular backbone, is completely similar to the pattern of tilt reorientations described by Y_5.

The harmonics Y_3, Y_4, and Y_5 transform as components of a second-rank tensor; therefore, every observable tensorial property will be described by its thermal averages. Quantitative relations for macroscopic strain components are derived at the end of this section. Here, we relate these averages to the mean cross section of a molecule. The cross section of a rigid rod that models a molecule can be represented as a 2-D second rank tensor similar to the nematic order parameter Eq. (1.1). Therefore, the statistical averages of the surface harmonics, $\langle Y_3 \rangle$, $\langle Y_4 \rangle$, and $\langle Y_5 \rangle$, can be a measure of an average cross section of the rigid rod. Furthermore, this cross section would correspond to an average orientation of a molecular backbone. Physically, this means that a rigid rod with circular cross section represents a molecule rotating around the long axis, and the rotation becomes hindered by a hexagonal lattice deformation and/or tilt of the molecule. This is reflected in nonzero orientational order parameters, with the tensorial parameters interpreted as indicating the angular distribution of the molecular backbones.

Finally, for instabilities without translational symmetry changes, we shall discuss an instability due to coupled (Y_3 (A_1) and Y_4 (E_2)) harmonics. The inverse susceptibility submatrix corresponding to this is

$$X^{-1}(q \Rightarrow 0; A_1, E_2) = \begin{pmatrix} X_{33}^{-1} & X_{34}^{-1} \\ X_{34}^{-1} & X_{44}^{-1} \end{pmatrix} \quad (3.24)$$

where

$$X_{33}^{-1} = \beta^{-1} \chi_3^{-1} - \delta V_{13}^2 + L_{33}^S + 6J_{33} \quad (3.25)$$

$$X_{44}^{-1} = \beta^{-1} \chi_4^{-1} - \delta V_{14}^2 + L_{44}^S + 3(J_{44} + J_{55}) \quad (3.26)$$

$$X_{34}^{-1}(q_x \Rightarrow 0) = \delta V_{13} V_{14} \quad (3.27)$$

$$X_{34}^{-1}(q_y \Rightarrow 0) = -\delta V_{13} V_{14} \quad (3.28)$$

where we note the difference in the off-diagonal terms for approaching the Γ-point ($\mathbf{q} \Rightarrow 0$) from the x and y directions. The instability condition (Eq. (3.16))

$$X_{33}^{-1}(T_0) X_{44}^{-1}(T_0) = (X_{34}^{-1})^2 \quad (3.29)$$

determines the instability temperature, T_0. At this temperature, the lowest eigenvalue of the matrix (Eq. (3.24)) becomes zero, and the system is unstable against orientational fluctuations described by the corresponding eigenvector,

$$\xi^x = Y_3 \cos \varphi + Y_4 \sin \varphi \quad (3.30)$$

when approaching the Γ-point from the x direction and

$$\xi^y = -Y_3 \sin \varphi + Y_4 \cos \varphi \quad (3.31)$$

for the y direction, where $\tan 2\varphi = 2|X_{34}^{-1}|(X_{33}^{-1} - X_{44}^{-1})^{-1}$. The eigenvectors express hybridization of the orientational fluctuations described by the Y_3 and Y_4 surface harmonics. The order parameters corresponding to the instability are defined as statistical averages of the eigenvectors,

$$\eta^x(A_1, E_2) = \langle \xi^x \rangle; \qquad \eta^y(A_1, E_2) = \langle \xi^y \rangle \quad (3.32)$$

Note that the order parameters are orthogonal. It is this indirect, lattice-mediated, rotational interaction that makes the difference in the ordering of molecules, depending on the direction of approach to the Γ-point (see Eqs. (3.27) and (3.28)). This is a well-known problem in solids with magnetic or electrical dipolar interactions. Here, the lattice-mediated, indirect, interaction described by the matrix $L(\mathbf{q})$ represents the interaction of elastic dipoles [66], and this aspect will be discussed subsequently [4]. Therefore, we can expect a shape dependence for the interaction. Because a real system breaks up into domains, no shape dependence can be observed for the system as a whole. However, the domains themselves would have the most favorable shapes, because in this way the system can lower its free energy.

Detailed analysis of the shape dependence problem would require lengthy considerations and derivations, but some insight into the preferred shapes of domains can be developed. Clearly, Eq. (3.32) defines order parameters for two domains of the same phase. The domain with the η^x order parameter is formed with relative dimensions $L_y \gg L_x$, and, conversely, the domain characterized by the order parameter η^y has relative dimensions $L_x \gg L_y$. In other words, the phase with the order parameter that is a hybrid of $\langle Y_3 \rangle$ and $\langle Y_4 \rangle$ will split into two kinds of domains with mutually perpendicular stripes as their favored shapes. Indeed, it has been found that monolayer crystallites are stripes elongated perpendicular to the molecular tilt direction [18].

Every domain will be characterized by C_{2v} symmetry because the $\langle Y_4 \rangle$ component is responsible for the symmetry breaking. Because of the lattice-mediated interactions that take part in the instability mechanism, the phase transition would be accompanied by a lattice distortion measured by a linear combination of ($e_{xx} + e_{yy}$) (A_1 symmetry) and ($e_{xx} + e_{yy}$) (E_2 symmetry) strain tensor components. Later, we shall elaborate this aspect in greater detail and show that the Langmuir mesophases can be viewed as elastically interacting domains [7].

Discussion of the orientational instabilities concludes with the case of broken translational symmetry. Consider the direction $\Sigma(0, q_y)$ with the end point $M(0, 2\pi/\sqrt{3}a)$ at the Brillouin zone boundary (see the Appendix). For this direction, the $\mathbf{J}(\mathbf{q})$ matrix, as well as the matrix of indirect couplings, $L(\mathbf{q})$, is specified in the Appendix. It follows from the symmetry of the matrix, $L(\mathbf{q})$ in the Σ direction that there is a coupling between Y_1 and Y_5 and between Y_2, Y_3, and Y_4. Consequently, the inverse susceptibility matrix, $X^{-1}(\Sigma)$, is decomposed into 2×2 and 3×3 submatrices. For practical reasons, we analyze the 2×2 matrix corresponding to hybridization between Y_1 and Y_5 in the Σ direction. A related orientational instability will be determined by the lowest eigenvalue of the matrix,

$$\lambda(\Sigma) = \frac{1}{2}\Big\{X_{11}^{-1} + X_{55}^{-1} \\ - \big[(X_{11}^{-1} - X_{55}^{-1})^2 + 4(X_{15}^{-1})^2\big]^{\frac{1}{2}}\Big\} \quad (3.33)$$

where the elements of the inverse susceptibility submatrix are

$$X_{11}^{-1} = \beta^{-1}\chi_1^{-1} - L_{11}(\Sigma) + J_{11}(\Sigma) \quad (3.34)$$

$$X_{15}^{-1} = -L_{15}(\Sigma) \quad (3.35)$$

$$X_{55}^{-1} = \beta^{-1}\chi_4^{-1} - L_{55}(\Sigma) + J_{55}(\Sigma) \quad (3.36)$$

One then finds that the minimum eigenvalue is obtained for the wave vector, $q_y = 2\pi/\sqrt{3}a$, e.g., for the **M**-point that is the end of the Brillouin zone in the Σ direction. At this point, Y_1 and Y_5 are decoupled. Moreover, the indirect interactions are nonzero only for the Y_1 and Y_2 fluctuations. Because the interaction is larger for the Y_1 harmonic, the instability temperature is found from

$$T_0(\mathbf{M}) = k_B^{-1}\chi_1(T_0)\big[48\delta V_{11}^2 + (3J_{22} - J_{11})\big] \quad (3.37)$$

If $(3J_{22} - J_{11}) < 0$, then the instability at the **M**-point will be preferred only when $48\delta V_{11}^2 > |(3J_{22} - J_{11})|$. In this case, the instability at the **M**-point is a result of successful competition between the indirect and direct rotational interaction. It would be favored for a sufficiently soft lattice, which gives a large δ coefficient.

In the case $(3J_{22} - J_{11}) > 0$, the instability related to the Y_1 orientational fluctuation will be preferred at the **M**-point, and the new phase will be characterized by the order parameter,

$$\eta(E_1, \mathbf{M}) = \langle Y_1(\mathbf{M})\rangle \quad (3.38)$$

(see Eq. (2.15)). This phase is characterized by the tilt of molecules in the direction of NN (x), with a modulation described by $\exp[i\mathbf{q}(m)\mathbf{R}(k)]$. The modulation term may be interpreted as equivalent to the "weak crystallization" order parameter [2, 62–64]. The transition to this phase results in a doubling of the unit cell and creation of a herringbone pattern of molecular backbones. The new phase can be assigned to L_2'' of the schematic phase diagram (Fig. 1). The symmetry change would be $p6m \Rightarrow pg$[72, 73].

3.3. Thermoelastic and Structural Instabilities

Thermoelasticity implies a temperature dependence of the elastic properties of a system, and, thus, we calculate effective (T-dependent) elastic constants for a Langmuir film. We define the translational susceptibility matrix,

$$D^{-1}(\mathbf{q}) = \beta\big[\langle\mathbf{u}(\mathbf{q})\mathbf{u}(\mathbf{q})\rangle - \langle\mathbf{u}(\mathbf{q})\rangle\langle\mathbf{u}(\mathbf{q})\rangle\big] \quad (3.39)$$

The thermal averages of the displacement variables are calculated with the total Hamiltonian of the system (Eq. (2.37)). On introducing Eqs. (3.1) and (3.2) into the above equation, we get

$$D^{-1}(\mathbf{q}) = M^{-1}(\mathbf{q})\big[I + V(\mathbf{q})X(\mathbf{q})V(-\mathbf{q})M^{-1}(\mathbf{q})\big] \quad (3.40)$$

where we have used

$$M^{-1}(\mathbf{q}) = \beta\langle\mathbf{w}(\mathbf{q})\mathbf{w}(\mathbf{q})\rangle \quad (3.41)$$

In the limit $\mathbf{q} \Rightarrow 0$, Eq. (3.40) gives a link between the rotational susceptibility, $\mathbf{X}(\mathbf{q})$, and the elastic constants in the presence of the translational–rotational coupling, $\mathbf{V}(\mathbf{q})$. Matrix $\mathbf{D}(\mathbf{q})$ has exactly the same form as matrix $\mathbf{M}(\mathbf{q})$, with bare elastic constants replaced by the effective constants. Without the central force assumption, the matrix takes the form

$$D^{-1}(\mathbf{q} \Rightarrow 0) = \begin{pmatrix} C_{11}q_x^2 + C_{66}q_y^2 & \frac{1}{2}(C_{11} + C_{22})q_xq_y \\ \frac{1}{2}(C_{11} + C_{22})q_xq_y & \frac{1}{2}C_{66}q_x^2 + C_{11}q_y^2 \end{pmatrix} \quad (3.42)$$

In principle, the effective elastic constants should be calculated from the exact equation, Eq. (3.40). However, this is tedious, and instead we use an approximate equation that gives insight into the renormalization of the elasticity due to the translational–rotational coupling. We use, as an approximation to (3.40),

$$D(\mathbf{q} \Rightarrow 0) \cong M(\mathbf{q} \Rightarrow 0) - V(\mathbf{q} \Rightarrow 0)X(\mathbf{q} = 0)V(\mathbf{q} \Rightarrow 0) \quad (3.43)$$

The rotational susceptibility, $\mathbf{X}(\mathbf{q} = 0)$, is determined by direct interactions and self-interactions only and is diagonal, as required by the symmetry of the C_{6v} point group. The result of calculations for the effective elastic constants is

$$C_{11} = C_{11}^0 - 9\big(X_{33}V_{13}^2 + X_{44}V_{14}^2\big) \quad (3.44)$$

$$C_{66} = C_{66}^0 - 18X_{44}V_{14}^2 \quad (3.45)$$

$$C_{11} + C_{12} = \big(C_{11}^0 + C_{12}^0\big) - 18X_{33}V_{13}^2 \quad (3.46)$$

where the rotational susceptibilities are

$$X_{33} = \big[\beta^{-1}\chi_3^{-1} + \big(J_{33}(0) + L_{33}^S\big)\big]^{-1} \quad (3.47)$$

$$X_{44} = \big[\beta^{-1}\chi_4^{-1} + \big(J_{44}(0) + L_{44}^S\big)\big]^{-1} \quad (3.48)$$

(3.48) The susceptibilities increase with decreasing temperature, consequently making the effective elastic constants smaller. The thermoelasticity makes the system softer with decreasing temperature, an effect totally due to translational–rotational coupling.

An elastic instability occurs when one of the eigenvalues of the effective dynamical matrix, $\mathbf{D}(\mathbf{q})$ (Eq. (3.42)), becomes zero. The diagonalization of the matrix for $\mathbf{q} = (q_x, 0)$ and $\mathbf{q} = (0, q_y)$ yields the same eigenvalues, and the lowest one corresponds to the C_{66} elastic constant. Thus, the elastic instability in the system will be determined by the condition, $C_{66} \Rightarrow 0$. This means that the system becomes unstable against a shear strain of the 2-D lattice, e.g., the $e_{xy} \equiv e_6$ component of the strain tensor. For the hexagonal lattice, the relation $C_{66} = (1/2)(C_{11} - C_{12})$ implies that the hexagonal lattice also becomes unstable with respect to the $(e_{xx} - e_{yy}) \equiv (e_1 - e_2)$ strain. Both strains transform according to the E_2 irreducible rerepresentation and will cause a corresponding lattice symmetry change: $C_{6v} \Rightarrow C_{2v}$. The elastic instability will take place simultaneously with the orientational instability, causing E_2 type symmetry changes as discussed above. This is due to the translational–rotational coupling effect, and, as a result, phases characterized by the orientational order parameters, as defined in Eqs. (3.23) and (3.32), will show lattice deformations measured by e_6 and $(e_1 - e_2)$ strains, respectively. They are equivalent to the orientational order parameters of E_2 symmetry.

4. FERROELASTICITY OF LANGMUIR MONOLAYERS

Ferroelasticity commonly refers to the interaction between anisotropic defects in elastic media that become correlated through the elastic interaction of a medium that is deformed by the presence of the defects. According to our model of the Langmuir monolayer, the molecular tails, characterized by orientational degrees of freedom, are "defects" grafted onto a 2-D elastic medium with translational order of head groups [3]. Ferroelasticity would therefore mean a correlation of orientations of molecular tails through the elastic interaction of the translationally ordered head groups. First, we shall illustrate this coupling by numerical calculations, then we shall introduce the concept of elastic dipoles.

4.1. Strain-State Calculations For Stearic Acid: An Illustration of Translational–Rotational Coupling

This section is devoted to illustrating, by detailed numerical calculations, the importance of translational–rotational coupling in a Langmuir monolayer of stearic acid [5]. We retain the model of orientationally free tails grafted to a 2-D net formed by the head groups of the amphiphilic molecules. Following the above microscopic derivation, the system energy can be described as

$$V(\{\varepsilon\}, \{\Omega\}) = V^R(\{\Omega\}) + V^{TR}(\{\varepsilon\}, \{\Omega\}) + V^T(\{\varepsilon\}) \quad (4.1)$$

where the first term is the contribution from the orientational degrees of freedom ($\{\Omega\}$), the third is the contribution from the translational degrees of freedom ($\{\varepsilon\}$), and the second term represents the coupling between these two types of variables. Although the appropriate dynamical orientational variables for these systems are spherical harmonics, in these calculations we consider small fluctuations, and so three orientational angles are used: $\{\Omega\} = \{R_x, R_y, R_z\}$, i.e., rotation about the orthogonal crystallographic x, y, and z axes, respectively. Thus, R_z is a rotation about the film normal and is analogous to the azimuthal angle, ϕ, when either R_x or R_y is negligible. It is convenient to describe the translational subsystem via strain variables ($\{\varepsilon\}$) appropriate to a given 2-D lattice. The translational contribution to the energy in Eq. (4.1) may be written as

$$V^T(\{\varepsilon\}) = V_H^T(\{\varepsilon\}) + V_T^T(\{\varepsilon\}) \quad (4.1a)$$

where the first term denotes contributions from the head groups, and the second denotes contributions from the tails. The translational energy contribution of the tails is crucial to the determination of the global energy minimum and to subsequent local minimizations with respect to the area. The head-group contribution is not directly calculated, but is instead attributed to a rescaling of the translational energy by a πA contribution to the total energy. Nonetheless, the translational variables remain two-dimensional, even when tail interactions are included; there are no translations allowed along the film normal. The total system may be considered as a connected stack of two-dimensional subsystems consisting of various cross sections of the film.

The orientational fluctuations are expected to follow, or quickly reach equilibrium, for a particular set of strains. The above energy may therefore be minimized with respect to Ω, where the braces indicating a set of variables have been dropped,

$$\left(\frac{\partial V(\varepsilon, \Omega)}{\partial \Omega}\right)_\varepsilon = V_c'(\varepsilon, \Omega^*) + V_\Omega'(\Omega^*) = 0 \quad (4.2)$$

and the resulting expression is solved for $\Omega^*(\varepsilon)$, the equilibrium orientation that minimizes $V(\Omega^*(\varepsilon))$ in Eq. (4.1). The solution $\Omega^*(\varepsilon)$, which is a type of stress–strain relation, shows the energetically most favorable path for reorientation of the molecules due to a two-dimensional strain. The energy along this path was investigated.

Intermolecular interactions between the tail groups are calculated as for rigid molecules, with the use of 6-exp atom–atom potentials [83]. This is a reasonable approximation for condensed phases that is supported by the fact that tilt of the molecule as a whole is shown experimentally to be preferred over conformational rearrangement for these systems [84]. Head-group and surface interactions are not directly calculated, but are not ignored. Surface interactions are implied, because the molecules are constrained to a plane [85], and head-group interactions are implicit in the value of the strain variables chosen. These may be rescaled by adding a surface pressure–area (πA) term to the potential energy [86]. With the use of geometrical parameters appropriate for a fatty acid molecule, it can be shown that the two-dimensional Gibbs free energy (πA) contribution from a surface tension of 1 mN/m is on the order of that exerted by a three-dimensional pressure of 1 kbar.

The calculative procedure consists of several steps:

1. The global minimum with respect to all strain and orientational variables is determined. All lattice parameters and orientational degrees of freedom are minimized. For example, this involves minimizing a structure with an assumed fatty acid molecular structure [87] in a two-dimensional crystal with either one or two molecules per unit cell.
2. Strain states relative to this global minimum are calculated. This is done by utilizing the dependence of the strain on the lattice parameters as described below. The states calculated depend on the symmetry of the particular system investigated.
3. The symmetries (phases) of the states from step 2 are determined. It was found in every case presented here that the phase space can be separated into various

partitions characterized by a specific symmetry, as determined by a set of orientational variables.

4. A complete set of strain states for each symmetry (phase) is calculated, meaning that the energy $V(\Omega^*(\varepsilon))$ is determined for the most important set of strain variables.

Once this procedure has been completed, the contribution to the strain state partition function [88] may be computed for each phase from step 3 according to $Q^\varepsilon = \sum_{\text{all }\varepsilon} \exp(-\beta V(\Omega^*(\varepsilon)))$. It is then straightforward to calculate the strain state contribution to the free energy for a given phase as $F^\varepsilon = -\frac{1}{\beta} \ln Q^\varepsilon$. A plot of this free energy against temperature gives an estimate of the transition temperature between phases.

Combinations of the three two-dimensional strains e_{xx}, e_{yy}, and e_{xy} are considered. The last is a pure shear strain and is denoted as ε_6, to be compatible with Voigt notation. $e_{xx} + e_{yy}$ is proportional to the change in area and is defined as ε_1. This leaves $\varepsilon_2 = e_{xx} - e_{yy}$ as the remaining strain variable. It is straightforward to express the strain variables in terms of the lattice parameters [89]. The third dimension is not relevant to the present model, and the equations reduce to

$$
\begin{aligned}
e_{xx} &= \frac{a_1 \sin \gamma_1^*}{a_0 \sin \gamma_0^*} - 1 \\
e_{yy} &= \frac{b_1}{b_0} - 1 \\
e_{xy} &= \frac{1}{2}\left[\frac{b_1 \cot \gamma_0^*}{b_0} - \frac{a_1 \cos \gamma_1^*}{a_0 \sin \gamma_0^*}\right]
\end{aligned}
\tag{4.3a}
$$

where a and b are the lengths of the lattice vectors and γ^* is the angle between the lattice vectors in reciprocal space. The subscripts 0 and 1 denote states before and after deformation, respectively.

The strain variables introduced above take the following form for a hexagonal (C_{6v}) reference cell:

$$
\begin{aligned}
\varepsilon_1 &= e_{xx} + e_{yy} = \frac{1}{a_0}\left[\frac{2}{\sqrt{3}}a_1 \sin \gamma_1^* + b_1\right] - 2 \\
\varepsilon_2 &= e_{xx} - e_{yy} = \frac{1}{a_0}\left[\frac{2}{\sqrt{3}}a_1 \sin \gamma_1^* - b_1\right] \\
\varepsilon_6 &= e_{xy} = \frac{1}{2\sqrt{3}a_0}\left[b_1 - 2a_1 \cos \gamma_1^*\right]
\end{aligned}
\tag{4.3b}
$$

For two molecules per unit cell, the reference state chosen is centered rectangular, and the strain variables are of the form

$$
\begin{aligned}
\varepsilon_1 &= e_{xx} + e_{yy} = \frac{1}{a_0}\left[a_1 \sin \gamma_1^* + b_1\right] - 2 \\
\varepsilon_2 &= e_{xx} - e_{yy} = \frac{1}{a_0}\left[a_1 \sin \gamma_1^* - b_1\right] \\
\varepsilon_6 &= e_{xy} = -\frac{1}{2}\left[\frac{a_1}{a_0} \cos \gamma_1^*\right]
\end{aligned}
\tag{4.3c}
$$

For either reference state, three variables must be specified. This is done by scanning ε_2 and ε_6 relative to the global minimum and minimizing the energy for these various states with respect to ε_1 (and the three orientational variables). In other words, the area of the minimized system provides the third necessary constraint. This is implemented by letting $a = a(\gamma^*)$ and $b = b(\gamma^*)$, according to the equations above, and by minimizing the lattice with respect to the one remaining independent variable, γ^*.

The system chosen for initial calculation is a monolayer composed of stearic acid amphiphiles in the conformation found in form C [87]. The calculation with one molecule per unit cell most clearly illustrates this method. The calculated minimum energy (Table II) is determined with respect to all degrees of freedom for an ordered lattice. A quasi-hexagonal (planar group $p6m$, point group C_{6v}) reference state may then be defined by appropriately distorting the lattice according to Eqs. (4.3b) above. This symmetry can apply to the reported high-pressure, high-temperature LS phase if one considers a vertical molecule averaged over all orientations about the film normal [3]. In a system of $p6m$ symmetry ε_1 is totally symmetric and ε_2 and ε_6 transform according to the irreducible representation E_2. The energy $V(\Omega^*(\varepsilon))$ is then identical with respect to a scan over either ε_2 or ε_6, because these orthogonal variables couple bilinearly with appropriate orthogonal spherical harmonics of identical E_2 symmetry [3]. Fatty acids in an ordered phase, however, are never of higher symmetry than C_σ, which leads to planar groups of lower symmetry. In general this necessitates consideration of a complete set of states, including ε_6 and ε_2 as well as ε_1. In practice, the lowest energy states for this system accompany changes in ε_6, and so this variable is studied in detail with ε_2 set to 0. If this reduced set contains the lowest energy states and the lowest energy path between them, it will give a good first-order prediction of the phase behavior of this system, as well as being less complicated to interpret and

Table II. Results for Calculated Phases of Stearic Acid Langmuir Film

Phase	Energy (kcal/mol)	Area/molecule (Å)	A (Å)	B (Å)	γ (degrees)
Vertical (1 mol/u.c.)	−26.31	18.28	4.303	4.581	112.00
Herringbone (*pg*)	−26.27	18.32	5.027	7.288	90.00
Tilted (1 mol/u.c.)	−26.01	19.09	4.695	5.289	129.76
Tilted antiparallel (*pm*)	−25.89	18.95	5.149	7.413	96.73

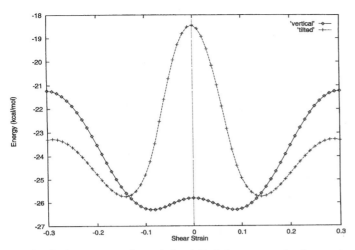

Fig. 3. Strain-state curves for vertical and tilted phases, one molecule per unit cell, plotting energy versus shear strain, ε_6. The vertical and tilted phase curves are found by fixing the tilt magnitude to 0 and 14.69°, respectively. The lower curve at every point is found when minimizing with respect to all degrees of freedom.

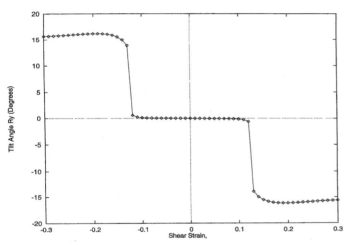

Fig. 4. Tilt angle R_y versus shear strain ε_6. The average magnitude of the two nonzero portions of the plot defines the tilted phase geometry.

much more efficient to calculate. The effect of ε_1 is included in the minimization of the area, which in practice changes only slightly along the minimum energy path.

By scanning the shear strain ε_6 over all states while minimizing the energy with respect to the orientational degrees of freedom and ε_1, we obtained the results in Figure 3. There is a sudden change in the tilt angle R_y at shear strain ≈ 0.19, from 0° to 14.69° (Fig. 4). Because R_x is nearly 0 for the entire set of states scanned (Fig. 5), R_y corresponds to the tilt angle and R_z to the azimuthal angle (Fig. 6). This allows the separation of phase space into two and the definition of a vertical and a tilted phase, with the magnitude of the tilt angle R_y fixed at 0° and 14.69°, respectively. The remaining energies for these two defined phases were then calculated; complete curves are shown in Figure 3.

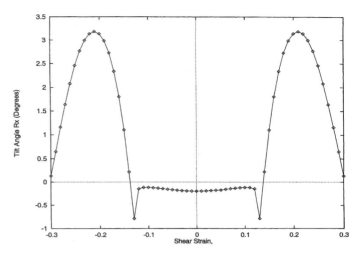

Fig. 5. Tilt angle R_x versus shear strain ε_6. Note the small magnitude of this angle.

The curves in Figure 3 were fit to a 16th-order polynomial by a χ^2 minimization. This was then used to calculate the strain state partition function as the integral $Q^\varepsilon = \int_{\text{one period}} \exp(-\beta V(\Omega^*(\varepsilon))) d\varepsilon$. Because the pure shear strain ε_6 results in a periodic function (Fig. 3), the integral is performed over one period. These functions are sufficiently simple that they may be integrated by Romberg integration [90]. These curves show that the shear elastic constant C_{66} is negative for the reference hexagonal system, and, therefore, the hexagonal phase is obviously unstable with respect to shear strain due to the coupling with the orientational fluctuations. Because the alkane chain has no more than C_σ symmetry, it cannot be accommodated in an ordered phase by a C_{6v} translational subsystem. Furthermore, any translational–rotational coupling will lead to a breaking of C_{6v} symmetry.

The one-molecule unit cell results have been introduced above. Surprisingly, tilting behavior is quite discontinuous with respect to shear strain, indicating nonlinear coupling between the tilt angle R_y and ε_6 (see Fig. 4). Because there are two phases found with respect to a scan of ε_6, vertical and tilted, these were treated separately to determine the ground state and possible transitions. There is considerably more linear coupling between R_z (the azimuthal angle) and ε_6, as shown in Figure 6. This coupling results in the continuum of states shown in Figure 3.

Free energy contributions were calculated as described above. Calculations including no numerically applied pressure indicate that the stable phase at all temperatures is the vertical phase (Fig. 7). A negative surface pressure must be introduced to produce a tilted phase. This rescaling is attributed to the effect of head-group and surface interactions that can cause the film to expand [91]. The magnitude of the surface pressure difference applied here is similar to that required to produce a vertical phase on the aqueous surface. Estimates of the transition temperature were calculated as shown in Figure 8, indicating a transition from a vertical to a tilted phase with increasing temperature [54]. Although this transition temperature

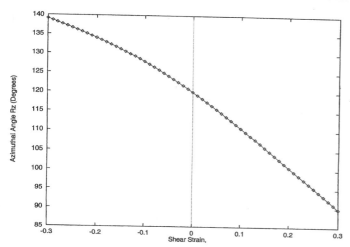

Fig. 6. Azimuthal angle R_x versus shear strain ε_6. This plot is considerably more linear than that of R_x and R_y.

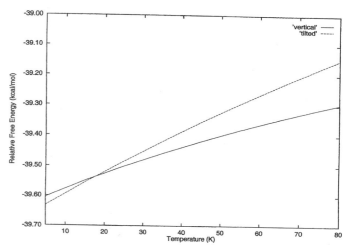

Fig. 8. Strain-state free energy contributions for vertical and tilted phases (one molecule per unit cell) at an applied surface pressure of −50. The sign of this pressure is attributed to a rescaling due to surface and head group interactions.

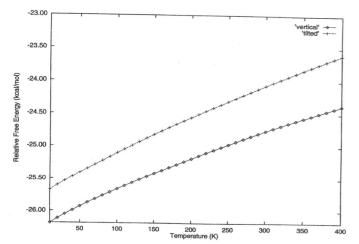

Fig. 7. Strain-state free energy contributions for vertical and tilted phases (one molecule per unit cell) at 0 applied surface pressure.

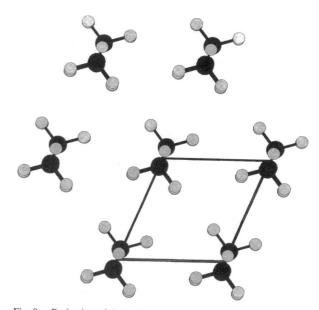

Fig. 9. Projection of the vertical phase (one molecule per unit cell).

is unobtainable in practice, it increases with decreasing applied pressure.

The calculated minimum for a two-molecule unit cell (as proposed for the CS phase [1, 84]) is essentially identical in energy to that found for a one-molecule unit cell (Table II, Figs. 9 and 10). Two phases are again found by scanning ε_6, a vertical herringbone phase of symmetry pg, and a tilted phase of symmetry pm. The herringbone pg phase contains a crystallographic glide plane between the molecules, whereas in the tilted pm phase this is replaced by a mirror plane coincident with the mirror plane containing the fatty acid chain. A twofold axis is found between the tilted molecules and parallel to their long axis, but not coincident with a crystallographic axis. The long axes of these molecules are therefore parallel, but the rotation about them differs by 180°. This phase is therefore denoted as the antiparallel phase.

A third phase of higher energy is also found that exhibits a crystallographic twofold axis between pairs of molecules of significant tilt magnitude. This nonuniformly tilted pm phase was not investigated in detail. The antiparallel pm phase is extremely similar to the original tilted phase of one molecule per unit cell (Figs. 11 and 12), as is clearly demonstrated by viewing these phases down the long axes of the molecules. The closest distance between molecular centers in the plane perpendicular to the long molecular axis is 4.11 and 4.14 Å for the tilted and antiparallel phases, respectively. The corresponding tilt angles are 74.2° and 76.6°. Because reorienting a given molecule in these two phases by 180° with respect to its surrounding molecules makes little energetic difference, it seems highly unlikely that such phases will exhibit long coherence lengths in the condensed phase.

The results from two-molecule per unit cell systems are shown in Figure 13. A clear transition between the pg her-

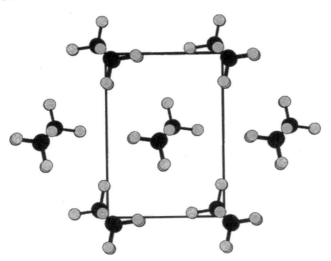

Fig. 10. Projection of the herring bone (*pg*) phase (two molecules per unit cell).

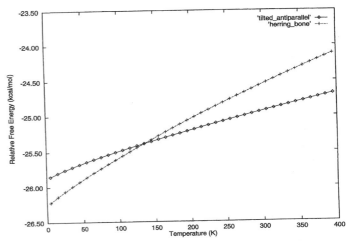

Fig. 13. Strain-state free energy contributions for herring bone (*pg*) and tilted antiparallel (*pm*) phases (two molecules per unit cell) at 0 applied surface pressure.

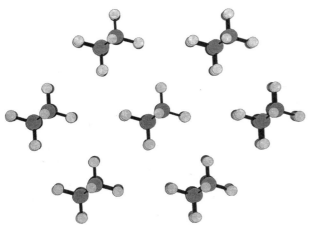

Fig. 11. Perspective projection along the long axis of the molecule of the tilted phase (one molecule per unit cell).

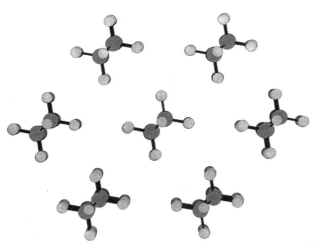

Fig. 12. Perspective projection along the long axis of the molecule of the tilted antiparallel (*pm*) phase (two molecules per unit cell). The orientation of the molecules depicted in the center row is rotated 180° about the long axis.

ringbone and the *pm* tilted antiparallel phase is found with increasing pressure. This may be compared to a CS–S′ transition [1]. The slopes of these curves are considerably different at the crossing point. This indicates a high entropy of transition (≈ 3 cal/mol·K) and, therefore, a first-order transition that is in agreement with phase coexistence reports in the literature [84].

Concluding this section, we stress that these calculations should be viewed as modeling Langmuir monolayers by a realistic potential while exploiting the underlying order and symmetry of these systems to arrive at reasonable predictions of the most likely phases. We intend to illustrate the feasibility of such an approach, which requires several extensions for quantitative agreement with experiment. These include a more detailed description of the interaction of the head groups with the substrate, especially their chemical nature and their relation to the surface pressure. Orientational disorder about the tail axis, as well as conformational disorder, should also be modeled [92, 93]. As such, conclusions from this work must still be expressed qualitatively, although they arise from potentials that are considered to be realistic.

The results for one molecule per unit cell are illustrative and clear but are not directly comparable to most π-T phase diagrams that describe ordered condensed phase unit cells as containing two molecules. Nonetheless, the high surface pressure calculations predict no transition between vertical and tilted phases with respect to temperature, and the qualitatively necessary result of tilt with decreasing surface pressure is obtained. Further data are necessary to determine the slope of the line separating vertical and tilted phases at intermediate pressures. There is nonlinear coupling between the tilt magnitude and the shear strain (Fig. 4), but this is a much more linear coupling with what, in practice, is the azimuthal angle of the amphiphile (Fig. 6). The graph of $V(\Omega^*(\varepsilon))$ in Figure 3 allows straightforward determination of the shear elastic constant C_{66} by this methodology. It also may facilitate the calculation of coupling constants involving ε_6 and orientational variables (Ω_i)

in the Landau expansion. By sequentially setting ε_i and then Ω_i to zero and performing further scans involving the remaining variables, one may quantitatively obtain all of the constants in the Landau expansion.

Systems of two molecules per unit cell exhibit a tilting transition with respect to temperature, and the considerable divergence of the two free energy contribution curves indicates that this is a first-order transition. It may be compared with the recently reported coexistence of similar phases [84]. Details of the potential are important, because the potential energy surface is, in general, quite flat with many close-lying local minima.

The curves in Figure 3 provide a unique view of fatty acid monolayers. They indicate that at low temperatures the monolayer is ferroelastic, as predicted [3]. Stress is caused by orientational fluctuations, although these calculations derive this necessary stress by first fixing the strain. Clearly, a tilted amphiphile introduces an elastic dipole, which is a local stress, into the system [5]. Domains represented by minima in the double-well potential decrease in size with increasing temperature until, finally, a phase that has average hexagonal symmetry results. This mesophase is reached when the barrier between the minima is overcome, although there remains a higher barrier to larger shear strain. At a high enough temperature, this barrier is also overcome, and a truly liquid state is obtained. Aside from this qualitative picture, this investigation of stress–strain relationships has reproduced several features of the fatty acid phase diagram. Extensions, including the chemical nature of the head groups, the disorder of the tail groups, and the explicit contributions of the other strain variables will further improve the quantitative predictions of these preliminary calculations.

This methodology may prove to be well suited to inclusion compounds such as alkane derivatives in urea, which have been analyzed in this context [94, 95]. Here the neglect of interalkane translational contributions may be justified, and the explicit calculation of contributions from the urea cage are not complicated by solvent interactions, as in the case of head groups considered here. This makes calculation of $V^{T}(\{\varepsilon\})$ straightforward. The remaining parts of the energy are similarly accessible computationally.

4.2. Elastic Dipoles as a Measure of Orientational Fluctuations

In this section the theoretical description of the translational–rotational coupling is reconsidered to introduce the concept of local stress. As shown in Section 3, Langmuir monolayer phases are characterized by both strain and orientational order parameters [3] that are related to each other through the elasticity of the system. This suggests that molecular tilt can be conveniently described by the concept of elastic dipoles embedded in an isotropic 2-D elastic medium. The elastic multipoles were originally introduced as a model for the interaction of defects and impurities in a bulk solid [96, 97]. They have been shown to be a very useful and elegant concept for orientational glasses [98, 99] and crystals with orientational disorder [66, 100]. Very recently elastic multipoles have been adopted

for a theory of solid-state reactions in a quantitative description of the mechanical characteristics of the reaction cavity [101]. The concept of elastic multipoles, which describe a force distribution around a molecule, is thus convenient for modeling molecular systems. Individual molecules can be treated as objects immersed in an isotropic elastic medium. The objects are characterized by a multipole moment that is a measure of the extent to which an object embedded in the elastic medium disturbs the medium by deviating from sphericity. The model is especially suitable for molecular systems where molecular shape [46] rather than chemical nature determines the predominant interaction and packing. It is reminiscent of the well-known "close-packing" principle commonly believed to govern the structures of simple molecular solids.

Langmuir monolayers are appropriate for modeling by elastic multipoles, and we are unaware of any other attempts to do so, although mechanical models are commonly used to mimic monolayer molecules (e.g., grafted rods). Here, we demonstrate how the elastic multipole concept models the Langmuir monolayer by representing the molecules as elastic multipoles and by taking the interaction between them as that between multipoles. The model can be extended, and molecular compressibility can be introduced, which, for the amphiphilic molecules, would be a measure of aliphatic chain flexibility. This will allow inclusion of coupling between molecular tilt and chain conformation.

Following the approach proposed in Section 2, we treat the LS phase as a hexagonal crystal-like phase, the equilibrium behavior of which is described by continuum elasticity theory. We rewrite Eq. (2.25) as a Fourier transform in the elastic limit,

$$V_i'(\mathbf{q}) = i V_{i\alpha}(\mathbf{q}) Y_\alpha(\mathbf{q}) \qquad (4.4)$$

It is possible to determine the form of the (2×5) coupling matrix $V(q)\{V_{i\alpha}\}$ from the requirement that V^{TR} must have full hexagonal symmetry and, in the elastic limit,

$$V(\mathbf{q}) = 3a \begin{pmatrix} 0 & 0 & Aq_x & Bq_x & Bq_y \\ 0 & 0 & Aq_y & -Bq_y & Bq_x \end{pmatrix} \qquad (4.5)$$

where a is the hexagonal lattice constant and A and B are coupling constants for nearest neighbor molecules located at $(0, 0)$ and $(a, 0)$. The fact that there is no translational–rotational coupling for harmonics Y_1 and Y_2 indicates that forces due to translational displacements along x and y are balanced at equilibrium by molecular torques associated with orientational fluctuations described by Y_1 and Y_2.

Equation (4.4) is now written in the form

$$V_i'(\mathbf{q}) = i N^{-\frac{1}{2}} \sum_k P_{ij}(k) q_j \exp[i\mathbf{q}\mathbf{R}(k)] \qquad (4.6)$$

where we have used the elastic limit of the translational–rotational coupling and the Fourier transform of the surface harmonics. We define

$$P_{ij}(k) = a_{ij}^\alpha A_\alpha Y_\alpha(k) \qquad (4.7)$$

as the elastic dipole of the kth molecule induced by orientational fluctuations specified by the surface harmonics. A_α equals A for $\alpha = 3$ and equals B for $\alpha = 4, 5$ and represents the translational–rotational coupling constants (see Eq. (2.17)). The a are elements of real, spherical unit tensors [97]. In writing Eq. (4.7), we have represented a force distribution around a reorienting molecule, k, by the elastic dipole.

With Eq. (4.7), we can write the translational–rotational part of the energy, Eq. (2.29), as

$$V^{TR} = i N^{-\frac{1}{2}} \sum_q u_i(\mathbf{q}) \sum_k P_{ij}(k) q_j \exp i\mathbf{q}\mathbf{R}(k) \quad (4.8)$$

with the strain tensor components defined as

$$e_{ij}(k) = \frac{1}{2}\left(\frac{\partial u_i(k)}{\partial R_j(k)} + \frac{\partial u_j(k)}{\partial R_i(k)}\right) \quad (4.9)$$

When calculated from the Fourier transform of the translational displacement, $u(\mathbf{q})$ can be used to write Eq. (4.8) in the compact form,

$$V^{TR} = \sum_k e_{ij}(k) P_{ij}(k) \quad (4.10)$$

This shows that the translational–rotational part of the energy is conveniently represented as a sum over sites of products: local strain times local stress. This gives a simple interpretation of an elastic dipole as a local stress generated by a molecular tilt.

As in elasticity theory, we define the elastic Green's function [102],

$$G_{ij}(kk') = \sum_q M_{ij}(\mathbf{q}) \exp\left(i\mathbf{q}[\mathbf{R}(k) - \mathbf{R}(k')]\right) \quad (4.11)$$

which satisfies the relation $G_{ij}(kk') M_{il}(k'k'') = \delta_{il}(kk'')$. We can express elastic displacement and local strain as

$$s_i^{el}(k) = G_{ijk}(kk') P_{jk}(k') \quad (4.12)$$

$$e_{ij}(k) = G_{ijkl}(kk') P_{kl}(k') \quad (4.13)$$

where the nonlocal response functions are corresponding derivatives of the Green's function, e.g.,

$$G_{ijkl}(kk') = \frac{\partial^2}{\partial \mathbf{R}_j(k) \partial \mathbf{R}_i(k')} G_{ik}(kk') \quad (4.14)$$

Relation (4.14) implies that the above response function has the meaning of nonlocal compliance. With this identification, we write

$$V_{ind} = \frac{1}{2} \sum_k \sum_{k'} P_{ij}(k) G_{ijkl}(kk') P_{kl}(k') \quad (4.15)$$

This indicates that the indirect rotational energy is the lattice deformation energy due to local stresses induced by the reorientation of molecules. In the equation, the summation runs over all molecular sites and includes the term $k = k'$, which is the energy of creation of an elastic dipole in the elastic medium, E. The energy is $E_C = P_{ij} S_{ijkl}^0 P_{kl}$, where $S^0 = (C^0)^{-1}$ is the bare elastic compliance tensor.

The elastic Green's function for a 2-D hexagonal (isotropic) system can be calculated following the procedure in [103] with

the result

$$G_{ij}(kk') = -[4\pi\mu(2\mu + \lambda)]^{-1}\big[(3\mu + \lambda)\delta_{ij}\ln R \\ - (\mu + \lambda)c_i c_j\big] \quad (4.16)$$

where $R = |R(k) - R(k')|/a$ and $c_i = R_i/R$. We calculate the elastic coupling between the dipoles,

$$G_{ijkl}(kk') = [4\pi\mu(2\mu + \lambda)]^{-1} R^{-2}\big\{(3\mu + \lambda)[\delta_{ik}\delta_{jl} \\ - c_j c_i \delta_{ik}] - (\mu + \lambda)[\delta_{il}\delta_{kj} + \delta_{ij}\delta_{kl} \\ - 2(\delta_{ij}c_k c_l + \delta_{kj}c_i c_l + \delta_{jl}c_i c_k + \delta_{il}c_k c_j \\ - \delta_{kl}c_i c_j) + 8c_i c_j c_k c_l]\big\} \quad (4.17)$$

It is convenient to represent the indirect interaction, Eq. (4.15), in spherical coordinates rather than Cartesian ones. We split the elastic dipoles into components, $P_{ij} = a_{ij}^\alpha P_\alpha$, where $\mathrm{P}_\alpha = A_\alpha Y_\alpha$ defines the transformed couplings between dipoles,

$$G_{\alpha\beta}(kk') = a_{ij}^\alpha G_{ijkl}(kk') a_{kl}^\beta \quad (4.18)$$

with the transformed elastic compliance,

$$S_{\alpha\beta}^0 = a_{ij}^\alpha S_{ijkl}^0(kk') a_{kl}^\beta \quad (4.19)$$

The indirect interaction is

$$V_{ind} = -\frac{1}{2} \sum_{k \neq k'} \sum P_\alpha(k)\big[N^{-1}S_{\alpha\beta}^0 + G_{\alpha\beta}(kk')\big]P_\beta(k') \quad (4.20)$$

This interaction is highly anisotropic. To see this, we consider the interaction of two elastic dipoles, separated by the vector, R, in the direction of the x axis. The result is

$$V_{ind}(\mathbf{R}||x) = -(4\mu)^{-1}\frac{2\mu + \lambda}{\mu + \lambda}\left[1 + \frac{2\mu(\mu + \lambda)}{\pi(2\mu + \lambda)^2 R^2}\right] \\ \times (A + B)^2[Y_3(0) + Y_4(0)][Y_3(R) + Y_4(R)] \\ - \frac{\lambda}{\mu + \lambda}\left[1 + \frac{(2\mu + 3\lambda)(\mu + \lambda)}{\pi(2\mu + \lambda)\lambda R^2}\right] \\ \times (A - B)^2[Y_3(0) - Y_4(0)][Y_3(R) - Y_4(R)] \\ + \left[1 + \frac{(5\mu + 3\lambda)}{4\pi(2\mu + \lambda)R^2}\right]B^2 Y_5(0) Y_5(R) \quad (4.21)$$

It is convenient to use the linear combination $[Y_3 + Y_4]/2$ for the (xx) component of the elastic dipole and $[Y_3 - Y_4]/2$ for the (yy) component of the elastic dipole. We conclude that a tilt of molecules in the direction perpendicular to the vector $\mathbf{R}$ is highly unlikely (viz. the second term), and the molecules will tend to tilt in the direction of the vector, which joins them. This supports the analogy to the interaction of electric dipoles, where a "head-to-tail" orientation is favored over a parallel arrangement of dipoles. This analogy can be further extended by considering elastic compliance as an analog of the macroscopic dielectric constant and the nonlocal compliance as corresponding to the dielectric function.

Now we can calculate the indirect potential for a system of hexagonal symmetry. The summation over interacting nearest neighbors, which are at distance R, gives

$$
\begin{aligned}
V_{\text{ind}}(R) = & -\frac{1}{2(\mu + \lambda)} \left[1 + \frac{6(\mu^2 - \lambda^2)}{\pi \mu (2\mu + \lambda) R^2} \right] A^2 \\
& \times \left[3 \cos^2 \Theta(0) - 1 \right] \left[3 \cos^2 \Theta(R) - 1 \right] \\
& - \frac{1}{2\mu} \left[1 + \frac{3(\mu - \lambda)}{4\pi (2\mu + \lambda) R} \right] B^2 \\
& \times \sin^2 \theta(0) \sin^2 \theta(R) \cos 2 \left[\phi(0) - \phi(R) \right]
\end{aligned}
$$

$$(4.22)$$

This equation gives two contributions to the total energy of a system (Eq. (4.1)). The indirect potential should be understood as an increase (decrease) in energy of the system with respect to the energy of the 2-D elastic medium, because surfactant molecules are long and are oriented at the interface. The first term in the indirect potential gives the interaction of the elastic dipole components (zz) along the long axes of molecules forming the Langmuir monolayer. Because molecules are oriented at the air–water interface, this term is nonzero and is largest for a phase with vertical molecules. The second term describes the interaction of (xx) and (yy) components of the elastic dipoles, which are related to the effective cross sections of the molecules. This term prefers the molecules to be tilted.

For $\mu = \lambda$, which corresponds to the Cauchy relation between elastic constants, and when there are only central forces between molecules, the indirect interactions in the hexagonal system are R-independent. Moreover, while the energy is independent of the tilt direction of molecules, it prefers them to be tilted, $\theta \neq 0$. For the general case, however, as long as $[(8\pi + 3) \cdot \mu + (4\pi - 3) \cdot \lambda] > 0$, the elastic dipole interaction prefers the molecules to be tilted in such a way as to maintain the smallest difference between tilted directions of neighboring molecules. This tendency may change, depending on the elastic properties of the system and, in particular, on the interrelation between the elastic constants, μ and λ. Notice that the interaction is critically dependent on the shear modulus μ and is important for soft systems. When the system reaches an elastic instability ($\mu \Rightarrow 0$), the interaction becomes extremely attractive, provided the translational–rotational coupling constants A and B, are nonzero. If the elastic instability drives the system toward melting, the couplings diminish (there is no static translational–rotational coupling in an isotropic liquid) as well, and the elastic dipole interaction does not influence the collective orientation of the molecules. However, the elastic instability, $\mu \Rightarrow 0$, might drive a transition toward a solid phase or mesophase where the coupling still exists. Then the elastic dipole interaction becomes attractive to the extent that it overcomes direct rotational interactions and leads to a collective tilt of the molecules. This creates a new phase with correlated (ordered) elastic dipoles, a ferroelastic phase.

Now, we analyze how results are modified when one considers mesophases. Instead of a system with long-range translational order where the continuum elastic limit can be used, we deal with a system with a mesoscopic range of the translational order. Following an approach adopted for the glassy

state, we use the microscopic Hamiltonian from Section 2 as a guideline for a phenomenological coarse-graining procedure. There are reasons why such a phenomenological procedure is needed. A mesophase is characterized by large fluctuations predominantly in translational displacements on microscopic length scales. Therefore, standard procedures for extracting elastic properties (e.g., by gradient expansions) cannot be used here. Second, the reorientation of a molecule in a translationally disordered system is a strongly anharmonic process on microscopic length scales. It is hard to estimate it from first principles.

In a mesophase we consider ranges of thermodynamic parameters for regions where the correlation length x of thermal orientational fluctuations is much larger than the NN distance, i.e., $x \gg a$. A coarse-graining procedure should lead to an effective Hamiltonian, H_{eff}, for long-wavelength fluctuations. The form of H_{eff} is restricted by the following simplifying assumptions: (i) The range of all direct intermolecular forces is short compared with the mesoscopic coarse-graining length ξ_0 ($a < \xi_0 < \xi$), with the possible exception of direct interactions between molecules due to electrical multipoles. This assumption is well satisfied for Langmuir monolayers where the predominant interaction is between aliphatic chains and is of short range. As a consequence of this assumption, H_{eff} should consist of terms and their derivatives that are local in a coarse-grained displacement field, $u(x)$. (ii) On mesoscopic length scales, the system behaves like an elastic medium, so that H_{eff} is at most quadratic in $u(x)$. This does not imply the harmonic approximation at shorter length scales, where a system with translational disorder, such as a mesophase, is anharmonic in molecular displacements. (iii) The coarse-graining procedure divides the system into mesoscopic grains of size x. Within a grain, the system is considered to possess hexagonal, C_{6v} point group symmetry.

Given these three simplifying assumptions, the simplest effective Hamiltonian takes the form

$$
\begin{aligned}
H_{\text{eff}} = & \sum_x \left(\frac{1}{2} e_{ij}(x) C^0_{ijkl} e_{kl}(x) + P_{ij} \left[\Omega(x) \right] e_{ij}(x) \right) \\
& + \frac{1}{2} \sum_x \sum_{x'} V^R \left[\Omega(x), \Omega(x') \right]
\end{aligned}
$$

$$(4.23)$$

The summation extends over the points of a coarse-grained 2-D net with hexagonal, C_{6v} point group symmetry. The three terms can be identified as corresponding to the microscopic terms. In particular, the second term corresponds to Eq. (3.11). The stress on the mesoscopic length scale, $P[\Omega(x)]$, depends on the orientation of molecules within the grain x and is expressed in terms of spherical harmonics. The elastic constants are those on the mesoscopic coarse-graining scale. The direct rotational potential can be decomposed into a potential that denotes local orientational anisotropy, $V^R_0[\Omega(x)]$, and couplings $x \neq x'$. The coupling between grains will, in general contain contributions from direct electrical multipole interactions as well as indirect interactions mediated by lattice distortions on the short length scales, $\lambda < \xi_0$. For Langmuir monolayers, the coupling in the effective Hamiltonian (Eq. (4.1)) is essentially the indirect interaction, V_{ind}, found in the previous section.

Thermodynamic properties of the model can be obtained from the partition function with the Hamiltonian, H_{eff}, by integrating over $e(x)$ and $\Omega(x)$. It is convenient to decompose the strain into a homogeneous part, e, and an inhomogeneous one, $e(x) = \epsilon + \delta e(x)$. Minimization of the partition function with respect to homogeneous strain gives

$$N^{-1} \sum_x \langle P_{ij}(x) \rangle = C^0_{ijkl} \langle \varepsilon_{kl} \rangle \qquad (4.24)$$

where, as follows from Eq. (3.8), $\langle P_{ij}(x) \rangle = a^\alpha_{ij} A_a \langle Y_a(x) \rangle$. The orientationally ordered phase ($\langle Y_\alpha \rangle = 0$) is always accompanied by a homogeneous deformation, which is characteristic of ferroelasticity. This is a deformation of the 2-D elastic medium formed by the head groups, and Eq. (4.24) tells us that Langmuir monolayers with orientationally ordered tails are always strained 2-D systems. In particular, for the phase with vertical tails, $\langle Y_\alpha \rangle \neq 0$, and consequently from Eq. (4.24), the totally symmetric strain ($e_{11} + e_{22}) \neq 0$. If this phase serves as a reference for a Landau free energy expansion, $\langle Y_\alpha \rangle$ should not be considered as the orientational order parameter. However, if the reference state is the 2-D elastic medium of the molecular head groups, Eq. (4.24) gives the linear relation between the orientational order parameter $\langle Y_\alpha \rangle$ and macroscopic strain.

If orientations within grains are frozen into random directions such as $\sum_x \langle P_{ij}(x) \rangle$, which corresponds to fluidized molecular tails whose head groups are frozen in a crystalline arrangement, the macroscopic deformation vanishes. But there are randomly frozen inhomogeneous strains, $\delta e(x)$, within every grain. This can be found by minimization of the partition function with respect to displacement, $u(x)$, yielding

$$\langle u_i(x) \rangle = \sum_{x'} G_{ij}(xx') \left(\frac{\partial \langle P_{jk}(x') \rangle}{\partial x'_k} \right) \qquad (4.25)$$

and inhomogeneous strain can be calculated from Eq. (4.9). The elastic Green's function is that calculated above, because the elastic 2-D medium is the crystalline arrangement of head groups, and the derivative of the elastic dipole represents an average force induced by tilted molecules within grain x', acting on the grain at x. The inhomogeneous strain vanishes when averaged over all grains of the system. Langmuir monolayers with frozen inhomogeneous strains are, thus, reminiscent of an orientational glassy state.

4.3. Elastic Domains in Mesophases

The mesophase, LS, also called the rotational phase, of global hexagonal structure with, on average, vertical molecules is assumed to be a mixture of mesoscopic local structures characterized by strains or elastic dipoles related to local structural stresses. This view is consistent with the absence of long-range translational order in 2-D systems where local equilibria compete with a global one (as indicated by kinetic processes observed in the systems), as well as with observations that mesophases are more crystalline than hexatic. Moreover, there

is growing evidence that microscopic properties of monolayers are different from their macroscopic properties, as clearly shown by their elasticity [31]. Obviously, for any renormalization of physical properties, one has to treat the mixture of domains as an interacting system [5, 6].

Numerical calculations for Langmuir monolayer structural optimization have already suggested that a mesophase can be considered as a mixture of domains [5, 6]. The potential energy landscape for a system with vertical molecules has minima corresponding to elastic domains with small energy barriers for reorientation [5]. Moreover, the calculated energy minima for a two-molecule unit cell (herringbone structure) is essentially identical to that found for a one-molecule unit cell, thereby making discrimination between these structures impossible. Results for the monolayers' structural minimization (in particular Figs. 7 and 8 of [5]), can be taken as optimum local structures of rectangular and oblique domains. Grafting these structures onto a global, hexagonal net, as in Figure 2, serves as a schematic illustration of our model.

We consider three types of domains that are compatible with a 2-D hexagonal net. These are characterized by local stress tensors or, equivalently, the elastic dipoles arising from orientational fluctuations as discussed above. The elastic dipole is represented as

$$\mathbf{P} = P_s + P_2 \mathbf{a}_2 + P_6 \mathbf{a}_6 \qquad (4.26)$$

where

$$P_s = 1/2(P_{xx} + P_{yy}) \qquad (4.27)$$
$$P_2 = 1/2(P_{xx} - P_{yy}) \equiv \langle Y_4 \rangle \qquad (4.28)$$
$$P_6 = P_{xy} \equiv \langle Y_5 \rangle \qquad (4.29)$$

and the unit matrices are

$$\mathbf{a}_2 = \begin{pmatrix} 1 & 0 \\ 0 & -1 \end{pmatrix} \qquad \mathbf{a}_6 = \begin{pmatrix} 0 & 1 \\ 1 & 0 \end{pmatrix} \qquad (4.30)$$

As the domains are determined by symmetry elements lost upon ordering, the local stress tensors for other domains are generated by applying six-fold and three-fold axis symmetry operations, e.g.,

$$\mathbf{P}^{(2)} = \mathbf{C}_6 \mathbf{P}^{(1)} \mathbf{C}_6^T \quad \text{and} \quad \mathbf{P}^{(3)} = \mathbf{C}_3 \mathbf{P}^{(1)} \mathbf{C}_3^T \qquad (4.31)$$

Three types of domains are considered because of the same symmetry of rectangular and oblique deformations, but every domain is, in principle, a combination of a totally symmetric strain (change in the surface of a unit cell) and rectangular and oblique deformations. Having specified the mesoscopic domains by elastic dipoles we must find fluctuations in their density.

The density of domains at point $\mathbf{x}$ in the system is taken as $\rho_0(\mathbf{x})$. $\mathbf{x}$ locates a center of local stress that is characteristic for a given domain. The mesophase, which is considered as a mixture of domains, is assumed to be in a global thermodynamic equilibrium (stress-free state) [96]. At equilibrium the density of domains of type α is

$$\rho_\alpha(\mathbf{x}) = \rho_0(\mathbf{x})\left[\exp(-\beta E_\alpha(\mathbf{x}))/V^{-1}\right.$$
$$\left.\times \int dV \sum_\alpha \exp\left(-\beta E_\alpha(\mathbf{x})\right)\right] \quad (4.32)$$

where V is the volume of the system and $E_\alpha(\mathbf{x})$ is the energy of interaction of the domain in orientation α with a surrounding field formed by a distribution of domains. The energy is assumed to be purely elastic,

$$E_\alpha(\mathbf{x}) = -\mathbf{P}^\alpha(\mathbf{x})\varepsilon(\mathbf{x}) = -P_{ij}^\alpha(\mathbf{x})\varepsilon_{ij}(\mathbf{x}) \quad (4.33)$$

where $\varepsilon(\mathbf{x})$ is the effective strain acting at site $\mathbf{x}$. $\varepsilon(\mathbf{x})$ is a sum of macroscopic uniform strain, ε, and the local, mesoscopic strain, $\delta\varepsilon(\mathbf{x})$, due to surrounding mesoscopic elastic dipoles. Thus,

$$\varepsilon_{ij}(\mathbf{x}) = \varepsilon_{ij} + \delta\varepsilon_{ij}(\mathbf{x}) \quad (4.34)$$

where

$$\delta\varepsilon_{ij}(\mathbf{x}) = \sum_{x \neq x'} K_{ijlm}(\mathbf{x}, \mathbf{x}')p_{lm}(\mathbf{x}') \quad (4.35)$$

The elastic strain field, $K_{ijlm}(\mathbf{x}, \mathbf{x}')$, is just a correlation function for the mesoscopic strains in the uniform, disordered system,

$$K_{ijlm}(\mathbf{x}, \mathbf{x}') = \beta\langle\delta\varepsilon_{ij}(\mathbf{x})\delta\varepsilon_{lm}(\mathbf{x}')\rangle_{\text{unif}} \quad (4.36)$$

The density of elastic dipoles at site $\mathbf{x}'$ is

$$p_{lm}(\mathbf{x}') = \sum_\alpha \delta\rho_\alpha(\mathbf{x}')P_{lm}^\alpha(\mathbf{x}') \quad (4.37)$$

The fluctuation in the elastic dipole density is

$$\delta\rho_\alpha(\mathbf{x}) = \rho_\alpha(\mathbf{x}) - \rho_0(\mathbf{x}) \quad (4.38)$$

and is the key parameter that determines local order. It may be viewed as equivalent to orientational probabilities, p_i. The difference is, however, that the elastic dipole density fluctuation measures an ordering of mesoscopic domains due to elastic interaction, while the orientational probabilities were determined for a homogeneous system (within a domain). Disorder in the inhomogeneous system means that $\delta\rho_\alpha(\mathbf{x}) = 0$ at every site, whereas in a state with order, $\delta\rho_\alpha(\mathbf{x}) \neq 0$.

The assumption that the elastic energy is less than the thermal energy can be described by making the external strain sufficiently small (see Eq. (4.33)). With this approximation and the global equilibrium condition (Eq. (4.32)), we find the density fluctuation,

$$\delta\rho_\alpha(\mathbf{x}) = \beta\left[\rho_0(\mathbf{x})/3\right]P_{ij}^\alpha(\mathbf{x})\varepsilon_{ij}(\mathbf{x}) \quad (4.39)$$

This relation is valid for a stress-free system, and, in the case of Langmuir monolayers, the global equilibrium will appear after a system relaxes to such a state. This process is known to be rather slow, and in our considerations it will be taken as the reference state.

If an experiment on Langmuir monolayers is performed without allowing the system to relax to global equilibrium, one

has to use the condition for local equilibrium. This yields the following result for the density fluctuation:

$$\delta\rho_\alpha(\mathbf{x}) = -\beta(\rho_0(\mathbf{x})/3)\left[P_{ij}^\alpha(\mathbf{x}) - (1/3)\sum_\alpha P_{ij}^\alpha(\mathbf{x})\right]\varepsilon_{ij}(\mathbf{x}) \quad (4.40)$$

This shows that when a system is in local thermodynamic equilibrium (within a domain), the density fluctuation of local stress is determined by the deviatoric part of the elastic dipole only. In further discussion, we assume global equilibrium, although the same considerations can be performed for the local equilibrium situation.

If an experiment on Langmuir monolayers is performed without allowing the system to relax to global equilibrium, one has to use the condition for local equilibrium. This yields the following result for the density fluctuation:

$$\delta\rho_\alpha(\mathbf{x}) = -(\beta\rho_0(\mathbf{x})/3)\left[P_{ij}^\alpha(\mathbf{x}) - (1/3)\sum_\alpha P_{ij}^\alpha(\mathbf{x})\right]\varepsilon_{ij}(\mathbf{x}) \quad (4.40a)$$

This shows that when a system is in local thermodynamic equilibrium (within a domain), the density fluctuation of local stress is determined by the deviatoric part of the elastic dipole only. In further discussion, we assume global equilibrium, although the same considerations can be performed for the local equilibrium situation.

4.4. Elastic Dipole Density Correlation

The density fluctuation of the elastic dipoles shows (see Eq. (4.40)) a self-consistent dependence via the local strain field coupled with the elastic dipole density. The key quantity that couples the fluctuations is the elastic strain correlation function, $K_{ijlm}(\mathbf{x}, \mathbf{x}')$, considered as a nonlocal compressibility tensor. A discussion of the function is given in the Appendix, and we treat it as known. Moreover, we assume the limit of large distances between domains and drop the site indices.

With the introduction of Eqs. (4.34), (4.35), and (4.36) into (4.39), the density fluctuation of elastic dipoles is given by the formula

$$\delta\rho_\alpha = \beta(\rho_0/3)\left[1 - \beta(\rho_0/3)\mathbf{J}\right]_{\alpha\beta}^{-1}P_{ij}^\beta\varepsilon_{ij} \quad (4.41)$$

where

$$J_{\alpha\beta} = P_{ij}^\alpha K_{ijlm}P_{lm}^\beta \quad (4.42)$$

is elastic energy of interaction between domains of types α and β. The matrix is composed of values $a = J_{\alpha\alpha}$ and $b = J_{\alpha\beta}(\alpha \neq \beta)$. The correlation function for the elastic dipoles' density fluctuations can be deduced from Eq. (4.42). The susceptibility is

$$\chi_{\alpha\beta} = \beta\langle\delta\rho_\alpha\delta\rho_\beta\rangle = \beta\rho_0/3\left[1 - \beta\rho_0/3\mathbf{J}\right]_{\alpha\beta}^{-1} \quad (4.43)$$

The symmetry of the susceptibility matrix implies that there will be three eigenvalues,

$$\chi_A = (\rho_0/3)\left[\beta^{-1} - \beta_c^{-1}(A_g)\right]^{-1} \quad (4.44)$$

for the totally symmetric representation and

$$\chi_E = (\rho_0/3)\big[\beta^{-1} - \beta_c^{-1}(E_2)\big]^{-1} \qquad (4.45)$$

for the doubly degenerate representation, E_2. The eigenvalues can be conveniently expressed in terms of critical temperatures,

$$\beta_c^{-1}(A_g) = (\rho_0/3)(a + 2b) \qquad (4.46)$$

$$\beta_c^{-1}(E_2) = (\rho_0/3)(a - b) \qquad (4.47)$$

which determine critical points for ordering of elastic dipoles of the corresponding symmetries. At the temperatures given by Eqs. (4.46) and (4.47), the compressibility and shear compliance of a system become infinite. The latter instability is related to ferroelastic ordering.

Now assume that the totally symmetric part of the elastic dipole is zero. Taking into account the symmetries of the elastic dipoles for the three domains, the elastic interaction energies are calculated (see Appendix) as

$$\begin{aligned} a &= (2\pi\mu_0)^{-1} R^{-2} \big(P_2^2 + P_6^2\big); \\ b &= -(4\pi\mu_0)^{-1} R^{-2} \big(P_2^2 + P_6^2\big) \end{aligned} \qquad (4.48)$$

where μ_0 is the shear elastic constant of the system in the disordered state. This result confirms our intuition that elastic domains in the same orientation repel each other and prefer orthogonal orientations in their neighborhood. Because of specific features of the elastic interaction for a 2-D net, the critical temperature for an ordering of elastic domains, Eq. (4.47), is distance dependent. Ordering, therefore, appears as a continuous process with decreasing temperature and with a corresponding increase of size of uniform domains. In terms of the E_2 eigenvalue, the density fluctuation of the elastic dipole type α is

$$\delta\rho_\alpha = \chi_E(R) P_{ij}^\alpha \varepsilon_{ij} \qquad (4.49)$$

which is also distance dependent in the sense as discussed above. The susceptibility shows an algebraic decay with the distance.

4.5. Ferroelasticity in a Quasi-Two-Dimensional System

As evident from the above discussion, Langmuir monolayers are excellent candidates for ferroelastic behavior. By this we mean a strain reversal process, which can be driven by a change in orientation of the molecular tails. The ferroelastic phase is characterized, therefore, by an average strain, $\langle\varepsilon\rangle$, and average orientational fluctuations, $\langle Y_\alpha\rangle$. These can be considered as the order parameters for the system if the reference state is the 2-D elastic medium of the head groups. Because of the strain-molecular reorientation coupling, the bare elastic constants, C_{ij}^0, of the medium are renormalized to their observed values, C_{ij}. The elastic compliance, $S = (C)^{-1}$, defined as the matrix of second derivatives of the free energy with respect to externally applied homogeneous stress, is

$$S_{ijkl} = \beta N^{-1} \big[\langle\varepsilon_{ij}\varepsilon_{kl}\rangle - \langle\varepsilon_{ij}\rangle\langle\varepsilon_{kl}\rangle\big] \qquad (4.50)$$

The thermodynamic averages are calculated with H_{eff} Eq. (4.23), with the use of the decomposition of strain into homogeneous and inhomogeneous parts and a shift of variables, $\varepsilon' = \varepsilon + (C^0)^{-1}\sum_x P(x)$. The result is

$$S = S^0 + [S^0 \mathbf{a}^\alpha A_\alpha]^T X_{\alpha\beta}[S^0 \mathbf{a}^\beta A_\beta] \qquad (4.51)$$

where S^0 is the bare compliance tensor and $\mathbf{a}^\alpha$ is the real spherical unit matrix. The rotational susceptibility is

$$X_{\alpha\beta} = N^{-1} \sum_x \sum_{x'} X_{\alpha\beta}(x, x') \qquad (4.52)$$

where

$$X_{\alpha\beta}(x, x') = \beta\langle Y_\alpha(x) Y_\beta(x')\rangle \qquad (4.53)$$

The susceptibility in Eq. (4.52) has to be calculated for every system for which the rotational part of the system Hamiltonian has been specified. This has been discussed in detail [66].

Now we consider the simplest approximation for the rotational susceptibility in two limits, $x = x'$ and $(x - x') \Rightarrow \infty$, to get a rough estimation of the ferroelasticity. Neglecting the direct rotational potential, the on-site susceptibility is simply $(x - x') \Rightarrow \infty$. In the limit of an infinite distance, the interaction of two orientational fluctuations is equal to the deformation energy. We write for the approximate rotational susceptibility of the system

$$X_{\alpha\alpha} = \big[\beta^{-1} - A_\alpha^2 S_{\alpha\alpha}^0\big]^{-1} \qquad (4.54)$$

As a property of the system, the susceptibility has to be invariant with respect to hexagonal symmetry. Thus, there are only diagonal elements in this representation. $S_{\alpha\alpha}^0$ is the elastic compliance expressed in coordinates of spherical harmonics (see Eq. (4.20)). Although it represents an oversimplified version of the rotational susceptibility, it provides a plausible interpretation of an incipient instability. For temperatures at which the thermal energy is greater than the elastic dipole deformation energy, the susceptibility is positive and finite, indicating that the system is stable with respect to orientational fluctuations described by Y_α. When the temperature decreases, however, the deformation energy overcomes the thermal energy and the system becomes orientationally unstable, $X_{\alpha\alpha} \Rightarrow \infty$. As long as there is sufficient thermal energy to create elastic dipoles, the system is stable. If, however, the thermal energy is insufficient to form the elastic dipoles, the system tends toward a structural change to gain enough energy for their creation (reoriented molecules). This structural change is driven by an elastic instability, as seen from the equation

$$S_{\alpha\alpha} = S_{\alpha\alpha}^0 \left[1 + \left\{\frac{k_B T}{E_\alpha} - 1\right\}^{-1}\right] \qquad (4.55)$$

where $E_\alpha = A_\alpha^0 S_{\alpha\alpha}^0$ represents the energy of deformation due to the interaction of two elastic dipoles of type α. In particular, we seek a renormalization of the shear modulus,

$$\mu = \mu^0 \big[1 + \{\mu^0 k_B T A_4^{-2} - 1\}^{-1}\big]^{-1} \qquad (4.56)$$

For $k_B T = B^2 (\mu^0)^{-1}$, μ vanishes, leading to a distorted system in which there is nonzero strain and the molecular tilt is expressed in terms of elastic dipoles,

$$\langle e_{12} \rangle = -(2\mu^0)^{-1} \langle P_{12} \rangle \propto \langle \sin 2\phi \rangle \quad (4.57)$$

$$\langle e_{11} - e_{22} \rangle = -(2\mu^0)^{-1} \langle P_{11} - P_{22} \rangle \propto \langle \cos 2\phi \rangle \quad (4.58)$$

The distorted hexagonal structures measured by the strains given by Eqs. (4.57) and (4.58) represent oblique and rectangular lattices, respectively. The sign of strains is determined by the microscopic translational–rotational constant, B, which is determined by the microscopic nature of intermolecular interactions and by the tilt angle, ϕ. For positive elastic dipoles, the deformation is reflected in contraction (negative strain) and vice versa.

The essence of ferroelasticity is that one can stimulate a strain reversal by changing the orientation of molecules. This implies that the opposite process is also possible. The strain reversal as a function of azimuthal tilt angle is seen from the angular dependence of the elastic dipoles. Phases with tilt toward NN and NNN are ferroelastic domains within a phase, which results from an orientational instability against Y fluctuations. It follows that, for example, the "swiveling" transition between NN and NNN phases [104] proceeds through the unstable state of vertical molecules on the hexagonal net. This mechanism allows the molecules to rotate around the long axes in the process, which is what has been observed [104]. Moreover, because the two strains, $\langle e_{12} \rangle$ and $\langle e_{11} - e_{22} \rangle$, are induced by the same elastic instability, a tilted phase may, in general, be a mixture of all of those domains if there is no external (uniaxial) stress applied to the system. Under the usual experimental conditions for Langmuir monolayers, this is difficult to achieve. Therefore, some preference for the existence of the ferroelastic domains will be forced. With the concept of elastic dipoles, this can be analyzed for a particular experiment.

The above analysis shows that it is both useful and helpful to model tilted molecules in Langmuir monolayers by elastic dipoles. The elastic dipoles model the orientational degrees of freedom of the molecular tails and are local stresses expressed by expectation values of spherical harmonics with $l = 2$, and their strength is given by translational–rotational coupling constants. The stresses interact with each other through the elastic field of the monolayer, which is a 2-D elastic medium of molecular head groups arranged on the water surface. The deformation energy is an indirect interaction between orientational fluctuations.

The interaction of reorienting molecules can be modeled by the interaction of elastic dipoles. The potential, with parameters directly dependent on the elastic constants, has been obtained for the 2-D hexagonal lattice. This may prove useful for computer experiments on these systems. The concept of elastic dipoles also allows representation of an amphiphilic molecule as a set of segments where an elastic dipole can be assigned to each. This approach allows modeling of intermolecular forces as interactions of segments-elastic dipoles, thereby including the molecular compressibility as well as the elastic properties of the system.

When the elastic dipoles' interaction is strong enough, they will order, leading to an orientational order of the molecules. The direct relation between the tilt and strain suggests that the ordered phase is ferroelastic where domains with different orientations are characterized by different strains. In particular, the angular distributions in terms of the tilt-azimuthal angle, ϕ, of the macroscopic shear strain, e_{12}, have the symmetry of the Y_5 harmonics, and that for $(e_{11} - e_{22})$ is of Y_4 symmetry. We conclude that the Langmuir monolayer phases with tilts toward NN and NNN are ferroelastic domains and that a tilted phase can in general be a mixture of such domains. Within this concept, the "swiveling" transition between NN and NNN phases is seen as a strain reversal process, which goes through the undeformed hexagonal lattice. This result strongly suggests that monolayer textures may be characterized quantitatively and, on a microscopic scale, by models based upon elastic multipoles and their mediation by the film itself.

These results demonstrate the efficacy of applying an analysis based on solid-state theory to understanding the phase behavior of molecular monolayers. In particular, the approach permits direct identification of processes that occur in Langmuir films with well-known phenomena observed in solids. The elastic properties of monolayers provide an extensive arena for investigating these commonalities and, coupled with an appropriate design of the amphiphiles based on these models, may lead to novel films with desirable material properties.

5. FREE ENERGY AND ORDER PARAMETERS

5.1. Generalized Free Energy Expansion

From the total Hamiltonian of the system (Eq. (2.37)), the free energy can be written as a sum,

$$F = F_0 + F^{RR} + F^{TR} + F^T \quad (5.1)$$

Here, F_0 is the single-molecule orientational free energy per molecule,

$$F_0 = -N^{-1}\beta^{-1} \ln Z_R. \quad (5.2)$$

where the partition function, Z_R, is calculated classically with the effective single-molecule orientational potential given by Eq. (3.7). The other parts of the free energy can be obtained by replacing the variables, $\mathbf{u}(\mathbf{q})$ and $\mathbf{Y}(\mathbf{q})$, in Eqs. (2.40)–(2.42) with their instantaneous thermal expectation values [77]. The result is

$$F = F_0 + \frac{1}{2}\sum_{\mathbf{q}} \{ \langle \mathbf{u}(\mathbf{q}) \rangle M(\mathbf{q}) \langle \mathbf{u}(-\mathbf{q}) \rangle$$
$$+ 2\langle \mathbf{u}(\mathbf{q}) \rangle V(\mathbf{q}) \langle \mathbf{Y}(-\mathbf{q}) \rangle$$
$$+ \langle \mathbf{Y}(\mathbf{q}) \rangle [(X^0)^{-1} + J(\mathbf{q}) + L^S] \langle \mathbf{Y}(-\mathbf{q}) \rangle \} \quad (5.3)$$

Recall that $\mathbf{X}^0$ is the single-molecule orientational susceptibility and $\mathbf{J}(\mathbf{q})$ is the direct rotational–rotational coupling matrix.

The average, $\langle \mathbf{Y}(\mathbf{q}) \rangle$, represents a vector of orientational order parameters.

For a given orientational configuration, $\langle Y(q) \rangle$, we minimize the free energy with respect to the displacements, $\langle \mathbf{u}(\mathbf{q}) \rangle$, and obtain

$$\langle \mathbf{u}(\mathbf{q}) \rangle = -M^{-1}(\mathbf{q}) V(-\mathbf{q}) \langle \mathbf{Y}(\mathbf{q}) \rangle \tag{5.4}$$

Substituting back into Eq. (3.51) yields

$$F = F_0 + \frac{1}{2} N^{-1} \sum_{\mathbf{q}} \langle \mathbf{Y}(\mathbf{q}) \rangle X^{-1}(\mathbf{q}) \langle \mathbf{Y}(-\mathbf{q}) \rangle \tag{5.5}$$

where $\mathbf{X}(\mathbf{q})$ is the coupled orientational susceptibility matrix for the system defined by Eqs. (3.9)–(3.14). Equation (5.5) gives the most general result: the free energy of the system is expressed in terms of primary orientational order parameters, $\langle \mathbf{Y}(\mathbf{q}) \rangle$. These order parameters are more general than those used in current theoretical developments. In particular, the statistical average of the Fourier transform of the spherical harmonics is equivalent to a product of the tilt order parameter and the so-called weak crystallization order parameter [62–64]. This is clear when one remembers that the set of $\mathbf{q}$ vectors is determined by the extent of crystallinity defined by the correlation lengths (see the Appendix). Equation (5.4) gives translational displacements (as a measure of lattice distortion) that accompany the change in orientation of molecules, e.g., it gives the elastic response of the lattice to the primary order parameter. Instabilities were analyzed using the condition $\det |X^{-1}(\mathbf{q})| = 0$, because the microscopic nature of intermolecular interactions and mutual competition is contained in this quantity.

Finally, we derive the free-energy contribution in the mesoscopic limit ($\mathbf{q} \Rightarrow 0$). The corresponding parts of the free energy are

$$F^{RR} = \frac{1}{2} \langle Y_\alpha \rangle \left[X^{-1}(0) \right]_{\alpha\beta} \langle Y_\beta \rangle \tag{5.6}$$

$$F^{TR} = 3 \{ V_{13} \langle Y_3 \rangle (e_1 + e_2) + V_{14} \left[\langle Y_4 \rangle (e_1 - e_2) + \langle Y_5 \rangle e_6 \right] \} \tag{5.7}$$

$$F^{T} = \frac{1}{2} \left[\frac{1}{2} (C_{11}^0 + C_{12}^0)(e_1 + e_2)^2 + \frac{1}{2} (C_{11}^0 - C_{12}^0)(e_1 - e_2)^2 + C_{66}^0 e_6^2 \right] \tag{5.8}$$

The minimization of the free energy given by the above equations with respect to the strain components yields macroscopic strain as a response to the orientational order parameters,

$$(e_1 + e_2) = -6 (C_{11}^0 + C_{12}^0)^{-1} V_{13} \langle Y_3 \rangle \tag{5.9}$$

$$(e_1 - e_2) = -6 (C_{11}^0 - C_{12}^0)^{-1} V_{14} \langle Y_4 \rangle \tag{5.10}$$

$$e_6 = -3 (C_{66}^0)^{-1} V_{14} \langle Y_5 \rangle \tag{5.11}$$

Note that Eqs. (5.9) and (5.10) are coupled as the averages; $\langle Y_3 \rangle$ and $\langle Y_4 \rangle$ are coupled in the "hybridized" order parameter, Eq. (3.32). Decoupling leads to the relations $e_1 \propto \eta^x(A_1, E_2)$ and $e_2 \propto \eta^y(A_1, E_2)$. Moreover, Eq. (5.9) gives important

information on the change of the area of the 2-D unit cell, $\Delta A = A_0(e_1 + e_2)$, where A_0 is the area of the unit cell in the high-symmetry (C_{6v}) phase. In analogy to Eqs. (5.9)–(5.11), one can derive qualitative relations between the effective cross section of a rigid rod and the orientational order parameters of a second-rank tensor. This analogy is used in representing the effective molecular shapes in the phase diagram (Fig. 1).

Introducing Eqs. (5.9)–(5.11) into (5.7) and (5.8), we obtain the free energy in terms of the orientational order parameters, $\langle Y \rangle$, which characterize phases of the system without translational symmetry change. In the presence of an external 2-D pressure, π, we consider the Gibbs free energy per molecule,

$$G = F \left[e, \langle \mathbf{Y} \rangle \right] + \pi(e_1 + e_2) \tag{5.12}$$

where the first term represents the free energy expressed by Eq. (5.6). The equilibrium condition, Eq. (5.9), changes to

$$\pi = -(C_{11}^0 + C_{12}^0)^{-1} (e_1 + e_2) - 6 V_{13} \langle Y_3 \rangle \tag{5.13}$$

This serves as an equation of state for the system. It reflects the physical behavior of the system because the externally applied pressure (a stress) comprises a macroscopic stress due to the bare elasticity of the lattice (the first term on the right-hand side) and local stress due to reorientation of the molecules. The second term forms the diagonal, isotropic part of the local stress tensor, conveniently expressed as an elastic dipole tensor. The equation of state shows that only totally symmetric reorientations of the molecules, measured by $\langle Y_3 \rangle$, are 2-D pressure dependent. Therefore, only transitions to phases characterized by $\langle Y_3 \rangle \neq 0$, e.g., due to instability with respect to hybridized (A_1, E_2) orientational fluctuations, will depend on the pressure. The conclusion is that phase transitions driven by E_1- and E_2-type tilt instabilities are weakly pressure dependent. When a nonuniform pressure is applied, it couples to e_1 and e_2 separately, and the local stress is due to reorientations expressed by the order parameters $\eta^x(A_1, E_2)$ and $\eta^y(A_1, E_2)$, respectively.

This approach to Langmuir monolayers from a solid-state point of view has been made to understand the microscopic mechanisms that cause the richness of the phase behavior of these systems. We have attempted to provide a uniform description of different phases through a logical choice of order parameters.

The key variables in our theory are symmetry-adapted surface harmonics that offer a guideline to yet unresolved questions: (i) What are the intermolecular interactions related to a possible phase (instability) transition? (ii) How should the phases of Langmuir monolayer systems be characterized? Starting from the highest symmetry phase, the LS phase of C_{6v} point group symmetry, we have analyzed orientational instabilities that can cause different phase transitions. The instabilities are possible responses of the model system of an assumed symmetry. We have shown that the instabilities taking place depend, however, on the details of the intermolecular interactions.

5.2. Natural Order Parameters: Dipolar and Quadrupolar

To better understand the phenomena involved with the phases of Langmuir films, we introduce the following terminology: *dipolar phases* (*instabilities*) are those characterized by the order parameters, $\langle Y_1 \rangle$ and $\langle Y_2 \rangle$, and which transform as vector components; *quadrupolar phases* (*instabilities*) are those characterized by order parameters related to $\langle Y_3 \rangle$, $\langle Y_4 \rangle$, and $\langle Y_5 \rangle$ and which transform as second rank tensor components. The order parameters of quadrupolar tilt phases correspond to the 2-D nematic order parameter tensor components (Eq. (1.12)).

Dipolar instabilities without translational symmetry breaking have been analyzed. The instabilities are caused by competition between the single-molecule orientational potential, which includes the molecule–surface interaction, and the direct rotational–rotational coupling. No lattice-mediated interaction is involved. Therefore, the instability does not cause any deformation of the hexagonal lattice. The dipolar tilt on the hexagonal lattice requires strong, attractive dipolar-like intermolecular interactions. This can be understood as the interaction of elastic counterparts of an electric dipole, e.g., as forces created by the tilted molecules. Thus, the dipolar tilt instability would be preferred for longer molecules. The proper order parameter for the transition is the linear combination of the expectation values of surface harmonics, $\langle Y_1 \rangle$ and $\langle Y_2 \rangle$, which transform as the x and y components of a vector, respectively. For symmetry reasons, the E_1-type, dipolar instability is isotropic in the hexagonal plane of the monolayer, and there is no preferred direction for a molecular tilting where a precession of molecules is expected. The experimentally observed transitions from the LS phase to the tilted, deformed phases, therefore, must be initiated by quadrupolar instabilities causing the lattice deformations.

Quadrupolar instabilities are of two types and are driven by both direct and indirect rotational interactions. The E_2-type instability is driven by a coupling between the xy components of molecular quadrupoles. This instability involves hexagonal lattice deformation measured by the strain component, e_6, and is due to the indirect interaction. The hybridized (A_1, E_2) type instability is driven by the coupling between diagonal, xx and yy, components of molecular quadrupoles. The instability involves hexagonal lattice deformations, $(e_1 + e_2)$ and $(e_1 - e_2)$. In fact, the deformations, e_6 and $(e_1 - e_2)$, can be considered as equivalent to primary order parameters for the quadrupolar phases.

We obtain the following mechanism for the tilt transitions from the LS phase. Phases Ov, L_1', and L_{2d} (Fig. 1) are assigned as dipolar phases. Phase L_1' may result from an instability with respect to Y_1 and Y_2 orientational fluctuations. The order parameter for the phase is given by Eq. (3.21), and molecules are tilted on the hexagonal lattice. Because there is no preferred tilting direction, the molecules may show random tilt directions. Phases Ov and L_{2d} have molecules tilted toward NNN and NN, respectively, and they are characterized by order parameters $\langle Y_2 \rangle$ and $\langle Y_1 \rangle$, respectively. The rotating molecules in all of the dipolar phases have circular effective cross sections that allow retention of hexagonal packing in the plane perpendicular to the average molecular long axis.

L_{2h} and L_2^* are deformed hexagonal phases driven by a structural (elastic) instability, $C_{66} \Rightarrow 0$ (see Eq. (3.48)), causing e_6 and $(e_1 - e_2)$ deformations. Because the transition to these phases seems to be pressure dependent, we expect that a quadrupolar instability of the hybridized (A_1, E_2) type drives the transitions. The deformation, $(e_1 - e_2) \neq 0$, a result of the instability, lowers the system's symmetry and causes coupling to the dipolar order parameters. Molecular backbones are ordered on a centered rectangular lattice, and the molecules are tilted. For $(e_1 - e_2) > 0$, which represents contraction of the lattice in the direction NN, the dipolar interaction between molecules tilted in that orientation would become more attractive and the system would gain more energy by tilting the molecules in this direction. As a result, the L_{2h} phase is established and characterized by the order parameters $(e_1 - e_2) > 0$ and $\langle Y_1 \rangle \neq 0$. The dipolar interaction between tilted molecules would favor the direction of lattice contraction along the NNN direction corresponding to the $(e_1 - e_2) < 0$ deformation. This generates the L_2^* phase, which is characterized by $(e_1 - e_2) < 0$ and $\langle Y_2 \rangle \neq 0$.

We posit that the L_{2h} and L_2^* phases appear as a result of ferroelastic domain formation, and the transitions from the LS phase to the L_2^* and the L_{2h} phases are ferroelastic phase transitions. This assignment of lattice strains to these phases is particularly well supported by recent studies [104] of structures and phase diagrams of Langmuir monolayers composed of mixtures of heneicosanic acid and heneicosanol.

We take the "swiveling transition" observed as a 2-D pressure-driven [104] transformation to be a "transition" between ferroelastic domains. From this perspective, it is interesting that fatty acids show this phenomenon, whereas corresponding alcohols prefer to retain a NNN tilt direction and $(e_1 - e_2) < 0$ deformation. We attribute this to details of the molecule–surface interaction and thus to the 2-D elasticity of the lattice.

The mechanism of the tilt transitions, therefore, can be understood by representing the tilted molecules by force vectors (zero-order elastic multipoles) in the system's plane. The force is proportional to $\langle Y_1 \rangle$ in the x direction (NN) and to $\langle Y_2 \rangle$ in the y direction (NNN). The forces directed along the vector joining interacting molecules attract each other in the same way as electric dipoles, with, however, a different distance dependence. In contrast, forces directed perpendicular to the joining vector repel each other. In the hexagonal, undeformed lattice, tilting is possible only when there are strong, dipolar-like interactions that overcome all other interactions, with the molecules subsequently undergoing precession. Lattice deformations suppress the precession, and the direction of lattice contraction determines the direction of collective tilt. Thus, ferroelastic domains are formed. There is strong correlation between the lattice deformation and the ordering of molecular backbones, so that every deformation results in molecular ordering with tilting occurring in the direction perpendicular to the backbone located on the centered rectangular lattice.

The phase transition LS $\Rightarrow$ S can be driven by a quadrupolar instability of the E_2 type that involves the lattice distortion, e_6

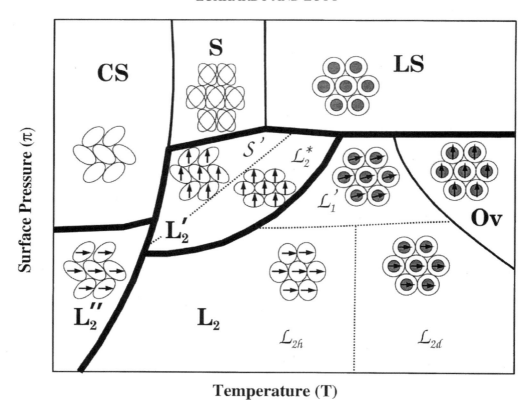

Fig. 14. Schematic phase diagram of *n*-alkanoic acids with a pictorial description of the orientational order parameters and symmetries of the phases. Vectorial (dipolar) order parameters are represented by arrows showing the tilt direction, and tensorial (quadrupolar) order parameters are designated by the effective molecular cross section.

Table III.   Characterization of Phases of the Phase Diagram Using Primary Orientational Order Parameters[b]

Phase	Order parameters	Instability	Symmetry[a]
Ov	$\langle Y_2 \rangle$	Dipolar	*cm*
L_{2d}	$\langle Y_1 \rangle$	Dipolar	*cm*
L_1'	$\langle Y_1 \rangle + \langle Y_2 \rangle$	Dipolar	*p1*
L_{2h}	$\langle Y_1 \rangle, [\langle Y_3 \rangle + \langle Y_4 \rangle)]$	Quadrupolar	*cm*
L_2^*	$\langle Y_2 \rangle, \langle Y_5(M) \rangle$	Quadrupolar	*cm*
S	$\langle Y_5 \rangle$	Quadrupolar	*cmm*
S'	$\langle Y_2 \rangle, \langle Y_5(M) \rangle$	Quadrupolar	*pg*
L_2''	$\langle Y_1 \rangle, \langle Y_5(M) \rangle$	Quadrupolar	*pg*
CS	$\langle Y_5(M) \rangle$	Quadrupolar	*pmg*

[a] Symmetry of the phases is given in symbols of 2-D space groups.

[b] For the phases appearing as a result of quadrupolar instabilities the lattice strain and effective cross section of molecules serve as equivalent order parameters.

and $\langle Y_5 \rangle \neq 0$. This instability is temperature driven and almost 2-D pressure independent. What is believed to be the phase with molecules in an orientationally disordered state [1] is mirrored here by the average value of the orientational fluctuations described by the Y_5 surface harmonic. The order parameter $\langle Y_5 \rangle$

describes the angular distribution of the orientational fluctuations at every site of the rectangular (deformed hexagonal) lattice. Finally, the phases believed to possess a herringbone pattern of molecular backbones, L_2'', S', and CS, are consequences of the corresponding instabilities at the *M*-point of the Brillouin zone.

The order parameters formulated in this study are more general than any used to date. In particular, these order parameters, defined as statistical averages of Fourier transforms of the surface harmonics, contain information about both the orientational fluctuations and crystallinity, because the wave vector **q** has its discrete values determined by the size of a crystalline sample. Thus, what has been suggested as an order parameter for "weak crystallization" [62–64] is obtained naturally in our formalism.

Figure 14 presents a pictorial assignment of phases to the schematic diagram of Fig. 1 according to our considerations. Table III gives our description of the phases in terms of the orientational order parameters and symmetries. There is a simple convention used to represent the order parameters. The vectorial parameters, $\langle Y_1 \rangle$ and $\langle Y_2 \rangle$, are represented by a vector (tilt direction), and the tensorial parameters, $\langle Y_3 \rangle$, $\langle Y_4 \rangle$, and $\langle Y_5 \rangle$, are designated by the effective cross section of molecules. The cross sections, of course, are related to the macroscopic strain.

5.3. Observations on the Microscopic Development of Order Parameters

We have formulated a new approach to understanding Langmuir monolayers from the viewpoint of the solid state and have used methods commonly associated with solid-state analysis. We have exploited surface harmonics for the description of orientational fluctuations of molecules in the context of earlier work on monolayers by Somoza and Desai [59]. This has provided a compact treatment of rotational degrees of freedom and allowed description of orientational fluctuations of E_1 as well as E_2 symmetry. We have included vector and second-rank tensor order parameters that are consistently defined by the average values of the surface harmonics. By explicitly treating translational–rotational coupling, we have been able to calculate indirect rotational coupling between the molecules and show that it gives rise to orientational instabilities. These instabilities are related to the nature of intermolecular interactions because of their representation by the surface harmonics. We have calculated the thermoelastic properties and have found their free energy in terms of the order parameters. The resulting expression serves as a microscopic foundation for a Landau free-energy expansion.

We have found that the structural (elastic) instability related to the softening of the C_{66} elastic constant most likely drives the phase transitions from the "superliquid" to tilted phases. The transitions are initiated by the instability, which induces lattice strain, and the resulting decrease in symmetry allows for subsequent coupling with the collective tilt of molecules. The mechanism of the coupling of the lattice strain to molecular tilt has also been outlined. The transitions are interpreted as ferroelastic, and the phases L_{2h} and L_2^* are found to be ferroelastic domains. These considerations can be used in the 2-D crystal engineering of ferroelectric Langmuir films.

6. APPLICATION OF THE GENERAL THEORY

As an application of the general theory we consider two specific transitions: the so-called swiveling transition and the LS-S transition. The ferroelastic aspect of the transformations is emphasized. The hexagonal, untilted phase (LS) has been chosen as the parent phase. We now derive a general free energy expansion in terms of the order parameters introduced in the microscopic theory.

6.1. Generalized Free Energy Expansion

The orientational fluctuations were expressed in terms of surface harmonics (composed of spherical harmonics) adapted for C_{6v} symmetry [3]:

$Y_1 \approx \eta \cos \phi$ and $Y_2 \approx \eta \sin \phi$ belong to the doubly degenerate E_1 representation,

$Y_3 \approx (3\cos^2 \phi - 1)$ belongs to the totally symmetric A_1 representation, and

$Y_4 \approx \eta^2 \cos 2\phi$ and $Y_5 \approx \eta^2 \sin 2\phi$ transform as components of the doubly degenerate E_2 representation.

This set of spherical harmonics describes orientational fluctuations of molecular tails in terms of tilt (θ) and azimuthal (ϕ) angles and provides the simplest rotational potential. The description is exact when the molecules are treated as rigid rods, e.g., when their effective cross section is well approximated by a circle. This is true for the rotational phases, of which the parent hexagonal LS phase is one. We have used such a set in our theory where the molecular tails have been averaged to a cylindrical shape. Strictly speaking, these functions describe orientational fluctuations of the tails, assuming nearly free rotation about the long axis. The functions Y_1 and Y_2 transform as x and y components of a vector, whereas Y_3, Y_4, and Y_5 transform as components of a second-rank tensor. The symmetry properties of the harmonics are very important because they allow the relation of every observable property, vectorial and/or tensorial, to statistical averages of the corresponding surface harmonics.

The translational part of the energy is conveniently described in terms of the strain tensor for the 2-D net. The variables $e_{xx} - e_{yy} = \varepsilon_2$ and $e_{xy} = \varepsilon_6$ transform as components of the E_2 representation, and $e_{xx} + e_{yy} = \varepsilon_1$ transforms as the totally symmetric A_1 representation. The nonsymmetric strain variables take the following form for the hexagonal C_{6v} reference cell [5],

$$\varepsilon_2 = (1/a_0)\big[(2/\sqrt{3})a_1 \sin \gamma_1^* - b_1\big] \quad \text{and}$$
$$\varepsilon_6 = \big[1/(2\sqrt{3}a_0)\big]\big[b_1 - 2a_1 \cos \gamma_1^*\big] \tag{6.1}$$

where a and b are the lengths of the 2-D lattice vectors and γ^* is the angle between the vectors in reciprocal space. Because the strains ε_2 and ε_6 transform as components of the doubly degenerate E_2 representation, they can be written in analogy to surface harmonics Y_4 and Y_5, as $\varepsilon_2 = \xi \cos 2\beta$ and $\varepsilon_6 = \xi \sin 2\beta$. The distortion amplitude may then be identified as $\xi = (2a_1/\sqrt{3}a_0)[\sin \gamma_1 - \sqrt{3} \cos \gamma_1]$, where a_0 is the lattice period before deformation. The angle β defines the direction of the lattice deformation, and for $\beta = 0, \pi, \ldots$, it is along the nearest neighbor (NN) direction, contracting for $\xi < 0$ and stretching for $\xi > 0$.

The strain $\varepsilon_2 \neq 0$ describes a deformation of the lattice from hexagonal to rectangular symmetry, and $\varepsilon_6 \neq 0$ defines distortion to an oblique lattice. *These strain parameters appear naturally from the symmetry analysis*, and the identification with order parameters as above is physically advantageous compared with other formulations [64] that suggest the formula for the distortion order parameter is $(a^2 - b^2)/(a^2 + b^2)$, where a and b are the major and minor axes of an ellipse drawn through the six nearest neighbors of a hexagonal net. This choice of deformation parameter lacks generality and shows an unnecessarily complicated relation to 2-D crystallography.

Having identified the symmetry properties of the orientational and strain variables, we can write the contributions to the free energy as follows:

$$F^R = a(\langle Y_4 \rangle^2 + \langle Y_5 \rangle^2)$$
$$+ b(\langle Y_4 \rangle^3 - 3\langle Y_4 \rangle \langle Y_5 \rangle^2) + c(\langle Y_4 \rangle^2 + \langle Y_5 \rangle^2)^2$$
$$+ f(\langle Y_1 \rangle, \langle Y_2 \rangle) \qquad (6.2)$$

$$F^T = (1/2)C_{66}^0(\varepsilon_2^2 + \varepsilon_6^2) + A(\varepsilon_2^3 - \varepsilon_2\varepsilon_6^2) + B(\varepsilon_2^2 + \varepsilon_6^2)^2 \,(6.3)$$

$$F^{TR} = \alpha(\varepsilon_2\langle Y_4 \rangle + \varepsilon_6\langle Y_5 \rangle) \qquad (6.4)$$

Thermal averages of the surface harmonics are taken with the total potential of the system given by Eq. (2.2). It is important to notice that this free energy function describes transitions from the LS, hexagonal parent phase to distorted and/or tilted phases without a translational symmetry change. As has been extensively discussed, transitions to herringbone ordering of the tilted molecules are conveniently described by the same orientational order parameters, but including wavevector dependence [3].

The most important feature of the free energy function is bilinear coupling between tensorial order parameters, orientational fluctuations, and strains. The fluctuations described by vectorial parameters $\langle Y_1 \rangle$ and $\langle Y_2 \rangle$ are responsible for transitions to tilted phases within hexagonal nets, and because they do not couple linearly with strains, we neglect their contribution. The bilinear coupling in the translational–rotational part of the free energy follows from the symmetry of the system and results in the relations $\langle Y_4 \rangle \propto \varepsilon_2$ and $\langle Y_5 \rangle \propto \varepsilon_6$. These linear dependencies have been found experimentally [64] (described as $\xi \approx \eta^2$) and are a simple consequence of the symmetry considerations and the more general relation between our natural order parameters. It is clear that the parameter sets $\{\langle Y_4 \rangle, \langle Y_5 \rangle\}$ and $\{\varepsilon_2, \varepsilon_6\}$ are equivalent, and it is a matter of taste which of them to use in characterizing Langmuir phases and their phase transitions. To stress the elastic aspects of a transition, as for ferroelastic transitions in monolayers, the free energy functional is expressed as a polynomial in terms of the strain order parameters [4]. Minimization of the free energy with respect to orientational fluctuations (assuming that the fluctuations follow a particular set of strains) gives

$$F = (1/2)C_{66}(T)[\varepsilon_2^2 + \varepsilon_6^2] + A(T)[\varepsilon_2^3 - 3\varepsilon_2\varepsilon_6^2]$$
$$+ B(T)[\varepsilon_2^4 + \varepsilon_6^4] + \cdots \qquad (6.5)$$

where

$$C_{66}(T) = C_{66}^0 - \alpha^2/a(T) \qquad (6.6)$$

is the renormalized elastic constant. This renormalization is due to orientational fluctuations at a molecular level and is the expected result of translational–rotational coupling [3]. The temperature dependence, which enters via the $A(T)$ coefficient, may arise from either or both of the following contributions: an effective orientational field experienced by a single molecular tail in the environment of its neighbors that is temperature dependent, and an orientational entropy contribution due to a change in the orientational distribution function. Thus, an orientational disorder at the molecular scale renormalizes the elasticity of a monolayer.

For Langmuir monolayers, it is much more common to stress the orientational order of the molecules and to express the free energy in terms of the thermal averages of the surface harmonics. From a minimization of the total free energy with respect to strain components (assuming the strain follows a particular set of orientational fluctuations), the free energy function is

$$F = a'(T)[\langle Y_4 \rangle^2 + \langle Y_5 \rangle^2] + b'(T)[\langle Y_4 \rangle^3 - 3\langle Y_4 \rangle \langle Y_5 \rangle^2]$$
$$+ c'(T)[\langle Y_4 \rangle^2 + \langle Y_5 \rangle^2]^2 \qquad (6.7)$$

The coefficients are renormalized with respect to those in Eq. (6.2), and they are, in general, temperature dependent. For the quadratic term, this dependence can be justified in a simple approximation by calculating the rotational susceptibility, $\chi = (kT)^{-1}\langle Y^2 \rangle$ [3].

6.2. The Orientational Entropy

We now show how the temperature dependence can follow from the orientational entropy by assuming a kind of orientational disorder within the monolayers. This seems quite reasonable for systems that we consider as disordered solids. The temperature dependencies of the parameters will determine the shape of the free energy function. We expect that for some temperature region the function represents a "sombrero," with three minima located away from the center point, which represents the LS hexagonal phase. The minima would then correspond to three domains of a tilted (and correspondingly deformed) ferroelastic phase. The swiveling transition can be subsequently identified as a transition between two ferroelastic phases.

The orientational order of the molecules at a temperature T can be described by a single-particle orientational distribution function, $P(\Omega)$. The probability that a molecule has its axis directed within the solid angle $d\Omega = \sin\theta d\theta d\phi$ about the direction $\Omega = (\theta, \phi)$ is given by $P(\Omega)d\Omega$. The orientational distribution function can be expressed in terms of the symmetry-adapted spherical harmonics, and, for highly delocalized orientations, higher order harmonics are needed to describe the function. For a system with C_{6v} symmetry, we expect six states ($\phi_i = i2\pi/6, i = 1, \ldots, 6$) that reflect localization of the orientations of the molecules. However, as is evident from the above symmetry considerations, only three of these are strongly coupled with the strains.

As has been shown in many experiments for Langmuir monolayers, the orientational fluctuations are strongly coupled with the strains, and for this reason we may assume that there are three pocket states that contribute to the orientational distribution function for the strain-orientation coupled system. This can be attributed to fluctuations in the tilt of cylindrical (rotating) molecules and/or fluctuations in molecular backbone orientation. For the tilt fluctuations, which can be visualized as precession-like fluctuations, the order parameters are directly related to tilt angle, via $\eta = \sin\theta$, $Y_4 \approx \eta^2 \cos 2\phi$, and $Y_5 \approx \eta^2 \sin 2\phi$. Fluctuations in backbone orientation may also be characterized by the same surface harmonics, say, $Y_4 \approx \delta \cos 2\phi$ and $Y_5 \approx \delta \sin 2\phi$. In this case δ is a measure of distortion of the effective cross section of a molecule from a circular (rotating) one. In a real system, such a separation of contributions to the orientational disorder is highly artificial, and, for this reason, it is more convenient to interpret the orientational order parameters, $\langle Y_4 \rangle$ and $\langle Y_5 \rangle$, as components of local stress, the elastic dipoles, which represent both contributions. Thus, the convenient functions are $Y_4 \approx \Delta \cos 2\phi$ and $Y_5 \approx \Delta \sin 2\phi$, with $\Delta = (\eta^2 + \delta)$ indicating the short axis of an ellipse along the $\psi = 0$ direction.

Thermal averages of the harmonics are determined by an orientational distribution function. The states are defined as $\Omega_j = (\theta, \phi_j)$, $j = 1, 2, 3$, with $\phi_1 = 0$, $\phi_2 = 2\pi/3$, and $\phi_3 = 4\pi/3$ (Fig. 2), and for a function approximated by three "pocket states" located at $\phi = 0, 2\pi/3, 4\pi/3$, they are

$$\langle Y_4 \rangle = \Delta \sum P_j \cos 2\phi_j, \qquad \langle Y_5 \rangle = \Delta \sum P_j \sin 2\phi_j \quad (6.8)$$

From these equations, we have the following relations:

$$P_1 = 1/3 + (2/3)\langle Y_4 \rangle \qquad (6.9a)$$

$$P_2 = 1/3 - (1/3)\langle Y_4 \rangle + \left(1/\sqrt{3}\right)\langle Y_5 \rangle \qquad (6.9b)$$

$$P_3 = 1/3 - (1/3)\langle Y_4 \rangle - \left(1/\sqrt{3}\right)\langle Y_5 \rangle \qquad (6.9c)$$

The contribution to the orientational entropy is therefore approximated by

$$S \cong -Nk \sum P_j \ln P_j \qquad (6.10)$$

where N is the number of molecules. Using the above relations for the orientational distribution and expanding for small values of the order parameters, the orientational entropy energy contribution per molecule is

$$-(1/N)TS = -\text{const}T + kT\left(\langle Y_4 \rangle^2 + \langle Y_5 \rangle^2\right) - kT\left(\langle Y_4 \rangle^3 - 3\langle Y_4 \rangle\langle Y_5 \rangle^2\right) + (3/2)kT\left(\langle Y_4 \rangle^2 + \langle Y_5 \rangle^2\right)^2 \qquad (6.11)$$

Within this approximation, which assumes a certain degree of disorder between the three states in the hexagonal parent phase, the free energy is described by Eq. (6.7), with coefficients linearly dependent on the temperature. For example, $a'(T) = a' + kT$, where $a' = a - \alpha^2/C_{66}^0$, but $b'(T) = b' - kT$, and the opposite effect of the temperature will decide the relative stability of different phases.

If the surface pressure effect is included, the Gibbs free energy is

$$G(T, \pi) = F(T) + \pi\varepsilon_1 \qquad (6.12)$$

assuming that the 2-D pressure is isotropic and couples with the totally symmetric strain $\varepsilon_1 = (e_{xx} + e_{yy})$ only. This might not be fulfilled by normal experimental conditions.

The free energy of the system expressed in terms of the orientational order parameters determines the three domains that are characterized by the molecular orientations,

$$\langle Y_4 \rangle; \qquad -(1/2)\langle Y_4 \rangle + \left(\sqrt{3}/2\right)\langle Y_5 \rangle; \\ -(1/2)\langle Y_4 \rangle - \left(\sqrt{3}/2\right)\langle Y_5 \rangle \qquad (6.13)$$

and the corresponding strains,

$$\varepsilon_2; \qquad -(1/2)\varepsilon_{2+}\left(\sqrt{3}/2\right)\varepsilon_6; \qquad -(1/2)\varepsilon_2 - \left(\sqrt{3}/2\right)\varepsilon_6 \qquad (6.14)$$

These domains are rectangular, 2-D lattices with tilted molecules. $\varepsilon_2 < 0$ indicates a compression of the hexagonal lattice along the (NN) direction, and $\varepsilon_2 > 0$ corresponds to a stretching along this direction or a compression along the orthogonal (NNN) direction. The tilt direction of molecules is, to a first approximation, linearly related to the strain, $\langle Y_4 \rangle = -(\alpha/C_{66}^0)\varepsilon_2$, and its direction depends on the sign of the coupling constant α and the bare, shear elastic constant. Because coupling usually decreases the free energy of the system, one may assume $\alpha < 0$, and because the hexagonal LS phase is a super liquid, one can certainly assume the shear modulus $C_{66}^0 < 0$. We conclude that the tilt of molecules is expected in the direction of the unit cell compression, an observation that has been experimentally observed but hitherto unexplained. This correlation can also be found from considerations involving the concept of elastic dipoles and their interactions [4]. We may therefore identify a domain characterized by $\langle Y_4 \rangle > 0$ (or, equivalently $\varepsilon_2 < 0$) as one of three domains of the L_2 (L_{2h}) phase.

6.3. The Swiveling Transition ($L_2 \rightarrow L_2'$)

The swiveling transition, which has recently been extensively studied [104, 105], is the transition between the L_2 (L_{2h}) phase (rectangular lattice with molecules tilted toward the (NN) direction), and the phase is denoted as $L_2'(L_2^*)$ (rectangular lattice, characterized by molecular tilt ordering toward the (NNN) direction. In terms of the order parameters introduced here, the transition corresponds to the change: $\langle Y_4 \rangle \Rightarrow -\langle Y_4 \rangle$, or, equivalently, $\varepsilon_2 \Rightarrow -\varepsilon_2$, for one domain of the phases and with corresponding relations for the two other domains. The domains of the L_{2h} and L_2^* phases can be conveniently represented on the $(\langle Y_4 \rangle, \langle Y_5 \rangle)$ plane of the orientational order parameters (Fig. 14). They correspond to (local) minima of the free energy function described by Eq. (6). The free energies of the two phases are different, and their relative stabilities are temperature and (strongly) pressure dependent, as shown by experimental studies [105].

Having identified and characterized the phases that take part in the swiveling transition, the easiest way to analyze the transition is to consider just one domain of a L_{2h} or L_2^* phase. The transition may be analyzed through the cross section of the free energy function for $\langle Y_5 \rangle = 0$. This implies, however, that the route for the swiveling transition corresponds to a "least motion" path, but this may not be true. The transition most likely follows a path along lowest energy barriers. In fact, what has been observed recently for fatty acids indicates that the transition goes via an intermediate "I" phase, where molecules are tilted to an intermediate direction between (NN) and (NNN), and the lattice has been identified as oblique [105]. In terms of the order parameters, this indicates that the transition does not go via the $\langle Y_4 \rangle \Rightarrow -\langle Y_4 \rangle$ route, but rather corresponds to a path $\langle Y_4 \rangle \Rightarrow \langle Y_4 \rangle + (\sqrt{3}/2)\langle Y_5 \rangle$, i.e., to the nearest domain of another phase. This can be equivalently represented as the transition $[-(1/2)\langle Y_4 \rangle + (\sqrt{3}/2)\langle Y_5 \rangle] \Rightarrow [(1/2)\langle Y_4 \rangle + (\sqrt{3}/2)\langle Y_5 \rangle]$, which has to go through an intermediate state where $\langle Y_4 \rangle = 0$. This implies that the intermediate state with $\langle Y_5 \rangle \neq 0$ and on an oblique lattice ($\varepsilon_6 \neq 0$) can be formed and identified with the "I" phase.

The swiveling transition is highly surface-pressure dependent. One may even view the transition as pressure-induced [105]. The question arises as to how isotropic (as is usually assumed) the experimentally applied surface pressure, π, is. If it is isotropic, then the pressure does not couple directly to nonsymmetric strain components (or orientational counterparts), but modifies the free energy function by changing the coefficients of the expansion. However, the manner in which this modification appears is not defined without numerical calculations for each specific system. An example of such calculations has been presented [5]. If, on the other hand, the surface pressure is not isotropic, it might be considered to be an anisotropic external stress, $\sigma_2 = \sigma_{xx} - \sigma_{yy}$ or $\sigma_6 = \sigma_{xy}$. The macroscopic stress then couples to macroscopic strain components. For example, considering only one ferroelastic domain characterized by $\varepsilon_2 \neq 0$, $\varepsilon_6 = 0$, the Gibbs free energy is

$$G(\varepsilon_2) = (1/2)C_{66}^{o}\varepsilon_2^2 + A'\varepsilon_2^3 + B'\varepsilon_2^4 + \sigma_2\varepsilon_2 \qquad (6.15)$$

and the stability of phases will be determined by the sign of the external stress. As indicated in Figure 14, the phase with $\varepsilon_2 < 0$ is stable under higher external stress.

6.4. An Internal Stress Effect

So far, we have considered a system in which molecules have been assumed to be cylindrical, e.g., in a rotational state. However, when the molecules experience a strong orientational potential, the rotational dynamics become so slow that the molecular cross sections no longer average to a circle, and the orientation of the molecular backbone has to be specified. Therefore, to characterize orientational fluctuations of a molecule in a nonrotational phase, it is helpful to introduce a "measure" of the distortion of the cross section from a circle. One way to do this is by introducing the so-called nematic order parameter, the tensor constructed from a director as mentioned above. However, it is more meaningful if the "measure" of the effective cross section is related to the elastic dipole concept as introduced into the solid-state theory of Langmuir monolayers [4]. With this concept, the orientational fluctuations of a molecule form a distribution of forces around the molecular site. A multipole expansion, where the first term corresponds to a net force, the second to an elastic dipole, and so on, conveniently represents this distribution.

The elastic dipole has a clear physical meaning and represents an internal (local) stress due to the orientational fluctuations. The internal stress is due to all orientational fluctuations, tilt of molecular tail, and backbone orientation and, in principle, should be expressed in terms of Wigner rotation matrices. However, as discussed above for Langmuir monolayers, it is useful

to formally separate the contributions of tilt from the contribution due to backbone orientation. The first contributions have been expressed in terms of surface harmonics as $\langle Y_3 \rangle$, $\langle Y_4 \rangle$, and $\langle Y_5 \rangle$ order parameters. The bilinear coupling between these elastic dipoles and their corresponding strains allows for the clear interpretation of the coupling as a stress–strain relation [4].

The contribution due to the backbone orientation of the molecules is expressed by the same harmonics due to the symmetry constraint. Therefore, a transition from a rotational phase to a nonrotational phase is modeled by introducing an internal stress into the system—the stress created by a frozen orientation of the molecular backbone. The internal stress, being a measure of the difference between the effective cross section and the circular one, can be described by the elastic dipole components $P_2 = P_{xx} - P_{yy}$ and $P_6 = P_{xy}$, which transform according to the E_2 representation of the C_{6v} group. Thus, the free energy of the system can be described by the equation

$$F(P, \langle Y \rangle) = F(\langle Y \rangle) + \{P_2\langle Y_4 \rangle + P_6\langle Y_5 \rangle\} \qquad (6.16)$$

and the fact that the elastic dipole components P_2 and P_6 are nonzero dictates the broken symmetry of the nonrotational phase. Taking into account the angular form of the orientational order parameters, the free energy is

$$F = B\eta^4 - D\eta^6 \cos 6\phi + J'\eta^2 \cos 2(\phi - \alpha) \qquad (6.17)$$

In the above expression, the elastic dipole components have been expressed as $P_2 \propto \cos 2\alpha$ and $P_6 \propto \sin 2\alpha$. If, in addition, twofold symmetry is required, $P_6 = 0$, and the formula becomes

$$F_{2-\text{fold}} = J\eta^2 \cos 2\phi + B\eta^4 \cos 4\phi - D\eta^6 \cos 6\phi \qquad (6.18)$$

This is the same formula used to analyze the swiveling transition. The important point is, as we have shown throughout the derivation, that the coefficient J has a clear physical meaning: it is the internal stress arising from a specific orientation of the backbone of the molecule ($J \propto \cos 2\alpha$). Thus, the elastic dipole determines the sign of the coefficient. Analysis of the function in Eq. (6.18) shows that the swiveling transition corresponds to a change in the J coefficient ($J > 0$, $\alpha = 0$; $J < 0$, $\alpha = \pi/2$). Moreover, if twofold symmetry is imposed on the system, the possibility of analyzing an oblique lattice is excluded, and, in fact, only transitions from rectangular (NN) to rectangular (NNN) lattices can be analyzed. Therefore, not surprisingly, the oblique distortion characteristic for the intermediate I phase could not be analyzed by the limited approach adopted in a previous study based on a liquid-crystal model of Langmuir monolayers.

6.5. Calculations

Simple calculations illustrate the utility of the theory presented here, as well as confirm its veracity. A full potential minimization was executed for a perfectly crystalline monolayer utilizing atom–atom potentials [5]. It should be noted that head-group

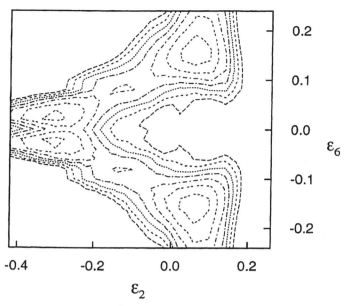

Fig. 15.   Potential energy contours of a NN (L_{2h}) phase, showing a threefold minimum distinct from that of the one for a NNN phase.

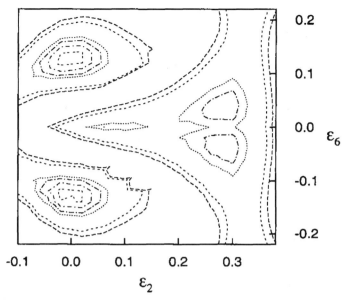

Fig. 16.   Potential energy contours of a NNM (L_{2h}) phase, showing a threefold minimum.

interactions and temperature effects have not been modeled. Furthermore, we have departed from the assumptions of previous authors by modeling these phases with only one molecule per unit cell [104]. These calculations were done at sufficiently low isotropic pressure to ensure a global minimum that exhibited tilt (-45 in our arbitrary pressure scale) [5]. After we located the potential minimum (with respect to lattice vectors **a** and **b** and the three orientational angles), ε_1 was set equal to 0, and a_0 of the parent hexagonal lattice was found by solving the equation

$$\varepsilon_1 = (1/a_0)\left[(2/\sqrt{3})a_1 \sin\gamma_1 + b_1\right] - 2 \qquad (6.19)$$

For any given ε_2 and ε_6 strains, Eqs. (6.1) may subsequently be solved for γ_1^*, a_1, and b_1. ε_2 and ε_6 were thus scanned, and the three orientational angles were minimized for each point in the (ε_2, ε_6) plane (Fig. 15). A gradient minimization was used with initial trial values for any point consisting of the final (minimum) values for an adjacent point. Thus, large deviations from the initial point are not likely to be found, because an energy barrier is likely to be encountered in such a case. Accordingly, the calculations prefer the "phase" of the global minimum. As can be seen from the energy contours in Figure 15, a threefold minimum is found for this approximately NNN (L_2^*) phase (actual tilt direction $\beta \approx 20°$).

A related triply degenerate minimum that resembles the NN (L_{2h}) phase must also be found. This was achieved by setting the lattice parameters to those values associated with the minimum near $\varepsilon_6 = 0$, $\varepsilon_2 = -0.3$, and minimizing all parameters from an initial orientation $90°$ from that of the previously found minimum. A new minimum was found nearby with $\varepsilon_1 = 0.1$, assuming the same a_0 as for the NNN phase. This is reflected by an increase in the area per molecule from 20.16 Å^2 to 22.11 Å^2.

Thus, the relative stabilities of these two phases should exhibit significant pressure dependence.

The results of a scan of the (ε_2, ε_6) plane are shown in Figure 16. Another triple degeneracy is found. In both cases, a slight splitting is observed in the minima located about $\varepsilon_6 = 0$. The barrier between these double minima is considerably less than $0.5kT$ at room temperature and will not be observable in practice. This splitting, as well as the departure from the idealized locations of the minima predicted in Section 2, is directly related to the departure of the molecular geometry from that of a perfect cylinder. Further minima that are also close (within 1 kcal/mol) are also found. Although these results demonstrate the concepts related above and are consistent with both present theoretical and published experimental results, further work is required to establish a more quantitative comparison.

The solid-state model demonstrates that it is the nature of Langmuir monolayers that the swiveling of molecules must be coupled with a reorientation of the cross sections arising from stress–strain (local) coupling. Thus, we have also shown why the change in the distortion direction tracks the change in tilt direction, thereby answering a question posed but unresolved by other models [105]. In addition, the phases taking part in the swiveling transition are characterized by the order parameters $\langle Y_4 \rangle$ and $-\langle Y_4 \rangle$, or, equivalently, ε_2 and $-\varepsilon_2$. It is clear that the cell dimensions in phases L_{2h} and L_2^* are the same as those observed experimentally [105]. The solid-state model also explains why the axes of the lattices are so insensitive to the cross section orientation. Expression of this model in terms of strain–stress couplings, by macroscopic or local elastic dipoles, is the easiest way to analyze the complicated phase behavior of Langmuir monolayers as demonstrated here for the swiveling transition.

7. EXTENSIONS OF THE SOLID-STATE THEORY OF LANGMUIR FILM PHASES

7.1. Implications for Elastic Properties

There are some consequences for properties of the system that follow from this model that treats a mesophase as a mixture of elastic domains. The elastic interaction between domains causes an ordering process in the system that may lead to a change in macroscopic elastic properties. First, we discuss static response to an applied stress. The formula for the elastic dipole density fluctuation (Eq. (4.42)), may be rewritten as

$$\delta\rho_\alpha = \beta(\rho_0/3)\chi_{\alpha\beta}(R)P_{ij}^\beta S_{ijlm}^0 \sigma_{lm} \qquad (7.1)$$

The macroscopic strain has been expressed in terms of the macroscopic stress tensor, σ, and the elastic compliance tensor for an isotropic medium,

$$S_{ijlm}^0 = (1/2\mu_0)\big[1/2(\delta_{il}\delta_{jm} + \delta_{im}\delta_{jl}) \\ - (\lambda_0/(3\lambda_0 + 2\mu_0))\delta_{ij}\delta_{lm}\big] \qquad (7.2)$$

The total strain experienced by the macroscopic system is the sum of the elastic ($S^0\sigma$) and plastic strains (the density of elastic dipoles). Written in tensor notation, this is

$$\varepsilon = S^0[\sigma + p] \qquad (7.3)$$

where the elastic dipole density is

$$p = P^\alpha \chi_{\alpha\beta}(R)P^\beta S^0 \sigma \qquad (7.4)$$

The total strain is then

$$\varepsilon = S^0\big[1 + P^\alpha \chi_{\alpha\beta}(R)P^\beta S^0\big]\sigma \qquad (7.5)$$

where the renormalized compliance is identified as

$$S = S^0 + (S^0 P^\alpha)\chi_{\alpha\beta}(R)(P^\beta S^0) \qquad (7.6)$$

The second term in Eq. (7.6) can be recognized as the correlation function for strain fluctuations at large distances for a system of interacting elastic domains,

$$\beta\langle\delta\varepsilon(R)\delta\varepsilon(0)\rangle = (S^0 P^\alpha)\chi_{\alpha\beta}(R)(P^\beta S^0) \qquad (7.7)$$

For the symmetry of the domains in this system, the shear elastic constant is renormalized according to the relation

$$[\mu(R)]^{-1} = \mu_0^{-1}\{1 + \mu_0^{-1}(P_2^2 + P_6^2)\chi_E(R)\}. \qquad (7.8)$$

The R-dependence of the susceptibility is explicitly a function of the distance,

$$\chi_E(R) = (\rho_0/3)\big[\beta^{-1} - (4\pi\mu_0)^{-1}R^{-2}(P_2^2 + P_6^2)\big] \qquad (7.9)$$

and correspondingly, the R-dependent shear elastic constant may be considered as the macroscopic counterpart of the microscopic elastic constant. The renormalization of the shear elasticity is due to orientational relaxation of the elastic domains, and such a relaxation will, obviously, influence the elastic response of a macroscopic system. One may expect that the difference between macroscopic and microscopic elasticity increases with decreasing temperature because of the temperature-induced ordering process. In a real system the temperature interval studied is usually quite small, and it will be difficult to detect such a change with temperature.

An important aspect of the elasticity of Langmuir monolayers is that the elastic response can be, and often is, a nonequilibrium one. This means that a system responds to an applied stress within a time during which elastic domains relax. What is the time evolution of the elastic domains and does the relaxation process influence the elastic response? The elastic dipole density fluctuation at a thermodynamic equilibrium is given by Eq. (4.50) for the rectangular and oblique domains. We denote this as the value at infinite time,

$$\delta\rho_\alpha(\infty) = \chi_E P_{ij}^\alpha \varepsilon_{ij} \qquad (7.10)$$

Time evolution of the density fluctuation can be found from the kinetic equation

$$d[\delta\rho_\alpha(t)]/dt = -(1/\tau)[\delta\rho_\alpha(t) - \delta\rho_\alpha(\infty)] \qquad (7.11)$$

where τ is the relaxation time for a process of ordering the elastic domains. It might also be considered as the residence time for a system in an orientational potential well. The time evolution of the elastic dipole density fluctuation is

$$\delta\rho_\alpha(t) = \delta\rho_\alpha(\infty)\big[1 - \exp(-t/\tau)\big] \qquad (7.12)$$

It is, in fact, an approximation that we use one relaxation time for the reorientation of the elastic domains. Because the domains are, in general, rectangular (P_2 elastic dipole) and/or oblique (P_6), it may be necessary to introduce two separate relaxation times for rectangular and oblique distortions, respectively.

The elastic dipole density (the local stress) will show the same time evolution, and, following the derivation for the static case, we find the following time dependence of the shear elastic constant:

$$[\mu(R,t)]^{-1} = \mu_0^{-1}\Big\{1 + \mu_0^{-1}(P_2^2 + P_6^2)\chi_E(R) \\ \times [1 - \exp(-t/\tau)]\Big\} \qquad (7.13)$$

It is evident that any instantaneous measurements (at $t = 0$) will not register an elastic response of the elastic dipole density fluctuations, and the response is insensitive to an ordering process of elastic domains. On the other hand, experiments within the time required for a system to reach thermodynamic equilibrium will measure the shear elastic constant of a system where an ordering process takes place. Equation (7.13) may be rewritten in terms of macroscopic and microscopic elastic constants:

$$[\mu_{\text{MACRO}}(R)]^{-1} = \mu_0^{-1}\{1 + \mu_0^{-1}(P_2^2 + P_6^2)\chi_E(R)\} \qquad (7.14)$$

It has been found that the ratio $(\mu_{\text{MACRO}}/\mu_0)$ is on the order of 10^{-2} [106]. Such a large difference indicates that the renormalization of the elasticity due to an ordering process of elastic domains in mesophases is very important. One may also represent the renormalized shear elastic constant by an approximate formula derived from Eq. (7.13),

$$\mu(R,t) = \mu_0 - (P_2^2 + P_6^2)\chi_E(R)\big[1 - \exp(-t/\tau)\big] \qquad (7.15)$$

It has been found [16] that such a formula describes well the time-dependent shear elastic response of monolayers and that the ratio $\mu(R, t \to \infty)/\mu_0$ is on the order of 10^{-2}. Within the model of elastic domains, it is not surprising that the ratio coincides with that for macroscopic/microscopic elastic constants, as the two are determined by the same renormalization factor.

For dynamical experiments performed within specific frequency domains, we introduce a frequency-dependent susceptibility,

$$\chi_E(R, \omega) = \chi_E(R)(1 + i\omega\eta)^{-1} \qquad (7.16)$$

that reflects the relaxation process of domain ordering. With this susceptibility, the renormalized elastic constant is made frequency dependent,

$$[\mu(R, \omega)]^{-1} = \mu_0^{-1}\{1 + \mu_0^{-1}(P_2^2 + P_6^2)\chi_E(R, \omega)\} \quad (7.17)$$

and, following the standard notation, it may be represented as

$$\mu(\omega) = \mu'(\omega) + i\omega\eta(\omega) \qquad (7.18)$$

The real part of the shear elastic constant is

$$\mu'(\omega) = \mu_0\left\{1 + \mu_0^{-1}(P_2^2 + P_6^2)\chi_E(R)(1 + \omega^2\tau^2)^{-1}\right\}^{-1} \quad (7.19)$$

where the shear elastic constant, μ_0, can be taken as a high-frequency property.

The imaginary part of the shear elasticity is expressed in terms of shear viscosity (viscoelasticity), which is

$$\eta(\omega) = [\mu'(\omega)]^2\mu_0^{-2}(P_2^2 + P_6^2)\chi_E(R)\iota(1 + \omega^2\tau^2)^{-1} \quad (7.20)$$

This expression shows that the viscoelasticity of the system changes in the same way as the real part of the shear elastic constant. Moreover, in the region of Langmuir monolayer mesophases where elastic domains possess orientational freedom, the viscoelasticity and shear elastic constant show minimum values [107]. This provides an explanation of observations of "viscosity" made by Copeland et al. [108], and of results from recent experiments on the shear elasticity of Langmuir monolayers [30, 31, 106].

7.2. Implications for Diffuse X-Ray Scattering

Clearly, the "anomalous" elastic properties of the mesophases are related to disorder stimulated by the existence of mesoscopic elastic domains. The disorder is seen in X-ray scattering experiments and manifests itself in anomalous broadening of diffraction peaks at temperatures around the LS/S transition [26]. This observation has been intuitively connected with elastic properties [109] and brought into close analogy with bulk alkanes [29]. However, even in a recent analysis of the positional disorder [26], it has not been shown how to quantitatively relate the X-ray scattering observations to the elastic properties. The model of elastic domains is a first step toward a consistent physical picture. We demonstrate how it can be used to analyze X-ray diffuse scattering and to calculate a contribution due to positional disorder induced by the elastic domains.

For the purpose of diffuse scattering, a Fourier transform of the density fluctuation, $\delta\rho_\alpha(\mathbf{q})$, is used. In fact, the wavevector ($\mathbf{q}$)-dependent density fluctuation can be considered as an analog of an order parameter in the Landau theory of "weak crystallization" [32, 63]. According to that theory, crystallization occurs when a system becomes unstable against the formation of a density fluctuation wave with a given wavevector. In the system of disordered elastic domains, an ordering process corresponds to "crystallization". It appears as a formation of a static distribution of elastic dipoles according to a pattern of a density fluctuation corresponding to the lowest eigenvalue of the $\mathbf{q}$-dependent susceptibility,

$$\chi_{\alpha\beta}(\mathbf{q}) = \beta\langle\delta\rho_\alpha(\mathbf{q})\delta\rho_\beta(\mathbf{q})\rangle \qquad (7.21)$$

Conceptually, the approach using the elastic dipole density fluctuations is similar to and consistent with "weak crystallization" theory. In our model, the "crystallization" of elastic domains may be considered as a kind of spinoidal decomposition process.

According to the kinematic theory of X-ray scattering, the intensity at a point $\mathbf{Q} = \mathbf{q} + 2\pi\mathbf{H}$, where $\mathbf{H}$ is a reciprocal lattice vector, is

$$I_{\text{dif}}(\mathbf{Q}) = \langle|\Delta\varphi(\mathbf{q})|^2\rangle \qquad (7.22)$$

where $\Delta\varphi(\mathbf{q})$ is a fluctuation in the scattering amplitude at point $\mathbf{q}$ of the reciprocal lattice. For the system composed of rectangular/oblique domains, the fluctuation is due to the elastic dipole density fluctuation, $\delta\rho_\alpha(\mathbf{q})$. The fluctuation causes inhomogeneous displacements $u(\mathbf{q})$, which may be expressed as a response to a fluctuating force, $\mathbf{V}^\alpha(\mathbf{q})$, produced by a given elastic domain,

$$u(\mathbf{q}) = \mathbf{G}(\mathbf{q})\sum_\alpha\mathbf{V}^\alpha(\mathbf{q})\delta\rho_\alpha(\mathbf{q}) \qquad (7.23)$$

$\mathbf{G}(\mathbf{q})$ is the elastic Green's function specified in the Appendix. Assuming that the displacements are small, the contribution to the X-ray diffuse scattering is

$$I_{\text{dif}}(\mathbf{Q}) = \beta^{-1}F_\alpha(\mathbf{q})\chi_{\alpha\beta}(\mathbf{q})F_\beta(-\mathbf{q}) \qquad (7.24)$$

where

$$F_\alpha(\mathbf{q}) = [\Delta f_\alpha + f\mathbf{Q}\mathbf{G}(\mathbf{q})\mathbf{V}^\alpha(\mathbf{q})] \qquad (7.25)$$

Δf_α is the difference between scattering factors for different types of elastic domains and the reference, disordered system. Simplifications may be imposed. First, we approximate the susceptibility by its value calculated for "large R." Second, within the same limit, the force can be expressed in terms of the elastic dipoles as

$$\mathbf{V}^\alpha(\mathbf{q}) \propto i\mathbf{v}\mathbf{P}^\alpha\mathbf{q} \qquad (7.26)$$

With these approximations, the contribution to X-ray diffuse scattering due to positional disorder caused by the rectangular/oblique elastic domains is

$$I_{\text{dif}}(\mathbf{Q}) = \beta^{-1}\chi_{\alpha\beta}[\Delta f_\alpha + f\mathbf{Q}\mathbf{G}(\mathbf{q})\mathbf{P}^\alpha\mathbf{q}][\Delta f_\beta + f\mathbf{Q}\mathbf{G}(\mathbf{q})\mathbf{P}^\beta\mathbf{q}]$$
$$(7.27)$$

This formula may be further rearranged, after some approximations, yielding

$$I_{\text{dif}}(\mathbf{Q}) \propto \beta^{-1}(\mathbf{S}^0\mathbf{P}^\alpha)\chi_{\alpha\beta}(\mathbf{P}^\beta\mathbf{S}^0) \qquad (7.28)$$

and showing that the contribution to the diffuse scattering is proportional to that which renormalizes elasticity of the system. Thus, observations of anomalous elasticity and anomalous scattering are closely related by the same mechanism of elastic domain ordering. It is evident that the scattering is proportional to the square of the elastic dipole, i.e., the local stress, which, in a product with the density fluctuation, may serve as an order parameter. This has been assumed intuitively in a recent analysis of positional disorder in LS/S phases [6]. The model of elastic domains will also help in understanding the X-ray scattering experiments for bulk alkanes and addresses the need for a model-based analysis of these systems [29].

8. COMPUTER SIMULATIONS

The ability of simulations to generate "pictures" of the evolution of molecular processes is of particular use in the study of the phase and phase transition behavior of Langmuir films. The number of such simulations is now legion, and it is not our purpose to review them. Our interest is in specific calculations that emphasize the solid-state nature of Langmuir monolayers and that reflect some of the developments of the theory developed above. Most researchers have not interpreted the results of their calculations in the light of this approach to the modeling of films, and, although our discussion is not inclusive, it describes a set of related computer simulations that represent an approach consistent with the model developed in previous sections. Many of the concepts, and indeed results, generated by the theory of Langmuir films based on a solid-state model are substantiated by these calculations.

In the ensuing parts of this section, we examine computer experiments based on idealized models that are conceptually imbued with the solid-state model of films of amphiphiles. In particular, the calculations demonstrate some of the conclusions or expectations generated by the solid-state theory for Langmuir films that is extensively developed in prior sections. The first calculation is a direct application of Lennard–Jones pair potentials to the problem of alkanoic films. These calculations are typical of those employed to determine the packing of 3-D crystals of molecular crystals. The next model examines the role of the cross sectional geometry of an amphiphile in determining the film packing. The model also serves as a representation of interacting 2-D elastic dipoles. The cross section potential is used to examine some phase behavior by employing it in Monte Carlo simulations. Finally, a realistic film is addressed by modeling amphiphiles as beads on a string. This breaks the pure 2-D nature of the model calculations but still provides a qualitative description of film phase behavior with additional insights into how amphiphiles should be designed to achieve these behaviors.

8.1. Calculation of Packing of Model Amphiphiles and Selected Fatty Acids

Langmuir's original premise was that the packing of organic amphiphiles on an aqueous subphase would be governed solely by the molecules' tails [110]. This was examined [85] with the use of a model of rigid amphiphiles anchored to the plane of the subphase. In this approach, only the contribution of the tail portions of the molecules is examined. This leads to the realization that only the cross section of the amphiphile determines the collective symmetry of the net.

A Buckingham potential, parameterized using values proposed by Williams [111], was used to characterize the interactions of molecular tails of representative, although sometimes unrealistic, conformations that are similar to those of fatty acid chains. The molecules were deemed vertical when their inertial axis from head to tip of tail was coincident with the surface normal, and the static contributions to the energy were minimized with respect to the unit cell axes, the angle between them, the translational offsets from the unit cell origin, and one Euler angle about the plane normal. Deviation from circular cross section was measure by an ellipticity parameter, ϵ, defined as the projection of the vertical molecule's cross section ratio of minor and major axes. Enantiomers were generated by creating alkane helices of opposite handedness but varying ellipticities.

Pure enantiomers with $\epsilon = 1.02$, essentially of circular cross section, were found to pack in a centered hexagonal lattice (*p3*), and their racemate gave the expected centered rectangular lattice (*pg*). For a more elliptical coil $\epsilon = 1.07$, an oblique lattice (p1) was found for a pure enantiomer but still centered rectangular (*pg*) for the racemate, although this lattice was of higher energy than the oblique net. It was further found that the energies of the more elliptical molecule had considerably larger lattice energies than the corresponding circular systems. These are conclusions that would be predicted by reasoning based on traditional solid-state concepts of molecular packing.

Calculations were also carried out on stearic acid itself, assuming an all-trans configuration. Calculations of lattices in the centered rectangular geometry were within 0.3 kcal/mol of the lower energy oblique net. This difference is negligible for films at normal ambient temperature. Because there is no ellipticity for stearic acid, comparison with the results of the helical molecules is not appropriate. Nevertheless, the potential energy surface displays similar close-lying local minima similar to that of the circular coil. They apparently determine the preference for the *pg* symmetry of the 2-D lattice. Essentially the same results, quite consistent with experiment [112–114], were found for two other alkane tails of fatty acid amphiphiles ($C_{17}H_{36}$ and $C_{22}H_{43}$).

These calculations, based on a rather naive geometry, support Langmuir's idea and further suggest that the cross section of the amphiphiles' tails may be determinative of the packing. These calculations for rigidified molecules yielded agreement with experimental finding. Perhaps the most interesting implication is that the packing is consistent with solid-state principles and thus is in line with the theoretical development given above.

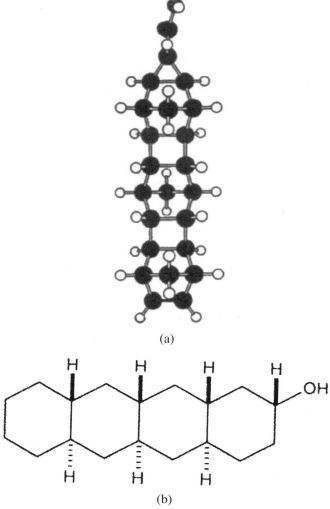

(a)

(b)

Fig. 17.   (a) TNBC amphiphile. (b) TCA amphiphile.

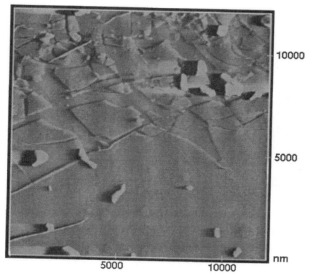

Fig. 18.   Unfiltered atomic force microscopy image of a 10 μm × 10 μm area of TNBC on atomically flat mica [116]. The bright structures are small fragments of matter that have been deposited on the film.

8.2. Control of Planar Packing by the Design of an Amphiphile's Cross Section

Based on the results of the above model calculation, several amphiphiles were synthesized that had cross sections designed to differ significantly from that of fatty acids. To further simplify behavior, the amphiphiles were made to be as rigid as possible. Two such molecules are of particular interest. One, trinorboranecarboxylic acid (TNBC) (Fig. 17a), has an average cross section well approximated by a lozenge formed by abutting two equilateral triangles at their bases. Atomic force microscopy studies of the TNBC monolayer film deposited on atomically flat mica showed nearly hexagonal static packing ($\gamma = 64.6° \pm 0.9°$) [115]. Of course, there is no guarantee this is the packing on the aqueous subphase, but it is certainly quite likely.

A startling observation was reported for atomic force microscopy images of TCNB obtained at lower, mesoscopic resolution [116]. Upon transfer to the mica, it was obvious that the film had shattered in some regions (Fig. 18). Crystallites could

be clearly observed that had distinct interlinear angles of 60° and 120° (within error). This clear demonstration of cleavage lines was complemented by observation of regions of the transferred film that showed conchoidal fractures typical of glasses. This suggests that the film was plastically deformed during the transfer to the mica to such a degree that it became amorphous. This emphatically underlines the importance of elasticity in the consideration of a film's physical behavior.

Another molecule, tetracyclohexane carboxylic acid (TCA) (Fig. 17b), was synthesized so that a rigid, chiral amphiphile with a pronounced ellipticity could be examined [117]. Three phases were observed on the aqueous subphase. Upon transfer of each phase to atomically flat mica, it was observed that the lowest pressure phase was amorphous, whereas the intermediate pressure phase was a rectangular centered lattice. This is exactly what would be expected from solid-state packing of a racemate of markedly elliptical molecules. A surprise came with the high-pressure phase, where two mirror image oblique lattices were seen to be accompanied by significant local disorder. This could only have occurred with a chiral phase separation. However, it is unlikely this happened on the aqueous subphase, because there is no easily envisaged mechanism that would allow two enantiomers of opposite chirality to diffuse past each other when the compression of the film is increasing. This suggests that this occurs when transfer to the mica subphase is achieved and again emphasizes the role of the elastic properties of the film in dictating its stability under mechanical perturbation. Thus, it is likely that the intermediate phase of the TCA Langmuir film is tilted and rectangular, whereas the high-pressure phase is vertical and rectangular. These experimental findings strongly support a solid-state view of Langmuir monolayers. This concept was pursued by performing additional computer experiments.

8.3. Cross Section Potentials

The results of both the atom–atom potential calculations for model amphiphiles and fatty acid chains and the experimental observations of real model compounds whose cross sections were based on the conclusions of these calculations, led to computer simulations that centered on the idea that packing could be studied by the development of so-called cross section potentials. The goal of these simulations would be to ascertain whether the phase behavior predicted by cross section potentials would track that of real systems.

Because untilted phases are universally observed for fatty acid amphiphiles, modeling of vertical molecular chains could be used to assist understanding of the vertical phases: CS, S, and LS. Recall that the first two form centered rectangular nets, with the S phase possessing more orientational disorder reminiscent of the rotator I phase in paraffin crystals in which molecular rotations are restricted. The LS phase is the rotator phase, which has a dynamic hexagonal packing but liquid-like viscosity. The S and LS phases and the transition between them were modeled with the use of cross section potentials in a Monte Carlo simulation [46]. Although there were other simulations of films of diatomic molecules [118] and rigid rods [56–58], they were not successful at describing transitions between the untilted phases. Although there are excellent simulations that provide for variation of all degrees of freedom of the system [44, 45, 119], their very complexity make the underlying physical picture difficult to see.

To address the efficacy of the importance of the cross section in determining amphiphile packing, a simple model was adopted [120]. A number of 2-D molecules were allowed to unrestrictedly rotate and translate in a plane under the influence of an anisotropic pair potential and applied external surface pressures. The pair potential is constructed to model the interaction of two rigid 2-D objects with the cross section of an amphiphile projected onto the plane. The forces that keep the molecules in the plane constrain these molecules to be rigid rotors. The translational freedom permits the system to display different space group symmetries. A Monte Carlo simulation of the isobaric–isothermal ensemble [46] was used to study the behavior of this model system, with a particular focus on the role of the anisotropy of the cross sections of the molecules.

The cross section of a plane figure is specified by $\rho(\gamma)$, which expresses the dependence of the radial distance from the center of the object, ρ, on the angle from the direction of the x-axis, γ. For a nearly circular profile, this relationship can be expressed as

$$\rho(\gamma) = \rho_0\big[1 + \alpha g(\gamma)\big] \tag{8.1}$$

where ρ_0 is the nominal radius of the molecule's cross section and α and $g(\gamma)$ determine the magnitude and type of deviation from a circular cross section. Because the cross section of a fatty acid amphiphile is reminiscent of an almost square rectangle with curved corners, values of $\alpha = 1/30$ and $g(\gamma) = \cos 4\gamma$ were chosen.

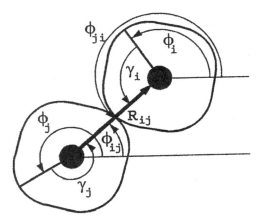

Fig. 19.   Cross section model system coordinates.

A modified Lennard–Jones pair potential of the form

$$\phi(r) = \frac{A}{r^{12}} - \frac{B}{r^6} \tag{8.2}$$

was used where r is the center-to-center distance of the molecules. The touching of the perimeters of the cross sections is associated with the separation $r = r_{\mathrm{pm}}$ that minimizes the potential. Radii of the two interacting molecules are taken to be additive, and the attractive (second) term of the potential is assumed to be isotropic, forcing all of the anisotropy into the repulsive term. These considerations produce an intermolecular cross section potential that depends on the relative positions of the centers of the molecules and their orientations:

$$\phi(r_{ij}, \gamma_i, \gamma_j) = \varepsilon\left[\left(\frac{2\rho_0}{r_{ij}}\right)^{12}\left(1 + \frac{\alpha}{2}\{g(\gamma_i) + g(\gamma_j)\}\right)^6 - 2\left(\frac{2\rho_0}{r_{ij}}\right)^6\right] \tag{8.3}$$

Using the $\rho(\gamma)$ that describes a rounded square, expanding in powers of α, and neglecting higher order terms gives

$$\phi(r_{ij}, \phi_{ij}, \phi_i, \phi_j) = \varepsilon\left[\left(\frac{2\rho_0}{r_{ij}}\right)^{12}\left(1 + 3\alpha\big[\cos 4(\phi_{ij} - \phi_i)\right.\right. \\ \left.\left. + \cos 4(\phi_{ij} - \phi_j)\big]\right) - 2\left(\frac{2\rho_0}{r_{ij}}\right)^6\right] \tag{8.4}$$

which is the cross section potential that was used in these calculations (Fig. 19).

A typical snapshot of the system is shown in Figure 20. The molecules are in a random hexagonal arrangement in Figure 20 (top), but at low temperature are oriented in an ordered centered rectangular structure (Fig. 20) (middle). The ratio of the NN to NNN (NN/NNN) distances was $\sqrt{3}$ for hexagonal lattices that always yielded one molecule per unit cell.

The calculation addressed the question of free rotation in the random or ordered phases. Because the calculation only evaluates anisotropy along the molecular line of centers there is little

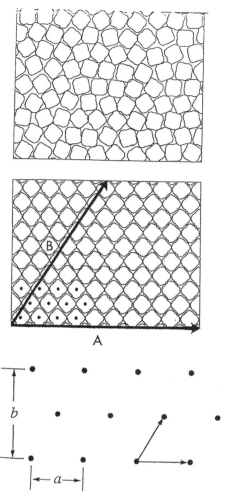

Fig. 20. *Top*: A snapshot of the high temperature ($T = 0.200$) phase. *Middle*: A snapshot of the low temperature ($T = 0.001$) phase. **A** and **B** indicate vectors used to generate the periodic space. *Bottom*: Diagram illustrating a and b which, for the systems studied, correspond to the next-nearest and nearest neighbor distances, respectively. Average primitive lattice vectors, shown as bold arrows, may be derived directly from the periodic space vectors **A** and **B**.

of the cross section potential and the saddle points between them.

The cross section model gives an extremely simple but qualitatively accurate picture of the S-to-LS transition in alkanoic acid amphiphiles. The anisotropy of the potential, essentially that of the elastic dipoles, dictates this behavior: the anisotropy must be thermally overcome so that the molecules may rotate. The success of this simple model, so directly related to the elastic dipole concept, indicates the importance of the latter in understanding Langmuir film packing and phase behavior.

8.4. Extension of the Cross Section Potential to Simulation of the S → LS Transition

In the prior simulation, the symmetry of the amphiphilic cross section was taken as fourfold, thereby ignoring the slight difference between the long and short axes of the projected cross section. Although the phase transition from the solid to the superliquid phase was found to be continuous, its order was open to further investigation. This requires both a larger system for simulation as well as the calculation of many more data points near the phase transition. To achieve meaningful statistical results, more Monte Carlo steps are needed because of the greater amplitude of the fluctuations near the phase transition.

Because weakly first-order processes can be difficult to distinguish from continuous ones, the effect of finite size must be examined. Of course, for macroscopic systems, essentially infinite, first-order transitions are discontinuous in the internal energy, leading to singularities in the heat capacity at the transition. This behavior of the heat capacity is not observed for "finite" systems, and the specific heat function is also finite. In such systems, however, first-order transitions have the maxima in the function proportional to the dimensional volume, L^D, with widths decreasing as L^{-D}.

For infinite systems, second-order or continuous transitions show the heat capacity as a λ-function proportional to $\tau^{-\alpha}$ for infinite systems where $\tau = (T - T_c)/T_c$, T_c is the critical temperature, and α is the critical exponent. In finite systems the specific heat function is broadened because the correlation length is size-limited. Scaling theory [121, 122] shows that the maximum in the specific heat should be proportional to $L^{\alpha/\nu}$, with width decreasing as $L^{-1/\nu}$, where ν is related to the correlation length by $\tau^{-\nu}$. The temperature at the specific heat maximum is also size-dependent, with $\tau \propto L^{-1/\nu}$. By extension from 2-D Ising models [123], a further effect of the size of the lattice can be ascertained. For a first-order transition the area of the specific heat curve should be proportional to the maximum's value, whereas for the second-order transition it will be independent of size.

The model is that of the cross section potential given in Eq. (10.4), with $\alpha = 1/30$ to allow for anisotropy. The potential was smoothly taken to zero at $r_{ij} = 1.25\rho_0$. Using system sizes ranging from 144 to 576 particles, was initiated a starting configuration of centered rectangular packing with particles at identical orientational angles at low temperature ($kT/\varepsilon =$

dependence on the external surface pressure, π, and α. The initial b/a ratio decreases with increasing α but increases only slightly with increasing π. This arises from the potential being more anisotropic near the minima than in the hard core region. Although not in quantitative agreement with experimental results, the qualitative transition behavior between the S and LS phases is mirrored.

The simulation indicates that the essential mechanism inducing the transition between the two phases in question arises from the anisotropy of the molecules. The prior theoretical discussion emphasizes the relation of such cross section potentials to the interactions of elastic dipoles representing the molecules. The simulations are also consistent with the purely theoretical result that the phase transition is continuous. It is also clear that the angular correlation coefficients decrease more rapidly for NNN than for NN. The thermal energy for random orientations is on the order of the energy difference between the minima

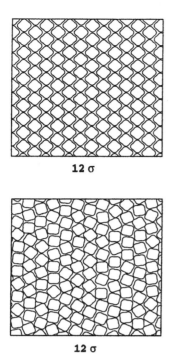

12 σ

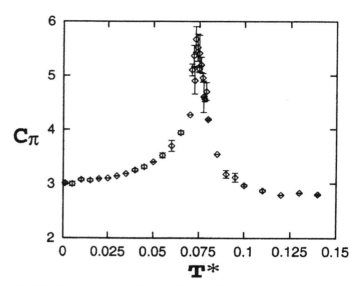

Fig. 22.   Temperature dependence of the specific heat shown in reduced units for $N = 100$. The shape of the curve is typical of those employed in a finite-size effect study. For larger sizes, calculations were concentrated around the transition temperature.

12 σ

Fig. 21.   *Top*: Monolayer amphiphiles represented by objects of four-fold symmetry in a plane. In the low temperature S-phase there is insufficient thermal energy for free rotation. The molecules pack in a centered rectangular net. *Bottom*: The same monolayer in the high temperature LS phase. Rotation of the molecules makes them essentially isotropic yielding a hexagonal lattice.

0.05), and thereafter it was incrementally increased, with the final configuration of the prior calculation as input. This process was repeated until the phase transition was observed (Fig. 21). The heat capacity was calculated from potential energy fluctuations, an average for each run of 250,000 moves per particle, according to

$$C_\pi = \frac{\langle V^2 \rangle - \langle V \rangle^2}{NkT^2} + \frac{3}{2}k \qquad (8.5)$$

The calculation clearly shows that the ratio of peak height in the calculated heat capacity curve is proportional to the area that is characteristic of a first-order transition (Fig. 22).

A clear bound-to-free rotor transition is observed in this highly idealized model that is consistent with the simpler one discussed above. The weakly first-order transition is consistent with experimental observations.

Of greater interest to the problem of hexatic character addressed in the theoretical section, the correlation function of the sixfold bond orientational order parameter was calculated according to

$$g_6(r) = \left\langle \psi_6(0)\psi_6(r)^* \right\rangle \qquad (8.6)$$

where ψ_6 is the sixfold bond orientational order parameter:

$$\psi_6(r) = \sum_{k=1}^{6} \exp(6i\theta_{jk}) \qquad (8.7)$$

Here θ_{jk} is the angle with an arbitrary axis made by an imaginary bond between nearest neighbors j and k with summation over the six nearest neighbors of particle j. For a hexatic phase, the envelope of the correlation function should decrease algebraically, and it is indeed found to do so. Because the cross section model is a strict 2-D system, any other outcome would have been surprising. The calculation also shows that the translational order is short range, which is also consistent with hexatic order. It appears that the hexatic order arises from the anisotropy of the cross sections rather than from free dislocations, as is the usual case. However, it may also arise from the fact that there is insufficient equilibration to remove the order of the original lattice.

Introduction of twofold instead of fourfold symmetry could influence the S phase packing and drive it to a herringbone ordering that is characteristic of S and CS phases.

8.5. Bead Potentials

Although the 2-D cross section potentials described in the previous section are certainly useful, their usefulness is effectively limited to those Langmuir film phases where the surface pressures are high or at low temperature and/or area per molecule. These are conditions where the molecules are normal to the plane of the subphase, but the abundance of phases compels some expansion of the cross section potential model. Tilt necessarily involves an angle of orientation of the molecule from the normal to the plane and thus increases the problem to a 3-D one.

Again, simulations of great complexity met early on with a great deal of success, especially in terms of their accuracy [39–45], but their complexity inhibits easy conceptualization. Simple models employing rigid rods [54, 56, 58] or centers with

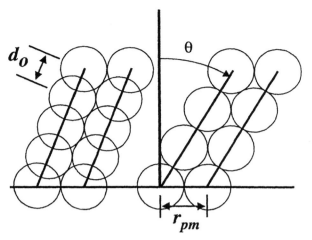

Fig. 23. Illustration of interbead distance's (d_0) effect on the degree of amphiphile tilt.

Lennard–Jones potentials [92, 124, 125] were also soon devised to assuage this difficulty and have provided a good qualitative picture of the mechanisms driving the phase behavior of Langmuir films. A particularly interesting model involved spherocylinders grafted onto a hard 2-D lattice [58]. This simulation found a first-order phase transition going from an untilted phase to a tilted NNN when the lattice spacing was increased. The lattice was seen to expand in the direction of the tilt but contracted perpendicular to that direction. Furthermore, a second tilt transition toward NN was seen when amphiphile–subphase interactions were modeled by the introduction of a surface potential.

The characteristics of the amphiphiles play no role in models involving tilting of rigid rods, where lattice spacing and length of the rods, are determinative, with the latter inversely proportional to the degree of tilt. The introduction of beaded string models [92, 124, 125] permitted some insight into the role of molecular properties. In these models, the methylene groups of the alkanoic acid amphiphiles are modeled by linked spherical potentials. These provided additional insight into the tilt behavior of the fatty acid amphiphiles, especially with reference to the role of temperature. The appearance of a uniform NNN tilt for seven bead molecules constrained to a triangular lattice was found [92]. Other studies dealt with the consequences of allowing the beads to change their conformations [125]. These calculations, which are very similar to those used in the analysis of packing problems in solids, left some problems that required investigation: (i) influence of string length, (ii) role of the interbead distance, and (iii) role of pressure. Uniformly tilted molecules were simulated because they are the most commonly observed packing.

The model employed here relies on the interbead distance, d_0, as the primary geometric parameter that characterizes the model amphiphile (Fig. 23). The beads are an averaged group of atoms, essentially the methylene groups. It is possible to have values of $d_0 \geq 1$ for amphiphiles such as staffanes [126] where the molecular beads may be essentially directly linked.

As seen in Figure 23, close packing principles can be seen as operative. Considering the beads as packing in a plane perpendicular to the subphase, it is clear that a hexagonal close packed array is achievable. Values of d_0 of 0.3 and 0.7 allow a comparison with previous results [92, 124, 125], whereas those of 1.0 and 1.4 are limiting geometries for van der Waals separation between the beads and hexagonal closest packing (to two significant figures), respectively. A NN tilt results in a non–physical geometry for $d_0 > \sqrt{2}$ because there would, realistically, be some chemical entity connecting the beads and sterically forbiding such an arrangement.

In the spirit of previous calculations, linear strings of identical, but finite, numbers of beads are attached to a perfect lattice. The only constraint on lattice symmetry was that there could be only one molecule in a unit cell, which is consistent with the assumption of uniform tilt for molecules of radial symmetry and previous studies of similar systems at low temperature [92, 124, 125].

The system's energy was minimized with respect to the lattice parameters (α, β, γ) and orientational angles (θ and φ), where φ is the azimuthal angle and θ is shown in Figure 21. The meaning of NN and NNN becomes unclear when the lattice distorts significantly from hexagonal symmetry, so care was taken to ensure that these labels remained appropriate for a particular minimum. This was achieved by using a gradient minimization that located the local minimum nearest to $\theta = 0$ in each of three directions: NN ($\varphi = 0$), NNN ($\varphi = 90°$), and an intermediate tilt with $\varphi = 15°$. This prevents the lattice from significant departures from hexagonal symmetry. A NN tilt is then equivalent to a tilt along the x axis, and a NNN tilt is equivalent to a tilt along the y axis.

The Lennard–Jones potential was written as a function of the distance to the potential minimum, r_{pm}:

$$\phi(r) = \varepsilon\left(\frac{r_{pm}^{12}}{r^{12}} - \frac{2r_{pm}^6}{r^6}\right) \qquad (8.8)$$

where r is the distance between the centers of the beads and $r_{pm} = 2^{1/4}\sigma$, where σ is the core diameter. Both ε and r_{pm} were set to unity, and the potential was smoothly truncated, beginning at the reduced distance 2.0. Energy differences between minima and saddle points were often small, making the details of the potential important.

Minimum energy configurations were sought, with particular emphasis on NN and NNN tilt directions. Because there is an energy barrier between these latter two directions, the two associated minima were compared to determine the thermodynamically favored tilt direction. Results of these calculations are summarized in Tables IV and V.

The picture that develops from the calculations is that a global energy minimum for short-length molecules is found for vertical ($d_0 = 1/3$) or NNN ($d_0 = 0.7$) configurations. For long lengths, however, the tilt is toward NN. However, substantial tilt was not observed until the model molecules had more than 20 beads, because at these lengths the amount of overlap increases, thereby allowing the repulsive part of the potential to become significant and drive the system toward a collective

Table IV.   Length Dependence of Preferred Tilt Direction: Zero Pressure

d_0	NN	NNN	Vertical	Intermediate
1/3	—	>20 beads	<20 beads	>20 beads (hex lattice)
0.7	>12 beads	<12 beads		
1.0	—	All lengths		
1.4	—	All lengths		

Table V.   Pressure Dependence of Preferred Tilt Direction: Length = 15 Beads

d_0	NN	NNN	Vertical	Intermediate
1/3	—	Pressure <100	Pressure ≥ 100	
0.7	—	Pressure ≥ 100	Pressure $\geq 10^4$	$10^3 \leq$ Pressure $\leq 10^4$
1.0	—	All pressures		
1.4	—	Pressure $\leq 10^3$		Pressure $\geq 10^3$

tilt. Because a pressure–volume term was not included in the energy, this tilt is not due to pressure. The only role isotropic pressure plays in this model is to require molecules to be longer for NNN tilt to occur.

The two special cases, $d_0 = 1/3$ and 0.7, are of interest. The first is straightforward where the associated hexagonal lattice had molecules consistently tilted in an intermediate direction ($\varphi \approx 130°$–$140°$) for systems with more than 20 beads. The other case was less simple, with the minimum energy configuration for molecules with less than 12 beads showing NNN tilt but longer ones displaying NN tilt. Because NN minima have a larger "area" per string than those for NNN, application of pressure will determine which tilt will occur for a given length of molecule. With large local pressure, the NN phase completely disappears.

In general, isotropic pressure affects the direction and magnitude of the tilt. The pressures required to produce vertical order are orders of magnitude larger than those required to produce changes in the tilt. At higher pressures the lattice approaches hexagonal symmetry, with responses mirrored by those where $d_0 = 1/3$, where tilt angles are again in intermediate directions. This latter observation is suggestive of the L'_1 phase of fatty acid monolayers. Thus, the $d_0 = 0.7$ system has the same progression of tilt phases as the fatty acids: NN at low pressure, NNN at higher pressure, and vertical at high pressure. The pressures do not scale similarly, however, although this could be greatly improved by the inclusion of temperature dependence.

For the $d_0 = 1.0$ system, NNN is the minimum energy tilt direction, and the system is that of layer spheres with hexagonal closest packing, which is stable for all tested reduced pressures. By increasing d_0 to 1.4, a tilt to NNN was found for reduced pressures up to 10^3, with intermediate tilts preferred at higher pressures.

This model provides a qualitative picture of three types of amphiphiles that may be designed to form films with predeter-mined properties consistent with the molecular properties. The amphiphiles may be characterized by ranges of d_0:

1. *Smooth amphiphiles* ($d_0 \leq 0.3$): rigid rods model such amphiphiles that form vertical phases with finite pressure and subsequently pack according to their cross section potentials (Section 8.3). A novel amphiphile that demonstrates the predicted behaviors has been reported [115].

2. *Discrete amphiphiles* ($d_0 \geq 1.0$): The packing of amphiphiles of this type is dictated by the disparity in the size of the "beads" and the sections linking them. These may be expected to produce stiffer films than usual. In fact, staffanes are reported to pack in a vertical phase [126]. Another unusual amphiphile with methylene group protuberances clearly models the discrete amphiphiles and shows the expected behavior [115, 116].

3. *Corrugated amphiphiles* ($d_0 \approx 0.7$): amphiphiles of this type, of which the fatty acids are prototypical, are expected to exhibit the richest phase behavior. They possess an overlapped geometry that requires significant but easily obtained surface pressures to achieve vertical packing. Their geometry is associated with a potential energy surface that is shallow enough to provide a multiplicity of states.

A characteristic of those systems for which $d_0 < 1$ is that tilt directions are determined by the surface pressure. The temperature of the film determines whether states with NN or NNN tilt form, depending on the nature of the rotational barrier that separates them. A resulting precessing phase would be analogous to the LS phase in the case of fatty acids, whereby the molecules would only be vertical on average. Experiments indicate that this would only be so for small tilt angles [116].

The role of temperature for these systems is only in qualitative agreement with the alkanoic acid phase diagram because it does not scale quantitatively. This behavior may indicate that temperature-induced nonuniform tilt and conformational lability are needed to accurately reproduce experiments. Given that the potential of this idealized amphiphile is quite flat at its bottom, the introduction of lattice distortion would be quite important in quantitatively reproducing experimental results.

9. CLOSING REMARKS ON THE SOLID-STATE MODEL FOR LANGMUIR FILMS

Guided by recent discussions on the ordering of Langmuir monolayer mesophases [26, 29], we have suggested a phenomenological model of the hexagonal, parent phase, LS. Within this picture, the phase is composed of clusters (domains), which have local equilibrium structures that are, rectangular and/or oblique. Orientational ordering is assumed to be at two levels, molecular and cluster. At the molecular level, the local structures are formed, and they are determined by an interplay of molecular translational–rotational coupling and direct van der Waals interaction between tails [3]. Orientational order of the clusters may come from elastic interaction between the domains which is more characteristic for lower symmetry than for the global symmetry of the LS phase. We chose the simplest model: three rectangular/oblique domains characterized by local stresses—the elastic dipoles. Correlation between the domains due to the elastic interaction has been considered.

The following consequences of the model have been obtained. The elastic shear response of a macroscopic system is expected to be smaller than the microscopic one because of the elastic orientational relaxation of the domains. This provides an explanation for an already observed effect [30, 31]. Moreover, we have considered the time evolution and frequency dependence of the elastic response and interpreted the difference between static and high-frequency shear elasticity [107], as well as anomalous shear viscosity [108]. We have shown that the positional order of the monolayers, as observed by X-ray scattering [26], can be influenced by the disorder of the clusters and their correlation. We have demonstrated how elastic domains cause positional disorder and contribute to X-ray diffuse scattering. The latter's intensity is proportional to the local strain correlation function and depends on the product of elastic dipoles, which are measures of local stresses. These considerations provide a model-based explanation for the observed anomalous broadening of the diffraction peaks [26]. For systems where sufficiently small clusters are formed and whose monolayers are either close to a solid–liquid transition or possess rather short molecular tails (vanishingly small elastic dipoles), positional disorder will occur at the molecular level with an exponential decay, as predicted for a purely 2-D net. However, for molecules with longer tails that are able to form larger clusters and mesoscopic domains, the scattering experiment must measure their correlation function, which decays algebraically and which is a symptom of a crystalline state for the monolayers.

The computer experiments serve to demonstrate the efficacy of the model. When the normal solid-state packing calculation is applied to model (indeed, idealized) amphiphiles, the results clearly indicate that behavior mirrors the packing of 3D systems. Leaving the question of the actual dimensionality aside, these results are sufficient to confirm the idea of Langmuir that the tails determine the packing and that amphiphiles can be designed by a "crystal engineering" approach to design films of desired properties.

In an effort to determine the 2-D nature of films and examine the role of the cross section of the amphiphile in packing, a potential related to the molecule's cross section was devised for an idealized alkanoic chain. This potential of fourfold symmetry can be viewed as representative of the $\langle Y_4 \pm Y_5 \rangle/2$ (xx, yy) components of the elastic constant multipole. This extremely simple potential successfully reproduced the nearly vertical phase boundary between the S and LS phases. In addition, it demonstrated that a representation of an amphiphile's cross section could produce a quite simple calculation that would be reasonably predictive of its packing in a film at high pressure.

Extension of the cross section model to Monte Carlo simulations further refined the picture of the S-to-LS transition that was consistent with experiment. At low temperatures the anisotropy of the potential prevents the molecules from rotating, but at higher temperatures this constraint weakens until, at the phase transition, free rotation occurs and the molecules pack in a hexagonal array. The transition is shown to be weakly first-order, and the simulation clearly shows hexatic structure for this pure 2D system, as may be expected.

Recognizing that the films are not actually 2D, the model was extended to consideration of clusters of atoms, chemical groups such as methylenes, as "beads" on a string. The interaction of beads on different strings was represented by a truncated Lennard–Jones potential. Again, high-pressure behavior was reproduced, and the influence of chain length was able so show both length and pressure dependence consistent with experiment. The model permits the beads to be arranged in several geometries of interest, and three general types were found: smooth, discrete, and corrugated. The tilt and packing behavior are related to closest packing concepts commonly employed in the solid state. Smooth amphiphiles exhibit vertical phases under finite pressure. Discrete amphiphiles are always tilted because of closest packing considerations. The corrugated amphiphiles are intermediate in behavior with the richest phase behavior qualitatively similar to that of a fatty acid phase diagram.

The calculational experiments demonstrate the applicability of the solid-state model of amphiphile packing in monolayers on liquid subphases. The limited results are quite consistent with both the theoretical structure given above and experimental results for alkanoic acid films.

Our phenomenological picture serves as a bridge between more rigorous theories for 2-D systems. These have been described and compared by Kats and Lejzerowicz [32], who conclude that the Landau "weak crystallization" concept is more suitable for molecules with long tails, which are expected to have a clear solid–liquid transition, whereas the theory of melting by a dislocation dissociation mechanism (predicting the hexatic phase) is more suitable for systems with molecules with shorter tails. Within our picture this might be mapped onto a scale of clusters that are dependent on direct tail–tail interactions. We have briefly indicated how the "weak crystallization" concept can be understood within our phenomenological model. In general terms, the model we offer treats Langmuir monolayers as frustrated systems, where a local structure (and equilibrium) competes with a global structure (and equilibrium). In terms of competing interactions, the effective rotational coupling between tails (van der Waals and translational–rotational coupling at a molecular level) that form clusters of lower symmetry competes with long-range elastic interactions between the elastic domains. This review demonstrates that the long advocated approach of treating mesophases of Langmuir monolayers as disordered solids [3, 5, 6] is not only a useful concept, but is the only model that is consistent with all present experimental evidence.

10. APPENDIX

10.1. Symmetry Aspects of the 2-D Hexagonal Close-Packed Lattice

The 2-D hexagonal, close-packed lattice is represented in Figure 4a. It belongs to the *p6m* planar space group. The basis vectors of the Bravais lattice, given in terms of the (x, y) Cartesian coordinates, are $\mathbf{t}_1 = ((1/2)a, (1/2)\sqrt{3}a)$ and $\mathbf{t}_2 = (-a, 0)$; a is the hexagonal lattice constant. The basis vectors, $\mathbf{g}_1$ and $\mathbf{g}_2$, of the reciprocal 2-D lattice are given in terms of Cartesian coordinates as $\mathbf{g}_1 = (2\pi/a)(-1, 1\sqrt{3})$ and $\mathbf{g}_2 = (2\pi/a)(0, 2/\sqrt{3})$. Figure 4b shows the vectors in the reciprocal lattice and the corresponding Brillouin zone. The high-symmetry points (Γ, Σ, and $\mathbf{M}$), expressed in terms of the reciprocal lattice vectors, are $\Gamma(0, 0)$, $\Sigma(0, 0 < \zeta < 1/2)$, and $\mathbf{M}(0, 1/2)$.

The wave vector, $\mathbf{q}$, is defined in the reciprocal lattice as

$$\mathbf{q} = \lambda_1 \mathbf{g}_1 + \lambda_2 \mathbf{g}_2 = |\mathbf{q}|(q_x, q_y) \tag{A.1}$$

where the discrete values of the vector are determined by the size of the crystalline sample, i.e., by the correlation lengths, L_x and L_y,

$$q_i = \frac{1}{NL_i} \tag{A.2}$$

with the number of the wavevectors given by

$$N = \left(L_x^{-2} + L_y^{-2}\right)^{1/2} \tag{A.3}$$

The mesoscopic limit, $\mathbf{q} \Rightarrow 0$, has to be understood in the context of the finite-size crystallinity of a sample.

Symmetries of the coupling matrices, $\mathbf{M}(\mathbf{q})$, $\mathbf{V}(\mathbf{q})$, and $\mathbf{J}(\mathbf{q})$, were determined with the use of symmetry operations of the point group C_{6v}. Only couplings between nearest neighbors are considered. The matrices of coupling constants for the pair of molecules $(0, 0)$ and $(a, 0)$ have to be invariant with respect to the σ_{v1} symmetry operation, e.g.,

$$R(\sigma_{v1})M(0, 0; a, 0)R^t(\sigma_{v1}) = M(0, 0; a, 0) \tag{A.4}$$

$$V(\sigma_{v1})V(0, 0; a, 0)S^t(\sigma_{v1}) = V(0, 0; a, 0) \tag{A.5}$$

$$S(\sigma_{v1})j(0, 0; a, 0)S^t(\sigma_{v1}) = J(0, 0; a, 0) \tag{A.6}$$

Transformation of Cartesian coordinates is given by the matrix

$$R(\sigma_{v1}) = \begin{pmatrix} 1 & 0 \\ 0 & -1 \end{pmatrix} \tag{A.7}$$

and the transformation of the functions, Y_α, $\alpha = 1, \ldots, 5$, is given by the matrix

$$S(\sigma_{v1}) = \begin{pmatrix} 1 & 0 & 0 & 0 & 0 \\ 0 & -1 & 0 & 0 & 0 \\ 0 & 0 & 1 & 0 & 0 \\ 0 & 0 & 0 & 1 & 0 \\ 0 & 0 & 0 & 0 & -1 \end{pmatrix} \tag{A.8}$$

The transformations (A.4)–(A.6) reduce the number of independent parameters of the coupling matrices. The coupling matrices for the other five pairs of molecules are generated by other symmetry operations of the point group. For example,

for the pair of molecules $(0, 0)$ and $(1/2a, (\sqrt{3}/2)a)$, the six-fold rotational symmetry operation can be used. Corresponding transformation matrices are

$$R(C_6) = \begin{pmatrix} \frac{1}{2} & -\frac{\sqrt{3}}{2} \\ -\frac{\sqrt{3}}{2} & \frac{1}{2} \end{pmatrix} \tag{A.9}$$

$$J(\Sigma) = \begin{pmatrix} 2J_{11} + (J_{11} + 3J_{33})\,f_1 & 0 & 0 \\ 0 & 2J_{22} + (J_{22} + 3J_{11})\,f & 0 \\ 0 & 2J_{33}f_2 & 2J_{34}f_3 \\ 0 & 0 & 2J_{34}f_3 \\ 0 & 0 & 0 \end{pmatrix}$$

$$\begin{matrix} 0 & 0 \\ 0 & 0 \\ 0 & 0 \\ 2J_{44} + (J_{44} + 3J_{55})\,f_1 & 0 \\ 0 & 2J_{55} + (J_{55} + 3J_{44})\,f_1 \end{matrix} \Bigg)$$

$$S(C_6) = \begin{pmatrix} \frac{1}{2} & -\frac{\sqrt{3}}{2} & 0 & 0 & 0 \\ \frac{\sqrt{3}}{2} & \frac{1}{2} & 0 & 0 & 0 \\ 0 & 0 & 0 & 0 & 0 \\ 0 & 0 & 0 & -\frac{1}{2} & -\frac{\sqrt{3}}{2} \\ 0 & 0 & 0 & \frac{\sqrt{3}}{2} & -\frac{1}{2} \end{pmatrix} \tag{A.10}$$

With the coupling matrices for all six pairs of nearest neighbors determined, the Fourier transforms of the coupling matrices $\mathbf{J}(\mathbf{q})$, $\mathbf{V}(\mathbf{q})$, and $\mathbf{M}(1)$ are calculated. In particular, the elements of the rotational–rotational coupling matrix are

$$J_{11}(\mathbf{q}) = 2J_{11} \cos 2\alpha + (J_{11} + 3J_{22}) \cos \alpha \cos \beta$$

$$J_{12}(\mathbf{q}) = i\sqrt{3}(J_{22} - J_{11}) \sin \alpha \cos \beta = -J_{21}(\mathbf{q})$$

$$J_{13}(\mathbf{q}) = i2J_{13}(\sin 2\alpha + \cos \beta \sin \alpha) = -J_{31}(\mathbf{q})$$

$$J_{14}(\mathbf{q}) = i2J_{14} \sin 2\alpha + i(3J_{25} - J_{14}) \sin \alpha \cos \beta = -J_{41}(\mathbf{q})$$

$$J_{15}(\mathbf{q}) = -i\sqrt{3}(J_{14} - J_{25}) \sin \alpha \cos \beta = -J_{51}(\mathbf{q}) = J_{42}(\mathbf{q})$$
$$= -J_{42}(\mathbf{q})$$

$$J_{22}(\mathbf{q}) = 2J_{22} \cos 2\alpha + (J_{22} + 3J_{11}) \cos \alpha \cos \beta$$

$$J_{23}(\mathbf{q}) = -i2\sqrt{3}J_{13} \sin \alpha \cos \beta = -J_{32}(\mathbf{q}) \tag{A.11}$$

$$J_{25}(\mathbf{q}) = i2J_{25} \sin 2\alpha + i(J_{14} - J_{25}) \sin \alpha \cos \beta = -J_{52}(\mathbf{q})$$

$$J_{33}(\mathbf{q}) = 2J_{33}(\cos 2\alpha + 2\cos \alpha \cos \beta)$$

$$J_{34}(\mathbf{q}) = 2J_{34}(\cos 2\alpha - \cos \alpha \cos \beta) = J_{43}(\mathbf{q})$$

$$J_{35}(\mathbf{q}) = i2\sqrt{3}J_{34} \sin \alpha \cos \beta = -J_{53}(\mathbf{q})$$

$$J_{44}(\mathbf{q}) = 2J_{44} \cos 2\alpha + (J_{44} + 3J_{55}) \cos \alpha \cos \beta$$

$$J_{45}(\mathbf{q}) = i\sqrt{3}(J_{44} - J_{55}) \cos \alpha \cos \beta = -J_{54}(\mathbf{q})$$

$$J_{55}(\mathbf{q}) = 2J_{55} \cos 2\alpha + (J_{55} + 3J_{44}) \cos \alpha \cos \beta$$

with $\alpha = (1/2)q_x$ and $\beta = (\sqrt{3}/2)q_y a$. The matrices, $\mathbf{V}(\mathbf{q})$ and $\mathbf{M}(\mathbf{q})$ are specified in Eqs. (2.30) and (2.33)–(2.35), respectively.

For the $\Sigma(0, q_y)$ direction, the $\mathbf{J}(\mathbf{q})$ matrix reduces to Eq. (A.12) with $f_1 = \cos \beta$, $f_2 = 1 + 2\cos \beta$, and $f_3 =$

$1 - \cos\beta$. For the indirect interaction matrix we get (A.12)

$$L(\Sigma) = \delta \times \begin{pmatrix} 24V_{11}^2 f_1 & 0 & 0 \\ 0 & 4V_{11}^2 & -V_{11}V_{13}f_3 \\ 0 & V_{11}V_{13}f_3 & V_{33}f_2 \\ 0 & 2V_{11}V_{14}f_3 & V_{13}V_{14}f_3 \\ -12V_{11}V_{14}f_3 & 0 & 0 \\ & 0 & 12V_{11}V_{14}f_3 \\ & -2V_{11}V_{14}f_3 & 0 \\ & -V_{13}V_{14}f_3 & 0 \\ & V_{14}f_2 & 0 \\ & 0 & 6V_{14}^2 f_2 \end{pmatrix}$$ (A.12)

with $f_1 = 1 - \cos\beta$, $f_2 = \sin^2\beta(1 - \cos\beta)^{-1}$, and $f_3 = -i\sin\beta$.

10.2. The Local Stress Correlation Function

Here we discuss the correlation function that couples local stresses—the elastic dipoles. The function can be written in terms of the wavevector ($\mathbf{q}$), transformation as

$$K_{ijlm}(\mathbf{x}, \mathbf{x}') = (1/N)\sum_q G_{ijlm}(\mathbf{q})\left[\exp\left(i\mathbf{q}(\mathbf{x} - \mathbf{x}')\right) - 1\right]$$ (A.13)

where

$$G_{ijlm}(\mathbf{q}) = \beta\langle\delta\varepsilon_{ij}(\mathbf{q})\delta\varepsilon_{lm}(-\mathbf{q})\rangle_{\text{unif}}$$ (A.14)

is the correlation function of strain fluctuations in a uniform (disordered) system. The second term in square brackets ensures that the self-term, $K_{ijlm}(\mathbf{x} = \mathbf{x}')$, is excluded from the elastic coupling. This term gives rise to the energy for creation of an elastic dipole in an elastic medium. The strain fluctuation correlation function is

$$G_{ijlm}(\mathbf{q}) = q_i q_l G_{jm}(\mathbf{q})$$ (A.15)

where $G_{jm}(\mathbf{q})$ is the displacement–displacement correlation function. For an isotropic elastic medium, an appropriate approximation for the hexagonal net in the limit of small $\mathbf{q}$ is

$$G_{jm}(\mathbf{q}) = \beta\langle u_j(\mathbf{q})u_m(-\mathbf{q})\rangle$$
$$= (1/q^2\mu_0)\left[\delta_{jm} - (\lambda_0 + \mu_0)/(\lambda_0 + 2\mu_0)n_j n_m\right]$$ (A.16)

where λ_0 and μ_0 are Lamé's constants and $n_i = q_i/q$.

Alternatively, the strain field can be calculated as

$$K_{ijlm}(\mathbf{x}, \mathbf{x}') = (\partial/\partial R_i)(\partial/\partial R_l)G_{jm}(\mathbf{R})$$ (A.17)

where $\mathbf{R} = \mathbf{x} - \mathbf{x}'$, and

$$G_{jm}(\mathbf{R}) = (1/N)\sum_q G_{jm}(\mathbf{q})\left[\exp\left(i\mathbf{q}(\mathbf{x} - \mathbf{x}')\right) - 1\right]$$ (A.18)

Summation over $\mathbf{q}$-space is done by integration over 2-D space. For the calculations, it is important to specify the elastic Green's function, the inverse of the dynamical matrix. At this point, one must decide how the translational subsystem, formed in our model by the head groups, is to be considered. If treated as a purely 2-D elastically isotropic net without an interaction with the water substrate, then one introduces (A.4) into (A.6). The result is the well-known displacement–displacement correlation function for a 2-D net,

$$\langle u_j(0)u_l(R)\rangle_{2D} \approx -(B/4\pi\beta)\delta_{jl}\ln(R/a)$$ (A.19)

in the limit of large R (a being of the order of lattice spacing). The constant B is expressed in terms of the elastic constants,

$$B = 1/\mu_0 + 1/(2\mu_0 + \lambda_0)$$ (A.20)

For our simple model, we also assume the limit of large distances. This can be justified with the model that a domain of a monolayer represents a cluster of tails that are tilted toward (or opposite) a center of the domain. This arises from a geometric frustration due to incompatibility between areas of tails and heads, as mentioned in the introduction. Interaction of domains is assumed to be elastic. Therefore, the coupling is between centers of the domain's stress, which is identified with the geometrical center of a domain. Thus, for reasonably sized domains, the limit of large distances between the stress centers can be justified. In the limit, the correlation function, $K_{ijlm}(\mathbf{x}, \mathbf{x}')$, is denoted as K_{ijlm} and, for the symmetries of the elastic dipoles considered, will be approximated by the formula

$$\beta\langle\delta\varepsilon_{ij}(\mathbf{x})\delta\varepsilon_{lm}(\mathbf{x}')\rangle_{\text{unif}} = K_{ijlm}$$
$$\approx (4\pi\mu_0)^{-1}R^{-2}\delta_{jm}\delta_{il}$$ (A.21)

This result, although approximate, shows that correlation between local strain fluctuations within a uniform 2-D net decays algebraically, as do orientational fluctuations. It is important to notice that the tensor, $\mathbf{K}$, although considered in the limit of large distances, does not correspond to the compliance tensor, $\mathbf{S}^0 = \mathbf{C}^{-1}$ (the inverse of the elastic constant tensor), as it would for a 3-D system.

Acknowledgments

This work was supported by the Chemical and Biological Sciences Division of the United States Army Research Office under grants DAAH04-93-0159, DAAL03-89-G-0094, DAAL03-92-G-0396, and DAAH04-96-1-0394. The authors thank Dr. Krzysztof Rohleder, Dr. David R. Swanson, James J. Haycraft, Rick Albro, and Christine Erickson for their assistance in the preparation of this chapter.

REFERENCES

1. C. M. Knobler and R. C. Desai, *Annu. Rev. Phys. Chem.* 43, 207 (1992).
2. V. M. Kaganer, H. Möhwald, and P. Dutta, *Rev. Mod. Phys.* 71, 779 (1999).
3. T. Luty and C. J. Eckhardt, *J. Phys. Chem.* 99, 8872 (1995).
4. T. Luty and C. J. Eckhardt, *J. Phys. Chem.* 100, 6793 (1996).
5. D. R. Swanson, T. Luty, and C. J. Eckhardt, *J. Chem. Phys.* 107, 4744 (1997).
6. T. Luty, D. R. Swanson, and C. J. Eckhardt, *J. Chem. Phys.* 110, 2606 (1999).

7. T. Luty, C. J. Eckhardt, and J. Lefebvre, *J. Chem. Phys.* 111, 10321 (1999).

8. D. K. Schwartz and C. M. Knobler, *J. Phys. Chem.* 97, 8849 (1993).

9. A. M. Bibo, C. M. Knobler, and I. R. Peterson, *J. Phys. Chem.* 95, 5591 (1991).

10. I. R. Peterson, R. M. Kenn, A. Goudot, P. Fontaine, F. Rondelez, W. G. Bouwman, and K. Kjaer, *Phys. Rev. E: Stat. Phys., Plasmas, Fluids, Relat. Interdiscip. Top.* 52, 667 (1996).

11. K. Kjaer, J. Als-Nielsen, C. A. Helm, L. A. Laxhuber, and H. Möhwald, *Phys. Rev. Lett.* 58, 2224 (1987).

12. S. W. Barton, F. B. Horn, S. A. Rice, B. Lin, J. B. Peng, J. B. Ketterson, and P. Dutta, *J. Chem. Phys.* 89, 2257 (1988).

13. N. M. Peachey and C. J. Eckhardt, *Micron* 25, 271 (1994).

14. T. Raising, Y. R. Shen, M. W. Kim and S. Grubb, *Phys. Rev. Lett.* 55, 2903 (1985).

15. G. A. Overbeck and D. Mobius, *J. Phys. Chem.* 97, 7999 (1993).

16. G. C. Nutting and W. D. Harkins, *J. Am. Chem. Soc.* 61, 2040 (1939).

17. S. Ramos and R. Castillo, *J. Chem. Phys.* 110, 7021 (1999).

18. D. Jacquemain, S. Grayer Wolf, F. Leveiller, M. Deutsch, K. Kjaer, J. Als-Nielsen, M. Lahav, and L. Leiserowitz, *Angew. Chem., Int. Ed. Engl.* 31, 130 (1992).

19. F. Leveiller, Ch. Bohrn, D. Jacquemain, H. Mohwald, L. Leiserowitz, K. Kjaer, and J. Als-Nielsen, *Langmuir* 10, 819 (1994).

20. C. Bohm, F. Leveiller, D. Jacquemain, H. Mohwald, K. Kjaer, J. Als-Nielsen, I. Weissbuch, and L. Leiserowitz, *Langmuir* 10, 830 (1994).

21. T. M. Fischer, R. F. Bruinsma, and C. M. Knobler, *Phys. Rev. E: Stat. Phys., Plasmas, Fluids, Relat. Interdiscip. Top.* 50, 413 (1994).

22. D.G. Dervichian, *J. Chem. Phys.* 7, 931 (1939).

23. D. Lacquemain, S. Grayer Wolf, F. Leveiller, M. Lahav, L. Leiserowitz, M. Deutsch, K. Kjaer, and J. Als-Nielsen, *J. Am. Chem. Soc.* 112, 7724 (1990).

24. S. W. Barton, A. Goudot, O. Bouloussa, F. Rondelez, B. Lin, F. Novak, A. Acero, and S. A. Rice, *J. Chem. Phys.* 96, 1343 (1992).

25. D. Jacquemain, S. Grayer Wolf, F. Leveiller, F. Frolow, M. Eisenstein, M. Lahav, and L. Leiserowitz, *J. Am. Chem. Soc.* 114, 9983 (1992).

26. V. M. Kaganer, G. Brezesinski, H. Mohwald, P. B. Howes, and K. Kjaer, *Phys. Rev. Lett.* 81, 5864 (1998).

27. I. R. Peterson and R. M. Kenn, *Langmuir* 10, 4645 (1994).

28. V. M. Kaganer and E. B. Loginov, *Phys. Rev. E: Stat. Phys., Plasmas, Fluids, Relat. Interdiscip. Top.* 51, 2237 (1995).

29. E. B. Sirota, *Langmuir* 13, 3847 (1997).

30. C. Zakri, A. Renault, J.-P. Rieu, M. Vallade, B. Berge, J.-F. Legrand, G. Vignault, and G. Grubel, *Phys. Rev. B: Solid State* 55, 14163 (1997).

31. C. Zakri, A. Renault, and B. Berge, *Physica B* 248, 208 (1998).

32. E. I. Kats and J. Lajzerowicz, *JETP Lett. (Engl. Transl.)* 83, 495 (1996).

33. S. Riviere, S. Henon, J. Meunier, D. K. Schwartz, M. W. Tsao, and C. M. Knobler, *J. Chem. Phys.* 101, 10045 (1994).

34. J. V. Selinger and D. R. Nelson, *Phys. Rev. A: At., Mol., Opt. Phys.* 39, 3135 (1989).

35. J. V. Selinger, Z. G. Wang, R. F. Bruinsma, and C. M. Knobler, *Phys. Rev. Lett.* 70, 1139 (1993).

36. C. Lautz, Th. M. Fisher, and J. Kildea, *J. Chem. Phys.* 106, 7448 (1997).

37. J. G. Harris and S. A. Rice, *J. Chem. Phys.* 89, 5898 (1988).

38. G. Cardini, J. P. Bareman, and M. L. Klein, *Chem. Phys. Lett.* 145, 493 (1988).

39. J. P. Bareman, G. Cardini, and M. L. Klein, *Phys. Rev. Lett.* 60, 2152 (1988).

40. S. Karaborni and S. Toxvaerd, *J. Chem. Phys.* 96, 5505 (1992).

41. S. Karaborni and S. Toxvaerd, *J. Phys. Chem.* 96, 4965 (1992).

42. S. Karaborni and S. Toxvaerd, *J. Chem. Phys.* 97, 5876 (1992).

43. S. Shin, N. Collazo, and S. A. Rice, *J. Chem. Phys.* 96, 1352 (1992).

44. N. Collazo, S. Shin, and S. A. Rice, *J. Chem. Phys.* 96, 4735 (1992).

45. S. Shin, N. Collazo, and S. A. Rice, *J. Chem. Phys.* 98, 3469 (1993).

46. D. R. Swanson, R. J. Hardy, and C. J. Eckhardt, *J. Chem. Phys.* 99, 8194 (1993).

47. M. Schoen, D. J. Diestler, and H. J. Cushman, *Mol. Phys.* 78, 1097 (1993).

48. U. Nilsson, B. Jonsson, and H. Wennerstrom, *J. Phys. Chem.* 97, 5654 (1993).

49. J. I. Siepmann, S. Karaborni, and M. L. Klein, *J. Phys. Chem.* 98, 6675 (1994).

50. Z. Cai and S. A. Rice, *Faraday Discuss. Chem. Soc.* 89, 211 (1990).

51. Z. Cai and S. A. Rice, *J. Chem. Phys.* 96, 6629 (1992).

52. S. Shin and S. A. Rice, *J. Chem. Phys.* 101, 2508 (1994).

53. M. Li, A. A. Acero, Z. Huang, and S. A. Rice, *Nature (London)* 367, 151 (1994).

54. Z. G. Wang, *J. Phys. (Paris)* 51, 1431 (1990).

55. M. E. Costas, Z. G. Wang, and W. M. Gelbart, *J. Chem. Phys.* 96, 2228 (1992).

56. D. Kramer, A. Ben-Shaul, Z. G. Chen, and W. M. Gelbart, *J. Chem. Phys.* 96, 2236 (1992).

57. M. Scheringer, R. Hilfer, and K. Binder, *J. Chem. Phys.* 96, 2269 (1992).

58. V. M. Kaganer, M. A. Osipow, and I. R. Petersen, *J. Chem. Phys.* 98, 3512 (1993).

59. A. M. Somoza and R. C. Desai, *J. Phys. Chem.*, 96, 1401 (1992).

60. M. Jing, F. Zhong, D. Y. Xing, and J. Dong, *J. Chem. Phys.* 110, 2660 (1999).

61. P. G. de Gennes, "The Physics of Liquid Crystals." Clarendon, Oxford, 1974.

62. V. M. Kaganer and V. L. Indenbom, *J. Phys. II* 3, 813 (1993).

63. V. M. Kaganer and E. B. Loginov, *Phys. Rev. Lett.* 71, 2599 (1993).

64. V. M. Kaganer and E. B. Loginov, *Phys. Rev. E: Stat. Phys., Plasmas, Fluids, Relat. Interdiscip. Top.* 51, 2237 (1995).

65. K. H. Michel and J. Naudts, *J. Chem. Phys.* 67, 547 (1977).

66. B. De Raedt, K. Binder, and K. H. Michel, *J. Chem. Phys.* 75, 2977 (1981).

67. J. C. Raich, H. Yasuda, and E.R. Bernstein, *J. Chem. Phys.* 78, 6209 (1983).

68. K. H. Michel, *Phys. Rev. B: Solid State* 35, 1405 (1987).

69. K. Binder and J. Reger, *Adv. Phys.* 41, 547 (1992).

70. T. H. M. van der Berg and A. van der Avoird, *Phys. Rev. B: Solid State* 43, 13926 (1991).

71. A. P. Cracknell, *Thin Solid Films* 21, 107 (1974).

72. D. M. Hatch and H. T. Stokes, *Phys. Rev. B: Solid State* 30, 5156 (1984).

73. C. J. Bradley and A. P. Cracknell, "The Mathematical Theory of Symmetry in Solids." Clarendon, Oxford, 1972.

74. H. M. James and T. A. Keenan, *J. Chem. Phys.* 31, 12 (1959).

75. W. Press and A. Huller, *Acta Crystallogr. Sect. A* 29, 252 (1973).

76. M. Yvinec and R. Pick, *J. Phys. (Paris)* 41, 1045 (1980).

77. K. H. Michel and K. Parlinski, *Phys. Rev.* B31, 1823 (1985).

78. J. C. Raich and N. S. Gillis, *J. Chem. Phys.* 65, 2088 (1976).

79. A. van der Avoird, P. E. S. Wormer, F. Mulder, and R. Berns, *Top. Curr. Chem.* 93, 1 (1980).

80. G. Venkataraman and V. C. Sahni, *Rev. Mod. Phys.* 42, 409 (1970).

81. Y. Yamada, M. Mori, and Y. Noda, *J. Phys. Soc. Jpn.* 32, 1565 (1972).

82. M. Born and K. Huang, "Dynamical Theory of Crystal Lattices." Oxford Univ. Press, London, 1985.

83. D. E. Williams, *J. Chem. Phys.* 47, 4680 (1967).

84. M. Li and S. A. Rice, *J. Chem. Phys.* 104, 6860 (1996).

85. C. J. Eckhardt and D. R. Swanson, Chem. Phys. Lett., 194, 370 (1992).

86. D. R. Swanson, R. J. Hardy, and C. J. Eckhardt, *J. Chem. Phys.* 105, 673 (1996).

87. V. Malta, G. Celotti, R. Zanneti, and A. Martelli, *J. Chem. Soc. B* 548 (1971).

88. A. Criado and T. Luty, *Phys. Rev. B: Solid State* 48, 12419 (1993).

89. J. L. Schlenker, G. V. Gibbs, and M. B. Boisen, *Acta Crystallogr., Sect. A* 34, 52 (1978).

90. W. H. Press, B. P. Flannery, S. A. Teukolsky, and W. T. Vetterling, "Numerical Recipes." Cambridge Univ. Press, Cambridge, UK, 1986.

91. B. Lin, T. M. Bohanon, and M. C. Shih, *Langmuir* 6, 1665 (1990).

92. F. M. Haas, R. Hilfer, and K. Binder, *J. Chem. Phys.* 102, 2960 (1995).

93. M. Tarek, D. J. Tobias, and M. L. Klein, *J. Phys. Chem.* 99, 1393 (1995).

94. M. E. Brown and M. D. Hollingsworth, *Nature (London)* 376, 323 (1995).

95. R. M. Lynden-Bell, *Mol. Phys.* 79, 313 (1993).

96. R. Siems, Phys. *Status. Solidi* 30, 645 (1968).

97. R. T. Shuey and V. Z. Beyerler, *Z. Angew. Math. Phys.* 66, 278 (1968).

98. K. H. Michel and J. M. Rowe, *Phys. Rev. B: Solid State* 22, 1417 (1980).

99. H. Vollmayer, R. Kree, and A. Zippelius, *Phys. Rev. B: Solid State* 44, 12238 (1991).

100. R. S. Pfeiffer and G. D. Mahan, *Phys. Rev. B: Solid State* 48, 669 (1993).

101. T. Luty and C. J. Eckhardt, *J. Am. Chem. Soc.* 117, 2441 (1995).

102. R. De Wit, *Solid State Phys.* 10, 249 (1960).

103. D. R. Nelson and B. I. Halperin, *Phys. Rev. B: Solid State* 19, 2457 (1979).

104. M. C. Shih, M. K. Dubin, A. Malik, P. Zschack, and P. Dutta, J. Chem. Phys., 101, 9132 (1994).

105. M. K. Durbin, A. Malik, A. G. Richter, R. Ghaskadari, T. Gog, and P. Dutta, *J. Chem. Phys.* 106, 8216 (1997).

106. M. Abraham, K. Miyano, J. B. Ketterson, and S. Q. Xu, *Phys. Rev. Lett.* 51, 1975 (1983).

107. V. M. Kaganer, G. Brezesinski, H. Mohwald, P. B. Howes, and K. Kjaer, *Phys. Rev. Lett.* 81, 5864 (1998).

108. L. E. Copeland, W. D. Harkins, and G. E. Boyd, *J. Chem. Phys.* 10, 357 (1942).

109. M.C. Shih, T.M. Bohanon, J.M. Mikrut, P. Zschak, and P. Dutta, *Phys. Rev. A: At., Mol., Opt. Phys.* 45, 5734 (1992).

110. I. Langmuir, *J. Am. Chem. Soc.* 39, 1848 (1917).

111. D. E. Williams, *J. Chem. Phys.* 47, 4680 (1967).

112. B. Lin, M. Shih, T. M. Bohanon, G. E. Ice, and P. Dutta, *Phys. Rev. Lett.* 65, 191 (1990).

113. L. A. Feigen, Y. M. Lvov, and V. I. Troitsky, *Sov. Sci. Rev. Sect. A* 11, 285 (1988).

114. M. L. Schlossman, D. K. Schwartz, P. S. Pershan, E. H. Kawamoto, G. J. Kellog, and S. Lee, Phys. Rev. Lett., 66, 1599 (1991).

115. C. J. Eckhardt, N. M. Peachey, D. R. Swanson, J.-H. Kim, J. Wang, R. A. Uphaus, G. P. Lutz, and P. Beak, *Langmuir* 8, 2591 (1992).

116. C. J. Eckhardt, N. M. Peachey, J. M. Takacs, and R. A. Uphaus, *Thin Solid Films* 242, 67 (1994).

117. C. J. Eckhardt, N. M. Peachey, D. R. Swanson, J. J. Takacs, M. A. Khan, X. Gong, J. H. Kim, J. Wang, and R. A. Uphaus, *Nature (London)* 362, 614 (1992).

118. R. M. J. Cotterill, *Biochem. Biophys. Acta* 433, 264 (1976).

119. J. P. Bareman and M. Klein, *J. Phys. Chem.* 94, 5202 (1990).

120. M. D. Gibson, D. R. Swanson, C. J. Eckhardt, and X. C. Zeng, *J. Chem. Phys.* 106, 1961 (1997).

121. M. S. Challa, D. P. Landau, and K. Binder, *Phys. Rev. B* 34, 1841 (1986).

122. M. E. Fisher, in "Critical Phenomena" (M. S. Green, Ed.) Academic Press, New York, 1971.

123. M. E. Fisher, in "Critical Phenomena" (F. J. W. Hahne, Ed.) Springer-Verlag, New York, 1983.

124. M. Kreer, K. Kremer, and K. Binder, *J. Chem. Phys.* 92, 6195 (1990).

125. M. Scheringer, R. Hilfer, and K. Binder, *J. Chem. Phys.* 96, 2269 (1992).

126. H. C. Yang, T. F. Magnera, C. Lee, A. J. Bard, and J. Michel, *Langmuir* 8, 2740 (1992).

Chapter 15

SOLID STATE NMR OF BIOMOLECULES

Akira Naito, Miya Kamihira

Department of Life Science, Himeji Institute of Technology, 3-2-1 Kouto, Kamigori, Hyogo 678-1297, Japan

Contents

1. INTRODUCTION

Nuclear magnetic resonance (NMR) spectroscopy is recognized as one of the most powerful means to elucidate structures and dynamics of a wide variety of molecules. In particular, in the biological sciences, it is now possible to determine the three-dimensional structure of proteins in solution. Solid state NMR has also become useful for obtaining information on the structure and dynamics of biologically important molecules such as membrane proteins and insoluble fibril proteins, since high-resolution solid state NMR spectroscopy was developed. The latter technique combines high-power proton decoupling with magic angle spinning (MAS) and cross polarization (CP) techniques. This allows one to observe high-resolution NMR signals of solid biomolecules. In this technique, anisotropic interactions are eliminated to achieve high-resolution signals, although these interactions contain important geometric information for elucidating the structure of biomolecules. To reintroduce the anisotropic interaction under a high-resolution condition, a number of useful techniques have been developed. Solid state NMR of anisotropic media such as lipid bilayer systems may serve as a magnetically anisotropic interaction to explore the molecular orientation in the membrane. Multidimensional NMR has been used to distinguish between interactions or to assign the connectivity of the targeted nuclei. These rapidly developing techniques in the field of high-resolution solid state NMR now promise to elucidate the detailed structure of biomolecules and thus the structure–function relationship in structural biology.

2. FUNDAMENTALS OF SOLID STATE NMR SPECTROSCOPY

2.1. Description of Spin Dynamics Using the Density Operator Formalism [1, 2]

NMR signals can be obtained by looking at the time evolution of the density operator $\rho(t)$, which is evaluated by solving the

Handbook of Thin Film Materials, edited by H.S. Nalwa
Volume 2: Characterization and Spectroscopy of Thin Films

ISBN 0-12-512910-6/$35.00

following Liouville–von Neumann equation:

$$d\rho(t)/dt = i[\rho(t), \mathcal{H}] \tag{1}$$

A formal solution of this equation can be given by

$$\rho(t) = U(t)\rho(0)U(t)^{-1} \tag{2}$$

where $U(t)$ is called the *propagator* and is given by

$$U(t) = \int_{-\infty}^{+\infty} T_d \exp(-i\mathcal{H}(t))\, dt \tag{3}$$

where T_d is Dyson's time-ordering operator that rearrange the order of operators to a standard time-ordered form. However it is not possible to apply this operator for the general case where $\mathcal{H}(t)$ is not commutative between different times. If $\mathcal{H}(t)$ is in fact commutative, T_d can be removed to give

$$U(t) = \int_{-\infty}^{+\infty} \exp(-i\mathcal{H}(t))\, dt \tag{4}$$

Once $\rho(t)$ is obtained, expectation values of any operator, such as the nuclear spin magnetization $\langle I_i \rangle(t)$ at time t, can be calculated from

$$\langle I_i \rangle(t) = \text{Tr}\{\rho(t), I_i\} \tag{5}$$

In the case of NMR, the observable operators are I_x and I_y, which provide an NMR spectrum by Fourier transforming the complex transverse magnetization as follows:

$$F(\omega) = \int_{-\infty}^{+\infty} (I_x + i I_y) \exp(i\omega t)\, dt \tag{6}$$

This mathematical transformation can be performed using a digital fast Fourier transform algorithm, which saves a large amount of time in computer calculation [3]. It is therefore important to take into account the applied rf pulse and internal interaction to evaluate the spin dynamics of system in the presence of a static magnetic field.

2.2. Internal Interaction

2.2.1. Chemical Shift Interaction

In the presence of a static magnetic field, the motion of electron spins induces electric current and hence a local magnetic field that shields the static magnetic field at the nuclei. This local field can be expressed as $-\sigma H$, where σ is called the shielding tensor. Since σ is much smaller than the static magnetic field, the chemical shielding interaction can be treated as a first-order correction (Zeeman term) and is given by

$$\mathcal{H}_{cs} = \gamma h \sigma_{zz} I_z H_0 \tag{7}$$

where σ_{zz} is the zz component of the chemical shielding tensor in the laboratory frame, z being the direction of the static magnetic field. The components of the chemical shielding tensor in the crystalline frame can be obtained from the experiments using a single crystal sample, knowledge of them allows one to

calculate the principal values of the chemical shielding tenor by using the following transformation:

$$\begin{pmatrix} \sigma_{11} & 0 & 0 \\ 0 & \sigma_{22} & 0 \\ 0 & 0 & \sigma_{33} \end{pmatrix} = L \cdot \begin{pmatrix} \sigma_{XX} & \sigma_{XY} & \sigma_{XZ} \\ \sigma_{YX} & \sigma_{YY} & \sigma_{YZ} \\ \sigma_{ZX} & \sigma_{ZY} & \sigma_{ZZ} \end{pmatrix} \cdot L^{-1} \tag{8}$$

where L is the matrix of transformation of the chemical shielding tensor from the crystall frame to the principal axes are frame. It is known that the principal axes are related to the molecular symmetry axes and hence these principal values and the principal axes are important parameters in elucidating the molecular structure of biomolecules. In the case of solution NMR, the chemical shielding tensor can be averaged to the isotropic value

$$\sigma_{iso} = \tfrac{1}{3}\text{Tr}(\sigma) = \tfrac{1}{3}(\sigma_{11} + \sigma_{22} + \sigma_{33}) \tag{9}$$

In solid state NMR, this value can usually be obtained by a MAS experiment.

Anisotropic components of the ^{13}C chemical shift tensor of a variety of biomolecules have been determined using either single crystals or polycrystalline samples as summarized in Table I. The direction of the principal axes reflects the local molecular symmetry. In the case of the carboxyl carbon of amino acids, the most shielded direction is perpendicular to the sp^2 plane, the least shielded direction is parallel to the bisector of the O−C−O bond angle, and the intermediate shielded direction is perpendicular to those axes. The principal values of the ^{13}C chemical shift tensor also reflect the secondary structure of polypeptide chains, such as the α-helix and β-sheet.

Therefore, anisotropic ^{13}C chemical shift values are important parameters for elucidating the structure of biomolecules.

The conformation dependence of isotropic ^{13}C chemical shifts of biomolecules is of practical importance. It is well recognized that all of the ^{13}C chemical shifts of amino acid residues that are more than two residues away from a chain end in peptides and proteins adopting an unfolded conformation in solution are effectively independent of all neighboring residues except proline [12]. Therefore, it is expected that ^{13}C chemical shifts of the backbone C_α and C=O and side chain C_β signals of peptides and protein are significantly displaced (up to 8 ppm) depending on their local secondary structure as defined by a set of torsion angles in the peptide unit (ϕ, ψ), irrespective of there being any of a variety of neighboring amino acid residues [13, 14]. The ^{13}C NMR signal of polypeptides in the solid state allows secondary structures (α-helix, β-sheet, 3_1-helix, etc.) to be recorded. In particular, the two major conformations, α-helix and β-sheet forms, are readily distinguished from the peak position of the ^{13}C NMR signals: the C_α and C=O ^{13}C NMR signal of the α-helix form are displaced upward in frequency by 3–8 ppm with respect to those of the β-sheet forms, whereas the C_β signals of the α-helix are displaced downward, as summarized in Table II. In addition, it has been demonstrated that seven conformations, including the random coil form, can be distinguished by the conformation-dependent displacements of peaks as manifested from those of Ala residues [13–16]. This means that the local conformation of particular amino acid residues from any polypeptides or structural, globular, or membrane protein is readily evaluated, in an empirical manner, by means of the conformation-dependent displacements of ^{13}C chemical shifts of the respective residues with reference to the database so far accumulated from a number of polypeptides, because the transferability of these parameters from a simple model system to more complicated proteins has proved to be excellent. For globular proteins in solution, specific displacements of ^{13}C NMR peaks with respect to those of the random coil have been utilized as a convenient probe for sequential assignment of the secondary structures of proteins [17, 18], as inspired by the mentioned success of the compilation of the conformation-dependent ^{13}C chemical shifts.

Table I. ^{13}C Chemical Shift Tensor of Solid Amino Acids and Peptides

Substance	Site	δ_{11}	δ_{22}	δ_{33}	δ_{iso}	Reference
Glycine	C_α	65.4	46.3	26.7	46.1	[4]
	C—O	247.3	181.4	102.6	177.1	
L-Alanine	C_α	65.1	56.5	31.3	51.0	[5]
	COO^-	242.9	183.5	106.7	177.7	
	CH_3	30.3	21.4	8.3	20.0	
L-Serine	C_α	70.0	60.6	44.2	58.3	[6]
	COO^-	238.9	180.9	106.9	175.6	
	C_β	85.5	65.7	34.1	61.8	
L-Threone	C_α	69.0	58.9	52.6	60.2	[7]
	COO^-	240.2	164.7	105.0	170.0	
	C_β	89.7	54.4	45.1	63.1	
	CH_3	32.1	23.1	1.4	18.9	
L-Asparagine	C_α	71.8	53.5	30.1	51.8	[8]
	COO^-	240.2	179.7	109.3	176.4	
	C_δ	246.2	196.6	88.5	177.1	
Glycylglycine	CONH	244.1	177.1	87.9	169.7	[9]
Glycine in collagen	CONH	224	173	121	173	[10]
Polyglycine (β-sheet)	CONH	243	174	88	168	[11]
Polyglycine (3_1 helix)	CONH	243	179	94	172	[11]
Glycine in poly-L-alanine (α-helix)	CONH	244	178	94	172	[11]
Glycine in poly (β-benzyl L-asparatic acid) (ω-helix)	CONH	242	178	93	171	[11]
Glycine in poly (L-valine) (β-sheet)	CONH	242	171	93	169	[11]

2.2.2. Magnetic Dipolar Interaction

The magnetic dipole–dipole coupling between two nuclear spins I and S can be expressed in angular velocity units as the Hamiltonian

$$\mathcal{H}_{IS} = \frac{\gamma_I \gamma_S h}{2\pi r^3}\left[\mathbf{I}\cdot\mathbf{S} - 3(\mathbf{I}\cdot\mathbf{r})(\mathbf{S}\cdot\mathbf{r})/r^2\right] \quad (10)$$

where $\mathbf{r}$ is the internuclear vector between the spins I and S. This interaction depends on the interatomic distance and the angle between the internuclear vector and the static magnetic field as illustrated in Figure 1. For the case of a homonuclear two-spin system, this interaction can be given in the presence of strong magnetic field $\mathbf{H}_0$ by

$$\begin{aligned}\mathcal{H}_{II} &= \frac{\gamma_I^2 h}{2\pi r^3}\frac{3\cos^2\theta - 1}{2}(\mathbf{I_1}\cdot\mathbf{I_2} - 3I_{1z}I_{2z})\\ &= A + B \end{aligned} \quad (11)$$

where

$$A = -\frac{\gamma_I^2 h}{2\pi r^3}\left(3\cos^2\theta - 1\right)I_{1z}I_{2z}$$

$$B = \frac{\gamma_I^2 h}{2\pi r^3}\left(3\cos^2\theta - 1\right)\frac{I_{1+}I_{2-} + I_{1-}I_{2+}}{4}$$

In Eq. (11), I_1 and I_2 are the like spins, γ_I is the gyromagnetic ratio of I-nuclei, h is Planck's constant, and r is the length of the I_1–I_2 internuclear vector. θ is the angle between the static magnetic field and $\mathbf{r}$. The term B is called the *flip-flop* term and causes mutual spin exchange when the energy levels of the states are very close to each other. Two dipolar precession frequencies in the rotating frame can be obtained from Eq. (11) as follows:

$$\omega_D = \omega_0 \pm \tfrac{3}{4}D\left(3\cos^2\theta - 1\right) \quad (12)$$

where $D = \gamma_I^2 h / 2\pi r^3$. D is called the dipolar coupling constant, and ω_0 is the resonance frequency of the observed nuclei.

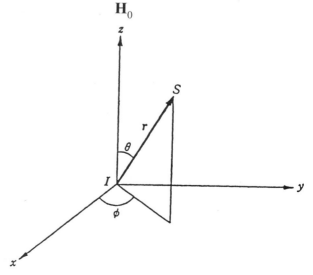

Fig. 1.   Orientation of I–S internuclear vector in spherical coordinates.

Table II. ^{13}C Chemical Shifts Characteristic of the α-Helix and β-Sheet Forms (from TMS) [14]

Amino acid residues in polypeptides	Chemical shift (ppm)								
	C_α			C_β			C=O		
	α-helix	β-sheet	Δ^a	α-helix	β-sheet	Δ^a	α-helix	β-sheet	Δ^a
Ala	52.4	48.2	4.2	14.9	19.9	−5.0	176.4	171.8	4.6
	52.3	48.7	3.6	14.8	20.0	−5.2	176.2	171.6	4.6
Leu	55.7	50.5	5.2	39.5	43.3	−3.8	175.7	170.5	5.2
	55.8	51.2	4.6	43.7^b	39.6	(4.1)	175.8	171.3	4.5
Glu(OBzl)	56.4	51.2	5.2	25.6	29.0	−3.4	175.6	171.0	4.6
	56.8	51.1	5.7	25.9	29.7	−3.8	175.4	172.2	3.2
Asp(OBzl)	53.4	49.2	4.2	33.8	38.1	−4.3	174.9	169.8	5.1
Val	65.5	58.4	7.1	28.7	32.4	−3.7	174.9	171.8	3.1
		58.2			32.4			171.5	
Ile	63.9	57.8	6.1	34.8	39.4	−4.6	174.9	172.7	2.2
Lys^c	57.4			29.9			176.5		
Lys(z)	57.6	51.4	6.2	29.3	28-5	−0.8	175.7	170.4	5.3
Arg^c	57.1			28.9			176.8		
Phe	61.3	53.2	8.1	35.0	39.3	−4.3	175.2	169.0	6.2
Met	57.2	52.2	5.0	30.2	34.8	−4.6	175.1	170.6	4.5
Gly		43.2						168.4	
							171.6^d	168.5	3.1

^{a}Difference in the ^{13}C chemical shifts of the α-helix form from those of the β-sheet form.

^{b}Mistyping or erroneous assignment. This assignment should be reversed.

^{c}Data uken from neutral aqueous solution.

^{d}Averaged values from the data of polypeptides containing ^{13}C-labeled glycine residues.

The heteronuclear dipolar Hamiltonian can be expressed as

$$\mathcal{H}_{IS} = -\frac{\gamma_I \gamma_S h}{2\pi r^3}\left(3\cos^2\theta - 1\right)I_z S_z \qquad (13)$$

where I and S are like and unlike nuclear spins, respectively. In this case, the difference of the Zeeman interaction between the I and S nuclei is quite large, and hence spin exchange cannot occur. The dipolar precession frequency in this case can be obtained from Eq. (13) as follows:

$$\omega_D = \omega_0 \pm \tfrac{1}{2}D\left(3\cos^2\theta - 1\right) \qquad (14)$$

It is therefore possible to determine internuclear distances from the dipolar interaction to evaluate molecular structure.

2.2.3. Nuclear Quadrupole Interaction

Quadrupolar nuclei with nuclear spin greater than 1/2 have a nuclear quadrupole moment, eQ. This quadrupole moment interacts with the electric field gradient around the nuclei to give a nuclear quadrupole interaction as expressed by the following Hamiltonian:

$$\mathcal{H}_Q = \frac{eQ}{2I(2I-1)}\, I \cdot V \cdot I \qquad (15)$$

The electric field gradient $V_{\alpha\beta} = \partial^2 V/\partial\alpha\partial\beta$ is a real symmetric tensor, and hence its trace is zero. Thus, when $|V_{zz}| > |V_{xx}| > |V_{yy}|$, $eq = V_{zz}$, and $\eta = (V_{xx} - V_{yy})/V_{zz}$ are defined, Eq. (15) can be written

$$\mathcal{H}_Q = \frac{e^2qQ}{4I(2I-1)}\left[3I_z^2 - I(I+1) + \tfrac{1}{2}\eta\left(I_+^2 + I_-^2\right)\right] \qquad (16)$$

where e^2qQ is the nuclear quadrupole coupling constant and η is the asymmetry parameter.

The nuclear quadrupolar interaction is not always much smaller than the Zeeman interaction. It is therefore necessary to take account of the second-order terms. Then the first-order

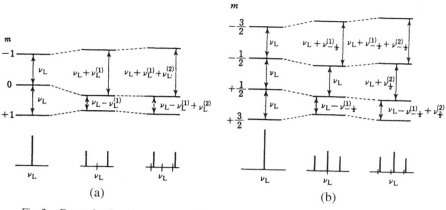

Fig. 2. Energy levels and resonance positions for the nuclei with (a) $I = 1$ and (b) $I = 3/2$.

energy correction can be expressed as

$$E_m^{(1)} = \frac{h\nu_Q}{4}\left(3\cos^2\theta - 1 + \eta\sin 2\theta\cos 2\phi\right)$$
$$\times\left[m^2 - \tfrac{1}{3}I(I+1)\right] \tag{17}$$

The second-order energy correction can be given by

$$E_m^{(2)} = -\frac{h\nu_Q^2}{12\nu_L}\frac{m}{24}$$
$$\times\left\{\left[3\sin 2\theta - \eta(\sin 2\theta\cos 2\phi + i2\sin\theta\sin 2\phi)\right]\right.$$
$$\times\left[3\sin 2\theta - \eta(\sin 2\theta\cos 2\phi - i2\sin\theta\sin 2\phi)\right]$$
$$\times\left[8m^2 - 4I(I+1) + 1\right]\right\}$$
$$-\left\{3\sin^2\theta + \eta\left[(1 + \cos^2\theta)\cos 2\phi\right.\right.$$
$$\left.+ i2\cos\theta\sin 2\phi\right]\right\}$$
$$\times\left\{3\sin^2\theta + \eta\left[(1 + \cos 2\theta)\cos 2\phi\right.\right.$$
$$\left.- i2\cos\theta\sin 2\phi\right]\right\}$$
$$\times\left[2m^2 - 2I(I+1) + 1\right] \tag{18}$$

where $\nu_Q = 3e^2qQ/2I(2I-1)h$, $\nu_L = \gamma H/2\pi$. The allowed NMR transition frequency $\nu_m = (E_{m-1} - E_m)/h$ for the m-to-$(m-1)$ transition can be expanded as

$$\nu_m^{(0)} = \nu_L \tag{19}$$
$$\nu_m^{(1)} = -(\nu_Q/2)\left(3\cos^2\theta - 1 + \eta\sin^2\theta\cos 2\phi\right)(m - 1/2) \tag{20}$$
$$\nu_m^{(2)} = (\nu_Q/12\nu_L)\left\{\left[3\sin 2\theta - \eta(\sin 2\theta\cos\phi\right.\right.$$
$$\left.+ i2\sin\theta\cos 2\phi)\right]$$
$$\times\left[3\sin 2\theta - \eta(\sin 2\theta\cos 2\phi - i2\sin\theta\sin 2\phi)\right]$$
$$\times\left[m(m-1) - I(I+1)/6 + 3/8]\right\}$$
$$-\left\{3\sin^2\theta + \eta\left[(1 + \cos^2\theta)\cos 2\phi + i2\cos\theta\sin 2\phi\right]\right\}$$
$$\times\left\{3\sin^2\theta + \eta\left[(1 + \cos^2\theta)\cos 2\phi - i2\cos\theta\sin 2\phi\right]\right\}$$
$$\times\left(m(m-1)/4 - I(I+1)/12 + 1/8\right)\right\} \tag{21}$$

for zero, first, and second-order terms, respectively. Energy levels and the NMR transitions for the nuclei with $I = 1$ and $3/2$ are shown in Figure 2.

In the case of $I = 1$ nuclei, the NMR lines will be symmetrically split from the center frequency ν_L by the first-order term, and the split line will be shifted in the same direction by the second-order term for single-crystal samples. In polycrystalline or powder samples, a broad line with width comparable to the magnitude of the quadrupole coupling interaction will be observed.

In the case of nuclei with $I = 3/2$, a triplet line with the intensity ratio $3:4:3$ can be observed. The center line is displaced from ν_L by the second-order frequency. It is therefore difficult to observe a whole line for the nuclei, because the quadrupole coupling constant is usually larger than the excitation width for RF pulses. One can observe a central ($1/2$ to $-1/2$) transition whose line width is determined only by the second-order term.

Deuterium nuclei have a relatively small quadrupole coupling constant, on the order of 250 kHz, and hence it is possible to observe the entire powder pattern to extract the motional information about the local deuterium-labeled site.

3. SOLID STATE NMR TECHNIQUES

3.1. Cross-Polarization and Magic Angle Spinning

Enhancement of sensitivity for the dilute spins can be achieved through a technique called cross-polarization [19]. This technique relies on polarization transfer from the abundant spins such as ^{1}H to the rare spins such as ^{13}C, through matching the H_1 RF fields of ^{1}H's and the ^{13}C's. This is known as matching the *Hartmann–Hahn condition*

$$\gamma_H H_{1H} = \gamma_C H_{1C} \tag{22}$$

where H_{1H} and H_{1C} are known as the RF spin-locking fields.

The pulse sequence for a cross-polarization (CP) is shown in Figure 3. In the first step, the proton magnetization is rotated through 90° to the x-axis and then locked there by a spin-locking y-pulse. The proton spins are kept locked for a time period t_1 known as the *contact time*. During this period, a strong on-resonance pulse is applied to the ^{13}C spins, which are also oriented along the x-axis. If the Hartmann–Hahn condition is matched, the Zeeman splittings in the rotating frame for ^{1}H and ^{13}C become equal, as shown in Figure 4. Thus spin exchange between the two spin reservoirs accelerates as characterized by the CP time T_{CH}. Consequently, polarization transfer occurs from the ^{1}H reservoir to the ^{13}C reservoir. Since the ^{1}H spin reservoir is much greater than the ^{13}C spin reservoir, the ^{13}C magnetization will be increased by a factor $\gamma_H/\gamma_C = 4$, which is related to the ratio of the Zeeman splittings of ^{1}H and ^{13}C nuclei in the rotating frame. After the carbon magnetization has built up during the contact time, the carbon field is switched off, and the free induction decay (FID) recorded. The proton field is kept on for high-power decoupling. One of the important consequences of this pulse sequence is that the carbon magnetization, which yields an FID, does not depend on the regrowth of the carbon magnetization in between scans, but arises entirely from contact with the proton spins. This means that the intensity of the carbon spectrum effectively depends on the relaxation of the proton spin system. The great advantage of this is that, generally, the ^{1}H spin longitudinal relaxation times are much shorter than the ^{13}C ones, so that successive scans may be recycled much faster than for normal ^{13}C acquisitions, yielding better signal-to-noise ratio in a given time period.

During an acquisition period, a high-power decoupling pulse is applied to the ^{1}H nuclei. This pulse eliminates the heteronuclear dipolar interaction. Because the directly bonded ^{13}C–^{1}H dipolar interaction has a magnitude as large as 20 kHz, much higher decoupling power for solids is required for liquids to decoupled the dipolar interaction. It is advised to use an on-resonance frequency for decoupling to give the best decoupling

efficiency. Instead of using cw decoupling, composite decoupling pulse is designed to achieve efficient decoupling for solids under the MAS condition. The two-pulse phase modulation (TPPM) pulse sequence is frequently used to give more effective decoupling than cw [20].

After high-power proton decoupling has greatly reduced the linewidth of dilute nuclei, the chemical shift anisotropy remains. This anisotropy can be eliminated when the rotation axis of the sample spinning is inclined at $\theta = 54°44'$, the so-called *magic angle*, to the static magnetic field. Not only the chemical shift anisotropy, but also all the anisotropic terms in the Hamiltonian, are minimized under the MAS condition. However, if the spinning speed is not sufficiently faster than the order of the anisotropic interaction, sidebands will appear on either side of the center line. It is useful to use sideband suppression pulse sequence such as *total sideband suppression* (TOSS), which removes the rotational sidebands without enhancing the intensity of the center band [21].

When CP is combined with MAS and high-power proton decoupling, spectra, called *CP-MAS high-resolution spectra*, comparable to solution state spectra are obtained [22]. This CP-MAS technique is routinely used to obtain high-resolution solid state NMR signals.

3.2. Quadrupole Echo Measurements

It is possible to obtain an entire spectrum for ^{2}H ($I = 1$) nuclei, because the quadrupole coupling constant is exceptionally small ($\sim$250 kHz) in comparison with other quadrupolar nuclei. However, it is still necessary to use a very short excitation pulse (of duration less than 3 μs) to obtain distortion-free spectra. Furthermore, it is required to use an echo method to avoid the dead time of the receiver. In the ^{2}H nuclei, a quadrupole echo pulse sequence ($90°x–\tau–90°y–\tau$–acquisition) is used, and the FID is acquired starting from the top of the echo signals [23, 25].

After Fourier-transforming the echo signals, a symmetric powder spectrum called a *Pake doublet* is obtained when the molecule is static. As the frequency of the molecular motion is increased, the powder pattern shows characteristic line shape, particularly for the frequency range from 10^4 to 10^7 kHz. Figure 5 shows the ^{2}H line shape of phenyl deuterons in

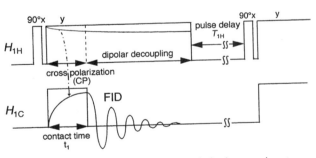

Fig. 3.   Pulse sequence for the cross-polarization experiment.

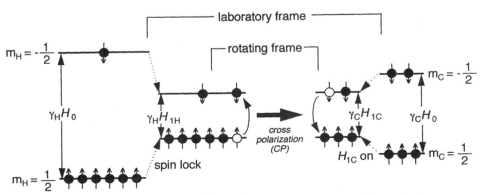

Fig. 4.   Energy level diagram for the ^{13}C and ^{1}H nuclei in the Zeeman and the rotating frame.

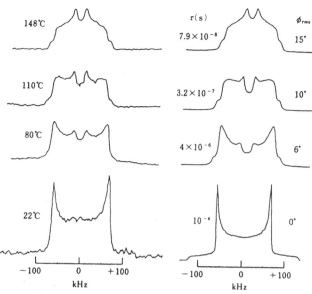

Fig. 5. ^{2}H NMR spectra of phenyl deuterons in *p*-fluorophenylalanine.

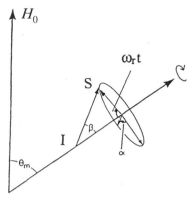

Fig. 6. Representation of I–S internuclear vector rotating about the MAS axis.

p-fluorophenylalanine [25]. It is important to point out that not only the frequency but also the type of molecular motion (such as jump or diffusion) can be identified by analyzing the ^{2}H powder line shape. It is therefore useful to characterize the local molecular motion of biomolecules from the analysis of the ^{2}H NMR line shape.

3.3. Interatomic Distance Measurements

3.3.1. Dipolar Interaction under MAS

When samples are rotated about an axis inclined at θ_m to the static magnetic field in the CP MAS experiment as shown in Figure 5, θ is time-dependent and the factor $3\cos^2\theta - 1$ in Eqs. (11) and (13) can be expressed as a function of time as follows:

$$3\cos^2\theta(t) - 1 = \frac{1}{2}\left(3\cos^2\theta_m - 1\right)\left(3\cos^2\beta - 1\right)$$
$$+ \frac{3}{2}\sin 2\theta_m \sin 2\beta \cos(\alpha + \omega_r t)$$
$$+ \frac{3}{2}\sin^2\theta_m \sin^2\beta \cos 2(\alpha + \omega_r t) \quad (23)$$

where α is the azimuthial angle and β is the polar angle of the internuclear vector with respect to the rotor axis. ω_r is the angular velocity of the rotor. When θ_m is the magic angle, the first term in Eq. (23) vanishes and Eq. (23) becomes

$$3\cos^2\theta(t) - 1$$
$$= \sqrt{2}\sin 2\beta \cos \omega_r t + \sin^2\beta \cos 2(\alpha + \omega_r t) \quad (24)$$

Two frequencies due to the I–S dipolar interaction are expressed in angular velocity units as follows:

$$\omega_D(\alpha, \beta, t)$$
$$= \pm\frac{D}{2}\left[\sin^2\beta \cos 2(\alpha + \omega_r t) - \sqrt{2}\sin 2\beta \cos(\alpha + \omega_r t)\right]$$
$$\quad (25)$$

Therefore, the dipolar interaction under the magic angle spinning condition is a function of time. It can be null after taking an average over the rotor period as follows:

$$\overline{\omega_D} = \frac{1}{T_r}\int_0^{T_r} \omega_D(t)\, dt$$
$$= 0 \quad (26)$$

This fact indicates that the dipolar interaction cannot affect the line shape of the center peak except for the intensities of the sidebands. For this reason, the dipolar interaction was thought to be difficult to obtain under the MAS condition.

3.3.2. Recoupling of the Dipolar Interaction

Accurate interatomic distances can be evaluated from dipolar interactions, which were normally sacrificed under the condition of high-power decoupling and magic angle spinning techniques [26, 27]. A considerable improvement has been established in recoupling the dipolar interaction by either introducing rf pulses synchronized with the MAS rotor period [28] or adjusting the rotor frequency according to the difference of the chemical shift values of two isotopically labeled homonuclei [29]. Rotational echo double resonance (REDOR) [26, 28] was explored to recouple the relatively weak heteronuclear dipolar interactions under the MAS condition by applying a π-pulse synchronously with the rotor period. Consequently, the transverse magnetization cannot be refocused completely at the end of the rotor cycle, leading to a reduction of the echo amplitude. The extent of the reduction of the echo amplitude as a function of the number of rotor periods depends on the strength of the heteronuclear dipolar interaction. This method is extensively used to determine relatively remote interatomic distances of 2–8 Å. When a number of isolated pairs are involved in a REDOR dephasing, the REDOR transformation can be useful in determining interatomic distances because it yields single peaks in the frequency domain for each heteronuclear coupling strength [30, 31]. This approach is, however, not applicable for the case where the observed nuclei in REDOR are coupled with multiple nuclei.

The *rotational resonance* (RR) phenomenon [26, 32] is a recoupling of the homonuclear dipolar interaction under the MAS condition. When the rotor frequency is adjusted to a multiple of the difference between the chemical shift values of two different resonance lines, line broadening and acceleration of the exchange rate of the longitudinal magnetization are observed. These effects depend strongly on the magnitude of the homonuclear dipolar interaction.

REDOR and RR methods have been most extensively explored, although several other approaches to determine the interatomic distances in solid molecules have been proposed: *Transferred echo double resonance* (TEDOR) [33] is a similar method used to determine heteronuclear dipolar interactions by observing the buildup of the echo amplitude. The magnetization in this method is transferred from one nucleus to the other through the heteronuclear dipolar interaction. It is therefore useful for eliminating natural isotope background signals. *Dipolar recovery at the magic angle* (DRAMA) [34] is used to recouple the homonuclear dipolar interaction, which is normally averaged out by MAS, by applying $90° + x$ and $90° - x$ pulses synchronously with the rotor period so that the distances between the two homonuclei can be determined. Since DRAMA strongly depends on the offset of the carrier frequency, Sun et al. developed *melding of spin locking and DRAMA* (MELODRAMA) [35] by combining DRAMA with a spin-lock technique. This technique reduced the offset effect. *Simple excitation for the dephasing of the rotational echo amplitude* (SEDRA) [36] and *rf-driven dipolar recoupling* (RFDR) [37] are techniques used to apply a π-pulse synchronously with the rotor period. *Dipolar decoupling with a windowless sequence* (DRAWS) is a pulse sequence method using phase-shifted, windowless irradiation applied synchronously with sample spinning [38]. Using this pulse sequence, accurate interatomic distances can be determined for the case where the coupled spins have large chemical shift anisotropies and large differences in isotropic chemical shifts, as well as the case of equal isotropic chemical shifts.

These techniques also apply to determine the homonuclear dipolar interaction under the MAS condition. Because they are not sensitive to the MAS frequency and offset effects, they will be useful for determining the dipolar interaction using multidimensional NMR for multiple-site labeled systems [35]. It has not been fully evaluated, however, how accurately interatomic distances can be determined by these methods.

As an alternative approach to evaluating molecular structure on the basis of interatomic distances, methods of determining torsion angles have been proposed using magnetization or coherence transfer through a particular bond [39–42] or spin diffusion between isotopically labeled nuclei [43]. This approach is expected to be useful for obtaining structural information, although much effort in spectral simulation has to be made to analyze the two-dimensional data to determine even one torsion angle or one pair of dihedral angles.

3.3.3. Simple Description of the REDOR Experiment [26]

The transverse magnetization, which precesses about the static magnetic field because of the dipolar interaction under the MAS condition, moves back to the same direction once every rotor period because the integral of ω_D over one rotor period is zero. Consequently, the rotational echo signals are refocused at every rotor period. When a π-pulse is applied to the S-nucleus, which is coupled with the I-nucleus, in one rotor period, this pulse serves to invert the precession direction of the magnetization of the observed I-nucleus. Consequently, the magnetization vector of the I-nucleus cannot move back to the same direction after one rotor period. Therefore the amplitude of the echo intensity decreases. The extent of the reduction of the rotational echo amplitude yields the interatomic distances. To evaluate the REDOR echo amplitude theoretically, one has to consider the average precession frequency in the presence of a π-pulse at the center of the rotor period over one rotor cycle as follows:

$$\overline{\omega_D(\alpha, \beta)} = \pm \frac{1}{T_r} \left[\int_0^{T_r/2} \omega_D \, dt - \int_{T_r/2}^{T_r} \omega_D \, dt \right]$$
$$= \pm \frac{D}{\pi} \sqrt{2} \sin 2\beta \sin \alpha \qquad (27)$$

Therefore, the phase angle $\Delta\Phi(\alpha, \beta)$ for the rotor cycle N_c is given by

$$\Delta\Phi(\alpha, \beta) = \overline{\omega_D(\alpha, \beta)} N_c T_r \qquad (28)$$

where T_r is the rotor period. Finally, the echo amplitude can be obtained by averaging over every orientations as follows:

$$S_f = \frac{1}{2\pi} \int_\alpha \int_\beta \cos \Delta\Phi(\alpha, \beta) \, d\alpha \sin \beta \, d\beta \qquad (29)$$

Therefore, the normalized echo difference, $\Delta S/S_0$, is given by

$$\Delta S/S_0 = (S_0 - S_f)/S_0$$
$$= 1 - S_f \qquad (30)$$

Experimentally, REDOR and full echo spectra are acquired for a variety of $N_c T_r$ values, and the respective REDOR (S_f) and full echo (S_0) amplitudes are evaluated.

3.3.4. Rotational Echo Amplitude Calculated by the Density Operator Approach

The REDOR echo amplitude can be evaluated more rigorously by using density matrix operators and a pulse sequence for the REDOR experiment shown in Figure 7 [44]. The time evolution of the density operator ρ_0 under the heteronuclear dipolar interaction during one rotor period can be considered by taking the pulse length into account. The average Hamiltonian in the rotating frame over one rotor period is given by

$$\overline{\mathcal{H}} = \frac{1}{T_r} \left[\overline{\mathcal{H}_1(t)} \tau + \overline{\mathcal{H}_2(t)} t_w + \overline{\mathcal{H}_3(t)} \tau \right]$$
$$= \frac{D}{4\pi} \left\{ \sin^2 \beta \left[\sin(2\alpha + \omega_r t_w) + \sin(2\alpha - \omega_r t_w) - 2 \sin 2\alpha \right] \right.$$

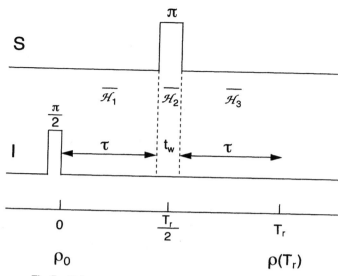

Fig. 7. Pulse sequence and timing chart of REDOR experiment.

$$-2\sqrt{2}\sin 2\beta\big[\sin\big(\alpha+\tfrac{1}{2}\omega_r t_w\big)$$
$$+\sin\big(\alpha-\tfrac{1}{2}\omega_r t_w\big)+2\sin 2\alpha\big]$$
$$-\sin^2\beta\big[\sin(2\alpha+\omega_r t_w)+\sin(2\alpha-\omega_r t_w)\big]$$
$$\times\frac{4\omega_r^2 t_w^2}{4\omega_r^2 t_w^2-\pi^2}$$
$$+\sqrt{2}\sin 2\beta\big[\sin\big(\alpha+\tfrac{1}{2}\omega_r t_w\big)+\sin\big(\alpha-\tfrac{1}{2}\omega_r t_w\big)\big]$$
$$\times\frac{2\omega_r^2 t_w^2}{\omega_r^2 t_w^2-\pi^2}\bigg\}I_z S_z$$
$$+\frac{D}{4\pi}\bigg\{\sin^2\beta\big[\cos(2\alpha+\omega_r t_w)+\cos(2\alpha-\omega_r t_w)\big]$$
$$\times\frac{2\pi\omega_r t_w}{4\omega_r^2 t_w-\pi^2}$$
$$-\sqrt{2}\sin 2\beta\big[\cos\big(\alpha+\tfrac{1}{2}\omega_r t_w\big)$$
$$+\cos\big(\alpha-\tfrac{1}{2}\omega_r t_w\big)\big]$$
$$\times\frac{2\pi\omega_r t_w}{\omega_r^2 t_w-\pi^2}\bigg\}I_z S_y$$
$$=a I_z S_z+b I_z S_y \tag{31}$$

where the same notation as in Eq. (23) is used. $\overline{\mathcal{H}_1(t)}$, $\overline{\mathcal{H}_2(t)}$, and $\overline{\mathcal{H}_3(t)}$ are the average Hamiltonians corresponding to the period shown in Figure 7. The pulse length t_w is also considered in the calculations for the analysis of the REDOR results. The density operator $\rho(T_r)$ at T_r after evolution under the average Hamiltonian can be calculated as

$$\rho(T_r)=\exp(-i\overline{\mathcal{H}}T_r)\rho_0\exp(i\overline{\mathcal{H}}T_r) \tag{32}$$

where ρ_0 is taken as I_y after the contact pulse. Then finally the transverse magnetization at T_r can be given by

$$\langle I_y(T_r)\rangle = \mathrm{Tr}\{\rho(T_r)I_y\}$$
$$=\cos\big(\tfrac{1}{2}\sqrt{a^2+b^2}\,T_r\big) \tag{33}$$

The echo amplitude in the powder sample can be calculated by averaging over every orientation as follows:

$$S_f=\frac{1}{2\pi}\int_\alpha\int_\beta\langle I_y(T_r)\rangle\sin\beta\,d\beta\,d\alpha \tag{34}$$

Therefore, the normalized echo difference, $\Delta S/S_0$, is given by Eq. (30). When t_w is zero, Eq. (33) can be simplified as follows:

$$\langle I_y(T_r)\rangle=\cos\Big(\frac{D}{\pi}\sqrt{2}\sin 2\beta\sin\alpha\,T_r\Big) \tag{35}$$

In this case, Eq. (34) is equivalent to Eq. (33) in the case of $N_c=1$.

3.3.5. Echo Amplitude in the Three-Spin System [45]

It is important to consider the case where the observed nucleus (I_1) is coupled with two other heteronuclei (S_1 and S_2). The Hamiltonian in the three-spin system is given by

$$\mathcal{H}(t)=-\frac{\gamma_I\gamma_S h}{2\pi r_1^3}\big[3\cos^2\theta_1(t)-1\big]I_{Z1}S_{Z1}$$
$$-\frac{\gamma_I\gamma_S h}{2\pi r_2^3}\big[3\cos^2\theta_2(t)-1\big]I_{Z1}S_{Z2} \tag{36}$$

where r_1 and r_2 are the I_1-S_1 and the I_1-S_2 interatomic distances, respectively. $\theta_1(t)$ and $\theta_2(t)$ are the angles between the magnetic field and the I_1-S_1 and the I_1-S_2 internuclear vectors, respectively. In the molecular coordinate system, the x-axis is along the I_1-S_1 internuclear vector, and the $S_1-I_1-S_2$ plane is the $x-y$ plane. The angle between I_1-S_1 and I_1-S_2 is denoted by ζ. The coordinate system is transformed from the molecular axis system to the MAS system by applying a rotation transformation matrix $R(\alpha,\beta,\gamma)$ with Euler angles α, β, γ, and then transformed from the MAS to the laboratory coordinate system by applying $R(\omega_r t,\theta_m,0)$. Finally, $\cos\theta_1(t)$ and $\cos\theta_2(t)$ are calculated as follows:

$$\cos\theta_1(t)=(\cos\gamma\cos\beta\cos\alpha-\sin\gamma\sin\alpha)\sin\theta_m\cos\omega t$$
$$-(\sin\gamma\cos\beta\cos\alpha+\cos\gamma\sin\alpha)\sin\theta_m\sin\omega t$$
$$+\sin\beta\cos\alpha\cos\theta_m$$

and

$$\cos\theta_2(t)=\big[(\cos\gamma\cos\beta\cos\alpha-\sin\gamma\sin\alpha)\cos\zeta\sin\theta_m$$
$$+(\cos\gamma\cos\beta\sin\alpha+\sin\gamma\cos\alpha)\sin\zeta\sin\theta_m\big]$$
$$\times\cos\omega t$$
$$-\big[(\sin\gamma\cos\beta\cos\alpha+\cos\gamma\sin\alpha)\cos\zeta\sin\theta_m$$
$$+(\sin\gamma\cos\beta\sin\alpha-\cos\gamma\cos\alpha)\sin\zeta\sin\theta_m\big]$$
$$\times\sin\omega t$$
$$+\sin\beta\sin\alpha\cos\theta_m\sin\zeta+\sin\beta\sin\alpha\cos\theta_m\sin\zeta \tag{37}$$

where θ_m is the magic angle between the spinner axis and the static magnetic field, and ω_r is the angular velocity of the spinner rotating about the magic angle axis. The four resonance frequencies in the system are given by

$$\omega_{D1} = (D_1 - D_2)/2$$
$$\omega_{D2} = (D_1 + D_2)/2$$
$$\omega_{D3} = -(D_1 + D_2)/2$$
$$\omega_{D4} = -(D_1 - D_2)/2$$

(38)

These dipolar transition frequencies are time-dependent and follow the cycle of the spinning.

In the REDOR pulse sequence, a π-pulse is applied at the center of the rotor period. In this case, the averaged angular velocity over one rotor cycle for each resonance is given by

$$\overline{\omega_i(\alpha, \beta, \gamma, T_r)} = \frac{1}{T_r}\left(\int_0^{T_r/2} \omega_{Di}\, dt - \int_{T_r/2}^{T_r} \omega_{Di}\, dt\right) \quad (39)$$

The phase accumulation after the cycle N_c is given by

$$\Delta\Phi_i(\alpha, \beta, \gamma, N_c, T_r) = \overline{\omega_i(\alpha, \beta, \gamma, T_r)}N_cT_r \quad (40)$$

Finally, the REDOR echo amplitude after averaging over all Euler angles was calculated as

$$S_f = \frac{1}{8\pi^2}\sum_{i=1}^{4}\int_\alpha\int_\beta\int_\gamma[\cos\Delta\Phi i(\alpha, \beta, \gamma, T_r)]\, d\alpha\,\sin\beta\, d\beta\, d\gamma$$

(41)

The normalized echo difference $\Delta S/S_0$ is given by Eq. (30). This relation strongly depends not only on the dipolar I_1-S_1 and I_1-S_2 couplings, but also on the angle $S_1-I_1-S_2$ [45].

3.3.6. Practical Aspects of the REDOR Experiment

It is emphasized that accurate interatomic distances are a prerequisite to achieving the three-dimensional structure of peptides, proteins, and macromolecules. A careful evaluation of the following several points is the most important step to obtaining reliable interatomic distances by the REDOR experiment, although they were not always seriously taken into account in the early papers. In practice, it is advisable to employ a standard sample such as [1-^{13}C, ^{15}N]glycine [44], whose C—N interatomic distance was determined to be 2.48 Å by a neutron diffraction study, to check that the instrumental conditions of a given spectrometer are correct, prior to an experiment on a new sample.

As described in Section 3.3.4, the finite pulse length may affect the REDOR factor. In fact, this effect is experimentally observed and calculated using Eq. (34) as shown in Figure 8. The REDOR parameter, $\Delta S/S_0$, as measured for 20% [1-^{13}C, ^{15}N]Gly, is plotted, for lengths of the ^{15}N π-pulse equal to 13.0 μs and 24.6 μs (chosen to satisfy the requirement of a 10% rotor cycle) for the experiment, as a function of N_cT_r, together with the calculated lines using the pulse length δ as well as 13.0 and 24.6 μs. It turns out, however, that the shortness of the ^{15}N π-pulse does not significantly affect the REDOR effect provided that the pulse length is less than 10% of the rotor cycle at the rotor frequency of 4000 Hz.

For most spectrometers, it is very difficult avoid *fluctuations of* rf *power* during REDOR experiments. It is therefore

very important for the rf power to be stabilized after waiting a certain time. If not, the π-pulse cannot be precisely controlled over the duration of the experiment. The REDOR factor is then greatly decreased, yielding spuriously long interatomic distances. Compensation of such instability of the rf power by the pulse sequence is therefore necessary. xy-4 and xy-8 pulse sequences have been developed for this purpose, and an xy-8 pulse is known to be the best sequence to compensate for the fluctuation of the rf power [46].

Since the early history of the REDOR experiment, contributions of natural isotopes have been considered as the major error source for the distance measurement [47]. It appears that the observed dipolar interaction can be modified by the presence of such neighboring nuclei. This effect was originally taken into account by simply calculating the $\Delta S/S_0$ value for isolated pairs and weighting them according to their natural abundances [47]. Careful analysis of the three-spin system, however, indicates that this sort of simple addition of contributions of two-spin systems may result in a serious overestimate

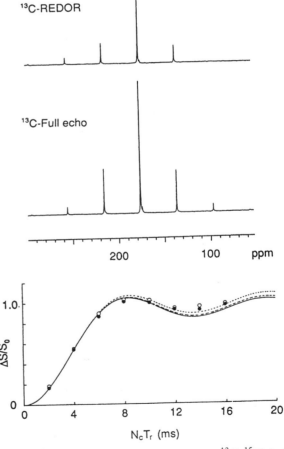

Fig. 8. Top: ^{13}C REDOR and full echo spectra of [1-^{13}C, ^{15}N]glycine as recorded at the rotor frequency of 4000 Hz and N_cT_r of 4 ms. Bottom: plots of $\Delta S/S_0$ vs N_cT_r. Solid and open circles denote the experimental points recorded using a ^{15}N π-pulse of 13.0 and 24.6 μs, rrespectively. Solid, broken, and dotted lines are calculated using π-pulses of δ, 13.0 μs, and 24.6 μs, respectively, with C—N interatomic distance of 2.48 Å. From [44].

of the effect, to yield shorter than actual distances [44]. The most accurate way to consider the natural isotope effect is therefore to treat the whole spin system as a three-spin system by taking into account the neighboring carbons in addition to the labeled pair. In practice, contributions from natural isotopes can be ignored [45] for ^{13}C REDOR but not for ^{15}N REDOR, because the proportion of natural ^{13}C nuclei is much higher than that of ^{15}N nuclei.

^{13}C, ^{15}N-doubly labeled samples are usually used in the REDOR experiment to determine the interatomic distances between the labeled nuclei. More importantly, the dipolar interaction with the *labeled ^{15}N nuclei in the neighboring molecules* should be taken into account as an additional contribution to the dipolar interaction of the observed pair under consideration. This contribution can be substantial when the observed distance is remote, because there are many contributions from nearby nuclei. It can be completely removed by diluting the labeled sample with a sample of natural-abundance molecules. The sensitivity of the signals, however, has to be sacrificed if one wants to remove the effect completely, as the sample is previously diluted to 1/49 [35]. Instead, it is advised to evaluate the REDOR factors in the infinitely diluted condition by extrapolating the data obtained by stepwise dilution of the sample (e.g., 60%, 30%, etc.) without losing sensitivity [45, Fig. 2.5], because a linear relationship between the REDOR factor and the dilution has been deduced theoretically. Alternatively, the observed plots of $\Delta S/S_0$ values against the corresponding $N_c T_r$ values for the sample without dilution can be fitted by a theoretical curve obtained from the dipolar interactions among three-spin systems, although the accuracy is not always improved to the level of the dilution experiment.

The transverse magnetization of the REDOR experiment decreases with increasing ^{1}H decoupling field [48, 49]. Dipolar decoupling may be strongly interfered with by molecular motion when the motional frequency is of the same order of magnitude as the decoupling field, and hence the transverse relaxation times T_2 are significantly shortened. In fact, it was found that the T_2-values of the carbonyl carbons in crystalline Leu-enkephalin are very short because of the presence of backbone motion [50]. This is a serious problem for the REDOR experiment, especially for the long distance pairs, because the S/N ratio is significantly deteriorated. In this case, it is worth considering measuring the ^{13}C REDOR signal under a strong decoupling field to prolong the transverse relaxation times. It is also useful to measure the distances at low temperature to be able to reduce the motional frequency. It is cautioned, however, that a crystalline phase transition may be associated with the freezing of the solvent molecule, as encountered for a variety of enkephalin samples [51, 81].

In a commercial spectrometer, the rotor is designed to allow the sample volume to be as large as possible in order to gain better sensitivity. Obviously, this arrangement causes inhomogeneity of H_1, which results in a broad distribution of the lengths of the 90° pulses. This problem is serious for the REDOR experiment, in which a number of π-pulses are applied. As a result, the pulse error can accumulate during the acqui-

sition to give serious error. Particularly, the samples located at the top or bottom of the sample rotor feel a quite weak rf field [44, 53]. This causes a great reduction of the REDOR factor for a sample that accupies the whole length of the sample rotor. This effect should be carefully taken into account prior to experiment with a commercial spectrometer. It is therefore strongly recommended to have the sample occupy only the middle of the coil, just as in a multiple-pulse experiment, so as to be able to acquire as accurate interatomic distances as possible by the REDOR method.

3.3.7. Natural Abundance ^{13}C REDOR Experiment [52]

Natural abundance REDOR experiments are free from both homonuclear dipolar and scalar interactions. This approach provides the interatomic distances of a number of isolated spin pairs simultaneously with high resolution. Further, this approach is free of the problem of shortened T_2-values due to homonuclear ^{13}C spin interaction. Figure 9 shows the REDOR and full echo spectra of natural abundance ^{13}C nuclei of singly ^{15}N-labeled crystalline ammonium[^{15}N] L-glutamate monohydrate (I) at $N_c T_r = 8$ ms. The assignment of peaks to the C_β, C_γ, C_α, C=O, and C_δ carbon nuclei, from high to low field, is also shown in Figure 9. The C=O (176.6 ppm) peak was distinguished from the C_δ (179.5 ppm) peak by its yielding a stronger REDOR effect, as shown in Figure 10. Similarly, the peak at 56.0 ppm was assigned to the C_α carbon nucleus, which gave the strongest REDOR effect among the

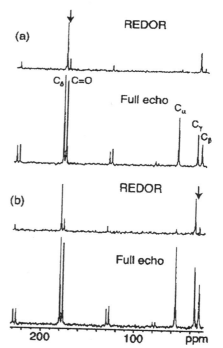

Fig. 9. Naturally abundant ^{13}C-REDOR and full echo spectra at $N_c T_r = 8$ ms with two different carrier frequencies for crystalline ammonium [^{15}N] L-glutamate monohydrate. Arrows indicate carrier frequency positions at the center of: (a) C=O and C_δ resonances, and (b) C_β and C_γ resonances.

three aliphatic ^{13}C peaks; and the peak at 30.5 ppm showed a stronger REDOR effect than that at 35.2 ppm. Thus, the peaks at 35.2 and 30.5 ppm were assigned to the C_γ and C_β carbon nuclei, respectively.

The interatomic C—N distances between ^{15}N and ^{13}C=O, $^{13}C_\alpha$, $^{13}C_\beta$, $^{13}C_\gamma$, and $^{13}C_\delta$ carbon nuclei for I were determined with a precision of 0.15 Å, after experimental conditions were carefully optimized. ^{13}C-REDOR factors for the three-spin system $(\Delta S/S_0)_{CN_1N_2}$, and the sum of the factors of two isolated two-spin systems, $(\Delta S/S_0)^* = (\Delta S/S_0)_{CN_1} + (\Delta S/S_0)_{CN_2}$, were further evaluated by REDOR measurements on isotopically diluted I in a controlled manner. Subsequently, the intra- and intermolecular C—N distances were separated by searching for the minima in a contour map of the root mean square deviation (RMSD) between the theoretically and experimentally obtained $(\Delta S/S_0)^*$ values against two interatomic distances, r_{C-N_1} and r_{C-N_2}. When the intermolecular C—N distance (r_{C-N_1}) of the particular carbon nucleus was substantially shorter than the intermolecular one (r_{C-N_2}), C—N distances between the molecule in question and the nearest neighboring molecules could also be obtained, although the accuracy was lower. On the contrary, it was difficult to determine the interatomic distances in the same molecule when the intermolecular dipolar contribution is larger than the intramolecular one, as in the case of the C_δ nucleus.

3.3.8. Simple Description of the RR Experiment [32]

In contrast to the REDOR experiment, homonuclear dipolar interactions can be recoupled in the RR experiment by adjusting the rotor frequency to be a multiple of the frequency difference between the isotropic chemical shift values of two chemically different homonuclear spins. This is called the *rotational resonance condition* ($\Delta\omega_{\text{iso}} = n\omega_r$). Under this condition, the energy level due to the sideband of one resonance becomes equal to that due to the centerband of the other, and therefore mixing of two spin states occurs, as indicated by the dotted arrows in Figure 10. Consequently, broadening of the line shape and acceleration of the exchange rate of the longitudinal magnetization are observed. These rotational resonance phenomena are strongly dependent on the interatomic distance. One can therefore determine the interhomonuclear distance by the RR experiment. When the distance between the labeled nuclei is short, a characteristic line shape due to the mixing of the spin states can be observed. In contrast, when the interatomic distance is long, a change of the signal intensity due to exchange of the longitudinal magnetizations can be observed.

The homonuclear dipolar interaction under the MAS condition has Fourier components $\omega_B^{(m)}$ at the frequency $m\omega_r$ as follows:

$$\omega_B(t) = \sum_{m=-2}^{2} \omega_B^{(m)} \exp(im\omega_r t) \qquad (42)$$

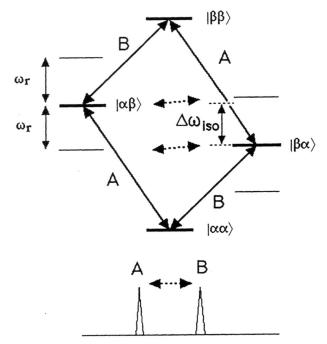

Fig. 10. Energy level diagram of the homonuclear two-spin system near the rotational resonance condition.

The Fourier components in this equation can be further expanded in the chemical shift anisotropy as follows:

$$\omega_B^{(n)} = \sum_{n=-2}^{2} \omega_B^{(m)} a_\Delta^{(m-n)} \qquad (43)$$

so that $\omega_B^{(n)}$ is expressed in terms of the dipolar frequencies $\omega_B^{(m)}$ and the chemical shift anisotropies $a_\Delta^{(n)}$. When the rotational resonance condition $\Delta\omega_{\text{iso}} = n\omega_r$ is fulfilled, $\omega_B^{(n)}$ is the Fourier component that is equal to the energy gap between two spin states (Fig. 10), and therefore the spin exchange (mixing of the wave functions) becomes efficient. At the rotational resonance condition, the exchange rate R can be expressed as

$$R^2 = r^2 - 4\left|\omega_B^{(n)}\right|^2 \qquad (44)$$

where $r = 1/T_2^{ZQ}$ is the zero quantum transverse magnetization rate, and $|\omega_B^{(n)}|$ the rotationally driven exchange rate. When $R^2 > 0$, the difference of the two longitudinal spins is given by

$$\langle I_z - S_z \rangle(t) = e^{-rt/2}\left[\cosh(Rt/2) + \frac{r}{R}\sinh(Rt/2)\right] \qquad (45)$$

When $R^2 < 0$,

$$\langle I_z - S_z \rangle(t) = e^{-rt/2}\left[\cos(iRt/2) + \frac{r}{iR}\sin(iRt/2)\right] \qquad (46)$$

As is expected, T_2^{ZQ}, the chemical shift anisotropy, and the dipolar interaction are involved in the exchange rate of the longitudinal magnetization under the MAS condition. Therefore, the separation of the dipolar interaction is quite complicated as compared with the REDOR experiment.

3.3.9. Practical Aspects of the RR Experiment [53]

Experimentally, homonuclear dipolar interactions can be determined by measuring the exchange rate of the longitudinal magnetization as a function of the mixing time τ_m. A selective inversion pulse is used to invert one of the two resonances after the mixing time. One of the advantages of the RR experiment is that it can be performed by an ordinary double-resonance spectrometer as long as the spinner speed can be controlled using a spinner frequency controller. It is advised to use $n = 1$ as the rotational resonance condition, because the chemical shift anisotropy is strongly affected when n is greater than 3. When a high-field spectrometer is used, the chemical shift anisotropy increases proportionally to the frequency used. In that case, one has to include the chemical shift anisotropy tensor for both nuclei in the analysis of the longitudinal magnetization. In the analysis of the RR experiment, one has to use T_2^{ZQ} as discussed in the previous sub-subsection. Its value is difficult to determine experimentally. In practice, it can be approximated from the single-quantum relaxation time (T_2) of the two spins by the expression [54]

$$\frac{1}{T_2^{ZQ}} = \frac{1}{T_2^{I_1}} + \frac{1}{T_2^{I_2}} \qquad (47)$$

or

$$T_2^{ZQ} = \frac{1}{\pi(\nu_{I_1} + \nu_{I_2})} \qquad (48)$$

When the chemical shift difference is very small, it is difficult to perform the RR experiment because of the overlap of the resonance lines of the dipolar-coupled nuclei, leading to difficulty in the analysis of the RR data. In this case, a rotating resonance experiment in the tilted rotating frame [55] can be used, because it allows a much higher spinning speed for small chemical shift differences.

The natural abundance background signal can also affect the apparent amount of RR magnetization exchange. The observed magnetization exchange rate then yields a smaller magnetization exchange rate than the observed one. This results in an overestimate for the interatomic distance. Incomplete proton decoupling prevents magnetization exchange between the coupled spins because of the H_1 field inhomogeneity. It is advised to irradiate with a strong proton decoupling field (>80 kHz) and used a small-size sample in the rf coil to avoid H_1 inhomogeneity.

3.4. Oriented Biomembranes

To obtain the information on the orientation of biomolecules bound to a membrane, it is necessary to align the samples uniformly in the NMR spectrometer with respect to the magnetic field direction. Because a number of peptides and membrane proteins are strongly bound to membrane, these biomolecules

•Lipid bilayers on glass plate

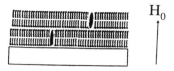

•Bicelle

•Elongated liposome

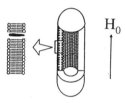

Fig. 11. Magnetically aligned lipid bilayers.

can be aligned with the magnetic field provided that the lipid bilayers are so aligned.

Ordering of the lipid bilayers with respect to the magnetic field can be achieved in one of two ways, as shown in Figure 11. First, lipid can be oriented macroscopically by pressing lipid–water dispersions between flat glass plates, thus orienting the membrane microdomains by mechanical shearing forces. Second, the lipid molecules themselves can align spontaneously with the magnetic field because of their diamagnetic anisotropy. The interaction of the diamagnetic anisotropy and the magnetic field is given by

$$F = -\tfrac{1}{2}H_0^2\{\chi_\perp - (\chi_\parallel - \chi_\perp)\cos 2\theta\} \qquad (49)$$

where θ is the angle between $\chi_\parallel$ and the magnetic field. Therefore, the orientation energy of a lipid bilayer with N lipids is expressed as

$$N\Delta F = -\tfrac{1}{2}N\,\Delta\chi\,H_0^2 \qquad (50)$$

where $\Delta F = F(\theta = 0°) - F(\theta = 90°)$ and $\Delta\chi = \chi_\parallel - \chi_\perp$. If N is small, the orientation energy is not large enough to align the macrodomain of the lipid bilayer, because the thermal energy kT tends to disturb the alignment. In the case of phospholipid bilayers, however, a macrodomain of lipid containing 10^6 lipid molecules can be spontaneously aligned with the magnetic field.

Magnetic ordering of lipid bilayers has been reported for pure and mixed phosphatidylcholine bilayers [56–60], including melittin–phospholipid systems [61–64]. Subsequently, such magnetic ordering has been reported in a detergent–lipid mixture called a *bicelle* [65, 66], which was shown to be oriented in the magnetic field by the negative magnetic anisotropy of the lipid acyl chain. Therefore, the acyl chain tends to orient perpendicular to the magnetic field if a large number of lipid molecules are ordered in the liquid crystalline phase and possess a sufficient degree of magnetic anisotropy to align the lipid bilayers along the magnetic field.

^{31}P NMR spectra clearly show that at moderately high concentrations of melittin incorporated into a DMPC bilayer (DMPC:melittin = 10:1 molar ratio), the bilayer shows lysis and fusion at temperatures both lower and higher than T_m. Above T_m, giant vesicles were observed for the melittin–DMPC bilayer systems. This lipid bilayer systems shows magnetic ordering at a temperature higher than T_m. Therefore, it is suggested that elongated bilayer vesicles rather than discoidal bilayers are formed above T_m in the melittin–DMPC bilayer system, in which most of the surface area of the bilayers is oriented parallel to the magnetic field, as shown in schematically in Figure 12. Thus a large magnetic anisotropy can be induced, because most of the phospholipids, which have negative magnetic anisotropy along the acyl chain axes, are aligned perpendicular to the magnetic field.

A mechanically aligned lipid bilayer system can be is established by casting the lipid bilayer on a glass plate and then hydrating under high humidity. In this system the acyl chains can be aligned perpendicular to the glass plate. Such a system offers the advantage that it is possible to align the lipid bilayer surface in any orientation with respect to the magnetic field. It has the disadvantage of a low filling factor because of the presence of plate in the NMR coil, leading to small NMR signals.

The orientation of peptides bound to the magnetically aligned lipid bilayer can be investigated by looking at the chemical shift anisotropy of the carbonyl carbon in the backbone in the peptide chain. In particular when the α-helix is rotating

about its axis, it is possible to determine the orientation angle of that axis with respect to the bilayer normal [64]. As we discuss in Section 4, the tilt angle of an α-helix with respect to the average can be determined by comparing the anisotropy patterns of carbonyl carbons of consecutive amino acid residues that form the α-helix with the chemical shift values of the corresponding magnetically aligned state. This is related to the inter-peptide-plane angle of 100° for consecutive peptide planes in the case of an ideal α-helix.

Similar information can be obtained by looking at ^{15}N–^{1}H dipolar interaction in the peptide backbone. Polarization inversion spin exchange at the magic angle (PISEMA) gives excellent resolution for the dipolar dimension in the correlation spectra between ^{15}N chemical shift values and the ^{15}N–^{1}H dipolar interaction [67]. It is of interest to note that a characteristic circular pattern, called the *polarity index slant angle* (PISA) wheel, is seen when an α-helix is formed in the membrane, and the shape of wheel is sensitive to the tilting angle of the α-helix with respect to the bilayer normal [68–70] (Fig. 13). When one observe this PISA wheel, one can identify the amino

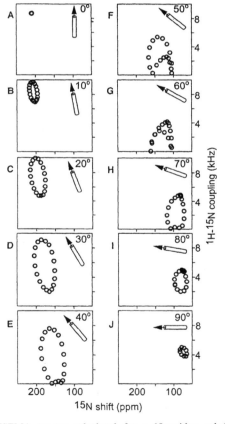

Fig. 13. PISEMA spectra calculated for a 19-residue α-helix with 3.6 residues per turn and uniform dihedral angles ($\phi = -65°$, $\psi = -40°$) at various helix tilt angles relative to the bilayer normal: A, 0°; B, 10°; C, 20°; D, 30°; E, 40°; F, 50°; G, 60°; H, 70°; I, 80°; J, 90°. Spectra were calculated on a Silicon Graphics O2 computer (Mountain View. CA), using the FORTRAN program FINGERPRINT [23, 26]. The principal values and molecular orientation of the ^{15}N chemical shift tensor ($\sigma_{11} = 64$ ppm, $\sigma_{22} = 77$ ppm, $\sigma_{33} = 217$ ppm; $\angle$NH = 17°) and the NH bond distance (1.07 Å) were as previously determined.

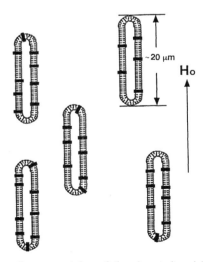

Fig. 12. Schematic representation of the elongated vesicles of melittin–DMPC bilayers in a strong magnetic field. The longer axis is parallel to the magnetic field, and so is most of the bilayer surface in the vesicles. The melittin molecule forms a transmembrane helix with the helical axis parallel to the bilayer normal. The average length of the longer axis is ~20 μm.

acid residues involved in the α-helix even if the amino acid sequence is not known.

4. APPLICATION OF SOLID STATE NMR TO BIOMOLECULES

4.1. Three-Dimensional Structure of Peptides and Proteins

Schaefer et al. [71] synthesized an emerimicine fragment (Ac–Phe–[1-^{13}C]MeA^2–MeA–MeA–Val–[^{15}N]Gly^6–Leu–MeA–MeA–OBz). The ^{13}C–^{15}N interatomic distance for a distance of four residues was determined to be 4.07 Å by the REDOR method. It was concluded that the structure is an α-helix, because the expected distances are 4.13 and 5.87 Å for the α-helix and the 3_{10}-helix, respectively. In a similar manner, the ^{19}F–^{13}C interatomic distance was measured for the ^{19}F, ^{13}C, and ^{15}N triply labeled fragment (^{19}FCH$_2$CO–Phe–MeA–MeA–[1-^{13}C]MeA–[^{15}N]Val–Gly–Leu–MeA–MeA–OBzl) [72] and found to be 7.8 Å by the TEDOR method [8] after transferring the magnetization from ^{15}N to ^{13}C. Because the TEDOR method makes it possible to eliminate background signals due to naturally abundant nuclei, quite remote interatomic distances can be determined. Schaefer et al. also tried to determine the C–N interatomic distances of an ion channel peptide Val^1–[1-^{13}C]Gly^2–[^{15}N]Ala^3–gramicidin A in a DMPC bilayer [73]. The dipolar interaction of the peptide in the lipid bilayer showed much smaller values than that in the powder state, because the helix motions significantly averaged the dipolar interactions. The extent of the scaling of the dipolar interaction shows that gramicidin A consists of a dimer with a single helix. A magainin analog in the membrane was investigated by ^{13}C, ^{31}P REDOR [74]. The result indicates that the α-helical Ala$_{19}$-magainin 2 amide is bound to the head group of the lipid bilayers. Incidentally, the data analysis described seems worth considering in general in order to improve the accuracy of REDOR data.

A complete three-dimensional structure can be determined by combining a variety of interatomic distances [44, 75–77].

Garbow et al. have synthesized three peptides that are labeled at different positions. These interatomic distances were converted to torsion angles to yield the β-turn II structure [75, 76]. Naito et al. systematically applied this technique to elucidate the three-dimensional structure of N-acetyl-Pro–Gly–Phe [44]. They proposed that the carbonyl carbon of the $(i-1)$th residue and the amino nitrogen of the $(i+1)$th residue should be labeled with ^{13}C and ^{15}N, respectively. Namely, [1-^{13}C]N-acetyl-Pro–[^{15}N]Gly–Phe (I), N-acetyl-[1-^{13}C]Pro–Gly–[^{15}N]Phe (II), and [1-^{13}C]N-acetyl-Pro–Gly–[^{15}N]Phe (III) were synthesized, and the resulting distances were determined to be 3.24, 3.43, and 4.07 Å, respectively, utilizing the REDOR factor obtained for the infinitely diluted state to prevent errors from the contributions of the neighboring labeled nuclei. No correction for the contribution of the natural isotopes turned out to be necessary. Surprisingly, these distances do not agree well with the values obtained from an X-ray diffraction study [78] available at that time. The maximum discrepancy between them is 0.5 Å, which is much larger than the expected error in the REDOR experiment (± 0.05 Å). The disagreement is explained by the fact that the crystal (orthorhombic) used for the REDOR experiments was different from that used in the X-ray diffraction study (monoclinic). To check the accuracy of the REDOR experiment, an X-ray diffraction study was performed on the same crystals used for the REDOR experiment. It is found that the distances from the new crystalline polymorph (orthorhombic) agree within an accuracy of 0.05 Å, as shown in Table III. Conformational maps based on the possible combinations of the torsion angles of the Pro and Gly residues were calculated. Further, the difference of the chemical shifts between the C_β and C_γ carbons of the Pro residues, $\Delta_{\beta\gamma}$, was used as a constraint to determine the ψ-value ($-13°$) [79]. The angle ϕ of the Pro residue is in many instances restricted to $-75°$, which yields the minimum energy in the residue. Therefore, the torsion angles of the Pro residue are uniquely determined to be ($-75°$, $-28°$). Using these torsion angles, conformational maps were calculated again. Finally, two pairs of torsion angles were selected: ($-112°$, $48°$) and ($-112°$, $-48°$). Energy minimization by molecular mechanics yielded

Table III. C–N Interatomic Distances Determined from REDOR Experiments as Compared with Those Determined by X-Ray Diffraction and MD [4]

	Distance (Å)					
	Experimental			Calculated		
Labelled peptide	Redor	X-ray			MD	
	Orthorhombic	Orthorhombic	Monoclinic	Energy-minimized[a]	Orthorhombic	Monoclinic
I	3.24 ± 0.05 (3.43 ± 0.05)[b]	3.19	3.76	3.17	3.22 ± 0.10	3.63 ± 0.10
II	3.43 ± 0.05 (3.66 ± 0.05)	3.35	3.21	3.57	3.22 ± 0.10	3.33 ± 0.10
III	4.07 ± 0.05 (4.45 ± 0.05)	3.99	3.91	4.17	3.92 ± 0.12	3.83 ± 0.10

[a] Energy-minimized structure based REDOR data.
[b] Based on fully packed 7.5-mm rotor system.

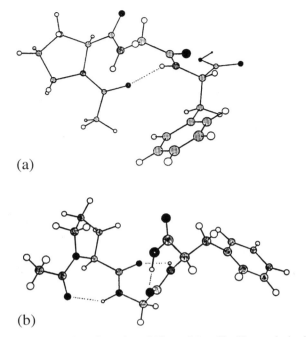

(a)

(b)

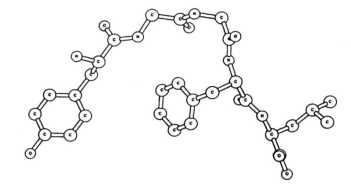

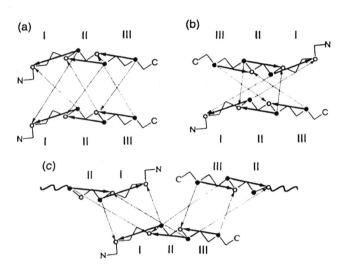

Fig. 14. Optimized conformation of *N*-acetyl-Pro–Gly–Phe as obtained by the minimization of energy from the initial form as deduced from the REDOR experiment (a), and a "snapshot" of the conformation deduced by MD simulation *in vacuo* (at 100 K) (b).

Fig. 15. Top: Three-dimensional structure of Leu–enkephalin dihydrate determined uniquely by the successive application of conformation maps as well as additional constraints on the conformation-dependent ^{13}C chemical shifts. Botom: Three kinds of models for putative intermolecular packing: interaction (a) through parallel β-sheets, (b) through antiparallel β-sheets, and (c) through interaction of three molecules. Solid and dotted arrows correspond with the dipolar interactions between intramolecular and intermolecular ^{13}C–^{15}N pairs, respectively.

the structure of the β-turn I structure, as shown in Figure 14. It is found that the three-dimensional structure of this peptide is well reproduced by a molecular dynamics simulation taking into account all of the intermolecular interactions in the crystals [44, 80].

Elucidation of the three dimensional structure of an opioid peptide Leu–enkephalin crystal, Tyr–Gly–Gly–Phe–Leu, grown from a mixed MeOH–H$_2$O solvent, was performed by the REDOR method [81]. It is a challenge for this technique alone to reveal the three-dimensional structure of such a complicated system. Six differently labeled Leu–enkephalin molecules were synthesized following the strategy described, and the resulting interatomic distances were accurately determined. It turns out, however, that the crystalline polymorph under consideration was very easily converted to another form. Therefore it is necessary to check whether the six differently labeled samples are all in the same crystalline polymorph by means of the ^{13}C chemical shifts. Meaningless data can be obtained without this precaution. When the distance data are converted to yield the necessary numbers of torsion angles, a unique combination of torsion angles in the corresponding conformational map is determined by using the chemical shift data as additional constraints. A three-dimensional structure was thus determined as shown in Figure 15. However, this structure is not the same as that previously determined by X-ray diffraction, because one is dealing with a crystalline polymorph that is not fully explored. It is possible to figure out the molecular packing in the crystals by looking at the intermolecular dipolar contribution. Since the intermolecular dipolar contributions in I and III are larger than that in II, we conclude that the

β-sheet is formed by the interaction of one peptide with two other peptides and is antiparallel.

Another important application of the REDOR and RR methods is to determine protein structure. It is, however, still difficult to determine the three-dimensional structure of a whole protein molecule using these methods. Instead, Schaefer et al. [82–85] have determined the structure of the ligand and the binding site of the ligand–protein complex. It is possible to determine the interatomic distance by the REDOR method in the case of the enzyme–analog complex, because the reaction will not proceed. It is, however, difficult to measure the interatomic distances of the enzyme–substrate complex, since they will react in a short time. Evans et al. [86, 87] froze the reaction instantaneously, and the structure of the intermediate state of the complex could be observed by the REDOR method. Griffin et al. [88–90] have used the RR method to determine retinal configurations in

SOLID STATE NMR SPECTROSCOPY   751

various states of photointermediates in the membrane protein bacteriorhodopsin.

4.2. Dynamics of Peptides and Proteins

It is well recognized that side chains of amino acid residues in proteins undergo several types of internal motions, such as rotational isomerism, ring puckering of proline or flip-flop motion of aromatic residues, both in aqueous solution and in the solid state. The presence or absence of such motions is considered to be a very convenient means to examine a manner of molecular packing of peptides in the solid state or in the interiors of globular proteins in aqueous solution. The rate constants for such motion have been determined over a wide range (10^8–10^2 s^{-1}) by means of spin–lattice relaxation times in the laboratory frame, line-shape analysis of quadrupolar interaction, etc. It is also possible to examine very slow motions, if any, whose rate constant is much less than 10^2 s^{-1}, by using two-dimensional exchange NMR spectroscopy. A detailed analysis of such motions for peptides in the solid state seems to be very valuable as a pertinent model for protein, because such local motions are readily separated from the contribution of the overall motion.

Local motion of Leu^5– and Met^5–enkephalin crystallized from a variety of solvent systems were examined using ^{13}C CP-MAS NMR spectroscopy [91]. It was demonstrated that ^{13}C NMR spectra of the trihydrate recorded below $-40°$C display an additional spectral change: the doubling of peaks in Tyr C$_\zeta$ and Phe C$_\delta$ caused by conformational isomerism about the C$_\alpha$–C$_\beta$. The spectral profile of Tyr C$_\varepsilon$ was well related to the presence or absence of the flip-flop motion of the tyrosin side chain, because the broad single peak was changed into a well-defined doublet peak when the sample was cooled down to $-80°$C. The rate constant for such flip-flop motion was estimated to be 1.3×10^2 s^{-1} at ambient temperature, based on a spectral simulation utilizing the two-site exchange model. One- or two-dimensional exchange spectra were further recorded to analyze similar flip-flop motions whose rate constants are much smaller than the limiting values as estimated from the simple analysis of the two-site exchange (10 s^{-1}). The rate constant for the flip-flop motion of Ac–Tyr–NH$_2$ and the extended form of Met^5–enkephalin are found to be 1.94 and 1.45 s^{-1} at ambient temperature, respectively. The rate constants of very slow flip-flop motion, of order to 10^{-3} to 1 s^{-1}, were determined for Tyr–OH and Tyr–NH$_2$.

The backbone and side-chain motions of enkephalin crystals were further examined by means of ^{13}C and proton spin–lattice relaxation times in the laboratory (T_{1C}) and the rotating frame ($T_{1\rho}^H$), respectively [92]. The β-bend structure of crystalline Leu–enkephalin trihydrate turned out to be very flexible, in view of the T_{1C} and $T_{1\rho}^H$ values, as compared with the other crystals taking the β-sheet form. In contrast, there appears no such flexibility in Leu–enkephalin dihydrate, despite a claimed similar β-bend structure as determined by X-ray diffraction. It was also demonstrated that the presence or absence of side-chain motions was conveniently monitored by the relative peak intensities of individual residues, which were strongly influenced by the manner in which local motions interfered with the proton decoupling frequency. It was further shown that these molecular motions were strongly affected by the bound solvent molecules.

Kamihira et al. examined the phenyl ring dynamics of [^{2}H$_5$]Phe^4-labeled Leu^5– and Met^5–enkephalin molecules in crystals grown from four solvents using solid state ^{2}H NMR spectroscopy [93]. The ^{2}H NMR powder pattern clearly indicated the presence of 180° flip motions about the C$_\beta$–C$_\gamma$ bond axis of the phenyl rings. The frequencies of the 180° flip motions were estimated to be 5.0×10^3, 3.0×10^4, and 2.4×10^6 Hz for Leu–enkephalin crystallized from H$_2$O, methanol–H$_2$O, and N,N-dimethylformamide (DMF)–H$_2$O, respectively, and 1.0×10^4 Hz for Met–enkephalin crystallized from ethanol–H$_2$O at ambient temperature. The difference of the frequencies for the motion was attributed to the manner of their molecular packing in the crystals as determined by X-ray diffraction. Because the correlation times determined from the ^{2}H spin–lattice relaxation times (T_{1D}-values) were much shorter than those of the 180° flip motions, it was shown that the phenyl rings of these four crystals have small-amplitude librations. Therefore, it was concluded that the T_{1D}-values were dominated by the librations, even for the ring deuterium. These motions became slower at lowered temperatures and caused changes of the peak intensities and increased quadrupole splittings, which were observed in each ^{2}H NMR spectrum. Isotropic sharp signals due to natural isotopes in solvent molecules were observed at the center of the ^{2}H NMR spectra. The stepwise loss of the signal intensity was interpreted in terms of different temperatures of freezing of motions of both bound water or organic solvent and mixed solvent as the temperature was lowered, consistent with the buildup of solvent peaks in the ^{13}C CP-MAS NMR spectra. It is suggested that there are a number of bound mobile solvent molecules in the crystals and that the freezing of the solvent causes considerable changes in the conformations and dynamics of enkephalin molecules.

4.3. Membrane-Bound Peptide

The structure of membrane-bound peptides has been extensively studied by using lipid bilayer media, oriented either magnetically, mechanically, or spontaneously as described in the previous section.

Melittin is a hexacosapeptide with a primary structure of Gly–Ile–Gly–Ala–Val–Leu–Lys–Val–Leu–Thr–Thr–Gly–Leu–Pro–Ala–Leu–Ile–Ser–Trp–Ile–Lys–Arg–Lys–Arg–Gln–Gln–NH$_2$ and is the main component of bee venom. Melittin has powerful hemolytic activity in addition to causing voltage-dependent ion conductance across planar lipid bilayers at low concentration. It is also causes selective micellization of bilayers as well as membrane fusion at high concentration [94]. It is important to determine the orientation of the melittin helix in lipid bilayers to understand the nature of the interaction of melittin with membranes. As is described in the previous section, the melittin–DMPC lipid bilayer aligns along a static

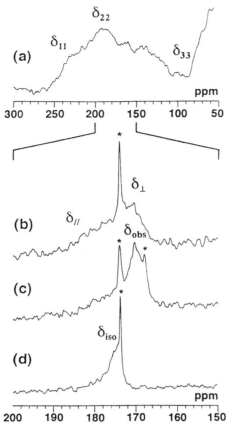

Fig. 16. Temperature variation of ^{13}C NMR spectrum of a DMPC bilayer in the presence of $(1\text{-}^{13}\text{C})\text{Ile}^{20}$-melittin in the static condition at $-60°$C (a) and hat the slow MAS (b), static (c), and fat MAS (d) conditions at $40°$C. The signals (marked by asterisks) appearing at 173 ppm in (b), 173 and 168 ppm in (c), and 173 ppm in (d) are assigned to the C=O groups of DMPC.

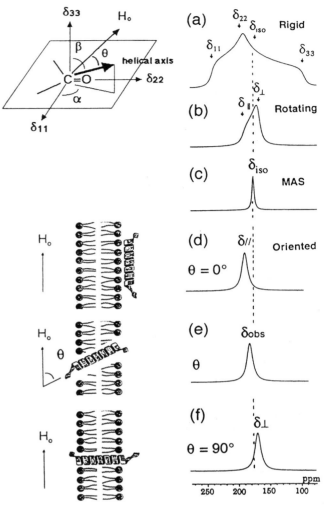

Fig. 17. Directions of the principal axes of the ^{13}C chemical shift tensor of the C=O group, the helical axis, and the static magnetic field (H_0), and ^{13}C NMR spectral patterns of the C=O carbons corresponding to the orientation of the α-helix with respect to the surface of the magnetically oriented lipid bilayers. Simulated spectra were calculated using $\delta_{11} = 241$, $\delta_{22} = 189$, and $\delta_{33} = 96$ ppm (principal values of Ile^{20} C=O at $-60°$C) for the rigid case (a), rotation about the helical axis (slow MAS) (b), fast MAS (c), magnetic orientation parallel to the magnetic field (d), magnetic orientation at an angle θ with the magnetic field (e), and magnetic orientation perpendicular to the magnetic field (f).

magnetic field by forming elongated vesicles with the long axis parallel to the field. Using this magnetic orientability of the membrane, the structure, orientation, and dynamics of melittin have been extensively studied [64].

Figure 16 shows the ^{13}C NMR spectra of $[1\text{-}^{13}\text{C}]\text{Ile}^{20}$–melittin bound to a DMPC bilayer hydrated with TRIS buffer. A broad asymmetrical powder pattern characterized by $\delta_{11} = 241$, $\delta_{22} = 189$, and $\delta_{33} = 96$ ppm appeared at $-60°$C (Fig. 16). The presence of this broad signal indicates that any motion of melittin bound to the DMPC bilayer is completely frozen at $-60°$C. A narrowed ^{13}C NMR signal was observed at 174.8 ppm for Ile^{20} C=O in a fast MAS experiment at $40°$C, and its position was displaced upfield by 4.6 ppm in the oriented bilayer at $40°$C, as observed in the magnetically oriented state. An axially symmetric powder pattern with an anisotropy of 14.9 ppm was recorded at $40°$C in a slow MAS experiment. Because the line width due to the anisotropy at $40°$C is not as broad as that at $-60°$C, it is expected that the α-helical segment undergoes rapid reorientation about the helical axis at $40°$C.

The secondary structure of melittin bound to DMPC bilayers can be determined in an empirical manner by utilizing the isotropic chemical shift values of ^{13}C-labeled amino acid residues with reference to those of model systems as dis-

cussed in the previous section. Because the isotropic ^{13}C chemical shifts of $[1\text{-}^{13}\text{C}]\text{Gly}^3$, $[1\text{-}^{13}\text{C}]\text{Val}^5$, $[1\text{-}^{13}\text{C}]\text{Gly}^{12}$, $[1\text{-}^{13}\text{C}]\text{Leu}^{16}$, and $[1\text{-}^{13}\text{C}]\text{Ile}^{20}$ residues in melittin are found to be 172.7, 175.2, 171.6, 175.6, and 174.8 ppm, respectively, all of the residues mentioned are involved in the α-helix, as summarized in Table IV. The ^{13}C chemical shift tensor of the carbonyl carbon forming the α-helix was used to reveal the molecular motion. It has been reported that the principal directions of δ_{22} and δ_{33} are nearly parallel to the C=O bond direction and the peptide plane normal, respectively, and δ_{11} is perpendicular to both δ_{22} and δ_{33} axes, as is schematically depicted in Figure 17. It is assumed that the C=O direction in an α-helix is nearly parallel to the helical axis, to form

Table IV. ^{13}C Chemical Shifts [ppm] Structure and Orientation of Melittin Bound to Magnetically Oriented Lipid Bilayers at 40°C

Residue	δ_{obs}	$\Delta\delta_{obs}{}^a$	δ_{iso}	$\Delta\delta_{iso}{}^a$	$\delta_\parallel - \delta_\perp$	$\delta_{11}{}^b$	$\delta_{22}{}^b$	$\delta_{33}{}^b$	Structure^c	θ^d
[1-^{13}C]Gly^3	179.5 ± 0.2	3.4	172.7 ± 0.2 (172.4)^c	2.4	−20.7 ± 2.0	237.6	186.7	92.9	α-Helix	86° ± 10°
[1-^{13}C]Val^5	177.0 ± 0.2	3.6	175.2 ± 0.2 (175.2)^c	3.4	−5.4 ± 0.8	237.6	191.1	96.9	α-Helix	90° ± 15°
[1-^{13}C]Gly^{12}	170.9 ± 0.3	4.1	171.6 ± 0.2 (171.6)^c	3.2	~0	240.6	180.7	93.5	α-Helix	—
[3-^{13}C]Ala^{15}	17.2 ± 0.1	1.4	16.1 ± 0.1 (15.8)^c	1.4	—	—	—	—	α-Helix	—
[1-^{13}C]Leu^{16}	177.2 ± 0.3	4.2	175.6 ± 0.2 (175.5)^c	3.2	−6.2 ± 1.0	240.6	192.8	93.1	α-Helix	74° ± 15°
[1-^{13}C]Ile^{20}	170.2 ± 0.3	4.3	174.8 ± 0.2 (175.1)^c	3.4	14.9 ± 2.0	240.6	188.9	96.1	α-Helix	81° ± 10°

^{a}Line width at half height.

^{b}Obtained at −60°C. The error range for the tensor elements are ±1.5 ppm.

^{c}Typical ^{13}C chemical shift values (ppm) of (δ_{iso} of α-helix, δ_{iso} of β-sheet) are (171.6, 168.5), (174.9, 171.8), (15.5, 20.3), (175.7, 170.5), and (174.9, 172.7) for Gly C=O, Val C=O, Ala CH$_3$, Leu C=O, and Ile C=O, respectively [14].

^{d}Angles between the average helical axis and the membrane surface.

C=O···H−N hydrogen bonds (both α and β are 90° in Figure 17). Under this condition, an axially symmetrical powder pattern characterized by $\delta_\parallel$ and $\delta_\perp$, corresponding to δ_{22} and $(\delta_{11} + \delta_{33})/2$, respectively, is obtained as shown in Figure 17, when the helix rotates rapidly about its axis. It was demonstrated that the principal values δ_{11}, δ_{22}, and δ_{33} of the chemical shift tensor of Ile^{20} C=O are 241, 189, and 96 ppm, respectively, in the low-temperature experiments. Therefore, the average ^{13}C chemical shift tensor becomes axially symmetrical to give $\delta_\perp = (\delta_{11} + \delta_{33})/2 = 168$ and $\delta_\parallel - \delta_{22} = 189$ ppm. Indeed, an axially symmetrical powder pattern with $\delta_\perp = 170$ and $\delta_\parallel = 185$ ppm was obtained for Ile^{20} C=O in the slow MAS experiment, which agree well with the values obtained from the motional model as mentioned above. In the case of Gly^3 C=O, the values $\delta_\perp = 165$ and $\delta_\parallel = 187$ ppm were evaluated from the principal values ($\delta_{11} = 238$, $\delta_{22} = 187$, and $\delta_{33} = 93$ ppm) obtained from the powder pattern recorded at −60°C, whereas $\delta_\perp = 180$ and $\delta_\parallel = 159$ ppm were obtained from the slow MAS experiment. The axially symmetrical powder pattern for Gly^3 C=O was reversed in shape as compared to that for Ile^{20} C=O. This observation suggest that the C=O direction of Gly^3 is not parallel to the rotation axis. Therefore, it is suggested that melittin molecules rotate rapidly about the averaged helical axis of the whole body of melittin, whose N- and C-terminal helical rods are inclined about 30° ± 12° and 10° ± 12° to the average axis, as estimated from the ^{13}C chemical shift values of [1-^{13}C]Gly^3-melittin and [1-^{13}C]Ile^{20}-melittin molecules, respectively. The error range was estimated by considering the possibility of deviation of the C=O axis from the helical axis over the range from 0 to 12°. The chemical shift values of [1-^{13}C]Gly^3 and [1-^{13}C]Ile^{20} were used in this discussion to determine the tilt angle, because clear axially symmetrical powder pattern were observed for these amino acid residues. Hence the two possible kink angles between the N- and C-terminal helical axes are estimated as 140° ± 24° or 160° ± 24°, which is considerably larger than 120°, as reported in an X-ray diffraction study of the static condition [95, 96]. On the other hand,

a large kink angle of 160° is obtained from monomeric melittin in methanol [97].

It is useful [98] to consider the ^{13}C=O chemical shift tensor to evaluate the orientation of averaged α-helical axis of the peptides bound to the magnetically oriented lipid bilayer in a case where the C=O axis is parallel to the helical axis. In this case the α-helical axis is defined by the polar angles α and β with respect to the principal axes of the ^{13}C=O chemical shift tensor, as shown in the top of Figure 17. When the average α-helix is inclined by θ to the static magnetic field, the observed ^{13}C chemical shift δ_{obs} for the rotating α-helix in the magnetically oriented state, obtained from the static experiment, is given by

$$\delta_{obs} = \delta_{iso} + \tfrac{1}{4}\big[(3\cos 2\theta - 1)(\delta_{33} - \delta_{iso}) + \sin 2\beta \cos 2\alpha(\delta_{11} - \delta_{22})\big] \quad (51)$$

When $\theta = 0°$, the α-helical axis is considered to be parallel to the static magnetic field. In that direction, δ_{obs} is denoted as $\delta_\parallel$ and is given by

$$\delta_\parallel = \delta_{iso} + \tfrac{1}{2}(2\cos 3\beta - 1)(\delta_{33} - \delta_{iso}) \quad (52)$$

When $\theta = 90°$, the α-helix is perpendicular to the static magnetic field, and then δ_{obs} corresponds to $\delta_\perp$ and is expressed as

$$\delta_\perp = \delta_{iso} - \tfrac{1}{4}\big[(3\cos 2\beta - 1)(\delta_{33} - \delta_{iso}) + \sin 2\beta \cos 2\alpha(\delta_{11} - \delta_{33})\big] \quad (53)$$

Using Eqs. (52) and (53), Eq. (51) can be written as

$$\delta_{obs} = \delta_{iso} + \tfrac{1}{3}(3\cos 2\theta - 1)(\delta_\parallel - \delta_\perp). \quad (54)$$

This relation allows one to predict that the α-helix is parallel to the magnetic field when δ_{obs} is displaced downfield, to $\delta_\parallel$, whereas the α-helix is perpendicular to the magnetic field when δ_{obs} is displaced upfield to $\delta_\perp$, as shown in Figure 17d and f, respectively, provided that $\delta_\parallel$ appears at a lower field than $\delta_\perp$ does. Generally, the orientation θ of the α-helical axis with respect to the lipid bilayer surface can be determine by using Eq. (54) after δ_{iso}, δ_{obs}, and $\delta_\parallel - \delta_\perp$ have been obtained from MAS, static, and slow MAS experiments, respectively.

Because it turns out that the lipid bilayer is oriented with respect to the magnetic field with the bilayer surface parallel to the magnetic field and the α-helical axis of melittin precessing about the average helical axis, θ reflects the direction of the average α-helix with respect to the surface of the lipid bilayer. Actually, the static ^{13}C chemical shift δ_{obs} of Ile^{20} C=O in the magnetically oriented state was displaced upfield by 4.6 ppm from the isotropic value δ_{obs}. This result allows one to determine that the average α-helical axis was inclined nearly 90° to the bilayer plane. On the other hand, δ_{obs} of Gly^3 C=O was displaced downfield by 6.8 ppm, whereas the axially symmetrical powder pattern was reversed in shape as compared to that of Ile^{20} C=O, and hence $\delta_{\parallel} - \delta_{\perp}$ is negative in this case. This result leads to the conclusion that the average axis of the α-helix is again inclined 90° to the bilayer plane. Therefore, it is concluded that transmembrane α-helices of melittin are formed in the lipid bilayer systems and both N- and C-terminal helices reorient about the average helical axis, which is parallel to the lipid bilayer normal. It is emphasized that the charged amino acid residues such as Lys^7 in the N-terminus and Lys^{21}, Arg^{22}, Lys^{23}, and Arg^{24} in the C-terminus may be closely located on opposite sides of the polar head groups of lipid bilayers, although melittin forms an amphiphilic helix in the lipid bilayers.

Solid state NMR results indicate that melittin forms a transmembrane α-helix in the lipid bilayer, and its average axis is parallel to the bilayer normal. It was also shown that the transmembrane helix is not static, but undergoes motion; namely, the N- and C-terminal α-helical rods rotate (reorient) rapidly about the average helical axis. Although the average direction of the α-helical axis is parallel to the bilayer normal, the local helical axis may precess about the bilayer normal, making angles of 30° and 10° with the N- and C-terminal helical rods, respectively, as shown in Figure 18.

It is of interest to relate the lytic activity of melittin to the molecular association in the lipid bilayer, as shown schematically in Figure 18. Although it is not possible to obtain detailed information on the molecular association from this NMR experiment alone, the dynamic behavior of melittin strongly suggests that it exists as monomers in the lipid bilayer at temperatures higher than T_m. When monomeric melittin adopts the transmembrane α-helical form in the lipid bilayer, lysis can be initiated by the association of melittin molecules in the lipid bilayer to separate the lipid bilayer surface, resulting in pore formation in the lipid bilayer. After growing a number of pores, small areas of the lipid bilayer are surrounded by melittin helices to form small discoidal bilayers to disperse in the solution. The monomeric transmembrane form of melittin is considered to be unstable in the lipid bilayers because of the amphiphilic nature of the melittin helix. These unstable helices might associate to make pores by aligning the hydrophobic side toward the lipid bilayer. This lytic activity might therefore be related to the lateral diffusion of lipid bilayers rather than the structural change of melittin, because this lytic behavior is altered on passing through T_m.

Gramicidin A (gA), a major synthetic product of *Bacillas brevis*, is a polypeptide of 15 amino acid residues hav-

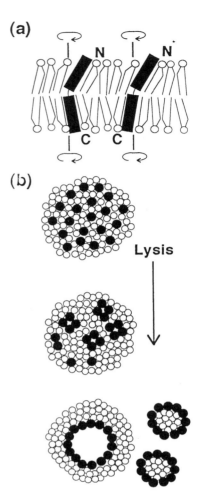

Fig. 18. (a) Schematic representation of the orientation of melittin helices bound to magnetically oriented lipid bilayers. N- and C-terminal helix axes make angles of 30° and 10°, respectively, with the average axis, which is perpendicular to the bilayers surface. Two kink angles (140° and 160°) can occur, but they cannot be distinguished by this NMR experiment. (b) The lytic process of lipid bilayers in the presence of melittin at temperatures below T_m. ○, ●, Lipid and melittin molecules, respectively.

ing the sequence formyl-L-Val–Gly–L-Ala–D-Leu–L-Ala–D-Val–L-Val–D-Val–L-Trp–D-Leu–L-Trp–D-Leu–L-Trp–D-Leu–L-Trp–etha-nolamine. All the alternating D- and L-amino acid side chains project on one side of the β-strand secondary structure, so as to force the strand to take on a helical conformation. In lipid bilayers, the polypeptide forms a monovalent cation selective channel that is dimeric, but single-stranded. The high-resolution structure of the channel monomer has been defined with 120 precise orientation constraints from solid-state NMR of uniformly aligned sample in bilayers [99–101]. It forms a single-stranded helix with a right-handed sense, 6.5 residues per turn, and β-strand torsion angles. The monomer–monomer geometry (amino terminal to amino terminal) has been characterized by solution NMR in sodium dodecyl sulfate (SDS) micelles [102] in which the monomer fold proved to be the same as in lipid bilayers. However, the refined solid state NMR constraints for gA in a lipid bilayer [101] are not consistent

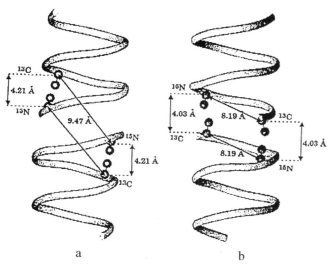

a   b

Fig. 19. Positions of the specifically labeled ^{13}C and ^{15}N sites in gramicidin A observed in hydrated phospholipid bilayers. Both ^{13}C and ^{15}N labels are indicated by solid balls at the ribbon. The solid balls, away from the ribbon, but linked to the ^{13}C and ^{15}N labels, represent the oxygen and the hydrogen nuclei, respectively, which are favorably oriented for hydrogen bonding. The distances indicated in the figures are derived from the coordinates of the high-resolution gA structure [100] (PDB accession no. #1 MAG). (a) ^{13}C$_1$-Val$_7$, ^{15}N-Gly$_2$ gA. For these labels, the distance between the intramonomer ^{13}C and ^{15}N sites is 4.21 Å, while the intermonomer ^{13}C and ^{15}N sites are 9.47 Å apart. (b) ^{13}C$_1$-Val$_1$, ^{15}N-Ala$_5$ gA. In this case, the separation between the intramonomer ^{13}C and ^{15}N sites is 8.19 Å, while the intermonomer ^{13}C and ^{15}N sites are modeled to be 4.03 Å apart, across the monomer–monomer junction.

with an X-ray crystallographic structure for gramicidine having a double-stranded, right-handed helix with 7.2 residues per turn [103].

The intermolecular distance measurements described here provide a straightforward approach for characterizing the dimeric structure of gA in lipid bilayers. Figure 19 present two choices for specific ^{13}C–^{15}N isotropic labeling of gA. The ^{13}C and ^{15}N labels are incorporated into each monomer, but in different sites. Figure 19a shows a model of ^{13}C$_1$-Val^7, ^{15}N-Gly^2 gA. Based on the high-resolution monomer structure [100], the intramonomer distance is 4.21 Å, and the orientation of the internuclear vector with respect to the motional axis is 35.3°, yielding a scaling factor of 0.5 for the dipolar interaction. The intermonomer distance between these ^{13}C and ^{15}N labels is 9.47 Å, too long to yield a detectable dipolar coupling, based on the model of the amino-terminal-to-amino-terminal hydrogen-bonded single-stranded dimer. In the ^{13}C$_1$-Val^1, ^{15}N-Ala^5 gA sample, the intramonomer distance is 8.19 Å with a scaling factor of 0.36, resulting from an orientation of 40.9° with respect to the global motion axis. Such a dipolar coupling is too weak to be detectable. On the other hand, the intermonomer distance between the labels is 4.03 Å with an angle of 2.2° between the internuclear vector and the motional axis. The scaling factor resulting from this small angle is negligible, 0.997. Such a distance constraint could be observable.

In the case of ^{13}C$_1$-Val^7, ^{15}N-Gly^2 gA in unoriented hydrated DMPC bilayers, the ^{13}C–^{15}N interatomic distance was

determined to be 4.2 ± 0.2 Å between the labeled sites. This result is in good agreement with the one calculated from the high-resolution structure of the monomer [100]. Similarly, the interatomic distance in ^{13}C$_1$-Val$_1$, ^{15}N-Ala gA in unoriented hydrated DMPC bilayers was determined to be 4.3 ± 0.1 Å. This result agrees well with the gA dimer structure in SDS micelles.

4.4. Membrane Protein

Membrane proteins are integral parts of a membrane and have at least one segment of peptide chain traversing the lipid bilayer. They are not soluble in ordinary solvents, owing to the presence of both hydrophilic and hydrophobic regions in the same molecule. Thus, solution NMR studies are very difficult because of the high-molecular-weight complex. Crystallization is extremely difficult as compared with soluble proteins: whole membrane lipids must first be solubilized and replaced by appropriate detergent molecules prior to crystallization. In this system solid state NMR is a suitable means to clarify the structure and dynamics of membrane proteins with the aid of isotropic labeling.

Bacteriorhodopsin (bR) is a light-driven proton pump in the purple membrane of *Halobacterium salinarum,* consisting of seven transmembrane α-helices, with a retinal chromophore linked to Lys-216. In addition to studies on the mechanism by ^{13}C NMR, this protein can also serve as an ideal model system to gain insight into the general aspects of the conformation and dynamics of membrane protein, because bR in PM is organized as two-dimensional crystals and a large-scale preparation of ^{13}C-labeled bR is exceptionally simple as compared with other membrane proteins. ^{13}C NMR signals of specifically ^{13}C-labeled bR are clearly distinguished from those of unlabeled preparation. It is preferable to utilize the ^{13}C-labeled Ala residue as the conformational probe, because the ^{13}C chemical shifts of Ala residues have been most thoroughly examined for a variety of local conformations. The ^{13}C NMR spectra of bR have been recorded under several conditions: lyophilized preparation, lyophilized preparation followed by hydration, and hydrated pellets of PM [104–106]. It was found that dehydration of bR by lyophilization resulted in substantial conformational distortion of the protein backbone as manifested from the obvious line broadening of ^{13}C NMR signals, although that distortion was partially alleviated in a subsequent hydration experiment [104, 106]. Therefore, it is preferable to use hydrated pellets to avoid ambiguity arising from such distortion.

It has been demonstrated that the spectral pattern of [1-^{13}C]Ala-bR is significantly different when obtained by CP-MAS and by DD-MAS NMR [105, 106], because the local flexibility of the peptide backbone is substantially different among the transmembrane α-helix, loop, and N- or C-terminus regions, as inferred from the primary structure of bR (Fig. 20). It was found that seven Ala residues located both at N- and C-terminus residues are missing in the ^{13}C CP-MAS NMR,

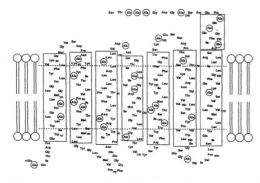

Fig. 20. Schematic representation of the secondary structure of bR, after Henderson et al. Residues enclosed by the boxes belong to transmembrane α-helices.

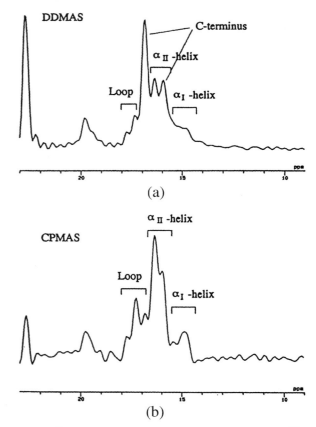

Fig. 21. (a) ^{13}C DD-MAS and (b) CP-MAS NMR spectra of [1-^{13}C]Ala-bR [16].

but they are fully recovered in the ^{13}C DD-MAS NMR spectra [105, 106]. This is because these Ala residues, being located in terminal regions, undergo rapid isotropic tumbling motions with correlation time of 10^{-8} s, which average out the dipolar interactions essential for cross polarization. This is consistent with the fact that the ^{13}C spectral pattern recorded by CP-MAS NMR was unchanged even if the C-terminus moiety containing six Ala residues was cleaved by papain. This is of course not true for the ^{13}C NMR spectra recorded by DD-MAS.

Clearly, both the ^{13}C CP-MAS and DD-MAS NMR spectra of [3-^{13}C]Ala-bR resonating at 14–18 ppm arise from at least seven resolved ^{13}C NMR signals (Fig. 21). The individual peaks are ascribed to the portion of transmembrane α_I-helix, α_{II}-helix, loop, or N- or C-terminus, with reference to the conformation-dependent ^{13}C isotropic chemical shift as summarized in Table II. The reference data for the α_I-helix and α_{II}-helix were taken from the ^{13}C NMR spectra of (Ala)$_n$ in the solid and in hexafluoroisopropanol (HFIP) solution. As indicated by the top trace, these types of α-helices are well distinguishable. The largest part of the intense peak at 16.9 ppm is ascribed to the Ala residues taking a random coil conformation as a result of undergoing rapid reorientation in the N- and C-terminal regions and is visible in the DD-MAS NMR experiment alone. A part of the peak at 16.9 ppm (three Ala residues involved in the CP-MAS experiment), however, is not due to the distorted α_{II}-helix of Ala residues located at the membrane surface, because this signal is displaced to lower frequency on removal of retinal [107]. The remaining two peaks at 17.3 and 17.9 ppm are then ascribed to the loop region connecting the transmembrane helices.

Proteolytic digestion is a very convenient means of locating signals that are strongly affected by a cleavage of the specific site by an enzyme. For instance, two peaks emerge in the difference spectrum between intact and papain-cleaved bRs (16.9 and 15.9 ppm), which are unequivocally ascribed to six Ala residues at the C-terminus: the peak at 16.9 ppm is assigned to Ala245–248 taking a random coil conformation at the terminal end, whereas 15.9 ppm is assigned to Ala228 and 233 taking the α_{II}-helix conformation [103, 108]. In particular, the signal from Ala246 and 247 can also be assigned on the basis

of the difference spectrum between carboxypeptidase A- and papain-cleaved bR.

An alternative means of site-specific assignment is to compare the ^{13}C NMR spectrum of the wild type with that of site-directed mutants, if the three-dimensional structure of mutant is not severely altered by such a site-directed mutagenisis. For instance, the assignment of [3-^{13}C]Ala^{53} is very easily done from the difference spectra using two mutants, A53G and A53V [106].

Solid state NMR is very convenient for evaluating the native conformation and dynamics of bR, which vary with environmental factors, such as temperature [109], pH [108], variety of cation [106], ionic strength [107], and in vivo or in vitro binding to bO [111]. The ^{13}C-NMR signals from the C-terminal residues are suppressed at temperatures below $-20°C$, due to interference of motional frequencies with proton decoupling frequencies. It was found that the well-resolved ^{13}C-NMR signals of bR in the presence of 10 mM NaCl at ambient temperature were broadened considerably at temperatures below $-40°C$, although no such change appears in the absence of NaCl [109]. This finding was interpreted in terms of chemical exchange among peptide chains taking several slightly different conformations, with a rate constant of 10^2 s^{-1} at ambient temperature. At lower temperatures, these peaks are naturally broad, even without such chemical exchange. That this exchange process was strongly influenced by the presence of sodium ion suggested that it plays an essential role in maintaining the secondary structure of bacteriorhodopsin, perhaps through reorganization of lipid molecules by partial screening of negative charge in the acidic head group of lipids. Obviously, this kind of screening effect would arise from all types of mono- or divalent cations.

The native purple membrane has a considerably changed photocycle when cations are completely removed to yield the *blue membrane*. As already demonstrated by ^{13}C-NMR spectra, the structure of bR is substantially modified in the blue membrane because of lowered surface pH [111]. Tuzi et al. therefore recorded ^{13}C-NMR spectra of [3-^{13}C]Ala-labeled sodium purple membranes prepared by neutralization of acid blue membrane through titration of NaOH alone, to search for the exclusive binding site of divalent cations to bR. They showed that there are high-affinity cation-binding sites in both the extracellular and Kytopläsmic regions, and one of the preferred binding sites is located at the F−G loop near Ala-196 on the exteracellular side. Furthermore, the bound cations undergo rather rapid exchange among various types of cation binding site [111].

It has been suggested that translocation of a proton from the cytoplasmic surface to the extracellular surface is associated with a protein conformational change. It is recognized that the first proton transfer in the photocycle of bR is from the protonated Schiff base to the anionic Asp-85, in the L−M reaction. This protonation induces proton release from the proton release group containing Glu-194 and -204 at the extracellular surface, and might be linked also to the subsequent deprotonation of Asp-96 in the cytoplasmic region, which causes the proton uptake. This means that the information of the protonation at Asp-85 should be transmitted to both the extracellular and the cytoplasmic regions, through specific interactions. Tanio et al. recorded ^{13}C-NMR spectra of a variety of site-directed mutations in order to clarify how such interactions might be modified by changes of electric charge or polarity in mutants [112, 113]. This is based on the expectation that such interaction should also exist in bR even in the unphotolyzed state, among backbone, side chains, bound water molecules, etc. They found that there is indeed a long-distance interaction between Asp-96 and the extracellular surface through Thr-46, Val-49, Asp-85, Arg-82, Glu-204, and Glu-194 in the unprotolyzed state. For instance, conformational changes were induced in the extracellular region through a reorientation of Arg-82 when Asp-85 was unchanged, as manifested in the recovery of the missing Ala-126 signal of D85N in the double mutant D85N/R82Q. The underlying spectral change might be interpreted in terms of the presence of perturbed Ala-126 mediated by Tyr-83, which is located between Arg-82 and Ala-126. It is possible that disruption of this interaction after protonation of Asp-85 in the photocycle could cause the same kind of conformational change as detected by the [3-^{13}C]Ala-labeled peaks of Ala-196 and Ala-126 and also the [1-^{13}C]Val-labeled peaks of Val-49 and -199 [107].

The transmembrane portion of the M2 protein from the influenza A virus has been studied in hydrated DMPC lipid bilayers with solid state NMR [114]. Orientational constraints were obtained from isotopically labeled peptide samples mechanically aligned between thin glass plates. ^{15}N chemical shifts from single-site-labeled samples constrain the molecular frame with respect to the magnetic field. When these constraints were applied to the peptides, modeled as a uniform α-helix, the tilt of the helix with respect to the bilayer normal was determined to be 33°. Furthermore, the orientation about the helix axis was also determined within an error of ±30°. These results imply that the packing of this tetrameric protein is in a left-handed four-helix bundle. Only with such a large tilt angle are the hydrophilic residues aligned with the channel axis.

The structure of functional peptides corresponding to the predicted channel-lining M2 segments of the nicotinic acetylcholine receptor (AChR) and of a glutamate receptor of the NMDA subtype (NMDAR) were determined from solution NMR experiments on micelle samples, and solid-state NMR experiments on bilayer samples [115]. Both M2 segments form straight transmembrane α-helices with no kinks. The AChR M2 peptide inserts in the lipid bilayer at an angle of 12° relative to the bilayer normal, with a rotation about the helix long axis such that the polar residues face the N-terminal side of the membrane, which is assigned to be intracellular. A model built from these solid state NMR data, assuming a symmetric pentameric arrangement of M2 helices, results in a funnel-like architecture for the channel, with the wide opening on the N-terminal intracellular side.

4.5. Fibril Structure of Amyloid and Related Biomolecules

Amyloid fibril formation—the process in which normally innocuous, soluble proteins or peptides polymerize to form insoluble fibrils—has now been seen in biochemically diverse conditions. A number of nonfibrillar proteins and peptides have been shown to form such fibrils, all of which exhibit similar morphologies in electron micrographs [116]. Some of these phenomena are related to the misfolding of proteins leading to severe diseases, such as the fibril deposit in the brain in the case of Alzheimer's disease, in the pancreas in the case of type II diabetes, etc. Therefore, the elucidation of the molecular structure of amyloid fibrils is important for understanding the mechanism of self-aggregation. However, it has been difficult to determine the molecular structures of fibrils at high resolution by ordinary spectroscopic methods, because the fibrils are heterogeneous solids. In the last decade, solid-state NMR spectroscopy has shown its advantage for the conformational determination of Alzheimer's amyloid β-peptides (Aβ), which are the main component of the amyloid plaques of Alzheimer's disease and comprise from the 39 to 43 amino acids. Not only the intrachain conformation of the Aβ molecule in the fibrils, but also the intermolecular alignment has been analyzed to explore the mechanism of molecular association to form fibrils.

The RR method has been used to characterize the structures of fragments of amyloid [117–119]. Griffin et al. have synthesized the β-amyloid fragment Aβ(34–42) (H$_2$N–Leu–Met–Val–Gly–Gly–Val–Val–Ile–Ala–CO$_2$H), which is the C-terminus of the β-amyloid protein. The fragment was chosen because the region is implicated in the initiation of amyloid formation. The structure of this molecule was determined by the $^{13}C-^{13}C$ interatomic distances and the ^{13}C chemical shift values using fibril samples of 10–20% of the doubly ^{13}C-labeled Aβ(34–42) diluted with the unlabeled peptide. The α-carbon of the ith residue and the carbonyl carbon of the $(i + 1)$th residue were doubly labeled, and the A[$\alpha i, i + 1$] interatomic distance was observed by using a rotational resonance method. Similarly, the B[$i, \alpha(i + 2)$] and C[$i, \alpha(i + 3)$] interatomic distances were also determined. Since the rotational resonance signal of A does not show a dilution effect, an intermolecular contribution does not exist. On the other hand, B and C show strong intermolecular contributions from B* and C*. Therefore, it turns out that the fragment forms an antiparallel β-sheet. Furthermore, the intermolecular contribution indicates that β-strands consist of antiparallel β-sheets forming hydrogen bonds at the position that is offset from the N-terminus position. It is interesting that the information on the intermolecular contribution made it possible to reveal the assembly of the amyloid molecules.

On the other hand, the DRAWS solid state NMR technique using peptides with ^{13}C in the carbonyl position was applied to characterize the peptide conformation and the supramolecular organization of fibrils formed by Aβ(10–35), which contains both hydrophobic and nonhydrophobic segments of Aβ. The peptide backbones in the fibrillated precipitation were confirmed to form an extended conformation in view of the ^{13}C chemical shifts of carbonyl carbon and the interatomic distance between I- and $I + 1$-labeled carbonyl carbons obtained by the DRAWS method. Interpeptide distances between the ^{13}C nuclei measured by DRAWS were 4.9–5.8 (± 0.4) Å, respectively, throughout the entire length of the peptide. These results indicated that an in-register parallel organization of β-sheets is formed in the fibril of Aβ(10–35) [120, 122].

Recently, the fibril structure of another fragment of Aβ, which comprises residues 16–22 of the Alzheimer's β-amyloid peptide [Aβ(16–22)] and is the shortest fibril forming one, was determined [123, 124]. 2D MAS exchange and constant-time double-quantum-filtered dipolar (CTDQFD) recoupling techniques revealed that the torsion angles of the peptide backbone showed the extended conformation. The ^{13}C chemical shifts determined from two-dimensional chemical shift correlation spectra also indicated that the entire hydrophobic segment forms a β-strand conformation. ^{13}C multiple quantum (MQ) NMR and $^{13}C-^{15}N$ REDOR data indicate an antiparallel organization of β-sheets in Aβ(16–22) fibrils. However, the MQ NMR data indicate an in-register, parallel organization for full-length β-amyloid [Aβ(1–40)] fibrils.

Calcitonin (CT) is a peptide hormone consisting of 32 amino acid residues that contains an intrachain disulfide bridge between Cys[1] and Cys[7] and a proline amide at the C-terminus. CT has been found to be a useful drug for various bone disorders such as Paget's disease and osteoporosis. However, human calcitonin (hCT) has a tendency to associate to form fibril precipitate in aqueous solution. This fibril is known to be of the same type as amyloid fibril, and hence its formation has been studied as a model of amyloid fibril formation.

Conformations of hCT in several solvents have been studied by solution NMR spectroscopy. In TFE/H$_2$O, hCT forms a helical structure between residues 9 and 21 [125]. A short double-stranded antiparallel β-sheet form, however, was observed in the central region made by residues 16–21 in DMSO/H$_2$O. It was suggested that hCT exhibit an amphiphilic nature when it forms an α-helix. However, the secondary structure of hCT in H$_2$O was shown to be a totally random coil, as determined from the ^1H chemical shift data. Subsequent studies by two-dimensional NOESY and CD measurements indicated that it adopts an extended conformation with high flexibility in aqueous solution [126]. Recently, an α-helical conformation was reported to be present in the central region of hCT in aqueous acidic solution, although it is shorter than in TFE/H$_2$O. Similarly, NMR studies showed that salmon calcitonin (sCT) also forms an amphiphilic α-helical structure in the central region in TFE/H$_2$O, methanol/H$_2$O, and SDS micelles. It has been reported that sCT has a high activity and stability as a drug, and its fibrillation is much slower than that of hCT.

Two-dimensional NMR measurements of the time course of the fibrillation process have shown that the peaks from residues in the N-terminal (Cys[1]–Cys[7]) and the central (Met[8]–Pro[23]) regions are broadened and disappear earlier than those in the C-terminal region [128]. These findings indicate that the α-helices are bundled together in the first homogeneous nucleation process. Subsequently, it appears that larger fibrils grow in a second, heterogeneous process. However, local structures and characteristics of hCT in the fibril have not been obtained directly by solution NMR spectroscopy, because the intrinsically broadened signals from the fibril components cannot be observed.

Conformational transition of hCT during fibril formation in the acidic and neutral conditions was investigated by high-resolution solid state ^{13}C NMR spectroscopy [129]. In aqueous acetic acid solution (pH 3.3), a local α-helical form is present around Gly[10], whereas a random coil form is dominant as viewed from Phe[22], Ala[26], and Ala[31] in the monomer form on the basis of the ^{13}C chemical shifts. On the other hand, a local β-sheet form as viewed from Gly[10] and Phe[22], and both β-sheet and random coil as viewed from Ala[26] and Ala[31], were detected in the fibril at pH 3.3. The results indicate that conformational transitions from α-helix to β-sheet and from random coil to β-sheet form occurred in the central and C-terminus regions, respectively, during fibril formation. The increased ^{13}C resonance intensities of fibrils after a certain delay suggest that the fibrillation can be explained by a two-step reaction mechanism in which the first step is a homogeneous association to form a nucleus, and second step is an autocatalytic heterogeneous fibrillation as shown in Figure 22. In contrast to the fibril at pH 3.3, the fibril at pH 7.5 formed a local β-sheet conformation in the central region and exhibited random coiling in the C-terminus region. Not only a hydrophobic interaction among the amphiphilic α-helices, but also an electrostatic interaction between charged side chains, can play an important role in fibril

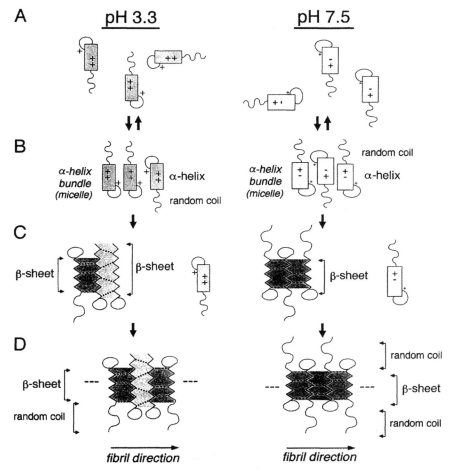

Fig. 22. Schematic representation of proposed models for fibril formation at pH 3.3 and 7.5: A hCT monomers in solution; B, a homogeneous association to form the α-helical bundle (micelle); C, a homogeneous nucleation process to form the β-sheet, and heterogeneous associating process; D, a heterogeneous fibrillation process to grow a large fibril.

formation at pH 7.5 and 3.3, respectively; the latter is a favorable interactions. These results suggest that hCT fibril is formed by stacking antiparallel β-sheets at pH 7.5 and a mixture of antiparallel and parallel β-sheets at pH 3.3.

5. CONCLUDING REMARKS

It has been demonstrated that high-resolution solid state NMR spectroscopy is a powerful means to elucidate the structure, orientation, molecular assembly, and dynamics of a variety of biomolecules that cannot be studied by other spectroscopic methods. Conformation-dependent chemical shifts are the starting point in studing the structure of membrane proteins, because it is now understood that they reflect the secondary structure of the protein. Accurate interatomic distances may provide high-resolution structure of solid peptides, fibril proteins, and (more importantly) molecular assemblies of amyloid peptides. Magnetically aligned lipid bilayer media provide detailed information on the orientation of membrane-bound peptide and membrane proteins. This information provides otherwise unavailable insight into the structure–function relationship for important solid biomolecules.

REFERENCES

1. M. Mehring, "Principles of High Resolution NMR in Solids." Springer-Verlag, Berlin, 1983.
2. C. P. Slichter, "Principles of Magnetic Resonance." Springer-Verlag, Berlin, 1990.
3. N. Ahmed and K. R. Rao, "Orthogonal Transforms for Digital Processing." Springer-Verlag, 1975.
4. R. A. Haberkorn, R. E. Stark, H. van Willigen, and R. G. Griffin, *J. Am. Chem. Soc.* 103, 2534 (1981).
5. A. Naito, S. Ganapathy, K. Akasaka, and C. A. McDowell, *J. Chem. Phys.* 74, 3190 (1981).
6. A. Naito, S. Ganapathy, P. Raghunathan, and C. A. McDowell, *J. Chem. Phys.* 79, 4173 (1983).
7. N. Janes, S. Ganapathy, and E. Oldfield, *J. Magn. Reson.* 54, 111 (1983).
8. A. Naito and C. A. McDowell, *J. Chem. Phys.* 81, 4795 (1984).
9. R. E. Stark, L. W. Jelinski, D. J. Ruben, D. A. Torchia, and R. G. Griffin, *J. Magn. Reson.* 55, 266 (1983).
10. L. W. Jelinski and D. A. Torchia, *J. Mol. Biol.* 133, 45 (1979).

11. S. Ando, T. Yamanobe, I. Ando, A. Shoji, T. Ozaki, R. Tabeta, and H. Saito, *J. Am. Chem. Soc.* 107, 7648 (1985).

12. O. W. Howarth and D. M. J. Lilley, *Prog. Nucl. Magn. Reson. Spectrosc.* 12, 1 (1978).

13. H. Saitô, *Magn. Reson. Chem.* 24, 835 (1983).

14. H. Saitô and I. Ando, *Annu. Rep. NMR Spectrosc.* 21, 209 (1989).

15. H. Saitô, S. Tuzi, and A. Naito, *Annu. Rep. NMR Spectrosc.* 35, 79 (1998).

16. H. Saito, S. Tuzi, S. Yamaguchi, S. Kimura, M. Tanio, M. Kamihira, K. Nishimura, and A. Naito, *J. Mol. Struct.* 441, 231 (1998).

17. S. Spera and A. Bax, *J. Am. Chem. Soc.* 113, 5490 (1981).

18. D. S. Wishart and B. D. Sykes, *Methods Enzymol.* 239, 363 (1994).

19. A. Pines, M. G. Gibby, and J. S. Waugh, *J. Chem. Phys.* 59, 569 (1973).

20. A. E. Bennett, C. M. Rienstra, M. Auger, K. V. Lakshmi, and R. G. Griffin, *J. Chem. Phys.* 103, 6951 (1995).

21. W. T. Dixon, *J. Chem. Phys.* 77, 1800 (1982).

22. J. Schaefer and E. O. Stejskal, *J. Am. Chem. Soc.* 98, 1031 (1976).

23. J. H. Davis, K. R. Jeffery, M. Bloom, M. I. Valio, and T. P. Higgs, *Chem. Phys. Lett.* 42, 390 (1976).

24. H. W. Spiess, *J. Chem. Phys.* 72, 6755 (1980).

25. Y. Hiyama, J. V. Silverton, D. A. Torchia, J. T. Gerig, and S. J. Hammond, *J. Am. Chem. Soc.* 108, 2715 (1986).

26. T. Gullion and J. Schaefer, *Adv. Magn. Reson.* 13, 57 (1989).

27. E. Bennett, R. G. Griffin, and S. Vega, *NMR* 33, 1 (1994).

28. T. Gullion and J. Schaefer, *J. Magn. Reson.* 81, 196 (1989).

29. D. P. Raleigh, M. H. Levitt, and R. G. Griffin, *Chem. Phys. Lett.* 146, 71 (1988).

30. K. T. Mueller, T. P. Javie, D. J. Aurentz, and B. W. Roberts, *Chem. Phys. Lett.* 242, 535 (1995).

31. T. P. Javie, G. T. Went, and K. T. Mueller, *J. Am. Chem. Soc.* 118, 5330 (1996).

32. M. H. Levitt, D. P. Raleigh, F. Creuzet, and R. G. Griffin, *J. Chem. Phys.* 92, 6347 (1990).

33. A. W. Hing, S. Vega, and J. Schaefer, *J. Magn. Reson.* 96, 205 (1992).

34. R. Tycko and G. Dabbagh, *Chem. Phys. Lett.* 173, 461 (1990).

35. B.-Q. Sun, P. R. Costa, D. Kocisko, P. T. Lansbury, Jr., and R. G. Griffin, *J. Chem. Phys.* 102, 702 (1995).

36. T. Gullion and S. Vega, *Chem. Phys. Lett.* 194, 423 (1992).

37. A. E. Bennett, J. H. Ok, R. G. Griffin, and S. Vega, *J. Chem. Phys.* 96, 8624 (1992).

38. D. M. Gregory, D. J. Mitchell, J. A. Stringer, S. Kiihne, J. C. Shiels, J. Callahan, M. A. Mehta, and G. P. Drobny, *Chem. Phys. Lett.* 246, 654 (1995).

39. G. J. Boender, J. Raap, S. Prytulla, H. Oschkinat, and H. J. M. de Groot, *Chem. Phys. Lett.* 237, 502 (1995).

40. T. Fujiwara, K. Sugase, M. Kainosho, A. Ono, and H. Akutsu, *J. Am. Chem. Soc.* 117, 11351 (1995).

41. K. Schmidt-Rohr, *Macromolecules* 29, 3975 (1996).

42. M. Baldus, R. J. Iuliucci, and B. M. Meier, *J. Am. Chem. Soc.* 119, 1121 (1997).

43. D. P. Weliky and R. Tyco, *J. Am. Chem. Soc.* 118, 8487 (1996).

44. A. Naito, K. Nishimura, S. Kimura, S. Tuzi, M. Aida, N. Yasuoka, and H. Saitô, *J. Phys. Chem.* 100, 14995 (1996).

45. A. Naito, K. Nishimura, S. Tuzi, and H. Saitô, *Chem. Phys. Lett.* 229, 506 (1994).

46. T. Gullion and J. Schaefer, *J. Magn. Reson.* 92, 439 (1991).

47. Y. Pan, T. Gullion, and J. Schaefer, *J. Magn. Reson.* 90, 330 (1990).

48. D. Suwelack, W. P. Rothwell, and J. S. Waugh, *J. Chem. Phys.* 73, 2559 (1980).

49. W. P. Rothwell and J. S. Waugh, *J. Chem. Phys.* 74, 2721 (1981).

50. A. Naito, A. Fukutani, M. Uitdehaag, S. Tuzi, and H. Saitô, *J. Mol. Struct.*, in press.

51. M. Kamihira, A. Naito, K. Nishimura, S. Tuzi, and H. Saitô, *J. Phys. Chem.*, in press.

52. K. Nishimura, K. Ebisawa, E. Suzuki, H. Saito, and A. Naito, *J. Mol. Struct.*, in press.

53. O. B. Peersen, M. Groesbeek, S. Aimoto, and S. O. Smith, *J. Am. Chem. Soc.* 117, 7728 (1995).

54. A. Kubo and C. A. McDowell, *J. Chem. Soc. Faraday Trans. 1* 84, 3713 (1988).

55. K. Takegoshi, K. Nomura, and T. Terao, *Chem. Phys. Lett.* 232, 424 (1995).

56. F. Scholz and W. Helfrich, *Biophys. J.* 45, 589 (1984).

57. J. Seelig, F. Borle, and T. A. Cross, *Biochim. Biophys. Acta* 814, 195 (1985).

58. J. B. Spryer, P. K. Spipada, S. K. Das Gupta, G. G. Shipley, and R. G. Griffin, *Biophys. J.* 51, 687 (1987).

59. T. Brumm, C. Mops, C. Dolainsky, S. Bruckner, and T. M. Bayerl, *Biophys. J.* 61, 1018 (1997).

60. X. Qin, P. A. Mirau, and C. Pidgeon, *Biochim. Biophys. Acta* 1147, 59 (1993).

61. C. E. Dempsey and A. Watts, *Biochemistry* 26, 5803 (1987).

62. C. E. Dempsey and B. Sternberg, *Biochim. Biophys. Acta* 1061, 175 (1991).

63. T. Pott and E. J. Dufourc, *Biophys. J.* 68, 965 (1995).

64. A. Naito, T, Nagao, K. Norisada, T. Mizuno, S. Tuzi, and H. Saitô, *Biophys. J.* 78, 2405 (2000).

65. C. R. Sanders and J. H. Prestegard, *Biophys. J.* 58, 447 (1990).

66. C. R. Sanders and J. P. Schwonek, *Biochemistry* 31, 8898 (1992).

67. C. H. Ho, A. Ramamoorthy, and S. J. Opella, *J. Magn. Reson.* A109, 270 (1994).

68. F. M. Marassi and S. J. Opella, *J. Magn. Reson.* 144, 150 (2000).

69. F. M. Marassi, C. Ma, J. J. Gesell, and S. J. Opella, *J. Magn. Reson.* 144, 156 (2000).

70. J. Wang, J. Denny, C. Tian, S. Kim, Y. Mo, F. Kovacs, Z. Song, K. Nishimura, Z. Gan, R. Fu, J. R. Quine, and T. A. Cross, *J. Magn. Reson.* 144, 162 (2000).

71. G. R. Marshall, D. P. Beusen, K. Kociolek, A. S. Redlinski, M. T. Leplawy, and J. Schaefer, *J. Am. Chem. Soc.* 112, 4963 (1990).

72. S. M. Holl, G. R. Marshall, D. P. Beusen, K. Kociolek, A. S. Redlinski, M. T. Leplway, R. Makey, S. Vega, and J. Schaefer, *J. Am. Chem. Soc.* 114, 4830 (1992).

73. A. W. Hing and J. Schaefer, *Biochemistry* 32, 7593 (1993).

74. D. J. Hirsh, J. Hammer, W. L. Maloy, J. Blazyk, and J. Schaefer, 35, 12733 (1996).

75. J. R. Garbow and C. A. McWherter, *J. Am. Chem. Soc.* 115, 238 (1993).

76. J. R. Garbow, M. Breslav, O. Antohi, and F. Naider, *Biochemistry* 33, 10094 (1994).

77. R. C. Anderson, T. Gullion, J. M. Joers, M. Shepiro, E. B. Villhauer, and H. P. Weber, *J. Am. Chem. Soc.* 117, 10546 (1995).

78. S. K. Brahmachari, T. N. Bhat, V. Sudhakar, M. Vijayan, R. S. Rapaka, R. S. Bhatnagar, and V. S. Aranthanarayanan, *J. Am. Chem. Soc.* 103, 1703 (1981).

79. I. Z. von Siemion, T. Wieland, and K.-H. Pook, *Angew. Chem.* 87, 712 (1975).

80. M. Aida, A. Naito, and H. Saitô, *J. Mol. Struct. Theory* 388, 187 (1996).

81. K. Nishimura, A. Naito, S. Tuzi, H. Saitô, C. Hashimoto, and M. Aida, *J. Phys. Chem.* B102, 7476 (1998).

82. A. W. Hing, N. Tjandra, P. F. Cottam, J. Schaefer, and C. Ho, *Biochemistry* 33, 8651 (1994).

83. A. M. Christensen and J. Schaefer, *Biochemistry* 32, 2868 (1993).

84. L. M. McDowell, A. Schmidt, E. R. Cohen, D. R. Studelska, and J. Schaefer, *J. Mol. Biol.* 256, 160 (1996).

85. L. M. McDowell, C. K. Klug, D. D. Beusen, and J. Schaefer, *Biochemistry* 35, 5396 (1996).

86. Y. Li, R. J. Appleyard, W. A. Shuttleworth, and J. N. S. Evans, *J. Am. Chem. Soc.* 116, 10799 (1994).

87. Y. Li, F. Krekel, C. A. Ramilo, N. Amrhein, and J. N. S. Evans, *FEBS Lett.* 377, 208 (1995).

88. L. K. Thompsom, A. E. McDermott, J. Raap, C. M. Van der Wielen, J. Lugtenburg, J. Herzfeld, and R. G. Griffin, *Biochemistry* 31, 7931 (1992).

89. K. V. Lakshmi, M. Auger, J. Raap, J. Lugtenburg, R. G. Griffin, and J. Hertzfeld, *J. Am. Chem. Soc.* 115, 8515 (1993).

90. K. V. Lakshmi, M. R. Farrar, J. Raap, J. Lugtenburg, R. G. Griffin, and J. Herzfeld, *Biochemistry* 33, 8854 (1994).
91. A. Naito, M. Kamihira, S. Tuzi, and H. Saitô, *J. Phys. Chem.* 99, 12041 (1995).
92. M Kamihira, A. Naito, K. Nishimura, S. Tuzi, and H. Saitô, *J. Phys. Chem.* 102, 2826 (1998).
93. M. Kamihira, A. Naito, S. Tuzi, and H. Saitô, *J. Phys. Chem.* 103, 3356 (1999).
94. C. E. Dempsey, *Biochim. Biophys. Acta* 1031 (1990).
95. T. C. Terwilliger and D. E. Senberg, *J. Biol. Chem.* 257, 6010 (1982).
96. T. C. Terwilliger, L. Weissman, and D. E. Senberg, *Biophys. J.* 37, 353 (1992).
97. R. Bazzo, M. J. Tappin, A. Pastore, T. S. Harrey, J. A. Carrer, and I. D. Campbell, *Eur. J. Biochem.* 173, 139 (1988).
98. R. Smith, F. Separovic, T. J. Milne, A. Whittaker, F. M. Bennett, B. A. Cornell, and A. Makriyannis, *J. Mol. Biol.* 241, 456 (1994).
99. R. R. Ketchem, W. Hu, and T. A. Cross, *Science* 261, 1457 (1993).
100. R. R. Ketchem, B. Ronx, and T. A. Cross, *Structure* 5, 1655 (1997).
101. F. Kovacs, J. Quine, and T. A. Cross, *Proc. Nat. Acad. Sci. U.S.A.* 96, 7910 (1999).
102. A. L. Lomize, V. Y. Urekhov, and A. S. Arseniev, *Bioorg. Khim.* 18, 182 (1992).
103. R. M. Burkhart, N. Li, D. A. Langs, W. A. Pangborn, and W. L. Duax, *Proc. Nat. Acad. Sci. U.S.A.* 95, 12950 (1998).
104. S. Tuzi, A. Naito, and H. Saitô, *Eur. J. Biochem.* 218, 837 (1993).
105. S. Tuzi, A. Naito, and H. Saitô, *Biochemistry* 33, 15046 (1994).
106. S. Tuzi, S. Yamaguchi, A. Naito, R. Needleman, J. K. Lanyi, and H. Saitô, *Biochemistry* 35, 7520 (1996).
107. M. Tanio, S. Tuzi, S. Yamaguchi, H. Konishi, A. Naito, R. Needleman, J. K. Lanyi, and H. Saitô, *Biochim. Biophys. Acta* 1375, 84 (1998).
108. S. Yamaguchi, S. Tuzi, T. Seki, M. Tanio, R. Needleman, J. K. Lanyi, A. Naito, and H. Saitô, *J. Biochem. (Tokyo)* 103, 78 (1998).
109. S. Tuzi, A. Naito, and H. Saitô, *Eur. J. Biochem.* 239, 294 (1996).
110. S. Yamaguchi, S. Tuzi, M. Tanio, A. Naito, J. K. Lanyi, R. Needleman, and H. Saitô, *J. Biochem. (Tokyo)* 127, 861 (2000).
111. S. Tuzi, S. Yamaguchi, M. Tanio, H. Konishi, and H. Saitô, *Biophys. J.* 76, 1523 (1999).
112. M. Tanio, S. Inoue, K. Yokota, T. Seki, S. Tuzi, R. Needleman, J. K. Lanyi, A. Naito, and H. Saitô, *Biophys. J.* 77, 431 (1999).
113. M. Tanio, S. Tuzi, S. Yamaguchi, R. Kawaminami, A. Naito, R. Needleman, J. K. Lanyi, and H. Saitô, *Biophys. J.* 77, 1577 (1999).
114. I. Song, F. A. Kovacs, J. Wang, K. Denny, S. C. Shekav, J. R. Quine, and T. A. Cross, *Biophys. J.* 79, 767 (2000).
115. S. J. Opella, F. M. Marassi, J. I. Gesell, A. P. Valente, Y. Kim, M. Oblatt-Montal, and M. Montal, *Nature Struct. Biol.* 6, 374 (1999).
116. J. D. Sipe, *Annu. Rev. Biochem.* 61, 947–975 (1992).
117. R. G. S. Spencer, K. J. Halverson, M. Auger, A. E. McDermott, and R. G. Griffin, *Biochemistry* 30, 10382 (1991).
118. P. T. Lansbury, Jr., P. R. Costa, J. M. Griffiths, E. J. Simon, M. Auger, K. J. Halverson, D. A. Kocisko, Z. S. Hendsch, T. T. Ashburn, R. G. S. Spencer, B. Tidor, and R. G. Griffin, *Nature Struct. Biol.* 2, 990 (1995).
119. J. M. Griffiths, T. T. Ashburn, M. Auger, P. R. Costa, R. G. Griffin, and P. T. Lansbury, Jr., *J. Am. Chem. Soc.* 117, 3539 (1995).
120. P. R. Costa, D. A. Kocisko, B. Q. Sun, P. T. Lansbury, Jr., and R. G. Griffin, *J. Am. Chem. Soc.* 119, 10487 (1997).
121. T. L. S. Benzinger, D. M. Gregory, T. S. Burkoth, H. Miller-Auer, D. G. Lynn, R. E. Botto, and S. C. Meredith, *Proc. Nat. Acad. Sci. U.S.A.* 95, 13407 (1998).
122. T. L. S. Benzinger, D. M. Gregory, T. S. Burkoth, H. Miller-Auer, D. G. Botto, and S. C. Meredith, *Biochemistry* 39, 3491 (2000).
123. T. S. Burkoth, T. L. S. Benzinger, V. Urban, D. M. Morgan, D. M. Gregory, P. Thiyagarajan, R. E. Botto, S. C. Meredith, and D. G. Lynn, *J. Am. Chem. Soc.* 122, 7883 (2000).
124. J. J. Balbach, Y. Ishii, O. N. Antzutkin, R. D. Leapman, N. W. Rizzo, F. Reed, and R. Tycko, *Biochemistry* 39, 13748 (2000).
125. O. N. Antzutkin, J. J. Balbach, R. D. Leapman, N. W. Rizzo, J. Reed, and R. Tycko, *Proc. Nat. Acad. Sci. U.S.A.* 97, 13045 (2000).
126. M. Doi, Y. Kobayashi, Y. Kyogoku, M. Takimito, and K. Gota, "Peptides: Chemistry, Structure & Biology," Proc. 11th Symp. July 9–14, 1989, pp. 165–167. La Jolla, CA, 1990.
127. A. Motta, P. A. Tenussi, E. Wunsch, and G. Bovermann, *Biochemistry* 30, 2364 (1991).
128. Y. H. Jeon, K. Kanaori, H. Takashima, T. Koshiba, and Y. A. Nosaka, "Proc. ICMRBS XVIII," p. 61. Tokyo, 1998.
129. K. Kanaori and A. Y. Nosaka, *Biochemistry* 34, 12138 (1995).
130. M. Kamihira, A. Naito, S. Tuzi, A. Y. Nosaka, and H. Saitô, *Protein Sci.* 9, 867 (2000).
131. D. M. Gregory, M. A. Methta, J. C. Shiels, and G. P. Drobny, *J. Chem. Phys.* 107, 28 (1997).

Index